GLOBAL ENVIRONMENT OUTLOOK
GEO-6
HEALTHY PLANET, HEALTHY PEOPLE

Edited by

Paul Ekins
Co-chair Global
Environment Outlook
University College
London

Joyeeta Gupta
Co-chair Global
Environment Outlook
University of
Amsterdam

Pierre Boileau
Senior Environmental
Affairs Officer

CAMBRIDGE
UNIVERSITY PRESS

University Printing House, Cambridge CB2 8BS, United Kingdom
One Liberty Plaza, 20th Floor, New York, NY 10006, USA
477 Williamstown Road, Port Melbourne, VIC 3207, Australia
314–321, 3rd Floor, Plot 3, Splendor Forum, Jasola District Centre, New Delhi – 110025, India
79 Anson Road, #06–04/06, Singapore 079906

Cambridge University Press is part of the University of Cambridge.
It furthers the University's mission by disseminating knowledge in the pursuit of education, learning and research at the highest international levels of excellence.

www.cambridge.org
Information on this title: www.cambridge.org/9781108707664
DOI: 10.1017/9781108627146

First published 2019

The designations employed and the presentation of the material in this publication do not imply the expression of any opinion whatsoever on the part of UN Environment concerning the legal status of any country, territory or city or its authorities, or concerning the delimitation of its frontiers or boundaries. For general guidance on matters relating to the use of maps in publications please go to http://www.un.org/Depts/Cartographic/english/htmain.htm

This document may be cited as: UN Environment (2019). Global Environment Outlook – GEO-6: Healthy Planet, Healthy People. Nairobi. DOI 10.1017/9781108627146.

Job Number: DEW/2214/NA
© Maps, photos, and illustrations as specified
Cover design: Joseph Shmidt-Klingenberg and Sebastian Obermeyer
Graphic Design: Joseph & Sebastian
Layout: Jennifer Odallo and Catherine Kimeu (UNON Publishing Services Section – ISO 14001-certified)

Printed in Singapore by Markono Print Media Pte Ltd

A catalogue record for this publication is available from the British Library.
ISBN 978-1-108-70766-4 Paperback

Disclaimers

Mention of a commercial company or product in this document does not imply endorsement by UN Environment or the authors. The use of information from this document for publicity or advertising is not permitted. Trademark names and symbols are used in an editorial fashion with no intention on infringement of trademark or copyright laws.

Cambridge University Press has no responsibility for the persistence or accuracy of URLs for external or third-party internet websites referred to in this publication, and does not guarantee that any content on such websites is, or will remain, accurate or appropriate.

UN Environment promotes environmentally sound practices globally and in its own activities. This report is printed on Forest Stewardship Council (FSC) certified paper from sustainable forests with non-toxic inks. Our distribution policy aims to reduce UN Environment's carbon footprint.

GLOBAL ENVIRONMENT OUTLOOK
GEO-6
HEALTHY PLANET, HEALTHY PEOPLE

A healthy planet is important for the health and well-being of all people.
It directly supports the lives and livelihoods of 70 per cent of the Earth's population
The top 10 emitting countries emit 45 per cent of global GHG emissions; the bottom 50 per cent emit only 13 per cent
Exposure to indoor/outdoor air and water pollution costs at least 9 million lives annually

Meat production currently uses 77 per cent of agricultural land. Agricultural production consumes 70 per cent of global water.
Current science justifies policy action now. More detailed knowledge can enable more refined and ambitious policy
Achieving the 2°C Paris Agreement target could save US$ 54.1 trillion for a global expenditure of $US 22.1 trillion. Achieving a 1.5°C target could lead to a health saving of $US 3.3-8.4 trillion for India and $US 0.3-2.3 trillion for China

Acknowledgements

The sixth Global Environment Outlook (GEO-6) assessment report is a product of the generous dedication and extraordinary investment of numerous individuals, whose knowledge, expertise and insight helped shape this important body of work. UN Environment acknowledges the contributions made by many governments, individuals and institutions to the preparation and publication of this report. A more complete list of names of individuals and institutions involved in the assessment process is included following the annexes of this publication. Special thanks are extended to:

High Level Inter-Governmental and Stakeholder Advisory Group (HLG)

Nassir S. Al-Amri, Hæge Andenæs, Juan Carlos Arredondo, Sara Baisai Feresu, Benon Bibbu Yassin, Simon Birkett, Gillian Bowser, Joji Carino, Fernando E.L.S. Coimbra, Marine Collignon (alternate), Victoria de Higa Rodriguez, Laksmi Dhewanthi, Noasilalaonomenjahary Ambinintsoa Lucie, Arturo Flores Martinez (alternate), Sascha Gabizon, Prudence Galega, Edgar Gutiérrez Espeleta, Keri Holland (alternate), Pascal Valentin Houénou (Vice-chair), Yi Huang (Co-chair), Ingeborg Mork-Knutsen (alternate), Melinda Kimble, Asdaporn Krairapanond, Yaseen M. Khayyat, Pierluigi Manzione, Veronica Marques (alternate), Jock Martin, John M. Matuszak, Megan Meaney, Naser Moghaddasi, Bedrich Moldan, Roger Roberge, Najib Saab, Mohammed Salahuddin, Jurgis Sapijanskas (alternate), Paolo Soprano (Co-chair), Xavier Sticker, Sibylle Vermont (Vice-chair), Andrea Vincent (alternate), Terry Yosie.

Scientific Advisory Panel

Asma Abahussain, John B.R Agard, Odeh Al-Jayyousi, Paulo Eduardo Artaxo Netto, Rosina M. Bierbaum, Enrico Giovannini, Sarah Green (Co-chair), Torkil Jønch Clausen, Ahmed Khater, Nicholas King (Co-chair), Paolo Laj, Byung-Kook Lee, Alastair Charles Lewis, Franklyn Lisk, Majid Shafiepour Motlagh, Carlos Afonso Nobre, Toral Patel-Weynand, Anand Patwardhan, N.H Ravindranath (Vice-chair), Wendelin Stark, Danling Tang, Maria del Mar Viana Rodriguez (Vice-chair), Naohiro Yoshida.

Members of the Assessment Methodologies, Data and Information Working Group

Maria Andrzejewska, Ousséni Arouna, Sandra De Carlo (Co-chair), Rosario Gomez, Wabi Marcos, Reza Maknoon, Graciela Metternicht, Thy Nguyen Van, Nicolas Perritaz (Co-chair), Qurat ul Ain Ahmad, Mathis Wackernagel, Fei Wang.

Co-chairs and Vice-chairs of the assessment

Co-chairs: Paul Ekins, Joyeeta Gupta.
Vice-chairs: Jane Bemigisha, Kejun Jiang.

Coordinating Lead Authors (CLAs)

Ghassem Asrar, Elaine Baker, Tariq Banuri, Graeme Clark, John Crump, Florence Mayocyoc-Daguitan, Jonathan Davies, Phillip Dickerson, Nicolai Dronin, Mark Elder, Erica Gaddis, Jia Gensuo, Anna Maria Grobicki, Cristina Guerreiro, Andrés Guhl, Peter Harris, Rowena Hay, Steve Hedden, Klaus Jacob, Mikiko Kainuma, Terry Keating, Peter King, Pali Lehohla, Christian Loewe, Paul Lucas, Diana Mangalagiu, Diego Martino, Shanna McClain, Catherine McMullen, Adelina Mensah, Indu K. Murthy, Charles Mwangi, John Muthama Nzioka, Jacob Park, Laura Pereira, Fernando Filgueira Prates, Walter Rast, Jake Rice, Joni Seager, William Sonntag, Peter Stoett, Michelle Tan, Detlef van Vuuren, Dimitri Alexis Zenghelis.

Review Editors

Amr Osama Abdel-Aziz, Ahmed Abdelrehim, Majdah Aburas, Mohammad Al Ahmad, Chandani Appadoo, Michael Brody, Louis Cassar, William W. Dougherty, Manal Elewah, Amr El-Sammak, Elsa Patricia, Galarza Contreras, Jose Holguin-Veras, Muhammad Ijaz, Joy Jadam, Emmanuel Dieudonné Kam Yogo, Yoon Lee, Clever Mafuta, Simone Maynard, Joan Momanyi, Jacques André Ndione, Washington Odongo Ochola, Renat Perelet, Linn Persson, Jan Plesnik, Ariana Rossen, Mayar Sabet, John Shilling, Binaya Raj Shivakoti, Asha Singh, Asha Sitati, Lawrence Surendra, Paul C. Sutton, Khulood Abdul Razzaq Tubaishat; Emma Archer van Garderen, Lei Yu, Samy Mohamed Zalat.

GEO-6 Funders

Producing an assessment of this scale requires many generous contributions. The following organizations provided funding directly or indirectly to the sixth *Global Environment Outlook*: The Government of Norway, the European Union, the Governments of Italy, Singapore, China, Mexico, Switzerland, Denmark, Egypt and Thailand. Together with UN Environment's Environment Fund and Regular Budget, these contributions allowed for the production of GEO-6 and its accompanying Summary for Policymakers, as well as subsequent outreach activities.

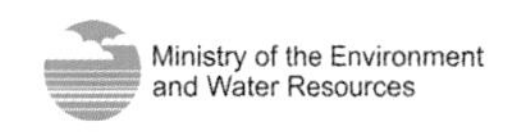

GEO-6 Partners

GEO-6 also benefited from the generous contributions of several partners, including: GRID-Arendal, World Conservation Monitoring Centre (WCMC), The Centre for Environment and Development in the Arab Region and Europe (CEDARE), The Big Earth Data Science Engineering Program (CASEarth), the European Space Agency (ESA), the Netherlands Environmental Assessment Agency (PBL), the Freie Universität Berlin and the Massachusetts Institute of Technology (MIT).

The sixth Global Environment Outlook Production Team

GEO-6 Core Team
Pierre Boileau (Head of GEO Unit), Minang Acharya, Yunting Duan, Sandor Frigyik, Orlane Jadeau, Caroline Kaimuru, Jian Liu, Caroline Mureithi, Franklin Odhiambo, Brigitte Ohanga, Adele Roccato, Sharif Shawky, Simone Targetti Ferri, Brennan Van Dyke, Edoardo Zandri.

Chapter Coordinators
Joana Akrofi, Hilary Allison, Jennifer Bailey, Pierre Boileau, Jillian Campbell, Kilian Christ, John Crump, Valentin Foltescu, Jason Jabbour, Hartwig Kremer, Maarten Kappelle, Patrick M'Mayi, Franklin Odhiambo, Mohamed Sessay, Michael Stanley-Jones, Simone Targetti Ferri, Kaisa Uusimaa, Clarice Wilson.

Production and Data Support
Misha Alberizzi, Matthew Billot, Alexandre Caldas, Jillian Campbell, Ludgarde Coppens, Ananda Dias, Angeline Djampou, Dany Ghafari, Virginia Gitari, Florence Kahiro, Isabell Kempf, Sera Kinoyan, Ian Magero, Nada Matta, Pascil Muchesia, Josephine Mule, Jane Mureithi, Theuri Mwangi, Immaculate Mwololo, Samuel Opiyo, Audrey Ringler, Jinhua Zhang.

Peer-review System and Database
Ahmed Abdelrehim (CEDARE), Tesfaye Demissie (UNECA), Clever Mafuta (GRID-Arendal), Mayar Sabet (CEDARE), Ayman Soliman (CEDARE), Viola Sawiris (CEDARE), Youssef Younis (CEDARE).

Maps and Graphics
Fang Chen (CASEarth/RADI), Pouran Ghaffarpour (UNON Publishing), Catherine Kimeu (UNON Publishing), Samuel Kinyanjui (UNON Publishing), Jie Liu (CASEarth/RADI), Jennifer Odallo (UNON Publishing), Audrey Ringler, Jinita Shah (UNON Publishing), Zeeshan Shirazi (CASEarth/RADI), Lei Wang (CASEarth/RADI).

Editorial Team
Markus MacGill (Green Ink Limited), Clare Pedrick (Green Ink Limited), Anne Sweetmore (Green Ink Limited), Tim Woods (Green Ink Limited).

Design and Layout
Jennifer Odallo, Catherine Kimeu (UNON Publishing Services Section), Audrey Ringler (UN Environment)

Table of contents

PART A: STATE OF THE GLOBAL ENVIRONMENT

PART B: POLICIES, GOALS, OBJECTIVES AND ENVIRONMENTAL GOVERNANCE: AN ASSESSMENT OF THEIR EFFECTIVENESS

PART C: OUTLOOKS AND PATHWAYS TO A HEALTHY PLANET WITH HEALTHY PEOPLE

Figures

Land and Soil

Freshwater

The Way Forward

Future Data and Knowledge Needs

Annexes

Tables

Systemic Policy Approaches for Cross-cutting Issues

A Long-Term Vision for 2050

Future Developments Without Targeted Policies

Pathways Toward Sustainable development

Bottom-up Initiatives and Participatory Approaches for Outlooks

Future Data and Knowledge Needs

Annexes

Boxes

Land and Soil Policy

Systemic Policy Approaches for Cross-cutting Issues

Future Developments Without Targeted Policies

Pathways Toward Sustainable Development

Bottom-up Initiatives and Participatory Approaches for Outlooks

The Way Forward

Future Data and Knowledge Needs

Foreword

The sixth *Global Environment Outlook* is an essential check-up for our planet. Like any good medical examination, there is a clear prognosis of what will happen if we continue with business as usual and a set of recommended actions to put things right. *GEO-6* provides both a statement of the problems and a how-to guide to advance us on the path set out in the 2030 Agenda and the 17 Sustainable Development Goals.

The theme, "Healthy Planet, Healthy People", highlights the inextricable link between the environment and our survival and progress. The challenges outlined are multiple. From climate change to the extinction of species, economies too dependent on the wasteful use of resources and unprecedented pressure on terrestrial and marine ecosystems, we are at a decisive moment in our role as custodians of the planet.

It is not all bad news. Many indicators point to progress on issues such as global hunger, access to clean water, sanitation and clean energy. We can also see some signs of the decoupling of environmental degradation and unsustainable resource use from economic growth, as well as unprecedented technological innovation.

The overall message, however, is that we need a significant shift in trajectory – indeed, the kind of transformational change prescribed by the Intergovernmental Panel on Climate Change in its recent report on limiting global warming to 1.5 degrees.

GEO-6 details both the perils of delaying action and the opportunities that exist to make sustainable development a reality. We have the necessary policy guidance and the science that underpins it. The only missing ingredient for success is our collective resolve.

António Guterres
Secretary-General of the United Nations

January 2019

Foreword

"Grow now, clean up later". That's sadly been the business model for much of the world since the industrial revolution. It's as if looking after environment is a needless distraction, but ultimately a nice add-on when economies are doing well, and when luxuries can be afforded.

The *Global Environment Outlook*, now in its sixth edition, has been a key driver of the shift in this mindset. Grounded in the best available science and real-world case studies, it underscores the fact that a healthy planet is a prerequisite for healthy people, and that is in turn the foundation of any healthy economy. And most importantly, it shows how it's possible to win on all fronts.

In this drive towards a green economy, greater sustainability and the hope that we can thrive rather than merely survive, there has never been a more critical moment than now. The science and data are crystal clear on the multitude of challenges we face, but also the small window of opportunity we have to turn things around.

The *Global Environment Outlook* is therefore a roadmap to achieving the United Nations' Agenda 2030, in which hunger and poverty are consigned to history, and where biodiversity, oceans, land and freshwater are protected and restored to health.

It makes it clear that achieving this requires a transformation in human lifestyles and productive activities: our industry, agriculture, buildings, transport and the energy system which powers them. It means renewables like wind and solar must be the new norm, as must energy efficient, green buildings and transport. At the same time, this work also opens up huge economic opportunities – a new, better industrial revolution.

The task may be enormous, but we should also be inspired. Global environmental actions like the Montreal Protocol, our innovative defence against the hole in the Ozone layer, prove that we have the institutions and capacity to come together. The issue of plastics pollution has shown how diverse communities around the world – school children in Bali, coastal residents in Mumbai or surfers from Cornwall – can come together. After all, making the world a better place and cleaning up our act is a non-partisan, unifying cause we can all get behind.

Joyce Msuya
Acting Executive Director, UN Environment

Co-Chairs' Foreword

What is the Outlook for humanity? This sixth *Global Environment Outlook (GEO-6)* shows clearly that our species now stands at a crossroads. It can choose a challenging but navigable path towards a new golden age of sustainable development as envisaged by the United Nations' Agenda 2030 in which human hunger and poverty are consigned to history through the sustainable use of Earth's resources and the natural environment that leaves no-one behind. Or it can continue with current trends and practices, which will lead to a losing struggle against environmental disruptions, which threaten to overwhelm large parts of the world.

GEO-6 clearly identifies the problems that have to be addressed if this latter outcome is to be avoided. But it also points to the solutions to these problems, to ways in which the aspirations of the Sustainable Development Goals (SDGs) can be realised and Earth's air, biodiversity, oceans, land and freshwater restored to health, to the incalculable benefit of Earth's people: Healthy Planet, Healthy People, the title of *GEO-6*.

GEO-6 makes clear that achieving the SDGs will require a transformation in human lifestyles and productive activities: our industry, agriculture, buildings, transport and the energy system which powers them. This necessary transformation over the coming decades represents an enormous economic opportunity to those countries, policy makers and businesses who show the enterprise and innovative spirit to put in place the technologies, social practices and institutions that can make sustainable development a reality.

As co-chairs of the sixth *Global Environment Outlook* we have overseen the work of the tireless authors and experts who have contributed to this analysis. The scientific integrity of the process has been monitored by the Scientific Advisory Panel. The High Level Group helped us to find the language that can communicate to policymakers. The Secretariat provided the staying power to ensure that the entire process moved smoothly. Some States provided the necessary funding, encouraged us and hosted some of our meetings. We feel that the *GEO-6* has gathered the evidence to show what needs to be done, and what can be done. We respectfully present it to the world's decision makers, and ask them to face and address these challenges, for all of our sakes and future generations.

Joyeeta Gupta

Paul Ekins

Co-Chairs' Message

UN Environment's sixth *Global Environmental Outlook (GEO-6)* has reviewed the state of the health of the environment and the related health of the people, and the prospects for meeting the Sustainable Development Goals (SDGs) of the UN's Agenda 2030. As co-chairs, we draw six key messages from the report:

First, a healthy planet supports healthy people: A healthy planet is important for the health and well-being of all people. It directly supports the lives and livelihoods of 70 per cent of the Earth's population living in poverty [SPM 2.2.2; 6, 6.3.4, 6.6.3; boxes 6.5, 13.2], in particular those who are very poor, and it provides the basis for the production of the goods and services that are necessary for the global formal economy, which had a global GDP value of $US 75 trillion in 2017. Overall the biosphere is essential for human survival and civilization and its value to humans is therefore effectively infinite. However, for some purposes it is useful to calculate the monetary value of ecosystem goods and services; as an example the total global ecosystem services have been valued at $US (2007) 125 trillion/year [1.3.1]. This number does not capture the benefits of, for example, a climate suitable for agriculture or how melting glaciers affect the water security of more than a billion people [4.2.2], and so is clearly an underestimate. The value of lost ecosystem services between 1995 and 2011 have been estimated at $US 4-20 trillion (Costanza *et al.* 2014). More particularly, the value of pollinators which provide crucial services for commercial and non-commercial food production, has been estimated at $US 351 Billion/year to the commercial sector (Lautenbach *et al.* 2012).

Second, an unhealthy planet leads to unhealthy people: The planet is becoming increasingly unhealthy through the negative impacts of biodiversity loss (including pollinators, coral reefs and mangroves), climate change and other air pollution, water pollution, ocean pollution and depletion, and land use change. An unhealthy planet has huge social costs in terms of human health and well-being as well as on the formal economy and livelihoods worldwide. As with ecosystem goods and services, these costs are difficult to express comprehensively in monetary or other terms. However, *GEO-6* provides data that illustrate the sort of costs involved. For example, exposure to indoor/outdoor air and water pollution costs at least 9 million lives annually [4.1.1] including 300,000 in the G7 countries in 2015 (Organisation for Economic Co-operation and Development [OECD] 2017). About 2.8 million people died in 2015 from indoor air pollution [5.3.1] and about 2.8 million depend on unclean traditional biomass [21.2.3]. Many more millions suffer from ill-health and loss of livelihoods. Pollution-related costs have been estimated at $US 4.6 trillion annually [1.3.1]. 29 per cent of land is degraded affecting the lives and livelihoods of 1.3-3.2 billion people [8.3.2] and slow onset disasters are triggering migration [9.3.4; 9.7.3]. In 2016, 24.2 million people were internally displaced in 118 countries as a result of sudden-onset disasters [4.1.2]. Such disasters affected not just the poor countries, but also rich countries like the USA and Japan. Between 1995-2015, 700,000 people died and 1.7 billion people were affected by extreme weather events costing $US 1.4 trillion [4.1.2; Figure 4.2] (Centre for Research on the Epidemiology of Disasters and United Nations Office for Disaster Risk Reduction 2015). Between 2010 and 2016, an average of around 700 extreme events each year cost an average of $US 127 billion per annum. While 90 per cent of the losses came from high and upper-middle income countries, the less than 1 per cent of the losses from low-income countries amounted to around 1.5 per cent of their GDP, a much higher proportion than in high-income countries, and was almost all uninsured (Watts *et al.* 2017). The damage of climate variability and change to some small island regions is in the order of 1-8 per cent of GDP averaged over 1970-2010 (United Nations Environment Programme [UNEP] 2016a); if average global warming is not limited to 1.5°C, small island states and coastal populations may face existential threats. Water-related health costs are estimated at about $US140 billion in lost earnings and $US 56 billion in health costs annually (LiXil, Water Aid and Oxford Economics 2016). Such impacts are likely to exacerbate inequalities within and between countries, as opposed to reducing them in line with SDG10.

Third, the drivers and pressures leading to an unhealthy planet need to be addressed: The drivers and pressures result from a continuing failure to internalize environmental and health impacts into economic growth processes, technologies and city design. The pressures arise from massive use of chemicals (many with toxic health and environmental implications), huge waste streams (many largely unmanaged), committed and intensifying climate change impacts, and inequality which contributes to demographic changes and other drivers and pressures. The environmental footprint of rich people is significantly higher than that of poorer people. For example, the monthly emissions per capita in rich countries are mostly higher than the yearly emissions per capita in poorer countries (Ritchie and Roser 2018). The wealthiest countries consume 10 times the materials per person compared to the poorest countries (UNEP 2016b). While ideas around a green, healthy and inclusive economy aim to address these challenges, these ideas have yet to be systematically reflected in existing national policies. The IPCC 1.5°C report highlights the very limited time left to reduce greenhouse gas emissions to the extent necessary to limit average global warming to this level, thereby avoiding the potentially very expensive adaptation costs that will otherwise be required (Intergovernmental Panel on Climate Change 2018).

Fourth, current science justifies policy action now, but more detailed knowledge can enable more refined and preemptive policy. Existing knowledge is sufficient to mobilize action now [1,2, 4-9]. New knowledge including disaggregated data from earth observation, in-situ data, citizen science, ground truthing and indigenous and local knowledge are necessary in national policy and accounting more broadly [3]. There are major benefits in accounting systems that register the details about who causes damage to the environment, how and why; what is the extent of nature's contributions to humans, the loss of ecosystem goods and services; and who is affected [Figure 3.6]. Statistics and accounting systems also need to recognize the realities of the predominantly poor people in the informal economy, who are often particularly dependent on nature's contributions to people, and hence more vulnerable to environmental degradation.

Fifth, environmental policy is necessary but inadequate by itself to address systemic ecological problems, solutions to which require a more holistic approach. Current (inter) national policies are not on track to address the key environmental challenges effectively and equitably, in line with the aspirations of the SDGs. Environmental considerations need to be integrated into all policy areas, such that the potential and actual implications for natural resources and the environment are robustly included in policies for economic growth, technological development and urban design, so that there is effective long-term decoupling between economic growth, resource use and environmental degradation. Climate mitigation needs to be accompanied by policy for the equitable adaptation to committed climate change. Policies will only be effective if they are well designed, involving clear goals and flexible mixes of policy, including monitoring, instruments aimed at achieving them [12-17] and when access to judicial remedies are available [23.3; 23.11; 24.2]. Such a holistic approach need not require additional economic costs. If 2 per cent of global GDP is invested in maintaining and restoring natural capital, it could deliver the same economic growth outcome as a similar investment along current lines [18.1]. The health benefits from reduced air pollution of achieving the 2°C target could be 1.4-2.5 times the cost of mitigation, the higher figure involving benefits of $US 54.1 trillion for a global expenditure of $US 22.1 trillion. Moving from a 2°C to a 1.5°C target would generate further substantial health benefits for China and India [Box 24.1]. Food security could be enhanced if food wastage, currently running at 33 per cent globally, is curtailed [SPM 2.2.4].

Sixth, healthy people, a healthy planet and a healthy economy can be mutually supportive: Healthy diets (less meat) and lifestyles, healthy cities with good waste management (2 out of five people lack access to waste disposal services [SPM 2.2.6; 4.4.1]) and the use of green infrastructure in built-up areas, and healthy mobility can increase labour productivity, reduce the need for land for agriculture (e.g. meat production currently uses 77 per cent of agricultural land [SPM 2.2.4; 8.5.1, 8.5.3]) and reduce the costs associated with urban congestion and transport-related pollution and address the potential trade-offs between land for food/biofuel and biodiversity protection (OECD 2017). Technological and social innovation that supports environmentally sound economic development provides a viable and attractive alternative to the 'grow now, clean up later' practices of the past. In addition, a healthy people approach requires implementation of the rights of access to clean water and food, tenure rights, and gender equality. Millions of lives could be saved and livelihoods improved by access to clean air, water, fuel and food. Secure tenure rights for poor and indigenous people would enhance their ability to protect biodiversity and the different ecosystems that sustain them – for example, indigenous and poor people live on 22 per cent of the land that supports 80 per cent of global biodiversity (Sobrevila 2008) generating billions of dollars' worth of carbon sequestration, reduced pollution, clean water, erosion control, etc. (SPM 2.2.4; 8.5.3]. If gender equality is promoted, including the right to inherit and own land, then food security and many health issues relating especially to women and children could be better addressed [4.1.12]. Embracing the urgent and transformative changes that are required to accelerate the transition to a more equitable and environmentally sustainable economy, and a healthier society, through top down policy guidance and bottom-up initiatives will underpin the well-being and prosperity of countries and their people now and in the future.

Joyeeta Gupta

Paul Ekins

References

Centre for Research on the Epidemiology of Disasters and United Nations Office for Disaster Risk Reduction (2015). *The Human Cost of Weather-Related Disasters 1995-2015.* https://www.unisdr.org/files/46796_cop21weatherdisastersreport2015.pdf.

Costanza, R., de Groot, R., Sutton, P., van der Ploeg, S., Anderson, S.J., Kubiszewski, I. *et al.* (2014). Changes in the global value of ecosystem services. *Global Environmental Change* 26, 152-158. https://doi.org/10.1016/j.gloenvcha.2014.04.002.

Intergovernmental Panel on Climate Change (2018). *Global Warming of 1.5 °C an IPCC Special Report on the Impacts of Global Warming of 1.5 °C Above Pre-Industrial Levels and Related Global Greenhouse Gas Emission Pathways, in the Context of Strengthening the Global Response to the Threat of Climate Change, Sustainable Development, and Efforts to Eradicate Poverty.* http://www.ipcc.ch/report/sr15/.

Lautenbach, S., Seppelt, R., Liebscher, J. and Dormann, C.F. (2012). Spatial and temporal trends of global pollination benefit. *PLOS ONE* 7(4), e35954. https://doi.org/10.1371/journal.pone.0035954.

LiXil, Water Aid and Oxford Economics (2016). *The True Cost of Poor Sanitation.* https://www.lixil.com/en/sustainability/pdf/the_true_cost_of_poor_sanitation_e.pdf.

Organisation for Economic Co-operation and Development (2017). *Healthy People, Healthy Planet: The Role of Health Systems in Promoting Healthier Lifestyles and a Greener Future.* Paris. https://www.oecd.org/health/health-systems/Healthy-people-healthy-planet.pdf.

Ritchie, H. and Roser, M. (2018). CO_2 and other greenhouse gas emissions. 23 November. Our World in Data https://ourworldindata.org/co2-and-other-greenhouse-gas-emissions.

Sobrevila, C. (2008). *The Role of Indigenous Peoples in Biodiversity Conservation: The Natural but Often Forgotten Partners.* Washington, D.C.: World Bank. https://siteresources.worldbank.org/INTBIODIVERSITY/Resources/RoleofIndigenousPeoplesinBiodiversityConservation.pdf.

United Nations Environment Programme (2016a). *GEO-6 Regional Assessment for Asia and the Pacific.* Nairobi. http://wedocs.unep.org/bitstream/handle/20.500.11822/7548/GEO_Asia_Pacific_201611.pdf?isAllowed=y&sequence=1.

United Nations Environment Programme (2016b). *Global Material Flows and Resource Productivity: Assessment Report for the UNEP International Resource Panel.* Schandl, H., Fischer-Kowalski, M., West, J., Giljum, S., Dittrich, M., Eisenmenger, N. *et al.* (eds.). http://wedocs.unep.org/bitstream/handle/20.500.11822/21557/global_material_flows_full_report_english.pdf?sequence=1&isAllowed=y.

Watts, N., Amann, M., Ayeb-Karlsson, S., Belesova, K., Bouley, T., Boykoff, M. *et al.* (2017). The lancet countdown on health and climate change: From 25 years of inaction to a global transformation for public health. *The Lancet* 391(10120), 581-630. https://doi.org/10.1016/S0140-6736(17)32464-9.

Setting the Stage

Chapter 1

Introduction and Context

Coordinating Lead Authors: Mark Elder (Institute for Global Environmental Strategies), Christian Loewe (German Environment Agency)

1.1 GEO-6: Healthy Planet, Healthy People – humanity's transformative challenge

Providing a decent life and well-being for nearly 10 billion[1] people by 2050, without further compromising the ecological limits of our planet and its benefits, is one of the most serious challenges and responsibilities humanity has ever faced. People worldwide rely on the smooth functioning of Earth's natural life-support systems, in different ways and in different contexts. A healthy planet is a necessary foundation for the overall well-being and further advancement of humanity (United Nations 2015a; Organisation for Economic Co-operation and Development [OECD] 2017a).

Under the theme of 'Healthy Planet, Healthy People,' the sixth Global Environment Outlook (GEO-6) is an integrated assessment which considers various scientific perspectives and inputs from across the world in a holistic manner. The assessment urges the world's decision makers and all citizens to apply the principles of sustainable development to help ensure that Earth's environment remains the foundation of society and of people's well-being and resilience.

GEO-6 aims to answer the following questions:

- What is the state of the global environment, how is it changing, and what are the major factors and drivers, both positive and negative, influencing these changes?
- How are people and their livelihoods affecting and affected by environmental change in terms of health, economic prosperity, social equity, food security and overall well-being?
- Are environmental benefits, responsibilities and risks distributed fairly across different regions, socioeconomic groups and genders?
- What are the main responses and policy measures that have been taken to strengthen environmental protection and governance at various levels? How effective have they been in terms of improving environmental quality, and resource efficiency?
- What are the possible pathways, critical opportunities and policies, including Multilateral Environmental Agreements (MEAs) and Sustainable Development Goals (SDGs), to transform the global human-environment system to become more sustainable and contribute to a healthy planet for healthy people? What are the likely consequences if no additional actions are taken?

The first three points above are addressed by the introductory chapters and those in Part A of this report. The chapters in Part B consider the fourth point, on policy effectiveness, and the final point, on the most promising future pathways, is covered in Part C.

GEO-6 comes at a time of great uncertainty about the current trajectory of global human development (United States National Intelligence Council 2017). One major reason is that over the last few decades, human activities, such as human-caused climate change and other human impacts on ecosystems, have transformed the Earth's natural systems, exceeding their capacity and disrupting their self-regulatory mechanisms, with irreversible consequences for global humanity (Intergovernmental Panel on Climate Change [IPCC] 2014). Humanity has already been seriously affected by ongoing systemic ecological changes, such as climate change and land use change (especially deforestation). These have reached the point that the ecological foundations of human society and natural systems that support other species and provide invaluable ecosystem services are in great danger (Millennium Ecosystem Assessment 2005).

Human activities are causing increasing amounts of pollution, to the extent that this is now recognised as the biggest single risk to human health worldwide (Landrigan *et al.* 2018). Continuing to live on the brink of or outside of ecological limits, from the global to the local, will make it dramatically more difficult to achieve prosperity, justice, equity and a healthy life for all (Crutzen and Stoermer 2000; Crutzen 2002; Steffen, Crutzen and McNeill 2007; Steffen *et al.* 2011; Steffen *et al.* 2015; Steffen *et al.* 2018). The need for humanity to remain within the planetary boundaries' safe operating space and the need to eradicate poverty and accelerate social and economic development are linked by the concept of "a safe and just space for humanity" (Raworth 2012).

To cope with this range of human-induced damages, including climate change, deforestation, desertification, loss of biodiversity, scarcity of natural resources, pollution, and the consequent natural and the associated environmental impacts, is a great challenge. While many old and new societal contradictions and conflicts have to be solved simultaneously (Beck 2009; Beck 2015; Raskin 2016), these accumulative and omnipresent challenges should be addressed as humanity's transformative challenge (Beck 2009), by creating opportunities for further human development which achieve human well-being. This would be, where the universally applied principles of sustainability govern the pathway towards 'Healthy Planet, Healthy People', with no one left behind and endeavouring to reach the furthest behind first (United Nations 2015a).

GEO-6 addresses this transformative challenge, which is taken up by the United Nations 2030 Agenda for Sustainable Development (2030 Agenda) and its 17 SDGs. Transforming human-environment interactions (and related human-human interactions), especially consumption and production patterns and lifestyles, towards sustainability requires a better information base and new, diversified knowledge of planetary systems (Steffen 2000; Schellnhuber *et al.* eds. 2004) and transformative processes within globalized social and economic systems (Schneidewind 2013). This includes the cultural dynamics and ethical foundation of human perceptions and understanding of 'nature and environmental sustainability' (Morton 2009; Lammel *et al.* 2013; Díaz *et al.* 2015; Intergovernmental Science-Policy Platform on Biodiversity and Ecosystem Services [IPBES] 2015; Pascual *et al.* 2017).

The increasing body of global environmental assessments undertaken by international organizations in cooperation with the global science community and UN Member States provides the knowledge to understand the vital inter-connections and accelerating dynamics of natural ecosystems, socio-ecological systems and the dependence of human life on healthy and natural ecosystems. Increasing use of Earth observation techniques, from outer space and on Earth, in combination

[1] Throughout this publication the term 'billion' refers to 1000 million.

with new tools for data analysis, disciplines like environmental accounting (e.g. Kim and Kim eds. 2016), and environmental economics (Siebert 2008; Wiesmeth 2012; Ghosh *et al.* eds. 2016), has revolutionized our ability to recognize patterns of what causes environmental change and how it impacts life (Chuvieco ed. 2008; Tomás and Li 2017; Mathieu and Aubrecht eds. 2018).

Integrated and systems-based approaches (i.e. those that consider multiple benefits at the same time) enable cross-linkages to be explored and system-wide effects to be managed, so that policies can effectively support a number of social, economic and environmental goals to support human well-being, ensuring that various preconditions for this well-being are in place. These new scientific approaches and methods, including the study of cross-cutting inter-relationships between many areas, facilitate the preparation of more appropriate, equitable and effective policy responses, including shifting investment, production, distribution and consumption towards more sustainable approaches, and the development of better governance capacities at multiple scales. The GEO-6 assessment endeavours to support the vision that equal opportunities for prosperity and well-being for all, within the Earth's ecological limits, will be possible through sustainable development pathways that are shared and pursued globally.

GEO-6 is intended to be solution-oriented, with these solutions drawing on facts and statistics. Based on multidisciplinary perspectives from various scientific fields, GEO-6 also provides an interpretative framework and tells stories, including successes, failures and aspirations, to help people, governments and the global community work to prevent and repair environmental damage and respond more effectively to environmental changes and opportunities. GEO-6 highlights existing evidence of these environmental changes and reflects on possible pathways and critical opportunities for transformation of the global human-environmental system to become more sustainable in the mid to long term (2030/2050).

GEO-6 is entitled 'Healthy Planet, Healthy People', a conceptual approach that considers the human dimensions for achieving a healthy planet. It underlines the importance of maintaining the integrity of ecosystems and recognizes their interlinkages with socioeconomic systems. It emphasizes that a healthy planet is a necessary foundation for human physical, psychological, social, economic and emotional health and well-being, and is therefore critical for achieving all the SDGs.

Raïssa Lara Lütolf/Unsplash

Figure 1.1: Choices to be made to achieve a healthy planet for healthy people

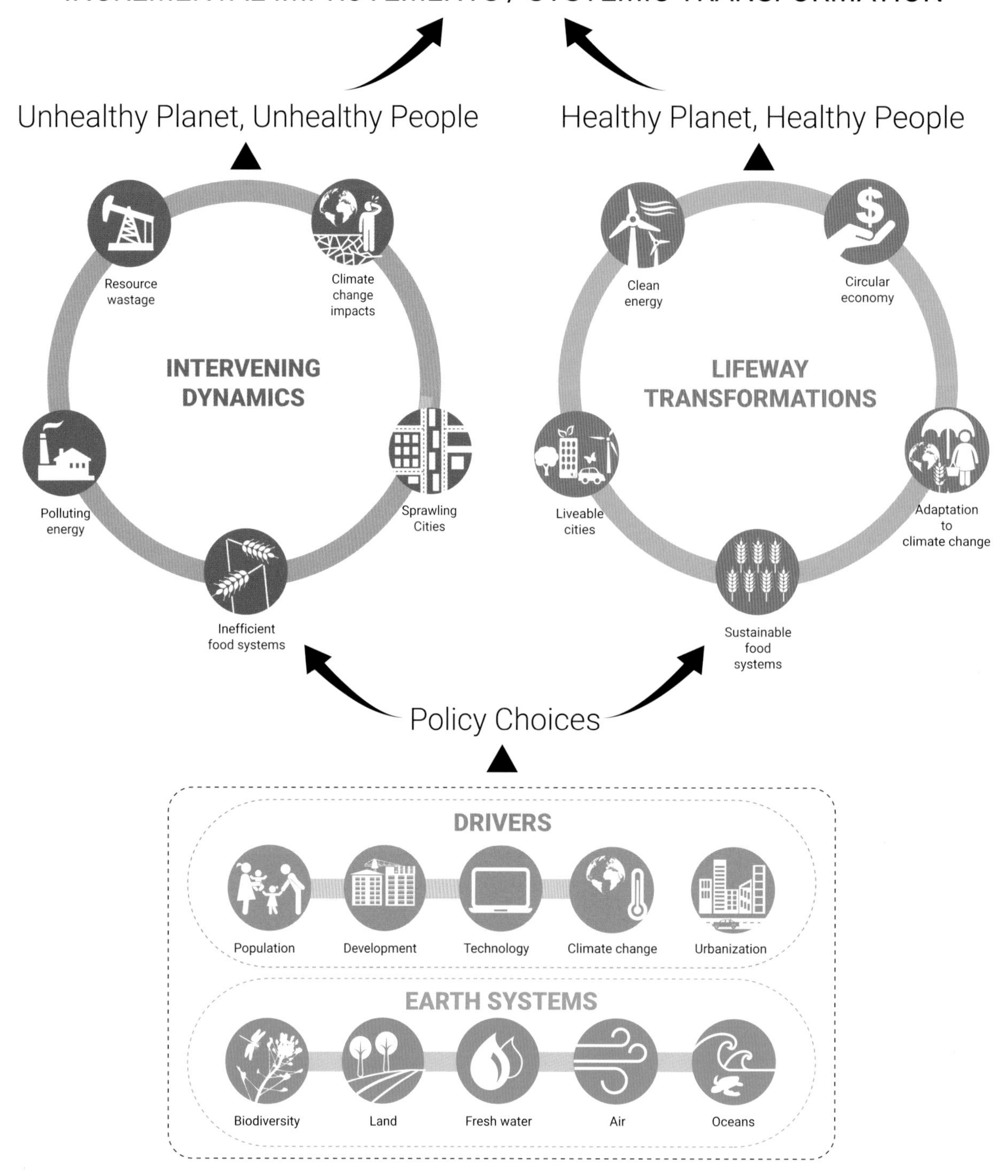

Figure 1.1 illustrates how a healthy planet contributes directly to healthier people by encouraging healthier lifestyles. Environmental degradation increases the burden of disease through exposure to harmful pollutants, as well as through reduced access to the ecosystem contributions from nature. Avoiding these problems will require protecting natural capital through detoxification, decarbonization, dematerialization and restoration of ecosystems to enhance planetary and human well-being.

A healthy planet requires protection and sustainable management of natural capital, in the form of nature's contributions to people, and human capital. People's opportunities in life are affected by humanity's ability to generate sustainable, long-term economic and social prosperity from human, physical and natural assets, the extent of environmental degradation and resource depletion, pollution and climate impacts, in addition to disparities in income and wealth.

This report recognizes that the environmental, economic and social equity dimensions are integrally linked, as they are in the SDGs with their overarching objective to 'Leave no one behind', and that all SDGs are rooted in human rights and dignity. Furthermore, many SDGs have environmental targets, some of which have equity components. Throughout GEO-6, evidence is presented of how fundamentally nature's contributions to people underpin human health and well-being. The SDGs recognize that inequality, including poverty and gender discrimination, results in a sizeable waste of human productivity and prosperity, and limits the scope for effective and accountable civic governance, quite apart from the ethical dimension of fairness and opportunity. Human resources are being underutilized and are not contributing to the sum total of human innovation required to help us live sustainably, demonstrated by the continued poverty in many parts of the world, which Agenda 2030 aims to eradicate (World Bank 2016a). The SDGs also recognize that disparities in access to resources, ecosystem services, income and wealth play an important role in shaping people's opportunities in life (Whitmee *et al.* 2015; OECD 2017), disproportionately affecting women and girls, as well as poor people.

1.2 UNEP's flagship assessment to deliver the environmental dimension of the 2030 agenda

Recognizing these important challenges, governments of the world have sought to better understand the interrelationships across the environmental dimension of the Sustainable Development Goals by requesting the preparation of a sixth edition of the Global Environment Outlook.

1.2.1 Mandate

Member States attending the first session of the United Nations Environment Assembly (UNEA-1) in Nairobi, June 2014, requested:

> ... the Executive Director, within the programme of work and budget, to undertake the preparation of the sixth Global Environment Outlook (GEO-6), supported by UNEP Live, with the scope, objectives and procedures of GEO-6 to be defined by a transparent global intergovernmental and multi-stakeholder consultation informed by document UNEP/EA.1/INF/14, resulting in a scientifically credible, peer-reviewed GEO-6 and its accompanying summary for policy makers, to be endorsed by the United Nations Environment Assembly no later than 2018.

As requested by Member States (UNEP/EA.1/4) and based on the decision (UNEP/IGMS.2 Rev.2) made by the Global Intergovernmental and Multi-stakeholder Consultation in Berlin, 21-23 October 2014, GEO-6 builds on six regional assessments that were conducted in a similar fashion to the global GEO-6 process and launched in May-June 2016. In addition, the main messages of GEO-6 are compiled in an accompanying Summary for Policymakers, which is drafted by the authors of the main report and negotiated by the governments. See Annex 1-1 for more details on UN Environment's mandate to produce the sixth Global Environment Outlook.

More recently, recognizing that the date of the fourth session of the UN Environment Assembly (UNEA-4) would be shifted to 11-15 March 2019, Member States decided at UNEA-3 to:

> ... [Request] the Executive Director to issue the sixth Global Environment Outlook report at least three months before the fourth session of the United Nations Environment Assembly;
>
> Also [request] the Executive Director to schedule the negotiations on the summary for policymakers at least six weeks in advance of the fourth session of the United Nations Environment Assembly and to present the sixth Global Environment Outlook report and its accompanying summary for policymakers for consideration and possible endorsement by the Environment Assembly at its fourth session.

With these decisions, the delivery date of the embargoed version of the main report is now the week of 5 December 2018, and the delivery date of the adopted and translated version of the Summary for Policymakers is 28 January 2019.

1.2.2 Role of GEO-6

GEO-6 comes at a critical time for global development, and it will build on the knowledge and experience gained from previous GEOs. Previous GEO editions have already presented substantial evidence that environmental degradation, even within the planetary limits of the Earth's ability to support human civilization, has undermined current and future development, and threatened different aspects of human well-being (United Nations Environment Programme [UNEP] 2007; UNEP 2012a).

GEO-6 explores some issues further, attempting to show the interlinkages across environmental challenges and geo-political, economic, industrial, social, technological and cultural issues, while considering potential transformative sustainable development pathways and policies for achieving the SDGs and other Internationally Agreed Environmental Goals (IAEG). In this respect, GEO-6 aims to apply a wider scope to the discussion of global environmental security (Matthew *et al.* 2010; UNEP *et al.* 2013)

Box 1.1: Concept of Well-being

Human well-being is assumed to have multiple constituents, including:

- the basic material for a good life, such as secure and adequate livelihoods,
- enough food at all times, shelter, clothing, and access to goods;
- health, including feeling well and having a healthy physical environment, such as clean air and access to clean water; good social relations, including social cohesion, mutual respect, and the ability to help others and provide for children;
- security, including secure access to natural and other resources, personal safety, and security from natural and human-made disasters; and freedom of choice and action, including the opportunity to achieve what an individual values doing and being.

Freedom of choice and action is influenced by other constituents of well-being (as well as by other factors, notably education) and is also a precondition for achieving other components of well-being, particularly with respect to equity and fairness.

Source: Millennium Ecosystem Assessment 2005

Additionally, GEO-6 attempts to further strengthen understanding of the macro perspective of socio-ecological systems (including economics), and also to use a more people-centred approach (UNEP 2016a). GEO-6 underlines that people are part of ecosystems and depend on them, emphasizing the importance of conserving nature not only for its intrinsic value, but also because it is crucial for the well-being of humanity. Such an approach is urgently needed to help address the vulnerability and different conditions and capabilities enabling people to react to hazards and disruptions in daily life (resilience) (Millennium Ecosystem Assessment 2005). With this knowledge, it is hoped that people will be encouraged to respond to the challenge by changing their behaviour as citizens, consumers, voters, politicians, religious leaders and business leaders (UNEP 2016b).

GEO-6 highlights an updated understanding of the relationship between the environment and the economy, which is a foundation of the people-centred approach. This emphasizes nature's contribution to people, the environmental functions that support human well-being (including the benefits of environmental investments, innovations and technologies), as well as the high costs of inaction, business as usual, and stranded assets.

Furthermore, this perspective within GEO-6 helps to better inform future policy decisions by addressing complex distributional impacts and conflicts as the new baseline to design sustainable development policies and governance systems associated with implementation of the 2030 Agenda (World Bank 2016b). Creating such knowledge and its evidence base through this assessment will help to better communicate possible policies, actions and investments that could be used by governments, as well as other stakeholders and citizens, to address current and future development challenges, as well as to explain the benefits of taking such actions. How this perspective is integrated into the GEO-6 assessment is further explained in Section 1.7.

1.3 GEO-6 in a changing global context

The world is facing a wide range of economic, social, cultural and political/military security challenges (World Economic Forum 2017). Despite significant global progress in economic development and poverty reduction in some regions, a large portion of the population in many areas suffers from poverty or extreme poverty, and many people who are not impoverished are still concerned about economic security and future life opportunities. Some areas are experiencing social friction, growing inequality, poor governance, cultural erosion, reactions against globalization, political instability, large numbers of refugees, large-scale migration and violent conflicts due to these economic and social insecurities, injustices and corruption.

Many of these global economic, social and political/military security challenges are related to the environment in terms of causes, impacts and possible solutions. Moreover, recent scientific concepts of environmental safeguards for society, f or example planetary boundaries (Rockström *et al.* 2009; Steffen *et al.* 2011; Steffen *et al.* 2015), explain that the environment is the foundation for human life on Earth. Current methods of generating material prosperity have undermined ecosystem health and caused massive environmental damage, crossing several of these planetary boundaries, to the point where the development of human societies and the 'safe operating space' for human life on Earth is at risk. In this planetary boundaries framework, environmental problems are considered to be inherent systemic problems of humans' deep-rooted transformation of nature and ongoing cultural dynamics, and are not seen only as collateral damage of societal development (Steffen 2000). Biodiversity is also critical for human well-being (Secretariat of the Convention on Biological Diversity [CBD] 2014), as are ecosystem services more broadly (Millennium Ecosystem Assessment 2005).

Clearly, the functions of environmental policy have expanded, and it now contributes to political/military security, economic and social policy and other development activities. Likewise, these other policy areas also have a major influence on the state of the environment. A key implication of these interlinkages is the need for an integrated approach to address environmental, economic and social problems holistically (United Nations 2015b; Jetzkowitz *et al.* 2018). GEO-6 aims to integrate the linkages between the environment, social and economic security, global justice and human well-being, to promote a new framework for sustainability to be an integral part of all aspects of global, regional and national development (United Nations Educational, Scientific and Cultural Organization [UNESCO] 2014a; Lehmann *et al.* 2015; UNEP 2016a; UNESCO 2016).

1.3.1 Environmental and economic challenges and opportunities

The environment is closely related, in both positive and negative ways, to key economic issues such as poverty, prosperity, jobs, production patterns, innovation, and resource availability/scarcity. On one hand, the economy is a major source of environmental problems, while environmental problems are increasingly causing economic losses. Recent articles have noted that "welfare losses due to pollution are estimated to amount to US$4.6 trillion per year," which is "about 6.2 per cent of global economic output" (Landrigan *et al.* 2018, p. 462). Economically, countries are often still guided by an approach of 'grow now, clean up later'. This report will show that this is simply not sustainable in a world already crossing planetary boundaries on a number of dimensions, a situation which threatens to undermine economic growth if not addressed. In addition, this option is likely to prove far more expensive for most countries, because it is often costlier to clean up later than prevent damage in the first place; it creates stranded assets which lose their value, and is now leading to irreversible negative impacts, including on human health. This renders an economy unproductive and uncompetitive compared with a flexible and proactive approach, capable of managing the transition to a sustainable, innovative and resource-efficient economy that can take advantage of domestic and export market opportunities in fast-growing, environmentally aware markets.

On the other hand, protecting the environment, as well as preventing and mitigating the impacts of pollution, are also major sources of economic opportunity, providing jobs, reducing poverty, driving innovation and addressing resource availability/scarcity and depletion. Positive synergies between the economy and the environment are now more widely recognized (Porter and van der Linde 1995; The Economics of Ecosystems and Biodiversity [TEEB] 2010; OECD 2011; UNEP 2011a; UNEP 2011b; Hepburn and Bowen 2012; United Nations Economic and Social Commission for Asia and the Pacific [UNESCAP] and Korea International Cooperation Agency [KOICA] 2012; Global Commission on the Economy and Climate 2014; Altenburg and Assmann 2017; OECD 2017b), compared with the view that trade-offs exist between the environment and the economy.

The global economic value of ecosystem services was estimated to be about US$ 125 trillion in 2011 (in 2007 US$[2]) (Costanza *et al.* 2014). Still, more effort is needed to communicate this message about positive synergies, as the perspective of the trade-off between the economy and the environment is still reinforced by current methods of calculating economic growth, which generally externalize environmental impacts and emphasize short-term, rather than long-term, perspectives. Especially in nations/regions where people have anxieties about jobs, wages and economic prosperity, there is a risk of weakening support for environmental protection and MEAs if the linkages among these concerns are not well understood. GEO-6 aims to contribute to a more thorough assessment of costs and benefits, as well as the cost-effectiveness of environmental policies and practices, and how they are distributed in society.

Many businesses around the world now understand that environmental problems pose major challenges to their operations, and that addressing them presents significant business opportunities, for example through circular economy business practices (see Chapter 17), in the context of sustainable consumption and production (Lacy and Rutqvist 2015; Ghisellini, Cialani and Ulgiati 2016; Murray, Skene and Haynes 2017; Hopkinson, Zils and Hawkins 2018, see section 17.5 of this report), as well as enabling increases in productivity and profitability (at least in the initial stages of waste reduction and efficiency improvements). It also avoids major liabilities and burdens for future generations. Prominent business groups, such as the World Business Council for Sustainable Development and UN Global Compact, promote environmental sustainability at all levels of society and decision-making.

Environmental protection and environmental business can also be major sources of jobs (International Labour Office [ILO] 2016). In the global energy sector, renewable energy sources are growing much faster than expected, and global annual investment in these systems is now greater than investment in fossil fuels (Renewable Energy Policy Network for the 21st Century [REN21] 2018). It is suggested that 'clean' energy (renewable and low-carbon energy) and energy efficiency may have more job creation potential than coal and natural gas (Wei, Patadia and Kammen 2010; Garrett-Peltier 2017; International Renewable Energy Agency [IRENA] 2018; Yihdego, Salem and Pudza 2017). Most recently, in the United States of America, the solar industry accounts for more than twice as many jobs as coal (United States Department of Energy 2017).

Still, many economic trends pose challenges for addressing environmental problems. Many governments face challenges in raising revenue, and deregulation initiatives often focus on weakening environmental standards/regulations (Castree 2008; Steinebach and Knill 2017). The Addis Ababa Action Agenda, which addresses the means of implementation for sustainable development in general, including the SDGs, suggests ways to help governments strengthen their domestic financing capacity (United Nations 2015c).

Globalization has been an overall trend for several decades, and its possible environmental effects have been a major research focus. However, the linkages between economic development and the environment are very complex and difficult to summarize. Some aspects of globalization may worsen environmental problems, while others may be beneficial (Boyce 2004; Gallagher 2009; Clapp and Dauvergne 2011; Newell and Roberts eds. 2016). Identifying such trade-offs and synergies is a major element of the GEO-6 assessment (see chapters 4 and 17).

1.3.2 Environment and social challenges and opportunities

Environmental issues are closely related to social issues such as hunger, consumption patterns, health, education, inequality, gender gaps, waste and sanitation, refugees, migration, conflicts and intolerance. For example, hunger and food, addressed in SDG 2, are linked to agriculture, which in turn is linked to the environment, especially SDG target 2.4 on sustainable agriculture. Environmental pollution harms agriculture, while a cleaner environment will help to improve agriculture, nutrition and health (Landrigan *et al.* 2018).

Education promotes a healthier environment and vice versa (UNESCO 2014b; UNEP 2017a). Environmental pollution, biodiversity loss and climate change are important causes of health problems and environmental diseases, which in turn can negatively affect education and learning, especially

2 Readers should assume that all values in this report are nominal market values, unless it is stated to the contrary.

among children; they can also be a hindrance to employment among adults (Mohai *et al.* 2011; Zhang and Zhang 2018). In contrast, cleaning up, avoiding pollution, and protecting and restoring habitats are major opportunities to improve health, which in turn helps people lead fuller and more productive lives. Diseases related to air pollution caused 9 million premature deaths in 2015, accounting for 16 per cent of all deaths globally (Landrigan *et al.* 2018) while in some countries, hazardous air pollution has forced schools to close (Sastry 2002; Li *et al.* 2014; British Broadcasting Corporation [BBC] 2016; Reuters 2017).

The environment is also related to growing social inequality, including gender inequality, in many ways that may put burdens on poor or socially disadvantaged people. These can include unequal access to resources (e.g. land, water, food, seeds), uneven distribution of the impacts of environmental degradation (e.g. the health impacts of climate change and waste), job creation and loss due to shifting consumption and production between geographic areas, and uneven distribution of responsibilities with respect to addressing environmental challenges. Children are particularly susceptible to the negative health impacts of chemicals, due to their rapid growth and development and greater exposure relative to body weight.

In many cases, people's environmental impacts are related to their income levels (Moser and Kleinhückelkotten 2017). Wealthier people are more able than poor people to insulate themselves from environmental problems, while they have more potential to contribute to solutions through their greater resources and scope for lifestyle changes (UNEP 2016b). This is also related to the geographic, economic and social distribution of areas affected by environmental problems.

The drivers and pressures of environmental change, as well as its state and impacts, have people-centred aspects that need to be taken into account in order to develop effective and just policies in an Agenda 2030 world. This approach is needed to help address the vulnerability and different conditions and capabilities of people to react to hazards and disruptions in daily life (resilience) (Millennium Ecosystem Assessment 2005). People – poor and rich, women and men – affect and are affected differently by environmental changes and related risks. These differences play a major role in related political decisions (Serret and Johnstone eds. 2006; UNEP 2016b). Using this perspective, GEO-6 attempts to interpret how environmental 'equity' will be experienced by different people, so it can inform future policy decisions by addressing complex distributional impacts and conflicts. This perspective provides a new baseline to design sustainable development policies and governance systems for implementing the 2030 Agenda (World Bank 2016).

In Agenda 2030, the sustainable use of the environmental and natural resources is now understood to be complementary and necessary to "end poverty in all its forms everywhere" (SDG 1). Approximately 70 per cent of the world's poor people depend directly on natural resources for all or part of their livelihoods, particularly women and girls, as well as other marginalized groups. Efforts to eradicate poverty and ensure prosperity are directly linked to improving the management of both the environment and natural resources in an integrated way (TEEB 2010).

1.3.3 Environment and political/military security challenges and opportunities

Environmental problems such as land degradation (United Nations Convention to Combat Desertification [UNCCD] 2017) and resource scarcity and depletion, especially water, energy, food and biodiversity, have the potential to be major sources of conflict, security problems and migration (Homer-Dixon 1991; Homer-Dixon 1999; Barnett and Adger 2007; Gupta, Dellapenna and Heuvel 2016). Political/military security problems may be amplified by climate change effects. Water security is being compromised by pollution and unsustainable use, as well as demand exceeding sustainable supply, climate variability, droughts, flooding, etc. Climate change, including related weather extremes, and environmental degradation are already having a range of complex effects, especially in fragile states and ecosystems. For example, they worsen the problems of migrants and refugees (both within and between countries), which in turn contribute to increasing political uncertainty and instability worldwide. Environmental refugees displaced by environmental degradation may also suffer from health problems and difficulties maintaining their livelihoods.

Wars and conflicts are major sources of pollution, especially air, water and soil pollution, waste, greenhouse gases and land degradation. Likewise, addressing environmental problems may provide important opportunities to help address political/military security problems (Brown, Hammill and McLeman 2007; UNEP *et al.* 2013), including by helping to secure livelihoods and reduce the necessity for migration. International funding to war-torn states may be productively aimed at addressing environmental problems through development of sustainable infrastructure, including natural infrastructure and ecosystem restoration, and services such as waste, wastewater and resource management.

1.3.4 Resource availability and scarcity

Resource availability and scarcity problems clearly illustrate the tight interlinkages between economic, social, human, political/military security and environmental issues (Qasem 2010; UNEP 2011a; Pereira 2015). Resources have significant negative environmental and social/health impacts in their production and/or use, for example through mining or other extraction processes.

At the same time, they are important inputs to environmental solutions. Resources are important economic inputs and sources of jobs, and are used in products and services supporting human well-being. This is not just related to the key resources of water, energy and food, which have been extensively researched as 'nexus' issues (UNESCAP 2013; Food and Agriculture Organization [FAO] 2014; International Resource Panel 2015). Phosphorus (Cordell and White 2015) is a key input for food production and other important scarce resources including materials such as rare earth metals (Gupta and Krishnamurthy 2004; Abraham 2015; Graedel *et al.* 2015), are used in many industrial applications including key environmental technologies such as wind and solar energy as well as advanced batteries. These materials, and the many products made from them, also have important military applications. On the negative side, in addition to the environmental damage caused by their production, these resources are also scarce (Calvo, Valero and Valero 2017), leading to political/military security concerns related to securing their availability.

1.4 Environmental governance

Environmental governance is increasingly important at all levels, including global, regional, national and subnational (local, provincial, etc.) governments, as well as business and civil society stakeholders (Biermann *et al.* 2012; Biermann 2014; United Nations Economic Commission for Europe [UNECE] 2014; Patterson *et al.* 2015; Mortensen and Petersen 2017). New environmental governance challenges are emerging, such as the opening of the Arctic and the advent of new materials, while many old challenges have not been adequately addressed. Greater multi-stakeholder participation in governance is a major global trend, but there is a need for greater synergies between governments and civil society organizations. Many efforts have been made to develop more effective facilitation methods to enable this collaboration (Ansell and Gash 2008; UNECE 2014; Pattberg and Widerberg 2016; Dodds, Donoghue and Leiva-Roesch 2017). This includes new technologies and social media, and citizen science, which engages citizens in scientific research (Kobori *et al.* 2016, see section 25.2) which may be the only way to obtain some kinds of data. Also, governance within the private sector has become an important innovation space.

Environmental problems have always been very complex and closely related with other policy areas (Jordan and Lenschow 2010), but efforts to overcome separate sector/silo boundaries have not made sufficient progress (Adelle and Nilsson 2015). It is now more important than ever to promote the integration and coordination of environmental concerns with other development areas, such as economy, trade, health, water, energy, education, food systems and urban planning (FAO 2014; Le Blanc 2015; OECD 2015; Elder, Bengtsson and Akenji 2016; United Nations 2016; Scheyvens *et al.* 2017).

Moreover, ecosystem boundaries often do not correspond to geopolitical boundaries, so many environmental problems, especially those related to pollution, are often transboundary in nature, such as air pollution, freshwater contamination (surface and groundwater), marine pollution, wastewater, leakages of pollutants, dumping of hazardous and nuclear wastes and species loss. Because many of these transboundary problems are interrelated, there are extensive opportunities to take advantage of co-benefits from policy solutions, but these require greater cooperation and coordination across political boundaries.

Many efforts have been made to develop ways to improve environmental governance, ranging from stronger regulation and enabling policies to support voluntary actions, to stakeholder self-governance. The state has an important role in strengthening environmental governance, including by ratifying and implementing environmental conventions, supporting environmental research and supporting vulnerable populations. Still, the best way forward is not always clear, and further efforts are needed (Ansell and Gash 2008; Jordan 2008; Newig and Fritsch 2009; Biermann *et al.* 2012; Galaz *et al.* 2012; Biermann 2014; United Nations Development Programme [UNDP] 2014; Kanie, Andresen and Haas eds. 2014; Pattberg and Widerberg 2015; Pattberg and Zelli eds. 2016; Biermann, Kanie and Kim 2017).

1.5 The environmental dimension of the sustainable development goals, global environmental governance and multilateral environmental agreements

Until recent years, global environmental governance has mainly focused on MEAs (Najam, Papa and Taiyab 2006; Environment Canada, University of Joensuu and UNEP 2007; Kanie 2007), along with many regional and bilateral agreements (Balsiger and VanDeveer 2012). It has been estimated that there are over 1,300 MEAs and 2,200 bilateral environmental agreements (Mitchell 2018).

Despite these MEAs and five previous Global Environment Outlooks, the state of the environment remains troubled and has continued to deteriorate in many respects (Susskind and Ali 2015; UNEP 2012b), to the point where the environmental foundation for human society is increasingly at risk (Rockstrom *et al.* 2009; Steffen *et al.* 2011; Steffen *et al.* 2015). Moreover, some environmental pollutants, such as plastic waste, marine pollution, military-related waste and pesticides, remain largely unregulated at the global level. There has been insufficient progress in achieving sustainable consumption and production patterns.

Some international agreements and frameworks are working to deal with global problems more comprehensively, rather than focusing narrowly on specific environmental issues. They combine political, economic, social and environmental dimensions, while strengthening the environmental elements. These include the SDGs, the Paris Agreement on climate change, the Sendai Framework for Disaster Risk Reduction, the New Urban Agenda - Habitat III, and the United Nations Convention to Combat Desertification.

The SDGs and the 2030 Agenda are at the vanguard of this trend, bringing an integrated, holistic perspective to sustainable development. They link the environment with other dimensions of sustainable development in order to take advantage of synergies and minimize trade-offs between them. They also represent a major change from the Millennium Development Goals (MDGs). Not only are the SDGs universal and challenging all countries, they also offer a broad sustainability agenda, giving equal attention to social, economic and environmental issues; by contrast, the MDGs had a greater focus on the social agenda, paying insufficient attention to economic and environment issues. Thus, the environment is incorporated into the SDGs more extensively than it was into the MDGs (UNEP 2016c).

Furthermore, where the MDGs mainly aimed at poverty reduction in developing countries (with developed countries committing to a Global Partnership for Development), the 2030 Agenda is a universal one, with goals and targets to be achieved by all countries. According to UNEA, the 2030 Agenda "represents a paradigm shift to replace today's growth-based economic model with a new model that aims to achieve sustainable and equitable economies and societies worldwide" (UNEA 2016 p.1), noting that "ecosystems and the services they provide, such as food, water, disease management, climate regulation, and spiritual fulfilment are preconditions" for sustainable development, while "unsustainable patterns of production and consumption threaten our ability to achieve sustainable development".

1

The environment is represented in all the SDGs. More than half have a direct environmental focus or address the sustainable use of natural resources (UNEP 2016d). Many goals are directly related to the quality of the physical environment, e.g. water (SDG 6), climate (SDG 13), oceans (SDG 14) and land and biodiversity (SDG 15). Other goals are more indirectly related to the physical environment, e.g. via natural disasters (SDGs 1 and 11), food, hunger and agriculture (SDG 2), human health (SDG 3), energy (SDG 7), economic growth and employment (SDG 8), industry (SDG 9) and cities (SDG 11) (International Resource Panel 2014; International Resource Panel 2015; OECD 2015; Lucas *et al.* 2016). For example, SDG 8 emphasizes sustainable economic growth and decent jobs, while one of its targets calls for decoupling economic growth from environmental degradation and improving global resource efficiency in consumption. SDG 12 on sustainable consumption and production, SDG 16 on peace, justice and strong institutions, and SDG 17 on means of implementation are cross-cutting goals that support all other goals and their environmental dimensions. Clearly, the SDGs cannot be achieved without fundamental environmental progress. This is recognized in the 2030 Agenda, which directly calls for an integrated approach to sustainable development (International Resource Panel 2015).

Although the SDGs link the environment much more closely to other development areas, they do not comprehensively represent the global environmental agenda (Wackernagel, Hanscom and Lin 2017). Some important environmental problems are not well reflected in the SDGs, such as mining and natural resource extraction, and the links between gender and the environment (e.g. indoor air pollution from cooking; Elder and Zusman 2016). The climate goal (SDG 13) does not have a target or indicator directly related to the state of the climate, although it references the Paris Agreement, which does have such a target. Moreover, the environmental indicators are not as well developed as those for other areas, and there is less data available to quantify their impacts and/or progress towards achieving the related targets. Many targets have several dimensions, and often the dimension related to the environment is not included in the indicator(s). The SDGs address the goals of many MEAs, although few of the many IAEGs are directly mentioned in the SDGs.

Similarly to the SDGs, other major recent United Nations agreements and frameworks, such as the Paris Agreement on climate change, the Sendai Framework for Disaster Risk Reduction and the New Urban Agenda - Habitat III, require substantial contributions from all sectors and actors, as well as significant transformation of economic and social practices. Thus, like the SDGs, these agreements have a broad scope and should be implemented using an integrated approach.

Similarly, major non-United Nations global forums (e.g. the Group of Twenty [G20], the Group of Seven [G7] and the World Economic Forum) focus increasingly on environmental issues and associated risks, especially in relation to the SDGs. In 2015, the leaders and heads of states of the G7 met in Elmau, Germany, and agreed to decarbonize the world economy by the end of this century (G7 2015); at the Ise-Shima Summit in Japan, 2016, the G7 agreed to make concerted efforts to fulfil their SDG and Paris Agreement commitments. At the Taormina Summit in 2017 in Italy, all the G7 members reaffirmed their strong commitment to swiftly implement the Paris Agreement (except the United States of America, which was in the process of reviewing its related policies). The G7 has been holding environment ministers' meetings regularly. The G20 also adopted an SDG Action Plan (G20 2016).

Other major meetings of environment ministers include the BRICS (Brazil, Russian Federation, India, China, South Africa), the Asia-Pacific Ministerial Summit on the Environment, the African Ministerial Conference on the Environment, and the Tripartite Environment Ministers Meeting among China, Japan and the Republic of Korea. The SDGs provide a framework and common language to bring all these agreements and actions together.

The target and indicator-based approach, which was a key innovation of the MDGs, was also used by the SDGs, as well as by the Strategic Plan for Biodiversity 2011-2020, including the Aichi Biodiversity Targets developed under the CBD (Kanie and Biermann eds. 2017). Many felt that this approach made an important contribution to the MDGs' relative success in mobilizing action and support, although it also has some disadvantages (Fukuda-Parr, Yamin and Greenstein 2014). If this approach is implemented broadly in line with the spirit and language of the SDGs, and not in a narrow instrumental manner, then implementation and accountability may be strengthened (Biermann, Kanie and Kim 2017). Another major innovation for the SDGs and the Paris Agreement is that each country agreed to translate the global goals and targets into national targets and indicators; however, this will introduce the challenge/opportunity of accounting for progress.

Therefore, it is very important for GEO-6 to continue to focus global attention on MEAs, IAEGs and its new focus on SDGs and non-United Nations global forums. Still, implementation of some traditional MEAs may also benefit from a more integrated approach, possibly through greater linkages with the SDGs.

One of the main tasks of GEO-6 is to assess progress on the Internationally Agreed Environmental Goals (IEAG) that have been established by MEAs, highlighting gaps between the commitments and achievements of these agreements. More importantly, it will help to inform the global response and institutional capacity-building needed to address the increasing complexity and uncertainties associated with environmental problems and addressing them through global development. Given the urgency of the challenges associated with environment and development, and the limited financial and human resources available to address them, GEO-6 is focused on a holistic and integrated approach to assessment in order to leverage synergies across issues and minimize trade-offs, and to communicate the resulting knowledge.

1.6 GEO-6 in the context of other environmental assessments

To address environmental challenges effectively, their wider impacts on people, economies, societies, markets, institutions, justice, security and culture must be well understood. The GEO-6 process recognizes a need for participatory and integrated environmental assessments (IEAs), and for institutionalized tools and platforms to empower people, organizations and decision makers by co-developing information and relevant knowledge on the state and trends of the environment to inform policy action and adequate responses (UNEP 2015).

GEO-6 is part of the growing body of global environmental assessments (Mitchell *et al.* 2006; Kowarsch *et al.* 2014; Jabbour and Flachsland 2017; Kowarsch *et al.* 2017). Some of these assessments are, or include, regional assessments (e.g. the European Environment Agency's State of the Environment Report) or country-level assessments, while others focus on specific themes, such as the Global Gender Environment Outlook (UNEP 2016e). These assessments are typically conducted by international organizations and programmes, like UNEP through its Environment Under Review sub-programme (UNEP 2018) and create the needed evidence base that brings clarity and transparency to the main concerns facing the planet and humanity. This evidence base includes successes and failures in addressing these issues and, most importantly, provides options for actions to make sure that current and anticipated problems are equitably and effectively addressed. This action-oriented and stakeholder-focused approach has the desired attributes of incorporating feedback from decision makers into the knowledge development process and shortening the time for implementing the information and knowledge. Annex 1-2 lists the IEAs from which GEO-6 draws.

An IEA (such as GEO assessments) follows a common methodology and procedures to ensure the consistent application of relevant quality standards, and links science to policy by:

- analysing and synthesizing existing environmental, social and economic data to determine the state of the environment using the Drivers, Pressures, State, Impact, Response (DPSIR) framework, taking into account all ecosystem components and processes **(see Figure 1.2)**;
- determining risk and uncertainty in the information;
- identifying and assessing past and potential policy and management actions;
- providing guidance for decision makers on the consequences of various policy and management actions, including not taking any action (UNEP 2017b).

1.7 GEO-6 approach, theory of change and structure

1.7.1 Approach

Historically, the GEO process was established as part of the follow up to the adoption of Agenda 21 in 1992, with the aim of placing the status of the environment under permanent review (UNEP 1995 - UNEP Governing Council in its decision 18/27). Since the first GEO in 1997, its approach and structure have undergone several changes and improvements.

Figure 1.2: The DPSIR approach used in GEO-6

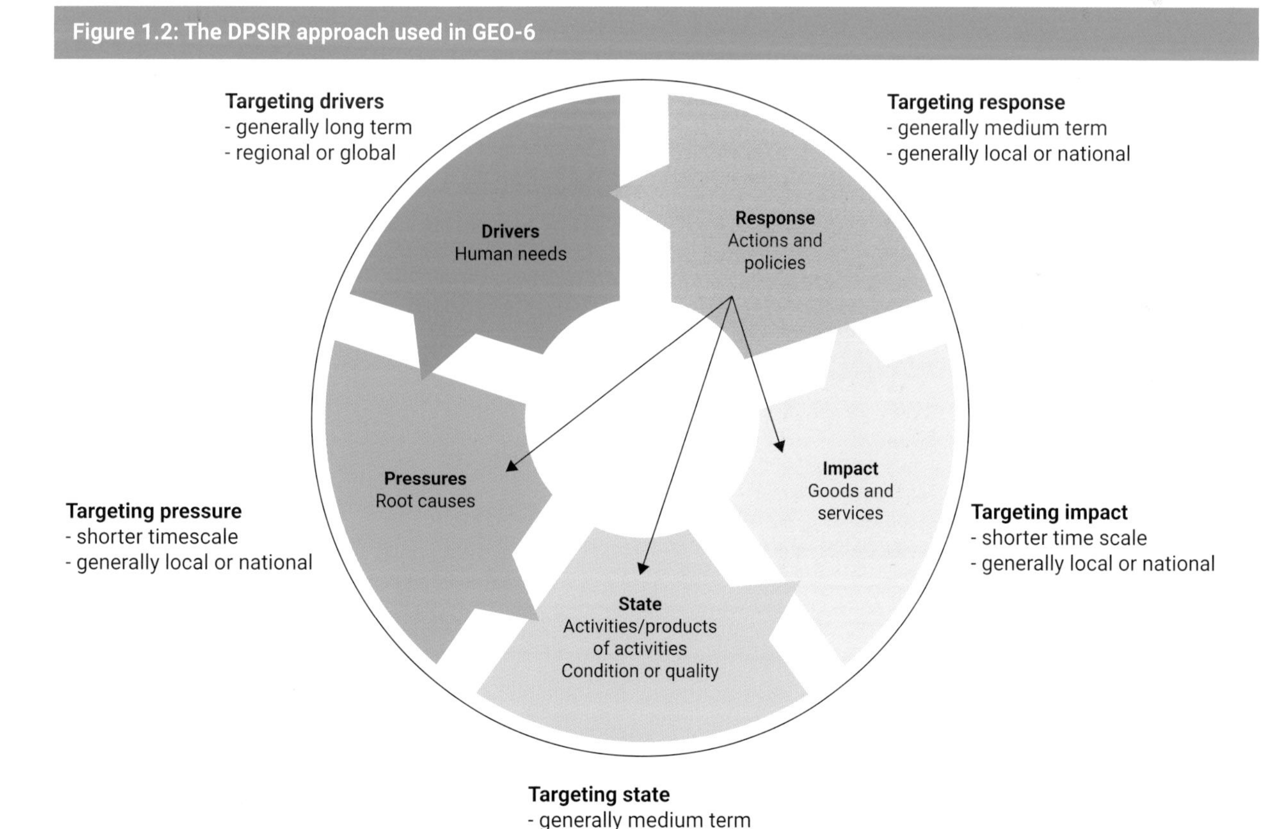

In 1995, UNEP adopted the DPSIR causal framework approach for the GEO assessments. This represents a systems-analysis view in which the driving forces of social and economic development exert pressures on the environment, which change the state of the environment. The changing state of the environment leads to impacts on, for example, human well-being and ecosystem health, which then produces human responses to remedy these impacts, such as social controls, redirecting investments, and/or policies and political interventions to influence human activity. Finally, these responses influence the state of the environment, either directly or indirectly, through the driving forces or the pressures. Existing policies increasingly need to be assessed in terms of how they address the drivers and impacts of environmental challenges.

Source: UNEP (2017b)

Box 1.2: Multidimensional aspects of the analysis

GEO assessments are multidimensional in scope, in an effort to incorporate environmental, social, economic, policy, geographic and temporal perspectives to form various threads of evidence to answer the overarching questions. The main approaches used in GEO-6 include the following.

- The 'Healthy Planet, Healthy People' theme emphasizes the foundational importance of healthy ecosystems and environment for human health. Human health is systematically covered throughout GEO-6 by focusing on the many direct and indirect health-related impacts (e.g. diseases, mortality) deriving from environmental change and deterioration (see Chapters 2 and 4). In addition, health-related objectives are recognized within transformative environmental policies and pathways (Part B). Where possible, health-related impacts are analysed related to social criteria such as age and gender.
- GEO-6 includes thematic dimensions that track the state and trends of air, fresh water, oceans, land and biodiversity, and constitute a 'state of the environment' report (Part A).
- GEO-6 presents more than 25 policy case studies that highlight the importance of evidence-based policymaking (Chapters 12-16). The case studies allow examination of how to design effective policies without being policy prescriptive.
- GEO-6 includes cross-cutting dimensions that combine social, economic and environmental aspects of complex challenges. GEO-6 covers 12 cross-cutting issues (e.g. food, energy, resource use, gender, health, disasters, etc.) throughout the assessment of the environmental themes (Chapter 4) and policy effectiveness (Chapter 17), adding a specific focus on the interlinkages between the environmental and other dimensions of the 2030 Agenda/SDGs.
- For the first time, GEO-6 examines climate change as both a driver (i.e. built-in climate change) and a cross-cutting issue (i.e. anticipated impacts) in the sustainable development context through its overarching relevance to all other aspects of GEO-6 (Chapters 2 and 4).
- GEO-6 considers the equity dimension systematically by considering distributional, representational and procedural issues in the various parts of the assessment, highlighting impacts and possible opportunities of environmental policies and future development pathways to overcome inequities.
- The outlook chapters of GEO-6 (Part C) combine traditional global and scenario-based analysis with local, participative and decision-based analysis. This aims to provide a solution-oriented perspective which considers relevance and efficacy.
- GEO-6 uses modern tools and platforms (e.g. crowdsourcing) to expand stakeholder engagement in the assessment process (Chapter 23).

Based on the core principles of developing integrated environmental assessments (UNEP 2017b), the scope of the GEO has evolved. A key new feature of GEO-6 is increased emphasis on the interactions and interlinkages between the environment and human health. The changing approach and structure of this GEO reflects the most recent scientific evidence and the new geopolitical context, particularly implementation of the 2030 Agenda. GEO-6 provides the evidence base for addressing the environmental dimension of the SDGs.

The GEO-6 process in itself is part of the effort to strengthen overall capacity-building within the global environmental governance system, in order to increase the level of science-based decision-making on multiple levels (United Nations General Assembly Resolution 2997 of 1972). The GEO is an independent, expert-led, participatory process created to facilitate the interaction between scientific understanding and policy development. Policymakers, as well as a wide range of scientists and stakeholders, are consulted on each edition's focus and methodology through the High-Level Intergovernmental and Stakeholder Advisory Group, the Scientific Advisory Panel and the Assessment Methodologies, Data and Information Group, which provide advice and guidance throughout the GEO process. This participatory and consultative process gives GEO assessments scientific credibility, accuracy and authority, as well as policy relevance.

In addition to producing GEOs, UNEP has a mandate for capacity-building. This is an integral part of the GEO process and works at different levels, using various mechanisms. GEO reports include contributions from leading international experts from a wide range of organizations worldwide, as well as a team of GEO Fellows who are early-career professionals or students.

The thematic dimensions (state and trends of air, fresh water, oceans, land and biodiversity) were also core elements of previous GEOs, but all the other approaches listed above are new to GEO-6.

Annex 1-3 contains information on the theory of change that GEO-6 is built upon, and Annex 1-4 provides information on how the authors of each chapter have established confidence statements for the main findings of each chapter. These confidence statements can be found in the Executive Summaries for each chapter and are expected to assist policymakers in understanding the extent of the evidence that exists on a subject, and how much of that evidence is in agreement on the findings presented in this assessment.

1.7.2 Structure

Based on this mandate and scope, the contents of GEO-6 are structured as shown in **Figure 1.3**.

Three chapters complement this introduction, Chapter 2: Drivers of Environmental Change, Chapter 3: The State of Our Data and Knowledge, and Chapter 4: Cross-cutting Issues. As information and data become more important in society, knowledge creation and use also become even more important within GEO-6, since the organization of data, information and knowledge form the foundation of scientifically sound assessments and informed policy decision-making. Therefore,

Figure 1.3: Structure of GEO-6, with a link to its Theory of Change (see Annex 1-3)

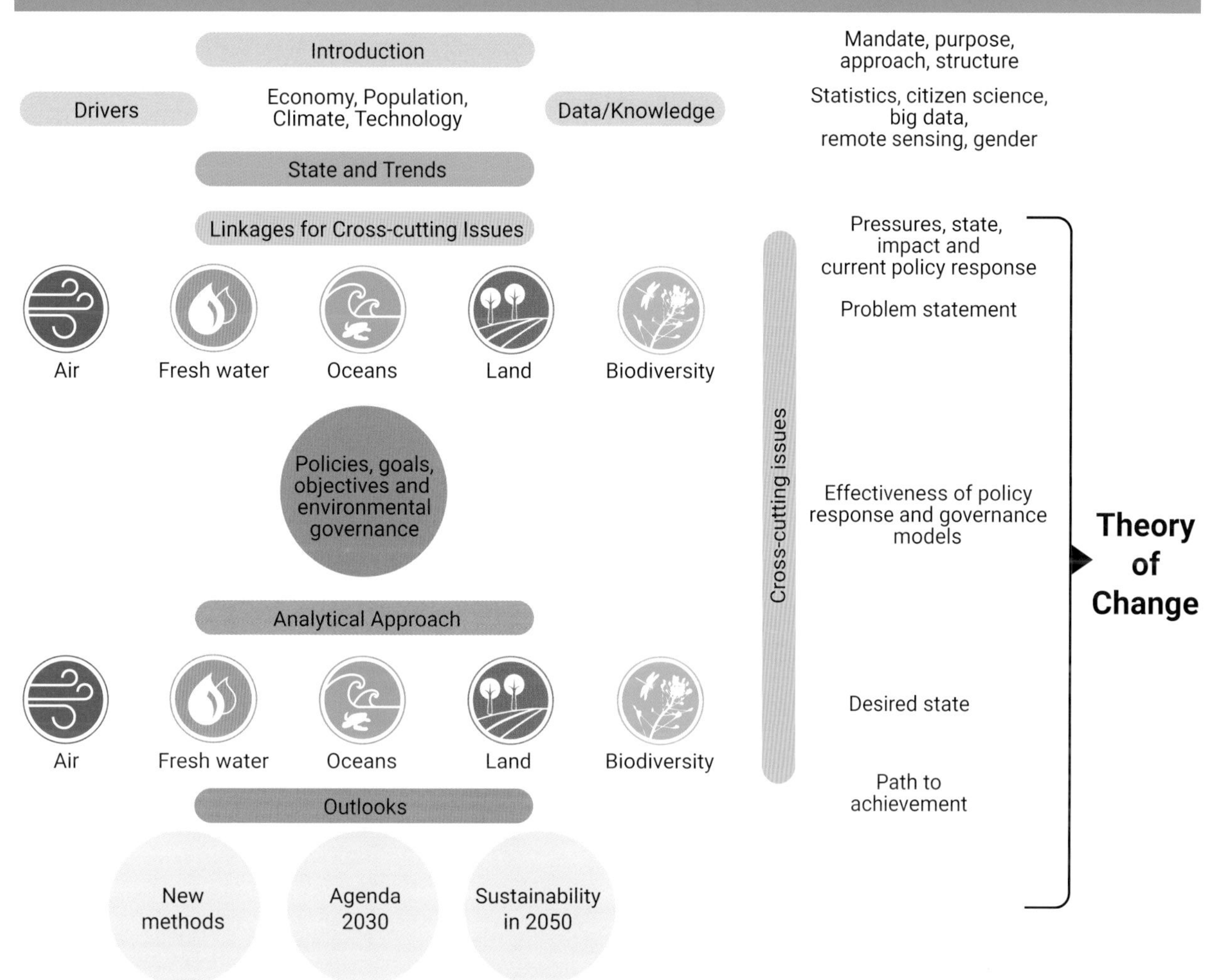

GEO-6 makes greater efforts to explain both changing needs and new opportunities around data, information and knowledge generation derived from UNEP's mandate to deliver the environmental dimension of the SDGs.

The global human system has many deep-rooted path dependencies, which have evolved over a long time. As society and civilization have evolved and developed, the interlinkages between human and environmental systems have become more complex and dynamic. To understand the most relevant structural elements of the human system, GEO-6 systematically examines the overall drivers, for example, population and demographic changes, including the causes of migration, current economic trends and technological developments.

One new element is Chapter 4 on cross-cutting issues, which presents the evidence explaining how the state and trends of the environment are already impacting human systems on various scales. The twelve cross-cutting issues addressed in GEO-6 are also important SDG issues: health, environmental disasters, gender, education, urbanization, climate change, polar regions and mountains, chemicals, waste and wastewater, resource use, energy and food systems). GEO-6 uses a matrix-approach to address these cross-cutting issues, considering each within the context of the five environmental themes (air, biodiversity, oceans, land, fresh water). This approach helps reflect the growing need to synthesize more effectively our knowledge on the environment's multidimensional functionality and how it already affects human systems.

The analysis in GEO-6 is divided into four parts:
Part A: State of the Global Environment features five thematic chapters providing the latest data and information on the state and trends of air, biodiversity, oceans, land, oceans and fresh water. Chapters 5-9 have a common structure using the DPSIR approach, and each includes information on related policy responses.

Part B: Policies, Goals, Objectives and Environmental Governance, an assessment of their effectiveness evaluates the effectiveness of the current policy landscape within the existing environmental governance structure at multiple scales, based on the policy responses identified in the thematic chapters in Part A, including the cross-cutting issues (Chapters 10-17). The methodology developed for this assessment is based on a combined top-down and bottom-up approach. The results are used to extract guidance for policymakers and to support the promising policy approaches addressed in the final

section of the report. Based on this analysis, Part B also identifies needs for further improvements to the global environmental governance system (Chapter 18).

Part C: Outlooks and Pathways to a Healthy Planet with Healthy People incorporates the most promising policy approaches from Part B into the pathways of transformation. It combines global and scenario-based analysis (Chapters 20-22) with local, participative analysis (Chapter 23) to identify possible pathways towards achieving the environmental dimension of the SDGs and other MEAs (up to 2030), and assesses long-term or mid-century strategies required for achieving long-term sustainability (to 2050) (Chapter 24). The outcomes and conclusions provide a baseline to guide policymaking and implementation of the SDGs, as well as the development of more transformative pathways to reach scientific targets over a longer time-horizon (to 2050), such as the objective to become a climate-neutral, resource-efficient society. This long-term perspective will help to guide the further development of global, regional and national governance systems to ensure future human development stays within the Earth's ecological limits, and helps to create a more equitable world with no one left behind. Where possible, Part C emphasizes the economic and social costs and benefits of various options for action and non-action.

1

Part D: Remaining Data and Knowledge Gaps (Chapter 25) provides an overview of the data/knowledge trends and issues and identifies the gaps that need to be filled in order to implement the SDGs and achieve the IAEGs established in MEAs. This is based on the premise that more data/knowledge leads to better and more effective actions/solutions in more places. A revolution in communications and information technology is creating significant new data and information opportunities beyond traditional environmental monitoring and assessment.

We hope readers – whether policymakers, researchers or citizens – find the analysis and assessment findings presented in the following chapters useful, helping to inform future efforts to address our collective environmental challenges.

References

Abraham, D.S. (2015). *The Elements of Power: Gadgets, Guns, and the Struggle for a Sustainable Future in the Rare Metal Age*. New Haven, CT: Yale University Press. https://yalebooks.yale.edu/book/9780300196795/elements-power.

Adelle, C. and Nilsson, M. (2015). Environmental Policy Integration. In *Encyclopedia of Global Environmental Governance and Politics*. Pattberg, P., H. and Zelli, F. (eds.). Cheltenham: Edward Elgar. chapter 58. 454-461. https://www.e-elgar.com/shop/eep/preview/book/isbn/9781782545798/.

Altenburg, T. and Assmann, C. (2017). *Green Industrial Policy: Concept, Policies, Country Experiences*. Geneva/Bonn: United Nations Environment Programme/German Development Institute. http://www.greengrowthknowledge.org/sites/default/files/downloads/resource/Green%20Industrial%20Policy_Concept%2C%20Policies%2C%20Country%20Experiences.pdf.

Ansell, C. and Gash, A. (2008). Collaborative governance in theory and practice. *Journal of Public Administration Research and Theory* 18(4), 543-571. https://doi.org/10.1093/jopart/mum032.

Balsiger, J. and VanDeveer, S.D. (2012). Navigating regional environmental governance. *Global Environmental Politics* 12(3), 1-17. https://doi.org/10.1162/GLEP_e_00120.

Barnett, J. and Adger, W.N. (2007). Climate change, human security and violent conflict. *Political Geography* 26(6), 639-655. https://doi.org/10.1016/j.polgeo.2007.03.003.

Beck, U. (2009). World risk society and manufactured uncertainties. *IRIS European Journal of Philosophy and Public Debate* 1(2), 291-299. http://www.fupress.net/index.php/iris/article/view/3304.

Beck, U. (2015). *The Metamorphosis of the World: How Climate Change is Transforming Our Concept of the World*. Cambridge: Polity Press. https://www.wiley.com/en-us/The+Metamorphosis+of+the+World%3A+How+Climate+Change+is+Transforming+Our+Concept+of+the+World-p-9780745690216.

Biermann, F. (2014). *Earth System Governance: World Politics in the Anthropocene*. Cambridge, MA: MIT Press. https://www.jstor.org/stable/j.ctt1287hkh.

Biermann, F., Abbott, K., Andresen, S., Backstrand, K., Bernstein, S., Betsill, M.M., Bulkeley, H., Cashore, B., Clapp, J., Folke, C. *et al.* (2012). Navigating the Anthropocene: Improving Earth system governance. *Science* 335(6074), 1306-1307. https://doi.org/10.1126/science.1217255.

Biermann, F., Kanie, N. and Kim, R.E. (2017). Global governance by goal-setting: The novel approach of the UN Sustainable Development Goals. *Current Opinion in Environmental Sustainability* 26-27, 26-31. https://doi.org/10.1016/j.cosust.2017.01.010.

Boyce, J.K. (2004). Green and brown? Globalization and the environment. *Oxford Review of Economic Policy* 20(1), 105-128. https://doi.org/10.1093/oxrep/grh007.

British Broadcasting Corporation (2016). Delhi smog: Schools closed for three days as pollution worsens. https://www.bbc.com/news/world-asia-india-37887937.

Brown, O., Hammill, A. and McLeman, R. (2007). Climate change as the 'new' security threat: Implications for Africa. *International Affairs* 83(6), 1141-1154. https://doi.org/10.1111/j.1468-2346.2007.00678.x.

Calvo, G., Valero, A. and Valero, A. (2017). Assessing maximum production peak and resource availability of non-fuel mineral resources: Analyzing the influence of extractable global resources. *Resources, Conservation and Recycling* 125, 208-217. https://doi.org/10.1016/j.resconrec.2017.06.009.

Castree, N. (2008). Neoliberalising nature: The logics of deregulation and reregulation. *Environment and Planning A: Economy and Space* 40(1), 131-152. https://doi.org/10.1068/a3999.

Chuvieco, E. (ed.) (2008). *Earth Observation of Global Change: The Role of Satellite Remote Sensing in Monitoring the Global Environment*. Dordrecht: Springer. https://www.springer.com/gp/book/9781402063572.

Clapp, J. and Dauvergne, P. (2011). *Paths to a Green World: The Political Economy of the Global Environment*. Cambridge, MA: MIT Press. https://mitpress.mit.edu/books/paths-green-world.

Cordell, D. and White, S. (2015). Tracking phosphorus security: Indicators of phosphorus vulnerability in the global food system. *Food Security* 7(2), 337-350. https://doi.org/10.1007/s12571-015-0442-0.

Costanza, R., de Groot, R., Sutton, P., van der Ploeg, S., Anderson, S.J., Kubiszewski, I., Farber, S. and Turner, R.K. (2014). Changes in the global value of ecosystem services. *Global Environmental Change - Human and Policy Dimensions* 26, 152-158. https://doi.org/10.1016/j.gloenvcha.2014.04.002.

Crutzen, P.J. (2002). Geology of mankind. *Nature* 415(6867), 23. https://doi.org/10.1038/415023a.

Crutzen, P.J. and Stoermer, E.F. (2000). The "Anthropocene". *Global Change Newsletter* 41, pp. 17-18. Stockholm: International Geosphere–Biosphere Programme. http://www.igbp.net/download/18.316f18321323470177580001401/1376383088452/NL41.pdf.

Díaz, S., Demissew, S., Carabias, J., Joly, C., Lonsdale, M., Ash, N., Larigauderie, A., Adhikari, J.R., Arico, S., Báldi, A. *et al.* (2015). The IPBES Conceptual Framework -- connecting nature and people. *Current Opinion in Environmental Sustainability* 14, 1-16. https://doi.org/10.1016/j.cosust.2014.11.002.

Dodds, F., Donoghue, D. and Leiva-Roesch, J. (2017). *Negotiating the Sustainable Development Goals. 1st edition*. London: Routledge. https://www.taylorfrancis.com/books/e/9781315527086.

Elder, M., Bengtsson, M. and Akenji, L. (2016). An optimistic analysis of the means of implementation for Sustainable Development Goals: Thinking about goals as means. *Sustainability* 8(9), 962-986. https://doi.org/10.3390/su8090962.

Elder, M. and Zusman, E. (2016). *Strengthening the Linkages Between Air Pollution and the Sustainable Development Goals*. IGES Policy BriefsInstitute for Global Environmental Strategies. https://www.jstor.org/stable/resrep02907.

Environment Canada, University of Joensuu and United Nations Environment Programme (2007). *Multilateral Environment Agreement Negotiator's Handbook. 2nd* edition. Joensuu: University of Joensuu. https://unfccc.int/resource/docs/publications/negotiators_handbook.pdf.

Food and Agriculture Organization of the United Nations (2014). *The Water-Energy-Food Nexus: A New Approach in Support of Food Security and Sustainable Agriculture*. Rome. http://www.fao.org/3/a-bl496e.pdf.

Fukuda-Parr, S., Yamin, A.E. and Greenstein, J. (2014). The power of numbers: A critical review of Millennium Development Goal targets for human development and human rights. *Journal of Human Development and Capabilities* 15(2-3), 105-117. https://doi.org/10.1080/19452829.2013.864622.

Group of Seven (2015). *Leaders' Declaration: G7 Summit*. 7-8 June 2015. https://www.bundesregierung.de/resource/blob/72444/436680/e077d51d67486b1df34e539f621aff8c/2015-06-08-g7-abschluss-eng-en-data.pdf?download=1.

Group of Twenty (2016). *G20 Action Plan on the 2030 Agenda for Sustainable Development*. https://www.b20germany.org/fileadmin/user_upload/G20_Action_Plan_on_the_2030_Agenda_for_Sustainable_Development.pdf.

Galaz, V., Biermann, F., Folke, C., Nilsson, M. and Olsson, P. (2012). Global environmental governance and planetary boundaries: An introduction. *Ecological Economics* 81, 1-3. https://doi.org/10.1016/j.ecolecon.2012.02.023.

Gallagher, K.P. (2009). Economic globalization and the environment. *Annual Review of Environment and Resources* 34(1), 279-304. https://doi.org/10.1146/annurev.environ.33.021407.092325.

Garrett-Peltier, H. (2017). Green versus brown: Comparing the employment impacts of energy efficiency, renewable energy, and fossil fuels using an input-output model. *Economic Modelling* 61, 439-447. https://doi.org/10.1016/j.econmod.2016.11.012.

Ghisellini, P., Cialani, C. and Ulgiati, S. (2016). A review on circular economy: The expected transition to a balanced interplay of environmental and economic systems. *Journal of Cleaner Production* 114, 11-32. https://doi.org/10.1016/J.JCLEPRO.2015.09.007.

Ghosh, N., Mukhopadhyay, P., Shah, A. and Panda, M. (eds.) (2016). *Nature, Economy and Society: Understanding the Linkages*. New Delhi: Springer. https://www.springer.com/gp/book/9788132224037.

Global Commission on the Economy and Climate (2014). *The New Climate Economy Report: Better Growth, Better Climate*. Washington, D.C.: World Resources Institute. http://newclimateeconomy.report/2014/wp-content/uploads/sites/2/2014/08/BetterGrowth-BetterClimate_NCE_Synthesis-Report_web.pdf.

Graedel, T.E., Harper, E.M., Nassar, N.T. and Reck, B.K. (2015). On the materials basis of modern society. *Proceedings of the National Academy of Sciences* 112(20), 6295–6300. https://doi.org/10.1073/pnas.1312752110.

Gupta, C.K. and Krishnamurthy, N. (2004). *Extractive Metallurgy of Rare Earths*. 2nd edn: CRC Press. https://www.crcpress.com/Extractive-Metallurgy-of-Rare-Earths-Second-Edition/Krishnamurthy-Gupta/p/book/9781466576346.

Gupta, J., Dellapenna, J.W. and van den Heuvel, M. (2016). Water sovereignty and security, high politics and hard power: The dangers of borrowing discourses. In *Handbook on Water Security*. Pahl-Wostl, C., Bhaduri, A. and Gupta, J. (eds.). Cheltenham: Edward Elgar. chapter 8. https://www.elgaronline.com/view/9781782548003.00014.xml.

Hepburn, C. and Bowen, A. (2012). *Prosperity with growth: Economic growth, climate change and environmental limits*. GRI Working Papers 93. London: Grantham Research Institute on Climate Change and the Environment. https://ideas.repec.org/p/lsg/lsgwps/wp93.html.

Homer-Dixon, T. (1991). On the threshold: Environmental change as causes of acute conflict. *International Security* 16(2), 76-116. https://doi.org/10.2307/2539061.

Homer-Dixon, T. (1999). *Environment, Scarcity, and Violence*. Princeton, NJ: Princeton University Press. https://press.princeton.edu/titles/6640.html.

Hopkinson, P., Zils, M. and Hawkins, P. (2018). Managing a complex global circular economy business model: Opportunities and challenges. *California Management Review* 60(3), 71-94. https://doi.org/10.1177/0008125618764692.

Intergovernmental Panel on Climate Change (2014). Fifth assessment report: Summary for policymakers. In *Climate Change 2014: Mitigation of Climate Change. Contribution of Working Group III to the Fifth Assessment Report of the Intergovernmental Panel on Climate Change*. Edenhofer, O., Pichs-Madruga, R., Sokona, Y., Farahani, E., Kadner, S., Seyboth, K., Adler, A., Baum, I., Brunner, S., Eickemeier, P. *et al.* (eds.). Cambridge: Cambridge University Press. https://www.ipcc.ch/pdf/assessment-report/ar5/wg3/ipcc_wg3_ar5_summary-for-policymakers.pdf.

Intergovernmental Science-Policy Platform on Biodiversity and Ecosystem Services (2015). *Preliminary Guide regarding Diverse Conceptualization of Multiple Values of Nature and its Benefits including Biodiversity and Ecosystem Functions and Services: Note by the Secretariat*. IPBES/4/INF/13. Plenary of the Intergovernmental Science-Policy Platform on Biodiversity and Ecosystem Services Fourth session. Kuala Lumpur, 22–28 February. https://www.ipbes.net/sites/default/files/downloads/IPBES-4-INF-13_EN.pdf.

Intergovernmental Science-Policy Platform on Biodiversity and Ecosystem Services (2017). *Progress Report on the guide on the production of assessments (Deliverable 2 (a)). Note by the Secretariat*. IPBES/5/INF/6. Plenary of the Intergo°ernmental Science-Policy Platform on Biodiversity and Ecosystem Services Fifth session. Bonn, 7-10 March. https://www.ipbes.net/system/tdf/downloads/doc/ipbes-5-inf-6.docx?file=1&type=node&id=13725.

International Labour Office (2016). *Green Jobs: Progress Report 2014-2015*. Geneva. http://www.ilo.org/wcmsp5/groups/public/---ed_emp/---emp_ent/documents/publication/wcms_502730.pdf.

International Renewable Energy Agency (2017). *Renewable Energy and Jobs: Annual Review 2017*. Abu Dhabi. http://www.irena.org/-/media/Files/IRENA/Agency/Publication/2017/May/IRENA_RE_Jobs_Annual_Review_2017.pdf.

International Resource Panel (2014). *Managing and Conserving the Natural Resource Base for Sustained Economic and Social Development*. Nairobi: United Nations Environment Programme. http://www.resourcepanel.org/file/244/download?token=OHRPH1MH.

International Resource Panel (2015). *Policy Coherence of the Sustainable Development Goals: A Natural Resource Perspective*. Nairobi: United Nations Environment Programme. http://www.resourcepanel.org/file/251/download?token=678P6Zys.

Jabbour, J. and Flachsland, C. (2017). 40 years of global environmental assessments: A retrospective analysis. *Environmental Science & Policy* 77, 193-202. https://doi.org/10.1016/j.envsci.2017.05.001.

Jetzkowitz, J., van Koppen, C.S.A., Lidskog, R., Ott, K., Voget-Kleschin, L. and Wong, C.M.L. (2018). The significance of meaning. Why IPBES needs the social sciences and humanities. *Innovation: The European Journal of Social Science Research* 31, S38-S60. https://doi.org/10.1080/13511610.2017.1348933.

Jordan, A. (2008). The governance of sustainable development: Taking stock and looking forwards. *Environment and Planning C: Government and Policy* 26(1), 17-33. https://doi.org/10.1068/cav6.

Jordan, A. and Lenschow, A. (2010). Environmental policy integration: A state of the art review. *Environmental Policy and Governance* 20(3), 147-158. https://doi.org/10.1002/eet.539.

Kanie, N. (2007). Governance with multilateral Agreements: A healthy or ill-equipped fragmentation. In *Global Environmental Governance: Perspectives on the Current Debate Center for UN reform*. Swart, L. and Perry, E. (eds.). New York, NY: Center for UN Reform Education. 67-86. http://www.centerforunreform.org/sites/default/files/GEG_Kanie.pdf.

Kanie, N., Andresen, S. and Haas, P.M. (eds.) (2014). *Improving Global Environmental Governance: Best Practices for Architecture and Agency*. New York, NY: Routledge. https://www.routledge.com/Improving-Global-Environmental-Governance-Best-Practices-for-Architecture/Kanie-Andresen-Haas/p/book/9780415811767.

Kanie, N. and Biermann, F. (eds.) (2017). *Governing Through Goals: Sustainable Development Goals as Governance Innovation*. Cambridge, MA: MIT Press. https://mitpress.mit.edu/books/governing-through-goals.

Kim, E. and Kim, B.H.S. (eds.) (2016). *Quantitative Regional Economic and Environmental Analysis for Sustainability in Korea*. Singapore: Springer. https://www.springer.com/gp/book/9789811002984.

Kobori, H., Dickinson, J.L., Washitani, I., Sakurai, R., Amano, T., Komatsu, N., Kitamura, W., Takagawa, S., Koyama, K., Ogawara, T. *et al.* (2016). Citizen science: A new approach to advance ecology, education, and conservation. *Ecological Research* 31(1), 1-19. https://doi.org/10.1007/s11284-015-1314-y.

Kowarsch, M., Flachsland, C., Jabbour, J., Garard, J. and Riousset, P. (2014). *The Future of Global Environmental Assessment Making (FOGEAM): Reflecting on Past Experiences to Inform Future Choices*. Berlin: Mercator Research Institute on Global Commons and Climate Change. https://www.mcc-berlin.net/fileadmin/data/C18_MCC_Publications/FOGEAM_Preliminary_Draft_Report_17102014.pdf.

Kowarsch, M., Jabbour, J., Flachsland, C., Kok, M.T.J., Watson, R., Haas, P.M., Minx, J.C., Alcamo, J., Garard, J., Riousset, P. *et al.* (2017). A road map for global environmental assessments. *Nature Climate Change* 7(6), 379-382. https://www.nature.com/articles/nclimate3307.pdf.

Lacy, P. and Rutqvist, J. (2015). *Waste to Wealth: The Circular Economy Advantage*. New York, NY: Palgrave Macmillan. https://www.palgrave.com/gp/book/9781137530684.

Lammel, A., Gutierez, E.G., Dugas, E. and Jamet, F. (2013). Cultural and environmental changes: Cognitive adaptation to global warning. In *Steering the Cultural Dynamics: Selected Papers from the 2010 Congress of the International Association for Cross-cultural Psychology*. Kashima, Y., Kashima, E. and Beatson, R. (eds.). Melbourne: International Association for Cross Cultural Psychology. http://iaccp.org/sites/default/files/melbourne_pdf/Lammel.pdf.

Landrigan, P.J., Fuller, R., Acosta, N.J.R., Adeyi, O., Arnold, R., Basu, N., Baldé, A.B., Bertollini, R., Bose-O'Reilly, S., Boufford, J.I. *et al.* (2018). The Lancet Commission on pollution and health. *The Lancet* 391(10119), 462-512. https://doi.org/10.1016/S0140-6736(17)32345-0.

Le Blanc, D. (2015). Towards integration at last? The sustainable development goals as a network of targets. *Sustainable Development* 23(3). https://doi.org/10.1002/sd.1582.

Lehmann, H., Rajan, S.C., Annavarapu, S., Kabel, C., Lowe, C. and Matthey, A. (2015). *Sustainable Lifestyles: Pathways and Choices for India and Germany*. Berlin: Deutsche Gesellschaft für Internationale Zusammenarbeit (GIZ) GmbH. https://www.giz.de/en/downloads/giz2015-en-IGEG_3_sustainable-lifestyles.pdf.

Li, F., Liu, Y., Lü, J., Liang, L. and Harmer, P. (2014). Ambient air pollution in China poses a multifaceted health threat to outdoor physical activity. *Journal of Epidemiology and Community Health* 69(3), 201-204. http://dx.doi.org/10.1136/jech-2014-203892.

Lucas, P.L., Ludwig, K., Kok, M.I.J. and Kruitwagen, S. (2016). *Sustainable Development Goals in the Netherlands: Building Blocks for Environmental Policy for 2030*. The Hague: PBL Netherlands Environmental Assessment Agency. http://www.pbl.nl/sites/default/files/cms/publicaties/pbl-2016-sustainable-development-in-the-Netherlands_1966.pdf.

Mastrandrea M.D., Field, C.B., Stocker, T.F., Edenhofer, O., Ebi, K.L., Frame, D.J. *et al.* (2010). *Guidance Note for Lead Authors of the IPCC Fifth Assessment Report on Consistent Treatment of Uncertainties*. Intergovernmental Panel on Climate Change. https://www.ipcc.ch/pdf/supporting-material/uncertainty-guidance-note.pdf.

Mathieu, P.-P. and Aubrecht, C. (eds.) (2018). *Earth Observation Open Science and Innovation*. New York, NY: Springer. https://www.springer.com/gb/book/9783319656328.

Matthew, R.A., Barnett, J., McDonald, B. and O'Brien, K.L. (2010). *Global Environment Change and Human Security*. Cambridge, MA: MIT Press. https://mitpress.mit.edu/books/global-environmental-change-and-human-security.

Millennium Ecosystem Assessment (2005). *Ecosystems and Human Well-being: Synthesis*. Washington, D.C.: Island Press. https://www.millenniumassessment.org/documents/document.356.aspx.pdf.

Mitchell, R.B., Clark, W.C., Cash, D.W. and Dickson, N.M. (eds.) (2006). *Global Environmental Assessments: Information and Influence*. Cambridge, MA: MIT Press. https://mitpress.mit.edu/books/global-environmental-assessments.

Mitchell, R.B. (2018). Data from Ronald B. Mitchell. 2002-2018. *International Environmental Agreements Database Project (Version 2018.1)*. Oregon: University of Oregon. https://iea.uoregon.edu/. (Accessed 13 February 2018).

Mohai P., Kweon B.S., Lee, S. and Ard, K. (2011). Air pollution around schools is linked to poorer student health and academic performance health affairs. *Health Affairs* 30(5), 852-862. https://doi.org/10.1377/hlthaff.2011.0077.

Mortensen, F.L. and Petersen, L.K. (2017). Extending the boundaries of policy coherence for sustainable development: Engaging business and civil society. *Solutions Journal* 8(3). https://www.thesolutionsjournal.com/article/extending-boundaries-policy-coherence-sustainable-development-engaging-business-civil-society/.

Morton, T. (2009). *Ecology Without Nature: Rethinking Environmental Aesthetics*. Cambridge, MA: Harvard University Press. http://www.hup.harvard.edu/catalog.php?isbn=9780674034853.

Moser, S. and Kleinhückelkotten, S. (2017). Good intents, but low impacts: Diverging importance of motivational and socioeconomic determinants explaining pro-environmental behavior, energy use, and carbon footprint. Environment and Behavior 50(6), 626-656. https://doi.org/10.1177/0013916517710685.

Murray, A., Skene, K. and Haynes, K. (2017). The circular economy: An interdisciplinary exploration of the concept and application in a global context. *Journal of Business Ethics* 140(3), 369–380. https://doi.org/10.1007/s10551-015-2693-2.

Najam, A., Papa, M. and Taiyab, N. (2006). *Global Environmental Governance: A Reform Agenda*. Winnipeg: International Institute for Sustainable Development. https://www.iisd.org/pdf/2006/geg.pdf.

Newell, P. and Roberts, J.T. (eds.) (2016). *The Globalization and Environment Reader*. Oxford: Wiley-Blackwell. https://www.wiley.com/en-us/The+Globalization+and+Environment+Reader-p-9781118964132.

Newig, J. and Fritsch, O. (2009). Environmental governance: Participatory, multi-level - and effective? *Environmental Policy and Governance* 19(3), 197-214. https://doi.org/10.1002/eet.509.

Organisation for Economic Co-operation and Development (2011). *Towards Green Growth*. Paris. https://www.oecd-ilibrary.org/docserver/9789264111318-en.pdf.

Organisation for Economic Co-operation and Development (2015a). *OECD Principles on Water Governance*. Paris. http://www.oecd.org/cfe/regional-policy/OECD-Principles-on-Water-Governance-brochure.pdf.

Organisation for Economic Co-operation and Development (2017a). *Healthy People, Healthy Planet: The Role of Health Systems in Promoting Healthier Lifestyles and a Greener Future*. Paris. https://www.oecd.org/health/health-systems/Healthy-people-healthy-planet.pdf.

Organisation for Economic Co-operation and Development (2017b). *Investing in Climate, Investing in Growth*. Paris. http://www.oecd.org/env/investing-in-climate-investing-in-growth-9789264273528-en.htm.

Pascual, U., Balvanera, P., Díaz, S., Pataki, G., Roth, E., Stenseke, M., Watson, R.T., Başak Dessane, E., Islar, M., Kelemen, E. *et al.* (2017). Valuing nature's contributions to people: The IPBES approach. *Current Opinion in Environmental Sustainability* 26-27, 7-16. https://doi.org/10.1016/j.cosust.2016.12.006.

Pattberg, P. and Widerberg, O. (2015). Theorising global environmental governance: Key findings and future questions. *Millennium: Journal of International Studies* 43(2), 684-705. https://doi.org/10.1177/0305829814561773.

Pattberg, P. and Widerberg, O. (2016). Transnational multistakeholder partnerships for sustainable development: Conditions for success. *Ambio* 45(1), 42-51. https://doi.org/10.1007/s13280-015-0684-2.

Pattberg, P. and Zelli, F. (eds.) (2016). *Environmental Politics and Governance in the Anthropocene*. London: Routledge. https://www.routledge.com/Environmental-Politics-and-Governance-in-the-Anthropocene-Institutions/Pattberg-Zelli/p/book/9781138902398.

Patterson, J., Schulz, K., Barau, A., Obani, P., Sethi, M., Hissen, N. *et al.* (2015). *Transformations Towards Sustainability: Emerging Approaches, Critical Reflections, and a Research Agenda*. Earth System Governance Working Paper No. 33. Lund: Earth System Governance Project. https://ueaeprints.uea.ac.uk/54624/1/ESG_WorkingPaper_34_Patterson_et_al.pdf.

Pereira, J.C. (2015). Environmental issues and international relations, a new global (dis)order - the role of international relations in promoting a concerted international system. *Revista Brasileira de Política Internacional* 58(1), 191–209. https://doi.org/10.1590/0034-7329201500110.

Porter, M. and van der Linde, C. (1995). Green and competitive: Ending the stalemate. *Harvard Business Review* September–October 1995. Boston, MA: Harvard Business Publishing. https://hbr.org/1995/09/green-and-competitive-ending-the-stalemate.

Qasem, I. (2010). *Resource Scarcity in the 21st Century: Conflict or Cooperation?* Den Haag: Hague Centre for Strategic Studies (HCSS) and TNO. https://hcss.nl/sites/default/files/files/reports/Strategy_Change_PAPER_03_web.pdf.

Raskin, P. (2016). *Journey to Earthland: The Great Transition to Planetary Civilization*. Boston, MA: Tellus Institute. http://www.tellus.org/pub/Journey-to-Earthland.pdf.

Raworth, K. (2012). *A Safe and Just Space for Humanity: Can We Live Within the Doughnut*. Oxford: Oxfam. https://www.oxfam.org/sites/www.oxfam.org/files/dp-a-safe-and-just-space-for-humanity-130212-en.pdf.

Renewable Energy Policy Network for the 21st Century (2018). *Renewables 2018 Global Status Report*. Paris. http://www.ren21.net/wp-content/uploads/2018/06/17-8652_GSR2018_FullReport_web_final_.pdf.

Reuters (2017). Schools shut in Iran capital, major cities due to high pollution. https://www.reuters.com/article/us-china-pollution/china-regions-accused-of-faking-pollution-compliance-in-new-probe-idUSKCN1MW0PM.

Rockström, J., Steffen, W., Noone, K., Persson, Å., Chapin III, F.S., Lambin, E. *et al.* (2009). Planetary boundaries: Exploring the safe operating space for humanity. *Ecology and society* 14(2). https://doi.org/10.5751/ES-03180-140232.

Sastry, N. (2002). Forest fires, air pollution, and mortality in southeast Asia. *Demography* 39(1), 1-23. https://www.jstor.org/stable/3088361.

Schellnhuber, H.J., Crutzen, P.J., Clark, W.C., Claussen, M. and Held, H. (eds.) (2004). *Earth System Analysis for Sustainability*. Cambridge, MA: MIT Press. https://mitpress.mit.edu/books/earth-system-analysis-sustainability.

Scheyvens, H., Shaw, R., Endo, I., Kawasaki, J., Ngoc Bao, P., Shivakoti, B.R., Samejima, H., Mitra, B.K. and Takahashi, Y. (2017). *Promoting the Landscape Approach in Asia-Pacific Developing Countries: Key Concepts and Ways Forward*. Hayama: Institute for Global Environmental Strategies. https://pub.iges.or.jp/pub_file/pb37efinalpdf/download.

Schneidewind, U. (2013). Wandel verstehen: Auf dem weg zu einer "Transformative Literacy". In *Wege aus der Wachstumsgesellschaft*. Welzer, H. and Wiegandt, K. (eds.). Frankfurt am Main: Fischer. 115-140. https://epub.wupperinst.org/frontdoor/deliver/index/docId/4935/file/4935_Schneidewind.pdf.

Secretariat of the Convention on Biological Diversity (2014). *Global Biodiversity Outlook 4*. Montreal. https://www.cbd.int/gbo/gbo4/publication/gbo4-en.pdf.

Serret, Y. and Johnstone, N. (eds.) (2006). *The Distributional Effects of Environmental Policy*. Paris/Cheltenham: Organisation for Economic Co-operation and Development/Edward Elgar. http://www.oecd.org/env/tools-evaluation/thedistributionaleffectsofenvironmentalpolicy.htm.

Siebert, H. (2008). *Economics of the Environment: Theory and Policy. 7th edition*. Berlin Heidelberg: Springer-Verlag. https://www.springer.com/gp/book/9783540737063.

Steffen, W. (2000). An integrated approach to understanding Earth's metabolism. *Global Change Newsletter* 41, 9-10. Stockholm: International Geosphere–Biosphere Programme. http://www.igbp.net/download/18.316f18321323470177580001401/1376383088452/NL41.pdf.

Steffen, W., Crutzen, P.J. and McNeill, J.R. (2007). The Anthropocene: Are humans now overwhelming the great forces of nature? *Ambio* 36(8), 614-621. https://doi.org/10.1579/0044-7447(2007)36[614:TAAHNO]2.0.CO;2.

Steffen, W., Persson, A., Deutsch, L., Zalasiewicz, J., Williams, M., Richardson, K., Crumley, C., Crutzen, P., Folke, C., Gordon, L. *et al.* (2011). The Anthropocene: From global change to planetary stewardship. *Ambio* 40(7), 739-761. https://doi.org/10.1007/s13280-011-0185-x.

Steffen, W., Richardson, K., Rockstrom, J., Cornell, S.E., Fetzer, I., Bennett, E.M., Biggs, R., Carpenter, S.R., de Vries, W., de Wit, C.A. *et al.* (2015). Planetary boundaries: Guiding human development on a changing planet. *Science* 347(6223), 1259855-1259855. https://doi.org/10.1126/science.1259855.

Steffen, W., Rockström, J., Richardson, K., Lenton, T.M., Folke, C., Liverman, D. *et al.* (2018). Trajectories of the earth system in the anthropocene. *Proceedings of the National Academy of Sciences* 115(33), 8252-8259. https://doi.org/10.1073/pnas.1810141115.

Steinebach, Y. and Knill, C. (2017). Still an entrepreneur? The changing role of the European Commission in EU environmental policy-making. *Journal of European Public Policy* 24(3), 429-446. https://doi.org/10.1080/13501763.2016.1149207.

Susskind, L.E. and Ali, S.H. (2015). *Environmental Diplomacy: Negotiating More Effective Global Agreements. 2nd edition*. New York, NY: Oxford University Press. https://global.oup.com/academic/product/environmental-diplomacy-9780199397990?cc=ke&lang=en&.

The Economics of Ecosystems and Biodiversity (2010). *The Economics of Ecosystems and Biodiversity: Mainstreaming the Economics of Nature: A Synthesis of the Approach, Conclusions and Recommendations of TEEB*. Geneva. http://doc.teebweb.org/wp-content/uploads/Study%20and%20Reports/Reports/Synthesis%20report/TEEB%20Synthesis%20Report%202010.pdf.

Tomás, R. and Li, Z. (2017). Earth observations for geohazards: Present and future challenges. *Remote Sensing* 9(3), 194. https://doi.org/10.3390/rs9030194.

United Nations (2015a). *Transforming Our World: The 2030 Agenda for Sustainable Development*. New York, NY. https://sustainabledevelopment.un.org/content/documents/21252030%20Agenda%20for%20Sustainable%20Development%20web.pdf.

United Nations (2015b). *Global Sustainable Development Report: 2015 Edition, Advance Unedited Version*. New York, NY. https://sustainabledevelopment.un.org/globalsdreport/2015.

United Nations (2015c). *Addis Ababa Action Agenda of the Third International Conference on Financing for Development*. New York, NY. http://www.un.org/esa/ffd/wp-content/uploads/2015/08/AAAA_Outcome.pdf.

United Nations (2016). *Concepts, Tools and Experiences in Policy Integration for Sustainable Development*. New York, NY. https://www.un.org/ecosoc/sites/www.un.org.ecosoc/files/publication/desa-policy-brief-policy-integration.pdf.

United Nations, General Assembly (1972). 2997 (XXVII). Institutional and Financial Arrangements for International Environmental Cooperation: Resolution adopted by the General Assembly. 15 December. A/RES/27/2997. http://www.un-documents.net/a27r2997.htm.

United Nations Convention to Combat Desertification (2017). *Global Land Outlook*. Bonn. https://www.unccd.int/sites/default/files/documents/2017-09/GLO_Full_Report_low_res.pdf.

United Nations Development Programme (2014). *Governance for Sustainable Development: Integrating Governance in the Post-2015 Development Framework*. New York, NY. http://www.undp.org/content/dam/undp/library/Democratic%20Governance/Discussion-Paper--Governance-for-Sustainable-Development.pdf.

United Nations Economic and Social Commission for Asia and the Pacific (2013). *Water, Food and Energy Nexus in Asia and the Pacific*. Bangkok. http://www.unescap.org/sites/default/files/Water-Food-Nexus%20Report.pdf.

United Nations Economic and Social Commission for Asia and the Pacific and Korea International Cooperation Agency (2012). *Low Carbon Green Growth Roadmap for Asia and the Pacific*. Bangkok/Seongnam. http://www.unescap.org/sites/default/files/Full-report.pdf.

United Nations Economic Commission for Europe (2014). *The Aarhus Convention: An Implementation Guide. Second edition*. Geneva. https://www.unece.org/fileadmin/DAM/env/pp/Publications/Aarhus_Implementation_Guide_interactive_eng.pdf.

United Nations Educational, Scientific and Cultural Organization (2014). Culture, creativity and sustainable development: Florence declaration. *Third UNESCO World Forum on Culture and Culture Industries*. Florence, 2-4 October 2014. http://www.lacult.unesco.org/docc/ENG_Florence_Declaration_4oct.pdf.

United Nations Educational, Scientific and Cultural Organization (2014). *Shaping the Future We Want: UN Decade of Education for Sustainable Development (2005-2014) Final Report*. Paris. http://unesdoc.unesco.org/images/0023/002303/230302e.pdf.

United Nations Educational Scientific and Cultural Organization (2016). *Culture Urban Future: Global Report on Culture for Sustainable Urban Development*. Paris. http://unesdoc.unesco.org/images/0024/002459/245999e.pdf.

United Nations Environment Assembly of the United Nations Environment Programme (2014). *Proposed Procedures for Enhancing Future Assessment Processes: Note by the Secretariat*. 16 May. UNEP/EA.1/INF/14. http://undocs.org/UNEP/EA.1/INF/14.

United Nations Environment Assembly of the United Nations Environment Programme (2014). *State of the Environment: Report of the Executive Director*. 11 April. UNEP/EA.1/4. http://wedocs.unep.org/bitstream/handle/20.500.11822/17682/K1400635.pdf?sequence=6&isAllowed=y.

United Nations Environment Assembly of the United Nations Environment Programme (2016). *UNEA's contribution to the global follow-up and review in the 2016 High Level Political Forum (HLPF) on the work of the United Nations Environment Programme*. Nairobi. https://sustainabledevelopment.un.org/content/documents/10554UNEA%20inputs%20to%20the%20HLPF%202016%20(Final).pdf.

United Nations Environment Programme (2007). *Global Environment Outlook 4: Environment for Development*. Nairobi. https://wedocs.unep.org/bitstream/handle/20.500.11822/7646/-Global%20Environment%20Outlook%20%204%20(GEO-4)-2007768.pdf.

United Nations Environment Programme (2011a). *Decoupling Natural Resource Use and Environmental Impacts from Economic Growth. A Report of the Working Group on Decoupling to the International Resource Panel*. Fischer-Kowalski, M., Swilling, M., von Weizsäcker, E.U., Ren, Y., Moriguchi, Y., Crane, W., Krausmann, F., Eisenmenger, N., Giljum, S., Hennicke, P. *et al.* (eds.). Nairobi. http://wedocs.unep.org/bitstream/handle/20.500.11822/9816/-Decoupling%3a%20natural%20resource%20use%20and%20environmental%20impacts%20from%20economic%20growth%20-2011Decoupling_1.pdf.

United Nations Environment Programme (2011b). *Towards a Green Economy: Pathways to Sustainable Development and Poverty Eradication*. Nairobi. https://reliefweb.int/sites/reliefweb.int/files/resources/Full_Report_2176.pdf.

United Nations Environment Programme (2012a). *Global Environment Outlook- 5: Environment for the Future We Want*. Nairobi. https://wedocs.unep.org/bitstream/handle/20.500.11822/8021/GEO5_report_full_en.pdf.

United Nations Environment Programme (2012b). *Global Environment Outlook- 5: Summary for Policymakers*. Nairobi. http://web.unep.org/geo/sites/unep.org.geo/files/documents/geo5_spm_english.pdf.

United Nations Environment Programme (2015a). *An Introduction to Environmental Assessment*. Nairobi. http://wedocs.unep.org/bitstream/handle/20.500.11822/7557/-An_introduction_To_environmental_Assessment-2015An_introduction_to_environmental_assessment_WEB_1.pdf.pdf.

United Nations Environment Programme (2015b). *The United Nations Environment Programme and the 2030 Agenda: Global Action for People and the Planet*. Nairobi. https://wedocs.unep.org/bitstream/handle/20.500.11822/9851/-The_United_Nations_Environment_Programme_and_the_2030_Agenda_Global_Action_for_People_and_the_Planet-2015EO_Brochure_WebV.pdf.pdf?sequence=3&isAllowed=y

United Nations Environment Programme (2016a). *GEO-6 Assessment for the Pan-European Region*. Nairobi. http://wedocs.unep.org/bitstream/handle/20.500.11822/7735/unep_geo_regional_assessments_europe_16-07513_hires.pdf.

United Nations Environment Programme (2016b). *A Framework for Shaping Sustainable Lifestyles: Determinants and Strategies*. Nairobi. http://www.oneplanetnetwork.org/sites/default/files/a_framework_for_shaping_sustainable_lifestyles_determinants_and_strategies_0.pdf.

United Nations Environment Programme (2016c). Report on the implementation of the integrated approach to financing the sound management of chemicals and waste. *The Second United Nations Environment Assembly* Nairobi, 23–27 May 2016. http://wedocs.unep.org/bitstream/handle/20.500.11822/17557/K1601889%20INF%2018.docx.

United Nations Environment Programme (2016d). *Delivering on the Environmental Dimension of the 2030 Agenda for Sustainable Development: A Concept Note*. Nairobi. http://sdgtoolkit.org/wp-content/uploads/2017/02/Delivering-on-the-Environmental-Dimension-of-the-2030-Agenda-for-Sustainable-Development-%E2%80%93-a-concept-note.pdf.

United Nations Environment Programme (2016e). *Global Gender and Environment Outlook*. Nairobi. https://wedocs.unep.org/bitstream/handle/20.500.11822/14764/Gender_and_environment_outlook_HIGH_res.pdf?sequence=1&isAllowed=y.

United Nations Environment Programme (2017a). *Towards a Pollution-Free Planet: Background Report*. Nairobi. http://wedocs.unep.org/bitstream/handle/20.500.11822/21800/UNEA_towardspollution_long%20version_Web.pdf?sequence=1&isAllowed=y.

United Nations Environment Programme (2017b). *Guidelines for Conducting Integrated Assessments*. Nairobi. https://wedocs.unep.org/bitstream/handle/20.500.11822/16775/IEA_Guidelines_Living_Document_v2.pdf.

United Nations Environment Programme (2018). *Environment Under Review*. Nairobi. https://www.unenvironment.org/explore-topics/environment-under-review.

United Nations Environment Programme, United Nations Development Programme, United Nations Economic Commission for Europe, Regional Environment Center, Organization for Security and Co-operation in Europe and North Atlantic Treaty Organization (2013). *Transforming Risks into Cooperation: The Environment and Security Initiative 2003-2013*. Geneva. http://documents.rec.org/publications/ENVSECTransformingRisks_FINAL_Web.pdf.

United States Department of Energy (2017). *U.S. Energy and Employment Report January 2017*. https://www.energy.gov/sites/prod/files/2017/01/f34/2017%20US%20Energy%20and%20Jobs%20Report_0.pdf.

United States National Intelligence Council (2017). *Global Trends: Paradox of Progress*. Washington, D.C. https://www.dni.gov/files/documents/nic/GT-Full-Report.pdf.

Wackernagel, M., Hanscom, L. and Lin, D. (2017). Making the Sustainable Development Goals Consistent with Sustainability. *Frontiers in Energy Research* 5(18). https://doi.org/10.3389/fenrg.2017.00018.

Wei, M., Patadia, S. and Kammen, D.M. (2010). Putting renewables and energy efficiency to work: How many jobs can the clean energy industry generate in the US? *Energy Policy* 38(2), 919-931. https://doi.org/10.1016/j.enpol.2009.10.044.

Whitmee, S., Haines, A., Beyrer, C., Boltz, F., Capon, A.G., de Souza Dias, B.F., Ezeh, A. *et al.* 2015. Safeguarding human health in the Anthropocene epoch: Report of the Rockefeller Foundation–Lancet Commission on planetary health. *Rockefeller Foundation–Lancet Commission on Planetary Health* 386(10007), 1973-2028. https://www.doi.org/10.1016/S0140-6736(15)60901-1

Wiesmeth, H. (2012). *Environmental Economics: Theory and Policy in Equilibrium*. Berlin Heidelberg: Springer-Verlag. https://www.springer.com/gp/book/9783642245138.

World Bank (2016a). *Poverty and Shared Prosperity 2016: Taking on Inequality*. Washington, D.C. https://openknowledge.worldbank.org/bitstream/handle/10986/25078/9781464809583.pdf.

World Bank (2016b). *The Global Monitoring Report 2015/2016: Development Goals in an Era of Demographic Change*. Washington, D.C. https://openknowledge.worldbank.org/bitstream/handle/10986/22547/9781464806698.pdf.

World Economic Forum (2017). *The Global Risks Report 2017: 12th Edition*. Geneva. http://www3.weforum.org/docs/GRR17_Report_web.pdf.

Yihdego, Y., Salem, H.S. and Pudza, M.Y. (2017). Renewable energy: Wind farm perspectives – The case of Africa. *Journal of Sustainable Energy Engineering* 5(4), 281-306. https://doi.org/10.7569/JSEE.2017.629521.

Zhang, X., Chen, X. and Zhang, X. (2018). The impact of exposure to air pollution on cognitive performance. *Proceedings of the National Academy of Sciences* 115(37), 9193-9197. https://doi.org/10.1073/pnas.1809474115.

Chapter 2

Drivers of Environmental Change

Coordinating Lead Authors: Tariq Banuri (University of Utah), Fernando Filgueira Prates (Centro de Informaciones y Estudios del Uruguay (CIESU), Diego Martino (AAE Asesoramiento Ambiental Estratégico and ORT University), Indu K Murthy (Indian Institute of Science), Jacob Park (Green Mountain College), Dimitri Alexis Zenghelis (London School of Economics)

Contributing Author: Matthew Kosko (University of Utah)

GEO Fellow: Maria Jesus Iraola Trambauer (University College London)

Executive summary

Population growth will be highest in countries that are very poor, have a low carbon footprint per capita and high gender inequity in terms of access to education, work, and sexual and reproductive rights *(well established)*[1]. It will also remain important in countries going through their early or late demographic dividend (most middle-income and upper middle-income countries). These are also the countries that have presented the highest increases in carbon footprints per capita – and in ecological footprints more broadly. {2.3.1}[2]

The world's population will become older, including in the global South, more urban and will live in smaller households *(well established)*. In a business-as-usual scenario, all these trends will contribute to higher levels of emissions. This is true even if, in some cases, urban milieux show a more efficient relationship between welfare improvement and environmental footprint. {2.3.3}

Between today and 2050, the global urban population will continue to increase *(well established)*. Around 90 per cent of the growth of cities will take place in low-income countries, mainly in Asia and Africa, which are the world's most rapidly urbanizing regions. {2.4}

Serious social and environmental challenges of urbanization remain unsolved in many urban areas, particularly but not solely, in the global South *(well established)*. These challenges can be exacerbated by climate change and rapid urban growth in regions and cities that currently lack the capacity to face these mounting pressures. {2.4.1, 2.4.2}

On the other hand, urban population growth can represent an opportunity to increase citizens' well-being while decreasing their ecological footprint *(established but incomplete)*. Urbanizing areas can therefore be seen as an opportunity for the reduction of greenhouse gas (GHG) emissions through the appropriate planning and design of urban form and infrastructure. {2.4.4}

Economic development in the past has been a driver of increased resource use and environmental damage *(well established)*. The production of internationally traded goods accounts for about 30 per cent of all CO_2 emissions. The household consumption, meanwhile, of goods and services over their life cycle, accounts for about 60 per cent of the total environmental impact from consumption (UNEP 2010). Economic development continues to be the number-one policy priority in most countries, because of its material benefits and its potentials for poverty eradication, for narrowing inequalities in income and wealth between and within countries, and for providing win-win scenarios that can facilitate collective action and global solidarity. At the same time, economic development must coincide with sustainable consumption and production. {2.5.1, 2.5.4}

Achieving the SDGs will require that the fruits of sustainable economic development are predominantly used to increase the capacity, capabilities and opportunities of the least-advantaged people in societies *(well established)*. Educating girls, improving the status and opportunities for women, and enabling poor people to achieve full participation in society will strengthen both sustainable economic growth and sustainable economic development, and reduce alienation and conflicts in society. {2.5.2, 2.5.3}

Technological advances have resulted in both positive as well as negative impacts *(well established)*. Oil and other fossil fuels have accelerated economic development and lifted the standard of living for billions of people in both industrialized and developing countries, but they have also contributed to climate change. At the same time, there are current and emerging technology business models, which are building a more circular economy, creating less resource-intensive processes, and accelerating more effective resource innovation cycles. {2.6.1, 2.6.2}

Technological advances have created unintended consequences that make it difficult to determine whether the advances have long-term positive and/or negative impacts *(established but incomplete)*. Scientific analyses of technology issues often fail to capture the important negative and rebound effects of technologies as well as the complex policy and market challenge of diffusing sustainable technologies to developing countries. {2.6.3, 2.6.4}

Climate change has become an independent driver of environmental change and poses a serious challenge to future economic development *(well established)*. Regardless of human action, or even human presence on the planet, impacts will continue to occur. Climate change thus poses a challenge to growth and development. {2.7.1, 2.7.2}

Climate change poses risks to human societies through impacts on food, and water security *(established but incomplete)*, and on human security, health, livelihoods and infrastructure. These risks are greatest for people dependent on natural resource sectors, such as coastal, agricultural, pastoral and forest communities; and those experiencing multiple forms of inequality, marginalization and poverty are most exposed to the impacts. {2.7.3}

Climate change will amplify existing risks and create new risks for natural and human systems *(well established)*. Risks are unevenly distributed and are generally greater for developing countries (mainly for SIDS) and for disadvantaged people and communities in countries at all levels of development. Risk of climate-related impacts results from the interaction of climate-related hazards with the vulnerability and exposure of human and natural systems, including their resilience and ability to adapt. {2.7.4}

1 This assessment uses confidence statements to better inform policy makers of the extent of evidence on a particular subject and the level of agreement across this evidence. The various confidence statements used include: "well established" (much evidence and high agreement), "unresolved" (much evidence but low agreement), "established but incomplete" (limited evidence but good agreement) and "inconclusive" (limited or no evidence and little agreement). Annex 1-4 provides more information on the use of confidence statements.

2 Statements in the Executive Summaries of different chapters are referend to the subsections of the chapter where the underlying analysis and evidence for the statement can be found.

There is an important need to limit the potential negative sustainability impacts of drivers of population, economic development and climate change *(established but incomplete)*. Whether these three drivers serve as catalysts of positive (rather than negative) transformative response in the form of social equity, environmental resilience, and poverty eradication is likely to be determined by uncertain long-term impacts of drivers of urbanization and technology. {2.8, **Figure 2.23**}

2

2.1 Introduction and context

The environmental movement has gone through many phases. Initially the movement consisted broadly of the conservation school, which emphasized husbanding of both renewable and non-renewable resources (especially forests) for future development, and the preservation school, which saw nature as intrinsically valuable (Eckersley 1992). In addition to these economic and aesthetic concerns, the modern environmental movement is now more about risk, the risk that environmental degradation poses to human health and well-being (Carson 1962; Rees 1995; Guha 1999; Lenton *et al.* 2008; Rockstrom *et al.* 2009a; Diamond 2011). Increasingly, there are concerns that the enormous gains in life expectancy and quality of life since the industrial revolution are in danger of being reversed (GBD 2015 Mortality and Causes of Death Collaborators; Harari 2017).

The five drivers reviewed in this chapter — population growth and demographics, urbanization, economic development, new technological forces, and climate change — have led to an unprecedented expansion of wealth for many but have also left many behind and could produce trouble for the future. If current trends in inequality continue, the top 0.1 per cent of the population will own more wealth than the global middle class by 2050 (WID 2018).

2.1.1 Overview of the Drivers

As noted in Section 1.6, the analysis conducted in the GEO-6 uses the DPSIR framework, where DPSIR stands for Drivers, Pressure, State (of the environment), Impact (on the environment and human well-being), and Response[3]. 'Drivers' are anthropogenic *inertial* forces – social, economic, ecological, technological, and political. They are inertial forces, in the sense that they have their own rules of motion and reversing them will require time and effort. GEO-5 referred to two drivers – population and economic development – to which GEO-6 adds three more, urbanization (previously covered under population), technology and climate change.

Three of these drivers – population, economic development, and technology – are ubiquitous in the DPSIR literature (Nelson 2005) and represent the disaggregation into three components of aggregate human consumption, and therefore of what is necessary for meeting survival as well as other welfare needs.

- *Population*: Other things being equal, more people will mean a proportionally higher pressure on the environment. In such a scenario, long-term sustainability is incompatible with growing populations, which the literature indicates will continue to grow at a global scale throughout this century. It is imperative in the present, therefore, to attend to how key population dynamics – including fertility rates, ageing populations, displacement and gender inequality – interact at multiple scales and impact environmental sustainability.
- *Economic development*: This refers to an increase in human welfare, which depends on material consumption and many other factors, including the environment. While economic development has been highly correlated with economic growth in the modern era, the two are quite distinct, empirically as well as conceptually. Per capita consumption is expected to continue increasing in the foreseeable future (because of the unfinished agenda of eradicating poverty, meeting survival needs and enabling individuals to pursue prosperity). To decouple growth from negative environmental impacts, resource-efficient, sustainable patterns of consumption are needed.
- *Technology*: Technological change is well understood as a driver of change, both negative and positive. Negatively, it provides an opportunity to accelerate, with incentives, the harnessing of natural resources for human ends; in times of crisis, incentives strongly favour adoption of riskier options and elimination or minimization of safeguards. Positively, technological progress also creates more efficient options, which can meet human needs at lower resource costs.

In this assessment, urbanization and climate change are added as independent drivers because of their importance in socioeconomic change.

Urbanization has been going on throughout history, but its pace, scale and impact have accelerated sharply in recent decades. As such, it is included independently as a fourth driver.

Likewise, climate change has been added as a fifth driver, even though, in principle it could be represented as an outcome of the other drivers. According to the Fifth Assessment Report (AR5) (Intergovernmental Panel on Climate Change [IPCC] 2014), the world is on the threshold of entering the era of 'committed climate change', namely that some impacts of climate change have now become irreversible (such as extinction of species and loss of biodiversity) and regardless of future mitigation or adaptation actions. In other words, even if all human activity were to cease, the impacts of climate change would continue to manifest themselves over the next few centuries.

Taken together, these five drivers are bringing about changes in natural as well as social systems. These impacts range from resource depletion to biodiversity loss, water scarcity, changes in the hydrological cycle, health impacts, and ecosystem degradation as well as pollution. In the absence of an adequate response, a changing climate could lead to a pre-modern world of famine, plague, war, and premature death.

2.2 Changes since the last assessment

A number of changes, summarized as follows, have taken place since the fifth Global Environmental Assessment (GEO-5).

- *Population*: With the 2018 world population estimated at 7.6 billion people, estimates by the United Nations indicate that the peak human population is likely to be higher than had been projected earlier. The world has also seen an increase in the number of migrants and refugees, in part as the result of heightened conflict and increased environmental degradation. Other demographic variables remain on track.
- *Urbanization*: Having passed the symbolic 50 per cent of population living in urban areas, trends indicate that rural-to-urban migration will continue, with acceleration in the global south. This represents both an increased driver of environmental pressure and an opportunity to enhance sustainability.

3 Note that The DPSIR framework has come under some criticism, especially on the elision over the interdependence between the drivers. In this assessment, we include an explicit examination of this interaction.

- *Economic development*: The global economy is coming through a slow recovery from the 2008 recession, and there are concerns about the persistent debt crisis, the increase in income inequality, and emerging instability due to trade wars. Offsetting factors include the increasing role and contribution of emerging economies, and the adoption of the Sustainable Development Goals (SDGs) as a new global aspiration and orientation for development (Section 2.5.1).
- *Technology*: The environmental crisis is creating perverse incentives for countries and businesses to resort to environmentally riskier technological options, including geo-engineering and nuclear technology. Yet it is also providing sound incentives for such technologies as renewable energy, energy efficiency, energy storage, and expanded application of information and communications technologies (ICTs).
- *Climate change*: The IPCC-AR5 states that 'the warming of the climate system is unequivocal, as evidenced by observations of increases in global temperatures, widespread melting of snow and ice and rising sea level'. IPCC also notes that human influence on the climate system is clear, and that 'many aspects of climate change and associated impacts will continue for centuries, even if anthropogenic emissions of greenhouse gases are stopped' (IPCC 2014, p. 16).

Besides the drivers themselves, various policy developments since GEO-5 also need mention. A number of global agreements were reached to address key issues pertinent to this assessment, including a new comprehensive treaty to address climate change, an agreement on the new development agenda, including the adoption of the SDGs, and agreements on mobilization of finance for development as well as climate action. In addition, several countries adopted national policies on disaster risk management, renewable energy, urbanization, transport, and water and sanitation.

Recent years have also seen increased interest in technologies that can accelerate social and environmental benefits and enable people, institutions, and communities to achieve their needs at lower resource costs. Section 2.7 focuses on the interactions across the five selected drivers and how actions on one driver may affect the others.

2.3 Population

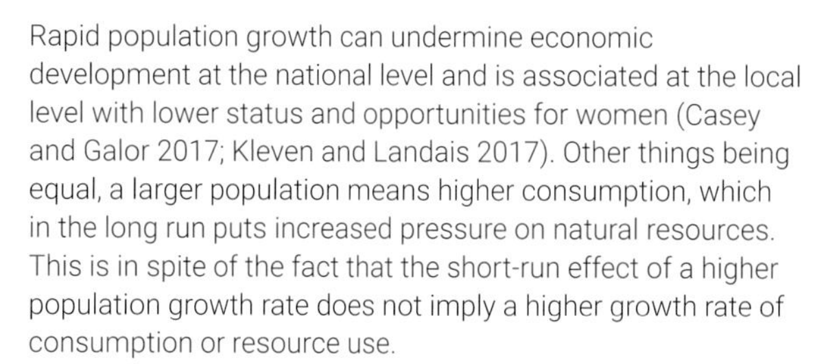

Rapid population growth can undermine economic development at the national level and is associated at the local level with lower status and opportunities for women (Casey and Galor 2017; Kleven and Landais 2017). Other things being equal, a larger population means higher consumption, which in the long run puts increased pressure on natural resources. This is in spite of the fact that the short-run effect of a higher population growth rate does not imply a higher growth rate of consumption or resource use.

Box 2.1: Relationship between higher population and growth rate of consumption and resource use

Countries with higher population growth rates are typically also poorer, have lower carbon footprints per capita and experience slower growths in income per capita. For this reason, increased population does not always lead to increased consumption or resource use. High inequality and population growth are also inextricably linked. Inequality is a root cause of both rapid population growth and environmental degradation. To moderate population growth in high-growth regions, people need access to voluntary family planning and other reproductive health services, as well as to educational and employment opportunities.

While the most important source of environmental pressure comes from the global North and its high carbon footprint per capita, high population growth in the global South is expected to– under current conditions – reinforce environmental pressures and enhance global inequality. Here, countries are transitioning to early and late demographic dividend stages.

Equally, high population growth rates constitute a drag on the development process. Whereas most countries that have been able to make the transition to developed status have seen massive reduction in their fertility rates (Sinding 2009), at the level of families and individuals, poverty, conversely, is typically associated with having many children (Gillespie *et al.* 2007).

Box 2.2: The demographic dividend

The demographic dividend takes place when the dependency ratio goes down – because of lower fertility and the fact that societies have not aged yet. Post-dividend societies are those that are already starting to increase their dependency ratio, led now by older people. Countries going through their demographic dividend – also called the window of demographic opportunity – benefit from increasing numbers of active-age population (15-64 years), decreasing numbers of young dependents (0-14 years) and small numbers of older people (64 years and over). In schematic terms, pre-dividend countries are the poorest, early dividend ones are the low- to middle-income countries, and late demographic dividend countries are mostly upper middle-income countries. Post-dividend countries are almost always rich countries with some upper middle-income countries from the former socialist block. Pre-dividend countries and those in early stages of the demographic dividend are expected to increase their population quite strongly, late-dividend societies are expected to grow still, but more moderately, and post-dividend societies will increase their populations in the years to come at a much slower rate or, in some cases, might even decrease their absolute population, and will continue to increase their older population. Pre-dividend countries and those in early stages of the demographic dividend have a smaller carbon footprint per capita and GDP. Yet as can be seen in this chapter, both early- and late-dividend countries (where both population and GDP should be expected to grow) have increased their carbon footprint per capita substantially.

Figure 2.1: World population, emissions and fertility

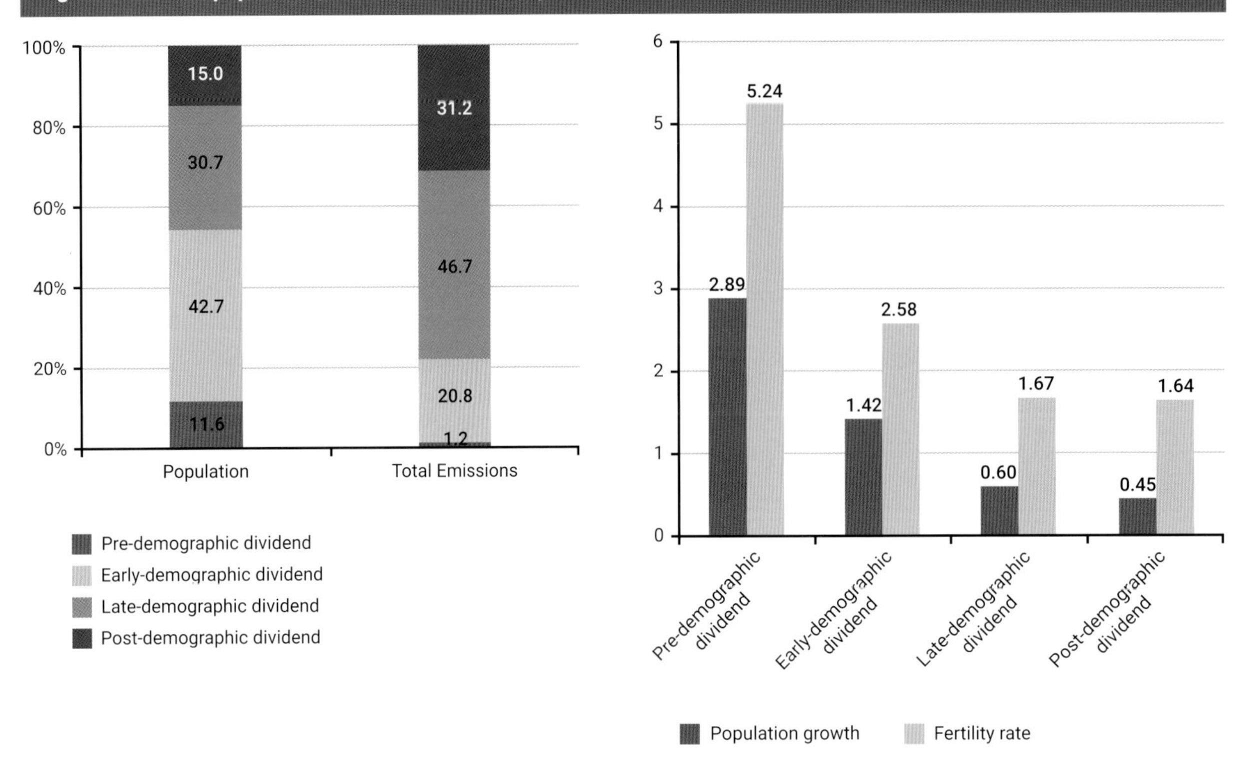

Source: Own elaboration based on World Development Indicators (2017 (https://data.worldbank.org/products/wdi)

Finally, countries with high population growth rates are often characterized by adverse conditions for women, including lack of access to education and health services, lower levels of literacy and life expectancy, higher rates of maternal and child mortality, significant barriers to participation in the labour force, and other discriminatory factors (Iversen and Rosenbluth 2010).

Sexual and reproductive health is often thought of as a universal right. While no single human right is framed in such terms, in the words of the United Nations Population Fund [UNFPA], "no country today – even those considered the wealthiest and most developed – can claim to be fully inclusive, where all people have equal opportunities and protections, and fully enjoy their human rights" [UNFPA 2017, p. 10.) Not only are sexual and reproductive inequalities and economic inequality strongly correlated, but the literature demonstrates they may be mutually reinforcing (UNFPA 2017). Poor women, particularly those who are less educated and live in rural areas, are often least able to access sexual and reproductive health services. Lack of access to these services, including contraception, places a woman at heightened risk of unintended pregnancy, which results in greater health risks and lifelong negative economic repercussions for herself and her children (UNFPA 2017).

Population growth can affect the environment not only through consumption and use of natural resources, but also through its impact on other factors. This includes the strain it can create on governance, its effects on the probability of conflict over limited resources, and its impact on rapid and unplanned urbanization (Organization for Economic Cooperation and Development [OECD] 2016).

As an example, consider the experience of Latin America. As one of the regions with the highest inequalities, it experienced rapid urbanization and the formation of megacities far too rapidly for governing systems to cope. The result was inequality within dysfunctional urban milieux, rendering them segregated, unsafe and violent, in turn starving them of public resources, dampening economic growth, shrinking civic spaces, weakening public and merit goods, and undermining the quality and availability of collective services (Filgueira 2014). This reinforces inequality by encouraging private, segregated solutions for leisure, education, security, transport and housing.

The following analysis focuses on global population trends and global effects on environmental sustainability with some discussion of impacts on a subregional, national and local level.

The expected trends show that global population growth rates will slow but will continue to be positive in all regions except Europe, at least until 2040 – even in the most conservative estimates (United Nations Department of Economic and Social Affairs [UN DESA] 2017). This means that population growth will remain quite strong in many developing regions. These regions will also rapidly increase gross domestic product (GDP) and consumption per capita given both historical trends and accepted projections. The rapid increase in the carbon footprint per capita of countries sitting in the middle of the demographic transition (early- and late-demographic dividend) clearly illustrates the likely effects of high population growth on CO_2 aggregate emissions under current circumstances **(Figure 2.2)**.

Migration will probably move a large part of the population born in areas of low carbon footprint per capita (rural areas, the

Figure 2.2: Emissions per capita according to demographics

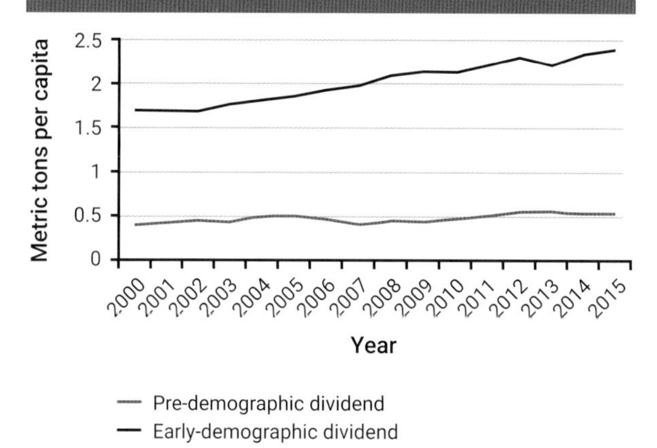

Source: Own elaboration based on World Development Indicators (2017 (https://data.worldbank.org/products/wdi)

global South) to areas of higher carbon footprint (O'Neill *et al.* 2012; OECD 2016). These are shifts that can increase the efficiency of carbon production per unit of output (technology or agglomeration reduces pressure for a given level of welfare). These shifts also increase consumption, however, and so increase aggregate CO_2 emissions in the process.

Finally, the still-growing world population will become older and is living and will live in smaller households (Dalton *et al.* 2008; O'Neill *et al.* 2012; UN DESA 2017).

These trends imply – on average, and again, with other things being equal – a higher carbon footprint per capita. In most cases, this simplified logic of population growth, dynamics and growing carbon emissions (assuming a business-as-usual scenario – see Chapter 21) also applies to the national and local levels and to other environmental variables such as water and air pollution, soil degradation, desertification and deforestation.

It should be stressed that population dynamics and population growth do not in themselves lead to an unsustainable environmental path. Rather, this path is the result of population growth happening with the current consumption and production patterns. Unsustainable consumption and production are each largely fuelled by heightened inequality. Both within and between countries, inequality remains one of the largest obstacles to environmental sustainability (Chancel and Piketty 2015; Oxfam 2015).

There are two detrimental effects against sustainability that are produced directly by heightened inequality:

1. because of the highly uneven distribution of resources, the level of growth required to lift people out of poverty is far larger than it would be in a more egalitarian distribution (Ravallion 2001; Bourguignon 2002; World Bank Group 2004). Put another way, the world would not have to grow at very high rates to improve the lives of those worse off if the distribution of those gains was more equally distributed.

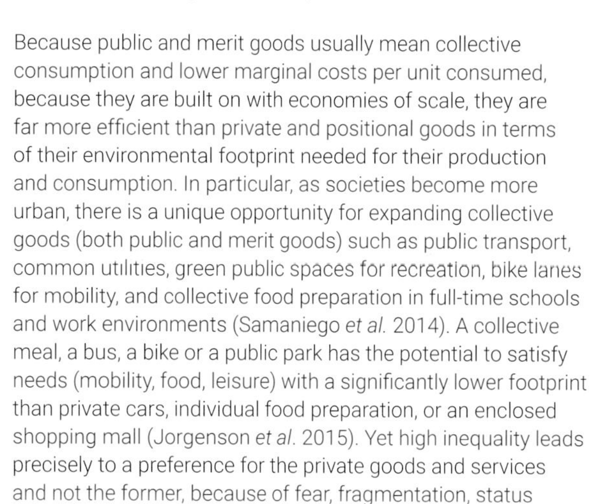

2. high inequality is associated with a preference for overconsumption of private and positional goods, weakening public and merit goods (López and Palacios 2014; Samaniego *et al.* 2014).

Because public and merit goods usually mean collective consumption and lower marginal costs per unit consumed, because they are built on with economies of scale, they are far more efficient than private and positional goods in terms of their environmental footprint needed for their production and consumption. In particular, as societies become more urban, there is a unique opportunity for expanding collective goods (both public and merit goods) such as public transport, common utilities, green public spaces for recreation, bike lanes for mobility, and collective food preparation in full-time schools and work environments (Samaniego *et al.* 2014). A collective meal, a bus, a bike or a public park has the potential to satisfy needs (mobility, food, leisure) with a significantly lower footprint than private cars, individual food preparation, or an enclosed shopping mall (Jorgenson *et al.* 2015). Yet high inequality leads precisely to a preference for the private goods and services and not the former, because of fear, fragmentation, status competition and segregation.

It is because of the inevitability of population growth and other demographic dynamics (urbanization, smaller households and ageing populations) that it is critical to decouple these trends from unsustainable environmental pressure, by changing current consumption and production patterns.

2.3.1 Global population growth and composition

Four trends can be predicted with confidence: the world population will continue to grow (until at least 2050; **Figure 2.3**), average age will increase, populations will become more urban, and household sizes will become smaller (United Nations 2015a). These trends are the inevitable results of underlying processes: industrialization, the agricultural technological revolution and resulting landholding patterns, the shift from extended households towards nuclear ones, and the dramatic drop in mortality due to the epidemiological transition (Lopez and Murray 1996; GBD 2015 Mortality and Causes of Death Collaborators 2016).

Policy and behavioural changes could moderate the rate at which these changes occur, but not reverse them. All other things being equal, smaller households, urbanization and ageing will generate more environmental impact per capita. Given that such trends are inevitable – to a larger or lesser extent – there are only three possible courses of action.

1. when possible and desirable, such trends can be moderated. For example, lower fertility (due to improved access to contraception and improved economic and social empowerment for women) is positive for economic development, moderating inequality, combating poverty and decreasing environmental pressure.
2. avoiding rapid surges in unplanned urbanization due to expulsion from rural areas provides a win-win scenario, allowing for national trajectories and urbanization processes that are more balanced and welfare-enhancing, which could enhance green cities and improve ecosystem connectivity. (rural expulsion is due in part to underinvestment in sustainable farming techniques and overexploitation in the depletion of natural resources, among other causes.)
3. patterns of consumption and production remain highly inefficient in terms of CO_2 production and other environmental pressures. Both hard and soft technological innovations (substitutes for fossil fuel energy sources, soil management, urban planning, collective care services in urban centres, public transportation, etc.) can drastically change the elasticity of consumption and production to units of environmental pressure.

2.3.2 Population growth estimates

In 2017 (UN DESA 2015a), the total world population was 7.55 billion, growing at 1.10 per cent annually, a decline from a decade earlier, when it was growing at 1.24 per cent. Under middle projections for fertility, there will be 8.55 billion people by 2030, and almost 10 billion by 2050 (9.77 billion). However, any forecast looking a century into the future comes with significant caveats. Depending on the rate of decline in fertility rates, the global population could rise as high as 13.2 billion by the end of this century or reach 9.4 billion by mid-century and stay around those levels until 2100 (see Section 21.3.1).

Figure 2.3: Projected world population

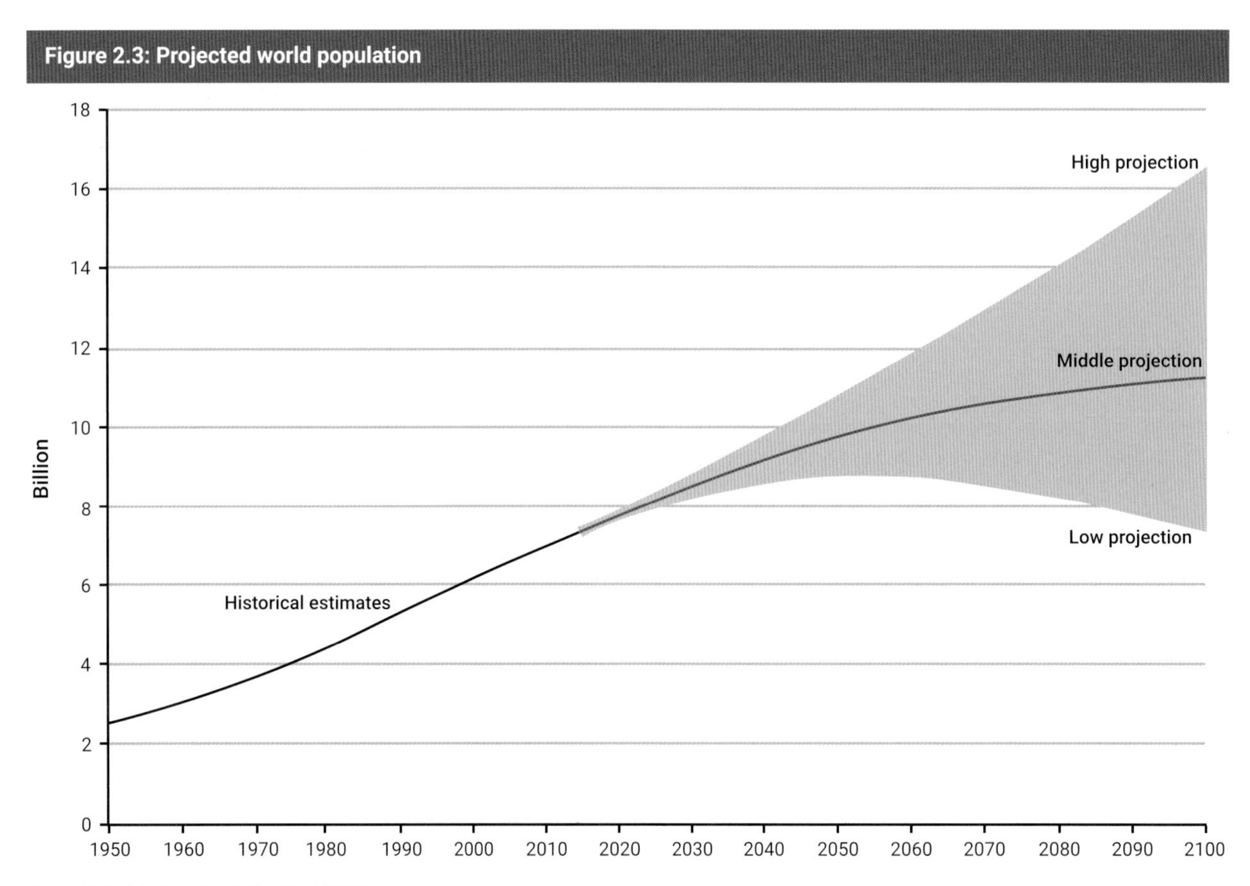

Source: United Nations Population Fund (2017)

Figure 2.4: Consumption and associated environmental pressures are unequally distributed between nations

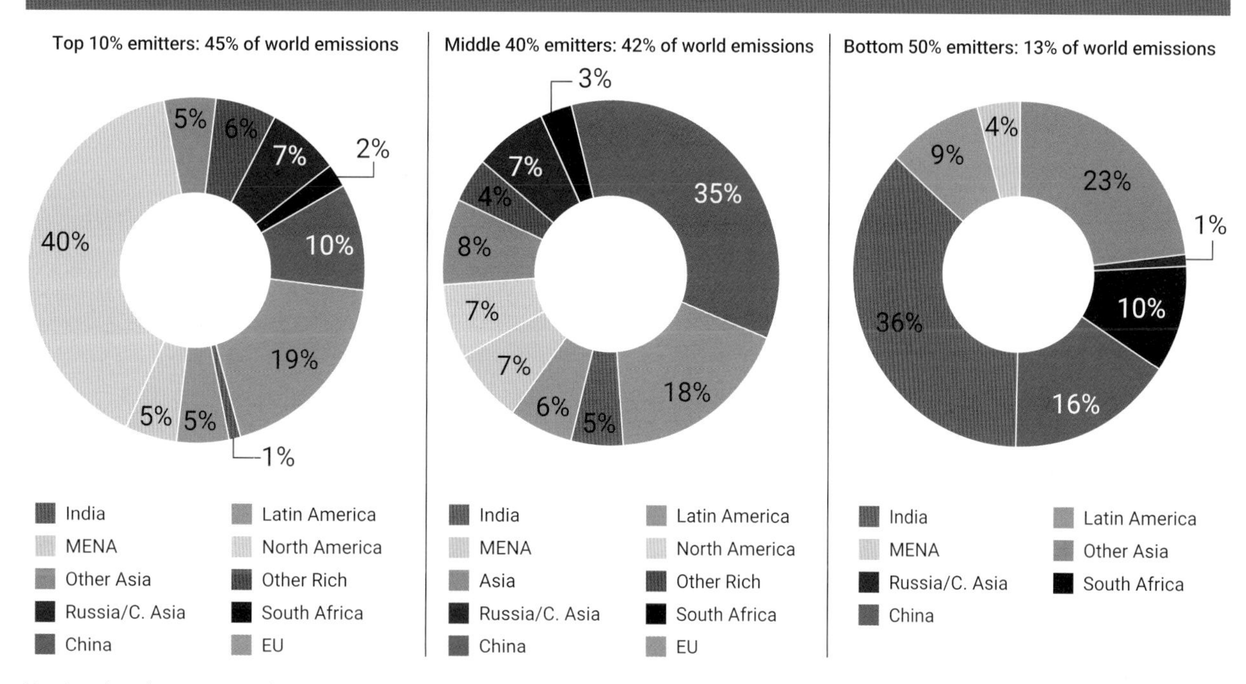

Note: In order to better represent the contribution of different groups of emitters to total CO_2 emissions, the charts split the world in three groups: top 10 per cent, middle 40 per cent and bottom 50 per cent CO_2 emitters in each country. For each of these groups, the chart presents the percentage of the group's emissions stemming from each region of the world.

Source: Chancel and Picketty 2015.

The key points to take away from this projection are:

a. that the population will continue to rise until at least the middle of the century, and perhaps longer,
b. that there are significant uncertainties about long-term trends, and
c. that population control is not responsive to direct policy intervention, but rather indirectly to policies that, for example, lower fertility rates through women's control over reproductive choices.

Population growth depends on the numbers of births and deaths in a given year, and these in turn depend on three interrelated factors – fertility, mortality and the age and sex structure of the population. These last three depend on human behaviour, health conditions and demographic inertia, respectively. While age and sex structures change slowly, there are uncertainties about the rate of decline of fertility rates, as well as future trends in mortality rates. Also, while changes in fertile behaviour result in a lower rate of population growth, they do so only eventually, after considerable lags.

Mortality rates are declining rapidly in almost all developing countries, but fertility rates remain high in the least developed group, where the average is above 4 children per woman, almost twice the replacement level of fertility of 2.1 children (UNFPA 2017). Fertility rates can respond to gender policies, but if emerging medical technologies result in a dramatic extension of lifespans, population growth would be closer to the higher-end estimates, and ageing of the world population would be far more pronounced.

2.3.3 Population composition and distribution

There is increasing evidence of the complex interactions between the environment and the distribution and composition of the population (age, urban/rural residence, and household structure) (see Jiang and O'Neill 2007; Dalton *et al.* 2008; O'Neill *et al.* 2012; Liddle 2014).

Population growth is distributed unevenly around the globe and within nations, as a result of differences in fertility patterns and migration trends. Countries with high fertility rates, young populations and steeply declining mortality rates will grow more rapidly than others. In the coming decades **(Figure 2.5)**, According to current trends Africa is projected to grow the fastest, followed by Asia, Latin America, North America, Oceania and Europe (United Nations 2015b, 2017).

The impact of natural population growth is partially mitigated by migratory patterns, which will lead to shifts in population from less developed regions to more developed ones, and from rural to urban areas (OECD 2016). The pace of migration has increased in the last 50 years and will continue to do so in the next 30 years (Massey and Taylor 2004; International Organization for Migration [IOM] 2015). This is driven by the persistence of the underlying push-and-pull causal factors:

- the push effects of global inequality, poverty, conflict-ridden regions, and
- the pull effects, such as already established migrant communities in more developed regions sometimes attracting others from less developed regions.

South-South international migration has also increased along the same patterns as South-to-North migration (Hujo and Piper 2010). In many cases, migration is actually fuelled by environmental degradation that makes life unsustainable in the original locations (Leighton 2006).

Migration tends to dampen population growth, as data show that migrants typically have lower fertility rates in their new contexts (Majelantle and Navaneetham 2013). The net impact on the environment can still be adverse, however, given that migrants access higher levels of income and consumption than they had in their previous milieux. Given that one of the objectives of development, as well as of migration, is less poverty, increased income and consumption are desirable outcomes.

Figure 2.5: World population distribution and composition

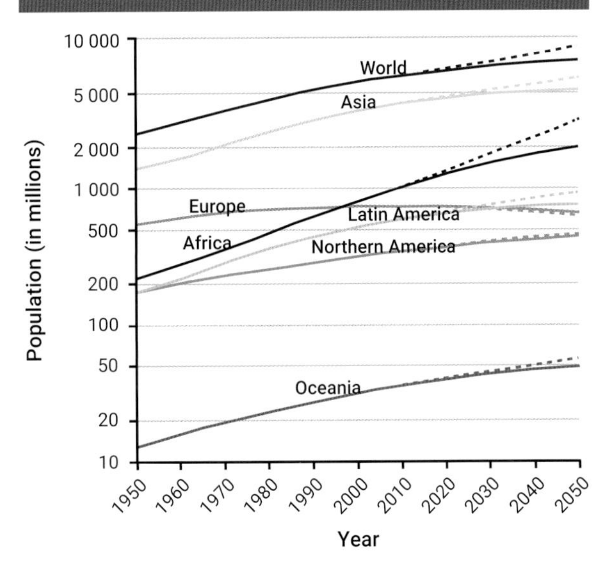

Source: United Nations Population Fund (2017)

Increased per capita consumption of resources may not be the only impact of migration on the environment and natural resources; resource efficiency may also change – for example, energy and materials use per unit of consumption may decline.

2.3.4 From population programmes to gender equality and women's empowerment

Population programmes, which were a major policy focus in the 1960s and 1970s, have since been discontinued in many countries, even though their benefits are widely recognized (UNFPA 2017). Part of the explanation for their decline was the systematic violation of basic rights that some of these programmes entailed through mass sterilization or forced and coercive policies limiting women's reproductive choices.

Figure 2.6: Contraceptive prevalence and total fertility

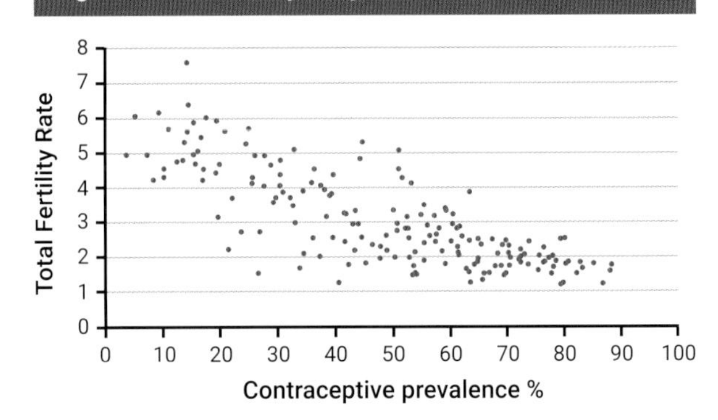

Source: Own elaboration based on World Development Indicators (2017) (https://data.worldbank.org/products/wdi)

The United Nations International Conference on Population and Development in Cairo in 1994 and the Women's Conference in Beijing in 1995 contributed to the view that population policies should respect the rights of women and their choices, moving from population targets to a rights-based approach that places reproductive control in the hands of women. There is little doubt that existing population policies in Africa, Asia and parts of Latin America can contribute markedly to moderating the rate of population growth while respecting gender equality and empowering women. In turn, this seems likely to contribute to more robust economic growth, through higher female labour-force participation in the market economy, and improved health for mothers as well as children (UNFPA 2017).

These policies comprise a suite of actions, including access to modern contraceptive methods **(Figure 2.6)**, improved access for women and men to voluntary family planning and other reproductive health services, investment in women's education, removal of barriers to female labour-force participation, institution of legal penalties for discriminatory practices associated with traditional patriarchal behaviour, and investment in the social and economic uplift of less developed areas within countries, and of developing countries more generally.

2.3.5 Gender and education

Placing reproductive choices as much as possible in the hands of women has proven to have a definite impact on timing and quantity of childbearing (UNFPA 2017; United Nations Entity for Gender Equality and the Empowerment of Women [UN Women] 2017). This is affected, in part, by access to education and employment. One of the main contributing factors to high fertility rates is lack of women's access to education and employment opportunities. In least developed countries, where fertility rates are highest, access to education for girls tends to be lowest. Causal relations run both ways. **(Figure 2.7)**.

Figure 2.7: Female secondary education and total fertility rates

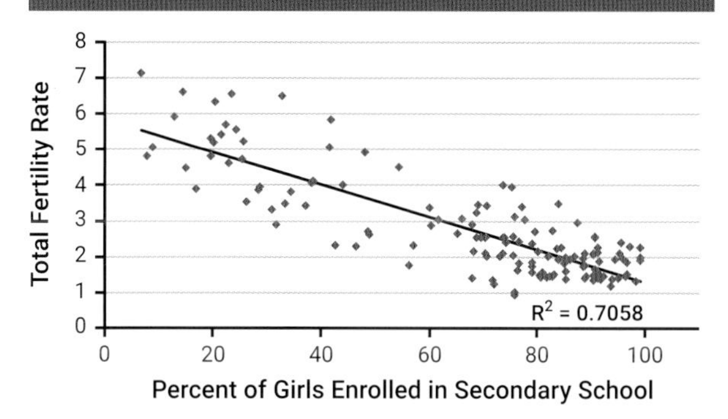

Source: Earth Policy Institute (2011)

2.3.6 Inequality, migration and cities

North-South inequality and international inequality in general is a major driver of migratory patterns. Closing international welfare gaps and promoting growth in the South has proven to help moderate migratory flows, which can allow for slower and eventually less CO_2 intensive welfare enhancing trajectories.

Similarly, within-country migration is driven by inequalities, especially between rural and urban areas, leading to rapid – and sometimes environmentally unmanageable – urbanization. Again, adequate developmental support for rural areas helps moderate such pressures. (IOM 2015).[4]

2.4 Urbanization

A distinct channel through which demographic trends affect environmental resources is through urbanization (also analysed as a cross-cutting issue in Section 4.2.5 of this report). The facts about urbanization are well known. Urban areas have higher incomes and consumption, greater access to political power, higher rates of economic growth, and, per capita, place a higher pressure on natural resources. On the other hand, cities exhibit greater efficiency in the use of resources per unit of income generated and better potential for energy efficiency (Dodman 2009; Bettencourt and West 2010; Barrera, Carreón and de Boer 2018; Cottineau *et al.* 2018). Cities are also the engines of economic growth. No country has made the transition from poverty to middle-income status without experiencing a period of rapid urbanization. Managed effectively, though, urbanization can help in the achievement of SDGs, efficiently and sustainably. Finally, urbanization is generally associated with a lowering of fertility rates (Martine, Alves and Cavenaghi 2013).

Slightly more than half of the world's population is currently living in urban areas, a share that is expected to rise to 60 per cent by 2030 and 66.4 per cent by 2050 (Brenner and Schmid 2014; United Nations 2014; Melchiorri *et al.* 2018). It should be noted that urban areas are defined in different ways worldwide, so UN DESA information is based on heterogeneous data sources. Using a globally harmonized definition of urban areas that combines demographic characteristics and density grids, Melchiorri *et al.* (2018) place global urban population at 85 per cent in 2015. Alternative understandings of the urban condition (Brenner and Schmid 2014) that can benefit from these new methodologies, and from an analysis of the transboundary ramifications of cities (Section 4.2.5), could represent an important tool for policy analysis and environmental governance.

Around 90 per cent of the growth of cities will take place in low-income countries (United Nations Human Settlements Programme [UN-Habitat] 2014). Africa is the world's most rapidly urbanizing region, while European cities grew the least in the 1995-2015 period (UN-Habitat 2016). The critical factor accounting for these trends is neither fertility nor age structure (which are, respectively, lower and older in urban areas), but migration (UN-Habitat 2016).

The coming decades are crucial. It took 200 years for the urban share of the world's population to rise from 3 per cent to 50 per cent, to 3.5 billion people in 2010 (United Nations 2014). This population is set to more than double over this century, but in all the centuries that follow, we may add, at most, another billion or so. This makes the current global urbanization era not just immense, but also brief (Fuller and Romer 2014). The choices around investment and design of new and existing cities are effectively determining the infrastructure, technologies, institutions and patterns of behaviour that will define the functioning of our cities and the future of the planet for the foreseeable future. This suggests there is a very narrow window of opportunity to help plan and design this future. The world's infrastructure will more than double in the next 20 years (Bhattacharya *et al.* 2016).

2.4.1 Cities of different sizes face different challenges

The pattern of urbanization is also relevant for understanding both the potential for growth and the impact on natural resources. At the high end of urbanization are megacities, defined by UN-Habitat as cities with more than 10 million people (UN-Habitat 2016, p. 7), most of which are located in the global South. In 1990, there were 10 megacities housing 153 million people, or 7 per cent of the total urban population; by 2014, there were 28 megacities, with 453 million people, or 12 per cent of the total (UN DESA 2014); in 2016 there were 31 megacities, 24 located in the less developed regions or the global South; of these, 6 were in China and 5 in India (UN DESA, Population Division 2016).

4 This support in rural areas is not an alternative to avoiding migration. Such a policy can still have detrimental effects on migrants and host areas.

However, while megacities might be economic powerhouses, they do not represent the majority of the urban population **(see Figure 2.8)**, and are not the fastest-growing urban centres **(see Figure 2.9)**. Small and medium cities now account for roughly 50 per cent of the world's urban population and are growing at the fastest rates (UN DESA 2014; United Nations Economic and Social Commission for Asia and the Pacific [UNESCAP] and UN-Habitat 2015). They will "deliver nearly 40 per cent of global growth by 2025, more than the entire developed world and emerging market megacities combined" (UN-Habitat 2015a, p. 2; Dobbs *et al.* 2011). Small and medium cities are also more vulnerable to natural hazards than big cities and megacities (Birkmann *et al.* 2016).

2.4.2 Urban agglomeration economies

Agglomeration economies reflect the advantage of people clustering to reduce transport costs for goods, people and ideas. Higher productivity attracts inflows of people, who in turn further increase productivity. Agglomeration economies thus generate a positive feedback loop and multiply the impact of external productivity factors, and so boost urban populations and wages (Glaeser and Gottlieb 2009; Zenghelis 2017).

Figure 2.8: Global urban population growth propelled by cities

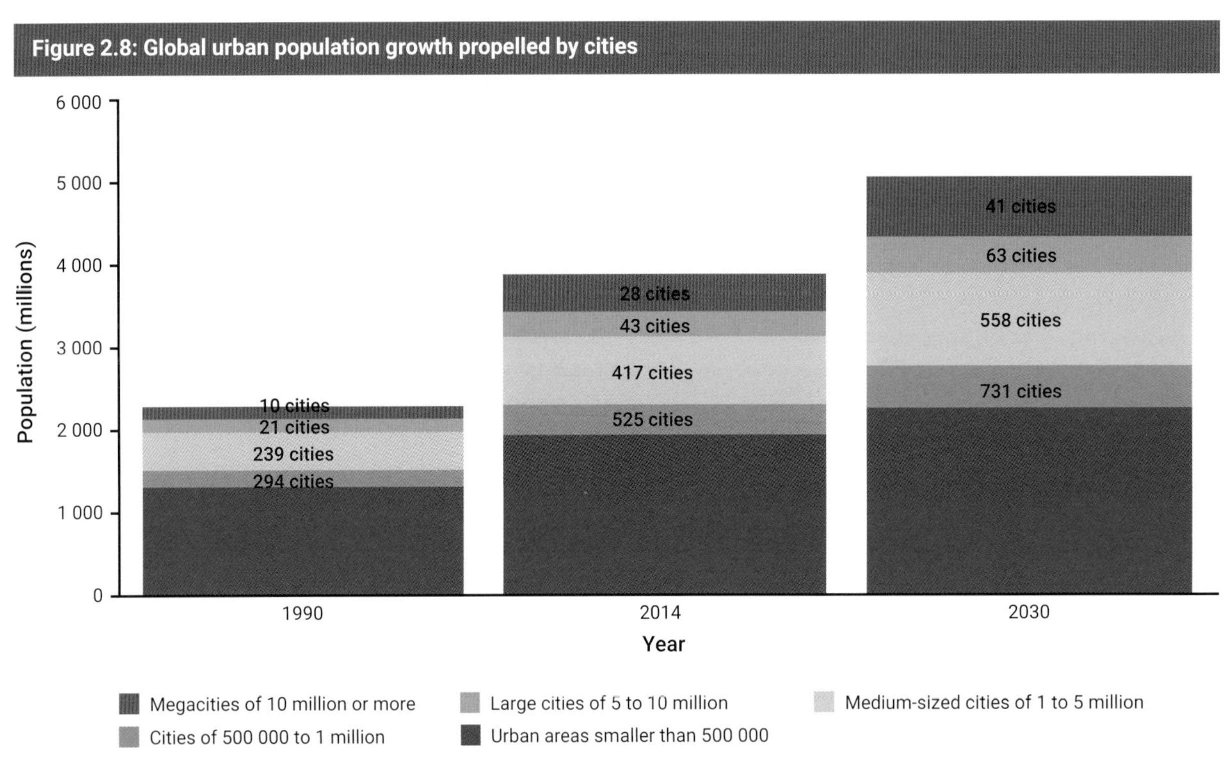

Source: United Nations (2014, p. 13)

Figure 2.9: City growth rates

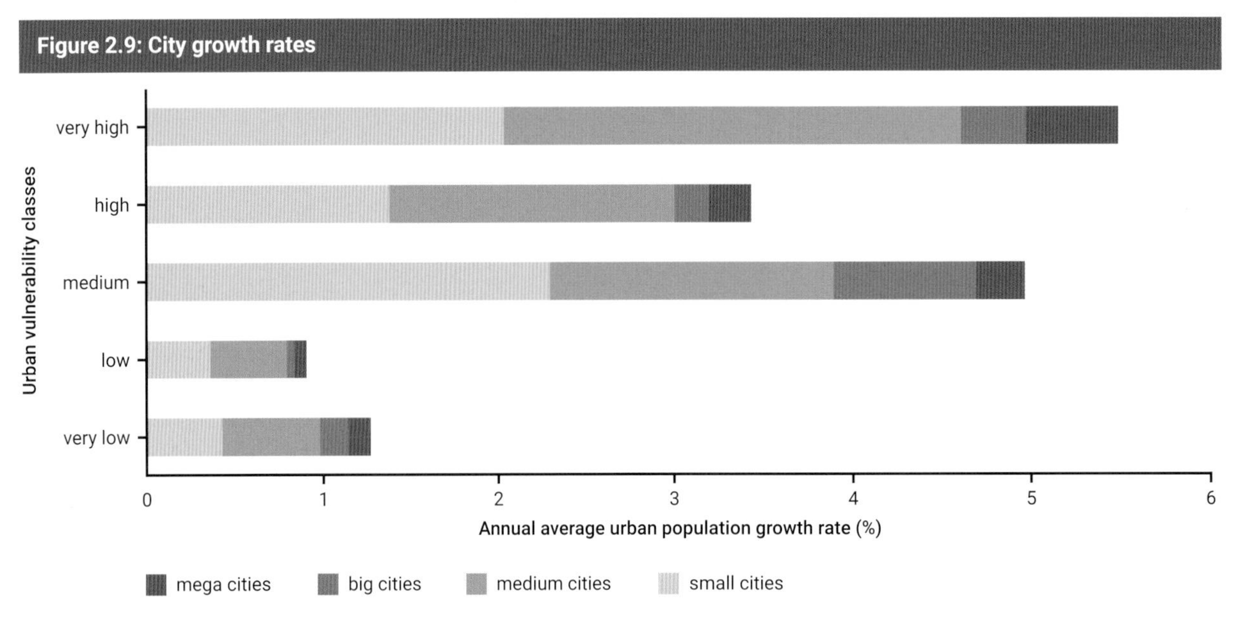

Source: Birkmann *et al.* (2016)

Cities are a source of wealth creation, where wealth is measured as the sum of natural, human, and physical assets (Hamilton and Hartwick 2017). Natural capital includes land, parks, green spaces, water and biodiversity. Human capital includes the population's education, knowledge and skills. Physical (or manufactured) capital includes such things as housing, infrastructure, industry and offices. Added to these is intangible capital – ideas and inspiration captured in forms that include research and development, patents, intellectual property rights, customer lists, brand equity, social capital and institutional governance. Intangible capital is perhaps the most important but feeds off and interacts with the other forms of capital. It also provides the source for innovation and investment necessary to decouple growth from resource use and CO_2 emissions, in absolute levels as well as in rate of growth terms.

A growing body of research supports the hypothesis that cities have the capacity to spread knowledge, so the key driver of wealth is now the ability to attract skilled and creative individuals and to nurture and spread ideas. Cities therefore appear to have a comparative advantage in more idea-intensive sectors. Unlike manufacturing, which is increasingly located outside cities, ideas-oriented industries tend to cluster in urban centres. It is unsurprising that much of the generation and distribution of ideas occurs in major cities given the role of close spatial proximity. The evidence clearly suggests that the direction of innovation is strongly influenced by urban and national planning and policy, and there is substantial scope for policy to direct cities towards resource-efficient low-carbon innovation.

Urbanization carries its own penalties of success, including pollution, congestion, urban heat effect, ill health, crime, informal settlements (slums), lack of affordability and waste. Unregulated, unplanned urban sprawl might appear to be the cheapest option in the short run, as it requires minimal institutional interference, infrastructure provision and urban planning. But the medium- and long-run costs to society, the economy and the environment can be dire. Unregulated cities will be less attractive, more polluted, congested and inefficient in the use of resources. About a third of the global urban population lives in slum-like conditions without basic services and social protection (United Nations Population Fund 2010/2011 cited in Urban Habitat III #1, p.3). Poor women living in slums are particularly vulnerable and face barriers to accessing some of the advantages of urban living (United Nations Population Fund 2014 cited in Urban Habitat III #1, p.2). Moreover, two thirds of urban dwellers live in cities where income inequalities increased between 1980 and 2010 Lopez Moreno 2012 cited in Urban Habitat III #1 p.1). Urban sprawl, poor public transport and a lack of access to basic services such as water, waste collection and energy offset the economic benefits of urban concentrations and increase costs. These growth penalties hinder opportunities to prosper and also exacerbate urban poverty. Unplanned urban growth also leads to excessive GHG emissions, alienation and social exclusion, as well as a range of other social, economic and environmental costs such as congestion, ill health and crime (Floater and Rode 2014).

These trends place an enormous burden on governance structures (Frank and Martinez-Vazquez 2014; UNESCAP and UN-Habitat 2015). In developing countries, local taxes, measured as a percentage of GDP, are three times lower than in industrialized countries (Bird and Bahl 2008).

Similarly, many of the small and medium-sized cities "lack the technical capacity to lead a major urban development process" (UN-Habitat 2012, p. XIV) and suffer from devolved responsibilities without corresponding resources, hampering their planning capacity (Frank and Martinez-Vazquez 2014). The result is that the capacity of urban governments to protect both natural resources and the rights of its citizens is severely circumscribed.

© Shutterstock/Harvey O. Stowe

Mass urbanization is not new in Europe, North America and richer parts of Asia, but the most recent wave is focused in developing regions, including Southern Asia and sub-Saharan Africa. This influx of people into cities can place great strain on urban institutional resources and infrastructure in growing cities. **Figure 2.10** shows that in countries with lower levels of urbanization and higher growth, urban citizens are highly vulnerable – vulnerability being "calculated by adding the urban susceptibility, the lack of coping capacities and the lack of urban adaptive capacities" (Garschagen *et al.* 2014, p. 46). If the relative change in the degree of urbanization is disaggregated by income class for the 1990-2015 period, it can be seen that in Asia, low-income countries (LICs) are urbanizing at the fastest rates (15.5 per cent) in comparison with low to middle-income countries (LMCs), at 1.2 per cent, and upper middle-income countries (UMCs), at 1.5 per cent. A similar pattern is seen in Africa – where the urbanization rates are 8 per cent for LICs, 3.6 per cent for LMCs and 5.7 per cent for UMCs – and in Latin America and the Caribbean. Globally, the pace of change in urbanization overall is 2.3 per cent (1990-2015), and the disaggregation by income class reveals the pace of change in LICs is 8 per cent while in LMCs it is 1.6 per cent (Melchiorri *et al.* 2018)

These rapidly urbanizing areas present a challenge but also represent "the largest opportunities for future urban GHG emissions reduction [... because their ...] urban form and infrastructure is not locked-in" (Seto *et al.* 2014, p. 928). As is presented below and in Part B of this report, there are positive and negative examples of rapidly urbanizing areas with regard to environmental effects. Cities exemplify the reality emphasized in this report, that when it comes to the creation of complex spatial networks, the future is not 'God-given', but is system and path-dependent. If new cities are built over the next two or three decades on a resource-hungry, carbon-intensive model, based on sprawling urbanization, all hope of meeting ambitious resource and climate-risk targets will be lost. This could leave cities and countries struggling to meet their

2

Figure 2.10: Where rapid growth faces high vulnerability

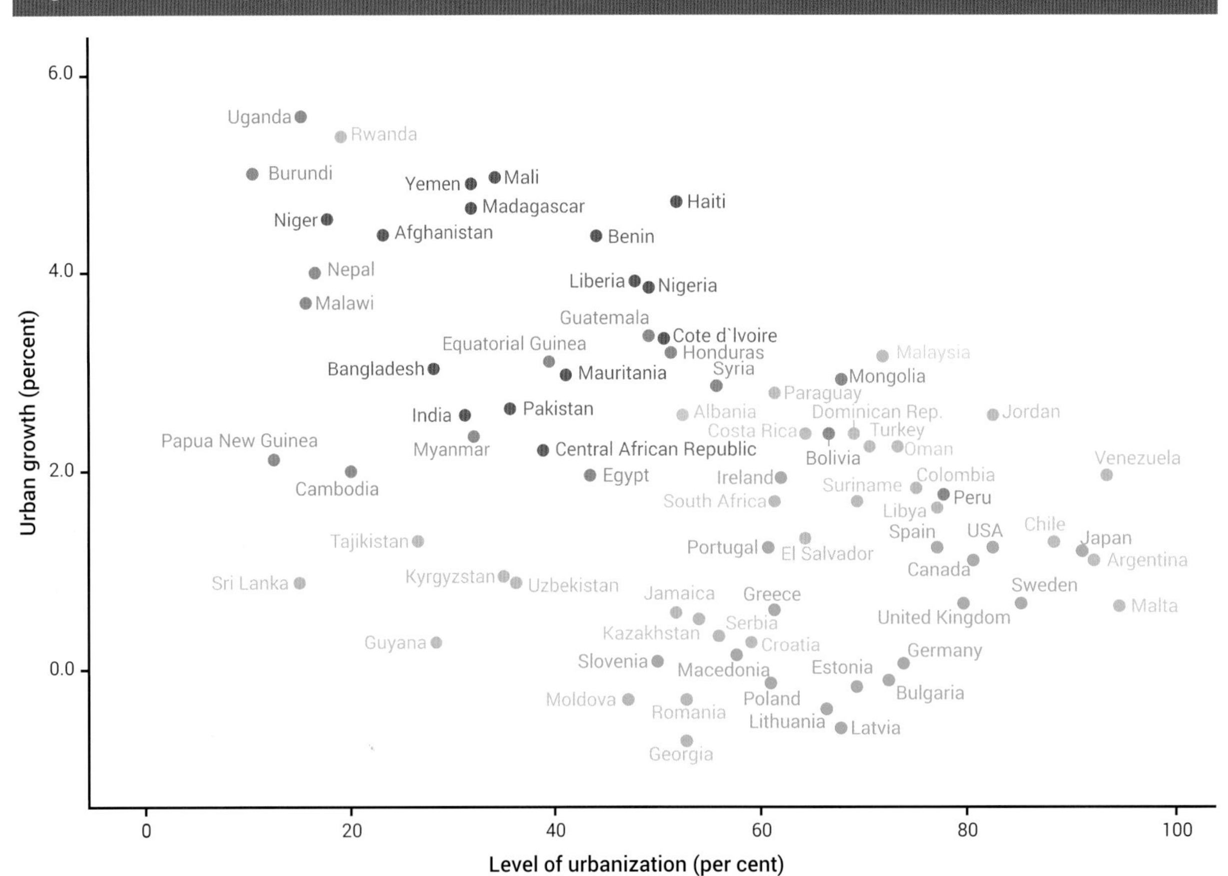

Classes of urban vulnerability	Level of urbanization	Growth rate 2000-2015
Very low	75.80	0.71
Low	69.19	0.92
Medium	56.07	2.36
High	43.51	2.89
Very high	38.59	3.71

Source: Garschagen *et al.* (2014)

resource needs and unable to compete in global markets, with the stranding of physical and human assets. Cities are also vulnerable to environmental and climate impacts such as heat, water stress, and floods; while coastal cities face sea level rise, saltwater incursion and storm surges.

2.4.3 Trends in urban expansion and density

Currently there are different views regarding territorial expansion of cities and population growth. In the absence of sustainable urban management, some studies show that cities are growing in size more than in population, reporting territorial expansion at double the population growth (Angel *et al.* 2011). Pesaresi *et al.* (2016) show that between 1975 and 2015, built-up areas increased 2.5 times while total population increased by a factor of 1.8 **(Figure 2.11)**, with the highest urban growth concentrated in India, China and countries in Africa. Urban land growth in these regions has also outpaced urban population growth rates, suggesting that urbanization has resulted in sprawled developments (Seto *et al.* 2011; Wolf, Haase and Haase 2018). Even in cities that are shrinking in population, sprawl still occurs (Schmidt 2011; Wolf, Haase and Haase 2018). Conversely, recent studies from Asia have shown that urban population has grown faster than urban land (in eastern South-East Asia, a 31 per cent population increase compares with a 22 per cent land increase) and that urban areas (in East Asia) are four times more dense than in land-rich developed countries: two times more than in Europe, 1.5 times more than in the Latin American and Caribbean region, and 1.3 times more dense than in the Middle East (Schneider *et al.* 2015; World Bank Group 2015).

Figure 2.11: Built-up area vs. Population (1975-2015)

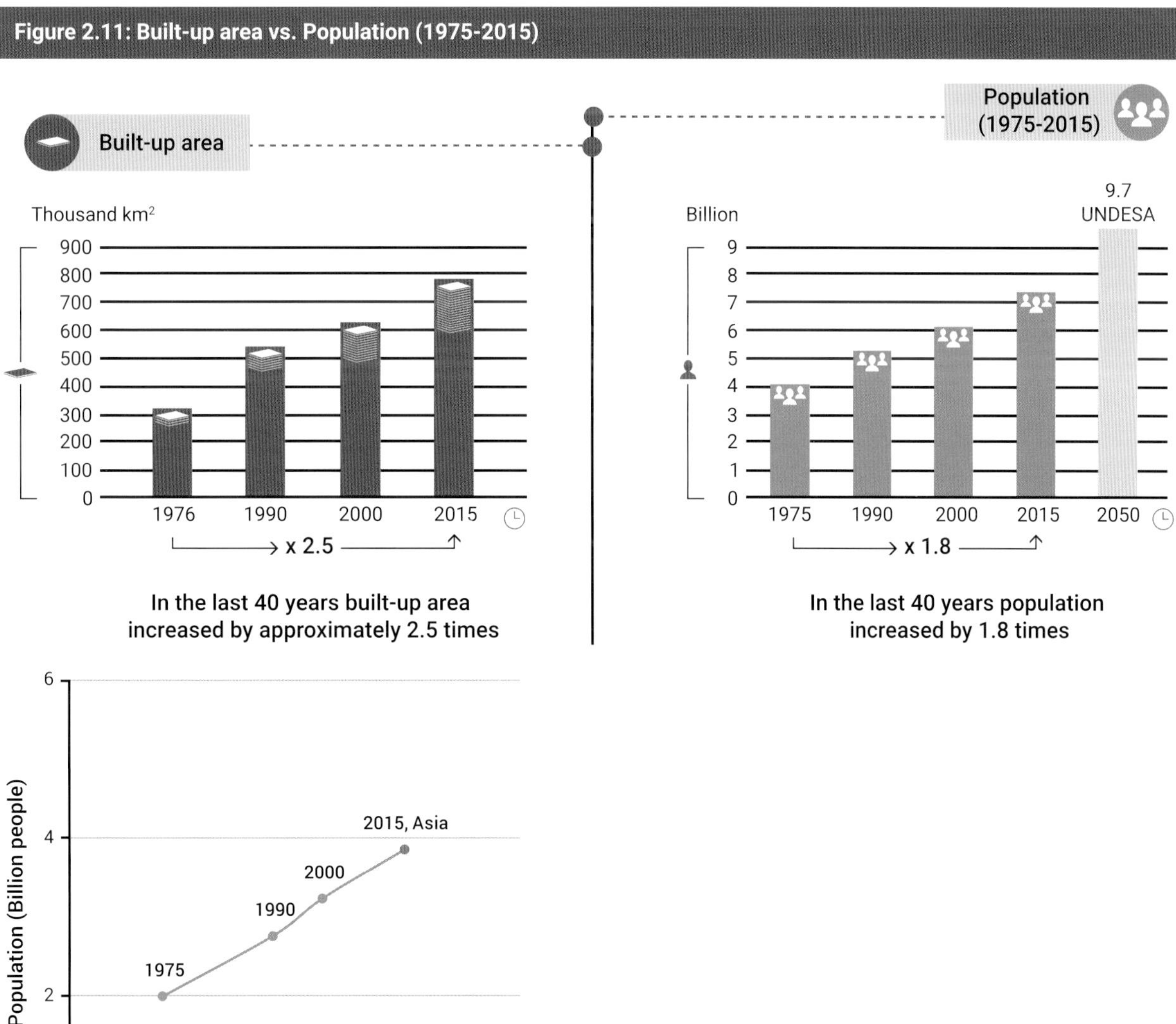

Source: Pesaresi *et al.* 2016

Increasing density alone is insufficient to make the transition to sustainable cities. Another factor that affects urban impact on the environment is urban form, namely the pattern of urban physical infrastructure, which cannot be easily modified, and determines land use, transportation and energy demand for long periods of time (Seto *et al.* 2016; Güneralp *et al.* 2017). Form patterns have implications for energy consumption, GHG emissions, biodiversity (Seto, Güneralp and Hutyra 2012; Salat, Chen and Liu 2014), water infrastructure (Farmani and Butler 2014) and land use and conversion of croplands (Bren d'Amour *et al.* 2016). Urban form, "infrastructure design and socio-spatial disparities within cities are emerging as critical determinants of human health and well-being" (Ramaswami *et al.* 2016, p. 940).

2.4.4 Urbanization as an opportunity

In a world where environmental limits are visibly closer, and with rural to urban migration expected to continue, urban population growth can represent an opportunity to increase citizens' well-being while decreasing their ecological footprint. This is made possible through lifestyle choices, improved governance, awareness and education programmes, the availability of infrastructure and services, and technological solutions. Small and medium cities have a particularly important role to play as they are usually a stepping stone between rural populations and urban centres (UN-Habitat 2015c, p. 3). In other words, urbanization can be positive, but will only amplify existing challenges if poorly managed. If cities could build technological solutions that took advantage of economies of scale not feasible in rural contexts, they could potentially hold the promise of limiting the negative environmental effects of population growth and increased consumption.

2.5 Economic development

The term economic development has been used in the literature to distinguish it from a one-dimensional measure of human welfare, which focused solely on economic growth (or, properly speaking, the growth in GDP). It includes, for example, social equity, poverty eradication, the meeting of basic human needs (access to health, education, and water and sanitation services), the provision of physical infrastructure (housing, energy, transport and communications), and the guarantee of essential political, economic, and social freedoms as elaborated by Sen (2011). Similarly, the term economic development highlights structural transformation, namely the changes in industrial structure (from an agriculture-based structure towards industry and services), social organization (from small-scale productive activities towards large-scale organizational structures), and the diversification of skills. The SDGs are derived from this broader concept of economic development.

2.5.1 The social role of economic growth

As the economy has moved from an 'empty world' to a 'full world' (Daly 1973), it has become clear that conventional growth cannot continue far into the future (United Nations Environment Programme [UNEP] 2011). Yet the social and political commitment to a vision of unending growth remains as strong as ever. The reasons are easy to see. Economic growth plays a number of vital roles in modern society, including poverty eradication, the pursuit of social justice, the building of social solidarity, the defence of civic peace and the establishment of good governance.

The most important of these is poverty eradication. Two and a half centuries after the advent of the Industrial Revolution, about 783 million people (10.7 per cent of the global population), still live on less than US$1.90 per day, and 48.7 per cent of the population lives on less than US$5.50 per day (World Bank Group 2013). Globally, about 22 per cent of children are stunted and 7.5 per cent are underweight (UNICEF 2018a) while 264 million children and adolescents are unable to enter or complete school (UNICEF 2018b), the majority of them girls. Nearly 2.1 billion lack access to safely managed water and 2.3 billion lack basic sanitation (UNICEF/WHO 2017).

This poverty is not because of a lack of economic resources. In 2017, the world's average income per capita was $16,906 per year (PPP, current international $), which is $46 per day (World Bank Group 2018) and about 24 times the poverty threshold. While redistributive policies and social security arrangements can help people to cope with poverty, the only reliable mechanism for eradicating poverty is to enable the poor to benefit from fast, steady growth.

Another argument for economic growth in developing countries is the need to narrow the huge income gap that separates them from developed countries. Indeed, this gap continued to widen well into the second half of the 20th century. Only in the 21st century was there evidence of a narrowing of the gap, as growth rates in developing countries began to outstrip those in developed countries **(Figure 2.12).**[5]

As such, even critics of the growth agenda agree that it is essential for developing countries (see, e.g. Jackson 2009, p. 4). Their main critique focuses on developed countries, where growth is, they argue, neither necessary nor desirable (see Daly 1973; Rees 1995; Victor 2008; Jackson 2009). Others (e.g. Friedman 2005) argue, however, that growth continues to play important political roles in developed countries, including supporting fairness, social mobility and social solidarity, while attracting popular support for civic and international peace (Benhabib and Rustichini 1996, p. 139; Weede 1996, p. 32; Gartzke 2007, p. 180).

In sum, then, the recent episodes of global economic growth are associated with:

a. a narrowing of the income gap between developed and developing countries, and
b. a huge dent in the incidence of poverty in the latter countries.

The danger is that if the growth engine slows down, these trends may not continue, and this could – as outlined in the report of the United Nations Secretary-General on climate

5 This was in large part due to the higher growth rate in the large populous economies, especially China and India, but was not restricted to them. Indeed, the first decade of this century witnessed the first occasion when sub-Saharan developing countries, as a group, grew by more than 5 per cent per year for 5 years. However, the global financial crisis has resulted in the slowing of average developing country growth rates by 2-3 per cent, and a widening in the variance in growth rates, as larger countries (e.g., Brazil, China, Germany and the United States) recovered more quickly than smaller economies.

Figure 2.12: How growth rates in developing countries began outstripping those in developed countries

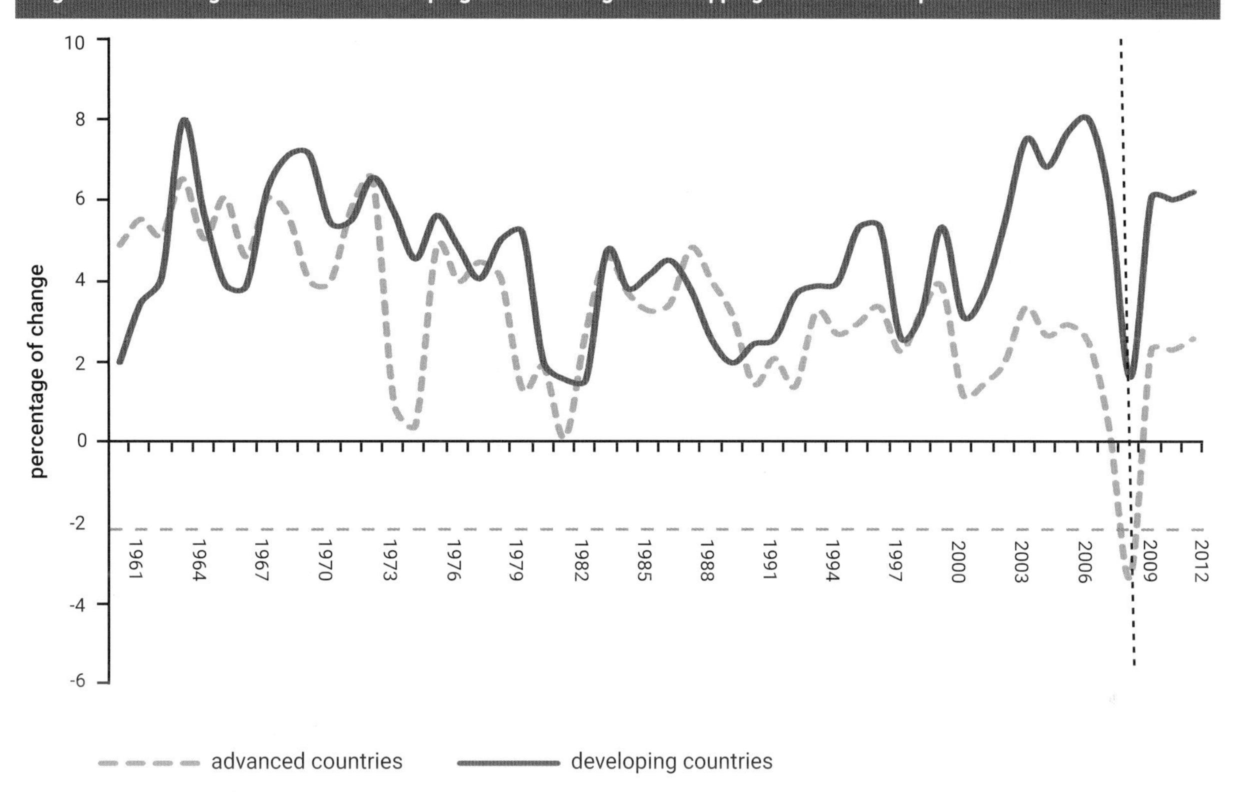

Source: Canuto (2010)

change and its possible security implications (A/64/350) – signal a reversion to a zero-sum world in which conflict and war would proliferate, governance systems atrophy and popular support diminish for social justice, solidarity and civic peace.

The question is not whether growth in developed countries is needed to meet their material aspirations, but whether it is an essential element in the quest of modern societies to meet their political, social, cultural and even moral and ethical goals. Ideally, economic growth and environmental sustainability are mutually reinforcing rather than in conflict.

2.5.2 From growth to development

Economic growth is only one of the factors contributing to human welfare, which also depends on social justice, poverty eradication, good governance (including anti-corruption efforts) and environmental health. The global policy process has sought to reflect this integrated approach in the form of the Millennium Development Goals (MDGs) and SDGs.

The structure of the SDGs provides an insight into the broader issues discussed in this section. The MDGs were motivated by a simple idea, namely the resolve of heads of state and government at the Millennium Summit of the United Nations General Assembly to halve poverty in 15 years (United Nations, 2000). The SDGs take it a step further and seek to eradicate poverty and hunger by 2030. In addition, the SDGs draw explicit attention to environmental and social factors, including climate change, terrestrial and marine biodiversity, sustainable consumption and production, inequality, industrialization and decent jobs, and peace and justice (United Nations Development Programme [UNDP] 2018).

In retrospect, the MDGs were a qualified success; they coincided with accelerated progress on poverty eradication, health and education, but lagged on nutrition and on access to water and sanitation (McArthur and Rasmussen 2017). The successes of the MDGs can be attributed to four factors, in descending order of significance, namely: high economic growth in developing countries, support for local programmes and community-based initiatives, large vertical programmes (especially in the health sector), and the enactment of legal rights and protections. Although it is difficult to assess the causal impact of MDGs (it is impossible to know what would have happened in their absence), some empirical research has found evidence of the MDGs accelerating progress in these areas (McArthur and Rasmussen 2017).

Although the SDGs seek to build on this success, the underlying context is very different. Their adoption was preceded by a major financial crisis, barely avoiding a full-fledged financial meltdown, a long-drawn out recession in industrialized countries, a potentially disastrous debt crisis, a dramatic rise in income inequality in the OECD countries, recurrent commodity-price volatility, significant political fallout from food price shocks, shrinking natural resources and biodiversity, growing evidence of adverse climate-change impacts, an increasing awareness that the global economy was coming up against planetary boundaries (Rockström *et al.* 2009b), and a dramatic rise in global conflicts.

The SDGs can be loosely grouped into three categories:

- *Human development*: tackling income poverty, hunger, lack of access to basic services (health, education, water and sanitation) and gender inequality (i.e. SDGs 1-6),
- *Economic development*: enabling conditions for poverty eradication, providing access to energy, providing economic growth, decent jobs, infrastructure and industry, declining inequality, housing, and peaceful societies (SDGs 7-11 and 16-17)
- *Environment*: ensuring that the agenda of poverty eradication (and by implication, of economic growth) is protected against ecological threats (SDGs 12-15).

This agenda is relevant to the Global Environment Outlook assessment. The poverty agenda remains unfinished, and the development consensus remains that its pursuit will require further economic growth in the world economy. There is a growing concern, however, that the prospects of development itself are increasingly threatened by the closing in of planetary boundaries, especially through the impacts of climate change. In the loop's other direction, there continue to be fears that the growth process entails increasing use of natural resources and sinks, thus increasing the pressure on the natural environment.

The poverty agenda remains the highest priority of the international policy community, as documented in almost every international agreement pertaining to economic development and the environment in the past quarter century. The reasons for this are not exclusively or even primarily altruistic. In the words of the founding principles of the International Labour Organization (ILO), as cited by its Director-General: "poverty anywhere is a threat to prosperity everywhere" (ILO 2011). The reasons thereby reflect an understanding that global peace cannot be built on conditions that condemn a significant segment of humanity to permanent deprivation and subservience.

2.5.3 Recent experience

The financial crisis was followed by a slowdown in global growth. The reasons for this were to do with stagnant international trade, revival of the spectre of trade wars, heightened policy uncertainty, and a dampening of the main engine of global growth, namely emerging economies (World Bank Group 2017, p. 3). From an average of about 6 per cent growth per year between 1992 and 2008 (and a height of 10 per cent per year in 2006-2008), growth in global trade has shrunk to about 1 per cent since 2010 **(see Figure 2.13)**. More recently, this appears to have resulted in renewed threats of trade wars.

A second notable trend is the rising inequality in industrialized countries. There is a paradox in the contrasting movements in international and intra-national inequality. For much of the 20th century, income inequality between countries widened (or, at best, was static), while income inequality within countries narrowed (or, at worst, remained static). Since 1980, however, both these trends have reversed.

One consequence is Milanovic's global elephant curve, so called for its shape as seen in **Figure 2.14** (Lakner and Milanovic 2013, p. 31; Weldon 2016). This shows that between 1988 and 2011, while the incomes in the top 1 per cent as well as those in the 40-70 percentiles (presumed to be in developing economies) were rising, incomes in the bottom 10 per cent and in the 80-90 percentiles (presumed to be in the middle class of developed countries) were growing more slowly.

Figure 2.13: World trade growth

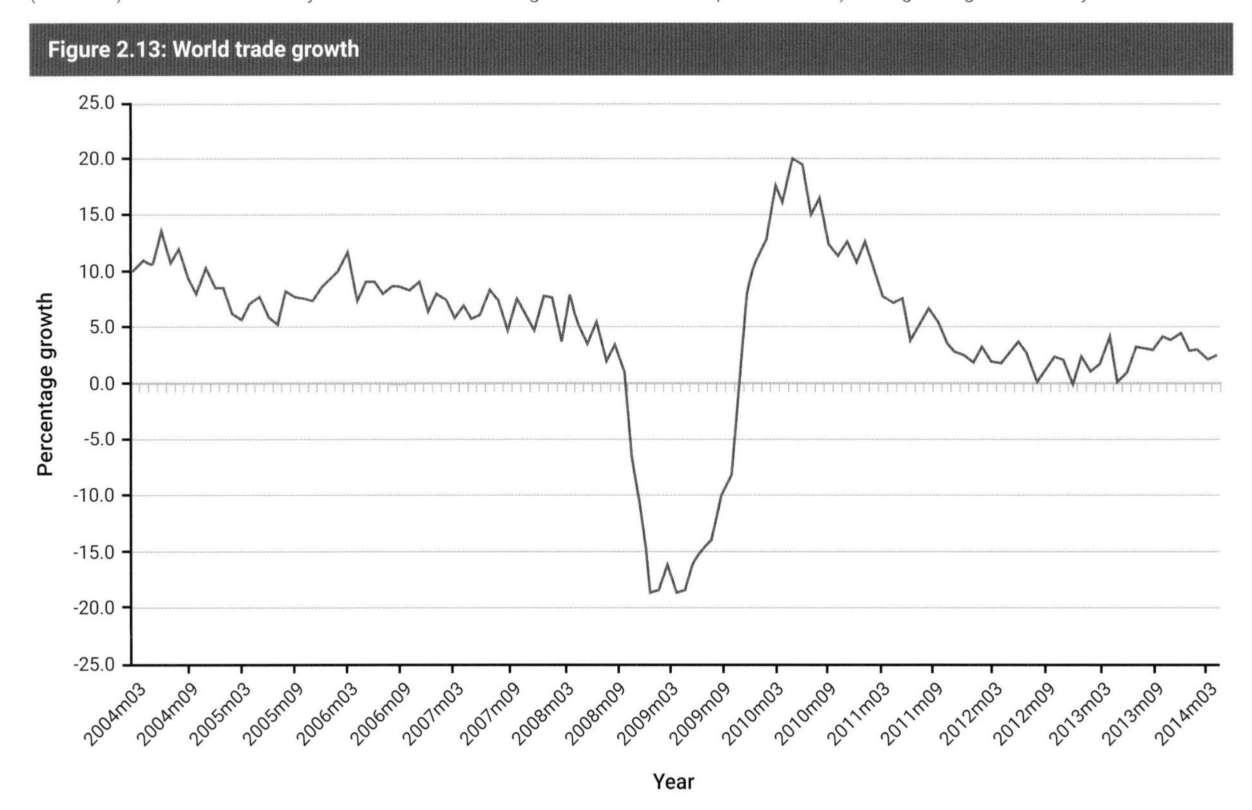

Source: Data from CPB Netherlands Bureau for Economic Policy Analysis (2018)

Figure 2.14: Milanovic´s elephant curve

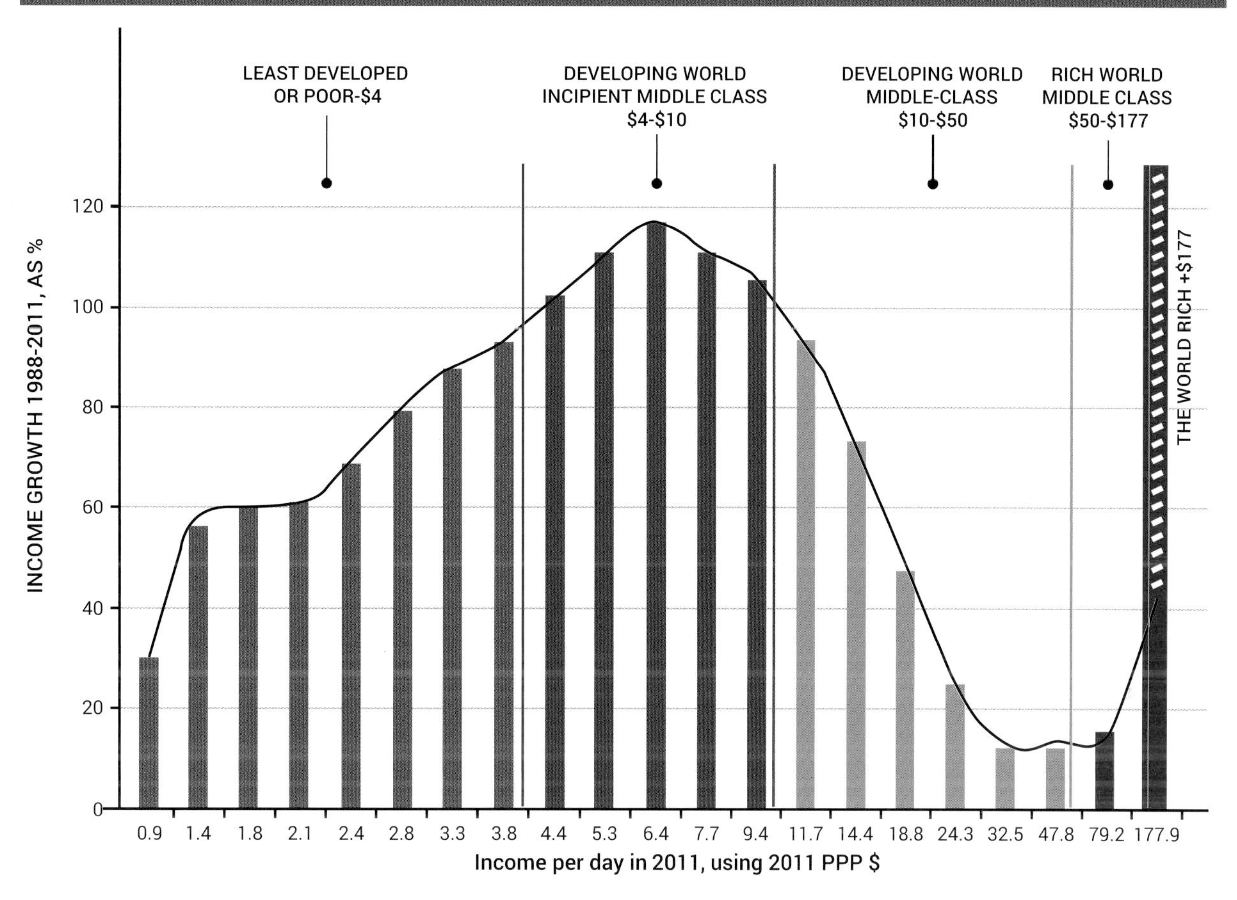

Source: Data from Lakner (2013) and artwork from da Costa (2017)

The relationship between income inequality and the use of natural resources is not straightforward. On the one hand, the classic economic argument is that the poor have a higher propensity to consume than the rich (Carroll *et al.* 2017), and transferring income from the former to the latter should therefore reduce the impact on the natural environment. On the other hand, heightened income inequality creates upward pressure on resources, both through the impact of conspicuous consumption and out of the squeezing of the middle class. More importantly, inequality has the potential to exacerbate conflict, which in turn has an adverse impact on the environment. Inequality's effects on the environment move through the consumption, investment and community channels (Islam 2015).

There are two main justifications for the global growth agenda: the material one and the political one. The former is for the role of growth in poverty eradication while the latter is for the pursuit of growth for its possible contribution to other needed political aims, such as social justice, fairness, solidarity, civic peace and democratic governance.

Recent trends show that a significant dent has been made in the twin agendas of poverty eradication and reduction of global, between-country inequality. On the other hand, the manner and pace at which this has happened has given rise to new tensions and fractures, both within and between countries. This may indicate that there has been a renewed urgency for reviving the growth momentum, not only in developing countries, but equally in developed ones too.

2.5.4 The role of energy

A key question is the relationship between two different dimensions of economic development, namely aggregate economic growth and resource consumption, especially the consumption of energy.

There is a large body of literature on decoupling economic growth from its impact on resource consumption (see, e.g. UNEP 2011; UNEP 2017; Hennicke 2014). A key distinction is between renewable and non-renewable resources. Since the latter are finite in nature, the only way to reduce depletion is to reduce, reuse and recycle (the three Rs). However, as has been noted prominently in the literature, this redirects the focus onto energy consumption (i.e. on the energy component embedded in resource use). The 3R strategies are fairly well known; their viability depends on the cost of energy used for recycling or reuse relative to the cost of new extraction.

Some renewable resources, such as solar and wind, are drawn on without concern that these resources will run out. Other resources are renewable as long as the encompassing ecosystems are not degraded. For renewable biomass

resources such as forests, the primary challenge is ensuring that their use does not exceed the rate of natural (or enhanced) regeneration. This, too, boils down to the rate with which these resources can regenerate themselves by harnessing the energy of the sun. Various techniques of increasing natural resource productivity are equivalent to enhancing their energy-harnessing potential.

In short, as noted by Hennicke (2014, p. 2), energy is the key to decoupling. Not surprisingly, environmental analysis has often used the concept of energy flows to frame these issues. Energy, construed broadly, is the motive force in both human and natural affairs. The miraculous transformation introduced by the Industrial Revolution is in essence the result of harnessing an enormous volume of readily available energy resources, namely fossil fuels (Smil 2010; Bithas and Kalimeris 2016). This is a major factor responsible for the idea of a permanently growing economy.

As we look to the future, the need for energy will continue to increase, not only to promote economic development in poor countries, but also to help reduce the unsustainable consumption of material resources. To avoid catastrophic climate change and a scarcity of resources, a major shift is needed by this increase, towards affordable and sustainable energy resources (see e.g., Yihdego, Salem and Pudza 2017; UNDP 2018, SDG 7).

2.6 Technology, innovation, and global sustainability

Technology can be a positive and a negative driver of environmental change. Technological innovation has been – and is likely to continue being – a critical driver of sustainability changes at a global level (Segars 2018). At the same time, technologies have often created unintended consequences that are far beyond the predictive ability of our best scientific analysis (e.g. the impact of effects of fossil fuel consumption on the climate system). Existing scientific analyses of technology issues often fail to capture the important negative/rebound effects (Chitnis *et al*. 2013) of the systematic impact of technologies as well as underplay the problem of technology diffusion, particularly in terms of agricultural technological innovation (Juma 2015). Motor vehicles and electricity are good examples of past scientific limitations. They represent two of the most important technological breakthroughs of the 20th century, but their negative environmental and resource impacts are likely to persist well through the 21st century.

2.6.1 Technological innovation and sustainable economic development

From an economic perspective, technological innovation has long been recognized as one of the core drivers of economic development, but in modern theories of growth it is given a pre-eminent role (see Romer 1994; Acemoglu and Daron 2009; Zenghelis 2011). Innovation in human capital, through investment in research and development and knowledge-sharing, is the key not only to productivity growth, but also to getting more out of the resources we have. This is crucial to solving many environmental problems.

Innovation offers the most important route out of many environmental problems. In an environmentally sustainable economy, economic growth and development would still occur, and humanity would continue to prosper. Economic growth and human well-being can be decoupled from material throughput and environmental impact, though the policy challenge of actually achieving this is considerable (Jacobs 1991; Hepburn and Bowen 2013).

Recognition of new opportunities, together with the falling cost of key low-carbon technologies (solar, wind, etc.), has proved game-changing in terms of driving global policy action. While the United Nations Framework Convention on Climate Change (UNFCCC) negotiations are often seen as moving slowly, 40 countries and 20 subnational regions have implemented or are planning to implement carbon pricing and other types of low-carbon technology-enabling policies (Global Commission on the Economy and Climate 2015). There are now over 1,200 climate change or climate change-relevant laws worldwide, which is a 20-fold increase over the past 20 years (Nachmany *et al*. 2017).

The Paris Agreement on climate change (United Nations Framework Convention on Climate Change [UNFCCC] 2015) itself can be seen as a consequence of accelerating momentum in countries, cities and businesses across the world to reduce GHGs. Falling technology costs of renewables and energy efficiency, growing market opportunities, changing behaviours, and a growing awareness of the co-benefits of lower emissions (such as less urban pollution and congestion, and fiscal opportunities from pricing scarce resources, carbon and pollution and from removing environmentally damaging subsidies) all helped to support the voluntary commitments signed into action after the Paris Agreement.

2.6.2 Cleaner and energy-efficient technologies

Rapid advances are occurring in the market development of cleaner and energy-efficient technologies, including renewables (solar, wind, advanced biomass, etc.), storage (batteries, pumped hydro, etc.), energy efficiency (e.g. demand-side management and dematerialization), decarbonized transport options (e.g. electric vehicles). Research and development advances are also emerging for cleaner technology options (e.g. carbon, capture and storage, second- and third-generation biofuels, decentralized electricity generation at small/micro scales, self-driving vehicles) (International Energy Agency [IEA] 2016b).

In the case of renewable energy, for instance, diffusion and scale-up become both feasible and affordable worldwide. By the year 2040, renewables will constitute two-thirds of the global investment in power generation, while solar energy will become the largest source of global low-carbon capacity, fuelled by growth in China and India. In the case of the European Union, renewables are expected to account for 80 per cent of new power-generating capacity, with wind energy becoming the leading source of regional electricity after 2030 (IEA 2017b).

Regionally, in the case of sub-Saharan Africa, where there are a number of public, private and cross-sector initiatives to address energy poverty, a rapidly developing cleaner and energy-efficient technology ecosystem is incubating early-stage off-grid solar technology companies, as well as helping to accelerate the overall market dynamics of sub-Saharan African countries (Park 2016; Yihdego, Salem, and Pudza 2017).

For instance, investments in off-grid solar companies in sub-Saharan Africa and other countries went up tenfold, to more than US$200 million between 2013 and 2016 (Bloomberg New Energy Finance 2017), although it should be stressed that this rapid growth still represents a small percentage of the investments that will be needed to make an impact on the regional energy marketplace.

Scalable solar-powered off-grid electrification solutions are important for sustainable development in many developing regions and represent a critical element in the case of the sub-Saharan Africa region (International Renewable Energy Agency [IRENA] 2013). Access to energy represents a critical economic, social and environmental issue in both industrialized and developing countries because energy access is linked to a wide range of economic and environmental benefits (IRENA 2016). Yet sub-Saharan Africa as a region consumes just 145 terawatt-hours of electricity a year – or one incandescent light bulb per person used three hours a day (Lucas 2015) – making it the most energy-poor region in the world (Park 2016).

There is substantial potential for the unit costs of resource-efficient and low-carbon technologies to continue to fall as these new technologies are developed and deployed, and as engineers learn how to connect and service them cheaply. This potential is far higher for new technologies than it is for long-established, high-carbon incumbents.[6] For example, price drops in renewable energy technologies have allowed new combinations of solar, wind, and energy storage to outcompete coal and gas on cost.[7]

Not only does the energy sector benefit from productivity improvements associated with a transition to low-carbon, there are also important economic spillovers from low-carbon innovations. Acemoglu *et al.* (2012) argue that sustainable growth can be achieved by adopting temporary policy levers such as a carbon tax that can redirect innovation towards clean inputs, while Dechezleprêtre, Martin and Mohnen (2014) conclude that economic spillovers from low-carbon innovation are consistently 40 per cent greater compared with conventional technologies, while information and communication technologies (ICTs) can, in theory, vastly increase productivity and energy efficiency, while reducing material consumption throughout the lifespan of a product (a mobile phone for instance). While ICTs may one day usher in a new era in which digital technologies play a key role in accelerating global environmental governance, it is not yet clear if the energy and materials savings are greater and outweigh the cumulative sustainability impact of the ICT product lifespan from resource extraction to waste disposal **(see Box 2.3 on** electronic waste).

Box 2.3: Electronic waste

Electronic waste (e-waste) – which can be defined as "items of electrical and electronic equipment and their parts that have been discarded by the owner as waste without the intention of re-use" – represents one of the fastest-growing waste streams in the world (Solving the E-waste Problem (StEP) Initiative 2014).

Fuelled by rapid global sales of computers and electronics, combined with shortening product life cycles, 44.7 million metric tons – the equivalent of 6.1 kg per inhabitant of e-waste were generated in 2016, while the overall e-waste stream is expected to increase to 52.2 million metric tons or 6.8 kg per inhabitant by 2021 (Baldé et al. 2017).

Some e-waste from industrialized countries is being shipped to the developing world, "where crude and inefficient techniques are often used to extract materials and components", a trend which is posing challenges to global sustainability governance (Baldé, Wang and Kuehr 2016).

Beyond the direct social and environmental impacts of ICTs, one emerging sustainability issue is the electricity use of data centres, which in the case of the United States is estimated to be around 2 percent of the country's total electricity consumption (Whitney and Kennedy 2012). With energy efficiency of computers reportedly doubling every 1.5 years (Koomey *et al.* 2011), the more important long-term sustainability question may be the use and application of ICTs in avoiding future energy use and lowering climate change impact.

Digital technologies such as smart meters are projected to link more than 1 billion households and 11 billion smart appliances in interconnected electricity systems by 2040. The use of digital technology innovations will enable individual homes to determine when and how much they draw electricity from the grid. They will also enable the design of environmentally friendly demand-side responses in the building, industry and transport sectors, resulting in US$270 billion of avoided new investments in new electricity infrastructure (IEA 2017a). Governments of cities ranging from Copenhagen to Addis Ababa are also investing in ICT-based smart technologies (e.g. open data stores, citizen engagement platforms) to help improve urban governance at lower financial and environmental cost (C40 Cities Climate Leadership Group 2015).

2.6.3 Food-agricultural technology

A number of global food-agricultural trends – population growth and increasing global affluence, among others – will require increased agricultural productivity (by as much as 60-120 per cent on 2005 levels), in direct conflict with the wider SDGs (Ort *et al.* 2015).

Moreover, there is a wide range of perspectives in terms of what the yield gap is likely to be - the difference between how much a crop could yield per hectare with enough water and nutrients, and how much is currently being harvested (White 2015) - and over what technology options are available to address it. Total agricultural production is projected to increase by 60 per cent by 2050 compared with 2005 (Alexandratos and Bruinsma 2012), due to an increase in global population and in the number of people from the developing world who can afford to eat more and better food. The emerging question confronting the international community is likely to be: will the global food supply be adequate to meet global food demand,

6 The so-called sailing ship effect (whereby the introduction of steam ships induced a leap forward in efficiency and design of sailing ships) suggests that incumbent industries can respond with competitive innovation when faced with existential competition.

7 Solar photovoltaic and onshore wind technologies are competitive with gas and coal in a number of global locations, even without a carbon price. The cost of solar photovoltaic modules fell by 60 per cent in the two years to the first half of 2017, and by a factor of five in the five years post-2008 (Bloomberg NEF 2017). Energy storage prices are falling even faster than solar photovoltaic and wind prices. A recent study found that research and development investments for energy storage projects have lowered lithium ion battery costs from US$10,000/kWh in the early 1990s to a trajectory set to reach US$100/kWh on or by 2018 (Kittner, Lill, and Kammen 2017).

and can this demand be met without adversely impacting land use, biodiversity, freshwater use and other natural resources? If not, can this demand be met, or reshaped, using alternative technologies beyond agriculture as we know it today?

Bijl *et al.* (2017) suggest that a sustainable balance between reducing global hunger and staying within, among others, the planetary boundaries of land and water use might be struck by changing dietary patterns and more effectively addressing food waste as a policy priority. In the case of agricultural water use, a Pacific Institute study (2014) concluded that the adoption of existing water technologies and management techniques could reduce agricultural water use in the state of California by 5.6 million to 6.6 million acre-feet (one acre-foot is 1,233.48 cubic metres) per year, or by between 17 and 22 per cent, while maintaining the same level of agricultural productivity.

The International Food Policy Research Institute (IFPRI) (2014) argues that certain agricultural technologies and practices (e.g. crop protection, drip irrigation, drought tolerance, heat tolerance, integrated soil fertility management, no-till farming, nutrient use efficiency, organic agriculture, precision agriculture **[see Box 2.4]**, sprinkler irrigation, water harvesting, and land conservation measures) might be scaled up to achieve the dual goal of increasing food production and reducing food insecurity in the developing world. No-till farming alone can increase maize yields by 20 per cent, while heat-tolerant varieties of wheat can lead to a 17 per cent rise in crop yields (IFPRI 2014).

With the livestock sector accounting for about half of food-system GHG emissions (Food and Agricultural Organization of the United Nations [FAO] 2017; Gerber *et al.* 2013), emerging food-agricultural technologies may have the potential to reshape demand for animal produce and increase the sustainability of the food system. Reducing overall meat consumption as well as providing alternatives to conventional livestock production systems (e.g. through the introduction of plant-based meat alternatives) would, for instance, substantially reduce the agricultural land use footprint from food production (Alexander *et al.* 2017). In another example, although there are uncertainties in terms of an increased energy-use rebound effect[8], production of cultured or *in vitro* meat requires smaller quantities of agricultural inputs and land compared with raising livestock (Mattick *et al.* 2015).

Box 2.4: Precision agricultural technologies

The world's population is expected to reach 9 billion by 2050, while climate change and income growth will drive food demand in the coming decades. Baseline scenarios show food prices for maize, rice and wheat would significantly increase between 2005 and 2050, and the number of people at risk of hunger in the developing world would grow from 881 million in 2005 to more than a billion people by 2050 (IFPRI 2014).

While no single technology can be offered as a solution to these global agricultural and food challenges, precision agriculture (GPS-assisted, machine-to-machine solutions that combine information collected by sensors with automated management) represents one of 11 agricultural innovations, which, in aggregate, might help by 2050 to improve global crop yields by up to 67 per cent while reducing food prices by nearly half (IFPRI 2014).

Other emerging technological advances are demonstrating the potential to decouple crop production from the vulnerability of land use and climate (Gilmont *et al.* 2018). Hydroponics employ nutrient-rich water rather than soils to grow crops, and aeroponics use nutrient-dense sprays to nourish plants suspended in the air. Both techniques permit precise application of nutrients to crops grown under controlled conditions, including in land-sparing indoor vertical farms that can be located in urban and degraded environments (Eigenbrod and Gruda 2015).

As the cost of decentralized renewable energy sources falls, the constraints to the broader deployment of these technologies, including broadening them to grow staple crops, will continue to decline as the environmental benefits increase (Kalantari *et al.* 2017). To truly accelerate innovative food and agricultural technologies on the global level, particularly in the developing world, it will also be critically important to have complementary sustainable policy initiatives, such as the FAO Global Agenda for Sustainable Livestock Initiative, to diffuse both technology-based and non-technology-based sustainable food and agricultural innovations.

2.6.4 Technology diffusion and global sustainability

While there is strong scientific consensus on the importance of technological innovation as a driver of global sustainability change, there is far less scientific consensus in terms of two issues: first, sustainable technology diffusion – particularly in terms of the adoption and deployment of what might be described as sustainable technologies – in the developing world, and, second, how to regulate and govern new and emerging technologies in terms of global sustainability (Juma 2015). For technological diffusion, in terms of the rates of both adoption and acceleration, a good place to start might be the market development of solar, wind and other renewable energy technologies in the developing world, particularly relating to cities and urbanization (IEA 2016a).

Although renewable energy sources accounted for 70 per cent of the net increase in the global power capacity in 2017 due to the rising economic competitiveness of solar and wind energy (REN21 2018), rising energy demand, particularly in the developing world, coupled with population growth, is likely to outpace the development of economically viable and scalable renewable-based solutions without additional technology breakthroughs in the energy sector (IRENA 2017).

To provide the necessary institutional and socioeconomic conditions for technological diffusion, there is a critical need to design the appropriate innovation scale-up conditions (Rogers 2003) and to implement new public and private measures to more effectively deal with incoherent policies, misalignments in electricity markets and cumbersome and risky investment conditions (Ang, Röttgers and Burli 2017) in both industrialized and developing countries.

8 The energy use rebound effect refers to the observation that people may begin to consume more energy as a result of increases in energy efficiency.

In terms of technology diffusion and sustainability pathways in OECD member countries, the emerging 'industry 4.0' model is likely to have a major impact on the nexus of technology diffusion, market development and sustainability. Industry 4.0 – which can best be described as a digital industry technology platform powered by sensors, machines and information technology systems **(see Figure 2.15)** – is regarded by many scientists, technology experts and business executives as the fourth wave of technological advancement (Rüßmann *et al.* 2015).

While the industry 4.0 model, particularly as a technological platform, has the short-term potential to produce more efficient processes and higher-quality goods at reduced costs, the long-term social, environmental and economic impacts, particularly in terms of employment and workforce development, remain, at best, unclear. The emerging industry 4.0 model, along with artificial intelligence, additive manufacturing, the Internet of things, and other disruptive technologies, reflects a deep uncertainty that lies at the technology-sustainability nexus: how can the international community properly weigh the sustainability risks and benefits, particularly with regard to short- and long-term impacts on employment and economic development?

Despite the growing visibility of the social, environmental and economic impacts of global climate change and environmental dilemmas, slow progress on a wide range of international environmental (e.g. climate change) and social (e.g. refugees) policy negotiations has limited the scope for so-called good public policy options and tilted the governance framework towards riskier forms of technology like climate geoengineering as a policy alternative. Whether a particular emerging technology should be adopted or actively promoted by public organizations or private companies is not the critical issue. Rather, it is how and to what degree the international community can make sure that proper oversight, monitoring and protection against the potential adverse effects are in place as we proceed with the complex task of identifying, developing and diffusing technologies that positively impact wealthier OECD as well as lesser developed countries.

Figure 2.15: Industry 4.0: technological transformation of future industrial production

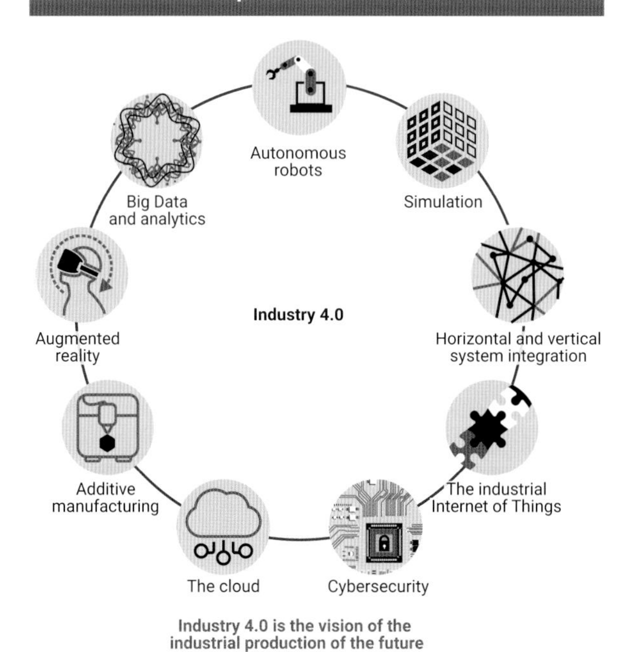

Source: Rüßmann *et al.* (2015)

2.7 Climate change

GEO-6 includes anthropogenic climate change as a driver of environmental change because it has acquired a momentum independent of future human activity; it is also analysed as a cross-cutting issue in Section 4.3.1 of this report. **Figure 2.16** demonstrates the increase in CO_2 concentration over the industrial period, charted on the same scale as the data for the transitions in CO_2 concentration between the glacial and interglacial periods over the past 20,000 years. Other GHGs such as methane and nitrous oxide have also been increasing consistently over the decades, as indicated by the National Oceanic and Atmospheric Administration's greenhouse gas index and shown by Hartmann *et al.* (2013). The impact of such changes demonstrates that climate change is now a major driver of environmental change – an inexorable force that can no longer be ignored.

According to the Fifth Assessment Report of the IPCC (2014), the world has entered an era of committed climate change. The concept of climate commitment, first introduced by Ramanathan (1988), refers to changes that are already in the pipeline, regardless of any further emissions or any future change in GHG concentrations in the atmosphere. "A large fraction of anthropogenic climate change resulting from CO_2 emissions is irreversible on a multi-century to millennial time scale, except in the case of a large net removal of CO_2 from the atmosphere over a sustained period" (IPCC 2013, p. 28).

Figure 2.16: Mean atmospheric CO_2 concentration

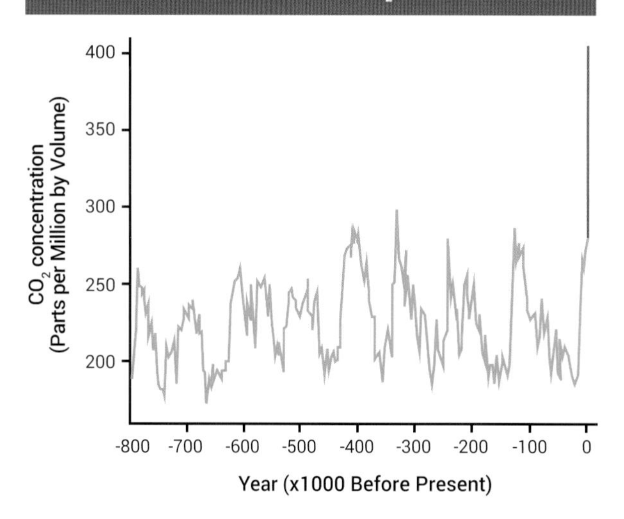

Source: Based on (in blue) NOAA data from http://www.ncdc.noaa.gov/paleo/metadata/noaa-icecore-6091.html and (in red) data provided by Pieter Tans, NOAA/ESRL (www.esrl.noaa.gov/gmd/ccgg/trends/ and scrippsco2.ucsd.edu/) [])

2

Surface temperatures will remain roughly constant at elevated levels for many centuries after a complete cessation of net anthropogenic CO_2 emissions. According to Mauritsen and Pincus (2017), "due to the lifetime of CO_2, the thermal inertia of the oceans [Wigley 2005] and the temporary impacts of short-lived aerosols [Hare and Meinshausen 2006] and reactive greenhouse gases, the earth's climate is not equilibrated with anthropogenic forcing. As a result, even if fossil-fuel emissions were to suddenly cease, some level of committed warming is expected, due to past emissions, as studied previously using climate models [Solomon *et al*. 2009; Gillett *et al*. 2011; Frölicher *et al*. 2014]."

Therefore, the current global temperature is controlled largely by past CO_2 emitted over past decades, a consequence of the inertia in the climate and carbon cycle. The climate is committed at the current concentration of GHGs. This means that climate change has now become an independent driver of environmental change. Regardless of human action, or even human presence on the planet, impacts will continue to occur through temperature change, fluctuations of precipitation, snow melt, sea level rise, drought and other climate variables, and through changes in the hydrological cycle (Salem 2011). Climate change thus poses a challenge to growth and development.

2.7.1 Greenhouse gas emissions and concentration

The emission trends in selected countries are illustrated in **Figure 2.17** and **Figure 2.18**. More than half of total cumulative emissions since the Industrial Revolution were emitted in the past four decades. Cumulative CO_2 emissions for the period 1750-1970 (220 years) are estimated at 910 gigatons[9], while those for the period 1970-2010 (just 40 years) are about 1,090 gigatons (IPCC 2014). This growth is despite the presence of a wide array of multilateral institutions as well as national policies aimed at mitigation. The 2007/2008 global economic crisis only temporarily reduced the GHG emissions growth rate, compared with the trend since 2000 (Peters *et al*. 2011).

There is an unequal distribution of GHG emissions, both in terms of individual emissions coming from varied lifestyle consumption patterns and in terms of country emissions. The richest 10 per cent of the population emits 50 per cent of total GHG emissions, while the poorest 50 per cent emit only 10 per cent (King 2015). At the same time, when the carbon budget for limiting global warming below 2°C is considered, a generational inequality arises, with future generations having a lower allowance to emit. If the current Nationally Determined Contributions (NDCs) are fully implemented, the carbon budget for limiting global warming below 2°C will be 80 per cent depleted by 2030 (UNEP 2017).

Atmospheric concentrations of GHGs have increased from around 277 parts per million (ppm) in 1750 to 403.3 ppm in 2016 (World Meteorological Organization 2016). Regional contributions to this global GHG concentration are detailed in the GEO-6 Regional Assessments (UNEP 2016). The growth in atmospheric CO_2 was 6.0 ± 0.2 gigatons in 2016 (2.85 ± 0.09 ppm), well above the 2007-2016 average of 4.7 ± 0.1 gigatons a year (Le Quéré *et al*. 2017).

Figure 2.17: Global growth in emissions of GHGs by economic region

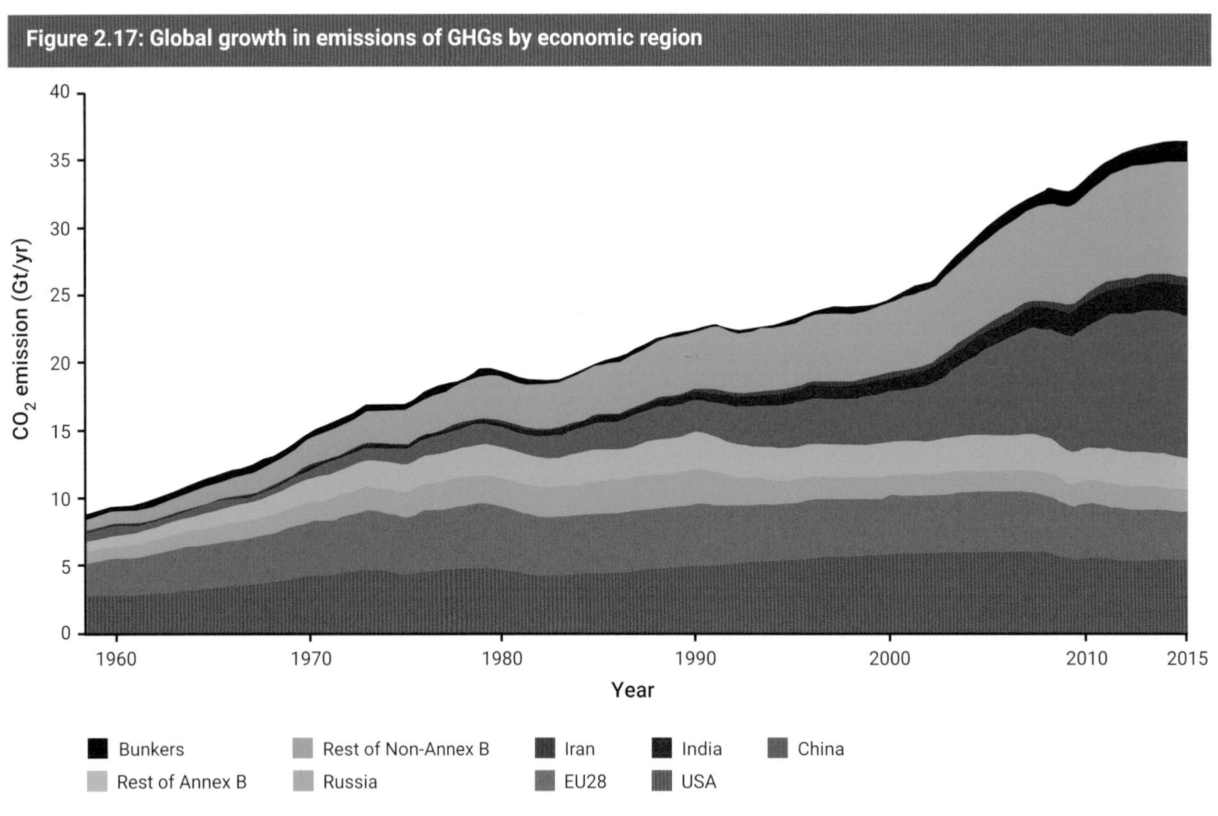

Source: Le Quéré *et al.* (2016)

9 Throughout this publication the term 'ton' refers to a metric ton or 1000 kilograms

Figure 2.18: Emission trends in different countries from 1990-2015. Orange lines highlight growth while blue ones show reductions

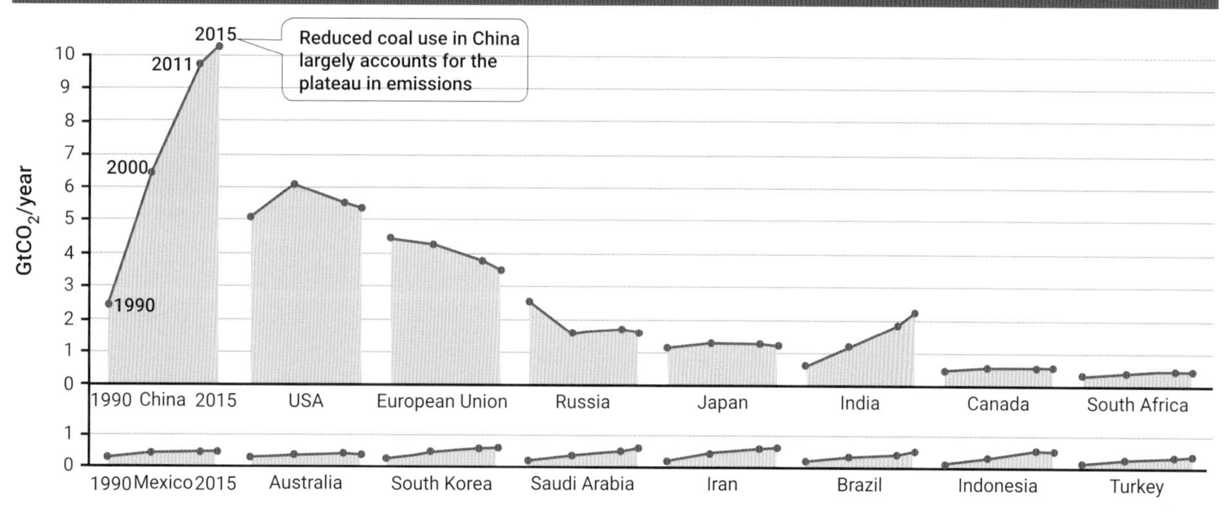

Source: Le Quéré *et al.* (2016)

2.7.2 The emissions budget

Cumulative total emissions of CO_2 and the response of the global mean surface temperature are approximately linearly related. Any given level of warming is associated with a range of cumulative CO_2 emissions. Therefore, a given temperature target (e.g. 2°C) will translate into a long-term emissions budget. Using this information, in the synthesis report of the Fifth Assessment, the IPCC (2014) estimated how much CO_2 we could emit and yet keep the global average temperature rise over pre-industrial levels to no more than 1.5°C, 2°C or even 3°C, which could be catastrophic.

The Paris Agreement's central aim is to strengthen a global response to the threat of climate change by keeping a global temperature rise this century well below 2°C above pre-industrial levels, and to pursue efforts to limit the temperature increase even further, to 1.5°C (UNFCCC 2015). To accomplish this, countries have submitted NDCs outlining their post-2020 climate action, which will undergo a global stocktake every five years to assess the collective progress and to inform further individual actions by parties (UNFCCC 2015).

In order to achieve the Paris temperature target, the carbon budget that remains after deducting past emissions is between 150 and 1,050 gigatons CO_2. At the current annual emission rates, the lower limit of this range will be crossed in four years and the midpoint (600 gigatons CO_2) in 15 years **(Figure 2.19)**. The emissions would have to drop to zero almost immediately after the budget is exhausted (Figueres *et al.* 2017).

Figure 2.19: The carbon crunch

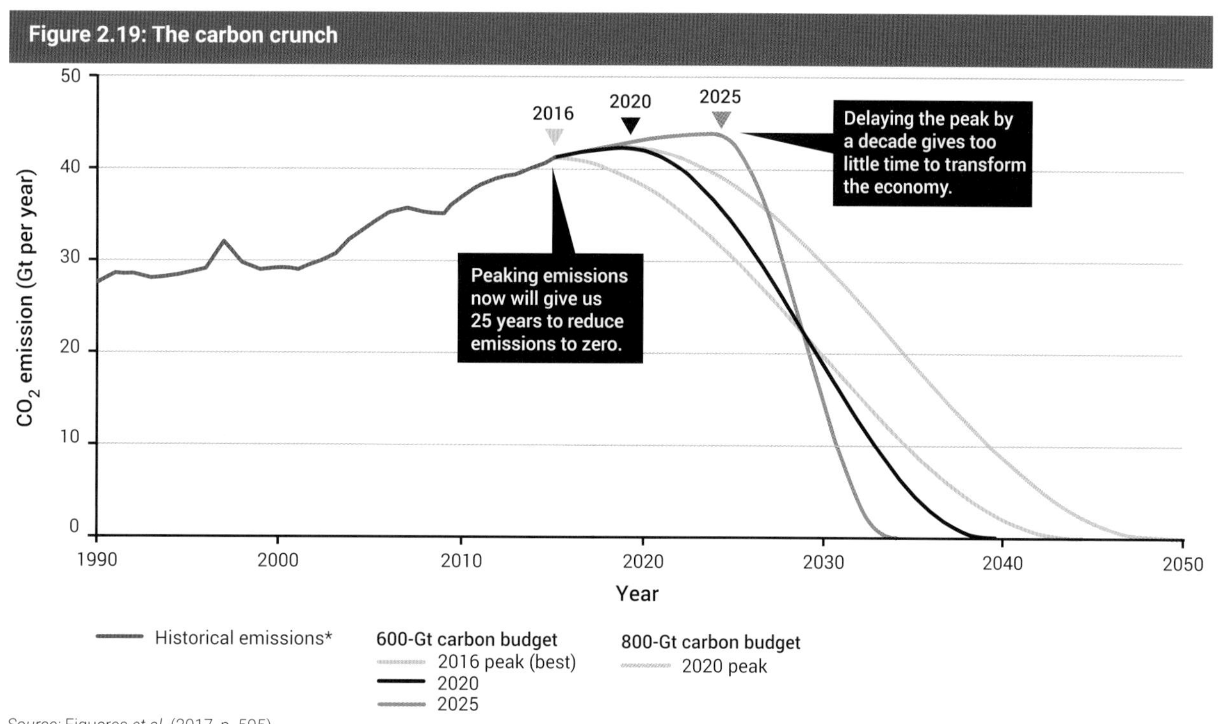

Source: Figueres *et al.* (2017, p. 595)

If the emission pledges in the Paris Agreement are fulfilled, the worst effects of climate change can be avoided, and studies suggest this could avoid a temperature increase of 3°C by 2100 (Le Quéré *et al.* 2016). The implications of the 2017 withdrawal of the United States, the second-largest emitter, from the Paris Agreement are mixed, because the withdrawal does not preclude individual American states' policies to support environmentally friendly innovation. It is still possible to meet the Paris temperature goals if global emissions begin to fall by 2020 (Figueres *et al.* 2017).

Under current and planned policies, the world would exhaust its energy-related carbon budget (CO_2) in under 20 years to keep the global temperature rise to well below 2°C. To meet the below 2°C goal, immediate action is crucial to reduce further cumulative emissions by 470 gigatons by 2050, compared with current and planned policy targets (IRENA, 2018)

2.7.3 Impacts of Climate Change

Climate change will amplify existing risks and create new risks for natural and human systems (IPCC 2014). The risks are not only unevenly distributed but are generally greater for disadvantaged people and communities. This is so in countries at all levels of development. The risk of climate-related impacts is a result of complex interactions between climate-related hazards and the vulnerability, exposure and adaptive capacity of human and natural systems. The rise in the rates and magnitudes of warming and other changes in the climate system, accompanied by ocean acidification, increase the risk of severe, pervasive and in some cases irreversible detrimental impacts. Already, the annual global mean surface temperature has increased at an average rate of 0.07°C per decade since 1880 and at an average rate of 0.17°C per decade since 1970 (National Oceanic and Atmospheric Administration [NOAA] 2015). The trends in sea surface temperature, marine air temperature, sea level, tropospheric temperature, ocean heat content and specific humidity are similar (IPCC 2014) **(Figure 2.20)**.

Beyond temperature increase, the impacts already observed include changes in the water cycle, warming of the oceans, shrinking of the Arctic ice cover, increase in the global mean sea level, and altering of the carbon and biogeochemical cycles (see more detail in Chapters 4 and 5). Further, there have been increases in the frequency and intensity of wildfires that in turn release GHGs. Observations and climate model simulations indicate polar warming amplification resulting from various feedbacks in the climate system – the positive ice-albedo feedback being the strongest (Taylor *et al.* 2013). The reduced extent of ice cover reveals a darker surface, which leads to a decreased albedo, in turn resulting in a stronger absorption of solar radiation and a further acceleration of warming. In response to the increased warming in the Arctic, sea-ice extent is strongly decreasing, especially in summer (Vaughan *et al.* 2013). However, recent literature has concluded that temperature feedbacks play a dominant role, making surface albedo feedback the second main contributor to Arctic amplification (Pithan and Mauritsen 2014).

The global water cycle has been affected, impacting on global-scale precipitation patterns over land, and on surface and subsurface ocean salinity, contributing to global-scale changes in frequency and intensity of daily temperature extremes since the mid-20th century. The global mean sea level rose by 0.19 metres (range, 0.17-0.21 metres) over the period 1901-2010, calculated using the mean rate over these 110 years, and based on tide gauge records plus, since 1993, satellite data (IPCC 2014).

Figure 2.20: Multiple independent indicators of a changing global climate

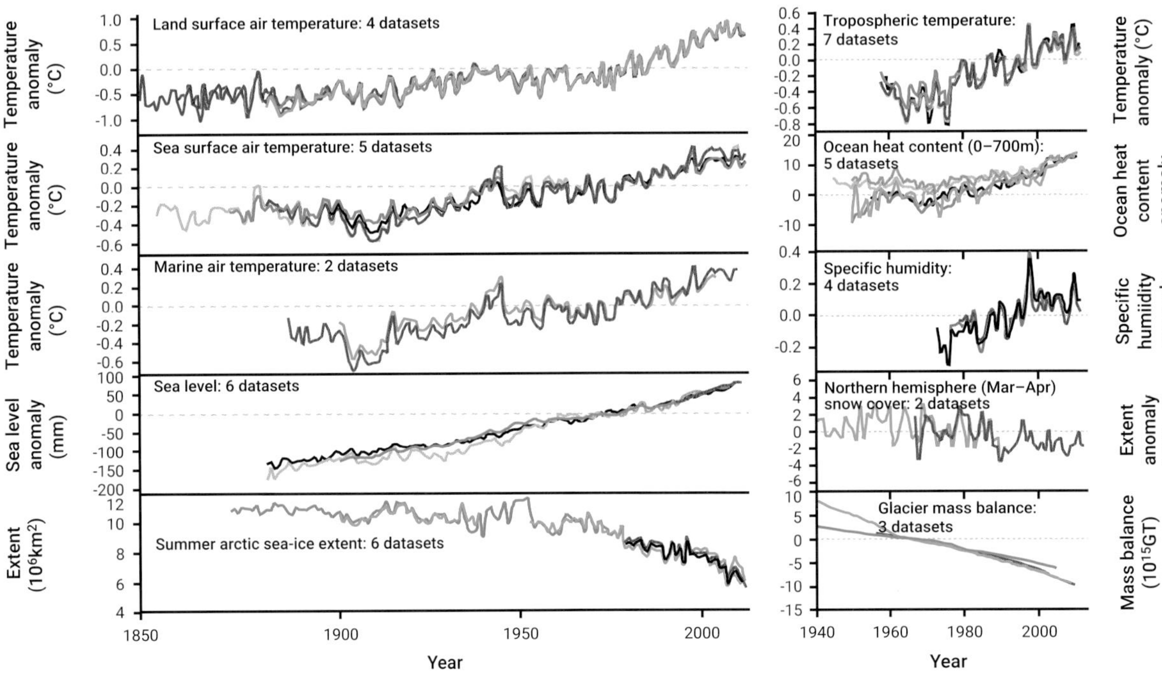

Source: IPCC (2014)

Changes in the climate system have had large-scale impacts on various ecosystems, as documented across the thematic chapters that follow in Part A. As a driver of environmental change, climate change is exacerbating current pressures on land, water, biodiversity and ecosystems. If atmospheric CO_2 concentration increases from the current levels of 406 ppm to 450-600 ppm, leading to greater than 2°C warming over the coming century, it will lead to several irreversible impacts, including sea level rise (Smith *et al.* 2011). O'Neill *et al.* (2017) have elaborated individual risks as well as overarching key risks, including risks to biodiversity, health, agriculture and so on, as well as risks of extreme events such as extreme precipitation and heat waves and risks to specific ecosystems such as mountain and Arctic, to name but a few **(see Figure 2.21)**.

Future climate will thus depend on the combination of committed warming caused by past anthropogenic emissions, the impact of future anthropogenic emissions, natural climate variability and climate sensitivity. There are regions (particularly at northern, mid- and high latitudes) already experiencing greater warming than the global average, with mean temperature rise exceeding 1.5°C in these regions.

These impacts have implications for the quality and quantity of ecosystem services, as well as for patterns of resource use, their distribution and access across regions and within countries.

Time is running out to prevent the irreversible and dangerous impacts of climate change. Unless GHG emissions are reduced radically, the world remains on a course to exceed the agreed temperature threshold of 2°C above pre-industrial levels, which would increase the risk of pervasive effects of climate change, beyond what is already seen. These effects include extreme events (including flooding, hurricanes and cyclones) leading to loss of lives and livelihoods, pervasive droughts leading to loss of agricultural productivity and food insecurity, severe heat waves, changes in disease vectors resulting in increases in morbidity and mortality, slowdowns in economic growth, and increased potentials for violent conflict (Salem 2011; SIDA 2018). The extent, distribution and acute nature of the impacts is different between countries, and several islands have faced multiple impacts in one season – Haiti in 2004, for example – or annually in multiple years, such as Dominica experiencing hurricanes Erika in 2015 and Maria in 2017. These impacts can undermine food security mechanisms and systems, as well as social and economic progress in health and other areas.

Figure 2.21: The enhanced burning embers diagram, providing a global perspective on climate-related risks

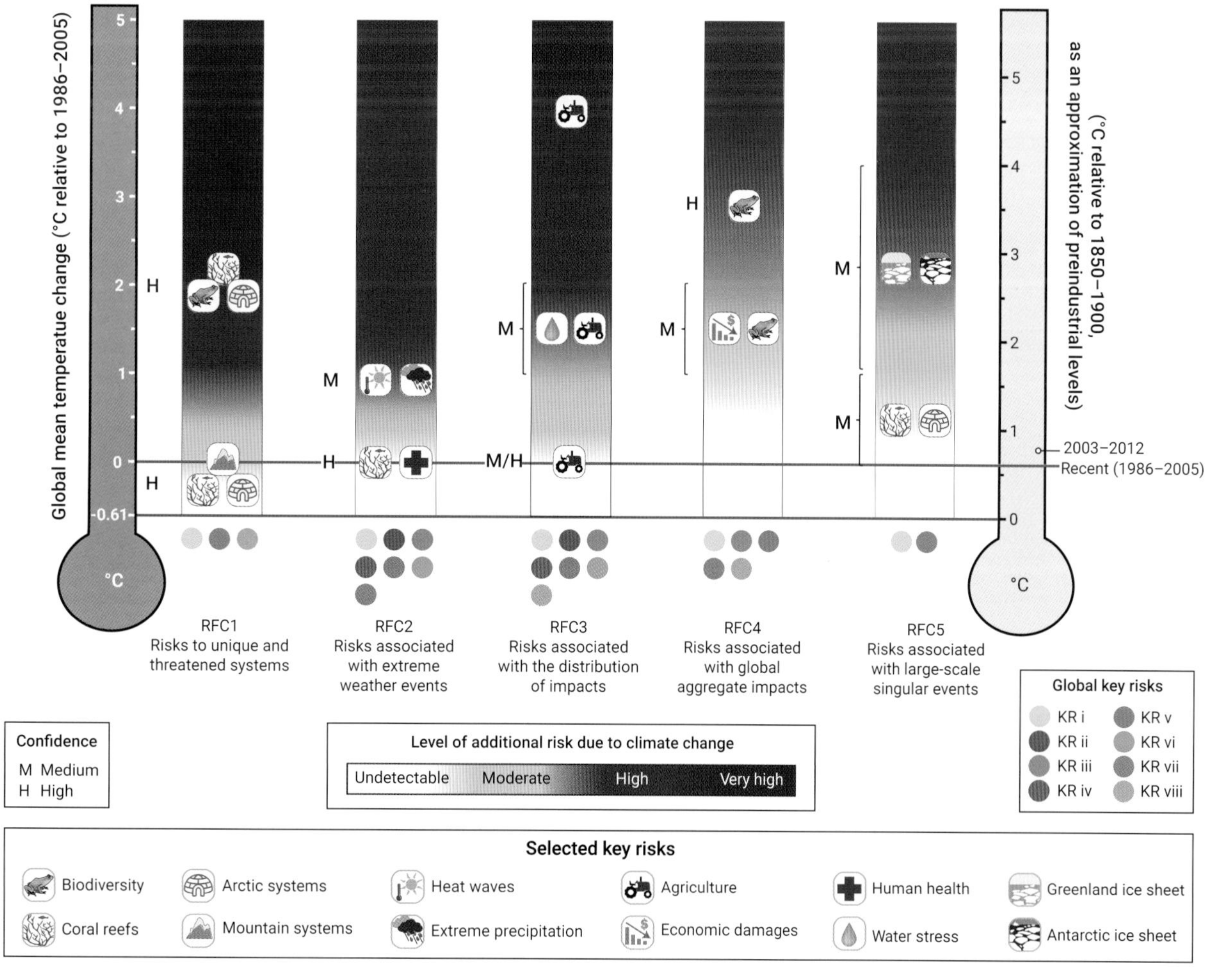

Source: O'Neill *et al.* (2017, p. 30)

Figure 2.22: Trends in numbers of loss-relevant natural events

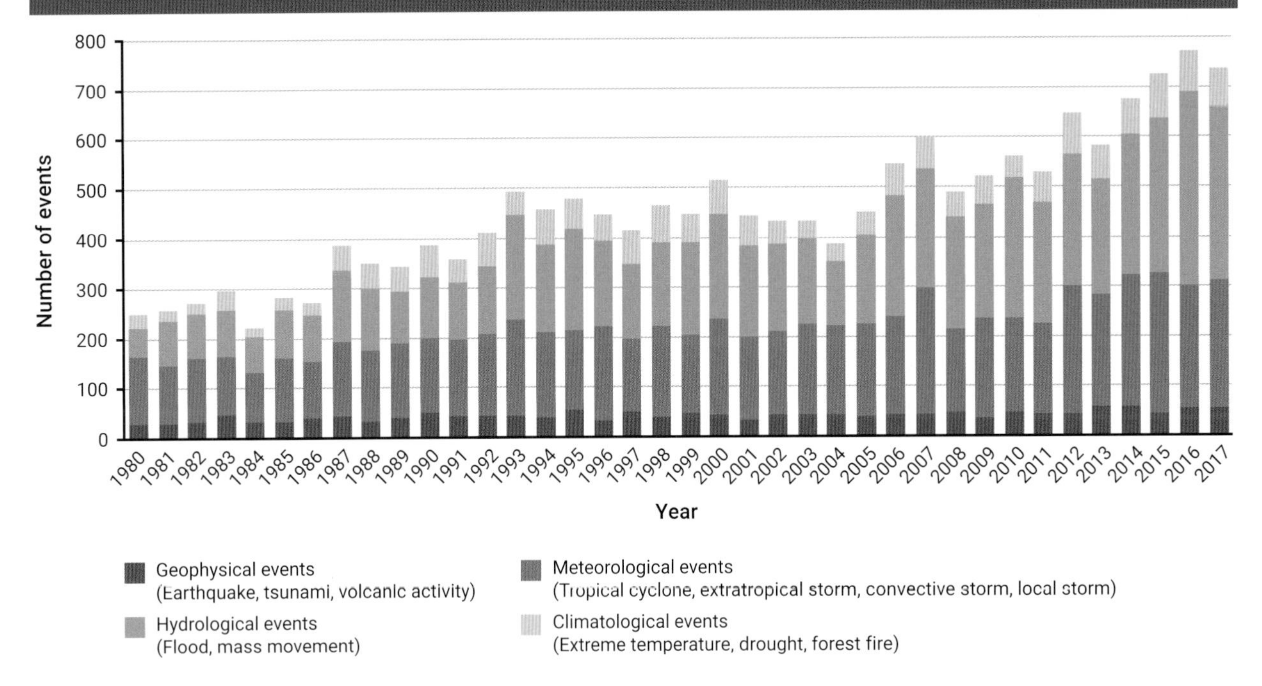

Source: Munich Re (2017)

One indication of the potential impacts is the doubling of the frequency of climate-related loss events **(Figure 2.22)** since 1980 (Hoeppe 2016). These events are already estimated to have resulted in the loss of 400,000 lives and the imposition of a cost of US$1.2 trillion annually on the global economy, wiping 1.6 per cent from global GDP.

These risks are greatest – currently as well as in the future – for people who are dependent on natural-resource sectors. Such people include coastal communities, people in agricultural and forest communities, and those experiencing multiple forms of inequality, marginalization and poverty, thereby amplifying existing risks and create new ones for natural and human systems. The scale of potential damage from climate change poses a major systemic risk to our future well-being and the ecosystems on which we depend, in particular for societies in less-developed, less-resilient countries (OECD 2017).

2.7.4 Implications

The Paris Agreement recognizes that limiting warming by the end of the century could help prevent more problems. It explicitly states the need for achieving a balance of emissions and removals in the second half of the century. The 2°C target is important to achieve, to reduce the likelihood of more intense storms, longer droughts, rising sea levels and other natural disasters that are being increasingly reported (Munich Re 2016). To keep a good chance of staying below 2°C, and at manageable costs, emissions should drop by 40-70 per cent globally between 2010 and 2050, falling to zero by 2100 (IPCC 2014; Kroeze and Pulles 2015). The current trajectory of global annual and cumulative emissions of GHGs is inconsistent with the widely discussed goals of limiting global warming to 1.5-2.0°C above pre-industrial levels. Should emissions continue to rise beyond 2020, or even remain level, the temperature goals set in Paris become almost unattainable. Delayed action or weak near-term policies increase the mitigation challenges in the long-term. There are risks associated with exceeding 1.5°C global warming by the end of the century (increases in the severity of projected impacts and in the adaptation needs), making the achievement of many SDGs much more difficult. The overall costs and risks of climate change include a prediction that some regions could see growth decline by as much as 6 per cent of GDP by 2050, according to a recent report from the World Bank Group (2016) on climate change, water and the economy. If the worst of the climate change-related risks are to be avoided, the pace and scale of the required economic transformation is unprecedented (OECD 2017).

2.8 Unravelling drivers and their interactions

The same driver of environmental change can exert both positive and negative forces on the environment, as described in the previous sections. Moreover, the five drivers highlighted in this chapter are mutually interdependent, and this interdependence can itself also be positive or negative. The cumulative effect the drivers can have on the environment has been extensively discussed in the literature (Wu *et al.* 2017).

Table 2.1 presents the interactions between the drivers covered in this chapter. These are first-order interactions (excluding interactions with other variables) at a global scale and under current conditions.

The aggregate effects of these interactions on climate change are negative. This is clear from the current trajectory of GHG emissions, which not only continue to increase, but at a rate that has accelerated in the last 15 years, compared with the 1980-2000 trajectory (Section 2.7). Thus, there is little

Table 2.1: Interrelationships between the drivers

	Population growth	Economic growth	Technological change	Climate change	Urbanization
Population growth	–	Negative impact due to delay in the demographic window of opportunity	Population growth fosters technological innovation, to accommodate the additional demands. Alternatively, it could lead to lower savings and investment due to high dependency rates	Population growth increases environmental pressure, and climate change	Increased pressure on urban areas, more people might move to urban areas
Economic growth	Higher GDP and development in general is associated with lower fertility rates	–	Economic growth is associated with increased investment and technological innovation	Increased economic output is associated with increased environmental pressure	Growth will push towards increased urbanization
Technological change	Technological innovation is associated with increased capacity to lower fertility rates	Innovation is associated with increased growth in GDP	–	Current trends show an increase in green technological innovation, thus lowering pressure per unit of output	Technological change can contribute to processes of urbanization or it can help to decrease the migration patterns through better access to technologies and communication
Climate change	Climate change increases mortality rates and negatively affects health	There are costs associated with climate change that limit economic growth	Climate change pressures foster adaptive technological innovation	–	Effects of climate change on rural communities puts pressure on migration towards urban areas
Urbanization	Urbanization is associated with lower fertility rates (due to access to better health care and education)	Urbanization is strongly associated with higher economic output	Urbanization will lead to intensification of technology use due to greater population density	There is no clear causal link, but there is an association between urbanization and higher emissions	–

doubt about the unsustainability of the current interaction and aggregated effects of population growth, economic development and technological innovation.

These aggregated effects are not the same for different regions. In developed countries (such as Canada, European Union countries, Japan and the United States of America), emissions have plateaued and in some cases diminished substantially. Moderate growth, stable populations, some change in consumption patterns, and technological innovations have allowed for a reduction in aggregate GHG emissions. At the same time, emerging economies that are moving from lower middle-income status to upper middle-income status have increased aggregate emissions (this is the case in most middle-income countries, including China and India).

On the other hand, both in per-capita terms and in aggregate, it is the richest and better off countries that contribute, by far, the most to emissions. This is true both for countries by income level (the developed world accounts for more than half of total emissions, with a far higher carbon footprint per capita) or for individuals by income level within countries (people in the world's richest quintiles, both from developed and developing countries, produce both higher carbon footprints per capita and greater aggregate emissions). Consumption patterns

and production functions in the developed world, and the lifestyle and consumption options of the world's elites and better-off, therefore have to change drastically to adjust GHG emissions for a more sustainable path. The pathway to growth in emerging economies cannot reiterate the carbon expansion and GHG emissions witnessed in the last 20 years. Both technology and urbanization provide a window of opportunity – but no guarantee – for emerging economies to follow a developmental path that will prove more sustainable, from both consumption and production perspectives.

The diagram below **(Figure 2.23)** shows another way to look at the interactions between the drivers, by focusing on how each domain relates to the other, and how that can change. The diagram is used to evaluate the obstacles that different paths will confront. Three of the drivers considered – economic growth, population growth and climate change – stand at the left, while the right of the diagram presents the desired or preferred outcomes – lower environmental pressure, human well-being and equity. In the middle are the mediating factors of technological change and urbanization (potentially enabling mechanisms, but also potentially negative forces). Economic growth, population growth and climate change are to the left because they reflect the fundamental realities of human aspiration, demographic momentum and climate change commitment. What can change the impacts of these processes is the nature of the two other drivers in the middle – technology and urbanization.

Limiting the negative effects of the various drivers described – and indeed reframing them as the catalysts of an urgently needed transformative response – is necessary to achieve sustainable development and equity, including poverty eradication. At the same time, it is important to ensure that efforts to address one driver do not undermine actions overall to promote sustainable development.

The chapters in Parts B, C and D present a comprehensive assessment that looks at development pathways more broadly, along with their policy implications.

Figure 2.23: Relationship across the drivers

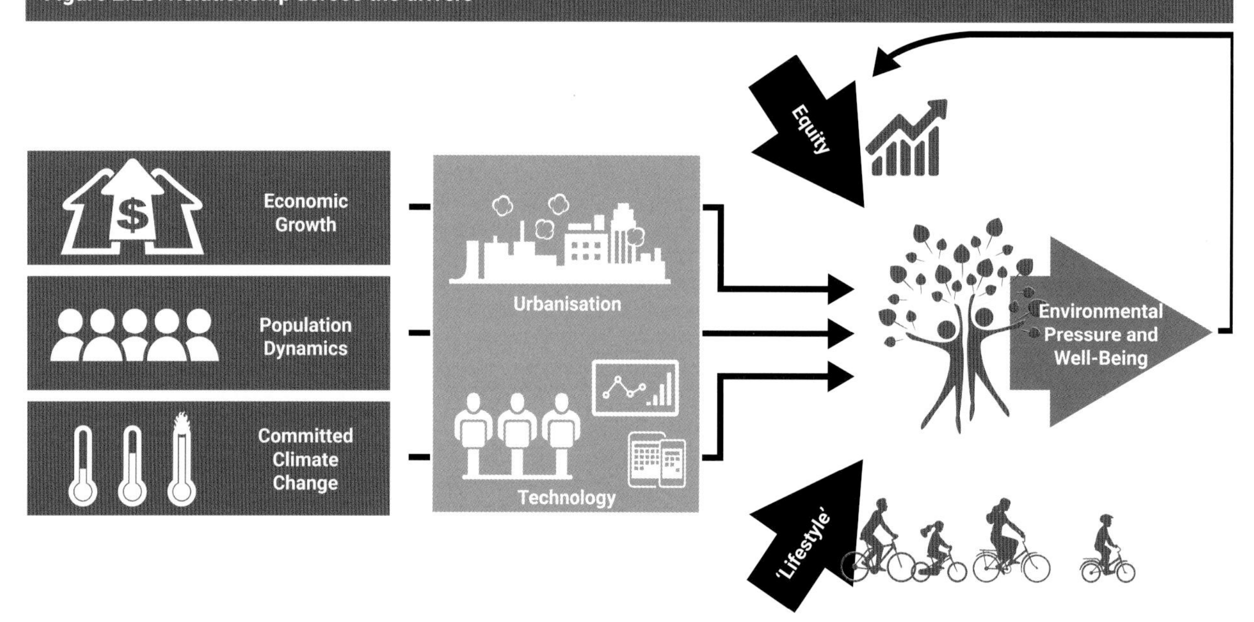

Box 2.5: IPAT identity

The IPAT identity makes a conceptual link between population, development, and technology, and different trajectories depend on the interactions between these determining factors. The IPAT identity has the following form:

I = P. A. T = P. (Y/P). (I/Y) = I, where (Eq. 1a)

- I = impact, i.e. use of natural resources or energy
- P = population
- A = affluence, an alternative term for per capita income, P/Y
- Y = national output or GDP
- T = technology, or the efficiency with which production takes place, generally interpreted as the amount of resource use (or resource impact) per dollar of output.

Eq. 1a suggests a simple multiplicative relationship between the three constituent factors, P, A and T. Indeed, population is viewed in some of the scientific and policy literature as simply a proportional factor or multiplier of the environmental impacts of the more 'substantive' factors of economic growth, technological change and regulatory restriction. Other things being fixed, a doubling of population will lead to double the consumption of natural resources and energy. We know that this does not happen, however – other things are not fixed – overall moderating population growth will improve economic growth in emerging and low-income economies thus limiting the positive effects that population moderation will have on aggregate emissions. On the other hand moderating economic growth can limit growth in lower-income economies, affecting also the rate of population moderation. So once again, what might be a gain on one side can be lost on the other. Further, growth is required to meet the other substantive SDGs. Thus a radical decoupling of emissions from both population and economic growth has to be achieved.

The Kaya Identity has often been used to analyse the various drivers of climate change.

C = P. A. e. c = P. (Y/P). (E/Y). (C/E) (Eq. 1b)

Where:

C = carbon emissions

P = population,

A = affluence = Y/P, where Y = Income (or consumption)

e = energy intensity (or energy consumption per dollar of output) = E/Y, where E is total energy consumption

c = carbon intensity (i.e. carbon emissions per unit of energy consumed) = C/E

What this suggests is that the reduction of emissions will occur only if one or more variables in Eq. 1b are reduced. Two inferences can be drawn from this relationship. First, while a marginal and gradual reduction in emissions can be achieved through marginal changes in one or more of the constituent factors (P, A, e, and c), radical reductions implied by the Paris Agreement (including, e.g., reducing emissions to zero by 2050) can be achieved only through some combination of rapid decarbonization of energy use (i.e. reducing C/E), reduction of the overall energy intensity (E/Y) of the economy, reduction of the consumption level (Y/P) of the world's rich and better off (both in the developed and the developing world), and reduction of the ultimate level of the population (P). All of these options present challenges.

References

Acemoglu, D. (2009). Endogenous technological change. In *Introduction to Modern Economic Growth*. Princeton, NJ: Princeton University Press. 411-533. https://www.theigc.org/wp-content/uploads/2016/06/acemoglu-2007.pdf

Acemoglu, D., Aghion, P., Bursztyn, L. and Hemous, D. (2012). The environment and directed technical change. *American Economic Review* 102(1), 131-166. http://dx.doi=10.1257/aer.102.1.131.

Alexander, P., Brown, C., Arneth, A., Dias, C., Finnigan, J., Moran, D. *et al.* (2017). Could consumption of insects, cultured meat or imitation meat reduce global agricultural land use? *Global Food Security* 15, 22-32. https://doi.org/10.1016/j.gfs.2017.04.001.

Alexandratos, N. and Bruinsma, J. (2012). *World Agriculture Towards 2030/2050: The 2012 Revision*. Rome: Food and Agriculture Organization of the United Nations. http://www.fao.org/docrep/016/ap106e/ap106e.pdf.

Alvaredo, F., Chancel, L., Piketti, T., Saez, E. and Zucman, G. (2017). *World Inequality Report 2018*. The World Inequality Lab. https://wir2018.wid.world/files/download/wir2018-full-report-english.pdf.

Ang, G., Röttgers, D. and Burli, P. (2017). *The Empirics of Enabling Investment and Innovation in Renewable Energy*. OECD Environment Working Papers. Paris: Organization for Economic Cooperation and Development. https://www.oecd-ilibrary.org/deliver/67d221b8-en.pdf?itemId=%2Fcontent%2Fpaper%2F67d221b8-en&mimeType=pdf.

Angel, S., Parent, J., Civco, D.L., Blei, A. and Potere, D. (2011). The dimensions of global urban expansion: Estimates and projections for all countries, 2000–2050. *Progress in Planning* 75(2), 53-107. https://doi.org/10.1016/j.progress.2011.04.001.

Baldé, C.P., Forti, V., Gray, V., Kuehr, R. and Stegmann, P. (2017). *The Global E-Waste Monitor 2017: Quantities, Flows and Resources*. Bonn: United Nations University, International Telecommunication Union and International Solid Waste Association. http://collections.unu.edu/view/UNU:6341.

Baldé, C.P., Wang, F. and Kuehr, R. (2016). *Transboundary Movements of Used and Waste Electronic and Electrical Equipment*. Bonn: United Nations University. http://www.step-initiative.org/files/_documents/other_publications/UNU-Transboundary-Movement-of-Used-EEE.pdf.

Barrera, P.P., Carreón, J.R. and de Boer, H.J. (2018). A multi-level framework for metabolism in urban energy systems from an ecological perspective. *Resources, Conservation and Recycling* 132, 230-238. https://doi.org/10.1016/j.resconrec.2017.05.005.

Benhabib, J. and Rustichini, A. (1996). Social conflict and growth. *Journal of Economic growth* 1(1), 125-142. https://doi.org/10.1007/BF00163345.

Bettencourt, L. and West, G. (2010). A unified theory of urban living. *Nature* 467(7318), 912-913. https://doi.org/10.1038/467912a.

Bhattacharya, A., Meltzer, J., Oppenheim, J., Qureshi, Z. and Stern, N. (2016). *Delivering on Sustainable Infrastructure for Better Development and Better Climate*. Washington, D.C: The Brookings Institution. https://www.brookings.edu/wp-content/uploads/2016/12/global_122316_delivering-on-sustainable-infrastructure.pdf.

Bijl, D.L., Bogaart, P.W., Dekker, S.C., Stehfest, E., de Vries, B.J.M. and van Vuuren, D.P. (2017). A physically-based model of long-term food demand. *Global Environmental Change* 45, 47-62. https://doi.org/10.1016/j.gloenvcha.2017.04.003.

Bird, R. and Bahl, R. (2008). *Subnational Taxes in Developing Countries: The Way Forward*. Institute for International Business. https://papers.ssrn.com/sol3/papers.cfm?abstract_id=1273753.

Birkmann, J., Welle, T., Solecki, W., Lwasa, S. and Garschagen, M. (2016). Boost resilience of small and mid-sized cities. *Nature* 537(7622), 605-608. http://dx.doi.org/10.1038/537605a.

Bithas, K. and Kalimeris, P. (2016). *Revisiting the Energy-Development Link: Evidence from the 20th Century for Knowledge-Based and Developing Economies*. Springer. https://books.google.co.ke/books/about/Revisiting_the_Energy_Development_Link.html?id=4O1irgEACAAJ&redir_esc=y.

Bloomberg New Energy Finance (2017). 'Off-grid and mini-grid: Q1 2017 market outlook'. 5 January 2017 https://about.bnef.com/blog/off-grid-mini-grid-q1-2017-market-outlook/

Bourguignon, F. (2002). The growth elasticity of poverty reduction: Explaining heterogeneity across countries and time periods. In Inequality and Growth. Eicher, T. and Turnovski, S. (eds.). Cambridge, MA: MIT Press. chapter 1. 3-26. http://documents.worldbank.org/curated/en/503161468780002293/pdf/28104.pdf

Bren d'Amour, C., Reitsma, F., Baiocchi, G., Barthel, S., Güneralp, B., Erb, K.-H. *et al.* (2016). Future urban land expansion and implications for global croplands. *Proceedings of the National Academy of Sciences* 114(34), 8939-8944. https://doi.org/10.1073/pnas.1606036114.

Brenner, B. and Schmid, C. (2014). The 'urban age' in question. *International Journal of Urban and Regional Research* 38(3), 731-755. https://doi.org/10.1111/1468-2427.12115.

C40 Cities Climate Leadership Group and Arup (2015). *Polisdigitocracy: Digital Technology, Citizen Engagement, and Climate Action*. London. http://www.c40.org/researches/polisdigitocracy-digital-technology-citizen-engagement-and-climate-action.

Canuto, O. (2010). *Towards a Switchover of Locomotives in the Global Economy*. Economic PremiseWorld Bank. http://documents.worldbank.org/curated/en/694391468174866224/pdf/568290BRI0EP330Box353739B01PUBLIC1.pdf.

Carroll, C., Slacalek, J., Tokuoka, K. and White, M.N. (2017). The distribution of wealth and the marginal propensity to consume. *Quantitative Economics* 8(3), 977-1020. https://doi.org/10.3982/QE694.

Carson, R. (1962). *Silent Spring*. Boston, MA: Houghton Mifflin. http://www.rachelcarson.org/SilentSpring.aspx.

Casey, G. and Galor, O. (2017). Is faster economic growth compatible with reductions in carbon emissions? The role of diminished population growth. *Environmental Research Letters* 12(1), 014003. https://doi.org/10.1088/1748-9326/12/1/014003.

Chancel, L. and Piketty, T. (2015). *Carbon and Inequality: From Kyoto to Paris. Trends in the Global Inequality of Carbon Emissions (1998-2013) and Prospects for an Equitable Adaptation Fund*. Paris: Paris School of Economics. http://piketty.pse.ens.fr/files/ChancelPiketty2015.pdf.

Chitnis, M., Sorrell, S., Druckman, A., Firth, S.K. and Jackson, T. (2013). Turning lights into flights: Estimating direct and indirect rebound effects for UK households. *Energy Policy* 55, 234-250. https://doi.org/10.1016/j.enpol.2012.12.008.

Cottineau, C., Finance, C., Hatna, E., Arcaute, E. and Batty, M. (2018). Defining urban agglomerations to detect agglomreation economies. *Environment and Planning B: Urban Analytics and City Science*. https://doi.org/10.1177/2399808318755146.

CPB Netherlands Bureau for Economic Policy Analysis (2018). *CPB world trade monitor*. https://www.cpb.nl/en/worldtrademonitor (Accessed: 5 November 2018).

da Costa, P.N. (2017). *There's a morsel of good news about inequality that's buried in all the scary headlines*. Business Insider UK. http://uk.businessinsider.com/inequality-rising-in-rich-countries-but-falling-in-many-poor-ones-2017-8?r=US&IR=T.

Dalton, M., O'Neill, B., Prskawetz, A., Jiang, L. and Pitkin, J. (2008). Population aging and future carbon emissions in the United States. *Energy economics* 30(2), 642-675. https://doi.org/10.1016/j.eneco.2006.07.002.

Daly, H.E. (1973). *Towards a Steady State Economy*. San Francisco, CA: W.H. Freeman. http://www.worldcat.org/title/toward-a-steady-state-economy/oclc/524050.

Dechezleprêtre, A., Martin, R. and Mohnen, M. (2014). *Knowledge Spillovers from Clean and Dirty Technologies*. CEP Discussion Paper. London: Centre for Economic Performance. http://eprints.lse.ac.uk/60501/.

Diamond, J.M. (2011). *Collapse: How Societies Choose to Fail or Succeed*. *New* edn. London: Penguin. http://www.e-reading.by/bookreader.php/133781/Collapse%253A_How_Societies_Choose_to_Fail_or_Master.pdf.

Dobbs, R., Smit, S., Remes, J., Manyika, J., Roxburgh, C. and Restrepo, A. (2011). *Urban World: Mapping the Economic Power of Cities*. New York, NY: McKinsey Global Institute. https://www.mckinsey.com/~/media/McKinsey/Featured%20Insights/Urbanization/Urban%20world/MGI_urban_world_mapping_economic_power_of_cities_full_report.ashx.

Dodman, D. (2009). Blaming cities for climate change? An analysis of urban greenhouse gas emissions inventories. *Environment and Urbanization* 21(1), 185-201. https://doi.org/10.1177/0956247809103016.

Earth Policy Institute (2011). Data highlights: Education leads to lower fertility and increased prosperity. Rutgers University http://www.earth-policy.org/data_highlights/2011/highlights13

Eckersley, R. (1992). *Environmentalism and Political Theory: Toward an Ecocentric Approach*. Albany, NY: Suny Press. http://www.sunypress.edu/p-1386-environmentalism-and-political-.aspx.

Eigenbrod, C. and Gruda, N. (2015). Urban vegetable for food security in cities: A review. *Agronomy for Sustainable Development* 35(2), 483-498. https://doi.org/10.1007/s13593-014-0273-y.

Farmani, R. and Butler, D. (2014). Implications of urban form on water distribution systems performance. *Water resources management* 28(1), 83-97. https://doi.org/10.1007/s11269-013-0472-3.

Figueres, C., Schellnhuber, H.J., Whiteman, G., Rockström, J., Hobley, A. and Rahmstorf, S. (2017). Three years to safeguard our climate. *Nature* 546(7660), 593-595. https://doi.org/10.1038/546593a.

Filgueira, F. (2014). The politics and policies of social incorporation in Latin America. *New Directions in Social Policy: Alternatives from and for the Global South*. Geneva, 7-8 April 2014. United Nations Research Institute for Social Development http://www.unrisd.org/80256B3C005BCCF9/(httpAuxPages)/BC8DABDEC1C728D3C1257D08003AD5EA/$file/Filgueira.pdf

Floater, G. and Rode, P. (2014). *Cities and The New Climate Economy: The Transformative Role of Global Urban Growth*. The New Climate Economy. https://files.lsecities.net/files/2014/11/LSE-Cities-2014-The-Transformative-Role-of-Global-Urban-Growth-NCE-Paper-01.pdf.

Food and Agriculture Organization of the United Nations (2017). *The Future of Food and Agriculture: Trends and Challenges*. Rome. http://www.fao.org/3/a-i6583e.pdf.

Frank, J. and Martinez-Vazquez, J. (2014). *Decentralization and Infrastructure in the Global Economy: From Gaps to Solutions*. Atlanta, GA: Georgia State University. http://icepp.gsu.edu/files/2015/03/ispwp1405.pdf.

Friedman, B.M. (2006). *The Moral Consequences of Economic Growth*. New York, NY: Vintage Books. https://www.penguinrandomhouse.com/books/56526/the-moral-consequences-of-economic-growth-by-benjamin-m-friedman/9781400095711/.

Frölicher, T.L., Winton, M. and Sarmiento, J.L. (2014). Continued global warming after CO_2 emissions stoppage. *Nature Climate Change* 4, 40–44. https://doi.org/10.1038/nclimate2060.

Fuller, B. and Romer, P. (2014). *Urbanization as Opportunity*. Policy Research Working Paper. Washington, D.C: World Bank. http://documents.worldbank.org/curated/en/775631468180872982/pdf/WPS6874.pdf.

Garschagen, M., Hagenlocher, M., Kloos, J., Pardoe, J., Lanzendörfer, M., Mucke, P. *et al.* (2014). *World Risk Report 2015*. Bündnis Entwicklung Hilft (Alliance Development Works) and United Nations University – Institute for Environment and Human Security (UNU-EHS). https://collections.unu.edu/eserv/UNU:3303/WRR_2015_engl_online.pdf.

Gartzke, E. (2007). The capitalist peace. *American journal of political science* 51(1), 166-191. https://doi.org/10.1111/j.1540-5907.2007.00244.x.

GBD 2015 Mortality and Causes of Death Collaborators (2016). Global, regional, and national life expectancy, all-cause mortality, and cause-specific mortality for 249 causes of death, 1980-2015: A systematic analysis for the Global Burden of Disease Study 2015. *The Lancet* 388(10053), 1459-1544. https://doi.org/10.1016/S0140-6736(16)31012-1.

Gerber, P.J., Steinfeld, H., Henderson, B., Mottet, A., Opio, C., Dijkman, J. *et al.* (2013). *Tackling Climate Change Through Livestock: A Global Assessment of Emissions and Mitigation Opportunities*. Rome: Food and Agriculture Organization of the United Nations. http://www.fao.org/3/a-i3437e.pdf.

Gillespie, D., Ahmed, S., Tsui, A. and Radloff, S. (2007). Unwanted fertility among the poor: An inequity? *Bulletin of the World Health Organization* 85(2), 100-107. http://www.who.int/bulletin/volumes/85/2/06-033829.pdf.

Gillett, N.P., Arora, V.K., Zickfeld, K., Marshall, S.J. and Merryfield, W.J. (2011). Ongoing climate change following a complete cessation of carbon dioxide emissions. *Nature Geoscience* 4, 83–87. https://doi.org/10.1038/ngeo1047.

Gilmont, M., Nassar, L., Rayner, S., Tal, N., Harper, E. and Salem, H.S. (2018). The potential for enhanced water decoupling in the Jordan Basin through regional agricultural best practice. *Land* 7(2), 63. https://doi.org/10.3390/land7020063.

Glaeser, E.L. and Gottlieb, J.D. (2009). The wealth of cities: Agglomeration economies and spatial equilibrium in the United States. *Journal of Economic Literature* 47(4), 983-1028. https://doi.org/10.1257/jel.47.4.983.

Global Commission on The Economy and Climate (2015). *Seizing the Global Opportunity: Partnerships for Better Growth and A Better Climate. The 2015 New Climate Economy Report*. Washington, D.C. http://newclimateeconomy.report/2015/wp-content/uploads/sites/3/2014/08/NCE-2015_Seizing-the-Global-Opportunity_web.pdf.

Guha, R. (1999). *Environmentalism: A Global History*. *1st* edn. London: Pearson.

Güneralp, B., Zhou, Y., Ürge-Vorsatz, D., Gupta, M., Yu, S., Patel, P.L. *et al.* (2017). Global scenarios of urban density and its impacts on building energy use through 2050. *Proceedings of the National Academy of Sciences* 114(34), 8945-8950. https://doi.org/10.1073/pnas.1606035114.

Hamilton, K. and Hartwick, J. (2017). Wealth and sustainability. In National *Wealth: What is Missing, Why it Matters*. Hamilton, K. and Hepburn, C. (eds.). Oxford: Oxford Universiy Press. chapter 15. http://www.oxfordscholarship.com/view/10.1093/oso/9780198803720.001.0001/oso-9780198803720-chapter-15.

Harari, Y.N. (2017). *Homo Deus: A Brief History of Tomorrow*. Harper. http://www.elboomeran.com/upload/ficheros/obras/z3_homo_deus.pdf.

Hare, B. and Meinshausen, M. (2006). How much warming are we committed to and how much can be avoided? *Climatic Change* 75(1-2), 111–149. https://doi.org/10.1007/s10584-005-9027-9.

Hartmann, D.L., Tank, A.M.G.K., Rusticucci, M., Alexander, L.V., Brönnimann, S., Charabi, Y.A.R. *et al.* (2013). Observations: Atmosphere and surface. In Climate *Change 2013 the Physical Science Basis: Working Group I Contribution to the Fifth Assessment Report of the Intergovernmental Panel on Climate Change.* Cambridge: Cambridge University Press. chapter 2. 159-254. https://www.ipcc.ch/pdf/assessment-report/ar5/wg1/WG1AR5_Chapter02_FINAL.pdf

Hennicke, P. and Hauptstock, D. (2014). Decoupling resource consumption and economic growth: Insights into an unsolved global challenge. In *Handbook of Research on Sustainable Consumption.* Reisch, L.A. (ed.). Cheltenham: Elgar. 377-393. https://epub.wupperinst.org/frontdoor/index/index/searchtype/authorsearch/author/Dorothea+Hauptstock/start/2/rows/10/nav/next/docId/6077

Hepburn, C. and Bowen, A. (2013). Prosperity with growth: Economic growth, climate change and environmental limits. In *Handbook of Energy and Climate Change.* Fouquet, R. (ed.). Edward Elgar. chapter 29. https://www.e-elgar.com/shop/handbook-on-energy-and-climate-change?__website=uk_warehouse

Hoeppe, P. (2016). Trends in weather related disasters–Consequences for insurers and society. *Weather and Climate Extremes* 11, 70-79. https://doi.org/10.1016/j.wace.2015.10.002.

Hujo, K. and Piper, N. (2010). Linking migration, social development and policy in the south–An introduction. In *South-South Migration: Implications for Social Policy and Development Series.* Hujo, K. and Piper, N. (eds.). London: Palgrave Macmillan. chapter 1. 1-45. https://link.springer.com/chapter/10.1057/9780230283374_1

Intergovernmental Panel on Climate Change (2013). *Climate Change 2013: The Physical Science Basis. Summary for Policymakers.* Cambridge: Cambridge University Press. http://www.ipcc.ch/pdf/assessment-report/ar5/wg1/WG1AR5_SPM_FINAL.pdf

Intergovernmental Panel on Climate Change (2014). *Climate Change 2014: Synthesis Report. Contributios of Working Groups I, II and III to the Fifth Assessment Report of the IPCC.* Pachauri, R.K. and Meyer, L.A. (eds.). Geneva. http://epic.awi.de/37530/1/IPCC_AR5_SYR_Final.pdf.

International Energy Agency (2016a). *Energy Technology Perspectives: Towards Sustainable Urban Energy Systems. Executive Summary.* Paris. https://www.iea.org/publications/freepublications/publication/EnergyTechnologyPerspectives2016_ExecutiveSummary_EnglishVersion.pdf.

International Energy Agency (2016b). *World Energy Outlook 2016: Exectuvie Summary.* Paris. https://www.iea.org/publications/freepublications/publication/WorldEnergyOutlook2016ExecutiveSummaryEnglish.pdf.

International Energy Agency (2017a). *World Energy Outlook 2017: Executive Summary.* Paris. http://www.iea.org/Textbase/npsum/weo2017SUM.pdf

International Energy Agency (2017b). *Digitalization & Energy.* Paris. http://www.iea.org/publications/freepublications/publication/DigitalizationandEnergy3.pdf.

International Food Policy Research Institute (2014). *Food Security in a World of Natural Resource Scarcity: The Role of Agricultural Technologies.* Washington, D.C. http://www.ifpri.org/cdmref/p15738coll2/id/128022/filename/128233.pdf.

International Labour Organization (2011). *ILO Director-General address to the European Parliament.* 14 September https://www.ilo.org/global/about-the-ilo/newsroom/statements-and-speeches/WCMS_162828/lang--nl/index.htm.

International Organization for Migration (2015). *World Migration Report 2015: Migrants ad Cities: New Partnerships to Manage Mobility.* Geneva. http://publications.iom.int/system/files/wmr2015_en.pdf.

International Renewable Energy Agency (2013). *Africa's Renewable Future: The Path to Sustainable Future.* Abu Dhabi. http://www.irena.org/-/media/Files/IRENA/Agency/Publication/2013/Africa_renewable_future.pdf.

International Renewable Energy Agency (2016). *REmap: Roadmap for a Renewable Energy Future.* Abu Dhabi. http://www.irena.org/DocumentDownloads/Publications/IRENA_REmap_2016_edition_report.pdf.

International Renewable Energy Agency (2017). *Accelerating the Energy Transition through Innovation: Working Paper Based on Global REmap Analysis.* Abu Dhabi. http://www.irena.org/DocumentDownloads/Publications/IRENA_Energy_Transition_Innovation_2017.pdf.

International Renewable Energy Agency (2018). *Global Energy Transformation: A roadmap to 2050.* Abu Dhabi. http://www.irena.org/-/media/Files/IRENA/Agency/Publication/2018/Apr/IRENA_Report_GET_2018.pdf?la=en&hash=9B1AF0354A2105A64CFD3C4C0E38ECCEE32AAB0C.

Islam, S.N. (2015). *Inequality and Environmental Sustainability.* United Nations. http://www.un.org/esa/desa/papers/2015/wp145_2015.pdf

Iversen, T. and Rosenbluth, F.M. (2010). *Women, Work, and Politics: The Political Economy of Gender Inequality.* New Haven, CT: Yale University Press. https://www.jstor.org/stable/j.ctt1nq33z.

Jackson, T. (2009). *Prosperity Without Growth? The Transition to a Sustainable Economy.* London: Sustainable Development Commission. http://www.sd-commission.org.uk/data/files/publications/prosperity_without_growth_report.pdf.

Jacobs, M. (1991). *The Green Economy: Environment, Sustainable Development and the Politics of the Future.* Chicago, IL: University of Chicago Press. http://press.uchicago.edu/ucp/books/book/distributed/G/bo21611742.html.

Jiang, L. and O'Neill, B.C. (2007). Impacts of demographic trends on US household size and structure. *Population and Development Review* 33(3), 567-591. https://doi.org/10.1111/j.1728-4457.2007.00186.x.

Jorgenson, A.K., Schor, J.B., Huang, X. and Fitzgerald, J. (2015). Income inequality and residential carbon emissions in the United States: A preliminary analysis. *Human Ecology Review* 22(1), 93-106. https://www.jstor.org/stable/24875150.

Juma, C. (2015). *The New Harvest: Agricultural Innovation in Africa.* 2nd edn. Oxford: Oxford University Press. https://global.oup.com/academic/product/the-new-harvest-9780190237233?cc=ke&lang=en&.

Kalantari, F., Mohd Tahir, O., Akbari Joni, R. and Fatemi, E. (2017). Opportunities and challenges in sustainability of vertical farming: A review. *Journal of Landscape Ecology* 11(1). https://doi.org/10.1515/jlecol-2017-0016.

King, R. (2015). *Carbon Emissions and Income Inequality.* Technical Note. Oxford: Oxfam International. https://oxfamilibrary.openrepository.com/oxfam/bitstream/10546/582545/2/tb-carbon-emissions-inequality-methodology-021215-en.pdf.

Kittner, N., Lill, F. and Kammen, D.M. (2017). Energy storage deployment and innovation for the clean energy transition. *Nature Energy* 2(17125). https://doi.org/10.1038/nenergy.2017.125.

Kleven, H. and Landais, C. (2017). Gender inequality and economic development: Fertility, education and norms. *Economica* 84(334), 180-209. https://doi.org/10.1111/ecca.12230.

Koomey, J., Berard, S., Sanchez, M. and Wong, H. (2011). Implications of historical trends in the electrical efficiency of computing. *IEEE Annals of the History of Computing* 33(3), 46-54. http://doi.ieeecomputersociety.org/10.1109/MAHC.2010.28.

Kroeze, C. and Pulles, T. (2015). The importance of non-CO_2 greenhouse gases. *Journal of Integrative Environmental Sciences* 12(1), 1-4. https://doi.org/10.1080/1943815X.2015.1118131.

Kuehr, R. (ed.) (2014). *Solving the E-waste Problem (Step) White Paper: One Global Definition of E-waste.* Bonn: United Nations University. http://collections.unu.edu/view/UNU:6120.

Lakner, C. and Milanovic, B. (2013). *Global Income Distribution: From the Fall of the Berlin Wall to the Great Recession.* Washington, D.C: World Bank. http://documents.worldbank.org/curated/en/914431468162277879/pdf/WPS6719.pdf.

Le Quéré, C., Andrew, R.M., Canadell, J.G., Sitch, S., Korsbakken, J.I., Peters, G.P. *et al.* (2016). Global carbon budget 2016. *Earth System Science Data* 8(2), 605. https://doi.org/10.5194/essd-8-605-2016.

Le Quéré, C., Andrew, R.M., Friedlingstein, P., Sitch, S., Pongratz, J., Manning, A.C. *et al.* (2017). Global carbon budget 2017. *Earth System Science Data Discussions*, 1-79. https://doi.org/10.5194/essd-2017-123.

Leighton, M. (2006). Desertification and migration. In Governing *Global Desertification: Linking Environmental Degradation, Poverty and Participation.* Johnson, P.M., Mayrand, K. and Paquin, M. (eds.). London: Routledge. chapter 4. 43-58. https://www.taylorfrancis.com/books/e/9781351932486

Lenton, T.M., Held, H., Kriegler, E., Hall, J.W., Lucht, W., Rahmstorf, S. *et al.* (2008). Tipping elements in the Earth's climate system. *Proceedings of the National Academy of Sciences* 105(6), 1786-1793. https://doi.org/10.1073/pnas.0705414105.

Liddle, B. (2014). Impact of population, age structure, and urbanization on carbon emissions/energy consumption: Evidence from macro-level, cross-country analyses. *Population and Environment* 35(3), 286-304. https://doi.org/10.1007/s11111-013-0198-4.

Lopez, A.D. and Murray, C.J.L. (eds.) (1996). *The Global Burden of Disease: A Comprehensive Assessment Of Mortality and Disability From Diseases, Injuries, and Risk Factors in 1990 and Projected to 2020: Summary.* Boston, MA: Harvard University Press. http://apps.who.int/iris/bitstream/handle/10665/41864/0965546608_eng.pdf?sequence=1&isAllowed=y.

López, R. and Palacios, A. (2014). Why has Europe become environmentally cleaner? Decomposing the roles of fiscal, trade and environmental policies. *Environmental and Resource Economics* 58(1), 91-108. https://doi.org/10.1007/s10640-013-9692-5.

Lucas, E. (2015). *Let There Be Light.* Special Report: Energy and Technology. The Economist. http://media.economist.com/sites/default/files/sponsorships/MCR75_20150117_Accenture/20150117_Energy.pdf.

Majelantle, R.G. and Navaneetham, K. (2013). Migration and fertility: A review of theories and evidences. *Journal of Global Economics* 1(101), 2. https://www.omicsonline.org/open-access/migration-and-fertility-a-review-of-theories-and-evidences-2375-4389.1000101.php?aid=18557.

Martine, G., Alves, J.E. and Cavenaghi, S. (2013). *Urbanization and Fertility Decline: Cashing in on Structural Change.* London: International Institute for Environment and Development. http://pubs.iied.org/pdfs/10653IIED.pdf.

Massey, D.S. and Taylor, J.E. (2004). *International Migration: Prospects and Policies in A Global Market.* Oxford: Oxford University Press. http://www.oxfordscholarship.com/view/10.1093/0199269009.001.0001/acprof-9780199269006.

Mattick, C., Landis, A., Allenby, B. and Genovese, N. (2015). Anticipatory life cycle analysis of in vitro biomass cultivation for cultured meat production in the United States. *Environmental science & technology* 49(19), 11941-11949. https://doi.org/10.1021/acs.est.5b01614.

Mauritsen, T. and Pincus, R. (2017). Committed warming inferred from observations. *Nature Climate Change* 7, 652–655. https://doi.org/10.1038/nclimate3357.

McArthur, J. and Rasmussen, K. (2017). 'How successful were the millenium development goals?'. *Future Development*, 11 January 2017 https://www.brookings.edu/blog/future-development/2017/01/11/how-successful-were-the-millennium-development-goals/

Melchiorri, M., Florczyk, A., Freire, S., Schiavina, M., Pesaresi, M. and Kemper, T. (2018). Unveiling 25 years of planetary urbanization with remote sensing: Perspectives from the global human settlement layer. *Remote Sensing* 10(5), 768. https://doi.org/10.3390/rs10050768.

Moreno, L.E. (2012). *Concept Paper for the World Urban Forum 7.* Nairobi: United Nations Human Settlement Programme.

Munich RE (2017). *Natural disasters: The year in figures.* https://natcatservice.munichre.com/events/1?filter=eyJ5ZWFyRnJvbSI6MTk4MCwieWVhclRvIjoyMDE3fQ%3D%3D&type=.

Nachmany, M., Fankhauser, S., Setzer, J. and Averchenkova, A. (2017). *Global Trends in Climate Change Legislation and Litigation: 2017 Update.* London Grantham Research Institute on Climate Change and the Environment. http://www.lse.ac.uk/GranthamInstitute/wp-content/uploads/2017/04/Global-trends-in-climate-change-legislation-and-litigation-WEB.pdf.

Nelson, G.C. (2005). Drivers of ecosystem change: Summary chapter. In The *Millenium Ecosystem Assesment.* Washington, D.C: Island Press. chapter 3. 74-76. https://millenniumassessment.org/documents/document.272.aspx.pdf

O'Neill, B.C., Liddle, B., Jiang, L., Smith, K.R., Pachauri, S., Dalton, M. *et al.* (2012). Demographic change and carbon dioxide emissions. *The Lancet* 380(9837), 157-164. https://doi.org/10.1016/S0140-6736(12)60958-1.

O'Neill, B.C., Oppenheimer, M., Warren, R., Hallegatte, S., Kopp, R.E., Pörtner, H.O. *et al.* (2017). IPCC reasons for concern regarding climate change risks. *Nature Climate Change* 7(1), 28-37. https://doi.org/10.1038/nclimate3179.

Organisation for Economic Co-operation and Development (2016). *International Migration Outlook 2016.* Paris. http://dx.doi.org/10.1787/migr_outlook-2016-en.

Organization for Economic Cooperation and Development (2017). *Investing in Climate, Investing in Growth.* Paris. https://www.oecd-ilibrary.org/docserver/9789264273528-en.pdf?expires=1540967063&id=id&accname=ocid195767&checksum=1570E9A640FBA6E18BC5D2D336A820BC.

Ort, D.R., Merchant, S.S., Alric, J., Barkan, A., Blankenship, R.E., Bock, R. *et al.* (2015). Redesigning photosynthesis to sustainably meet global food and bioenergy demand. *Proceedings of the National Academy of Sciences* 112(28), 8529-8536. https://doi.org/10.1073/pnas.1424031112.

Oxfam (2015). *Extreme carbon inequality.* 2 December https://www.oxfam.org/sites/www.oxfam.org/files/file_attachments/mb-extreme-carbon-inequality-021215-en.pdf.

Pacific Institute (2014). *Agricultural Water Conservation and Efficiency Potential in California* Oakland, CA: Pacific Institute. https://www.nrdc.org/sites/default/files/ca-water-supply-solutions-ag-efficiency-IB.pdf.

Park, J. (2016). Clean energy entrepreneurship in sub-saharan Africa. In Global *Entrepreneurship: Past, Present & Future.* Devinney, T.M., Markman, G., Pedersen, T. and Tihanyi, L. (eds.). Emerald Group Publishing Limited. 257-277. https://www.emeraldinsight.com/doi/pdfplus/10.1108/S1571-502720160000029015

Pesaresi, M., Melchiorri, M., Siragusa, A. and Kemper, T. (eds.) (2016). *Atlas of the Human Planet 2016* Mapping Human Presence on Earth with the Global Human Settlement Layer. Brussels: European Union. http://ghsl.jrc.ec.europa.eu/atlas2016Overview.php.

Peters, G.P., Minx, J.C., Weber, C.L. and Edenhofer, O. (2011). Growth in emission transfers via international trade from 1990 to 2008. *Proceedings of the National Academy of Sciences* 108(21), 8903-8908. https://doi.org/10.1073/pnas.1006388108.

Pithan, F. and Mauritsen, T. (2014). Arctic amplification dominated by temperature feedbacks in contemporary climate models. *Nature Geoscience* 7, 181-184. https://doi.org/10.1038/ngeo2071.

Ramanathan, V. (1988). The greenhouse theory of climate change: A test by an inadvertent global experiment. *Science* 240(4850), 293-299. http://science.sciencemag.org/content/240/4850/293.

Ramaswami, A., Russell, A.G., Culligan, P.J., Sharma, K.R. and Kumar, E. (2016). Meta-principles for developing smart, sustainable, and healthy cities. *Science* 352(6288), 940-943. http://science.sciencemag.org/content/352/6288/940.

Ravallion, M. (2001). Growth, inequality and poverty: Looking beyond averages. *World development* 29(11), 1803-1815. http://siteresources.worldbank.org/INTPGI/Resources/13996_MR2.pdf.

Rees, W.E. (1995). Achieving sustainability: Reform or transformation? *Journal of Planning Literature* 9(4), 343-361. https://doi.org/10.1177/088541229500900402.

Renewable Energy Policy Network for the 21st Century (2018). *Renewables 2018 Global Status Report*. Paris. http://www.ren21.net/wp-content/uploads/2018/06/17-8652_GSR2018_FullReport_web_final_.pdf

Rockström, J., Steffen, W., Noone, K., Persson, Å., Chapin, F.S., Lambin, E.F. *et al.* (2009b). A safe operating space for humanity. *Nature* 461(7263), 472-475. https://doi.org/10.1038/461472a.

Rockström, J., Steffen, W., Noone, K., Persson, Å., Chapin III, F.S., Lambin, E. *et al.* (2009a). Planetary boundaries: Exploring the safe operating space for humanity. *Ecology and society* 14(2). https://www.ecologyandsociety.org/vol14/iss2/art32/.

Rogers, E. (2003). *Diffusion of Innovations*. 5th edn. New York, NY: Free Press. http://www.simonandschuster.com/books/Diffusion-of-Innovations-5th-Edition/Everett-M-Rogers/9780743222099.

Romer, P.M. (1994). The origins of endogenous growth. *Journal of Economic perspectives* 8(1), 3-22. www.jstor.org/stable/2138148.

Rüßmann, M., Gerbert, P., Lorenz, M., Waldner, M., Justus, J., Engel, P. *et al.* (2015). *Industry 4.0: The Future of Productivity and Growth in Manufacturing Industries*. Boston, MA: Boston Consulting Group. https://www.bcg.com/publications/2015/engineered_products_project_business_industry_4_future_productivity_growth_manufacturing_industries.aspx.

Salat, S., Chen, M. and Liu. F. (2014). *Planning Energy Efficient and Livable Cities: Energy Efficient Cities*. Energy Sector Management Assistance Program. https://www.esmap.org/sites/esmap.org/files/DocumentLibrary/ESMAP_CEETI_MayoralNote_6_PlanningEE%20Livable%20Cities_optimized.pdf.

Salem, H.S. (2011). Social, environmental and security impacts of climate change of the eastern Mediterranean. In *Coping with Global Environmental Change, Disasters and Security – Threats, Challenges, Vulnerabilities and Risks*. Brauch, H.S., Spring, U.O., Mesjasz, C., Grin, J., Kameri-Mbote, P., Chourou, B. *et al.* (eds.). Springer. 421-445. https://www.researchgate.net/publication/299562984_Social_Environmental_and_Security_Impacts_of_Climate_Change_of_the_Eastern_Mideterranian.

Samaniego, J., Galindo. Luis Miguel., Alatorre, J.E., Ferrer, J., Gómez, J.J., Lennox, J. *et al.* (2014). *The Economics of Climate Change in Latin America and the Caribbean: Paradoxes and Challenges. Overview For 2014*. United Nations, Economic Commission for Latin America and the Caribbean. http://repositorio.cepal.org/bitstream/handle/11362/37056/4/S1420806_en.pdf.

Schmidt, S. (2011). Sprawl without growth in eastern Germany. *Urban Geography* 32(1), 105-128. https://pdfs.semanticscholar.org/6599/5ccbec31100240fceadb0af3856ecad85db1.pdf.

Schneider, A., Mertes, C.M., Tatem, A.J., Tan, B., Sulla-Menashe, D., Graves, S.J. *et al.* (2015). A new urban landscape in East–Southeast Asia, 2000–2010. *Environmental Research Letters* 10(3), 034002. https://doi.org/10.1088/1748-9326/10/3/034002.

Segars, A.H. (2018). Seven technologies remaking the world: An MIT SMR executive guide. *MIT Sloan Management Review*, Massachusetts Institute of Technology https://sloanreview.mit.edu/projects/seven-technologies-remaking-the-world/?switch_view=PDF.

Sen, A. (2000). *Development as Freedom*. New York, NY: Anchor Books. https://www.penguinrandomhouse.com/books/163962/development-as-freedom-by-amartya-sen/9780385720274/.

Seto, K.C., Davis, S.J., Mitchell, R.B., Stokes, E.C., Unruh, G. and Ürge-Vorsatz, D. (2016). Carbon lock-in: Types, causes, and policy implications. *Annual Review of Environment and Resources* 41, 425-452. https://doi.org/10.1146/annurev-environ-110615-085934.

Seto, K.C., Dhakal, S., Bigio, A., Blanco, H., Delgado, G.C., Dewar, D. *et al.* (2014). Human settlements, infrastructure and spatial planning. In *Climate Change 2014: Mitigation of Climate Change. Contribution of Working Group III to the Fifth Assessment Report of the Intergovernmental Panel on Climate Change*. Edenhofer, O., Pichs-Madruga, R., Sokona, Y., Farahani, E., Kadner, S., Seyboth, K. *et al.* (eds.). Cambridge: Cambridge University Press. chapter 12. http://www.ipcc.ch/pdf/assessment-report/ar5/wg3/ipcc_wg3_ar5_chapter12.pdf

Seto, K.C., Fragkias, M., Güneralp, B. and Reilly, M.K. (2011). A meta-analysis of global urban land expansion. *PloS one* 6(8), e23777. https://doi.org/10.1371/journal.pone.0023777.

Seto, K.C., Güneralp, B. and Hutyra, L.R. (2012). Global forecasts of urban expansion to 2030 and direct impacts on biodiversity and carbon pools. *Proceedings of the National Academy of Sciences* 109(40), 16083-16088. https://doi.org/10.1073/pnas.1211658109.

Sinding, S.W. (2009). Population, poverty and economic development. *Philosophical Transactions of the Royal Society of London B: Biological Sciences* 364(1532), 3023-3030. https://doi.org/10.1098/rstb.2009.0145.

Smil, V. (2010). Science, energy, ethics, and civilization. In *Visions of Discovery: New Light on Physics, Cosmology, and Consciousness*. Chiao, R.Y., Cohen, M.L., Leggett, A.J., Phillips, W.D. and Harper, J.C.L. (eds.). Cambridge: Cambridge University Press. chapter 35. 709-729. http://vaclavsmil.com/wp-content/uploads/docs/smil-articles-science-energy-ethics-civilization.pdf

Smith, D.E., Harrison, S. and Jordan, J.T. (2011). The early Holocene sea level rise. *Quaternary Science Reviews* 30(15-16), 1846-1860. https://doi.org/10.1016/j.quascirev.2011.04.019.

Solomon, S., Plattner, G.-K., Knutti, R. and Friedlingstein, P. (2009). Irreversible climate change due to carbon dioxide emissions. *Proceedings of the National Academy of Sciences* 106(6), 1704-1709. https://doi.org/10.1073/pnas.0812721106.

Swedish International Development Cooperation Agency (2018). *The Relationship Between Climate Change and Violent Conflict*. https://www.sida.se/contentassets/c571800e01e448ac9dce2d097ba125a1/working-paper---climate-change-and-conflict.pdf.

Taylor, P.C., Cai, M., Hu, A., Meehl, J., Washington, W. and Zhang, G.J. (2013). A decomposition of feedback contributions to polar warming amplification. *Journal of Climate* 26(18), 7023-7043. https://doi.org/10.1175/JCLI-D-12-00696.1.

United Nations (2000). *Millennium summit (6-8 September 2000)*. http://www.un.org/en/events/pastevents/millennium_summit.shtml

United Nations, General Assembly (2009). *Climate Change and Its Possible Security Implications: Report of the Security-General*. Sixty-fourth session, 11 September. A/64/350. https://undocs.org/A/64/350.

United Nations (2014). *World Urbanization Prospects: 2014 Revision, Highlights*. New York, NY. https://esa.un.org/unpd/wup/Publications/Files/WUP2014-Highlights.pdf.

United Nations (2015a). *World Population Prospects: Key Findings and Advanced Tables. The 2015 Revision*. New York, NY. https://esa.un.org/Unpd/wpp/Publications/Files/Key_Findings_WPP_2015.pdf.

United Nations (2015b). *International Migration Report 2015: Highlights*. New York, NY. http://www.un.org/en/development/desa/population/migration/publications/migrationreport/docs/MigrationReport2015_Highlights.pdf.

United Nations (2016). *The World's Cities in 2016: Data Booklet*. New York, NY. http://www.un.org/en/development/desa/population/publications/pdf/urbanization/the_worlds_cities_in_2016_data_booklet.pdf.

United Nations (2017a). *World Population Prospects: Key Findings and Advance Tables - The 2017 Revision*. https://population.un.org/wpp/Publications/Files/WPP2017_KeyFindings.pdf.

United Nations (2017b). *Household Size and Composition Around the World 2017: Data Booklet*. New York, NY. http://www.un.org/en/development/desa/population/publications/pdf/ageing/household_size_and_composition_around_the_world_2017_data_booklet.pdf.

United Nations Children's Fund (2018a). *Malnutrition*. https://data.unicef.org/topic/nutrition/malnutrition/.

United Nations Children's Fund (2018b). *Education*. https://www.unicef.org/education.

United Nations Children's Fund and World Health Organization (2017). *Progress on Drinking Water, Sanitation and Hygiene: 2017 Update and Sustainable Development Goal Baselines*. Geneva. https://www.unicef.org/publications/files/Progress_on_Drinking_Water_Sanitation_and_Hygiene_2017.pdf.

United Nations Development Programme (2018). *Sustainable Development Goals*. http://www.undp.org/content/undp/en/home/sustainable-development-goals.html (Accessed: 27 September 2018).

United Nations Economic and Social Commission for Asia and the Pacific and United Nations Human Settlement Programme (2015). *The State of Asian and Pacific Cities 2015: Urban Transformations. Shifting from Quantity to Quality*. https://www.unescap.org/sites/default/files/The%20State%20of%20Asian%20and%20Pacific%20Cities%202015.pdf.

United Nations Entity for Gender Equality and the Empowerment of Women (2017). *Annual Report 2016-2017*. New York, NY. http://www2.unwomen.org/-/media/annual%20report/attachments/sections/library/un-women-annual-report-2016-2017-en.pdf?vs=5634.

United Nations Environment Programme (2010). *Assessing the Environmental Impacts of Consumption and Production: Priority Products and Materials. A Report of the Working Group on the Environmental Impacts of Products and Materials to the International Panel for Sustainable Resource Management*. Hertwich, E., van der Voet, E., Suh, S., Tukker, A., Huijbregts M., Kazmierczyk, P. *et al.* (eds.). http://www.unep.fr/shared/publications/pdf/dtix1262xpa-priorityproductsandmaterials_report.pdf.

United Nations Environment Programme (2011). *Towards A Green Economy: Pathways to Sustainable Development and Poverty Eradication*. Nairobi. https://wedocs.unep.org/bitstream/handle/20.500.11822/22025/green_economyreport_final_dec2011.pdf?amp%3BisAllowed=&sequence=1.

United Nations Environment Programme (2016). *GEO-6 Regional Assesment for North America*. Nairobi. http://wedocs.unep.org/bitstream/handle/20.500.11822/7611/GEO_North_America_201611.pdf?sequence=1&isAllowed=y.

United Nations Environment Programme (2017). *The Emissions Gap Report 2017: A UN Environment Synthesis Report*. Nairobi. https://wedocs.unep.org/bitstream/handle/20.500.11822/22070/EGR_2017.pdf?isAllowed=y&sequence=1.

United Nations Framework Convention on Climate Change (2015). Paris Agreement. Paris https://unfccc.int/files/meetings/paris_nov_2015/application/pdf/paris_agreement_english_.pdf

United Nations Human Settlement Programme (2010). *State of the World's Cities 2010/2011: Bridging the Urban Divide*. Nairobi. https://unhabitat.org/books/state-of-the-worlds-cities-20102011-cities-for-all-bridging-the-urban-divide/#.

United Nations Human Settlement Programme (2012). *State of Latin America and Caribbean cities 2012: Towards a New Urban Transition*. Nairobi. http://mirror.unhabitat.org/pmss/getElectronicVersion.aspx?nr=3386&alt=1.

United Nations Human Settlement Programme (2014). Urbanization and Sustainable Development: Towards A New United Nations Urban Agenda. CEB/2014/HLCP-28/CRP.5. New York, NY Agenda Item 6: New UN Urban Agenda https://habnet.unhabitat.org/sites/default/files/oo/urbanization-and-sustainable-development.pdf

United Nations Human Settlement Programme (2015b). *Habitat III Issue Papers 10 – Urban-Rural Linkages*. Nairobi. http://habitat3.org/wp-content/uploads/Habitat-III-Issue-Paper-10_Urban-Rural-Linkages-2.0.pdf.

United Nations Human Settlement Programme (2016). *World Cities Report 2016: Urbanization and Development - Emerging Futures*. Nairobi. http://wcr.unhabitat.org/wp-content/uploads/2017/02/WCR-2016-Full-Report.pdf.

United Nations Human Settlements Programme (2015a). *Habitat III Issue Papers 1 - Inclusive Cities*. Nairobi. https://www.alnap.org/system/files/content/resource/files/main/habitat-iii-issue-paper-1-inclusive-cities.pdf.

United Nations Population Fund (2014). *Framework of Actions for the Follow-Up to the Programme of Action of the International Conference of Population and Development Beyond 2014*. New York, NY. https://www.unfpa.org/sites/default/files/pub-pdf/ICPD_beyond2014_EN.pdf.

United Nations Population Fund (2017). *State of World Population: Reproductive Health and Rights in an Age of Inequality*. New York, NY. https://www.unfpa.org/sites/default/files/sowp/downloads/UNFPA_PUB_2017_EN_SWOP.pdf.

United States National Oceanic and Atmospheric Administration (2015). *State of the Climate: Global Climate Report for Annual 2015*. https://www.ncdc.noaa.gov/sotc/global/201513.

United States National Oceanic and Atmospheric Administration (2018). *Paleoclimatology Data*. https://www.ncdc.noaa.gov/data-access/paleoclimatology-data

Vaughan, D.G., Comiso, J.C., Allison, I., Carrasco, J., Kaser, G., Kwok, R. *et al.* (2013). Observations: Cryosphere. In *Climate Change 2013: The Physical Science Basis. Contribution of Working Group I to the Fifth Assessment Report of the Intergovernmental Panel on Climate Change* Stocker, T.F., Qin, D., Plattner, G.-K., Tignor, M., Allen, S.K., Boschung, J. *et al.* (eds.). Cambridge: Cambridge University Press. chapter 4. 317-382. https://www.ipcc.ch/pdf/assessment-report/ar5/wg1/WG1AR5_Chapter04_FINAL.pdf

Victor, P.A. (2008). *Managing Without Growth: Slower by Design, Not Disaster*. Cheltenham: Edward Elgar Publishing. https://www.e-elgar.com/shop/eep/preview/book/isbn/9781848442993/.

Weede, E. (1996). *Economic Development, Social Order, and World Politics*. Boulder, CO: Lynne Rienner Publishers. https://www.rienner.com/title/Economic_Development_Social_Order_and_World_Politics.

Weldon, D. (2016). Globalisation is fraying: Look under the Elephant Trunk. *Bull Market*. 13 June 2016. https://medium.com/bull-market/globalisation-is-fraying-look-under-the-elephant-trunk-e79f76e9754b.

White, M. (2015). Redesigning crops for the 21st century. *Pacific Standard Magazine*. 5 August. https://psmag.com/environment/redesigning-crops-for-21st-century-gmos.

Whitney, J. and Kennedy, J. (2012). *The Carbon Emissions of Server Computing for Small-to Medium-sized Organization: A Performance Study of On-Premise vs. The Cloud*. Natural Resources Defense Council. https://www.nrdc.org/sites/default/files/NRDC_WSP_Cloud_Computing_White_Paper.pdf.

Wigley, T.M.L. (2005). The climate change commitment. *Science* 307(5716), 1766-1769. https://doi.org/10.1126/science.1103934.

Wolf, M., Haase, D. and Haase, A. (2018). Compact or spread? A quantitative spatial model of urban areas in Europe since 1990. *PloS one* 13(2), e0192326. https://doi.org/10.1371/journal.pone.0192326.

World Bank (2004). *The Poverty-Growth-Inequality Triangle*. Washington, D.C. http://documents.worldbank.org/curated/en/449711468762020101/pdf/28102.pdf.

World Bank (2013). *Poverty and Equity*. http://databank.worldbank.org/data/reports.aspx?source=poverty-and-equity-database.

World Bank (2015). *East Asia's Changing Urban Landscape: Measuring a Decade of Spatial Growth*. Washington, D.C. https://www.worldbank.org/content/dam/Worldbank/Publications/Urban%20Development/EAP_Urban_Expansion_full_report_web.pdf.

World Bank (2016). *High and Dry: Climate Change, Water, and the Economy*. Washington, D.C. https://openknowledge.worldbank.org/bitstream/handle/10986/23665/K8517.pdf?sequence=3&isAllowed=y.

World Bank (2017). *World Development Indicators*. http://datatopics.worldbank.org/world-development-indicators/.

World Meteorological Organization 12 (2016). WMO Greenhouse Gas Bulletin: The State of Greenhouse Gases in the Atmosphere Based on Global Observations in the Atmosphere through 2015. https://library.wmo.int/doc_num.php?explnum_id=3084.

Wu, Y., Shen, J., Zhang, X., Skitmore, M. and Lu, W. (2017). Reprint of: The impact of urbanization on carbon emissions in developing countries: A Chinese study based on the U-Kaya method. *Journal of Cleaner Production* 163(1), S284-S298. https://doi.org/10.1016/j.jclepro.2017.05.144.

Yihdego, Y., Salem, H.S. and Pudza, M.Y. (2017). Renewable energy: Wind farm perspectives–the case of Africa. *Journal of Sustainable Energy Engineering* 5(4), 281-306. https://doi.org/10.7569/JSEE.2017.629521.

Zenghelis, D. (2011). *The Economics of Network-Powered Growth*. San Jose, CA: Cisco Internet Business Solutions Group. https://www.cisco.com/c/dam/en_us/about/ac79/docs/Economics_NPG_FINALFINAL.pdf.

Zenghelis, D. (2017). Cities, wealth and the era of urbanisation. In National *Wealth: What is Missing, Why It Matters*. Hamilton, K. and Hepburn, C. (eds.). Oxford University Press. chapter 14. http://www.oxfordscholarship.com/view/10.1093/oso/9780198803720.001.0001/oso-9780198803720-chapter-14.

2

Chapter 3

The Current State of our Data and Knowledge

© Shutterstock/Sergey Nivens

Coordinating Lead Authors: Florence Daguitan (Tebtebba, Indigenous Peoples' International Centre for Policy Research and Education), Pali Lehohla (Pan African Institute for Evidence - PIE), Charles Mwangi (GLOBE Programme), Joni Seager (Bentley University), William Sonntag (Group on Earth Observation Secretariat), Graeme Clark (University of New South Wales)

Lead Authors: James M. Donovan (ADEC Innovations), Sheryl Joy Anne S. Gutierrez (ADEC Innovations), Michelle G. Tan (ADEC Innovations)

Contributing Authors: Amit R. Patel (Planned Systems International, Inc.)

Executive summary

There is a growing demand for environmental indicators and analysis, particularly analysis that addresses interlinkages across different environmental domains and between the environment, society and the economy (*well established*). There have been advancements in terms of collecting official statistics related to the environment, including geospatial statistics, particularly in terms of promoting environmental economic accounting and building geospatial information systems, which contribute to environmental monitoring. However, there are still methodological gaps in measuring some aspects of the environment, there is very limited information which links people and the environment, and there are capacity gaps in countries attempting to build their environmental information systems. {3.2}

Measuring the nexus between gender and the environment has been identified as a high priority, as women and men, in many contexts, have differing rights over and access to the environment (*well established*). Women and men have different vulnerabilities to environmental degradation and hazards, and often play different roles in environmental management decision-making. Currently, only limited time series data and statistics are available on the gender-environment nexus. {3.5}

Much environmental data collection is part of one-off studies or projects, limiting their usefulness (*well established*). Through the Sustainable Development Goals (SDGs) there has been a global recognition that monitoring the environmental dimension of development will require regular, standardized data collection, which can translate into time series statistics and indicators, including time series for geospatial data products. This will increase the emphasis on compiling high-quality information based on international best practices. {3.7}

Transforming the provisioning of environmental data and statistics will require new and innovative means of data collection, (*well established*) including new partnerships with the private sector, multilateral institutions, space agencies, non-governmental organizations and other partners. {3.8}

3.1 Introduction

This section provides an introduction to environmental statistics and data and covers the state of existing data and knowledge that contribute to any environmental assessment, including national-, regional- and global-level assessments. It attempts to elaborate the state of data collection and the use of data to compile statistics and produce indicators. Emerging areas of statistics, such as big data, citizen science and traditional knowledge – which are currently underutilized, but present tremendous opportunities for better measuring – are discussed in Chapter 25 of this report.

3.2 The demand for environmental statistics and data

Knowledge and data are essential bedrocks of environmental assessment. Without an evidence base to work from, conducting and publishing an accurate assessment is impossible. But what is an evidence base, and how do we generate it?

'The Environment' was traditionally considered to refer only to biophysical earth systems. But this paradigm is shifting. It is important not only to measure the state of the environment, but also to determine how environmental problems, which manifest in the biophysical environment, arise from social systems and economic arrangements, and how economic development and social well-being depend on the environment.

The GEO-5 report chapter on the *Review of Data Needs* presents the deficiencies in scientifically credible data on the environment; in particular, the report notes the need for time series on freshwater quantity and quality, groundwater depletion, ecosystem services, loss of natural habitat, land degradation, chemicals and waste, and other issues (United Nations Environment Programme [UNEP] 2012). It also acknowledges that the factual and scientific quality of an assessment rely on the quality and availability of data on the environment (UNEP 2012). Further, it indicates that more systematic data-collection can help governments, as well as regional and international bodies, to assess their progress towards international goals.

In his 2015 Millennium Development Goals (MDG) Report, Ban Ki Moon **(Box 3.1)** (United Nations 2015a) called for urgent and rapid improvements in data for the post-2015 agenda, especially its availability, reliability and timeliness. He urged governments to make substantial investments in their national statistics offices and systems, as well as to scale up the capacity and capability for producing high-quality data.

The Drivers, Pressures, State, Impact, Response (DPSIR) conceptual framework (see Section 1.6) is a useful framework for environmental monitoring and assessment. Many of the drivers and pressures of environmental change are located in the social realm, and so are many of the impacts. Many environmental challenges are the result of inequalities in access to resources and institutions of power, as well as along the axes of gender, age, race, ethnicity, income and other social status.

Box 3.1: Statement from Ban Ki Moon, 2015

"Strong political commitment and significantly increased resources will be needed to meet the data demand for the new development agenda."

Ban Ki Moon, 2015 (United Nations 2015)

As highlighted in the GEO-5 report (UNEP 2012), there is a need not only for regular monitoring data, but also for harmonization of data-collection approaches and methodologies. Governments rely on national statistical systems to provide the necessary data for national policy; however, historically, many national statistical systems have not considered environmental statistics to be within their purview.

3.3 History of environmental statistics

Historically, official statistics have risen in response to a clear demand from governments for information. The first Roman census was justified by the need for accountability in terms of taxation and military service (Hin 2007). National accounts were born out of the stock market crash of 1929 and the need for wartime statistics, which would allow countries to avoid economic catastrophe and provide information on how to pay for World War II (Stone 1947; Vanoli 2005). In 1947, the United Nations established the United Nations Statistical Commission (UNSC) to develop and promote statistical guidelines which could be used by countries for national monitoring. The scope of the Commission's work covers statistical methodologies for keeping stock of the economy, and for policy on global macroeconomic stability, including economic growth, price movements and population dynamics, migration, mortality, births and longevity – but not the environment.

The Brundtland Commission of 1983 led to the Framework for the Development of Environment Statistics which was first adopted by the UN Statistical Commission in 1984. Later, the UNSC worked on environmental economic accounting which arose from the 1992 Earth Summit. There have been three revisions of the System of Environmental Economic Accounts (SEEA) these include the SEEA 1993, the SEEA 2003 and the SEEA 2012 – the latest was adopted as a statistical standard in 2012 (United Nations 1993; United Nations *et al*. 2003; United Nations 2012). Additionally, the Experimental Ecosystem Accounts were adopted in 2013. The link between these two statistical frameworks forms the basis for monitoring progress towards sustainable development and focuses on the effects of life on the environment and that of the environment on life.

There was a 91 per cent participation by countries and territories in the 2010 census round, and a 95 per cent submission rate of national accounts to the United Nations Statistics Division (United Nations 2015b; United Nations 2017a). However, for the first six decades of the UNSC, progress in official statistics was mostly related to demography and economic statistics. The adoption of the MDGs, which included goals focused mostly on social development, and the desire to track progress as measured by the MDG indicators was transformational in terms of increasing investment in statistics. The MDG implementation efforts resulted in increased statistical capacity of countries to produce and use statistics on poverty, education, health, gender, environment and governance (World Bank 2002; Organisation for Economic Co-operation and Development [OECD] 2015; United Nations 2016a).

Environmental statistics and statistics disaggregated by location, gender, age, poverty and other factors were not a focus of MDG monitoring and therefore received less investment. These areas are a focus of the SDGs; however, many challenges remain in terms of measuring different aspects of the environment and also in creating disaggregated statistics.

3.4 Better data for a healthy planet with healthy people

Improved environmental data and statistics are required for many levels of decision-making, for environmental assessments at the local, national, regional and international levels, and for analysis of the interaction between the environment and the economy and society. A robust environmental statistics system, which is geospatially disaggregated, would ideally provide information that could be used for different purposes and at different levels.

3.4.1 Measuring the environmental dimension of sustainable development

The context within which this report is produced is one where the MDGs have run their course. In September 2015, the United Nations General Assembly endorsed *Transforming Our World: The 2030 Agenda for Sustainable Development*, a global development agenda which captures goals and targets needed to achieve economic, social, and environmental development (A/RES/70/1). The Sustainable Development Goals (SDGs) represent a move away from treating social development in isolation towards an approach aimed at sustainable prosperity, dignity for people, and a healthy planet through national action and partnerships.

In the quest for achieving these ambitious goals, the SDGs are defined around 17 goals, 169 targets and 244 indicators (inclusive of duplication) (United Nations 2017b). *Transforming Our World...* clearly notes that data requirements for the global indicators present a tremendous challenge to all countries. One study estimated that an investment of US$ 1 billion per annum will be needed in order for lower-income countries to monitor the SDGs (Sustainable Development Solutions Network 2017). Thus, as highlighted in the 2016 SDG report, tracking progress on the SDGs will require a shift in how data are collected, processed, analysed and disseminated, including using data from new and innovative data sources (United Nations 2016b).

Although the SDG framework creates monitoring challenges, it also creates opportunities. It represents the first time that there has been an attempt to holistically include environment-related indicators in a global monitoring framework. Although the SDG framework has set out indicators for measuring across all 17 SDG goals, many of the indicators lack a statistical methodology. This is recognized in the framework by assigning each indicator to one of three tiers **(see Figure 3.2)**. The inclusion of a broad range of environment-related SDG indicators can be used to leverage increased investment in environmental statistics and to promote their use.

Figure 3.1: SDGs data and knowledge framework

Figure 3.2: SDG indicator status

Tier 1 Indicator — Indicator has an internationally established methodology and data are regularly produced by 50 per cent of countries.

Tier 2 Indicator — Indicator is conceptually clear, but data are not regularly produced by countries.

Tier 3 Indicator — No internationally established methodology or data collection.

Source: United Nations (2018, p.3)

Figure 3.3: Environment-related SDG indicators by goal and tier

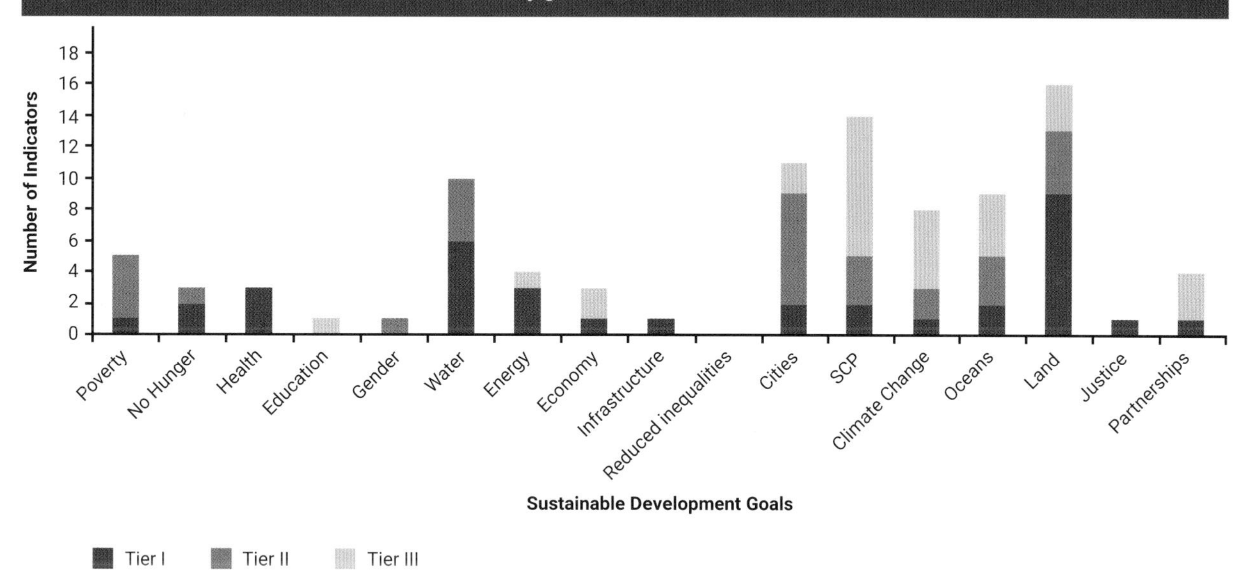

Source: United Nations (2018).

There are 93 SDG indicators directly related to the environment **(Figure 3.3)**. There are also a number of additional indicators that are indirectly related to the environment (e.g. poverty, zoonotic disease, nutrition and life expectancy, economic growth, inclusive societies and policy processes which are not included in **Figure 3.3**). The environment-related SDG indicators are spread across all of the SDGs, with at least one environmentally relevant SDG indicator for each, except Goal 10 – which reflects the cross-cutting nature of the SDGs and the interactions between people, the environment and the economy. However, of the 93 environment-related SDG indicators, only 34 currently have an existing agreed methodology and data that are available from most countries (Tier I). The other indicators have either been given a Tier II or III status (27 and 34 indicators, respectively) by the Inter-Agency and Expert Group on the SDG indicators **(Figure 3.4)**.

Monitoring the environmental dimension of the SDGs will not only require research and development in terms of statistical methodologies but will also require investing in environmental statistics and utilizing new data sources to achieve a data revolution. Traditional data collection by national statistical offices cannot be the only source of data, but countries will need integrated data systems which bring together official statistics, earth observation, citizen science, big data and traditional knowledge. Integrated data systems can bring many sources of information together to provide a more complete picture. Environmental data integration includes:

- bringing together ethnographic information about environmental changes as experienced on the ground;
- a participatory understanding of personal experience; indigenous and traditional knowledge; geospatial information about people and the environment;
- combined information on the environment and women, the poor, and other vulnerable groups in order to reveal patterns and challenges hidden in other systems of knowledge;
- knowledge from Big Data on sustainable consumption and production patterns; and
- poverty dynamics within and between contries, some dimensions of which can be revealed by satellite observations of 'the earth at night' or deforestation boundaries.

Figure 3.4: GEO-6 major data gaps organized by respective chapter

DRIVERS

Population

Urbanization

AIR

Air Quality

Health impact

FRESHWATER

Water consumption

Groundwater

Water withdrawals

Wastewater

BIODIVERSITY

Genomic data

Economic valuation

LAND

Biofuels and LIDAR

Forest plantations

Soil degradation

Land use and ownership

Stockpiling and pesticides

OCEANS

Polar region forecasting

Environmental disasters

Planetary boundaries

Survey systems

A new approach to data and knowledge systems with increased emphasis on an evidence-based approach to decision-making is crucial to meet the various SDGs. By placing these systems at the forefront for all end users, cross collaboration can be innately encouraged to foster new skills, technologies, and sources of data. In turn, our knowledge of sustainable development will improve, alongside our understanding of the SDGs. However, organizational and methodological challenges will arise regarding data privacy, ownership and use (Sustainable Development Solutions Network 2017).

3.4.2 Thematic data gaps

In almost all thematic areas (including biodiversity, land, air, water and oceans), available data are lacking (**Figure 3.4**), particularly in developing countries. Environmental indicators linked to industrial activities are easier to measure and monitor, for example energy consumption or water use. Land cover and ecosystem extent can be assessed on a broad scale using satellite remote sensing, but not always with the necessary resolution. The effects of environmental change, air and water pollution, and other environmental conditions are particularly difficult to measure (UNEP 2012); hence the need to explore a paradigm shift on environmental monitoring approaches – depicting the social orientation and complementing the approach with physical attributions.

The following is a brief description of some of the major data gaps from the thematic chapters in Part A of this report.

Drivers (Chapter 2)
National-level population data are relatively sound for most countries due to government census requirements, but weaknesses arise from aggregation across sectors of the population. National census data are generally insufficient to answer important intra-household questions, such as contraception use and access, fertility, household decision making and family structure (e.g. age of marriage). To be properly understood, these and other variables should be disaggregated by age, gender, race and other socioeconomic factors. Urbanization data are plagued by similar issues of national aggregation. There is a lack of information on small and medium-sized cities, and inconsistency in the scale of reporting. For both population and urbanization, there need to be standard agreements on statistics at a global scale, and greater consistency and coverage. Other significant data gaps include rural to urban migration, the role of nuclear households, the distribution of benefits provided by technology, and patterns of production and consumption. Uncertainty also exists in the myriad factors affecting economic development, and dependencies between this and other drivers. For example, financial estimates of the cost of unsustainable practices and impacts of climate change require greater accuracy and transparency, given that, while numbers exist, there is low confidence in their accuracy.

In addition to gaps in raw data, gaps exist in the mechanistic understanding of the driver processes. Future technologies and events will alter the global landscape and qualitatively change the roles of other drivers. For example, automation may change the nature of transportation, which would have flow-on effects to many other areas. The impact of climate change on human health requires better analyses and understanding of current and future links between these factors. More data are needed on the effect of climate change on human demographics, including migration estimates at finer scales (McMichael, Barnett and McMichael 2012). There is sectoral imbalance in knowledge on the effects of climate change, with impacts on the energy sector being well understood, while impacts on land use, ecosystem processes and functions, and intersectoral issues are not.

Air (Chapter 5)
An overarching issue with air quality data is that, unlike meteorological variables, few air pollutant concentrations are measured with sufficient spatial and temporal coverage. Therefore, the effects of most chemicals are estimated by using other (measured) chemicals as proxies, which is likely to be inaccurate in many cases. For example, only a few persistent, bioaccumulative and toxic substances are measured, and their data are globally patchy. Where monitoring does exist, it is biased towards developed countries, compromising analyses of air pollution versus human health in developing countries. There is a general need for capacity-building to facilitate the measurement of air pollution in developing countries, both for national benefit and to complete global coverage. Bias in air quality sampling also exists within countries, and there is a need for more sampling in areas of low socioeconomic status (e.g. informal or slum dwellings).

Impacts of air quality on human health gained attention in the Global Burden of Disease Study – a global study of factors influencing human health (World Health Organization [WHO] 2018), which elevated air pollution to a top priority. Instead of relying only on cities that have air quality monitoring, satellite data and modelling were used to estimate air pollution at large scales (Brauer *et al.* 2016). Additionally, there are currently few consistent global emissions inventories. Inventories are gathered or modelled in some regions, but data quality and sources vary. There are, however, consistent inventories available at a European level and at international level for a selection of pollutants (e.g. under the United Nations Economic Commission for Europe Convention on Long-range Transboundary Air Pollution).

Efforts should focus on greater sampling coverage and/or modelling, potentially with sensors or satellites. The Copernicus programme aims to measure a number of air quality variables on a regular basis and provide data for all countries. Another European initiative for data reporting is the Air Quality Directive, which provides for statistical data that are produced annually, and an online map of air quality that is updated every 6 hours.

Biodiversity (Chapter 6)
Biological data and knowledge are sparse compared with the complexity and diversity of biological systems. In general, data paucity increases at finer spatial scales, and at higher taxonomic resolutions. Estimates of the total number of species vary between 2 million and 13 million (Costello, Wilson and Houlding 2012; Scheffers *et al.* 2012), with the majority (86 per cent of terrestrial species, and 91 per cent of oceanic species) believed to be undescribed (Mora *et al.* 2011). Invertebrates and deep-sea ecosystems are particularly poorly described. Biologists increasingly use genetic information to identify species (a technique known as DNA barcoding) (Hosein *et al.* 2017), but more traditional taxonomy is still needed to describe morphological traits.

Gaps in data of ecological processes and ecosystem and community structure are even greater than gaps in species information. Examples include ecosystem function and services, which are understood conceptually but are often difficult to measure. A consequence of this is an inability to effectively prevent species invasions, which is considered by some to be the second greatest threat to global biodiversity (Doherty *et al.* 2016).

There is substantial uncertainty in the extent of climate change impacts on biodiversity, and bioinformatic challenges in processing the volume of earth observation data relevant to climate-driven biological change (e.g. in forest cover). Current solutions to such big data problems include change-detection software, which minimizes the need to store data for every fly-over, and multidimensional data structures such as 'data cubes', which manipulate large amounts of raster data efficiently.

Global initiatives to advance biological data include the Convention on Biological Diversity's Global Taxonomy Initiative (Siebenhüner 2006) and the Global Biodiversity Information Facility (GBIF) (Yesson *et al.* 2007). GBIF species occurrence records now cover all parts of the globe (1 billion records referring to 1.7 million species); (GBIF 2018), and its taxonomy follows the Catalogue of Life (http://www.catalogueoflife.org) using established Biodiversity Information Standards (TDWG) for data transfer (http://www.tdwg.org/).

Much indigenous ecological knowledge (e.g. medicinal plants) is translated by word of mouth and risks being lost if undocumented (McCarter *et al.* 2014). However, a framework has recently been developed for connecting indigenous knowledge with other knowledge systems (Tengö *et al.* 2013), such as international assessments (Sutherland *et al.* 2014), and some indigenous knowledge is now captured digitally (Liebenberg *et al.* 1999; Stevens *et al.* 2014).

In addition to data gaps, there are deficiencies in data sharing and access. Some biological problems are inherently regional or global and require coordinated multinational management. A field where this is a major problem is transnational environmental crime (White ed. 2017), which includes harvesting, transporting and tracing trade of endangered species, illegal mining, fishing and deforestation. Improvements in shared data infrastructure are essential for effective regulation in this area.

Oceans (Chapter 7)

Ocean data have many gaps, which is unsurprising since satellite observations cannot penetrate below surface waters. Most oceanic data are collected by direct measurement or modelling, so it is difficult to obtain good coverage for a vast environment that extends over 70 per cent of the earth's surface. Some issues exist through lack of global coordination, as both coral reefs and marine litter lack global databases. The National Oceanic and Atmospheric Administration (NOAA) maintains the largest coral reef database, but it does not draw upon all sources globally. Similarly, marine litter data are collected by different countries with different protocols and have not been globally consolidated. In addition to litter abundance and distribution, significant knowledge gaps exist regarding the ecological impacts of marine litter, including the toxicity of ingestion, impacts of nanoparticles, microplastics, and how plastics ingested by fish impact human consumption.

Global fish catch data are maintained by the Food and Agriculture Organization of the United Nations (FAO), to which all countries report national catch and yield. Commercial fishing catches are well monitored in developed countries, but are almost certainly underestimated since illegal fishing constitutes as much as 40 per cent of all catch in some areas (Agnew *et al.* 2009). In countries with fewer resources to devote to reporting, fishing estimates are often based on a small number of samples and are therefore less reliable. Research vessel costs are a major impediment to obtaining fisheries-independent data, particularly in developing countries where even catch monitoring in ports may not be economically viable.

Land (Chapter 8)

Land is one of the most data-rich domains due to the effectiveness of earth observation in monitoring land surfaces, but there are still notable data gaps and quality issues. Earth observation generally measures the quantity rather than the quality of change, and is unable to measure certain processes. For example, there is agreement that land degradation has increased, but it is not done often and is inconsistently measured. The interrelationships between the Normalized Difference Vegetation Index (NDVI) and land degradation are often difficult to generalize and transfer, since land use and biophysical conditions are changing regionally. While forest cover data have improved since the mid-1990s and some broad-scale data are maintained by FAO, other data exist in multiple databases that are not always comparable. Soil erosion, salinization, desertification and change in ecosystem services are all difficult to measure from satellite images, and there are questions as to the appropriate scale of observation. There is no global database or standardized measurements of soil erosion, preventing a globally coherent or comprehensive assessment. Other difficult areas are land tenure and cadastral (map-based) information, since there is no global standard for defining land use, and systems are not comparable across countries.

Freshwater (Chapter 9)

Data on fresh water suffer from spatial and temporal patchiness, and a divide between variables than can be remotely sensed by earth observation versus those that cannot (Lawford *et al.* 2013). Data-deficient areas at all scales include water quality, water consumption, groundwater quantity, water withdrawals and wastewater. The SDGs require monitoring of ambient water quality, but not all countries have the capacity or will to meet these reporting requirements. There are better data for surface-water quality than for groundwater, but these are still patchy. Earth observation systems measure optical qualities of water (chlorophyll, salinity, turbidity), but cannot measure nitrogen or phosphorous concentrations. In recent years, progress has been made in using satellite data from the GRACE mission to estimate changes in groundwater storage (depletion), but assessing groundwater resources requires the collection of direct data which are relatively expensive as they require access to groundwater through wells or boreholes. There are also gaps in glacier, snow and ice data, and uncertainty around impacts of climate change (Salzmann *et al.* 2014), though the Copernicus programme may address this on a global scale with a satellite dedicated to monitoring snow/ice cover. Some other variables are difficult to measure by any

means, such as groundwater and saltwater intrusion, which are mostly understood by modelling rather than observation. These models are in urgent need of reliable on-the-ground data for calibration and verification. Geopolitical issues of water use, such as transboundary water sharing, are another area requiring more data, particularly at times of water scarcity.

Citizen science may offer some solutions to issues of freshwater sampling coverage and basic monitoring of groundwater levels. Examples include the use of mobile applications to monitor water quality (Lemmens *et al.* 2017) and the use of testing kits in EarthWatch Freshwater Watch (http://www.freshwaterwatch.thewaterhub.org/) and other volunteer groups (Overdevest *et al.* 2004). An early form of citizen science has successfully been deployed for many decades in Netherlands where volunteers from across the country measure groundwater levels in piezometers bimonthly, contributing to the building up of long-term time series of groundwater data in the country. However, citizen science initiatives usually involve simple water monitoring and do not measure the suite of modern pollutants such as antibiotics, persistent organic pollutants, current use pesticides, microplastics, nanoparticles and endocrine disruptors.

3.5 Gender and social-environment intersectionality

The paradigm shift that is bringing social analysis into the heart of environmental assessment has developed since the mid-1990s with the emergence of gender-disaggregated environmental analysis and analysis focused on other vulnerable groups. This section will focus on the gender-environment nexus; however, many of the issues presented could be applied to other vulnerable groups. Broader equity issues, including, importantly, North-South inequalities in environmental footprints and impacts – which are themselves gendered – are addressed elsewhere in this report.

The role of gender in environmental analysis will accelerate as the social equity and equality commitments of Agenda 2030 shape global policymaking **(Box 3.2)**.

At the heart of gender analysis is the understanding that virtually all environmental relationships, including drivers and impacts, are 'gendered'. Socially constructed gender roles and norms position men and women differently in relation to the environment. Men and women are often exposed to different environmental problems and risks; in turn, this may mean that men and women have different perspectives on the extent and seriousness of environmental problems, and on what solutions might best be attempted or deployed. Further, because of the social construction of gender roles, men and women are often positioned differently in terms of being able to take action or being taken seriously as agents of environmental interpretation and change.

Gender analysis requires new approaches to the structure of environmental inquiry. Analysing the environment through a gender lens requires new and different questions, brings to the foreground different dimensions of human-environment relationships, and requires different methodological tools and approaches. Gender analytical lenses encompass 'the environment' in both its physical and social aspects, and in the interactions of these. Gendered commitments to "lift the roof off the household" in data collection reveal intra-household dynamics of resource utilization and decision-making, which are often critically important in understanding local environmental behaviour and environmental outcomes (Seager 2014).

Gender analysis also brings to the fore intersectionality – an understanding that social relationships with the environment are seldom shaped by a single social identity, but rather by a combination of gender identities and norms, as well as other social identities such as race, sexuality and class.

The UNEP *Guidelines for Conducting Integrated Environmental Assessment* (UNEP 2017) reflect these new approaches by bringing to the fore gender-informed questions that should be integrated into environmental assessment from the earliest planning stages **(Box 3.3)**.

Data availability and statistical systems have not kept pace with the interest in and demand for gender-disaggregated analysis in environmental assessment. The GEO-5 assessment notes the lack of – and need for – gender-disaggregated environmental data (UNEP 2012). One of the most consistent messages in the field of gender-disaggregated environment analysis is that this information is crucial to a comprehensive analysis (United Nations 2015a; UNEP 2016). Some progress has been made since the GEO-5 assessment, and UNEP (2016) synthesizes the data and analytical approaches that are now

Box 3.2: Gender statistics

"Gender statistics are defined as statistics that adequately reflect differences and inequalities in the situation of women and men in all areas of life...First, gender statistics have to reflect gender issues, that is, questions, problems and concerns related to all aspects of women's and men's lives, including their specific needs, opportunities and contributions to society. In every society, there are differences between what is expected, allowed and valued in a woman and what is expected, allowed and valued in a man. These differences have a specific impact on women's and men's lives throughout all life stages and determine, for example, differences in health, education, work, family life or general well-being. Producing gender statistics entails disaggregating data by sex and other characteristics to reveal those differences or inequalities and collecting data on specific issues that affect one sex more than the other or relate to gender relations between women and men. Second, gender statistics should adequately reflect differences and inequalities in the situation of women and men. In other words, concepts and definitions used in data collection must be developed in such a way as to ensure that the diversity of various groups of women and men and their specific activities and challenges are captured. In addition, data collection methods that induce gender bias in data collection, such as underreporting of women's economic activity, underreporting of violence against women and undercounting of girls, their births and their deaths should be avoided..."

Source: UNSD (2015)

available. Nonetheless, very little information is available about the different needs of men and women, their different use of resources, and their different responsibilities in contributing to conservation and sustainable development.

Even less information is available to support intersectional analysis of gender with age, race, caste or class dynamics. Existing data on gender and the environment are fragmented and scattered among small and often grey-literature sources or across hard-to-access scholarly reports. There are almost no common standards or complementarities across countries, making it almost impossible to aggregate and compare issues across regions. The lack of sufficient long-term data further impedes gender-disaggregated environmental assessment because relationships between gender and the environment may only become evident over long time periods.

Box 3.3: Gender-informed questions

- What are the geographic locations and subject areas, sectors and activities in which gender difference and social class impact one's relationship with the environment?
- Are there any other intersectional issues that might need to be considered (e.g. how different cultural/ethnic/class groups use, imagine and/or relate to place and are there any conflicts between these groups)?
- How do general differences between socioeconomic classes, in relation to the environment (as mapped in reports such as the *Global Gender and Environment Outlook*, UNEP 2016) apply to the environmental issues undergoing assessment?
- What are the differences in behaviour of men, women, boys and girls in relation to the environmental issues undergoing assessment (as mapped in reports such as the *Global Gender and Environment Outlook*?
- Are gender-disaggregated data available to understand that relationship or will it need to be collected?

The absence of gender data undercuts the momentum towards further gender-environmental analysis – 'what's not counted is assumed to not count'. In the absence of data, environmental assessments remain partial; establishing baselines, monitoring progress and assessing outcomes are almost impossible. Progress towards SDG commitments to gender equity and equality in all domains, including the environment, will be impossible to measure without substantial improvement in gendered data.

Even simple gender-disaggregated data-based analysis, such as that on average time spent in unpaid work by men and women **(Figure 3.5)**, can reveal important gender dynamics. The burden of unpaid work restricts women, more than men, from undertaking paid work and from participating fully in civil and economic spheres. Figure 3.5 illustrates the uneven burden of unpaid work between men and women. Many hours of women's unpaid work, especially in poorer countries, are spent in directly managing local environmental resources to meet the needs of household water, fuel and food. At the same time, 'time poverty', which is produced by the burden of unpaid work, means that women are less likely than men to be available for environmentally relevant training, nor are they available to participate in formal processes relating to environmental use, management and decision-making.

Figure 3.5: Unpaid care work

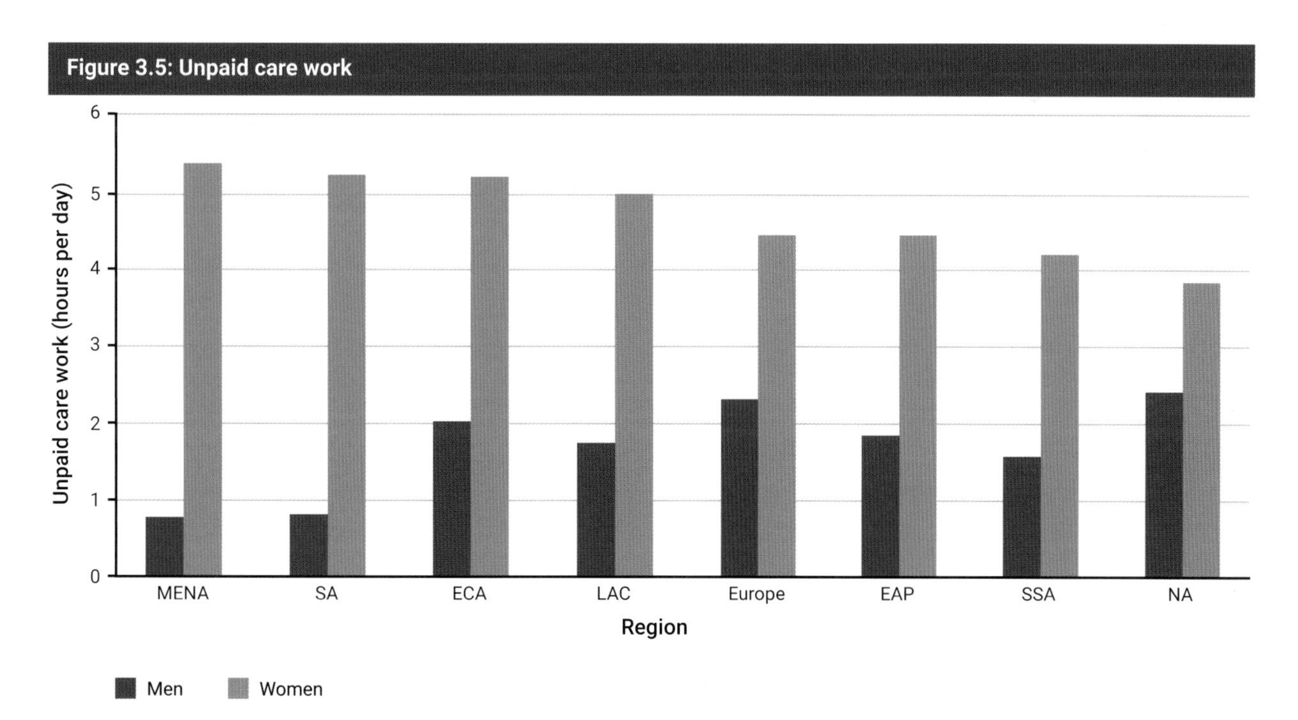

MENA, Middle East and North Africa; SA, South Asia; ECA, Eastern and Central Africa; LAC, Latin America and the Caribbean; EAP, East Asia and the Pacific; SSA, sub-Saharan Africa; NA, North Africa.

Source: Ferrant, Pesando and Nowacka (2014, p. 2).

The expectation that environmental assessments will include gender analysis and data is achieving mainstream acceptance. In 2016, UNEP produced the *Global Gender and Environmental Outlook* (GGEO) entirely through a gender lens. The GGEO report concluded that the effectiveness of environmental decision-making would be enhanced by "Strengthening the focus on developing, collecting and analysing gender-disaggregated data, indicators and other information, including at the intra-household level." (UNEP 2016, p. 201).

The SDG target 17.18 specifically calls for improved collection and availability of gender-disaggregated data: "By 2020, enhance capacity-building support to developing countries, including for least developed countries and small island developing States, to increase significantly the availability of high-quality, timely and reliable data disaggregated by income, gender, age, race, ethnicity, migratory status, disability, geographic location and other characteristics relevant in national contexts" (A/RES/70/1).

The GGEO provides a summary of the most complete gender-disaggregated data sets available as of 2016. These include several gender disaggregated agricultural indices (from FAO) on indicators such as agricultural employment and landholders; cross-national comparative information on access to and ownership of land (from FAO, the Organization for Economic Co-operation and Development [OECD] and the World Bank); and sex-disaggregated burden-of-disease data for a few environmental factors (Prüss-Ustün *et. al.* 2017).

Additional large-scale efforts are under way to collect and analyse environment-related gender-disaggregated data:

- in 2014, the United Nations Educational, Scientific and Cultural Organization (UNESCO) launched a project to identify gender and water priority indicators (UNESCO 2014);
- the FAO Gender and Land Rights Database "was launched in 2010 to highlight the major political, legal and cultural factors that influence the realisation of women's land rights" (FAO 2018). By 2018, the FAO database had data from more than 80 countries, and the FAO 'Legal Assessment Tool' maps the intricacies of men's and women's access to land.

The prospects for improving gender-disaggregated environmental data are promising and the expectations for data collection for the SDGs should accelerate efforts to systematically collect both sex-disaggregated (indicators specifically related to biologically rooted activities, roles and impacts), as well as gender-disaggregated (related to social roles and impacts) environmental data. There remains, however, a considerable gap between demand and supply.

3.6 Equity and human-environment interactions

Assessing human-environment interactions requires data, knowledge and integrated approaches as outlined in Chapter 1 of this report. A balanced evaluation of existing data and scientific results can lead to balanced policy choices. However, is the knowledge base able to provide a balanced story about human-environment interactions? This leads to three key questions, as shown in **Figure 3.6**.

Figure 3.6: Equity questions in data and knowledge

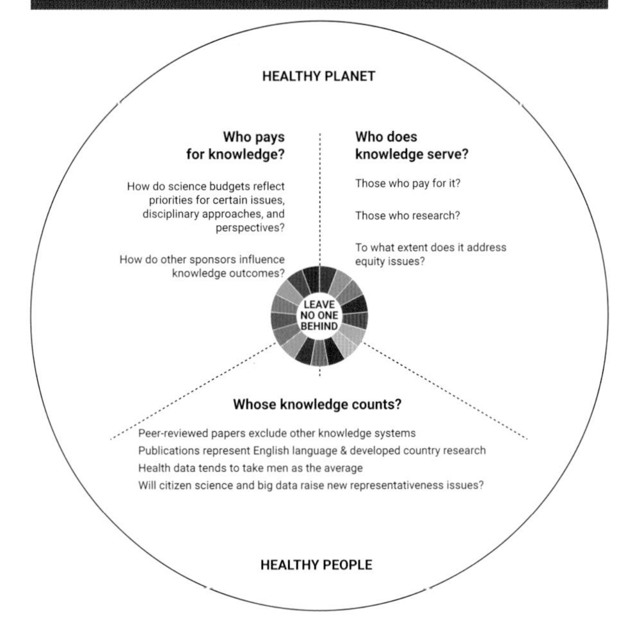

Who pays for data and knowledge and for what sorts of data and knowledge? All data and research are funded by specific actors – the state, but also non-state actors such as civil society, industry and philanthropists. There is clear evidence that states invest large sums of money in natural science and technology research, but there is significantly less invested in environment- and resource-related social science and equity-related research. For example, a study of funding in the United States of America shows that between 1970 and 2015 social sciences received very little funding in comparison with other fields of study (National Science Foundation 2017).

Whose interests do the existing data and knowledge serve? Research questions and data tend to serve dominant interests, for example those identified by the funding agencies. They may also serve disciplinary interests rather than human-environment interactions more integratively (McMichael, Butler and Folke 2003). Furthermore, although there is need for data and knowledge on the causes and impacts of internally displaced people, such data are not yet available (Bennett *et al.* 2017, p. 11). The need for disaggregated data is vital to address issues of equity, but such data and knowledge are limited.

Whose data and knowledge counts and why? In international assessments there is increasing evidence that researchers come from the richer 'developed' world, rather than the non-English speaking and/or developing world. For example, 87 per cent of the world's researchers, 92 per cent of the research budget and 94 per cent of scientific publications come from the G20 countries (UNESCO 2015). In 2015, the majority of authors in the Intergovernmental Panel on Climate Change (IPCC) came from developed countries with significantly fewer from developing countries (Schulte-Uebbing *et al.* 2015).

There is little assessment in any of the environmental literature of the politics of data and knowledge, and this is challenging.

3.6.1 Environment and economy

A number of SDGs depend for their realization on understanding and properly taking into account the costs and benefits of environment-economy relationships. Most importantly, the SDGs and natural capital accounting indicators provide insight into the value of 'nature's contributions to people', human societies more broadly, and of the cost of residuals such as pollution and waste. The economics of nature or natural capital accounting involve the assessment, measurement, aggregation and valuation of these contributions, to help policymakers ensure that this value is reflected in the economic activities of production, consumption, trade and investment through such instruments as pricing, costing and regulation. Evaluating the economic dimension of the environmental impacts of economic activities helps policymakers realize synergies between economic and environmental issues, gain efficiency in the allocation of limited resources and avoid trade-offs (or minimize them where they are inevitable). Any such evaluation should take into account that economic activities are increasingly characterized by global chains (e.g. investment, trade), and the role of such 'teleconnections' is crucial in determining overall impacts. Therefore, what we do to sustain environmental resources in one place may be at the expense of resources or environmental quality elsewhere. The System of Environmental Economic Accounts provides a framework for analysing the interactions between the environment and the economy. It includes information on four policy quadrants, namely: access to services and resources; managing environmental resources supply and demand; the state of the environment; and risks and extreme events (United Nations 2014).

When considering the benefits of nature, a fundamental issue is whether these values are comparable with and substitutable by other economic benefits. Most conventional economic analysis assumes substitutability of factors of production, called 'weak sustainability', when applied to natural capital (Solow 1974; Hartwick 1977). But there are many instances when the contributions of nature to human life (e.g. the regulation of the climate) cannot be provided by other human activities. These situations of 'strong sustainability', often related to the planetary boundaries, need to be revealed through robust analysis. Such analysis will need to rely on methodological diversity, using insights from ecology, economics, social and cultural studies, and recognizing their dynamic evolution.

In natural capital accounting and in the System of Environmental Economic Accounting, methodologies may be used to give monetary value to environmental benefits and costs, so that they may be compared with other economic activities and costs. Alternatively, in some cases, only the stocks and flows of environmental resources and residuals are measured using an accounting framework as opposed to including a valuation. An accounting framework provides information on the use of environmental resources, such as water and energy, and the residual to the environment, such as emissions and wastes, by industrial classification.

For economic valuation, the valuation should always be applied in a way that it can capture trade-offs and the demand for resources for competing uses. It should also be recognized that there are numerous environmental situations in which, due either to lack of data or absence of credible science or methodological agreement, economic analysis has limited scope.

Economic analysis of the environment can be oriented towards the wider goals of the United Nations system, and the SDGs related to peace, human rights, equity and security, as well as sustainability. It needs to recognize the complexity of environmental-economic interactions and highlight uncertainties, through clear and simple communication.

Sustainability, and the policies necessary to achieve it, should focus on trends in per capita wealth, as well as flows of income and non-monetary benefits. It is the natural capital stock that generates nature's contribution to people, and correct wealth accounting in relation to the environment and resources, avoids the mixing up of income and wealth.

Macro-models are required to assess national and global outcomes of policies for the use of resources and the environment. Recent results from use of these models suggest that the conventional perception of the economy and environment having a trade-off relationship may be incorrect. Increasingly, 'green economy' analyses seem to suggest that natural resources are an essential input to sustainable economic growth. From this perspective, an appropriate 'economics of nature' could be a great enabler of both conservation and development. Such messages need to be transmitted with clarity and confidence.

3.6.2 Environment and health

The environments in which we live are a key determinant of human health and well-being. The physical environment provides us with the air we breathe, the food and water required for sustenance, solar radiation that provides heat and light, and more. These are direct effects, but indirect effects are also important in supporting healthy ecosystems, which in turn provide food security and other ecosystem services. The social environment also has a strong influence on health and well-being, as clearly shown through socioeconomic gradients in health, whereby social disadvantage is associated with poor health and well-being across a wide range of diseases and health-risk behaviours (Friel and Marmot 2011). Degradation of our environment (e.g. air pollution, contamination of food and/or water, insufficient or excessive sun exposure, excessive noise, conflict and war) adversely affects food and water security, health and well-being.

Exploring the links between the environment, in its broadest sense, and human health and well-being requires measurement of the 'exposure' (the environmental factor of interest) and the 'outcome' (some measure of health and/or well-being). The next step is to assess whether there is a causal relationship between the exposure (e.g. air pollutants, conflict, green space, noise) and the outcome, which typically requires good study design, appropriate statistical methods, and causal analysis. The size of the effect, coupled with an understanding of the prevalence of the exposure in the population, can be used to provide an attributable effect (i.e. what proportion of the health outcome is caused by exposure to the environmental risk factor) (Prüss-Ustün *et al.* 2017). In addition to analysing exposure to certain contaminants, analysis of environmental

conditions and health and well-being can also reveal underlying relationships between health and the environment. For example, data on underweight children, malnutrition and other food security indicators can be analysed through an environmental lens to better understand the relationship between climate change and food security, health and well-being.

Environmental exposure can be directly measured at the individual level (usually only for relatively small numbers of people) or inferred at the individual level or at an ecological level using data from routine monitoring (e.g. of air and water quality, levels of solar radiation, or modelled, for example using combinations of atmospheric variables to estimate climate change-related exposures). These methods can also be combined, for example, where data from multiple weather stations are used to calculate individual exposures at different locations within an area (Miranda *et al.* 2016). Exposure measurement is more precise for some environmental factors (e.g. blood lead levels) than for others (e.g. lifetime exposure to noise pollution) (Klompmaker *et al.* 2018), and for short-term rather than long-term (e.g. lifetime) exposures. Here, large sample sizes ('big data'), plus innovative study designs and data analysis, are required, but there must also be recognition of the potential biases within these 'noisy' data sets (Ehrenstein *et al.* 2017).

Data to assess the burden of the health outcome with environmental factors are available at the individual level through epidemiological studies, and from administrative databases (e.g. hospital separations data, where modern data linkage methods can allow examination of individual-level data). However, considerable challenges remain to using administrative data due to ethical issues around protection of individual privacy. Administrative data can also be used in ecological studies (e.g. of the effect of air pollution on hospital admissions). For some health outcomes in some countries, surveillance through disease registries provides comprehensive and accurate incidence and mortality data. These can be linked to other data sets to derive associations at an individual level (Korda *et al.* 2017), or used in ecological studies to assess relationships between disease and environmental parameters (Adams *et al.* 2016). The Global Burden of Disease (GBD) Study is a valuable data set for disease-specific incidence and mortality (GBD 2016 Causes of Death Collaborators 2017; GBD 2016 Disease and Injury Incidence and Prevalence Collaborators 2017). The GBD is now updated annually and seeks to collect the best possible health (typically disease) data from all countries to provide comprehensive estimates at the global, country and, for some countries, regional levels. In addition, the GBD Study estimates health loss through morbidity, as well as disability adjusted life years (DALYs) and health adjusted life expectancy (HALEs) (GBD 2016 DALYs and HALE Collaborators 2017). However, additional disaggregated information on who is impacted and on location, which would be necessary for a comprehensive assessment, is typically not available. Recent developments in 'omics' technologies – genomics, metabolomics, exposomics, epigenomics and others – deliver a huge amount of data that may allow assessment of the effects of environmental exposures on human health and well-being. However, challenges remain in separating out effects of specific exposure (e.g. the various components of 'air pollution') and accurately quantifying effects attributable to exposures:

a) that are difficult to measure precisely,
b) have non-linear dose-response or threshold effects,
c) when exposure levels change over life, or
d) have both risks and benefits to human health.

3.7 Existing data systems

Official statistics, national geospatial data, and Earth observation monitoring data often are not part of a single data system at the national level, and there is a need for better integration of data from these sources in assessments. Although gaps remain in official statistics, national geospatial data and Earth observation data, these data sources are currently being used for environmental assessment and are better developed globally than the emerging tools for environmental assessment presented in Chapter 1.

3.7.1 Official statistics

The disciplines of official statistics and Earth observation have developed independently and manifestations of their interconnectedness have been sporadic. The relationship has benefited from guidance emanating from the national statistical systems through the following developments: adoption of the System of Environmental Economic Accounting (SEEA) under the Central Framework in 2012, adoption of the SEEA Experimental Ecosystem Accounts in 2013 and the revised Framework for the Development of Environment Statistics in 2013. These three statistical frameworks provide an increased methodological basis for statistics; however, there is still a need to scale up statistical production and to involve more actors in the production of environment statistics, including local-level actors. Additionally, there remains a need for methodological guidance on the interactions between society and the environment, including the gender dimension.

Technological change – including better satellite data, monitoring stations and personal electronic devices – is changing the data landscape, including through citizen science. The data revolution and its technological derivatives, namely big data and citizen science, unleashed new possibilities for measurement, potentially disrupting existing organizational and institutional relationships in the management of measurement and production of scientific knowledge. The response to these new manifestations of technology-inspired measurement have been led by, among others, the United Nations Committee of Experts on Global Geospatial Information Management (UN-GGIM). However, it is likely to be long before the urgent need of integration is achieved. There remains a need to better utilize technologies, including mobile applications, smart devices and other tools, to make data accessible to populations and to provide an interface for making citizen science data discoverable.

The imperative for statistics and data

The injunction of 'leaving no one behind' imposes a high premium on the production and delivery of disaggregated data by all attributes possible, including (importantly) by local area. In so doing, the SDGs bring to bear the importance of geospatial data and statistics. A geospatial statistical approach to measurement provides a transformative infrastructure that improves the information necessary to 'leave no one behind' through analysis of interactions and causality at the local level and for particular populations.

The SDG indicator framework

While the SDG agenda is bold and ambitious, it is not possible to cover everything at the same time. Therefore, the ability to prioritize and sequence is strategic for success in the delivery of measurement to the global agenda. So, which data, which statistics and which indicators?

An attempt to answer the questions cannot be made without historical experiences of global development measurement exercises, of which the MDGs represented the most enlightening. In his 2015 MDG report, the United Nations Secretary-General notes with regret that, first, statistical information is collected with a major temporal lag. Yet, today's world is a fast moving one requiring real-time data; second, that the information is highly aggregated and lacks locational specificity for use in directing interventions; and third, there is minimal resource allocation to countries and institutions that require data the most for development (for their people and environment).

That the nature of the problem has been defined does not imply that the questions the problem raises should not be answered. The benefit of defining the problem is in identifying with a greater level of clarity what needs to be done in prioritizing and sequencing.

With regard to indicators, official statisticians, under the guidance of the UNSC, have worked hard to identify an indicator framework and the feasibility of indicators that would feed into the framework. However, the design of the framework has, in practice, been directed towards the number of indicators, rather than to an architecture that would determine the indicators. The existence of the framework and the ability to identify indicators feeding into the framework is a commendable start.

In relation to the Global Environment Outlook, it is more important to note that the task becomes even more serious and politically challenging given that tens of the goals in the SDGs relate directly to or are closely linked to the environment. Perhaps this will lead to a different GEO outcome.

As defined in Section 3.1, less than a quarter of the environment-related SDG indicators are Tier I. This gives some idea of the difficulty of measurement, including resourcing of the statistics systems, in some countries.

The former United Nations Secretary General has recognized the need for clear coordination mechanisms for data and statistics. In this regard, the Secretary-General called on countries to recognize the significance of coordination among national agencies **(Box 3.4)**, including national statistical institutions, in providing, encouraging and enforcing compliance with statistical standards through principles, legislation and practice notes.

Box 3.4: Statement from the United Nations Secretary-General

"National statistical offices should have a clear mandate to lead the coordination among national agencies involved and to become the data hub for monitoring."

Ban Ki Moon, United Nations Secretary-General 2007-2016, (United Nations 2015a)

Measuring the environment in the context of the SDGs

Accurately assessing the interaction between people and the environment will require new data sources and new tools for environmental assessment. For example, geospatial information can be incorporated with population maps to determine the regional environmental issues that affect people (e.g. where poor people live and where water quality issues are).

The key driver to the exponential growth of access to and use of technology has been the ability of technology to create and push towards common standards. Through this innovation, a movement has emerged towards standardized forms of data to be collected at a much lower cost. This has made collection of larger amounts of data a lot more attractive. More importantly, technology has unleashed possibilities for the use of geospatial statistics and a greater ability to observe changes in the environment.

Environmental data, statistics and knowledge are the foundations of successful environmental assessments. Remote technologies, Earth observation systems and national statistical offices remain the leading generators of environmental data. New and emerging knowledge frameworks and data capacities in database management, citizen science, disaggregated social and gender analysis, big data, data visualization tools, spatial modelling, social media and the Internet offer opportunities to collect and disseminate information. Collectively, data aggregated from these approaches improve capacity to support strategic decision-making processes that are based on wide-ranging and multidisciplinary knowledge. Effective monitoring of environmental trends is critical to clean up environmental damage.

The disaggregated and location-based information needed to 'leave no one behind' is believed to be achievable, and this meets the requirements for effective monitoring of environmental trends.

If we are, however, true to the notion of 'leaving no one behind' as prescribed in the SDGs, then multiple methods need to be handled by information management systems. These include the well-established traditions of statistical standards and the future of statistics is enhanced with the availability and analysis potential of land information systems. Furthermore, new technologies and their capabilities in the data and geographic space create new ways for citizens to participate in science and also to increase the possibilities of environmental data integration.

The challenge, however, for these new knowledge platforms to be useful, is if they are supported by an institution. First, can they be seen as systems of today and tomorrow that attract reasonable resources, and that enhance the well-being of people and the planet? The 2015 MDG report of the United Nations Secretary-General argues the need for coordination and involvement of national agencies in monitoring (see Box 3.5). Second, can these data and information systems work together across space and time? That is, can they be trusted to help social, economic and political discussions, and also withstand times of transition? Third, are they auditable? Will they stand up to scrutiny? Fourth, knowledge, statistics and data are inherently political and can create challenges to governments in the knowledge discourse.

Underlying challenges

Major data gaps, across the globe and across environmental domains, limit our ability to identify trends and manage unwanted outcomes. In many countries, official statistics on the environment are rarely generated, are difficult to access, are scattered across different institutions, and reporting is fragmented (UNEP 2016). Across many environmental topics, data availability is geographically unbalanced, being scarcer for rural areas and developing countries. Monitoring systems from global to regional scales are fragmented, lack coverage and are often not updated on a regular basis (UNEP 2012, p. 129). There is a pressing need to create regular monitoring that follows commonly agreed international standards that are best enacted through international cooperation. There is also a need for increased sharing of data in a standardized format, for example, data that is compliant with Statistical Data and Metadata e-change standards.

The United Nations SDG report of 2016 explains that data requirements for the global indicators are almost as unprecedented as the SDGs themselves, and constitute a challenge for many countries. Tracking progress on the SDGs would require the collection, processing, analysis and dissemination of an unprecedented amount of data and statistics at the subnational, national, regional and global levels, including those derived from official statistical systems, as well as from new and innovative data sources (United Nations 2016b).

While knowledge systems often cross national boundaries, the creation, custodianship, distribution and use of knowledge have historically and politically been associated with governments. Knowledge does not exist in a geopolitical, social or economic vacuum. Will these new systems be able to inform political decision making and acceptance of environmental development and management?

3.7.2 Geospatial information

Environmental monitoring and forecasting systems have been growing rapidly. However, combining information from multiple systems to generate statistics and indicators remains a major challenge. Earth observation is defined by the global Group on Earth Observations as both surface observations (*in situ*) and those collected by aircraft and remote sensing, including from satellites and other space missions. Similarly, a data set collected for one purpose can often be used for multiple purposes. For example, agricultural land cover could be useful for understanding natural disaster risk, examining the migration of people, the nature of informal settlements, urban infrastructure and their relationship with biodiversity and ecosystems.

Earth observations and environmental monitoring are being transformed through integration of administrative data from national statistical agencies, including economic data, and open data policies for Earth observations that benefit both emerging economies and developed countries. Open Earth observations, citizen science, social media, and digital platform or big data access can stimulate a transformation to a new model for creating data which results in more inclusive, social, robust knowledge for decision-making, where there is broader understanding and access to policy-relevant knowledge.

For example, the first *Atlas of the Human Planet* (Pesaresi *et al.* 2017), derived from the Global Human Settlement Layer (GHSL), provides a validated source of information on human habitations, from villages to megacities. The baseline data, spatial metrics and indicators related to population and settlements, developed in the frame of the Group on Earth Observations Human Planet initiative, provide users with a baseline data platform for monitoring and analysis. The GHSL resource is an example of the potential of public data to support global, national and local analyses of human settlements and, in particular, support policy and decision-making. This application of Earth observations is essential for evidence-based modelling of human and physical exposure to environmental contamination and degradation, as monitored through multilateral environmental agreements; disasters as encompassed by the Sendai Framework for Disaster Risk Reduction; the impact of human activities on ecosystems, as measured by the Convention on Biological Diversity; and human access to resources, assessed by the SDGs (European Commission 2018).

In September 2015, the United Nations General Assembly endorsed *Transforming Our World: The 2030 Agenda for Sustainable Development*, a global development agenda to use to monitor progress on economic, social and environmental aspects of sustainability, as stipulated in Article 76 **(Box 3.5)** (A/RES/70/1).

Within the United Nations system, agencies including the UNSC InterAgency Expert Working Group (IAEG-SDG) and the United Nations custodial agencies taking a lead in developing monitoring methodologies are examining, and in some cases preparing to incorporate, Earth observation and geospatial data for support of the SDGs, its targets and indicators. A 2016 analysis by the Group on Earth Observations estimated that at least 98 targets and indicators could benefit from and use

Box 3.5: Article 76 of the 2030 Agenda

"We will promote transparent and accountable scaling-up of appropriate public-private cooperation to exploit the contribution to be made by a wide range of data, including Earth observation and geo-spatial information, while ensuring national ownership in supporting and tracking progress."

– United Nations, General Assembly (2015)

Earth observations data (United Nations 2016c). The Earth observations global community is fully engaged and ready to provide expertise to all United Nations members, particularly developing countries, with regional and specific national capacity-building.

3.8 Conclusion

Gender and social-environment intersectionality
The differences in exposure to environmental problems and risks result in different perspectives for men and women, thereby reflecting unequal reaction to and interpretation of opportunities for development and sustainability. Since the environment is shaped by a blend of social identities and norms, improved collection and strengthened analysis of high-quality and timely disaggregated data by gender, age, race and other characteristics in the national contexts are required to establish a holistic baseline, and for monitoring and assessment. Such data should also be spatially disaggregated and geographically sensitive to capture local variations.

Equity and the human-environment interactions
Collection, disaggregation and analysis of data for the most vulnerable communities remain a challenge. More work in this area would better capture issues of inequality (United Nations 2012, p. 12). Industry generally funds research that helps improve industrial processes and increase shareholder value, while philanthropists may cover a range of issues including equity issues. It is important to promote data and knowledge on how "to overcome barriers to political and social participation and to accessing services and proactive policies and sustained social communication to influence social norms that perpetuate discrimination and exclusion" (United Nations 2012, p. 9). Furthermore, in terms of regional concentration, research is concentrated geographically in the United States of America, China, Japan and Germany, which collectively account for 63 per cent of the global research and development expenditures, mostly funded by the business sector (National Science Board 2016, pp. 41-46). Businesses as funders of research have overtaken government-led funding, which has moved the balance towards more applied research than basic research (United States National Science Board 2016). This issue raises the question of who is reaping the benefits of research and if a greater good is achieved by it.

Environment and economy
Economic evaluation of environmental impacts involves the overall assessment of nature's contributions to the lives of people; the accounting of global economic activities, investment and trade to people and the environment; and the comprehensive institutional issues affecting equity and market operations. Specific findings on sustainability can only be revealed through a robust analysis, covering ecological, social and cultural factors, and their interaction over time. Valuation attributes a monetary values to environmental benefits and costs, as well as trade-offs and competition. Economic analysis of the environment should be oriented towards the wider scope of the SDGs, including peace, equity and security. The monetary and non-monetary values in relation to the environment and resources, as well as models reflecting the economics of nature, can only be generated through timely and reliable data and information from statistical surveys and other new data sources such as big data.

Environment and health
Combined physical and social environments have strong influences, both direct and indirect, on human health and well-being. With this, the measurement of linkages between the 'exposure' and 'outcome', the assessment of causal relationships, and the exposure to populations need strong statistical bases and large sample sizes (i.e. big data). Challenges facing epidemiological studies include data protection, reliability and disparity when using administrative databases. Recent developments include the use of big data to allow assessment of long-term environmental exposures. It is necessary to explore the use of other sources of information to validate the long-term effects of human activities and natural disturbance such as climate change on health, with the use of new forms of data and knowledge (i.e. citizen science and traditional knowledge).

3.8.1 Better data for a better planet and better lives

The United Nations 2030 Agenda serves as the global framework for assessing economic, social and environmental development, focusing on building a healthier planet and fostering better lives through national engagement and partnerships. Monitoring the progress on the SDGs requires shifts in data collection, analysis and dissemination, including using environmental statistics, geospatial data, Earth observation and new data sources (i.e. citizen science, big data, traditional knowledge).

A new and innovative approach to data and knowledge systems with an aligned focus on evidence-based information gathering is essential for achieving the ambitious SDG framework. However, monitoring the entire SDG framework over the 2016-2030 period is estimated to cost as much as a quarter of a trillion dollars (Jerven 2014). So, in addition to improving data systems, there is also a need for priority setting to target data collection and improve efficiencies.

Environmental change is difficult to measure, and the effects of environmental change are even more complicated to measure, especially in relation to identifying causes. A shift from focusing solely on the physical dimensions to including social orientation, economic value and impacts on health and well-being is crucial but is a challenge for even well-developed statistical systems.

References

Adams, S., Lin, J., Brown, D., Shriver, C.D. and Zhu, K. (2016). Ultraviolet radiation exposure and the incidence of oral, pharyngeal and cervical cancer and melanoma: An analysis of the SEER data. *Anticancer research* 36(1), 233-238. https://www.ncbi.nlm.nih.gov/pubmed/26722048.

Agnew, D.J., Pearce, J., Pramod, G., Peatman, T., Watson, R., Beddington, J.R. *et al.* (2009). Estimating the worldwide extent of illegal fishing. *PLoS ONE* 4(2), e4570. https://doi.org/10.1371/journal.pone.0004570.

Bennett, K., Bilak, A., Bullock, N., Cakaj, L., Clarey, M., Desai, B. *et al.* (2017). *Global Report on Internal Displacement*. International Displacement Monitoring Centre and Norwegian Refugee Council. http://www.internal-displacement.org/global-report/grid2017/pdfs/2017-GRID.pdf.

Brauer, M., Freedman, G., Frostad, J., van Donkelaar, A., Martin, R.V., Dentener, F. *et al.* (2016). Ambient air pollution exposure estimation for the global burden of disease 2013. *Environmental Science & Technology* 50(1), 79–88. https://doi.org/10.1021/acs.est.5b03709.

Costello, M.J., Wilson, S. and Houlding, B. (2012). Predicting total global species richness using rates of species description and estimates of taxonomic effort. *Systematic Biology* 61(5). https://doi.org/10.1093/sysbio/syr080.

Doherty, T.S., Glen, A.S., Nimmo, D.G., Ritchie, E.G. and Dickman, C.R. (2016). Invasive predators and global biodiversity loss. *Proceedings of the National Academy of Sciences* 113(40), 11261-11265. https://doi.org/10.1073/pnas.1602480113.

Ehrenstein, V., Nielsen, H., Pedersen, A.B., Johnsen, S.P. and Pedersen, L. (2017). Clinical epidemiology in the era of big data: New opportunities, familiar challenges. *Clinical epidemiology* 9, 245-250. https://doi.org/10.2147/CLEP.S129779.

European Commission (2018). *GHSL - Global Human Settlement Layer*. http://ghsl.jrc.ec.europa.eu/.

Ferrant, G., Pesando, L.M. and Nowacka, K. (2014). *Unpaid Care Work: The Missing Link in the Analysis of Gender Gaps in Labour Outcomes*. Centro de Desarrollo de la OCDE. Paris: Organisation for Economic Co-operation and Development. https://www.oecd.org/dev/development-gender/Unpaid_care_work.pdf.

Food and Agriculture Organization of the United Nations (2018). *What is the GLRD?* http://www.fao.org/gender-landrights-database/background/en/.

Friel, S. and Marmot, M.G. (2011). Action on the social determinants of health and health inequities goes global. *Annual review of public health* 32, 225-236. https://doi.org/10.1146/annurev-publhealth-031210-101220.

GBD 2016 Causes of Death Collaborators (2017). Global, regional, and national age-sex specific mortality for 264 causes of death, 1980-2016: A systematic analysis for the Global Burden of Disease Study 2016. *Lancet Global Health Metrics* 390(10100), 1151-1210. https://doi.org/10.1016/S0140-6736(17)32152-9.

GBD 2016 DALYs and HALE Collaborators (2017). Global, regional, and national disability-adjusted life-years (DALYs) for 333 diseases and injuries and healthy life expectancy (HALE) for 195 countries and territories, 1990-2016: A systematic analysis for the Global Burden of Disease Study 2016. *Lancet Global Health Metrics* 390(10100), 1260-1344. https://doi.org/10.1016/S0140-6736(17)32130-X.

Global Biodiversity Information Facility (2018). *Free and open access to biodiversity data*. https://www.gbif.org/.

Hartwick, J.M. (1977). Intergenerational equity and the investment of rents from exhaustible resources. *American Economic Review* 67(5), 972-974. https://www.jstor.org/stable/1828079.

Hin, S. (2007). *Counting Romans*. Princeton/Stanford Working Papers in Classic Stanford University. https://www.princeton.edu/~pswpc/pdfs/hin/110703.pdf.

Hin, S. (2007). *Counting Romans*. Princeton/Stanford Working Papers in Classic Stanford University. https://www.princeton.edu/~pswpc/pdfs/hin/110703.pdf.

Hosein, F.N., Austin, N., Maharaj, S., Johnson, W., Rostant, L., Ramdass, A.C. *et al.* (2017). Utility of DNA barcoding to identify rare endemic vascular plant species in Trinidad. *Ecology and Evolution* 7(18), 7311–7333. https://doi.org/10.1002/ece3.3220.

Jerven, M. (2014). *Data for Development Assessment Paper: Benefits and Costs of the Data for Development Targets for the Post-2015 Development Agenda*. Copenhagen Consensus Center. https://www.copenhagenconsensus.com/sites/default/files/data_assessment_-_jerven.pdf.

Klompmaker, J.O., Hoek, G., Bloemsma, L.D., Gehring, U., Strak, M., Wijga, A.H. *et al.* (2018). Green space definition affects associations of green space with overweight and physical activity. *Environmental research* 160, 531-540. https://doi.org/10.1016/j.envres.2017.10.027.

Korda, R.J., Clements, M.S., Armstrong, B., K., Di Law, H., Guiver, T., Anderson, P.R. *et al.* (2017). Risk of cancer associated with residential exposure to asbestos insulation: A whole-population cohort study. *The Lancet Public Health* 2(11), e522-e528. https://doi.org/10.1016/S2468-2667(17)30192-5.

Lawford, R., Strauch, A., Toll, D., Fekete, B. and Cripe, D. (2013). Earth observations for global water security. *Current Opinion in Environmental Sustainability* 5(6), 633–643. https://doi.org/10.1016/j.cosust.2013.11.009.

Lemmens, R., Lungo, J., Georgiadou, Y. and Verplanke, J. (2017). Monitoring rural water points in Tanzania with mobile phones: The evolution of the SEMA App. *ISPRS International Journal of Geo-Information* 6(10), 316. https://doi.org/10.3390/ijgi6100316.

Liebenberg, L., Steventon, L., Benadie, K. and Minye, J. (1999). Rhino tracking with the CyberTracker field computer. *Pachyderm* 27, 59-61. https://www.iucn.org/backup_iucn/cmsdata.iucn.org/downloads/pachy27.pdf#page=60.

McCarter, J., Gavin, M.C., Baereleo, S. and Love, M. (2014). The challenges of maintaining indigenous ecological knowledge. *Ecology and Society* 19(3), 39. http://dx.doi.org/10.5751/ES-06741-190339.

McMichael, A.J., Butler, C.D. and Folke, C. (2003). New visions for addressing sustainability. *Science* 302(5652), 1919-1920. https://doi.org/10.1126/science.1090001.

McMichael, C., Barnett, J. and McMichael, A.J. (2012). An ill wind? climate change, migration, and health. *Environmental health perspectives* 120(5), 646–654. https://doi.org/10.1289/ehp.1104375.

Miranda, A.I., Ferreira, J., Silveira, C., Relvas, H., Duque, L., Roebeling, P. *et al.* (2016). A cost-efficiency and health benefit approach to improve urban air quality. *Science of the Total Environment* 569-570, 342-351. https://doi.org/10.1016/j.scitotenv.2016.06.102.

Mora, C., Tittensor, D.P., Adl, S., Simpson, A.G.B. and Worm, B. (2011). How many species are there on earth and in the ocean? *PLoS Biology* 9(8), e1001127. https://doi.org/10.1371/journal.pbio.1001127.

Organisation for Economic Co-operation and Development (2015). *Strengthening National Statistical Systems to Monitor Global Goals*. OECD and Post-2015 Reflections: Element 5, Paper 1. Paris. https://www.oecd.org/dac/POST-2015%20P21.pdf.

Overdevest, C., Orr, C.H. and Stepenuck, K. (2004). Volunteer stream monitoring and local participation in natural resource issues. *Human Ecology Review* 11(2), 177-185. https://www.humanecologyreview.org/pastissues/her112/overdevestorrstepenuck.pdf.

Pesaresi, M., Ehrlich, D., Kemper, T., Siragusa, A., Florczyk, A., Freire, S. *et al.* (2017). *Atlas of the Human Planet: Global Exposure to Natural Hazards*. European Union. http://publications.jrc.ec.europa.eu/repository/bitstream/JRC106292/atlas2017_online.pdf.

Prüss-Ustün, A., Wolf, J., Corvalán, C., Neville, T., Bos, R. and Neira, M. (2017). Diseases due to unhealthy environments: An updated estimate of the global burden of disease attributable to environmental determinants of health. *Journal of public health* 39(3), 464-475. https://doi.org/10.1093/pubmed/fdw085.

Salzmann, N., Huggel, C., Rohrer, M. and Stoffel, M. (2014). Data and knowledge gaps in glacier, snow and related runoff research – A climate change adaptation perspective. *Journal of Hydrology* 518, 225–234. https://doi.org/10.1016/j.jhydrol.2014.05.058.

Scheffers, B.R., Joppa, L.N., Pimm, S.L. and Laurance, W.F. (2012). What we know and don't know about Earth's missing biodiversity. *Trends in Ecology & Evolution* 27(9), 501-510. https://doi.org/10.1016/j.tree.2012.05.008.

Schulte-Uebbing, L., Hansen, G., Hernandez, A.M. and Winter, M. (2015). Chapter scientists in the IPCC AR5—experience and lessons learned. *Current Opinion in Environmental Sustainability* 14, 250-256. https://doi.org/10.1016/j.cosust.2015.06.012.

Seager, J. (2014). *Background and Methodology for Gender Global Environmental Outlook*. Nairobi: United Nations Environment Programme. https://web.unep.org/sites/default/files/ggeo/documents/GGEO_Multi-stakehoder_consultation_Background_document_final.pdf.

Siebenhüner, B. (2006). Administrator of global biodiversity: The secretariat of the convention on biological diversity. *Biodiversity and Conservation* 16(1), 259–274. https://doi.org/10.1007/s10531-006-9043-8.

Solow, R.M. (1974). Intergenerational equity and exhaustible resources. *The Review of Economic Studies* 41(5), 29-46. https://doi.org/10.2307/2296370.

Stevens, M., Vitos, M., Altenbuchner, J., Conquest, G., Lewis, J. and Haklay, M. (2014). Taking participatory citizen science to extremes. *IEEE Pervasive Computing* 13(2). https://doi.org/10.1109/MPRV.2014.37.

Stone, R. (1947). *Measurement of national income and the construction of social accounts: Report of the Sub-committee on National Income Statistics of the League of Nations Committee of Statistical Experts*. United Nations. http://www.worldcat.org/title/measurement-of-national-income-and-the-construction-of-social-accounts-report/oclc/610219052?referer=di&ht=edition.

Sustainable Development Solutions Network (2017). *Counting on the World: Building Modern Data Systems for Sustainable Development*. New York, NY. http://unsdsn.org/wp-content/uploads/2017/09/sdsn-trends-counting-on-the-world-1.pdf.

Sutherland, W.J., Gardner, T.A., Haider, L.J. and Dicks, L.V. (2014). How can local and traditional knowledge be effectively incorporated into international assessments? *Oryx* 48(1), 1-2. https://doi.org/10.1017/S0030605313001543.

Tengö, M., Malmer, P., Brondizio, E., Elmqvist, T. and Spierenburg, M. (2013). *The Multiple Evidence Base as a Framework for Connecting Diverse Knowledge Systems in the IPBES*. Stockholm: Stockholm Resilience Centre. http://www.stockholmresilience.org/download/18.416c425f13e06f977b11277/Multiple+Evidence+Base+for+IPBES+2013-06-05.pdf.

United Nations (1993). *Handbook of National Accounting: Integrated Environmental and Economic Accounting*. New York, NY. http://unstats.un.org/unsd/publication/SeriesF/SeriesF_61E.pdf.

United Nations (2012). *Addressing Inequalities: The Heart of the Post-2015 Agenda and the Future We Want for All*. http://www.un.org/millenniumgoals/pdf/Think%20Pieces/10_inequalities.pdf.

United Nations (2014). *System of Environmental-Economic Accounting 2012: Central Framework*. New York, NY. https://unstats.un.org/unsd/envaccounting/seearev/seea_cf_final_en.pdf.

United Nations, General Assembly (2015). *70/1. Transforming Our World: The 2030 Agenda for Sustainable Development. Resolution adopted by the General Assembly on 25 September 2015*. 21 October. A/RES/70/1. http://www.un.org/ga/search/view_doc.asp?symbol=A/RES/70/1&Lang=E.

United Nations (2015a). *The Millennium Development Goals Report*. New York, NY. http://www.un.org/millenniumgoals/2015_MDG_Report/pdf/MDG%202015%20rev%20(July%201).pdf.

United Nations (2015b). *World's Women 2015: Trends and Statistics*. https://unstats.un.org/unsd/gender/downloads/worldswomen2015_report.pdf.

United Nations (2016a). *World Economic and Social Survey 2014/2015*. New York, NY. http://www.un.org/en/development/desa/policy/wess/wess_archive/2015wess_full_en.pdf.

United Nations (2016b). *Sustainable Development Goals Report 2016*. New York, NY. http://unstats.un.org/sdgs/report/2016/.

United Nations (2016c). Geospatial information and earth observations: Supporting official statistics in monitoring the SDGs. *47th Session of the United Nations Statistical Commission Statistical-Geospatial Integration Forum*. New York, NY, 7 March. United Nations Statistical Commission https://www.fgdc.gov/organization/working-groups-subcommittees/unggim-wg/unggim-meeting-march-2016.pdf.

United Nations (2017a). Report of the Inter-Secretariat Working Group on National Accounts: Supplement to the Report of the Inter-Secretariat Working Group on National Accounts. https://unstats.un.org/unsd/statcom/48th-session/documents/BG-NationalAccounts-Supplement-E.pdf.

United Nations (2017b). *Report of the Inter-agency and Expert Group on Sustainable Development Goal Indicators*. https://unstats.un.org/unsd/statcom/48th-session/documents/2017-2-IAEG-SDGs-E.pdf.

United Nations (2018). *Tier Classification for Global SDG Indicators*. https://unstats.un.org/sdgs/files/Tier%20Classification%20of%20SDG%20Indicators_11%20May%202018_web.pdf.

United Nations, European Commission, International Monetary Fund, Organisation for Economic Co-operation and Development and World Bank (2003). *Handbook of National Accounting: Integrated Environmental and Economic Accounting 2003*. http://unstats.un.org/unsd/EconStatKB/Attachment60.aspx?AttachmentType=1.

United Nations Educational Scientific and Cultural Organization (2014). *Water and gender*. http://www.unesco.org/new/en/natural-sciences/environment/water/wwap/water-and-gender/.

United Nations Educational Scientific and Cultural Organization (2015). *UNESCO Science Report: Towards 2030*. Institutions and Economies. Paris: United Nations Educational, Scientific and Cultural Organization. http://unesdoc.unesco.org/images/0023/002354/235406e.pdf.

United Nations Environment Programme (2012). *Global Environment Outlook-5: Environment for the Future We Want*. Nairobi. https://wedocs.unep.org/bitstream/handle/20.500.11822/8021/GEO5_report_full_en.pdf?isAllowed=y&sequence=5.

United Nations Environment Programme (2016). *Global Gender and Environment Outlook*. Nairobi. https://wedocs.unep.org/bitstream/handle/20.500.11822/14764/Gender_and_environment_outlook_HIGH_res.pdf?sequence=1&isAllowed=y.

United Nations Environment Programme (2017). *Guidelines for Conducting Integrated Environmental Assessment* Nairobi. https://wedocs.unep.org/bitstream/handle/20.500.11822/16775/IEA_Guidelines_Living_Document_v2.pdf?sequence=1&isAllowed=y.

United States National Science Board (2016). Research and development: National trends and international comparisons. In Science *and Engineering Indicators*. Arlington, VA: National Science Foundation. chapter 4. https://www.nsf.gov/statistics/2016/nsb20161/uploads/1/7/chapter-4.pdf.

United States National Science Foundation (2017). *Federal funds for research and development*. http://www.nsf.gov/statistics/fedfunds/.

Vanoli, A. (2005). *National Accounting at the beginning of the 21st century: Where from? Where to?* European Commission. https://ec.europa.eu/eurostat/cros/system/files/p1-national_accounting_at_the_beginning_of_the_21st_century.pdf.

White, R. (ed.) (2017). *Transnational Environmental Crime*. London: Routledge. https://www.taylorfrancis.com/books/9781409447856.

World Bank (2002). *Building Statistical Capacity to Monitor Development Progress*. Washington, D.C. http://siteresources.worldbank.org/SCBINTRANET/Resources/Building_Statistical_Capacity_to_Monitor_Development_Progress.pdf.

World Health Organization (2018). *Global Health Observatory (GHO) Data: Mortality and Global Health Estimates*. http://www.who.int/gho/mortality_burden_disease/en/.

Yesson, C., Brewer, P.W., Sutton, T., Caithness, N., Pahwa, J.S., Burgess, M. *et al.* (2007). How global is the global biodiversity information facility? *PLoS ONE* 2(11), e1124. https://doi.org/10.1371/journal.pone.0001124.

Chapter 4

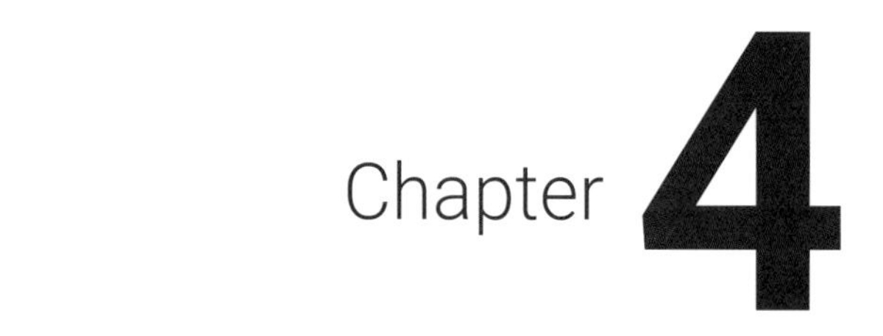

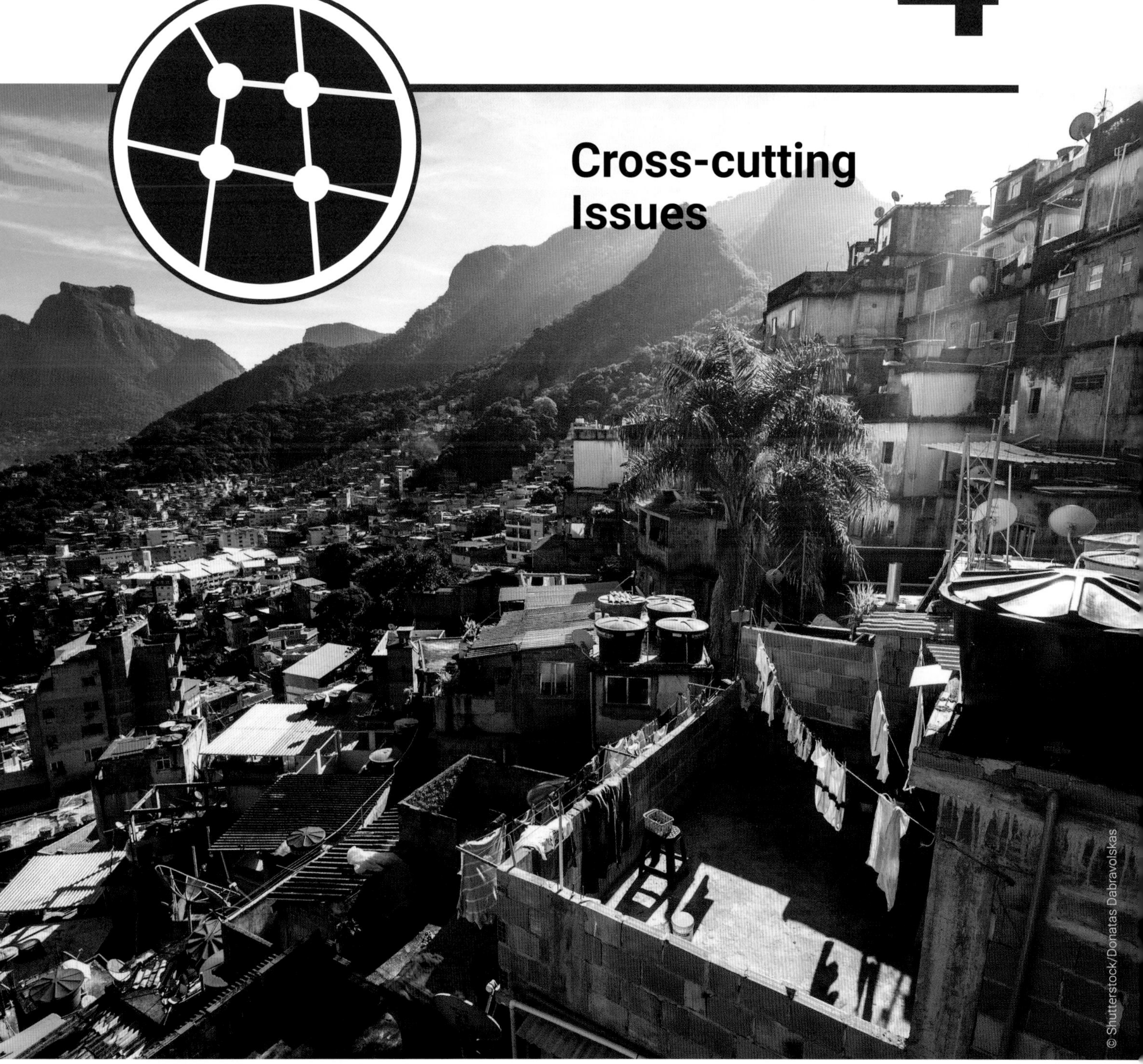

Cross-cutting Issues

© Shutterstock/Donatas Dabravolskas

Coordinating Lead Authors: Shanna N. McClain (Environmental Law Institute), Catherine P. McMullen (Stockholm Environment Institute – Asia)

Lead Authors: Babatunde Joseph Abiodun (University of Cape Town), Giovanna Armiento (Italian National Agency for New Technologies, Energy and Economic Sustainable Development), Rob Bailey (Chatham House –The Royal Institute of International Affairs), Rajasekhar Balasubramanian (National University of Singapore), Ricardo Barra (University of Concepción), Kathryn Jennifer Bowen (Australian National University), John Crump (Grid-Arendal), Irene Dankelman (Radboud University), Kari DePryck (Sciences Po University), Riyanti Djalante (United Nations University), Monica Dutta (The Energy Research Institute), Francois Gemenne (Université de Liège), Linda Godfrey (Council of Scientific & Industrial Research, South Africa), James Grellier (University of Exeter), Maha Halalsheh (University of Jordan), Fintan Hurley (Institute of Occupational Medicine), Maria Jesus Iraola (University of Uruguay), Richard King (Chatham House – The Royal Institute of International Affairs), Andrei Kirilenko (University of Florida), Shi Lei (Tsinghua University), Peter Lemke (Alfred-Wegener-Institut), Daniela Liggett (University of Canterbury), Robyn Lucas (National Centre for Epidemiology and Population Health, The Australian National University), Oswaldo dos Santos Lucon (Sao Paulo State Environment Secretariat), Katrina Lyne (James Cook University), Diego Martino (AAE Asesoramiento Ambiental Estratégico and ORT University), Ritu Mathur (TERI), Emma Gaalaas Mullaney (Bucknell University), Leisa N. Perch (SAEDI Consulting), Marco Rieckmann (University of Vechta), Fülöp Sándor (National University of Public Services), Atilio Savino (ARS), Heinz Schandl (Commonwealth Scientific and Industrial Research Organisation [CSIRO]), Joeri Scholtens (University of Amsterdam), Patricia Nayna Schwerdtle (Monash University), Joni Seager (Bentley University), Frank Thomalla (Stockholm Environment Institute – Asia), Laura Wellesley (Chatham House – The Royal Institute of International Affairs), Caradee Y. Wright (Medical Research Council of South Africa), Dimitri Alexis Zenghelis (London School of Economics), Caroline Zickgraf (Université de Liège)

Executive summary

Environmental pollution is still a major source of damage to the health of the planet (*well established*), **human health** (*well established*), **equity** (*well established*) **and economic sustainability** (*established but incomplete*). The risks, however, are systemic and wide-ranging, including climate change, ecosystem and biodiversity loss, wildlife damage, systemic change and other major issues. Sustainable development is possible if 'Healthy Planet, Healthy People' becomes central to our understanding of genuine progress. Solutions need to be both evidence-based and systemic, tackling sources of pollution, aiming for co-benefits and checking for unintended consequences. {4.2.1}

The number of people affected by both slow and sudden-onset environmental disasters is increasing due to compounding effects of multiple and interacting drivers (*well established*)**.** These drivers include climate change and environmental degradation, poverty and social inequality, demographic change and settlement patterns, increasing population density in urban areas, unplanned urbanization, unsustainable use of natural resources, weak institutional arrangements, and policies which do not consider disaster risk. Disasters undermine human security and well-being, resulting in loss and damage to ecosystems, property, infrastructure, livelihoods, economies and places of cultural significance while forcing millions of people each year to flee their homes. {4.2.2}

Gender equality and women's empowerment are multipliers of sustainability (*well established*). Ensuring gender-equal representation in environmental assessments, resource management and environmental decision-making ensures that diverse experiences and knowledge systems about the environment are integrated and ecosystem conservation and sustainable use of natural resources are enhanced. In this way, increasing gender equality and women's empowerment contribute to achieving the environmental dimension of the Sustainable Development Goals (SDGs). {4.2.3}

Significant progress has been made around the world with implementing education for sustainable development (ESD) in all educational sectors (*well established*). However, upscaling of ESD is still needed in order to include it as a core element in the structures of educational systems globally. In this way, education will contribute to achieving the SDGs. Policies are needed that eliminate economic and gender barriers to accessing education. {4.2.4}

Urban footprints have transboundary ramifications (*well established*). The magnitude, scale and scope of contemporary urbanization is now so large as to be affecting global resource flows and planetary cycles. At the same time, the current urbanization process and its prospects represent not only a challenge, they also represent an opportunity to improve human well-being with potentially decreasing environmental impacts per capita and per unit of production. {4.2.5}

Climate change is one of the most pressing issues affecting natural (*well established*) **and human systems** (*established but incomplete*) **(SDG 13).** The evidence of current global climate change is unequivocal. Worldwide, the average surface temperature has gone up by about 1.0°C since the 1850-1879 period; if the current rate of greenhouse gas emission persists by the 2040s warming will exceed 1.5°C. Eight of the ten warmest years on record have occurred within the past ten years. The impacts of climate change are much wider than temperature increase, affecting water availability, ecosystems, energy demand and production, transportation and other sectors. Shifts in weather patterns, extreme events (e.g. heat waves and droughts) and environmental disruptions (e.g. crop failures) result in greater risks to human health and well-being, and livelihoods, especially among the poorest and most vulnerable groups. {4.3.1}

Current observations and climate model experiments indicate that polar surface temperatures increases exceed twice the mean global temperature rise (*well established*). This amplified warming has cascading effects on other components of the polar-climate system, with sea ice in the Arctic retreating; permafrost thawing; snow cover extent decreasing; ice sheets decaying; and ice sheets, ice shelves and mountain glaciers continuing to lose mass, contributing substantially to sea level rise. {4.3.2}

Modern society is living in the most chemical-intensive era in human history, the pace of production of new chemicals largely surpasses the capacity to fully assess their potential adverse impacts on human health and ecosystems (*well established*). The risks to human health and ecosystem integrity produced by the combined effects of certain currently used chemicals, including in products, given their occurrence in the environment as a complex mixture, even in remote areas, are poorly understood and need further evaluation. Regulations, assessment and monitoring as well as industry and consumer responsibility, in informing and substituting the use of chemicals of global concern with safer alternatives are needed. Sustainable and green chemistry is aiming to achieve the sustainable design, production, use and disposal of chemicals throughout their life cycle, while taking into account the three dimensions of sustainable development. {4.3.3}

The disposal and discharge of waste to receiving environments is negatively impacting ecosystem and human health (*well established*). Issues of global concern include: increasing distribution and impact of marine litter, in particular plastic, in the world's oceans; the loss and wastage of approximately one-third of the food produced for human consumption; and increased trafficking of waste from developed to developing countries. While developed countries transition to reduced waste generation and greater resource efficiency, developing countries grapple with basic waste management challenges, including uncontrolled dumping, open burning, and inadequate access to waste services. {4.3.4}

The use of resources and the environmental impacts of resource extraction and use are growing despite a large potential for resource efficiency through circular economy and sustainable consumption and production approaches (*well established*)**.** Global resource use has accelerated since the year 2000 and reached 90 billion tons in 2017; high-income countries consume ten times the amount of resources that low-income countries consume; resource efficiency has been stagnant and the environmental impacts of resource use have been growing at a rate commensurate with overall resource use; there are many economically attractive opportunities for resource efficiency in the short term; in the medium and long term resource efficiency creates better economic outcomes compared with business as usual; there are considerable co-benefits of resource efficiency for climate mitigation.{4.4.1}

Coupled with efficiency improvements, transition to low-carbon energy sources has been accelerating globally over the last decade but it is still not sufficient to achieve the 2°C target of the Paris Agreement (*well established*), warranting bolder action in terms of technology innovation. Meanwhile the access of billions of poorer people to electricity and other modern energy services remains a challenge. {4.4.2}

The food system is increasing local to global pressures on ecosystems and the climate (*well established*). Farming is the most expansive human activity in the world and the principal user of fresh water. Food production is the main driver of biodiversity loss, a major polluter of air, fresh water and seawater, a leading source of soil degradation, and a significant source of greenhouse gas emissions. Changing consumption patterns are both increasing these pressures and presenting new food security challenges resulting in malnourishment, including overnourishment, as well as undernourishment. Climate change, natural resource constraints, and demographic trends suggest that the challenge of producing and distributing nourishing and sustainable food for all continues to escalate and will necessitate significant changes in food production and consumption. {4.4.3}

4.1 Introduction

As understanding of the interdependence between a healthy planet and healthy people becomes more developed, complex issues that thread through systems and societies gain new importance. Beyond the traditional Global Environment Outlook (GEO) themes addressing air, biodiversity, oceans, land and fresh water, this GEO-6 assessment addresses cross-cutting issues worthy of further examination. Using a systems approach, these cross-cutting issues offer entry points allowing another dimension for analysing GEO-6 themes as well as understanding the network of interconnections throughout earth and human systems. These cross-cutting issues are grouped according to shared characteristics: health, environmental disasters, gender, education and urbanization are grouped as 'people and livelihoods'; climate change, polar and mountain regions, chemicals and waste and wastewater are grouped as 'changing environments'; and resource use, energy and food systems are considered as 'resources and materials'. While each issue provides useful entry points into GEO-6 themes, it is important to discuss the state of the environment and policy context for each one.

As the deficiencies in our traditional issues-based approach to environmental assessment limit our ability to consider truly transformative pathways, cross-cutting and more integrated approaches are essential and must ultimately displace those based on single-issue analyses. Therefore, this chapter initiates a new approach in the GEO assessment process through an analysis of selected cross-cutting issues that illustrate the pressing need for more integrated and transformative policy responses. Given the global scale of the GEO-6 assessment, the chapter can address only a few cross-cutting issues, threads and influences among the myriad possible combinations. The cross-cutting issues selected for this assessment are chosen because of their close alignment with the SDGs and the fact that the scope and influence of these different issues vary dramatically over time, scale and region.

Given the obvious intersections among these cross-cutting issues, a number of emerging issues arose in regard to taking a 'Healthy Planet, Healthy People' perspective. This chapter addresses the health of the environment, the consequences for human health from pollution of all kinds, climate change impacts, environmental disasters and unsustainable consumption of natural resources, as well as the longer-term health effects of rapid and intense changes to lives, livelihoods and the environment, which require a wider focus.

The policy implications of addressing these cross-cutting issues converge on four particular human and economic systems that could accomplish the required transformation into a healthy planet supporting healthy people. Contributions from all 12-issue teams, including insights from at least 50 issue specialists from around the world, developed into system studies on climate change adaptation, sustainable food, clean energy systems and a more circular economy. The products of these collaborative efforts are presented in Chapter 17 (Part B) of this report.

4.2 People and livelihoods

4.2.1 Health

The public health community has two long-established ways of reflecting the complex web of relationships between healthy planet and healthy people that is central to GEO-6. One way is to define human health inclusively as "a state of complete physical, mental, and social well-being and not merely the absence of disease or infirmity" (World Health Organization [WHO] 1948), and then use 'well-being' (Glatzer *et al.* 2015; Maggino 2015) together with 'health' to incorporate the psychological, emotional and social dimensions. The second way focuses on the determinants of health: it recognizes that human health is mediated by multiple factors in the natural, social and built environments, including our senses of equity and safety as well as equitable access to environmental resources and human contact with nature (WHO 2008). So, while human health is the direct focus of Sustainable Development Goal (SDG) 3, this complexity links health and well-being directly and indirectly to all the SDGs (e.g. Section 20.3.1) and to issues throughout GEO-6, including the thematic chapters and other cross-cutting topics.

Buse *et al.* (2018) identify six frameworks developed from late 20th century onward to show and deal with this complexity: political ecology of health, environmental justice, Ecohealth, One Health, Ecological Public Health, and Planetary Health. These frameworks represent a shift towards a more sophisticated understanding of the implicit, complex and systemic links between human health and well-being and the natural environment. They build on an older tradition (from the mid-19th century), of 'occupational and environmental health'. This is narrower (e.g. Ayres *et al.* eds. 2010) than the more recent frameworks in two ways. First, health is often interpreted as risk of death and disease or illness, referred to as mortality and morbidity, rather than as the more holistic health and well-being. Second, it focuses on the physical, chemical and biological spheres, rather than on the social as well as determinants of health.

Within this traditional but narrow framework of pollution and disease, this report shows numerous examples of how health is damaged by environmental changes including air, water and land pollution; heat waves, flooding and other weather extremes; toxic chemicals; pathogens; ultraviolet and other radiation; desertification; reduced biodiversity; melting of polar ice; and destruction of coral reefs. Overall, "natural systems are being degraded to an extent unprecedented in human history" (Whitmee *et al.* 2015, p. 1,974) and the damage to human health is already severe. For example, the Lancet Commission on pollution and health (Landrigan *et al.* 2017) estimated that diseases caused by environmental pollution resulted in 9 million premature deaths in 2015. The biggest effects are from exposure to outdoor and indoor air pollution, which together caused 6.4 million deaths in 2015 (Cohen *et al.* 2017). More generally, the incidence of non-communicable diseases is on the rise globally and will continue to be affected by the state of the environment in relation to pollution, diet and physical (in)activity. However, human health depends on much more than a healthy planet.

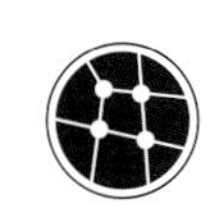

Similarly, Prüss-Ustün *et al.* (2016) estimated that in 2012 modifiable environmental health risks caused 12.6 million deaths globally, representing 23 per cent (13-34 per cent, 95 per cent confidence interval [CI]) of all deaths. These are big impacts, but nevertheless they show that even if it were desirable and feasible to attain a healthy, sustainable planet without addressing socioeconomic issues and associated determinants of health, it would still leave humanity far short of the goal of 'healthy people' (see also Section 20.3.1).

Environmental pressures and their impacts on health and well-being are not equitably distributed. They fall especially on groups that are already vulnerable or disadvantaged, such as young people and elders, women, poor people, those with chronic health conditions, indigenous peoples and people targeted by racism (Solomon *et al.* 2016; Landrigan *et al.* 2017, pp. 27-31). For example, unsafe food and water can cause diarrhoeal diseases (Mills and Cumming 2016), with children under five in sub-Saharan Africa and South Asia being the most affected (Walker *et al.* 2013; Prüss-Ustün *et al.* 2014) (SDG 3 notes that four out of every five deaths of children under age five occur in these regions).

New challenges (which may be countered by relevant, sound, scientific research) include the growth of resistance of pathogens to antibiotics (antimicrobial resistance) that have been, and are, used heavily in agriculture and aquaculture (Finley *et al.* 2013; Wallinga, Rayner and Lang 2015); the multitude of industrial chemicals (though not all are widely used) that challenges our ability to meaningfully test their potential impacts on environmental and human health, including for future generations (The American Society of Human Genetics *et al.* 2011; Sharma *et al.* 2014; Landrigan *et al.* 2017); the cumulative effect (both social and environmental) of multiple exposures, including those of chemical mixtures (Solomon *et al.* 2016); emergence and re-emergence of infections originating in birds and animals (Ostfeld 2009; Lindahl and Grace 2015; Hassell *et al.* 2017); increased physical inactivity associated with new technology for work and leisure; and others including some whose effects on human health are currently unclear (e.g. the presence of microplastics in fish and marine biological resources).

Solutions to the degradation of natural systems, including the management of environmental pollution at its sources, should take account of the complex interactions between planet and health (Whitmee *et al.* 2015) and consider environment-health as a complex system, seeking co-benefits (Haines 2017), and where practicable avoiding trade-offs or win-lose situations or unintended adverse consequences (von Schneidemesser *et al.* 2015). There are now many examples of health co-benefits, especially of greenhouse gas reductions (Chang *et al.* 2017; Quam *et al.* 2017; Deng *et al.* 2018). For example, the unfolding transition to cleaner energy improves air quality and slows climate change effects, each of which greatly benefits health and well-being (Smith *et al.* 2014a; Haines 2017; see also Section 4.2.1). Active travel, such as walking and bicycling, can have multiple benefits for health and well-being (Saunders *et al.* 2013; Smith *et al.* 2014a); however, benefits will vary with (for example) climate and pollution levels. Reducing red meat intake per capita where there is high consumption, especially of processed meat, will improve human health (McMichael *et al.* 2007; Wolk 2017), while reducing pressure on biodiversity and greenhouse gas emissions, including methane. The benefits to human health and well-being of access to safe and biodiverse natural environments, green and blue spaces, are being recognized (Coutts and Hahn 2015; Wolf and Robbins 2015; Wall, Derham and O'Mahony eds. 2016; Grellier *et al.* 2017).

Rigorous incorporation and integration of human health considerations within health-determining sectoral plans (e.g. agriculture, water, disaster management, urban design) can support responses that address human health impacts, with a focus on prevention activities. Initiatives to reduce environmental risks, focusing on benefits across sectors, are consistent with the World Health Organization's (WHO) call for Health in All Policies (WHO 2014) and the development of tools for integrated environmental and health assessment (Fehr *et al.* 2016). The health sector must rapidly strengthen the way that it articulates messages on human health and emphasize that the majority of environmental pressures will ultimately have human health impacts.

More fundamental changes may be needed, for example "the redefinition of prosperity to focus on the enhancement of quality of life and delivery of improved health for all, together with respect for the integrity of natural systems" (Whitmee *et al.* 2015). This view resonates with intentions to keep the GEO-6 goal of Healthy Planet, Healthy People central to our understanding of genuine progress.

4.2.2 Environmental disasters

Hazards become disasters when they disrupt human communities. Therefore, the consequences of these disasters are as much a part of where and how people live as the presence of the hazard itself (Sun 2016, p. 30). This includes anthropogenic effects on the climate, but also disasters directly caused by human activities such as oil spills, accidents at nuclear power stations or other hazardous installations, and even earthquakes triggered by fracking and the building of large dams (Legere 2016). Sudden-onset disasters, such as earthquakes, tsunamis, landslides, flash floods and severe storms, are distinguished from slow-onset events, experienced as drought, desertification, sea level rise and coastal erosion. Slow-onset events comprise as much as 90 per cent of disasters worldwide and threaten growth, development and livelihoods (Lucard, Jaquemet and Carpentier 2011). Development and disaster risk are closely linked; decisions regarding the management of natural resources and development pathways determine patterns of vulnerability and exposure to a range of environmental hazards. Disasters, in turn, can set back development gains by years or even decades, at immense social and economic cost. Over the long or short term, these decisions and their management can act as drivers of migration and displacement (United Kingdom Government Office for Science 2011). They can also affect peace and security (Schilling *et al.* 2017).

Environmental disasters are affecting an increasing number of people globally and taking an ever-larger toll on societies and economies, particularly in the poorest communities and countries. Between 2005 and 2015, they affected more than 3 billion people (Centre for Research on the Epidemiology of Disasters 2017). This is partly due to an increase in frequency and magnitude of climate and hydrometeorological hazards

Figure 4.1: The economic and human impact of disasters in the last ten years

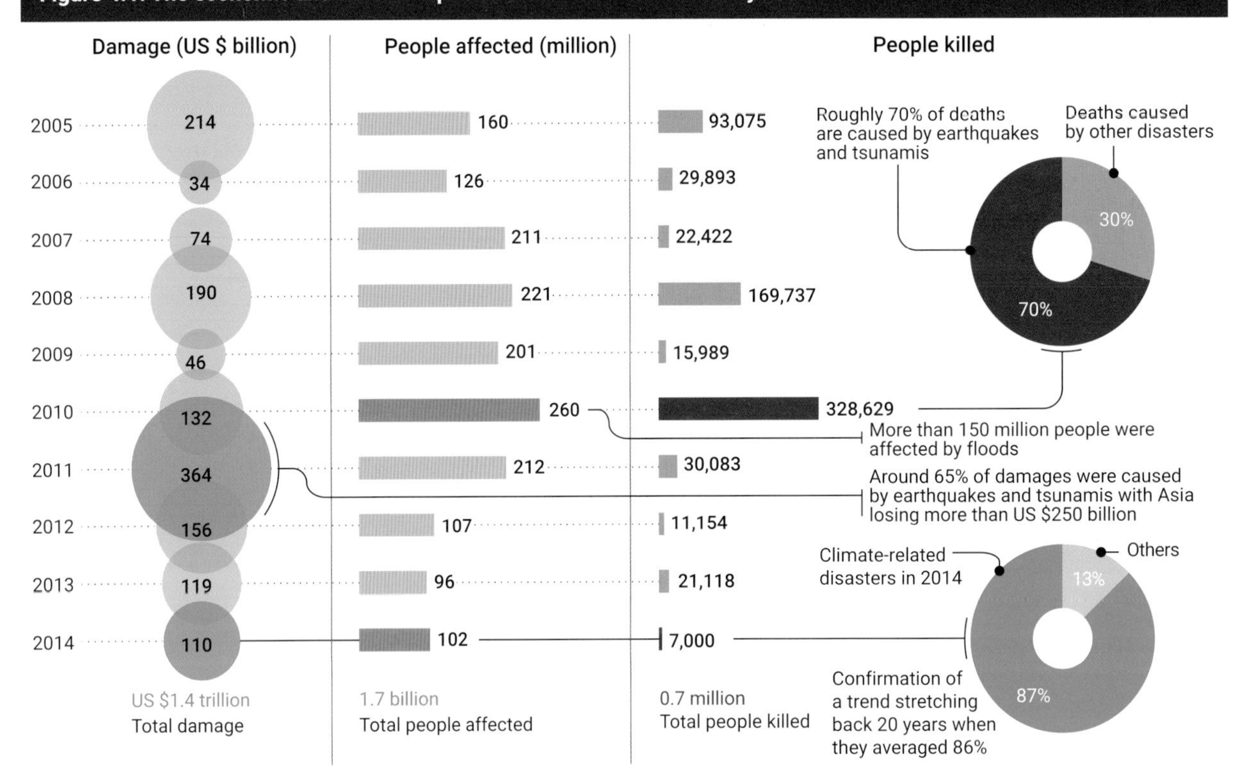

Source: United Nations Office for Disaster Risk Reduction (UNISDR) 2014

such as tropical cyclones, fires and floods. However, social and economic processes that increase exposure to hazards by placing more people, infrastructure and economic activities in harm's way significantly escalate disaster risk. For example, migration away from rural drought to overcrowded, poorly planned, coastal megacities in flood-prone zones can increase mortality, displacement, health and disaster risks in urban areas.

In some cases, disasters result from the combined effect of several interacting hazard events. The 2011 Tohoku disaster in Japan exemplified such a case when a sequence of cascading events occurred, including an earthquake, a tsunami and a nuclear power plant accident, all contributing to 15,893 casualties. The disaster forced more than 350,000 people into protracted displacement (i.e. displacement of more than one year) and cost an estimated US$ 210 billion in direct damage. Disasters also disproportionately affect some of the most vulnerable populations; 54 per cent of fatalities from the Tohoku disaster were women and girls, and 56 per cent were above age 65 (Leoni 2012). To date, it remains the most expensive environmental disaster in history (Ranghiere and Ishiwatari eds. 2014, pp. 2, 269, 284).

The consequences of disasters are far-reaching and long lasting. In 2016 alone, 24.2 million people in 118 countries became newly internally displaced by sudden-onset disasters (Internal Displacement Monitoring Centre [IDMC] 2017, p. 10). They outnumbered those who were newly displaced by conflict and violence three to one (IDMC 2017). Precipitation shocks, droughts, floods and storms in Philippines, for example, correspond with significant intensifications of conflict (Eastin 2016, p. 12). The Protection Agenda of the Nansen Initiative, endorsed by 109 governments in 2015, is a key instrument to foster the protection of the rights of those displaced across borders by disasters. The Platform on Disaster Displacement, established in 2016, is tasked with supervising implementation of the Agenda and following up on the work carried out by the Nansen Initiative between 2012 and 2015 (Disaster Displacement 2017). In many cases, drivers of displacement are difficult to disentangle from other destabilizing factors. The African Union's Kampala Convention, a legally binding protection instrument shielding those displaced by conflict, violence and human rights abuses alongside disasters, is an important step in recognizing these interactions (African Union 2009).

Learning from past disasters and shifting from a culture of disaster response to one of prevention, preparedness and resilience is imperative. While initiatives such as disaster response and recovery strategies have been formulated in many countries following disaster events, the number of countries that have incorporated prevention, mitigation and preparedness as part of a comprehensive disaster risk reduction strategy remains quite low (Ranghiere and Ishiwatari eds. 2014, p. xv). The Sendai Framework for Disaster Risk Reduction 2015-2030 (UNISDR 2015) represents a new opportunity to further improve disaster risk reduction efforts. Improvements can be achieved by mobilizing and prioritizing investments, enhancing policy and institutional coherence, promoting innovation and technological development, increasing collaboration and cooperation, and mainstreaming disaster risk reduction in development and climate change adaptation efforts.

4.2.3 Gender

A gender approach redefines the environmental situation through the lens of social relationships and their reflection in human-environment interactions, instead of defining the state of the environment primarily in its physical or ecological forms. Gender analysis reveals that while systemic environmental problems typically manifest in physical landscapes and ecosystems, the state of the environment can only be explained by examining social, cultural and economic systems and arrangements. Those structures are 'gendered': they are shaped by socially constructed roles and relationships between women and men. For example, in *The State of Food and Agriculture 2010-11* paragraph 4.3.3 on 'Food systems' the role of women in agriculture is underlined (Food and Agriculture Organization of the United Nations [FAO] 2011).

Figure 4.2 shows that women's and girls' responsibilities in collecting water is much larger than that of men and boys (United Nations Entity for Gender Equality and the Empowerment of Women [UN-Women] UN Women 2015; Sagrario and Willoughby 2016; United Nations Environment Programme [UNEP] 2016a; WHO 2017).

Assessments of the economic value of environment-related sectors are often seriously distorted because women's contributions are overlooked (see also Section 4.1.3). For example, the economic work of women in fisheries continues to be undercounted, partly because fishing is often defined only as catching fish at sea with specialized equipment. This type of fishing is highly masculinized (Harper *et al.* 2013; UNEP 2016a; Harper *et al.* 2017). Women's tasks in the fishing sector focus on coastal fishing, fish processing and trade, and are often neglected (Lambeth *et al.* 2014). Throughout this publication, some other examples of the gender-environment relationship are included.

Figure 4.2: Percentage distribution of the water collection burden across 61 countries

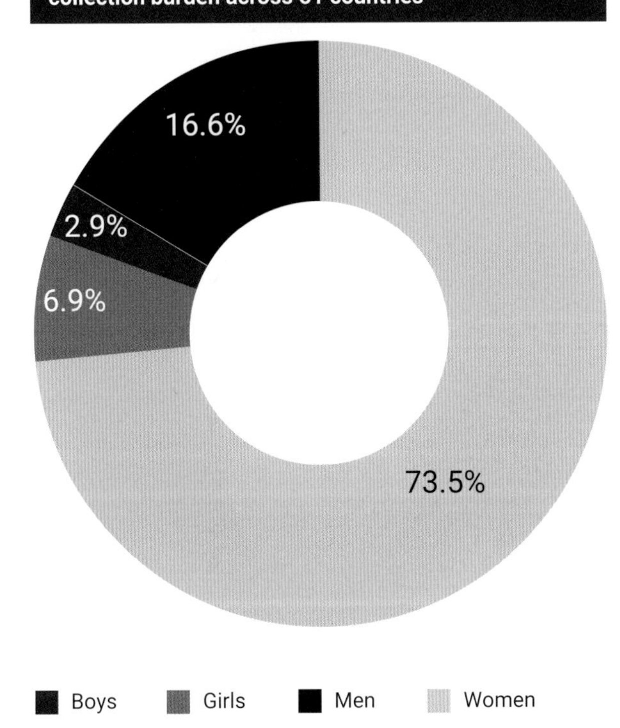

Source: UNICEF and WHO (2017, p. 30).

The scholarly and practitioner field of gender and environment has been developing since the 1980s and is now a large and robust domain of analysis and assessment (Skinner 2011; Aguilar, Granat and Owren 2015). Early directions in this field focused on identifying the gender-differentiated impacts of environmental change (Dankelman and Davidson 1988). Now, an emerging focus is examining the ways in which the drivers of environmental change are also gendered, rooted in socially constructed norms of masculinity and femininity, including in our economies, sciences and technologies (Harcourt and Nelson eds. 2015; UNEP 2016a). Revealing the gendered dimensions of environmental dynamics illuminates new aspects of environmental states and trends, as well as pointing out pathways for transformations and policy solutions that are sustainable. The Global Gender and Environment Outlook, which elaborates on the importance of gender in most environmental areas, provides the first comprehensive global assessment of the gender-environment nexus and offers a channel for gender analysis in GEO-6 (UNEP 2016a). Applying a gender lens to environmental assessment also creates awareness of the relevance of additional social dimensions and intersections in environmental use and management, such as differentiation by class, race or ethnicity, caste and age (Harris 2011).

Recent studies recognize the diverse roles of men and women in collecting forest products and their related diverse knowledge systems (Sunderland *et al.* 2014; Chiwona-Karltun *et al.* 2017). Evidence from studies on community forest management point to the understanding that women's participation in environmental assessment and resource management can enhance ecosystem conservation and sustainable use of natural resources (Agarwal 2010; Agarwal 2015).

Other evidence suggests that when women are accorded equal voice in environmental decision-making, public resources are more likely to be directed towards human development priorities and investments (Chattopadhyay and Duflo 2004; UN-Women 2014). Women's enhanced access to and control over productive agricultural resources helps create food security and sustainable livelihoods (FAO 2011; UN-Women 2014). The use of gender budgeting is another important approach to promote gender-responsive financing. The SDG framework reveals that sustainable development will not evolve, nor will environmental policies and initiatives be effective, if gender equality and women's empowerment are not enhanced (United Nations 2015a). Environmental sustainability and justice contribute significantly to SDG 5: achieving gender equality and empowering all women and girls, and to the gender targets of SDGs 1, 4, 8 and 10 (Agarwal 2010; UNEP *et al.* 2013; Agarwal 2015; United Nations 2015b; Dankelman 2016; UNEP 2016a). While gender equality can be tacitly read in all the other SDG goals, there are almost no explicit gender targets and indicators included in the environment-related SDGs.

Bringing gender perspectives to bear on environmental frameworks is not a matter of simply adding 'women' into environmental analyses. Approaching the environment through a gender lens means new and different questions in environmental assessment, emphasizing different dimensions of human-environment relationships and requiring gender-

responsive methodological tools and approaches, as well as gender-disaggregated data (Patt, Dazé and Suarez 2009; Doss 2014; Seager 2014; Bradshaw and Fordham 2015; Harcourt and Nelson eds. 2015; Jerneck 2018). Given the difficult state of the environment, the persistence of drivers of environmental change, and the severity of societal and ecological consequences that societies face, a gender-integrative approach is a precondition for more effective and transformative environmental policies and interventions.

4.2.4 Education

Education for Sustainable Development (ESD), a key area of education, reaching gender equality, developing healthier and more sustainable lifestyles, and creating more peaceful societies. However, this requires access to education for all and a high quality of education (United Nations Development Programme[UNDP] 2016; United Nations Educational, Scientific and Cultural Organization [UNESCO] 2017a). Despite all efforts to provide all children worldwide with access to education, this is still not a reality for all children. "Worldwide, 91 per cent of primary-school-age children were enrolled in school in 2015" (UNICEF 2018). "In 2015, there were 264 million primary and secondary age children and youth out of school: 61 million children of primary school age (9% of the age group), 62 million adolescents of lower secondary school age (16%), and 141 million youth of upper secondary school age (37%)" (UNESCO 2017a, p. 118). Also gender equality is still a major challenge: "While there is gender parity in education participation, global averages mask gaps between countries: only 66% have achieved gender parity in primary education, 45% in lower secondary and 25% in upper secondary" (UNESCO 2017a, p. 182). Education for Sustainable Development, a key area of education, aims to enable individuals to contribute to fostering sustainable development. Instead of promoting certain behaviours and ways of thinking (instrumental approach), an emancipatory concept of ESD concentrates in particular on the critical reflection on expert opinions, testing possibilities of sustainable development and exploring the trade-offs of a sustainable lifestyle (Wals 2015; UNESCO 2017b; Rieckmann 2018). It aims to empower individuals to act responsibly in order to contribute to the creation of sustainable societies, and to prepare them for disruptive thinking and the co-creation of new knowledge (Lotz-Sisitka *et al.* 2015; UNESCO 2017b), but also for exploring and using traditional and indigenous knowledge.

With the overall aim to develop cross-cutting sustainability competencies within learners (Wiek, Withycombe and Redman 2011; Rieckmann 2018), ESD is an important contribution to achieving the SDGs: it enables all people to contribute to achieving the SDGs by providing them, not only with the knowledge to understand what the SDGs are all about, but also the competencies to make a difference towards a more sustainable society (UNESCO 2017b).

The emancipatory ESD approach asks which key competencies are needed for learners to be 'sustainability citizens' (Wals and Lenglet 2016). Various key competencies essential to sustainable development have been outlined (e.g. Wiek, Withycombe and Redman 2011; Rieckmann 2012; Glasser and Hirsh 2016; Wiek *et al.* 2016) – describing what individuals need to be able to do to transform their own individual lifestyles to more sustainable ones and to contribute to societal transformation towards sustainability. In the international ESD discourse, there is agreement that the following key competencies are of particular importance for thinking and acting in favour of sustainable development (UNESCO 2017b; Rieckmann 2018):

- Systems thinking competency
- Anticipatory competency
- Normative competency
- Strategic competency
- Collaboration competency
- Critical thinking competency
- Self-awareness competency
- Integrated problem-solving competency

However, while competencies describe the capacity or disposition of acting, they do not necessarily imply that an individual will act in a certain way in a specific situation. Sustainability-oriented performance depends on the interplay of knowledge and skills, values and motivational drivers, and opportunities (Biberhofer *et al.* 2018). The interrelation of these dimensions influences personal behaviour **(Figure 4.3)**.

ESD is directly related to the other cross-cutting issues. It enables people, for example,

- "to act in favour of people threatened by climate change", and "to promote climate protecting public policies" (UNESCO 2017b, p. 36);
- "to develop a vision of a reliable, sustainable energy production, supply and usage in their country", and "to apply and evaluate measures in order to increase energy efficiency and sufficiency in their personal sphere and to increase the share of renewable energy in their local energy mix" (UNESCO 2017b, p. 24);
- "to communicate the need for sustainable practices in production and consumption", and "to challenge cultural and societal orientations" (UNESCO 2017b, p. 34);
- "to reflect on their own gender identity and gender roles", and "to plan, implement, support and evaluate strategies

Figure 4.3: Key competencies and performance of sustainability citizens

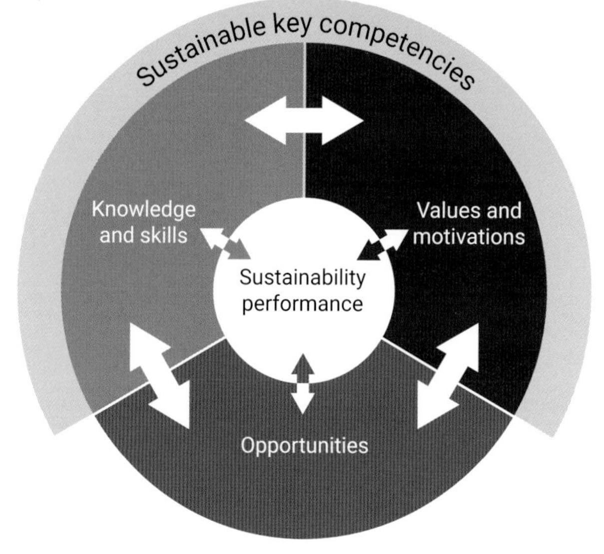

Source: Rieckmann (2018).

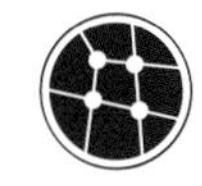

for gender equality" (UNESCO 2017b, p. 20); and

- "to encourage others to decide and act in favour of promoting health and well-being for all", and "to include health promoting behaviours in their daily routines" (UNESCO 2017b, p. 16).

ESD is at the heart of teaching and learning and should not be seen as a complement to the existing curriculum. "Mainstreaming ESD requires integrating sustainability topics into the curricula, but also sustainability-related intended learning outcomes" (UNESCO 2017b, p. 49). Since sustainability competencies cannot be taught or conveyed, but can only be developed by the learners themselves, an action-oriented transformative pedagogy is required (Mindt and Rieckmann 2017; UNESCO 2017b; Rieckmann 2018). In addition to the formal education curricula, ESD should also be promoted by non-formal and informal education. Community engagement and local learning can also play an important role, especially for involving traditional and indigenous knowledge into the learning process.

During the United Nations Decade for Education for Sustainable Development (2005-2014) (DESD) significant progress was made around the world with implementing ESD in all educational sectors (e.g. McKeown 2015; Watson 2015). Monitoring and evaluation of the DESD has shown many good examples of integrating ESD in curricula. Reviews of official curriculum documents show that "many countries now include sustainability and/or environmental themes as one of the general goals of education" (UNESCO 2014, p. 30). Most progress has been made in developing curricula towards ESD in primary and secondary education. "Close to 40% of Member States indicate that their greatest achievement over the DESD has been the integration of ESD into formal curricula, with another fifth describing specific school projects as being their most important contributions to ESD" (UNESCO 2014, p. 82). There has also been good progress with the implementation of ESD in higher education (Karatzoglou 2013; Lozano *et al.* 2015). This is particularly the case in Europe, where there has been a stronger interest in the integration of sustainable development in higher education institutions than in other parts of the world (Lozano *et al.* 2015; Barth and Rieckmann 2016).

However, upscaling of ESD is still needed in order to include it as a core element in the structures of educational systems (Singer-Brodowski *et al.* 2018). The Global Action Programme on Education for Sustainable Development, which was launched in 2014 at the UNESCO World Conference on ESD in Aichi-Nagoya, Japan, has five priority areas:

1. advancing policy;
2. transforming learning and training environments;
3. building capacities of educators and trainers;
4. empowering and mobilizing youth; and
5. accelerating sustainable solutions at local level.

It strives to scale up ESD, building on the DESD (Hopkins 2015; Mickelsson, Kronlid and Lotz-Sisitka 2018). Of particular importance in this context is the increased integration of ESD into (pre-service and in-service) teacher education. "Efforts to prepare teachers to implement ESD have not advanced sufficiently. More work still needs to be done to reorient teacher education to approach ESD in its content and its teaching and learning methods" (UNESCO 2017b, p. 51). For achieving this reorientation of teacher education towards sustainable development, it is necessary to form strategic institutional alliances among national, regional and local governments, non-governmental organizations, universities and other educational institutions involved in teacher education. Further challenges for scaling up ESD are:

- integrating ESD in policies, strategies and programmes;
- integrating ESD in curricula and textbooks;
- delivering ESD in the classroom and other learning settings;
- and changing the ways ESD learning outcomes and the quality of ESD programmes are assessed (UNESCO 2017b).

In order for all learners to benefit from ESD and to develop sustainability competencies, policies are needed that eliminate economic and gender barriers to access to education.

4.2.5 Urbanization

As explained in Section 2.3, urbanization is a major driver shaping the economy, the environment, the planet and human well-being worldwide. About 54 per cent of the world's population lives in urban areas that collectively generate more than 80 per cent of the world's gross domestic product (GDP) (United Nations Human Settlements Programme [UN-Habitat] 2011; UN-Habitat 2016a). By the year 2050, about 6.7 billion people – some 66 per cent of the world total population of 9.7 billion – are expected to be living in cities, adding 3.1 billion to cities' populations over the short span of about 40 years (United Nations 2018). While all world regions (except polar regions) will continue to urbanize, 90 per cent of future urban population growth is expected to occur in Africa and Asia (UN-Habitat 2014).

4

Cities are centres of innovation and historically they experience economies of scale with GDP increasing linearly with city population numbers (Bettencourt 2013). This capacity for innovation and wealth-generation, enabled by proximity and activity-intensity, is one of the features that attracts migrants to cities (International Organization for Migration [IOM] 2015), and will lead to an expansion of urban population by 2050 **(Figure 4.4)**. However, the wealth of cities is not distributed equally across the globe, with only 600 cities contributing more than 62 per cent of the global GDP (UN-Habitat 2011).

There is also significant inequality within cities, with a staggering 2 to 3 billion people –35 to 50 per cent of the urban population in 2050 – expected to be living in informal settlements (UN-Habitat 2014; UN-Habitat-2016a: UN-Habitat 2016b). Urbanization is associated with lower fertility rates, longer life expectancy, and better access to basic physical infrastructure and social amenities such as education and health care. However, inequality, crime and social exclusion are becoming characteristics of many urban areas, where living conditions are deteriorating in relation to the rural origins of many migrants (United Nations 2014).

Cities face huge challenges regarding social inclusion and improved provisioning of basic physical services. Energy, water, buildings, transportation and communication, food, public spaces and waste management emerge as key factors that shape the effect of cities on people, the environment and the planet.

The magnitude, scale and scope of contemporary urbanization is now so large as to be affecting global resource flows and planetary cycles. Urbanization is affecting the entire planet, not solely the areas defined as urban. Through networks of trade, migration and infrastructure, cities are influencing the natural environment well beyond their administrative boundaries (Wiggington *et al.* 2016). For example, although directly occupying only 3 per cent of the world's land area, energy supply to cities contributes more than 70 per cent of the world's energy-related carbon emissions (Seto *et al.* 2014). Direct water supply to cities puts pressure on 42 per cent of the world's watersheds (McDonald *et al.* 2014). In addition, water embodied in food supplied to cities exceeds direct water requirements in urban areas by more than a factor of ten (Ramaswami *et al.* 2017).

Figure 4.4: World urbanization trends

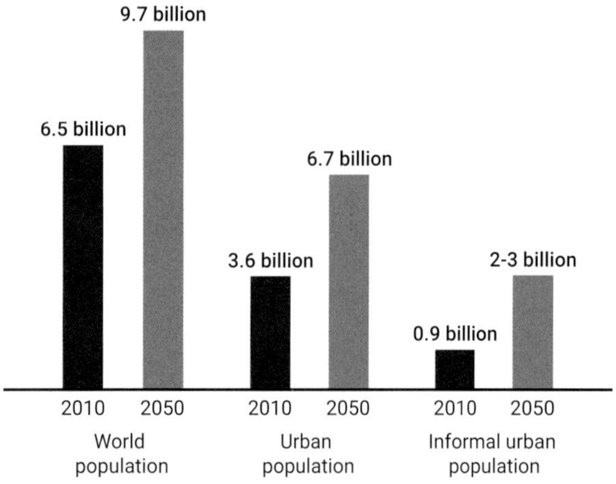

Source: Own elaboration based on (UN-Habitat 2014; UN-Habitat 2016a; UN-Habitat 2016b; United Nations 2018)

Urban footprints that represent both the bounded and transboundary ramifications that cities have on natural resources and the environment are essential to characterize the consequences of different urban activities, such as household consumption, production and community-wide infrastructure provisioning, and to chart pathways towards a sustainable future. In some regions, urban areas are de-densifying: urban population growth at declining densities leads to urban land expansion, which, in ecologically sensitive regions, can cause habitat fragmentation and contribute to large-scale biodiversity loss (Seto, Guneralp and Hutyra 2012).

Cities also face management and technological transformative opportunities. Around 60 per cent of the urban area required to accommodate the urban population of 2050 is yet to be built (Secretariat of the Convention on Biological Diversity [SCBD] 2012). Once built, it will last for at least the next 40 years. The bases of urban structures (e.g. street networks, blocks) "can affect and lock in energy demand for long time periods" (Seto *et al.* 2016).

At the same time, existing cities in advanced economies are repairing or replacing ageing infrastructures. Several infrastructural innovations are on the horizon in cities of both developed and developing countries that can enhance equity, resource efficiency and environmental sustainability. These innovations include new strategies for shared mobility, *in situ* slum rehabilitation, a One-Water approach to urban water management, urban-industrial symbiosis based on sustainable production and consumption through a circular economy, electric and autonomous vehicles for mass transit and private trips, and distributed renewable energy to achieve a decarbonized and resilient grid. Cities around the world are experimenting with infrastructure involving technology, human behaviour, financing and novel governance arrangements. This provides a historic opportunity and the imperative to build inclusive and sustainable infrastructure (UNEP 2013a). Successful urbanization relies on human as well as infrastructural assets.

Urban areas will continue to act as generators of economic growth and, through fertility and migration, they will continue growing in population and size. This can result in increased impacts of cities, but also in potential decreases in impacts per unit of production and per capita. As stated in the Section 2.3 of this report, there are clear challenges and opportunities that urgently need to be understood and addressed. These are related as much to governance as to technology, as is highlighted in Part B of this report (UNEP 2017).

4.3 Changing environments

4.3.1 Climate change

As explained in Section 2.7, climate change is driven by modifications in atmospheric composition due to land-use change, primarily deforestation, and to greenhouse gas (GHG) emissions, such as CO_2 emitted through fossil fuel burning and methane released from agriculture and other sources, as well as the emissions of aerosol particles (Vaughan *et al.* 2013). The evidence of current global climate change is unequivocal (Vaughan *et al.* 2013).

Eight of the ten warmest years on record have occurred within the past decade (United States National Oceanic and Atmospheric Administration [NOAA] 2018). Within this period, 2016 was the warmest year in the history of instrumental observation (NOAA 2017), and 2017 was the warmest year without an El Niño influence (NOAA 2018). As a result, global warming has reached approximately 1.0±0.2°C above the pre-industrial level (**Figure 4.5,** Haustein *et al.* 2017; Yin *et al.* 2017).

The current GHG emission rate, if it persists, will result in continuation of the current rate of global temperature increase of ~0.2°C per decade (e.g. Haustein *et al.* 2017), crossing the 1.5°C Paris Agreement target by the 2040s (Leach *et al.* 2018). While not unattainable, the goal of limiting warming to 1.5°C requires transformational changes leading to radical reduction of GHG emissions and expedited transition to carbon neutrality (Schellnhuber, Rahmstorf and Winkelmann 2016), that requires balancing of remaining anthropogenic CO_2 emissions with anthropogenic CO_2 removals.

Climate change modifies the water cycle by altering precipitation patterns and seasons. In general, dry areas are becoming drier, and wet areas are becoming wetter (Trenberth 2011; Intergovernmental Panel on Climate Change [IPCC] 2014; Feng and Zhang 2015), but numerous exceptions exist. Additionally, the increased water-holding capacity of warmer air leads to more extreme rainstorms that arrive less frequently (Trenberth 2011). Higher temperatures increase evapotranspiration rates and shift precipitation from snow to rain. A warmer atmosphere also governs the growth, melt and discharge of glaciers (Bliss, Hock and Radić 2014). These hydrological modifications determine river flows and the risks of early spring flooding and summer drought (Seneviratne *et al.* 2012; Cook *et al.* 2014; Kundzewicz *et al.* 2014). Changes in flow patterns alter water availability and, at the same time, higher temperatures increase demands from and competition among agricultural, industrial and domestic users (Hanjra and Qureshi 2010; Jiménez-Cisneros *et al.* 2014).

Oceans play an important role in climate regulation, having stored 93 per cent of the additional heat absorbed by the earth system since 1955. During that period, land has taken up 3 per cent of the heat absorbed, ice another 3 per cent, and the atmosphere only 1 per cent (IPCC 2013; Levitus *et al.* 2012). Heat-induced expansion of ocean water contributes to the observed sea level rise that has been accelerating over the past two decades; this trend will continue into the future even if the warming is limited to 1.5°C (Schewe, Levermann and Meinshausen 2011). Higher sea levels increase risks from storm surges for vulnerable small islands, coastal communities and exposed infrastructure. Oceans also absorb CO_2 from the atmosphere. Estimates suggest that, of all the CO_2 released to the atmosphere from human activities since the beginning of the industrial era, approximately 40 per cent has been absorbed by oceans (IPCC 2013; Khatiwala *et al.* 2013), resulting in a reduction of seawater pH (acidification), referred to as 'the other CO_2 problem' (Caldeira and Wickett 2003; Doney *et al.* 2009). This ocean acidification combines with warmer water temperatures and de-oxygenation processes to alter ocean

Figure 4.5: Global annual average temperature anomalies (relative to the long-term average for 1981-2010). Labelling designates different data sets; for explanation refer to the source

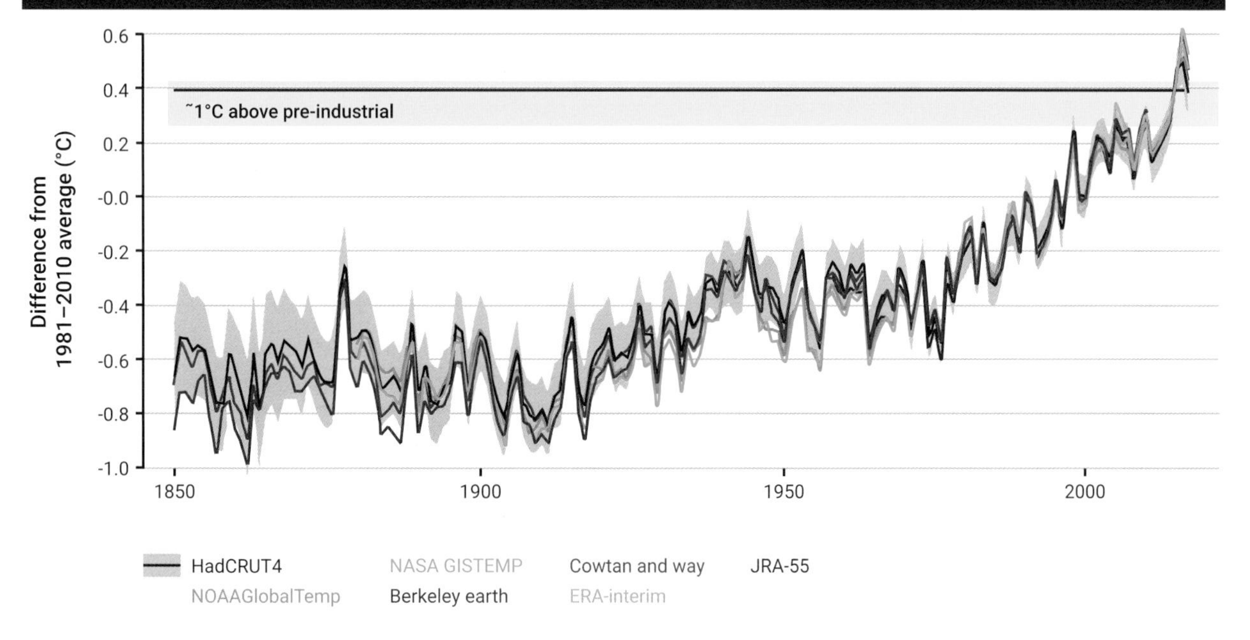

Source: United Kingdom Government Met Office (2018)

ecosystems (Achterberg 2014), most visibly as coral bleaching (see Chapter 7) when symbiotic algae are expelled from the reefs, reducing or ending their productivity (Fabry *et al.* 2008).

Estimates suggest that approximately 20 per cent of fossil-fuel CO_2 emissions are absorbed by land ecosystems (Arneth *et al.* 2017). Increased concentrations of CO_2 in the atmosphere may eventually benefit some C_3 crops[1], a category that includes wheat and beans, through carbon fertilization (McGrath and Lobell 2013). Warmer temperatures could bring yield gains in high-latitude regions, if soil and precipitation characteristics are suitable (IPCC 2014). Seventy per cent of global agriculture is rain-fed, and shifting rainfall patterns may benefit certain regions, but higher temperatures generally cause water stress that limits yields (Lobell, Schlenker and Costa-Roberts 2011; Challinor *et al.* 2014). Despite potential local yield increase, at a global level, yields are expected to suffer due to elevated risks from droughts and heat stress (Schlenker and Roberts 2009; Lobell and Gourdji 2012; Jiménez-Cisneros *et al.* 2014; Porter *et al.* 2014). Additionally, climate change, together with direct effects of rising atmospheric CO_2 concentration, has also been demonstrated to benefit invasive plant species (Ziska and Dukes eds. 2014).

Climate change also affects forest productivity, including increased stress from droughts, wildfires, insects, pathogens and windstorms (Williams *et al.* 2013; IPCC 2014). However, the influence of carbon fertilization on forest productivity is not well understood given the complexity of contributing factors (Norby *et al.* 2016). In combination with other human pressures, such as habitat destruction, climate change affects biodiversity at genetic, species and ecosystem levels. Seasonal changes can disrupt the timing of gestation, birth, hibernation, resource availability and optimal productivity. Species that are able are shifting their ranges, patterns and interactions on land, in fresh water and in oceans (IPCC 2014). There are possible shifts in infectious disease distributions in flora, fauna and humans (Lafferty 2009).

The shifts in weather patterns and extreme events, such as heat waves and droughts, and environmental disruptions, including crop failures, result in greater risks to human health and survival, especially among the poor and most vulnerable groups (Smith *et al.* 2014b). Climate change is also affecting the toxicity, environmental fate and behaviour of chemical toxicants by modifying physical, chemical and biological drivers of partitioning between the atmosphere, water, soil/sediment and biota, wet/dry deposition, and reaction rates with a potential of adverse impacts on biodiversity and human health (Noyes *et al.* 2009). Recent studies have examined the link between climate change and poverty in developing countries. In general, rural households in developing countries depend on crops, forest extraction and other income sources for their livelihoods, which tend to be extremely sensitive to climate change (Wunder, Noack and Angelsen 2018). The poor are more exposed to extreme climate conditions and experience greater rainfall fluctuations, while the poorest in dry regions experience the greatest forest loss (Angelsen and Dokken 2018). Poor people are often disproportionately exposed to droughts and floods, particularly in urban areas, and in many countries in Africa (Winsemius *et al.* 2018). Poorer households

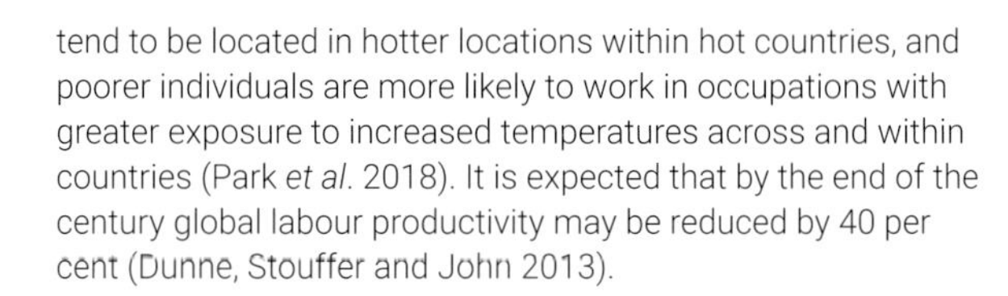

tend to be located in hotter locations within hot countries, and poorer individuals are more likely to work in occupations with greater exposure to increased temperatures across and within countries (Park *et al.* 2018). It is expected that by the end of the century global labour productivity may be reduced by 40 per cent (Dunne, Stouffer and John 2013).

The climate continues to change and the impacts on the natural and human system are increasingly recognized. Social responses such as population migration and displacement exacerbate health risks and threats to geopolitical stability (Adger *et al.* 2014); these risks increase with continuing warming beyond 1.5°C as detailed in chapters 3 and 5 of the IPCC 1.5°C report (IPCC 2018). Limiting the observed warming trend to 1.5°C requires transformational changes in policies, technologies and societal goals.

4.3.2 Polar regions and mountains

Covering approximately 20 per cent of the Earth's surface and containing the ice sheets of Greenland and Antarctica, the polar regions play a significant role in the global climate system. Land and sea ice not only regulate the energy balance of the climate system due to their high albedo, or reflectivity, but also store a record of climate information. In addition to their role as engines of global climate processes, the Arctic and Antarctic act as bellwethers of climate change because warming is amplified at their high latitudes (Taylor *et al.* 2013). Warming is also amplified at high altitudes, so mountain regions can be included in this discussion as a 'third pole' (Pepin *et al.* 2015).

Amplified warming affects all components of the polar climate system. Arctic Sea ice is shrinking in area and volume **(Figure 4.6)**. Permafrost is thawing resulting in a release of greenhouse gases, including CO_2, and snow cover extent is decreasing. Ice sheets and mountain glaciers continue to lose mass, contributing significantly to sea level rise that threatens coastal regions at every latitude (Vaughan *et al.* 2013). These transformations have consequences for polar and high-altitude ecosystems and for the people who live there. Shifting environmental and socioeconomic conditions in the Arctic in particular are delivering consequences to environments and populations further south through teleconnections within the climate system (Francis, Vavrus and Cohen 2017) and through close geopolitical connections. In fact, polar regions are gaining politico-strategic importance. The Arctic has already been subjected to resource extraction and exploitation, from hydrocarbons to diamonds (Dodds 2010; Ruel 2011), and the Antarctic is becoming an area of strategic interest for countries looking at potential resource extraction in the future. At the same time, the Arctic and particularly the Antarctic, which has a treaty devoting the continent to peace and scientific cooperation, are regions of peaceful international coordination and enhanced environmental cooperation, exhibiting governance systems that can be exemplars for environmental protection in other regions.

The ecosystem services of the polar regions that relate to global climate regulation are further enhanced by the formation of super-dense Antarctic bottom water, and to a lesser extent of North Atlantic deep water, which are significant contributors to the thermohaline circulation. The cooler ocean waters of higher latitudes, especially the Southern Ocean, also represent important carbon sinks and areas of high marine productivity.

1 The plants that utilize C3 photosynthesis (85% of all plants) have disadvantage in hot, dry conditions. C3 crops include wheat, rice, soybeans, and many others.

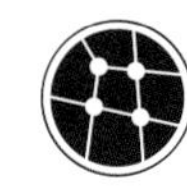

Figure 4.6: Arctic sea ice age and extent

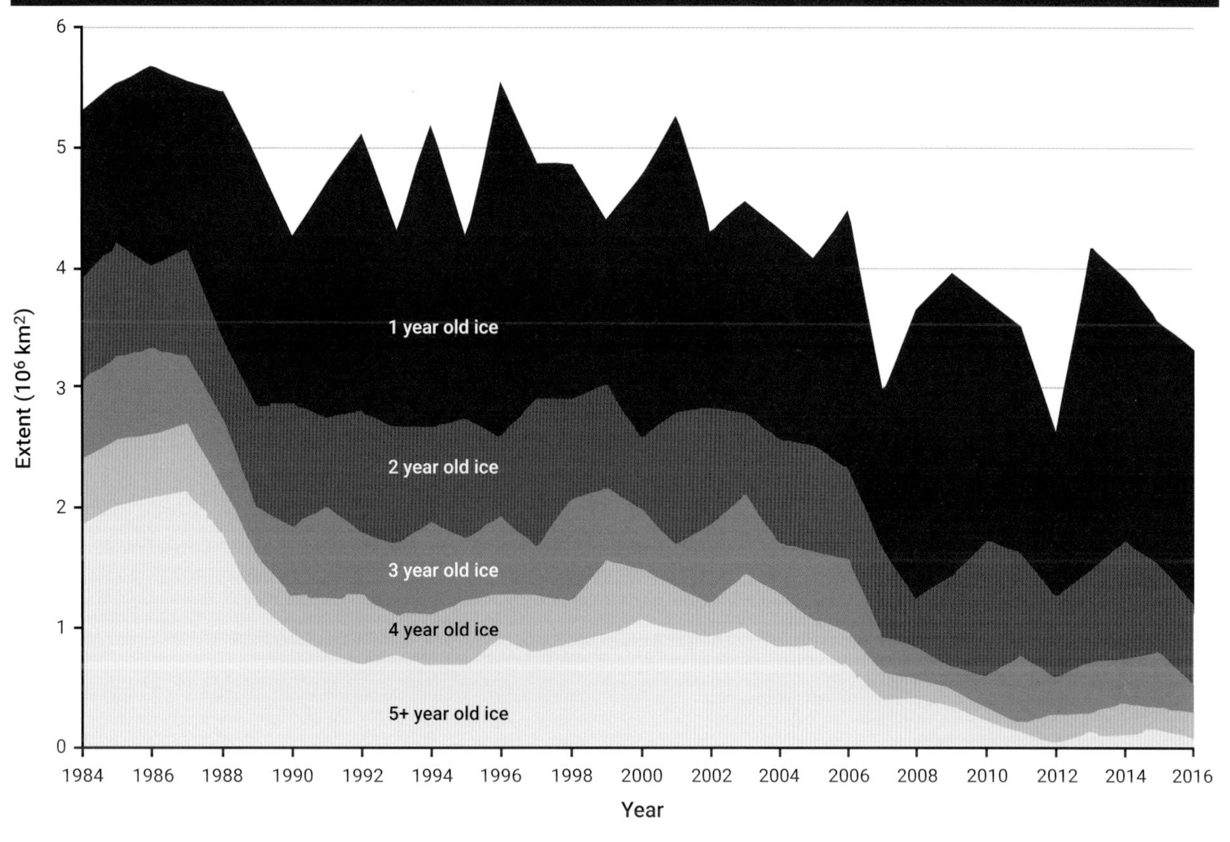

Source: United States National Snow and Ice Data Center (2017).

They play a significant role in food production in the high latitudes and require careful management through agencies such as the North Atlantic Fisheries Organization and the Commission for the Conservation of Antarctic Marine Living Resources. Some high-latitude fisheries have been significantly affected by fishing activities in the last century as highlighted in the collapse of the Atlantic cod fishery (Villasante *et al.* 2011).

More than 70 per cent of the planet's fresh water is locked up in ice in the polar regions. If released, the water stored in the Greenland Ice Sheet would result in a 7.4 metre rise in sea level, the water in the Antarctic Ice Sheet would result in a 58.3 metre rise, and the water stored in all mountain glaciers would yield a 0.4 metre rise (Vaughan *et al.* 2013). In a scenario limiting temperature increase to below 2°C, the world would still see a mean rise of global sea levels by 0.4 to 0.6 metres. A business-as-usual scenario produces an average sea level rise of 0.7 to 1.2 metres by the end of the 21st century (Horton *et al.* 2014). As the latest IPCC report and multiple independent scientific studies indicate, mountain glaciers and polar ice sheets are already losing mass and are contributing on average the equivalent of 1.85 mm of sea level rise per year (Bamber *et al.* 2018).

As more fresh water is transported to the ocean from seasonal permafrost thaw, iceberg calving, glacier and ice sheet melt, and other fluvial discharge, the increase of silt, carbon and other nutrients will affect the polar regions' primary productivity in the marine food chain. The source and quality of food for higher organisms will shift, with much less primary productivity originating from ice-related algae, so that species at higher trophic levels, such as krill and fish, will be challenged (Alsos *et al.* 2016; Frey *et al.* 2016). This, combined with invasive species shifting into newly tolerable conditions and their potential threats, requires humans to adapt to new economic and cultural livelihoods and may result in conflicts, especially with regard to resource use, governance, cultural concerns and marine protected areas (Conservation of Arctic Flora and Fauna [CAFF] and Protection of the Arctic Marine Environment [PAME] 2017).Nearly all of the world's glaciers are losing mass and some will vanish in the coming decades (Kaltenborn, Nellemann and Vistnes eds. 2010; Vaughan *et al.* 2013). More than a billion people rely on mountain glaciers for water, with the majority of these people living in Asia, which has around 100,000 km² of glaciers (Yao *et al.* 2012). Over 200 million people rely on water from the Hindu Kush Himalayan mountains with hundreds of millions more people downstream who are affected by reduced reliability of local water sources and increased hazards, including glacial lake outburst floods. Run-off is expected to decrease until 2050 in the Ganges, Brahmaputra and Mekong basins. At the same time, the Hindu Kush Himalaya region can expect higher variability in water flows and more water in pre-monsoon months leading to more floods and droughts. The Andes are already experiencing less run-off. Changes in temperature and precipitation will affect agriculture, water resources and health (Shrestha *et al.* eds. 2015).

Further adjustment to new realities will warrant responses to increasing levels of contaminants that have been transported long distances and accumulate in the polar regions. Despite few local industrial sources, persistent environmental contaminants were detected decades ago in these remote locations and pose significant threats to local people and environments through polar food chains (Andrew 2014). Sea-ice melting will result in air-water exchange of persistent organic pollutants in areas of the Arctic that are no longer covered with ice. Likewise, melting of polar and alpine glaciers, ice sheets and shelves, and permafrost will also release persistent organic pollutants and mercury, enabling further air-soil exchange of these pernicious compounds (Arctic Monitoring and Assessment Programme [AMAP] 2015; Sun *et al.* 2017). Due to new regulations, the levels of many persistent organic pollutants are now declining, but new chemicals are a cause for increased concern, such as organophosphate-based flame retardants, phthalates, some siloxanes, and some currently used pesticides (AMAP 2017). Equally, microplastics have now been detected in all of the world's oceans (Thompson *et al.* 2004; Browne *et al.* 2011), including in deep-sea sediments (Barnes, Walters and Gonçalves 2010) and even in Arctic sea ice (Thompson *et al.* 2004, Browne *et al.* 2011, Ivar do Sul and Costa 2014; Obbard *et al.* 2014; Isobe *et al.* 2017; Waller *et al.* 2017). More research is needed to trace the distribution and impact of microplastics in the Antarctic, but their existence in the Southern Ocean (Isobe *et al.* 2017; Waller *et al.* 2017) and in the Ross Sea (Cincinelli *et al.* 2017) has already been confirmed.

Those who live at high latitudes and in mountain regions are vulnerable to the compounding effects of air pollution, and-use changes and other factors, as well as the threats from climate change. However, people in these areas, especially the indigenous peoples who have inhabited the Arctic and mountain regions for millennia, have a rich knowledge about their environment that provides crucial insights for effective adaptation strategies (Magga *et al.* eds. 2009; Nakashima *et al.* 2012).

4.3.3 Chemicals

Modern societies produce and inhabit the most chemical-intensive environment humans have ever experienced – today, it is estimated that there are more than 100,000 chemicals on the market of modern society (European Chemicals Agency [ECHA] 2018) – and now chemical pollution is considered a global threat (Barrows, Cathey and Petersen 2018). Common categories of chemicals include pharmaceutical and veterinary chemicals, pesticides, antibiotics, flame-retardants, plasticizers and nanomaterials (Tijani *et al.* 2016). Even the more familiar chemicals, used for generations in agriculture and industry, are now used so intensively and in such concentrations as to require responsible monitoring and evaluation programmes **(Figure 4.7)** (Bernhardt, Rossi and Gessner 2017).

Global chemical pollution has been raised as a problem that needs urgent action: calls for more active involvement of governments and industry and for more research are included

Figure 4.7: Chemical intensification, 1955-2015

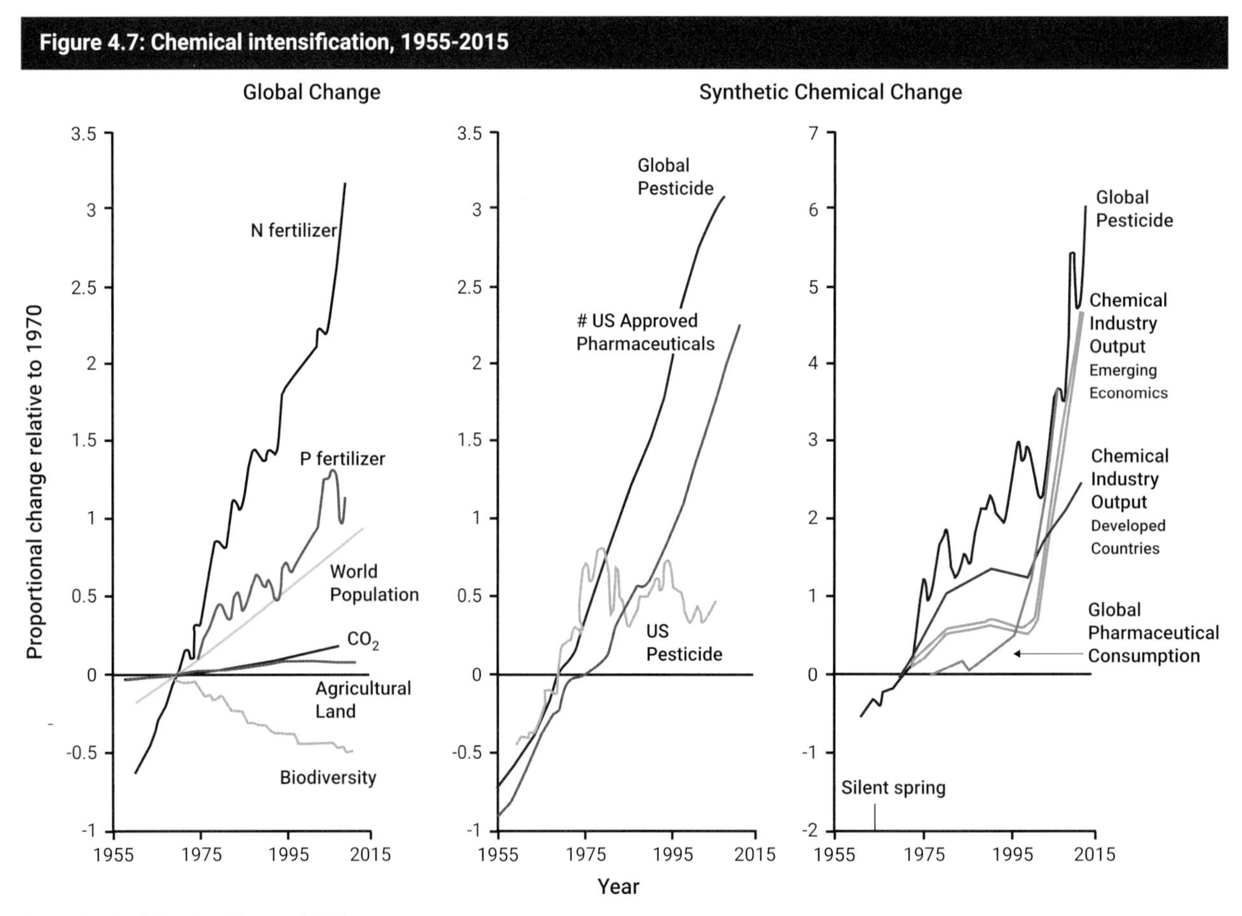

Source: Bernhardt, Rossi and Gessner (2017).

in all relevant studies and the existence of the problem is admitted in global change assessments and clarion calls (UNEP 2012; Stehle and Schulz 2015; Bernhardt, Rossi and Gessner 2017). However, the assimilative capacities for chemical burdens are largely categorized as undetermined and then ignored, even in efforts to inspire concern about planetary environmental issues (Diamond *et al.* 2015; Steffen *et al.* 2015). The global dimension of chemical pollution manifests as these substances spread to the most remote environments on the planet, including the polar regions (Andrew 2014), high mountain peaks (Ferrario, Finizio and Villa 2017) and the deepest oceans: persistent organic pollutants were detected in fauna found at more than 10,000 metres depth in the Pacific Ocean's Mariana Trench (Jamieson *et al.* 2017). However, there are currently ongoing efforts in developed countries to carry out regular monitoring programmes to mitigate the impact of chemicals, especially pesticides, on human and environmental health (Brouwer 2018).

Some chemicals that are persistent, toxic and bioaccumulating, and may travel long distances, are listed under international conventions, such as the Stockholm Convention (persistent organic pollutants) and Minimata Convention (mercury), but scientific evidence shows that more chemicals regularly made available for commercial use display the same properties as the regulated persistent organic pollutants (Strempel *et al.* 2012). Countless new chemicals, as well as old chemicals that were not well understood, are not regulated at all even though they are suspected of causing adverse effects (Petrie, Barden and Kasprzyk-Hordern 2015; Ferrario, Finizio and Villa 2017).

Pharmaceuticals are commonly mishandled 'from cradle to grave' with over 200 different substances reported in river waters globally (Petrie, Barden and Kasprzyk-Hordern 2015). Antibiotic-resistant bacteria have evolved and spread due to mismanagement of antibacterial drugs (Marti, Variatza and Balcazar 2014; Grenni, Ancona and Caracciolo 2017). Recent research indicates that the development of antimicrobial resistance in pathogens is accelerated and achieved at lower exposure concentrations, in the presence of heavy metals and other contaminants that are commonly found in the same contaminated reservoirs (The Lancet Planetary Health 2018). The presence of such contaminants in the natural environment results from the discharge of wastewater from treatment plants that are unequipped to effectively remove these dangerous compounds (Petrie, Barden and Kasprzyk-Hordern 2015) and from mismanagement of their use for agricultural production, particularly in livestock (Hamscher and Bachour 2018).

The effects of some endocrine-disrupting chemicals are of particular concern because of potential multigenerational effects on the health of humans and wildlife (Gore *et al.* 2015). Endocrine activity or disruption has been associated with a wide variety of compounds, including some persistent organic pollutants (Kabir, Rahman and Rahman 2015) and industrial chemicals (UNEP and WHO 2013). They are present in many pesticides that are designed to interfere with the life cycles of organisms and are highly valued for those abilities (Gore *et al.* 2015). Endocrine disruption potential has also been attributed to certain chemicals present in manufactured plastics (Schug *et al.* 2016).

Products used in everyday life may contain toxic compounds that interfere with human and environmental health, spanning cosmetics, plastic containers, and household cleaners and pesticides. Addressing the issue of chemicals in products may offer new opportunities in terms of innovation through green and sustainable chemistry efforts and could represent a valuable opportunity to improve sustainable consumption and production patterns and life cycle thinking. Application of the circular economy model to chemical production and consumption could establish some measure of control from the extraction of primary materials, through the design, formulation, production, use and final disposal of the substances and products that people use (Roschangar, Sheldon and Senanayake 2015). Chemicals in everyday products, as well as endocrine disruptors and nanomaterials, have been identified as emerging policy areas under the Strategic Approach to International Chemicals Management (SAICM) (UNEP 2013b). Highly hazardous pesticides, used in agricultural practices in developing countries, are another issue addressed by SAICM: alternative approaches rely on agroecological practices to promote substitution of hazardous pesticides by pest management approaches and products that pose less risk (FAO and WHO 2016), as well as demand reduction and non-chemical alternatives.

Nanotechnology, by decreasing the particle size of materials and increasing its reactivity, may give a material some interesting properties, but these may be toxic (Schulte *et al.* 2016). There remain a number of questions about the toxicity of nanoparticles to humans and the environment, but comparison of nanomaterials of certain size and shape with asbestos indicates similar toxicological potential (Nagai and Toyokuni 2012; Allegri *et al.* 2016).

Even those substances considered under control in some regions may be distributed in developing countries with no guidance on health and safety issues and proper use. The Global Chemical Outlook (UNEP 2013b; UNEP 2013c) estimates total health-related pesticide costs – the costs of inaction – for agricultural smallholders in sub-Saharan Africa from 2015 to 2020 at US$90 billion, assuming a continued scenario of inadequate capacity for pesticide management.

Further studies evaluating the combined effects of chemical mixtures are critical, in addition to understanding the cumulative effects of chemicals over time. Equally, more information is needed on causal linkages between exposures to certain chemicals and related health effects (The Lancet Planetary Health 2018). Promoting safer and sustainable alternatives to chemicals, especially biodegradable replacements for plastics, and sound cradle-to-cradle chemicals management is essential. Institutions and instruments are available and coordination through United Nations agencies is an objective of SAICM. The costs of inaction to global society is high if measures are not taken to detoxify the environment and to create a safe-chemical future in coming decades (UNEP 2013c).

4.3.4 Waste and wastewater

The Global Waste Management Outlook (UNEP 2015) estimates the total 'urban' waste generation, including municipal solid waste, commercial and industrial waste, and construction and demolition waste, at around 7-10 billion tons per year. Waste generation rates are stabilizing in developed regions. However, Asia and Africa are expected to contribute significant amounts to global waste generation over the next century (UNEP 2015).

GEO-6 highlights key global waste management challenges consistent across the regional assessments prepared for it and prioritized in the Global Waste Management Outlook (UNEP 2015). These include food waste, marine litter, waste trafficking and crime, and the growing disparity in waste management between developed and developing countries.

Approximately one-third of the food produced for human consumption is wasted or lost annually, at a financial cost of US$750 billion to US$1 trillion (FAO 2013; FAO 2015; UNEP 2015). This wasted food could feed over 2 billion people, more than twice the number of undernourished people estimated globally (FAO 2013). Food losses and waste result in unnecessary greenhouse gas emissions, estimated at 3.3 gigatons of CO_2 equivalent in 2007, or around 9 per cent of total global GHG emissions that year (UNEP 2015). This estimate does not take into account GHG emissions as a result of land-use changes. Considering land-use changes, GHG emissions from food waste would be 25-40 per cent higher. Even without counting land-use change, if food losses and waste all occurred in one country, it would rank as the third largest country in the world in terms of CO_2 emissions (FAO 2013).

With increasing global demand for resources, the waste market has become a viable economic sector, estimated at US$ 410 billion a year, from collection to recycling. In a context of increasing costs for the safe disposal of hazardous waste, weak environmental regulations and enforcement, and increasing resource scarcity, this market creates opportunities for waste trafficking and illegal activities. This is evident in large quantities of often hazardous waste being unlawfully exported to developing countries, with the potential to cause significant, and displaced, impacts **(Figure 4.8)** (Rucevska *et al.* 2015). The illegal trafficking of end-of-life electrical and electronic equipment has become an issue of global concern (UNEP 2015; UNEP 2016b).

Figure 4.8: Global illegal waste traffic

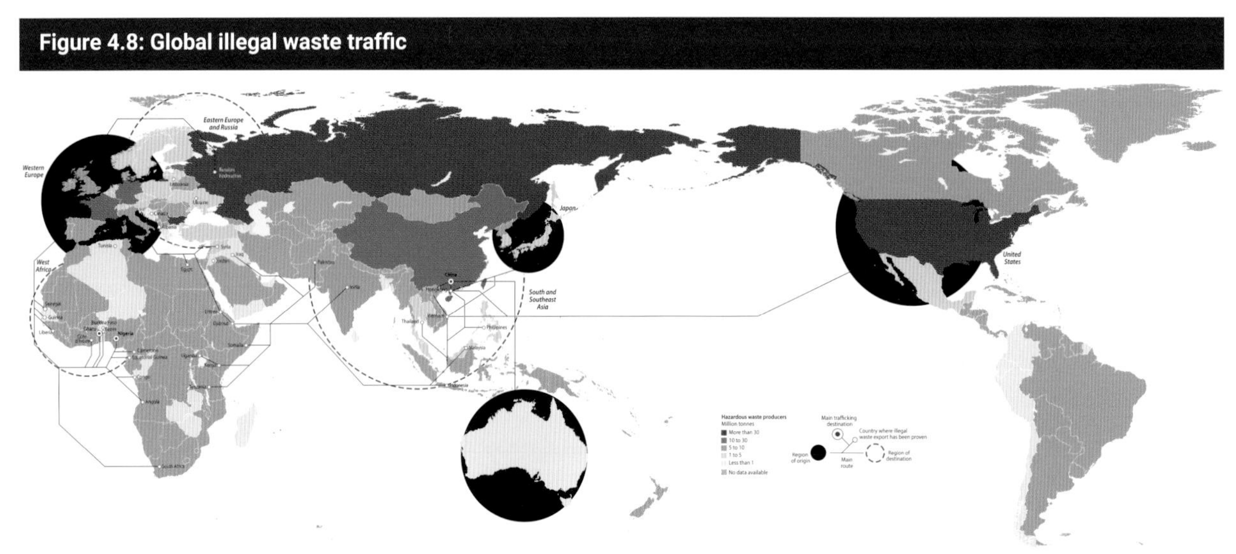

Source: Pravettoni (2015).

Developed countries have advanced their waste management systems to the point where they can consider strategies for integrating new and complex waste types; driving sustainable consumption and production; moving towards near zero waste schemes and a circular economy; and the adoption of emerging and potentially disruptive technologies on waste management. Developing countries are still grappling with basic waste management challenges, including uncontrolled dumping, open burning and inadequate access to waste services. Globally, 3 billion people lack access to controlled waste disposal facilities, according to United Nations estimates, with the potential to cause significant environmental, social and economic impacts from poor waste management (UNEP 2015). In the first seven months of 2016, an estimated 750 people died due to poor waste management at dumpsites (International Solid Waste Association [ISWA] 2016). In early 2017, some 115 people were killed in a waste landslide in Addis Ababa, Ethiopia (Gardner 2017) and 16 people were killed in the collapse of the Hulene Garbage Landfill in February 2018 in Maputo, Mozambique. A high percentage of the fatalities were women. Such dumpsites in developing countries are often home to millions of informal waste pickers (ISWA 2016; Duan, Li and Liu 2017). While developed countries chase the ideals of reduced waste, a circular economy and greater resource efficiency, developing countries must not be left behind.

Any circular economy plan incorporates wastewater in its design. This includes human sewage, industrial effluent and both agricultural and urban run-off (Mateo-Sagasta *et al.* 2013). Agriculture is the main contributor, accounting for 79 per cent of wastewater produced in arid West Asia, where it is discharged straight into the environment **(Figure 4.9)**

Figure 4.9: West Asia non-conventional annual water resources

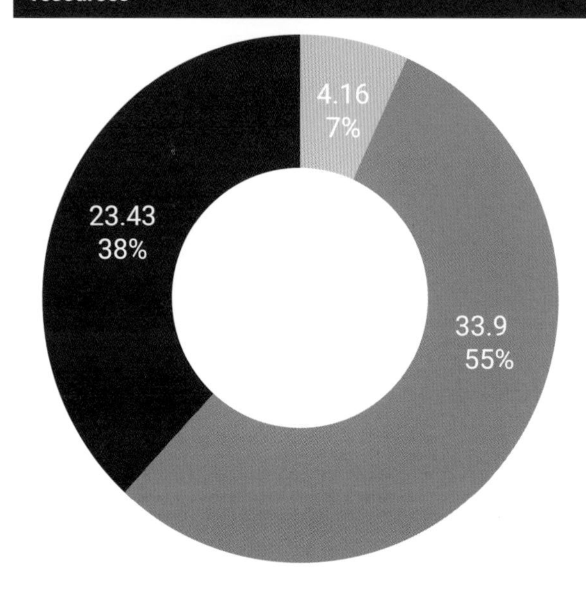

Source: Abuzeid *et al.* (2014).

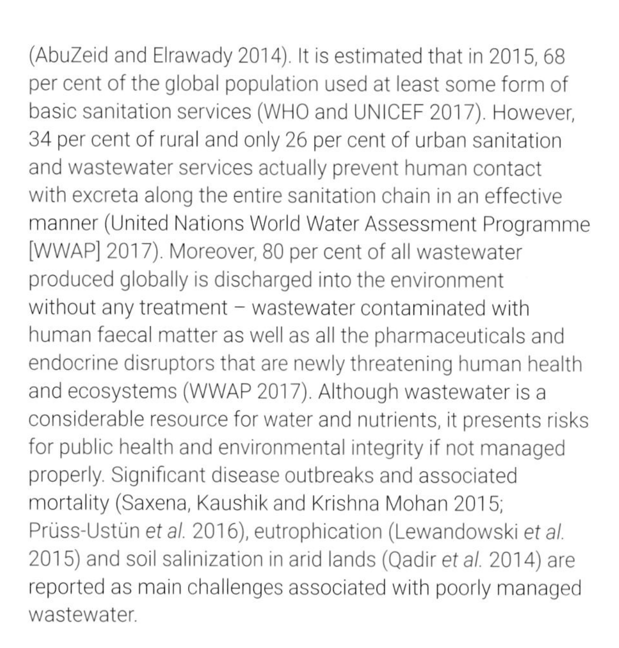

(AbuZeid and Elrawady 2014). It is estimated that in 2015, 68 per cent of the global population used at least some form of basic sanitation services (WHO and UNICEF 2017). However, 34 per cent of rural and only 26 per cent of urban sanitation and wastewater services actually prevent human contact with excreta along the entire sanitation chain in an effective manner (United Nations World Water Assessment Programme [WWAP] 2017). Moreover, 80 per cent of all wastewater produced globally is discharged into the environment without any treatment – wastewater contaminated with human faecal matter as well as all the pharmaceuticals and endocrine disruptors that are newly threatening human health and ecosystems (WWAP 2017). Although wastewater is a considerable resource for water and nutrients, it presents risks for public health and environmental integrity if not managed properly. Significant disease outbreaks and associated mortality (Saxena, Kaushik and Krishna Mohan 2015; Prüss-Ustün *et al.* 2016), eutrophication (Lewandowski *et al.* 2015) and soil salinization in arid lands (Qadir *et al.* 2014) are reported as main challenges associated with poorly managed wastewater.

4.4 Resources and materials

4.4.1 Resource use

Sustainable resource use requires sound management of renewable resources and aims to recycle non-renewable resources, leading to the concept of a circular economy in which a waste, the by-product of a process, becomes a raw material for another process. In a circular economy, efficient use of resources across their entire life cycle is critical: from extraction to manufacturing, through consumption and use, to recycling and reuse (Ellen MacArthur Foundation 2012; European Commission 2015).

From the 20th century, resource exploitation has grown considerably, especially of metals, such as iron and copper, and of minerals, such as sand and limestone for cement. Fossil fuel exploration and extraction, and its consumption, exemplify modern society's great advances, according to one narrative. However, fossil fuel exploitation has also created great challenges. The momentum of consumption has led to ever increasing scales of resource exploitation, leading to concerns over the cumulative and global consequences of such activities, as well as over local damage (Rockström *et al.* 2009).

Traditionally, the discovery of new and accessible deposits of non-renewable resources has kept pace with or even outpaced growing extraction, so concern over the depletion of such resources would not be considered highly important (Mudd, Weng and Jowitt 2013; Mudd and Jowitt 2014; Weng *et al.* 2015; Mudd and Jowitt 2017). However, as a measure of their quality, the grades of most mined ores are in gradual decline, meaning that the most easily and economically refined ores have already been exploited (Ruth 1995; Mudd 2010). Larger amounts of lower grade ore have to be extracted and processed to meet global demands, as can be shown by tracking exploitation of copper ore deposits **(Figure 4.10)**.

When declining ore grades are combined with the larger project scales needed to extract enough ore to supply market demand, greater risks threaten the natural environment. More land is cleared, or simply removed and shipped away, as mountain-top removal illustrates. Larger volumes of mine waste accumulate, with heavy metals and reactive agents recombining into noxious compounds. Water pollution risks, especially from acid and metalliferous drainage, increase. Threats to biodiversity become more complex. Energy demand intensifies, along with associated greenhouse gas emissions (Norgate and Haque 2010). To meet global demands in 2014, the global metals and mining industry produced around 90 billion tons of mine waste, excluding construction materials (Mudd and Jowitt 2016). This massive mining scale requires an acute focus on environmental assessment, monitoring and management for primary resource extraction (Hudson-Edwards 2016; Mudd and Jewitt 2016). Currently, much of the mine waste is stored, exposed to changing environmental and management conditions. The 2015 Samarco tailings dam failure in Brazil, among other events, demonstrated how long-term storage strategies are not solutions (Philips 2016; Roche, Thygesen and Baker 2017).

Some mined resources are widely distributed around the world, including sand, gold, copper and lead-zinc; other resources, such as nickel, rare earth elements and phosphorous, are concentrated in a small number of countries. Given the fundamental contribution of mineral resources to modern social systems, technologies and infrastructure, these materials need to be assessed for their role in modern society. This analytical approach is known as criticality – examination of the potential implications of supply disruption, resource substitution, recyclability and environmental impacts (Graedel *et al.* 2015) For example, many metals such as iron, copper, gold and lead are recyclable. Other minerals, such

Figure 4.10: Example of ore grade decline over time for copper mining, showing world annual copper production and estimated tailings generated annually

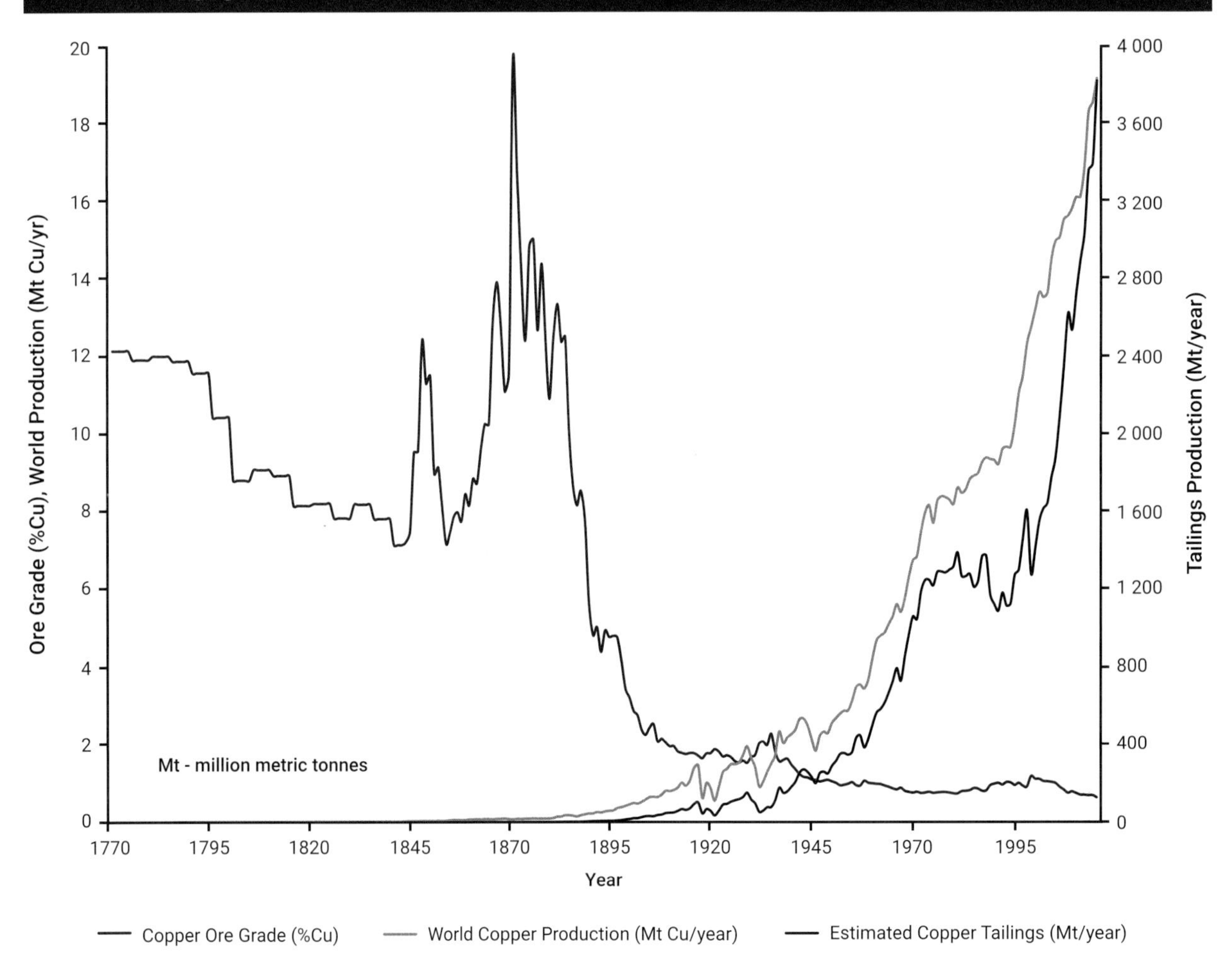

Source: Ruth (1995); Crowson (2012); Mudd, Weng and Jowitt (2013); Mudd and Jowitt (2016).

4

as phosphorous, are dispersed in soils and water bodies, ultimately washing away and being effectively lost to any further use. That kind of material dissipation raises alarms over the eventual depletion of the essential resource (Ciacci *et al.* 2015; Nassar, Graedel and Harper 2015).

In contrast, when a metal is recycled, the environmental risks are typically much lower. For instance, fabricating a product from recycled aluminium uses one-twentieth of the energy than production from primary aluminium does. For the circular economy, this means that recycling should lead to reduced environmental pressures and risks, mainly due to lower energy and raw material needs (Wernick *et al.* 1996; Wernick and Ausubel 1997; Balke *et al.* 2017). The focus of a circular economy concentrates on sound product or infrastructure design, as well as on the systems in place to monitor resource use, waste and environmental repercussions (Ghisellini, Cialani and Ulgiati 2016). Other strategies may include variations of upcycling or recycling: refuse, rethink, reduce, reuse, repair, refurbish, remanufacture and repurpose. Here, environmental and sustainability education is crucial.

An important issue arising from resource use is that the environmental and social costs are typically greatest during extraction when land is cleared, or populations displaced, while the greatest benefits accrue at the other end of the supply chain. To fully appreciate the cost-benefit ratio and the actual value of a product, it is important to consider the environmental consequences of global trade in resources, including the repercussions for local communities in areas of resource extraction. Interest is growing in tracing the origins and added values of supplied resources through sustainable supply chain management. This traceability supports action on issues such as conflict minerals, chemical and pharmaceutical waste, food contamination and illegal trade in endangered species (Mundy and Sant 2015; Paunescu, Stark and Grass 2016; Tijani *et al.* 2016; Sauer and Seuring 2017). The availability and distribution of this type of information defines a connection between supplier and consumer and encourages more sustainable resource use choices. Recent research indicates, however, that humanity has overshot the safe operating space for certain planetary systems, specifically climate change, the rate of biodiversity loss and the biogeochemical flow of the nitrogen cycle (Rockström *et al.* 2009; Steffen *et al.* 2015). Some updated analyses would add phosphorus to that overshoot list (Carpenter and Bennett 2011; Cordell and Neset 2014).

The pressures upon our planet have therefore brought global society to a decisive crossroads: the continuation of a conventional process model to 'extract-make-use-discard' through a linear economy or the transformation into a circular economy with society focused on the entire life cycle of resource use and management. Some thinkers consider that it may already be too late (Urry 2010; Scheffer 2016). Others suggest the transition from a linear economy with wasteful resource management to a circular economy with sustainable resource management can be accomplished but requires new concepts of de-growth and a post-capitalistic economic vision (Jackson and Senker 2011; Kosoy *et al.* 2012; Krausmann *et al.* 2017).

The transition to a circular economy will provide many opportunities for technology innovation and deployment that also present many new business prospects. At heart, a circular economy will require sound policies for resource accounting and waste management that create the demand for recycled resources and deliver a resource efficient and sustainable economy (Ghisellini, Cialani and Ulgiati 2016; Balke *et al.* 2017). Resource use is also intimately connected to energy technologies and policies, such as the materials required for various renewable energy technologies, highlighting the need to consider the links among material resources, energy and environmental outcomes (Akenji *et al.* 2016; McLellan 2017).

All 17 of the Sustainable Development Goals involve competition for natural resources, with many requiring efficient and sustainable use of resources and minimizing associated impacts – especially the metals considered critical for renewable energy and, consequently, for progress on climate change solutions (Arrobas *et al.* 2017; International Resource Panel 2017).

4.4.2 Energy

By 2015, global energy consumption reached around 13.5 billion tons of oil equivalent (International Energy Agency [IEA] 2018). That is expected to increase to around 19 billion tons by 2040 (IEA 2016). Much of this increase is attributed to consumption expected in developing economies that currently depend largely on fossil-based energy sources. This makes accelerated efficiency a crucial strategy to mitigate energy-related impacts. At the same time, nearly 1.2 billion people remain without access to electricity and 2.7 billion still resort to traditional fuels for cooking and heat, facing exposure to concentrated indoor air pollution (IEA 2016). Improved access to modern energy services is not only closely connected to all Sustainable Development Goals and indicators, including food security, health and quality education, but shifting to clean and efficient forms of energy also empowers women and other marginalised groups responsible for the collection and burning of primitive solid fuels (World Energy Council 2016).

Energy demand also leads to competition for water, land and even atmospheric limits; to inequitable distribution of these and other sets of natural capital, such as mineral resources and access to sensitive ecosystems; and to processes involving different approaches that often cause disputes and conflicts at several levels and magnitudes (Rodriguez *et al.* 2013; Jägerskog *et al.* 2014; McLellan 2017).

The competition between biofuels and food re-emphasizes the need to understand the nexus of energy, food, water and land use (see Chapter 8). Popp *et al.* (2014) discuss the impact of biofuel production on food supply, environmental health and land requirements, and highlight the need for integrated policies to manage the various components of the energy, food, water and land-use nexus.

The rise in water demand, while usable water reserves decline, accentuates the need to examine water-energy linkages against the backdrop of growing energy demand. Jägerskog *et al.* (2014) discuss the energy and environmental trade-offs related to hydropower. Rodriguez *et al.* (2013) also provide an overview of water requirements for generating power, particularly in the case of thermal power plants. Copeland and Carter (2017) address the energy requirement for delivering water to end users and for the disposal of wastewater in the United States of America.

At the global scale, greenhouse gas emissions amounted to 33 gigatons of CO_2 equivalent in 2014 and may reach 38 gigatons in 2040, due mostly to the burning of fossil fuels (IEA 2015). Historical data demonstrate trends in decoupling through decarbonization and improved efficiency, but the current trend still indicates a global temperature increase beyond the 2°C threshold target of the Paris Agreement **(Figure 4.11)** (IEA 2015; United Nations 2015b; IEA 2016). This likely overshoot warrants bolder action.

The economics of transition to low-carbon energy sources have been greatly assisted by a dramatic reduction in the cost of renewables, especially wind and solar photovoltaic systems. Solar photovoltaic systems experienced a price decline of 23 per cent for each cumulative doubling of production over the last 35 years. In many cases these costs are now lower than those of conventional fossil fuel electricity generation technologies (International Renewable Energy Agency [IRENA] 2015). Further reductions are expected making them possibly the best economic-environmental option in practically every country in the world before 2025 **(Figure 4.12)**.

Figure 4.11: Technology wedges to achieve the 2°C pathway

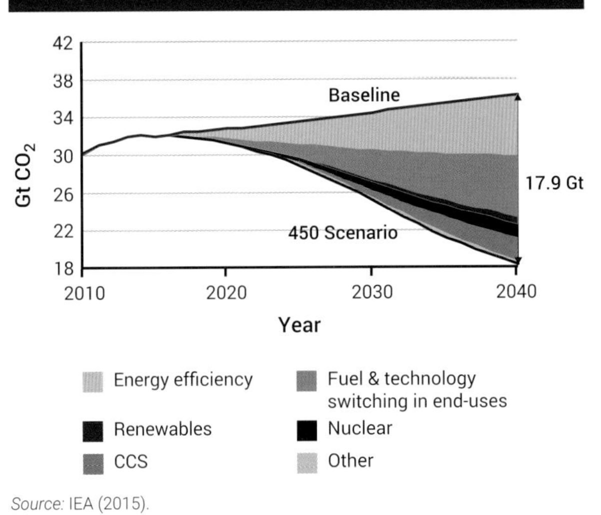

Source: IEA (2015).

Figure 4.12: Ranges of levelized cost of electricity for different renewable power generation technologies, 2014 and 2025

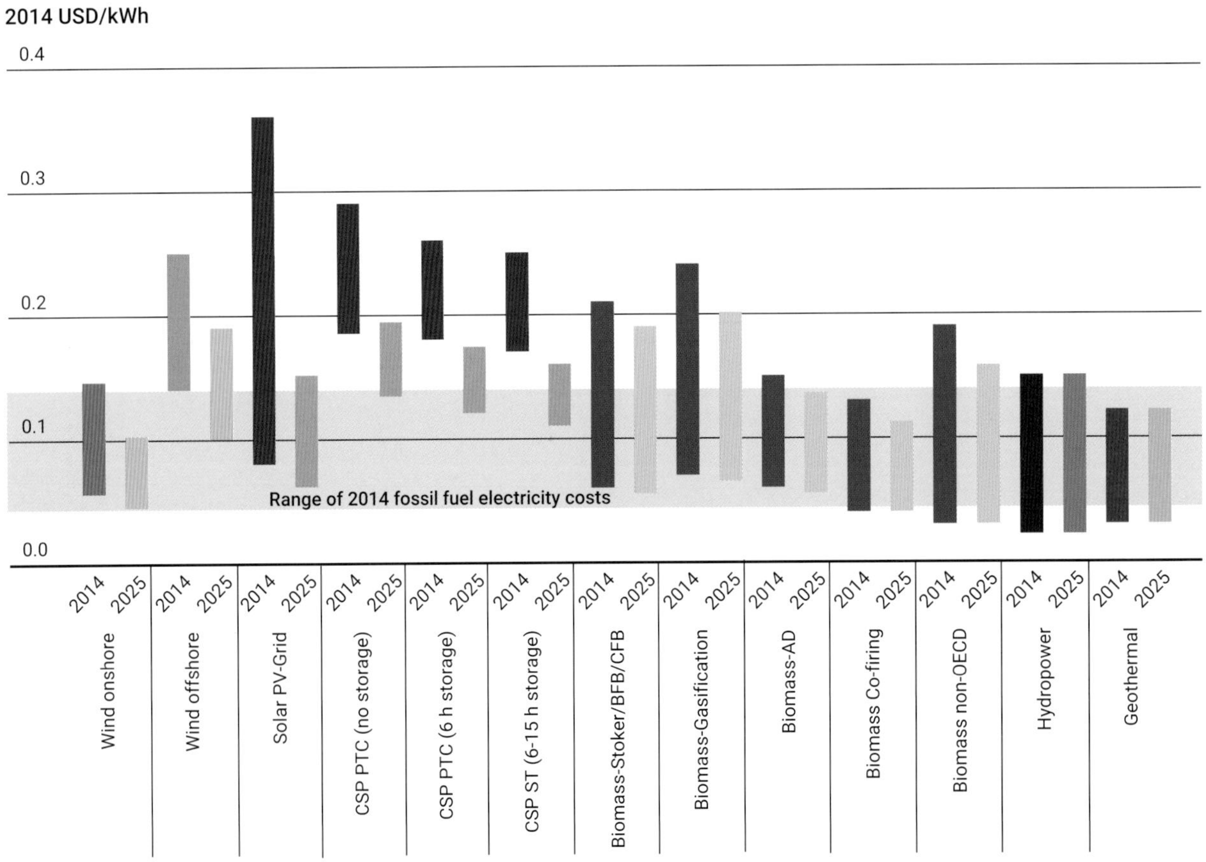

2014 USD/kWh

0.4

0.3

0.2

0.1

0.0

Range of 2014 fossil fuel electricity costs

Wind onshore 2014 2025 | Wind offshore 2014 2025 | Solar PV-Grid 2014 2025 | CSP PTC (no storage) 2014 2025 | CSP PTC (6 h storage) 2014 2025 | CSP ST (6-15 h storage) 2014 2025 | Biomass-Stoker/BFB/CFB 2014 2025 | Biomass-Gasification 2014 2025 | Biomass-AD 2014 2025 | Biomass Co-firing 2014 2025 | Biomass non-OECD 2014 2025 | Hydropower 2014 2025 | Geothermal 2014 2025

Source: IRENA (2015).

Education is crucial for developing energy literacy. Seen from the perspective of the SDGs, it enables individuals to apply and evaluate measures to increase energy efficiency and sufficiency in their own lives. It also influences public policies related to energy production, supply and usage (Aguirre-Bielschowsky *et al.* 2015; UNESCO 2017a).

4.4.3 Food systems

The global food system is central to sustainable development and to many of the SDGs. Across the complex interactions of activities including farming, fishing, food processing, retailing, preparing and consuming, and the multiple actors who perform them, the food system both significantly affects and is affected by environmental and social-economic dynamics (UNEP 2016c). Agriculture provides jobs for over 30 per cent of the global workforce, the majority in developing countries where 40 per cent of smallholder farmers and laborers are women (FAO 2011; FAO 2017a). Smallholder-dominated systems in developing countries produce more than half of all global food calories (Samberg *et al.* 2016) and contribute significantly to micronutrient production (Herrero *et al.* 2017). Fifty-seven million people work in fisheries and aquaculture, where women's roles are often invisible and underrecognized (Koralagama, Gupta and Pouw 2017), with many more in food manufacturing and retail (FAO 2016). A great number of these women and men live in poverty.

While the food system produces more than enough to feed the world's population adequately, it does not distribute it well. Over 800 million people are undernourished (FAO 2017a) and more than 2 billion suffer from micronutrient deficiencies (Global Panel on Agriculture and Food Systems for Nutrition 2016). However, over 2.3 billion people – about one-third of the human population – are obese or overweight (Abarca-Gómez *et al.* 2017). Diet-related diseases are globally pervasive, and many are associated with overconsumption of saturated fats and processed foods, such as type 2 diabetes, colorectal cancer and cardiovascular disease (Monteiro *et al.* 2013; Tilman and Clark 2014; UNEP 2016c). These diseases are becoming increasingly prevalent in low-income and middle-income countries, as animal protein and products high in fats and sugars become more widely available (Popkin 2006; McMichael *et al.* 2007).

The environmental footprint of the global food system is immense. It is estimated to account for 19-29 per cent of global greenhouse gas emissions (Vermeulen, Campbell and Ingram 2012). Farming is the most expansive human activity in the world, accounting for 38 per cent of global land area, and it is the principal user of fresh water, responsible for 70 per cent of withdrawals (FAO 2017a; FAO 2017b). Food production is the main driver of biodiversity loss (Kok *et al.* 2014). It is a major polluter of air, fresh water and seawater, particularly in farming systems that make heavy or poorly managed use of chemical pesticides and fertilizers (Popp, Petö and Nagy 2013; Sutton *et al.* 2013; Zhang, Zeiss and Geng 2015). Food production systems are also a leading source of soil degradation and deforestation (Amundson *et al.* 2015; Vanwalleghem *et al.* 2017; FAO 2017a). Yet the global food system is estimated to convert only 38 per cent of harvested energy and 28 per cent of harvested protein into required food consumption after accounting for losses from food waste, trophic losses from livestock and human overconsumption (Alexander *et al.* 2017).

Within the global food system's environmental footprint, the consequences of livestock raising are disproportionately large. While supplying only 18 per cent of calories and 40 per cent of protein to the world's food supply, the livestock sector accounts for about half of agriculture's greenhouse gas emissions (Gerber *et al.* 2013; FAO 2017a) and almost 80 per cent of agricultural land use – a third of all cropland is used to produce feed crops (FAO 2009). Due to the livestock sector, food production is the principal cause of habitat destruction (Machovina, Feeley and Ripple 2015) and the main disrupter of the nitrogen and phosphorous cycles that produce most of agriculture's pollution (Bouwman *et al.* 2013; Sutton *et al.* 2013). As with many resource extraction activities, the environmental burden of food production is localized, and often spatially dislocated from the consumption that drives demand. Around 20 per cent of cropland area and agricultural water use is devoted to agricultural commodities consumed in other countries (MacDonald *et al.* 2015). Similarly, overexploitation of wild fish stocks and intensive aquaculture have detrimental effects on marine and terrestrial ecosystems (see Chapter 7).

© Shutterstock/Magdalena Paluchowska

4

Current environmental pressures from the global food system cannot be sustained, yet to meet projected demand in 2050, with current efficiencies, world agricultural production would need to increase by 50 per cent from 2013 levels (FAO 2017a) with global crop demand forecast to increase 100-110 per cent over the same period (Tilman *et al.* 2011). Flows of nitrogen and phosphorous into the biosphere and oceans already exceed globally sustainable levels **(Figure 4.13)** (Steffen *et al.* 2015). On current trajectories, agricultural emissions are incompatible with a 2°C pathway. Action to reduce the volume and intensity of agricultural emissions, the amount of food waste and, most importantly, the share of animal products in diets will be necessary if the Paris Agreement's goal is to be achieved (Bajželj *et al.* 2014; Hedenus, Wirsenius and Johansson 2014; United Nations 2015b). On a global basis, diets with lower levels of animal products and higher levels of fruit, vegetables, pulses, whole grains and nuts are necessary to meet environmental and nutritional goals (Springmann *et al.* 2018), although particular requirements for dietary change will vary according to national context.

The food system is highly vulnerable to the pressures it is exerting on ecosystem services. Habitat loss is degrading pollinator services, with implications for crops important to human nutrition (Vanbergen 2013; Intergovernmental Science-Policy Platform on Biodiversity and Ecosystem Services 2016). Land degradation decreases crop yields, and abandonment rates of agricultural land due to that degradation appear to have increased (Gibbs and Salmon 2015; United Nations Convention to Combat Desertification 2017). Rising temperatures are thought to be diminishing crop yields rather than enhancing them in certain regions, especially for wheat and maize (Asseng *et al.* 2014; Porter *et al.* 2014; Moore and Lobell 2015; Schauberger *et al.* 2017). This trend is likely to have an increasingly detrimental effect on agriculture, particularly in low-latitude developing countries, although some temperate regions may benefit from warmer temperatures and longer growing seasons in the medium term, if soil and water characteristics are right (Deryng *et al.* 2014; Porter *et al.* 2014; Zhao *et al.* 2017). Water scarcity may limit the extent to which irrigation expansion can counter climate threats to crop yields;

Figure 4.13: The subglobal distributions and current status of the control variables for (A) biogeochemical flows of phosphorus; (B) biogeochemical flows of nitrogen

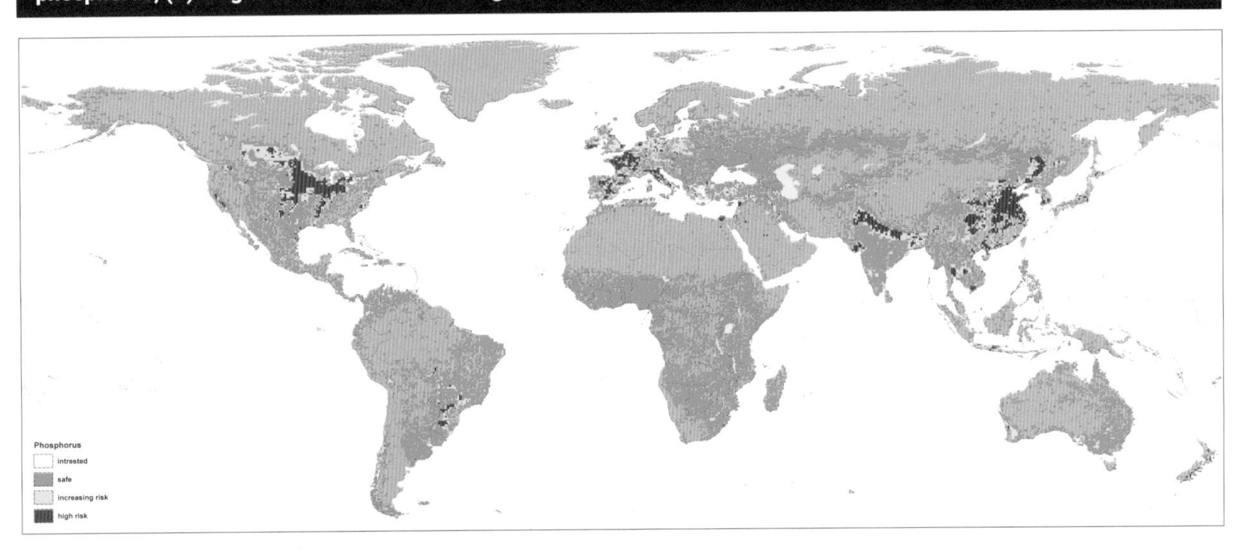

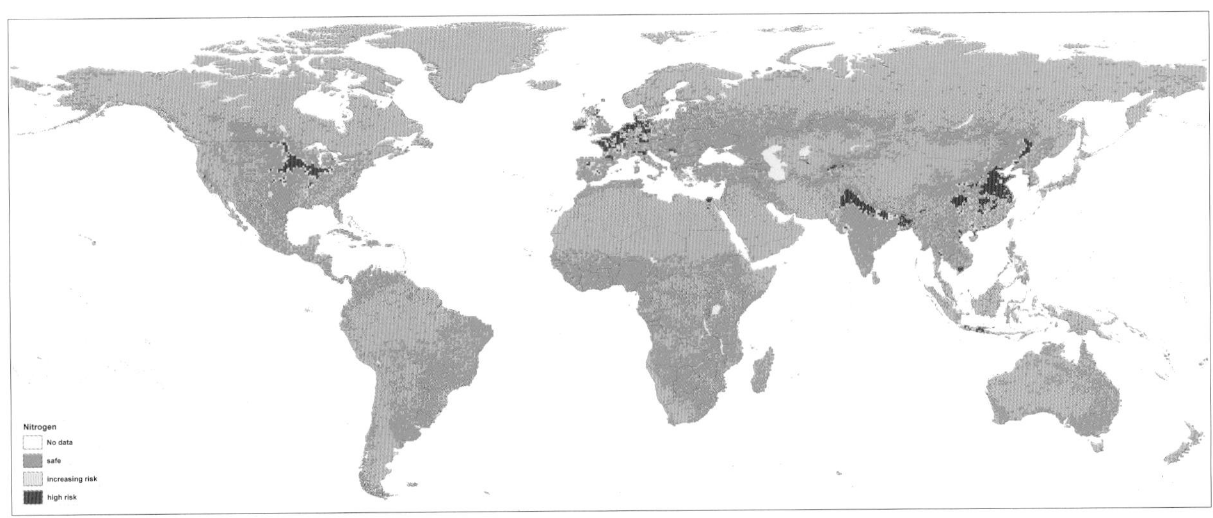

Source: Steffen *et al.* (2015).

4

in fact, it may force reversion to rain-fed agriculture in a number of important crop-producing regions by the end of this century, with further consequences for crop production (Elliott *et al.* 2013). Overexploitation is already compromising groundwater in several large aquifers critical to agriculture (Gleeson *et al.* 2012).

4.5 Conclusions

This GEO-6 assessment offers opportunities to identify cross-cutting issues as entry points for further understanding the state of the global environment. By exploring the 12 cross-cutting issues and how they relate to the Earth system topics, GEO can demonstrate where intersections and nexus issues will need synergistic solutions with the objective of achieving true transformative change.

References

Abarca-Gómez, L., Abdeen, Z.A., Hamid, Z.A., Abu-Rmeileh, N.M., Acosta-Cazares, B., Acuin, C. et al. (2017). Worldwide trends in body-mass index, underweight, overweight, and obesity from 1975 to 2016: A pooled analysis of 2416 population-based measurement studies in 128·9 million children, adolescents, and adults. The Lancet 390(10113), 2627–2642. https://doi.org/10.1016/S0140-6736(17)32129-3.

AbuZeid, K., Elrawady, M., CEDARE and Arab Water Council (2014). 2nd Arab State of the Water report 2012. http://www.arabwatercouncil.org/images/Publications/Arab_state/2nd_Arab_State_of_the_Water_Report.pdf.

Achterberg, E.P. (2014). Grand challenges in marine biogeochemistry. Frontiers in Marine Science 1(7), 1-5. https://doi.org/10.3389/fmars.2014.00007.

Adger, W.N., Pulhin, J.M., Barnett, J., Dabelko, G.D., Hovelsrud, G.K., Levy, M. *et al.* (2014). Human security. In *Climate Change 2014: Impacts, Adaptation, and Vulnerability. Part A: Global and Sectoral Aspects. Contribution of Working Group II to the Fifth Assessment Report of the Intergovernmental Panel on Climate Change*. Field, C.B., Dokken, D.J., Mastrandrea, M.D., Mach, K.J., Bilir, T.E., Chatterjee, M. *et al.* (eds.). Cambridge: Cambridge University Press. chapter 12. 755-791. https://www.ipcc.ch/pdf/assessment-report/ar5/wg2/WGIIAR5-Chap12_FINAL.pdf.

African Union (2009). African Union Convention for the Protection and Assistance of Internally Displaced Persons in Africa (Kampala Convention). Kampala https://au.int/en/treaties/african-union-convention-protection-and-assistance-internally-displaced-persons-africa.

Agarwal, B. (2010). *Gender and Green Governance: The Political Economy of Women's Presence*. Oxford: Oxford University Press. https://global.oup.com/academic/product/gender-and-green-governance-9780199569687?cc=ke&lang=en&#.

Agarwal, B. (2015). The power of numbers in gender dynamics: Illustrations from community forestry groups. *The Journal of Peasant Studies* 42(1), 1-20. https://doi.org/10.1080/03066150.2014.936007.

Aguilar, L., Granat, M. and Owren, C. (2015). *Roots for The Future: The Landscape and Way Forward on Gender and Climate Change*. Washington, D.C: International Union for Conservation of Nature and Global Gender Office. https://portals.iucn.org/library/sites/library/files/documents/2015-039.pdf.

Aguirre-Bielschowsky, I., Lawson, R., Stephenson, J. and Todd, S. (2015). Energy literacy and agency of New Zealand children. *Environmental Education Research* 23(6), 832–854. https://doi.org/10.1080/13504622.2015.1054267.

Akenji, L., Bengtsson, M., Bleischwitz, R., Tukker, A. and Schandl, H. (2016). Ossified materialism: Introduction to the special volume on absolute reductions in materials throughput and emissions. *Journal of Cleaner Production* 132, 1-12. https://doi.org/10.1016/j.jclepro.2016.03.071.

Alexander, P., Brown, C., Arneth, A., Finnigan, J., Moran, D. and Rounsevell, M.D.A. (2017). Losses, inefficiencies and waste in the global food system. *Agricultural Systems* 153, 190-200. https://doi.org/10.1016/j.agsy.2017.01.014.

Allegri, M., Bianchi, M.G., Chiu, M., Varet, J., Costa, A.L., Ortelli, S. *et al.* (2016). Shape-related toxicity of titanium dioxide nanofibres. *PLoS One* 11(3), e0151365. https://doi.org/10.1371/journal.pone.0151365.

Alsos, I.G., Ehrich, D., Seidenkrantz, M.S., Bennike, O., Kirchhefer, A.J. and Geirsdottir, A. (2016). The role of sea ice for vascular plant dispersal in the Arctic. *Biological Letters* 12(9). http://dx.doi.org/10.1098/rsbl.2016.0264.

Amundson, R., Berhe, A.A., Hopmans, J.W., Olson, C., Sztein, A.E. and Sparks, D.L. (2015). Soil and human security in the 21st century. *Science* 348(6235), 1261071. https://doi.org/10.1126/science.1261071.

Andrew, R. (2014). *Socio-Economic Drivers of Change in the Arctic. AMAP Technical Report No. 9*. Oslo: Arctic Monitoring and Assessment Programme. https://www.amap.no/documents/download/3011.

Angelsen, A. and Dokken, T. (2018). Climate exposure, vulnerability and environmental reliance: A cross-section analysis of structural and stochastic poverty. *Environment and Development Economics* 23(3), 257-278. https://doi.org/10.1017/S1355770X18000013.

Arctic Monitoring and Assessment Programme (2015). *Summary for Policy-makers: Arctic Pollution Issues 2015. Persistent Organic Pollutants; Radioactivity in the Arctic; Human Health in the Arctic*. Oslo. https://www.amap.no/documents/download/2222.

Arctic Monitoring and Assessment Programme (2017). *AMAP Assessment 2016: Chemicals of Emerging Arctic Concern*. Oslo. https://www.amap.no/documents/download/3003.

Arneth, A., Sitch, S., Pongratz, J., Stocker, B.D., Ciais, P., Poulter, B. *et al.* (2017). Historical carbon dioxide emissions caused by land-use changes are possibly larger than assumed. *Nature Geoscience* 10(2), 79–84. https://doi.org/10.1038/ngeo2882.

Arrobas, D., Hund, K., Mccormick, M., Ningthoujam, J. and Drexhage, J. (2017). *The Growing Role of Minerals and Metals for a Low Carbon Future*. Washington, D.C: World Bank. http://documents.banquemondiale.org/curated/fr/207371500386458722/pdf/117581-WP-P159838-PUBLIC-ClimateSmartMiningJuly.pdf.

Asseng, S., Ewert, F., Martre, P., Rötter, R.P., Lobell, D.B., Cammarano, D. *et al.* (2014). Rising temperatures reduce global wheat production. *Nature Climate Change* 5(2), 143-147. 10.1038/nclimate2470.

Ayres, J.G., Harrison, R.M., Nichols, G.L. and Maynard, R.L. (eds.) (2010). *Environmental Medicine*. London: CRC Press. https://www.crcpress.com/Environmental-Medicine/Ayres-Harrison-Nichols-Maynard-CBE/p/book/9780340946565.

Bajželj, B., Richards, K.S., Allwood, J.M., Smith, P., Dennis, J.S., Curmi, E. *et al.* (2014). Importance of food-demand management for climate mitigation. *Nature Climate Change* 4( ), 924-929. https://doi.org/10.1038/nclimate2353.

Balke, V., Evans, S., Rabbiosi, L. and Monnery, S.A. (2017). Promoting circular economies. In *Green Industrial Policy: Concept, Policies, Country Experiences*. Altenburg, T. and Assmann, C. (eds.). Nairobi: United Environment Environment Programme. chapter 8. 120-133. http://wedocs.unep.org/bitstream/handle/20.500.11822/22277/Green_industrial_policy.pdf?sequence=1&isAllowed=y.

Bamber, J.L., Westaway, R.M., Marzeion, B. and Wouters, B. (2018). The land ice contribution to sea level during the satellite era. *Environmental Research Letters* 13(6), 063008. https://doi.org/10.1088/1748-9326/aac2f0.

Barnes, D.K., Walters, A. and Gonçalves, L. (2010). Macroplastics at sea around Antarctica. *Marine Environmental Research* 70(2), 250-252. https://www.doi.org/10.1016/j.marenvres.2010.05.006.

Barrows, A.P.W., Cathey, S.E. and Petersen, C.W. (2018). Marine environment microfiber contamination: Global patterns and the diversity of microparticle origins. *Environmental Pollution* 237, 275-284. https://doi.org/10.1016/j.envpol.2018.02.062.

Barth, M. and Rieckmann, M. (2016). State of the art in research on higher education for sustainable development. In *Routledge Handbook of Higher Education for Sustainable Development*. Barth, M., Michelsen, G., Thomas, I. and Rieckmann, R. (eds.). London: Routledge. chapter 7. 100-113. https://www.taylorfrancis.com/books/e/9781317918110/chapters/10.4324%2F9781315852249-18

Bernhardt, E.S., Rossi, E.J. and Gessner, M.O. (2017). Synthetic chemicals as agents of global change. *Frontiers in Ecology and the Environment* 15 (2), 84–90. https://doi.org/10.1002/fee.1450.

Bettencourt, L. (2013). The origins of scaling in cities. *Science* 340(6139), 1438-1441. https://doi.org/10.1126/science.1235823.

Biberhofer, P., Lintner, C., Bernhardt, J. and Rieckmann, M. (2018). Facilitating work performance of sustainability-driven entrepreneurs through higher education. *The International Journal of Entrepreneurship and Innovation*. https://doi.org/10.1177/1465750318755881.

Bliss, A., Hock, R. and Radić, V. (2014). Global response of glacier runoff to twenty-first century climate change. *Journal of Geophysical Research: Earth Surface* 119(4), 717-730. https://doi.org/10.1002/2013JF002931.

Bouwman, L., Goldewijk, K.K., Van Der Hoek, K.W., Beusen, A.H.W., Van Vuuren, D.P., Willems, J. *et al.* (2013). Exploring global changes in nitrogen and phosphorus cycles in agriculture induced by livestock production over the 1900–2050 period. *Proceedings of the National Academy of Sciences* 110(52), 21195-21196. https://doi.org/10.1073/pnas.1012878108.

Bradshaw, S. and Fordham, M. (2015). Double disaster: Disaster through a gender lens. In *Hazards, Risks and Disasters in Society*. London: Elsevier. 233-251. http://nrl.northumbria.ac.uk/21075/

Brouwer M. (2018). *Progress in Pesticide Exposure Assessment: The Case of Parkinson's Disease in the Netherlands*. Utrecht University https://www.ris.uu.nl/ws/files/41629133/Brouwer.pdf

Browne, M.A., Crump, P., Niven, S.J., Teuten, E., Tonkin, A., Galloway, T. *et al.* (2011). Accumulation of microplastic on shorelines woldwide: Sources and sinks. *Environmental Science & Technology* 45(21), 9175-9179. https://www.doi.org/10.1021/es201811s.

Buse, C.G., Oestreicher, J.S., Ellis, N.R., Patrick, R., Brisbois, B., Jenkins, A.P. *et al.* (2018). Public health guide to field developments linking ecosystems, environments and health in the Anthropocene. *Journal of Epidemiology and Community Health* 72(5), 420-425. http://dx.doi.org/10.1136/jech-2017-210082.

Caldeira, K. and Wickett, M.E. (2003). Oceanography: Anthropogenic carbon and ocean pH. *Nature* 425(6956), 365-365. https://doi.org/10.1038/425365a.

Carpenter, S.R. and Bennett, E.M. (2011). Reconsideration of the planetary boundary for phosphorus. *Environmental Research Letters* 6(1), 014009. https://doi.org/10.1088/1748-9326/6/1/014009.

Centre for Research on the Epidemiology of Disasters (2017). *EM-DAT: The International Disaster Database*. Centre for Research on the Epidemiology of Disaster. https://www.emdat.be/database (Accessed: 20 August 2017).

Challinor, A.J., Watson, J., Lobell, D.B., Howden, S.M., Smith, D.R. and Chhetri, N. (2014). A meta-analysis of crop yield under climate change and adaptation. *Nature Climate Change* 4, 287-291. https://doi.org/10.1038/nclimate2153.

Chang, K., M., Hess, J., J.,, Balbus, J.M., Buonocore, J.J., Cleveland, D.A., Grabow, M.L. *et al.* (2017). Ancillary health effects of climate mitigation scenarios as drivers of policy uptake: A review of air quality, transportation and diet co-benefits modeling studies. *Environmental Research Letters* 12(11), 113001. https://doi.org/10.1088/1748-9326/aa8f7b.

Chattopadhyay, R. and Duflo, E. (2004). Women as policy makers: Evidence from a randomized policy experiment in India. *Econometrica* 72(5), 1409-1443. https://doi.org/10.1111/j.1468-0262.2004.00539.x.

Chiwona-Karltun, L., Kimanzu, N., Clendenning, J., Lodin, J.B., Ellingson, C., Lidestav, G. *et al.* (2017). What is the evidence that gender affects access to and use of forest assets for food security? A systematic map protocol. *Environmental Evidence* 6(2). https://doi.org/10.1186/s13750-016-0080-9.

Ciacci, L., Reck, B.K., Nassar, N.T. and Graedel, T.E. (2015). Lost by design. *Environmental Science & Technology* 49(16), 9443-9451. https://doi.org/10.1021/es505515z.

Cincinelli, A., Scopetani, C., Chelazzi, D., Lombardini, E., Martellini, T., Katsoyiannis, A. *et al.* (2017). Microplastic in the surface waters of the Ross Sea (Antarctica): Occurrence, distribution and characterization by FTIR. *Chemosphere* 175, 391-400. https://doi.org/10.1016/j.chemosphere.2017.02.024.

Cohen, A.J., Brauer, M., Burnett, R., Anderson, H.R., Frostad, J., Estep, K. *et al.* (2017). Estimates and 25-year trends of the global burden of disease attributable to ambient air pollution: An analysis of data from the Global Burden of Diseases Study 2015. *The Lancet* 389(10082), 1907-1918. https://doi.org/10.1016/S0140-6736(17)30505-6.

Conservation of Arctic Flora and Fauna and Protection of the Arctic Marine Environment (2017). *Arctic Invasive Alien Species: Strategy and Action Plan 2017*. Akureyri. https://www.doi.gov/sites/doi.gov/files/uploads/arias-27april2017_web.pdf.

Cook, B.I., Smerdon, J.E., Seager, R. and Coats, S. (2014). Global warming and 21st century drying. *Climate Dynamics* 43(9-10), 2607-2627. https://doi.org/10.1007/s00382-014-2075-y.

Copeland, C. and Carter, N.T. (2017). *Energy-Water Nexus: The Water Sector's Energy Use*. Congressional Research Service. https://fas.org/sgp/crs/misc/R43200.pdf.

Cordell, D. and Neset, T. (2014). Phosphorus vulnerability: A qualitative framework for assessing the vulnerability of national and regional food systems to the multi-dimensional stressors of phosphorus scarcity. *Global Environmental Change* 24, 108-122. https://doi.org/10.1016/j.gloenvcha.2013.11.005.

Coutts, C. and Hahn, M. (2015). Green infrastructure, ecosystem services, and human health. *International Journal of Environmental Research and Public Health* 12(8), 9768-9798. https://doi.org/10.3390/ijerph120809768.

Crowson, P. (2012). Some observations on copper yields and ore grades. *Resources Policy* 37(1), 59-72. https://doi.org/10.1016/j.resourpol.2011.12.004.

Dankelman, I. (2016). *Action Not Words: Confronting Gender Equality Through Climate Change Action and Disaster Risk Reduction in Asia. An overview of progress in Asia with Evidence from Bangladesh, Cambodia and Viet Nam*. Aipira, C., Kidd, A., Reggers, A., Fordham, M., Shreve, C. and Burnett, A. (eds.) United Nations Entity for Gender Equality and the Empowerment of Women. http://www2.unwomen.org/-/media/field%20office%20eseasia/docs/publications/2017/04/ccdrr_130317-s.pdf?la=en&vs=5239.

Dankelman, I. and Davidson, J. (1988). *Women and the Environment in the Third World: Alliance for the Future*. 1st edn. London: Earthscan. https://www.popline.org/node/381129.

Deng, H.-M., Liang, Q.-M., Liu, L.-J. and Anadon, L.D. (2018). Co-benefits of greenhouse gas mitigation: a review and classification by type, mitigation sector, and geography. *Environmental Research Letters* 12(12), 123001. https://doi.org/10.1088/1748-9326/aa98d2.

Deryng, D., Conway, D., Ramankutty, N., Price, J. and Warren, R. (2014). Global crop yield response to extreme heat stress under multiple climate change futures. *Environmental Research Letters* 9(3). https://doi.org/10.1088/1748-9326/9/3/034011.

Diamond, M.L., de Wit, C.A., Molander, S., Scheringer, M., Backhaus, T., Lohmann, R. *et al.* (2015). Exploring the planetary boundary for chemical pollution. *Environment International* 78, 8-15. https://doi.org/10.1016/j.envint.2015.02.001.

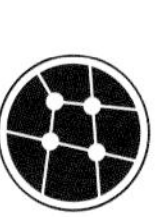

Disaster Displacement (2017). *Platform on disaster displacement*. [Disaster Displacement https://disasterdisplacement.org/ (Accessed: 27 September 2018).

Dodds, K. (2010). A polar Mediterranean? Accessibility, resources and sovereignty in the Arctic Ocean. *Global Policy* 1(3), 303-311. https://doi.org/10.1111/j.1758-5899.2010.00038.x.

Doney, S.C., Fabry, V.J., Feely, R.A. and Kleypas, J.A. (2009). Ocean acidification: The other CO^2 problem. *Annual review of marine science* 1, 169-192. https://doi.org/10.1146/annurev.marine.010908.163834.

Doss, C. (2014). Data needs for gender analysis in agriculture. In *Gender in agriculture: Closing the knowledge gap*. Quisumbing, A.R., Meinzen-Dick, R., Raney, T.L., Croppenstedt, A., Behrman, J.A. and Peterman, A. (eds.). Dordrecht: Springer. 55-68. https://link.springer.com/book/10.1007/978-94-017-8616-4

Duan, H., Li, J. and Liu, G. (2017). Developing countries: Growing threat of urban waste dumps. *Nature* 546(7660), 599-599. https://doi.org/10.1038/546599b.

Dunne, J., Stouffer, R. and John, J. (2013). Reductions in labour capacity from heat stress under climate warming. *Nature Climate Change* 3(3), 563-566. https://doi.org/10.1038/nclimate1827.

Eastin, J. (2016). Hell or high water: Precipitation shocks and conflict violence in the Philippines. *Political Geography* 63, 116-134. https://doi.org/10.1016/j.polgeo.2016.12.001.

Ellen MacArthur Foundation (2012). *Towards the Circular Economy: Economic and Business Rationale for an Accelerated Transition*. https://www.ellenmacarthurfoundation.org/assets/downloads/publications/Ellen-MacArthur-Foundation-Towards-the-Circular-Economy-vol.1.pdf.

Elliott, J., Deryng, D., Müller, C., Frieler, K., Konzmann, M., Gerten, D. *et al.* (2013). Constraints and potentials of future irrigation water availability on agricultural production under climate change. *Proceedings of the National Academy of Sciences* 111(9), 3239-3244. https://doi.org/10.1073/pnas.1222474110.

European Chemicals Agency (2018). *Sostanze registrate*. https://echa.europa.eu/it/information-on-chemicals/registered-substances.

European Commission (2015). Communication from the Commission to the European Parliament, the Council, the European Economic and Social Committee and the Committee of the Regions. Closing the Loop - An EU Action Plan for the Circular Economy *COM/2015/0614 final*. Brussels https://eur-lex.europa.eu/legal-content/EN/TXT/?uri=CELEX:52015DC0614

Fabry, V.J., Seibel, B.A., Feely, R.A. and Orr, J.C. (2008). Impacts of ocean acidification on marine fauna and ecosystem processes. *ICES Journal of Marine Science* 65(3), 414-432. https://doi.org/10.1093/icesjms/fsn048.

Fehr, R., Mekel O.C.L., Hurley, J.F., Mekel, O.C. and Mackenbach, J.P. (2016). Health impact assessment: A survey of quantifying tools. *Environmental Impact Assessment Review* 57, 178-186. http://dx.doi.org/10.1016/j.eiar.2016.01.001.

Feng, H. and Zhang, M. (2015). Global land moisture trends: Drier in dry and wetter in wet over land. *Scientific Reports* 5(18018), 1-6. https://doi.org/10.1038/srep18018.

Ferrario, C., Finizio, A. and Villa, S. (2017). Legacy and emerging contaminants in meltwater of three Alpine glaciers. *Science of The Total Environment* 574, 350-357. https://www.doi.org/10.1016/j.scitotenv.2016.09.067.

Finley, R.L., Collignon, P., Larsson, D.G.J., Mcewen, S.A., Li, X.Z., Gaze, W.H. *et al.* (2013). The scourge of antibiotic resistance: The important role of the environment. *Clinical Infectious Diseases* 57(5), 704–710. https://doi.org/10.1093/cid/cit355.

Food and Agriculture Organization of the United Nations (2009). *The State of Food and Agriculture 2009: Livestock in the Balance*. Rome. http://www.fao.org/docrep/012/i0680e/i0680e.pdf.

Food and Agriculture Organization of the United Nations (2011). *The State of Food and Agriculture 2010–11. Women in Agriculture: Closing the Gender Gap for Development*. Rome. http://www.fao.org/docrep/013/i2050e/i2050e.pdf.

Food and Agriculture Organization of the United Nations (2013). *Food Wastage Footprint. Impacts on Natural Resources*. Rome. http://www.fao.org/docrep/018/i3347e/i3347e.pdf.

Food and Agriculture Organization of the United Nations (2015). *Global Initiative on Food Loss and Waste Reduction*. Rome. http://www.fao.org/3/a-i4068e.pdf.

Food and Agriculture Organization of the United Nations (2016). *The State of World Fisheries and Aquaculture 2016: Contributing to Food Security and Nutrition for All*. Rome. http://www.fao.org/3/a-i5555e.pdf.

Food and Agriculture Organization of the United Nations (2017a). *The Future of Food and Agriculture: Trends and Challenges*. Rome. http://www.fao.org/3/a-i6583e.pdf.

Food and Agriculture Organization of the United Nations (2017b). *FAOStat*. http://www.faostat.org/en (Accessed: 20 July 2017).

Food and Agriculture Organization of the United Nations and World Health Organization (2016). *International Code of Conduct on Pesticide Management: Guidelines on Highly Hazardous Pesticides*. Rome. http://apps.who.int/iris/bitstream/handle/10665/205561/9789241510417_eng.pdf?sequence=1&isAllowed=y.

Francis, J.A., Vavrus, S.J. and Cohen, J. (2017). Amplified Arctic warming and mid-latitude weather: New perspectives on emerging connections. *Wiley Interdisciplinary Reviews: Climate Change* 8(5), e474. https://doi.org/10.1002/wcc.474.

Frey, K.E., Comiso, J.C., Cooper, L.W., Gradinger, R.R., Grebmeier, J.M. and Tremblay, J.-É. (2016). *Arctic Ocean Primary Productivity*. Arctic Report Card 2016Arctic Program. http://arctic.noaa.gov/Report-Card/Report-Card-2016/ArtMID/5022/ArticleID/284/Arctic-Ocean-Primary-Productivity.

Gardner, T. (2017). 'Ethiopia's deadly rubbish dump landslide sparks landrights battle'. *Reuters* 3 May 2017 https://www.reuters.com/article/us-ethiopia-landslide-landrights/ethiopias-deadly-rubbish-dump-landslide-sparks-landrights-battle-idUSKBN17Z106.

Gerber, P.J., Steinfeld, H., Henderson, B., Mottet, A., Opio, C., Dijkman, J. *et al.* (2013). *Tackling Climate Change Through Livestock: A Global Assessment of Emissions and Mitigation Opportunities*. Rome: Food and Agriculture Organization of the United Nations. http://www.fao.org/3/a-i3437e.pdf.

Ghisellini, P., Cialani, C. and Ulgiati, S. (2016). A review on circular economy: The expected transition to a balanced interplay of environmental and economic systems. *Journal of Cleaner Production* 114(11-32). https://doi.org/10.1016/j.jclepro.2015.09.007.

Gibbs, H.K. and Salmon, J.M. (2015). Mapping the world's degraded lands. *Applied Geography* 57, 12-21. https://doi.org/10.1016/j.apgeog.2014.11.024.

Glasser, H. and Hirsh, J. (2016). Toward the development of robust learning for sustainability core competencies. *Sustainability: The Journal of Record* 9(3), 121-134. https://doi.org/10.1089/SUS.2016.29054.hg.

Glatzer, W., Camfield, L., Møller, V. and Rojas, M. (eds.) (2015). *Global Handbook of Quality of Life: Exploration of Well-Being of Nations and Continents*. Dordrecht: Springer. https://www.springer.com/gp/book/9789401791779.

Gleeson, T., Wada, Y., Bierkens, M.F.P. and van Beek, L.P.H. (2012). Water balance of global aquifers revealed by groundwater footprint. *Nature* 488(7410), 197–200. https://doi.org/10.1038/nature11295.

Global Panel on Agriculture and Food Systems for Nutrition (2016). *Food Systems and Diets: Facing the Challenges of the 21st Century*. London. http://glopan.org/sites/default/files/ForesightReport.pdf.

Gore, A.C., Chappell, V.A., Fenton, S.E., Flaws, J.A., Nadal, A., Prins, G.S. *et al.* (2015). Executive summary to EDC-2: The endocrine society's second scientific statement on endocrine disrupting chemicals. *Endocrine Reviews* 36(6), 593–602. https://doi.org/10.1210/er.2015-1093.

Graedel, T.E.H., Harper, E.M., Nassar, N.T. and Reck, B.K. (2015). On the materials basis of modern society. *Proceedings of the National Academy of Sciences* 112(20), 6295-6300. https://www.doi.org/10.1073/pnas.1312752110.

Grellier, J., White, M.P., Albin, M., Bell, S., Elliott, L.R., Gascon, M. *et al.* (2017). Bluehealth: A study programme protocol for mapping and quantifying the potential benefits to public health and well-being from Europe's blue spaces. *BMJ Open* 7(6), e016188. http://dx.doi.org/10.1136/bmjopen-2017-016188.

Grenni, P., Ancona, V. and Caracciolo, A.B. (2018). Ecological effects of antibiotics on natural ecosystems: A review. *Microchemical Journal* 136, 25-39. https://www.doi.org/10.1016/j.microc.2017.02.006.

Haines, A. (2017). Health co-benefits of climate action. *The Lancet Planetary Health* 1(1), e4-e5. https://doi.org/10.1016/S2542-5196(17)30003-7.

Hamscher, G. and Bachour, G. (2018). Veterinary drugs in the environment: Current knowledge and challenges for the future. *Journal of Agricultural and Food Chemistry* 66(4), 751-752. https://www.doi.org/10.1021/acs.jafc.7b05601.

Hanjra, M.A. and Qureshi, M.E. (2010). Global water crisis and future food security in an era of climate change. *Food Policy* 35(5), 365-377. https://doi.org/10.1016/j.foodpol.2010.05.006.

Harcourt, W. and Nelson, I.L. (eds.) (2015). *Practising Feminist Political Ecologies: Moving Beyond the Green Economy*. Chicago, IL: University of Chicago Press. http://press.uchicago.edu/ucp/books/book/distributed/P/bo20504936.html.

Harper, S., Grubb, C., Stiles, M. and Sumaila, U.R. (2017). Contributions by women to fisheries economies: Insights from five maritime countries. *Coastal Management* 45(2), 91-106. https://doi.org/10.1080/08920753.2017.1278143.

Harper, S., Zeller, D., Hauzer, M., Pauly, D. and Sumaila, U.R. (2013). Women and fisheries: Contribution to food security and local economies. *Marine Policy* 39, 56-63. https://doi.org/10.1016/j.marpol.2012.10.018.

Harris, G.L.A. (2011). The quest for gender equity. *Public Administration Review* 71(1), 123-126. https://doi.org/10.1111/j.1540-6210.2010.02315.x.

Hassell, J.M., Begon, M., Ward, M.J. and Fèvre, E.M. (2017). Urbanization and disease emergence: Dynamics at the wildlife–livestock–human interface. *Trends in Ecology & Evolution* 32(1), 55-67. https://doi.org/10.1016/j.tree.2016.09.012.

Haustein, K., Allen, M.R., Forster, P.M., Otto, F.E.L., Mitchell, D.M., Matthews, H.D. *et al.* (2017). A real-time global warming index. *Scientific Reports* 7(15417). https://doi.org/10.1038/s41598-017-14828-5.

Hedenus, F., Wirsenius, S. and Johansson, D.J.A. (2014). The importance of reduced meat and dairy consumption for meeting stringent climate change targets. *Climatic Change* 124(1-2), 79-91. https://doi.org/10.1007/s10584-014-1104-5.

Herrero, M., Thornton, P.K., Power, B., Bogard, J.R., Remans, R., Fritz, S. *et al.* (2017). Farming and the geography of nutrient production for human use: A transdisciplinary analysis. *The Lancet Planetary Health* 1(1), e33-e42. https://doi.org/10.1016/S2542-5196(17)30007-4.

Hopkins, C. (2015). Beyond the decade: The global action program for education for sustainable development. *Applied Environmental Education & Communication* 14(2), 132-136. https://doi.org/10.1080/1533015X.2015.1016860.

Horton, B.P., Rahmstorf, S., Engelhart, S.E. and Kemp, A.C. (2014). Expert assessment of sea-level rise by AD 2100 and AD 2300. *Quaternary Science Reviews* 84, 1-6. https://doi.org/10.1016/j.quascirev.2013.11.002.

Hudson-Edwards, K.A. (2016). Tackling mine wastes. *Science Magazine* 35(6283), 288-290. https://doi.org/10.1126/science.aaf3354.

Intergovernmental Panel on Climate Change (2013). *Climate Change 2013: The Physical Science Basis. Contribution of Working Group I to the Fifth Assessment Report of the Intergovernmental Panel on Climate Change*. Stocker, T.F., Qin, D., Plattner, G.-K., Tignor, M., Allen, S.K., Boschung, J. *et al.* (eds.). Cambridge, MA: Cambridge University Press. http://www.climatechange2013.org/images/report/WG1AR5_ALL_FINAL.pdf.

Intergovernmental Panel on Climate Change (2014). *Climate Change 2014: Impacts, Adaptation, and Vulnerability. Part B: Regional Aspects. Contribution of Working Group II to the Fifth Assessment Report of the Intergovernmental Panel on Climate Change*. Field, C.B., Dokken, D.J., Mastrandrea, M.D., Mach, K.J., Bilir, T.E., Chatterjee, M. *et al.* (eds.). Cambridge: Cambridge University Press. http://www.ipcc.ch/report/ar5/wg2/.

Intergovernmental Panel on Climate Change (2018). *Global Warming of 1.5°C*. http://www.ipcc.ch/report/sr15/.

Intergovernmental Science-Policy Platform on Biodiversity and Ecosystem Services (2016). *The Assessment Report of the Intergovernmental Science-Policy Platform on Biodiversity and Ecosystem Services on Pollinators, Pollination and Food Production*. Potts, S.G., Imperatriz-Fonseca, V.L. and Ngo, H.T. (eds.). Bonn. https://www.ipbes.net/sites/default/files/downloads/pdf/individual_chapters_pollination_20170305.pdf.

Internal Displacement Monitoring Centre (2017). *GRID 2017: Global Report on Internal Displacement*. Internal Displacement Monitoring Centre and Norwegian Refugee Council. http://www.internal-displacement.org/global-report/grid2017/pdfs/2017-GRID.pdf.

International Energy Agency (2015). *World Energy Outlook 2015*. Paris. https://www.iea.org/publications/freepublications/publication/WEO2015.pdf.

International Energy Agency (2016). *World Energy Outlook 2016*. Paris. https://webstore.iea.org/world-energy-outlook-2016.

International Energy Agency (2018). *Statistics*. https://www.iea.org/statistics/?country=WORLD&year=2016&category=Key%20indicators&indicator=TPESbySource&mode=chart&categoryBrowse=false&dataTable=BALANCES&showDataTable=false (Accessed: 17 August 2018).

International Organization for Migration (2015). *World Migration Report 2015. Migrants and Cities: New Partnerships to Manage Mobility*. Geneva. http://publications.iom.int/system/files/wmr2015_en.pdf.

International Renewable Energy Agency (2015). *Renewable Power Generation Costs in 2014*. Abu Dhabi. http://www.irena.org/DocumentDownloads/Publications/IRENA_RE_Power_Costs_2014_report.pdf.

International Resource Panel (2017). *Green Technology Choices: The Environmental and Resource Implications of Low-Carbon Technologies. A Report of the International Resource Panel*. Suh, S., Bergesen, J., Gibon, T.J., Hertwich, E. and Taptich M. (eds.). Nairobi: United Nations Environment Programme. http://www.resourcepanel.org/reports/green-technology-choices.

International Solid Waste Association (2016). *A Roadmap for Closing Waste Dumpsites: The World's most Polluted Places*. Vienna. https://www.iswa.org/fileadmin/galleries/About%20ISWA/ISWA_Roadmap_Report.pdf.

Isobe, A., Uchiyama-Matsumoto, K., Uchida, K. and Tokai, T. (2017). Microplastics in the southern ocean. *Marine Pollution Bulletin* 114(1), 623-626. https://doi.org/10.1016/j.marpolbul.2016.09.037.

Ivar do Sul, J.A. and Costa, M.F. (2013). The present and future of microplastic pollution in the marine environment. *Environmental Pollution* 185, 352-364. https://doi.org/10.1016/j.envpol.2013.10.036.

Jackson, T. and Senker, P. (2011). Prosperity without growth: Economics for a finite planet. *Energy & Environment* 22(7), 1013-1016. https://doi.org/10.1260/0958-305X.22.7.1013.

Jägerskog, A., Clausen, T.J., Holmgren, T. and Lexén, K. (2014). *Energy and Water: The Vital Link for a Sustainable Future*. Stockholm: Stockholm International Water Institute. http://www.worldwaterweek.org/wp-content/uploads/2014/08/2014_WWW_Report_web-2.pdf.

Jamieson, A.J., Malkocs, T., Piertney, S.B., Fujii, T. and Zhang, Z. (2017). Bioaccumulation of persistent organic pollutants in the deepest ocean fauna. *Nature Ecology & Evolution* 1. https://doi.org/10.1038/s41559-016-0051.

Jerneck, A. (2018). What about gender in climate change? Twelve feminist lessons from development. *Sustainability* 10(3), 627. https://doi.org/10.3390/su10030627.

Jiménez-Cisneros, B.E., Oki, T., Arnell, N.W., Benito, G., Cogley, J.G., Döll, P. *et al.* (2014). Freshwater resources. In *Climate Change 2014: Impacts,Adaptation, and Vulnerability. Part A: Global and Sectoral Aspects. Contribution of Working Group II to the Fifth Assessment Report of the Intergovernmental Panel on Climate Change*. Field, C.B., Dokken, D.J., Mastrandrea, M.D., Mach, K.J., Bilir, T.E., Chatterjee, M. *et al.* (eds.). Cambridge: Cambridge University Press. chapter 3. 229-269. https://www.ipcc.ch/pdf/assessment-report/ar5/wg2/WGIIAR5-Chap3_FINAL.pdf.

Kabir, E.R., Rahman, M.S. and Rahman, I. (2015). A review on endocrine disruptors and their possible impacts on human health. *Environmental toxicology and pharmacology* 40(1), 241-258. https://doi.org/10.1016/j.etap.2015.06.009.

Kaltenborn, B.P., Nellemann, C. and Vistnes, I.I. (eds.) (2010). *High mountain glaciers and climate change – Challenges to human livelihoods and adaptation*: United Nations Environment Programme and GRID-Arendal. http://wedocs.unep.org/bitstream/handle/20.500.11822/8101/-High%20mountain%20glaciers%20and%20climate%20change%20-%20Challenges%20to%20human%20livelihoods%20and%20adaptation-20101128.pdf?sequence=2&isAllowed=y.

Karatzoglou, B. (2013). An in-depth literature review of the evolving roles and contributions of universities to education for sustainable development. *Journal of Cleaner Production* 49, 44-53. https://doi.org/10.1016/j.jclepro.2012.07.043.

Khatiwala, S., Tanhua, T., Fletcher, S.M., Gerber, M., Doney, S.C., Graven, H.D. *et al.* (2013). Global ocean storage of anthropogenic carbon. *Biogeosciences* 10 (4), 2169-2191. https://doi.org/10.5194/bg-10-2169-2013.

Kok, M., Alkemade, R., Bakkenes, M., Boelee, E., Christensen, V., van Eerdt, M. *et al.* (2014). *How sectors can contribute to sustainable use and conservation of biodiversity*. CBD Technical Series No 79PBL Netherlands Environmental Assessment Agency. https://www.cbd.int/doc/publications/cbd-ts-79-en.pdf.

Koralagama, D., Gupta, J. and Pouw, N. (2017). Inclusive development from a gender perspective in small scale fisheries. *Current Opinion in Environmental Sustainability* 24, 1-6. https://doi.org/10.1016/j.cosust.2016.09.002.

Kosoy, N., Brown, P.G., Bosselmann, K., Duraiappah, A., Mackey, B., Martinez-Alier, J. *et al.* (2012). Pillars for a flourishing earth: Planetary boundaries, economic growth delusion and green economy. *Current Opinion in Environmental Sustainability* 4(1), 74–79. https://doi.org/10.1016/j.cosust.2012.02.002.

Krausmann, F., Schandl, H., Eisenmenger, N., Giljum, S. and Jackson, T.D. (2017). Material flow accounting: Global material use and sustainable development. *Annual Review of Environment and Resources* 42, 647-675. https://doi.org/10.1146/annurev-environ-102016-060726.

Kundzewicz, Z.W., Kanae, S., Seneviratne, S.I., Handmer, J., Nicholls, N., Peduzzi, P. *et al.* (2014). Flood risk and climate change: Global and regional perspectives. *Hydrological Sciences Journal* 59(1), 1–28. https://doi.org/10.1080/02626667.2013.857411.

Lafferty, K.D. (2009). The ecology of climate change and infectious diseases. *Ecology* 90(4), 888-900. https://doi.org/10.1890/08-0079.1.

Lambeth, L., Hanchard, B., Aslin, H., Fay-Sauni, L., Tuara, P., Rochers, K.D. *et al.* (2014). An overview of the Involvement of women in fisheries activities in Oceania. In *Global Symposium on Women in Fisheries*. Williams, M.J., Chao, N.H., Choo, P.S., Matics, K., Nandeesha, M.C., Shariff, M. *et al.* (eds.). Penang: ICLARM – The World Fish Center. 127-142. https://www.researchgate.net/profile/Heather_Aslin/publication/23550943_An_overview_of_the_involvement_of_women_in_fisheries_activities_in_Oceania/links/00463525cdb0453fc7000000/An-overview-of-the-involvement-of-women-in-fisheries-activities-in-Oceania.pdf

Landrigan, P.J., Fuller, R., Acosta, N.J.R., Adeyi, O., Arnold, N.N., Baldé, A.B. *et al.* (2017). The Lancet Commission on pollution and health. *The Lancet* 391(10119), 1-57. https://doi.org/10.1016/S0140-6736(17)32345-0.

Leach, N.J., Millar, R.J., Haustein, K., Jenkins, S., Graham, E. and Allen, M.R. (2018). Current level and rate of warming determine emissions budgets under ambitious mitigation. *Nature Geoscience* 11(8), 574-579. https://doi.org/10.1038/s41561-018-0156-y.

Legere, L. (2016). 'State seismic network helps tell fracking quakes from natural ones'. *Pittsburgh Post Gazette* June 25 2016 http://powersource.post-gazette.com/powersource/policy-powersource/2016/06/26/State-seismic-network-helps-tell-fracking-quakes-from-natural-ones/stories/201606210014.

Leoni, B. (2012). *Japan quake took toll on women and elderly*. [United Nations Office for Disaster Risk Reduction https://www.unisdr.org/archive/25598.

Levitus, S., Antonov, J.I., Boyer, T.P., Baranova, O.K., Garcia, H.E., Locarnini, R.A. *et al.* (2012). World ocean heat content and thermosteric sea level change (0-2000 m), 1955-2010. *Geophysical Research Letters* 39(10), 1-5. https://doi.org/10.1029/2012GL051106.

Lewandowski, J., Meinikmann, K., Nützmann, G. and Rosenberry, O. (2015). Groundwater the disregarded component in lake water and nutrient budgets. Part 2: Effects of groundwater on nutrients. *Hydrological processes* 29(13), 2922-2955. https://doi.org/10.1002/hyp.10384.

Lindahl, J.F. and Grace, D. (2015). The consequences of human actions on risks for infectious diseases: A review. *Infection Ecology & Epidemiology* 5(1). https://doi.org/10.3402/iee.v5.30048.

Lobell, D.B. and Gourdji, S.M. (2012). The influence of climate change on global crop productivity. *Plant Physiology* 160(4), 1686-1697. https://doi.org/10.1104/pp.112.208298.

Lobell, D.B., Schlenker, W. and Costa-Roberts, J. (2011). Climate trends and global crop production since 1980. *Science* 333(6042), 616-620. https://doi.org/10.1126/science.1204531.

Lotz-Sisitka, H., Wals, A.E.J., Kronlid, D. and McGarry, D. (2015). Transformative, transgressive social learning: Rethinking higher education pedagogy in times of systemic global dysfunction. *Current Opinion in Environmental Sustainability* 16, 73-80. https://doi.org/10.1016/j.cosust.2015.07.018.

Lozano, R., Ceulemans, K., Alonso-Almeida, M., Huisingh, D., Lozano, F.J., Waas, T. *et al.* (2015). A review of commitment and implementation of sustainable development in higher education: Results from a worldwide survey. *Journal of Cleaner Production* 108(Part A), 1-18. https://doi.org/10.1016/j.jclepro.2014.09.048.

Lucard, M., Jaquemet, I. and Carpentier, B. (2011). Out of sight, out of mind. *The Magazine of the International Red Cross and Red Crescent Movement*. http://www.redcross.int/EN/mag/magazine2011_2/18-23.html

MacDonald, G.K., Brauman, K.A., Sun, S., Carlson, K.M., Cassidy, E.S., Gerber, J.S. *et al.* (2015). Rethinking agricultural trade relationships in an era of globalization. *BioScience* 65(3), 275-289. https://doi.org/10.1093/biosci/biu225.

Machovina, B.K., Feeley, J. and Ripple, W.J. (2015). Biodiversity conservation: The key is reducing meat consumption. *Science of the Total Environment* 536, 419-431. https://doi.org/10.1016/j.scitotenv.2015.07.022.

Magga, O.H., Mathiesen, S.D., Corell, R.W. and Oskal, A. (2009). *Reindeer Herding, Traditional Knowledge, Adaptation to Climate Change and Loss of Grazing Land*. Arctic Council. http://reindeerherding.org/wp-content/uploads/2013/06/EALAT-Final-Report.pdf.

Maggino, F. (2015). Assessing the subjective wellbeing of nations. In *Global Handbook of Quality of Life*. Glatzer, W., Camfield, L., Møller, V. and Rojas M. (eds.). Springer. 803–822. https://link.springer.com/chapter/10.1007/978-94-017-9178-6_37

Marti, E., Variatza, E. and Balcazar, J.L. (2014). The role of aquatic ecosystems as reservoirs of antibiotic resistance. *Trends in Microbiology* 22(1), 36-41. https://www.doi.org/10.1016/j.tim.2013.11.001.

Mateo-Sagasta, J., Medlicott, K., Manzoor, Q., Raschid-Sally, L., Drechsel, P. and Liebe, J. (2013). *Proceedings of the UN-Water project on the Safe use of wastewater in agriculture: Proceedings of the UN–Water project*. Liebe, J. and Ardakanian, R. (eds.). Bonn: UN-Water Decade Programme on Capacity Development (UNW-DPC). http://www.ais.unwater.org/ais/pluginfile.php/62/course/section/29/proceedings-no-11_WEB.pdf.

McDonald, R.I., Weber, K., Padowski, J., Flörke, M., Schneider, C., Green, P.A. *et al.* (2014). Water on an urban planet: Urbanization and the reach of urban water infrastructure. *Global Environmental Change* 27, 96-105. https://doi.org/10.1016/j.gloenvcha.2014.04.022.

McGrath, J.M. and Lobell, D.B. (2013). Regional disparities in the CO_2 fertilization effect and implications for crop yields. *Environmental Research Letters* 8(1), 1-9. https://doi.org/10.1088/1748-9326/8/1/014054.

McKeown, R. (2015). What happened during the UN decade of education for sustainable development? *Applied Environmental Education & Communication* 14(2), 67-69. https://doi.org/10.1080/1533015X.2014.971979.

McLellan, B.C. (2017). The minerals-energy nexus: Past, present and future. In *Sustainability Through Innovation in Product Life Cycle Design*. Matsumoto, M., Masui, K., Fukushige, S. and Kondoh, S. (eds.). Singapore: Springer. 619-631. https://link.springer.com/chapter/10.1007/978-981-10-0471-1_42

McMichael, A.J., Powles, J.W., Butler, C.D. and Uauy, R. (2007). Food, livestock prodution, energy, climate change, and health. *The Lancet* 370(9594), 1253-1263. https://doi.org/10.1016/S0140-6736(07)61256-2.

Mickelsson, M., Kronlid, D.O. and Lotz-Sisitka, H. (2018). Consider the unexpected: Scaling ESD as a matter of learning. *Environmental Education Research*, 1-16. https://doi.org/10.1080/13504622.2018.1429572.

Mills, J.E. and Cumming, O. (2016). *The Impact of Water, Sanitation and Hygiene on Key Health and Social Outcomes: Review of Evidence*. Barahman, V. and Poirier, P. (eds.)Sanitation and Hygiene Applied Research for Equity. https://www.unicef.org/wash/files/The_Impact_of_WASH_on_Key_Social_and_Health_Outcomes_Review_of_Evidence.pdf.

Mindt, L. and Rieckmann, M. (2017). Developing competencies for sustainability-driven entrepreneurship in higher education: A literature review on teaching and learning methods. *Teoría de la Educación. Revista Interuniversitaria* 29(1), 129-159. https://doi.org/10.14201/teoredu291129159.

Monteiro, C.A., Moubarac, J.-C., Cannon, G., Ng, S.W. and Popkin, B. (2013). Ultra-processed products are becoming dominant in the global food system. *Obesity Reviews* 14(2), 21-28. https://doi.org/10.1111/obr.12107.

Moore, F.C. and Lobell, D.B. (2015). The fingerprint of climate trends on European crop yields. *Proceedings of the National Academy of Sciences of the United States of America* 112(9), 2670-2675. https://doi.org/10.1073/pnas.1409606112.

Mudd, G.M. (2010). The environmental sustainability of mining in Australia: Key mega-trends and looming constraints. *Resources Policy* 35(2), 98-115. https://doi.org/10.1016/j.resourpol.2009.12.001.

Mudd, G.M. and Jowitt, S.M. (2014). A detailed assessment of global nickel resource trends and endowments. *Economic Geology* 109(7), 1813-1841. https://doi.org/10.2113/econgeo.109.7.1813.

Mudd, G.M. and Jowitt, S.M. (2016). From mineral resources to sustainable mining - The key trends to unlock the holy grail? In *Proceedings The Third AusIMM International Geometallurgy Conference (GeoMet) 2016*. Melbourne: The Australasian Institute of Mining and Metallurgy. 37–54. https://www.ausimm.com.au/publications/epublication.aspx?ID=16949

Mudd, G.M. and Jowitt, S.M. (2017). Global resource assessments of primary metals: An optimistic reality check. *Natural Resources Research* 27(2), 229-240. https://doi.org/10.1007/s11053-017-9349-0.

Mudd, G.M., Weng, Z. and Jowitt, S.M. (2013). A detailed assessment of global Cu resource trends and endowments. *Economic Geology* 108(5), 1163-1183. http://dx.doi.org/10.2113/econgeo.108.5.1163.

Mundy, V. and Sant, G. (2015). *Traceability Systems in the CITES context: A Review of Experiences, Best Practices and Lessons Learned for the Traceability of Commodities of CITES-listed Shark Species*. Geneva: Secretariat of the Convention on International Trade in Endangered Species of Wild Fauna and Flora. https://cites.org/sites/default/files/eng/prog/shark/docs/BodyofInf12.pdf.

Nagai, H. and Toyokuni, S. (2012). Differences and similarities between carbon nanotubes and asbestos fibers during mesothelial carcinogenesis: Shedding light on fiber entry mechanism. *Cancer Science* 103(8), 1378-1390. https://www.doi.org/10.1111/j.1349-7006.2012.02326.x.

Nakashima, D.J., Galloway McLean, K., Thulstrup, H.D., Ramos Castillo, A. and Rubis, J.T. (2012). *Weathering Uncertainty: Traditional Knowledge for Climate Change Assessment and Adaptation*. Paris: United Nations Educational, Scientific and Cultural Organization and United Nations University. http://unesdoc.unesco.org/images/0021/002166/216613e.pdf.

Nassar, N.T., Graedel, T.E. and Harper, E.M. (2015). By-product metals are technologically essential but have problematic supply. *Science Advances* 1(3), e1400180. https://doi.org/10.1126/sciadv.1400180.

Norby, R.J., De Kauwe, M.G., Domingues, T.F., Duursma, R.A., Ellsworth, D.S., Goll, D.S. *et al.* (2016). Model–data synthesis for the next generation of forest free-air CO_2 enrichment (FACE) experiments. *New Phytologist* 209(1), 17-28. https://doi.org/10.1111/nph.13593.

Norgate, T.E. and Haque, N. (2010). Energy and greenhouse gas impacts of mining and mineral processing operations. *Journal of Cleaner Production* 18(3), 266-274. https://doi.org/10.1016/j.jclepro.2009.09.020.

Noyes, P.D., McElwee, M.K., Miller, H.D., Clark, B.W., Van Tiem, L.A., Walcott, K.C. *et al.* (2009). The toxicology of climate change: Environmental contaminants in a warming world. *Environment International* 35(6), 971-986. https://doi.org/10.1016/j.envint.2009.02.006.

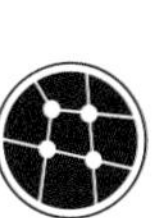

Obbard, R.W., Sadri, S., Wong, Y.Q., Khitun, A.A., Baker, I. and Thompson, R.C. (2014). Global warming releases microplastic legacy frozen in Arctic Sea ice. *Earth's Future* 2(6), 315-320. https://doi.org/10.1002/2014EF000240.

Ostfeld, R.S. (2009). Biodiversity loss and the rise of zoonotic pathogens. *Clinical Microbiology and Infection* 15(1), 40-43. https://doi.org/10.1111/j.1469-0691.2008.02691.x.

Park, J., Bangalore, M., Hallegatte, S. and Sandhoefner, E. (2018). Households and heat stress: Estimating the distributional consequences of climate change. *Environment and Development Economics* 23(3), 349-368. https://doi.org/10.1017/S1355770X1800013X.

Patt, A., Dazé, A. and Suarez, P. (2009). Gender and climate change vulnerability: What's the problem, what's the solution? In *Distributional Impacts of Climate Change and Disasters: Concepts and Cases*. Ruth, M. and Ibarraran, M.E. (eds.). Cheltenham: Edward Elgar. chapter 5. https://www.e-elgar.com/shop/distributional-impacts-of-climate-change-and-disasters

Paunescu, D., Stark, W.J. and Grass, R.N. (2016). Particles with an identity: Tracking and tracing in commodity products. *Powder Technology* 291, 344-350. https://www.doi.org/10.1016/j.powtec.2015.12.035.

Pepin, N., Bradley, R.S., Diaz, H.F., Baraër, M., Caceres, E.B., Forsythe, N. *et al.* (2015). Elevation-dependent warming in mountain regions of the world. *Nature Climate Change* 5(5), 424-430. https://doi.org/10.1038/nclimate2563.

Petrie, B., Barden. R. and Kasprzyk-Hordern, B. (2015). A review on emerging contaminants in wastewaters and the environment: Current knowledge, understudied areas and recommendations for future monitoring. *Water Research* 72, 3-27. http://doi.org/10.1016/j.watres.2014.08.053.

Phillips, D. (2016). Samarco dam collapse: One year on from Brazil's worst environmental disaster. *The Guardian*, Guardian News and Media Limited https://www.theguardian.com/sustainable-business/2016/oct/15/samarco-dam-collapse-brazil-worst-environmental-disaster-bhp-billiton-vale-mining.

Popkin, B.M. (2006). Global nutrition dynamics: The world is shifting rapidly toward a diet linked with noncommunicable diseases. *American Journal of Clinical Nutrition* 84(2), 289-298. https://doi.org/10.1093/ajcn/84.2.289.

Popp, J., Lakner, Z., Harangi-Rakos, M. and Fari, M. (2014). The effect of bioenergy expansion: Food, energy, and environment. *Renewable and Sustainable Energy Reviews* 32. https://doi.org/10.1016/j.rser.2014.01.056.

Popp, J., Petö, K. and Nagy, J. (2013). Pesticide productivity and food security: A review. *Agronomy for Sustainable Development* 33(1), 243-255. https://doi.org/10.1007/s13593-012-0105-x.

Porter, J.R., Xie, L., Challinor, A.J., Cochrane, K., Howden, S.M., Iqbal, M.M. *et al.* (2014). Food security and food production systems. In *Climate Change 2014: Impacts, Adaptation, and Vulnerability. Part A: Global and Sectoral Aspects. Contribution of Working Group II to the Fifth Assessment Report of the Intergovernmental Panel on Climate Change*. Field, C.B., Dokken, D.J., Mastrandrea, M.D., Mach, K.J., Bilir, T.E., Chatterjee, M. *et al.* (eds.). Cambridge: Cambridge University Press. 485-533. http://www.ipcc.ch/report/ar5/wg2/

Pravettoni, R. (2015). *Global Illegal Waste Traffic*. GRID-Arendal, Arendal. http://www.grida.no/resources/8061.

Prüss-Ustün, A., Bartram, J., Clasen, T., Colford, J.M., Cumming, O., Curtis, V. *et al.* (2014). Burden of disease from inadequate water, sanitation and hygiene in low- and middle-income settings: A retrospective analysis of data from 145 countries. *Tropical Medicine & International Health* 19(8), 894-905. https://doi.org/10.1111/tmi.12329.

Prüss-Ustün, A., Wolf, J., Corvalán, C., Bos, R. and Neira, M. (2016). *Preventing Disease Through Healthy Environments: A Global Assessment of the Burden of Disease from Environmental Risks*. Geneva: World Health Organization. http://apps.who.int/iris/bitstream/10665/204585/1/9789241565196_eng.pdf?ua=1.

Qadir, M., Quillerou, E., Nangia, V., Murtaza, G., Singh, M., Thomas, R.J. *et al.* (2014). Economics of slat-induced land degradation and restoration. *Natural Resources Forum* 38 (4), 282-295. https://doi.org/10.1111/1477-8947.12054.

Quam, V., Rocklöv, J., Quam, M. and Lucas, R. (2017). Assessing greenhouse gas emissions and health co-benefits: A structured review of lifestyle-related climate change mitigation strategies. *International Journal of Environmental Research and Public Health* 14(5), 468. https://doi.org/10.3390/ijerph14050468.

Ramaswami, A., Boyer, D., Nagpure, A., Fang, A., Bogra, S., Bakshi, B. *et al.* (2017). An urban systems framework to assess the transboundary food-energy-water nexus: Implementation in Delhi, India. *Environmental Research Letters* 12(2), 1-14. https://doi.org/10.1088/1748-9326/aa5556.

Ranghiere, F. and Ishiwatari, M. (eds.) (2014). *Learning from Megadisasters: Lessons from the Great East Japan Earthquake*. Washington, D.C: World Bank. https://openknowledge.worldbank.org/bitstream/handle/10986/18864/9781464801532.pdf?sequence=1&isAllowed=y.

Rieckmann, M. (2012). Future-oriented higher education: Which key competencies should be fostered through university teaching and learning? *Futures* 44(2), 127-135. https://doi.org/10.1016/j.futures.2011.09.005.

Rieckmann, M. (2018). Learning to transform the world: Key competencies in ESD. In *Issues and Trends in Education for Sustainable Development: Education on the Move*. Leicht, A., Heiss, J. and Byun, W.J. (eds.). Paris: United Nations Educational, Scientific and Cultural Organization. chapter 2. 39-59. http://unesdoc.unesco.org/images/0026/002614/261445E.pdf

Roche, C., Thygesen, K. and Baker, E. (eds.) (2017). *Mine Tailings Storage: Safety is no Accident. A UNEP Rapid Response Assessment*. Nairobi: United Nations Environment Programme and GRID-Arendal. https://gridarendal-website-live.s3.amazonaws.com/production/documents/:s_document/371/original/RRA_MineTailings_lores.pdf?1510660693.

Rockström, J., Steffen, W., Noone, K., Persson, Å., Chapin, F.S., Lambin, E.F. *et al.* (2009). A safe operating space for humanity. *Nature* 461(7263), 472-475. https://doi.org/10.1038/461472a.

Rodriguez, D., Delgado, A., DeLaquil, P. and Sohns, A. (2013). *Thirsty Energy: Securing Energy in a Water-Constrained World*. World Bank working paper. Washington, D.C: World Bank. http://documents.worldbank.org/curated/en/835051468168842442/pdf/789230REPLACEM0sty0Energy0204014web.pdf.

Roschangar, F., Sheldon, R.A. and Senanayake, C.H. (2015). Overcoming barriers to green chemistry in the pharmaceutical industry–the Green Aspiration Level™ concept. *Green Chemistry* 17(2), 752-768. https://doi.org/10.1039/C4GC01563K.

Rucevska, I., Nellemann, C., Isarin, N., Yang, W., Liu, N., Yu, K. *et al.* (2015). *Waste Crime – Waste Risks: Gaps in Meeting the Global Waste Challenge. A UNEP Rapid Response Assessment*. Nairobi: United Nations Environment Programme. http://apps.unep.org/publications/index.php?option=com_pub&task=download&file=011703_en.

Ruel, G.K. (2011). The (Arctic) show must go on. *International Journal: Canada's Journal of Global Policy Analysis* 66(4), 825-833. https://doi.org/10.1177/002070201106600411.

Ruth, M. (1995). Thermodynamic constraints on optimal depletion of copper and aluminum in the United States: A dynamic model of substitution and technical change. *Ecological Economics* 15(3), 197-213. https://doi.org/10.1016/0921-8009(95)00053-4.

Sagrario, M.F. and Willoughby, J. (2016). Feminist economics and the analysis of the global economy: The challenge that awaits us. *The Fletcher Forum of World Affairs* 40(2), 15-27. https://static1.squarespace.com/static/579fc2ad725e253a86230610/t/57ec6a1d5016e1636a21dcad/1475111454239/FletcherForum_Sum16_40-2_15-27_FLORO_WILLOUGHBY.pdf.

Samberg, L.H., Gerber, J.S., Ramankutty, N., Herrero, M. and West P.C. (2016). Subnational distribution of average farm size and smallholder contributions to global food production. *Environment Research Letters* 11(12). https://doi.org/10.1088/1748-9326/11/12/124010.

Sauer, P.C. and Seuring, S. (2017). Sustainable supply chain management for minerals. *Journal of Cleaner Production* 151, 235-249. https://doi.org/10.1016/j.jclepro.2017.03.049.

Saunders, L.E., Green, J.M., Petticrew, M.P., Steinbach, R. and Roberts, H. (2013). What are the health benefits of active travel? A systematic review of trials and cohort studies. *PLoS One* 8(8), e69912. https://doi.org/10.1371/journal.pone.0069912.

Saxena, T., Kaushik, P. and Krishna Mohan, M. (2015). Prevalence of E colid O157:H7 in water sources: An overview on associated diseases, outbreaks and detection methods. *Diagnostic Microbiology and Infectious Disease* 82(3), 249-264. https://doi.org/10.1016/j.diagmicrobio.2015.03.015.

Schauberger, B., Archontoulis, S., Arneth, A., Balkovic, J., Ciais, P., Deryng, D. *et al.* (2017). Consistent negative response of US crops to high temperatures in observations and crop models. *Nature Communications* 8(13931). https://doi.org/10.1038/ncomms13931.

Scheffer, M. (2016). Anticipating societal collapse. Hints from the stone age. *Proceedings of the National Academy of Sciences* 113(39), 10733-10735. https://doi.org/10.1073/pnas.1612728113.

Schellnhuber, H.J., Rahmstorf, S. and Winkelmann, R. (2016). Why the right climate target was agreed in Paris. *Nature Climate Change* 6, 649-653. https://doi.org/10.1038/nclimate3013.

Schewe, J., Levermann, A. and Meinshausen, M. (2011). Climate change under a scenario near 1.5 C of global warming: Monsoon intensification, ocean warming and steric sea level rise. *Earth System Dynamics* 2, 25-35. https://doi.org/10.5194/esd-2-25-2011.

Schilling, J., Ide, T., Scheffran, J. and Froese, R. (2017). Resilience and environmental security: Towards joint application in peacebuilding. *Global Change, Peace & Security* 29(2), 1-30. https://doi.org/10.1080/14781158.2017.1305347.

Schlenker, W. and Roberts, M.J. (2009). Nonlinear temperature effects indicate severe damages to US crop yields under climate change. *Proceedings of the National Academy of sciences* 106(37), 15594-15598. https://doi.org/10.1073/pnas.0906865106.

Schug, T.T., Johnson, A.F., Birnbaum, L.S., Colborn, T., Guillette, L.J., Jr., Crews, D.P. *et al.* (2016). Minireview: Endocrine disruptors: Past lessons and future directions. *Molecular Endocrinology* 30(8), 833-847. https://doi.org/10.1210/me.2016-1096.

Schulte, P.A., Roth, G., Hodson, L.L., Murashov, V., Hoover, M.D., Zumwalde, R. *et al.* (2016). Taking stock of the occupational safety and health challenges of nanotechnology: 2000–2015. *Journal of Nanoparticle Research* 18(159), 1–21. https://doi.org/10.1007/s11051-016-3459-1.

Seager, J. (2014) *Background and Methodology for Gender Global Environmental Outlook*. http://www.unep.org/sites/default/files/ggeo/documents/GGEO_Multi-stakehoder_consultation_Background_document_final.pdf

Secretariat of the Convention on Biological Diversity (2012). *Global Biodiversity Outlook 4: A Mid-Term Assessment of Progress Towards the Implementation of the Strategic Plan for Biodiversity 2011-2020*. Montréal https://www.cbd.int/gbo/gbo4/publication/gbo4-en.pdf.

Seneviratne, S.I., Nicholls, N., Easterling, D., Goodess, C.M., Kanae, S., Kossin, J. *et al.* (2012). Changes in climate extremes and their impacts on the natural physical environment: An overview of the IPCC SREX report. *EGU General Assembly*. Vienna, 22-27 April 2012. 12566 http://adsabs.harvard.edu/abs/2012EGUGA..1412566S.

Seto, K., Guneralp, B. and Hutyra, L. (2012). Global forecasts of urban expansion to 2030 and direct impacts on biodiversity and carbon pools. *Proceedings of the National Academy of Sciences of the United States of America* 109(40), 16083-16088. https://doi.org/10.1073/pnas.1211658109.

Seto, K., Shakal, S., Bigio, A., Blanco, H., Delgado, G.C., Dewar, D. *et al.* (2014). Human settlements, infrastructure and spatial planning. In *Climate Change 2014: Mitigation of Climate Change. Contribution of Working Group III to the Fifth Assessment Report of the Intergovernmental Panel on Climate Change*. Edenhofer, O., Pichs-Madruga, R., Sokona, Y., Farahani, E., Kadner, S., Seyboth, K. *et al.* (eds.). Cambridge. chapter 12. 923-1000. http://www.ipcc.ch/pdf/assessment-report/ar5/wg3/ipcc_wg3_ar5_chapter12.pdf

Seto, K.C., Davis, S.J., Mitchell, R.B., Stokes, E.C., Unruh, G. and Ürge-Vorsatz, D. (2016). Carbon lock-in: Types, causes, and policy implications. *Annual Review of Environment and Resources* 41(1), 425-452. https://doi.org/10.1146/annurev-environ-110615-085934.

Sharma, B.M., Bharat, G.K., Tayal, S., Nizzetto, L., Cupr, P. and Larssen, T. (2014). Environment and human exposure to persistent organic pollutants (POPs) in India: A systematic review of recent and historical data. *Environmental International* 66, 48-64. https://doi.org/10.1016/j.envint.2014.01.022.

Shrestha, A.B., Agrawal, N.K., Alfthan, B., Bajracharya, S.R., Maréchal, J. and van Oort, B. (eds.) (2015). *The Himalayan Climate and Water Atlas: Impact of Climate Change on Water Resources in Five of Asia's Major River Basins*. https://gridarendal-website-live.s3.amazonaws.com/production/documents/:s_document/20/original/HKHwateratlas2016_screen.pdf?1483646266.

Singer-Brodowski, M., Brock, A., Etzkorn, N. and Otte, I. (2018). Monitoring of education for sustainable development in Germany – insights from early childhood education, school and higher education. *Environmental Education Research*, 1-16. https://doi.org/10.1080/13504622.2018.1440380.

Skinner, E. (2011). *Gender and Climate Change: Overview Report*. Brighton: Institute of Development Studies. http://docs.bridge.ids.ac.uk/vfile/upload/4/document/1211/Gender_and_CC_for_web.pdf.

Smith, K.R., Bruce, N., Balakrishnan, K., Adair-Rohani, H., Balmes, J., Chafe, Z. *et al.* (2014b). Millions dead: How do we know and what does it mean? Methods used in the comparative risk assessment of household air pollution. *Annual Review of Public Health* 35, 185-206. https://doi.org/10.1146/annurev-publhealth-032013-182356.

Smith, K.R., Woodward, A., Campbell-Lendrum, D., Chadee, D.D., Honda, Y., Liu, Q. *et al.* (2014a). Human health: Impacts, adaptation, and co-benefits. In *Climate Change 2014: Impacts, Adaptation, and Vulnerability. Part A: Global and Sectoral Aspects. Contribution of Working Group II to the Fifth Assessment Report of the Intergovernmental Panel on Climate Change*. Field, C.B., Dokken, D.J., Mastrandrea, M.D., Mach, K.J., Bilir, T.E., Chatterjee, M. *et al.* (eds.). Cambridge: Cambridge University Press. 709-754. http://www.ipcc.ch/pdf/assessment-report/ar5/wg2/WGIIAR5-Chap11_FINAL.pdf

Solomon, G.M., Morello-Frosch, R., Zeise, L. and Faust, J.B. (2016). Cumulative environmental impacts: Science and policy to protect communities. *Annual Review of Public Health* 37, 83-96. https://doi.org/10.1146/annurev-publhealth-032315-021807.

Springmann, M., Clark, M., Mason-D'Croz, D., Wiebe, K., Bodirsky, B.L., Lassaletta, L. *et al.* (2018). Options for keeping the food system within environmental limits. *Nature* 562, 519–525. https://doi.org/10.1038/s41586-018-0594-0.

Steffen, W., Richardson, K., Rockström, J., Cornell, S.E., Fetzer, I., Bennett, E.M. *et al.* (2015). Planetary boundaries: Guiding human development on a changing planet. *Science* 347(6223), 1259855-1259810. https://www.doi.org/10.1126/science.1259855.

Stehle, S. and Schulz, R. (2015). Agricultural insecticides threaten surface waters at the global scale. *Proceedings of the National Academy of Sciences* 112(18), 5750-5755. https://doi.org/10.1073/pnas.1500232112.

Strempel, S., Scheringer, M., Ng, C.A. and Hungerbuhler, K. (2012). Screening for PBT chemicals among the "existing" and "new" chemicals of the EU. *Environmental Science and Technology* 46(11), 5680-5687. https://doi.org/10.1021/es3002713.

Sun, L.G. (2016). Climate change and the narrative of disaster. In *The Role of International Environmental Law in Disaster Risk Reduction*. Peel, J. and Fisher, D. (eds.). Leiden: Brill. 27-48. http://booksandjournals.brillonline.com/content/books/b9789004318816_003

Sun, X., Wang, K., Kang, S., Guo, J., Zhang, G., Huang, J. *et al.* (2017). The role of melting alpine glaciers in mercury export and transport: An intensive sampling campaign in the Qugaqie Basin, inland Tibetan Plateau. *Environmental Pollution* 220, 936-945. https://doi.org/10.1016/j.envpol.2016.10.079.

Sunderland, T., Achdiawan, R., Angelsen, A., Babigumira, R., Ickowitz, A., Paumgarten, F. *et al.* (2014). Challenging perceptions about men, women, and forest product use: A global comparative study. *World Development* 64, S56-S66. http://dx.doi.org/10.1016/j.worlddev.2014.03.003.

Sutton, M.A., Bleeker, A., Howard, C.M., Bekunda, M., Grizzetii, B., de Vries, W. *et al.* (2013). *Our Nutrient World: The Challenge to Produce More Food and Energy with Less Pollution*. Edinburgh: Centre for Ecology and Hydrology. http://nora.nerc.ac.uk/id/eprint/500700/1/N500700BK.pdf.

Taylor, P., Cai, M., Hu, A., Meehl, J., Washington, W. and Zhang, G.J. (2013). A decomposition of feedback contributions to polar warming amplification. *American Meteorological Society* 26, 7023-7043. https://doi.org/10.1175/JCLI-D-12-00696.1.

The American Society of Human Genetics, The American Society for Reproductive Medicine, The Endocrine Society, The Genetics Society of America, The Society for Developmental Biology, The Society for Pediatric Urology *et al.* (2011). Assessing chemical risk: Societies offer expertise. *Science* 331(6021), 1136. https://doi.org/10.1126/science.331.6021.1136-a.

The Lancet Planetary Health (2018). The natural environment and emergence of antibiotic resistance. *The Lancet Planetary Health* 2(1). https://doi.org/10.1016/S2542-5196(17)30182-1.

Thompson, R.C., Olsen, Y., Mitchell, R.P., Davis, A., Rowland, S.J., John, A.W.G. *et al.* (2004). Lost at Sea: Where is all the Plastic? *Science* 304(5672), 838. https://doi.org/10.1126/science.1094559.

Tijani, J.O., Fatoba, O.O., Babajide, O.O. and Petrik, L.F. (2016). Pharmaceuticals, endocrine disruptors, personal care products, nanomaterials and perfluorinated pollutants: A review. *Environmental Chemistry Letters* 14(1), 27-49. https://doi.org/10.1007/s10311-015-0537-z.

Tilman, D., Balzer, C., Hill, J. and Befort, B.L. (2011). Global food demand and the sustainable intensification of agriculture. *Proceedings of the National Academy of Sciences* 108(50), 20260-20264. https://doi.org/10.1073/pnas.1116437108.

Tilman, D. and Clark, M. (2014). Global diets link environmental sustainability and human health. *Nature* 515(7528), 518-522. https://doi.org/10.1038/nature13959.

Trenberth, K.E. (2011). Changes in precipitation with climate change. *Climate Research* 47(1/2), 123-138. https://doi.org/10.3354/cr00953.

United Kingdom Government Met Office (2018). *An overview of global surface temperatures in 2017*. https://www.metoffice.gov.uk/research/news/2018/global-surface-temperatures-in-2017.

United Kingdom Government Office for Science (2011). *Migration and Global Environmental Change: Future Scenarios*. London. https://www.gov.uk/government/uploads/system/uploads/attachment_data/file/288793/11-1117-migration-global-environmental-change-scenarios.pdf.

United Nations (2014). *World Urbanization Prospects: The 2014 Revision, Highlights*. ST/ESA/SER.A/352. New York, NY. https://esa.un.org/unpd/wup/publications/files/wup2014-highlights.pdf.

United Nations (2015a). *Transforming Our World: The 2030 Agenda For Sustainable Development*. New York, NY: United Nations. http://www.un.org/ga/search/view_doc.asp?symbol=A/RES/70/1&Lang=E.

United Nations (2015b). *The World's Women 2015: Trends and Statistics*. New York, NY. https://unstats.un.org/unsd/gender/downloads/worldswomen2015_report.pdf.

United Nations (2018). *World Urbanization Prospects: The 2018 Revision*. New York, NY. https://population.un.org/wup/Publications/Files/WUP2018-KeyFacts.pdf.

United Nations Children's Fund (2018). *Primary education*. https://data.unicef.org/topic/education/primary-education/

United Nations Children's Fund and World Health Organization (2017). *Safely Managed Drinking Water*. Geneva. https://data.unicef.org/wp-content/uploads/2017/03/safely-managed-drinking-water-JMP-2017-1.pdf.

United Nations Convention to Combat Desertification (2017). *Global Land Outlook*. Bonn. https://knowledge.unccd.int/sites/default/files/2018-06/GLO%20English_Full_Report_rev1.pdf.

United Nations Development Programme (2016). *Human Development Report 2016. Human Development for Everyone*. New York, NY. http://hdr.undp.org/sites/default/files/2016_human_development_report.pdf.

United Nations Educational Scientific and Cultural Organization (2014). *Shaping the Future We Want: UN Decade of Education for Sustainable Development (2005-2014)*. Paris. https://sustainabledevelopment.un.org/content/documents/1682Shaping%20the%20future%20we%20want.pdf.

United Nations Educational Scientific and Cultural Organization (2017a). *Global Education Monitoring Report 2017/8. Accountability in Education: Meeting our Commitments*. Paris. http://unesdoc.unesco.org/images/0025/002593/259338e.pdf.

United Nations Educational Scientific and Cultural Organization (2017b). *Education for Sustainable Development Goals: Learning Objectives*. Paris. https://europa.eu/capacity4dev/file/69191/download?token=Jw3cLunL.

United Nations Entity for Gender Equality and the Empowerment of Women (2014). *The World Survey on The Role Of Women In Development 2014: Gender Equality and Sustainable Development*. New York, NY. https://gest.unu.edu/static/files/world-survey-on-the-role-of-women-in-development-2014.pdf.

United Nations Entity for Gender Equality and the Empowerment of Women (2015). *Progress of The World's Women 2015–2016: Transforming Economies, Realizing Rights*. New York, NY. http://progress.unwomen.org/en/2015/pdf/UNW_progressreport.pdf.

United Nations Environment Programme (2012). *Progress Report on the Chemicals in Products Project, including Proposed Recommendations for Further International Cooperative Action*. International Conference on Chemicals Management. Nairobi. http://www.saicm.org/Portals/12/documents/meetings/ICCM3/doc/SAICM_ICCM3_15_EN.pdf.

United Nations Environment Programme (2013a). *City-Level Decoupling: Urban Resource Flows and the Governance of Infrastructure Transitions. A Report of the International Resource Panel*. Swilling, M., Robinson, B., Marvin, S. and Hodson, M. (eds.). Nairobi. http://wedocs.unep.org/handle/20.500.11822/8488.

United Nations Environment Programme (2013b). *Global Chemicals Outlook: Towards Sound Management of Chemicals*. Nairobi. https://sustainabledevelopment.un.org/content/documents/1966Global%20Chemical.pdf.

United Nations Environment Programme (2013c). *Costs of Inaction on the Sound Management of Chemicals*. Nairobi. http://wedocs.unep.org/bitstream/handle/20.500.11822/8412/-Costs%20of%20inaction%20on%20the%20sound%20management%20of%20chemicals-2013Report_Cost_of_Inaction_Feb2013.pdf?sequence=3&isAllowed=y.

United Nations Environment Programme (2015). *Global Waste Management Outlook*. Nairobi. http://apps.unep.org/publications/index.php?option=com_pub&task=download&file=011782_en.

United Nations Environment Programme (2016a). *Global Gender and Environment Outlook*. Nairobi. http://wedocs.unep.org/bitstream/handle/20.500.11822/14764/GLOBAL%20GENDER%20AND%20ENVIRONMENT%20OUTLOOK.pdf?sequence=1&isAllowed=y.

United Nations Environment Programme (2016b). *GEO-6 Regional Assessment for West Asia*. Nairobi. http://wedocs.unep.org/bitstream/handle/20.500.11822/7668/GEO_West_Asia_201611.pdf?isAllowed=y&sequence=1.

United Nations Environment Programme (2016c). *Food Systems and Natural Resources: A Report of the Working Group on Food Systems of the International Resource Panel*. Westhoek, H., Ingram, J., Van Berkum, S., Özay, L. and Hajer M. (eds.). Nairobi. http://www.resourcepanel.org/file/133/download?token=6dSyNtuV.

United Nations Environment Programme (2017). *Towards a Pollution-Free Planet: Background Report*. Nairobi. https://wedocs.unep.org/bitstream/handle/20.500.11822/21800/UNEA_towardspollution_long%20version_Web.pdf?sequence=1&isAllowed=y.

United Nations Environment Programme, United Nations Entity for Gender Equality and the Empowerment of Women, United Nations Peacebuilding Support Office and United Nations Development Programme (2013). *Women and Natural Resources: Unlocking the Peacebuilding Potential*. Nairobi. https://reliefweb.int/sites/reliefweb.int/files/resources/UNEP_UN-Women_PBSO_UNDP_gender_NRM_peacebuilding_report%20pdf.pdf.

United Nations Environment Programme and World Health Organization (2013). *State of the Science of Endocrine Disrupting Chemicals - 2012*. Bergman, Å., Jerrold J. Heindel., Jobling, S., Kidd, K.A. and Zoeller, R.T. (eds.). https://www.who.int/iris/bitstream/10665/78101/1/9789241505031_eng.pdf?ua=1.

United Nations Human Settlements Programme (2011). *The Economic Role of Cities*. The Global Urban Economic Dialogue Series. Nairobi. https://unhabitat.org/books/economic-role-of-cities/#.

United Nations Human Settlements Programme (2014). *Sustainable Urban Development and Agenda 2030: UN-Habitat's Programme Framework: PSUP, Transforming the Lives of One Billion Slum Dwellers*. Nairobi. https://unhabitat.org/sustainable-urban-development-and-agenda-2030-un-habitats-programme-framework-psup-transforming-the-lives-of-one-billion-slum-dwellers/#

United Nations Human Settlements Programme (2016a). *Urbanization and Development: Emerging Future. World Cities Report 2016*. Nairobi. https://unhabitat.org/wp-content/uploads/2014/03/WCR-%20Full-Report-2016.pdf.

United Nations Human Settlements Programme (2016b). *Pretoria Declaration on Informal Settlements*. Nairobi. https://unhabitat.org/pretoria-declaration-on-informal-settlements/#.

United Nations Office for Disaster Risk Reduction (2014). The Economic and human impact of disasters in the last ten years. https://www.unisdr.org/files/42862_economichumanimpact20052014unisdr.pdf

United Nations Office for Disaster Risk Reduction (2015). *Sendai Framework for Disaster Risk Reduction 2015-2030*. http://www.unisdr.org/files/43291_sendaiframeworkfordrren.pdf.

United Nations World Water Assessment Programme (2017). *The United Nations World Water Development Report 2017. Wastewater: The Untapped Resource*. Paris: United Nations Educational, Scientific and Cultural Organization. http://unesdoc.unesco.org/images/0024/002471/247153e.pdf.

United States National Oceanic and Atmospheric Administration (2017). *2016 marks three consecutive years of record warmth for the globe*. National Oceanic and Atmospheric Administration. http://www.noaa.gov/stories/2016-marks-three-consecutive-years-of-record-warmth-for-globe

United States National Oceanic and Atmospheric Administration (2018). *2017 was 3rd warmest year on record for the globe*. http://www.noaa.gov/news/noaa-2017-was-3rd-warmest-year-on-record-for-globe.

United States National Snow and Ice Data Center (2017). *Arctic sea ice 2017: Tapping the brakes in September*. [National Snow and Ice Data Center http://nsidc.org/arcticseaicenews/2017/10/ (Accessed: 1 November 2018).

Urry, J. (2010). Consuming the planet to excess. *Theory, Culture & Society* 27(2-3), 191-212. https://doi.org/10.1177/0263276409355999.

Vanbergen, A.J. (2013). Threats to an ecosystem service: Pressures on pollinators. *Frontiers in Ecology and the Environment* 11(5), 251-259. https://doi.org/10.1890/120126.

Vanwalleghem, T., Gómez, J.A., Infante Amate, J., González de Molina, M., Vanderlinden, K., Guzmán, G. *et al.* (2017). Impact of historical land use and soil management change on soil erosion and agricultural sustainability during the Anthropocene. *Anthropocene* 17, 13-29. https://www.doi.org/10.1016/j.ancene.2017.01.002.

Vaughan, D.G., Comiso, J.C., Allison, I., Carrasco, J., Kaser, G., Kwok, R. *et al.* (2013). Observations: Cryosphere. In *Climate Change 2013: The Physical Science Basis. Contribution of Working Group I to the Fifth Assessment Report of the Intergovernmental Panel on Climate Change*. Stocker, T.F., Qin, D., Plattner, G.-K., Tignor, M., Allen, S.K., Boschung, J. *et al.* (eds.). Cambridge: Cambridge University Press. chapter 4. 317-382. http://www.ipcc.ch/pdf/assessment-report/ar5/wg1/WG1AR5_Chapter04_FINAL.pdf.

Vermeulen, S.J., Campbell, B.M. and Ingram, J.S.I. (2012). Climate change and food systems. *Annual Review of Environment and Resources* 37, 195-222. https://doi.org/10.1146/annurev-environ-020411-130608.

Villasante, S., do Carme García-Negro, M., González-Laxe, F. and Rodríguez, G.R. (2011). Overfishing and the common fisheries policy: (Un)successful results from TAC regulation? *Fish and Fisheries* 12(1), 34-50. https://doi.org/10.1111/j.1467-2979.2010.00373.x.

von Schneidemesser, E., Monks, P.S., Allan, J.D., Bruhwiler, L., Forster, P., Fowler, D. *et al.* (2015). Chemistry and the linkages between air quality and climate change. *Chemical Reviews* 115(10), 3856-3897. https://doi.org/10.1021/acs.chemrev.5b00089.

Walker, C.L., Rudan, I., Liu, L., Nair, H., Theodoratou, E., Bhutta, Z.A. *et al.* (2013). Global burden of childhood pneumonia and diarrhoea. *Lancet* 381(9875), 1405-1416. https://doi.org/10.1016/S0140-6736(13)60222-6.

Wall, B., Derham, J. and O'Mahony, T. (eds.) (2016). *Ireland's Environment 2016: An Assessment*. Wexford: Ireland Environmental Protection Agency. http://www.epa.ie/pubs/reports/indicators/SoE_Report_2016.pdf.

Waller, C.L., Griffiths, H.J., Waluda, C.M., Thorpe, S.E., Loaiza, I., Moreno, B. *et al.* (2017). Microplastics in the Antarctic marine system: An emerging area of research. *Science of The Total Environment* 598, 220-227. https://www.doi.org/10.1016/j.scitotenv.2017.03.283.

Wallinga, D., Rayner, G. and Lang, T. (2015). Antimicrobial resistance and biological governance: Explanations for policy failure. *Public Health* 129(10), 1314-1325. https://doi.org/10.1016/j.puhe.2015.08.012.

Wals, A.E.J. (2015). *Beyond Unreasonable Doubt. Education and Learning For Socio-Ecological Sustainability in the Anthropocene*. Wageningen: Wageningen University. https://arjenwals.files.wordpress.com/2016/02/8412100972_rvb_inauguratie-wals_oratieboekje_v02.pdf.

Wals, A.E.J. and Lenglet, F. (2016). Sustainability citizens: Collaborative and disruptive social learning. In *Sustainability Citizenship in Cities: Theory and Practice*. Horne, R., Fien, J., Beza, B. and Nelson, A. (eds.). London: Routledge. chapter 5. https://www.taylorfrancis.com/books/e/9781317391081/chapters/10.4324%2F9781315678405-14

4

Watson, M. (2015). The UN decade of ESD: What was achieved in Scotland 2005–2014. *Applied Environmental Education & Communication* 14(2), 90-96. https://doi.org/10.1080/153301 5X.2014.971980.

Weng, Z., Jowitt, S.M., Mudd, G.M. and Haque, N. (2015). A detailed assessment of global rare earth element resources: Opportunities and challenges. *Economic Geology* 110(8), 1925-1952. http://dx.doi.org/10.2113/econgeo.110.8.1925.

Wernick, I.K. and Ausubel, J.H. (1997). *Industrial Ecology: Some Directions for Research*. New York, NY: The Rockefeller University. https://phe.rockefeller.edu/ie_agenda/.

Wernick, I.K., Herman, R., Govind, S. and Ausubel, J.H. (1996). Materialization and dematerialization: Measures and trends. *Daedalus* 125(3), 171-198. https://phe.rockefeller.edu/Daedalus/Demat/.

Whitmee, S., Haines, A., Beyrer, C., Boltz, F., Capon, A.G., de Souza Dias, B.F. *et al.* (2015). Safeguarding human health in the Anthropocene epoch: Report of The Rockefeller Foundation–Lancet Commission on planetary health. *The Lancet* 386(10007), 1973-2028. https://doi.org/10.1016/S0140-6736(15)60901-1.

Wiek, A., Bernstein, M.J., Foley, R.W., Cohen, M., Forrest, N., Kuzdas, C. *et al.* (2016). Operationalising competencies in higher education for sustainable development. In *Routledge Handbook of Higher Education for Sustainable Development*. Barth, M., Michelsen, G., Thomas, I. and Rieckmann, M. (eds.). London: Routledge. chapter 16. 241-260. https://www.routledgehandbooks.com/doi/10.4324/9781315852249.ch16

Wiek, A., Withycombe, L. and Redman, C.L. (2011). Key competencies in sustainability: A reference framework for academic program development. *Sustainability Science* 6(2), 203-218. https://doi.org/10.1007/s11625-011-0132-6.

Wiggington, N., Fahrenkamp-Uppenbrink, J., Wible, B. and Malakoff, D. (2016). Cities are the future. *Science* 352(6288), 904-905. https://doi.org/10.1126/science.352.6288.904.

Williams, A.P., Allen, C.D., Macalady, A.K., Griffin, D., Woodhouse, C.A., Meko, D.M. *et al.* (2013). Temperature as a potent driver of regional forest drought stress and tree mortality. *Nature Climate Change* 3(3), 292-297. https://doi.org/10.1038/nclimate1693.

Winsemius, H.C., Jongman, B., Veldkamp, T.I.E., Hallegatte, S., Bangalore, M. and Ward, P.J. (2018). Disaster risk, climate change, and poverty: Assessing the global exposure of poor people to floods and droughts. *Environment and Development Economics* 23(3), 328-348. https://doi.org/10.1017/S1355770X17000444.

Wolf, K.L. and Robbins, A.S.T. (2015). Metro nature, environmental health, and economic value. *Environmental Health Perspectives* 123(5), 390-398. https://doi.org/10.1289/ehp.1408216.

Wolk, A. (2017). Potential health hazards of eating red meat. *Journal of Internal Medicine* 281(2), 106-122. https://doi.org/10.1111/joim.12543.

World Energy Council (2016). *World Energy Trilemma: Defining measures to accelerate the energy transition*. London. https://www.worldenergy.org/wp-content/uploads/2016/05/World-Energy-Trilemma_full-report_2016_web.pdf.

World Health Organization (1948). Constitution of the World Health Organization. Geneva http://www.who.int/governance/eb/who_constitution_en.pdf

World Health Organization (2008). *Closing the Gap in a Generation: Health Equity Through Action on the Social Determinants of Health*. Geneva. http://apps.who.int/iris/bitstream/10665/43943/1/9789241563703_eng.pdf.

World Health Organization (2014). Health in All Policies: Helsinki Statement. *The 8th Global Conference on Health Promotion*. Helsinki, 10-14 June 2013. World Health Organization, Geneva http://apps.who.int/iris/bitstream/10665/112636/1/9789241506908_eng.pdf?ua=1

World Health Organization and United Nations Children's Fund (2017). *Progress on Drinking Water, Sanitation and Hygiene: 2017 Update and SDG Baselines*. Geneva. https://www.unicef.org/publications/files/Progress_on_Drinking_Water_Sanitation_and_Hygiene_2017.pdf.

Wunder, S., Noack, F. and Angelsen, A. (2018). Climate, crops, and forests: A pan-tropical analysis of household income generation. *Environment and Development Economics* 23(3), 279-297. https://doi.org/10.1017/S1355770X18000116.

Yao, T., Thompson, L., Yang, W., Yu, W., Gao, Y., Guo, X. *et al.* (2012). Different glacier status with atmospheric circulations in Tibetan Plateau and surroundings. *Nature Climate Change* 2, 663-667. https://doi.org/10.1038/nclimate1580.

Yin, J., Overpeck, J., Peyser, C. and Stouffer, R. (2018). Big jump of record warm global mean surface temperature in 2014-2016 related to unusually large oceanic heat releases. *Geophysical Research Letters* 45(2), 1069-1078. https://doi.org/10.1002/2017GL076500.

Zhang, M., Zeiss, M.R. and Geng, S. (2015). Agricultural pesticide use and food safety: California's model. *Journal of Integrative Agriculture* 14(11), 2340-2357. https://doi.org/10.1016/S2095-3119(15)61126-1.

Zhao, C., Liu, B., Piao, S., Wang, X., Lobell, D.B., Huang, Y. *et al.* (2017). Temperature increase reduces global yields of major crops in four independent estimates. *Proceedings of the National Academy of Sciences of the United States of America* 114(35), 9326-9331. https://doi.org/10.1073/pnas.1701762114.

Ziska, L.H. and Dukes, J.S. (eds.) (2014). *Invasive Species and Global Climate Change* CABI Invasives Series. https://www.cabi.org/bookshop/book/9781780641645.

PART A

State of the Global Environment

Chapter 5

Air

© Shutterstock/Adam Radosavljevic

Coordinating Lead Authors: Phillip Dickerson (United States Environmental Protection Agency), Cristina Guerreiro (Norwegian Institute for Air Research), Terry Keating (United States Environmental Protection Agency), John Muthama Nzioka (University of Nairobi)

Lead Authors: Serena H. Chung (United States Environmental Protection Agency), Stefan Reis (Centre for Ecology and Hydrology)

Contributing Authors: Babatunde Joseph Abiodun (University of Cape Town), Kathryn Jennifer Bowen (Australian National University), Riyanti Djalante (United Nations University – Institute for the Advanced Study of Sustainability), James Grellier (European Centre for Environment and Human Health, University of Exeter), Fintan Hurley (Institute of Occupational Medicine), Andrei Kirilenko (University of Florida), Robyn Lucas (National Centre for Epidemiology and Population Health – The Australian National University), Caradee Y. Wright (Medical Research Council of South Africa)

GEO Fellow: HE Chenmin (Peking University)

Executive summary

Concentrations of CO_2 and other long-lived greenhouse gases (GHGs) continue to increase, driven mainly by people consuming fossil fuels to satisfy ever-increasing demands for energy (*well established*). {5.2.4}

Given the current concentrations of GHGs and their lifetime in the atmosphere, significant changes in climate and sea levels are unavoidable, with widespread consequences for people and the environment (*well established*). There is robust evidence that climate change and increased climate variability worsen existing poverty, exacerbate inequalities and trigger new vulnerabilities. However, even greater changes are expected in the future if action is not taken soon to halt GHG emissions. {5.3.4}

Climate change impacts include increased frequency and magnitude of heatwaves and storms (*established but incomplete*); changes in the distribution of disease vectors, exacerbation of air pollution episodes, and decreases in water supply and impacts on crop yields and food prices. {5.3.4}

Efforts to decrease emissions of short-lived climate pollutants (SLCP), specifically black carbon (BC), methane (CH_4), tropospheric ozone (O_3) and hydrofluorocarbons (HFCs), are a critical component of an integrated climate change mitigation and air quality management programme (*well established*). Along with rapid mitigation of long-lived GHG emissions, decreases in SLCP emissions achieve the objectives of the United Nations Framework Convention on Climate Change (UNFCCC). {5.2.4}

Air pollution is the most important environmental contributor to the global burden of disease, leading to an estimated 6 million to 7 million premature deaths annually and large economic losses (*established but incomplete*). Of those deaths, 2.6 million to 3.8 million deaths have been attributed to burning wood, coal, crop residue, dung and kerosene for cooking, heating and lighting. Another 3.2 million to 3.5 million deaths have been attributed to other sources of ambient air pollution. The monetary value of the global welfare losses has been estimated at US$5.1 trillion (or 6.6 per cent of global world product). {5.3.1}

People who are elderly, very young, sick and poor are more susceptible to air pollution, which can exacerbate pre-existing illnesses or conditions (*well established*). Exposures are highest for people living in urban areas in low- and middle-income countries and for the approximately 3 billion people who depend on burning solid fuels or kerosene to meet household energy needs. {5.3.1}

Globally, decreasing emission trends in some sectors and regions have been offset by increasing emission trends in rapidly developing and emerging economies and areas of rapid urbanization (*well established*). {5.2}

East and South Asia have the highest total number of deaths attributable to air pollution, due to large populations and cities with high levels of pollution (*well established*). These regions also bear the largest health burden caused by the production of goods consumed in other regions of the world, primarily Western Europe and North America. {5.3.1}

As controls have been placed on power plants, large industrial facilities and vehicles, the relative contributions of other sources have grown in importance *(well established)*. Sources of pollution that are increasingly relevant to achieving air quality objectives include agriculture, domestic fuel burning, construction and other portable equipment, artisanal manufacturing and fires. The relative contributions of these sources to air quality problems differs from region to region, such that priorities for air pollution control may vary in different locations. {5.2.1}

Emissions of ozone-depleting substances (ODSs) have decreased dramatically as a result of the Montreal Protocol (*well established*). New studies provide robust evidence that stratospheric ozone over Antarctica has started to recover. Although stratospheric ozone concentrations in other regions have increased since 2000, the expected increase in total atmospheric column ozone and decrease in ultraviolet (UV) radiation reaching the Earth's surface have not been observed outside Antarctica due to natural variability, increases in GHGs, and changes in attenuation of the UV radiation by tropospheric ozone, clouds and aerosols. {5.2.3}

International agreements have been successful in addressing specific chemicals, but new chemical risks are emerging (*established but incomplete*). Environmental concentrations of persistent organic pollutants (POPs) have been reduced in Europe, North America, Asia and the Pacific, and the Arctic. {5.2.2}

Rapid development and urbanization combined with insufficient environmental governance in many regions suggest that climate change and air pollution are likely to worsen before they improve without additional policy interventions *(well established)*. However, future policy efforts can build upon renewed attention to these issues in international forums and several decades of experience with various governance strategies in different countries. {5.4}

5.1 Introduction

Emissions generated by human activity have changed the composition of the Earth's atmosphere, with consequences for the health of people and the planet. The impacts of human activity on the atmosphere are often framed in terms of four separate challenges: air pollution; climate change; stratospheric ozone depletion; and persistent, bioaccumulative, toxic substances (PBT) (Abelkop, Graham and Royer 2017). The causes of these four challenges, their effects on atmospheric composition and meteorological processes, and their impacts on humans and ecosystems are closely intertwined **(see Figure 5.1)**. Solutions to these challenges are also interrelated, as changes in lifestyle, technology and policy alter emissions of multiple pollutants simultaneously with a variety of interrelated implications. This chapter describes these four challenges together following the Drivers, Pressures, State, Impact, Response (DPSIR) framework (see Section 1.6).

Since the fifth Global Environment Outlook (GEO-5) was published in 2012, a number of developments have focused international attention on changing atmospheric composition. Estimates of the global burden of disease contributed by air pollution have doubled (comparing assessments published in 2004, 2012 and 2017) primarily due to new exposure estimates informed by satellite-borne instruments (Lim *et al.* 2012; Cohen *et al.* 2017). The United Nations Environment Assembly of the United Nations Environment Programme (UNEA) (2014; 2017) and World Health Assembly of the World Health Organization (WHO) (2015) have responded with resolutions to encourage national-level actions to address air pollution. Concentrations of major GHGs are still growing strongly (World Meteorological Organization [WMO] 2017a) and indicators of climate change

Table 5.1: Some atmospheric chemical components

BC	black carbon
CFCs	chlorofluorocarbons
CH_4	methane
CO	carbon monoxide
CO_2	carbon dioxide
GHGs	greenhouse gases
HCFCs	hydrochlorofluorocarbons
HFCs	hydrofluorocarbons
Hg	mercury
N_2O	nitrous oxide
NH_3	ammonia
NMVOC	non-methane volatile organic compounds
NO	nitrogen oxide
NO_2	nitrogen dioxide
NO_x	nitrogen oxides
O_3	ozone, tropospheric and stratospheric
OC	organic carbon
ODS	ozone-depleting substances
PAHs	polycyclic aromatic hydrocarbons
Pb	lead
PBDE	polybrominated diphenyl ethers
PBTs	persistent, bioaccumulative, toxic chemicals (includes POPs, metals)
PCB	polychlorinated biphenyl
PFAS	per- and polyfluoroalkyl substances
PM	particulate matter
PM_{10}	PM less than 10 µm in diameter
$PM_{2.5}$	PM less than 2.5 µm in diameter
POPs	persistent organic pollutants (as defined by international agreements)
SO_2	sulphur dioxide

Figure 5.1: Primary linkages between pressures, state and impacts of atmospheric change

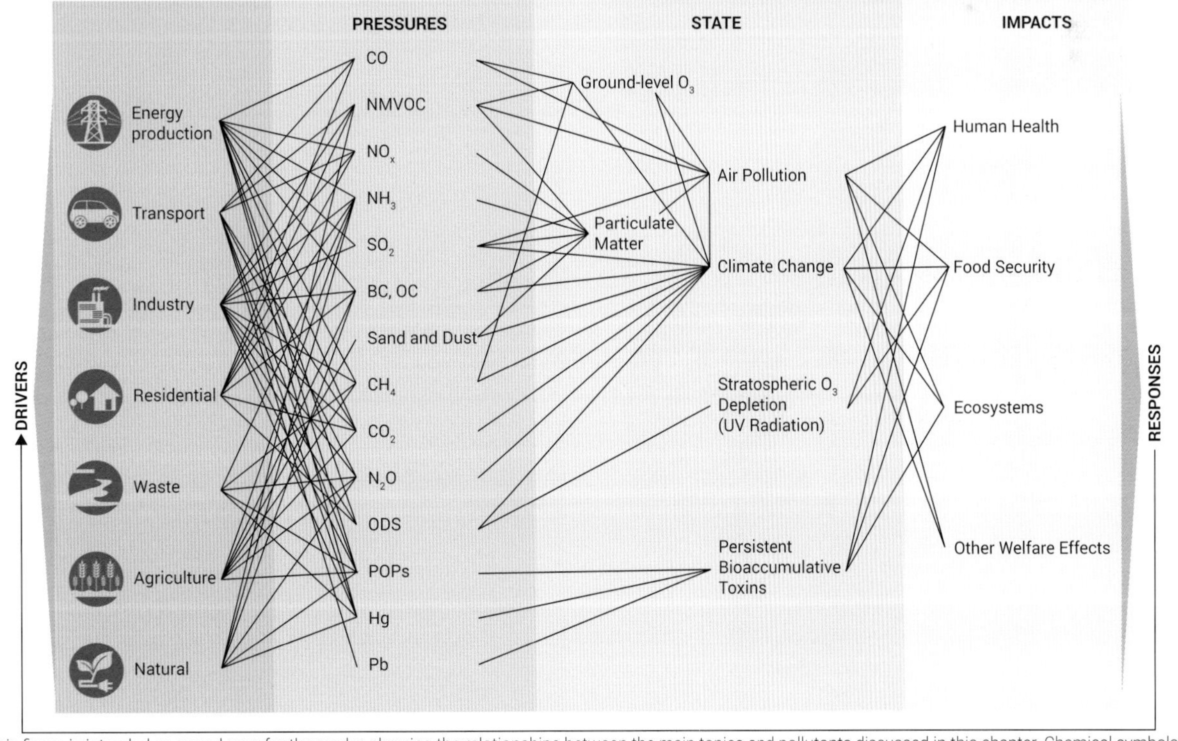

This figure is intended as a road map for the reader, showing the relationships between the main topics and pollutants discussed in this chapter. Chemical symbols and abbreviations are defined in Table 5.1.

have continued to accumulate. Targets in the Kyoto Protocol of the United Nations Framework Convention on Climate Change (UNFCCC) expired but were replaced by new ones under the Doha Amendment and new commitments under the Paris Agreement (UNFCCC 2016). Complementing the work of the UNFCCC, new efforts have targeted reductions of short lived climate pollutants (SLCPs) from specific sectors with benefits for climate change mitigation and human health (Climate and Clean Air Coalition [CCAC] 2015). As stratospheric ozone (O_3) has continued its recovery, the Kigali Amendment to the Montreal Protocol (United Nations 2016a) has harnessed this successful international agreement to help mitigate the climate impacts of hydrofluorocarbons (HFCs), originally introduced as substitutes for ozone-depleting substances (ODS). Emissions of mercury (Hg) have declined in some regions and increased in others. Emissions of some banned persistent organic pollutants (POPs) have declined due to the implementation of international agreements. However, atmospheric burdens of other POPs and PBTs remain at levels of concern, and new chemical risks have been identified (United Nations Environment Programme [UNEP] 2017a).

Efforts to achieve each of the Sustainable Development Goals (SDGs) are linked directly or indirectly to mitigating air emissions and changes to atmospheric composition, as shown in **Figure 5.2**.

In the GEO-6 regional assessments, air pollution, climate change and energy development, as well as the intersection of these three issues, were identified as top priorities in every region. Growing cities, energy, and transportation demand were consistently identified as issues of concern. Indoor air pollution and access to clean household energy were priorities in Africa and Asia. Other regional priorities highlight differences in the institutional capacities of governments in different regions: improving observational networks (Africa, Latin America and the Caribbean, West Asia), strengthening governance (Asia, Latin America and the Caribbean), and understanding costs and benefits of mitigation measures (Asia). The following sections build upon the GEO-6 regional assessments to explore the state of these challenges from a global perspective.

5.2 Pressures: emissions

People alter the atmosphere primarily by generating emissions. Trends in human-caused emissions are driven by changes in population, urbanization, economic activity, technology and climate ('the drivers'), as well as by behavioural choices, including lifestyle, and conflict. In turn, these drivers are influenced by policies ('responses'). Natural emission sources, including emissions from vegetation, soils, wildfires, and windblown sand and dust, also contribute to emissions, but can be affected by people (e.g. through land-use change).

Although an increasing amount of emissions information in some GEO regions is publicly available, there is no global reporting programme applicable to all sources and pollutants and no comprehensive emissions data repository. The Aarhus Convention and its Protocol on Pollutant Release and Transfer Registers (PRTR) aspires to establish a global network, building on the work of the United Nations Economic Commission for Europe (UNECE) and the Organisation for Economic Co-operation and Development (OECD) (see http://prtr.net). Currently, compiling a consistent global emissions inventory requires research effort. This assessment uses the latest anthropogenic emissions data developed using the Community Emissions Data System (CEDS), an open source, global emissions inventory data system that was developed

Figure 5.2: Linkages between changes in atmospheric composition and achievement of the Sustainable Development Goals

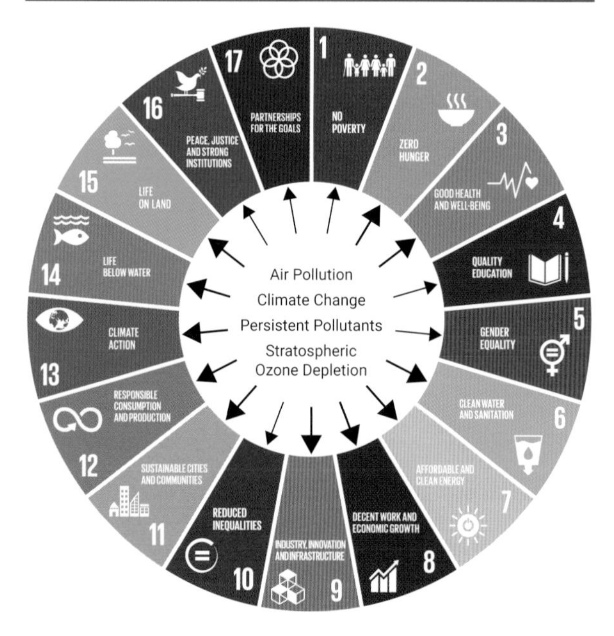

Direct linkages are shown with bold arrows, indirect linkages with light arrows.

to provide consistent long-term emission trends for use in global atmospheric modelling efforts, such as those supporting the preparation of Intergovernmental Panel on Climate Change (IPCC) 6th Assessment Report (Hoesly *et al.* 2018). Open biomass burning emissions, whether anthropogenic or natural, are drawn from a separate inventory created for global modelling efforts by merging information from satellite-based estimates, sedimentary charcoal records, historical visibility records and multiple fire models (van Marle *et al.* 2017). Together, these data sets provide an up-to-date and consistent basis to examine trends for most air pollutants and greenhouse gases (GHGs) **(see Figure 5.3)**.

Globally, anthropogenic carbon dioxide (CO_2) emissions increased by more than 40 per cent over the period 1990-2014, driven by large increases in Asia and counteracted by small declines in North America and Europe. Sulphur dioxide (SO_2) emissions are the only ones to have declined globally during this period, with increases of more than 50 per cent in Asia offset by a more than 75 per cent decrease in North America and Europe. In recent years, emissions of SO_2 and nitrogen oxides (NO_X) have begun to decline in East Asia. The inclusion of wild and agricultural fires significantly increases the inter-annual variability of emissions of non-methane volatile organic compounds (NMVOC), carbon monoxide (CO), black carbon (BC) and organic carbon (OC).

The emissions data presented here are best estimates with different degrees of uncertainty depending on pollutant, sector, region and time period. Hoesly *et al.* (2018) found that CEDS estimates are slightly higher than previous global inventories (e.g. Lamarque *et al.* 2010; European Commission 2016). In general, estimates of CO_2 and SO_2 emissions have uncertainties on the order of ±10 per cent for a 5-95 per cent confidence interval, whereas BC and OC emissions have uncertainties on the order of a factor of two. Uncertainties for CO, NO_X, NMVOC and ammonia (NH_3) emissions lie in between these two endpoints (Hoesly *et al.* 2018). Uncertainty also varies by sector: emissions from large electricity generation plants are well characterized, whereas emissions generated by military conflicts are not well understood or commonly included in inventories.

Figure 5.3: Annual emission trends from 1990 to 2014 in kilotons by pollutant, region and sector

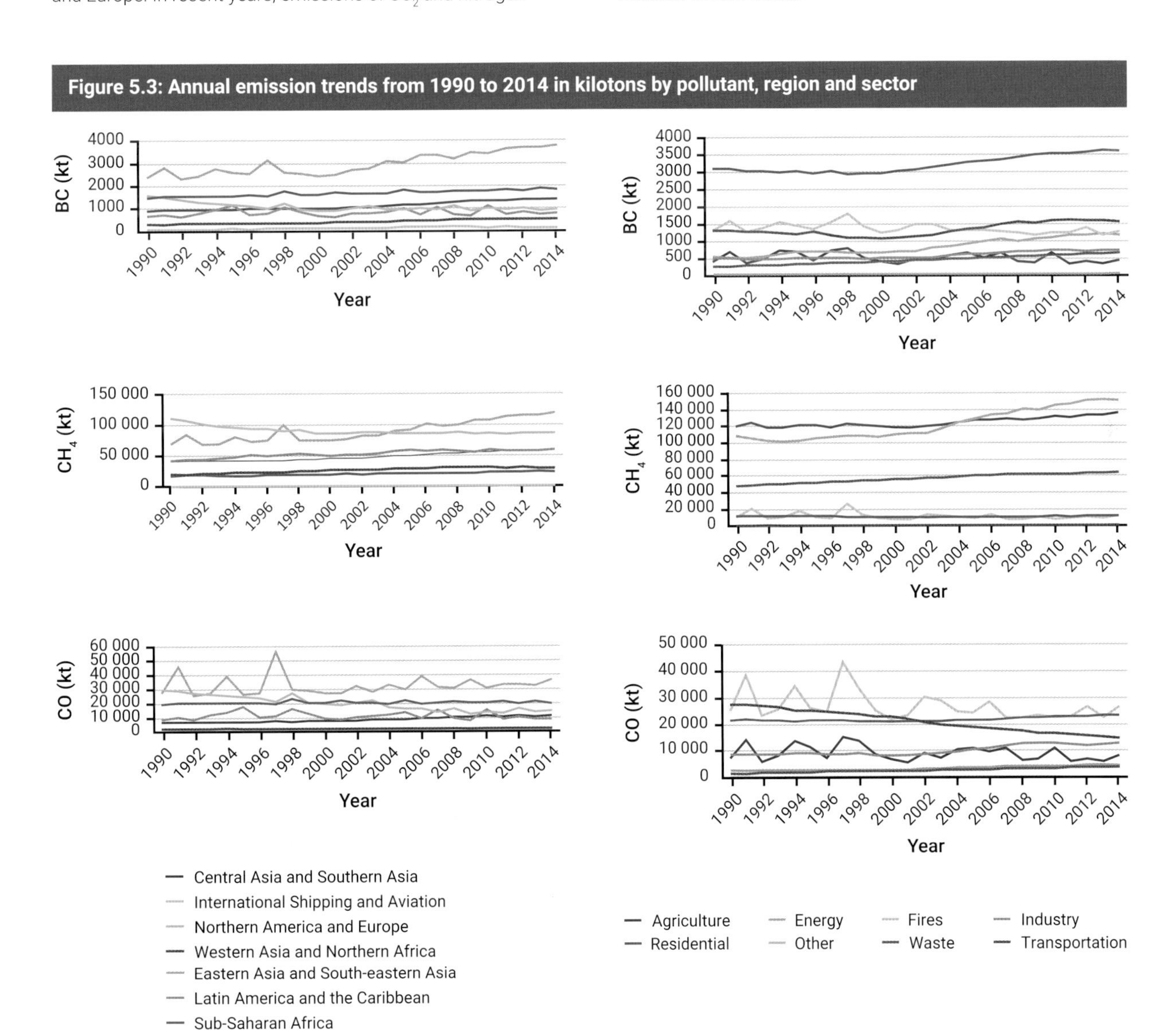

Source: Hoesly *et al.* (2018).

Figure 5.3 (continued): Annual emission trends from 1990 to 2014 in kilotons by pollutant, region and sector

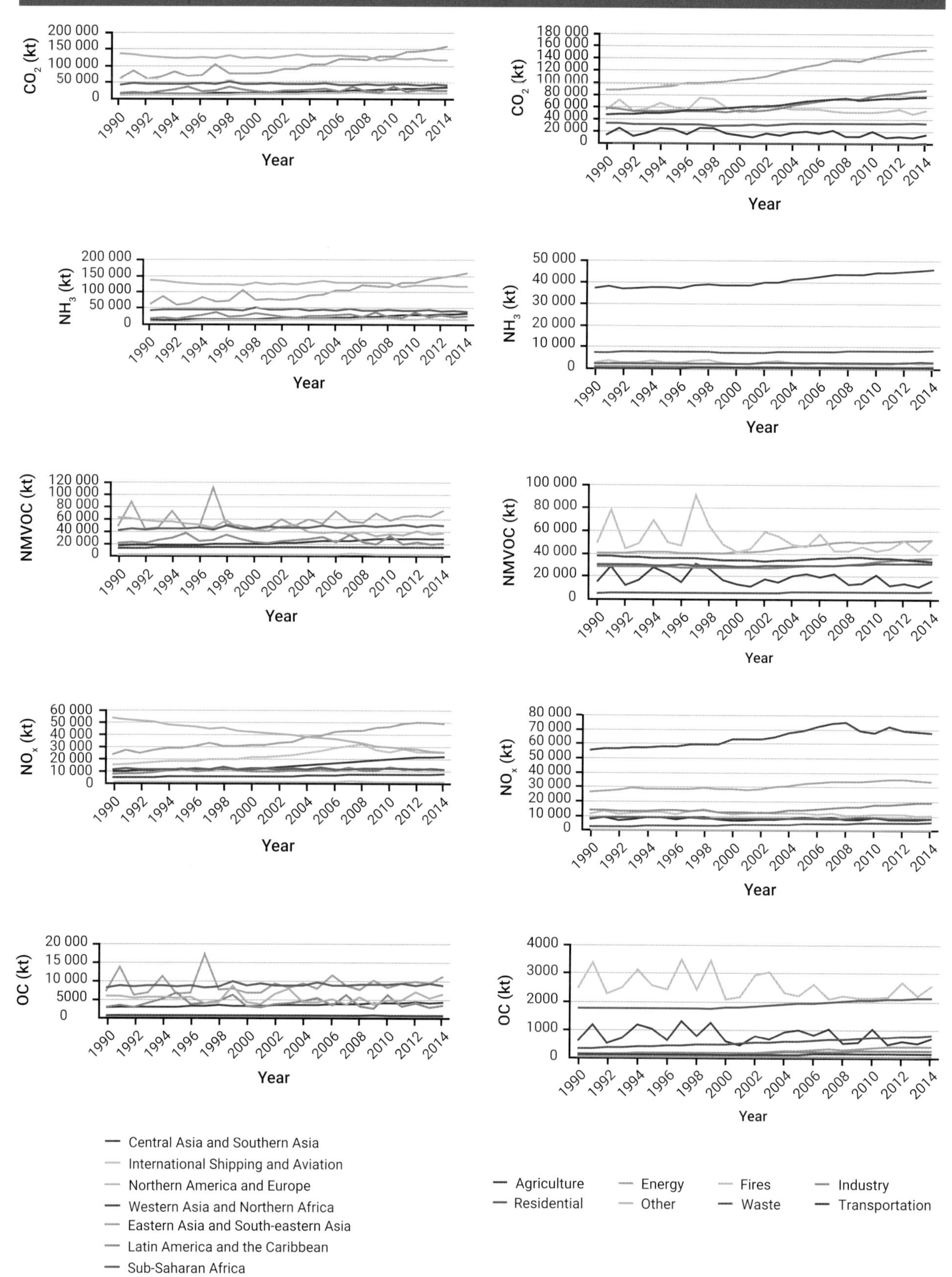

Source: Hoesly *et al.* (2018).

Figure 5.3 (continued): Annual emission trends from 1990 to 2014 in kilotons by pollutant, region and sector

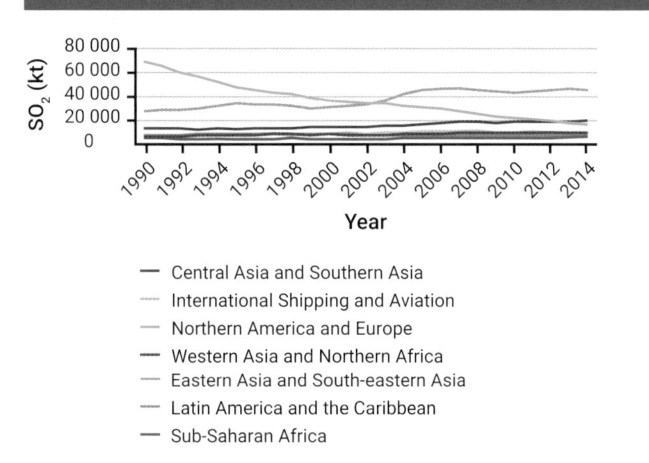

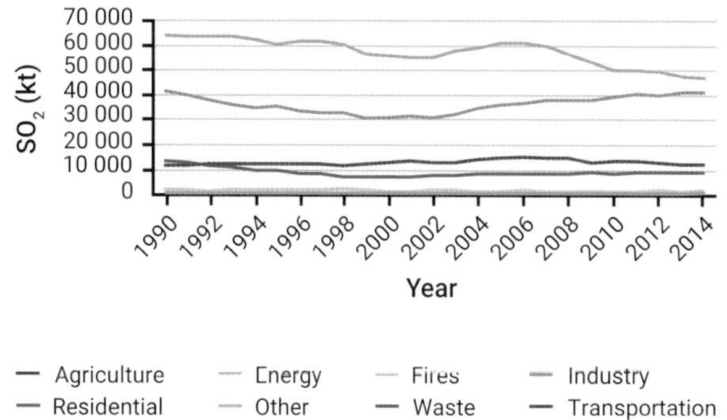

Source: Hoesly *et al.* (2018).

There are considerable gaps in available emissions data for POPs, which include pesticides, industrial chemicals and products of incomplete combustion or chemical reactions. Available data in Europe, America and Central Asia indicate that emissions decreased significantly between 1990 and 2012 for the most studied POPs, due to regulations, including the Stockholm Convention (UNEP 2014a; UNEP 2014b; UNEP 2015a; UNEP 2015b). Nevertheless, alongside the growing number of listed POPs and candidate substances, unregulated POPs emissions may be increasing. Many commercial products contain unknown quantities and types of unregulated POPs, often with unknown effects (see also Section 4.3.3).

The UNEP Global Mercury Assessment estimated that anthropogenic Hg emissions to air were 2,220 (2,000-2,820) (metric) tons/year for 2015 (UNEP 2013a). Globally, artisanal and small-scale gold mining (ASGM) was responsible for about 38 per cent of total anthropogenic Hg emissions to air in 2015, followed by coal combustion (about 21 per cent), non-ferrous metal production (about 15 per cent) and cement production (about 11 per cent). Asia is the main source region, contributing about 49 per cent of 2015 global anthropogenic Hg emissions, followed by South America (18 per cent) and sub-Saharan Africa (16 per cent). Current anthropogenic sources contribute about 30 per cent of annual Hg emissions to air, while natural geological sources contribute about 10 per cent. The remaining 60 per cent comes from 're-emissions' of previously released Hg from soils and oceans, mostly from anthropogenic sources (UNEP 2013a).

Globally, both the production and consumption of ODS, and thus ODS emissions, declined by more than 99 per cent between 1990 and 2016 (UNEP 2017b). Chlorofluorocarbons (CFCs) and halons, the most potent ozone depleters, have been replaced by shorter-lived hydrochlorofluorocarbons (HCFCs) and hydrofluorocarbons (HFCs), although recent measurements suggest that new emissions of trichlorofluoromethane (CFC-11) may be occurring (Montzka *et al.* 2018). The less-depleting HCFCs are now being phased out in favour of chemicals that do not contribute to ozone depletion. Concerns about the potential future contribution of HFCs to climate change led to the 2016 Kigali Amendment to the Montreal Protocol, which will limit future HFC emissions.

5.2.1 Electricity and fuel production

The electricity and fuel production sector (labelled 'energy' in **Figure 5.3**) is the largest anthropogenic emitting sector of CO_2, methane (CH_4), SO_2 and NMVOC, and the main emitting sector of other air pollutants. Within the sector, electricity generation contributed around 70 per cent of CO_2, 71 per cent of SO_2 and 72 per cent of NO_x in 2014 (Hoesly *et al.* 2018).

Figure 5.4: Global fuel shares of electricity generation in 2015[1]

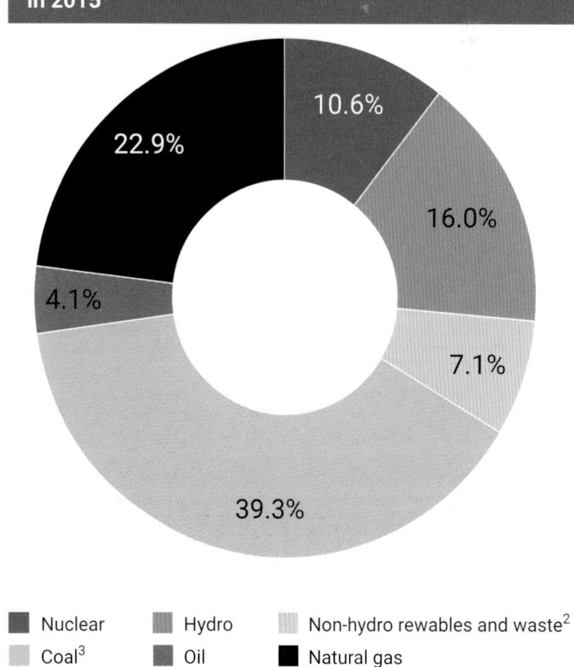

Notes: [1] Excludes electricity generation from pumped storage. [2] Includes geothermal, solar, wind, heat, etc. [3] Peat and oil shale are aggregated with coal.

Source: IEA (2017).

Despite increases in renewable energy capacity, fossil fuels still dominate the global power system (see **Figure 5.5**). Three-quarters of the sector's SO_2 emissions, 70 per cent of its NO_X emissions and over 90 per cent of those of primary particulate matter less than 2.5 μm in diameter ($PM_{2.5}$) are from coal-fired plants. Coal combustion is also the second most important anthropogenic source of global Hg emissions (International Energy Agency [IEA] 2016a). In 2015, gas-fired generation emitted close to 20 per cent of NO_X from power generation, but barely any SO_2 or primary $PM_{2.5}$ (IEA 2016a).

From 1990 to 2015, global petroleum fuel production saw slow but sustained growth (see **Figure 5.5**). CH_4 and NMVOC emissions from fuel production showed a corresponding increase (**Figure 5.3**). However, for electricity generation, production doubled between 1990 and 2015 (**Figure 5.6**),

Figure 5.5: World petroleum refinery output by-product (million tons)

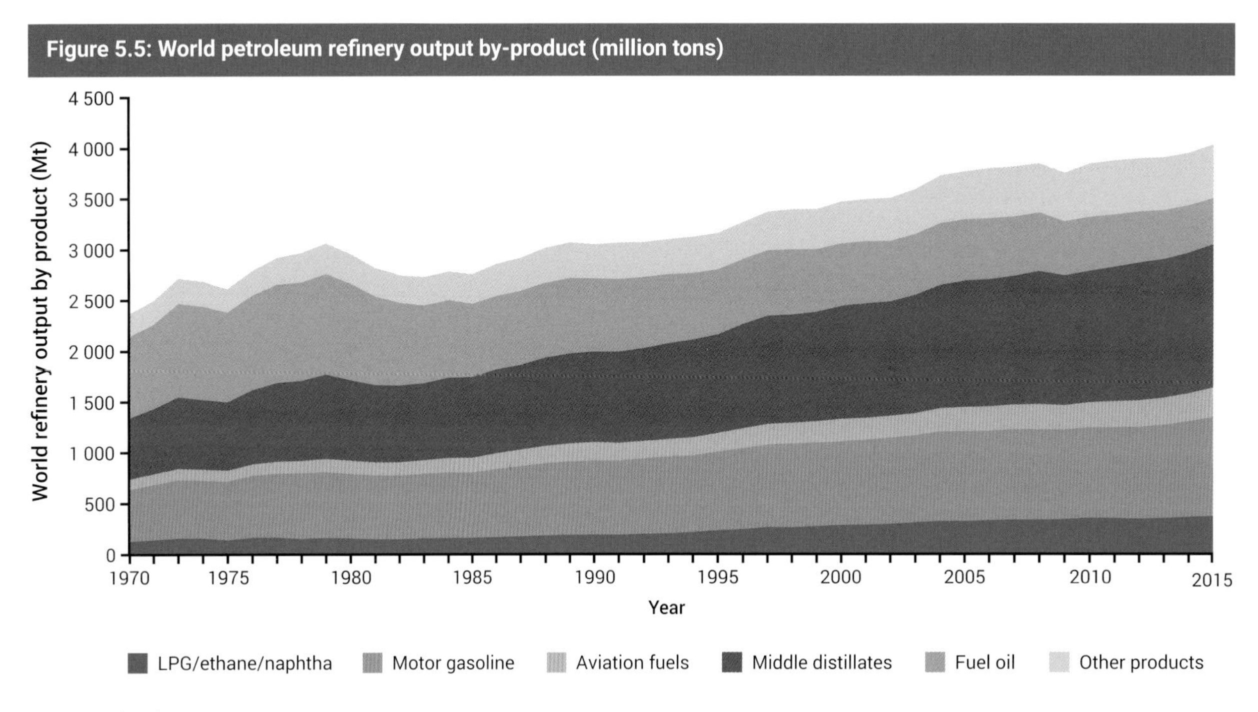

Source: IEA (2017).

Figure 5.6: World electricity generation by fuel (terawatt hours)[1]

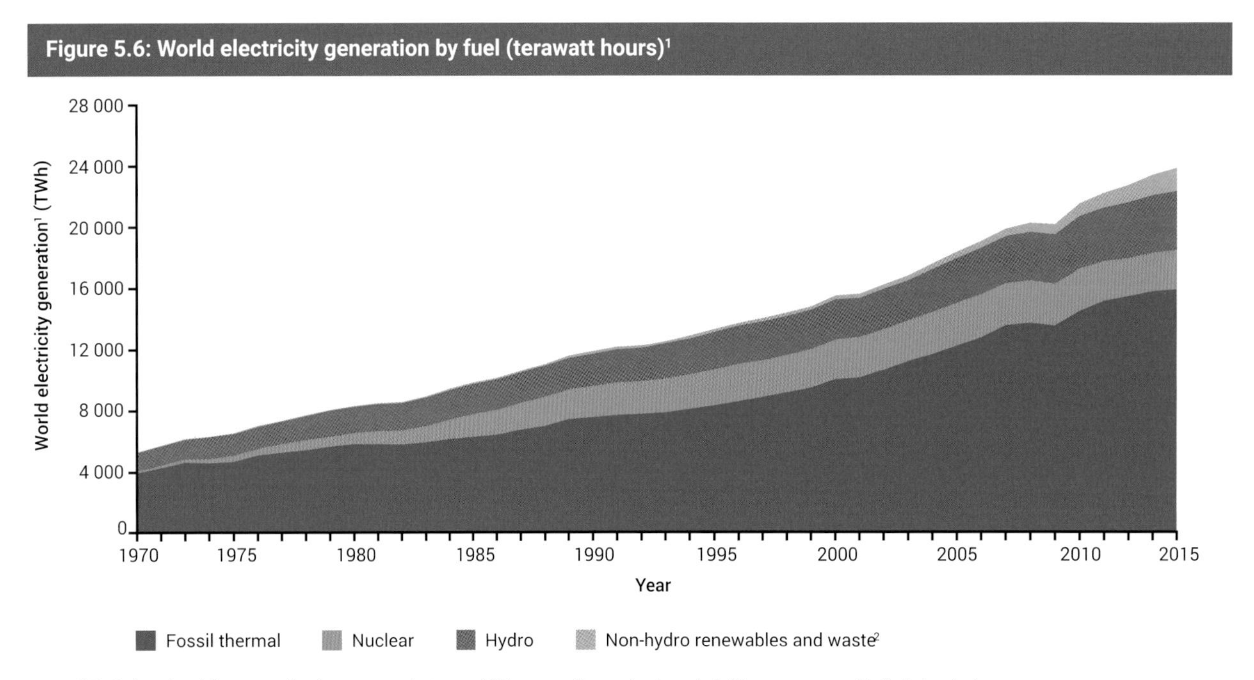

Notes: [1] Excludes electricity generation from pumped storage. [2] Other = geothermal, solar, wind, tide, wave, ocean, biofuels, heat, etc.
Source: IEA (2017).

but emissions of air pollutants did not increase at the same rate. Most importantly, SO_2 emissions from electricity generation declined after 2006 (see **Figure 5.3**). The main reasons for this decoupling include:

1. improvement of energy efficiency;
2. tighter emission standards for power plants and progress of end-of-pipe control technologies;
3. development of natural gas, renewable and nuclear power (Renewable Energy Policy Network for the 21st Century [REN21] 2016).

However, despite existing policies and the announced aims, targets and intentions, electricity demand is expected to increase by two-thirds by 2040 (IEA 2016b). Both end-of-pipe pollution control technologies and coal with low sulphur content may be used to achieve lower air pollutant emissions.

5.2.2 Transportation

In all regions of the world, the transportation of people and goods are significant sources of emissions of air pollutants, GHGs, ODS (from automobile air-conditioning units) and PBTs (including lead [Pb] and other metals). Road transport, including petrol- (gasoline-) and diesel-fuelled passenger cars and heavy-duty trucks, account for a dominant fraction of NO_x emissions, and a significant fraction of CO_2, CO, NMVOC and BC emissions (see **Figure 5.3**; Hoesly *et al.* 2018). Road traffic also contributes to emissions of primary PM from tyre and brake wear and entrained road dust (not included in **Figure 5.3**). Because cars and trucks operate and emit pollutants near where people live and work, they have a larger impact on air pollution exposures and associated health impacts than is proportional to their fraction of total emissions.

Total road transport activity is higher in North America and Europe than in other regions and is therefore responsible for greater CO_2 emissions, but those emissions have held steady for the last decade, with improvements in fuel efficiency keeping pace with increasing transport demand (Hoesly *et al.* 2018). The emissions of other transportation-related pollutants in North America and Europe have declined due to the introduction of vehicle emissions and fuel standards (see Section 12.2).

In developing countries, road transport emissions continue to rise as vehicle use is increasing faster than technological improvements, despite the introduction of emissions and fuel standards, which lag behind those in North America and Europe. Implementation of cleaner technologies is slowed by the trade in used vehicles from richer countries (UNECE and UNEP 2017). However, continued progress towards decreasing the sulphur content of fuel will enable the use of advanced emission control systems in all countries.

As emission standards are more widely applied to road vehicles, the relative fraction of emissions from non-road vehicles, such as heavy-duty construction equipment, is becoming increasingly important. Often running on diesel fuel, and with long lifetimes, such vehicles can be good candidates for retrofit control technologies or alternative fuels.

Maritime shipping is used to transport 80 per cent of global trade measured by volume (International Transport Forum 2017) and grew by more than 300 per cent between 1990 and 2015 when measured by ton-miles (United Nations Conference on Trade and Development [UNCTAD] 1997; UNCTAD 2017). Typically burning the heaviest petroleum products, ships are a significant source of SO_2 and CO_2 emissions globally and a source of SO_2, NO_x and BC emissions in coastal regions and port cities. Emission Control Areas have been established under international law (e.g. covering the North and Baltic seas and North American coastal waters) and national laws (e.g. covering Chinese ports and inland waters). The International Maritime Organization has announced new emission and fuel standards that are expected to dramatically decrease shipping emissions starting in 2020.

Aviation is a small but growing contributor to global emissions, accounting for less than about 2 per cent of global anthropogenic CO_2 emissions from fuel combustion (IEA 2017). Between 2000 and 2016, global air passenger travel increased 235 per cent (measured in passenger-km) and airfreight increased 174 per cent (measured in tons-km) (International Civil Aviation Organization [ICAO] 2016a). Aircraft emit pollutants directly into the upper atmosphere where their impact on ozone formation and climate forcing is larger than if emitted near the surface. The contribution of aviation CO_2 to radiative forcing is well quantified, but planes also emit water vapour, other gases and aerosols at high altitudes that trigger cloud formation and modify natural clouds and alter ozone and methane concentrations in the upper troposphere and lower stratosphere. The effects of these changes on climate forcing are not well quantified (Brasseur *et al.* 2016; Fahey *et al.* 2016). In 2016, the International Civil Aviation Organization (ICAO) adopted the Carbon Offsetting and Reduction Scheme for International Aviation (CORSIA), with a goal of capping the net CO_2 emissions from international aviation at 2020 levels (ICAO 2016b).

5.2.3 Industrial

Industry includes both manufacturing and mining sectors. The industrial sector emits air pollutants, GHGs, ODSs and PBTs, providing opportunities for multi-pollutant controls. Emissions and emission controls are often industry and process specific, or even regionally specific for some industries.

Nearly two-thirds of historic CO_2 and CH_4 emissions can be attributed to 90 investor- or government-owned businesses involved in the production of fossil fuels and cement (Heede 2014). Global emissions from industry increased for all pollutants between 1990 and 2014, except for SO_2 **(Figure 5.3)**, due to decreases in Europe's and North America's emissions being smaller than increases in other continents. Global SO_2 industrial emissions declined by 26 per cent from 1990 to 1999, due to the decrease in European and North American emissions, and increased after 1999, due to a considerable increase in China's (up to 2012 and reduced thereafter; Zheng *et al.* 2018) and other Asian countries' emissions (Hoesly *et al.* 2018).

The creation of many new industrial products, nanomaterials and chemicals poses a considerable challenge in terms of regulation and control. Their emissions are often neither regulated nor quantified, leading to unknown effects on the environment and health.

Technological innovation, technology transfer and tighter emission regulations to improve energy efficiency in

manufacturing and mining sectors are key to reducing emissions. Examples include cleaner brick kiln technology, piloted in Asia and Latin America (Maithel *et al.* 2012; Center for Human Rights and Environment 2015); cleaner technologies and approaches to reduce or eliminate mercury use in ASGM piloted in several countries (United States Environmental Protection Agency [US EPA] 2018a); and Perform-Achieve-Trade schemes for energy intensive industry in India (Kumar and Agarwala 2013; Bhandari and Shrimali 2018).

5.2.4 Residential and commercial

Around 3.1 billion people, about 43 per cent of the global population in 2014, depend on burning fuels such as wood, crop residue, dung, coal and kerosene to cook their food and heat and light their homes (World Health Organization [WHO] 2016a). These fuels are the dominant source of BC and OC emissions globally and a major source of primary PM, polycyclic aromatic hydrocarbons (PAHs), CO and CO_2 emissions (Hoesly *et al.* 2018). Globally, exposure to residential smoke is one of the largest environmental health risk factors (Cohen *et al.* 2017). Lack of access to clean household energy is most severe in low- and middle-income countries, but the use of polluting fuels takes place in high-income countries and in urban as well as rural areas. Women and children are the most exposed to household air pollution, and also bear the greatest burden of gathering or procuring the fuels (WHO 2016b). Improving access to cleaner stoves and fuels (including wood pellets, liquid petroleum gas, natural gas, and sources of electricity) has been identified as a global priority, and although progress is being made, many challenges remain (Global Alliance for Clean Cookstoves 2014; WHO 2016a) (see Section 12.2.3).

The energy demands of the built environment (primarily the construction, heating, cooling, and lighting of residential and commercial buildings) account for a large fraction of GHG emissions in countries with developed economies and some cities in developing economies. Improving the energy efficiency of buildings and cities is necessary to meet global goals for GHG mitigation and to achieve co-benefits for air quality. These improvements require policy approaches such as building standards, labelling and rating systems, land-use planning, tax incentives, financing, voluntary commitments, awareness and education.

5.2.5 Waste management

While most developed countries have shifted towards cleaner and more efficient technologies for waste management, developing countries are still grappling with basic challenges in this area. Open dumping and burning of solid waste remain predominant in low-income countries and continues to be practised in many cities in lower-middle and upper-middle income countries. An estimated 2 billion people worldwide lack access to solid waste collection services, while 3 billion people lack access to adequate waste disposal facilities (UNEP and International Solid Waste Association 2015). Approximately 64 million people are directly affected by uncontrolled dumping and open burning at the world's 50 largest dumpsites, 42 of which are within 2 km of settlements (Waste Atlas Partnership 2014).

Open waste burning emits CO_2, CH_4, NMVOC and PM, and is a major source of POPs, including dioxins and furans, in many developing countries (UNEP 2014a; UNEP 2014b; UNEP 2015a; UNEP 2015b). In developed countries, the waste sector is also an important source of CH_4, metals and POPs. The illicit export of discarded electrical and electronic equipment (ewaste) from industrialized to developing countries (Rucevska *et al.* 2015) leads to significant emissions of POPs as well as other semivolatile organic contaminants (e.g. other halogenated flame retardants) in the informal e-waste receiving and processing areas (Breivik *et al.* 2016).

5.2.6 Agricultural and forestry

A broad array of agricultural and livestock farming practices alter the nitrogen cycle and GHG emissions, and increase pollution by fertilizers and pesticides, promoting biodiversity loss and soil degradation (DeLonge, Miles and Carlisle 2016). Agriculture, forestry and other land uses contribute 25 per cent to global GHG emissions (Seto *et al.* 2014). In developed countries, agriculture forms about 10 per cent of national GHG inventories (European Environment Agency 2017; US EPA 2017), while in developing countries the contribution is much higher.

Meat and dairy production, distribution and consumption have large environmental impacts on scales ranging from local to global (Leip *et al.* 2015). Industrial meat production and livestock operations are significant sources of GHGs, NH_3, dust and bioaerosols (Cole and McCoskey 2013). GHG emissions from livestock farming increased by 51 per cent globally between 1961 and 2010, mostly due to a 117 per cent increase in developing countries, moderated by a 23 per cent decrease in the developed countries (Caro *et al.* 2014; Pagano *et al.* 2017). Livestock production is responsible for 9 per cent of total GHG emissions (Caro *et al.* 2014). The largest source of these emissions (74 per cent) are dairy and beef cattle. N_2O and CH_4 emissions, which emanate from manure left on pasture, manure management and fermentation, increased by 57 per cent globally in the same period. However, rotational livestock grazing and other pasture management techniques are available to decrease the production of GHGs by the very same cattle, and at the same time preserve biodiversity (Nordborg and Röös 2016).

Along with livestock farming, fertilizer use results in significant emissions of NH_3, accounting for about 75 per cent of anthropogenic and about 60 per cent of total NH_3 emissions globally (Ciais *et al.* 2013) and contributing to regional PM formation and detrimental effects on terrestrial, freshwater
and marine ecosystems (Galloway *et al.* 2003).

Irrigation and fertilization practices for crops in general, as well as pasture management, can alter soil respiration rates, changing the amount of CO_2 emitted by soils to the atmosphere (UNEP 2017c). Pesticides used in agricultural applications are a major source of unregulated POPs into the environment and food chain, with various detrimental effects on health (see Section 4.3.4).

Biomass burning – including natural wildfires, prescribed burning of crop and forest residues, and prescribed burning of forests and savannah for land clearing – contributes significantly to air pollution by emitting CO, OC, BC, NO_x and NH_3, as well as GHGs, CO_2 and CH_4. Dominant types of biomass burned are savannah in Africa; boreal forest in the former Soviet Union, savannah and tropical forest in Latin America; and savannah, peat and tropical forest in eastern Asia. Biomass burning in South-East Asia, the drought triggered by the 2015-2016 El-Niño, coupled with anthropogenically induced deforestation over peat swamps and effects of previous widespread fires, have all led to severe regional air pollution events (Wooster, Perry and Zoumas 2012; Koplitz *et al.* 2016; Parker *et al.* 2016) (see Section 12.2.5).

5.2.7 Natural emissions and land-use change

Natural sources also contribute to emissions, but people have a strong influence on these in some regions through land-use change, especially cropland expansion (Pacifico *et al.* 2012; Ciais *et al.* 2013). Wind-blown dust from natural landscapes and unprotected cropland in arid and semi-arid regions is the largest source of atmospheric PM and the dominant fraction of coarse PM in many regions, such as northern Africa and the Middle East (Ginoux *et al.* 2012; Albani *et al.* 2014). Sustainable land and water management practices can decrease sand and dust storms, while contributing to reduced desertification, preserving biodiversity and mitigating climate change. Regional and national action plans, including those developed under the United Nations Convention to Combat Desertification (UNCCD), have the potential to address the underlying causes of sand and dust storms (UNEP, WMO and UNCCD 2016).

Globally, terrestrial vegetation is the dominant source of atmospheric NMVOCs, outweighing anthropogenic sources by a factor of ten (Guenther *et al.* 2012; Sindelarova *et al.* 2014). Biogenic NMVOCs tend to be highly reactive and can contribute significantly to O_3 and PM formation even in urban areas (Chameides *et al.* 1988). Soil microbial processes are an important part of the nitrogen cycle and can be a significant source of NO_x emissions outside urban areas and the dominant source of nitrous oxide (N_2O), a potent GHG, on a global basis (Ciais *et al.* 2013). Soil NO_x emissions are highest in croplands due to increased soil nitrogen content from fertilizer application (Vinken *et al.* 2014). Deforestation associated with expansion of croplands and pasturelands is estimated to have reduced global annual biogenic NMVOC emissions by 10-35 per cent and increased soil NO_x emissions by about 50 per cent since the 1850s, except in parts of the eastern United States and Western Europe where reforestation has taken place (Unger 2014; Heald and Geddes 2016). Bouwman *et al.* (2013) estimated that agricultural soil N_2O emissions increased by a factor of three during the 20th century.

Soil respiration is a major source of CO_2 to the atmosphere at a global scale (Hashimoto *et al.* 2015) that in recent decades has increased its contribution (Bond-Lamberty *et al.* 2018).

5.3 State: atmospheric composition and climate

For meteorology and climate variables, a well-developed global observation system with spatial coverage adequate to monitor regional patterns is coordinated by WMO. For atmospheric composition, however, the amount of information available varies significantly by pollutant and region. Countries in North America, Europe and East Asia, have well-developed in-situ ground-based monitoring networks for ground-level O_3 and PM, as well as SO_2, CO and, in some areas, NO and NO_2. For other pollutants, observations tend to be relatively sparse. There is a need for a global catalogue of monitoring station metadata, currently being pursued through expansion of the WMO Global Atmosphere Watch Station Information System (GAWSIS, https://gawsis.meteoswiss.ch) and Observing Systems Capability Analysis and Review (OSCAR, https://oscar.wmo.int) tool. For many regions of the world, however, ground-based networks do not have sufficient density and coverage to characterize spatially representative trends. Observations from satellites, aircraft and other platforms, as well as atmospheric chemistry and transport models, are needed to complement traditional networks.

Existing polar-orbiting satellite instruments provide global observations of a number of important air pollutants (including PM, O_3, CO, SO_2, NO_2, NH_3, formaldehyde and CH_4) albeit with relatively coarse temporal, spatial and vertical resolution (Duncan *et al.* 2014; Duncan *et al.* 2016). In some parts of the world, however, monthly average total column observations from satellites provide the only information available. Current efforts to improve understanding of the relationship between space-based and ground-based observations should help to fill data gaps in areas with sparse monitoring (e.g. Snider *et al.* 2015).

Space agencies in the Republic of Korea, the United States of America and Europe are working to deploy a constellation of geostationary satellites over East Asia, North America, Europe, North Africa and the Mediterranean to measure O_3, PM and their precursors. In geostationary orbit, these instruments will have much finer temporal and spatial resolution than current polar-orbiting satellites, providing a wealth of information about air pollution over these regions in near real-time (Committee on Earth Observing Satellites 2011).

At the other end of the spectrum of cost and complexity, inexpensive electronic sensors for measuring different pollutants are being developed, marketed to governments, businesses and even individuals, and deployed in a variety of mobile and stationary settings (e.g. Apte *et al.* 2017). The quality of information varies significantly and is currently low, but efforts are in place to better understand the performance of different sensors, and to develop standardized tests and guidance on how to deploy and use the observations gathered (UNEP 2016; Lewis *et al.* 2017; US EPA 2018b).

Increasingly, air quality information from ground-based networks as well as air quality forecasts are being made available publicly. The United States of America pioneered such systems with AirNow.gov starting in 1998, and similar information is now available in countries and cities worldwide, as well as through open source platforms (e.g. OpenAQ.org) (see Section 12.2.4).

5.3.1 Air pollution: urban to global scales

From a global public health perspective, the two most important air pollutants are PM and its components and ground-level O_3. Ambient PM may be emitted directly as a fine particle (e.g. BC, OC and soil dust) or formed in the atmosphere from emissions of gaseous precursors (e.g. SO_2, NO_X, NH_3 and NMVOC). Ground-level O_3 is not directly emitted but is formed in the atmosphere from reactions of NO_X, NMVOC, CH_4 and CO (Seinfeld and Pandis 2016). Globally, the highest annual average concentrations of $PM_{2.5}$ are seen in areas affected by windblown sand and dust (e.g. northern Africa and west Asia), fires (e.g. Central Africa and Latin America) and anthropogenic pollution (e.g. South and East Asia) (Cohen *et al.* 2017; Shaddick *et al.* 2018) (see **Figure 5.7**). From 1998 to 2012, satellite observations suggest that $PM_{2.5}$ decreased significantly over eastern North America, and increased over west Asia, South Asia and East Asia (Boys *et al.* 2014). Ground-based measurements suggest that the trends over North America, South Asia and East Asia are associated with changes in anthropogenic pollution, but the changes over west Asia are due to changes in windblown sand and dust (Boys *et al.* 2014).

Ground-level O_3 is highest in the northern mid latitudes and tropics, and peaks in the warm season. North America, the Mediterranean, South Asia and East Asia are hotspots for O_3 pollution (see **Figure 5.8**). However, high population weighted O_3 concentrations are also estimated in Central Africa, west Asia and South-East Asia (Health Effects Institute 2017).

Satellite observations have identified rapid changes in the ground-level concentrations of SO_2 and NO_2 over the last 10-15 years, with declining trends in Europe and North America, and increasing trends in some regions in East Asia, South Asia, Africa and South America (Schneider, Lahoz and van der A 2015; Geddes *et al.* 2016; Krotkov *et al.* 2016).

Figure 5.7: Annual average $PM_{2.5}$ concentrations in 2016 compared with the WHO Air Quality guideline and interim targets

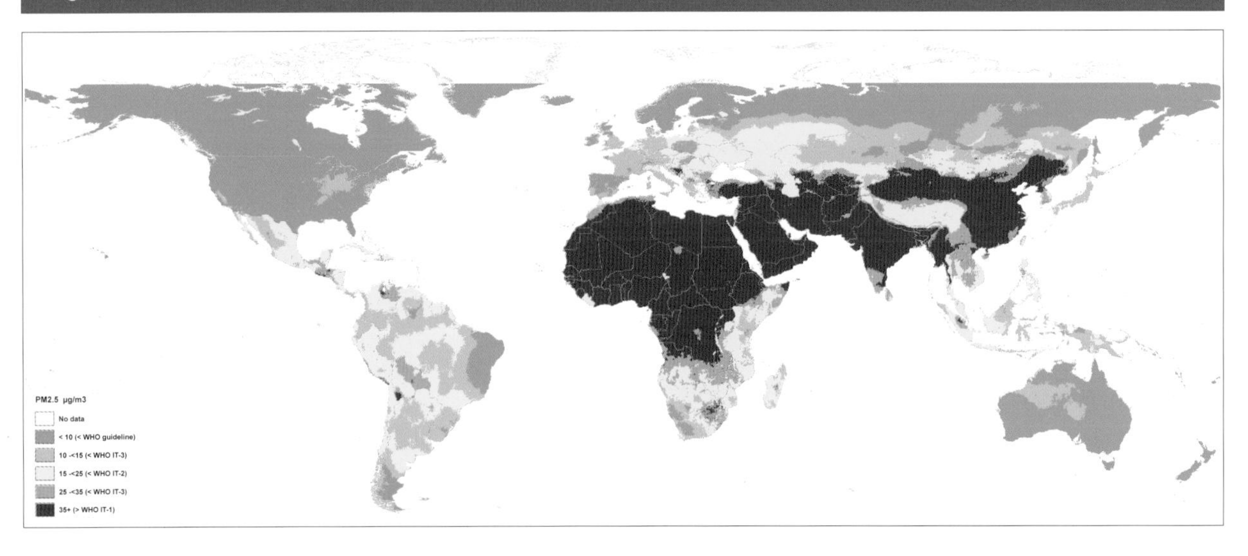

This map combines data from satellite observations, surface monitors and an atmospheric chemistry and transport model. IT = Interim Target.

Source: Shaddick *et al.* (2018).

Figure 5.8: Seasonal average population-weighted O_3 concentration in 2016 for season with maximum ozone levels by country

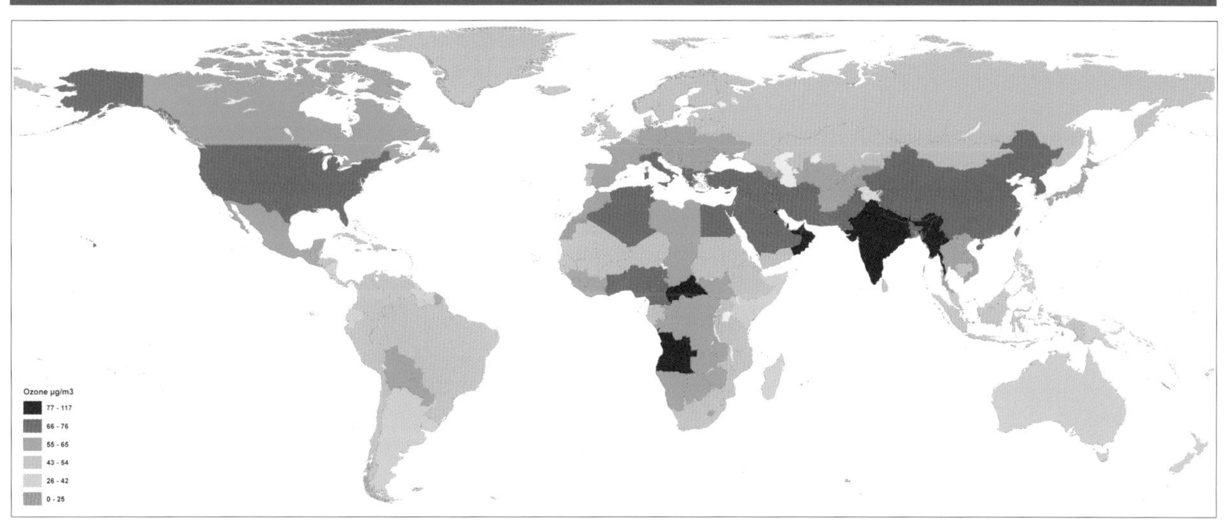

Source: HEI (2018).

Urban areas, which are home to over half of the world's population, have higher overall levels of air pollution. A review of published $PM_{2.5}$ observations for 71 megacities (over 5 million people) for 2013 found that, of the 45 megacities with available observations, only 4 attained the WHO guideline for annual average concentrations (Cheng *et al.* 2016) **(Figure 5.9)**. Cities with the highest levels were clustered in east-central China and the Indo-Gangetic Plain. Many cities in low- and middle-income countries lack available measurements, but where data is available, 98 per cent of cities exceed the WHO guidelines for $PM_{2.5}$ or PM_{10}, compared with the 56 per cent of cities in high-income countries with available data (WHO 2016b).

Figure 5.9: Annual average PM_{10} levels for megacities of more than 14 million inhabitants with available data for the period 2011-2015

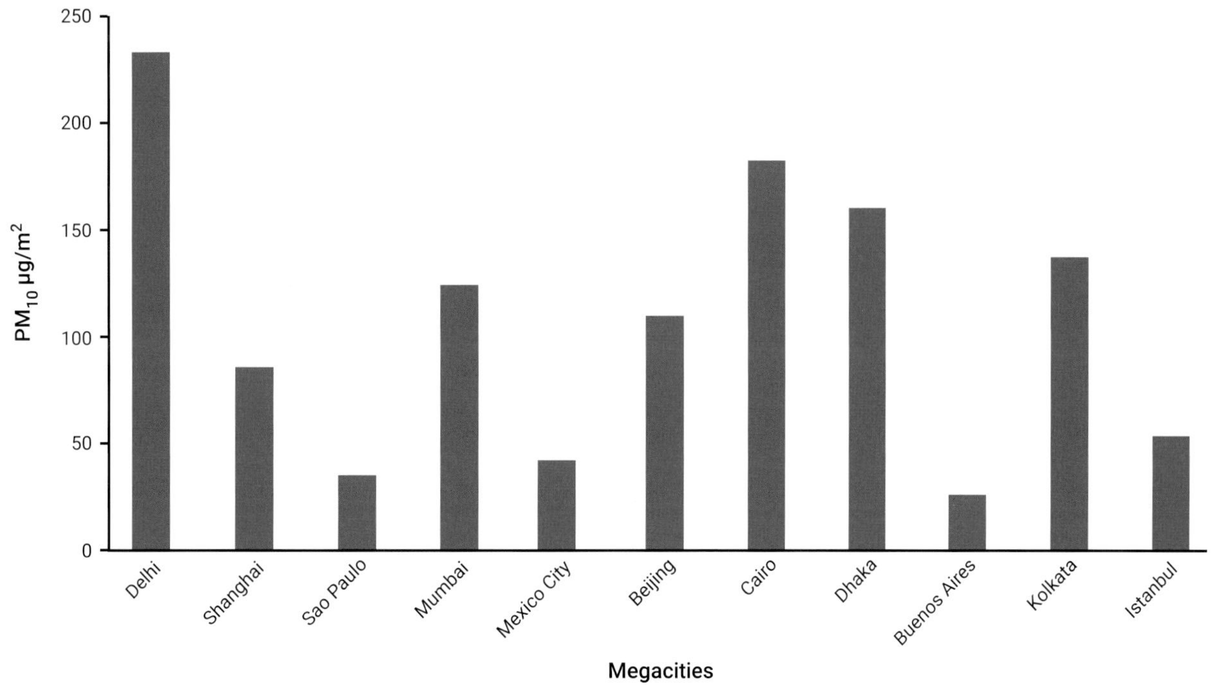

Source: WHO (2016b).

5

Traffic, residential fuel burning, electricity generation, industry and agriculture all contribute to urban air pollution, although the contribution of different sectors in individual cities can vary significantly (Karagulian *et al.* 2015). In growing cities across Africa, Asia and other developing regions, there has been an unprecedented rapid increase in the number of vehicles, driven by population growth and economic development (e.g. Adiang *et al.* 2017). It is projected that by 2030 there will be 41 megacities (population greater than 10 million), the majority in developing countries (United Nations 2016b). Impacts of the pollution from megacities extends far beyond the urban area with effects at local, regional and global scales (Ang'u, Nzioka and Mutai 2016; WHO 2016b).

Air pollution observed in any given location may be comprised of contributions from local, regional and even global sources **(Figure 5.10)**.

Better global models, additional monitoring and field studies, and accumulated observations from satellite-based instruments have improved our understanding of the processes and trends that drive such long-range transport of pollution. However, quantifying the absolute contributions of distant sources to observed values on a given day remains challenging. Data assembled for the Tropospheric O_3 Assessment Report (TOAR) demonstrates that recent trends in peak values upon which most health-based standards are founded are strongly decreasing in North America and Europe, and strongly increasing in parts of East Asia. However, for summer daytime average O_3 concentrations, the trends are more mixed in North America and Western Europe, with some sites showing significant increases (Chang *et al.* 2017; Schultz *et al.* 2017). This finding is consistent with observations of increasing 'background' O_3 above the boundary layer throughout the Northern Hemisphere (Task Force on Hemispheric Transport of Air Pollution 2010; Parrish *et al.* 2014). The observed increasing trend in global tropospheric O_3 from 1980 to 2010 may be due primarily to an equatorward shift in the distribution of global precursor emissions, the effect of which is larger than the increase in global methane and the total mass of other precursor emissions combined (Zhang *et al.* 2016).

The largest source of particulate matter in the atmosphere globally, on an annual basis, is windblown sand and dust. A 'dust belt' extends from the west coast of North Africa, over the Mediterranean Basin, the Middle East, Central and South Asia, to Mongolia and China (see **Figure 5.11**). This encompasses both natural areas, such as the Sahara and Taklamakan deserts, as well as agricultural areas. Outside the dust belt, sand and dust storms (SDS) are less prevalent; however, SDS can have important local impacts in central Australia, Southern Africa (Botswana and Namibia), the Atacama in South America, and the North America Great Basin (UNEP, WMO and UNCCD 2016). People influence dust sources through land clearing and land management practices and other influences on

Figure 5.10: Model estimates of the sources of $PM_{2.5}$ observed in several cities in each of three countries shows local $PM_{2.5}$ concentrations are strongly influenced by secondary particles from transboundary sources. The source of emissions is divided into natural, international (emitted outside the country), national (emitted within the country but outside the urban area), urban (emitted within the city) and street (emitted within the immediate vicinity of the observation) and interim targets

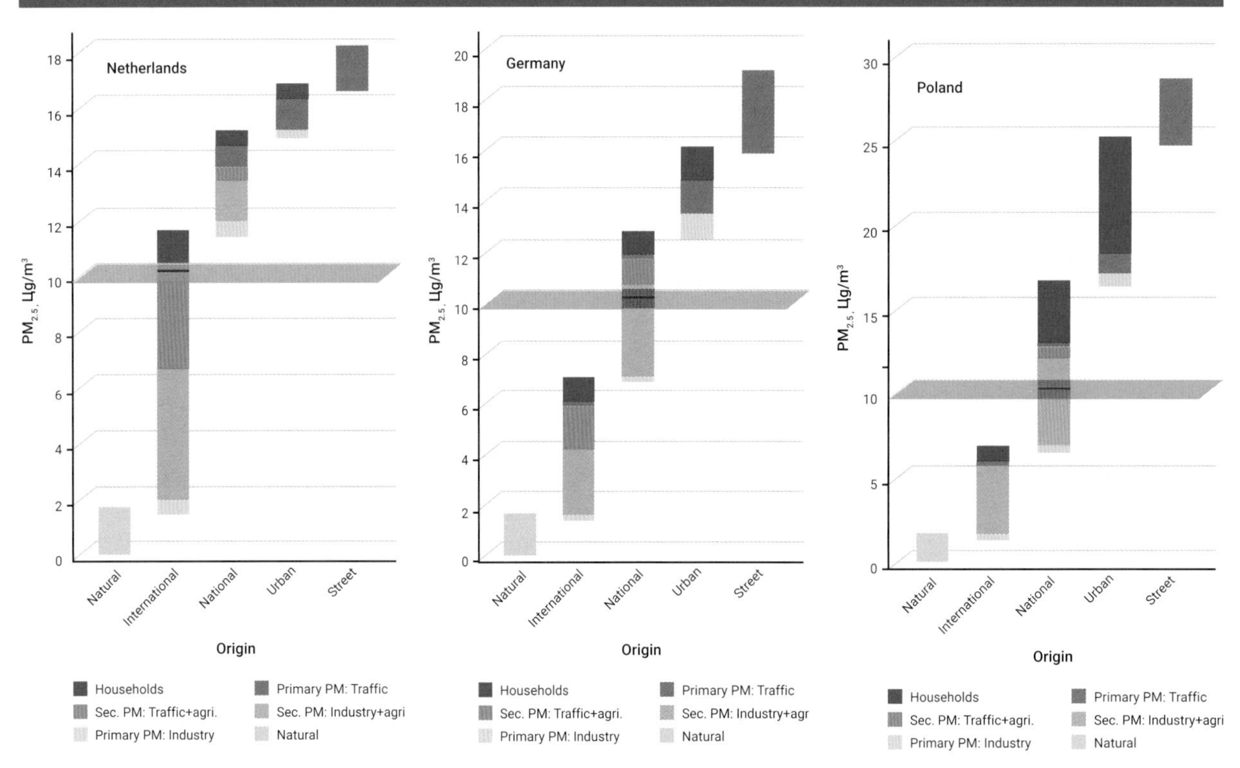

Source: Reprinted from UNEP/UNECE (2016), based on (Kiesewetter and Amann 2014).

Figure 5.11: The Dust Belt

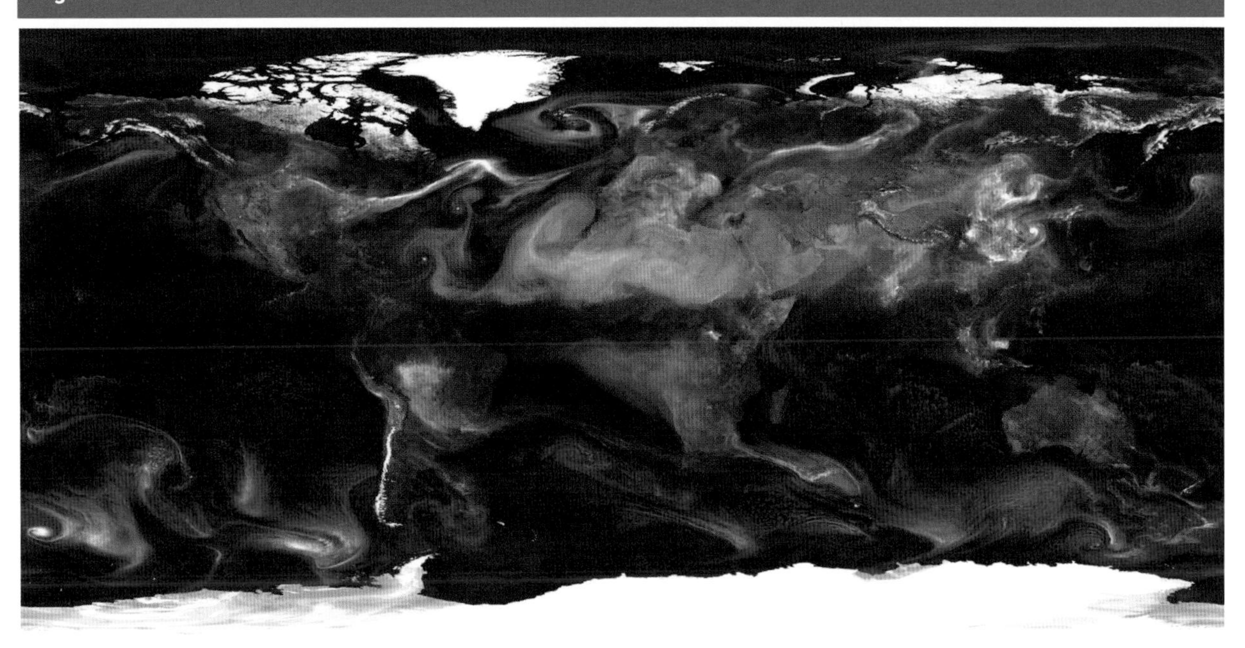

Using a global model, the aerosol optical depths attributable to different types of particulate matter are shown in different colours: dust (red and yellow), black and organic carbon (green), sulphate (white); and sea salt (blue). An animated version of this image is available at https://gmao.gsfc.nasa.gov/research/aerosol/modeling/nr1_movie/

Source: Puttman and da Silva (2013).

desertification (see Section 8.4.2). UNEP, WMO and UNCCD (2016) concluded that there has been little change in the frequency and severity of SDS in North Africa, the Middle East and South America over the last 30 years, but significant increases have been observed in North America, Central Asia and Australia. Klingmuller *et al.* (2016) found an increasing trend in dust over large parts of the Middle East during the period 2001 to 2012 that is correlated with climatic changes.

Transported dust contributes to a wide range of impacts: it affects climate and precipitation patterns; fertilizes distant forests and oceans; contributes to human respiratory ailments; and spreads human, animal and plant pathogens far downwind of the source region. Within the source region, dust storms may damage infrastructure, interrupt transportation and communication systems, and cause air and road traffic accidents. To better understand, forecast and mitigate these impacts, WMO has established a global Sand and Dust Storm Warning Advisory and Assessment System (SDS-WAS) (UNEP, WMO and UNCCD 2016; WMO 2017b).

Fires, primarily associated with land clearing or lightning, are another large contributor to transboundary pollution. In South-East Asia, perennial forest and peatland fires associated, primarily with slash-and-burn agriculture, intensify during dry seasons (Page and Hooijer 2016; Wijedasa *et al.* 2017). In 2015, fires blanketed the region with smoke, leading to an estimated 100,000 premature deaths associated with air pollution exposure, mostly in Indonesia (Koplitz *et al.* 2016) (see Section 12.2.5). Boreal forest fires in Siberia, Canada and Alaska contribute to the deposition of BC and other particles in the Arctic, darkening the surface of snow and ice and accelerating melting (Arctic Monitoring and Assessment Programme [AMAP] 2011; AMAP 2015).

5.3.2 Persistent bioaccumulative toxic substances

Gaseous elemental Hg is a global pollutant with the highest concentrations in East, South and South-East Asia, and in the artisanal gold mining regions of Equatorial Africa and South America (see **Figure 5.12**) (UNEP and AMAP 2018).

Concentrations of POPs that are regulated and monitored under the Stockholm Convention have been reduced in Europe, North America, and Asia and the Pacific (UNEP 2014a; UNEP 2014b; UNEP 2015a; UNEP 2015b).

Measurements of regulated POPs in Arctic air and biota show predominantly downward trends for substances that have been banned for more than 20-30 years in developed countries, but the rate of their decrease has slowed (Hung *et al.* 2016). Trends of POPs in the Arctic appear to be sensitive to changes in climate, due to increased volatilization from sources (AMAP 2014, Ma *et al.* 2011) and to changes in Arctic land-use and emission patterns, such as increases in mining and shipping (UNEP and AMAP 2011) (see also Sections 4.3.2 and 4.3.3). Although Antarctica is the Earth's continent least subject to direct human impact, low but sometimes significant contamination levels can be found there (Vecchiato *et al.* 2015). Concentrations of PAHs and PCBs in Antarctic snow have decreased over recent decades (Vecchiato *et al.* 2015).

Trends for many new PBTs, however, are not yet established, although baseline data have become available in some regions, such as Europe (UNEP 2015a). As some POPs have been regulated or banned, other unregulated PBTs have emerged as substitutes and are widely used in consumer and household items (e.g. furniture and electronics) and construction materials (Lee *et al.* 2016; Rauert *et al.* 2016). The growing

Figure 5.12: Global distribution of annual mean gaseous elemental mercury concentration in near-surface air (top) and wet-deposition flux (bottom) in 2015 simulated by a model ensemble

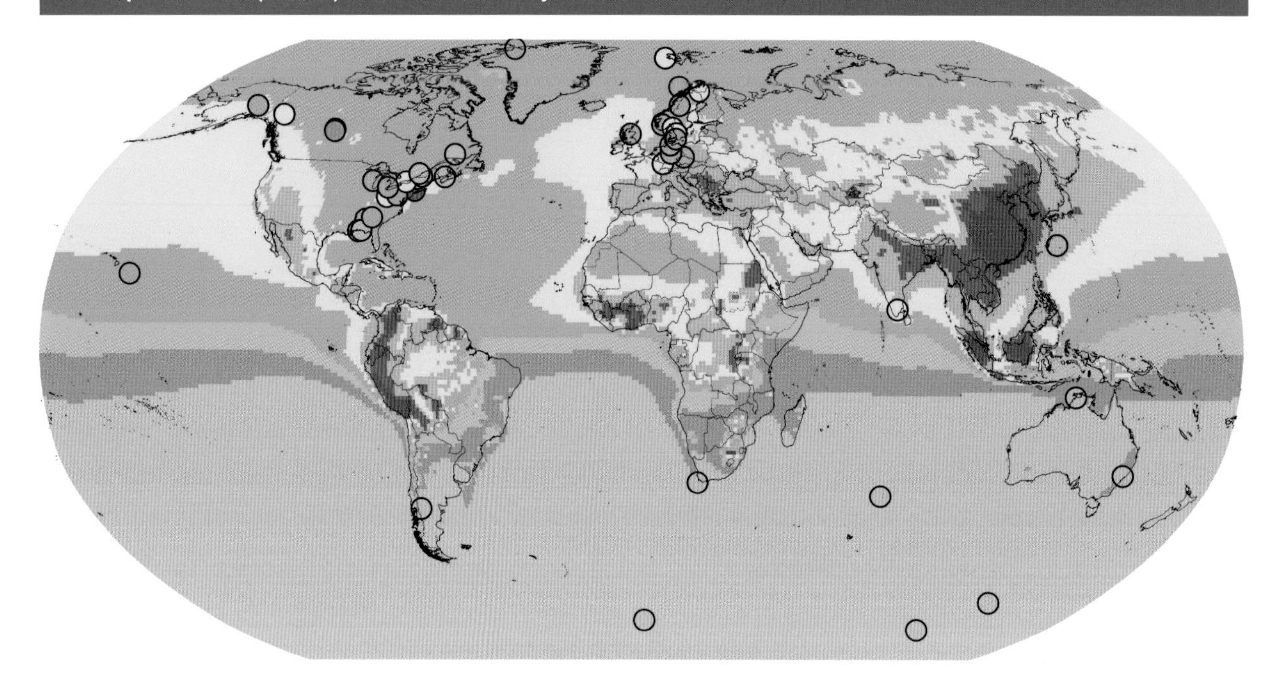

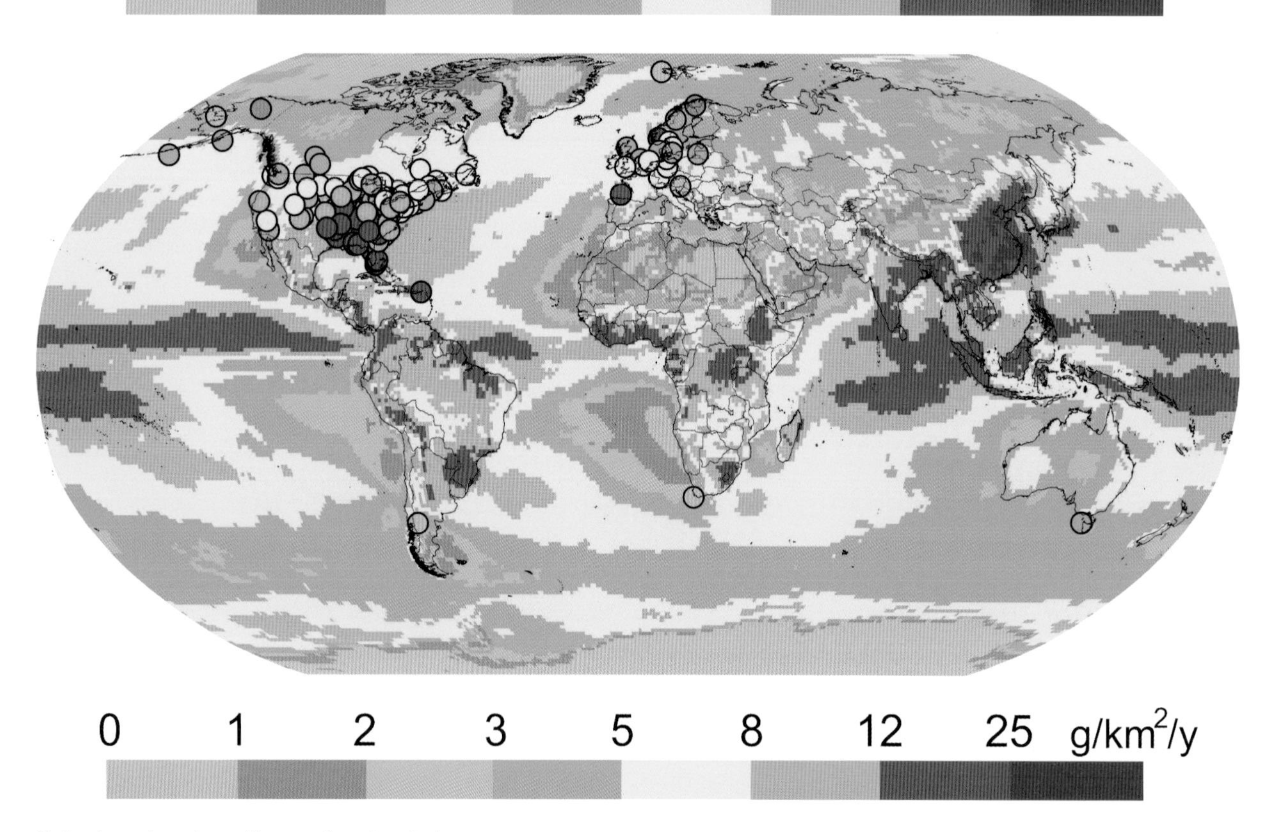

Circles show values observed in ground-based monitoring

Source: UNEP and AMAP (2018).

number of listed POPs and candidate substances presents a resource pressure for existing monitoring programmes (UNEP 2015a). The emission, transport and environmental fate of new unregulated PBTs differs from regulated POPs, further challenging their assessment.

5.3.3 Stratospheric O_3 and ultra-violet radiation

Perennial ground-based *in situ* observations of ODS show a clear decline since the implementation of the Montreal Protocol (Newman *et al.* 2007; Engel *et al.* 2018). However, the decreasing trend slowed down by about 50 per cent after 2012 for trichlorofluoromethane (CFC-11) (Montzka *et al.* 2018). There are indicators that the stratospheric O_3 layer is starting to recover. Total atmospheric column O_3 declined over most of the globe during the 1980s and early 1990s, but has remained stable since 2000, and there are indications of an increase in global-mean total column O_3 over 2000-2013 **(Figure 5.13)** (WMO 2014). Since around 2000, measured concentrations of O_3 in the upper stratosphere show an increasing trend, and modelling results indicate that decreasing ODSs and increasing GHGs, which increases stratospheric ozone by cooling the stratosphere, contributed equally to the increase in upper stratospheric ozone (WMO 2014; Harris *et al.* 2015; Chipperfield *et al.* 2017). Over Antarctica, positive trends for 2001-2013 were found for O_3 concentrations in the lower stratosphere (about 10-20 km) for austral summer and for total column O_3 for spring and summer (Kuttippurath and Nair 2017; Solomon *et al.* 2017). For the mid-latitudes (60°S and 60°N), there is no clear indication of O_3 recovery for reasons that are not clear (Ball *et al.* 2018). As ODS concentrations continue to decline throughout the 21st century, stratospheric O_3 concentrations are expected to rise, though the trends will be increasingly dominated by effects from rising GHG concentrations; thus, the time frame for stratospheric O_3 to recover to 1960 levels is uncertain (Chipperfield *et al.* 2017).

Changes in ultraviolet (UV) radiation at the Earth's surface in response to the recovery of stratospheric O_3 have not yet been documented, because such changes are still masked by varying attenuation of UV radiation by O_3, clouds, aerosols and other factors (Bais *et al.* 2018).

5.3.4 Climate change

In 2016, global averaged concentrations of CO_2, CH_4 and N_2O reached 403.3±0.1 ppm, 1853±2 ppb and 328.9±0.1 ppb, respectively, corresponding to 145 per cent, 257 per cent and 122 per cent above pre-industrial levels (WMO 2017c). The global CO_2 growth rate from 2015 to 2016 was the largest of the last 30 years (partly driven by El Niño) and the CO_2 concentration was the highest in at least the last 800,000 years. CH_4 concentrations plateaued during 1999-2006 but have been increasing since then. Studies point to a variety of different processes driving the change in CH_4, mainly changes in anthropogenic sources, permafrost melting or wetland emissions (Dean *et al.* 2018). N_2O concentrations have been increasing steadily since the mid-1980s. Concentrations of CFC

Figure 5.13: Vertical profiles of annual mean O_3 trends over 35°-60°N averaged over all available observations (black) for the periods of stratospheric ODS decline (left) and ODS increase (right), with the corresponding modelled trends for ODS changes only (red), GHG changes only (blue) and both together (grey)

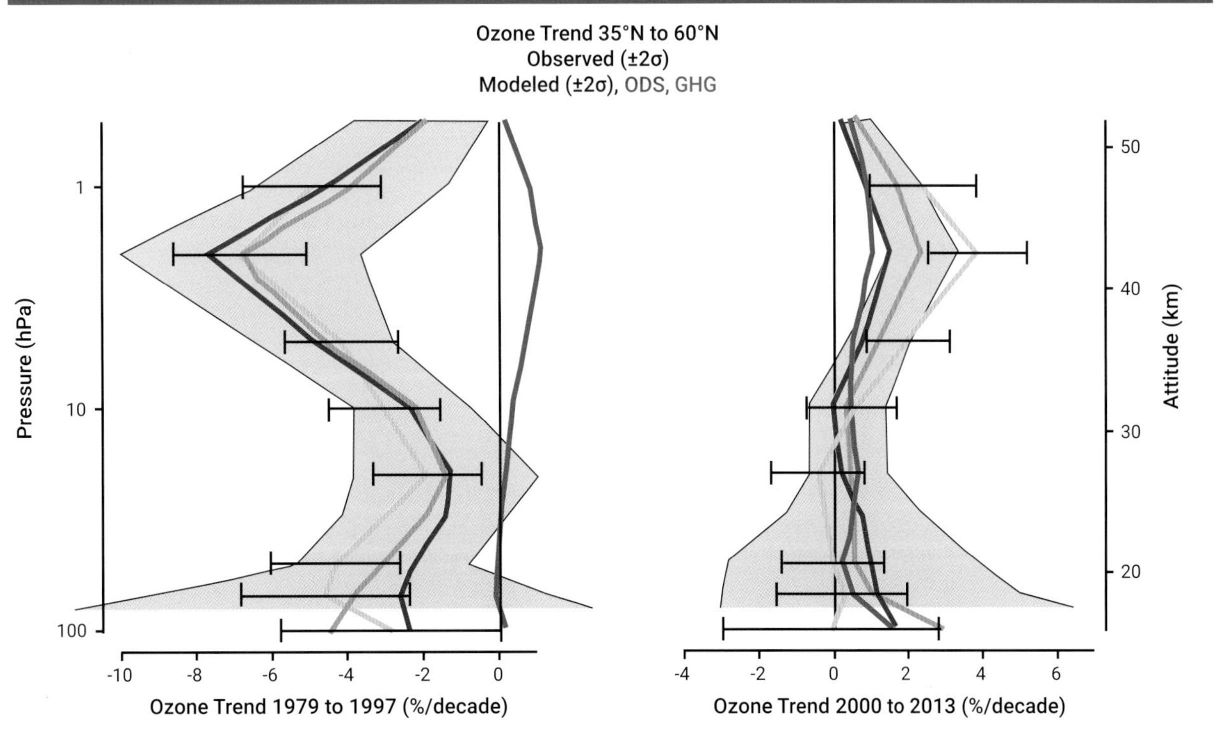

Note: The ±2 standard error uncertainty range for the trends is shown by the horizontal bars for the observations and by the grey shading for the all-changes modelled trend.

Source: WMO (2014).

replacements, HCFCs and HFCs, which are potent GHGs, have been increasing exponentially since 2005, though these remain low overall and currently contribute to less than 4 per cent combined of the radiative forcing due to all GHGs. According to the National Oceanic and Atmospheric Administration Annual Greenhouse Gas Index (AGGI), radiative forcing by long-lived GHGs increased by 78 per cent between 1979 and 2016, with CO_2 accounting for about 72 per cent of this increase.

Since 1901 almost the whole globe has experienced surface warming, and it is extremely likely that anthropogenic activities caused more than half the observed increase in global mean surface temperature since the mid-20th century (Bindoff *et al.* 2013). The global mean surface temperature increase over the 1901-2012 period (see **Figure 4.2**) was approximately 0.89°C, but some regions experienced warming of greater than 2°C (Hartmann *et al.* 2013).

Trends in precipitation are less clear and differ by locations. In general, dry areas are becoming drier, and wet areas are becoming wetter, but multiple exceptions exist (Trenberth 2011; IPCC 2014; Feng and Zhang 2015). For tropical land areas, observations show a decreasing trend from the mid-1970s to mid-1990s and an increasing trend the following decade, resulting in no significant overall trend from 1951 to 2008 (Hartmann *et al.* 2013). A statistically significant increase in precipitation occurred from 1901 to 2008 for the northern mid-latitudes (30°N to 60°N) land areas; in contrast, there is only limited evidence of a long-term increase in the southern mid-latitudes (Hartmann *et al.* 2013). Observed changes in the latitudinal distribution of precipitation over land are suggestive of human influence; however, the results are still inconclusive, due to incomplete data and model uncertainties (Bindoff *et al.* 2013).

Climate change can also impact atmospheric circulations and features at global and regional levels. Observations indicate a widening of the tropical belt, a poleward shift of storm tracks and jet streams, and a contraction of the northern polar vortex since the 1970s are likely (Hartmann *et al.* 2013). Stratospheric O_3 depletion and GHG warming may have contributed to the poleward shift of the southern Hadley cell and positive trend in the Southern Annular Mode, which characterizes the north-south movement of the belt of westerly winds that circles Antarctica, during the austral summer (Bindoff *et al.* 2013). Attribution of anthropogenic influence on the poleward shift of the Hadley cell in the Northern Hemisphere is less certain (Bindoff *et al.* 2013). While many studies have indicated changes in the El Niño-Southern Oscillation (ENSO) and monsoon circulations, there are large observational and modelling uncertainties such that there is low confidence that changes, if observed, can be attributed to anthropogenic activities (Bindoff *et al.* 2013).

There is increasing evidence that climate change has led to changes in the frequency and intensity of extreme events since the mid-20th century (Trenberth 2011; Hartmann *et al.* 2013; Alexander 2016). It is likely that the frequency of extreme warm days has increased in North America, Central America, Europe, Southern Africa, Asia and Australia, and the frequency of heat waves has increased in Europe, Australia and across large parts of Asia (Hartmann *et al.* 2013). Observations have shown a general increase in heavy precipitation at the global scale (Trenberth 2011; Hartmann *et al.* 2013). Regionally, it is likely that the frequency or intensity of heavy precipitation events has increased in North America, Central America and Europe, and it is virtually certain that there has been an increase in the frequency and intensity of the strongest tropical cyclones in the North Atlantic basin since the 1970s (Hartmann *et al.* 2013). For drought, the frequency and intensity likely have increased in the Mediterranean and West Africa, and likely have decreased in central North America and north-west Australia (Hartmann *et al.* 2013).

Air pollution, stratospheric O_3 depletion, persistent pollutants and climate change are interlinked problems (see **Figure 5.1**). Climate warming agents such as BC, tropospheric O_3, CH_4 and HFCs have a relatively short lifetime in the atmosphere compared with long-lived GHGs and are referred to as short-lived climate pollutants (SLCPs) (Haines *et al.* 2017). Tropospheric O_3 contributes to warming directly as a GHG. However, O_3 also contributes to warming by impairing vegetation growth and decreasing plant uptake of CO_2 (Ainsworth *et al.* 2012). BC has a warming effect both in the atmosphere and when deposited on snow and ice. Decreasing emissions of SLCPs can decrease warming in the near term, which may be essential for achieving near-term climate targets or avoiding climate tipping points (Shindell *et al.* 2017). However, decreasing emissions of SLCPs in the near term needs to be combined with mitigation of long-lived GHGs, which dominate climate forcing over the long term (UNEP 2017c).

Other PM constituents (e.g. sulphates and nitrates) also affect climate and may cool the climate by scattering solar radiation. PM also affects climate indirectly by affecting cloud formation, leading to changes in cloud reflectivity, cloud distribution and precipitation patterns. There is still a significant amount of uncertainty on the net radiative effects of aerosols (Fuzzi *et al.* 2015).

Through its impact on synoptic and local-scale meteorology, climate change impacts air pollution and PBT concentrations in multiple, non-linear ways (UNEP and AMAP 2011; Fiore, Naik and Leibensperger 2015). Higher temperatures can increase the chemical reaction rates involving O_3 formation or reduce PM concentrations as components volatilize (Megaritis *et al.* 2013; Czernecki *et al.* 2016). Higher temperatures also increase primary emissions of POPs that can volatilize and secondary emissions by revolatilizing previously deposited POPs (Ma *et al.* 2011). Because particle-bound POPs are more efficiently removed from the atmosphere via deposition, semi-volatile POPs may last longer in the atmosphere at higher temperatures and be transported further from source regions. Higher temperatures may also increase degradation of POPs (Ma *et al.* 2011). Reduced cloud cover promotes the formation of O_3 by increasing photolysis rates (Na, Moon and Kim 2005). Higher temperatures and light intensity can also increase emissions of biogenic NMVOC (Guenther *et al.* 2012), which are O_3 and PM precursors. At the same time, higher temperatures and water stress lower stomatal uptake of O_3 and thus reduce O_3 deposition (Solberg *et al.* 2008; Huang *et al.* 2016). More rain reduces pollution by washing out PM and other pollutants. Extreme events such as heat waves and drought increase risks of high PM pollution associated with wildfires (Bowman *et al.* 2017) and dust (Achakulwisut, Mickley and Anenberg 2018). Extreme events such as floods and storms can also impact the remobilization and bioavailability of POPs (Ma *et al.* 2011).

5

Meteorological parameters that affect air quality often co-vary with and depend on synoptic-scale or other larger-scale phenomena. For example, surface O_3 and PM concentrations are strongly influenced by ventilation and dilution, which are governed by winds and boundary-layer height and are often correlated with temperature and humidity. A decline in the number of summertime mid-latitude cyclones travelling across North America since 1980 has been associated with increases in stagnation and O_3 pollution episodes in the eastern United States of America, offsetting some of the air quality improvement in the north-eastern United States of America from reductions in anthropogenic emissions (Leibensperger, Mickley and Jacob 2008). Extreme wintertime stagnation and pollution episodes in eastern China have been associated with melting sea ice in the Arctic during the preceding autumn and increased snowfall across Siberia during early winter (Zou *et al.* 2017).

5.4 Impacts

Activities that generate emissions threaten human health and well-being, food security and ecosystems. This section focuses on the direct impacts of changing atmospheric composition.

5.4.1 Human health

Exposure to air pollution outdoors and indoors, temperature extremes, airborne pathogens and allergens, and ultraviolet radiation directly affect human health. The following focuses on air pollution effects due to anthropogenic emissions.

Air Pollution

Exposure to indoor and outdoor air pollution was responsible for 6 million (Global Burden of Disease [GBD] Risk Factor Collaborators 2017) to 7 million (WHO 2018) premature deaths in 2016. The GBD Study estimated that long-term exposure to ambient PM was responsible for between 3.6 and 4.6 millions of those premature deaths and between 95 and 118 million years of healthy life lost from heart disease, stroke, lung cancer, chronic lung disease and respiratory infections (Cohen *et al.* 2017; GBD Risk Factor Collaborators 2017; HEI 2018). Consequently, exposure to ambient $PM_{2.5}$ is the highest environmental risk factor for the global burden of disease and sixth among all risk factors in terms of disability-adjusted life years lost, behind high blood pressure, smoking, low birth weight, high levels of blood sugar and high body mass index (GBD Cancer Collaboration 2017). The estimates of premature deaths underestimate the total number of individuals affected, because air pollution has potential effects on everyone who breathes the air, rather than being the sole reason for early death in a small subset of the population (Committee on the Medical Effects of Air Pollutants [COMEAP] 2010).

Even brief periods (minutes to hours) of exposure to high concentrations of pollutants can have significant health impacts (WHO 2006), and episodes of unusually high air pollution attract public concern (e.g. Vidal 2016; Safi 2017). However, the greatest damage to public health is associated with long-term exposure – living in areas of high annual average exposure (HEI 2017). Importantly, there is no known safe level of annual average $PM_{2.5}$ exposure (WHO 2013).

About 43 per cent of the world's population, primarily in low-income countries, uses biomass for heating and cooking. The resulting indoor and outdoor air pollution contributes to acute lower respiratory infections (ALRTI) and pneumonia among children, and chronic obstructive pulmonary disease (COPD) and lung cancer among adults (WHO 2007; Sumpter and Chandramohan 2013; WHO 2018). The GBD Study attributed between 66 and 88 million disability-adjusted life years (DALYs) lost, and between 2.2 and 3.0 million premature deaths in 2016 to household air pollution (GBD Risk Factor Collaborators 2017), whereas WHO estimated the burden to be approximately 3.8 million premature deaths (WHO 2018).

An additional 0.09 to 0.38 million deaths in 2016 from chronic lung disease were attributed to ambient ground-level O_3 exposure (GBD Risk Factor Collaborators 2017). Associations of mortality with other gases are well established, notably NO_2 (a marker of traffic pollution) and SO_2 (a marker of industrial pollution) (WHO 2013). Because these are markers of mixtures, it is unclear to what extent effects associated with them are caused by the gases themselves or by correlated pollutants (WHO 2013; COMEAP 2018).

The number of deaths attributable to air pollution varies widely among countries, reflecting different pollution levels as well as differences in population size, demographics, underlying rates of disease and other socioeconomic characteristics **(Figure 5.14)**

Between 2010 and 2016, deaths attributable to ambient $PM_{2.5}$ exposure increased by 11% per cent globally, due to increased air pollution, as well as growth and ageing of the population. In 2016, 95 per cent of the world's population lived in areas with levels of $PM_{2.5}$ exceeding the WHO air quality guideline (HEI 2018). While mortality attributable to $PM_{2.5}$ has declined in Western Europe and North America, many other regions have seen sharp increases. Deaths attributable to ground-level O_3, though much fewer, have increased nearly 60 per cent globally between 1990 and 2015, with increases in some countries as high as 250-400 per cent (HEI 2017).

In addition to premature mortality, air pollution contributes to a wide range of chronic and acute diseases, especially cardiovascular (Brook *et al.* 2010; McCracken *et al.* 2012) and respiratory disease (American Thoracic Society 2000). Studies suggest associations between air pollution and other diseases such as diabetes (Eze *et al.* 2015); adverse birth outcomes (Stieb *et al.* 2012; Li *et al.* 2017) including premature births, low birth weight (Fleischer *et al.* 2014) and birth defects (Farhi *et al.* 2014); and neurological ailments, including dementia (Calderon-Garciduenas and Villarreal-Rios 2017). Emerging research highlights the potential interactions between air pollution and airborne pathogens and allergens (Hussey *et al.* 2017; Liu *et al.* 2018).

People who are elderly, very young, with pre-existing cardio-respiratory diseases or of low socioeconomic status are most susceptible to air pollution (Sacks *et al.* 2011). Women and children have higher exposures to air pollution indoors, where cooking and heating with solid fuels is the major source (Smith *et al.* 2014). There is increasing evidence that indoor smoke contributes to cataracts, the leading cause of blindness worldwide (Clougherty 2010; Sacks *et al.* 2011; Global Alliance for Clean Cookstoves 2014; Villeneuve *et al.* 2015; WHO 2016b).

The economic impacts of life years lost, increased health care and lost worker productivity due to air pollution are considerable. Premature mortality due to ambient and household air pollution in 2013 was estimated to cost the world's economy US$ 5.1 trillion in welfare losses (World Bank and Institute for Health Metrics and Evaluation 2016). This is equivalent to the 2013 gross domestic product (GDP) of Japan. WHO (2015) estimated that air pollution in Europe in 2010 cost US$ 1.575 trillion per year. In 2011, the US EPA estimated emission controls implemented as a result of the 1990 Clean Air Act Amendments avoided US$ 1.3 trillion in damages in 2010 (US EPA 2011). The impact of $PM_{2.5}$ air pollution on the labour force in China in 2007 was estimated to create economic losses of 346 billion yuan (approximately 1.1 per cent of GDP) (Xia *et al.* 2016). A recent OECD analysis estimated the combined cost of ambient and household air pollution in Africa to be US$ 450 billion in 2013 (Roy 2016).

Asia had the highest absolute number of deaths in 2016 attributable to $PM_{2.5}$ exposure, due to its large populations and high levels of industrial activity. However, $PM_{2.5}$ exposures have begun to decline in China but are increasing in parts of South Asia (HEI 2018). Asian countries also bear the largest burden of air pollution caused by the production of goods consumed in other regions of the world, primarily Western Europe and North America. For example, 97 per cent of $PM_{2.5}$ related deaths in East Asia were associated with emissions in East Asia, but only 80 per cent were associated with goods or services consumed in East Asia. Consumption in Europe and Russia and in North America of goods made in East Asia were estimated to contribute 7 per cent and 6 per cent, respectively, to the $PM_{2.5}$ mortality burden in East Asia (Zhang *et al.* 2017) **(Figure 5.15)**.

Figure 5.14: Deaths per 100,000 people in 2016 attributable to ambient $PM_{2.5}$ air pollution; age-standardized data

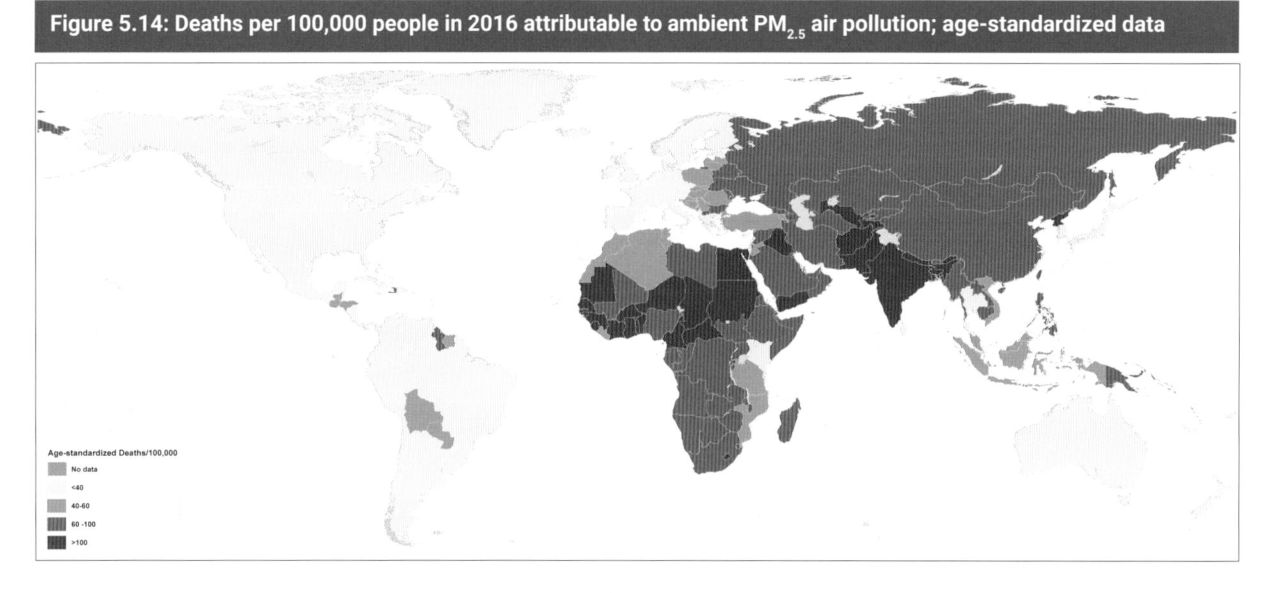

Age standardization allows the estimates to be compared for countries with different age distributions. Note that these estimates do not include deaths attributable to exposure to household air pollution

Source: Adapted from HEI (2018).

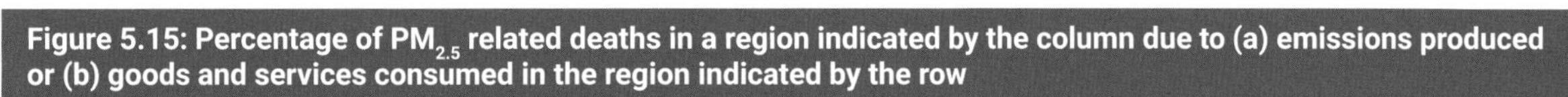

Figure 5.15: Percentage of $PM_{2.5}$ related deaths in a region indicated by the column due to (a) emissions produced or (b) goods and services consumed in the region indicated by the row

		China and East Asia	India and Rest of Asia	Europe and Russia	Middle East and North America	North America	Latin America	Sub Saharan Africa and Rest of World
Where air pollution was emitted	China and East Asia	97%	3%	1%	1%	2%	1%	0%
	India and Rest of Asia	1%	93%	1%	2%	0%	0%	2%
	Europe and Russia	1%	0%	94%	18%	1%	0%	1%
	Middle East and North Africa	0%	3%	2%	78%	0%	0%	5%
	North America	0%	0%	1%	1%	95%	2%	0%
	Latin America	0%	0%	0%	0%	1%	97%	0%
	Sub Saharan Africa and Rest of World	0%	0%	0%	0%	0%	0%	93%
Where associated goods were consumed	China and East Asia	80%	4%	3%	3%	6%	4%	2%
	India and Rest of Asia	3%	84%	2%	3%	1%	1%	2%
	Europe and Russia	7%	4%	86%	24%	5%	6%	4%
	Middle East and North Africa	2%	3%	4%	64%	2%	1%	4%
	North America	6%	3%	3%	4%	82%	12%	2%
	Latin America	1%	0%	1%	1%	4%	75%	1%
	Sub Saharan Africa and rest of World	1%	1%	1%	1%	1%	1%	84%

Source: Based on Zhang *et al.* (2017).

Stratospheric ozone depletion

The health risks of stratospheric O_3 depletion occur as a result of increased levels of biologically damaging wavelengths of UV radiation reaching the Earth's surface. Although some exposure to UV is necessary, too much exposure damages the skin and eyes and can cause immune suppression. Impacts include sunburn, keratinocyte (previously called non-melanoma) cancers, cutaneous malignant melanoma (CMM), Merkel cell carcinoma, photoconjunctivitis. photokeratitis (e.g. snow blindness), cataracts, pterygium and conjunctival melanoma.

In recent decades, most countries with predominantly fair-skinned populations have experienced a steady increase in the incidence rates of CMM which is responsible for about 80 per cent of the deaths due to skin cancer (Lucas *et al.* 2015). Excessive exposure to UV radiation accounts for 60-90 per cent of the risk for CMM (Olsen, Carroll and Whiteman 2010; WHO 2004). Increasing incidence rates of CMM and other UV-related adverse health impacts are unlikely to be due to changes in UV exposure due to stratospheric O_3 depletion, but rather to increases in risky sun exposure behaviour (Lucas *et al.* 2015). However, without the Montreal Protocol, incidence of skin cancer may have been 14 per cent greater, affecting 2 million people by 2030 (van Dijk *et al.* 2013).

Climate change

Over the coming decades to centuries, adverse health effects from climate change are forecast to greatly exceed any potential health benefits (Smith *et al.* 2014; Watts *et al.* 2017). The effects of climate change on human health can be classified as direct (e.g. heat waves, storms), less direct (e.g. changes in disease-vector ecology, reductions in water supply, or exacerbation of air pollution episodes) and diffuse (Butler 2014; Melillo, Richmond and Yohe 2014). The category of diffuse effects could have the largest burden of disease through means such as conflict (Kelley *et al.* 2015), migration (Piguet, Pecoud and de Guchteneire eds. 2011) and famine. Mental health effects arise from all three categories (e.g. post-traumatic stress disorder).

The health impacts of a changing climate will be inequitably distributed globally. Climate change and increasing climate variability "worsen existing poverty, exacerbate inequalities, and trigger both new vulnerabilities and some opportunities for individuals and communities" (IPCC 2014, p. 796).

Buildings and roads retain heat more than rural landscapes and depress humidity, creating urban heat islands. In northern mid-latitudes and subtropics, nights are up to 4°C warmer and

10-15 per cent drier in urban areas compared with surrounding rural areas. In northern Africa, the number of nights with exceptional heat stress is around ten times higher in urban areas than in rural areas (Fischer, Oleson and Lawrence 2012).

5.4.2 Food security

The Food and Agriculture Organization of the United Nations (FAO 2008) describes four dimensions of food security: availability, related to quantity; access, including affordability; utilization, related to meeting nutritional needs and food safety; and stability, related to the temporal variation in the other dimensions.

Availability: Current levels of ground-level O_3 decrease yields of key staple crops – including wheat, soybean, maize and rice – by 2-15 per cent depending on crop types and locations (Feng and Kobayashi 2009; Van Dingenen *et al.* 2009; Fishman *et al.* 2010; Avnery *et al.* 2011). Global estimates of damage are uncertain because different cultivars of crops have different sensitivities and not all crops have been studied. The economic implications of loss of crop productivity are substantial. For example, elevated O_3 concentrations in the United States of America reduce maize and soybean production by about 10 per cent and 5 per cent, respectively, at a cost of US$9 billion annually (McGrath *et al.* 2015).

Climate change already affects crop production through changes in average and extreme temperatures and precipitation, the spread and impacts of invasive weeds and pests and deforestation. Although increased CO_2 fertilization (see Section 4.4.3) is thought to offset negative impacts, the interactions between changes in CO_2, O_3, nitrogen, water availability and temperature are still not well understood (Schlenker and Roberts 2009; Porter *et al.* 2014).

Yields in tropical countries are expected to suffer the most serious impacts, while some temperate regions may benefit from higher yields, expansion of productive areas and longer growing seasons (though these benefits may be offset by increasingly frequent extreme events, temperature and water stresses and ineffective adaptations) (Schmidhuber and Tubiello 2007; Gornall *et al.* 2010; Porter *et al.* 2014). In short, the impact of climate change on crop production will be felt most heavily in developing countries where large numbers of people depend on agriculture for their livelihoods, food insecurity is high and adaptive capacity low. Climate change impacts on the availability and distribution of aquatic species are also expected to disproportionately affect developing countries (see Section 7.3.2).

Higher temperatures are likely to adversely affect livestock productivity by changing the availability of pasture, fodder crops and water (Andre *et al.* 2011; Renaudeau *et al.* 2011; Porter *et al.* 2014). The impacts of climate change on livestock diseases remain difficult to predict and highly uncertain (Mills, Gage and Khan 2010; Tabachnick 2010).

Access: Climate change exerts upward pressure on global food prices (Porter *et al.* 2014), disproportionately affecting poor consumers who may spend a significant proportion of their income on food, with implications for health and nutrition (Springmann *et al.* 2016). Women and girls disproportionately suffer from both the health consequences of nutritional deficiencies and the greater burdens of caregiving for others who are ill (WHO 2014; FAO 2016).

Utilization: Higher temperatures and higher CO_2 levels are associated with lower protein content of grains (Porter *et al.* 2014; Feng *et al.* 2015) and reduced micronutrient content of grains and legumes (Myers *et al.* 2014).

The nutritional content and safety of food supply is affected by pollution, primarily by PBTs, including Hg and POPs. Hg can travel long distances in the air and water, bioaccumulate and biomagnify up food chains, reaching levels that can be dangerous to the health of ecosystems and humans (Gibb and O'Leary 2014; Sundseth *et al.* 2017). Concentrations of methylmercury in the blood of populations that consume top marine predators, such as indigenous Arctic people, are among the highest recorded globally, giving rise to serious health concerns (UNEP 2013a; UNEP 2013b). Hg is toxic to the central nervous system (CNS) leading to cognitive and motor dysfunction (Karagas *et al.* 2012; Antunes dos Santos *et al.* 2016; Sundseth *et al.* 2017). Hg exposure also increases the risk of cardiovascular diseases, causes kidney damage, adversely affects the reproductive, endocrine and immune systems, and leads to premature death (Rae and Graham 2004; AMAP 2009; Rice *et al.* 2014).

Similarly, POPs and other PBTs can travel long distances and bioaccumulate up food chains (e.g. Gibson *et al.* 2016; Ma, Hung, and Macdonald 2016). A wide range of health effects has been associated with exposure to POPs, including changes to the reproductive, endocrine, immunologic and neurologic systems, cancer, dermal and ocular changes, and reduced birth weight (Damstra 2002; El-Shahawi *et al.* 2010; Fry and Power 2017). The exposure of pregnant and breastfeeding women to POPs is of particular concern, as POPs can cross the placenta and the blood-milk barrier, which may increase the risk of adverse developmental outcomes in children (Vizcaino *et al.* 2014; Women in Europe for a Common Future and Women International for a Common Future 2016).

Little is known about the potential health effects of some chemicals that have substituted for banned POPs, such as non-polybrominated diphenyl ether (PBDE) organophosphate flame retardants. Human exposure to such flame retardants in the United States of America has been observed to be increasing over the last decade (Hoffman *et al.* 2017).

Stability: The increasing frequency and severity of extreme weather caused by climate change will have serious consequences for the stability of food prices and food supply, such as the wheat harvest failure and price spike experienced following the 2010 Russian heat wave (Otto *et al.* 2012; Porter *et al.* 2014). Droughts, floods and other weather-related disasters can lead to acute, localized food crises, particularly in countries with pre-existing vulnerabilities such as high levels of poverty and undernutrition. For example, climate change contributed to the drought that led to the 2011 East African food crisis and ultimately contributed to famine in Somalia (Bailey 2013; Lott, Christidis and Stott 2013; Coghlan *et al.* 2014). If transport infrastructure supporting exports from major crop-producing regions is disrupted by acute weather shocks, the impacts on food security could be more widespread (Bailey and Wellesley 2017).

5.4.3 Ecosystems

Air pollution, climate change, UV radiation and PBTs all have effects on the health of natural ecosystems and wildlife. These adverse impacts in turn affect the services provided to humans by those ecosystems, or 'nature's contribution to people' (NCP) (Diaz *et al.* 2018).

Since the 1970s, international attention has focused on air pollution in the form of wet and dry deposition of sulphur and nitrogen, often referred to as 'acid rain', which led to acidification of soils and fresh water, and damage to vegetation and fish kills. In Asia and Africa, significant increases and decreases in sulphur deposition have been observed depending on location (Vet *et al.* 2014). In Western Europe and eastern North America, after decades of declining sulphur emissions and deposition levels, acidification is declining or slowing, and some forests and lakes are showing signs of recovery (Maas and Grennfelt eds. 2016). As sulphur emissions have decreased due to the implementation of emission controls, recent assessments have focused attention on the effect that humans have had on the global nitrogen cycle and its implications.

Human activity, mainly through combustion and fertilizer production, are responsible for as much nitrogen fixation as natural and unmanaged ecosystems, significantly altering the nitrogen cycle from its pre-industrial state (Fowler *et al.* 2015). Since 2000, nitrogen deposition has decreased in North America and Europe and increased in Africa and Asia, directly corresponding to decreases of NO_x and increases in NH_3 continent-wide emissions (Zhao *et al.* 2017). Nitrogen deposition exceeds critical loads over large parts of Europe and the area of exceedance has shown little change in recent decades (Hettelingh *et al.* 2015). High levels of nitrogen deposition contribute to the eutrophication of aquatic ecosystems and can affect terrestrial plant communities, possibly favouring dominant species, which in turn affects insects, birds and other animals. The loss of biodiversity due to excess nitrogen deposition is very likely to be occurring in many parts of the world, although the impacts have not been well quantified. Changes in climate, land use and other global changes will continue to alter the nitrogen cycle in the future, with consequences for ecosystems and human health (Fowler *et al.* 2015).

Marine ecosystems are also affected by air pollution, climate change and PBT pollution, for instance through the distribution of oceanic dissolved nutrients and oxygen (York 2018). Human activity is now increasing the inputs of all fixed nitrogen to the oceans by about 50 per cent (more in local hotspots near high emission regions in South-East Asia, Europe and North America) and atmospheric transport is now the dominant route contributing anthropogenic nitrogen into the open ocean beyond the continental shelf (Joint Group of Experts on the Scientific Aspects of Marine Environmental Protection [GESAMP] 2018). Harmful algal blooms in turn can contribute to respiratory health impacts through airborne transmission of aerosols (Centers for Disease Control and Prevention 2017).

Ozone exposure can affect plant growth, flowering, pollination and susceptibility to pathogens, with impacts on species composition and biodiversity (Fuhrer *et al.* 2016). Critical load thresholds have been identified for some terrestrial ecosystems (International Cooperative Programme on Effects of Air Pollution on Natural Vegetation and Crops 2017), but there are many ecosystems for which O_3 sensitivity is poorly understood.

The full extent of PBT exposure and their biological effects on wildlife and natural ecosystems is still not well known and is an area of active research (AMAP 2017). However, given the widespread presence of PBTs in the environment, the potential exists for long-term damage to food chains and ecosystem functions especially in sensitive areas, such as the Arctic (AMAP 2011; AMAP 2016; AMAP 2017).

5.4.4 Social well-being

Beyond the impacts on human and ecosystem health and food security, changes in the atmosphere have negative impacts on social well-being, or welfare.

Air pollution degrades materials and coatings, decreasing their useful life and generating costs for cleaning, repair and replacement. When the materials affected are structures or objects of cultural significance, the damage can be priceless (Watt *et al.* eds. 2009). In Europe, visible pollution damage to cultural heritage sites and artworks was highlighted as a justification for air pollution control policies (Di Turo *et al.* 2016; Maas and Grennfelt eds. 2016). In India, the government has taken steps to protect, in addition to public health, the white marble Taj Mahal, which has become discoloured over time due to high levels of PM, possibly from the open burning of municipal solid waste (Bergin *et al.* 2015; Raj *et al.* 2016).

Sand and dust storms, fires and extreme weather events all create disruptions to society, transportation and economic activity. Such events can be a drag on a local economy and may also drive dislocations and migration (Hanlon 2016). In the short term, increased pollution levels affect worker productivity. These effects are not limited to outdoor workers or to extreme pollution levels (Chang *et al.* 2016; Zivin and Neidell 2018). In the longer term, elevated pollution exposures have been associated with poor educational and labour-market performance, creating a long-term human capital deficit (Zivin and Neidell 2018).

5.5 Response: policies and governance

A wide variety of governance approaches and policy instruments have been used to help mitigate the sources and impacts of air pollution, climate change, stratospheric O_3 depletion and PBTs, including the following.

- **Planning regimes,** strategies or action plans designed to achieve ambient air quality standards or objectives or attain emission ceilings, combined with analyses and environmental impact assessments.
- **Command and control**, including technology, emissions or ecosystem restoration standards; record-keeping and reporting requirements, or limits on manufacture, trade or use of specific chemicals or products; each of which are implemented through permitting and enforcement programmes.
- **Market interventions**, including economic instruments, such as taxes, fees or markets for tradable emission rights, as well as loans and subsidies.

- **Public information**, including product labelling, air quality forecasting, near real-time observations and training.
- **Cooperative frameworks**, including international agreements and voluntary sectoral standards or initiatives.

The effectiveness of specific examples of these policies is explored further in Chapter 12.

Different governance approaches have been adopted at local, provincial, country and international scales depending on the specific institutional, economic, technological and political contexts. Often multiple complementary approaches are deployed simultaneously to address a single issue or source. Different mixes of approaches may be used to address similar issues, even in a single jurisdiction.

The existence and extent of implementation of air-related policies also vary widely based on differences in institutional capacity and culture in different regions of the world and at different spatial scales. In some regions, such as North America and Europe, there are well-developed, federated systems of national, provincial and local policies and enforcement programmes designed to achieve common policy objectives. In other regions, international agreements or national legislation may exist, but implementation and enforcement are weak due to a lack of institutional capacity at the national or subnational scale. In some regions, city governments are developing the primary policy response to these issues, with simultaneous benefits for other parts of their countries.

Climate change, stratospheric O_3 depletion and PBTs have been recognized as shared global problems. **Table 5.2** lists some global environmental agreements that have been developed to motivate, enable and coordinate ongoing efforts to address these challenges. These set out common objectives and obligations, which are implemented through different policies developed at national to local levels. One of the most successful global agreements is the Vienna Convention and Montreal Protocol to address stratospheric O_3 depletion, which in 2009 became the first United Nations convention to be ratified by all United Nations member states. The most recent amendment to the Montreal Protocol, the 2016 Kigali Amendment, is designed to limit the impact of ODS substitutes on climate change.

Adopted in 1992, the United Nations Framework Convention on Climate Change (UNFCCC) has led to the negotiation of a series of protocols and agreements on "common but differentiated responsibilities" to address GHG emissions (United Nations 1992). The UNFCCC divides countries into developed (Annex I) and developing countries. This differentiation has been key to the design of mechanisms to transfer between countries the technology and resources needed to mitigate emissions (including Activities Implemented Jointly, Clean Development Mechanism and Joint Implementation). Under the Kyoto Protocol and Doha Amendment, Annex I countries agreed to specific emission reduction commitments. The second commitment period (2013-2020) of the 1997 Kyoto Protocol has yet to be approved by a quorum of 144 nations. The 2015 Paris Agreement set the goal of limiting the global average temperature increase to well below 2°C above pre-industrial levels by 2100, with ambition to limit the increase to less than 1.5°C. All countries are required to present periodically to the Convention Secretariat national GHG inventories and Nationally Determined Contributions (NDCs), or emission reduction commitments. To achieve the 1.5°C goal, GHG emissions need to be decreased significantly in the coming years and be brought to net zero by around mid-century (see Chapters 21 and 22). Studies have suggested that there is a greater than 90 per cent chance of exceeding

Table 5.2: Global environmental agreements relevant to climate change, stratospheric O_3 depletion and PBTs

Climate change

- 1992 United Nations Framework Convention on Climate Change (UNFCCC)
- 1997 Kyoto Protocol
    - 2012 Doha Amendment
- 2016 Paris Agreement

Stratospheric O_3 depletion

- 1985 Vienna Convention for the Protection of the Ozone Layer
- 1987 Montreal Protocol on Substances that Deplete the Ozone Layer
    - 1990 London Amendment
    - 1992 Copenhagen Amendment
    - 1997 Montreal Amendment
    - 1999 Beijing Amendment
    - 2016 Kigali Amendment

Persistent bioaccumulative toxic chemicals (e.g. POPs and Hg)

- 1989 Basel Convention on the Control of Transboundary Movements of Hazardous Wastes and their Disposal
- 1998 Rotterdam Convention on the Prior Informed Consent Procedure for Certain Hazardous Chemicals and Pesticides in International Trade
- 2001 Stockholm Convention on Persistent Organic Pollutants
- 2013 Minamata Convention on Mercury

2°C warming under the current pledges submitted by national governments, which achieve only a third of the mitigation required to be on a least cost path to stay below that threshold. However, pathways towards staying below 1.5°C and 2°C are still technically feasible (Xu and Ramanathan 2017).

Although air pollution travels around the world, there is no single global agreement addressing air pollution; rather there is a patchwork of regional intergovernmental agreements **(Figure 5.16)**. In general, this patchwork has good geographic coverage, but is uneven in terms of the coverage of pollutants, sources and capabilities. Furthermore, this patchwork does not encourage the transfer of experience and resources from richer to poorer countries. The oldest and most-developed among these is the 1979 Convention on Long-Range Transboundary Air Pollution (CLRTAP) organized under the United Nations Economic Commission for Europe (Sliggers and Kakebeeke eds. 2004; Maas and Grennfelt eds. 2016). In the Russian Federation and Central Asia, the CLRTAP overlaps with the grouping of agreements under the umbrella of the Asia and the Pacific Clean Air Partnership. There are three regional agreements on air pollution in Africa which overlap each other and have a few members in common with the Council of Arab Ministers Responsible for the Environment.

Figure 5.16: Map of groupings of selected regional multilateral air pollution agreements

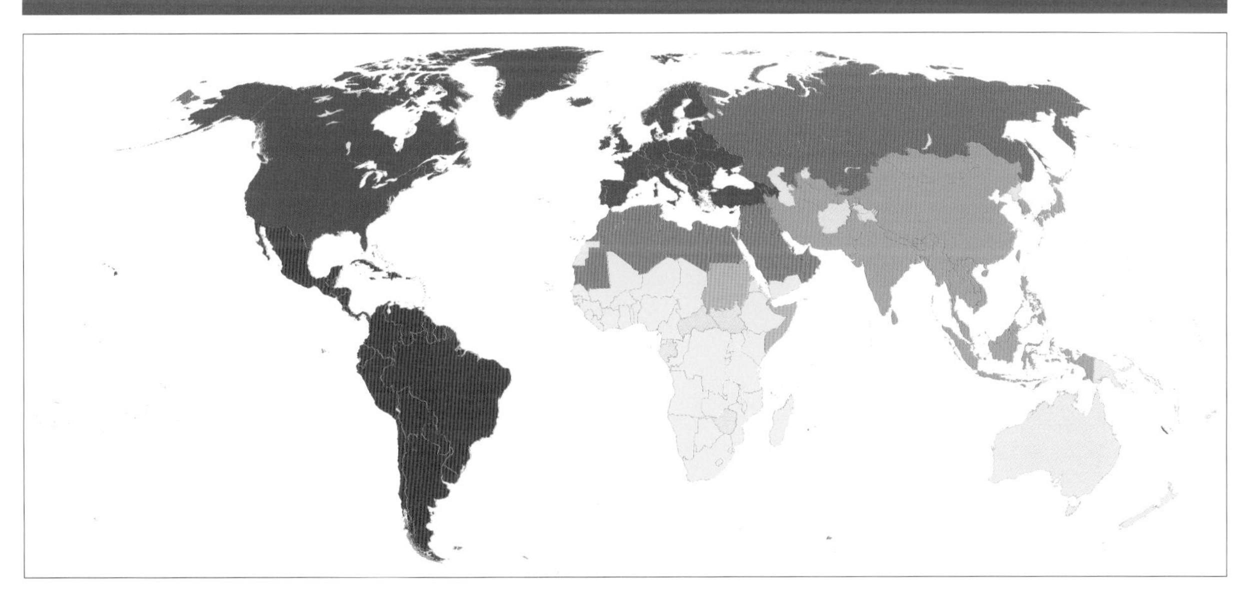

1979 United Nations Economic Commission for Europe (UNECE) Convention on Long-range Transboundary Air Pollution (Geneva)
1998 Acid Deposition Monitoring Network in East Asia (EANET)
1998 Malé Declaration on Control and Prevention of Air Pollution and its likely Transboundary Effects for South Asia
2002 Association of Southeast Asian Nations (ASEAN) Agreement on Transboundary Haze Pollution
2006 Framework Convention on Environmental Protection for Sustainable Development in Central Asia (Ashkhabad)
2015 Asia and the Pacific Clean Air Partnership
2008 Eastern Africa Regional Framework Agreement on Air Pollution (Nairobi)
2008 Southern African Development Community Regional Policy Framework on Air Pollution (Lusaka)
2009 West and Central Africa Regional Framework Agreement on Air Pollution (Abidjan)
1986 Council of Arab Ministers Responsible for the Environment (CAMRE)
2008 Intergovernmental Network on Air Pollution for Latin America and the Caribbean
No agreements

To guide their air pollution policies, many countries have developed national ambient air quality standards, or guidelines for a number of common pollutants (Kutlar Joss *et al.* 2017). These can differ with respect to the pollutant targeted, concentration level, averaging time, frequency of occurrence and measurement protocols, making comparisons of stringency difficult. In 2005, a WHO expert panel developed a set of air quality guidelines that are intended to be globally applicable for general population exposure and a set of recommended interim targets for some pollutants for areas that exceed the guidelines (WHO 2006; see **Table 5.3**). The interim targets were suggested for use by highly polluted areas as incremental steps towards achieving the guideline values. Each interim target is associated with a specified decrease in mortality risk (WHO 2006).

The ability of governments and the public to compare air quality monitoring data to such guidelines and standards and associated information about health benefits has been important in developing awareness and motivating mitigation. Thus, improving air quality monitoring infrastructure and the use of air quality and health effects information in benefit-cost analyses of mitigation measures were identified as priorities in the GEO-6 regional assessments.

Significant successes have been achieved through national and international policy and regulatory structures that have been developed over recent decades, as evidenced by the declining trends in emissions and increasing trends in activity and production (see Section 5.2). However, past policy responses may not be well suited to addressing the problems and sources

Table 5.3: WHO Air Quality Guidelines and Interim Targets

Pollutant	Averaging time	Unit	Interim targets			Air quality
			1	2	3	Guideline
PM_{10}	Annual	μg/m³	70	50	30	20
	24 hours	μg/m³	150	100	75	50
$PM_{2.5}$	Annual	μg/m³	35	25	15	10
	24 hours	μg/m³	75	50	37.5	25
NO_2	Annual	μg/m³	–	–	–	40
	1 hour	μg/m³	–	–	–	200
SO_2	24 hours	μg/m³	125	50	–	20
O_3	8 hours	μg/m³	160	–	–	100
CO	1 hour	mg/m³	–	–	–	30

Source: WHO (2006).

Box 5.1: UNEA 3/8 Resolution

Preventing and Reducing Air Pollution to Improve Air Quality Globally

The resolution urges Member States to:

- Take action to decrease all forms of air pollution
- Establish systems to monitor air quality and emissions
- Set ambitious air quality standards
- Address short lived climate pollutants as part of national action plans
- Integrate air pollution management into national development planning
- Create awareness of air pollution costs and benefits of air pollution control
- Strengthen national and sub-national capacity for air quality management

In addition, it calls for strengthened cooperation to address air pollution at the local, national, regional and global levels. The resolution also requests UN Environment to undertake additional technical support, capacity building and analysis to support Member States in improving air quality.

that remain or that are emerging, particularly in the near term. Particularly if government capacity or regulatory structures are lacking, responses that engage a broad mix of stakeholders to integrate air-related concerns into broader policy and investment decisions (e.g. transportation planning, land-use planning, economic development investments, behavioural change) may be more capable of addressing diffuse sources of emissions and promoting innovation.

Cities have been important centres of policy innovation and policy integration and continue to provide important opportunities for progress. The non-governmental organization Clean Air Asia is a leading example of efforts in this arena, bringing together city governments, national ministries, industry and other stakeholder groups from more than 1,000 cities across Asia to share lessons in developing air pollution, climate change, transportation, land-use and energy policies (Clean Air Asia 2017). The C40 Cities Climate Leadership Group is another example, which connects officials in cities to their peers in cities around the world to exchange information as they face common challenges associated with climate change mitigation and adaptation (Day *et al.* 2018).

At both international and local levels, coalitions and initiatives have formed between governments, industry and other groups to facilitate specific actions. The Climate and Clean Air Coalition for Reduction of Short-Lived Climate Pollutants (CCAC) is an example of a coordinated effort to make near-term progress focused on specific pollutants and sectors (CCAC 2015).

References

Abelkop, A.D.K., Graham, J.D. and Royer, T.V. (2017). *Persistent, Bioaccumulative, and Toxic (PBT) Chemicals: Technical Aspects, Policies, and Practices*. Boca Raton, FL: CRC Press. https://www.crcpress.com/Persistent-Bioaccumulative-and-Toxic-PBT-Chemicals-Technical-Aspects/Abelkop-Graham-Royer/p/book/9781138792944.

Achakulwisut, P., Mickley, L.J. and Anenberg, S.C. (2018). Drought-sensitivity of fine dust in the US Southwest: Implications for air quality and public health under future climate change. *Environmental Research Letters* 13(5), 054025. https://doi.org/10.1088/1748-9326/aabf20.

Adiang, C.M., Monkam, D., Njeugna, E. and Gokhale, S. (2017). Projecting impacts of two-wheelers on urban air quality of Douala, Cameroon. *Transportation Research Part D: Transport and Environment* 52, Part A, 49-63. https://doi.org/10.1016/j.trd.2017.02.010.

Ainsworth, E.A., Yendrek, C.R., Sitch, S., Collins, W.J. and Emberson, L.D. (2012). The effects of tropospheric ozone on net primary productivity and implications for climate change. *Annual Review of Plant Biology* 63(1), 637-661. https://doi.org/10.1146/annurev-arplant-042110-103829.

Albani, S., Mahowald, N.M., Perry, A.T., Scanza, R.A., Zender, C.S., Heavens, N.G. *et al.* (2014). Improved dust representation in the Community Atmosphere Model. *Journal of Advances in Modeling Earth Systems* 6(3), 541-570. https://doi.org/10.1002/2013MS000279.

Alexander, L.V. (2016). Global observed long-term changes in temperature and precipitation extremes: A review of progress and limitations in IPCC assessments and beyond. *Weather and Climate Extremes* 11, 4-16. https://doi.org/10.1016/j.wace.2015.10.007.

American Thoracic Society (2000). What constitutes an adverse health effect of air pollution? Official statement of the American Thoracic Society. *American Journal of Respiratory and Critical Care Medicine* 161(2), 665-673. https://doi.org/10.1164/ajrccm.161.2.ats4-00.

Andre, G., Engel, B., Berentsen, P.B., Vellinga, T.V. and Lansink, A.G. (2011). Quantifying the effect of heat stress on daily milk yield and monitoring dynamic changes using an adaptive dynamic model. *Journal of Dairy Science* 94(9), 4502-4513. https://doi.org/10.3168/jds.2010-4139.

Ang'u, C., Nzioka, J.M. and Mutai, B., K., (2016). Aerosol optical depth patterns associated with urbanization and weather in Nairobi and Lamu. *Journal of Meteorology and Related Sciences (Kenya Meteorological Society)* 8(3). https://doi.org/10.20987/jmrs.2016.02.803.

Antunes Dos Santos, A., Appel Hort, M., Culbreth, M., Lopez-Granero, C., Farina, M., Rocha, J.B. *et al.* (2016). Methylmercury and brain development: A review of recent literature. *Journal of Trace Elements in Medicine and Biology* 38, 99-107. https://doi.org/10.1016/j.jtemb.2016.03.001.

Apte, J.S., Messier, K.P., Gani, S., Brauer, M., Kirchstetter, T.W., Lunden, M.M. *et al.* (2017). High-Resolution Air Pollution Mapping with Google Street View Cars: Exploiting Big Data. *Environmental Science & Technology* 51(12), 6999-7008. https://doi.org/10.1021/acs.est.7b00891.

Arctic Monitoring and Assessment Programme (2009). *AMAP Assessment 2009: Human Health in the Arctic*. Oslo: Arctic Monitoring and Assessment Programme (AMAP). https://www.amap.no/documents/doc/amap-assessment-2009-human-health-in-the-arctic/98.

Arctic Monitoring and Assessment Programme (2011). *The Impact of Black Carbon on Arctic Climate*. Oslo: Arctic Monitoring and Assessment Programme. https://www.amap.no/documents/download/977.

Arctic Monitoring and Assessment Programme (2014). *Trends in Stockholm Convention Persistent Organic Pollutants (POPs) in Arctic Air, Human media and Biota*. Wilson, S., Hung, H., Katsoyiannis, A., Kong, D., van Oostdam, J., Riget, F. and Bignert, A. (eds.). Oslo: Arctic Monitoring and Asessment Programme (AMAP). http://www.amap.no/documents/download/1972.

Arctic Monitoring and Assessment Programme (2015). *AMAP Assessment 2015: Black Carbon and Ozone as Arctic Climate Forcers*. Oslo: Arctic Monitoring and Assessment Programme. https://www.amap.no/documents/download/2506.

Arctic Monitoring and Assessment Programme (2016). *AMAP Assessment 2015: Temporal Trends in Persistent Organic Pollutants in the Arctic*. Oslo: Arctic Monitoring and Assessment Programme. https://www.amap.no/documents/doc/amap-assessment-2015-temporal-trends-in-persistent-organic-pollutants-in-the-arctic/1521.

Arctic Monitoring and Assessment Programme (2017). *AMAP Assessment 2016: Chemicals of Emerging Arctic Concern*. Oslo: Arctic Monitoring and Assessment Programme. https://www.amap.no/documents/doc/amap-assessment-2016-chemicals-of-emerging-arctic-concern/1624.

Avnery, S., Mauzerall, D.L., Liu, J.F. and Horowitz, L.W. (2011). Global crop yield reductions due to surface ozone exposure: 1. Year 2000 crop production losses and economic damage. *Atmospheric Environment* 45(13), 2284-2296. https://doi.org/10.1016/j.atmosenv.2010.11.045.

Bailey, R. (2013). *Managing Famine Risk: Linking Early Warning to Early Action*. A Chatham House Report. London: Royal Institute of International Affairs. https://www.chathamhouse.org/sites/files/chathamhouse/public/Research/Energy%2C%20Environment%20and%20Development/0413r_earlywarnings.pdf.

Bailey, R. and Wellesley, L. (2017). *Chokepoints and Vulnerabilities in Global Food Trade*. London: Chatham House. https://www.chathamhouse.org/sites/files/chathamhouse/publications/research/2017-06-27-chokepoints-vulnerabilities-global-food-trade-bailey-wellesley-final.pdf.

Bais, A.F., Lucas, R.M., Bornman, J.F., Williamson, C.E., Sulzberger, B., Austin, A.T. *et al.* (2018). Environmental effects of ozone depletion, UV radiation and interactions with climate change: UNEP environmental effects assessment panel, update 2017. *Photochemical & Photobiological Sciences* 17(2), 127-179. https://doi.org/10.1039/c7pp90043k.

Ball, W.T., Alsing, J., Mortlock, D.J., Staehelin, J., Haigh, J.D., Peter, T. *et al.* (2018). Evidence for a continuous decline in lower stratospheric ozone offsetting ozone layer recovery. *Atmospheric Chemistry and Physics* 18, 1379-1394. https://doi.org/10.5194/acp-18-1379-2018.

Bergin, M.H., Tripathi, S.N., Devi, J.J., Gupta, T., Mckenzie, M., Rana, K.S. *et al.* (2015). The discoloration of the Taj Mahal due to particulate carbon and dust deposition. *Environmental Science & Technology* 49(2), 808-812. https://doi.org/10.1021/es504005q.

Bhandari, D. and Shrimali, G. (2018). The perform, achieve and trade scheme in India: An effectiveness analysis. *Renewable and Sustainable Energy Reviews* 81(Part 1), 1286-1295. https://doi.org/10.1016/j.rser.2017.05.074.

Bindoff, N.L., Stott, P.A., AchutaRao, K.M., Allen, M.R., Gillett, N., Gutzler, D. *et al.* (2013). Detection and attribution of climate change: From global to regional.In *Climate Change 2013: The Physical Science Basis. Contribution of Working Group I to the Fifth Assessment Report of the Intergovernmental Panel on Climate Change*. Stocker, T.F., Qin, D., Plattner, G.-K., Tignor, M., Allen, S.K., Boschung, J. *et al.* (eds.). Cambridge: Cambridge University Press. chapter 10. https://www.ipcc.ch/pdf/assessment-report/ar5/wg1/WG1AR5_Chapter10_FINAL.pdf.

Bond-Lamberty, B., Bailey, V.L., Chen, M., Gough, C.M. and Vargas, R. (2018). Globally rising soil heterotrophic respiration over recent decades. *Nature* 560(7716), 80-83. https://doi.org/10.1038/s41586-018-0358-x.

Bouwman, A.F., Beusen, A.H.W., Griffioen, J., Van Groenigen, J.W., Hefting, M.M., Oenema, O. *et al.* (2013). Global trends and uncertainties in terrestrial denitrification and N(2)O emissions. *Philosophical Transactions of the Royal Society B: Biological Sciences* 368(1621), 20130112. https://doi.org/10.1098/rstb.2013.0112.

Bowman, D.M.J.S., Williamson, G.J., Abatzoglou, J.T., Kolden, C.A., Cochrane, M.A. and Smith, A.M.S. (2017). Human exposure and sensitivity to globally extreme wildfire events. *Nature Ecology & Evolution* 1, 0058. https://doi.org/10.1038/s41559-016-0058.

Boys, B.L., Martin, R.V., van Donkelaar, A., MacDonell, R.J., Hsu, N.C., Cooper, M.J. *et al.* (2014). Fifteen-year global time series of satellite-derived fine particulate matter. *Environmental Science & Technology* 48(19), 11109-11118. https://doi.org/10.1021/es502113p.

Brasseur, G.P., Gupta, M., Anderson, B.E., Balasubramanian, S., Barrett, S., Duda, D. *et al.* (2016). Impact of aviation on climate: FAA's Aviation Climate Change Research Initiative (ACCRI) phase II. *Bulletin of the American Meteorological Society* 97(4), 561-583. https://doi.org/10.1175/bams-d-13-00089.1.

Breivik, K., Armitage, J.M., Wania, F., Sweetman, A.J. and Jones, K.C. (2016). Tracking the global distribution of persistent organic pollutants accounting for e-waste exports to developing regions. *Environmental Science & Technology* 50(2), 798-805. https://doi.org/10.1021/acs.est.5b04226.

Brook, R.D., Rajagopalan, S., Pope, C.A., 3rd, Brook, J.R., Bhatnagar, A., Diez-Roux, A.V. *et al.* (2010). Particulate matter air pollution and cardiovascular disease: An update to the scientific statement from the American Heart Association. *Circulation* 121(21), 2331-2378. https://doi.org/10.1161/CIR.0b013e3181dbece1.

Butler, C. (ed.) (2014). *Climate Change and Global Health*. Oxfordshire: Centre for Agriculture and Biosciences International. https://www.cabi.org/bookshop/book/9781780642659.

Calderon-Garciduenas, L. and Villarreal-Rios, R. (2017). Living close to heavy traffic roads, air pollution, and dementia. *Lancet* 389(10070), 675-677. https://doi.org/10.1016/S0140-6736(16)32596-X.

Caro, D., Davis, S.J., Bastianoni, S. and Caldeira, K. (2014). Global and regional trends in greenhouse gas emissions from livestock. *Climatic Change* 126(1-2), 203-216. https://doi.org/10.1007/s10584-014-1197-x.

Center for Human Rights and Environment (2015). *Policy Advocacy Network for Latin America for Clean Brick Production: Compilation of Existing Policy Frameworks*. Paris: Climate and Clean Air Coalition. http://ccacoalition.org/en/file/2170/download?token=cogeNCXF.

Centers for Disease Control and Prevention (2017). *Harmful Algal Bloom (HAG)-associated illness – Illness & symptoms*. [https://www.cdc.gov/habs/illness.html (Accessed: 5 October 2018).

Chameides, W., Lindsay, R., Richardson, J. and Kiang, C. (1988). The role of biogenic hydrocarbons in urban photochemical smog: Atlanta as a case study. *Science* 241(4872), 1473-1475. https://doi.org/10.1126/science.3420404.

Chang, K.-L., Petropavlovskikh, I., Copper, O.R., Schultz, M.G. and Wang, T. (2017). Regional trend analysis of surface ozone observations from monitoring networks in eastern North America, Europe and East Asia. *Elem Sci Anth* 5(0), 50. https://doi.org/10.1525/elementa.243.

Chang, T., Zivin, J.G., Gross, T. and Neidell, M. (2016). *The Effect of Pollution on Worker Productivity: Evidence from Call-Center Workers in China*. National Bureau of Economic Research Working Paper Series. Cambridge, MA: National Bureau of Economic Research. http://www.nber.org/papers/w22328.

Cheng, Z., Luo, L., Wang, S., Wang, Y., Sharma, S., Shimadera, H. *et al.* (2016). Status and characteristics of ambient PM2. 5 pollution in global megacities. *Environment International* 89, 212-221. https://doi.org/10.1016/j.envint.2016.02.003.

Chipperfield, M.P., Bekki, S., Dhomse, S., Harris, N.R.P., Hassler, B., Hossaini, R. *et al.* (2017). Detecting recovery of the stratospheric ozone layer. *Nature* 549(7671), 211-218. https://doi.org/10.1038/nature23681.

Ciais, P., Sabine, C., Bala, G., Bopp, L., Brovkin, V., Canadell, J. *et al.* (2013). Carbon and other biogeochemical cycles.In *Climate Change 2013: The Physical Science Basis. Contribution of Working Group I to the Fifth Assessment Report of the Intergovernmental Panel on Climate Change*. Stocker, T.F., Qin, D., Plattner, G.-K., Tignor, M., Allen, S.K., Boschung, J. *et al.* (eds.). Cambridge: Cambridge University Press. chapter 6. 465–570. http://www.ipcc.ch/pdf/assessment-report/ar5/wg1/drafts/WG1AR5_SOD_Ch06_All_Final.pdf.

Clean Air Asia (2017). *Clean air Asia summary*. [Clean Air Asia http://cleanairasia.org/clean-air-asia-summary/2017).

Climate and Clean Air Coalition (2015). *CCAC Five-Year Strategic Plan*. Paris: Climate and Clean Air Coalition. http://www.ccacoalition.org/en/resources/ccac-five-year-strategic-plan.

Clougherty, J.E. (2010). A growing role for gender analysis in air pollution epidemiology. *Environmental Health Perspectives* 118(2), 167-176. https://doi.org/10.1289/ehp.0900994.

Coghlan, C., Muzammil, M., Ingram, J., Vervoort, J., Otto, F. and James, R. (2014). *A Sign of Things to Come? Examining Four Major Climate-related Disasters, 2010-2013, and their Impacts on Food Security*. Oxford: Oxfam International. http://policy-practice.oxfam.org.uk/publications/a-sign-of-things-to-come-examining-four-major-climate-related-disasters-2010-20-326092.

Cohen, A.J., Brauer, M., Burnett, R., Anderson, H.R., Frostad, J., Estep, K. *et al.* (2017). Estimates and 25-year trends of the global burden of disease attributable to ambient air pollution: An analysis of data from the Global Burden of Diseases Study 2015. *Lancet* 389(10082), 1907-1918. https://doi.org/10.1016/S0140-6736(17)30505-6.

Cole, J.R. and McCoskey, S. (2013). Does global meat consumption follow an environmental Kuznets curve? *Sustainability: Science, Practice & Policy* 8(2), 26-36. https://doi.org/10.1080/15487733.2013.11908112.

Committee on Earth Observing Satellites (2011). *A Geostationary Satellite Constellation for Observing Global Air Quality: An International Path Forward*. Committee on Earth Observing Satellites. http://ceos.org/document_management/Virtual_Constellations/ACC/Documents/AC-VC_Geostationary-Cx-for-Global-AQ-final_Apr2011.pdf.

Committee on the Medical Effects of Air Pollutants (2010). *The Mortality Effects of Long-Term Exposure to Particulate Air Pollution in the United Kingdom*. Oxfordshire. https://www.gov.uk/government/publications/comeap-mortality-effects-of-long-term-exposure-to-particulate-air-pollution-in-the-uk.

Committee on the Medical Effects of Air Pollutants (2018). *Associations of Long-Term Average Concentrations of Nitrogen Dioxide with Mortality: A Report by the Committee on the Medical Effects of Air Pollutants*. Chilton. https://assets.publishing.service.gov.uk/government/uploads/system/uploads/attachment_data/file/734799/COMEAP_NO2_Report.pdf.

Czernecki, B., Półrolniczak, M., Kolendowicz, L., Marosz, M., Kendzierski, S. and Pilguj, N. (2016). Influence of the atmospheric conditions on PM_{10} concentrations in Poznań, Poland. *Journal of Atmospheric Chemistry* 74(1), 115-139. https://doi.org/10.1007/s10874-016-9345-5.

Damstra, T. (2002). Potential effects of certain persistent organic pollutants and endocrine disrupting chemicals on the health of children. *Journal of Toxicology: Clinical Toxicology* 40(4), 457-465. https://doi.org/10.1081/clt-120006748.

Day, T., Gonzales-Zuñiga, S., Nascimento, L., Höhne, N., Fekete, H. and Sterl, S. (2018). *Climate Opportunity: More Jobs; Better Health; Liveable Cities*. Cologne: New Climate Institute. https://newclimate.org/wp-content/uploads/2018/09/ClimateOpportunity_Full.pdf.

Dean, J.F., Middelburg, J.J., Röckmann, T., Aerts, R., Blauw, L.G., Egger, M. *et al.* (2018). Methane feedbacks to the global climate system in a warmer world. *Reviews of Geophysics* 56(1), 207-250. https://doi.org/10.1002/2017rg000559.

DeLonge, M.S., Miles, A. and Carlisle, L. (2016). Investing in the transition to sustainable agriculture. *Environmental Science & Policy* 55(Part 1), 266-273. https://doi.org/10.1016/j.envsci.2015.09.013.

Di Turo, F., Proietti, C., Screpanti, A., Fornasier, M.F., Cionni, I., Favero, G. *et al.* (2016). Impacts of air pollution on cultural heritage corrosion at European level: What has been achieved and what are the future scenarios. *Environmental pollution* 218, 586-594. https://doi.org/10.1016/j.envpol.2016.07.042.

Díaz, S., Pascual, U., Stenseke, M., Martín-López, B., Watson, R.T., Molnár, Z. *et al.* (2018). Assessing nature's contributions to people. *Science* 359(6373), 270-272. https://doi.org/10.1126/science.aap8826.

Duncan, B.N., Lamsal, L.N., Thompson, A.M., Yoshida, Y., Lu, Z., Streets, D.G. *et al.* (2016). A space-based, high-resolution view of notable changes in urban NOx pollution around the world (2005–2014). *Journal of Geophysical Research: Atmospheres* 121(2), 976-996. https://doi.org/10.1002/2015JD024121.

Duncan, B.N., Prados, A.I., Lamsal, L.N., Liu, Y., Streets, D.G., Gupta, P. *et al.* (2014). Satellite data of atmospheric pollution for U.S. air quality applications: Examples of applications, summary of data end-user resources, answers to FAQs, and common mistakes to avoid. *Atmospheric Environment* 94, 647-662. https://doi.org/10.1016/j.atmosenv.2014.05.061.

El-Shahawi, M.S., Hamza, A., Bashammakh, A.S. and Al-Saggaf, W.T. (2010). An overview on the accumulation, distribution, transformations, toxicity and analytical methods for the monitoring of persistent organic pollutants. *Talanta* 80(5), 1587-1597. https://doi.org/10.1016/j.talanta.2009.09.055.

Engel, A., Bönisch, H., Ostermöller, J., Chipperfield, M.P., Dhomse, S. and Jöckel, P. (2018). A refined method for calculating equivalent effective stratospheric chlorine. *Atmospheric Chemistry and Physics* 18(2), 601-619. https://doi.org/10.5194/acp-18-601-2018.

European Commission (2016). *Emissions database for global atmospheric research (EDGAR) v4.3.1*. http://edgar.jrc.ec.europa.eu/overview.php?v=431.

European Environment Agency (2017). *Annual European Union Greenhouse Gas Inventory 1990-2015 and Inventory Report 2017*. Copenhagen: European Environment Agency. https://www.eea.europa.eu/publications/european-union-greenhouse-gas-inventory-2017.

Eze, I.C., Hemkens, L.G., Bucher, H.C., Hoffmann, B., Schindler, C., Kunzli, N. *et al.* (2015). Association between ambient air pollution and diabetes mellitus in Europe and North America: systematic review and meta-analysis. *Environmental Health Perspectives* 123(5), 381-389. https://doi.org/10.1289/ehp.1307823.

Fahey, D.W., Baughcum, S.L., Fuglestvedt, J., Gupta, M., Lee, D.S., Sausen, R. *et al.* (2016). White paper on climate change aviation impacts on climate: State of the science.In *On Board A Sustainable Future: ICAO 2016 Environmental Report, Aviation and Climate Change*. Montreal: International Civil Aviation Organization. 99-107. https://www.icao.int/environmental-protection/Documents/ICAO%20Environmental%20Report%202016.pdf

Farhi, A., Boyko, V., Almagor, J., Benenson, I., Segre, E., Rudich, Y. *et al.* (2014). The possible association between exposure to air pollution and the risk for congenital malformations. *Environmental Research* 135, 173-180. https://doi.org/10.1016/j.envres.2014.08.024.

Feng, H. and Zhang, M. (2015). Global land moisture trends: Drier in dry and wetter in wet over land. *Scientific Reports* 5(18018). https://doi.org/10.1038/srep18018.

Feng, Z., Rütting, T., Pleijel, H., Wallin, G., Reich, P.B., Kammann, C.I. *et al.* (2015). Constraints to nitrogen acquisition of terrestrial plants under elevated CO_2. *Global Change Biology* 21(8), 3152-3168. https://doi.org/10.1111/gcb.12938.

Feng, Z.Z. and Kobayashi, K. (2009). Assessing the impacts of current and future concentrations of surface ozone on crop yield with meta-analysis. *Atmospheric Environment* 43(8), 1510-1519. https://doi.org/10.1016/j.atmosenv.2008.11.033.

Fiore, A.M., Naik, V. and Leibensperger, E.M. (2015). Air quality and climate connections. *Journal of the Air & Waste Management Association* 65(6), 645-685. https://doi.org/10.1080/10962247.2015.1040526.

Fischer, E.M., Oleson, K.W. and Lawrence, D.M. (2012). Contrasting urban and rural heat stress responses to climate change. *Geophysical Research Letters* 39(3). https://doi.org/10.1029/2011GL050576.

Fishman, J., Creilson, J.K., Parker, P.A., Ainsworth, E.A., Vining, G.G., Szarka, J. *et al.* (2010). An investigation of widespread ozone damage to the soybean crop in the upper Midwest determined from ground-based and satellite measurements. *Atmospheric Environment* 44(18), 2248-2256. https://doi.org/10.1016/j.atmosenv.2010.01.015.

Fleischer, N.L., Merialdi, M., van Donkelaar, A., Vadillo-Ortega, F., Martin, R.V., Betran, A.P. *et al.* (2014). Outdoor air pollution, preterm birth, and low birth weight: Analysis of the world health organization global survey on maternal and perinatal health. *Environmental Health Perspectives* 122(4), 425-430. https://doi.org/10.1289/ehp.1306837.

Food and Agriculture Organization of the United Nations (2008). *The State of Food Insecurity in the World 2008*. Rome. http://www.fao.org/3/a-i4646e.pdf.

Food and Agriculture Organization of the United Nations (2016). *Climate Change and Food Security: Risks and Responses*. Rome. http://www.fao.org/3/a-i5188e.pdf.

Fowler, D., Steadman, C.E., Stevenson, D., Coyle, M., Rees, R.M., Skiba, U.M. *et al.* (2015). Effects of global change during the 21st century on the nitrogen cycle. *Atmospheric, Chemistry and Physics* 15(24), 13849-13893. https://doi.org/10.5194/acp-15-13849-2015.

Fry, K. and Power, M.C. (2017). Persistent organic pollutants and mortality in the United States, NHANES 1999-2011. *Environmental Health* 16(1), 105. https://doi.org/10.1186/s12940-017-0313-6.

Fuhrer, J., Val Martin, M., Mills, G., Heald, C.L., Harmens, H., Hayes, F. *et al.* (2016). Current and future ozone risks to global terrestrial biodiversity and ecosystem processes. *Ecology and Evolution* 6(24), 8785-8799. https://doi.org/10.1002/ece3.2568.

Fuzzi, S., Baltensperger, U., Carslaw, K., Decesari, S., van der Gon, H.D., Facchini, M.C. *et al.* (2015). Particulate matter, air quality and climate: Lessons learned and future needs. *Atmospheric Chemistry and Physics* 15(14), 8217–8299. https://doi.org/10.5194/acp-15-8217-2015.

Galloway, J.N., Aber, J.D., Erisman, J.W., Seitzinger, S.P., Howarth, R.W., Cowling, E.B. *et al.* (2003). The nitrogen cascade. *BioScience* 53(4), 341-356. https://doi.org/10.1641/0006-3568(2003)053[0341:TNC]2.0.CO;2.

Geddes, J.A., Martin, R.V., Boys, B.L. and van Donkelaar, A. (2016). Long-term trends worldwide in ambient NO2 concentrations inferred from satellite observations. *Environmental Health Perspectives* 124(3), 281-289. https://doi.org/10.1289/ehp.1409567.

Gibb, H. and O'Leary, K.G. (2014). Mercury exposure and health impacts among individuals in the artisanal and small-scale gold mining community: A comprehensive review. *Environmental Health Perspectives* 122(7), 667-672. https://doi.org/10.1289/ehp.1307864.

Gibson, J., Adlard, B., Olafsdottir, K., Sandanger, T.M. and Odland, J.Ø. (2016). Levels and trends of contaminants in humans of the Arctic. *International Journal of Circumpolar Health* 75(1), 33804. https://doi.org/10.3402/ijch.v75.33804.

Ginoux, P., Prospero, J.M., Gill, T.E., Hsu, N.C. and Zhao, M. (2012). Global-scale attribution of anthropogenic and natural dust sources and their emission rates based on modis deep blue aerosol products. *Reviews of Geophysics* 50. https://doi.org/10.1029/2012rg000388.

Global Alliance for Clean Cookstoves (2014). *Results Report 2014: Sharing Progress on the Path to Adoption of Clean and Efficient Cooking Solutions*. Washington, DC.: Global Alliance for Clean Cookstoves. http://cleancookstoves.org/binary-data/RESOURCE/file/000/000/414-1.pdf.

Global Burden of Disease 2016 Risk Factor Collaborators (2017). Global, regional, and national comparative risk assessment of 84 behavioural, environmental and occupational, and metabolic risks or clusters of risks, 1990–2016: A systematic analysis for the Global Burden of Disease Study 2016. *The Lancet* 390(10100), 1345-1422. https://doi.org/10.1016/S0140-6736(17)32366-8.

Global Burden of Disease Cancer Collaboration, Fitzmaurice, C., Allen, C., Barber, R.M., Barregard, L., Bhutta, Z.A. *et al.* (2017). Global, regional, and national cancer incidence, mortality, years of life lost, years lived with disability, and disability-adjusted life-years for 32 cancer groups, 1990 to 2015: A systematic analysis for the global burden of disease study. *JAMA Oncology* 3(4), 524-548. https://doi.org/10.1001/jamaoncol.2016.5688.

Cornell, J., Betts, R., Burke, E., Clark, R., Camp, J., Willett, K. *et al.* (2010). Implications of climate change for agricultural productivity in early twenty first century. *Philosophical Transactions of the Royal Society B* 365(1554), 2973–2989. https://doi.org/10.1098/rstb.2010.0158.

Guenther, A.B., Jiang, X., Heald, C.L., Sakulyanontvittaya, T., Duhl, T., Emmons, L.K. *et al.* (2012). The model of emissions of gases and aerosols from nature version 2.1 (MEGAN2.1): An extended and updated framework for modeling biogenic emissions. *Geoscientific Model Development* 5(6), 1471-1492. https://doi.org/10.5194/gmd-5-1471-2012.

Haines, A., Amann, M., Borgford-Parnell, N., Leonard, S., Kuylenstierna, J. and Shindell, D. (2017). Short-lived climate pollutant mitigation and the Sustainable Development Goals. *Nature Climate Change* 7(12), 863-869. https://doi.org/10.1038/s41558-017-0012-x.

Hanlon, W.W. (2016). *Coal Smoke and the Costs of the Industrial Revolution*. National Bureau of Economic Research Working Paper Series. Cambridge, MA: National Bureau of Economic Research. http://www.nber.org/papers/w22921.pdf.

Harris, N.R.P., Hassler, B., Tummon, F., Bodeker, G.E., Hubert, D., Petropavlovskikh, I. *et al.* (2015). Past changes in the vertical distribution of ozone – Part 3: Analysis and interpretation of trends. *Atmospheric Chemistry and Physics* 15(17), 9965-9982. https://doi.org/10.5194/acp-15-9965-2015.

Hartmann, D.L., Tank, A.M.G.K., Rusticucci, M., Alexander, L.V., Brönnimann, S., Charabi, Y.A.R. *et al.* (2013). Observations: Atmosphere and surface.In *Climate Change 2013: The Physical Science Basis. Contribution of Working Group I to the Fifth Assessment Report of the Intergovernmental Panel on Climate Change*. Cambridge, MA: Cambridge University Press. chapter 2. 159-254. https://www.ipcc.ch/pdf/assessment-report/ar5/wg1/WG1AR5_Chapter02_FINAL.pdf

Hashimoto, S., Carvalhais, N., Ito, A., Migliavacca, M., Nishina, K. and Reichstein, M. (2015). Global spatiotemporal distribution of soil respiration modeled using a global database. *Biogeosciences* 12, 4121-4132. https://doi.org/10.5194/bg-12-4121-2015.

Heald, C.L. and Geddes, J., A., (2016). The impact of historical land use change from 1850 to 2000 on secondary particulate matter and ozone. *Atmospheric Chemistry and Physics* 16(23), 14997-15010. https://doi.org/10.5194/acp-16-14997-2016.

Health Effects Institute (2017). *State of Global Air 2017: A Special Report on Global Exposure to Air Pollution and Its Disease Burden*. Boston, MA. https://www.stateofglobalair.org/sites/default/files/SoGA2017_report.pdf.

Health Effects Institute (2018). *State of Global Air 2018 Special Report: A Special Report on Global Exposure to Air Pollution And Its Disease Burden*. Boston, MA. https://www.stateofglobalair.org/sites/default/files/soga-2018-report.pdf.

Heede, R. (2014). Tracing anthropogenic carbon dioxide and methane emissions to fossil fuel and cement producers, 1854-2010. *Climatic Change* 122(1-2), 229–241. https://doi.org/10.1007/s10584-013-0986-y.

Hettelingh, J.-P., Stevens, C., Posch, M., Bobbink, R. and de Vries, W. (2015). Assessing the impacts of nitrogen deposition on plant species richness in Europe.In *Critical Loads and Dynamic Risk Assessments: Nitrogen, Acidity and Metals in Terrestrial and Aquatic Ecosystems*. de Vries, W., Hettelingh, J. and Posch, M. (eds.). Dordrecht: Springer. 573-586. https://link.springer.com/chapter/10.1007/978-94-017-9508-1_23

Hoesly, R.M., Smith, S.J., Feng, L., Klimont, Z., Janssens-Maenhout, G., Pitkanen, T. *et al.* (2018). Historical (1750–2014) anthropogenic emissions of reactive gases and aerosols from the Community Emission Data System (CEDS). *Geoscientific Model Development* 11, 369-408. https://doi.org/10.5194/gmd-2017-43.

Hoffman, K., Butt, C.M., Webster, T.F., Preston, E.V., Hammel, S.C., Makey, C. *et al.* (2017). Temporal trends in exposure to organophosphate flame retardants in the United States. *Environmental Science & Technology Letters* 4(3), 112-118. https://doi.org/10.1021/acs.estlett.6b00475.

Huang, L., McDonald-Buller, E.C., McGaughey, G., Kimura, Y. and Allen, D., T., (2016). The impact of drought on ozone dry deposition over eastern Texas. *Atmospheric Environment* 127, 176-186. https://doi.org/10.1016/j.atmosenv.2015.12.022.

Hung, H., Katsoyiannis, A.A., Brorstrom-Lunden, E., Olafsdottir, K., Aas, W., Breivik, K. *et al.* (2016). Temporal trends of Persistent Organic Pollutants (POPs) in arctic air: 20 years of monitoring under the Arctic Monitoring and Assessment Programme (AMAP). *Environmental Pollution* 217, 52-61. https://doi.org/10.1016/j.envpol.2016.01.079.

Hussey, S.J.K., Purves, J., Allcock, N., Fernandes, V.E., Monks, P.S., Ketley, J.M. *et al.* (2017). Air pollution alters Staphylococcus aureus and Streptococcus pneumoniae biofilms, antibiotic tolerance and colonisation. *Environmental Microbiology* 19(5), 1868-1880. https://doi.org/10.1111/1462-2920.13686.

Intergovernmental Panel on Climate Change (2014). *Climate Change 2014: Impacts, Adaptation, and Vulnerability: Part B: Regional Aspects, Contribution of Working Group II to the Fifth Assessment Report*. Barros, V.R., Field, C.B., Dokken, D.J., Mastrandrea, M.D., Mach, K.J., Bilir, T.E. *et al.* (eds.). Cambridge, MA: Cambridge University Press. http://www.ipcc.ch/report/ar5/wg2/.

International Civil Aviation Organization (2016a). *Presentation of 2016 Air Transport Statistical Results*. Annual Report of the Council. Montreal: International Civil Aviation Organization. https://www.icao.int/annual-report-2016/Documents/ARC_2016_Air%20Transport%20Statistics.pdf.

International Civil Aviation Organization (2016b). *Assembly Resolution A39-3*. Montreal. https://www.icao.int/Meetings/a39/Documents/WP/wp_530_en.pdf.

International Cooperative Programme on Effects of Air Pollution on Natural Vegetation and Crops (2017). Mapping critical levels for vegetation.In *Manual On Methodologies and Criteria For Modelling and Mapping Critical Loads and Levels and Air Pollution Effects, Risks and Trends*. Geneva: United Nations Economic Commission for Europe. chapter 3. https://icpvegetation.ceh.ac.uk/publications/documents/FinalnewChapter3v4Oct2017_000.pdf

International Energy Agency (2016a). *Energy and Air Pollution*. Paris: International Energy Agency. https://www.iea.org/publications/freepublications/publication/WorldEnergyOutlookSpecialReport2016EnergyandAirPollution.pdf.

International Energy Agency (2016b). *World Energy Outlook*. Paris: International Energy Agency. https://www.iea.org/publications/freepublications/publication/WorldEnergyOutlook2016ExecutiveSummaryEnglish.pdf.

International Energy Agency (2017). *CO_2 Emissions from Fuel Combustion: 2017 Overview*. Paris: International Energy Agency. http://www.iea.org/publications/freepublications/publication/CO2EmissionsFromFuelCombustion2017Overview.pdf.

International Transport Forum (2017). *ITF Tranport Outlook 2017*. Paris: Organization for Economic Co-operation and Development. https://www.oecd-ilibrary.org/transport/itf-transport-outlook_25202367.

Joint Group of Experts on the Scientific Aspects of Marine Environmental Protection (GESAMP) (2018). *The Magnitude and Impacts of Anthropogenic Atmospheric Nitrogen Inputs to the Ocean*. Geneva: World Meteorological Organization. http://www.gesamp.org/publications/the-magnitude-and-impacts-of-anthropogenic-atmospheric-nitrogen-inputs-to-the-ocean.

Karagas, M.R., Choi, A.L., Oken, E., Horvat, M., Schoeny, R., Kamai, E. *et al.* (2012). Evidence on the human health effects of low-level methylmercury exposure. *Environmental Health Perspectives* 120(6), 799-806. https://doi.org/10.1289/ehp.1104494.

Karagulian, F., Belis, C.A., Dora, C.F., C., Prüss-Ustün, A., M., Bonjour, S., Adair-Rohani, H. *et al.* (2015). Contributions to cities' ambient particulate matter (PM): A systematic review of local source contributions at global level. *Atmospheric Environment* 120, 475-483. https://doi.org/10.1016/j.atmosenv.2015.08.087.

Kelley, C.P., Mohtadi, S., Cane, M.A., Seager, R. and Kushnir, Y. (2015). Climate change in the fertile crescent and implications of the recent Syrian drought. *Proceedings of the National Academy of Sciences* 112(11), 3241-3246. https://doi.org/10.1073/pnas.1421533112.

Kiesewetter, G. and Amann, M. (2014). *Urban PM2.5 levels under the EU Clean Air Policy Package*. Laxenburg: International Institute for Applied Systems Analysis. http://ec.europa.eu/environment/air/pdf/TSAP_12.pdf.

Klingmuller, K., Pozzer, A., Metzger, S., Stenchikov, G.L. and Lelieveld, J. (2016). Aerosol optical depth trend over the Middle East. *Atmospheric Chemistry and Physics* 16(8), 5063-5073. https://doi.org/10.5194/acp-16-5063-2016.

Koplitz, S.N., Mickley, L.J., Marlier, M.E., Buonocore, J.J., Kim, P.S., Liu, T. *et al.* (2016). Public health impacts of the severe haze in Equatorial Asia in September–October 2015: Demonstration of a new framework for informing fire management strategies to reduce downwind smoke exposure. *Environmental Research Letters* 11(9), 094023. https://doi.org/10.1088/1748-9326/11/9/094023.

Krotkov, N.A., McLinden, C.A., Li, C., Lamsal, L.N., Celarier, E.A., Marchenko, S.V. *et al.* (2016). Aura OMI observations of regional SO2 and NO2 pollution changes from 2005 to 2015. *Atmospheric Chemistry and Physics* 16(7), 4605-4629. https://doi.org/10.5194/acp-16-4605-2016.

Kumar, R. and Agarwala, A. (2013). Renewable energy certificate and perform, achieve, trade mechanisms to enhance the energy security for India. *Energy Policy* 55, 669-676. https://doi.org/10.1016/j.enpol.2012.12.072.

Kutlar Joss, M., Eeftens, M., Gintowt, E., Kappeler, R. and Künzli, N. (2017). Time to harmonize national ambient air quality standards. *International Journal of Public Health* 62(4), 453-462. https://doi.org/10.1007/s00038-017-0952-y.

Kuttippurath, J. and Nair, P.J. (2017). The signs of Antarctic ozone hole recovery. *Scientific Reports* 7(585). https://doi.org/10.1038/s41598-017-00722-7.

Lamarque, J.F., Bond, T.C., Eyring, V., Granier, C., Heil, A., Klimont, Z. *et al.* (2010). Historical (1850–2000) gridded anthropogenic and biomass burning emissions of reactive gases and aerosols: Methodology and application. *Atmospheric Chemistry and Physics* 10(15), 7017-7039. https://doi.org/10.5194/acp-10-7017-2010.

Lee, S.C., Sverko, E., Harner, T., Pozo, K., Barresi, E., Schachtschneider, J. *et al.* (2016). Retrospective analysis of "new" flame retardants in the global atmosphere under the GAPS Network. *Environmental Pollution* 217, 62-69. https://doi.org/10.1016/j.envpol.2016.01.080.

Leibensperger, E.M., Mickley, L.J. and Jacob, D.J. (2008). Sensitivity of US air quality to mid-latitude cyclone frequency and implications of 1980–2006 climate change. *Atmospheric Chemistry and Physics* 8(23), 7075-7086. https://doi.org/10.5194/acp-8-7075-2008.

Leip, A., Billen, G., Garnier, J., Grizzetti, B., Lassaletta, L., Reis, S. *et al.* (2015). Impacts of European livestock production: nitrogen, sulphur, phosphorus and greenhouse gas emissions, land-use, water eutrophication and biodiversity. *Environmental Research Letters* 10(11), 115004. https://doi.org/10.1088/1748-9326/10/11/115004.

Lewis, A.C., Zellweger, C., Schultz, M.G., Tarasova, O.A. and Reactive Gases Science Advisory Group (2017). *Technical Advice Note On Lower Cost Air Pollution Sensors*. World Meteorological Organization http://www.wmo.int/pages/prog/arep/gaw/documents/GAW_Sensors_advice.pdf.

Li, X., Huang, S., Jiao, A., Yang, X., Yun, J., Wang, Y. *et al.* (2017). Association between ambient fine particulate matter and preterm birth or term low birth weight: An updated systematic review and meta-analysis. *Environmental Pollution* 227, 596-605. https://doi.org/10.1016/j.envpol.2017.03.055.

Lim, S.S., Vos, T., Flaxman, A.D., Danaei, G., Shibuya, K., Adair-Rohani, H. *et al.* (2012). A comparative risk assessment of burden of disease and injury attributable to 67 risk factors and risk factor clusters in 21 regions, 1990–2010: a systematic analysis for the Global Burden of Disease Study 2010. *The Lancet* 380(9859), 2224-2260. https://doi.org/10.1016/s0140-6736(12)61766-8.

Liu, H., Zhang, X., Zhang, H., Yao, X., Zhou, M., Wang, J. *et al.* (2018). Effect of air pollution on the total bacteria and pathogenic bacteria in different sizes of particulate matter. *Environmental Pollution* 233, 483-493. https://doi.org/10.1016/j.envpol.2017.10.070.

Lott, F.C., Christidis, N. and Stott, P.A. (2013). Can the 2011 East African drought be attributed to human-induced climate change? *Geophysical Research Letters* 40(6), 1177-1181. https://doi.org/10.1002/grl.50235.

Lucas, R.M., Norval, M., Neale, R.E., Young, A.R., de Gruijl, F.R., Takizawa, Y. *et al.* (2015). The consequences for human health of stratospheric ozone depletion in association with other environmental factors. *Photochemical and Photobiological Sciences* 14(1), 53-87. https://doi.org/10.1039/c4pp90033b.

Ma, J., Hung, H. and Macdonald, R.W. (2016). The influence of global climate change on the environmental fate of persistent organic pollutants: A review with emphasis on the Northern Hemisphere and the Arctic as a receptor. *Global and Planetary Change* 146, 89-108. https://doi.org/10.1016/j.gloplacha.2016.09.011.

Ma, J.M., Hung, H.L., Tian, C. and Kallenborn, R. (2011). Revolatilization of persistent organic pollutants in the Arctic induced by climate change. *Nature Climate Change* 1(5), 255-260. https://doi.org/10.1038/Nclimate1167.

Maas, R. and Grennfelt, P. (eds.) (2016). *Towards Cleaner Air: Scientific Assessment Report 2016*. Oslo: United Nations Economic Commission for Europe. https://www.unece.org/fileadmin/DAM/env/lrtap/ExecutiveBody/35th_session/CLRTAP_Scientific_Assessment_Report_-_Final_20-5-2016.pdf.

Maithel, S., Lalchandami, D., Malhotra, G., Bhanware, P., Uma, R., Ragavan, S. *et al.* (2012). *Brick Kilns Performance Assessment: A Roadmap for Cleaner Brick Production in India*. New Delhi: Greentech Knowledge Solutions Pvt. Ltd. http://ccacoalition.org/en/file/575/download?token=wpSU0X2x.

McCracken, J.P., Wellenius, G.A., Bloomfield, G.S., Brook, R.D., Tolunay, H.E., Dockery, D.W. *et al.* (2012). Household air pollution from solid fuel use: Evidence for links to CVD. *Global Heart* 7(3), 223-234. https://doi.org/10.1016/j.gheart.2012.06.010.

McGrath, J.M., Betzelberger, A.M., Wang, S., Shook, E., Zhu, X.-G., Long, S.P. *et al.* (2015). An analysis of ozone damage to historical maize and soybean yields in the United States. *Proceedings of the National Academy of Sciences* 112(46), 14390-14395. https://doi.org/10.1073/pnas.1509777112.

Megaritis, A.G., Fountoukis, C., Charalampidis, P.E., Pilinis, C. and Pandis, S.N. (2013). Response of fine particulate matter concentrations to changes of emissions and temperature in Europe. *Atmospheric Chemistry and Physics* 13(6), 3423-3443. https://doi.org/10.5194/acp-13-3423-2013.

Melillo, J.M., Richmond, T.C. and Yohe, G.W. (2014). *Climate Change Impacts in the United States: The Third National Climate Assessment*. Washington, D.C: U.S. Global Change Research Program. https://nca2014.globalchange.gov/report.

Mills, J.N., Gage, K.L. and Khan, A.S. (2010). Potential influence of climate change on vector-borne and zoonotic diseases: A review and proposed research plan. *Environmental Health Perspectives* 118(11), 1507-1514. https://doi.org/10.1289/ehp.0901389.

Montzka, S.A., Dutton, G.S., Yu, P., Ray, E., Portmann, R.W., Daniel, J.S. *et al.* (2018). An unexpected and persistent increase in global emissions of ozone-depleting CFC-11. *Nature* 557(7705), 413-417. https://doi.org/10.1038/s41586-018-0106-2.

Myers, S.S., Zanobetti, A., Kloog, I., Huybers, P., Leakey, A.D.B., Bloom, A.J. *et al.* (2014). Increasing CO_2 threatens human nutrition. *Nature* 510(7503), 139-142. https://doi.org/10.1038/nature13179.

Na, K., Moon, K.C. and Kim, Y.P. (2005). Source contribution to aromatic VOC concentration and ozone formation potential in the atmosphere of Seoul. *Atmospheric Environment* 39(30), 5517-5524. https://doi.org/10.1016/j.atmosenv.2005.06.005.

Newman, P.A., J. S. Daniel., Waugh, D.W. and Nash, E.R. (2007). A new formulation of equivalent effective stratospheric chlorine (EESC). *Atmospheric Chemistry and Physics* 7, 4537-4552. https://doi.org/10.5194/acp-7-4537-2007.

Nordborg, M. and Röös, E. (2016). *Holistic Management – A Critical Review Of Allan Savory's Grazing Method*. Uppsala: SLU/EPOK - Centre for Organic Food and Farming & Chalmers. https://www.fcrn.org.uk/research-library/holistic-management-%E2%80%93-critical-review-allan-savory%E2%80%99s-grazing-method.

Olsen, C.M., Carroll, H.J. and Whiteman, D.C. (2010). Estimating the attributable fraction for melanoma: a meta-analysis of pigmentary characteristics and freckling. *International Journal of Cancer* 127(10), 2430-2445. https://doi.org/10.1002/ijc.25243.

Otto, F.E.L., Massey, N., van Oldenborgh, G.J., Jones, R.G. and Allen, M.R. (2012). Reconciling two approaches to attribution of the 2010 Russian heat wave. *Geophysical Research Letters* 39(4). https://doi.org/10.1029/2011GL050422.

Pacifico, F., Folberth, G.A., Jones, C.D., Harrison, S.P. and Collins, W.J. (2012). Sensitivity of biogenic isoprene emissions to past, present, and future environmental conditions and implications for atmospheric chemistry. *Journal of Geophysical Research* 117(D22302). https://doi.org/10.1029/2012jd018276.

Pagano, M.C., Correa, E.J.A., Duarte, N.F., Yelikbayev, B., O'Donovan, A. and Gupta, V.K. (2017). Advances in eco-efficient agriculture: The plant-soil mycobiome. *Agriculture* 7(2). https://doi.org/10.3390/agriculture7020014.

Page, S.E. and Hooijer, A. (2016). In the line of fire: The peatlands of Southeast Asia. *Philosophical Transactions of the Royal Society B* 371(1696). https://doi.org/10.1098/rstb.2015.0176.

Parker, R.J., Boesch, H., Wooster, M.J., Moore, D.P., Webb, A.J., Gaveau, D. *et al.* (2016). Atmospheric CH4 and CO2 enhancements and biomass burning emission ratios derived from satellite observations of the 2015 Indonesian fire plumes. *Atmospheric Chemistry and Physics* 16(15), 10111-10131. https://doi.org/10.5194/acp-16-10111-2016.

Parrish, D.D., Lamarque, J.-F., Naik, V., Horowitz, L., Shindell, D.T., Staehelin, J. *et al.* (2014). Long-term changes in lower tropospheric baseline ozone concentrations: Comparing chemistry-climate models and observations at northern midlatitudes. *Journal of Geophysical Research* 119(9), 5719-5736. https://doi.org/10.1002/2013JD021435.

Piguet, E., Pecoud, A. and de Guchteneire, P. (eds.) (2011). *Migration and Climate Change*. Cambridge, MA: Cambridge University Press. http://www.cambridge.org/gb/academic/subjects/politics-international-relations/international-relations-and-international-organisations/migration-and-climate-change?format=PB&isbn=9781107662254#KhdrPz8VSzmAvqFf.97.

Porter, J.R., L. Xie., Challinor, A.J., Cochrane, K., Howden, S.M., Iqbal, M.M. *et al.* (2014). Food security and food production systems.In *Climate Change 2014: Impacts, Adaptation, and Vulnerability. Part A: Global and Sectoral Aspects. Contribution of Working Group II to the Fifth Assessment Report of the Intergovernmental Panel of Climate Change*. Field, C.B., Barros, V.R., Dokken, D.J., Mach, K.J., Mastrandrea, M.D., Bilir, T.E. *et al.* (eds.). Cambridge, MA: Cambridge University Press. https://www.cambridge.org/core/books/climate-change-2014-impacts-adaptation-and-vulnerability-part-a-global-and-sectoral-aspects/1BE4ED76F97CF3A75C64487E6274783A

Puttman, W. and da Silva, A. (2013). *Simulating the transport of aerosols with GEOS-5*. [National Aeronautics and Space Administration https://gmao.gsfc.nasa.gov/research/aerosol/modeling/nr1_movie/index.php.

Rae, D. and Graham, L. (2004). *Benefits of Reducing Mercury in Saltwater Ecosystems: A Case Study*. Washington, DC: United States Environmental Protection Agency http://nepis.epa.gov/Exe/ZyPURL.cgi?Dockey=901K0B00.TXT.

Raj, M.L., Ajay, S.N., Lina, L., Sachchida, N.T., Anu, R., Michael, H.B. *et al.* (2016). Municipal solid waste and dung cake burning: Discoloring the Taj Mahal and human health impacts in Agra. *Environmental Research Letters* 11(10), 104009. https://doi.org/10.1088/1748-9326/11/10/104009.

Rauert, C., Harner, T., Schuster, J.K., Quinto, K., Fillmann, G., Castillo, L.E. *et al.* (2016). Towards a regional passive air sampling network and strategy for new POPs in the GRULAC region: Perspectives from the GAPS Network and first results for organophosphorus flame retardants. *Science of the Total Environment* 573, 1294-1302. https://doi.org/10.1016/j.scitotenv.2016.06.229.

Renaudeau, D., Collin, A., Yahav, S., de Basilio, V., Gourdine, J.L. and Collier, R.J. (2011). Adaptation to hot climate and strategies to alleviate heat stress in livestock production. *Animal* 6(5), 707-728. https://doi.org/10.1017/S1751731111002448.

Renewable Energy Policy Network for the 21st Century (2016). *Renewables 2016 Global Status Report*. Paris: Renewable Energy Policy Network for the 21st Century. http://www.ren21.net/wp-content/uploads/2016/06/GSR_2016_Full_Report.pdf.

Rice, K.M., Walker, E.M., Jr., Wu, M., Gillette, C. and Blough, E.R. (2014). Environmental mercury and its toxic effects. *Journal of Preventive Medicine & Public Health* 47(2), 74-83. https://doi.org/10.3961/jpmph.2014.47.2.74.

Roy, R. (2016). *The Cost of Air Pollution in Africa*. OECD Development Centre Working Papers. Paris: Organization for Economic Co-operation and Development. https://www.oecd-ilibrary.org/development/the-cost-of-air-pollution-in-africa_5jlqzq77x6f8-en.

Rucevska, I., Nelleman, C., Isarin, N., Yang, W., Liu, N., Yu, K. *et al.* (2015). *Waste crime - waste risks. Gaps in meeting the global waste challenge*. Nairobi: United Nations Environment Programme and GRID-Arendal. https://europa.eu/capacity4dev/file/25575/download?token=WAWKTk7p.

Sacks, J.D., Stanek, L.W., Luben, T.J., Johns, D.O., Buckley, B.J., Brown, J.S. *et al.* (2011). Particulate matter-induced health effects: who is susceptible? *Environmental Health Perspectives* 119(4), 446-454. https://doi.org/10.1289/ehp.1002255.

Safi, M. (2017). 'Delhi doctors declare pollution emergency as smog chokes city'. *The Guardian* 7 November 2017 https://www.theguardian.com/world/2017/nov/07/delhi-india-declares-pollution-emergency-as-smog-chokes-city.

Schlenker, W. and Roberts, M.J. (2009). Nonlinear temperature effects indicate severe damages to U.S. crop yields under climate change. *Proceedings of the National Academy of Sciences* 106(37), 15594-15598. https://doi.org/10.1073/pnas.0906865106.

Schmidhuber, J. and Tubiello, F.N. (2007). Global food security under climate change. *Proceedings of the National Acacemy of Sciences* 104(50), 19703-19708. https://doi.org/10.1073/pnas.0701976104.

Schneider, P., Lahoz, W.A. and van der, A., R., (2015). Recent satellite-based trends of tropospheric nitrogen dioxide over large urban agglomerations worldwide. *Atmospheric Chemistry and Physics* 15(3), 1205-1220. https://doi.org/10.5194/acp-15-1205-2015.

Schultz, M.G., Schröder, S., Lyapina, O., Cooper, O., Galbally, I., Petropavlovskikh, I. *et al.* (2017). Tropospheric ozone assessment report: Database and metrics data of global surface ozone observations. *Elementa: Science of the Anthropocene* 5(58). https://doi.org/10.1525/elementa.244.

Seinfeld, J.H. and Pandis, S.N. (2016). *Atmospheric Chemistry and Physics: From Air Pollution to Climate Change.* 3rd edn. Hoboken, NJ: John Wiley & Sons, Inc. https://www.wiley.com/en-us/Atmospheric+Chemistry+and+Physics%3A+From+Air+Pollution+to+Climate+Change%2C+3rd+Edition-p-9781118947401.

Seto, K.C.-Y., Dhakal, S., Bigio, A., Blanco, H., Delgado, G.C., Dewar, D. *et al.* (2014). Agriculture, forestry and other land use (AFOLU). In *Climate Change 2014: Mitigation of Climate Change. Contribution of Working Group III to the Fifth Assessment Report of the Intergovernmental Panel on Climate Change.* Edenhofer, O., Pichs-Madruga, R., Sokona, Y., Farahani, E., Kadner, S., Seyboth, K. *et al.* (eds.). Cambridge: Cambridge University Press. chapter 11. 811-922. https://www.ipcc.ch/pdf/assessment-report/ar5/wg3/ipcc_wg3_ar5_chapter11.pdf

Shaddick, G., Thomas, M.L., Amini, H., Broday, D., Cohen, A., Frostad, J. *et al.* (2018). Data integration for the assessment of population exposure to ambient air pollution for global burden of disease assessment. *Environmental Science & Technology* 52(16), 9069-9078. https://doi.org/10.1021/acs.est.8b02864.

Shindell, D., Borgford-Parnell, N., Brauer, M., Haines, A., Kuylenstierna, J.C.I., Leonard, S.A. *et al.* (2017). A climate policy pathway for near- and long-term benefits. *Science* 356(6337), 493. https://doi.org/10.1126/science.aak9521.

Sindelarova, K., Granier, C., Bouarar, I., Guenther, A., Tilmes, S., Stavrakou, T. *et al.* (2014). Global data set of biogenic VOC emissions calculated by the MEGAN model over the last 30 years. *Atmospheric Chemistry and Physics* 14(17), 9317-9341. https://doi.org/10.5194/acp-14-9317-2014.

Sliggers, J. and Kakebeeke, W. (eds.) (2004). *Clearing the Air: 25 years of the Convention on Long-range Transboundary Air Pollution.* Geneva: United Nations Economic Commission for Europe. http://www.unece.org/index.php?id=10091.

Smith, K.R., Woodward, A., Campbell-Lendrum, D., Chadee, D.D., Honda, Y., Liu, Q. *et al.* (2014). Human health: Impacts, adaptation, and co-benefits.In *Climate Change 2014: Impacts, Adaptation, and Vulnerability. Part A: Global and Sectoral Aspects. Contribution of Working Group II to the Fifth Assessment Report of the Intergovernmental Panel of Climate Change.* Field, C.B., Barros, V.R., Dokken, D.J., Mach, K.J., Mastrandrea, M.D., Bilir, T.E. *et al.* (eds.). Cambridge: Cambridge University Press. chapter 11. 709-754. https://www.ipcc.ch/pdf/assessment-report/ar5/wg2/WGIIAR5-Chap11_FINAL.pdf

Snider, G., Weagle, C.L., Martin, R.V., van Donkelaar, A., Conrad, K., Cunningham, D. *et al.* (2015). SPARTAN: A global network to evaluate and enhance satellite-based estimates of ground-level particulate matter for global health applications. *Atmospheric Measurement Techniques* 8(1), 505-521. https://doi.org/10.5194/amt-8-505-2015.

Solberg, S., Hov, Ø., Søvde, A., Isaksen, I.S.A., Coddeville, P., De Backer, H. *et al.* (2008). European surface ozone in the extreme summer 2003. *Journal of Geophysical Research* 113(D7). https://doi.org/10.1029/2007jd009098.

Solomon, S., Ivy, D., Gupta, M., Bandoro, J., Santer, B., Fu, Q. *et al.* (2017). Mirrored changes in Antarctic ozone and stratospheric temperature in the late 20th versus early 21st centuries. *Journal of Geophysical Research: Atmospheres* 122(16), 8940-8950. https://doi.org/10.1002/2017jd026719.

Springmann, M., Mason-D'Croz, D., Robinson, S., Garnett, T., Godfray, H.C.J., Gollin, D. *et al.* (2016). Global and regional health effects of future food production under climate change: A modelling study. *The Lancet* 387(10031), 1937-1946. https://doi.org/10.1016/S0140-6736(15)01156-3.

Stieb, D.M., Chen, L., Eshoul, M. and Judek, S. (2012). Ambient air pollution, birth weight and preterm birth: a systematic review and meta-analysis. *Environmental Research* 117, 100-111. https://doi.org/10.1016/j.envres.2012.05.007.

Sumpter, C. and Chandramohan, D. (2013). Systematic review and meta-analysis of the associations between indoor air pollution and tuberculosis. *Tropical Medicine & International Health* 18(1), 101-108. https://doi.org/10.1111/tmi.12013.

Sundseth, K., Pacyna, J.M., Pacyna, E.G., Pirrone, N. and Thorne, R.J. (2017). Global sources and pathways of mercury in the context of human health. *International Journal of Environmental Research and Public Health* 14(1). https://doi.org/10.3390/ijerph14010105.

Tabachnick, W.J. (2010). Challenges in predicting climate and environmental effects on vector-borne disease episystems in a changing world. *The Journal of experimental biology* 213(6), 946-954. https://doi.org/10.1242/jeb.037564.

Task Force on Hemispheric Transport of Air Pollution (2013). Answers to policy-relevant science questions.In *Hemispheric Transport of Air Pollution 2010: Part D - Answers to Policy-Relevant Science Questions.* New York: United Nations. 1-42. https://www.un-ilibrary.org/hemispheric-transport-of-air-pollution-2010_2edceeff-en.pdf?itemId=%2Fcontent%2Fpublication%2F2edceeff-en&mimeType=pdf

Trenberth, K.E. (2011). Changes in precipitation with climate change. *Climate Research* 47(1), 123-138. https://doi.org/10.3354/cr00953.

Unger, N. (2014). Human land-use-driven reduction of forest volatiles cools global climate. *Nature Climate Change* 4(10), 907-910. https://doi.org/10.1038/nclimate2347.

United Nations (1992). United Nations Framework Convention on Climate Change. New York, NY http://unfccc.int/files/essential_background/background_publications_htmlpdf/application/pdf/conveng.pdf

United Nations (2016a). Amendment to the Montreal Protocol on Substances that Deplete the Ozone Layer, Kigali, 15 October 2016. United Nations, New York, NY https://treaties.un.org/doc/Publication/CN/2016/CN.872.2016-Eng.pdf

United Nations (2016b). *The World's Cities in 2016 – Data Booklet.* New York, NY. http://www.un.org/en/development/desa/population/publications/pdf/urbanization/the_worlds_cities_in_2016_data_booklet.pdf.

United Nations Conference on Trade and Development (1997). *Review of Maritime Transport.* Geneva. http://unctad.org/en/Docs/rmt1997_en.pdf.

United Nations Conference on Trade and Development (2017). *Review of Maritime Transport.* Geneva. http://unctad.org/en/PublicationsLibrary/rmt2017_en.pdf.

United Nations Economic Commission for Europe and United Nations Environment Programme (2017). Used vehicles: A global overview. Background Paper. *Ensuring Better Air Quality and Reduced Climate Emissions Through Cleaner Used Vehicles.* Geneva, 20-24 February 2017. United Nations Economic Commission for Europe https://www.unece.org/fileadmin/DAM/trans/doc/2017/itc/UNEP-ITC_Background_Paper-Used_Vehicle_Global_Overview.pdf

United Nations Environment Assembly of the United Nations Environment Programme (2014). 1/7. Strengthening the Role of the United Nations Environment Programme in Promoting Air Quality. In *Proceedings of the United Nations Environment Assembly of the United Nations Environment Programme at its First Session. UNEP/EA.1/10.* Nairobi. 38-39. http://undocs.org/UNEP/EA.1/10

United Nations Environment Assembly of the United Nations Environment Programme (2017). *3/8. Preventing and Reducing Air Pollution to Improve Air Quality Globally. UNEP/EA.3/Res.8.* https://papersmart.unon.org/resolution/uploads/k1800222.english.pdf.

United Nations Environment Programme (2013a). *Global Mercury Assessment 2013: Sources, Emissions, Releases and Environmental Transport.* Nairobi: United Nations Environment Programme. http://wedocs.unep.org/bitstream/handle/20.500.11822/7984/-Global%20Mercury%20Assessment-201367.pdf?sequence=3&isAllowed=y.

United Nations Environment Programme (2013b). *Minamata Convention on Mercury, Text and Annexes.* Nairobi: United Nations Environment Programme. http://mercuryconvention.org/Portals/11/documents/Booklets/Minamata%20Convention%20on%20Mercury_booklet_English.pdf.

United Nations Environment Programme (2014a). *Global Monitoring Plan for Persistent Organic Pollutants under the Stockholm Convention Article 16 on Effectiveness Evaluation: Second Regional Monitoring Report of the Central, Eastern European and Central Asian Region.* http://chm.pops.int/portals/0/download.aspx?d=UNEP-POPS-GMP-RMR-CEE-2015.English.pdf.

United Nations Environment Programme (2014b). *Global Monitoring Plan for Persistent Organic Pollutants.* Nairobi: United Nations Environment Programme. http://chm.pops.int/portals/0/download.aspx?d=UNEP-POPS-GMP-RMR-GRULAC-2015.English.pdf.

United Nations Environment Programme (2015a). *Global Monitoring Plan for Persistent Organic Pollutants under the Stockholm Convention Article 16 on Effectiveness Evaluation - 2nd Regional Monitoring Report Western Europe and Others Group (WEOG) Region.* Nairobi: United Nations Environment Programme. http://wedocs.unep.org/bitstream/handle/20.500.11822/19330/WEOG_Report_FINAL_2015_03_31.pdf?sequence=1&isAllowed=y.

United Nations Environment Programme (2015b). *Global Monitoring Plan for Persistent Organic Pollutants. Under the Stockholm Convention Article 16 on Effectiveness Evaluation. Second Regional Monitoring Report Asia-Pacific Region.* Nairobi: United Nations Environment Programme. https://www.informea.org/sites/default/files/imported-documents/UNEP-POPS-GMP-RMR-ASIAPACIFIC-ANNEX-2015.English.pdf.

United Nations Environment Programme (2016). UNEP Air Quality Monitoring System. Nairobi http://pre-uneplive.unep.org/media/docs/news_ticker/Air_Quality_Leaflet_Letter_size.pdf

United Nations Environment Programme (2017a). *The new POPs under the Stockholm Convention.* [http://chm.pops.int/TheConvention/ThePOPs/TheNewPOPs/tabid/2511/Default.aspx (Accessed: 30 June 2017.

United Nations Environment Programme (2017b). *Ozone secretariat data access centre.* [http://ozone.unep.org/en/data-reporting/data-centre.

United Nations Environment Programme (2017c). *The Emissions Gap Report 2017.* Nairobi: United Nations Environment Programme. http://www.unenvironment.org/resources/emissions-gap-report.

United Nations Environment Programme and Arctic Monitoring and Assessment Programme (2011). *Climate Change and POPs: Predicting the Impacts. Report of the UNEP/AMAP Expert Group.* http://www.amap.no/documents/doc/climate-change-and-pops-predicting-the-impacts/753.

United Nations Environment Programme and Arctic Monitoring and Assessment Programme (2018). *Global Mercury Assessment 2018 - Draft Technical Background Document.* https://www.unenvironment.org/explore-topics/chemicals-waste/what-we-do/mercury/global-mercury-assessment.

United Nations Environment Programme and International Solid Waste Association (2015). *Global Waste Management Outlook.* Nairobi. http://web.unep.org/ietc/what-we-do/global-waste-management-outlook-gwmo.

United Nations Environment Programme and United Nations Economic Commission for Europe (2016). *GEO-6 Assessment for the Pan-European Region.* Nairobi. http://wedocs.unep.org/bitstream/handle/20.500.11822/7735/unep_geo_regional_assessments_europe_16-07513_hires.pdf?sequence=1&isAllowed=y.

United Nations Environment Programme, World Meteorological Organization and United Nations Convention to Combat Desertification (2016). *Global Assessment of Sand and Dust Storms.* Nairobi: United Nations Environment Programme. https://wedocs.unep.org/bitstream/handle/20.500.11822/7681/Global_Assessment_of_sand_and_dust_storms_2016.pdf?sequence=1&isAllowed=y.

United Nations Framework Convention on Climate Change (2016). *Decision 1/CP.21 Adoption of the Paris Agreement* Bonn: United Nations Framework Convention on Climate Change. https://unfccc.int/resource/docs/2015/cop21/eng/10a01.pdf.

United States Environmental Protection Agency (2011). *The Benefits and Costs of the Clean Air Act from 1990 to 2020: Final Report, Rev. A.* US Environmental Protection Agency, Office of Air and Radiation. https://www.epa.gov/sites/production/files/2015-07/documents/fullreport_rev_a.pdf.

United States Environmental Protection Agency (2017). *Inventory of U.S. Greenhouse Gas Emissions and Sinks: 1990-2015.* Washington, D.C: United States Environmental Protection Agency. https://www.epa.gov/sites/production/files/2017-02/documents/2017_complete_report.pdf.

United States Environmental Protection Agency (2018a). *Artisanal and small-scale gold mining without mercury.* [United States Environmental Protection Agency https://www.epa.gov/air-sensor-toolbox (Accessed: 8 October 2018).

United States Environmental Protection Agency (2018b). *Air sensor toolbox for citizen scientists, researchers and developers.* [United States Environmental Protection Agency https://www.epa.gov/air-sensor-toolbox.

van Dijk, A., Slaper, H., den Outer, P.N., Morgenstern, O., Braesicke, P., Pyle, J.A. *et al.* (2013). Skin cancer risks avoided by the Montreal Protocol—worldwide modeling integrating coupled climate-chemistry models with a risk model for UV. *Photochemistry and Photobiology* 89(1), 234-246. https://doi.org/10.1111/j.1751-1097.2012.01223.x.

Van Dingenen, R., Dentener, F.J., Raes, F., Krol, M.C., Emberson, L. and Cofala, J. (2009). The global impact of ozone on agricultural crop yields under current and future air quality legislation. *Atmospheric Environment* 43(3), 604-618. https://doi.org/10.1016/j.atmosenv.2008.10.033.

van Marle, M.J.E., Kloster, S., Magi, B.I., Marlon, J.R., Daniau, A.L., Field, R.D. *et al.* (2017). Historic global biomass burning emissions based on merging satellite observations with proxies and fire models (1750-2015). *Geoscientific Model Development* 2017, 1-56. https://doi.org/10.5194/gmd-2017-32.

Vecchiato, M., Argiriadis, E., Zambon, S., Barbante, C., Toscano, G., Gambaro, A. *et al.* (2015). Persistent Organic Pollutants (POPs) in Antarctica: Occurrence in continental and coastal surface snow. *Microchemical Journal* 119, 75-82. https://doi.org/10.1016/j.microc.2014.10.010.

Vet, R., Artz, R.S., Carou, S., Shaw, M., Ro, C.-U., Aas, W. *et al.* (2014). A global assessment of precipitation chemistry and deposition of sulfur, nitrogen, sea salt, base cations, organic acids, acidity and pH, and phosphorus. *Atmospheric Environment* 93, 3-100. https://doi.org/10.1016/j.atmosenv.2013.10.060.

Vidal, J. (2016). 'Clouds of filth envelop Asian cities: 'you can't escape''. *The Guardian* 22 November 2017 https://www.theguardian.com/global-development/2016/nov/22/cloud-filth-envelope-asian-cities-urban-smog-air-pollution-india-china.

Villeneuve, P.J., Weichenthal, S.A., Crouse, D., Miller, A.B., To, T., Martin, R.V. *et al.* (2015). Long-term exposure to fine particulate matter air pollution and mortality among Canadian women. *Epidemiology* 26(4), 536-545. https://doi.org/10.1097/ede.0000000000000294.

Vinken, G.C.M., Boersma, K.F., Maasakkers, J.D., Adon, M. and Martin, R.V. (2014). Worldwide biogenic soil NOx emissions inferred from OMI NO2 observations. *Atmospheric Chemistry and Physics* 14(18), 10363-10381. https://doi.org/10.5194/acp-14-10363-2014.

Vizcaino, E., Grimalt, J.O., Fernandez-Somoano, A. and Tardon, A. (2014). Transport of persistent organic pollutants across the human placenta. *Environment International* 65, 107-115. https://doi.org/10.1016/j.envint.2014.01.004.

Waste Atlas Partnership (2014). *Waste Atlas: The World's 50 Biggest Dumpsites: 2014 Report*. http://www.atlas.d-waste.com/Documents/Waste-Atlas-report-2014-webEdition.pdf.

Watt, J., Tidblad, J., Kucera, V. and Hamilton, R. (eds.) (2009). *The Effects of Air Pollution on Cultural Heritage*. New York, NY: Springer. http://www.springer.com/gp/book/9780387848921.

Watts, N., Amann, M., Ayeb-Karlsson, S., Belesova, K., Bouley, T., Boykoff, M. *et al.* (2017). The Lancet Countdown on health and climate change: from 25 years of inaction to a global transformation for public health. *The Lancet*. https://doi.org/10.1016/S0140-6736(17)32464-9.

Wijedasa, L.S., Jauhiainen, J., Kononen, M., Lampela, M., Vasander, H., Leblanc, M.C. *et al.* (2017). Denial of long-term issues with agriculture on tropical peatlands will have devastating consequences. *Global Change Biology* 23(3), 977-982. https://doi.org/10.1111/gcb.13516.

Women in Europe for a Common Future and Women International for a Common Future (2016). *Women And Chemicals - The Impact Of Hazardous Chemicals On Women*. Women in Europe for a Common Future (WECF) and Women International for a Common Future (WICF). http://www.wecf.eu/download/2016/March/WomenAndChemicals_PublicationIWD2016.pdf.

Wooster, M.J., Perry, G.L.W. and Zoumas, A. (2012). Fire, drought and El Niño relationships on Borneo (Southeast Asia) in the pre-MODIS era (1980–2000). *Biogeosciences* 9, 317-340. https://doi.org/10.5194/bg-9-317-2012.

World Bank and Institute for Health Metrics and Evaluation (2016). *The Cost of Air Pollution: Strengthening the Economic Case for Action*. Washington, DC.: World Bank. http://documents.worldbank.org/curated/en/781521473177013155/pdf/108141-REVISED-Cost-of-PollutionWebCORRECTEDfile.pdf.

World Health Assembly of the World Health Organization (2015). *Health and the Environment: Addressing the Health Impact of Air Pollution. WHA68.8*. http://apps.who.int/iris/bitstream/handle/10665/253237/A68_R8-en.pdf?sequence=1&isAllowed=y.

World Health Organization (2004). *Global Health Risks: Mortality and Burden of Disease Attributable To Selected Major Risks*. Geneva: World Health Organization. http://apps.who.int/iris/bitstream/handle/10665/44203/9789241563871_eng.pdf?sequence=1&isAllowed=y.

World Health Organization (2006). *Air Quality Guidelines: Global Update 2005*. Copenhagen: World Health Organization. http://www.euro.who.int/__data/assets/pdf_file/0005/78638/E90038.pdf?ua=1.

World Health Organization (2007). *Indoor Air Pollution: National Burden of Disease Estimates*. Geneva: World Health Organization. http://www.who.int/airpollution/publications/nationalburden/en/.

World Health Organization (2013). *Review of Evidence on Health Aspects of Air Pollution – REVIHAAP Project: Technical Report*. Copenhagen: World Health Organization http://www.euro.who.int/__data/assets/pdf_file/0004/193108/REVIHAAP-Final-technical-report-final-version.pdf?ua=1.

World Health Organization (2014). *Gender, Climate Change and Health*. Geneva: World Health Organization. http://www.who.int/globalchange/publications/reports/gender_climate_change/en/.

World Health Organization (2015). *Economic Cost Of The Health Impact Of Air Pollution In Europe: Clean Air, Health And Wealth*. Copenhagen: World Health Organization. http://www.euro.who.int/__data/assets/pdf_file/0004/276772/Economic-cost-health-impact-air-pollution-en.pdf.

World Health Organization (2016a). *Ambient Air Pollution: A Global Assessment of Exposure and Burden of Disease*. Geneva. http://www.who.int/phe/publications/air-pollution-global-assessment/en/.

World Health Organization (2016b). *Burning Opportunity: Clean Household Energy for Health, Sustainable Development and Well-Being of Women and Children*. Geneva. http://apps.who.int/iris/bitstream/handle/10665/204717/9789241565233_eng.pdf?sequence=1&isAllowed=y.

World Health Organization (2018). *Health Topics: Air Pollution*. [http://www.who.int/airpollution/en/ (Accessed: 4 October 2018).

World Meteorological Organization (2014). *Scientific Assessment of Ozone Depletion: 2014*. Geneva. http://www.wmo.int/pages/prog/arep/gaw/ozone_2014/documents/Full_report_2014_Ozone_Assessment.pdf.

World Meteorological Organization (2017a). *WMO Greenhouse Gas Bulletin: The State of Greenhouse Gases in the Atmosphere Based on Global Observations through 2016*. Geneva: World Meteorological Organization. https://reliefweb.int/sites/reliefweb.int/files/resources/GHG_Bulletin_12_EN_web_JN161640.pdf.

World Meteorological Organization (2017b). *Sand and Dust Storm Warning Advisory and Assessment System*. WMO Airborne Dust BulletinWorld Meteorological Organization. https://library.wmo.int/doc_num.php?explnum_id=3416.

World Meteorological Organization (2017c). *WMO Statement on the State of the Global Climate*. Geneva: World Meteorological Organization. https://library.wmo.int/doc_num.php?explnum_id=4453.

Xia, Y., Guan, D., Jiang, X., Peng, L., Schroeder, H. and Zhang, Q. (2016). Assessment of socioeconomic costs to China's air pollution. *Atmospheric Environment* 139, 147-156. https://doi.org/10.1016/j.atmosenv.2016.05.036.

Xu, Y. and Ramanathan, V. (2017). Well below 2 °C: Mitigation strategies for avoiding dangerous to catastrophic climate changes. *Proceedings of the National Acacemy of Sciences* 114(39), 10315-10323. https://doi.org/10.1073/pnas.1618481114.

York, A. (2010). Marine biogeochemical cycles in a changing world. *Nature Reviews Microbiology* 16, 259. https://doi.org/10.1038/nrmicro.2018.40.

Zhang, Q., Jiang, X., Tong, D., Davis, S.J., Zhao, H., Geng, G. *et al.* (2017). Transboundary health impacts of transported global air pollution and international trade. *Nature* 543(7647), 705-709. https://doi.org/10.1038/nature21712.

Zhang, Y., Cooper, O.R., Gaudel, A., Thompson, A.M., Nedelec, P., Ogino, S.-Y. *et al.* (2016). Tropospheric ozone change from 1980 to 2010 dominated by equatorward redistribution of emissions. *Nature Geoscience* 9(12), 875-879. https://doi.org/10.1038/ngeo2827.

Zhao, Y., Zhang, L., Chen, Y., Liu, X., Xu, W., Pan, Y. *et al.* (2017). Atmospheric nitrogen deposition to China: A model analysis on nitrogen budget and critical load exceedance. *Atmospheric Environment* 153, 32-40. https://doi.org/10.1016/j.atmosenv.2017.01.018.

Zheng, B., Tong, D., Li, M., Liu, F., Hong, C., Geng, G. *et al.* (2018). Trends in China's anthropogenic emissions since 2010 as the consequence of clean air actions. *Atmospheric Chemistry and Physics* 18(19), 14095-14111. https://doi.org/10.5194/acp-18-14095-2018.

Zivin, J.G. and Neidell, M. (2018). Air pollution's hidden impacts. *Science* 359(6371), 39-40. https://doi.org/10.1126/science.aap7711.

Zou, Y., Wang, Y., Zhang, Y. and Koo, J.H. (2017). Arctic sea ice, Eurasia snow, and extreme winter haze in China. *Science Advances* 3(3), e1602751. https://doi.org/10.1126/sciadv.1602751.

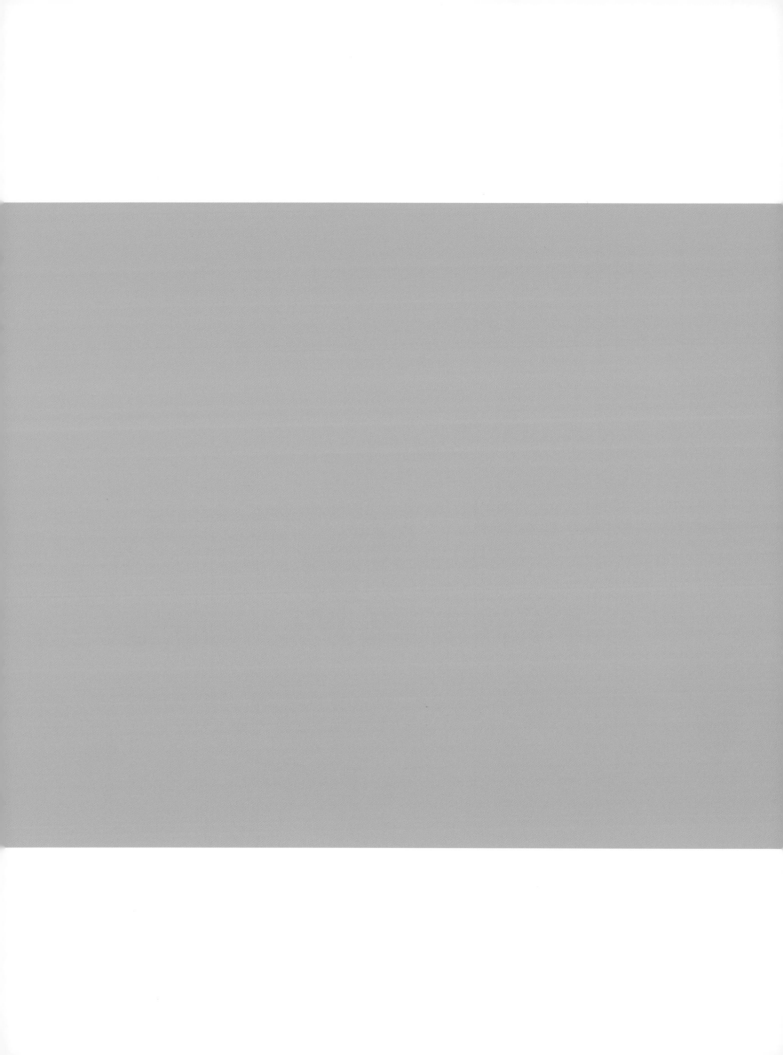

Biodiversity

Coordinating Lead Authors: Peter Stoett (University of Ontario Institute of Technology), Jonathan Davies (University of British Columbia)

Lead Authors: Dolors Armenteras (Universidad Nacional de Colombia), Jeremy Hills (University of the South Pacific), Louise McRae (Zoological Society of London), Carol Zastavniouk (Golder Associates)

Contributing Authors: Rob Bailey (Chatham House – The Royal Institute of International Affairs), Colin Butler (University of Canberra), Irene Dankelman (Radboud University), Keisha Garcia (University of the West Indies), Linda Godfrey (Council for Scientific and Industrial Research, South Africa), Andrei Kirilenko (University of Florida), Peter Lemke (Alfred Wegener Institute), Daniela Liggett (University of Canterbury), Gavin Mudd (RMIT University), Joni Seager (Bentley University), Caradee Y. Wright (Medical Research Council of South Africa), Caroline Zickgraf (Université de Liège)

Executive summary

Biodiversity is in crisis. There is *well-established* evidence indicating an irrevocable and continuing decline of genetic and species diversity, and degradation of ecosystems at local and global scales. Scientists are increasingly concerned that, if anthropogenic pressures on Biodiversity continue unabated, we risk precipitating a sixth mass extinction event in Earth history, with profound impacts on human health and equity. {6.1}

Biodiversity provides many valuable goods and services – nature's contributions to people (*well established*). Biodiversity helps regulate climate through carbon storage and control of local rainfall, filters air and water, and mitigates the impact of natural disasters such as landslides and coastal storms. Direct benefits include timber from forests, fish from oceans and freshwater systems, crops and medicines from plants, cultural identity, and the health benefits gained from access to nature. {6.1}

Biodiversity loss has consequences for human health and equity (*well established*). Biodiversity contributes positively to human health and well-being. The livelihoods of more than 70 per cent of the world's population living in poverty depend on natural resources to some extent and over 80 per cent of global biodiversity is found in the traditional territories of indigenous peoples. Depleting this natural capital will therefore disproportionately affect the people least able to offset losses and reduce options for future generations. {6.1}

The loss of biodiversity reduces ecosystem resilience and increases vulnerability to threats including negative impacts of climate change (*well established*). At local scales, it is *likely* that ecosystems with greater biodiversity are more productive and more stable through time. {6.5.4, 6.5.6}

The critical pressures on Biodiversity are well recognized (*well established*). Biodiversity is being eroded by land-use change, direct exploitation, climate change, pollution and invasive alien species. While habitat loss and transformation is likely the most significant present pressure, climate change may be the most significant future pressure. {6.3.1, 6.3.2, 6.3.3, 6.3.4, 6.3.5}

Pressures often overlap and there are positive feedback loops between many of them (*well established*). Habitat changes may increase exposure to pollutants, pests, exotic pathogens and emerging infectious diseases harmful to humans, livestock and wildlife, and exacerbate human-wildlife conflicts. Forests are experiencing alteration due to multiple land-use changes such as logging, mining, road building and agricultural expansion; the resulting habitat fragmentation and loss of biodiversity can lower forest resilience to climate change impacts and the introduction of invasive species. {6.3.1}

Newly recognized and aggravating factors add to pressures on biodiversity (*well established*). Energy production, resource extraction, wildlife trade and poaching, chemical waste and plastics in the marine environment are exacerbating factors that contribute to biodiversity decline. {6.3.1, 6.3.3, 6.3.4}

Genetic diversity is the vital raw material allowing adaptation (*well established*). The decline in the population size of many species represents a loss in genetic diversity. Genetic diversity of crops, crop wild relatives and livestock provides resilience of agricultural systems to changing environments. The ongoing long-term loss of crop and livestock genetic diversity is a threat to food security. {6.4.1}

There is no slowing in the rate of species population decline globally (*well established*). The increase in species extinction risks through time is *well established*, and there is no slowing in the rate of population declines globally. Freshwater species have the highest rates of population declines, whereas amphibians, reef-forming corals and cycads are the taxa with the highest proportion of species currently considered at risk of extinction. There is less data on invertebrate groups, but recent evidence indicates large declines in local abundance. The loss of invertebrate pollinators has been highlighted as a growing problem, with major consequences for agricultural production, ecosystem functioning and human well-being. {6.4.2}

There is no global overview of ecosystem health (*well established*)., The status of many habitat types is very likely in decline. While global monitoring is challenging, across terrestrial habitats 10 out of 14 have seen a decrease in vegetation productivity, and just under half of all terrestrial ecoregions are classified as having an unfavourable status. Natural wetland areas and marine habitats, such as deep-sea ecosystems and coral reefs, are highlighted as of particular concern globally. {6.4.3}

Biodiversity loss is being experienced across all Earth's major biomes (*well established*). In the oceans, overexploitation of fish stocks is leading to fisheries collapse, warming is destroying coral reefs, and habitat destruction of coastal systems, such as mangrove forests, exposes communities to greater risks from erosion and extreme weather events. Marine plastic pollution is a major and growing threat to biodiversity. In freshwater systems, agricultural and chemical pollution, including increased nitrogen input, results in toxic algal blooms and a decline in drinking-water quality; invasive species are spreading through waterways; and freshwater species are declining at a faster rate than those in any other biome. In the terrestrial environment, rising temperatures are converting grasslands into deserts, and unsustainable irrigation has turned drylands into inhospitable, toxic landscapes unsuitable for wildlife or agriculture. Mountain ecosystems and polar regions are especially vulnerable to climate change, and extinctions may be likely for species at the upper limits of their thermal ranges and those dependent on sea ice. Tropical forests represent some of the most biodiverse terrestrial ecosystems, yet deforestation and forest degradation continue in many regions, often in response to demands for wood, fibre, food and fuel products such as palm oil, as well as external drivers. {6.5.1, 6.5.2, 6.5.3, 6.5.4, 6.5.5, 6.5.6, 6.5.7, 6.5.8}

A range of national and international instruments work to conserve biodiversity (*well established*). These include National Biodiversity Strategies and Actions Plans (NBSAPS)

under the Convention on Biological Diversity (CBD), the Strategic Plan for Biodiversity 2011-2020 (encompassing the Aichi targets), the Cartagena Protocol on Biosafety, the Nagoya Protocol, and the Intergovernmental Platform for Biodiversity and Ecosystem Services (IPBES). {6.6.1, 6.6.2}

Species and ecosystems are most effectively safeguarded through the conservation of natural habitats (*well established*). There has been significant progress in expanding the global network of protected areas, but the total area under protection remains insufficient, and habitats within protected areas are often degraded. {6.6.3}

Ex-situ conservation of biological material can contribute to conserving genetic diversity (*well established*). Seed banks and gene banks, aided by the use of these new genomic tools, have contributed to the conservation of the genetic diversity of crops and their wild relatives. Advances in technology allow cheaper and faster genome sequencing, however, genetic data for most wild species are still lacking. {6.4.1}

At a local scale indigenous people and local communities (IPLC) play a key role in protecting biodiversity (*well established*). IPLCs can offer bottom-up, self-driven, cost-effective and innovative solutions, and have potential to be scaled up and inform national and international practice. Such solutions provide a practical governance approach as an alternative to top-down policy-setting. This is essential to achieve many of the Sustainable Development Goals. {Box 6.6, 6.6.3}

Biodiversity policy responses are visible and operating at international, national and local levels, but they have been insufficient to slow or reverse the decline in global biodiversity (*well established*). There is an urgent need to bolster current policy responses. There are additional opportunities to maintain biodiversity and the contributions of nature through addressing distribution, access and governance, and by recognizing the role of IPLCs in biodiversity conservation. {6.6.3, 6.7}

The cost of inaction is large and escalating (*well established*)**.** The full cost of inaction is rarely quantified; however, failure to act now will impose much higher costs in the future as shown by many examples, such as the spread of invasive species, and extinctions have immeasurable costs for future generations. {6.3.2}

6.1 Introduction

Biodiversity – the "variability among living organisms from all sources including … diversity within species, between species and of ecosystems" (United Nations 1992, Article 2) – helps regulate climate through carbon sequestration and control of local rainfall, filters air and water, and mitigates the impact of natural disasters such as landslides and coastal storms. Direct benefits include food and fibres from natural vegetation, wood and non-wood products from forests, fish from oceans and freshwater systems, pollination of crops, medicines from plants, and psychological health (Clark *et al.* 2014; Harrison *et al.* 2014; World Health Organization [WHO] and Secretariat of the Convention on Biological Diversity [SCBD] 2015, p. 200; Pascual *et al.* 2017). Never before have we known so much about the biodiversity that enables ecosystems to function (Cardinale *et al.* 2012), yet biodiversity loss and habitat decline continues to accelerate, potentially beyond planetary boundaries (Tittensor *et al.* 2014; Steffen *et al.* 2015).

Current rates of species loss are estimated to be 1,000-fold greater than background rates (Pimm *et al.* 2014), sparking debate among scientists over whether we have already entered into a sixth mass extinction event (Barnosky *et al.* 2011; Ceballos, Ehrlich and Dirzo 2017). For many species, populations are in decline globally (Ceballos, Ehrlich and Dirzo 2017; McRae, Deinet and Freeman 2017), and genetic diversity – vital for future adaptation to global change – is eroding (Food and Agriculture Organization of the United Nations [FAO] 2015a). Natural communities of plants and animals are being reshaped through climate change and human-mediated movement of species (Pacifici *et al.* 2015); some displaced species are invasive, posing risks to human health, genetic diversity, and food and water security. These changes seem likely to reduce the efficiency by which ecosystems are able to capture essential resources, produce biomass, decompose and recycle nutrients (Cardinale *et al.* 2012), and decrease the resilience of ecosystems (MacDougall *et al.* 2013). The restoration and maintenance of biodiversity will enhance

Figure 6.1: Schematic from the Intergovernmental Science-Policy Platform on Biodiversity and Ecosystem Services describing the main elements and relationships linking nature, biodiversity and ecosystem services, human well-being and sustainable development. (In this diagram, anthropogenic drivers equate to the pressures as described in Section 6.3)

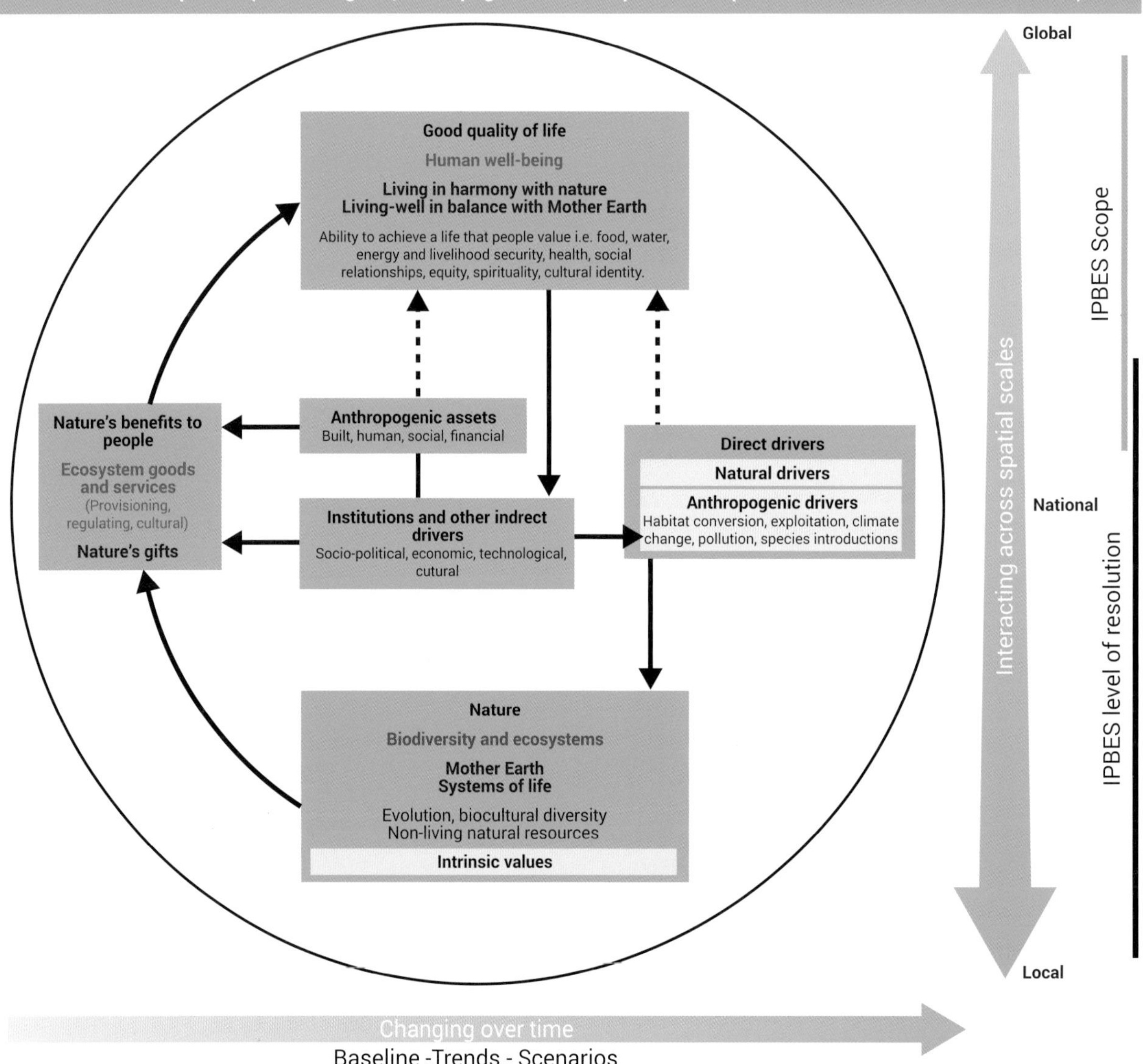

Source: IPBES (2013, p. 2).

adaptive potential, and help sustain nature's contributions to people's livelihoods, health and well-being (Intergovernmental Science-Policy Platform on Biodiversity and Ecosystem Services [IPBES] 2016). These critical services are frequently neglected as they largely bypass the market and there are no clear price signals for them (e.g. Foale *et al.* 2013; Seddon *et al.* 2016; Costanza *et al.* 2017). The loss of biodiversity is also a significant equity issue: the livelihoods of 70 per cent of people living in poverty rely to some extent on natural resources (Green Economy Coalition 2012, p. 4); 80 per cent of global biodiversity is found in the traditional territories of indigenous peoples (Sobrevila 2008, p. xii); and future generations will experience relatively impoverished lives if losses continue (Naeem *et al.* 2016).

6.2 Further assessments since the fifth Global Environmental Outlook (GEO-5)

GEO-5 (United Nations Environment Programme [UNEP] 2012) concluded that pressure on biodiversity continues to increase through habitat loss, degradation from agriculture and infrastructure development, overexploitation, pollution, invasive alien species and climate disruption, as well as interactions between these pressures, and that the state of global biodiversity is continuing to decline with substantial ongoing losses of populations, species and habitats. Since GEO-5, a midterm assessment of progress towards the Aichi Biodiversity Targets concluded that while progress has been made, this was insufficient to achieve them by 2020 (SCBD 2014). A series of GEO regional assessments (UNEP 2016a; UNEP 2016b; UNEP 2016c; UNEP 2016d; UNEP 2016e; UNEP 2016f), State of Biodiversity reports looking at regional progress towards the Aichi Biodiversity Targets (United Nations Environment Programme World Conservation Monitoring Centre [UNEP-WCMC] 2016a; UNEP-WCMC 2016b; UNEP-WCMC 2016c; UNEP-WCMC 2016d), and regional assessments on biodiversity and ecosystem services from the Intergovernmental Science-Policy Platform on Biodiversity and Ecosystem Services (IPBES) (https://www.ipbes.net/outcomes), have summarized evidence for declines in the state of biodiversity from different parts of the world while highlighting variation in responses to regional pressures. Among many other developments encouraged by these assessments, the gradual acceptance of the numerous benefits of biodiversity conservation for human health has been recognized (WHO and SCBD, 2015; see also **Box 6.1**).

© Shutterstock/Photobyt

Box 6.1: Biodiversity, disease and One Health

Several dimensions of global change, including shifts in urbanization, agricultural practices, land use and biodiversity, are altering ecological dynamics and in some cases facilitating human-animal contact that exacerbates the risk of zoonotic disease emergence and spread. Zoonotic diseases are transmissible from domestic or wild animals to humans through direct contact or through water, food and the environment (WHO and SCBD 2015; Centers for Disease Control and Prevention [CDC] 2017).

One Health is an approach that recognizes the opportunities and challenges related to these interconnections at the human-animal-ecosystem interface, and aims for optimal health outcomes for all; it is particularly relevant in the prevention and control of zoonoses, which account for more than 60 per cent of human infectious diseases (Karesh *et al.* 2012; WHO and SCBD 2015; CDC 2017).

The United States Agency for International Development (USAID) Emerging Pandemic Threats PREDICT project is expanding the detection and discovery of zoonotic viruses with pandemic potential through surveillance in 'hotspots' for emerging infectious diseases (EIDs), such as Ebola, to help track their circulation and understand factors driving their emergence (Kelly *et al.* 2017; Marlow 2017). Using the One Health approach, the project considers the behaviours, practices, and ecological and biological factors driving disease emergence, transmission and spread. Through enhanced understanding of EID risks, countries can be better equipped to prevent, prepare for and respond to the threat of an outbreak, ideally through taking preventive measures before major disease outbreaks. PREDICT partners include the University of California Davis One Health Institute, USAID, EcoHealth Alliance, Metabiota, Wildlife Conservation Society, and Smithsonian Institution.

6.3 Drivers

Drivers of environmental change – population demography, urbanization, economic development, technology and innovation, and climate change (see Chapter 2) – impose multiple negative impacts on biodiversity, leading to loss of genetic diversity, population declines that have pushed some species towards a heightened risk of extinction, and the reshaping of natural communities, with ramifications for the stability and functioning of ecosystems **(Figure 6.2)**. While most drivers are projected to increase, climate change is likely to become the dominant driver of biodiversity change in the next few decades (Leadley *et al.* 2014; Newbold *et al.* 2015). Ultimately, reducing pressures on biodiversity will require addressing these drivers of change.

6.4 Pressures

The main direct pressures on global biodiversity are habitat stress and land-use change, invasive species, pollution, unsustainable use/overexploitation and climate change (mainly as a consequence of higher temperatures, changes in precipitation patterns and increasing frequency and severity of extreme weather events and wildfires) (UNEP 2012). The spatial distribution and combination of these pressures varies across the globe **(Figure 6.3)** and affects species groups in different ways **(Figure 6.4),** although detailed data for invertebrates, which comprise most of the diversity of life, are lacking (Collen *et al.* 2012).

Figure 6.2: Interconnections between people, biodiversity, ecosystem health and provision of ecosystem services showing drivers and pressures

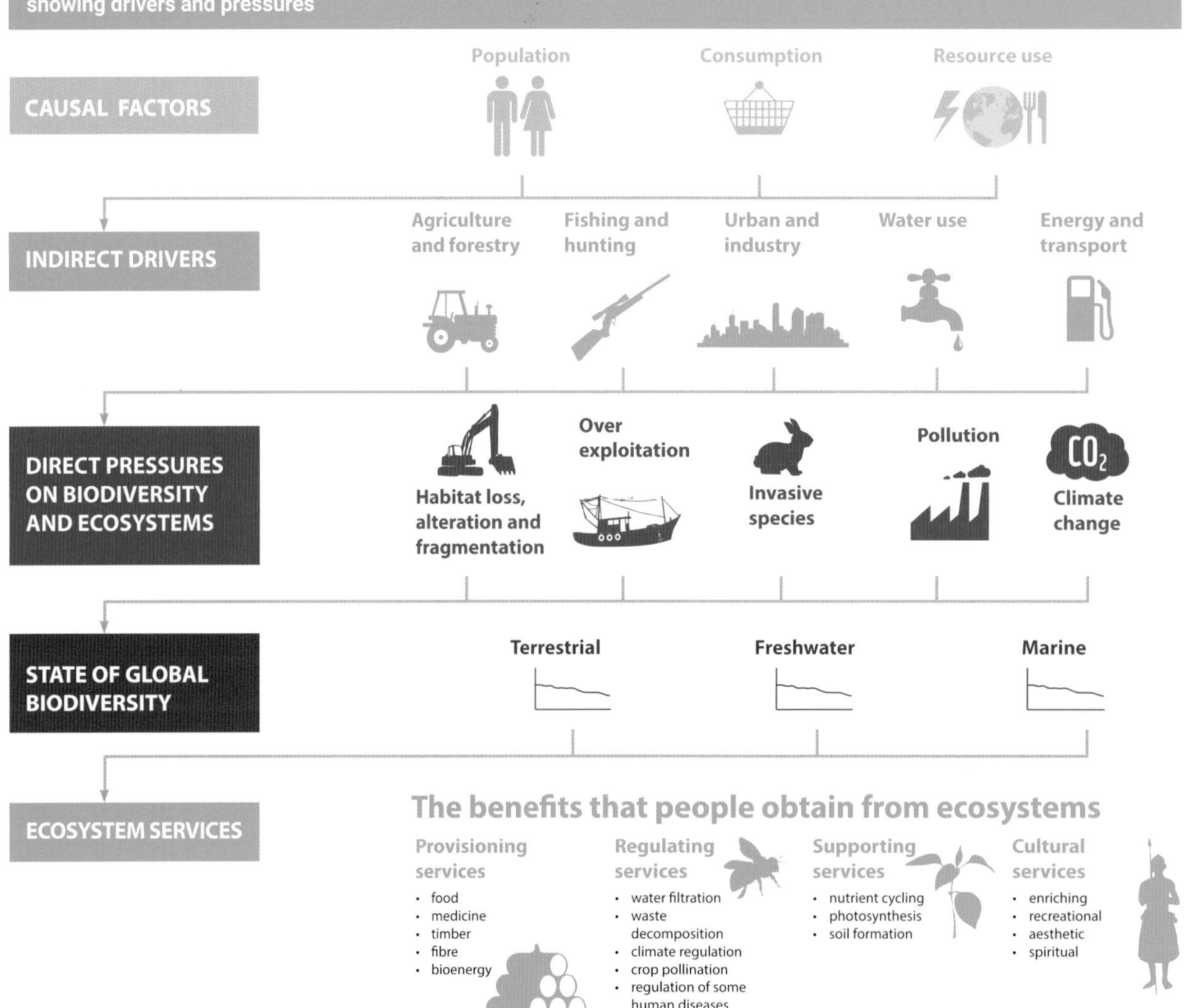

Source: World Wide Fund for Nature (WWF) *et al.* (2012).

Figure 6.3: Examples of global distribution of pressures on (a) threat intensity (H: high; L: low; M: medium; VH: very high; VL: very low) from terrestrial invasive alien species and (b) cumulative fisheries by-catch intensity for seabirds, sea mammals and sea turtles, by all gear types (gillnet, longline and trawl)

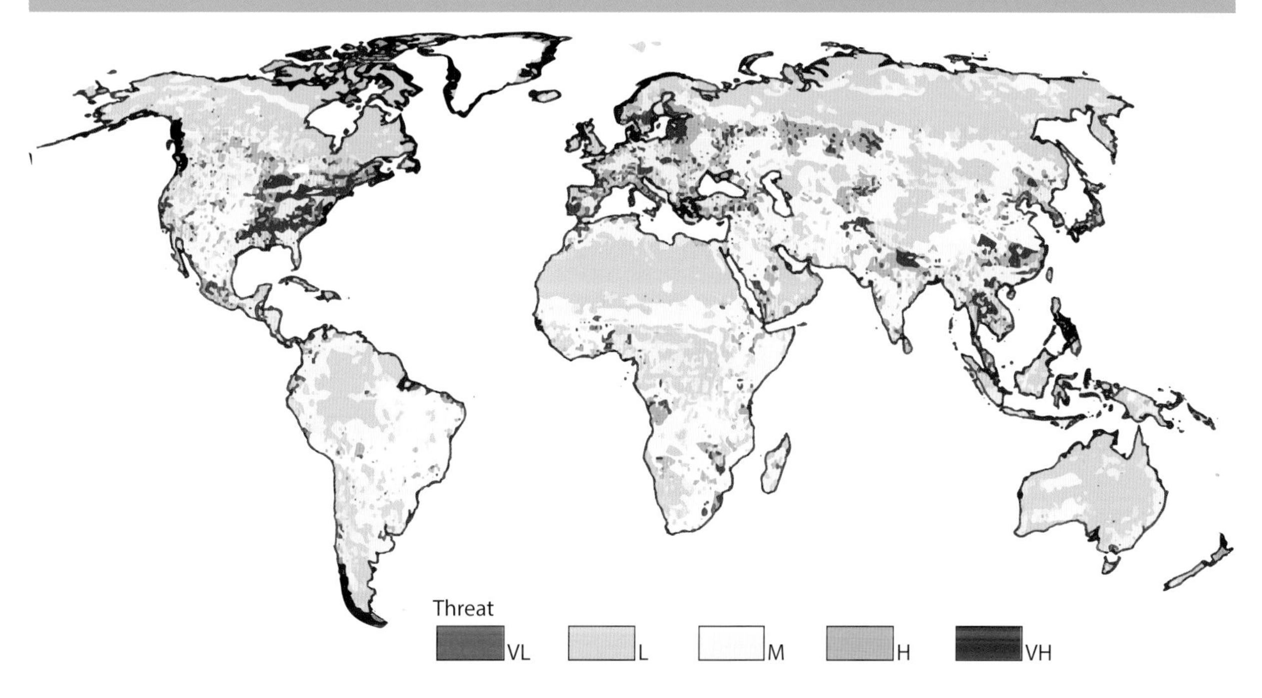

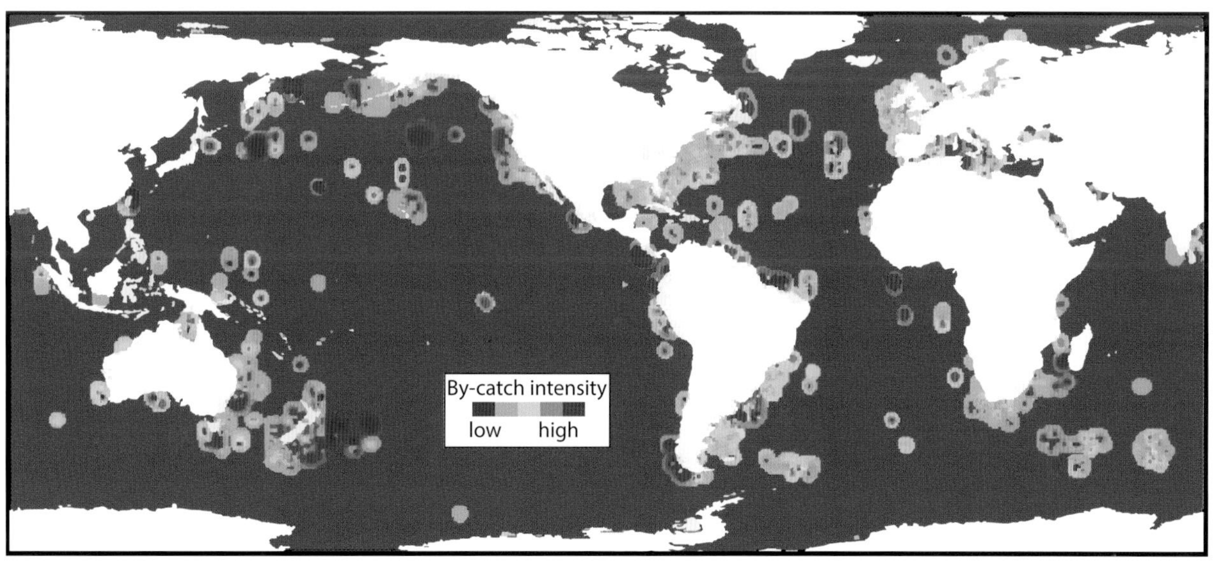

Sources (a) Early *et al.* 2016 (b) Lewison *et al.* (2014).

Figure 6.4: Percentage of threatened (critically endangered, endangered and vulnerable) and near threatened amphibian, bird and mammal species by major threat class

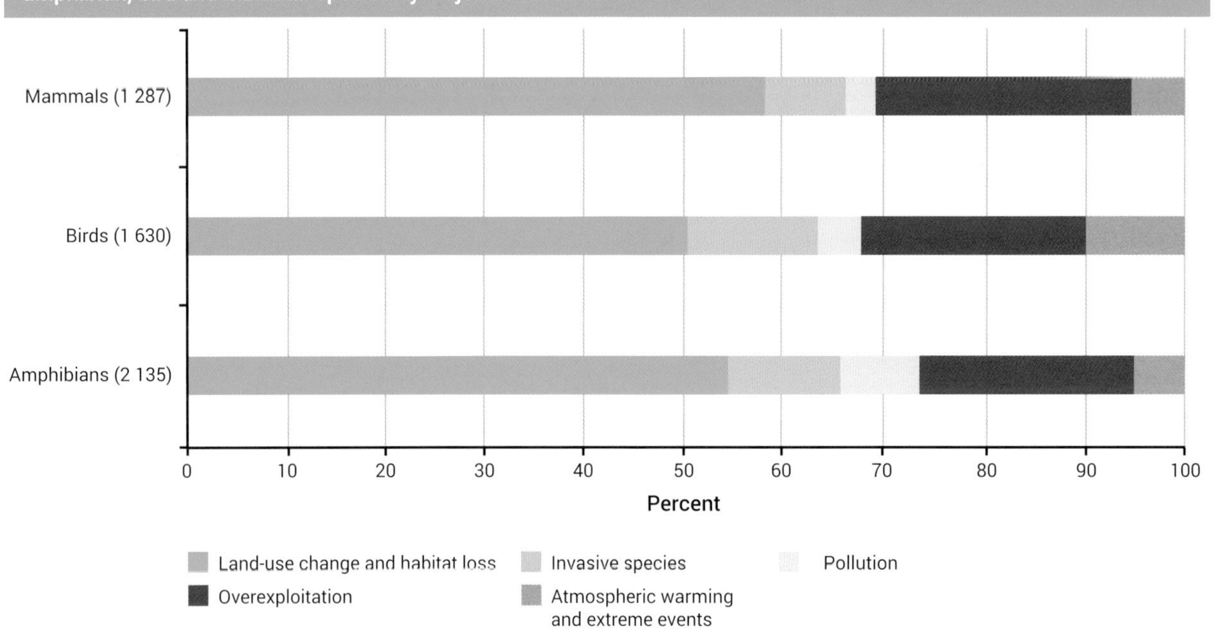

Number of threatened species in each taxonomic class in parentheses. Threat classes were aggregated as follows: 1 = Residential and commercial development, Agriculture and aquaculture, Energy production and mining, Transportation and service corridors, Human intrusions and disturbance, Natural system modifications; 2 = Invasive and other problematic species, genes and disease; 3 = Pollution; 4 = Biological resource use; 5 = Geological events, Climate change and severe weather.

Source: Maxwell *et al.* (2016) updated with International Union for Conservation of Nature [IUCN] (2018).

6.4.1 Land-use change and habitat loss

The global human footprint – infrastructure, land cover and human access into natural areas – is expanding **(Figure 6.5)** (Venter *et al.* 2016). Economic drivers and demographic pressures are the primary sources of accelerating land-use change. These drive agricultural expansion – the largest contributor to land-use change – for food, commodities, fodder and biofuels (Alexander *et al.* 2015), demand for extraction of mineral, metal and energy resources (Mudd and Jowitt 2017), urbanization, road building, land-take and deforestation, land degradation, desertification and habitat fragmentation.

Urban growth is a major driver of land-use change and habitat loss through deforestation. In developing countries, the establishment and expansion of urban areas (many of which lack adequate planning) and the growth of infrastructure can coincide with biodiversity hotspots (UNEP 2016d). Road construction facilitates the spread of invasive species, and allows for easier access into previously intact habitats, exposing them to threats from hunting and resource exploitation (Alamgir *et al.* 2017). Additional land-use practices, such as burning (or the suppression of natural fire) (Smith *et al.* 2016) and livestock grazing, impose further pressures on already degraded systems (Royal Botanic Gardens Kew2010). The marine environment is equally affected and heavily impacted by commercial fishing practices, such as bottom trawling, coastal development and dredging (Ocean Health Index 2017) (see Chapter 7). International trade can export threats to biodiversity, resulting from demand in developed countries, to developing countries (Lenzen *et al.* 2012). Many of the causes of habitat destruction also contribute to human population pressure and movement, which further compound threats to biodiversity (Black *et al.* 2011) (see Chapter 2).

Pressure from agricultural land use is widely expected to increase (Kehoe *et al.* 2017). Global food production is forecast to rise by between 60 and 100 per cent by 2050 as a result of population growth and economic development, with an accompanying minimum net increase in land under crop production of 70 million ha (Tilman *et al.* 2011; Alexandratos and Bruinsma 2012) (see Chapter 8). Large-scale industrial agriculture has many unfavourable environmental and social effects, such as land degradation, albedo changes, increase in methane emissions and loss of carbon sequestration capacities (Laurance, Sayer and Cassman2014; Dangal *et al.* 2017; Houspanossian *et al.* 2017). Agricultural intensification can reduce pressure on non-agricultural lands (Phalan *et al.* 2016), but may have detrimental impacts on wild plant and animal species that cohabit within diverse agroecosystems (Emmerson *et al.* 2016).

Rapid development-induced impacts result from the construction of dams, mines and other hard infrastructure developments, including those associated with energy production (Butt *et al.* 2013).

Climate warming and increasing frequency of extreme weather events contribute to habitat loss and degradation (see Chapter 2). Warming seas are reducing sea ice extent (critical hunting habitat for polar bears, seals and fishing birds) (Intergovernmental Panel on Climate Change [IPCC] 2014, p. 80) and, in conjunction with elevated atmospheric CO_2, acidifying ocean habitats (Hoegh-Guldberg *et al.* 2017). Extreme weather events, such as flooding, drought and fire, can accelerate the degradation of already vulnerable habitats (IPCC 2014, p. 294).

Land-use change, which may impact both aquatic and terrestrial environments, can result in:

- exposure to pollutants, exotic pathogens and emerging infectious diseases harmful to humans, livestock and wildlife (WHO and SCBD 2015, pp. 1-19);
- increased human conflict (Ghazi, Muniruzzaman and Singh 2016, p. ii);
- loss of habitat for wild species and the ecosystem services they provide, such as pollinators and predators of agricultural pests (Potts *et al.* 2016; Woodcock *et al.* 2016); and
- loss of human access to nature (see Chapter 8), with disproportionate impacts on vulnerable and indigenous communities (Haines-Young and Potschin 2010).

Figure 6.5: Map of the global human footprint for 2009 (combined pressures of infrastructure, land cover and human access into natural areas, using a 0-50 on a cool to hot colour scales) (a), and absolute change in average human footprint from 1993 to 2009 at the ecoregion scale (b)

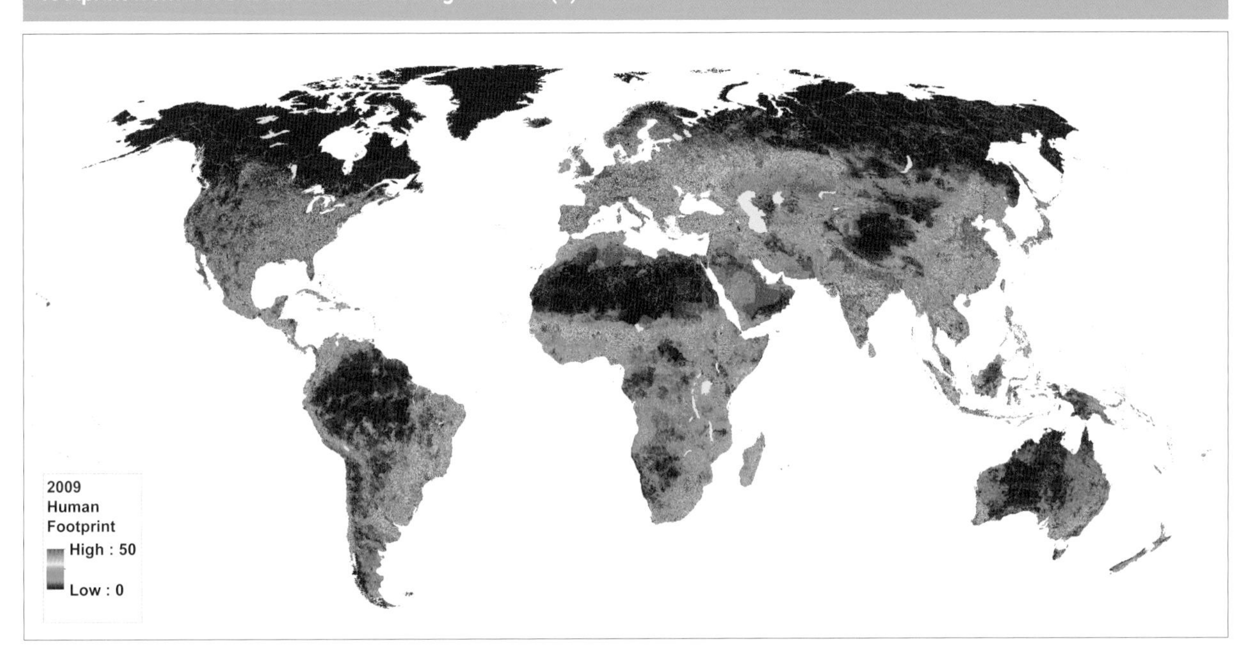

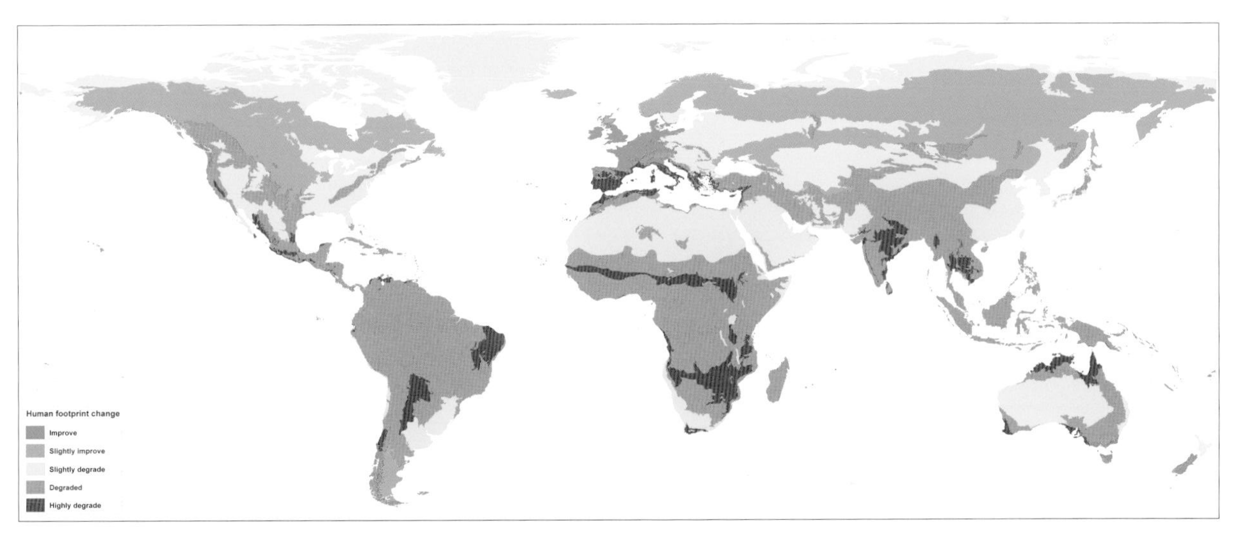

Source: Venter *et al.* (2016).

6.4.2 Invasive species

Invasive species threaten ecosystems, habitats and other species (Bellard, Cassey and Blackburn 2016). They are usually non-native (invasive alien species) but can also include expanding native populations (Nackley *et al.* 2017). The annual rate of first records of non-native species has increased during the last 200 years and the increase in numbers does not show any sign of saturation, meaning that efforts to mitigate invasions have not been effective (Seebens *et al.* 2017). The ecological impacts of invasive species are felt through direct and indirect competition, predation, habitat degradation, hybridization, and their role as disease agents and vectors – also a threat to human health and food security **(Figure 6.6)** (Strayer 2010; Paini *et al.* 2016).

Figure 6.6: Impact mechanism of invasive alien species on threatened species in Europe

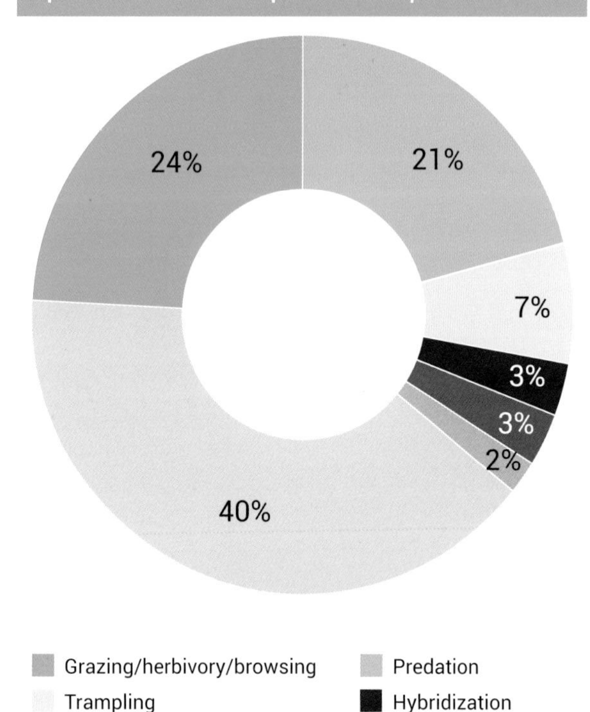

Source: Genovesi, Carnevali and Scalera (2015).

Invasive plants can impact the provisioning of key ecosystem services, such as access to clean water, by the congestion and eutrophication of waterways, degradation of catchment areas, and viability of pasture and rangeland (Packer *et al.* 2017). Invertebrate species that have become invasive may pose an even greater risk. The population expansion of the invasive zebra mussel in the North American Great Lakes was so great that it impeded water flow of municipal water supplies and hydroelectric companies (Rapai 2016). Invasive pests, such as the gypsy moth, emerald ash borer and hemlock woolly adelgid in North America, have both large biodiversity and economic impacts (Aukema *et al.* 2011). Invasive insect vectors can also facilitate the spread of parasites and emerging infectious diseases (Rabitsch, Essl and Schindler 2017), including chikungunya, dengue and Zika, which are vectored by mosquitoes (Akiner *et al.* 2016). Invasive vertebrates present grave danger on islands (Spatz *et al.* 2017), where they may be the major driver of biodiversity loss (Leadley *et al.* 2014; Doherty *et al.* 2016).

The economic costs, both direct and indirect (e.g. costs of control efforts), amount to many billions of dollars annually (for regional estimates see Kettunen *et al.* 2008; Pejchar and Mooney 2009; van Wilgen *et al.* 2012). The cost of restoring lost ecosystem services following invasion of the Laurentian Great Lakes by the spiny water flea was estimated to be between US$86.5 million and US$163 million (Walsh, Carpenter and Vander Zanden 2016). These costs do not reflect the additional environmental and societal/cultural impacts of invasive species.

Major routes for species invasion include deliberate release, escape and accidental introductions via trade, tourism and ship ballast water (CBD 2014; Early *et al.* 2016). Good governance may decrease invasion risk from trade (Brenton-Rule, Barbieri and Lester 2016), whereas climate change may facilitate increased spread by opening up new niche space (Wolkovich *et al.* 2013) and lowering barriers to establishment, especially in more extreme environments (Duffy *et al.* 2017). Loss of native biodiversity is likely to enhance invasion risk, while rising temperatures in cold regions increase the likelihood of establishment (Molina-Montenegro *et al.* 2012; Cuba-Díaz *et al.* 2013; Chown *et al.* 2017). Future threats are posed by increased transport in the Arctic with the decrease in sea ice, commercial use of microbes in crop production, horizontal gene transfer from genetically modified organisms, and the emergence of invasive microbial pathogens (Ricciardi *et al.* 2017).

6.4.3 Pollution

Pollution can take many forms (e.g. waste and chemical products deliberately or accidentally released into the environment, but also light, noise, heat and microbes); major emitters include transport, industry, agriculture (Landrigan *et al.* 2017) and aquaculture (Klinger and Naylor 2012; Bouwman *et al.* 2013). Emerging pollutants include a wide range of synthetic chemicals, pesticides, cosmetics, personal and household care products, and pharmaceuticals (Gavrilescu *et al.* 2015; Landrigan *et al.* 2017).

On land, open waste dumps have local impacts on plants and animals (see Chapter 8), and soil pollution can affect the microbial population and reduce important ecosystem functioning (Wall, Nielson and Six 2015). Pesticides, fertilizers and other chemicals used in agricultural processes can harm pollinators and natural predators of pests (Woodcock *et al.* 2016), with surface run-off also impacting freshwater and coastal biodiversity (see Chapters 7 and 9). Bioaccumulation of toxins, including heavy metals (Araújo and Cedeño-Macias 2016), may have cascading impacts across the entire food chain, including humans. In marine and freshwater environments, the accumulation of microplastic and nanoplastic pollution (see Chapter 7 and **Box 6.2**) has been identified as an emerging issue (SCBD 2016).

The accumulation of endocrine-disrupting chemicals (EDCs) and persistent organic pollutants (POPs) in natural ecosystems pose additional threats to wildlife (Bergman *et al.* eds. 2013), particularly in aquatic systems (Wang and Zhou 2013) (see Chapter 9).

Box 6.2: The threats to biodiversity from marine litter and microplastics

Marine litter, including marine plastic litter and microplastics, is considered a major threat to biodiversity, with serious impacts reported over the last four decades (SCBD 2012). Recent research shows that more than 800 marine and coastal species are now affected through ingestion, entanglement, ghost fishing or dispersal by rafting (SCBD 2016). Between 2012 and 2016, aquatic mammal and seabird species known to be affected by marine litter ingestion increased from 26 per cent and 38 per cent to 40 per cent and 44 per cent, respectively (SCBD 2016). Plastics, which constitute 75 per cent of marine litter, have been shown to act as carriers for persistent bioaccumulative and toxic substances (PBTs); provide habitats for unique microbial communities; act as a potential vector for disease; and provide a means to transport invasive alien species across oceans and lakes (Rochman *et al.* 2013; SCBD 2016). Research on the physical and toxicological effects of microplastic provides evidence of trophic transfer in planktonic food chains as well as the direct uptake of microplastics by marine invertebrates (Wright, Thompson and Galloway 2013; SCBD 2016). Ingestion of microplastic by fish has been shown to cause physiological stress, liver cancer and endocrine dysfunction, affecting female fertility and the growth of reproductive tissue in male fish (Joint Group of Experts on the Scientific Aspects of Marine Environmental Protection [GESAMP] 2015). According to the United Nations, 51 trillion microplastic particles, 500 times more than stars in our galaxy, litter our seas, seriously threatening marine wildlife (van Sebille *et al.* 2015).

Air pollution contributes to the acidification and eutrophication of terrestrial ecosystems, lakes, estuaries and coastal waters (O'Dea *et al.* 2017; Payne *et al.* 2017), and to mercury bioaccumulation in aquatic food webs (Lavoie *et al.* 2013) (see Chapter 5).

6.4.4 Overexploitation

Overexploitation includes illegal, unreported and unregulated fishing, illegal and unsustainable logging, overgrazing, unregulated bushmeat consumption, wildlife poaching and illegal killing (often for foreign markets). It also includes legal but ecologically unsustainable harvesting as a consequence of poorly designed quotas, lack of knowledge of the resource base or new advances in technology that allow more efficient resource exploitation. Direct exploitation has resulted in threats to iconic land and marine species alike, such as the beluga sturgeon prized for caviar (He *et al.* 2017), sharks harvested for their fins (Worm *et al.* 2013), rhinoceros species targeted by poachers for their horns **(Figure 6.7)**, African elephants hunted for their ivory (Maxwell *et al.* 2016), the Andean condor of South America hunted for feathers and bones (Williams *et al.* 2011), and agarwood (*Thymelaeaceae*) harvested for perfume and incense (United Nations Office on Drugs and Crime [UNODC] 2016, p. 59).

Illegal trade in wildlife, fisheries and forest products is extensive, with estimates of their combined value between US$90- 270 billion per year, and links to transnational organized crime (UNEP 2014; Stimson Center 2016; Stoett 2018; see also 'Project Predator' case study in Section 13.3.2). Poverty provides a strong incentive for poaching, while economic development can improve infrastructure that facilitates access to wildlife-rich areas and fuels demand for wildlife products (UNODC 2016, p. 19). However, legal but unsustainable exploitation of wildlife is likely an even greater threat to biodiversity than currently illegal practices (FAO 2018a). The impact of mismanaged harvesting is perhaps most clearly evident in marine fisheries (see Section 6.6.1, and Chapter 7), although future projections are less certain (Costello *et al.* 2016).

Figure 6.7: Recorded number of rhinoceros poached in South Africa, 2007-2015. In 2011, the rhino population in South Africa numbered just over 20,000

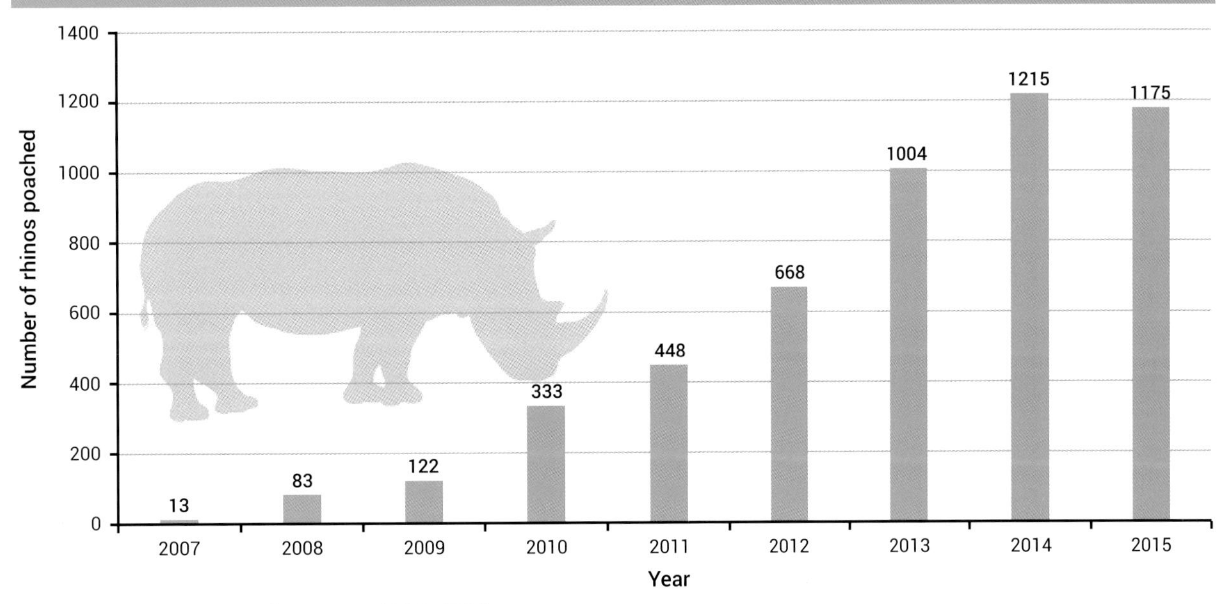

Source: South Africa Department of Environmental Affairs (2016).

The overexploitation of wildlife has implications for equity as it deprives poor and vulnerable local communities and indigenous peoples of sustenance, traditional medicines, tourist income and other ecosystem benefits (Haines-Young and Potschin 2010; O'Neill *et al.* 2017). Conversely, increased regulation of wildlife harvesting can have positive societal consequences, such as strengthening women's leadership roles, which may feed back into biodiversity conservation policy designs (FAO 2016).

6.4.5 Climatic warming and extreme events

The impacts of anthropogenic climate change on biodiversity are most evident in natural systems (IPCC 2014, p. 40), and manifest as changes in both average climate and frequency of extreme weather events **(see Box 6.3)**. One estimate suggests that up to one in six species could be threatened with extinction by 2050 if current warming trends continue (Urban 2015). However, known impacts are not distributed evenly and our knowledge of impacts remains incomplete **(Figure 6.8)**.

In response to rising temperatures, species may move to cooler locations or alter their phenology to flower, breed or migrate sooner (Parmesan 2006; Scheffers *et al.* 2016). Evidence suggests they are doing both: species are moving, on average, 16.9 km per decade to higher latitudes or 11 m per decade upward in elevation (Chen *et al.* 2011), and advances in flowering phenology are suggested to be between 2.3 and 5.1 days per decade (Wolkovich *et al.* 2012; IPCC 2014). There is increasing speculation that such climate-induced shifts in distributions and phenologies might cascade through trophic interactions, resulting in species asynchronies, such as between flowers and their pollinators. An analysis of over 10,000 time series suggests climate sensitivity (i.e. phenological shift in response to climate change) differs among trophic groups (Thackeray *et al.* 2016), but data on interacting species remains sparse (Kharouba *et al.* 2018).

Box 6.3: Extreme events – further pressures on biodiversity

Natural disasters, such as earthquakes and tsunamis, or floods, landslides, wildfires and droughts following extreme weather events kill and injure hundreds of thousands of people a year, cause widespread destruction to ecological habitats, and threaten wildlife populations with local extinction. Following the 2011 Great East Japan earthquake and tsunami, there was an overall decline in local species diversity, and coastal forests and other vegetation on sandy beaches and low-lying coastal areas were severely damaged (Miura, Sasaki and Chiba 2012; Hara *et al.* 2016). The loss of natural coastal habitat, such as mangrove forest and coral reefs, through pollution, habitat transformation and increased sea surface temperatures, can further undermine protection of coastlines from waves, storm surges and coastal erosion. When communities are rapidly rebuilt post-disaster, building material is often gathered unsustainably, posing an additional threat to local habitats, and communities can be relocated to environmentally sensitive areas.

In the marine environment, warming and acidifying oceans are associated with coral bleaching events, with unprecedented pan-tropical bleaching recorded during 2015-2016 (Hughes *et al.* 2017) (see Section 7.3.1). Ocean acidification may also have negative impacts on other marine systems, including mussel beds and some macroalgal habitats (Sunday *et al.* 2017). Warmer waters additionally impose direct metabolic costs on reef fish, reducing swimming capacity and increasing mortality rates (Johansen and Jones 2011). In polar regions, decrease in sea ice and greater surface run-off may increase primary and secondary productivity, altering food-web dynamics (Post *et al.* 2013), and increase the probability of the establishment of invasive species (Duffy *et al.* 2017) (see Section 4.4.2).

Figure 6.8: Global map showing species vulnerable to climate change

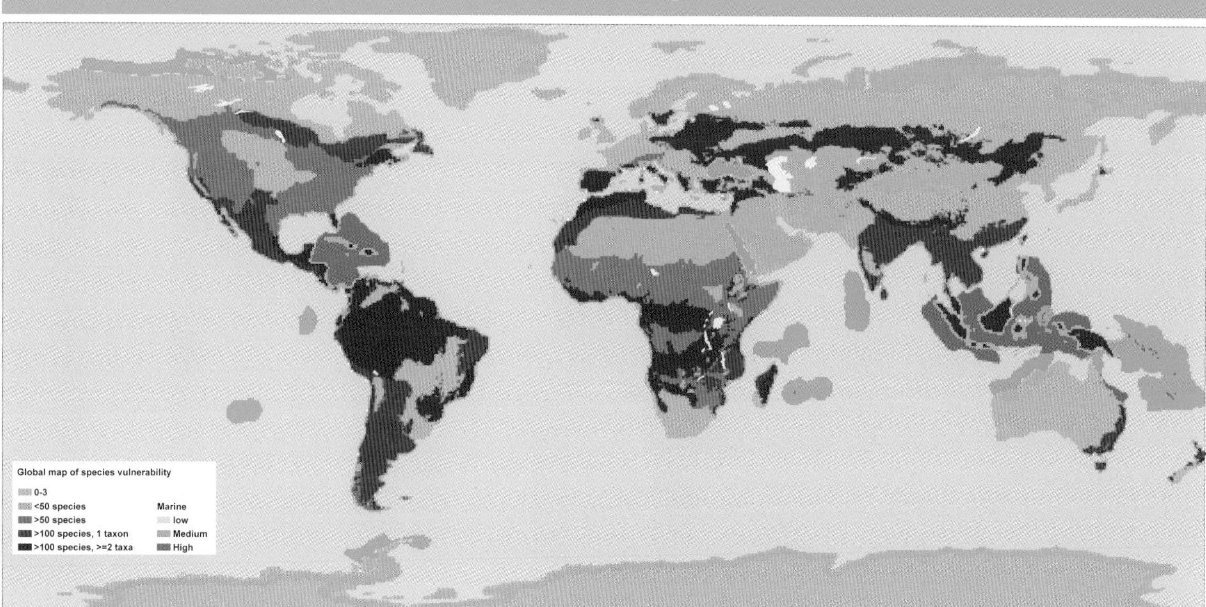

Terrestrial areas with high numbers of vulnerable species were identified on the basis of the number of species assessed and the taxonomic ranks higher than species considered.

Source: Pacifici *et al.* (2015).

6.5 Global state and trends of biodiversity

Global change is having negative impacts across all dimensions of biodiversity, from genes to ecosystems. However, the genetic diversity of most natural populations remains unmeasured, population baseline data is often lacking, and the status of ecosystems is under evaluated. More data and science-based targets for evaluation are needed urgently.

6.5.1 State and trends in genetic diversity

Genetic diversity is of fundamental importance not only as the raw material for continued adaptation of wild species by natural selection, but also in maintaining and enhancing the diversity of cultivated plants and breeds of livestock underpinning the resilience of agricultural systems and food security (Khoury *et al.* 2014; FAO 2015a; Bruford *et al.* 2017). Conservation of genetic diversity can be implemented *in situ* in the wild or crop fields, or increasingly *ex situ* in gene banks and seed collections maintained at local and national levels (see Section 13.2.4).

Long-term declines in the number of varieties of crops and breeds of livestock continue, and much of this diversity, alongside that of wild relatives and lesser used species, still lacks sufficient protection (FAO 2015a). More than 35 species of birds and mammals have been domesticated for use in agriculture and food production, and there are about 8,800 recognized breeds (FAO 2018a). An assessment of extinction risk for existing local animal breeds found 65 per cent are classified as 'status unknown' because of missing population data or lack of recent updates, 20 per cent as 'at risk' and only 16 per cent as 'not at risk' (FAO 2018a). These proportions vary regionally, particularly with respect to the availability of data **(Figure 6.9).**

Figure 6.9: Proportions of local animal breeds, classified as being at risk, not at risk or unknown level of risk of extinction

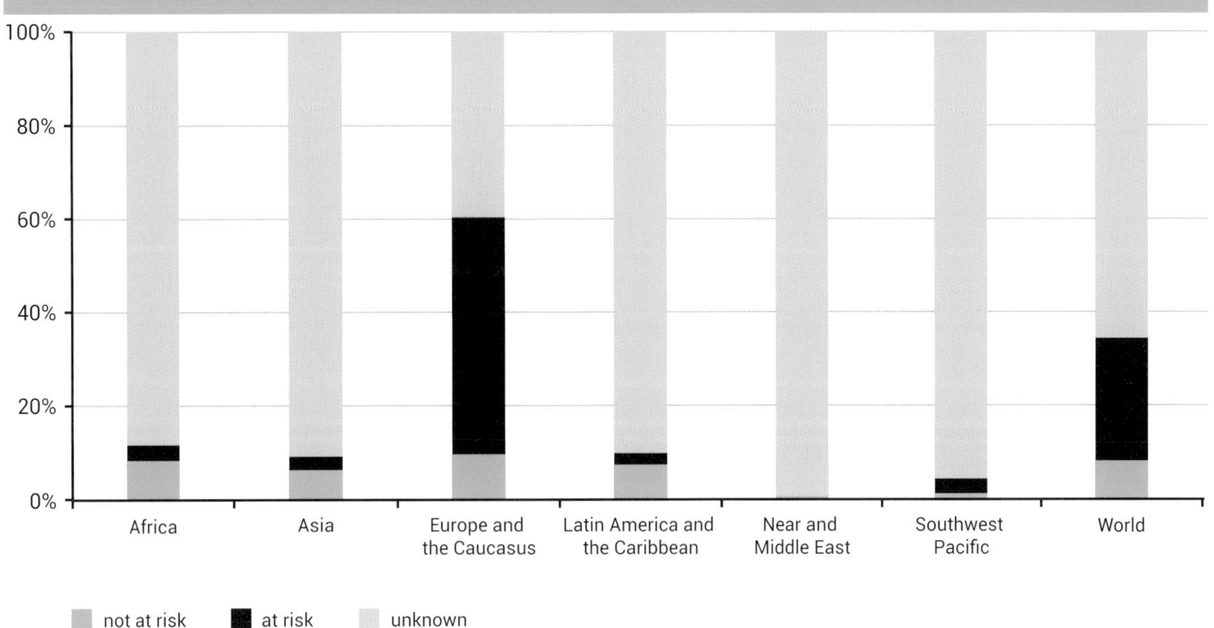

Source: FAO (2018a).

New genomic tools that allow rapid and increasingly low-cost DNA sequencing have become an integral part of conserving genetic diversity *ex situ*, helping us to understand the genetic potential of crop wild relatives for enhancing productivity, nutritional content and resilience to environmental change (Royal Botanic Gardens Kew 2016). As of 2017, some 225 species of plants, mostly crops, had complete genome sequences (Royal Botanic Gardens Kew 2017; see **Figure 6.10**). However, this remains an expensive enterprise and there is an ongoing need to share related information with those whose livelihoods are dependent on biodiversity but lack the resources to access such data.

Traditional approaches to breeding-enhanced varieties of plants and breeds of livestock still predominate; however, genetically modified (GM) organisms continue to draw attention and new advances, such as the CRISPR/Cas genome editing techniques, are advancing synthetic biology (SCBD 2015; CBD 2016). There is evidence of the positive contribution of genome-editing techniques through the control of invasive species (Webber, Raghu and Edwards 2015) due to the lessened need for insecticides that are harmful to non-target organisms (e.g. Li *et al.* 2015). However, the propagation of genome-edited crops may also contribute to negative biodiversity and environmental outcomes, such as facilitating the spread of herbicide-resistant weeds (Rótolo *et al.* 2015) and reduced insect diversity (Schütte *et al.* 2017; Tsatsakis *et al.* 2017), and the natural adaptation of ecosystems to GM traits may ultimately require further technological innovation and increased use of herbicides and insecticides (Rótolo *et al.* 2015).

The conservation status of genetic diversity for most wild species unrelated to agricultural crops and livestock remains poorly documented (although there are concerted efforts to close this gap, see http://www.genomicobservatories.org/). Yet population declines are increasingly commonplace (Ceballos, Ehrlich and Dirzo 2017; McRae, Deinet and Freeman 2017). A loss in population size, particularly when persisting over several generations, frequently translates into a loss in genetic diversity. Thus, the drivers that threaten species and populations also likely erode the genetic diversity within them.

6.5.2 Global state and trends in species

The global decline in biodiversity as illustrated by trends in species remains striking (Dirzo *et al.* 2014). Many observers have suggested that we are witnessing a new mass extinction event (Ceballos *et al.* 2015), although there is as yet no scientific consensus. The International Union for the Conservation of Nature's (IUCN) **(Box 6.4)** Red List of Threatened Species (http://www.iucnredlist.org/) provides the most comprehensive inventory of the global conservation status of plant, animal and fungi species. The status of vertebrates has been relatively well studied (Rodrigues *et al.* 2014), but fewer than 1 per cent of described invertebrates (Collen *et al.* 2012) and only about 5 per cent of vascular plants (Royal Botanical Gardens Kew 2016) have been assessed for extinction risk.

According to IUCN's latest estimates, cycad species face the greatest risk of extinction with 63 per cent of species in this plant group considered threatened **(Figure 6.11)**. The most threatened group of vertebrates are amphibians (41 per cent). Of the few invertebrate species assessments completed, 42 per cent of terrestrial, 34 per cent of freshwater and 25 per cent of marine species are considered at risk of extinction (Collen *et al.* 2012). Among well sampled invertebrate groups, reef-forming corals have the highest proportion (33 per cent) of species under threat.

Figure 6.10: Cumulative number of species with whole genome sequences (2000-2016)

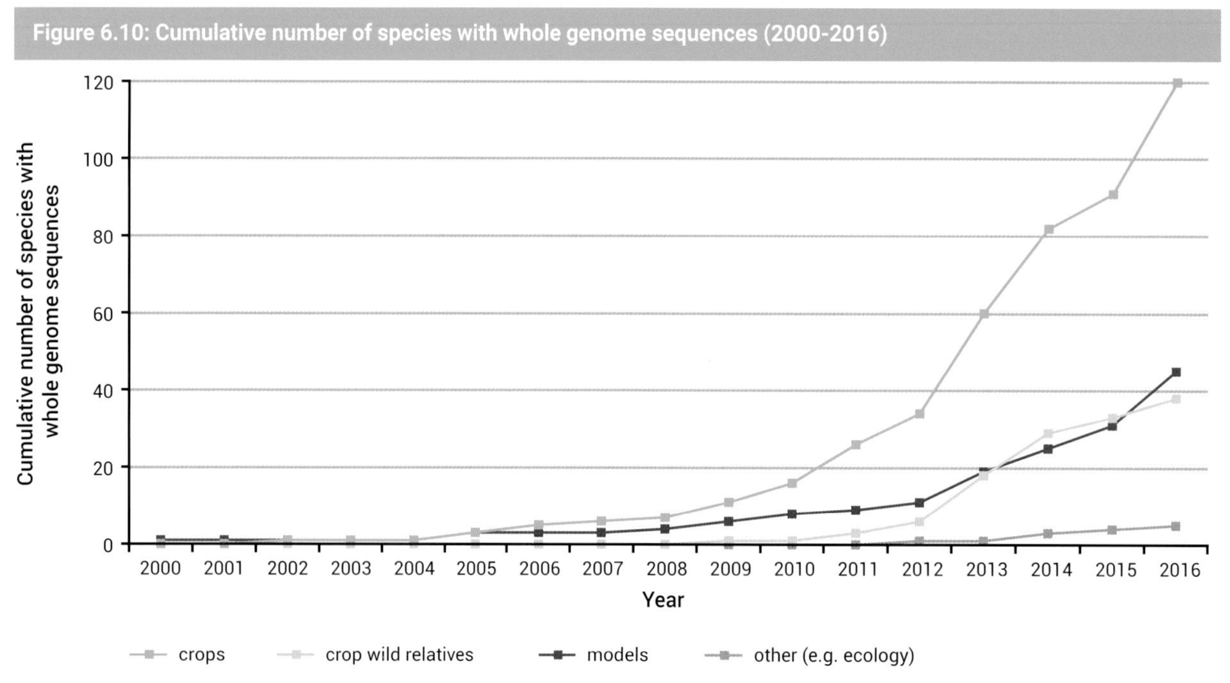

Colours denote the type of species: crops, usually for food; crop wild relatives; model species to help understand plant ecology or evolution; other species, e.g. dominant species in an ecosystem.

Source: Royal Botanic Gardens Kew (2017).

Figure 6.11: The proportion of species in each extinction risk category of the IUCN Red List of Threatened Species

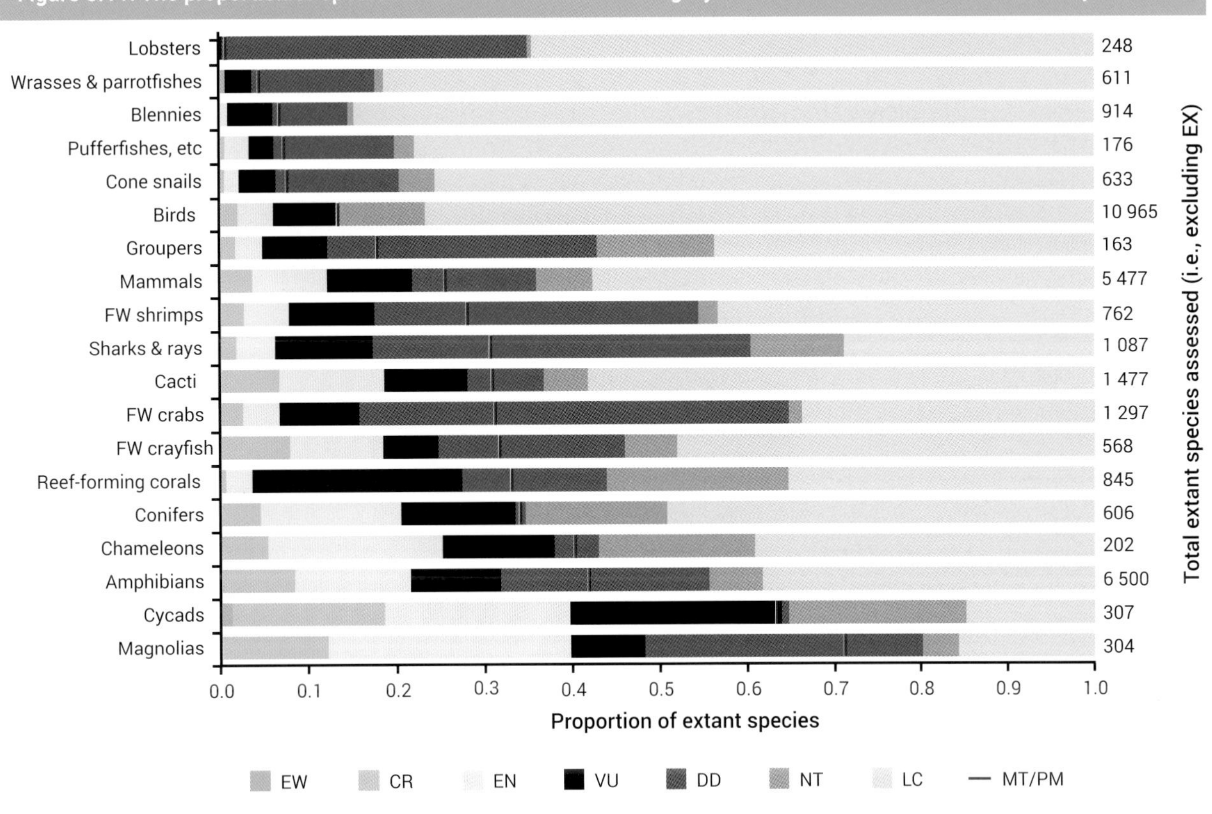

The numbers to the right of each bar represent the total number of existing species assessed for each group. EW: Extinct in the wild; CR: Critically endangered; EN: Endangered; VU: Vulnerable; NT: Near threatened; DD: Data deficient; LC: Least concern.

Source: IUCN 2018 (Red List Version 2018-1).

Box 6.4: International Union for the Conservation of Nature (IUCN)

The International Union for Conservation of Nature (IUCN) has, since 1948, served as a science-policy interface for biodiversity and ecosystem services. IUCN has a membership of which the governance weight is exactly 50 per cent intergovernmental (with over 200 state and government agency members) and exactly 50 per cent civil society and indigenous peoples' organizations (over 1,000 civil society members). The Union mobilizes independent commissions to provide expert input into pressing challenges of nature conservation; there are currently six commissions (Ecosystem Management, Education and Communication, Environmental Economic and Social Policy, Species Survival Commission, World Commission on Environmental Law, and World Commission on Protected Areas), comprising over 10,000 specialists in total. The IUCN Red List of Threatened Species, initiated in 1964, remains the most authoritative global inventory of endangered species today **(Figure 6.11)**.

For those groups that have been comprehensively assessed more than once, changes in extinction risk through time have been examined using the IUCN Red List Index. The evidence suggests an increase in risk of extinction for all groups individually and as an aggregate from 1993 to 2017 **(Figure 6.12)**.

Figure 6.12: Red List Index of species survival for birds, mammals, amphibians, corals and cycads, and an aggregate (in light green) for all species

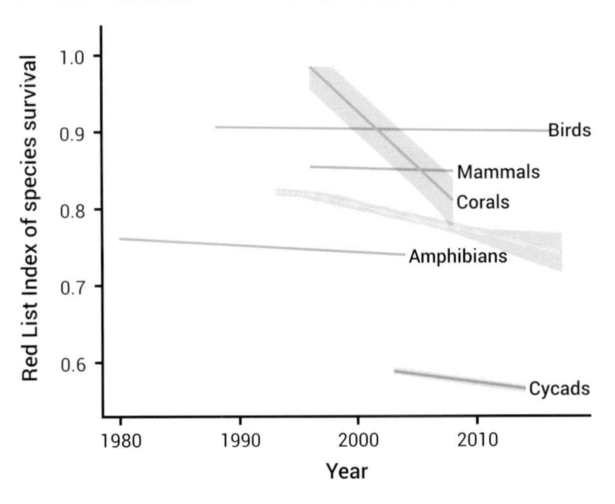

A decline in the trend line indicates either more species have become at risk of extinction over time or there has been an increase in the level of extinction risk over time for some species. The shading denotes 95 per cent confidence intervals.

Sources: IUCN (2017a), Hoffman *et al.* (2018).

Figure 6.13: Global Living Planet Index

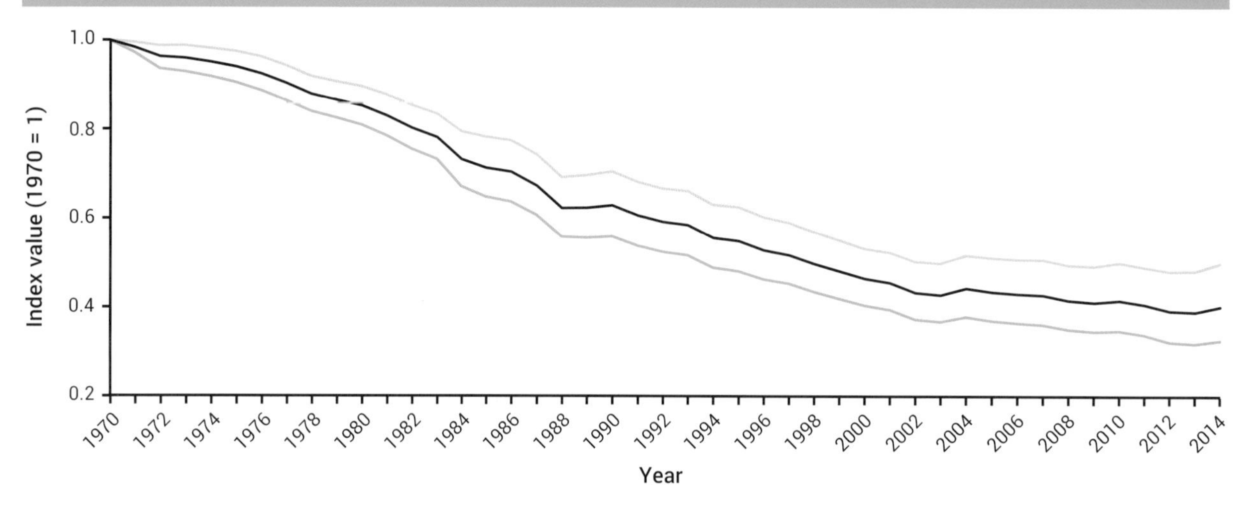

The centre line shows the index values indicating a 60 per cent decline between 1970 and 2014 and the upper and lower lines represent the 95 per cent confidence limits surrounding the trend. This is the average change in population size of 4,005 vertebrate species, based on data from 16,704 time series from terrestrial, freshwater and marine habitats.

Source: WWF (2018).

Monitoring the abundance of species provides a complementary indicator of status and trends. Although lacking the comprehensive coverage of many taxonomic groups found in the IUCN Red List Index, these indicators provide finer spatial and temporal resolution. Trends in global vertebrate species population abundances as measured by the Living Planet Index **(Figure 6.13)** show an average decline of 60 per cent between 1970 and 2014 (McRae, Deinet and Freeman 2017; WWF 2018). Freshwater species have higher rates of population declines than either terrestrial or marine species (McRae, Deinet and Freeman 2017). Globally, average local abundance of terrestrial species is estimated to have fallen to 85 per cent of modelled abundances in the absence of anthropogenic land-use change (Newbold *et al.* 2016), although the intactness of biodiversity varies spatially (Newbold *et al.* 2015; Newbold *et al.* 2016; **Figure 6.14**), and data on species population trends of both flora and fauna are sparse.

Trends in invertebrates may well echo those observed in vertebrates. A global index sampling populations of 452 invertebrate species revealed an average 45 per cent decline in abundance over 40 years (Dirzo *et al.* 2014) and recent reports of declines greater than 75 per cent in biomass of flying insects has been found in protected areas in Germany (Hallmann *et al.* 2017), with similar findings emerging elsewhere in Western Europe (Vogel 2017) and central Europe (Hussain *et al.* 2017;

Figure 6.14: Terrestrial Biodiversity Intactness Index

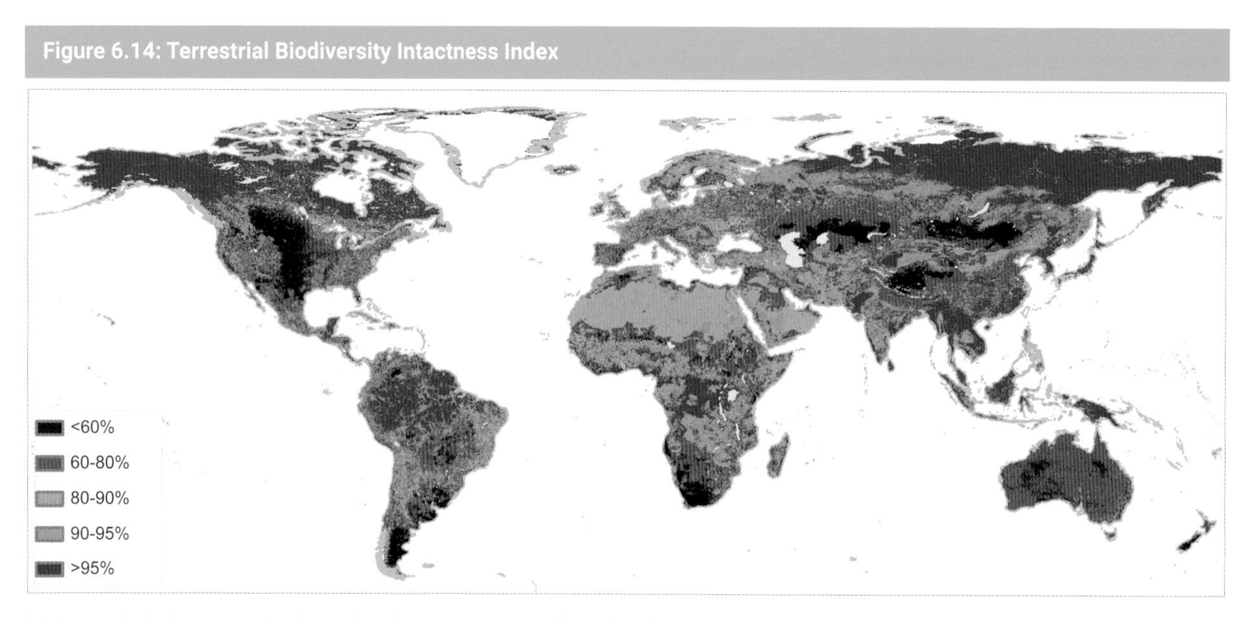

Intactness value is the average abundance of species as a percentage of the modelled abundance in an undisturbed habitat.

Source: Newbold *et al.* (2016).

Hussain *et al.* 2018). Particularly steep declines were observed in hoverflies, which are important pollinators (Vogel 2017). Declines in pollinator abundance have also been documented elsewhere, for example, bumble bee species in North America (Bartomeus *et al.* 2013).

The Living Planet Index **(Figure 6.13)** and the Biodiversity Intactness Index **(Figure 6.14)** both indicate that terrestrial species abundance has declined as a result of anthropogenic land-use change, and that the trend of population decline in the last 44 years has shown no sign of slowing (McRae, Deinet and Freeman 2017; WWF 2018). It has been suggested from the Biodiversity Intactness Index that a terrestrial planetary boundary has been crossed (based on a reduction of 10 per cent in Biodiversity Intactness); from this, it is inferred that ecosystem function may be impaired (Newbold *et al.* 2016).

6.5.3 Global state and trends in ecosystems

There is a pressing need to expand ecosystem assessments. The IUCN has begun to issue a Red List for Ecosystems to complement its global species-based assessment (Keith *et al.* 2015), and a few ecosystems have been assessed by global and regional criteria. One ecosystem, the Aral Sea, has been assessed as 'collapsed' **(Figure 6.15)** (Sehring and Diebold 2012; Keith *et al.* 2013), and several others, such as the gnarled mossy cloud forest on Lord Howe Island of Australia, and the Gonakier forests of the Senegal river floodplain shared by Senegal and Mauritania, have been listed as 'critically endangered' (see Red List of Ecosystems; IUCN 2017b).

Collapse may be reversible if all the component parts of the collapsed ecosystem still exist in other ecosystems (Rodríguez *et al.* 2015). However, shifts to alternative stable states, such as that documented in coral reef systems, from coral dominated to algal dominated, with human-induced eutrophication, cannot be simply reversed (Hughes *et al.* 2017).

Figure 6.15: Mechanisms of ecosystem collapse, and symptoms of the risk of collapse

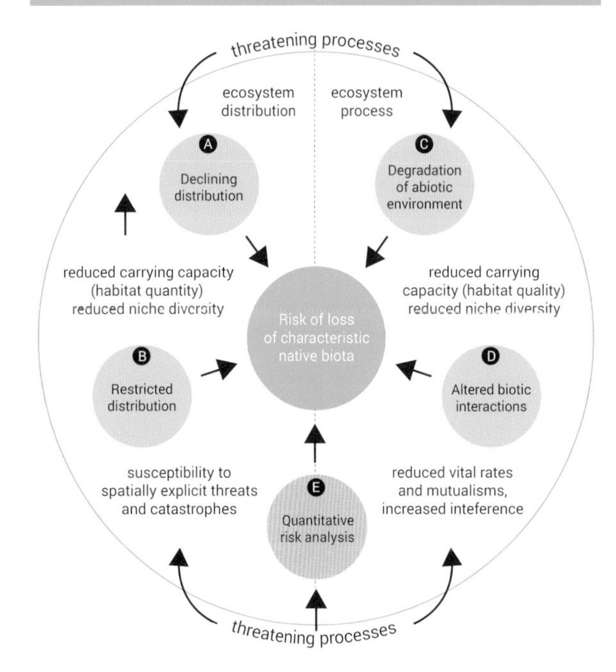

Source: Keith *et al.* (2013).

Figure 6.16: Mean percentage change in each broad habitat type based on satellite imagery: (a) change from original land-cover type between 2001 and 2012; (b) vegetation productivity as measured using the Enhanced Vegetation Index between the years 2000-2004 and 2009-2013

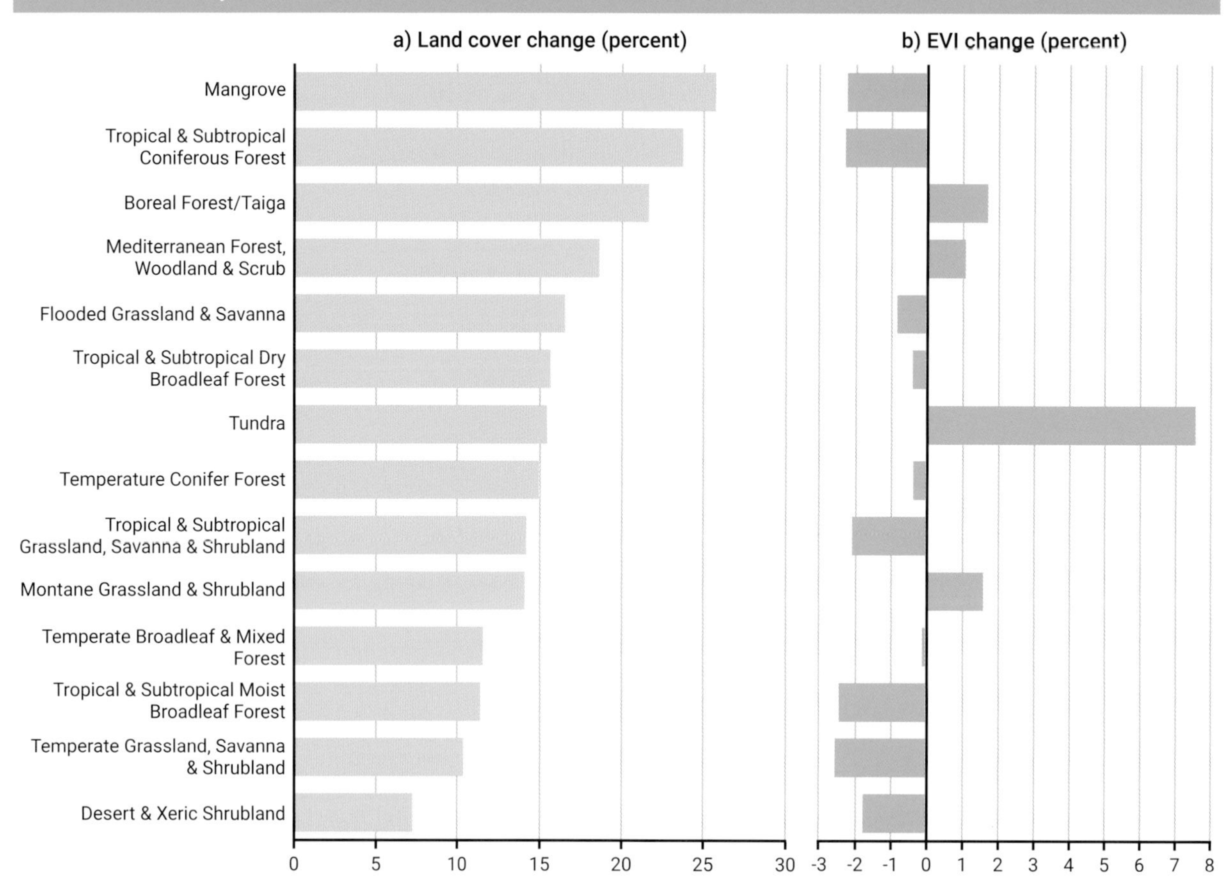

Source: Royal Botanical Gardens Kew (2016).

Some information is available at a large scale for broad terrestrial habitat types, and it is estimated that 10 out of 14 experienced a decrease in vegetation productivity between 2000 and 2013, while 4 increased in productivity **(Figure 6.16)**, with anthropogenic factors thought to be driving these trends (Royal Botanical Gardens Kew 2016). At a finer scale, 24 per cent of terrestrial ecoregions have been classified as 'Nature imperilled' (Dinerstein *et al.* 2017).

More is known about the status of terrestrial species and ecosystems than their aquatic counterparts. However, an average decline in natural wetland area of about 30 per cent between 1970 and 2008 was observed globally (Dixon *et al.* 2016), varying from a 50 per cent decline in Europe to 17 per cent in Oceania. While the spatial extent of anthropogenic impacts on marine ecosystems has been estimated (Jones *et al.* 2018), relatively little is known about their current status. Nonetheless, the impact of pressures on the marine environment is thought to be increasing, as evidenced by marine wildlife loss (McCauley *et al.* 2015) and the current critical status of coral reefs (Hughes *et al.* 2017). The deep-sea ecosystem is probably one of the least well studied and is expected to be particularly vulnerable to habitat loss and climate change (Barbier *et al.* 2014).

The status of biodiversity that explicitly underpins nature's contribution to people has not yet been comprehensively assessed, although a global assessment of biodiversity and ecosystem services will be published by IPBES in 2019. However, many of these ecosystem processes are thought to be under threat as a consequence of observed wildlife declines and ongoing threats to biodiversity (Cardinale *et al.* 2012; Mace, Norris and Fitter 2012). Mammal and bird species that are used for food and/or medicine are at greater risk of extinction than those not used; the opposite was found for the same assessment of amphibian species (Almond *et al.* 2013). The perceived value of a species may impose an additional pressure on biodiversity conservation: of the 28,187 plant species that are recorded as being of medicinal use, there are controls on international trade for 1,280 to reduce threats from overexploitation (Royal Botanical Gardens Kew 2017).

6.6 Impacts on the world's biomes

A biome is defined as a major ecological community of organisms adapted to a particular climatic or environmental condition across a large geographic area. Within biomes, several ecosystems may coexist. This section examines eight broadly defined biomes that encompass most of Earth's biodiversity.

6.6.1 Oceans and coasts

The primary pressures on open ocean biodiversity are overexploitation, pollution from land-based activities and climate change; coastal ecosystems have additional pressures associated with habitat destruction, aquaculture and invasive species (see Section 7.2). Although data are limited, these pressures affect the state of marine biodiversity from populations to ecosystems.

Coastal systems are particularly vulnerable; for example, between 20 and 35 per cent of mangrove area has been lost since 1980 (Innis and Simcock eds. 2016) and the current annual rate of seagrass habitat destruction is about 8 per cent (Innis and Simcock eds. 2016). Coral reefs are among the most biodiverse marine ecosystems, yet they are also among the most fragile (see Section 7.3.1).

The decline in the health of marine ecosystems and biodiversity is increasingly affecting people (WWF 2015). Marine capture fisheries provide healthy food and support livelihoods (see Section 7.3.2). However, overexploitation is leading to population declines in marine fisheries with the percentage of global stocks fished at biologically unsustainable levels increasing from 10 per cent in 1975 to 33 per cent in 2015, with the largest increases in the late 1970s and 1980s (FAO 2018b; **Figure 6.17**). In 2015, over 50 per cent of the stocks in the Mediterranean, Black Sea, the Pacific Southwest and the Atlantic Southwest were fished at biologically unsustainable levels (FAO 2018b).

Exploitation of target species is coupled with additional negative biodiversity impacts from by-catch and damage to benthic environments from trawling, although some seabird populations have increased through feeding on discards (Foster, Swann and Furness 2017). The rise of aquaculture can reduce pressures of exploitation for some wild species, but can also lead to invasive species, inter-species breeding, eutrophication and disease spread (Ottinger, Clauss and Kuenzer 2016) (see Section 7.4.3).

Pollution, including marine plastic litter and microplastics **(see Box 6.2)**, and loss and degradation of habitat leads to further reduced contributions from natural systems, such as declining fish nursery grounds or mangrove wood supply (Nordlund *et al.* 2016; Quinn *et al.* 2017), as well as increases in vulnerability to extreme events **(see Box 6.3)** through reduced coastal protection.

6.6.2 Freshwater

Freshwater systems are exposed to the full gamut of multiple pressures with changes in land use, habitat loss, invasive species, use of watercourses for development of hydroelectric power, and pollution creating widespread and significant impacts (see Section 9.2). Wetland loss has been long term and extensive, and freshwater species, especially in tropical ecosystems, have declined at a faster rate than those in any other biome (see Section 6.4.1).

Figure 6.17: Global trends in the state of the world's marine stocks 1975-2015

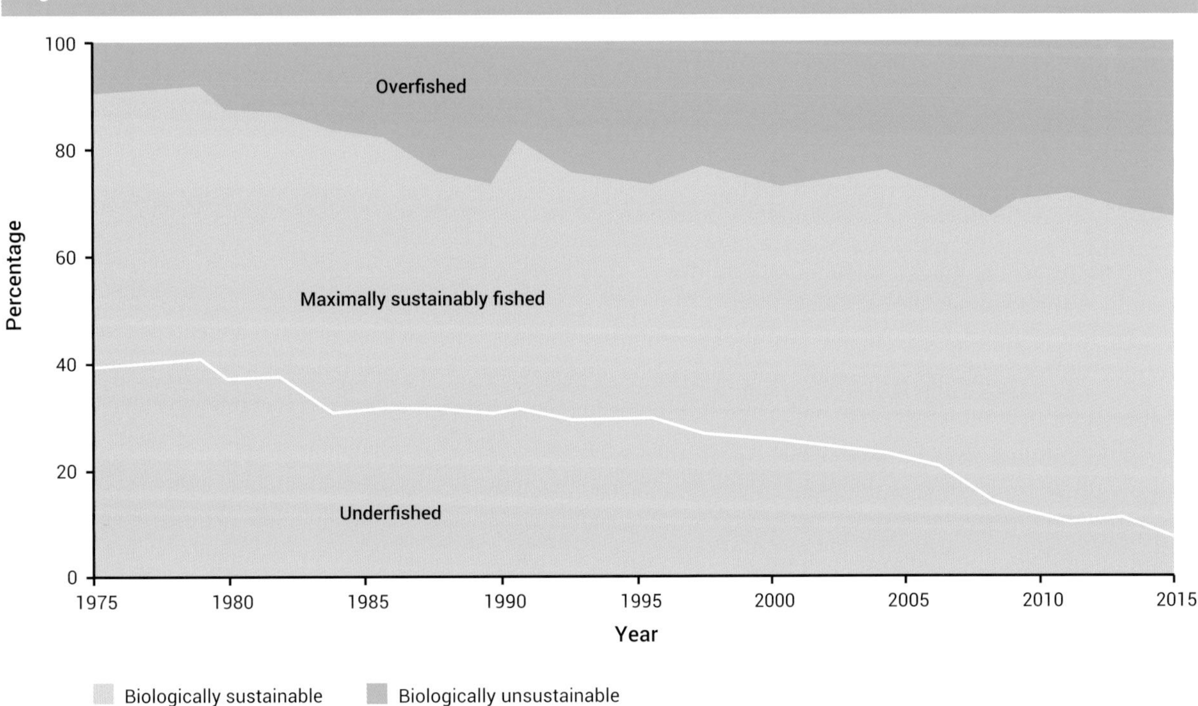

Source: FAO (2018b).

The abundance of monitored populations of freshwater vertebrate species declined an average of 81 per cent over the past 42 years (WWF 2016). A summary of extinction risk of global freshwater fauna indicates that reptiles have the highest estimated risk among the six groups assessed **(Figure 6.18)**. About a third of the more than 7,000 freshwater invertebrate species on the IUCN Red List are considered threatened, with gastropods being the most threatened group (Collen *et al.* 2012). These species combine to provide a wide range of critical services for humans, such as flood protection, food, water filtration and carbon sequestration (Collen *et al.* 2014).

Industrial-era agriculture results in nitrogen- and phosphorous-driven eutrophication of terrestrial, freshwater and nearshore marine ecosystems, and pesticide use can further degrade freshwater ecosystems (Malaj *et al.* 2014; Mekonnen and Hoekstra 2015). Globally, it is estimated that the number of lakes with harmful algal blooms will increase at least 20 per cent by 2050 (United Nations Educational, Scientific and Cultural Organization [UNESCO] 2014). Cyanobacterial algal blooms can result in lowered value for recreational uses, reduced aesthetics, lower dissolved oxygen concentrations, decline in drinking water quality and the production of toxins, which can impact both wildlife and human health (Brooks *et al.* 2016).

6.6.3 Grasslands

Grasslands cover about 8 per cent of total land area and were once home to some of the largest wildlife assemblages on Earth (IUCN 2017c). They are now considered the most altered terrestrial ecosystem worldwide and the most endangered ecosystem on most continents, facing multiple pressures including land-use change, overgrazing, fragmentation, invasive species, suppression of natural fire, climate change and afforestation (IUCN 2017c).

Though grasslands contain high plant diversity, agricultural expansion is causing habitat destruction and fragmentation; for example, soybean production has replaced traditional livestock subsistence on natural pastures in much of the cerrado, a woodland savanna ecosystem, of South America (Aide *et al.* 2013). The Brazilian Cerrado holds roughly five per cent of global biodiversity and has lost close to 50 per cent of its original range (Brazil, Ministério de Meio Ambiente 2015). Rising temperatures are associated with woody encroachment and desertification across Africa (Midgley and Bond 2015; Engelbrecht and Engelbrecht 2016), South America and, to a lesser extent, Australia (Stevens *et al.* 2017).

It is estimated that 49 per cent of grassland ecosystems experienced degradation over a ten-year period (2000-2010), with nearly 5 per cent experiencing strong to extreme degradation (Gang *et al.* 2014), greatly decreasing the ability of these ecosystems to support biodiversity. Currently, 4.5 per cent of global grasslands have protected status (IUCN 2017c).

The strong relationship between grassland biodiversity and biomass (Cardinale *et al.* 2012), which is often used for animal fodder, agricultural products and raw textile materials for local populations, suggests that reductions in biodiversity will have negative implications for small-scale economic productivity and livelihoods.

Figure 6.18: Extinction risk of global freshwater fauna by taxonomic group

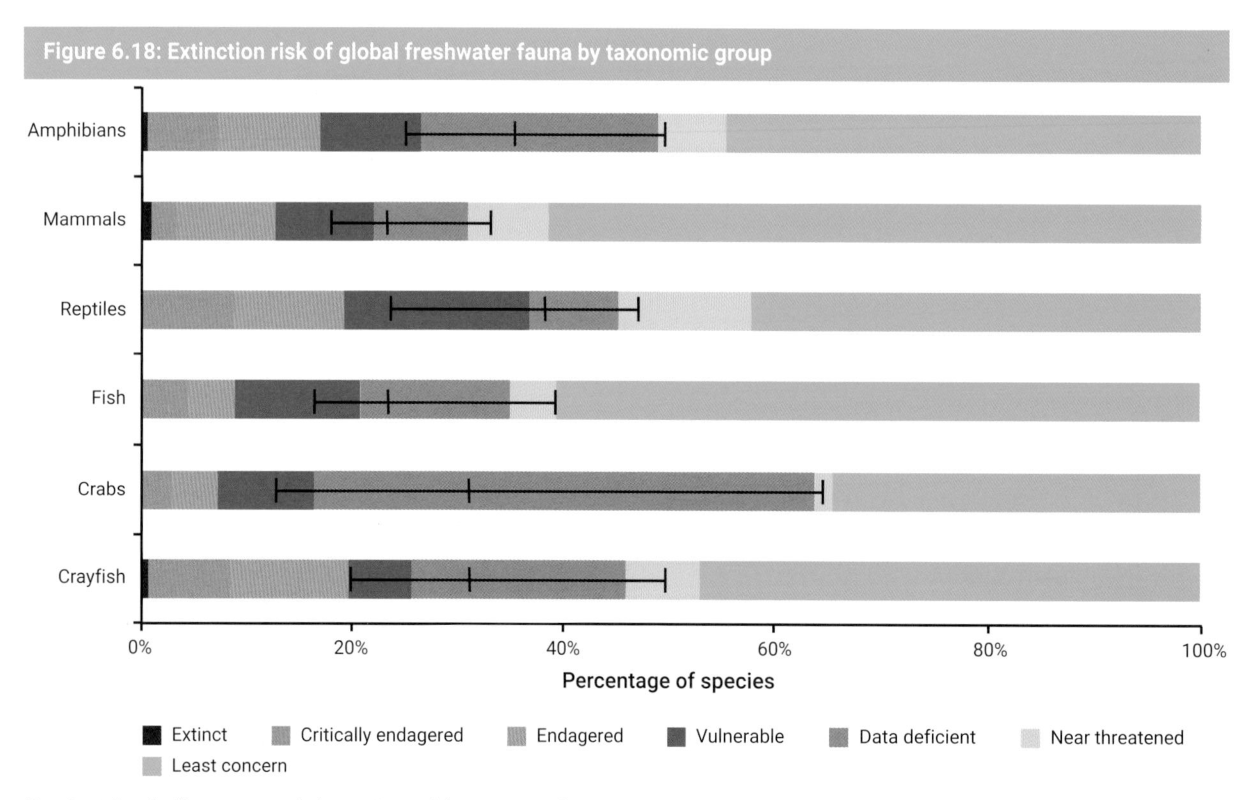

Note: Central vertical lines represent the best estimate of the proportion of species threatened with extinction, with whiskers showing confidence limits. Data for fish and reptiles are samples from the respective group; all other data are comprehensive assessments of all species (n = 568 crayfish, 1191 crabs, 630 fish, 57 reptiles, 490 mammals and 4147 amphibians).

Source: Collen et al. (2014).

Box 6.5: Agrobiodiversity and gender

In many societies, women have traditionally been the keepers of deep knowledge of the plants, animals and ecological processes around them. The use of hybrid seed varieties (to which there has been a widespread shift in recent decades) can prevent women collecting seeds, undermining their status as seed collectors, as well as food security, especially in developing countries (Bhutani 2013). The erosion of biodiversity driven by industrial agriculture has therefore had specific impacts for women, including a loss of knowledge related to seeds, food processing and cooking (International Panel of Experts on Sustainable Food Systems 2016). In recent years, community seed banks that preserve local seeds have been re-established in some areas and are frequently managed by women, including through local seed exchanges. Participatory plant-breeding schemes to improve seeds further enhance women's status in farming (Galiè *et al.* 2017).

6.6.4 Agricultural landscapes

Beginning about 8,000 years ago, agricultural expansion and intensification has led to biodiversity loss in many biomes (United Nations Convention to Combat Desertification [UNCCD] 2017). Global demand and supply chains concentrate production in 'breadbasket' regions (Khoury *et al.* 2014), where landscape transformation reduces and fragments natural habitat, and yield-enhancing inputs (fertilizers and pest control) can impact non-cropped areas, watercourses and air quality. Recent decades are notable for marked land-use change in tropical regions associated with increasing oilseed production, in particular for soya and oil palm, much of which has come at the expense of highly biodiverse biomes (Foley *et al.* 2011). A dramatic decline in animal populations both inside and outside protected areas (Keesing and Young 2014) is associated with increased risk of predators attacking livestock (Zheng and Cao 2015; Malhi *et al.* 2016), negatively impacting agricultural livelihoods. Agricultural practices, such as tillage, crop combinations, and application of fertilizers and pesticides, also have impacts on below-ground biodiversity. (FAO and the Platform for AgroBiodiversity Research 2011, p. ix). Importantly, agricultural landscapes can sometimes maintain rare species in semi-natural habitats, while abandonment of agricultural practices may even lead to biodiversity decline (Plieninger *et al.* 2014).

Loss of diversity in agroecosystems increases their vulnerability and thus reduces the sustainability of many production systems. Reduction in the provisioning of regulating and support services can drive additional chemical use and may create harmful feedback loops (WHO and SCBD 2015, p. 5). There is some evidence that farmers in homogeneous landscapes have higher incomes than farmers in heterogeneous landscapes (Watts and Williamson 2015), but their resilience to pressures such as climate change is often lower and income variability is greater (Abson, Fraser and Benton 2013). In addition, the homogenization of crop production has health impacts, contributing to the homogenization of diets and increasing consumption of processed foods associated with obesity and diet-related non-communicable diseases (Khoury *et al.* 2014). In contrast, production diversity is strongly associated with dietary and nutrition diversity among smallholder farmers whose market participation is limited (Sibhatu, Krishna and Qaim 2015) and local knowledge about seed varieties is often held by women farmers **(see Box 6.5)**.

In some cases, intensive agriculture might also increase the prevalence of infectious diseases (Cable *et al.* 2017). For example, oil palm plantations in South America appear to increase the risk of Chagas disease (Rendón *et al.* 2015), and in Kalimantan, Indonesia, the burning of forests to plant oil palm may have contributed to the migration of bats, known to carry Nipah virus (Pulliam *et al.* 2011).

Biodiversity in agricultural landscapes is key to food and nutrition security **(see Box 6.6)**. Pollination by about 100,000 species of insects, birds and mammals accounts for 35 per cent of global crop production (SCBD 2013; IPBES 2016), and up to 15 per cent of the value of economies based on cash crops (IPBES 2016, p. 209). Production is declining at local scales in places where the diversity of pollinators has been declining (IPBES 2016, pp. 154,185-186). Maintaining remnant patches within a few hundred metres of farms can help support pollinator populations and increase crop yield (Pywell *et al.* 2015; IPBES 2016, p. 394).

Box 6.6: Importance of traditional practices and knowledge in pollinator conservation

Indigenous and local knowledge has been recognized as an important source of expertise in finding solutions to declines in animal pollinators – wild species such as birds, bats, bumblebees and hoverflies, and managed species such as honeybees (Lyver *et al.* 2015; IPBES 2016, p. xxii). In 2013, the Indigenous Pollinators Network was established with a view to combining traditional knowledge of indigenous peoples with modern science for the benefit of conserving pollinators and their vital services (Platform for AgroBiodiversity Research 2013). As well as conserving pollinators, traditional practices of beekeeping may have wider benefits for biodiversity, for example strengthening watershed conservation in the face of climate change (Kumsa and Gorfu 2014) and in forest conservation (Wiersum, Humphries and van Bommel 2013).

Ethiopia is the largest producer of honey and beeswax in Africa (Begna 2015). These products are used for making candles and Tej or honey wine (an important drink in cultural life), and white honey from the Bale mountain region is used medicinally (IPBES 2016, pp. 312-314). Women contribute to this value chain, usually by manufacturing honey products rather than beekeeping itself. However, there is potential for beekeeping to provide income generation and empowerment for women in rural areas of Ethiopia (Ejigu, Adgaba and Bekele 2008; Serda *et al.* 2015).

6.6.5 Drylands

Though drylands are less diverse than other ecosystems, they contain thousands of species that are highly adapted to the dryland environment yet often neglected in conservation efforts. Arid and semi-arid rangeland ecosystems have seasonal climatic extremes and unpredictable rainfall patterns, but dryland species have evolved to be highly resilient by recovering quickly from drought, fire and herbivore pressure. Desertification (also known as land degradation in drylands) is a worldwide phenomenon (see Section 8.4.2).

Dryland degradation has many causes, including human conflicts. Large amounts of waste, garbage and toxic material were dumped and burned in desert ecosystems due to the Islamic Republic of Iran-Iraq war (UNEP 2016f). Drought, overgrazing, overuse of groundwater and unsustainable agricultural practices impose additional pressures (O'Connor and Ford 2014; Southern Africa Development Community 2014), though the extent of human versus natural causes are often difficult to disentangle.

The degradation of semi-arid and arid landscapes reduces capacity in terms of freshwater supply and food production, decreases wild food availability, and presents a threat to emblematic species and genetic resources (Low ed. 2013). Desertification has a damaging effect on soil health and vegetation, leading to adverse impacts that cascade through the food chain (Assan, Caminade and Obeng 2009). Salinization, mostly due to unsustainable irrigation systems, irrigated areas with poor drainage and poor quality of irrigation water, is a major problem in arid and semi-arid regions (see Section 9.5.6). The almost complete desiccation of the Aral Sea has led to the creation of the Aral Kum desert, which has caused degradation of riparian forests, pastures and other vegetation cover (Kulmatov 2008).

6.6.6 Forests

Forests provide habitat for large numbers of animal and plant species, and deforestation is one of the top threats to species diversity (FAO 2015b; Alroy 2017). Deforestation and forest degradation continue in many regions, often in response to demands for biomass as well as drivers outside the forest sector, such as urban expansion and agriculture, energy, mining and transportation development (see Section 8.4.2). Recent estimates show that tree cover loss is high across all forest types but differs across regions (Leadley *et al.* 2014). Tree cover density is associated with both losses and gains, but losses are especially high in the tropics and boreal forests; tropical rainforest accounted for 32 per cent of global tree cover loss over the period 2000-2012, with half of this loss occurring in South America (Hansen *et al.* 2013). Rates of forest gains approach or exceed rates of tree cover loss in some areas, particularly in temperate regions, reflecting forestry-dominated land management.

Recent work suggests that more biodiverse forests contribute a greater range of ecosystem services (Gamfeldt *et al.* 2013). Forests supply essential regulating services, including carbon sequestration, important for the regulation of climate, and protection of soil and water (Foley *et al.* 2007; Brockerhoff *et al.* 2017). With increasing deforestation and forest degradation, however, forest ecosystems can transform from net carbon sinks to carbon sources (Baccini *et al.* 2017).

The total number of people deriving benefits from forests — in the form of food, forest products, employment, and direct or indirect contributions to livelihoods and incomes — is estimated to be between 1 billion and 1.5 billion (Agrawal *et al.* 2013). In Africa, approximately 80 per cent of people are dependent on fuelwood (including charcoal) as their sole source of energy (UNEP 2016a, p. 76). Global exports of forest products were worth US$226 billion in 2015, with wood fuel comprising 9 million m^3 and industrial roundwood 122 million m^3 (FAO 2015b). Non-wood forest products, including wild plant resources, typically contribute less to local economies, but can have high global market value. Contributions of forests to economies of the developing world are estimated at over US$250 billion (Agrawal *et al.* 2013). These economic benefits can only be maintained if forests are managed sustainably (FAO 2015a).

Though there are short-term employment gains from deforestation, the loss of forests translates into a loss of livelihoods: over 13 million people are employed in the formal forest sector, and another 40-60 million people may be employed in informal small and medium-sized forest operations (Agrawal *et al.* 2013; FAO 2018c). A well-documented gender gap in access to forest resources suggests that poor management or loss of forest ecosystems may have different impacts on women and men (WWF 2013; Djoudi *et al.* 2015).

The direct health consequences of deforestation are complex: there is some evidence that forests can promote physical and mental well-being (Oh *et al.* 2017), while forest loss may increase exposure to infectious diseases, including malaria (Guerra, Snow and Hay 2006; Fornace *et al.* 2016) and other vector-borne parasites (Plowright *et al.* 2015; Hunt *et al.* 2017; Olivero *et al.* 2017).

6.6.7 Mountains

Mountain ranges cover around 22 per cent of the terrestrial space of the planet and provide multiple ecosystem services. At lower elevations, mountain habitats, especially those in tropical regions, are often more biodiverse and have higher levels of endemism than adjacent lowlands. However, habitat degradation and fragmentation has impacted many mountain ecosystems (Shrestha, Gautam and Bawa 2012; Chettri 2015; Venter *et al.* 2016) (see Section 4.3.2).

Mountain ecosystems are especially vulnerable to climate change: effects include shifts in species ranges and composition, with notable impacts on those organisms whose dispersal might be limited, or which are restricted to high altitudes, and local extinctions can occur for species in the upper margins of elevation gradients (Pauli *et al.* 2012; Khan *et al.* 2013; Grytnes *et al.* 2014; Knapp *et al.* 2017). Climate-induced warming can change ecosystem functioning, advance spring phrenology, and increase productivity and carbon uptake (Piao *et al.* 2012; Shen *et al.* 2016). Localised pressures include road construction, deforestation, mining, tourism, grazing of domestic livestock, burning and armed conflict (see Epple and Dunning 2014; Young 2014).

Figure 6.19: Capacity of mountains to provide ecosystem services

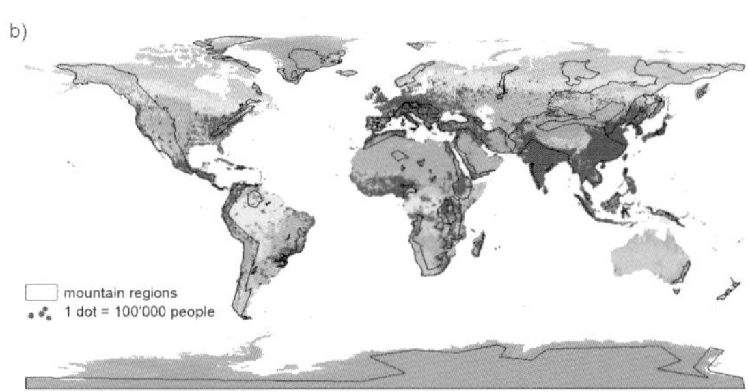

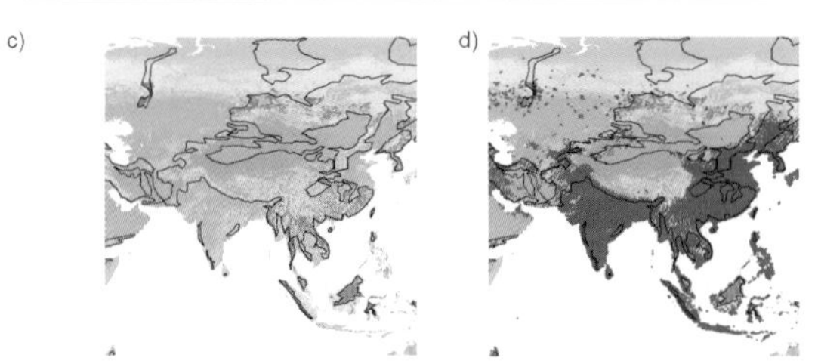

The maps display the proxy capacity of land to provide ecosystem services, measuring to what degree 15 selected ecosystem services are supported by the underlying land characteristics: (a) global analysis; (b) population density data highlighting regions of high demand for ecosystem services; (c) and (d) high supply of and high demand for ecosystem services in the Himalayas.

Source: Grêt-Regamey, Brunner and Kienast (2012).

Most mountain areas today are under high human pressure, including the Tropical Andes and Central Asian Mountain biodiversity hotspots. The Himalayas, with approximately 19,000 species (Khan *et al.* 2013), have been documented as highly vulnerable to climate change (Shrestha, Gautam and Bawa 2012). In Europe, warming has driven many species upward, resulting in local increases of boreal and temperate mountaintop diversity; but the opposite effect has been noted for Mediterranean mountains, which have lost some species (Pauli *et al.* 2012). In some areas, the abandonment of agricultural land in mountain ranges has also led to decreases in biodiversity, especially among bird populations (Hussain *et al.* 2018).

Loss of biodiversity reduces nature's contributions to people in both mountains and lowlands **(Figure 6.19)** (Grêt-Regamey, Brunner and Kienast 2012). Degradation in mountain ecosystems will result in changes in air quality and climate regulation, such as the reduction of greenhouse gas sequestration (Ward *et al.* 2014). Threats to local communities include loss of food security, medicinal plants, and water quality and provision, and increased exposure to risks associated with landslides, sedimentation of rivers and flooding modifying their livelihoods and land cover (Eriksson *et al.* 2009; Khan *et al.* 2013; Young 2014). A few mountain areas still maintain the traditional use of species (e.g. Andes, Himalayas), while ethnobotanical knowledge in the Alps has been lost due to changes in land-use patterns (Khan *et al.* 2013). Glacier loss impacts water security, with some populations in South Asian countries dependent upon the flow of rivers from the western but also central and eastern Himalayas (Khan *et al.* 2013; see **Box 6.7**). Economic costs of land-use change may also be high; for example, a 75 per cent reduction in economic benefits from nature-based recreation has been reported following replacement of mountain forest with crops in Nepal (Thapa *et al.* 2016).

6.6.8 Polar regions

Biodiversity in the Arctic and Antarctic regions is under particular stress (Bennett *et al.* 2015) (see Section 4.3.2). Many native species are in decline; rising temperatures and invasive species, especially in the sub-Antarctic and Antarctic Peninsula, are major pressures (Hughes, Cowan and Wilmotte 2015; Amesbury *et al.* 2017). Industrial development, pollution and local disturbances present additional pressures (Conservation of Arctic Flora and Fauna [CAFF] 2013), with polar regions acting as a sink for many anthropogenic pollutants such as persistent organic pollutants (POPs) and other synthetic organic chemicals (Alava *et al.* 2017).

Box 6.7: Climate change and the need for ecosystem-based adaptation: the Hindu Kush Himalayas

While climate change may bring some benefits to mountain regions (e.g. longer growing seasons), the preponderance of impact is negative. Increased variability in precipitation patterns (including variability in monsoon and more frequent extreme rainfall) coupled with glacial ice melt, is predicted to increase risks of floods (carrying rock, sediments and debris), landslides, fire, soil erosion and spread of water-related and vector-borne diseases (Ebi *et al.* 2007; Armstrong 2010; Ahmed and Suphachalasai 2014). Of particular concern are the potentially devastating impacts from glacial lake outburst floods which have become more frequent since the middle of the 20th century (Armstrong 2010; International Centre for Integrated Mountain Development 2011).

The Hindu Kush Himalayas, the greater Himalayan region extending from eastern Nepal and Bhutan to northern Afghanistan, are among the most extensive areas covered by glaciers and permafrost on the planet. They contain water resources that drain through ten of the largest rivers in Asia, from which over 1.3 billion people derive their livelihoods and upon which many more depend for water and other resources (Eriksson *et al.* 2009). The region has been recognized as a unique biodiversity-rich area with equally unique topographic characteristics and socioeconomic and environmental challenges. The accelerated rate of warming, glacier ice melt and related implications on the hydrological systems are among the most pressing challenges to this unique mountain ecosystem (Gerlitz *et al.* 2017). It is essential that these macro-climatic effects are integrated into plans to conserve the fragile biodiversity of the region.

Substantial changes expected to Antarctic ice sheets before the turn of the century may have considerable global consequences (Chown *et al.* 2017) (see Section 4.3.2). Under most climate scenarios, the Arctic is projected to be ice-free in summer by 2050 (IPCC 2013, p. 1090), although remnants of multi-year ice will remain off the coasts of Canada and Alaska. The retreat of sea ice is likely to result in major ecological shifts linked to:

a) an increase in primary productivity as a result of more open water and greater freshwater flow carrying nutrients;
b) a comparable shift in the source and quality of food for species at higher trophic levels such as krill, fish and marine mammals (Frey *et al.* 2016; Alsos *et al.* 2016); and
c) an influx of new species into the polar regions with productivity and food web relationships changing as coastal and sea ice systems of polar regions experience earlier spring bloom and longer growing periods for microalgae (Potts *et al.* 2016).

Average abundance of Arctic vertebrates increased from 1970 until 1990 and then remained fairly stable through to 2007, as measured by the Arctic Species Trend Index (McRae *et al.* 2012; CAFF 2013). However, some food resources are being lost in areas of diminishing sea ice, posing health risks to species such as the walrus, ivory gull, polar bear and Barents Sea harp seal (CAFF 2017). Penguins are one of the more regularly monitored species groups in Antarctica, and populations have been changing over the last century with recorded declines in some colonies of macaroni, Adélie and chinstrap penguins (Trathan, Lynch and Fraser 2016).

It is likely that, due to higher productivity, the availability of some natural resources will increase for circumpolar peoples and communities (Arrigo 2014), but changes in hunting conditions will have a detrimental impact on the Inuit and other groups that have relied on seal hunting and other traditional food sources for which sea ice provides access. Some negative impacts are already being felt; for example, a significant die-off of seals and walruses in the Pacific Arctic in 2011 affected food sources for indigenous communities in the United States of America, Canada and Russian Federation (CAFF 2017). Breaks in the dormancy of pathogenic bacteria and viruses in thawing permafrost are a direct threat to human health (Sutherland *et al.* 2018).

The opening of potential new fishing zones, oil and gas development and shipping may result in future conflicts, especially with regard to economic use, governance, cultural interests and marine protected areas. As the Antarctic has no indigenous people or local communities and is outside the range of the Convention on Biological Diversity's Nagoya Protocol, the equitable sharing of benefits from biodiversity to people, including those benefits derived from bioprospecting, represents a particular challenge not completely addressed by the Antarctic Treaty System (Chown *et al.* 2017).

6.7 Responses

A broad spectrum of governance approaches and policy instruments are used to help address biodiversity loss. Their effectiveness and specific examples are explored in Chapter 13.

6.7.1 The Convention on Biological Diversity (CBD)

The CBD has been the key global convention on biodiversity in recent decades and it has three central goals: the conservation of biological diversity, the sustainable use of its components, and the fair and equitable sharing of the benefits arising out of the utilization of genetic resources. With 196 Parties in 2018, it establishes international norms and provides a forum for states to cooperate and share information and coordinate policy. In 2010 member states adopted the Strategic Plan for Biodiversity 2011-2020, as well as the more specific Aichi Biodiversity Targets, a comprehensive and ambitious array of goals subsequently reflected in many of the United Nations Sustainable Development Goals (SDGs). The midterm assessment of progress towards the Aichi Biodiversity targets concluded that, while progress has been made, it was insufficient to achieve them by 2020 (SCBD 2014).

The CBD's Cartegena Protocol on Biosafety deals with the international transfer of living modified organisms (LMOs), demanding advanced and 'informed' agreement from the importing country prior to the exchange of any LMOs, which includes genetically modified organisms (GMOs) such as seeds. The Nagoya Protocol on Access to Genetic Resources and the Fair and Equitable Sharing of Benefits Arising from their Utilization to the Convention on Biological Diversity establishes a framework for access to genetic resources and the sharing of benefits arising from their utilization, including the transfer of relevant technologies, which directly aims to curb biopiracy and promote equity in future bioprospecting agreements. It has been ratified by 105 countries as of May 2018. The Secretariat of the CBD plays a key role in raising awareness and organizing regional workshops and other capacity-building exercises.

An important mandatory requirement of Parties to the CBD is a commitment to produce National Biodiversity Strategies and Action Plans (NBSAPs) with associated targets (see Chapter 13.1). The Global Environment Facility (GEF), through its enabling activities window, provides support to eligible Parties which focuses on revising/updating their NBSAPs considering the CBD Strategic Plan for Biodiversity 2011-2020 and the Aichi Biodiversity Targets. This support is routed through the United Nations Development Programme (UNDP) and UN Environment (UNEP) as the key implementing agencies (Pisupati and Prip 2015). The CBD also supports the creation of subnational biodiversity strategies and action plans and regional (supranational) plans, and collaborates with the other key multilateral environmental agreements that have biodiversity-related mandates such as the Convention on International Trade in Endangered Species of Wild Fauna and Flora (CITES) (see **Box 6.8** and Annex 6-1).

6.7.2 Intergovernmental Science-Policy Platform on Biodiversity and Ecosystem Services (IPBES)

In 2012, IPBES was officially established with a stated mission "to strengthen the science-policy interface for biodiversity and ecosystem services/nature's contributions to people for the conservation and sustainable use of biodiversity, long-term human well-being and sustainable development." IPBES is organized under the auspices of four United Nations agencies – UNEP, United Nations Educational, Scientific

Box 6.8: The international wildlife trade and CITES

The Convention on International Trade in Endangered Species of Wild Flora and Fauna (CITES) came into force in 1975 and had 183 Parties by 2018. International trade of flora and fauna is worth billions of dollars and includes hundreds of millions of species and species parts, including food products, artistic ornaments and many traditional medicines (Broad, Mulliken and Roe 2003; Rosen and Smith 2010). Today, the agreement assigns various degrees of protection to over 35,000 species of plants and animals (CITES 2018).

Species listed in CITES that are traded across borders are subject to controls through a licensing system managed by member countries. CITES species are listed in three Appendices attached to the Convention: Appendix I provides the highest degree of protection, effectively banning all commercial trade in wild-taken alive or dead specimens of the species; trade in specimens on Appendix II is strictly regulated; Appendix III indicates a country has unilaterally asked for the help of other Parties in controlling trade in the species, subject to regulation within its jurisdiction.

The CITES agenda is ambitious, and the Convention is not self-executing: parties must implement and enforce its provisions under national law. This is a difficult task requiring significant educational and enforcement resources, and corruption can be problematic (Bennett 2015).

and Cultural Organization (UNESCO), Food and Agriculture Organization of the United Nations (FAO) and UNDP – and is administered by UNEP. By June 2018, its membership comprised 130 governments as well as a number of major stakeholder groups.

6.7.3 Protected areas

Protected areas have been successful in reducing habitat loss (Aichi Biodiversity Target 5) and have helped in lowering extinction risk for some target species (Aichi Target 12) (UNEP-WCMC and IUCN 2018). However, despite clear evidence that investment in conservation can help reduce biodiversity loss (Geldmann *et al.* 2013; Waldron *et al.* 2017), less than 15 per cent of the world's terrestrial and inland waters, less than 11 per cent of the coastal and marine areas within national jurisdiction, and less than 4 per cent of the global ocean is covered by protected areas **(Figure 6.20)** (UNEP-WCMC and IUCN 2018; Sala *et al.* 2018). In addition, a third of the land area within protected area boundaries is already degraded by human impacts (Jones *et al.* 2018).

While providing biodiversity benefits, protected areas can have potentially negative effects on livelihoods in local communities due to decreased access to natural resources or the lack of support for the development of cultural, social, financial, natural, human, physical and political capital assets (Bennett and Dearden 2014). This can result in ineffective management,

Figure 6.20: Protected areas of the world

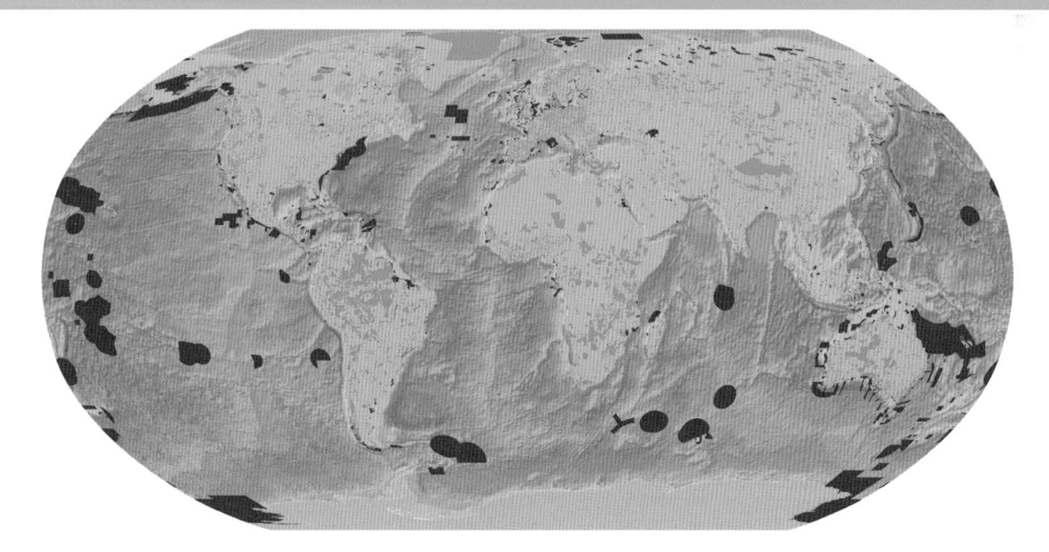

Protected Areas of the World
- Terrestrial protected areas
- Marine and coastal protected areas

Source: UNEP-WCMC and IUCN (2018).

Box 6.9: Biodiversity conservation and poverty

It is increasingly accepted that biodiversity loss and poverty are closely coupled problems, though seeking to solve one does not automatically address the other (SCBD 2010; Suich, Howe and Mace 2015). Indeed, some approaches to protecting particular species or natural areas have exacerbated existing uneven access to natural resources and placed disproportionate burdens on already-vulnerable populations (Dowie 2009; Sylvester, Segura and Davidson-Hunt 2016). Intergenerational justice is also an important theme, since loss of biodiversity will impoverish future generations in a variety of ways, including reducing their ability to rely upon and connect with a biodiverse natural world.

Biodiversity conservation is likely to be more effective in programmes that successfully integrate social and ecological support, and the benefits from conservation are more likely to be directly accessible by local human populations (Figurel, Durán and Bray 2011; Persha, Agrawal and Chhatre 2011; Fischer *et al.* 2017).

equity issues, lack of accountability or conflict (Halpern *et al.* 2014; Watson *et al.* 2014; Di Minin and Toivonen 2015; Eklund and Cabeza 2017; see also **Box 6.9**). The active engagement of indigenous and local communities in the decision-making process has proven highly effective at addressing these imbalances **(see Box 6.10)**. Analysis of deforestation rates indicate that these can be significantly lower in community-managed forests in comparison to strictly protected areas (Porter-Bolland *et al.* 2012). The development of a more inclusive and integrated approach linking communities with national, divisional and provincial governments for sustainable development has proved highly efficient (see Locally Managed Marine Areas case study in Fiji in Section 13.2.1). Increasingly, indigenous and local communities' contributions and collective actions have the potential to be scaled up and to inform national and international practice and provide a practical governance approach as an alternative to top-down policy-setting.

6.7.4 Other approaches

Many other approaches have evolved to confront biodiversity loss and respond to related drivers. Biodiversity offsets create biodiversity benefits to compensate for losses (Gordon *et al.* 2015; Apostolopoulou and Adams 2017). Controversially based on the monetization of nature (Adams 2014; Costanza *et al.* 2017), offset programmes have been developed in numerous countries within the last ten years. Monetary valuation can serve as a useful tool in underpinning policy instruments such as socioeconomic assessments of public policies and investments, and economic incentives such as payment for ecosystem services, permits and taxation schemes (Bateman *et al.* 2013; Gaworecki 2017). Another economic instrument is the United Nations System of Environmental-Economic Accounting (Experimental Ecosystem Accounting), developed in 2012. Examples of ecosystem accounting have been prepared (e.g. Victoria in Australia, Uganda, and the United Kingdom of Great Britain and Northern Ireland; Eigenraam, Chua and Hasker 2013; UNEP-WCMC and Institute for Development of Environmental-Economic Accounting [IDEEA] 2017; United Kingdom Office for National Statistics 2018), and initiatives to encourage its use in planning have been launched (see https://www.wavespartnership.org and https://naturalcapitalcoalition.org/).

Efforts to address deforestation and forest degradation in developing countries culminated in international agreement under the United Nations Framework Convention on Climate Change (UNFCCC) on methodological guidance for implementing activities relating to reducing emissions from deforestation and forest degradation (REDD) and the role of conservation, sustainable management of forests and enhancement of forest carbon stocks in developing countries – known as REDD+ (UNFCCC 2018). Forest certification, such as that promoted by the Forest Stewardship Council (https://www.fsc.org/) and the Programme for the Endorsement of Forest Certification (https://www.pefc.org/) provides greater information flow to consumers, encompassing not just logging and extraction but also the social and economic well-being of workers and local communities (e.g. forest management certification

Box 6.10: Female rangers in South Africa

In 2015, a South African ranger group consisting mostly of women, the Black Mamba Anti-Poaching Unit, was one of the winners of the top United Nations environmental prize. The unit was formed in a bid to engage local communities outside conservation parks in protecting biodiversity inside the fences. Initially comprising 26 unemployed female high-school graduates, the unit has reduced snaring by 76 per cent since its launch in 2013, removed more than 1,000 snares, and put five poachers' camps and two bushmeat kitchens out of action (United Nations 2015).

http://www.blackmambas.org/uploads/8/3/5/5/83556980/screen-shot-2016-07-18-at-4-34-38-pm_orig.png

in Indonesia; Miteva, Loucks and Pattanayak 2015), and transparency and inclusiveness in decision-making. In the European Union (EU) Common Agricultural Policy, some mechanisms have been developed to address environmental problems through protecting and promoting biodiversity in the European countryside.

Within urban settings, a movement towards 'green cities' is gathering pace, especially, but not only, within developed countries (Hegazy, Seddik and Ibrahim 2017), which highlights the protection and expansion of urban forests and green spaces and parks, and the recreational and air quality benefits they provide to people (Salbitano *et al.* 2016), including increased exposure to microbial biodiversity, important for healthy immune responses (Lax, Nagler and Gilbert 2015). Public engagement in urban agriculture, and specific programmes on beekeeping and bird conservation can facilitate human contact with nature in an urban setting. Urban and peri-urban agriculture, when guided by principles of agroecology, with wastes (or by-products) reused as raw materials, promotes self-sufficiency, gender equality, disaster resilience, water and soil conservation and environmental sustainability (FAO 2001; van Veenhuizen 2012).

More generally, ecosystem-based adaptation (EbA) promotes the conservation, sustainable management and restoration of natural ecosystems to help people and communities adapt to climate change (Cohen-Shacham *et al.*2016). However, the effective integration of EbA is challenged by scientific uncertainty at the international scale and disputes over criteria for prioritization (Ojea 2015; Bourne *et al.* 2016).

Ocean governance is particularly complex. Current efforts are focused on the elaboration of the text of an international legally binding instrument under the United Nations Convention on the Law of the Sea on the conservation and sustainable use of marine biological diversity of areas beyond national jurisdiction (ABNJ).

6.8 Conclusion

Our understanding of the natural world and the threats posed to its integrity has never been greater. New technologies have allowed us unparalleled insight into the different dimensions of biodiversity, from genomes to biomes. The major pressures on biodiversity are increasingly well-understood – habitat transformation/land-use change, invasive species, pollution, overexploitation including the illegal wildlife trade and climate change – though each of the world's biomes faces distinct challenges, reflecting particular geographic, ecological and socioeconomic contexts. Biodiversity loss is exacerbated where there is significant inequality in wealth and is a major threat to intergenerational justice. But the political and social will necessary to preserve biological diversity has been lacking. While certain policy responses have demonstrated effectiveness in promoting biodiversity conservation, persistent negative trends in almost every aspect of biodiversity indicate the need for more concerted action. Wildlife populations are thinning, reducing their adaptive potential; current rates of species extinctions are estimated to be orders of magnitude greater than background rates, with some scientists suggesting that we may be entering a sixth mass extinction event, and ecosystems are becoming increasingly degraded.

Increased investment in conservation on a global scale is urgently required. Greater focus on strengthening governance systems; improving policy frameworks through research; integration, implementation and effective enforcement; and encouraging partnerships and participation, are all measures that have the potential to address the greatest pressures on biodiversity. Efforts to combat biodiversity loss must also address poverty eradication, gender inequality, systemic corruption in governance structures and other social variables. The path to conserving global biodiversity and to finding solutions for sustainable use is a long but critical journey; humankind depends on it to support nature's contributions to people and the flourishing of health and development.

References

Abson, D.J., Fraser, E.D.G. and Benton, T.G. (2013). Landscape diversity and the resilience of agricultural returns: A portfolio analysis of land-use patterns and economic returns from lowland agriculture. *Agriculture & Food Security* 2(2). https://doi.org/10.1186/2048-7010-2-2.

Adams, W.M. (2014). The value of valuing nature. *Science* 346(6209), 549. https://doi.org/10.1126/science.1255997.

Agrawal, A., Cashore, B., Hardin, R., Shepherd, G., Benson, C. and Miller, D. (2013). Economic contributions of forests. *United Nations Forum on Forests Tenth Session*. Istanbul, 8-19 April 2013. United Nations Forum on Forests http://www.un.org/esa/forests/pdf/session_documents/unff10/EcoContrForests.pdf

Ahmed, M. and Suphachalasai, S. (2014). *Assessing the costs of climate change and adaptation in South Asia*. Manila: Asian Development Bank. https://think-asia.org/bitstream/handle/11540/46/assessing-costs-climate-change-and-adaptation-south-asia.pdf?sequence=1.

Aide, T.M., Clark, M.L., Grau, H.R., López-Carr, D., Levy, M.A., Redo, D. *et al.* (2013). Deforestation and reforestation of Latin America and the Caribbean (2001–2010). *Biotropica* 45(2), 262-271. https://doi.org/10.1111/j.1744-7429.2012.00908.x.

Akiner, M.M., Demirci, B., Babuadze, G., Robert, V. and Schaffner, F. (2016). Spread of the Invasive Mosquitoes Aedes aegypti and Aedes albopictus in the Black Sea Region Increases Risk of Chikungunya, Dengue, and Zika Outbreaks in Europe. *PLOS Neglected Tropical Diseases* 10(4), e0004664. https://doi.org/10.1371/journal.pntd.0004664.

Alamgir, M., Campbell, M.J., Sloan, S., Goosem, M., Clements, G.R., Mahmoud, M.I. *et al.* (2017). Economic, Socio-Political and Environmental Risks of Road Development in the Tropics. *Current Biology* 27(20), R1130-R1140. https://doi.org/10.1016/j.cub.2017.08.067.

Alava, J.J., Cheung, W.W.L., Ross, P.S. and Sumaila, U.R. (2017). Climate change–contaminant interactions in marine food webs: Toward a conceptual framework. *Global Change Biology* 23(10), 3984–4001. https://doi.org/10.1111/gcb.13667.

Alexander, P., Rounsevell, M.D.A., Dislich, C., Dodson, J.R., Engström, K. and Moran, D. (2015). Drivers for global agricultural land use change: The nexus of diet, population, yield and bioenergy. *Global Environmental Change* 35, 138-147. https://doi.org/10.1016/j.gloenvcha.2015.08.011.

Alexandratos, N. and Bruinsma, J. (2012). *World Agriculture Towards 2030/2050: The 2012 Revision*. ESA Working Paper No. 12-03. Rome: Food and Agriculture Organization. http://www.fao.org/docrep/016/ap106e/ap106e.pdf.

Almond, R.E.A., Butchart, S.H.M., Oldfield, T.E.E., McRae, L. and de Bie, S. (2013). Exploitation indices: Developing global and national metrics of wildlife use and trade.In *Biodiversity Monitoring and Conservation: Bridging the gap between global commitment and local action*. Collen, B., Pettorelli, N., Baillie, J.E.M. and Durant, S.M. (eds.). Oxford: Wiley-Blackwell. chapter 8. 159-188. https://onlinelibrary.wiley.com/doi/pdf/10.1002/9781118490747.ch8

Alroy, J. (2017). Effects of habitat disturbance on tropical forest biodiversity. *Proceedings of the National Academy of Sciences* 114(23), 6056-6061. https://doi.org/10.1073/pnas.1611855114.

Alsos, I.G., Ehrich, D., Seidenkrantz, M.-S., Bennike, O., Kirchhefer, A.J. and Geirsdottir, A. (2016). The role of sea ice for vascular plant dispersal in the Arctic. *Biology letters* 12(9). https://doi.org/10.1098/rsbl.2016.0264.

Amesbury, M.J., Roland, T.P., Royles, J., Hodgson, D.A., Convey, P., Griffiths, H. *et al.* (2017). Widespread biological response to rapid warming on the Antarctic Peninsula. *Current Biology* 27(11), 1616-1622. https://doi.org/10.1016/j.cub.2017.04.034.

Apostolopoulou, E. and Adams, W.M. (2017). Biodiversity offsetting and conservation: Reframing nature to save it. *Oryx* 51(1), 23-31. https://doi.org/10.1017/S0030605315000782.

Araújo, C.V.M. and Cedeño-Macias, L.A. (2016). Heavy metals in yellowfin tuna (Thunnus albacares) and common dolphinfish (Coryphaena hippurus) landed on the Ecuadorian coast. *Science of the Total Environment* 541, 149-154. https://doi.org/10.1016/j.scitotenv.2015.09.090.

Armstrong, R.L. (2010). *The Glaciers of the Hindu Kush-Himalayan Region: A Summary of the Science Regarding Glacier Melt/Retreat in the Himalayan, Hindu Kush, Karakoram, Pamir, and Tien Shan Mountain Ranges*. Kathmandu: International Centre for Integrated Mountain Development. http://lib.icimod.org/record/26917/files/attachment_734.pdf.

Arrigo, K.R. (2014). Sea ice ecosystems. *Annual review of marine science* 6, 439-467. https://doi.org/10.1146/annurev-marine-010213-135103.

Assan, J.K., Caminade, C. and Obeng, F. (2009). Environmental variability and vulnerable livelihoods: Minimising risks and optimising opportunities for poverty alleviation. *Journal of International Development* 21(3), 403-418. https://doi.org/10.1002/jid.1563.

Aukema, J.E., Leung, B., Kovacs, K., Chivers, C., Britton, K.O., Englin, J. *et al.* (2011). Economic Impacts of Non-Native Forest Insects in the Continental United States. *PloS one* 6(9), e24587. https://doi.org/10.1371/journal.pone.0024587.

Baccini, A., Walker, W., Carvalho, L., Farina, M., Sulla-Menashe, D. and Houghton, R.A. (2017). Tropical forests are a net carbon source based on aboveground measurements of gain and loss. *Science* 358(6360), 230-234. https://doi.org/10.1126/science.aam5962.

Barbier, E.B., Moreno-Mateos, D., Rogers, A.D., Aronson, J., Pendleton, L., Danovaro, R. *et al.* (2014). Protect the deep sea. *Nature* 505(7484), 475-477. https://www.nature.com/news/ecology-protect-the-deep-sea-1.14547.

Barnosky, A.D., Matzke, N., Tomiya, S., Wogan, G.O.U., Swartz, B., Quental, T.B. *et al.* (2011). Has the Earth's sixth mass extinction already arrived? *Nature* 471(7336), 51-57. https://doi.org/10.1038/nature09678.

Bartomeus, I., Ascher, J.S., Gibbs, J., Danforth, B.N., Wagner, D.L., Hedtke, S.M. *et al.* (2013). Historical changes in northeastern US bee pollinators related to shared ecological traits. *Proceedings of the National Academy of Sciences* 110(12), 4656-4660. https://doi.org/10.1073/pnas.1218503110.

Bateman, I.J., Harwood, A.R., Mace, G.M., Watson, R.T., Abson, D.J., Andrews, B. *et al.* (2013). Bringing ecosystem services into economic decision-making: Land use in the United Kingdom. *Science* 341(6141), 45. https://doi.org/10.1126/science.1234379.

Begna, D. (2015). Assessment of pesticides use and its economic impact on the apiculture subsector in selected districts of Amhara Region, Ethiopia. *Journal of Environmental & Analytical Toxicology* 5(3), 267. https://doi.org/10.4172/2161-0525.1000267.

Bellard, C., Cassey, P. and Blackburn, T.M. (2016). Alien species as a driver of recent extinctions. *Biology letters* 12(2). https://doi.org/10.1098/rsbl.2015.0623.

Bennett, E.L. (2015). Legal ivory trade in a corrupt world and its impact on African elephant populations. *Conservation Biology* 29(1), 54-60. https://doi.org/10.1111/cobi.12377.

Bennett, J.R., Shaw, J.D., Terauds, A., Smol, J.P., Aerts, R., Bergstrom, D.M. *et al.* (2015). Polar lessons learned: Long-term management based on shared threats in Arctic and Antarctic environments. *Frontiers in Ecology and the Environment* 13(6), 316-324. https://doi.org/10.1890/140315.

Bennett, N.J. and Dearden, P. (2014). Why local people do not support conservation: Community perceptions of marine protected area livelihood impacts, governance and management in Thailand. *Marine Policy* 44, 107-116. https://doi.org/10.1016/j.marpol.2013.08.017.

Bergman, Å., Heindel, J.J., Jobling, S., Kidd, K.A. and Zoeller, R.T. (eds.) (2013). *State of the Science of Endocrine Disrupting Chemicals - 2012*. Geneva: United Nations Environment Programme and the World Health Organization. http://www.who.int/iris/bitstream/10665/78101/1/9789241505031_eng.pdf?ua=1.

Bhutani, S. (2013). *Researching Agriculture in South Asia: The Law and Policy Context For Agricultural Research and Development and Its Impact on Smallholder Farmers*. London: International Institute for Environment and Development. http://re.indiaenvironmentportal.org.in/files/file/ReSearchingAgricultureJune2013.pdf.

Black, R., Adger, W.N., Arnell, N.W., Dercon, S., Geddes, A. and Thomas, D. (2011). The effect of environmental change on human migration. *Global Environmental Change* 21(Supplement 1), S3-S11. https://doi.org/10.1016/j.gloenvcha.2011.10.001.

Bourne, A., Holness, S., Holden, P., Scorgie, S., Donatti, C.I. and Midgley, G. (2016). A socio-ecological approach for identifying and contextualising spatial ecosystem-based adaptation priorities at the sub-national level. *PloS one* 11(5), e0155235. https://doi.org/10.1371/journal.pone.0155235.

Bouwman, A.F., Beusen, A.H.W., Overbeek, C.C., Bureau, D.P., Pawlowski, M. and Glibert, P.M. (2013). Hindcasts and Future Projections of Global Inland and Coastal Nitrogen and Phosphorus Loads Due to Finfish Aquaculture. *Reviews in Fisheries Science* 21(2), 112-156. https://doi.org/10.1080/10641262.2013.790340.

Brazil Ministério de Meio Ambiente (2015). *Terraclass: Projeto terraclass cerrado mapeamento do uso e cobertura vegetal do cerrado* [http://www.dpi.inpe.br/tccerrado/index.php?mais=1.

Brenton-Rule, E.C., Barbieri, R.F. and Lester, P.J. (2016). Corruption, development and governance indicators predict invasive species risk from trade. *Proceedings of the Royal Society B: Biological Sciences* 283(1832). https://doi.org/10.1098/rspb.2016.0901.

Broad, S., Mulliken, T. and Roe, D. (2003). The nature and extent of legal and illegal trade in wildlife. In *The trade in wildlife: regulation for conservation*. Oldfield, S. (ed.). London: Earthscan Publications. chapter 1. 3-22. http://dlib.scu.ac.ir/bitstream/Hannan/462459/2/185383954X.pdf

Brockerhoff, E.G., Barbaro, L., Castagneyrol, B., Forrester, D.I., Gardiner, B., González-Olabarria, J.R. *et al.* (2017). Forest biodiversity, ecosystem functioning and the provision of ecosystem services. *Biodiversity and Conservation* 26(13), 3005-3035. https://doi.org/10.1007/s10531-017-1453-2.

Brooks, B.W., Lazorchak, J.M., Howard, M.D.A., Johnson, M.-V.V., Morton, S.L., Perkins, D.A.K. *et al.* (2016). Are harmful algal blooms becoming the greatest inland water quality threat to public health and aquatic ecosystems? *Environmental Toxicology and Chemistry* 35(1), 6-13. https://doi.org/10.1002/etc.3220.

Bruford, M.W., Davies, N., Dulloo, M.E., Faith, D.P. and Walters, M. (2017). Monitoring changes in genetic diversity.In *The GEO Handbook on Biodiversity Observation Networks*. Walters, M. and Scholes, R. (eds.). Cham: Springer. 107-128. https://link.springer.com/chapter/10.1007/978-3-319-27288-7_5#citeas

Butt, N., Beyer, H.L., Bennett, J.R., Biggs, D., Maggini, R., Mills, M. *et al.* (2013). Biodiversity risks from fossil fuel extraction. *Science* 342(6157), 425-426. https://doi.org/10.1126/science.1237261.

Cable, J., Barber, I., Boag, B., Ellison, A.R., Morgan, E.R., Murray, K. *et al.* (2017). Global change, parasite transmission and disease control: Lessons from ecology. *Philosophical Transactions of the Royal Society B: Biological Sciences* 372(1719). https://doi.org/10.1098/rstb.2016.0088.

Cardinale, B.J., Duffy, J.E., Gonzalez, A., Hooper, D.U., Perrings, C., Venail, P. *et al.* (2012). Biodiversity loss and its impact on humanity. *Nature* 486(7401), 59-67. https://doi.org/10.1038/nature11148.

Ceballos, G., Ehrlich, P.R., Barnosky, A.D., García, A., Pringle, R.M. and Palmer, T.M. (2015). Accelerated modern human–induced species losses: Entering the sixth mass extinction. *Science advances* 1(5), e1400253. https://doi.org/10.1126/sciadv.1400253.

Ceballos, G., Ehrlich, P.R. and Dirzo, R. (2017). Biological annihilation via the ongoing sixth mass extinction signaled by vertebrate population losses and declines. *Proceedings of the National Academy of Sciences* 114(30), E6089-E6096. https://doi.org/10.1073/pnas.1704949114.

Centers for Disease Control and Prevention (CDC) (2017). *Zoonotic Diseases*. https://www.cdc.gov/onehealth/basics/zoonotic-diseases.html (Accessed: 2017 1 December).

Chen, I.C., Hill, J.K., Ohlemüller, R., Roy, D.B. and Thomas, C.D. (2011). Rapid range shifts of species associated with high levels of climate warming. *Science* 333(6045), https://doi.org/10.1126/science.1206432.

Chettri, N. (2015). Reconciling mountain biodiversity conservation in a changing climate: A Hindu Kush-Himalayan perspective. *Conservation Science* 2(1), 17-27. https://doi.org/10.3126/cs.v2i1.13766.

Chown, S.L., Brooks, C.M., Terauds, A., Le Bohec, C., van Klaveren-Impagliazzo, C., Whittington, J.D. *et al.* (2017). Antarctica and the strategic plan for biodiversity. *PLoS biology* 15(3), e2001656. https://doi.org/10.1371/journal.pbio.2001656.

Clark, N.E., Lovell, R., Wheeler, B.W., Higgins, S.L., Depledge, M.H. and Norris, K. (2014). Biodiversity, cultural pathways, and human health: A framework. *Trends in Ecology & Evolution* 29(4), 198-204. https://www.doi.org/10.1016/j.tree.2014.01.009.

Cohen-Shacham, E., Walters, G., Janzen, C. and Maginnis, S. (2016). *Nature-based Solutions to address global societal challenges*. Gland: IUCN. https://www.researchgate.net/profile/Emmanuelle_Cohen-Shacham/publication/307608144_Nature-based_Solutions_to_address_global_societal_challenges/links/57cd67f408ae59825189ca7a.pdf.

Collen, B., Böhm, M., Kemp, R. and Baillie, J.E. (2012). *Spineless: status and trends of the world's invertebrates*. London: Zoological Society of London. https://www.zsl.org/sites/default/files/media/2014-02/spineless-report.pdf.

Collen, B., Whitton, F., Dyer, E.E., Baillie, J.E.M., Cumberlidge, N., Darwall, W.R.T. *et al.* (2014). Global patterns of freshwater species diversity, threat and endemism. *Global Ecology and Biogeography* 23(1), 40-51. https://doi.org/10.1111/geb.12096.

Conservation of Arctic Flora and Fauna (2013). *Arctic Biodiversity Assessment: Status and Trends in Arctic Biodiversity*. Akureyri. http://arcticlcc.org/assets/resources/ABA2013Science.pdf.

Conservation of Arctic Flora and Fauna (2017). *State of The Arctic Marine Biodiversity: Key Findings and Advice For Monitoring*. Akureyri: Conservation of Arctic Flora and Fauna. https://oaarchive.arctic-council.org/bitstream/handle/11374/1955/SAMBR_Summary_April_2017_LR.pdf?sequence=1&isAllowed=y.

Convention on Biological Diversity (2014). *Pathways of introductions of invasive species, their prioritization and management*. Subsidiary Body on Scientific, Technical and Technological Advice (SBSTTA) of the Convention on Biological Diversity. https://www.cbd.int/doc/meetings/sbstta/sbstta-18/official/sbstta-18-09-add1-en.pdf.

Convention on Biological Diversity (2016). Decision Adopted by the Conference of the Parties to the Convention on Biological Diversity XIII/17. Synthetic Biology. CBD/COP/DEC/XIII/17. 4 https://www.cbd.int/doc/decisions/cop-13/cop-13-dec-17-en.pdf

Convention on International Trade in Endangered Species of Wild Fauna and Flora (2018). *What is CITES?* https://www.cites.org/eng/disc/what.php (Accessed: 5 June 2017).

Costanza, R., de Groot, R., Braat, L., Kubiszewski, I., Fioramonti, L., Sutton, P. *et al.* (2017). Twenty years of ecosystem services: How far have we come and how far do we still need to go? *Ecosystem Services* 28, 1-16. https://doi.org/10.1016/j.ecoser.2017.09.008.

Costello, C., Ovando, D., Clavelle, T., Strauss, C.K., Hilborn, R., Melnychuk, M.C. *et al.* (2016). Global fishery prospects under contrasting management regimes. *Proceedings of the National Academy of Sciences* 113(18), 5125-5129. https://doi.org/10.1073/pnas.1520420113.

Cuba-Díaz, M., Troncoso, J.M., Cordero, C., Finot, V.L. and Rondanelli-Reyes, M. (2013). Juncus bufonius, a new non-native vascular plant in King George Island, South Shetland Islands. *Antarctic Science* 25(3), 385-386. https://doi.org/10.1017/S0954102012000958.

Dangal, S.R.S., Tian, H., Zhang, B., Pan, S., Lu, C. and Yang, J. (2017). Methane emission from global livestock sector during 1890–2014: Magnitude, trends and spatiotemporal patterns. *Global Change Biology* 23(10), 4147-4161. https://doi.org/10.1111/gcb.13709.

Di Minin, E. and Toivonen, T. (2015). Global protected area expansion: Creating more than paper parks. *BioScience* 65(7), 637-638. https://doi.org/10.1093/biosci/biv064.

Dinerstein, E., Olson, D., Joshi, A., Vynne, C., Burgess, N.D., Wikramanayake, E. *et al.* (2017). An ecoregion-based approach to protecting half the terrestrial realm. *BioScience* 67(6), 534-545. https://doi.org/10.1093/biosci/bix014.

Dirzo, R., Young, H.S., Galetti, M., Ceballos, G., Isaac, N.J.B. and Collen, B. (2014). Defaunation in the anthropocene. *Science* 345(6195), 401-406. https://doi.org/10.1126/science.1251817.

Dixon, M.J.R., Loh, J., Davidson, N.C., Beltrame, C., Freeman, R. and Walpole, M. (2016). Tracking global change in ecosystem area: The Wetland Extent Trends index. *Biological Conservation* 193, 27-35. https://doi.org/10.1016/j.biocon.2015.10.023.

Djoudi, H., Vergles, E., Blackie, R.R., Koame, C.K. and Gautier, D. (2015). Dry forests, livelihoods and poverty alleviation: understanding current trends. *International Forestry Review* 17, 54-69. https://doi.org/10.1505/146554815815834868.

Doherty, T.S., Glen, A.S., Nimmo, D.G., Ritchie, E.G. and Dickman, C.R. (2016). Invasive predators and global biodiversity loss. *Proceedings of the National Academy of Sciences* 113(40), 11261-11265. https://doi.org/10.1073/pnas.1602480113.

Dowie, M. (2009). *Conservation Refugees: The Hundred-Year Conflict Between Global Conservation and Native Peoples.* Cambridge, MA: MIT Press. http://web.mnstate.edu/robertsb/307/Articles/Conservation_Refugees_Intro.pdf.

Duffy, G.A., Coetzee, B.W.T., Latombe, G., Akerman, A.H., McGeoch, M.A. and Chown, S.L. (2017). Barriers to globally invasive species are weakening across the Antarctic. *Diversity and Distributions* 23(9), 982-996. https://doi.org/10.1111/ddi.12593.

Early, R., Bradley, B.A., Dukes, J.S., Lawler, J.J., Olden, J.D., Blumenthal, D.M. *et al.* (2016). Global threats from invasive alien species in the twenty-first century and national response capacities. *Nature Communications* 7, 12485. https://doi.org/10.1038/ncomms12485.

Ebi, K.L., Woodruff, R., von Hildebrand, A. and Corvalan, C. (2007). Climate change-related health impacts in the Hindu Kush–Himalayas. *EcoHealth* 4(3), 264-270. https://doi.org/10.1007/s10393-007-0119-z.

Eigenraam, M., Chua, J. and hasker, J. (2013). *Environmental-Economic Accounting: Victorian Experimental Ecosystem Accounts, Version 1.0.* [Department of Sustainability and Environment, State of Victoria https://www.researchgate.net/profile/Mark_Eigenraam2/publication/273692801_Environmental-Economic_Accounting_Victorian_Experimental_Ecosystem_Accounts_Version_10/links/550881190cf2d7a28129f415/Environmental-Economic-Accounting-Victorian-Experimental-Ecosystem-Accounts-Version-10.pdf.

Ejigu, K., Adgaba, N. and Bekele, W. (2008). The role of women and indigenous knowledge in Ethiopian beekeeping. *Bees for Development* 86. http://www.beesfordevelopment.org/media/2656/bfdj86-women-ethiopia008.pdf.

Eklund, J. and Cabeza, M. (2017). Quality of governance and effectiveness of protected areas: Crucial concepts for conservation planning. *Annals of the New York Academy of Sciences* 1399(1), 27-41. https://doi.org/10.1111/nyas.13284.

Emmerson, M., Morales, M.B., Oñate, J.J., Batáry, P., Berendse, F., Liira, J. *et al.* (2016). How Agricultural Intensification Affects Biodiversity and Ecosystem Services. In *Advances in Ecological Research.* Dumbrell, A.J., Kordas, R.L. and Woodward, G. (eds.). Academic Press. 43-97. http://www.sciencedirect.com/science/article/pii/S0065250416300204

Engelbrecht, C.J. and Engelbrecht, F.A. (2016). Shifts in Köppen-Geiger climate zones over southern Africa in relation to key global temperature goals. *Theoretical and applied climatology* 123(1 2), 247-261. https://doi.org/10.1007/s00704-014-1354-1.

Epple, C. and Dunning, E. (2014). *Ecosystem Resilience to Climate Change: What is it and How Can it be Addressed in the Context of Climate Change Adaptation?* Cambridge: United Nations Environment Programme World Conservation Monitoring Centre. https://www.unep-wcmc.org/system/dataset_file_fields/files/000/000/288/original/Ecosystem_resilience_to_climate_change_formatted_20141219.pdf?1419260116.

Eriksson, M., Xu, J., Shrestha, A.B., Vaidya, R.A., Santosh, N. and Sandström, K. (2009). *The Changing Himalayas: Impact of Climate Change on Water Resources and Livelihoods in The Greater Himalayas.* Kathmandu: International Centre for Integrated Mountain Development. https://www.cabdirect.org/cabdirect/abstract/20093086376.

Figurel, J.J., Durán, E. and Bray, D.B. (2011). Conservation of the jaguar *Panthera onca* in a community-dominated landscape in montane forests in Oaxaca, Mexico. *Oryx* 45(4), 554-560. https://doi.org/10.1017/S0030605310001353.

Fischer, J., Abson, D.J., Bergsten, A., Collier, N.F., Dorresteijn, I., Hanspach, J. *et al.* (2017). Reframing the food–biodiversity challenge. *Trends in Ecology & Evolution* 32(5), 335-345. https://doi.org/10.1016/j.tree.2017.02.009.

Foale, S., Adhuri, D., Aliño, P., Allison, E.H., Andrew, N., Cohen, P. *et al.* (2013). Food security and the coral triangle initiative. *Marine Policy* 38, 174-183. https://doi.org/10.1016/j.marpol.2012.05.033.

Foley, J.A., Asner, G.P., Costa, M.H., Coe, M.T., DeFries, R., Gibbs, H.K. *et al.* (2007). Amazonia revealed: Forest degradation and loss of ecosystem goods and services in the Amazon Basin. *Frontiers in Ecology and the Environment* 5(1), 25-32. https://doi.org/10.1890/1540-9295(2007)5[25:ARFDAL]2.0.CO;2.

Foley, J.A., Ramankutty, N., Brauman, K.A., Cassidy, E.S., Gerber, J.S., Johnston, M. *et al.* (2011). Solutions for a cultivated planet. *Nature* 478(7369), 337-342. https://doi.org/10.1038/nature10452.

Food and Agriculture Organization of the United Nations (2001). *Urban and Peri-Urban Agriculture: A Briefing Guide for the Successful Implementation of Urban and Peri-urban Agriculture in Developing Countries and Countries of Transition.* Handbook Series. Rome. http://www.fao.org/fileadmin/templates/FCIT/PDF/briefing_guide.pdf.

Food and Agriculture Organization of the United Nations (2015a). *Coping with Climate Change: The Roles of Genetic Resources for Food and Agriculture.* Rome. http://www.fao.org/3/a-i3866e.pdf.

Food and Agriculture Organization of the United Nations (2015b). *FAOSTAT-Forestry Database: Global Production and Trade of Forest Products in 2015.* Rome http://www.fao.org/forestry/statistics/80938/en/ (Accessed April 2, 2017)

Food and Agriculture Organization of the United Nations (2016). *Sustainable Wildlife Management and Gender.* Rome. http://www.fao.org/3/a-i6574e.pdf.

Food and Agriculture Organization of the United Nations (2018a). *Sustainable development goals: SDG Indicator 2.5.2 - Risk status of livestock breeds.* Food and Agriculture Organization http://www.fao.org/sustainable-development-goals/indicators/252/en/ (Accessed: 1 June 2017).

Food and Agriculture Organization of the United Nations (2018b). *The State of World fisheries and Aquaculture 2018: Meeting the Sustainable Development Goals.* Rome. http://www.fao.org/3/i9540en/I9540EN.pdf.

Food and Agriculture Organization of the United Nations (2018c). *The State of the World's Forest 2018: Forest Pathways To Sustainable Development.* Rome. http://www.fao.org/3/ca0188en/ca0188en.pdf.

Food and Agriculture Organization of the United Nations and Platform for AgroBiodiversity Research (2011). *Biodiversity for Food and Agriculture: Contributing to Food Security and Sustainability in A Changing World.* Rome. http://www.fao.org/fileadmin/templates/biodiversity_paia/PAR-FAO-book_lr.pdf.

Fornace, K.M., Abidin, T.R., Alexander, N., Brock, P., Grigg, M.J., Murphy, A. *et al.* (2016). Association between landscape factors and spatial patterns of Plasmodium knowlesi infections in Sabah, Malaysia. *Emerging infectious diseases* 22(2), 201-209. https://doi.org/10.3201/eid2202.150656.

Foster, S., Swann, R.L. and Furness, R.W. (2017). Can changes in fishery landings explain long-term population trends in gulls? *Bird Study* 64(1), 90-97. https://doi.org/10.1080/00063657.2016.1274287.

Frey, K.E., Comiso J.C., Cooper, L.W., Gradinger, R.R., Grebmeier, J.M. and Tremblay, J.É. (2016). Arctic ocean primary productivity.In *Arctic Report Card 2016.* ftp://ftp.oar.noaa.gov/arctic/documents/ArcticReportCard_full_report2016.pdf

Galiè, A., Jiggins, J., Struik, P.C., Grando, S. and Ceccarelli, S. (2017). Women's empowerment through seed improvement and seed governance: Evidence from participatory barley breeding in pre-war Syria. *NJAS-Wageningen Journal of Life Sciences* 81, 1-8. https://doi.org/10.1016/j.njas.2017.01.002.

Gamfeldt, L., Snäll, T., Bagchi, R., Jonsson, M., Gustafsson, L., Kjellander, P. *et al.* (2013). Higher levels of multiple ecosystem services are found in forests with more tree species. *Nature Communications* 4(1340). https://doi.org/10.1038/ncomms2328.

Gang, C., Zhou, W., Chen, Y., Wang, Z., Sun, Z., Li, J. *et al.* (2014). Quantitative assessment of the contributions of climate change and human activities on global grassland degradation. *Environmental Earth Sciences* 72(11), 4273-4282. https://doi.org/10.1007/s12665-014-3322-6.

Gavrilescu, M., Demnerová, K., Aamand, J., Agathos, S. and Fava, F. (2015). Emerging pollutants in the environment: present and future challenges in biomonitoring, ecological risks and bioremediation. *New Biotechnology* 32(1), 147-156. https://doi.org/10.1016/j.nbt.2014.01.001.

Gaworecki, M. (2017). Cash for conservation: Do payments for ecosystem services work? *Mongabay Series: Conservation Effectiveness*, Mongabay https://news.mongabay.com/2017/10/cash-for-conservation-do-payments-for-ecosystem-services-work/.

Geldmann, J., Barnes, M., Coad, L., Craigie, I.D., Hockings, M. and Burgess, N.D. (2013). Effectiveness of terrestrial protected areas in reducing habitat loss and population declines. *Biological conservation* 161, 230-238. https://doi.org/10.1016/j.biocon.2013.02.018.

Genovesi, P., Carnevali, L. and Scalera, R. (2015). *The Impact of Invasive Alien Species on Native Threatened Species in Europe.* Technical report for the European Commission. Rome: Institute for Environmental Protection and Research. https://wedocs.unep.org/bitstream/handle/20.500.11822/19388/ISSG_report_impact_of_IAS_on_biodiversity_in_E.pdf?sequence=1.

Gerlitz, J.-Y., Macchi, M., Brooks, N., Pandey, R., Banerjee, S. and Jha, S.K. (2017). The multidimensional livelihood vulnerability index – an instrument to measure livelihood vulnerability to change in the Hindu Kush Himalayas. *Climate and Development* 9(2), 124-140. https://doi.org/10.1080/17565529.2016.1145099.

Ghazi, W.T., Muniruzzaman, A.N.M. and Singh, A.K. (2016). *Climate Change and Security in South Asia: Cooperating for Peace.* Global Military Advisory Council on Climate Change. http://gmaccc.org/wp-content/uploads/2016/05/Climate_Change_and_Security_in_South_Asia.pdf.

Gordon, A., Bull, J.W., Wilcox, C. and Maron, M. (2015). FORUM: Perverse incentives risk undermining biodiversity offset policies. *Journal of Applied Ecology* 52(2), 532-537. https://doi.org/10.1111/1365-2664.12398.

Green Economy Coalition (2012). *The Green Economy Pocketbook: The Case For Action.* London. http://www.greengrowthknowledge.org/sites/default/files/downloads/resource/The_GE_Pocketbook_The_case_for_action_GEC.pdf.

Grêt-Regamey, A., Brunner, S.H. and Kienast, F. (2012). Mountain ecosystem services: Who cares? *Mountain Research and Development* 32, S23-S34. https://doi.org/10.1659/MRD-JOURNAL-D-10-00115.S1.

Grytnes, J.A., Kapfer, J., Jurasinski, G., Birks, H.H., Henriksen, H., Klanderud, K. *et al.* (2014). Identifying the driving factors behind observed elevational range shifts on European mountains. *Global Ecology and Biogeography* 23(8), 876-884. https://doi.org/10.1111/geb.12170.

Guerra, C.A., Snow, R.W. and Hay, S.I. (2006). A global assessment of closed forests, deforestation and malaria risk. *Annals of tropical medicine and parasitology* 100(3), 189-204. https://www.ncbi.nlm.nih.gov/pmc/articles/PMC3204444/.

Haines-Young, R. and Potschin, M. (2010). The links between biodiversity, ecosystem services and human well-being.In *Ecosystem Ecology: a new synthesis.* Raffaelli, D.G. and Frid, C.L.J. (eds.). Cambridge: Cambridge University Press. chapter 6. 110-139. https://www.nottingham.ac.uk/cem/pdf/Haines-Young&Potschin_2010.pdf

Hallmann, C.A., Sorg, M., Jongejans, E., Siepel, H., Hofland, N., Schwan, H. *et al.* (2017). More than 75 per cent decline over 27 years in total flying insect biomass in protected areas. *PloS one* 12(10), e0185809. https://doi.org/10.1371/journal.pone.0185809.

Halpern, B.S. (2014). Making marine protected areas work. *Nature* 506, 167-168. https://doi.org/10.1038/nature13053.

Hansen, M.C., Potapov, P.V., Moore, R., Hancher, M., Turubanova, S.A., Tyukavina, A. *et al.* (2013). High-resolution global maps of 21st-century forest cover change. *Science* 342(6160), 850-853. https://doi.org/10.1126/science.1244693.

Hara, K., Zhao, Y., Tomita, M., Kamagata, N. and Li, Y. (2016). Impact of the Great East Japan Earthquake and Tsunami on coastal vegetation and landscapes in northeast Japan: Findings based on remotely sensed data analysis.In *Ecological Impacts of Tsunamis on Coastal Ecosystems.* Urabe J. and Nakashizuka, T. (eds.). Tokyo: Springer. 253-269. https://link.springer.com/chapter/10.1007/978-4-431-56448-5_16

Harrison, P.A., Berry, P.M., Simpson, G., Haslett, J.R., Blicharska, M., Bucur, M. *et al.* (2014). Linkages between biodiversity attributes and ecosystem services: A systematic review. *Ecosystem Services* 9, 191-203. https://doi.org/10.1016/j.ecoser.2014.05.006.

He, F., Zarfl, C., Bremerich, V., Henshaw, A., Darwall, W., Tockner, K. *et al.* (2017). Disappearing giants: A review of threats to freshwater megafauna. *Wiley Interdisciplinary Reviews: Water* 4(3), e1208. https://doi.org/10.1002/wat2.1208.

Hegazy, I., Seddik, W. and Ibrahim, H. (2017). Towards green cities in developing countries: Egyptian new cities as a case study. *International Journal of Low-Carbon Technologies* 12(4), 358-368. https://doi.org/10.1093/ijlct/ctx009.

Hoegh-Guldberg, O., Mumby, P.J., Hooten, A.J., Steneck, R.S., Greenfield, P., Gomez, E. *et al.* (2007). Coral reefs under rapid climate change and ocean acidification. *Science* 318(5857), 1737-1742. https://doi.org/10.1126/science.1152509.

Hoffmann, M., Brooks, T.M., Butchart, S.H.M., Gregory, R.D. and McRae, L. (2018). Trends in biodiversity: Vertebrates. *Encyclopedia of the Anthropocene* 3, 175-184. https://doi.org/10.1016/B978-0-12-809665-9.09963-8.

Houspanossian, J., Giménez, R., Jobbágy, E. and Nosetto, M. (2017). Surface albedo raise in the South American Chaco: Combined effects of deforestation and agricultural changes. *Agricultural and Forest Meteorology* 232, 118-127. https://doi.org/10.1016/j.agrformet.2016.08.015.

Hughes, K.A., Cowan, D.A. and Wilmotte, A. (2015). Protection of Antarctic microbial communities–'out of sight, out of mind'. *Frontiers in microbiology* 6(151). https://doi.org/10.3389/fmicb.2015.00151.

Hughes, T.P., Kerry, J.T., Álvarez-Noriega, M., Álvarez-Romero, J.G., Anderson, K.D., Baird, A.H. *et al.* (2017). Global warming and recurrent mass bleaching of corals. *Nature* 543(7645), https://doi.org/10.1038/nature21707.

Hunt, S.K., Galatowitsch, M.L. and McIntosh, A.R. (2017). Interactive effects of land use, temperature, and predators determine native and invasive mosquito distributions. *Freshwater Biology* 62(9), 1564-1577. https://doi.org/10.1111/fwb.12967.

Hussain, R.I., Walcher, R., Brandl, D., Arnberger, A., Zaller, J.G. and Frank, T. (2018). Efficiency of two methods of sampling used to assess the abundance and species diversity of adult Syrphidae (Diptera) in mountainous meadows in the Austrian and Swiss Alps. *European Journal of Entomology* 115, 150-156. https://doi.org/10.14411/eje.2018.014.

Hussain, R.I., Walcher, R., Brandl, D., Jernej, I., Arnberger, A., Zaller, J.G. *et al.* (2017). Influence of abandonment on syrphid assemblages in mountainous meadows. *Journal of Applied Entomology* 142(4), 450-456. https://doi.org/10.1111/jen.12482.

Innis, L. and Simcock, A. (eds.) (2016). *The First Global Integrated Marine Assessment: World Ocean Assessment I*. New York, NY. http://www.un.org/depts/los/global_reporting/WOA_RegProcess.htm.

Intergovernmental Panel on Climate Change (2013). *Climate Change 2013: The Physical Science Basis. Contribution of Working Group I to the Fifth Assessment Report of the Intergovernmental Panel on Climate Change.* Stocker, T.F., Qin, D., Plattner, G.-K., Tignor, M., Allen, S.K., Boschung, J. *et al.* (eds.). Cambridge. https://www.ipcc.ch/report/ar5/wg1/.

Intergovernmental Panel on Climate Change (2014). *Climate Change 2014: Mitigation of Climate Change. Contribution of Working Group III to the Fifth Assessment Report of the Intergovernmental Panel on Climate Change.* Edenhofer, O., Pichs-Madruga, R., Sokona, Y., Farahani, E., Kadner, S., Seyboth, K. *et al.* (eds.). Cambridge. https://www.ipcc.ch/pdf/assessment-report/ar5/wg3/ipcc_wg3_ar5_frontmatter.pdf.

Intergovernmental Science-Policy Platform on Biodiversity and Ecosystem Services (2013). Decision IPBES-2/4: Conceptual framework for the Intergovernmental Science-Policy Platform on Biodiversity and Ecosystem Services. Intergovernmental Science-Policy Platform on Biodiversity and Ecosystem Services, 9 https://www.ipbes.net/sites/default/files/downloads/Decision%20IPBES_2_4.pdf.

Intergovernmental Science-Policy Platform on Biodiversity and Ecosystem Services (2016). *The Assessment Report of The Intergovernmental Science-Policy Platform on Biodiversity and Ecosystem Services on Pollinators, Pollination and Food Production.* Potts, S.G., Imperatriz-Fonseca, V.L. and Ngo, H.T. (eds.). Bonn. https://www.researchgate.net/profile/Jean_Michel_Salles/publication/311486448_The_assessment_report_of_the_Intergovernmental_Science-Policy_Platform_on_Biodiversity_and_Ecosystem_Services_on_pollinators_pollination_and_food_production/links/58c27ef145851538eb7e6958/The-assessment-report-of-the-Intergovernmental-Science-Policy-Platform-on-Biodiversity-and-Ecosystem-Services-on-pollinators-pollination-and-food-production.pdf?origin=publication_detail.

International Centre for Integrated Mountain Development (2011). *Glacial Lakes and Glacial Lake Outburst Floods in Nepal.* Kathmandu. http://www.icimod.org/dvds/201104_GLOF/reports/final_report.pdf.

International Panel of Experts on Sustainable Food Systems (2016). *From Uniformity to Diversity: A Paradigm Shift from Industrial Agriculture to Diversified Agroecological Systems.* http://www.ipes-food.org/images/Reports/UniformityToDiversity_FullReport.pdf.

International Union for Conservation of Nature (2010). *Plants Under Pressure, A Global Assessment. The First Report of the IUCN Sampled Red List.* Kew: Royal Botanical Gardens, Natural History Museum and International Union for Conservation of Nature. https://www.kew.org/sites/default/files/kppcont_027304.pdf.

International Union for Conservation of Nature (2017a). *The Red List Index.* https://www.iucn.org/theme/species/our-work/iucn-red-list-threatened-species/red-list-index.

International Union for Conservation of Nature (2017b). *IUCN Red List of Ecosystems.* [https://iucnrle.org/ (Accessed: October 2 2017).

International Union for Conservation of Nature (2017c). *Grasslands.* [https://www.iucn.org/theme/protected-areas/wcpa/what-we-do/grasslands (Accessed: 12 June 2017).

International Union for Conservation of Nature (2018). *The IUCN Red List of Threatened Species. Version 2018-1.* http://www.iucnredlist.org (Accessed: June 9 2017).

Johansen, J.L. and Jones, G.P. (2011). Increasing ocean temperature reduces the metabolic performance and swimming ability of coral reef damselfishes. *Global Change Biology* 17(9), 2971-2979. https://doi.org/10.1111/j.1365-2486.2011.02436.x.

Joint Group of Experts on the Scientific Aspects of Marine Environmental Protection (2015). *Sources, Fate and Effects of Microplastics in The Marine Environment: A Global Assessment.* Kershaw, P.J. (ed.). London: International Maritime Organization. http://ec.europa.eu/environment/marine/good-environmental-status/descriptor-10/pdf/GESAMP_microplastics%20full%20study.pdf.

Jones, K.R., Venter, O., Fuller, R.A., Allan, J.R., Maxwell, S.L., Negret, P.J. *et al.* (2018). One-third of global protected land is under intense human pressure. *Science* 360(6390), 788. https://doi.org/10.1126/science.aap9565.

Karesh, W.B., Dobson, A., Lloyd-Smith, J.O., Lubroth, J., Dixon, M.A., Bennett, M. *et al.* (2012). Ecology of zoonoses: natural and unnatural histories. *The Lancet* 380(9857), 1936-1945. https://doi.org/10.1016/S0140-6736(12)61678-X.

Keesing, F. and Young, T.P. (2014). Cascading consequences of the loss of large mammals in an African savanna. *BioScience* 64(6), 487-495. https://doi.org/10.1093/biosci/biu059.

Kehoe, L., Romero-Muñoz, A., Polaina, E., Estes, L., Kreft, H. and Kuemmerle, T. (2017). Biodiversity at risk under future cropland expansion and intensification. *Nature Ecology & Evolution* 1(8), 1129-1135. https://doi.org/10.1038/s41559-017-0234-3.

Keith, D.A., Rodríguez, J.P., Brooks, T.M., Burgman, M.A., Barrow, E.G., Bland, L. *et al.* (2015). The IUCN red list of ecosystems: motivations, challenges, and applications. *Conservation Letters* 8(3), 214-226. https://doi.org/10.1111/conl.12167.

Keith, D.A., Rodriguez, J.P., Rodríguez-Clark, K.M., Nicholson, E., Aapala, K., Alonso, A. *et al.* (2013). Scientific foundations for an IUCN red list of ecosystems. *PloS one* 8(5), e62111. https://doi.org/10.1371/journal.pone.0062111.

Kelly, T.R., Karesh, W.B., Johnson, C.K., Gilardi, K.V.K., Anthony, S.J., Goldstein, T. *et al.* (2017). One Health proof of concept: Bringing a transdisciplinary approach to surveillance for zoonotic viruses at the human-wild animal interface. *Preventive Veterinary Medicine* 137, 112-118. https://doi.org/10.1016/j.prevetmed.2016.11.023.

Kettunen, M., Genovesi, P., Gollasch, S., Pagad, S., Starfinger, U., ten Brink, P. *et al.* (2008). *Technical support to EU strategy on invasive species (IS) - Assessment of the impacts of IS in Europe and the EU (final module report for the European Commission).* Brussels: Institute for European Environmental Policy (IEEP). http://ec.europa.eu/environment/nature/invasivealien/docs/Kettunen2009_IAS_Task%201.pdf.

Khan, S.M., Page, S.E., Ahmad, H. and Harper, D.M. (2013). Sustainable utilization and conservation of plant biodiversity in montane ecosystems: The western Himalayas as a case study. *Annals of botany* 112(3), 479-501. https://doi.org/10.1093/aob/mct125.

Kharouba, H.M., Ehrlén, J., Gelman, A., Bolmgren, K., Allen, J.M., Travers, S.E. *et al.* (2018). Global shifts in the phenological synchrony of species interactions over recent decades. *Proceedings of the National Academy of Sciences* 115(20), 5211-5216. https://doi.org/10.1073/pnas.1714511115.

Khoury, C.K., Bjorkman, A.D., Dempewolf, H., Ramirez-Villegas, J., Guarino, L., Jarvis, A. *et al.* (2014). Increasing homogeneity in global food supplies and the implications for food security. *Proceedings of the National Academy of Sciences* 111(11), 4001-4006. https://doi.org/10.1073/pnas.1313490111.

Klinger, D. and Naylor, R. (2012). Searching for Solutions in Aquaculture: Charting a Sustainable Course. *Annual Review of Environment and Resources* 37(1), 247-276. https://doi.org/10.1146/annurev-environ-021111-161531.

Knapp, S., Schweiger, O., Kraberg, A., Asmus, H., Asmus, R., Brey, T. *et al.* (2017). Do drivers of biodiversity change differ in importance across marine and terrestrial systems—Or is it just different research communities' perspectives? *Science of the Total Environment* 574, 191-203. https://doi.org/10.1016/j.scitotenv.2016.09.002.

Kulmatov, R. (2008). Modern problems in using, protecting, and managing water and land resources of the Aral Sea Basin.In *Environmental Problems of Central Asia and their Economic, Social and Security Impacts.* Qi J. and Evered K.T. (eds.). Dordrecht: Springer. 15-30. https://link.springer.com/chapter/10.1007%2F978-1-4020-8960-2_2?LI=true.

Kumsa, T. and Gorfu, B. (2014). Beekeeping as integrated watershed conservation and climatic change adaptation: An action research in Boredo watershed. *Journal of Earth Science and Climate Changes* 5(7), 213. https://doi.org/10.4172/2157-7617.1000213.

Landrigan, P.J., Fuller, R., Acosta, N.J.R., Adeyi, O., Arnold, R., Basu, N. *et al.* (2017). The Lancet Commission on pollution and health. *The Lancet.* https://doi.org/10.1016/S0140-6736(17)32345-0.

Laurance, W.F., Sayer, J. and Cassman, K.G. (2014). Agricultural expansion and its impacts on tropical nature. *Trends in Ecology & Evolution* 29(2), 107-116. https://doi.org/10.1016/j.tree.2013.12.001.

Lavoie, R.A., Jardine, T.D., Chumchal, M.M., Kidd, K.A. and Campbell, L.M. (2013). Biomagnification of mercury in aquatic food webs: A worldwide meta-analysis. *Environmental science & technology* 47(23), 13385-13394. https://doi.org/10.1021/es403103t.

Lax, S., Nagler, C.R. and Gilbert, J.A. (2015). Our interface with the built environment: Immunity and the indoor microbiota. *Trends in immunology* 36(3), 121-123. https://doi.org/10.1016/j.it.2015.01.001.

Leadley, P.W., Krug, C.B., Alkemade, R., Pereira, H.M., Sumaila, U.R., Walpole, M. *et al.* (2014). *Progress towards the Aichi Biodiversity Targets: An Assessment of Biodiversity Trends, Policy Scenarios and Key Actions.* Technical Series 78. Montreal: Secretariat of the Convention on Biological Diversity. https://www.cbd.int/doc/publications/cbd-ts-78-en.pdf.

Lenzen, M., Moran, D., Kanemoto, K., Foran, B., Lobefaro, L. and Geschke, A. (2012). International trade drives biodiversity threats in developing nations. *Nature* 486, 109. https://doi.org/10.1038/nature11145.

Lewison, R.L., Crowder, L.B., Wallace, B.P., Moore, J.E., Cox, T., Zydelis, R. *et al.* (2014). Global patterns of marine mammal, seabird, and sea turtle bycatch reveal taxa-specific and cumulative megafauna hotspots. *Proceedings of the National Academy of Sciences.* https://doi.org/10.1073/pnas.1318960111.

Li, Y., Hallerman, E.M., Liu, Q., Wu, K. and Peng, Y. (2015). The development and status of Bt rice in China. *Plant Biotechnology Journal* 14(3), 839-848. https://doi.org/10.1111/pbi.12464.

Low, P.S. (2013). Economic and social impacts of desertification, land degradation and drought: White paper i. *United Nations Convention to Combat Desertification 2nd Scientific Conference.* Bonn, 9-12 April 2013. United Nations Convention to Combat Desertification https://profiles.uonbi.ac.ke/jmariara/files/unccd_white_paper_1.pdf.

Lyver, P., Perez, E., Carneiro da Cunha, M. and Roué, M. (eds.) (2015). *Indigenous and Local Knowledge About Pollination and Pollinators Associated With Food Production: Outcomes From the Global Dialogue Workshop.* http://www.unesco.org/fileadmin/MULTIMEDIA/HQ/SC/pdf/IPBES_Pollination-Pollinators_Workshop.pdf.

MacDougall, A.S., McCann, K.S., Gellner, G. and Turkington, R. (2013). Diversity loss with persistent human disturbance increases vulnerability to ecosystem collapse. *Nature* 494(7435), 86-89. https://doi.org/10.1038/nature11869.

Mace, G.M., Norris, K. and Fitter, A.H. (2012). Biodiversity and ecosystem services: A multilayered relationship. *Trends in Ecology & Evolution* 27(1), 19-26. https://doi.org/10.1016/j.tree.2011.08.006.

Malaj, E., von der Ohe, P.C., Grote, M., Kühne, R., Mondy, C.P., Usseglio-Polatera, P. *et al.* (2014). Organic chemicals jeopardize the health of freshwater ecosystems on the continental scale. *Proceedings of the National Academy of Sciences* 111(26), 9549-9554. https://doi.org/10.1073/pnas.1321082111.

Malhi, Y., Doughty, C.E., Galetti, M., Smith, F.A., Svenning, J.-C. and Terborgh, J.W. (2016). Megafauna and ecosystem function from the pleistocene to the anthropocene. *Proceedings of the National Academy of Sciences* 113(4), 838-846. https://doi.org/10.1073/pnas.1502540113.

Marlow, J. (2017). *The Virus Hunters.* Undark Magazine https://undark.org/article/virus-hunters-ebola-usaid-predict/ (Accessed: 2017 5 December).

Maxwell, S.L., Fuller, R.A., Brooks, T.M. and Watson, J.E.M. (2016). Biodiversity: The ravages of guns, nets and bulldozers. *Nature* 536(7615), 143-145. https://doi.org/10.1038/536143a.

McCauley, D.J., Pinsky, M.L., Palumbi, S.R., Estes, J.A., Joyce, F.H. and Warner, R.R. (2015). Marine defaunation: Animal loss in the global ocean. *Science* 347(6219), 1255641. https://doi.org/10.1126/science.1255641.

McRae, L., Böhm, M., Deinet, S., Gill, M. and Collen, B. (2012). The Arctic Species Trend Index: using vertebrate population trends to monitor the health of a rapidly changing ecosystem. *Biodiversity* 13(3-4), 144-156. https://doi.org/10.1080/14888386.2012.705085.

McRae, L., Deinet, S. and Freeman, R. (2017). The diversity-weighted Living Planet Index: controlling for taxonomic bias in a global biodiversity indicator. *PloS one* 12(1), e0169156. https://doi.org/10.1371/journal.pone.0169156.

Mekonnen, M.M. and Hoekstra, A.Y. (2015). Global gray water footprint and water pollution levels related to anthropogenic nitrogen loads to fresh water. *Environmental science & technology* 49(21), 12860-12868. https://doi.org/10.1021/acs.est.5b03191.

Midgley, G.F. and Bond, W.J. (2015). Future of African terrestrial biodiversity and ecosystems under anthropogenic climate change. *Nature Climate Change* 5(9), 823-829. https://doi.org/10.1038/nclimate2753.

Miteva, D.A., Loucks, C.J. and Pattanayak, S.K. (2015). Social and environmental impacts of forest management certification in Indonesia. *PloS one* 10(7), e0129675. https://doi.org/10.1371/journal.pone.0129675.

Miura, O., Sasaki, Y. and Chiba, S. (2012). Destruction of populations of Batillaria attramentaria (Caenogastropoda: Batillariidae) by tsunami waves of the 2011 Tohoku earthquake. *Journal of Molluscan Studies* 78(4), 377-380. https://doi.org/10.1093/mollus/eys025.

Molina-Montenegro, M.A., Carrasco-Urra, F., Rodrigo, C., Convey, P., Valladares, F. and Gianoli, E. (2012). Occurrence of the non-native annual bluegrass on the Antarctic mainland and its negative effects on native plants. *Conservation Biology* 26(4), 717-723. https://doi.org/10.1111/j.1523-1739.2012.01865.x.

Mudd, G.M. (2010). The environmental sustainability of mining in Australia: key mega-trends and looming constraints. *Resources Policy* 35(2), 98-115. https://doi.org/10.1016/j.resourpol.2009.12.001.

Nackley, L.L., West, A.G., Skowno, A.L. and Bond, W.J. (2017). The nebulous ecology of native invasions. *Trends in Ecology & Evolution* 32(11), 814-824. https://doi.org/10.1016/j.tree.2017.08.003.

Naeem, S., Chazdon, R., Duffy, J.E., Prager, C. and Worm, B. (2016). Biodiversity and human well-being: an essential link for sustainable development. *Proceedings of the Royal Society B: Biological Sciences* 283(1844). https://doi.org/10.1098/rspb.2016.2091.

Newbold, T., Hudson, L.N., Hill, S.L.L., Contu, S., Lysenko, I., Senior, R.A. *et al.* (2015). Global effects of land use on local terrestrial biodiversity. *Nature* 520(7545), 45-50. https://doi.org/10.1038/nature14324.

Newbold, T., Hudson, L.N., Arnell, A.P., Contu, S., De Palma, A., Ferrier, S. *et al.* (2016). Has land use pushed terrestrial biodiversity beyond the planetary boundary? A global assessment. *Science* 353(6296), 288-291. https://doi.org/10.1126/science.aaf2201.

Nordlund, L.M., Koch, E.W., Barbier, E.B. and Creed, J.C. (2016). Seagrass ecosystem services and their variability across Genera and geographical regions. *PloS one* 12(1), e0169942. https://doi.org/10.1371/journal.pone.0163091.

O'Neill, A.R., Badola, H.K., Dhyani, P.P. and Rana, S.K. (2017). Integrating ethnobiological knowledge into biodiversity conservation in the Eastern Himalayas. *Journal of Ethnobiology and Ethnomedicine* 13(21), 1-14. https://doi.org/10.1186/s13002-017-0148-9.

Ocean Health Index (2017). *Habitat Destruction*. [http://www.oceanhealthindex.org/methodology/components/habitat-destruction (Accessed: October 7 2017).

O'Connor, D. and Ford, J. (2014). Increasing the effectiveness of the "Great Green Wall" as an adaptation to the effects of climate change and desertification in the Sahel. *Sustainability* 6(10), 7142-7154. https://doi.org/10.3390/su6107142.

O'Dea, C.B., Anderson, S., Sullivan, T., Landers, D. and Casey, C.F. (2017). Impacts to ecosystem services from aquatic acidification: Using FEGS-CS to understand the impacts of air pollution. *Ecosphere* 8(5), e01807. https://doi.org/10.1002/ecs2.1807.

Oh, B., Lee, K.J., Zaslawski, C., Yeung, A., Rosenthal, D., Larkey, L. *et al.* (2017). Health and well-being benefits of spending time in forests: Systematic review. *Environmental health and preventive medicine* 22(71), 1-11. https://doi.org/10.1186/s12199-017-0677-9.

Ojea, E. (2015). Challenges for mainstreaming ecosystem-based adaptation into the international climate agenda. *Current Opinion in Environmental Sustainability* 14, 41-48. https://doi.org/10.1016/j.cosust.2015.03.006.

Olivero, J., Fa, J.E., Real, R., Márquez, A.L., Farfán, M.A., Vargas, J.M. *et al.* (2017). Recent loss of closed forests is associated with Ebola virus disease outbreaks. *Scientific Reports* 7(1), 14291. https://doi.org/10.1038/s41598-017-14727-9.

Ottinger, M., Clauss, K. and Kuenzer, C. (2016). Aquaculture: Relevance, distribution, impacts and spatial assessments – A review. *Ocean & Coastal Management* 119, 244-266. https://doi.org/10.1016/j.ocecoaman.2015.10.015.

Pacifici, M., Foden, W.B., Visconti, P., Watson, J.E.M., Butchart, S.H.M., Kovacs, K.M. *et al.* (2015). Assessing species vulnerability to climate change. *Nature Climate Change* 5(3), 215-224. https://doi.org/10.1038/nclimate2448.

Packer, J.G., Meyerson, L.A., Richardson, D.M., Brundu, G., Allen, W.J., Bhattarai, G.P. *et al.* (2017). Global networks for invasion science: benefits, challenges and guidelines. *Biological invasions* 19(4), 1081-1096. https://doi.org/10.1007/s10530-016-1302-3.

Paini, D.R., Sheppard, A.W., Cook, D.C., De Barro, P.J., Worner, S.P. and Thomas, M.B. (2016). Global threat to agriculture from invasive species. *Proceedings of the National Academy of Sciences* 113(27), 7575. https://doi.org/10.1073/pnas.1602205113

Parmesan, C. (2006). Ecological and evolutionary responses to recent climate change. *Annual Review Ecology, Evolution, and Systematics* 37, 637-669. https://doi.org/10.1146/annurev.ecolsys.37.091305.110100.

Pascual, U., Balvanera, P., Díaz, S., Pataki, G., Roth, E., Stenseke, M. *et al.* (2017). Valuing nature's contributions to people: The IPBES approach. *Current Opinion in Environmental Sustainability* 26-27, 7-16. https://doi.org/10.1016/j.cosust.2016.12.006.

Pauli, H., Gottfried, M., Dullinger, S., Abdaladze, O., Akhalkatsi, M., Alonso, J.L.B. *et al.* (2012). Recent plant diversity changes on Europe's mountain summits. *Science* 336(6079), 353-355. https://doi.org/10.1126/science.1219033.

Payne, R.J., Dise, N.B., Field, C.D., Dore, A.J., Caporn, S.J.M. and Stevens, C.J. (2017). Nitrogen deposition and plant biodiversity: Past, present, and future. *Frontiers in Ecology and the Environment* 15(8), 431-436. https://doi.org/10.1002/fee.1528.

Pejchar, L. and Mooney, H.A. (2009). Invasive species, ecosystem services and human well-being. *Trends in Ecology & Evolution* 24(9), 497-504. https://doi.org/10.1016/j.tree.2009.03.016.

Persha, L., Agrawal, A. and Chhatre, A. (2011). Social and ecological synergy: Local rulemaking, forest livelihoods, and biodiversity conservation. *Science* 331(6024), 1606-1608. https://doi.org/10.1126/science.1199343.

Phalan, B., Green, R.E., Dicks, L.V., Dotta, G., Feniuk, C., Lamb, A. *et al.* (2016). How can higher-yield farming help to spare nature? *Science* 351(6272), 450-451. https://doi.org/10.1126/science.aad0055.

Piao, S., Tan, K., Nan, H., Ciais, P., Fang, J., Wang, T. *et al.* (2012). Impacts of climate and CO_2 changes on the vegetation growth and carbon balance of Qinghai–Tibetan grasslands over the past five decades. *Global and Planetary Change* 90-99, 73 80. https://doi.org/10.1016/j.gloplacha.2012.08.009.

Pimm, S.L., Jenkins, C.N., Abell, R., Brooks, T.M., Gittleman, J.L., Joppa, L.N. *et al.* (2014). The biodiversity of species and their rates of extinction, distribution, and protection. *Science* 344(6187), 1246752. https://doi.org/10.1126/science.1246752.

Pisupati, B. and Prip, C. (2015). *Interim Assessment of Revised National Biodiversity Strategies and Action Plans (NBSAPs)*. Cambridge: United Nations Environment Programme World Conservation Monitoring Centre. https://www.cbd.int/doc/nbsap/Interim-Assessment-of-NBSAPs.pdf.

Platform for AgroBiodiversity Research (2013). *The indigenous pollinators network*. [http://agrobiodiversityplatform.org/par/2013/12/24/the-indigenous-pollinators-network/.

Plieninger, T., van der Horst, D., Schleyer, C. and Bieling, C. (2014). Sustaining ecosystem services in cultural landscapes. *Ecology and Society* 19(2), 59. https://doi.org/10.5751/ES-06159-190259.

Plowright, R.K., Eby, P., Hudson, P.J., Smith, I.L., Westcott, D., Bryden, W.L. *et al.* (2015). Ecological dynamics of emerging bat virus spillover. *Proceedings of the Royal Society B* 282(1798), 20142124. https://doi.org/10.1098/rspb.2014.2124.

Porter-Bolland, L., Ellis, E.A., Guariguata, M.R., Ruiz-Mallén, I., Negrete-Yankelevich, S. and Reyes-García, V. (2012). Community managed forests and forest protected areas: An assessment of their conservation effectiveness across the tropics. *Forest Ecology and Management* 268, 6-17. https://doi.org/10.1016/j.foreco.2011.05.034.

Post, E., Bhatt, U.S., Bitz, C.M., Brodie, J.F., Fulton, T.L., Hebblewhite, M. *et al.* (2013). Ecological consequences of sea-ice decline. *Science* 341(6145), 519-524. https://doi.org/10.1126/science.1235225.

Potts, S.G., Imperatriz-Fonseca, V., Ngo, H.T., Aizen, M.A., Biesmeijer, J.C., Breeze, T.D. *et al.* (2016). Safeguarding pollinators and their values to human well-being. *Nature* 540, 220-229. https://doi.org/10.1038/nature20588.

Pulliam, J.R.C., Epstein, J.H., Dushoff, J., Rahman, S.A., Bunning, M., Jamaluddin, A.A. *et al.* (2011). Agricultural intensification, priming for persistence and the emergence of Nipah virus: A lethal bat-borne zoonosis. *Journal of The Royal Society Interface* 9(66). https://doi.org/10.1098/rsif.2011.0223.

Pywell, R.F., Heard, M.S., Woodcock, B.A., Hinsley, S., Ridding, L., Nowakowski, M. *et al.* (2015). Wildlife-friendly farming increases crop yield: evidence for ecological intensification. *Proceedings of the Royal Society B: Biological Sciences* 282(1816). https://doi.org/10.1098/rspb.2015.1740.

Quinn, C.H., Stringer, L.C., Berman, R.J., Le, H.T.V., Msuya, F.E., Pezzuti, J.C.B. *et al.* (2017). Unpacking changes in mangrove social-ecological systems: Lessons from Brazil, Zanzibar, and Vietnam. *Resources* 6(1), 14. https://doi.org/10.3390/resources6010014.

Rabitsch, W., Essl, F. and Schindler, S. (2017). The rise of non-native vectors and reservoirs of human diseases.In *Impact of Biological Invasions on Ecosystems Services*. M., V. and P., H. (eds.). Cham: Springer. 263-275. https://doi.org/10.1007/978-3-319-45121-3_17

Rapai, W. (2016). *Lake Invaders: Invasive Species and the Battle for the Future of the Great Lake*. Detroit, MI: Wayne State University Press. https://www.wsupress.wayne.edu/books/detail/lake-invaders.

Rendón, L.M., Guhl, F., Cordovez, J.M. and Erazo, D. (2015). New scenarios of *Trypanosoma cruzi* transmission in the Orinoco region of Colombia. *Memórias do Instituto Oswaldo Cruz* 110(3), 283-288. https://doi.org/10.1590/0074-02760140403.

Ricciardi, A., Blackburn, T.M., Carlton, J.T., Dick, J.T.A., Hulme, P.E., Iacarella, J.C. *et al.* (2017). Invasion Science: A Horizon Scan of Emerging Challenges and Opportunities. *Trends in Ecology & Evolution* 32(6), 464-474. https://doi.org/10.1016/j.tree.2017.03.007.

Rochman, C.M., Hoh, E., Kurobe, T. and Teh, S.J. (2013). Ingested plastic transfers hazardous chemicals to fish and induces hepatic stress. *Scientific Reports* 3(3263). https://doi.org/10.1038/srep03263.

Rodrigues, A.S.L., Brooks, T.M., Butchart, S.H.M., Chanson, J., Cox, N., Hoffmann, M. *et al.* (2014). Spatially explicit trends in the global conservation status of vertebrates. *PloS one* 10(3), e0121040. https://doi.org/10.1371/journal.pone.0113934.

Rodríguez, J.P., Keith, D.A., Rodríguez-Clark, K.M., Murray, N.J., Nicholson, E., Regan, T.J. *et al.* (2015). A practical guide to the application of the IUCN Red List of Ecosystems criteria. *Philosophic Transactions of the Royal Society B* 370(1662), 20140003. https://doi.org/10.1098/rstb.2014.0003.

Rosen, G.E. and Smith, K.F. (2010). Summarizing the evidence on the international trade in illegal wildlife. *EcoHealth* 7(1), 24-32. https://doi.org/10.1007/s10393-010-0317-y.

Rótolo, G.C., Francis, C., Craviotto, R.M., Viglia, S., Pereyra, A. and Ulgiati, S. (2015). Time to re-think the GMO revolution in agriculture. *Ecological Informatics* 26, 35-49. https://doi.org/10.1016/j.ecoinf.2014.05.002.

Royal Botanic Gardens Kew (2010). *Plants Under Pressure, a Global Assessment. The First Report of the IUCN Sampled Red List*. Kew and London: Royal Botanical Gardens, Natural History Museum and IUCN. https://www.kew.org/sites/default/files/kppcont_027304.pdf.

Royal Botanical Gardens Kew (2016). *The State of the World's Plants 2016*. https://stateoftheworldsplants.org/2016/report/sotwp_2016.pdf.

Royal Botanic Gardens Kew (2017). *The State of the World's Plants 2017*. https://stateoftheworldsplants.com/2017/report/SOTWP_2017.pdf.

Sala, E., Lubchenco, J., Grorud-Colvert, K., Novelli, C., Roberts, C. and Sumaila, U.R. (2018). Assessing real progress towards effective ocean protection. *Marine Policy* 91, 11-13. https://doi.org/10.1016/j.marpol.2018.02.004.

Salbitano, F., Borelli, S., Conigliaro, M. and Chen, Y. (2016). *Guidelines on Urban and Peri-Urban Forestry*. FAO Forestry Paper No.178. Rome: Food and Agriculture Organisation of the United Nations. http://www.fao.org/3/a-i6210e.pdf.

Scheffers, B.R., De Meester, L., Bridge, T.C.L., Hoffmann, A.A., Pandolfi, J.M., Corlett, R.T. *et al.* (2016). The broad footprint of climate change from genes to biomes to people. *Science* 354(6313). https://doi.org/10.1126/science.aaf7671.

Schütte, G., Eckerstorfer, M., Rastelli, V., Reichenbecher, W., Restrepo-Vassalli, S., Ruohonen-Lehto, M. *et al.* (2017). Herbicide resistance and biodiversity: Agronomic and environmental aspects of genetically modified herbicide-resistant plants. *Environmental Sciences Europe* 29(5). https://doi.org/10.1186/s12302-016-0100-y.

Secretariat of the Convention on Biological Diversity (2010). *Linking Biodiversity Conservation and Poverty Alleviation: A State of Knowledge Review*. Montreal. https://www.cbd.int/doc/publications/cbd-ts-55-en.pdf.

Secretariat of the Convention on Biological Diversity (2012). *Cities and Biodiversity Outlook: Action and Policy. A Global Assessment of the Links between Action and Policy Urbanization, Biodiversity, and Ecosystem Services*. Montreal. https://www.cbd.int/doc/health/cbo-action-policy-en.pdf.

Secretariat of the Convention on Biological Diversity (2013). Biodiversity is key to sustainable, efficient, resilient and nutritious food production. *Biodiversity for Food Security and Nutrition*, 5. July 2013. https://www.cbd.int/doc/newsletters/development/news-dev-2015-2013-07-en.pdf.

Secretariat of the Convention on Biological Diversity (2014). *Global Biodiversity Outlook 4: A Mid-Term Assessment of Progress Towards the Implementation of the Strategic Plan for Biodiversity 2011-2020*. Montréal. https://www.cbd.int/gbo/gbo4/publication/gbo4-en-hr.pdf.

Secretariat of the Convention on Biological Diversity (2015). *Synthetic Biology*. CBD Technical Series No. 82. Montreal. https://www.cbd.int/ts/cbd-ts-82-en.pdf.

Secretariat of the Convention on Biological Diversity (2016). *Marine Debris: Understanding, Preventing and Mitigating the Significant Adverse Impacts on Marine and Coastal Biodiversity*. CBD Technical Series No. 83. Montreal. https://www.cbd.int/doc/publications/cbd-ts-83-en.pdf.

Seddon, N., Mace, G.M., Naeem, S., Tobias, J.A., Pigot, A.L., Cavanagh, R. *et al.* (2016). Biodiversity in the Anthropocene: Prospects and policy. *Proceedings of the Royal Society B: Biological Sciences* 283(1844). https://doi.org/10.1098/rspb.2016.2094.

Seebens, H., Blackburn, T.M., Dyer, E.E., Genovesi, P., Hulme, P.E., Jeschke, J.M. *et al.* (2017). No saturation in the accumulation of alien species worldwide. *Nature communications* 8, 14435. https://doi.org/10.1038/ncomms14435.

Sehring, J. and Diebold, A. (2012). *From The Glaciers To The Aral Sea-Water Unites*. 1st edn: Trescher Verlag. https://www.researchgate.net/publication/319112234_From_the_Glaciers_to_the_Aral_Sea_Water_Unites.

Serda, B., Zewudu, T., Dereje, M. and Aman, M. (2015). Beekeeping practices, production potential and challenges of bee keeping among beekeepers in Haramaya District, Eastern Ethiopia. *Journal of Veterinary Science and Technology* 6(5), 255. https://doi.org/10.4172/2157-7579.1000255.

Shen, M., Piao, S., Chen, X., An, S., Fu, Y.H., Wang, S. *et al.* (2016). Strong impacts of daily minimum temperature on the green-up date and summer greenness of the Tibetan Plateau. *Global Change Biology* 22(9), 3057-3066. https://doi.org/10.1111/gcb.13301.

Shrestha, U.B., Gautam, S. and Bawa, K.S. (2012). Widespread climate change in the Himalayas and associated changes in local ecosystems. *PloS one* 7(5), e36741. https://doi.org/10.1371/journal.pone.0036741.

Sibhatu, K.T., Krishna, V.V. and Qaim, M. (2015). Production diversity and dietary diversity in smallholder farm households. *Proceedings of the National Academy of Sciences* 112(34), 10657-10662. https://doi.org/10.1073/pnas.1510982112.

Smith, P., House, J.I., Bustamante, M., Sobocká, J., Harper, R., Pan, G. *et al.* (2016). Global change pressures on soils from land use and management. *Global Change Biology* 22(3), 1008-1028. https://doi.org/10.1111/gcb.13068.

Sobrevila, C. (2008). *The Role of Indigenous Peoples in Biodiversity Conservation*. World Bank. https://siteresources.worldbank.org/INTBIODIVERSITY/Resources/RoleofIndigenousPeoplesinBiodiversityConservation.pdf.

South Africa, Department of Environmental Affairs (2016). *Rhino poaching statistics update 2007-2015*. [https://www.environment.gov.za/projectsprogrammes/rhinodialogues/poaching_statistics#2015 (Accessed: 2 April 2017).

Southern Africa Development Community (2014). *Livestock information management system*. Southern Africa Development Community. http://gisportal.sadc.int/lims-db/

Spatz, D.R., Zilliacus, K.M., Holmes, N.D., Butchart, S.H.M., Genovesi, P., Ceballos, G. *et al.* (2017). Globally threatened vertebrates on islands with invasive species. *Science advances* 3(10). https://doi.org/10.1126/sciadv.1603080.

Steffen, W., Richardson, K., Rockström, J., Cornell, S.E., Fetzer, I., Bennett, E.M. *et al.* (2015). Planetary boundaries: Guiding human development on a changing planet. *Science* 347(6223), 1259855. https://doi.org/10.1126/science.1259855.

Stevens, N., Lehmann, C.E.R., Murphy, B.P. and Durigan, G. (2017). Savanna woody encroachment is widespread across three continents. *Global Change Biology* 23(1), 235-244. https://doi.org/10.1111/gcb.13409.

Stimson Center (2016). *Environmental crime: Defining the challenge as a global security issue and setting the stage for integrated collaborative solutions*. [Stimson https://www.stimson.org/enviro-crime/ (Accessed: 12 April 2017).

Stoett, P. (2018). Unearthing under-governed territory: Transnational environmental crime.In *Just Security in an Undergoverned World*. Larik, J., Ponzio, R. and Durch, W. (eds.). Oxford: Oxford University Press. 238-263. https://global.oup.com/academic/product/just-security-in-an-undergoverned-world-9780198805373

Strayer, D.L. (2010). Alien species in fresh waters: ecological effects, interactions with other stressors, and prospects for the future. *Freshwater Biology* 55(1), 152-174. https://www.doi.org/10.1111/j.1365-2427.2009.02380.x.

Suich, H., Howe, C. and Mace, G. (2015). Ecosystem services and poverty alleviation: A review of the empirical links. *Ecosystem Services* 12, 137-147. https://doi.org/10.1016/j.ecoser.2015.02.005.

Sunday, J.M., Fabricius, K.E., Kroeker, K.J., Anderson, K.M., Brown, N.E., Barry, J.P. *et al.* (2017). Ocean acidification can mediate biodiversity shifts by changing biogenic habitat. *Nature Climate Change* 7, 81-85. https://doi.org/10.1038/nclimate3161.

Sutherland, W.J., Butchart, S.H.M., Connor, B., Culshaw, C., Dicks, L.V., Dinsdale, J. *et al.* (2018). A 2018 horizon scan of emerging issues for global conservation and biological diversity. *Trends in Ecology & Evolution* 33(1), 47-58. https://doi.org/10.1016/j.tree.2017.11.006.

Sylvester, O., Segura, A.G. and Davidson-Hunt, I.J. (2016). The protection of forest biodiversity can conflict with food access for indigenous people. *Conservation and society* 14(3), 279-290. https://doi.org/10.4103/0972-4923.191157.

Thackeray, S.J., Henrys, P.A., Hemming, D., Bell, J.R., Botham, M.S., Burthe, S. *et al.* (2016). Phenological sensitivity to climate across taxa and trophic levels. *Nature* 535(7611), 241-245. https://doi.org/10.1038/nature18608.

Thapa, I., Butchart, S.H.M., Gurung, H., Stattersfield, A.J., Thomas, D.H.L. and Birch, J.C. (2016). Using information on ecosystem services in Nepal to inform biodiversity conservation and local to national decision-making. *Oryx* 50(1), 147-155. https://doi.org/10.1017/S0030605314000088.

Tilman, D., Balzer, C., Hill, J. and Befort, B.L. (2011). Global food demand and the sustainable intensification of agriculture. *Proceedings of the National Academy of Sciences* 108(50), 20260-20264. https://doi.org/10.1073/pnas.1116437108.

Tittensor, D.P., Walpole, M., Hill, S.L.L., Boyce, D.G., Britten, G.L., Burgess, N.D. *et al.* (2014). A mid-term analysis of progress toward international biodiversity targets. *Science* 346(6206), 241. https://doi.org/10.1126/science.1257484.

Trathan, P.N., Lynch, H.J. and Fraser, W.R. (2016). *Changes in penguin distribution over the Antarctic Peninsula and Scotia Arc*. [Antarctic Environments Portal https://doi.org/10.18124/D43019 (Accessed: 18 May 2017).

Tsatsakis, A.M., Nawaz, M.A., Tutelyan, V.A., Golokhvast, K.S., Kalantzi, O.-I., Chung, D.H. *et al.* (2017). Impact on environment, ecosystem, diversity and health from culturing and using GMOs as feed and food. *Food and Chemical Toxicology* 107, 108-121. https://doi.org/10.1016/j.fct.2017.06.033.

United Kingdom Office for National Statistics (2018). *UK natural capital: Ecosystem service accounts, 1997 to 2015*. https://www.ons.gov.uk/economy/environmentalaccounts/bulletins/uknaturalcapital/ecosystemserviceaccounts1997to2015 (Accessed: 10 April 2018).

United Nations (1992). *Convention on Biological Diversity*, 1992. https://www.cbd.int/doc/legal/cbd-en.pdf.

United Nations (2015). *Majority female ranger unit from South Africa wins top UN environmental prize*. http://www.un.org/sustainabledevelopment/blog/2015/09/majority-female-ranger-unit-from-south-africa-wins-top-un-environmental-prize-2/.

United Nations Convention to Combat Desertification (2017). *Global Land Outlook*. Bonn. https://knowledge.unccd.int/sites/default/files/2018-06/GLO%20English_Full_Report_rev1.pdf.

United Nations Educational Scientific and Cultural Organization (2014). *Addressing the impacts of harmful algal blooms on water security*. [http://www.unesco.org/new/en/natural-sciences/environment/water/wwap/display-single-publication/news/addressing_the_impacts_of_harmful_algal_blooms_on_water_secu/ (Accessed: 15 January 2018).

United Nations Environment Programme (2012). *Global Environmental Outlook-5: Environment for the Future We Want*. Nairobi. https://wedocs.unep.org/bitstream/handle/20.500.11822/8021/GEO5_report_full_en.pdf?isAllowed=y&sequence=5.

United Nations Environment Programme (2014). *UNEP Year Book 2014 Emerging Issues Update: Illegal Trade in Wildlife*. Nairobi. https://wedocs.unep.org/bitstream/handle/20.500.11822/18380/UNEP_Year_Book_2014_Emerging_issues_update_l.pdf?sequence=1&isAllowed=y.

United Nations Environment Programme (2016a). *GEO-6 Regional Assessment for Africa*. Nairobi. http://wedocs.unep.org/bitstream/handle/20.500.11822/7595/GEO_Africa_201611.pdf?sequence=1&isAllowed=y.

United Nations Environment Programme (2016b). *GEO-6. Regional Assessment for Asia and the Pacific*. Nairobi,: United Nations Environment Programme. http://web.unep.org/geo/assessments/regional-assessments/regional-assessment-asia-and-pacific.

United Nations Environment Programme (2016c). *GEO-6. Regional Assessment for Latin America and the Caribbean*. Nairobi: United Nations Environment Programme. http://web.unep.org/geo/assessments/regional-assessments/regional-assessment-latin-america-and-caribbean.

United Nations Environment Programme (2016d). *GEO-6. Regional Assessment for North America*. Nairobi: United Nations Environment Programme. http://web.unep.org/geo/assessments/regional-assessments/regional-assessment-north-america.

United Nations Environment Programme (2016e). *GEO-6. Regional Assessment for Pan European Region*. Nairobi: United Nations Environment Programme. http://web.unep.org/geo/assessments/regional-assessments/regional-assessment-pan-european-region.

United Nations Environment Programme (2016f). *GEO-6. Regional Assessment for West Asia*. Nairobi: United Nations Environment Programme. http://web.unep.org/geo/assessments/regional-assessments/regional-assessment-west-asia.

United Nations Environment Programme World Conservation Monitoring Centre (2016a). *The State of Biodiversity in Africa: A mid-term review of progress towards the Aichi Biodiversity Targets*. Cambridge: UNEP-WCMC.

United Nations Environment Programme World Conservation Monitoring Centre (2016b). *The State of Biodiversity in Asia and the Pacific: A mid-term review of progress towards the Aichi Biodiversity Targets*. Cambridge: UNEP-WCMC.

United Nations Environment Programme World Conservation Monitoring Centre (2016c). *The State of Biodiversity in Latin America and the Caribbean: A mid-term review of progress towards the Aichi Biodiversity Targets*. Cambridge. UNEP-WCMC.

United Nations Environment Programme World Conservation Monitoring Centre (2016d). *The State of Biodiversity in West Asia: A mid-term review of progress towards the Aichi Biodiversity Targets*. Cambridge: UNEP-WCMC.

United Nations Environment Programme -World Conservation Monitoring Centre, International Union for Conservation of Nature and National Geographic Society (2018). Protected Planet Report 2018. Gland.https://livereport.protectedplanet.net/pdf/Protected_Planet_Report_2018.pdf

United Nations Environment Programme World Conservation Monitoring Centre and Institute for Development of Environmental-Economic Accounting (2017). *Experimental Ecosystem Accounts for Uganda*. Cambridge. https://www.unep-wcmc.org/resources-and-data/experimental-ecosystem-accounts-for-uganda.

United Nations Framework Convention on Climate Change (2018). *Reducing emissions from deforestation and forest degradation and the role of conservation, sustainable management of forests and enhancement of forest carbon stocks in developing countries (REDD-plus)*. https://unfccc.int/topics/land-use/workstreams/reddplus (Accessed: 3 June 2017).

United Nations Office on Drugs and Crime (2016). *World Wildlife Crime Report: Trafficking in Protected Species*. Vienna. https://www.unodc.org/documents/data-and-analysis/wildlife/World_Wildlife_Crime_Report_2016_final.pdf.

Urban, M.C. (2015). Accelerating extinction risk from climate change. *Science* 348(6234), 571-573. https://doi.org/10.1126/science.aaa4984.

van Sebille, E., Wilcox, C., Lebreton, L., Maximenko, N., Hardesty, B.D., Van Franeker, J.A. *et al.* (2015). A global inventory of small floating plastic debris. *Environmental Research Letters* 10(12), 124006. https://doi.org/10.1088/1748-9326/10/12/124006.

van Veenhuizen, R. (2012). Urban and Peri-Urban Agriculture and Forestry (UPAF): An Important Strategy to Building Resilient Cities? The Role of Urban Agriculture in Building Resilient Cities *Webinar ICLEI*. 18 October 2012. Resource Centres on Urban Agriculture and Food Security Foundation, http://resilient-cities.iclei.org/fileadmin/sites/resilient-cities/files/Resilient_Cities_2012/Digital_Congress_Proceedings/RUAF_RvV_ICLEI_181012.pdf

van Wilgen, B.W., Cowling, R.M., Marais, C., Esler, K.J., McConnachie, M. and Sharp, D. (2012). Challenges in invasive alien plant control in South Africa. *South African Journal of Science* 108(11-12), 8-11. http://ref.scielo.org/ksrrpx.

Venter, O., Sanderson, E.W., Magrach, A., Allan, J.R., Beher, J., Jones, K.R. *et al.* (2016). Sixteen years of change in the global terrestrial human footprint and implications for biodiversity conservation. *Nature communications* 7(12558). https://doi.org/10.1038/ncomms12558.

Vogel, G. (2017). Where have all the insects gone? *Science* 356(6338), 576-579. https://doi.org/10.1126/science.356.6338.576.

Waldron, A., Miller, D.C., Redding, D., Mooers, A., Kuhn, T.S., Nibbelink, N. *et al.* (2017). Reductions in global biodiversity loss predicted from conservation spending. *Nature* 551, 364. https://doi.org/10.1038/nature24295.

Wall, D.H., Nielson, U.N. and Six, J. (2015). Soil biodiversity and human health. *Nature* 528(7580), 69-78. https://doi.org/10.1038/nature15744.

Walsh, J.R., Carpenter, S.R. and Vander Zanden, M.J. (2016). Invasive species triggers a massive loss of ecosystem services through a trophic cascade. *Proceedings of the National Academy of Sciences* 113(15), 4081-4085. https://doi.org/10.1073/pnas.1600366113.

Wang, Y. and Zhou, J. (2013). Endocrine disrupting chemicals in aquatic environments: A potential reason for organism extinction? *Aquatic ecosystem health & management* 16(1), 88-93. https://doi.org/10.1080/14634988.2013.759073.

Ward, A., Dargusch, P., Thomas, S., Liu, Y. and Fulton, E.A. (2014). A global estimate of carbon stored in the world's mountain grasslands and shrublands, and the implications for climate policy. *Global Environmental Change* 28, 14-24. https://doi.org/10.1016/j.gloenvcha.2014.05.008.

Watson, J.E.M., Dudley, N., Segan, D.B. and Hockings, M. (2014). The performance and potential of protected areas. *Nature* 515(7525), 67-73. https://doi.org/10.1038/nature13947.

Watts, M. and Williamson, S. (2015). *Replacing Chemicals with Biology: Phasing Out Highly Hazardous Pesticides with Agroecology*. Penang: Pesticide Action Network Asia and the Pacific. https://www.panna.org/sites/default/files/Phasing-Out-HHPs-with-Agroecology.pdf.

Webber, B.L., Raghu, S. and Edwards, O.R. (2015). Opinion: Is CRISPR-based gene drive a biocontrol silver bullet or global conservation threat? *Proceedings of the National Academy of Sciences* 112(34), 10565-10567. https://doi.org/10.1073/pnas.1514258112.

Wiersum, K.F., Humphries, S. and van Bommel, S. (2013). Certification of community forestry enterprises: Experiences with incorporating community forestry in a global system for forest governance. *Small-scale Forestry* 12(1), 15-31. https://doi.org/10.1007/s11842-011-9190-y.

Williams, R.S.R., Jara, J.L., Matsufuiji, D. and Plenge, A. (2011). Trade in Andean condor Vulture gryphus feathers and body parts in the city of Cusco and the Sacred Valley, Cusco region, Peru. *Vulture News* 61, 16-26. http://dx.doi.org/10.4314/vulnew.v61i1.2.

Wolkovich, E.M., Cook, B.I., Allen, J.M., Crimmins, T.M., Betancourt, J.L., Travers, S.E. *et al.* (2012). Warming experiments underpredict plant phenological responses to climate change. *Nature* 485(7399), 494-497. https://doi.org/10.1038/nature11014.

Wolkovich, E.M., Davies, T.J., Schaefer, H., Cleland, E.E., Cook, B.I., Travers, S.E. *et al.* (2013). Temperature-dependent shifts in phenology contribute to the success of exotic species with climate change. *American Journal of Botany* 100(7), 1407-1421. https://doi.org/10.3732/ajb.1200478.

Woodcock, B.A., Isaac, N.J.B., Bullock, J.M., Roy, D.B., Garthwaite, D.G., Crowe, A. *et al.* (2016). Impacts of neonicotinoid use on long-term population changes in wild bees in England. *Nature communications* 7(12459). https://doi.org/10.1038/ncomms12459.

World Health Organization and Secretariat of the Convention on Biodiversity (2015). *Connecting Global Priorities: Biodiversity and Human Health. Summary of the State of Knowledge Review.* Geneva. https://www.cbd.int/health/SOK-biodiversity-en.pdf.

World Wide Fund for Nature (2015). *Living Blue Planet Report: Species, Habitats and Human Well-being.* Gland. https://www.wwf.or.jp/activities/data/20150831LBPT.pdf.

World Wide Fund for Nature (2016). *Living Planet Report 2016: Risk and Resilience in a New Era.* Gland. http://awsassets.panda.org/downloads/lpr_living_planet_report_2016.pdf.

World Wide Fund for Nature (2018). *Living Planet Report 2018: Aiming Higher.* Gland. https://c402277.ssl.cf1.rackcdn.com/publications/1187/files/original/LPR2018_Full_Report_Spreads.pdf.

World Wide Fund for Nature, Zoological Society of London, Global Footprint Network and European Space Agency (2012). *Living Planet Report 2012. Biodiversity, Biocapacity and Better Choices.* Gland: WWF International. https://portals.iucn.org/library/node/29018.

World Wildlife Fund for Nature (2013). *Chitwan Annapurna Landscape (CHAL): A Rapid Assessment.* Nepal. http://pdf.usaid.gov/pdf_docs/PA00K357.pdf.

Worm, B., Davis, B., Kettemer, L., Ward-Paige, C.A., Chapman, D., Heithaus, M.R. *et al.* (2013). Global catches, exploitation rates, and rebuilding options for sharks. *Marine Policy* 40, 194-204. https://doi.org/10.1016/j.marpol.2012.12.034.

Wright, S.L., Thompson, R.C. and Galloway, T.S. (2013). The physical impacts of microplastics on marine organisms: A review. *Environmental Pollution* 178, 483-492. https://doi.org/10.1016/j.envpol.2013.02.031.

Young, K.R. (2014). Ecology of land cover change in glaciated tropical mountains. *Revista peruana de biología* 21(3), 259-270. http://dx.doi.org/10.15381/rpb.v21i3.10900.

Zheng, H. and Cao, S. (2015). Threats to China's biodiversity by contradictions policy. *Ambio* 44(1), 23-33. https://doi.org/10.1007/s13280-014-0526-7.

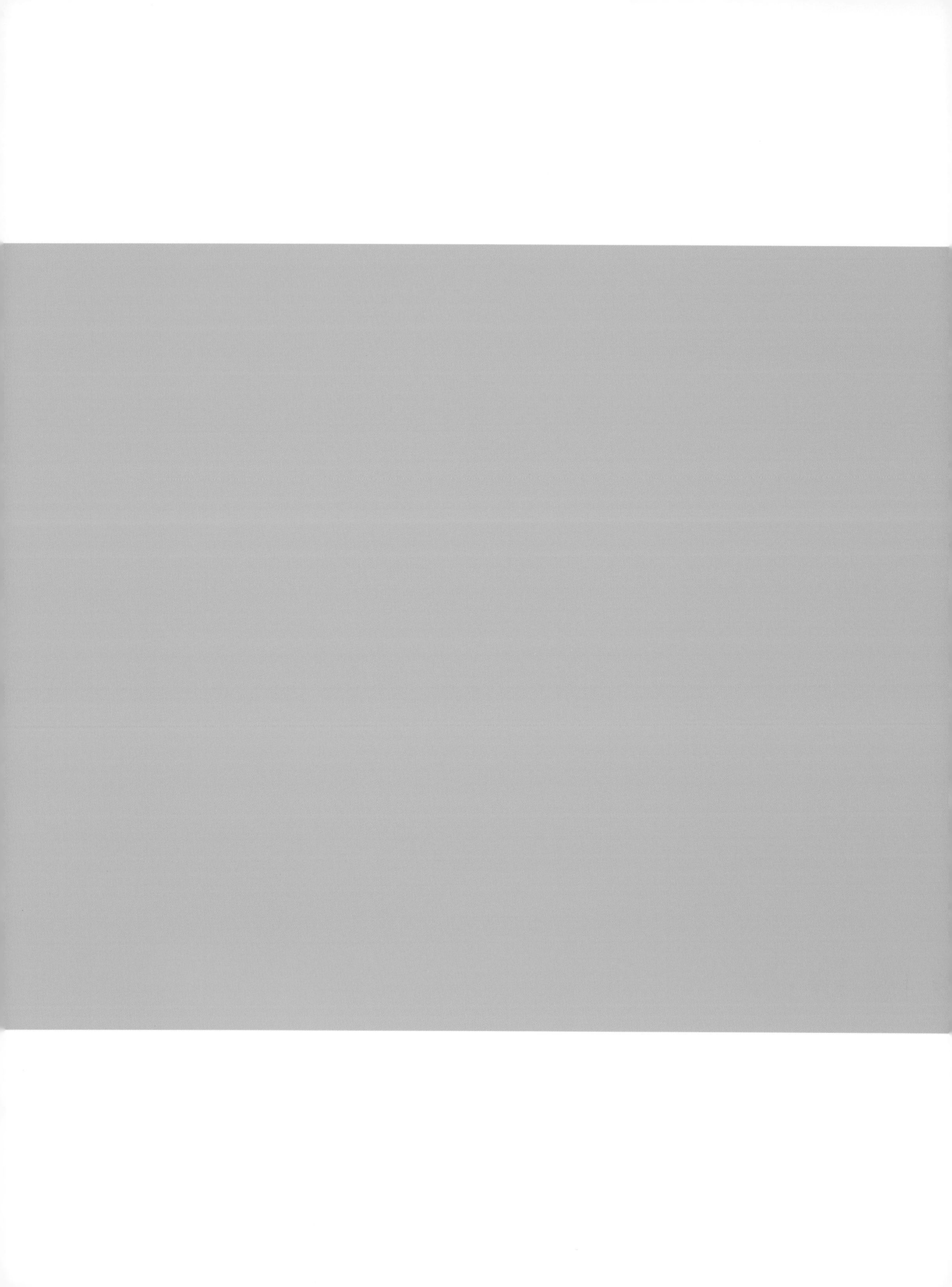

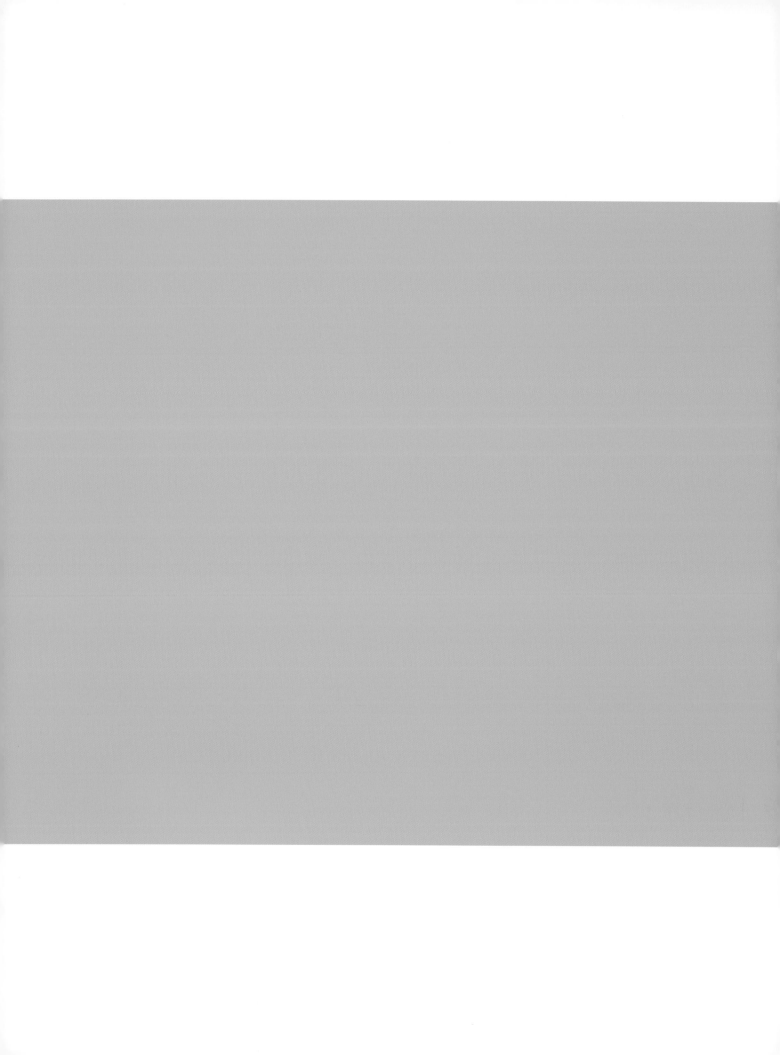

Chapter 7

Oceans and Coasts

Coordinating Lead Authors: Elaine Baker (GRID-Arendal at the University of Sydney), Peter Harris (GRID-Arendal), Adelina Mensah (University of Ghana), Jake Rice (Department of Fisheries and Oceans, Canada)

Contributing Author: James Grellier (European Centre for Environment and Human Health, University of Exeter)

GEO Fellow: Al Anoud Alkhatlan (Arabian Gulf University)

Executive summary

Human pressures on the health of the oceans have continued to increase over the last decade, in concert with the growing human population and the expanded use of ocean resources (*well established*). Multiple stressors give rise to cumulative impacts that affect the health of marine ecosystems and diminish nature's benefits to humans. However, there has been success in the management of some pressures, with concomitant improvements in ocean health, and these provide lessons on which to build. Out of numerous existing pressures we have selected three for particular attention in this Global Environment Outlook (GEO-6) assessment: bleaching of coral reefs; marine litter; and challenges to achieving sustainable fisheries in the world's oceans. {7.1}

Tropical coral reefs have passed a tipping point whereby chronic bleaching has killed many reefs that are unlikely to recover even over century-long timescales (*well established*). Coral bleaching is due to warming of the oceans, which is in turn, attributed to anthropogenic emissions of green house gases (GHGs; especially CO_2) since the industrial revolution. Ocean warming lags behind GHG emissions by several decades, such that the tipping point for coral reef bleaching was passed in the 1980s when atmospheric concentration of CO_2 exceeded about 350 parts per million (ppm). {7.3.1}

Reef bleaching events now have a recurrence interval of about six years, while reef recovery rates are known to exceed ten years (*established but incomplete*). This means that, on average, reefs will not have sufficient time to recover between bleaching events and so a steady downward spiral in reef health is to be expected in coming decades. The oceans SDG target 14.2 "by 2020, sustainably manage and protect marine and coastal ecosystems to avoid significant adverse impacts, including by strengthening their resilience, and take action for their restoration in order to achieve healthy and productive oceans" may not be attainable for most tropical coral reef ecosystems. {7.3.1}.

There is evidence that reef death will be followed by loss in fisheries, tourism, livelihoods and habitats (*inconclusive*). The demise of tropical coral reef ecosystems will be a disaster for many dependent communities and industries, and governments should, over the next decade, prepare for the eventual collapse of reef-based industries. The contributions provided by coral reefs have collectively been valued at US$29 billion, which includes their value to tourism, fisheries and coastal protection. Losses to these sectors have not yet been documented but there is significant risk that losses will occur over the next decade. {7.4.1}.

Fisheries and aquaculture are estimated to be worth US$362 billion in 2016, with aquaculture contributing US$232 billion (*established but incomplete*). Mariculture is expanding but most of the increase is in aquaculture, especially inland aquaculture (*established*). Aquaculture provides more than 10 per cent of the total tonnage of fish production and this proportion is increasing. Together fisheries and aquaculture support between 58-120 million livelihoods, depending on how part-time employment and employment in secondary processing is counted. The large majority of livelihoods are provided by small-scale fisheries and this has been stable for over a decade, yet commercial harvesting accounts for the large majority of commodity value, including more than US$80 billion per year exported from developing countries to international markets. {**Table 7.1**, 7.3.2}.

Fish, high in protein and micronutrients important for health, currently provide 3.1 billion people with over 20 per cent of their dietary protein, with higher proportions in many areas of the world where food insecurity is widespread (*established but incomplete*). To meet future challenges of food security and healthy populations, in addition to using all natural products harvested for food more efficiently, more fish, invertebrates and marine plants will have to be taken as food from the oceans and coasts, so both capture fisheries and aquaculture are expected to expand. {7.5.2}.

It is possible to keep capture fisheries sustainable, but this requires significant investments in monitoring, assessment and management and strong local community-based approaches (*established but incomplete*). Likewise, sustainable aquaculture requires knowledge and care in management of operations. {7.6}.

Reviews show wide variation among countries in the sustainability of their fisheries and aquaculture, with factors such as overall wealth to invest in fisheries research and management, while avoiding capacity-enhancing subsidies, strongly affecting the ability to keep large-scale fisheries sustainable (*established but incomplete*). For small-scale fisheries coherence of the social structures and cultural practices that promote effective community self-regulation strongly affect sustainability. {7.5.2}

The ecosystem approach to fisheries has been widely adopted in national and regional policies and operational guidance on actions to manage the footprint of fisheries has been provided by the Food and Agriculture Organization of the United Nations (FAO) (*inconclusive*). Despite the acknowledgement of the large footprint of fisheries on marine ecosystems and its full uptake in policy, measures to minimize the ecosystem effects of fishing have had mixed success. However, as with sustainability of exploitation of target species, in general the ecosystem footprint of by-catches, discards and negative habitat impacts of fishing gear is declining in the parts of world with sufficient economic resources to invest in fisheries monitoring and gear technologies that improve selectivity of harvest and reduce habitat impacts. This approach is also being applied in aquaculture, with comparable objectives and rapid uptake by the industry. {7.4.2}

The amount of marine litter continues to increase – an estimated 8 million tons (Mt) of plastics enters the ocean each year, as a result of the mismanagment of domesic waste in coastal areas (*established but incomplete*). Marine litter has been found at all ocean depths. Without intervention, the quantity of plastic in the ocean is expected to increase to 100-250 Mt by 2025. {7.3.3}.

Plastic particles are increasingly being found in the digestive systems of marine organisms including fish and shellfish consumed by humans (*established but incomplete*). The human health risks of ingesting seafood contaminated with plastic are unclear. There is well-documented evidence of physical damage to marine organisms from both entanglement in marine litter and ingestion of plastic. Some plastic contains potential toxins and can also adsorb and concentrate toxic substances from the surrounding seawater. However, there is currently no evidence of serious toxic effects to marine biota from these pollutants. Marine litter can also provide a means of transport for the spread of pathogens and invasive species (*well established*). {7.4.4}.

The economic, social and environmental costs of marine litter are continually increasing and include the direct economic costs of clean-up and loss of revenue from industries such as tourism and fishing (*unresolved*). Social and health costs are more difficult to quantify beyond local scales, as are environmental costs such as reduction in ecosystem function and services. {7.4.4}.

7.1 Introduction

The world's oceans comprise more than 70 per cent of the Earth's surface. More than 1.9 billion people lived in coastal areas in 2010, and the number is expected to reach 2.4 billion by 2050 (Kummu *et al.* 2016). Twenty of the 30 megacities[1] are located on coasts, and these megacities are expected to increase in population faster than non-urban areas (Kummu *et al.* 2016). The three fastest-growing coastal megacities are Lagos, Nigeria (4.17 per cent population growth rate), Guangzhou, China (3.94 per cent) and Dhaka, Bangladesh (3.52 per cent) (Grimm and Tulloch eds. 2015).

7.1.1 Welcome to the ocean

The health and livelihoods of many people are directly linked to the ocean through its resources and the important aesthetic, cultural and religious benefits it provides. Seafood provides at least 20 per cent of the animal protein supply for 3.1 billion people globally (Food and Agriculture Organization of the United Nations [FAO] 2016a). This is particularly important for economically disadvantaged coastal areas and communities. Coastal ecosystems also provide numerous benefits not readily monetized, such as coastal stabilization, regulation of coastal water quality and quantity, biodiversity and spawning habitats for many important species. The ocean is an integral part of the global climate system (Intergovernmental Panel on Climate Change [IPCC] 2013), contributing to the transport of heat, which influences temperature and rainfall across the planet. About 50 per cent of global primary production occurs in the ocean (Mathis *et al.* 2016). The ocean also provides a reservoir of additional economically important resources such as aggregates and sand, renewable energy and biopharmaceuticals. However, people, their livelihoods and the many indirect benefits the ocean provides are being affected by the deteriorating health of marine and coastal ecosystems, from causes including pollution, climate change, overfishing, and habitat and biodiversity loss.

By definition a healthy ocean would be one in which the basic ecosystem function and structure are intact, thereby:

- able to support livelihoods and contribute to human well-being;
- resilient to current and future change.

The full range of benefits can only continue to be enjoyed if marine and coastal ecosystems are functioning and used within environmental limits, in a way that does not cause severe or irreversible harm. However, sustainable use of marine and coastal ecosystems is challenged by many drivers of change (see Chapter 2), and by the competing pressure on natural resources and the complexities of governance and multiple, often conflicting, uses **(Figure 7.1).** Coastal states have rights and obligations within their marine jurisdiction (United Nations 1982). However, the ocean imposes special challenges on the exercise of jurisdiction. Ocean currents can carry chemicals, waste, emerging organic pollutants and pathogens beyond areas under national maritime boundaries, and marine organisms and seabirds may not stay within an area under the jurisdiction of a state. Coordination of governance measures is particularly difficult in areas beyond national jurisdiction, where a large number of institutions and agreements regulate sectoral issues such as shipping, fishing and seabed mining.

Not only must states cooperate across borders, they must also integrate decision-making across the various uses of marine and coastal ecosystems. The interlinkages between ocean conditions and marine life, and the spatially dynamic ocean processes mean that the activities of any single industry sector may have far-reaching impacts. These may disrupt the livelihoods of people who have received no benefits from the industry that has caused the impact. Similarly, benefits expected from conservation measures taken in one sector or jurisdiction may be reduced or negated by lack of action in other sectors or jurisdictions.

Global challenges such as climate change and ocean acidification must also be addressed. Climate change impacts ocean temperature, sea-ice extent and thickness, salinity, sea level rise and extreme weather events. Although climate change impacts vary at regional levels and therefore require adaptive management actions at local and regional scales (Von Schuckmann *et al.* 2016), these efforts need to be coordinated at larger scales, and lessons and best practices shared efficiently.

7.1.2 Focus of this chapter

Oceans have many uses, and there are too many linkages among marine ecosystems and between the land and adjacent seas to review them all in this chapter. *The First Global Integrated Marine Assessment* (A/RES/70/235; Inniss and Simcock eds. 2016) and reports of the Intergovernmental Panel on Climate Change (IPCC 2013) have provided recent comprehensive reviews of the state of the ocean. Therefore, three topics have been selected here that warrant particular attention – tropical coral reefs, fishing and debris entering the marine environment. Several topics of emerging or particular interest – mercury, sand mining, deep sea mining and ocean noise – are also briefly considered.

The rationale for selecting the three main topics stems from resolutions adopted by the United Nations Environmental Assembly (UNEA) at its second session in May 2016, which included specific mention of coral reefs in Resolution UNEP/EA.2/Res.12 (UNEA 2016a), and marine litter in Resolution UNEP/EA.2/Res.11 (UNEA 2016b). Marine litter was also included in a special Decision CBD/COP/DEC/XIII/10 of the Conference of the Parties to the Convention on Biological Diversity (CBD) (CBD 2016) and in Decision BC 13/17 of the Conference of the Parties to the Basel Convention (2017) . Fisheries have linkages to multiple Sustainable Development Goals (SDGs) and they also intersect the cross-cutting themes identified in Chapter 4 (notably gender, health, food systems, climate change, polar regions, and chemicals and waste).

1 Cities with populations of more than 10 million.

Figure 7.1: Generalized schematic showing the drivers and pressures relevant to the marine environment

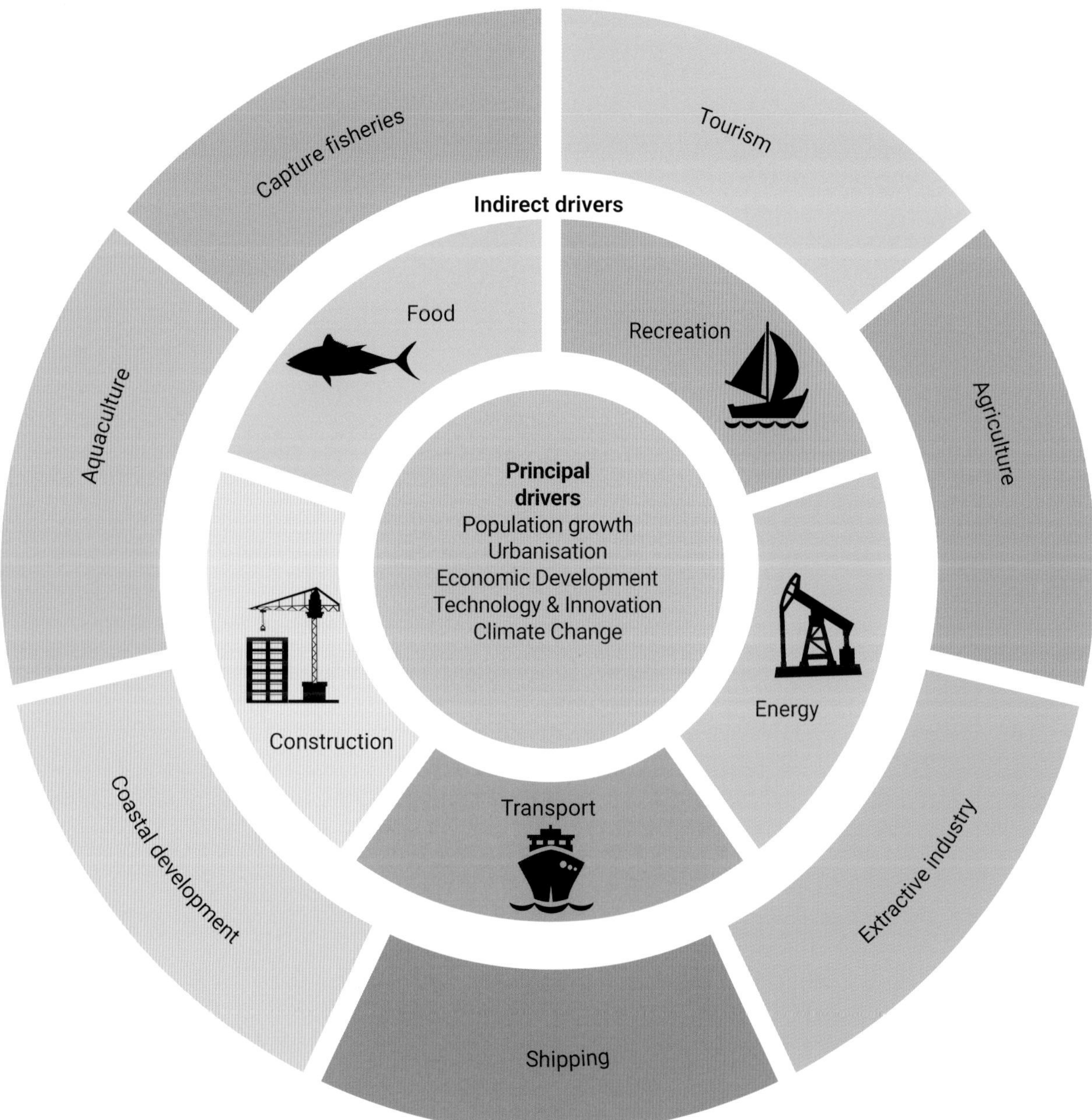

The central circle represents major high-level drivers of change in human demands on the ocean. The inner ring represents the types of societal needs promoted by the drivers, and the outer ring represents the industry sectors addressing the needs, for which policies are commonly established. The needs expressed through sector actions are the relevant pressures.

7.2 Pressures

Human activities can alter the ocean and its resources in many ways, particularly through activities that are land-based. Part V of the *First Global Integrated Marine Assessment* (Inniss and Simcock eds. 2016) describes both the societal benefits and major impacts of human activities, whether directly through resource extraction (e.g. fish, hydrocarbons, sand) or indirectly (e.g. seabed impacts of fishing gear or mining operations). The report also documents the economic value and number of livelihoods supported by each industry sector **(Table 7.1)**

The footprints of many ocean industries overlap (**Table 7.1**: column 4) and sometimes multiple sectors use the same resource for different purposes (e.g. fish for ecotourism, versus food for a coastal community; see also Halpern *et al.* 2012).

Table 7.1: Estimates of economic value, employment and major environmental impacts of the major ocean-related industries

Sector [and World Ocean Assessment chapter]	Economic value or scale of operation	Employment/ livelihoods	Major environmental impacts if inadequately regulated
Fishing [9,11,12]	US$362 billion (includes mariculture and freshwater aquaculture – approx. US$28 billion but accounting not fully separated) Competent IGOs	58-120 million (depending on how part-time employment and secondary processing employment are counted)	Changes of food web structure and function if top predators or key forage species are depleted or fishing is highly selective. By-catches of non-targeted species, some of which can sustain only very low mortality rates (e.g. sea turtles, many seabirds and small cetaceans). Gear impacts on seabed habitats and benthos, especially structurally fragile habitats (e.g. corals, sponges). Continued fishing of lost fishing gear.
Shipping [17]	50,500 billion ton-miles of cargo; 2.05 billion passenger trips Competent IGO – and conventions – IMO and MARPOL	> 1.25 million seafarers	Shipping disasters and accidents that may result in release of cargos, fuel and loss of life. Toxicity of cargos ranges from nil to severe. Chronic and episodic release of fuel and other hydrocarbons. Infrequent loss of containers with toxic contents. Discharge of sewage, waste and 'grey water'. Transmission of invasive species through ballast water and bilge water. Use of anti-fouling paints. Noise from ships. Maritime transport responsible for about 3 per cent of global greenhouse gas emissions.
Ports [18]	5.09 billion tons of bulk cargo Competent IGO – IMO and MARPOL conventon, but mostly local jurisdiction	Technology development has made consistent dockworker statistics unavailable	Concentration of shipping and potential environmental impacts of shipping. Need for dredging and access to deep water passages. Impacts on seabed and coastline from construction of infrastructure. Noise.
Offshore hydrocarbon industries [21]	US$500 billion (at US$50 per barrel)	200,000 workers in offshore production	Release of hydrocarbons particularly during blowouts or platform disasters, with potential for very large volumes to enter marine systems, with high persistence impacting on tourism and aesthetic and cultural values. Oiling of marine and coastal organisms and habitats. Contaminants entering food webs and potential human food sources Chronic release of chemicals used in operations. Episodic release of dispersants during spill clean-up. Local smothering of benthos. Noise from seismic surveys and shipping. Disturbances of biota during decommissioning.
Other marine-based energy industries [2]	7.36 MW (megawatts) produced Competent IGO – primarily local jurisdiction	7-11 job-years per MW generated	Competition for space for infrastructure and displacement of biota. Localized mortality of benthos due to infrastructure. Mortality of birds, fish in energy turbines and windmills. Noise and physical disturbance during construction and decommissioning of infrastructure.
Marine-based mining [23]	US$5.0-5.4 billion Competent IGO – ISA	7,100–12,000 (incomplete)	Mortality, displacement or extinction of marine species, particularly benthos. Destruction of seabed habitat, esp. if fragile or sensitive. Creation of sediment plumes and deposition of sediments. Noise. Potential contamination of food chains from deep-sea mining. Creation of microhabitats vulnerable to sediment concentration and anoxia [23.3].
Marine-based tourism [27]	US$2.3 trillion (35 per cent of coarse estimate of all tourism, including multiplier effects) Competent IGO – none	Not estimated due to lack of common treatment of multiplier effects. Overall tourism considered to comprise 3.3 per cent of global workforce, but breakout of marine and not-marine not consistent.	Construction of coastal infrastructure changing habitats, increasing erosion, mortality and displacement of biota, noise. Contamination of coastal waters by waste and sewage. Disturbance of organisms by increased presence of people, especially diving in high-diversity habitats, and watching marine megafauna. Increased mortality due to recreational fishing. Increases boating with all the impacts of shipping on local scales.

IGO: Intergovernmental organisations; IMO: International Maritime Organization; ISA: International Seabed Authority; MARPOL: the International Convention for the Prevention of Pollution from Ships.

Sources: Unless indicated otherwise, all information is taken from the First Global Integrated Marine Assessment (United Nations 2016), with chapter(s) indicated in first column. For some industries, economic value is recorded so differently by different countries that global economic value cannot be estimated meaningfully, and other indicators of scale of the industry are used. Reporting year also not standardized for all rows, but all estimates are 2012 or later. Table entries should be taken as indicative of global scale with large variation regionally and nationally. IMO (2015).

Developing effective management strategies therefore requires policies that can address cumulative impacts and not just separate sectoral footprints (Halpern *et al.* 2008).

7.3 State

7.3.1 Coral bleaching crisis 2015-17

Tropical coral reefs[2] are among the most biodiverse ecosystems on earth, hosting approximately 30 per cent of all marine biodiversity (Burke *et al.* 2012). The 'Coral Triangle' region, which includes Indonesia, Malaysia, Philippines, Timor-Leste, Papua New Guinea and Solomon Islands, is the area of greatest biodiversity, hosting more than 550 species of hard corals (c.f. 65 coral species in the Caribbean and Atlantic region). Globally, coral reefs cover an area of around 250,000 km^2. Due to multiple human pressures, including pollution, fishing and coral bleaching, the current state of reef health is very poor at many sites.

Coral bleaching occurs when corals are stressed by changes in conditions such as temperature, light or nutrients, causing them to expel symbiotic algae living in their tissues, revealing their white skeltons. Large-scale coral reef bleaching events attributed to warmer surface ocean temperatures have been regularly reported over the last two decades and climate research reveals that the recurrence interval between events is now about six years (Hughes *et al.* 2018). The 2015 northern hemisphere and 2015-2016 southern hemisphere summers were the hottest ever recorded and caused the worst coral bleaching on record. The United States National Oceanic and Atmospheric Administration (NOAA) declared 2015 as the beginning of the third global coral bleaching event, following similar events in 1998 and 2010. Still ongoing, this third event is the longest and most damaging recorded, to date affecting 70 per cent of the world's reefs, with some areas experiencing annual bleaching **(Figure 7.2)**. Australia's Great Barrier Reef has been particularly hard hit, with more than 50 per cent of the reef impacted since 2016 (Australia, Great Barrier Reef Marine Park Authority [GBRMPA] 2017).

The severity of bleaching varies both within reefs and between regions, and some areas that have not previously experienced bleaching have been impacted in this latest event. A recent initiative to identify the 50 reef areas most likely to survive beyond the year 2050 has been announced, with the goal of encouraging governments to set these areas aside for protection and conservation (https://50reefs.org).

The recently published summary of IPCC Fifth Assessment Report, O'Neill *et al.* (2017) concluded that there "is robust evidence (from recent coral bleaching) of early warning signals that a biophysical regime shift already may be underway". Veron *et al.* (2009) predicted the coral reef bleaching tipping point (an abrupt change in state that occurs when a threshold value is exceeded) would occur once global atmospheric CO_2 reached 350 ppm. This value was reached in about 1988, but because ocean warming lags behind global atmospheric CO_2 levels (Hansen *et al.* 2005) it has taken almost 30 years for the impact of this level of CO_2 to be revealed. The lag effect is due to the slow rate of global ocean circulation compared with the rapid rate of rising CO_2 levels. In effect, the ocean is currently responding to CO_2 levels of decades ago and the balance of evidence indicates that a tipping point for coral bleaching has now been passed (Hoegh-Guldberg *et al.* 2007; Frieler *et al.* 2013). The Veron *et al.* (2009) 350 ppm tipping point, reached 29 years ago, may have been the death sentence for many corals. And given that global atmospheric CO_2 levels are now in excess of 400 ppm, there are serious implications for the very survival of coral reefs. Recent modelling suggests more than 75 per cent of reefs will experience annual severe bleaching before 2070, even if pledges made following the 2015 Paris Climate Change Conference (COP 21) become reality (van Hooidonk *et al.* 2016; UNEP 2017). Experts agree that the coral reefs that survive to the end of the 21st century will bear little resemblance to those we are familiar with today (Hughes et al. 2017).

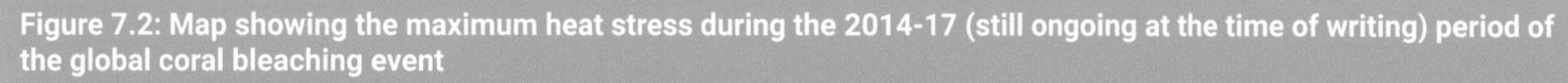

Figure 7.2: Map showing the maximum heat stress during the 2014-17 (still ongoing at the time of writing) period of the global coral bleaching event

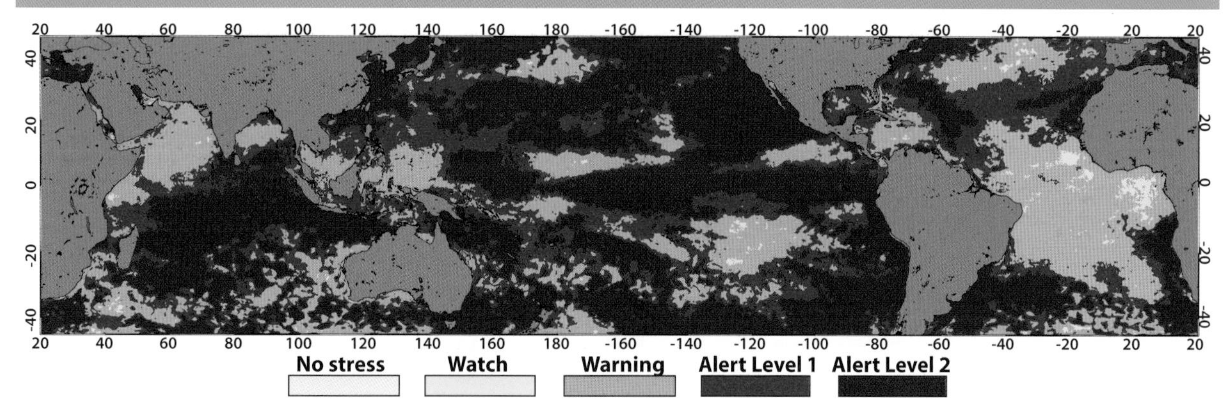

Alert Level 2 heat stress indicates widespread coral bleaching and significant mortality. Level 1 heat stress indicates significant coral bleaching. Lower levels of stress may have caused some bleaching as well.

Source: United States National Oceanic and Atmospheric Administration (NOAA) (2017).

2 Tropical coral reefs do not include deep, cold-water reefs or temperate rocky reefs.

7.3.2 Fisheries

Capture fisheries

In addition to changes in ocean status due to natural variation and climate change, people change the state of the ocean by removing resources from it. Most widespread and largest in magnitude is the harvesting of fish and other marine organisms for human consumption and some industrial uses (e.g. feed for aquaculture).

The ocean is an increasingly important source of food (International Labour Organisation [ILO] 2014). Total production from capture fisheries and mariculture[3] exceeded 170 million (metric) tons by 2017 and the mariculture contribution continues to grow (FAO 2018a). Fish provide more than 20 per cent of dietary protein to over 3.1 billion people, with this percentage high in coastal areas where food security concerns are also high. Moreover, the micronutrients in fish are an important contribution to human health, and are difficult to replace in areas where availability of fish is declining (Roos *et al.* 2007; FAO and World Health Organization [WHO] 2014; Thilsted *et al.* 2014).

Capture fisheries have been stable at around 90 million tons for over 15 years, whereas production from culture facilities has continued to increase **(Figure 7.3)** There are debates about the sustainability of present levels of fishing, with disagreements about many fundamental points regarding stock status, causes of trends and effectiveness of management measures (Worm *et al.* 2009; Froese *et al.* 2013; Melnychuk *et al.* 2016). Some fishing crises have become textbook stories of harm from diverse combinations of overexpansion of fishing capacity and effort, unmanaged technological innovation, politicized or non-precautionary decision-making, and ineffective science, management and governance. In addition, interactions of environmental change and stock dynamics in the face of inertia in management decisions played central roles in the collapse of the cod fisheries in eastern Canada (Rose 2007; Rice 2018), and fisheries for Pacific small pelagic species off Peru and Chile (Chavez *et al.* 2008).

The large volume of literature on fisheries sustainability contains many cases of both unsustainable expansion, and successes in managing exploitation rates and rebuilding previously depleted stocks. For countries where capacity and political will exist to assess stock status and fishing mortality, and implement monitoring, control and surveillance measures, trends from 1990 to the present indicate that overfishing is usually avoided (Hilborn and Ovando 2014; Melnychuk *et al.* 2016). However, the reviews also show wide variation among countries, with factors such as overall wealth to invest in fisheries research and management while avoiding capacity-enhancing subsidies, strongly affecting the ability to keep fisheries sustainable. In the large majority of cases where jurisdictions have resources for sufficient research and management, and have implemented effective governance, fishing mortality has been constrained or reduced to sustainable rates, and stocks are assessed as either healthy or recovering from historical overfishing **(Figure 7.4)**. However, where significant funding for resource assessments and monitoring, control and surveillance measures are not made available, overfishing, illegal, unreported or unregulated (IUU)[4] fishing and resource depletion continue and may be expanding.

Figure 7.3: World capture fisheries and aquaculture production

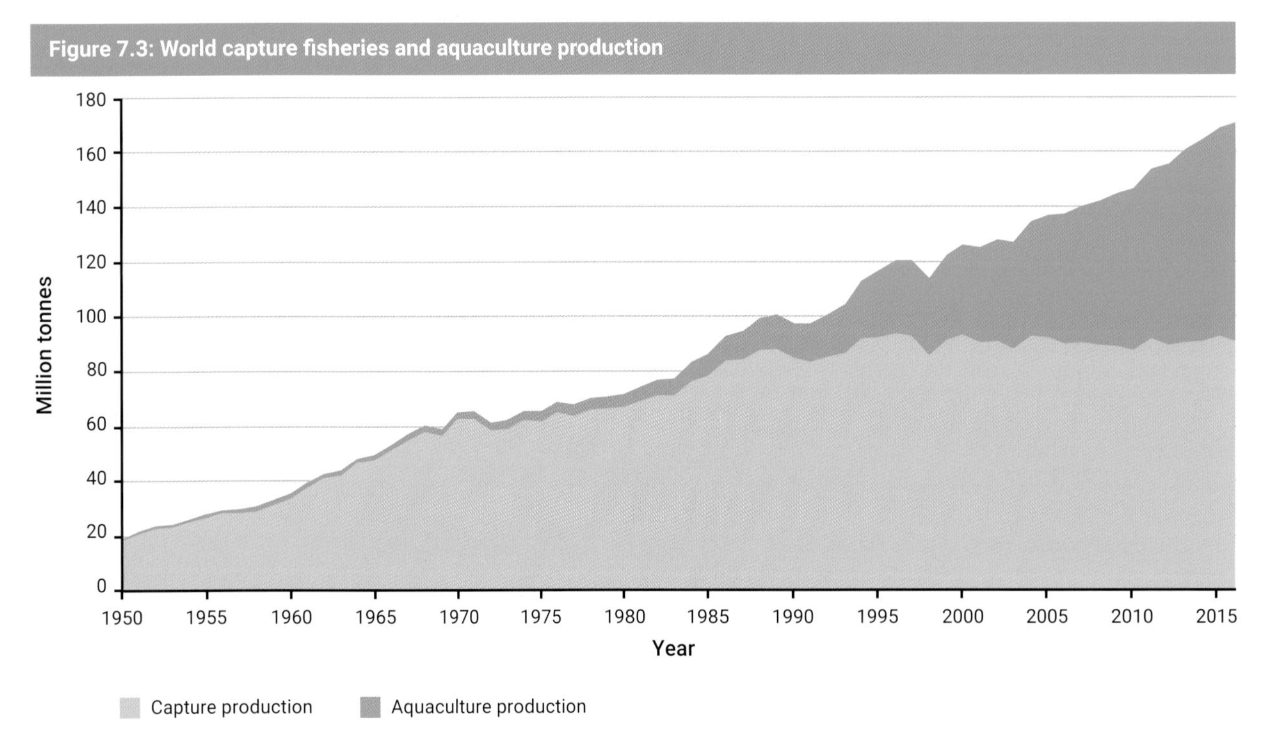

Source: FAO (2018a).

3 For this report 'aquaculture' is a general term used for raising fish and shellfish in captivity for eventual human consumption, whereas 'mariculture' is the portion of aquaculture practised in marine, coastal and estuarine areas.

4 Illegal, unreported and unregulated (IUU) fishing is a broad term which includes: fishing and fishing-related activities conducted in contravention of national, regional and international laws; non-reporting, misreporting or under-reporting of information on fishing operations and their catches.

Figure 7.4: Status of fish stocks and fishing mortality as influenced by various factors of science, management and governance. Higher relative scores on vertical axis reflect better stock status relative to theoretically 'ideal' management

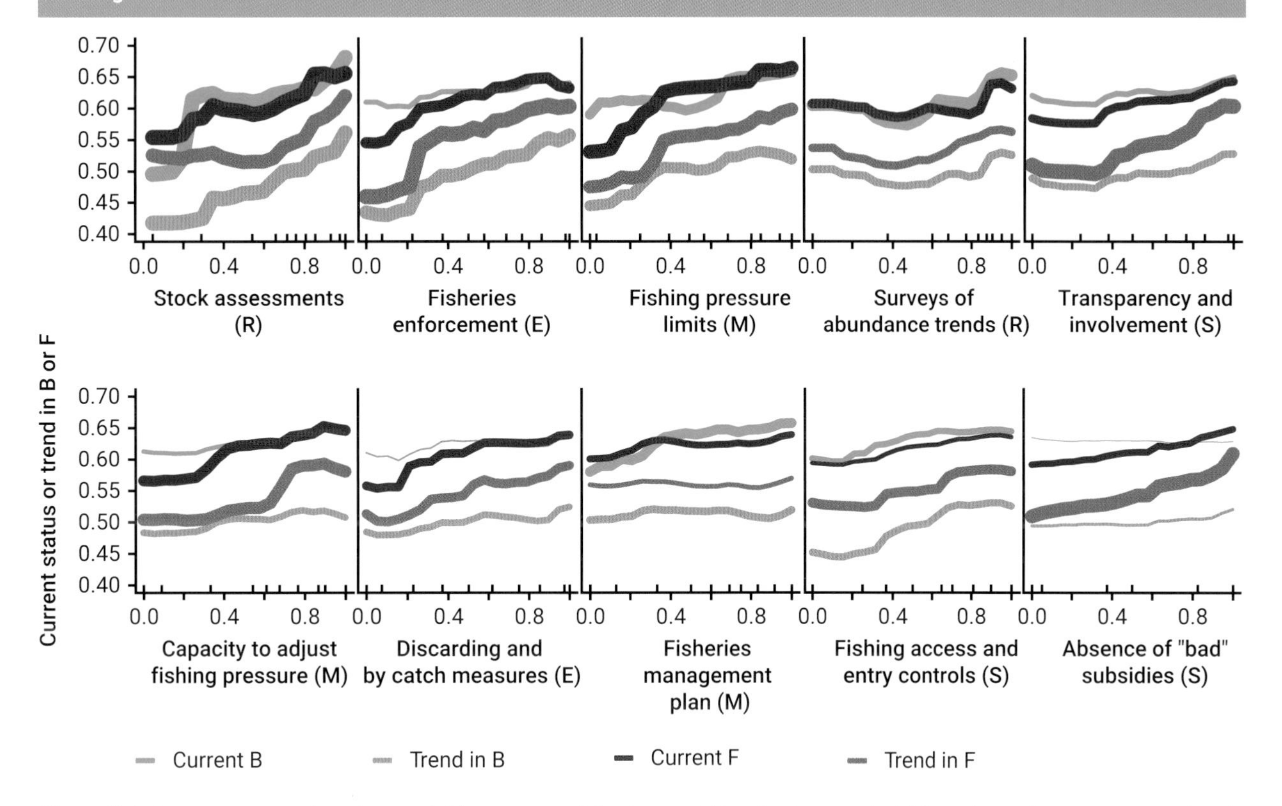

Effects of fisheries management attributes in research (R), management (M), enforcement (E), and socioeconomics (S) dimensions on the current status and trends of biomass (B) and fishing mortality (F). Line thickness reflects the different importance of each dimension on the relationship of the x and y variable.

Source: Melnychuk *et al.* 2016

In addition, fisheries are still expanding geographically, with management jurisdictions scrambling to keep pace. Causes include:

- effort displaced from jurisdictions trying to reduce exploitation on stocks within their authority,
- a continued increase in fishing capacity of fleets based in Asia (although fleet capacity of other jurisdictions is decreasing), and
- overall increases in efficiency of fishing on global scales (Bell, Watson and Ye 2017; Jacobsen, Burgess and Andersen 2017).

Spatial realignment of fishing effort will occur as stocks move in response to changes in ocean conditions due to anthropogenic global warming (Cheung, Watson and Pauly 2013), but the details of species' redistributions is uncertain (Barange *et al.* 2014; Johnson *et al.* 2016; Salinger *et al.* 2016) and management strategies appropriate for such dynamics are in the early stages of development (Schindler and Hilborn 2015; Creighton *et al.* 2016).

Fisheries have expanded to many oceanic seamounts, where accumulated biomass of long-lived, slow-growing fishes, such as orangy roughy and oreos, are often depleted even before the regional fisheries management organizations/bodies can collect sufficient information to assess sustainable harvest levels (FAO 2009a; Koslow *et al.* 2016). As fish stocks in polar

Box 7.1: Fisheries in the polar oceans

The polar oceans were not identified as a GEO-6 Region, but many of the sectors listed in **Table 7.1** are also present in one or both polar regions. Estimates of economic value and livelihoods supported are incomplete, but marine resources remain essential to the livelihoods of over 150,000 Inuit in the North American Arctic (Inuit Circumpolar Council 2011). Commercial fishing in the Arctic Ocean is under moratorium by the United States of America and Canada within their national jurisdictions, and in the international Arctic waters the initial Canada–Russian Federation–United States of America moratorium was recently joined by China, Denmark (for Greenland), the European Union, Iceland, Japan and Republic of Korea.[5] For the polar areas under Norwegian and Russian jurisdiction, fisheries are managed by the national authorities and regularly assessed by the International Council for Exploration of the Seas (ICES).

In the Southern Ocean, commercial fisheries for toothfish, icefish and krill have been prosecuted under Commission for the Conservation of Antarctic Marine Living Resources' (CCAMLR) regulatory framework since 1982. The toothfish and krill fisheries expanded rapidly, with krill catches less than a third of the precautionary catch limit (Commission for the Conservation of Antarctic Marine Living Resources [CCAMLR] 2016). Toothfish and icefish fisheries have been certified as sustainable (by the Marine Stewardship Council, an independent body), with substantial progress in deterring IUU (Österblom and Bodin 2012). The legal fisheries produced annual revenue of over US$200 million (toothfish) and US$70 million krill over five years (Hoshino and Jennings 2016). CCAMLR has periodic independent reviews of its performance (e.g. CCAMLR 2016). Polar oceans are experiencing the most rapid climate change and northern livelihoods are being impacted in many detrimental ways (Inuit Circumpolar Council 2011). For example, seasonal access of indigenous fishers to sea-ice fisheries has become problematic as sea ice thins and disappears. Opportunities for mining seabed, hydrocarbon resources and commercial shipping will require development of appropriate policies to ensure any benefits flow to local inhabitants.

latitudes become more available to commercial fisheries through a combination of melting sea ice and improved technologies for harvesting, overfishing could be a particular threat, if not carefully regulated **(Box 7.1)**. Such fisheries can expand rapidly, challenging the capabilities of management jurisdictions (Swan and Gréboval 2005), with regional fisheries management organizations/bodies playing a major role as fisheries expand in areas beyond national jurisdiction.

Where overfishing has been reduced or eliminated, or new fisheries have been constrained within sustainable levels, a wide mix of measures have been used (Melnychuk *et al.* 2016; Garcia *et al.* 2018). Efforts to constrain total catches (number and sizes of fishing vessels, days fishing, etc.) are almost universally present and technological innovation is at least monitored if not managed. Where science and management resources allow, the regulatory measures are usually informed by biologically based management reference points and harvest control rules (Inniss and Simcock eds. 2016). However, top down management based on scientific assessments and advice is not essential in all types of fisheries. In small-scale community-based fisheries community management is often effective, as long as the coherence with traditional cultural practices is high (FAO 2015). In all scales of fisheries, co-management and inclusiveness of industry participants in management can pay off in greater compliance and lower management costs (Gray 2005; Dichmont *et al.* 2016; Leite and Pita 2016).

Small-scale fisheries have been a cornerstone of livelihoods and food security in many parts of the world for centuries but only recently have been recognized as a major consideration in fisheries status and trends. (FAO 2005; SDG 14.b.a; FAO 2018b). Providing nearly 80 per cent of the employment in fisheries globally (FAO 2016a) they often operate in circumstances where centralized top-down managment would be both very expensive and culturally intrusive (FAO 2015;FAO 2016b). After extensive consultation globally, guidelines for the performance

Box 7.2: Mercury in the marine environment

The World Health Organization places mercury in the top ten chemicals of major public health concern (WHO 2017). This is because mercury, especially in the form of methylmercury, is a powerful neurotoxin, which even at low concentrations can affect fetal and childhood development and cause neurological damage (Karagas *et al.* 2012; Ha *et al.* 2017). Epidemiological studies of elevated prenatal methylmercury exposure in populations from the Faroe Islands and New Zealand have found some adverse developmental impacts (Grandjean *et al.* 1997; Crump *et al.* 1998). However, studies in the Seychelles and the United Kingdom of Great Britain and Northern Ireland found that the regular consumption of ocean fish during pregnancy did not pose a developmental risk (Myers *et al.* 2003; Daniels *et al.* 2004; van Wijngaarden *et al.* 2017). Further research on the United Kingdom cohort found that seafood intake during pregnancy (>340 g per week) improved developmental, behavioural and cognitive outcomes (Hibbeln *et al.* 2007), suggesting other nutrients present in fish such as long-chain polyunsaturated fatty acids (Strain *et al.* 2008) or selenium (Ralston and Raymond 2010) may obscure or counteract the negative effects of the methylmercury.

The health benefits of eating fish are well established (FAO and WHO 2011; FAO and WHO 2014); however, due to high methylmercury levels in some seafood and the uncertainty regarding risk, many countries have advisories suggesting that pregnant women should limit their intake of fish to species that record low concentrations of mercury (Taylor *et al.* 2018). Generally, the fish to be avoided are predatory species such as shark, tuna and swordfish and long-lived fish such as orange roughy due to the processes of biomagnification and bioaccumulation (United States Food and Drug Administration 2017).

[5] 2017 Agreement to Prevent Unregulated High Seas Fisheries in the Central Arctic Ocean.

Figure 7.5: Biomagnification and bioaccumulation of methylmercury in the food chain

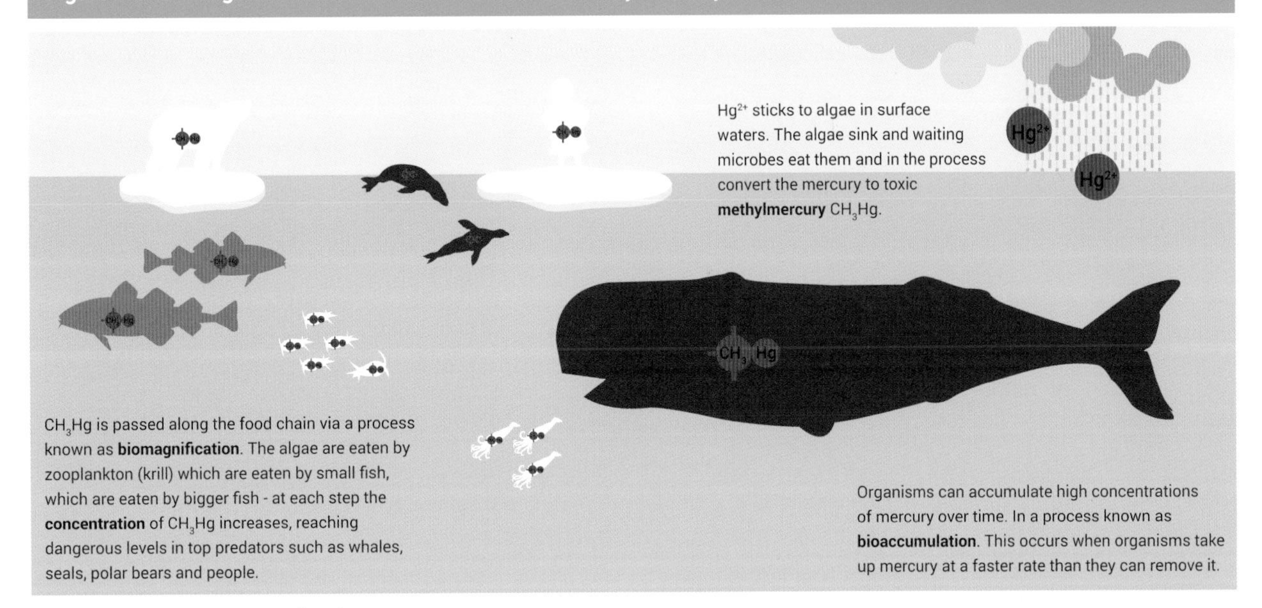

Source: Baker, Thygesen and Roche (2017).

and governance of small-scale fisheries are already leading to improvements in these fisheries (FAO 2015; FAO 2016b).

Emergence of mariculture

Although capture fisheries plateaued in the early 2000s, mariculture continues to expand and, if current trends continue, will soon surpass them (**Figure 7.4**; FAO 2018a). Large-scale mariculture of market-oriented, high-value fish and shellfish such as tuna, salmon, mussels, oysters and other bivalves, now contributes significantly to the economies of most coastal developed countries. Small-scale mariculture is also expanding through less-developed countries and economies in transition. Freshwater and marine culture which use fish-processing by-products and low-value fish as feed, create both new markets for low-value fisheries products and some potential for market competition as mariculture demand for feedstocks increases. Data on production from small-scale operations are incomplete, especially for community consumption, as these products do not enter the market.

Populations reliant on marine organisms for nutrition may have particularly high exposures to methylmercury and persistent organic pollutants and these risks are highest in areas where food security is not assured (Gribble *et al.* 2016).

In addition, climate change may lead to changes in emissions of mercury, for instance through its release from long-term storage in the frozen peatlands of the northern hemisphere (UNEP 2013; Schuster *et al.* 2018). This has the potential to increase input of mercury into the oceans.

7.3.3 Marine litter

Marine litter is a growing problem, that has serious impacts on marine organisms, habitats and ecosystems (Secretariat of the Convention on Biological Diversity [SCBD] 2016). Litter has been found at all ocean depths and on the ocean floor (Pham *et al.* 2014) and on the shores of even the most remote Pacific islands (Lavers and Bond 2017). Three-quarters of all marine

litter is composed of plastic. This includes microplastics of less than 5 mm in size, which are either purposefully manufactured (primary microplastics) for use in various industrial and commercial products (e.g. pellets, microbeads in cosmetics), or are the result of weathering of plastic products and synthetic fibres that can produce micro- and nanoplastic particles (Joint Group of Experts on the Scientific Aspects of Marine Environmental Protection [GESAMP] 2015; Gigault *et al.* 2016). Weathering can also release the chemical additives that are used in plastic manufacture (Jahnke *et al.* 2017).

Based on global solid waste data, population density and economic status, Jambeck *et al.* (2015) estimate that 275 million tons of plastic waste were generated in 192 coastal countries in 2010, of which 4.8 to 12.7 (8) million tons may have washed into the ocean **(Figure 7.6)**. They calculate that without global intervention, the quantity of plastic in the ocean could increase to 100-250 million tons by 2025. Sources of marine litter can generally be correlated with the efficiency of solid waste management and wastewater treatment (Schmidt *et al.* 2017).

It is generally accepted that a large proportion of the plastic entering the ocean originates on land. It makes its way into the marine environment via storm water run-off, rivers or is directly discharged into coastal waters (Cozar *et al.* 2014; Wang *et al.* 2016). Uncollected waste is thought to be the major source, with lesser amounts coming from collected waste re-entering the system from poorly operated or located formal and informal dumpsites (see 5.2.5). There is less information on the percentage of plastic coming from ocean-based sources, but we do know that lost fishing gear is a problem. This includes gear that is lost as a result of fishing method, washed overboard during storms or is intentionally discarded (Macfadyen, Huntington and Cappell 2009).

7.4 Impacts

7.4.1 Social and economic consequences of death of coral reefs

Coral reefs are of major importance for 275 million people located in 79 countries who depend on reef-associated fisheries as their major source of animal protein (Wilkinson *et al.* 2016). The contributions provided by coral reefs have collectively been valued at US$29 billion per annum, in the form of tourism (US$11.5 billion), fisheries (US$6.5 billion) and coastal protection (US$10.7 billion) (Burke *et al.* 2012). Bleaching of corals in the Great Barrier Reef alone could cost the Australian economy US$1 billion pa in lost tourism revenue (Willacy 2016). The total annual economic value of coral reefs in the United States of America has been valued at US$3.4 billion (Brander and Van Beukering 2013).

Coral reefs that have been degraded by the compounding effects of pollution from land or repeated bleaching events, are less able to provide the benefits on which local communities depend (Cinner *et al.* 2016). Once corals have died, they no longer grow vertically upwards, so the reefs gradually erode. Dead reefs become submerged under rising sea level and are less effective in providing shoreline protection from wave attack during storms. Dead corals not only lack the aesthetic appeal that is fundamental to reef tourism, they also sustain a less biodiverse fish community (Jones *et al.* 2004). This results in reduced tourist activity and reduced income from fisheries, which can threaten the livelihoods of local communities. Living coral reefs are also important religious symbols for some communities (Wilkinson *et al.* 2016).

Figure 7.6: Global map of potential marine plastic input to the oceans based on human activities and watershed characteristics

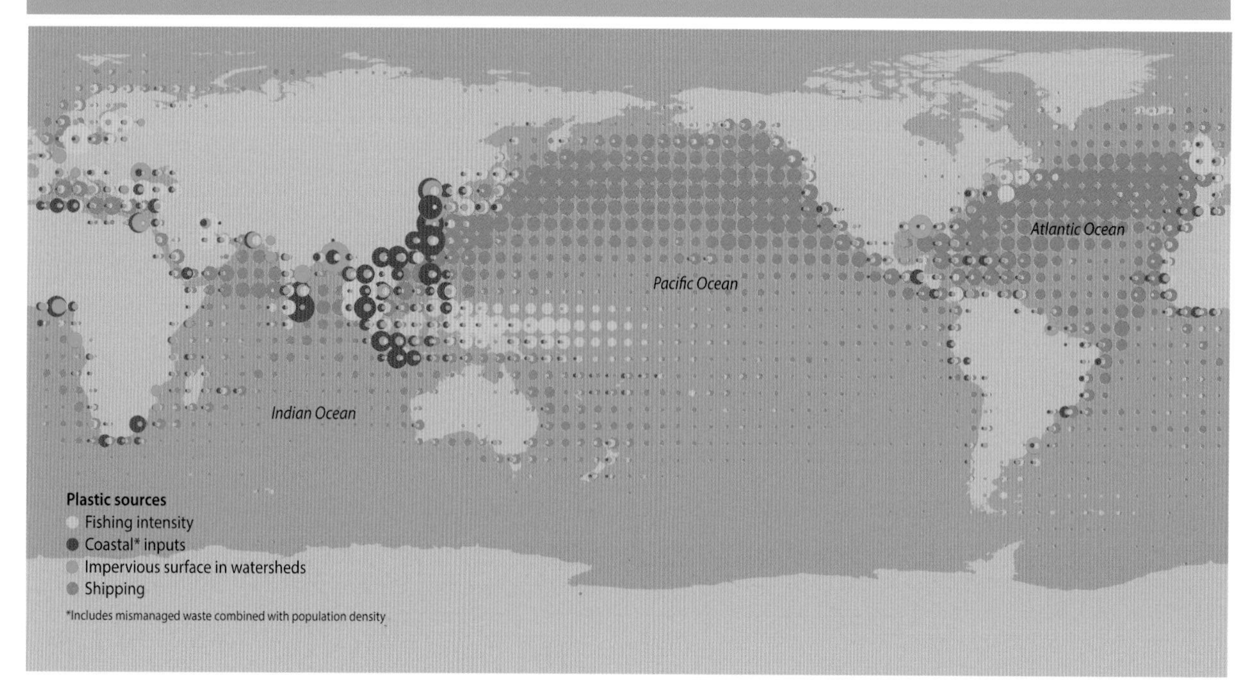

Source: Map produced by GRID-Arendal (2016a) based on data from Halpern *et al.* (2008), Watson *et al.* (2012) and Jambeck *et al.* (2015).

7.4.2 Capture fisheries

The initial impact of fishing on the target species is to reduce abundance from the unfished level. This reduction, in turn, is expected to produce increases in population productivity as density-dependence pressures are reduced, so both growth and energy reserves are available for spawning increase. This reasoning underpins basic fisheries science (Beverton and Holt 1957; Ricker 1975) and the concept of a Maximum Sustainable Yield (MSY) is entrenched in the United Nations Convention on the Law of the Sea (UNCLOS). This concept is a global norm for fisheries management, when the rate of removals by fisheries has maximized productivity without depleting the size of the spawning population sufficiently to impair production of recruits. If the exploitation rate increases beyond this level, spawning potential is diminished faster than productivity is enhanced, and overfishing occurs. The current global outcomes of fishing on target species were summarized in Section 7.3.2.

The impacts of fishing on marine ecosystems are well documented and have been studied for several decades (Jennings and Kaiser 1998; Gislason and Sinclair 2000). Major impacts include:

- by-catches of non-target species in fishing operations
- impacts of fishing gear on seabed habitats and sedentary benthic communities
- alteration of food webs through reduction in abundance of either top predators potentially allowing release of prey populations, or depletion of prey populations leading to decreased productivity of predator populations.

The pathways of these impacts are well described, and have been central in the development of the ecosystem approach to fisheries. This was entrenched in the United Nations Fish Stocks Agreement and has been widely adopted in national and regional policies (Rice 2014). FAO has provided operational guidance on actions to manage fisheries' footprint (FAO 2003) and updates, and it has been taken into the Code of Conduct on Responsible Fishing (FAO 2005; FAO 2011).

Despite acknowledgement of fisheries' large footprint on marine ecosystems, and the full uptake in policy, measures to minimize the ecosystem effects of fishing have had mixed success. There appears to be overall progress, as two global reviews a decade apart found estimates of global annual discards from fisheries to have declined from 27 million tons in 1994 to 7.3 million tons in 2004 (Alverson *et al.* 1994; Kelleher 2005). However, substantial discarding remains in many fisheries, particularly small mesh fisheries for species such as shrimp in less-developed countries, where incentives for reduction of discards and by-catch are absent or ineffective (FAO 2016a; FAO 2016b). Moreover, even where by-catches of highly vulnerable species have been reduced, levels still present population concerns for some sharks and seabirds (Campana 2016; Northridge *et al.* 2017).

Similarly, the footprint of fishing gear on sea floor habitat and benthic communities is being taken seriously by fisheries management organizations at national and regional scales. This concern has increased, prompting the adoption in the United Nations General Assembly of Resolution 61/105 in 2007, which required all regional fisheries management organizations (RFMOs) to identify marine ecosystems in their jurisdiction that would be vulnerable to bottom-contacting gear and to either protect them from harm or close them to such fishing. The evidence for policy effectiveness of this approach is examined in Chapter 14. However, despite all relevant RFMOs acting to comply with this requirement (Rice 2014), regional studies find that well over 50 per cent of fishable seabed has been impacted by fishing gear more often than benthic communities can recover fully from the disturbance, and repeated impacts remain common (Eigaard *et al.* 2017).

7.4.3 Mariculture

Mariculture has a substantial impact on the marine ecosystem, and documentation of these effects is growing. Conversion of mangroves for mariculture has resulted in widespread habitat loss with far-reaching implications for dependent species. In open, dense culture facilities, antibiotics and other medications used to prevent disease are carried by currents and tides well outside the waters in the culture area. Excessive feed sinking through the cages can accumulate on the sea floor, decompose and reduce oxygen levels. These and other effects, such as being vectors or resources for parasites and diseases, or increasing risks of non-adaptive gene-flow and invasive species, can be managed through careful, albeit sometimes costly operations (Bernal and Oliva 2016). However, the ecosystem approach is also being applied in aquaculture, with comparable objectives and rapid uptake by industry (FAO 2010).

7.4.4 Marine litter

Although the greatest accumulation of marine litter is in coastal environments (Derraik 2002), plastic (including microplastic) is distributed worldwide in the ocean, with increased accumulation in the convergence zones of each of the five subtropical gyres (Cozar *et al.* 2014; Van Sebille *et al.* 2015; Yang *et al.* 2015; see **Figure 7.7**).

Plastic pollution has been recognized for decades as a threat to marine biodiversity (Gray 1997). One of the most visible impacts is death or injury of marine life from entanglement with derelict fishing gear and plastic packaging. Many animals also ingest litter, either accidently or intentionally when it is mistaken for food. This can cause starvation due to intestinal blockage or lack of nutritian (UNEP and GRID-Arendal 2016). Recent reviews have found that a growing number of turtles, marine mammals and seabirds are endangered or killed by floating litter (Thiel *et al.* 2018; O'Hanlon *et al.* 2017).

Figure 7.7: Plastic litter in the open ocean

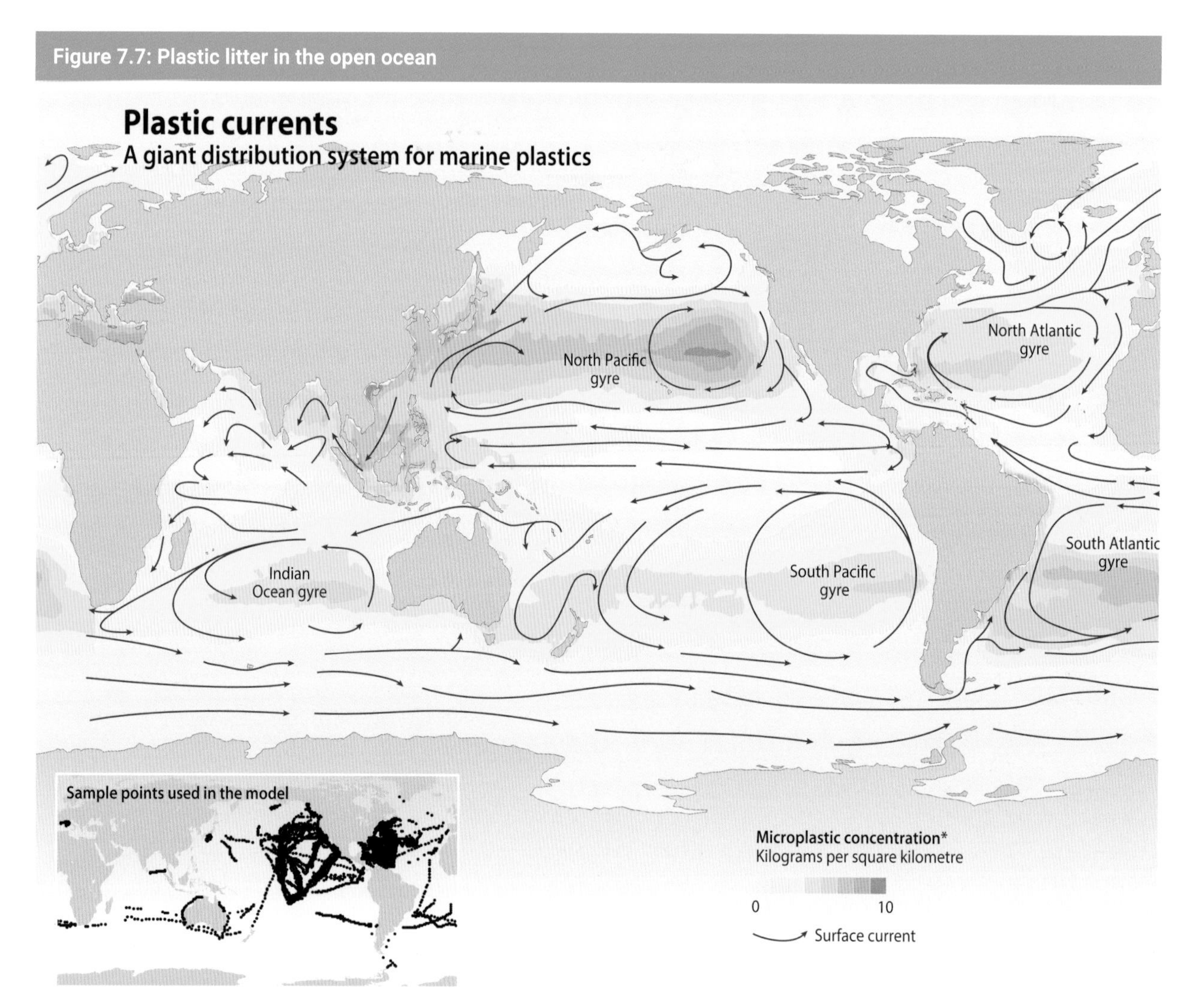

Source: GRID-Arendal (2016b), based on data from Van Sebille *et al.* (2015)

Microplastics are now appearing in food consumed by humans; however, the impact on human health is uncertain (GESAMP 2015; Halden 2015). Plastic particles have been found in the intestines of fish from all oceans and in products such as sea salt (e.g. Yang *et al.* 2015; Güven *et al.* 2017). There are currently no standard methods for assessing the health risks of ingesting plastic particles. For fish at least, people do not generally consume their digestive tract where plastic accumulates, so intake is probably limited. In instances where people consume whole organisms, such as mussels and oysters, ingestion rates could be higher (Van Cauwenberghe and Janssen 2014; Li *et al.*, 2018). Moreover, the aesthetic and restorative value of the ocean for people is well known, but there is evidence that the presence of marine litter can undermine the psychological benefits generally provided (Wyles *et al.* 2015).

Some plastic products contain dangerous chemicals (e.g. fire retardants) and plastic marine litter can also attract chemicals from the surrounding seawater (e.g. UNEP 2016; UNEP and GRID-Arendal 2016). However, the fraction of chemicals contained in plastic or sorbed to plastic in the ocean, is currently considered to be small compared to the chemicals found in seawater and organic particles that originate from other land-based sources of pollution (Koelmans *et al.* 2016). There are currently no proven toxic effects of chemicals sorbed by plastic particles found across a range of marine biota, but more data are needed to fully understand the relative importance of exposure to sorbed chemicals from microplastics compared with other exposure pathways (Ziccardi *et al.* 2016).

The economic and social costs of marine litter include indirect effects such as interfering with small-scale fishing opportunities, tourism and recreation (Watkins *et al.* 2017). These costs are generally unquantified but may fall disproportionately on those with livelihoods most closely tied to coastal activities. Some direct economic costs include the cost of beach cleanup and accidents related to navigation hazards or fouling (UNEP 2016). The European Union has estimated that every year up to €62 million are lost to the fishing industry from damage to vessels and gear and reduced catch due to ghost fishing (abandoned gear that continues to catch marine organisms as it drifts) and up to €630 million is spent on beach cleaning (Acoleyen *et al.* 2013).

7.4.5 Emerging Issues for the Ocean

Exploitation of the ocean is expanding and a number of key emerging issues will need to be addressed by policy makers as this exploitation continues.

Box 7.3: Coastal sand mining

Around the globe, coastal and nearshore areas are being mined for construction sand and gravel. These are non-renewable resources, although deposits are replenished by a number of processes including erosion of the coast, riverine transport of sediments and biological production (Woodroffe *et al.* 2016) and landward sediment transport. Sand and gravel are the second most-used natural resource on our planet, after water. Annual sand and gravel consumption is estimated at around 40-50 billion tons (5.2-6.6 tons per person per year, or c.20 kg per person per day), 26 billion tons of which is used for making concrete (Peduzzi 2014).

Most sand comes from the erosion of mountains by rivers and glaciers. It is estimated that all the Earth's rivers deliver around 12.6 billion tons of sediment to the sea each year (Syvitski *et al.* 2005). Consequently, humans are currently using sand at a rate four-times that at which it is being produced by nature. Desert sand cannot be used as an aggregate because the grains are too smooth and rounded from constant motion over desert dunes.

Many European countries have been mining sand from offshore sand banks for several decades (Baker *et al.* 2016). The practice is expanding rapidly in other parts of the world, but the exact volume mined is currently uncertain. The act of dredging the seabed kills organisms in the mined area and the plume of disturbed mud can blanket the seabed and smother sea life in surrounding areas. Illegal and poorly regulated sand mining on beaches (and in rivers) is causing major damage to ecosystems and landscapes (Larson 2018). For example, in Kiribati, beach mining has increased vulnerability to coastal inundation (Ellison 2018) and in central Indonesia, sand mining is one of the identified threats to seagrass beds (Unsworth *et al.* 2018).

Actions to reduce the global 'sand mining footprint' include conserving existing buildings and substituting recycled material for sand and gravel in new projects. It is also possible to replace sand in concrete with 15-70 per cent of incinerator ash, depending on the use (Rosenberg 2010). Research into developing desert-sand-based concrete is expanding and new products are currently being trialled (Material District 2018).

Improved knowledge of sandy environments and their dependent ecosystems is needed in order to make the wisest use of remaining sand and gravel resources (Peduzzi 2014). There is no mention of seabed mining or coastal erosion in the SDG indicators.

Box 7.4: Deep sea mining

Commercial deep sea mining has not yet begun, but the International Seabed Authority (ISA) has currently entered into 15-year contracts with companies for exploration of polymetallic nodules (the Clarion Clipperton Fracture Zone and the Central Indian Basin), polymetallic sulphides (South West Indian Ridge, Central Indian Ridge and the Mid-Atlantic Ridge) and cobalt-rich ferromanganese crusts (Western Pacific Ocean). In addition, a number of Pacific Island nations with potential deep sea mineral resources have issued exploration licences or are updating relevant policies before doing so.

Globally, deep sea mineral deposits are becoming more attractive to mining companies as they search for higher grade ore bodies (Secretariat of the Pacific Community [SPC] 2013a; SPC 2013b). These include: (1) manganese nodules that exist as cobble- to boulder-sized rocks scattered over broad areas of the abyssal ocean floor at depths exceeding 5,000 m; (2) cobalt-rich crusts formed on the flanks of seamounts and other volcanic sea floor features; and (3) massive sulphide deposits that are formed in association with hydrothermal vents found along sea floor spreading ridges, back arc-basins and submarine volcanic arcs. Benthic communities inhabiting these environments are globally unique and host many endemic species (Beaudoin and Smith 2012). Interest in mining these deposits is most advanced in relation to massive sulphide deposits located in the south-west Pacific, but many unanswered questions remain about the environmental impacts (Boschen *et al.* 2013).

Potential impacts of deep sea mining are poorly studied, but are generally assumed to include (1) direct impacts on the benthic communities where nodules/ore deposits are removed; (2) impacts on the benthos due to mobilization, transport and redeposition of sediment over potentially broad areas; and (3) impacts in the water column in cases where mining vessels discharge a plume of sediment near the sea surface, thus affecting photosynthesizing biota and pelagic fish (Morgan, Odunton and Jones 1999; Sharma 2001). A seabed disturbance experiment in the Peru Basin found very little recovery of benthic fauna 26 years after mimicking mining operations (Marcon *et al.* 2016). Lack of knowledge and understanding has been argued as one reason for countries to proceed with caution in developing these resources (Van Dover 2011; Van Dover *et al.* 2017). In the context of deep sea mining, the world has a unique opportunity to make wise decisions about an industry before it has started.

The ISA is responsible for ensuring effective protection of the marine environment from harmful effects of deep sea mining in areas beyond national jurisdiction (in accordance with Part XI of the United Nations Convention on the Law of the Sea). The Authority is in the process of developing the Mining Code, which contains rules, regulations and procedures to regulate prospecting, exploration and exploitation of marine minerals in the area (International Seabed Authority [ISA] 2017).

Many states with potential deep sea minerals have developed or are developing policies to regulate this new industry. These include a range of initiatives – for example, the Secretariat of the Pacific Community Regional Legislative and Regulatory Framework for Deep Sea Minerals Exploration and Exploitation (SPC 2013b), Cook Islands National Seabed Minerals Policy (Cook Islands Seabed Minerals Authority 2014) and the Tuvalu Seabed Mining Act 2014 (Tuvalu 2014).

Box 7.5: Anthropogenic ocean noise

There is increasing concern regarding the potential impact of anthropogenic acoustic noise on marine life. This is noise generated by a range of activities including shipping, seismic surveys, military operations, wind farms, channel dredging and aggregate extraction (Inger *et al.* 2009). Large commercial ships generate noise in the frequency range from 10 to 1,000 Hz, which coincides with frequencies used by marine mammals for communication and navigation (Richardson *et al.* 1995). There is evidence that low-frequency noise has increased significantly in the deep ocean since the 1950s (Andrew *et al.* 2002; McDonald *et al.* 2006; Chapman and Price 2011). However, some recent observations have shown a constant level or slightly decreasing trend in low-frequency noise (Andrew *et al.* 2011; Miksis-Olds and Nichols 2016). There is limited information on noise levels in the shallower water of the continental shelf (Harris *et al.* 2016).

Evolutionary adaptations that have allowed many marine species to detect sound may now make them vulnerable to noise pollution (Popper and Hastings 2009). Sound energy dissipates as a function of the distance squared, so proximity to the sound source is a major factor for calculating impact. Early research on noise and marine mammals focused on high-frequency sound, such as ship sonar, which had been implicated in whale strandings (e.g. Fernández *et al.* 2005). More recently, researchers have tried to determine the impacts of common, low-frequency sounds on marine mammals. Although it is difficult to determine the impact of anthropogenic noise on marine mammals, there is general consensus that it can cause adverse effects, from behavioural changes to strandings (Götz *et al.* 2009). A review by Cox *et al.* (2016) on the impact of ocean noise on fish behaviour and physiology determined that certain sounds can disrupt communication and interfere with predator-prey interactions. Low-frequency noise has also been found to impact crustaceans, producing changes in behaviour and ecological function (Tidau and Briffa 2016).

There are increasing concerns about the long-term and cumulative effects of noise on marine biodiversity (CBD 2012). The CBD (operational paragraph 3 of Decision XIII/10) calls for improved assessment of noise levels in the ocean, further research, development and transfer of technologies and capacity-building and mitigation (CBD 2016). The European Union Marine Strategy Framework Directive 2017/848 (European Commission 2017) has recently provided criteria and methodological standards to ensure that introduced noise does not adversely affect the marine environment and proposed standardized methods for monitoring and assessment.

The United Nations Convention on the Law of the Sea makes no specific mention of anthropogenic noise, but if the introduction of noise into the marine environment is likely to have a negative impact on the environment, it may be considered a form of pollution under UNCLOS. Delegates at the United Nations Open-ended Informal Consultative Process on Oceans and the Law of the Sea (ICP-19, 2018) disussed recognizing underwater noise as a form of transboundary pollution to be mitigated and addressed through an United Nations General Assembly resolution.

7.5 Response

Governance approaches and policy instruments that address impacts on the marine environment are quite varied. General discussion of these policy approaches is provided here while the effectiveness of specific examples is explored in Chapter 14 (Part B).

7.5.1 Coral reefs

Since the increased frequency of coral bleaching is attributed to global anthropogenic climate change, only a global policy response can address the root cause of the problem. The term 'coral reefs' is not mentioned in the SDG indicators, including SDG 14 "Conserve and sustainably use the oceans, seas and marine resources for sustainable development". Aichi Target 10 is related to coral reefs conservation: "By 2015, the multiple anthropogenic pressures on coral reefs, and other vulnerable ecosystems impacted by climate change or ocean acidification are minimized, so as to maintain their integrity and functioning." The oceans SDG target 14.2 – "by 2020, sustainably manage and protect marine and coastal ecosystems to avoid significant adverse impacts, including by strengthening their resilience, and take action for their restoration in order to achieve healthy and productive oceans" – may not be attainable for most tropical coral reef ecosystems. The resilience of coral reefs is affected by cumulative human impacts (e.g. fishing, coastal pollution, sediment run-off, invasive species), hence these impacts must be curbed to sustain reefs into the future.

Nations dependent upon reef-based fisheries, tourism and other sectors will need to develop policies for a transition to post-reef economies within the next decade, including dealing with associated cultural trauma, especially in cases where reef degradation is most rapid and spatially widespread. In addition, low-lying coral atoll countries will need to develop policies for a transition to environments where the natural benefits of coral reefs to people are much reduced or no longer available. Given that some reef habitat may be in locations where the impacts of climate change will be less severe, and where corals might survive, reef-owning nations should consider taking immediate action to protect all known coral reef habitat from any non-subsistence uses (i.e. establish all reefs as total no-take, no-go conservation zones) until such time as the location of reefs that are most likely to survive becomes known (Beyer *et al.* 2018). Studies show that where 'no-take' MPAs have been established, reef ecosystem resilience is improved (Steneck *et al.* 2018).

The challenge is to evolve from local management and monitoring towards the multiscale governance of addressing drivers, thresholds and feedbacks at relevant scales. Coral reef management must adapt to embrace new approaches such as resilience and ecosystem-based management, including the manipulation of ecosystems, bio-engineering of heat-resistant coral species as well as building new international institutions and partnerships to tackle the global aspects of the decline in coral reefs (Hughes *et al.* 2017).

7.5.2 Fishing

Policies and measures to manage fisheries impacts on ecosystems

The impacts of fisheries on species not delivered to markets (collectively called by-catch) on the sea floor and its biota, and on marine ecosystem structure and function, have been studied since before the 1980s. Measures to manage all these types of impacts are known and feasible, and can keep them within safe ecological limits (FAO 2009a). These include technologies and practices that make fishing gear more selective for target species, discourage by-catches of marine birds, mammals and reptiles, and avoid or reduce impacts of fishing gear on the sea floor (FAO 2009a; FAO 2009b). Guidance on how, and under what conditions, to apply all these measures has been available for well over a decade (FAO 2003), and has been expanded and updated regularly (e.g. FAO and World Bank 2015). Significant global policy commitments have been made to avoid or mitigate such ecosystem effects of fishing (Rice 2014).

Spatial measures have had a role in fisheries management for over a century and the growing establishment of marine protected areas (MPAs) has accelerated the interest in spatial management approaches. Many ecological and governance factors appear to influence the effectiveness of MPAs and their incremental value to other measures (Rice *et al.* 2012). Overall there is growing awareness that they can help to keep fisheries sustainable, particularly with regard to protection of sensitive habitat features or contributing to improving the status of fish stocks when conventional fisheries management measures are not being implemented effectively. However, MPAs also have a wide range of social and economic impacts that need to be considered on a case-by-case basis (FAO 2007). In addition, conflicting results are found with regard to MPA benefits such as 'spillover effects', and studies of their impacts on coastal livelihoods and implications for food security have produced mixed results (FAO 2016b).

Fisheries are being impacted by climate change in many ways, well documented in IPCC's Fifth Aseessment Report, Working Group I (IPCC 2013), and the subject of an upcoming IPCC special report on oceans and cryosphere, expected late in 2019. As temperature and salinity profiles change with global warming, the distribution and productivity of important target species is already being reflected in changes in distribution of fishery catches. Moreover, environmental changes are impacting stock productivity of fish and making them available at different places and/or at different times of the year, with impacts on large-scale mobile fisheries (which may have to fish in different places or at different times) and small-scale fisheries with lower mobility (which may have to adapt to changing species available for harvest). Depending on the cultural practices associated with fishing, these challenges may be disruptive to address.

Ocean acidification is a potential threat to many species, particularly in early life stages, including many shellfish, as calcium carbonate for shell formation is less available in seawater of higher acidity. Estimates of losses from ocean acidification are highly variable, but some projections suggest losses over US$100 billion by 2100 (Narita, Redhanz and Tol 2012; Lemasson *et al.* 2017). Acidification is considered a particularly serious threat in polar areas (Tarling *et al.* 2016), and should be an important consideration.

Box 7.6: Examples of existing global policy commitments to sustainable fisheries using an ecosystem approach (dates of agreements in brackets)

United Nations Convention on the Law of the Sea Articles 61(4) and 119(1) both make explicit reference to sustainability of associated and dependent species, and many articles in parts V, VI and VII refer to sustainable fisheries [1982].

United Nations Fish Stocks Agreement Article 5.3.d: Develop data-collection and research programmes to assess the impact of fishing on non-target and associated or dependent species and their environment, and adopt plans necessary to ensure the conservation of such species and to protect habitats of special concern [1995].

Aichi Target 6: By 2020, all fish and invertebrate stocks and aquatic plants are managed and harvested sustainably, legally and applying ecosystem-based approaches, so that overfishing is avoided; recovery plans and measures are in place for all depleted species; fisheries have no significant adverse impacts on threatened species and vulnerable ecosystems; and the impacts of fisheries on stocks, species and ecosystems are within safe ecological limits [2010].

United Nations General Assembly 61/105 Paragraph 80: Calls upon states to take action immediately, individually and through regional fisheries management organizations and arrangements, and – consistent with the precautionary approach and ecosystem approaches – to sustainably manage fish stocks and protect vulnerable marine ecosystems, including seamounts, hydrothermal vents and cold-water corals, from destructive fishing practices, recognizing the immense importance and value of deep sea ecosystems and the biodiversity they contain [2006]. This resolution has been followed by several updates.

SDG Target 14.4: By 2020, effectively regulate harvesting and end over-fishing, illegal, unreported and unregulated fishing and destructive fishing practices and implement science-based management plans, in order to restore fish stocks in the shortest time feasible, at least to levels that can produce maximum sustainable yield as determined by their biological characteristics [2016].

Social and economic benefits of fishing

The benefits and opportunities for development presented by fisheries is important to different large-scale and small-scale fisheries (LSF and SSF). Some SSF have severely depleted the stocks they exploit, as have some LSF, and some of the most destructive fishing practices, including fishing with dynamite and poisons, are restricted to SSF. The geographic scale of LSF means that even modest by-catch rates or habitat impacts of fishing gear can result in substantial pressure on species taken as by-catch and seabed features (FAO 2009a; FAO 2018a).

SSF and LSF differ in the magnitude of the market value of their catches, and in the employment created, livelihoods supported and social distribution of the benefits provided from fishing. As a generalization with occasional exceptions, LSF provide greater direct economic revenues, but also require much greater capital investment in fishing vessels, gear and processing capacity. On the other hand, employment for the same volume of catch is usually much greater in SSF, especially since significant additional jobs are created in shore-based small-scale market and processing, with sometimes multiple layers of these secondary employment opportunities. These multiplication factors also apply to LSF, which can create substantial coastal employment in rural areas, but data are rarely collected systematically, so total employment created in all types of fisheries is probably underestimated.

Gender roles also differ between LSF and SSF. Most open ocean fishers are men. Women generally fish on shallow reefs and tidal flats, and in mangroves and coastal estuaries (Lambeth *et al.* 2014). Women often predominate in the post-harvest processing, marketing and trading of fish. These roles are often omitted from data-collection efforts, and overlooked in conventional government or aid programmes that support fishing and fishers (Siason *et al.* 2010). However, when all of the industry workforce is counted, women make up nearly 50 per cent (World Bank 2012; **Table 7.2**).

These issues of magnitude and distribution of revenue and employment created by LSF and SSF present complex choices to policymakers. In developing countries, SSF potentially contribute substantially to development and equitable distribution of livelihoods from fishing. This does not mean that earnings from fishing alone are sufficient to sustain households at a level above the poverty line or above a country's minimum wage (FAO 2016a), and these fisheries are particularly vulnerable to outside threats from factors such as climate

Table 7.2: Global capture fisheries employment

	Small-scale fisheries			Large-scale fisheries			Total
	Marine	Inland	Total	Marine	Inland	Total	
Number of fishers (millions)	13	18	31	2	1	3	34
Number of post-harvest jobs (millions)	37	38	75	7	0.5	7.5	82.5
Total	50	56	106	9	1.5	10.5	116.5
Percentage of women	36%	54%	46%	66%	28%	62%	47%

Source: World Bank (2012).

change (Barange *et al.* 2014; Guillotreau, Campling and Robinson 2012). LSF have greater opportunity to generate revenues for participants and governments (World Bank 2012), but are at greater risk of concentrating the wealth and opportunity generated among a small number of individuals (Olson 2011). How available fish harvests are distributed between SSF and LSF consequently has major consequences for development, employment and revenue generation, which need to be considered fully in any comprehensive fisheries policies.

Fisheries and SDGs and the Aichi Targets

Fisheries have important roles in meeting both SDGs 1 and 2 (end poverty and hunger) as well as SDG 14 (conserve and sustainably use the ocean and its resources). To meet global food security needs, dietary protein from marine sources will have to increase by 50 per cent and likely much more (Rice and Garcia 2011). Some combination of innovative harvest strategies that increase harvest of food sources with presently low market value and ensure their distribution to appropriate markets (e.g. Garcia *et al.* (2012) and expansion of mariculture production will be essential to meeting SDG 2, and can contribute to improving employment and livelihoods supported by-production of marine food (SDG 1). These needs pose challenges for SDG 14, as plans for advancing this goal usually involve discussions of reducing the pressure from fisheries on marine ecosystems, rebuilding depleted stocks, ending over- and IUU fishing, and greatly expanding the coverage of no-take MPAs. These goals can be pursued in unison, but only if planning for expanded catches and mariculture production, including its offshore expansion, is done very carefully, with full ecosystem impacts considered in each case. If the 'conserved' part of SDG 14 is interpreted as complementary with 'sustainably used', systems altered from their pristine state are considered 'conserved' as long as major structural properties and functional processes are not altered beyond safe ecological limits as specified in Aichi Target 6. Such careful planning for expansion of food production from the sea could also contribute to SDGs 3 (health and well-being), 5 (gender equity) and 12 (sustainable consumption and production patterns), as long as these factors are part of the benefits sought from the increased food production.

Aichi Target 6 also focuses directly on fishing. In much more detail than SDG 14, it spells out all the ecological factors related to fishing that need to be made sustainable by 2020, including catch levels of all stocks, commitments to rebuilding depleted stocks, management of by-catches and habitat impacts of fishing gear, and establishing resilient ecosystem structure and function.

7.5.3 Marine litter

Policy responses to marine plastics are growing and range from global instruments such as MARPOL, UNCLOS and the Honolulu Commitment and Strategy, through regional action plans such as the Regional Plan on Marine Litter Management in the Mediterranean (UNEP/MAP 2015), and specific product bans (e.g. single-use plastic bags) at municipal or national levels. Marine litter has been incorporated into SDG target 14.1 indicator 14.1.1 as a composite indicator that includes (i) the index of coastal eutrophication and (ii) floating plastic litter density. The third United Nations Assembly (UNEA-3) adopted resolution UNEP/EA.3/Res.7 which includes the establishment of an open-ended ad hoc expert group to further examine the barriers to and options for combating marine plastic litter and microplastics from all sources, especially land-based sources (UNEA 2017). The first meeting of the expert group was held in Nairobi, Kenya from 29 to 31 May 2018.

Cleaning up coasts and beaches can provide environmental and economic benefits (e.g. Orange County California estimated an economic benefit of more than US$140 million could be generated annually from the increased number of visitors attracted to cleaner beaches (Leggett *et al.* 2014). However, cleaning up the open ocean does not currently appear to be a practical solution to marine litter. The cost of the ship-time alone needed to clean the litter concentrated in 1 per cent (approximately one million km^2) of the Central Pacific Gyre is estimated to be between US$122 million and US$489 million (NOAA Office of Response and Restoration 2012). Large-scale booms may be effective at trapping surface litter in small areas. The trail of a 600 m long boom by the NGO Ocean Cleanup recently began offshore California. If succcessful, the boom will be deployed in the open ocean of the North Pacific gyre (Stokstad 2018).

Research suggests that up to 95 per cent of the plastic entering the ocean does not remain in the surface waters (Eriksen *et al.* 2014). However, there is a major knowledge gap in understanding the behaviour and breakdown of plastic in the ocean and where it eventually ends up (Cozar *et al.* 2014). Therefore, efforts to address marine litter should focus primarily on its prevention at source through sustainable consumption and production patterns, sound waste management, wastewater treatment and resource recovery using the priciples of a circular economy (Eriksen *et al.* 2014; UNEP 2016).

7.6 Conclusions

The oceans are impacted by numerous human activities and the most serious impacts are related to climate change, land-based pollution and fishing. Within the impacts of climate change, our assessment has mentioned several issues: ocean acidification; sea level rise; changes to bottom water formation; the distribution of many fish and invertebrate species; and ocean circulation. The most dramatic and immediate impact of climate change on the oceans in recent years (GEO-6 cycle) is the bleaching and death of coral reefs. Pollution, particularly from plastic, is a major concern for many marine and coastal ecosystems. In relation to the fisheries sector, the chapter highlights concerns of overfishing, climate change impacts on species distribution patterns and the rise of aquaculture. We therefore summarize some key findings:

1. Tropical coral reefs have passed a tipping point whereby chronic bleaching has killed many reefs that are unlikely to recover even over centuries-long timescales. Reef death will be followed by loss of fisheries, tourism livelihoods and habitats. The demise of tropical coral reef ecosystems will be a disaster for many dependent communities and industries. Even if reef-owning nations take immediate action to protect their coral reefs from non-subsistence uses, there is a major risk that many reef-based industries will collapse over the next decade.

2. Marine litter has been found across all oceans and at all depths. Micro- and nano-plastics are now documented in the food web, including in seafoods consumed by humans. Marine litter has increased, with an estimated 8 million tons per year of plastics entering the ocean, mainly from land-based sources. If nations do not take action to prevent litter from entering the ocean, it will continue to accumulate and compromise ecosystem health and human food security. Prevention involves ensuring recovery and recycling of all used plastic products, encouraging communities to reduce the volume of rubbish generated, and improving solid waste management and wastewater treatment. Cleaning up the oceans is not a sustainable option without action to stop litter from entering the oceans.

3. To meet future challenges of food security and healthy populations, in addition to using all natural products harvested for food more efficiently, more fish, invertebrates and marine plants will have to be taken as food from the oceans and coasts, so both capture fisheries and mariculture must expand while preserving sustainability. It is possible to keep capture fisheries sustainable, but this requires significant investments in monitoring, assessment and management (at national, regional and international levels) and/or strong local community-based approaches. Sustainable mariculture requires knowledge and care in management of operations. Without sound bases in knowledge and governance of fisheries and mariculture, patterns of overexploitation, environmental damage and resource depletion are likely, and neither food security nor health goals will be met.

References

Acoleyen, M., Laureysens, I., Lambert, S., Raport, L., van Sluis, C., Kater, B. *et al.* (2013). *Final Report: Marine Litter Study To Support The Establishment of an Initial Quantitative Headline Reduction Target - SFRA0025.* European Commission. http://ec.europa.eu/environment/marine/good-environmental-status/descriptor-10/pdf/final_report.pdf.

Alverson, D.L., Freeberg, M.H., Murawaski, S.A. and Pope, J.G. (1994). *A Global Assessment of Fisheries Bycatch and Discards.* FAO Fisheries Technical Paper. Rome. http://www.fao.org/docrep/003/t4890e/t4890e00.htm.

Andrew, R.K., Howe, B.M. and Mercer, J.A. (2011). Long-time trends in ship traffic noise for four sites off the North American west coast. *The Journal of the Acoustical Society of America* 129(2), 642-651. https://doi.org/10.1121/1.3518770.

Andrew, R.K., Howe, B.M., Mercer, J.A. and Dzieciuch, M.A. (2002). Ocean ambient sound: Comparing the 1960s with the 1990s for a receiver off the California coast. *Acoustics Research Letters Online* 3(2), 65-70. https://doi.org/10.1121/1.1461915.

Australia, Great Barrier Reef Marine Park Authority (2017). *Reef health.* http://www.gbrmpa.gov.au/about-the-reef/reef-health.

Baker E., Gaill F., Karageorgis A., Lamarche G., Narayanaswamy B., Parr J. *et al.* (2016). Offshore mining industries.In *The First Global Integrated Marine Assessment - World Ocean Assessment I.* United Nations. chapter 23. http://www.un.org/Depts/los/global_reporting/WOA_RPROC/Chapter_23.pdf

Baker, E.K., Thygesen, K. and Roche, C. (2017). Why we need action on mercury now. [Grid-Arendal https://news.grida.no/why-we-need-action-on-mercury-now (Accessed: June 2018).

Barange, M., Merino, G., Blanchard, J.L., Scholtens, J., Harle, J., Allison, E.H. *et al.* (2014). Impacts of climate change on marine ecosystem production in societies dependent on fisheries. *Nature Climate Change* 4(3), 211-216. https://doi.org/10.1038/nclimate2119.

Basel Convention (2017). *BC-13/17: Work Programme and Operations of the Open-ended Working Group for the biennium 2018–2019.* Basel Convention. https://www.informea.org/en/decision/work-programme-and-operations-open-ended-working-group-biennium-2018–2019.

Beaudoin, Y.C. and Smith, S. (2012). Habitats of the Su Su Knolls hydrothermal site, eastern Manus Basin, Papua New Guinea.In *Seafloor Geomorphology as Benthic Habitat.* Harris, P. and Baker, E. (eds.). Elsevier. 843-852. https://www.researchgate.net/publication/284781274_Habitats_of_the_Su_Su_Knolls_Hydrothermal_Site_Eastern_Manus_Basin_Papua_New_Guinea

Bell, J.D., Watson, R.A. and Ye, Y. (2017). Global fishing capacity and fishing effort from 1950 to 2012. *Fish and Fisheries* 18(3), 489-505. https://doi.org/10.1111/faf.12187.

Bernal, P. and Olivia, D. (2016). Aquaculture.In *The First Global Integrated Marine Assessment - World Ocean Assessment I.* Innis, L. and Simcock, A. (eds.). United Nations. chapter 12. http://www.un.org/Depts/los/global_reporting/WOA_RPROC/Chapter_12.pdf

Beverton, R.J.H. and Holt, S.J. (1957). *On The Dynamics of Exploited Fish Populations.* 1st edn. London: Her Majesty's Stationery Office. https://trove.nla.gov.au/work/13338365?q&sort=holdings+desc&_=1539166792028&versionId=25601182.

Beyer, H.L., Kennedy, E.V., Beger, M., Chen, C.A., Cinner, J.E., Darling, E.S. *et al.* (2018). Risk-sensitive planning for conserving coral reefs under rapid climate change. *Conservation Letters*, e12587. https://doi.org/10.1111/conl.12587.

Boschen, R.E., Rowden, A.A., Clark, M.R. and Gardner, J.P.A. (2013). Mining of deep-sea seafloor massive sulfides: A review of the deposits, their benthic communities, impacts from mining, regulatory frameworks and management strategies. *Ocean & Coastal Management* 84, 54-67. https://doi.org/10.1016/j.ocecoaman.2013.07.005.

Brander, L. and Van Beukering, P. (2013). *The Total Economic Value of U.S. Coral Reefs: A Review of The Literature.* Silver Spring, MD: National Oceanographic and Atmospheric Administration (NOAA) Coral Reef Conservation Programme (CRCP). https://data.nodc.noaa.gov/coris/library/NOAA/CRCP/other/other_crcp_publications/TEV_US_Coral_Reefs_Literature_Review_2013.pdf.

Burke, L., Reytar, K., Spalding, M. and Perry, A. (2012). *Reefs at Risk Revisited in the Coral Triangle.* Washington, DC: World Resources Institute. http://pdf.wri.org/reefs_at_risk_revisited_coral_triangle.pdf

Campana, S.E. (2016). Transboundary movements, unmonitored fishing mortality, and ineffective international fisheries management pose risks for pelagic sharks in the Northwest Atlantic. *Canadian Journal of Fisheries and Aquatic Sciences* 73(10), 1599-1607. https://doi.org/10.1139/cjfas-2015-0502.

Chapman, N.R. and Price, A. (2011). Low frequency deep ocean ambient noise trend in the Northeast Pacific Ocean. *The Journal of the Acoustical Society of America* 129(5), EL161-EL165. https://doi.org/10.1121/1.3567084.

Chavez, F.P., Bertrand, A., Guevara-Carrasco, R., Soler, P. and Csirke, J. (2008). The northern Humboldt Current System: Brief history, present status and a view towards the future. *Progress in Oceanography* 79(2–4), 95-105. https://doi.org/10.1016/j.pocean.2008.10.012.

Cheung, W.W., Watson, R. and Pauly, D. (2013). Signature of ocean warming in global fisheries catch. *Nature* 497(7449), 365-368. https://doi.org/10.1038/nature12156.

Cinner, J.E., Pratchett, M.S., Graham, N.A.J., Messmer, V., Fuentes, M.M.P.B., Ainsworth, T. *et al.* (2016). A framework for understanding climate change impacts on coral reef social–ecological systems. *Regional environmental change* 16(4), 1133-1146. https://doi.org/10.1007/s10113-015-0832-z.

Commission for the Conservation of Antarctic Marine Living Resources (2016). *Toothfish fisheries.* https://www.ccamlr.org/en/fisheries/toothfish-fisheries.

Convention on Biodiversity (2012). Scientific Synthesis on the Impacts of Underwater Noise on Marine and Coastal Biodiversity and Habitats- Note by the Executive Secretary. UNEP/CBD/SBSTTA/16/INF/12 *19th ASCOBANS Advisory Committee Meeting.* 20-22 March 2012. https://www.cbd.int/doc/meetings/sbstta/sbstta-16/information/sbstta-16-inf-12-en.doc

Convention on Biological Diversity (2016). *XIII/10. Addressing impacts of marine debris and anthropogenic underwater noise on marine and coastal biodiversity: Decision adopted by the Conference of the Parties to the Convention on Biological Diversity. CBD/COP/DEC/XIII/10.* Cancun. https://www.cbd.int/doc/decisions/cop-13/cop-13-dec-10-en.pdf.

Cook Islands Seabed Minerals Authority (2014). *Cook Islands National Seabed Minerals Policy.* https://www.seabedmineralsauthority.gov.ck/PicsHotel/SeabedMinerals/Brochure/Cook%20Islands%20Seabed%20Minerals%20Policy%20.pdf.

Cox, K.D., Brennan, L.P., Dudas, S.E. and Juanes, F. (2016). Assessing the effect of aquatic noise on fish behavior and physiology: A meta-analysis approach. *Proceedings of Meetings on Acoustics* 27(1). https://doi.org/10.1121/2.0000291.

Cozar, A., Echevarria, F., Gonzalez-Gordillo, J.I., Irigoien, X., Ubeda, B., Hernandez-Leon, S. *et al.* (2014). Plastic debris in the open ocean. *Proceedings of the National Academy of Sciences* 111(28), 10239-10244. https://doi.org/10.1073/pnas.1314705111.

Creighton, C., Hobday, A.J., Lockwood, M. and Pecl, G.T. (2016). Adapting management of marine environments to a changing climate: a checklist to guide reform and assess progress. *Ecosystems* 19(2), 187-219. https://doi.org/10.1007/s1002.

Crump, K.S., Kjellström, T., Shipp, A.M., Silvers, A. and Stewart, A. (1998). Influence of prenatal mercury exposure upon scholastic and psychological test performance: Benchmark analysis of a New Zealand cohort. *Risk Analysis* 18(6), 701-713. https://doi.org/10.1111/j.1539-6924.1998.tb01114.x.

Daniels, J.L., Longnecker, M.P., Rowland, A.S., Golding, J. and ALSPAC Study Team-University of Bristol Institute of Child Health (2004). Fish intake during pregnancy and early cognitive development of offspring. *Epidemiology* 15(4), 394-402. https://doi.org/10.1097/01.ede.0000129514.46451.ce.

Derraik, J.G.B. (2002). The pollution of the marine environment by plastic debris: A review. *Marine pollution bulletin* 44(9), 842-852. https://doi.org/10.1016/S0025-326X(02)00220-5.

Dichmont, C.M., Dutra, L.X.C., Owens, R., Jebreen, E., Thompson, C., Deng, R.A. *et al.* (2016). A generic method of engagement to elicit regional coastal management options. *Ocean & Coastal Management* 124, 22-32. https://doi.org/10.1016/j.ocecoaman.2016.02.003.

Eigaard, O.R., Bastardie, F., Hintzen, N.T., Buhl-Mortensen, L., Buhl-Mortensen, P., Catarino, R. *et al.* (2017). The footprint of bottom trawling in European waters: Distribution, intensity, and seabed integrity. *ICES Journal of Marine Science* 74(3), 847-865. https://doi.org/10.1093/icesjms/fsw194.

Ellison, J.C. (2018). Pacific Island beaches: Values, threats and rehabilitation.In *Beach Management Tools-Concepts, Methodologies and Case Studies.* Botero C., Cervantes O. and Finkl, C. (eds.). Cham: Springer. 679-700. https://link.springer.com/chapter/10.1007/978-3-319-58304-4_34

Eriksen, M., Lebreton, L.C.M., Carson, H.S., Thiel, M., Moore, C.J., Borerro, J.C. *et al.* (2014). Plastic pollution in the world's oceans: More than 5 trillion plastic pieces weighing over 250,000 tons afloat at sea. *PLoS One* 9(12), e111913. https://doi.org/10.1371/journal.pone.0111913.

European Commission (2017). *Commission Decision (EU) 2017/848 of 17 May 2017 laying down Criteria and Methodological Standards on Good Environmental Status of Marine Waters and Specifications and Standardised Methods for Monitoring and Assessment, and repealing Decision 2010/477/EU.* European Union. https://publications.europa.eu/en/publication-detail/-/publication/a7523a58-3b91-11e7-a08e-01aa75ed71a1/language-en.

Fernández, A., Edwards, J.F., Rodriguez, F., De Los Monteros, A.E., Herraez, P., Castro, P. *et al.* (2005). "Gas and fat embolic syndrome" involving a mass stranding of beaked whales (family Ziphiidae) exposed to anthropogenic sonar signals. *Veterinary Pathology* 42(4), 446-457. https://doi.org/10.1354/vp.42-4-446.

Food and Agriculture Organization of the United Nations (2003). *Fisheries Management 2: The Ecosystem Approach to Fisheries.* FAO Technical Guidelines for Responsible Fisheries. Rome. http://www.fao.org/3/a-y4470e.pdf.

Food and Agriculture Organization of the United Nations (2005). *Increasing The Contribution of Small-Scale Fisheries To Poverty Alleviation and Food Security.* FAO Technical Guidelines For Responsible Fisheries. Rome. http://www.fao.org/tempref/docrep/fao/008/a0237e/a0237e00.pdf.

Food and Agriculture Organization of the United Nations (2007). *Marine Protected Areas as as a Tool for Fisheries Management (MPAs).* Rome. http://sih.ifremer.fr/content/download/5924/43589/file/MPA_FAO_website_Sep_2007.pdf.

Food and Agriculture Organization of the United Nations (2009a). *International guidelines for the management of deep-sea fisheries in the high seas.* http://www.fao.org/docrep/011/i0816t/i0816t00.htm.

Food and Agriculture Organization of the United Nations (2009b). *FAO/UNEP Expert Meeting on Impacts of Destructive Fishing Practices, Unsustainable Fishing, and Illegal, Unreported and Unregulated (IUU) Fishing on Marine Biodiversity and Habitats.* FAO Fisheries and Aquaculture. Rome. http://www.fao.org/docrep/012/i1490e/i1490e00.pdf.

Food and Agriculture Organization of the United Nations (2010). *Aquaculture Development 4. Ecosystem Approach to Aquaculture.* FAO Technical Guidelines for Responsible Fisheries Rome. http://www.fao.org/docrep/013/i1750e/i1750e.pdf.

Food and Agriculture Organization of the United Nations (2011). *Code of Conduct for Responsible Fisheries.* Rome. http://www.fao.org/3/a-v9878e.pdf.

Food and Agriculture Organization of the United Nations (2015). *Voluntary Guidelines for Securing Sustainable Small-Scale Fisheries in the Context of Food Security and Poverty Eradication.* Rome: Food and Agriculture Organization. http://www.fao.org/policy-support/resources/resources-details/en/c/418453/.

Food and Agriculture Organization of the United Nations (2016a). *The State of World Fisheries and Aquaculture 2016: Contributing to Food Security and Nutrition for All.* Rome. http://www.fao.org/3/a-i5555e.pdf.

Food and Agriculture Organization of the United Nations (2016b). *Technical and Socio-Economic Characteristics of Small-Scale Coastal Fishing Communities, and Opportunities for Poverty Alleviation and Empowerment.* FAO Fisheries and Aquaculture. Rome. http://www.fao.org/3/a-i5651e.pdf.

Food and Agriculture Organization of the United Nations (2018a). *The State of World Fisheries and Aquaculture: Meeting the Sustainable Development Goals.* Rome. http://www.fao.org/3/i9540en/I9540EN.pdf.

Food and Agriculture Organization of the United Nations (2018b). *Policy support and governance: Sustainable small-scale fisheries.* http://www.fao.org/policy-support/policy-themes/sustainable-small-scale-fisheries/en/ (Accessed: 1 October 2018).

Food and Agriculture Organization of the United Nations and World Bank (2015). *Aquaculture Zoning, Site Selection and Area Management under the Ecosystem Approach To Aquaculture.* http://www.fao.org/3/a-i5004e.pdf.

Food and Agriculture Organization of the United Nations and World Health Organisation (2011). *Report of the Joint FAO/WHO Expert Consultation on the Risks and Benefits of Fish Consumption.* Rome, 25–29 January. http://www.fao.org/docrep/014/ba0136e/ba0136e00.pdf

Food and Agriculture Organization of the United Nations and World Health Organization (2014). *Conference Outcome Document: Rome Declaration on Nutrition.* Second International Conference on Nutrition. Rome. www.fao.org/3/a-ml542e.pdf.

Frieler, K., Meinshausen, M., Golly, A., Mengel, M., Lebek, K., Donner, S.D. *et al.* (2013). Limiting global warming to 2 °C is unlikely to save most coral reefs. *Nature Climate Change* 3(2), 165-170. https://doi.org/10.1038/nclimate1674.

Froese, R., Zeller, D., Kleisner, K. and Pauly, D. (2013). Worrisome trends in global stock status continue unabated: a response to a comment by RM Cook on "What catch data can tell us about the status of global fisheries". *Marine Biology* 160(9), 2531-2533. https://doi/org/10.1007/s00227-013-2185-9.

Garcia, S.M., Kolding, J., Rice, J., Rochet, M.-J., Zhou, S., Arimoto, T. *et al.* (2012). Reconsidering the consequences of selective fisheries. *Science* 335(6072), 1045-1047. https://doi.org/10.1126/science.1214594.

Garcia, S.M., Ye, Y., Rice, J. and Charles, A.T. (2018). *Rebuilding of Marine FisheriesPart 1: Global Review.* FAO Fisheries and Aquaculture Technical Paper 630/1. Rome: Food and Agriculture Organization of the United Nations. http://www.fao.org/3/ca0161en/CA0161EN.pdf.

Gigault, J., Pedrono, B., Maxit, B. and Ter Halle, A. (2016). Marine plastic litter: The unanalyzed nano-fraction. *Environmental science: Nano* 3(2), 346-350. https://doi.org/10.1039/C6EN00008H.

Gislason, H. and Sinclair, M.M. (2000). Ecosystem effects of fishing. *ICES Journal of Marine Science* 57(3), 466–467. https://doi.org/10.1006/jmsc.2000.0742.

Götz, T., Hastie, G., Hatch, L.T., Raustein, O., Southall, B.L., Tasker, M. *et al.* (2009). *Overview of the Impacts of Anthropogenic Underwater Sound in the Marine Environment.* OSPAR Biodiversity Series. London: OSPAR Commission. https://tethys.pnnl.gov/sites/default/files/publications/Anthropogenic_Underwater_Sound_in_the_Marine_Environment.pdf.

Grandjean, P., Weihe, P., White, R.F., Debes, F., Araki, S., Yokoyama, K. *et al.* (1997). Cognitive deficit in 7-year-old children with prenatal exposure to methylmercury. *Neurotoxicology and teratology* 19(6), 417-428. https://doi.org/10.1016/S0892-0362(97)00097-4.

Gray, J.S. (1997). Marine biodiversity: Patterns, threats and conservation needs. *Biodiversity and Conservation* 6(1), 153-175. https://doi.org/10.1023/A:1018335901847.

Gray, T.S. (ed.) (2005). *Participation in Fisheries Governance.* Dordrecht: Springer. https://www.springer.com/gp/book/9781402037771.

Gribble, M.O., Karimi, R., Feingold, B.J., Nyland, J.F., O'Hara, T.M., Gladyshev, M.I. *et al.* (2016). Mercury, selenium and fish oils in marine food webs and implications for human health. *Journal of the Marine Biological Association of the United Kingdom* 96(1), 43-59. https://doi.org/10.1017/S0025315415001356.

GRID-Arendal (2016a). *Plastic input into the ocean.* http://www.grida.no/resources/6906.

GRID-Arendal (2016b). *Plastic currents.* Grid-Arendal http://www.grida.no/resources/6913.

Grimm, M. and Tulloch, J. (eds.) (2015). *The megacity state:The world's biggest cities shaping our future.* Munich: Allianz SE. https://www.allianz.com/content/dam/onemarketing/azcom/Allianz_com/migration/media/press/document/Allianz_Risk_Pulse_Megacities_20151130-EN.pdf.

Guillotreau, P., Campling, L. and Robinson, J. (2012). Vulnerability of small island fishery economies to climate and institutional changes. *Current Opinion in Environmental Sustainability* 4(3), 287-291. https://doi.org/10.1016/j.cosust.2012.06.003.

Güven, O., Gökdağ, K., Jovanović, B. and Kıdeyş, A.E. (2017). Microplastic litter composition of the Turkish territorial waters of the Mediterranean Sea, and its occurrence in the gastrointestinal tract of fish. *Environmental Pollution* 223, 286-294. https://doi.org/10.1016/j.envpol.2017.01.025.

Ha, E., Basu, N., Bose-O'Reilly, S., Dórea, J.G., McSorley, E., Sakamoto, M. *et al.* (2017). Current progress on understanding the impact of mercury on human health. *Environmental Research* 152, 419-433. https://doi.org/10.1016/j.envres.2016.06.042.

Halden, R.U. (2015). Epistemology of contaminants of emerging concern and literature meta-analysis. *Journal of Hazardous Materials* 282, 2-9. https://doi.org/10.1016/j.jhazmat.2014.08.074.

Halpern, B.S., Longo, C., Hardy, D., McLeod, K.L., Samhouri, J.F., Katona, S.K. *et al.* (2012). An index to assess the health and benefits of the global ocean. *Nature* 488(7413), 615-620. https://doi.org/10.1038/nature11397.

Halpern, B.S., Walbridge, S., Selkoe, K.A., Kappel, C.V., Micheli, F., D'agrosa, C. *et al.* (2008). A global map of human impact on marine ecosystems. *Science* 319(5865), 948-952. https://doi.org/10.1126/science.1149345.

Hansen, J., Nazarenko, L., Ruedy, R., Sato, M., Willis, J., Del Genio, A. *et al.* (2005). Earth's energy imbalance: Confirmation and implications. *Science* 308(5727), 1431-1435. https://doi.org/10.1126/science.1110252.

Harris, P., Philip, R., Robinson, S. and Wang, L. (2016). Monitoring anthropogenic ocean sound from shipping using an acoustic sensor network and a compressive sensing approach. *Sensors* 16(3), 415. https://doi.org/10.3390/s16030415.

Hibbeln, J.R., Davis, J.M., Steer, C., Emmett, P., Rogers, I., Williams, C. *et al.* (2007). Maternal seafood consumption in pregnancy and neurodevelopmental outcomes in childhood (ALSPAC study): An observational cohort study. *The Lancet* 369(9561), 578-585. https://doi.org/10.1016/S0140-6736(07)60277-3.

Hilborn, R. and Ovando, D. (2014). Reflections on the success of traditional fisheries management. *ICES Journal of Marine Science: Journal du Conseil* 71(5), 1040-1046. https://doi/org/10.1093/icesjms/fsu034.

Hoegh-Guldberg, O., Mumby, P.J., Hooten, A.J., Steneck, R.S., Greenfield, P., Gomez, E. *et al.* (2007). Coral reefs under rapid climate change and ocean acidification. *Science* 318(5857), 1737-1742. https://doi.org/10.1126/science.1152509.

Hoshino, E. and Jennings. S. (2016). *The value of marine resources harvested in the CCAMLR Convention Area – an assessment of GVP.* Tasmania: Conservation of Antarctic Marine Living Resources. https://www.ccamlr.org/en/ccamlr-xxxv/10.

Hughes, T.P., Anderson, K.D., Connolly, S.R., Heron, S.F., Kerry, J.T., Lough, J.M. *et al.* (2018). Spatial and temporal patterns of mass bleaching of corals in the Anthropocene. *Science* 359(6371), 80-83. https://doi.org/10.1126/science.aan8048.

Hughes, T.P., Barnes, M.L., Bellwood, D.R., Cinner, J.E., Cumming, G.S., Jackson, J.B. *et al.* (2017). Coral reefs in the Anthropocene. *Nature* 546(7656), 82-90. https://doi.org/10.1038/nature22901

Hughes, T.P., Barnes, M.L., Bellwood, D.R., Cinner, J.E., Cumming, G.S., Jackson, J.B. *et al.* (2017). Coral reefs in the Anthropocene. *Nature* 546(7656), 82-90. https://doi.org/10.1038/nature22901.

Inger, R., Attrill, M.J., Bearhop, S., Broderick, A.C., James Grecian, W., Hodgson, D.J. *et al.* (2009). Marine renewable energy: Potential benefits to biodiversity? An urgent call for research. *Journal of Applied Ecology* 46(6), 1145-1153. https://doi.org/10.1111/j.1365-2664.2009.01697.x.

Inniss, L. and Simcock, A. (eds.) (2016). *The First Global Integrated Marine Assessment: World Ocean Assessment I.* New York, NY: United Nations. http://www.un.org/depts/los/global_reporting/WOA_RegProcess.htm.

Intergovernmental Panel on Climate Change (2013). *Climate Change 2013: The Physical Science Basis. Contribution of Working Group I to the Fifth Assessment Report of the Intergovernmental Panel on Climate Change.* Stocker, T.F., Qin, D., Plattner, G.-K., Tignor, M., Allen, S.K., Boschung, J. *et al.* (eds.). Cambridge, MA: Cambridge University Press. http://www.climatechange2013.org/images/report/WG1AR5_ALL_FINAL.pdf.

International Labour Organization (2014). *Report of the UN Secretary-General on Oceans and Law of the Sea, 2014.* Geneva. http://www.un.org/depts/los/general_assembly/contributions_2014/ILO.pdf.

International Seabed Authority (2017). *Selected Decisions and Documents of The Twenty-Third Session.* Kingston: International Seabed Authority. https://www.isa.org.jm/sites/default/files/files/documents/en_3.pdf.

Inuit Circumpolar Council (2011). *A Circumpolar Inuit Declaration on Resource Development Principles in Inuit Unaat.* http://www.inuitcircumpolar.com/uploads/3/0/5/4/30542564/declaration_on_resource_development_a3_final.pdf (Accessed: 27 July 2016).

Jacobsen, N.S., Burgess, M.G. and Andersen, K.H. (2017). Efficiency of fisheries is increasing at the ecosystem level. *Fish and Fisheries* 18(2), 199-211. https://doi.org/10.1111/faf.12171.

Jahnke, A., Arp, H.P.H., Escher, B.I., Gewert, B., Gorokhova, E., Kühnel, D. *et al.* (2017). Reducing uncertainty and confronting ignorance about the possible impacts of weathering plastic in the marine environment. *Environmental Science & Technology Letters* 4(3), 85-90. https://doi.org/10.1021/acs.estlett.7b00008.

Jambeck, J.R., Geyer, R., Wilcox, C., Siegler, T.R., Perryman, M., Andrady, A. *et al.* (2015). Plastic waste inputs from land into the ocean. *Science* 347(6223), 768-771. https://doi.org/10.1126/science.1260352.

Jennings, S. and Kaiser, M.J. (1998). The effects of fishing on marine ecosystems. *Advances in Marine Biology* 34, 201-352. https://doi.org/10.1016/S0065-2881(08)60212-6.

Johnson, J.E., Welch, D.J., Maynard, J.A., Bell, J.D., Pecl, G., Robins, J. *et al.* (2016). Assessing and reducing vulnerability to climate change: Moving from theory to practical decision-support. *Marine Policy* 74, 220-229. https://doi.org/10.1016/j.marpol.2016.09.024.

Joint Group of Experts on the Scientific Aspects of Marine Environmental Protection (GESAMP) (2015). *Sources, Fate and Effects of Microplastics in the Marine Environment: A Global Assessment.* Kershaw, P.J. (ed.). London: International Maritime Organization. http://ec.europa.eu/environment/marine/good-environmental-status/descriptor-10/pdf/GESAMP_microplastics%20full%20study.pdf.

Jones, G.P., McCormick, M.I., Srinivasan, M. and Eagle, J.V. (2004). Coral decline threatens fish biodiversity in marine reserves. *Proceedings of the National Academy of Sciences* 101(21), 8251-8253. https://doi.org/10.1073/pnas.0401277101.

Karagas, M.R., Choi, A.L., Oken, E., Horvat, M., Schoeny, R., Kamai, E. *et al.* (2012). Evidence on the human health effects of low-level methylmercury exposure. *Environmental Health Perspectives* 120(6), 799-806. https://doi.org/10.1289/ehp.1104494.

Kelleher, K. (2005). *Discards in The World's Marine Fisheries: An Update.* FAO Fisheries Technical Paper. Rome: Food and Agriculture Organization of the United Nations. http://www.fao.org/docrep/008/y5936e/y5936e00.htm.

Koelmans, A.A., Bakir, A., Burton, G.A. and Janssen, C.R. (2016). Microplastic as a vector for chemicals in the aquatic environment: Critical review and model-supported reinterpretation of empirical studies. *Environmental Science & Technology* 50(7), 3315-3326. https://doi.org/10.1021/acs.est.5b06069.

Koslow, J.A., Auster, P., Bergstad, O.A., Roberts, J.M., Rogers, A., Vecchione, M. *et al.* (2016). Biological communities on seamounts and other submarine features potentially threatened by disturbance. In *The First Global Integrated Marine Assessment: World Ocean Assessment I.* Inniss, L. and Simcock, A. (eds.). United Nations. chapter 51. http://www.un.org/depts/los/global_reporting/WOA_RPROC/Chapter_51.pdf

Kummu, M., De Moel, H., Salvucci, G., Viviroli, D., Ward, P.J. and Varis, O. (2016). Over the hills and further away from coast: Global geospatial patterns of human and environment over the 20th–21st centuries. *Environmental Research Letters* 11(3), https://doi.org/10.1088/1748-9326/11/3/034010.

Lambeth, L., Hanchard, B., Aslin, H., Fay-Sauni, L., Tuara, P., Rochers, K.D. *et al.* (2014). An overview of the involvement of women in fisheries activities in Oceania.In *Global Symposium on Women in Fisheries.* Williams M.J., Chao N.H., Choo P.S., Matics K., Nandeesha M.C., Shariff M. *et al.* (eds.). Penang: ICLARM – The World Fish Center. 21-33.

Larson, C. (2018). Asia's hunger for sand takes toll on ecology. *Science* 359(6379), 964-965. https://doi.org/10.1126/science.359.6379.964

Lavers, J.L. and Bond, A.L. (2017). Exceptional and rapid accumulation of anthropogenic debris on one of the world's most remote and pristine islands. *Proceedings of the National Academy of Sciences* 114(23), 6052-6055. https://doi.org/10.1073/pnas.1619818114.

Leggett, C., Scherer, N., Curry, M., Bailey, R. and Haab, T. (2014). *Final Report: Assessing the economic Benefits of Reductions in Marine Debris: A Pilot Study of Beach Recreation in Orange County, California.* https://marinedebris.noaa.gov/sites/default/files/publications-files/MarineDebrisEconomicStudy_0.pdf.

Leite, L. and Pita, C. (2016). Review of participatory fisheries management arrangements in the European Union. *Marine Policy* 74, 268-278. https://doi.org/10.1016/j.marpol.2016.08.003.

Lemasson, A.J., Fletcher, S., Hall-Spencer, J.M. and Knights, A.M. (2017). Linking the biological impacts of ocean acidification on oysters to changes in ecosystem services: A review. *Journal of Experimental Marine Biology and Ecology* 492, 49-62. https://doi.org/10.1016/j.jembe.2017.01.019.

Li, H.-X., Ma, L.-S., Lin, L., Ni, Z.-X., Xu, X.-R., Shi, H.-H. *et al.* (2018). Microplastics in oysters Saccostrea Cucullata along the Pearl River Estuary, China. 236, 619-625. https://doi.org/10.1016/j.envpol.2018.01.083.

Macfadyen, G., Huntington, T. and Cappell, R. (2009). *Abandoned, Lost Or Otherwise Discarded Fishing Gear.* UNEP Regional Seas Reports and Studies No.185; FAO Fisheries and Aquaculture Technical Paper, No. 523. Rome: Food and Agriculture Organization of the United Nations and the United Nations Environment Programme. http://www.fao.org/docrep/011/i0620e/i0620e00.htm.

Marcon, Y., Purser, A., Janssen, F., Lins, L., Brown, A. and Boetius, A. (2016). Megabenthic community structure within and surrounding the DISCOL Experimental Area 26 years after simulated manganese nodule mining disturbance. *EU FP7 MIDAS Final Meeting.* Gent, 3-7 October 2016. http://epic.awi.de/44161/

Material District (2018). *Finite: A more sustainable alternative to concrete made from desert sand.* [Material District https://materia.nl/article/finite-concrete-desert-sand/ (Accessed: October 2018).

McDonald, M.A., Hildebrand, J.A. and Wiggins, S.M. (2006). Increases in deep ocean ambient noise in the Northeast Pacific west of San Nicolas Island, California. *The Journal of the Acoustical Society of America* 120(2), 711-718. https://doi.org/10.1121/1.2216565.

Melnychuk, M.C., Peterson, E., Elliott, M. and Hilborn, R. (2016). Fisheries management impacts on target species status. *Proceedings of the National Academy of Sciences* 114(1), 178-183. https://doi.org/10.1073/pnas.1609915114.

Miksis-Olds, J.L. and Nichols, S.M. (2016). Is low frequency ocean sound increasing globally? *The Journal of the Acoustical Society of America* 139(1), 501-511. https://doi.org/10.1121/1.4938237.

Morgan, C.L., Odunton, N.A. and Jones, A.T. (1999). Synthesis of environmental impacts of deep seabed mining. *Marine Georesources and Geotechnology* 17(4), 307-356. https://doi.org/10.1080/106411999273666.

Myers, G.J., Davidson, P.W., Cox, C., Shamlaye, C.F., Palumbo, D., Cernichiari, E. *et al.* (2003). Prenatal methylmercury exposure from ocean fish consumption in the Seychelles child development study. *The Lancet* 361(9370), 1686-1692. https://doi.org/10.1016/S0140-6736(03)13371-5.

Narita, D., Rehdanz, K. and Tol, R.S.J. (2012). Economic costs of ocean acidification: A look into the impacts on global shellfish production. *Climatic Change* 113(3-4), 1049-1063. https://doi.org/10.1007/s10584-011-0383-3.

Northridge, S., Coram, A., Kingston, A. and Crawford, R. (2017). Disentangling the causes of protected-species bycatch in gillnet fisheries. *Conservation Biology* 31(3), 686-695. https://doi.org/10.1111/cobi.12741.

7

O'Hanlon, N.J., James, N.A., Masden, E.A. and Bond, A.L. (2017). Seabirds and Marine Plastic Debris in the Northeastern Atlantic: A Synthesis and Recommendations for Monitoring and Research. *Environmental Pollution* 22, 1291-1301. https://doi.org/10.1016/j.envpol.2017.08.101.

O'Neill, B.C., Oppenheimer, M., Warren, R., Hallegatte, S., Kopp, R.E., Pörtner, H.O. *et al.* (2017). IPCC reasons for concern regarding climate change risks. *Nature Climate Change* 7(1), 28-37. https://doi.org/10.1038/nclimate3179.

Olson, J. (2011). Understanding and contextualizing social impacts from the privatization of fisheries: An overview. *Ocean & Coastal Management* 54(5), 353-363. https://doi.org/10.1016/j.ocecoaman.2011.02.002.

Österblom, H. and Bodin, Ö. (2012). Global cooperation among diverse organizations to reduce illegal fishing in the Southern Ocean. *Conservation Biology* 26(4), 638-648. https://doi.org/10.1111/j.1523-1739.2012.01850.x.

Peduzzi, P. (2014). Sand, rarer than one thinks. *Environmental Development* 11, 208-218. https://doi.org/10.1016/j.envdev.2014.04.001.

Pham, C.K., Ramirez-Llodra, E., Alt, C.H.S., Amaro, T., Bergmann, M., Canals, M. *et al.* (2014). Marine litter distribution and density in European seas, from the shelves to deep basins. *PLoS One* 9(4), e95839. https://doi.org/10.1371/journal.pone.0095839.

Popper, A.N. and Hastings, M.C. (2009). The effects of anthropogenic sources of sound on fishes. *Journal of fish biology* 75(3), 455-489. https://doi.org/10.1111/j.1095-8649.2009.02319.x

Ralston, N.V. and Raymond, L.J. (2010). Dietary selenium's protective effects against methylmercury toxicity. *Toxicology* 278(1), 112-123. https://doi.org/10.1016/j.tox.2010.06.004.

Rice, J. (2014). Evolution of international commitments for fisheries sustainability. *ICES Journal of Marine Science* 71(2), 157-165. https://doi.org/10.1093/icesjms/fst078.

Rice, J., Moksness, E., Attwood, C., Brown, S.K., Dahle, G., Gjerde, K.M. *et al.* (2012). The role of MPAs in reconciling fisheries management with conservation of biological diversity. *Ocean & Coastal Management* 69, 217-230. https://doi.org/10.1016/j.ocecoaman.2012.08.001.

Rice, J.C. and Garcia, S.M. (2011). Fisheries, food security, climate change, and biodiversity: characteristics of the sector and perspectives on emerging issues. *ICES Journal of Marine Science* 68(6), 1343-1353. https://doi.org/10.1093/icesjms/fsr041.

Richardson, W.J., Greene, C.R., Malme, C.I. and Thomson, D.H. (1995). *Marine Mammals and Noise*. San Diego, CA: Academic Press. https://www.elsevier.com/books/marine-mammals-and-noise/richardson/978-0-08-057303-8.

Ricker, W.E. (1975). *Computation and Interpretation of Biological Statistics of Fish Populations*. Bulletin of the Fisheries Research Board of CanadaEnvironment Canada. http://www.dfo-mpo.gc.ca/Library/1485.pdf.

Roos, N., Wahab, M.A., Chamnan, C. and Thilsted, S.H. (2007). The role of fish in food-based strategies to combat vitamin A and mineral deficiencies in developing countries. *The Journal of Nutrition* 137(4), 1106-1109. https://doi.org/10.1093/jn/137.4.1106.

Rose, G.A. (2007). *Cod: The Ecological History of the North Atlantic Fisheries*. St. John's, Newfoundland: Breakwater Books. http://www.breakwaterbooks.com/books/cod-the-ecological-history-of-the-north-atlantic-fisheries/.

Rosenberg, A. (2010). *Using fly ash in concrete*. [National Precast Concrete Association https://precast.org/2010/05/using-fly-ash-in-concrete/.

Salinger, J., Hobday, A., Matear, R., O'Kane, T., Risbey, J., Dunstan, P. *et al.* (2016). Chapter one-decadal-scale forecasting of climate drivers for marine applications. *Advances in Marine Biology* 74, 1-68. https://doi.org/10.1016/bs.amb.2016.04.002.

Schindler, D.E. and Hilborn, R. (2015). Prediction, precaution, and policy under global change. *Science* 347(6225), 953-954. https://doi.org/10.1126/science.1261824.

Schmidt, C., Krauth, T. and Wagner, S. (2017). Export of plastic debris by rivers into the sea. *Environmental Science & Technology* 51(21), 12246-12253. https://doi.org/10.1021/acs.est.7b02368.

Schuster, P.F., Schaefer, K.M., Aiken, G.R., Antweiler, R.C., Dewild, J.F., Gryziec, J.D. *et al.* (2018). Permafrost stores a globally significant amount of mercury. *Geophysical Research Letters* 45(3), 1463-1471. https://doi.org/10.1002/2017GL075571.

Secretariat of the Convention on Biological Diversity (2016). *Marine Debris: Understanding, Preventing and Mitigating the Significant Adverse Impacts on Marine and Coastal Biodiversity*. CBD Technical Series No. 83. Montreal. https://www.cbd.int/doc/publications/cbd-ts-83-en.pdf.

Secretariat of the Pacific Community (ed.) (2013a). *Deep Sea Minerals: Sea Floor Massive Sulphides, a Physical, Biological, Environmental, and Technical Review*. http://dsm.gsd.spc.int/public/files/meetings/TrainingWorkshop4/UNEP_vol1A.pdf.

Secretariat of the Pacific Community (2013b). *Deep Sea Minerals: Deep Sea Minerals and the Green Economy*. Baker, E. and Beaudoin, Y. (eds.). https://www.researchgate.net/publication/260596769_Deep_Sea_Minerals_and_the_Green_Economy.

Sharma, R. (2001). Indian Deep-sea Environment Experiment (INDEX): An appraisal. *Deep Sea Research Part II: Topical Studies in Oceanography* 48(16), 3295-3307. https://doi.org/10.1016/S0967-0645(01)00041-8.

Steneck, R.S., Mumby, P.J., MacDonald, C., Rasher, D.B. and Stoyle, G. (2018). Attenuating effects of ecosystem management on coral reefs. *Science Advances* 4(5), eaao5493. https://doi.org/10.1126/sciadv.aao5493.

Stokstad, E. (2018). Controversial plastic trash collector begins maiden ocean voyage. Science Magazine http://www.sciencemag.org/news/2018/09/still-controversial-plastic-trash-collector-ocean-begins-maiden-voyage.

Strain, J.J., Davidson, P.W., Bonham, M.P., Duffy, E.M., Stokes-Riner, A., Thurston, S.W. *et al.* (2008). Associations of maternal long-chain polyunsaturated fatty acids, methyl mercury, and infant development in the Seychelles child development nutrition study. *NeuroToxicology* 29(5), 776-782. https://doi.org/10.1016/j.neuro.2008.06.002.

Swan, J. and Gréboval, D. (2005). *Overcoming Factors of Unsustainability and Overexploitation in Fisheries: Selected Papers on Issues and Approaches*. Rome: Food and Agriculture Organization of the United Nations. http://www.fao.org/docrep/009/a0312e/A0312E00.htm.

Syvitski, J.P.M., Vörösmarty, C.J., Kettner, A.J. and Green, P. (2005). Impact of humans on the flux of terrestrial sediment to the global coastal ocean. *Science* 308(5720), 376-380. https://doi.org/10.1126/science.1109454.

Tarling, G.A., Peck, V.L., Ward, P., Ensor, N.S., Achterberg, E., Tynan, E. *et al.* (2016). Effects of acute ocean acidification on spatially-diverse polar pelagic foodwebs: Insights from on-deck microcosms. *Deep Sea Research Part II: Topical Studies in Oceanography* 127, 75-92. https://doi.org/10.1016/j.dsr2.2016.02.008.

Taylor, C.M., Emmett, P.M., Emond, A.M. and Golding, J. (2018). A review of guidance on fish consumption in pregnancy: Is it fit for purpose? *Public Health Nutrition* 21(11), 2149-2159. https://doi.org/10.1017/S1368980018000599.

Thiel, M., Luna-Jorquera, G., Álvarez-Varas, R., Gallardo, C., Hinojosa, I.A., Luna, N. *et al.* (2018). Impacts of Marine Plastic Pollution from Continental Coasts to Subtropical Gyres—Fish, Seabirds, and Other Vertebrates in the SE Pacific. *Frontiers in Marine Science* 5(238). https://doi.org/10.3389/fmars.2018.00238.

Thilsted, S.H., James, D., Toppe, J., Subasinghe, R. and Karunasagar, I. (2014). Maximizing the contribution of fish to human nutrition. *ICN2 Second International Conference on Nutrition*. Food Agriculture Organizationof the United Nations and World Health Organization http://www.fao.org/3/a-i3963e.pdf

Tidau, S. and Briffa, M. (2016). Review on behavioral impacts of aquatic noise on crustaceans. *Proceedings of Meetings on Acoustics* 27(010028). https://doi.org/10.1121/2.0000302.

Tuvalu Seabed Mining Act 2014, 2014 (Tuvalu, P.o.). https://www.tuvalu-legislation.tv/cms/images/LEGISLATION/PRINCIPAL/2014/2014-0014/TuvaluSeabedMineralsAct_1.pdf.

United Nation Environment Programme (2013). *Global Mercury Assessment 2013: Sources, Emissions, Release and Environmental Transport*. Nairobi. http://wedocs.unep.org/bitstream/handle/20.500.11822/7984/-Global%20Mercury%20Assessment-201367.pdf?sequence=3&isAllowed=y.

United Nations (1982). *United Nations Convention on the Law of the Sea (LOSC)*. http://www.un.org/Depts/los/convention_agreements/texts/unclos/closindx.htm.

United Nations, General Assembly (2016). *70/235. Oceans and the law of the sea: Resolution adopted by the General Assembly on 23 December 2015*. https://undocs.org/A/RES/70/235.

United Nations Environment Assembly of the United Nations Environment Programme (2016a). *2/12. Sustainable Coral Reefs Management. UNEP/EA.2/Res.12*. http://wedocs.unep.org/bitstream/handle/20.500.11822/11187/K1607234_UNEPEA2_RES12E.pdf?sequence=1&isAllowed=y.

United Nations Environment Assembly of the United Nations Environment Programme (2016b). *2/11. Marine Plastic Litter and Microplastics. UNEP/EA.2/Res.11*. http://wedocs.unep.org/bitstream/handle/20.500.11822/11186/K1607228_UNEPEA2_RES11E.pdf?sequence=1&isAllowed=y.

United Nations Environment Assembly of the United Nations Environment Programme (2017). *3/7. Marine Litter and Microplastics. UNEP/EA.3/Res.7*. https://papersmart.unon.org/resolution/uploads/k1800210.english.pdf.

United Nations Environment Programme (2016). *Marine Plastic Debris and Microplastics: Global Lessons and Research To Inspire Action and Guide Policy Change*. Nairobi. https://wedocs.unep.org/rest/bitstreams/11700/retrieve.

United Nations Environment Programme (2017). *Coral Bleaching Futures: Downscaled Projections of Bleaching Conditions for the World's Coral Reefs, Implications of Climate Policy and Management Responses*. Nairobi. http://wedocs.unep.org/bitstream/handle/20.500.11822/22048/Coral_Bleaching_Futures.pdf?sequence=1&isAllowed=y.

United Nations Environment Programme and GRID-Arendal (2016). *Marine Litter Vital Graphics*. Nairobi: United Nations Environment Programme and GRID-Arendal. https://gridarendal-website-live.s3.amazonaws.com/production/documents/:s_document/11/original/MarineLitterVG.pdf?1488455779.

United Nations Environment Programme World Conservation Monitoring Centre (2015). *Marine Litter Assesment in the Mediterranean 2015. A Report of the Mediterranean Action Plan*. https://wedocs.unep.org/bitstream/handle/20.500.11822/7098/MarineLitterEng.pdf?sequence=1&isAllowed=y.

United States Food and Drug Administration (2017). *Eating fish: What pregnant women and parents should know*. https://www.fda.gov/Food/ResourcesForYou/Consumers/ucm393070.htm (Accessed: June 2018.

United States National Oceanic and Atmospheric Administration (2012). *How much would it cost to clean up the pacific garbage patches?* NOAA Coral Reef Watch https://response.restoration.noaa.gov/about/media/how-much-would-it-cost-clean-pacific-garbage-patches.html.

United States National Oceanic and Atmospheric Administration (2017). *Global warming and recurrent mass bleaching of corals. Coral Reef Watch*. https://coralreefwatch.noaa.gov/satellite/publications_hughes-etal_nature_20170316.php (Accessed: June 2017).

Unsworth, R.K.F., Ambo-Rappe, R., Jones, B.L., La Nafie, Y.A., Irawan, A., Hernawan, U.E. *et al.* (2018). Indonesia's globally significant seagrass meadows are under widespread threat. *Science of the Total Environment* 634, 279-286. https://doi.org/10.1016/j.scitotenv.2018.03.315.

Van Cauwenberghe, L. and Janssen, C.R. (2014). Microplastics in bivalves cultured for human consumption. *Environmental Pollution* 193, 65-70. https://doi.org/10.1016/j.envpol.2014.06.010.

Van Dover, C.L. (2011). Tighten regulations on deep-sea mining. *Nature* 470(7332), 31-33. https://doi.org/10.1038/470031a.

Van Dover, C.L., Ardron, J.A., Escobar, E., Gianni, M., Gjerde, K.M., Jaeckel, A. *et al.* (2017). Biodiversity loss from deep-sea mining. *Nature Geoscience* 10, 464–465. https://doi.org/10.1038/ngeo2983.

Van Hooidonk, R., Maynard, J., Tamelander, J., Gove, J., Ahmadia, G., Raymundo, L. *et al.* (2016). Local-scale projections of coral reef futures and implications of the Paris Agreement. *Scientific reports* 6(39666). https://doi.org/10.1038/srep39666.

Van Sebille, E., Wilcox, C., Lebreton, L., Maximenko, N., Hardesty, B.D., Van Franeker, J.A. *et al.* (2015). A global inventory of small floating plastic debris. *Environmental Research Letters* 10(12), 124006. https://doi.org/10.1088/1748-9326/10/12/124006.

Van Wijngaarden, E., Thurston, S.W., Myers, G.J., Harrington, D., Cory-Slechta, D.A., Strain, J.J. *et al.* (2017). Methyl mercury exposure and neurodevelopmental outcomes in the Seychelles Child Development Study Main cohort at age 22 and 24 years. *Neurotoxicology and teratology* 59, 35-42. https://doi.org/10.1016/j.ntt.2016.10.011.

Veron, J., Hoegh-Guldberg, O., Lenton, T., Lough, J., Obura, D., Pearce-Kelly, P. *et al.* (2009). The coral reef crisis: The critical importance of< 350ppm CO_2. *Marine Pollution Bulletin* 58(10), 1428-1436. https://doi.org/10.1016/j.marpolbul.2009.09.009.

Von Schuckmann, K., Palmer, M.D., Trenberth, K.E., Cazenave, A., Chambers, D., Champollion, N. *et al.* (2016). An imperative to monitor Earth's energy imbalance. *Nature Climate Change* 6(2), 138-144. https://www.nature.com/articles/nclimate2876.

Wang, J., Kiho, K., Ofiara, D., Zhao, Y., Bera, A., Lohmann, R. *et al.* (2016). Marine debris.In *The First Global Integrated Marine Assessment - World Ocean Assessment I*. Innis, L. and Simcock, A. (eds.). chapter 25. http://www.un.org/depts/los/global_reporting/WOA_RPROC/Chapter_25.pdf

Watkins, E., ten Brink, P., Sirini Withana, M.K., Russi, D., Mutafoglu, K., Schweitzer, J.-P. *et al.* (2017). The socio-economic impacts of marine litter, including the costs of policy inaction and action.In *Handbook on the Economics and Management of Sustainable Oceans*. Nunes, P.A.L.D., Svensson, L.E. and Markandya, A. (eds.). Edward Elgar Publishing. chapter 14. 296-319. https://www.elgaronline.com/view/9781786430717.00024.xml

Watson, R.A., Cheung, W.W.L., Anticamara, J.A., Sumaila, R.U., Zeller, D. and Pauly, D. (2012). Global marine yield halved as fishing intensity redoubles. *Fish and Fisheries* 14(4), 493-503. https://doi.org/10.1111/j.1467-2979.2012.00483.x.

Wilkinson, C., Salvat, B., Eakin, C.M., Brathwaite, A., Francini-Filho, R., Webster, N. *et al.* (2016). Tropical and sub-tropical coral reefs.In *The First Global Integrated Marine Assessment: World Ocean Assessment I*. Simcock, A. and Innis, L. (eds.). chapter 43. http://www.un.org/depts/los/global_reporting/WOA_RPROC/Chapter_43.pdf

Woodroffe, C.D., Hall, F.R., Farrell , J.W. and Harris, P.T. (2016). Calcium carbonate production and contribution to coastal sediments.In *The First Global Integrated Marine Assessment: World Ocean Assessment I*. Innis, L. and Simcock, A. (eds.). Cambridge, MA: Cambridge University Press. chapter 7. http://www.un.org/Depts/los/global_reporting/WOA_RPROC/Chapter_07.pdf

World Bank (2012). *Hidden Harvest.The Global Contribution of Capture Fisheries*. Washington, DC: World Bank. http://documents.worldbank.org/curated/en/515701468152718292/pdf/664690ESW0P1210120HiddenHarvest0web.pdf.

World Health Organization (2017). *Mercury and health: Key facts*. http://www.who.int/news-room/fact-sheets/detail/mercury-and-health (Accessed: June 2018).

Worm, B., Hilborn, R., Baum, J.K., Branch, T.A., Collie, J.S., Costello, C. *et al.* (2009). Rebuilding global fisheries. *Science* 325(5940), 578-585. https://doi/org/10.1126/science.1173146.

Wyles, K.J., Pahl, S., Thomas, K. and Thompson, R.C. (2016). Factors that can undermine the psychological benefits of coastal environments: Exploring the effect of tidal state, presence, and type of litter. *Environment and behavior* 48(9), 1095-1126. https://doi.org/10.1177/0013916515592177.

Yang, D., Shi, H., Li, L., Li, J., Jabeen, K. and Kolandhasamy, P. (2015). Microplastic pollution in table salts from China. *Environmental science & technology* 49(22), 13622-13627. https://doi.org/10.1021/acs.est.5b03163.

Ziccardi, L.M., Edgington, A., Hentz, K., Kulacki, K.J. and Kane Driscoll, S. (2016). Microplastics as vectors for bioaccumulation of hydrophobic organic chemicals in the marine environment: A state-of-the-science review. *Environmental toxicology and chemistry* 35(7), 1667-1676. https://doi.org/10.1002/etc.3461.

Chapter 8

Land and Soil

Coordinating Lead Authors: Nicolai Dronin (Moscow State University), Andres Guhl (Universidad de los Andes), Jia Gensuo (Chinese Academy of Sciences)
Lead Authors: Javier Ñaupari (Universidad Nacional Agraria La Molina)
GEO Fellows: Darshini Ravindranath (University College London), Hung Vo (Harvard University), Ying (Grace) Wang (Tongji University)

Executive summary

Land resources are essential for achieving 10 of the 17 Sustainable Development Goals (SDGs). Agricultural and food production are still responsible for most of the changes of land, including forests and other types of ecosystems, while human-induced land degradation remains a fundamental environmental problem affecting food security, livelihoods and lives of the people on this planet. Globalization, population growth, urbanization and shifting dietary preferences are responsible for some of the changes in our food system over the past 50 years and have increased food imports and teleconnections. There is also a growing concern over land grabbing and speculation throughout the world. Clear property rights and land-resource stewardship are crucial for ensuring sustainable production of food while preserving the ability of land ecosystems to continue providing a wide variety of other benefits to people (e.g. hydrological regulation, pollination). Rural inhabitants play a fundamental role in land conservation. The main findings regarding land can be summarized as follows.

Current trends, based on technological optimism, improved seeds, machinery and fertilizers, are not likely to supply future demands for food, energy, timber and other ecosystem services and values taking into consideration even moderate projections for land-resource availability (*well established*). By 2050, the world needs to produce at least 50 per cent more food to feed the projected global population of 10 billion people. Current land management cannot achieve this while preserving ecosystem services, the loss of natural capital, combating climate change, addressing energy and water security, and promoting gender and social equality. {8.5.1}

Food production is the largest anthropogenic use of land, accounting for 50 per cent of habitable land (*well established*). Livestock production uses 77 per cent of agricultural land for feed production, pasture and grazing land. The livestock sector provides only 17 per cent of dietary energy and 33 per cent of dietary protein demands. Therefore, using about 80 per cent of agricultural land for livestock is inefficient. {8.4.1}

The expansion of agricultural area has been slowed by increasing productivity (*established but incomplete*). Although there are regional variations, globally, the harvested crop area increased by 23 per cent between 1984 and 2015, while global crop production rose by 87 per cent. On average, per capita daily food supply in the world increased 10 per cent between 1993 and 2013. However, monocultural farming systems, sometimes assumed to be more productive and profitable, are often associated with environmental degradation and biodiversity loss. Grasslands in southern South America have been converted into soybean fields mostly for export. The expansion of oil palm in South-East Asia has been at the expense of forests and peatlands. {8.4.1}

Global food supply has become dependent on the growing trade of a small number of crops grown in a few regions with increasing crop specialization (*well established*). The share of production traded internationally in 2014 was 24, 11 and 60 per cent of global wheat, maize and soybean production, respectively. This leads to lower food prices and food-deficit countries benefit from these food imports. However, the geographic concentration of production increases systemic risk, as illustrated by recent spikes in international commodity prices due to poor harvests in certain regions. Furthermore, the growing prevalence of certain crops in global food supplies has contributed to the increasing consumption of nutritionally poor, highly processed foods, with potentially serious consequences for population health. {8.5.1}

The linkages between different places (teleconnections) are strengthening worldwide (*well established*). Demand in some places generates land transformations in others. The distance between producers and consumers may obscure ecosystem degradation in production areas. For example, demand for land resources in many urban areas is affecting land use in rural and other urban areas, both within national boundaries and internationally. {8.3.2}

Approximately one-third of food produced globally for human consumption is lost or wasted (*well established*). Approximately 56 per cent of total food loss and food waste occurs in industrialized countries, while 44 per cent originates from developing countries. {8.5.1}

Deforestation rates differ among regions, and while the global trend is continuing forest loss, many regions, especially in more developed countries, are showing an increase in forest cover (mostly in plantations) (*well established*). In the 1990s, about 10.6 million ha of natural forests were lost per year. For the period 2010-2015, this rate had dropped to 6.5 million ha/year. Simultaneously, the growth rate of planted forests is about 3.2 million ha/year, and by 2015 they accounted for 7 per cent of the global forest area mostly concentrated in high-income countries. Plantations do not provide the same diversity of ecosystem services as natural forests. {8.4.1}

Although built-up areas represent only a relatively small fraction of land, their impacts extend beyond built areas (*well established*). Since 1975 urban settlements have grown approximately 2.5 times, accounting for 7.6 per cent of the global land area in 2015. Cities and infrastructure expand differently across regions. By covering the ground with impervious surfaces, cities affect the hydrological cycle and soil function, as well as generating urban heat islands. About 3 billion urban dwellers lack access to adequate waste disposal facilities, which poses health risks (infections, exposure to chemicals, dust, others) and generates environmental impacts (soil and water pollution, greenhouse gas [GHG] emissions, others) and land-use competition. {8.4.1; 8.5.2}

Land is the most important asset for people in large sections of the world and secure rights can help turn these assets into development opportunities (*well established*). Indigenous populations, the poor, landless and women are among the groups most vulnerable to the implications of unequal landownership and access. Estimates suggest that only about 10 per cent of formal land rights are registered or recorded worldwide. Without formal recognition and protection of their land rights, communities in some countries face loss of land

due to land acquisition, land grabbing and land leasing amid fear of food scarcity and rising food prices. Around the world, 26.7 million ha of agricultural land have been transferred into the hands of foreign investors since 2000. {8.5.3, 8.5.4}

Unequal tenure of land resources is a critical challenge for sustainable land management (*well established*). Tenure-security of indigenous peoples' lands can generate billions of dollars' worth of benefits (carbon sequestration, reduced pollution, clean water, erosion control) and a suite of other local, regional and global 'ecosystem services'. These benefits far outweigh the costs of securing land tenure. {8.5.3}

Continuing on the current track, it will be difficult to achieve the land degradation neutrality target adopted in the United Nations Conference on Sustainable Development (Rio+20) (*well established*). Assessments based on satellite data show that land degradation hotpots cover about 29 per cent of global land area. However, there is variance between different data sets and disagreement between methods. About 3.2 billion people live in these degrading areas. Investing in avoiding land degradation and the restoration of degraded land makes sound economic sense; the benefits generally far exceed the cost. Innovative technologies, land management strategies and land-resource stewardship at different scales (e.g. good agricultural practices, sustainable forest management, agro-silvopastoral production systems, agricultural innovation, payment for ecosystem services, land restoration, land titling) need to be more effectively promoted and adopted at local, regional, international and national levels. These alternatives also contribute to climate change resilience. Existing multilateral environmental agreements provide a platform of unprecedented scope and ambition for action to avoid and reduce land degradation and promote restoration. {8.6.1; 8.6.3}

Decreasing the gender gap in access to information and technology, and access to and control over production inputs and land, could increase agricultural productivity and reduce hunger and poverty (*well established*). New policies should explicitly target indigenous peoples, women, family farmers, pastoralists and fishers, so these groups can have secure and equitable access to land, inputs, knowledge, resources, markets, financial services, opportunities for adding value and non-farm employment. {8.6}

Minimizing food losses and waste will have significant environmental, social and economic benefits in supporting global food security (*well established*). Where waste cannot be prevented, opportunities to recover value from this waste stream, such as conversion to compost, liquid fertilizers, biogas or higher value end-use products such as animal feed protein or biochemicals, should be pursued. {8.6}

8.1 Land resources and the Sustainable Development Goals

Land is complex to define as it has multiple interconnected dimensions (e.g. land as a provider of resources and services, as shelter, as property, as a key to cultural identity) (United Nations Convention to Combat Desertification [UNCCD] 2017). In this chapter, we emphasize land as a provider of food, fodder, fibre and forest products. Its ability to provide ecosystem services that regulate ecological processes is treated in Chapter 6 and the latest Intergovernmental Science-Policy Platform on Biodiversity and Ecosystem Services (IPBES) assessment reports (see below). Land is where a large proportion of food is produced, therefore it is closely related to Sustainable Development Goal (SDG) 2: End hunger, achieve food security and improved nutrition and promote sustainable agriculture. Specific targets for this goal include ensuring access to sufficient, healthy and nutritious food, especially for the most vulnerable groups. Furthermore, SDG 2 is closely related to increasing productivity through sustainable food production systems that are more resilient under increasing threats of climate change, and for maintaining and improving soil quality for future generations. Sustainable and more resilient food production systems require working towards gender equality and reducing other forms of inequality (SDG 10) since men and women do not have equal access to land resources in many parts of the world.

Land is the home of terrestrial biodiversity, is associated with food production, is where people live and where most economic activities take place. Over 54 per cent of the global population lives in urban areas (United Nations 2015a) and this poses additional challenges for land management: how to deal with hazardous pollutants and chemicals and their impacts on people and the environment. Pollution on land is becoming an important pressure, and human-generated waste and chemicals are impacting the health of people and the functioning of many ecosystem processes (SDGs 3, 15).

Additionally, human use of land is exerting enormous pressure on land resources, privileging short-term gains over long-term sustainability (UNCCD 2017), decreasing the supply of many ecosystem services (nature's contributions to people). The Millennium Ecosystem Assessment presented evidence that we are living beyond our means (Millennium Ecosystem Assessment 2004) and that ecosystems' abilities to provide us with food, fibre, forest resources, fodder and other biodiversity-related benefits are threatened. The recent IPBES report on land degradation and restoration reinforces this critical message (Intergovernmental Science-Policy Platform on Biodiversity and Ecosystem Services [IPBES] 2018). A healthy planet is the basis for development, and sustainable land-resource management is at the core of this challenge.

8.2 Setting the stage for GEO-6: the GEO-5 legacy

The main messages of the fifth report of the *Global Environment Outlook* (GEO-5) could be extrapolated to GEO 6. Perhaps the most important difference is the recognition of climate change as a driver of environmental change, and how it has the potential for altering land resources on its own (see Chapter 2). Climate change usually exacerbates ecosystem degradation and a more variable climate degrades ecosystems more strongly.

Another difference is the increasing recognition of the critical function that clear property rights play for land-resource stewardship and the crucial role of rural inhabitants in land conservation. The Land Rights Now initiative (http://www.landrightsnow.org) states that 2.5 billion people depend on land resources that are held, managed or used collectively. These people manage and protect 50 per cent of land, but only have legal ownership of 10 per cent. Clear property rights usually result in better management and stewardship of land resources (Lawry *et al.* 2017). Without them, these people are vulnerable to land dispossession in the hands of powerful actors (e.g. multinationals, governments).

Finally, there is increased concern over how land resource degradation is leading to widespread migration and even conflict. Since recording of these instances began in 2015, the Environmental Justice Atlas (https://ejatlas.org/) has listed more than 2,000 cases of socioenvironmental conflicts across the globe where land mismanagement, largely due to poor governance, has led to land degradation, conflict and/or dispossession of resources.

8.3 Drivers and pressures

8.3.1 Population

As chapter 2 notes, population growth is a key driver of land-use transformation with its associated environmental impacts. In the developing world, particularly Africa, there will be a doubling or tripling of population by the mid-21st century (United Nations 2014). In contrast, by 2050 developed countries will experience only small increases or even decreases in their population (United Nations 2015). Since the developed world has already entered a post-industrial society based increasingly on the tertiary sector, it is expected to be more stable in terms of land use, while developing countries are currently experiencing a rapid transition from agrarian societies to the industrial regime, with consequent radical change in land- and resource-use patterns (Haberl *et al.* 2011).

© Shutterstock/Tsering

Population growth can present a serious threat to the inherent limits of land to provide food, shelter and appropriate nutrition for local communities. However, impacts depend on specific socioeconomic contexts and are present mostly in developing countries. For example, a study of land-use change in north-western Ethiopia (1972-2010) shows conversion of 62 per cent of woodland into cropland, with high environmental costs (dust storms, droughts, severe soil erosion), due to population growth, but also because of attractive subsidies to farmers (Zewdie and Csaplovies 2015). Most studies on the subject recognize the importance of rural-to-urban migration for mitigating some of the negative impacts of population growth on land resources in rural areas. Some natural increase in population in rural areas can now be absorbed outside the country due to intraregional infrastructure improvements, as observed in Africa where a majority of migrants circulate within the continent looking for economic opportunities (Awumbila 2017).

8.3.2 Urbanization

Urban and rural areas are interconnected in terms of people, resources and services. Rural areas are connected to urban regions through networks of roads, information technology, electricity and trade. Meanwhile, urban areas are increasingly reliant on land-based resources yielding nature's contributions to people such as clean water, food and fibre. Urbanization can both positively and negatively impact these flows and functions and influence the economy and development of peri-urban and rural areas (Brenner and Schmid 2014). Cities operate within ecosystems that usually extend beyond jurisdictional boundaries (Solecki and Marcotullio 2013), requiring new methods to accurately measure the extent of urbanization to aid decision makers and civil society in responding to existing and emerging challenges (United Nations 2016). Urban demands for food, water, fibre and construction materials have established strong linkages between cities, rural areas and even regions in other countries. These linkages, also known as teleconnections, mean that land use in rural areas increasingly depends on demands from distant, urban agglomerations (Seto *et al.* 2012; Bergmann and Holmberg 2016). Urban infrastructure (energy, water, buildings and transportation) and food supply are particularly reliant on transboundary supplies (Kennedy and Hoornweg 2012; Ramaswami *et al.* 2012; Ramaswami *et al.* 2017).

Rural-to-urban migration continues and it has multifaceted impacts on land use through changing diets and demands on infrastructure and housing, as well as the ability of land to continue providing nature's contributions to people (UNCCD 2017). Much of the increase in population in built-up areas has taken place in disaster-prone regions such as within 10 metres (above sea level) of low elevation coastal zones (Seto *et al.* 2011; Paresi *et al.* 2016).

Figure 8.1: Different perspectives on the globalization of lands in 2007 (Eckert IV projection)

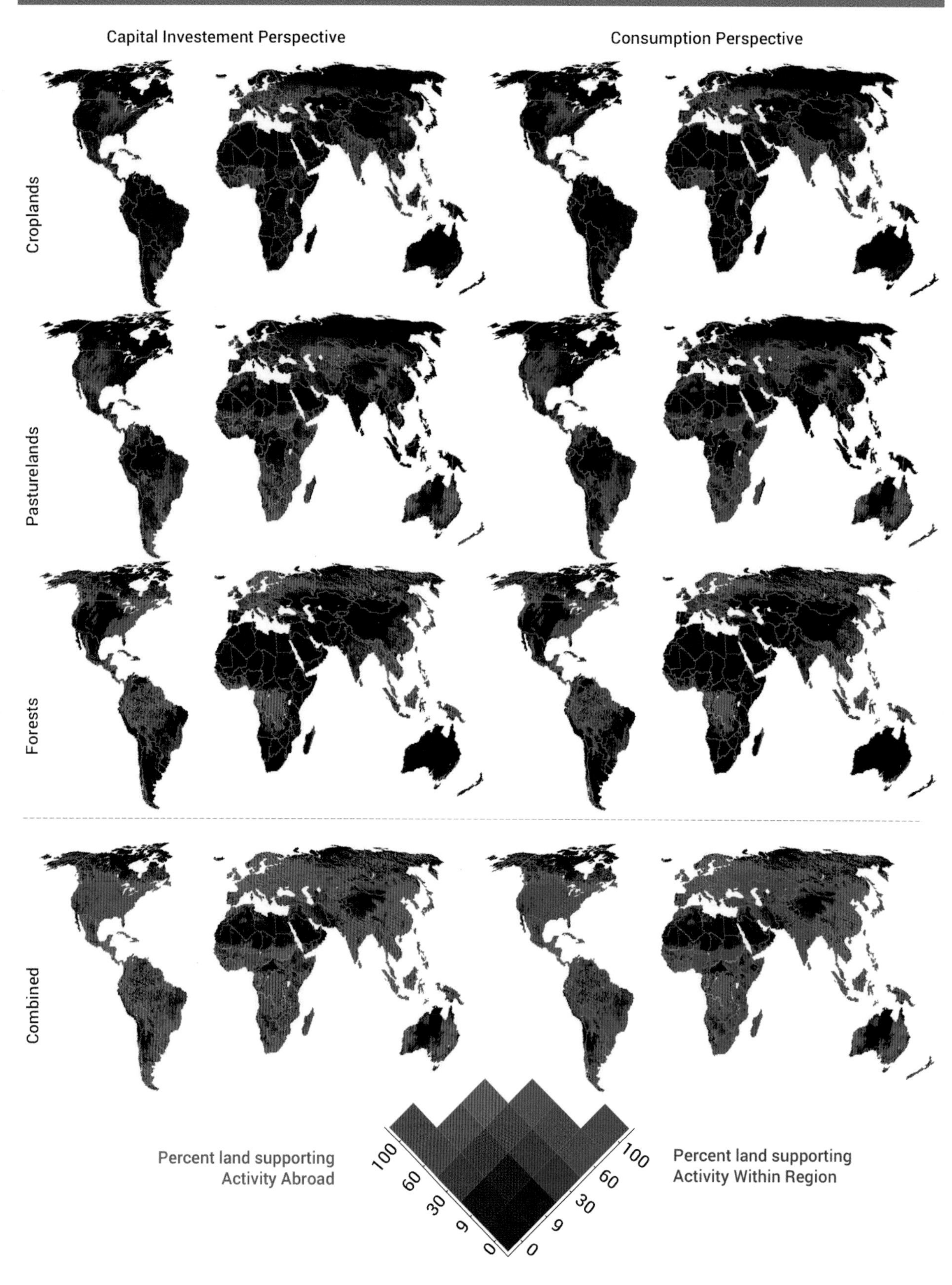

The figure illustrates how capital and consumption are linked regionally and globally for different lands' economic activities.

Source: Bergmann and Holmberg (2016).

8.3.3 Economic development

Globalization forces exert increasing pressures on land systems and their functions, leading to landscape change (Fischer-Kowalski and Haberl 2007; Henders and Ostwald 2014; Schaffartzik *et al.* 2015). Global trade and capital flows influence land use (e.g. agriculture, forestry) in developing countries (Bergmann and Holmberg 2016) (**Figure 8.1**, **Figure 8.2**). These flows of agricultural goods require transport and storage, which may increase economic and environmental costs and may also lead to the deterioration of the nutritional value of food, increase risks of disease transmission and generate food waste (UNEP 2016a). The significance of pressures on land tenure and land access is discussed in further detail in Section 8.5.3.

8.3.4 Technology and innovation

Around the globe, fast advancing technologies shape production and consumption, and drive patterns of land use and terrestrial ecosystems at various scales. Earth's big data and citizen science improve environmental monitoring and assessment, while allowing more public involvement (see Chapter 25).

Although it still has some limitations, satellite-based Earth observation has been combined with big data to track forest changes worldwide (e.g. Global Forest Watch, www.globalforestwatch.org; Terra-i, www.terra-i.org). Drones, powered by mobile technology, are becoming widely used to monitor biomass burning and unauthorized land-use conversion. The global explosion of cell phone access, and especially smartphones, can be used to democratize data access. Technological developments such as precision agriculture and drip irrigation are examples of more efficient agrochemical and water use.

Mobile communication and the Internet enable critical environmental information to spread within seconds to any corner of the world, rich or poor. Rural inhabitants in many parts of the developing world can use these technologies to improve land management with potential impacts on biodiversity conservation and land use (Chin 2018).

Figure 8.2: Relative roles played by agricultural commodities versus manufactures and services in globalizing lands (Eckert IV projections)

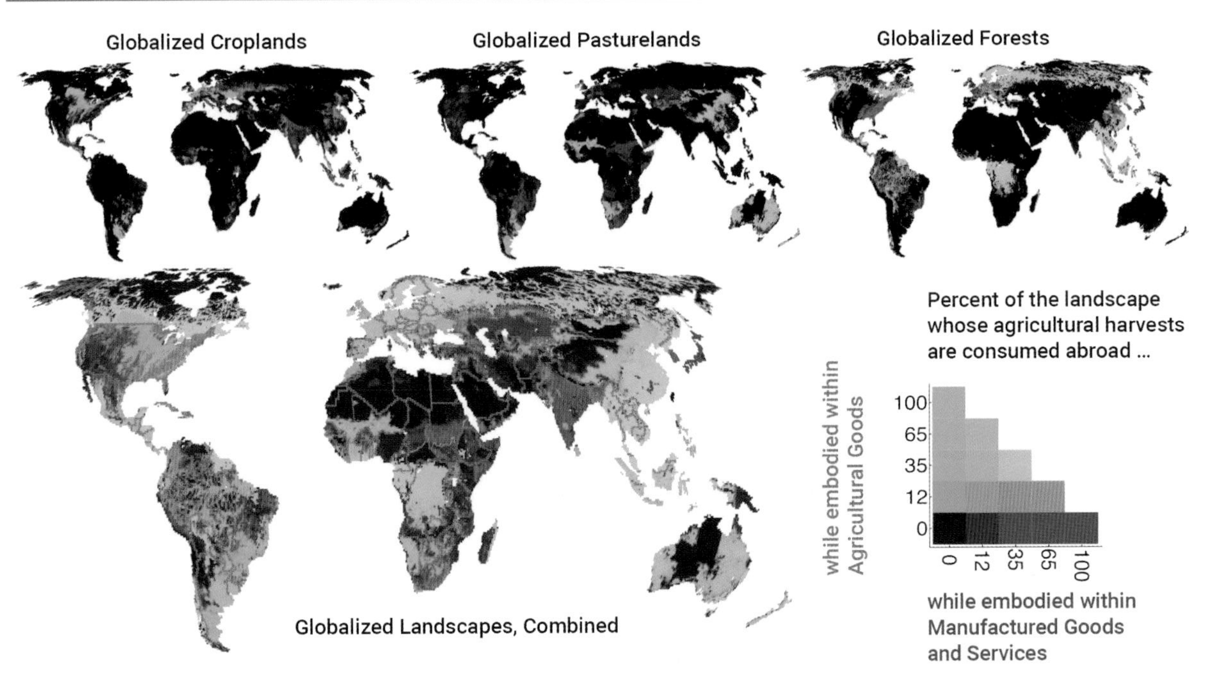

Source: Bergmann and Holmberg (2016).

8.3.5 Climate change

Rising global temperature and changing rainfall patterns have already impacted terrestrial ecosystems and crop yields (see **Figure 8.3**) In tropical regions, the effects of higher temperatures will likely be greater than in temperate zones (Intergovernmental Panel on Climate Change [IPCC] 2014). Shifting rainfall patterns may benefit certain regions, but greater variability in precipitation (more frequent droughts) poses a risk to 70 per cent of global agriculture that is rain-fed (IPCC 2014). As the growing seasons change, yield growth has slowed (Lobell, Schlenker and Costa-Roberts 2011; Lobell and Gourdji 2012). Rising sea level due to climate change generates risks of coastal area loss and subsidence (IPCC 2014), threatening the livelihoods of many coastal inhabitants (Paresi *et al.* 2016) (see Section 8.3.5).

Increased concentrations of CO_2 in the atmosphere may benefit crop yields in certain regions through greater CO_2 fertilization (McGrath and Lobell 2013), while warmer temperatures could bring yield gains in high-latitude regions (IPCC 2014). At a global level, however, yields are expected to suffer as average temperatures and ozone concentrations in the troposphere continue to rise (Schlenker and Roberts 2009; IPCC 2014). Higher temperatures have led to increased distribution of certain weeds and pests (Pautasso *et al.* 2012) and have exacerbated existing stresses during certain growing periods (Gourdji, Sibley and Lobell 2013).

On the other hand, climate-smart agricultural practices such as minimum tillage and energy-efficient crops and practices present an opportunity for increasing the atmospheric carbon sink in soils and hence contribute to mitigation of climate change (Han *et al.* 2018). Similarly, efforts to reduce deforestation and forest degradation, conserve and enhance forest carbon stocks, and sustainably managed forests globally can contribute significantly to reducing greenhouse gas (GHG) emissions and to carbon sequestration in living biomass and forest products.

Figure 8.3: Estimated net impact of climate trends for 1980-2008 on crop yields by country

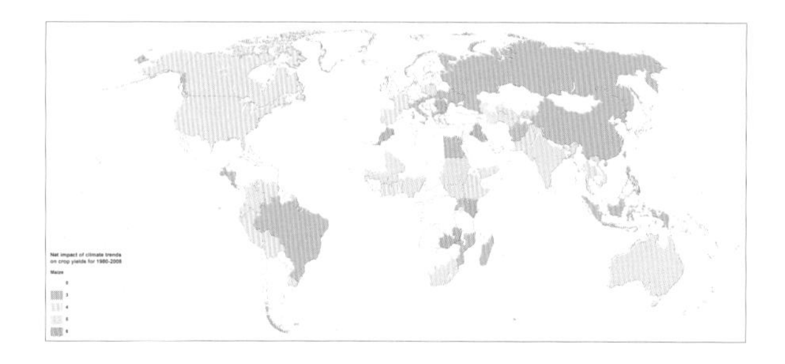

Source: Lobell, Schlenker and Costa-Roberts (2011).

8.4 Key state and trends

8.4.1 Land-use dynamics

Land-cover change

Land is extremely dynamic and land cover changes due to climatic, geologic or ecological processes. However, human land use, mostly agriculture, is currently responsible for most of the changes of land cover and its condition (Haberl 2015; de Ruiter *et al.* 2017; **Figure 8.4**).

Agricultural production needs to nearly double in the period 2012-2050 to meet increasing food, feed and biofuel demand (Food and Agriculture Organization of the United Nations [FAO] 2017a). Although the Food and Agriculture Organization of the United Nations (FAO) estimates that 1,400 million ha are available for expansion (Alexandratos *et al.* 2012), these are mostly in forests and other ecosystems with little disturbance, where nature's contributions to people such as clean water and climate regulation are generated (Machovina, Feeley and Ripple 2015). When possible, people abandon degraded land and expand production elsewhere. As land becomes abandoned, it may slowly start to regenerate: vegetation and wildlife begin to reclaim the spaces left by the abandoned land use, as the spontaneous regrowth of 362,430 km^2 of woody vegetation in Latin America (2000-2010) illustrates (Aide *et al.* 2013).

Global economic forces are shaping local land-use patterns. For example, modern mining is growing in scale due to increased global demand. This is compounded by declining ore grades, which means more ore needs to be processed to meet demand, with extensive use of open cast mining and its associated waste rock. Mining presents cumulative environmental impacts, especially in intensively mined regions,

Figure 8.4. Changes of global forests (top) and cropland (bottom) 1992-2015 based on European Space Agency land cover data time series

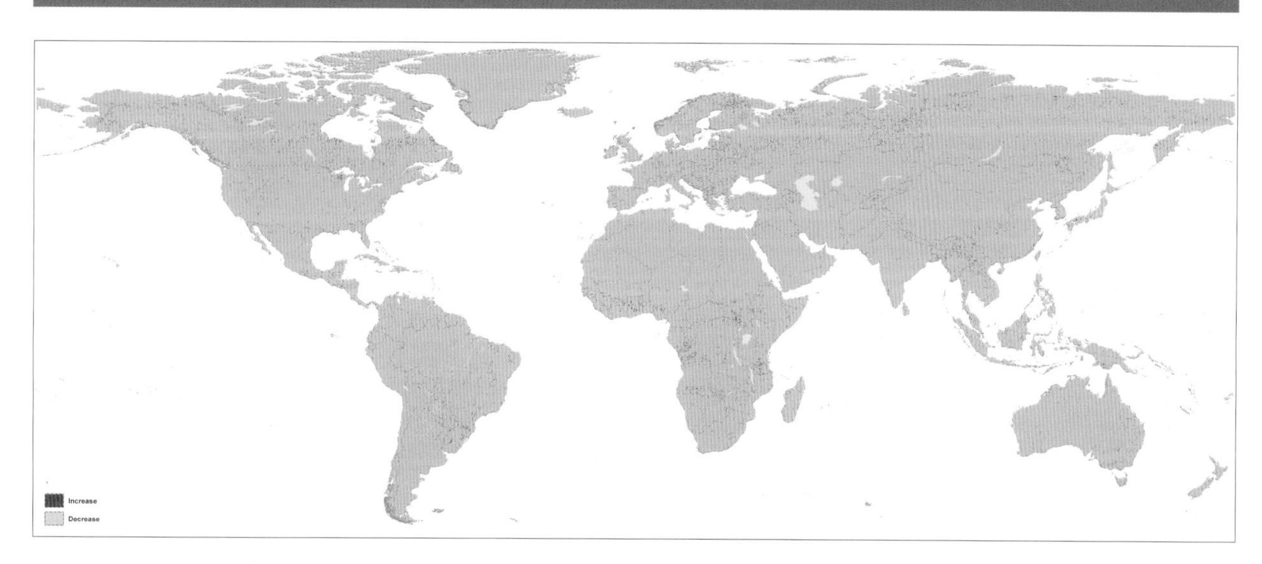

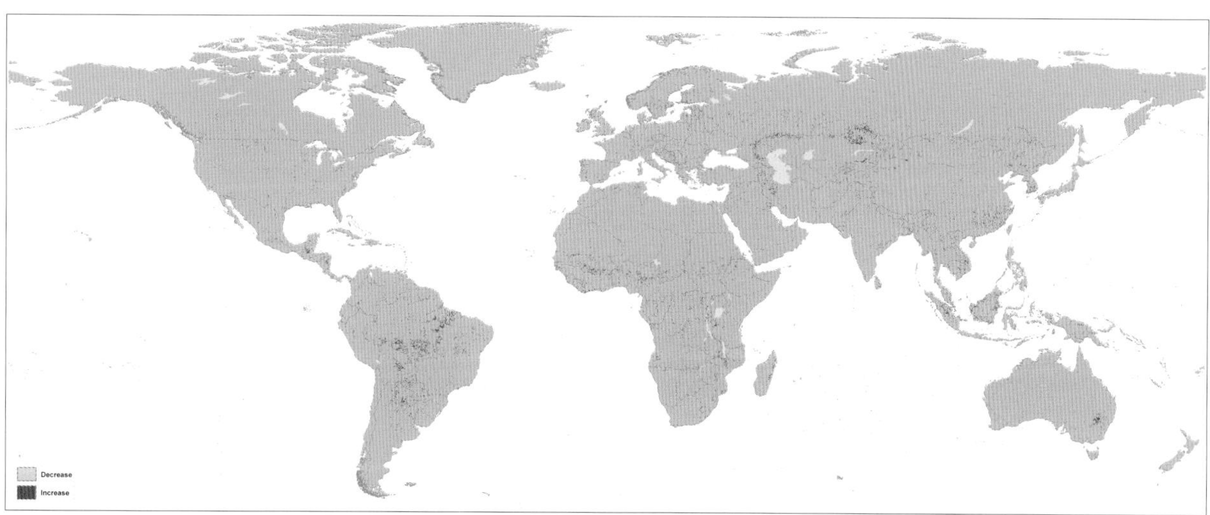

Source: Adapted from European Space Agency (2015).

including areas subject to hydraulic fracturing for oil. A map of areas in Colombia, Ecuador, Peru and Bolivia (**Figure 8.5**) shows land areas that are or have the potential to be exploited for mining, gas and oil highlights the conflict that can emerge from land-use competition (Asociación Pro Derechos Humanos [Aprodeh] *et al.* 2018).

Agricultural dynamics

Food production accounts for the largest anthropogenic use of land – 38 per cent of ice-free land (Holmes *et al.* 2013) or 50 per cent of habitable land (Roser and Ritchie 2018). Within this, the livestock sector dominates, using more than three-quarters of agricultural land for feed production, pasture and grazing (Foley *et al.* 2011; Roser and Ritchie 2018) (**Figure 8.6**).

Primary food production accounts for about 23 per cent of agricultural land use (Figure 8.6), although in recent years a growing proportion of land has been used to grow crops for biofuel production (Cassidy *et al.* 2013). By 2009, biofuel production accounted for 2 per cent of total ice-free land use and is expected to increase to 4 per cent by 2030 (FAO 2009). Agricultural area has decreased by about 1 per cent since 2000 (**Figure 8.7**; FAO 2017b). Although a small drop, this figure does not consider land degradation (see below) or how, despite the reduction in the total agricultural area, this may mask the abandonment of degraded lands and the expansion of the agricultural frontier elsewhere.

While the global harvested crop area increased by 23 per cent between 1984 and 2015, global crop production rose by 87 per cent (FAO 2017b), mostly through monoculture farming. However, these food production systems might be associated with environmental degradation and biodiversity loss (Benton, Vickery and Wilson 2003; Foley *et al.* 2011; UNCCD 2017).

Figure 8.5: Areas designated for extractive activities in the Andean region (South America)

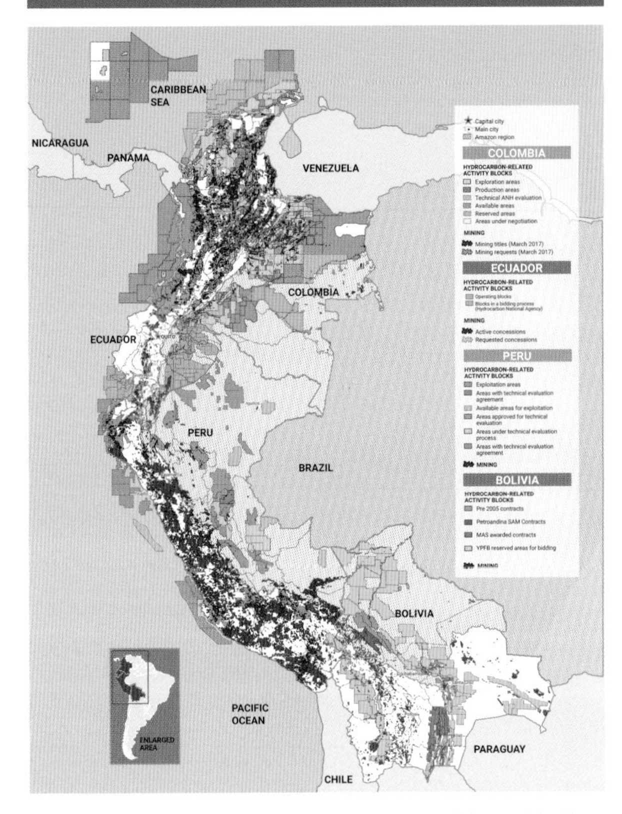

Source: Aprodeh et al. (2018). Adapted from a map compiled in 2018 by the Bolivian Information and Documentation Center (Cedib) from official country sources available on the internet

Figure 8.6: Global area allocation for food production

The breakdown of the surface of the Earth by functional and allocated uses, down to agricultural land allocation for livestock and food crop production, measured in millions of square kilometres. The area for livestock farming includes land for animals, and arable land used for animal feed production.

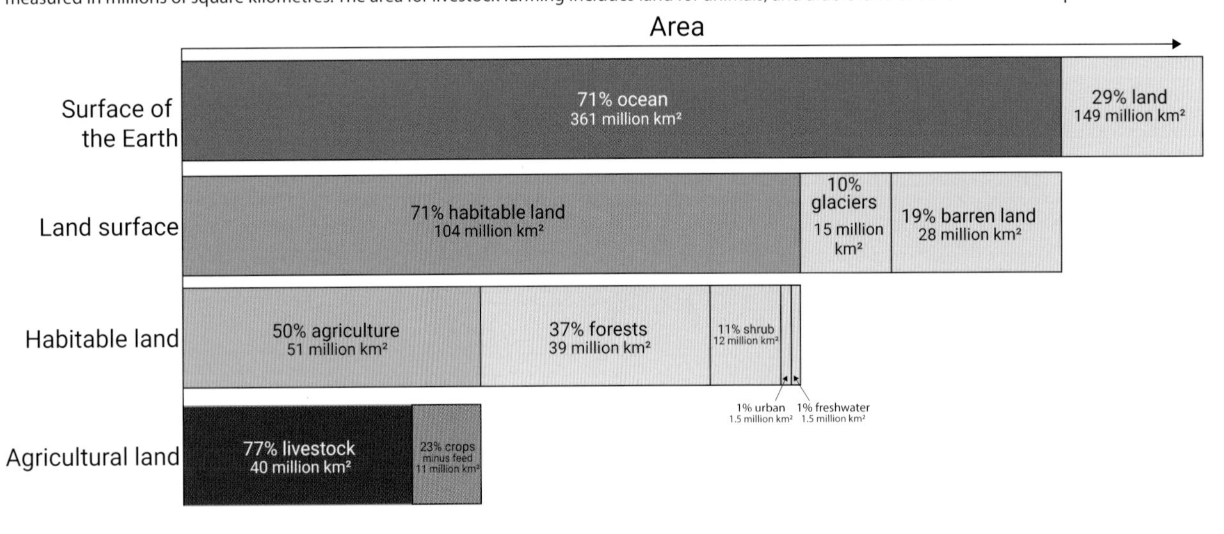

Source: FAO (2017b); Roser and Ritchie (2018).

Figure 8.7: Agricultural area 2000-2014

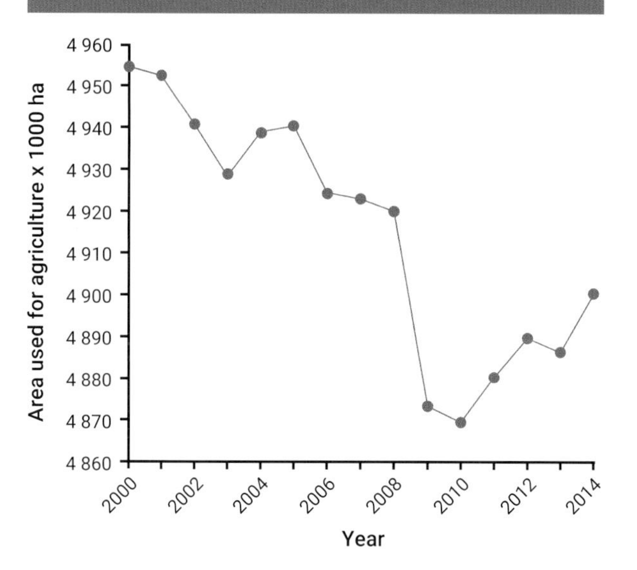

Agricultural area includes the area under agriculture (arable land), permanent crops, and pasture and meadows in a given year.

Source: FAO (2017b).

Figure 8.9: Soybean production in South America 2000–2014

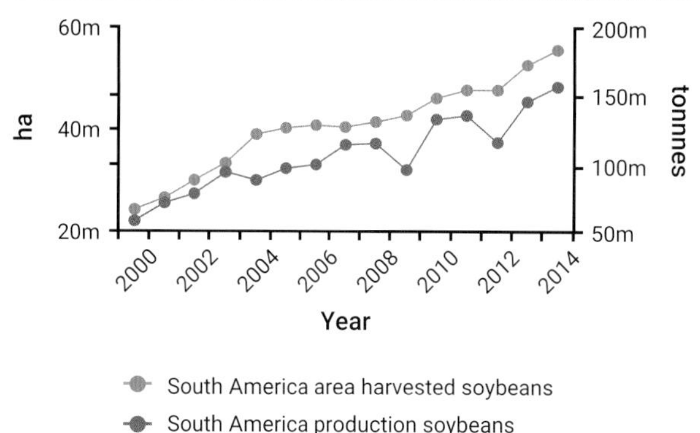

Source: FAO (2017b).

Similarly, per capita daily food supply in the world increased 10 per cent between 1993 and 2013 (**Figure 8.8**; FAO 2017b). Many areas have been converted to cropland as the demand for flexible crops increases (Borras *et al.* 2012). Grasslands in Argentina, Bolivia, Brazil, Paraguay and Uruguay have been converted into soybean fields mostly for export (Graesser *et al.* 2015). Soybean area has more than doubled since 2000 (**Figure 8.9**). The areas harvested in South America and North America account for approximately 47 per cent and 30 per cent, respectively, of the soybean area worldwide (FAO 2017b).

A similar process occurs with oil palm production in South-East Asia. The area planted with this crop has increased since 2000 (**Figure 8.10**). In 2014, more than 68 per cent of total oil palm crop area was in this region and 85 per cent was in Asia (FAO 2017b).

The expansion of oil palm plantations in South-East Asia has been at the expense of forests. This increase has been the result of the rising demand for biofuels and edible oil. In Kalimantan, Indonesia, from 1990 to 2010, some 90 per cent of land converted to oil palm plantations were forested (Carlson *et al.* 2012). From 2001 to 2015, more than 9.5 million ha were deforested on Borneo (World Resources Institute [WRI] 2018). In the oil-palm plantations in the lowlands of peninsular Malaysia (2 million ha), Borneo (2.4 million ha) and Sumatra (3.9 million ha), Koh *et al.* (2011) found that about 880,000 ha of tropical peatlands in the region had been converted to oil palm plantations by the early 2000s. By 2010, some 2.3 million ha of peat-swamp forests were deforested but were not yet converted to oil palm plantations.

Figure 8.8: Food supply in the world (kcal/capita per day)

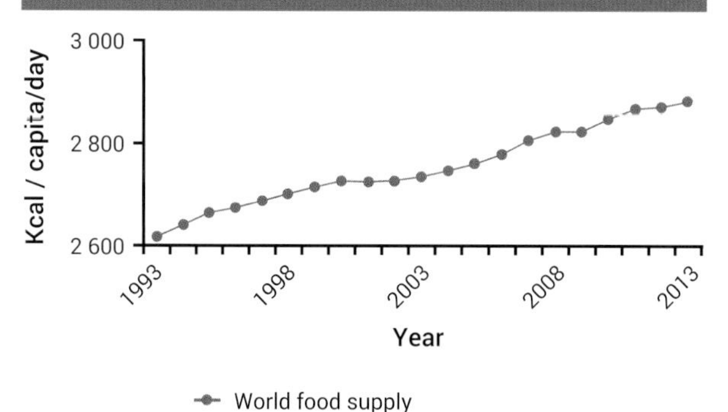

Source: FAO (2017b).

Figure 8.10: Production of oil palm fruit in South-East Asia

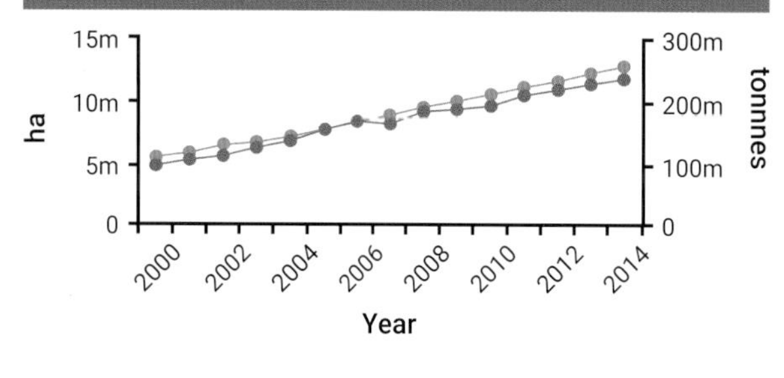

Source: FAO (2017b).

Global livestock populations increased between 2000 and 2014 (**Figure 8.11**, **Figure 8.12**). While human population grew by nearly 19 per cent, numbers of cattle and buffalo, goat and sheep, poultry birds and pigs grew by 13.8 per cent, 21.9 per cent, 45.4 per cent and 15.1 per cent respectively. However, the increase in livestock numbers has been accompanied by a decrease in pasture and permanent meadows (**Figure 8.13**). These high growth rates are mostly associated with more intensive livestock production systems that rely on the efficient use of animal feed (Mottet *et al.* 2017).

Figure 8.11: Numbers of herbivores and poultry

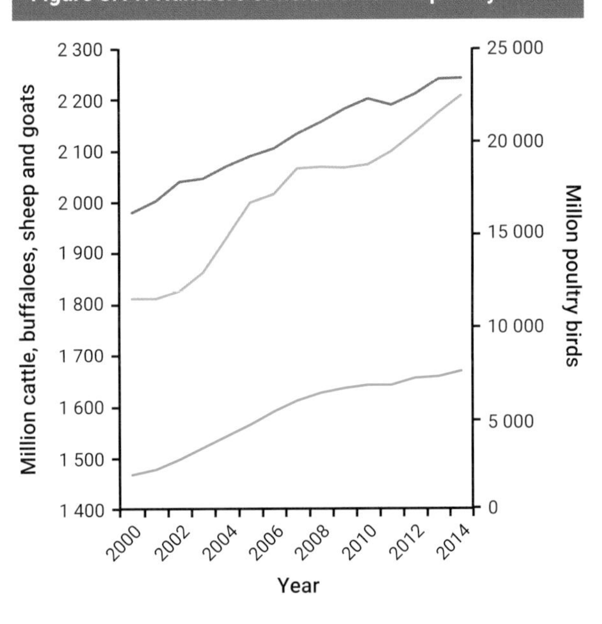

Source: FAO (2017b).

Figure 8.12: Numbers of pigs, 2000-2014

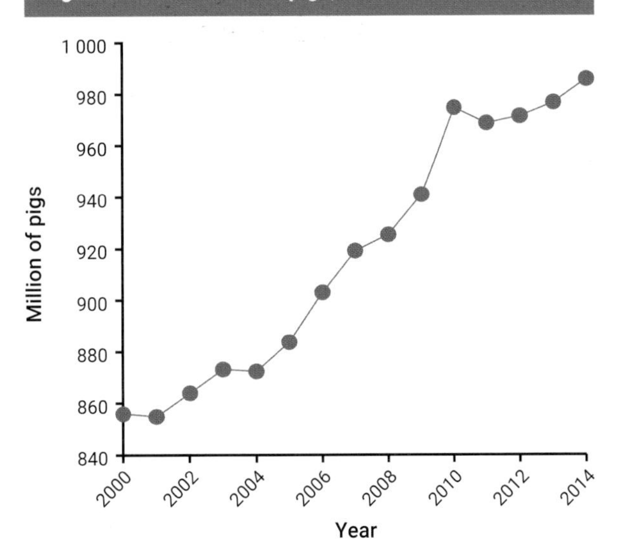

Source: FAO (2017b).

Figure 8.13: Permanent meadows and pastures (1,000 ha)

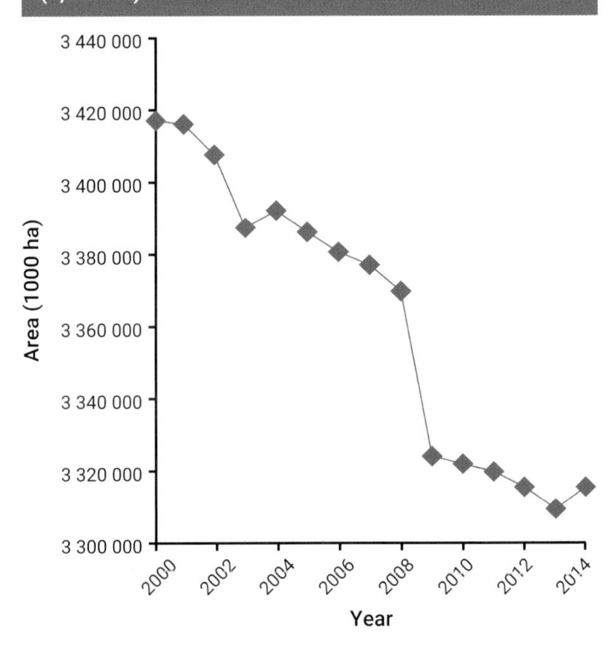

Source: FAO (2017b).

Forest dynamics

Forests continue to decline (**Figure 8.14**). In 1990, they represented 31.6 per cent of the planet's land area. This decreased to 30.6 per cent in 2015 (FAO 2015a), but forest loss rates are declining. In the 1990s, about 10.6 million ha of natural forests were lost each year. For the period 2010-2015, this rate had dropped to 6.5 million ha/year. At the same time, the increase in planted forests was about 3.2 million ha/year; by 2015 they accounted for 7 per cent of the global forest area mostly concentrated in high-income countries (FAO 2015a; **Figure 8.15**). Forest loss rates differ among regions and, while the global trend is towards forest loss, many regions, especially

Figure 8.14: Forest land in the world, 2000-2015

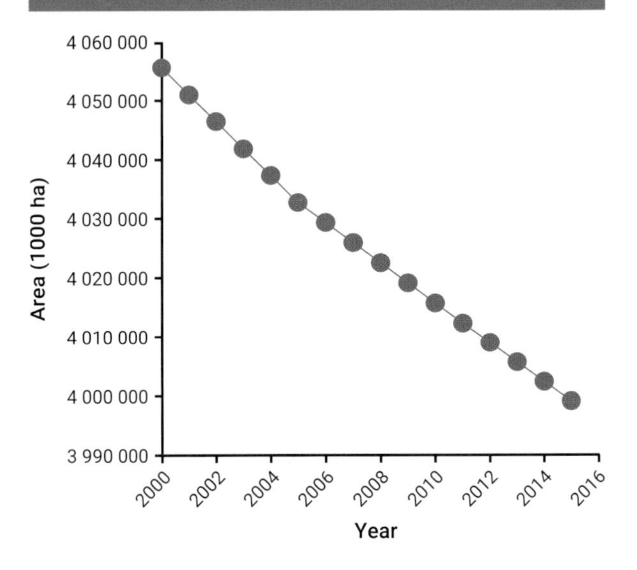

Source: FAO (2017b)

Figure 8.15: Forest area annual net change, (1990-2000, 2000-2010, 2010-2015)

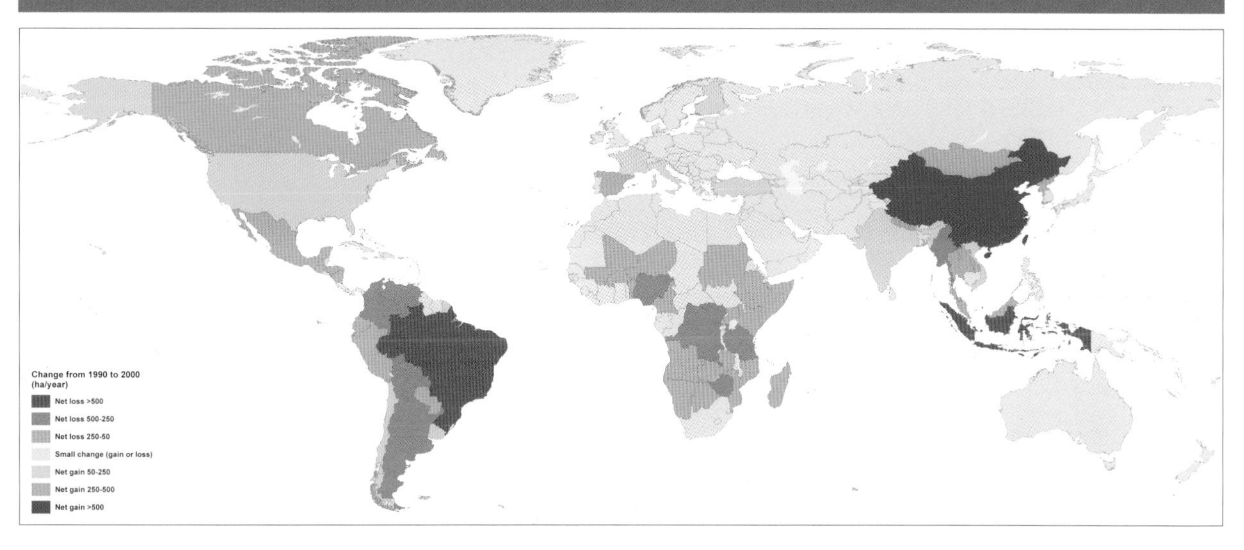

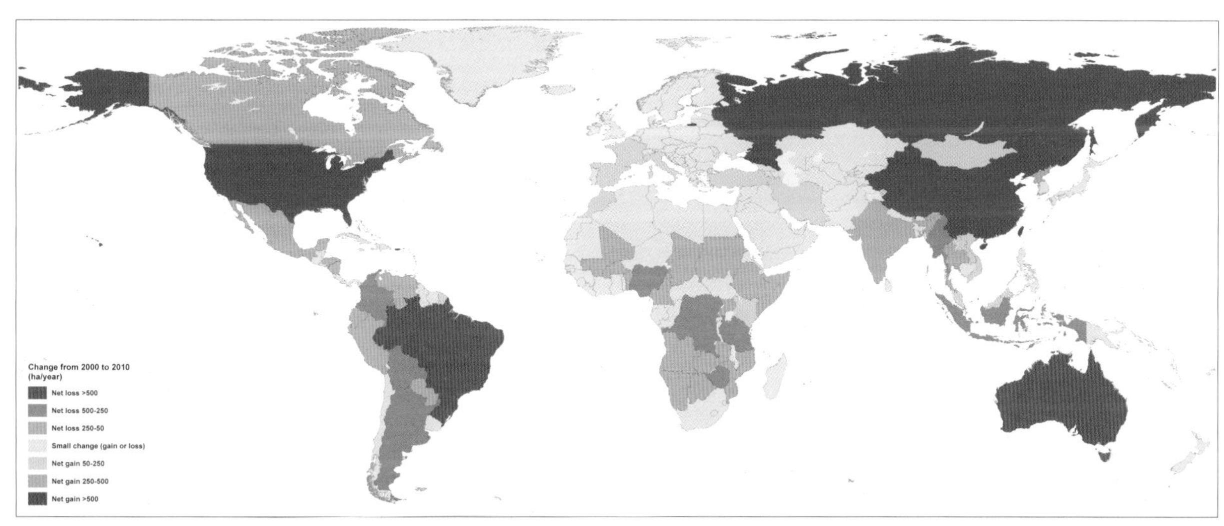

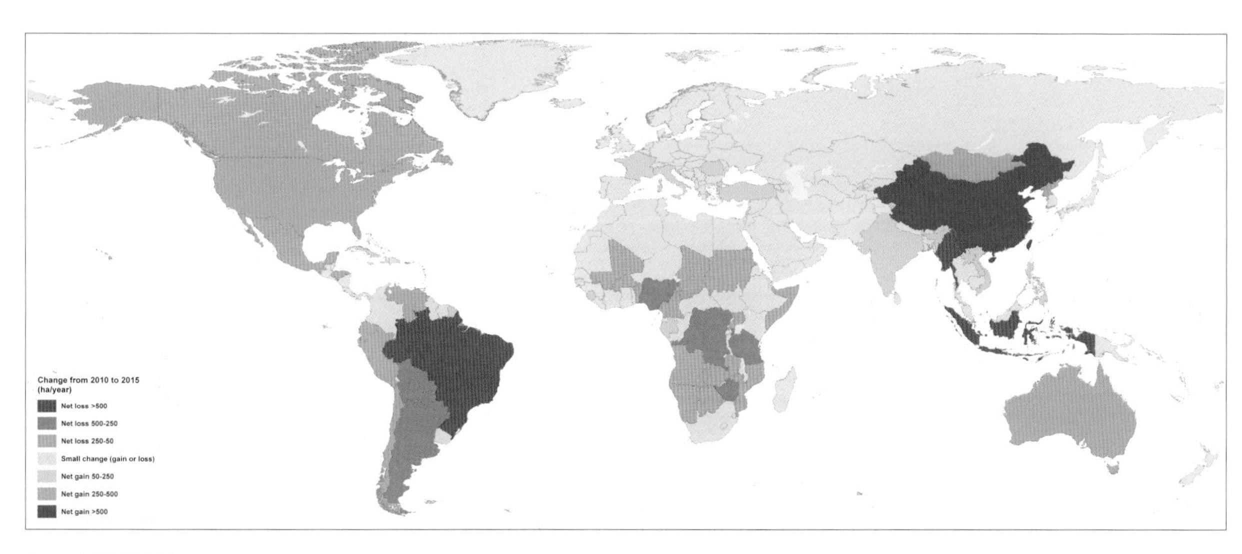

Source: FAO (2015a).

in more developed countries, are showing an increase in forest cover, though some of this forest is as plantations. Natural forests continue to decline in most areas of the world (**Figure 8.15**), threatening the supply of essential benefits to people. For example, as deforestation increases in the Amazon rainforest, rainfall has been decreasing. Recent estimates indicate that a critical tipping point for the hydrological cycle in this part of South America will be reached if deforestation reaches 20-25 per cent of the original forest cover in the Amazon basin (Lovejoy and Nobre 2018). In the last 50 years, 17 per cent of the original extent of the Amazon rainforest has been deforested (World Wide Fund for Nature [WWF] 2018) and the forest cover continues to decrease (Butler 2017; WRI 2018; WWF 2018).

Urban expansion

Built-up areas occupy a very small fraction of land. However, since 1975 urban clusters (i.e. urban centers as well as surrounding suburbs) have expanded approximately 2.5 times, accounting for 7.6 per cent of global land area (Paresi *et al.* 2016). Between 1975 and 2015, built-up areas doubled in size in Europe, while in Africa they grew approximately fourfold. Cities have grown in both regions, but urban population remained relatively constant in Europe while it tripled in Africa. This means that the built-up area per-capita is different across the world (Paresi *et al.* 2016). In addition, urban expansion leads to landscape fragmentation and urban sprawl. As cities expand, urban land uses usually take over agricultural lands (van Vliet, Eitelberg and Verburg 2017), and the demand for food, fibre and minerals can transform previously unconnected locations (Seto *et al.* 2012; van Vliet, Eitelberg and Verburg 2017). In Latin America, a pervasive spatial expansion (almost 84 per cent of the population lives in cities) has been observed leading to less compactness (Inostroza, Baur and Csaplovics 2013).

By covering the ground with impervious surfaces, cities affect the hydrological cycle and soil function. They also generate what are called urban heat islands. But they can also be more efficient in providing access to education, housing, clean water and electricity. Since 2000, cities have incorporated more green spaces and trees (Paresi *et al.* 2016).

While cities are expanding into hinterlands, there is increasing recognition of the value of preserving natural systems (e.g. lakes and natural water bodies) as well as constructing enhanced-engineered urban green infrastructures (e.g. parks, urban farms, bioswales). These have potential to offer multiple benefits that can enhance biodiversity and human well-being, including water management, flood risk mitigation; heat island mitigation (Pataki *et al.* 2011); emotional well-being, health (Groenewegen *et al.* 2006; Pataki *et al.* 2011; White *et al.* 2013; Sturm and Cohen 2014; World Health Organization [WHO] 2017); pollution capture; and cultural amenities.

In 2015, some 52 per cent of people lived in high-density urban centres, 33 per cent in towns and suburbs and 15 per cent in rural areas (Paresi *et al.* 2016). While many cities continue to grow in population and expand, others experience population decline. Shrinking cities leave behind vacant parcels as part of a cycle of growth and decline, whose management offers new opportunities to enhance the environment.

8.4.2 Land quality dynamics

Land degradation and crop production

Land degradation involves the decline or disruption of land ecosystem services, including net primary production (NPP) (Le, Nkonya and Mirzabaev 2016). It results from different processes: soil erosion, salinization, compaction and contamination, organic matter decline, forest fires and

Figure 8.16: Natural forest area by region, 1990-2015

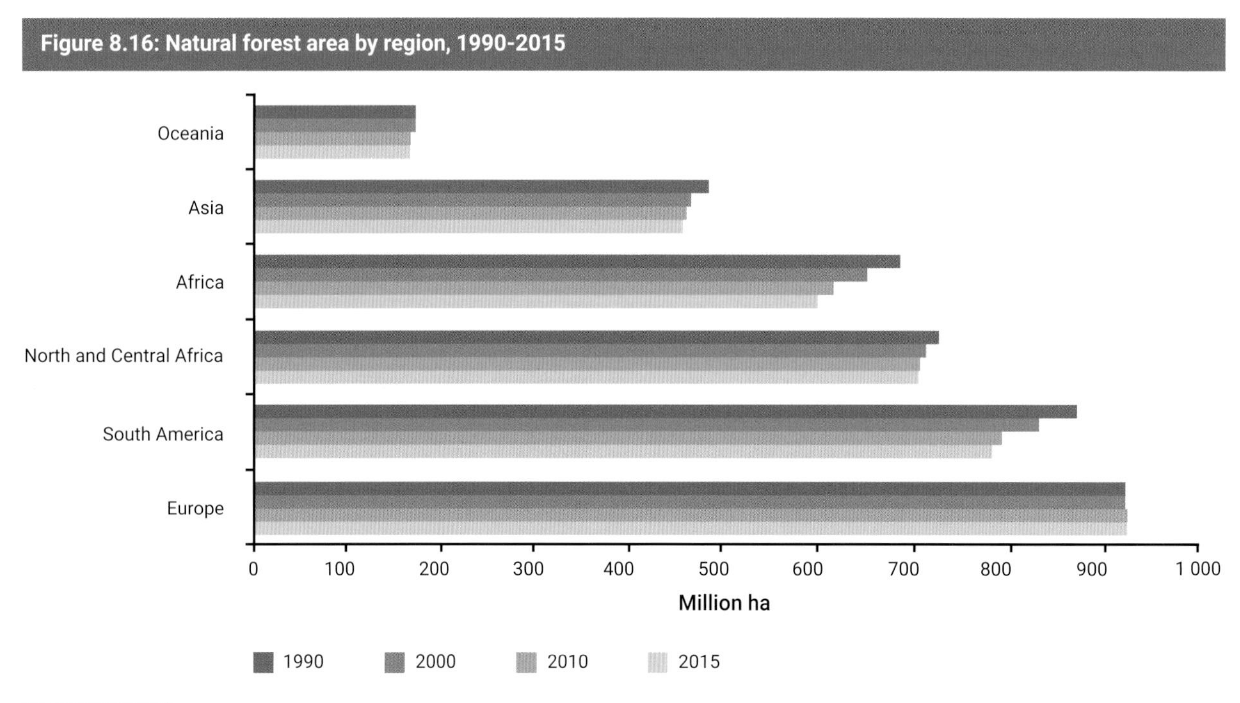

Source: FAO (2015a).

overgrazing (Jones *et al.* 2012; Kosmas *et al.* 2014). Decline of NPP is also a reduction in microbiological activity and water retention capacity, lower hydraulic conductivity, and decreasing soil resistance, among others (Soane *et al.* 2012). FAO (2015b) estimates current land degradation at 12 million ha/year. It is estimated that annual losses from ecosystem services resulting from land degradation range between US$6.3 trillion and 10.6 trillion (The Economics of Land Degradation [ELD] 2015). While degradation could be a biophysical phenomenon, the causes and implications are also economic and social. Many efforts attempt to assess observable land degradation trends, scales and consequences. However, different definitions of degradation and methods used to measure them lead to differing results regarding its magnitude, where it takes place, its effects and its costs (FAO 2018). A recent estimate using satellite imagery estimates that 29 per cent of global land area is degraded, while improvement has occurred in 2.7 per cent of global land area in the last three decades, and about 3.2 billion people live in the degrading areas (Le, Nkonya and Mirzabaev 2016). Reducing land degradation and increasing land restoration are critical for providing necessary ecosystem services that contribute to life on Earth and human well-being (IPBES 2018).

Desertification

The United Nations Convention to Combat Desertification (UNCCD) defines desertification as "land degradation in arid, semi-arid, and dry sub-humid areas resulting from various factors, including climatic variations and human activities" (UNCCD 1994). However, desertification is still a highly controversial issue usually leading to expert disagreement (Reynolds and Smith 2002; Bestelmeyer *et al.* 2015). The extent of desertification ranges from 15 per cent to 63 per cent globally as well as 4 per cent to 74 per cent for drylands (Safriel 2007), and can be equally variable within a country like Mongolia, where degradation estimates range from 9 per cent to 90 per cent (Addison *et al.* 2012).

Recent research (Global Assessment of Soil Degradation [GLASOD]) shows that previous generalizations claiming that land degradation is occurring in semiarid areas worldwide is not supported by satellite-based observations (de Jong *et al.* 2011; Fensholt *et al.* 2012; Cherlet *et al.* 2018). Desertification and drought research in the Sahel indicate that the first process is not taking place (Behnke and Mortimore 2016). This trend may be explained by increasing precipitation, as well as by lower pressure on land due to outmigration (Olsson, Eklundh and Ardö 2005). However, current climatic conditions in the Sahel appear to be still below the more humid conditions of 1930-1965 (Anyamba and Tucker 2005; Nicholson 2013).

A positive trend is also observed in semi-arid areas of China where human actions might explain the 'expansion of desertification' between 1980 to 1990, although conservation activities have begun to reverse these trends (1990-2000) (Xu *et al.* 2009). Recent modelling results indicate that global greening might also be caused by CO_2 fertilization, nitrogen deposition and climate change (Zhu *et al.* 2016).

Recognizing the inherent complexity underlying land degradation, the recent edition of the World Atlas of Desertification (WAD) (Cherlet *et al.* 2018) presents several global data sets of biophysical and socioeconomic processes that, individually or combined, can contribute to land degradation (Reynolds *et al.* 2011; Bisaro *et al.* 2014).

Soil salinization

In arid and semi-arid regions, lack of adequate drainage in irrigated areas triggers salt accumulation in the root zone, negatively affecting crop productivity and soil properties (Qadir *et al.* 2014). In some countries, soil salinization affects half of irrigated land (Metternicht and Zinck 2003). Other sources suggest that about 33 per cent of the globally irrigated area has declining productivity due to inadequate irrigation, causing waterlogging and salinization (Khan and Hanjra 2008). Several studies of grain yield losses due to salinization indicate grain yield losses of 32-48 per cent on average (Murtaza 2013). The global annual losses in irrigated crops caused by salt-induced land degradation could be about US$27.3 billion due to lost crop production (Qadir *et al.* 2014). The costs of inaction on these lands may result in 15-69 per cent revenue losses depending on the type and intensity of land degradation, crop variety and irrigation water quality and management (Qadir *et al.* 2014). Additional losses, which are not included in these estimates, cover a wide range of issues – from deterioration of animal health to decline in property values of affected farms, among others (Qadir *et al.* 2014).

Permafrost thawing

Due to various feedbacks in the climate system, warming in the Arctic currently exceeds twice the mean global temperature rise (Taylor *et al.* 2013a). Sea ice is retreating, permafrost is thawing, and the ice-free season is lengthening, such that waves and warm air are increasingly degrading the thawing permafrost in the interior, as well coastal areas. The thawing of permafrost releases GHGs and alters the landscape. Thaw reduces soil and landform stability, increases erosion and affects arctic habitat, albedo and hydrology.

By far, the largest fraction of the Arctic coastline consists of thawing permafrost (**Box 8.1**). Arctic permafrost coasts represent 34 per cent of all coasts on Earth. Coastal erosion rates have increased in recent years with values ranging around 1 metres /year. Erosion rates are highest along the Alaskan and Siberian coastlines, with maxima as high as 25 metre/year (**Figure 8.17, Figure 8.18**) (Günther *et al.* 2013; Overduin *et al.* 2014; Fritz, Vonk and Lantuit 2017). Therefore, increasing fluxes of organic carbon are released into the shelf seas. In some locations (Alaska), villages have had to be relocated further inland.

Box 8.1: Livelihood impacts in the Arctic

Reindeer (caribou) herds are an important part of Arctic ecosystems and integral to the livelihoods of indigenous peoples in Alaska, Arctic Canada, Scandinavia and the Russian Federation. Reindeer-herding communities depend on access to seasonal pastures. The seasonality and extent of pastures is changing as a result of climate change, impacting these pastoral communities.

Mining and resource extraction are also important in the Arctic. Changing Arctic conditions have made the construction and operation of the winter ice roads that supply mining outposts problematic. A warming climate has delayed freeze-up in the autumn (fall) and produced an earlier spring melt as well as thinner ice during the winter. This has led to shorter winter-road seasons. As the Arctic climate continues to warm, co-management institutions will find themselves increasingly dealing with trade-offs between sustainable development and sociocultural and ecological integrity of Arctic lands and livelihoods.

Figure 8.17: Coastal erosion rates at selected sites in the Arctic

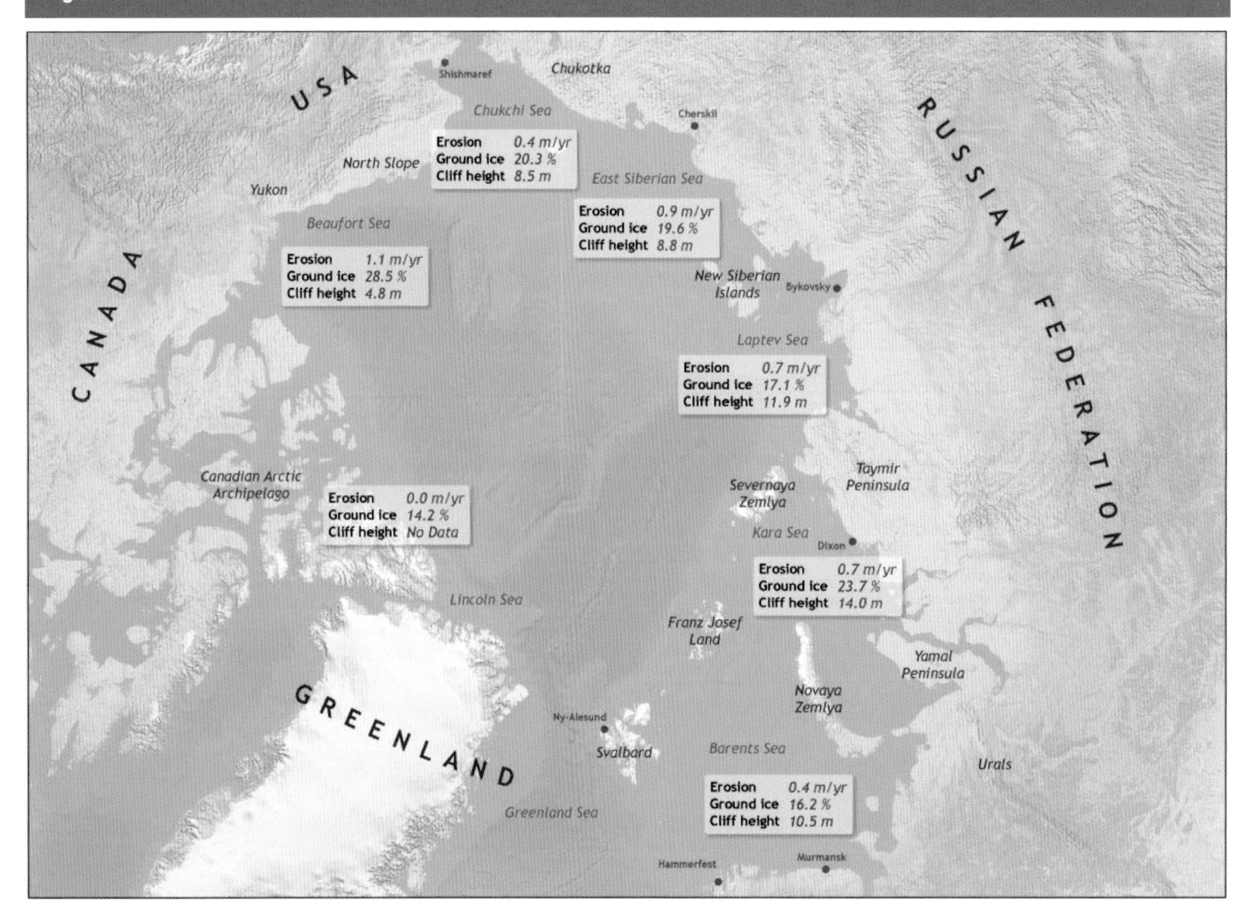

Source: Overduin *et al.* (2014).

Figure 8.18: Estimated coastal erosion threat in the Arctic

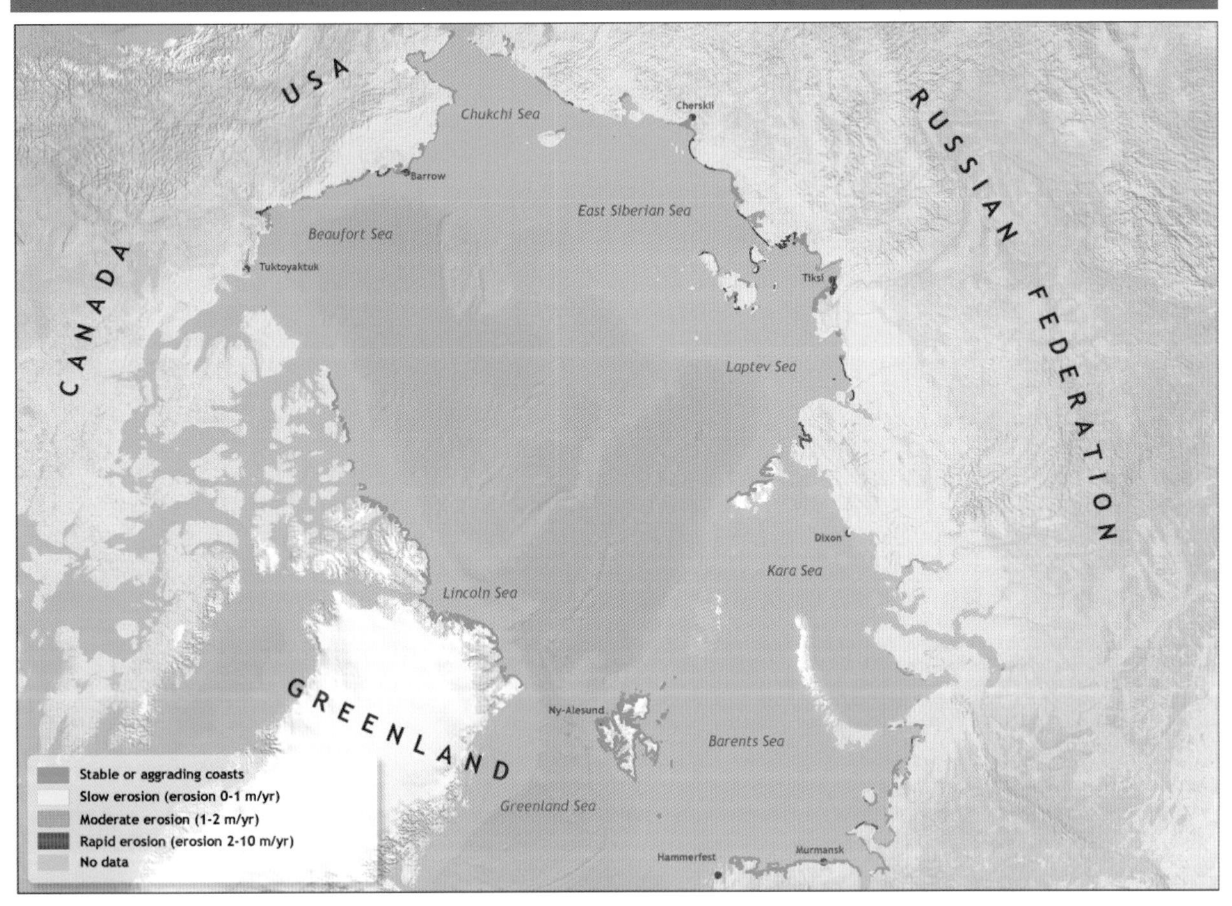

Source: Lantuit, Overduin and Wetterich (2012).

8.5 Key impacts

8.5.1 Food security

People are considered food secure when they always have availability of and adequate access to sufficient, safe, nutritious food to maintain a healthy and active life (FAO *et al.* 2017). The discussions in this section cover three critical issues–food availability, food access and food utilization.

Hunger and malnourishment

A sizeable proportion of the worlds' seven billion people are hungry and malnourished. Roughly one billion people have energy-deficient diets, and about one billion people suffer from diseases of energy surplus (called the 'hidden hunger' of micronutrient deficiencies) (Godfray and Garnett 2014). Although undernutrition is slowly declining, 155 million children under five years old, mostly in sub-Saharan Africa and South Asia, still suffer from stunted growth. Simultaneously, increasing numbers of people are suffering from overnutrition: more than 2 billion adults are overweight and 500 million are obese. Moreover, 88 per cent of countries face two or three forms of malnutrition (Development Initiatives 2017), and undernutrition and obesity increasingly coexist in the same households (FAO *et al.* 2017).

Malnutrition and changing consumption patterns put greater pressure on land resources making land-use decisions more important than ever before. Most food is sourced from terrestrial environments, though 17 per cent of global animal protein and 6.7 per cent of all protein consumption is from fish (FAO 2016). While food costs have fallen since 2008, this trend has not been constant (FAO 2017c), with volatility attributed to increased demand from rapidly developing countries and competition among first-generation biofuel producers

(The Royal Society 2008; Godfray *et al.* 2010). **Figure 8.19** shows vulnerability to food security using meteorological data for the period 1981-2010 and socioeconomic data representative of the year 2010. The results indicate that disasters such as floods and droughts are already having a strong impact on food security, and their frequency and intensity may increase as a result of climate change (Met Office Hadley Centre and World Food Programme 2018). In developing countries, agriculture absorbs about 22 per cent of the total damage and losses caused by natural hazards (FAO 2015b). Although disasters may impact rural livelihoods directly, the disruption to agricultural production and development can have negative repercussions across national economies, with devastating effects on food security, including in urban areas (**Box 8.2**).

Sustainable food production and efficient use
Approximately one-third of the food produced globally for human consumption is lost or wasted (Lipinski *et al.* 2013; United Nations Environment Programme [UNEP] 2015), together with the resources used in its production (land, energy, water, etc.) with the associated environmental impacts. Food losses and waste in 2007 utilized almost 1.4 billion ha of land, equivalent to about 28 per cent of the world's agricultural land area (FAO 2013). Based on food crop data for the period 2005-2007, food losses and waste consumed 23 per cent of total global fertilizer use (28 million tons/year) and 24 per cent of total freshwater resource use (Kummu *et al.* 2012). Furthermore, an estimated 99 per cent of food wastage at the agricultural production stage is produced in areas where soils are facing medium to strong land degradation, placing further stresses on these areas (FAO 2013, p. 47).

Approximately 56 per cent of total food loss and food waste occurs in developed countries, while 44 per cent originates from developing countries (Lipinski *et al.* 2013). This wastage generates GHGs. If food wastage were a country, it would be the third largest emitting country in the world (FAO 2015c). In the global South, losses are mainly due to the absence of food-chain infrastructure and lack of knowledge or investment in storage techniques. In the global North, pre-retail losses are lower but those arising from retail, food service and home stages of the food chain have grown dramatically in recent years (Godfray *et al.* 2010; **Figure 8.20**).

Box 8.2: The Syrian crisis: droughts and land degradation as factors

The Syrian conflict has sometimes been labelled a 'climate conflict', since some of the root causes could be traced to the drought that affected the country between 2007 and 2010 (Kelley *et al.* 2015), the worst drought on record, causing widespread crop failure in the region. In Syrian Arab Republic, some 1.5 million people from rural farming areas migrated to the peripheries of urban centres, leading to a spike in food prices and eventually to the upheaval of the population (Kelley *et al.* 2015). The government could not provide migrants with housing, jobs and economic opportunities. This combination of factors contributed to a war that has now lasted several years and left the country in ruins, with about two-thirds of its 22 million population displaced.

Sustainable intensification (e.g. agroecology-based production, agricultural innovation) is promoted as a sustainable land management strategy. Besides a sustainable food supply, it maintains nature's contributions to people, promotes human health and nutrition (Pretty, Toulmin and Williams 2011; Robinson *et al.* 2015).

Food security and food trade
International trade is increasingly important to meeting global food demand (Nelson *et al.* 2010; MacDonald *et al.* 2015). Population growth, urbanization and shifting dietary preferences have increased dependency on food imports (Msangi and Rosegrant 2011; Alexandratos *et al.* 2012; Porkka *et al.* 2013). The proportion of the global population living in food-deficit countries rose from 72 per cent in 1965 to 80 per cent in 2005 (Porkka *et al.* 2013).

Just under one-quarter of all food produced for human consumption is traded on international markets (D'Odorico *et al.* 2014; **Figure 8.21**).

Figure 8.19: Potential impacts of climate change on food security

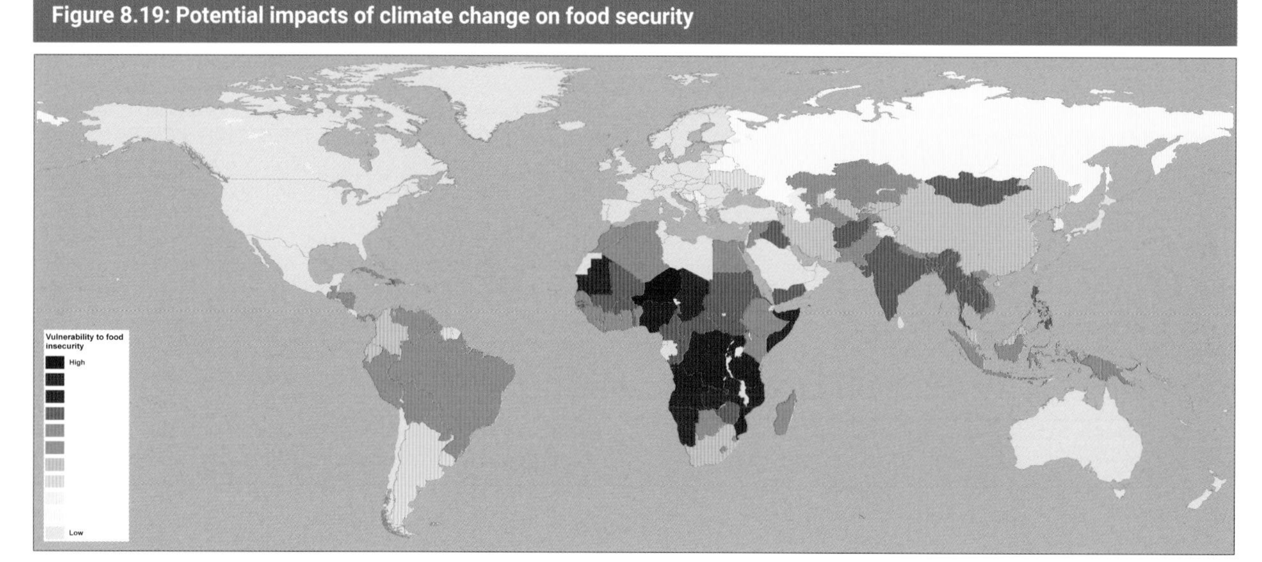

Source: Met Office Hadley Centre and World Food Programme (2018).

Figure 8.20: Make-up of total food waste in developed and developing countries

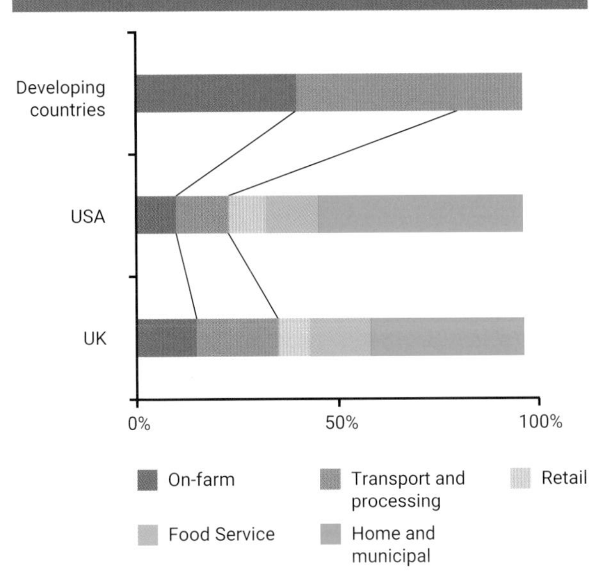

Retail, food service and home and municipal (subnational government sphere) categories are presented together for developing countries.

Source: Godfray *et al.* (2010).

Some low-income food-deficit countries have capacity to increase food productivity. But in others, including those where food insecurity is high – for example, Eritrea, Burundi and Somalia – food availability from domestic production is falling and the capacity to increase production is limited (Fader *et al.* 2016). Most developing countries have become increasingly reliant on imports to meet domestic demand, a trend that will likely continue through to 2050 (Alexandratos *et al.* 2012; **Figure 8.22**).

Figure 8.21: Share of global production volumes traded internationally in 2014

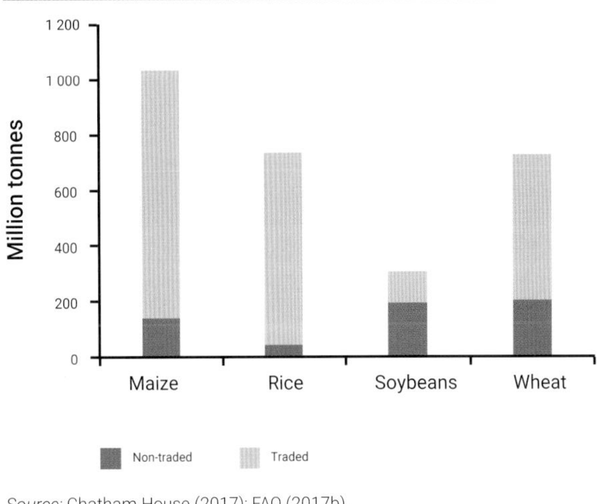

Source: Chatham House (2017); FAO (2017b).

Global food supply has become dependent on the growing trade of a small number of crops grown in a few 'breadbasket' regions with increasing specialization (Khoury *et al.* 2014). This has led to lower food prices, with food-deficit countries benefiting from these food imports. However, the geographic

Figure 8.22: Developing countries: net cereals trade (million tons)

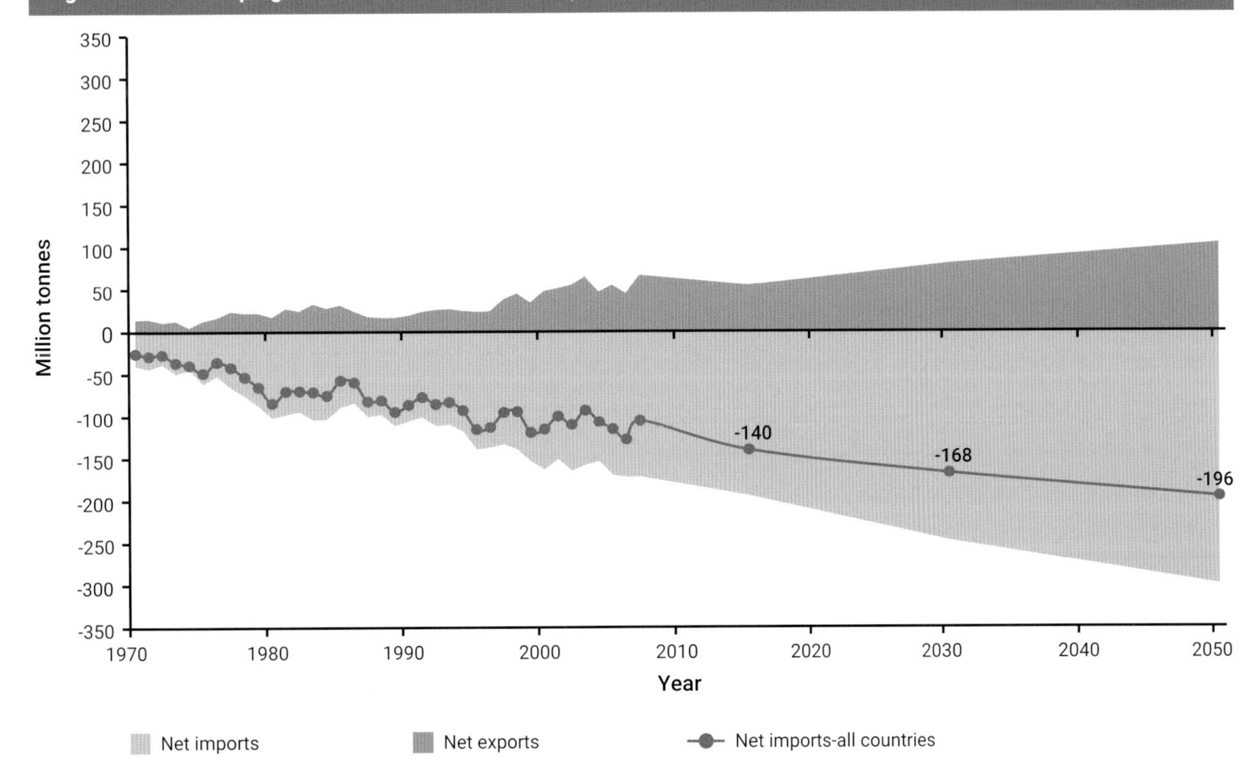

Net cereal imports have increased since 1970 and are expected to rise.

Source: Alexandratos *et al.* (2012).

8

concentration of production increases systemic risk, as illustrated by recent spikes in international commodity prices due to poor harvests in certain regions (Puma *et al.* 2015; The Global Food Security Programme 2015). Due to climate change, such events may become more likely (Porter *et al.* 2014). Furthermore, the growing prevalence of certain crops in global food supplies has contributed to the increasing consumption of nutritionally poor food, some of which is highly processed (processed in a nutrient-poor manner), with serious consequences for human health (Khoury *et al.* 2014).

8.5.2 Human Health and land management

Health effects from mining

Adverse human health issues are also associated with mining and ore processing. While such operations generate employment and provide essential fuels and raw materials, residues such as lead affect air quality, posing a hazard especially to children, who are more likely to ingest such dust (Taylor *et al.* 2013b). The mining of some rare minerals, such as tantalum, often involves exploitation and even slavery (Gold, Trautrims and Trodd 2015).

Mining waste is one of the world's largest waste streams by volume, with the potential to cause significant environmental impacts, including abrupt and extensive land use change (Sonter *et al.* 2014; Murguía 2015; Hudson-Edwards 2016; Sonter *et al.* 2017). The Global Waste Management Outlook (UNEP 2015) estimates mining waste to be in the order of 10-20 billion tons per year. Mining waste will probably continue to grow, since companies are now turning to lower-grade ores, which typically generate more waste per unit extracted. However, mining waste should also be regarded as a potential resource within a circular economy (Lèbre and Corder 2015). Mining activities generate impacts on ecosystems and lead to soil contamination. Toxic and radioactive dust emissions from mining waste are a relevant health issue in many parts of the world (see Chapter 5). Water pollution also results from mining (acid metalliferous drainage and leakages from tailing management facilities) (see Chapter 9) (Hudson-Edwards 2016). In many parts of Latin America, mining activities have an important impact. For example, artisanal gold mining in the Amazon basin deposited an estimated 3,000-4,000 tonnes of mercury during the late 1980s and early 1990s (Lacerda 2003). Although gold mining has shifted to different parts of the region, mercury contamination is still present in many soils and rivers as a result of land-use change (Lacerda, Bastos and Almeida 2012). This mercury also contributes to atmospheric pollution.

Waste and human health

The Global Waste Management Outlook indicates that cities generate between 7 and 10 billion tonnes of waste per year, figures that are expected to rise, even double, in lower-income African and Asian cities by 2030 (UNEP 2015). It also estimates that 3 billion people lack access to adequate waste disposal facilities, which poses health risks (infections, exposure to chemicals, dust) and generates environmental impacts (soil and water pollution, GHG emissions). An estimated 15 million people are operating globally as informal recyclers, many of them in dump sites (Binion and Gutberlet 2012). Identified health risks for these workers include exposure to chemical hazards, infections, musculoskeletal damage and poor mental health (Binion and Gutberlet 2012). Working in organized groups, such as recycling cooperatives in developing countries (e.g. Bolivia and Colombia), has helped to reduce the domestic waste flow to landfills and improved the livelihoods of the recyclers (UNEP 2015). A key step towards reducing the environmental and health impacts of domestic waste is to shift from regarding waste as a health and environmental threat to including a resource management perspective, using waste as a source of raw materials (UNEP 2015).

Soil contamination

Soil health is essential for life, food security and the ecosystems services provided by soils. Many chemicals coming from industrial, urban and agricultural sources end up contaminating soils. In most developed countries, the main direct causes of site contamination are industrial and commercial activity. The extent of these sites can vary considerably, from small parcels of land to large industrial facilities or agricultural areas. Governments in the developed world maintain an inventory of contaminated and remediated sites. More than 2.5 million potentially contaminated sites are located in Europe, of which 342,000 are thought to be actually contaminated. About one-third of these have been identified, and more than 50,000 sites had been successfully remediated by 2014 (van Liedekerke *et al.* 2014). In the United States of America, the Superfund National Priorities list includes the sites contaminated with complex hazardous substances and pollutants (1,342 in 2016) that impact soil groundwater or surface water and that pose the greatest potential risks to public health and the environment (United States Environmental Protection Agency 2016). In Canada, more than 23,000 contaminated or suspected sites have been identified (Government of Canada 2017).

Developing countries are undergoing significant industrialization and urbanization. In large urban areas, provision of sanitation and drainage is needed as well as adequate governance so that urban waste is disposed of adequately (FAO and Intergovernmental Technical Panel on Soils [ITPS] 2015). Trace elements contaminate agricultural soil and crops in many Asian countries (Thangavel and Sridevi 2017). In many parts of Latin America intensive use of agricultural inputs contributes to soil contamination (UNEP 2010). In Africa, agrochemicals, mining, spills and improper handling of waste have contaminated soils (Gzik *et al.* 2003; Kneebone and Short 2010). In the Near East and North Africa, soil contamination is primarily the result of oil production and heavy mining.

Soil and human health

The burden of disease of soil-transmitted helminths – a group of parasitic worms including hookworm, ascariasis and trichuriais/whipworm – is substantial, affecting human development and cognitive potential (Bartsch *et al.* 2016). These are generally acquired by walking barefoot on soil that has been contaminated by human faeces. High-intensity hookworm infection commonly affects both children and adults (Bartsch *et al.* 2016).

Land contains many trace elements, which enter the human food chain through the raising of crops and animals. Some are essential for good health (e.g. iodine, iron, selenium and zinc), while others are harmful in large quantities (e.g. arsenic and fluoride) (Oliver and Gregory 2015). Soils in mountainous areas often have reduced levels of iodine, and human populations in such areas can face higher health risks, as they are likely to have reduced access to iodine-rich marine foods. Fertilizers are often contaminated by cadmium, which is not essential to human health and is harmful in high doses (Newbigging, Yan and Le 2015).

Positive effects of healthy soils in human health are related to nature's available benefits to people (FAO 2015d). For example, some valuable antibiotics have been derived from soil microorganisms (Oliver and Gregory 2015).

Food, chemicals and human health

Pesticides (defined here as also including herbicides) have generated an almost universal human exposure to synthetic chemicals, many of which are harmful and even fatal at high doses (Nicolopoulou-Stamati *et al.* 2016). However, there is much uncertainty concerning the health effects of chronic exposure to pesticides at lower doses. While human exposure to some chemicals, such as organochlorines, has reduced in recent years due to regulation, other synthetic compounds have entered the human food chain, such as other pesticides, artificial sweeteners and colorants. The health effects of these substances, whether in isolation or combination, are very difficult to determine for reasons including uncertainty concerning exposure, varying rates and times of the accumulation of these compounds and their release from human tissue, and the lag between exposure and disease. In 1990, the World Health Organization (WHO) estimated an annual 735,000 cases of specific chronic effects linked to pesticides globally (WHO and UNEP 1990), but pesticide use has increased dramatically since then, especially in developing countries where lax regulations and an absence of compliance mechanisms expose millions of farmers and workers to pesticides capable of causing chronic effects that include cancers; reproductive, respiratory, immune and neurological effects; and much more (Watts and Williamson 2015).

There is good evidence from high-income countries that groups occupationally exposed to pesticides, such as farmers, have higher rates of non-Hodgkins lymphoma, attributed to pesticides (Schinasi and Leon 2014). Higher than expected rates of Parkinson's disease have also been related to occupational exposure to pesticides (Liew *et al.* 2014). Other factors that influence health, such as age, undernutrition and impaired immune status, may also interact with the health effects of pesticides, but this issue is currently under-studied. The health effects of chronic pesticide exposures vary considerably on women and men due to their different physiologies. Data on pesticide use (and protection) by women and men in food production are incomplete and inconsistent. Overall, men are less sensitive than women to many pesticides (Hardell 2003; Watts 2007; Watts 2013). Pesticides and breast cancer rates have a strong connection (Watts 2007; Watts 2013) and women are more vulnerable than men to endocrine disruption from pesticides (Howard 2003). On the other hand, men are more sensitive to some (other) pesticides (Alavanja *et al.* 2003).

Food quality can also be impaired through biotic contamination, both microbiological and fungal (Gnonlonfin *et al.* 2013). Mycotoxins, including aflatoxins, can be generated when cereals are damaged by rain, both pre-harvest and through poor storage and are an important cause of liver cancer in many low-income settings (Wild and Gong 2010).

8.5.3 Tenure security

Land tenure, land deals

Despite heavy reliance on land resources, communities, especially in the global South, frequently lack ownership of the land they farm or hold in common. While high-impact scientific studies on the causal linkages between tenure security and food security are lacking (Ghebru and Stein 2013; Holden and Ghebru 2016; Lawry *et al.* 2017), there is sufficient evidence to show that food and energy security of local communities is profoundly diminished when they lose reliable access to their land resources (Godfray *et al.* 2010; Muchomba 2017; Tomei and Ravindranath 2018). Land and housing are the most important assets in large sections of the world. Secure rights, for both men and women, can help turn these assets into economic opportunities (Doss, Kieran and Kilic 2017). It also allows communities to tap into the benefits of institutional support and regulation (Dekker 2016). Indigenous populations, the poor, landless and women are among the most vulnerable to the repercussions of unequal landownership and access (Narh *et al.* 2016).

While the precise amount of community land in the world is unknown, estimates suggest that only approximately 10 per cent of formal land rights are registered or recorded worldwide (Veit and Reytar 2017). Estimates indicate that local communities and indigenous people depend on and manage 50-65 per cent of the world's land area (Alden Wily 2011; Pearce 2016), yet many governments still recognize their rights over only a fraction of these lands (Rights and Resources Initiative [RRI] 2015) **(Figure 8.23)**.

Figure 8.23: Global forest ownership, 2002-2013 (%)

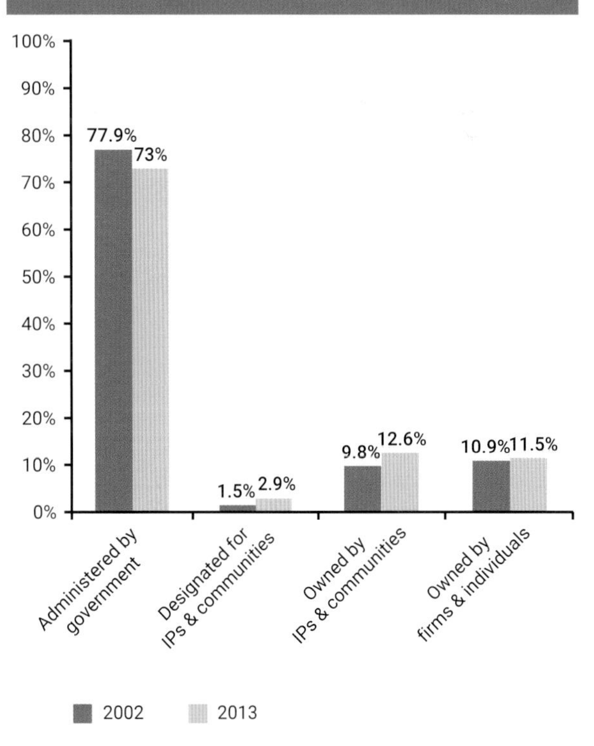

IP: indigenous peoples.
Source: RRI (2015).

As industrial agriculture and monoculture plantations have expanded, competition for land between industry, governments and communities has increased, putting pressure on forests and drylands, threatening local peoples' livelihoods in some parts of the world (UNCCD 2017). Without formal recognition and protection of their land rights, communities in some countries lack legal recourse following infringement of those rights. In the recent past, stories of poor governance have been under a global spotlight due to issues of land acquisition, land grabbing and land leasing amid fears of food scarcity and rising food prices. Although estimates vary, since 2000, between 26.7 million ha (Nolte, Chamberlain and Giger 2016) and 42 million ha (UNCCD 2017) of agricultural land around the world have become controlled by foreign investors. As of April 2016, Africa remains the most significant target area, with 42 per cent of all deals and 10 million ha (37 per cent) (**Figure 8.24**). Most deals involve the private sector, whose focus is on flexible crops. Importantly, food and biofuels produced on such land are unlikely to reach local communities. Most acquisitions do not include domestic shareholders or local community negotiations, despite often targeting relatively highly populated areas dominated by croplands.

Studies have shown that lack of tenure security among local communities can translate into reduced investments in human capital (Dekker 2016), negative effects on land improvements (Eskander and Barbier 2017), reduced agricultural productivity (Place 2009; Lawry *et al.* 2014) and lower resilience in times of disaster risk (Unger, Zevenbergen and Bennett 2017).

There is increasing evidence of local indigenous communities successfully managing and conserving lands (**Box 8.3**). The World Resources Institute (Ding *et al.* 2016; Veit and Reytar 2017) indicates that 'tenure-secure' indigenous lands generate billions and sometimes trillions of dollars' worth of benefits in the form of clean water, erosion control, carbon sequestration, reduced pollution, and a suite of other local, regional and global ecosystem services (**Figure 8.25**).

Figure 8.24: Global maps of land deals, number of land deals per country (top), land deal area per country (bottom)

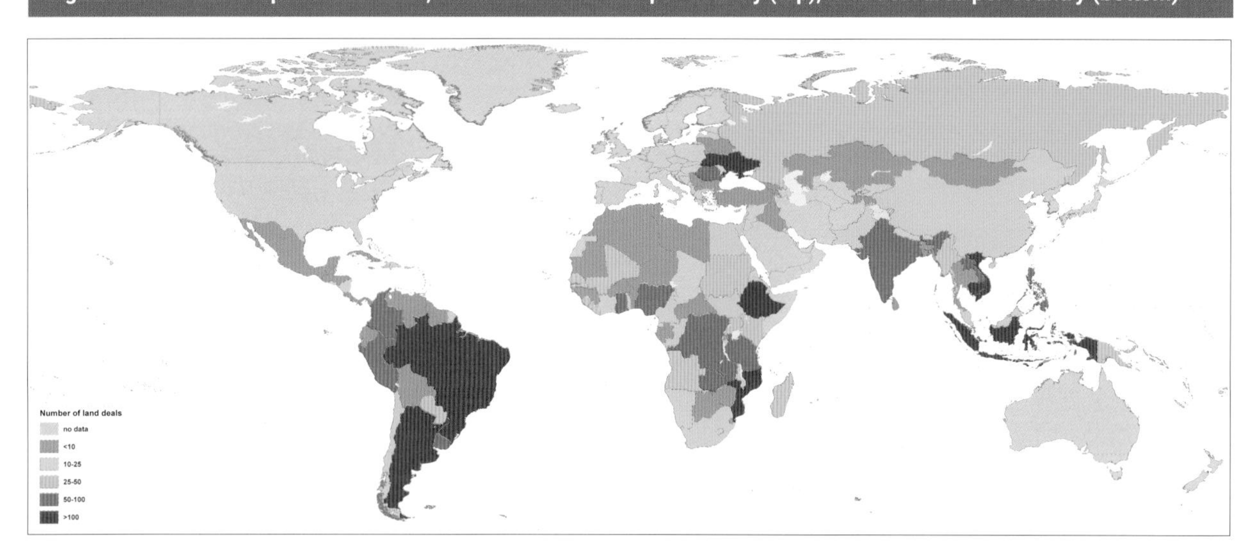

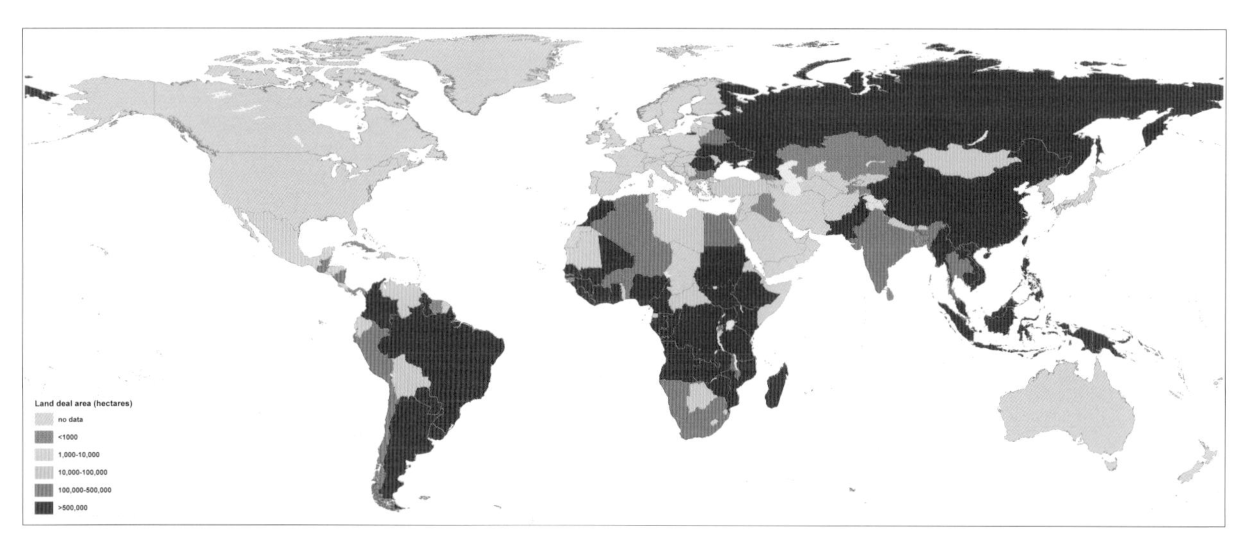

Source: Alexandratos and Bruinsma (2012).

Box 8.3: Cultural values and conservation in Bhutan

Sustainable land management can be promoted by strengthening environment-friendly cultural values and customary institutions. In Bhutan, cultural values play a role in protecting ecosystem services. Mahayana Buddhism places strong significance on the peaceful coexistence of people with nature and the sanctity of life and compassion for others. This explains in large part the high share (71 per cent) of land area under forests in Bhutan and the fact that 25 per cent of Bhutan's population lives within protected areas (Nkonya, Mirzabaev and Von Braun 2016). Many of Bhutan's Buddhist monasteries are located within the forested landscapes of the country.

© Darshini Ravindranath

Figure 8.25: Benefits of tenure-secure lands outweigh the costs in three Latin American countries

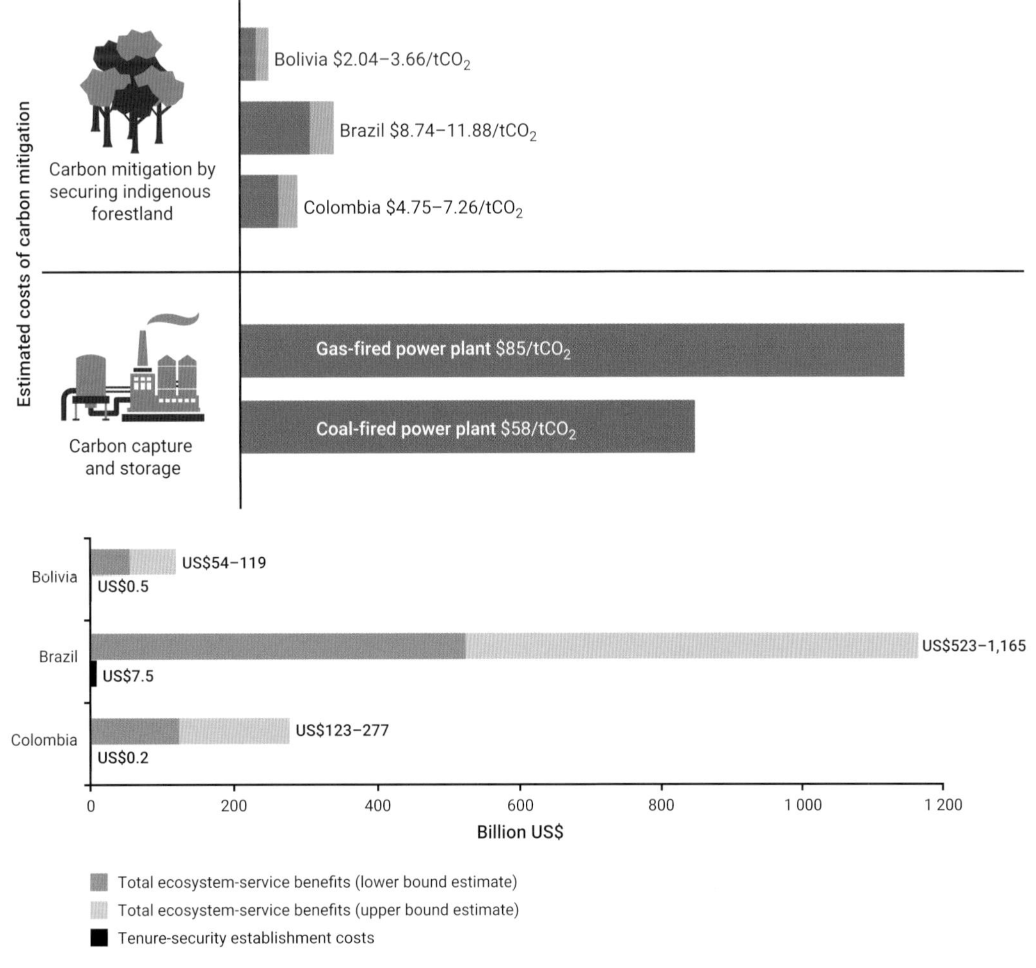

Source: Ding *et al*. (2016).

Both the benefits and impacts can, however, vary by region and context due to the complex nature of defining and measuring land tenure. For instance, Eskander and Barbier (2017) find that, in Bangladesh, secure land tenure is associated with improvements in topsoil conservation. However, it is also related to lower human capital investments (e.g. lower spending on educational and recreational activities). Such heterogeneities in findings suggest that adequate attention needs to be given to the broader macro and sector conditions in addition to the local context within which tenure systems are governed.

At a global level, recommendations for stronger land governance in countries that are the targets of large-scale investments are becoming a priority. The rights of indigenous people to their lands and territories are explicitly mentioned in the United Nations Declaration on the Rights of Indigenous People (Article 25 and Article 26) (United Nations 2007).

The FAO *Voluntary Guidelines on the Responsible Governance of Tenure* (VGGT) also seek to improve the governance of land tenure with respect to all forms: public, private communal, indigenous, customary and informal (FAO 2012).

Land and sociocultural services

Land provides a variety of sociocultural and aesthetic benefits to people that are essential for sustainable, healthy livelihoods. Land degradation, deforestation and desertification lead to increases in land abandonment, outmigration and changes in rural power structures (due to increasing demand for intensification), among others. One of the key impacts of these changes has been a loss of critical sociocultural services provided by land, leading to a lowering of overall community resilience (Wilson *et al.* 2016; Wilson *et al.* 2017).

In many developing countries, most people reside in rural areas and are heavily dependent on land resources for their livelihoods. They grow crops for food and to sell in local markets; collect fodder for their livestock; gather wood for their stoves; and collect tree products for their health and well-being (Tomei and Ravindranath 2018). Here, the value of land is often an assertion of their long-standing sociocultural identity, place and heritage (Tomei and Ravindranath 2018). Kelly *et al.* (2015) show that ancient traditions such as festivals related to the preservation of timber, food and fuel resources reveal a deeply embedded relationship between land, culture and identity. In the European Union (EU), the recreational and cultural significance of land is incorporated, to an extent, through national and regional policies on management of ecosystem services. The EU 2020 Biodiversity Strategy, currently being implemented throughout Europe, predominantly covers 'cultural landscapes' (European Commission 2011; Plieninger *et al.* 2013).

Despite progress in recognizing these challenges, land-use trends and impact research continue to be dominated by the study of land-use change from the perspective of productivity, seldom acknowledging and documenting trends in the deep-rooted need for conservation from the perspective of communities (Sharmina *et al.* 2016).

8.5.4 Gender inequality: land, health and food

Existing gender inequality may contribute to increased poverty, people displacement, resource scarcity and other conflicts (Behrman, Meinzen-Dick and Quisumbing 2012; Verma 2014; White, Park and Mi Yong 2015). While progress has been made on the importance of incorporating women to sustain land productivity, it has often been at a superficial level (e.g. to meet certain global targets). Furthermore, women in agrarian societies often have a strategic role in reducing hunger, malnutrition and poverty as they play a central role in household food security, dietary diversity and children's health. Evidence suggests that women are much more likely than men are to spend income from these resources on their children's nutritional and educational needs (Malapit *et al.* 2015; Komatsu, Malapit and Theis 2018).

Agricultural contributions by women tend to be underestimated or not considered in official statistics since their focus is usually on formal employment in agriculture and on commercial agriculture. Women are usually engaged in subsistence agriculture, they tend home gardens and collect wild foods, and all these contributions are essential to food security (UNEP 2016a). In 2011, women represented 43 per cent of those economically active in agriculture (FAO 2011). However, they hold titles to less than 20 per cent of agricultural land (FAO 2010). In Africa, only Cape Verde can report that women own over half of agricultural holdings (50.5 per cent) (Doss *et al.* 2017). Few statistics show improvements in land tenure of women during the current decade, especially in countries of the global South (**Figure 8.26**).

Closing the gender gap in access to information and technology, and access to and control over production inputs and land, could increase agricultural productivity and reduce hunger and poverty (Croppenstedt, Goldstein and Rosas 2013).

8.6 Policy responses

Countless policies and actions attempt to address environmental degradation on land. Some strategies have been successful or are promising (e.g. restoration of degraded lands in specific locations such as the Great Green Wall Project in China – see chapter 15, sustainable management strategies such as no-tillage cultivation in Australia, payment for ecosystem services such as Mexico's National Program), while the benefits of others are not necessarily clear (e.g. the expansion of agricultural lands for flexible crop and biofuel production). However, most of these approaches do not consider the variety of benefits people obtain from land and focus only on its productive potential. Globally, land is becoming a scarce resource and is increasingly traded instead of being treated as a global common good due to its importance in the provision of basic services such as food production (Creutzig 2017). This section reviews this undesirable trend, while chapter 15 in Part B discusses in detail alternative land-use policies that could change this unsustainable trajectory.

Figure 8.26: Distribution of agricultural land holdings: females

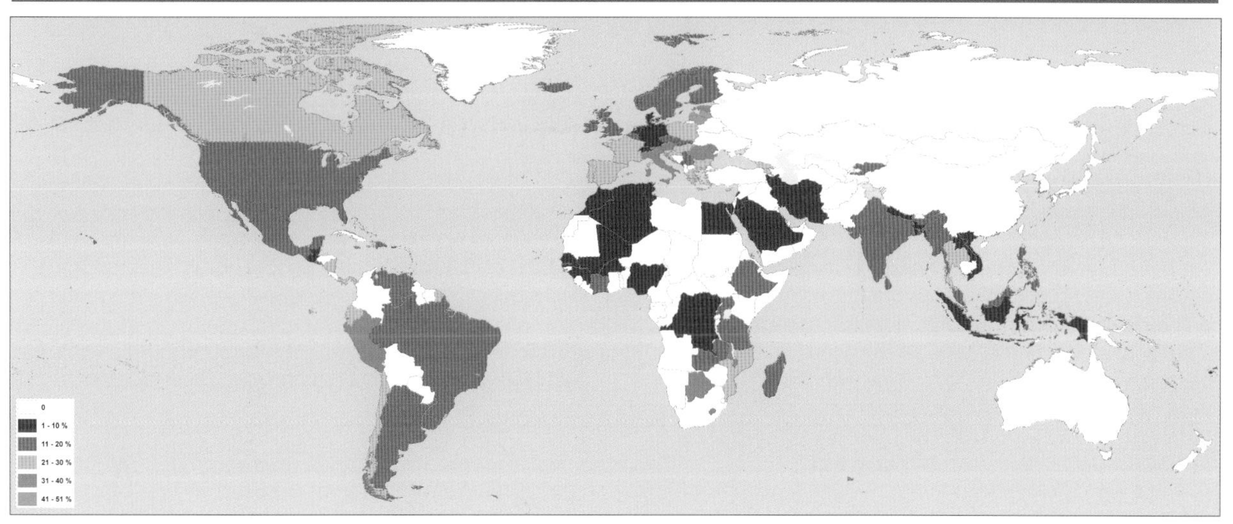

Source: FAO (2017d).

8.6.1 Economic optimism and land degradation

Land degradation is a key global issue due to its adverse impacts on the environment, agricultural productivity and human welfare. The current paradigm of land management usually considers that the losses caused by land degradation and mismanagement can be compensated by increasing inputs in agriculture, expanding to new areas, and managing lands through command and control strategies such as replacing native forests with plantations (e.g. Chile, Indonesia). This approach also considers that nutritional and other associated social problems will gradually disappear as agricultural production expands (Rosegrant *et al.* 2001). However, social and environmental scientists warn that constantly improving agrotechnology may offer agricultural managers a false sense of security (Eswaran, Lal and Reich 2001).

Current trends are unlikely to supply future demands for food, energy, timber and other ecosystem services taking into consideration even moderate projections for land resources availability. By 2050, demand for food across all categories is likely to be 50 per cent more than today due to dietary changes associated with increasing incomes and population growth (Tilman *et al.* 2011; Alexandratos *et al.* 2012). At the aggregate level, yields are not increasing fast enough to meet demand without significant expansion of the agricultural area (Ray *et al.* 2012; Ray *et al.* 2013; Bajželj *et al.* 2014). This would be difficult to reconcile with large-scale afforestation or deployment of BioEnergy with Carbon Capture and Storage (BECCS) at the levels thought necessary to limit global warming to less than 2°C. For example, Smith *et al.* (2015) estimate that BECCS could require 380-700 million ha by the end of the century, representing up to 14 per cent of global agricultural land, for a 2°C pathway.

Continuing the current track it will be difficult to achieve the land degradation neutrality target adopted at the United Nations Conference on Sustainable Development (Rio+20) in 2012. Land degradation neutrality (LDN) is captured in SDG 15.3. Achieving land degradation neutrality by 2030 is regarded as critical for attaining other key international goals related to reducing biodiversity loss and deforestation, improving human welfare, and climate change adaptation and mitigation. Land-use change, a warmer climate, stagnating yields and unsustainable agricultural practices continue to lead to a reduced stock of organic soil carbon (Wiesmeier *et al.* 2016).

While scientists provide alarming estimates for the decline of productivity of lands globally and regionally due to soil erosion and desertification (Nkonya, Mirzabaev and Von Braun eds. 2016), many economists still believe that if land degradation were a severe issue, market forces would have taken account of it (Utuk and Daniel 2015). In other words, agricultural managers would not let their lands degrade to the point that it affects their incomes (Wiebe 2003). Cumulative productivity losses due to land degradation appear economically acceptable for most agricultural actors. In many instances, farmers can rely on government agricultural policies (e.g. subsidies for inputs and machinery) to curb losses associated with land degradation (Jat, Sahrawat and Kassam 2013).

However, these policies are not sustainable in either developing or developed countries. Market fluctuations of agricultural inputs could be more volatile than output prices. From 2005 to 2008, fertilizer prices rose much faster (by 400 per cent) than maize prices (by 100 per cent) and reached record high levels in 2008. In this case an input subsidy would be inefficient as

it would encourage unprofitable use of inputs (**Figure 8.27**) (Baltzer and Hansen 2011). The same study indicates that, in Malawi, the subsidy ratio jumped from 79 per cent to 91 per cent or from 3.4 per cent to 6.6 per cent of GDP in 2008-2009.

In sub-Saharan Africa (SSA), the contribution of fertilizer subsidies to national food security strategies remains highly controversial (Druilhe and Barreiro-Hurlé 2012). Success in the Asian Green Revolution was based on two main food crops grown under irrigation, wheat and rice. In SSA countries, yield response to fertilizer application is observed for some crops (e.g. maize), but not for most other staple crops grown in rain-fed areas (e.g. cassava, plantain, yam). In these contexts, fertilizer use is not profitable under market conditions, especially in some remote areas where output prices are too low. In order to be effective, agricultural programmes should be complemented with other government investments in infrastructure, education, health and rural development (Druilhe and Barreiro-Hurlé 2012) (**Figure 8.28**).

Reducing farm subsidies in rich countries would be positive for poor countries, although the effect will depend on their economic, trade and poverty characteristics (Boysen, Jensen and Matthews 2016). Meanwhile, the availability of subsidies in rich countries does not provide an incentive to adopt innovative soil conservation strategies.

For a long time, the market price of crops has been the standard for determining land-use policy. However, a new trend is being observed in growing competition between the financial and economic values of land. Land speculation and land grabbing can distort the actual economic value generated by land. With increasing land scarcity, the trend to consider land as a 'commodity' is only strengthened (ELD 2013). As land prices increase, more farmland will be sold to outsiders purely for speculative purposes. Consequently, lands might be left idle for some time, leading to less agricultural production, exacting a significant social cost if the practice becomes widespread.

In the EU, inflationary pressures are fueling land speculation and the acquisition of farmland. This rapid inflation has been attributed to the rise of 'new investors' in farmland, some with little connection to agriculture or farming. This process has been termed by French activists as one of 'land artificialization': the loss of prime agricultural land, the expansion of cities, urban development, tourism and other commercial undertakings (Borras, Franco and van der Ploeg 2013). Land speculation and land 'artificialization' contribute to farmland concentration in the EU by raising the stakes and increasing the barriers for prospective farmers to take up farming (Kay, Peuch and Franco 2015).

One of the indicators of ever-increasing commoditization and commercialization of land was a recent boom for biofuel production. The relative abundance of cheap and suitable land in poor countries and increasingly liberalized trade and

Figure 8.27: Fertilizer and maize prices, 2000-2010

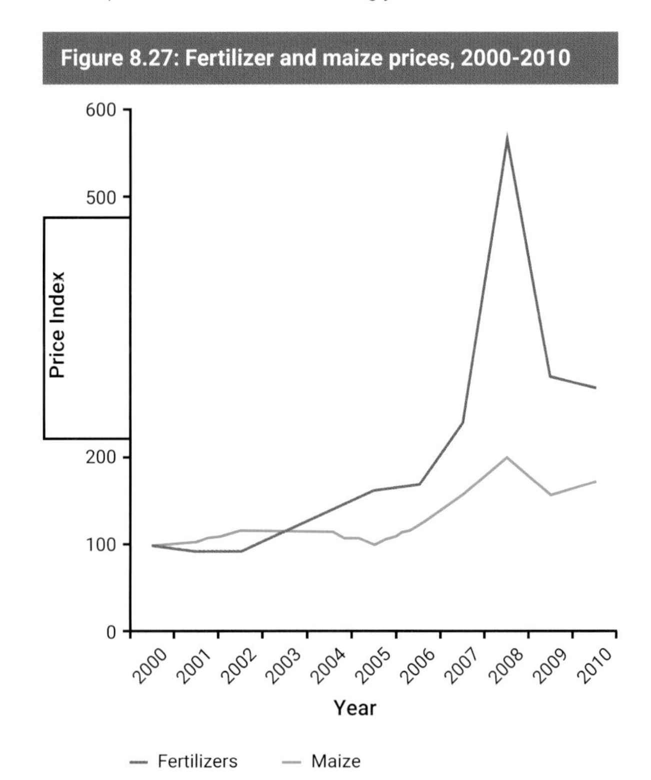

Prices are real US$ indices of world market prices.
Source: Baltzer and Hansen (2011).

Figure 8.28: Where should subsidies fit?

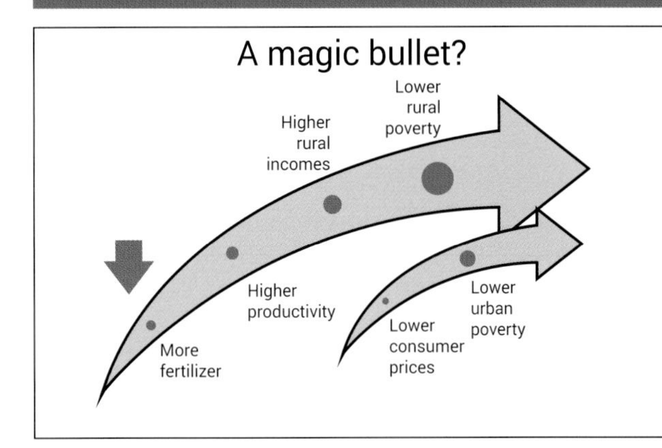

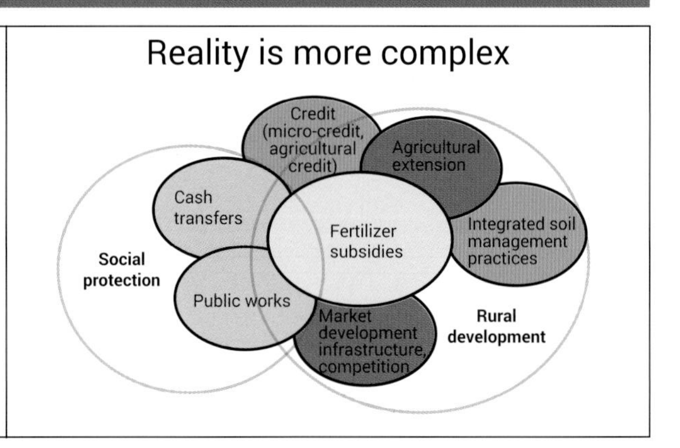

Source: Druilhe and Barreiro-Hurlé (2012).

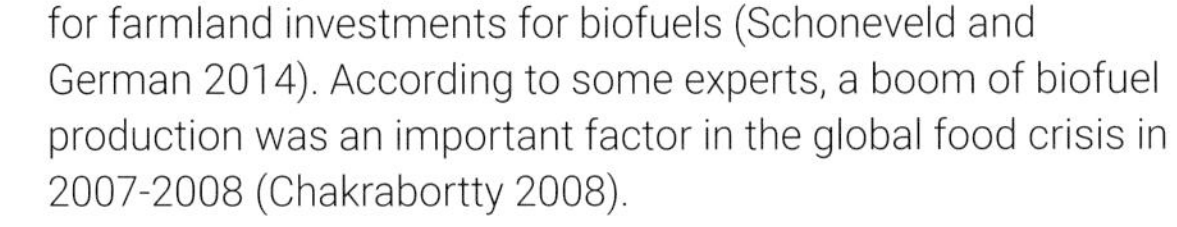

investment regimes made them an attractive destination for farmland investments for biofuels (Schoneveld and German 2014). According to some experts, a boom of biofuel production was an important factor in the global food crisis in 2007-2008 (Chakrabortty 2008).

8.6.2 Challenges for achieving the SDGs

Estimating the full economic benefits of land is neither easy nor straightforward (UNEP 2016b). The ecosystem services framework can contribute to comprehensive ecosystem assessments by dividing ecosystem services provided by land into categories that are interdependent and can be valued separately (**Figure 8.29**).

Current land management cannot prevent loss of natural capital while preserving ecosystem services (e.g. moisture retention, nutrient cycling), combating climate change (e.g. carbon sequestration), providing sustainable food production, addressing energy and water security, and promoting fair access to land (ELD 2013).

Intergenerational equity is not necessarily considered in current land management strategies, and present productivity gains are valued more than sustainable production for the future. Furthermore, land-use policy may not reflect the teleconnections that link production and consumption across the globe. According to current land policy approaches, most issues which cannot be addressed by increasing inputs are automatically dropped outside the land-policy domain. However, this approach is inappropriate as many social, gender, poverty and health issues are directly or indirectly associated with conventional ways of managing land resources and trading them across the globe.

Economic optimism plays in favour of enlarging farms due to their economic effectiveness and the difficulty of incorporating the economic impacts of the degradation of land resources. However, maximizing smallholders' potential, including women and indigenous peoples, is essential for food security and proper nutrition, and for reaching many SDGs. There are about 570 million farms in the world, and 84 per cent operate on less than 2 ha of land (International Food Policy Research Institute

Figure 8.29: The provision of ecosystem services from natural capital: linkages between ecosystem services and human well-being

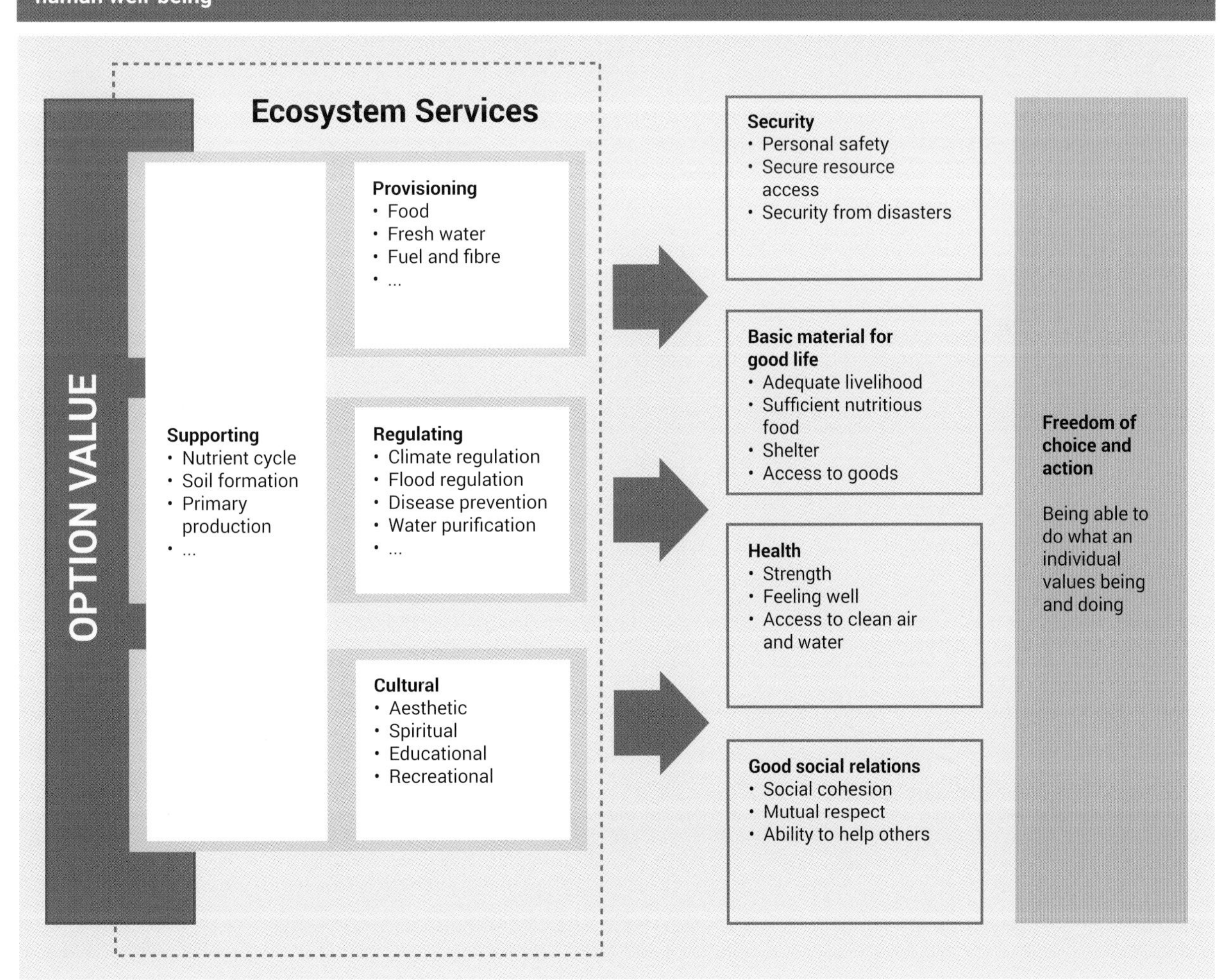

Source: Millennium Ecosystem Asessment (2003).

8

[IFPRI] 2016). Small farms play different roles: billions of people get their income, employment and food from these lands. They are also home to most of the world's undernourished population. FAO estimates that if gender inequality in access to land resources is eliminated, agricultural output could increase by 2.5-4.0 per cent. Additionally, it would lead to a reduction of 12-17 per cent reduction in the number of undernourished people in developing countries (IFPRI 2016). In low-income agrarian societies, agricultural growth is more effective for reducing hunger and poverty than promoting any other sector of the economy (FAO 2015e). If SDG Target 2.3 is to be achieved by 2030, agricultural productivity of small farms should increase simultaneously with the incomes of their farmers. Policies should especially target the most vulnerable small-scale food producers (e.g. women, indigenous peoples), so they can have guaranteed access to market and other production means, including their material, informational and financial needs.

It is clear that minimizing food losses and waste will have significant environmental, social and economic benefits in supporting global food security (UNEP 2015). Where waste cannot be prevented, opportunities to recover value, such as conversion to compost, liquid fertilizers, biogas or higher value end-use products such as animal feed protein or biochemicals, should be explored (Jayathilakan *et al.* 2012; Nguyen, Tomberlin and Vanlaerhoven 2015; UNEP 2015). Achieving SDG Target 12.3 of halving per capita global food losses and waste at the retail and consumer levels and reducing food losses along production and supply chains, including post-harvest losses, by 2030, will require significant intervention and commitment, but also diverse strategies, since the reasons for food losses and waste, and the area within the food supply chain where losses and waste occur, differ between developed and developing countries (FAO 2015c).

References

Addison, J., Friedel, M., Brown, C., Davies, J. and Waldron, S. (2012). A critical review of degradation assumptions applied to Mongolia's Gobi Desert. *The Rangeland Journal* 34(2), 125-137. https://doi.org/10.1071/RJ11013.

Aide, T.M., Clark, M.L., Grau, H.R., López-Carr, D., Levy, M.A., Redo, D. *et al.* (2013). Deforestation and reforestation of Latin America and the Caribbean (2001-2010). *Biotropica* 45(2), 262-271. https://doi.org/10.1111/j.1744-7429.2012.00908.x.

Alavanja, M.C.R., Samanic, C., Dosemeci, M., Lubin, J., Tarone, R., Lynch, C.F. *et al.* (2003). Use of agricultural pesticides and prostate cancer risk in the Agricultural Health Study cohort. *American Journal of Epidemiology* 157(9), 800-814. https://doi.org/10.1093/aje/kwg040.

Alden Wily, L. (2011). *The Tragedy of Public Lands: The Fate of The Commons Under Global Commercial Pressure*. International Land Coalition. http://www.landcoalition.org/sites/default/files/documents/resources/WILY_Commons_web_11.03.11.pdf.

Alexandratos, N. and Bruinsma, J., (2012). World agriculture: Towards 2030/2050. The 2012 Revision. Rome: Food and Agriculture Organization. http:// www.fao.org/docrep/016/ap106e/ap106e.pdf.

Anyamba, A. and Tucker, C.J. (2005). Analysis of Sahelian vegetation dynamics using NOAA-AVHRR NDVI data from 1981–2003. *Journal of Arid Environments* 63(3), 596-614. https://doi.org/10.1016/j.jaridenv.2005.03.007.

Asociación Pro Derechos Humanos (Aprodeh), Broederlijk Delen, Colectivo de Abogados José Álvaro Restrepo, Centro de Documentación e Información Bolivia and Comisión Ecuménica de Derechos Humanos (2018). *Abusos De Poder Contra Defensores Y Defensoras De Los Derechos Humanos, Del Territorio Y Del Ambiente: Informe Sobre Extractivismo Y Derechos En La Región Andina*. Bogota. http://www.broederlijkdelen.be/sites/default/files/downloads/andesrapport_2018_lr.pdf.

Awumbila, M. (2017). *Drivers of Migration and Urbanization in Africa: Key Trends and Issues*. New York, NY: United Nations. http://www.un.org/en/development/desa/population/events/pdf/expert/27/presentations/III/presentation-Awunbila-final.pdf.

Bajželj, B., Richards, K.S., Allwood, J.M., Smith, P., Dennis, J.S., Curmi, E. *et al.* (2014). Importance of food-demand management for climate mitigation. *Nature Climate Change* 4, 924-929. https://doi.org/10.1038/nclimate2353.

Baltzer, K. and Hansen, H. (2011). *Evaluation Study Agricultural input subsidies in Sub-Saharan Africa*. Copenhagen: Institute of Food and Resource Economics. https://www.oecd.org/derec/49231998.pdf.

Bartsch, S.M., Hotez, P.J., Asti, L., Zapf, K.M., Bottazzi, M.E., Diemert, D.J. *et al.* (2016). The global economic and health burden of human hookworm infection. *PLOS Neglected Tropical Diseases* 10(9), e0004922. https://doi.org/10.1371/journal.pntd.0004922.

Behnke, R. and Mortimore, M. (eds.) (2016). *The End of Desertification? Disputing Environmental Change in the Drylands*. Heidelberg: Springer Berlin. https://www.springer.com/gp/book/9783642160134.

Behrman, J., Meinzen-Dick, R. and Quisumbing, A. (2012). The gender implications of large-scale land deals. *Journal of Peasant Studies* 39(1), 49-79. https://doi.org/10.1080/03066150.2011.652621.

Benton, T.G., Vickery, J.A. and Wilson, J.D. (2003). Farmland biodiversity: Is habitat heterogeneity the key? *Trends in Ecology and Evolution* 18(4), 182-188. https://doi.org/10.1016/S0169-5347(03)00011-9.

Bergmann, L. and Holmberg, M. (2016). Land in motion. *Annals of the American Association of Geographers* 106(4), 932-956. https://doi.org/10.1080/24694452.2016.1145537.

Bestelmeyer, B.T., Okin, G.S., Duniway, M.C., Archer, S.R., Sayre, N.F., Williamson, J.C. *et al.* (2015). Desertification, land use, and the transformation of global drylands. *Frontiers in Ecology and the Environment* 13(1), 28-36. https://doi.org/10.1890/140162.

Binion, E. and Gutberlet, J. (2012). The effects of handling solid waste on the wellbeing of informal and organized recyclers: A review of the literature. *International Journal of Occupational and Environmental Health* 18(1), 43-52. https://doi.org/10.1179/1077352512Z.0000000001.

Bisaro, A., Kirk, M., Zdruli, P. and Zimmermann, W. (2014). Global drivers setting desertification research priorities: Insights from a stakeholder consultation forum. *Land Degradation & Development* 25(1), 5-16. https://doi.org/10.1002/ldr.2220.

Borras, S., Franco, J. and van der Ploeg, J. (2013). Land concentration, land grabbing and people's struggles in Europe: Introduction to the collection of studies. In *Land Concentration, Land Grabbing and People's Struggles in Europe*. Amsterdam: Transnational Institute. chapter 1. 6-30. https://www.tni.org/files/download/land_in_europe-jun2013.pdf.

Borras, S.M., Franco, J.C., Gómez, S., Kay, C. and Spoor, M. (2012). Land grabbing in Latin America and the Caribbean. *The Journal of Peasant Studies* 39(3-4), 845-872. https://doi.org/10.1080/03066150.2012.679931.

Boysen, O., Jensen, H.G. and Matthews, A. (2016). Impact of EU agricultural policy on developing countries: A Uganda case study. *The Journal of International Trade & Economic Development* 25(3), 377-402. https://doi.org/10.1080/09638199.2015.1069884.

Brenner, N. and Schmid, C. (2014). The 'Urban Age' in Question. *International Journal of Urban and Regional Research* 38(3), 731-755. https://doi.org/10.1111/1468-2427.12115.

Butler, R. (2017). *Amazon destruction*. [https://rainforests.mongabay.com/amazon/amazon_destruction.html.

Carlson, K.M., Curran, L.M., Asner, G.P., Pittman, A.M., Trigg, S.N. and Marion Adeney, J. (2012). Carbon emissions from forest conversion by Kalimantan oil palm plantations. *Nature Climate Change* 3, 283-287. https://doi.org/10.1038/nclimate1702.

Cassidy, E.S., West, P.C., Gerber, J.S. and Foley, J.A. (2013). Redefining agricultural yields: From tonnes to people nourished per hectare. *Environmental Research Letters* 8(3). https://doi.org/10.1088/1748-9326/8/3/034015.

Chakrabortty, A. (2008). 'Secret report: Biofuel caused food crisis'. *The Guardian* 3 July 2008 https://www.theguardian.com/environment/2008/jul/03/biofuels.renewableenergy.

Chatham House (2017). *Exploring interdependencies in global resource trade*. [Chatham House https://resourcetrade.earth/.

Cherlet, M., Reynolds, J., Hutchinson, C., Hill, J. and von Maltitz, G. (eds.) (2018). *World Atlas of Desertification: Rethinking Land Degradation and Sustainable Land Management*. 3rd edn. Luxembourg. https://wad.jrc.ec.europa.eu/sites/default/files/atlas_pdf/JRC_WAD_fullVersion.pdf.

Chin, A. (2018). Notes from the field: The value of observational data and natural history. *Pacific Conservation Biology* 24(1). https://doi.org/10.1071/PCv24n1_ED.

Creutzig, F. (2017). Govern land as a global commons. *Nature* 546(7656), 28-29. https://doi.org/10.1038/546028a.

Croppenstedt, A., Goldstein, M. and Rosas, N. (2013). Gender and agriculture inefficiencies, segregation, and low productivity traps. *World Bank Research Observe* 28(1), 79-109. http://hdl.handle.net/10986/19493.

de Jong, R., de Bruin, S., de Wit, A., Schaepman, M.E. and Dent, D.L. (2011). Analysis of monotonic greening and browning trends from global NDVI time-series. *Remote Sensing of Environment* 115(2), 692-702. https://doi.org/10.1016/j.rse.2010.10.011.

de Ruiter, H., Macdiarmid, J.I., Matthews, R.B., Kastner, T., Lynd, L.R. and Smith, P. (2017). Total global agricultural land footprint associated with UK food supply 1986–2011. *Global Environmental Change* 43, 72-81. https://doi.org/10.1016/j.gloenvcha.2017.01.007.

Dekker, H.A.L. (2016). *The Invisible Line: Land Reform, Land Tenure Security, and Land Registration*. Routledge. https://www.crcpress.com/The-Invisible-Line-Land-Reform-Land-Tenure-Security-and-Land-Registration/Dekker/p/book/9781138258709.

Development Initiatives (2017). *Global Nutrition Report 2017: Nourishing the SDGs*. Bristol. https://www.gainhealth.org/wp-content/uploads/2017/11/GNR-Report_2017.pdf.

Ding, H., Veit, P.G., Blackman, A., Gray, E., Reytar, K., Altamirano, J.C. *et al.* (2016). *Climate Benefits, Tenure Costs. The Economic Case for Securing Indigenous Land Rights in the Amazon*. World Resources Institute. Washington, DC. http://wriorg.s3.amazonaws.com/s3fs-public/Climate_Benefits_Tenure_Costs.pdf.

D'Odorico, P., Carr, J.A., Laio, F., Ridolfi, L. and Vandoni, S. (2014). Feeding humanity through global food trade. *Earth's Future* 2, 458-469. https://doi.org/10.1002/2014EF000250.

Doss, C., Kieran, C. and Kilic, T. (2017). *Measuring Ownership, Control, and Use of Assets*. World Bank Policy Research Working Paper. Washington, DC: World Bank. http://documents.worldbank.org/curated/en/934731500383137028/pdf/WPS8146.pdf.

Druilhe, Z. and Barreiro hurlé, J. (2012). *Fertilizer Subsidies in Sub-Saharan Africa*. Rome: Food and Agriculture Organization. http://www.fao.org/3/a-ap077e.pdf.

Eskander, S.M.S.U. and Barbier, E.B. (2017). Tenure security, human capital and soil conservation in an overlapping generation rural economy. *Ecological Economics* 135, 176-185. https://doi.org/10.1016/J.ECOLECON.2017.01.015.

Eswaran, H., Lal, R. and Reich, P.F. (2001). *Land degradation: An overview*. [United States Department of Agriculture https://www.nrcs.usda.gov/wps/portal/nrcs/detail/soils/use/?cid=nrcs142p2_054028.

European Commission (2011). *Our Life Insurance, Our Natural Capital: An EU Biodiversity Strategy To 2020*. Brussels. http://ec.europa.eu/environment/nature/biodiversity/comm2006/pdf/EP_resolution_april2012.pdf.

European Space Agency (2015). *Land cover*. [https://www.esa-landcover-cci.org/ (Accessed: 7 November 2018).

Fader, M., Rulli, M.C., Carr, J., Dell'Angelo, J., D'Odorico, P., Gephart, J.A. *et al.* (2016). Past and present biophysical redundancy of countries as a buffer to changes in food supply. *Environmental Research Letters* 11, 055008. https://doi.org/10.1088/1748-9326/11/5/055008.

Fensholt, R., Langanke, T., Rasmussen, K., Reenberg, A., Prince, S.D., Tucker, C. *et al.* (2012). Greenness in semi-arid areas across the globe 1981–2007 – an Earth Observing Satellite based analysis of trends and drivers. *Remote Sensing of Environment* 121, 144-158. https://doi.org/10.1016/j.rse.2012.01.017.

Fischer-Kowalski, M. and Haberl, H. (2007). Socioecological transitions and global change :Trajectories of social metabolism and land use. *Journal of Industrial Ecology* 12(5-6), 806-807. https://doi.org/10.1111/j.1530-9290.2008.00091_4.x.

Foley, J.A., Ramankutty, N., Brauman, K.A., Cassidy, E.S., Gerber, J.S., Johnston, M. *et al.* (2011). Solutions for a cultivated planet. *Nature* 478, 337-342. https://doi.org/10.1038/nature10452.

Food and Agriculture Organization and Intergovernmental Technical Panel on Soils (2015). *Status of the World's Soil Resources*. Rome: Food and Agriculture Organization. http://www.fao.org/3/a-i5199e.pdf.

Food and Agriculture Organization, International Fund for Agricultural Development, United Nations Children's Fund, World Food Programme and World Health Organization (2017). *The State of Food Security and Nutrition in the World*. Rome. http://www.fao.org/3/a-I7695e.pdf.

Food and Agriculture Organization of the United Nations (2009). *How to Feed the World in 2050*. Rome. http://www.fao.org/fileadmin/templates/wsfs/docs/expert_paper/How_to_Feed_the_World_in_2050.pdf.

Food and Agriculture Organization of the United Nations (2010). *Gender and Land Rights: Understanding Complexities, Adjusting Policies*. Rome. http://www.fao.org/docrep/012/al059e/al059e00.pdf.

Food and Agriculture Organization of the United Nations (2011). *The Role of Women in Agriculture*. ESA Working Paper. Rome: Food and Agriculture Organization. http://www.fao.org/docrep/013/am307e/am307e00.pdf.

Food and Agriculture Organization of the United Nations (2012). *Voluntary Guidelines on the Responsible Governance of Tenure of Land, Fisheries and Forests in the Context of National Food Security of Tenure*. Rome. http://www.fao.org/docrep/016/i2801e/i2801e.pdf.

Food and Agriculture Organization of the United Nations (2013). *Food Wastage Footprint: Impacts on Natural Resources. Summary Report*. Rome. http://www.fao.org/docrep/018/i3347e/i3347e.pdf.

Food and Agriculture Organization of the United Nations (2015a). *Global Forest Resources Assessment 2015*. Rome: Food and Agriculture Organization. http://www.fao.org/3/a-i4808e.pdf.

Food and Agriculture Organization of the United Nations (2015b). *Global Initiative on Food Loss and Waste Reduction*. Rome: Food and Agriculture Organization. http://www.fao.org/3/a-i4068e.pdf.

Food and Agriculture Organization of the United Nations (2015c). *Food Wastage Footprint and Climate Change*. Rome. http://www.fao.org/3/a-bb144e.pdf.

Food and Agriculture Organization of the United Nations (2015d). *Healthy Soils are the Basis for Healthy Food Production*. Rome. http://www.fao.org/3/a-i4405e.pdf.

Food and Agriculture Organization of the United Nations (2015e). *FAO and the 17 Sustainable Development Goals*. Rome. http://www.fao.org/3/a-i4997e.pdf.

Food and Agriculture Organization of the United Nations (2016). *The State of World Fisheries and Aquaculture: Contributing to Food Security and Nutrition for all*. Rome. http://www.fao.org/3/a-i5555e.pdf.

Food and Agriculture Organization of the United Nations (2017a). *The Future of Food and Agriculture: Trends and Challenges*. Rome. http://www.fao.org/3/a-i6583e.pdf.

Food and Agriculture Organization of the United Nations (2017b). *Food and agriculture data*. http://www.fao.org/faostat/en/#home.

Food and Agriculture Organization of the United Nations (2017c). *FAO Food Price Index*. [Food and Agriculture Organization http://www.fao.org/worldfoodsituation/foodpricesindex/en/ (Accessed: 19 December 2017).

Food and Agriculture Organization of the United Nations (2017d). *Gender and land rights database*. http://www.fao.org/gender-landrights-database/en/ (Accessed: 11 April 2018).

Food and Agriculture Organization of the United Nations (2018). *Land & water*. [http://www.fao.org/land-water/databases-and-software/gladis/en/.

Fritz, M., Vonk, J.E. and Lantuit, H. (2017). Collapsing Arctic coastlines. *Nature Climate Change* 7, 6-7. https://doi.org/10.1038/nclimate3188.

Ghebru, H. and Stein, H. (2013). *Links Between Tenure Security and Food Security: Evidence from Ethiopia*. Washington, DC: International Food Policy Research Institute. http://www.ifpri.org/cdmref/p15738coll2/id/127861/filename/128072.pdf.

Gnonlonfin, G.J.B., Hell, K., Adjovi, Y., Fandohan, P., Koudande, D.O., Mensah, G.A. *et al.* (2013). A review on aflatoxin contamination and its implications in the developing world: A sub-saharan African perspective. *Critical Reviews in Food Science and Nutrition* 53(4), 349-365. https://doi.org/10.1080/10408398.2010.535718.

Godfray, H.C.J., Beddington, J.R., Crute, I.R., Haddad, L., Lawrence, D., Muir, J.F. *et al.* (2010). Food security: The challenge of feeding 9 billion people. *Science* 327(5967), 812-818. https://doi.org/10.1126/science.1185383.

Godfray, H.C.J. and Garnett, T. (2014). Food security and sustainable intensification. *Philosophical Transactions of the Royal Society B: Biological Sciences* 369(1639), 20120273-20120273. https://doi.org/10.1098/rstb.2012.0273.

Gold, S., Trautrims, A. and Trodd, Z. (2015). Modern slavery challenges to supply chain management. *Supply Chain Management: An International Journal* 20(5), 485-494. https://doi.org/10.1108/SCM-02-2015-0046.

Gourdji, S.M., Sibley, A.M. and Lobell, D.B. (2013). Global crop exposure to critical high temperatures in the reproductive period: Historical trends and future projections. *Environmental Research Letters* 8(2). https://doi.org/10.1088/1748-9326/8/2/024041.

Government of Canada (2017). *Federal contaminated sites inventory*. [https://www.tbs-sct.gc.ca/fcsi-rscf/home-accueil-eng.aspx.

Graesser, J., Aide, T.M., Grau, H.R. and Ramankutty, N. (2015). Cropland/pastureland dynamics and the slowdown of deforestation in Latin America. *Environmental Research Letters* 10(3), 034017. https://doi.org/10.1088/1748-9326/10/3/034017.

Groenewegen, P.P., van den Berg, A.E., de Vries, S. and Verheij, R.A. (2006). Vitamin G: effects of green space on health, well-being, and social safety. *BMC Public Health* 6(149), 149. https://doi.org/10.1186/1471-2458-6-149.

Günther, F., Overduin, P.P., Sandakov, A.V., Grosse, G. and Grigoriev, M.N. (2013). Short- and long-term thermo-erosion of ice-rich permafrost coasts in the Laptev Sea region. *Biogeosciences* 10, 4297-4318. https://doi.org/10.5194/bg-10-4297-2013.

Gzik, A., Kuehling, M., Schneider, I. and Tschochner, B. (2003). Heavy metal contamination of soils in a mining area in South Africa and its impact on some biotic systems. *Journal of Soils and Sediments* 3(1), 29-34. https://doi.org/10.1007/BF02989466.

Haberl, H. (2015). Competition for land: A sociometabolic perspective. *Ecological Economics* 119, 424-431. https://doi.org/10.1016/j.ecolecon.2014.10.002.

Haberl, H., Fischer-Kowalski, M., Krausmann, F., Martinez-Alier, J. and Winiwarter, V. (2011). A socio-metabolic transition towards sustainability? Challenges for another Great Transformation. *Sustainable Development* 19(1), 1-14. https://doi.org/10.1002/sd.410.

Han, D., Wiesmeier, M., Conant, R.T., Kühnel, A., Sun, Z., Kögel-Knabner, I. *et al.* (2018). Large soil organic carbon increase due to improved agronomic management in the North China Plain from 1980s to 2010s. *Global Change Biology* 24, 987-1000. https://doi.org/10.1111/gcb.13898.

Hardell, L. (2003). Environmental Organochlorine Exposure and the Risk for Breast Cancer. In *Silent Invaders : Pesticides, Livelihoods, and Women's Health*. Jacobs, M. and Dinham, B. (eds.). London: Zed Books. chapter 16. 342. http://press.uchicago.edu/ucp/books/book/distributed/S/bo20852234.html

Henders, S. and Ostwald, M. (2014). Accounting methods for international land-related leakage and distant deforestation drivers. *Ecological Economics* 99, 21-28. https://doi.org/10.1016/j.ecolecon.2014.01.005.

Holden, S.T. and Ghebru, H. (2016). Land tenure reforms, tenure security and food security in poor agrarian economies: Causal linkages and research gaps. *Global Food Security* 10, 21-28. https://doi.org/10.1016/J.GFS.2016.07.002.

Holmes, M., Hughes, R., Jones, G., Sturman, V., Whiting, M., Wiltshire, J. *et al.* (2013). *A 2020 Vision for the Global Food System*. World Wide Fund for Nature. https://www.wwf.org.uk/sites/default/files/2013-04/2020vision_food_report_feb2013.pdf.

Howard, J.M. (2003). Measuring Gender Differences in Response to Pesticide Exposure. In *Silent Invaders : Pesticides, Livelihoods, and Women's Health*. Jacobs, M. and Dinham, B. (eds.). London: Zed Books. chapter 13. http://press.uchicago.edu/ucp/books/book/distributed/S/bo20852234.html

Hudson-Edwards, K. (2016). Tackling mine wastes. *Science* 352(6283), 288-290. https://doi.org/10.1126/science.aaf3354.

Inostroza, L., Baur, R. and Csaplovics, E. (2013). Urban sprawl and fragmentation in Latin America: A dynamic quantification and characterization of spatial patterns. *Journal of Environmental Management* 115, 87-97. https://doi.org/10.1016/j.jenvman.2012.11.007.

Intergovernmental Panel on Climate Change (2014). Climate Change 2014: Impacts, Adaptation, and Vulnerability. Part A: Global and Sectoral Aspects. In *Climate Change 2014: Impacts, Adaptation, and Vulnerability*. Intergovernmental Panel on Climate Change. 1132. http://www.ipcc.ch/pdf/assessment-report/ar5/wg2/WGIIAR5-PartA_FINAL.pdf

Intergovernmental Science-Policy Platform on Biodiversity and Ecosystem Services (2018). *Summary for Policymakers of the Assessment Report on Land Degradation and Restoration of the Intergovernmental Science- Policy Platform on Biodiversity and Ecosystem Services*. Scholes, R.J., Montanarella, L., Brainich, E., Brainich, E., Barger, N., ten Brink, B. *et al.* (eds.). Bonn: Intergovernmental Science-Policy Platform on Biodiversity and Ecosystem Services. https://www.ipbes.net/system/tdf/spm_3bi_ldr_digital.pdf?file=1&type=node&id=28335.

International Food Policy Research Institute (2016). *Global Food Policy Report*. Washington, DC: International Food Policy Research Institute. http://www.ifpri.org/cdmref/p15738coll2/id/130207/filename/130418.pdf.

Jat, R., Sahrawat, K. and Kassam, A. (eds.) (2013). *Conservation Agriculture: Global Prospects and Challenges*. Wallingford: CABI. https://www.cabi.org/cabebooks/ebook/20133423246.

Jayathilakan, K., Sultana, K., Radhakrishna, K. and Bawa, A.S. (2012). Utilization of byproducts and waste materials from meat, poultry and fish processing industries: A review. *Journal of food science and technology* 49(3), 278-293. https://doi.org/10.1007/s13197-011-0290-7.

Jones, A., Panagos, P., Barcelo, S., Bouraoui, F., Bosco, C., Dewitte, O. *et al.* (2012). *The State of Soil in Europe*. Copenhagen: European Environment Agency. http://publications.jrc.ec.europa.eu/repository/bitstream/JRC68418/lbna25186enn.pdf.

Kay, S., Peuch, J. and Franco, J. (2015). *Extent of Farmland Grabbing in the EU*. Brussels: European Parliament. http://www.europarl.europa.eu/RegData/etudes/STUD/2015/540369/IPOL_STU(2015)540369_EN.pdf.

Kelley, C.P., Mohtadi, S., Cane, M.A., Seager, R. and Kushnir, Y. (2015). Climate change in the Fertile Crescent and implications of the recent Syrian drought. *Proceedings of the National Academy of Sciences* 112(11), 3241-3246. https://doi.org/10.1073/pnas.1421533112.

Kelly, C., Ferrara, A., Wilson, G.A., Ripullone, F., Nolè, A., Harmer, N. *et al.* (2015). Community resilience and land degradation in forest and shrubland socio-ecological systems: Evidence from Gorgoglione, Basilicata, Italy. *Land Use Policy* 46, 11-20. https://doi.org/10.1016/J.LANDUSEPOL.2015.01.026.

Kennedy, C. and Hoornweg, D. (2012). Mainstreaming urban metabolism. *Journal of Industrial Ecology* 16(6), 780-782. https://doi.org/10.1111/j.1530-9290.2012.00548.x.

Khan, S. and Hanjra, M.A. (2008). Sustainable land and water management policies and practices: A pathway to environmental sustainability in large irrigation systems. *Land Degradation & Development* 19(5), 469-487. https://doi.org/10.1002/ldr.852.

Khoury, C.K., Bjorkman, A.D., Dempewolf, H., Ramirez-Villegas, J., Guarino, L., Jarvis, A. *et al.* (2014). Increasing homogeneity in global food supplies and the implications for food security. *Proceedings of the National Academy of Sciences of the United States of America* 111(11), 4001-4006. https://doi.org/10.1073/pnas.1313490111.

Kneebone, P. and Short, D. (2010). *Soil Contamination in West Africa*. London: Shift Soil Remediation. https://www.scribd.com/doc/71599035/Soil-Contamination-in-West-Africa.

Koh, L.P., Miettinen, J., Liew, S.C. and Ghazoul, J. (2011). Remotely sensed evidence of tropical peatland conversion to oil palm. *Proceedings of the National Academy of Sciences* 108(12), 5127-5132. https://doi.org/10.1073/pnas.1018776108.

Komatsu, H., Malapit, H.J.L. and Theis, S. (2018). Does women's time in domestic work and agriculture affect women's and children's dietary diversity? Evidence from Bangladesh, Nepal, Cambodia, Ghana, and Mozambique. *Food Policy* 79, 256–270. https://doi.org/10.1016/J.FOODPOL.2018.07.002.

Kosmas, C., Kairis, O., Karavitis, C., Ritsema, C., Salvati, L., Acikalin, S. *et al.* (2014). Evaluation and selection of indicators for land degradation and desertification monitoring: Methodological approach. *Environmental Management* 54(5), 951-970. https://doi.org/10.1007/s00267-013-0109-6.

Kummu, M., de Moel, H., Porkka, M., Siebert, S., Varis, O. and Ward, P.J. (2012). Lost food, wasted resources: Global food supply chain losses and their impacts on freshwater, cropland, and fertiliser use. *Science of The Total Environment* 438, 477-489. https://doi.org/10.1016/j.scitotenv.2012.08.092.

Lacerda, L.D., Bastos, W.R. and Almeida, M.D. (2012). The impacts of land use changes in the mercury flux in the Madeira River, Western Amazon. *Anais Da Academia Brasileira De Ciencias* 84(1), 69-78. https://doi.org/10.1590/S0001-37652012000100007.

Lacerda, L.D.D. (2003). Updating global Hg emissions from small-scale gold mining and assessing its environmental impacts. *Environmental Geology* 43(3), 308-314. https://doi.org/10.1007/s00254-002-0627-7.

Lantuit, H., Overduin, P.P. and Wetterich, S. (2012). Arctic Coastal erosion: A review. *Tenth International Conference on Permafrost, Salekhard, Russia*. Salekhard, 25 June - 29 June 2012. http://epic.awi.de/30700/

Lawry, S., Samii, C., Hall, R., Leopold, A., Hornby, D. and Mtero, F. (2014). The impact of land property rights interventions on investment and agricultural productivity in developing countries: A systematic review *Campbell Systematic Reviews* 2014(1). https://doi.org/10.4073/csr.2014.1.

Lawry, S., Samii, C., Hall, R., Leopold, A., Hornby, D. and Mtero, F. (2017). The impact of land property rights interventions on investment and agricultural productivity in developing countries: A systematic review. *Journal of Development Effectiveness* 9(1), 107. https://doi.org/10.1080/19439342.2016.1160947.

Le, Q.B., Nkonya, E. and Mirzabaev, A. (2016). Biomass Productivity-Based Mapping of Global Land Degradation Hotspots. In *Economics of Land Degradation and Improvement – A Global Assessment for Sustainable Development*. Nkonya, E., Mirzabaev, A. and von Braun, J. (eds.). Cham: Springer International Publishing. chapter 4. 55-84. https://link.springer.com/content/pdf/10.1007%2F978-3-319-19168-3_4.pdf

Lèbre, É. and Corder, G. (2015). Integrating industrial ecology thinking into the management of mining waste. *Resources* 4(4), 765-786. https://doi.org/10.3390/resources4040765.

Liew, Z., Wang, A., Bronstein, J. and Ritz, B. (2014). Job exposure matrix (jem)-derived estimates of lifetime occupational pesticide exposure and the risk of parkinson's disease. *Archives of Environmental and Occupational Health* 69(4), 241-251. https://doi.org/10.1080/19338244.2013.778808.

Lipinski, B., Hanson, C., Lomax, J., Kitinoja, L., Waite, R. and Searchinger, T. (2013). *Reducing Food Loss and Waste*. Washington, DC: World Resources Institute. http://wriorg.s3.amazonaws.com/s3fs-public/reducing_food_loss_and_waste.pdf.

Lobell, D.B. and Gourdji, S.M. (2012). The influence of climate change on global crop productivity. *Plant Physiology* 160(4), 1686-1697. https://doi.org/10.1104/pp.112.208298.

Lobell, D.B., Schlenker, W. and Costa-Roberts, J. (2011). Climate trends and global crop production since 1980. *Science* 333(5042), 616-620. https://doi.org/10.1126/science.1204531.

Lovejoy, T.E. and Nobre, C. (2018). Amazon tipping point. *Science Advances* 4(2). https://doi.org/10.1126/sciadv.aat2340.

MacDonald, G.K., Brauman, K.A., Sun, S., Carlson, K.M., Cassidy, E.S., Gerber, J.S. *et al.* (2015). Rethinking agricultural trade relationships in an era of globalization. *BioScience* 65(3), 275-289. https://doi.org/10.1093/biosci/biu225.

Machovina, B., Feeley, K.J. and Ripple, W.J. (2015). Biodiversity conservation: The key is reducing meat consumption. *Science of The Total Environment* 536, 419-431. https://doi.org/10.1016/j.scitotenv.2015.07.022.

Malapit, H.J.L., Kadiyala, S., Quisumbing, A.R., Cunningham, K. and Tyagi, P. (2015). Women's empowerment mitigates the negative effects of low production diversity on maternal and child nutrition in Nepal. *The Journal of Development Studies* 51(8), 1097–1123. https://doi.org/10.1080/00220388.2015.1018904.

McGrath, J.M. and Lobell, D.B. (2013). Regional disparities in the CO_2 fertilization effect and implications for crop yields. *Environmental Research Letters* 8(1). https://doi.org/10.1088/1748-9326/8/1/014054.

Met Office Hadley Centre and World Food Program (2018). *Food insecurity: Climate change – met office*. [https://www.metoffice.gov.uk/food-insecurity-index/ (Accessed: 11 April 2018).

Metternicht, G.I. and Zinck, J.A. (2003). Remote sensing of soil salinity: Potentials and constraints. *Remote Sensing of Environment* 85(1), 1-20. https://doi.org/10.1016/S0034-4257(02)00188-8.

Millennium Ecosystem Assessment (2003). *Ecosystems and Human Well-being: A Framework for Assessment*. Washington, DC: Island Press. http://pdf.wri.org/ecosystems_human_wellbeing.pdf.

Millennium Ecosystem Assessment (2004). *Living Beyond Our Means: Natural Assets and Human Well-being*. Washington, DC. https://www.millenniumassessment.org/documents/document.429.aspx.pdf.

Mottet, A., de Haan, C., Falcucci, A., Tempio, G., Opio, C. and Gerber, P. (2017). Livestock: On our plates or eating at our table? A new analysis of the feed/food debate. *Global Food Security* 14, 1-8. https://doi.org/10.1016/j.gfs.2017.01.001.

Msangi, S. and Rosegrant, M. (2011). World Agriculture in a Dynamically Changing Environment: IFPRI's Long-Term Outlook for Food and Agriculture. In *Looking Ahead in World Food and Agriculture: Perspectives to 2050*. Conforti, P. (ed.). Rome: Food and Agriculture Organization. 57-94. http://www.fao.org/docrep/pdf/012/ak542e/ak542e05.pdf

Muchomba, F.M. (2017). Women's land tenure security and household human capital: Evidence from Ethiopia's land certification. *World Development* 98, 310-324. https://doi.org/10.1016/j.worlddev.2017.04.034.

Murguía, D.I. (2015). *Global Area Disturbed and Pressures on Biodiversity by Large-Scale Metal Mining*. Kassel: Kassel University Press. http://www.uni-kassel.de/upress/online/OpenAccess/978-3-7376-0040-8.OpenAccess.pdf.

Murtaza, G. (2013) *Economic aspects of growing rice and wheat crops on salt-affected soils in the Indus Basin of Pakistan*. Institute of Soil and Environmental Sciences, University of Agriculture

Narh, P., Lambini, C., Sabbi, M., Pham, V. and Nguyen, T. (2016). Land sector reforms in Ghana, Kenya and Vietnam: A comparative analysis of their effectiveness. *Land* 5(2), 8. https://doi.org/10.3390/land5020008.

Nelson, G.C., Rosegrant, M.W., Palazzo, A., Gray, I., Ingersoll, C., Robertson, R. *et al.* (2010). *Food Security, Farming, and Climate Change to 2050: Scenarios, Results, Policy Options*. Research reports IFPRI. Washington, DC: International Food Policy Research Institute. http://www.ifpri.org/cdmref/p15738coll2/id/127066/filename/127277.pdf.

Newbigging, A.M., Yan, X. and Le, X.C. (2015). Cadmium in soybeans and the relevance to human exposure. *Journal of Environmental Sciences (China)* 37, 157-162. https://doi.org/10.1016/j.jes.2015.09.001.

Nguyen, T.T.X., Tomberlin, J.K. and Vanlaerhoven, S. (2015). Ability of black soldier fly (Diptera: Stratiomyidae) larvae to recycle food waste. *Environmental Entomology* 44(2), 406-410. https://doi.org/10.1093/ee/nvv002.

Nicholson, S.E. (2013). The West African Sahel: A review of recent studies on the rainfall regime and its interannual variability. *International Scholarly Research Notices*(453521). https://doi.org/10.1155/2013/453521.

Nicolopoulou-Stamati, P., Maipas, S., Kotampasi, C., Stamatis, P. and Hens, L. (2016). Chemical pesticides and human health: The urgent need for a new concept in agriculture. *Frontiers in Public Health* 4(148). https://doi.org/10.3389/fpubh.2016.00148.

Nkonya, E., Mirzabaev, A. and Von Braun, J. (eds.) (2016). *Economics of Land Degradation and Improvement – A Global Assessment for Sustainable Development*: Springer. https://link.springer.com/book/10.1007/978-3-319-19168-3#about.

Nolte, K., Chamberlain, W. and Giger, M. (2016). *International Land Deals for Agriculture: Fresh Insights from the Land Matrix; Analytical Report II*. Bern: Centre for Development and Environment, University of Bern. https://landmatrix.org/media/filer_public/ab/c8/abc8b563-9d74-4a47-9548-cb59e4809b4e/land_matrix_2016_analytical_report_draft_ii.pdf.

Oliver, M.A. and Gregory, P.J. (2015). Soil, food security and human health: A review. *European Journal of Soil Science* 66(2), 257-276. https://doi.org/10.1111/ejss.12216.

Olsson, L., Eklundh, L. and Ardö, J. (2005). A recent greening of the Sahel - Trends, patterns and potential causes. *Journal of Arid Environments* 63(3), 556-566. https://doi.org/10.1016/j.jaridenv.2005.03.008.

Overduin, P.P., Strzelecki, M.C., Grigoriev, M.N., Couture, N., Lantuit, H., St-Hilaire-Gravel, D. *et al.* (2014). Coastal changes in the Arctic. *Geological Society, London, Special Publications* 388(1), 103-129. https://doi.org/10.1144/SP388.13.

Paresi, M., Melchiorri, M., Siragusa, A. and Kemper, T. (2016). *Atlas of the Human Planet 2016: Mapping Human Presence on Earth with the Global Human Settlement Layer*. European Commission. http://publications.jrc.ec.europa.eu/repository/bitstream/JRC103150/atlas%20of%20the%20human%20planet_2016_online.pdf.

Pataki, D.E., Carreiro, M.M., Cherrier, J., Grulke, N.E., Jennings, V., Pincetl, S. *et al.* (2011). Coupling biogeochemical cycles in urban environments: Ecosystem services, green solutions, and misconceptions. *Frontiers in Ecology and the Environment* 9, 27-36. https://doi.org/10.1890/090220.

Pautasso, M., Döring, T.F., Garbelotto, M., Pellis, L. and Jeger, M.J. (2012). Impacts of climate change on plant diseases-opinions and trends. *European Journal of Plant Pathology* 133(1), 295-313. https://doi.org/10.1007/s10658-012-9936-1.

Pearce, F. (2016). *Common Ground: Securing Land Rights and Safeguarding the Earth*. Oxford: Oxfam, International Land Coalition and Rights and Resources Initiative. https://www.oxfamamerica.org/static/media/files/GCA_REPORT_EN_FINAL.pdf.

Place, F. (2009). Land tenure and agricultural productivity in Africa: A comparative analysis of the economics literature and recent policy strategies and reforms. *World Development* 37(8), 1326-1336. https://doi.org/10.1016/J.WORLDDEV.2008.08.020.

Plieninger, T., Dijks, S., Oteros-Rozas, E. and Bieling, C. (2013). Assessing, mapping, and quantifying cultural ecosystem services at community level. *Land Use Policy* 33, 118-129. https://doi.org/10.1016/J.LANDUSEPOL.2012.12.013.

Porkka, M., Kummu, M., Siebert, S. and Varis, O. (2013). From food insufficiency towards trade dependency: A historical analysis of global food availability. *PLoS ONE* 8(12), e82714. https://doi.org/10.1371/journal.pone.0082714.

Porter, J.R., Xie, L., Challinor, A.J., Cochrane, K., Howden, S.M., Iqbal, M.M. *et al.* (2014). Food Security and Food Production Systems. In *Climate Change 2014: Impacts, Adaptation, and Vulnerability. Part A: Global and Sectoral Aspects. Contribution of Working Group II to the Fifth Assessment Report of the Intergovernmental Panel on Climate Change*. Cambridge, MA: Cambridge University Press. chapter 7. 485-533. https://www.ipcc.ch/pdf/assessment-report/ar5/wg2/WGIIAR5-Chap7_FINAL.pdf

Pretty, J., Toulmin, C. and Williams, S. (2011). Sustainable intensification in African agriculture. *International Journal of Agricultural Sustainability* 9(1), 5-24. https://doi.org/10.3763/ijas.2010.0583.

Puma, M.J., Bose, S., Chon, S.Y. and Cook, B.I. (2015). Assessing the evolving fragility of the global food system. *Environmental Research Letters* 10, 024007. https://doi.org/10.1088/1748-9326/10/2/024007.

Qadir, M., Quillérou, E., Nangia, V., Murtaza, G., Singh, M., Thomas, R.J. *et al.* (2014). Economics of salt-induced land degradation and restoration. *Natural Resources Forum* 38(4), 282-295. https://doi.org/10.1111/1477-8947.12054.

Ramaswami, A., Boyer, D., Nagpure, A.S., Fang, A., Bogra, S., Bakshi, B. *et al.* (2017). An urban systems framework to assess the trans-boundary food-energy-water nexus: Implementation in Delhi, India. *Environmental Research Letters* 12(2), 025008. https://doi.org/10.1088/1748-9326/aa5556.

Ramaswami, A., Weible, C., Main, D., Heikkila, T., Siddiki, S., Duvall, A. *et al.* (2012). A social-ecological-infrastructural systems framework for interdisciplinary study of sustainable city systems: An integrative curriculum across seven major disciplines. *Journal of Industrial Ecology* 16(6), 801-813. https://doi.org/10.1111/j.1530-9290.2012.00566.x.

Ray, D.K., Mueller, N.D., West, P.C., Foley, J.A. and Meybeck, A. (2013). Yield trends are insufficient to double global crop production by 2050. *PLoS ONE* 8(6), e66428. https://doi.org/10.1371/journal.pone.0066428.

Ray, D.K., Ramankutty, N., Mueller, N.D., West, P.C. and Foley, J.A. (2012). Recent patterns of crop yield growth and stagnation. *Nature Communications* 3(1293), 1293. https://doi.org/10.1038/ncomms2296.

Reynolds, J.F., Grainger, A., Stafford Smith, D.M., Bastin, G., Garcia-Barrios, L., Fernández, R.J. *et al.* (2011). Scientific concepts for an integrated analysis of desertification. *Land Degradation & Development* 22(2), 166-183. https://doi.org/10.1002/ldr.1104.

Reynolds, J.F. and Smith, D.M. (eds.) (2002). *Global Desertification: Do Humans Cause Deserts?* 1st edn: Dahlem University Press. https://imedea.uib-csic.es/master/cambioglobal/Modulo_II_cod101606/M%C3%B3dulo%201.%20Presentaci%C3%B3n%20de%20la%20asignatura/reynolds%20do%20humans%20cause%20deserts.pdf.

Rights and Resources Initiative (2015). *Who Owns the World's Land? A Global Baseline of Formally Recognized Indigenous and Community Land Rights*. Washington, D.C.: Rights and Resources Initiative. https://rightsandresources.org/wp-content/uploads/GlobalBaseline_web.pdf.

Robinson, L.W., Ericksen, P.J., Chesterman, S. and Worden, J.S. (2015). Sustainable intensification in drylands: What resilience and vulnerability can tell us. *Agricultural Systems* 135, 133-140. https://doi.org/10.1016/j.agsy.2015.01.005.

Rosegrant, M.W., Paisner, M.S., Siet, M. and Witcover, J. (2001). *2020 Global Food Outlook*. Washington, DC: International Food Policy Research Institute. http://ebrary.ifpri.org/cdm/ref/collection/p15738coll2/id/57811#img_view_container.

Roser, M. and Ritchie, H. (2018). Yields and land use in agriculture. OurWorldInData.org https://ourworldindata.org/yields-and-land-use-in-agriculture.

Safriel, U.N. (2007). The Assessment of Global Trends in Land Degradation. In *Climate and Land Degradation*. Sivakumar M.V.K. and Ndiang'ui, N. (eds.). Heidelberg: Springer Berlin. chapter 1. 1-38. https://link.springer.com/chapter/10.1007/978-3-540-72438-4_1

Schaffartzik, A., Haberl, H., Kastner, T., Wiedenhofer, D., Eisenmenger, N. and Erb, K.H. (2015). Trading land: A review of approaches to accounting for upstream land requirements of traded products. *Journal of Industrial Ecology* 19(5), 703-714. https://doi.org/10.1111/jiec.12258.

Schinasi, L. and Leon, M. (2014). Non-hodgkin lymphoma and occupational exposure to agricultural pesticide chemical groups and active ingredients: A systematic review and meta-analysis. *International Journal of Environmental Research and Public Health* 11(4), 4449-4527. https://doi.org/10.3390/ijerph110404449.

Schlenker, W. and Roberts, M.J. (2009). Nonlinear temperature effects indicate severe damages to U.S. crop yields under climate change. *Proceedings of the National Academy of Sciences* 106(37), 15594-15598. https://doi.org/10.1073/pnas.0906865106.

Schoneveld, G.C. and German, L. (2014). Translating legal rights into tenure security: Lessons from the new commercial pressures on land in Ghana. *The Journal of Development Studies* 50(2), 187-203. https://doi.org/10.1080/00220388.2013.858129.

Seto, K.C., Fragkias, M., Güneralp, B. and Reilly, M.K. (2011). A meta-analysis of global urban land expansion. *PLoS ONE* 6, e23777. https://doi.org/10.1371/journal.pone.0023777.

Seto, K.C., Reenberg, A., Boone, C.G., Fragkias, M., Haase, D., Langanke, T. *et al.* (2012). Urban land teleconnections and sustainability. *Proceedings of the National Academy of Sciences of the United States of America* 109(20), 7687-7692. https://doi.org/10.1073/pnas.1117622109.

Sharmina, M., Hoolohan, C., Bows-Larkin, A., Burgess, P.J., Colwill, J., Gilbert, P. *et al.* (2016). A nexus perspective on competing land demands: Wider lessons from a UK policy case study. *Environmental Science & Policy* 59, 74–84. https://doi.org/10.1016/j.envsci.2016.02.008.

Smith, P., Davis, S.J., Creutzig, F., Fuss, S., Minx, J., Gabrielle, B. *et al.* (2015). Biophysical and economic limits to negative CO2 emissions. *Nature Climate Change* 6, 42-50. https://doi.org/10.1038/nclimate2870.

Soane, B.D., Ball, B.C., Arvidsson, J., Basch, G., Moreno, F. and Roger-Estrade, J. (2012). No-till in northern, western and south-western Europe: A review of problems and opportunities for crop production and the environment. *Soil and Tillage Research* 118, 66-87. https://doi.org/10.1016/j.still.2011.10.015.

Solecki, W. and Marcotullio, P.J. (2013). Climate Change and Urban Biodiversity Vulnerability. In *Urbanization, Biodiversity and Ecosystem Services: Challenges and Opportunities: A Global Assessment*. Dordrecht: Springer Netherlands. chapter 25. 485–504. https://link.springer.com/content/pdf/10.1007%2F978-94-007-7088-1_25.pdf

Sonter, L.J., Herrera, D., Barrett, D.J., Galford, G.L., Moran, C.J. and Soares-Filho, B.S. (2017). Mining drives extensive deforestation in the Brazilian Amazon. *Nature Communications* 8, 1013. https://doi.org/10.1038/s41467-017-00557-w.

Sonter, L.J., Moran, C.J., Barrett, D.J. and Soares-Filho, B.S. (2014). Processes of land use change in mining regions. *Journal of Cleaner Production* 84, 494-501. https://doi.org/10.1016/j.jclepro.2014.03.084.

Sturm, R. and Cohen, D. (2014). Proximity to urban parks and mental health. *Journal of Mental Health Policy and Economics* 17(1), 19-24. https://www.ncbi.nlm.nih.gov/pmc/articles/PMC4049158/.

Taylor, M.P., Camenzuli, D., Kristensen, L.J., Forbes, M. and Zahran, S. (2013b). Environmental lead exposure risks associated with children's outdoor playgrounds. *Environmental Pollution* 178, 447-454. https://doi.org/10.1016/j.envpol.2013.03.054.

Taylor, P.C., Cai, M., Hu, A., Meehl, J., Washington, W., Zhang, G.J. *et al.* (2013a). A decomposition of feedback contributions to polar warming amplification. *Journal of Climate* 26(21), 7023-7043. https://doi.org/10.1175/JCLI-D-12-00696.1.

Thangavel, P. and Sridevi, G. (2017). Soil Security: A Key Role for Sustainable Food Productivity. In *Sustainable Agriculture towards Food Security*. Dhanarajan, A. (ed.). Singapore: Springer Singapore. 309-325. https://link.springer.com/chapter/10.1007/978-981-10-6647-4_16

The Economics of Land Degradation (2013). *Economics of Land Degradation Initiative: A Global Strategy for Sustainable Land Management; The Rewards of Investing in Sustainable Land Management* Bonn. https://macsphere.mcmaster.ca/bitstream/11375/15701/1/ELD%20Initiative_2013%20-%20The%20rewards%20of%20investing%20in%20sustainable%20land%20management%20%20Interim%20Report_Web-Version4.pdf.

The Economics of Land Degradation (2015). *Report for Policy Makers - Key Facts and Figures*. Bonn. http://www.eld-initiative.org/fileadmin/pdf/Key_facts_and_figures_-_Report_for_policy_and_decision_makers2015.pdf.

The Global Food Security Programme (2015). *Final Project Report from the UK-US Taskforce on Extreme Weather and Global Food System Resilience*. Wiltshire. https://www.foodsecurity.ac.uk/publications/extreme-weather-resilience-global-food-system.pdf.

The Land Matrix Global Observatory (2018). *Land Matrix*. https://landmatrix.org/en/.

The Royal Society (2008). *Sustainable Biofuels: Prospects and Challenges*. Policy Document. London: The Royal Society. https://royalsociety.org/~/media/Royal_Society_Content/policy/publications/2008/7980.pdf.

Tilman, D., Balzer, C., Hill, J. and Befort, B.L. (2011). Global food demand and the sustainable intensification of agriculture. *Proceedings of the National Academy of Sciences of the United States of America* 108(50), 20260-20264. https://doi.org/10.1073/pnas.1116437108.

Tomei, J. and Ravindranath, D. (2018). Governing Land in the Global South. In *Routledge Handbook of the Resource Nexus*. Bleischwitz, R., Hoff, H., Spataru, C., van der Voet, E. and VanDeveer, S.D. (eds.). Routledge. https://www.routledgehandbooks.com/doi/10.4324/9781315560625-22

Unger, E.-M., Zevenbergen, J. and Bennett, R. (2017). On the need for pro-poor land administration in disaster risk management. *Survey Review* 49(357), 437-448. https://doi.org/10.1080/00396265.2016.1212160.

United Nations (2007). *United Nations Declaration on the Rights of Indigenous People. (A/RES/61/295)*. [https://www.un.org/development/desa/indigenouspeoples/declaration-on-the-rights-of-indigenous-peoples.html.

United Nations (2014). *World Urbanization Prospects: The 2014 Revision*. United Nations. New York, NY: United Nations. https://esa.un.org/Unpd/Wup/Publications/Files/WUP2014-Highlights.pdf.

United Nations (2015). *World Population Prospects: the 2015 Revision, Key Findings and Advance Tables*. United Nations. New York, NY: United Nations. https://esa.un.org/Unpd/wpp/Publications/Files/Key_Findings_WPP_2015.pdf.

United Nations (2016). *Global Sustainable Development Report*. New York, NY. https://sustainabledevelopment.un.org/content/documents/2328Global%20Sustainable%20development%20report%202016%20(final).pdf.

United Nations Convention to Combat Desertification (1994). *United Nations Convention to Combat Desertification: In Those Countries Experiencing Serious Drought and/or Desertification Particularly in Africa*. Bonn https://www.unccd.int/sites/default/files/relevant-links/2017-01/An%20explanatory%20leaflet.pdf

United Nations Convention to Combat Desertification (2017). *Global Land Outlook*. Bonn: United Nations Convention to Combat Desertification. https://knowledge.unccd.int/sites/default/files/2018-06/GLO%20English_Full_Report_rev1.pdf.

United Nations Environment Programme (2010). *Latin America and the Caribbean: Environmental Outlook*. Nairobi. http://wedocs.unep.org/bitstream/handle/20.500.11822/8663/-Global_environment_outlook_Latin_America_and_the_Caribbean_GEO_LAC_3-2010Latinin_America_and_the_Caribbean_-_Environment_Outlook_3.pdf.pdf?sequence=3&isAllowed=y.

United Nations Environment Programme (2015). *Global Waste Management Outlook*. Nairobi. http://apps.unep.org/publications/index.php?option=com_pub&task=download&file=011782_en.

United Nations environment Programme (2016a). *Global Gender and Environment Outlook: The Critical Issues*. Nairobi. http://web.unep.org/sites/default/files/ggeo/ggeo_summary_report_final.pdf.

United Nations Environment Programme (2016b). *Unlocking the Sustainable Potential of Land Resources: Evaluation Systems, Strategies and Tools*. Nairobi. https://wedocs.unep.org/bitstream/handle/20.500.11822/7710/-Unlocking_the_sustainable_potential_of_land_resources_Evaluating_systems_strategies_and_tools-2016Unlocking_Land_Resources_full_report.pdf.pdf?sequence=3&isAllowed=y.

United States Environmental Protection Agency (2016). *Superfund: National priorities list (NPL)*. [https://www.epa.gov/superfund/superfund-national-priorities-list-npl.

Utuk, I.O. and Daniel, E.E. (2015). Land degradation : A threat to food security : A global assessment. *Journal of Environment and Earth Science* 5(8), 13-22. https://www.iiste.org/Journals/index.php/JEES/article/view/22020/22057.

Van Liedekerke, M., Prokop, G., Rabl-Berger, S., Kibblewhite, M. and Louwagie, G. (2014). *Progress in Management of Contaminated Sites in Europe*. European Commission. http://publications.jrc.ec.europa.eu/repository/bitstream/JRC85913/lbna26376enn.pdf.

van Vliet, J., Eitelberg, D.A. and Verburg, P.H. (2017). A global analysis of land take in cropland areas and production displacement from urbanization. *Global Environmental Change* 43, 107-115. https://doi.org/10.1016/j.gloenvcha.2017.02.001.

Veit, P. and Reytar, K. (2017). 'By the Numbers: Indigenous and Community Land Rights'. 20 March 2017 https://www.wri.org/blog/2017/03/numbers-indigenous-and-community-land-rights

Verma, R. (2014). Land grabs, power, and gender in east and southern Africa: So, what's new? *Feminist Economics* 20, 52-75. https://doi.org/10.1080/13545701.2014.897739.

Watts, M. (2007). *Pesticides and Breast Cancer: A Wakeup Call*. Penang: Pesticide Action Network Asia and the Pacific. http://files.panap.net/resources/Pesticides-and-Breast-Cancer-A-Wake-Up-Call.pdf.

Watts, M. (2013). *Breast Cancer, Pesticides and You*. Penang: Pesticide Action Network Asia and the Pacific http://files.panap.net/resources/Breast-cancer-pesticides-and-you.pdf.

Watts, M. and Williamson, S. (2015). *Replacing Chemicals with Biology*. Penang: Pesticide Action Network Asia and the Pacific. https://www.panna.org/sites/default/files/Phasing-Out-HHPs-with-Agroecology.pdf.

White, B., Park, C. and Mi Young, J. (2015). The Gendered Political Ecology of Agrofuels Expansion. In *The Political Ecology of Agrofuels*. Engels, D. and Pye, O. (eds.). London: Routledge. chapter 4. 53-69. https://www.taylorfrancis.com/books/e/9781317747444/chapters/10.4324%2F9781315795409-4

White, M.P., Alcock, I., Wheeler, B.W. and Depledge, M.H. (2013). Would you be happier living in a greener urban area? A fixed-effects analysis of panel data. *Psychological Science* 24(6), 920-928. https://doi.org/10.1177/0956797612464659.

Wiebe, K. (2003). *Linking Land Quality, Agricultural Productivity, and Food Security*. Agricultural Economic Report. Washington, DC: United States Department of Agriculture. https://www.ers.usda.gov/webdocs/publications/41563/18547_aer823fm_1_.pdf?v=41061.

Wiesmeier, M., Poeplau, C., Sierra, C.A., Maier, H., Frühauf, C., Hübner, R. *et al.* (2016). Projected loss of soil organic carbon in temperate agricultural soils in the 21st century: Effects of climate change and carbon input trends. *Scientific Reports* 6(32525). https://doi.org/10.1038/srep32525.

Wild, C.P. and Gong, Y.Y. (2010). Mycotoxins and human disease: a largely ignored global health issue. *Carcinogenesis* 31(1), 71-82. https://doi.org/10.1093/carcin/bgp264.

Wilson, G., Quaranta, G., Kelly, C. and Salvia, R. (2016). Community resilience, land degradation and endogenous lock-in effects: Evidence from the Alento region, Campania, Italy. *Journal of Environmental Planning and Management* 59(3), 518-537. https://doi.org/10.1080/09640568.2015.1024306.

Wilson, G.A., Kelly, C.L., Briassoulis, H., Ferrara, A., Quaranta, G., Salvia, R. *et al.* (2017). Social memory and the resilience of communities affected by land degradation. *Land Degradation & Development* 28(2), 383-400. https://doi.org/10.1002/ldr.2669.

World Health Organization (2017). *Urban Green Space Interventions and Health*. Copenhagen: World Health Organization. http://www.euro.who.int/__data/assets/pdf_file/0010/337690/FULL-REPORT-for-LLP.pdf?ua=1.

World Health Organization and United Nations Environment Programme (1990). *Public Health Impact of Pesticides used in Agriculture*. Geneva: World Health Organization. http://apps.who.int/iris/bitstream/handle/10665/39772/9241561394.pdf?sequence=1&isAllowed=y.

World Resources Institute (2018). *Global forest watch*. [https://www.globalforestwatch.org/ (Accessed: 10 June 2018).

World Wildlife Fund (2018). *Deforestation: Overview*. [World Wildlife Fund https://www.worldwildlife.org/threats/deforestation.

Xu, D., Kang, X., Qiu, D., Zhuang, D. and Pan, J. (2009). Quantitative assessment of desertification using Landsat data on a regional scale - a case study in the Ordos Plateau, China. *Sensors* 9(3), 1738-1753. https://doi.org/10.3390/s90301738.

Zewdie, W. and Csaplovies, E. (2015). Remote Sensing based multi-temporal land cover classification and change detection in northwestern Ethiopia. *European Journal of Remote Sensing* 48(1), 121-139. https://doi.org/10.5721/EuJRS20154808.

Zhu, Z., Piao, S., Myneni, R.B., Huang, M., Zeng, Z., Canadell, J.G. *et al.* (2016). Greening of the earth and its drivers. *Nature Climate Change* 6, 791-795. https://doi.org/10.1038/nclimate3004.

Chapter 9

Freshwater

© Shutterstock/Mikadun

Coordinating Lead Authors: Erica Gaddis (Utah Department of Environmental Quality), Anna Maria Grobicki (Food and Agriculture Organization of the United Nations [FAO]), Rowena Hay (Umvoto), Gavin Mudd (RMIT University), Walter Rast (Meadows Center for Water and the Environment, Texas State University)

GEO Fellows: Beatriz Rodríguez-Labajos (Universitat Autònoma de Barcelona), Jaee Sanjay Nikam (Arizona State University)

Executive summary

Freshwater mobilizes and amplifies the risks to human health and the environment associated with human activities (*established but incomplete*). The global water cycle integrates the impacts of population growth, agriculture, economic development, urbanization, industrialization, deforestation and climate change. All of these impacts affect freshwater quality and quantity. Hence, freshwater is now simultaneously a public good and a risk multiplier, affecting human and ecosystem health through pollutants and through climate change, which is intensifying storms, floods, droughts and desertification of land. Improved governance of every aspect of the water cycle is urgently needed in order to prevent, mitigate and manage these increasing risks. {9.2}

The per capita availability of freshwater in the global water cycle is decreasing with population growth, coupled with the associated agricultural, industrial and energy requirements (*established but incomplete*), while the continents are becoming drier in many places due to climate change impacts. {9.2}

Increasing numbers of people are at risk of 'slow-onset disasters' such as water scarcity, droughts and famine. Such events sometimes lead to increased migration and social conflicts (*well established*) {4.2}. The rising severity and frequency of water-related disasters pose growing risks to social and economic stability, as well as to ecosystems and their life-supporting ecosystem goods and services. There is evidence that water scarcity drives greater competition for available resources, reflected in food insecurity, prices and trade (*established but incomplete*). {9.2}

Groundwater comprises a much larger freshwater volume than surface water. It is increasingly important for water security in many countries and regions (*established but incomplete*). Some major aquifers at subregional and regional levels are threatened by poor management, resulting in unsustainable abstraction levels, groundwater pollution and issues of saline intrusion. {9.4}

Approximately 1.4 million people die annually from diseases associated with pathogen-polluted drinking water and inadequate sanitation, with many millions more becoming ill (*well established*)**.** Some 2.3 billion people still do not have access to safe sanitation. The total global disease burden could be cut by up to an estimated 10 per cent with improved drinking water quality and access, sanitation, hygiene and integrated water resources management. {9.5}

Human illnesses and deaths due to antibiotic- and antimicrobial-resistant infections are increasing rapidly and are projected to become a main cause of death worldwide by 2050 (*well established*). Antibiotics reach the aquatic environment from a wide range of sources, including treated and untreated human waste, agriculture, animal husbandry and aquaculture. Antibiotic-resistant bacteria are now found in both source water and treated drinking water worldwide. {9.5}

New pollutants not easily removed by current wastewater treatment technologies are of emerging concern, including certain veterinary and human pharmaceuticals, pesticides, antimicrobial disinfectants, flame retardants, detergent metabolites and microplastics (*well established*). Endocrine-disrupting chemicals are of particular concern as they are now widely distributed through the freshwater system on all continents. Their long-term impacts on human health include fetal underdevelopment, child neurodevelopment and male infertility. {9.7}

Freshwater ecosystems are disappearing rapidly, representing a high rate of loss of biodiversity and ecosystem services (*well-established*). Wetlands are the natural areas most affected by increasing urbanization, agricultural expansion and deforestation. Approximately 40 per cent of the world's wetlands were lost between 1997 and 2011, and this rate of loss continues. This is linked to an 81 per cent freshwater species population decline over the same period, the highest for any type of habitat (*likely*). The annual economic cost of wetland ecosystem losses between 1996 and 2011 was estimated at US$2.7 trillion. {9.6}

Peatlands (one type of wetland) store more carbon than all the world's forests combined (*established but incomplete*)**.** Climate change is thawing permafrost in boreal peatlands in and around the Arctic Circle, causing increased carbon emissions. Increased drainage and agricultural use of tropical peatlands cause wildfires and release significant quantities of carbon dioxide and methane as greenhouse gases. Altogether, about 15 per cent of peatlands worldwide had been drained by 2015, and currently contribute approximately 5 per cent of annual global carbon emissions. {9.6}

SDG 6 water targets can be realized through engaging the public, private and non-governmental sectors, civil society and local actors in practising effective, efficient and transparent water resources governance (*well established*). {9.9}

Promoting water-use efficiency, water recycling and rainwater harvesting is becoming increasingly important to ensure greater water security and more equitable water allocation for different users and uses (*well established*). {9.9}

The agricultural sector, the largest consumer of freshwater globally, needs substantial improvements in water-use efficiency and productivity (*well established*). The industrial and mining sectors also have strong potential for increasing water-use efficiency, recycling and reuse, as well as limiting water pollution. {9.9}

Limited capacity currently exists to control long-term impacts of aquifer overabstraction and pollution in many locations (*established but incomplete*). Monitoring, modelling and managing aquifer systems are essential to implement sound aquifer and integrated water resources management. Salinization of aquifers resulting from subsidence in river deltas is a complex catchment and coastal urbanization issue (SDG 11), but saline intrusion into coastal aquifers can be controlled by managed aquifer recharge (*well established*). {9.9}

Efficient water use requires water-sensitive urban design of water infrastructure, including conjunctive surface and groundwater development and the promotion of managed aquifer recharge (*well established*). Together with investment in wastewater treatment and recycling, these approaches support water quantity and quality management, and promote drought risk reduction and resilient urban water supplies. At the same time, the provision of drinking water supply and sanitation services to all, as well as leakage control from bulk water supplies, are still challenges in many cities worldwide. {9.9}

9.1 Introduction and priority issues

Freshwater is essential for the health and well-being of people, animals, plants, and aquatic and terrestrial ecosystems. The global water cycle is the most important component of weather and climate systems and it is accelerating because of climate change (**Figure 9.1**) (Stocker and Raible 2005; Huntington 2006; Organisation for Economic Co-operation and Development [OECD] 2016, pp. 5-6). The proportion of total freshwater that is readily available as surface water in rivers, lakes and wetlands is 0.4 per cent and decreasing dramatically. Increased floods and droughts (Huntington 2006) and the loss of glaciers (Gao *et al.* 2011; Yao *et al.* 2012; Rodell *et al.* 2018) result in direct and indirect impacts to human and ecosystem health (e.g. Holloway 2003, p. 2; Liu *et al.* 2005; Wang, Wang and Tong 2016; Liu *et al.* 2018).

Water is implicated in most of the Sustainable Development Goals (SDGs), being crucial to food security (SDG 2), health and well-being (SDG 3), energy security (SDG 7), sustainable cities (SDG 11), responsible consumption and production (SDG 12), climate impacts (SDG 13), life below water (SDG 14) and terrestrial biodiversity (SDG 15). Most other SDGs are not achievable without adequate supplies of good quality freshwater (United Nations Water [UN-Water] 2016, p. 9). The sixth Global Environment Outlook (GEO-6) highlights links between water (SDG 6) and health (SDG 3). Degradation of water quality impacts human and ecosystem health (United Nations Environment Programme [UNEP] 2017). Nearly 1.7 million people die annually from preventable diarrhoeal diseases (Lozano *et al.* 2013; Sevilimedu *et al.* 2016, p. 637).

9.2 Pressures on freshwater

Multiple pressures on water resulting from the global drivers of environmental change (see chapter 2) are evident in the rapid deterioration in freshwater quantity and quality in different regions. This is exacerbated in certain regions by pressures from ongoing conflicts, human migration and the cumulative impacts of increasing frequency and severity of droughts,

Figure 9.1: Global hydrological fluxes and storages (expressed in 1,000 km³ per year), illustrating natural and anthropogenic cycles

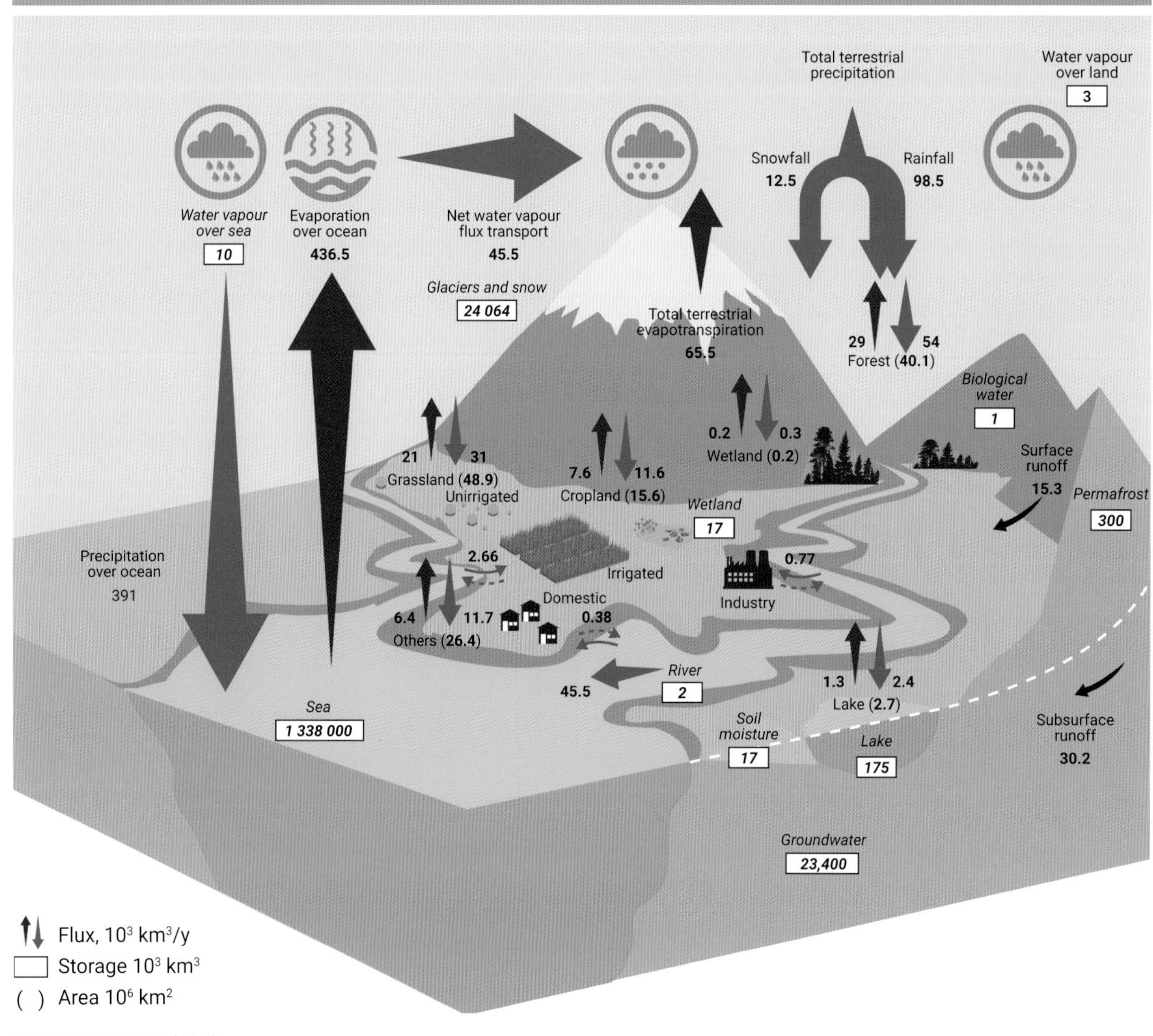

Source: Oki and Kanae (2006).

floods and storm surges (Intergovernmental Panel on Climate Change [IPCC] 2014). Impacts of natural and human-made disasters are compounded by unsustainable use of freshwater and related ecosystems, which reduces the resilience of the ecosystems (Sheffer *et al.* 2001; Holling and Gunderson 2002). Recent satellite data show that freshwater bodies are rapidly disappearing in many irrigated agriculture areas due to this combination of climate change and overabstraction (Rodell *et al.* 2018).

9.2.1 Climate change

The global water cycle is intimately tied to our changing climate. As the planet warms, the water cycle accelerates, with multiple changes in precipitation patterns putting pressure on freshwater ecosystems (Oki and Kanae 2006). The quantity of salt water is now increasing relative to freshwater due to global warming, land-use changes, melting ice and snow reserves, pumping of groundwater, drying of the continents, and rising sea levels (Bates *et al.* eds. 2008).

Many areas now receive less precipitation than in the past, while others receive more, with most regions experiencing increasingly unpredictable and variable temperature and precipitation patterns. Polar regions and high mountain regions are warming much faster than other parts of the world, with unforeseeable consequences (see Section 4.3.2). A 12 per cent increase in record-breaking high rainfall events occurred globally during 1981-2010 (Lehmann, Coumou and Frieler 2015). By contrast, there is evidence of increasing drought severity in Europe (Vicente-Serrano *et al.* 2014), with historical records indicating increased aridity over many areas since the 1950s (Dai 2011).

Global climate change interacts with weather and local-scale climate effects, as well as unsustainable water uses and diversions, leading to dramatic impacts such as shrinking freshwater bodies (e.g. Lake Chad, see **Box 9.1**; the Aral Sea; the disappearing wetlands of Islamic Republic of Iran [e.g. Lake Urmia] and the Iraqi Marshes; and even the Caspian Sea (Rodell *et al.* 2018)).

Box 9.1: Impacts of climate change on disappearing lakes and wetlands

Lakes and wetlands are important in regulating water cycles, for example by creating more moderate local climates (Kodama, Eaton and Wendler 1983; Laird *et al.* 2001; Saaroni and Ziv 2003; McInnes 2016; Dai *et al.* 2018). They warm up during the day and lose heat more slowly at night than the land surface, reducing temperature extremes in their basins. Through evaporation, they provide water vapour and precipitation during winter, and they cool and stabilize the local climate in summer. Urban wetlands have been shown to provide a local cooling effect of at least 1-3°C (Filho *et al.* 2017).

Climate change alters water cycles over lakes, wetlands and other standing (lentic) water systems, reducing the quantity of fresh water and waterbody surface area. A warmer climate increases evaporation over the waterbody and adjacent land, but a warmer atmosphere also takes more time to become saturated with water to subsequently produce rainfall. Thus, moisture evaporated from a waterbody may blow away before it can fall as rainfall in its own basin. The basin then becomes drier, with less run-off into the waterbody and associated rivers and wetlands, increasing the need for agricultural irrigation water. These factors collectively accelerate the shrinking of a waterbody, as illustrated in the case of Lake Chad (below), which has lost 90 per cent of its surface area, with an enormous loss of its associated biodiversity, especially fish, and loss of livelihoods for the millions of people dependent upon the lake. Human water use is estimated to account for 50 per cent of the shrinkage and climate change for the remainder (Coe and Foley 2001; Gao *et al.* 2011). The resulting change in microclimate establishes a cycle that further contributes to the drying and desertification of the continent and intensifies the impacts of global climate change.

Figure 9.2: Shrinkage of Lake Chad

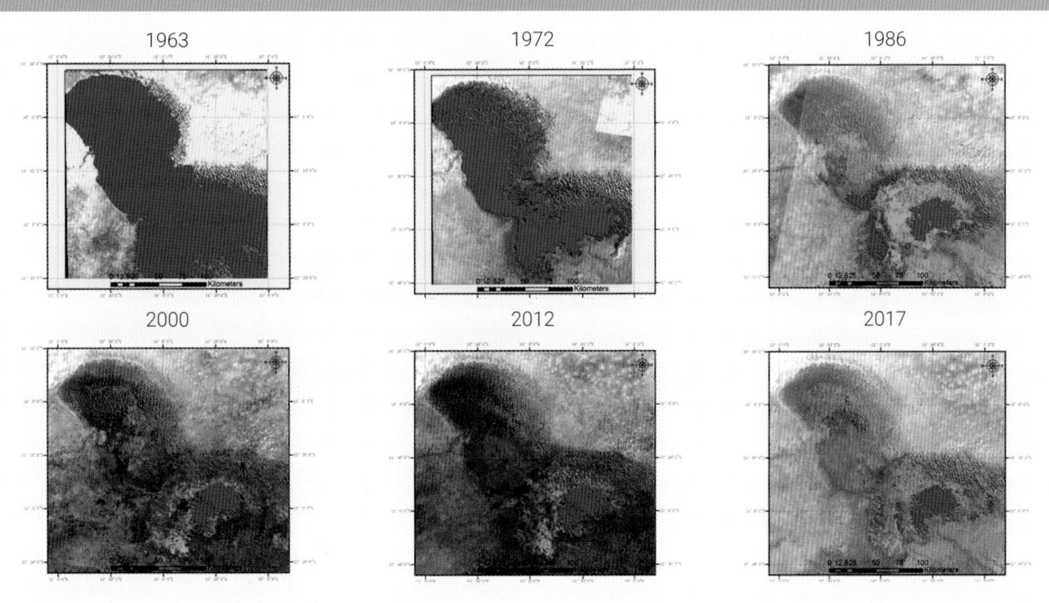

Sources: Hansen *et al.* (2013); Guzinski *et al.* (2014).

Too much rainfall brings pollution, soil erosion, avalanches and mud slides which, together with floods, tornadoes and cyclones, are responsible for much physical damage to infrastructure, loss of life and injury. Too little rainfall causes drought, extreme wildfires, sandstorms, soil degradation and increased competition over water sources, often leading to the accelerated shrinkage and loss of these goods. Collectively, these realities and risks have grave socio-political, economic, environmental and ecological implications, making better management and governance of freshwater resources an imperative.

9.3 Water and land use

Growing cities and agricultural intensification are increasingly depleting both surface water and aquifers. Wetlands are being drained, and many rivers, lakes and ponds are vanishing in water-scarce regions. Land-use changes result in surface hardening of natural areas, reducing infiltration and aquifer recharge, while increasing water run-off and pollution. Land degradation and deforestation also cause increased run-off, carrying eroded sediment through rivers into oceans (see Section 8.4.2). In areas experiencing large-scale deforestation, the likelihood of precipitation events is decreasing and soil erosion is increasing (Birkinshaw *et al.* 2011; Ellison, Futter and Bishop 2012).

Agriculture is responsible for an average 70 per cent of global water withdrawals (UN-Water 2017). Industrial processes and energy generation increasingly compete with agriculture and cities for available water. However, much energy water demand is for non-consumptive uses (e.g. cooling) (UNEP 2012a).

The interconnections between water, energy security and food security have identified tensions and trade-offs between them requiring careful scrutiny and consideration (Rosengrant *et al.* 2009). This nexus becomes especially important when considering drivers such as urbanization, population, economic growth, technology and innovation (Bleischwitz *et al.* 2018).

9.4 Global state and trends of freshwater

9.4.1 Water quantity

Geographic variations, coupled with climate change, result in uneven distribution of rainfall and freshwater sources, with deserts and rainforests highlighting these water availability extremes (**Figure 9.1** and **Figure 9.4**). Groundwater is the major drinking water source for the majority of people globally, particularly in arid regions and during drought. The estimated available renewable groundwater resource in Africa is more than 100 times that of total annual renewable surface-water resources (MacDonald *et al.* 2012, p. 5). However, deeper aquifer water is constrained by exploration and abstraction costs. Abstraction of very ancient 'fossil groundwater' is unsustainable, because this is not a renewable resource.

9.4.2 Water withdrawals

Human and environmental water demands vary spatially and culturally across rural and urban areas. While an average of 70 per cent of water withdrawals worldwide are for the agricultural sector, this varies widely across regions and countries (Hoekstra and Mekonnen 2012, p. 3232; Food and Agriculture Organization of the United Nations [FAO] 2016; UN-Water 2017). South-East Asia uses more than 80 per cent of its available freshwater for agriculture (FAO 2016).

The North American region has the highest per capita freshwater use (Hoekstra and Mekonnen 2012, p. 3232; UNEP 2016a, p. 71), although increased water-use efficiency is helping to lower demand, despite population and economic growth (UNEP 2016a, p. 71). Water withdrawals by all sectors in the United States of America (**Figure 9.3**) illustrate high water usage for cooling in electricity production.

Groundwater is increasingly important globally, representing estimated withdrawals of about 982 km^3 (Margat and van der Gun 2013), equivalent to nearly 33 per cent of total water withdrawals (Seibert *et al.* 2010, p. 1863; Famiglietti 2014, p. 945). Since conventional groundwater withdrawal technology is easily accessible to landowners, extraction is highly decentralized. Groundwater in confined artesian basins (Bundesanstalt für Geowissenschaften und Rohstoffe [BGR] 2008) can be accessed at depths of up to 2 km, and often provides a strategic water resource, especially during droughts (e.g. Great Artesian Basin, Australia [GABCC] 2016); Table Mountain Group, South Africa) (Hay and Hartnady *et al.* 2001; Weaver *et al.* 2002; Blake *et al.* 2010).

Industries that abstract from aquifers include industrial agriculture, mining, geothermal energy and ground-source heat pumps, disposal and/or storage of hazardous wastes (e.g. landfills, nuclear waste), fluid injection (e.g. oil and gas extraction through hydraulic fracturing or 'fracking' and associated wastewater reinjection), and underground construction activities. Such pressures are leading inexorably to stronger competition/interactions between the different industries, with sometimes unforeseen consequences.

Figure 9.3: United States water withdrawals from all sources (1950-2010)

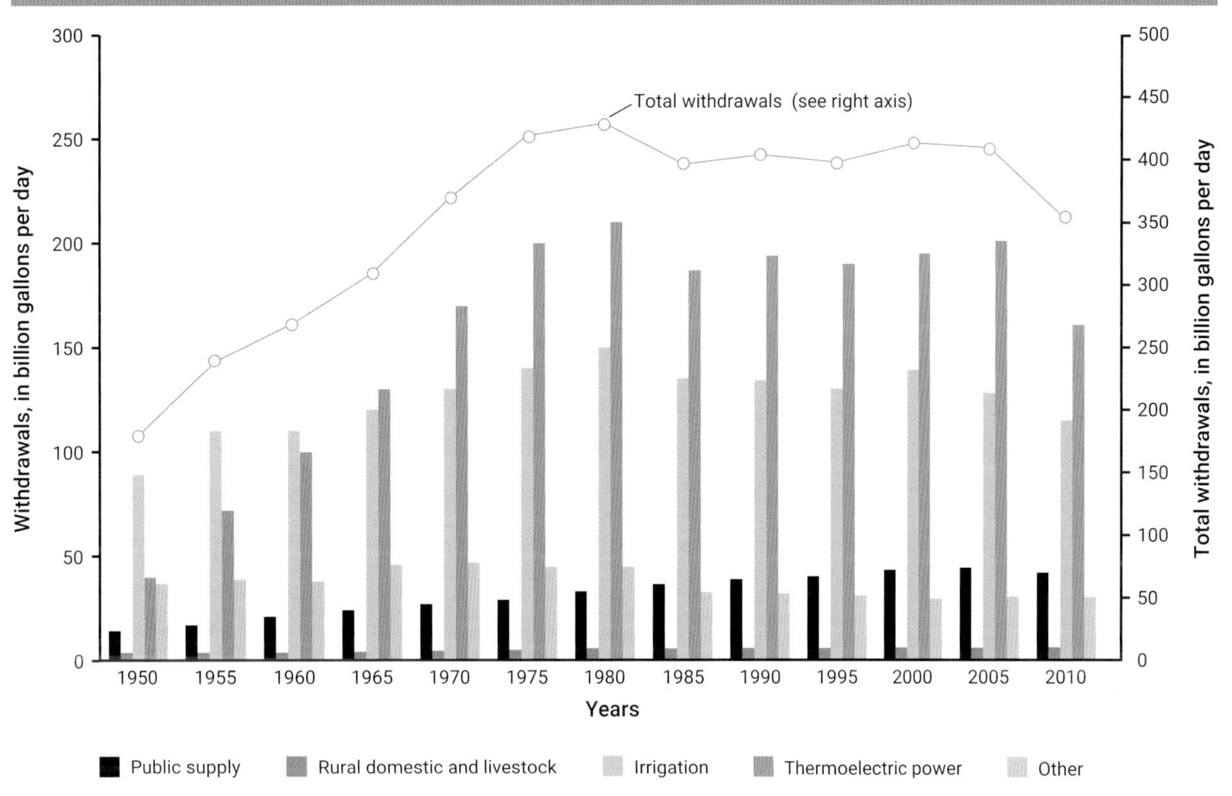

Note: 1 billion gallons = 3.8 million m³.

Source: Maupin *et al*. (2014, p. 46).

Figure 9.4: Global hydrogeological map illustrating various aquifers and groundwater resources

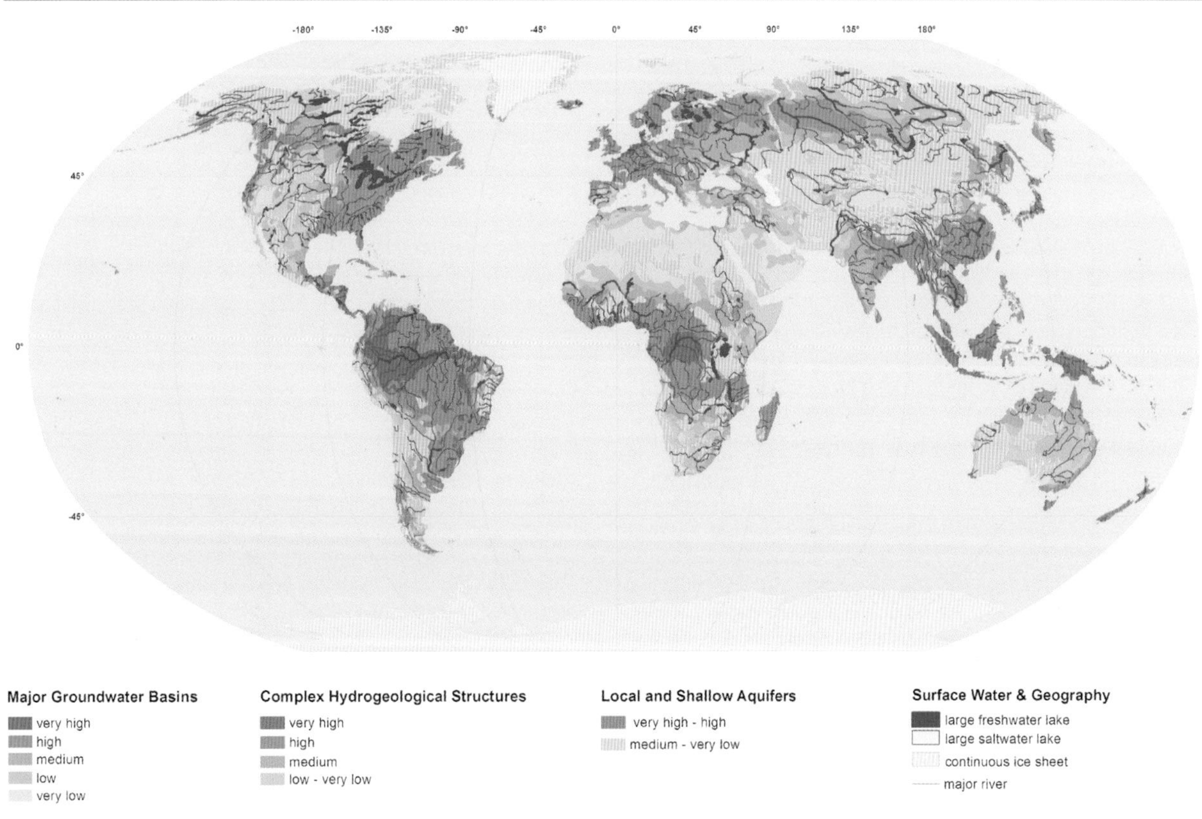

Source: BGR and United Nations Educational, Scientific and Cultural Organization [UNESCO] (2008).

Figure 9.5: Global trends in increasing groundwater use

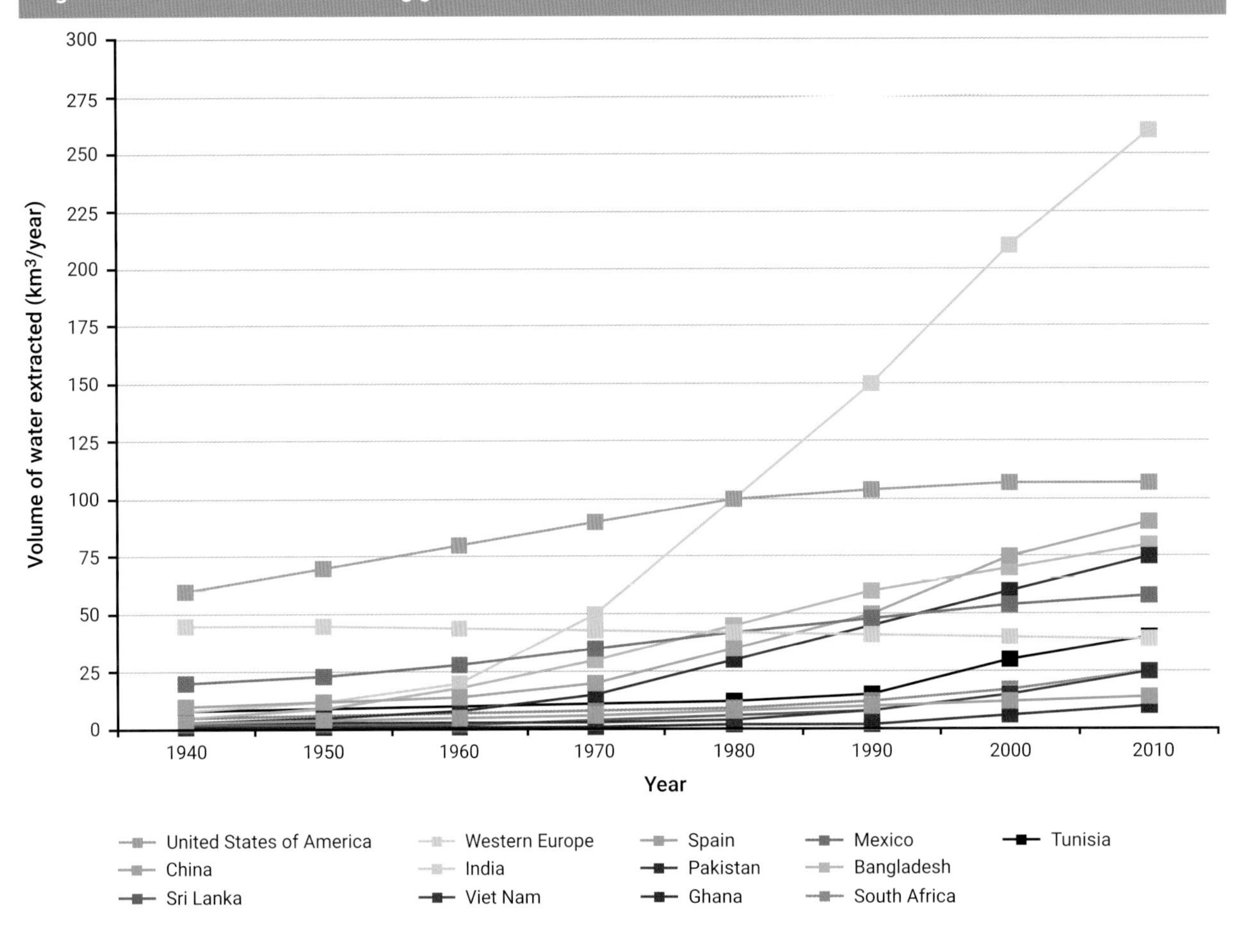

Source: Shah (2014, p. 12).

Groundwater use has plateaued in some regions but is increasing elsewhere (**Figure 9.5**), such as in Asia and the Pacific and West Asia (e.g. about two-thirds of freshwater utilized in West Asia). About 75 per cent of European Union (EU) inhabitants rely on groundwater for drinking (European Commission 2008, p. 7), and groundwater use, compared with surface water, has increased substantially to 1.3 trillion m^3 per year across North America (Famiglietti and Rodell 2013, p. 1301). Groundwater accounts for 30 per cent of water withdrawals in Latin America (Campuzano *et al.* 2014, p. 38) and an estimated 75 per cent of the African population depends on it (Altchenko and Villholth 2013, p. 1498). It must be noted, however, that estimates of groundwater withdrawals and use vary widely, constituting a critical data gap.

Increased agricultural groundwater use has led to rising depletion rates in major aquifers in arid and semi-arid zones (UNEP 2012b). Pumping rates that for decades have exceeded long-term natural recharge have resulted in some larger aquifers being 'mined' unsustainably (Famiglietti 2014, p. 946). Five of the world's seven largest aquifers are in Asia and the Pacific, and are overstressed (UNEP 2016b, p. 84).

Excessive groundwater abstraction has caused land subsidence in some coastal cities (e.g. Bangkok; Ho Chi Minh City; Jakarta; Manila) (UNEP 2016b, p. 87). Overexploitation of an aquifer can also impact wetland ecosystems. Hydraulic fracturing (fracking) for oil and gas extraction merits concern for its groundwater impacts (see **Box 9.2**). Groundwater is often underexplored on some islands due to surface-water availability, while other islands can be wholly reliant on it. Climate change impacts may lead to a greater reliance on and pose a threat to ground water because of sea level rise. Further studies are needed since islands are experiencing increasing freshwater shortages (Famiglietti 2014, p. 946).

9.4.3 Glacial retreat

Climate change is affecting regional water availability around the world, especially in areas reliant on glacial meltwater. Rivers originating in the Hindu-Kush Himalayas are among the most meltwater-dependent systems, and the source of ten large Asian river systems (Amu Darya, Brahmaputra, Ganges, Indus, Irrawaddy, Mekong, Salween, Tarim, Yangtse, Yellow), providing water for 20 per cent of the world's population (UNEP 2016b, p. 81) (**Figure 9.7**).

Box 9.2: Water quality impacts of mining

Modern mining generates large volumes of tailings (finely ground rock remaining after extracting ore) and waste rock (non-mineralized rock; low-grade ore), often containing iron sulphide minerals (e.g. pyrite). Exposed to the surface environment, these can react with water and oxygen to form sulphuric acid, producing acid metalliferous drainage (AMD). AMD can degrade water quality and impact aquatic biodiversity. Recent tailings dam failures (e.g. Mount Polley, Canada; Samarco, Brazil) demonstrate that mine wastes escaping into the environment can also significantly impact aquatic ecosystems and biodiversity, with tailings particles smothering riverbeds, reducing light penetration and oxygen levels, and affecting river geomorphology (Mudd *et al.* 2013).

Figure 9.6: Examples of surface streams affected by acid and metalliferous drainage (AMD) and/or tailings discharges: (left) Urban stream severely affected by AMD in western Witwatersrand Basin, Johannesburg, South Africa; (right) Tailings sediment from Samarco Dam

Figure 9.7: Rivers originating in the Hindu-Kush Himalayas are among the most meltwater-dependent systems

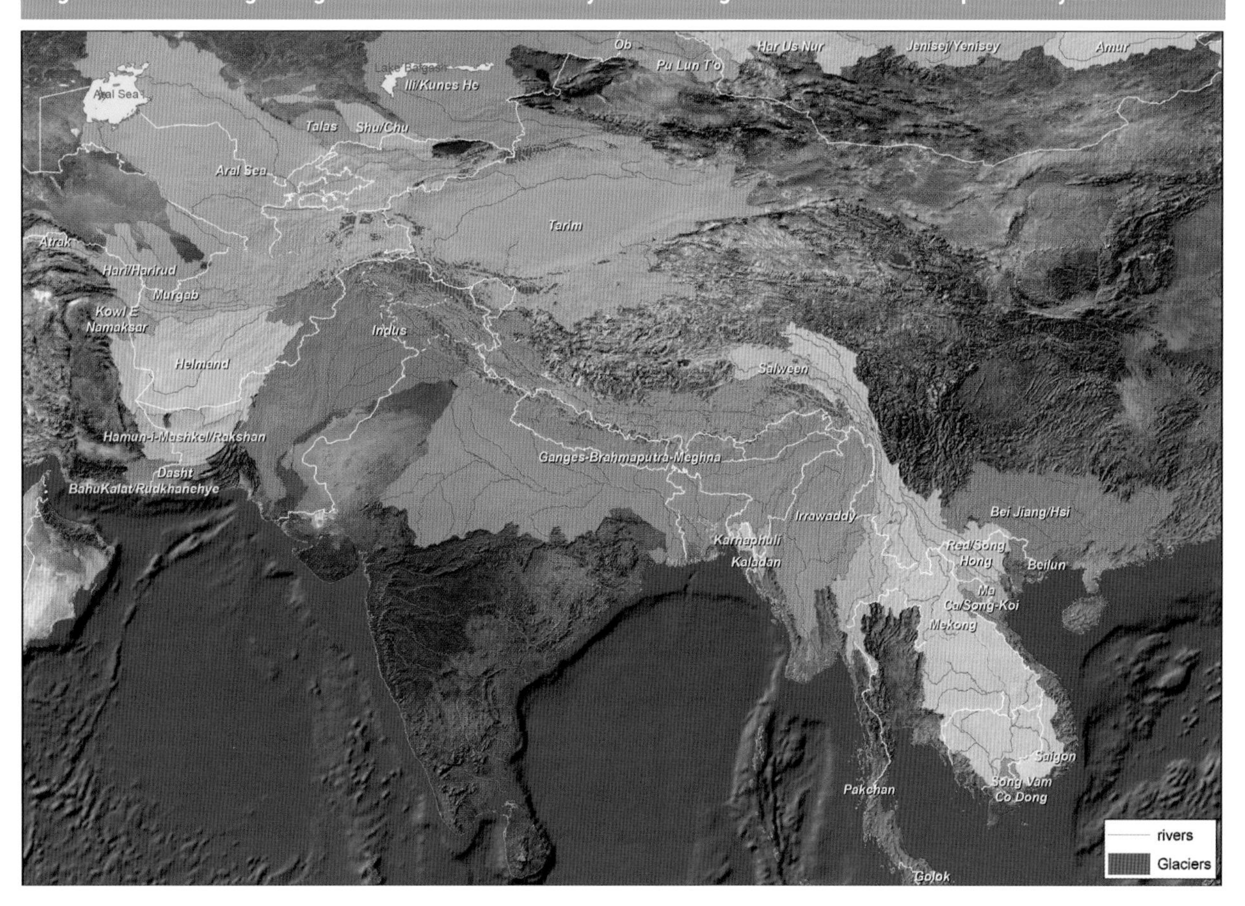

Sources: UNEP and Global Environment Facility [GEF] 2018; Global Land Ice Measurements from Space [GLIMS] 2018.

Figure 9.8: Retreat of Quelccaya ice cap in Peru between 1988 (left) to 2010 (right)

Source: Schoolmeester *et al.* (2018).

Tropical glaciers in the Andes account for more than 80 per cent of available freshwater for downstream populations and ecosystems in Latin American semi-arid tropic regions (UNEP 2013, p. 1). These are melting at an accelerating rate attributed to climate change (Chevallier *et al.* 2011; Rabatel *et al.* 2013), raising concerns about sustainable water supplies (**Figure 9.8**). Glacial retreat in the European Alps has accelerated over the last two decades (Huss 2012, p. 1132), while Central Asian glaciers have lost 27 per cent of their mass and 18 per cent of their area (Farinotti *et al.* 2015, p. 720; Yao *et al.* 2012).

9.4.4 Water scarcity

Water scarcity is defined as less than 1,000 m^3 per capita of available, renewable freshwater per year (United Nations World Water Assessment Programme [WWAP] 2012, p. 124). The differentiation between areas of economic water scarcity (where storage, treatment and conveyance infrastructure are lacking) and absolute or physical water scarcity is illustrated in **Figure 9.9** (WWAP 2012).

Sustainable freshwater supplies from surface and groundwater sources are critical for human and ecosystem needs, and for achieving the SDGs. Excessive withdrawals are often the cause of water scarcity. Lack of infrastructure, combined with rapid population growth, can lead to economic water scarcity, although there is not always agreement about the cause of water scarcity being physical, economic or indeed political in nature. Water of appropriate volume and quality is not always available at the right time or in the right place for a specific use.

Water scarcity is common throughout West Asia and the Asia and the Pacific region, and in arid parts of Africa, Latin America, the western United States of America and the Middle East. Factors that typically stress water resources include large populations, agricultural expansion and intensification, rainfall variability, rapid development, increasing urbanization, industrialization and climate change. The desiccation of the Aral Sea in Central Asia remains one of the most dramatic water-related environmental disasters of the 20th century. Most global climate model projections predict a 20 per cent rainfall decrease over the next 50 years in West Asia, with increased temperatures, evaporation and relative humidity all influencing water availability (UNEP 2016c, p. 12).

Desertification is a pressing problem in Africa's sub-Saharan region, arising from climate change and internal migration (UNEP 2016d). Although physical and economic water scarcity prevails across Africa, its surface- and groundwater resources are considered underdeveloped, in terms of meeting human livelihood and development needs (UNEP 2016d). In this context, many small- to medium-scale water infrastructure projects are well suited to local water demand.

In parts of the developed world (e.g. Europe, North America, Australia), water scarcity is a challenge that is commonly addressed through large water infrastructure projects, such as dams, long-distance pipelines and desalinization plants. Given expected population growth trends, regions such as the Middle East, Africa and Asia need to address water scarcity in innovative and scale-appropriate ways, including water governance, rainwater harvesting and wastewater recycling, leapfrogging the conventional solutions of the past.

9.5 Water quality

Although natural processes also generate water pollutants, human activities related to population growth, urbanization, agricultural expansion, transportation, and human and industrial waste discharges are typically the main sources of water pollution (UNEP 2016e). They include pathogens, nutrients, heavy metals and organic chemicals (Annex 9-1) from point sources (domestic, industrial or sewage pipeline discharges; septic tank leakage) and/or catchment non-point sources (land surface run-off from extensive diffuse agricultural use and urban areas following rainfall and snowmelt events).

Water quality in many Latin American, African, Asian and Pacific rivers has generally decreased since the 1990s, although the majority are still in relatively good condition (UNEP 2016e). Water quality in many European rivers has improved since adoption of the EU Water Framework Directive

Figure 9.9: Global physical and economic water scarcity

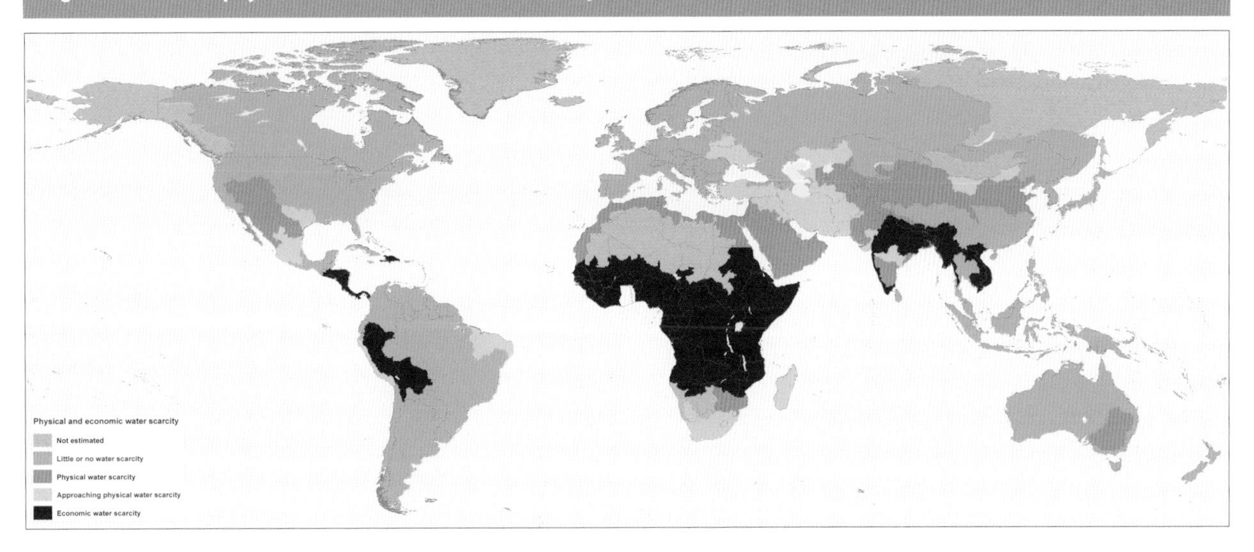

Source: WWAP (2012, p. 125).

in 2000. About half of United States waters do not meet standards protective of aquatic life, with more than 40 per cent not meeting recreational-use standards (UNEP 2016a). The water quality of many lakes and reservoirs is particularly endangered worldwide because of their long water residence times and tendency to accumulate pollutants (International Lake Environment Committee Foundation [ILEC] and UNEP 2016).

Groundwater pollution sources include non-point agricultural and urban run-off, on-site wastewater treatment, oil and gas extraction and fracking activities, mining, and industrial sources (Foster *et al.* 2016). Natural contamination occurs in some cases (e.g. sodium-chloride salinity, arsenic, fluoride, radioactivity in fossil groundwater aquifers). Human health impacts from untreated groundwater are of particular concern (Morris *et al.* 2003; UNEP 2016e).

9.5.1 Pathogens

Water-borne diseases remain major challenges in many African, Asian, Pacific and Latin American cities and rural communities (Annex 9-1). Although collection and treatment of human excrement has largely curtailed the problem in developed countries, sewage outfalls still generate large pathogen loads. However, parasites can survive waterbody conditions for many weeks, while viruses may survive drinking water treatment.

Pathogens remain a major cause of human death and illness, particularly in developing countries (http://www.who.int/water_sanitation_health/takingcharge.html). High child mortality, for example, is associated with diarrheal diseases from contaminated water in Africa, Asia, the Pacific and Latin America (Annex 9-1). Principal pathogen sources include inadequately treated human and livestock wastes, and combined sewer overflows and leaks. **(Figure 9.10)**

Irrigation with inadequately treated or diluted wastewater occurs in many developing countries, increasing agricultural productivity for many poor communities, but often at the expense of human health and environmental risks. Comparison studies highlight environmental degradation and higher water-borne disease rates in wastewater-irrigated areas (75 per cent prevalence rate for gastroenteritis in children 8-12 years of age, compared with 13 per cent in freshwater-irrigated areas) (Grangier, Qadir and Singh 2012).

Antibiotic and antimicrobial resistance is a major global health concern, with the spread of resistant bacteria and resistance genes in the environment being a critical component of integrated control efforts (Berendonk *et al.* 2015). The major source remains human and animal excreta, with aquaculture increasingly adding to the levels in the water environment (Kümmerer 2009). The World Health Organization (WHO) predicts antimicrobial resistance to become a major cause of deaths globally by 2050 (Annex 9-1). Wastewater-treatment plants have diverse abilities to remove antibiotic-resistant bacteria, and limited capacity to remove antibiotic drugs (Pruden *et al.* 2013; Berendonk *et al.* 2015).

9.5.2 Nutrients

Eutrophication represents the natural ageing process of lakes and wetlands, wherein they become enriched with nutrients and sediments, becoming more biologically productive, usually over a long period (Annex 9-1). Human activities can greatly increase these nutrient loads, accelerating this process, with detrimental effects on the whole ecosystem. The resulting algal blooms and aquatic plant growths interfere with many human water uses and can greatly affect the balance and diversity of aquatic flora, fauna and algal species (OECD 1982; Research Center for Sustainability and Environment-Shiga University and ILEC 2014). Major nutrient sources include inadequately treated domestic sewage discharges, urban and agricultural run-off, aquaculture and mariculture. Algal blooms can turn lakes, reservoirs and slow-moving rivers turbid and green in colour, depleting the water's oxygen content when algae die and undergo decomposition. Some blue-green algal species are toxic to fish and livestock (O'Neil *et al.* 2012) and affect human health. A clear relationship between climate change and eutrophication of lakes has also been reported (Jeppersen *et al.* 2010).

More than half of the total phosphorus loads in the five UN Environment regions originate from inorganic agricultural fertilizer run-off **(Figure 9.11)**. Livestock waste used as fertilizer can also be problematic because its nitrogen-phosphorus ratio is higher than that needed by crops, thereby potentially saturating soils with phosphorus, which can then reach waterbodies via non-point source run-off. River nutrient contributions to coastal areas almost doubled during 1970-2000 (Annex 9-1). The Gulf of Mexico exhibits a 'dead zone' typically covering nearly 13,800 km^2 attributed to nitrogen from grain fields in the midwestern United States of America carried down the Mississippi River, with eventual decay of the algal growth consuming the oxygen in the water, suffocating marine life. There are nearly four times as many dead zones (400) in the oceans now as there were in 1950, including in the Mediterranean Sea (Pearce 2018).

Some major urban areas in Asia and the Pacific experience groundwater nitrate concerns from sewer and septic tank leaks (Umezawa *et al.* 2009), and rural areas in many countries are affected by excessive chemical fertilizer application (Novotny *et al.* 2010). The effects of nitrates in groundwater has long been a public health concern, particularly as a causative factor in methemoglobinemia in infants ('blue baby' syndrome).

Figure 9.10: Model estimates of trends in faecal coliform bacteria levels in rivers during 1990-1992 and 2008-2010

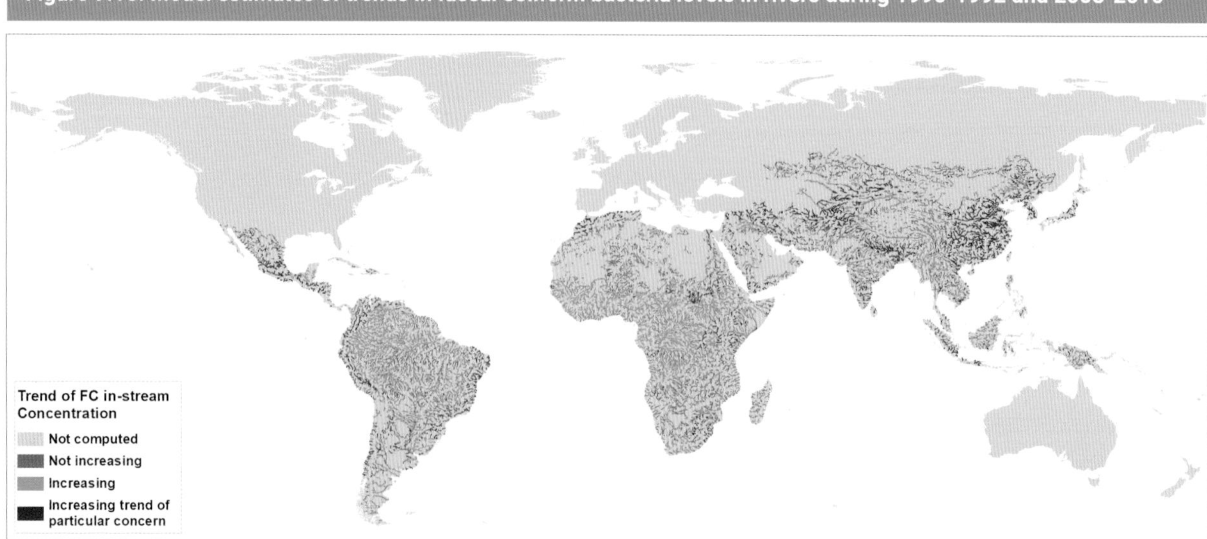

Orange or red river stretches indicate increasing concentrations between these periods; red river stretches indicate increasing trends of particular concern.

Source: UNEP (2016e).

Although over half of EU surface waters improved during 1992-2010 (average river phosphate and nitrate levels decreasing by 57 per cent and 20 per cent, respectively), many still do not meet European Water Framework Directive environmental objectives (European Union 2000).

9.5.3 Sediments

Sediments result from erosion of exposed soil surfaces, with much eroded soil being deposited in basins throughout the world, including in Africa, Asia and Latin America. Deforestation, poor agricultural and livestock practices, intensive fuelwood harvesting, mining, urbanization and unplanned settlements are major causes of soil vulnerability to erosion, and storm-generated run-off carries soil into downstream waterbodies (Annex 9-1). Sediment-associated pollutants can have human health impacts, and interfere with water uses and aquatic organism metabolism and habitats (UNEP 2017). Artificial channels from dams and urban development can change sediment flow paths, lead to erosion and reduce sediment available to build up banks, river deltas and beaches along coastlines, causing aquatic ecosystem changes (Blum and Roberts 2009, Syvitski *et al.* 2009; Yang *et al.* 2011; Cloern and Jassby 2012; Adams *et al.* 2016; Yihdego, Khalil and Salem 2017).

9.5.4 Organic pollutants

Biodegradation of organic pollutants such as liquid manure, sewage effluents and sewage treatment sludge can deplete oxygen concentrations in waterbodies, causing fish kills and releasing heavy metals from bottom sediments back into the water column, a process characterized by a high biochemical oxygen demand (BOD) from microbial decomposition of these pollutants. Algal bloom decomposition also can deplete the oxygen content in eutrophic waterbodies, particularly lakes and wetlands.

Based on model analyses, BOD concentrations increased in many parts of Africa, Asia and the Pacific, and Latin America during 1990-2010 from industrial and domestic wastewater discharges, and agricultural and urban run-off, with highest increases in rapidly urbanizing and industrializing countries (Annex 9-1). BOD pollution in most developed countries has significantly decreased with enhanced wastewater treatment (e.g. implementation of the 1991 EU Urban Wastewater Treatment Directive).

Synthetic organic pollutants include pesticides, industrial chemicals and solvents, and personal care and pharmaceutical products. Persistent organic pollutants (POPs) are particularly problematic because they do not readily biodegrade in the aquatic environment. Used in many industrial and agricultural applications, they can impact human health and aquatic ecosystems, persisting in fatty tissues of humans, fish and other organisms, and accumulating in sediments. DDT has human carcinogenic and teratogenic risks, for example, but is still used in many regions to control malaria (Annex 9-1). Other synthetic compounds, including non-POPs, continue to enter the ecological food chain globally, while others, such as endocrine-disrupting chemicals, are considered contaminants of emerging concern (see Section 9.5.7). Neonicotinoid and fipronil systemic insecticides, for example, are water soluble and can leach into freshwater and marine systems. Neonicotinoid insecticides are toxic to most arthropods and invertebrates, while fipronil is toxic to fish and some bird species (Annex 9-1; van Lexmond *et al.* 2015; Intergovernmental Science-Policy Platform on Biodiversity and Ecosystem Services 2017).

Figure 9.11: Sources of anthropogenic total phosphorus loadings to lakes (five largest lakes by surface area in each of the five UN Environment regions), showing average percentage contributions in 2008-2010 annual loads

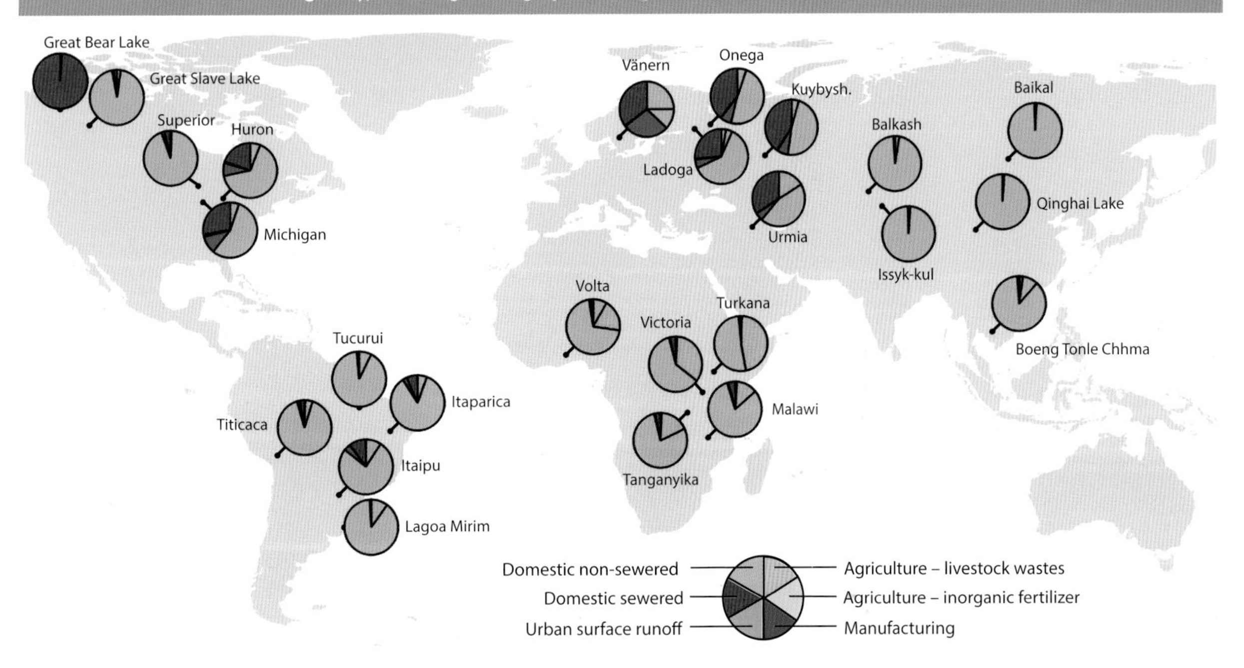

Source: UNEP (2016e).

9.5.5 Heavy metals

Used in a range of industrial and agricultural sectors, heavy metals exhibit widespread environmental distribution. Heavy metals from industrial activities, and large-scale and artisanal mining, have seriously degraded water in some Asian, Pacific and South American countries (Da Rosa *et al.* 1997; Spitz and Trudinger 2008; Sikder *et al.* 2013; Annex 9-1). They can bioaccumulate in plants grown with contaminated irrigation water (Arunakumara, Walpola and Yoon 2013; Lu *et al.* 2015). Many (mercury, lead, chromium, cadmium) are toxic to humans and aquatic organisms (Kim *et al.* 2017).

Heavy metals associated with water-intensive mining are problematic in Africa and Latin America (Annex 9-1). Water drainage from active and abandoned mines can cause significant water degradation (e.g. mercury and arsenic used in gold mining can pollute surface and groundwater). Examples of untreated mine-water discharging into streams and rivers include Mount Morgan (Australia) and Tisza River (Hungary), where reservoirs, agricultural irrigation water and aquatic ecosystem biodiversity have all been degraded. Groundwater pollution also has been reported to have occurred in Alberta, Canada because of the tar sands industry (Timoney and Lee 2009).

Groundwater contamination with naturally occurring arsenic occurs in South Asia and other countries in Asia and the Pacific (Rahman, Ng and Naidu 2009; Annex 9-1). Arsenic mobilization can also be facilitated or worsened through such human activities as metal mining and groundwater abstraction and, in some cases, through use of arsenic-based pesticides in agriculture and wood preservation. Although some problems remain, heavy metal contamination has generally diminished in EU countries since 2000. A dramatic example of heavy metal contamination involved Flint, Michigan (United States of America). A decision to switch the city's drinking water supply in 2014 from Lake Huron to the Flint River, containing more corrosive water, released lead from leaded pipes in the city's water distribution system, with significant human health impacts (Masten *et al.* 2016).

9.5.6 Salinity

Increases in salinity, a measure of the quantity of dissolved minerals in freshwater, result from land-use changes, agricultural irrigation drainage, lake evaporation and seawater intrusion, usually most severe in arid and semi-arid regions (Vengosh 2003). Excess salinity renders the water unfit for many human uses, and most freshwater organisms have limited salinity tolerance (UNEP 2016e).

Salinity problems persist at various degrees in rivers throughout Africa, Asia and the Pacific, and Latin America, affecting agricultural irrigation as a result of accumulation of naturally occurring minerals in irrigation water, as well as industrial water uses (Foster *et al. 2018*; Annex 9-1), with surface-water salinization being a major issue in Central Asia. Saline water intrusion into coastal aquifers can result from over-abstraction and mismanagement, as well as sea level rise. Apart from sodium, waters with elevated levels of magnesium are emerging examples of water quality deterioration leading to environmental and food security constraints in several irrigation schemes (Qadir *et al.* 2018).

9.5.7 Contaminants of emerging concern

Water contaminants of emerging concern include certain human and veterinary pharmaceuticals, personal care products, insect repellents, antimicrobial disinfectants, flame retardants, detergent metabolites, microplastics and manufactured nanomaterials ('nanoparticles') (**Figure 9.13**; Kolpin *et al.* 2002; UNESCO 2016; Yuan *et al.* 2018). The United States Geological Survey detected such contaminants in a

Figure 9.12: Model estimates of trends in biochemical oxygen demand (BOD) concentrations in rivers between 1990-1992 and 2008-2010

Orange and red river stretches: increasing concentrations between these two periods; red river stretches: increasing trend of concern.

Source: UNEP (2016e).

Figure 9.13: Source and pathways of pharmaceutical and personal care products (PPCPs) entering surface and groundwater, highlighting need for improved detection of commonly found PPCPs and their transformative products

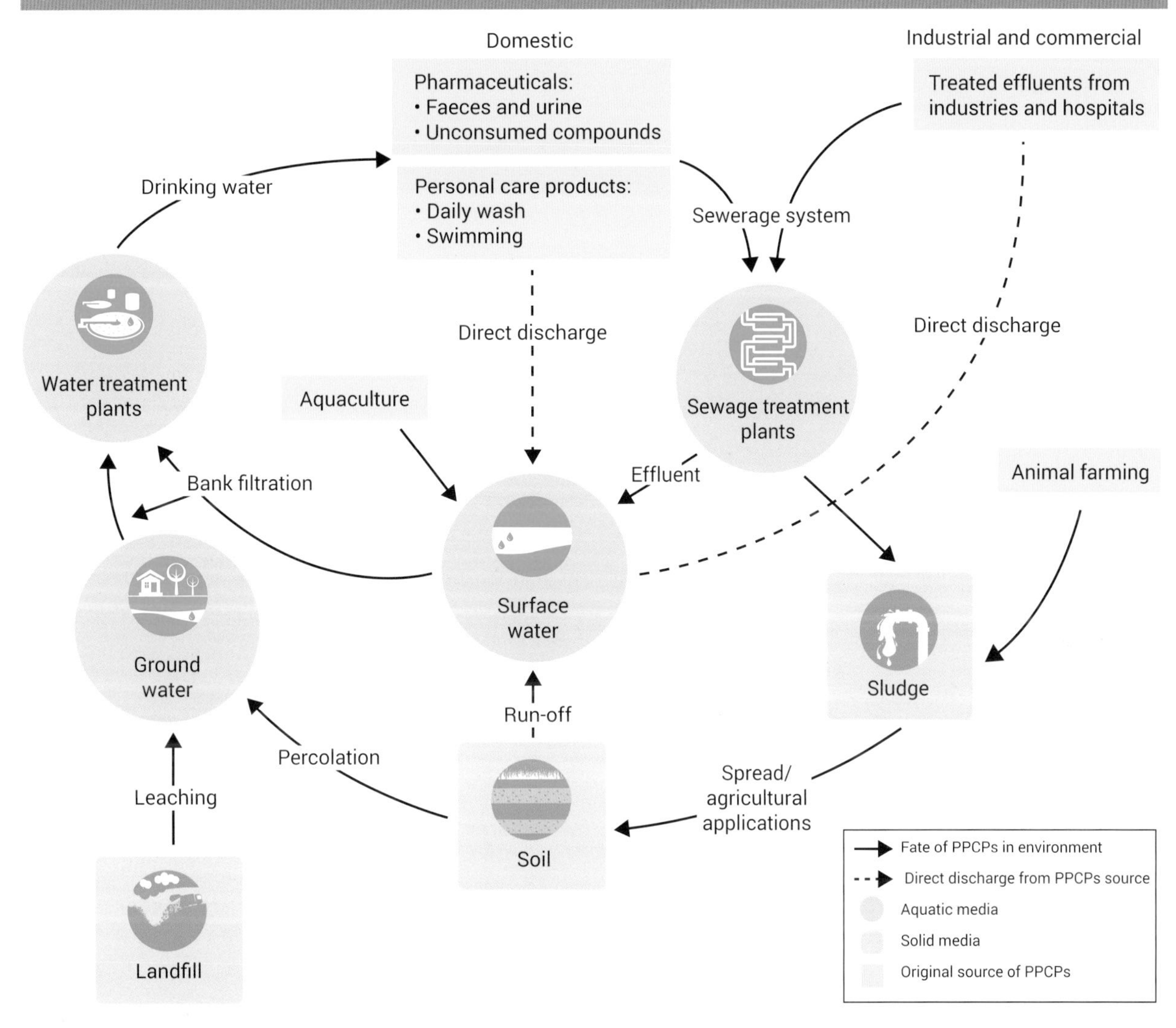

Source: Adapted from Petrović *et al.* (2003); Mompelat *et al.* (2009); Yang *et al.* (2017).

majority of sampled streams in the United States of America (Annex 9-1; Kolpin *et al.* 2002). They also were detected in all pan-European seas, as well as groundwater (Sui *et al.* 2015; Corada-Fernández *et al.* 2017). Used in fire retardants, oil and water repellents, furniture, waterproof clothes, takeaway containers and non-stick cookware, poly- and perfluoroalkyl substances (PFAS) were found in water systems serving 16 million people in 33 states in the United States of America between 2013 and 2015 (INTJ Input 2017). Conventional wastewater treatment is not effective in removing most of these contaminants from domestic and industrial wastewaters.

Many are endocrine-disrupting chemicals (EDCs), attributed partly to wastewater treatment plant overflows, particularly those with combined sewer systems. Being found in site-specific studies in Europe, the Asia and the Pacific region, Canada and the United States of America (Annex 9-1; Sui *et al.* 2015), their long-term human health impacts include fetal underdevelopment, child neurodevelopment and male infertility (Meeker 2012).

Micro- and nanoplastics (manufactured nanomaterials) resulting from microplastics in cosmetics, fragmentation of large plastic waste, tyre wear particles and laundering of synthetically based clothes are increasingly affecting freshwater and marine ecosystems (Annex 9-1; Horton *et al.* 2017). Of the 275 million (metric) tons of plastic waste generated by 192 countries in 2010, an estimated 4.8-12.7 million tons ended up in the oceans because of inadequate solid waste management. They are found worldwide in fresh and ocean waters, river and delta sediments, and in the stomachs of organisms ranging from zooplankton to whales (UNEP 2016g). Microplastics also can contain and absorb toxic chemicals. Electronic wastes also are of increasing concern because of their widespread abundance and unknown risks to surface- and groundwater quality.

Figure 9.14: Status and trends of the world's wetlands disaggregated by region

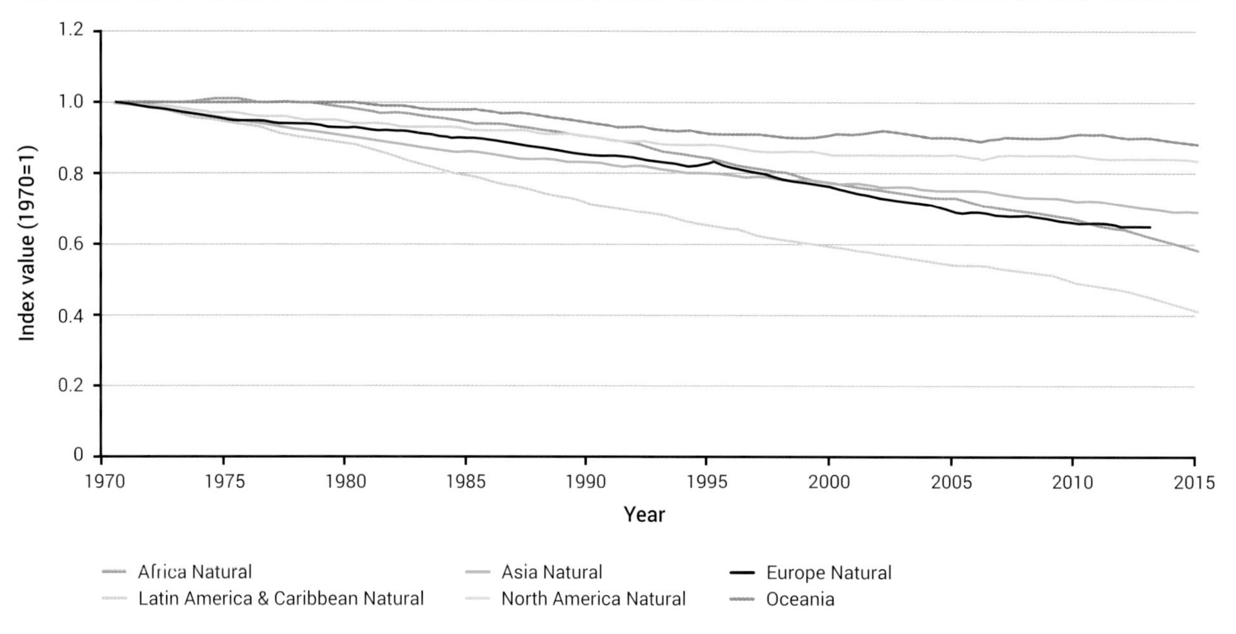

An index taking the total extent of wetland area in 1970 = 1, based upon a literature search.
Source: UNEP World Conservation Monitoring Centre (UNEP-WCMC) (2017).

9.5.8 Other water quality concerns

Groundwater pollution from oil and gas fracking activities, which use large quantities of chemicals and discharge large volumes of 'produced water', is problematic in the Americas (Osborn *et al.* 2011; Vengosh *et al.* 2014; Annex 9-1). Heavy metals, particulate matter, various organic chemicals, and EDCs are widely used in, or become by-products of, these oil and natural gas operations (Webb *et al.* 2017).

Lake acidification from atmospheric deposition of fossil fuel emissions causes problems in areas lacking soils or bedrock capable of buffering the emissions, including the north-east United States of America, south-east Canada and some Scandinavian regions. The situation is improving for affected lakes in the Adirondack Mountains region of the north-eastern United States of America, where sulphur and nitrogen oxide emissions have decreased since the 1970s (Annex 9-1; Driscoll *et al.* 2016).

Thermal pollution and radionuclides also represent water quality concerns. Thermal pollution, often resulting from using freshwater as a coolant in power plants and industrial manufacturing activities, can degrade water quality by changing ambient water temperature. The impacts can be multiple, including reducing the dissolved oxygen concentration while at the same time increasing the respiration rates of aquatic organisms using it. Some aquatic species populations may decrease because they cannot thrive or reproduce in waters at higher temperature, while others may increase, potentially changing the overall ecosystem dynamics of a waterbody. Radioactive contamination typically in the form of accidental releases of radionucleotides from nuclear activities have polluted inland freshwater systems in some areas, with negative implications for aquatic and other organisms, including humans, using these waterbodies (Echols, Meadows and Orazion 2009).

9.6 Freshwater ecosystems

9.6.1 Continuing loss of wetlands

Freshwater ecosystems (or inland wetlands) include marshes, swamps, peatlands, wetland forests, rivers, lakes, ponds and headwaters. They provide a range of provisioning, regulatory and supporting ecosystem services, including water and food supply, fodder and building materials, carbon and nutrient sequestration, unique habitats for endangered species (including migratory birds), flood- and drought-buffering capacity, ecotourism and cultural services (WWAP 2018). Although freshwater ecosystems only cover 0.8 per cent of Earth's surface, they support approximately 10 per cent of all known species (World Wide Fund For Nature [WWF] 2016), and are among the world's most biodiverse habitats. They are the ecosystems most affected by changing land use, particularly increasing urbanization and agricultural expansion.

Between 69 per cent and 75 per cent of wetlands worldwide are estimated to have been lost since 1900 due to rapid population growth, urbanization and agricultural expansion (Davidson 2014). The extent of the loss since 1970 differs notably across regions, the slowest loss rate being apparent in Oceania and North America. Levelling off of the loss rate in North America is due partly to the current national policy of "no net loss of wetlands" in the United States of America (United States Fish and Wildlife Service 1994). Although constructed wetlands can compensate to some degree for some natural wetland removal, they cannot typically provide the same level of ecosystem functioning, resilience and biodiversity, emphasizing the need for natural wetland protection and conservation (see Section 9.4).

Ecosystem services for all wetland types have been valued financially across a very wide range from US$300-US$887,828 per hectare per year, with a median value of US$12,163 (de

Groot *et al.* 2012). More specific assessments are needed. A recent valuation of swamp and floodplain ecosystem services attributed an average annual global value of US$25,000 per hectare per year, excluding the value of the land itself (Costanza *et al.* 2014). The estimated annual loss to the global economy from diminishing swamp and floodplain areas from 165 to 60 million ha between 1997-2011 is US$2.7 trillion (Costanza *et al.* 2014).

Although covering only 3 per cent of the planet's land surface, peatlands have a high carbon-sequestration value, hence they contain more carbon than all global forest biomass combined (Joosten 2015). The world's largest tropical peatland (Cuvette Centrale) covers an area of 145,500 km^2 in the Congo River basin, containing an estimated 30 gigatons of carbon accumulated over the past 11,000 years (Dargie *et al.* 2017). Draining peatland areas for agriculture (e.g. the large Indonesian and Malaysian palm oil plantations) breaks down the peat, rapidly emitting carbon as CO_2 and methane. About 15 per cent of peatlands worldwide have been drained in the last 40 years, contributing approximately 5 per cent of global carbon emissions (Joosten 2015).

As the drained peat decomposes quickly, it dries out, shrinks and subsides. Tropical coastal peatlands are subsiding by an average 5-7 cm/year, and thus become vulnerable to salinization during storm surges. During hot, dry periods, the fire hazard in peatlands is high (Jayachandran 2009), an example being extensive peat fires in Indonesia exacerbating brown-haze pollution of the whole Asian region in the summer of 2015 (Carmenta, Zabala and Phelps 2015).

The permafrost in boreal peatlands in and around the Arctic Circle is thawing and draining due to climate change, with effects on local and global carbon fluxes (Joosten 2015; Couture *et al.* 2018). Apart from additional emitted carbon, the permafrost thawing is damaging infrastructure and housing, affecting Arctic people's quality of life. For both tropical and boreal peatlands, the straightforward technical solution to addressing carbon emissions from drained peatlands is to rewet the peatland, bringing the water table back to the soil surface, as is currently being done at a large scale in Indonesia, Canada, Sweden and Switzerland (Zerbe *et al.* 2013).

9.6.2 Biodiversity loss

There is evidence of significantly reduced abundance of populations of flora and fauna attributed to wetland loss and pollution impacts, particularly eutrophication, chemical and metal toxicity, and the hazards of plastic and other wastes (WWF 2016). Although wetlands have the capacity to filter and improve water quality, continuous breaking down of organic matter and other nutrients can lead to a tipping point in water quality, beyond which a wetland can no longer regenerate itself, with species assemblages potentially changing markedly.

Fragmentation of rivers through dam building and water diversion, with resultant wetland habitat losses and degradation, has a significant impact on fish populations, especially migratory and endemic fish species. Fish populations are also being overexploited for food. Amphibian species are experiencing dramatic declines through habitat loss, invasive species, disease and pollution, followed by climate change (WWF 2016) **(see Figure 9.15).** Reptiles and many bird species are deeply affected by loss of wetlands, while aquatic mammals such as otters also suffer local extinctions from habitat loss and overexploitation.

The Living Planet Index (LPI) measures population abundance trends of 881 freshwater species monitored worldwide across

Figure 9.15: Taxonomic differences in threat frequency for 449 declining freshwater populations in Living Planet Index (LPI) database

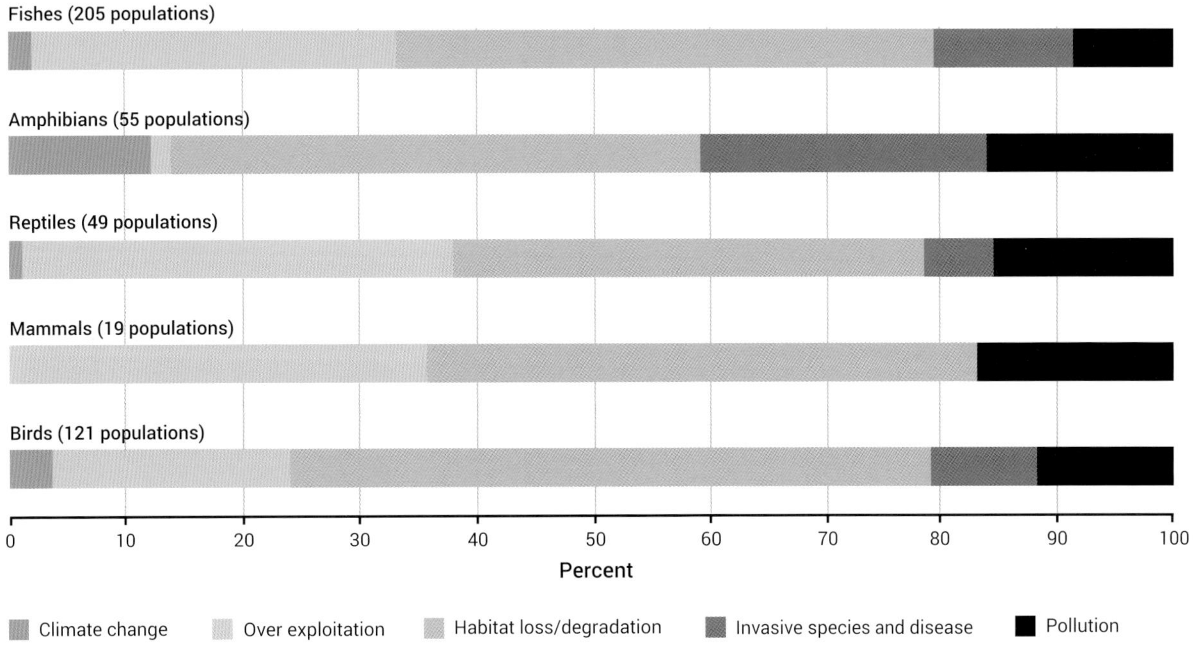

Source: WWF (2016).

Figure 9.16: Migratory fish from the Living Planet Index (LPI) exhibiting a decline of 41 per cent between 1970 and 2012, with a recent upturn, and freshwater LPI for 881 monitored freshwater species exhibiting an 81 per cent decline

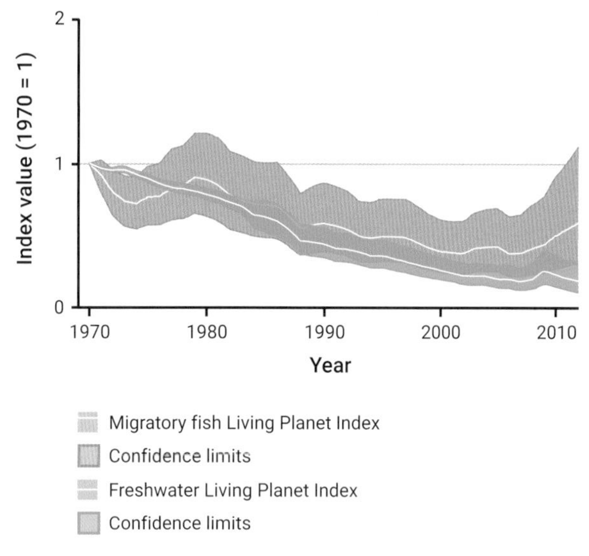

Migratory fish Living Planet Index
Confidence limits
Freshwater Living Planet Index
Confidence limits

Source: WWF (2016).

3,324 different populations (see chapter 6). Recent analyses indicate an 81 per cent decline in LPI in freshwater ecosystems between 1970 and 2012, the highest of any habitat type monitored using this index (WWF 2016).

The LPI declined by 41 per cent over this period for migratory fish, based on measurement of over 162 fish species (**Figure 9.16**). Some improvement is evident from 2008 onwards, in response to removing weirs, installing fish ladders, and improving the up- and downstream passage of migratory fish in many places. Migratory species of birds and mammal populations in certain managed wetlands are also starting to recover, in response to habitat conservation and restocking. In contrast, the decline in amphibian and invertebrate wetland species, including insects, is much higher (WWF 2016).

9.7 Water infrastructure

9.7.1 Drinking water supply: treatment and distribution

Provision of safe, reliable drinking water is a continuing goal of development institutions, requiring modernization of ageing infrastructure and construction of new infrastructure. In many parts of the world, the original infrastructure is deteriorating, requiring upgrading. Provision of drinking water services has not kept pace with the rate of urbanization in many Latin American and Caribbean cities (UNEP 2016f; World Health Organization [WHO] and United Nations Children's Fund [UNICEF] 2016). The Millennium Development Goal (MDG) of halving the number of people without access to safe drinking water by 2015 was achieved by 2010, although progress was uneven across urban and rural populations in all regions (WHO and UNICEF 2015) (**Figure 9.17**).

Figure 9.17: Variations in trends in drinking water supply coverage across regions

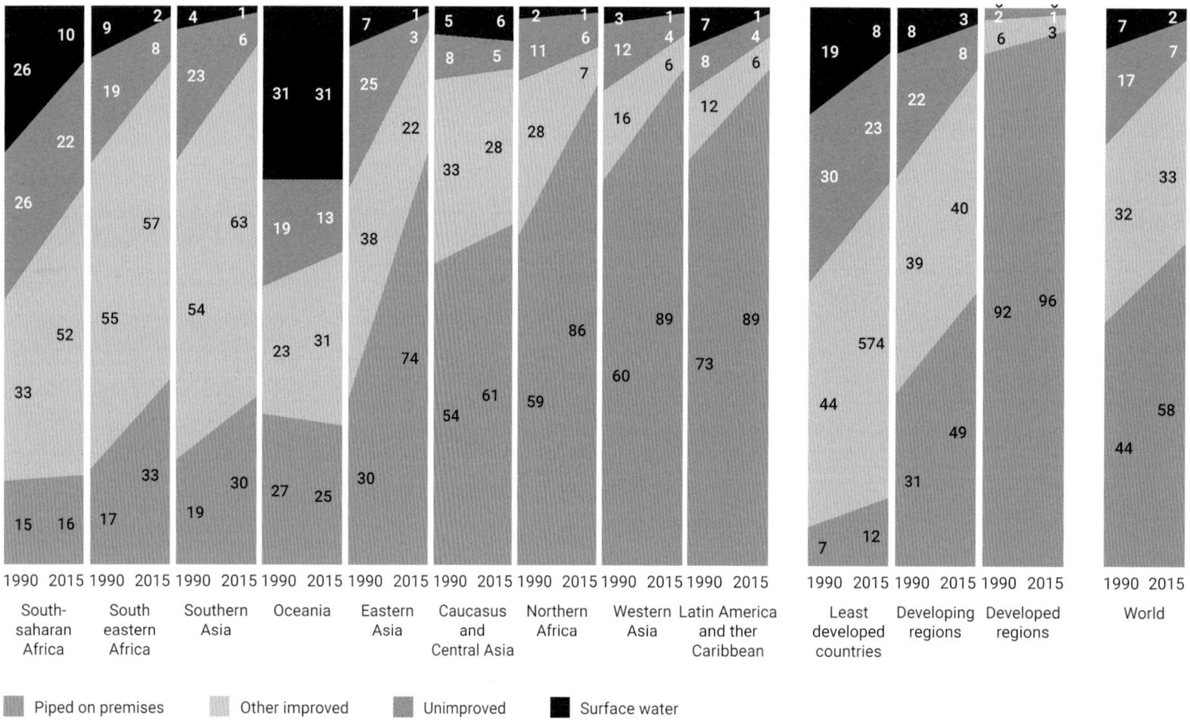

1990 2015 — South-saharan Africa; South eastern Africa; Southern Asia; Oceania; Eastern Asia; Caucasus and Central Asia; Northern Africa; Western Asia; Latin America and ther Caribbean; Least developed countries; Developing regions; Developed regions; World

Piped on premises · Other improved · Unimproved · Surface water

Source: WHO and UNICEF (2015).

In Asia and the Pacific, progress in provision of drinking water was significant, with 90 per cent of the population having access to improved water supply by 2015. Drinking water supply in Africa increased from 56 per cent in 1990 to 65 per cent in 2013, albeit mostly in urban areas, with 90 per cent of the urban population using improved water sources (UNEP 2016d).

There were significant improvements in access to drinking water in West Asia, with 89 per cent of the population having access to improved water supply by 2015. The reliability and continuity of service remains challenging, however, especially in conflict zones (UNEP 2016c; WHO and UNICEF 2016).

The relatively high quality of North American and Western European drinking water contributes to good public health, with these regions having some of the lowest rates of water-borne disease in the world. Most of the countries in these regions are party to the United Nations Economic Commission for Europe/ World Health Organization (WHO) Regional Office for Europe Protocol on Water and Health, and to the 1992 Convention on the Protection and Use of Transboundary Watercourses and International Lakes, a multilateral agreement committing countries to work actively to lower water-borne disease outbreaks (UNEP 2016a; UNEP 2016h).

Gender is a significant factor in water supply, with women and (mostly female) children continuing to carry the major burden and safety risks of acquiring and physically transporting water from source to place of use, particularly in developing countries, despite 1.5 billion more people gaining access to clean water since 2000 (WHO and UNICEF 2017; WHO and UNICEF 2012) **(Figure 9.18)**.

The water collection time burden and the physical labour involved have implications for the livelihoods and safety of women and girls. Time spent by school-age girls collecting water competes with schooling (UNICEF and WHO 2012). It also reduces the ability of women to participate in other pursuits. It represents a substantial economic drain. Women in India spend an estimated 150 million workdays per year collecting and carrying water, the equivalent of a national income loss of 10 billion rupees (approximately US$160 million per year). The positive impacts of women being able to spend time on other activities should be widely acknowledged, since economic surveys indicate they typically reinvest up to 90 per cent of their income in their families, improving family health and nutrition, and increasing access to schooling for their children (Unilever *et al.* 2015).

Figure 9.18: Summary of global progress in providing basic drinking water services and disproportionate impact on women in areas still lacking access to basic drinking water services

1.5 billion people gained access to basic drinking water services from 2000 to 2015

But women still spend
16 million hours per day
collecting water in 25 sub-Saharan countries

WOMEN	MEN	CHILDREN
16 MILLION HOURS	6 MILLION HOURS	4 MILLION HOURS

Source: UNICEF and WHO (2012); WHO and UNICEF (2017).

9.7.2 Sanitation and wastewater treatment

Improved sanitation, including proper human waste treatment and disposal, is one of the most effective measures for improving public health globally (Sedlak 2014). It remains a challenge, however, in many parts of the world (**Figure 9.19**). Growing megacities, especially in Africa and Asia, do not have adequate sanitation services to accommodate population growth, contributing to open defecation and poor or non-existent wastewater treatment and disposal (UNEP 2016b; UNEP 2016d). Even in areas with improved sanitation, large-scale septic tank and leachfield use in many expanding urban centres affects downstream water supplies as well as groundwater quality.

Approximately 1.4 million people still die annually from treatable diseases associated with pathogen-polluted drinking water and inadequate sanitation, with many millions of others becoming ill (Lozano *et al.* 2013). An estimated 2.3 billion people still lacked access to improved sanitation in 2015. While almost all developed countries had achieved 'universal sanitation coverage' by 2015, only four of the nine developing regions met the sanitation target (Caucasus and Central Asia, East Asia, North Africa, West Asia). The population proportion served by improved sanitation was particularly low in parts of Oceania, sub-Saharan Africa and South Asia (WHO and UNICEF 2015).

There are significant inequalities in access to improved sanitation between rural and urban areas. About 82 per cent of the global urban population has access to improved sanitation, compared with only 51 per cent of the rural global population (WHO and UNICEF 2015). Public sanitary facilities tend to be regulated at local level in most countries. Where facilities are inadequate, they are often especially so for women and girls, including those located in markets, public transport stations and public event venues. Inadequate sanitation in schools has a deleterious effect on education, especially for girls. The problem is compounded for people living in slums and informal settlements lacking access to adequate drinking water and sanitation facilities, or to durable housing, sufficient living area and security of tenure.

9.7.3 Dams and reservoirs for water storage and hydroelectric power

Many developing countries continue to construct dams to secure domestic water supply for communities, agricultural irrigation and hydroelectric power generation. Such multifunctional dams can also be operated to provide flood protection to downstream communities, as well as being sensitive to downstream ecological flow requirements (e.g. providing flow pulses to support fish spawning). Hydroelectric power is a key energy source, often critical to provide energy for drinking-water pumps, with additional growth potential evident in Latin America, Africa and Asia (Campuzano *et al.* 2014; UNEP 2016e). In addition, the use of reservoirs for pumped hydro-energy storage systems is increasingly being used to offset the fluctuating nature of other renewable sources of energy (Rehman, Al-Hadhrami and Alam 2015; Barbour *et al.* 2016) On the other hand, efforts to employ run-of-the-river hydroelectric power technologies have shown promise in the Amazon region in supplying electricity to rural communities, exhibiting fewer environmental impacts than traditional dams (Sánchez, Torres and Kalid 2015).

In recent years, dam construction in industrialized countries has slowed considerably. Many older dams are being decommissioned for economic (e.g. high dam operation and maintenance costs) and environmental reasons (e.g. effects on migratory fish, downstream ecosystems and sediment patterns) (O'Connor, Duda and Grant 2015; UNEP 2016e). Dam density nevertheless remains highest in industrialized countries (**Figure 9.20**).

More than 1,270 dams have now been constructed across Africa for irrigation, hydroelectric power production and domestic water supply purposes, although only about 20 per cent of the potential to generate hydroelectric power is currently being utilized and lack of resources to properly maintain dams has resulted in reduced power generating capacity in some places. Increased dam construction in some locations (e.g. Ghana, Benin, Burkina Faso) has caused

Figure 9.19: Proportion of population using improved sanitation facilities in 2015

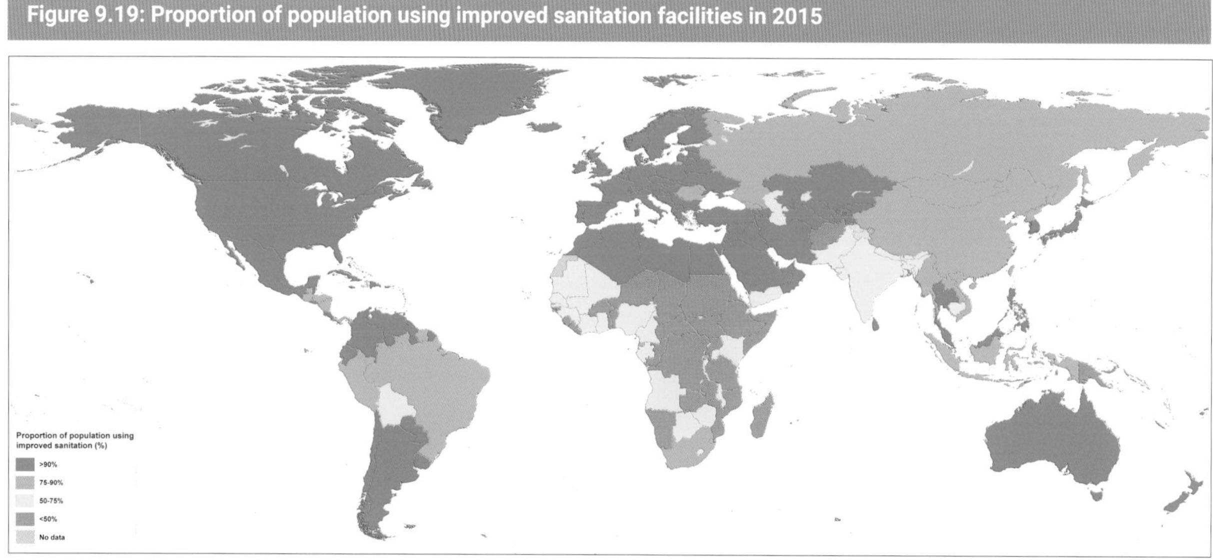

Source: WHO and UNICEF (2015).

Figure 9.20: Location of dams and reservoirs around the world. Data include dams associated with reservoirs that have a storage capacity of more than 0.1 km³ and may not represent large dams and reservoirs that have been constructed in more recent years

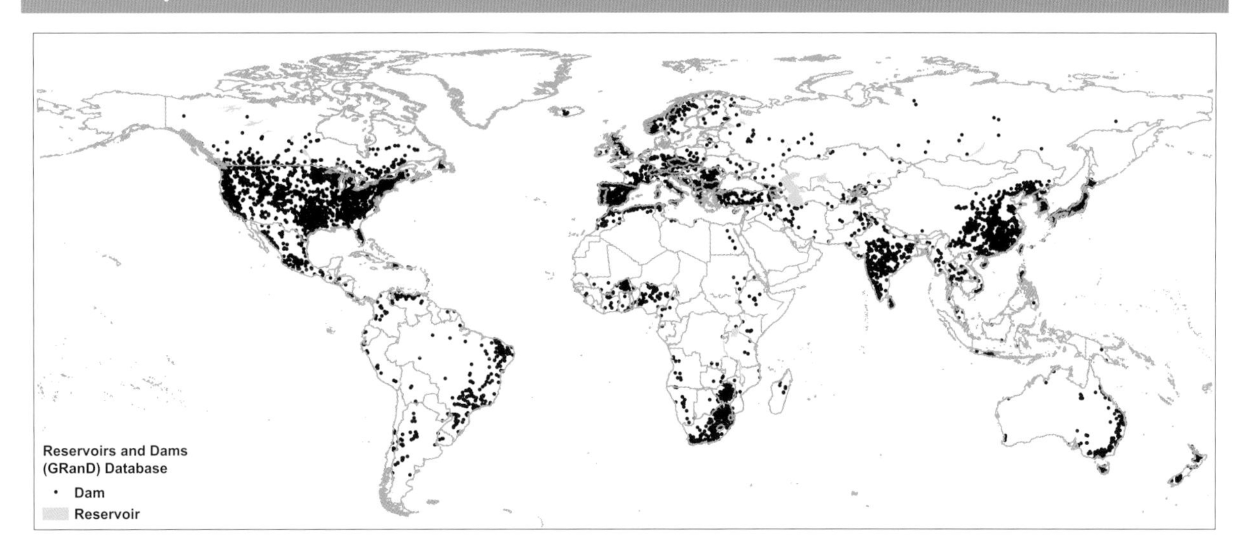

Source: Lehner *et al.* (2011).

water contamination, irregular flows, methane emissions and degraded ecosystems, including increased sedimentation, and invasive aquatic plant and animal species (Zarfl *et al.* 2014; UNEP 2016d). Increased sediment trapping associated with reservoirs has been linked to subsidence in deltaic areas and reduced soil fertility, for example the White Volta delta (Boateng, Bray and Hooke 2012; Anthony 2015). Recent construction of large dams has generated significant controversy in many areas, including Africa (e.g. Grand Ethiopian Renaissance Dam; Yihdego, Khalil and Salem 2017), Turkey, the Balkans and the Mekong basin.

Drought is a major risk for hydroelectric power generation. Brazil experienced one of its most debilitating droughts ever in 2015, resulting in decreased reservoir water levels and flows, with many hydroelectric power facilities nearing zero capacity, and causing water shortages to several major Brazilian cities, including São Paulo (Poindexter 2015). The Brazilian example demonstrates the need to foresee conditions that must be dealt with by engineering systems while striving to find an even balance between guaranteeing water supplies and minimizing social or environmental costs.

9.8 Impacts

9.8.1 Human health

Contamination of water and food by faecal material as a result of poor sanitation and poor hygiene, leading to unsafe drinking water, is a major cause of gastrointestinal illness, particularly diarrhoea. Diseases and organisms associated with diarrhoea include cholera, typhoid, hepatitis A, giardia and cryptosporidium (Lozano *et al.* 2013).

The most important known viral cause of diarrhoea (rotavirus) is being reduced by vaccination programmes (Burnett *et al.* 2017). Open defecation also causes important parasitic diseases transmitted via contact with soil and water (e.g. ascaris, hookworm, water snails) (McCarty, Turkeltaub and Hotez 2014; Lo *et al.* 2017).

While hygiene-related diseases have diminished greatly, deaths from diarrhoea still constituted the second most prevalent cause of death (about 13 per cent) in children aged one to four years in 2010 (Lozano *et al.* 2013). Sub-Saharan Africa and South Asia have the highest mortality rates associated with water, sanitation and hygiene (WHO 2017a). Chronic gastrointestinal infections, including those from parasites, cause disability, economic loss and cognitive impairment (Pinkerton *et al.* 2016; Lo *et al.* 2017). Because freshwater provides a habitat for mosquito-breeding, malaria and dengue fever exhibit an even higher disease burden for individuals residing near such habitats, although the situation is slowly improving (e.g. through widespread bed net use) (Ebi *et al.* 2016; Hemingway *et al.* 2016). Recent estimates of the burden of disease due to diarrheal diseases are summarized in **Figure 9.21**.

Health effects from water and sanitation-related diseases appear to vary by gender. Women may have less access to sanitation compared with men and spend more time in environments where open defecation has occurred, thus incurring a greater risk of parasite exposure. Gendered roles of fetching water and caring for young children, including disposing of their faeces, may further increase the exposure of women to sources of infection. Nevertheless, a recent systematic review and meta-analysis found that, overall, infectious diarrhoea was more common in males. Schistosomiasis was also more common in men, but cholera was more frequent in women (Sevilimedu *et al.* 2016).

Predicted hydrologic cycle changes associated with climate change may exacerbate the environmental health-related diseases, particularly diarrhoea (GBD 2015 DALYs and HALE Collaborators 2016; Mukabutera *et al.* 2016; Musengimana *et al.* 2016; Thiam *et al.* 2017).

9.8.2 Food security

Agricultural uses, primarily irrigation, account for 70 per cent of global water withdrawals (FAO 2016). Irrigated land, which accounted for 25 per cent of total cropland in 2012 (FAO 2016),

Figure 9.21: Morbidity (total disability-adjusted life years, DALYs) from diarrheal diseases (all ages) for females (upper graphic) and males (lower graphic), globally

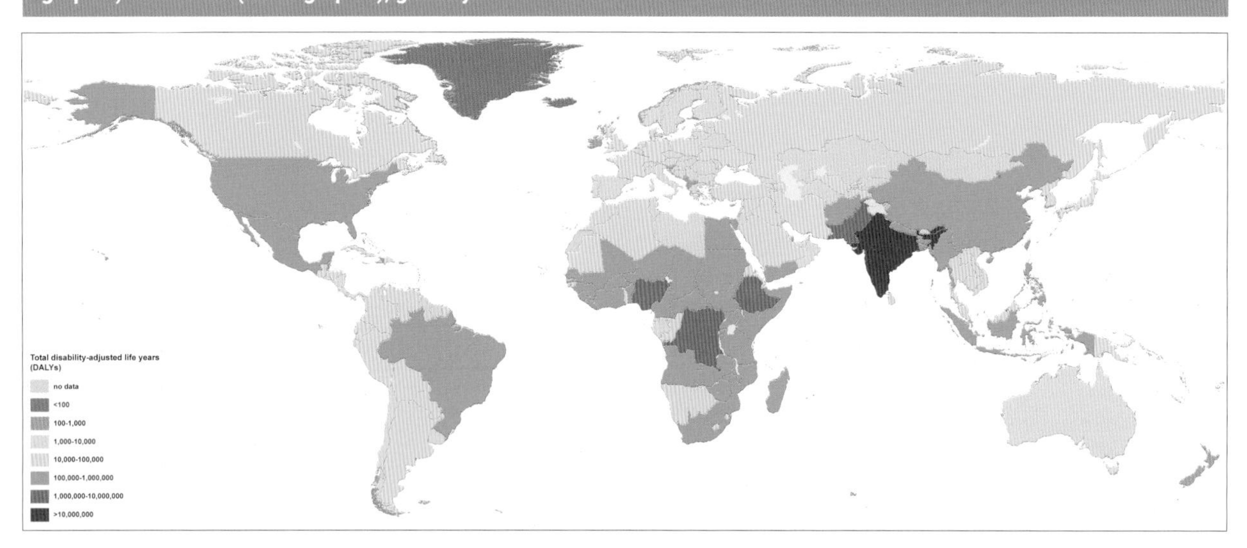

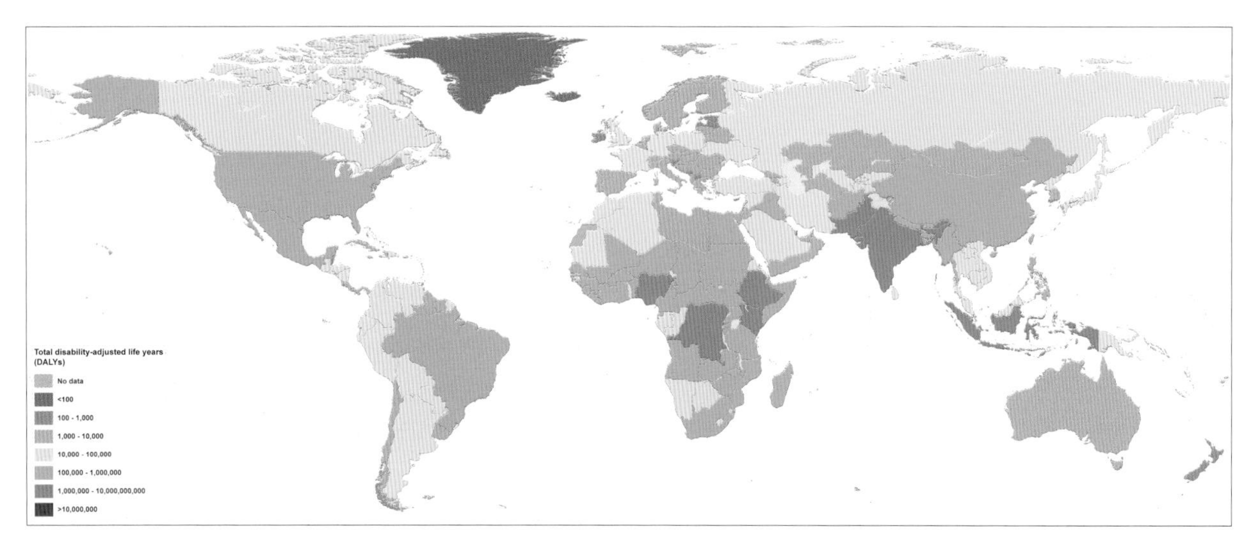

Source: GBD 2015 DALYs and HALE Collaborators 2016 (2016).

nevertheless represented half of global crop production (FAO 2016). Climate change effects on temperature and rainfall patterns may drive additional irrigation demands, with water scarcity in many parts possibly limiting crop yields by 2070 (Elliott *et al.* 2014). Efforts are under way worldwide to address predicted hydrologic changes, including shifts to more water-efficient irrigation technologies, while trade of agricultural products provides opportunities for improving food security and adjusting to water scarcity through food imports (United Nations 2017).

The quality and availability of irrigation water and irrigated land are projected to decrease concomitantly, with potential negative effects on food security and human health. About 34.2 million ha of irrigated area has been affected by salinization (Mateo-Sagasta and Burke 2012), representing 10 per cent of total irrigated area globally (324 million ha) (FAO 2017). About 60 per cent of irrigation water does not reach crops due to leakage, spillage and evaporation (FAO 2017), with losses being especially high in developing countries with poor irrigation infrastructure. Improved irrigation efficiency could make a substantial difference. The Mediterranean region could save 35 per cent of its irrigation water through efficiency improvements (Fader *et al.* 2016).

Food security and associated water demands are and will be further stressed by a growing population (FAO 2016). Changing food preferences with rising incomes also increases water demands, with livestock products being more water-intensive than crops. Global meat and dairy consumption are projected to increase by 89 per cent and 81 per cent, respectively, during 2002-2050, with higher growth rates in developing countries (Thornton 2010). However, use of drought-tolerant or flood-tolerant crops will be critical to improving the productivity of the agricultural industry with changing water supply conditions (Zandalinas *et al.* 2018).

The virtual water trade concept (i.e. water embedded in traded products ranging from crops to manufactured goods) illustrates the comparative advantages of certain water uses, including agriculture and energy, in particular regions (Gilmont *et al.* 2018). If water is appropriately priced and allocated, market forces can lead to overall efficiency by capitalizing on these advantages, with virtual water trade redistributing water efficiently, and partially helping to address the disconnect between consumption and production impacts (Mekonnen and Hoekstra 2011; Vörösmarty *et al.* 2015). However, water is not always priced and valued appropriately: water embedded in food commodities is controlled by supply chain corporations and international trade that neither account for ecosystem services nor costs of watershed degradation. The problem lies in the lack of accounting systems for water stewardship in market systems and the practice of subsidies and taxes to keep food prices low (Allan *et al.* 2015; Allan and Matthews 2016). Farmers are faced with the resulting pressures on food prices, further disempowering them from managing and sustaining water and ecosystems (Allan and Matthews 2016).

9.8.3 Human safety and security

Degraded water quality, physical and economic water scarcity, and loss of freshwater ecosystem services have significant impacts on human safety and security. Floods and droughts affect ever-larger numbers of vulnerable people (IPCC 2014), with security and migration implications magnified in transboundary basins.

Transboundary cooperation in addressing water scarcity, floods and droughts is challenging, but can enable more effective, efficient management and adaptation by pooling available data, models, scenarios and resources, and enlarging the planning space for locating adaptation measures, including transboundary basins (United Nations Economic Commission for Europe and International Network of Basin Organizations 2015). Transboundary water management creates benefits in international trade, climate change adaptation, economic growth, food security, and improved governance and regional integration.

About 286 international transboundary river basins involving 151 countries pose challenging management problems (UNEP-DHI Partnership and UNEP 2016), as do transboundary lakes and reservoirs. Further, there are currently 366 identified transboundary aquifers and 226 transboundary 'groundwater bodies' underlying almost every nation (International Groundwater Resources Assessment Centre and United Nations Educational Scientific and Cultural Organization - International Hydrological Programme 2015). Even within federated countries (e.g. Australia, India, the United States of America), transboundary problems may be no less acute at a state/provincial level. Although water management has historically led to cooperative, rather than conflicting, outcomes, significant conflicts between stakeholders can still occur over the implementation of international and inter-state agreements. Intensification of water pollution and water scarcity can cause tensions within and between nations, though rarely being the sole trigger of conflict, since a complex mix of social and political conflicts, economic, demographic and environmental factors, and military occupation and water wars (hegemony) is typically the origin of such conflicts.

9.9 Policy responses

Human activities now dominate changes in the biosphere and functioning of the Earth system (Green *et al.* 2015; Vörösmarty, Meybeck and Pastore 2015; Vörösmarty *et al.* 2015), "causing complex, and frequently unwanted outcomes including unprecedented changes to global water circulation" (Bhaduri *et al.* 2016).

Box 9.3: Jordan faces a combined refugee and water crisis

Jordan is one of the world's most water-scarce countries, providing only 150 m^3 of water annually per person, much lower than the 1,000 m^3/capita level denoting water scarcity. Jordan also currently hosts over 717,000 Syrian and Iraqi refugees, adding to freshwater supply pressures. The formerly permanent lush Azraq Oasis in Jordan used to cover more than 6,000 ha, supporting a variety of plant and animal life, including migratory birds, as well as being the main water source for Jordan's capital city, Amman. However, it was almost completely dried out by 1990, due to overexploitation of the underlying aquifer. By 2017, there were over 35,000 refugees living in the Azraq refugee camp in the oasis (United Nations High Commissioner for Refugees [UNHCR] 2017), an unsustainable situation contributing to further water stress (Alhajahmad and Lockhart 2017).

In addition to many other challenges facing sustainability (Yihdego and Salem 2017), the 'Future We Want' adopted by the Member States of the United Nations in 2012 (Rio +20) recognizes that "water is at the core of sustainable development" (United Nations General Assembly 2012 [66/288]; UNESCO and WWAP 2015). Urgent local-scale actions to meet human water needs, however, may trigger increased regional and global environmental stress, and trade-offs (Bhaduri *et al.* 2016).

With the adoption of the 2030 Agenda for Sustainable Development, the United Nations General Assembly adopted a dedicated water goal (SDG 6), incorporating eight targets, in a holistic framework linking drinking water provision, sanitation, water-use efficiency, water quality and sustainability. The framework includes targets on integrated water resource management and transboundary cooperation, and this section examines a variety of global and regional governance approaches and policy responses to achieve them. The effectiveness of specific examples is explored in Chapter 16.

9.9.1 Expanding access to safe drinking water and sanitation (SDG Targets 6.1 and 6.2)

Many policy tools and responses proved successful in increasing access to potable water and improved sanitation during the MDG period (2000-2015). Though water-related investments and appropriate policy tools remain a top global priority, there are significant differences across nations and between rural and urban areas. Innovative technology has played an important role, with the introduction of ventilated improved pit latrines by the United Nations Children's Fund (UNICEF) in the 1980s, and small-bore hand pumps proving effective in many parts of Africa. An array of technologies will be needed to meet the unique circumstances of individual communities and the aggressive goals of SDG targets 6.1 and 6.2. Construction of water-harvesting cisterns in Brazilian rural areas reduced time spent collecting water by 90 per cent (Gomes and Heller 2016). Nonetheless, much still remains to be done to narrow the gaps in access identified in Section 9.7.1.

Funding mechanisms
Drinking water and sanitation are recognized as basic human rights with considerable economic benefits realized through investing in water and sanitation provisions. These benefits can be quantified as an overall estimated gain of 1.5 per cent of global gross domestic product (GDP) and a US$4.3 return for every dollar invested. This is attributable to reduced human health-care costs, greater workplace productivity and involvement through better access to relevant facilities (WHO 2017b).

Inadequate funding, corruption and rapid population growth still limit the achievement of SDG water and sanitation targets in African, Latin American and West Asian countries (UNEP 2016c; UNEP 2016d; UNEP 2016f). The funding gap is partly being addressed with allocation of domestic funds, for example in the 2003 Pan African Implementation and Partnership Conference on Water Declaration (African Union [AU] 2015; UNEP 2016d). Latin American governments have provided wider access to safe potable water for vulnerable populations using public funds (United Nations Economic Commission for Latin America and the Caribbean 2017; UNEP 2016f). WHO launched 'TrackFin', a methodology to track financing of water, sanitation and hygiene at national level, enabling more evidence-based policymaking (UN-Water and WHO 2015).

Market approaches as policy tools
Drinking water access and sanitation are generally considered public goods financed and/or provided through governmental or quasi-governmental entities. However, water pricing for users that reflect water treatment costs (both capital and operations) and incentivize water conservation (Giannakis *et al.* 2016), as well as private investment in water, have become more common in parts of the world, while remaining controversial in others (Harris *et al.* 2015).

Regulatory programmes
Regulatory programmes throughout North America, Europe and many parts of Asia rely on enforceable regulations at multiple governmental levels, focusing on delivery of safe drinking water through public utilities and appropriate wastewater treatment before discharging. Drinking-water standards protect public health, especially for vulnerable communities. A robust regulatory programme, focusing on enforceable municipal and specifically industrial discharge permits, could improve the policy approach in many parts of Africa, Asia and Latin America (Masson, Walter and Priester 2013; Aguilar-Barajas *et al.* 2015; UNEP 2016f).

9.9.2 Improving water quality (SDG Targets 6.3 and 15.1)

This target focuses on reducing pollution, halving the proportion of untreated wastewater, and increasing water recycling and reuse globally, as a means of improving water quality both for human uses, addressed by the WHO Drinking Water Safety Plan, and aquatic ecosystem health.

In the pan-European region, the basis for wastewater discharge limits, and wastewater collection, discharge and treatment, was set by regional legal instruments, including the 1992 Convention on the Protection and Use of Transboundary Watercourses and International Lakes (which has been open to accession to all United Nations member states since 2016), including its Protocol on Water and Health, and the European Union's Urban Wastewater Treatment Directive (WWAP 2017). The implementation of these at national level has achieved water quality benefits beyond the implementing countries.

Knowledge about the quantity and quality of pollutants, and where they are released into water, remains a prerequisite for addressing water pollution, and its impacts on human and environmental health (Sustainable Facilities Tool 2017). Some countries (or regions) address this goal by Pollutant Release and Transfer Registers (see United Nations Economic Commission for Europe [UNECE] 1998). On a pathway to a circular economy (SDG 12), however, full 'life cycle analysis' and management should be considered.

9.9.3 Water-use efficiency and responses to water scarcity (SDG target 6.4)

Addressing water scarcity requires reduction of use and improved water-use efficiency. This includes water reuse, shifts to less demanding crops and industries, water rationing, improved agricultural practices, and use of virtual water trade accounting for embedded water costs. However, even higher water-use efficiency sometimes does not meet community needs, requiring development of additional water sources (e.g. rainwater harvesting, desalinization, fog interception). Water is transferred across large distances and even between drainage basins in arid regions (e.g. Salem 2009). Management strategies and technological improvements outlined here address water scarcity and stress.

Water efficiency

Improved water efficiency is central to the water-food-energy nexus, considering factors such as climate change, population and land use (Fader *et al.* 2016). Water efficiency refers to reducing water wastage, in contrast to water conservation, which focuses on reducing water use. To this end, growing food demands require increased productivity per litre of water. Increased water efficiency could also result in reduced water use for energy production, assuming a gradual transition to non-fossil fuel energy sources. Rapid urbanization requires protection of water sources, reduction of reticulation losses and increased water in storage.

Efficiency gains across sectors and regions have been realized through technology and management improvements. As the largest global water user, agriculture represents the greatest potential in water-use efficiency. However, inadequate global data exist to accurately evaluate the overall state and trends of industrial and domestic water-use efficiency. The UN-Water Integrated Monitoring Initiative, initiated in 2014, attempts to address the water-related global monitoring gaps (UN-Water 2017). Existing data are informing the transition from the MDGs to the SDGs, but spatial distribution and frequency of measurements need to improve to strengthen water-resource monitoring, modelling and management.

Desalinization

Desalinization addresses water scarcity in arid regions and large coastal cities such as the Gaza Strip on the Mediterranean Sea (United Nations Office for the Coordination of Humanitarian Affairs [OCHA] 2017). About 60 per cent of global desalinization occurs in arid West Asian countries (e.g. Bahrain, Kuwait, Oman, Qatar, Saudi Arabia, United Arab Emirates) (Abuzeid 2014; Abuzeid *et al.* 2014; UNEP 2016c). It is also becoming more common in California, United States of America and eastern Australia, which are prone to recurrent drought years (Little 2015; UNEP 2016a).

Impacts of desalinization include large energy demands, associated greenhouse gas emission risks, the effects of heavy brine releases into coastal ecosystems (Jenkins *et al.* 2012), and entrainment of marine organisms in infrastructure (Dawoud and Al Mulla 2012). The desalinization industry is working to mitigate these impacts and advances in membrane efficiency and energy efficiency may reduce the cost of doing so by 20 per cent over the next five years, and up to 60 per cent over the next 20 years (Voutchkov 2016).

Water rationing

In water scarcity conditions, water authorities and governments must prioritize water allocations to specific sectors and users. While rationing mechanisms are usually determined by legal water rights, there may also be emergency measures protecting the public and the economy (see also **Box 9.4**).

Water reuse

Water reuse or reclamation is the concept of treating wastewater as a resource, rather than as variably contaminated waste discharged to the environment (UNESCO and WWAP 2015). Reclaimed water is most commonly used in developed countries for non-potable purposes (e.g. agriculture, landscape and park irrigation), thermal power plant cooling, industrial processes, and enhancing natural or artificial lakes and wetlands (UNEP 2016a; UNEP 2016c). Singapore uses recycled water for indirect potable use and for direct non-potable use. Windhoek (Namibia) uses it to recharge aquifers which thereafter feed water into the bulk water supply. Recycling treated wastewater provides multiple benefits by decreasing water diversions from sensitive ecosystems and reducing wastewater discharges to surface waters, in addition to being a dependable, locally controlled water supply and an opportunity to create green jobs.

Box 9.4: How cities face water scarcity

In late February 2018, Cape Town faced the prospect of 'Day Zero', a term coined for the date – then estimated to be 9 July – when the city was expected to run out of water, taps would run dry, and all municipal supply would be rerouted to emergency pickup points (Poplak 2018). This severe urban water scarcity in Cape Town is significant because it could have been the first major modern city to literally run out of municipal water if Day Zero was not averted by sufficient rainfall in the early winter season. There have been past cases of other cities such as Barcelona, regional capital of Catalonia, suffering its worst drought in 2008 since records began 60 years ago, with reservoirs down to a quarter of normal capacity (Keeley 2008). In 2015, Brazil's financial capital, São Paulo, one of the world's most populated cities (over 21.7 million inhabitants) experienced an ordeal similar to that of Cape Town when its main reservoir fell below 4 per cent capacity (Gerberg 2015).

The situation in Cape Town was caused by a three-year drought, considered to be a roughly 1-in-400-year hydrological event, resulting in the levels of the largest storage reservoir (Theewaterskloof Dam) to drop to 11 per cent of capacity (Poplak 2018). However, this proximate cause needs to be understood within a context of efforts to redress historical inequities and overcome institutional divides, and the need to innovate in the face of climate change.

Analysis of water consumption data from 400,000 households (Visser and Brühl 2018) illustrates how Capetonians rallied to avert Day Zero. Over four years of water consumption data indicate that usage by all domestic consumption brackets converged, with 63 per cent of households reaching the recommended target (under 10.5 kilolitres per month) in July 2017, and 30 per cent of households reaching the lower target of 6 kilolitres per household per month even before it came into effect in February 2018. Hence, Cape Town succeeded in halving its water consumption within three years, through a common vision and commitment by its people. A take-home message for Cape Town, and possibly for the world, is that "people's faith in each other's ability to safeguard the remaining water as part of a common pool resource, is critical" (Visser and Brühl 2018).

Using treated wastewater for agricultural irrigation can fertilize crops and benefit production while preventing nutrients and organic matter from entering freshwater systems. Insufficiently treated wastewater, however, can introduce pathogens, metals, excessive nutrients, POPs and emerging contaminants, and pose grave risks to workers and surrounding communities. Increased regulation, investments in treatment and risk assessments are essential for safe wastewater reuse (WHO 2006).

In West Asia, the United Arab Emirates currently reuses all treated wastewaters (290 million m^3 per annum), while Saudi Arabia reuses 166 million m^3. This reclaimed water is reused for agricultural production in Saudi Arabia's Al Hassa Oasis, after being mixed with groundwater (UNEP 2016c).

Effective management considers an entire watershed or basin as a socio-ecological system integrating across agriculture, forestry, industry, domestic and commercial uses in the ecosystem context. This has improved water availability, sanitation and wastewater treatment in many countries (SDG 6.5 and 6.6) (UNEP 2016a; UNEP 2016f; UNEP 2016h). European river basin management identifies various pressures, classifies monitoring results and enforces environmental objectives (e.g. International Commission for the Protection of the Danube River 2008). There has also been substantial progress in transboundary river basin management (e.g. European Commission 1992; European Commission 2000). Furthering surface and groundwater governance requires cooperation from multinational to local levels, supported by real-time data and information management (Cross *et al.* 2016).

9.9.4 Water governance (SDG Target 6.5)

The commonly accepted definition of integrated water resources management (IWRM) is "a process which promotes the coordinated development and management of water, land and related resources, to maximise economic and social welfare in an equitable manner, without compromising the sustainability of vital ecosystems" (Global Water Partnership 2000). IWRM recognizes water both as a natural resource critical to society and economy and as an integral component of all ecosystems. While discussions over the merits of the IRWM approach continue (e.g. Jeffrey and Gearey 2006; Mukhtarov and Gerlak 2014), it is the major policy concept in place in over one hundred countries (Conca 2006; UNEP 2012a). IWRM is a progressive tool for reform, requiring strong political will for change, and contextual embedding in specific policy problems. However, it is not a panacea for all complexities of water governance (Ingram 2013). The Organisation for Economic Co-operation and Development (OECD) Principles of Water Governance have relevance to IWRM, emphasizing trust between stakeholders. A complementary approach recognizing the buffering capacity of lakes, wetlands and standing water systems is integrated lake basin management (ILBM), which focuses on "gradual, continuous and holistic improvement of basin governance by basin stakeholders" (Research Center for Sustainability and Environment-Shiga University and ILEC 2014).

The SDG 6.5 target calls upon all countries to implement IWRM at all levels by 2030, including through transboundary cooperation. Likely transboundary impacts on water resources are also often addressed in the procedures under the Espoo Convention and its Strategic Environmental Assessment (SEA) Protocol. To facilitate transboundary water system assessments and management, UN Environment, in collaboration with the Global Environment Facility (GEF) and partners, prepared a global assessment of the status of transboundary lakes, rivers, aquifers and small island groundwater systems, large marine ecosystems and open oceans, Transboundary Waters Assessment Programme (TWAP), (UNEP 2011). The International Groundwater Resources Assessment Centre (IGRAC), a TWAP partner, developed a groundwater information management system to tackle the paucity of standardized quantitative real-time data on key groundwater parameters, and underlined the lack of adequate groundwater governance at all levels.

Recent developments in international water law have significantly strengthened the legal basis regarding shared still (lentic) and flowing (lotic) surface waters and groundwaters. The 1997 Convention on the Law of the Non-navigational Uses of International Watercourses (UN Watercourses Convention) entered into force; the 1992 Convention on the Protection and Use of Transboundary Watercourses and International Lakes (UNECE Water Convention, as amended 2013) was opened to all United Nations member states; and the International Law Commission's 2008 Draft Articles on the Law of Transboundary Aquifers were commended to governments by the United Nations General Assembly. The two conventions, now operating in tandem at the global level, act as an important catalyst for the revision of existing agreements and negotiation of new river, lake and aquifer agreements at basin scale. Financing to support implementation of existing agreements remains a challenge. They are complemented by the United Nations Framework Convention on Climate Change, the Convention on Wetlands of International Importance especially as waterfowl habitat (Ramsar 2016) and the Convention on Biological Diversity (Convention on Biological Diversity [CBD] 1992), which address the protection of water-related ecosystems. Regional-level instruments for water management include the EU Water Framework Directive (European Union 2000).

9.9.5 Surface water-groundwater conjunctive management

Groundwater depletion can lead to streamflow depletion (Hunt 1999; Kendy and Bredehoeft 2006), while streamflow diversions can limit groundwater recharge. Managing these two sources as separate entities arises from limited knowledge of groundwater systems and their spatial and temporal relationships with surface waters, a situation that is no longer justifiable (Famiglietti 2014; McNutt 2014). Current experiences highlight the value of conjunctively managing and using surface and groundwater as 'one water' (Sticklor 2014), thereby buffering against both droughts and floods. Sound management would consider the potential long-term impact of sustained groundwater abstraction on groundwater-dependent ecosystems in arid or semi-arid areas.

Aquifer storage and recovery recharge (Pyne 1995) or managed aquifer recharge (Dillon *et al.* 2009) are becoming important tools to battle chronic water scarcity (e.g. in the state of Arizona, United States of America) (Lacher *et al.* 2014; Scanlon *et al.* 2016; Stefan and Ansems 2017). Underground water storage could play a significant role in semi-arid and arid parts of Africa (e.g. Botswana, South Africa) during episodic heavy rain events, and/or where surface-water storage and transfer options are exhausted (Tredoux, van der Merwe and Peters 2009; Bugan *et al.* 2016). Capture and storage of monsoon

rains in depleted aquifers is being piloted in India (International Water Management Institute 2016).

Water sensitive urban design principles are critical to water-use efficiency, reuse (Wong 2011; Fisher-Jeffes, Carden and Armitage 2017) and flood management (Dai *et al.* 2018); for example, storing reclaimed storm and wastewater from the urban environment in aquifer(s). This approach is especially effective in mitigating subsidence of, and saline intrusion into, coastal city aquifers (Ortuño *et al.* 2010; Bugan *et al.* 2016).

Box 9.5 illustrates conjunctive development of surface and groundwater for Hermanus, a coastal town in South Africa, without inducing saline intrusion – a case aimed at mitigation of drought risks by balancing surface- and groundwater storage.

Box 9.5: Hermanus, near Cape Town, Western Cape Province, South Africa: A case study for conjunctive surface- and groundwater development and management

Groundwater was used for private housing developments and garden irrigation between 1971 and 2001. The greater Hermanus area water demand was met by the DeBos Dam (blue line, running concurrently with the purple total supply line in **Figure 9.22**). During 2002, 7,750 kilolitres/year of groundwater (green line) came online, with 24,191 kl/year added in 2009, as illustrated by the total supply (purple line) separating from the DeBos Dam inflow (blue line), with the groundwater addition (green line) keeping the supply line above the red demand line.

Groundwater augmentation was particularly effective in keeping supply above demand in 2010, when the DeBos Dam supply (blue line) could not meet it (red line). The water demand was met by surface-water supply from the dam, augmented by three well fields. Water restrictions were introduced in 2009 in anticipation of reduced surface-water supply and later lifted.

In contrast to other towns throughout the Western Cape Province suffering from severe drought, residents in the Greater Hermanus area were only advised on 27 February 2018 that it would be necessary to introduce Level 1B water restrictions from 1 March 2018; although water tariffs would only be increased once the dam had dropped to 40 per cent full level. The DeBos Dam was 46.5 per cent full on this date. (Overstrand Municipality 2018).

Figure 9.22: Hermanus Conjunctive Use

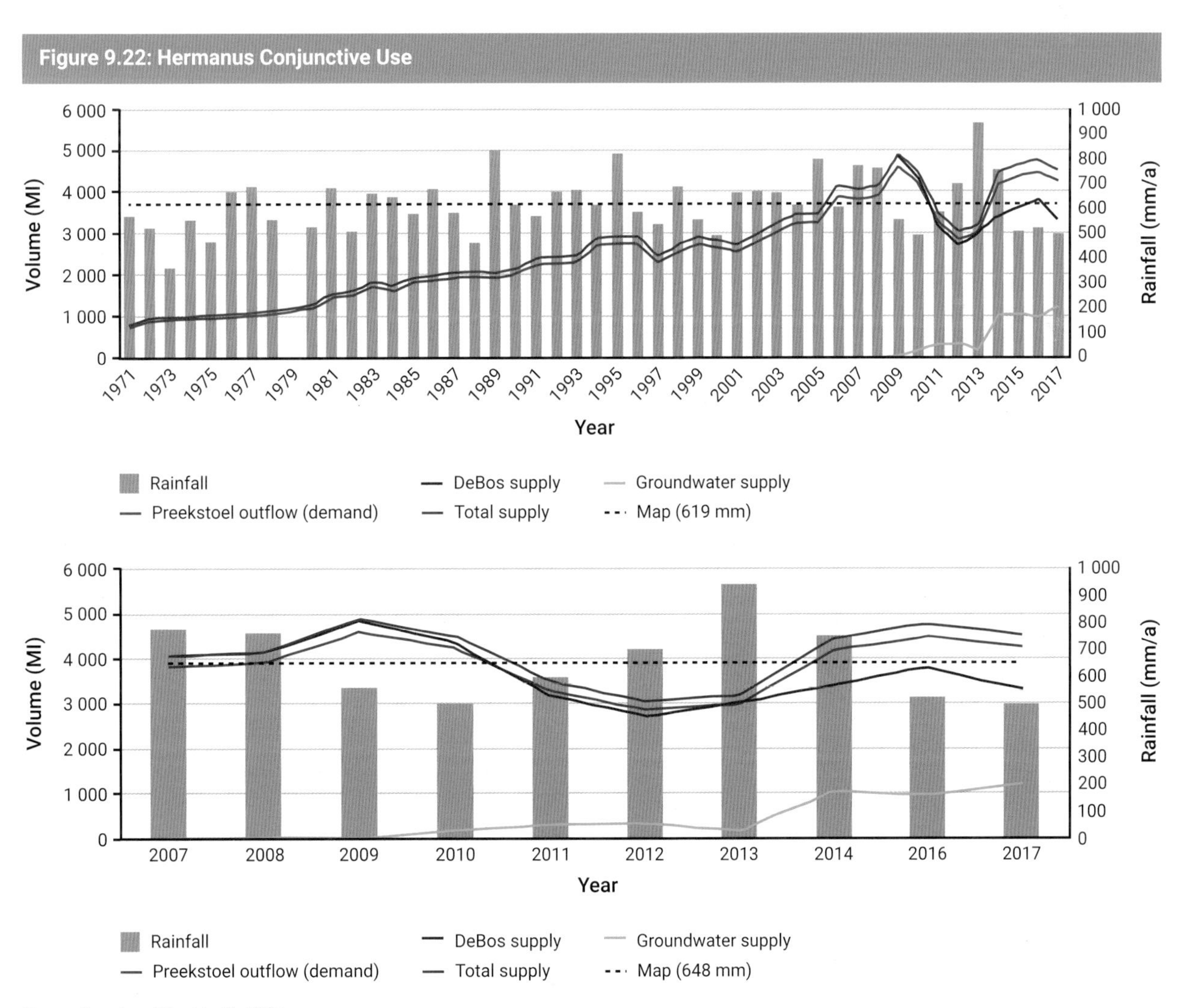

Source: Overstrand Municipality 2018.

Figure 9.23: Supply of and demand for water, Greater Hermanus, 1971-2001 (a) and 2002-2017 (b)

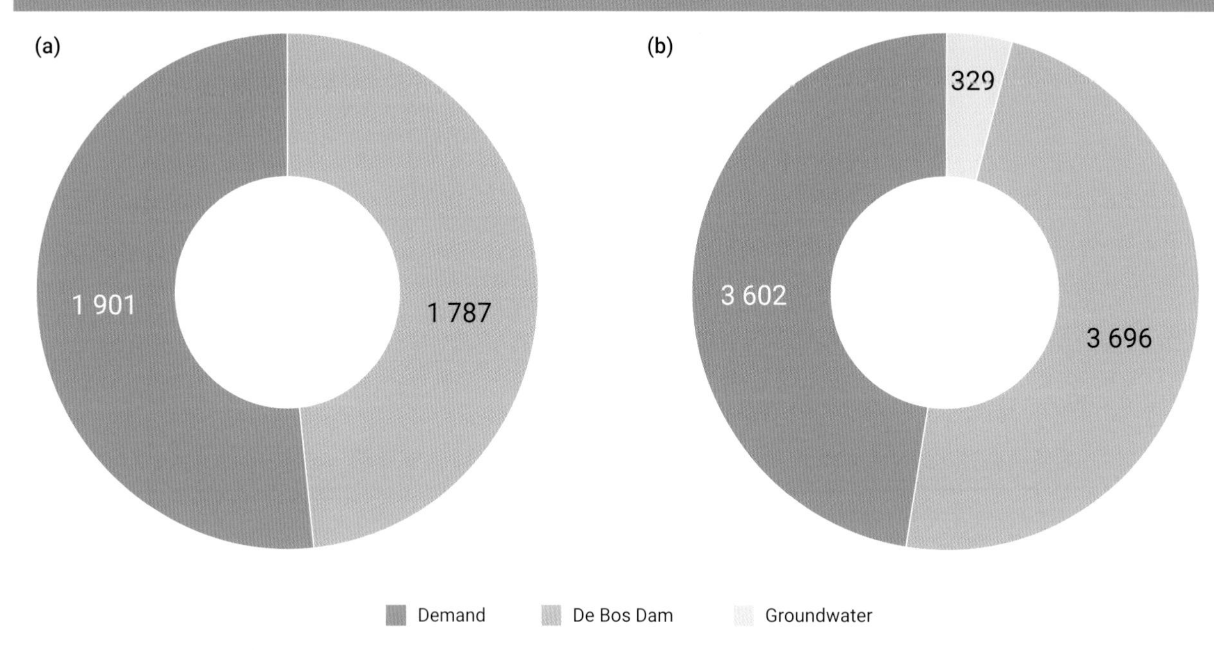

Source: Overstrand Municipality (2018).

Underlying successful conjunctive use is comprehensive monitoring, modelling and risk assessment of the aquifer and surface-water source(s), associated catchment areas and social systems in a learn-by-doing approach (Bidwell 2003). Managing aquifer resources requires a land-use zoning system based on aquifer vulnerability and constraints, to enable adequate abstraction rates and natural recharge (Cross *et al.* 2016).

Monitoring and management of the full water-use cycle by the private sector (e.g. agriculture and mining) is gaining recognition. Examples of stewardship programmes include Woolworths in South Africa (in partnership with WWF-South Africa, WWF-UK, the Alliance for Water Stewardship and Marks and Spencer); Coca-Cola and the United States Agency for International Development (USAID) Water and Development Alliance, H&M and WaterAid (Workers' Need Project in India), Unilever and Nestlé in Europe. The CEO Water Mandate (https://ceowatermandate.org) was instrumental in promoting the business benefits of water stewardship. This intersection between water governance, use, users, real-time monitoring, and modelling to inform evidence-based resource development and management is gaining momentum. In the fast-growing city of Bangalore, where 40 per cent of the water entering the system is lost to leakages, Water Supply and Sewerage Works formed an alliance with IBM and installed flow meters at several critical points in the water reticulation system. Data is being transmitted via GSM (Global System for Mobile communications) technology to a central Supervisory Control and Data Acquisition (SCADA) server to be transformed, aggregated and presented on a web interface and mobile application for end users.

9.9.6 Protecting and restoring water ecosystems (SDG target 6.6)

The importance of water-related ecosystems is specifically reflected in the water goal (SDG 6) and the terrestrial biodiversity goal (SDG 15). Target 6.6 aims "to protect and restore water-related ecosystems, including mountains, forests, wetlands, rivers, aquifers and lakes," emphasizing their crucial role in water cycle functions and watershed management.

SDG 6.6 monitors changes in the spatial extent of water-related ecosystems. Given wetland losses and associated biodiversity declines, many countries respond with natural wetland protection and management programmes and environmental flow requirements (e.g. Mexico's water reserves; South African National Water Act 1998 [Government of South Africa 1998]). River and wetland restoration and construction efforts are proceeding, including constructed wetlands for storm water treatment in Australia, recapturing floodplain areas in The Netherlands, and reconnecting wetlands and lakes to the main stem of the Yangtze River in China. Improved Earth observation data, combined with a classification methodology, enable countries to gain accurate pictures of their water-related ecosystems. However, there is a pressing need to extend on-the-ground monitoring of water cycle components and harmonize observations.

The Ramsar Convention on Wetlands (1971) is a multinational environmental agreement, devoted specifically to the conservation and wise use of wetlands. Each signatory country must designate and protect one or more "Wetlands of International Importance" (known as 'Ramsar sites'). As Contracting Parties to the Convention, 170 countries had designated 2,326 Ramsar sites by early 2018. The total wetland area protected by the Ramsar designation has increased from 81 million ha to almost 250 million ha since 2000 (**Figure 9.24**). New Ramsar sites designated in recent years tend to follow hydrological boundaries, to protect whole catchments and river basins (Ramsar 2018).

Figure 9.24: Ramsar sites designated by year and by region

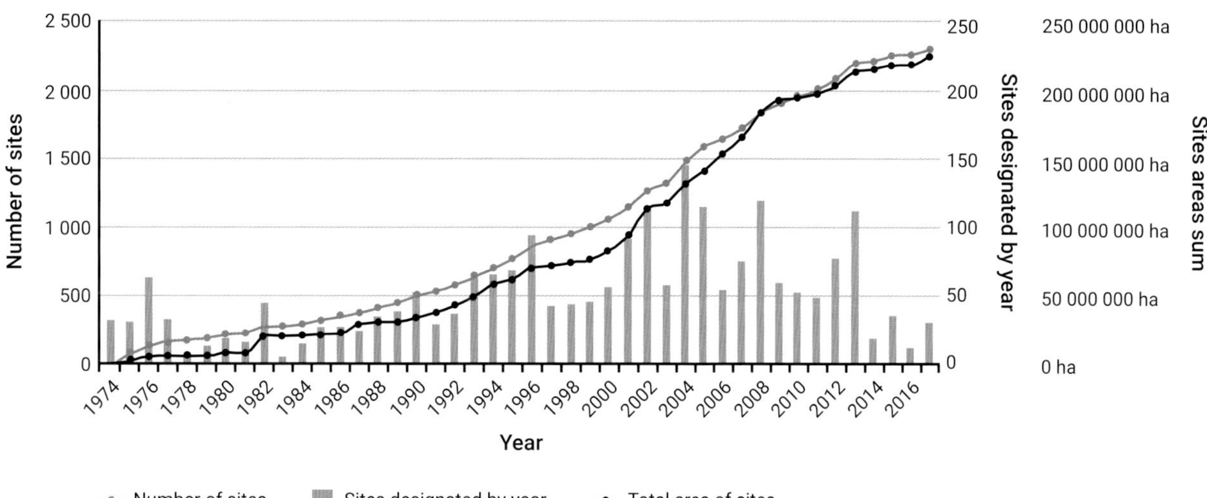

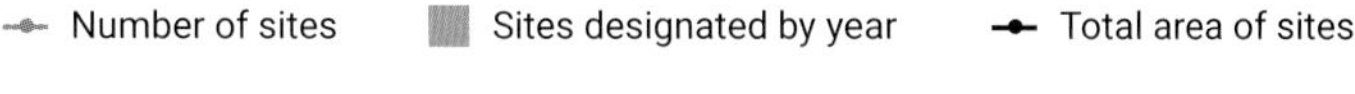

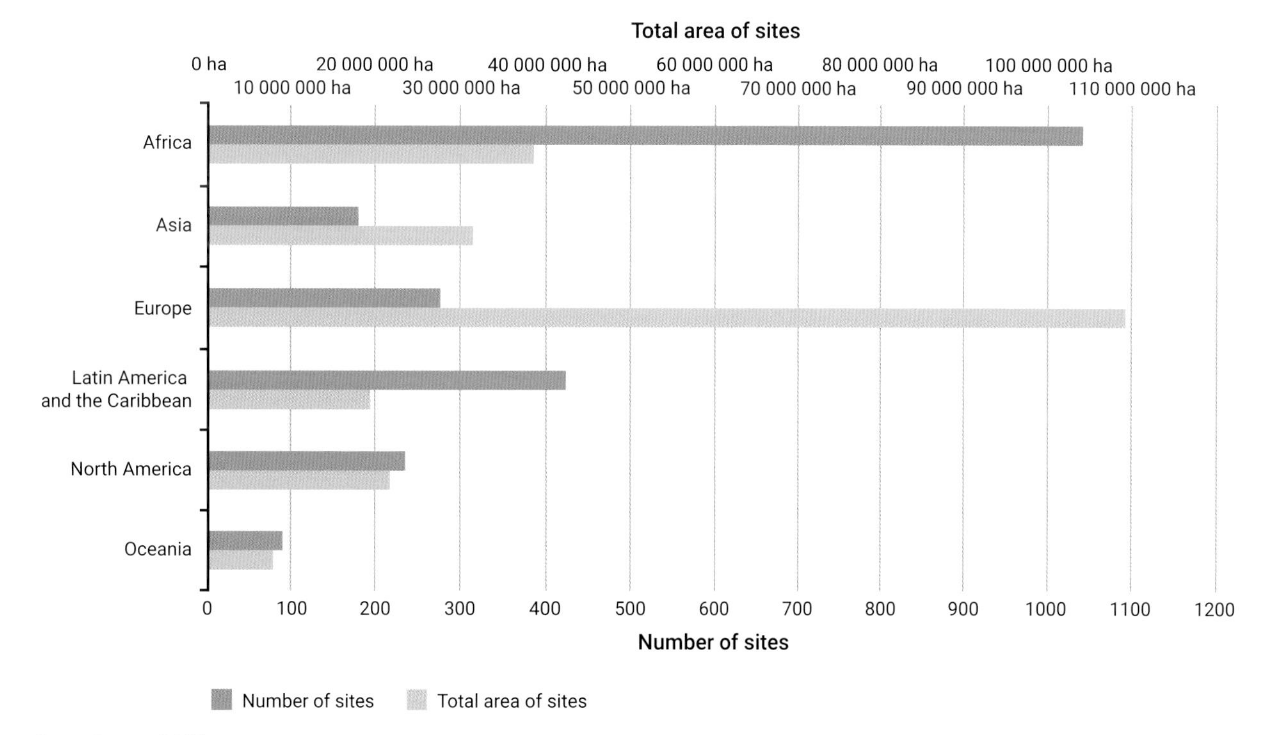

Source: Ramsar (2018).

9.10 Conclusions

Widespread water scarcity is now an outcome of the connections and linkages between the hydrological cycle, unsustainable agriculture, and energy systems. At a local level, water is contested and plays a part in social conflicts and human migration decisions, against this backdrop of complex interlinkages. At a global level, the water cycle integrates the impacts of human activities, population growth and climate change. The deterioration of water quality across regions and continents threatens the health of people and ecosystems, while climate change is accelerating the water cycle and causing increased impacts on communities through storms, floods and droughts, extreme wildfires and landslides, as well as increasing dust and sandstorms in the most arid areas. Hence water, in addition to being a public good, is now becoming a risk multiplier for the health of people and of the planet.

However, the realization of SDG 6 (water) targets can be achieved through engaging public, private and non-governmental sectors, civil society and local actors, and by mutual reinforcement or trade-offs that also consider other interlinked SDG goals focused on poverty eradication (SDG 1), food security (SDG 2), health (SDG 3), gender equity (SDG 5), sustainable cities (SDG 11) and protection of biodiversity (SDGs 14 and 15).

Multinational environmental agreements (MEAs) governing water resources and water-related ecosystem management and climate change can support the embedding of integrated water resources management in the rules of law – through national and local legislation.

Effective, efficient and transparent water resources governance is required that includes improved collaboration and coordination between governments, technical institutions, non-governmental organizations and civil society towards improved monitoring and data quality, culminating in better hydrological and hydrogeological services, as discussed in the recent WMO conference held in May 2018 (World Meteorological Organization 2018). Increased investment in the scope and rigor of standardized water data is essential to improve policy and governance for sound water management.

References

AbuZeid, K. (2014). An Arab perspective on the applicability of the water convention in the Arab region: Key aspects and opportunities for the Arab countries. *Key Aspects and Opportunities for the Arab Countries*. Tunis, 11-12 June 2014. http://gwp.dev.sublime.se/globalassets/global/gwp-med-files/news-and-activities/mena/tunis-workshop_june2014/5.2.k.abuzeid_waterconvention__arab_region.pdf

AbuZeid, K., Elrawady, M., CEDARE and Arab Water Council (2014). *2nd Arab State of the Water Report 2012*. Water Resources Management Program-Center for Environment and Development for the Arab Region & Europe (CEDARE) and Arab Water Council http://www.arabwatercouncil.org/images/Publications/Arab_state/2nd_Arab_State_of_the_Water_Report.pdf.

Adams, M.P., Hovey, R.K., Hipsey, M.R., Bruce, L.C., Ghisalberti, M., Lowe, R.J. *et al.* (2016). Feedback between sediment and light for seagrass: Where is it important? *Limnology and Oceanography* 61(6), 1937-1955. https://doi.org/10.1002/lno.10319.

African Union (2015). *Annual Report of the Commission on the Implementation of the July 2008 Assembly Declaration on the SharmEl-Sheikh Commitments for Accelerating the Achievement of Water and Sanitation Goals in Africa. Assembly Decision (Assembly/AU/DEC.1(XI) of July 2008. The 2015 Africa Water and Sanitation Sector Report: Reviewing Progress and Positioning Africa for 2030 SDGs & Agenda 2063 on Sustainable Water Management and Sanitation*. http://www.amcow-online.org/images/docs/monitoring_evaluation_report_2015_eng.pdf.

Aguilar-Barajas, I., Mahlknecht, J., Kaledin, J., Kjellén, M. and Abel Mejía, B. (eds.) (2015). *Water and Cities in Latin-America: Challenges for Sustainable Development*. 1st Edition. Water Policy. London: Routledge. https://www.routledge.com/Water-and-Cities-in-Latin-America-Challenges-for-Sustainable-Development/Aguilar-Barajas-Mahlknecht-Kaledin-Kjellen-Mejia-Betancourt/p/book/9780415730976.

Alhajahmad, S. and Lockhart, D. (2017). *Jordan's Recent Economic Performance: Implications for Future Growth, Investment, Refugee Policy, and Refugees*. Amman: West Asia–North Africa (WANA) Institute. http://wanainstitute.org/sites/default/files/publications/Publication_JordansRecentEconomicPerformance_English.pdf.

Allan, T., Keulertz, M. and Colman, T. (2015). The complexity and urgency of water: Time for the accountancy profession to step up. *Global Knowledge Gateway, International Federation of Accountants*, International Federation of Accountants https://www.ifac.org/global-knowledge-gateway/viewpoints/complexity-and-urgency-water-time-accountancy-profession-step.

Allan, T. and Matthews, N. (2016). The water, energy and food nexus and ecosystems: The political economy of food non-food supply chains. In *The Water, Food, Energy and Climate Nexus: Challenges and an Agenda for Action*. Dodds, F. and Bartram, J. (eds.). Oxon: Routledge. 78-89. https://cgspace.cgiar.org/handle/10568/78373.

Altchenko, Y. and Villholth, K.G. (2013). Transboundary aquifer mapping and management in Africa: A harmonised approach. *Hydrogeology Journal* 21(7), 1497-1517. https://doi.org/10.1007/s10040-013-1002-3.

Arunakumara, K.K.I.U., Walpola, B.C. and Yoon, M.H. (2013). Current status of heavy metal contamination in Asia's rice lands. *Reviews in Environmental Science and Bio-Technology* 12(4), 355-377. https://doi.org/10.1007/s11157-013-9323-1.

Barbour, E., Wilson, G., Radcliffe, J., Ding, Y. and Li, Y. (2016). A review of pumped hydro energy storage development in significant international electricity markets. *Renewable and Sustainable Energy Reviews* 61, 421-432. https://doi.org/10.1016/j.rser.2016.04.019.

Bates, B.C., Kundzewicz, Z.W., Wu, S. and Palutikof, J.P. (eds.) (2008). *Climate Change and Water: IPCC Technical Paper VI* Technical Paper of the Intergovernmental Panel on Climate Change. Geneva: Intergovernmental Panel on Climate Change. https://www.ipcc.ch/pdf/technical-papers/climate-change-water-en.pdf.

Berendonk, T.U., Manaia, C.M., Merlin, C., Fatta-Kassinos, D., Cytryn, E., Walsh, F. *et al.* (2015). Tackling antibiotic resistance: The environmental framework. *Nature Reviews Microbiology* 13, 310-317. https://doi.org/10.1038/nrmicro3439.

Bergeron, S., Boopathy, R., Nathaniel, R., Corbin, A. and LaFleur, G. (2015). Presence of antibiotic resistant bacteria and antibiotic resistance genes in raw source water and treated drinking water. *International Biodeterioration & Biodegradation* 102, 370-374. https://doi.org/10.1016/j.ibiod.2015.04.017.

Bhaduri, A., Bogardi, J., Siddiqi, A., Voigt, H., Vörösmarty, C., Pahl-Wostl , C. *et al.* (2016). Achieving sustainable development goals from a water perspective. *Frontiers in Environmental Science* 4(64), 1-13. https://doi.org/10.3389/fenvs.2016.00064.

Bidwell, V. (2003). *Groundwater Management Tools: Analytical Procedure and Case Studies*. http://citeseerx.ist.psu.edu/viewdoc/download?doi=10.1.1.75.1401&rep=rep1&type=pdf.

Birkinshaw, S.J., Bathurst, J.C., Iroume, A. and Palacios, H. (2011). The effect of forest cover on peak flow and sediment discharge-an integrated field and modelling study in central-southern Chile. *Hydrological Processes* 25(8), 1284-1297. https://doi.org/10.1002/hyp.7900.

Blake, D., Mlisa, A. and Hartnady, C. (2010). Large scale quantification of aquifer storage and volumes from the Peninsula and Skurweberg Formations in the southwestern Cape. *Water SA* 36(2), 177-184. http://www.scielo.org.za/pdf/wsa/v36n2/a05v36n2.pdf.

Bleischwitz, R., Hoff, H., Spataru, C., van der Voet, E. and vanDeveer, S.D. (eds.) (2018). *Routledge Handbook of the Resource Nexus*. https://www.routledge.com/Routledge-Handbook-of-the-Resource-Nexus/Bleischwitz-Hoff-Spataru-Voet-VanDeveer/p/book/9781138675490.

Blum, M.D. and Roberts, H.H. (2009). Drowning of the Mississippi Delta due to insufficient sediment supply and global sea-level rise. *Nature Geoscience* 2, 488-491. https://doi.org/10.1038/ngeo553.

Boateng, I., Bray, M. and Hooke, J. (2012). Estimating the fluvial sediment input to the coastal sediment budget: A case study of Ghana. *Geomorphology* 138(1), 100-110. https://doi.org/10.1016/j.geomorph.2011.08.028.

Bugan, R.D.H., Jovanovic, N., Israel, S., Tredoux, G., Genthe, B., Steyn, M. *et al.* (2016). Four decades of water recycling in Atlantis (Western Cape, South Africa): Past, present and future. *Water SA* 42(4), 577-594. https://doi.org/10.4314/wsa.v42i4.08.

Bundesanstalt für Geowissenschaften und Rohstoffe and United Nations Educational Scientific and Cultural Organization (2008). *Groundwater resources of the world*. Bundesanstalt für Geowissenschaften und Rohstoffe https://www.whymap.org/whymap/EN/Home/gw_world_g.html;jsessionid=FED2359D86966F2836EC2516AC762173.1_cid292?nn=1577094.

Burnett, E., Jonesteller, C.L., Tate, J.E., Yen, C. and Parashar, U.D. (2017). Global impact of rotavirus vaccination on childhood hospitalizations and mortality from diarrhea. *Journal of Infectious Diseases* 215(11), 1666-1672. https://doi.org/10.1093/infdis/jix186.

Campuzano, C., Hansen, A.M., De Stefano, L., Martínez-Santos, P., Torrente, D. and Willaarts, B.A. (2014). Water resources assessment. In *Water for Food and Wellbeing in Latin America and the Caribbean: Social and Environmental Implications for a Globalized Economy*. Garrido, A., Willaarts, B.A. and Llamas, M.R. (eds.). Oxon: Routledge. chapter 2. 27-53. http://www.fundacionbotin.org/89dguuytdfr276ed_uploads/Observatorio%20Tendencias/PUBLICACIONES/LIBROS%20SEM%20INTERN/water%20for%20food%20security/capitulo2.pdf.

Carmenta, R., Zabala, A. and Phelps, J. (2015). *Indonesian Peatland Fires: Perceptions of Solutions*. Bogor: Center for International Forestry Research. http://www.cifor.org/publications/pdf_files/flyer/5882-flyer.pdf.

Chevallier, P., Pouyaud, B., Suarez, W. and Condom, T. (2011). Climate change threats to environment in the tropic Andes: Glaciers and water resources. *Regional Environmental Change* 11 179-187. https://doi.org/10.1007/s10113-010-0177-6.

Cloern, J. and Jassby, A. (2012). Drivers of change in estuarine-coastal ecosystems: Discoveries from four decades of study in San Francisco Bay. *Reviews of Geophysics* 50(4), 1-33. https://doi.org/10.1029/2012RG000397.

Coe, M.T. and Foley, J.A. (2001). Human and natural impacts on the water resources of the Lake Chad basin. *Journal of Geophysical Research* 106(D4), 3349-3356. https://doi.org/10.1029/2000JD900587.

Conca, K. (2006). *Governing Water: Contentious Transnational Politics and Global Institution Building*. Cambridge, MA: The MIT Press. https://mitpress.mit.edu/books/governing-water.

Corada-Fernández, C., Candela, L., Torres-Fuentes, N., Pintado-Herrera, M.G., Paniw, M. and González-Mazo, E. (2017). Effects of extreme rainfall events on the distribution of selected emerging contaminants in surface and groundwater: The Guadalete River basin (SW, Spain). *Science of the Total Environment* 605-606, 770-783. https://doi.org/10.1016/j.scitotenv.2017.06.049.

Costanza, R., de Groot, R., Sutton, P., van der Ploeg, S., Anderson, S.J., Kubiszewski, I. *et al.* (2014). Changes in the global value of ecosystem services. *Global Environmental Change-Human and Policy Dimensions* 26, 152-158. https://doi.org/10.1016/j.gloenvcha.2014.04.002.

Couture, N.J., Irrgang, A., Pollard, W., Lantuit, H. and Fritz, M. (2018). Coastal erosion of permafrost soils along the Yukon Coastal Plain and fluxes of organic carbon to the Canadian Beaufort Sea. *Journal of Geophysical Research: Biogeosciences* 123(2), 406-422. https://doi.org/10.1002/2017JG004166.

Cross, K., Laban, P., Paden, M. and Smith, M. (2016). *Spring: Managing Groundwater Sustainably*. Gland: International Union for Conservation of Nature. https://portals.iucn.org/library/sites/library/files/documents/2016-039.pdf.

Da Rosa, C.D., Lyon, J.S. and Hocker, P.M. (1997). *Golden Dreams, Poisoned Streams: How Reckless Mining Pollutes America's Waters, and How We Can Stop It*. Hocker, P.M. and Aley, T.J. (eds.). Washington, D.C: Mineral Policy Center. https://trove.nla.gov.au/work/29007642?q&versionId=35310352.

Dai, A. (2011). Drought under global warming: A review. *Wiley Interdisciplinary Reviews-Climate Change* 2(1), 45-65. https://doi.org/10.1002/wcc.81.

Dai, Y., Wang, L., Yao, T., Li, X., Zhu, L. and Zhang, X. (2018). Observed and simulated lake effect precipitation over the Tibetian Plateau: An initial study at Nam Co Lake. *Journal of Geophysical Research-Atmospheres* 123(13), 6746-6759. https://doi.org/10.1029/2018JD028330.

Dargie, G.C., Lewis, S.L., Lawson, I.T., Mitchard, E.T.A., Page, S.E., Bocko, Y.E. *et al.* (2017). Age, extent and carbon storage of the central Congo Basin peatland complex. *Nature* 542(7639), 86–90. https://doi.org/10.1038/nature21048.

Davidson, N.C. (2014). How much wetland has the world lost? Long-term and recent trends in global wetland area. *Marine and Freshwater Research* 65(10), 934-941. https://doi.org/10.1071/mf14173.

Dawoud, M.A. and Al Mulla, M.M. (2012). Environmental impacts of seawater desalination: Arabian gulf case study. *International Journal of Environment and Sustainability* 1(3), 22-37. https://www.sciencetarget.com/Journal/index.php/IJES/article/view/96/25.

de Groot, R., Brander, L., van der Ploeg, S., Costanza, R., Bernard, F., Braat, L. *et al.* (2012). Global estimates of the value of ecosystems and their services in monetary units. *Ecosystem Services* 1(1), 50-61. https://doi.org/10.1016/j.ecoser.2012.07.005.

Dillon, P.J., Pavelic, P., Page, D., Beringen, H. and Ward, J. (2009). *Managed Aquifer Recharge: An Introduction*. Waterlines Report Series. Canberra: National Water Commission. https://www.researchgate.net/publication/304620744_Managed_aquifer_recharge_an_introduction_Waterlines_Report_Series_no_13_February_2009_National_Water_Commisssion_Canberra.

Dris, R., Gasperi, J., Rocher, V., Saad, M., Renault, N. and Tassin, B. (2015). Microplastic contamination in an urban area: A case study in Greater Paris. *Environmental Chemistry* 12(5), 592-599. https://doi.org/10.1071/EN14167.

Driscoll, C.T., Driscoll, K.M., Fakhraei, H. and Civerolo, K. (2016). Long-term temporal trends and spatial patterns in the acid-base chemistry of lakes in the Adirondack region of New York in response to decreases in acidic deposition. *Atmospheric Environment* 146, 5-14. https://doi.org/10.1016/j.atmosenv.2016.08.034.

Ebi, K.L. and Nealon, J. (2016). Dengue in a changing climate. *Environmental Research* 151, 115-123. https://doi.org/10.1016/j.envres.2016.07.026.

Echols, K.R., Meadows, J.C. and Orazion, C.E. (2009). Pollution of aquatic ecosystems II: Hydrocarbons, synthetic organics, radionuclides, heavy metals, acids and thermal pollution. In *Encyclopedia of Inland Waters*. Likens, G.E. and Tochner, K. (eds.). Waltham, MA: Academic Press. 120-128. https://www.researchgate.net/publication/284820061_Pollution_of_Aquatic_Ecosystems_II_Hydrocarbons_Synthetic_Organics_Radionuclides_Heavy_Metals_Acids_and_Thermal_Pollution.

Elliott, J., Deryng, D., Müller, C., Frieler, K., Konzmann, M., Gerten, D. *et al.* (2014). Constraints and potentials of future irrigation water availability on agricultural production under climate change. *Proceedings of the National Academy of Sciences* 111(9), 3239-3244. https://doi.org/10.1073/pnas.1222474110.

Ellison, D., Futter, M.N. and Bishop K. (2012). On the forest cover-water yield debate: From demand to supply-side thinking. *Global Change Biology* 18(3), 806-820. https://doi.org/10.1111/j.1365-2486.2011.02589.x.

European Commission (1992). *Commission Decision of 27 July 1992 concerning Questionnaires relating to directives in the water Sector (92/446/EEC)*. https://eur-lex.europa.eu/legal-content/EN/TXT/PDF/?uri=CELEX:31992D0446&from=EN.

European Commission (2000). *Directive 2000/60/EC of the European Parliament and of the Council of 23 October 2000 establishing A Framework for Community Action in the Field of Water Policy*. https://eur-lex.europa.eu/legal-content/EN/TXT/PDF/?uri=CELEX:32000L0060&from=EN.

European Commission (2008). *Groundwater Protection in Europe: The New Groundwater Directive – Consolidating The EU Regulatory Framework*. Brussels: European Commission. http://ec.europa.eu/environment/water/water-framework/groundwater/pdf/brochure/en.pdf.

Fader, M., Shi, S., von Bloh, W., Bondeau, A. and Cramer, W. (2016). Mediterranean irrigation under climate change: More efficient irrigation needed to compensate for increases in irrigation water requirements. *Hydrology and Earth System Sciences* 20(2), 953-973. https://doi.org/10.5194/hess-20-953-2016.

Famiglietti, J.S. (2014). The global groundwater crisis. *Nature Climate Change* 4(11), 945-948. https://doi.org/10.1038/nclimate2425.

Famiglietti, J.S. and Rodell, M. (2013). Water in the balance. *Science* 340(6138), 1300-1301. https://doi.org/10.1126/science.1236460.

Farinotti, D., Longuevergne, L., Moholdt, G., Duethmann, D., Moelg, T., Bolch, T. *et al.* (2015). Substantial glacier mass loss in the Tien Shan over the past 50 years. *Nature Geoscience* 8(9), 716-722. https://doi.org/10.1038/ngeo2513.

Filho, L.W., Echevarria Icaza, L., Emanche, V.O. and Quasem Al-Amin, A. (2017). An evidence-based review of impacts, strategies and tools to mitigate urban heat islands. *International Journal of Environmental Research and Public Health* 14(12), 1600. https://doi.org/10.3390/ijerph14121600.

Fisher-Jeffes, L., Kirsty C. and Armitage, N. (2017). A water sensitive urban design framework for South Africa. *Journal of Town and Regional Planning* 71, 1-10. https://doi.org/10.18820/2415-0495/trp71i1.1.

Food and Agriculture Organization of the United Nations (2016). *Water Withdrawal by Sector, Around 2010*. Rome http://www.fao.org/nr/water/aquastat/tables/WorldData-Withdrawal_eng.pdf.

Food and Agriculture Organization of the United Nations (2017). *AQUASTAT website*. http://www.fao.org/nr/water/aquastat/main/index.stm (Accessed: 12 January 2018).

Foster, S., Pulido-Bosch, A., Vallejos, Á., Molina, L., Llop, A. and MacDonald, A.M. (2018). Impact of irrigated agriculture on groundwater-recharge salinity: A major sustainability concern in semi-arid regions. *Hydrogeology Journal*. https://doi.org/10.1007/s10040-018-1830-2.

Foster, S., Tyson, G., Colvin, C., Wireman, M., Manzano, M., Kreamer, D. *et al.* (2016). *Ecosystem Conservation and Groundwater*. International Association of Hydrogeologists. https://www.researchgate.net/publication/297698654_Ecosystem_Conservation_Groundwater.

Gao, H., Bohn, T.J., Podest, E., McDonald, K.C. and Lettenmaier, D.P. (2011). On the causes of the shrinking of Lake Chad. *Environmental Research Letters* 6(3). https://doi.org/10.1088/1748-9326/6/3/034021.

GBD 2015 DALYs and HALE Collaborators (2016). Global, regional, and national disability-adjusted life-years (DALYs) for 315 diseases and injuries and healthy life expectancy (HALE), 1990–2015: A systematic analysis for the Global Burden of Disease study 2015. *Lancet* 388, 1603 – 1658. https://doi.org/10.1016/S0140-6736(16)31460-X.

Gerberg, J. (2015). A megacity without water: São Paulo's drought. *Time Magazine*. 13 October 2015. http://time.com/4054262/drought-brazil-video/.

Giannakis, E., Bruggeman, A., Djuma, H., Kozyra, J. and Hammer, J. (2016). Water pricing and irrigation across Europe: Opportunities and constraints for adopting irrigation scheduling decision support systems. *Water Science and Technology: Water Supply* 16(1), 245-252. https://doi.org/10.2166/ws.2015.136.

Gilmont, M., Nassar, L., Rayner, S., Tal, N., Harper, E. and Salem, H. (2018). The potential for enhanced water decoupling in the Jordan Basin through regional agricultural best practice. *Land* 7(2), 63. https://doi.org/10.3390/land7020063.

Global Land Ice Measurements from Space (2018). *Monitoring the World's Changing Glaciers*. Global Land Ice Measurements from Space https://www.glims.org/.

Global Water Partnership (2000). *Integrated Water Resources Management*. Technical Advisory Committee Background Paper. Stockholm. https://www.gwp.org/globalassets/global/gwp-cacena_files/en/pdf/tec04.pdf.

Gomes, U.A.F. and Heller, L. (2016). Acesso à água proporcionado pelo Programa de Formação e Mobilização Social para Convivência com o Semiárido: Um Milhão de Cisternas Rurais: combate à seca ou ruptura da vulnerabilidade? *Engenharia Sanitária e Ambiental* 21(3), 623-633. https://doi.org/10.1590/S1413-41522016128417.

Government of South Africa Act No 36 of 1998 (1998). South Africa National Water Act 1998. 20 August 1998. http://portal.unesco.org/en/files/47385/12670886571NWA_1998.pdf/NWA%2B1998.pdf.

Grangier, C., Qadir, M. and Singh, M. (2012). Health implications for children in wastewater-irrigated peri-urban Aleppo, Syria. *Water Quality, Exposure and Health* 4(4), 187-195. https://doi.org/10.1007/s12403-012-0078-7.

Great Artesian Basin Coordinating Committee (2016). *Great Artesian Basin: Resource Study 2014*. Canberra. http://www.gabcc.gov.au/sitecollectionimages/resources/66540f98-c828-4268-8b8b-b37f8193cde7/files/resource-study-2016.pdf.

Green, P.A., Voeroesmarty, C.J., Harrison, I., Farrell, T., Saenz, L. and Fekete, B.M. (2015). Freshwater ecosystem services supporting humans: Pivoting from water crisis to water solutions. *Global Environmental Change-Human and Policy Dimensions* 34, 108-118. https://doi.org/10.1016/j.gloenvcha.2015.06.007.

Gross-Sorokin, M.Y., Roast, S.D. and Brighty, G.C. (2006). Assessment of feminization of male fish in English rivers by the Environment Agency of England and Wales. *Environmental health perspectives* 114, 147-151. https://doi.org/10.1289/ehp.8068.

Guzinski, R., Kass, S., Huber, S., Bauer-Gottwein, P., Jensen, I., Naeimi, V. *et al.* (2014). Enabling the use of earth observation data for integrated water resource management in Africa with the water observation and information system. *Remote Sensing* 6(8), 7819-7839. https://doi.org/10.3390/rs6087819.

Hansen, M.C., Potapov, P.V., Moore, R., Hancher, M., Turubanova, S.A.A., Tyukavina, A. *et al.* (2013). High-resolution global maps of 21st-century forest cover change. *Science* 342(6160), 850-853. http://doi.org/10.1126/science.1244693. Data available on-line from: http://earthenginepartners.appspot.com/science-2013-global-forest.

Harris, L.M., Phartiyal, J., Scott, D.N. and Peloso, M. (2015). Women Talking about Water: Feminist Subjectivities and Intersectional Understandings. *Canadian Women's Studies Journal* 30(2-3), 15-24. https://doi.org/10.14288/1.0366125.

Hay, E.R. and Hartnady, C.J.H. (2001). Development of deep groundwater reserve of strategic importance. *Journal of the South African Institution of Civil Engineering* 9(5), 13-16.

Hemingway, J., Ranson, H., Magill, A., Kolaczinski, J., Fornadel, C., Gimnig, J. *et al.* (2016). Averting a malaria disaster: Will insecticide resistance derail malaria control? *Lancet* 387(10029), 1785-1788. https://doi.org/10.1016/s0140-6736(15)00417-1.

Hoekstra, A.Y. and Mekonnen, M.M. (2012). The water footprint of humanity. *Proceedings of the National Academy of Sciences* 109(9), 3232 – 3237. https://doi.org/10.1073/pnas.1109936109.

Holling, C.S. and Gunderson, L.H. (2002). Resilience and adaptive cycles. In *Panarchy: Understanding Transformations In Human and Ecological Systems*. Gunderson, L.H. and Holling, C.S. (eds.). Washington, D.C.: Island Press. 25-62. https://vtechworks.lib.vt.edu/handle/10919/67621?show=full.

Holloway, A. (2003). Disaster risk reduction in southern Africa: Hot rhetoric—cold reality. *African Security Review* 12(1), 29-38. https://doi.org/10.1080/10246029.2003.9627568.

Horton, A.A., Walton, A., Spurgeon, D.J., Lahive, E. and Svendsen, C. (2017). Microplastics in freshwater and terrestrial environments: Evaluating the current understanding to identify the knowledge gaps and future research priorities. *Science of the Total Environment* 586, 127-141. https://doi.org/10.1016/j.scitotenv.2017.01.190.

Hunt, B. (1999). Unsteady stream depletion from ground water pumping. *Ground Water* 37(1), 98-102. https://doi.org/10.1111/j.1745-6584.1999.tb00962.x.

Huntington, T.G. (2006). Evidence for intensification of the global water cycle: Review and synthesis. *Journal of Hydrology* 319, 83-95. https://doi.org/10.1016/j.jhydrol.2005.07.003.

Huss, M. (2012). Extrapolating glacier mass balance to the mountain-range scale: The European Alps 1900-2100. *Cryosphere* 6(4), 713-727. https://doi.org/10.5194/tc-6-713-2012.

Ingram, H. (2013). No universal remedies: Design for contexts. *Water International* 38(1), 6-11. https://doi.org/10.1080/02508060.2012.739076.

Intergovernmental Panel on Climate Change (2014). *Climate Change 2014: Synthesis Report. Contribution of Working Groups I, II and III to the Fifth Assessment Report of the Intergovernmental Panel on Climate Change*. Pachauri, R.K. and Meyer, L.A. (eds.). Geneva. https://www.ipcc.ch/pdf/assessment-report/ar5/syr/AR5_SYR_FINAL_All_Topics.pdf.

Intergovernmental Science-Policy Platform on Biodiversity and Ecosystem Services (2017). *The Assessment Report on Pollinators, Pollination and Food Production*. Potts, S.G., Imperatriz-Fonseca, V.L. and Ngo, H.T. (eds.). Bonn. https://www.ipbes.net/sites/default/files/downloads/pdf/individual_chapters_pollination_20170305.pdf.

International Commission for the Protection of the Danube River (2008). *Analysis of the Tisza River Basin 2007: Initial Step Toward the Tisza River Basin Management Plan – 2009*. Vienna. http://www.icpdr.org/main/sites/default/files/Tisza_RB_Analysis_2007.pdf.

International Groundwater Resources Assessment Centre and United Nations Educational Scientific and Cultural Organization - International Hydrological Programme (2015). *Transboundary aquifers of the world map 2015*. International Groundwater Resources Assessment Centre, Delft. https://www.un-igrac.org/resource/transboundary-aquifers-world-map-2015.

International Lake Environment Committee Foundation and United Nation Environment Programme (2016). *Transboundary Lakes and Reservoirs: Status and Trends. Volume 2: Lake Basins and Reservoirs*. Nairobi. http://geftwap.org/publications/TWAPVOLUME2TRANSBOUNDARYLAKESANDRESERVOIRS.pdf.

International Water Management Institute (2016). *Managing the monsoon*. [International Water Management Institute http://www.iwmi.cgiar.org/2016/05/managing-the-monsoon/ (Accessed: 24 June 2017).

INTJ Input (2017). PFAS: new biohazards identified in fast food wrappers. 19 April https://intjinput.wordpress.com/tag/environmental-science-and-technology-letters/.

Jayachandran, S. (2009). Air quality and early-life mortality: Evidence from Indonesia's wildfires. *The Journal of Human Resources* 44(4), 916-954. https://doi.org/10.3386/w14011.

Jeffrey, P. and Gearey, M. (2006). Integrated water resources management: Lost on the road from ambition to realisation? *Water Science and Technology* 53(1), 1-8. https://doi.org/10.2166/wst.2006.001.

Jenkins, S., Paduan, J., Roberts, P., Schlenk, D. and Weis, J. (2012). *Management of Brine Discharges to Coastal Waters: Recommendations of a Science Advisory Panel*. Technical Report. Costa Mesa, CA: State Water Resources Control Board. http://wedocs.unep.org/bitstream/handle/20.500.11822/20069/jenkins_management.pdf?sequence=1&isAllowed=y.

Jeppersen, E., Moss, B., Bennion, H., Carvalho, L., DeMeester, L., Feuchtmayr, H. *et al.* (2010). Interaction of climate change and eutrophication. In *Climate Change Impacts on Freshwater Ecosystems*. Kernan, M., Battarbee, R. and Moss. B. (eds.). Blackwell Publishing Ltd. chapter 6. 119-151. https://doi.org/10.1002/9781444327397.Chapter 6.

Joosten, H. (2015). *Peatlands, Climate Change Mitigation and Biodiversity Conservation: An Issue Brief on the Importance of Peatlands for Carbon and Biodiversity Conservation and the Role of Drained Peatlands as Greenhouse Gas Emission Hotspots*. Copenhagen: Nordic Council of Ministers. https://www.ramsar.org/sites/default/files/documents/library/ny_2_korrektur_anp_peatland.pdf.

Keeley, G. (2008). Barcelona forced to import emergency water. *The Guardian* 14 May 2008 https://www.theguardian.com/world/2008/may/14/spain.water.

Kendy, E. and Bredehoeft, J.D. (2006). Transient effects of groundwater pumping and surface-water-irrigation returns on streamflow. *Water Resources Research* 42(8). https://doi.org/10.1029/2005wr004792.

Kim, S., De Jonghe, J., Kulesa, A.B., Feldman, D., Vatanen, T., Bhattacharyya, R.P. *et al.* (2017). High-throughput automated microfluidic sample preparation for accurate microbial genomics. *Nature Communications* 8(13919). https://doi.org/10.1038/ncomms13919.

Kodama, Y., Eaton, F. and Wendler, G. (1983). The influence of Lake Minchumina, interior Alaska, on its surroundings. *Archives for Meteorology Geophysics and Bioclimatology Series B* 33(3), 199-218. https://doi.org/10.1007/bf02275094.

Kolpin, D.W., Furlong, E.T., Meyer, M.T., Thurman, E.M., Zaugg, S.D., Barber, L.B. *et al.* (2002). Pharmaceuticals, hormones, and other organic wastewater contaminants in US streams, 1999-2000: A national reconnaissance. *Environmental Science & Technology* 36(6), 1202-1211. https://doi.org/10.1021/es011055j.

Kümmerer, K. (2009). Antibiotics in the aquatic environment – A review – Part II. *Chemosphere* 75(4), 435-441. https://doi.org/10.1016/j.chemosphere.2008.12.006.

Lacher, L.J., Turner, D.S., Gungle, B., Bushman, B.M. and Richter, H.E. (2014). Application of hydrological tools and monitoring to support managed aquifer recharge decision making in the upper San Pedro River, Arizona, USA. *Water Resources Research* 6(11), 3495 – 3527. https://doi.org/10.3390/w6113495.

Laird, N.F., Kristovich, D.A.R., Liang, X.Z., Arritt, R.W. and Labas, K. (2001). Lake Michigan lake breezes: Climatology, local forcing, and synoptic environment. *Journal of Applied Meteorology* 40(3), 409-424. https://doi.org/10.1175/1520-0450(2001)040<0409:lmlbcl>2.0.co;2.

Lehmann, J., Coumou, D. and Frieler, K. (2015). Increased record-breaking precipitation events under global warming. *Climatic Change* 132(4), 501-515. https://doi.org/10.1007/s10584-015-1434-y.

Lehner, B., Liermann, C.R., Revenga, C., Voeroesmarty, C., Fekete, B., Crouzet, P. *et al.* (2011). High-resolution mapping of the world's reservoirs and dams for sustainable river-flow management. *Frontiers in Ecology and the Environment* 9(9), 494-502. https://doi.org/10.1890/100125.

Little, A. (2015). Can desalination counter the drought? *The New Yorker*. 22 July 2015. http://www.newyorker.com/tech/elements/can-desalination-counter-the-drought.

Liu, P.L.F., Lynett, P., Fernando, H., Jaffe, B.E., Fritz, H., Higman, B. *et al.* (2005). Observations by the international tsunami survey team in Sri Lanka. *Science* 308(5728), 1595. https://doi.org/10.1126/science.1110730.

Liu, W., Sun, F., Lim, W.H., Zhang, J., Wang, H., Shiogama, H. *et al.* (2018). Global drought and severe drought-affected popullations in 1.5 and 2°C warmer worlds. *Earth System Dynamics* 9(1), 267-283. https://doi.org/10.5194/esd-9-267-2018.

Lo, N.C., Addiss, D.G., Hotez, P.J., King, C.H., Stothard, J.R., Evans, D.S. *et al.* (2017). A call to strengthen the global strategy against schistosomiasis and soil-transmitted helminthiasis: The time is now. *The Lancet Infectious Diseases* 17(2), e64-e69. https://doi.org/10.1016/S1473-3099(16)30535-7.

Lozano, R., Naghavi, M., Foreman, K., Lim, S., Shibuya, K., Aboyans, V. *et al.* (2013). Global and regional mortality from 235 causes of death for 20 age groups in 1990 and 2010: A systematic analysis for the global burden of disease study 2010. *Lancet* 380(9859), 2095-2128. https://doi.org/10.1016/S0140-6736(12)61728-0.

Lu, Y., Song, S., Wang, R., Liu, Z., Meng, J., Sweetman, A.J. *et al.* (2015). Impacts of soil and water pollution on food safety and health risks in China. *Environment International* 77, 5-15. https://doi.org/10.1016/j.envint.2014.12.010.

MacDonald, A.M., Bonsor, H.C., Dochartaigh, B.E.O. and Taylor, R.G. (2012). Quantitative maps of groundwater resources in Africa. *Environmental Research Letters* 7(2). https://doi.org/10.1088/1748-9326/7/2/024009.

Margat, J. and van der Gun, J. (2013). *Groundwater Around the World: A Geographic Synopsis*. 1st edn. London: CRC Press. https://www.crcpress.com/Groundwater-around-the-World-A-Geographic-Synopsis/Margat-Gun/p/book/9781138000346.

Masson, M., Walter, M. and Priester, M. (2013). *Incentivizing Clean Technology in the Mining Sector in Latin America and the Caribbean: The Role of Public Mining Institutions*. IDB Technical Note. Washington, D.C.: Inter-American Development Bank. https://publications.iadb.org/bitstream/handle/11319/6018/Incentivizing%20Clean%20Technology%20in%20the%20Mining%20Sector%20in%20Latin%20America%20and%20the%20Caribbean.pdf?sequence=1&isAllowed=y.

Masten, S.J., Davies, S.H. and McElmurry, S.P. (2016). Flint water crisis: What happened and why? *Journal of the American Water Works Association* 108(12), 22-34. https://doi.org/10.5942/jawwa.2016.108.0195.

Mateo-Sagasta, J. and Burke, J. (2012). *Agriculture and Water Quality Interactions: A Global Overview*. SOLAW Background Thematic Report. Rome: Food and Agricultural Organization of the United Nations. http://www.fao.org/3/a-bl092e.pdf.

Maupin, M.A., Kenny, J.F., Hutson, S.S., Lovelace, J.K., Barber, N.L. and Linsey, K.S. (2014). *Estimated Use of Water in the United States in 2010*. United States Geological Survey. https://pubs.usgs.gov/circ/1405/pdf/circ1405.pdf.

McCarty, T.R., Turkeltaub, J.A. and Hotez, P.J. (2014). Global progress towards eliminating gastrointestinal helminth infections. *Current Opinion in Gastroenterology* 30(1), 18-24. https://doi.org/10.1097/mog.0000000000000025.

McInnes, R.J., Simpson, M., Lopez, B., Hawkins, R. and Shore, R. (2016). Wetland ecosystem services and the Ramsar convention: An assessment of needs. *Wetlands* 37(1), 123-134. https://doi.org/10.1007/s13157-016-0849-1.

McNutt, M. (2014). The drought you can't see. *Science* 345(6204), 1543. https://doi.org/10.1126/science.1260795.

Meeker, J.D. (2012). Exposure to environmental endocrine disruptors and child development. *Archives of Pediatrics Adolescent Medicine* 166(10), 952-958. https://doi.org/10.1001/archpediatrics.2012.241.

Mekonnen, M.M. and Hoekstra, A.Y. (2011). *National Water Footprint Accounts: The Green, Blue and Grey Water Footprint of Production and Consumption*. Value of Water Research Report Series. Delft: United Nations Educational, Scientific and Cultural Organization-IHE Institute for Water Education. http://waterfootprint.org/media/downloads/Report50-NationalWaterFootprints-Vol1.pdf.

Mitsch, W.J. and Gosselink, J.G. (2015). *Wetlands, 5th Edition*. 5th edn: Wiley. https://www.wiley.com/en-us/Wetlands%2C+5th+Edition-p-9781118676820.

Mompelat, S., Le Bot, B. and Thomas, O. (2009). Occurrence and fate of pharmaceutical products and by-products, from resource to drinking water. *Environment International* 35(5), 803-814. https://doi.org/10.1016/j.envint.2008.10.008.

Morris, B.L., Lawrence, A.R.L., Chilton, P.J.C., Adams, B., Calow, R.C. and Klinck, B.A. (2003). *Groundwater and its Susceptibility to Degradation: A Global Assessment of the Problem and Options for Management*. Early Warning and Assessment Report Series. Nairobi: United Nations Environment Programme. https://www.ircwash.org/sites/default/files/BGS-2003-Groundwater.pdf.

Mudd, G.M., Weng, Z. and Jowitt, S.M. (2013). A detailed assessment of global Cu resource trends and endowments. *Economic Geology* 108(5), 1163–1183. https://doi.org/10.2113/econgeo.109.7.1813.

Mukabutera, A., Thomson, D., Murray, M., Basinga, P., Nyirazinyoye, L., Atwood, S. *et al.* (2016). Rainfall variation and child health: Effect of rainfall on diarrhea among under 5 children in Rwanda, 2010. *BMC Public Health* 16(1), 731. https://doi.org/10.1186/s12889-016-3435-9.

Mukhtarov, F. and Gerlak, A. (2014). Epistemic forms of integrated water resources management: Towards knowledge versatility. *Policy Sciences* 47(2), 101-120. https://doi.org/10.1007/s11077-013-9193-y.

Musengimana, G., Mukinda, F.K., Machekano, R. and Mahomed, H. (2016). Temperature variability and occurrence of diarrhoea in children under five-years-old in Cape Town metropolitan sub-Districts. *International Journal of Environmental Research and Public Health* 13(9), 859. https://doi.org/10.3390/ijerph13090859.

Novotny, V., Wang, X., Englande, A.J.J., Bedoya, D., Promakasikorn, L. and Tirado, R. (2010). Comparative assessment of pollution by the use of industrial agricultural fertilizers in four rapidly developing Asian countries. *Environment Development and Sustainability* 12(4), 491-509. https://doi.org/10.1007/s10668-009-9207-2.

O'Connor, J.E., Duda, J.J. and Grant, G.E. (2015). 1000 dams down and counting. *Science* 348(6234), 496-497. https://doi.org/10.1126/science.aaa9204.

Oki, T. and Kanae, S. (2006). Global hydrological cycles and world water resources. *Science* 313(5790), 1068-1072. https://doi.org/10.1126/science.1128845.

O'Neil, J.M., Davis, T.W., Burford, M.A. and Gobler, C.J. (2012). The rise of harmful cyanobacteria blooms: The potential roles of eutrophication and climate change. *Harmful Algae* 14, 313-334. https://doi.org/10.1016/j.hal.2011.10.027.

O'Neill Commission (2014). *Antimicrobial Resistance: Tackling A Crisis for The Health and Wealth of Nations*. London. https://amr-review.org/sites/default/files/AMR%20Review%20Paper%20-%20Tackling%20a%20crisis%20for%20the%20health%20and%20wealth%20of%20nations_1.pdf.

Organization for Economic Cooperation and Development (1982). *Eutrophication of Waters: Monitoring, Assessment and Control*. Paris. https://catalogue.nla.gov.au/Record/1904210.

Organization for Economic Co-operation and Development (2016). *OECD Council: Recommendations on Water*. Paris. https://www.oecd.org/environment/resources/Council-Recommendation-on-water.pdf.

Ortuño, F., Molinero, J., Custodio, E., Juárez, I., Garrido, T. and Fraile, J. (2010). Seawater intrusion barrier in the deltaic Llobregat aquifer (Barcelona, Spain): Performance and pilot phase results. *21st Salt Water Intrusion Meeting*. Azores, 21-26 June 2010. http://www.swim-site.nl/pdf/swim21/pages_135_138.pdf.

Osborn, S.G., Vengosh, A., Warner, N.R. and Jackson, R.B. (2011). Methane contamination of drinking water accompanying gas-well drilling and hydraulic fracturing. *Proceedings of the National Academy of Sciences* 108(20), 8172-8176. https://doi.org/10.1073/pnas.1100682108.

Overstrand Municipality (2018). *Overstrand implements Level 1B water restrictions from 1 March*. https://www.overstrand.gov.za/en/media-section/news/330-overstrand-implements-level-1b-water-restrictions-from-1-march (Accessed: 9 October 2019).

Pearce, F. (2018). *Can the world find solutions to the nitrogen pollution crisis?* YaleEnvironment360. https://e360.yale.edu/features/can-the-world-find-solutions-to-the-nitrogen-pollution-crisis (Accessed: 8 October 2018).

Petrović, M., Gonzalez, S. and Barceló, D. (2003). Analysis and removal of emerging contaminants in wastewater and drinking water. *TrAC Trends in Analytical Chemistry* 22(10), 685-696. https://doi.org/10.1016/S0165-9936(03)01105-1.

Pinkerton, R., Oria, R.B., Lima, A.A.M., Rogawski, E.T., Oria, M.O.B., Patrick, P.D. *et al.* (2016). Early childhood diarrhea predicts cognitive delays in later childhood independently of malnutrition. *American Journal of Tropical Medicine and Hygiene* 95(5), 1004-1010. https://doi.org/10.4269/ajtmh.16-0150.

Poindexter, G.F. (2015). Brazil's drought brings water supply to near zero capacity at hydroelectric facilities. Hydro Review, 28 January. http://www.hydroworld.com/articles/2015/01/brazil-s-drought-brings-water-supply-to-near-zero-capacity-at-hydroelectric-facilities.html (Accessed: 24 June 2017).

Poplak, R. (2018). What's actually behind Cape Town's water crisis. *The Atlantic*. 15 February 2018. https://www.theatlantic.com/international/archive/2018/02/cape-town-water-crisis/553076/.

Pruden, A., Larsson, D.G.J., Amézquita, A., Collignon, P., Brandt, K.K., Graham, D.W. *et al.* (2013). Management options for reducing the release of antibiotics and antibiotic resistance genes to the environment. *Environmental Health Perspectives* 121(8), 878-885. https://doi.org/10.1289/ehp.1206446.

Pyne, R.D.G. (1995). *Groundwater Recharge and Wells: A Guide to Aquifer Storage Recovery*. 1st edn: CRC Press. https://www.crcpress.com/Groundwater-Recharge-and-Wells-A-Guide-to-Aquifer-Storage-Recovery/Pyne/p/book/9781566700979#googlePreviewContainer.

Qadir, M., Schubert. S., Oster, J.D., Sposito, G., Minhas, P.S., Cheraghi, S.M. *et al.* (2018). High-magnesium waters and soils: Emerging environmental and food security constraints. *Science of the Total Environment* 642, 1108-1117. https://doi.org/10.1016/j.scitotenv.2018.06.090.

Qu, J. and Fan, M. (2010). The Current State of Water Quality and Technology Development for Water Pollution Control in China. *Critical Reviews in Environmental Science and Technology* 40(6), 519-560. https://doi.org/10.1080/10643380802451953.

Rabatel, A., Francou, B., Soruco, A., Gomez, J., Caceres, B., Ceballos, J.L. *et al.* (2013). Current state of glaciers in the tropical Andes: A multi-century perspective on glacier evolution and climate change. *Cryosphere* 7(1), 81-102. https://doi.org/10.5194/tc-7-81-2013.

Rahman, M.M., Ng, J.C. and Naidu, R. (2009). Chronic exposure of arsenic via drinking water and its adverse health impacts on humans. *Environmental Geochemistry and Health* 31, 189-200. https://doi.org/10.1007/s10653-008-9235-0.

Ramsar Convention Secretariat (2016). *Introduction to the Ramsar Convention*. Gland. https://www.ramsar.org/sites/default/files/documents/library/handbook1_5ed_introductiontoconvention_e.pdf.

Ramsar Convention Secretariat (2018). *Ramsar sites information service*. https://rsis.ramsar.org/ to https://rsis.ramsar.org/?pagetab=2 (Accessed: 26 January 2018).

Rehman, S., Al-Hadhrami, L.M. and Alam, M.M. (2015). Pumped hydro energy storage system: A technological review. *Renewable Sustainable Energy Reviews* 44, 586-598. https://doi.org/10.1016/j.rser.2014.12.040.

Research Center for Sustainability and Environment-Shiga University and International Lake Environment Committee Foundation (2014). *Development of ILBM Platform Process: Evolving Guidelines through Participatory Improvement*. Kusatsu: International Lake Environment Committee Foundation. http://www.ilec.or.jp/en/wp/wp-content/uploads/2013/02/Development-of-ILBM-Platform-Process_2nd_Edition11.pdf.

Rodell, M., Famiglietti, J.S., Wiese, D.N., Reager, J.T., Beaudoing, H.K., Landerer, F.W. *et al.* (2018). Emerging trends in global freshwater availability. *Nature* 557(7707), 651-659. https://doi.org/10.1038/s41586-018-0123-1.

Rosegrant, M.W., Ringler, C., Sulser, T.B., Ewing, M., Palazzo, A., Zhu, T. *et al.* (2009). *Agriculture and Food Security Under Global Change: Prospects for 2025/2050. Prepared for the Strategy Committee of the CGIAR*. Washington, D.C.: International Food Policy Research Institute.

Saaroni, H. and Ziv, B. (2003). The impact of a small lake on heat stress in a Mediterranean urban park: The case of Tel Aviv, Israel. *International Journal of Biometeorology* 47(3), 156-165. https://doi.org/10.1007/s00484-003-0161-7.

Salem, H.S. (2009). The Red Sea-Dead Sea Conveyance (RSDS) Project: A solution for some problems or a cause for many problems. In *The Second International Conference: Water: Values and Rights*. Messerschmid, C., El-Jazairi, L., Khatib, I. and Daoud, A.A.H. (eds.). Ramallah: Palestine Academy Press. https://www.researchgate.net/profile/Hilmi_Salem/publication/299563326_The_Red_Sea-Dead_Sea_Conveyance_RSDSC_Project_A_Solution_for_Some_Problems_or_A_Cause_for_Many_Problems/links/56ff7ebf08aee995dde744b9/The-Red-Sea-Dead-Sea-Conveyance-RSDSC-Project-A-Solution-for-Some-Problems-or-A-Cause-for-Many-Problems.pdf?origin=publication_detail.

Sánchez, A.S., Torres, E.A. and Kalid, R.A. (2015). Renewable energy generation for the rural electrification of isolated communities in the Amazon Region. *Renewable and Sustainable Energy Reviews* 49, 278-290. https://doi.org/10.1016/j.rser.2015.04.075.

Scanlon, B.R., Reedy, R.C., Faunt, C.C., Pool, D. and Uhlman, K. (2016). Enhancing drought resilience with conjunctive use and managed aquifer recharge in California and Arizona. *Environmental Research Letters* 11(3). https://doi.org/10.1088/1748-9326/11/4/049501.

Scheffer, M., Carpenter, S., Foley, J.A., Folke, C. and Walker, B. (2001). Catastrophic shifts in ecosystems. *Nature* 413, 591-596. https://doi.org/10.1038/35098000.

Schoolmeester, T., Johansen, K.S., Alfthan, B., Baker, E., Hesping, M. and Verbist, K. (2018). *The Andean Glacier and Water Atlas – The Impact of Glacier Retreat on Water Resources*. United Nations Educational, Scientific and Cultural Organization and GRID-Arendal

Secretariat of the Convention on Biological Diversity (1992). Convention on Biological Diversity. https://www.cbd.int/doc/legal/cbd-en.pdf.

Sedlak, D. (2014). *Water 4.0: The Past, Present, and Future of the World's Most Vital Resource*. New Haven, CT: Yale University Press. https://books.google.co.ke/books/yup?vid=ISBN9780300212679&redir_esc=y.

Seibert, S., Burke, J., Faures, J.M., Krenken, K., Hoogeveen, J., Döll, P. *et al.* (2010). Groundwater use for irrigation – a global inventory. *Hydrology and Earth System Sciences* 14, 1863-1880. https://doi.org/10.5194/hess-14-1863-2010.

Sevilimedu, V., Pressley, K.D., Snook, K.R., Hogges, J.V., Politis, M.D., Sexton, J.K. *et al.* (2016). Gender-based differences in water, sanitation and hygiene-related diarrheal disease and helminthic infections: A systematic review and meta-analysis. *Transactions of the Royal Society of Tropical Medicine and Hygiene* 110(11), 637-648. https://doi.org/10.1093/trstmh/trw080.

Shah, T. (2014). *Groundwater Governance and Irrigated Agriculture*. Stockholm: Global Water Partnership. http://www.gwp.org/globalassets/global/toolbox/publications/background-papers/gwp_tec_19_web.pdf.

Sikder, M.T., Kihara, Y., Yasuda, M., Yustiawati., Mihara, Y., Tanaka, S. *et al.* (2013). River water pollution in developed and developing countries: Judge and assessment of physicochemical characteristics and selected dissolved metal concentration. *Clean-Soil Air Water* 41(1), 60-68. https://doi.org/10.1002/clen.201100320.

Spitz, K. and Trudinger, J. (2008). *Mining and the Environment: From Ore to Metal*. 1st edn: CRC Press. https://www.crcpress.com/Mining-and-the-Environment-From-Ore-to-Metal/Spitz-Trudinger/p/book/9780415465106.

Stefan, C. and Ansems, N. (2017). Web-based global inventory of managed aquifer recharge applications. *Sustainable Water Resources Management* 4, 153–162 https://doi.org/10.1007/s40899-017-0212-6.

Sticklor, R. (2014). Is underground water storage the answer to water security? 22 April. https://wle.cgiar.org/thrive/2014/04/22/underground-water-storage-answer-water-security

Stocker, T.F. and Raible, C.C. (2005). Water cycle shifts gear. *Nature* 434(7035), 830–833. https://doi.org/10.1038/434830a.

Sui, Q., Cao, X., Lu, S., Zhao, W., Qiu, Z. and Yu, G. (2015). Occurrence, sources and fate of pharmaceuticals and personal care products in the groundwater: A review. *Emerging Contaminants* 1(1), 14-24. https://doi.org/10.1016/j.emcon.2015.07.001.

Sustainable Facilities Tool (2017). *Life Cycle Assessment (LCA) overview*. https://sftool.gov/plan/400/life-cycle-assessment-lca-overview (Accessed: 24 June 2017).

Syvitski, J.P.M., Kettner, A.J., Overeem, I., Hutton, E.W.H., Hannon, M.T., Brakenridge, G.R. *et al.* (2009). Sinking deltas due to human activities. *Nature Geoscience* 2, 681-686. https://doi.org/10.1038/ngeo629.

Thiam, S., Diène, A.N., Sy, I., Winkler, M.S., Schindler, C., Ndione, J.A. *et al.* (2017). Association between childhood diarrhoeal incidence and climatic factors in urban and rural settings in the health district of Mbour, Senegal. *International Journal of Environmental Research and Public Health* 14(9), 1049. https://doi.org/10.3390/ijerph14091049.

Thornton, P.K. (2010). Livestock production: Recent trends, future prospects. *Philosophical Transactions of the Royal Society B-Biological Sciences* 365(1554), 2853-2867. https://doi.org/10.1098/rstb.2010.0134.

Timoney, K.P. and Lee, P. (2009). Does the Alberta tar sands industry pollute? The scientific evidence. *The Open Conservation Biology Journal* 3(1), 65-81. https://doi.org/10.2174/1874839200903010065.

Tredoux, G., van der Merwe, B. and Peters, I. (2009). Artificial recharge of the Windhoek aquifer, Namibia: Water quality considerations. *Boletin Geologico y Minero* 120(2), 269 – 278. http://aguas.igme.es/Boletin/2009/120_2_2009/269-278.pdf.

Umezawa, Y., Hosono, T., Onodera, S., Siringan, F., Buapeng, S., Delinom, R. *et al.* (2009). Erratum to "Sources of nitrate and ammonium contamination in groundwater under developing Asian megacities". *Science of the Total Environment* 407(9), 3219-3231. https://doi.org/10.1016/j.scitotenv.2009.01.048.

UNEP-DHI and United Nations Environment Programme (2016). *Transboundary River Basins: Status and Trends*. WAP RB Technical Assessment Report. Nairobi. http://twap-rivers.org/assets/GEF_TWAPRB_FullTechnicalReport_compressed.pdf.

Unilever, Sunlight, Oxfam, NextDrop and WaterAid (2015). *Every Woman Counts, Every Second Counts: Water for Women*. https://www.unilever.com/Images/slp_water-for-women-march-2015_tcm244-423659_en.pdf.

United Nations (2017). *Reconciling Resource Uses in Transboundary Basins: Assessment of the Water-Food-Energy-Ecosystems Nexus in the Sava River Basin*. New York, NY. http://www.unece.org/fileadmin/DAM/env/water/publications/GUIDELINES/2017/nexus_in_Sava_River_Basin/Nexus-SavaRiverBasin_ECE-MP.WAT-NONE-3_WEB_final_corrected_for_gDoc.pdf.

United Nations Children's Fund and World Health Organization (2012). *Progress on Drinking Water and Sanitation: 2012 Update*. United Nations Children's Fund and World Health Organization. https://www.unicef.org/publications/files/JMPreport2012(1).pdf.

United Nations Commission for Europe and International Network of Basin Organizations (2015). *Water and Climate Change Adaptation in Transboundary Basins: Lessons Learned and Good Practices*. Geneva and Paris: UNECE, INBO. http://www.unece.org/fileadmin/DAM/env/water/publications/WAT_Good_practices/ece.mp.wat.45.pdf.

United Nations Economic Commission for Europe (1998). Convention on Access to Information, Public Participation in Decision-Making and Access to Justice in Environmental Matters. http://www.unece.org/fileadmin/DAM/env/pp/documents/cep43e.pdf

United Nations Economic Commission for Latin America and the Caribbean (2017). *CEPALSTAT. Bases de Datos y Publicaciones Estadísticas de la Comisión Económica para América Latina y el Caribe*. http://estadisticas.cepal.org/cepalstat/ (Accessed: 26 January 2017).

United Nations Educational Scientific and Cultural Organization and World Water Assessment Programme (2015). *The United Nations World Water Development Report: Water for a Sustainable Development*. Paris. http://unesdoc.unesco.org/images/0023/002318/231823E.pdf.

United Nations Educational Scientific and Cultural Organization-IHP and United Nations Environment Programme (2016). *Transboundary Aquifers and Groundwater Systems of Small Island Developing States: Status and Trends, Summary for Policy Makers. Volume 1: Groundwater*. Nairobi. http://unesdoc.unesco.org/images/0024/002449/244912e.pdf.

United Nations Environment Programme (2011). *Methodology for the GEF Transboundary Waters Assessment Programme: Volume 1. Methodology of the Assessment of Transboundary Aquifers, Lake Basins, River Basins, Large Marine Ecosystems, and the Open Oceans*. Jeftic, L., Glennie, P., Talaue-McManu, L. and Thornton, J.A. (eds.). Nairobi. http://www.geftwap.org/publications/methodologies-for-the-gef-transboundary-assessment-programme-1/volume-1.

United Nations Environment Programme (2012a). *UN-Water Report: Status Report on the Application of Integrated Approaches to Water Resources Management*. Nairobi. http://www.unwater.org/app/uploads/2017/05/UNW_status_report_Rio2012.pdf.

United Nations Environment Programme (2012b). *Measuring Water Use in a Green Economy. A Report of the Working Group on Water Efficiency to the International Resource Panel*. McGlade, J., Werner, B., Young, M., Matlock, M., Jefferies, D., Sonnemann, G. *et al.* (eds.). Nairobi. https://waterfootprint.org/media/downloads/UNEP-2012-MeasuringWaterUse_1.pdf.

United Nations Environment Programme (2013). *Where Will the Water Go? Impacts of Accelerated Glacier Melt in the Tropical Andes*. Nairobi. https://europa.eu/capacity4dev/file/15905/download?token=yowiV9D7.

United Nations Environment Programme (2016a). *GEO-6 Regional Assessment for North America*. Nairobi. http://wedocs.unep.org/bitstream/handle/20.500.11822/7611/GEO_North_America_201611.pdf?isAllowed=y&sequence=1.

United Nations Environment Programme (2016b). *GEO-6 Regional Assessment for Asia and the Pacific*. Nairobi. http://wedocs.unep.org/bitstream/handle/20.500.11822/7548/GEO_Asia_Pacific_201611.pdf?isAllowed=y&sequence=1.

United Nations Environment Programme (2016c). *GEO-6 Regional Assessment for West Asia*. Nairobi. http://wedocs.unep.org/bitstream/handle/20.500.11822/7668/GEO_West_Asia_201611.pdf?isAllowed=y&sequence=1.

United Nations Environment Programme (2016d). *GEO-6 Regional Assessment for Africa*. Nairobi. http://wedocs.unep.org/bitstream/handle/20.500.11822/7595/GEO_Africa_201611.pdf?isAllowed=y&sequence=1.

United Nations Environment Programme (2016e). *A Snapshot of the World's Water Quality: Towards a Global Assessment*. Nairobi. https://wedocs.unep.org/bitstream/handle/20.500.11822/19524/UNEP_WWQA_report_03052016.pdf?sequence=1&isAllowed=y.

United Nations Environment Programme (2016f). *GEO-6 Regional Assessment for Latin America and the Caribbean*. Nairobi. http://apps.unep.org/publications/index.php?file=012096_en&option=com_pub&task=download.

United Nations Environment Programme (2016g). Microplastics: Trouble in the food chain. In *UNEP Frontiers 2016 Report: Emerging Issues of Environmental Concern*. 11. https://wedocs.unep.org/bitstream/handle/20.500.11822/7664/Frontiers_2016.pdf?sequence=1&isAllowed=y

United Nations Environment Programme (2016h). *GEO-6 Regional Assessment for the Pan-European Region*. Nairobi. http://wedocs.unep.org/bitstream/handle/20.500.11822/7735/unep_geo_regional_assessments_europe_16-07513_hires.pdf?isAllowed=y&sequence=1.

United Nations Environment Programme (2017). *Towards a Pollution-Free Planet: Background Report*. Nairobi. https://wedocs.unep.org/bitstream/handle/20.500.11822/21800/UNEA_towardspollution_long%20version_Web.pdf?sequence=1&isAllowed=y.

United Nations Environment Programme and Global Environment Facility (2018). *Transboundary Waters Assessment Programme: River Basins*. United Nations Environment Programme Global Environment Facility. http://twap-rivers.org/indicators/.

United Nations Environment Programme World Conservation Monitoring Centre (2017). Natural wetlands have declined and artificial wetlands increased. In *Global Wetland Outlook: State of the World's Wetlands and their Services to People*. https://medwet.org/wp-content/uploads/2018/09/ramsar_gwo_english_web.pdf.

United Nations General Assembly (2012). Resolution Adopted by the General Assembly on 27 July 2012. 66/288. The Future We Want. A/RES/66/288. http://www.un.org/en/development/desa/population/migration/generalassembly/docs/globalcompact/A_RES_66_288.pdf.

United Nations High Commissioner for Refugees (2017). *Azraq Camp Fact Sheet*. https://data2.unhcr.org/fr/documents/download/53299.

United Nations Office for the Coordination of Humanitarian Affairs (2017). *Largest seawater desalination plant opened in Gaza*. https://www.ochaopt.org/content/largest-seawater-desalination-plant-opened-gaza (Accessed: 11 October 2018).

United Nations Water (2016). *Water and Sanitation Interlinkages across the 2030 Agenda for Sustainable Development*. Geneva. http://www.unwater.org/app/uploads/2016/08/Water-and-Sanitation-Interlinkages.pdf.

United Nations Water (2017). *What we do: Monitor and report*. http://www.unwater.org/what-we-do/monitoring-and-report/ (Accessed: 26 June 2017).

United Nations Water and World Health Organization (2015). *UN-Water GLAAS TrackFin Initiative: Tracking Financing to Sanitation, Hygiene and Drinking-Water at the National Level* Geneva. http://apps.who.int/iris/bitstream/10665/204861/1/WHO_FWC_WSH_15.23_eng.pdf?ua=1.

United Nations World Water Assessment Programme (2012). *Managing Water Under Uncertainty and Risk: The United Nations World Water Development Report 4*. Paris: United Nations Educational, Scientific and Cultural Organization. http://unesdoc.unesco.org/images/0021/002156/215644e.pdf.

United Nations World Water Assessment Programme (2017). *The United Nations World Water Development Report 2017: Wastewater, the Untapped Resource*. Paris: United Nations Educational, Scientific and Cultural Organization. http://unesdoc.unesco.org/images/0024/002471/247153e.pdf.

United Nations World Water Assessment Programme (2018). *The United Nations World Water Report 2018: Nature-Based Solutions for Water*. Paris: United Nations Educational, Scientific and Cultural Organization. http://unesdoc.unesco.org/images/0026/002614/261424e.pdf.

United States Fish and Wildlife Service (1994). *660 FW 1, Wetlands Policy and Action Plan*. https://www.fws.gov/policy/660fw1.html (Accessed: 26 January 2018).

van Lexmond M, B., Bonmatin, I.M., Goulson, D. and Noome, D.A. (2015). Worldwide integrated assessment on systemic pesticides. *Environmental Science and Pollution Research* 22(1), 1-4. https://doi.org/10.1007/s11356-014-3220-1.

Vengosh, A. (2003). Salinization and saline environments. In *Treatise on Geochemistry*. Lollar, B.S. (ed.). Elsevier Science. chapter 9.09. 325-378. https://doi.org/10.1016/B0-08-043751-6/09051-4.

Vengosh, A., Jackson, R.B., Warner, N., Darrah, T.H. and Kondash, A. (2014). A critical review of the risks to water resources from unconventional shale gas development and hydraulic fracturing in the United States. *Environmental Science & Technology* 48(15), 8334-8348. https://doi.org/10.1021/es405118y.

Vicente-Serrano, S.M., Lopez-Moreno, J.-I., Beguería, S., Lorenzo-Lacruz, J., Sanchez-Lorenzo, A., García-Ruiz, J.M. *et al.* (2014). Evidence of increasing drought severity caused by temperature rise in southern Europe. *Environmental Research Letters* 9(4). https://doi.org/10.1088/1748-9326/9/4/044001.

Visser, M. and Brühl, J. (2018). *Op-Ed: A drought-stricken Cape Town did come together to save water*. Daily Maverick. https://www.dailymaverick.co.za/article/2018-03-01-op-ed-a-drought-stricken-cape-town-did-come-together-to-save-water/#.WpluhWpublV (Accessed: 26 March 2018).

Vörösmarty, C.J., Hoekstra, A.Y., Bunn, S.E., Conway, D. and Gupta, J. (2015). Freshwater goes global. *Science* 349(6247), 478-479. https://doi.org/10.1126/science.aac6009.

Vörösmarty, C.J., Meybeck, M. and Pastore, C.L. (2015). Impair-then-repair: A brief history & global-scale hypothesis regarding human-water interactions in the Anthropocene. *Daedalus* 144(3), 94-109. https://doi.org/10.1162/DAED_a_00345.

Voutchkov, N. (2016). *Desalination – past, present and future*. [International Water Association http://www.iwa-network.org/desalination-past-present-future/ (Accessed: 23 February 2018).

Wang, X., Wang, W. and Tong, C. (2016). A review on impact of typhoons and hurricanes on coastal wetland ecosystems. *Acta Ecologica Sinica* 36(1), 23-29. https://doi.org/10.1016/j.chnaes.2015.12.006.

Weaver, J.M.C., Rosewarne, P., Hartnady, C.J.H. and Hay, E.R. (2002). Potential of table mountain group aquifers and integration into catchment water management. In *A Synthesis of the Hydrogeology of the Table Mountain Group – Formation of a Research Strategy*. Pietersen, K. and Parsons, R. (eds.). chapter 7. 241-255. http://www.wrc.org.za/Knowledge%20Hub%20Documents/Research%20Reports/TT-158-01.pdf.

Webb, E., Moon, J., Dyrszka, L., Rodriguez, B., Cox, C., Patisaul, H. *et al.* (2017). Neurodevelopmental and neurological effects of chemicals associated with unconventional oil and natural gas operations and their potential effects on infants and children. *Reviews on Environmental Health* 33(1), 3–29. https://doi.org/10.1515/reveh-2017-0008.

Wong, T.H.F. (2011). Framework for stormwater quality management in Singapore. *12th International Conference on Urban Drainage*. Association, I.W. (ed.). Porto Alegre, 11-16 September 2011. https://www.scribd.com/document/327618575/Framework-for-Stormwater-Quality-Management-in-Singapore.

World Health Organisation and United Nations Children's Fund (2017). *Safely Managed Drinking Water: Thematic Report on Drinking Water 2017*. Geneva, Switzerland. https://washdata.org/report/jmp-2017-tr-smdw.

World Health Organization (2006). *Guidelines for the Safe Use of Wastewater, Excreta and Greywater: Volume 2. Wastewater Use in Agriculture*. Geneva. http://www.who.int/water_sanitation_health/wastewater/wwuvol2intro.pdf.

World Health Organization (2017a). *Global Health Observatory Data: Mortality and Burden of Disease from Water and Sanitation*. Geneva http://www.who.int/gho/phe/water_sanitation/burden/en/.

World Health Organization (2017b). *UN-Water Global Analysis and Assessment of Sanitation and Drinking-water (GLAAS)*. http://www.who.int/water_sanitation_health/monitoring/investments/glaas/en/ (Accessed: 24 June 2017).

World Health Organization and United Nations Children's Fund (2012). *WHO/UNICEF Joint Monitoring Programme for Water Supply and Sanitation (JMP). 2012 Annual Report* https://washdata.org/report/jmp-2012-annual-report.

World Health Organization and United Nations Children's Fund (2015). *2015 Annual Report WHO/UNICEF Joint Monitoring Programme for Water Supply and Sanitation*. https://d26p6gt0m19hor.cloudfront.net/whywater/JMP-2015-Annual-Report.pdf.

World Health Organization and United Nations Children's Fund (2016). *2016 Annual Report WHO/UNICEF Joint Monitoring Programme for Water Supply, Sanitation and Hygiene (JMP)*. https://washdata.org/sites/default/files/documents/reports/2017-07/JMP-2016-annual-report.pdf.

World Meteorological Organization (2018). *What are hydrological services?* http://hydroconference.wmo.int/en/about.

World Wildlife Fund (2016). *Living Planet Report 2016: Risk and Resilience in a New Era*. Gland. http://awsassets.panda.org/downloads/lpr_2016_full_report_low_res.pdf.

Yang, S.L., Milliman, J.D., Li, P. and Xu, K. (2011). 50,000 dams later: Erosion of the Yangtze River and its delta. *Global and Planetary Change* 75(1), 14-20. https://doi.org/10.1016/j.gloplacha.2010.09.006.

Yang, Y., Ok, Y.S., Kim, K.-H., Kwon, E.E. and Tsang, Y.F. (2017). Occurrences and removal of pharmaceuticals and personal care products (PPCPs) in drinking water and water/sewage treatment plants: A review. *Science of the Total Environment* 596-597, 303-320. https://doi.org/10.1016/j.scitotenv.2017.04.102.

Yao, T., Thompson, L., Yang, W., Yu, W., Gao, Y., Guo, X. *et al.* (2012). Different glacier status with atmospheric circulations in Tibetan Plateau and surroundings. *Nature Climate Change* 2, 663-667. https://doi.org/10.1038/nclimate1580.

Yihdego, Y., Khalil, A. and Salem, H.S. (2017). Nile rivers basin dispute: Perspectives of the grand Ethiopian Renaissance Dam (GERD). *Global Journal of Human-Social Science: B - Geography, Geo-Sciences, Environmental Science & Disaster Management* 17(2), 1-21. https://socialscienceresearch.org/index.php/GJHSS/article/view/2239.

Yihdego, Y. and Salem, H.S. (2017). The challenges of sustainability: Perspective of ecology. *Journal of Sustainable Energy Engineering* 5(4), 22. https://doi.org/10.7569/JSEE.2017.629519.

Yuan, L., Richardson, C.J., Ho, M., Willis, C.W., Colman, B.P. and Wiesner, M.R. (2018). Stress responses of aquatic plants to Silver Nanoparticles. *Environmental Science & Technology* 52(5), 2558-2565. https://doi.org/10.1021/acs.est.7b05837.

Zandalinas, S., Mittler, R., Balfagón, D., Arbona, V. and Gómez-Cadenas, A. (2018). Plant adaptations to the combination of drought and high temperatures. *Physiologia Plantarum* 162(1), 2-12. https://doi.org/10.1111/ppl.12540.

Zarfl, C., Lumsdon, A., Berlekamp, J., Tydeck, L. and Tockner, K. (2014). A global boom in hydropower dam construction. *Aquatic Sciences* 77, 161–170. https://doi.org/10.1007/s00027-014-0377-0.

Zerbe, S., Steffenhagen, P., Parakenings, K., Timmermann, T., Frick, A., Gelbrecht, J. *et al.* (2013). Ecosystem service restoration after 10 years of rewetting peatlands in NE Germany. *Environmental Management* 51(6), 1194-1209. https://doi.org/10.1007/s00267-013-0048-2.

PART B

Policies, Goals, Objectives and Environmental Governance: An assessment of their effectiveness

Chapter 10

Approach to Assessment of Policy Effectiveness

Coordinating Lead Authors: Klaus Jacob (Freie Universität Berlin), Peter King (Institute for Global Environmental Strategies), Diana Mangalagiu (University of Oxford and Neoma Business School)

Contributing Author: Beatriz Rodríguez-Labajos (Universitat Autònoma de Barcelona)

10.1 The Context

Policies are crucial in determining and improving the state of our environment. A simple way to think about policy and policy instruments is that a policy is a statement of intent to change behaviour in a positive way, while an instrument is the means or a specific measure to translate that intent into action (Mees *et al.* 2014). Therefore, discussion of effectiveness of environmental policies means addressing both aspects. Goal setting (including targets, indicators and time frames) is an important step towards legitimization of environmental policies. Execution of the policy instruments is through effective governance. Governance is "the process whereby societies or organizations make important decisions, determine whom they involve and how they render account" (United Nations Economic and Social Council 2006). The recently adopted Sustainable Development Goals (SDGs) give a new impetus to 'governing through goals' (Yoshida and Zusman 2015).

Strong environmental policies form an integral component of UN Environment's theory of change, which posits alternative pathways to global sustainable development. UN Environment defines a theory of change as when "an intervention depicts the causal pathways from outputs through outcomes via intermediate states towards impact" (United Nations Environment Programme [UNEP] 2017). The theory of change further defines the external factors that influence change along the major pathways – that is factors that affect whether one result can lead to the next. These contributing factors are called drivers and assumptions.

The theory of change for the fifth Global Environment Outlook (GEO-5) showed an expectation that GEO should be policy relevant and draw from a good understanding of global and regional policy issues (UNEP 2012). In GEO-6, however, policy effectiveness is seen to be more central in the theory of change, as shown in Annex 1-3. Reflecting on the mandate of UN Environment's High Level Intergovernmental and Stakeholder Advisory Group, it is no longer sufficient to be merely policy relevant. Member governments want to know which policies are most effective in dealing with seemingly intransigent and insurmountable environmental problems. Using the Drivers, Pressures, State, Impact, Response (DPSIR) framework (see Figure 1.2, Chapter 1), current responses to environmental problems are discussed in the thematic chapters in Part A of this report, while Part B addresses the question of when these policies are effective, and Part C incorporates the most promising policy approaches into the pathways of transformation. While GEO-6 is not policy prescriptive, it offers guidance to governments and policymakers who would like to know which policies have worked best in which circumstances, under what governance arrangements and whether that experience is transferable to other contexts.

10.2 Environmental policy and governance

Environmental policies are pursued through a multitude of modes of governance and are designed to promote desirable behaviours by a defined set of actors and to overcome a range of challenges that impede effective environmental management. Policy objectives are to be achieved via policy measures or instruments – structured activities targeted at changing other activities in society towards achieving environmental goals. Not all effective policy instruments are for environmental policy alone, other instruments (e.g. in energy and transport policy) may include environmental policy goals, often as secondary goals to the prime non-environmental goal (e.g. reducing congestion). This is the usual case now in most integrated policies (as discussed in Chapter 11). Accordingly, environmental governance extends well beyond environmental ministries.

Governments are often thought of as the primary domain for development and implementation of policies. While governments are often the most important actors in formulating, implementing and enforcing policy instruments, they do not act alone, and various governance arrangements are needed. Effective policies usually involve a wide range of stakeholder inputs throughout the policy cycle. Governments at all levels are active in policy formulation and implementation, as are private sector and civil society actors. Roles and responsibilities are spread not only between governmental and non-governmental institutions, but also across all levels vof governance.

Politicians, policy think tanks, education and research institutions, non-governmental organizations (NGOs), civil society organizations (CSOs), lobbyists, communities and companies all have roles to play in influencing policy outcomes in different contexts. At the regional and global levels, policy instruments are created and implemented by global, regional or national institutions in multilayered governance arrangements. There is also a growing number of 'public-private partnerships' and 'corporate sustainability initiatives', including the emergence of 'business-NGO interactions' aimed at stimulating responsible and sustainable behaviour in specific sectors (Forsyth 2005; van Tulder *et al.* 2016). Such partnerships (e.g. UN Environment's Clean Seas Initiative) emerged in the design and production of goods, risk assessments, due diligence, training, monitoring, reporting and mediation, transparency in supply chains and more. In many countries, citizens and communities are also contributing to the realization of collective environmental goals. These are often framed as 'citizen co-production' and/or 'community-based initiatives' (Mees, Crabbé and Driessen 2017).

The challenge is for all these actors, layers and levels to mesh together and provide a coherent mix of policies appropriate for the scale and period of application and consistent with the national social, cultural, historical and political context (European Environment Agency [EEA] 2001a; EEA 2001b; European Commission [EC] 2012; Niles and Lubell 2012; EEA 2017).

Polycentric governance is a source of innovation and by enabling the competition of ideas, collaboration and alignment, it creates momentum for environmental policies (Jordan and Huitema 2014). However, dispersion of responsibility may lead to fragmented policies, poorly defined roles and responsibilities, weak follow-up and monitoring mechanisms, lack of accountability for results or a stalemate in decision-making.

10.3 Policy instruments

Policy instruments come in multiple forms and can be implemented by multiple actors (not only governments) at multiple levels of governance (Mees *et al.* 2014; Keskitalo *et al.* 2016). Policy instruments can be aimed at various mechanisms:

i. available alternatives can be amended;
ii. impacts of alternatives can be changed; and
iii. evaluation of outcomes can be influenced (Boersema and Reijnders eds. 2009).

These forms vary between traditional top-down steering by governments to self-regulation of business organizations. Some forms are more, and others less, successful in fulfilling their policy objectives. It is often stated that collaborative modes of governance, which rely on stakeholder participation, are needed to address the complex, multi scalar and cross sectoral dimensions of environmental problems (Challies *et al.* 2017; Kochskämper *et al.* eds. 2018). Moreover, increasing understanding of environmental challenges have changed policy approaches and instruments from targeted policies and single-use instruments to policy integration and raising of public awareness to policy coherence and systematic approaches (e.g. green economy) (EEA 2017).

One policy tool to address transboundary environmental problems and maintain 'the commons' is robust and legally binding international agreements. However, given the structure and legal basis of international law-making, such agreements often fall short in meeting the ambitions of the front-running or most affected countries (Sandler 2017). Therefore, coalitions or clubs of countries may step in and develop more ambitious environmental policies (Hovi *et al.* 2016).

An often-stated understanding of environmental policy instruments is that they can be 'carrots, sticks or sermons,' although this is only a partial characterization of the full range (Niles and Lubell 2012). Some common types of policy instrument include legislation and regulatory policies, financial incentives/disincentives, voluntary approaches, treaties and agreements, and international soft law (Hildén, Jordan and Rayner 2014). GEO-4 used the following traditional structure: regulations and standards, market-based instruments, voluntary agreements, research and development, and information instruments (UNEP 2007). In GEO-5, common threads were traced between and across different world regions, emphasizing particular policy approaches that had proven successful in a number of cases. Successful policy responses in several regions were assumed to be more likely to accelerate achievement of internationally agreed goals. Within the DPSIR framework, policies that address 'drivers' tend to be preferred, as they intend to address the roots of environmental problems rather than treating the symptoms (UNEP 2012). For GEO-6, the typology in Table 10.1 has been used to provide guidance on the selection of types of policies instruments and governance approaches and associated case studies. Note, however, that the typology is not intended to be an exhaustive coverage of all possible environmental policies or policy instruments.

Environmental policies ultimately aim to preserve a state of the environment that protects habitats, safeguards ecosystem services and minimizes risks for human health from pollution. Therefore, environmental policies have been typically developed to protect different environmental media (air, water, land), to influence the state of the environment outlined in Part A of this report, and usually under the control of an environment ministry. However, effective environmental policies not only address the state of the environment, but also the drivers and pressures originating from social and economic activities (outlined in Chapter 2). Accordingly, governments across the world have developed institutions and policies that address the most important polluting sectors, such as energy, mobility, industry and agriculture (Organisation for Economic Co-operation and Development [OECD] 2016).

In Part B of this report, a selection of the most commonly adopted policies, cutting-edge policies and policy clusters that show real promise in each thematic area and cross-cutting issues are analysed, recognizing that there are literally thousands of policies and policy instruments and it is not possible to cover them all. The selected policies represent a sample from different types of policy instruments and governance approaches (see **Table 10.1**) from different regions of the world.

10.4 Policy mixes and coherence

Given the multiple actors and factors causing environmental deterioration and the various types of barriers to environmental innovation, a single policy instrument is rather unlikely to be sufficient for achieving the desired goals. Against this background and the multiple challenges to address when developing effective environmental policies, adopting policy mixes rather than a single policy is claimed to be more effective (Jänicke *et al.* 2000; Mees *et al.* 2014; Kivimaa and Kern 2016).

However, different policies may not always complement each other, but could impair each other (e.g. economic incentives may undercut intrinsic motivation). A policy package, portfolio, mix or cluster is a collection of policy instruments all designed to achieve a common goal or set of intentions (Lay *et al.* 2017). When developing policy mixes, their coherence has to be ensured in order to achieve optimal results (Howlett and Rayner 2007; Huttunen, Kivimaa and Virkamäki 2014).

Policy coherence is the systematic promotion of mutually reinforcing policies that can accumulate synergies in attempting to achieve agreed objectives (OECD 2016). Policy coherence can be sectoral, transnational, across governance regimes, multi-level (from global to local) and/or from policy objective through to instrument design and implementation practice (Hood 2011). Policy coherence occurs when the balance of policies is aligned with that common goal or set of intentions.

In addition to policy coherence, policy synergy is also necessary. In order to realize environmental objectives, environmental concerns need to be incorporated in other policy sectors. This is often referred to as policy synergy or environmental policy integration and contributes to policy coherence (Hood 2011; Lay *et al.* 2017). Policy synergy occurs when successive policy instruments have a cumulative or reinforcing impact in achieving the common goal or overarching policy aspiration (OECD *et al.* 2015).

Table 10.1: Policy typology

Policy instrument / governance approach	Point of intervention[a]	Assumed causal mechanism	Barriers to effectiveness	Potential costs	Typical policy instruments	Contextual requirements
Command and Control	Industrial processes and products Technologies End of pipe or smokestack pollution control	Prohibition of environmentally harmful technologies (products and processes) or demanding environmentally friendly technologies as part of permitting => reduced emissions/resource use => if emissions/resource use cannot be reduced enough by upstream controls, then improved waste management is needed. Can also address enabling issues such as property rights and access issues.	Lack of monitoring Corruption.	Costs for firms, harmful for competitiveness. No incentives to innovate.	Permitting processes/ fines. Discharge/emission standards. Total pollutant load caps Real-time monitoring systems. Legal reviews.	Administrative capacities. Self-reporting by industry.
Promotion of innovation	Green innovation	Incentivizing R&D in green technologies => introducing green technologies to markets (=> cost savings + exports)	Spill over.	Additional costs.	Subsidies, tax relief for R&D, green procurement.	Public and private budgets.
Market-based/ economic incentives	Pricing of products or processes	Change of relative prices between environmentally harmful and green technologies => increased markets for green services; => incentives to innovate and disincentives to cause environmental harm.	Lack of significance within the overall context of consumer budgets and/or industrial capital plans.	Distributional/ regressive impacts, harmful for competitiveness. Retrofitting costs for existing industries.	Cap and trade, public procurement, removal of subsidies.	Private investment in new products or processes Effective tax regimes.
Convincing consumers, employees and stockholders	Public information, education, knowledge, awareness, advocacy	Knowledgeable consumers and producers will voluntarily choose environmentally sound products and processes.	Cost disincentives Inadequate supply of alternative products and processes.	Power imbalances cause poor choices by consumers.	Campaigning. Labelling. Certifying. Nudging. Stockholder voting blocs.	Media and stock markets Education system.
Enabling actors	Environmental actors	Strengthening participation of governmental and non-governmental actors in decision-making on policies or projects leading to improved project design and implementation.	Environmentally disinclined actors may wield more power.	Access to all relevant information, skill levels.	Access to information, requirements on transparency, enabling participation, requesting evidence base (EIA, SEA, IA), legal review.	Environmental and social safeguards.
Supporting investments	Infrastructure and technologies	'Green' infrastructure (waste management, electricity grids for renewables, railways, etc.) => enabling market access for green technologies => demand for increased access.	Incremental costs.	Uncertainty in relation to future impacts such as climate change could cause maladaptation.	Climate-proofing infrastructure Green investment funds.	Public-private partnerships.

[a] Point of intervention refers to the issues considered as key for environmental degradation, especially its improvement (e.g. technologies, innovation, infrastructures, actors, behaviour).
EIA: Environmental Impact Assessment; IA: Impact Assessment; R&D: research and development; SEA: Strategic Environmental Assessment.

Policy conflict, on the other hand, sees the impact of one set of policies, often in unrelated sectors or from an external actor, undermining the intended outcomes of the desired environmental policies. For example, providing a subsidy to first car buyers may conflict with policies to reduce air pollution from transportation. Accordingly, any analysis of environmental policy effectiveness also needs to address the influence of economic and social policies in other domains (Perrels 2001; Interwies, Görrlach and Newcombe 2007; Lambin *et al.* 2014).

10.5 Methodology adopted to assess policy effectiveness

The assessment of policy effectiveness in the remaining chapters of Part B serves three main purposes:

1. To showcase policies and governance approaches at all levels that have demonstrated an impact and that can potentially be applied elsewhere.
2. To identify needs for further action by improving the effectiveness of policies. The analysis builds as much as possible on quantification of policy effectiveness (i.e. indication of how much/how often policies do have an effect, not only how and why).
3. To establish methods and best available knowledge for assessing policy effectiveness that can be used beyond GEO-6 for improving the evidence base of policymaking and thereby strengthen environmental policies.

The gold standard to evaluate and quantify the effectiveness of policies is the comparison of empirical observations with a control group in an experimental design or with a counterfactual scenario. However, constructing such experiments or scenarios is in many cases costly if not impossible as the objects of policy interventions are complex social systems. For example, it is not possible to predict the reactions of markets with or without policy interventions. Furthermore, in many cases control groups cannot be identified and it may be unethical to deliberately withhold the benefits of a policy.

Evaluating policy effectiveness is still possible based on theoretical assumptions and empirical observations of policy impact. Theory-based evaluation uses an explicit theory of change throughout the causal chain from policy outputs to outcomes and final impacts (Blamey and Mackenzie 2007; Rogers and Weiss 2007).

Attributing causality to policies in often extensive and complex causal chains from policy, through its implementation, to behavioural changes and processes that are triggered, to impacts, indirect and induced impacts, is a particular challenge for policy evaluation (Forss, Marra and Schwartz eds. 2011). A conceptual approach was adopted in Part B of this report which aims to minimize the problem of attribution by combining a top-down and a bottom-up perspective (Sabatier 1986). The top-down perspective shown in **Figure 10.1** starts with the policy and traces the causal chains that are expected from the implementation of the policy. The bottom-up perspective starts from the observed outcomes and uses policy-relevant indicators to trace the causalities back to the policy interventions. This helps analysts to evaluate the effects of policy mixes. Both perspectives have their shortcomings – the top-down perspective tends to overemphasize the impacts of policies compared to other factors, the bottom-up perspective tends to overemphasize the impacts of contextual factors.

Figure 10.1: Methodological approach for assessing policy effectiveness: top-down and bottom-up approach

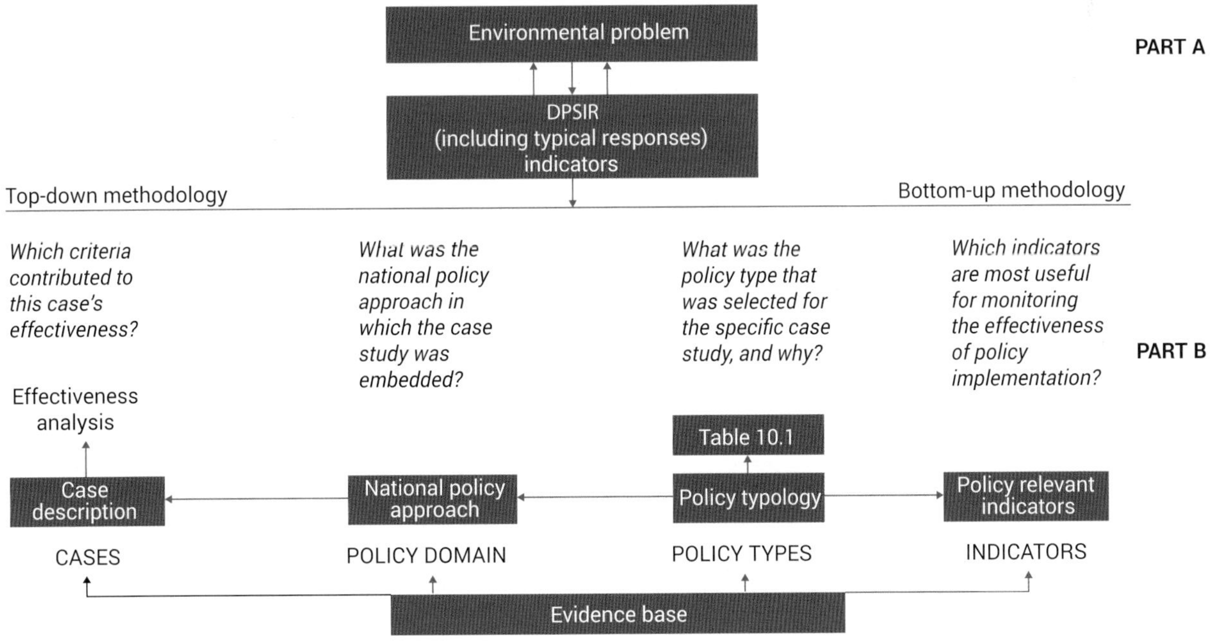

DPSIR: Drivers, Pressures, State, Impact, Response.

The conceptual approach taken in relation to policy effectiveness in this section follows this dual perspective, combining a theory-based top-down evaluation and a bottom-up, observed outcomes-based evaluation. Despite the recognized shortcomings, this dual approach is the best available option for assessing policy effectiveness.

The top-down approach is particularly suitable for identifying policies that may serve as examples of good practice that can be applied elsewhere (the primary goal of such evaluations). Chapters 12-17 provide a narrative description of the most commonly implemented or most important policy instruments in the five thematic areas, as well as the cross-cutting policies that have positive or negative impacts on the themes. In addition, they have identified policy instruments at the cutting edge that appear to have considerable potential but have not been widely adopted to date, through a series of case studies, evaluated against the criteria listed in Section 10.6. However, a quantification of aggregate effects of this mix of policies will not be possible due to a lack of representative sampling and the limited number of policies reviewed.

The bottom-up evaluation, based on policy-relevant indicators, complements the analysis and contributes in particular to the quantification and the identification of needs for further action (goal 2 of the evaluation). The methods for each approach are further elaborated in the following sections. Figure 10.1 graphically summarizes how to assess policy effectiveness through these top-down and bottom-up evaluations.

10.6 Top-down evaluation methodology

The top-down evaluation of policy effectiveness in GEO-6 starts with the selection of policies and governance arrangements and associated cases. For each thematic area, up to five promising policy types or governance arrangements are selected by considering the coverage of the variety of policy types and governance arrangements, geographical diversity and the availability of data.

These policy types and governance arrangements are then evaluated using the available knowledge based on peer-reviewed publications, official reports and statistics.

Next, for each policy type or governance arrangement selected and evaluated, a case exemplifying the implementation of the policy is chosen and assessed in terms of policy effectiveness according to a common research protocol covering the achievement of stated objectives or improvement of relevant indicators, the quality criteria of policy formulation and implementation (e.g. level of participation), *ex ante* or *ex post* assessments and the contextual requirements for the effectiveness.

Note that 'policy effectiveness' is not merely a matter of achieving the policy goals at any cost. For example, an island nation may decide on a policy of carbon neutrality and attempt to achieve it by banning import of gasoline and fuel oil. If the local fishers are unable, however, to power their boats then there could be widespread malnutrition as fish disappears from the diet. Crime might also increase in order to meet the unsatisfied demand for fuel.

The criteria for assessing the cases are derived from the literature on policy design and effectiveness. They are not prescriptive in terms of methods, data, policy instruments or causal chains, but on each aspect the relevant knowledge from the literature is presented. As all case studies are based on secondary data analysis, the research protocol necessarily leaves discretion to be adapted. For example, evaluation studies may be based on measuring effectiveness against the stated goal of policymakers, against an indicator, against a control group or against a counterfactual scenario. The research protocol does not prescribe one or the other method for assessment, but provides transparency on the underlying methods, theories and data sources of the individual case study.

The evaluation criteria and associated guiding questions for the case studies are the following:

1. Effectiveness/goal achievement – What effects did the policy have on the targeted problem?
2. Unintended effects – What were the unintended effects of this policy?
3. Baseline – Was the baseline defined at the policy design stage?
4. Coherence/convergence/synergy – How does the policy intersect with other related policies?
5. Co-benefits – Did the policy design provide for co-benefits?
6. Equity/winners and losers – What are the effects of this policy on different population groups?
7. Enabling/constraining factors – What external factors are likely to influence the intended policy effects?
8. Cost/cost-effectiveness – What were the financial/economic costs and benefits of this policy? Is it the most cost-effective or the least-cost approach?
9. Time frame – Was the policy implemented within the expected time frame?
10. Feasibility/implementability – Is the policy technically feasible in the institutional context?
11. Acceptability – Do the relevant policy stakeholders view the policy as generally acceptable?
12. Stakeholder involvement – To what extent were affected stakeholders actively involved in implementation?
13. Any other factors – such as transformative potential, intergenerational effects, transboundary impacts, sociocultural concerns, political interference, enforcement issues, compliance with legal standards (e.g. national/international human rights).

As a caveat, there is abundant evidence from the environmental policy and governance literature that policy effectiveness is highly context dependent (Jordan and Huitema 2014). Therefore, effective policies from one region or country cannot simply be transferred to another context. Social, cultural, historical and political differences do matter.

This top-down evaluation is complemented by a bottom-up evaluation described in the next section.

10.7 Bottom-up evaluation methodology

An indicator-based assessment of policy-sensitive/policy-relevant indicators for each thematic area and cross-cutting issue complements the top-down evaluation and provides evidence on the quantification of policy effectiveness (Hezri and Dovers 2006; Bauler 2012; Moldan, Janoušková and Hák 2012). Indicators are constructed to measure the state of complex systems which may not be observed directly or comprehensively. They measure certain aspects and based on theoretical considerations and/or evidence, conclusions can be drawn regarding the state of the overall system. For the purpose of measuring policy effectiveness, it is necessary to be explicit on the theory regarding how policies and the selected indicator interact.

Indicators that provide insights to the state of ecological or economic systems and their environmental performance are in many cases not directly influenced by policies. Instead, cultural, structural, political, geographical and other factors may intervene. Measuring the policy outputs (e.g. adoption of policy instruments) would not adequately capture the preferences of different countries for one or another instrument. For example, for mitigating greenhouse gas (GHG) emissions, one country may regulate emissions, another imposes taxes or emission trading schemes, a third implements information campaigns or subsidizes climate-friendly technologies. In each of these cases, the expected impact will be reducing emissions of GHGs. The indicator is influenced, however, by the industrial structure, natural conditions, level of income and other factors that are not, or not directly, impacted by (environmental) policies.

Therefore, an indicator-based assessment must have a transparent underlying theory on how policies would impact on the selected indicator and what other factors may also have an influence. **Figure 10.2** shows the underlying rationale for developing the theory on the relationship between policies and relevant indicators.

Figure 10.2: Approach of assessing policy effectiveness from the bottom-up

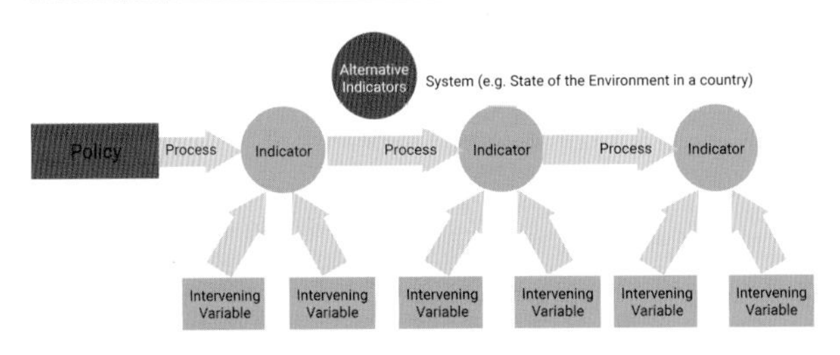

SoE: State of Environment

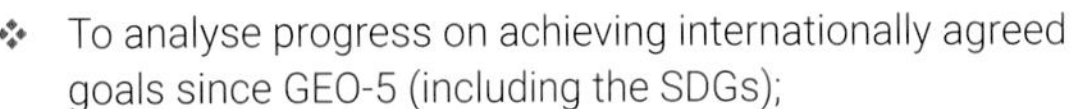

The analysis of data for the selected indicators has multiple objectives:

- To analyse progress on achieving internationally agreed goals since GEO-5 (including the SDGs);
- To identify countries or groups of countries that demonstrate – with their policy approaches and implementation experience – a high level of effectiveness; and
- To quantify policy impacts and thus identify where further action may be needed.

The selection of indicators is based on the following rationale:

- There is a causal relationship determining the variation of the indicator to policy instrument (preferably, different types of policy instruments) and their implementation can be demonstrated;
- The indicator has a relationship to a multilateral environmental agreement (MEA) and/or SDGs to guarantee the alignment of the analysis with the future global agenda;
- Data need to be available (at least at country level, and possibly at global scale, and also in a time series);
- The indicator should be relevant for the thematic area, i.e. it would provide insights into the state of the environment for the respective thematic area; the indicator should ideally consider the policy responses discussed in Part A of this report.

For each indicator, the following aspects are considered, based on peer-reviewed literature.

1. Scope and measurement: the indicator should provide insight into the state of the environment for the respective thematic area. The argument behind selecting each indicator in a thematic area is made transparent.

2. Policy relevance: the causal relationship between policy measures and instruments and the indicator is specified. Not all indicators are policy sensitive but the following questions can be asked of the indicators. Through which mechanisms would policies impact on the indicator? What triggers (e.g. prices, command and control, persuasion) are used by policy instruments that would impact on the respective indicator? Which actors change their behaviour as a result of these policies and how does this impact on the indicator? How does this indicator relate to the state of the environment (ideally at the country level)? What processes are triggered by changes in the indicator and what are the impacts on the environment?

3. Causal relations/causal chain(s): policy-sensitive indicators are those for which a causal relationship to policies and their implementation can be demonstrated. While attribution of causality is challenging, indicators can be selected for their responsiveness to policies as compared to other intervening factors such as sociodemographic factors, infrastructures, natural conditions and culture. Is there any evidence showing that indicators can be associated with these causal output-outcome-impact chains? 'Outputs' and 'outcomes' are processes that are triggered by the policy and 'impacts' are the ultimate effects of a policy. Initial impact may again trigger other processes and have indirect or secondary impacts as well.

4. The analysis must consider other factors impacting on the indicator.
   Is there any evidence showing that other factors not directly or immediately affected by policies (natural conditions, infrastructure, cultural, natural disasters) have demonstrated any impact on the indicator? Are there uncertainties on causal relationships that affect the indicator?

5. Graphic representation and visualization: for each indicator, data are presented on progress towards achieving the relevant international goals as well as the development of each indicator at the country level (cross-longitudinal and cross-sectional analysis).
   Are there outstanding countries in terms of best performer or poor performer? Based on the previous steps a critical reflection should be undertaken: Is it possible to attribute this to policies and other factors? What are the uncertainties?

6. Possible alternative indicators: in case there are suggestions in the literature for other indicators to measure policy effectiveness for the given thematic area, these are discussed.
   Is the suitability or relevance of the indicator disputed? Are there other, alternative indicators proposed to measure policy effectiveness? Why were these not considered in the analysis?
   The indicator-based assessment of policy-sensitive indicators described above does not aim for comprehensiveness: it is certainly not possible to cover all indicators and all aspects for all the thematic areas and cross-cutting issues. This very partial coverage is acknowledged, and further efforts are needed in future GEOs to improve the coverage.

10.8 Content of Part B

The remainder of Part B reflects these top-down and bottom-up approaches, with Chapter 11, based on literature, focusing on issues of policy design, spatial and temporal policy diffusion and evolution, and the effectiveness of international and multi-level governance. Chapters 12-17 cover the key policy approaches mentioned in Part A of the report, under the 'responses' section of the DPSIR framework. For each of these key policy approaches, a case study is used to exemplify the application of the policy approach in a specific context. Policy-sensitive indicators for each thematic area and cross-cutting issue are also included in these six chapters. Chapter 18 draws conclusions and key messages for Part B and provides guidance for policymakers and the link to Part C of the report.

References

Bauler, T. (2012). An analytical framework to discuss the usability of (environmental) indicators for policy. *Ecological Indicators* 17, 38-45. https://doi.org/10.1016/j.ecolind.2011.05.013.

Blamey, A. and Mackenzie, M. (2007). Theories of change and realistic evaluation: Peas in a pod or apples and oranges? *Evaluation* 13(4), 439-455. https://doi.org/10.1177/1356389007082129.

Boersema, J. and Reijnders, L. (eds.) (2009). *Principles of Environmental Sciences*. Dordrecht: Springer. https://link.springer.com/book/10.1007/978-1-4020-9158-2#about.

Challies, E., Newig, J., Kochskämper, E. and Jager, N.W. (2017). Governance change and governance learning in Europe: Stakeholder participation in environmental policy implementation. *Policy and Society* 36(2), 288-303. https://doi.org/10.1080/14494035.2017.1320854.

European Commision (2012). *Assessing and Strengthening the Science and EU Environment Policy Interface*. European Commision https://publications.europa.eu/en/publication-detail/-/publication/a7123d5f-52ee-4f12-9d82-a59eb7d93a26 (Downloaded: 30 October 2017).

European Environment Agency (2001a). *Reporting on Environmental Measures: Are We Being Effective?* Environmental Issue Report. Copenhagen. https://www.eea.europa.eu/publications/rem.

European Environment Agency (2001b). *Paper 1: Defining Criteria for Evaluating the Effectiveness of EU Environmental Measures*. Towards a New EU Framework for Reporting on Environmental Policies and Measures. Copenhagen. https://www.eea.europa.eu/publications/rem/defining.pdf.

European Environment Agency (2017). *Perspectives on Transitions to Sustainability*. Copenhagen. https://www.eea.europa.eu/publications/perspectives-on-transitions-to-sustainability/at_download/file.

Forss, K., Marra, M. and Schwartz, R. (eds.) (2011). *Evaluating the Complex: Attribution, Contribution, and Beyond*. 1st edn Comparative Policy Evaluation. New Brunswick, NJ: Transaction Publishers. https://www.routledge.com/Evaluating-the-Complex-Attribution-Contribution-and-Beyond/Marra/p/book/9781138509832.

Forsyth, T. (2005). Building deliberative public–private partnerships for waste management in Asia. *Geoforum* 36(4), 429-439. https://doi.org/10.1016/j.geoforum.2004.07.007.

Hezri, A.A. and Dovers, S.R. (2006). Sustainability indicators, policy and governance: Issues for ecological economics. *Ecological Economics* 60(1), 86-99. https://doi.org/10.1016/J.ECOLECON.2005.11.019.

Hildén, M., Jordan, A. and Rayner, T. (2014). Climate policy innovation: Developing an evaluation perspective. *Environmental Politics* 23(5), 884-905. https://doi.org/10.1080/09644016.2014.924205.

Hood, C. (2011). *Summing up the Parts: Combining Policy Instruments for Least-Cost Climate Mitigation Strategies*. Paris: International Energy Agency. https://www.iea.org/publications/freepublications/publication/Summing_Up.pdf.

Hovi, J., Sprinz, D.F., Sælen, H. and Underdal, A. (2016). Climate change mitigation: A role for climate clubs? *Palgrave Communications* 2(16020). https://doi.org/10.1057/palcomms.2016.20.

Howlett, M. and Rayner, J. (2007). Design principles for policy mixes: Cohesion and coherence in 'new governance arrangements'. *Policy and Society* 26(4), 1-18. https://doi.org/10.1016/S1449-4035(07)70118-2.

Huttunen, S., Kivimaa, P. and Virkamäki, V. (2014). The need for policy coherence to trigger a transition to biogas production. *Environmental Innovation and Societal Transitions* 12, 14-30. https://doi.org/10.1016/J.EIST.2014.04.002.

Interwies, E., Görlach, B. and Newcomb, J. (2007). Evaluating the cost-effectiveness of environmental policies: Theoretical aspirations and lessons from european practice for global governance. *Amsterdam Conference on the Human Dimensions of Global Environmental Change: Earth System Governance: Theories and Strategies for Sustainability*. Amsterdam, 24-26 May 2007. http://citeseerx.ist.psu.edu/viewdoc/download?doi=10.1.1.664.6369&rep=rep1&type=pdf.

Jänicke, M., Blazejczak, J., Edler, D. and Hemmelskamp, J. (2000). Environmental policy and innovation: An international comparison of policy frameworks and innovation effects. In *Innovation-Oriented Environmental Regulation*. Hemmelskamp, J., Rennings, K. and Leone, F. (eds.). Heidelberg: Springer. 125-152. https://link.springer.com/chapter/10.1007/978-3-662-12069-9_7

Jordan, A. and Huitema, D. (2014). Innovations in climate policy: The politics of invention, diffusion, and evaluation. *Environmental Politics* 23(5), 715-734. https://doi.org/10.1080/09644016.2014.923614.

Keskitalo, E.C.H., Juhola, S., Baron, N., Fyhn, H. and Klein, J. (2016). Implementing local climate change adaptation and mitigation actions: The role of various policy instruments in a multi-level governance context. *Climate* 4(1), 7. https://doi.org/10.3390/cli4010007.

Kivimaa, P. and Kern, F. (2016). Creative destruction or mere niche support? Innovation policy mixes for sustainability transitions. *Research Policy* 45(1), 205-217. https://doi.org/10.1016/j.respol.2015.09.008.

Kochskämper, E., Challies, E., Jager, N.W. and Newig, J. (eds.) (2018). *Participation for Effective Environmental Governance: Evidence from European Water Framework Directive Implementation*. London: Routledge. https://www.routledge.com/Participation-for-Effective-Environmental-Governance-Evidence-from-European/Kochskamper-Challies-Jager-Newig/p/book/9781138713291.

Lambin, E.F., Meyfroidt, P., Rueda, X., Blackman, A., Börner, J., Cerutti, P.O. *et al.* (2014). Effectiveness and synergies of policy instruments for land use governance in tropical regions. *Global Environmental Change* 28, 129-140. https://doi.org/10.1016/j.gloenvcha.2014.06.007.

Lay, J., Brandi, C., Upendra Das, R., Klein, R., Thiele, R., Alexander, N. *et al.* (2017). *Coherent G20 Policies Towards the 2030 Agenda for Sustainable Development*. G20 Insights. http://www.g20-insights.org/wp-content/uploads/2017/03/TF_2030_Agenda_PolicyCoherence.pdf.

Mees, H., Crabbé, A. and Driessen, P.P.J. (2017). Conditions for citizen co-production in a resilient, efficient and legitimate flood risk governance arrangement. A tentative framework. *Journal of Environmental Policy & Planning* 19(6), 827-842. https://doi.org/10.1080/1523908X.2017.1299623.

Mees, H., Dijk, J., van Soest, D., Driessen, P., van Rijswick, M. and Runhaar, H. (2014). A method for the deliberate and deliberative selection of policy instrument mixes for climate change adaptation. *Ecology and Society* 19(2). http://dx.doi.org/10.5751/ES-06639-190258.

Moldan, B., Janoušková, S. and Hák, T. (2012). How to understand and measure environmental sustainability: Indicators and targets. *Ecological Indicators* 17, 4-13. https://doi.org/10.1016/j.ecolind.2011.04.033.

Niles, M.T. and Lubell, M. (2012). Integrative frontiers in environmental policy theory and research. *Policy Studies Journal* 40, 41-64. https://doi.org/10.1111/j.1541-0072.2012.00445.x.

Organisation for Economic Co-operation and Development (2016). *Better Policies for Sustainable Development 2016: A New Framework for Policy Coherence*. Paris. http://dx.doi.org/10.1787/9789264256996-en.

Organisation for Economic Co-operation and Development, International Energy Agency, International Transport Forum and Nuclear Energy Agency (2015). *Aligning Policies for a Low-Carbon Economy*. Paris. https://www.oecd-ilibrary.org/aligning-policies-for-a-low-carbon-economy_5js4lch2tsjj.pdf?itemId=%2Fcontent%2Fpublication%2F9789264233294-en&mimeType=pdf.

Perrels, A. (2001). Efficiency and effectiveness of policy instruments: Concepts and practice. *Workshop on Good Practices in Policies and Measures*. Copenhagen, 8-10 October 2001. 10 https://unfccc.int/files/meetings/workshops/other_meetings/application/pdf/perrels.pdf

Rogers, P.J. and Weiss, C.H. (2007). Theory-based evaluation: Reflections ten years on: Theory-based evaluation: Past, present, and future. *New directions for evaluation* (114), 63-81. https://doi.org/10.1002/ev.225

Sabatier, P.A. (1986). Top-down and bottom-up approaches to implementation research: A critical analysis and suggested synthesis. *Journal of Public Policy* 6(1), 21-48. https://doi.org/10.1017/S0143814X00003846.

Sandler, T. (2017). Environmental cooperation: Contrasting international environmental agreements. *Oxford Economic Papers* 69(2), 345-364. https://doi.org/10.1093/oep/gpw062.

United Nations Economic and Social Council (2006). *Definition of Basic Concepts and Terminologies in Governance and Public Administration. Note by the Secretariat**. E/C.16/2006/4. http://unpan1.un.org/intradoc/groups/public/documents/un/unpan022332.pdf

United Nations Environment Programme (2007). *Global Environment Outlook-4: Environment for Development*. Nairobi. https://wedocs.unep.org/bitstream/handle/20.500.11822/7646/-Global%20Environment%20Outlook%20%204%20(GEO-4)-2007768.pdf?isAllowed=y&sequence=3.

United Nations Environment Programme (2012). *Global Environment Outlook-5: Environment for the Future We Want*. Nairobi. http://wedocs.unep.org/bitstream/handle/20.500.11822/8021/GEO5_report_full_en.pdf?sequence=5&isAllowed=y.

United Nations Environment Programme (2017). *Use of Theory of Change in Project Evaluations: Introduction*. Nairobi. http://wedocs.unep.org/bitstream/handle/20.500.11822/7116/11_Use_of_Theory_of_Change_in_Project_Evaluation_26.10.17.pdf?sequence=6&isAllowed=y.

van Tulder, R., Seitanidi, M.M., Crane, A. and Brammer, S. (2016). Enhancing the impact of cross-sector partnerships. *Journal of Business Ethics* 135(1), 1-17. https://doi.org/10.1007/s10551-015-2756-4.

Yoshida, T. and Zusman, E. (2015). *Can the Sustainable Development Goals Strengthen Existing Legal Instruments? The Case of Biodiversity and Forests*. Global Environmental ResearchInstitute for Global Environmental Strategies. http://www.airies.or.jp/attach.php/6a6f75726e616c5f476c6f62616c456e7669726f6e6d656e74616c52657365617263685f6736737748764e6a/save/0/0/19_2-13.pdf.

Chapter 11

Policy Theory and Practice

Coordinating Lead Authors: Klaus Jacob (Freie universität Berlin), Peter King (Institute for Global Environmental Strategies)

Lead Authors: Pedro Fidelman (Centre for Policy Futures, University of Queensland), Leandra Regina Gonçalves (University of Campinas/Centre for Environmental Studies and Research [NEPAM]), James Hollway (Graduate Institute of International and Development Studies), Sebastian Sewerin (Swiss Federal Institute of Technology Zurich [ETH Zurich])

Fellow: He Chenmin (Peking University)

Executive summary

Environmental policy struggles with some conceptual and empirical challenges, so a good starting point for analysis is what constitutes 'good' policy design. Within the definition of 'good' policy design, ecosystem properties and problems, the performance of existing policies, practices and actors need to be considered common elements. Analysts and policymakers should better understand the temporal dynamics of policy change, how and why specific policies work (or not) and how policy choices interact in increasingly complex policy mixes. {11.2}

In the field of environmental policy research, diffusion across borders has featured prominently, especially in relation to renewable energy policies such as feed-in tariffs, renewable portfolio standards and emissions trading. Among four possible mechanisms of renewable energy policy diffusion (emulation, suasion, learning and competition), suasion and emulation were found to be more common than learning and competition. There has been little research on post-adoption dynamics of diffused policies, with a risk of undermining other sector policies and policy coherence. {11.2.1}

In relation to how policies change over time and what factors drive these changes, two approaches are dominant in the literature: policy stability—or lock-in—on the one hand and punctuated equilibrium on the other. Punctuations may be driven by external shocks that rock the otherwise stable policy environment until a new equilibrium is established. Such punctuations may open opportunities for new environmental policies (e.g. the impact of Japan's Fukushima nuclear disaster on Germany's policy decision to phase out nuclear power). Often, to avoid risks from taking decisions with unwanted side effects, policymakers tend to delay decisive action as long as possible and, confronted with external shocks, choose symbolic action rather than effective policymaking. {11.2.2}

Policy innovation can be regarded as a mix of invention (new or novel approaches), diffusion (transfer and adoption) and monitoring of effects (outcomes, impacts and possibly disruption). Good practice suggests that multiple innovative policies should be implemented as a form of quasi-experiment, with best practices emerging from the monitored effects. {11.2.2}

Evaluation of policy effectiveness often comes down to expert judgement, as there is no commonly agreed approach to assessing effectiveness. Ideally, a combination of quantitative and qualitative assessments will be most reliable in assessing policy effectiveness. Some policy design tools that can supplement expert judgement are: (i) cost–benefit analysis and cost-effectiveness analysis, both of which can be used *ex ante* (before implementation) or *ex post* (after implementation); (ii) regulatory impact analysis; (iii) benchmarking or distance to goal or target; and (iv) content analysis or pattern matching. {11.2.3}

A key lesson learned is to carefully craft a mix of policies that are well aligned with an overall policy objective followed by monitoring of the actual effects to determine best practice and likely contributions of different policies. For example, climate change mitigation policy objectives will need a comprehensive mix of carbon pricing, support for energy efficiency and renewable energy, phasing out fossil fuel subsidies, innovation policies, preventing lock-in of certain technologies and changes in consumer behaviour, among others. {11.2.4}

Environmental objectives cannot be realized by environmental policies alone, but need to be incorporated in non-environmental policy sectors too. Environmental ambitions often clash with other sectoral goals, so environmental policy integration should be used to address conflicts between environmental and other policy objectives. A corollary to policy integration is policy coherence: the promotion of mutually reinforcing policy actions creating synergies towards achieving objectives in multiple sectors. {11.3}

An important argument in favour of environmental policy integration is the economic and social co-benefits that can be expected or demonstrated from implementing environmental policies. These may include additional economic growth from innovation, savings from more efficient use of natural resources and avoiding the costs of environmental damage. {11.3.3}

Involving alliances, clubs and non-State actors in policy design may provide opportunities for peer pressure to overcome institutional reluctance. Hybrid governance, combining different modes and instruments of governance, can help mutually strengthen institutional responses. There is no 'one size fits all' governance structure, however. As for policy effectiveness, different approaches have been proposed to gauge institutional effectiveness, involving both qualitative and quantitative methods. Increasingly important in international environmental policy discourse is the role of non-State actors such as local governments, cities and civil society organizations. As the 17 SDGs are intended to be fully integrated and universal, several countries are now grappling with the task of devising the most effective institutional arrangements to address the desired vertical and horizontal policy integration. {11.4.2}

Finally, the importance of good policy design cannot be overstressed. Some common elements are: (i) setting a long-term vision and avoiding crisis-mode policy decisions, through inclusive, participatory design processes; (ii) establishing a baseline, as well as quantified targets and milestones; (iii) conducting *ex ante* and *ex post* cost–benefit or cost-effectiveness analysis to ensure that public funds are being used most efficiently and effectively; (iv) building in policy monitoring regimes during implementation, preferably involving affected stakeholders; and (v) conducting post-intervention evaluation of the policy outcomes and impacts, to close the loop for future policy design improvements. {11.5}

11.1 Introduction

Academic and practical or policy advice-related research asks important questions about environmental policy. However, the literature still struggles with common conceptual and empirical challenges, including:

i. diverse conceptualizations and measurements of policy, which frustrate generalization;
ii. poorly understood dynamics of policy change and stability; and
iii. complex effects of policy design choices.

Overall, these common challenges impede the comparability of findings across environmental policy fields, risking somewhat narrow approaches and perspectives.

One potential starting point for overcoming this narrow focus is a renewed interest in policy design. Are there common elements of 'good' policy design that are transferable across problem domains? How do policymakers search for effective policy precedents, and how does this search lead to policy diffusion across countries and across problem domains? How does the theoretical understanding of policy design contribute to policy practice?

This section examines the literature and teases out some lessons learned for validation by the policy domains/ instruments and governance arrangements and associated case studies in Chapters 12–17 of Part B **(Figure 11.1)**. Essentially, it addresses the top portion of Figure 11.1, while Chapters 12–17 explore the lower half. Figure 11.1 illustrates the importance of going beyond an analysis of individual policy instruments when trying to determine policy effectiveness. At the policy design stage, policymakers should examine how an environmental policy will either support or conflict with policies in other sectors, and vice versa. Various policy integration tools are available to contribute to this design requirement. Within the environmental policy mix, policymakers should ensure that any new policy is coherent with and supportive of the intended policy outcomes. Policymakers at multiple levels of governance may also examine experience from other countries, subnational areas or corporations, leading to policy diffusion across borders and over time. In subsequent chapters, this experience is examined from the perspective of multiple case studies, to tease out explanations of why particular policy approaches appear to have been more or less effective, as well as to analyse policy-sensitive indicators.

11.2 Policy design

The challenges described above, relate both to the spatial and temporal dynamics of policy change and to the complexity of how policy instruments interplay within a policy mix. These have led to a renewed interest in questions of policy design (Howlett and Lejano 2013). The literature acknowledges the complexities of design elements in dense policy mixes (Howlett and Cashore 2009; Howlett and del Rio 2015; Young 2017), and it also recognizes that the compliance system highly influences it (Grey and Shimshack 2011). However, as the research agenda develops, approaches for institutional diagnosis (Young 2008; Ostrom 2009), empirical assessment and analysis of policy change (Jabbour *et al.* 2012; Knill, Schulze and Tosun 2012; Schaffrin, Sewerin and Seubert 2015) have been increasingly applied to systematic, more quantitative analysis of policy (mix) dynamics (Voigt 2013). Young (2008) argues that ecosystem properties and problems, policies, practices and players need to be considered as elements for regime design under a diagnostic analytical approach that tries to match institutional arrangements to those properties and structures. Ostrom (2007) takes a similar approach and builds a framework that contains many types of resource properties across multiple scales, including local.

Regarding policymakers' decision-making, there is an emerging consensus that policy design is at least as important as policy instrument choice for both individual policy effectiveness and for the effectiveness of the overall policy mix (Mitchell 2002; Yin and Powers 2010; Flanagan *et al.* 2011; Kemp and Pontoglio 2011). For example, technology-specificity—i.e. where a policy

Figure 11.1: Conceptual outline of policy effectiveness analysis

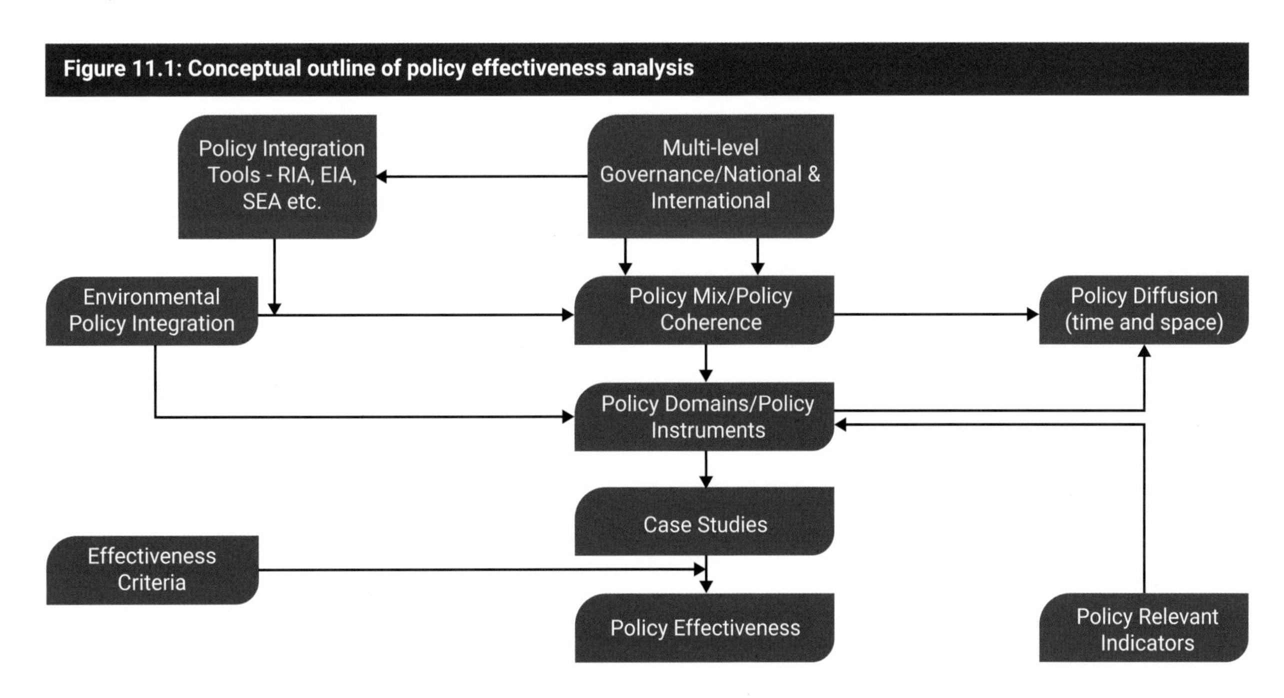

explicitly differentiates between different available technologies that address a given policy problem—is increasingly applied in the analysis of low-carbon technology deployment policies (e.g. Schmidt *et al.* 2016). Yet, integrating the various perspectives on policy design (e.g. from implementation research or governance studies) remains challenging. Still, literature on policy dynamics and policy mixes, on policy design and policy effectiveness and on long-term intervention logics is increasingly available (Young 1999; Miles *et al.* 2002; Howlett and Rayner 2013; Mees *et al.* 2014).

When trying to tackle the various pressing environmental problems outlined in Part A, a policy design perspective would be very helpful. Analysts and policymakers alike need to better understand the temporal dynamics of policy change, how and why specific policies work (or not) and how policy choices interact in an increasingly complex mix of policies. Ideally, this would help to improve the design of policies that create positive feedback loops (including *ex post* impact assessment), eventually changing the underlying instrumental logic of a policy mix **(Figure 11.2)**.

11.2.1 Spatial dynamics: policy diffusion across borders

Policy diffusion research aims to understand how and why policies spread across borders and are adopted and designed by different jurisdictions. The academic literature, primarily from the field of political science, tends to focus on drivers of the spatial diffusion of policies (Tews, Busch and Jorgens 2003; Holzinger, Knill and Sommerer 2011; Graham, Shipan and Volden 2013; Matisoff and Edwards 2014), and much less so on what policies have actually diffused and when (Jordan and Huitema 2014a). In the environmental policy research literature, policy diffusion featured prominently in the 2000s and early 2010s. The main focus of this research was on environmental regulations (Knill, Schulze and Tosun 2012) and renewable energy policy (Alizada 2017). The former was helped by the availability of a large data set of regulatory standards covering primarily emissions (Heichel *et al.* 2008), while the latter was driven by the debate about the effectiveness of feed-in tariffs and renewable portfolio standards as renewable energy deployment policy tools. Generally, these studies focused either on the macro-level characteristics of policies (e.g. policy instrument types), as in Stadelmann and Castro (2014), or on very specific regulatory standards (e.g. NOx or SO_2 emissions standards for large combustion plants) as in, for example, Liefferink *et al.* (2009); Holzinger, Knill and Sommerer (2011). In relation to renewable energy, four possible mechanisms of policy diffusion—emulation, suasion, learning and competition—were examined (Alizada 2017). Suasion and emulation were found to be more common than learning and competition.

More recently, there has been an explosion in climate change-related policies in both developed and developing countries. Over the period 1998–2010, there was a fivefold increase in national climate laws, and by 2012 these laws covered 67 per cent of all emissions (Jordan and Huitema 2014b).

There is little research, however, on the post-adoption dynamics of diffused policies. Only isolated studies (e.g. Biesenbender and Tosun 2014) analyse how these policies are adapted in new jurisdictions—i.e. how they are modified subsequent to the initial diffusion. The European Union's emissions trading scheme is a good example of the difficulties in adjusting policy from one jurisdiction to another (Cass 2005). Post-adoption dynamics may even undermine the intended policy impact and policy coherence (Jordan and Huitema 2014a).

Figure 11.2: The policy cycle

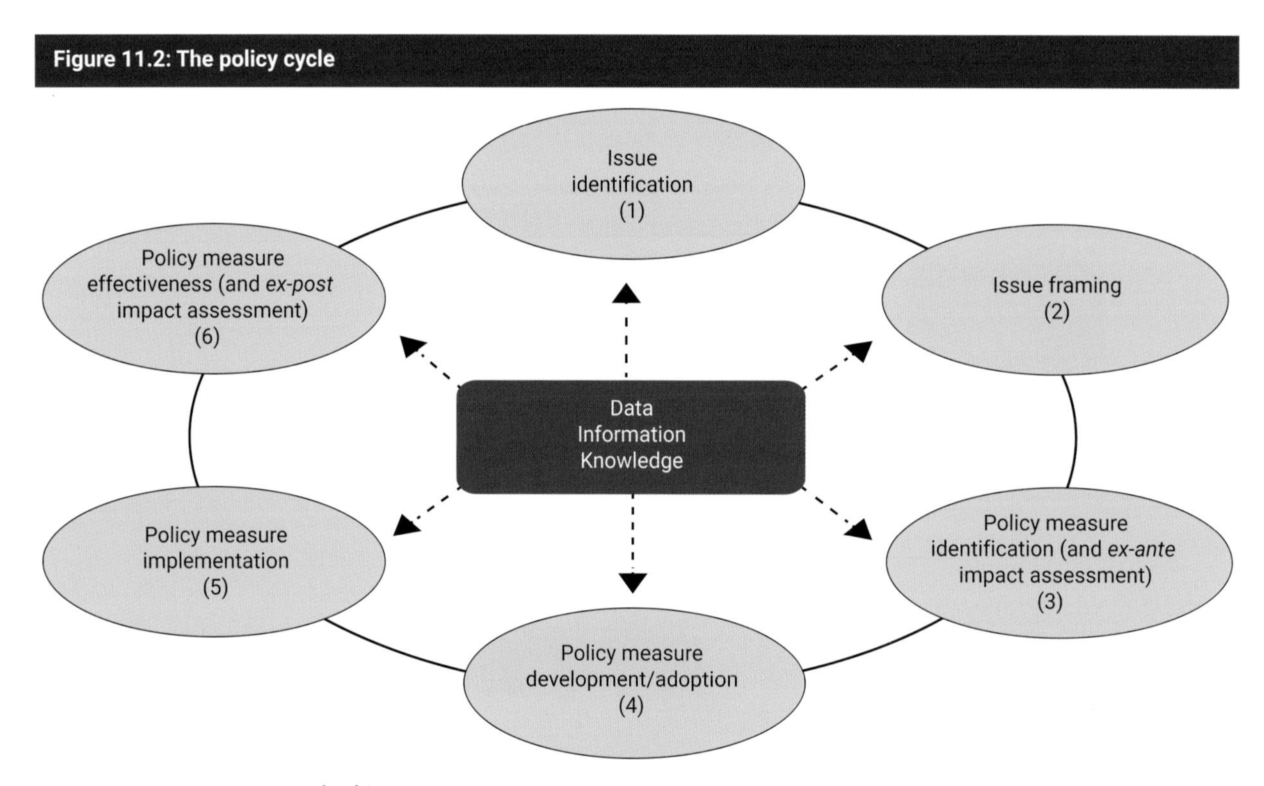

Source: European Environment Agency [EEA] (2006)

Some publicly available data sets aim to help chart the diffusion of environmental policies, particularly related to climate change and, more specifically, to renewable energy policy. The London School of Economics' Climate Change Laws of the World Database (Nachmany *et al.* 2017), for example, compiles information on national-level climate policies ranging from adaptation to mitigation to transport. Similarly, REN21's Global Status Report charts the use of renewable energy policies across a large sample of national and subnational jurisdictions. International organizations, such as the International Energy Agency (IEA), also collect information on renewable energy-related policies in use across a large sample of jurisdictions. The quality of all these data sets, though, varies, as does the method of collection, categorization of policies and the level of detailed policy information included. This problem relates to what has been coined the "dependent variable problem in the study of policy change" (Howlett and Cashore 2009)—i.e. the underlying challenge of how to assess policy output systematically across cases. While efforts have been made to develop a common methodology for measuring policy output in a comparable way (e.g. Knill, Schulze and Tosun 2012; Schaffrin, Sewerin and Seubert 2015), these approaches are only slowly being taken up, and most policy dynamics analysis continues to apply diverse or ad hoc concepts and measurements of policy output. Thus, despite prolonged interest in the topic and efforts to provide systematic policy information, knowledge of the spatial diffusion of environmental policies, especially outside the specific policy field of renewable energy, remains limited.

11.2.2 Temporal dynamics: policy change over time

If and how policies change over time and what factors drive these changes are important topics in the academic literature. The different approaches for understanding policy change can be categorized, on the one hand, in path-dependency literature, which stresses that early policy decisions lock in policy choices and that most policies only change incrementally after they are implemented (Pierson 2000). The main reason for such stability is thought to be positive feedback, through for example, policy learning that creates and sustains self-reinforcing processes around a policy. On the other hand, the punctuated equilibrium approach seeks to explain how otherwise stable policies can unravel in a sudden burst of abrupt, non-incremental change (Baumgartner and Jones 2009; Colgan, Keohane and Van de Graaf 2012). The main driver of these punctuations is thought to be external shocks that tilt the otherwise stable balance of positive and negative feedbacks towards a new equilibrium. One example of such a shock could be a legal case that challenges the legitimacy of the environmental policy. A punctuation could also create opportunities for environmental policies; for example, the Fukushima disaster in Japan may have led to Germany's policies to phase out nuclear energy (Wittneben 2012).

Both approaches have been applied in the analysis of environmental policy change (e.g. Daugbjerg 2003; Repetto 2006), although the applications have mostly concentrated on large programmes in particular policy areas, such as agriculture. Recent literature has argued that focusing exclusively on positive feedback or on the catalytic effect of external shocks is not very helpful for intentionally designing policies that can both create positive feedback and withstand sustained negative feedback and external shocks (Jordan and Matt 2014).

The complexity of environmental problems can also increase the risk of 'policy under-reaction' by decision-makers, since it is difficult for policymakers to accurately estimate risks (Maor 2014). To avoid risks from taking decisions with unwanted side effects, policymakers tend to delay decisive action as long as possible and, confronted with external shocks, choose symbolic action rather than effective policymaking (Howlett 2014). There are a number of suggestions for strengthening the importance of the environment within States to overcome such shortcomings (Kloepfer 1989; Calliess 2001; Eckersley 2005; Jänicke 2007); however, this has not happened so far: the importance of the environment is not institutionalized as a priority but competes with other goals of governments.

Against this background, research is increasingly turning to policy design (Howlett and Lejano 2013) and seeks to understand how policy design choices can create policy change—i.e. how steps of incremental policy change can, over time, build up to create transformational change. Policies that are 'sticky' (i.e. persistent) but not 'stuck' (i.e. unresponsive to changing conditions) and that create positive feedbacks are seen as a potential way to increase the effectiveness of environmental policies (Jordan and Matt 2014). The Paris Agreement on climate change and its ratcheting-up mechanism is a prominent example of this concept (Falkner 2016). The need for such a forward-looking approach to policy design can be seen in policy fields that are troubled with complexity, as is the case for most environmental problems (Levin *et al.* 2013). The design of international regimes heavily influences their effectiveness—even more importantly than the type of underlying problem. In other words: an easy problem is not solved if an international regime is poorly designed (Young 2011). Given the context dependency of policies and regimes, a careful diagnosis of the appropriateness of their design is essential (Young 2011).

Policy innovation can be regarded as a mix of invention (new or novel approaches), diffusion (transfer and adoption) and monitoring of their effects (outcomes, impacts and possibly disruption) (Jordan and Huitema 2014b). The literature on polycentric governance suggests that multiple innovative policies should be implemented as a form of quasi-experiment, with best practices emerging from the monitored effects. It has been argued that governance at the lowest possible level minimizes free-riding as a motivation, and that monitoring is easier in smaller entities, e.g. communities (Marshall 2009). However, on a global scale, polycentric governance could lead to free-riding by governments, for example in the absence of a global regime, governments could be tempted to avoid actions while benefiting from mitigation efforts by others (Ostrom 2010). However, the role of policy entrepreneurship, and the contribution of civil society, in motivating policy shifts should not be underestimated.

Policy innovation, however, is not necessarily the most effective pathway to policy packaging, as tried and true command and control and economic incentive policies may deliver most of the impact (Hildén, Jordan and Rayner 2014; Jordan and Huitema 2014a). Greater focus on policy coherence, successful implementation and compliance may prove that traditional policy approaches still work effectively. Innovative policies may bring new implementation and compliance challenges which existing institutions are not well equipped to handle.

11.2.3 Policy effectiveness through improved design

In the past, attempts to measure policy effectiveness assumed there was a one-to-one correlation between an environmental policy and its outcomes (Weber, Driessen and Runhaar 2013). In some cases, this may be warranted, such as a government command and control policy to remove lead from petrol or from paint (see the subsequent discussion of policy-sensitive indicators). However, in most of the case studies in Chapters 12-17, attributing environmental outcomes to specific policies is shown to be challenging. Counterfactual scenarios (i.e. without policy) cannot be implemented experimentally, for practical and ethical reasons among others, as it is not justifiable to expose one group to a policy against a harmful pollutant and not others (Niles and Lubell 2012).

For these reasons, evaluation of policy effectiveness often comes down to the use of expert judgement **(Figure 11.3)** (EEA 2001; Egger *et al.* 2015). Figure 11.3 shows that not all energy efficiency policies are ranked equally, and a significant proportion of experts believe some policies are not effective at all. Although there are some limitations to measuring policy effectiveness, some important insights have emerged, not only from the use of statistical procedures to separate the effects of individual variables but also from the application of alternative techniques, such as Qualitative Comparative Analysis (QCA), designed to identify combinations of factors that operate together, to determine the effectiveness of policies (Breitmeier, Underdal and Young 2011). Ideally, a combination of quantitative and qualitative assessments will be most reliable in assessing policy effectiveness (Egger *et al.* 2015).

Some policy design tools that can supplement expert judgement are:

i. cost–benefit analysis and cost-effectiveness analysis, both of which can be used *ex ante* or *ex post* (Interwies, Gorlach and Newcombe 2007);
ii. regulatory impact analysis (Organisation for Economic Co-operation and Development [OECD] 2008);
iii. benchmarking or distance to target (Uslu, Mozzaffarian and Stralen 2016); and
iv. content analysis or pattern matching (Di Gregorio *et al.* 2017).

Of course, environmental problems are essentially social constructs, and what appears as an environmental problem to one group may not be seen as a problem by another group with different interests. Therefore, in designing effective policies, framing of the 'problem' is extremely important. For example, framing climate change as an issue involving employment and security, as exercised in Europe, may be more effective than simply discussing it as a technical or scientific issue. Changing the approach to economic development from 'grow now, clean up later' may be an important shift in environmental policy design in several developing countries.

11.2.4 Policy mixes

As indicated in Chapter 10, within the environmental domain a key lesson learned is to carefully craft a mix of policies that are well aligned with the overall policy objective (OECD, IEA, Nuclear Energy Agency [NEA] and International Transport Forum

Figure 11.3: Results of expert perspectives on European energy efficiency policies

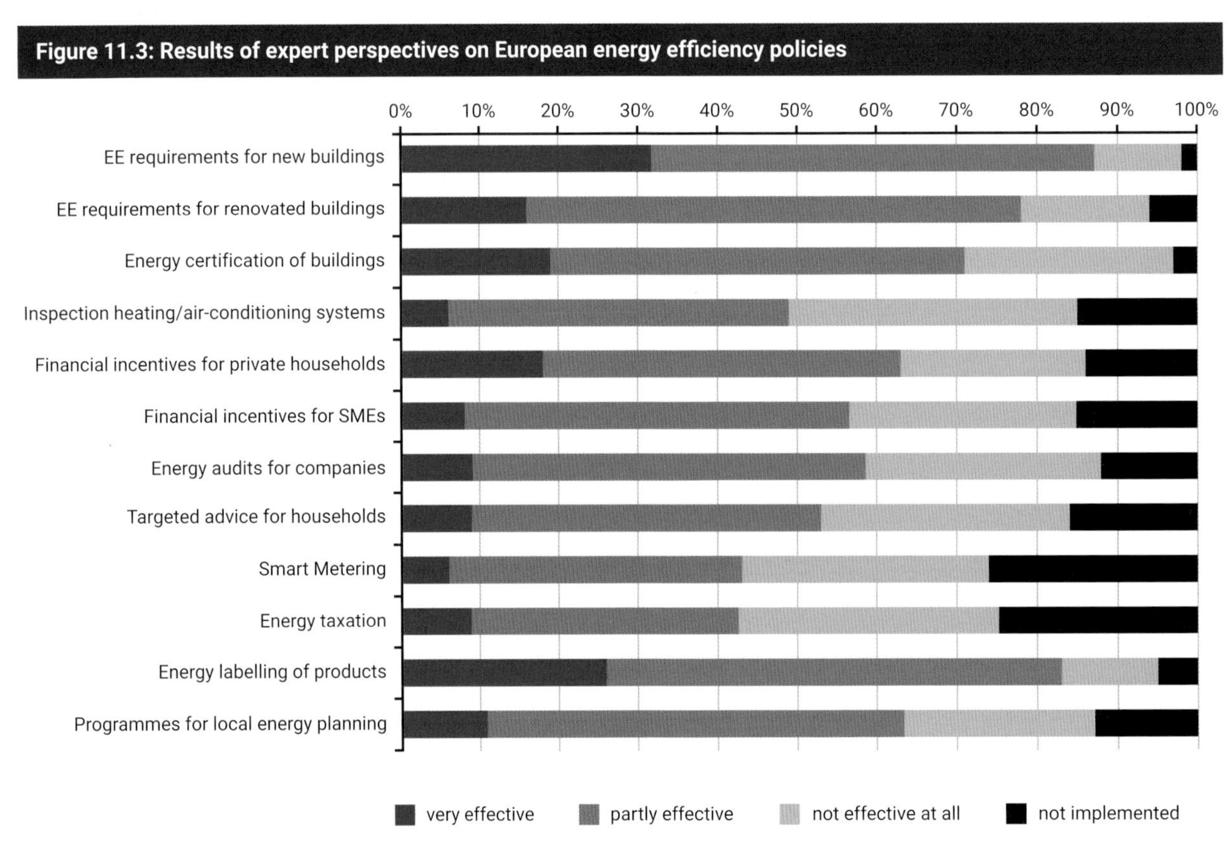

Source: Egger *et al.* (2015)

[ITF] 2015) and then monitor the actual effects to determine best practice. For example, climate change mitigation policy objectives will need a comprehensive mix of carbon pricing, support for energy efficiency and renewable energy, phasing out fossil fuel subsidies, innovation policies, preventing locked-in technology and changes in consumer behaviour, among others (Hood 2011). Too often policies are not well aligned and may even be in conflict, so solving misalignment and ensuring that policies are mutually reinforcing may be one of the best ways of achieving environmental improvement (OECD *et al.* 2015).

Policy analysis literature increasingly recognizes the need to view policies not in isolation but to consider each policy in its wider context among a mix of other policies. Policy mixes are generally thought of in relation to specific policy fields (e.g. the renewable energy policy mix). Their quality is traditionally assessed in qualitative terms and draws on a set of concepts, such as

i. consistency of multiple policy instruments (i.e. instruments' ability to reinforce rather than undermine each other),
ii. coherence of multiple policy goals (i.e. goals not contradicting each other) and
iii. congruence of multiple policy goals and instruments (i.e. their ability to work together in a unidirectional fashion) (Howlett and Rayner 2013; Kern, Kivimaa and Martiskainen 2017).

While these concepts are widely used, they are neither consistently defined nor is there an accepted common practice of assessment, leading to a lack of empirical analysis of patterns of policy mixes (Howlett and del Rio 2015). Other than a rather broad understanding that some types of policy instruments do not necessarily work well together, the interplay of policy instruments in a mix is not well understood. One main reason for this is the persistent challenge to adequately and systematically assess and evaluate individual policies (Capano and Howlett 2009; Howlett and Cashore 2009)—i.e. the building blocks of complex policy mixes—let alone how they mix with other policies. Only recently has research begun to tackle these important challenges, either by conducting policy mix analysis based on comprehensive data sets (Schmidt and Sewerin 2018), or by modelling the interplay of policy instruments. These models, however, are generally limited to the interplay of two or three policies, whereas the real-world mix, consisting of many more policies, is usually more complex.

11.3 Policy integration

Environmental objectives cannot be realized by implementing a mix of environmental policies alone; they need to be integrated in non-environmental policy sectors too. This is underpinned in the integrated approach of Agenda 2030 and the Sustainable Development Goals (SDGs). However, environmental ambitions often clash with other sectoral goals—for example, when the utilization of natural resources by the energy sector, agriculture industries or for building infrastructures clashes with efforts to conserve those natural resources. A concept that promotes the environment in other policy sectors, and recognized in previous GEO assessments, is 'environmental policy integration' (EPI) (Lay *et al.* 2017). EPI aims to settle such conflicts between environmental and other policy objectives and has been discussed in the scientific literature (Nilsson *et al.* 2012; Runhaar, Driessen and Uittenbroek 2014) and debated in policy contexts from early on (Mullally and Dunphy 2015). It led to the introduction of a wide range of institutions, processes and instruments for its implementation (Jacob *et al.* 2008). The principle of EPI also contributed to change in policy discussions which again affected policy outcomes (Scarse and Ockwell 2010; Espinosa *et al.* 2017). Others suggest that EPI needs to go further and demand the "incorporation of environmental objectives into all stages of policy-making in non-environmental policy sectors, with a specific recognition of its role as a guiding principle for the planning and execution of policy" (Lafferty 2004, p. 201). This level of ambition is not achieved in reality, as policy incoherence and competition typically prevail, pointing to the limitation of the institutions, instruments and processes that have been introduced to promote EPI.

It is not always clear how institutional and socio-economic conditions associated with other policy domains degrade the environment, making it difficult to understand which EPI strategies would work to mitigate this degradation (Runhaar, Driessen and Uittenbroek 2014). Systematic evidence is lacking on how sectors such as agriculture, transport, urban planning or water management incorporate environmental concerns and standards to prevent, reduce or mitigate any harmful environmental effects. Nonetheless, one necessary but insufficient condition for policy integration is political leadership and the acknowledgement of co-benefits across multiple policy domains (Persson 2007; Jordan and Lenschow 2010).

Policy coherence

A closely related concept to policy integration is policy coherence: the promotion of mutually reinforcing policy actions creating synergies towards achieving objectives in multiple sectors. Attempts at better coherence include the development of national sustainable development strategies and various road maps, such as those developed by the European Commission or the 'better regulation' agenda pursued by both the European Union (EU) and the OECD (European Commission 2010). Here too, policy leadership and co-benefits are necessary but insufficient. In 2016, the OECD elaborated a framework of policy coherence for sustainable development based on eight building blocks:

i. political commitment and leadership at the highest level;
ii. integrated approaches to implementation;
iii. an intergenerational time frame;
iv. analysis and assessments of potential transboundary effects;
v. policy and institutional coordination;
vi. local and regional involvement;
vii. stakeholder participation; and
viii. monitoring and reporting (OECD 2017).

A major example of an attempt at environmental integration and coherence within the United Nations environmental umbrella are the SDGs (United Nations General Assembly [UNGA] 2015). The SDGs encompass major environmental areas such as climate change, chemical pollution, waste, and marine and terrestrial ecosystems, but they also include social, economic and institutional development objectives applicable to both low-income and high-income countries, such as access to food, water, sanitation, energy, health,

education and justice, and the development of infrastructure, cities, employment and growth (Nilsson and Persson 2017). The SDGs mark a historic shift for the United Nations towards a unique 'sustainable' development agenda after a long history of trying to integrate economic and social development with environmental sustainability. Before the SDGs, international agreements were more fragmented and sectoral, and while environmental integration was regularly mentioned on paper, it was rarely translated into practice. Of course, the effectiveness of the holistic, indivisible approach recommended for the SDGs remains to be seen.

Chapters 12-17 identify examples of regulatory and other policy instruments that have demonstrated some utility for policy integration and coherence. Regulatory policies have emerged in most countries over the last two decades under the umbrella of so-called 'better regulation policies' (Turnpenny *et al.* 2009; De Francesco 2012; Adelle *et al.* 2015). In many countries these have led to the establishment of institutions and the adoption of instruments and processes, such as regulatory impact assessments, participatory approaches and *ex post* evaluation. Initially, this was motivated by concerns to cut the costs of regulation and deregulation. However, in some countries these instruments were broadened in their scope and used to promote the integration of concerns about sustainable development into policymaking (Bäcklund 2009; Adelle and Weiland 2012; Renda 2017).

The SDGs (e.g. target 17.4) emphasize the need to "enhance policy coherence for sustainable development" (OECD 2016). A key lesson learned is to ensure policy coherence between different levels of governance and across economic, social and environmental domains. Accordingly, it is not sufficient to consider environmental policies as separate from macroeconomic or sectoral policies, which often act as drivers of environmental degradation, or from social policies that attempt to address the human impacts of environmental damage. It is possible to see in the case studies in Chapters 12-17 that these other policies often act as enabling or constraining factors in achieving environmental policy effectiveness. When analysing policy coherence, policymakers also need to consider long-term as well as short-term impacts (OECD 2016).

11.3.1 Integration of environmental aspects in regulatory policies

Integrating environmental concerns and policy objectives into different policy domains comprises the cornerstone of EPI (Runhaar, Driessen and Uittenbroek 2014). Ensuring that such concerns and policy objectives are incorporated in the development of legislation may prove critical in promoting EPI (Jacob *et al.* 2011). Many countries have adopted approaches/instruments to assess the potential impacts of proposed legislation on stakeholders, economic sectors and the environment (Radaelli 2009; Jacob *et al.* 2011; Adelle and Weiland 2012; Adelle *et al.* 2016). In OECD countries, *ex ante* assessment of regulatory policies has become a standard administrative procedure. For example, the Netherlands demands such an assessment for all new laws, orders in councils and proposed amendments; in Canada, a key part of the regulatory process is describing how government actions affect citizens; and in Australia, it is mandatory for proposed legislation and tax-related reforms to include regulatory impact statements and assessment (Jacob *et al.* 2011).

To effectively address climate change, Di Gregorio *et al.* (2017) suggest that climate policy integration needs:

i. policy coherence between mitigation and adaptation;
ii. policy coherence between climate change and development objectives;
iii. vertical policy integration by mainstreaming climate change into sector policies; and
iv. horizontal policy integration through cross-sectoral coordination.

These multiple dimensions and governance arrangements make effective policy design particularly difficult.

Assessment requirements and practices vary across countries, as do the extent to which environmental aspects are considered (Jacob, Volkery and Lenschow 2008; Jacob *et al.* 2011). However, among OECD countries, assessments of regulatory policy share four key aspects or objectives:

i. assessment of impacts;
ii. integration of policies;
iii. promotion of transparency; and
iv. the improvement of accountability (Ritzka 2016).

Further, regulatory impact assessment is believed to improve policy coherence and minimize policy conflicts, ultimately contributing to regulatory quality and good governance (Hertin *et al.* 2009).

Usually, regulatory impact assessment involves several stages, with environmental aspects being relevant to all of them **(Table 11.1)**. Integrating environmental aspects in assessments of regulatory policies involves tools for gathering and analysing data about the likely outcomes and impacts of policy options. These tools are used to generate and analyse data on specific impact areas (e.g. socio-economic, biophysical models and integrated models), to integrate and aggregate data, such as scenario tools and cost–benefit analysis, and to involve stakeholders in policy development (Jacob *et al.* 2011).

Table 11.1: Typical stages of regulatory impact assessment
❖ Selection of policy proposals to be subject to assessment
❖ Description of the problem and the objective of the proposed regulation
❖ Description of the baseline scenario
❖ Identification of policy options to be assessed
❖ Assessment of options, including the anticipated impacts in the different areas as well as the weighting and aggregation of different impacts
❖ Consultation of stakeholders and other interested parties on the assessment results
❖ Review of the quality of the assessment

Source: Jacob *et al.* (2011)

Box 11.1: Carbon valuation as part of United Kingdom of Great Britain and Northern Ireland's policy assessment

In 2002, the United Kingdom of Great Britain and Northern Ireland Ministry of Economics and Finance (HM Treasury) and the Department of Environment, Food and Rural Affairs (DEFRA) jointly published a report on how to integrate the social costs of carbon emissions into policy decisions. Since 2003 the Greenhouse Gas (GHG) Impact Assessment has been mandatory as part of the broader policy assessment process (United Kingdom, Department for Business Innovation and Skills [BIS] 2010, p. 73). It uses cost–benefit analysis and requires assessment of all policy initiatives.

The rationale for estimating GHG emissions that arise from potential government policies is "to inform key climate change policy decisions". Policies shall be developed to meet United Kingdom of Great Britain and Northern Ireland short and long-term CO_2 reduction targets, which establish choices between competing objectives (BIS 2010.). GHG tests are applied within the overall cost–benefit analysis to assess whether a policy is cost-effective in comparison with other alternatives (*ibid.*, p. 91).

The approach of estimating the social cost of carbon was reviewed in 2007. As a result, it was replaced by the shadow price of carbon to allow for consideration of then more recent evidence drawn from the Stern Review. In 2009, the shadow price of carbon was in turn revised and replaced by a target-consistent approach (United Kingdom, Department of Energy and Climate Change [DECC] 2009, p. 5).

Source: Jacob *et al.* (2011).

The assessment of the social costs of carbon in the United Kingdom illustrates the approach and the associated tools described above **(Box 11.1).** The United Kingdom of Great Britain and Northern Ireland is considered to be the country with the longest experience with climate impact assessment, featuring one of the most elaborate approaches to policy assessment, as well as specific legislation to support this work (the United Kingdom of Great Britain and Northern Ireland Climate Change Act) (Jacob *et al.* 2011).

Considering the variety of approaches and tools for assessing regulatory policies, Adelle *et al.* (2011) suggest that that there is no 'one way' or 'best way' of conducting these assessments. However, some jurisdictions may be regarded as exemplars, such as the EU, for the high level of integration of its approach and consideration of social, economic and environmental dimensions (Adelle *et al.* 2016). Based on a review of regulatory policy assessments in selected OECD countries, Jacob *et al.* (2011) suggest how to better consider environmental aspects in assessments of regulatory policies, including:

- taking into consideration environmental costs and benefits when regulating economic activities;
- undertaking early assessment, notification and participatory approaches to minimize conflicts between departments and with stakeholders and increase the social robustness of proposals;
- using well-established models, such as those available for climate change, emission of harmful substances, and land use;
- introducing institutional requirements, including mechanisms for quality control, transparency and consultation; and
- building capacity of environmental departments and agencies to perform or support regulatory impact assessment.

11.3.2 Other policy integration tools

There are other tools for policy integration besides the tools of regulatory policies such as regulatory impact assessments. Environmental impact assessments (EIAs) are used around the world, especially on major projects such as dams and other infrastructure (Morgan 2012). EIA has been used steadily over the last two decades and is recognized in a large number of international agreements (for example, the Espoo Convention, Ramsar Convention, Aarhus Convention, UNFCCC, UNCLOS and the Protocol on Environmental Protection to the Antarctic Treaty (Madrid Protocol)).

Strategic environmental assessment (SEA) was elaborated as an extension of EIA's principles, procedures and methods to higher levels of decision-making (Lee and Walsh 1992). SEA is seen as a tool able to evaluate a set of policies with broader lenses and within a more systematic and comprehensive process of evaluating the environmental impacts of a policy, plan or programme and its alternatives. SEA is the process by which environmental considerations are fully integrated into the preparation of plans and programmes prior to their final adoption. The objectives of the SEA process are to "provide for a high level of protection of the environment and to promote sustainable development by contributing to the integration of environmental considerations into the preparation and adoption of specified Plans and Programmes" (United Kingdom Environmental Protection Agency [UK EPA] 2018).

For the European Commission, the SEA procedure can be summarized as follows: "an environmental report is prepared in which the likely significant effects on the environment and the reasonable alternatives of the proposed plan or programme are identified. The public and the environmental authorities are informed and consulted on the draft plan or programme and the environmental report prepared." The EU ratified the Protocol on Strategic Environmental Assessment in November 2008. The SEA Directive (Directive 2001/42/EC) transposes the Protocol in EU legislation (European Commission 2001).

There is little evidence on the actual outcomes of applying the various tools of policy integration and attempts to measure the level of policy integration, and most examples of evaluations were undertaken in the context of Europe. For example, in countries where environmental liability is weakly developed, EIA/SEA tools may have limited effectiveness (United Nations Economic Commission for Africa 2005; Kotze and Plessis 2006; Gitari *et al.* 2016).

Among the few exceptions is the Publishing and the Ecology of European Research (PEER) project (Mickwitz *et al.* 2009). The study assessed the extent of climate policy integration in different European countries, policy sectors and, in some

cases, regions and municipalities. The assessment is based on five criteria: inclusion, consistency, weighting, reporting and resources. The report also analyses measures and means for enhancing climate policy integration and improving policy coherence in each country of many policies, but mostly centred on one or two, and on some regions and municipalities. With this work it was possible to draw some conclusions, such as the fact that reducing emissions has become a key political issue, and climate change is widely integrated into government programmes. The selected countries recognized the need for climate policy integration if the more ambitious climate change mitigation commitments are to be achieved.

In terms of policy integration, one lesson from PEER is that cities and municipalities have also integrated climate aims in their strategies and in specific measures, and their goals are sometimes more ambitious than those of their respective countries. The study also highlights that effective climate policy integration will require sufficient resources in the form of knowledge and money. Without these resources, there will be no realistic possibility of truly recognizing the links between general or sectoral policies and climate change or of finding alternatives and implementing them. According to the PEER project, given the great complexity of the socio-economic processes that result in GHG emissions, as well as those of adapting to a changing climate, policies need to be based on learning.

The increasing role of non-State actors (e.g. cities, civil society groups, etc.) in global climate governance is contributing considerably to the advancement of mitigation efforts. The example of the announcement of the United States pulling out of the Paris Agreement demonstrates that climate policy integration is not a one-way path but is reversible. Although the announcement possibly had no adverse impacts on the activities of US communities and firms to reduce GHG emissions, it still points to the need for more robust institutionalization.

Another example of a policy integration tool is the EU evaluation of its Common Agricultural Policy (CAP) compared to its energy policies. The CAP dates back to 1962 and is one of the oldest policies with the aim of providing price support and food security. In 2013, CAP reforms placed sustainable development as a core objective of the programme. Policy integration thus evolved from a traditional position that assumed agricultural and environmental objectives were intrinsically aligned, to a broader recognition that explicit environmental policy integration is necessary. Still, climate change considerations are conspicuously absent from the agricultural sector policy efforts.

By contrast, energy policy development efforts have introduced environmental considerations because of explicit environmental concerns, and more recently, the growing awareness of climate change has intensified efforts to integrate environment and energy policies. The policy integration approach for energy has notably shifted from one of sustainable development in the late 1990s to one where the climate change agenda has all but captured the environmental dimension of the sector, leading to such apparent anomalies as the promotion of 'sustainable nuclear energy' and a possible overemphasis on the need for biofuels. This lack of consistency across policy boundaries makes successful environmental policy integration more difficult and may lead to conflicting policy instruments where the domains intersect— for example, biofuels in the case of energy and agriculture (Mullally and Dunphy 2015).

Another notable example is the Nepal initiative to include climate change not only in environmental matters but as a major consideration in all development planning. The Climate Public Expenditure and Institutional Review (Government of Nepal *et al.* 2011) reviews the financial management systems as well as the institutional arrangements and policy directives for allocating and spending climate change-related finance. This study examined the early emphasis being given to climate change programming within Nepal and acknowledges the role played by communities in the entire process, including civil society, the private sector and international support. The main findings include the lack of institutional collaboration and capacity-building to integrate policies across the different ministries. In addition, the fragmentation of budget implementation frustrates the coordination of expenditures to facilitate and promote the best outputs and outcomes, leading to an attempt to build climate change expenditures into the national chart of accounts.

A final example of EPI relates to the global trade regime. The EU attempted to explicitly integrate environmental concerns into its trade agreement strategy in 2010, with the Communication on Trade, Growth and World Affairs, a part of the EU's 'Europe 2020' strategy. The EU also tried again in 2012, with the Communication on Trade, Growth and Development (Morin, Pauwelyn and Hollway 2017). As the controversies on the recent negotiations on trade agreements with Canada, the United States of America and Japan demonstrate, their effectiveness is under dispute.

At the multilateral level, the EU is actively involved in advancing the mandate of paragraph 31 of the Doha Declaration on the liberalization of environmental goods and services in the

regular and special sessions of the World Trade Organization (WTO) Committee on Trade and Environment, albeit with little progress so far. In 2014, 14 WTO members (including the EU), representing more than 80 per cent of world trade in environmental goods, launched the Green Goods Initiative which, as a first step, aims to eliminate tariffs on a broad list of green goods. The objective of the ongoing Environmental Goods Agreement (EGA) negotiations is to make high-quality environmental goods and technologies available at cheaper cost.

The EU, for example, also incorporates environmental provisions into bilateral and regional preferential trade agreements in the form of Trade and Sustainable Development (TSD) chapters. These provisions *inter alia* commit EU trade partners to ratify and implement key multilateral environmental agreements (MEAs) domestically and enforce them effectively. They are integrated into the agreement negotiation process through sustainability impact assessments (SIAs). SIAs are independent assessments carried out by external consultants but rely on input from stakeholders. Both the EU and civil society then closely monitor partners' implementation of TSD environmental provisions. Since such provisions may represent a costly commitment, partners may then demand that similar environmental provisions are included in their subsequent trade agreements with third parties (Milewicz *et al.* 2016). The EU has unilaterally established the Generalized Scheme of Preferences (GSP) to allow developing-country exporters to pay lower duties on exports to the EU. The GSP+ arrangement is intended to build the capacity of vulnerable countries to integrate environmental concerns into their sustainable development plans by offering them additional trade preferences.

This relationship between trade and the environment is apparent in other ways. For example, environmental policies may impede some undesirable forms of trade, and trade policies may water down potentially stronger international environmental policies. Trade policy measures appear in a range of environmental instruments, such as the restrictions on trade in endangered animal and plant species, illegal timber, illegal, unreported and unregulated (IUU) fishing, chemicals of regional or global concern and ozone-depleting substances. Generally, the influence of environmental non-governmental organizations (NGOs) and their concerns over trade policies remains limited (Dür and De Bièvre 2007).

In summary, EPI and associated tools have been used by governments trying to include environmental concerns in other sectoral policies of interest. However, there is a lack of evaluation of actual outcomes and impacts of EPI and major challenges persist: institutional fragmentation, lack of capacity-building, the difficulty of stakeholder participation and even integration with other environmental issues beyond climate change.

11.3.3 Co-benefits: findings on the impacts of environmental policies on economic growth, innovation and employment

An important argument in favour of EPI is the economic and social co-benefits that can be expected or demonstrated as a result of implementing environmental policies. These may include additional economic growth from innovation, savings from more efficient use of natural resources and avoiding the costs of environmental damage. However, the concept of co-benefits is contested because it mostly ignores political and 'North–South' aspects (Mayrhofer and Gupta 2016).

More specifically, policies that integrate environmental aspects in key economic sectors benefit from synergies and promote long-term growth by mitigating scarcities. In this regard, it is estimated that a green investment of 2 per cent of global gross domestic product (GDP) would deliver long-term growth over 2011–2050 that could be at least as high as an optimistic business-as-usual scenario, while minimizing the adverse impacts of climate change, water scarcity and the loss of ecosystem services (United Nations Environment Programme [UNEP] 2011).

Well-crafted policies that integrate environmental concerns can in many cases promote innovation (especially technological innovation, but also policy and institutional innovation) (Ambec and Barla 2002; Ambec *et al.* 2013). This is based on the following premises (Porter and van der Linde 1995, pp. 99–100, cited in Ambec *et al.* 2013):

- "...[R]egulation signals companies about likely resource inefficiencies and potential technological improvements."
- "... [R]egulation focused on information gathering can achieve major benefits by raising corporate awareness."
- "... [R]egulation reduces the uncertainty that investments to address the environment will be valuable."
- "... [R]egulation creates pressure that motivates innovation and progress."
- "... [R]egulation levels the transitional playing field. During the transition period to innovation-based solutions, regulation ensures that one company cannot opportunistically gain position by avoiding environmental investments."

In this context, market-based and flexible instruments such as environmental taxes and tradable emissions are believed to be more conducive to innovation by allowing business to determine the best ways to achieve compliance (Ambec *et al.* 2013). Further, there is an increasing tendency for over-compliance by businesses seeking to gain competitive advantage and/or maintain their social licence to operate (Ford, Steen and Verreyne 2014). Market-based instruments are, therefore, essential for triggering the efficiency-based green economy process (EEA 2014). Nevertheless, a green economy approach and market-based instruments focusing on efficiency are frequently criticized on the grounds of poor consideration of social equity—for example, by having distributional effects that disadvantage poor people.

Environmental policies can also have a positive impact on employment, particularly in the context of economic activities integrating the environmental dimension; these include renewable energy, construction, transport, agriculture, forestry and recycling and waste management (UNEP 2011; OECD 2017). Renewable energy is a critical source of employment growth; in 2016, it was estimated that this sector was responsible for 8.1 million jobs globally. Projections indicate that this figure may reach up to 20 million jobs by 2030: 2.1 million jobs in wind energy production, 6.3 million in solar photovoltaics and 12 million in biofuels-related agriculture and industry (OECD 2017). Other sectors, such as agriculture, buildings, forestry and transport are predicted to see job growth in the short, medium and long term exceeding their comparable

business-as-usual scenarios, as a result of a more resource-efficient and low-carbon economy (UNEP 2011). For example, China, which leads global employment in renewable energy, is predicted to generate at least 4.5 million jobs as a result of greening in sectors such as transport and forestry (Pan, Ma and Zhang 2011). Other studies, which take into account possible losses in other sectors and calculate the net effects of jobs created from environmental policies, are less optimistic, but overall evidence suggests that the net effects are at least not negative (Telli *et al.* 2008; Lin and Jiang 2011; Willenbockel 2011; Jacob, Quitzow and Bär 2015).

11.4 Effectiveness of international and multilevel governance

11.4.1 Enduring conceptual challenges of institutional effectiveness

Of course, proper framing of an environmental problem and good policy design form only part of the policy effectiveness analysis (as shown in the policy cycle Figure 11.2). Effective institutions are needed for designing, integrating and implementing successful environmental policy. There are several key challenges when conceiving of institutional effectiveness. One is to disentangle effectiveness from adjacent concepts such as compliance and enforcement (Chayes and Chayes 1993). This is important because an institution may see regular compliance from participants without being effective at all. Formal compliance with a regulatory instrument is an example of first-order effectiveness, addressing the identified problem but not necessarily addressing second- and third-order issues (other impacts and side effects).

If an institution relies on voluntary participation to solve an environmental problem (as is often the case internationally), then participants may be predisposed to comply with (or without) the institution because they are driven by the same reasons to join the institution in the first place. Thus, some institutions may not change behaviour so much as screen those that are not willing to comply in the first place (Downs, Rocke and Barsoom 1996; Von Stein 2005; Simmons 2010). Alliances and clubs may provide opportunities for peer pressure to overcome institutional reluctance. Hybrid governance—i.e. combining different modes and instruments of governance—can help in mutual strengthening (e.g. information bases and regulatory approaches), as shown for the European chemical regulation REACH (Hey *et al.* 2007).

Another issue is to disentangle effectiveness from performance (Gutner and Thompson 2010). In relation to biodiversity, Le Prestre (2002) distinguishes between uses of effectiveness in problem-solving (see also Young 2011), goal attainment, implementation, compliance, behaviour change, cooperation and normative gains (justice).

Different approaches have been proposed to gauge institutional effectiveness, involving both qualitative and quantitative methods. For example, the Oslo-Potsdam solution (Hovi, Sprinz and Underdal 2003) proposes an analytical approach in which institutional effectiveness is measured against both a no-regime counterfactual (i.e. what would happen if there were no responsible institution) and an analytically derived collective optimum. The approach has been challenged—for example, on the grounds of failing to propose a consistent baseline (Young 2003). A common alternative is an approach that relies on a well-specified statistical model to capture the no-regime counterfactual by offering an estimate of an 'institutions effect', controlling for other plausible effects on the behavioural variable of interest (Bernauer 1995).

11.4.2 Determinants of institutional effectiveness

What is important for strengthening existing international environmental institutions and/or creating new ones is the understanding of the effectiveness of these institutions (Young 2011). Increasingly important in international environmental policy discussions is the role of non-State actors such as local governments, cities and civil society organizations (Nasiritousi *et al.* 2016). In the absence of national government support for internationally agreed environmental goals, individual states and cities may carve out their own implementation agendas, such as in the Paris Agreement.

A major determinant of institutional effectiveness is the structure of the problem that the institution is trying to tackle (Mitchell and Keilbach 2001). These contextual factors include the distribution and enforcement problems faced, as well as various types of uncertainty (Koremenos, Lipson and Snidal 2001). It is also important that actors recognize that there is a problem (Mitchell 2009; Breitmeier, Underdal and Young 2011) and provide the necessary environmental leadership (Sprinz and Vaahtoranta 1994).

Next are the specific actors involved in the policy problem. In some cases, the support of a powerful actor can be important for institutional success; however, this is not a necessary condition (Young 2011). Some institutions rely on a powerful coalition of willing actors to establish and run an effective institution (Sebenius 1991). These 'pushers' can be frustrated by 'laggards', however (Sprinz and Vaahtoranta 1994; Haas, Keohane and Levy 1993). Often in complex negotiations, the lowest level of ambition that can be accepted by all becomes a significant barrier to progress (Underdal 1983).

One important mechanism for the efficacy of international institutions is domestic leverage. By providing the information resources, international institutions can induce change in national policies via domestic constituents that are empowered through that information provided by these resources (Dai 2005).

Another key determinant is institutional design. Young (2011) argues that design is often more significant than problem structure in determining an institution's effectiveness. The depth and density of regime rules is important (Breitmeier, Underdal and Young 2011). Moreover, the 'deepest' institutions do not necessarily scare off potential participants (Bernauer *et al.* 2013). Many actors are attracted to institutions that promise results (Hollway and Koskinen 2016).

However, the effects of an institution's design reach beyond what is strictly regulatory (Young 2011), especially where international organizations are established. An organization's design can foster certain institutional cultures and enable that organization to play a role in orchestrating various governance actors active in an issue area, such as private governance or public–private governance (Abbott and Snidal 2010; Andonova 2017).

Additional considerations of 'cutting-edge' concern, may hold significant interest for policymakers. These include: (i) the deep structure of international society in which environmental institutions are embedded, and the need to align the regime to this structure, most notably the power structure and norms; and (ii) the non-linear nature of human interactions with the environment (Young 2011).

11.4.3 Vertical and horizontal interplay in multilevel governance

International environmental institutions interact among themselves and with institutions from other policy areas such as trade, energy and finance (Stokke 2001; Gehring and Oberthür 2008; Oberthür 2009; Oberthür and Stokke 2011). In general, MEAs support environmental decision-making at the national level; however, coherence and interaction remain a challenge. Institutional interaction may be distinguished in terms of horizontal (i.e. across agencies at the same level) and vertical (i.e. from international down to local government levels) interplay (Young 2002; Young 2006). It can also be distinguished in terms of functional interactions, when problems addressed by two or more institutions are linked in spatial, bio-geophysical or socio-economic terms. In this case, the operation of one institution directly influences the effectiveness of another (Adger, Brown and Tompkins 2005; Young 2002; Young 2006). Interplay can also be influenced by political linkages, when actors create links between institutions to advance individual or collective goals (Young 2002; Young 2006). It also opens options for forum shopping (i.e. trying to find an institutional arrangement that gives maximum benefit to an individual or collective) (Gehring and Oberthür 2009).

Interplay is likely to produce tensions between and among institutions. However, it is equally likely to result in positive or synergistic interaction. In case of tensions, these may be resolved through negotiation entailing compromises ensuring, however, that the institutions involved can operate without disproportionately affecting each other's ability to address the problems they were designed to address (Young 2011). The notion of interplay may provide relevant entry points to efforts aiming to improve horizontal and vertical integration.

As the 17 SDGs are intended to be fully integrated and universal, several countries are now grappling with the task of devising the most effective institutional arrangements to address the desired vertical and horizontal integration. The 2017 synthesis of the Voluntary National Reports submitted to date found that only about one third of countries were addressing all the SDGs (United Nations 2017), but almost all had put in place some relevant institutional arrangements.

Some examples of institutional approaches for horizontal integration include the following:

- Mongolia initially created a Ministry of Environment and Green Development, recently amended to Ministry of Environment and Tourism. The Ministry chairs a coordination committee for green development.
- Sri Lanka placed responsibility for the SDGs under the Office of the President, who chairs the National Council on Sustainable Development.
- Afghanistan has an existing High Council of Ministers which now supervises the nationalization of the SDGs and allocation of budgets against the targets and indicators.
- Costa Rica established a High-level SDG Council, jointly chaired by the President and three key ministers.
- Nigeria established a Presidential Committee on the SDGs and created the post of Senior Special Assistant to the President on the SDGs.
- Bangladesh formed an inter-ministerial SDG monitoring and implementation committee, involving 21 ministries.
- Belarus has a National Coordinator for the Achievement of the SDGs, chairing the National Council for Sustainable Development, comprising 30 agencies.
- Botswana has a National Steering Committee that includes the United Nations and all stakeholder groups.
- The Czech Republic has a Government Council for Sustainable Development, which includes nine thematic committees.
- Japan established the SDG Promotion Headquarters as a cabinet-level body headed by the Prime Minister.
- Denmark has an inter-ministerial SDG working group coordinated by the Ministry of Finance.

Examples of vertical integration include the following:

- Brazil's National Commission for the SDGs comprises 27 representatives from federal, state, district and municipal governments and civil society.
- Belgium's Inter-Ministerial Conference for Sustainable Development comprises federal, regional and community ministers responsible for sustainable development.
- India has created a National Institution for Transforming India, chaired by the Prime Minister.
- The Local Government Authority of the Maldives has aligned its five-year development plan, implemented by island councils, with the SDGs.
- Ethiopia has a Growth and Transformation Plan for implementation of the SDGs, with annual reports to a Standing Committee of Parliament.

Among others, Afghanistan, Argentina, Bangladesh, Belize, Benin, Botswana, Brazil, Chile, Costa Rica, the Czech Republic, Honduras, Kenya, Malaysia, Maldives, Nepal, Peru, Thailand and Zimbabwe have explicitly incorporated stakeholder engagement in their SDG institutional arrangements.

A pertinent question, given this wide range of institutional arrangements, is whether the lessons learned from previous attempts at institutional integration arrangements have been learned and incorporated into the current approaches. This should become more evident as more countries submit their Voluntary National Reviews to the High-level Political Forum on sustainable development.

An earlier form of horizontal integration, National Councils for Sustainable Development (NCSDs), came into vogue following the 1992 United Nations Conference on Environment and Development and were strengthened by the Johannesburg Summit on Sustainable Development in 2002. Their forms, functions and effectiveness vary considerably across countries (Osbourn, Cornforth and Ullah 2014). Following some progress in implementation of the Johannesburg Plan

of Implementation, the 2012 United Nations Conference on Sustainable Development (Rio+20) launched the process that led to the SDGs (also called the post-2015 development agenda).

To feed into this process, the Global Network of National Councils for Sustainable Development and similar bodies reviewed 25 years of attempts at integration, to draw out good practice and success factors (Osborn, Cornforth and Ullah 2014). The Global Network and local governments responsible for Local Agenda 21 plans also illustrate the importance of vertical integration.

As with many Environment ministries, many NCSDs found it difficult to get their recommendations accepted, so they often resorted to disseminating key recommendations through non-traditional media channels. A sufficient arms-length relationship from the normal silo-like government bureaucracy may facilitate such non-traditional communication mechanisms and an ability to reach out to a wider group of stakeholders.

The composition of the NCSDs has usually reflected the national political context, and no clear preference has emerged. With only government agencies as members, there is a higher risk of being influenced by political interests and possibly create lower levels of ambition. Mixed memberships had difficulty in avoiding the dominance of government views and keeping track of the larger picture. Councils dominated by NGOs and other members outside government had difficulty in influencing decision-makers and often had long-term funding issues. A key factor in the success of NCSDs, however, has been the status and engagement of the Chair, with an independent Chair or co-Chair appearing to offer the best results.

Despite the drawbacks listed above, the following conclusion has emerged: "Where NCSDs exist, they should be nourished. Where they do not yet exist, careful consideration should be given to establishing them. Where they have been discontinued for essentially short-term reasons, consideration should be given to re-establishing them, possibly in a new format" (Osborn, Cornforth and Ullah 2014).

11.5 Conclusions

The importance of good policy design cannot be overstressed. Some common elements are:

i. setting a long-term vision and avoiding crisis-mode policy decisions, through inclusive, participatory design processes;
ii. establishing a baseline, quantified targets and milestones;
iii. conducting *ex ante* and *ex post* cost–benefit or cost-effectiveness analysis to ensure that public funds are being used most efficiently and effectively;
iv. building in monitoring regimes during implementation, preferably involving affected stakeholders; and
v. conducting post-intervention evaluation of the policy outcomes and impacts to close the loop for future policy design improvements (Mickwitz *et al.* 2009, p.12).

Focus also needs to be on ensuring that regulatory arrangements and policy instruments and tools take local conditions into account. The need for appropriate design applies also to international regimes (Young 2011). Policy design complexity increases when an effective mix of policies is required, often in areas under the control of different sectoral priorities. Policy coherence and environmental policy integration are critical considerations to ensure that policies are synergistic and do not undermine each other. Institutional effectiveness often springs from collaborative and participatory arrangements, involving both horizontal and vertical integration.

Policy diffusion is generally positive but can be misused if: (i) the policy adopted is not truly effective in the new context; and (ii) the transferability of the policy is merely assumed and not tested under different conditions. While it is human nature to want to copy, there is no substitute for evidence-based policy.

References

Abbott, K.W. and Snidal, D. (2010). International regulation without international government: Improving IO performance through orchestration. *The Review of International Organizations* 5(3), 315-344. https://doi.org/10.1007/s11558-010-9092-3.

Adelle, C., Jordan, A., Turnpenny, J., Bartke, S., Bournaris, T., Kuittinen, H. *et al.* (2011). *Impact Assessment Practices in Europe*. Berlin: LIAISE Innovation. https://refubium.fu-berlin.de/bitstream/handle/fub188/18658/LIAISE_innovation_report_n02.pdf?sequence=1&isAllowed=y.

Adelle, C., Macrae, D., Marusic, A. and Naru, F. (2015). New development: Regulatory impact assessment in developing countries—tales from the road to good governance. *Public Money & Management* 35(3), 233-238. https://doi.org/10.1080/09540962.2015.1027500.

Adelle, C. and Weiland, S. (2012). Policy assessment: the state of the art. *Impact Assessment and Project Appraisal* 30(1), 25-33. https://doi.org/10.1080/14615517.2012.663256.

Adelle, C., Weiland, S., Dick, J., Olivo, D.G., Marquardt, J., Rots, G. *et al.* (2016). Regulatory impact assessment: A survey of selected developing and emerging economies. *Public Money & Management* 36(2), 89-96. https://doi.org/10.1080/09540962.2016.1118930.

Adger, W.N., Brown, K. and Tompkins, E.L. (2005). The political economy of cross-scale networks in resource co-management. *Ecology and Society* 10(2). https://doi.org/10.5751/ES-01465-100209.

Alizada, K. (2017). *Diffusion of Renewable Energy Policies*. Doctor of Philosophy (PhD), Old Dominion University https://digitalcommons.odu.edu/cgi/viewcontent.cgi?article=1015&context=gpis_etds

Ambec, S. and Barla, P. (2002). A theoretical foundation of the Porter hypothesis. *Economics Letters* 75(3), 355-360. https://doi.org/10.1016/s0165-1765(02)00005-8.

Ambec, S., Cohen, M.A., Elgie, S. and Lanoie, P. (2013). The Porter Hypothesis at 20: Can environmental regulation enhance innovation and competitiveness? *Review of Environmental Economics and Policy* 7(1), 2-22. https://doi.org/10.1093/reep/res016.

Andonova, L.B. (2017). *Governance Entrepreneurs: International Organizations and the Rise of Global Public-Private Partnerships*. New York, NY: Cambridge University Press. https://www.cambridge.org/core/books/governance-entrepreneurs/70BCBC8857B11FB1593A6767993AC62B.

Bäcklund, A.-K. (2009). Impact assessment in the European Commission- a system with multiple objectives. *Environmental Science & Policy* 12(8), 1077-1087. https://doi.org/10.1016/j.envsci.2009.04.003.

Baumgartner, F.R. and Jones, B.D. (2009). *Agendas and Instability in American Politics*. 2nd edn. American Politics and Political Economy Series. Chicago, IL: The University of Chicago Press. http://press.uchicago.edu/ucp/books/book/chicago/A/bo6763995.html.

Bernauer, T. (1995). The effect of international environmental institutions: How we might learn more. *International Organization* 49(2), 351-377. https://doi.org/10.1017/s0020818300028423.

Bernauer, T., Kalbhenn, A., Koubi, V. and Spilker, G. (2013). Is there a "Depth versus Participation" dilemma in international cooperation? *The Review of International Organizations* 8(4), 477-497. https://doi.org/10.1007/s11558-013-9165-1.

Biesenbender, S. and Tosun, J. (2014). Domestic politics and the diffusion of international policy innovations: How does accommodation happen? *Global Environmental Change-Human and Policy Dimensions* 29, 424-433. https://doi.org/10.1016/j.gloenvcha.2014.04.001.

Breitmeier, H., Underdal, A. and Young, O.R. (2011). The effectiveness of international environmental regimes: Comparing and contrasting findings from quantitative research. *International Studies Review* 13(4), 579-605. https://doi.org/10.1111/j.1468-2486.2011.01045.x.

Calliess, C. (2001). *Rechtsstaat und Umweltstaat: Zugleich ein Beitrag zur Grundrechtsdogmatik im Rahmen mehrpoliger Verfassungsverhältnisse*. Tübingen: Mohr Siebeck. https://www.mohr.de/buch/rechtsstaat-und-umweltstaat-9783161475788.

Capano, G. and Howlett, M. (2009). Introduction: The determinants of policy change: Advancing the debate. *Journal of Comparative Policy Analysis: Research and Practice* 11(1), 1-5. https://doi.org/10.1080/13876980802648227.

Cass, L. (2005). Norm entrapment and preference change: The evolution of the European Union position on international emissions trading. *Global Environmental Politics* 5(2), 38-60. https://doi.org/10.1162/1526380054127736.

Chayes, A. and Chayes, A.H. (1993). On compliance. *International Organization* 47(2), 175-205. https://doi.org/10.1017/S0020818300027910.

Colgan, J.D., Keohane, R.O. and Van de Graaf, T. (2012). Punctuated equilibrium in the energy regime complex. *Review of International Organizations* 7(2), 117-143. https://doi.org/10.1007/s11558-011-9130-9.

Dai, X. (2005). Why comply? The domestic constituency mechanism. *International Organization* 59(2), 363-398. https://doi.org/10.1017/s0020818305050125.

Daugbjerg, C. (2003). Policy feedback and paradigm shift in EU agricultural policy: The effects of the MacSharry reform on future reform. *Journal of European Public Policy* 10(3), 421-437. https://doi.org/10.1080/1350176032000085388.

De Francesco, F. (2012). Diffusion of regulatory impact analysis among OECD and EU member states. *Comparative Political Studies* 45(10), 1277-1305. https://doi.org/10.1177/0010414011434297.

Di Gregorio, M., Nurrochmat, D.R., Paavola, J., Sari, I.M., Fatorelli, L., Pramova, E. *et al.* (2017). Climate policy integration in the land use sector: Mitigation, adaptation and sustainable development linkages. *Environmental Science & Policy* 67, 35-43. https://doi.org/10.1016/j.envsci.2016.11.004.

Directive 2001/42/EC of the European Parliament and of the Council of 27 June 2001 on the Assessment of the Effects of Certain Plans and Programmes on the Environment, 2001. https://eur-lex.europa.eu/legal-content/EN/TXT/PDF/?uri=CELEX:32001L0042&from=EN.

Downs, G.W., Rocke, D.M. and Barsoom, P.N. (1996). Is the good news about compliance good news about cooperation? *International Organization* 50(3), 379-406. https://doi.org/10.1017/S0020818300033427.

Dür, A. and De Bièvre, D. (2007). Inclusion without influence? NGOs in European trade policy. *Journal of Public Policy* 27(1), 79-101. https://doi.org/10.1017/S0143814X0700061X.

Eckersley, R. (2005). *The Green State: Rethinking Democracy and Sovereignty*. Cambridge, MA: MIT Press. https://mitpress.mit.edu/books/green-state.

Egger, C., Priewasser, R., Rumpl, J., Gignac, M., European Federation of Agencies and Regions for Energy and the Environment, European Council for an Energy Efficient Economy and Energy Cities (2015). *Survey Report 2015: Progress in Energy Efficiency Policies in the EU Member States - The Experts Perspective: Findings from the Energy Efficiency Watch Project*. Linz: OÖ Energiesparverband. http://www.energy-efficiency-watch.org/fileadmin/eew_documents/EEW3/Survey_Summary_EEW3/EEW3-Survey-Report-fin.pdf.

Espinosa, C., Pregernig, M. and Fischer, C. (2017). *Narrative und Diskurse in der Umweltpolitik: Möglichkeiten und Grenzen ihrer strategischen Nutzung*. Umweltbundesamt. https://www.umweltbundesamt.de/sites/default/files/medien/1410/publikationen/2017-09-27_texte_86-2017_narrative_0.pdf.

European Commission (2010). *Communication from the Commission to the European Parliament, the Council, the European Economic and Social Committee and the Committee of the Regions: Smart Regulation in the European Union* Brussels. https://eur-lex.europa.eu/legal-content/EN/TXT/PDF/?uri=CELEX:52010DC0543&from=EN.

European Environment Agency (2001). *Reporting on Environmental Measures: Are we Being Effective?* Environmental Issue Report. Copenhagen: European Environment Agency. https://www.eea.europa.eu/publications/rem/issue25.pdf.

European Environment Agency (2006). *Policy Effectiveness Evaluation: The Effectiveness of Urban Wastewater Treatment and Packaging Waste Management Systems*. Copenhagen K, Denmark. https://publications.europa.eu/en/publication-detail/-/publication/7c46337b-cfd7-4a13-8c83-082bcd5f2d13.

European Environment Agency (2014). *Resource-Efficient Green Economy and EU Policies*. Copenhagen. https://www.eea.europa.eu/publications/resourceefficient-green-economy-and-eu/at_download/file.

Falkner, R. (2016). The Paris Agreement and the new logic of international climate politics. *International Affairs* 92(5), 1107-1125. https://doi.org/10.1111/1468-2346.12708.

Flanagan, K., Uyarra, E. and Laranja, M. (2011). Reconceptualising the 'policy mix' for innovation. *Research Policy* 40(5), 702-713. https://doi.org/10.1016/j.respol.2011.02.005.

Ford, J.A., Steen, J. and Verreynne, M.-L. (2014). How environmental regulations affect innovation in the Australian oil and gas industry: Going beyond the Porter Hypothesis. *Journal of Cleaner Production* 84, 204-213. https://doi.org/10.1016/j.jclepro.2013.12.062.

Gehring, T. and Oberthür, S. (2009). The causal mechanisms of interaction between international institutions. *European Journal of International Relations* 15(1), 125–156. https://doi.org/10.1177/1354066108100055.

Gitari, E., Kahumbu, P., Jayanathan, S., Karani, J., Maranga, W., Muliro, N. *et al.* (2016). *Outcome of Court Trials in the First Two Years of Implementation of the Wildlife Conservation and Management Act, 2013: Courtroom Monitoring Report, 2014 and 2015*. Nairobi, Kenya: Wildlife Direct. https://wildlifedirect.org/wp-content/uploads/2017/03/WildlifeDirect-Courtroom-Monitoring-Report1.pdf.

Government of Nepal, National Planning Commission, United Nations Development Programme, United Nations Environment Programme and Capacity Development for Development Effectiveness Facility for Asia Pacific (2011). *Nepal Climate Public Expenditure and Institutional Review*. Kathmandu: National Planning Commission, Government of Nepal. https://www.climatefinance-developmenteffectiveness.org/sites/default/files/documents/05_02_15/Nepal_CPEIR_Report_2011.pdf.

Graham, E.R., Shipan, C.R. and Volden, C. (2013). The diffusion of policy diffusion research in political science. *British Journal of Political Science* 43(3), 673-701. https://doi.org/10.1017/s0007123412000415.

Gray, W.B. and Shimshack, J.P. (2011). The effectiveness of environmental monitoring and enforcement: A review of the empirical evidence. *Review of Environmental Economics and Policy* 5(1), 3-24. https://doi.org/10.1093/reep/req017.

Gutner, T. and Thompson, A. (2010). The politics of IO performance: A framework. *The Review of International Organizations* 5(3), 227-248. https://doi.org/10.1007/s11558-010-9096-z.

Haas, P.M., Keohane, R.O. and Levy, M.A. (eds.) (1993). *Institutions for the Earth: Sources of Effective International Environmental Protection*. Cambridge, MA: MIT Press. https://mitpress.mit.edu/books/institutions-earth.

Heichel, S., Holzinger, K., Sommerer, T., Liefferink, D., Pape, J. and Veenman, S. (2008). Research design, variables and data. In *Environmental Policy Convergence in Europe: The Impact of International Institutions and Trade*. Holzinger, K., Knill, C. and Arts, B. (eds.). Cambridge, MA: Cambridge University Press. chapter 4. 64-97. https://www.cambridge.org/core/books/environmental-policy-convergence-in-europe/research-design-variables-and-data/F6DB3BADF226FC81481C102CD4CEA471.

Hertin, J., Jacob, K., Pesch, U. and Pacchi, C. (2009). The production and use of knowledge in regulatory impact assessment – An empirical analysis. *Forest Policy and Economics* 11(5-6), 413-421. https://doi.org/10.1016/j.forpol.2009.01.004.

Hey, C., Jacob, K. and Volkery, A. (2007). Better regulation by new governance hybrids? Governance models and the reform of European chemicals policy. *Journal of Cleaner Production* 15(18), 1859-1874. https://doi.org/10.1016/j.jclepro.2006.11.001.

Hildén, M., Jordan, A. and Rayner, T. (2014). Climate policy innovation: Developing an evaluation perspective. *Environmental Politics* 23(5), 884-905. https://doi.org/10.1080/09644016.2014.924205.

Hollway, J. and Koskinen, J. (2016). Multilevel bilateralism and multilateralism: States' bilateral and multilateral fisheries treaties and their secretariats. In *Multilevel Network Analysis for the Social Sciences*. Lazega, E. and Snijders, T.A.B. (eds.). Cham: Springer. 315-332. https://link.springer.com/chapter/10.1007/978-3-319-24520-1_13

Holzinger, K., Knill, C. and Sommerer, T. (2011). Is there convergence of national environmental policies? An analysis of policy outputs in 24 OECD countries. *Environmental Politics* 20(1), 20-41. https://doi.org/10.1080/09644016.2011.538163.

Hood, C. (2011). *Summing Up the Parts: Combining Policy Instruments for Least-Cost Climate Mitigation Strategies*. International Energy Agency Information Paper. Paris: International Energy Agency. www.iea.org/publications/freepublications/publication/Summing_Up.pdf.

Hovi, J., Sprinz, D.F. and Underdal, A. (2003). The Oslo-Potsdam solution to measuring regime effectiveness: Critique, response, and the road ahead. *Global Environmental Politics* 3(3), 74-96. https://doi.org/10.1162/152638003322469286.

Howlett, M. (2014). Why are policy innovations rare and so often negative? Blame avoidance and problem denial in climate change policy-making. *Global Environmental Change* 29, 395-403. https://doi.org/10.1016/j.gloenvcha.2013.12.009.

Howlett, M. and Cashore, B. (2009). The dependent variable problem in the study of policy change: Understanding policy change as a methodological problem. *Journal of Comparative Policy Analysis: Research and Practice* 11(1), 33-46. https://doi.org/10.1080/13876980802648144.

Howlett, M. and del Rio, P. (2015). The parameters of policy portfolios: Verticality and horizontality in design spaces and their consequences for policy mix formulation. *Environment and Planning C: Government and Policy* 33(5), 1233-1245. https://doi.org/10.1177/0263774x15610059.

Howlett, M. and Lejano, R.P. (2013). Tales from the crypt: The rise and fall (and rebirth?) of policy design. *Administration & Society* 45(3), 357-381. https://doi.org/10.1177/0095399712459725.

Howlett, M. and Rayner, J. (2013). Patching vs packaging in policy formulation: Assessing policy portfolio design. *Politics and Governance* 1(2), 170. https://doi.org/10.17645/pag.v1i2.95.

Interwies, E., Görlach, B. and Newcombe, J. (2007). Evaluating the cost-effectiveness of environmental policies: Theoretical aspirations and lessons from European practice for global governance. *Amsterdam Conference on the Human Dimensions of Global Environmental Change*. Amsterdam, 24-26 May 2007. http://citeseerx.ist.psu.edu/viewdoc/download?doi=10.1.1.664.6369&rep=rep1&type=pdf

Jabbour, J., Keita-Ouane, F., Hunsberger, C., Sánchez-Rodríguez, R., Gilruth, P., Patel, N. *et al.* (2012). Internationally agreed environmental goals: A critical evaluation of progress. *Environmental Development* 3, 5-24. https://doi.org/10.1016/j.envdev.2012.05.002.

Jacob, K., Quitzow, R. and Bär, H. (2015). *Green Jobs: Impacts of a Green Economy on Employment*. Berlin: Deutsche Gesellschaft für Internationale Zusammenarbeit. http://www.greengrowthknowledge.org/sites/default/files/downloads/resource/Jacob,%20Quitzow,%20B%C3%A4r%202014%20Green%20Jobs_ENGLISH.pdf.

Jacob, K., Volkery, A. and Lenschow, A. (2008). Instruments for environmental policy integration in 30 OECD countries. In *Innovation in Environmental Policy? Integrating the Environment for Sustainability*. A. Jordan and A. Lenschow (eds.). Cheltenham: Edward Elgar Publishing. chapter 2. 24–46. https://www.elgaronline.com/view/9781847204905.00013.xml

Jacob, K., Weiland, S., Ferretti, J., Wascher, D. and Chodorowska, D. (2011). *Integrating the Environment in Regulatory Impact Assessments*. Paris: Organization for Economic Co-operation and Development. https://www.oecd.org/gov/regulatory-policy/Integrating%20RIA%20in%20Decision%20Making.pdf.

Jänicke, M. (2007). Umweltstaat – eine neue Basisfunktion des Regierens. Umweltintegration am Beispiel Deutschlands. In *Politik und Umwelt*. Jacob, K., Biermann, F., Busch, P.-O. and Feindt, P.H. (eds.). 342-259. http://www.polsoz.fu-berlin.de/polwiss/forschung/systeme/ffu/aktuelle-publikationen/martin_jaenicke__2007_d/index.html

Jordan, A. and Huitema, D. (2014a). Innovations in climate policy: The politics of invention, diffusion, and evaluation. *Environmental Politics* 23(5), 715-734. https://doi.org/10.1080/09644016.2014.923614.

Jordan, A. and Huitema, D. (2014b). Policy innovation in a changing climate: Sources, patterns and effects. *Global Environmental Change* 29, 387-394. https://doi.org/10.1016/j.gloenvcha.2014.09.005.

Jordan, A. and Lenschow, A. (2010). Environmental policy integration: A state of the art review. *Environmental Policy and Governance* 20(3), 147-158. https://doi.org/10.1002/eet.539.

Jordan, A. and Matt, E. (2014). Designing policies that intentionally stick: Policy feedback in a changing climate. *Policy Sciences* 47(3), 227-247. https://doi.org/10.1007/s11077-014-9201-x.

Kemp, R. and Pontoglio, S. (2011). The innovation effects of environmental policy instruments — A typical case of the blind men and the elephant? *Ecological Economics* 72, 28-36. https://doi.org/10.1016/j.ecolecon.2011.09.014.

Kern, F., Kivimaa, P. and Martiskainen, M. (2017). Policy packaging or policy patching? The development of complex energy efficiency policy mixes. *Energy Research & Social Science* 23, 11-25. https://doi.org/10.1016/j.erss.2016.11.002.

Kloepfer, M. (1989). Der begriff umweltstaat. In *Umweltstaat*. Kloepfer, M. (ed.). Berlin: Springer. 43-44. https://link.springer.com/chapter/10.1007/978-3-642-95596-9_6

Knill, C., Schulze, K. and Tosun, J. (2012). Regulatory policy outputs and impacts: Exploring a complex relationship. *Regulation & Governance* 6(4), 427-444. https://doi.org/10.1111/j.1748-5991.2012.01150.x.

Koremenos, B., Lipson, C. and Snidal, D. (2001). The rational design of international institutions. *International Organization* 55(4), 761-799. https://doi.org/10.1162/002081801317193592.

Kotze, L.J. and Plessis, A.D. (2006). The inception and role of international environmental law in domestic biodiversity conservation efforts: The South African experience. *Queensland University of Technology Law and Justice Journal* 6(1), 30-53. https://doi.org/10.5204/qutlr.v6i1.191.

Lafferty, W.M. (ed.) (2004). *Governance for Sustainable Development: The Challenge of Adapting Form to Function*. Cheltenham: Edward Elgar. https://www.e-elgar.com/shop/governance-for-sustainable-development.

Lay, J., Brandi, C., Das, R.U., Klein, R., Thiele, R., Alexander, N. *et al.* (2017). *Coherent G20 Policies Towards the 2030 Agenda for Sustainable Development*. G20 Insights. http://www.g20-insights.org/wp-content/uploads/2017/03/TF_2030_Agenda_PolicyCoherence.pdf.

Le Prestre, P.G. (2002). Studying the effectiveness of the CBD. In *Governing global biodiversity: The evolution and implementation of the convention on biological diversity*. Le Prestre, P.G. (ed.). London: Routledge. chapter 3. https://www.taylorfrancis.com/books/e/9781351932547

Lee, N. and Walsh, F. (1992). Strategic environmental assessment: An overview. *Project Appraisal* 7(3), 126-136. https://doi.org/10.1080/02688867.1992.9726853.

Levin, S., Xepapadeas, T., Crepin, A.S., Norberg, J., De Zeeuw, A., Folke, C. *et al.* (2013). Social-ecological systems as complex adaptive systems: Modeling and policy implications. *Environment and Development Economics* 18(2), 111-132. https://doi.org/10.1017/S1355770x12000460.

Liefferink, D., Arts, B., Kamstra, J. and Ooijevaar, J. (2009). Leaders and laggards in environmental policy: A quantitative analysis of domestic policy outputs. *Journal of European Public Policy* 16(5), 677-700. https://doi.org/10.1080/13501760902983283.

Lin, B. and Jiang, Z. (2011). Estimates of energy subsidies in China and impact of energy subsidy reform. *Energy Economics* 33(2), 273-283. https://doi.org/10.1016/j.eneco.2010.07.005.

Maor, M. (2014). Policy bubbles: Policy overreaction and positive feedback. *Governance* 27(3), 469-487. https://doi.org/10.1111/gove.12048.

Marshall, G.R. (2009). Polycentricity, reciprocity, and farmer adoption of conservation practices under community-based governance. *Ecological Economics* 68(5), 1507-1520. https://doi.org/10.1016/J.ECOLECON.2008.10.008.

Matisoff, D.C. and Edwards, J. (2014). Kindred spirits or intergovernmental competition? The innovation and diffusion of energy policies in the American states (1990–2008). *Environmental Politics* 23(5), 795-817. https://doi.org/10.1080/09644016.2014.923639.

Mayrhofer, J.P. and Gupta, J. (2016). The science and politics of co-benefits in climate policy. *Environmental Science & Policy* 57, 22-30. https://doi.org/10.1016/j.envsci.2015.11.005.

Mees, H.L.P., Dijk, J., van Soest, D., Driessen, P.P.J., van Rijswick, M.H.F.M.W. and Runhaar, H. (2014). A method for the deliberate and deliberative selection of policy instrument mixes for climate change adaptation. *Ecology and Society* 19(2). https://doi.org/10.5751/ES-06639-190258.

Mickwitz, F.A., Beck, S., Carss, D., Ferrand, N., Görg, C., Jensen, A. *et al.* (2009). *Climate Policy Integration, Coherence and Governance*. Leipzig: Partnership for European Environmental Research. http://www.peer.eu/publications/climate-policy-integration-coherence-and-governance/.

Miles, E.L., Underdal, A., Andresen, S., Wettestad, J., Skjærseth, J.B. and Carlin, E.M. (2002). *Environmental Regime Effectiveness: Confronting Theory with Evidence*. Cambridge: The MIT Press. https://mitpress.mit.edu/books/environmental-regime-effectiveness.

Milewicz, K., Hollway, J., Peacock, C. and Snidal, D. (2016). Beyond trade: The expanding scope of the nontrade agenda in trade agreements. *Journal of Conflict Resolution* 62(4), 743-773. https://doi.org/10.1177/0022002716662687.

Mitchell, R.B. (2002). Of course international institutions matter: But when and how? In *How Institutions Change: Perspectives on Social Learning in Global and Local Environmental Contexts*. Breit, H., Engels, A., Moss, T. and Troja, M. (eds.). Potsdam: VS Verlag für Sozialwissenschaften. 35-52. https://link.springer.com/chapter/10.1007/978-3-322-80936-0_5

Mitchell, R.B. (2009). *International Politics and the Environment*. Carlsnaes, W. and Checkel, J.T. (eds.). London: SAGE Publications. http://uk.sagepub.com/en-gb/eur/international-politics-and-the-environment/book228800.

Mitchell, R.B. and Keilbach, P.M. (2001). Situation structure and institutional design: Reciprocity, coercion, and exchange. *International Organization* 55(4), 891-917. https://doi.org/10.1162/002081801317193637.

Morgan, R.K. (2012). Environmental impact assessment: The state of the art. *Impact Assessment and Project Appraisal* 30(1), 5-14. https://doi.org/10.1080/14615517.2012.661557.

Morin, J.-F., Pauwelyn, J. and Hollway, J. (2017). The trade regime as a complex adaptive system: Exploration and exploitation of environmental norms in trade agreements. *Journal of International Economic Law* 20(2), 365-390. https://doi.org/10.2139/ssrn.3008543.

Mullally, G. and Dunphy, N.P. (2015). *State of Play: Review of Environmental Policy Integration Literature*. National Economic and Social Council (NESC). http://files.nesc.ie/nesc_research_series/Research_Series_Paper_7_UCC.pdf.

Nachmany, M., Fankhauser, S., Setzer, J. and Averchenkova, A. (2017). *Global Trends in Climate Change Legislation and Litigation: 2017 Update*. London: Grantham Research Institute on Climate Change and the Environment. http://www.lse.ac.uk/GranthamInstitute/wp-content/uploads/2017/04/Global-trends-in-climate-change-legislation-and-litigation-WEB.pdf.

Nasiritousi, N., Hjerpe, M. and Linnér, B.O. (2016). The roles of non-state actors in climate change governance: Understanding agency through governance profiles. *International Environmental Agreements: Politics, Law and Economics* 16(1), 109-126. https://doi.org/10.1007/s10784-014-9243-8.

Niles, M.T. and Lubell, M. (2012). Integrative frontiers in environmental policy theory and research. *Policy Studies Journal* 40(s1), 41-64. https://doi.org/10.1111/j.1541-0072.2012.00445.x.

Nilsson, M. and Persson, Å. (2017). Policy note: Lessons from environmental policy integration for the implementation of the 2030 Agenda. *Environmental Science & Policy* 78, 36-39. https://doi.org/10.1016/j.envsci.2017.09.003.

Nilsson, M., Zamparutti, T., Petersen, J.E., Nykvist, B., Rudberg, P. and McGuinn, J. (2012). Understanding policy coherence: Analytical framework and examples of sector-environment policy interactions in the EU. *Environmental Policy and Governance* 22(6), 395-423. https://doi.org/10.1002/eet.1589.

Oberthür, S. (2009). Interplay management: Enhancing environmental policy integration among international institutions. *International Environmental Agreements: Politics, Law and Economics* 9(4), 371-391. https://doi.org/10.1007/s10784-009-9109-7.

Oberthür, S. and Stokke, O.S. (eds.) (2011). *Managing Institutional Complexity: Regime Interplay and Global Environmental Change*. Cambridge, MA: MIT Press. https://mitpress.mit.edu/books/managing-institutional-complexity.

Organization for Economic Co-operation and Development (2008). *Introductory Handbook for Undertaking Regulatory Impact Analysis (RIA)*. Paris: Organization for Economic Co-operation and Development. https://www.oecd.org/gov/regulatory-policy/44789472.pdf

Organization for Economic Co-operation and Development (2016). *Better Policies for Sustainable Development 2016: A New Framework for Policy Coherence*. Paris. https://sustainabledevelopment.un.org/content/documents/commitments/493_12066_commitment_Better%20Policies%20for%20Sustainable%20Development%202016.pdf.

Organization for Economic Co-operation and Development (2017). *Policy Coherence for Sustainable Development 2017: Eradicating Poverty and Promoting Prosperity*. Paris: Organization for Economic Co-operation and Development. https://www.oecd-ilibrary.org/policy-coherence-for-sustainable-development-2017_5jg03xm8f9lw.pdf?itemId=%2Fcontent%2Fpublication%2F9789264272576-en&mimeType=pdf.

Organization for Economic Co-operation and Development, International Energy Agency, International Transport Forum and Nuclear Energy Agency (2015). *Aligning Policies for a Low-carbon Economy*. Paris: Organization for Economic Co-operation and Development. https://www.oecd-ilibrary.org/environment/aligning-policies-for-a-low-carbon-economy_9789264233294-en.

Osborn, D., Cornforth, J. and Ullah, F. (2014). *National Councils for Sustainable Development: Lessons from the Past and Present*. Ottawa: SDplanNet. https://www.iisd.org/sites/default/files/publications/sdplannet_lessons_from_the_past.pdf.

Ostrom, E. (2007). A diagnostic approach for going beyond panaceas. *Proceedings of the National Academy of Sciences* 104(39), 15181-15187. https://doi.org/10.1073/pnas.0702288104.

Ostrom, E. (2009). A general framework for analyzing sustainability of social-ecological systems. *Science* 325(5939), 419-422. https://doi.org/10.1126/science.1172133.

Ostrom, E. (2010). Polycentric systems for coping with collective action and global environmental change. *Global Environmental Change* 20(4), 550-557. https://doi.org/10.1016/J.GLOENVCHA.2010.07.004.

Pan, J., Ma, H. and Zhang, Y. (2011). *Green Economy and Green Jobs in China: Current Status and Potentials for 2020*. Mastny, L. (ed.). Washington, D.C.: Worldwatch Institute. http://www.worldwatch.org/system/files/185%20Green%20China.pdf.

Persson, Å. (2007). Different perspectives on EPI. In *Environmental Policy Integration in Practice: Shaping Institutions for Learning*. Nilsson, M. and Eckerberg, K. (eds.). London: Routledge. chapter 2. 25-48. https://www.taylorfrancis.com/books/e/9781136548185/chapters/10.4324%2F9781849773843-10

Pierson, P. (2000). Increasing returns, path dependence, and the study of politics. *American Political Science Review* 94(2), 251-267. https://doi.org/10.2307/2586011.

Porter, M.E. and van der Linde, C. (1995). Toward a new conception of the environment-competitiveness relationship. *The Journal of Economic Perspectives* 9(4), 97-118. https://doi.org/10.1257/jep.9.4.97.

Radaelli, C.M. (2009). Measuring policy learning: Regulatory impact assessment in Europe. *Journal of European Public Policy* 16(8), 1145-1164. https://doi.org/10.1080/13501760903332647.

Renda, A. (2017). *How can Sustainable Development Goals be 'mainstreamed' in the EU's Better Regulation Agenda?* CEPS Policy Insights. Brussels: Centre for European Policy Studies. https://www.ceps.eu/system/files/Better%20regulation%20and%20sustainable%20development_CEPS%20Policy%20Insights_%20A_Renda.pdf.

Repetto, R. (ed.) (2006). *Punctuated Equilibrium and the Dynamics of US Environmental Policy*. New Haven, CT: Yale University Press. https://yalebooks.yale.edu/book/9780300110760/punctuated-equilibrium-and-dynamics-us-environmental-policy.

Ritzka, M.S. (2016). *Incorporation of Sustainable Development Concerns in Regulatory Impact Assessments.* Masters in Sustainable Development, Uppsala University http://www.diva-portal.se/smash/get/diva2:945065/FULLTEXT01.pdf

Runhaar, H., Driessen, P. and Uittenbroek, C. (2014). Towards a systematic framework for the analysis of environmental policy integration. *Environmental Policy and Governance* 24(4), 233-246. https://doi.org/10.1002/eet.1647.

Schaffrin, A., Sewerin, S. and Seubert, S. (2015). Toward a comparative measure of climate policy output. *Policy Studies Journal* 43(2), 257-282. https://doi.org/10.1111/psj.12095.

Schmidt, T.S., Battke, B., Grosspietsch, D. and Hoffmann, V.H. (2016). Do deployment policies pick technologies by (not) picking applications? - A simulation of investment decisions in technologies with multiple applications. *Research Policy* 45(10), 1965-1983. https://doi.org/10.1016/j.respol.2016.07.001.

Schmidt, T.S. and Sewerin, S. (2018). Measuring the temporal dynamics of policy mixes – An empirical analysis of renewable energy policy mixes' balance and design features in nine countries. *Research Policy*. https://doi.org/10.1016/j.respol.2018.03.012.

Scrase, J.I. and Ockwell, D.G. (2010). The role of discourse and linguistic framing effects in sustaining high carbon energy policy—An accessible introduction. *Energy Policy* 38(5), 2225-2233. https://doi.org/10.1016/j.enpol.2009.12.010.

Sebenius, J.K. (1991). Designing negotiations toward a new regime: The case of global warming. *International Security* 15(4), 110-148. https://doi.org/10.2307/2539013.

Simmons, B. (2010). Treaty compliance and violation. *Annual Review of Political Science* 13(1), 273-296. https://doi.org/10.1146/annurev.polisci.12.040907.132713.

Sprinz, D. and Vaahtoranta, T. (1994). The interest-based explanation of international environmental policy. *International Organization* 48(1), 77-105. https://doi.org/10.1017/S0020818300000825.

Stadelmann, M. and Castro, P. (2014). Climate policy innovation in the South – Domestic and international determinants of renewable energy policies in developing and emerging countries. *Global Environmental Change* 29, 413-423. https://doi.org/10.1016/j.gloenvcha.2014.04.011.

Stokke, O.S. (ed.) (2001). *Governing High Seas Fisheries: The Interplay of Global and Regional Regimes*. Oxford: Oxford University Press. https://global.oup.com/academic/product/governing-high-seas-fisheries-9780198299493?cc=ke&lang=en&.

Telli, Ç., Voyvoda, E. and Yeldan, E. (2008). Economics of environmental policy in Turkey: A general equilibrium investigation of the economic evaluation of sectoral emission reduction policies for climate change. *Journal of Policy Modeling* 30(2), 321-340. https://doi.org/10.1016/j.jpolmod.2007.03.001.

Tews, K., Busch, P.O. and Jörgens, H. (2003). The diffusion of new environmental policy instruments. *European Journal of Political Research* 42(4), 569-600. https://doi.org/10.1111/1475-6765.00096.

Turnpenny, J., Radaelli, C.M., Jordan, A. and Jacob, K. (2009). The policy and politics of policy appraisal: Emerging trends and new directions. *Journal of European Public Policy* 16(4), 640-653. https://doi.org/10.1080/13501760902872783.

Underdal, A. (1983). Causes of negotiation 'failure'. *European Journal of Political Research* 11, 183-195. https://doi.org/10.1111/j.1475-6765.1983.tb00055.x

United Kingdom, Department of Energy and Climate Change, (2009). *Carbon Valuation in UK Policy Appraisal: A Revised Approach*. London. https://assets.publishing.service.gov.uk/government/uploads/system/uploads/attachment_data/file/245334/1_20090715105804_e____carbonvaluationinukpolicyappraisal.pdf.

United Kingdom, Department for Business Innovation and Skills, (2010). *Impact Assessment Toolkit. A Guide to Undertaking an Impact Assessment and Completing the IA Template*. London. http://webarchive.nationalarchives.gov.uk/20110120023243/http://www.bis.gov.uk/assets/biscore/better-regulation/docs/10-901-impact-assessment-toolkit.pdf.

United Kingdom Environmental Protection Agency (2018). *Strategic environmental assessment*. http://www.epa.ie/monitoringassessment/assessment/sea/.

United Nations (2005). *Review of the Application of Environmental Impact Assessment in Selected African Countries*. Addis Ababa. http://repository.uneca.org/bitstream/handle/10855/5607/bib.%2041846_I.pdf?sequence=1.

United Nations (2017). *Voluntary National Reviews 2017 Synthesis Report*. New York, NY. https://sustainabledevelopment.un.org/content/documents/17109Synthesis_Report_VNRs_2017.pdf.

United Nations Environment Programme (2011). *Towards a Green Economy: Pathways to Sustainable Development and Poverty Eradication: A Synthesis for Policy Makers*. Nairobi: United Nations Environment Programme. https://sustainabledevelopment.un.org/content/documents/126GER_synthesis_en.pdf.

United Nations General Assembly (2015). *Transforming Our World: The 2030 Agenda for Sustainable Development. A/RES/70/1*. New York, NY. http://www.un.org/ga/search/view_doc.asp?symbol=A/RES/70/1&Lang=E.

Uslu, A., Mozaffarian, H. and van Stralen, J. (2016). *Deliverable 3.2: Benchmarking Bioenergy Policies in Europe*. Strategic Initiative for Resource Efficient Biomass Policies. Brussels: European Union. https://www.ecn.nl/publicaties/PdfFetch.aspx?nr=ECN-O--16-009.

Voigt, S. (2013). How (Not) to measure institutions. *Journal of Institutional Economics* 9(1), 1-26. https://doi.org/10.1017/s1744137412000148.

Von Stein, J. (2005). Do treaties constrain or screen? Selection bias and treaty compliance. *American Political Science Review* 99(4), 611-622. https://doi.org/10.1017/s0003055405051919.

Weber, M., Driessen, P.P.J. and Runhaar, H.A.C. (2013). Evaluating environmental policy instruments mixes, a methodology illustrated by noise policy in the Netherlands. *Journal of Environmental Planning and Management* 57(9), 1381-1397. https://doi.org/10.1080/09640568.2013.808609.

Willenbockel, D. (2011). *Environmental Tax Reform in Vietnam: An Ex Ante General Equilibrium Assessment*. Institute of Development Studies at the University of Sussex. https://mpra.ub.uni-muenchen.de/44411/1/MPRA_paper_44411.pdf.

Wittneben, B.B.F. (2012). The impact of the Fukushima nuclear accident on European energy policy. *Environmental Science & Policy* 15(1), 1-3. https://doi.org/10.1016/J.ENVSCI.2011.09.002.

Yin, H.T. and Powers, N. (2010). Do state renewable portfolio standards promote in-state renewable generation? *Energy Policy* 38(2), 1140-1149. https://doi.org/10.1016/j.enpol.2009.10.067.

Young, O.R. (1999). *The Effectiveness of International Environmental Regimes: Causal Connections and Behavioral Mechanisms*. Cambridge, MA: The MIT Press. https://mitpress.mit.edu/books/effectiveness-international-environmental-regimes.

Young, O.R. (2002). *The Institutional Dimensions of Environmental Change: Fit, Interplay, and Scale*. Cambridge, MA: MIT Press. https://mitpress.mit.edu/books/institutional-dimensions-environmental-change.

Young, O.R. (2003). Determining regime effectiveness: A commentary on the Oslo-Potsdam solution. *Global Environmental Politics* 3(3), 97–104. https://doi.org/10.1162/152638003322469295.

Young, O.R. (2006). Vertical interplay among scale-dependent environmental and resource regimes. *Ecology and Society* 11(1), 27. https://doi.org/10.5751/ES-01519-110127.

Young, O.R. (2008). Building regimes for socio-ecological systems: Institutional diagnosis. In *Institutions and Environmental Change: Principal Findings, Applications, and Research Frontiers*. Young, O.R., King, L.A. and Schroeder, H. (eds.). The MIT Press. chapter 4. http://mitpress.universitypressscholarship.com/view/10.7551/mitpress/9780262240574.001.0001/upso-9780262240574-chapter-4

Young, O.R. (2011). Effectiveness of international environmental regimes: Existing knowledge, cutting-edge themes, and research strategies. *Proceedings of the National Academy of Sciences* 108(50), 19853-19860. https://doi.org/10.1073/pnas.1111690108.

Young, O.R. (2017). *Governing Complex Systems: Social Capital for The Anthropocene*. Cambridge, MA: The MIT Press. https://mitpress.mit.edu/books/governing-complex-systems.

Chapter 12

Air Policy

Coordinating Lead Author: Peter King (Institute for Global Environmental Strategies)

Lead Authors: Frederick Ato Armah (University of Cape Coast), Phillip Dickerson (United States Environmental Protection Agency), Cristina Guerreiro (Norwegian Institute for Air Research), Terry Keating (United States Environmental Protection Agency), Oswaldo dos Santos Lucon (São Paulo State Environment Secretariat, Brazil), Asami Miyazaki (Kumamoto Gakuen University), Amit Patel (Planned Systems International, Inc.), Stefan Reis (Centre for Ecology and Hydrology, Natural Environment Research Council,)

GEO Fellow: Kari DePryck (Institut d'etudes politiques de Paris)

Executive summary

Institutional capacity to manage air pollution, climate change, stratospheric ozone depletion and persistent bio-accumulative toxic substances varies significantly across the world (*well established*). In some regions and countries (e.g. North America, Western Europe, East Asia), there are well-developed federated systems of national, provincial and local policies and enforcement programmes. In other regions, international agreements or national legislation may exist, but implementation and enforcement are weak due to a lack of institutional capacity at the national or subnational scale (*established but incomplete*). In some regions, city governments are leading the way with benefits for other parts of their countries {5.5, 12.2}.

Different investments are needed to improve management capacity in different regions. For example, the GEO-6 regional assessments identified improving air quality monitoring infrastructure as a priority for Africa and Latin America and improving the use of benefit cost analyses of climate change and air pollution mitigation measures as a priority for Asia and the Pacific {5.1}.

Traditional regulatory approaches, including the use of emissions and technology standards, have been successful in addressing some pollution sources (*established but incomplete*)**.** Successes are evident in declining trends in emissions and increasing trends in economic activity and production. However, such approaches rely on adequate human resources and effective enforcement and legal systems, which may not exist {12.2.1, 12.2.2}.

There is no single global agreement addressing air pollution, but there is a patchwork of regional intergovernmental agreements and initiatives focused on public–private partnerships (*unresolved*)**.** Global agreements have been adopted to address climate change, stratospheric O_3 depletion, persistent organic pollutants, and mercury {5.5, 12.2.5}.

National commitments on climate change under the United Nations Framework Convention on Climate Change (UNFCCC) processes are still insufficient to meet the agreed global temperature stabilization goals, and options are foreclosing (*established but incomplete*)**.** The 2015 Paris Agreement on climate change set a limit to the average global temperature rise in this century well below 2 degrees Celsius, with the ambition to achieve 1.5 degrees or less, as a means of transiting towards a low-carbon and resilient future. To date, the set of national commitments and their implementation is not on track to avoid dangerous to catastrophic climate change, and delayed ambitions will lead to greater risks to the economy and to overall planetary health {5.5}.

12.1 Introduction

The composition of Earth's atmosphere is one of the major determinants of a healthy planet, influencing the climate, ecosystems and human health. This is highlighted by the existence of a direct or indirect link between the challenges of air pollution, climate change, stratospheric ozone depletion, persistent bioaccumulative and toxic (PBT) chemicals and each of the Sustainable Development Goals (SDGs) (United Nations 2015).

A plethora of international, national, subnational and regional policies have been deployed to address these challenges. The fundamental forms include:

i. technology or emissions standards, commonly referred to as 'command and control';
ii. planning regimes;
iii. market interventions;
iv. public information; and
v. cooperative forums, including international agreements.

The various policy instruments that are used in each of these approaches are discussed in the following sections along with a case study to illustrate each approach. Key features of each case are highlighted using the methodology described in section 10.6. The case studies are selected from a diverse range of geographical contexts, spatial scales and implementation time frames. The case studies are not intended to be all encompassing, but highlight the context-specific nuances, generic patterns and issues that require attention from relevant stakeholders to elicit better policy outcomes. They are not intended to be replicable without considering the local context.

Policies enacted to address air pollution, climate change, stratospheric ozone depletion and PBTs should account for the mix of emissions, the atmospheric or environmental lifetime of the pollutant and its associated benefits and trade-offs (Melamed *et al.* 2016). Pertinent questions include:

i. how can goals for affordable, clean and reliable energy across spatial scales (local, national, regional, global) be achieved by considering possible synergies and trade-offs?;
ii. what synergies and co-benefits between climate policy and air pollution control can be identified?; and
iii. how will emissions of greenhouse gases (GHGs) and air pollutants co-evolve in scenarios with and without policy interventions, such as context-specific regulation of PBTs, climate policy and air pollution control?

From a systems perspective, the existence of various policy instruments and regimes at a range of spatial and temporal scales brings into sharp focus the complexity of addressing air quality challenges in an integrated and comprehensive manner. This makes it imperative to take an integrated approach to address potential conflicts and define trade-offs between environmental policy objectives, as well as to isolate and consolidate policies with possible co-benefits such as improved energy security, urban air quality and human health (see Section 11.3).

12.2 Key policies and governance approaches

12.2.1 Planning regimes

Planning regimes (or frameworks) establish ambient targets (e.g. concentration standards, total pollutant loads or a change in global mean temperature) and emissions budgets (or ceilings). Clusters of policies are then developed and implemented to meet the targets or budgets. Progress towards these is monitored, and if necessary, additional policies are developed or existing policies are revised. Planning regimes are often considered to be fundamental to managing air pollution, climate change, ozone-depleting substances (ODSs) and PBTs, as they provide a strategic policy framework within which specific actions can be integrated.

Ambient concentration standards or other environmental targets define the desired state of the environment, often linked to cause–effect relationships in relation to human health. Emissions budgets, pollutant loads or ceilings are the estimated levels of pressure that still enable achievement of the desired state, or present a no-effect threshold (e.g. critical loads/levels). As an example, the US National Ambient Air Quality Standards (NAAQS) are standards for harmful pollutants established by the US Environmental Protection Agency (EPA) under authority of the Clean Air Act (42 U.S.C. 7,401 et seq.). These standards are applied for outdoor air throughout the country (United States Environmental Protection Agency 2016) and aim to protect human health from harmful air pollution.

Emissions budgets can be related to environmental targets using quantitative (e.g. atmospheric or climate) models properly evaluated against field observations. In the case of secondary pollutants (which are created in the atmosphere) or pollutants that have long lifetimes in the environment, the relationship between emissions and ambient concentrations may not be linear. Making a causal linkage between emissions

Table 12.1: Typology of policy and governance approaches described in this chapter

Governance approach	Policy instrument(s)	Case study
Planning regimes	Ambient standards, emissions budgets	United Kingdom of Great Britain and Northern Ireland Climate Act and carbon budgets
Technology and emissions standards	Emissions standards, fuel quality standards, efficiency standards, best available control technology	Diesel emissions standards in Europe
Market interventions	Subsidies, tax policy, tradeable credits/allowances	Improved cookstoves in Kenya (Global Alliance for Clean Cookstoves)
Public information	Information, forecasts, labelling and branding	Provision of real-time air quality data and forecasts
International cooperation	Multilateral and bilateral binding agreements, voluntary organizations	ASEAN Agreement on Transboundary Haze

and the desired state is often important to justify the extent or costs of emissions controls or other policies and requires a significant amount of information (as inputs to the model) and expertise. However, in cases where emissions are high and the government's technical planning capacity is low, it is often not necessary to quantify the linkage between policies and the desired state to justify some control measures on the largest sources of emissions. It may be sufficient to qualitatively demonstrate that the sources contribute to adverse impacts and that there would be benefits from controls, with progress towards long-term objectives being achieved through implementation.

Linked to emissions budgets, emissions trading schemes have been introduced in particular for GHGs, where the location of emissions matters less in contrast to air pollutants. In December 2017, China launched its emissions trading system (United Nations Framework Convention on Climate Change [UNFCCC] 2017a), which will initially cover only the electricity sector. However, it is set to be the world's largest system of its sort and accounts for approximately 3 billion tons of traded CO_2. A similar system is in place in Europe, with approximately 2 billion tons traded (European Commission 2018). Similar trading schemes exist in several countries and regions (Carbon Market Data 2018). By setting emissions budgets, which may gradually be constrained further, prices for traded emissions may be adjusted.

As noted in section 5.5, countries have identified different ambient concentration standards based on their own interpretation of the epidemiological evidence on relationships between environmental state and health effects; their existing levels of air pollution; and their own perceptions about their ability to achieve decreases in air pollution. The World Health Organization (WHO) has established guidelines and interim targets (WHO 2018) that countries can use to establish their own standards and targets. As WHO and individual countries establish or revise their ambient standards based on new information, other countries often take note of the changes and consider whether to make similar changes to their own standards. For the protection of sensitive habitats and ecosystems, critical loads and critical levels have been set in the context of the United Nations Economic Commission for Europe (UNECE) Convention on Long-range Transboundary Air Pollution (CLRTAP), initially focusing on acidification in the early 1980s, and later extended to include nutrient critical loads and ammonia (NH_3) critical levels based on growing evidence of ecosystem damage and biodiversity losses even at lower concentration and deposition levels, stemming, for instance, from agricultural NH_3 emissions (Sutton, Reis and Baker eds. 2009).

Emissions budgets and ceilings not only provide a way of evaluating whether a suite of policies would be expected to achieve the relevant environmental targets; they also provide a way of apportioning responsibility for achieving environmental targets among regions, jurisdictions, sectors or individual sources. For example, emissions budgets have been applied at the national and state level in air pollution planning and at the national level in international agreements to mitigate ODS emissions. Within Europe and the UNECE region, the 1999 Gothenburg Protocol to Abate Acidification, Eutrophication and Ground-level Ozone (as amended) under CLRTAP (UNECE 2018) as well as the European Union (EU) National Emission Ceilings Directive (European Environment Agency [EEA] 2016a) present recent examples of national emissions ceilings agreed for pollutants known to contribute to a range of environmental and human health effects.

In the following case study, emissions budgets are applied to long-lived GHG emissions in the United Kingdom.

Case study: The United Kingdom of Great Britain and Northern Ireland's energy and climate policies

The 2008 Climate Change Act (United Kingdom of Great Britain and Northern Ireland [UK] Government 2008a; UK Government 2008b) is an example of how a national policy can be established within an international framework to tackle climate change (including targets and timeframes, carbon prices and emissions trading). Legally binding targets aim to progressively reduce emissions through five-year carbon budgets up to 2050, with significant benefits such as international market competitiveness, resource conservation, cost-effective removal of barriers, support to low-carbon technologies, promotion of carbon capture and storage and considerations of social implications such as fuel poverty. To achieve such aims, practical measures included mandatory cap-and-trade schemes, standards, financing and taxing, innovation strategies and technology deployment (UK, Department of Trade and Industry 2007).

Table 12.2: Summary of assessment criteria: United Kingdom of Great Britain and Northern Ireland's energy and climate policies

Criterion	Description	References
Success or failure	GHG emissions reduced by 5 per cent/year since 2012, reaching 42 per cent below 1990 levels in 2016, with the economy growing by 60 per cent. Between 2008 and 2015 GHG per capita emissions fell from 8.22 to 5.99 tCO_2, GHG per GDP (0.20 to 0.15 tCO_2/2010 US$, or 0.22 to 0.16 tCO_2/2010 US$ PPP); per capita energy use as total primary energy supply in tons of oil equivalent (TPES 3.37 to 2.78 toe/person), TPES per GDP (0.08 to 0.07 toe/1,000 2010 US$) and electricity consumption per capita (6.01 to 5.08 MWh/person).	UK, Department for Business, Energy & Industrial Strategy 2016; International Energy Agency [IEA] 2017; UK, Committee on Climate Change 2017
Independence of evaluation	Official Report to Parliament, supported by statistics.	
Key actors	Mainly government bodies (including devolved administrations).	
Baseline	1990 economy-wide GHG emissions, plus other associated baselines such as shares of renewable energy.	
Time frame	GHG cuts of 80 per cent by 2050, with other interim targets (50 per cent by 2025), based on five-year carbon budgets, set 12 years in advance to allow preparation; other goals in the energy sector (renewable electricity and biofuels, transport efficiency, phase-in of electric vehicles, and others, including carbon capture and storage).	
Constraining factors	Slow GHG curbs in transport and buildings sectors, very limited carbon capture and storage.	
Enabling factors	The main driver was a reduction in the use of coal by 75 per cent in the power sector. The Act had strong support due to lower energy bills, salience of scientific evidence, public awareness, political responses following innovation outside government, technology improvements, value of institutional innovation, use of evidence reframing climate change as an economic issue and the importance of leadership.	
Cost-effectiveness	Costs will be around 1–2 per cent of GDP in 2050, with significant business opportunities from a low-carbon economy (in 2009 a market worth GBP112 billion, with over 900,000 jobs).	
Equity	Contraction and convergence approach.	
Co-benefits	Improved market access, innovation, infrastructure resilience, energy supply security and system flexibility (storage, interconnection), take-up of new technologies, quality of life (air, water, health and well-being, land use).	
Transboundary issues	Necessary policy realignment depending on exit from the EU and more ambition to fulfil the Paris Agreement goals.	
Possible improvements	Scale-up low-carbon power generation, accelerated uptake of electric vehicles, more low-carbon heat alongside energy efficiency, restart work on carbon capture and storage, address land management practices, improve and clarify combinations of policy instruments (carbon pricing, standards and regulations, research and development funding, subsidies, market design and taxation).	

The Climate Change Act of the United Kingdom of Great Britain and Northern Ireland was the world's first legal instrument to set a long-range and significant carbon reduction target with a legally binding framework. Its approach considers a socio-technical transition and equity under a contraction and convergence model (Lovell, Bulkeley and Owens 2009; UK, Department for Business, Energy and Industrial Strategy 2016; Global Commons Institute 2018), with carbon budgets (caps in GHG emissions) used as an umbrella policy with strong direction for all the main economic sectors.

The importance of carbon budgets in guiding climate policies is illustrated by **Figure 12.1**, which allocates limits to cumulative emissions by country. A global carbon budget translates the atmospheric carrying capacity to withstand anthropogenic GHG emissions within the goals set by the UNFCCC 2015 Paris Agreement (1.5-2°C), based on representative concentration pathways (RCPs). According to the Intergovernmental Panel on Climate Change [IPCC] (2014), for the period 2011–2100, limits range around 1,000 $GtCO_2$ (750–1,400 for 2°C with >66 per cent probability, or 550–600 Gt CO_2 for the 1.5°C goal with a >50 per cent chance).

Despite being an important lesson for other countries, the United Kingdom is still only partly on track to achieve its GHG emissions targets. The cluster of policies that have been implemented have been moderately successful in some sectors, such as transport and buildings. Entrenched interests,

Figure 12.1: Regional allocation of cumulative CO_2 emissions

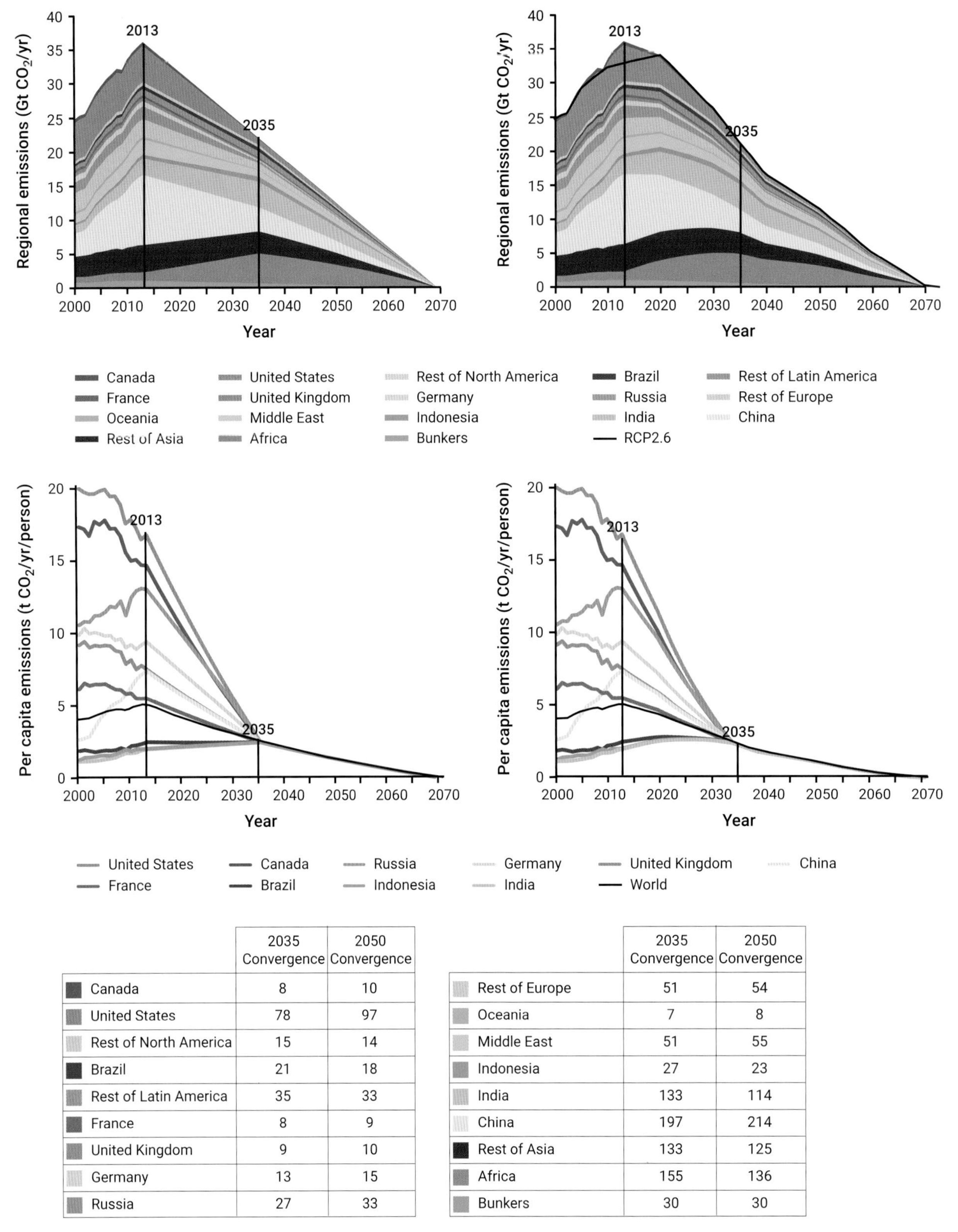

	2035 Convergence	2050 Convergence
Canada	8	10
United States	78	97
Rest of North America	15	14
Brazil	21	18
Rest of Latin America	35	33
France	8	9
United Kingdom	9	10
Germany	13	15
Russia	27	33

	2035 Convergence	2050 Convergence
Rest of Europe	51	54
Oceania	7	8
Middle East	51	55
Indonesia	27	23
India	133	114
China	197	214
Rest of Asia	133	125
Africa	155	136
Bunkers	30	30

Note: The regional allocation of cumulative CO_2 emissions following a linear emissions decrease to zero (left) and the RCP 2.6 global emission scenario (right). Per capita convergence occurs in 2035, and total cumulative emissions after 2013 are equal to 1,000 Gt CO_2 for both scenarios. The side table compares regional cumulative emissions allocation (values in Gt CO_2 from 2014 onwards) from the RCP 2.6 scenario, for per capita convergence in 2035 and 2050 (Gignac and Matthews 2015).

such as coal-fired power plants and the transport sector, are not changing as quickly as needed to achieve the targets and additional policies to accelerate change will be needed to achieve the mandatory emissions cuts. Finally, since much of the country's existing environmental legislation arises from membership of the EU, a wide range of new policies and programmes would be needed after the British exit (UK, Committee on Climate Change 2017).

12.2.2 Technology and emissions standards

One of the most common approaches to address challenges related to air quality and climate change is to define emissions standards or other performance standards for specific industrial processes, equipment or products. Such standards may mandate that a process, piece of equipment or product should not emit more than a specified mass of emissions of a given pollutant per unit of time, input or output. For example, boilers used to generate electricity may be limited to a mass of emissions per kilowatt-hour generated; vehicles are typically limited to certain emissions per kilometre (km) travelled. Alternatively, a standard may require the application of a specific type of technology or a specific operational practice. For example, to limit emissions of fugitive dust from a construction site, vehicles leaving the site may be required to go through a wheel wash station. In most cases, standards are designed to be neutral with respect to the choice of fuel or to the choice of a control manufacturer, but the general principle is that the polluter is responsible for attaining the standard and hence should bear the cost of emissions control. For example, polluters are responsible for retrofitting existing technologies, where specific equipment needs to be installed to achieve existing or new emissions standards. Where emissions control technologies are integrated into new equipment—for example, in the case of vehicles that comply with Euro emissions standards, which must pass type approval at production—the costs of emissions control are included in the unit price.

The implementation and enforcement of technology and emissions standards present a direct measure affecting 'pressures' and contribute to the attainment of the emissions targets and budgets, the desired 'state' established in a planning regime (see Figure 1.2, Chapter 1). Often referred to as 'command and control' regulations, technology and emissions standards may be implemented and enforced through permit programmes, type approval schemes, inspections and audits, and reinforced by emissions monitoring and reporting requirements. Technology and emissions standards, however, may also be developed and applied voluntarily by industry groups, self-policed or subject to third-party verification, and associated with branding programmes (see section 12.2.4). Lifestyle choices and consumption patterns may play a vital role in determining the effectiveness of voluntary approaches.

Different standards in jurisdictions in the same geographical market increase the costs to manufacturers of products and equipment, which are required to meet the various standards. Although regulatory compliance costs are only one factor that affects business location decisions, a lack of harmonization of standards can lead to shifts in economic activity and associated emissions between jurisdictions as businesses seek locations with lower overall compliance costs. Therefore, standards are typically defined at the national level, taking into account standards in other countries, jurisdictions and markets.

Under the US Clean Air Act, New Source Performance Standards (NSPS) define a minimum level of air pollution control for all major new industrial emissions sources across the country, but more stringent standards may be applied depending on the level of existing pollution. To prevent the deterioration of air quality in areas already meeting the National Ambient Air Quality Standards (NAAQS) short-term exposure to air pollution, new major sources are required to use Best Available Control Technology (BACT), which is determined on a case-by-case basis as part of a permitting process and is at least as stringent as the NSPS. In polluted areas that already exceed the NAAQS, major new sources are required to meet the Lowest Achievable Emission Rate (LAER), which reflects the most stringent requirements in practice. Existing sources in such areas are required to meet the less stringent standard of Reasonably Available Control Technology (RACT). The USEPA maintains a database of these standards and case-by-case determinations in what is known as the RACT/BACT/LAER Clearinghouse (United States Environmental Protection Agency 2018b). As technology evolves, so do the RACT/BACT/LAER determinations, but by making the information public, the standards in different jurisdictions can evolve together. Likewise, the Industrial Emissions Directive (2010/75/EU) regulates industrial activities in Europe and also uses a legally binding concept of 'Best Available Techniques' (BAT) to set environmental performance levels which industry must achieve (European Commission 2018b). The definition of a BAT is done through a transparent exchange of information between industry, non-governmental organizations (NGOs) and regulators, and is recognized beyond the EU.

For marine shipping emissions, the International Maritime Organization (IMO) and the International Convention for the Prevention of Pollution from Ships (MARPOL, Annex VI) (International Maritime Organization [IMO] 2018) set standards for a global limit for sulphur in fuel oil used on board ships of 0.50 per cent m/m (mass by mass), to enter into force on 1 January 2020, representing a fuel-based standard. Compliance can be achieved by ships using low-sulphur compliant fuel oil. In contrast, a nitrogen oxide (NO_x) Emission Control Area (NECA), which will enter into force for the North Sea and the Baltic Sea from 1 January 2021, will require all vessels built after 2021 to comply with emissions standards aimed to reduce NO_x emissions by 80 per cent compared to present levels. Compliance can only be achieved in practice by equipping vessels with catalysts or using liquefied natural gas (LNG) fuels. Existing emission control areas for sulphur oxide (SO_x) cover the Baltic Sea (19 May 2006) and the North Sea (22 November 2007), and for SO_x and NO_x the North American east and west coasts (1 August 2012) and the US Caribbean Sea (1 January 2014).

The effectiveness of technology and emissions standards depends on the level of compliance, affordability and the extent to which the standards reflect the real-world impact of the emissions sources. The level of compliance, in turn, depends on the level of education, as well as monitoring and enforcement, among other factors. However, in countries or jurisdictions where the government has little capacity to enforce standards through inspections and audits, compliance with standards can be low. Even in countries where sophisticated inspections and audits are routine, some businesses and individuals may violate standards to gain a competitive economic advantage.

Cost–benefit analyses, such as that conducted by Åström *et al.* (2018), can provide an insight into the cost-effectiveness and the distributional effects of setting standards, for both *ex ante* and *ex post* evaluations.

Finally, the use of solid fuels, including biomass, in residential boilers is gaining more attention recently, as other key sectors have been regulated in the past decades. For example, the European Commission has established performance standards for solid-fuel space heaters with a nominal heat output of 50 kW or less, which have so far been mostly unregulated and may contribute to local air quality problems due to emissions of NO_x and particulate matter less than 2.5 μm ($PM_{2.5}$) (European Union 2015). Similarly, the EU Ecodesign-Directive (Directive 2009/125/EC) establishes a framework for setting ecodesign requirements for energy-related products, household appliances, and information and communications technologies.

The following case study illustrates the effectiveness of emissions standards through a type approval scheme through the example of the Euro 6 vehicle emissions standards, the difference between how compliance is measured and real-world performance, and the efforts of some businesses to thwart compliance testing.

Case study: Excess diesel emissions in Europe

Since the 1970s, the key mechanism for regulating vehicle air pollutant emissions in Europe has been a regulatory/command and control policy that has progressively set increasingly stringent standards for emissions of air pollutants and GHGs (since 2009). A series of directives, known as Euro standards, define the acceptable limits for exhaust emissions of new road vehicles sold in the EU. The Euro standards, starting with the release of Euro 1, which entered into force in 1993, have since been amended regularly. The most recent Euro standard is Euro 6 for light passenger and commercial vehicles, which came into force in 2014 (Commission Regulation (EU) No 582/2011).

The aim of this policy is to contribute to reductions in actual or real-world emissions from road transport, an important source of GHG and air pollutant emissions. However, road transport enables people to access employment, education, goods and services. Manufacturers of road vehicles in the EU also contribute to the economy and employment. Hence, the policy aims to achieve a balance across these objectives.

Table 12.3: Summary of assessment criteria: Excess diesel emissions in Europe

Criterion	Assessment	Reference
Success or failure	The Euro standards have been successful in decreasing emissions of air pollutants and GHGs per unit of travel and decreasing measurable air pollution concentrations, especially close to roads. However, real-world reductions fall short of the potential reductions measured in laboratory testing.	EEA 2015; EEA 2016a; EEA 2016b; EEA 2017
Independence of evaluation	Independent evaluations of overall progress have been conducted.	
Key actors	Associations of car owners and vehicles manufacturers, NGOs etc. were highly involved. It was a long process with a high level of stakeholder involvement.	
Baseline	Quantitative baselines were established and updated with each amendment.	
Timeframe	Policies were implemented on time, but some intended targets were missed because of the inadequacy of the test cycle and wide-ranging exceptions enabling manufacturers to switch off emissions control technologies under certain ambient conditions.	
Constraining factors	Lobbying from industry has led to delays and weakening of the policy. In some cases, manufacturers have circumvented the standards designing cars to have lower emissions during testing than on the road.	Grice *et al.* 2009; Guerreiro *et al.* 2010, p. 3; Hotten 2015;
Enabling factors	The European regulatory and governance structure was key to enabling policy implementation. Some countries lacked the resources to independently verify emissions reported by manufacturers. The level of participation in policy development was high, leading to high levels of public approval.	
Cost-effectiveness	Costs and benefits, including impacts on health, agriculture and ecosystems, are considered carefully in the process of European policy development, aiming at high cost-effectiveness.	European Commission 2018
Equity	The impacts can be considered positive for everyone, but they particularly benefit people living close to roads, who may be economically disadvantaged.	
Co-benefits	The Euro standards set an example for the world. Industry is driven to innovate. Fuel efficiency increased, improving energy security and decreasing GHG emissions. Some national policies providing incentives to purchase diesel vehicles led to increases in NO_2 emissions. Transport subsidies may offset the reduction of emissions.	Franco *et al.* 2014; European Commission 2015; EEA 2015
Transboundary effects	As the Euro standards progressively penetrated the international second-hand vehicle market, vehicle emissions were lowered in countries without similar regulations.	
Possible improvements	Reduce the margins of technical uncertainty in testing, eliminate test flexibilities, and increase emissions checks of cars in circulation.	

The main objective of the original European standards was the reduction of emissions of NO_x, carbon monoxide (CO), $PM_{2.5}$ and volatile organic compounds (VOCs), as well as carbon dioxide (CO_2) and other pollutants, and subsequent amendments have implemented further limits. Furthermore, increasingly strict Euro standards require cleaner petrol and diesel fuels (the quality of which was regulated by Directive 2003/17/EC, e.g. low sulphur content), leading to, for example, lower PM emissions. These regulations have been driven primarily by air pollution effects on human health and—to a lesser extent—natural and semi-natural ecosystems, in addition to climate change. These increasingly stringent emissions standards have achieved positive results. They have led to the introduction of new vehicle technologies, which have achieved significant reductions in vehicle emissions over recent decades in Europe. For example, the latest standard (Euro 6) for diesel cars requires a reduction of almost 97 per cent of PM emissions compared to the Euro 1 standard for a 20-year-older vehicle.

The Euro standards have been successful in decreasing emissions of air pollutants and GHGs per unit of travel, as well as reducing overall transportation emissions even while transportation activity has increased (EEA 2015, pp. 25, 30, 32, 33, 37, 38, 46; EEA 2017, p. 19). They have also led to a measurable decrease in air pollution concentrations, especially close to roads (EEA 2016b, pp. 31, 43, 77–79, 82). However, these reductions fall short of the targets outlined in the policy, due to differences between real-world behaviour and emissions under laboratory testing conditions (EEA 2015, p. 46; EEA 2016a, pp. 27–37). These differences have increased considerably since 2000, especially for CO_2 and NO_x emissions in diesel cars, mainly due to:

i. the test procedure, which did not reflect real-world driving conditions;
ii. flexibilities in the procedures, which allowed manufacturers to achieve lower fuel consumption and emissions values during testing;
iii. several in-use factors, such as driving style and environmental conditions; and
iv. the use of 'defeat devices' designed to lower emissions measured during vehicle testing in the laboratory but not on the road.

Only after the diesel-gate scandal in 2015 (Hotten 2015) was there enough political awareness and will to change the laboratory test cycle to better reflect real-world emissions (European Union 2016).

In view of the shortcomings of the implemented standards, the test cycles were reviewed, and new and more reliable emissions tests in real driving conditions, as well as improved laboratory tests, were introduced for new car models sold in Europe in 2017. In addition, the European Commission presented a new proposal in 2018, aiming to reduce the margins of technical uncertainty in testing, eliminate test flexibilities and increase emissions checks of cars in circulation.

12.2.3 Market interventions

As a potential alternative to mandating lifestyle and technology choices, governments can also guide lifestyle and choices by creating economic incentives (e.g. subsidies, tax credits, loans or price guarantees) or disincentives (e.g. fees or taxes) in existing markets or by creating new markets for rights or commodities that have not been traded (e.g. emissions reduction credits or renewable energy credits). All these types of market interventions have been used to some extent to mitigate air pollution, climate change, ODSs or PBTs.

In the DPSIR framework (Figure 1.2, Chapter 1), market interventions affect the 'drivers' of environmental issues, which in turn affect the 'pressures' and the environmental 'state'.

Although market interventions do not directly decrease emissions or ambient concentrations, they provide regulated individuals and businesses with flexibility and can create an incentive to improve performance and lower costs. Thus, a properly designed and adjusted market intervention may achieve emissions reductions more efficiently than 'command and control' approaches.

Markets are affected by many factors which are beyond government control. Therefore, market interventions must be adjusted periodically to reflect changing conditions. It is helpful if those adjustments can be made within the context of a planning regime (see section 11.3) so that progress towards a desired state can be evaluated and tracked.

In some cases, government interventions are needed to bring new technologies to a given market. Once the technology is introduced to the market and provided with a foothold, such as through initial subsidies or loans, it should be able to compete with other technologies without government assistance. The following case study explores how the Government of Kenya helped introduce clean cookstoves and fuels to reduce residential air pollution.

Case study: Improved cookstoves in Kenya

Household air pollution from the use of traditional fuel sources (wood, dung and charcoal) for cooking is a leading contributory factor to the global burden of disease (Lim *et al.* 2012; Cepeda *et al.* 2017; Landrigan 2017, see Section 5.2.4). A range of chronic illnesses such as cataracts, lung cancer and bronchitis are associated with smoke from cooking, with women and children most affected (Cepeda *et al.* 2017). Beyond this, black carbon (a household air pollutant) has been identified as the second most important anthropogenic emission that significantly affects the world's climate (Bond *et al.* 2013; Myhre *et al.* 2013).

The adverse effects of the dependence of the developing world on traditional renewable energy for cooking has necessitated the timely intervention by the Global Alliance for Clean Cookstoves (the Alliance), a public–private partnership geared towards the creation of a global market for improved, clean and efficient household cooking solutions in a bid to reduce the carbon footprint of highly polluting traditional stoves. The Alliance was launched in 2010 with an ambitious 10-year goal to foster the adoption of clean cookstoves and fuels in 100 million households by 2020. It partnered with NGOs, foundations, women's cooperatives, trade associations, academic institutions, investors and entrepreneurs to expand markets for clean cookstoves. Given the extensive use of unprocessed fuelwood in rural sub-Saharan Africa in particular, the Alliance is operational in a number of countries in this region, including Kenya, where 84 per cent of the population

use solid fuel for cooking and 16,500 deaths annually have been attributed to exposure to indoor air pollutants (Global Alliance for Clean Cookstoves 2012a).

Kenya's development, marketing and distribution of clean cookstoves is the most advanced in the sub-Saharan African region, having emerged in the 1980s led by development of the Kenyan Ceramic Jiko (United States Agency for International Development and Winrock International 2011). Yet by 2007, the penetration rate in the Kenyan market for cookstoves was approximately 36 per cent, and adoption in rural areas was quite low. Since then the Alliance has made notable inroads in the Kenyan cookstoves market through its partnership with the Clean Cookstoves Association of Kenya (CCAK) to encourage government officials to adopt market incentives—for example, abolishing or minimizing taxes and tariffs that impede the growth of the clean cooking sector. A notable accomplishment was the reduction in the import duty on efficient cookstoves from 25 per cent to 10 per cent by the Kenyan Government in 2016. The Alliance has also provided grants to boost brand-building and marketing efforts and has supported two women-owned businesses through its Women's Empowerment Fund.

Carbon financing for clean cookstoves has not only been beneficial to Kenyans but has also enabled international companies to achieve their emissions reduction goals through carbon trading, which allows for carbon credits to be used to comply with emissions reduction obligations under cap-and-trade schemes or for voluntary reduction schemes (Lambe *et al.* 2015).

An evaluation of six types of biomass stove highlighted the need to address key factors that contribute to reducing emissions—namely: the cookstove design and performance; other potential sources of emissions; the availability and cost of cookstoves; ventilation; and the strategies to ensure the adoption and use of clean cookstoves (Pilishvili *et al.* 2016). The results of the evaluation showed that the biomass stoves did reduce emissions compared with the three-stone traditional stove baseline. However, the reduction in emissions did not reach thresholds where public health benefits could be maximized.

Table 12.4: Summary of assessment criteria: Improved cookstoves in Kenya

Criterion	Description	Reference
Success or failure	Approximately 37 per cent (3.5 million) households use improved biomass cookstoves (ICSs), while over 50 per cent (approximately 5 million) households still use traditional biomass cookstoves. Between 240,000 and 300,000 ICSs are sold to new customers annually. The Kenyan Government aims to achieve 50 per cent of abatement potential (i.e. approx. 2.6 $MtCO_2e$) by 2030.	Kenya, Ministry of Energy and Petroleum and Sustainable Energy for All 2016; Kenya, Ministry of Environment and Natural Resources 2017
Independence of evaluation	Green Climate Fund, a UNFCCC funding mechanism.	
Key actors	Over 80 per cent of the market share for biomass ICSs is dominated by artisanal fabricated stoves.	Green Climate Fund 2018
Baseline	The project started in 2010, and by April 2016 the Global Alliance for Clean Cookstoves project had sold approximately 251,000 cookstoves across Kenya.	Natural Capital Partners 2018
Timeframe	GHG emissions mitigation objective of 30 per cent by 2030 in relation to the business-as-usual (BAU) scenario of 143 $MtCO_2e$. Of this, ICS interventions are considered to have an abatement potential in 2030 of 5.6 Mt CO_2e.	Global Alliance for Clean Cookstoves2014: Green Climate Fund 2018
Constraining factors	Underdeveloped ICS supply chain; communities that collect wood for free; costs and risks associated with investment to cover remote rural areas; weak consumer awareness; regulatory constraints (i.e. import duties, taxes and poorly targeted subsidies); limited product testing capacity to enforce standards; and insufficient investment into product improvement.	Global Alliance for Clean Cookstoves 2012b; Green Climate Fund 2018
Enabling factors	Kenya removed the 16 per cent value added tax (VAT) on LPG, reduced the import duty on efficient cookstoves from 25 per cent to 10 per cent, and placed a zero-rating VAT on improved cookstoves, raw materials and their accessories.	Global Alliance for Clean Cookstoves 2016
Cost-effectiveness	Design of cookstoves does not offer the full benefit of fuel savings and reduced emissions and high quality cookstoves are not easily distinguished from their competitors.	
Equity	Households in the poorest quintile, women in younger age groups and people living in remote areas are less likely to adopt and install improved cookstoves.	Silk *et al.* 2012; Kapfudzaruwa, Fay and Hart 2017
Co-benefits	Livelihood improvements, social impacts (including gender), reductions in co-pollutants (ozone damage to crops), among others.	
Transboundary effects	Lessons learned can be applied to other sub-Saharan African countries.	
Possible improvements	Customer segmentation studies are needed to understand customer needs and tailor financial products for purchasers. Existing non-cookstove distribution and wholesale networks need to be used to improve consumer access and affordability.	(Global Alliance for Clean Cookstoves 2013; Kenya, Ministry of Energy and Petroleum 2015)

Overall the clean cookstove market in Kenya shows promise, but there is a strong need for policy intervention in rural areas to scale up the adoption and use of cookstoves. Human health could be better integrated within this policy in the future by clearly delineating the health indicators of local and global significance that need to be monitored periodically to effectively track progress in health benefits spatially and temporally.

12.2.4 Information policies

In addition to using regulatory mandates and market incentives, governments can, at times, aim to support a change in lifestyle to either reduce emissions or exposure to harmful levels of air pollution by providing the public with better information. This goes in line with an improved understanding of the hazards and risks of exposure to air pollution in recent years (Kelly and Fussel 2015), although the importance of public education for air pollution control was raised more than 50 years ago by Auerbach and Flieger (1967). One example of such an approach is the provision of near real-time air quality observations and forecasts. The posting of near real-time air quality data online is increasingly common in many nations and major cities around the world. Some locations also provide air quality forecasts, predicting air quality levels for one or more days into the future. This information is often further circulated via other websites, social media, smart phone apps, newspapers, local radio and television. These dynamic media may be complemented by educational posters and pamphlets. The objective of providing such information is to encourage citizens to change their lifestyle to:

- decrease their exposure to pollution and, consequently, lower the risk of adverse health effects (e.g. by avoiding exercise outdoors during peak concentrations, or for especially vulnerable individuals to stay indoors); and
- decrease their emissions (e.g. commute via public transportation instead of personal transport or curtail use of wood fires or other biomass burning during forecast pollution episodes).

Air quality forecasts and real-time values may be used to complement other air pollution control policies. For example, some localities in the United States impose wood-burning restrictions when a pollution episode alert has been issued, while in Europe several large cities, such as Paris, have issued restrictions for private car use during high pollutant concentration episodes in recent years. For the severest of episodes, governments may shutdown factories and other non-essential activities. In China, several cities impose a ban on trucks and other high emission vehicles during the day to manage air quality.

Other examples of providing information to guide behaviour are labelling and branding. In this context, labelling refers to providing information on a product about its environmental impacts, such as its relative emissions or energy consumption, to inform consumers' choices. Such labels may be required by government regulations (e.g. emissions and fuel economy estimates are required on the labels for new cars in the United States) or may be voluntarily provided by the manufacturer. Branding, in this context, involves associating a logo or symbol with a product or service that indicates to consumers that the product or service has met some set criteria for environmental performance (e.g. Energy Star). Such brands have been established by governments, industry associations or public advocacy groups as voluntary programmes, and often involve independent testing.

These approaches need to reach a threshold level of awareness and recognition across their target populations before they can have much effect on environmental pressures (consumption and other emissions-generating behaviour) and impacts (exposure-related behaviour). However, once the threshold is reached, citizens and consumers may begin to expect and demand such information. Moreover, increasing awareness of the sources and impacts of pollution may increase public demand for cleaner air, lower emitting products and services, and more stringent policies.

Access to information also promotes innovation. In the past, most of the information on ambient air quality originated from regulatory bodies. With the curiosity of the public to know more about the ambient air quality and its linkages to human health, there is now a new market of non-regulatory air quality monitors, which are providing the same information as the regulatory monitors at lower cost. While the data quality produced by (often referred to as low-cost) air quality sensors for crowdsourcing of air quality information is at times questionable (Lewis and Edwards 2016; Thompson 2016), the empowerment of citizens to take ownership of environmental data should not be underestimated (see Section 25.2). This serves society with information and lets the public make informed decisions on when to go out, when to spend more time outdoors etc., and is further supported by activities to make air quality information openly available and easily accessible—for instance, through the OpenAQ initiative ((https://openaq.org), which aims to *"enable previously impossible science, impact policy and empower the public to fight air pollution through open data, open-source tools, and cooperation"*.

The need for more information has led to the establishment of private organizations as well as regulatory agencies taking charge of experimenting with novel ways to generate information most suitable for consumption. Experimentation is feasible to expand, and there is room for the regulatory bodies to step in and endorse the emerging technologies for further expansion.

With the wide use of smart phones and open applications, this information is now serving everyone with access to a phone, in addition to traditional dissemination channels. Due to the wide spread use of mobile applications, citizens' awareness of and access to information is increasing rapidly. As a consequence, the public has generated a demand for change. The availability of targeted information can further empower individuals to be prepared and manage health risks in response to environmental hazards, illustrated, for example, by the Know and Respond information system in Scotland, or the United Kingdom of Great Britain and Northern Ireland-wide air pollution forecast provided by the UK, Department for Environment, Food and Rural Affairs (2018). Know and Respond is a free service to subscribers in Scotland that sends registered users an alert message if air pollution in their area is forecast to be moderate, high or very high (Air Quality in Scotland 2018).

Case study: AirNow, real-time air quality data and forecasts

In 1995, the American Lung Association of Maryland, a non-governmental advocacy group, began to create a daily map of ozone observations from the monitors operated by the State of Maryland (state and local governments are responsible for operating air quality monitoring stations in the United States). In 1997, the daily ozone maps were expanded to 14 north-eastern states. In 1998, the USEPA took over the operation of the central data system, added seven more states and named the system AirNow (http://airnow.gov/).

Over the next 10 years, AirNow grew as more states and local agencies contributed their data. By 2008, agencies in all 50 US states, 4 Canadian provinces and the national government, and 2 Mexican states were submitting data. Over time, the data system and product delivery were improved incrementally, experimenting with new products and services. In 2009, the system architecture and software tools were overhauled to allow the software to be implemented in different settings. The new software, branded AirNow-International, was deployed in Shanghai, China, for the 2010 World Expo, and in Monterrey, Mexico in 2012.

AirNow currently gathers and distributes observations and forecasts from more than 130 federal, state and local agencies. Data for the United States is provided to the public and the media using the Air Quality Index (AQI), a colour-coded and numerical scale based on the US National Ambient Air Quality Standards. An applications programming interface (API) has opened the data system to smart phone applications, which have proliferated. AirNow-International has increased its scope geographically, and the US Department of State has begun providing air quality observations from selected US embassies and consulates around the world.

As an information policy, the US EPA AirNow programme provides a low-cost but high-benefit example. By building on existing structures, such as monitoring networks and state or local air quality agencies, the programme leverages infrastructure in a new way. The primary benefit of this information policy comes by helping individuals reduce or avoid exposure to high levels of pollution. Timely information encourages citizens to take mitigative actions to reduce their contributions to pollution. In addition, the provision of information creates an awareness of air pollution, a demand for cleaner air and greater acceptance of other regulatory and market policies to decrease pollution. While no formal evaluation of the programme exists, various studies show that air quality information has an impact on awareness and lifestyle.

Table 12.5: Summary of assessment criteria: AirNow, real-time air quality data and forecasts

Criterion	Assessment	Reference
Success or failure	Studies have demonstrated increased awareness due to alerts, but changes in lifestyle (e.g. decreased driving, energy consumption) are more difficult to quantify.	Blanken, Dillon and Wismann 2001; Henry and Gordon 2003; Mansfield and Corey 2003; Kansas Department of Health and Environment 2006; Mansfield, Johnson and von Houtven 2006; McDermott, Srivastava and Croskell 2006; Semenza *et al.* 2008; Mansfield, Sinha and Henrion 2009; Neidell 2008
Independence of evaluation	Various studies.	
Key actors	National, state/provincial and local governments.	
Baseline	Depends on the location.	
Timeframe	Between 1995 and 2008, the system evolved from an effort by one state to a system involving agencies in all 50 US states, 4 Canadian provinces and the national government, and 2 Mexican states. Currently, more than 130 federal, state and local agencies are participating.	
Constraining factors	Air quality information can generate anxiety about potential health effects and may encourage some people to seek unnecessary medical attention. Moreover, once information is made freely available, a variety of uses may be devised that were not envisioned when the system was created. Policies and attitudes about data transparency and openness can be important constraints.	
Enabling factors	A prerequisite is an effective air quality monitoring infrastructure and data collection and dissemination programme.	
Cost-effectiveness	Additional costs of the data management and dissemination are small compared to the costs of the monitoring activity and the value of the media coverage.	
Equity	Access to air quality information is not uniform. People who are poorly educated or without access to the Internet may be excluded.	
Co-benefits	Creates a greater demand for air quality and improves acceptance of air quality management policies.	
Transboundary effects	The programme is expanding internationally under the auspices of the Group on Earth Observations.	
Possible improvements	As access to various types of information increases, expectations for access to air quality information also rise.	

12.2.5 International cooperation

No one can live without air, and its quality is indispensable. However, 90 per cent of the global population is now forced to live with unhealthy air, particularly in Asia and Africa (WHO 2018). Air pollution and PBTs are of particular concern, as they travel locally, internationally, regionally and globally. International cooperation plays an important role when air pollution crosses borders or needs to be addressed across borders.

International cooperation can take many forms, ranging from formal to informal, bilateral to multilateral diplomacy. Governments are one of the key actors to align their actions—negotiating and concluding multilateral environmental agreements (MEAs) in a tangled web of national interests, providing international aid, conducting capacity-building/technical assistance under or beyond the agreement, monitoring and modelling air pollution to improve scientific knowledge with help from a community of experts, sharing information with the public, raising awareness and engaging in voluntary efforts for further cooperation to reduce air pollution. Formal training, technology demonstration, and cooperative research and assessment activities provide more effective knowledge-sharing and capacity-building opportunities. These activities may have the most significant long-term influence on environmental outcomes, but their immediate impact is difficult to quantify. Local governments are important actors to implement national environmental policies. Cooperation from business and industry is crucial to increase the effectiveness of the policies. Considerable progress on policy cooperation, emissions control and reporting, and ecosystem recovery has been achieved under the eight legally binding CLRTAP Protocols. The application of the effects-oriented critical load concept with regional cost minimization of science-based mitigation measures, technical as well as structural, has offered a sophisticated but successful way forward for participating countries.

A relatively recent approach to international cooperation on air-related issues has been the development of public–private initiatives, such as the Partnership for Clean Fuels and Vehicles, the Global Alliance for Clean Cookstoves (see Section 12.2.3), the Global Research Alliance on Agricultural Greenhouse Gases, and the Climate and Clean Air Coalition for the Reduction of Short-Lived Climate Pollutants (CCAC). These initiatives bring together interested national governments, intergovernmental organizations, private-sector companies, civil society organizations and philanthropic foundations to advance specific pollution mitigation efforts. For example, CCAC was founded in 2012 by the United Nations Environment Programme (UNEP) and the governments of Bangladesh, Canada, Ghana, Mexico, Sweden and the United States, to catalyse action to decrease emissions of black carbon, methane and hydrofluorocarbons. CCAC now has more than 100 state and non-state partner organizations participating in 11 different initiatives. This could contribute to cleaner fuels and technologies in the homes of 3 billion vulnerable people suffering from household air pollution (Apte and Salvi *et al.* 2016).

International financial institutions such as the World Bank, the Asian Development Bank, the African Development Bank, the Global Environmental Facility and the Green Climate Fund play major roles in project funding. Financial assistance and cooperative implementation of control measures can have a clear and demonstrable effect on decreasing emissions in the short term, but the long-term impacts may be much larger if the control measures are replicated.

Regional organizations can function in two ways. One, like the EU, is taking the leading role in negotiations as a global actor and an increasing role in environmental politics, while the other, like the Association of South East Asian Nations (ASEAN), is to function as an international forum. Both provide opportunities to set regional agendas, learn new knowledge and perspectives, share information and discuss common issues. Treaty secretariats could influence the negotiation process among States under the accords. Like-minded groups, alliances and friends of the Chairs could also lead, mediate or slow down negotiations. Some of these cooperation processes could also be activated by environmental NGOs, green parties and citizens as well as international organizations such as WHO, the World Meteorological Organization (WMO), the Organisation for Economic Co-operation and Development (OECD), the International Energy Agency (IEA), the International Maritime Organization (IMO), the International Civil Aviation Organization (ICAO) and UNEP, helping to set the environmental agenda, framing environmental issues or providing resources for international cooperation.

As introduced in Section 5.5, global MEAs which have linkages to air pollution are those targeting climate change (UNFCCC), stratospheric ozone depletion (Vienna Convention and Montreal Protocol), mercury (Minamata Convention) and persistent organic pollutants (Stockholm Convention). Although there is no global convention on air pollution, several regional MEAs and bilateral agreements exist. One of the oldest regional MEAs is the 1979 CLRTAP negotiated under UNECE. Considerable progress on policy cooperation, monitoring and modelling, emissions control and reporting, and ecosystems recovery has been achieved under the eight legally binding CLRTAP Protocols.

It is difficult to evaluate the impact of international agreements. Compliance with legal commitments can be evaluated, but it is not always clear that emission decreases that occur are a result of an international agreement or if they would have occurred in the absence of the agreement. Furthermore, perfect compliance may be an indication of unambitious targets that require little more than business as usual efforts.

None of the MEAs identified above have an effective international enforcement mechanism. Depending on a country's own laws, a national government may be taken to court in its own country for not abiding by its international treaties. However, this kind of action is rare, and ensuring compliance with international commitments mostly relies on diplomatic or peer pressure.

The following case study explores the progress made under the regional agreement on transboundary haze negotiated under ASEAN in 2002. This case provides valuable lessons on the challenges of international cooperation.

Case study: ASEAN Agreement on Transboundary Haze Pollution

Haze or smoke-haze as used in Southeast Asia is synonymous with air pollution from land and forest fires and agriculture-waste burning from rural populations. Transboundary haze pollution refers to haze pollution whose physical origin is situation wholly or in part within an area under the national jurisdiction of one country and which is transported into the area under the jurisdiction of another country (ASEAN Agreement on Transboundary Haze Pollution). The worst haze episodes occur in years affected simultaneously by the climatological anomalies of *El Niño*—Southern Oscillation and Indian Ocean Dipole. The most recent severe haze event in 2015 was caused in part by the dryness triggered by both anomalies in the region, particularly on the Indonesian islands of Sumatra and Kalimantan (Koplitz *et al*. 2016).

Table 12.6: Summary of assessment criteria: ASEAN Agreement on Transboundary Haze Pollution

Criterion	Description	References
Success or failure	Mixed view on effectiveness and achievement of goals.	
Independence of evaluation	No consensus in evaluations	
Key actors	Governments, the ASEAN Secretariat and its agencies, NGOs, foreign governments, industries and international/regional organizations.	
Baseline	The Agreement does not provide a specific baseline.	
Timeframe	The Agreement is expected to remain in force perpetually with the active participation of all ASEAN member states, with the goal of achieving a haze-free ASEAN by 2020.	Haze Action Online 2018
Constraining factors	Logging and clearing land by burning creates employment for less fortunate people due to limited employment in other sectors. Oil palm and pulpwood industries have thus largely flourished under weak law enforcement by local authorities. Once cleared to make room for plantations, peat forests tend to smoulder underground for weeks even after surface fires are fully extinguished.	
Enabling factors	Singapore collaborated with Jambi in Indonesia between 2007 and 2011 to implement action programmes to prevent and mitigate land and forest fires. Malaysia cooperated with an NGO, the Global Environmental Centre, to reduce the risks associated with peatland fires. The 2005 cloud-seeding project in Riau and West Kalimantan conducted by Indonesia, Malaysia and Singapore indicated cooperation. An environmental partnership between Singapore and Indonesia from 2009 to 2011 enables fire management and extinction in Riau. The institutional arrangements and collaboration have been key enabling factors, as was the willingness of the Parties to commence action even prior to Indonesia's ratification. Programmes to improve peatland management and control haze pollution have been welcomed by member states.	Haze Action Online 2017
Cost-effectiveness	The cost of 2010–2014 ASEAN Peatland Forests Project was US$5.9million. The project successfully scaled up the management and rehabilitation of critical sites in Philippines and Viet Nam. An estimate of the cost to control Indonesian haze is put at US$5.7 billion, and another to address forest fires is US$1.2 billion, but this does not include losses from peatland burning, which the World Bank puts at US$16 billion for the 2015 fires alone.	World Bank 2016; Nazeer and Furuoka 2017; Hans 2018
Equity	Haze contributes to respiration difficulties and other haze-related illness, particularly in children and in poor people.	Malaysian Nature Society Science and Conservation Unit 1997; Gordon, Mackay and Rehfuess 2004
Co-benefits	Associated impacts from haze include CO_2 emissions, losses of wildlife and potential benefits from nature preservation and biodiversity; impacts are linked to the Sustainable Development Goals such as goals 3, 11 and 12. The ASEAN peatland management strategy and the SEA peat project include contributions to climate change mitigation (2006–2020). The Agreement is relevant to the achievement of national air quality targets. E.g., in Singapore, where targets for 2020 have been adopted, effective management of peatlands and the control of haze pollution in neighbouring countries have a direct impact on the quality of air.	Asia Pacific Clean Air Partnership [APCAP] 2015, p. 2; World Wide Fund for Nature [WWF] 2018
Transboundary effects	Smoke from wildfires and burning agricultural waste travels to and adversely affects other countries.	
Possible improvements	Development of a zero-burning policy, accomplishment of ASEAN Community Vision 2025, and raising political will for further cooperation; could gain potential benefit through collaboration with other initiatives such as the Acid Deposition Monitoring Network in East Asia, the Climate and Clean Air Coalition and the Asian Co-benefits Partnership.	

Haze is a health issue, worsening existing heart and lung conditions (WHO 2016). It caused acute respiratory infections in over 500,000 people and 19 deaths in 2015. The economic cost was estimated at US$16 billion, affecting the transportation (US$372 million), tourism (US$399 million), trade (US$1.3 billion), manufacturing and mining, and mass education sectors (World Bank 2016, pp. 1–2, 4–8). Emissions from burning peatlands contained 90 gases with high levels of toxicity from formaldehyde, acrolein, benzene, carbon monoxide and nitrogen dioxide (Stockwell *et al*. 2016).

The first severe haze episode in 1997-1998 called for regional action. Following the Asian Development Bank's major damage assessment (Qadri ed. 2001), the ASEAN Agreement on Transboundary Haze Pollution was adopted in 2002 and entered into force in 2003 (Haze Action Online 2018; UNEP 2010).

The Agreement attempts to change the governance arrangements towards an ASEAN haze-free region by preventing and monitoring transboundary haze pollution as a result of land/forest fires mitigated through concerted national efforts and intensified regional and international cooperation. (ASEAN Agreement on Transboundary Haze Pollution). The ASEAN Specialised Meteorological Centre has been mandated as the regional centre to monitor and assess the weather and smoke haze situation and provide early warning on the occurrence of transboundary haze.Malaysian Meteorological Department provides information on fire risk through a Fire Danger Rating System. National Monitoring Centres are designated within each country to monitor the fire and haze situation locally (Velasco and Rastan 2015).

There are mixed views on the effectiveness of the Agreement in achieving its goals. In 2003, it became the first regional environmental agreement under ASEAN to enter into force, but not all ASEAN countries ratified the agreement until 2014. Some international cooperative activities and national policies were initiated following entry into force, but major haze events have continued to occur, with notable episodes in 2013 and 2015. Like many MEAs, the Agreement has no sanctions clause on failure to meet stipulated obligations.

Some progress has been observed since 2015. Brunei has developed its own transboundary haze forecasting system. Singapore has enacted a law to enhance the country's ability to deal with transboundary haze by holding companies accountable for haze pollution in Singapore due to activity which had occurred within or outside of Singapore. Indonesia implemented regulations to protect primary forests and peatland, building on a moratorium on new concessions started in 2011. This policy is expected to avoid 7.8 Gt of CO_2 emissions by 2030 (Republic of Indonesia 2015, pp. 5–6).

Activities under the Agreement can also support other cooperation in the region, not least the coordination of monitoring networks, such as through the Asia Pacific Clean Air Partnership (APCAP). APCAP brings together the ASEAN Haze Agreement, the Malé Declaration, the East Asia Network for Acid Deposition and the Atmospheric Brown Cloud initiatives.

12.3 Indicators

As discussed in Chapter 5, atmospheric change (including air pollution, climate change, stratospheric ozone depletion and PBT pollution) is directly or indirectly related to each of the SDGs (see Figure 5.2). These issues are also addressed by a collection of global and regional MEAs, as well as policies at the national, provincial and municipal level. To evaluate the effectiveness or sufficiency of any given policy or suite of policies, it is useful to have measurable indicators. Ideally, such indicators should be sensitive to the policy changes of interest but not confounded by other influences. However, the indicator should also be relatable to impacts of the policy change that society values. These two objectives point to either end of the DPSIR framework (Figure 1.2) used in the preceding chapters, suggesting that selecting one best indicator is often a compromise. The compromise might be different if the intent is to evaluate the effectiveness of a specific policy (such as in the case studies above) versus the sufficiency of a suite of policies (including policies at different geographical scales).

This section describes three indicators of atmospheric change that are intended to track progress towards the SDGs and compare the sufficiency of existing policies at the national scale. The indicators chosen focus on air pollution, stratospheric ozone depletion and climate change. The data available for emissions and concentrations of PBTs are not sufficient for evaluating the effectiveness of policies globally or at the national level.

12.3.1 Indicator 1: Population-weighted annual mean concentration of $PM_{2.5}$

Figure 12.2: Population-weighted annual country-wide mean concentration of $PM_{2.5}$ in 2016

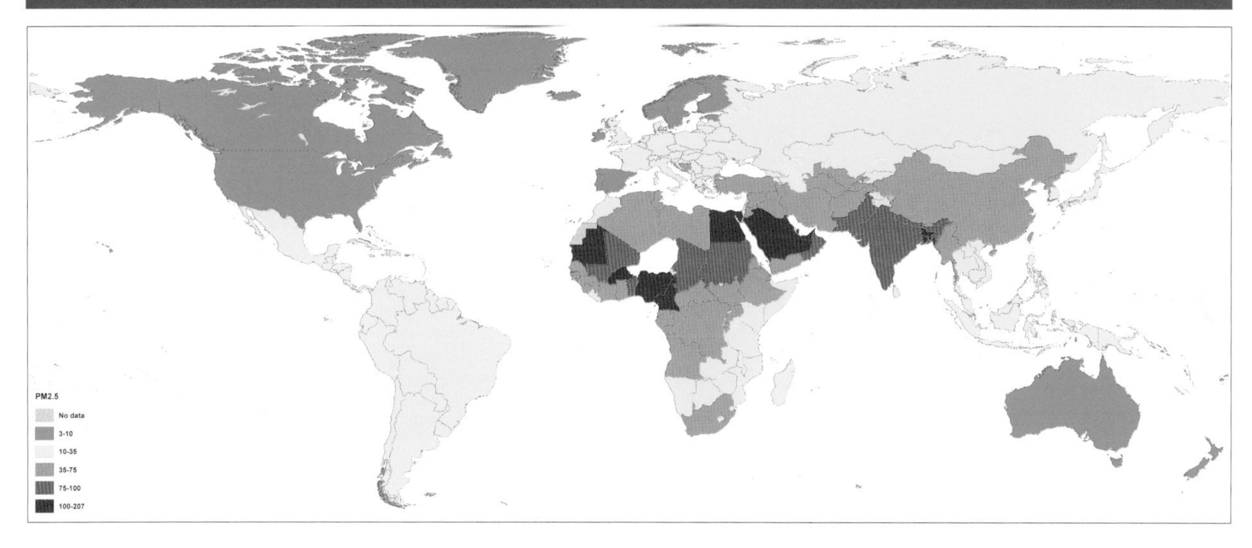

This map combines data from satellite observations, surface monitors and an atmospheric chemistry and transport model.

Source: Adapted from Health Effects Institute (2017); Shaddick *et al.* (2018).

Policy relevance/causal chain

$PM_{2.5}$ concentrations are state variables that are driven by emissions of pollutants, but also by meteorology and climate. $PM_{2.5}$ is directly emitted but also formed in the atmosphere from emissions of precursor gases. Policies impact $PM_{2.5}$ concentrations by changing emissions and, over the long term, climate. $PM_{2.5}$ concentrations are related to exposures and impacts on human health, ecosystems, visibility and short-term climate forcing.

Ambient $PM_{2.5}$ concentrations can be monitored at surface locations and also estimated from observations from satellite-based instruments. The best characterization of the distribution of $PM_{2.5}$ concentrations is developed by combining information from surface observations, satellite observations and computer models. The resulting concentration field can be combined with the population distribution to estimate the population-weighted annual average.

Exposure to $PM_{2.5}$ concentrations leads to a variety of human health impacts, including premature mortality. Weighting the average concentration by population distribution creates an indicator that is focused on the concentrations to which people are exposed.

Emissions of $PM_{2.5}$ and its precursors come from a wide variety of anthropogenic sources, including electricity generation, transportation, residential combustion, industrial processes and agricultural burning. These sources can be managed using a wide variety of policy approaches.

Other factors

$PM_{2.5}$ concentrations are also affected by weather, wildfires, windblown dust and volcanoes. Weather is subject to significant inter-annual variability, decadal cycles and long-term trends. The contributions of wildfires, windblown dust and volcanoes also vary from year to year. These influences must be accounted for when attributing observed trends to the impact of control policies.

Observed trends can be quantitatively apportioned to changes in anthropogenic emissions, natural emissions and weather, using computer models of atmospheric chemistry and dynamics. However, the uncertainty in model estimates can be as large as the impact of a change in policy.

Possible alternatives

$PM_{2.5}$ concentrations are well-accepted metrics for air pollution. WHO has established a guideline value and interim targets for maximum annual average $PM_{2.5}$ concentrations. Under the SDGs, the population-weighted annual mean level of $PM_{2.5}$ or PM_{10} was selected as an indicator of progress towards sustainable cities. Weighting concentrations by the population exposed focuses the metric on the overall impact on the population. However, a population weighted average may mask the number of people exposed to the most extreme air pollution levels. Thus, absolute concentrations provide another alternative metric.

Direct emissions of $PM_{2.5}$ and its gaseous precursors (including SO_2, NO_X and NH_3) are alternative indicators. Emissions are more directly affected by policy changes, but more distantly related to health impacts. For many sources, measurement of emissions is impractical, and estimates are highly uncertain.

12.3.2 Indicator 2: Emissions of ozone-depleting substances

Figure 12.3: Ozone-depleting substance consumption in ozone depletion tons in 2016

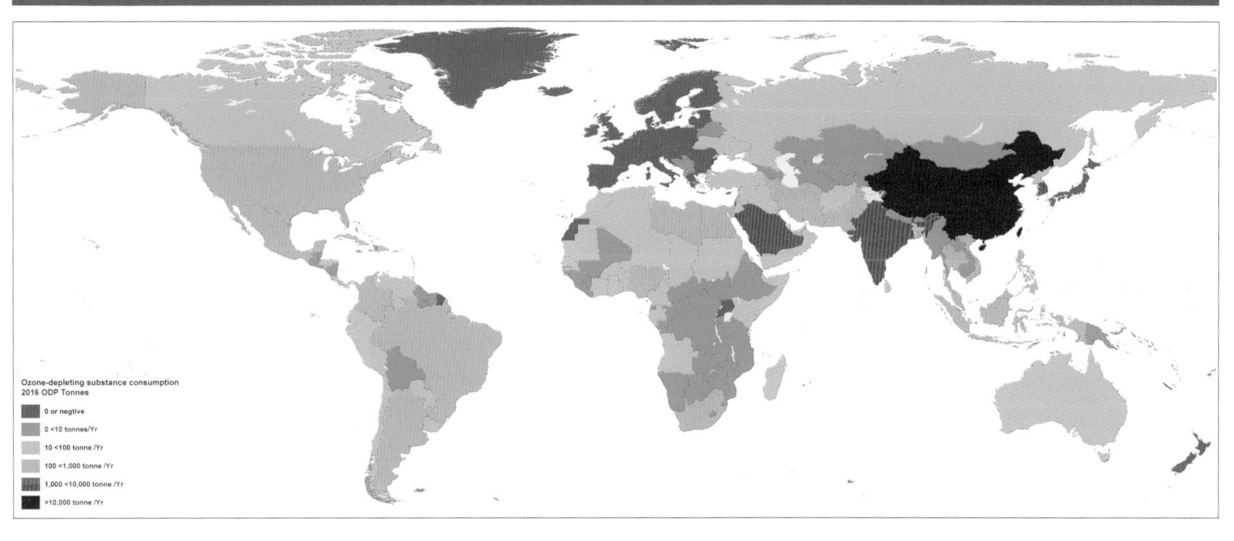

Source: UNEP (2017)

Policy relevance/causal chain

The manufacture and use of ODSs lead to emissions that affect the concentration of ozone in the stratosphere and, as a result, the level of ultraviolet (UV) radiation reaching the Earth's surface. Increases in UV exposure adversely affect human health and ecosystems. ODSs vary in the extent to which they deplete stratospheric ozone. The effectiveness of policies to limit the manufacture and use of ODSs can be measured using ODS emissions estimates weighted by the ozone-depleting potential (ODP) of each compound (2014).ODP-weighted ODS emissions are reported to the Secretariat of the Vienna Convention and Montreal Protocol.

Other factors

ODS emissions are directly linked to the effectiveness of policies to eliminate their manufacture, use and improper disposal.

The relationship between ODS emissions, stratospheric ozone concentrations, UV exposures and health effects can be estimated using models which account for atmospheric chemistry and dynamics, exposure behaviours and population characteristics. These model estimates can be evaluated by comparing them to observed ozone concentrations, UV levels and disease incidence rates.

Possible alternatives

Other metrics have been used to gauge the success of efforts to protect stratospheric ozone, including the minimum observed stratospheric ozone concentration, changes in UV radiation levels and the spatial extent of the Antarctic ozone 'hole'. These metrics are influenced by ODS policies over the long term, but also by inter-annual variability, decadal climate cycles and long-term climate change.

ODS emissions could be compared on a per capita or per gross domestic product (GDP) basis, each of which implies a different assumption about what constitutes an equitable distribution of the burden of additional controls.

12.3.3 Indicator 3: Anthropogenic emissions of long-lived greenhouse gases (CO_2 equivalents)

Figure 12.4: National total GHG emissions in 2014 in $MtCO_2e$, including land-use change and forestry sources and sinks

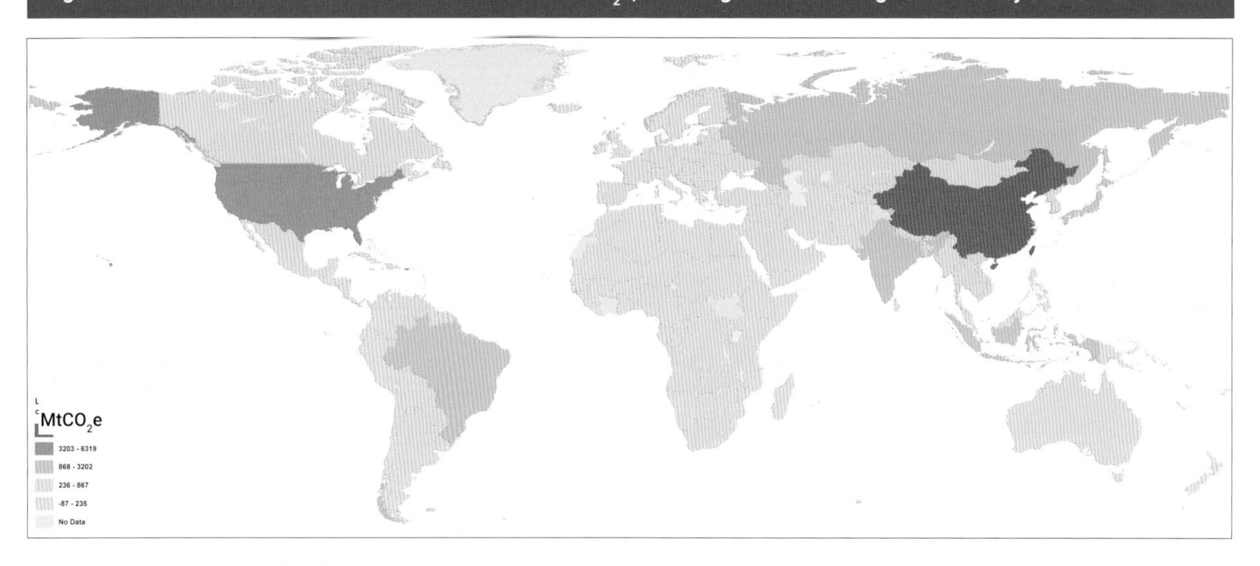

Source: World Resources Institute (2017). Climate Analysis Indicators Tool: WRI's Climate Data Explorer. Washington, DC: World Resources Institute. Available at http://cait.wri.org/historical/. Note these are derived from several sources including Food and Agriculture Organization of the United Nations [FAO] 2016; IEA 2016; UNFCCC 2017b).

Policy relevance/causal chain

Long-term climate change is primarily driven by emissions of GHGs, including CO_2, methane (CH_4), nitrous oxide (N_2O) and fluorinated gases. The contribution of each of these pollutants to climate change can be compared in terms of CO_2 equivalents using the Global Warming Potential (GWP) index, which accounts for the pollutants' different atmospheric lifetimes. GWPs are calculated for a specific time-horizon, with 100 years being the most common for comparing the impact of long-lived GHGs.

Anthropogenic emissions of GHGs come from a wide range of sources but are primarily associated with the production and consumption of energy, land-use change and deforestation. Different types of policies can directly or indirectly affect energy production and consumption behaviours and technologies and thus the generation of GHG emissions. Anthropogenic emissions are quantified and reported at the national level to the UNFCCC and provide an aggregate indicator of the effectiveness of policies. Annual emissions data from 1990 to the present are available from the UNFCCC for industrialized countries or economies in transition (called Annex I Parties under the Convention). Under the Paris Agreement, all countries are required to submit emissions inventories every two years.

Other factors

In addition to policies, anthropogenic GHG emissions are also affected by economic, social and technological trends. GHG emissions are also generated by natural sources. Both natural and anthropogenic sources are affected by natural meteorological and climate variability.

The relationship between GHG emissions, radiative forcing, changes in climate and climate variability and downstream impacts on human health, ecosystems and infrastructure can be estimated using Earth system models that must account for a significant number of processes and feedbacks.

GHG emission estimates can be compared to observations of GHG concentrations from *in situ* measurements and satellite-borne instruments, although it can be difficult to attribute small changes in specific anthropogenic sources to changes in the observed distribution.

Possible alternatives

GWP-weighted long-lived GHG emissions are well-accepted metrics of climate policy effectiveness under the UNFCCC.

There are long lag times between changes in emissions and changes in atmospheric concentrations, radiative forcing or climate state variables (temperatures, precipitation etc.), which make these state variables less useful for assessing policy effectiveness.

Emissions of short-lived climate pollutants (such as black carbon and hydrofluorocarbons) could be included in the aggregation, and other indices (such as a Global Temperature Potential (GTP) or a GWP over a 20-year horizon) could be used to weight the contribution of different pollutants.

GHG emissions could be compared on a per capita or per GDP basis, each of which implies a different assumption about what constitutes an equitable distribution of the burden of additional controls.

12.4 Discussion and conclusions

A wide variety of policy approaches, including but not limited to planning regimes, emissions and technology standards, market interventions, public information and international cooperation, have been applied to the problems of air pollution, climate change, ODSs and PBTs. Lessons can be learned about each type of policy approach from applications to the four different problems at different geographical scales.

One lesson is that policy approaches must be adapted to specific contexts. There is no single model policy that is most appropriate for all settings. High-income countries rely on information-rich planning regimes and regulatory approaches backed by government enforcement capacity. These approaches may not be the most appropriate for settings where information is poor and enforcement capacity is lacking. In such settings, voluntary standards, market interventions and public information may prove more effective in decreasing emissions and hazardous exposures. To improve the effectiveness of such attempts to strengthen climate finance and reduce air pollution, development assistance will play a crucial role in capacity-building and green economy development. Capacity-building should focus on strengthening the technical and planning capabilities at local and national levels that are most relevant for anticipating the potential impacts of climate change and developing appropriate policy responses. Air quality measures need to be combined with climate and energy measures, agricultural policy, transport policy and urban planning, with a focus on improving health and biodiversity. A key message and challenge is how to ensure that climate policies do not increase health risks (e.g. from biomass burning and diesel) and that air quality policy is climate neutral. Also, it is imperative to consolidate a multi-scale governance approach that aligns international, national and local actions (Maas and Grennfelt eds. 2016).

References

Air Quality in Scotland (2018). *Know & respond - Scotland, the free air pollution alert messaging system*. http://www.scottishairquality.co.uk/know-and-respond/ (Accessed: 4 November 2018).

Apte, K. and Salvia, S. (2016). Household air pollution and its effects on health. *F1000Research* 5(2593). https://doi.org/10.12688/f1000research.7552.1.

Asia Pacific Clean Air Partnership (2015). *Asia Pacific Clean Air Partnership: Promoting Better Air Quality in Asia Pacific*. http://staging.unep.org/documents/APCAP%20Brochure%208%20October%20 2015.pdf.

Åström, S., Yaramenka, K., Winnes, H., Fridell, E. and Holland, M. (2018). The costs and benefits of a nitrogen emission control area in the Baltic and North Seas. *Transportation Research Part D: Transport and Environment* 59, 223-236. https://doi.org/10.1016/j.trd.2017.12.014.

Auerbach, I.L. and Flieger, K. (1967). The importance of public education in air pollution control. *Journal of the Air Pollution Control Association* 17(2), 102-104. https://doi.org/10.1080/00022470.1 967.10468947.

Blanken, P.D., Dillon, J. and Wismann, G. (2001). The impact of an air quality advisory program on voluntary mobile source air pollution reduction. *Atmospheric Environment* 35(13), 2417-2421. https://doi.org/10.1016/S1352-2310(00)00523-9.

Bond, T.C., Doherty, S.J., Fahey, D.W., Forster, P.M., Berntsen, T., DeAngelo, B.J. *et al.* (2013). Bounding the role of black carbon in the climate system: A scientific assessment. 118(11), 5380-5552. https://doi.org/10.1002/jgrd.50171.

Carbon Market Data (2018). *World carbon market database: A unique data platform covering the world's carbon trading markets*. https://www.carbonmarketdata.com/en/products/world-ets-database/presentation (Accessed: 4 November 2018).

Cepeda, M., Schoufour, J., Freak-Poli, R., Koolhaas, C.M., Dhana, K., Bramer, W.M. *et al.* (2017). Levels of ambient air pollution according to mode of transport: A systematic review. *The Lancet Public Health* 2(1), e23-e34. https://doi.org/10.1016/S2468-2667(16)30021-4.

European Commission (2015a). *Cost-benefit analysis of the CAFE programme*. http://ec.europa.eu/environment/archives/cafe/activities/cba.htm (Accessed: 4 November 2018).

European Commission (2015b). *Evaluation of Regulations 443/2009 and 510/2011 on CO_2 Emissions from Light-Duty Vehicles. Final Report*. https://ec.europa.eu/clima/sites/clima/files/transport/vehicles/docs/evaluation_ldv_co2_regs_en.pdf.

European Commission (2018a). *EU emissions trading system (EU ETS)*. https://ec.europa.eu/clima/policies/ets_en (Accessed: 10 October 2018).

European Commission (2018b). *The industrial emissions directive: Summary of directive 2010/75/EU on industrial emissions (integrated pollution prevention and control)*. http://ec.europa.eu/environment/industry/stationary/ied/legislation.htm (Accessed: 4 November 2018).

European Environment Agency (2015). *Evaluating 15 Years of Transport and Environmental Policy Integration Term 2015: Transport Indicators Tracking Progress Towards Environmental Targets in Europe*. Copenhagen. https://www.eea.europa.eu/publications/term-report-2015/download.

European Environment Agency (2016a). *Air Quality in Europe: 2016 Report*. Copenhagen. https://www.eea.europa.eu/publications/air-quality-in-europe-2016/at_download/file.

European Environment Agency (2016b). *Explaining Road Transport Emissions: A NonTechnical Guide*. Copenhagen. https://www.eea.europa.eu/publications/explaining-road-transport-emissions/at_download/file.

European Environment Agency (2017). *Air quality in Europe: 2017 report*. Copenhagen. https://www.eea.europa.eu/publications/air-quality-in-europe-2017/at_download/file.

European Union 2015/1185 (2015). Directive 2009/125/EC of the European Parliament and of the Council with Regard to Ecodesign Requirements for Solid Fuel Local Space Heaters Union, E., 1-19 https://eur-lex.europa.eu/legal-content/EN/TXT/PDF/?uri=CELEX:32015R1185&from=EN

European Union (2016). Commission Regulation (EU) 2016/427 of 10 March 2016 amending Regulation (EC) No 692/2008 as Regards Emissions from Light Passenger and Commercial Vehicles (Euro 6) (Text with EEA relevance). https://eur-lex.europa.eu/legal-content/EN/TXT/?uri=CELEX%3A32016R0427

Food and Agriculture Organization of the United Nations (2016). *FAOSTAT: Food and Agriculture Data*. http://www.fao.org/faostat/en/#data.

Franco, V., Sánchez, F.P., German, J. and Mock, P. (2014). *Real-World Exhaust Emissions from Modern Diesel Cars - A Meta-Analysis of PEMS Emissions Data from EU (Euro 6) and US (Tier 2 Bin 5/ULEV II) Diesel Passenger Cars. Part 1: Aggregated Results*. Berlin: International Council on Clean Transportation. https://www.theicct.org/sites/default/files/publications/ICCT_PEMS-study_diesel-cars_20141010.pdf.

Gignac, R. and Matthews, H.D. (2015). Allocating a 2 °C cumulative carbon budget to countries. *Environmental Research Letters* 10(7), 075004. https://doi.org/10.1088/1748-9326/10/7/075004.

Global Alliance for Clean Cookstoves (2012a). *East Africa Regional Analysis*. Washington, D.C: Global Alliance for Clean Cookstoves. http://cleancookstoves.org/binary-data/RESOURCE/file/000/000/158-1.pdf.

Global Alliance for Clean Cookstoves (2012b). *Global Alliance for Clean Cookstoves: Kenya Market Assessment. Sector Mapping*. http://cleancookstoves.org/resources_files/kenya-market-assessment-mapping.pdf.

Global Alliance for Clean Cookstoves (2013). *Kenya Country Action Plan (CAP) 2013*. http://cleancookstoves.org/resources_files/kenya-country-action-plan.pdf.

Global Alliance for Clean Cookstoves (2016). *Kenya: Overview and country statistics*. http://cleancookstoves.org/country-profiles/focus-countries/4-kenya.html. (Accessed: 4 November 2018).

Global Commons Institute (2018). *Contraction and convergence: Climate truth and reconciliation*. http://www.gci.org.uk/endorsements_UK_Climate_Act.html (Accessed: 10 October 2018).

Gordon, B.A., Mackay, R. and Rehfuess, E. (2004). *Inheriting the World: The Atlas of Children's Health and the Environment*. Geneva: World Health Organization. http://www.who.int/ceh/publications/en/atlas.pdf?ua=1.

Green Climate Fund (2018). *Concept Note: Promotion of Climate-Friendly Cooking: Bangladesh, Kenya and Senegal*. https://www.greenclimate.fund/documents/20182/893456/19770_-_Promotion_of_Climate-Friendly_Cooking_Bangladesh__Kenya_and_Senegal.pdf.

Grice, S., Stedman, J., Kent, A., Hobson, M., Norris, J., Abbott, J. *et al.* (2009). Recent trends and projections of primary NO_2 emissions in Europe. *Atmospheric Environment* 43(13), 2154-2167. https://doi.org/10.1016/j.atmosenv.2009.01.019.

Guerreiro, C., Horálek, J., de Leeuw, F., Hak, C., Nagl, C., Kurfürst, P. *et al.* (2010). *Status and Trends of NO_2 Ambient Concentrations in Europe*. ETC/ACC Technical Paper. Bilthoven: European Topic Centre on Air and Climate Change. https://www.researchgate.net/profile/Cristina_Guerreiro/publication/260600537_Status_and_trends_of_NO2_ambient_concentrations_in_Europe/links/5433a1070cf22395f29e1e57/Status-and-trends-of-NO2-ambient-concentrations-in-Europe.pdf.

Hans, N.J. (2018). *Study on economic loss from Indonesia's peat policies criticized*. Mongabay https://news.mongabay.com/2018/01/study-on-economic-loss-from-indonesias-peat-policies-criticized/ (Accessed: 4 November 2018).

Haze Action Online (2017). *Information on fire and haze*. Association of South East Asian Nations http://haze.asean.org/%20about-us/information-on-fire-and-haze/ (Accessed: 4 November 2018).

Haze Action Online (2018). *Haze hotspot map today*. Association of South East Asian Nations https://haze.asean.org/ (Accessed: 4 November 2018).

Health Effects Institute (2017). *State of Global Air 2017: A Special Report on Global Exposure to Air Pollution and its Disease Burden*. Boston, MA. https://www.stateofglobalair.org/sites/default/files/SoGA2017_report.pdf.

Henry, G.T. and Gordon, C.S. (2003). Driving less for better air: Impacts of a public information campaign. *Journal of Policy Analysis and Management* 22(1), 45-63. https://doi.org/10.1002/pam.10095

Hotten, R. (2015). Volkswagen: The scandal explained. British Broadcasting Corporation https://www.bbc.com/news/business-34324772.

Intergovernmental Panel on Climate Change (2014). *Climate Change 2014: Synthesis Report. Contribution of Working Groups I, II and III to the Fifth Assessment Report of the Intergovernmental Panel on Climate Change*. Pachauri, R.K., Meyer, L.A. Geneva. https://www.ipcc.ch/pdf/assessment-report/ar5/syr/AR5_SYR_FINAL_All_Topics.pdf.

International Energy Agency (2016). *CO_2 Emissions from Fuel Combustion 2016*. Paris. https://www.oecd-ilibrary.org/docserver/co2_fuel-2016-en.pdf?expires=1541330155&id=id&accname=ocid195767&checksum=B4C10B7629C3B96202B183CE7A93E533.

International Energy Agency (2017). *Energy Balances of OECD Countries*. https://www.iea.org/statistics/relateddatabases/energybalancesofoecdcountries/.

International Maritime Organization (2018). *Prevention of air pollution from ships*. http://www.imo.org/en/OurWork/Environment/PollutionPrevention/AirPollution/Pages/Air-Pollution.aspx (Accessed: 4 November 2018).

Kansas Department of Health and Environment (2006). *Environmental Factors, Outdoor Air Quality, and Activity Level: Results from 2005 Kansas Behavioral Risk Factor Surveillance System*. Topeka, KS. http://www.kdheks.gov/brfss/PDF/cste_report_final.pdf.

Kapfudzaruwa, F., Fay, J. and Hart, T. (2017). Improved cookstoves in Africa: Explaining adoption patterns. *Development Southern Africa* 34(5), 548-563. https://doi.org/10.1080/037683 5X.2017.1335592.

Kelly, F.J. and Fussell, J.C. (2015). Air pollution and public health: Emerging hazards and improved understanding of risk. *Environmental geochemistry and health* 37(4), 631-649. https://doi.org/10.1007/s10653-015-9720-1.

Kenya, Ministry of Environment andv Natural Resources, (2017). *Nationally Determined Contribution Sector Analysis Report: Evidence Base for Updating Kenya's National Climate Change Action Plan*. Nairobi: Ministry of Environment and Natural Resources. http://www.starckplus.com/documents/ta/ndc/NDC%20Sector%20Analysis%20Report%202017.pdf.

Kenya, Ministry of Energy and Petroleum and Sustainable Energy for All (2016). *Kenya Action Agenda*. https://www.seforall.org/sites/default/files/Kenya_AA_EN_Released.pdf.

Kenya, Ministry of Energy and Petroleum (2015). *Draft Strategy and Action Plan for Bioenergy and LPG Development in Kenya*. Nairobi. https://kcpsa.or.ke/download/draft-strategy-and-action-plan-for-bioenergy-and-lpg-development-in-kenya/?wpdmdl=12841.

Koplitz, S.N., Mickley, L.J., Marlier, M.E., Buonocore, J.J., Kim, P.S., Liu, T. *et al.* (2016). Public health impacts of the severe haze in Equatorial Asia in September-October 2015: Demonstration of a new framework for informing fire management strategies to reduce downwind smoke exposure. *Environmental Research Letters* 11(9). https://doi.org/10.1088/1748-9326/11/9/094023.

Lambe, F., Jürisoo, M., Lee, C. and Johnson, O. (2015). Can carbon finance transform household energy markets? A review of cookstove projects and programs in Kenya. *Energy Research & Social Science* 5, 55-66. https://doi.org/10.1016/j.erss.2014.12.012.

Landrigan, P.J. (2017). Air pollution and health. *The Lancet Public Health* 2(1), e4-e5. https://doi.org/10.1016/S2468-2667(16)30023-8.

Lewis, A. and Edwards, P. (2016). Validate personal air-pollution sensors. *Nature* 535(7610). https://doi.org/10.1038/535029a.

Lim, S.S., Vos, T., Flaxman, A.D., Danaei, G., Shibuya, K., Adair-Rohani, H. *et al.* (2012). A comparative risk assessment of burden of disease and injury attributable to 67 risk factors and risk factor clusters in 21 regions, 1990–2010: A systematic analysis for the global burden of disease study 2010. *The Lancet* 380(9859), 2224-2260. https://doi.org/10.1016/S0140-6736(12)61766-8.

Lovell, H., Bulkeley, H. and Owens, S. (2009). Converging agendas? Energy and climate change policies in the UK. *Environment and Planning C: Government and Policy* 27(1), 90-109. https://doi.org/10.1068/c0797j.

Maas, R. and Grennfelt, P. (2016). *Towards Cleaner Air. Scientific Assessment Report 2016. EMEP Steering Body and Working Group on Effects of the Convention on Long-Range Transboundary Air Pollution*. Oslo. http://www.unece.org/fileadmin/DAM/env/documents/2016/AIR/Publications/CLRTAP_Scientific_Assessment_Report_-_Final_20-5-2016.pdf.

Malaysian Nature Society Science and Conservation Unit (1997). The haze thing. *Malaysian Naturalist* 51(2), 18-21.

Mansfield, C. and Corey, C. (2003). *Task 4: Analysis of Survey Data on Ozone Alert Days: Final Report*. Research Triangle Park, NC: United States Environmental Protection Agency. https://www.rti.org/sites/default/files/resources/rti-publication-file-83190b2b-f906-414b-9709-8bf84f88b367.pdf.

Mansfield, C., Johnson, F.R. and van Houtven, G. (2006). The missing piece: Valuing averting behavior for children's ozone exposures. *Resource and Energy Economics* 28(3), 215-228. https://doi.org/10.1016/j.reseneeco.2006.02.002.

Mansfield, C., Sinha, P. and Henrion, M. (2009). *Influence Analysis in Support of Characterizing Uncertainty in Human Health Benefits Analysis*. United States Environmental Protection Agency. https://www3.epa.gov/ttnecas1/regdata/Benefits/influence_analysis_final_report_psg.pdf.

McDermott, M., Srivastava, R. and Croskell, S. (2006). Awareness of and compliance with air pollution advisories: A comparison of parents of asthmatics with other parents. *Journal of Asthma* 43(3), 235-239. https://doi.org/10.1080/02770900600567114.

Melamed, J.R., Riley, R.S., Valcourt, D.M. and Day, E.S. (2016). Using gold nanoparticles to disrupt the tumor microenvironment: An emerging therapeutic strategy. *ACS Nano* 10(12), 10631-10635. https://doi.org/10.1021/acsnano.6b07673.

Myhre, G., Samset, B.H., Schulz, M., Balkanski, Y., Bauer, S., Berntsen, T.K. *et al.* (2013). Radiative forcing of the direct aerosol effect from AeroCom Phase II simulations. *Atmospheric Chemistry and Physics* 13, 1853–1877. https://doi.org/10.5194/acp-13-1853-2013.

Natural Capital Partners (2018). *Kenya improved cookstoves*. Natural Capital Partners https://www.naturalcapitalpartners.com/projects/project/kenya-improved-cookstoves (Accessed: 1 May 2018).

Nazeer, N. and Furuoka, F. (2017). Overview of ASEAN environment transboundary haze pollution agreement and public health. *International Journal of Asia Pacific Studies* 13(1), 73-94. https://doi.org/10.21315/ijaps2017.13.1.4.

Neidell, M.J. (2008). *Information, Avoidance Behavior, and Health: The Effect of Ozone on Asthma hospitalizations*. Cambridge, MA National Bureau of Economic Research. https://www.nber.org/papers/w14209.pdf.

Pilishvili, T., Loo, J.D., Schrag, S., Stanistreet, D., Christensen, B., Yip, F. *et al.* (2016). Effectiveness of six improved cookstoves in reducing household air pollution and their acceptability in rural Western Kenya. *PloS one* 11(11), e0165529. https://doi.org/10.1371/journal.pone.0165529.

Qadri, S.T. (ed.) (2001). *Fire, Smoke, and Haze: The ASEAN Response Strategy*. Manila: Asian Development Bank. https://www.adb.org/sites/default/files/publication/28035/fire-smoke-haze.pdf.

Republic of Indonesia (2015). *Intended Nationally Determined Contribution: Republic of Indonesia*. Jakarta. http://www4.unfccc.int/Submissions/INDC/Published%20Documents/Indonesia/1/INDC_REPUBLIC%20OF%20INDONESIA.pdf.

Semenza, J.C., Hall, D.E., Wilson, D.J., Bontempo, B.D., Sailor, D.J. and George, L.A. (2008). Public perception of climate change. *American Journal of Preventive Medicine* 35(5), 479-487. https://doi.org/10.1016/j.amepre.2008.08.020.

Shaddick, G., Thomas, M.L., Green, A., Brauer, M., Donkelaar, A., Burnett, R. *et al.* (2018). Data integration model for air quality: A hierarchical approach to the global estimation of exposures to ambient air pollution. *Journal of the Royal Statistical Society* 67(1), 231-253. http://doi.org/10.1111/rssc.12227.

Silk, B.J., Sadumah, I., Patel, M.K., Were, V., Person, B., Harris, J. *et al.* (2012). A strategy to increase adoption of locally-produced, ceramic cookstoves in rural Kenyan households. *BMC Public Health* 12(359), 1-10. https://doi.org/10.1186/1471-2458-12-359.

Stockwell, C.E., Jayarathne, T., Cochrane, M.A., Ryan, K.C., Putra, E.I., Saharjo, B.H. *et al.* (2016). Field measurements of trace gases and aerosols emitted by peat fires in central Kalimantan, Indonesia. *Atmospheric Chemistry and Physics* 16, 11711-11732. https://doi.org/10.5194/acp-16-11711-2016.

Sutton, M., Reis, S. and Baker, S. (eds.) (2009). *Atmospheric Ammonia: Detecting Emission Changes and Environmental Impacts. Results of an Expert Workshop Under the Convention on Long-Range Transboundary Air Pollution*. 1st edn: Springer. https://www.springer.com/gp/book/9781402091209.

Thompson, J.E. (2016). Crowd-sourced air quality studies: A review of the literature & portable sensors. *Trends in Environmental Analytical Chemistry* 11, 23-34. https://doi.org/10.1016/j.teac.2016.06.001.

United Kingdom, Department of Trade and Industry, (2007). *Meeting the Energy Challenge: A White Paper on Energy*. London. https://assets.publishing.service.gov.uk/government/uploads/system/uploads/attachment_data/file/243268/7124.pdf.

United Kingdom, Department for Business Energy & Industrial Strategy, (2016). *Guidance: Carbon budgets*. https://www.gov.uk/guidance/carbon-budgets (Accessed: 4 November 2018).

United Kingdom, Committee on Climate Change, (2017). *Meeting Carbon Budgets: Closing the Policy Gap. 2017 Report to Parliament*. London. https://www.theccc.org.uk/wp-content/uploads/2017/06/2017-Report-to-Parliament-Meeting-Carbon-Budgets-Closing-the-policy-gap.pdf.

United Kingdom, Department of Environment Food and Environmental Affairs, (2018). *Pollution forecast*. https://uk-air.defra.gov.uk/forecasting/ (Accessed: 5 November 2018).

United Kingdom Government (2008a). *Carbon Plan*. London. https://www.gov.uk/government/uploads/system/uploads/attachment_data/file/47621/1358-the-carbon-plan.pdf.

United Kingdom Government (2008b). Climate Change Act 2008. 108 1 http://www.legislation.gov.uk/ukpga/2008/27/pdfs/ukpga_20080027_en.pdf

United Nations (2015). *Sustainable development goals: 17 goals to transform our world*. https://www.un.org/sustainabledevelopment/home/ (Accessed: 4 November 2018).

United Nations Economic Commission for Europe (2018). *Gothenburg protocol: Guidance documents and other methodological materials for the implementation of the 1999 protocol to abate acidification, eutrophication and ground-level ozone*. http://www.unece.org/environmental-policy/conventions/envlrtapwelcome/guidance-documents-and-other-methodological-materials/gothenburg-protocol.html (Accessed: 4 November 2018).

United Nations Environment Programme (2010). *Air Pollution: Promoting Regional Cooperation*. Nairobi. http://www.rrcap.ait.asia/Publications/Air_pollution_book.pdf.

United Nations Environment Programme (2017). *Ozone-Depleting Substance Consumption: Country Data*. Ozone Secretariat, Nairobi. http://ozone.unep.org/countries/data

United Nations Framework Convention on Climate Change (2017a). *China to launch world's largest emissions trading system*. https://unfccc.int/news/china-to-launch-world-s-largest-emissions-trading-system (Accessed: 10 October).

United Nations Framework Convention on Climate Change (2017b). *GHG data from UNFCCC*. https://unfccc.int/process/transparency-and-reporting/greenhouse-gas-data/ghg-data-unfccc (Accessed: 4 November 2018).

United States Agency for International Development and Winrock International (2011). *Clean and Efficient Cooking Technologies and Fuels: The Fuel-Efficient Cookstoves and Clean Fuels Sector is Evolving Rapidly, Enabling Cross-Cutting Solutions that can Achieve Greater Development Impacts*. United States Agency for International Development and Winrock International. https://www.winrock.org/wp-content/uploads/2017/09/Winrock_Cookstove_final_reduced.pdf.

United States Environmental Protection Agency (2016). *NAAQS table*. https://www.epa.gov/criteria-air-pollutants/naaqs-table (Accessed: 10 October 2018).

United States Environmental Protection Agency (2017). *RACT/BACT/LAER clearinghouse (RBLC)*. https://cfpub.epa.gov/RBLC/ (Accessed: 10 October 2018).

Velasco, E. and Rastan, S. (2015). Air quality in Singapore during the 2013 smoke-haze episode over the strait of Malacca: Lessons learned *Sustainable Cities and Society* 17, 122-131. https://doi.org/10.1016/j.scs.2015.04.006.

World Bank (2016). *The Cost of Fire: An Economic Analysis of Indonesia's 2015 Fire Crisis*. Washington, D.C. http://documents.worldbank.org/curated/en/776101467990969768/pdf/103668-BRI-Cost-of-Fires-Knowledge-Note-PUBLIC-ADD-NEW-SERIES-Indonesia-Sustainable-Landscapes-Knowledge-Note.pdf.

World Health Organization (2016). *Ambient Air Pollution: A Global Assessment of Exposure and Burden of Disease*. Geneva. http://www.who.int/iris/bitstream/10665/250141/1/9789241511353-eng.pdf?ua=1.

World Health Organization (2018). *WHO outdoor air quality guidelines*. http://www.euro.who.int/en/health-topics/environment-and-health/air-quality/policy/who-outdoor-air-quality-guidelines (Accessed: 10 October 2018).

World Resources Institute (2017). *Climate analysis indicators tool: WRI's climate data explorer*. http://cait.wri.org/historical/Country%20GHG%20Emissions?indicator[]=Total%20GHG%20Emissions%20Excluding%20Land-Use%20Change%20and%20Forestry&indicator[]=Total%20GHG%20Emissions%20Including%20Land-Use%20Change%20and%20Forestry&year[]=2014&sortIdx=NaN&chartType=geo (Accessed: 4 November 2018).

World Wide Fund for Nature (2018). *Forests and Sustainable Development: The Role of SDG 15 in Delivering the 2030 Agenda*. Gland. http://d2ouvy59p0dg6k.cloudfront.net/downloads/wwf_forest_practice_report_hlpf_2018__forests_and_sustainable_development___the_role_of_.pdf.

Biodiversity Policy

Coordinating Lead Author: Diana Mangalagiu (University of Oxford and Neoma Business School)

Lead Authors: Nibedita Mukherjee (University of Cambridge and University of Exeter), Dolors Armenteras Pascual (Universidad Nacional de Colombia), Jonathan Davies (University of British Columbia), Leandra Regina Gonçalves (University of Campinas), Jeremy Hills (University of the South Pacific), Louise McRae (Zoological Society of London), Peter Stoett (University of Ontario Institute of Technology)

Contributing Authors: Irene Dankelman (Radboud University), Souhir Hammami (Freie universität Berlin), Caradee Wright (South African Medical Research Council), Carol Zastavniouk (Golder Associates Ltd)

Executive summary

Biodiversity is a key component of a healthy planet with healthy people (*well established*)**.** Though evidence regarding the importance of biodiversity for economic output, health and security has grown significantly in the last two decades, it is certain that existing measures to conserve and sustainably manage biodiversity are inadequate {Box 13.1, Section 6.1, 13.1}.

Policy instruments working in silos are insufficient to stem biodiversity loss *(well established)*. Instead, multiple approaches that embrace a diversity of instruments and scales, including platforms for encouraging behaviour change, are vital {13.1, 13.2.3}.

The cost of inaction (societal and economic) for biodiversity conservation and restoration is extremely high, as biodiversity loss is largely irreversible *(established but incomplete)* {13.1, 13.2.1, 13.2.4}.

There is an urgent need to act now and strengthen policy responses for conserving biodiversity and invest in capacity-building and institutional infrastructure to reach the Aichi targets and Sustainable Development Goals *(well established)* {13.3, 13.4.2}.

Current valuation methods are not adequate to account for the negative impacts of biodiversity loss *(well established)*. Developing appropriate valuation metrics and methods to make the multiple values of biodiversity understandable to decision makers is urgently needed (e.g. natural capital accounts). Such valuation techniques should consider the full natural capital value at the national level and integrate it into their National Biodiversity Strategic Action Plans {13.2.4}.

Mainstreaming biodiversity should be promoted by all stakeholders, including governments and the private sector, across themes such as health, agriculture, social security, trade and education *(well established)* {13.2.2, 13.2.3, 13.2.4}.

There is a lack of baseline information to measure the success or failure of most biodiversity policy and governance interventions *(well established)*. Investing in long-term research programmes would be useful, particularly in biodiverse developing nations, to develop effective baselines. In addition, a well-defined time frame to turn goals into actions will be very likely to be useful for effective conservation policy implementation {13.2}.

Investing in independent monitoring and cost–benefit analysis could help in measuring policy effectiveness *(well established)*. Countries could integrate autonomous monitoring and evaluation in the implementation of programmes to improve effectiveness. As a start, building an evidence base of what works in conservation could be prioritized at a national level {13.2}.

Conservation problems require long-term solutions, while conservation and research funding is usually short term *(well established)*. Addressing this **timescale mismatch** is urgently needed in the design phase of policy interventions {13.2.3}.

Policies and mechanisms need to be in place to support innovative measures to strengthen biodiversity protection. For example, while traditional approaches such as protected areas have been the norm to secure tenure, other forms of arrangements such as community-based protected areas (e.g. Locally Managed Marine Areas) are needed to supplement protected areas for conserving biodiversity in the long term {13.1}.

Economic development is commonly perceived as a threat to biodiversity conservation, but sustainable growth and development of the green economy (low in carbon, resource-efficient and socially inclusive) can also promote and enhance biodiversity *(well established)* {13.2.3, 13.2.5}.

In the policy design phase, adequate attention needs to be paid to equity, gender and health aspects *(well established)*. To deliver desired co-benefits between enhanced biodiversity and other environmental and societal goals, there is a need for scale-up, further innovation and transformation in the approach to biodiversity management. It would also help other sectors achieve their goals through biodiversity conservation. This reflects Sustainable Development Goal (SDG) 17, which calls for the building of partnerships to achieve the SDGs. {4.2, 13.1, 13.2.4}

The astounding wealth of biodiversity that we collectively share is on loan from future generations. To create the future we want, member states, community-based organizations, non-governmental organizations and corporations are urged to create financial and social incentives that enable individuals and policymakers to make decisions that favour the protection and promotion of biodiversity {13.2.2, 13.2.3, 13.2.4}.

13.1 Introduction

Biodiversity is an integral facet for achieving a healthy planet and human well-being (Cardinale *et al.* 2012; World Health Organization [WHO] and Secretariat of the Convention on Biological Diversity [SCBD] 2015). However, the rate of biodiversity loss continues unabated, and it is well known that species extinction risks are increasing over time (see Section 6.1). The estimated annual cost to the global economy from biodiversity loss and loss of ecosystem functions is up to €14 trillion by 2050; this is equivalent to 7 per cent of projected global gross domestic product (GDP) (Braat and ten Brink (eds.) 2008). Another estimate places the global cost from the loss of ecosystem services solely from land-use change at US$4.3–20.2 trillion (in 2007 valuation) per year between 1997 and 2011 (Costanza *et al.* 2014). Though it is impossible to be precise, quantifying the costs of inaction motivates the need for policy action (Braat and ten Brink (eds.) 2008; Oliver *et al.* 2015). In addition, the importance of biodiversity to health in all its dimensions (WHO and SCBD 2015) has emerged in initiatives such as ecosystem approaches to health, Ecohealth, One Health and Planetary Health (see Section 4.2.1). There is a growing focus on interrelationships between the health of humans, domesticated and wild animals and other species in the context of complex social-ecological systems (Charron 2012; Wilcox, Aguirre and Horwitz 2012; WHO and SCBD 2015) **(Box 13.1)**.

Biodiversity loss is a complex issue (see Section 6.1), and biodiversity conservation relies on strategies involving a wide range of policy approaches such as regulatory command and control, economic incentives, supporting investment, the promotion of innovation, enabling actors, capacity-building and goal-setting, among others. The major policy and governance responses include the Convention for Biological Diversity (CBD) (CBD 1992), the Intergovernmental Science-Policy Platform on Biodiversity and Ecosystem Services (IPBES) and protected areas (see Section 6.7). While there are variations in the effectiveness and perceived legitimacy of international environmental agreements (IEAs) and multilateral environmental agreements (MEAs) (see Annex 6-1 for a list of MEAs relevant to biodiversity at the end of this document), they form the basis of global environmental governance and continue to shape governmental behaviour and expectations (Stoett 2012). Biodiversity has more MEAs in place than other environmental policy domains (see Annex 13.1).

Box 13.1: Global recognition of the link between human health and biodiversity.

A joint work programme between CBD and WHO was formally established in 2012 (CBD 2012 (Decision XI/6)). Health was identified as a priority mainstreaming sector at the 13th CBD Conference of the Parties (CoP) in Mexico in December 2016 (CBD 2016a (Decision XIII/3)); a comprehensive decision to integrate biodiversity and health linkages in national policies was also adopted (CBD 2016b (Decision XIII/6)). The United Nations Environment Programme (UNEP) GEO-6, 'Healthy Planet, Healthy People' and the joint WHO-CBD publication 'State of Knowledge Review Connecting Global Priorities' (WHO and SCBD 2015) recognize that human health and biodiversity are inextricably linked. At the 71st World Health Assembly in 2018, biodiversity loss was recognized as a significant human health issue by many Member States. Increasingly, the medical, public health, biodiversity conservation and policy communities are forging new networks and breaking traditional silos, and One Health, Ecohealth and Planetary Health have emerged as animating approaches.

Over the last 10 years, and particularly since the last GEO, awareness about the loss of biodiversity has risen significantly in international policy, health and economic discourse (WHO and SCBD 2015; Jabbour and Flaschsland 2017; World Economic Forum 2018). The most recent developments in the global biodiversity policy and governance landscape are described in Annex 13.2 at the end of this document.

The 196 CBD member states are required to develop National Biodiversity Strategies and Action Plans (NBSAPs) according to Article 6. To date, 190 of the 196 parties (96 per cent) have developed NBSAPs (SCBD 2018a) **(Figure 13.1)**.

Figure 13.1: Cumulative number of countries that have adopted the NBSAPs as of 2018

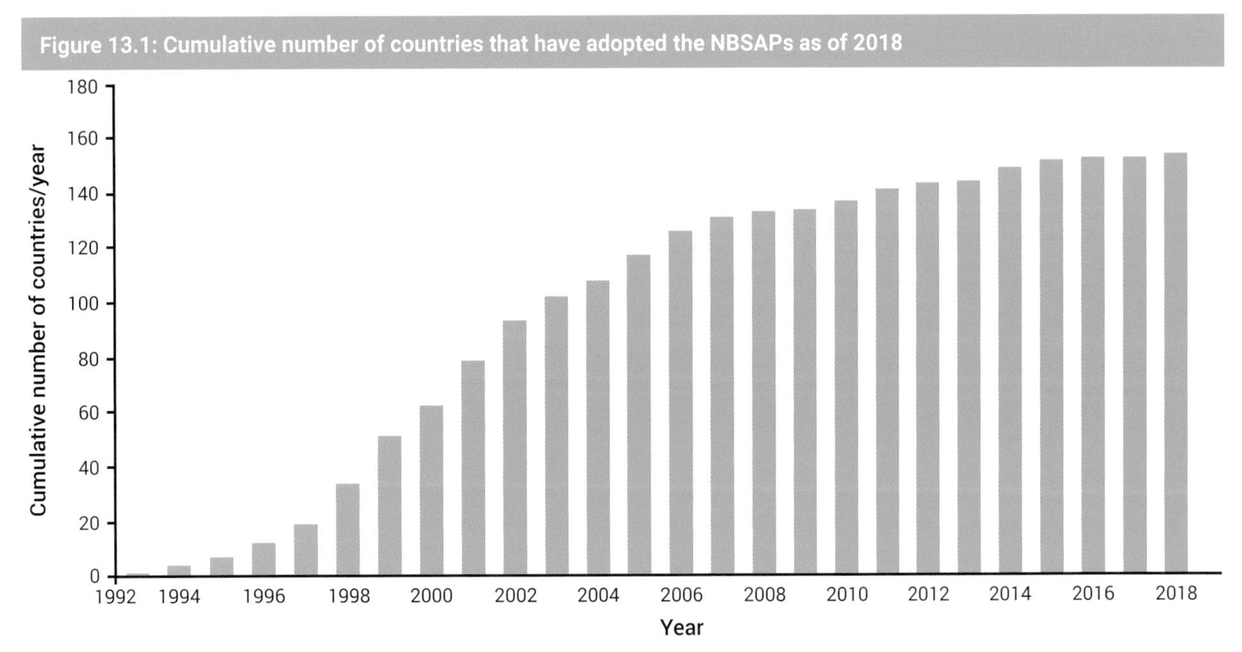

Source: SCBD (2018a)

Moreover, similar to other environmental issues, biodiversity conservation and restoration require the involvement of a range of different stakeholders with often conflicting value positions (Mukherjee *et al.* 2018). Over time, there has been greater recognition of the gender and equity dimensions of conservation policies and their implementation **(Box 13.2)**.

This chapter follows the twofold (top down and bottom up) policy effectiveness assessment framework outlined in Chapter 10.[1]

Key policies and governance approaches (top down): A set of policy clusters and five specific policy instruments pertaining to these clusters is elaborated. Five case studies are drawn as illustrative examples of these policy instruments **(Table 13.1)**. The case studies are not intended to be representative in any form. They were selected to cover the three dimensions of biodiversity (ecosystems, species and genetics), a range of approaches within the typology, geographical spread of examples and varying degrees of success (see Section 10.6). The case studies are drawn from North America, South Asia, Europe, the Pacific and Africa.

Indicators (bottom up): The case studies are followed by a review of three policy-relevant indicators (see Section 10.7), which map the progress towards internationally agreed goals and targets, complementing the policy and governance approaches above.

Box 13.2: Highlights of the gender and equity dimensions in biodiversity policies

Paragraph 13 of the CBD Preamble recognizes gender issues in conservation, and the Subsidiary Body on Scientific, Technical and Technological Advice (SBSTTA) mentions women's practices, knowledge and gender roles in food production (SCBD 2018b). The need for the full and active involvement of relevant stakeholders, including indigenous and local communities, youth, non-governmental organizations (NGOs), women and the business community, is underlined in the Convention.

- The Nagoya Protocol on Access to Genetic Resources and the Fair and Equitable Sharing of Benefits Arising from their Utilization created an international framework that provides concrete measures, rules and procedures.
- Out of the 254 NBSAP reports from 174 countries, 143 reports (56 per cent of all documents) from 107 countries (61 per cent of all countries examined) contain at least one gender keyword; 145 of the 174 countries (83 per cent) identify gender equality as a guiding principle; and 12 per cent have gender equality or women's empowerment as an objective or goal (International Union for Conservation of Nature [IUCN] 2016; SCBD and IUCN 2018).

13.2 Key policies and governance approaches

13.2.1 Enabling actors: Community-based conservation

Engaging local stakeholders through community-based conservation is a central feature of many biodiversity conservation and natural resource management efforts globally to make conservation more effective. Within the Drivers, Pressures, State, Impact, Response (DPSIR) framework (see Figure 1.2 Chapter 1), community-based conservation as a policy approach addresses the drivers, as it counterbalances external resource users who do not have the same cultural and historical attachment to the area.

Protected areas are a key tool for biodiversity conservation. There has been a shift over the last few decades away from exclusive protected areas, where humans were not welcome, towards more people-centred or community-based conservation (Brown 2003; Oldekop *et al.* 2015) and integrated landscape management (Food and Agriculture Organization [FAO] 2018). A nuanced understanding of governance and sociocultural context plays an important role in all types of stakeholder engagement efforts for biodiversity conservation (Bennett *et al.* 2017; Mukherjee *et al.* 2018) and makes those efforts more legitimate, salient, robust and effective (Sterling *et al.* 2017).

Communities are the major players in decision-making in indigenous peoples' and community-conserved territories and areas (ICCAs). ICCAs play a key role in conserving traditional ecological knowledge, cultures and languages, which are often inextricably linked to conservation of biodiversity (Corrigan and Hay-Edie 2013). This role helps in addressing CBD Aichi Biodiversity Target 18, which is aimed at preserving traditional knowledge, innovations and practices of indigenous peoples and local communities and integrating them into biodiversity conservation interventions (ICCA Registry 2018).

Table 13.1: Typology of policy and governance approaches described in this chapter

Policy type /governance approach	Policy instrument(s)	Case study	Spatial scale
Enabling actors	Community-based conservation	Locally managed marine areas, Fiji	National
Command and control	Policing of illegal wildlife trade	Wildlife trafficking and Project Predator, South Asia	Regional
Economic incentive	Payment for ecosystem services	Working for Water programme, South Africa	Subnational
Supporting investment	Banking of genetic material	Svalbard global seed vault, Norway	Global
Enabling actors[1]	Strategic environmental planning	Urban biodiversity in Edmonton, Canada	City

1 The policy type 'enabling actors' has been showcased through two different examples of associated policy instruments.

The case study below on Locally Managed Marine Areas (LMMAs) elaborates one such type of community-based sustainable management in the marine realm.

Case study: Locally Managed Marine Areas in Fiji

Fiji LMMAs are defined as "areas of nearshore waters and coastal resources that are primarily managed at a local level by the coastal communities, land-owning groups, partner organizations, and/or collaborative government representatives who reside or are based in the immediate area" **(Figure 13.2)**. They cover 145 traditionally defined fishing areas (79 per cent of Fiji's inshore fishing areas); the remaining areas permit comparison of the effectiveness of the approach. The LMMA approach is signified by empowered local actors acting at a community scale to sustainably manage inshore resources for mutual community-wide benefit, most commonly through periodically harvested closures (Jupiter *et al.* 2017). After gaining traction in Fiji, the approach was extended further to Melanesia and Polynesia and into Asia through the LMMA network.

Community-based approaches have garnered support because of their adaptability to different contexts and focus on locally identified objectives, negotiated and implemented by stakeholders. Rather than promoting new and alternative visions for serving short-term human needs, community-based approaches such as the LMMA approach are built on refreshing and revitalizing long-standing traditional systems. The non-prescriptive nature of the approach, however, leads to multiple objectives that confound simple measurement of natural resource and biodiversity outcomes (Jupiter *et al.* 2014). Further benefits of the approach could extend to human health through improved food and nutrition security and community integrity, though this has not been documented to date.

The organic expansion of alternative approaches is a positive indication of their effectiveness. Some management tools used with the LMMA approach, such as periodically harvested closures, are not consistently effective for fish biomass and biodiversity outcomes (Jupiter *et al.* 2017). While the approach has spatially expanded dramatically in the absence of any alternative and is currently heavily relied on to achieve conservation and fisheries management outcomes in Fiji, there is no unequivocal evidence at present that it has been completely effective in terms of site-based biological outcomes. The approach has transformative potential, through promoting benefits based on a long-established community system strengthened by coherent resource management approaches.

Figure 13.2: Inshore fishing is an important source of food in Fiji, and many of these inshore areas are under traditional tenure by local communities

As the costs of both inaction and action are predominately borne by the local community, the incentive for progressive transformation is apparent. Attention needs to be paid to the causality between community-based governance arrangements and the effectiveness of conservation efforts (Eklund and Cabeza 2017). Existing analyses of policy effectiveness, such as the 'Protected Planet' report (Bertzky *et al.* 2012; Juffe-Bignoli *et al.* 2014; United Nations Environment Programme World Conservation Monitoring Centre [UNEP-WCMC] and IUCN 2016), could be consulted to identify gaps between policy intent and governance effectiveness.

13.2.2 Command and control policies: Policing of the illegal wildlife trade

Command and control policies (CCPs) are characterized by centralized, often hierarchical and bureaucratic, decision-making structures that have defined jurisdictional authority and less flexibility in implementation compared to economic incentive policies (Cox 2016). CCPs are the most traditional form of regulatory instruments seeking to 'control' activities that could negatively affect biodiversity through penalties, prohibitive rules, enforcement and compliance checks. Typically, national or subnational governments are the decision-making authorities which create the rules and decide how, when and by whom the rules will be implemented (Holling and Meffe 1996). Due to their centralized structure and institutional support, it is easier to evaluate the policy effectiveness of CCPs, especially when the policies have clearly stated objectives and time frames (Gunningham and Young 1997). Therefore, they may be well suited to complex, non-linear issues such as biodiversity loss (e.g. due to ecological tipping points). However, top-down approaches can also present issues of legitimacy, equity and sustainability for local communities (Redpath *et al.* 2017).

Integrating the views of local stakeholders in the decision-making and implementation phases is often key to the success of CCPs (Mukherjee *et al.* 2018). For example, though the European Union (EU)-wide Birds Directive 79/409/EEC (European Council 1979) and the Habitats Directive 92/43/EEC (European Council 1992) engaged several actors in the policy design phase, they are often implemented in an inflexible way at the national level in EU member states (Primmer *et al.* 2014). In Greece, local communities were rarely engaged in the effective implementation and enforcement of EU directives; this led to limited representation of species endemic to Greece in Natura 2000 appendices and inadequate responses (Apostolopoulou and Pantis 2009), conflicts and a lack of trust (Primmer *et al.* 2014). Furthermore, the effectiveness of CCPs (e.g. protected areas) is directly proportional to the capacity and resources available to manage them (Geldmann *et al.* 2018).

Table 13.2: Summary of assessment criteria: Locally Managed Marine Areas in Fiji case study

Criterion	Description	References
Success or failure	The Fiji LMMA approach can increase fish and invertebrate size and abundance. Three of the eight LMMAs studied had fish biomass benefits, and one had biodiversity benefits. This improves the potential for the sustainable use of coastal fisheries resources and, therefore, supports national policy agendas (e.g. Roadmap for Democracy and Sustainable Socio-economic Development and Green Growth Framework) as well as international obligations such as UNCLOS and CBD.	Jupiter *et al.* 2017
Independence of evaluation	Expert-based assessments involved several organizations, including: UN Environment World Conservation Monitoring Centre (UNEP-WCMC), the Centre for Sustainable Development and Environment (CENESTA), the International Union for Conservation of Nature (IUCN), the United Nations Development Programme (UNDP), the Secretariat of the Pacific Regional Environment Programme (SPREP), the South Pacific Community (SPC), the World Wide Fund for Nature (WWF), WorldFish and Reefbase.	Govan 2009; Jupiter *et al.* 2014; Jupiter *et al.* 2017,
Key actors	There is external input from the NGO and community-based organization (CBO) sector guided by local norms. The Government plays no direct role in management approaches and has a passive administrative role such as collection of dues for fisheries permits from non-customary (non-community) fisheries, which are then returned to the community.	
Baseline	No baseline data were collected at the start of initiatives. The selection of target areas was based on interest from the community, rather than any particular biological or societal status.	
Time frame	No time frame was established for the initiative. LMMAs emerged in the 1980s, coupled with a realization of the ineffectiveness of Western approaches to conservation in countries with local tenure and with little ability to enforce conservation measures. Fiji's LMMAs expanded through the 1990s and 2000s to their present coverage of 145 traditional fishing areas covering 79 per cent of Fiji's inshore fishing area.	
Constraining factors	While the community may be able to manage local resources, strategies must be implemented that improve management of threats operating at larger scales and across boundaries (e.g. provincial scale, land–sea interactions, climate change).	
Enabling factors	The openness of the Fiji LMMA network allowed a wide range of entities to participate in the expansion, including NGOs, universities and CBOs, and facilitated the associated financial support. Working within existing sociocultural norms allowed this inclusive and integrated approach linking communities to natural resources to flourish.	
Cost-effectiveness	No cost-effectiveness assessment has been conducted.	
Equity	Equity gains from the approach include: increased fish and invertebrate size and abundance, which improve diet and also improvement of the potential for the sustainable use of coastal fisheries resources when under LMMA management. However, these biological gains are not guaranteed, and other co-benefits may be more important, such as reinforcing customs and asserting access and tenure rights. This is assumed to be relatively equitable in the way the gains are spread across the communities involved; however, there are sections of the population who are not bestowed with traditional customary rights.	Jupiter *et al.* 2014; Jupiter *et al.* 2017
Co-benefits	The extension to other Pacific countries and Asian countries was not an intended consequence at the start but can be perceived as a co-benefit, additional to those identified at the community level.	
Transboundary issues	There are no intra- or international transboundary issues.	
Possible improvements	There is a need for increased engagement and alignment with government. Government 'ratification' and sustainable financing of support for customary systems delivering natural resource management would help stabilize the approach. Increased clarity in the costs of such approaches and improved monitoring to assess resource management and biodiversity outcomes would be useful.	

The case study below examines the effectiveness of CCPs in the context of addressing the global illegal wildlife trade. The estimated value of illegal wildlife trade ranges between US$50 billion and US$150 billion per year (illegal fisheries are estimated at between US$10 billion and US$23.5 billion, and illegal logging at between US$30 billion and US$100 billion (Nellemann and International Criminal Police Organization [INTERPOL] Environmental Crime eds. 2012; Higgins and White 2016). The corrupt engagement of some government officials, including customs officials and local police, in addition to a chronic lack of resources, make effective monitoring and enforcement difficult. Even in countries with relatively advanced technological and criminological infrastructures, wildlife crime lags behind other aspects of law enforcement (Wellsmith 2011). Violence is not uncommon either, as poaching involves weapons, and anti-poaching efforts can be lethal; armed rebel groups also use the trade to finance their military campaigns.

Within the DPSIR framework (see Section 1.6), this policy approach is mostly aimed at the pressure of overexploitation, by tackling related biodiversity loss or by protecting endemic species and traditional human practices (see Section 6.4.4). However, the development of effective CCPs to constrain undesirable human activities demands significant capacity-building efforts.

Case study: Project Predator and policing the global illegal wildlife trade

Project Predator was launched in 2011, at the 80th INTERPOL General Assembly in Hanoi, Viet Nam, and is focused on building law enforcement capacity for the conservation of Asian big cats, most notably the tiger. Wild tiger populations are falling at a precipitous rate, down from over 100,000 at the start of the 20th century to less than 4,000 today (Goodrich *et al.* 2015). The main threat to big cats is habitat destruction, but poaching remains a major problem throughout their range. As reported by the IUCN Red List, the largest market is for tiger bone used in Asian traditional medicines, but other illegal markets for tiger products, especially skins, teeth and claws (particularly in Sumatra), contribute to poaching pressure. Tigers killed by farmers or villagers who believe their livestock or human inhabitants are at risk of tiger attacks can also feed into the illegal trade.

The specific objectives of Project Predator include:

i. encouragement of the creation of National Environmental Security Taskforces **(Figure 13.3)** and strengthening the South Asia Wildlife Enforcement Network;
ii. information and intelligence management, and enhancement of investigative skills;
iii. capacity-building and international integration; and
iv. intelligence-led anti-poaching activity.

Tiger range States include Bangladesh, Bhutan, Cambodia, China, India, Indonesia, Lao PDR, Malaysia, Myanmar, Nepal, Russian Federation, Thailand and Viet Nam. The collaboration between INTERPOL, national governments and legal systems is a relatively new development in global environmental governance and supports the Convention on International Trade in Endangered Species of Wild Fauna and Flora (CITES) and other international conventions. Similar programmes have been implemented for the ivory trade, hazardous waste, illegal logging, and illegal fishing.

Figure 13.3: National Environmental Security Taskforces are direct liaisons between national bureaucracies and the INTERPOL National Central Bureau; image showing seizure of 114kg of tiger bones

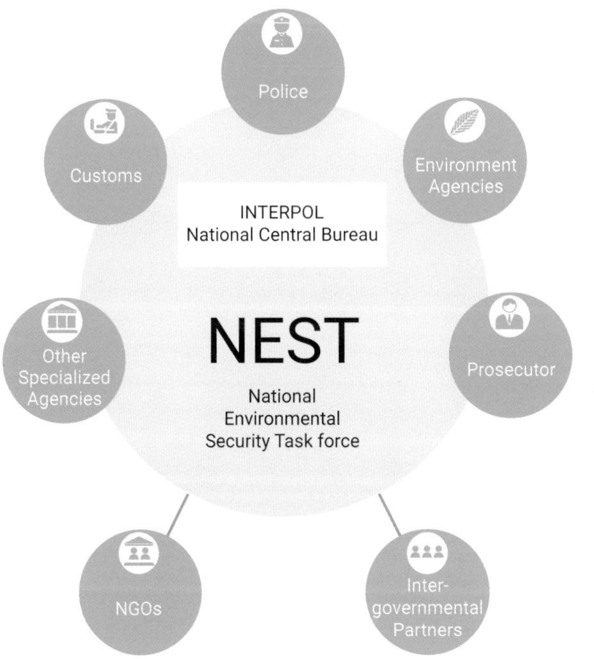

Source: https://greennews.ie/wanted-wildlife-trafficker-arrested-nepal-this-month/

Table 13.3: Summary of assessment criteria: Project Predator case study

Criterion	Description	References
Success or failure	Success refers to empirical evidence of animal parts seized. In 2015, officials organized Operation PAWS (Protection of Asian Wildlife Species) across 17 countries. This led to the arrest of more than 300 wildlife criminals and revealed the location of four wildlife crime fugitives. Officers seized 6 tiger skins and parts, more than 150 common and clouded leopard skins and parts, including 12 big-cat skins, more than 9 tonnes of ivory, 37 rhino horns, more than 2,000 turtles and reptiles, 282 pangolins, 5 tonnes of pangolin meat, and 275 kg of pangolin scales.	INTERPOL 2015
Independence of evaluation	To our knowledge, no formal evaluation of Project Predator has taken place. However, a recent study by the independent wildlife trade monitoring network TRAFFIC emphasizes the need to share intelligence among range States and the potentially helpful role of INTERPOL.	Stoner *et al.* 2016
Key actors	Project Predator's main funders include the United Kingdom of Great Britain and Northern Ireland Government, Environment Canada, the International Fund for Animal Welfare (IFAW), the Smithsonian, USAID and the Global Tiger Initiative. The latter is an umbrella organization formed in 2008 by the World Bank, the Global Environment Facility, the Smithsonian and the Save the Tiger Fund. It is related in turn to the International Tiger Coalition, which comprises some 40 NGOs in 13 tiger range countries. The CITES Secretariat is a formal partner.	United States Agency for International Development [USAID] 2016
Baseline	Wild tiger populations have fallen from over 100,000 at the start of the 20th century to less than 4,000 today.	Goodrich *et al.* 2015
Time frame	Operation Predator was established in 2011. Funding is expected to continue into the 2020s.	
Constraining factors	Corruption at all levels continues to be a problem, as does the inability to establish environmental crime as a punishable offence in many countries. The transnational environmental crime networks involved in wildlife trafficking are powerful, and their crossover illicit activities are believed to include human trafficking, drug and arms smuggling, money-laundering and extortion.	
Enabling factors	International outrage over the fate of wild tiger and snow leopard populations related to the charismatic nature of these iconic species was a motivating factor. Intelligent policing and the introduction of new tracking technology was essential. Since establishing an Environmental Crime Committee in 1992, INTERPOL has become an active agent in efforts to curb and punish transnational environmental crime.	
Cost-effectiveness	Not conducted yet	
Equity	Problematically, the low-income poacher often assumes the brunt of legal prosecution, while the enriched 'middle man' or purchaser of illicit wildlife trade escapes (including developed nations (United Kingdom of Great Britain and Northern Ireland, United States of America) which continue to trade in 'legal' wildlife when sources are often hard to identify) (Nelson 2017).	
Co-benefits	Big cats are central to ecosystem resilience and biodiversity, so their protection is beneficial to everyone who relies on related ecosystem services. The enhancement of judicial systems through National Environmental Security Taskforces is another main co-benefit.	Thinley *et al.* 2018
Transboundary issues	Wildlife trafficking involves a wide variety of international actors, and INTERPOL is unable to monitor them all. Ultimately, the success of anti-poaching efforts will depend on the capacity of national governments to monitor their own borders in a corruption-free context, and to impose serious punishment on offenders.	
Possible improvements	More information is needed on the impact of INTERPOL's interventions and National Environmental Security Taskforces. More accurate tracking of big cat populations would be helpful across the range States. More local community involvement is needed.	

Command and control strategies have historically dominated efforts to promote environmental protection. However, they face difficulties in terms of a lack of human resources and local participation (Harrington, Morgenstern and Sterner 2004; Laitos and Wolongevicz 2014). Though CCPs have their fair share of demerits, they may be highly pertinent in situations where critically endangered species and habitats are at stake and their loss is imminent (see Section 6.4.4). For instance, the relaxation of land clearing regulations and enforcement has led to increased forest loss, particularly in remnant forests (Marcos-Martinez *et al.* 2018). The challenge lies in greater integration of local communities in both the design and implementation phases (Paavola, Gouldson and Kluvánková-Oravská 2009). Adequate, power-neutral consultation of different stakeholders during policy design, and regular monitoring and adaptation could help improve the effectiveness of CCPs for biodiversity conservation. In the United Kingdom of Great Britain and Northern Ireland,

Box 13.3: The centrality of indigenous peoples and local communities

UN Environment Assembly Resolution 2/14 asked for a review of "best practices in engaging rural communities in wildlife management" (Cooney *et al.* 2018), focused specifically on engaging indigenous peoples and local communities (IPLCs) in combating the illegal trade in wildlife. The report prepared by IUCN and the International Institute for Environment and Development (IIED) concludes that local communities must be central actors in stemming illegal trade and be viewed as stakeholders and not just passive victims or witnesses. Though policing activities are important, they can also be seen as militarized responses that alienate the local communities that have the most to gain from enhanced biodiversity conservation. As the report states, partly as a "result of an increased militarization of poaching, the response [has included] the resurgence of a top-down protectionist approach emphasizing fences and fines, guns and boots. However, unless accompanied by strengthened accountability measures, this can lead to—and has led to—human rights abuses, restricted livelihood options, and hardship for IPLCs [and can drive] disenfranchisement, resentment and anger" (Cooney *et al.* 2018, p. 5).

Community-based approaches demand patience, as local stakeholders need to organize and build their own capacity. Building robust opportunities for IPLCs to be heard and to exercise their rights at all levels is critical in promoting more effective and equitable wildlife conservation strategies.

implementation of the Natura 2000 sites was carried out in an integrated manner, leading to wider acceptance (Primmer *et al.* 2014). If the INTERPOL approaches described above prove successful, they could serve as models for further initiatives aimed at stemming international crime and environmental destruction. In addition, setting up randomized control trials and regularly measuring and reporting on the success or failure of conservation interventions can help monitor effectiveness (Schwartz *et al.* 2017).

13.2.3 Economic incentive policies: Payment for ecosystem services

Economic incentive policies (EIPs) and market-based instruments are arguably more flexible than CCPs and regulatory policies, allowing the development of innovative approaches that reframe the relationship between people and the environment. EIPs are generally a response measure in the DPSIR framework (see Section 1.6) and are based on the assumption that economic incentives can account for market externalities by facilitating pro-conservation behaviour and disincentivizing negative behaviour. Such economic tools can also be used to compensate stakeholders who are negatively affected by biodiversity conservation.

EIPs are, therefore, able to address scale mismatches in biodiversity conservation—for example, where the benefits of conservation are felt at a regional or national scale, while the cost is borne by local communities at a smaller scale. Examples of EIPs include schemes related to REDD+; species enhancement; eco-certification; setting aside agricultural land; or purchasing public or grant-aided land. Others include conservation easements, incentive payments for organic farming, fiscal/taxation measures and payment for ecosystem services (PES) (UNDP 2017). For instance, municipalities located in the core area of a national park in France now receive an 'ecological allocation' for the protection of these areas (General Code for Local Authorities, article L2,334-7). There is also a 20 per cent reduction in the property tax rate for all wetlands in the country (Primmer *et al.* 2014).

PES captures many of the important elements of EIPs. As a policy instrument, PES was first widely implemented on a national scale in Costa Rica (Porras *et al.* 2013), where it has been operative since 1996, but has since spread to many countries in different forms. Typically, PES rewards local stewards of an ecosystem so that they maintain the natural resources on which they (and often downstream users) depend. Farmers on steep slopes, for example, can be incentivized to return their land to forest cover, so that an important water supply can be protected. In one well-documented example, the city of New York paid landholders in the Catskill Mountains to protect the landscape and thus avoid the greater cost of a new water treatment plant (Appleton 2013). By providing economic incentives to encourage better stewardship of the land, PES enables new actors in biodiversity conservation and simultaneously promotes a more sustainable relationship between people and nature by emphasizing the value of the ecosystem services that biodiversity supports.

However, the effectiveness of PES schemes is an area of current debate, as there are few randomized control studies to evaluate its success (Börner *et al.* 2017). A recent analysis of 38 peer-reviewed articles found that evidence of effectiveness was weak (Gaworecki 2017). Most studies had not compared areas where PES had been implemented with a relevant non-PES control area (Gaworecki 2017), and the more rigorously designed studies showed reductions in deforestation of just a few percentage points. Payments were often too low compared to the opportunity costs of other land uses – for instance, agricultural development – although this does not take into account potential co-benefits. The following case study explores an example of a PES scheme that had dual goals of reducing invasive species (one of the major pressures on biodiversity) and generating employment.

Case study: Working for Water programme, South Africa

In South Africa, a major pressure on water resources is imposed by non-native plants, both terrestrial (e.g. *Pinus*, *Acacia* and *Eucalyptus* species that have escaped from commercial cultivation) and aquatic (e.g. water hyacinth [*Eichhornia crassipes*], also a threat to the African Great Lakes) biomes (Chamier *et al.* 2012). In 1995, the South African Government established the Working for Water (WfW) programme to clear invasive species from environmentally degraded water catchments and address social equity issues and unemployment among low-skilled people. WfW focussed mainly on rural women, youth and people with disabilities, by providing them with employment opportunities associated with the removal and control of invasive plants (McQueen, Noemdoe and Jezile 2001). WfW provides one of the longest-running examples of the PES approach linked to employment generation.

Table 13.4: Summary of assessment criteria: Working for Water case study

Criterion	Description	References
Success or failure	The aims of the WfW programme were to enhance water security, improve ecological integrity and restore the productive potential of land, promote sustainable use of natural resources and invest in the most marginalized sectors of South African society. Today, over 3 million hectares have been cleared of alien species (30 per cent of the total affected area in South Africa), showing some success and promise for the future of the policy. Stream flows were increased, although benefits decline over time as vegetation regrows.	Barnes *et al.* 2007; Bonnardeaux 2012; Jarmain and Meijninger 2012; Le Maitre, Gush and Dzikiti 2015; Scott-Shaw, Everson and Clulow 2017
Independence of evaluation	Extensively evaluated in the peer-reviewed scientific literature	Hobbs 2004; Turpie, Marais and Blignaut 2008; Buch and Dixon 2009; Meijninger and Jarmain 2014
Key actors	WfW's framework comprises the following: ❖ Inter-ministerial Board (Cabinet ministers chaired by the Minister of Water Affairs and Forestry) ❖ Inter-departmental Steering Committee ❖ Provincial Steering Committees and Project Steering Committees of relevant stakeholders at local level	McQueen, Noemdoe and Jezile 2001
Baseline	The report 'Water for Growth and Development in South Africa, Version 6' was the baseline. It reported that 10.1 million hectares (6.8 per cent) of South Africa and Lesotho were invaded by alien plants in 1997, reducing mean annual water flow by 3.3 billion m^3 and resulting in wastage of about 7 per cent of South Africa's water annually.	Barnes *et al.* 2007
Time frame	WfW has been operational for over two decades. Measurable ecosystem gains were reported in the few years immediately following implementation.	
Constraining factors	The short-term employment and low wage provided by WfW has been suggested as providing only a temporary solution to the chronic problems of unemployment and the skills gap in South Africa.	Buch and Dixon 2009
Enabling factors	Effective legislation used in the programme includes the Agricultural Pests Act, Conservation of Agricultural Resources Act, National Environmental Management Act, Environment Conservation Act, National Water Act, and National Veld and Forest Fire Act. WfW maintains a research unit as part of its commitment to the management of invasive alien plants.	Venter 2005
Cost-effectiveness	There have been several cost–benefit analyses, with differing results, but overall leaning towards this being a cost-effective policy. An important aspect is the high cost of doing nothing. In 1998, the South African Department of Environmental Affairs estimated the cost of controlling invasive plants at R600 million (US$100 million) a year over 20 years but indicated that this could double within 15 years if appropriate action is not taken.	South African National Biodiversity Institute 2008; Turpie, Marais and Blignaut 2008; South Africa, Department of Environmental Affairs 2010; South Africa, Department of Water Affairs [DWAF] 2010a; DWAF 2010b; McConnachie *et al.* 2012; van Wilgen *et al.* 2012
Equity	Landowners clearing invasive species through the WfW programme were eligible for tax breaks. The employees clearing invasive species from the landscape (mostly women and disadvantaged people) benefit the most.	Turpie, Marais and Blignaut 2008; Buch and Dixon 2009
Co-benefits	WfW provides more than 20,000 temporary jobs each year for the most marginalized people who might not have access to any other employment (52 per cent of beneficiaries are women), educates and trains unskilled labourers and assists in community development programmes (http://www.dwaf.gov.za/wfw/). With a particular emphasis on HIV/AIDS, WfW aimed to provide support for those with a positive diagnosis, and education and training to reduce the risk of transmission.	Magadlela and Mdzeke 2004
Transboundary issues	Not applicable	
Possible improvements	Recommendations include: a) robust ecological indicators to evaluate: (i) the extent of the area treated; (ii) the reduction in the degree of invasion; (iii) the impact on water resources; and (iv) the rate of ecosystem recovery (Levendal *et al.* 2008); and b) further integration of social development more fully with the programme's environmental goals.	Levendal *et al.* 2008

When implemented well, EIPs allow cross-sector integration (e.g. facilitating women's empowerment by controlling invasive species, as shown in the WfW case study), greater stakeholder engagement and multi-level governance. However, a drawback of EIPs stems from the assumption that financial incentives alone will influence the actors to change their behaviour towards a pro-conservation stance. This assumption may lead to further questions on the sustainability of such policies when funding is exhausted. Finding the correct financial tipping point to prevent biodiversity loss and improve human well-being (e.g. correct level of compensation) by matching projected opportunity costs may be challenging. Cost-effectiveness analysis can help find the optimal solution when multiple conservation interventions are possible (Bryan 2010). A further gap in the implementation of EIPs lies in the treatment of landowners as independent and individual decision makers (e.g. in the Finnish 'Nature Values Trading' PES experiment) (Paloniemi and Vilja 2009). However, landowners may be influenced by professional advisers and a range of group-based factors in their decision-making (Mukherjee *et al*. 2016). In addition, the focus should remain on biodiversity conservation rather than simply the benefits derived from it.

Considerable progress has been made in the last couple of years towards mainstreaming the value of nature (e.g. IPBES 2016). A cautionary note though would be to retain the focus on biodiversity. The essence of biodiversity conservation should not be lost in the enthusiasm to value its benefits and services since biodiversity underpins all the services (see **Figure 13.4** below from Kusmanoff 2017, which shows that while the use of economic language has risen, the use of the term 'biodiversity' has declined in Australia).

Figure 13.4: Usage of the terms containing 'biodiversity', 'econo' and 'ecosystem services' over time in Australian Government environment portfolio media releases (n= 3,553). Error bars indicate 95 per cent confidence intervals based on the ecosystem services framing subsample (n = 516)

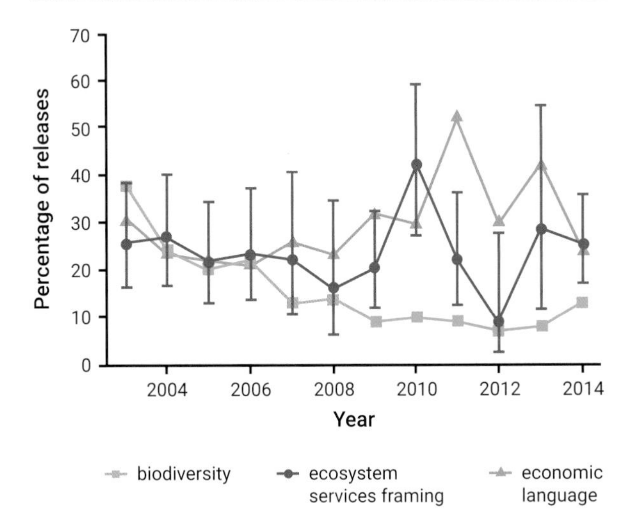

Source: Kusmanoff (2017)

13.2.4 Supporting investment: Banking of genetic material

Currently only a tiny fraction (~0.002 per cent) of global GDP is invested in the conservation of biodiversity (Sumaila *et al.* 2017). Yet sustaining natural capital by meeting the 2020 Aichi Biodiversity Targets would provide monetary and non-monetary gains that far outweigh the costs of achieving these goals (Sumaila *et al.* 2017).

Though progress is slow, some governments are warming to the cause. For example, the New South Wales Government in Australia has set up a Biodiversity Conservation Trust to deliver a comprehensive conservation programme on private land in its 2017-2037 strategy (New South Wales, Office for Environment and Heritage 2017). Government investment of A$240 million over five years, with A$70 million in ongoing annual funding, has been earmarked for this project targeting private landholders.

The EU has estimated that the cost of managing the Natura 2000 sites, its protected area network, would amount to €5.8 billion annually, while the benefits range from €200 billion to €300 billion annually and could create 180,000 jobs (Bourguignon 2015). EU LIFE funding (the Financial Instrument for the Environment), launched in 1992, and its successor LIFE+ have co-financed site management, capacity-building, and species action plans. Between 2014 and 2020, €2.6 billion has been earmarked under LIFE for environmental protection, half of which is for nature and biodiversity conservation (Bourguignon 2015). The United Kingdom of Great Britain and Northern Ireland Government recently announced that it would set up a Green Business Council to support environmental entrepreneurialism in its 25-year Environment Plan (United Kingdom [UK], Department for Environment, Food and Rural Affairs 2018, p. 150). The United Kingdom of Great Britain and Northern Ireland also plans to create a Natural Environment Impact Fund to issue a variety of loans and grants at submarket rates that could be repaid on a long-term basis. This is aimed at addressing potential market failures that might have limited the uptake of return-generating natural environment projects in the past (UK Department for Environment, Food and Rural Affairs 2018, p. 149).

Sources of financing for investment in biodiversity can come from multiple sources (SCBD 2012), including core national biodiversity funding sources, national government financing, international flows of Official Development Assistance and multilateral funding. In addition, tax breaks for green infrastructure, conservation agreements, carbon offsets, green fiscal policies and green bonds, as well as private- and third-sector investment are also in the toolkit available to policymakers to support investment in biodiversity conservation.

The Green Bond principles of 2016 explicitly recognize biodiversity conservation as one of the categories eligible for funding (GreenInvest 2017). Green Bonds are one of the fastest growing fixed-income market segments, with US$81 billion in 2016. These Green Bonds could be used strategically by governments and corporations to tap international capital to support investment in biodiversity conservation (GreenInvest 2017). Green Bonds could also provide a platform for interactions between financial and investment policymaking,

which are often institutionally separate in some countries (GreenInvest 2017, p. 40).

An innovative example of supporting investment is the Svalbard Global Seed Vault (SGSV), which is a gene bank representing the largest collection of crop diversity in the world. Within the DPSIR framework (see Section 1.6), this serves as a policy response focused on *ex situ* conservation of seeds to improve the status of plant species important for food and agriculture.

Case study: Svalbard Global Seed Vault

FAO estimates that 75 per cent of plant genetic diversity was lost in the last century (FAO 2010). A primary form of conservation for plant genetic material is *ex situ* in the form of gene banks (currently over 1,750 worldwide, collectively maintaining an estimated 7.4 million accessions) (FAO 2010).

The SGSV **(Figure 13.5)** was established in 2008 with the primary goal of providing a backup for plant genetic resources for food and agriculture. The priority is on preserving intraspecific diversity of crop species and crop wild relatives. The risks from natural disaster, war and the mismanagement of some national gene banks demand backup storage for globally important crops (Fowler 2008).

The construction was funded by the Norwegian Government at a cost of US$8.8 million (Hopkin 2008), and the operating costs for the SGSV are around US$300,000 annually, shared by the Norwegian Government and Global Crop Diversity Trust. The latter provides long-term grants from an endowment fund built by public and private donations.

Figure 13.5: The SGSV is located 100m inside a mountain on a remote island in the Svalbard archipelago, midway between mainland Norway and the North Pole, and the samples are stored at -18°C

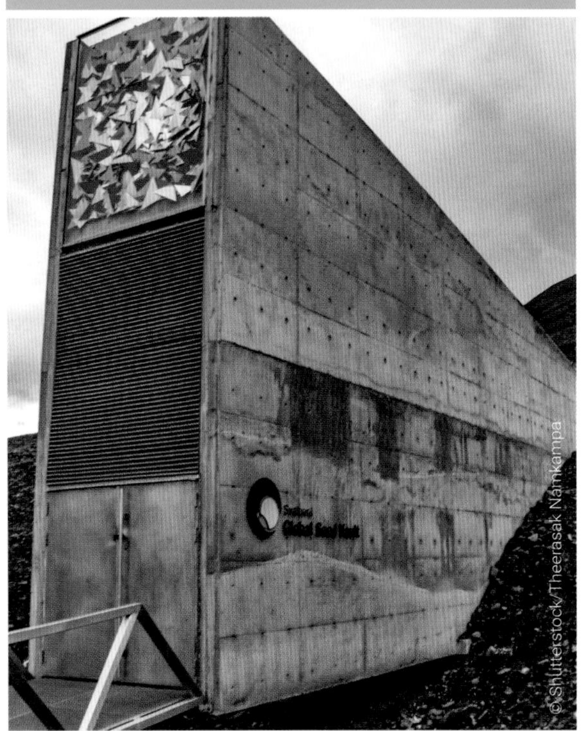

Supporting investment policies are urgently needed to complement CCPs, EIPs and enabling actors in stemming rates of biodiversity loss (see Section 6.5). Similar to EIPs, the supporting investment policies are also more flexible and adaptable in their approach. They also allow for unique and innovative solutions as shown in the SGSV case study. Foreign direct investment to developing tropical countries could be directed for biodiversity conservation through supporting investment policies, such as Green Bonds (GreenInvest 2017). Initiatives such as the SGSV are in line with Sustainable Development Goal (SDG) 16, as the outputs of such investments are accountable, transparent and inclusive. One concern, however, is the power structure inherent in the decision-making and implementation of supporting investment policies. Who invests and who benefits in the long term are key questions to be asked in *ex ante* analysis of such policies.

In terms of wider biodiversity conservation, the SGSV is a backup, and it does not seek to maintain the traditional knowledge for harvesting crops that could be lost as agriculture evolves, whereas *in situ* conservation could sustain these skills and also allow species to adapt to changes in their environment. *Ex situ* conservation also faces the issue of genetic erosion (van de Wouw *et al.* 2010), whereby the seeds being conserved may not be viable in perpetuity. Protection of genetic resources requires a range of actors to be involved, as there are political, ethical and technical challenges to be overcome in the conservation of crop genetic resources (Esquinas-Alcázar 2005).

In addition, the contribution of biodiversity to food security needs to be mainstreamed. The Ecosystem Based Adaptation for Food Security (EBAFOSA) initiative was launched in 2015. It aims to reconcile the sustainable management of ecosystems (including the conservation of biodiversity) with adaptations to climate change to ensure food security in Africa.[2]

13.2.5 Enabling actors: Strategic environmental planning

The enhancement of the quality of urban environments for ecological and social benefits is becoming widely accepted as a critical component of urban planning. The United Nations General Assembly (A/71/266 of 1 August 2016) has discussed the 'Mother Earth' concept under 'harmony with Nature', seeking to inspire citizens and societies and change the way they interact with the natural world. This links closely to the concept of green infrastructure, green spaces and the recognition of the vital connections between the ecosystem services and biodiversity. These include benefits linked to water quality, flood attenuation, improved air quality, physical and mental health and noise reduction, all of which are important in reducing problems posed by urban living (Carrus *et al.* 2015; Ürge-Vorsatz *et al.* 2018) and in contributing to climate change mitigation and adaptation (Rosenzweig *et al.* 2018). Biodiversity's role in cities has also been recognized by other international forums, such as the Intergovernmental Panel on Climate Change (IPCC) cities conference in March 2018; experiencing biodiversity has been proven to improve life quality, human health and environmental consciousness in urban areas (WHO and SCBD 2015; Ürge-Vorsatz *et al.* 2018).

2 https://www.ebafosa.org

Table 13.5: Summary of assessment criteria: Svalbard Global Seed Vault case study

Criterion	Description	References
Success or failure	The target is to hold 4.5 million varieties of crops, with each variety having 500 seeds on average (a total of 2.5 billion seeds). In the first five years of operation, 53 of the world's gene banks had deposited a substantial part of their collections, and the vault currently contains over 960,000 samples.	Westengen, Jeppson and Guarino 2013; Asdal 2018
Independence of evaluation	The formal assessment was published in a peer-reviewed journal.	Westengen, Jeppson and Guarino 2013; Asdal 2018
Key actors	Actors include the FAO Commission on Plant Genetic Resources, the Norwegian Government, the Nordic Genetic Resource Centre, the Global Crop Diversity Trust and the International Advisory Council (technical and policy experts representing FAO, national gene banks, the Consultative Group on International Agricultural Research and the International Treaty on Plant Genetic Resources for Food and Agriculture).	Westengen, Jeppson and Guarino 2013; Asdal 2018
Baseline	Five benchmarks assessed the duplication covered by the collection in the SGSV. This assessment aimed to quantify how far the SGSV is away from its target of duplicating all the distinct accessions (unique sample of seeds) of plant genetic resources for food and agriculture conserved as orthodox seeds (those that can survive drying or freezing) globally.	Westengen, Jeppson and Guarino 2013; Asdal 2018
Time frame	The vault theoretically has a permanent lifetime. Currently, there are a third of globally distinct accessions of 156 crop genera.	Westengen, Jeppson and Guarino 2013; Asdal 2018
Constraining factors	The willingness of countries to sign up (e.g. China and Japan) was identified as a constraint, although new seed samples were deposited from countries including India, Peru and Kenya in 2018. Changes in climate could be seen as a future constraint to the facility.	
Enabling factors	Signing of the International Treaty on Plant Genetic Resources for Food and Agriculture; permafrost offers natural freezing in case the cooling equipment breaks down; geopolitical stability and a supportive local government (military activity is prohibited under the International Treaty of Svalbard)	
Cost-effectiveness	None conducted so far	
Equity	Currently, plants for traditional use and their cultivation practices are not prioritized, and they might also be vulnerable to loss. The Global Crop Diversity Trust provides funding for developing countries to assist in the logistics of transporting accessions to the SGSV.	Eastwood *et al.* 2015
Co-benefits	The SGSV has helped raise public awareness (particularly promoted by the media) of the importance of conserving genetic diversity—especially plants—for future food security.	Friel and Ford 2015; Westengen, Jeppson and Guarino 2013
Transboundary issues	The SGSV's Standard Deposit Agreement ensures that the legal ownership of accessions cannot be transferred and that accessions can only be returned to the gene banks that originally supplied them.	Westengen, Jeppson and Guarino 2013
Possible improvements	1) Gaps in accessions from other gene banks which have no backup collection. 2) The importance of *in situ* conservation to complement *ex situ* approaches has also been highlighted, as stored genetic material is evolutionarily static and cannot adapt to changes in climate and habitat. 3) Another form of *ex situ* conservation—DNA banks—could be a complementary approach to plant genetic conservation.	Dulloo 2015; Hopkin 2008; Hodkinson *et al.* 2007

Engaging communities in effective land use and management of natural ecosystems in urban areas can be beneficial to both residents and biodiversity and promote inclusive city governance. The involvement of different stakeholders at different scales and partnerships between experts from various disciplines (e.g. ecologists, urban designers, landscape architects) is also considered important for biodiversity conservation (Felson and Pickett 2005; Colding 2007). Progress is measurable: for example, the City Biodiversity Index, which "provides a monitoring tool to assist local authorities to evaluate their progress in urban biodiversity conservation, which can be further included in national reports" (CBD 2014).

Various institutional arrangements and approaches take into account the importance of biodiversity in green areas. For example, in Italy health and well-being aspects (Carrus *et al.* 2015), in Brazil restoring Atlantic Forest in urban areas through municipal plans (Sansevero *et al.* 2017), and in Finland preservation of ecosystem services (Niemelä *et al.* 2010) are considered. Mainstreaming biodiversity requires the integration of biodiversity and environmental components and norms into sectoral policies, enabling stakeholders' involvement. Within the DPSIR framework (Section 1.6), mainstreaming is a response made by a group of actors to address pressures and drivers such as habitat loss and fragmentation and human population pressure (Section 2.2). The Edmonton Natural Area System Policy shows how to engage local actors to mainstream biodiversity into the urban environment.

Case study: Edmonton Natural Area Systems Policy

The City of Edmonton has made biodiversity protection a priority by integrating biodiversity considerations into urban planning. In 2006 it approved its Environmental Policy to promote the development of environmentally sustainable communities. In 2007, the city approved its Natural Area Systems Policy with a clear goal to "conserve, protect, and restore Edmonton's biodiversity, and to balance ecological and environmental considerations with economic and social considerations in its decision-making". As an outcome of this policy, a strategic plan emerged for the conservation and restoration of Edmonton's natural systems and the biodiversity they contain **(Figure 13.6).**

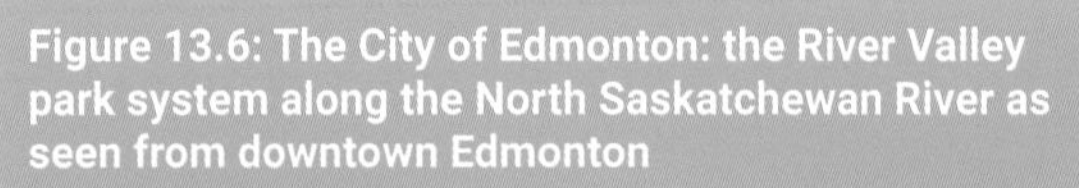
Figure 13.6: The City of Edmonton: the River Valley park system along the North Saskatchewan River as seen from downtown Edmonton

Enabling actors and institutional arrangements in local and urban biodiversity conservation has been proven in certain cases to be successful when governments collaborate across different levels to enhance the quality of urban environments for ecological and social benefits. Extensive stakeholder participation on environmental management "may seem very risky, but there is growing evidence that if well designed, these perceived risks may be well worth taking" (Reed 2008). However, fiscal and budget prioritization remain serious challenges for the public administration.

The Edmonton case study illustrates a successful implementation of the Protected Areas System Policy, securing 110ha/year of priority natural areas. Although Edmonton's Ecological Footprint has decreased, it is still 7.45ha per capita, far above the global average of 2.71ha per capita, and 4.5ha per capita higher than the sustainability indicator of global capacity; this is largely driven by consumption of resources from outside city boundaries.

Table 13.6: Summary of assessment criteria: Edmonton Natural Area Systems Policy

Criterion	Description	References
Success or failure	The Edmonton 'Report on the Environment 2015' includes several ecological indicators, including naturalization of turf, priority natural areas secured, land secured for natural areas and wetland expansion, and the number of trees managed and cared for by the City Council. Time series data indicate that most indices show positive trends, with increases in the number of trees maintained by the city, land secured for natural areas and reconstructed wetland.	
Independence of evaluation	Policy success has been self-assessed with oversight from a City Environmental Management Steering Committee.	
Key actors	Key actors include the Edmonton City Council and the departments responsible for initiating best practices for biodiversity protection. The Office of Natural Areas coordinates the city's corporate strategic efforts to protect the network. Local communities participate in programmes such as the Master Naturalist, which exchanges knowledge and education for volunteering for stewardship of the natural areas within the city, monitoring of invasive species via citizen science, and participation in governance of a not-for-profit land trust.	
Baseline	The findings of the City of Edmonton's 2006 'State of Natural Areas' report revealed that its business-as-usual land use would result, over time, in the loss of more than half the area of existing natural systems in Edmonton's tablelands.	City of Edmonton 2009
Time frame	'The Way We Green Vision: 2040' set out the City of Edmonton's 30-year environmental strategic plan, with an emphasis on resilience and sustainability, and defined 12 goals that need to be reached for Edmonton to achieve a sustainable and resilient future.	
Constraining factors	The city continues to experience significant losses of natural areas as new residents move to Edmonton in unprecedented numbers. Responses to this have been the purchase by the city of valuable lands to protect them from development pressures (see below).	City of Edmonton 2009
Enabling factors	Leadership within the City Council seems to be strong and sustained in driving through both policy and implementation. Edmonton's City Council authorized a Can$20 million fund allocation and permit borrowing land trust for the acquisition of forests and wetlands in new neighbourhoods and, as part of a separate initiative, a Can$1 million per year agreement to purchase wetlands. A strong international profile and reputation may also help in continuing to focus attention on sustaining successes.	City of Edmonton, 2009; Local Governments for Sustainability 2013
Cost-effectiveness	The City of Edmonton evaluated the environmental effects, value and structure of Edmonton's urban forest, considering three ecosystem services: cleansing the air; sequestering carbon; and reducing storm water. The average benefit per tree in Edmonton's urban forest was US$74.73, whereas the cost of caring for each tree is US$18.38.	City of Edmonton 2009
Equity	The project has contributed to the social integration of immigrants into the life of the city. Land developers have to comply with environmental regulations, and new suburban areas are designed with new green spaces, natural areas and parks for the benefit of communities. However, the increase in the value of land means that buying land for conservation purposes is prohibitively costly for the City trust, especially because landowners are more reluctant to sell.	
Co-benefits	Increasing green spaces in urban settings provides additional benefits, including reducing stress, crime and violence and increasing neighbourhood social cohesion. They support a range of benefits associated with psychological, cognitive and physiological health (WHO and SCBD 2015). There are some indications of increased opportunities for renewable energy businesses (Alberta Canada 2017).	Maas *et al.* 2009; Garvin, Cannuscio and Branas 2013; Roe *et al.* 2013
Transboundary issues	None identified or recorded in reviewing the progress reports	
Possible improvements	Some long-term tracking of a wider range of social as well as environmental benefits would be useful, as would a more formal evaluation by independent peers. There is also a need to incorporate the trade-offs, such as increased land costs, and conflicts between priorities in a city with a population that has increased over the last 25 years.	

13.3 Indicators: Biodiversity policy

Policy-sensitive indicators provide an interesting way to understand policy implementation (see Chapter 10). Both IPBES and CBD have produced global assessments using a wide variety of indicators; for example, Global Biodiversity Outlook-4 used 55 biodiversity indicators (SCBD 2014; Tittensor *et al*. 2014). For the purposes of the sixth Global Environment Outlook (GEO-6), three global indicators were selected based on their linkages with the SDGs, national disaggregation and continuity with previous GEOs **(see Table 13.7).**

Currently, there is a lack of indicators which can adequately capture the links between biodiversity and human health, though ways to improve biodiversity health indicators have been described previously (Huynen, Martens and De Groot 2004; Hough 2014; Sandifer, Sutton-Grier and Ward 2015).

Table 13.7: Policy-sensitive indicators

Indicator	Rationale for selection	Addressed in Part A	Addressed in the case studies	Connection with the SDGs or MEAs	Data sources
1) Proportion of countries adopting relevant national legislation and adequately resourcing the prevention or control of invasive alien species	Links to the Convention on Biological Diversity as an indicator for Aichi Biodiversity Target 9. Indicator is policy-responsive and relevant and was designed as a response indicator. It was used in the fifth Global Environment Outlook (GEO-5) and is a confirmed SDG indicator.	**Yes:** invasive species are dealt with as one of the five main pressures on biodiversity (Section 6.4.2).	**Yes:** invasives are the subject of the WfW case study from South Africa (Section 13.2.3), which uses PES as a means of tackling invasives.	**Aichi Biodiversity Target 9.** This is also the indicator for SDG Target 15.8	IUCN, IUCN SSC, IUCN ISSG, Monash University; bipindicators.net for factsheets, graphs, meta-data
2) Red List Index (impacts of utilization)	Links to CBD as an indicator for Aichi Biodiversity Target 4. This is a response indicator. It was used in GEO-5 and is relevant to the SDGs. It has global coverage, can be disaggregated, is a quantitative measure based on scientific assessment and has a long data series. Red List (impacts of utilization) was also chosen to demonstrate the degree to which species of direct relevance to human livelihoods and culture are responding to measures to ensure their sustainable use over time.	**Yes:** subsets of the Red List are used throughout Chapter 6, particularly in Section 6.5 in species section. The Red List Index is the leading global source on species extinction status.	No	**Aichi Biodiversity Target 4**. Also related to Aichi Targets 3, 6, 7 and 12. Relates to SDGs 8.4, 12.2, 14 and 15.	IUCN Red List Index bipindicators.net for factsheets, graphs, meta-data
3) Global Ecological Footprint	Links to CBD as an indicator for Aichi Biodiversity Target 4. The indicator tracks pressures. It was used in GEO-5 and is relevant to the SDGs. It is global, based on a long data series and can be disaggregated. This indicator was chosen because an increase in a nation's Ecological Footprint would mean an increase in its population's pressure on biodiversity and an increased risk of biodiversity loss.	**Yes:** in Section 6.4.1, as a leading driver of biodiversity loss.	**Yes**: the Ecological Footprint of Edmonton is quoted in the policy effectiveness assessment Section 13.2.5.	**Aichi Biodiversity Target 4**. Related to SDG targets 8.4 and 12.2.	Global Footprint Network bipindicators.net for factsheets, graphs, meta-data

13.3.1 Indicator 1: Proportion of countries adopting relevant national legislation and adequately resourcing the prevention or control of invasive alien species (SDG Indicator 15.8.1)

Invasive alien species (IAS) may threaten local biodiversity through direct and indirect competition, predation and habitat degradation, and as disease agents and vectors (Pejchar and Mooney 2009; Strayer 2010). They are considered the second greatest threat to biodiversity after land-use change and habitat loss (Section 6.4.2) (Wilcove *et al.* 1998; Bellard, Cassey and Blackburn 2016).

This indicator evaluates the "trends in policy responses, legislation and management plans to control and prevent spread of invasive alien species" (species that have been introduced to an area and have spread beyond the area of introduction) and the "proportion of countries adopting relevant national legislation and adequately resourcing the prevention or control of invasive alien species" (see methodology in Biodiversity Indicators Partnership 2018a) (**Figure 13.7** and **Figure 13.8**).

Policy relevance

This indicator directly tracks progress towards global multilateral environmental agreements, and in particular Target 9 of the Aichi Biodiversity Targets. It is also relevant to Aichi Targets 5, 11, 12 and 17 and Goal 15 (Target 15.8) of the SDGs ('Life on Land') (UNEP 2015).

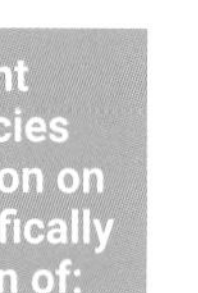

Figure 13.7: Trends in national legislation relevant to the prevention or control of invasive alien species (IAS) for 196 countries reporting to the Convention on Biological Diversity (1967–2016), showing specifically the percentage of countries having a combination of: (i) IAS legislation; (ii) NBSAP targets on IAS; and (iii) IAS targets aligned with Aichi Target 9

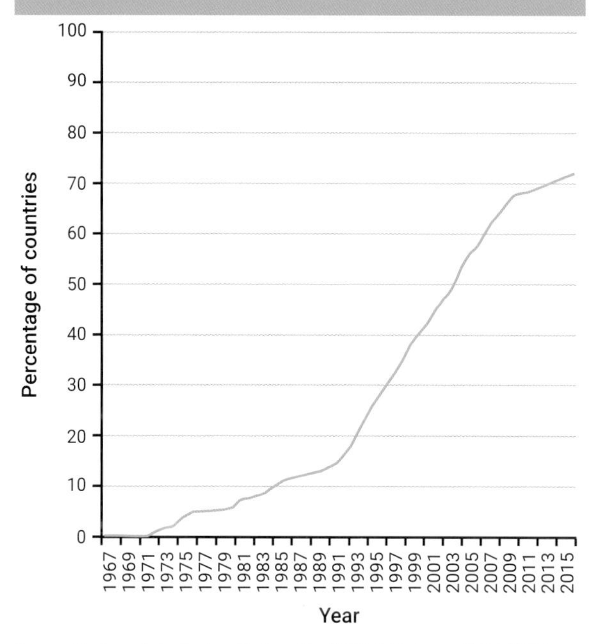

Source: Biodiversity Indicators Partnership (2018a). Partners: IUCN, IUCN Species Survival Commission and IUCN Invasive Species Specialist Group, Monash University

Figure 13.8: Percentage of countries whose institutions have a clear mandate and/or legal authority to manage IAS (a positive result is given by a Yes and is included in the overall percentage)

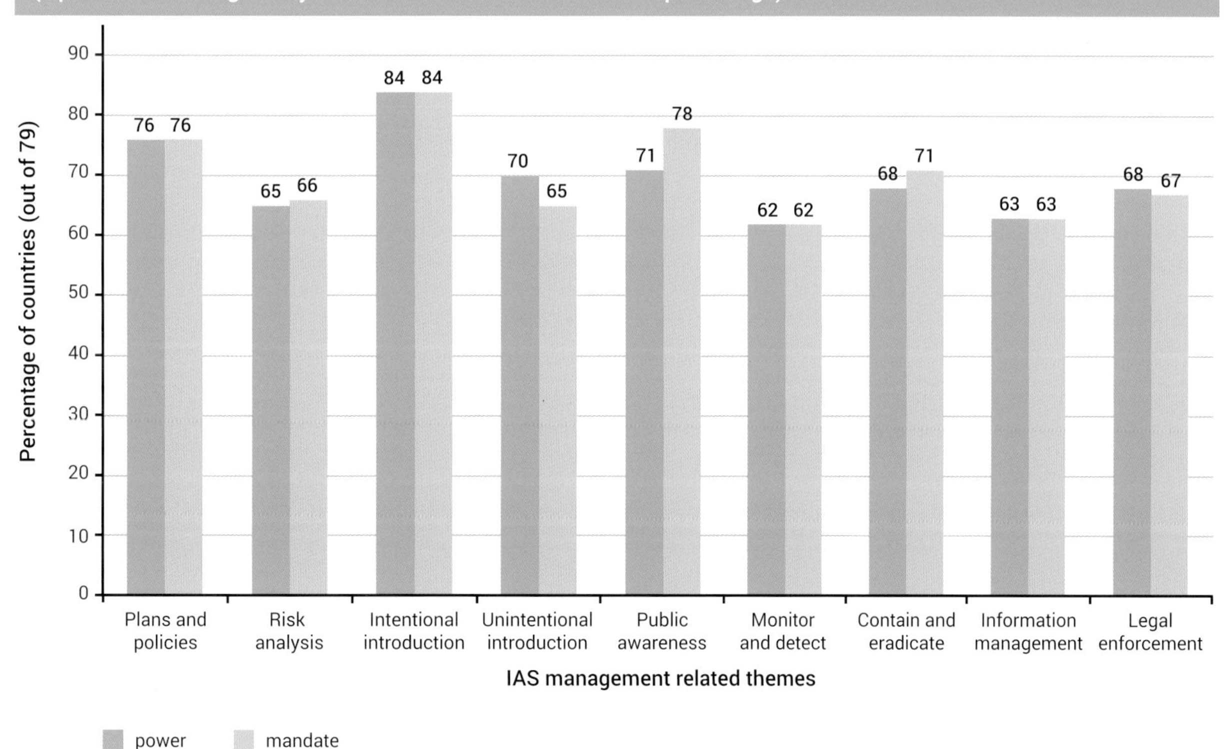

Source: Biodiversity Indicators Partnership (2018a). Partners: IUCN, IUCN Species Survival Commission and IUCN Invasive Species Specialist Group, Monash University

Causal relations
As more multilateral international agreements relevant to IAS are introduced (such as the Cartagena Protocol, the International Plant Protection Convention and the Agreement on Sanitary and Phytosanitary Measures of the World Trade Organization), the level of national commitment to related policies increases. This, in turn, reflects a greater global commitment to controlling IAS (Biodiversity Indicators Partnership 2018a). Those countries that are party to the CBD have agreed to Aichi Target 17, and policies related to the control of IAS should be addressed in their NBSAPs. This is an example of an international policy trigger and a top-down approach leading to the creation of national IAS regulations. A bottom-up causal relation (the creation of an IAS policy due to an increase in IAS within a country) is more difficult to demonstrate.

Within national IAS-relevant policies, governments may use several policy instruments to reduce IAS. These responses can be quite varied and specific. The WfW programme in South Africa (Section 13.2.3) uses PES to encourage the removal of IAS from waterways by giving monetary incentives to local communities (Buch and Dixon 2009). Other nations may use CCPs, such as the United Kingdom of Great Britain and Northern Ireland plant health policy that imposes strict regulations on and certifications for the import and moving of certain plants, seeds, organic matter and plant products to prevent the introduction and spread of harmful plant pathogens (UK Department for Environmental and Rural Affairs 2014), as well as Australia's well-developed strategic plan (Australia, Invasive Species Council 2015). In addition, island nations may have stronger IAS policies, reflecting a higher presence of endemic species, and ports can be subjected to stronger regulation, such as the recent international Ships Ballast Water and Sediments policy (International Maritime Organization 2017).

Other international and national policies may influence this indicator, especially trade policies. As globalization progresses and international commerce creates new trade routes and markets, new opportunities are created for alien species to establish themselves in new areas (Meyerson and Mooney 2007; Seebens *et al.* 2015). A direct positive link has been shown between the degree of international trade by a nation and the number of IAS (Westphal *et al.* 2008; Hulme 2009; Liebhold *et al.* 2012; Brockerhoff *et al.* 2014).

Other factors
Climate change, especially in colder regions, poses an IAS risk, as it may lower the barrier to establishment by creating new niche space (Wolkovich *et al.* 2013; Duffy *et al.* 2017). Emerging economies with increasing economic development in tourism, the exotic pet trade and infrastructure projects are also at greater risk of IAS (Hulme 2015).

13.3.2 Indicator 2: Red List Index (impacts of utilization)

Humans depend on biodiversity and the use of wildlife in a range of different ways (e.g. hunting, trapping and collecting wild birds for food, sport or feathers). The Red List Index (RLI) (impacts of utilization) shows trends in the status of mammals, amphibians and birds driven by two factors: the negative impacts of utilization (i.e. the use of wildlife leading to a decrease in status) or the positive impacts of measures taken (i.e. controlling or managing the utilization of wildlife towards sustainability) (Biodiversity Indicators Partnership 2018b, see Section 6.5.2). **Figure 13.9** shows the RLI (impacts of utilization) for birds, mammals and amphibians from 1980 to 2017.

Figure 13.9: The Red List Index (RLI) for 1980–2017 for mammals, birds and amphibians, showing the trends driven only by utilization (by only including utilized species)

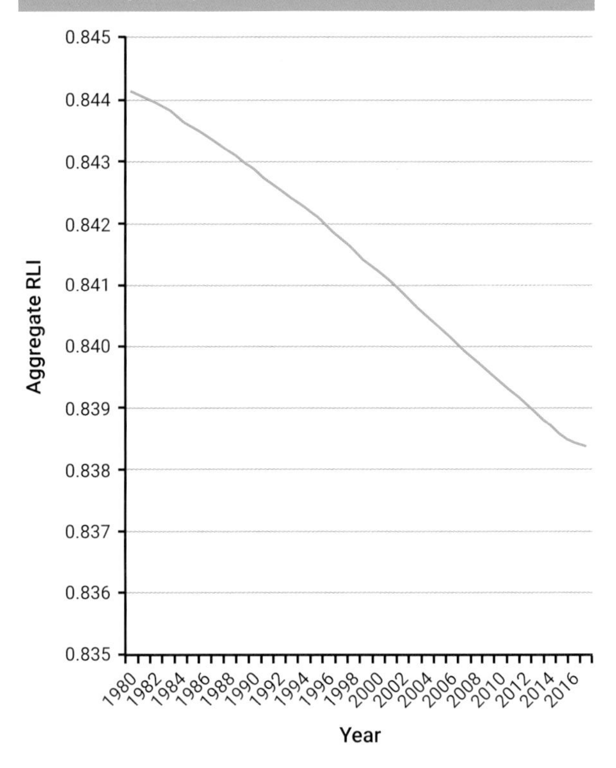

Source: Biodiversity Indicators Partnership (2018b). Partners: Birdlife International, Kew Gardens, ZSL, IUCN

Scope and measurement
The indicator is determined from species-level data which may be analysed on several scales (country, region and/or global). The IUCN Red List assigns species to seven categories of relative extinction risk (Extinct to Least Concern, or 'Data Deficient' for poorly known species). This is done using quantitative criteria for species based on population size, area of distribution and rate of decline (Bubb *et al.* 2009). In the 2012 update, the IUCN Red List included assessments for 63,837 species, of which 19,817 were threatened with extinction (SCBD and IUCN 2018). An RLI of 1 means all species in that group are categorized as Least Concern, while an RLI of 0 means that all species in the group are Extinct (Bubb *et al.* 2009). Currently, an RLI can be calculated for birds, mammals, amphibians, corals and gymnosperms. To assess taxonomic groups that are poorly known and/or have a very large number of species, a sampling approach was developed in which 1,500 species are randomly chosen and assumed to represent the larger group (Baillie *et al.* 2008).

For the RLI (impacts of utilization), only species that are utilized by humans (as pets, for food, medicine, materials or other uses) are included. Utilization categories are defined by the IUCN Use and Trade Classification Scheme (version 3.2) (IUCN

2006; Almond *et al.* 2013). The resulting trend can be used to indicate the degree to which consumption is sustainable and the impact of natural resource use is within safe ecological limits. A declining trend indicates that current utilization is unsustainable (negative impact of utilization), while an upward trend means that human use of this group of species is sustainable (positive impact of utilization through measures to control or manage sustainably) (Birdlife International 2012).

Policy relevance
The RLI (impacts of utilization) is directly related to Target 4 of the Aichi Biodiversity Targets. It is also directly related to several targets within SDGs 8, 12, 14 and 15 (Biodiversity Indicators Partnership 2018b; UNEP 2015).

Causal relations
Policies that limit the utilization or promote sustainable management of species have the potential to directly impact this indicator, though there is little to no published literature demonstrating policy effectiveness. The lack of evidence for policy impact may be partly because the average time for species status to improve by one Red List category is 16 years (Young *et al.* 2014). However, this indicator should be sensitive to economic changes or policies that increase or decrease the price of a species-derived product. For example, a higher market price creates an incentive for greater use of a species by the manufacturer or hunter and, therefore, can put that group of species at greater risk of extinction, reflected in a lower RLI (Ayling 2013). It has been shown that CCPs, such as CITES international trade bans and regulations on poaching products from endangered species, can fail when there are strong economic incentives to continue poaching (Rivalan *et al.* 2007; Conrad 2012). Policies that instead focus on incentivizing and building capacity within communities to sustainably manage wildlife (e.g. as showcased in the Project Predator case study, Section 13.2.2) can decrease the long-term use of and demand for species (Challender and MacMillan 2014), effectively increasing the RLI (impact of utilization). Similarly, modelling has shown that more effective management of protected areas (i.e. design of protected areas, adequacy and appropriateness of management, delivery of objectives; SCBD 2018c) can have a greater positive impact on the RLI than only expanding protected areas (Costelloe *et al.* 2016).

Other factors
Other factors include cultural and marketplace trends, such as people not buying items of clothing made using animals (fur, leather, feather down) and the increase in vegetarian/vegan diets in Western countries (Newport 2012; Saner 2016). Both these trends can result in a decrease in the use of species, and an increase in the RLI. Advocacy groups and consumer policies that push for decreases in the use of threatened species play a large role in marketplace trends; for example, consumer awareness campaigns, an increase in the number of organizations certifying environmental sustainability, and government restrictions have combined to dramatically reduce the consumption of shark fins in China in recent years (Fabinyi 2016).

Caveats
Empirical data supporting policy effectiveness remain scarce. One study showed that the efforts of a local conservation trust resulted in improving the status of a small set of 17 threatened vertebrate species in Brazil, India, Madagascar, Mauritius and Spain (Young *et al.* 2014). However, other studies have shown that the RLI has the potential to exhibit a shifting baseline over the long term. This is because the Red List measures declines in abundance over species-specific time frames, so if populations stabilize, a species may return to a low-risk category despite being at very low population levels (Costelloe *et al.* 2016; Nicholson, Fulton and Collen 2017).

© Dr Nibedita Mukherjee

13.3.3 Indicator 3: Ecological footprint

The Ecological Footprint, or Ecological Footprint Accounting, "compares human demand on nature against biocapacity, or nature's supply" and capacity to regenerate (Rees and Wackernagel 1996). "Demand is measured by the biologically productive area a human population uses for producing the natural resources it consumes and absorbing its waste." Biocapacity is measured in surface area (Biodiversity Indicators Partnership 2018c). The Ecological Footprint is measured "by taking the amount of biologically productive land and water area, or biocapacity that is required to produce the food, fibre and renewable raw materials an individual, population or activity consumes". It also takes into account the materials needed to absorb carbon dioxide emissions generated (Global Footprint Network 2018). The Ecological Footprint uses an area-equivalent unit called global hectares (gha); 1 gha represents a biologically productive hectare with world average productivity (Galli 2015). The Ecological Footprint encompasses production and consumption, and each of these comprises the cropland, grazing, forest product, carbon and fish footprints, as well as built-up land (Global Footprint Network 2018). As a population's pressure on biodiversity grows, so does its Ecological Footprint (see Section 13.2.5 Edmonton case study). The world Ecological

Footprint by component (land type) between 1961 and 2013 is shown in **Figure 13.10**.

Policy relevance
The Ecological Footprint indicator is directly relevant to Target 4 of the Aichi Biodiversity Targets and to several targets within SDG 8 (8.4) and SDG 12 (12.2).

Causal relations
There have been many studies on Ecological Footprint Accounting and how it can guide policy creation (e.g. the global Ecological Footprint aiding in the recent adoption of a National Strategy for Sustainable Development in Montenegro (Galli 2015; Galli *et al.* 2018), but few empirical examples of how policy changes have affected the global or national Ecological Footprint. Any policies that reduce or promote sustainable management of resource consumption, land use or carbon emissions will lower the Ecological Footprint, while those policies that directly or indirectly promote increases in these parameters raise it. One study has found that economic globalization drives the Ecological Footprint of consumption, production, imports and exports, while social globalization increases the Ecological Footprint of imports and exports but lowers the Ecological Footprint of consumption and production (Rudolph and Figuree 2017).

Other factors
Other factors that can influence the Ecological Footprint are environmental events that change the biocapacity of a region (e.g. climate change causing a previously unproductive area to become productive or vice versa), technological advancements that increase the biocapacity of a region (e.g. heat-resistant genetically modified crops that increase the productivity of an area) or cultural consumer choices that increase or decrease resource consumption (e.g. opting for public transit, walking or biking instead of using motor vehicles).

Caveats
Although the Ecological Footprint has been widely embraced due to its clear depiction for policymakers of the overuse of ecosystem services (Galli 2015), it has also been criticized because it fails to track human-induced depletion of natural capital stocks. However, the methodology is actively being improved by the Global Footprint Network (Mancini *et al.* 2017).

13.4 Conclusions

It is well established that biodiversity is in a crisis and that existing policy and governance measures to conserve biodiversity have not been adequate (see Chapter 6, Executive Summary). This may be because policy responses may be insufficient to counteract the growth of drivers of loss (SCBD 2014).

Evidence suggests that inadequate economic incentives and investments in ensuring effective compliance and enforcement of legal instruments at the national level could lead to ineffective policies and governance (Ambalam 2014). A qualitative study assessing the United Nations Convention to Combat Desertification in Africa identified additional challenges, including a lack of adequate baseline data on desertification, poor monitoring mechanisms and ill-defined

Figure 13.10: The world Ecological Footprint by component (land type) between 1961 and 2013, measured by number of Earths

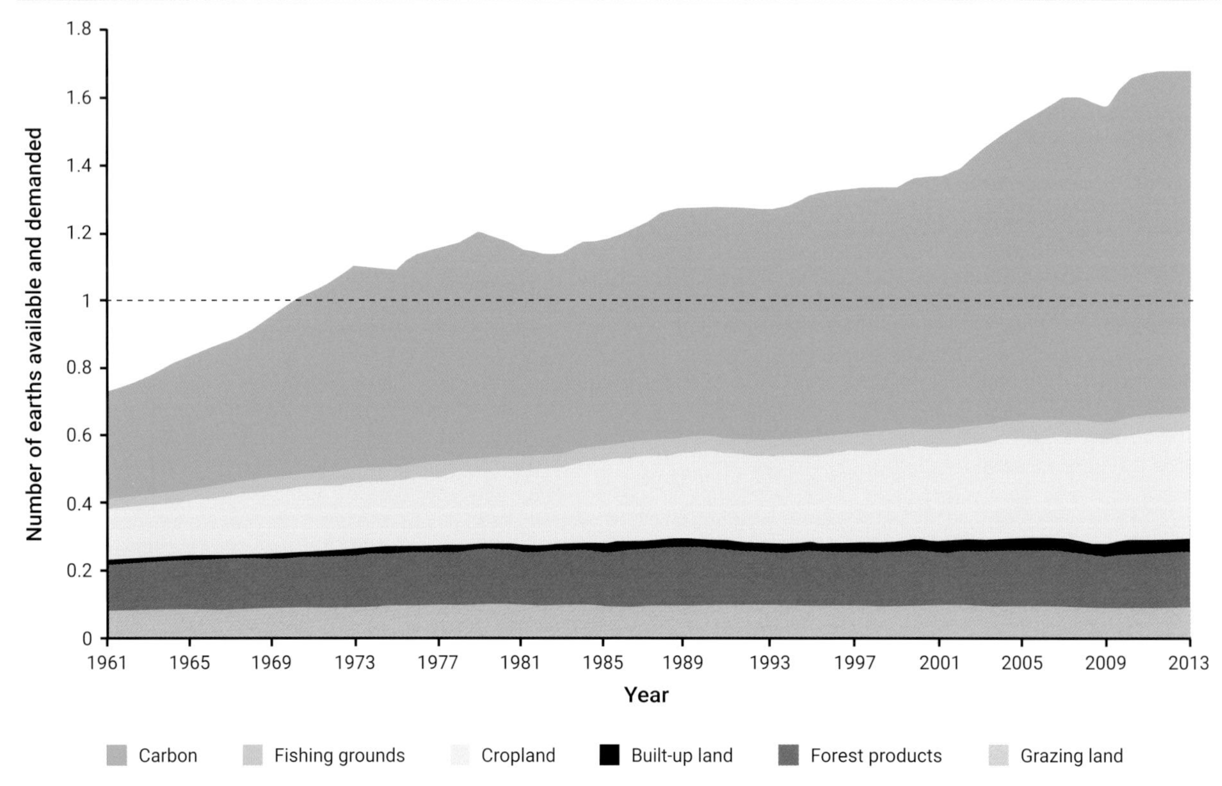

Source: Global Footprint Network (2018).

policy objectives, which hindered compliance (Ambalam 2014). An analysis of the Finnish NBSAP revealed how a range of different forms of responsibility (liability, accountability, responsiveness and care) in different policy sectors could be constructed by introducing new knowledge, providing better process design and building institutional networks (Sarkki *et al.* 2016). However, there remained a lack of intersectoral dialogue despite pro-biodiversity outcomes in other targeted policy sectors, and the responsibilities did not percolate from the environmental administration to other policy sectors. Addressing this cross-sectoral 'responsibility gap' remains a major challenge for effective environmental policies (Mukherjee *et al.* 2015; Sarkki *et al.* 2016). In addition, International Environmental Agreements, in particular, seldom go beyond business-as-usual outcomes (Kellenberg and Levinson 2014). Diffuse language and the lack of quantitative or measurable goals in many International Environmental Agreements leave signatory countries' actions open to interpretation and prevent rigorous appraisal of their performance in improving the quality of ecosystems.

Biodiversity conservation policy is inherently multifaceted, and it is more vital than ever that a 'big picture' perspective emerges among practitioners and governments. Integrating climate, health and equity issues into efforts to mainstream biodiversity, and developing awareness across sectors of policy commitments, are key to the overall success of the SDGs. Many of the policy initiatives discussed in this chapter can serve as models for scaling up efforts to the global level with appropriate and sustained support from governments.

References

Allison, H. and Brown, C. (2017). A review of recent developments in ecosystem assessment and its role in policy evolution. *Current Opinion in Environmental Sustainability* 29, 57-62. https://doi.org/10.1016/j.cosust.2017.11.006.

AlbertaCanada (2017). *Opportunities in Alberta's renewable energy sector*. https://www.albertacanada.com/business/industries/re-opportunities.aspx (Accessed: 2 February 2018).

Almond, R.E.A., Butchart, S.H.M., Oldfield, T.E.E., McRae, L. and de Bie, S. (2013). Exploitation indices: Developing global and national metrics of wildlife use and trade. In *Biodiversity Monitoring and Conservation: Bridging the Gap Between Global Commitment and Local Action*. Collen, B., Pettorelli, N., Baillie, J.E.M. and Durant, S.M. (eds.). Oxford: John Wiley & Sons. chapter 8. 159-188. http://doi.org/10.1002/9781118490747.ch8

Ambalam, K. (2014). Challenges of compliance with multilateral environmental agreements: The case of the United Nations Convention to Combat Desertification in Africa. *Journal of Sustainable Development Studies* 5(2), 145-168. http://infinitypress.info/index.php/jsds/article/download/552/276.

Apostolopoulou, E. and Pantis, J.D. (2009). Conceptual gaps in the national strategy for the implementation of the European natura 2000 conservation policy in Greece. *Biological Conservation* 142(1), 221-237. https://doi.org/10.1016/j.biocon.2008.10.021.

Appleton, A.F. (2013). *How New York City Used an Ecosystem Services Strategy Carried out Through an Urban-Rural Partnership to Preserve the Pristine Quality of its Drinking Water and Save Billions of Dollars*. Water Commons, Water Citizenship and Water Security: Revolutionizing Water Management and Governance for Rio + 20 and Beyond. Minneapolis, MN: Our Water Commons. http://www.ourwatercommons.org/sites/default/files/New-York-preserving-the-pristine-quality-of-its-drinking-water.pdf.

Asdal, Å. (2018). *One million seed samples deposited*. Norwegian Ministry of Agriculture and Food https://www.seedvault.no/news/one-million-seed-samples-deposited/ (Accessed: 5 October 2018).

Australia, Invasive Species Council (2015). *Strategic Plan 2016-2022*. https://invasives.org.au/wp-content/uploads/2015/02/Strategic-Plan-Report-2016-2022.pdf.

Ayling, J. (2013). What sustains wildlife crime? Rhino horn trading and the resilience of criminal networks. *Journal of International Wildlife Law & Policy* 16(1), 57-80. https://doi.org/10.1080/13880292.2013.764776.

Baillie, J.E.M., Collen, B., Amin, R., Akcakaya, H.R., Butchart, S.H.M., Brummitt, N. *et al.* (2008). Toward monitoring global biodiversity. *Conservation Letters* 1(1), 18-26. https://doi.org/10.1111/j.1755-263X.2008.00009.x.

Barnes, A., Ebright, M., Gaskin, E. and Strain, W. (2007). *Working for Water: Addressing Social and Environmental Problems with Payments for Ecosystem Services in South Africa*. https://rmportal.net/library/content/translinks/translinks-2007/earth-institute/WorkingForWaterSouthAfrica_CaseStudy_Translinks_2007.pdf/at_download/file.

Bellard, C., Cassey, P. and Blackburn, T.M. (2016). Alien species as a driver of recent extinctions. *Biology Letters* 12(2). https://doi.org/10.1098/rsbl.2015.0623.

Bennett, N.J., Roth, R., Klain, S.C., Chan, K., Christie, P., Clark, D.A. *et al.* (2017). Conservation social science: Understanding and integrating human dimensions to improve conservation. *Biological Conservation* 205, 93-108. https://doi.org/10.1016/j.biocon.2016.10.006.

Bertzky, B., Corrigan, C., Kemsey, J., Kenney, S., Ravilious, C., Besançon, C. *et al.* (2012). *Protected Planet Report 2012: Tracking Progress Towards Global Targets for Protected Areas*. Gland: International Union for Conservation of Nature and United Nations Environment Programme World Conservation Monitoring Centre. https://cmsdata.iucn.org/downloads/protected_planet_report.pdf.

Biodiversity Indicators Partnership (2018a). *Legislation for prevention and control of invasive alien species (IAS), encompassing "Trends in policy responses, legislation and management plans to control and prevent spread of invasive alien species" and "Proportion of countries adopting relevant national legislation and adequately resourcing the prevention or control of invasive alien species"*. https://www.bipindicators.net/indicators/adoption-of-national-legislation-relevant-to-the-prevention-or-control-of-invasive-alien-species (Accessed: 2 January 2018).

Biodiversity Indicators Partnership (2018b). *Red list index (impacts of utilisation)*. https://www.bipindicators.net/indicators/red-list-index/red-list-index-impacts-of-utilisation (Accessed: 5 February 2018).

Biodiversity Indicators Partnership (2018c). *Ecological footprint*. [United Nations Environment Programme World Conservation Monitoring Centre https://www.bipindicators.net/indicators/ecological-footprint (Accessed: 13 February 2018).

Birdlife International (2012). *Developing and Implementing National Biodiversity Strategies and Action Plans: How to Set, Meet and Track the Aichi Biodiversity Targets*. Cambridge. http://www.birdlife.org/datazone/userfiles/file/sowb/pubs/NBSAP_booklet_Sep_2012.pdf.

Bonnardeaux, D. (2012). *Linking Biodiversity Conservation and Water, Sanitation, and Hygiene: Experiences from Sub-Saharan Africa*. https://www.conservation.org/publications/Documents/ABCG-CI_LinkingBiodiversityConservationWASH.pdf.

Börner, J., Baylis, K., Corbera, E., Ezzine-de-Blas, D., Honey-Rosés, J., Persson, U.M. *et al.* (2017). The effectiveness of payments for environmental services. *World Development* 96, 359-374. https://doi.org/10.1016/j.worlddev.2017.03.020.

Börner, J., Baylis, K., Corbera, E., Ezzine-de-Blas, D., Honey-Rosés, J., Persson, U.M. *et al.* (2017). The effectiveness of payments for environmental services. *World Development* 96, 359-374. https://doi.org/10.1016/j.worlddev.2017.03.020.

Bourguignon, D. (2015). *Safeguarding Biological Diversity - EU Policy and International Agreements*. Brussels: European Union. http://www.europarl.europa.eu/RegData/etudes/IDAN/2015/554175/EPRS_IDA(2015)554175_EN.pdf.

Braat, L.C. and ten Brink, P. (2008). *The Cost of Policy Inaction: The Case of not Meeting the 2010 Biodiversity Target*. Wageningen: Alterra. http://edepot.wur.nl/152014.

Brockerhoff, E.G., Kimberley, M., Liebhold, A.M., Haack, R.A. and Cavey, J.F. (2014). Predicting how altering propagule pressure changes establishment rates of biological invaders across species pools. *Ecology* 95(3), 594-601. https://doi.org/10.1890/13-0465.1.

Brown, K. (2003). Three challenges for a real people-centred conservation. *Global Ecology and Biogeography* 12(2), 89-92. https://doi.org/10.1046/j.1466-822X.2003.00327.x.

Bryan, B.A. (2010). Development and application of a model for robust, cost-effective investment in natural capital and ecosystem services. *Biological Conservation* 143(7), 1737-1750. https://doi.org/10.1016/j.biocon.2010.04.022.

Bubb, P., Butchart, S.H.M., Collen, B., Dublin, H., Kapos, V., Pollock, C. *et al.* (2009). *IUCN Red List Index: Guidance for National and Regional Use. Version 1.1*. Gland. https://portals.iucn.org/library/sites/library/files/documents/2009-001.pdf.

Buch, A. and Dixon, A.B. (2009). South Africa's working for water programme: Searching for win–win outcomes for people and the environment. *Sustainable Development* 17(3), 129-141. https://doi.org/10.1002/sd.370.

Cardinale, B.J., Duffy, J.E., Gonzalez, A., Hooper, D.U., Perrings, C., Venail, P. *et al.* (2012). Biodiversity loss and its impact on humanity. *Nature* 486(7401), 59-67. https://doi.org/10.1038/nature11148.

Carrus, G., Scopelliti, M., Lafortezza, R., Colangelo, G., Ferrini, F., Salbitano, F. *et al.* (2015). Go greener, feel better? The positive effects of biodiversity on the well-being of individuals visiting urban and peri-urban green areas. *Landscape and Urban Planning* 134, 221-228. https://doi.org/10.1016/j.landurbplan.2014.10.022.

Challender, D.W.S. and MacMillan, D.C. (2014). Poaching is more than an enforcement problem. *Conservation Letters* 7(5), 484-494. https://doi.org/10.1111/conl.12082.

Chamier, J., Schachtschneider, K., le Maitre, D.C., Ashton, P.J. and van Wilgen, B.W. (2012). Impacts of invasive alien plants on water quality, with particular emphasis on South Africa. *Water SA* 38(2), 345-356. https://doi.org/10.4314/wsa.v38i2.19

Charron, D.F. (2012). Ecohealth research in practice. In *Ecohealth Research in Practice: Innovative Applications of an Ecosystem Approach to Health*. Charron, D.F. (ed.). New York, NY: Springer. 255-271. https://doi.org/10.1007/978-1-4614-0517-7_22

City of Edmonton (2009). *Natural Connections: Biodiversity Action Plan*. Edmonton. https://www.edmonton.ca/city_government/documents/PDF/Edmonton_Biodiversity_Action_Plan_Final.PDF.

Colding, J. (2007). 'Ecological land-use complementation' for building resilience in urban ecosystems. *Landscape and Urban Planning* 81(1-2), 46-55. https://doi.org/10.1016/j.landurbplan.2006.10.016.

Conrad, K. (2012). Trade bans: A perfect storm for poaching? *Tropical Conservation Science* 5(3), 245-254. https://doi.org/10.1177/194008291200500302.

Cooney, R., Roe, D., Dublin, H. and Booker, F. (2018). *Wild Life, Wild Livelihoods: Involving Communities in Sustainable Wildlife Management and Combatting the Illegal Wildlife Trade*. Nairobi: United Nations Environment Programme. http://wedocs.unep.org/bitstream/handle/20.500.11822/22864/WLWL_Report_web.pdf.

Convention on Biological Diversity (1992). *Convention on Biological Diversity*. https://www.cbd.int/doc/legal/cbd-en.pdf.

Convention on Biological Diversity (2012. XI/6. *Cooperation with other Conventions, International Organizations, and initiatives. COP 11 Decision XI/6*. https://www.cbd.int/decision/cop/?id=13167.

Convention on Biological Diversity (2016a). XIII/3. *Strategic Actions to Enhance the Implementation of the Strategic Plan for Biodiversity 2011-2020 and the Achievement of the Aichi Biodiversity Targets, including with respect to Mainstreaming and the Integration of Biodiversity within and across Sectors. Decision adopted by the Conference of Parties to the Convention on Biological Diversity*. https://www.cbd.int/doc/decisions/cop-13/cop-13-dec-03-en.pdf.

Convention on Biological Diversity (2016b). XIII/6. Biodiversity and Human Health. *Decision adopted by the Conference of Parties to the Convention on Biological Diversity*. https://www.cbd.int/doc/decisions/cop-13/cop-13-dec-06-en.pdf.

Corrigan, C. and Hay-Edie, T. (2013). *A Toolkit to Support Conservation by Indigenous Peoples and Local Communities: Building Capacity and Sharing Knowledge for Indigenous Peoples' and Community Conserved Territories and Areas (ICCAs)*. Cambridge: United Nations Environment Programme World Conservation Monitoring Centre. http://www.silene.es/documentos/Toolkit_conservation_indigenous_peoples.pdf.

Costanza, R., de Groot, R., Sutton, P., van der Ploeg, S., Anderson, S.J., Kubiszewski, I. *et al.* (2014). Changes in the global value of ecosystem services. *Global Environmental Change* 26, 152-158. https://doi.org/10.1016/j.gloenvcha.2014.04.002.

Costelloe, B., Collen, B., Milner-Gulland, E.J., Craigie, I.D., McRae, L., Rondinini, C. *et al.* (2016). Global biodiversity indicators reflect the modeled impacts of protected area policy change. *Conservation Letters* 9(1), 14-20. https://doi.org/10.1111/conl.12163.

Cox, M. (2016). The pathology of command and control: A formal synthesis. *Ecology and Society* 21(3), 33. https://doi.org/10.5751/ES-08698-210333.

Díaz, S., Demissew, S., Carabias, J., Joly, C., Lonsdale, M., Ash, N. *et al.* (2015). The IPBES Conceptual Framework — connecting nature and people. *Current Opinion in Environmental Sustainability* 14, 1-16. https://doi.org/10.1016/j.cosust.2014.11.002

Duffy, G.A., Coetzee, B.W.T., Latombe, G., Akerman, A.H., McGeoch, M.A. and Chown, S.L. (2017). Barriers to globally invasive species are weakening across the Antarctic. *Diversity and Distributions* 23(9), 982-996. https://doi.org/10.1111/ddi.12593.

Dulloo, M.E. (2015). Conservation and availability of plant genetic diversity: Innovative strategies and technologies. *Acta Horticulturae* 1101, 1-8. https://doi.org/10.17660/ActaHortic.2015.1101.1.

Eastwood, R.J., Cody, S., Westengen, O.T. and Bothmer, R. (2015). Conservation roles of the millennium seed bank and the svalbard global seed vault. In *Crop Wild Relatives and Climate Change*. Redden, R., Yadav, S.S., Maxted, N., Dulloo, M.E., Guarino, L. and Smith, P. (eds.). John Wiley & Sons, Inc. chapter 10. 173-186. https://onlinelibrary.wiley.com/doi/pdf/10.1002/9781118854396.ch10

Eklund, J. and Cabeza, M. (2017). Quality of governance and effectiveness of protected areas: Crucial concepts for conservation planning. *Annals of the New York Academy of Sciences* 1399(1), 27-41. https://doi.org/10.1111/nyas.13284.

Esquinas-Alcázar, J. (2005). Protecting crop genetic diversity for food security: Political, ethical and technical challenges. *Nature Reviews Genetics* 6, 946-953. https://doi.org/10.1038/nrg1729.

European Council (1979). *Council Directive of 2 April 1979 on the Conservation of Wild Birds (79/409/EEC)*. https://eur-lex.europa.eu/LexUriServ/LexUriServ.do?uri=CELEX:31979L0409:EN:PDF.

European Council (1992). *Council Directive 92/43/EEC of 21 May 1992 on the Conservation of Natural Habitats and of Wild Fauna and Flora*. https://eur-lex.europa.eu/legal-content/EN/TXT/PDF/?uri=CELEX:31992L0043&from=EN

Fabinyi, M. (2016). Sustainable seafood consumption in China. *Marine Policy* 74, 85-87. https://doi.org/10.1016/j.marpol.2016.09.020.

Felson, A.J. and Pickett, S.T.A. (2005). Designed experiments: New approaches to studying urban ecosystems. *Frontiers in Ecology and the Environment* 3(10), 549-556. https://doi.org/10.1890/1540-9295(2005)003[0549:DENATS]2.0.CO;2.

Food and Agriculture Organization of the United Nations (2010). *The Second Report on the State of the World's Plant Genetic Resources for Food and Agriculture*. Rome. http://www.fao.org/docrep/013/i1500e/i1500e.pdf.

Food and Agriculture Organization of the United Nations (2018). *Integrated landscape management*. http://www.fao.org/land-water/overview/integrated-landscape-management/en/ (Accessed: 21 June 2018).

Fowler, C. (2008). The svalbard seed vault and crop security. *BioScience* 58(3), 190-191. https://doi.org/10.1641/B580302.

Friel, S. and Ford, L. (2015). Systems, food security and human health. *Food Security* 7(2), 437-451. https://doi.org/10.1007/s12571-015-0433-1.

Galli, A. (2015). On the rationale and policy usefulness of ecological footprint accounting: The case of Morocco. *Environmental Science & Policy* 48, 210-224. https://doi.org/10.1016/j.envsci.2015.01.008.

Galli, A., Đurović, G., Hanscom, L. and Knežević, J. (2018). Think globally, act locally: Implementing the sustainable development goals in Montenegro. *Environmental Science & Policy* 84, 159-169. https://doi.org/10.1016/j.envsci.2018.03.012.

Garvin, E.C., Cannuscio, C.C. and Branas, C.C. (2013). Greening vacant lots to reduce violent crime: A randomised controlled trial. *Injury Prevention* 19(3), 198. https://doi.org/10.1136/injuryprev-2012-040439.

Gaworecki, M. (2017). Cash for conservation: Do payments for ecosystem services work? *Mongabay Series: Conservation Effectiveness*, Mongabay. https://news.mongabay.com/2017/10/cash-for-conservation-do-payments-for-ecosystem-services-work/.

Geldmann, J., Coad, L., Barnes, M.D., Craigie, I.D., Woodley, S., Balmford, A. *et al.* (2018). A global analysis of management capacity and ecological outcomes in terrestrial protected areas. *Conservation Letters* 11(3), e12434. https://doi.org/10.1111/conl.12434.

Global Footprint Network (2018). *Ecological footprint*. [https://www.footprintnetwork.org/our-work/ecological-footprint/ (Accessed: 20 May 2018).

Goodrich, J., Lynam, A., Miquelle, D., Wibisono, H., Kawanishi, K., Pattanavibool, A. *et al.* (2015). *Panthera Tigris: The IUCN Red List of Threatened Species 2015: e.T15955A50659951*. Gland: International Union for Conservation of Nature. https://doi.org/10.2305/IUCN.UK.2015-2.RLTS.T15955A50659951.en.

Govan, H. (2009). *Status and Potential of Locally-Managed Marine Areas in the Pacific Island Region: Meeting Nature Conservation and Sustainable Livelihood Targets Through Wide-Spread Implementation of LMMAs*. https://mpra.ub.uni-muenchen.de/23828/1/MPRA_paper_23828.pdf.

GreenInvest (2017). *Green Foreign Direct Investment in Developing Countries*. Nairobi: United Nations Environment Programme. http://wedocs.unep.org/bitstream/handle/20.500.11822/22280/Green_Invest_Dev_Countries.pdf?sequence=1.

Gunningham, N. and Young, M.D. (1997). Toward optimal environmental policy: The case of biodiversity conservation. *Ecology Law Quarterly* 24(2), 243-298. https://doi.org/10.15779/Z38BN7K.

Harrington, W., Morgenstern, R.D. and Sterner, T. (2004). Overview: Comparing instrument choices. In *Choosing Environmental Policy: Comparing Instruments and Outcomes in the United States and Europe*. Harrington, W., Morgensterm, R.D. and Sterner, T. (eds.). New York, NY: Routledge. https://www.taylorfrancis.com/books/e/9781136524943/chapters/10.4324%2F9781936331468-6

Higgins, D. and White, R. (2016). Collaboration at the front line: INTERPOL and NGOs in the same NEST. In *Environmental Crime and Collaborative State Intervention*. Pink, G. and White, R. (eds.). London: Palgrave Macmillan. chapter 6. 101-116. https://link.springer.com/chapter/10.1007/978-1-137-56257-9_6

Hobbs, R.J. (2004). The working for water programme in South Africa: The science behind the success. *Diversity and Distributions* 10(5-6), 501-503. https://doi.org/10.1111/j.1366-9516.2004.00115.x.

Hodkinson, T.R., Waldren, S., Parnell, J.A.N., Kelleher, C.T., Salamin, K. and Salamin, N. (2007). DNA banking for plant breeding, biotechnology and biodiversity evaluation. *Journal of Plant Research* 120(1), 17-29. https://doi.org/10.1007/s10265-006-0059-7.

Holling, C.S. and Meffe, G.K. (1996). Command and control and the pathology of natural resource management: Comando-y-control y la patología del manejo de los recursos naturales. *Conservation Biology* 10(2), 328-337. https://doi.org/10.1046/j.1523-1739.1996.10020328.x.

Hopkin, R. (2008). Biodiversity: Frozen futures. *Nature* 452, 404-405. https://doi.org/10.1038/452404a.

Hough, R.L. (2014). Biodiversity and human health: Evidence for causality? *Biodiversity and Conservation* 23(2), 267-288. https://doi.org/10.1007/s10531-013-0614-1.

Hulme, P.E. (2009). Trade, transport and trouble: Managing invasive species pathways in an era of globalization. *Journal of Applied Ecology* 46(1), 10-18. https://doi.org/10.1111/j.1365-2664.2008.01600.x.

Hulme, P.E. (2015). Invasion pathways at a crossroad: Policy and research challenges for managing alien species introductions. *Journal of Applied Ecology* 52(6), 1418-1424. https://doi.org/10.1111/1365-2664.12470.

Huynen, M.M., Martens, P. and De Groot, R.S. (2004). Linkages between biodiversity loss and human health: A global indicator analysis. *International Journal of Environmental Research and Public Health* 14(1), 13-30. https://doi.org/10.1080/09603120310001633895.

ICCA Registry (2018). *International conservation and targets*. [United Nations Environment Programme World Conservation Monitoring Centre http://www.iccaregistry.org/en/about/international-conservation-and-targets (Accessed: 23 March 2018).

Intergovernmental Science-Policy Platform on Biodiversity and Ecosystem Services (2016). *The Assessment Report of the Intergovernmental Science-Policy Platform on Biodiversity and Ecosystem Services on Pollinators, Pollination and Food Production*. Potts, S.G., Imperatriz-Fonseca, V.L. and Ngo, H.T. (eds.). Bonn. https://www.ipbes.net/sites/default/files/downloads/pdf/spm_deliverable_3a_pollination_20170222.pdf.

International Maritime Organization (2017). *International convention for the control and management of ships' ballast water and sediments (BWM)*. [http://www.imo.org/en/About/Conventions/ListOfConventions/Pages/International-Convention-for-the-Control-and-Management-of-Ships'-Ballast-Water-and-Sediments-(BWM).aspx.

International Union for Conservation of Nature (2006). *Unified Classification of Conservation Actions: Version 1.0*. Gland. http://www.conservationmeasures.org/wp-content/uploads/2010/04/IUCN-CMP_Unified_Actions_Classification_2006_06_01.pdf.

International Union for Conservation of Nature (2016). *Inclusion and Characterization of Women and Gender Equality Considerations in National Biodiversity Strategies and Action Plans (NBSAPs)*. Washington, D.C. https://www.cbd.int/gender/doc/gender-nbsaps-factsheet.pdf.

International Union for Conservation of Nature (2017). *Gender and Biodiversity: Analysis of Women and Gender Equality Considerations in National Biodiversity Strategies and Action Plans (NBSAPs)*. Environment & Gender Information. Washington DC: International Union for Conservation of Nature. https://www.cbd.int/gender/doc/gender-biodiversity-nbsaps-report-final.pdf.

International Union for Conservation of Nature (2018a). *Red list index*. https://www.iucnredlist.org/assessment/red-list-index (Accessed: 6 November 2018).

International Union for Conservation of Nature (2018b). *About: What is a protected area?* https://www.iucn.org/theme/protected-areas/about (Accessed: 6 November 2018).

International Criminal Police Organization (2015). *Protection of Asian wildlife species: Operation PAWS II (2015)*. https://cites.org/sites/default/files/eng/com/sc/66/E-SC66-44-01-A3.pdf.

Jabbour, J. and Flachsland, C. (2017). 40 years of global environmental assessments: A retrospective analysis. *Environmental Science & Policy* 77, 193-202. https://doi.org/10.1016/j.envsci.2017.05.001.

Jarmain, C. and Meijninger, W.M.L. (2012). Assessing the impact of invasive alien plants on South African water resources using remote sensing techniques. In *Remote Sensing and Hydrology*. Neale, C.M.U. and Cosh, M.H. (eds.). IAHS Press. 388-392. http://library.wur.nl/WebQuery/wurpubs/508172

Juffe-Bignoli, D., Burgess, N.D., Bingham, H., Belle, E.M.S., de Lima, M.G., Deguignet, M. *et al.* (2014). *Protected Planet Report 2014: Tracking Progress Towards Global Targets for Protected Areas*. Cambridge: United Nations Environment Programme World Conservation Monitoring Centre. https://www.unep-wcmc.org/system/dataset_file_fields/files/000/000/289/original/Protected_Planet_Report_2014_01122014_EN_web.pdf?1420549522.

Jupiter, S.D., Cohen, P.J., Weeks, R., Tawake, A. and Govan, H. (2014). Locally-managed marine areas: Multiple objectives and diverse strategies. *Pacific Conservation Biology* 20(2), 165-179. https://doi.org/10.1071/PC140165.

Jupiter, S.D., Epstein, G., Ban, N.C., Mangubhai, S., Fox, M. and Cox, M. (2017). A social–ecological systems approach to assessing conservation and fisheries outcomes in Fijian locally managed marine areas. *Society & Natural Resources* 30(9), 1096-1111. https://doi.org/10.1080/08941920.2017.1315654.

Kellenberg, D. and Levinson, A. (2014). Waste of effort? International environmental agreements. *Journal of the Association of Environmental and Resource Economists* 1(1/2), 135-169. https://doi.org/10.1086/676037.

Kusmanoff, A. (2017). *Framing the Conservation Conversation: An Investigation into Framing Techniques for Communicating Biodiversity Conservation*. Doctor of Philosophy (PhD), RMIT University https://researchbank.rmit.edu.au/eserv/rmit:162021/Kusmanoff.pdf

Laitos, J.G. and Wolongevicz, L.J. (2014). Why Environmental Laws Fail. *William & Mary Environmental Law and Policy Review* 39(1). https://scholarship.law.wm.edu/wmelpr/vol39/iss1/2.

Le Maitre, D.C., Gush, M.B. and Dzikiti, S. (2015). Impacts of invading alien plant species on water flows at stand and catchment scales. *AoB Plants* 7(1), 1-21. https://doi.org/10.1093/aobpla/plv043.

Levendal, M., Le Maitre, D.C., van Wilgen, B.W. and Ntshotso, P. (2008). *The Development of Protocals for the Monitoring and Evaluation of Benefits Arising from the Working for Water Programme*. Monitoring & Evaluation Frameworks. Pretoria: Council for Scientific and Industrial Research. http://www.dwaf.gov.za/wfw/docs/Levendaletal.,2008.pdf.

Liebhold, A.M., Brockerhoff, E.G., Garrett, L.J., Parke, J.L. and Britton, K.O. (2012). Live plant imports: The major pathway for forest insect and pathogen invasions of the US. *Frontiers in Ecology and the Environment* 10(3), 135-143. https://doi.org/10.1890/110198.

Local Governments for Sustainability (2013). *Cities and Biodiversity: Exploring how Edmonton and Montreal are Mainstreaming the Urban Biodiversity Movement*. Toronto. http://www.biopolis.ca/wp-content/uploads/2013/01/Cities-and-Biodiversity-Exploring-how-Edmonton-and-Montreal-are-Mainstreaming-the-Urban-Biodiversity-Movement.pdf.

Maas, J., van Dillen, S.M.E., Verheij, R.A. and Groenewegen, P.P. (2009). Social contacts as a possible mechanism behind the relation between green space and health. *Health & Place* 15(2), 586-595. https://doi.org/10.1016/j.healthplace.2008.09.006.

Magadlela, D. and Mdzeke, N. (2004). Social benefits in the Working for Water programme as a public works initiative: Working for water. *South African Journal of Science* 100(1-2), 94-96. https://hdl.handle.net/10520/EJC96206.

Mancini, M.S., Galli, A., Niccolucci, V., Lin, D., Hanscom, L., Wackernagel, M. *et al.* (2017). Stocks and flows of natural capital: Implications for ecological footprint. *Ecological Indicators* 77, 123-128. https://doi.org/10.1016/j.ecolind.2017.01.033.

Marcos-Martinez, R., Bryan, B.A., Schwabe, K.A., Connor, J.D. and Law, E.A. (2018). Forest transition in developed agricultural regions needs efficient regulatory policy. *Forest Policy and Economics* 86, 67-75. https://doi.org/10.1016/j.forpol.2017.10.021.

McConnachie, M.M., Cowling, R.M., van Wilgen, B.W. and McConnachie, D.A. (2012). Evaluating the cost-effectiveness of invasive alien plant clearing: A case study from South Africa. *Biological Conservation* 155, 128-135. https://doi.org/10.1016/j.biocon.2012.06.006.

McQueen, C., Noemdoe, S. and Jezile, N. (2001). The working for water programme. *Land Use and Water Resources Research* 1(4), 1-4. https://core.ac.uk/download/pdf/6569950.pdf.

Meijninger, W.M.L. and Jarmain, C. (2014). Satellite-based annual evaporation estimates of invasive alien plant species and native vegetation in South Africa. *Water SA* 40(1), 95-107. https://doi.org/10.4314/wsa.v40i1.12.

Meyerson, L.A. and Mooney, H.A. (2007). Invasive alien species in an era of globalization. *Frontiers in Ecology and the Environment* 5(4), 199-208. https://doi.org/10.1890/1540-9295(2007)5[199:IASIAE]2.0.CO;2.

Mukherjee, N., Dahdouh-Guebas, F., Koedam, N. and Shanker, K. (2015). An interdisciplinary framework to evaluate bioshield plantations: Insights from peninsular India. *Acta Oecologica* 63, 91-100. https://doi.org/10.1016/j.actao.2014.01.005.

Mukherjee, N., Dicks, L.V., Shackelford, G.E., Vira, B. and Sutherland, W.J. (2016). Comparing groups versus individuals in decision making: A systematic review protocol. *Environmental Evidence* 5(19). https://doi.org/10.1186/s13750-016-0066-7.

Mukherjee, N., Zabala, A., Huge, J., Nyumba, T.O., Adem Esmail, B. and Sutherland, W.J. (2018). Comparison of techniques for eliciting views and judgements in decision-making. *Methods in Ecology and Evolution* 9(1), 54-63. https://doi.org/10.1111/2041-210X.12940.

Nellemann, C. and INTERPOL Environmental Crime (eds.) (2012). *Green Carbon, Black Trade: Illegal Logging, Tax Fraud and Laundering in the Worlds Tropical Forests. A Rapid Response Assessment*. https://gridarendal-website-live.s3.amazonaws.com/production/documents/:s_document/148/original/RRAlogging_english_scr.pdf?1483646716.

Nelson, A. (2017). UK named as world's largest legal ivory exporter. *The Guardian* 15 October 2017 https://www.theguardian.com/environment/2017/aug/10/uk-named-as-worlds-largest-legal-ivory-exporter.

New South Wales, Office of Environment and Heritage, (2017). *Draft Biodiversity Conservation Investment Strategy 2017-2037: A Strategy to Guide Investment in Private Land Conservation*. Sydney. http://www.environment.nsw.gov.au/resources/biodiversity/strategy/draft-biodiversity-conservation-investment-strategy-170450.pdf.

Newport, F. (2012). In U.S., 5% consider themselves vegetarians: Even smaller 2% say they are vegans. Gallup https://news.gallup.com/poll/156215/consider-themselves-vegetarians.aspx.

Nicholson, E., Fulton, E.A. and Collen, B. (2017). Linking biodiversity indicators with global conservation policy. In *Decision-Making in Conservation and Natural Resource Management: Models for Interdisciplinary Approaches*. Bunnefeld, N., Nicholson, E. and Milner-Gulland, E.J. (eds.). Cambridge: Cambridge University Press. chapter 9. 196-212. https://www.cambridge.org/core/books/decisionmaking-in-conservation-and-natural-resource-management/linking-biodiversity-indicators-with-global-conservation-policy/50F2977DF871CEE99AD8E04C84086449

Niemelä, J., Saarela, S.-R., Söderman, T., Kopperoinen, L., Yli-Pelkonen, V., Väre, S. *et al.* (2010). Using the ecosystem services approach for better planning and conservation of urban green spaces: A Finland case study. *Biodiversity and Conservation* 19(11), 3225-3243. https://doi.org/10.1007/s10531-010-9888-8.

Oldekop, J.A., Holmes, G., Harris, W.E. and Evans, K.L. (2015). A global assessment of the social and conservation outcomes of protected areas. *Conservation Biology* 30(1), 133-141. https://doi.org/10.1111/cobi.12568.

Oliver, T.H., Heard, M.S., Isaac, N.J.B., Roy, D.B., Procter, D., Eigenbrod, F. *et al.* (2015). Biodiversity and resilience of ecosystem functions. *Trends in Ecology & Evolution* 30(11), 673-684. https://doi.org/10.1016/j.tree.2015.08.009.

Paavola, J., Gouldson, A. and Kluvánková-Oravská, T. (2009). Interplay of actors, scales, frameworks and regimes in the governance of biodiversity. *Environmental Policy and Governance* 19(3), 148-158. https://doi.org/10.1002/eet.505.

Paloniemi, R. and Vilja, V. (2009). Changing ecological and cultural states and preferences of nature conservation policy: The case of nature values trade in South-Western Finland. *Journal of Rural Studies* 25(1), 87-97. https://doi.org/10.1016/j.jrurstud.2008.06.004.

Pejchar, L. and Mooney, H.A. (2009). Invasive species, ecosystem services and human well-being. *Trends in Ecology & Evolution* 24(9), 497-504. https://doi.org/10.1016/j.tree.2009.03.016.

Porras, I., Barton, D.N., Chacón-Cascante, A. and Miranda, M. (2013). *Learning from 20 Years of Payments for Ecosystem Services in Costa Rica*. London: International Institute for Environment and Development. http://pubs.iied.org/pdfs/16514IIED.pdf.

Primmer, E., Paloniemi, R., Mathevet, R., Apostolopoulou, E., Tzanopoulos, J., Ring, I. *et al.* (2014). An approach to analysing scale-sensitivity and scale-effectiveness of governance in biodiversity conservation. In *Scale-sensitive Governance of the Environment*. Padt, F., Opdam, P., Polman, N. and Termeer, C. (eds.). chapter 15. https://onlinelibrary.wiley.com/doi/pdf/10.1002/9781118567135.ch15

Redpath Steve, M., Linnell John, D.C., Festa-Bianchet, M., Boitani, L., Bunnefeld, N., Dickman, A. *et al.* (2017). Don't forget to look down – collaborative approaches to predator conservation. *Biological Reviews* 92(4), 2157-2163. https://doi.org/10.1111/brv.12326.

Reed, M.S. (2008). Stakeholder participation for environmental management: A literature review. *Biological Conservation* 141(10), 2417-2431. https://doi.org/10.1016/j.biocon.2008.07.014.

Rees, W. and Wackernagel, M. (1996). Urban ecological footprints: Why cities cannot be sustainable—And why they are a key to sustainability. *Environmental Impact Assessment Review* 16(4-6), 223-248. https://doi.org/10.1016/S0195-9255(96)00022-4.

Rivalan, P., Delmas, V., Angulo, E., Bull, L.S., Hall, R.J., Courchamp, F. *et al.* (2007). Can bans stimulate wildlife trade? *Nature* 447, 529-530. https://doi.org/10.1038/447529a.

Roe, J.J., Thompson, W.C., Aspinall, A.P., Brewer, J.M., Duff, I.E., Miller, D. *et al.* (2013). Green space and stress: Evidence from cortisol measures in deprived urban communities. *International Journal of Environmental Research and Public Health* 10(9), 4086–4103. https://doi.org/10.3390/ijerph10094086.

Rosenzweig, C., Solecki, W.D., Romero-Lankao, P., Mehrotra, S., Dhakal, S. and Ibrahim, S.A. (eds.) (2018). *Climate Change and Cities: Second Assessment Report of the Urban Climate Change Research Network*. Cambridge, MA: Cambridge University Press. https://www.cambridge.org/core/books/climate-change-and-cities/climate-change-and-cities-second-assessment-report-of-the-urban-climate-change-research-network/BE242A59BEA99C3DB5E663BAF5FD480F.

Rudolph, A. and Figuree, L. (2017). Determinants of ecological footprints: What is the role of globalization? *Ecological Indicators* 81, 348-361. https://doi.org/10.1016/j.ecolind.2017.04.060.

Sandifer, P.A., Sutton-Grier, A.E. and Ward, B.P. (2015). Exploring connections among nature, biodiversity, ecosystem services, and human health and well-being: Opportunities to enhance health and biodiversity conservation. *Ecosystem Services* 12, 1-15. https://doi.org/10.1016/j.ecoser.2014.12.007.

Saner, E. (2016). Fit, macho, sexy: The reinvention of vegans. *The Guardian* 18 May 2016 https://www.theguardian.com/lifeandstyle/2016/may/18/vegans-veganism-fit-macho-sexy-beyonce-ufc-fighters-wellness-bloggers.

Sansevero, J.B.B., Prieto, P.V., Sánchez-Tapia, A., Braga, J.M.A. and Rodrigues, P.J.F.P. (2017). Past land-use and ecological resilience in a lowland Brazilian Atlantic forest: Implications for passive restoration. *New Forests* 48(5), 573-586. https://doi.org/10.1007/s11056-017-9586-4.

Sarkki, S., Ficko, A., Grunewald, K. and Nijnik, M. (2016). Benefits from and threats to European treeline ecosystem services: An exploratory study of stakeholders and governance. *Regional Environmental Change* 16(7), 2019-2032. https://doi.org/10.1007/s10113-015-0812-3.

Schwartz, M., W.,, Cook, C., N.,, Pressey, R., L.,, Pullin, A., S.,, Runge, M., C.,, Salafsky, N. *et al.* (2017). Decision support frameworks and tools for conservation. *Conservation Letters* 11(2), e12385. https://doi.org/10.1111/conl.12385.

Scott-Shaw, B.C., Everson, C.S. and Clulow, A.D. (2017). Water-use dynamics of an alien-invaded riparian forest within the Mediterranean climate zone of the Western Cape, South Africa. *Hydology and Earth System Sciences* 21(9), 4551-4562. https://doi.org/10.5194/hess-21-4551-2017.

Secretariat of the Convention on Biological Diversity (2012). *Resourcing the Aichi Biodiversity Targets: A First Assessment of the Resources Required for Implementing the Strategic Plan for Biodiversity 2011-2020*. Montreal. https://www.cbd.int/doc/meetings/fin/hlpgar-sp-01/official/hlpgar-sp-01-01-report-en.pdf.

Secretariat of the Convention on Biological Diversity (2014). *Global Biodiversity Outlook 4: A Mid-Term Assessment of Progress Towards the Implementation of the Strategic Plan for Biodiversity 2011-2020*. Montreal. https://www.cbd.int/gbo/gbo4/publication/gbo4-en-hr.pdf.

Secretariat of the Convention on Biological Diversity (2018a). *Latest NBSAPs*. https://www.cbd.int/nbsap/about/latest/default.shtml (Accessed: 6 May 2018).

Secretariat of the Convention on Biological Diversity (2018b). *Background: Gender Mainstreaming in International Agreements*. https://www.cbd.int/gender/background/ (Accessed: 28 March 2018).

Secretariat of the Convention on Biological Diversity (2018c). *Protected Areas Management Effectiveness*. https://www.cbd.int/protected-old/PAME.shtml (Accessed: 2 May 2018).

Secretariat of the Convention on Biological Diversity and International Union for Conservation of Nature (2018). *Gender and Access and Benefit Sharing of Genetic Resources (ABS)*. Gland. https://portals.iucn.org/union/sites/union/files/doc/gender_and_access_and_benefits_sharing_of_genetic_resources.pdf.

Seebens, H., Essl, F., Dawson, W., Fuentes, N., Moser, D., Pergl, J. *et al.* (2015). Global trade will accelerate plant invasions in emerging economies under climate change. *Global Change Biology* 21(11), 4128-4140. https://doi.org/10.1111/gcb.13021.

South Africa, Department of Environmental Affairs (2010). *Value added industries and wetlands projects*. https://www.environment.gov.za/projectsprogrammes/wfw/valueadded_industries_wetlands (Accessed: 1 October 201).

South Africa, Department of Water Affairs (2010a). *Research*. http://www.dwaf.gov.za/wfw/problem.aspx (Accessed: 1 October 2017).

South Africa, Department of Water Affairs (2010b). *Welcome to the working for water webpage*. http://www.dwaf.gov.za/wfw/ (Accessed: 1 October 2017).

South African National Biodiversity Institute (2008). *Rietvlei Rehabilitation Project Aids in Water Purification*. Cape Town.

Sterling, E.J., Betley, E., Sigouin, A., Gomez, A., Toomey, A., Cullman, G. *et al.* (2017). Assessing the evidence for stakeholder engagement in biodiversity conservation. *Biological Conservation* 209, 159-171. https://doi.org/10.1016/j.biocon.2017.02.008.

Stoett, P. (2012). *Global Ecopolitics: Crisis, Governance and Justice*. Toronto: University of Toronto Press. https://books.google.ca/books?id=KvByBgAAQBAJ&dq=global+ecopolitics+crisis+governance+and+justice&lr=.

Stoner, S., Krishnasamy, K., Wittmann, T., Delean, S. and Cassey, P. (2016). *Reduced to Skin and Bones Re-Examined: Full Analysis. An Analysis of Tiger Seizures from 13 Range Countries from 2000-2015*. Selangor: TRAFFIC. http://tigers.panda.org/wp-content/uploads/Reduced-to-Skin-and-Bones-Re-examined-Full-Analysis.pdf.

Strayer, D.L. (2010). Alien species in fresh waters: Ecological effects, interactions with other stressors, and prospects for the future. *Freshwater Biology* 55, 152-174. https://doi.org/10.1111/j.1365-2427.2009.02380.x.

Sumaila, U.R., Rodriguez, C.M., Schultz, M., Sharma, R., Tyrrell, T.D., Masundire, H. *et al.* (2017). Investments to reverse biodiversity loss are economically beneficial. *Current Opinion in Environmental Sustainability* 29, 82-88. https://doi.org/10.1016/j.cosust.2018.01.007.

Thinley, P., Rajaratnam, R., Lassoie, J.P., Morreale, S.J., Curtis, P.D., Vernes, K. *et al.* (2018). The ecological benefit of tigers (Panthera tigris) to farmers in reducing crop and livestock losses in the eastern Himalayas: Implications for conservation of large apex predators. *Biological Conservation* 219, 119-125. https://doi.org/10.1016/j.biocon.2018.01.015.

Tittensor, D.P., Walpole, M., Hill, S.L.L., Boyce, D.G., Britten, G.L., Burgess, N.D. *et al.* (2014). A mid-term analysis of progress toward international biodiversity targets. *Science* 346(6206), 241-244. https://doi.org/10.1126/science.1257484.

Turpie, J.K., Marais, C. and Blignaut, J.N. (2008). The working for water programme: Evolution of a payments for ecosystem services mechanism that addresses both poverty and ecosystem service delivery in South Africa. *Ecological Economics* 65(4), 788-798. https://doi.org/10.1016/j.ecolecon.2007.12.024.

United Kingdom, Department for Environmental and Rural Affairs (2014). *Protecting Plant Health: A Plant Biosecurity Strategy for Great Britain* London. https://assets.publishing.service.gov.uk/government/uploads/system/uploads/attachment_data/file/307355/pb14168-plant-health-strategy.pdf.

United Kingdom, Department for Environment Food & Rural Affairs (2018). *A Green Future: Our 25 Year Plan to Improve the Environment*. London. https://assets.publishing.service.gov.uk/government/uploads/system/uploads/attachment_data/file/693158/25-year-environment-plan.pdf.

United Nations Development Programme (2017). *Payments for ecosystem services*. http://www.undp.org/content/sdfinance/en/home/solutions/payments-for-ecosystem-services.html (Accessed: 2 October 2017).

United Nations Environment Programme (2015). *Annual Report 2015*. Nairobi. http://wedocs.unep.org/bitstream/handle/20.500.11822/7544/-UNEP_2015_Annual_Report-2016UNEP-AnnualReport-2015-EN.pdf.pdf?sequence=8&isAllowed=y.

United Nations Environment Programme World Conservation Monitoring Centre and International Union for Conservation of Nature (2016). *Protected Planet Report 2016: How Protected Areas Contribute to Achieving Global Targets for Biodiversity*. Cambridge: United Nations Environment Programme - World Conservation Monitoring Centre and International Union for Conservation of Nature. http://wcmc.io/protectedplanetreport_2016.

United States Agency for International Development (2016). *Protecting Tigers with Project Predator*. Washington, D.C. http://pdf.usaid.gov/pdf_docs/PA00MFT9.pdf.

Ürge-Vorsatz, D., Rosenzweig, C., Dawson, R.J., Sanchez Rodriguez, R., Bai, X., Barau, A.S. *et al.* (2018). Locking in positive climate responses in cities. *Nature Climate Change* 8(3), 174-177. https://doi.org/10.1038/s41558-018-0100-6.

van de Wouw, M., Kik, C., van Hintum, T., van Treuren, R. and Visser, B. (2010). Genetic erosion in crops: Concept, research results and challenges. *Plant Genetic Resources* 8(1), 1-15. https://doi.org/10.1017/S1479262109990062.

van Wilgen, B.W., Forsyth, G.G., Le Maitre, D.C., Wannenburgh, A., Kotzé, J.D.F., van den Berg, E. *et al.* (2012). An assessment of the effectiveness of a large, national-scale invasive alien plant control strategy in South Africa. *Biological Conservation* 148(1), 28-38. https://doi.org/10.1016/j.biocon.2011.12.035.

Venter, I. (2005). Back to basics. *Engineering News*. 21 October 2005.

Wellsmith, M. (2011). Wildlife crime: The problems of enforcement. *European Journal on Criminal Policy and Research* 17(2), 125-148. https://doi.org/10.1007/s10610-011-9140-4.

Westengen, O.T., Jeppson, S. and Guarino, L. (2013). Global ex-situ crop diversity conservation and the svalbard global seed vault: Assessing the current status. *PLOS ONE* 8(5), e64146. https://doi.org/10.1371/journal.pone.0064146.

Westphal, M.I., Browne, M., MacKinnon, K. and Noble, I. (2008). The link between international trade and the global distribution of invasive alien species. *Biological Invasions* 10(4), 391-398. https://doi.org/10.1007/s10530-007-9138-5.

Wilcove, D.S., Rothstein, D., Dubow, J., Phillips, A. and Losos, E. (1998). Quantifying threats to imperiled species in the United States. *BioScience* 48(8), 607-615. https://doi.org/10.2307/1313420.

Wilcox, B.A., Aguirre, A.A. and Horwitz, P. (2012). Ecohealth: Connecting ecology, health and sustainability. In *New Directions in Conservation Medicine: Applied Cases of Ecological Health*. Aguirre, A.A., Ostfeld, R. and Daszak, P. (eds.). New York, NY: Oxford University Press. 17-32. https://ro.ecu.edu.au/ecuworks2012/48/.

Wolkovich, E.M., Davies, T.J., Schaefer, H., Cleland, E.E., Cook, B.I., Travers, S.E. *et al.* (2013). Temperature-dependent shifts in phenology contribute to the success of exotic species with climate change. *American Journal of Botany* 100(7), 1407-1421. https://doi.org/10.3732/ajb.1200478.

World Economic Forum (2018). *The Global Risks Report 2018: 13th Edition*. Geneva. http://www3.weforum.org/docs/WEF_GRR18_Report.pdf.

World Health Organization and Secretariat of the Convention on Biological Diversity (2015). *Connecting Global Priorities: Biodiversity and Human Health: A State of Knowledge Review*. Geneva. https://www.cbd.int/health/SOK-biodiversity-en.pdf.

Young, R.P., Hudson, M.A., Terry, A.M.R., Jones, C.G., Lewis, R.E., Tatayah, V. *et al.* (2014). Accounting for conservation: Using the IUCN Red List Index to evaluate the impact of a conservation organization. *Biological Conservation* 180, 84-96. https://doi.org/10.1016/j.biocon.2014.09.039.

Chapter 14

Oceans and Coastal Policy

Coordinating Lead Author: Diana Mangalagiu (University of Oxford and Neoma Business School)

Lead Authors: Elaine Baker (GRID-Arendal at the University of Sydney), Pedro Fidelman (Centre for Policy Futures, The University of Queensland), Leandra Regina Gonçalves (University of Campinas), Peter Harris (GRID-Arendal), James Hollway (Graduate Institute of International and Development Studies), Rakhyun E. Kim (Utrecht University), Jake Rice (Department of Fisheries and Oceans – Canada)

GEO Fellow: AlAnoud Alkhatlan (Arabian Gulf University)

Executive summary

Responding to key drivers and pressures facing the oceans (e.g. climate change, pollution and overfishing; see Chapter 7 of this report) requires diverse policy instruments and governance approaches *(well established)*. These instruments and approaches include command and control, stakeholder partnerships, economic incentives and approaches to enable actors. {14.2}.

Policy coherence and integration are important in addressing cumulative impacts of local and regional threats to support the resilience of marine ecosystems (e.g. coral reefs) to climate change *(inconclusive)*. However, without international policies to curb carbon emissions, the effectiveness of resilience-based management is likely to be very limited, given the limits to the capacity of marine species to adapt to warmer ocean waters *(well established)*. {14.2.1}.

Problems involving numerous activities, sectors and sources (e.g. marine litter) may require policies involving comprehensive and coordinated measures *(well established but incomplete)*. When such problems involve multiple jurisdictions, governance approaches to engage neighbouring countries (e.g. the Regional Seas Programme) may be appropriate *(well established but incomplete)*. {14.2.2}.

Promoting more sustainable fisheries may require several policy instruments, given the range of contexts in which problems in this sector arise *(well established)*. Territorial use rights for fishing (TURF) programmes are a good fit for fisheries with relatively sedentary stocks, high exclusionary potential, and governments keen to devolve costly management and enforcement functions *(well established)*. Individual transferable quotas (ITQs) work best for relatively high-value stocks when supported by strong, independent, scientifically set quotas, strong monitoring, control and surveillance. Regulation of access and resource use rights may be successful when effective enforcement and compliance mechanisms are in place *(well established)*. {14.2.3}.

Some problems may be best addressed by policy instruments that entail community and stakeholder engagement *(well established)*. These include enabling local communities to develop and adopt measures tailored to their context and partnerships with the private sector *(well established)*. {14.2.3}.

Policy-sensitive indicators may be used to track progress in addressing key pressures and drivers *(well established)*. These include area-based indicators, such as the coverage of marine protected areas and of vulnerable marine ecosystems. Protected areas under national jurisdiction or in the high seas have the potential to address several pressures relating to marine biodiversity, including overfishing and habitat destruction *(well established but incomplete)*. {14.3.1}.

Many indicators may not entirely capture the multiple dimensions of different pressures and drivers *(well established)*. Area-based approaches alone do not gurantee effective area management; nor can they guard against the impact of climate change or pollution *(well established)*. Efforts to develop methods for evaluating the effectiveness of protected areas are, therefore, critical *(well established)*. {14.3.2}.

A lack of standardization may make it difficult to track progress towards marine conservation *(well established)*. This is the case of beach litter used as an indicator of litter in the marine environment. The lack of standardization and compatibility between methods used and results obtained in various bottom-up projects makes it difficult to reach an overall assessment of the status of marine litter over large geographical areas {14.3.2}.

14.1 Introduction

The impacts of human activities on the oceans have serious social and economic implications, which directly and indirectly affect human health and well-being. As noted in chapter 7 of this report, impacts of great concern include those associated with climate change, pollution and overfishing. Coral bleaching is perhaps one of the most dramatic and immediate impacts of climate change on oceans in recent years; marine litter and plastic pollution are rising to the forefront of pollution issues; and the depletion of fish stocks from overfishing remains critical. Drawing on selected policy typologies and related case studies, this chapter examines key approaches and instruments employed in response to these issues **(Table 14.1)**. In addition, case studies are used to illustrate responses in different governance (subnational, regional and global) and geographical contexts, and highlight challenges and opportunities for policy design and implementation.

This chapter also provides valuable insights into the effectiveness of policies at regional and global levels by drawing on selected policy-sensitive indicators, such as the coverage of marine protected areas, beach litter assessment and representation of vulnerable marine ecosystems in regional fisheries management organizations.

14.2 Key policies and governance approaches

14.2.1 Resilience-based management (climate change adaptation policy)

Resilience-based management (RBM) of coral reefs is an emerging concept in the context of very limited alternatives (van Oppen *et al.* 2015; van Oppen *et al.* 2017), given that the root cause of coral bleaching is the increasing level of atmospheric carbon dioxide (CO_2). RBM refers to strategic policy interventions at local and regional levels to support ecological resilience (i.e. the capacity to resist disturbances and recover from these disturbances) (Anthony 2016). It is believed to help offset to some extent the increasing effects of climate change (Anthony *et al.* 2015; Anthony 2016).

The basic premise underlying RBM is that the resilience of coral reefs can be enhanced by addressing the cumulative impacts of local and regional threats (e.g. pollution, sedimentation and overfishing) (Marshall and Schuttenberg 2006; Keller *et al.* 2009; Anthony *et al.* 2015; Anthony 2016). RBM may involve a mix of policy instruments and management actions (e.g. regulation, incentives and education) (Anthony *et al.* 2015, p. 53) relating to, for example, land use controls to improve water quality entering the reef system and spatial planning of marine protected areas, including no-take zones (Anthony *et al.* 2015; Anthony 2016). In terms of the DPSIR framework (Section 1.6), RBM aims to address a range of 'pressures' on the reefs, such as land use in adjacent catchments, coastal development and fisheries.

As an emerging concept, RBM is yet to be addressed in the policy literature. Elsewhere, in the case of coral reefs, there has not been much discussion beyond the suggested need for RBM and strategies to support its implementation.

Internationally, there has been considerable interest in resilience-based approaches to coral reef management. For example, the Coral Triangle Initiative – an intergovernmental effort involving Indonesia, Malaysia, Papua New Guinea, Philippines and Timor-Leste – incorporates resilience principles and multi-issue management (Coral Triangle Initiative Secretariat 2009). Further, the International Union for the Conservation of Nature and Natural Resources (IUCN) has adopted an agenda for action on coral reefs, climate change and resilience, which urges the development of policies to support RBM at national and international levels (Obura and Grimsditch 2009).

Case study: The Great Barrier Reef Climate Action Plan 2007-2012

Australia's Great Barrier Reef (GBR) Marine Park is one of the world's pioneers in coral reef management (Day 2016). It is an exemplar of approaches aiming to restore and maintain the resilience of coral reefs in the face of multiple threats, including climate change (Great Barrier Reef Marine Park Authority [GBRMPA] 2009; GBRMPA 2014). In 2007, the Great Barrier Reef Marine Park Authority (GBRMPA)[1] launched the GBR Climate

Table 14.1: Example of governance approaches and policy instruments to address coral bleaching, marine litter and overfishing

Governance approach	Policy instrument	Case study
Enabling actors	Production of knowledge, awareness-raising	Great Barrier Reef Climate Change Action Plan 2007-2012
Command and control and partnership with the private sector	Legally binding measures and voluntary approaches by business and other stakeholders	Regional Plan on Marine Litter Management in the Mediterranean
Enabling actors and economic incentives	Territorial use rights for fishing	Chilean Abalone Traditional User Rights Fishery
Economic incentives	Individual transferable quotas	British Columbia Groundfish Fishery Individual Transferrable Quotas
Command and control	Regulation of access and resource use rights	United Nations General Assembly Resolution 61/105 on Vulnerable Marine Ecosystems

1 GBRMPA is a federal statutory authority established under the Great Barrier Reef Marine Park Act 1975 with powers to prepare and publish plans and policies relating to the protection and management of the GBR (Commonwealth of Australia 1975).

Change Action Plan 2007-2012, which identified strategies and actions to enhance the reef's resilience and support adaptation by reef-dependent industries and communities (GBRMPA 2007). Once situated in the Council of Australian Governments' National Climate Change Adaptation Framework as a specific action item (Council of Australian Governments 2007), the Action Plan is regarded as the first of its kind, representing a relevant national and international case study on an adaptation policy cluster applied to the threat of climate change on a world heritage reef system (GBRMPA 2012). Further, the case exemplifies the enabling actors' governance approach; it involves actions for improving understanding of climate change vulnerability and adaptation and raising awareness among reef-dependent communities and industries.

Table 14.2: Australia's Great Barrier Reef

Criterion	Description	References
Success or failure	The overall goal of the GBR Climate Change Action Plan 2007-2012 was to maximize the resilience of the GBR to climate change. This involved four objectives: (i) targeted science; (ii) resilient ecosystems; (iii) adaptation of industries and communities; and (iv) reduced climate footprints. The review of the Action Plan highlights the delivery of over 250 individual projects or activities, generation of a diverse range of knowledge resources, including more than 150 reports and papers, and creation of scientific knowledge underpinning new decision-making tools and processes (e.g. developing and refining remote sensing tools that forecast coral bleaching and risks of outbreaks of coral disease). On the other hand, the GBR Outlook Report 2014 recognizes that, despite sound regional-scale management for climate change and other threats, the condition of the reef is still declining.	GBRMPA 2012; GBRMPA 2014
Independence of evaluation	A review of the Action Plan outcomes was undertaken by the GBRMPA (i.e. self-evaluation).	GBRMPA 2012
Key actors	Alongside the GBRMPA, implementation involved specific stakeholder groups, including traditional owners, tourism operators and the seafood industry, and is believed to have built stronger ongoing relationships across the public, private, community and research sectors.	Commonwealth of Australia 2016
Baseline	A comprehensive vulnerability assessment, including social and economic dimensions, undertaken in 2007 evaluated the threats posed by climate change to the GBR.	Johnson and Marshall 2007
Time frame	The Action Plan was implemented over a five-year period, between 2007 and 2012. The report *Climate Change Adaptation: Outcomes from the Great Barrier Reef Climate Change Action Plan 2007-2012* was released in 2012.	GBRMPA 2012
Constraining factors	Responding to climate change in the GBR involves cross-sectoral coordination involving several policy sectors and agencies spanning local, state and federal levels of government. Further challenges include compounding multiple spatial and temporal scales, uncertainty, and interlinkages between climate and non-climate drivers (see Chapter 2). Importantly, addressing major threats to the resilience of the reef, such as poor water quality from adjacent catchments and coastal development, are beyond the limits of the GBR Marine Park, therefore outside the jurisdiction of the GBRMPA and the application of the Action Plan.	Fidelman, Leitch and Nelson 2013
Enabling factors	The federal government allocated about A$9 million to implement the Action Plan. Further, the GBRMPA has provided leadership in managing the GBR since the mid-1970s. It also had capacity and the ability to mobilize additional expertise and partners.	Commonwealth of Australia 2016
Cost-effectiveness	Cost-effectiveness information is not available.	
Equity	The Action Plan did not involve fundamental equity issues. However, commentators point out the need to develop a user pays system for stakeholders impacting the GBR, including those responsible for shipping and port- and land-based activities.	Morrison and Hughes 2016 National Climate Change Adaptation Research Facility 2016
Co-benefits	Given the inherent nature of RBM, which involves addressing cumulative impacts of local and regional threats, the Action Plan had the potential to benefit existing policies relating to conservation, fisheries and tourism.	GBRMPA 2012
Transboundary issues	Many of the issues in the GBR span multiple administrative and ecological boundaries and involve multiple policy sectors (climate change, agriculture, coastal development and fishing). These pose significant challenges to RBM efforts.	Fidelman, Leitch and Nelson 2013; GBRMPA 2014
Possible improvements	The Action Plan focused mostly on actions within the GBR Marine Park. Major threats to the resilience of the reef, such as poor water quality from adjacent catchments and coastal development, lie beyond the Marine Park's boundaries. RBM efforts addressing external factors would be highly beneficial; they may require some level of coherence and integration with existing policies targeted at these factors.	

While RBM does not prevent coral bleaching, it may improve the prospect of recovery following bleaching events. However, without global action to curb carbon emissions, RBM alone is unlikely to be effective, given the limits to the capacity of corals to adapt to warmer ocean waters (Anthony 2016; Hughes *et al.* 2017).

The case of the GBR suggests that RBM will need to navigate complex governance settings involving multiple geographical and jurisdictional scales, levels of social and administrative organization, and policy and resource sectors (Fidelman, Leitch and Nelson 2013). Implementation of RBM may, therefore, involve fostering integration and coherence among existing policies addressing local and regional threats. In this regard, RBM has the potential to enhance overall governance across land–marine jurisdictional boundaries. Expanding the scope of RBM to incorporate the institutional and governance dimensions is critical – as addressing social resilience as part of RBM efforts is – since climate change has significant implications for reef-dependent communities and industries, including their well-being and health (Cinner *et al.* 2016).

14.2.2 Marine litter (regional cooperation policy)

Established in 1974, the Regional Seas Programme is one of the United Nations Environmental Programme's (UNEP) main efforts to address coastal and marine environmental issues. The programme illustrates regional cooperation approaches to coastal and marine management. It focuses on engaging neighbouring countries in regional action plans to address problems in shared marine environments. In many cases, these plans are underpinned by a legal framework in the form of a regional convention and associated protocols on specific issues.

There are currently 18 different Regional Seas Programmes, involving over 140 countries. These include the Mediterranean Action Plan with 22 contracting parties (Albania, Algeria, Bosnia and Herzegovina, Croatia, Cyprus, Egypt, France, Greece, Israel, Italy, Lebanon, Libya, Malta, Monaco, Montenegro, Morocco, Slovenia, Spain, Syrian Arab Republic, Tunisia, Turkey and the European Union).

Marine litter and debris in the Mediterranean are a well-recognized problem with environmental, economic, health and safety and cultural impacts (e.g. Galgani *et al.* 1995; Stefatos *et al.* 1999; Tomás *et al.* 2002; Campani *et al.* 2013; Pasquini *et al.* 2016). This has prompted the adoption of action plans to reduce pollution.

Case study: Regional Plan on Marine Litter Management in the Mediterranean

The densely populated coastline, fisheries, extensive tourism and maritime traffic, including the riverine inputs, have contributed to a continuous increase in marine litter over past decades (e.g. Santos, Friedrich and Barretto; Galgani *et al.* 2014; Rech *et al.* 2014; Unger and Harrison 2016). According to the International Coastal Cleanup Report (Ocean Conservancy 2017), cigarette butts are the most common item found at sea (see also Munari *et al.* 2016), but plastics, especially fragmented consumer products, make up by far the biggest type of marine litter (Li *et al.* 2016).

With the Regional Plan on Marine Litter Management in the Mediterranean (the Plan), the UNEP Mediterranean Action Plan (MAP) was the first Regional Seas Programme and Convention to develop legally binding measures to prevent and reduce the adverse effects of marine litter on marine and coastal environments. Adopted in 2013, the entry into force of the Plan coincided with the update of national action plans of the Mediterranean countries to combat pollution from land-based sources and activities.

The Plan involves some key principles on pollution control and prevention, including the integration of marine litter management into solid waste management and the reduction

© Shutterstock/ Fram Groomee

Table 14.3: Regional Plan on Marine Litter Management in the Mediterranean

Criterion	Description	References
Success or failure	The Plan contains 42 specific tasks, a timetable, lead authorities, verification indicators, costs and financial sources. The targets set for 2017 have been largely achieved, as many were conditional with "explore and implement to the extent possible". However, many of the aims have passed the explore stage to implementation.	
Independence of evaluation	It is the responsibility of the Contracting Parties to assess the state of marine litter, the impact of marine litter on the marine and coastal environment and human health as well as the socioeconomic aspects of marine litter management. The assessment will be conducted based on common agreed methodologies, national monitoring programmes and surveys.	
Key actors	The Plan was adopted by the Contracting Parties to the Convention for the Protection of the Marine Environment and the Coastal Region of the Mediterranean (Barcelona Convention), which includes 21 Mediterranean countries and the European Union (EU).	UNEP/MAP 2013
Baseline	An assessment of the status of marine litter in the Mediterranean was undertaken in 2008 and used as a basis for the development of the Plan. EU member states undertook a baseline evaluation of marine litter in accordance with the EU Marine Strategy Framework Directive (MSFD 2008). However, the 2015 marine litter assessment recommended a better definition of baselines and targets. Common baseline values for marine litter indicators (beaches, sea surface, sea floor, microplastics, ingested litter) should be proposed at the level of the entire Mediterranean Sea rather than at the subregional level.	European Parliament and European Council (2008) UNEP/MAP (2016); UNEP/MAP (2015a); UNEP (2016)
Time frame	The Plan is to be implemented between 2016 and 2025, with the majority of measures to be implemented, where possible, by 2020.	
Constraining factors	The behaviour of consumers remains a challenge; reducing marine litter will require changes in public perceptions, attitudes and behaviour. Compliance and improved detection and enforcement may prove challenging for effective legislation. Some States have inadequate waste management systems due to a lack of funding and poor governance. Furthermore, there has been a lack of consistency in methods used to tackle the marine litter problem. Responses include regional guidelines and the implementation of pilots such as Fishing-for-Litter and Adopt-a-Beach at a regional level, but there is still room for improvement.	UNEP/MAP 2013
Enabling factors	The aims of the Plan are also supported by the EU MSFD and synergistic policies, which include: the European Strategy on Plastic Waste in the Environment, which addresses plastic marine litter and ways to reduce it; a Directive to reduce the use of plastic bags; and the Port Reception Facility Directive, which addresses waste generated by ships at EU ports. The Plan is also supported by the G7 and G20 Action Plans on Marine Litter. Non-governmental organizations (NGOs) have been very active in awareness-raising and education activities. They have made a major contribution to data collection and cleanup operations, mobilizing thousands of volunteers in support of a litter-free Mediterranean. The Plan includes strong provisions on the effective coordination and important role of the various marine litter actors and stakeholders.	
Cost-effectiveness	Marine litter can cause significant socioeconomic damage, including losses for coastal communities, tourism, shipping and fishing. However, the costs of implementing measures necessary to meet the requirements of the Regional Plan through the National Action Plans are also significant. For example, the cost of coastal and beach cleaning in the EU has been assessed at almost €630 million per year, while the cost to the fishing industry could amount to almost €60 million. UNEP/MAP has carried out work to assist countries with estimating the costs for the Regional Plan and legally binding measures in the region. Furthermore, socioeconomic assessment of the costs and benefits of selected potential new measures has been conducted, including fishing-for-litter, port reception facilities and banning single-use plastic bags.	Ballance, Ryan and Turpie(2000); Williams *et al.* (2016); Brouwer *et al.* (2017); European Commission] (2017); UNEP/MAP (2015b)
Equity	Both the people and the environment benefit from a reduction in marine litter. Mitigation measures such as deposit return schemes, plastic bag levies and enforcement activities come at a cost, which is unevenly distributed among society.	
Co-benefits	Co-benefits include increased energy generation from recycling solid waste, and reduced demand for plastic packaging by awareness-raising. Reduced marine litter is also beneficial to marine species, ecosystems and biodiversity.	
Transboundary issues	Marine litter can be generated in many jurisdictions and migrates across boundaries. Mediterranean marine litter can even enter the sea from the Atlantic through the Strait of Gibraltar or via the Suez Canal, though the larger transboundary origins and effects of Mediterranean marine litter are from Mediterranean coastal States. Marine litter accumulates in hot spot areas. Preliminary work is currently being undertaken at regional level by the UNEP/MAP and other organizations and initiatives to identify where these areas are located.	
Possible improvements	The national data on marine litter show inconsistencies between reporting years and between countries with differing reporting systems. Therefore, the variations within the scope of the reporting, different methods of calculation and lack of data validation hinder identification of trends. The 2015 assessment recommended that countries develop more coherent monitoring programmes that include more data collection on sources of marine litter and regular monitoring of microparticles. Stronger enforcement measures need to be introduced to combat illegal discharge or dumping of marine litter, both from land-based sources and at sea, in accordance with national legislation.	UNEP/MAP 2014; UNEP/MAP 2016

of litter through a focus on promoting sustainable consumption and production practices. A key component of the Plan is collaboration with the private sector to reduce plastic consumption.

The Plan provides a legally binding set of actions and timelines to reduce marine litter in the Mediterranean. The targets set for 2017 have been largely achieved, as many were conditional with "explore and implement to the extent possible". However, many of the aims have passed the explore stage to implementation.

Some progress has been made in the use of recycled plastic and in reducing the use of single-use plastic bags. Some Mediterranean countries such as France and Morocco have a total ban on plastic bags. Other countries such as Croatia, Malta and Israel and some municipalities and districts of Spain and Greece have introduced a tax on single-use plastic bags. Tunisia has banned non-biodegradable plastic bags in large chain supermarkets (Legambiente ONLUS 2017).

On the other hand, the fishing sector has lagged in implementing litter reduction strategies. Although guidelines for the litter scheme have been developed, and the majority of Mediterranean fishermen have indicated a willingness to participate, country surveys indicate that vessels do not have bins or bags on board to store litter items. Fishermen continue to discard unwanted fishing gear into the sea (UNEP 2016). In this regard, a wide range of technologies for marking ownership of fishing gear are available. In fact, Moroccan and EU fisheries laws provide for the marking of both the vessel and the fishing gear carried on board (Food and Agriculture Organization [FAO] 2005), and the Food and Agriculture Organization of the United Nations adopted the Guidelines on Marking Fishing Gear in 2018.

14.2.3 Territorial use rights for fishing

An attractive policy for some countries seeking to manage small-scale fisheries is to (re-)enable the traditional users of the resource by allowing (or granting) them exclusive rights to collectively (or occasionally individually – Hauck and Gallardo-Fernández 2013) manage stocks in specific areas themselves. The logic behind these Territorial Use Rights for Fishing programmes (TURFs) (Christy 1992), stem from common property theory and the literature on community or local-scale governance (Ostrom 2002). TURFs are considered to ameliorate overfishing by stimulating resource stewardship among fishers and offering communities various sanctioning mechanisms to hold them accountable (Castilla and Fernández 1998; Wilen, Cancino and Uchida 2012). By engaging the community in the scientific, economic and political decision-making surrounding the setting of limits and the sanctioning of transgressions, TURFs are thought to promote equity and empower and encourage reinvestment in local communities (Villanueva-Poot *et al.* 2017).

TURFs are touted as a good fit for fisheries with relatively sedentary stocks and high exclusionary potential, and are valuable in locations where governance resources are limited (Fernández and Castilla 2005). Hybrid policy designs can extend their applicability though (Barner *et al.* 2015). For example, more mobile species or fishers can be addressed by establishing broader TURF networks (Aceves-Bueno *et al.* 2017), and some policies combine classic TURFs with marine reserves (so-called TURF-reserve systems – Afflerbach *et al.* 2014; Oyanedel *et al.* 2017). These TURF-reserves serve an important goal of restoring a healthy balance among competing species in the same ecosystem (Loot, Aldana and Navarrete 2005; Oyanedel *et al.* 2017), though studies have found that even classic TURFs may improve the abundance of non-target species through trophic interactions (Gelcich *et al.* 2008; Giacaman-Smith, Neira and Arancibia 2016). Indeed, TURFs could be targeted by private conservation actors (Costello and Kaffine 2017). Lastly, the literature shows that it is important to establish TURFs at an appropriate scale for the target species. TURFs for highly variable species subject to boom-and-bust dynamics should be established at a wide enough geographical scale to allow fishers to maintain their livelihood (Aburto, Stotz and Cundill 2014), while being attentive to interdependencies across individually managed areas due to larval dispersion or governance structures (Garavelli *et al.* 2014; Garavelli *et al.* 2016; Aceves-Bueno *et al.* 2017).

TURFs have proven popular with governments keen to devolve costly management and enforcement functions, but because TURFs can operate based on tradition and without formal establishment, it is unclear exactly how many exist or when they were first introduced (Christy 1992; Afflerbach *et al.* 2014). There are several ways to establish TURFs. In some cases (e.g. in Japan, Palau, Papua New Guinea, Samoa, Solomon Islands and Vanuatu), TURFs are based on centuries-old traditions granting local users exclusive access to nearshore fishing grounds (Le Cornu *et al.* 2017; Nomura *et al.* 2017; Yoshino 2017). In others (e.g. Chile and South Africa), TURFs have been initiated by the government as part of a national or regional co-management framework or were driven by local communities, with the regional or national government providing legal, operational or financial support (Charles 2002; Hauck and Gallardo-Fernández 2013).

A major challenge to TURFs continues to be the persistence of poaching (Andreu-Cazenave, Subida and Fernandez 2017; Oyanedel *et al.* 2018). One option often advocated is to complement local community management with some governmental resources for monitoring, enforcement and centralized dispute resolution (Hauck and Gallardo-Fernández 2013). The literature stresses though that even such co-management arrangements should be context-dependent (Defeo *et al.* 2016). The mix of formal and informal enforcement mechanisms deployed will depend on the biological productivity of the resource (Santis and Chávez 2015), and how well supported the regime is by fishers' social networks (Rosas *et al.* 2014; Crona, Gelcich and Bodin 2017). The importance of the underlying social network to the success of TURFs highlights how demographic changes and intergenerational shifts may ultimately undermine even successful TURFs (Tam *et al.* 2018). Lastly, another major challenge is that the integration of seafood markets continues to put global pressures even on the type of local, small-scale fisheries often governed by TURFs, with varying effects (Crona *et al.* 2015; Castilla *et al.* 2016; Crona *et al.* 2016), which may only be improved by transforming the coastal communities themselves (Saunders *et al.* 2016).

Case study: Chilean abalone TURFs

Despite some resemblance to abalone, 'Chilean abalone' is a different high-value species of sea snail, known locally as loco, and has been part of the local diet for at least 6,000 years (Reyes 1986; Santoro *et al.* 2017). Historically, the fishery had been open access, but as international '*loco* fever' (Meltzoff *et al.* 2002) demanded unsustainable catches, the Government experimented with a series of different policy instruments: seasonal closures from 1981 to 1984; a global national quota from 1985 to 1989; and then total closure from 1989 (Castilla 1995; Castilla and Fernández 1998; González *et al.* 2006; Gelcich *et al.* 2008; Hauck and Gallardo-Fernández 2013). All failed to stem widespread poaching. A 1991 fishing law then outlined area-based rights management schemes that evolved into the first TURFs being implemented in 1997 (Meltzoff *et al.* 2002). The Government banned *loco* fishing outside these TURFs and subsidized their establishment through a four-year tax deferment and contributions of up to 75 per cent on any baseline or follow-up assessments (Hauck and Gallardo-Fernández 2013). TURFs subsequently proliferated to other areas and other (relatively sedentary invertebrate) species (Gelcich *et al.* 2017), ultimately encompassing 80 per cent of the Chilean catch and 40 per cent of registered fishers in over 400 TURFs (Fernández and Castilla 2005; González *et al.* 2006; Hauck and Gallardo-Fernández 2013). This case was chosen as a relatively successful attempt to hand over governance to local communities and is a detailed illustration of how scale- and context-dependent different policy instruments are.

Table 14.4: Chilean fisheries

Criterion	Description	References
Success or failure	The Chilean Fisheries Department required a policy solution that reduced unsustainable pressure on a highly vulnerable species, returned all fishing access to adjacent community fisheries, and excluded mobile non-resident fishers who were poaching extensively.	Hauck and Gallardo-Fernández 2013
Independence of evaluation	Chilean TURFs have been evaluated several times, including by third parties and environmental NGOs.	Gonzalez *et al.* 2006; Earth Justice 2015
Key actors	Local communities developed and implemented their TURFs. Processing and marketing sectors were supportive throughout. Most environmental NGOs came late to the process but bought their way in through financial liaisons with individual communities.	
Baseline	Data to support quantitative baselines and targets were scarce, weak and ad hoc. However, all agreed that *loco* was severely depleted along much of the coastline and that individual transferable quotas (ITQs) had failed to control extensive illegal fishing.	Reyes 1986; Ruano-Chamarro, Subida and Fernández 2017
Time frame	The first TURFs were established over a two-year transition period and took another decade to spread, but numbers seem to have plateaued since.	
Constraining factors	Communities with high in-migration and fewer resources for surveillance and enforcement faced greater challenges.	
Enabling factors	The sedentary nature and high market value of the target species was essential to success. Community management relied on communities' cultural and social integrity and the law banning *loco* fishing outside TURFs.	Liu *et al.* 2016
Cost-effectiveness	Costs of TURFs to the Government were low, as it transferred monitoring and surveillance costs to the communities, which were willing to undertake them, given the large financial and political returns and some governmental support for their establishment.	Gutiérrez *et al.* 2011
Equity	'Communities' were self-defined and overlapped more than anticipated, so the first to obtain TURF authorization could marginalize and disempower others. Communities that struggled with adapting to the new system saw increased crime and poaching. Lastly, a 2008 law gave preferential rights to indigenous peoples, which some people considered inequitable.	Van Holt 2012; Hauck and Gallardo-Fernández 2013
Co-benefits	Chilean TURFs integrated and empowered local communities, facilitated policy experimentation and provided sustainable ecosystem services and tourism.	Hauck and Gallardo-Fernández 2013; Gelcich *et al.* 2017; Defeo and Castilla 2005, p. 275; de Juan *et al.* 2015; Biggs *et al.* 2016
Transboundary issues	TURFs increased fishing pressure on non-TURF areas and species once the fishing programme adopted by a community was fulfilled for the season or year.	Van Holt 2012
Possible improvements	Potential improvements include more stable funding for surveillance and enforcement, stronger integration across scales and better provision for those displaced from the fishery. Innovative business models and municipal conservation areas have been discussed and, in some cases, trialed, but it is too soon to tell whether these will address persistent poaching issues.	González *et al.* 2006; Gelcich and Donlan 2015; Gelcich *et al.* 2015

The Chilean abalone TURFs have been regarded as role models (González *et al.* 2006; Gelcich *et al.* 2017). They led to improved catch per unit of effort and sometimes substantial (as much as five-fold) improvements in economic returns. These successes were due to empowering local communities to develop and adopt instruments tailored to their geography and culture. However, illegal fishing continues (Andreu-Cazenave, Subida and Fernandez 2017), in some cases by fishers who abide by rules but fish illegally beyond their own TURFs, undermining ecological outcomes (González *et al.* 2006; Hauck and Gallardo-Fernández 2013; Oyanedel *et al.* 2018). Moreover, the sustainability of economic benefits from the system has seen competitive challenges from other markets and fishery products, and in one region only 5 out of 30 TURFs did well economically (Zuñiga *et al.* 2008). However, despite complaining that TURFs had not always provided significant financial returns and that monitoring costs had been increasing, Chilean fishers were reluctant to relinquish their TURFs, recognizing that they provided benefits across multiple dimensions, including ecological and economic empowerment (Gelcich *et al.* 2017). The transferability of this policy depends on having sedentary species, stable markets, and settled communities with an ability to exclude mobile, non-local fishers.

14.2.4 Individual transferable quotas

ITQs are a type of market-system approach that some governments use to manage fisheries (Selig *et al.* 2016). Typically, ITQs grant their owners exclusive and transferable rights to a given portion of the total allowable catch (TAC) from a fishery each season or year, which can then be bought, sold or leased in an open market. The logic is that because these quotas are individual and not collective, fishers cannot maximize earnings by racing to catch more fish from a common total quota or resource than other license-holders before depletion. Rather, income can only be increased by more strategically catching and marketing their share (for example, through more efficient fishing practices or timing the catch to market opportunities) and through resource stewardship by supporting stock growth so that their fixed percentage applies to a larger total quota. ITQs can thus generate substantial economic returns for society (Hoshino *et al.* 2017), promote economic efficiency by incentivizing reductions in fishing capacity (Blomquist and Waldo 2018) and create an economic incentive for the industry to value stock growth as well as present catch.

ITQs were first introduced on individual fish species in the late 1970s (Chu 2009) by the Netherlands (Hoefnagel and de Vos 2017), Iceland (Chambers and Kokorsch 2017; Kokorsch 2017) and Canada (Rice 2004; Pinkerton 2013; Edinger and Baek 2015; Gibson and Sumaila 2017). They have since been implemented at a range of scales, being first implemented as a national fisheries policy by New Zealand in 1986 (Mace, Sullivan and Cryer 2014) and Iceland in 1990 (Arnason 1993). ITQs have also been proposed as a potential reform option for the European Common Fisheries Policy (Waldo and Paulrud 2012; van Hoof 2013) and for international fisheries management (Pintassilgo and Costa Duarte 2000; Thøgersen *et al.* 2015), but they have not yet seen agreement at these scales.

A comprehensive review in 2009 found that 18 countries managed several hundred different fish stocks with ITQs (Chu 2009). They have been most vigorously adopted in tandem with the privatization of other common assets as a part of broader neoliberalist trends (Pinkerton 2017) – for example, in the United States of America (Porcelli 2017), Australia (Steer and Besley 2016; Emery *et al.* 2017), Argentina (Bertolotti *et al.* 2016) and Chile (Wiff *et al.* 2016), in addition to other countries listed above. Norway (Hannesson 2013; Hannesson 2017), Sweden (Waldo *et al.* 2013; Stage *et al.* 2016; Blomquist and Waldo 2018) and Denmark (Merayo *et al.* 2018) have seen more cautious adoption of ITQs, and other jurisdictions, such as France (Frangoudes and Bellanger 2017), have seen marked opposition. While several developing countries have shown interest in ITQs, they have not seen widespread adoption there, for various reasons that include concerns about economic participation, a backlash against 'privatizing nature', or the recognition that ITQs require sound stock assessment and reliable catch monitoring (see below).

Where conditions are favourable, ITQs are recognized as an excellent instrument for promoting economic efficiency in fisheries. However, their mixed record elsewhere has prompted the literature on marine policy and environmental economics to investigate the conditions for policy effectiveness. These conditions relate to scale, technology and capacity, as identified in Section 7.5.

First, ITQs operate best for relatively high-value stocks. Nonetheless, fishers' high-grading (discarding less valuable species or sizes into the sea to maximize quota value) can still produce negative ecological impacts and can only be deterred by on-board surveillance (as with any quota-based harvesting system). ITQs may have positive ecosystem effects through a variety of indirect mechanisms (Gibbs 2010), but, ultimately, ITQs are a relatively targeted policy instrument that should be well considered.

Second, successful ITQ programmes require strong, independent, scientifically set TACs (Sumaila 2010); otherwise, scientific uncertainty or political interference may erode quota owners' trust that the quotas are sustainable, restoring incentives to race for fish. For example, Nordic countries offer strong monitoring capabilities and high levels of trust in public institutions (Hannesson 2013; Merayo *et al.* 2018; Blomquist and Waldo 2018).

Third, the economic incentive value of ITQs is especially vulnerable to free-riding illegal, unreported and unregulated (IUU) fishing (Costello *et al.* 2010). Again, strong monitoring, control and surveillance (MCS) is required or target stock status will be undermined.

It should be acknowledged though that ITQs are a policy instrument for promoting economic efficiency in fisheries and not necessarily social equity (Costello, Gains and Lynham 2008; Høst 2015). Issues of social equity can arise during the initial allocation of ITQs or, later, upon their consolidation. Basing allocation on historical usage can exacerbate existing social inequities, particularly if the time frame used favours one group. The New Zealand Government spent considerable sums purchasing ITQs from the initial allocation to satisfy

Maori claims (Dewees 1998). Auctions are an alternative (Bromley 2009), but this may exacerbate pre-existing inequities too if not all parties have sufficient resources to buy in. Even if begun equitably, consolidation of ITQs can concentrate fishing gains and power (Pinkerton and Edwards 2009). Similar to other industries, the economic incentives of ITQs may promote further capitalization and ultimately 'armchair fishing', where corporate owners dissociated from coastal communities absorb harvesting profits. Where processing is also consolidated, small coastal communities may be left to slide into economic depression. To guard against this, many quota management systems limit how great a share each owner may collect. Initiatives such as licence banks may deter such consolidation of fishing opportunities (Edwards and Edwards 2017), but they have not been in place long enough for their social, economic and ecological consequences to be fully evaluated.

Lastly, by reducing the race to fish, ITQs are thought to considerably improve occupational health and safety. Generally, occupational injuries are more prevalent in fisheries than in other professions (Chauvin and Le Bouar 2007; Håvold 2010). But fishers in an ITQ can fill their quota at any time over the season, rather than compete for a total quota with other fishers, so they do not need to venture out in inclement weather, overload their vessels with gear or neglect vessel maintenance (Pfeiffer and Gratz 2016). However, these health benefits only accrue for quota owners; quota lessees or contract workers may still be subjected to pressures to take risks (Windle *et al.* 2008; Emery *et al.* 2014). Occupational safety can also affect how fishers perceive regulation. According to Håvold (2010), while serious fishing accidents justify regulatory frameworks to fishers, minor accidents undermine their impressions. Further research is required to determine how best to ensure the health and safety of those involved in the fishing industry (Lucas *et al.* 2014).

Case study: British Columbia groundfish fishery ITQs

The groundfish fishery of British Columbia, Canada, is a complex, multi-species commercial capture fishery. Species such as rockfish, hake, Pacific cod and pollock live and feed near the sea bottom, requiring large trawlers to catch them which results in a heavily capitalized and technologically advanced industry. From 1980 to 1995, Canada's Department of Fisheries and Oceans (DFO) managed the fishery through limits on the number of vessel licences and species- and season-specific TACs. However, this drove unsustainable capitalization, as fishers competed to catch as large a share of the quota as possible before it was exhausted (University of British Columbia [UBC] 2017), and several TACs were repeatedly exceeded (Turris 2000). In 1995, DFO closed the fishery and began consultations (see also Koolman *et al.* 2007). While relations between the industry and DFO were adversarial, all agreed that the fishery was heading towards an economic and environmental crash and that policy tweaks would be insufficient (Rice 2004). In 1997, the fishery reopened as an ITQ system. While not the first ITQ management system used in Canadian fisheries (Casey *et al.* 1995; Turris 2000), this was the broadest in terms of number of species governed (eventually over two dozen) and fleet impact (around 130 vessels at the start), and the first to tackle stocks that were already overfished. Ultimately, the ITQ scheme proved successful in improving the fishery's economics (Rice 2004; Branch 2006) and is thus illustrative of how ITQs can work well under the right conditions.

The ITQ system reversed the decline in status of many key stocks, secured the financial viability of the processing sector and reduced fleet capacity. Moreover, all four major stakeholders eventually supported the programme. DFO Science overcame its distrust of market incentives to reach conservation goals, and DFO Management came to recognize that making industry management partners somewhat relieved budgetary pressures associated with monitoring and enforcement. The processing sector enjoyed greater market stability and value, and licence holders (even those who ended up leaving the fishery) recognized alternatives as untenable and the market as ultimately safer and more stable. The British Columbia groundfish case is, therefore, instructive as a model for rationalizing a complex, larger-scale, multi-species and heavily capitalized fishery. Indeed, it refutes common wisdom that cooperation requires few parties (there are at least 30 independent players in the fishery) or should be localized (the fleet operates along the whole British Columbia coastline). Still, it is not a strategy for small-scale, livelihood-oriented fisheries and is usually expensive to set up, if not maintain. This case's success depended on strong science and management support, high product value and a reasonably strong economy. It should also be noted that, even if financially sustainable, the policy may not be ecologically sustainable, though more research is required.

Table 14.5: British Columbia fisheries

Criterion	Description	References
Success or failure	Two main policy goals were established by the formal Groundfish Advisory Committee (GAC): stopping the decline of many key stocks and securing financial viability for the processing sector. A subsidiary goal was to downsize fleet capacity to support positive revenue for each participant. These goals were met.	Turris 2000
Independence of evaluation	DFO evaluates all fisheries management plans periodically, and more detailed evaluations of several aspects of the ITQ have been conducted by external academics.	Fisheries and Oceans Canada 2017; Wallace *et al.* 2015
Key actors	The policy itself was developed behind closed doors by a subset of the GAC, which brought together all four main interests: harvesting, processing, science and management.	Rice 2004
Baseline	Baselines were based on historical records of stock status, and plant operating costs and revenues going back at least 15 years.	Richards 1994; Ainsworth *et al.* 2008
Time frame	The ITQ was successfully implemented within one year. Rockfish prices increased six-fold in six months, principally due to better matching of supply and demand. The number of vessels nearly halved within 18 months.	UBC 2017
Constraining factors	Funding to establish the allocations and monitoring and information systems and to buy out those willing to leave the fishery until fleet capacity adapted was the major constraining factor.	
Enabling factors	An important enabling factor was the economic status of British Columbia at the time, which enabled fishers who left the fishery to find alternative work.	
Cost-effectiveness	Setting up the ITQ system involved large upfront costs, especially from licence buy-outs. DFO had accurate estimates of these costs, though no ex-post cost-effectiveness analysis was done, since the only alternatives recognized were fishery closure or depletion.	
Equity	The policy eliminated both especially large vessels, which could no longer fill their holds, and smaller vessels, which could not bear the observer costs, from the fleet. While there was a licence buyback programme, no provision for employment transition was offered. More consistent supply also made for more consistent work for fish-cutters, mostly women. Overall, although the extension of the fishing season increased industry costs, these were largely in the form of wages, which may have improved social equity.	Stainsby 1994; Matulich, Mittelhammer and Reberte 1996; Dolan *et al.* 2005
Co-benefits	A major co-benefit was an improvement in workplace safety and occupational health. Whereas, under the previous regime, fishers might go out in hazardous conditions just because the fishery happened to be open, to catch as much as possible before the full quota was taken, now they could manage their own share over time, going out when it was safer to do so.	Dolan *et al.* 2005
Transboundary issues	Most international transboundary issues (with the United States) related not to groundfish but salmon, Pacific halibut or hake.	Ianson and Flostrand 2010
Possible improvements	Though financially sustainable, in the mid-2000s environmental NGOs protested about the ecological sustainability of bottom-trawling on marine habitats. They engaged the fishery industry and proposed by-catch limits to DFO that relied on the same quota and observer system for implementation.	Branch 2009; Wallace *et al.* 2015

14.2.5 Command and control approaches for the high seas

Command and control policies are a type of norm or policy arrangement that regulates activity by combining legal instruments detailing rules and obligations and 'control' mechanisms, such as sanctions, penalties or fees, that deter actors from infringing those rules. It is associated with the concept of legalization (Abbot *et al.* 2000) and includes three main characteristics: obligation, precision and delegation. Obligation means that actors (state and non-state) are legally bound by a set of rules. Precision means that rules unambiguously define the conduct required by a given actor or set of actors; and delegation means that third parties are granted authority to implement the rules, monitor compliance and apply sanctions for non-compliance.

Despite not being command and control, as defined above, many of the United Nations conventions and resolutions are translated, at the national level, into command and control approaches. Examples are the 1982 United Nations Convention on the Law of the Sea, which sets out the legal framework within which all activities in the oceans must be undertaken, and United Nations General Assembly (UNGA) Resolution 61/105 (UNGA 2006) on vulnerable marine ecosystems.

The United Nations Convention on the Law of the Sea contains a comprehensive set of rules for regulating the use and management of ocean spaces and their resources. It includes provisions on:

i. the extent and delimitation of the maritime zones;
ii. coastal States' sovereignty, sovereign rights and jurisdiction in the areas under national jurisdiction;
iii. flag States' rights and duties;
iv. the protection and preservation of the marine environment;
v. the conservation and management of living marine resources;
vi. the legal status of resources on the seabed, ocean floor and subsoil beyond the limits of national jurisdiction and activities therein; and
vii. marine scientific research; development and transfer of marine technology; and the settlement of disputes.

Many fish stocks have been overexploited at an unprecedented rate (Levin *et al.* 2016), particularly due to the effectiveness and intensification of modern vessels and technology to explore the oceans, and the difficulties of monitoring, control and surveillance (FAO 2016). Several rules have been implemented over the years, from local to global (Bigagli 2016), under the oceans' complex regime (Keohane and Nye 1977; Keohane and Victor 2011), to regulate resource use and protect biodiversity. However, the lack of enforcement mechanisms is worrisome, as only a fraction of treaties applying to oceans have specific enforcement mechanisms (Al-Abdulrazzak *et al.* 2017).

Within the DPSIR framework (Section 1.6), command and control policy instruments mostly address 'pressures' (e.g. fishing, mining and pollution). Command and control approaches applied to the high seas have been implemented at a regional and sectoral level, with multiple authorities managing parts of the same regions, and extensive areas without governance arrangements. Further, attempts to coordinate activities, mitigate conflicts, address cumulative impacts or facilitate communication have been inadequate (Ban *et al.* 2014). One of the reasons highlighted by Al-Abdulrazzak *et al.* (2017) for such a state of affairs is that States with small environmental budgets may be unable to participate effectively in the many distinct agreements. Further, the lack of coordination among these treaties risks turning the years of government negotiations into 'empty treaties' with no accomplishments. Ultimately, the success of command and control policy depends on the political will of national governments (Englender *et al.* 2014).

Case study: UNGA Resolution on Vulnerable Marine Ecosystems

Within the context of sustainable fisheries, UNGA adopted Resolution 61/105 (UNGA 2006), which calls on regional fisheries management organizations (RFMOs) and States to adopt and implement measures, in accordance with the precautionary approach, ecosystem approaches and international law, as a matter of priority. According to paragraph 83 of the Resolution, regional fisheries management organizations or arrangements (RFMO/As) with the competence to regulate bottom fisheries are called on to adopt and implement measures, such as:

- *"Conduct impact assessments of individual high seas bottom fisheries to ensure that 'significant' adverse impacts on vulnerable marine ecosystems (VMEs) would be prevented or else prohibit bottom fishing;*
- *Close areas of the high seas to bottom fishing where VMEs are known or likely to occur unless bottom fisheries can be managed in these areas to prevent significant adverse impacts on VMEs;*
- *Ensure the long-term sustainability of deep-sea fish stocks; and*
- *Require bottom-fishing vessels to move out of an area of the high seas where 'unexpected' encounters with VMEs occur" (UNGA 2004).*

The remoteness and extent of the high seas provide real challenges to law enforcement, and to command and control approaches more generally. Alternatives to these approaches are less likely to succeed, given the low social coherence among global actors participating in high-seas fisheries. Still, UNGA Resolution 61/105 (UNGA 2006) on VMEs has begun a process of addressing the problem and has engaged different stakeholders to protect marine ecosystems. It triggered subsequent actions, including further policy developments regarding implementation, and action at the RFMO level. Major gaps include shortcomings in the design and capacities of RFMOs and the political will of countries to enforce regulations. If fully implemented, the Resolution will provide a good basis for protecting VMEs from significant adverse impacts resulting from bottom fishing and ensuring the long-term sustainability of deep-sea fish stocks.

Table 14.6: International cooperation resolutions

Criterion	Description	References
Success of failure	Where VMEs have been identified and fishing vessels with bottom-contacting gears have been effectively excluded, the outcome of no further damage of the VMEs from fishing is likely to be occurring.	Rogers and Gianni 2010
Independence of evaluation	UNGA adopts resolutions on sustainable fisheries annually. As part of this process, following the adoption of Resolution 61/105 in 2006 (UNGA 2006), UNGA conducted dedicated reviews of the implementation of the provisions of the Resolution, as well as subsequent resolutions, addressing the impacts of bottom fishing on VMEs and the long-term sustainability of deep-sea fish stocks in 2008, 2011, 2014 and 2016. Each of these reviews resulted in the adoption of additional provisions in UNGA Resolutions 63/112, 66/68, 69/109 and 71/123. A further review is planned for 2020. In 2014 and 2016, the reviews were preceded by two-day informal multi-stakeholder workshops. In addition, bottom fishing is also addressed in the context of the Review Conference on the United Nations Fish Stocks Agreement, which was held in 2006 and resumed in 2010 and 2016.	
Key actors	Other than States, FAO and RFMO/As are the principal bodies involved in the implementation of the provisions of Resolution 61/105 et seq. Discussions regarding the implementation of the resolutions have involved representatives of these intergovernmental organizations, as well as representatives of environmental NGOs, the fishing industry and academia.	
Baseline	The Resolution was based on historical records of stock status and fish-processing plants' operating costs and revenues. Both sources are reliable for the last 15 years.	FAO (2009; FAO 2010)
Time frame	It took two years for the VME identification criteria to be developed by FAO, and another two years for some RFMOs to identify their VMEs. Most RFMOs identified their VMEs within the time frame established in the Resolution.	
Constraining factors	The capacity of some RFMOs to identify VMEs and develop protective measures is limited – for example, in parts of the Pacific and Indian oceans.	
Enabling factors	Protecting biodiversity in the high seas had been part of UNGA's agenda for several years, and it had accordingly adopted a series of pre-resolutions (e.g. Resolution 59/25). Improved technologies for distant-water surveillance, such as vessel monitoring systems and satellite tracking, have made the detection of illegal fishing more feasible. Video technologies also allow the automated and less costly monitoring of on-board operations. Increased scientific study of deep-sea habitats and the use of on-board observers also seem to be important factors.	UNGA 2004
Cost-effectiveness	No information on cost-effectiveness is available.	
Equity	The Resolution affects national and corporate interests large enough to have the technology to fish the high seas. However, it may entail a uniform burden on countries with different capacity to comply.	
Co-benefits	There is potential for improved fishing practices beyond VMEs; more active collaboration between RFMOs and other authorities (seabed mining, shipping and the Convention on Biodiversity) to coordinate conservation efforts; and increased participation of scientific experts in RFMOs and national assessment and advisory bodies.	
Transboundary issues	The Resolution applies to multiple jurisdictions and overlaps with other international efforts such as the Convention for Biological Diversity's (CBD) Ecologically or Biologically Significant Marine Areas. In this regard, CBD and FAO cooperate to harmonize the outcomes of these efforts. Cooperation between Canada and the United States, where federal fisheries management agencies identify VMEs or ecologically or biologically significant areas (EBSAs) that straddle national boundaries, illustrates such bilateral efforts. Regional seas conventions also engage in identifying transboundary and/or high-seas EBSAs and can be considered a transboundary issue within a multilateral effort.	
Possible improvements	Disseminating this type of policy at the national level would be important, given the role of the Resolution as a springboard for more meaningful negotiations in the context of the Marine Biological Diversity of Areas Beyond National Jurisdiction (BBNJ) process. The Secretary-General, in his 2016 report (A/71/351), concluded that "[o]verall, while a number of actions have been taken, implementation of resolutions 64/72 and 66/68 on a global scale continues to be uneven and further efforts are needed (UNGA 2016). Unless timely actions are taken by all the stakeholders concerned, overfishing of deep-sea species is likely to continue and some VMEs will not be adequately protected from significant adverse impacts".	

14

14.3 Indicators

The case studies analysed above provided insights into challenges and opportunities for policy design and implementation in responding to key contemporary threats to the oceans. Further insights may be gained by examining policy-sensitive indicators relating to these threats.

14.3.1 Indicator 1: Coverage of marine protected areas

Marine protected areas (MPAs) are defined as "a clearly defined geographical space, recognised, dedicated and managed, through legal or other effective means, to achieve the long-term conservation of nature with associated ecosystem services and cultural values" (Dudley 2008). The coverage of MPAs is calculated for each country using the World Database on Protected Areas, managed by the UNEP World Conservation Monitoring Centre (WCMC) and IUCN. It is expressed as the percentage of MPAs within waters under national jurisdictions.

Current projections indicate that 7.3 per cent of the world's oceans have been designed as MPAs (UNEP-WCMC and IUCN 2018). Sala *et al.* (2018) argue that these projections are overestimated, given that they include areas that are yet to exclude significant extractive activities. Their projection indicates that the actual coverage of MPAs is 3.6 per cent, and only 2 per cent is being strongly or fully protected (Sala *et al.* 2018). In any case, while MPA coverage has been increasing **(Figure 14.1)**, additional efforts are required to meet the internationally agreed targets.

Policy relevance

MPAs and other area-based management tools have been promoted thorough international conventions and agreements, including the Convention on Biological Diversity (CBD) and policy instruments, such as the annual UNGA resolutions and the Sustainable Development Goals (SDGs).[2] Protected areas are also essential in achieving the CBD Aichi Targets 5 and 12, which aim to prevent or reduce the rate of habitat and species loss, respectively. Further, some coastal MPAs are also recognized as wetlands of international importance under the Ramsar Convention.

Figure 14.1: Coverage of Marine Protected Areas

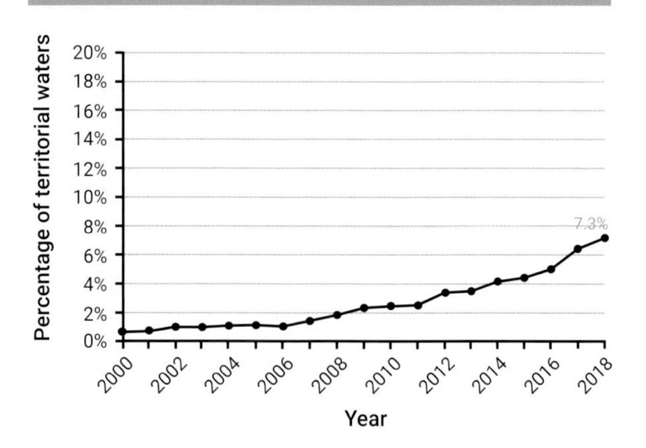

Source: UNEP-WCMC and IUCN (2018)

2 CBD Aichi Target 11 states: "[b]y 2020, at least... 10 per cent of coastal and marine areas, especially areas of particular importance for biodiversity and ecosystem services, are conserved through... systems of protected areas...". SDG 14.5 states: "[b]y 2020, conserve at least 10 per cent of coastal and marine areas, consistent with national and international law and based on the best available scientific information".

Casual relation

MPAs are a key conservation and management tool, particularly in the context of biodiversity and fisheries. They are part of area-based approaches, such as integrated coastal zone management and marine spatial planning. MPAs have the potential to address several pressures relating to marine biodiversity, including overfishing and habitat destruction. They help protect areas of ecological importance and ensure the provision of ecosystem services (e.g. fisheries, coastal protection, tourism and recreation) (Organisation for Economic Co-operation and Development [OECD] 2017), with important implications for human health and well-being (Kareiva and Mavier 2012). Further, MPAs have increasingly been promoted as a strategy to enhance the resilience of ecosystems to climate change (McLeod *et al.* 2009; Simard *et al.* 2016). Accordingly, the MPAs indicator addresses multiple issues identified in Chapter 7 of this report, particularly those relating to fisheries and climate change. Chapter 7 also recommends that, in the case of coral bleaching, reef-owning nations should consider taking immediate action (including establishing MPAs) to protect all known coral reef habitat from any non-subsistence uses (see Section 7.5.2).

Other influencing factors

National and subnational efforts are required to enhance the design and implementation of MPAs to ensure they meet their intended objectives. Evidence suggests that many nations are yet to meet key challenges such as:

i. strategically designing MPAs to maximize environmental and socioeconomic benefits;
ii. preparing and implementing adequate management plans;
iii. establishing robust monitoring and reporting frameworks;
iv. ensuring compliance and enforcement;
v. mobilizing finance to enable sustainable management; and
vi. embedding MPAs in policy mixes to address multiple pressures (OECD 2017).

Caveats

MPAs vary according to their management objective; they range from wholly biodiversity-focused MPAs to those incorporating human use (Dudley 2008). Accordingly, their contribution to achieving ocean conservation targets may vary. Further, the coverage of MPAs alone does not indicate that such areas are effectively and equitably managed. Efforts to develop methods for evaluating the effectiveness of MPAs are, therefore, critical. Examples of these methods include Protected Area Management Effectiveness and the Management Effectiveness Tracking Tool (Stolton *et al.* 2007; Coad *et al.* 2015).

14.3.2 Indicator 2: Beach litter assessment

Being relatively simple and cost-effective to monitor compared to other forms of marine litter (see Section 7.5.3), beach litter surveys are a common assessment method (e.g. Gabrielides *et al.* 1991; Madzena and Lasiak 1997; Willoughby *et al.* 1997; Velander and Mocogni 1999; Ballance, Ryan and Turpie 2000; Santos, Friedrich and Barretto 2005; Jayasiri *et al.* 2013; Hong

14

et al. 2014; Munari et al. 2016; Williams et al. 2016; Botero et al. 2017; Brouwer et al. 2017; Nelms et al. 2017; Rangel-Buitrago et al. 2017; Syakti et al. 2017). The key purpose is to assess trends in the volume, composition and spatial and temporal distribution of marine litter washed ashore or deposited on coastlines. The scope of the survey is limited to what is defined as a beach, which precludes very shallow tidal mudflat areas that may be many kilometres wide at low tide (Cheshire et al. 2009). The Northwest Pacific Action Plan (NOWPAP) and Convention for the Protection of the Marine Environment of the North-East Atlantic (OSPAR) selection criteria specify that sites should not be in close proximity to rivers, harbours or ports (NOWPAP 2008; OSPAR 2007). Buried litter is usually not sampled, though it may be a considerable proportion of beach litter.

Policy relevance

Although 'floating plastic debris density' was chosen as one of the indicators for SDG target 14.1: "By 2025, prevent and significantly reduce marine pollution of all kinds, in particular from land-based activities, including marine debris and nutrient pollution", it has been argued by many that beach litter should complement it. Many of the Regional Seas Conventions and Action Plans have agreed on beach litter as their core indicator for marine litter.

Various protocols outline the basic structure of the survey, the analysis of sampling units, the frequency and timing of surveys, the systems used for litter classification and the underpinning framework for facilitation and management of logistics. The data on beach litter generated through such standardized methodology can be useful for setting and achieving policy targets.

Causal relation

Beach litter originates from various sources; beach cleanup and monitoring programmes (such as Clean up Australia and the United Kingdom's Marine Conservation Society campaigns) have defined 'item indicators' to address the sources of litter. Some beaches will better indicate specific sources of litter than others due to their location (remote beaches or urban beaches tracking ship and urban pollution, respectively). Many studies dedicated to local beaches surveys and litter collection provide information on litter and tourism (UNEP/MAP 2015c). However, seasonal variations are common. While beach users were the main source of summer debris, litter in the tourist low season was primarily attributed to drainage and outfall systems. Other sources include floating litter washed ashore, coastal urbanization, wind-borne litter and illegal dumping. Changes in oceanographic (e.g. currents) and weather (e.g. storms) conditions may affect quantities of beach litter washed ashore.

Other influencing factors

The benefits of using beach litter as an indicator include the possibility to include citizen science (the participation of non-professional scientists in a scientific project). Because the technique is relatively simple, volunteers are able to participate in the quantification and monitoring of seasonal and site-specific beach litter (Rosevelt et al. 2013; van der Velde et al. 2017; Vincent et al. 2017). Furthermore, beach surveys provide a mechanism for education and building community understanding and awareness. For example, public participation in the cleaning campaigns is strong in the Mediterranean Sea. Comprehensive and regular surveys of marine litter on beaches have been made in many areas, often over a number of years, by various NGOs in the region (UNEP/MAP 2015c).

Caveats

It has been repeatedly emphasized that there is a need to develop and implement a standardized marine litter sampling protocol. A standardized method would allow quantification and understanding of the amount of litter within our seas and oceans through long-term, broad-scale, comparative studies (Cheshire et al. 2009; Besley et al. 2017). The lack of standardization and compatibility between methods used and results obtained in various bottom-up projects has made it difficult to compare data from different regions and to make an overall assessment of marine litter pollution for the entire region. Some regions have recently adopted a regional framework, such as the Regional Plan on Marine Litter Management in the Mediterranean, to coordinate and harmonize monitoring. Furthermore, it would help to make the categories for reporting compatible across different survey types (beaches, sea surface, sea floor), so that outcomes are comparable.

It can be difficult to draw conclusions regarding the overall increase or decrease of beach litter if variables change every year, including the number of volunteers participating in beach cleanups. More fundamentally, beach surveys may not relate to true marine pollution; because they may be affected by weather, the stranded debris may not necessarily provide a good indicator of changes in overall abundance (UNEP/MAP 2015c).

14.3.3 Indicator 3: Number of Vulnerable Marine Ecosystems identified by Regional Fisheries Management Organizations and closed to fishing or otherwise protected (1,000/934/934)

This indicator measures the number of marine ecosystems that have been identified as vulnerable to impacts from fishing activities and are protected by RFMOs **(Figure 14.2)**. This indicator serves as a complement to the stock status indicators (e.g. references to FAO State of World Fisheries and Aquaculture (SOFIA) reports) used in Chapter 7. It relates to a debate in wider policy literature on how to protect biodiversity. Although some approaches prefer sectoral regulation, such as on fisheries, mining or shipping, others (such as that underlying this indicator) advocate complete protection of biodiversity and habitats from all threats regardless of sector. VMEs are identified by an internationally agreed process that can be found in paragraph 42 of the International Guidelines for the Management of Deep-sea Fisheries in the High Seas (FAO 2009) and entail a management response that is generally embedded in the management process of RFMOs.

Policy relevance

As described in Section 14.2.5, 14.2.6, the concept of a VME gained momentum following UNGA Resolution 61/105. It stems from the *Rio +20* commitment to enhance actions to protect VMEs, such as impact assessments, but is most recently established in SDG 14 on oceans, particularly targets 14.2 and 14.4. VME protection also appears in CBD Aichi Target 6.

Causal relation

UNGA has identified a number of marine habitats with vulnerable ecosystem features **(Figure 14.2)**, including coastal lagoons, mangroves, estuaries, wetlands, seagrass beds and coral reefs, but also areas further from shore and sometimes beyond national jurisdiction, such as spawning and nursery grounds, cold-water corals, seamounts, various features associated with polar regions, hydrothermal vents, deep-sea trenches, submarine canyons and oceanic ridges (UNGA 2004). Here we concentrate on the identification and protection of VMEs by RFMOs, showing the areas of coverage through maps, as numbers were not available.

RFMOs have been required to protect VMEs since 2008, with specific requirements laid out under UNGA Resolutions 59/25, 61/105 and 64/72. VME protection typically consists of banning or otherwise restricting bottom-trawling in VMEs. Bottom-trawling consists of vessels dragging nets along or near the bottom of the sea; it is considered especially destructive because it is both indiscriminate, including considerable by-catch beyond the target species, and operates at the same part of the water column as many of the most vulnerable species and much oceanic habitat. RFMOs are expected to help identify VMEs within their regulatory areas, which are often beyond areas of national jurisdiction, and protect them against destructive practices.

Other influencing factors

Despite some early adoption, RFMO implementation has been variable. While recently established RFMOs such as the South Pacific (SP) RFMO and the Southern Indian Ocean Fisheries Agreement (SIOFA) expand the marine area beyond national jurisdiction subjected to regulatory opportunities, they may not yet provide adequate stock assessment and leave some

Shutterstock/Load dragon

VMEs open to bottom-trawling unless environmental impact assessments (see Section 11.3.2) highlight that further restrictions are necessary (Currie 2016). Other RFMOs, such as the North East Atlantic Fisheries Commission (NEAFC), the Northwest Atlantic Fisheries Organization (NAFO) and the South East Atlantic Fisheries Organisation (SEAFO), have closed substantial areas that are likely to contain VMEs, and the Convention on the Conservation of Antarctic Marine Living Resources (CCAMLR) has banned bottom-trawling in some areas. NAFO has identified 20 areas as being vulnerable to bottom-trawling and subsequently closed them **(Figure 14.3)**. The General Fisheries Commission for the Mediterranean (GFCM) lags behind other RFMOs in fulfilling these obligations. GFCM measures to protect VMEs are limited to three fisheries restricted areas (FRAs) and a prohibition on trawling below 1,000 metres. Most VMEs in the Mediterranean are, therefore, entirely unprotected (Oceana 2016).

Caveats

Banning destructive fishing practices in VMEs may not guarantee their preservation. Lost driftnets, marine litter, ocean acidification and eutrophication can all affect VMEs, even if they are protected from destructive fishing practices. Further, relying on protected VMEs as an indicator potentially disregards important, unprotected VMEs. However, compared to terrestrial ecosystems, data on marine biodiversity remain limited (Martin *et al.* 2015). IUCN's Red List of Ecosystems (IUCN 2017) provides a third-party attempt to catalogue ecosystems, including marine ecosystems, that are most vulnerable. The goal is to have all ecosystems assessed by 2025. This and further indicators should be developed.

Figure 14.2: Areas of predicted deep-sea vulnerable marine ecosystems

Note: Areas in red illustrate the extent of deep-sea (>200m) bottom-trawling on VMEs predicted from published habitat suitability models and binary predicted presence maps.

Source: Pham *et al.* 2014

Figure 14.3: Bottom-trawling and closed VMEs from 2006 to 2016

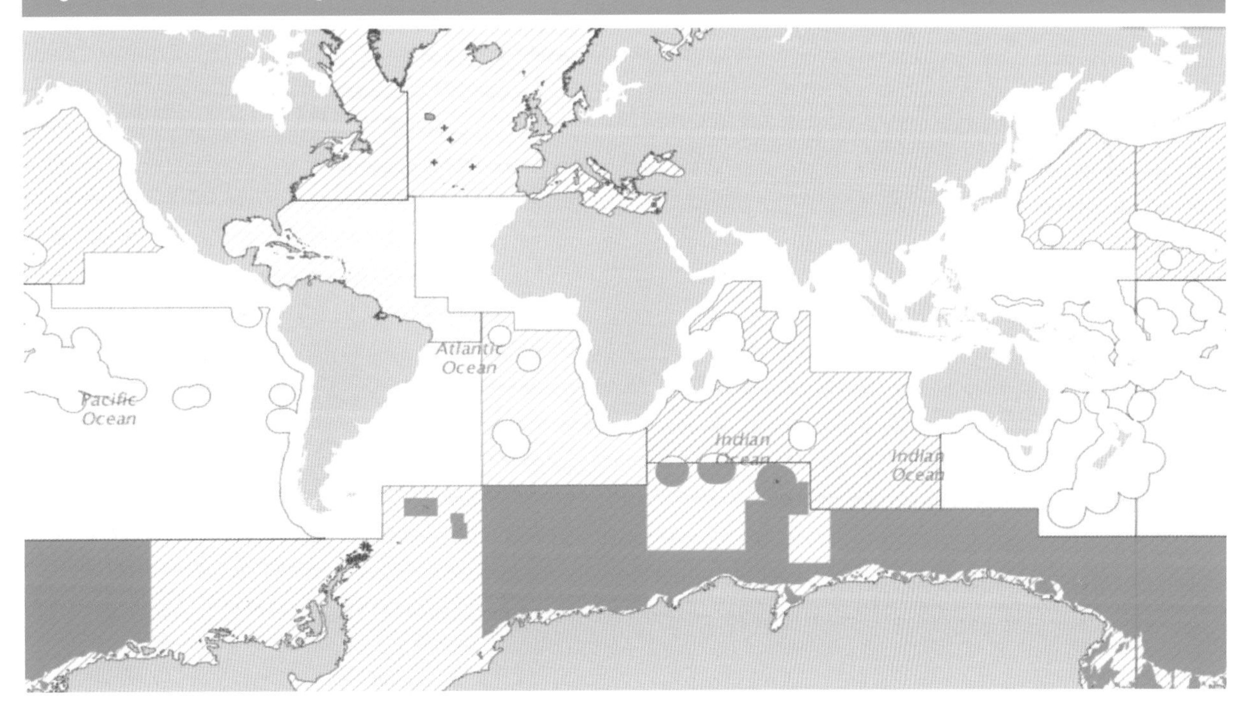

Note: Green-filled areas are bottom-fishing areas, and red-filled areas are VME closed areas. Diagonally shaded areas represent the regulatory areas of key regional fisheries bodies.

Source: FAO (2017).

14.4 Discussion and conclusions

Diverse governance approaches and policy instruments have been used in response to the impacts of climate change, pollution and overfishing on the ocean. These approaches and instruments have achieved different levels of success. For example, while RBM has only had a limited impact in minimizing coral bleaching in the GBR, ITQs reversed the decline in status of many key fish stocks and secured the financial viability of the processing sector in British Columbia.

The cases examined in this chapter provide useful insights into policy design and implementation. For example, the success of the Chilean abalone TURF is due to meaningful community involvement in developing and implementing a range of management arrangements. In the case of the Regional Plan on Marine Litter Management in the Mediterranean, stakeholder collaboration to reduce plastic consumption is a key component of the Plan. However, more diverse stakeholders were only included in the VME process after the UNGA Resolution was adopted. Common to most of the cases was the involvement of relevant stakeholders, including resource users, businesses, experts, environmental NGOs and government, at some point in the policy process.

Another feature common to most of the cases was the use of baseline information. For example, a comprehensive assessment of the threat posed by climate change to the GBR informed the RBM initiative; an assessment of the status of marine litter in the Mediterranean was used as a basis for the development of the Regional Plan; and historical records of stock status and plant operating costs and revenues supported the establishment of the ITQs in British Columbia. In addition to informing policy design, baselines establish the preconditions against which progress towards achieving desired goals can be measured, and additional interventions to improve implementation can be made. For example, in the case of UNGA Resolution 61/105 concerning VMEs, additional provisions were adopted (in Resolution 64/72) to improve implementation once it was recognized that adoption was not proceeding rapidly enough. Despite its importance, baseline information is not always reliable or available; though this should not prevent policy interventions. In the case of the Chilean abalone TURF, existing baseline data were weak and ad hoc. TURFs were established based on common knowledge of the severe levels of stock depletion and failed attempts to control extensive illegal fishing using ITQs.

Another important insight from the case studies is that policy effectiveness is context-dependent. That is, a policy is more likely to prove effective where favourable conditions exist. These enabling factors include leadership, expertise, funding and stakeholder support. For example, the relative implementation success of UNGA Resolution 61/105 in the North Atlantic is associated with existing scientific support and strong surveillance and enforcement capabilities. Conversely, conditions for implementing UNGA Resolution 61/105 are still to be developed in parts of the Pacific and Indian oceans. Strong governance capabilities have been key to successful ITQ implementation. Further, policy interventions need to be tailored to the circumstances where they apply. For instance, the success of Chilean abalone TURFs is attributed to management arrangements being adapted according to geographical and community characteristics.

Last, there is an apparent lack of explicit consideration of the policies and indicators examined regarding human health and well-being. For example, the establishment of MPAs which might restrict access rights of traditional coastal populations may have negative impacts on their livelihood, food security and health. Likewise, the impact of increasingly warmer oceans may result in more frequent phytoplankton blooms, some of which relate to shellfish and fish poisoning and conditions conducive to cholera outbreaks (Cinner *et al.* 2016). These and other health and well-being implications need to be considered as part of ocean policies, if the goal of a 'healthy planet, healthy people' is to be achieved.

References

Abbott, K.W., Keohane, R.O., Moravcsik, A., Slaughter, A.M. and Snidal, D. (2000). The concept of legalization. *International organization* 54(3), 401-419. https://doi.org/10.1162/002081800551271.

Aburto, J.A., Stotz, W.B. and Cundill, G. (2014). Social-ecological collapse: TURF governance in the context of highly variable resources in Chile. *Ecology and Society* 19(1). https://doi.org/10.5751/ES-06145-190102.

Aceves-Bueno, E., Cornejo-Donoso, J., Miller, S.J. and Gaines, S.D. (2017). Are territorial use rights in fisheries (TURFs) sufficiently large? *Marine Policy* 78, 189-195. https://doi.org/10.1016/j.marpol.2017.01.024.

Afflerbach, J.C., Lester, S.E., Dougherty, D.T. and Poon, S.E. (2014). A global survey of "TURF-reserves", Territorial use rights for fisheries coupled with marine reserves. *Global Ecology and Conservation* 2, 97-106. https://doi.org/10.1016/j.gecco.2014.08.001.

Ainsworth, C.H., Pitcher, T.J., Heymans, J.J. and Vasconcellos, M. (2008). Reconstructing historical marine ecosystems using food web models: Northern British Columbia from pre-European contact to present. *Ecological Modelling* 216(3-4), 354-368. https://doi.org/10.1016/j.ecolmodel.2008.05.005.

Al-Abdulrazzak, D., Galland, G.R., McClenachan, L. and Hocevar, J. (2017). Opportunities for improving global marine conservation through multilateral treaties. *Marine Policy* 86, 247-252. https://doi.org/10.1016/j.marpol.2017.09.036.

Andreu-Cazenave, M., Subida, M.D. and Fernandez, M. (2017). Correction: Exploitation rates of two benthic resources across management regimes in central Chile: Evidence of illegal fishing in artisanal fisheries operating in open access areas. *PLoS One* 13(1), e0191166. https://doi.org/10.1371/journal.pone.0180012.

Anthony, K., Marshall, P.A., Abdulla, A., Beeden, R., Bergh, C., Black, R. *et al.* (2015). Operationalizing resilience for adaptive coral reef management under global environmental change. *Global Change Biology* 21(1), 48-61. https://doi.org/10.1111/gcb.12700.

Anthony, K.R.N. (2016). Coral reefs under climate change and ocean acidification: challenges and opportunities for management and policy. *Annual Review of Environment and Resources* 41, 59-81. https://doi.org/10.1146/annurev-environ-110615-085610.

Arnason, R. (1993). The Icelandic individual transferable quota system: A descriptive account. *Marine Resource Economics* 8(3), 201-218. https://doi.org/10.1086/mre.8.3.42629066.

Ballance, A., Ryan, P.G. and Turpie, J.K. (2000). How much is a clean beach worth? The impact of litter on beach users in the Cape Peninsula, South Africa. *South African Journal of Science* 96(5), 210-230. https://journals.co.za/content/sajsci/96/5/AJA00382353_8975.

Ban, N.C., Maxwell, S.M., Dunn, D.C., Hobday, A.J., Bax, N.J., Ardron, J. *et al.* (2014). Better integration of sectoral planning and management approaches for the interlinked ecology of the open oceans. *Marine Policy* 49, 127-136. https://doi.org/10.1016/j.marpol.2013.11.024.

Barner, A.K., Lubchenco, J., Costello, C., Gaines, S.D., Leland, A., Jenks, B. *et al.* (2015). Solutions for recovering and sustaining the bounty of the ocean: Combining fishery reforms, rights-based fisheries management, and marine reserves. *Oceanography* 28(2), 252-263. https://doi.org/10.5670/oceanog.2015.51.

Bertolotti, M.I., Baltar, F., Gualdoni, P., Pagani, A. and Rotta, L. (2016). Individual transferable quotas in Argentina: Policy and performance. *Marine Policy* 71, 132-137. https://doi.org/10.1016/j.marpol.2016.05.024.

Besley, A., Vijver, M.G., Behrens, P. and Bosker, T. (2017). A standardized method for sampling and extraction methods for quantifying microplastics in beach sand. *Marine Pollution Bulletin* 114(1), 77-83. https://doi.org/10.1016/j.marpolbul.2016.08.055.

Bigagli, E. (2016). The international legal framework for the management of the global oceans social-ecological system. *Marine Policy* 68, 155-164. https://doi.org/10.1016/j.marpol.2016.03.005.

Biggs, D., Amar, F., Valdebenito, A. and Gelcich, S. (2016). Potential synergies between nature-based tourism and sustainable use of marine resources: Insights from dive tourism in territorial user rights for fisheries in Chile. *PLoS One* 11(3), e0148862. https://doi.org/10.1371/journal.pone.0148862.

Blomquist, J. and Waldo, S. (2018). Scrapping programmes and ITQs: Labour market outcomes and spill-over effects on non-targeted fisheries in Sweden. *Marine Policy* 88, 41-47. https://doi.org/10.1016/j.marpol.2017.11.004.

Botero, C.M., Anfuso, G., Milanes, C., Cabrera, A., Casas, G., Pranzini, E. *et al.* (2017). Litter assessment on 99 Cuban beaches: A baseline to identify sources of pollution and impacts for tourism and recreation. *Marine Pollution Bulletin* 118(1-2), 437-441. https://doi.org/10.1016/j.marpolbul.2017.02.061.

Branch, T.A. (2006). Discards and revenues in multispecies groundfish trawl fisheries managed by trip limits on the US west coast and by ITQs in British Columbia. *Bulletin of Marine Science* 78(3), 669-689. https://www.ingentaconnect.com/content/umrsmas/bullmar/2006/00000078/00000003/art00016#.

Branch, T.A. (2009). How do individual transferable quotas affect marine ecosystems? *Fish and Fisheries* 10(1), 39-57. https://doi.org/10.1111/j.1467-2979.2008.00294.x.

Bromley, D.W. (2009). Abdicating responsibility: The deceits of fisheries policy. *Fisheries* 34(6), 280-290. https://doi.org/10.1577/1548-8446-34.6.280.

Brouwer, R., Hadzhiyska, D., Ioakeimidis, C. and Ouderdorp, H. (2017). The social costs of marine litter along European coasts. *Ocean & Coastal Management* 138, 38-49. https://doi.org/10.1016/j.ocecoaman.2017.01.011.

Campani, T., Baini, M., Giannetti, M., Cancelli, F., Mancusi, C., Serena, F. *et al.* (2013). Presence of plastic debris in loggerhead turtle stranded along the Tuscany coasts of the Pelagos Sanctuary for Mediterranean marine mammals (Italy). *Marine Pollution Bulletin* 74(1), 225-230. https://doi.org/10.1016/j.marpolbul.2013.06.053.

Casey, K.E., Dewees, C.M., Turris, B.R. and Wilen, J.E. (1995). The effects of individual vessel quotas in the British Columbia halibut fishery. *Marine Resource Economics* 10(3), 211-230. https://doi.org/10.1086/mre.10.3.42629588.

Castilla, J.C. (1995). The sustainability of natural renewable resources as viewed by an ecologist and exemplified the fishery of the Mollusc *Concholepas concholepas* in Chile. In *Defining and Measuring Sustainability: The Biogeophysical Foundations*. Munasinghe, M. and Shearer, W. (eds.). Washington, D.C: World Bank. 153-159. http://documents.worldbank.org/curated/en/328001468764998700/pdf/multi0page.pdf

Castilla, J.C., Espinosa, J., Yamashiro, C., Melo, O. and Gelcich, S. (2016). Telecoupling between catch, farming, and international trade for the gastropods Concholepas concholepas (loco) and Haliotis spp. (abalone). *Journal of Shellfish Research* 35(2), 499-506. https://doi.org/10.2983/035.035.0223.

Castilla, J.C. and Fernandez, M. (1998). Small-scale benthic fisheries in Chile: On co-management and sustainable use of benthic invertebrates. *Ecological Applications* 8(sp1), S124-S132. https://doi.org/10.1890/1051-0761(1998)8[S124:SBFICO]2.0.CO;2.

Chambers, C. and Kokorsch, M. (2017). The social dimension in icelandic fisheries governance. *Coastal Management* 45(4), 330-337. https://doi.org/10.1080/08920753.2017.1327346.

Charles, A.T. (2002). Use rights and responsible fisheries: Limiting access and harvesting through rights-based management. In *A Fishery Manager's Guidebook - Management Measures and Their Application*. Cochrane, K.L. (ed.). Rome: Food and Agriculture Organization of the United Nations. chapter 6. 131-158. http://www.fao.org/tempref/docrep/fao/004/y3427e/y3427e06.pdf

Chauvin, C. and Le Bouar, G. (2007). Occupational injury in the French sea fishing industry: A comparative study between the 1980s and today. *Accident Analysis & Prevention* 39(1), 79-85. https://doi.org/10.1016/j.aap.2006.06.006.

Cheshire, A.C., Adler, E., Barbière, J., Cohen, Y., Evans, S., Jarayabhand, S. *et al.* (2009). *UNEP/IOC Guidelines on Survey and Monitoring of Marine Litter*. Nairobi: United Nations Environment Programme. http://wedocs.unep.org/xmlui/bitstream/handle/20.500.11822/13604/rsrs186.pdf?sequence=1&isAllowed=y.

Christy, F.T. (1992). *Territorial Use Rights in Marine Fisheries: Definitions and Conditions*. Rome: Food and Agriculture Organization of the United Nations. http://www.fao.org/docrep/003/t0507e/t0507e00.htm.

Chu, C. (2009). Thirty years later: The global growth of ITQs and their influence on stock status in marine fisheries. *Fish and Fisheries* 10(2), 217-230. https://doi.org/10.1111/j.1467-2979.2008.00313.x.

Cinner, J.E., Pratchett, M.S., Graham, N.A.J., Messmer, V., Fuentes, M.M.P.B., Ainsworth, T. *et al.* (2016). A framework for understanding climate change impacts on coral reef social–ecological systems. *Regional Environmental Change* 16(4), 1133–1146. https://doi.org/10.1007/s10113-015-0832-z.

Coad, L., Leverington, F., Knights, K., Geldmann, J., Eassom, A., Kapos, V. *et al.* (2015). Measuring impact of protected area management interventions: Current and future use of the global database of protected area management effectiveness. *Philosophical Transactions of the Royal Society B: Biological Sciences* 370(1681). https://doi.org/10.1098/rstb.2014.0281.

Commonwealth of Australia (2016). *Climate Change Adaptation Good Practice, Case Study: The Great Barrier Reef Climate Change Action Plan 2007-2012*. Canberra. https://www.nccarf.edu.au/localgov/sites/nccarf.edu.au.localgov/files/casestudies/pdf/Case%20Study_The%20Great%20Barrier%20Reef%20Climate%20Change%20Action%20Plan%202007_2012.pdf.

Coral Triangle Initiative Secretariat (2009). *Regional Plan of Action*. Coral Triangle Initiative on Coral Reefs, Fisheries and Food Security. https://www.environment.gov.au/system/files/pages/f072279b-828c-4743-b08e-c039270aa7b2/files/cti-rpoa.pdf.

Costello, C., Gaines, S.D. and Lynham, J. (2008). Can catch shares prevent fisheries collapse? *Science* 321(5896), 1678-1681. https://doi.org/10.1126/science.1159478.

Costello, C. and Kaffine, D. (2017). Private conservation in TURF-managed fisheries. *Natural Resource Modeling* 30(1), 30-51. https://doi.org/10.1111/nrm.12103.

Costello, C., Lynham, J., Lester, S.E. and Gaines, S.D. (2010). Economic incentives and global fisheries sustainability. *Annual Review of Resource Economics* 2(September), 299–318. https://doi.org/10.1146/annurev.resource.012809.103923.

Council of Australian Governments (2007). *National Climate Change Adaptation Framework*. Canberra: Council of Australian Governments. https://www.nccarf.edu.au/sites/default/files/Australian-Government-2007a.pdf.

Crona, B., Gelcich, S. and Bodin, Ö. (2017). The importance of interplay between leadership and social capital in shaping outcomes of rights-based fisheries governance. *World Development* 91, 70-83. https://doi.org/10.1016/j.worlddev.2016.10.006.

Crona, B.I., Basurto, X., Squires, D., Gelcich, S., Daw, T.M., Khan, A. *et al.* (2016). Towards a typology of interactions between small-scale fisheries and global seafood trade. *Marine Policy* 65, 1-10. https://doi.org/10.1016/j.marpol.2015.11.016.

Crona, B.I., Van Holt, T., Petersson, M., Daw, T.M. and Buchary, E. (2015). Using social–ecological syndromes to understand impacts of international seafood trade on small-scale fisheries. *Global Environmental Change* 35, 162-175. https://doi.org/10.1016/j.gloenvcha.2015.07.006.

Currie, D. (2016). Opportunities and challenges in further addressing the impacts of bottom fisheries on vulnerable marine ecosystems and long-term sustainability of deep-sea fish stocks. *Workshop to Discuss Implementation of Paragraphs 113, 117 and 119 to 124 of Resolution 64/72 and Paragraphs 121, 126, 129, 130 and 132 to 134 of Resolution 66/68 on Sustainable Fisheries, Addressing the Impacts of Bottom Fishing on Vulnerable Marine Ecosystems and the Long-Term Sustainability of Deep-Sea Fish Stocks*. United Nations Headquarters, 1-2 August 2016. http://www.un.org/depts/los/reference_files/Presentations/PPT/Segment5/DC.pdf

Day, J. (2016). The great barrier reef marine park: The grandfather of modern MPAs. In *Big, Bold and Blue: Lessons from Australia's Marine Protected Areas*. Fitzsimons, J. and Wescott, G. (eds.). Clayton: CSIRO Publishing. chapter 5. 65-97. https://www.researchgate.net/profile/Jon_Day/publication/304677045_The_Great_Barrier_Reef_Marine_Park_-_the_grandfather_of_modern_MPAs/links/578722ee08aec5c2e4e2fcba/The-Great-Barrier-Reef-Marine-Park-the-grandfather-of-modern-MPAs.pdf?origin=publication_detail

de Juan, S., Gelcich, S., Ospina-Alvarez, A., Perez-Matus, A. and Fernandez, M. (2015). Applying an ecosystem service approach to unravel links between ecosystems and society in the coast of central Chile. *Science of the Total Environment* 533, 122-132. https://doi.org/10.1016/j.scitotenv.2015.06.094.

Defeo, O. and Castilla, J.C. (2005). More than one bag for the world fishery crisis and keys for co-management successes in selected artisanal Latin American shellfisheries. *Reviews in Fish Biology and Fisheries* 15(3), 265-283. https://doi.org/10.1007/s11160-005-4865-0.

Defeo, O., Castrejón, M., Pérez-Castañeda, R., Castilla, J.C., Gutiérrez, N.L., Essington, T.E. *et al.* (2016). Co-management in Latin American small-scale shellfisheries: Assessment from long-term case studies. *Fish and Fisheries* 17(1), 176-192. https://doi.org/10.1111/faf.12101.

Dewees, C.M. (1998). Effects of individual quota systems on New Zealand and British Columbia fisheries. *Ecological Applications* 8(sp1), 133-138. https://doi.org/10.2307/2641371.

Dolan, A.H., Taylor, M., Neis, B., Ommer, R., Eyles, J., Schneider, D. *et al.* (2005). Restructuring and health in Canadian coastal communities. *EcoHealth* 2(3), 195-208. https://doi.org/10.1007/s10393-005-6333-7.

Dudley, N. (ed.) (2008). *Guidelines for Applying Protected Area Management Categories*. Gland: International Union for Conservation of Nature. https://cmsdata.iucn.org/downloads/guidelines_for_applying_protected_area_management_categories.pdf.

Earth Justice (2015). *Traditional use rights in fisheries in Chile: area-based management and conservation of benthic resources: Smithsonian ocean portal*. https://earthjustice.org/blog/2015-october/earthjustice-and-the-smithsonian-team-up-to-enhance-ocean-conservation-efforts (Accessed: 25 September.

Edinger, T. and Baek, J. (2015). The role of property rights in bycatch reduction: Evidence from the British Columbia groundfish fishery. *Fisheries Research* 168, 100-104. https://doi.org/10.1016/j.fishres.2015.04.001.

Edwards, D.N. and Edwards, D.G. (2017). Licence Banks as a tool to mitigate corporate control of fisheries: A British Columbia groundfish example. *Marine Policy* 80, 141-146. https://doi.org/10.1016/j.marpol.2016.11.006.

Emery, T.J., Gardner, C., Hartmann, K. and Cartwright, I. (2017). Incorporating economics into fisheries management frameworks in Australia *Marine Policy* 77, 136-143. https://doi.org/10.1016/j.marpol.2016.12.018.

Emery, T.J., Hartmann, K., Green, B.S., Gardner, C. and Tisdell, J. (2014). Fishing for revenue: How leasing quota can be hazardous to your health. *ICES Journal of Marine Science* 71(7), 1854-1865. https://doi.org/10.1093/icesjms/fsu019.

Englender, D., Kirschey, J., Stöfen, A. and Zink, A. (2014). Cooperation and compliance control in areas beyond national jurisdiction. *Marine Policy* 49, 186-194. https://doi.org/10.1016/j.marpol.2013.11.022.

European Commission (2017). *Our Ocean Seas and Coasts*. http://ec.europa.eu/environment/marine/good-environmental-status/descriptor-10/index_en.htm.

European Parliament and European Council (2008). *Directive 2008/56/EC of the European Parliament and of the Council of 17 June 2008 establishing a Framework for Community Action in the Field of Marine Environmental Policy (Marine Strategy Framework Directive)*. https://eur-lex.europa.eu/legal-content/EN/TXT/PDF/?uri=CELEX:32008L0056&from=EN.

Fernández, M. and Castilla, J.C. (2005). Marine conservation in Chile: Historical perspective, lessons, and challenges. *Conservation Biology* 19(6), 1752-1762. https://doi.org/10.1111/j.1523-1739.2005.00277.x.

Fidelman, P.I.J., Leitch, A.M. and Nelson, D.R. (2013). Unpacking multilevel adaptation to climate change in the Great Barrier Reef, Australia. *Global Environmental Change* 23(4), 800-812. https://doi.org/10.1016/j.gloenvcha.2013.02.016.

Fisheries and Oceans Canada (2017). *Groundfish, Pacific Region 2017: Integrated Fisheries Management Plan Summary*. Ottawa: Fisheries and Oceans Canada. http://www.pac.dfo-mpo.gc.ca/fm-gp/mplans/2017/ground-fond-sm-2017-eng.pdf.

Food and Agriculture Organization of the United Nations (2005). *Fisheries Laws and Regulations in the Mediterranean: A Comparative Study*. Rome: Food and Agriculture Organization of the United Nations. http://www.fao.org/3/a-y5880e.pdf.

Food and Agriculture Organization of the United Nations (2009). *International Guidelines for the Management of Deep-Sea Fisheries in the High Seas*. Rome. http://www.fao.org/docrep/011/i0816t/i0816t00.HTM.

Food and Agriculture Organization of the United Nations (2010). *Report of the FAO Workshop on the Implementation of the International Guidelines for the Management of Deep-sea Fisheries in the High Seas: Challenges and Ways Forward*. Rome. http://www.fao.org/docrep/014/i2135e/i2135e00.pdf.

Food and Agriculture Organization of the United Nations (2016). *The State of World Fisheries and Aquaculture 2016: Contributing to Food Security and Nutrition for All*. Rome: Food and Agriculture Organization of the United Nations. http://www.fao.org/3/a-i5555e.pdf.

Food and Agriculture Organization of the United Nations (2017). *Vulnerable marine ecosystems database*. http://www.fao.org/in-action/vulnerable-marine-ecosystems/vme-database/en/vme.html.

Frangoudes, K. and Bellanger, M. (2017). Fishers' opinions on marketization of property rights and the quota system in France. *Marine Policy* 80, 107-112. https://doi.org/10.1016/j.marpol.2017.01.010.

Gabrielides, G.P., Golik, A., Loizides, L., Marino, M.G., Bingel, F. and Torregrossa, M.V. (1991). Man-made garbage pollution on the Mediterranean coastline. *Marine Pollution Bulletin* 23, 437-441. https://doi.org/10.1016/0025-326X(91)90713-3.

Galgani, F., Claro, F., Depledge, M. and Fossi, C. (2014). Monitoring the impact of litter in large vertebrates in the Mediterranean Sea within the European Marine Strategy Framework Directive (MSFD): Constraints, specificities and recommendations. *Marine Environmental Research* 100, 3-9. https://doi.org/10.1016/j.marenvres.2014.02.003.

Galgani, F., Jaunet, S., Campillo, A., Guenegen, X. and His, E. (1995). Distribution and abundance of debris on the continental shelf of the north-western Mediterranean Sea. *Marine Pollution Bulletin* 30(11), 713–717. https://doi.org/10.1016/0025-326X(95)00055-R.

Garavelli, L., Colas, F., Verley, P., Kaplan, D.M., Yannicelli, B. and Lett, C. (2016). Influence of biological factors on connectivity patterns for Concholepas concholepas (loco) in Chile. *PLoS One* 11(1), e0146418. https://doi.org/10.1371/journal.pone.0146418.

Garavelli, L., Kaplan, D.M., Colas, F., Stotz, W., Yannicelli, B. and Lett, C. (2014). Identifying appropriate spatial scales for marine conservation and management using a larval dispersal model: the case of Concholepas concholepas (loco) in Chile. *Progress in Oceanography* 124, 42-53. https://doi.org/10.1016/j.pocean.2014.03.011.

Gelcich, S., Cinner, J., Donlan, C.J., Tapia-Lewin, S., Godoy, N. and Castilla, J.C. (2017). Fishers' perceptions on the Chilean coastal TURF system after two decades: Problems, benefits, and emerging needs. *Bulletin of Marine Science* 93(1), 53-67. https://doi.org/10.5343/bms.2015.1082.

Gelcich, S. and Donlan, C.J. (2015). Incentivizing biodiversity conservation in artisanal fishing communities through territorial user rights and business model innovation. *Conservation Biology* 29(4), 1076-1085. https://doi.org/10.1111/cobi.12477.

Gelcich, S., Godoy, N., Prado, L. and Castilla, J.C. (2008). Add-on conservation benefits of marine territorial user rights fishery policies in central Chile. *Ecological Applications* 18(1), 273-281. https://doi.org/10.1890/06-1896.1.

Gelcich, S., Peralta, L., Donlan, C.J., Godoy, N., Ortiz, V., Tapia-Lewin, S. *et al.* (2015). Alternative strategies for scaling up marine coastal biodiversity conservation in Chile. *Maritime Studies* 1(4), 1–13. https://doi.org/10.1186/s40152-015-0022-0.

Giacaman-Smith, J., Neira, S. and Arancibia, H. (2016). Community structure and trophic interactions in a coastal management and exploitation area for benthic resources in central Chile. *Ocean & Coastal Management* 119, 155-163. https://doi.org/10.1016/j.ocecoaman.2015.10.003.

Gibbs, M.T. (2010). Why ITQs on target species are inefficient at achieving ecosystem based fisheries management outcomes. *Marine Policy* 34(3), 708-709. https://doi.org/10.1016/j.marpol.2009.09.005.

Gibson, D. and Sumaila, U.R. (2017). Determining the degree of 'small-scaleness' using fisheries in British Columbia as an example. *Marine Policy* 86, 121-126. https://doi.org/10.1016/j.marpol.2017.09.015.

González, J., Stotz, W., Garrido, J., Orensanz, J.M., Parma, A.M., Tapia, C. *et al.* (2006). The Chilean TURF system: How is it performing in the case of the loco fishery? *Bulletin of Marine Science* 78(3), 499-527. https://www.ingentaconnect.com/content/umrsmas/bullmar/2006/00000078/00000003/art00007#.

Great Barrier Reef Marine Park Authority (2007). *Great Barrier Reef Climate Change Action Plan 2007-2012*. Townsville: Great Barrier Reef Marine Park Authority. http://www.gbrmpa.gov.au/__data/assets/pdf_file/0020/4493/climate-change-action-plan-2007-2012.pdf.

Great Barrier Reef Marine Park Authority (2009). *Great Barrier Reef Outlook Report 2009*. Townsville: Great Barrier Reef Marine Park Authority. http://elibrary.gbrmpa.gov.au/jspui/bitstream/11017/199/1/Great-Barrier-Reef-Outlook-Report-2009.pdf.

Great Barrier Reef Marine Park Authority (2012). *Climate Change Adaptation: Outcomes from the Great Barrier Reef Climate Change Action Plan 2007–2012*. Townsville: Great Barrier Reef Marine Park Authority. http://elibrary.gbrmpa.gov.au/jspui/bitstream/11017/1139/1/Climate%20Change%20Adaptation.pdf.

Great Barrier Reef Marine Park Authority (2014). *Great Barrier Reef Outlook Report 2014*. Townsville: Great Barrier Reef Marine Park Authority. http://elibrary.gbrmpa.gov.au/jspui/bitstream/11017/2855/1/GBR%20Outlook%20Report%202014_Web280714.pdf.

Gutiérrez, N.L., Hilborn, R. and Defeo, O. (2011). Leadership, social capital and incentives promote successful fisheries. *Nature* 470, 386-389. https://doi.org/10.1038/nature09689.

Hannesson, R. (2013). Norway's experience with ITQs. *Marine Policy* 37, 264-269. https://doi.org/10.1016/j.marpol.2012.05.008.

Hannesson, R. (2017). Return on capital and management reforms in Norway's fisheries. *Land Economics* 93(4), 710-720. https://doi.org/10.3368/le.93.4.710.

Hauck, M. and Gallardo-Fernández, G.L. (2013). Crises in the South African abalone and Chilean loco fisheries: Shared challenges and prospects. *Maritime Studies* 12(3). https://doi.org/10.1186/2212-9790-12-3.

Håvold, J.I. (2010). Safety culture aboard fishing vessels. *Safety Science* 48(8), 1054-1061. https://doi.org/10.1016/j.ssci.2009.11.004.

Hoefnagel, E. and de Vos, B. (2017). Social and economic consequences of 40 years of Dutch quota management. *Marine Policy* 80, 81-87. https://doi.org/10.1016/j.marpol.2016.09.019.

Hong, S., Lee, J., Kang, D., Choi, H.W. and Ko, S.H. (2014). Quantities, composition, and sources of beach debris in Korea from the results of nationwide monitoring. *Marine Pollution Bulletin* 84(1-2), 27-34. https://doi.org/10.1016/j.marpolbul.2014.05.051.

Hoshino, E., Pascoe, S., Hutton, T., Kompas, T. and Yamazaki, S. (2017). Estimating maximum economic yield in multispecies fisheries: A review. *Reviews in Fish Biology and Fisheries* 28(2), 261–276. https://doi.org/10.1007/s11160-017-9508-8.

Høst, J. (2015). *Market-Based Fisheries Management: Private Fish and Captains of Finance*. MARE Publication Series: Springer. https://www.springer.com/gp/book/9783319164311.

Hughes, T.P., Kerry, J.T., Álvarez-Noriega, M., Álvarez-Romero, J.G., Anderson, K.D., Baird, A.H. *et al.* (2017). Global warming and recurrent mass bleaching of corals. *Nature* 543(7645), 373-377. https://doi.org/10.1038/nature21707.

Ianson, D. and Flostrand, L. (2010). *Ecosystem Status and Trends Report: Coastal Waters off the West Coast of Vancouver Island, British Columbia*. Canadian Science Advisory Secretariat. http://publications.gc.ca/collections/collection_2011/mpo-dfo/Fs70-5-2010-046.pdf.

International Union for Conservation of Nature (2017). *Red list of ecosystems*. https://www.iucn.org/theme/ecosystem-management/our-work/red-list-ecosystems.

Jayasiri, H.B., Purushothaman, C.S. and Vennila, A. (2013). Plastic litter accumulation on high-water strandline of urban beaches in Mumbai, India. *Environmental Monitoring and Assessment* 185(9), 7709-7719. https://doi.org/10.1007/s10661-013-3129-z.

Johnson, J.E. and Marshall, P.A. (2007). *Climate Change and The Great Barrier Reef: A Vulnerability Assessment*. Great Barrier Reef Marine Park Authority. http://elibrary.gbrmpa.gov.au/jspui/bitstream/11017/3033/1/Johnson_Marshall_2007_CC_and_GBR_Vulnerability_Assessment.pdf.

Kareiva, P. and Marvier, M. (2012). What is conservation science? *BioScience* 62(11), 962–969. https://doi.org/10.1525/bio.2012.62.11.5.

Keller, B.D., Gleason, D.F., McLeod, E., Woodley, C.M., Airamé, S., Causey, B.D. *et al.* (2009). Climate change, coral reef ecosystems, and management options for marine protected areas. *Environmental Management* 44(6), 1069-1088. https://doi.org/10.1007/s00267-009-9346-0.

Keohane, R.O. and Nye, J.S. (1977). *Power and Interdependence: World Politics in Transition*. Boston, MA: Little, Brown. http://www.worldcat.org/title/power-and-interdependence-world-politics-in-transition/oclc/2748258.

Keohane, R.O. and Victor, D.G. (2011). The regime complex for climate change. *Perspectives on Politics* 9(1), 7-23. https://doi.org/10.1017/S1537592710004068.

Kokorsch, M. (2017). The tides they are a changin': Resources, regulation, and resilience in an Icelandic coastal community. *Journal of Rural and Community Development* 12(2-3), 59-73. http://journals.brandonu.ca/jrcd/article/view/1400/320.

Koolman, J., Mose, B., Stanley, R.D., Trager, D., Heifetz, J., DiCosimo, J. *et al.* (2007). Developing an integrated commercial groundfish strategy for British Columbia: Insights gained about participatory management. In *Biology, assessment, and management of North Pacific rockfishes*. Heifetz, J., DiCosimo, J., Gharrett, A.J., Love, M.S., O'Connell, V.M. and Stanley, R.D. (eds.). Fairbanks: University of Alaska. 353-366. https://seagrant.uaf.edu/bookstore/ssl/checkout2.php?step=1&bypass=TRUE&id=11251.

Le Cornu, E., Doerr, A.N., Finkbeiner, E.M., Gourlie, D. and Crowder, L.B. (2017). Spatial management in small-scale fisheries: A potential approach for climate change adaptation in Pacific Islands. *Marine Policy* 88, 350-358. https://doi.org/10.1016/j.marpol.2017.09.030.

Legambiente ONLUS (2017). Stop plastic bags in the Mediterranean Area. Our Oceans, Our Future: Partnering for the Implementation of Sustainable Development Goal 14. New York: United Nations. 5-9 June. https://oceanconference.un.org/commitments/?id=15599.

Levin, L.A., Mengerink, K., Gjerde, K.M., Rowden, A.A., Van Dover, C.L., Clark, M.R. *et al.* (2016). Defining "serious harm" to the marine environment in the context of deep-seabed mining. *Marine Policy* 74, 245-259. https://doi.org/10.1016/j.marpol.2016.09.032.

Li, W.C., Tse, H.F. and Fok, L. (2016). Plastic waste in the marine environment: A review of sources, occurrence and effects. *Science of the Total Environment* 566, 333-349. https://doi.org/10.1016/j.scitotenv.2016.05.084.

Liu, O.R., Thomas, L.R., Clemence, M., Fujita, R., Kritzer, J.P., McDonald, G. *et al.* (2016). An evaluation of harvest control methods for fishery management. *Reviews in Fisheries Science & Aquaculture* 24(3), 244-263. https://doi.org/10.1080/23308249.2016.1161002.

Loot, G., Aldana, M. and Navarrete, S.A. (2005). Effects of human exclusion on parasitism in intertidal food webs of central Chile. *Conservation Biology* 19(1), 203-212. https://doi.org/10.1111/j.1523-1739.2005.00396.x.

Lucas, D.L., Kincl, L.D., Bovbjerg, V.E. and Lincoln, J.M. (2014). Application of a translational research model to assess the progress of occupational safety research in the international commercial fishing industry. *Safety Science* 64, 71-81. https://doi.org/10.1016/j.ssci.2013.11.023.

Mace, P.M., Sullivan, K.J. and Cryer, M. (2014). The evolution of New Zealand's fisheries science and management systems under ITQs. *ICES Journal of Marine Science* 71(2), 204-215. https://doi.org/10.1093/icesjms/fst159.

Madzena, A. and Lasiak, T. (1997). Spatial and temporal variations in beach litter on the Transkei coast of South Africa. *Marine Pollution Bulletin* 34(11), 900-907. https://doi.org/10.1016/S0025-326X(97)00052-0.

Marshall, P. and Schuttenberg, H. (2006). Adapting coral reef management in the face of climate change. In *Coral reefs and climate change: Science and management*. Phinney, J.T., Hoegh-Guldberg, O., Kleypas, J., Skirving, W. and Strong, A. (eds.). Washington D.C.: American Geophysical Union. chapter 12. 223-241. https://agupubs.onlinelibrary.wiley.com/doi/pdf/10.1029/61CE13

Martin, C.S., Tolley, M.J., Farmer, E., McOwen, C.J., Geffert, J.L., Scharlemann, J.P.W. *et al.* (2015). A global map to aid the identification and screening of critical habitat for marine industries. *Marine Policy* 53, 45-53. https://doi.org/10.1016/j.marpol.2014.11.007.

Matulich, S.C., Mittelhammer, R.C. and Reberte, C. (1996). Toward a more complete model of individual transferable fishing quotas: Implications of incorporating the processing sector. *Journal of Environmental Economics and Management* 31(1), 112-128. https://doi.org/10.1006/jeem.1996.0035.

McLeod, E., Salm, R., Green, A. and Almany, J. (2009). Designing marine protected area networks to address the impacts of climate change. *Frontiers in Ecology and the Environment* 7(7), 362-370. https://doi.org/10.1890/070211.

Meltzoff, S.K., Lichtensztajn, Y.G. and Stotz, W. (2002). Competing visions for marine tenure and co-management: genesis of a marine management area system in Chile. *Coastal Management* 30(1), 85-99. https://doi.org/10.1080/08920750252692634.

Merayo, E., Nielsen, R., Hoff, A. and Nielsen, M. (2018). Are individual transferable quotas an adequate solution to overfishing and overcapacity? Evidence from Danish fisheries. *Marine Policy* 87, 167-176. https://doi.org/10.1016/j.marpol.2017.08.032.

Morrison T. and Hughes T. (2016). *Policy Information Brief 1: Climate Change and the Great Barrier Reef.* Gold Coast: National Climate Change Adaptation Research Facility. https://www.nccarf.edu.au/synthesis/policy-information-brief-1-climate-change-and-great-barrier-reef.

Munari, C., Corbau, C., Simeoni, U. and Mistri, M. (2016). Marine litter on Mediterranean shores: Analysis of composition, spatial distribution and sources in north-western Adriatic beaches. *Waste Management* 49, 483-490. https://doi.org/10.1016/j.wasman.2015.12.010.

Nelms, S.E., Coombes, C., Foster, L.C., Galloway, T.S., Godley, B.J., Lindeque, P.K. *et al.* (2017). Marine anthropogenic litter on British beaches: A 10-year nationwide assessment using citizen science data. *Science of the Total Environment* 579, 1399-1409. https://doi.org/10.1016/j.scitotenv.2016.11.137.

Nomura, K.J., Kaplan, D.M., Beckensteiner, J. and Scheld, A.M. (2017). Comparative analysis of factors influencing spatial distributions of marine protected areas and territorial use rights for fisheries in Japan. *Marine Policy* 82, 59-67. https://doi.org/10.1016/j.marpol.2017.05.005.

Northwest Pacific Action Plan (2008). *NOWPAP Regional Action Plan on Marine Litter.* Northwest Pacific Action Plan https://www.cbd.int/doc/meetings/mar/mcbem-2014-03/other/mcbem-2014-03-130-en.pdf.

Obura, D. and Grimsditch, G. (2009). *Coral Reefs, Climate Change and Resilience: An Agenda for Action from the IUCN World Conservation Congress in Barcelona, Spain.* Gland: International Union for the Conservation of Nature. https://portals.iucn.org/library/sites/library/files/documents/2009-022.pdf.

Ocean Conservancy (2017). *Together for Our Ocean: International Coastal Cleanup 2017 Report.* https://oceanconservancy.org/wp-content/uploads/2017/06/International-Coastal-Cleanup_2017-Report.pdf.

Oceana (2016). *VME Protection: What is Needed? 40th Session of the General Fisheries Commission for the Mediterranean.* Madrid. https://eu.oceana.org/sites/default/files/factsheet_gfcm_vme_eng.pdf.

Organization for Economic Co-operation and Development (2017). *Marine Protected Areas: Economics, Management and Effective Policy Mixes.* Paris. http://www.oecd.org/env/marine-protected-areas-9789264276208-en.htm.

OSPAR Commission (2010). *Guideline for Monitoring Marine Litter on the Beaches in the OSPAR Maritime Area.* London: OSPAR Commission. https://www.ospar.org/ospar-data/10-02e_beachlitter%20guideline_english%20only.pdf.

Ostrom, E. (2002). Managing resources in the global commons. *Journal of Business Administration and Policy Analysis* 30-31, 401-413. https://ostromworkshop.indiana.edu/library/node/49853.

Oyanedel, R., Keim, A., Castilla, J.C. and Gelcich, S. (2018). Illegal fishing and territorial user rights in Chile. *Conservation Biology* 32(3). https://doi.org/10.1111/cobi.13048.

Oyanedel, R., Macy Humberstone, J., Shattenkirk, K., Rodriguez Van-Dyck, S., Joye Moyer, K., Poon, S. *et al.* (2017). A decision support tool for designing TURF-reserves. *Bulletin of Marine Science* 93(1), 155-172. https://doi.org/10.5343/bms.2015.1095.

Pasquini, G., Ronchi, F., Strafella, P., Scarcella, G. and Fortibuoni, T. (2016). Seabed litter composition, distribution and sources in the Northern and Central Adriatic Sea (Mediterranean). *Waste Management* 58, 41-51. https://doi.org/10.1016/j.wasman.2016.08.038.

Pfeiffer, L. and Gratz, T. (2016). The effect of rights-based fisheries management on risk taking and fishing safety. *Proceedings of the National Academy of Sciences* 113(10), 2615-2620. https://doi.org/10.1073/pnas.1509456113.

Pham, C.K., Diogo, H., Menezes, G., Porteiro, F., Braga-Henriques, A., Vandeperre, F. *et al.* (2014). Deep-water longline fishing has reduced impact on vulnerable marine ecosystems. *Scientific reports* 4(4837). https://doi.org/10.1038/srep04837.

Pinkerton, E. and Edwards, D.N. (2009). The elephant in the room: The hidden costs of leasing individual transferable fishing quotas. *Marine Policy* 33(4), 707-713. https://doi.org/10.1016/j.marpol.2009.02.004.

Pinkerton, E.W. (2013). Alternatives to ITQs in equity-efficiency-effectiveness trade-offs: How the lay-up system spread effort in the BC Halibut Fishery. *Marine Policy* 42, 5-13. https://doi.org/10.1016/j.marpol.2013.01.010.

Pinkerton, E.W. (2017). Hegemony and resistance: Disturbing patterns and hopeful signs in the impact of neoliberal policies on small-scale fisheries around the world. *Marine Policy* 80, 1-9. https://doi.org/10.1016/j.marpol.2016.11.012.

Pintassilgo, P. and Duarte, C.C. (2000). The new-member problem in the cooperative management of high seas fisheries. *Marine Resource Economics* 15(4), 361-378. https://doi.org/10.1086/mre.15.4.42629331.

Porcelli, A.M. (2017). Comparing bonding capital in New England groundfish and scallop fisheries: differing effects of privatization. *Marine Policy* 84, 244-250. https://doi.org/10.1016/j.marpol.2017.07.023

Rangel-Buitrago, N., Williams, A., Anfuso, G., Arias, M. and Gracia, A. (2017). Magnitudes, sources, and management of beach litter along the Atlantico department coastline, Caribbean coast of Colombia. *Ocean & Coastal Management* 138, 142-157. https://doi.org/10.1016/j.ocecoaman.2017.01.021.

Rech, S., Macaya-Caquilpán, V., Pantoja, J.F., Rivadeneira, M.M., Madariaga, D.J. and Thiel, M. (2014). Rivers as a source of marine litter–a study from the SE Pacific. *Marine Pollution Bulletin* 82(1-2), 66-75. https://doi.org/10.1016/j.marpolbul.2014.03.019.

Reyes, F.E. (1986). ¿ Que pasó con el loco? Crónica de un colapso anunciado. *Revista Chile Pesquero*, 36. http://www.cipmachile.com/web/200.75.6.169/RAD/1986/2_Reyes.pdf.

Rice, J.C. (2004). The British Columbia rockfish trawl fishery. In *Report and Documentation of the International Workshop on the Implementation of International Fisheries Instruments and Factors of Unsustainability and Overexploitation in Fisheries.* Swan, J. and Greboval, D. (eds.). Rome: Food and Agriculture Organization of the United Nations. 161-187. http://www.fao.org/tempref/docrep/fao/007/y5242e/y5242e01c.pdf

Richards, L.J. (1994). Trip limits, catch, and effort in the British Columbia rockfish trawl fishery. *North American Journal of Fisheries Management* 14(4), 742-750. https://doi.org/10.1577/1548-8675(1994)014<0742:TLCAEI>2.3.CO;2.

Rogers, A.D. and Gianni, M. (2010). *The Implementation of UNGA Resolutions 61/105 and 64/72 in the Management of Deep-Sea Fisheries on the High Seas: A Report Prepared for the Deep-Sea Conservation Coalition* London: International Programme on the State of the Ocean. http://www.savethehighseas.org/publicdocs/61105-Implemention-finalreport.pdf.

Rosas, J., Dresdner, J., Chávez, C. and Quiroga, M. (2014). Effect of social networks on the economic performance of TURFs: The case of the artisanal fishermen organizations in Southern Chile. *Ocean & Coastal Management* 88, 43-52. https://doi.org/10.1016/j.ocecoaman.2013.11.012.

Rosevelt, C., Los Huertos, M., Garza, C. and Nevins, H.M. (2013). Marine debris in central California: Quantifying type and abundance of beach litter in Monterey Bay, CA. *Marine Pollution Bulletin* 71(1-2), 299-306. https://doi.org/10.1016/j.marpolbul.2013.01.015.

Ruano-Chamorro, C., Subida, M.D. and Fernández, M. (2017). Fishers' perception: An alternative source of information to assess the data-poor benthic small-scale artisanal fisheries of central Chile. *Ocean & Coastal Management* 146, 67-76. https://doi.org/10.1016/j.ocecoaman.2017.06.007.

Sala, E., Lubchenco, J., Grorud-Colvert, K., Novelli, C., Roberts, C. and Sumaila, U.R. (2018). Assessing real progress towards effective ocean protection. *Marine Policy* 91(1), 11-13. https://doi.org/10.1016/j.marpol.2018.02.004.

Santis, O. and Chávez, C. (2015). Quota compliance in TURFs: An experimental analysis on complementarities of formal and informal enforcement with changes in abundance. *Ecological Economics* 120, 440-450. https://doi.org/10.1016/j.ecolecon.2015.11.017.

Santoro, C.M., Gayo, E.M., Carter, C., Standen, V., Castro, V., Valenzuela, D. *et al.* (2017). Loco or no loco? Holocene climatic fluctuations, human demography and community base management of coastal resources in northern Chile. *Frontiers in Earth Science* 5 (October), 474–416. https://doi.org/10.3389/feart.2017.00077.

Santos, I.R., Friedrich, A.C. and Barretto, F.P. (2005). Overseas garbage pollution on beaches of northeast Brazil. *Marine Pollution Bulletin* 50(7), 783-786. https://doi.org/10.1016/j.marpolbul.2005.04.044.

Saunders, F.P., Gallardo-Fernández, G.L., Van Tuyen, T., Raemaekers, S., Marciniak, B. and Plá, R.D. (2016). Transformation of small-scale fisheries—critical transdisciplinary challenges and possibilities. *Current Opinion in Environmental Sustainability* 20, 26-31. https://doi.org/10.1016/j.cosust.2016.04.005.

Selig, E.R., Kleisner, K.M., Ahoobim, O., Arocha, F., Cruz-Trinidad, A., Fujita, R. *et al.* (2016). A typology of fisheries management tools: Using experience to catalyse greater success. *Fish and Fisheries* 18(3), 543–570. https://doi.org/10.1111/faf.12192.

Simard, F., Laffoley, D. and Baxter, J.M. (eds.) (2016). *Marine Protected Areas and Climate Change: Adaptation and Mitigation Synergies, Opportunities and Challenges.* Gland: International Union for Conservation of Nature. https://portals.iucn.org/library/sites/library/files/documents/2016-067.pdf.

Stage, J., Christiernsson, A. and Söderholm, P. (2016). The economics of the Swedish individual transferable quota system: Experiences and policy implications. *Marine Policy* 66, 15-20. https://doi.org/10.1016/j.marpol.2016.01.001.

Stainsby, J. (1994). " It's the Smell of Money": Women shoreworkers of British Columbia. *BC Studies: The British Columbian Quarterly* 103, 59-81. https://doi.org/10.14288/bcs.v0i103.931.

Steer, M. and Besley, M. (2016). The licence amalgamation scheme: Taming South Australia's complex multi-species, multi-gear marine scalefish fishery. *Fisheries Research* 183, 625–633. https://doi.org/10.1016/j.fishres.2016.05.008.

Stefatos, A., Charalampakis, M., Papatheodorou, G. and Ferentinos, G. (1999). Marine debris on the seafloor of the Mediterranean Sea: Examples from two enclosed gulfs in Western Greece. *Marine Pollution Bulletin* 38(5), 389-393. https://doi.org/10.1016/S0025-326X(98)00141-6.

Stolton, S., Hockings, M., Dudley, N., MacKinnon, K., Whitten, T. and Leverington, F. (2007). *Management Effectiveness Tracking Tool: Reporting Progress at Protected Area Sites.* Gland: WWF International http://assets.panda.org/downloads/mett2_final_version_july_2007.pdf.

Sumaila, U.R. (2010). A cautionary note on individual transferable quotas. *Ecology and Society* 15(3). https://doi.org/10.5751/ES-03391-150336.

Syakti, A.D., Bouhroum, R., Hidayati, N.V., Koenawan, C.J., Boulkamh, A., Sulistyo, I. *et al.* (2017). Beach macro-litter monitoring and floating microplastic in a coastal area of Indonesia. *Marine Pollution Bulletin* 122(1-2), 217-225. https://doi.org/10.1016/j.marpolbul.2017.06.046.

Tam, J., Chan, K.M.A., Satterfield, T., Singh, G.G. and Gelcich, S. (2018). Gone fishing? Intergenerational cultural shifts can undermine common property co-managed fisheries. *Marine Policy* 90, 1-5. https://doi.org/10.1016/j.marpol.2018.01.025.

Thøgersen, T., Eigaard, O.R., Fitzpatrick, M., Mardle, S., Andersen, J.L. and Haraldsson, G. (2015). Economic gains from introducing international ITQs—The case of the Mackerel and Herring fisheries in the Northeast Atlantic. *Marine Policy* 59, 85-93. https://doi.org/10.1016/j.marpol.2015.05.002.

Tomás, J., Guitart, R., Mateo, R. and Raga, J.A. (2002). Marine debris ingestion in loggerhead sea turtles, Caretta caretta, from the Western Mediterranean. *Marine Pollution Bulletin* 44(3), 211-216. https://doi.org/10.1016/S0025-326X(01)00236-3.

Turris, B.R. (2000). A comparison of British Columbia's ITQ fisheries for groundfish trawl and sablefish: Similar results from programmes with differing objectives, designs and processes. In *Use of Property Rights in Fisheries Management.* Shotton, R. (ed.). Fremantle, Western Austraria: Food and Agriculture Organization of the United Nations. 254-261. http://www.fao.org/3/a-x7579e.pdf

Unger, A. and Harrison, N. (2016). Fisheries as a source of marine debris on beaches in the United Kingdom. *Marine Pollution Bulletin* 107(1), 52-58. https://doi.org/10.1016/j.marpolbul.2016.04.024.

United Nations, General Assembly (2004). *Oceans and the Law of the Sea: Report of the Secretary-General – Addendum.* A/59/62/Add.1. http://www.un.org/Docs/journal/asp/ws.asp?m=A/59/62/Add.1.

United Nations, General Assembly (2006). *61/105. Sustainable Fisheries, including through the 1995 Agreement for the Implementation of the Provisions of the United Nations Convention on the Law of the Sea of 10 December 1982 relating to the Conservation and Management of Straddling Fish Stocks and Highly Migratory Fish Stocks, and related instruments.* A/RES/61/105. http://undocs.org/A/RES/61/105

United Nations, General Assembly (2016). *Actions taken by States and Regional Fisheries Management* Organizations and Arrangements in response to Paragraphs 113, 117 and 119 to 124 of General Assembly Resolution 64/72 and paragraphs 121, 126, 129, 130 and 132 to 134 of General Assembly resolution 66/68 on Sustainable Fisheries, addressing the Impacts of Bottom Fishing on Vulnerable Marine Ecosystems and the Long-term Sustainability of Deep-sea Fish Stocks:

United Nations Environment Programme (2016). *Implementing the Marine Litter Regional Plan in the Mediterranean.* United Nations Environment Programme. http://wedocs.unep.org/bitstream/handle/20.500.11822/6072/16ig22_28_22_10_eng.pdf?sequence=1.

United Nations Environment Programme World Conservation Monitoring Centre and International Union for Conservation of Nature (2018). *Protected Planet: The World Database on Protected Areas (WDPA)* https://www.protectedplanet.net/.

United Nations Environment Programme/Mediterranean Action Plan (2013). *Regional Plan on Marine Litter in the Mediterranean.* Athens: United Nations Environment Programme/Mediterranean Action Plan. https://www.cbd.int/doc/meetings/mar/mcbem-2014-03/other/mcbem-2014-03-120-en.pdf.

United Nations Environment Programme/Mediterranean Action Plan (2014). Mid-term evaluation of SAP/NAP implementation. *Meeting of MED POL Focal Points on LBS NAP update*. Athens, Greece, 26-28 March 2014. https://wedocs.unep.org/rest/bitstreams/45678/retrieve

United Nations Environment Programme/Mediterranean Action Plan (2015a). 1st report of the informal online Working Group on marine litter. *Meeting of the Integrated Monitoring Correspondence Group*. Athens, Greece, 30 March - 1 April 2015. https://wedocs.unep.org/bitstream/handle/20.500.11822/17302/15wg411_inf10_eng.pdf?sequence=1&isAllowed=y

United Nations Environment Programme/Mediterranean Action Plan (2015b). Approaches to estimating the costs for the Regional Plans/ legally binding measures adopted by the Contracting Parties. *Regional Meeting on Applying Methodology for Programmes of Measures and Economic Analysis in the NAP Update*. Athens, Greece, 11-13 May 2015. http://web.unep.org/unepmap/726869

United Nations Environment Programme/Mediterranean Action Plan (2015c). *Marine Litter Assessment in the Mediterranean*. https://wedocs.unep.org/rest/bitstreams/9739/retrieve.

United Nations Environment Programme/Mediterranean Action Plan (2016). *Integrated Monitoring and Assessment Programme of the Mediterranean Sea and Coast and Related Assessment Criteria*. Athens: United Nations Environment Programme - Mediterranean Action Plan (UNEP/MAP). https://wedocs.unep.org/bitstream/handle/20.500.11822/10576/IMAP_Publication_2016.pdf?sequence=1&isAllowed=y.

University of British Columbia (2017). *Canada's pacific groundfish trawl fishery: Ecosystem conflicts*. The University of British Columbia http://cases.open.ubc.ca/canada-pacific-groundfish-trawl/ (Accessed: 20 February 2018).

van der Velde, T., Milton, D.A., Lawson, T.J., Wilcox, C., Lansdell, M., Davis, G. *et al.* (2017). Comparison of marine debris data collected by researchers and citizen scientists: Is citizen science data worth the effort? *Biological conservation* 208, 127-138. https://doi.org/10.1016/j.biocon.2016.05.025.

Van Holt, T. (2012). Landscape influences on fisher success: Adaptation strategies in closed and open access fisheries in southern Chile. *Ecology and Society* 17(1), 28. https://doi.org/10.5751/ES-04608-170128.

van Hoof, L. (2013). Design or pragmatic evolution: Applying ITQs in EU fisheries management. *ICES Journal of Marine Science* 70(2), 462-470. https://doi.org/10.1093/icesjms/fss189.

Van Oppen, M.J.H., Gates, R.D., Blackall, L.L., Cantin, N., Chakravarti, L.J., Chan, W.Y. *et al.* (2017). Shifting paradigms in restoration of the world's coral reefs. *Global Change Biology* 23(9), 3437-3448. https://doi.org/10.1111/gcb.13647.

van Oppen, M.J.H., Oliver, J.K., Putnam, H.M. and Gates, R.D. (2015). Building coral reef resilience through assisted evolution. *Proceedings of the National Academy of Sciences* 112(8), 2307-2313. https://doi.org/10.1073/pnas.1422301112.

Velander, K. and Mocogni, M. (1999). Beach litter sampling strategies: Is there a 'best' method? *Marine Pollution Bulletin* 38(12), 1134-1140. https://doi.org/10.1016/S0025-326X(99)00143-5.

Villanueva-Poot, R., Seijo, J.C., Headley, M., Arce, A.M., Sosa-Cordero, E. and Lluch-Cota, D.B. (2017). Distributional performance of a territorial use rights and co-managed small-scale fishery. *Fisheries Research* 194, 135-145. https://doi.org/10.1016/j.fishres.2017.06.005.

Vincent, A., Drag, N., Lyandres, O., Neville, S. and Hoellein, T. (2017). Citizen science datasets reveal drivers of spatial and temporal variation for anthropogenic litter on Great Lakes beaches. *Science of the Total Environment* 577, 105-112. https://doi.org/10.1016/j.scitotenv.2016.10.113.

Waldo, S., Berndt, K., Hammarlund, C., Lindegren, M., Nilsson, A. and Persson, A. (2013). Swedish coastal herring fisheries in the wake of an ITQ system. *Marine Policy* 38, 321-324. https://doi.org/10.1016/j.marpol.2012.06.008.

Waldo, S. and Paulrud, A. (2012). ITQs in Swedish demersal fisheries. *ICES Journal of Marine Science* 70(1), 68-77. https://doi.org/10.1093/icesjms/fss141.

Wallace, S., Turris, B., Driscoll, J., Bodtker, K., Mose, B. and Munro, G. (2015). Canada's Pacific groundfish trawl habitat agreement: A global first in an ecosystem approach to bottom trawl impacts. *Marine Policy* 60, 240-248. https://doi.org/10.1016/j.marpol.2015.06.028.

Wiff, R., Quiroz, J.C., Neira, S., Gacitúa, S. and Barrientos, M.A. (2016). Chilean fishing law, maximum sustainable yield, and the stock recruitment relationship. *Latin American Journal of Aquatic Research* 44(2), 380-391. https://doi.org10.3856/vol44-issue2-fulltext-19.

Wilen, J.E., Cancino, J. and Uchida, H. (2012). The economics of territorial use rights fisheries, or TURFs. *Review of Environmental Economics and Policy* 6(2), 237-257. https://doi.org/10.1093/reep/res012.

Williams, A.T., Randerson, P., Di Giacomo, C., Anfuso, G., Macias, A. and Perales, J.A. (2016). Distribution of beach litter along the coastline of Cádiz, Spain. *Marine Pollution Bulletin* 107(1), 77-87. https://doi.org/10.1016/j.marpolbul.2016.04.015.

Willoughby, N.G., Sangkoyo, H. and Lakaseru, B.O. (1997). Beach litter: An increasing and changing problem for Indonesia. *Marine Pollution Bulletin* 34(6), 469-478. https://doi.org/10.1016/S0025-326X(96)00141-5.

Windle, M.J.S., Neis, B., Bornstein, S., Binkley, M. and Navarro, P. (2008). Fishing occupational health and safety: A comparison of regulatory regimes and safety outcomes in six countries. *Marine Policy* 32(4), 701-710. https://doi.org/10.1016/j.marpol.2007.12.003.

Yoshino, K. (2017). TURFs in the post-quake recovery: Case studies in Sanriku fishing communities, Japan. *Marine Policy* 86, 47-55. https://doi.org/10.1016/j.marpol.2017.08.029.

Zúñiga, S., Ramírez, P. and Valdebenito, M. (2000). Situación socioeconómica de las áreas de manejo en la región de Coquimbo, Chile. *Latin American Journal of Aquatic Research* 36(1), 63-81. https://doi.org/10.3856/vol36-issue1-fulltext-5.

Chapter **15**

Land and Soil Policy

Coordinating Lead Authors: Klaus Jacob (Freie universität Berlin), Peter King (Institute for Global Environmental Strategies), Diana Mangalagiu (University of Oxford and Neoma Business School)

Lead Authors: Pandi Zdruli (Mediterranean Agronomic Institute of Bari [CIHEAM-Bari]), Katharina Helming (Leibniz Centre for Agricultural Landscape Research [ZALF], Andrew Onwuemele (Nigerian Institute of Social and Economic Research [NISER]), Leila Zamani (Department of Environment of the Islamic Republic of Iran)

GEO Fellows: Darshini Ravindranath (University College London), Hung Vo (Harvard University)

Executive summary

Land protection policies differ between regions and countries from barely existent to well defined (*established but incomplete*). However, their implementation has many shortcomings. Often, national policies addressing socioeconomic development (e.g. economic incentives for agricultural, bioenergy and urban development) have overlooked land degradation side effects. As long as economic growth is not decoupled from environmental degradation, sustainable use and management of land requires policy frameworks that better integrate land management governance across sectors, especially in developing countries. This chapter analyses the effectiveness of policies and policy approaches addressing land quality dynamics in five case studies having different socioeconomic and physical contexts and management approaches. The intention is to draw messages for policy and decision makers when dealing with complex land issues in a context of competing interests and scarcity of resources. {15.4}

Land degradation is likely to be aggravated as long as effective land and soil management policy frameworks are not established at national and international levels (*established but incomplete*)**.** Global trade and land acquisitions, including land grabbing, have direct consequences on the livelihoods of local people and international food trade markets. {15.1, 15.2}

Land degradation and lack of policy action may accelerate migration in some regions (*well established*)**.** The Intergovernmental Science-Policy Platform on Biodiversity and Ecosystem Services notes "by 2050, an estimated 4 billion people will live in the drylands and until then it is likely that land degradation and climate change will have forced 50-700 million people to migrate". This will result in increased hardship for most areas in Africa, South Asia, the Middle East, and North Africa that will be impacted by rapid population growth, low per capita gross domestic product (GDP), limited options for agricultural expansion, increased water stress and high biodiversity losses. {15.1}

Land degradation and desertification could be prevented within the context of local social, economic and political conditions (*well established*). Sustainable land management practices can reverse even severe desertification processes. But the implementation of such practices necessitates policy frameworks that support the involvement and compensation of local people with public money or through public-private partnerships, including direct financing from the private sector alone. Such incentives, however, are country specific and depend on financial resources available. {15.2.1}

Land is a key source of ecosystem functions and services. Consequently, land-use change is the major direct driver of the loss of ecosystems services and biodiversity (*well established*). In 2017, the estimated global ecosystem services losses due to land degradation were between US$6.3 trillion and US$10.6 trillion per year. This estimated loss is equal to 10-17 per cent of global GDP (US$63 trillion in 2010), while halting and reversing current trends of land degradation could generate up to US$1.4 trillion per year in economic benefits if sustainable land management policies would be implemented. {15.2; 15.2.2}

Land policy frameworks to tackle risk to human health from soil contamination are scattered and incomplete (*established but incomplete*)**.** The 'polluter pays' principle is not widely applied and the cost of remediation is very high, preventing its implementation even in developed countries. Lack of knowledge and data gaps further hinder its implementation. Hence a reconsideration of that principle, or otherwise a strong commitment from the government (local, regional, national) to act, is needed to safeguard public health. The contamination of land with heavy metals, pesticides, organic pollutants and other toxic substances severely threatens humans as they are taken up by plants used for food. These effects are even more severe in developing countries that are faced with lack of financial resources and expertise to tackle soil pollution. {15.2.3, 15.2.4}

Sustainable land management is a major instrument for climate change mitigation because it improves carbon sequestration in the soil (*well established*). This is why land and soil policy gained increasing international recognition with respect to climate change negotiations of the 23rd Conference of the Parties to the United Nations Framework Convention on Climate Change (UNFCCC COP 23) held in Paris in 2015 when the '4 per 1000 Initiative' was launched by the French Government. The initiative promotes enhancement of soil quality, carbon sequestration and soil conservation through improved agricultural practices that mitigate climate change. {15.3.1, 15.4}

High-consumption lifestyles, especially in developed economies, aggravated by food waste, and rapid population growth rates have negative consequences for land and its resources (*established but incomplete*). The increased demand for food and biofuels will trigger agricultural intensification such that biomass production is expected to double by 2050. Policies need to steer sustainable intensification through conservation agriculture practices to mitigate negative effects on soil health and on the environment. {15.2.4}

Land-use planning, sustainable use of land resources and sustainable land management are the answers to balance production with environmental protection (*well established*). Sustainable intensification practices attempt to integrate increasing crop yields with maintaining soil fertility and improving water-use efficiency. Annually, US$75.6 trillion can be gained from implementing global policies that enable sustainable land management. Among many other sustainable land management practices, conservation agriculture that includes also zero tillage provides a good example of technologies that maintain land quality, enhance soil carbon sequestration, mitigate climate change, protect biodiversity and sustain productivity. Policies, economic analysis, science and farming incentives, however, are all necessary to support implementation of these technologies, especially for small landowners, in particular those in the developing countries. {15.2.4}

Land management, restoration and policies need to be tailored according to local conditions (*well established*). Experience has shown that 'one-size-fits-all' is not an option to promote sustainable land management worldwide. Success of policy implementation depends on a number of factors that consider an integrated landscape approach well-matched to socioeconomic and natural characteristics supported by good levels of governance and stakeholder engagement. {15.2.4, 15.2.5}

Implementing the right actions to combat land degradation and support sustainable land management policies has a direct effect on the livelihoods of millions of people across the planet (*well established*)**.** This imperative will become more difficult and costlier if no action is taken urgently. Unfortunately, there is still a disconnect between consumers and the ecosystems that provide the food and other commodities they depend upon. {15.5}

Land is a finite resource that is under human pressure, especially from urban sprawl (*established but incomplete*). Chaotic urban expansion has been observed worldwide mostly on fertile and productive lands and, by 2050, about 80 per cent of the productive soils are at risk of being lost as each year about 20 million ha of agricultural land is converted into urban and industrial developments. The situation along the coastal areas is worst. It is therefore imperative that land-use policies should define a proper allocation of land resources between competing interests. Cities play a major role in land-use changes, so municipal and city planners need to coordinate their actions with a large number of stakeholders, including civil society and establishing public-private partnerships, to ensure sustainable spatial planning, policy coherence, implementation and conflict resolution for both urban settlements and responsible food systems. {4.2.5, 15.3.3}

15.1 Introduction

As noted in chapter 8, land plays a crucial role within the 'Healthy Planet, Healthy People' theme and underpins global efforts towards sustainable development. Consequently, sustainable land management (SLM) is not only essential to promoting and maintaining the great diversity of nature's contribution to people but also in tackling poverty and hunger.

At the international level, the Sustainable Development Goals (SDGs) emphasize the need for SLM among stakeholders for the protection of natural ecosystems that are on the verge of collapse, including increased climate-induced natural disasters. SDG 15 is directly related to the analysis in Chapter 8. Furthermore, SDG target 15.3 focuses on land by demanding action against land degradation and efforts to achieve a land degradation-neutral world. Although land management is explicitly targeted in SDG 15.3, it is paramount for food security (SDG 2), climate action (SDG 13) and also has many interconnections with SDGs 1, 3, 6, 7, 11 and 12 **(Figure 15.1)**.

The driving forces and pressures (see Chapter 2) on land and its resources emanating from population growth, urbanization, economic development, technology and innovation, and climate change have elicited responses at global, regional and

Box 15.1: The Concepts of Land and Soil

The concepts land and soil are not synonyms. Land represents the terrestrial solid part of the earth that is not permanently under water and offers an endless set of services and functions from biomass production to urban living habitats. It comprises soil, vegetation, other biota, and the ecological and hydrological processes operating on it. Soil is the unconsolidated material on the land surface that has been formed by mineral particles, organic matter, water, air and living organisms simultaneously interacting over time. Ecological processes in the soil ensure biomass production, nutrient cycling, carbon sequestration, water filtration and buffering, cooling and hosting biodiversity.

Figure 15.1: Linkage between the land-related SDG target 15.3 and other SDGs

1.1 Eradicate extreme poverty
1.2 Halve % people in poverty
1.4 Ensure equal rights to resources, ownership over land
1.5 Build resilience, reduce vulnerability

2.1 End hunger, ensure access to food
2.2 End all forms of malnutrition
2.3 Double agricultural productivity and incomes
2.4 Ensure sustainable food production systems

15.1 Ensure conservation of ecosystems and their services
15.2 Promote sustainable management of forests
15.4 Ensure conservation of mountain ecosystems
15.5 Reduce degradation of natural habitats
15.8 Reduce impact of invasive alien species
15.9 Integrate ecosystem and biodiversity values in policy

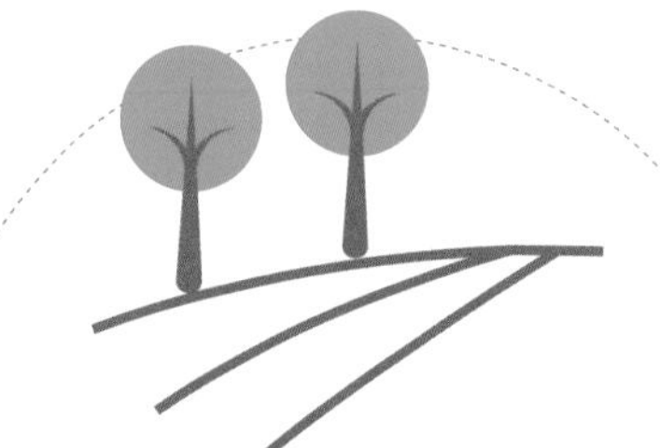

15.3 Achieve land degradation neutrality

6.1 Achieve access to safe drinking water for all
6.4 Increase water-use efficiency
6.5 Implement integrated water resources management
6.6 Protect and restore water-related ecosystems

13.1 Strengthen resilience to climate-related hazards
13.2 Integrate climate change measures in policy

12.3 Halve per capita global food waste

7.2 Increase share of renewable energy

Source: Akhtar-Schuster *et al.* (2017).

national levels. At the global level, several responses directly or indirectly related to sustainable land and soil management have been initiated as shown in **Table 15.1**.

This chapter provides a comprehensive overview of the current Land and Soil policy framework and the related shortcomings, especially in terms of policy cohesion and implementation, as well as overall effectiveness. The subsequent sections present case studies of different sets of legal and policy instruments, economic tools and incentives, as well as policies and programmes implemented across different countries. The case studies were selected based on the criteria of regional balance, different spatial scales, type of policies and/or governance arrangements, plus their relevance to state and trends of land resources as detailed in Chapter 8 of this report.

15.2 Key policies and governance approaches

An overview of key policies and respective case studies is provided in **Table 15.2**. The cases reflect the variety of driving forces, economic sectors and processes affecting land degradation, and are used as illustrative examples of policy instruments covering a diverse range of spatial scales and time frames of implementation. Despite addressing important drivers and respective policy approaches, the cases here do not

Table 15.1: Recent milestones in land governance and sustainable development

Year	Milestone
1981	2015 FAO World Soil Charter
1988	Intergovernmental Panel on Climate Change (IPCC)
1992	United Nations Conference on Environment and Development
	Rio Declaration
	Agenda 21
	Global Environment Facility
	United Nations Convention to Combat Desertification (UNCCD)
	United Nations Framework Convention on Climate Change (UNFCCC)
	Convention on Biological Diversity (CBD)
1997	Kyoto Protocol
2000	Millennium Development Goals (MDGs)
2005	Millennium Ecosystem Assessment
2008	UNCCD's Zero Net Land Degradation and Land Degradation Neutrality Initiative
2011	Global Soil Partnership initiated (FAO/European Union)
2012	Rio+20
2015	Sustainable Development Goals (SDGs) and Post-2015 Development Agenda
	Intergovernmental Technical Panel on Soils (ITPS) of the Global Soil Partnership (GSP)
	Land and Soils integrated in the Open Working Group of the Sustainable Development Goals
	Regional Soil Partnerships of the GSP
	International Year of Soils declared by the United Nations General Assembly
	The Economics of Land Degradation
	UNFCCC Paris Agreement
2017	(United Nations Economic and Social Council) United Nations Strategic Plan for Forests 2017-2030
	FAO Voluntary Guidelines for Sustainable Soil Management
2018	UNCCD's Land Degradation Neutrality Fund a public-private partnership for blended finance

Table 15.2: Typology of policy and governance approaches described in this chapter

Governance approach	Policy instrument(s)	Case studies
Policy mix: command and control, and economic incentives.	Funding programmes and standard setting for best management practices.	Strengthening foreign direct investment management and social and environmental safeguards in Lao People's Democratic Republic
Policy mix: Command and control, plus economic incentives.	Planning and compensation for halting desertification.	The Great Green Wall Project in China.
Command and control.	Setting threshold values for policy on remediation of contaminated sites.	Remedial treatment of Agent Orange-contaminated land in Viet Nam.
Promotion of innovation.	Provision of consultancy and networking for agricultural innovation.	Conservation agriculture and no-tillage cultivation in Australia.
Enabling actors.	Stakeholder network creation for responsible food systems and minimizing food waste.	Milan Urban Food Policy Pact.

address two further key aspects of land degradation that were pointed out in Chapter 8, namely insecure land tenure systems and land grabbing issues on the one side (Section 8.5.3), and teleconnections and spillover effects of consumption of land-based products (food, bioenergy) in one country on land resources depletion in other countries (Section 8.4.1). Both these issues severely affect the social dimension of land degradation impacts (IPBES 2018a).

15.2.1 Funding programmes and standard setting for best management practices

Sustainable land management is strongly influenced by policy frameworks that differ between countries and regions. A main driver is land-use change that has distinct economic and environmental consequences. Economic gains often are responsible for environmental degradation and, in many cases, not all stakeholders benefit from such gains (Castella *et al.* 2013). The case study below looks at these dynamics.

Case study: Strengthening foreign direct investment management and social and environmental safeguards in Lao People's Democratic Republic (Lao PDR).

Foreign investments in Lao People's Democratic Republic have a direct impact on the economic growth of the country and account for more than 50 per cent of gross domestic product (GDP), but they also raise serious environmental issues as the country has been experiencing significant forest depletion since the 1980s. Forests covered nearly 50 per cent of the country in 1982 (Phompila *et al.* 2017), but had declined to 41.5 per cent by 2013.

There are a number of factors that have influenced forest decline (Food and Agriculture Organization of the United Nations [FAO] 2010). They derive mostly from economic activities, such as forest land conversion for agriculture, mostly cash crops, and urban sprawl associated with infrastructure expansion and hydropower production. Other driving forces accelerating deforestation include uncontrolled hunting and logging as well as cleared forest areas converted to livestock grazing (United Nations Development Programme [UNDP] 2014). Economic activities emanating from forest land conversion for agriculture have triggered an increase in the number of land deals in Lao PDR 50-fold from 2000 to 2009 (Schonweger *et al.* 2012). The value of approved foreign and domestic investment projects exceeded US$29 billion by March 2018. All except for US$3.9 billion of this total was generated through foreign investment (US$8.5 billion) or joint ventures (US$16.6 billion) (UNDP-United Nations Environment Programme (UNEP) Poverty-Environment Initiative, UNDP and UNEP 2018).

The second largest type of land concession is related to agricultural investments. Between 1990 and 2007, the area of plantations, especially rubber plantations, increased dramatically from 1,000 ha to over 200,000 ha (Phimmavong *et al.* 2009). As of 2012, these covered more than 330,000 ha. An early study found that 85 per cent of all investment in agricultural concessions came from foreign investors, the five most important being from China (about 50 per cent), Thailand, Viet Nam, Republic of Korea and India (Wellmann 2012). Estimates of the area given to land concessions alone vary between 330,000 ha and 3.5 million ha, but the government has reported that 1.1 million ha was a conservative estimate. This is equivalent to 5 per cent of national territory or 18 per cent more than the total arable land in Lao People's Democratic Republic (Global Witness 2013). Thirteen per cent of all villages in Lao PDR have at least one concession within their village boundaries (Wellmann 2012).

Vast parts of communal lands that lack tenure titles are the target of big foreign companies, which have expanded their land acquisitions in Lao People's Democratic Republic. Unfortunately, this process is accompanied with a new poverty type that affects local people who become dependent on new investors for all of their basic livelihood needs (Messerli *et al.* 2015). In one case, 25 villages were displaced due to a land concession to a Vietnamese rubber company and local communities were prevented from accessing the natural resources upon which they based their livelihood (UNDP and UNEP 2013).

In response to the challenges presented by sustainable development of the environment and natural resources, and at the invitation of the Government of Lao People's Democratic Republic, the joint United Nations Development Programme-United Nations Environment Programme (UNDP-UNEP) Poverty-Environment Initiative supported the government from 2009 to 2015 in developing tools to guide promotion, screening, approval, monitoring and compliance enforcement of investments, and helped build the capacity of institutions to engage with impacted communities and respond to unintended social and environmental impacts of investments in key natural resource sectors. **Table 15.3** gives a summary of the assessment criteria.

Table 15.3: Summary of the assessment criteria for foreign investments

Criterion	Description	References
Success or failure	Success criteria include: benefits derived from the Poverty-Environment Initiative: a first green and pro-poor, quality investment management system; the assessment of development options; greater understanding of positive and negative investment implications; awareness of investors' degree of compliance; improved accountability; and the introduction of the 'green-growth' concept. Although training events and capacity-building took place, several relevant actors, at national and local levels, are not empowered to understand and enforce an equitable investment management system. Notwithstanding capacity-building events, uneven governmental capacity remains a challenge, but officials are showing commitment and appreciation. The lack of clear assignment of responsibilities in investment management remains unaddressed.	Tavera (2015)
Independence of evaluation	The assessment of this Lao PDR experience is part of the Initiative's independent midterm evaluation. Phase I (2009-2012) and Phase II (2012-2015) have been evaluated by an independent consultant at the request of the United Nations Development Programme Country Office.	
Key actors	The Poverty-Environment Initiative country team worked closely with the Lao PDR National Assembly, the Ministry of Planning and Investment, the Department of Environment and Social Impact Assessment, the Ministry of Natural Resources and Environment, the Investment Promotion Department and the National Economic Research Institute.	
Baseline	Before the commencement of the programme in 2011, annual average GDP growth was 7.9 per cent in the preceding decade (Lao People's Democratic Republic, Ministry of Planning and Investment 2011), while the poverty level was 27 per cent in 2007 (World Bank 2010). In 2010, Lao PDR had a GDP per capita of US$1,101.	
Time frame	Phase I took place from 2009 to 2012, Phase II from 2012 to 2015. Phase III (2016-2018) is not included in this case study.	
Constraining factors	There is an urgent need to build poverty-environment awareness and capacity within the National Assembly to make its normative work effective. Although technical staff of the National Assembly received training on compliance, the capacity-building efforts of the Initiative are limited and need to be sustained and expanded. Capacity needs to be strengthened also at the ministerial and, especially, local levels – fundamental for implementation and often hard to reach. Inter-ministerial and vertical coordination (especially with provinces) should be improved to achieve integrated development and equitable investment management.	
Enabling factors	Governmental commitment and involvement are major enabling factors, in particular with respect to the National Assembly, but also with the Investment Promotion Department and development-related ministries (the participation of the Ministry of Planning and Investment was fundamental). Shared and increased poverty-environment awareness by reference managers and staff resulted from the Initiative's efforts and allowed for the prioritization of equitable investment management. Improved assessments and investment data management had started to inform decision-making and create understanding of whether investments were economically beneficial for communities.	
Cost-effectiveness	Foreign direct investments are forecast to lead the country's development and comprise a significant share of its GDP. Their proper management is an important contribution to sustainable and equitable economic growth.	
Equity	Management of foreign direct investments directly addresses equity with regard to impacted communities. Unregulated investing led to cases of village displacement, land grabbing or segregation of resources vital to the livelihood of inhabitants, without necessarily contributing to the country's development (creating local jobs or providing significant revenues to the national government). With the support of the Poverty-Environment Initiative, Lao PDR was able to provide a legal framework and safety net to bind investments to more equitable conditions. This work also contributed to creating awareness among decision makers about equitable sustainable development and to orient the National Assembly and future national development strategies towards this aim. This was already clear during Phase I, which saw the 7th Five-Year National Socio-Economic Development Plan (2011-2015) integrate poverty-environment issues and highlight the objectives of quality growth and equity.	Lao People's Democratic Republic, Ministry of Planning and Investment (2011)
Co-benefits	Tackling the equity dimension in investment management also has external positive implications on the environmental side. It promotes more inclusive and sustainable land management and prevents the depletion of natural resources and biodiversity loss. These practices lead to economic benefits being more fairly distributed among local communities. These points highlight how all three dimensions of sustainable development (economic, social and environmental) are addressed. The Lao PDR case also provides a relevant example of foreign direct investment management for sustainable development that can be shared through South-South learning. The Lao PDR experience has further informed the work of the Poverty-Environment Initiative country projects in Myanmar, Mongolia and Philippines, in particular on investments in extractive industries.	Choi and Gankhuyag (2016)

Criterion	Description	References
Transboundary issues	A more binding investment regulation system might lead investors to flee to other countries with a laxer framework and where standards for compliance are lower. To avoid this, investment management performance should be enhanced, and working with the whole region could harmonize standards. The Lao PDR experience could serve countries beyond Asia and the Pacific, as it is relevant to global efforts to promote inclusive and greener economic growth.	
Possible improvements	❖ Realize a financial sustainability strategy for investment management, owned by the Investment Promotion Department and the Ministry of Finance, to make investment management sustainable. ❖ Enhance data sharing among relevant ministries to promote coordination for management of investments. ❖ Strengthen enforcement for compliance. ❖ Intensify training and ownership of tools for all actors involved, especially at the local level. ❖ Improve management of foreign direct investments in the whole region to avoid investors fleeing to contexts with lower standards. ❖ Strengthen communities' participation.	

An independent evaluation of the joint UNDP-UNEP Poverty-Environment Initiative programme conducted in 2016 rated its performance as "highly satisfactory". The effectiveness of this equitable and comprehensive legal framework is, however, subject to implementation and enforcement that go beyond the programme. The main obstacles to enforcement are lack of institutional capacity, tools and funds for investment monitoring. There are efforts to fill the information gap including improving database management for monitoring compliance. Yet, poor institutional coordination still prevents this data from generating consistent action on compliance and enforcement. There is also limited opportunity for community inputs into the national decision-making process (Tavera 2015; Tavera, Alderman and Nordin 2016). On the other hand, the conditions for equitable and sustainable growth have been laid. The policy effort was successful in providing comprehensive and fair tools and processes to enable quality investments and safeguard communities. There is now a legal framework to empower these actors to take part in development processes, and the country is moving one step closer to ensuring that investment is judged by the quality of its social and environmental benefits – and not just its benefits in economic terms. Community engagement was also enabled (UNDP 2016). These efforts in the investment sector also contributed to increasing decision makers' awareness and political prioritization of sustainable development.

15.2.2 Planning and compensation for halting desertification

The success of any strategy for combating desertification depends on the implementation of sustainable land and water management practices adapted to the specific local geo-biophysical and socioeconomic situation. Well-managed soils slow down the process of land degradation, regulate the water cycle, safeguard biodiversity, conserve landscape multifunctionality and improve the provision of ecosystem services (United Nations Convention to Combat Desertification [UNCCD] 2017a; Zdruli and Zucca 2018).

The general policy approach for combating desertification in terms of the DPSIR framework (section 1.6) needs to tackle the pressure derived from land cover losses, which in many cases are driven by economic incentives to increase agricultural production. Effective policy approaches generally combine command-and-control policies (in extreme cases, forcing farmers to stop farming) and offering incentives, such as subsidized tree planting. The Land Degradation Neutrality approach, included in SDG 15.3 and endorsed by the UNCCD, and the Economics of Land Degradation Initiative have become the mainstream strategic instruments to reduce net losses of land resources and ensure their sustainable management (Akhtar-Schuster *et al.* 2017; UNCCD 2017a); they also address climate change (Sanz *et al.* 2017). A set of management practices – including sustainable soil/land and water management, afforestation and reforestation, agroecology, pasture improvement and controlled grazing, watershed management, water harvesting and sustainable agricultural practices – have been implemented in support of this goal (Rojo *et al.* 2012; Schwilch, Liniger and Hurni 2014; Teshome *et al.* 2015; Marques *et al.* 2016).

Land restoration projects are usually funded with public money and follow top-down approaches (Marques *et al.* 2016). The top-down approach is traditionally applied by national governments, and as such these instruments are often called 'command-and-control' or regulatory instruments (King and Mori 2007; Weber, Driessen and Runhaar 2014). The government defines the rules and norms, and has the right to apply sanctions in those cases where rules are not implemented. Examples of regulatory instruments include standards, bans, permits, zoning and use restrictions (Lambin *et al.* 2014; Weber Driessen and Runhaar 2014). The 'command-and-control' approach is often implemented especially in the developing countries, as can be seen in the following case study.

Case study: The Great Green Wall to effectively decrease dust storm intensity in China

The Chinese Great Green Wall (GGW) is one of the most ambitious projects to combat desertification and control dust storms, similar to the Sahara Great Green Wall stretching from Dakar to Djibouti. The project was originally named the Three North Shelterbelt Forest programme, as launched in 1978; it still retains that same name but is also called the GGW. It is expected to continue until 2050. The programme name has become a common term in China, as the largest afforestation project in the country (Huang *et al.* 2016). It was designed to cover a total area of 4.1 million km^2 (or 42.7 per cent of the total land area of the country **Figure 15.2** (Wang *et al.* 2010).

Figure 15.2: The extent of the Great Green Wall in northern China

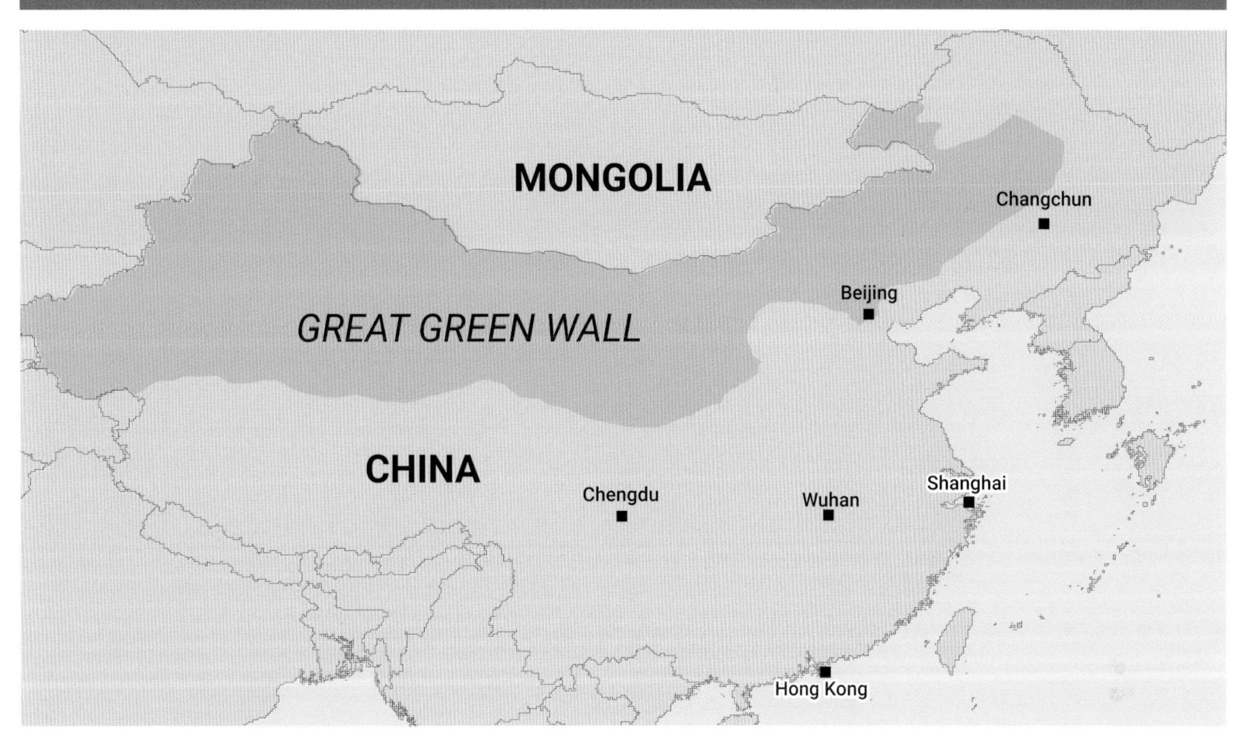

Source: Source: O'Callaghan (2014).

Table 15.4: Summary of the assessment criteria for desertification and dust control in China

Criterion	Description	References
Success or failure	Chinese researchers and government officials have reported that afforestation has successfully combated desertification, accelerated soil carbon sequestration and decreased soil erosion. By 2012, the programme had reportedly increased the tree cover from 5 per cent to 12.4 per cent in the programme area, with the cumulative tree planting area reaching 26.47 million ha. The policy showed success in that it reversed the trend of land degradation such that in many places soil carbon sinks are starting to increase. Vegetation cover increased in the GGW region compared to non-GGW regions, which led to a reduction of the dust storm intensity in northern China. During the project, desert expansion has been reduced to about 10 km² per year. In addition, about 1,060 km² of desert per year is transformed in a good condition. Critics of the programme, however, claim that it may have failed to meet its goals to date. One reason is that the observed decrease of dust storm intensity may simply be a consequence of climatic variability. Second, the programme was implemented only on a small portion of the affected areas and even not those that are known as core areas for dust storm sources. Third, not all of the planted trees and shrubs survived beyond the lifetime of the programme because of mismanagement and/or lack of water. A further point of critique was related to the overuse of groundwater by planting varieties that were not well suited to the arid areas.	Piao *et al.* (2007); Wang *et al.* (2010); State Forestry Administration (2011); Deng, Liu and Shangguan (2014); Sternberg, Rueff and Middleton (2015); Tan and Li (2015); Feng *et al.* (2016); Jiang (2016); Ahrends *et al.* (2017)
Independence of evaluation	There has been no independent evaluation, other than those conducted by Chinese researchers reported above.	
Key Actors	The key actor and investor is the Government of China, with 18 ministries and agencies of the central government and local authorities involved in various aspects of policy formulation concerning desertification control. The State Forestry Administration (SFA) is responsible for coordinating activities among these ministries. The local residents and communities have been involved in afforestation, grassland establishment and related activities. The programme additionally looks for the dynamic cooperation of non-governmental domestic and international entities, including the broad involvement of the private sector.	Lu and Wang (2003); State Forestry Administration (2011); Yin and Yin (2010)
Baseline	Since 1980, dust storm events were classified into ten grades (0-9) according to a visibility range that was being monitored at meteorological stations. The so-called Dust Storm Index (DSI) is equal to the mid value of the visibility for each grade. Despite considerable annual variation, a strong decrease of DSI has been observed since 1985. Also, the number of sandstorm days decreased and reached a low level in 1996. However, because of the complex interactions with vegetation, it was not possible to draw a clear, one-to-one cause-effect relationship between the implementation of the policy and the decrease of dust storm events.	Tan and Li (2015); Xiaoming *et al.* (2016)
Time frame	1978-2050	

15

Criterion	Description	References
Constraining factors	The excessive population growth of human and livestock is a challenge to the limited ecological carrying capacity in desertification-prone areas around the world (Pan *et al.* 2013) including in China. In the past decade alone, China invested more than US$100 billion into six key forestry programmes. However, returns for the large-scale tree planting investment in marginal areas may be low or take a long time to materialize. Another shortcoming is related to lack of interest from farmers after the trees have been planted, and lack of knowledge in forest management. Less supervision from local governmental offices and the size of subsidy levels were issues that limited the policy impact effect. Overall, the lower subsidies did not offer a strong motivation for farmers to participate in the programme despite the large amounts of money invested in ecosystem payment schemes to meet Chinese Government ambitions.	Ahrends *et al.* (2017) ; Xu, Song and Song (2017)
Enabling factors	A large number of institutional and administrative capacities have underpinned success to date. Since 1997, the SFA has established several institutions dealing with desertification, including the National Bureau to Combat Desertification, the National Desertification Monitoring Centre, a National Training Centre for Combating Desertification, and a National Research and Development Centre for Combating Desertification – all to conduct research and implement policy programmes on desertification issues. Moreover, in June 2009, the Institute of Desertification was established by the Chinese Forest Academy, which is also under the SFA. The government strongly supports desertification mitigation programmes by allocating significant funding (US$4 billion total during the initial 28 years). Several compensatory measures to increase vegetation have been implemented, including cash incentives to farmers willing to plant trees and shrubs.	Jiang (2016)
Cost effectiveness	"For the period from 2002 to 2006 the Three North Shelterbelt Forest Project has used 4,147 million yuan (US$545.6 million) of investment, created 2,840 million yuan (US$373.7 million) of ecological benefits, and 8,060 million yuan (US$1,060.5 million) of economic benefits." Direct costs of desertification are estimated at 64.2 billion yuan (Chinese Yuan Renminbi - CNY) annually (US$7.7 billion), while indirect costs of desertification are estimated at 288.9 billion yuan annually. Finally, the analysis shows that the costs of the rehabilitation of the lands degraded due to land use cover change are significantly lower than the costs of inaction, with returns of up to 4.7 times for every yuan invested over a 30-year period.	Deng and Li (2016); Jiang (2016)
Equity	Desertification control has mostly depended on the administration, specialists and other social elites for decision-making, while the local people are often inactively participating in the decision-making process. For instance, the local communities did not have the right to decide on control measures of the 'Sand Control Law'. In order to enhance land restoration of degraded areas, the government allocated land-use rights to local people for up to 70 years. This type of policy improved the land tenure issues and increased interest from local people. Government resettled farmers and herders on degraded lands and provided subsidies and compensations for those who participated in restoration activities. However, there is no systematic compensation method or proper regulations in land desertification control to support local people.	(Jiang 2016)
Co-benefits	Polycyclic aromatic hydrocarbons, also known as polyaromatic hydrocarbons, are organic compounds containing only carbon and hydrogen and can be dangerous to human health with cancer being the primary health risk from exposure to them. The implementation of the Three North Shelterbelt Forest Project resulted in atmospheric removal and long-term reduction trends of two polycyclic aromatic hydrocarbon species, phenanthrene and benzo[a]pyrene. A series of studies on the health effects of dust storms in north-western China show that dust events were significantly associated with respiratory and cardiovascular hospitalization (after adjusted the effect of SO_2 and/or NO_2). Events like this are also recorded in India, where in May 2018 it turned to be deadly. Based on published research (e.g. Tan and Li 2015; Wang *et al.* 2012), the frequency and severity of dust storm events have diminished over time thanks to the interventions of the GGW project that, according to the statistical data from National Meteorological Information Center, China (e.g. Tan and Li 2015), has positively affected the health status of the people living in the region and beyond. The other social-economic benefit of the project has been the development of tourism and increasing employment opportunities for local people.	Aunan and Pan (2004); Li and Huntsinger (2011); Huang *et al.* (2016)
Transboundary issues	From the global assessment report on sand and dust storms, it is clear that sand and dust storms from the desert areas of China and Mongolia affect the air and ocean quality as far as Korea, Japan, Pacific Islands and North America (e.g. https://youtu.be/jGPuCeEILeM). Furthermore, there is a Regional Master Plan for the Prevention and Control of Dust and Sandstorms in Northeast Asia, a project involving the governments of China, Japan, Mongolia and Republic of Korea. The goal is to mitigate health effects deriving from dust storms from this region outside north-west China (e.g. in Japan, Korea), emphasizing the long-range transport and the transboundary nature of these events and the need for regional cooperation. Desert dust also plays an integral role in the Earth system affecting air chemistry and climate processes, soil characteristics and water quality, nutrient dynamics and biogeochemical cycling in both oceanic and terrestrial environments.	Goudie and Middleton (2006); Abiodun *et al.* (2012); UNEP, World Meteorological Organization [WMO] and UNCCD (2016)
Possible improvements	Research data from similar ecological areas in the Loess Plateau in China showed that there is competition for water resources between the afforested vegetation and the human water needs. Hence government decisions and policies to combat desertification must be compliant with ecological and people's socioeconomic demands without disturbing the water balance in these areas. This could be achieved by protecting local vegetation in desertification-prone lands and planting suitable vegetation according to local conditions or in specific cases leaving the land to recover without human disturbance. While considering multiple ecosystem services and their potential consequences rather than focusing only on a few services and ignoring other influences, GGW could be a role model for other regions with similar natural conditions. However, the reduction of production of agricultural goods coming along with GGW implementation triggers an increase in agricultural production elsewhere, either within the country or abroad. Respective spillover effects on potential land degradation associated with intensified agricultural production have not been analysed adequately to date.	Xiaoming *et al.* (2016) Ahrends *et al.* (2017)

GGW is not the only project in China dealing with afforestation. For instance, the Grain for Green programme is being implemented over millions of hectares of crop- and grassland that are degraded or were at high risk of degradation had the farmland practices in use continued (Shuai *et al.* 2015; Xiaoming *et al.* 2016). Another example is the Beijing-Tianjin Sand Source Control programme that looks at ecological restoration and implementation of different types of management practices ranging from controlled grazing and associated restrictions to cropland conversion to forest or grassland not being used for grazing (Middleton and Kang 2017). The experience gained from the Three North Shelterbelt Forest programme was crucial for drafting the National Action Plan (NAP) to combat desertification in China. The first NAP was prepared in 1996, coinciding with the creation of the UNCCD, and was revised in 2003. It was the first NAP in the world to have a monitoring follow-up system to measure the trends of desertification.

The public-private partnership mode, especially with the Elion Resources Group Foundation, is proving very successful. In 2015, the UNCCD awarded the Elion Foundation with the prestigious Land for Life Award for improving the livelihoods of 100,000 farmers and herders in the Kubuqi Desert in Inner Mongolia and for recovering 11,000 km^2 of degraded land into productive areas and promoting the production of green energy.

The GGW's trees provide a barrier against desert winds and help hold moisture in the air and soil, allowing plants to grow. In spite of the very high costs of its implementation, this programme is cost-effective especially for improving human health, biodiversity and livelihoods. Therefore, the Chinese Government plans to expand the reforestation programmes.

15.2.3 Setting threshold values for policy and overall governance on remediation of contaminated land

Pollution is the world's leading environmental cause of disease and premature death (Landrigan *et al.* 2018), and increasing land contamination globally is affecting the sustainability of the land resources and their ability to support life systems (Plant *et al.* 2001; Ballabio *et al.* 2018; Rodríguez-Eugenio, McLaughlin and Pennock 2018). Approximately 342,000 contaminated sites were identified in Western Europe (European Environment Agency [EEA] 2014), while in the United States the Office of Land and Emergency Management (OLEM) oversees 540,000 contaminated sites, impacting 9.3 million ha (Pierzynski and Brajendra eds. 2017) and the Environmental Protection Agency manages approximately 1,400 highly contaminated sites (United States Environmental Protection Agency 2014). Contaminated land containing substances that are potentially hazardous to public health and the environment is found in many places around the world (Tóth *et al.* 2016). Land contamination results from mining, industrial activities, military action, farming, chemical and oil spills and waste disposal (Rodríguez-Eugenio, McLaughlin and Pennock 2018). Secondary soil salinization through excess or unsuitable irrigation is a further, yet unexplored process that threatens human health (Hamidov, Helmin and Balla 2016).

The Stockholm Convention on Persistent Organic Pollutants (POPs), which entered into force in 2004, is one of the Multilateral Environmental Agreements dealing with global policies and treaties to protect human health and the environment. It asks its Parties to adopt measures to eliminate POP releases (UNDP 2009). Many countries have already ratified this convention and are implementing various land remediation policies in compliance with the global treaty. Other conventions dealing with hazardous waste movements between countries and imports of hazardous chemicals include the Basel and Rotterdam Conventions, which have also been ratified by a number of Parties, but have significantly different obligations (Secretariat of the Basel, Rotterdam and Stockholm Conventions 2018).

Generally, land remediation policies adopt mandatory command-and-control approaches, mainly utilizing the 'polluter pays' principle (Rodrigues *et al.* 2009). Nevertheless, in most cases public financial resources are required to clean already polluted areas for the benefit to the common good. Within the DPSIR framework (Section 1.6), this policy approach targets the state of the environment, intending to reduce the quantity of pollutants in the soil. Several national governments have taken concrete steps, including the establishment of relevant policies and institutional frameworks, for the remediation of contaminated lands (Rodrigues *et al.* 2009; EEA 2014). A good example comes from China, which in 2018 approved a new law on soil pollution prevention due to enter into force in January 2019 (Xinhua 2018). The case study below examines the remedial treatment of Agent Orange-contaminated land in Viet Nam as adequate data are available to evaluate the policy outcomes, in contrast to many other sites where such data are missing.

Case study: Viet Nam remedial treatment of Agent Orange-contaminated land

Viet Nam has some of the worst contaminated lands in the world (Lupi and Hoa 2015), which occurred as a result of 2,3,7,8-Tetrachlorodibenzodioxin (TCDD or Agent Orange) contamination as a result of the Viet Nam (or Second Indochina) War (1955-1975). During this war (1961-1972), the United States army used herbicides (Agent Orange) against Vietnamese military installations and this eventually resulted in land contamination, and destruction of vegetation and crops. Decades after the conflict, the Government of Viet Nam initiated the dioxin decontamination programme (Environmental remediation of dioxin contaminated hotspots in Viet Nam) as part of its National Implementation Plan (NIP), developed in line with the regulations of the Stockholm Convention. The programme aims to decontaminate the most heavily polluted areas, to plant trees on 300,000 ha of contaminated land, to help all dioxin victims, offer allowances and health insurance for people with disabilities and enhance research into the effects of toxic chemicals. **Table 15.5** provides a summary of the assessment criteria.

Viet Nam has implemented the land remediation policy backed by a complete framework of laws and regulations with supports from the United States of America, UNDP and some philanthropic foundations. Overall, the policy is positive and effective in meeting its initial objectives, but this is a process that should be based on a long-term strategic planning and monitoring programme. Note that this is a particular case, so policies that should address contaminated lands must reflect local conditions, national regulatory frameworks accomplished by internationally agreed conventions. Budgetary constraints and limitations should not be the justification for non-action when public health and the well-being of entire communities and ecosystems are at stake.

Table 15.5: Summary of the assessment criteria for land decontamination in Viet Nam		
Criterion	**Description**	**References**
Success or failure	There are very few cases where dioxin-polluted soil has been effectively decontaminated and it appeared that the only viable solution in most countries has been land filling. The United States Environmental Protection Agency (EPA) states that, "remediation technologies for the clean-up of dioxin-contaminated soils and sediments are still being developed, and many of the accepted techniques rely on thermal destruction." The Viet Nam case shows some sort of success, as the spreading of TCDD in the environment was minimized thanks to the correct implementation of the NIP plan.	United Nations Industrial Development Organization [UNIDO] (2012); Lupi and Hoa (2015)
Independence of evaluation	Most of the evaluation work has been carried out by independent assessments commissioned by UNDP, USAID and UNIDO.	United States Agency for International Development [USAID] (2010); UNIDO (2012); Lupi and Hoa (2015)
Key actors	The key actors in the implementation of the policy include the Ministries of Defence, Environment and Natural Resources, Office 33 Committee, and the Department of Health. Stakeholders that provided the technical and financial assistance were also considered as key actors and included USAID, Czech experts, Bill & Melinda Gates Foundation and the Ford Foundation.	Lupi and Hoa (2015)
Baseline	About 45.000 m^3 of Agent Orange were sprayed by the United States military in about 10 per cent of the then South Viet Nam territory. About 4.8 million Vietnamese people were impacted by the contamination. Over 3 million Vietnamese were exposed to health challenges as a result of the contamination. In response to this, the United States Congress made a financial commitment of US$59.5 million for decontamination of the affected lands and related health-care activities in Viet Nam between 2007 and 2012. The estimated volume of dioxin in hotspots released to the environment was 1,736 g I-TEQ, while the volumes of soil remediated at Bien Hoa, Da Nang and Phu Cat were at least 100,000 m^3, 70,000 m^3 and 2,500 m^3, respectively.	Lupi and Hoa (2015)
Time frame	The process of remedial treatment of the Agent Orange-contaminated land commenced in 1999 with the issuing of Decision 33 which established the National Steering Committee 33 charged with the responsibility of coordination of all Agent Orange-related matters. This was followed by the ratification of the Stockholm Convention, which targeted the phasing out POPs. The time frame for the evaluation of success or failure of the Agent Orange remedial activities was five years.	
Constraining factors	The constraining factors include poor planning and absence of a robust regulatory framework regarding dioxin contamination, inadequate data on dioxin-contaminated lands and weak technological capacities. Other factors are the weak capacities of government ministries and departments for coordination of remediation activities and limited funds available.	UNDP (2009)
Enabling factors	There are several enabling factors, including the political will on the part of the government, which promoted the establishment of the relevant policy and institutional framework for the implementation and coordination of remedial activities. The support of the United States Government and the philanthropic support from the Bill & Melinda Gates Foundation and the Ford Foundation were critical enabling factors that facilitated the remedial treatment of contaminated lands.	
Cost-effectiveness	There is little information on the cost-effectiveness of the land remediation programmes, both in Viet Nam and elsewhere. However, an evaluation of cost-effectiveness of a land remediation programme in the Dominican Republic in a lead-contaminated site indicates that remediation activities reduced health burden associated with land contamination to an acceptable cost and thresholds in line with World Health Organization (WHO) standards.	Ericson *et al.* (2018)
Equity	The involvement of the local communities both in designing project-supported activities and their implementation promoted local participation and ownership, which helped to promote the equity dimension of the policy. Another policy equity dimension is the promotion of access to land that has become usable after the implementation of the remediation programmes.	Lupi and Hoa (2015)
Co-benefits	People and communities affected by dioxin contamination may benefit from employment opportunities during remediation activities. Also, business activities around the airport have benefited from remediation as more viable lands were made available. The project generated considerable health benefits for the country. Without action, the dioxin contamination would have spread, posing severe risk to human health and the environment. Apart from neutralizing the dioxin contamination, a considerable part of the project also focused on health education and risk reduction activities among the communities in the vicinity of the contaminated hotspots. This promoted positive health among the people.	University of the West of England, Science Communication Unit (2013)
Transboundary issues	There are no potential transboundary issues in the implementation of the policy, despite the fact that neighbouring countries were also affected during the Viet Nam war by the same form of contamination.	
Possible improvements	Viet Nam has demonstrated a strong commitment to land remediation policy, but it is not clear what quantity of the contaminated lands have been remediated. This is an important variable needed to accurately assess the policy effectiveness of the decontamination programme.	

15

15.2.4 Provision of consultancy and networking for agricultural innovation

One generic policy type is the promotion of innovation by land users and farmers alike. It uses instruments such as incentives and provision of education and extension services. Some of the innovation technologies that have expanded over the last two decades include no-tillage (NT) and conservation agriculture (CA). In fact, the two practices are complementary to each other and accomplish similar goals. While NT is not a primary practice in CA, CA is based on two other principles – the introduction of cover crops and crop rotations (Kassam and Friedrich 2011). NT and CA were initially developed to combat soil erosion; however, they can also optimize crop production, promote soil health by keeping soil organic matter and nutrients in place, and by improving water and air quality. NT and CA are both seen as community-driven development processes of acceptance of new principles of agriculture. Yet their global share is limited. NT is primarily practised in North America (32 per cent of the global area under NT) and South America (45 per cent). CA, on the other hand, covers about 11 per cent of the total cropland globally and like NT is most widely located in North and South America that together have 76.6 per cent of the total CA area. Europe is lagging behind with only about 7 million ha, largely found in Russian Federation, France, Spain and Italy (FAO 2016).

NT and CA farming are seen as very promising in terms of soil quality, carbon sequestration and environmental benefits (Reicosky 2015; Haddaway *et al.* 2017), though perhaps less economically beneficial at least in the first years of farming as yields tend to be lower than with conventional agriculture (Vastola *et al.* 2017) – but with time this gap can narrow. The drawback of NT and CA is an increased use of herbicides that goes along with reduced tillage.

Examples of policy instruments include decreased fertilizer taxes and governmental subsidies to farmers that have adopted NT (Lankoski, Ollikainen and Uusitalo 2004). Within the DPSIR framework (Section 1.6), this policy approach is mostly aimed at the pressure, to implement new tillage technologies that cause minimum soil disturbance, improve soil water retention capacity and provide erosion control (Dumanski *et al.* 2006; Serraj and Sidique 2012). The long-term application of NT and CA practices depends greatly on economic viability for farmers, especially those in developing countries (Krueger 2012). They rely largely on government subsides, in particular for the acquisition of agricultural machinery that is suitable for such farming. The International Food Policy Research Institute (IFPRI) projects a big gain associated with adoption of NT cultivation in South Asia and the Pacific as a whole – with up to 32 per cent higher yields of maize and 47 per cent for wheat as compared with baseline scenario (Rosegrant *et al.* 2014).

Case study: no-tillage cultivation in Western Australia

By the late 1970s, arable farming was severely challenged in Western Australia because of drought and soil compaction. Between the 1980s and early 1990s, Australian farmers attempted to identify ways of overcoming the negative consequences of the drought by implementing NT systems (Bellotti and Rochecouste 2014). With the seeming benefits of NT, the adoption rate among other farmers increased reaching around 80-90 per cent by 2008 (Bellotti and Rochecouste 2014). **Table 15.6** gives a summary of the assessment criteria.

The Australian NT implementation was rated effective in soil and water conservation, pest, diseases and weed control, as well as in plant nutrient availability. This is demonstrated in the New South Wales NT programme, where NT contributed to improvement in soil fertility, stabilization of soil acidity, as well as increase in soil organic carbon content (Bellotti and Rochecouste 2014).

Table 15.6: Summary of the assessment criteria for NT implementation in Australia

Criterion	Description	References
Success or failure	It is recognized that in Australia NT implementation leads to a profound improvement in soil water conservation, increasing the amount of soil water available for crop growth and nutrient uptake. New South Wales farmers adopted an NT programme as a strategy to stem soil nutrient losses from long-term conventional farming practices. The fertility of the soil at the sites improved considerably while soil acidity was stabilized. In addition, there was marked improvement in the organic carbon content of the soil. All these critical positive contributions of NT are key indicators for measuring success or failure.	Bellotti and Rochecouste (2014)
Independence of evaluation	Bellotti and Rochecouste (2014) implemented independent evaluations of the NT programme from Australian No-Till farming associations in 2014.	Bellotti and Rochecouste (2014)
Key actors	The key actors of the Australia NT programme include farmers, Western Australian No Tillage Farmers Association and the Australian Government, which offered tax credit incentives to farmers practising NT agriculture as well as the farmers advisers	
Baseline	The International Food Policy Research Institute (IFPRI) projects a big gain associated with adoption of NT cultivation in South Asia and the Pacific. The Economics of Land Degradation Initiative reports positive data of NT technology implementation in Tajikistan. Nevertheless, for Australia these figures might be lower taking into account the adoption rate of NT technology. Therefore, the potential for expansion of NT in Australia is considered negligible.	Rakhmon (2016)
Time frame	The introduction of the NT commenced between early 1960 and 1980, which was described as the awareness period. Subsequently, farmers experimented with NT techniques resulting in its rapid adoption and diffusion.	
Constraining factors	NT cultivation requires more nitrogen fertilizers for use, especially during the first 2-3 years of NT, which constitutes a serious constraint to farmers with poor access to inputs. The use of herbicides is also a principal requirement for any NT system. In many cases, the use of non-selective herbicides such as glyphosate is associated with NT systems. Extensive use of such herbicides may cause negative impacts on biodiversity and human health. For example, in Europe the future permission to use glyphosate as herbicide in agriculture is currently under heavy scrutiny because of its implications on biodiversity and even possible adverse effects on human health. If permission is not to be extended in the future, the adoption of NT systems might decrease. Future developments in precision agriculture as a new farming system that optimizes returns by reducing inputs enabled by big data technologies and new sensors may, however, allow for dramatic reductions of herbicide needs.	Trigo *et al.* (2009)
Enabling factors	There are several enabling factors for NT adoption in Australia. These include the perceived need for change and the changing complexity of farming. This has helped farmers to quickly understand the skill requirements for the successful practice of NT and they were quick to respond to the skill requirements in the context of NT in Australia. NT systems are also understood as promising instruments for climate change adaptation in agriculture. This thinking has promoted the widespread adoption and support for NT farming among stakeholders.	Bellotti and Rochecouste (2014); Lal (2014); Rosegrant *et al.* (2014)
Cost-effectiveness	Evidence shows that NT is cost-effective, with several soil improvement and agronomic advantages. It is noted that NT improves farm operating budgets. This can vary across locations. There is also some evidence in many areas that, under conditions of good management, NT and CA positively lessen yield variability across seasons, especially in areas with poor rainfall.	Serraj and Sidique 2012; Swella *et al.* 2015
Equity	As far as introduction of NT farming requires large capital investments in new equipment, poor farmers are unlikely to be able to afford this technology. However, statistics show no difference in extent of NT implementation between states in Australia, thus proving affordability of the system for most (80-90 per cent) farmers. There are no losers in NT agriculture in the real sense of it. However, farmers are the main gainers. Under NT agriculture, farmers have the opportunity to intensify cropping as much as possible due to the absence of fallowing in the NT as a CA practice.	Bellotti and Rochecouste (2014)
Co-benefits	NT practices enhance easy management of crop, soil and water conservation and improvements in crop yield, as well as saving of energy, cost and time, and therefore generally contribute to intensification of agriculture. The co-benefits of NT in this context include the farmers, governments and the general public that benefits from increased food production and sustainability of land resources and the environment.	Giller *et al.* (2015)
Transboundary issues	NT transboundary issues is in the areas of its contribution to reduction of global warming. NT contributes to reduction in albedo in cropland areas and thus has great potential for global cooling and reduction of global warming. NT also reduces the emission of nitrous oxide (N_2O), which is a potent greenhouse gas, by as much as 40-70 per cent, depending on rotation.	Wallheimer (2010); Omonode *et al.* (2011)
Possible improvements	NT farming is suggested to be compatible with other technological innovations. IFPRI projections for 2050 suggests testing the following combinations of schemes: (i) NT + water harvesting; (ii) NT + precision agriculture; (iii) NT + heat tolerance; and (iv) NT + drought tolerance. In wetter regions such combinations could compensate some decline of yields reported for areas under NT.	Rosegrant *et al.* (2014)

15.2.5 Responsible food policy systems to minimize food waste and promote stakeholder networking

The current food system is causing the potentially irreversible depletion of soils, water and biodiversity towards an irrecoverable degree (UNCCD 2017a, see Section 4.4.3). This also brings increased inequalities in accessing sufficient, fresh and healthy food, as well as a growing epidemic of (mal) nutrition-related illnesses, such as obesity, diabetes and heart disease (Rush and Yan 2017).

One attempt to address these challenges brought about the development of urban food policies aimed at integrating food issues and waste (Campoy-Muñoz, Cardenete and Delgado 2017), which in turn could reduce the pressure on land. It is estimated that 30 per cent of all food produced is wasted (FAO 2018); in the European Union alone, 88 million (metric) tons of food are lost annually (Stenmarck *et al.* 2016), amounting to a cost of €143 billion. Much of this loss comes from heavily populated urban areas. Hence, if losses were diminished more land would be available for environmentally friendly agricultural systems such as organic farming or agroecology with minor damages to the environment (Muller *et al.* 2017; Blakemore 2018). Moreover, almost all the biggest cities in Europe (Zdruli 2014) and around the world have expanded at the expense of the best soils suitable for crop production (Bren d'Amour *et al.* 2017).

The formal framing of urban food policy instruments was developed mostly in the last two decades. In the early 1990s, a few pioneering cities began to develop food strategies and food policy councils. Urban food policies represent actions on the part of city governments to deal with food-related challenges that require coordination between departments and policy areas, and the establishment of novel governance structures (International Panel of Experts on Sustainable Food Systems [IPESA-Food] 2017a).

Within the DPSIR framework (Section 1.6), this policy approach targets a set of natural resources drivers as well as rapid urbanization – to reduce unsustainable use of resources. Canada was one of the pioneering countries to develop an urban food policy in 1991 through the establishment of the Toronto Food Policy Council to advise the city on food policy issues, to serve as an advocate for community food security strategies, and to foster dialogue between stakeholders. Other urban food policies schemes are now being implemented around some European cities, such as Amsterdam, Ghent, Bristol, Edinburgh and London. A more detailed description is made in this chapter of the Milan Urban Food Policy Pact (De Cunto *et al.* 2017).

Box 15.2: UNCCD Statement on food system

"Our food system has put the focus on short-term production and profit rather than long-term environmental sustainability. The modern agricultural system has resulted in huge increases while soil, the basis for global food security, is being contaminated, degraded, and eroded in many areas, resulting in long-term declines in productivity".

(UNCCD 2017a)

Case study: City collaboration on urban food good practices (Milan Urban Food Policy Pact)

A typical example of the city collaboration on urban food good practices is the Milan Urban Food Policy Pact (MUFPP), which came about in October in 2015. This is an international policy pact signed by a number of cities around the world committed to improving the sustainability of food systems and agricultural land uses in urban areas (Clinton *et al.* 2018). The current food regime may no longer be sustainable given its negative and potentially irreversible impacts on natural resources, which are currently on course to reach an irrecoverable degree (UNCCD 2017a). A transformation in food production distribution and consumption patterns is necessary to accomplish the needed changes in the current food regime and diminish its negative impacts on land and public health. Therefore, urban food policies have been conceived as having the potential to effect the needed changes in the global food sector both in terms of food safety and food security and natural resources use and management (Milan Urban Food Policy Pact 2018).

By 2030, the United Nations goal is to reach zero hunger (SDG 2). SDG 15 (life on land) is relevant to meeting the zero-hunger goal, but this can only happen with the active support of cities. The MUFPP, currently signed by 167 cities, commits signatories to develop sustainable policies, programmes and initiatives across all sectors that impact urban food systems in six thematic clusters: "(i) governance or ensuring an enabling environment for effective action; (ii) sustainable diets and nutrition; (iii) social and economic equity; (iv) food production including urban rural linkages; (v) food supply and distribution; and (vi) food waste prevention, reduction and management" (Forster *et al.* eds. 2015). The MUFPP Framework for Action is voluntary and aims to accelerate city collaboration and enhance sustainable food systems while recognizing cities' diversity in terms of objectives and targets (Forster *et al.* eds. 2015). All these objectives are closely linked to environmental protection and biodiversity conservation **(Table 15.7)**. A typical example of MUFPP outcomes is the case study in Mexico on sugar-sweetened beverages tax (Colchero *et al.* 2016), described below.

Table 15.7: Summary of the assessment criteria on Milan Urban Food Policy Pact and it impacts in Mexico

Criterion	Description	References
Success or failure	The MUFPP consolidates the role of cities as key actors in the global food system and promotes collaboration linkages among them. Two years after its launch at EXPO 2015 Milan Universal Exposition 'Feeding the Planet Energy for Life', as part of the landmark document Carta di Milano signed by United Nations Secretary-General Ban Ki-Moon on 16 October 2015, the pact is proving to be a useful instrument to promote collaboration among cities on food policies and helping them better implement land-use planning and enhanced environmental sustainability.	European Association for the Study of Obesity (2015)
Independence of evaluation	Although there is no evidence yet of the implementation of an impact assessment of the Milan Urban Food Policy Pact, one of the foremost independent impact evaluations of it was implemented in Mexico.	Colchero *et al.* (2016)
Key actors	The key actors in the MUFPP include the city mayors who signed the pact, and the civil society organization in the cities, private sector and research communities.	
Baseline	Available data shows that as much as 43 gallons of soft drinks are consumed per capita per year in Mexico. In addition, Mexican schoolchildren between the ages of 5 and 11 years consumed 20.7 per cent of energy drinks with about half of them sugar-sweetened, while the majority (64 per cent) of Mexican adults are overweight, 28 per cent obese and 11 per cent of Mexicans have type 2 diabetes.	Flores *et al.* (2010); Bronwell *et al.* (2011); WHO (2015)
Time frame	This initiative was launched in January 2014 by the Mexican Government and the independent evaluation was conducted in 2016.	
Constraining factors	The constraining factors as identified by the Pact include the governance system within some cities with weak capacities of institutions and government departments, as well as poor stakeholder participation at the city level. Another constraining factor is the divergent cities policies that affect municipal authority or jurisdiction.	Forster *et al.* eds. (2015)
Enabling factors	The key enabling factors of the MUFPP implementation in Mexico are the evidence-based results framework of the Pact as well as availability of funding.	IPES-Food (2017b)
Cost effectiveness	Thavorncharoensap (2017) examined the cost-effectiveness of obesity prevention and control through beverages taxes. Results showed that beverages taxes are a relevant and cost-effective measure for prevention and control of obesity. A few city cases reveal some techniques or actions that would result in cost-effective policies. These include microgardens, multiple cooperative start-ups, mobile apps, family shops, popular restaurant programmes and the promotion of urban agriculture.	Forster *et al.* eds. (2015); FAO (2017); Thavorncharoensap (2017)
Equity	Six recommended actions are promoted for social and economic equity: (i) the use of social protection mechanisms such as cash and food transfers to vulnerable populations to increase access to food; (ii) reorientation of school feeding programmes to provide healthy food; (iii) promotion of decent employment; (iv) encouragement and support for social and solidarity economic activities; (v) promotion of networks and support for grass-roots activities; and (vi) promotion of participatory education, training and research.	Forster *et al.* eds. (2015)
Co-benefits	The co-benefits of the policy include local residents who receive support from the city government in many aspects of everyday life, including increased green areas and biodiversity, promotion of local economy that creates jobs, solidarity among inhabitants, better quality food, upgrading of abandoned areas, waste recycling and management, and the creation of a diverse urban landscape for recreation. Furthermore, the policy also dealt successfully with the heat waves and islands inside the city.	Forster *et al.* eds. (2015)
Transboundary issues	This network has the capacity to convene local governments and enhance their role in a multilevel governance structure providing a multi stakeholder platform for communication and exchange of successful implemented policies.	Cinzia and Licomati (2017)
Possible improvements	There are gaps in certain critical areas, including the need for improvement in the level of collaboration among key government departments in the implementation of the Pact, policy coherence and the inclusion of all critical stakeholders in the implementation of the food policy.	Forster *et al.* eds. (2015)

The MUFPP has implications for the environment, economy, social equity and health of urban populations, and their linkage with rural and urban agriculture. An increasing number of cities are involved in this initiative and many more are expected to join the effort. The third Annual Gathering and Mayors' Summit of MUFPP was held in Valencia, Spain in October 2017 involving more than 400 mayors, experts and city delegates. They called on United Nation agencies to recognize their role in shaping a sustainable food system and create a better living environment inside and outside the cities. The policy efficiency of the Pact in Mexico is reflected with increased awareness among the local people on the health consequences of excess use of sweetened soft drinks and the need to return to traditional food systems.

15.3 Indicators

Land management is explicitly targeted in SDG 15.3 and also has many interconnections with other SDGs, namely SDGs 1, 2, 3, 6, 7, 11, 12 and 13. The SDGs include a total of 244 indicators for which general agreement was reached. Based on data availability and relevance to land and soil policies, three SDG indicators stand out as most relevant for this chapter **(Table 15.8)**:

1. Proportion of land degraded over total land area (Indicator 15.3.1),
2. Terrestrial protected areas as a percentage of total land area (Indicator 15.3.2), and
3. Ratio of land consumption rate to population growth rate (Indicator 11.3.1).

Table 15.8: Indicators for assessing land policy effectiveness and for measuring the progress towards the achievement of global environmental goals

Indicator	Rationale for selection	Addressed in Part A: Yes/No How:	Addressed in the case studies: Yes/No Which	Connection with the SDGs or MEAs	Data sources	Causal chains to policies and other variables impacting the indicator
1. Proportion of land degraded over total land area	SDG indicator 15.3.1: There is scientific and political consensus as well as precedence and agreement.	Yes. Section 4.4.3 on the food system; a 'box' on the Syrian Crises **Box 9.4**: Jordan faces a combined refugee and water crisis.	Yes 2.2 Setting threshold values 2.3 Planning and compensation 2.4 Funding programme plus setting standards for best practice management 2.4 Provision of consultancy	It is an indicator for SDG 15. 'Land degradation' is defined and discussed in UNCCD.	See following sources on land degradation: Gibbs and Salmon 2015; Le, Nkonya and Mizarbaev 2016. FAO has the Global Land Degradation Information System (GLADIS); Sustainable Development Solutions Network [UNSDSN] (2014) noted that data on degraded land has been 'patchy'. Also see UNCCD's metadata.	IPBES (2018a) indicates that proportion of degraded land over total land continues to increase, mostly due to lack of policies or poor implementation.
2. Terrestrial protected areas as percentage of total land area	SDG indicators 15.1.2 and 15.4.1: There is scientific and political consensus as well as precedence and agreement.	No	No	SDG indicators 15.1.2 and 15.4.1.	The United Nations List of Protected Areas (International Union for Conservation of Nature [IUCN] 1994; IUCN 1998; Chape *et al.* 2003) are available online. There is also the World Database on Protected Areas. UNEP-WCMC has the Protected Planet Report. See (United Nations 2018).	Policies for the protected areas had overall positive impacts, especially in developed countries, with less pronounced results in the rest of the world.
3. Ratio of land consumption rate to population growth rate	Since land consumption is the strongest and mostly irreversible form of land degradation, its decoupling from population growth is the core step in maintaining land, also in relation to the nexus to the other SDGs.	Yes. Section 2.2	Yes. 2.1 Stakeholder Network creation	SDG indicator 11.3.1	UN-Habitat for all countries of the world. Data for more than 300 cities are monitored by The City Prosperity Initiative, Lincoln Institute, University of New York and UN-Habitat.	Globally, land cover today is altered principally by direct human use. Evidence from a study on 120 cities revealed a three-times faster growth of urban land cover compared to the growth of urban population. Other variables affecting the indicator are land degradation and crop production.

These were used in the assessment of policy effectiveness and for measuring progress towards internationally agreed environmental goals with special reference to land and soil.

15.3.1 Indicator 1: Proportion of land degraded over total land

The expansion of human economic activities and the competing interest for land resources have dramatically increased the pressure on land and on terrestrial ecosystems **(Figure 15.3)** and in some cases accentuated political conflicts at local level (Organisation for Economic Co-operation and Development [OECD] 2017). Estimates of global land degradation show that about 25 per cent of all land is degraded, 36 per cent is slightly or moderately degraded, while 10 per cent is improving (FAO 2011; IPBES 2018b). The unit of measurement of this indicator is the area (ha or km^2) of land that is degraded divided by the total land (UNCCD 2017b). This indicator is measured by adding all those areas that are subject to change, and whose conditions are considered as negative by national authorities when measuring and evaluating each of the following three sub-indicators:

- land cover and land cover change,
- land productivity, and
- carbon stocks above and below ground.

The indicator is linked to several targets and commitments that have been agreed by global, regional and national governments to halt and reverse land degradation and restore degraded land (IPBES 2018b). These include, for instance, the Aichi Biodiversity Targets, the Bonn Challenge and related regional initiatives (e.g. 20x20, African Forest Landscape Restoration Initiative [AFR100] 2018), and SDG target 15.3.

The indicator addresses the nature of land degradation, which is expressed as "the reduction in the capacity of land to provide ecosystem goods and services over a period of time for its beneficiaries" (Nachtergaele *et al.* 2011; Zdruli 2014). Land degradation has direct impacts on the capacity for net biomass primary production, but socioeconomic factors play a major role in its occurrence – such as the link between urbanization and its related soil and air pollution (Prasad and Badarinth 2004; Seto, Güneralpa and Hutyra 2012). In other cases, socioeconomic factors have hindered efforts to cope with land degradation (Lubwama 1999; Chasek *et al.* 2011). This is the case of urban sprawl or the expansion of solar panels promoted by renewable energy policies and lack of well-defined land-use planning guidelines that have accelerated these types of land-use changes at the expense of fertile soils that otherwise should be used for food production or preservation of nature-based contributions to people (Diaz *et al.* 2018).

Policy effectiveness in either halting or reversing the expansion of degraded areas over total land areas have produced limited results; globally, land degradation remains one of the most important degradation processes with huge consequences for food security, environmental consequences and threats to livelihoods (IPBES 2018b).

Figure 15.3 emphasizes the role of soil carbon sequestration as an important indicator directly linked to soil fertility and climate change mitigation. The 4 per 1000 Initiative is promoting carbon accumulation in the world soil at a rate of 0.4 per cent

Figure 15.3: Trends in land degradation and restoration worldwide

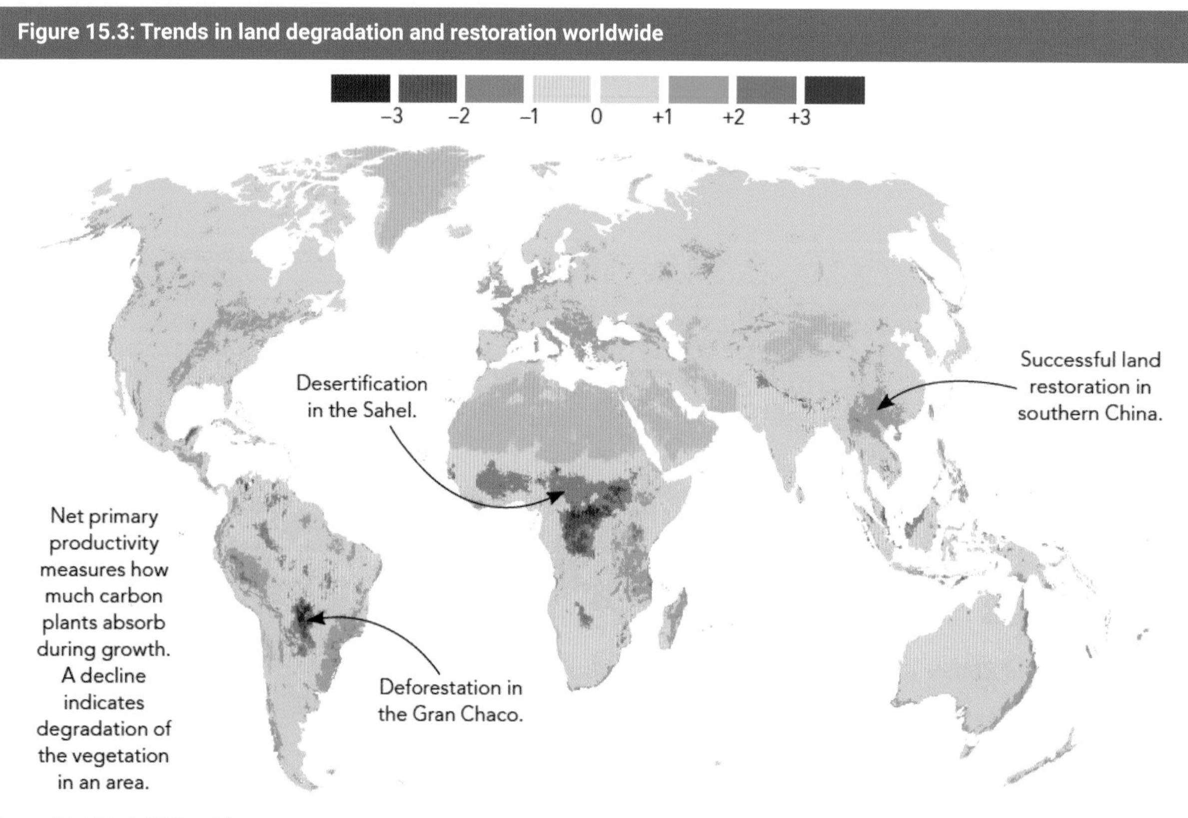

Source: World Bank (2018, p. 59).

per year to stop CO_2 increase in the atmosphere which is major contributor to climate change. The initiative intends to reach its goals by implementing the principles of conservation agriculture and agroecology. The figure shows expansion of desertification in the Sahel, deforestation in Latin America and positive results in land restoration in southern China.

15.3.2 Indicator 2: Terrestrial protected areas as a percentage of total land area

Since the mid-1990s, growing concerns over environmental degradation have led to the current emphasis on the roles that nature plays in maintaining societies (Butchart *et al.* 2015; Diaz *et al.* 2018). In recognition of the significance of this, 193 Parties of the Convention on Biological Diversity (CBD) adopted the 20 Aichi Targets to be met by 2020. Aichi biodiversity target 11 is of particular relevance and commits to a 1.6 per cent increase (from 15.4 per cent to at least 17 per cent) in protected areas by 2020 (CBD 2010). This could also contribute to reducing the loss of natural habitats (target 5), reducing human-induced species decline and extinction (target 12), and maintaining global carbon stocks (target 15). Countries have since agreed on the SDGs' targets (United Nations 2015) for 2020 and beyond, and this is likely to drive the agenda on protected terrestrial areas in the coming decades (Allen *et al.* 2016). SDG 15 specifically refers to land resources and their management.

This indicator measures the proportion of terrestrial protected areas as a share of the total land area expressed as a percentage (United Nations 2015). The purpose of the indicator is to represent the extent to which terrestrial areas that offer ecosystem value in terms of conserving biodiversity, cultural heritage, scientific research, recreation and other valuable uses are protected, in their diversity and integrity, from unsustainable land uses (United Nations 2015).

The indicator is calculated using all the nationally designated protected areas recorded in the World Database on Protected Areas (WDPA) (United Nations Environment World Conservation Monitoring Centre 2018). World Bank (2017) data show that the surface area of protected areas for the period 1990-2014 increased from 8.2 per cent to 14.8 per cent, indicating a positive trend reflecting the implementation of national and international policies for them **(Figure 15.4)**.

Governance of protected areas, in particular primary forests, is particularly relevant because evidence points to the impacts of agricultural output prices on deforestation rates both inside and outside of protected areas (Deiro and Escobar 2012). Assunção, Gadour and Rocha (2015), in a study in the Amazon, found high correlations between deforestation rates and agricultural output prices, while Deiro and Escobar (2012) point out that "between 1981 and 2010 an area of 45,000,000 hectares was downgraded or lost with almost 70 per cent of cases occurring since 2008". The authors (Deiro and Escobar 2012), however, conclude that changes in conservation policies implemented between 2004 and 2008 significantly contributed to the curbing of deforestation rates.

Location is another key influencing factor affecting protected areas. Joppa and Pfaff (2009) note that "the positioning of protected areas is not random; they are often located in areas that are inaccessible or unsuitable for agriculture, in remote and topographically challenging areas without transport links, such that they are unlikely to be under pressure from the developmental drivers of land use change".

In general, there is scientific and political consensus, as well as precedence and agreement, on this indicator. However, evidence of the impacts of market prices, management effectiveness and factors specific to other sites have since led to the proposal to incorporate indicators that aid in the measurement of protected area conditions and/or management effectiveness, including more equitable management and representative indicators of spatial coverage (e.g. forest area as a percentage of land area).

Figure 15.4: Terrestrial protected area as a percentage of total land area per country (1990-2014)

Source: World Bank (2018, p. 59).

15.3.3 Indicator 3: Ratio of land consumption rate to population growth

In 2016, about 54.5 per cent of the global (human) population lived in urban areas and by 2030, the United Nations predicts that 60 per cent of the global population will be urban (United Nations 2016a). The total increase in urban population between 2000 and 2020 is estimated at 1.48 billion, of which 1.35 billion will be concentrated in less-developed regions (United Nations 2012). As population growth increasingly consumes available land, cities expand far beyond their formal administrative boundaries and urban sprawl is visible around them (United Nations Human Settlements Programme [UN-Habitat] 2017).

Land consumption rate is computed as a function of: (a) "The expansion of built-up area which can be directly measured; (b) the absolute extent of land that is subject to exploitation by agriculture, forestry or other economic activities; and (c) the over-intensive exploitation of land that is used for agriculture and forestry" (United Nations 2015). Population growth rate shows the increase of population in a country during a certain period, typically one year, expressed as a percentage of the population at the start of that period (World Bank 2017). **(Figure 15.5)**.

The ratio of land consumption rate to population growth is a critical indicator that is closely connected with multiple SDGs. More directly, it is tied in with SDG 11, "Make cities and human settlements inclusive, safe, resilient and sustainable".

The changes in land consumption are largely driven by increases in transport infrastructure, poor urban and regional planning, and land speculation (UN-Habitat 2015). This in turn negatively impacts on the environment (per capita resource use and greenhouse gas emissions). For example, for every 10 per cent increase in urban sprawl, there is a 5.7 per cent increase in per capita CO_2 emissions and a 9.6 per cent increase in per capita hazardous pollution. It also increases socioeconomic and spatial inequalities. For instance, 30 per cent of the global urban population (880 million people) lived in slum-like conditions in 2014; in sub-Saharan Africa, that proportion was 55 per cent (United Nations 2016b).

While evidence shows that poor spatial planning is one of the main factors leading to urban sprawl, effective policymaking is central to managing land consumption (Rosni and Noor 2016). Many governments rely on policies such as land-use restrictions (e.g. urban growth boundaries and minimum-lot zoning); price-based policies such as property taxes (Gyourko and Molloy 2015; Glaeser and Gyourko 2017); and other regulatory systems composed of zoning ordinances, subdivision regulations and building codes, for controlling urban sprawl. Feng *et al.* (2016) conclude that the implementation of land-use planning policy in China played a major role in ensuring the lowest effective rate of change of sprawl. The potential of restoring and reutilizing former industrial and otherwise used land (brownfields) for mitigating land consumption is still underexplored in land planning and policies (Tobias *et al.* 2018).

Figure 15.5: Ratio of land consumption rate to population growth rate by region and period (1990-2015)

Sub-Saharan Africa
Land-Rich Developed Countries
East Asia and the Pacific
Southeast Asia
South and Central Asia
Latin America and the Caribbean
Western Asia and North Africa
Europe and Japan

-6.0 -4.0 -2.0 0.0 2.0 4.0 6.0 8.0 10.0 12.0

2000-2015 1990-2000

Source: UN-Habitat (2015).

Given the high rates of land consumption in the European Union – 275 ha of agricultural and natural habitats converted to urban sprawl and other forms of land take per day – the EU has endorsed a No Net Land Take by 2050 policy that intends to reduce land consumption throughout the Union, giving priority to greening areas and ecological corridors (University of the West of England, Science Communication Unit 2013).

Seto, Güneralpa and Hutyra (2012) state that varying causal factors make it difficult to observe this indicator on an international scale. The lack of standardized procedures in the delineation of spatial units and recognition of administrative boundaries lead to spatial inconsistencies (United Nations 2018). UN-Habitat therefore proposed a minor revision of indicator 11.3.1 to "Ratio of land consumption rate to population growth rate, including the term *Efficient land use*", where if the ratio is equal to or smaller than 1 it is qualified as efficient.

Other alternate indicators proposed include:

- "Resources per capita invested in human settlement per km^2" (by UNCDF)
- "Percentage of cities with direct participation structure of civil society in urban planning and management, which operate regularly and democratically" (by United Nations Statistical System Organizations)
- "Ratio of land consumption rate to urban population growth rate at comparable scale" (by UNFPA).

15.4 Conclusions

Across the globe, while different land policies and initiatives have been adopted and implemented, it is however difficult to attribute progress in the thematic area to specific policy approaches for several reasons.

Firstly, the transboundary nature of land and its resources (Sikor *et al.* 2013) hinders the assessment of policy effectiveness. Many land resources such as forests cannot be managed at state level alone because they straddle international borders. Activities in one country often have effects on neighbouring countries' land policies and initiatives. This hinders attribution of progress to a specific policy approach with respect to sustainable management of land and its resources (Creutzig 2017). Land tenure is also a constraining factor and global land acquisitions, or 'land grabbing', amounts to more than 42 million ha, mostly in Africa. Food-importing countries have accelerated their acquisitions to enhance food security globally.

Closely related to the above are the challenge of teleconnections. Demand for food in some places generates land uptake in others; for instance, Africa is a net contributor to the food needs of Europe (Bergmann and Holmberg 2016). In this context, sustainable land management policies in a country can be positively (or otherwise) impacted by demand from another country, which also makes it difficult to attribute progress to a specific policy approach. The regenerative capacity of land resources is another major obstacle to attribution of land policies. Food, water, forests and wildlife are all renewable resources. With or without any policy framework, some land resources such as forest systems can regenerate themselves, making attribution to specific forest conservation policy difficult.

The World Bank (2006) provides a set of principles, "where land and resource management policies have been successful" there is:

1. Local community participation in all aspects of the programme
2. Public support for private investment in soil and water conservation
3. Improvement and maintenance of roads
4. Sound macroeconomic management that does not discriminate against agriculture and natural resources
5. Robust local capacity-building by non-governmental organizations and other cooperative-type projects, and
6. Consistent efforts over at least a decade by concerned governments to increase not only land productivity but also awareness of environmental problems and possible solutions at local levels.

Some of these are conditions addressed in the case studies of this chapter. However, there are two emerging policy approaches which hold out promise for the future. The first policy approach involves the use of economic incentives to deal with environmental issues related to land, as the China case study (Section 15.2.2) demonstrates.

The second approach is the Sustainable Intensification of Land Use and Integrated Resource Management (Garnett *et al.* 2013), despite the criticism about the overall benefits of the agricultural intensification concept (Rasmussen *et al.* 2018). This approach is best described by technological advancements that ensure increases in crop production through implementation of sustainable land and water practices, such as conservation agriculture and no-tillage cultivation, as described in the Australia case study (Section 15.2.4) as well as combined cropping systems such as legumes and cereals; agroforestry, agroecology (World Bank 2006) and regenerative agriculture. One of the key lessons learned from the case studies is the importance of a robust institutional framework for policy implementation.

Across the case studies, the establishment of institutional and administrative capacities for policy implementation underpins the success of most of the key policies. The indicators are relevant to key interconnected international goals such as the SDGs and provide evidence of progress towards meeting the policy objectives. For instance, Indicator 2 on terrestrial protected areas as a percentage of total land area is connected to Aichi Biodiversity Targets 13, 11 and 5 and is also relevant to SDG 15 and its respective targets.

The case studies also indicate that evaluation of policy effectiveness in most cases has been commissioned by external stakeholders, while national governments that are often involved in policy design have not shown serious interest in policy evaluation. Land policy evaluation is important as it will provide significant lessons that can be useful in the refinement of policies and implementation of strategies.

One obvious gap is the fact that most national land policies are not linked to international goals. This is important, especially from the point of view of the SDGs, when viewed against the background that the implementation of such policies will have little or no contribution to the attainment of international goals. The Kyoto Protocol did not even mention relation to the role of land and soil in climate change dynamics. When national governments ratify international conventions, it is important that these are backed up with relevant national policies accompanied by baseline indicators to track progress towards reaching policy goals.

References

Abiodun, B.J., Adeyewa, Z.D., Oguntunde, P.G., Salami, A.T. and Ajayi, V.O. (2012). Modeling the impacts of reforestation on future climate in West Africa. *Theoretical and applied climatology* 110(1-2), 77-96. https://doi.org/10.1007/s00704-012-0614-1.

African Forest Landscape Restoration Initiative (2018). *Afr100*. http://afr100.org/ (Accessed: 6 November 2018).

Ahrends, A., Hollingsworth, P.M., Beckschäfer, P., Chen, H., Zomer, R.J., Zhang, L. *et al.* (2017). China's fight to halt tree cover loss. *Proceedings of the Royal Society B* 284(1854). https://doi.org/10.1098/rspb.2016.2559.

Akhtar-Schuster, M., Stringer, L.C., Erlewein, A., Metternicht, G., Minelli, S., Safriel, U. *et al.* (2017). Unpacking the concept of land degradation neutrality and addressing its operation through the Rio conventions. *Journal of Environmental Management* 195, 4-15. https://doi.org/10.1016/j.jenvman.2016.09.044.

Allen, C., Metternicht, G. and Wiedmann, T. (2016). National pathways to the global Sustainable Development Goals (SDGs): A comparative review of scenario modelling tools. *Environmental Science and Policy* 66, 199-207. https://doi.org/10.1016/j.envsci.2016.09.008.

American Museum of Natural History (2008). *China's great green wall: A dust antidote?* https://www.amnh.org/explore/science-bulletins/(watch)/bio/news/china-s-great-green-wall-a-dust-antidote (Accessed: 6 November 2018).

Assunção, J., Gandour, C. and Rocha, R. (2015). Deforestation slowdown in the Brazilian Amazon: Prices or policies? *Environment and Development Economics* 20(6), 697-722. https://doi.org/10.1017/S1355770X15000078.

Aunan, K. and Pan, X.-C. (2004). Exposure-response functions for health effects of ambient air pollution applicable for China – a meta-analysis. *Science of the Total Environment* 329(1–3), 3-16. https://doi.org/10.1016/j.scitotenv.2004.03.008.

Ballabio, C., Panagos, P., Lugato, E., Huang, J.-H., Orgiazzi, A., Jones, A. *et al.* (2018). Copper distribution in European topsoils: An assessment based on LUCAS soil survey. *Science of the Total Environment* 636, 282-298. https://doi.org/10.1016/j.scitotenv.2018.04.268.

Bellotti, B. and Rochecouste, J.F. (2014). The development of conservation agriculture in Australia—Farmers as innovators. *International Soil and Water Conservation Research* 2(1), 21-34. https://doi.org/10.1016/S2095-6339(15)30011-3.

Bergmann, L. and Holmberg, M. (2016). Land in Motion. *Annals of the Association of American Geographers* 106(4), 932-956. https://doi.org/10.1080/24694452.2016.1145537.

Blakemore, R.J. (2018). Critical decline of earthworms from organic origins under intensive, humic SOM-depleting agriculture. *Soil Systems* 2(2), 33. https://doi.org/10.3390/soilsystems2020033.

Bren d'Amour, C., Reitsma, F., Baiocchi, G., Barthel, S., Güneralp, B., Erb, K.-H. *et al.* (2017). Future urban land expansion and implications for global croplands. *Proceedings of the National Academy of Sciences* 114(34), 8939-8944. https://doi.org/10.1073/pnas.1606036114.

Bronwell, K., Farley, T., Willet, W.C., Popkin, B.M., Chaloupka, F., Thompson, J.W. *et al.* (2011). The public health and economic benefits of taxing sugar-sweetened beverages. *The New England Journal of Medicine* 361(16), 1599–1605 https://doi.org/10.1056/NEJMhpr0905723.

Butchart, S.H.M., Clarke, M., Smith, R.J., Sykes, R.E., Scharlemann, J.P.W., Harfoot, M. *et al.* (2015). Shortfalls and solutions for meeting national and global conservation area targets. *Conservation Letters: Journal of the Society for Conservation Biology* 8(5), 329-337. https://doi.org/10.1111/conl.12158.

Campoy-Muñoz, P., Cardenete, M.A. and Delgado, M.C. (2017). Economic impact assessment of food waste reduction on European countries through social accounting matrices. *Resources, Conservation and Recycling* 122, 202–209. https://doi.org/10.1016/j.resconrec.2017.02.010.

Castella, J.-C., Lestrelin, G., Hett, C., Bourgoin, J., Fitriana Y.R., Heinimann, A. *et al.* (2013). Effects of landscape segregation on livelihood vulnerability: Moving from extensive shifting cultivation to rotational agriculture and natural forests in northern Laos. *Human Ecology* 41(1), 63–76. https://doi.org/10.1007/s10745-012-9538-8.

Chape, S., Blyth, S., Fish, L., Fox, P. and Spalding, M.D. (2003). *2003 United Nations List of Protected Areas*. Gland: International Union for Conservation of Nature. https://archive.org/details/2003unitednation03chap/page/n3.

Chasek, P., Essahli, W., Akthar-Schuster, M., Stringer, L.C. and Thomas, R. (2011). Integrated land degradation monitoring and assessment: Horizontal knowledge management at the national and international levels. *Land Degradation and Development* 22(2), 272-284. https://doi.org/10.1002/ldr.1096.

Choi. S. and Gankhuyag, U. (2016). Mining and sustainable development in the Asia Pacific region. *Financing for the Sustainable Development Goals: The Role of Fiscal Reforms, Revenue Management and Sovereign Wealth Funds in the Extractives Sector*. Bangkok, 7-8 December 2016. http://www.greengrowthknowledge.org/event/financing-sustainable-development-goals-sdgs-role-fiscal-reforms-revenue-management-and

Cinzia, T. and Licomati, S. (2017). The Milan urban food policy pact: The potential of food and the key role of cities in localizing SDGS. *Journal of Universities and International Development Cooperation* 1, 372-378. http://www.ojs.unito.it/index.php/junco/article/view/2173.

Clinton, N., Stuhlmacher, M., Miles, A., Uludere Aragon, N., Wagner, M., Georgescu, M. *et al.* (2018). A global geospatial ecosystem services estimate of urban agriculture. *Earth's Future* 6(1), 40-60. https://doi.org/10.1002/2017EF000536.

Colchero, M.A., Popkin, B.M., Rivera, J.A. and Ng, S.W. (2016). Beverage purchases from stores in Mexico under the excise tax on sugar sweetened beverages: Observational study. *BMJ* 352, 6704. https://doi.org/10.1136/bmj.h6704.

Convention on Biological Diversity (2010). X/2. The Strategic Plan for Biodiversity 2011-2020 and the Aichi Biodiversity Targets. Decision adopted by the Conference of Parties to the Convention on Biological Diversity at its Tenth Meeting. 29 October. *UNEP/CBD/COP/DEC/X/2*. https://www.cbd.int/doc/decisions/cop-10/cop-10-dec-02-en.doc.

Creutzig, F. (2017). Govern land as a global commons. *Nature* 546(7656), 28-29. https://doi.org/10.1038/546028a.

De Cunto, A., Tegoni, C., Sonnino, R., Michel, C. and Lajili-Djalaï, F. (2017). *Food in Cities: Study on Innovation for A Sustainable and Healthy Production, Delivery, and Consumption of Food in Cities*. Brussels: European Commission. https://ec.europa.eu/research/openvision/pdf/rise/food_in_cities.pdf.

Deiro B. and Escobar, H. (2012). *Brasil perdeu um RJ de áreas protegidas*. Universidade Federal de Pernambuco http://www.estadao.com.br/noticias/impresso,brasil-perdeu-um-rj-de-areas-protegidas,975519.

Deng, L., Liu, G.b. and Shangguan, Z.P. (2014). Land-use conversion and changing soil carbon stocks in China's 'Grain-for-Green' program: A synthesis. *Global Change Biology* 20(11), 3544–3556. https://doi.org/10.1111/gcb.12508.

Deng, X. and Li, Z. (2016). Economics of land degradation in China. In *Economics of Land Degradation and Improvement: A Global Assessment for Sustainable Development*. Nkonya, E., Mirzabaev, A. and von Braun, J. (eds.). chapter 13. 385-399. https://link.springer.com/content/pdf/10.1007%2F978-3-319-19168-3_13.pdf

Diaz, S., Pascual, U., Stenseke, M., Martín-López, B., Watson, R.T., Molnár, Z. *et al.* (2018). Assessing nature's contribution to people. *Science* 359(6373), 270-272. https://doi.org/10.1126/science.aap8826.

Dumanski, J., Peiretti, R., Benites, J., McGarry, D. and Pieri, C. (2006). The paradigm of conservation tillage. *Proceedings of World Association of Soil and Water Conservation* P1(7), 58-64. http://www.unapcaem.org/publication/ConservationAgri/ParaOfCA.pdf.

Ericson, B., Caravanos, J., Depratt, C., Santos, C., Cabral, M.G., Fuller, R. *et al.* (2018). Cost effectiveness of environmental lead risk mitigation in low- and middle-income countries. *GeoHealth* 2(2), 87-101. https://doi.org/10.1002/2017GH000109.

European Association for the Study of Obesity (2015). *Carta Di Milano: 2015 Milan Declaration: A Call to Action on Obesity*. Teddington. http://carta.milano.it/wp-content/uploads/2015/11/112.pdf.

European Environment Agency (2014). *Progress in Management of Contaminated Sites*. Copenhagen. https://www.eea.europa.eu/downloads/a29faf166f9e45f78e3ae107e72d957c/1441389583/assessment.pdf.

Feng, L., Du, P., Zhu, L., Luo, J. and Adaku, E. (2016). Investigating sprawl along China's urban fringe from a spatio-temporal perspective. *Applied Spatial Analysis and Policy* 9(2), 233-250. https://doi.org/10.1007/s12061-015-9149-z.

Flores, M., Macias, N., Rivera, M., Lozada, A., Barquera, S., Rivera-Dommarco, J. *et al.* (2010). Dietary patterns in Mexican adults are associated with risk of being overweight or obese. *The Journal of Nutrition* 140(10), 1869–1873. https://doi.org/10.3945/jn.110.121533.

Food and Agriculture Organization of the United Nations (2010). *Global Forests Resources Assessment 2010: Country Report. Lao People's Democratic Republic*. Rome. http://www.fao.org/forestry/20366-06af02af6c37e155d6de871dafdf77bbf.pdf.

Food and Agriculture Organization of the United Nations (2011). *The State of the World's Land and Water Resources for Food and Agriculture: Managing Systems at Risk*. Rome. http://www.fao.org/docrep/017/i1688e/i1688e.pdf.

Food and Agriculture Organization of the United Nations (2016). *Conservation agriculture*. http://www.fao.org/ag/ca/6c.html (Accessed: 6 November 2018).

Food and Agriculture Organization of the United Nations (2017). *Voluntary Guidelines for Sustainable Soil Management*. Rome. http://www.fao.org/3/a-bl813e.pdf.

Food and Agriculture Organization of the United Nations (2018). *Food loss and food waste*. http://www.fao.org/food-loss-and-food-waste/en/ (Accessed: 6 November 2018).

Food and Agriculture Organization of the United Nations and Intergovernmental Technical Panel on Soils (2015). *Status of the World's Soil Resources: Main Report*. Rome. http://www.fao.org/3/a-i5199e.pdf.

Forster, T., Egal, F., Escudero, A.G., Dubbeling, M. and Renting, H. (eds.) (2015). *Milan Urban Food Policy Pact. Selected Good Practices from Cities*. Milano: Fondazione Giangiacomo Feltrinelli. https://www.ruaf.org/sites/default/files/MUFPP_SelectedGoodPracticesfromCities.pdf.

Garnett, T., Appleby, M.C., Balmford, A., Bateman, I.J., Benton, T.G., Bloomer, P. *et al.* (2013). Sustainable intensification in agriculture: Premises and policies. *Science* 341(6141), 33-34. https://doi.org/10.1126/science.1234485.

Gibbs, H.K. and Salmon, J.M. (2015). Mapping the world's degraded land. *Applied Geography* 57, 12-21. https://doi.org/10.1016/j.apgeog.2014.11.024.

Giller, K.E., Andersson, J.A., Corbeels, M., Kirkegaard, J., Mortensen, D., Erenstein, O. *et al.* (2015). Beyond conservation agriculture. *Frontiers in Plant Science* 6(872). https://doi.org/10.3389/fpls.2015.00870.

Glaeser, E.L. and Gyourko, J. (2017). *The Economic Implications of Housing Supply*. Cambridge, MA: National Bureau OF Economic Research. http://www.nber.org/papers/w23833.pdf.

Global Witness (2013). *Rubber Barons. How Vietnamese Companies and International Financiers are Driving a Land Grabbing Crisis in Cambodia and Laos*. https://www.globalwitness.org/documents/10525/rubber_barons_lores_0_1.pdf.

Goudie, A.S. and Middleton, N.J. (2006). *Desert Dust in the Global System*. Springer. https://www.springer.com/gp/book/9783540323549.

Gyourko, J. and Molloy, R. (2015). Regulation and housing supply. In *Handbook of Regional and Urban Economics*. Duranton, G., Henderson, J.V. and Strange, W. (eds.). Amsterdam: Elsevier Science Publishers. chapter 19. 1289-1337. https://econpapers.repec.org/bookchap/eeeregchp/5-1289.htm

Haddaway, N.R., Hedlund, K., Jackson, L.E., Kätterer, T., Lugato, E., Thomsen, I.K. *et al.* (2017). How does tillage intensity affect soil organic carbon? A systematic review. *Environmental Evidence* 6(30). https://doi.org/10.1186/s13750-017-0108-9.

Hamidov, A., Helming, K. and Balla, D. (2016). Impact of agricultural land use in Central Asia: a review. *Agronomy for Sustainable Development* 36(1), 6. https://doi.org/10.1007/s13593-015-0337-7.

Henríquez-Hernández, A.L., González-Antuña, A., Boada, L.D., Carranza, C., Pérez-Arellano, J.L., Almeida-González, M. *et al.* (2018). Pattern of blood concentrations of 47 elements in two populations from the same geographical area but with different geological origin and lifestyles: Canary Islands (Spain) vs. Morocco. *Science of the Total Environment* 636, 709-716. https://doi.org/10.1016/j.scitotenv.2018.04.311.

Huang, T., Zhang, X., Ling, Z., Zhang, L., Gao, H., Tian, C. *et al.* (2016). Impacts of large-scale land-use change on the uptake of polycyclic aromatic hydrocarbons in the artificial three northern regions shelter forest across northern China. *Environmental Science & Technology* 50(23), 12885-12893. https://doi.org/10.1021/acs.est.6b04835.

Intergovernmental Science-Policy Platform on Biodiversity and Ecosystem Services (2018a). *Summary for Policymakers of the Regional Assessment Report on Biodiversity and Ecosystem Services for Europe and Central Asia of the Intergovernmental Science-Policy Platform on Biodiversity and Ecosystem Services*. Fischer, M., Rounsevell, M., Torre-Marin Rando, A., Mader, A., Church, A., Elbakidze, M. *et al.* (eds.). Bonn. https://www.ipbes.net/system/tdf/spm_2b_eca_digital_0.pdf?file=1&type=node&id=28318.

Intergovernmental Science-Policy Platform on Biodiversity and Ecosystem Services (2018b). *Summary for Policymakers of the Assessment Report on Land Degradation and Restoration of the Intergovernmental Science-Policy Platform on Biodiversity and Ecosystem Services*. Scholes, R., Montanarella, L., Brainich, A., Barger, N., ten Brink, B., Cantele, M. *et al.* (eds.). Bonn. https://www.ipbes.net/system/tdf/spm_3bi_ldr_digital.pdf?file=1&type=node&id=28335.

International Panel of Experts on Sustainable Food Systems (2017a). *What Makes Urban Food Policy Happen? Insights from Five Case Studies: Executive Summary*. London. http://www.ipes-food.org/images/Reports/Cities_execsummary.pdf.

International Panel of Experts on Sustainable Food Systems (2017b). *What Makes Urban Food Policy Happen? Insights from Five Case Studies.* London. http://www.ipes-food.org/images/Reports/Cities_full.pdf.

International Union for Conservation of Nature (1994). *1993 United Nations List of National Parks and Protected Areas.* Gland. http://wedocs.unep.org/bitstream/handle/20.500.11822/22735/1993_UN_parks_protected_areas.pdf?sequence=1&isAllowed=y.

International Union for Conservation of Nature (1998). *1997 United Nations List of Protected Areas.* Gland. https://archive.org/details/1997unitednation97wcmc/page/n9.

Jiang, H. (2016). Taking down the "Great Green Wall": The science and policy discourse of desertification and its control in China. In *The End of Desertification? Disputing Environmental Change in the Drylands.* Behnke, R. and Mortimore, M. (eds.). Berlin: Springer. 513-536. https://link.springer.com/chapter/10.1007/978-3-642-16014-1_19

Joppa, L.N. and Pfaff, A. (2009). High and far: Biases in the location of protected areas. 4(12), e8273. https://doi.org/10.1371/journal.pone.0008273.

Kassam, A. and Fridriech, T.H. (2011). Conservation agriculture: Principles, sustainable land management and ecosystem services. *Società Italiana de Agronomia XL Convegno Nazionale, Università degli Studi Teramo.* Rome, 7-9 September. http://www.fao.org/ag/ca/CA-Publications/CA_Teramo_Kassam_Friedrich.pdf

King, P.N. and Mori, H. (2007). Policy selection and diffusion theory. In *International Review for Environmental Strategies: Best Practice on Environmental Policy in Asia and the Pacific.* Hayama: Institute for Global Environmental Strategies. chapter 2. 17-38. https://pub.iges.or.jp/pub_file/iresvol7-117pdf/download

Krueger, S. (2012). *Conservation crusader: Paraguayan Rolf Derpsch helped expand no-till across globe.* Corn and Soy Bean Digest http://www.cornandsoybeandigest.com/conservation/conservation-crusader-paraguayan-rolf-derpsch-helped-expand-no-till-across-globe (Accessed: 6 November 2018).

Lal, R. (2014). Soil conservation and ecosystem services. *International Soil and Water Conservation Research Journal* 2(3), 36–47. https://doi.org/10.1016/S2095-6339(15)30021-6.

Lambin, E.F., Meyfroidt, P., Rueda, X., Blackman, A., Börner, J., Cerutti, P.O. *et al.* (2014). Effectiveness and synergies of policy instruments for land use governance in tropical regions. *Global Environmental Change* 28, 129-140. https://doi.org/10.1016/j.gloenvcha.2014.06.007.

Landrigan, P.J., Fuller, R., Acosta, N.J.R., Adeyi, O., Arnold, R., Basu, N. *et al.* (2018). The Lancet Commission on pollution and health. *The Lancet* 391(10119), 462-512. https://doi.org/10.1016/S0140-6736(17)32345-0.

Lankoski, J., Ollikainen, M. and Uusitalo, P. (2004). *No-Till Technology: Benefits to Farmers and the Environment?* Helsinki: University of Helsinki. https://helda.helsinki.fi/bitstream/handle/1975/635/Discuss1.pdf?sequence=1.

Lao People's Democratic Republic, Ministry of Planning and Investment (2011). *The 7th Five-Year National Socio-Economic Development Plan (2011-2015).* Vientiane. http://www.la.undp.org/content/dam/laopdr/docs/Reports%20and%20publications/LA_7th%20NSEDP_Eng.pdf.

Le Q.B., Nkonya E. and Mirzabaev A. (2016). Biomass productivity- based mapping of global land degradation hotspots. In *Economics of Land Degradation and Improvement – A Global Assessment for Sustainable Development.* Nkonya, E., Mirzabaev, A. and von Braun, J. (eds.). Springer. https://link.springer.com/chapter/10.1007/978-3-319-19168-3_4#citeas

Li, W. and Huntsinger, L. (2011). China's grassland contract policy and its impacts on herder ability to benefit in inner Mongolia: Tragic feedbacks. *Ecology and Society* 16(2). https://www.ecologyandsociety.org/vol16/iss2/art1/.

Lu, Q. and Wang, S. (2003). Dust-Sand Storms in China: Disastrous effects and mitigation Strategies. *The XII World Forestry Congress.* Quebec City, 21-28 September. http://www.fao.org/docrep/ARTICLE/WFC/XII/0859-B5.HTM

Lubwama, F.B. (1999). Socio-economic and gender issues affecting the adoption of conservation tillage practices. In *Conservation Tillage with Animal Traction.* Kaumbutho, P.G. and Simalenga, T.E. (eds.). Kampala: Animal Traction Network for Eastern and Southern Africa. 155-162. https://pdfs.semanticscholar.org/27b8/6afc8c3669f5e4afa448bb376946714bea3b.pdf

Lupi, C. and Hoa, N.K. (2015). *GEF/UNDP project Environmental Remediation of Dioxin Contaminated Hotspots in Viet Nam Terminal Evaluation Report.* https://erc.undp.org/evaluation/documents/download/8716.

Marques, M.J., Schwilch, G., Lauterburg, N., Crittenden, S., Tesfai, M., Stolte, J. *et al.* (2016). Multifaceted impacts of sustainable land management in drylands: A review. *Sustainability* 8(2), 177. https://doi.org/10.3390/su8020177.

Messerli, P., Bader, C., Hett, C., Epprecht, M. and Heinimann, A. (2015). Towards a spatial understanding of trade-offs in sustainable development: A meso-scale analysis of the nexus between land use, poverty, and environment in the Lao PDR. *PloS one* 10(7), e0133418. https://doi.org/10.1371/journal.pone.0133418.

Middleton, N. and Kang, U. (2017). Sand and dust storms: Impact mitigation. *Sustainability* 9(6), 1053. https://doi.org/10.3390/su9061053

Milan Urban Food Policy Pact (2018). *Milan urban food policy pact.* https://www.milanurbanfoodpolicypact.org/ (Accessed: 6 November 2018).

Muller, A., Schader, C., El Haga Scialaba, N., Bruggemann, J., Isensee, A., Erb, K.H. *et al.* (2017). Strategies for feeding the world more sustainably with organic agriculture. *Nature Communications* 8(1290). https://doi.org/10.1038/s41467-017-01410-w.

Nachtergaele, F., Petri, M., Biancalani, R., van Lynden, G., van Velthuizen, H. and Bloise, M. (2011). *Global Land Degradation Information System (GLADIS). Version 1.0. An Information Database for Land Degradation Assessment at Global Level.* Rome: Food and Agriculture Organization of the United Nations.

O'Callaghan, J. (2014). Will China's Great GREEN Wall save the country from dust storms? 100 billion tree project could halt advancing Gobi Desert. Daily Mail, https://www.dailymail.co.uk/sciencetech/article-2874368/Will-China-s-Great-GREEN-Wall-save-country-dust-storms-100-billion-tree-project-halt-advancing-Gobi-Desert.html.

Omonode, R.A., Smith, D.R., Gál, A. and Vyn, T.J. (2011). Soil nitrous oxide emissions in corn following three decades of tillage and rotation treatments. *Soil Science Society of America Journal* 75(1), 152-163. https://doi.org/10.2136/sssaj2009.0147.

Organisation for Economic Co-operation and Development (2017). *The Governance of Land Use in OECD Countries: Policy Analyses and Recommendations.* Paris. https://www.oecd-ilibrary.org/urban-rural-and-regional-development/the-governance-of-land-use-in-oecd-countries_9789264268609-en.

Phimmavong, S., Ozarska, B., Midgley, S. and Keenan, R. (2009). Forest and plantation development in Laos: History, development and impact for rural communities. *The International Forestry Review* 11(4), 501-513. https://www.jstor.org/stable/43739828?seq=1#metadata_info_tab_contents.

Phompila, C., Lewis, M., Ostendorf, B. and Clarke, K. (2017). Forest cover changes in lao tropical forests: Physical and socio-economic factors are the most important drivers. *Land Contamination & Reclamation* 6(2), 23. https://doi.org/10.3390/land6020023.

Piao, S., Fang, J., Friedlingstein, P., Ciais, P., Viovy, N. and Demarty, J. (2007). Growing season extension and its impact on terrestrial carbon cycle in the northern hemisphere over the past 2 decades. *Global Biogeochemical Cycles* 21(3). https://doi.org/10.1029/2006GB002888

Pierzynski, G. and Brajendra (eds.) (2017). *Threats to Soils: Global Trends and Perspectives: A Contribution from the Intergovernmental Technical Panel on Soils, Global Soil Partnership Food and Agriculture Organization of the United Nations* Global Land Outlook Working Paper. Bonn: United Nations Convention to Combat Desertification. https://static1.squarespace.com/static/5694c48bd82d5e9597570999/t/5931752920099eabdb9b6a7a/1496413492935/Threats+to+Soils__Pierzynski_Brajendra.pdf.

Plant, J., Smith, D., Smith, B. and Williams, L. (2001). Environmental geochemistry at the global scale. *Applied Geochemistry* 16(11-12), 1291 –1308. https://doi.org/10.1016/S0883-2927(01)00036-1.

Prasad, V.K. and Badarinth, K.V.S. (2004). Land use changes and trends in human appropriation of above ground net primary production (HANPP) in India (1961–98). *The Geographical Journal* 170(1), 51-63. https://doi.org/10.1111/j.0016-7398.2004.05015.x

Rakhmon, S. (2016). *Tajikistan Case Study Policy Brief.* Bonn: Economics of Land Degradation Initiative. http://repo.mel.cgiar.org/handle/20.500.11766/5107.

Rasmussen, L.V., Coolsaet, B., Martin, A., Mertz, O., Pascual, U., Corbera, E. *et al.* (2018). Social-ecological outcomes of agricultural intensification. *Nature Sustainability* 1, 275–282. https://doi.org/10.1038/s41893-018-0070-8.

Reicosky, D.C. (2015). Conservation tillage is not conservation agriculture. *Journal of Soil and Water Conservation* 70(5), 103A–108A. https://doi.org/10.2489/jswc.70.5.103A.

Rodrigues, S.M., Pereira, M.E., da Silva, F., Hursthouse, A.S. and Duarte, A.C. (2009). A review of regulatory decisions for environmental protection: Part I - challenges in the implementation of national soil policies. *Environment Internantional* 35(1), 202-213. https://doi.org/10.1016/j.envint.2008.08.007.

Rodríguez-Eugenio, N., McLaughlin, M. and Pennock, D. (2018). *Soil Pollution: A Hidden Reality.* Rome: Food and Agriculture Organization of the United Nations. http://www.fao.org/3/I9183EN/i9183en.pdf.

Rojo, L., Bautista, S., Orr, B.J., Vallejo, R., Cortina, J. and Derak, M. (2012). Prevention and restoration actions to combat desertification. An integrated assessment: The PRACTICE Project. *Science et Changements Planetaires - Secheresse* 23(3), 219-226. https://doi.org/10.1684/sec.2012.0351.

Rosegrant, M.W., Koo, J., Cenacchi, N., Ringler, C., Robertson, R., Fisher, M. *et al.* (2014). *Food Security in a World of Natural Resource Scarcity: The Role of Agricultural Technologies.* Washington, D.C.: International Food Policy Research Institute. http://ebrary.ifpri.org/utils/getdownloaditem/collection/p15738coll2/id/128022/filename/128233.pdf/mapsto/pdf/type/singleitem.

Rosni, N.A. and Noor, A.P.D.N.M. (2016). A review of literature on urban sprawl: Assessment of factors and causes. *Journal of Architecture, Planning and Construction Management* 6(1), 12-35. http://journals.iium.edu.my/kaed/index.php/japcm.

Rush, E.C. and Yan, M.R. (2017). Evolution not revolution: Nutrition and obesity. *Nutrients* 9(5), 519. https://doi.org/10.3390/nu9050519.

Sanz, M.J., de Vente, J., Chotte, J.L., Bernoux, M., Kust, G., Ruiz, I. *et al.* (2017). *Sustainable Land Management Contribution to Successful Land-Based Climate Change Adaptation and Mitigation. A Report of the Science-Policy Interface.* Bonn: United Nations Convention to Combat Desertification. https://www.unccd.int/sites/default/files/documents/2017-09/UNCCD_Report_SLM.pdf.

Schonweger, O., Heiniman, A., Epprecht, M., Lu, J. and Thalongsengchanh, P. (2012). *Concessions and Leases in The Lao PDR: Taking Stock of Land Investments.* Bern: University of Bern. https://catalogue.nla.gov.au/Record/6571317.

Schwilch, G., Liniger, H.P. and Hurni, H. (2014). Sustainable land management (SLM) practices in drylands: How do they address desertification threats? *Journal of Environmental Management* 54(5), 983-1004. https://doi.org/10.1007/s00267-013-0071-3.

Secretariat of the Basel, Rotterdam and Stockholm Conventions (2018). *Synergies.* http://www.brsmeas.org/ (Accessed: 6 November 2018).

Serraj, R. and Siddique, K. (2012). Conservation agriculture in dry areas. *Field Crops Research* 132, 1-6. https://doi.org/10.1016/j.fcr.2012.03.002.

Seto, K.C., Güneralpa, B. and Hutyra, L.R. (2012). Global forecasts of urban expansion to 2030 and direct impacts on biodiversity and carbon pools. *Proceedings of the National Academy of Sciences* 109(40), 16083-16088. https://doi.org/10.1073/pnas.1211658109.

Shuai, W., Fu, B., Piao, S., Lü, Y., Ciais, P., Feng, X. *et al.* (2015). Reduced sediment transport in the Yellow River due to anthropogenic changes. *Nature Geoscience* 9, 38-41. https://doi.org/10.1038/NGEO2602.

Sikor, T., Auld, G., Bebbington, A.J., Benjaminsen, T.A., Gentry, B.S., Hunsberger, C. *et al.* (2013). Global land governance: From territory to flow? *Current Opinion in Environmental Sustainability* 5(5), 522-527. https://doi.org/10.1016/j.cosust.2013.06.006.

State Forestry Administration (2011). *A Bulletin of Status Quo of Desertification and Sandification in China.* Beijing. https://www.documentcloud.org/documents/1237947-state-forestry-administration-desertification.html.

Stenmarck, Å., Jensen, C., Quested, T. and Moates, G. (2016). *Estimates of European Food Waste Levels.* Brussles: European Union. https://www.eu-fusions.org/phocadownload/Publications/Estimates%20of%20European%20food%20waste%20levels.pdf.

Sternberg T., Rueff, H. and Middleton, N. (2015). Contraction of the Gobi Desert, 2000–2012. *Remote Sensing* 7(2), 1346–1358. https://doi.org/10.3390/rs70201346.

Sustainable Development Solutions Network (2014). *Indicators for Sustainable Development Goals: A Report by the Leadership Council of the Sustainable Development Solutions Network.* http://unsdsn.org/wp-content/uploads/2014/05/140522-SDSN-Indicator-Report.pdf.

Swella, G.B., Ward, P.R., Siddique, K.H.M. and Flower, K.C. (2015). Combinations of tall standing and horizontal residue affect soil water dynamics in rainfed conservation agriculture systems. *Soil and Tillage Research* 147, 30-38. https://doi.org/10.1016/j.still.2014.11.004.

Tan, M. and Li, X. (2015). Does the Green Great Wall effectively decrease dust storm intensity in China? A study based on NOAA NDVI and weather station data. *Land Use Policy* 43, 42-47. https://doi.org/10.1016/j.landusepol.2014.10.017.

Tavera, C. (2015). *Lao PDR Country Study Report for the Independent Evaluation of the Scale-up Phase (2008-2013) of the UNDP-UNEP Poverty – Environment Initiative and Mid-term Evaluation of the Second Phase (2012–2014) of the Lao PDR PEI Country Programme.* United Nations Development Programme and United Nations Environment Programme. http://www.unpei.org/sites/default/files/dmdocuments/PEI%20Evaluation%20Final%20Report.pdf.

Tavera, C., Alderman, C. and Nordin, N. (2016). *Independent Evaluation of the Scale-up Phase (2008-2013) of the UNDP-UNEP Poverty – Environment Initiative.* http://www.unpei.org/sites/default/files/dmdocuments/PEI%20Evaluation%20Final%20Report.pdf.

Teshome, A., de Graaff, J., Ritsema, C. and Kassie, M. (2015). Farmers' perceptions about the influence of land quality, land fragmentation and tenure systems on sustainable land management in the north western Ethiopian Highlands. *Land degradation & development* 27(4), 884-898. https://doi.org/10.1002/ldr.2298.

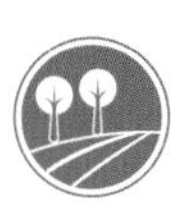

Thavorncharoensap, M. (2017). *Effectiveness of Obesity Prevention and Control*. Tokyo: Asian Development Bank Institute. https://www.adb.org/sites/default/files/publication/226281/adbi-wp654.pdf.

Tobias, S., Conen, F., Duss, A., Wenzel, L.M., Buser, C. and Alewell, C. (2018). Soil sealing and unsealing: State of the art and examples. *Land degradation & development* 29(6), 2015-2024. https://doi.org/10.1002/ldr.2919.

Tóth, G., Hermann, T., Szatmári, G. and Pásztor, L. (2016). Maps of heavy metals in the soils of the European Union and proposed priority areas for detailed assessment. *Science of Total Environment* 565, 1054-1062. https://doi.org/10.1016/j.scitotenv.2016.05.115.

Trigo, E., Cap, E., Malach, V. and Villarreal, F. (2009). *The Case of Zero-Tillage Technology in Argentina*. IFPRI Discussion Paper. Washington, D.C.: International Food Policy Research Institute. http://www.ifpri.org/cdmref/p15738coll2/id/29503/filename/29504.pdf.

United Nations (2012). *World Urbanization Prospects: The 2011 Revision*. New York, NY. http://www.un.org/en/development/desa/population/publications/pdf/urbanization/WUP2011_Report.pdf.

United Nations (2015). *17 goals to transform our world*. https://www.un.org/sustainabledevelopment/development-agenda/ (Accessed: 6 November 2018).

United Nations (2016a). *The World's Cities in 2016 – Data Booklet (ST/ESA/ SER.A/392)*. New York, NY. http://www.un.org/en/development/desa/population/publications/pdf/urbanization/the_worlds_cities_in_2016_data_booklet.pdf.

United Nations (2016b). *Sustainable development goal 11: Make cities and human settlements inclusive, safe, resilient and sustainable: Progress and Info (2016)*. https://sustainabledevelopment.un.org/sdg11 (Accessed: 6 November 2018 2018).

United Nations (2018). *Metadata & Reference*. http://data.un.org/DataMartInfo.aspx (Accessed: 6 November).

United Nations Convention to Combat Desertification (2017a). *Global Land Outlook*. Bonn. https://www.unccd.int/sites/default/files/documents/2017-09/GLO_Full_Report_low_res.pdf.

United Nations Convention to Combat Desertification (2017b). *Proportion of land that is degraded over total land area-indicator 15.3.1*. https://knowledge.unccd.int/publications/proportion-land-degraded-over-total-land-area-indicator-1531 (Accessed: 6 November 2018).

United Nations Development Programme (2009). *Environmental remediation of dioxin contaminated hotspots in Viet Nam*. http://www.vn.undp.org/content/vietnam/en/home/operations/projects/closed-projects/environment_climate/Environmental-Remediation-of-Dioxin-Contaminated-Hotspots-in-Vietnam.html (Accessed: 6 November 2018).

United Nations Development Programme (2014). *Sustainable Forest and Land Management in the Dry Dipterocarp Forest Ecosystems of Southern Lao PDR*. New York, NY. https://www.thegef.org/project/sustainable-forest-and-land-management-dry-dipterocarp-forest-ecosystems-southern-lao-pdr.

United Nations Development Programme (2016). *Managing investment through a poverty and environment lens*. http://www.la.undp.org/content/lao_pdr/en/home/presscenter/pressreleases/2016/04/27/managing-investment-through-a-poverty-and-environment-lens/ (Accessed: 6 November 2018).

United Nations Development Programme and United Nations Environment Programme (2013). *Stories of Change from the Joint UNDP-UNEP Poverty-Environment Initiative*. Nairobi. https://europa.eu/capacity4dev/file/14433/download?token=4l8JgbBb.

United Nations Development Programme and United Nations Environment Programme (2018). *Lao PDR*. http://www.unpei.org/what-we-do/pei-countries/lao-pdr (Accessed: 6 November 2018).

United Nations Environment Programme, World Meteorological Organization and United Nations Convention to Combat Desertification (2016). *Global Assessment of Sand and Dust Storms*. Shepherd, G. (ed.). http://wedocs.unep.org/bitstream/handle/20.500.11822/7681/Global_Assessment_of_sand_and_dust_storms_2016.pdf?sequence=1&isAllowed=y.

United Nations Environment World Conservation Monitoring Centre (2018). *World Database on Protected Areas*. https://www.protectedplanet.net/c/world-database-on-protected-areas (Accessed: 6 November 2018).

United Nations Human Settlements Programme (2015). *11.3 Sustainable urbanization*. https://unhabitat.org/un-habitat-for-the-sustainable-development-goals/11-3-sustainable-urbanization/ (Accessed: 2018 6 November).

United Nations Human Settlements Programme (2017). *UN Habitat Global Activities Report 2017. Strengthening Partnerships in Support of the New Urban Agenda and the Sustainable Development Goals*. Nairobi. https://unhabitat.org/wp-content/uploads/2017/02/GAR2017-FINAL_web.pdf.

United Nations Industrial Development Organization (2012). *Introduction of BAT/BEP Methodology to Demonstrate Reduction or Elimination of Unintentionally Produced Persistent Organic Pollutants (UPOPs) Releases from the Industry in Vietnam*. Vienna. https://vncpc.org/en/project/ap-dung-batbep-trong-giam-phat-thai-upop-2/.

United States Agency for International Development (2010). *Environmental Remediation at Da Nang Airport Environmental Assessment in Compliance with 22 CFR 216*. http://www.agentorangerecord.com/images/uploads/modules/EA%20DNG.pdf.

United States Environmental Protection Agency (2014). *Protection & Restoring Land, Making a visible difference in communities, OSWER FY14 End of Year Accomplishments Report*. https://www.epa.gov/sites/production/files/2014-03/documents/oswer_fy13_accomplishment.pdf.

University of the West of England Bristol, Science Communication Unit (2013). *Science for Environment Policy In-Depth Report: Soil Contamination: Impacts on Human Health*. European Commission. http://ec.europa.eu/environment/integration/research/newsalert/pdf/IR5_en.pdf.

Vastola, A., Zdruli, P., D'Amico, M., Pappalardo, G., Viccaro, M., Di Napoli, F. *et al.* (2017). A comparative multidimensional evaluation of conservation agriculture systems: A case study from a Mediterranean area of Southern Italy. *Land Use Policy* 68, 326–333. https://doi.org/10.1016/j.landusepol.2017.07.034.

Wallheimer, B. (2010). *No-till, rotation can limit greenhouse gas emissions from farm fields*. Purdue University https://www.purdue.edu/newsroom/research/2010/101220VynNitrous.html (Accessed: 6 November 2018).

Wang, X.M., Zhang, C.X., Hasi, E. and Dong, Z.B. (2010). Has the three Norths Forest Shelterbelt Program solved the desertification and dust storm problems in arid and semiarid China? *Journal of Arid Environments* 74(1), 13-22. https://doi.org/10.1016/j.jaridenv.2009.08.001.

Weber, M., Driessen, P.P. and Runhaar, H.A. (2014). Evaluating environmental policy instruments mixes: A methodology illustrated by noise policy in the Netherlands. *Journal of Environmental Planning and Management* 57(9), 1381-1397. https://doi.org/10.1080/09640568.2013.808609.

Wellmann, D. (2012). *The Legal Framework of State Land Leases and Concessions in the Lao PDR*. Integrated Rural Development in Poverty Regions of Laos. http://www.laolandissues.org/wp-content/uploads/2012/03/Legal-Framework-of-Concessions-in-the-Lao-PDR-Discussion-paper-GIZ-Wellmann.pdf.

World Bank (2006). *Sustainable Land Management: Challenges, Opportunities, and Trade-offs*. Washington, D.C. https://openknowledge.worldbank.org/bitstream/handle/10986/7132/366540PAPER0Su11PUBLIC0as0of0July71.pdf?sequence=1&isAllowed=y.

World Bank (2010). *Lao PDR Development Report 2010: Natural Resource Management for Sustainable Development: Hydropower and Mining*. Washington, D.C. http://siteresources.worldbank.org/LAOPRDEXTN/Resources/293683-1301084874098/LDR2010_Full_Report.pdf.

World Bank (2017). *World Development Indicators*. Washington, D.C. http://databank.worldbank.org/data/reports.aspx?source=2&type=metadata&series=ER.LND.PTLD.ZS

World Bank (2018). *Atlas of Sustainable Development Goals: World Development Indicators*. Washington, D.C. https://openknowledge.worldbank.org/bitstream/handle/10986/29788/9781464812507.pdf?sequence=5&isAllowed=y.

World Health Organization (2015). *Global health observatory data repository*. http://apps.who.int/gho/data/node.main.%20A897A?lang=en (Accessed: 6 November 2018).

World Overview of Conservation Approaches and Technologies (2016). *World overview of conservation approaches and technologies*. https://www.wocat.net/en/ (Accessed: 6 November 2018).

Xiaoming, F., Fu, B., Piao, S., Wang, S., Ciais, P., Zeng, Z. *et al.* (2016). Revegetation in China's Loess Plateau is approaching sustainable water resource limits. *Nature Climate Change* 6, 1019-1022. https://doi.org/10.1038/NCLIMATE3092.

Xinhua (2018). *China focus: China adopts new law on soil pollution prevention*. http://www.xinhuanet.com/english/2018-09/01/c_137434559.htm (Accessed: 6 November 2018).

Xu, D., Song, A. and Song, X. (2017). Assessing the effect of desertification controlling projects and policies in northern Shaanxi Province, China by integrating remote sensing and farmer investigation data. *Frontiers of Earth Science* 11(4), 689-701. https://doi.org/10.1007/s11707-016-0601-4.

Yin, R. and Yin, G. (2010). China's primary programs of terrestrial ecosystem restoration: Initiation, implementation, and challenges. *Environmental management* 45(3), 429-441. https://doi.org/10.1007/s00267-009-9373-x.

Zdruli, P. (2014). Land resources of the Mediterranean: Status, pressures, trends and impacts on future regional development. *Land Degradation and Development* 25(4), 373-384. https://doi.org/10.1002/ldr.2150

Zdruli, P. and Zucca, C. (2018). Maintaining soil health in dryland areas. In *Managing Soil Health for Sustainable Agriculture*. Cambridge Burleigh and Dods Science Publishing. https://shop.bdspublishing.com/checkout/Store/bds/Detail/WorkGroup/3-190-56261

Chapter 16

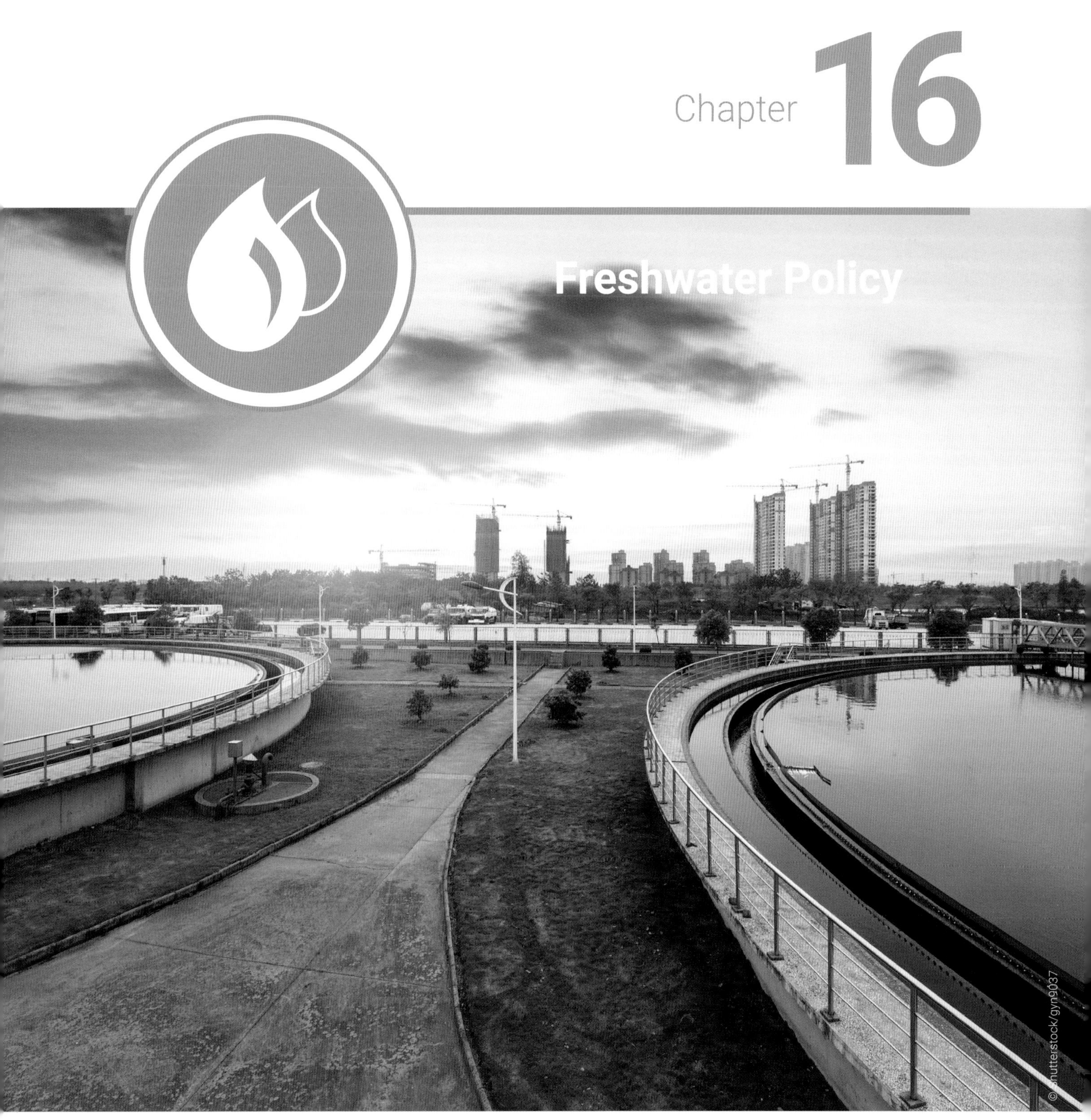

Freshwater Policy

Coordinating Lead Author: Peter King (Institute for Global Environmental Strategies)

Lead Authors: Erica Gaddis (Utah Department of Environmental Quality), James Grellier (European Centre for Environment and Human Health, University of Exeter), Anna Maria Grobicki (Food and Agriculture Organization of the United Nations [FAO]), Rowena Hay (Umvoto), Naho Mirumachi (King's College London), Gavin Mudd (RMIT University), Farhad Mukhtarov (Erasmus University Rotterdam), Walter Rast (Meadows Center for Water and the Environment, Texas State University),

GEO Fellows: Beatriz Rodríguez-Labajos (Universitat Autònoma de Barcelona), Jaee Sanjay Nikam (Arizona State University), Patricia Nayna Schwerdtle (Monash University)

Executive summary

Addressing drivers and pressures is the key to making effective freshwater policy. This can be achieved through regulatory command-and-control mechanisms, subsidies, supporting investments and enabling actors, but there is also value in process-based innovative approaches such as experimentation, learning and voluntary reporting. {16.2.1, 16.2.4}

Policy coherence and synergy are needed to address the water-food-energy-health-ecosystems nexus. Policy mixes are typically adopted to meet demands across multiple sectors and to manage implications outside the freshwater policy sphere. Intricate linkages among water quality and quantity, agriculture, human health, ecosystems and energy systems require that freshwater policy is developed with this nexus placed centre-stage. Achieving policy coherence and synergy are important benefits of this integrated thinking, as water policies influence policies in other sectors, especially agriculture and energy. {16.2.1, 16.2.2}

Much freshwater policy is highly context dependent, yet a variety of freshwater policy types and governance approaches can diffuse to fit diverse local contexts. Governance approaches and policy types are diverse. The design, implementation and evaluation of these policies require that institutional structures, economic resources and other enabling factors are in place. {16.1, 16.2.3, 16.2.5}

There is scope for freshwater policy to better consider co-benefits to ecosystems and human health. Changes to water quality and quantity through interventions such as infrastructure investment and natural hazards requires consideration of direct threats to human health but capitalizing on potential co-benefits is not yet widely practised. {16.1, 16.2.2, 16.3}

Policy effectiveness draws attention to the role of citizens, the private sector and non-governmental bodies, in particular through participatory processes. Implementing integrated water resources management (IWRM) is a participatory process, based upon intersectoral coordination and greater engagement of non-governmental actors. Collaborative efforts are required to involve the private sector and non-governmental organizations, or local governments and citizens. Stakeholder engagement is a long-term process and requires investment in supporting stakeholder relationships. Institutions should be designed to enable inputs into decision-making from these relationships rather than treating them on an ad hoc basis. Devolution of water governance requires supporting investments, capacity-building and sustained long-term efforts in raising awareness. Exchanging knowledge at the subnational level enables effective stakeholder involvement. {16.1, 16.2.1, 16.2.2, 16.2.5}

Evaluating policy effectiveness is enhanced by consistent and transparent reporting and systematic monitoring. For policy effectiveness, defining baseline conditions is needed prior to implementation for comparison and lesson learning. Standardization of sustainability reporting, development of national reporting mechanisms, and the use of knowledge hubs for scientific reporting have proven useful {Sections 16.2}. Reporting and monitoring helps tracking of Sustainable Development Goal (SDG) progress at both national and global levels and helps identify causal relations of specific policy interventions. {16.3.1, 16.3.2}

While policy approaches become further integrated and complex, there is an ongoing need to address basic environmental clean-up and the reversal of damaging legacies. Even in developed economies, regulation, technical fixes and investments are required to continually improve practices of water use and prevent water quality degradation {16.1}. Policies may need to be revised to change the direction of trends in water use. {16.6.1}

Environmental and freshwater policies can be effectively driven by the consideration of social issues, especially equity and health. Disparities within a country or between developed and developing countries can motivate national as well as global efforts for addressing access to water and sanitation services, underpinning the human right to water and sanitation. {16.2.3, 16.2.5}

Transformative potential can be seen in effective and innovative freshwater policies that benefit both people and planet. The environmental flows approach carries transformative potential, as it is a way of assessing quantitatively the water needs of the river as a living system, and of balancing these water requirements against the water requirements of various economic sectors. As more rivers are assessed in this way, environmental flows become a fundamental building block of river basin management and governance, leading to the integration of management of water and landscapes through the entire catchment. {16.2.1}

16.1 Introduction

The Millennium Development Goals (MDGs) motivated countries to tackle issues relating to sustainable access to safe drinking water and basic sanitation. The Sustainable Development Goals (SDGs) now present an even more ambitious global framework within which the multidimensional concerns of availability, quality, use and governance of water can be addressed. Chapter 9 in Part A identifies a broad set of policies that have been used across the world to address specific aspects of targets defined in SDG 6. These include generic policy approaches such as market instruments, regulatory programmes, monitoring, capacity-building, as well as water specific interventions such as desalination and conjunctive use of surface and groundwater.

Policies highlighted in Section 9.9 demonstrate the increasing attention given to larger spatial scales, including considerations beyond the scale of the river basin, as exemplified in the case of virtual water trading, and the incorporation of multiple institutional scales not confined to the national level. Accordingly, this analysis of freshwater policy effectiveness begins by focusing on multiple water uses within policies and the multi-sectoral considerations of given policy approaches and highlights the strengths and weaknesses of policies addressing the nexus that connects water, food, energy, climate, ecosystems and health.

16.2 Key policies and governance approaches

There is traction in policy communities to address the water-food-energy nexus so that freshwater policy approaches can be sensitive to the ways in which the hydrological cycle, ecosystems, food and energy systems are connected. Efforts to meet this need are, however, relatively new and must tackle the challenge of engaging multiple spatial, temporal and governance scales. Consideration of equity aspects, so that injustices in procedure and outcomes are averted, plays an important part in the discussion of effectiveness of policies addressing this nexus. Through case studies and an analysis of indicators related to the SDGs, this chapter shows how the nexus concept matters in relation to policy effectiveness and cost-effectiveness. The case studies collectively demonstrate national and transboundary water policies from around the world that have had moderate successes and implementation challenges in dealing with the following:

- Empowerment of local water managers while maintaining consistent protections across countries and transboundary water basins;
- Design and operation of dams to minimize impacts on ecosystems while providing benefits to human health, agriculture and energy as well as considering environmental flows and the use of adaptive management;
- Reform of flood risk management policy in line with integrated water resources management (IWRM) with greater responsibilities given to local authorities;
- Provision of basic water services to poor communities in water-scarce regions; and
- Improvement of the consistency and transparency of sustainability reporting conducted by the private sector.

In addition, three policy-relevant indicators on access to water and sanitation and on water withdrawals are reviewed. These indicators represent another way in which global water policy can be assessed.

Collectively, the case studies and indicators demonstrate the mix of policy instruments and clusters that have evolved to manage nexus concerns in an integrated way, which represents a shift from decision-making by a singular governmental authority to governance through sets of rules, principles and procedures involving various stakeholders.

Policy approaches and case studies addressed in this chapter **(Table 16.1)** are linked to the policy typology of Chapter 10.

16.2.1 Regulatory frameworks for transboundary water quality management

Transboundary water bodies are shared by two or more States. The management of these shared rivers, lakes or aquifers relies on multilateral coordination and institutional development. International agreements between States are formal arrangements for transboundary water governance.

Table 16.1: Policy approaches and case studies

Governance approach	Policy instrument(s)	Case study
Command and control; enabling actors; supporting investments	Water quality goals coordinated through a binational transboundary agreement	North American Great Lakes Water Quality Agreement
Enabling actors; command and control	Environmental flow	Adaptive management of Glen Canyon Dam
Economic incentives; command and control	Collaborative institutional design	Disaster Risk Reduction Flood Risk Management, United Kingdom of Great Britain and Northern Ireland – 'Making Space for Water' and Flood Risk Management Policy
Command and control; economic incentives; supporting investments	Water pricing and free provision of basic water supply	Free basic water policy, South Africa
Promotion of innovation; enabling actors; convincing consumers, employers and stockholders	Standardization of sustainability reporting	Mining – sustainable water

In particular for transboundary rivers, agreements have become more comprehensive and more numerous over time, reflecting an integrated approach to managing shared rivers and lakes (Giordano *et al.* 2014). By 2007, there were 250 freshwater treaties and 30 additional treaties are established every decade (Giordano *et al.* 2014), mostly focusing on water quality and the environment. However, obligations, responsibilities and enforcement mechanisms to address water quality have typically been left undefined (Giordano *et al.* 2014). Regulatory agreements have tended to exclude direct data- and information-exchange mechanisms (Gerlak, Lautze and Giordano 2011). These trends combined indicate that while water quality may be regarded as important, specific policy interventions have been hard to establish.

River basin organizations (RBOs) for transboundary water bodies can be vehicles of treaty implementation. In general, the main functions of RBOs are (i) data gathering, monitoring and regulation; (ii) river basin planning; and (iii) development of infrastructure and facilities (Global Water Partnership [GWP] 2017). Many RBOs are guided by IWRM principles that seek to achieve efficiency, equity and ecological sustainability, while also addressing water quality and quantity issues. Within the Drivers, Pressures, State, Impact, Response (DPSIR) framework (Section 1.6), this institutional approach is aimed at identifying pressures that cause water quality degradation, reasonable and equitable water use and ecosystem concerns.

The success of agreements and frameworks requires scrutiny as institutional development does not guarantee water quality improvement and prevent free-riding (Bernauer and Kuhn 2010), and effectiveness of cooperation can be questioned (Mirumachi 2015).

Case study: The North American Great Lakes Water Quality Agreement

In response to pollution of the Great Lakes Basin (Thornton *et al.* 1999), the United States of America and Canada, under the umbrella of the Boundary Waters Treaty (*Boundary Waters Treaty* 1909), signed the Great Lakes Water Quality Agreement in 1972 (*Great Lakes Water Quality Agreement* 2012).

With a population of more than 30 million people (United States Environmental Protection Agency [US EPA] 2017), the Great Lakes Basin receives substantial inputs of point source and non-point source pollution from a large range of industrial, agricultural, forestry and urban sources (Marvin, Painter and Rossmann 2004). Pollutants of particular concern in terms of impact on ecosystems and human health include biomagnifying metals such as mercury and persistent organic pollutants (POPs), including polychlorinated biphenyls (PCBs), polycyclic aromatic hydrocarbons (PAHs), polychlorinated naphthalene (PCNs), organochlorine pesticides (OCPs), polybrominated diphenylethers (PBDEs) and perfluorinated chemicals (PFCs) (Helm *et al.* 2011). Another danger to the ecosystem comes from invasive species and harmful algal blooms and eutrophication (Smith *et al.* 2015).

The current agreement comprises annexes addressing a range of Great Lakes water quality issues, including areas of concern, lake-wide management, pollution control, ecosystem maintenance and climate change impacts. It encompasses a range of policy clusters involving federal, state and local institutions, facilitation of cooperative actions (both regulated and voluntary), with each country contributing actions from their domestic programmes, policies and resources.

The International Joint Commission (IJC) is a permanent, binational institution for dispute resolution. Under the Treaty from 1909, the IJC was given powers to apply governing principles for water use and the arbitrational power to resolve disputes (Krantzberg and De Boer 2008). Additionally, the federal governments of the two countries periodically request IJC to investigate specific boundary water issues (Findlay and Telford 2006; McLaughlin and Krantzberg 2012). Accordingly, the IJC conducts semi-annual meetings under the Boundary Waters Treaty, with the scope of these meetings covering a full range of boundary issues across the Canada-United States of America boundary (http://www.ijc.org/en_/meetings_minutes). Under the Great Lakes Water Quality Agreement, the IJC was given a reference unique to the Great Lakes – namely to provide advice and recommendations to government, and to report on progress in implementing the agreement. To this end, the Parties to the Great Lakes Water Quality Agreement (i.e. national governments) conduct semi-annual meetings specific to the implementation of the agreement (IJC 1980; IJC 1981; IJC 2001; IJC 2017).

Pursuant to the 1987 Protocol in the Great Lakes Water Quality Agreement, 43 'areas of concern' were identified. These areas were found to exhibit severely degraded water quality and ecosystem health (12 in Canada, 26 in the United States of America and 5 shared). The environmental degradation is primarily a legacy of the past, attributable to industrial activities, agriculture, urban and rural run-off, municipal wastewater effluents, land-use planning and practices on urban and rural lands, all contributing to degraded water quality, contaminated river and lake sediments, and severely impacted fish and wildlife populations and habitats. **Table 16.2** presents our evaluation of the effectiveness of the Great Lakes Water quality Agreement.

Table 16.2: Evaluation of the effectiveness of the Great Lakes Water Quality Agreement

Criterion	Description	References
Success or failure	A total of seven areas of concern have been delisted (three in Canada; four in the United States of America). There are others considered areas of recovery, where actions have been completed and these areas are expected to be delisted soon.	(US EPA 2017)
Independence of evaluation	Progress is typically reported by the Parties and assessed by the International Joint Commission (IJC) on the basis of input from two major advisory boards (Great Lakes Water Quality Board and Science Advisory Board). The Water Quality Board provides policy advice and evaluation, and the Science Advisory Board provides scientific advice and evaluation. The IJC also publishes a triennial assessment report that reviews the progress of the Parties, summarizes public input on the Parties' progress report, and includes an assessment of the degree to which programmes are achieving the agreement's general and specific objectives.	United States National Research Council (1985); IJC (2017)
Key actors	The key actors are the governments of the United States of America and Canada, in collaboration with other jurisdictions that support implementation of the agreement. A key role of the IJC, in assisting the governments, is its assessment role.	IJC (2017)
Baseline	The Parties have adopted nine General Objectives under the agreement that outline high-level ecosystem objectives towards which they are working. The Parties have also established a suite of nine indicators of ecosystem health, supported by 44 sub-indicators, to assess the state of the Great Lakes, and whether or not progress is being made towards achieving the General Objectives.	IJC (2017)
Time frame	The agreement became effective with its adoption by both governments in 1972, with the most recent amendment in 2012. It has provided the binational framework for both countries to work towards the restoration and protection of the Great Lakes for over 45 years.	US EPA (2017)
Constraining factors	Although public scrutiny of the progress of the Parties provides a powerful oversight role (e.g. through binational public webinars on substantive issues), there have been calls for a more inclusive discourse and an increased role for citizen engagement, particularly during renegotiation periods of the agreement. A shift from an ad hoc problem-resolution mindset to a more imaginative and strategic thinking approach has also been advocated.	Krantzberg (2012)
Enabling factors	Canada and the United States of America have the capabilities to meet the substantial policy, institutional, technical, financial and personnel obligations inherent in carrying out the objectives of the agreement. They have an ongoing system of plans for remedial action in the areas of concern and target an adaptive system of experimentation and learning in pursuing remedial work in regard to addressing the General Objectives of the agreement.	Hall, O'Connor and Ranieri (2006)
Cost-effectiveness	Both countries depend on lake-derived ecosystem services that amount to US$7 billion annually in economic activity related to recreational and commercial fishing alone. The basin-wide approach resulted in doubling the habitat of fish species at a cost of US$70 million, whereas a less integrated approach of addressing only dams or only road crossings resulted in striking inefficiencies of 24 per cent and 88 per cent less habitat, respectively.	Southwick Associates (2008); Neeson *et al.* (2015)
Equity	The governance of the Great Lakes includes approximately 120 Native American, First Nation and Metis rights holders, as well as low-income and minority people, which provides opportunities for collective management. Traditional knowledge and input from the community of First Nations in Swan Lake Marsh, Canada, was used for a project on wetland restoration and to plan guided activities in the past few years.	Hildebrand, Pebbles and Fraser (2002); Jetoo (2017)
Co-benefits	The agreement, in enhancing the quality of the water, created wealth in several forms, including increased recreational and commercial fishing (see 'Cost-effectiveness'), as well as increased recreational use and tourism. The main beneficiaries were riparian inhabitants living around the lakes and their basins, water sports and fishing enthusiasts, tourists and visitors.	
Transboundary issues	Great Lakes water quality degradation, unless limited to restricted bays or similar settings, typically has transboundary implications; hence, the development of the agreement in the first place. The agreement fostered the creation of additional transboundary governance initiatives in the area, such as the Cities Initiative (see 'Possible improvements').	Jetoo (2017)
Possible improvements	In 2003, the Great Lakes and St. Lawrence Cities Initiative (the Cities Initiative) formalized a network of over 130 cities that participate in measures for Great Lakes restoration and protection. The Cities Initiative relies on government funding, associate fees and private foundation funding. Therefore, access to supranational funding would be an improvement. IJC may also pay more attention to indicators of progress, so that all those engaged in protecting and restoring the resource are up to date on the progress achieved.	Jetoo (2017)

The degree to which the 43 originally identified 'areas of concern' (AOC) have been addressed over time can indicate policy progress (**Figure 16.1**). The removal of 7 of the original AOC indicates a degree of success, although the 36 remaining areas highlight the difficulty of such remedial actions. As biophysical change may take decades to achieve in an AOC, a more immediate measure of policy progress is the adoption and implementation of policies to protect the Great Lakes in the areas in which specific remediation measures are undertaken.

The governments of Canada and the United States of America report on progress achieved under the agreement every three years through the 'Progress Report of the Parties', as well as through other means. The enabling factors for cooperation include the appreciation of the vast range of ecosystem services provided by the Great Lakes to the two countries. Furthermore, the two countries share similar visions, expectations and reliance on the Great Lakes. Finally, compared to many other areas in the world, they have the capabilities to meet the substantial policy, institutional, technical, financial and personnel obligations inherent in carrying out the objectives of the agreement.

The costs of the extensive monitoring, analysis and remediation of Great Lakes water quality issues are substantial. For instance, the Great Lakes Restoration Initiative, launched in 2010 to accelerate efforts to protect and restore the Great Lakes, has provided funds for projects amounting to more than US$2.3 billion for the future functioning of the initiative (US EPA 2017). The efforts to enhance habitat for the fisheries include the removal of hundreds of small dams and culverts that partially or fully impede fish movement for spawning (Kemp and O'Hanley 2010). While economically a relatively minor activity, dam and culvert removal have an important impact on the aquatic life of the lakes and contribute to recreational value to the lake-side population. On the other hand, the removal of dams and culverts allows for spawning of some of the most aggressive invasive species in the lake, a side effect of the policy that requires a scientific assessment. The optimization models used by Neeson *et al.* (2015) indicate that the most cost-effective way of managing the Great Lakes restoration is when dams and road crossings are removed across the whole basin. The basin-wide approach results in doubling the habitat of fish species at a cost of US$70 million, whereas a less integrated approach of addressing only dams or only road

Figure 16.1: Map showing location and status of all United States of America and Canadian Great Lakes Areas of Concern

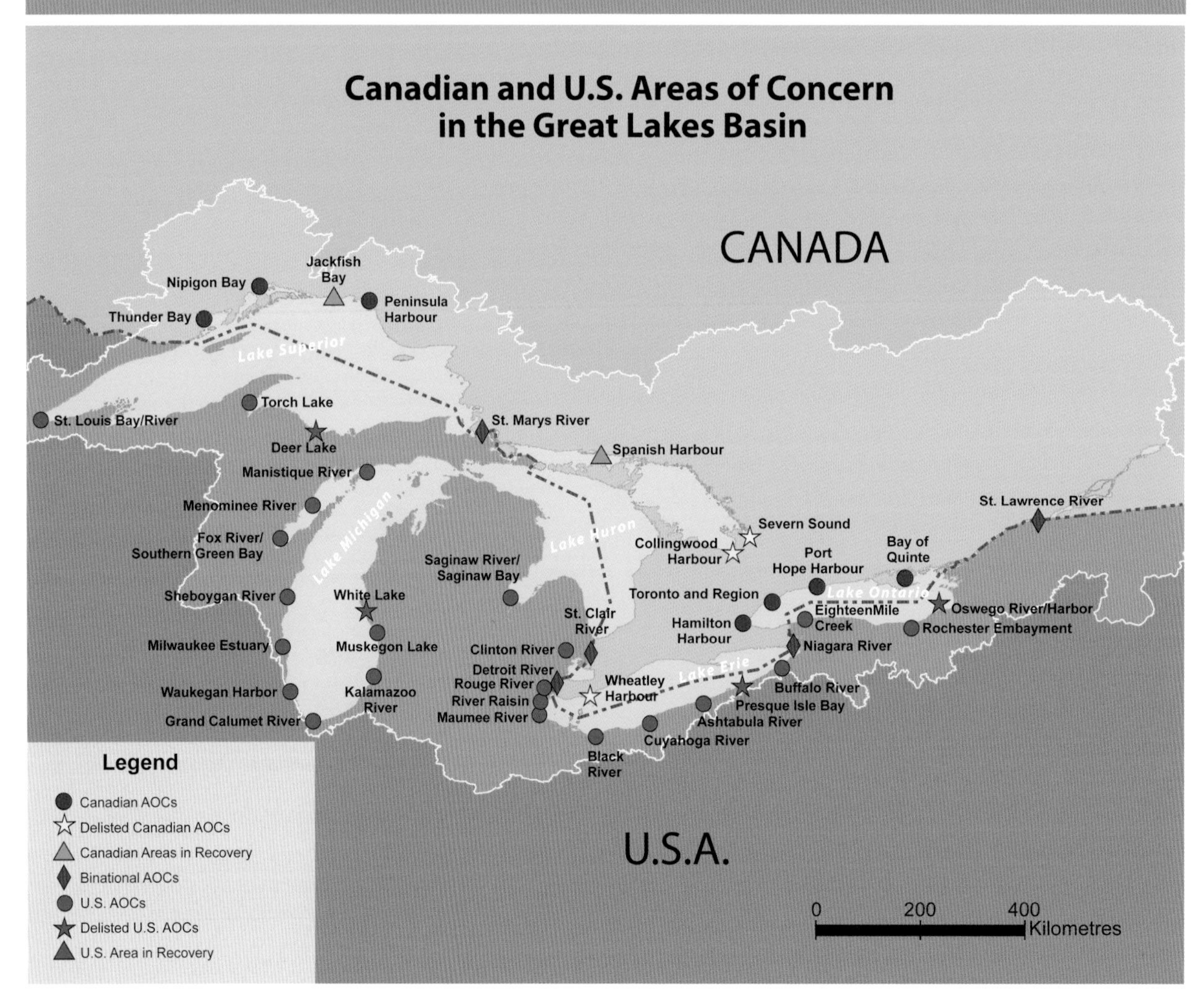

Source: Binational.net 2018.

crossings results in striking inefficiencies of 24 per cent and 88 per cent less habitat, respectively. These results provide a cost-effectiveness argument for the ecosystem-wide approach taken by the IJC.

16.2.2 Adaptive management of environmental flows in the water and energy sectors

Concerns arising in response to degradation of river ecosystems due to diversion and impoundment of water have led to the widespread recognition of the importance of environmental flows (Poff *et al.* 1997; Arthington *et al.* 2006; World Bank 2018). They are defined as the "quantity, timing and quality of water flows required to sustain freshwater and estuarine ecosystems and the human livelihoods and wellbeing that depend on these ecosystems" (International River Foundation 2007). As a 'master variable' for the sustainability of aquatic ecosystems, environmental flows can be incorporated into national-level legislation on water resources management as well as river basin planning (Poff *et al.* 1997; Speed *et al.* 2013). For example, the South African National Water Act (1998) requires water reserves to maintain river health as well as basic human needs. The environmental flow concept is particularly useful in considering the nexus between environmental development and human demands.

One way to influence and secure environmental flows is to adjust the timings and volumes of water released from dams in an adaptive manner. This approach attempts to influence the water and energy (i.e. hydropower) nexus, as well as the water-food nexus in cases where irrigation is required for agricultural production in water-scarce areas. Adaptive management utilizes experimental data derived from large-scale flow-release experiments designed to test hypotheses on physical and biological responses to streamflow in rivers, floodplains or estuaries (Konrad *et al.* 2011, p. 949). High flow release experiments are complex interventions and affect a range of factors beyond flow variability, and can result in more efficient attainment of a wide range of ecological, social and economic benefits (Olden *et al.* 2014, p. 179).

Adaptive management is considered to provide more flexibility than traditional management approaches because it has means available to account for and test uncertainties. Adaptive management focusing on environmental flows can simulate how the natural hydrological regime affects the sediment, water and habitat regimes downstream and can be modified over time as new information becomes available (Richter *et al.* 2006, p. 299). However, adaptive management is often constrained by complex institutional settings and lack of financing (Kingsford, Biggs and Pollard 2011; Allan and Watts 2017). In addition, environmental flow experiments have been constrained so far to large dams in the United States of America, Australia and South Africa, with little reporting from other regions such as South-East Asia, South America and parts of Europe where a significant number of dams exist or are being planned (Olden *et al.* 2014, p. 178). While adaptive management would enable procedural justice to be built into the process through instruments such as public participation, there are equity and ethical concerns because experiments differentiate groups within society (Huitema *et al.* 2009).

Large-scale flow experiments are not without contention and their success or failure is contested based on stakeholder perspectives (Olden *et al.* 2014, p. 177). The complexity and uncertainty of using flow experiments to inform adaptive management require a process for reflexive learning and incremental understanding (Sabatier *et al.* 2005). Active sharing of knowledge and collection of a diverse range of evidence regarding learning could further support the effectiveness of using environmental flows in adaptive management (Allan and Watts 2017). This feature of the policy approach addresses how responses can be influenced within the DPSIR framework (Section 1.6).

The following case study on the Colorado River below the Glen Canyon Dam in the United States of America highlights an example of a long-term commitment to experimentation and informed adaptive management used for the benefit of a spatially large area beyond the immediate dam catchment and for national conservation areas.

Case study: flow experiments and adaptive management of Glen Canyon Dam on the Colorado River, United States of America

Constructed in 1963, the Glen Canyon Dam impounds 300 kilometres of the Colorado River just upstream of Grand Canyon National Park, creating Lake Powell. The Colorado River carries a heavy sediment load that is integral to the habitat and ecology of the system. The dam had the effect of regulating river flow so that moderate flows are more frequent with less variance between high and low flows (Melis 2011, p. 8). Adaptive management was introduced as the negative impacts on aquatic and terrestrial species from the modified flow were observed; impacts such as riparian habitat loss and fish species endangerment (Collier, Web and Andrews 1997). Dam operating strategies began to take environmental flows into account with the Record of Decision of 1996 by the Secretary of the Interior setting up a flow experiment. The scheduled release of water from the dam aimed to artificially recreate conditions similar to pre-dam seasonal flows. The flow experiment addressed the water-energy nexus to "find an alternative dam operating plan that would permit recovery and long-term sustainability of downstream resources while limiting hydropower capability and flexibility only to the extent necessary to achieve recovery and long-term sustainability" (United States Department of the Interior [US DOI] 1996, p. G-11).

© Shutterstock/Alex Papp

Table 16.3: Evaluation of the effectiveness of adaptive management of the Glen Canyon Dam

Criterion	Description	References
Success or failure	The first experiment is considered to have been successful, paving the way for further experiments. The 2016 Record of Decision is a concrete example of an outcome that demonstrates the incremental nature of increased understanding of environmental flows.	US DOI (1996, p. G-11)
Independence of evaluation	The experiments and adaptive management approach have been evaluated by the United States Department of the Interior and United States Geological Survey, as well as extensively in peer-reviewed scientific literature.	Collier, Webb and Andrews (1997, p. 83); Webb *et al.* (1999); Meretsky, Patten and Stevens (2000); Hazel *et al.* (2006); US DOI (2008); Korman, Kaplinski and Melis (2010); Melis (2011); US DOI (2016)
Key actors	The Bureau of Reclamation and the National Park Services lead on setting out the adaptive management plan. These bodies engage with 15 stakeholder groups including other government agencies, river commissions, energy users and native tribes. The USGS Grand Canyon Monitoring and Research Centre plays a particularly important role in providing technical advice to government agencies and facilitates information exchange between these actors as well as civil society organizations.	US DOI (2016)
Baseline	The experiments were designed to mimic pre-dam conditions. Natural floods occur at a higher frequency and scale, during which the mean velocity of the river is estimated to be five times greater than base flows	Melis (2011, p. 7)
Time frame	Multiple experiments have occurred over the span of two decades and the *Glen Canyon Dam Long-term Experimental and Management Plan* (US DOI 2016) is designed to inform the next 20 years of dam operation.	US DOI (2016)
Constraining factors	An extensive list of laws, regulations and treaties constrains the alternatives for operation of Glen Canyon Dam and consideration must be given to a range of factors relating to the environment, cultural resources, tribal consultation, power marketing, and water allocation and delivery.	
Enabling factors	Multiple pieces of legislation work in tandem, in other words policy coherence, across different scales and sectors, has been ensured as the legislation has evolved.	e.g. United States Congress 1973; US DOI (1992); US DOI (2018)
Cost-effectiveness	The adaptive management decision of the Record of Decision 2016 used a comparison of seven options of dam operation and flow levels to assess costs and impacts. This enabled the finding that the net present value of adaptive management interventions compares favourably with the net present value of no action (status quo).	US DOI (2016)
Equity	Consultation with stakeholders in the form of public participation has sought to identify different ways that various stakeholders engage in processes. However, it has been pointed out that some native tribes have experienced challenges in expressing their cultural values, which do not fit the mould of scientific inquiry and assessments, highlighting some issues of capacity to engage in public participation processes, as well as the confrontation of scientific knowledge with traditional knowledge.	Austin and Drye (2011)
Co-benefits	The experiments have informed a set of co-benefits or 'resources goals' from the project site to downstream areas, ranging from cultural resources to recreational experience. Effects on human health from the experiments are not considered to be substantial (either as a benefit or harm), although it has been suggested that negative effects to health through degradation of water quality would be one criterion for terminating the experiments.	Valdez *et al.* (2000); Melis (2011); US DOI (2016)
Transboundary issues	The experiments are required to meet the allocation specified in the Water Treaty between the United States of America and Mexico, as well as in the Code of Federal Regulations Title XVIII-Grand Canyon Protection, Section 1801.	
Possible improvements	There are critiques that dispute-resolution mechanisms need to be further strengthened within the adaptive management approach.	Camacho, Susskind and Schenk (2010)

The first high flow experiment was conducted in 1996. This was considered to be the first large-scale international flow experiment (Collier, Webb and Andrews 1997, p. 83; Meretsky, Wegner and Stevens 2000, p. 583). Further experiments were conducted in 2004 and 2008 (Melis 2011, p. 9) and additionally in 2012, 2013, 2016 and 2017. The first 16 years of high flow experiments have provided the basis for the high-flow experiment protocol (US DOI 2011) that provides for adaptive management of the Glen Canyon Dam. The effects of these experiments are analysed as part of the Environmental Impact Statement and adaptive management plans required under the National Environmental Policy Act. **Table 16.3** presents our evaluation of the effectiveness of adaptive management of the Glen Canyon Dam.

Through these experiments, scientific understanding and the policy approach for adaptive management have been incrementally modified to balance hydropower generation with ecological concerns (Gunderson 2015). The high flow release protocol has been successful in increasing the size of sand bars, with benefits to the endangered humpback chub, re-establishing of riparian vegetation and increasing recreation. The programme was successful in accomplishing these improvements within the bounds of existing agreements for water allocation and supply and integrating water management with hydroelectric demand.

Adaptive management was enabled through several policy elements. The mandate of the Bureau of Reclamation, an agency within the Department of Interior, is charged with balancing environmental and economic consideration when developing a dam (US DOI 2016). In addition, legislation such as the Endangered Species Act enables conservation of endangered species and the Grand Canyon Protection Act of 1992 recommends adaptive management (Meretsky, Wegner and Stevens 2000, p. 580). In addition, water supply to downstream states in the United States of America needs to be considered, not to mention in Mexico, as determined by the Water Treaty of 1944 between the two countries. Adaptive management therefore does not operate in an institutional void and multiple institutions cut across different scales, causing interdependence. Consequently, adaptive management requires comprehensive understanding of the set of institutions that can affect this policy approach.

The use of data as well as knowledge generation are also important where uncertainty is inherent in the flow experiments (Konrad *et al.* 2011, p. 955). In this regard, the United States Geological Survey (USGS) Grand Canyon Monitoring and Research Centre acts as a knowledge hub to facilitate experimentation and learning. Such uncertainty is both an enabling and constraining factor to making environmental flows and adaptive management approaches effective. Continuing experimentation and monitoring are vital in helping to modify strategies (Melis 2011, pp. 141-142).

16.2.3 A new approach to water-related disaster risk reduction

Disaster risk reduction is the "concept and practice of reducing disaster risks through systematic efforts to analyse and reduce the causal factors of disasters" (United Nations Office for Disaster Reduction [UNISDR] 2017). Disaster risk reduction aims to reduce the severity of a disaster, considering that the occurrence of a natural hazard itself does not inevitably result in a disaster. Disaster risk reduction is therefore a preventative policy approach which includes objectives such as limiting exposure to hazards; reducing communities' vulnerability to property loss and damage, displacement, mortality and other negative outcomes of disasters. Benefits include: better managing and monitoring of land, environment and resources; and improving preparedness, for example through early warning systems and evacuation plans (UNISDR 2017). The key point of disaster risk reduction is that, through appropriate policy choices and implementing such preventative actions, countries and States can reduce the scale of environmental disasters. Disaster risk reduction frameworks have evolved to provide more effort to preventively limit the size of the disasters, consider the extended time frames (currently 2015-2030, aligned to other global frameworks), and to place the emphasis on implementation, rather than on final aims (Inter-Agency Regional Analysts Network [IARAN] 2016, p. 4).

The impacts of disasters are unique to each region and they impact disaster risk governance arrangements. Often this leads to high investments in infrastructure, often accompanied by strongly institutionalized arrangements (e.g. Poland, Netherlands, Singapore). Yet moderate (Belgium, France) or highly (United Kingdom of Great Britain and Northern Ireland) diversified strategies emerge from both the need and will to change, and from a mix of forces pushing for this change (Wiering *et al.* 2017, pp. 20-24).

Improvements in national disaster risk reduction efforts and individual preparedness often emerge after major disaster events such as the 2004 Indian Ocean tsunami (Hoffmann and Muttarak 2017, p. 32). Disaster risk reduction is gaining a higher profile on political agendas through developments such as the targets of the Sendai Framework of Action (UNISDR 2015). At the same time, a person-centred approach is being promoted through greater engagement of women, children and older people. There are stark differences on the level of disaster preparedness between the developed and developing world, and this raises serious equity concerns in terms of the capacity to deal with disasters, and the subsequent loss of lives in the developing world (Al-Nammari and Alzaghal 2015).

Policies may address disaster risk reduction by improving their effectiveness, that, for example, may affect availability of potable water, depending on local/regional vulnerability and levels of preparedness. In particular, floods and storms represent direct pressures on water quality, while droughts affect both water quantity and quality. Effective policies can also enhance disaster risk reduction responses by addressing threats to the availability of safely managed drinking water and the affected sewerage systems which can have impacts on human health.

Case study: Flood risk management policy in England and Wales (United Kingdom of Great Britain and Northern Ireland)
Floods cause great financial losses and health impacts in England and Wales; the floods of 2007 caused £3.2 billion damage (Penning-Rowsell 2015). As of 2014, the number of households on floodplain areas as designated by the Environment Agency (only England and Wales) "constitute 8.5 per cent of all properties, with one quarter of these at significant risk" (Penning-Rowsell and Pardoe 2015, p. 5). As a result, flood risk management policy has seen major shifts in the last 15 years. Most notably, there is an emphasis

Table 16.4: Evaluation of the effectiveness of the flood risk management policy in England

Criterion	Description	References
Success or failure	The United Kingdom Government has claimed that the flood risk management system has been effective and "has achieved notable successes including securing better protection for more than 500,000 properties since 2005". There are, however, concerns about the equity implications of the new cost-benefit analysis based allocation of funds and the remaining preference for structural methods of flood risk management	United Kingdom Parliament (2017)
Independence of evaluation	Sir Michael Pitt (2008) conducted an independent review following widespread flooding in summer 2007. The detailed assessment and recommendations of the review initiated a range of reforms in flood risk management as well as progress reports produced by DEFRA (2012). Chatterton *et al.* (2016) prepared an assessment of the costs and impacts of the 2013-2014 floods based on the categories of the 2007 assessment. The DEFRA report and the United Kingdom Government response in 2017 provide further material for assessing the policy.	Pitt (2008) ; DEFRA (2012); Chatterton *et al.* (2016)
Key actors	The stakeholders include the United Kingdom Department of Environment, Food and Rural Affairs (DEFRA), the Environment Agency, local authorities, water companies, flood wardens, National Flood Forum, consultants. The main policy reform to-date consisted in giving the local authorities greater responsibilities for flood risk management.	UK Government (2010); Laakso, Heiskanen and Matschoss (2016)
Baseline	Adjusted for inflation, the average damages from floods in the United Kingdom of Great Britain and Northern Ireland in the last 23 years are approximately £250 million per year (Penning-Rowsell 2015). Especially notable are the floods of the summer of 2007 with an estimated damage of £3.2 billion, which was a catalyst for the accelerated flood risk management reform.	Penning-Rowsell (2015)
Time frame	The Floods and Water Act (2010) is based on an earlier strategy, 'Making Space for Water', introduced in 2004 (DEFRA 2004) and a more recent government strategy 'Future Water' (DEFRA 2008), as well as the influential report from Sir Michael Pitt (DEFRA 2008).	DEFRA (2004); DEFRA (2008); Pitt (2008); UK Government (2010)
Constraining factors	The new policy framework has given local authorities the lead in preparing and responding to surface-water flooding without really equipping them with the necessary financial, human and technical capacity to deal with the new challenges. Local authorities across the country struggle with the high expectations for flood risk management as they remain underfunded and under-resourced.	Penning-Rowsell (2015)
Enabling factors	The record funds allocated to flood risk management for hard, soft and natural solutions amounted to £2.5 billion in the period 2015-2021.	Penning-Rowsell (2015)
Cost-effectiveness	The fact that the devastating floods of winter 2013/14 have caused economic damage that is within the average annual damage in the last two decades (ca. £250 million) may indicate that the flood risk management measures passed since 2007 have been effective (Thorne 2014).	Thorne (2014)
Equity	Some areas in the United Kingdom of Great Britain and Northern Ireland are more prone to flooding from surface water than others. The property insurance market is liberalized in the United Kingdom of Great Britain and Northern Ireland, which means that insurers may ask for higher premiums for houses located in areas under high risk of flooding. This puts some households at a disadvantage compared to others, raising concerns about the equity of flood risk management policy (Penning-Rowsell and Pardoe 2015; Begg, Walker and Kuhlicke 2015).	Begg, Walker and Kuhlicke. (2015); Penning-Rowsell and Pardoe (2015)
Co-benefits	Natural flood management, based on land-use planning and change, can help mitigate non-point pollution from agricultural land, and reduce soil erosion impacts on lake ecosystems (Dadson *et al.* 2017), illustrating how system scale management can address a complex nexus dynamic. The restoration of terrestrial and aquatic habitats would provide additional carbon storage services (Keesstra *et al.* 2018). Potential co-benefits may include retention of water upstream in ponds and aquifers that can supplement scarce water resources during droughts, as well as help mitigate the adverse ecological impacts of heat.	Dadson *et al.* (2017); Keesstra *et al.* (2018)
Transboundary issues	None	
Possible improvements	A future reform towards a stricter regulatory framework for sustainable drainage would be an improvement.	Begg, Walker and Kuhlicke (2015)

on the 'softer' measures of flood prevention, nature-based solutions, and citizen preparedness, as outlined in the strategy 'Making Space for Water' (United Kingdom, Department for Environment, Food and Rural Affairs [DEFRA] 2004; Mukhtarov 2009) and included in the Flood and Water Management Act (United Kingdom Parliament 2010). This is as opposed to a heavier reliance on infrastructure (e.g. Wiering *et al.* 2017).

Another governance approach to flood risk management uses stakeholder collaboration at multiple levels and implements integrated water resources management (IWRM) through cross-sector coordination and greater engagement of citizens (Mukhtarov 2009). Local authorities now have a number of responsibilities in planning for and responding to surface water (e.g. flood waters) as a result of devolution of responsibilities from the United Kingdom Department of Environment, Food and Rural Affairs (DEFRA) and the Environment Agency as well as the outcomes of policy reviews (Pitt 2008) and the above-mentioned Flood and Water Management Act.

Policy has also encouraged that disaster risk reduction strategies consider biodiversity, human health and water quality benefits. The Pitt Review (Pitt 2007; Pitt 2008) called for sustainable drainage systems to be installed in new buildings and urban land-use change to reduce run-off and improve water retention. Currently, voluntary measures address sustainable drainage. **Table 16.4** presents our evaluation of the effectiveness of the flood risk management policy in England.

The United Kingdom of Great Britain and Northern Ireland case of flood risk management is a comprehensive reform of flood policy including the overhaul of surface-water management in England and Wales. Time will tell whether the new system is more effective than the previous one. However, major positive outcomes are already clear in terms of the large number of properties with better protection against flood risk, 500,000 since 2005 (United Kingdom Parliament 2017).

The continued success of the new flood risk management policy and its focus on surface-water management seems to depend on the ability of the national and regional governments to coordinate alleviation schemes and natural flood risk management with the local authorities. While more research into the outcomes of local management in flood risk strategies is necessary for a better understanding of its impacts, it seems reasonable to state that successful devolution of these responsibilities should be accompanied by increasing the budgets and coordinating powers of local authorities, together with a greater supporting role for the national bodies, such as the Environment Agency and DEFRA. While each local authority is responsible for leading in preparing for and responding to flooding, they require the necessary financial, human and technical capacity to deal with the new challenges. Local authorities across the country struggle with the high expectations for flood risk management as they remain underfunded (Begg, Walker and Kuhlicke 2015; Penning-Rowsell and Johnson 2015). Furthermore, the multiple actors and responsibilities involved in the policy create a further challenge for coordinated implementation (Begg, Walker and Kuhlicke 2015).

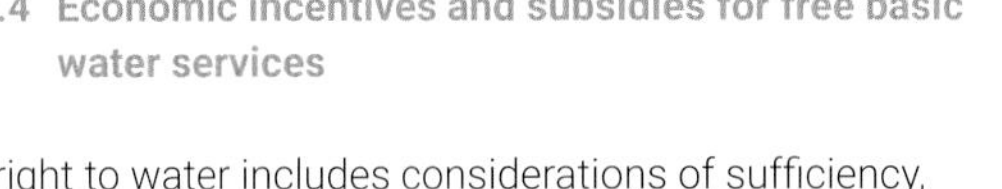

16.2.4 Economic incentives and subsidies for free basic water services

The right to water includes considerations of sufficiency, safety, acceptability, physical accessibility and affordability for personal and domestic uses (UN General Assembly Resolution A/RES/64/292, see United Nations, General Assembly [UNGA] 2010). A growing number of countries are now formally recognizing this human right following the 2010 UN General Assembly Resolution A/RES/64/292. The 2014 Global Analysis and Assessment of Sanitation and Drinking-Water reported that 70 out of 94 countries recognized the right to water (World Health Organization [WHO] 2014, p. 14).

Constitutions and legislation recognizing the right to water must be supported by policy instruments that target financing and budgeting. These instruments are important because financing and budgetary considerations are regarded as an obstacle to realizing the right to water and sanitation (de Albuquerque and Roaf 2012). States have often faced cost recovery and transaction cost issues (Obani and Gupta 2016, p. 679). Subsidies are part of water pricing efforts which provide economic incentives to realize the right to water. Subsidies are often used to support affordability of water and sanitation services, which include mechanisms such as income supplements, cross-subsidies, increasing block tariffs, universal price with rebate and free basic water (de Albuquerque and Roaf 2012, p. 54, 83).

The development of the MDGs enabled the human rights approach and treating water as an economic good to coexist, though there are tensions (Obani and Gupta 2014). Water pricing, if efficient, includes all of the economic costs required to provide water (Grafton 2017, pp. 30-31) and underscores that the right to water does not necessarily call for free water provision (United Nations Office of the United Nations High Commissioner for Human Rights [OHCHR] 2010, pp. 11-12). Moreover, the right to water needs to be considered within the wider context of multiple water uses, for example for food security, particularly at the household level. In general, the human right to water and sanitation does not consider how, for example, agricultural water requirements and virtual water transfer through trade could affect rich and poor users differently (Obani and Gupta 2016, p. 685). The right to water is thus defined by local contexts. The implication of rolling out this right indirectly illustrates the nexus of water and food security and the potential inequalities arising from differences in local contexts.

Within the DPSIR framework (Section 1.6), this policy approach is mostly aimed at guaranteeing a basic condition for safe and clean drinking water. Pricing and subsidies often operate within a broader governance context, including creating minimum requirements for water quality as well as establishing organizations as regulators (de Albuquerque and Roaf 2012). South Africa's free basic water policy is an early example of the constitutional recognition of the right to water, providing insight into over two decades of policy experience (**see Table 16.6**). It also exemplifies the need for economic instruments, such as increasing block tariffs, to be used in a way that considers local hydrological and socioeconomic contexts (von Hirschhausen *et al.* 2017).

Case study: Free Basic Water Policy, South Africa

The South African Government launched the Free Basic Water Policy (FBWP) in 2001. Its purpose was to address public health concerns of lack of access to safe water and sanitation, and to provide subsidized water services to the country's population. The policy targets the poor in particular, and allows for the provision of 6,000 litres (6 kilolitres) of safe water per household per month (Department of Water Affairs and Forestry [DWAF] 2002, p. 7). By extension, the policy aims to alleviate poverty through the provision of basic services (DWAF 2002, p. 1) in a country that had experienced historic inequalities within the population.

The FBWP requires that approaches to restrict water consumption ensure effective free provision of a basic level of water supply. Recognizing that municipalities are not homogeneous, the FBWP suggests that mixed service levels are offered according to the consumer's ability to pay. Services include hand pumps, communal taps, and regulated yard and roof tanks as well as house connections (DWAF 2002). In addition, several types of economic incentives have been put forward to meet the variety of consumers within a municipality (**see Table 16.5**). **Table 16.6** presents our evaluation of the effectiveness of economic incentives in the Free Basic Water Policy in South Africa.

The FBWP is an important first step to implementing the human right to water and brings together a set of legislation and policy instruments to consolidate the importance of this right. Municipalities are required to use a tariff system according to Section 74 of the Municipal Systems Act. This tariff system reflects the 'user pays' principle so that water consumption above a basic level is charged (DWAF 2003, p. 29). Metering is one way to measure or control the amount of water supplied without charge (DWAF 2002, p. 29). However, the figure of 6,000 litres/household/month has been controversial and it was recognized in 2007 by DWAF (2007, p. 5) that the amount of 25 litres per person per day might not be enough for many households and needs to be incrementally increased.

The economics of implementation are not negligible because there is high regional as well as socioeconomic variation across the country, with implications for equity. For rural water supply, the cost recovery is very low (WSP 2011). Urban water supply has achieved 96 per cent coverage, but maintaining assets has received low priority, risking deterioration in the future (WSP 2011) and with possible impacts on cost-effectiveness.

Efficient cost recovery by water service authorities (i.e. municipalities) also comes with issues of equity. Problems with cost recovery impinge on the levels of service provision. The free basic minimum has become the maximum amount for households in places like Durban (Loftus 2006). Cost recovery is necessary to provide benefits to extend coverage and address geographical unevenness of burden so that FBWP is not exclusive to those that already have infrastructure and thus benefit from the subsidy easily (Balfour *et al.* 2005, p. 16).

Table 16.5: Three options for free basic water supply

	Option 1 **Rising block tariffs**	**Option 2** **Targeted credits**	**Option 3** **Service-level targeting**
Description	Rising block tariff is applied to all residential consumers, with the first block typically set from 0 to 6 kilolitres with a zero tariff. No fixed monthly charge applicable to those using below poverty relief consumption limit.	Each consumer who is selected for poverty relief gets a credit on their water account which would typically be sufficient to cover the charge for the poverty relief amount (often 6 kilolitres per month) free.	Those service levels which provide a restricted flow, (below the poverty relief consumption level) are provided at no charge. Those with higher service levels pay the normal tariffs, except for poor consumers who historically have high service levels.
Targeting method	No targeting (first 6 kilolitres free to all households). However, targeted fixed monthly charge may be necessary for holiday areas.	Requires a system for identifying those who require poverty relief. Typically, this is based on a benchmark poverty indicator (household income or household expenditure).	Targeting takes place through selection of service level by the consumer (or authority in some cases).
Applicability	Mainly larger urban municipalities. Not suited to situations where there is a high proportion of holiday homes unless it is supplemented with a targeted fixed monthly charge.	Can be used in large municipalities but more typical for middle to small sized, largely urban municipalities. Requires a billing system to be in place for all consumers.	Best suited to municipalities which are largely rural in character.

Source: DWAF (2002, p. 27-29).

Table 16.6: Evaluation of the effectiveness of economic incentives through the Free Basic Water Policy in South Africa

Criterion	Description	References
Success or failure	The Department of Water Affairs and Forestry (DWAF) reported that there was a good track record of implementation particularly in urban areas during the first 22 months of implementing the provision of free basic water. In 2007, DWAF further reported that over 75 per cent of the population was provided with free basic water and the majority of those (69 per cent) were poor households. However, this success is uneven between urban and rural areas as provision of water supply in remote, rural locations has continued to lag behind. Moreover, it has been reported that drinking water provision decreased by 8 per cent from 2012 to 2014.	DWAF (2002); Muller (2008, p. 79); Water and Sanitation Program [WSP] (2011a, p. 2); Department of Water and Sanitation [DWS] (2014, p. 7)
Independence of evaluation	The Free Basic Water Policy (FBWP) has been evaluated internally through review by DWAF, the Water Research Commission and other related government agencies, and extensively in the peer-reviewed scientific and grey literature.	DWAF (2002); Mehta and Ntshona (2004); Balfour *et al.* (2005); Loftus (2006); DWAF (2007); Loftus (2007); Muller (2008); von Schnitzler (2008); Dugard (2008); WSP (2011a); Naidoo *et al.* (2012); DWS (2014); Statistics South Africa (2016)
Key actors	DWAF (now the Department of Water and Sanitation [DWS]) is the ministry responsible for overseeing the FBWP. Central government has the role of regulator in this decentralized process. Municipalities, water boards and private service providers are involved in local implementation.	WSP (2011a)
Baseline	There was no recognition of a right to water prior to policy implementation. When FBWP was introduced in 2001, it was reported that out of 44.8 million people, "5 million (11 per cent) had no access to safe water supply and a further 6.5 million (15%) did not have a defined basic service level" (DWAF 2003, p. 1).	DWAF (2003, p. 1)
Time frame	The policy developed out of a wider political process of post-apartheid democratization after 1994. In addition, the FBWP has also been implemented and monitored during the period 2000-2015 to achieve the MDG Target 7C.	
Constraining factors	Physical constraints of water availability in a dry region challenge the provision of water supply. The policy has not defined a 'poor' household, despite it targeting such water users.	Muller (2008); Naidoo *et al.* (2012)
Enabling factors	This policy enacts Section 27 of the Constitution, which states the right to water and is governed by the 1997 Water Services Act and the 1998 National Water Act. In addition, regulatory frameworks such as the 2003 Strategic Framework for Water Services guide the implementation of FBWP and are complemented by national standards on service levels such as DWS (2017). A mix of economic instruments is supported by the policy to help address provision in a situation where water has been bound up in social inequalities from apartheid. This policy is also part of an effort towards decentralization, and thus can be seen as part of a broader governance shift.	Muller (2008); DWS (2017)
Cost-effectiveness	The average per capita water supply investment in South Africa is relatively high (urban water supply US$385 per capita; rural water supply US$278 per capita). The uneven nature of cost-effectiveness is exemplified by the fact that across municipalities, the viability of cross-subsidization depends on a number of factors, including the level of wealth of consumers, and type and ratio of users.	DWAF (2007); WSP (2011)
Equity	According to one study, the disease burden attributable to unsafe drinking water and sanitation in South Africa in 2000 was estimated at 13,434 deaths, among which children were disproportionately highly represented (Lewin *et al.* 2007). The policy is a first step towards addressing these health implications. However, the use of prepaid water meters for cost recovery of water services has brought about a problem where consumption in excess of the free basic minimum becomes costly for some. The use of households as a unit of provision gives little attention to those in informal settlements and backyard dwellings.	Bond and Dugard (2008); McDonald (2008)
Co-benefits	The FBWP was set up to have co-benefits in public health, welfare and gender equity. Mehta and Ntshona (2004, p. 19) reported some evidence in this regard. However, published results and data on this aspect are not available in the public domain.	DWAF (2002); Mehta and Ntshona 2004, p. 19
Trans-boundary issues	While South Africa has several transboundary river basins and aquifers, the policy pertains to national aims and implementation, which do not seem to have direct or explicit implications for the exercise of the human right to water in other riparian states.	
Possible improvements	Since the introduction of FBWP, the Free Basic Sanitation Policy was established in 2009. Co-benefits of the latter's implementation to FBWP could be analysed in detail in the future. Cost-effectiveness could incorporate health costs and deal with the efficacy of the Free Basic Sanitation Policy. Policy coherence could be further enhanced by integrative approaches involving better institutional interplay. A more specific focus on the needs of informal settlements would improve equity.	

16.2.5 Voluntary sustainability reporting on water in the mining sector

Mining requires significant amounts of water and presents considerable short- and long-term risks to water resources (Spitz and Trudinger 2008) (see also Section 9.5.5). The potential impacts on existing users and values of water resources are a common concern for local communities faced with both large- and small-scale mining projects. Such concerns stem from experiences of mines that have caused (or continue to cause) pollution or other impacts on water resources (e.g. reductions in stream flows, declines in groundwater levels, river diversions, undesirable changes in quality). Governments, companies and communities have recognized the fundamental need for the mining industry to manage water resource-related risks effectively (e.g. Norgate and Lovel 2006; Rankin 2011).

The main protocol for sustainability reporting is the Global Reporting Initiative (GRI), which began in 1997 as a coalition of government, community and corporate stakeholders, and aimed to make sustainability reporting as commonplace and important as corporate reporting. The current GRI standard includes a wide range of indicators across social, economic, environmental and local-community health aspects, and was designed to be used not only by the mining sector, but by any company or organization. It addresses equity issues through providing guidance on reporting on management approaches affecting vulnerable groups, the means by which local stakeholders are identified and engaged with, and the means by which companies address risks to and impacts on local communities. Since the Johannesburg Earth Summit in 2002, the global mining industry, through the International Council of Mining and Metals, now requires their corporate members to publish annual sustainability reports.

Case study: Australian mining industry's Water Accounting Framework

Early research into the water data reported by mines found that data in sustainability reporting could be changed from year to year without explanation, that different mines interpreted terms such as 'consumed water' or 'recycled water' inconsistently, and that water quality issues were poorly addressed (Mudd 2008; Northey *et al.* 2016). This led the Minerals Council of Australia (MCA) to develop the Water Accounting Framework (MCA 2012), which allows a mine's water balance to be quantified and the specific Global Reporting Initiative indicators to be reported through sustainability reports. The Water Accounting Framework was a major step forward in providing a consistent reporting approach to water management for mines. The 49-member companies of the MCA represent 85 per cent of Australia's mining activity and more than 90 per cent of mineral exports (MCA 2017a).

Growing interest in corporate responsibility has been a strong enabling factor for sustainability reporting, with pressures from investors and shareholders in mining companies as well as local communities affected by mining. The main constraining factor for water risks in mining is the technical capacity of an individual company and its mines. For example, a mine may not be equipped with the necessary monitoring, technical (especially water-balance modelling) expertise and reporting systems to ensure accurate and timely sustainability reporting. Efficient management of water use and associated costs requires monitoring in any case, meaning that it is beneficial for a mine to invest in such systems to help it reduce operational costs, ensure transparency and improve its reputation, as well as minimizing water-resource-related risks. In terms of cost-effectiveness, the value gained from conducting good sustainability reporting compared to taking no action can be significant, from positive investor sentiment, a social licence to operate from a local community, reduction in operating costs from water efficiencies or recognition from regulators of successful environmental management – as noted by the MCA in its business case for the Water Accounting Framework (see MCA 2017b). **Table 16.7** below presents our evaluation of the effectiveness of the Australian mining industry's Water Accounting Framework.

The growing number of companies having adopted sustainability reporting is a sign of a successful policy initiative and approach. The fact that the MCA and now the International Council of Mining and Metals have mandated water reporting by their members also demonstrates success. However, four major weaknesses in the Water Accounting Framework and International Council of Mining and Metals protocols are:

i. the issue of water quality of the water sources used in mining;
ii. the links between detailed monitoring of potentially affected water resources, especially water quality and flows, and Global Reporting Initiative metrics;
iii. links between regulatory requirements for water resources and sustainability reporting; and
iv. improving the catchment and climate context of water data so that mining's use of water and risks to water resources can be more readily interpreted and understood.

Furthermore, there are very few formal evaluations of water data and information published in sustainability reports, except for a limited number of academic studies. With the Global Reporting Initiative moving to a standards framework rather than a guideline structure, independent auditing and assurance are now more prominent, as well as being important for responsible investors, regulators and interested community stakeholders.

The effectiveness of the policy in terms of the impact of mining operations on water resources generally has not yet been rigorously evaluated. However, the large proportion of Australian mining companies publishing sustainability reports incorporating the Water Accounting Framework suggests that the approach is useful as a management tool.

16.3 Indicators (link to SDGs and MEAs)

The following indicators on access to drinking water, sanitation and water withdrawal further examine the variety of policies used in managing freshwater resources, contributing to improving human health through various pathways. These indicators were selected for being policy sensitive and for being widely recognized for their importance under the current SDG targets and established multilateral environmental agreements. For the purposes of this chapter, the indicators are analysed in order to present policies influencing global

Table 16.7: Evaluation of the effectiveness of the Australian mining industry's Water Accounting Framework

Criterion	Description	Reference(s)
Success or failure	The growing number of companies that have now adopted sustainability reporting in Australia, as well as the fact that both the Minerals Council of Australia (MCA) and International Council of Mining and Metals (ICMM) have mandated water reporting, signals successful diffusion of the policy approach.	Mudd (2008); Northey, Haque and Mudd (2013)
Independence of evaluation	With the Global Reporting Initiative (GRI) moving to a standards framework rather than a guideline structure, external assurance auditing is being increasingly conducted, although the extent of such auditing is variable. Very few formal evaluations of water data reporting have been done.	Mudd (2008); Northey *et al.* (2016)
Key actors	Individual mining companies, their membership associations, local communities, interested stakeholders (e.g. environmental groups), government regulators, financial stakeholders.	Franks *et al.* (2014)
Baseline	There was no formal baseline. A tacit baseline could be the lack of water reporting prior to the mid-1990s.	Mudd (2008)
Time frame	The process of sustainability reporting and the data it contains have evolved over the past 20 years. From 2016, the GRI has been a formal standard rather than a guideline.	Mudd (2008); Northey, Haque and Mudd (2013); Northey *et al.* (2016)
Constraining factors	Companies and mines are constrained by their technical capacity to monitor and record water-related processes and impacts.	Mudd (2008); Northey, Haque and Mudd (2013); Northey *et al.* (2016)
Enabling factors	The growing interest in demonstrating corporate responsibility, with pressures from investors and shareholders in mining companies as well as communities affected by mining.	Mudd (2008); Franks *et al.* (2014); MCA (2017a)
Cost-effectiveness	It is logical for companies to engage in self-reporting to avoid project-delaying conflicts, expensive litigation and brand damage. Also, good sustainability reporting may lend companies a social licence to operate from local communities.	Mudd (2008); Franks *et al.* (2014);
Equity	Although sustainability reporting may result in win-win situations for mining companies, communities and government in Australia, it is unclear how sustainability reporting might impact on equity in other parts of the world.	Franks *et al.* (2014)
Co-benefits	Sustainability reporting results in data availability to researchers, enabling quantification of the life cycle costs of specific metals and minerals, innovation that may benefit the sustainability of processes, and evaluation of impacts of new mining technologies on water resources.	MCA (2017a)
Transboundary issues	The global uptake of standardized sustainability reporting may foster improved transnational management of mining water-related issues.	International Council of Mining and Metals (ICMM) (2017)
Possible improvements	Detailed study of water-resource-related sustainability reporting by companies should be conducted to assess its extent, quality and effectiveness. There are major weaknesses in the Water Accounting Framework and ICMM protocols. Online databases of pooled water-resource data would foster usability of water data and improve transparency.	Mudd (2008); Northey, Haque and Mudd (2013); Northey *et al.* (2016)

trends in drinking water and sanitation and water withdrawal. There is a considerable diversity of policies and our analysis underscores the importance of policy mixes in further achieving global targets such as the SDGs and facilitating implementation at the local level.

16.3.1 Indicator 1: Proportion of population using safely managed drinking water services

SDG indicator 6.1.1 is defined as the proportion of the population worldwide using safely managed drinking water services, in support of public health. 'Safely managed' refers to water from an improved water source located on premises, available when needed and free of faecal and priority chemical contamination (WHO and the United Nations Children's Fund [UNICEF] 2017), wherein 'improved water source' (the MDG indicator) includes rainwater, water that is piped, made available from taps, standpipes, boreholes, wells or springs, or is packaged or delivered. 'Drinking water services' refers to the accessibility, availability and quality of the main source used by households for drinking, cooking, personal hygiene and other domestic uses (WHO and UNICEF 2017). Priority chemical contaminants vary by country, but arsenic and fluoride are assigned as priority contaminants globally due to their potential impacts on human health.

Scope and measurement
From 2000 to 2015, the World Health Organization and United Nations Children's Fund (WHO/UNICEF) Joint Monitoring Programme for Water Supply and Sanitation (JMP) used a binary classification of improved/unimproved sources of drinking water as an indicator for monitoring and evaluation purposes. In order to monitor SDG target 6.1, the JMP further developed this indicator to facilitate further differentiation between service levels and assessment of safe management of supplies (WHO and UNICEF 2017). Corresponding updates were made to the JMP drinking water service ladders with 'safely managed' occupying a new rung positioned at the top **(Table 16.8)**.

According to the JMP, 2.6 billion people worldwide gained access to an improved source of drinking water in the period between 1990 and 2015 (UNICEF and WHO 2015) **(Figure 16.2).** This brought the proportion of the global population using piped water supplies on premises to approximately 75 per cent.

Policy relevance
This indicator is a modification of the MDG indicator 7.8 (proportion of population using an improved drinking water source) and directly relates to SDG target 6.1, which aims to achieve universal and equitable access to safe and affordable drinking water for all by 2030. This indicator also relates to long-standing global policy efforts addressing water and human health, including multilateral environmental agreements such as the 1999 Protocol on Water and Health.

Table 16.8: The JMP Service Ladder for drinking water

Service level	Definition
Safely managed 	Drinking water from an improved water source which is located on premises, available when needed and free of faecal and priority chemical contamination
Basic 	Drinking water from an improved water source provided collection time is not more than 30 minutes for a roundtrip including queuing
Limited 	Drinking water from an improved source where collection time exceeds over 30 minutes for a roundtrip to collect water including queuing
Unimproved	Drinking water from an unprotected dug well or unprotected spring
No service 	Drinking water collected directly from a river, dam, lake, pond, stream, canal or irrigation channel

Source: Adapted from WHO and UNICEF (2017).

Causal relations
The gradual shift towards water governance can be broadly attributed to changes in the provision of safe drinking water services. The initial intervention involves installing physical infrastructure for safe water supply. For example, efforts that upgrade water services to piped supplies typically reduce microbial contamination of both source and household-stored water quality (Shields *et al.* 2015). While technical solutions are still seen in the sanitation sector (WSP 2011b), there is increasing use of participatory approaches to complement them (see also Section 16.2.2). In India, for example, the national water policy aiming to provide safe water adopts a socio-technological approach (Khurana and Sen 2008).

Target setting at the national level also appears to encourage an increase in the size of populations with access to safe drinking water. In a recent assessment of water access in 97 countries, approximately half had established or were working towards universal access as a target between 1980 and 2013 (Luh *et al.* 2017). The MDG on safe drinking water halved the proportion of people requiring access by 2012, three years before the MDG deadline. This early success has been followed up with national targets motivated by the ambitious global goals of the SDGs: countries with the appropriate capacity can be expected to meet ambitious targets, leading to greater coverage than is found in countries lacking such ambitious targets (Luh *et al.* 2017).

Other influencing factors
Universal access may be hampered not only by hydrological factors such as rainfall, which may contribute to water scarcity, and water-related hazards such as microbial contamination, but also by economic factors. In rapidly developing countries such as India, pollution and overexploitation of water are linked to industrialization and agricultural expansion, which in turn influence water quality (Khurana and Sen 2008). The pace of population growth also challenges drinking water and sanitation coverage, especially in sub-Saharan Africa and Oceania (UNICEF and WHO 2015).

A lack of awareness or understanding of water quality problems may hinder safety of drinking water services in both developed and developing countries. In Bangladesh, while tube wells have increased, water quality testing is not commonly practised (Fischer 2017). This contributes to poor understanding of the health risks posed by both microbial and non-microbial contaminants. This can have serious consequences in terms of public health. For example, measures taken in the 1970s to reduce the health impacts of microbial disease from surface-water use resulted in the widespread installation of tube wells, themselves a source of water with high levels of inorganic arsenic (Flanagan, Johnston and Zheng 2012). Populations using these sources for drinking water have experienced severe health consequences ranging from skin lesions to cancer and cognitive effects (Abdul *et al.* 2015), resulting in stigmatization and other serious social impacts (Kabir *et al.* 2015).

Possible alternative indicators
A useful alternative indicator to understand the population benefiting from safely managed drinking water services might focus on disparities between rural and urban populations combined with wealth quintiles. The JMP has been able to track coverage between 1995 and 2012 (UNICEF and WHO 2015) but could benefit from comprehensive data and rigorous reporting.

Graphical representation

Figure 16.2: Change in global population by drinking water source, 1990-2015 (billions)

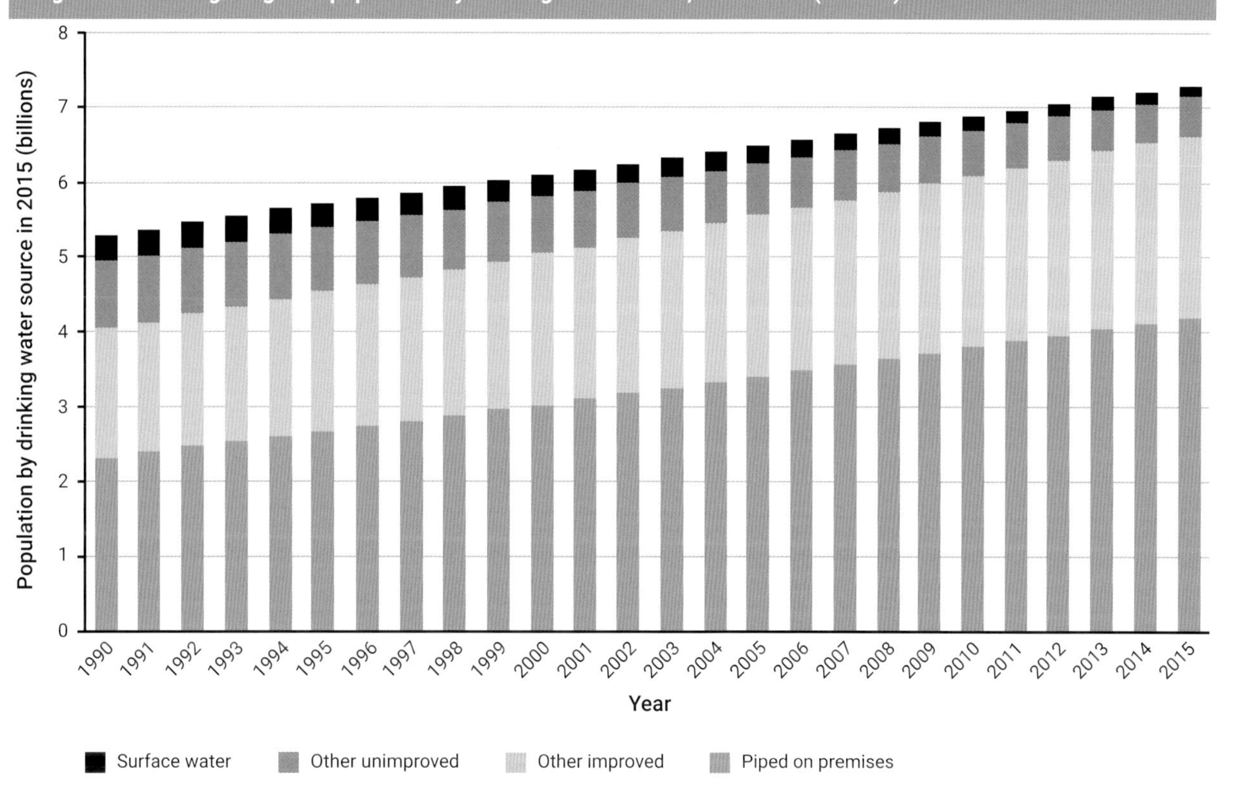

Source: Adapted from WHO and UNICEF (2017).

16.3.2 Indicator 2: Proportion of population using safely managed sanitation services, including a hand-washing facility with soap and water

SDG indicator 6.2.1 refers to the proportion of the population using safely managed sanitation services, including a hand-washing facility with soap and water, wherein 'safely managed' is defined as "an improved sanitation facility which is not shared with other households and where: excreta is safely disposed of *in situ*, or excreta is transported and treated off-site" (WHO 2017, p. 1).

Scope and measurement

The levels of sanitation services vary from safely managed, through basic, limited and unimproved, to no service according to the JMP. These levels focus on whether excreta is separated and disposed of safely, avoiding human contact. In addition, the levels depend on whether sanitation facilities are shared or private (WHO 2017).

Graphical representation

There has been great progress made in decreasing the number of people without access to safe sanitation services. As **Figure 16.3** shows, between 2000 and 2015 the number of people practising open defecation declined from 1,229 million to 892 million, which is an average reduction of 22 million people per year. Furthermore, all regions have made progress in decreasing this indicator apart from sub-Saharan Africa and Oceania.

Figure 16.3: Regional trends in proportion of national population practising open defecation, 2000-2015

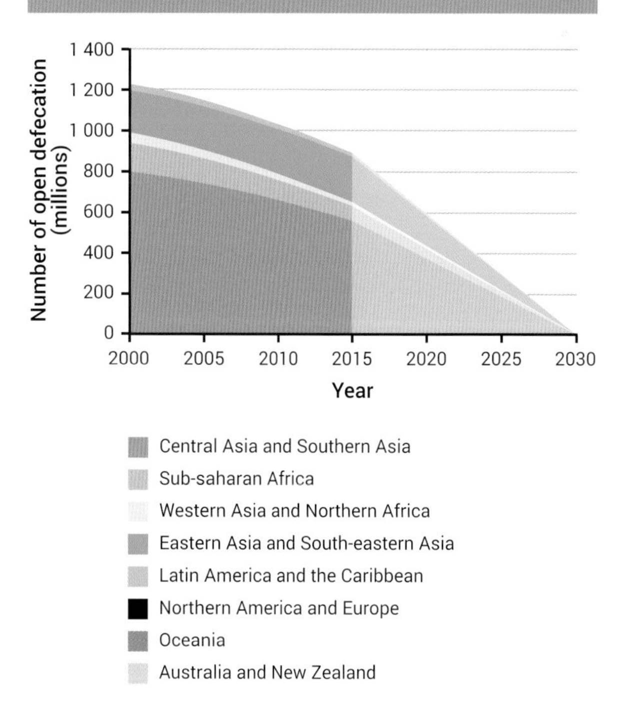

Source: WHO and UNICEF (2017).

Another notable trend is the rate of change in various countries in the world. While 14 countries have shown progress sufficient to be on track for universal basic sanitation by 2030, the majority either need to accelerate progress or to reverse a negative trend of increasing number of people with no access to safe sanitation **(Figure 16.4)**.

Policy relevance

The proportion of the population using safely managed sanitation services directly relates to the SDG target 6.2: "by 2030, achieve access to adequate and equitable sanitation and hygiene for all, and end open defecation, paying special attention to the needs of women and girls and those in vulnerable situations" (United Nations 2018). SDG indicator 6.2.1 increases recognition of these relationships and furthers ongoing global efforts to address water and sanitation, including the MDGs preceding this SDG target, as well as the Plan of Implementation of the World Summit on Sustainable Development.

Causal relations

Improved water supply and sanitation are the most fundamental indicators related to water, sanitation and hygiene (WASH) interventions. Policy interventions have aimed to provide and maintain infrastructure such as wells, water transport and distribution networks, and water-treatment facilities (Hunt 2011). Water quality interventions and hygiene promotion such as handwashing have also been effective in prevention of disease (Peletz *et al.* 2013).

Community-led total sanitation (CLTS) has been actively taken up in many parts of the world to improve the number of people using improved sanitation services. CLTS is the main policy used to tackle open defecation in rural areas in developing countries (Bateman and Engel 2017). The uptake has been rapid with 60 countries implementing CLTS since 2000 (Crocker *et al.* 2017). CLTS is a participatory and bottom-up approach that incorporates awareness-raising at the community level.

One reason for the spread of CLTS is its perceived low cost, even given the relative scarcity of studies examining its true costs (Crocker *et al.* 2017).

Other influencing factors

As with access to safe drinking water services, sanitation is a focus of global ambitions as reflected in the SDGs. However, rather than attempting to assess WASH and preventative health interventions at the global level, it is more effective to decentralize policy so as to better understand those factors that serve to enable WASH in local contexts (Whittington *et al.* 2012).

Possible alternative indicators

The concomitant rise in pit-latrine sanitation and groundwater use has led to increasing concerns about the potential impact of resulting contamination of drinking water on health. In order to measure the robustness of sanitation service hygienically separating excreta from human contact, consideration may be required to not simply measure the sanitation service provision but also any secondary or knock-on effects. Indicators based on integrated data to identify and mitigate risk could be useful and it has been suggested that water supply and pit-latrine mapping is effective, as well as the monitoring of key groundwater contamination indicators (Back *et al.* 2018).

16.3.3 Indicator 3: Level of water stress: freshwater withdrawal as a proportion of available freshwater resources

SDG indicator 6.4.2 refers to level of water stress (freshwater withdrawal as a proportion of available freshwater resources). Water withdrawal can be defined as the amount of freshwater resources removed from rivers or aquifers for agricultural, industrial and domestic uses (Food and Agriculture Organization of the United Nations [FAO] 2016). Agricultural water use makes up the majority of global water withdrawal, underscoring a major dimension of the water-food nexus

Figure 16.4: Progress towards universal basic sanitation services (2000-2015) among countries where at least 5 per cent of the population did not have basic services in 2015

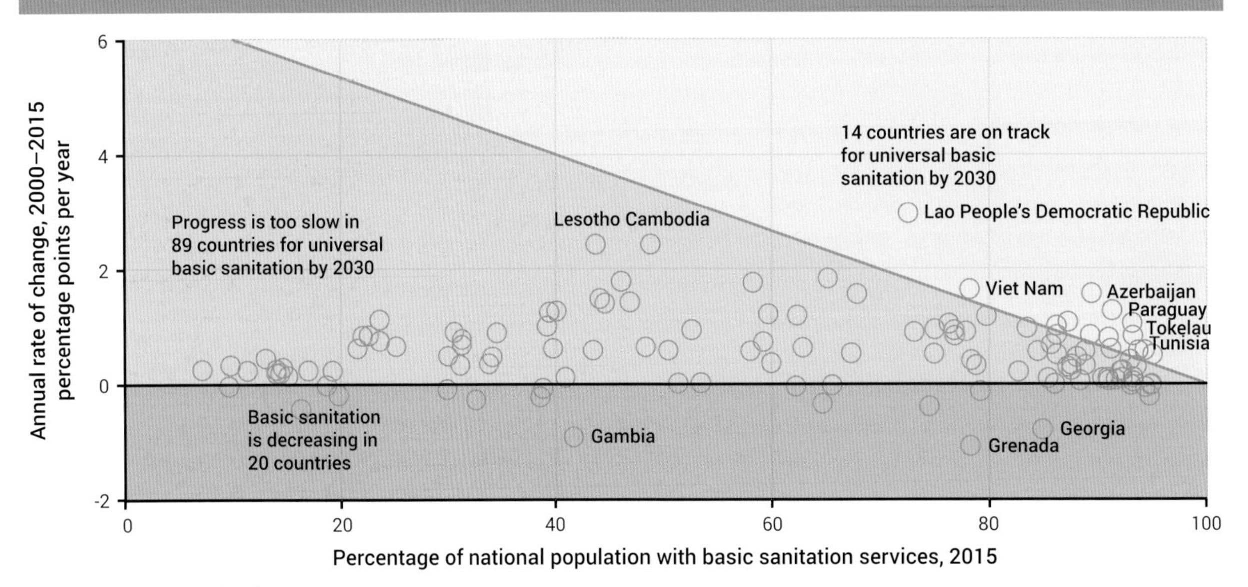

Source: WHO and UNICEF (2017).

(see Sections 4.4.3 and 9.8.2) with consequences for livelihoods, nutrition, public health and well-being. Agricultural water withdrawal is used for irrigation, livestock and aquaculture (FAO 2016). In particular, irrigation makes up the majority of total water withdrawal (67 per cent) (United Nations World Water Assessment Programme [WWAP] 2016).

Scope and measurement

Water withdrawal trends indicate how human use of fresh water has changed over time. At the global level, over the last century water withdrawal has increased **(Figure 16.5)**. The changes to blue water withdrawal suggest how irrigation has increased over time. The ratio of agricultural water withdrawal to total water withdrawal within a country varies across the globe with factors such as climate and priority given to agricultural activity **(Figure 16.6)**. The development of dams has contributed to anthropogenic water use and evaporation from storage of water in lakes or reservoirs. However, this type of water withdrawal is not currently reflected in the indicator discussed in this section (FAO 2016).

Graphical representation

Figure 16.5: Trends in global water withdrawal by sector between 1900 and 2010 (km³ per year)

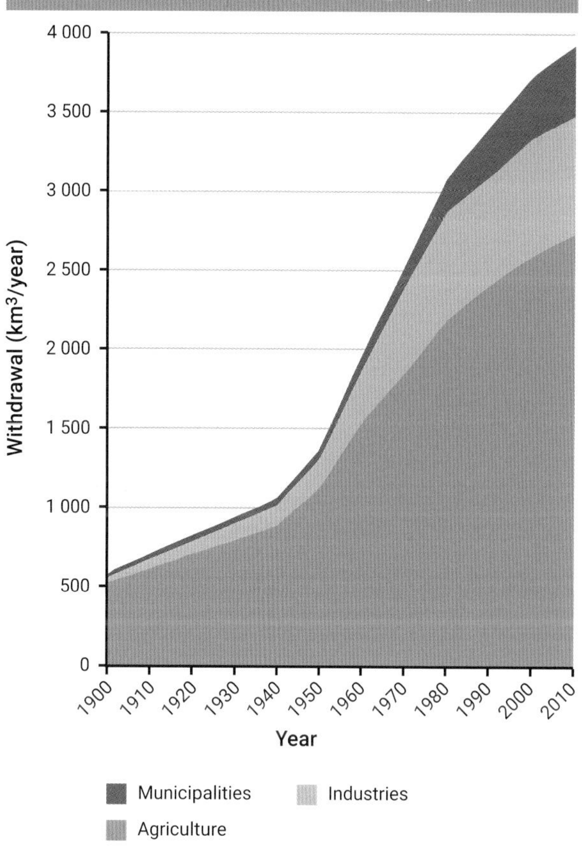

Source: Adapted from FAO (2016).

Figure 16.6: Proportion of total water withdrawn for agriculture

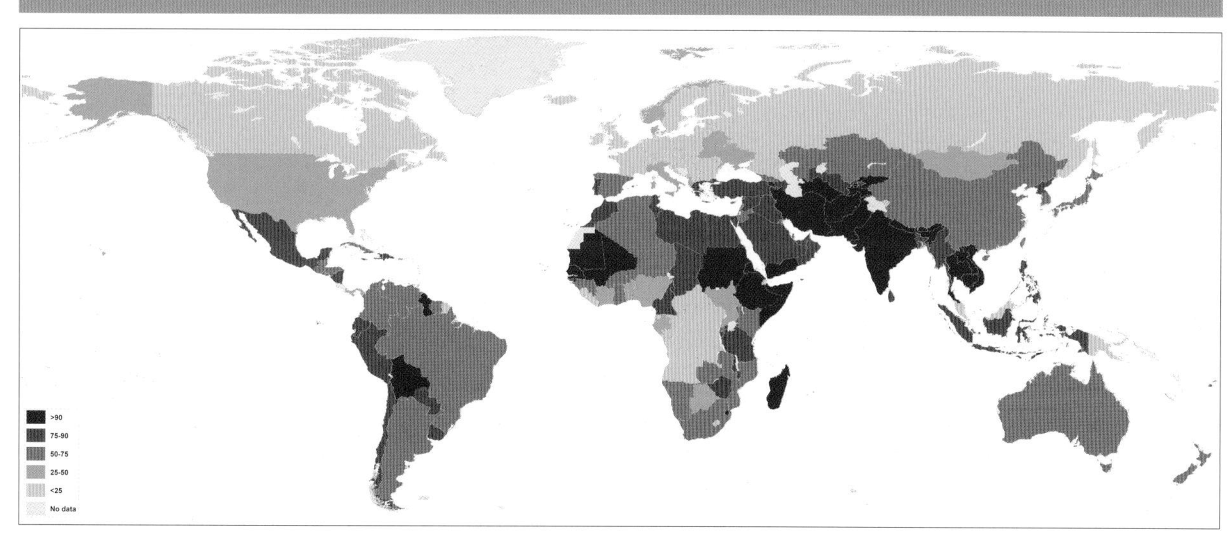

Source: Adapted from FAO (2015).

Figure 16.7: Changes in global gross crop water demand over time

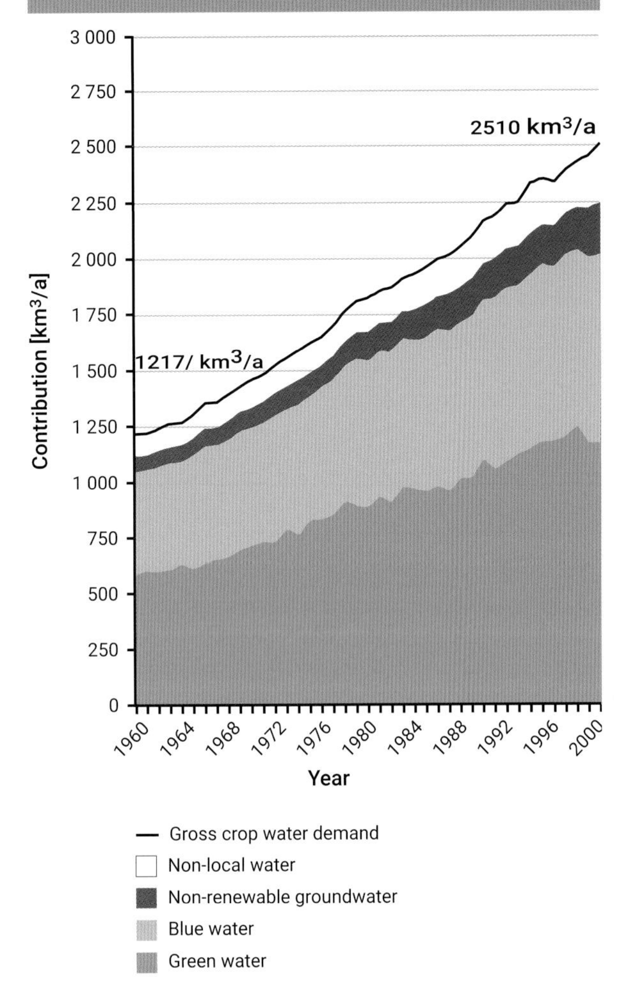

Source: Wada, van Beek and Bierkens (2012, p. 14)

Policy relevance

This indicator is directly relevant to SDG target 6.4: By 2030, substantially increase water-use efficiency across all sectors and ensure sustainable withdrawals and supply of fresh water to address water scarcity and substantially reduce the number of people suffering from water scarcity. Concerns over water quantity have been repeatedly raised in global policies and numerous multinational environmental agreements such as the 1977 Mar del Plata Action Plan, 1992 Dublin Statement of Water and Sustainable Development, the 1997 Convention on the Law of the Non-navigational Uses of International Watercourses (United Nations Watercourses Convention), the 1992 Convention on the Protection and Use of Transboundary Watercourses and International Lakes (UNECE Water Convention), and the International Law Commission's 2008 Draft Articles on the Law of Transboundary Aquifers. In addition, this indicator also draws attention to the balance between water for agriculture and water for industrial, household and ecosystem needs, which is addressed specifically in SDG target 6.5 advocating integrated water resources management (IWRM).

Causal relations

Subsidies are a major contributor to the expansion of irrigation agriculture. Full cost recovery rarely happens in developed countries. In developing countries, water user associations have been set up to decrease use of subsidies and charge for water use. However, such charges are not sufficient to meet full cost recovery (Toan 2016). Consequently, the price of irrigation undermines supply cost and disregards impacts on the environment. It has been suggested that the 'polluter pays' principle should be included in the price of irrigation (Howarth 2009).

While large-scale public investment in irrigation has been made in the past, it is unlikely for investments at such scale to be made in future. Instead, participatory irrigation management and irrigation management transfer are providing investment at the local scale and proving very popular (Turral *et al.* 2010).

Groundwater is increasingly used for agricultural purposes **(Figure 16.7)**. In particular, private groundwater wells and abstraction have become the main method for irrigation in India and are used widely in other developing countries such as China, Pakistan and Thailand (Turral *et al.* 2010). Here, the water-energy nexus is evident as cheaper pumping technology and easier energy access has enabled extraction, often at the individual level (Shah 2014). However, groundwater governance, especially for transboundary aquifers, has yet to be well established (Albrecht *et al.* 2017). There are also reported cases where efforts to improve irrigation efficiency have not contributed to the reduction of groundwater use, but rather the opposite (Pfeiffer and Lin 2014).

Other influencing factors

Molden *et al.* (2010) suggest supply management focusing on allocation has had a bigger impact on water efficiency than pricing to influence the behaviour of farmers. However, in large river systems, supply management through dams can lead to increased irrigation activity whereby dam-impacted catchments have 25 times more economic activity per unit of water compared with non-impacted catchments (Nilsson *et al.* 2005).

Possible alternative indicators

Vörösmarty *et al.* (2010) examined the ways human water security and biodiversity threats intersect globally. An indicator of these composite factors shows effects not only water withdrawals but also downstream and on ecosystems, beyond meeting water needs for agricultural output. Alternative indicators could provide insight into water scarcity at the subnational level. Considering that water scarcity is experienced at local level, alternative indicators could cover the spatial variation of water scarcity within countries. There are some emerging concepts and methodologies, for example the World Resources Institute Aqueduct water risk mapping makes detailed data accessible to a range of users including investors and companies (https://www.wri.org/our-work/project/aqueduct).

16.4 Discussion and conclusions

Various policy approaches show that water quantity and quality have serious implications for human and ecosystem health, and that these interactions are driven by changes in multiple sectors. Governance is increasingly opened up to non-State actors, such as the private sector and civil society. Decision-making thus needs to consider the full range of sectors and actors so that drivers and pressures (see Chapters 2 and 9) are addressed in an integrated fashion, considering economic, social and environmental issues. Achieving policy coherence and synergy are important features of the nexus interactions between fresh water and other sectors. Policy interventions should be designed to exceed purely technical fixes. This does not diminish the importance of provision of infrastructure such as wells, latrines and dams, but such provision should be considered within the complexity of a policy mix and with coherence in mind. In several case studies, public participation and stakeholder engagement have been implemented. However, the distribution of burdens and benefits of policies could be improved to address issues of equity and environmental justice.

The governance approaches and policy types examined in this chapter were not assessed in terms of evaluating non-monetary values. Where economic evaluations were conducted, trade-offs were mainly captured in monetary terms, and typically failed to assess impacts on human health or ecosystems. Negative impacts of policies on health have typically focused on natural hazards or infectious disease, and little has been done to capitalize on the potential co-benefits on human health (Grellier *et al.* 2017) or ecosystems.

Effective policies may be sought through active involvement of stakeholders. However, devolution of water governance does not necessarily result in better stakeholder engagement, as illustrated in the disaster risk reduction policy in England and Wales (Section 16.2.3); capacity-building and long term efforts of awareness-raising and knowledge use are also required to enable effective stakeholder involvement.

Monitoring thresholds and baseline conditions are a key component in the implementation of policy as well as for ensuring its overall effectiveness. Baseline conditions should be defined at implementation and subsequently monitored, causal relationships should be hypothesized and tested, and counterfactual thinking used to avoid misattribution of policy effectiveness due to confounding factors (Ferraro 2009). This is particularly true of access to safe drinking water and sanitation.

The selection of case studies in this chapter was guided by a number of requirements, in particular that case studies are described in depth in peer-reviewed literature, and as such the cases were drawn chiefly from developed economies. Developed economies are often equipped with resources and structures that allow for experimentation and innovation; accordingly, lessons learned from developed world case studies are not intended to be applied globally. On the contrary, a cautious approach should be taken with problems considered on a case-by-case basis, when embedded in their own specific context (Ingram 2013; Mukhtarov *et al.* 2015).

References

Abdul, K.S.M., Jayasinghe, S.S., Chandana, E.P.S., Jayasumana, C. and De Silva, P.M.C.S. (2015). Arsenic and human health effects: A review. *Environmental Toxicology and Pharmacology* 40(3), 828-846. https://doi.org/10.1016/j.etap.2015.09.016.

Albrecht, T.R., Varady, R.G., Zuniga-Teran, A.A., Gerlak, A.K. and Staddon, C. (2017). Governing a shared hidden resource: A review of governance mechanisms for transboundary groundwater security. *Water Security* 2, 43-56. https://doi.org/10.1016/j.wasec.2017.11.002.

Allan, C. and Watts, R.J. (2017). Revealing adaptive management of environmental flows. *Environmental Management* 61(3), 520-533. https://doi.org/10.1007/s00267-017-0931-3.

Al-Nammari, F. and Alzaghal, M. (2015). Towards local disaster risk reduction in developing countries: Challenges from Jordan. *International Journal of Disaster Risk Reduction* 12, 34-41. https://doi.org/10.1016/j.ijdrr.2014.11.005.

Arthington, A.H., Bunn, S.E., Poff, N.L. and Naiman, R.J. (2006). The challenge of providing environmental flow rules to sustain river ecosystems. *Ecological Applications* 16(4), 1311-1318. https://doi.org/10.1890/1051-0761(2006)016[1311:tcopef]2.0.co;2.

Austin, D. and Drye, B. (2011). The water that cannot be stopped: Southern Paiute perspectives on the Colorado River and the operations of Glen Canyon Dam. *Policy and Society* 30(4), 285-300. https://doi.org/10.1016/j.polsoc.2011.10.003.

Back, J.O., Rivett, M.O., Hinz, L.B., Mackay, N., Wanangwa, G.J., Phiri, O.L. *et al.* (2018). Risk assessment to groundwater of pit latrine rural sanitation policy in developing country settings. *Science of The Total Environment* 613-614, 592-610. https://doi.org/10.1016/j.scitotenv.2017.09.071.

Balfour, A., Wilson, I., de Jager, J., Still, D.A. and Louw, S. (2005). *Development of Models to Facilitate the Provision of Free Basic Water in Rural Areas.* WRC ReportSouth African Water Reseach Commission and The Mvula Trust. http://www.wrc.org.za/Knowledge%20Hub%20Documents/Research%20Reports/1379-1-05.pdf.

Bateman, M. and Engel, S. (2017). To shame or not to shame-that is the sanitation question. *Development Policy Review* 36(2), 155-173. https://doi.org/10.1111/dpr.12317.

Begg, C., Walker, G. and Kuhlicke, C. (2015). Localism and flood risk management in england: The creation of new inequalities? *Environment and Planning C-Government and Policy* 33(4), 685-702. https://doi.org/10.1068/c12216.

Bernauer, T. and Kuhn, P.M. (2010). Is there an environmental version of the Kantian peace? Insights from water pollution in Europe. *European Journal of International Relations* 16(1), 77-102. https://doi.org/10.1177/1354066109344662.

Binational.net (2012). *Areas of concern (annex 1).* https://binational.net/annexes/a1/.

Boundary Waters Treaty (1909). Treaty between Great Britain and the United States signed on 1 November. http://digitalcommons.unl.edu/cgi/viewcontent.cgi?article=1005&context=lawwater

Bond, P. and Dugard, J. (2008). Water, human rights and social conflict: South African experiences. *Law, Social Justice & Global Development* 1, 1-21. http://go.galegroup.com/ps/anonymous?id=GALE%7CA187844300&sid=googleScholar&v=2.1&it=r&linkaccess=abs&issn=14670437&p=AONE&sw=w.

Camacho, A.E., Susskind, L. and Schenk, T. (2010). Collaborative planning and adaptive management in Glen Canyon: A cautionary tale. *Columbia Journal of Environmental Law* 35, 1. https://papers.ssrn.com/sol3/papers.cfm?abstract_id=1572720.

Chatterton, J., Clarke, C., Daly, E., Dawks, S., Elding, C., Fenn, T. *et al.* (2016). *Delivering Benefits Through Evidence: The Costs and Impacts of the Winter 2013 to 2014 Floods.* Bristol: Environment Agency. http://rpaltd.co.uk/uploads/report_files/the-costs-and-impacts-of-the-winter-2013-to-2014-floods-report.pdf.

Collier, M.P., Webb, R.H. and Andrews, E.D. (1997). Experimental flooding in Grand Canyon. *Scientific American* 276(1), 82-89. www.jstor.org/stable/24993568.

Crocker, J., Saywell, D., Shields, K.F., Kolsky, P. and Bartram, J. (2017). The true costs of participatory sanitation: Evidence from community-led total sanitation studies in Ghana and Ethiopia. *Science of The Total Environment* 601-602, 1075-1083. https://doi.org/10.1016/j.scitotenv.2017.05.279.

Dadson, S., Hall, J., Murgatroyd, A., Acreman, M., Bates, P., Beven, K. *et al.* (2017). A restatement of the natural science evidence concerning catchment-based 'natural' flood management in the UK. *Proceedings of the Royal Society A: Mathematical, Physical and Engineering Sciences* 473(2199). https://doi.org/10.1098/rspa.2016.0706.

de Albuquerque, C. and Roaf, V. (2012). *On the Right Track: Good Practices in Realising the Rights to Water and Sanitation.* http://www.ohchr.org/Documents/Issues/Water/BookonGoodPractices_en.pdf.

Dugard, J. (2008). Rights, regulation and resistance: The Phiri Water Campaign. *South African Journal on Human Rights* 24(3), 593-611. https://doi.org/10.1080/19962126.2008.11864972.

Ferraro, P.J. (2009). Counterfactual thinking and impact evaluation in environmental policy. *New Directions for Evaluation* (122), 75-84. https://doi.org/10.1002/ev.297.

Findlay, R. and Telford, P. (2006). *The International Joint Commission and the Great Lakes Water Quality Agreement: Lessons for Canada-United States Regulatory Co-operation.* Toronto: Government of Canada. http://www.bvsde.paho.org/bvsacd/cd57/greatlakes.pdf.

Fischer, A. (2017). *Achieving and Sustaining Safely Managed Drinking Water in Bangladesh: Findings from a Water Audit.* Policy Brief. http://reachwater.org.uk/wp-content/uploads/2017/08/17_08-Matlab-Policy-brief-1.pdf.

Flanagan, S.V., Johnston, R.B. and Zheng, Y. (2012). Arsenic in tube well water in Bangladesh: Health and economic impacts and implications for arsenic mitigation. *Bulletin of the World Health Organization* 90(11), 839-846. https://doi.org/10.2471/BLT.11.101253.

Food and Agriculture Organization of the United Nations (2015). *Proportion of total water withdrawal withdrawn for agriculture.* Rome (http://www.fao.org/nr/water/aquastat/maps/WithA.WithT_eng.pdf.

Food and Agriculture Organization of the United Nations (2016). *Water uses.* http://www.fao.org/nr/water/aquastat/water_use/index.stm#time (Accessed: 19 October 2018).

Franks, D.M., Davis, R., Bebbington, A.J., Ali, S.H., Kemp, D. and Scurrah, M. (2014). Conflict translates environmental and social risk into business costs. *Proceedings of the National Academy of Sciences* 111(21), 7576-7581. https://doi.org/10.1073/pnas.1405135111.

Gerlak, A.K., Lautze, J. and Giordano, M. (2011). Water resources data and information exchange in transboundary water treaties. *International Environmental Agreements: Politics, Law and Economics* 11(2), 179-199. https://doi.org/10.1007/s10784-010-9144-4.

Giordano, M., Drieschova, A., Duncan, J.A., Sayama, Y., De Stefano, L. and Wolf, A.T. (2014). A review of the evolution and state of transboundary freshwater treaties. *International Environmental Agreements: Politics, Law and Economics* 14(3), 245-264. https://doi.org/10.1007/s10784-013-9211-8.

Global Water Partnership (2017). *What is the IWRM toolbox?* https://www.gwp.org/en/learn/iwrm-toolbox/About_IWRM_ToolBox/What_is_the_IWRM_ToolBox/ (Accessed: 18 October 2018).

Great Lakes Water Quality Agreement (2012). Protocol Amending the Agreement Between Canada and the United States of America on Great Lakes Water Quality, 1978, as Amended on October 16, 1983, and on November 18, 1987. Signed September 7, 2012. Entered into force February 12, 2013. https://binational.net//wp-content/uploads/2014/05/1094_Canada-USA-GLWQA-_e.pdf

Grafton, R.Q. (2017). Responding to the 'wicked problem' of water insecurity. *Water Resources Management* 31(10), 3023-3041. https://doi.org/10.1007/s11269-017-1606-9.

Grellier, J., White, M.P., Albin, M., Bell, S., Elliott, L.R., Gascón, M. *et al.* (2017). BlueHealth: A study programme protocol for mapping and quantifying the potential benefits to public health and well-being from Europe's blue spaces. *BMJ open* 7(6), e016188. https://doi.org/10.1136/bmjopen-2017-016188.

Gunderson, L. (2015). Lessons from adaptive management: Obstacles and outcomes. In *Adaptive Management of Social-Ecological Systems.* Allen, C.R. and Garmestani, A.S. (eds.). Dordrecht: Springer. 27-38. https://link.springer.com/chapter/10.1007%2F978-94-017-9682-8_3 (Downloaded: 03/12/2017)

Hall, J.D., O'Connor, K. and Ranieri, J. (2006). Progress toward delisting a great lakes area of concern: The role of integrated research and monitoring in the hamilton harbour remedial action plan. *Environmental Monitoring Assessment* 113(1-3), 227-243. https://doi.org/10.1007/s10661-005-9082-8.

Hazel, J.E., Topping, D.J., Schmidt, J.C. and Kaplinski, M. (2006). Influence of a dam on fine-sediment storage in a canyon river. *Journal of Geophysical Research-Earth Surface* 111(F1). https://doi.org/10.1029/2004jf000193.

Helm, P.A., Milne, J., Hiriart-Baer, V., Crozier, P., Kolic, T., Lega, R. *et al.* (2011). Lake-wide distribution and depositional history of current- and past-use persistent organic pollutants in Lake Simcoe, Ontario, Canada. *Journal of Great Lakes Research* 37, 132-141. https://doi.org/10.1016/j.jglr.2011.03.016.

Hildebrand, L.P., Pebbles, V. and Fraser, D.A. (2002). Cooperative ecosystem management across the Canada–US border: Approaches and experiences of transboundary programs in the Gulf of Maine, Great Lakes and Georgia Basin/Puget Sound. *Ocean & Coastal Management* 45(6), 421-457. https://doi.org/10.1016/S0964-5691(02)00078-9.

Hoffmann, R. and Muttarak, R. (2017). Learn from the past, prepare for the future: Impacts of education and experience on disaster preparedness in the philippines and Thailand. *World Development* 96, 32-51. https://doi.org/10.1016/j.worlddev.2017.02.016.

Howarth, W. (2009). Cost recovery for water services and the polluter pays principle. *ERA Forum* 10(4), 565-587. https://doi.org/10.1007/s12027-009-0134-3.

Huitema, D., Mostert, E., Egas, W., Moellenkamp, E., Pahl-Wostl, C. and Yalcin, R. (2009). Adaptive water governance: Assessing the institutional prescriptions of adaptive (co-)management from a governance perspective and defining a research agenda. *Ecology and Society* 14(1). http://www.jstor.org/stable/26268026.

Hunt, A. (2011). *Policy Interventions to Address Health Impacts Associated with Air Pollution, Unsafe Water Supply and Sanitation, and Hazardous Chemicals.* OECD Environment Working Papers. Paris: Organisation for Economic Co-operation and Development. https://www.oecd-ilibrary.org/docserver/5kg9qx8dsx43-en.pdf?expires=1533200166&id=id&accname=guest&checksum=00D5FDE049958B5C405E744BB0C5E186.

Ingram, H. (2013). No universal remedies: Design for contexts. *Water International* 38(1), 6-11. https://doi.org/10.1080/02508060.2012.739076.

Inter-Agency Regional Analysts Network (2016). *The Sendai Framework for Disaster Risk Reduction. A Three Year Outlook (2016-2018) at a Global Shift. Asia Report.* http://www.iris-france.org/wp-content/uploads/2017/01/IARAN-Sendai-Framework-F%C3%A9v-2016.pdf.

International Council on Mining and Metals (2017). *A Practical Guide to Consistent Water Reporting.* London. https://www.icmm.com/website/publications/pdfs/water/170315_water-reporting-guidance_en.pdf.

International Joint Commission (1980). *Pollution in the Great Lakes Basin from Land Use Activities: Summary.* https://scholar.uwindsor.ca/cgi/viewcontent.cgi?referer=https://www.google.com/&httpsredir=1&article=1251&context=ijcarchive.

International Joint Commission (1981). *Supplemental Report Under the Reference on Pollution in the Great Lakes System from Land Use Activities on Phosphorus Management Strategies.* https://scholar.uwindsor.ca/cgi/viewcontent.cgi?referer=&httpsredir=1&article=1250&context=ijcarchive.

International Joint Commission (2001). *Great Lakes Science and Policy Symposium, November 6-8, 2001. Discussion Papers.* https://scholar.uwindsor.ca/cgi/viewcontent.cgi?article=1559&context=ijcarchive.

International Joint Commission (2017). *First Triennial Assessment of Progress on Great Lakes Water Quality.* Washington, D.C. http://ijc.org/files/tinymce/uploaded/GLWQA/TAP.pdf.

International River Foundation (2007). The Brisbane Declaration. http://riverfoundation.org.au/wp-content/uploads/2017/02/THE-BRISBANE-DECLARATION.pdf (Accessed: 3 December 2018)

Jetoo, S. (2017). The role of transnational municipal networks in transboundary water governance. *Water* 9(1), 40. https://doi.org/10.3390/w9010040.

Kabir, R., Titus Muurlink, O. and Hossain, M.A. (2015). Arsenicosis and stigmatisation. *Global Public Health* 10(8), 968-979. https://doi.org/10.1080/17441692.2015.1015435.

Keesstra, S., Nunes, J., Novara, A., Finger, D., Avelar, D., Kalantari, Z. *et al.* (2018). The superior effect of nature-based solutions in land management for enhancing ecosystem services. *Science of The Total Environment* 610-611, 997-1009. https://doi.org/10.1016/j.scitotenv.2017.08.077.

Kemp, P.S. and O'Hanley, J. (2010). Procedures for evaluating and prioritising the removal of fish passage barriers: A synthesis. *Fisheries Management and Ecology* 17(4), 297-322. https://doi.org/10.1111/j.1365-2400.2010.00751.x.

Khurana, I. and Sen, R. (2008). *Drinking Water Quality in Rural India: Issues and Approaches. Background Paper.* London: WaterAid. http://www.indiawaterportal.org/sites/indiawaterportal.org/files/DrinkingWaterQuality_0.pdf.

Kingsford, R.T., Biggs, H.C. and Pollard, S.R. (2011). Strategic adaptive management in freshwater protected areas and their rivers. *Biological Conservation* 144(4), 1194-1203. https://doi.org/10.1016/j.biocon.2010.09.022.

Konrad, C.P., Olden, J.D., Lytle, D.A., Melis, T.S., Schmidt, J.C., Bray, E.N. *et al.* (2011). Large-scale flow experiments for managing river systems. *Bioscience* 61(12), 948-959. https://doi.org/10.1525/bio.2011.61.12.5.

Korman, J., Kaplinski, M. and Melis, T.S. (2010). *Effects of High-Flow Experiments From Glen Canyon Dam on Abundance, Growth, and Survival Rates of Early Life Stages of Rainbow Trout in The Lees Ferry Reach of The Colorado River.* Open–File Report. Reston, VA: United States Department of the Interior and U.S. Geological Survey. https://pubs.usgs.gov/of/2010/1034/of2010-1034.pdf.

Krantzberg, G. (2012). Renegotiation of the 1987 great lakes water quality agreement: From confusion to promise. *Sustainability* 4(6), 1239-1255. https://doi.org/10.3390/su4061239.

Krantzberg, G. and De Boer, C. (2008). A valuation of ecological services in the Laurentian Great Lakes Basin with an emphasis on Canada. *American Water Works Association* 100(6), 100-111. https://doi.org/10.1002/j.1551-8833.2008.tb09657.x.

Laakso, S., Heiskanen, E. and Matschoss, K. (2016). *Deliverable 3.2. ENERGISE Living Labs Background Report*. http://www.energise-project.eu/sites/default/files/content/ENERGISE_D3.2_141117_FINAL_0.pdf.

Lewin, S., Norman, R., Nannan, N., Thomas, E. and Bradshaw, D. (2007). Estimating the burden of disease attributable to unsafe water and lack of sanitation and hygiene in South Africa in 2000. *South African Medical Journal* 97(8), 755-762. https://www.ncbi.nlm.nih.gov/pubmed/17952234.

Loftus, A. (2006). Reification and the dictatorship of the water meter. *Antipode* 38(5), 1023-1045. https://doi.org/10.1111/j.1467-8330.2006.00491.x.

Loftus, A. (2007). Working the socio-natural relations of the urban waterscape in South Africa. *International Journal of Urban and Regional Research* 31(1), 41-59. https://doi.org/10.1111/j.1468-2427.2007.00708.x.

Luh, J., Ojomo, E., Evans, B. and Bartram, J. (2017). National drinking water targets-trends and factors associated with target-setting. *Water Policy* 19(5), 851-866. https://doi.org/10.2166/wp.2017.108.

Marvin, C., Painter, S. and Rossmann, R. (2004). Spatial and temporal patterns in mercury contamination in sediments of the Laurentian Great Lakes. *Environmental research* 95(3), 351-362. https://doi.org/10.1016/j.envres.2003.09.007.

McDonald, D.A. (2008). *World City Syndrome: Neoliberalism and Inequality in Cape Town*. 1st edn. New York, NY: Routledge. https://www.taylorfrancis.com/books/9781135903374.

McLaughlin, C. and Krantzberg, G. (2012). An appraisal of management pathologies in the Great Lakes. *The Science of the total environment* 416, 40-47. https://doi.org/10.1016/j.scitotenv.2011.12.015.

Mehta, L. and Ntshona, Z.M. (2004). *Dancing to Two Tunes? Rights and Market-Based Approaches in South Africa's Water Domain*. Sustainable Livelihoods in Southern Africa Research Paper. Brighton: Institute of Development Studies. https://www.ircwash.org/sites/default/files/Mehta-2004-Dancing.pdf.

Melis, T.S. (2011). *Effects of Three High-Flow Experiments on the Colorado River Ecosystem Downstream from Glen Canyon Dam, Arizona*. Circular. Reston, VA: U.S. Geological Survey. https://pubs.usgs.gov/circ/1366/c1366.pdf.

Meretsky, V.J., Wegner, D.L. and Stevens, L.E. (2000). Balancing endangered species and ecosystems: A case study of adaptive management in Grand Canyon. *Environmental Management* 25(6), 579-586. https://doi.org/10.1007/s002670010045.

Minerals Council of Australia (2012). *Water Accounting Framework for the Minerals Industry - User Guide Version 1.2*.

Minerals Council of Australia (2017a). *Annual Report 2016*. Canberra. https://www.minerals.org.au/sites/default/files/17%20732%20%20MCA%20Annual%20Report%202016%20to%20be%20released%207%20Jun%202017.pdf.

Minerals Council of Australia (2017b). *Water accounting framework for the Australian minerals industry*. https://www.minerals.org.au/water-accounting-framework-australian-minerals-industry.

Mirumachi, N. (2015). *Transboundary Water Politics in the Developing World*. 1st edn. Oxon: Routledge. https://www.routledge.com/Transboundary-Water-Politics-in-the-Developing-World/Mirumachi/p/book/9780415812955.

Molden, D., Oweis, T., Steduto, P., Bindraban, P., Hanjra, M.A. and Kijne, J. (2010). Improving agricultural water productivity: Between optimism and caution. *Agricultural Water Management* 97(4), 528-535. https://doi.org/10.1016/j.agwat.2009.03.023.

Mudd, G.M. (2008). Sustainability reporting and water resources: A preliminary assessment of embodied water and sustainable mining. *Mine Water and the Environment* 27(3), 136-144. https://doi.org/10.1007/s10230-008-0037-5.

Mukhtarov, F. (2009). *The Hegemony of Integrated Water Resources Management: A Study of Policy Translation in England, Turkey and Kazakhstan*. Master of Science in Environmental Sciences and Policy, Central European University http://www.cps.ceu.hu/theses/2/2005/the-hegemony-of-integrated-water-resources-management-a-study-of-policy-translation-in.

Mukhtarov, F., Fox, S., Mukhamedova, N. and Wegerich, K. (2015). Interactive institutional design and contextual relevance: Water user groups in Turkey, Azerbaijan and Uzbekistan. *Environmental Science & Policy* 53, 206-214. https://doi.org/10.1016/j.envsci.2014.10.006.

Muller, M. (2008). Free basic water - a sustainable instrument for a sustainable future in South Africa. *Environment and Urbanization* 20(1), 67-87. https://doi.org/10.1177/0956247808089149.

Naidoo, N., Longondjo, C., Rawatlal, T. and Brueton, V. (2012). *The Provision of Free Basic Water to Backyard Dwellers and/or More Than One Household Per Stand*. WRC Report. Pretoria: South African Water Research Commission. http://www.wrc.org.za/Knowledge%20Hub%20Documents/Research%20Reports/1987-1-12.pdf.

Neeson, T.M., Ferris, M.C., Diebel, M.W., Doran, P.J., O'Hanley, J.R. and McIntyre, P.B. (2015). Enhancing ecosystem restoration efficiency through spatial and temporal coordination. *Proceedings of the National Academy of Sciences* 112(19), 6236-6241. https://doi.org/10.1073/pnas.1423812112.

Nilsson, C., Reidy, C.A., Dynesius, M. and Revenga, C. (2005). Fragmentation and flow regulation of the world's large river systems. *Science* 308(5720), 405-408. https://www.doi.org/10.1126/science.1107887.

Norgate, T.E. and Lovel, R.R. (2006). Sustainable Water Use in Minerals and Metal Production. *Green Processing 2006: Third International Conference on Sustainable Processing of Minerals and Metals*. Newcastle, 5-6 June 2006. Australasian Institute of Mining & Metallurgy 133-141 https://publications.csiro.au/rpr/pub?list=BRO&pid=procite:3994a205-8750-4e3e-b214-9699562d42af

Northey, S., Haque, N. and Mudd, G. (2013). Using sustainability reporting to assess the environmental footprint of copper mining. *Journal of Cleaner Production* 40, 118-128. https://doi.org/10.1016/j.jclepro.2012.09.027.

Northey, S.A., Mudd, G.M., Saairvuori, E., Wessman-Jääskeläinen, H. and Haque, N. (2016). Water footprinting and mining: Where are the limitations and opportunities? *Journal of Cleaner Production* 135, 1098-1116. https://doi.org/10.1016/j.jclepro.2016.07.024.

Obani, P. and Gupta, J. (2014). Legal pluralism in the area of human rights: Water and sanitation. *A current Opinion in Environmental Sustainability* 11, 63-70. https://doi.org/10.1016/j.cosust.2014.09.014.

Obani, P. and Gupta, J. (2016). Human right to sanitation in the legal and non-legal literature: The need for greater synergy. *WIREs Water* 3(5), 678-691. https://doi.org/10.1002/wat2.1162.

Olden, J.D., Konrad, C.P., Melis, T.S., Kennard, M.J., Freeman, M.C., Mims, M.C. *et al.* (2014). Are large-scale flow experiments informing the science and management of freshwater ecosystems? *Frontiers in Ecology and the Environment* 12(3), 176-185. https://doi.org/10.1890/130076.

Patten, D.T. and Stevens, L.E. (2001). Restoration of the Colorado river ecosystem using planned flooding. *Ecological Applications* 11(3), 633-634. https://doi.org/10.1890/1051-0761(2001)011[0633:ROTCRE]2.0.CO;2.

Peletz, R., Mahin, T., Elliott, M., Harris, M.S., Chan, K.S., Cohen, M.S. *et al.* (2013). Water, sanitation, and hygiene interventions to improve health among people living with HIV/AIDS: A systematic review. *AIDS* 27(16), 2593-2601. https://doi.org/10.1097/QAD.0b013e3283633a5f.

Penning-Rowsell, E. and Pardoe, J. (2015). The distributional consequences of future flood risk management in England and Wales. *Environment and Planning C-Government and Policy* 33(5), 1301-1321. https://doi.org/10.1068/c13241.

Penning-Rowsell, E.C. (2015). A realistic assessment of fluvial and coastal flood risk in England and Wales. *Transactions of the Institute of British Geographers* 40(1), 44-61. https://doi.org/10.1111/tran.12053.

Penning-Rowsell, E.C. and Johnson, C. (2015). The ebb and flow of power: British flood risk management and the politics of scale. *Geoforum* 62, 131-142. https://doi.org/10.1016/j.geoforum.2015.03.019.

Pfeiffer, L. and Lin, C.Y.C. (2014). Does efficient irrigation technology lead to reduced groundwater extraction? Empirical evidence. *Journal of Environmental Economics and Management* 67(2), 189-208. https://doi.org/10.1016/j.jeem.2013.12.002.

Pitt, M. (2007). *Learning Lessons from the 2007 Floods. An Independent Review by Sir Michael Pitt*. London. http://webarchive.nationalarchives.gov.uk/20100702222546/http://archive.cabinetoffice.gov.uk/pittreview/_/media/assets/www.cabinetoffice.gov.uk/flooding_review/flood_report_lowres%20pdf.pdf.

Pitt, M. (2008). *The Pitt Review: Learning Lessons from the 2007 Floods. Final Report*. London. http://webarchive.nationalarchives.gov.uk/20100812084907/http://archive.cabinetoffice.gov.uk/pittreview/_/media/assets/www.cabinetoffice.gov.uk/flooding_review/pitt_review_full%20pdf.pdf.

Poff, N.L., Allan, J.D., Bain, M.B., Karr, J.R., Prestegaard, K.L., Richter, B.D. *et al.* (1997). The natural flow regime. *Bioscience* 47(11), 769-784. https://doi.org/10.2307/1313099.

Rankin, W.J. (2011). *Minerals, Metals and Sustainability: Meeting Future Material Needs*. 1st edn. Melbourne: CRC Press. https://www.crcpress.com/Minerals-Metals-and-Sustainability-Meeting-Future-Material-Needs/Rankin/p/book/9780415684590.

Richter, B.D., Warner, A.T., Meyer, J.L. and Lutz, K. (2006). A collaborative and adaptive process for developing environmental flow recommendations. *River Research and Applications* 22(3), 297-318. https://doi.org/10.1002/rra.892.

Sabatier, P. (2005). Linking science and public learning: An advocacy coalition perspective. In *Adaptive Governance and Water Conflict: New Institutions for Collaborative Planning*. Scholz, J.T. and Stiftel, B. (eds.). Washington, D.C. chapter 19. 196-203. https://www.routledge.com/Adaptive-Governance-and-Water-Conflict-New-Institutions-for-Collaborative/Scholz-Stiftel/p/book/9781936331475

Shah, T. (2014). *Groundwater Governance and Irrigated Agriculture*. TEC Background Papers. Stockholm: Global Water Partnership. https://www.gwp.org/globalassets/global/toolbox/publications/background-papers/gwp_tec_19_web.pdf.

Shields, K.F., Bain, R.E.S., Cronk, R., Wright, J.A. and Bartram, J. (2015). Association of supply type with fecal contamination of source water and household stored drinking water in developing countries: A bivariate meta-analysis. *Environmental Health Perspectives* 123(12), 1222-1231. https://doi.org/10.1289/ehp.1409002.

Smith, S.D.P., McIntyre, P.B., Halpern, B.S., Cooke, R.M., Marino, A.L., Boyer, G.L. *et al.* (2015). Rating impacts in a multi-stressor world: A quantitative assessment of 50 stressors affecting the Great Lakes. *Ecological Applications* 25(3), 717-728. https://doi.org/10.1890/14-0366.1.

South Africa, Department of Water Affairs and Forestry (2002). *Free Basic Water. Implementation Strategy*. Pretoria. http://www.dwaf.gov.za/Documents/FBW/FBWImplementStrategyAug2002.pdf.

South Africa, Department of Water Affairs and Forestry (2003). *Strategic Framework for Water Services: Water is Life, Sanitation is Dignity*. Pretoria. http://www.dwaf.gov.za/Documents/Policies/Strategic%20Framework%20approved.pdf.

South Africa, Department of Water Affairs and Forestry (2007). *Free Basic Water. Implementation Strategy 2007: Consolidating and Maintaining*. Pretoria. https://www.gov.za/sites/default/files/FBW%20strategy%20-%20Version%204%20final%2020070402%20mk_0.pdf.

South Africa, Department of Water and Sanitation (2014). *Blue Drop Report*. Pretoria. https://www.green-cape.co.za/assets/Water-Sector-Desk-Content/DWS-2014-Blue-Drop-report-national-overview-part-1-of-2-2016.pdf.

South Africa, Department of Water and Sanitation (2017). *National Norms and Standards for Domestic Water and Sanitation Services. Version 3- Final*. Pretoria. https://cer.org.za/wp-content/uploads/1997/12/National-norms-and-standards-for-domenstic-water-and-sanitation-services.pdf.

Southwick Associates (2008). *Sportfishing in America: An Economic Engine and Conservation Powerhouse*. Alexandria, VA: American Sportfishing Association. http://www.southwickassociates.com/wp-content/uploads/2011/10/sportfishiginamerica_2007.pdf.

Speed, R., Yuanyuan, L., Zhiwei, Z., Le Quesne, T. and Pegram, G. (2013). *Basin Water Allocation Planning: Principles, Procedures and Approaches for Basin Allocation Planning*. Paris: United Nations Educational, Scientific and Cultural Organization. https://think-asia.org/bitstream/handle/11540/82/basic-water-allocation-planning.pdf?sequence=1.

Spitz, K. and Trudinger, J. (2008). *Mining and the Environment: From Ore to Metal*. 1st edn: CRC Press. https://www.crcpress.com/Mining-and-the-Environment-From-Ore-to-Metal/Spitz-Trudinger/p/book/9780415465106.

Statistics South Africa (2016). *GHS Series Volume VIII. Water and Sanitation: In-depth analysis of the General Household Survey 2002–2015 and Community Survey 2016 Data*. Pretoria: Statistics South Africa. http://www.statssa.gov.za/publications/03-18-07/03-18-072015.pdf.

Thorne, C. (2014). Geographies of UK flooding in 2013/4. *Geographical Journal* 180(4), 297-309. https://doi.org/10.1111/geoj.12122.

Thornton, J.A., Rast, W., Holland, M.M., Jolánkai, G. and Ryding, S.O. (1999). *Assessment and Control of Nonpoint Source Pollution of Aquatic Ecosystems: A Practical Approach*. MAB Series. New York, NY: Parthenon. http://unesdoc.unesco.org/Ulis/cgi-bin/ulis.pl?catno=116965&set=005A57358C_1_376&gp=&lin=1&ll=c.

Toan, T.D. (2016). Water pricing policy and subsidies to irrigation: A review. *Journal of Environmental Processes* 3(4), 1081-1098. https://doi.org/10.1007/s40710-016-0187-6.

Turral, H., Svendsen, M. and Faures, J.M. (2010). Investing in irrigation: Reviewing the past and looking to the future. *Agricultural Water Management* 97(4), 551-560. https://doi.org/10.1016/j.agwat.2009.07.012.

United Kingdom, Department for Environment Food and Rural Affairs (2004). *Making Space for Water: Developing a New Government Strategy for Flood and Coastal Erosion Risk Management in England*. London. http://www.look-up.org.uk/2013/wp-content/uploads/2014/02/Making-space-for-water.pdf.

United Kingdom, Department for Environment Food and Rural Affairs (2008). *Improving Surface Water Drainage: Consultation to Accompany Proposals Set Out in the Government's Water Strategy*. London. https://uk.practicallaw.thomsonreuters.com/6-380-6771?transitionType=Default&contextData=(sc.Default)&firstPage=true&comp=pluk&bhcp=1.

United Kingdom, Department for Environment Food and Rural Affairs (2012). *UK Climate Change Risk Assessment: Government Report*. London. https://assets.publishing.service.gov.uk/government/uploads/system/uploads/attachment_data/file/69487/pb13698-climate-risk-assessment.pdf.

United Kingdom Parliament (2010). *Flood and Water Management Act 2010*. https://www.legislation.gov.uk/ukpga/2010/29/pdfs/ukpga_20100029_en.pdf.

United Kingdom Parliament (2017). *Future flood prevention: Government's response*. https://publications.parliament.uk/pa/cm201617/cmselect/cmenvfru/926/92605.htm.

United Nations, General Assembly (2010). *The Human Right to Water and Sanitation*. 3 August. A/RES/64/292. http://www.un.org/en/ga/search/view_doc.asp?symbol=A/RES/64/292

United Nations (2018). *Sustainable development goal 6: Ensure availability and sustainable management of water and sanitation for all*. https://sustainabledevelopment.un.org/sdg6 (Accessed: 19 October 2018).

United Nations Children's Fund and World Health Organization (2015). *Progress on Sanitation and Drinking Water: 2015 Update and MDG Assessment*. New York and Geneva. http://files.unicef.org/publications/files/Progress_on_Sanitation_and_Drinking_Water_2015_Update_.pdf.

United Nations Office for Disaster Risk Reduction (2015). *Sendai Framework for Disaster Risk Reduction 2015-2030*. Geneva. https://www.unisdr.org/files/43291_sendaiframeworkfordrren.pdf.

United Nations Office for Disaster Risk Reduction (2017). *What is disaster risk reduction?* https://www.unisdr.org/who-we-are/what-is-drr (Accessed: 21 November 2017).

United Nations Office of the High Commissioner for Human Rights (2010). *The Right to Water: Fact Sheet No. 35*. http://www.refworld.org/docid/4ca45fed2.html

United Nations World Water Assessment Programme (2016). *The United Nations World Water Development Report 2016: Water and Jobs*. Paris: United Nations Educational, Scientific and Cultural Organization. http://unesdoc.unesco.org/images/0024/002439/243938e.pdf.

United States, Department of the Interior (1992). *Grand Canyon Protection Act of 1992*. https://www.usbr.gov/lc/phoenix/AZ100/1990/grand_canyon_protection_act_1992.html (Accessed: 7 November 2018).

United States, Department of the Interior (1996). *Record of Decision, Operation of Glen Canyon Dam: Final Environmental Impact Statement*. Washington, D.C. http://www.usbr.gov/uc/rm/amp/pdfs/sp_appndxG_ROD.pdf.

United States, Department of the Interior (2008). *Final Environmental Assessment: Experimental Releases from Glen Canyon Dam, Arizona, 2008 Through 2012*. https://www.usbr.gov/uc/envdocs/ea/gc/2008hfe/GCD-finalEA2-29-08.pdf.

United States, Department of the Interior (2011). *Environmental Assessment: Development and Implementation of a Protocol for High-flow Experimental Releases from Glen Canyon Dam, Arizona, 2011 - 2020*. Salt Lake City, UT. https://www.usbr.gov/uc/envdocs/ea/gc/HFEProtocol/HFE-EA.pdf.

United States, Department of the Interior (2016). *Record of Decision for the Glen Canyon Dam Long-Term Experimental and Management Plan Final Environmental Impact Statement*. Salt Lake City, UT. http://ltempeis.anl.gov/documents/docs/LTEMP_ROD.pdf.

United States, Department of the Interior (2018). *Colorado river storage project*. https://www.usbr.gov/uc/rm/crsp/index.html (Accessed: 7 November 2018).

United States Congress (1973). *Endangered Species Act 1973*. https://history.house.gov/HistoricalHighlight/Detail/35155?ret=True.

United States Environmental Protection Agency (2017). *Great Lakes Restoration Initiative Report to Congress and the President: Fiscal Year 2016*. Washington, D.C. https://nepis.epa.gov/Exe/ZyPDF.cgi/P100UQE0.PDF?Dockey=P100UQE0.PDF.

United States National Research Council (1985). *The Great Lakes Water Quality Agreement: An Evolving Instrument for Ecosystem Management*. Washington, D.C.: National Academy press. https://doi.org/10.17226/18933.

Valdez, R.A., Carothers, S.W., House, D.A., Douglas, M.E., Douglas, M., Ryel, R.J. *et al.* (2000). *A Program of Experimental Flows for Endangered and Native Fishes of the Colorado River in Grand Canyon*. Flagstaff, AR. http://www.riversimulator.org/Resources/GCMRC/Aquatic/Valdez2000ExpFlow.pdf.

von Hirschhausen, C., Flekstad, M., Meran, G. and Sundermann, G. (2017). *Clean Drinking Water as a Sustainable Development Goal: Fair, Universal Access with Increasing Block Tariffs*. DIW Economic Bulletin. https://www.diw.de/documents/publikationen/73/diw_01.c.561626.de/diw_econ_bull_2017-28-3.pdf.

von Schnitzler, A. (2008). Citizenship prepaid: Water, calculability, and techno-politics in South Africa*. *Journal of Southern African Studies* 34(4), 899-917. https://doi.org/10.1080/03057070802456821.

Vörösmarty, C.J., McIntyre, P.B., Gessner, M.O., Dudgeon, D., Prusevich, A., Green, P. *et al.* (2010). Global threats to human water security and river biodiversity. *Nature* 467, 555-561. https://doi.org/10.1038/nature09440.

Wada, Y., van Beek, L.P.H. and Bierkens, M.F.P. (2012). Nonsustainable groundwater sustaining irrigation: A global assessment. *Water Resources Research* 48(6). https://doi.org/10.1029/2011WR010562.

Water and Sanitation Program (2011a). *Water Supply and Sanitation in South Africa: Turning Finance into Services for 2015 and Beyond*. https://openknowledge.worldbank.org/bitstream/handle/10986/17752/6992300REPLACE0LIC00CS00SouthAfrica.pdf?sequence=1&isAllowed=y.

Water and Sanitation Program (2011b). *The Political Economy of Sanitation: How Can We Increase Investment and Improve Service for The Poor? Operational Experiences from Case Studies in Brazil, India, Indonesia, and Senegal*. Washington, D.C. https://www.zaragoza.es/contenidos/medioambiente/onu/768-eng.pdf.

Webb, R.H., Schmidt, J.C., Marzolf, G.R. and Valdez, R.A. (1999). *The Controlled Flood in Grand Canyon*. Geophysical Monograph Series. Washington, D.C.: American Geophysical Union. http://adsabs.harvard.edu/abs/1999GMS...110.....W.

Whittington, D., Jeuland, M., Barker, K. and Yuen, Y. (2012). Setting priorities, targeting subsidies among water, sanitation, and preventive health interventions in developing countries. *World Development* 40(8), 1546-1568. https://doi.org/10.1016/j.worlddev.2012.03.004.

Wiering, M., Kaufmann, M., Mees, H., Schellenberger, T., Ganzevoort, W., Hegger, D.L.T. *et al.* (2017). Varieties of flood risk governance in Europe: How do countries respond to driving forces and what explains institutional change? *Global Environmental Change* 44, 15-26. https://doi.org/10.1016/j.gloenvcha.2017.02.006.

World Bank (2018). *Environmental Flows for Hydropower Projects: Guidance for the Private Sector in Emerging Markets*. Washington, D.C. http://documents.worldbank.org/curated/en/372731520945251027/pdf/124234-WP-Eflows-for-Hydropower-Projects-PUBLIC.pdf.

World Health Organization (2014). *Investing in Water and Sanitation: Increasing Access, Reducing Inequalities. UN-Water Global Analysis and Assessment of Sanitation and Drinking-Water (GLAAS) 2014 Report*. Geneva. http://apps.who.int/iris/bitstream/10665/139735/1/9789241508087_eng.pdf.

World Health Organization (2017). *Annex 2: Safely Managed Sanitation Services*. Geneva. http://www.who.int/water_sanitation_health/monitoring/coverage/indicator-6-2-1-safely-managed-sanitation-services-and-hygiene.pdf.

World Health Organization and United Nations Children's Fund (2017). *Progress on Drinking Water, Sanitation and Hygiene: 2017 Update and SDG Baselines*. Geneva. https://www.unicef.org/publications/files/Progress_on_Drinking_Water_Sanitation_and_Hygiene_2017.pdf.

Chapter 17

Systemic Policy Approaches for Cross-cutting Issues

© Shutterstock/aquatarkus

Coordinating Lead Authors: John Crump (GRID-Arendal), Klaus Jacob (Freie Universität Berlin), Peter King (Institute for Global Environmental Strategies), Diana Mangalagiu (University of Oxford and Neoma Business School), Caroline Zickgraf (Université de Liège)

Lead Authors: Babatunde Joseph Abiodun (University of Cape Town), Giovanna Armiento (Agenzia nazionale per le nuove tecnologie, l'energia e lo sviluppo economico sostenibile), Rob Bailey (Chatham House, the Royal Institute of International Affairs), Elaine Baker (GRID-Arendal at the University of Sydney), Kathryn Jennifer Bowen (Australian National University), Irene Dankelman (Radboud University), Riyanti Djalante (United Nations University – Institute for the Advanced Study of Sustainability), Monica Dutta (TERI), Fintan Hurley (Institute of Occupational Medicine), Maria Jesus Iraola (University of Uruguay), Rakhyun E. Kim (Utrecht University), Richard King (Chatham House, the Royal Institute of International Affairs), Andrei Kirilenko (University of Florida), Oswaldo dos Santos Lucon (Sao Paulo State Environment Secretariat), Katrina Lyne (James Cook University), Diego Martino (AAE Asesoramiento Ambiental Estratégico and ORT University), Ritu Mathur (TERI), Gavin Mudd (Royal Melbourne Institute of Technology), Sebastian Sewerin (Swiss Federal Institute of Technology Zurich (ETH Zurich), Tim Stephens (University of Sydney), Patricia Schwerdtle (Monash University), Joni Seager (Bentley University), Laura Wellesley (Chatham House, the Royal Institute of International Affairs), Caradee Y. Wright (Medical Research Council of South Africa)

GEO Fellow: Souhir Hammami (Freie Universität Berlin)

Executive summary

The physical, social, economic and health impacts of climate change, especially on the most vulnerable communities, require urgent adaptation approaches that are systemic, multidimensional and transformative *(established but incomplete).* Climate change adaptation is a complex process and needs to occur in all regions and sectors, at multiple temporal and geographical scales. It must consider the complex and interacting elements and feedback mechanisms of the human-environment system. {17.3.1}

Climate adaptation in coastal cities and small island developing states (SIDS) is generally categorized as 'protect', 'accommodate', 'retreat' *(established but incomplete)***.** Adaptation needs to deal with multiple slow and rapid onset hazards such as coastal erosion, sea level rise, tropical cyclones, floods or drought. Climate adaptation in coastal cities is still insufficient and may lead to increased risks in the future. Many low-lying SIDS are experiencing intensified flooding and coastal erosion and the area may become uninhabitable in the long term. {17.3.1}

A transformative approach for climate adaptation needs to deal with uncertainties and complexities arising from climate change impacts, address the drivers of risks and deal with the underlying factors of vulnerability, reduce inequality, address gender empowerment, and build resilience and adaptive capacity *(established but incomplete)*. {17.3.3}

The agrifood system is responsible for significant environmental externalities, including greenhouse gas emissions, and is highly inefficient on an energy basis (*well established*). Achieving the Sustainable Development Goals (SDGs) requires urgent action to reduce the agrifood system's environmental footprint and increase its overall efficiency. {17.4.1}

Agriculture is responsible for the majority of environmental consequences associated with food production (*well established*)**.** The two broad policy approaches for addressing this are: (1) incorporating the cost of negative environmental externalities into market prices via the 'polluter pays' principle; and (2) incentivizing farmers to minimize negative externalities or create positive externalities through payments for ecosystem services, which might be considered as the 'beneficiary pays' principle. {17.4.2}

Without a change in global dietary trends, food system emissions growth may mean that the Paris Agreement goal of limiting warming to well below 2°C is unlikely to be reached (*established but incomplete*). Most environmental policies in this area are oriented towards addressing the sustainability of food production, with less attention paid to waste and consumption. Several governments have introduced economic policy measures to encourage environmentally sensitive farming practices. There are nascent signs of sustainability criteria being incorporated into dietary guidelines to convince consumers to adjust their consumption patterns to optimize nutritional outcomes and to reduce the environmental burden of doing so. {17.4.3, 17.4.4}

Long-term planetary sustainability requires policy and technological interventions across energy systems to bring about choice of fuels, the way they are produced and consumed, and the way in which resources are affected systemically at every stage of the energy system (*established but incomplete*). {17.5.1, 17.5.2}

Mechanisms to address these challenges include carbon pricing (cap and trade systems, carbon taxes and other economic instruments such as fuel taxes and different subsidies to renewable energy), regulatory approaches (energy efficiency standards, command-and-control, mandatory decommissioning of old plants), information programmes (addressing behaviour, lifestyle and culture), and addressing administrative or political barriers (including through international cooperation) (*established but incomplete*). {17.5.3}

Decarbonizing supply and improving demand efficiency are two key policy elements that have been applied successfully (*well established*). Nevertheless, they need to be scaled up rapidly, together with the phasing in of new policies. {17.5.4}

The global economy currently operates predominantly in a linear mode whereby resources are extracted, converted through manufacturing to products and then disposed of (*well established*). {17.6.1}

The use of natural resources has grown rapidly over the last two decades and the global supply chains of resources have become more complex, resulting in growing environmental pressures and impacts (*well established*). {17.6.1}

A global shift is needed to a circular economy in which resource efficiency contributes to economic growth and human well-being, with reduced environmental pressures and impacts (*established but incomplete*). This would have substantial co-benefits for greenhouse gas abatement and waste and pollution minimization. {17.6.2}

A circular economy is a systems approach to industrial processes and economic activity that enables resource to maintain their highest value for as long as possible (*well established*). Key considerations in implementing a circular economy are reducing and rethinking resource use, and the pursuit of longevity, renewability, reusability, reparability, replaceability and upgradability for resources and products that are used.

Resource efficiency contributes to economic resilience by increasing the supply security of primary materials and closing of resource loops through remanufacturing and recycling, thereby reducing the pressures of resource exploitation, climate change, accumulation of toxic substances in ecosystems, and biodiversity loss (*well established*). {17.6.2}

Resource efficiency does not always happen spontaneously but requires well-designed policies that facilitate a change to sustainable systems of production and consumption and sustainable infrastructure (*established but incomplete*). {17.6.4}

Systems, product and service design that reduce demand and increase efficiency in resource use are key to bringing about the circular economy (*inconclusive*). Cross-sector and cross-disciplinary collaboration that empowers consumers as citizens is also key. {17.6.4}

Resource efficiency, greenhouse gas abatement and waste minimization policies, implemented together, will enable the decoupling of economic development and human well-being from global environmental degradation and resource exploitation (*inconclusive*). {17.6.4}

17

17.1 Cross-cutting policy issues and systemic change

The 2030 Agenda for Sustainable Development affirms the determination of governments to "take the bold and transformative steps which are urgently needed to shift the world on to a sustainable and resilient path." Achieving this transformation requires urgent and dramatic change in cross-cutting sustainable development policy areas which have closely intertwined social, economic and environmental dimensions.

Chapter 4 of this report identifies 12 cross-cutting issues of immediate concern for policymakers: health, environmental disasters, gender, education, urbanization, climate change, polar regions and mountains, chemicals, waste and wastewater, resource use, energy, and the food system. Because of their link to key economic, social and environmental systems, four of these 12 cross-cutting issues – climate change, food, energy and resource use – are selected for further analysis here.

This chapter evaluates the capacity of environmental policies to achieve transformational change in addressing cross-cutting global sustainable development challenges. To this end, the chapter addresses the major challenges of adapting socioeconomic systems to climate change, creating a sustainable agricultural and food production system, decarbonizing energy systems, and creating a circular economy. The world's pressing environmental challenges are the consequence of deeply rooted socioeconomic systems that reach across multiple policy areas. If global human needs are to be met within planetary boundaries there must be a transformation in the operation of these systems to reduce biophysical resource use and achieve just social outcomes (Raworth 2012; O'Neill *et al.* 2018). Systemic transformation will be very challenging for some communities but will provide a range of benefits and opportunities. Some of these opportunities can be realized in the short term, others over a longer period. In order to achieve a transformation which attracts widespread support, the opportunities and challenges will need clear communication, the expectations of affected groups and sectors will need to be considered, while those who suffer dislocation or negative distributional impacts from change will need to be compensated, retooled and reskilled.

17.1.1 A safe operating space

Transforming global systems towards a sustainable and resilient path is a major challenge because of the legacy of past policies, knowledge systems and cultural norms (Economic and Social Commission for Asia and the Pacific [ESCAP], Asian Development Bank [ADB] and United Nations Development Programme [UNDP] 2018) and because of the inherent complexity in policy arenas, involving many issue areas and actors. Climate change, for instance, has been described as a "diabolical policy problem" because its solution requires high levels of cooperation among governments and the implementation of policy measures across many economic sectors (Garnaut 2008).

In the Anthropocene, cross-cutting policy challenges involve a tightly coupled interdependency between the biophysical and socioeconomic elements of the Earth system (Liu *et al.* 2007; Biermann 2014; Young 2017). The central challenge for environmental policy in this new era is meeting human needs in a way that does not overstep planetary boundaries, and stay within a safe operating space for humanity (Rockström *et al.* 2009). For this objective to be reached there must be a radical reduction in biophysical resource use and a transformation in physical and social provisioning systems which connect resource use to just social processes and outcomes (Raworth 2012; O'Neill *et al.* 2018).

In pursuing transformation, it is vital that policymaking is strategic, coordinated and directed to the achievement of a clear vision. Environmental policies that address only one aspect of a systemic, cross-cutting, sustainable development challenge are unlikely to achieve the change necessary to shift the earth's socio-ecological systems to a pathway towards sustainability. For example, an isolated policy for reducing greenhouse gas emissions for one product may provide an economic incentive for production to shift to another, unregulated, product with the result that there is no net or economy-wide emissions reduction (Yang *et al.* 2012; van den Bergh *et al.* 2015). This is why in some contexts general regulation is preferable to technology-specific policies that 'pick winners'. Cross-cutting environmental issues must therefore be approached holistically, with policy interventions implemented with the objective of transforming the relevant system as a whole, including shifting collective behaviour and changing unsustainable social practices and norms.

However, setting the necessary and ambitious goal of transforming socio-economic or socio-technical systems does not always mean that the environmental policies directed to achieve this goal must be all-encompassing. An effective strategy for transformation that pursues a clear and overarching vision can be given operational effect through environmental policies applicable at macro, medium and micro scales. In some policy contexts, small-scale targeted interventions that can create innovation will be more effective than expansive policies. From this perspective, promotion of specific technological or social innovation can in some circumstances be justified. There is evidence that transformation of some socio-ecological systems can begin from change made within niches that can lead to technical and other innovations that result in more sustainable patterns of resource use (Doyon 2018). While small changes to one system may lead rapidly to a tipping point and a transformation of the system, other systems are more entrenched and robust and not easily shifted to a sustainable mode. Breaking through this path dependency requires a suite of policies and approaches at multiple scales.

17.2 Key actors, policies and governance approaches

Globalization has resulted in the emergence of complex global socio-ecological systems that do not operate in a predictable way and can give rise to nonlinear change. This means that policymaking and implementation occurs under conditions of uncertainty and there is an increasing premium on environmental governance that can respond in an agile way to rapid and unanticipated change (Young 2017). In this context, governments retain a central role in achieving successful transformation of socioeconomic systems. Governments continue to have the capacity to adopt a collection of policies from command-and-control regulations through to market-based measures in response to environmental problems. There are many examples where decisive government intervention has delivered major environmental benefit and transformed existing systems (e.g. the phase out of ozone-depleting substances, and the control of oil pollution from ships in the marine environment).

However, sometimes traditional governance approaches have their limits, including when what is needed is transformative change. Socio-ecological systems are increasingly complex in the variety of their components and their interactions so that it is not always possible to predict in advance what impact policy measures may have (Young 2017). Therefore, in addressing cross-cutting challenges, requiring whole-of-system change, there needs to be a willingness on the part of governments to engage in a reflective and experimental process of 'learning by doing', including regulatory experiments to test the feasibility of various approaches (e.g. Ostrom *et al.* 2007; Dryzek 2014).

This process of 'transformative learning' (ESCAP, ADB and UNDP 2018) can promote innovation by enabling experimentation through:

i. creating and highlighting opportunities for communities to embrace new and alternative visions for serving human needs in a sustainable way;
ii. enabling the participation of new actors that can provide more sustainable resources and services; and
iii. transparently phasing out existing unsustainable structures.

Government has an important role in this process but there is a broader dynamic at play in which it is possible to achieve 'governance without government' (Ostrom 1990). Key to this process is social mobilization around shared values and a vision for just and sustainable systems.

17.2.1 Evaluating the effectiveness of policies for cross-cutting issues

On the basis of our continually improving understanding of environmental policymaking, it is possible to evaluate the effectiveness of environmental policies that address cross-cutting issues and their systemic drivers. This not only refers to their immediate or short-term performance in achieving their specific targets, but also to their potential to engender systemic transformation. There are two key criteria in this respect, namely the objective of the policy and the outcome of the transformation.

This chapter focuses on four cross-cutting global-scale sustainable development challenges and asks:

i. What are the most urgent changes required in the system?
ii. Which elements of the system do policies seek to address?
iii. What has been done to date and how effective have these measures been?
iv. What is the transformative potential of the policy approaches discussed?

In undertaking this assessment, four sustainability challenges are examined through the lens of specific case studies which illustrate policy responses in a range of different settings and highlight challenges and opportunities for policy design and implementation. This chapter also provides broader insights on the effectiveness of cross-cutting environmental policies by examining policy-sensitive indicators.

17.3 Adapting socioeconomic systems to be more resilient to climate change

Climate change adaptation is a critical issue for coastal cities and Small Island Developing States (SIDS), as these are places where exposure to climate change impacts is increasing dramatically because of sea level rise. This is combined with dense populations and infrastructure along the coasts, rapid and often unplanned urbanization of low-lying areas, loss of ecosystems and environmental degradation, unsustainable management of natural resources, and lack of existing adaptive capacities.

Climate change adaptation needs to address both natural and human systems. Natural systems such as beaches, wetlands and coral reefs need to be protected by maintaining coastal ecosystems and processes and preventing erosion and flooding. Human systems – including settlements, industry, infrastructure, agriculture, fisheries, tourism, recreation and health – must be strengthened to become more climate-resilient. Adaptation strategies have recognized the special importance of safeguarding the most vulnerable groups, including Indigenous Peoples, women, children, those living with disabilities, and economically disadvantaged communities.

17.3.1 What are the most urgent changes required in the system?

The impacts of climate change differ across geographical locations, sectors and social groups. It particularly affects the lives, livelihoods and psychological well-being of the poor, vulnerable communities and people affected by disasters (Davis 2015; Dankelman 2016). Primary impacts include health risks related to temperature stress and extreme events leading to increased mortality and injury, internal and cross-border displacement, and infrastructure and economic loss and damages (Watts *et al.* 2015; Grimmins *et al.* 2016; Internal Displacement Monitoring Centre [IDMC] and Norwegian Refugee Council 2017). The secondary health impacts are mediated via the environment, including increased risk of climate-sensitive disease, which can be vector-, water- or food-borne. Tertiary impacts are socially mediated and include migration and conflicts (Watts *et al.* 2015). This requires adaptive responses to protect, preserve and promote human health and well-being.

What elements of the system do the policies seek to address?
Adaptation to sea level rise in coastal cities and SIDS seeks to address vulnerability to the following climate change impacts: coastal erosion, sea level rise, floods, and extreme events. They are generally categorized as 'protect', 'accommodate' or 'retreat':

- **Protection** of people and property by building higher seawalls, improving land-use management, developing new building codes to raise dwellings and infrastructure and reducing coastal erosion;
- **Accommodation** by changing the existing practices to make them more resilient to sea level rise, improving

infrastructure to increase absorption capacity of water bodies and wetlands, regulating water flow, introducing insurance; and

- **Retreat** by abandoning high-risk areas and relocating people away from the hazard.

Climate adaptation in coastal cities is still insufficient and may lead to increased risks in the future. Protecting existing populations and infrastructure has often led to even more development in high-risk areas, resulting in the accumulation of risk (Hallegatte *et al.* 2013). Climate adaptation programmes have not effectively dealt with multiple slow and rapid onset hazards, such as floods, droughts, tropical cyclones and sea level rise. They are often undertaken through sectoral programmes in, for example, agriculture, health and disaster management, rather than addressing the underlying causes of vulnerability. This has implications for human rights since persistent inequalities in terms of access to assets, opportunities, voice and participation, or discrimination mean poor and vulnerable communities lack adaptive capacity and are disproportionately exposed, and highly sensitive, to climatic hazards (United Nations 2016).

Some low-lying SIDS have experienced increasing flooding and significant coastal erosion and are expected to eventually become uninhabitable. Affected populations will be displaced and will need to migrate to other places or countries, with accompanying implications for their health and well-being (Schwerdtle, Bowen and McMichael 2018). Policy responses need to strengthen health systems to make them both climate-resilient and migrant-inclusive (Schwerdtle, Bowen and McMichael 2018). They also have to be integrated with other policy areas, such as border and labour market policies, and social and human rights protection.

17.3.2 What has been done to date and how effective have these measures been?

SDG 13 recognizes climate change as a critical issue and calls for urgent actions through strengthening resilience and adaptive capacity, mainstreaming it into policies and planning, education and capacity-building. The 2015 Paris Agreement on climate change seeks to strengthen the capacity of countries to deal with the impacts of climate change and support action by developing nations and the most vulnerable countries. Support provided through strategies and mechanisms under the Paris Agreement include climate adaptation funds, technology transfer and climate insurance.

Global climate finance is US$410 billion on average annually (Buchner *et al.* 2017). However, 93 per cent of this is spent on mitigation, while less than 5 per cent (US$22 billion) is spent on adaptation (Buchner *et al.* 2017). Looking deeper into adaptation finance, less than US$4 billion is spent on coastal protection, infrastructure and disaster risk management (Buchner *et al.* 2017). These are areas in greatest need if adaptation is to be strengthened in coastal cities and SIDS **(Figure 17.1)**.

The United Nations Office for Disaster Risk Reduction (UNISDR), through the Sendai Framework for Disaster Risk Reduction (UNISDR 2015), recognizes the need for better integration of disaster risk reduction (DRR) and adaptation, since climate change increases the severity, intensity and frequency of disasters. Strengthened and more coherent actions towards the 2030 Agenda, the Paris Agreement and the Sendai Framework are being developed (UNISDR 2017). The focus in this area has moved from emergency management and response to reducing disaster risks and mainstreaming it into development.

Deaths from disasters have been dramatically reduced through early warning systems and better disaster preparedness and planning, while the current challenge is that the number of people affected and economic loss continues to increase (UNISDR 2017). The New Urban Agenda (United Nations Human Settlements Programme [UN-Habitat] 2016), coordinated by the United Nations Human Settlements Programme, is a global framework on sustainable urbanization to "make cities and human settlements inclusive, safe, resilient,

Figure 17.1: Climate finance on adaptation

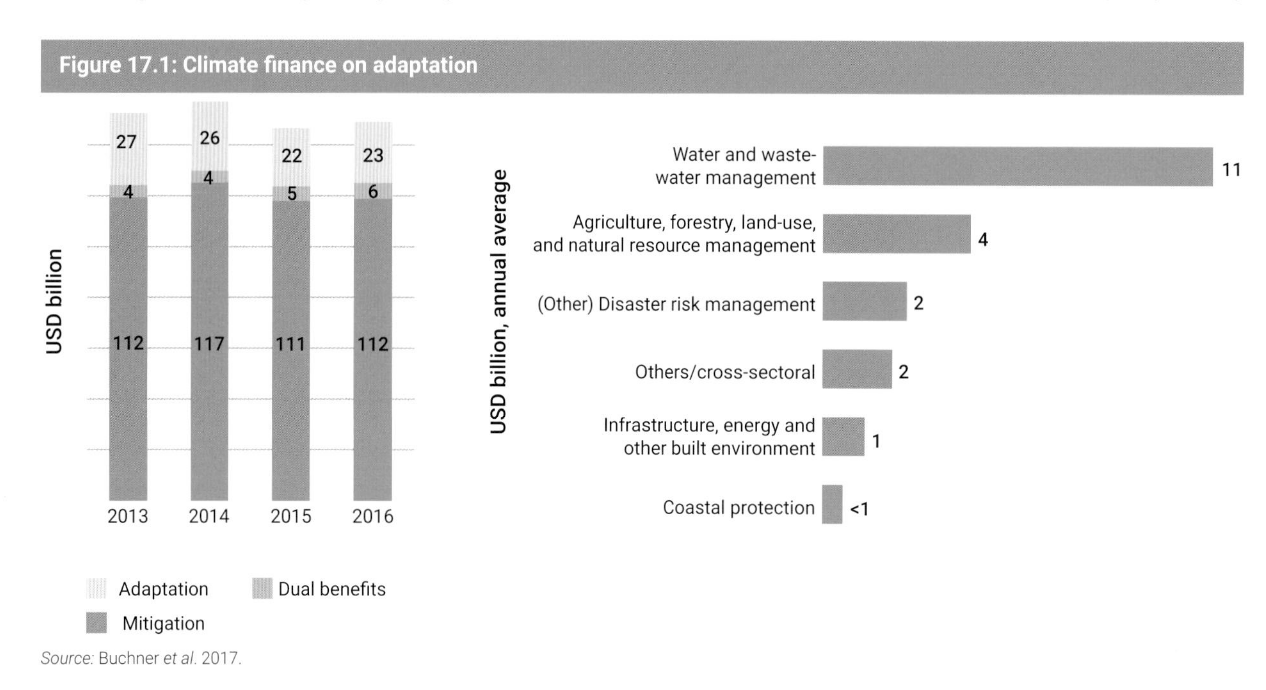

Source: Buchner *et al.* 2017.

and sustainable". It is clearly recognized that cities, especially those on coastlines, are where some of the most vulnerable places and infrastructure are located (World Bank 2013). Within the framework of the Global Action Programme (GAP) on Education for Sustainable Development, United Nations Educational, Scientific and Cultural Organization (UNESCO) implements Climate Change Education (CCE) alongside Education for Sustainable Development (ESD) programmes (UNESCO 2014). CCE includes, among other issues, the science of climate change, social and human aspects, policy responses and sustainable lifestyles (UNESCO 2010). To ensure effectiveness of this policy, research shows that educational interventions are most successful when they focus on local, tangible and actionable aspects of sustainable development and climate change, especially those that can be addressed by individual behaviour (Anderson 2013).

The Pacific Adaptation to Climate Change (PACC) Programme is the first major climate change adaptation initiative in the Pacific region and is a partnership between several key regional and national agencies and communities in 14 Pacific island countries. It is coordinated by the Secretariat of the Pacific Regional Environment Programme (SPREP). An assessment of this programme calls for a more integrated approach to climate change, disasters and climate mitigation, and better management of information and data (Hay 2009). Policies related to climate-related migration are only just emerging. There are policy frameworks that have sought to integrate migration with border protection, livelihoods and social and human rights protection such as those developed by the Asian Development Bank (ADB 2012) and the Protection Agenda of the Nansen Initiative (2015), but implementation remains rare at the local level.

17.3.3 What is the transformative potential of the policy approaches discussed?

The policy approaches and case study presented **(Box 17.1)** reinforce the need for better identification of governance for adaptation to address the complexities of the processes leading to and resulting from climate change impacts, and the underlying factors of vulnerability, and to build resilience and adaptive capacity. Governance of adaptation refers to the pattern that emerges from the processes of governing the social, political and administrative actors involved (Huitema *et al.* 2016). Successful adaptation requires consideration of effectiveness, efficiency, equity and legitimacy to ensure the sustainability of development pathways into an uncertain future (Adger, Arnell and Tompkins 2005).

A transformative approach towards adaptation to climate change is increasingly proposed as an approach to deal with the impacts of climate change and can potentially offer changes in the way current adaptation is governed and implemented towards being resilient and sustainable. It is an approach that has the potential to mediate complexity, uncertainty and rapid change. Its identified characteristics include adaptive management, particularly in allowing learning and self-organization; addressing scale to increase a governance 'fit' between social and ecological aspects; and a polycentric governance system allowing redundancy and diversity through participation and collaboration (Brunner *et al.* 2005; Folke 2006; Brunner and Lynch 2010; Djalante, Holley and Thomalla. 2011; Chaffin, Gosnell and Cosens 2014). The transformative potential is reflected through innovation, experimentation, vision and space for new actors. Learning allows for experimentation to take place, visions to be generated and innovations to flourish (Taylor 2017). Actions are taken based on the best available knowledge and allowing for learning from mistakes and innovation to take place. Climate change education also contributes to capacity-building for decision makers and empowers people to implement their own adaptation strategies – for example, by equipping them to understand complexity, perceive risks and take into account indigenous knowledge (Nakashima *et al.* 2012; Blum *et al.* 2013; Monroe *et al.* 2017; UNESCO 2017; UNESCO 2018). Overall, the learning by different stakeholders increases transformation capacity.

Box 17.1: Case study: 'Living With Floods' programme in Viet Nam

This case study provides an example of policy approaches towards achieving effective adaptation, despite vast complexities in setting targets for achieving policy effectiveness (i.e. social equity/human rights, community participation, economic variability, differing capacities, and multilevel policy fragmentation). The Vietnamese approach could be considered as transformative because flood risk management policy changed from control to 'living with floods'. However, the effectiveness of the approach is limited in the face of increasing hazard risks in the Mekong Delta.

The 'Living with Floods' programme is part of the National Strategy for Natural Disaster Prevention, Response and Mitigation to 2020. It aims to accommodate rather than control floods through the use of semi-dykes that allow occasional and controlled floods, which in turn lead to better soil management. Residential clusters are protected from flooding by a full- and semi-dyke system. There are permanent residences and access to basic public services and facilities, such as schools and health clinics (Central Committee for Flood and Storm Control [CCFSC] 2012). Up to 150,000 households are involved with the programme. These households are chosen directly by local authorities in the Mekong Delta. Poor households are eligible for a long-term, low-interest government loan to fund the acquisition of their new home, while wealthier households can purchase housing plots directly. A weakness of the process is a lack of transparency with regard to the selection of households and the allocation of funds. The sustainability of the funding from the national government is uncertain. There is no community participation or consultation in the selection of relocation sites (Chun 2015).

Overall, though the programme moved communities out of harm's way, it has resulted in an increase in the economic vulnerability of most households due to loss of livelihoods. Economic and solidarity networks have been dislocated in the process, and most households report decreased income, as well as difficulties in repaying their debts (Chun 2015; Entzinger and Scholten 2016). Such detrimental outcomes are largely the result of lack of integration of environmental with other policy objectives, such as long-term economic sustainability. As a result, though the programme contains many positive aspects and intentions, it has led to decreased community resilience.

17.3.4 Indicators

Indicators play a critical role in the monitoring and evaluation of climate change adaptation. The indicators for SDG 13, 'Climate action', do not provide the most direct measurement of adaptation effectiveness. The level to which the CCA action contributes to achieving SDG 5 (achieving gender equality and empowering women and girls) and SDG 10 (reducing inequality within and among countries) are also important indicators of success.

Scientific frameworks for measuring vulnerability, resilience and adaptive capacities along with indicators have been developed (e.g. Cutter, Boruff and Shirley 2003; Turner *et al.* 2003; Wisner *et al.* 2004; Hinkel *et al.* 2012; Taylor 2017). Examples of indicators to measure effective adaptation efforts for coastal cities can include identifying the amount of land area known to have (in)sufficient infrastructure, reducing the number of residents living in floodplains or low-elevation coastal zones, or developing a network of communication channels in times of crisis or disaster. For SIDS, indicators for adaptation include measures to respond to decreases in available fresh water (drought-resistant vegetation, water-saving devices, establishing buffer zones to protect catchment areas), prevention and removal of maladaptive practices (amend policies that lead to destruction of mangroves, laws preventing recycling of water, or allowing building in vulnerable areas), and address impacts of climate change on biodiversity and land degradation (land-use models for efficient farming, sustainable fishing practices, raising community awareness) (United Nations 2015).

Some considerations for achieving more transformational change – and ensuring effective adaptation measures – include consideration of scale (using a landscape- or basin-scale approach, and distinguishing between short-, medium- and long-term strategies), community participation, novel approaches to adaptation (e.g. the use of crop insurance in developing countries, building market resilience to climate change), and those that transform places or shift locations (artificial islands combined with relocation, and new institutions and funding mechanisms for reduced vulnerability) (Kates, Travis and Wilbanks 2012). Vulnerability and capacity assessments (VCA) together with climate risk screening and assessment are necessary to ensure that future development programmes consider impacts of climate change – see, for example, the International Federation of Red Cross and Red Crescent Societies VCA (International Federation of Red Cross and Red Crescent Societies n.d.), and Climate and the Disaster Risk Screening tool from the World Bank (World Bank 2018), or UNDP Report on stocktake of climate risks screening tool (Olhoff and Schaer 2010).

17.4 Creating a sustainable agrifood system

One of the best illustrations of the need to reduce uncertainties in the face of climate change is found in the agrifood system. The following section looks at some of the possibilities for transformation in this sector.

17.4.1 What are the most urgent changes required in the system?

The agrifood system is responsible for significant environmental impacts including greenhouse gas emissions, habitat destruction and biodiversity loss, and pollution of air and water resources. These environmental costs are compounded by the inefficiency of the agrifood system. According to one study, 62 per cent of the energy (in terms of kcal) harvested as crops and other biomass, is lost or wasted after accounting for losses from food waste, trophic losses from livestock, and human overconsumption (Alexander *et al.* 2017). Achieving the SDGs requires urgent action to reduce the system's environmental footprint and increase its overall efficiency and resilience. A whole-system approach is needed, including action to intensify agriculture sustainably, reduce food losses and greenhouse gas emissions along supply chains, and tackle wasteful consumption patterns including high consumer food waste and overconsumption of animal products.

Policies that shape the agrifood system can be broadly categorized in terms of production, processing and distribution, and consumption. Agricultural policies are typically focused on supporting farmers rather than on providing incentives for improved environmental outcomes. Moreover, reforming subsidy regimes often presents governments with significant political challenges. To the extent that they encourage production without accounting for environmental impacts, many agricultural policies exacerbate environmental problems (e.g. subsidies for fertilizer, water or energy use). Few governments have developed strategies for reducing greenhouse gas emissions from the agriculture and land-use sector (with the notable exception of forests); to date, no national government has fully included agriculture in a carbon pricing scheme.

Trade policies for agricultural commodities typically avoid explicit environmental criteria in order not to contravene World Trade Organization (WTO) rules that prevent governments from distinguishing between 'like' products, while regulations are concerned primarily with human health. Incentives to reduce food waste and losses have been eroded by low and declining real food prices (Benton and Bailey in press) and, despite increasing government intervention to shape consumption patterns for public health reasons (e.g. to reduce consumption of sugar, salt and trans fats), there is little policymaking that encourages sustainable diets (Garnet *et al.* 2015).

In sum, transforming the agrifood system to achieve the SDGs requires that the environmental footprint of agriculture is dramatically reduced, food losses and waste are drastically curtailed, and populations adopt healthier and more sustainable diets. This in turn requires a shift in policymaking to:

i. incentivize farmers to reduce negative environmental externalities, including greenhouse gas emissions, and create positive externalities, such as enhanced biodiversity or other ecosystem services;
ii. tackle food losses and waste along the entire value chain; **(Box 17.2)** and
iii. encourage the adoption of healthy and sustainable dietary patterns.

17.4.2 Which elements of the system do policies seek to address?

The polluter pays principle

Environmental impacts are a common symptom of agricultural policies that support farmers to maximize food production. Policy reforms designed to eliminate these impacts can take different forms, but essentially seek to ensure that the 'polluter pays'. Examples include taxes on fertilizer and pesticide use (rather than subsidies), water pricing schemes and regulations requiring farmers to build and maintain storage infrastructure for animal slurry.

While there is considerable national experience in applying the polluter pays principle to carbon emissions in the energy sector via emissions trading schemes and carbon taxes, agriculture remains excluded from such initiatives. Monitoring, reporting and verification of emissions in agriculture is considerably more complex and costly than for energy, because greenhouse gas emissions occur at the landscape scale according to farming practices and agroecological context. Nevertheless, this does not necessarily present an insurmountable barrier. For example, in New Zealand, the agricultural sector reports its greenhouse gas emissions without being part of the national emissions trading scheme, indicating that it is possible to quantify and account for emissions from agriculture.

The beneficiary pays principle: payments for ecosystem services (PES)

Several governments have introduced economic policy measures to encourage environmentally sensitive farming practices. The basic intention is to incentivize and reward those agricultural producers who take steps to minimize their environmental impacts or to deliver non-productive outputs (often termed 'payment for ecosystem services' [PES]), and to disincentivize and penalize those who do not (Meyer *et al.* 2014; Tanentzap *et al.* 2015). One such example is agricultural producers' participation in carbon markets by selling offset credits generated by specific projects to reduce emissions (Garnett 2012). In this case, rather than being penalized for emitting greenhouse gases as regulated entities under a carbon pricing scheme, farmers are paid for avoiding emissions.

The market for PES is growing and is now estimated at between US$36 billion and US$42 billion a year, including payments from non-governmental and private buyers. The largest areas include payments for watershed management and biodiversity, with the vast majority of payments for emissions reductions coming from forest projects (Salzman *et al.* 2018). Although by no means a negligible sum, these transfers are modest compared with conventional agricultural support, which totalled just under US$230 billion in 2017 in Organisation for Economic Co-operation and Development (OECD) countries and a similar amount in China (US$204 billion) (Organisation for Economic Co-operation and Development [OECD] 2018).

Consumer education

Consumer education, based on the concept of education for sustainable development, can enable consumers to understand how their individual dietary choices and habits influence social, economic and environmental development, to envision sustainable dietary choices and habits, and to adopt them (Fischer and Barth 2014; UNESCO 2017). For example, education can make meat consumers more aware of their own unsustainable consumption (Spannring and Grušovnik 2018).

Dietary guidelines

Governments typically use national guidelines to inform populations about good nutrition and healthy eating. In recent years, a small number of governments have begun to include environmental considerations in the guidelines they publish (see below for a discussion). National guidelines are unlikely to lead to widespread changes in eating habits on their own, but they can provide a basis for subsequent policymaking, and as such may constitute an important first step on the path to more concerted policy action (Bailey and Harper 2015, Garnet *et al.* 2015).

Labelling and certification

Schemes that provide consumers with assurance that a particular food meets certain environmental criteria have become increasingly common in developed country markets. These initiatives tend to be multi-stakeholder in their origins rather than policy led, often emerging from cooperation between the private sector and civil society; however, where sufficiently robust they can provide a basis for subsequent policymaking.

Public procurement

In many countries, public procurement of food can represent an appreciable share of market demand, hence public procurement policies in this area require suppliers to meet certain environmental standards and have the potential to drive wider change in the food system.

Consumption taxes

The costs of negative environmental impacts can also be incorporated at the point of consumption. To date, consumption taxes have been used to address health externalities associated with overconsumption of foodstuffs such as sugar. However, applying an emissions tax on foods at the point of consumption may be preferable to pricing emissions at the point of production. Although the latter approach may more accurately internalize the impact, consumption taxes may still be a better option because:

i. the costs of monitoring emissions in agriculture are high;
ii. the mitigation opportunities beyond reducing output of emissions-intensive foods are limited; and
iii. the opportunities for consumers to switch from foods of high emissions intensity to low emissions intensity are high (Wirsenius, Hedenus and Mohlin 2011).

Nonetheless, consumption taxes do not need to be blunt instruments with blanket rates applied indiscriminately across a product category. Differentiation between production and supply practices within a product category (e.g. by using disaggregated life cycle analyses) would allow for more nuanced reflection of externalities and incentivize the adoption of more sustainable practices, as well as consumer-switching to more sustainable products, within – as well as across – food categories. The transformative potential of consumption taxes could be high. It is estimated that worldwide emissions taxes on foods could save around 1 gigaton of CO_2 equivalent per year in 2020 and result in net health benefits at the global level due to reduced consumption of meat, although this would entail distributional impacts that governments would need to

manage with compensating policies (Springmann *et al.* 2016). No government has yet imposed an emissions tax on food, although some have implemented consumption taxes on certain foods for public health reasons.

17.4.3 What has been done to date and how effective have these measures been?

Production: economic incentives for ecosystem services
Payments for ecosystem services may pertain to additional conservation or sustainability practices to which agricultural producers commit voluntarily, or they may offer financial compensation to farmers whose income or production capacity is limited by the requirements of existing regulation (often referred to as 'cross-compliance') (Meyer *et al.* 2014).

In the European Union (EU), both approaches have been used under the Common Agricultural Policy (CAP). Agri-environment measures (AEMs) under Pillar II of the CAP are area-based mechanisms that occupy a middle ground between entirely voluntary schemes and direct compensation for cross-compliance. Funded jointly by the CAP and national authorities, AEMs are intended to encourage farmers to improve soil quality, use water resources more efficiently, reduce polluting inputs, and increase agricultural biodiversity. The majority of AEMs are action-based, compensating farmers for the activities they undertake, but more recently results-based AEMs have been introduced, with increased conditions and payments dependent on achieving desired environmental outcomes. These AEMs are less prescriptive with regard to management practices, are more cost-effective, and can encourage innovation (Illes *et al.* 2017). Generally, PES programmes applied nationally or internationally will be better able to maximize these benefits if they are flexible enough to be tailored to the unique conditions of local institutional and environmental contexts (de Blas *et al.* 2017).

As part of the CAP 2014-2020 Reform, the EU introduced a new form of direct payment support in 2015. The 'Greening Payment' was introduced under Pillar I of the CAP to supplement existing cross-compliance rules and oblige farmers who receive the direct payment support to meet three ecosystem service criteria. Initially the greening approach would provide "simple, generalized, annual and non-contractual payments" (European Commission 2011) that would create climatic and environmental benefits and permit Pillar II financial resources to be better spent on increasing the ambition of the agri-environment schemes (AESs). Relative to the original proposal, however, greening – as implemented – has affected a reduced area of farmland and encouraged fewer farmers to change their farming practices (Hart, Buckwell and Baldock 2016). Its effectiveness is also uncertain because ecosystem services usually need to be provided at a larger scale than permitted by agricultural management, requiring coordination across landowners (Benton 2012).

Although it is too early for a full end-of-project evaluation, there are a number of analyses that point to the greening programme having a limited impact and poor cost-effectiveness, given that it accounts for a sizeable proportion of the overall CAP budget (European Commission 2016; Gocht *et al.* 2016; Hart, Buckwell and Baldock 2016; Buckwell *et al.* 2017; OECD 2017).

Consumption: convincing stakeholders
There are early signs of sustainability criteria being incorporated into dietary guidelines, in an effort to convince consumers to adjust their consumption patterns to improve nutritional outcomes and to reduce the environmental burden. A recent global review of national dietary guidelines (Fischer and Garnett 2016) found that only four countries had so far included sustainability concerns into their food-based dietary guidelines (Brazil, Germany, Qatar and Sweden). Although most sustainability guidelines to date are health-oriented, reflecting the fact that their creation tends to be led by health ministries, and the link between behavioural change and influence from guidelines is challenging to demonstrate, more widespread inclusion of sustainability concerns in nutritional guidelines could serve to encourage policies that transform consumer demand.

17.4.4 What is the transformative potential of the policy approaches discussed?

Table 17.1 shows the transformative potential of some of the policy approaches discussed above as 'high', 'medium' or 'low'. These qualitative categories are posed as questions to

Box 17.2: Case study: Food losses and waste – multiple policy approaches in Japan

In Japan, multiple policy approaches are used to reduce food waste and losses, such as legislative targets, providing information to educate stakeholders, voluntary codes of conduct, and enabling new institutional arrangements. Those discussed here are primarily concerned with reducing waste in downstream sectors of the supply chain (processors, retailers, hospitality, consumers), but policy approaches are equally required to tackle upstream post-harvest losses. Policies to control and recycle food loss and waste have been implemented since 2000 under the Food Recycling Law, which obligates food manufacture, distribution and catering businesses to recycle waste materials and requires all businesses generating more than 100 tons of food waste annually to report on their waste generation and recycling activities (OECD 2014).

Following generally successful implementation – the majority of food waste associated with business activities is now recycled (as high as 95 per cent in the food manufacturing industry in 2011, though only 23 per cent in the catering industry in the same year [OECD 2014]) – food waste reduction is now a priority over reuse and recycling. Target values for controlling food waste generation have been established for 26 industry groups over the period of 2014-2019. Where unilateral action is challenging for businesses, such as waste resulting from returned goods and excess inventory, the Japanese food industry has formed a working group to address business practices such as changing delivery deadlines, best before date use standards, and labelling methods.

Levels of consumer food waste have changed little in recent years and this is now a priority area; it features prominently in the campaign introduced as a collaboration between six government ministries in 2013, 'No-Food-loss Project', aimed at increasing awareness and changing behaviour related to food losses at all stages of the supply chain (Food and Agriculture Organization of the United Nations [FAO] 2014).

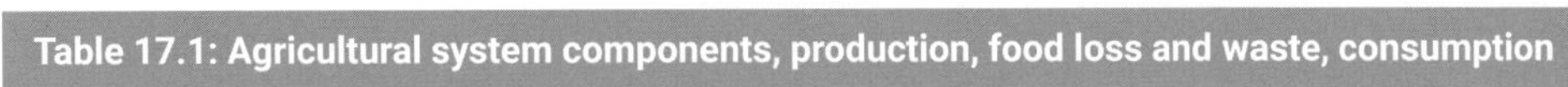

Table 17.1: Agricultural system components, production, food loss and waste, consumption

Agrifood system component	Production	Food losses and waste	Consumption
Policy approach	Economic incentives for ecosystem services: payments for ecosystem services	Various policy approaches, including food recycling laws	Convincing stakeholders: guidelines for sustainable, healthy diets
Promote innovation, including social and institutional innovation, that will not only improve existing approaches, but also entail completely new approaches to meet the needs of society?	Medium Payments for ecosystem services (PES) can be implemented in ways that are innovative – e.g. reforming existing subsidy regimes or developing new market mechanisms. Depending on the design and context, PES may create incentives for actors to develop new approaches.	Medium to high Mandatory targets encourage innovation across businesses in the supply chain to meet the requirements	Low May inform innovation in wider policymaking, but not intrinsically or in isolation. For example, regulatory nutritional labelling may encourage food manufacturers to reformulate products, but in isolation of wider regulation or policy, guidelines, are unlikely to promote innovation (Bailey and Harper 2015).
Enable experimentation, including regulatory experiments to test and demonstrate the feasibility of alternative configurations?	Medium The Common Agricultural Policy (CAP) example demonstrates that PES can enable regulatory experimentation through changing of subsidies. Outcome-based payments are more likely to encourage experimentation among the recipients, rather than activity-based payments.	Low to medium Regulation is unlikely to drive experimentation. Does not proactively drive businesses to experiment, but may create conditions in which they are encouraged to make changes.	Low Does not always enable experimentation, but could be the basis for subsequent regulatory experimentation.
Facilitate new and alternative visions for serving human needs in a sustainable manner?	High Depending on implementation, this approach could pay farmers to deliver new visions for agriculture and landscapes.	High Provides clarity on a low food waste future with expectations of high recycling rates.	Medium Shows a vision but does not aid the delivery of such a vision.
Create and enable new actors or new entrants that provide services to society in a more sustainable way?	Low Generally only works for existing businesses rather than encouraging new businesses to act.	Medium Enables new linkages and new opportunities for resource partnership to be realized by existing businesses, but does not necessarily encourage the entry of new businesses.	Low Creates few enabling conditions for new people to enter the business.
Organize the phase out of existing unsustainable structures?	Low Focuses on reforming the existing businesses.	Medium Reduces the volume of material going to landfill and the viability of existing waste chains, but does not necessarily fundamentally reorganize these structures.	Low Guidelines alone do little to reorganize existing structures – it requires accompanying policy measures to do so.

show the potential of the approach rather than the specific implementation of the instruments. How, and under what circumstances, each approach is implemented in any given situation will largely determine how transformative the outcomes are in that particular instance.

17.4.5 Indicators

Many existing indicators – such as agricultural emissions from different farming sectors – provide valuable information on the environmental sustainability of different parts of the food system, and others are still under development (e.g. SDG indicator 2.4.1 'Proportion of agricultural area under productive and sustainable agriculture'). However, these indicators are usually focused on productive aspects of the food system and tend not to show the efficiency or transformation of the system as a whole. To achieve this we propose a new policy-sensitive national-level indicator for the sustainability and nutrient efficiency of national dietary outcomes: dietary health and sustainability. The dietary health and sustainability indicator would be based on existing annual data series and measure the gap between national consumption patterns and national healthy and sustainable nutritional guidelines. However, as already noted, very few countries currently have nationally defined guidelines on the composition of healthy and sustainable diets. In the absence of such guidelines, alternative global values could be derived from the forthcoming recommendations of the EAT-Lancet Commission on Food,

Planet, Health, which intends to reach scientific consensus on what defines a healthy and sustainable diet (EAT-Lancet Commission on Food, Planet, Health 2018; Springmann *et al.* 2018).). The EAT-Lancet Commission recommendations could also be the basis of an aggregate global indicator.

If reliable national data on consumption are unavailable, the dietary health and sustainability indicator would use existing FAO Food Balance Sheet (FBS) data that include annual estimates of national food supplies per capita for each primary commodity and a number of processed commodities that are potentially available for human consumption. However, the FBS data are somewhat crude. The categories are summed to a high level, limiting the level of detail at which analysis can be conducted. Nor do they capture the nature of the food consumed, including whether it is heavily processed – which can have important health implications. Given these shortcomings, governments would be encouraged to gather more accurate data on national consumption patterns as well as to develop nationally appropriate guidelines for healthy and sustainable diets that better reflect the national context.

Table 17.2: Recommended intake for a healthy and sustainable diet

Hypothetical national (or EAT-Lancet Commission) recommended intake for a healthy and sustainable diet	Dietary Health and Sustainability indicator value (annual, national value)
X g/capita per day of fruit and vegetables	Vegetable intake: (+/−) Y per cent of healthy and sustainable levels
X kcal/capita per day of cereals and starches	Cereal and starch intake: (+/−) Y per cent of healthy and sustainable levels
X kcal/capita per day of oils and fats	Oil and fat intake: (+/−) Y per cent of healthy and sustainable levels
X g/capita per day of meat	Meat intake: (+/−) Y per cent of healthy and sustainable levels
X g/capita per day of dairy	Dairy intake: (+/−) Y per cent of healthy and sustainable levels

Food-groupings of national food intake or supply data would be measured to show the proportion by which they exceed or fall short of national guidelines or EAT-Lancet Commission recommended daily intakes for corresponding food groups:

$$[(\text{intake value} / \text{recommended intake}) - 1] \times 100$$

A value of zero represents 'ideal' consumption, negative values represent underconsumption and positive values show overconsumption. For example, if there were recommended intake values for the food groups in **Table 17.2**, the dietary health and sustainability indicator would express each country's supply in relative terms **(Figure 17.2)**.

The FBS data show the quantities of food available to the population after accounting for exports and imports, other uses (livestock feed, seed, non-food uses), and losses during storage and transportation. Therefore, the dietary health and sustainability indicator would provide a useful high-level picture of the performance of policies and measures across the entire food system, including actions to reorient agricultural production, trade measures, actions to reduce pre-household waste, and nutritional policies. It would provide an integrated measure of the agrifood system's contributions to progress against multiple SDGs.

Since this proposed dietary health and sustainability indicator is consumption-based, it would not fully reflect the impact of agricultural policies in countries that are large net exporters of agricultural goods, or which produce significant proportions of non-food agricultural products. For example, a country's consumption may appear to be healthy and sustainable, but if consumption is largely based on imported foods, this provides no indication of the sustainability of the agricultural system in that country. On a global basis, however, the dietary health and sustainability indicator would provide an aggregate indication of the sustainability of food production.

17.5 Decarbonizing energy systems

The previous section discussed how agricultural policies tend to focus on supporting farmers rather than providing incentives for improved environmental outcomes. In the complex agrifood system reducing energy use will also play an important role. This section explores the transformative potential that will come from decarbonizing all energy systems.

Figure 17.2: Health and sustainability of country X's dietary intake

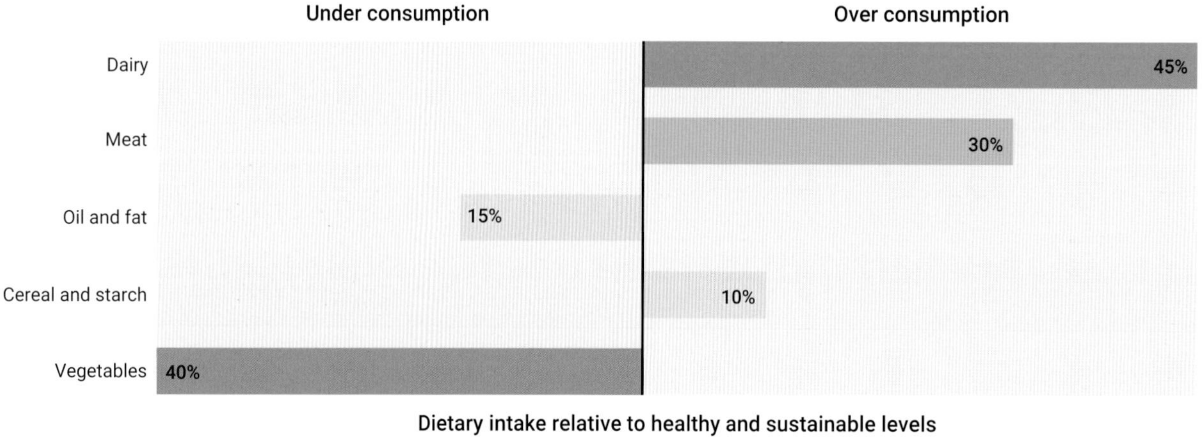

17.5.1 What are the most urgent changes required in the system?

Greenhouse gas emissions generated from energy use are a major driver of global climate change. Reducing the carbon footprint of global energy use requires integrated approaches that combine measures to:

i. reduce energy use;
ii. lower the greenhouse gas intensity of end-use sectors;
iii. decarbonize energy supply; and
iv. reduce net emissions and enhance carbon sinks.

There are important co-benefits of these measures, including:

i. reduced costs;
ii. greater energy security; and
iii. human and ecosystem health.

Near-term reductions in energy demand are cost-effective climate mitigation strategies, giving more flexibility for reducing carbon intensity in the energy supply sector, protecting against supply-side risks, and avoiding lock-in to carbon-intensive infrastructures. Delayed scaling up of low-carbon energy systems would make limiting warming over the 21st century to below 2°C very difficult to achieve, and will require much bolder actions such as a larger reliance on carbon dioxide removal in the long term (Intergovernmental Panel on Climate Change [IPCC] 2014).

17.5.2 What elements of the system do the policies seek to address?

Long-term planetary sustainability requires both policy and technological innovations to bring about changes in the choice of fuels, the way they are produced and consumed, and the way in which resources are impacted systemically at every stage of the energy system **(Figure 17.3)**.

Major areas of policy intervention in energy systems, which relate to the SDGs (especially SDG 7) are decarbonization measures that aim to substitute fossil fuels with clean(er) or renewable alternatives, implement efficiency measures that can provide the same service while using fewer resources, enhance access to other energy forms and services, apply land-use and urban planning which considers energy integration (e.g. distributed energy, smart grids, electric vehicle charging networks), and minimizes waste and lock-in of particular technologies by existing systems based on fossil fuels.

17.5.3 What has been done to date and how effective have these measures been?

Mechanisms to address these challenges include carbon pricing (cap and trade systems, carbon taxes and other economic instruments such as fuel taxes, different subventions to renewable energy), regulatory approaches (energy efficiency standards, command-and-control, mandatory decommissioning of old plants), information programmes (addressing behaviour, lifestyle and culture), and addressing administrative and political barriers (including through international cooperation) (IPCC 2014). Policy interventions also include research, development and demonstration (academic funding, grants, incubation support, research centres, public-private partnership, prizes, tax credits, voucher schemes, venture capital, soft and convertible loans), fiscal incentives (grants, energy production payments, rebates, tax credits and reductions, changes in depreciation), public finance (investments, guarantees, loans, procurement), regulations (quantity or quality driven, e.g. renewable portfolio standards, tendering and bidding, feed-in tariffs, green purchasing and labelling, net metering, priority to access to networks or dispatch) (Mitchell *et al.* 2011; International Renewable Energy Agency[IRENA] 2016; International Council for Science 2017; United Nations Industrial Development Organization [UNIDO] 2017).

Figure 17.3: An illustrative energy system

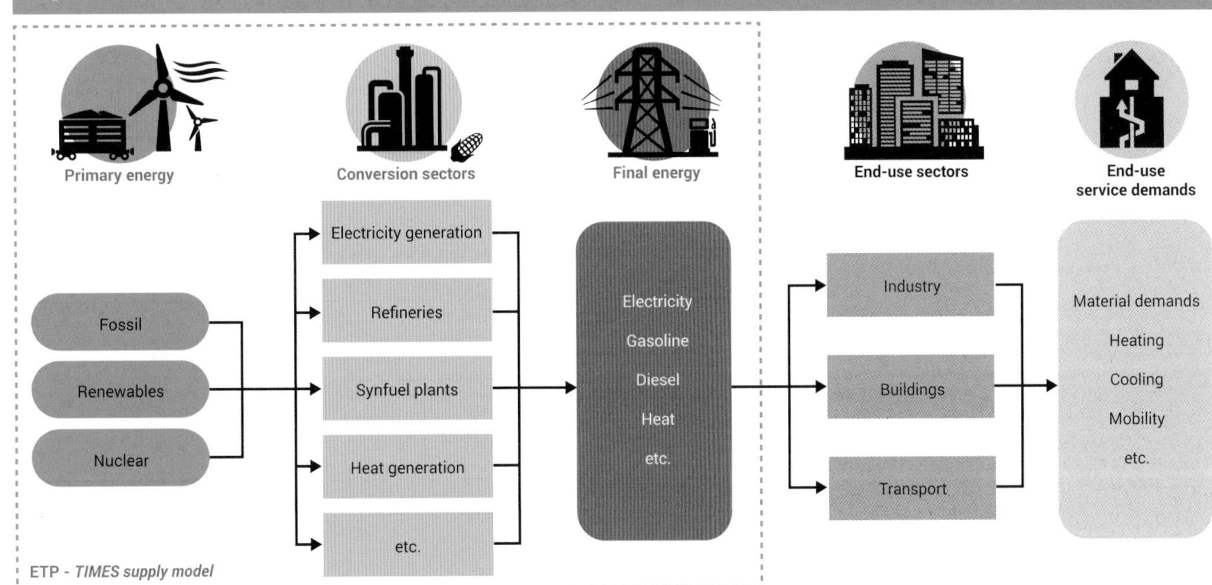

Source: Adapted from International Energy Agency (IEA) (2017)

Combined policies for renewable energy and efficiency form the basis of a low-carbon transformation for the global energy matrix. The diffusion, penetration and integration of these policies determine how effective this change can be. The effectiveness of these policy innovations depends on national capacities for action, on the demand for appropriate approaches applied by 'front-runner' countries, on the international policy transfer process, on the enabling conditions for such transfer, and whether policy models are developed at an early stage of the diffusion process to guide other countries (Kern, Jorgens and Jänicke 2001).

17.5.4 What is the transformative potential of the policy approaches discussed?

Building on the momentum created by the 2015 Paris Agreement, a total of 117 Nationally Determined Contributions were submitted, of which 55 included targets for increasing the use of renewable energy, while 89 made reference to renewable energy more broadly (Renewable Energy Policy Network for the 21st Century [REN21] 2017). In 176 countries, targets for renewable energy were a primary means by which governments expressed their commitments. As of 2016, nearly all countries directly supported renewable energy technology development and deployment through some mix of policies.

The other pillar of sustainable energy is efficiency. As shown in **Box 17.4**, improving energy efficiency can generate energy savings and mitigation of associated carbon emissions, encouraging large-scale investment in a competitive and innovative manufacturing industry.

Policy support for renewable energy has been focused mostly on power generation (as in the case in **Box 17.3)**, although implementation for such policies has slowed in recent years in response to tightening fiscal budgets and/or falling technology costs, with auction-based procurement now being a preferred policy approach. In 2014-2016, no new renewable portfolio standards or feed-in (tariffs and premiums) policies were introduced at the national level. However, support for new technologies is still an important driver for transformational change, and lessons from the past can be learned to allow an urgently needed scaling up to address climate change and other socio-ecological challenges.

On the demand side, electric efficiency tackles the purpose of environmental impact mitigation benefits along with improved energy access to cleaner energy. The India case **(Box 17.4)** resulted in spurring large-scale investment in manufacturing, improved standards, raised consumer awareness, generated employment and improved prospects for education, enhanced livelihoods and health.

17.5.5 Indicators

Energy production and consumption are one of the most tracked indicators, due to the heavy cost implications and the geopolitical implications of the energy sector. Because of this it is known that in 2015 the world consumed 13.65 billion metric tonnes of oil equivalent, with energy demand having doubled over the previous 40 years. Of this energy, 81.4 per cent was provided by fossil fuels (coal, oil and natural gas) emitting 32.3 billion tons of carbon dioxide (IEA 2017).

Box 17.3: Case study: Support for renewables in Germany: feed-in tariffs

The German Feed-in Tariff (FIT) policy under the 2000 Renewable Sources Act (Erneuerbare-Energien-Gesetz, EEG) was a remarkable intervention towards low-carbon technology (LCT) diffusion. The main policy design elements were: (i) guaranteed access to the grid for LCTs (purchase obligation); (ii) stable and long-term power purchase agreements (long payment duration); (iii) prices reflecting the varying costs of different LCTs (fixed tariffs with some particularly strong incentives for given technologies such as solar photovoltaics [PV] and onshore wind); and, more recently, (iv) expansion corridors for specific LCTs, limiting capacity additions and household costs. As a proxy for technology diffusion, installed capacity (2016) was 45.4 GW for onshore wind, 4.2 GW for offshore wind and 41.3 GW for solar PV (IRENA 2016). A 2016 amendment to the Act shifted the focus to large investors, with an auctioning scheme according to energy source, plant size and plant location. Design elements proved remarkably stable while flexible. Fixed tariffs led to a surge in deployment and the formation of a domestic solar industry.

In combination with the uptake of onshore wind, and farmers and house-owners profiting from the EEG's conditions, a powerful group of advocates evolved. Driven by a discussion about 'affordability' of continued LCT support schemes, the 2016 amendment replaced the FITs with an auctioning scheme, still technology-specific but aimed at existing large investors rather than at small ones that previously played a large role in the EEG. It was a blueprint for other countries which led to policy diffusion and learning (by doing and by using), ultimately driving down costs on a global scale faster than anticipated. Success was based on long-term guaranteed support and inter-technology differentiation, plus a relatively stable basic policy rationale, adjusted to changing conditions (e.g. cost changes) and minimizing windfall profits. Policy predecessors (1991 onwards) were already established in a highly regulated sector, ensuring fast decision-making, strong support and positive feedback loops. The Fukushima disaster and resulting commitment to nuclear phase-out also helped in creating long-term security in terms of LCT business models. Small decentralized project stakeholders were empowered, as was the domestic industry, in clusters around specific LCTs (wind, solar PV and others).

Household affordability was addressed through the introduction of caps for specific LCTs and factoring in social and environmental costs. Key actors were utilities and industry associations, environmental groups, political parties, and ministries. Some constraining complexity of the energy policies, the existing locked-in technology (fossil fuels, energy consumption), and badly designed policies (e.g. carbon pricing under the EU Emissions Trading System). Later, Fukushima changed the politics in the energy sector, and opponents criticized incentives due to cost inefficiency. However, even with the latest amendment replacing FITs by auctions, technology-specificity remained as a design element. The general public considered the policy necessary and effective (in terms of job and value creation, achieved technology innovation, disruption of incumbent systems, stable investment environment for LCTs) particularly during the first years that the policies were in effect.

Box 17.4: Case study: Demand-side management in India: affordable LED lights for all

The 2013 UJALA (Unnat Jyoti by Affordable LEDs for All) programme in India focused on the demand-side management of residential electricity. Implemented by Energy Efficiency Services Limited (EESL) with support from the Ministry of Power and local manufacturers, efficient LED lamps were distributed to domestic consumers at on-third of the market price. Having demonstrated success within 2-3 years, it covered high upfront costs for a large consumer base: the poorer sections of society. More than 260 million LEDs were sold, with annual savings of over 30 GWh of electricity, mitigation of around 3 million tons of CO_2 (2015) and one of the world's fastest reductions in LED retail market prices (US$12.28/bulb to US$3.07/bulb over 2012-2016).

The sale of new appliances provided energy savings, improved access to modern energy services, growth of domestic manufacturing to an internationally competitive business, better efficiency standards, and a growth in accredited testing laboratories and better consumer awareness. It was an example of low-carbon technology deployment, which created a large market (LED bulbs emerging as the preferred lighting option) using a bulk procurement model, with a technological advancement based on the idea of encouraging business models that could help in meeting the low-carbon emission targets at a faster rate. Domestic manufacturing has increased, and efficiency standards improved with market confidence in the product. Accredited testing laboratories have grown and consumer awareness has increased.

Empowered families had substantial money savings (over US$0.25 billion/year; household electricity bills fell 15 per cent), plus resource savings, emissions mitigation (about 3 million tons CO_2/year), improving quality of life, promoting productivity and local prosperity, and expanding energy access. Such a bulk procurement model allowed for a massive technology advancement. UJALA is an international demand side management showcase, being applied in the second largest world market (worth US$0.33 billion/year and growing) and more recently replicated in Malaysia, also with attempts to cover more appliances, sectors, companies and regions (Chunekar, Mulay and Kelkar 2014; ET Energy World 2017; Energy Efficiency Services Limited (EESL) and IEA 2016; Sundaramoorthy and Walia 2017; India, Ministry of Power 2018a; India, Ministry of Power 2018b).

Despite a slowing trend, global energy demand may still expand by 30 per cent between 2017 and 2040 according to the International Energy Agency (IEA 2018). This amount is the equivalent of adding another China and India to today's global energy demand. At the same time, universal access to electricity remains a challenge. Large-scale shifts in global energy systems are due to the rapid deployment and falling costs of clean energy technologies (chiefly renewables but also natural gas), the growing electrification of energy, and the shift to a more services-oriented economy. Renewable energies are expected to meet 40 per cent of the increase in primary demand, capturing two-thirds of global investment in power plants to 2040 as their costs drop, enabling policies to continue to support them, and the transformation of the power sector is amplified by millions of electricity end users investing directly in distributed solar photovoltaics, with an increasing share of smart connected devices and other digital technologies. Electrified transport will grow, pushing the global electric car fleet to 280 million by 2040, from the present 2 million. Global investment in electricity overtook oil and gas investment but the challenge of decarbonizing the global power supply remains. Natural gas plays an important role in replacing oil and coal, with 80 per cent of the projected growth in demand for natural gas taking place in developing economies and the shift towards a more flexible, liquid, global market (IEA 2018).

17.6 Towards a more circular economy

The three previous sections of this chapter illustrate some of the effects of a linear economic system on the global environment. In this section, we analyse the use of materials/resources throughout the value chain from extraction to waste in the prevailing economic systems and examine approaches for developing a circular economy.

17.6.1 What are the most urgent changes required in the system?

For several centuries, most societies have pursued development using a linear economy model, where the majority of resources are extracted, processed, converted to products (some of which have a very short lifespan) and are then disposed of after use (commonly referred to, as the "take, make, waste" process). Within this economic model, only a small percentage of materials is reused or recycled (the exception being commodities like iron and gold). Instead, at the end of life they are considered waste and there is often a high price, financially, socially and environmentally to dispose of this waste.

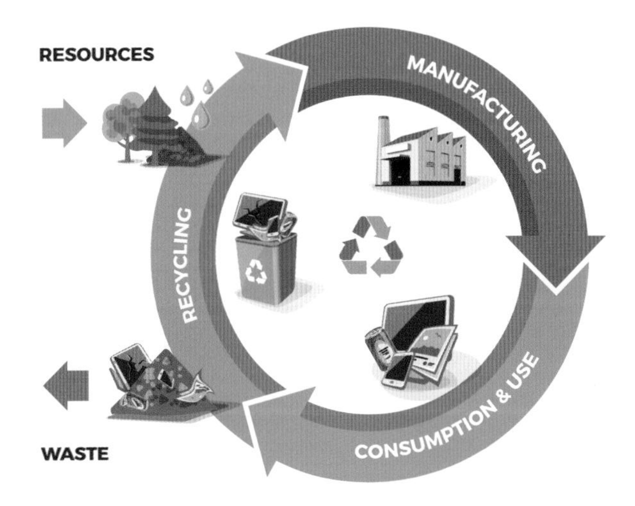

© Shutterstock/petovarga

The linear economy assumes that there will always be an abundant supply of raw materials and unlimited capacity to dispose of waste in the natural environment. However, as can be seen from Part A of this report, human societies cannot continue to operate in this way if we want to meet the demands of a growing population, preserve the health of the planet, and ensure that future generations are able to prosper. Continuing to extract natural resources such as minerals using this model implies an increasing environmental impact to extract ever diminishing ore grades. The example of fossil fuel resources shows that the capacity of ecosystems to absorb emissions is limited. Within the sustainability framework, some resources are finite and current levels of consumption are not compatible with reaching the SDGs. An alternative is to build sustainable economies that recognise the value of natural resources through a 'circular economy' **(Figure 17.4)**.

The components and strategies of a circular economy model were first identified in the early 1980s and refined in following decades (Stahel and Reday-Mulvey 1981; Ayres 1994). These earlier models referred only to waste management – collection, separation, recycling, reuse. Today, there are many circular economy strategies being applied by individuals, businesses and governments. These can go beyond dealing with waste to include better product design, reduced consumption and sustainable materials management. The common aim is to use resources in the most efficient way for the longest possible time. The resources circulate through various processes, being reused, repaired, redesigned or remanufactured, which reduces the need for new raw materials and minimizes waste **(Figure 17.4)**. When faced with persistent environmental problems such as climate change, resource scarcity and biodiversity loss, adopting resource circularity makes sense; however, society has been slow to adopt this model or has simply failed to take the actions necessary for large-scale change.

Speeding up the transition to a circular economy involves a large shift in business and consumer thinking, demanding the adoption of sustainable production and consumption processes. Fuenfschilling and Truffner (2014) identify breaking down long-standing rigid and interdependent system structures as the main challenge. The difficulty stems from having to enact large-scale socio-institutional change, which may require radical new ways of thinking and adjustment to normal customs and beliefs (Potting *et al.* 2017). Moving from

Figure 17.4: Building a circular economy

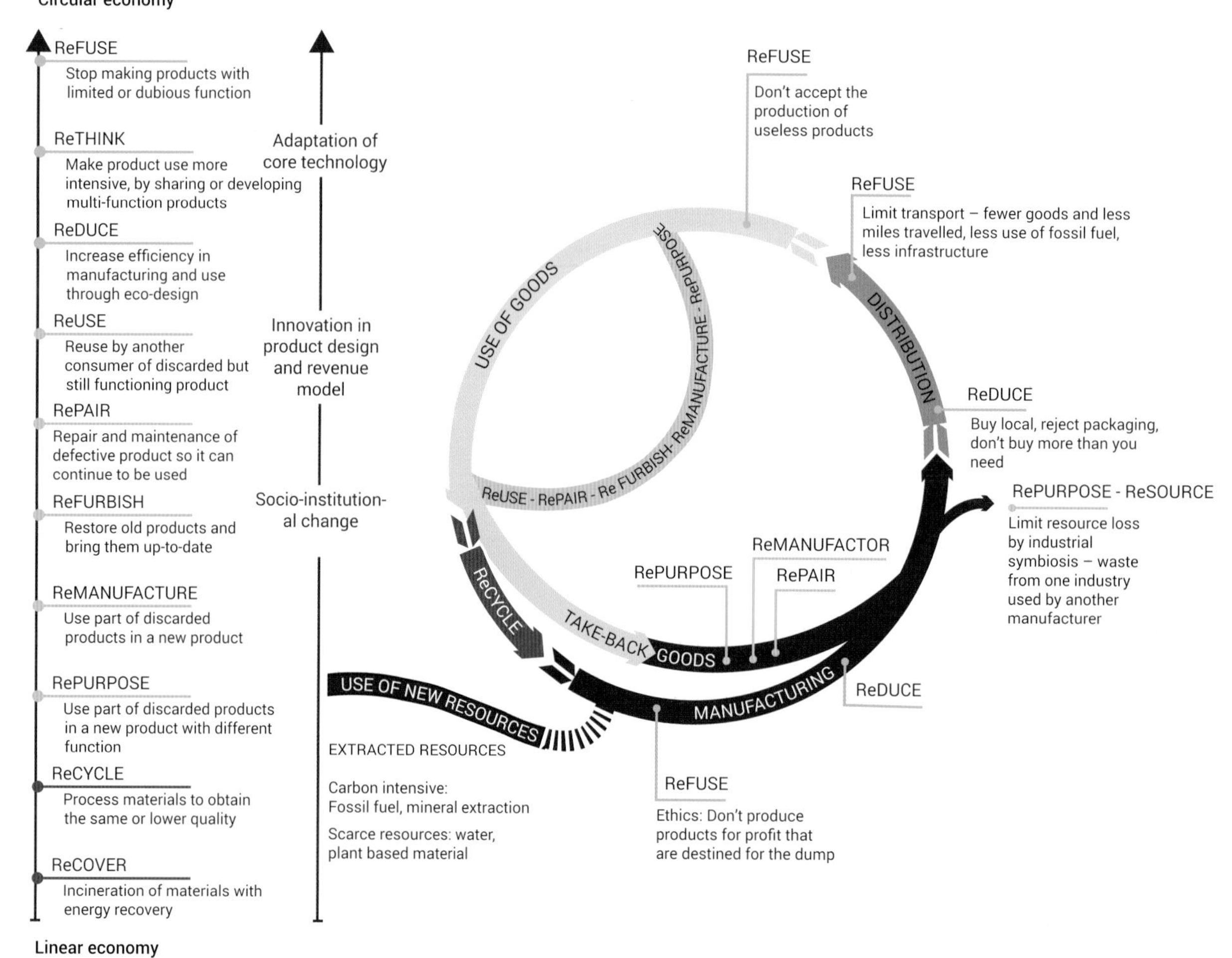

Source: Based on Stahel (2016) and Potting *et al.* (2017).

the established way of thinking involves the development of new laws and policies, which need revised, redesigned or new business models that integrate industries and incorporate a longer-term perspective, the internalization of the environmental and social costs of extraction, production and disposal, innovative technologies, and changes in consumer use patterns. Actions that can contribute to accelerated transformation have been outlined by the Government of the Netherlands (2016) and include the following.

- Decreasing demand for raw materials by increasing the efficiency of raw material use in the supply chain.
- In instances where raw materials are required, replacing fossil-based, scarce and non-sustainably produced raw materials with sustainably produced, renewable and readily available raw materials.
- Developing new innovative low-carbon production methods and smart product design.
- Promoting thoughtful consumption (e.g. reuse, smart design, extension of product life through design and repair, use of secondary and recycled materials, sharing economy).

Circular economy strategies have also been developed by Germany, Finland, Denmark and Slovenia. France, Italy and Spain have their road maps developed as well.

The circular economy promotes a production and consumption model that includes restoration and regeneration where possible (Ellen MacArthur Foundation 2015; Smol, Kulczycka and Avdiushchenko 2017). It ensures that the worth of products, materials, chemicals and resources is maintained in the economy at their highest utility and value for as long as possible (European Commission 2015; Stahel 2016). The circular economy, therefore, means reducing waste during production, ensuring asset recovery including waste utilization, and developing obsolescence prevention pathways in product and urban system designs through sustainable materials management **(Box 17.5)**. It also means ensuring product and service delivery with energy and materials from renewable sources, while changing business models to match these objectives (Ghisellini, Cialani and Ulgiati 2015; Rizos, Tuokko and Behrens 2017).

The circular economy preserves raw materials, thereby decoupling economic growth from the use of resources and its associated environmental externalities, including carbon emissions. However, in some cases the appearance of growth decoupling in one sector or territory can mask a continued environmental and social impact somewhere else (details in Ward *et al.* 2017). Ward *et al.* (2017) cite substituting one non-renewable resource for another (e.g. the cleaner energy systems that replace fossil fuels still require non-renewable resources) and shifting the cost somewhere else (e.g. importing resource-intensive consumer goods from developing countries).

17.6.2 What are the elements of the system that the policies seek to address?

Policies that support the transition to circularity are being developed and implemented in many places and involve a range of different approaches. Early examples include the German Closed Substance Cycle and Waste Management Act introduced in 1996 to recover materials from municipal and production waste (Germany, Federal Ministry for the Environment, Nature Conservation and Nuclear Safety [BMU] 2011) and the Japanese recycling initiative Basic Law for Establishing a Recycling-Based Society (Environment Agency Japan 2000). These actions are examples of what has become known as the 3Rs of reduce, reuse and recycle, and are the foundation of green manufacturing and consumption (Jawahir and Bradley 2016). However, in the last decade the focus has expanded from 'green' to sustainable manufacturing – for example, the 6Rs of manufacturing, which in addition to reduce, reuse and recycle, include recover (for a subsequent life cycle), redesign (the next generation of products) and remanufacture (meaning restoration to an 'as new' form) (Jawahir and Bradley 2016; **Figure 17. 5)**.

Box 17.5: Sustainable materials management

Sustainable materials management (SMM) is a policy approach that expands the focus of waste management to the whole life cycle of a material – from extraction to end of life. It seeks to maintain the availability of products and services by conserving valuable resources and keeping them in circulation indefinitely. One of the key aims of the holistic management approach is to reduce impacts on the environment across the whole life cycle of a resource. Producers and manufacturers need to extend sustainability across the value chain – this involves ensuring the sustainability standards of all suppliers, integrating sustainability into the design process, and identifying and addressing any negative social and environmental impacts.

Reducing the volume of waste produced and increasing material recovery are essential components of SMM (United States Environmental Protection Agency [US EPA] 2015). SMM promotes resource efficiency, which includes minimizing the economic, environmental and social costs of a production process and resource productivity, defined as the effectiveness with which natural resources are used (OECD 2012).

Figure 17.5: Closed-loop material flow diagram of 6R elements and the four life cycle stages

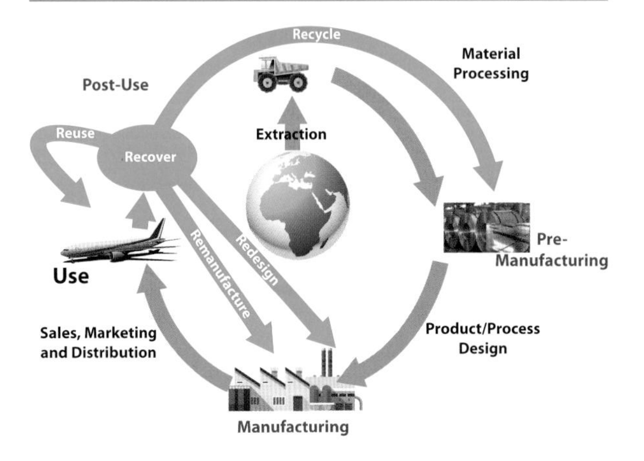

Source: Jawahir and Bradley (2016).

17

Table 17.3: Examples of policy focus to achieve key elements of the circular economy

Key elements of the circular economy	Policy examples	Result examples
Design for the future	EU Ecodesign Directive – ensures energy efficiency of products, such as household appliances, by setting minimum efficiency requirements (EU 2009).	It is estimated that the Ecodesign Directive will deliver a 16 per cent reduction in the primary energy consumption of 35 product groups compared with the consumption of these products in 2010. For example, the energy efficiency of televisions, under the Ecodesign scenario it is predicted to improve by a factor of 25 (measured from 1990) by 2030 (European Commission 2017).
Market-based instruments – green taxation	Taxes on virgin materials such as sand, gravel and rock used in the construction industry have been introduced by 16 EU states.	The United Kingdom of Great Britain and Northern Ireland introduced a tax on aggregates in 2002. Since the introduction of the tax, primary aggregates use has reduced by approximately 40 per cent per unit of construction (Ettlinger 2017).
Incorporate digital technology	Republic of Korea has some of the world's fastest internet speeds, with connections to more than 90 per cent of the population. The government has provided economic support for broadband infrastructure development, subsidies to ensure connectivity, and measures to stimulate information technology literacy (Falch and Henten 2018).	Streaming music reduces resource use and costs 80 per cent less than the cost of producing and distributing of compact disks (CDs) (Lacy 2015). The Republic of Korea was the sixth top music market in 2017 and has the largest number of paid music subscribers (International Federation of Phonographic Industry [IFPI] 2018).
Collaborate	In Sydney, Australia, the city council introduced policies to promote car sharing, including the provision of designated car-share parking spaces; and online listing of private vehicles participating in peer-to-peer sharing schemes (City of Sydney 2016).	GoGet is an Australian car-sharing company, operating in large cities. Members have access to a range of vehicles including cars and vans (GoGet https://www.goget.com.au).
Use waste as a resource	In 1997, Denmark introduced legislation that banned sending waste that could be recycled or incinerated to landfill. In 2015, a new law was introduced, the Environmental Technology Development and Demonstration Programme (MUDP). This includes a subsidy scheme, innovation partnerships and international cooperation to find resource-efficient solutions to environmental problems (Denmark, Ministry of Environment and Food n.d.).	The Kalundborg Symbiosis in Denmark is a network of businesses that was the first industry group to fully develop industrial symbiosis. The collaboration includes a coal-fired power plant, fish farming, fertilizer production and a host of other manufacturing and industrial operations (Kalundborg Symbiosis 2018).
Rethink the business model	New business models that utilize technologies are emerging, such as blockchain. Estonia, for example, has established an e-residency scheme to encourage entrepreneurs. E-residency provides anyone with a digital ID that allows them to access Estonia's e-services for online business development and management from anywhere in the world.	The China Construction Bank Corporation (CCB) is using the IBM Blockchain platform to improve procedures for the sale of its insurance products.
Preserve and extend existing products	The right to repair – the EU is preparing legislation making it mandatory for companies to provide spare parts and diagnostic tools that would make it cheaper and easier to repair products (European Parliament, Committee on the Internal Market and Consumer Protection 2017)	Inrego, a Swedish firm, is refurbishing electronic equipment such as laptops, personal computers, monitors and phones (European Remanufacturing Network 2018).
Prioritize regenerative resources	Norwegian policies to support battery electric vehicles (BEV): zero annual road tax (2018); 40 per cent reduced company car tax (2018); 50 per cent price reduction on ferries (2018); zero re-registration tax for used zero-emission cars (2018); free municipal parking in many cities (Norsk elbilforening 2018).	In Norway, incentive programmes to encourage use of BEVs began in the early 1990s. Norway currently leads the world with 21 per cent BEV market share (cf. Australia, where there is limited incentive and BEVs have 0.2 per cent of the market (ClimateWorks Australia 2018).

China adopted the circular economy as a development strategy in 2002, and this was given legal effect in 2009 through the Circular Economy Promotion Law (China, National People's Congress 2008). The European Commission released a 'Roadmap to a Resource Efficient Europe' in 2011, which was replaced in 2015 by 'Closing the Loop: An EU Action Plan for the Circular Economy' (McDowall *et al.* 2017). Both Europe and China were following earlier research and policy work in the United States of America, Japan and Europe that focused on waste management.

Box 17.6: Case study: Ellen MacArthur Foundation – A toolkit for policymakers in delivering the circular economy

The Ellen MacArthur Foundation, a UK-based non-governmental organization, has been a leading proponent of the circular economy, funding extensive research and education programmes. In 2015, the foundation partnered with the Danish Business Authority to develop a toolkit for policymakers (Ellen MacArthur Foundation 2015). In the development of the toolkit and subsequent pilot studies, the authors identified seven key insights which provide evidence of the potential economic, environmental and social benefits of moving towards a circular economy.

- A circular economy fosters more innovation, resilience and productivity, resulting in increased gross domestic product (GDP) and jobs, and reduces greenhouse gas emissions and virgin non-renewable resource consumption.
- Policymakers can break down the non-financial barriers that challenge the circular economy.
- There is no overarching solution that will instigate a circular economy – each sector must be analysed and tailored policies should be instituted.
- An overhaul of financial systems and the way we measure economic performance (i.e. currently excluding externalities such as environmental damage or social dislocation) will help illuminate the real value in transitioning towards a circular economy and the real cost of business as usual.
- Business needs to lead the way in identifying circular economy opportunities.
- Even developed countries that are moving towards a circular economy can increase the rate of change by scaling up and fostering enabling conditions across all sectors.

There needs to be policy coordination across countries as value chains extend across borders.
The policy environment is expanding, with states and other stakeholders such as the Ellen MacArthur Foundation playing an important role in promoting the circular economy transition to business and industry **(see Figure 17.6)**.

Figure 17.6: Outline of a circular economy

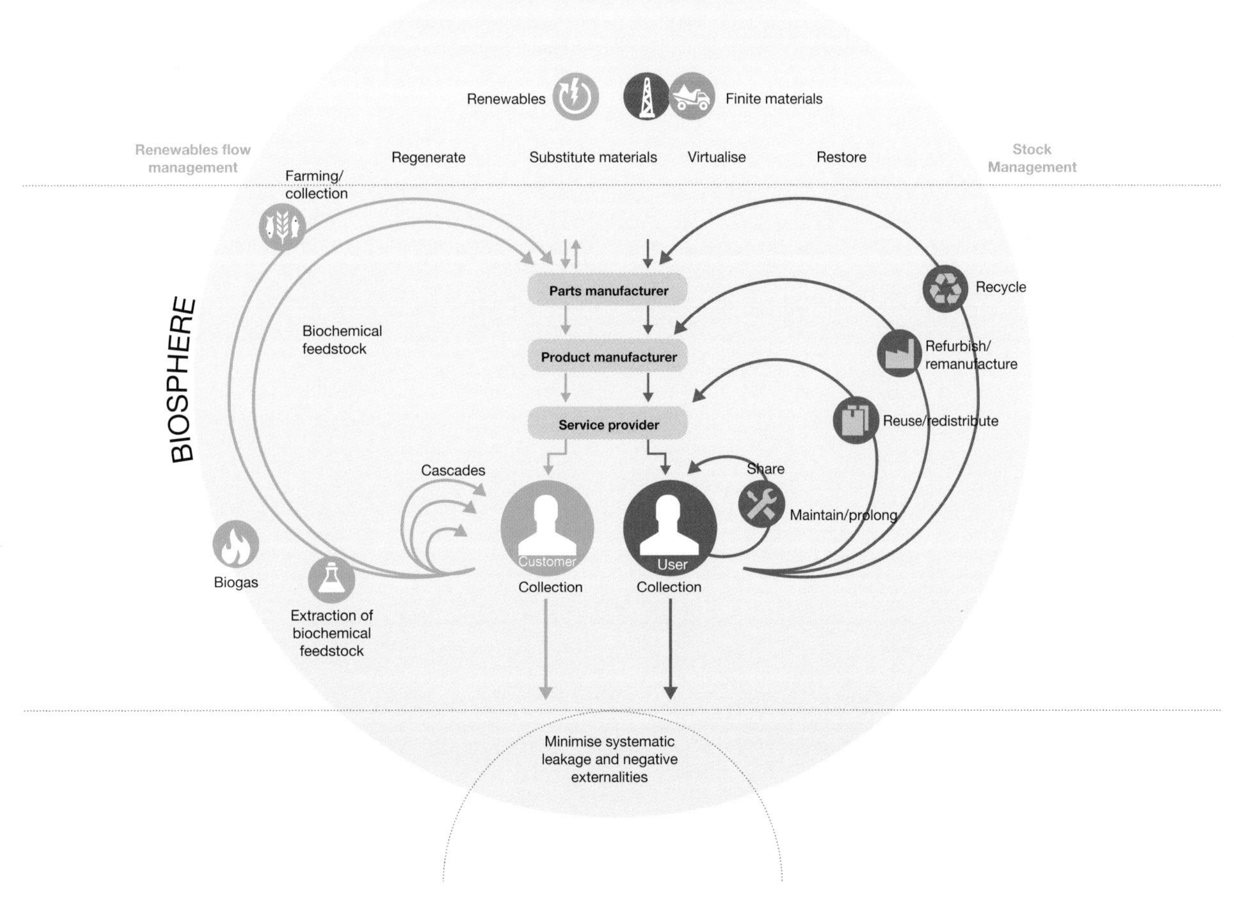

Source: Adapted from Cirular Norway (n.d).

17.6.3 What has been done to date and how effective have these measures been?

Many governments have introduced policies and regulations that address aspects of the circular economy. Policies supporting the circular economy can focus on one or more elements of the 'take, make and waste' process. While many policies have tended to address waste through recycling and resource recovery, there are significant gains to be made at the earliest stages of product design and manufacture. For example, products can be designed using eco-design principles, to use less material and last longer. They can be refurbished or repaired and made of non-toxic materials that are simple to recycle.

Policies that encourage eco-design also need to consider the potential adverse health, gender and developmental impacts of poorly planned policies (e.g. toxic exposures for women and children from recycling electronic waste). About 15 million people are involved in informal waste recycling of plastics, glass, metals and paper where these activities are a risk both to the environment and to the people performing the tasks (Yang *et al.* 2018). Individuals performing resource recovery, especially e-waste pickers in developing countries, risk considerable occupational and environmental health threats (Velis 2017). Women and children are among the vulnerable groups working in this informal sector who face exposure to hazardous chemicals and heavy metals (Heacock *et al.* 2016), with few to no measures for prevention or treatment (Han *et al.* 2018).

Policies can also support the move from managing waste to more environmentally sustainable outcomes, by focusing on behaviour change. These policies, which are often developed from grass-roots initiatives, aim to limit the amount of waste produced and increase material recovery (Silva *et al.* 2017).

Europe has established policies for implementing the circular economy, while in other areas this has happened at a national or subnational level. There have also been some international policy initiatives that align with, or promote, a circular economy approach, especially with regard to waste minimization (e.g. the Basel and Stockholm conventions). The new approach of green (or sustainable) chemistry is working to develop alternative solutions aimed at eliminating or at least significantly reducing hazardous chemicals and eventually their presence in the environment (Weber, Lissner and Fantke 2016). One of the challenges in relation to chemicals and the circular economy is increasing recycling and reuse, while making sure consumers are not at risk from exposure to substances of concern that may be present in products and passed on through waste (European Commission 2015). For some chemicals and toxic metals, such as persistent organic pollutants (POPs) and mercury, final disposal may be a better option than recycling and reuse.

17.6.4 What is the transformative potential of the policy approaches discussed?

A transition to a circular economy will be required in order to achieve the SDGs. There are insufficient natural resources to sustain the continued expansion of a global economy based upon a linear economic model. The circular economy offers opportunities not only to address fundamental resource constraints but also to create a more just and inclusive economic system (Raworth 2012). Circular economy policies therefore carry major transformative potential to address cross-cutting policy challenges.

17.6.5 Indicators

No indicator provides a single measure of progress towards a circular economy. However, there are several existing indicators of performance in areas that directly or indirectly contribute to the achievement of a circular economic system. Sustainable resource management, societal behaviour, business operations, material flow accounting or analysis are among a number of measures that have been proposed (Geng *et al.* 2012; Wiedmann *et al.* 2015; United Nations Environment Programme [UNEP] 2016). Taking into consideration linkages with the SDGs, we identify two policy-relevant indicators of circularity.

Indicator 1: Domestic material consumption (DMC) (SDG indicators 8.4.1, 8.4.2 and 12.2.2)

Domestic material consumption measures the territorial consumption of primary materials used in the economy, whether these are domestically sourced or imported. This indicator allows a comparison to be drawn between regions and states in per capita material consumption over time. DMC can also be used to estimate the amount of waste that may be generated in a given region. Domestic extraction (DE) is the amount of materials extracted in a given territory. DMC is higher than DE in net material importing countries and lower than DE in net material-exporting countries **(Figure 17.7).**

© Shutterstock/Wanna Thongpao

Figure 17.7: Domestic extraction and Domestic material consumption

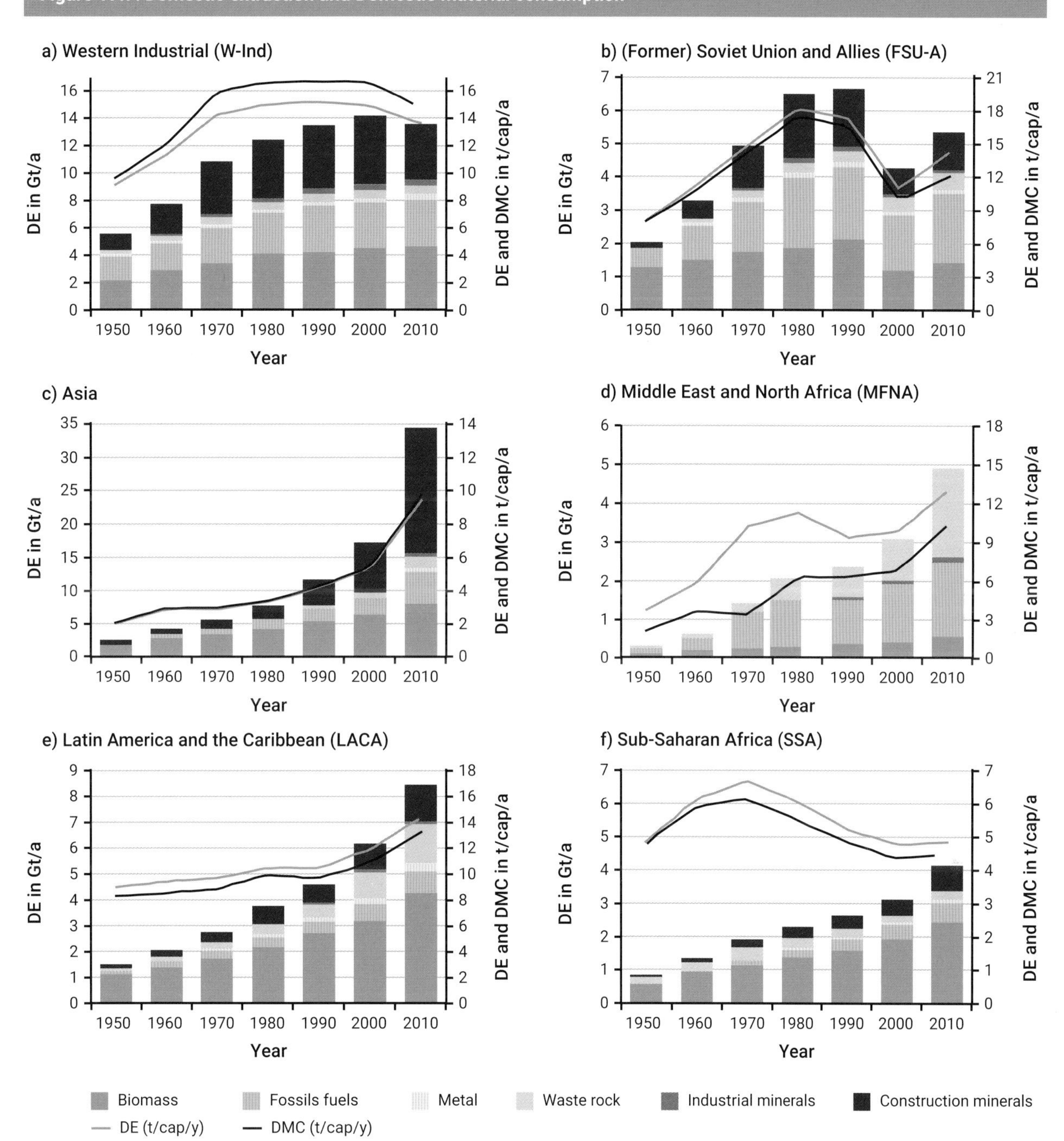

The figure shows data on extraction, trade, and apparent consumption of materials for six regions in Gigatonnes per year (Gt/a) and in per capita values per year (t/cap/a). DE: domestic extraction; DMC: domestic material consumption.

Source: Schaffartzik *et al.* (2014).

17

Indicator 2: Societal behaviour (SDG indicators 12.2.1 and 12.2.2)

In addition, developing a circular economy will involve people changing their consumption behaviour and choosing products and services that conserve resources. Sharing resources, a common strategy in many subsistence economies, is being increasingly adopted around the world. Sharing of expensive or infrequently used products such as cars, bikes, holiday houses, camping and other recreation equipment may be organized as formal schemes or informal agreements within communities **(Figure 17.8)**.

17.7 Conclusions

The cross-cutting nature of sustainability issues is well illustrated by the interactions across SDGs (Nilsson, Griggs and Visbeck 2016; Biermann, Kanie and Kim 2017; International Council for Science 2017). Meeting one goal or target will not guarantee that other SDGs will be achieved, just as some Millennium Development Goals (MDGs) were achieved in some parts of the world but not in others (Boas, Biermann and Kanie 2016; Kim 2016; Underdal and Kim 2017; Young 2017). This lesson is not new, but the overdue shift towards systemic policy approaches is beginning to occur. Some go as far as to argue that "the single most important [environmental] problem is our misguided focus on identifying the single most important problem" (Diamond 2005). The systems approach to environmental policy development and implementation, discussed in this chapter, can address multiple global goals and is no longer an option but is the only way forward for societal transformation to achieve global sustainability.

This chapter highlights the complex linkages between sustainability issues and the ways in which these present both challenges and opportunities. They are challenges in the sense that cross-cutting issues are difficult to address individually through incremental steps and in isolation from one another. As concluded in the thematic chapters of Part B (Chapters 12-16), many well-intended environmental policies and measures have had limited success. Policy improvements are visible, but they have not been made at a sufficient rate or scale. New sustainability issues have emerged that have greater complexity, often due to the unanticipated ways through which existing issues have interacted with one another. Some of the unwanted outcomes of interaction between global drivers in turn act as drivers for further suboptimal outcomes (Walker *et al.* 2009).

As the analyses in this chapter have shown, however, systems policy approaches with transformative potential do exist. If key leverage points can be identified in a system and the right policy interventions are made (Meadows 2008), transformative change leading to innovations will lead to net positive effects. Even small-scale interventions can sow the seeds for the larger systemic change that is required to deliver the SDGs. This chapter chose four socioeconomic systems to illustrate the transformative potential of the systemic approach to environmental policy intervention.

Figure 17.8: Citizen engagement in sharing: the percentage of 2013 survey respondents who had engaged in a sharing scheme, either formal or informal in the previous 12 months

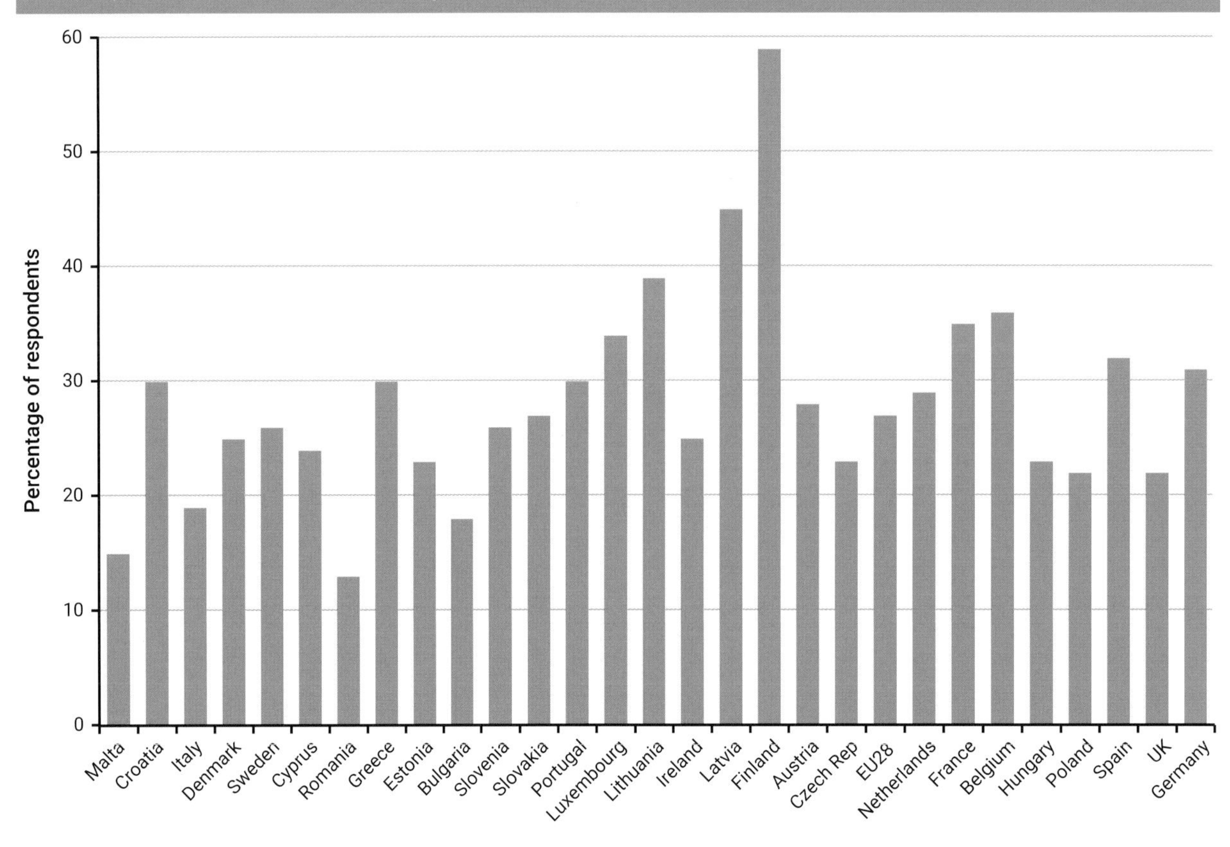

Source: Flash Eurobarometer 388 (2013)

Environmental, social and economic systems need to be understood and analysed by appreciating their complexity. Some understanding of a system is a prerequisite to identifying leverage points, that is, where 'seeds can be sown'. Acknowledging that there is no policy panacea (Ostrom 2007), various clusters of policies can then be deployed, and some degree of redundancy can be helpful as a policy safety net (Low *et al.* 2003). It is very difficult to predict whether a policy will work effectively to solve a cross-cutting issue without producing significantly perverse and unintended consequences. Attention to one element of a cross-cutting issue can lead to environmental problem shifting – both transboundary and trans-sectoral, or trade-offs and spillovers (Kim and van Asselt 2016). Adaptive governance or management approaches are therefore required that use experimentation (Hoffmann 2011) to build on lessons learned rather than 'reinventing the wheel'.

An effective response to cross-cutting environmental policy challenges requires cooperation and collaboration among a multitude of actors and institutions across issues, sectors, levels and jurisdictions. The transformation pathway for achieving human dignity and environmental sustainability this century requires a whole-of-system approach that can catalyse rapid technological innovation and economic and cultural paradigm shifts.

References

Adger, W.N., Arnell, N.W. and Tompkins, E.L. (2005). Successful adaptation to climate change across scales. *Global Environmental Change* 15(2), 77-86. https://doi.org/10.1016/j.gloenvcha.2004.12.005.

Alexander, P., Brown, C., Arneth, A., Finnigan, J., Moran, D. and Rounsevell, M. (2017). Losses, inefficiencies and waste in the global food system. *Agricultural Systems* 153, 190-200. https://doi.org/10.1016/j.agsy.2017.01.014.

Anderson, A. (2013). *Learning to Be Resilient Global Citizens for a Sustainable World*. Paper Commissioned for the EFA Global Monitoring Report 2013/4, Teaching and Learning: Achieving Quality for All. United Nations Educational, Scientific and Cultural Organization. http://unesdoc.unesco.org/images/0022/002259/225940e.pdf

Ayres, R.U. (1994). Industrial metabolism: theory and policy. In *The Greening of Industrial Ecosystems*. Allenby, B.R. and Richards, D.J. (eds.). Washington, DC: National Academy Press. 22-37. https://www.nap.edu/read/2129/chapter/4.

Asian Development Bank (2012). *Addressing Climate Change and Migration in Asia and the Pacific*. Manila. https://www.adb.org/sites/default/files/publication/29662/addressing-climate-change-migration.pdf

Bailey, R. and Harper, R. (2015). *Reviewing Interventions for Healthy and Sustainable Diets*. London: The Royal Institute of International Affairs. https://www.chathamhouse.org/sites/default/files/field/field_document/20150529HealthySustainableDietsBaileyHarperFinal.pdf

Benton, T.G. (2012). Managing agricultural landscapes for production of multiple services: the policy challenge. *Politica Agricola Internazionale [International Agricultural Policy]* 1, 7-17. http://ageconsearch.umn.edu/bitstream/130373/2/Benton.pdf

Benton, T.G. and Bailey, R. (forthcoming). The paradox of efficiency: agricultural efficiency promotes food system inefficiency. *Global Sustainability*.

Biermann, F. (2014). *Earth System Governance: World Politics in the Anthropocene*. Cambridge, MA: MIT Press. https://mitpress.mit.edu/books/earth-system-governance

Biermann, F., Kanie, N. and Kim, R.E. (2017). Global governance by goal-setting: the novel approach of the UN Sustainable Development Goals. *Current Opinion in Environmental Sustainability* 26-27, 26-31. https://doi.org/10.1016/j.cosust.2017.01.010

Boas, I., Biermann, F. and Kanie, N. (2016). Cross-sectoral strategies in global sustainability governance: Towards a nexus approach. *International Environmental Agreements: Politics, Law and Economics* 16(3), 449–464. https://doi.org/10.1007/s10784-016-9321-1

Blum, N., Nazir, J., Breiting, S., Goh, K.C. and Pedretti, E. (2013). Balancing the tensions and meeting the conceptual challenges of education for sustainable development and climate change. *Environmental Education Research* 19(2), 206-217. https://doi.org/10.1080/13504622.2013.780588

Brunner, R.D. and Lynch, A.H. (2010). *Adaptive Governance and Climate Change*. Chicago, IL: University of Chicago Press. https://www.press.uchicago.edu/ucp/books/book/distributed/A/bo8917780.html

Brunner, R.D., Steelman, T.A., Coe-Juell, L., Cromley, C., M., Edwards, C.M. and Tucker, D.W. (2005). *Adaptive Governance: Integrating Science, Policy, and Decision-Making*. New York, NY: Columbia University Press. https://cup.columbia.edu/book/adaptive-governance/9780231136259

Buchner, B.K., Oliver, P., Wang, X., Carswell, C., Meattle, C. and Mazza, F. (2017). *Global Landscape of Climate Finance 2017*. Climate Policy Initiative. https://climatepolicyinitiative.org/wp-content/uploads/2017/10/2017-Global-Landscape-of-Climate-Finance.pdf

Buckwell, A., Matthews, A., Baldock, D. and Mathijs, E. (2017). *Cap: Thinking Out of the Box: Further Modernisation of the Cap – Why, What and How?* RISE Foundation. http://www.risefoundation.eu/images/files/2017/2017_RISE_CAP_Full_Report.pdf

Burch, S., Mitchell, C., Berbes-Blazquez, M. and Wandel, J. (2017). Tipping toward transformation: progress, patterns and potential for climate change adaptation in the global south. *Journal of Extreme Events* 4(1). https://doi.org/10.1142/S2345737617500038

Chaffin, B.C., Gosnell, H. and Cosens, B.A. (2014). A decade of adaptive governance scholarship: Synthesis and future directions. *Ecology and Society* 19(3). http://dx.doi.org/10.5751/ES-06824-190356

China, National People's Congress (2008). *Circular Economy Promotion Law of the People's Republic of China*. Adopted at the 4th Meeting of the Standing Committee of the 11th National People's Congress. Beijing. http://www.fdi.gov.cn/1800000121_39_597_0_7.html

Chun, J.M. (2015). *Planned Relocations in the Mekong Delta, Vietnam: A Successful Model for Climate Change Adaptation, A Cautionary Tale, or Both?* Washington, DC: Brookings Institution. https://www.brookings.edu/wp-content/uploads/2016/06/Brookings-Planned-Relocations-Case-StudyJane-Chun-Vietnam-case-study-June-2015.pdf

Chunekar, A., Mulary, S. and Kelkar, M. (2014). *Understanding the Impacts of India's LED Bulb Programme, "UJALA"*. Kothrud: Prayas Energy Group. http://shaktifoundation.in/wp-content/uploads/2014/02/02-PEG-Report-on-impacts-of-UJALA.pdf

Circular Norway (n.d). Outlines of a circular economy. https://www.circularnorway.no/modeller/outlines-of-a-circular-economy (Accessed 19 October 2018).

City of Sydney (2016). *Car Sharing Policy*. http://www.cityofsydney.nsw.gov.au/__data/assets/pdf_file/0010/109099/2016-631840-Car-Sharing-Policy-2016-accessible.pdf

ClimateWorks Australia (2018). *Australia's Electric Vehicle Industry Gains Momentum: Report*. Sydney, Australia: Electric Vehicle Council. http://electricvehiclecouncil.com.au/australias-electric-vehicle-industry-gains-momentum-report/

Central Committee for Flood and Storm Control (2012). *Living with Floods Program*. Hanoi: Central Committe for Flood and Storm Control.

Cutter, S.L., Boruff, B.J. and Shirley, W.L. (2003). Social Vulnerability to Environmental Hazards. *Social Science Quarterly* 84(2), 242-261. https://doi.org/10.1111/1540-6237.8402002

Dankelman, I. (2016). *Action Not Words: Confronting Gender Inequality through Climate Change Action and Disaster Risk Reduction in Asia*. Aipira, C., Kidd, A., Reggers, A., Fordham, M., Shreve, C. and Burnett, A. (eds.). Bangkok: UN Women. http://www2.unwomen.org/-/media/field%20office%20eseasia/docs/publications/2017/04/ccdrr_130317-s.pdf

Davis, I. (2015). *Disaster Risk Management in Asia and the Pacific*. New York (NY): Routledge. https://www.adb.org/sites/default/files/publication/159311/adbi-disaster-risk-management-asia-pacific.pdf.

de Blas, F., Kettunen, M., Russi, D., Illes, A., Lara-Pulido, J.A., Arias, C. and Guevara, A. (2017). Innovative mechanisms for financing biodiversity conservation: a comparative summary of experiences from Mexico and Europe. Paris: La Recherche Agronomique pour le Développement (CIRAD). https://ieep.eu/uploads/articles/attachments/76fa5531-8333-4464-83bc-fef6b0317c77/IFMs_for_biodiversity_SYNTHESIS_Ezzine_de_Blas_et_al_2017_.pdf.

Denmark, Ministry of Environment and Food (n.d.). Danish lesson – waste management. http://eng.ecoinnovation.dk/the-danish-eco-innovation-program/results-and-cases/danish-lessons/waste-management/. (Accessed 21 October 2018).

Diamond, J. (2005). *Collapse: How Societies Choose to Fail or Succeed*. New York, NY: Penguin Group. https://pdfs.semanticscholar.org/8f2e/4df7d90c9744cef12c967a8755f89673d088.pdf.

Djalante, R., Holley, C. and Thomalla, F. (2011). Adaptive governance and managing resilience to natural hazards. *International Journal of Disaster Risk Science* 2(4), 1-14. https://doi.org/10.1007/s13753-011-0015-6.

Doyon, A. (2018). Niches: small scale interventions or radical innovations to build up internal momentum. In *Enabling Eco-Cities*. Hes, D. and Bush J. (eds.). Singapore: Palgrave Pivot. Chapter 10. 65-87. https://link.springer.com/chapter/10.1007/978-981-10-7320-5_5.

Dryzek, J.S. (2014). Institutions for the anthropocene: governance in a changing earth system. *British Journal of Political Science* 46(4), 937–956. https://doi.org/10.1017/S0007123414000453.

Economic and Social Commission for Asia and the Pacific, Asian Development Bank and United Nations Development Programme (2018). *Transformation Towards Sustainable and Resilient Societies in Asia and the Pacific*. Bangkok. https://www.unescap.org/sites/default/files/publications/SDG_Resilience_Report.pdf.

Ellen MacArthur Foundation (2015). *Delivering the Circular Economy: A Toolkit for Policymakers*. https://www.ellenmacarthurfoundation.org/publications/delivering-the-circular-economy-a-toolkit-for-policymakers.

Entzinger, H. and Scholten, P. (2016). *Adapting to Climate Change through Migration: A Case Study of the Vietnamese Mekong River Delta*. Geneva: International Organization for Migration. https://publications.iom.int/system/files/vietnam_survey_report_0.pdf.

Environment Agency Japan (2000). *The Basic Law for Establishing the Recycling-based Society*. http://www.env.go.jp/recycle/low-e.html. (Accessed 21 October 2018).

EAT-Lancet Commission for Food, Planet and Health (2018). The Report. https://foodplanethealth.org/the-report. (Forthcoming – late 2018).

Energy Efficiency Services Limited and International Energy Agency (2016). *India's UJALA Story – Energy Efficient Prosperity*. Noida, Uttar Pradesh and New Delhi: EESL. https://www.eeslindia.org/DMS/UJALA%20Case%20Studies.pdf.

Ettlinger, S. (2017). *Aggregates Levy in the United Kingdom*. Institute for European Environmental Policy. https://ieep.eu/uploads/articles/attachments/5337d500-9960-473f-8a90-3c59c5c81917/UK%20Aggregates%20Levy%20final.pdf?v=63680923242.

ET Energy World (2017). Eveready Industries turns to Ujala for its LED vertical, 14 August. https://energy.economictimes.indiatimes.com/news/power/eveready-industries-turns-to-ujala-for-its-led-vertical/60060082. (Accessed 19 October 2018).

European Commission (2011). *Proposal for a Regulation of the European Parliament and of the Council establishing Rules for Direct Payments to Farmers under Support Schemes within the Framework of the Common Agricultural Policy*, 19 October. Brussels. https://ec.europa.eu/agriculture/sites/agriculture/files/cap-post-2013/legal-proposals/com625/625_en.pdf.

European Commission (2013). Societal behaviours. In *Environment: Eco-Innovation Action Plan* [website]. https://ec.europa.eu/environment/ecoap/indicators/societal-behaviours_en. (Accessed 19 October 2018).

European Commission (2015). *Closing the Loop – An EU Action Plan for the Circular Economy*. Communication from the Commission to the European Parliament, the Council, the European Economic and Social Committee and the Committee of the Regions. COM(2015) 614 Final. Brussels. http://eur-lex.europa.eu/legal-content/EN/TXT/?uri=CELEX:52015DC0614.

European Commission (2016). *Review of Greening After One Year – Part 1/6*. Brussels. https://ec.europa.eu/agriculture/sites/agriculture/files/direct-support/pdf/2016-staff-working-document-greening_en.pdf.

European Commission (2017). *Ecodesign Impact Accounting. Status Report 2017*. Prepared by Van Holsteijn en Kemna B.V. (VHK) for the European Commission. https://ec.europa.eu/energy/sites/ener/files/documents/eia_status_report_2017_-_v20171222.pdf.

European Parliament, Committee on the Internal Market and Consumer Protection (2017). *Report on a Longer Lifetime for Products: Benefits for Consumers and Companies* (2016/2272(INI)). http://www.europarl.europa.eu/sides/getDoc.do?pubRef=-//EP//TEXT+REPORT+A8-2017-0214+0+DOC+XML+V0//EN.

European Remanufacturing Network (2018). *Business Model Case Study Description – Inrego Computers and Smart Phones*. Aylesbury, UK. http://www.remanufacturing.eu/studies/f6a18b15473d6fa8400e.pdf

European Union (2009). Directive 2009/125/EC of the European Parliament and of the Council of 21 October 2009 establishing a framework for the setting of ecodesign requirements for energy-related products. *Official Journal of the European Union* L 285, 10-35. https://eur-lex.europa.eu/legal-content/EN/ALL/?uri=CELEX%3A32009L0125. Accessed 19 October 2018.

Falch, M. and Henten, A. (2018). Dimensions of broadband policies and developments. *Telecommunications Policy* 42(9), 715-725. https://doi.org/10.1016/j.telpol.2017.11.004.

Folke, C. (2006). Resilience: The emergence of a perspective for social–ecological systems analyses. *Global Environmental Change* 16(3), 253-267. http://www.sciencedirect.com/science/article/pii/S0959378006000379.

Fischer, C.G. and Garnett, T. (2016). *Plates, Pyramids, Planet: Developments in National Healthy and Sustainable Dietary Guidelines: A State of Play Assessment*. Rome: Food and Agriculture Organization of the United Nations. http://www.fao.org/3/a-i5640e.pdf.

Fischer, D. and Barth, M. (2014). Key competencies for and beyond sustainable consumption: an educational contribution to the debate. *GAIA – Ecological Perspectives for Science and Society* 23(1), 193-200. https://doi.org/10.14512/gaia.23.S1.7.

Flash Eurobarometer 388 (2013). Attitudes of Europeans towards Waste Management and Resource Efficiency. http://ec.europa.eu/commfrontoffice/publicopinion/flash/fl_388_en.pdf.

Food and Agriculture Organization of the United Nations (2014). The No-Foodloss Project in Japan. http://www.fao.org/save-food/news-and-multimedia/news/news-details/en/c/242644/.

Fuenfschilling, L. and Truffer, B. (2014). The structuration of socio-technical regimes – Conceptual foundations from institutional theory. *Research Policy* 43(4), 772-791. https://doi.org/10.1016/j.respol.2013.10.010

Garnaut, R. (2008). *The Garnaut Climate Change Review: Final Report*. Cambridge: Cambridge University Press. https://trove.nla.gov.au/work/3576521?selectedversion=NBD43604049

Garnett, T. (2012). *Climate Change and Agriculture: Can Market Governance Mechanisms Reduce Emissions from the Food System Fairly and Effectively?* London: International Institute for Environment and Development. http://pubs.iied.org/pdfs/16512IIED.pdf

Garnett, T., Mathewson, S., Angelides, P. and Borthwick, F. (2015). *Policies and Actions to Shift Eating Patterns: What Works? A Review of The Evidence of the Effectiveness of Interventions Aimed at Shifting Diets in More Sustainable and Healthy Directions.* Oxford: Food Climate Research Network. https://www.fcrn.org.uk/sites/default/files/fcrn_chatham_house_0.pdf

Geng, Y., Fu, J., Sarkis, J. and Xue, B. (2012). Towards a national circular economy indicator system in China: an evaluation and critical analysis. *Journal of Cleaner Production* 23(1), 216–224. https://doi.org/10.1016/j.jclepro.2011.07.005

Germany, Federal Ministry for the Environment, Nature Conservation and Nuclear Safety (BMU) (2011). *Closed-loop Waste Management: Recovering Wastes – Conserving Resources.* Friederich, R., Jaron, A. and Schulz, J. (eds.). Berlin : BMU Public Relations Division. https://gnse.files.wordpress.com/2012/10/waste-management.pdf

Ghisellini P., Cialani, C. and Ulgiati, C. (2015). A review on circular economy: the expected transition to a balanced interplay of environmental and economic systems. *Journal of Cleaner Production* 114, 11-32. https://doi.org/10.1016/j.jclepro.2015.09.007

Gocht, A., Ciaian, P., Bielza, M., Terres, J.-M., Röder, N., Himics, M. *et al.* (2016). *Economic and Environmental Impacts of CAP Greening: CAPRI Simulation Results.* Brussels: European Commission. http://publications.jrc.ec.europa.eu/repository/bitstream/JRC102519/jrc%20report_cap%20greening-capri%20v12.pdf

Government of the Netherlands (2016). *A Circular Economy in the Netherlands by 2050:* Government-wide Programme for a Circular Economy. https://www.government.nl/binaries/government/documents/policy-notes/2016/09/14/a-circular-economy-in-the-netherlands-by-2050/17037+Circulaire+Economie_EN.PDF

Grimmins, A., Balbus, J., Gamble, J.L., Beard, C.B., Bell, J.E., Dodgen, D. *et al.* (2016). *The Impacts of Climate Change on Human Health in The United States: A Scientific Assessment.* Washington, D.C.: Global Change Research Program. https://s3.amazonaws.com/climatehealth2016/high/ClimateHealth2016_FullReport.pdf

Hallegatte, S., Green, C., Nicholls, R.J. and Corfee-Morlot, J. (2013). Future flood losses in major coastal cities. *Nature Climate Change* 3, 802. https://doi.org/10.1038/nclimate1979

Han, W., Gao, G., Geng, J., Li, Y. and Wang, Y. (2018). Ecological and health risks assessment and spatial distribution of residual heavy metals in the soil of an e-waste circular economy park in Tianjin, China. *Chemosphere* 197, 325-335. https://doi.org/10.1016/j.chemosphere.2018.01.043.

Hart, K., Buckwell, A. and Baldock, D. (2016). *Learning the Lessons of the Greening of the Cap.* Brussels: Institute for European Environmental Policy. https://www.nature.scot/sites/default/files/2017-06/A1943384.pdf.

Hay, J.E. (2009). *Assessment of Implementation of the Pacific Islands Framework for Action on Climate Change (PIFACC).* Apia: Secretariat of the Pacific Regional Environment. https://www.sprep.org/climate_change/PYCC/documents/HayReport_to_PCCR_2009.pdf.

Heacock, M., Kelly, C.B., Asante, K.A., Birnbaum, L.S., Bergman, Å.L., Bruné, M.N. *et al.* (2015). E-waste and harm to vulnerable populations: a growing global problem. *Environmental Health Perspectives* 124(5), 550-555. http://dx.doi.org/10.1289/ehp.1509699.

Hinkel, J., Brown, S., Exner, L., Nicholls, R.J., Vafeidis, A.T. and Kebede, A.S. (2012). Sea-level rise impacts on Africa and the effects of mitigation and adaptation: An application of DIVA. *Regional Environmental Change* 12(1), 207-224. https://doi.org/10.1007/s10113-011-0249-2.

Hoffmann, M.J. (2011). *Climate Governance at the Crossroads: Experimenting with a Global Response after Kyoto.* Oxford: Oxford University Press. https://doi.org/10.1080/09644016.2013.769805.

Huitema, D., Adger, W.N., Berkhout, F., Massey, E., Mazmanian, D., Munaretto, S. *et al.* (2016). The governance of adaptation: Choices, reasons, and effects. Introduction to the Special Feature. *Ecology and Society* 21(3). http://dx.doi.org/10.5751/ES-08797-210337.

Illes, A., Russi, D., Kettunen, M. and Robertson M. (2017). *Innovative Mechanisms for Financing Biodiversity Conservation: Experiences from Europe.* Final Report in the Context of the Project "Innovative Financing Mechanisms for Biodiversity in Mexico/ N°2015/368378". Brussels, Belgium. London: Institute for European Environmental Policy (IEEP). https://ieep.eu/uploads/articles/attachments/dcc74b53-6750-4ccd-99b9-dc9e9d659dd4/IFMs_for_biodiversity_EUROPE_Illes_et_al_2017.pdf.

India, Ministry of Power (2018a). *Energy Efficiency.* https://powermin.nic.in/en/content/energy-efficiency. (Accessed 19 October 2018).

India, Ministry of Power (2018b). *National UJALA Dashboard.* http://www.ujala.gov.in/. (Accessed 19 October 2018).

Intergovernmental Panel on Climate Change (2014). *Climate Change 2014: Synthesis Report. Contribution of Working Groups I, II and III to the Fifth Assessment Report of the Intergovernmental Panel on Climate Change.* Pachauri, R.K. and Meyer, L.A. (eds.). Geneva. http://www.ipcc.ch/pdf/assessment-report/ar5/syr/SYR_AR5_FINAL_full_wcover.pdf.

Internal Displacement Monitoring Centre and Norwegian Refugee Council (2017). *Global Report on Internal Displacement.* Geneva. http://www.internal-displacement.org/global-report/grid2017/pdfs/2017-GRID.pdf

International Council for Science (2017). *A Guide to SDG Interactions: From Science to Implementation.* Paris. https://council.science/cms/2017/05/SDGs-Guide-to-Interactions.pdf.

International Energy Agency (2017). *Energy Technology Perspectives 2017.* Paris. http://www.iea.org/etp/.

International Energy Agency (2018). *United Kingdom: Indicators for 2015.* Paris. https://www.iea.org/statistics/statisticssearch/report/?product=Indicators&country=UK.

International Federation of Phonographic Industry (2018). *Global Music Report 2018: Annual State of the Industry.* London. https://www.ifpi.org/downloads/GMR2018.pdf.

International Federation of Red Cross and Red Crescent Societies (n.d.) Vulnerability and capacity assessment (VCA), [no date]. https://www.ifrc.org/vca. (Accessed 22 October 2018).

International Renewable Energy Agency (2016). Renewable Energy and the UN Sustainable Development Goals (SDGs). IRENA Twelfth meeting of the Council. Abu Dhabi. http://www.irena.org/-/media/Files/IRENA/Agency/About-IRENA/Council/Twelth-Council/C_12_DN_4_RE-and-SDGs.pdf.

Jawahir, I.S. and Bradley, R. (2016). Technological elements of circular economy and the principles of 6R-based closed-loop material flow in sustainable manufacturing. *Procedia CIRP* 40, 103-108. https://doi.org/10.1016/j.procir.2016.01.067.

Kalundborg Symbiosis (2018). Kalundborg Symbiosis [website]. Kalundborg: SymbiosisCenter Denmark. http://www.symbiosis.dk/en/. (Accessed 19 October 2018).

Kates, R., Travis, W.R.T. and Wilbanks, T. (2012). Transformational adaptation when incremental adaptations to climate change are insufficient. *Proceedings of the National Academy of Sciences* 109(19), 7156-7161. https://doi.org/10.1073/pnas.1115521109.

Kern, K., Jorgens, H. and Jänicke, M. (2001). *The Diffusion of Environmental Policy Innovations: A Contribution to the Globalisation of Environmental Policy.* Berlin: Social Science Research Center. https://www.econstor.eu/obitstream/10419/48976/1/329601059.pdf.

Kim, R.E. (2016). The nexus between international law and the Sustainable Development Goals. *Review of European, Comparative & International Environmental Law* 25, 15–26. https://doi.org/10.1111/reel.12148.

Kim, R.E. and Bosselmann, K. (2013). International environmental law in the Anthropocene: Towards a purposive system of multilateral environmental agreements. *Transnational Environmental Law* 2, 285–309. https://doi.org/10.1017/S2047102513000149.

Kim, R.E. and van Asselt, H. (2016). International governance: dealing with problem shifting in the Anthropocene – the limits of international law. In *Research Handbook on International Law and Natural Resources.* Morgera, E. and Kulovesi, K. (eds.). Cheltenham: Edward Elgar. https://www.elgaronline.com/view/9781783478323.00039.xml.

Lacy P. 2015. Why the circular economy is a digital revolution, 17 August. *World Economic Forum.* https://www.weforum.org/agenda/2015/08/why-the-circular-economy-is-a-digital-revolution/. (Accessed 19 October 2018).

Low, B., Ostrom, E., Simon, C. and Wilson, J. (2003). Redundancy and diversity: do they influence optimal management? In *Navigating Social-Ecological Systems: Building Resilience for Complexity and Change.* Berkes, F., Colding, J. and Folke, C. (eds.). Cambridge: Cambridge University Press. Chapter 4. 53-83. http://assets.cambridge.org/052181/5924/sample/0521815924ws.pdf.

Liu, J., Dietz, T., Carpenter, S.R., Alberti, M., Folke, C., Moran, E. *et al.* (2007). Complexity of coupled human and natural systems. *Science* 317(5844), 1513–1516. https://doi.org/10.1126/science.1144004.

McDowall, W., Geng, Y., Huang, B., Barteková, E., Bleischwitz, R., Türkeli, S. *et al.* (2017). Circular economy policies in China and Europe. *Journal of Industrial Ecology* 21(3), 651-661. https://doi.org/10.1111/jiec.12597.

Meadows, D.H. (2008). *Thinking in Systems: A Primer.* Wright, D. (ed.): Chelsea Green Publishing. https://www.chelseagreen.com/product/thinking-in-systems/.

Meyer, C., Matzdorf, B., Muller, K. and Schleyer, C. (2014). Cross compliance as payment for public goods? Understanding EU and US agricultural policies. *Ecological Economics* 107, 185-194. https://doi.org/10.1016/j.ecolecon.2014.08.010.

Mitchell, C., Sawin, J.L., Pokharel, G.R., Kammen, D., Wang, Z., Fifita, S. *et al.* (2011). Policy, financing and implementation. In *IPCC Special Report on Renewable Energy Sources and Climate Change Mitigation.* Edenhofer, O., Pichs-Madruga, R., Sokona, Y., Seyboth, K., Matschoss, P., Kadner, S. *et al.* (eds.). Cambridge: Intergovernmental Panel on Climate Change. Chapter 11. https://www.ipcc.ch/pdf/special-reports/srren/drafts/SRREN-FOD-Ch11.pdf.

Monroe, M.C., Plate, R.R., Oxarart, A., Bowers, A. and Chaves., A. (2017). Identifying effective climate change education strategies: a systematic review of the research. *Environmental Education Research.* https://doi.org/10.1080/13504622.2017.1360842.

Nakashima, D.J., Galloway McLean, K., Thulstrup, H.D., Ramos Castillo, A. and Rubis, J.T. (2012). *Weathering Uncertainty Traditional Knowledge for Climate Change Assessment and Adaptation.* Paris: United Nations Educational, Scientific and Cultural Organization and United Nations University. http://unesdoc.unesco.org/images/0021/002166/216613e.pdf.

Nansen Initiative (2015) *Agenda for the Protection of Cross-Border Displaced Persons in the Context of Disasters and Climate Change Volume I.* The Nansen Initiative. https://nanseninitiative.org/wp-content/uploads/2015/02/PROTECTION-AGENDA-VOLUME-1.pdf.

Nilsson, M., Griggs, D. and Visbeck, M. (2016). Policy: map the interactions between Sustainable Development Goals. *Nature* 534(7607), 320–322. https://doi.org/10.1038/534320a.

Norsk elbilforening (2018). *Norwegian Electric Vehicle Association.* https://elbil.no. (Accessed 19 October 2018.)

Olhoff, A. and C. Schaer (2010). *Screening Tools and Guidelines to Support the Mainstreaming of Climate Change Adaptation into Development Assistance – A Stocktaking Report.* New York, NY: United Nations Development Programme. http://content-ext.undp.org/aplaws_publications/2386693/UNDP%20Stocktaking%20Report%20CC%20mainstreaming%20tools.pdf.

O'Neill, D.W., Fanning, A.L., Lamb, W.F. and Steinberger, J.K. (2018). A good life for all within planetary boundaries. *Nature Sustainability* 1, 88-95. https://doi.org/10.1038/s41893-018-0021-4.

Organisation for Economic Co-operation and Development (2012). *Sustainable Materials Management: Making Better Use of Resources.* Paris. https://www.oecd-ilibrary.org/docserver/9789264174269-en.pdf.

Organisation for Economic Co-operation and Development (2014). *Preventing Food Waste: Case Studies of Japan and the United Kingdom.* Trade and Agriculture Directorate, Committee for Agriculture, Working Party on Agricultural Policies and Markets. http://www.oecd.org/officialdocuments/publicdisplaydocumentpdf/?cote=TAD/CA/APM/WP(2014)25/FINAL&docLanguage=En.

Organisation for Economic Co-operation and Development (2017). *Working Party on Agricultural Policies and Markets: Evaluation of the EU Common Agricultural Policy (CAP) 2014-20.* Paris. https://ec.europa.eu/agriculture/sites/agriculture/files/direct-support/pdf/2016-staff-working-document-greening_en.pdf.

Organisation for Economic Co-operation and Development (2018). Producer and Consumer Support Estimates database. http://www.oecd.org/agriculture/pse. Accessed 21 October 2018.

Ostrom, E. (1990). *Governing the Commons: The Evolution of Institutions for Collective Action.* Cambridge: Cambridge University Press. http://www.cambridge.org/core_title/gb/478715.

Ostrom, E. (2007). A diagnostic approach for going beyond panaceas. *Proceedings of the National Academy of Sciences* 104(39), 15181–15187. https://doi.org/10.1073/pnas.0702288104.

Ostrom, E., Janssen, M.A. and Anderies, J.M. (2007). Going beyond panaceas. *Proceedings of the National Academy of Sciences* 104, 15176–15178. https://doi.org/10.1073/pnas.0701886104.

Potting, J., Hekkert, M.P., Worrell, E. and Hanemaaijer, A. (2017). *Circular Economy: Measuring Innovation in the Product Chain.* The Hague: PBL Netherlands Environmental Assessment Agency. http://www.pbl.nl/sites/default/files/cms/publicaties/pbl-2016-circular-economy-measuring-innovation-in-product-chains-2544.pdf.

Raworth, K. (2012). *A Safe and Just Space for Humanity: Can We Live Within the Doughnut?* Oxford: Oxfam. https://www.oxfam.org/sites/www.oxfam.org/files/dp-a-safe-and-just-space-for-humanity-130212-en.pdf.

Renewable Energy Policy Network for the 21st Century (2017). *Renewables 2017 Global Status Report.* Paris. http://www.ren21.net/wp-content/uploads/2017/06/17-8399_GSR_2017_Full_Report_0621_Opt.pdf.

Rizos, V., Tuokko, K. and Behrens, A. (2017). *The Circular Economy: A Review of Definitions, Processes and Impacts.* Brussels: Centre for European Policy Studies. https://www.ceps.eu/publications/circular-economy-review-definitions-processes-and-impacts#.

Rockström, J., Steffen, W., Noone, K., Persson, Å., Chapin, F. S. IIIrd, Lambin, E. *et al.* (2009). Planetary boundaries: exploring the safe operating space for humanity. *Ecology and Society* 14(2), 32. https://www.jstor.org/stable/26268316.

Salzman, J., Bennett, G., Carroll, N., Goldstein, A. and Jenkins, M. (2018). The global status and trends of payments for ecosystem services. *Nature Sustainability* 1, 136–144. https://doi.org/10.1038/s41893-018-0033-0.

Schwerdtle, P., Bowen, K. and McMichael, C. (2018). The health impacts of climate-related migration. *BMC Medicine* 16(1), 1. https://doi.org/10.1186/s12916-017-0981-7.

Schaffartzik A., Mayer, A., Gingrich, S., Eisenmenger, N., Loy, C. and Krausmann, F. (2014). The global metabolic transition: regional patterns and trends of global material flows, 1950–2010. *Global Environmental Change* 26, 87–97. https://doi.org/10.1016/j.gloenvcha.2014.03.013.

Silva, A., Rosano, M., Stocker, L. and Gorissen, L. (2017). From waste to sustainable materials management: three case studies of the transition journey. *Waste Management* 61, 547-557. https://doi.org/10.1016/j.wasman.2016.11.038.

Smol, M., Kulczycka, J. and Avdiushchenko, A. (2017). Circular economy indicators in relation to eco-innovation in European regions. *Clean Technologies and Environmental Policy* 19(3), 669-678. https://doi.org/10.1007/s10098-016-1323-8.

Spannring, R. and Grušovnik, T. (2018). Leaving the Meatrix? Transformative learning and denialism in the case of meat consumption. *Environmental Education Research*. https://doi.org/10.1080/13504622.2018.1455076.

Springmann, M., Clark, M., Mason-D'Croz, D., Wiebe, K., Bodirsky, B.L., Lassaletta, L., *et al.* (2018). Options for keeping the food system within environmental limits. Nature. https://doi.org/10.1038/s41586-018-0594-0.

Springmann, M., Mason-D'Croz, D., Robinson, S., Wiebe, K., Godfray, H., Rayner, M. and Scarborough, P. (2016). Mitigation potential and global health impacts from emissions pricing of food commodities. *Nature Climate Change* 7(1), 69-74. https://doi.org/10.1038/nclimate3155.

Stahel, W.R. and Reday-Mulvey, G. (1981). *Jobs for Tomorrow: The Potential for Substituting Manpower for Energy*. New York, NY: Vantage Press. https://searchworks.stanford.edu/view/9965259.

Stahel, W.R. (2016). The circular economy. *Nature* 531(7595), 435. https://doi.org/10.1038/531435a.

Sundaramoorthy, S. and Walia, A. (2017). India's experience in implementing strategic schemes to enhance appliance energy efficiency & futuristic integrated policy approaches to adopt most efficient technologies. New Delhi: CLASP. http://www.ieppec.org/wp-content/uploads/2017/10/walia_paper.pdf.

Tanentzap, A.J., Lamb, A., Walker, S. and Farmer, A. (2015). Resolving conflicts between agriculture and the natural environment. *PLoS Biology* 13(9). https://doi.org/10.1371/journal.pbio.1002242.

Taylor, E.W. (2017). Transformative learning theory. In *Transformative Learning Meets Bildung*. Laros, A., Fuhr, T. and Taylor, E.W. (eds.). Rotterdam: SensePublishers. chapter 2. 17-29. https://link.springer.com/chapter/10.1007/978-94-6300-797-9_2.

Turner, B.L., Kasperson, R.E., Matson, P.A., McCarthy, J.J., Corell, R.W., Christensen, L. *et al.* (2003). A framework for vulnerability analysis in sustainability science. *Proceedings of the National Academy of Sciences* 100(14), 8074. https://doi.org/10.1073/pnas.1231335100.

Underdal, A. and Kim, R.E. (2017). The sustainable development goals and multilateral agreements. In *Governing Through Goals: Sustainable Development Goals as Governance Innovation*. Kanie, N. and Biermann, F. (eds.). Cambridge, MA: The MIT Press. Chapter 10. http://mitpress.universitypressscholarship.com/view/10.7551/mitpress/9780262035620.001.0001/upso-9780262035620-chapter-010.

United Nations (2015). *Small Island Developing States in Numbers: Climate Change edition*. New York, NY. https://sustainabledevelopment.un.org/content/documents/2189SIDS-IN-NUMBERS-CLIMATE-CHANGE-EDITION_2015.pdf.

United Nations (2016) *World Economic and Social Survey 2016, Climate Change Resilience: An Opportunity for Reducing Inequalities*. New York. https://www.un.org/development/desa/dpad/wp-content/uploads/sites/45/publication/WESS_2016_Report.pdf

United Nations Educational, Scientific and Cultural Organization (2010). *Climate Change Education for Sustainable Development*. Paris. http://unesdoc.unesco.org/images/0019/001901/190101E.pdf.

United Nations Educational, Scientific and Cultural Organization (2014). *Shaping the Future We Want: UN Decade of Education for Sustainable Development (2005-2014). Final Report*. Paris. http://unesdoc.unesco.org/images/0023/002301/230171e.pdf.

United Nations Educational, Scientific and Cultural Organization (2017). *Education for Sustainable Development Goals: Learning Objectives*. Paris. http://unesdoc.unesco.org/images/0024/002474/247444e.pdf.

United Nations Educational, Scientific and Cultural Organization (2018). *Issues and Trends in Education for Sustainable Development*. Paris. http://unesdoc.unesco.org/images/0026/002614/261445E.pdf.

United Nations Environment Programme (2016). *Global Material Flows and Resource Productivity. Assessment Report for the UNEP International Resource Panel*. Schandl, H., Fischer-Kowalski.M., West, J., Giljum, S., Dittrich, M., Eisenmenger, N. *et al.* (eds.). Nairobi. http://wedocs.unep.org/bitstream/handle/20.500.11822/21557/global_material_flows_full_report_english.pdf?sequence=1&isAllowed=y.

United Nations Human Settlements Programme (2016). *New Urban Agenda adopted at Habitat III*. https://unhabitat.org/new-urban-agenda-adopted-at-habitat-iii/.

United Nations Industrial Development Organization (2017). *Vienna Energy Forum 2017*. Vienna. https://www.unido.org/sites/default/files/2017-08/VEF_REPORT_0.pdf.

United Nations Office for Disaster Risk Reduction (2015). *Sendai Framework for Disaster Risk Reduction 2015-2030*. https://www.unisdr.org/files/43291_sendaiframeworkfordrren.pdf.

United Nations Office for Disaster Risk Reduction (2017). *Disaster-related Data for Sustainable Development. Sendai Framework Data Readiness Review 2017: Global Summary Report*. https://www.unisdr.org/files/53080_entrybgpaperglobalsummaryreportdisa.pdf.

United States Environmental Protection Agency (2015). *Fiscal Year 2017-2022: U.S. EPA Sustainable Materials Management Program Strategic Plan*. Washington, DC. https://www.epa.gov/sites/production/files/2016-03/documents/smm_strategic_plan_october_2015.pdf.

van den Bergh, J., Folke, C., Polasky, S., Scheffer, M. and Steffen, W. (2015). What if solar energy becomes really cheap? A thought experiment on environmental problem shifting. *Current Opinion in Environmental Sustainability* 14, 170-179. https://doi.org/10.1016/j.cosust.2015.05.007.

Velis, C. (2017). Waste pickers in Global South: informal recycling sector in a circular economy era. *Waste Management & Research* 35(4), 329–331. https://doi.org/10.1177/0734242X17702024.

Walker, B., Barrett, S., Polasky, S., Galaz, V., Folke, C., Engström, G. *et al.* (2009). Looming global-scale failures and missing institutions. *Science* 325(5946), 1345–1346. https://doi.org/10.1126/science.1175325.

Ward, J., Chiveralls, K., Fioramonti, L., Sutton, P. and Costanza, R. (2017). The decoupling delusion: rethinking growth and sustainability, 12 March. *The Conversation*. http://theconversation.com/the-decoupling-delusion-rethinking-growth-and-sustainability-71996. (Accessed 19 October 2018).

Watts, N., Adger, W.N., Agnolucci, P., Blackstock, J., Byass, P., Cai, W. *et al.* (2015). Health and climate change: Policy responses to protect public health. *Lancet* 386(10006), 1861-1914. https://doi.org/10.1016/S0140-6736(15)60854-6.

Weber, R., Lissner, L., and Fantke, P. (2016). The substitution of hazardous chemicals in the international context - Opportunity for promoting sustainable chemistry. *Abstract from 1st Green and Sustainable Chemistry Conference, Berlin, Germany*. http://orbit.dtu.dk/files/126997510/Weber_2016a.pdf.

Wiedmann, T.O., Schandl, H., Lenzen, M., Moran, D., Suh, S., West, J. and Kanemoto, K. (2015). The material footprint of nations. *Proceedings of the National Academy of Sciences* 112(20), 6271-6276. https://doi.org/10.1073/pnas.1220362110

Wirsenius, S., Hedenus, F. and Mohlin, K. (2011). Greenhouse gas taxes on animal food products: rationale, tax scheme and climate mitigation effects. *Climatic Change* 108(1-2), 159-184. https://doi.org/10.1007/s10584-010-9971-x

Wise, R.M., Fazey, I., Smith, M.S., Park, S.E., Eakin, H.C., Van Garderen, E.A. and Campbell, B. (2014). Reconceptualising adaptation to climate change as part of pathways of change and response. *Global Environmental Change* 28, 325-336. https://doi.org/10.1016/j.gloenvcha.2013.12.002

Wisner, B., Blaikie, P.M., Cannon, T. and Davis, I. (2004). *At Risk: Natural Hazards, People's Vulnerability and Disasters*. 2nd edn. London: Routledge. https://trove.nla.gov.au/work/23504995?selectedversion=NBD24609117

World Bank (2013). Which coastal cities are at highest risk of damaging floods? New study crunches the numbers. http://www.worldbank.org/en/news/feature/2013/08/19/coastal-cities-at-highest-risk-floods

World Bank (2018) Climate & disaster risk screening tools. https://climatescreeningtools.worldbank.org/. (Accessed 22 October 2018).

Yang, Y., Bae, J., Kim, J. and Suh, S. (2012). Replacing gasoline with corn ethanol results in significant environmental problem-shifting. *Environmental Science & Technology* 46(7), 3671–3678. https://www.doi.org/10.1021/es203641p.

Yang, H., Ma, M., Thompson, J.R. and Flower, R.J. (2018). Waste management, informal recycling, environmental pollution and public health. *Journal of Epidemiology & Community Health* 72(3), 237-243. http://dx.doi.org/10.1136/jech-2016-208597.

Young, O.R. (2017). *Governing Complex Systems: Social Capital for the Anthropocene*. Cambridge, MA: MIT Press. https://mitpress.mit.edu/books/governing-complex-systems.

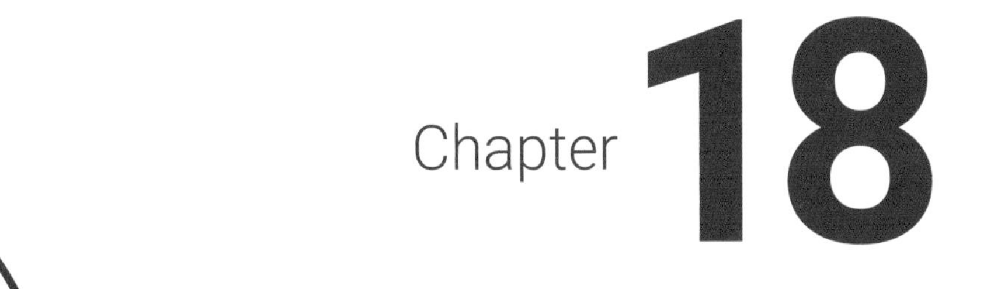

Chapter 18

Conclusions on Policy Effectiveness

© Shutterstock/PopTika

Coordinating Lead Authors: Klaus Jacob (Freie Universität Berlin), Peter King (Institute for Global Environmental Strategies), Diana Mangalagiu (University of Oxford and Neoma Business School)

Contributing Author: Beatriz Rodríguez-Labajos (Universitat Autònoma de Barcelona)

18.1 Overview of the outcomes

This chapter presents a set of conclusions for Part B of the sixth Global Environment Outlook (GEO-6), reached through the findings of the previous chapters about policy effectiveness (Chapters 10-17). It summarizes for policymakers what is known to work best and why, including a synthesized discussion of the limitations of the evidence available to date for policy effectiveness. We also make reference to Part C (Outlooks), which will examine the promising emerging policies for the future.

There is considerable innovation in policy approaches and instruments across all the environmental themes covered by the sixth Global Environment Outlook (GEO-6) (Chapters 12-17). New institutions, policies and policy instruments have been developed and introduced all over the world. Environmental policy innovation takes place not only in Western industrialized countries, but also in emerging and developing economies. Policies are developed that go beyond technical fixes by increasingly addressing social and economic practices.

Environmental policy innovation also takes place to address issues of equity and environmental protection at the same time. Examples of this include the territorial rights for fishing in Chile, or the free basic water allocation in South Africa, both of which are measures to secure access to natural resources for low-income communities while at the same time promoting sustainable management.

Environmental policies aim to reduce emissions and depletion of resources by encouraging behavioural change or limiting the choices of consumers, enterprises and communities. Different modes of intervention are being used: persuasion, economic incentives and regulation.

There is no single instrument for complex environmental problems, and policy mixes are more effective, often combining different modes of governance that mutually reinforce each other (referred to as 'hybrid governance'). Combining measures on the demand side, for taxing and labelling environmentally harmful consumption, with measures on the production side, to limit emissions, is one example that can mutually reinforce environmental innovation, and create markets for it.

Environmental policies are also defining the processes that enable and encourage actors to reflect on their environmental performance – environmental impact assessments, planning procedures and environmental management systems, for example.

Chapters 12-17 also show that environmental actors within and beyond governments are being established or strengthened by many environmental policies, showing an unfolding of effects on environmental performance. Environmental policies and institutions do not determine resource use and emissions on their own – there is also the role of policies in sectors such as housing, infrastructure, agriculture, industry, energy, and so on. A further mechanism that promotes effective environmental policy – albeit a difficult one to achieve –lies in the integration of environmental concerns into other sectoral policies.

While policy integration promises to settle conflicts between environmental and other objectives (Nilsson *et al.* 2012; Runhaar, Driessen and Uittenbroek 2014; Mullally and Dunphy 2015), the analysis in the previous chapters demonstrates that this has rarely been achieved in practice. There is a lack of systematic evidence on how sectors such as agriculture, transport, urban planning and water management can incorporate environmental standards to prevent, reduce or mitigate harmful environmental effects. Changes in policy mixes are often compelled by pressure from different groups and sectors that have opposing stakes on a resource, environmental asset or ecosystem service.

Many countries (and some international organizations) have begun to adopt integrated approaches or instruments to assess the potential impacts of proposed legislation on stakeholders and their well-being, economic sectors, and the environment (Radaelli 2009; Jacob *et al.* 2011; Adelle and Weiland 2012; Adelle *et al.* 2016; European Environment Agency [EEA] 2017). Such integrated policies may help to achieve the broader set of Sustainable Development Goals (SDGs) in a cost-efficient way, overcoming existing barriers and trade-offs.

Environmental policy integration tools include regulatory impact assessment, environmental and health impact assessment, and strategic environmental assessment. These evidence-based policymaking tools are increasingly being adopted to demonstrate the need for improved environmental policies. Considerable experience is emerging in the use of these tools, particularly in the European Union.

To date, however, there is little evidence to measure the level of policy integration or the actual outcomes from applying various tools. Among the few exceptions is the Partnership for European Environmental Research (Mickwitz *et al.* 2009), which assessed climate policy integration in Europe, at multiple scales. A key lesson from the project is that cities and municipalities have begun to integrate climate aims into their strategies and plans, and that such authorities sometimes have more ambitious goals than national governments.

An important argument in favour of environmental policy integration is the expected economic and social co-benefits from implementing environmental policies. These co-benefits include additional economic growth spurred by innovation, savings from the conservation of natural resources, and the avoided costs of environmental damage. The United Nations Environment Programme (UNEP) estimates that two per cent of global gross domestic product (GDP) in green investment would deliver long-term economic growth while minimizing the adverse impacts of climate change, water scarcity and loss of ecosystem services (UNEP 2011).

The analysis in GEO-6 of environmental policies and their integration demonstrates the diversity of institutional and cultural frameworks in which policymaking takes place. The roles of law, values, administrative capacities, socioeconomic conditions, and so on, are important in how effective policies can be. The design of policies that reflect on this set of conditions is important.

The effectiveness of the thousands of policy innovations cannot be assessed comprehensively; a case-by-case approach is needed to make evaluations. The effectiveness of different policy instruments cannot be compared given the multiple market failures, including, among others, the lack of price signals, lack of information and network effects. There is no evidence, for example, to support claims of a general superiority of market-based instruments over regulatory or persuasive ones. The analysis presented here by GEO-6 does show evidence, however, for the need to combine different policies into complementary policy mixes of clusters. Despite the recognition that coherent policy mixes are often more effective than stand-alone policies, the interplay of instruments within mixes is not well understood, aside from the rather broad understanding that some policy instrument types do not necessarily work well with others.

Effective and ambitious environmental policies are often contested by the sectors affected. Their design, and the level of ambition, is usually the subject of negotiation in the policy process, during which environmental actors usually need to find compromise. Second-best environmental policies are often adopted as a result. For many issues and in many countries, environmental policy does not make use of potentially powerful mixes of price signalling and hard regulation. Instead, mechanisms of persuasion, self-regulation or subsidy are introduced. Chapters 12-17 also find, finally, that vested rights and interests are often not touched on, with environmental policies instead focusing on new products or sites, by having permitting procedures for development projects, for example.

Once environmental policies have been established successfully, their scaling up has been observed. Moreover, new opportunities and capacities for advancing policies, and for raising the level of ambition over time have been observed once the technical, social and economic feasibility has been demonstrated and markets for environmentally friendly alternatives have been created. In a few cases, these opportunities were built into the policy design from the very beginning. The commitment for a continuous improvement of policies over time could be applied much more often than it is today, in the manner of the so-called ratchet mechanism of the Paris Agreement on climate change (United Nations Framework Convention on Climate Change [UNFCCC] 2015).

In view of the challenges outlined above, there is an emerging consensus that the design of policy instruments is at least as important as the choice of the instrument, for the effectiveness of individual policies and policy mixes (Yin and Powers 2010; Flanagan, Uyarra and Laranja 2011; Kemp and Pontoglio 2011). The temporal dynamics of policy change, how and why specific policies stick (or fail to stick) and how policy choices interact in an increasingly complex policy mix all need to be better understood. As these lessons are learned over time, the level of ambition is expected to increase – especially, as the GEO-6 finds, if environmental policies prove to have economic and social co-benefits.

Added to the observation that environmental policies are being scaled up within national borders, they are also diffusing across jurisdictions. Other countries, regions or communities are taking up and adapting the examples of pioneering countries. Some publicly available data sets aim to facilitate the charting of this diffusion, particularly for policies on climate change and renewable energy. The Climate Change Laws of the World database from the London School of Economics (Grantham Research Institute on Climate Change and the Environment 2017), for example, compiles information on national-level climate policies ranging from transport policy to adaptation and mitigation. Similarly, REN21's Global Status Report (Renewable Energy Policy Network for the 21st Century [REN21] 2018) charts the use of renewable energy policies across a large sample of national and subnational jurisdictions. InforMEA, finally, is the United Nation's portal for information on multilateral environmental agreements (UNEP, Food and Agriculture Organization of the United Nations [FAO] and United Nations Educational, Scientific and Cultural Organization [UNESCO] 2018).

Despite the prolonged interest in the spatial diffusion of environmental policies, and efforts to provide systematic policy information on it, knowledge about this particular aspect of policy development remains limited, especially outside the specific policy field of renewable energy. There is also a lack of research on the role of local contexts on the effectiveness of policies adopted from abroad. There is some evidence that less ambitious policies (e.g. distributional rather than re-distributional policies) are the subject of policy diffusion more often than cross-jurisdictional policies. This is in spite of the fact that policy diffusion may be considered a positive mechanism for learning across different jurisdictions – often facilitated by international regimes and multilevel governance.

The GEO-6 analysis finds that the importance of good policy design for the effectiveness of environmental policies cannot be overstressed. Mickwitz *et al.* (2009, p. 12) list some common elements of good design:

- set a long-term vision and avoid crisis-mode policy decisions, through inclusive, participatory design processes;
- establish a baseline, quantified targets and milestones;
- conduct *ex ante* (before implementation) and *ex post* (after implementation) analyses of cost-benefit or cost-effectiveness to ensure the best use of public funds;
- build in monitoring regimes during implementation, preferably involving affected stakeholders; and
- evaluate the policy outcomes and impacts, to close the loop for improving future policy design.

Despite this comprehensive list, assessments of policy effectiveness both *ex ante* and *ex post* against a baseline are usually missing, even for well-designed policies. The analysis finds that policy evaluation tools are rarely used. An evidence base for measuring policy effectiveness is therefore lacking because it is difficult in many cases to attribute effects to environmental policies, and whether these effects would have taken place without the policies. Impact assessments and policy evaluations are not being applied in a systematic way. Therefore, while the analysis of indicators, and the distance still to go to reach the goals, suggests that environmental policies are not yet sufficiently effective to achieve sustainable development, the analysis cannot reveal which policies and policy instruments are more effective or efficient than others.

There is no universally accepted methodology that can show causal relationships between the effects and the policies adopted, and unequivocal answers on policy effectiveness

unfortunately cannot be provided. It is rarely feasible or ethical in the environmental domain to conduct policy experiments that show the counterfactual – that is, what would have happened had there been no policy?

Further, the literature shows the importance of various constraining and enabling factors, such as institutional capacity and political will. Policies also rarely stand alone, and the importance is stressed, as discussed earlier, of coherent, synergistic policies, or policy mixes. It is important, too, to recognize co-benefits and unintended side effects. Finally, spillover effects need to be recognized, especially where these involve transboundary concerns.

Accordingly, a two-track process was adopted for the assessment of policy effectiveness (Chapter 10) in GEO-6. For the top-down perspective, the author teams identified typical policy approaches that have been employed to solve key environmental problems in the areas of air, biodiversity, oceans, land, fresh water (surface and groundwater), and cross-cutting issues (Chapters 12-17). To illustrate experience in the implementation of these policy approaches in greater detail, specific case studies were selected, and effectiveness criteria derived from the literature were used to provide a qualitative assessment of policy effectiveness.

The second track, bottom-up, was to identify policy-sensitive indicators, meaning that one should be able to construct, again from the literature, a plausible story around why each indicator appears to be improving in response to a policy or policy mix. Within Chapters 12-17, the subsections on indicators therefore cover:

- their descriptions and their relation to SDGs or other multilateral environmental agreements;
- how data are collected for each indicator;
- a plausible line of argument for how an observed improvement in the indicator across multiple countries could be due, at least partly, to one or more policies;
- what other factors might explain the improvement; and
- what alternative indicators could verify the role of policies.

The narrative is interspersed with infographics. Depending on the availability of data in the literature, these help to show: correlations between the adoption of certain policies by countries and improvements in the indicators; trend analysis showing improvement in the indicator; or the numbers of countries reporting on the indicator over time.

From the limited number of case studies that could be addressed in GEO-6, it is apparent that there are very few cases where all the effectiveness criteria have been comprehensively covered at the policy design, implementation or post-evaluation stages of the policy cycle. In many cases, no quantifiable baseline was established, making it difficult to show quantitative evidence that the policy was improving environmental outcomes as intended. In most cases, there was no *ex ante* cost-benefit or cost-effectiveness analysis, making it uncertain that the best policy choice had been made. While co-benefits were often identified, in most cases through hindsight, there was no evidence of a deliberate, prospective attempt to ensure policy coherence and synergies. While most policies specifying a timeframe had been conducted within that period, a surprising number of case studies appeared to be open ended, with no specific time for closure, evaluation or renewal. Many of the case studies were linked to global processes and agreements, which suggests that comprehensive environmental agreements like the Paris Agreement and the SDGs do provide an overarching policy framework that guides national policy processes.

The findings from GEO-6's assessment of policy effectiveness, as well as from its assessment of the evaluation methods used, have the potential to help develop a baseline for future research and global assessments. Continued efforts on policy evaluation would also help to close these gaps in data.

18.2 Connections to future policy

The analysis above of policy effectiveness inevitably comes after a lag in time because policymakers do not know if a policy has been effective until some years after its initial implementation, especially if part of the indication of effectiveness is viewed to be implementation across multiple countries. This means that Part B has not been able to showcase new emerging, promising policy approaches, which are instead addressed in Chapter 24 (The Way Forward). Future editions of GEO will need to assess the eventual effectiveness of these policy approaches following their implementation. Policymakers have the opportunity meanwhile to examine the effectiveness criteria selected in GEO-6 and to use these when designing the new generation of policies and planning their evaluation.

Improved policies and governance arrangements will form an essential part of crafting pathways towards sustainability. It is likely that the emerging and promising policies covered in Part C (Outlooks) will come into this picture – because the current set of policy approaches are unlikely, with the required urgency, to achieve the SDGs, Paris Agreement and other multilateral environmental agreements. One example of the need for new innovative policy is that the setting of national standards, as part of the normal command-and-control policy approach to combating pollution, is too slow and unwieldly to keep up with the thousands of new chemicals, materials, genetically modified organisms and nanotechnologies being released into the environment every day.

18.3 Gaps in knowledge

The policy-effectiveness analysis conducted for GEO-6 has struck out into a new direction for UN Environment. Policymakers want to know which policies work and why, but assessments should not stray too far into policy advocacy. The costs of inaction and inordinate delays in policy implementation also need to be studied, as well as the effectiveness of policy action. The key gap, surprising many of the authors, was the paucity of well-documented evaluations of the selected case studies that illustrate the importance of the science-policy interface. It appears that in most countries it is either not the practice to conduct post-evaluations of policies, or if such evaluations are conducted then the results are not in the public domain.

We suggest that UN Environment works with member countries to extract those policy evaluations not currently in the public domain to create a section for policy effectiveness in the online data portal, Environment Live. Researchers and

policy think tanks should also be encouraged to conduct policy-effectiveness studies to provide the independent analysis that appears to be in short supply.

The need for a universally acceptable methodology to assess policy effectiveness represents another critical gap. The challenge for researchers and policy think tanks is to conduct such analyses regularly, and for policymakers to apply this information in advancing policymaking.

Another gap in knowledge relates to the policy-sensitivity of indicators. Among the hundreds of indicators selected for the SDGs, which of these are policy-sensitive? Of the indicators that are policy-sensitive, what are the corresponding policies that they are sensitive to? Which of these should governments be considering to achieve accelerated progress on the indicators? Of the SDG targets that UN Environment is responsible for, which are policy-sensitive, and what should be the role of UN Environment in not only tracking these but also analysing policy effectiveness?

Finally, there is also a gap in relation to the analysis of social and economic policies, such as sectoral policies, that have important effects on environmental conditions. It is not sufficient to analyse only environmental policies. Tools such as Environmental Impact Assessments (EIA) and Strategic Environmental Assessments (SEA) can be used to examine the environmental consequences of projects, policies, plans and programmes. More importantly, however, sectoral agencies should engage with environmental experts for help with avoiding adverse environmental consequences to planned activities.

18.4 Key lessons from the analysis

Consider policy effectiveness at the design stage. Most of the weaknesses identified in environmental policy approaches stem from inadequate analysis at the design stage. The empirical analysis in GEO-6 demonstrates that, too often, environmental policy decisions are knee-jerk reactions to environmental crises rather than part of a deliberative, long-term process of policy selection and design to avoid environmental damage.

Establish a verifiable, quantitative baseline. A quantitative baseline that is science-based and verifiable, with firm targets, is an essential component of effective policies. For policies that will take a long time to reach fruition, quantitative milestones will also help to ensure that policy implementation is on the right track.

Conduct cost-benefit or cost-effectiveness analysis at the policy design stage. For most environmental problems there are multiple alternative policies that could achieve the desired outcomes. Water pollution, for example, could be controlled by regulations to change production processes, by the establishment of ambient or discharge standards, by imposing discharge fees, or through dilution from upstream reservoirs – or some combination of these. An examination needs to be made at the policy-design stage on what is the most effective use of public and private funding. This needs to be checked subsequently to ensure that the right policy choice has been made.

Ensure policy coherence and synergy. There is strength in numbers. Generally, a single, stand-alone policy will not be as effective as a mix of policies that work together towards the same policy goal. Equally important, however, is to examine policies that might adversely impact or conflict with the policy objective. For example, policies to promote renewable energy may be undermined by continued subsidies to thermal power plants.

Conduct independent post-evaluation studies. The literature examined for this report found that there were few independent post-evaluation studies of policies. Evaluation by governments is necessary and important, but greater confidence may be achieved by unbiased, independent studies. A crucial role for funding agencies, therefore, is to provide the necessary resources for more of these, particularly in developing countries, where there is a dearth of good policy assessment.

Engage key actors in all aspects of the policy cycle. A complex web of stakeholders who need to be involved in each part of the policy cycle is revealed by the case studies. This implies a need for transparency in the policy process. Inclusive policy processes will generally be more effective than those which exclude some of the actors. There may be a cost and additional time constraint in being inclusive, but this tends to be compensated during implementation, whereas protests or legal challenges could delay implementation plans.

Identify the indicators that are policy-sensitive and can demonstrate causal links. Establish a clear linkage from the indicators in the SDGs and other multilateral environmental agreements to known effective policies. The case studies showed the importance of multilateral agreements in the specific areas of air, biodiversity, oceans, land, fresh water (surface and groundwater), and in several cross-cutting areas. While such agreements are often unenforceable at the global level, they do carry moral suasion and provide peer pressure to embody the agreed approach in national and subnational policies, plans and programmes. Only weak links have been made so far between indicators and effective policy, however, and additional work needs to be done in establishing these connections.

Conduct additional research on policy effectiveness. Researchers often finish their assessments with a call for more research, but in this case, it is justified. There are remarkably few well-documented case studies of policy effectiveness that follow the policy decisions throughout the policy cycle, from design to post-evaluation. The future work of UN Environment needs to strengthen the link between policies and environmental outcomes, particularly since the indicators for SDGs are being monitored nationally and reported on globally. Further studies are needed on the political ecologies and stakeholder dynamics that underpin, drive or constrain the formation and movement of policies designed for environmental issues and sustainable development.

References

Adelle, C. and Weiland, S. (2012). Policy assessment: The state of the art. *Impact assessment and project appraisal* 30(1), 25-33. https://doi.org/10.1080/14615517.2012.663256

Adelle, C., Weiland, S., Dick, J., González Olivo, D., Marquardt, J., Rots, G. *et al.* (2016). Regulatory impact assessment: A survey of selected developing and emerging economies. *Public money & management* 36(2), 89-96. https://doi.org/10.1080/09540962.2016.1118930

European Environment Agency (2017). *Perspectives on Transitions to Sustainability.* Copenhagen. https://www.eea.europa.eu/publications/perspectives-on-transitions-to-sustainability/at_download/file.

Flanagan, K., Uyarra, E. and Laranja, M. (2011). Reconceptualising the 'policy mix' for innovation. *Research policy* 40(5), 702-713. https://doi.org/10.1016/j.respol.2011.02.005.

Grantham Research Institute on Climate Change and the Environment (2018). *Climate change laws of the world.* [London School of Economics http://www.lse.ac.uk/GranthamInstitute/climate-change-laws-of-the-world.

Jacob, K., Weiland, S., Ferretti, J., Wascher, D. and Chodorowska, D. (2011). *Integrating the Environment in Regulatory Impact Assessments.* Paris: Organisation for Economic Co-operation and Development. https://www.oecd.org/gov/regulatory-policy/Integrating%20RIA%20in%20Decision%20Making.pdf.

Kemp, R. and Pontoglio, S. (2011). The innovation effects of environmental policy instruments—A typical case of the blind men and the elephant? *Ecological Economics* 72, 28-36. https://doi.org/10.1016/j.ecolecon.2011.09.014.

Mickwitz, P., Aix, F., Beck, S., Carss, D., Ferrand, N., Görg, C. *et al.* (2009). *Climate Policy Integration, Coherence and Governance.* PEER Report. Helsinki: Partnership for European Environmental Research. http://library.wur.nl/WebQuery/wurpubs/fulltext/3987.

Mullally, G. and Dunphy, N. (2015). *State of Play: Review of Environmental Policy Integration Literature.* Research Series Paper. Dublin: National Economic and Social Council. https://cora.ucc.ie/handle/10468/3029.

Nilsson, M., Zamparutti, T., Petersen, J.E., Nykvist, B., Rudberg, P. and McGuinn, J. (2012). Understanding policy coherence: Analytical framework and examples of sector–environment policy interactions in the EU. *Environmental Policy and Governance* 22(6), 395-423. https://doi.org/10.1002/eet.1589.

Radaelli, C.M. (2009). Measuring policy learning: Regulatory impact assessment in Europe. *Journal of European Public Policy* 16(8), 1145-1164. https://doi.org/10.1080/13501760903332647.

Renewable Energy Policy Network for the 21st Century (2018). *Renewables 2018 Global Status Report. A Comprehensive Annual Overview of the State of Renewable Energy.* Paris. http://www.ren21.net/wp-content/uploads/2018/06/17-8652_GSR2018_FullReport_web_final_.pdf.

Runhaar, H., Driessen, P. and Uittenbroek, C. (2014). Towards a systematic framework for the analysis of environmental policy integration. *Environmental Policy and Governance* 24(4), 233-246. https://doi.org/10.1002/eet.1647.

United Nations Environment Programme (2011). Supporting the transition to a global green economy. In *Towards A Green Economy: Pathways to Sustainable Development and Poverty Eradication.* Nairobi. chapter Part III. https://wedocs.unep.org/bitstream/handle/20.500.11822/22025/green_economyreport_final_dec2011.pdf?sequence=1&isAllowed=

United Nations Environment Programme, Food and Agriculture Organization of the United Nations and United Nations Educational, Scientific and Cultural Organization (2018). *Access information on multilateral environmental agreements.* https://www.informea.org/ (Accessed: 17 October 2018.

United Nations Framework Convention on Climate Change (2015). *Paris Agreement.* Paris. https://unfccc.int/files/meetings/paris_nov_2015/application/pdf/paris_agreement_english_.pdf.

Yin, H. and Powers, N. (2010). Do state renewable portfolio standards promote in-state renewable generation? *Energy Policy* 38(2), 1140-1149. https://doi.org/10.1016/j.enpol.2009.10.067.

PART C

Outlooks and Pathways to a Healthy Planet with Healthy People

Chapter 19

Outlooks in GEO-6

Coordinating Lead Author: Ghassem R. Asrar (Pacific Northwest National Laboratory, United States of America)

Lead Authors: Paul Lucas (PBL Netherlands Environmental Assessment Agency), Detlef van Vuuren (PBL Netherlands Environmental Assessment Agency), Laura Pereira (Stellenbosch University Centre for Complex Systems in Transition), Joost Vervoort (Utrecht University)

Contributing Author: Rohan Bhargava (Utrecht University)

GEO Fellow: Semie Sama (Centre for International Governance Innovation)

Executive summary

For environmental assessments to be useful to decision makers, they should account for the interactions, interdependencies and co-evolutionary pathways of human-Earth systems in proposed policy options and scientific and technological solutions – including the direct effects and the co-benefits and/or trade-offs. (*established, but incomplete*). Global Environmental Assessments generally rely on model-based quantitative scenarios. While these models capture many important linkages, the social dimension is not very well represented. Moreover, it is difficult in global assessments to capture important details that are pertinent for local-level decision-making. A systemic and integrated approach is needed in scientifically based environmental assessments and future outlooks, in support of policy and investment decisions, to account for the highly complex, interdependent and continuously changing factors in assessing the human-Earth system changes {19.1}.

Global assessments have to shift the focus from what is happening and what could be done, to how trends and trajectories in development can be changed (*well established*). The use of top-down and bottom-up methods in Part C of GEO-6 – Outlooks and Pathways to a Healthy Planet for Healthy People is intended to provide science-based information for this purpose. The combined quantitative scenarios and participatory approaches also offer great potential to be more responsive to meeting the sector- and region-specific information required by decision makers. Therefore, the GEO-6 outlook analysis/assessment use: (1) model-based scenario analysis (generally referred to as the top-down approach); (2) information and knowledge from past and present initiatives, opportunities and trends (i.e. seeds of change); and (3) information resulting from integrative decision-making and participatory activities that are usually conducted at the local to regional levels (generally referred to as bottom-up approaches). This will ensure greater engagement of stakeholders in knowledge development and dissemination, and implementation of resulting policies and practices in a timely manner for greater success {19.2, 19.3}.

19.1 Introduction

Parts A and B of the sixth Global Environmental Outlook (GEO-6) indicate that the current global development trends and their future trajectories are not sustainable. At the same time, nations worldwide have agreed on a set of ambitious goals as part of the United Nations 2030 Agenda for Sustainable Development, including the Sustainable Development Goals (SDGs), a broad range of multilateral environmental agreements and other frameworks. Together, they aim for halting environmental degradation and aim to enable better development pathways that can benefit both humans and the ecosystems that support human well-being.

The key questions now are whether future trends would lead to the achievement of these ambitious goals and thereby a more sustainable future – and, if not, what would be required, both in policy and practice, to bend the trend towards positive and sustainable development pathways. Part C of GEO-6 aims to provide an integrated and holistic view of the scientific information to address these questions. It presents new approaches to developing science-based information for decision makers, by combining scenario-based quantitative projections (defined here as top-down approaches) with grass-roots and participatory methods (bottom-up) approaches.

19.2 Important elements of future-oriented environmental outlooks

Changes in Earth-human systems, and their cascading effects, transcend a wide range of scales of space (i.e. local, national, regional, global) and time (seasons, years, decades and longer), and vary significantly in different sectors (agriculture and food, water resources, energy systems, fisheries, etc.). Such complexities need consideration, through the active engagement of stakeholders and decision makers, in the design, development and implementation of environmental assessments and outlooks. This is because of the interrelationships among, for example, the following:

i. the choices for addressing legacy issues and current pressures (e.g. food-water-energy security),
ii. the development of management approaches that are responsive to a changing environment (e.g. SDGs and other targets) and,
iii. the extent to which emerging issues and future pressures are anticipated and prepared for.

Environmental assessments and outlooks should also consider the potential impact of proposed plans, policies and practices, and the need for improved communication between policymakers and the public. This requires consideration of the decision makers' needs much earlier in the assessment process. The expanding role of public, private and non-governmental organizations in the assessment process allows an intrinsic connection between environmental sustainability and equity, and enables the promotion of sustainability goals through this engagement (Simson 2012; Ho *et al.* 2013). The process of environmental assessment often has difficulty taking account of the socioeconomic impacts of development activities and issues associated with, for example, biodiversity, human health and cultural norms. These were not usually taken into account in the past (Mahboubi, Parkes and Chan 2015; Reid and Mooney 2016; Kok *et al.* 2017). The effectiveness of environmental assessments should be evaluated against the ability to raise the level of the environmental values that are considered important by stakeholders, such as stewardship, services and socioeconomic factors (Arts *et al.* 2012). Furthermore, investigation of effectiveness should explore whether the assessment process and products have resulted in better decision-making and the achievement of the desired outcomes (Fischer and He 2009).

Recent agreement and frameworks – including the 2030 Agenda for Sustainable Development and the Paris Climate Agreement – recognize the need for a change in the current trends and direction, and the need to promote a systemic and integrated approach to assessing the highly complex, interdependent and continuously changing factors that underpin these trends, including the states and dynamics of human-Earth systems. To achieve goals related to biodiversity, for example, one also needs to take account of goals related to food production, water availability, climate change and air pollution. Decision makers can benefit from science-based assessments of the outlooks that include the direct effects, co-benefits and consequences of the available responses, to avoid unsustainable and risk-prone development pathways (Kowarsch *et al.* 2017). Thus, the emerging global architecture for sustainable development and its governance requires environmental outlooks to take into account the complexities and interlinkages of Earth-human systems for developing a diverse range of policies and pragmatic solutions.

The SDGs offer a framework for such a holistic approach to identifying innovative ways and means for advancing human well-being and health together with environmental stewardship (Dye 2018). This framework requires interdisciplinary as well as multidisciplinary scientific research and assessments to be the norm; for the urgency of short-term needs and actions to be balanced strategically with the long-term risks in resource planning and allocation; and for more collaborative and participatory approaches to be promoted, to engage governments, businesses and citizens to reconsider their roles, responsibilities and contributions to the implementation of multilateral environmental agreements (Simson 2012; Ho *et al.* 2013). Stakeholder engagement could, for example, be an integral part of business development to bring the three aspects of sustainability – environment, society and economy – to the heart of societal value creation.

Specifically, environmental assessments and outlooks should identify the transformative interventions needed to achieve sustainable development pathways towards the stated goals/targets (e.g. SDGs), to ensure a healthy planet for healthy people. Such transformations must consider the role of humans in the form of socioeconomic development, the roles of the perturbations of natural systems and built systems, such as infrastructure, and also the interactions and interplay among these roles.

19.3 A new framework for combining top-down and bottom-up analysis methods

Various methods have been developed over the past decades to conduct environmental assessments and outlooks in support of decision-making. Model-based scenario analysis, for example, has been used as a method to define plausible future conditions, in relation to the current state and trends in socioeconomics, technologies, environmental conditions and policies (van Vuuren *et al.* 2012). In this approach, an envisioned scenario is a plausible and often simplified description of how the future may take shape, based on a coherent and internally consistent set of assumptions about key driving forces and their relationships (Millennium Ecosystem Assessment 2005). As such, scenarios are powerful tools that can help to conceptualize how alternate futures might unfold. They provide insight on where alternative pathways for sustainable development might lead us by taking into account the many interrelations between different subsystems (e.g. energy, agriculture, cities, etc.) and societal concerns (health, economy, climate, air, freshwater, biodiversity, etc.), thereby specifically addressing synergies and trade-offs between different developments and aspirations. Given the inherent uncertainty about the future, scenarios and associated analyses are also helpful for both assessing the future implications of different problems and inspiring the narratives around which decisions are made (with due consideration given to the level of confidence in their certainty and their likelihood of success). This approach is also referred to as the top-down approach and it usually starts with the consideration of a given policy and traces the causal chains expected by its implementation (see Chapters 21 and 22). It offers the opportunity to evaluate the potential effectiveness of the policies under consideration by evaluating the quantitative representation of the various systems involved, the interlinkages between them as much as possible, and the creativity used to represent these complex systems, often based on the current state of knowledge about them.

In contrast, the increasingly prominent participatory approaches, also known as bottom-up approaches, begin from the observed outcomes and trace the causality back to the policy interventions (see Chapter 23). Most of these approaches are based on the active engagement of stakeholders and citizens through workshops, crowdsourcing and competitions to identify innovative ideas, practices and solutions. The identified needs being answered are often sector- and region-specific, resulting in an evolution in the diversity of participatory approaches over the past decades. The greatest advantages of bottom-up approaches are threefold:

i. they focus on specific local and regional development challenges,
ii. they engage stakeholders and users in planning the intended analysis, and in the resulting knowledge for its design, development and implementation, and,
iii. they provide the ability to develop sector- and/or region-specific analysis and information.

The bottom-up approaches do have some limitations, such as their limited ability to be extended to larger scales, and their limited sustainability over time. They nonetheless offer significant potential considering the rapidly increasing needs for information for decision makers at local and subnational levels, and the desire to engage stakeholders actively in the knowledge-development process (Jabbour and Flachsland 2017; Kowarsch *et al.* 2017).

GEO-6 therefore uses both top-down and bottom-up approaches and methods towards target-seeking scenarios. This builds on the assessments in previous GEOs and pre-SDG pathways analysis and is based on quantitative scenarios (van Vuuren *et al.* 2015; see also Chapters 21 and 22) and on the participatory and grass-roots analysis that has been conducted through stakeholder workshops and crowdsourcing approaches (see Chapter 23). This opportunity to combine the desirable attributes of different approaches offers great potential for assessments and outlooks to capture the increased complexities of Earth-human systems and their interlinkages, and to be responsive to decision makers' needs for sector- and region-specific information. Some new features of the outlook analysis provided by GEO-6 are:

- A combination of top-down (e.g. pathways and trajectories) and bottom-up (or participatory) approaches (e.g. game changers, effective seeds and crowdsourcing) to ensure the efficacy and effectiveness of the resulting analysis.
- A focus on the 'how' question in integrated scenario analysis, by explicitly discussing target-seeking scenarios and linking them to the evolution of the pathway experience in the literature, with specific attention given to the synergies and trade-offs in simultaneously achieving well-being and environmental goals.
- The engagement of stakeholders in knowledge development, implementation and dissemination – through regional and sectoral stakeholder workshops and crowdsourcing platforms, for input into analysing, testing and refining the outcomes.
- Communication with decision makers (e.g. policy experts) throughout the knowledge-development process, not just in the final product, and using innovative means for communicating assessment outcomes, to increase their uptake in policy and practical decision-making.

The global top-down pathways considered in GEO-6 are based on a review of existing work, and can be grouped in three potential pathways that can drive change (PBL Netherlands Environmental Assessment Agency 2012; van Vuuren *et al.* 2012), namely:

i. technological innovations, which can serve as the dominant reason for change,
ii. shifts in consumer choices and behaviour, and,
iii. decentralized innovation in favour of more localized and community-level activities.

The bottom-up approaches evaluated in GEO-6 capture the richness of practices, ideas and visions for desirable global futures using a variety methods –examples include the Climate CoLab platform from the MIT Center for Collective Intelligence (Malone *et al.* 2017), initiatives dubbed the "seeds of a good Anthropocene" (Bennett *et al.* 2016), and the pathways projects sponsored by the European Commission's Seventh Framework Programme (Kok *et al.* 2015; see Chapter 23). The combined approach offers the potential to develop the required science-based analysis for the successful implementation of the SDGs, together with that of other multilateral environmental agreements (European Commission 2016; Patterson *et al.* 2017).

19.4 The role of scale

Scales plays a key role in environmental assessments (Gibson, Ostrom and Ahn 2000; Cash *et al.* 2006; Vervoort *et al.* 2012) because most environmental problems transcend a wide range of levels (i.e. across local, national, regional and global). The idea in that broad societal changes -which is described as the landscape level- can create opportunities for non-mainstream, radical practices and technologies at the niche level to replace the old social and technological mainstream practices at the regime level – the status quo in a specific domain of social and technological activity (Geels and Schot 2007). Many other theories of transformation applied to social-ecological systems share this general idea that some interplay between bottom-up change created by niche practices and top-down changes created by broad societal shifts, by changes in policies and economic activities, lead to transformation (Feola 2015; Patterson *et al.* 2017).

This can be illustrated for climate change and its impacts. First, the biophysical process plays out at different levels: global (CO_2 concentration), continental (weather patterns) and local (land-climate interactions). Second, levels also play a key role in terms of solutions. While, for example, international climate policy is negotiated at the global level, it needs to be implemented at the national and local levels. Connections across levels should therefore be a major consideration in environmental assessments and outlooks, and in recommended policies and actions (Zurek and Henrichs 2007). Mismatches between the levels of human-built and natural systems can lead to negative environmental impacts, for example, when a river basin falls under competing national jurisdictions (Cumming, Cumming and Redman 2006).

Integration across scales plays a significant role towards identifying synergistic and effective policies and actions (Palazzo *et al.* 2017). For example, the identification of concrete policy recommendations should consider what policy conditions have to be created (by governments, the private sector, civil society and others) to allow innovative bottom-up processes to flourish by scaling up to higher levels and deeply in future assessments (Moore, Riddell and Vocisano 2015; Mason-D'Croz *et al.* 2016). This implies that decision makers in different sectors receive useful information for formulating and implementing policies, strategies and investments that facilitate transformative change in their sector and area of interest.

The complementary features the top-down and bottom-up approaches being used in Part C allow the consideration of scales and their interactions in the evaluation of scenarios and strategies, by maximizing the synergies and, as much as possible, minimizing the trade-offs among them for potential pathways to achieving SDGs and other multilateral environmental agreements.

19.5 Roadmap for Part C of GEO-6

Building on previous assessments, particularly on GEO-5 (United Nations Environment Programme [UNEP] 2012), the focus of GEO-6 has shifted to a combination of the 'what' and 'how' questions, and the required approaches, to assess the state of scientific knowledge on the challenges and opportunities associated with global goals and targets. A universal, transformative and integrated agenda for sustainable development is now available in the form of the 2030 Agenda for Sustainable Development (without it being explicit about this), to allow the goals of a broad range of multilateral frameworks and agreements to be brought together in a more coherent manner. Part C seeks to address the synergies (co-benefits) and trade-offs (competing aspects) in achieving the multiple goals and targets of these frameworks and agreements (e.g. SDGs, Nationally Determined Contributions, Aichi biodiversity targets), rather than analysing how to achieve their many individual indicators separately. The guiding questions are:

- How can we achieve the environmental dimension of the SDGs and related multilateral environmental agreements?
- What mid- to long-term strategies are needed to achieve lasting sustainability?

The aim of the outlook chapters of GEO-6 (Part C of this report) is to address these questions by combining top-down, model-based scenario analysis with information resulting from bottom-up and participatory initiatives (see Section 19.2). The purpose is to illustrate how these can be used together towards meeting the information needs of decision makers at national and subnational as well as regional and global levels.

The following key elements are addressed in subsequent chapters:

- Formulating a quantitative long-term vision for 2050, consisting of key environmental targets from the SDGs and related multilateral environmental agreements (Chapter 20)
- Assessing long-term trends and discussing the potential implementation gaps (Chapter 21)
- Identifying potential pathways for achieving the long-term vision, with a specific focus on the many interrelations across the broad range of targets assessed (Chapter 22)
- Assessing innovative initiatives and game-changers in the context of future pathways (Chapter 23)
- Discussing the way forward for moving towards the theme of GEO-6 of healthy planet and healthy people (Chapter 24).

Chapter 20 translates the mid- to long-term vision of the SDGs and a few related multilateral environmental agreements into a more concise and quantitative formulation of targets, focusing on the food-water-energy nexus. This includes extracting the available information from these frameworks and agreements, selecting some key environment-related priorities in relation to healthy planet, healthy people and identifying indicators and quantitative target levels to track progress. Chapters 21 and 22 assess the model-based scenario literature (i.e. the top-down approach) to discuss current trends in Earth-human systems, and pathways towards achieving the long-term vision, respectively. No new scenarios were developed, and the analysis and assessment are based on existing scenarios (e.g. shared socioeconomic pathways). In Chapter 21, the scenario analysis focuses on current trends and identifies the potential implementation gaps between these and the targets identified in Chapter 20. Chapter 22, in contrast, identifies pathways that can achieve the selection of targets in a complementary and holistic way. Together, the three chapters provide a solutions-based perspective, including possible trade-offs and synergies for the identified pathways.

Chapter 23 focuses on the gap between current trends and the sustainable pathways based on grass-roots and participatory approaches that engage stakeholders and citizens (i.e. the bottom-up approach). Similar to the model-based scenarios, a combination of existing and future initiatives and best practices is identified that could help in achieving specific and combined SDGs and their targets. A major strength of this approach is that it takes into account the role of different actors. This type of analysis can be carried out by using the top-down scenarios to frame the bottom-up initiatives. Such framing will help to overcome the major challenges relating to the so-called game-changing and bottom-up strategies that are often specific to geographical areas and/or sectors, to evaluate their feasibility and benefits at the global level.

Finally, Chapter 24 presents the information resulting from the proposed integrative and holistic approaches examined across Part C that can contribute to the development and implementation of effective policies and practices towards achieving the SDGs and multilateral environmental agreements synergistically. In short, how they can contribute to transformative development pathways for a healthy planet, healthy people.

Figure 19.1: Conceptual framing of the chapters in Part C of GEO-6, how they are related, and how they contribute to a holistic analysis and assessment of human-Earth systems that identifies transformative development pathways

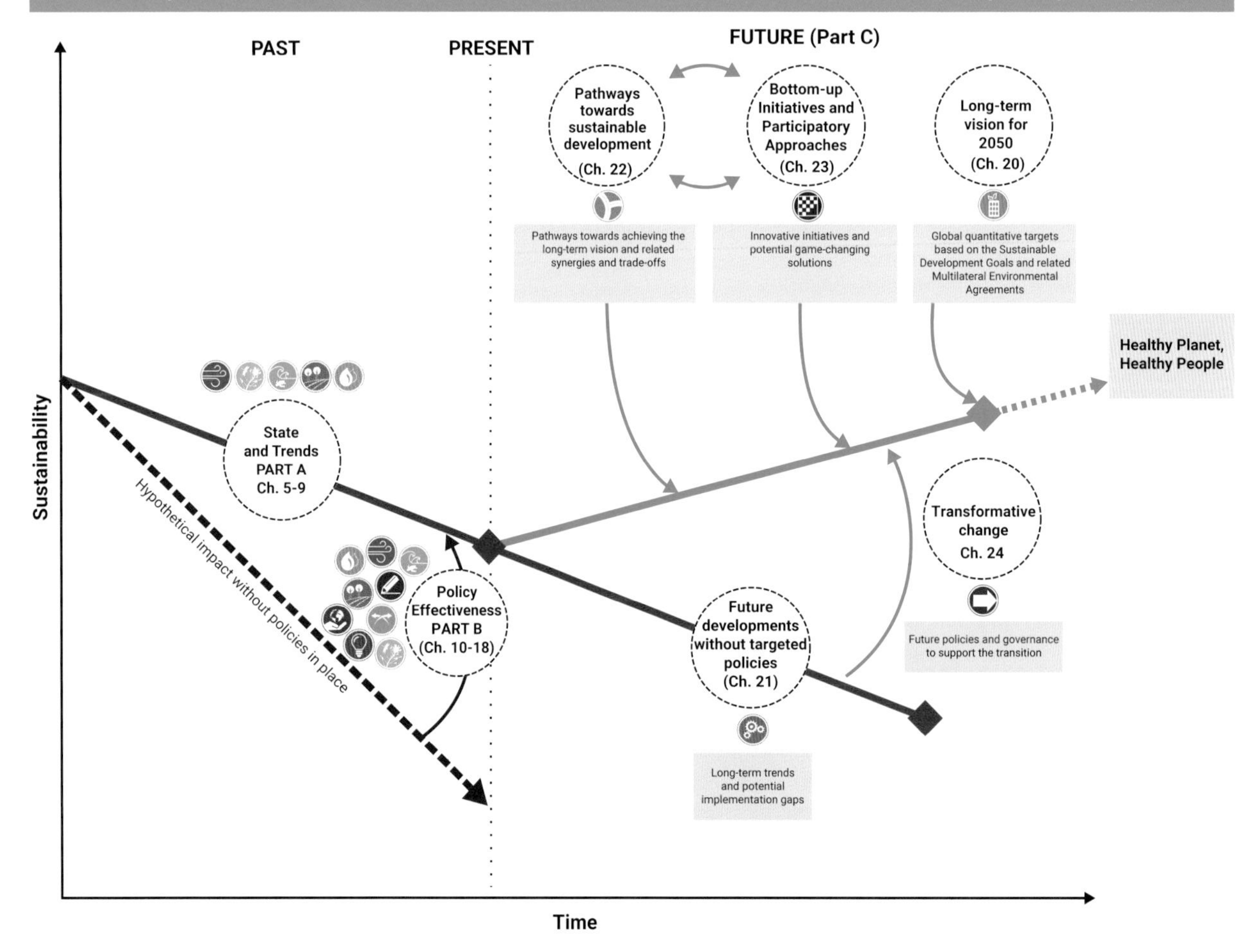

Source: Adapted from van Vuuren *et al.* 2015.

References

Arts, J., Runhaar, H.A.C., Fischer, T.B., Jha-Thakur, U., Laerhoven, F.V., Driessen, P.P.J. *et al.* (2012). The effectiveness of EIA as an instrument for environmental governance reflecting on 25 years of EIA practice in the Netherlands and the UK. *Journal of Environmental Assessment, Policy and Management* 14(4). https://doi.org/10.1142/S1464333212500251.

Bennett, E.M., Solan, M., Biggs, R., McPhearson, T., Norström, A.V., Olsson, P. *et al.* (2016). Bright spots: Seeds of a good Anthropocene. *Frontiers in Ecology and the Environment* 14(8), 441-448. https://doi.org/10.1002/fee.1309.

Cash, D.W., Adger, W.N., Berkes, F., Garden, P., Lebel, L., Olsson, P. *et al.* (2006). Scale and cross-scale dynamics: Governance and information in a multilevel world. *Ecology and Society* 11(2). http://www.ecologyandsociety.org/vol11/iss2/art8/.

Cumming, G.S., Cumming, D.H.M. and Redman, C.L. (2006). Scale mismatches in social-ecological systems: Causes, consequences, and solutions. *Ecology and Society* 11(1), 14. http://www.ecologyandsociety.org/vol11/iss1/art14/.

Dye, C. (2018). Expanded health system for sustainable development: Advance transformative research for 2030 agenda. *Science* 359(6382), 1337-1339. https://doi.org/10.1126/science.aaq1081.

European Commission (2016). *Exploring Transition Pathways to Sustainable, Low Carbon Societies.* https://cordis.europa.eu/project/rcn/111082_en.pdf.

Feola, G. (2015). Societal transformation in response to global environmental change: A review of emerging concepts. *Ambio: A Journal of the Human Environment* 44(5), 376-390. https://doi.org/10.1007/s13280-014-0582-z.

Fischer, T.B. and He, X. (2009). Differences in perceptions of effective sea in the UK and China. *Journal of Environmental Assessment Policy and Management* 11(4), 471-485. https://doi.org/10.1142/S1464333209003452.

Geels, F.W. and Schot, J. (2007). Typology of sociotechnical transition pathways. *Research Policy* 36(3), 399-417. https://doi.org/10.1016/j.respol.2007.01.003.

Gibson, C.C., Ostrom, E. and Ahn, T.K. (2000). The concept of scale and the human dimensions of global change: A survey. *Ecological Economics* 32(2), 217-239. https://doi.org/10.1016/S0921-8009(99)00092-0.

Ho, C.S., Matsuyoka, Y., Chau, L.W., Teh, B.T., Simson, J.J. and Gomi, K. (2013). Blueprint for the development of low carbon society scenarios for Asian regions- case study of Iskandar Malaysia. *IOP Conference Series: Earth and Environmental Science* 16(012125). https://doi.org/10.1088/1755-1315/16/1/012125.

Jabbour, J. and Flachsland, C. (2017). 40 years of global environmental assessments: A retrospective analysis. *Environmental Science and Policy* 77 (193-202). https://doi.org/10.1016/j.envsci.2017.05.001.

Kok, K., Pedde, S., Jäger, J. and Harrison, P. (2015). *European Shared Socioeconomic Pathways.* http://impressions-project.eu/getatt.php?filename=IMPRESSIONSReport_EuropeanSSPs_13773.pdf.

Kok, M.T.J., Kok, K., Peterson, G.D., Hill, R., Agard, J. and Carpenter, S.R. (2017). Biodiversity and ecosystem services require IPBES to take novel approach to scenarios. *Sustainability Science* 12(1), 177-181. https://doi.org/10.1007/s11625-016-0354-8.

Kowarsch, M., Jabbour, J., Flachsland, C., Kok, M.T.J., Watson, S.R., Haas, P.M. *et al.* (2017). A road map for global environmental assessments. *Nature Climate Change* 7(6), 379-382. https://doi.org/10.1038/nclimate3307

Mahboubi, P., Parkes, M.W. and Chan, H.M. (2015). Challenges and opportunities of integrating human health into the environmental assessment process: The Canadian experience contextualized in international efforts. *Journal of Environmental Assessment, Policy and Management* 17(4). https://doi.org/10.1142/S1464333215500349.

Malone, T.W., Nickerson, J.V., Laubacher, R.J., Fisher, L.H., de Boer, P., Han, Y. *et al.* (2017). Putting the pieces back together again: Contest webs for large-scale problem solving. *ACM Conference on Computer-Supported Cooperative Work and Social Computing*. Portland, OR, 25 February-1 March 2017 http://mitsloan.mit.edu/shared/ods/documents/?DocumentID=2711.

Mason-D'Croza, D., Vervoort, J., Palazzo, A., Islam, S., Lord, S., Helfgott, A. *et al.* (2016). Multi-factor, multi-state, multi-model scenarios: Exploring food and climate futures for Southeast Asia. *Environmental Modelling and Software* 83, 255-270. https://doi.org/10.1016/j.envsoft.2016.05.008.

Millennium Ecosystem Assessment (2005). *Ecosystems and Human Well-Being: Synthesis.* Washington D.C.: Island Press. https://www.millenniumassessment.org/documents/document.356.aspx.pdf.

Moore, M., Riddell, D. and Vocisano, D. (2015). Scaling out, scaling up, scaling deep: Strategies of non-profits in advancing systemic social innovation. *Journal of Corporate Citizenship* (58), 67-84. https://www.ingentaconnect.com/contentone/glbj/jcc/2015/00002015/00000058/art00009.

Palazzo, A., Vervoort, J.M., Mason-D'Croz, D., Rutting, L., Havlík, P., Islam, S. *et al.* (2017). Linking regional stakeholder scenarios and shared socioeconomic pathways: Quantified West African food and climate futures in a global context. *Global Environmental Change* 45, 227-242. https://doi.org/10.1016/j.gloenvcha.2016.12.002

Patterson, J., Schulz, K., Vervoort, J., van der Hel, S., Widerberg, O., Adler, C. *et al.* (2017). Exploring the governance and politics of transformations towards sustainability. *Environmental Innovation and Societal Transitions* 24, 1-16. https://doi.org/10.1016/j.eist.2016.09.001.

PBL Netherlands Environmental Assessment Agency (2012). *Roads from Rio+20: Pathways to Achieve Global Sustainability Goals By 2050.* The Hague. https://goo.gl/1vX3FC.

Reid, W.V. and Mooney, H.A. (2016). The millennium ecosystem assessment: Testing the limits of interdisciplinary and multi-scale science. *Current Opinion in Environmental Sustainability* 19, 40-46. https://doi.org/10.1016/j.cosust.2015.11.009.

Simson, J.J. (2012). *Study on Sustainable Low Carbon Society in Malaysian Regional Development.* Kyoto University https://repository.kulib.kyoto-u.ac.jp/dspace/bitstream/2433/157521/2/D_SIMSON_Janice%20Jeevamalar.pdf

United Nations Environment Programme (2012). *Global Environment Outlook 5: Environment for the Future We Want.* Nairobi. http://wedocs.unep.org/bitstream/handle/20.500.11822/8021/GEO5_report_full_en.pdf?sequence=5&isAllowed=y.

van Vuuren, D.P., Kok, M., Lucas, P.L., Prins, A.G., Alkemade, R., van den Berg, M. *et al.* (2015). Pathways to achieve a set of ambitious global sustainability objectives by 2050: Explorations using the IMAGE integrated assessment model. *Technological Forecasting and Social Change* 98, 303-323. https://doi.org/10.1016/j.techfore.2015.03.005.

van Vuuren, D.P., Kok, M.T.J., Girod, B., Lucas, P.L. and de Vries, B. (2012). Scenarios in global environmental assessments: Key characteristics and lessons for future use. *Global Environmental Change* 22(4), 884-895. https://doi.org/10.1016/j.gloenvcha.2012.06.001.

Vervoort, J.M., Rutting, L., Kok, K., Hermans, F.L.P., Veldkamp, T., Bregt, A.K. *et al.* (2012). Exploring dimensions, scales, and cross-scale dynamics from the perspectives of change agents in social–ecological systems. *Ecology and Society* 17(4), 24. https://doi.org/10.5751/ES-05098-170424.

Zurek, M.B. and Henrichs, T. (2007). Linking scenarios across geographical scales in international environmental assessments. *Technological Forecasting and Social Change* 74(8), 1282-1295. https://doi.org/10.1016/j.techfore.2006.11.005.

Chapter 20

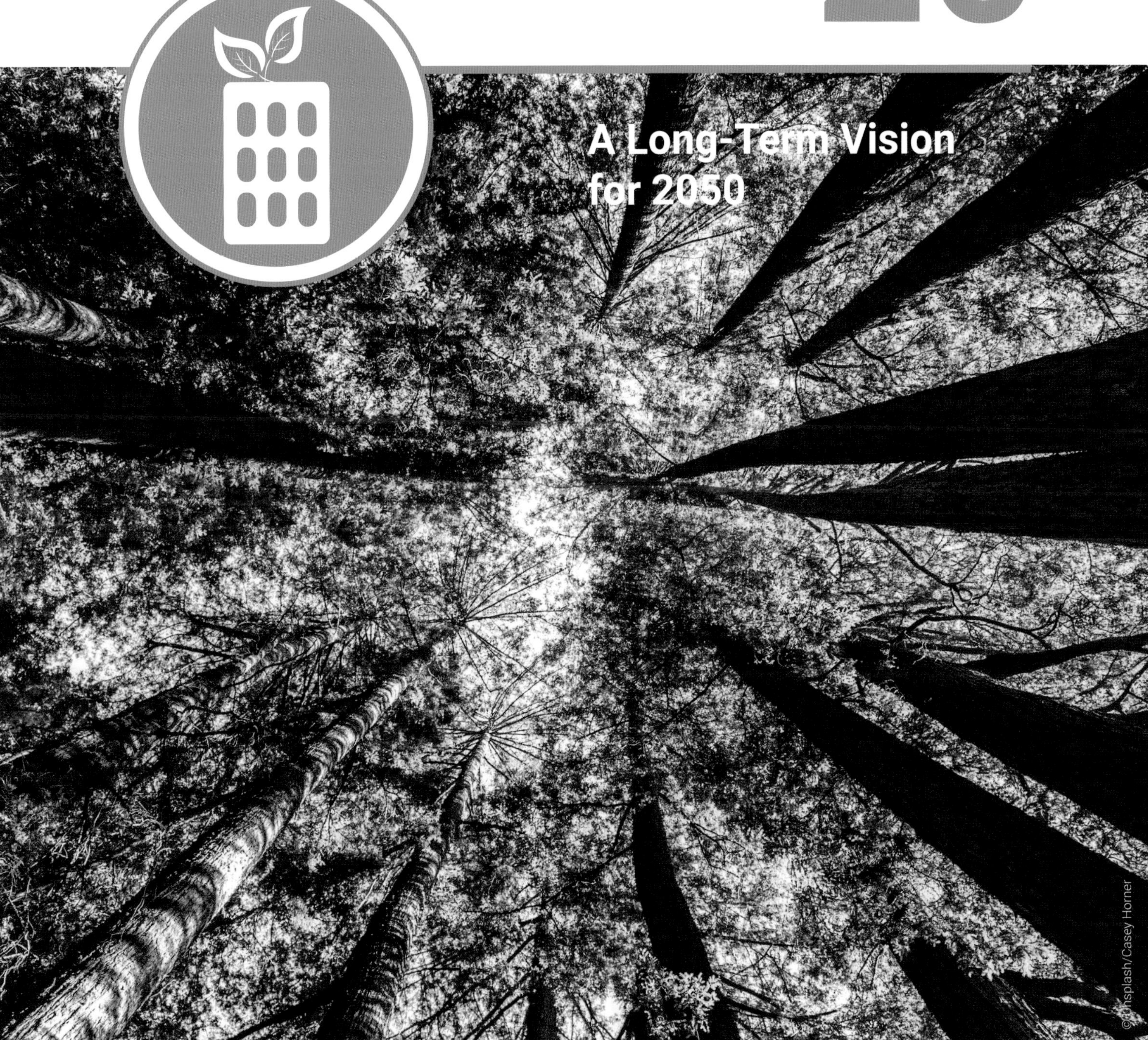

A Long-Term Vision for 2050

Coordinating Lead Author: Paul Lucas (PBL Netherlands Environmental Assessment Agency)
Lead Authors: Mark Elder (Institute for Global Environmental Strategies), Detlef van Vuuren (PBL Netherlands Environmental Assessment Agency),
Contributing Authors: Fintan Hurley (Institute of Occupational Medicine)

Executive summary

To assess future progress towards achieving the Sustainable Development Goals (SDGs) and related Multilateral Environmental Agreements (MEAs), their underlying targets need to be translated into a more concise and quantitative set of targets (*well established*). The SDGs cover a wide range of issues, with the environment represented in most of them, including in non-environmental goals. A range of challenges exists when interpreting the targets of the SDGs and related MEAs for an assessment of future progress. First, to make the assessment focused, a selection of the targets needs to be identified. Next, these targets need to be quantitative, using clear indicators accompanied by target values. {20.2; 20.3}

SDGs can be structured into groups based on how they address human well-being, sustainable consumption and production, and the natural resource base (*established, but incomplete*). The 2030 Agenda for Sustainable Development emphasizes that the goals and targets are integrated and indivisible and aimed at contributing to coherent sustainability policies. How the SDGs can be implemented synergistically is not apparent from the 2030 Agenda. To reveal potential trade-offs and synergies between achieving multiple SDGs and to point to ways these interlinkages can be governed, they can be grouped into goals focusing on social objectives or more broadly human well-being, goals addressing sustainable consumption and production with respect to resource use and access, and goals that address the protection and management of natural resources. These groups are bidirectionally connected in the sense that the environment provides the natural resource base on which human development and ultimately human well-being, including human health, are built. Unsustainable resource use can adversely impact both people and the planet, calling for policies that specifically focus on sustainable consumption and production as well as equitable distribution of natural resources and their benefits. The benefits to human health thus depend on the SDGs as a whole, not just those explicitly relating to health or well-being. {20.3}

Environment-related SDG targets can be further quantified based on internationally agreed targets from Multilateral Environmental Agreements (MEAs) and the scientific literature (*established, but incomplete*). Although many SDG targets have been formulated in clear and quantitative terms, for many environment-related ones, this is much less the case, both quantitatively and qualitatively. For several issues, such as climate change and biodiversity loss, the targets in MEAs are more concrete. Quantification of SDG targets can thus build on related MEAs. When internationally agreed environmental targets are lacking, so-called science-based targets can be used that are based on biophysical limits established in the scientific literature. {20.4}

The scenario assessment of GEO-6 centres around the food-water-energy nexus, linked to the five GEO-6 environmental themes and related multidimensional poverty and health. The selection of targets for the GEO-6 scenarios assessment, puts the use of natural resources central, focusing on the challenges addressed by, and linked to, the SDGs on food and agriculture (SDG 2), water (SDG 6) and energy (SDG 7). The use of natural resources is on the one hand linked to social objectives for people's access to food, water and energy, and subsequently to related health impacts (SDG 3). On the other hand, the use of natural resources is linked to the quality and quantity of the natural resource base that is required for, or impacted by, this use (SDG 13, SDG 14 and SDG 15). This focus and further quantification provide an integrated perspective on the environmental dimension of the SDGs and related agreements. The resulting set of targets is not an alternative to what is globally agreed, but a subset, and sometimes a proxy, to be used for the analysis in subsequent chapters. {20.4}

20.1 Introduction

An analysis of pathways towards sustainable development needs a long-term vision. Ideally, such a vision is summarized in a quantitative set of globally agreed key objectives or targets. The 2030 Agenda for Sustainable Development, adopted in September 2015, conceptualizes sustainable development through 17 SDGs and is further operationalized through 169 targets and 232 indicators (United Nations 2015a; see Annex 4-1). The Agenda formulates an ambitious and transformational vision for 2030. The SDGs address a broad range of issues, including eradicating poverty, transforming towards sustainable and resilient societies, and protecting and managing the natural resource base. Furthermore, the SDGs and its targets are related to several Multilateral Environmental Agreements (MEA). Together, the SDGs and related MEAs provide a globally agreed set of targets to guide the transformation towards long-term sustainability.

In this chapter, we define the vision used for the scenario analysis of GEO-6, taking a long-term perspective, beyond 2030. This vision uses the SDGs as an overarching, integrated set of goals and targets to start from, selecting key environment-related targets, and, where relevant, further quantifying these with targets from related MEAs or the scientific literature. This long-term vision addresses the theme of GEO-6 – Healthy Planet, Healthy People – by focusing on global environmental targets linked to the five themes discussed in Part A (State of the Global Environment), and related multidimensional poverty and health targets. In Chapters 21 and 22 the existing scenario literature is assessed, to allow discussion of future developments with respect to these targets and the pathways towards achieving them.

20.2 The environmental dimension of the SDGs

In 1972, as part of the United Nations Conference on the Human Environment, countries worldwide agreed that natural resources should be safeguarded, and pollution should not exceed the environment's capacity to clean itself (United Nations 1972). Since 1972, a proliferation of United Nations conferences, summits and international agreements have set targets for environmental protection and sustainable human development (Jabbour *et al.* 2012). The years 2015 and 2016 were a landmark for environmental multilateralism, thanks in large part to the formulation and adoption of five global frameworks, including the Paris Agreement (United Nations Framework Convention on Climate Change [UNFCCC] 2015) and the 2030 Agenda for Sustainable Development (United Nations 2015a).

The 2030 Agenda for Sustainable Development is explicit about the integrated nature of its goals and targets. The SDGs cover a wide range of environmental issues (see Section 1.5), with the environment represented in most of them (Organisation for Economic Co-operation and Development [OECD] 2015; United Nations Environment Programme [UNEP] 2015; Lucas *et al.* 2016; Reid *et al.* 2017) including their non-environmental goals (Elder and Olsen 2019).

More than half of the SDGs have an environmental focus and/or address the sustainability of natural resource use, while most include at least one target concerning environmental sustainability (United Nations Environment Assembly of the United Nations Environment Programme [UNEA] 2016). These targets link to the quality of the physical environment either directly (i.e. air, climate, biodiversity, oceans, land and freshwater) or indirectly (e.g. via health, education, agriculture, drinking water and sanitation, energy, and governance and institutions). Twelve SDGs promote human well-being through the sustainable use of natural resources, and ten can be achieved only if the efficiency of natural resource use is substantially improved (UNEP 2015). However, although many SDG targets have been formulated in clear and quantitative terms, many environment-related ones are not (Gupta and Vegelin 2016; Elder and Olsen 2018). This makes it more difficult to define a set of environment-related targets to be assessed in a quantitative scenario analysis.

20.3 An integrated view on the SDGs

The 2030 Agenda for Sustainable Development emphasizes that the goals and targets are integrated and indivisible, and aim to contribute to coherent sustainability policies (United Nations 2015a) – meaning that they depend on each other in different ways (Nilsson, Griggs and Visbeck 2016). Greater focus on the interlinkages and synergies among the SDGs could enhance the effectiveness of their implementation and reduce the total burden and cost of pursuing the goals and targets individually (Elder, Bengtsson and Akenji 2016; UNEA 2016). However, while the SDGs and their associated targets are fairly straightforward, how they can be integrated is not apparent from the 2030 Agenda (International Council for Science and International Social Science Council 2015). The systemic properties of an integrated and holistic approach are also poorly understood (Weitz *et al.* 2017). Nevertheless, the scientific community at large has called for an integrated approach for SDG implementation (Weitz, Nilsson and Davis 2014; United Nations 2015b; Boas, Biermann and Kanie 2016; Lucas *et al.* 2016; Yillia 2016; Stafford-Smith *et al.* 2017). Frameworks have been proposed that allow the interactions between the goals and targets to be mapped and scored (Nilsson *et al.* 2016; Nilsson, Griggs and Visbeck 2016; International Council for Science 2017; Weitz *et al.* 2017; Zhou, Moinuddin and Xu 2017; Singh *et al.* 2018).

Various studies have analysed the interlinkages between the goals and targets, from different perspectives and using different methodologies, for example, by looking at the goals and targets as a network in which links among goals exist through targets that refer to multiple goals (International Science Council and International Social Science Council 2015; Le Blanc 2015; UNEP 2015; Zhou and Moinuddin 2017) and are based on quantitative modelling (United Nations 2015b; van Vuuren *et al.* 2015; Collste, Pedercini and Cornell 2017). Furthermore, researchers have created frameworks to structure the goals, to reveal potential trade-offs and synergies, and to point to ways their interactions might be governed (Griggs *et al.* 2013; Nilsson, Lucas and Yoshida 2013; Lucas *et al.* 2014; Waage *et al.* 2015a; Waage *et al.* 2015b; Elder, Bengtsson and Akenji 2016; Folke *et al.* 2016; Gupta and Vegelin 2016; Reid *et al.* 2017).

Overall, these frameworks reveal a nested structure of goals **(Figure 20.1)**. Some focus on social objectives, related to lives and livelihoods or *human well-being* (SDGs 1, 3, 4, 5, 10); others address *sustainable consumption and production* from a resource use or security perspective (SDGs 2, 6, 7) or more broadly, such as in the context of industry or cities (SDGs 8, 9, 11, 12); and some goals address global public goods from an environmental perspective or the *natural resource base* (SDGs 13, 14, 15). Finally, these goals are supported by a goal on governance (SDG 16) and one addressing means of implementation (SDG 17).

The way of structuring links to the central theme of GEO-6, with Healthy People at the top (being part of human well-being) and Healthy Planet at the bottom (natural resource base). The groups of SDGs are bidirectionally connected in the sense that a healthy planet is the foundation for the economy, human development and, ultimately, human well-being, including healthy people. Unsustainable resource use, waste and pollution can impact adversely on both the natural resource base and on human well-being. A key role is thus played by the goals in the middle, addressing sustainable production and consumption and the equitable distribution of goods and services.

Figure 20.1: A framework for the classification and grouping of the SDGs

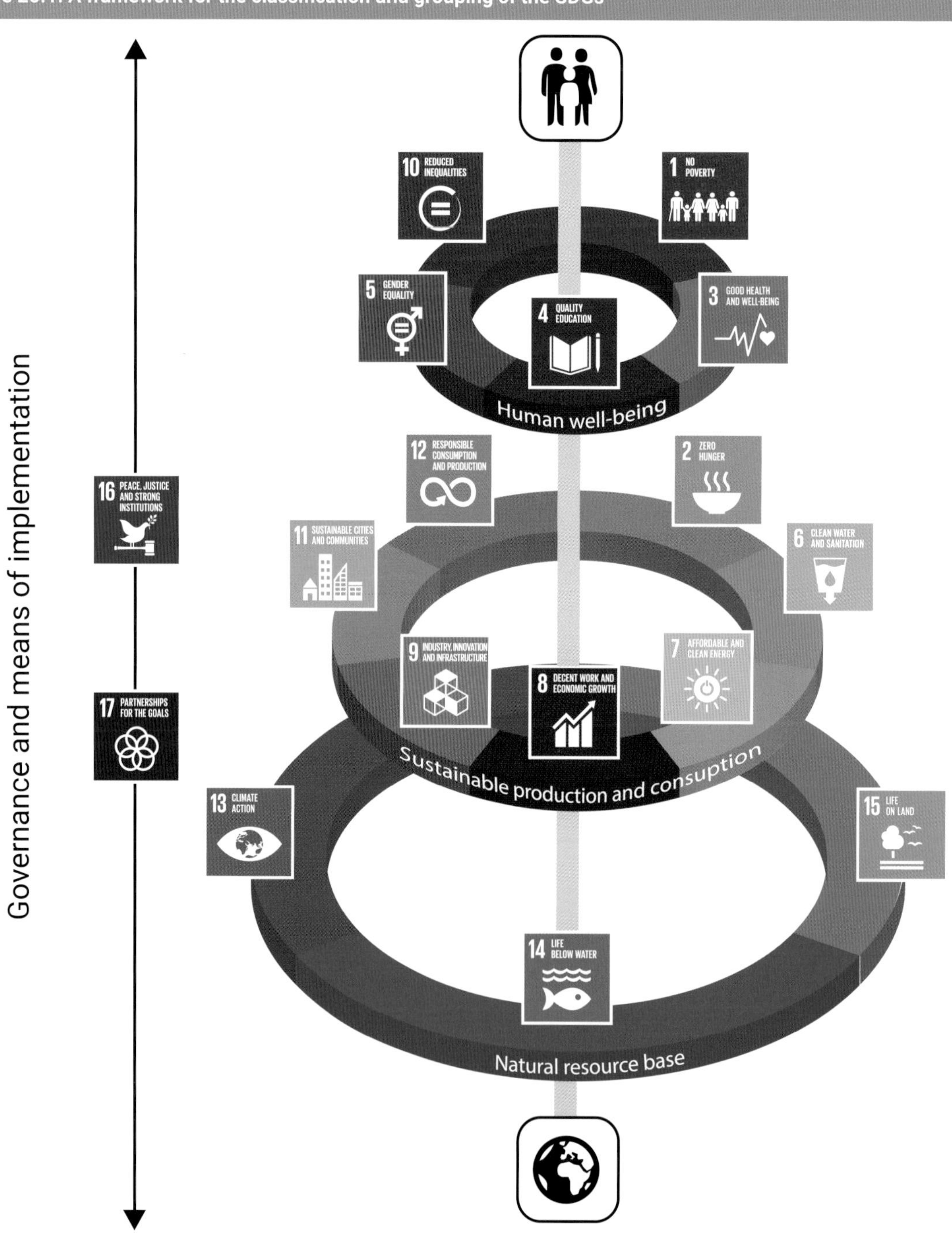

Source: PBL Netherlands Environmental Assessment Agency (2017).

The structure in **Figure 20.1** loosely follows the five areas of critical importance mentioned in the preamble of the 2030 Agenda for Sustainable Development – people, prosperity and planet, underpinned by peace and partnership (United Nations 2015a). It also shows similarities with the doughnut model proposed by Raworth (2012; 2017); a doughnut-shaped area between two boundaries: a social floor (human well-being) and an environmental boundary (the natural resource base). The doughnut model highlights the dependence of human well-being on a healthy environment and stresses the need for improved equity in incomes and resource use, and greater efficiency in the latter (Raworth 2017). Finally, the structuring also relates to the triangle or pyramid originally proposed by Herman Daly, which moves from a base of ultimate means to a tip of ultimate ends, and integrates human well-being, economic development and the state of natural resources into a holistic framework (Daly 1973; Meadows 1998; Pinter *et al.* 2014). According to this framework, ultimate means refer to the underlying natural resource base and the life-support system of the planet (equivalent to the bottom circle in **Figure 20.1**, the natural resource base); intermediate means involve the material economy (middle circle, sustainable consumption and production); intermediate ends represent the capacities of individuals and the condition and functioning of institutions (top circle; human well-being); and ultimate ends indicate human well-being or happiness (Pinter *et al.* 2014).

It should be noted that most SDGs can be classified within different groups, since each SDG is operationalized by multiple targets. SDG 2, for example, includes targets related to human well-being (such as reducing hunger and malnutrition), to sustainable resource use (such as promoting sustainable agriculture), and to the natural resource base (such as maintaining agricultural biodiversity). The structuring of the SDGs in **Figure 20.1** follows from an interpretation from the environmental perspective. In the case of SDG 2, this is sustainable agriculture.

Although it is not stated explicitly, the 2030 Agenda for Sustainable Development and the SDGs suggest sustainable development to be the overarching goal, while emphasizing poverty eradication (Elder, Bengtsson and Akenji 2016). As such, many SDGs are means or intermediate steps towards achieving the goal of poverty eradication (i.e. human health, well-being and security). The top circle therefore contains the people-centred or social goals that aim to deliver individual and collective well-being through improved health and education, ensuring equitable distribution within and between individuals and countries (Waage *at al.* 2015a). These goals can be considered minimum standards for human well-being, while there are also synergistic opportunities for implementation, between education, health and gender equality for example.

Achieving these people-centred goals depends strongly on the realization of goals that address sustainable consumption and production, and equitable distribution of goods and services, including food, water and energy, and more broadly the economy, infrastructure, cities and industries. Food, water and energy security are important resources needed to achieve social objectives such as poverty reduction and good health. The goals addressing these resource needs encompass two distinct resource aspects: (i) access to resources, relevant for poverty reduction, and (ii) sustainable use of resources, relevant for the long-term security of supply. At the same time, production of food, water and energy is highly interlinked. Water is needed for food and energy production, for example, and energy is needed to produce water and food. This is the so-called food-water-energy nexus (Hoff 2011; (Food and Agriculture Organization of the United Nations [FAO] 2014)). These resource goals are accompanied by economy-focused goals that address the production of goods and services more broadly for achieving social objectives. These latter goals focus on the economic system (economic growth and jobs), infrastructure and sustainable industrialization, human settlements, and sustainable consumption and production in general. From an environmental perspective, these goals address the decoupling of efforts to improve human well-being from negative effects on the natural resource based in different contexts.

Realization of these second-level resource and economy goals depends on conditions in the biophysical systems or the natural resource base, including climate, oceans, land and biodiversity (parts of SDG 6 on freshwater also fit here). These goals address protection, conservation, restoration and sustainable use of critical parts of the Earth system. They directly relate to the biophysical limits to ensuring long-term environmental sustainability, or planetary boundaries (Rockström *et al.* 2009; Steffen *et al.* 2015).

The goals in the middle circle connect environmental issues (such as biodiversity loss, climate change and ocean acidification) and social themes (such as health, equal opportunities and labour conditions) to economic activities, products and markets. The challenge of these goals is to seize the synergies and reduce the potential trade-offs between those goals aiming to eradicate poverty and improve human well-being versus those addressing the natural resource base. In other words, improving human well-being should not be achieved at the expense of the natural resource base, while safeguarding the planet should benefit all people and not interfere with poverty eradication. In addition, these goals in the middle are faced with the competition for resources needed to serve multiple goals, e.g. land, water and energy resources. A major transformation to more sustainable consumption and production is needed to address these challenges. From a production perspective, this requires a decoupling of economic growth from environmental degradation, including cleaner production processes, and improved resource efficiency and corporate responsibility. From a consumption perspective it requires changes in lifestyles, consumption preferences and consumer behaviour (Bizikova *et al.* 2015).

20.4 A long-term vision: selected targets and indicators

A range of challenges exist when interpreting SDG targets and related indicators with regard to their values. Assessing future developments and potential pathways for achieving all SDG targets is not possible because of limited data and time. Furthermore, such an analysis is limited by the scope of the existing scenario literature and the integrated assessment models that these studies are built on (see Chapter 21). A selection of targets should thus be made. Next to the challenge of selecting targets, many environment related SDG targets are broadly defined and/or phrased in non-quantitative terms (Lucas *et al.* 2016). In order to assess pathways towards achieving the environmental dimension of the SDGs, the selected targets need to be quantitative, requiring clear indicators accompanied by target values.

The grouping of SDGs in **Figure 20.1** was used to select and organize the SDG targets. Quantitative targets from related MEAs and the scientific literature (science-based targets) were used to quantify these targets, where relevant. The selection is centred around the so-called food-water-energy nexus, focusing on the challenges addressed by, and linked to, the SDGs on food and agriculture (SDG 2), water (SDG 6) and energy (SDG 7). The selection puts natural resource use at the centre (*sustainable consumption and production*), linked with social objectives concerned with people's access to these resources and related health impacts (*human well-being*), and environmental objectives related to the quality and quantity of environmental resources required for or impacted by human use (*natural resource base*). The selected SDG targets for *human well-being* **(Table 20.1)** and the *natural resource base* **(Table 20.2)** are endpoint targets, aiming for a healthy planet with healthy people. The selected SDG targets for *sustainable consumption and production* **(Table 20.3)** are effort- or activity-related targets that are relevant to achieving the endpoint targets.

The selected targets addressing the natural resource base link to the five environmental themes discussed in Part A of GEO-6 (air, biodiversity, oceans, land and freshwater), supplemented by climate change. Furthermore, the targets link to a range of GEO-6 cross-cutting issues (see Chapter 4), most prominently health, climate change, energy, and food systems. Chemicals, and waste and wastewater are two other GEO-6 cross-cutting issues, identified as issues of global concern and addressed under multiple SDGs. There is a general lack, however, of future chemicals and waste flow studies and scenarios in the scientific literature **(see Box 21.1)**. Therefore, chemicals, and waste and wastewater are not discussed as separate issues. More in-depth analysis of these two issues in the context of the SDGs can be expected in UNEP's upcoming Global Chemicals Outlook II and Global Waste Management Outlook, to be released in 2019.

For each target selected, one indicator (and where relevant, two) is selected to track progress. In the context of the SDGs, the United Nations General Assembly adopted an SDG indicator framework that consists of 232 indicators (United Nations 2017). Each indicator is being developed in order to provide accurate and reliable data from now until at least 2030. UNEP is the custodian agency for several SDG indicators related to water (SDG 6), sustainable consumption and production (SDG 12), conservation and sustainable use of ocean resources (SDG 14) and of terrestrial ecosystems (SDG 15) (United Nations 2018). In addition to being custodian agency for these SDG indicators, UNEP is involved in most other SDG indicators that have an environmental dimension. The selected indicators link as much as possible to these globally agreed indicators.

It should be noted that the selected indicators are meant to track progress at the global level and that they are not always relevant at the national or subnational scale. Moreover, many indicators, especially those related to *sustainable consumption and production* and the *natural resource base*, cover only part of what the goals and targets try to accomplish. For air quality in cities, for example, the proposed indicator tracks progress for only one kind of air pollutant (i.e. fine particulate matter [PM] of diameter less than 2.5 μm and 10 μm; $PM_{2.5}$ and PM_{10}) – yet there are several others, with some interacting with each other (e.g. ozone, volatile organic compounds, sulphur dioxide etc.). With respect to health, only one indicator was selected (the under-five mortality rate), which only partly reflects the interconnectedness of planet, society and human health that the SDGs, and GEO-6, are trying to represent. Focusing on a single indicator to track progress for such targets should thus be done with care. To keep the analysis focused however, a limited set of targets is selected to cover, as much as possible, the food-water-energy nexus, while the selected indicators are based mostly on the official SDG indicator set.

Next to the indicator and target levels presented in **Tables 20.1, 20.2** and **20.3**, additional indicators are used in Chapters 21 and 22 to discuss future development of the respective targets, including relevant underlying developments, as well to discuss the potential of specific measures and important synergies and trade-offs across these measures and the selected targets.

20.4.1 Human well-being

For *human well-being*, five SDG targets are selected **(Table 20.1)**. Overall, the SDGs express a strong commitment, both quantitatively and qualitatively, to eradicating poverty and improving human well-being. Among other relevant issues, they aim to end all forms of poverty, including ending hunger, and to achieve access for all to safe drinking water, adequate sanitation, modern energy services, health care, education, work, housing and more.

Despite the centrality of human health to the GEO-6 theme of Healthy Planet, Healthy People, only one target (3.2) and one indicator (3.2.1, under-five mortality rate) has been selected for the scenario analysis. Under-five mortality is generally seen as a good indicator of quality of life, is influenced by numerous environmental determinants, is strongly related to other targets selected for human well-being. And the SDGs set a quantitative target for 2030. Scenario projections, although limited, also exist in the scientific literature that link future developments in under-five mortality to underlying environmental risk factors (see Section 21.3.6). The under-five mortality rate also has

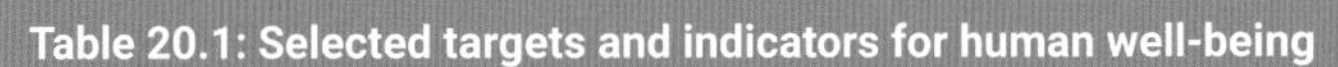

Table 20.1: Selected targets and indicators for human well-being

SDG target	Target for GEO-6	Related MEA	Indicator *	Target level	Based on	Cluster in Chapters 21 and 22
2.1 By 2030, end hunger and ensure access by all people, in particular the poor and people in vulnerable situations, including infants, to safe, nutritious and sufficient food all year round	End hunger	–	2.1.1 Prevalence of undernourishment	0 in 2030	SDGs	Agriculture, food, land and biodiversity
3.2 By 2030, end preventable deaths of newborns and children under 5 years of age, with all countries aiming to reduce neonatal mortality to at least as low as 12 per 1,000 live births and under-five mortality to at least as low as 25 per 1,000 live births	End preventable deaths of children under 5	–	3.2.1 Under-five mortality rate	< 25 in 2030	SDGs	Human health
6.1 By 2030, achieve universal and equitable access to safe and affordable drinking water for all	Achieve universal access to safe drinking water and adequate sanitation	–	6.1.1 Proportion of population using safely managed drinking water services	100 per cent in 2030	SDGs	Freshwater
6.2 By 2030, achieve access to adequate and equitable sanitation and hygiene for all and end open defecation, paying special attention to the needs of women and girls and those in vulnerable situations		–	Proportion of population using safely managed sanitation services (6.2.1)	100 per cent in 2030	SDGs	Freshwater
7.1 By 2030, ensure universal access to affordable, reliable and modern energy services	Achieve universal access to modern energy services	–	7.1.1 Proportion of population with access to electricity	100 per cent in 2030	SDGs	Energy, air and climate
		–	7.1.2 Proportion of population with primary reliance on clean fuels and technology	100 per cent in 2030	SDGs	Energy, air and climate

* Indicators different from official SDG indicator are in italics, with related SDG indicators shown in brackets
MEA = multilateral environmental agreement

significant limitations, however. It excludes measures of morbidity, for example, including psychosocial aspects of childhood well-being (e.g. a sense of safety and of being loved) and other aspects of childhood health that may affect health and survival in later life (see Section 4.2.1). However, these latter data could not be gathered routinely and reliably worldwide, let alone for inclusion in a scenario context. Finally, by being age-limited, the under-five mortality rate does not account for other vulnerable populations, such as older people or pregnant women. As a result, child mortality only partly represents the effect on human health of the many and varied policies and measures, whether business-as-usual ones or transformative scenarios, that are discussed in the following chapters.

SDG target 3.9 is more specific on particular environment-related health risk factors, targeting a substantial reduction in the number of deaths and illnesses from hazardous chemicals and from air, water and soil pollution and contamination. The specific indicators associated with this target involve the mortality rate attributable to household and ambient air pollution, deaths due to unsafe water, sanitation and hygiene and mortality from unintentional poisoning. These rates are largely a reflection of the pollution levels themselves. In effect, control of mortality implies control of pollution itself, which is the focus of several of the targets here, as well as that of several of the SDG targets selected for the *natural resource base*. And so, achievement of all of these targets is an essential part of the GEO-6 vision of Healthy Planet, Healthy People.

20.4.2 Natural resource base

For the *natural resource base*, nine SDG targets are selected that relate to the quality and quantity of environmental resources (i.e. air, climate, biodiversity, oceans, land and freshwater) **(Table 20.2)**. Compared with human well-being, none of these targets has clear quantitative target levels that could be used in a scenario analysis. Each aims to "halt" or "combat" a specific type of environmental degradation and to "restore" the natural system as much as possible.

Several natural resource targets link explicitly or implicitly to specific MEAs that have more explicit quantitative targets and/or take a longer-term perspective, beyond 2030. For these targets, the target levels can be based on quantitative measures given by the agreements. SDG 13 on climate change includes only process- or activity-based targets, but explicitly refers to the United Nations Framework Convention on Climate Change (UNFCCC) as the primary international intergovernmental forum for negotiating the global response to climate change. The target for climate change is thus based on the globally agreed target of the Paris Agreement: that is, "Holding the increase in the global average temperature to well below 2°C above pre-industrial levels and pursuing efforts to limit the temperature increase to 1.5°C above pre-industrial levels" (UNFCCC 2015). The World Health Organization (WHO) established an air-quality guideline of 10 $\mu g/m^3$ for annual mean $PM_{2.5}$ concentrations (WHO 2006), but also defined interim targets of 15 $\mu g/m^3$, 25 $\mu g/m^3$ and 35 $\mu g/m^3$. Here, we focus on the long-term effects of $PM_{2.5}$ and use the percentage of the population exposed to annual mean $PM_{2.5}$ concentrations above the highest interim target of 35 $\mu g/m^3$ as the indicator for achieving the target for air quality. For biodiversity loss, the SDG target does not include a target year for ending biodiversity loss. We derived the target from the Convention on Biological Diversity's strategic plan for biodiversity 2011-2020, and more specifically Aichi biodiversity target five: "By 2020, the rate of loss of all natural habitats, including forests, is at least halved and where feasible brought close to zero..." (Convention on Biological Diversity 2010). This target was translated by Kok *et al.* (2018, p. 138) into halting "biodiversity loss by 2020 for developed countries and from 2030 onwards for developing countries". Kok *et al.* (2018) use mean species abundance as a biodiversity-impact indicator to track progress on this target (Alkemade *et al.* 2009). Mean species abundance is a measure of the intactness of an ecosystem relative to its undisturbed state. Specifically, it is the mean change in the abundance of the species that were present in the original, undisturbed state. Although it is different from the Living Planet Index (the SDG indicator for target 15.5), it shares some important conceptual similarities.

When related MEAs do not exist or lack quantitative target levels, target levels could also be based on scientific literature. The planetary boundaries framework is one example (Hoff and Alva 2017; Lucas and Wilting 2018) – it proposes global quantitative limits for human disturbance of nine critical Earth-system processes (Rockström *et al.* 2009; Steffen *et al.* 2015). Crossing any of the boundaries at the global scale increases the risk of large-scale, and possibly abrupt or irreversible, environmental change. The planetary boundary framework thus provides a quantification of safe levels of global environmental change, based on Earth-system science.

The global limits from the planetary boundaries literature are used for targets related to freshwater quality (de Vries *et al.* 2013; Steffen *et al.* 2015), and ocean acidification (Steffen *et al.* 2015). As already noted, studies and scenarios for chemicals and waste flow are largely missing from the scientific literature. The selected targets for freshwater quality and marine pollution therefore focus on nutrient losses (of nitrogen and phosphorus), for which scenario literature, although limited, is available. Excessive nutrient losses through run-off and erosion can cause the eutrophication of freshwater and coastal ecosystems (de Vries *et al.* 2013; Steffen *et al.* 2015). While recognizing that regional distribution is critical to impacts, the two targets for freshwater quality are global averages. There are several other limitations in using these target levels. They do not account for future trends in the efficiency of nutrient use and do not include other relevant sources of pollution, primarily untreated sewage. Ocean acidification lowers the saturation state of aragonite, a form of calcium carbonate, making it more difficult for marine organisms to form shells and skeletons, which can also run the risk of dissolving as a result of the acidification. Taking into account geographic heterogeneity, the global target for ocean acidification is set for the average global surface aragonite saturation level (Steffen *et al.* 2015).

It should be stressed that the planetary boundaries are not politically endorsed and are subject to ongoing scientific debate. In the end, defining safe levels of global environmental changes and getting consensus on them is a political process, involving subjective elements such as risk acceptance, solidarity and precaution (Lucas and Wilting 2018). Here, the global limits defined by the planetary boundaries framework are used as a set of science-based targets. It should further be noted that there is large geographic heterogeneity underpinning these Earth-system processes that should also be monitored.

For the selected targets for water scarcity, marine nutrient pollution, ocean resources and land degradation, no globally agreed or scientific quantitative target level is available. Therefore, for these targets no quantitative target level is set. The SDG indicator for water scarcity is freshwater withdrawal as a proportion of available freshwater resources (SDG indicator 6.4.2). As this indicator is only relevant at the local level, the total global population living in water scarce areas is used as an indicator. For marine nutrient pollution, the SDG indicator is an index of Coastal Eutrophication (ICEP) and Floating Plastic Debris Density. The indicator is still under development. Here, the focus is on coastal eutrophication, using nutrient runoff into oceans (N and P) as indicator to track progress. For sustainable management of ocean resources, trends in the proportion of fish stocks within biologically sustainable levels, the SDG indicator, are used to track progress. The SDG indicator for land degradation is the proportion of land that is degraded over total land area, based on three sub-indicators, namely trends in land cover, land productivity and carbon stocks (United Nations Convention to Combat Desertification [UNCCD] 2017; van der Esch *et al.* 2017). Also this indicator is still under development. Recognizing that all three sub-indicators are important for assessing land degradation, trends in soil organic carbon stock are selected to track progress.

Table 20.2: Selected targets and indicators for the natural resource base

6.3 By 2030, improve water quality by reducing pollution, eliminating dumping and minimizing release of hazardous chemicals and materials, halving the proportion of untreated wastewater and substantially increasing recycling and safe reuse globally	Improve water quality	–	Nitrogen fertilizer use and biological nitrogen fixation	62 TgN/yr	(de Vries *et al.* 2013)	Freshwater
			Fertilizer use with phosphorus	6.2TgP/yr	(Steffen *et al.* 2015)	Freshwater
6.4 By 2030, substantially increase water-use efficiency across all sectors and ensure sustainable withdrawals and supply of freshwater to address water scarcity and substantially reduce the number of people suffering from water scarcity	Reduce water scarcity	–	*Population living in water scarce areas (6.4.2)*	Not quantified	-	Freshwater
11.6 By 2030, reduce the adverse per capita environmental impact of cities, including by paying special attention to air quality and municipal and other waste management	Improve air quality in cities	WHO guidelines	*Percentage population exposed to* $PM_{2.5}$ *above* 35 $\mu g/m^3$ *(11.6.2)*	0 per cent in 2050	(World Health Organization [WHO] 2006)	Energy, air and climate
SDG13 Take urgent action to combat climate change and its impacts	Limit global warming	Paris Agreement	*Global mean temperature increase*	2.0/1.5°C warming by 2100	(UNFCCC 2015)	Energy, air and climate
14.1 By 2025, prevent and significantly reduce marine pollution of all kinds, in particular from land-based activities, including marine debris and nutrient pollution	Reduce marine nutrient pollution	Aichi biodiversity targets	*N and P flow from freshwater systems into oceans (14.1.1)*	Not quantified	-	Oceans
14.3 Minimize and address the impacts of ocean acidification, including through enhanced scientific cooperation at all levels	Minimize ocean acidification	–	*Average global surface aragonite saturation level (14.3.1)*	Stay above 2.75 Ωarg	(Steffen *et al.* 2015)	Oceans
14.4 By 2020, effectively regulate harvesting and end overfishing, illegal, unreported and unregulated fishing and destructive fishing practices and implement science-based management plans, in order to restore fish stocks in the shortest time feasible, at least to levels that can produce maximum sustainable yield as determined by their biological characteristics	Sustainably manage ocean resources	Aichi biodiversity targets	14.4.1 Proportion of fish stocks within biologically sustainable levels	Not quantified	–	Oceans
15.3 By 2030, combat desertification, restore degraded land and soil, including land affected by desertification, drought and floods, and strive to achieve a land degradation-neutral world	Achieve land degradation neutrality	UNCCD and Aichi biodiversity targets	*Loss in soil organic carbon (15.3.1)*	Not quantified	–	Agriculture, food, land and biodiversity
15.5 Take urgent and significant action to reduce the degradation of natural habitats, halt the loss of biodiversity and, by 2020, protect and prevent the extinction of threatened species	Halt biodiversity loss	Aichi biodiversity targets	*Loss in Mean Species Abundance (MSA) (15.5.1)*	Less than 36 per cent from 2030 onwards	(Kok *et al.* 2018)	Agriculture, food, land and biodiversity

* Indicators different from official SDG indicators are in italics, with related SDG indicators shown in brackets
MEA = multilateral environmental agreement

20.4.3 Sustainable consumption and production

For *sustainable consumption and production*, five SDG targets are selected that address the decoupling of economic growth from environmental degradation (see **Table 20.3**: Selected targets and indicators for sustainable production and consumption). These SDG targets are mostly not quantitative – aiming to increase efficiency substantially without defining a specific target level. They address efforts or activities that help to achieve the endpoint targets. Their absolute level depends on specific overarching objectives. Yield improvements, for example, are important for achieving targets on hunger and biodiversity (SDG targets 2.1 and 15.5). Improvements in water-use efficiency are important for achieving the target on water stress targets (SDG target 6.4). And improvements in energy efficiency and the renewable energy share are important for achieving the target on climate change (SDG 13). The level of decoupling required thus depends on these endpoint targets. Therefore, for the selected SDG targets for *sustainable consumption and production*, no quantitative target levels are defined. Also not for SDG target 7.3, to "double the global rate of improvement in energy efficiency". Instead, the pathways analysis of Chapter 22 provides ranges for the efforts required to achieve the selected SDG targets that address human well-being and the natural resource base, taking into account the interdependencies across these efforts.

Table 20.3: Selected targets and indicators for sustainable consumption and production

SDG target	Target for GEO6	Related MEA	Indicator *	Target level	Based on	Cluster in Chapters 21 and 22
2.3 By 2030, double the agricultural productivity and incomes of small-scale food producers, in particular women, indigenous peoples, family farmers, pastoralists and fishers, including through secure and equal access to land, other productive resources and inputs, knowledge, financial services, markets and opportunities for value addition and non-farm employment	Increase agricultural productivity	–	Yield improvement	required effort results from the pathways analysis in Chapter 22		Agriculture, food, land and biodiversity
2.4 By 2030, ensure sustainable food production systems and implement resilient agricultural practices that increase productivity and production, that help maintain ecosystems, that strengthen capacity for adaptation to climate change, extreme weather, drought, flooding and other disasters and that progressively improve land and soil quality	Increase nutrient use efficiency	–	*Total N inputs to crop N yields (2.4.1)*	required effort results from the pathways analysis in Chapter 22		Agriculture, food, land and biodiversity
6.4 By 2030, substantially increase water-use efficiency across all sectors and ensure sustainable withdrawals and supply of freshwater to address water scarcity and substantially reduce the number of people suffering from water scarcity	Increase water-use efficiency	–	6.4.1 Change in water-use efficiency over time	required effort results from the pathways analysis in Chapter 22		Freshwater
7.2 By 2030, increase substantially the share of renewable energy in the global energy mix	Increase the share of renewable energy	–	7.2.1 Renewable energy share in the total final energy consumption	required effort results from the pathways analysis in Chapter 22		Energy, air and climate
7.3 By 2030, double the global rate of improvement in energy efficiency	Increase energy efficiency	–	7.3.1 Energy intensity measured in terms of primary energy and GDP	required effort results from the pathways analysis in Chapter 22		Energy, air and climate

* Indicators are different from official SDG indicators in are italics, with related SDG indicator shown in brackets
MEA = multilateral environmental agreement

20.5 Conclusions

The SDGs and related MEAs provide a long-term vision for sustainable development to influence policies at the global, regional, national and local levels. This chapter makes a selection of SDG targets, linked to targets from related MEAs and the scientific literature (science-based targets) where relevant, accompanied by clear indicators and quantitative target values, at the global level. The resulting target set provides an integrated perspective on the environmental dimension of the SDGs, focusing on the GEO-6 environmental themes in Part A (air, biodiversity, oceans, land and fresh water) and related multidimensional poverty (access to food, water and energy, and under-five mortality). Unlike the SDGs and related MEAs, the science-based targets selected when there are no globally agreed quantitative targets - are not politically endorsed. They provide a proxy for the related SDG ambitions. Finally, for some selected targets no quantitative globally agreed or science-based target is currently available.

The selection of targets is analysed further in subsequent chapters: Chapter 21 discusses the implementation gap if no new policies are formulated, and Chapter 22 discusses pathways towards achieving the targets, including relevant interrelations (synergies and trade-offs) between different measures and targets. The two chapters do not address regional, national or local differences in developments for these targets and the implementation of measures for achieving them. Chapter 23 discusses implementation from a bottom-up perspective, thereby explicitly taking into account the local situation, different actors and cultural perspectives.

References

Alkemade, R., van Oorschot, M., Miles, L., Nellemann, C., Bakkenes, M. and ten Brink, B. (2009). GLOBIO3: A framework to investigate options for reducing global terrestrial biodiversity loss. *Ecosystems* 12(3), 374-390. https://doi.org/10.1007/s10021-009-9229-5.

Bizikova, L., Pinter, L., Huppé, G.A., Schandl, H., Arden-Clarke, C. and Averous, S. (2015). *Sustainable Consumption and Production Indicators for the Future SDGs*. Nairobi: United Nations Environment Programme. https://wedocs.unep.org/bitstream/handle/20.500.11822/8966/-Sustainable_consumption_and_production_indicators_for_the_future_SDGs_UNEP_discussion_paper%2c_March_2015-2015Sustainable-consumption-and-production-.pdf?sequence=3&isAllowed=y.

Boas, I., Biermann, F. and Kanie, N. (2016). Cross-sectoral strategies in global sustainability governance: Towards a nexus approach. *International Environmental Agreements: Politics, Law and Economics* 16(3), 449-464. https://doi.org/10.1007/s10784-016-9321-1.

Collste, D., Pedercini, M. and Cornell, S.E. (2017). Policy coherence to achieve the SDGs: Using integrated simulation models to assess effective policies. *Sustainability Science* 12(6), 921-931. https://doi.org/10.1007/s11625-017-0457-x.

Convention on Biological Diversity (2010). *Decision Adopted by the Conference of the Parties to the Convention on Biological Diversity at its tenth meeting X/2. The Strategic Plan for Biodiversity 2011-2020 and the Aichi Biodiversity Targets*. 29 October. UNEP/CBD/COP/DEC/X/2. 18. https://www.cbd.int/doc/decisions/cop-10/cop-10-dec-02-en.pdf

Daly, H.E. (1973). *Toward a Steady-State Economy*. San Francisco, CA: W.H. Freeman and Co Ltd. http://www.worldcat.org/title/toward-a-steady-state-economy/oclc/524050.

de Vries, W., Kros, J., Kroeze, C. and Seitzinger, S.P. (2013). Assessing planetary and regional nitrogen boundaries related to food security and adverse environmental impacts. *Current Opinion in Environmental Sustainability* 5(3-4), 392-402. https://doi.org/10.1016/j.cosust.2013.07.004.

Elder, M., Bengtsson, M. and Akenji, L. (2016). An optimistic analysis of the means of implementation for sustainable development goals: Thinking about goals as means. *Sustainability* 8(9), 962. https://doi.org/10.3390/su8090962.

Elder M. and Olsen S.H. (2019). The Design of Environmental Priorities in the SDGs. *Global Policy*.

Folke, C., Biggs, R., Norström, A.V., Reyers, B. and Rockström, J. (2016). Social-ecological resilience and biosphere-based sustainability science. *Ecology and Society* 21(3), 41. http://dx.doi.org/10.5751/ES-08748-210341.

Food and Agriculture Organization of the United Nations (2014). *The Water-Energy-Food Nexus: A New Approach in Support of Food Security and Sustainable Agriculture*. Rome. http://www.fao.org/3/a-bl496e.pdf.

Gerten, D., Hoff, H., Rockström, J., Jägermeyr, J., Kummu, M. and Pastor, A. (2013). Towards a revised planetary boundary for consumptive freshwater use: Role of environmental flow requirements. *Current Opinion in Environmental Sustainability* 5(6), 551-558. https://doi.org/10.1016/j.cosust.2013.11.001.

Griggs, D., Stafford-Smith, M., Gaffney, O., Rockström, J., Öhman, M.C., Shyamsundar, P. *et al.* (2013). Sustainable development goals for people and planet. *Nature* 495, 305-307. http://dx.doi.org/10.1038/495305a.

Gupta, J. and Vegelin, C. (2016). Sustainable development goals and inclusive development. *International Environmental Agreements: Politics, Law and Economics* 16(3), 433-448. http://dx.doi.org/10.1007/s10784-016-9323-z.

Hoff, H. (2011). Understanding the nexus: Background paper for the Bonn 2011 Nexus Conference. *Bonn 2011 Conference the Water, Energy and Food Security Nexus: Solutions for the Green Economy*. Bonn, 16-18 November 2011. Stockholm Environment Institute http://wef-conference.gwsp.org/fileadmin/documents_news/understanding_the_nexus.pdf

Hoff, H. and Alva, I.L. (2017). *How the Planetary Boundaries Framework Can Support National Implementation of the 2030 Agenda*. SEI Policy Brief. Stockholm: Stockholm Environment Institute. https://www.sei.org/mediamanager/documents/Publications/SEI-2017-PB-Hoff-HowthePlanetary.pdf.

International Council for Science (2017). *A Guide to SDG Interactions: From Science to Implementation*. Griggs, D.J., Nilsson, M. and Stevance, A. (eds.). Paris: International Council for Science. https://www.icsu.org/cms/2017/05/SDGs-Guide-to-Interactions.pdf.

International Council for Science and International Social Science Council (2015). *Review of Targets for the Sustainable Development Goals: The Science Perspective*. Paris: International Council for Science. https://www.icsu.org/cms/2017/05/SDG-Report.pdf.

Jabbour, J., Keita-Ouane, F., Hunsberger, C., Sanchez-Rodriguez, R., Gilruth, P., Levy, M.A. *et al.* (2012). Internationally agreed environmental goals: A critical evaluation of progress. *Environmental Development* 3, 5-24. http://dx.doi.org/10.1016/j.envdev.2012.05.002.

Kok, M.T.J., Alkemade, R., Bakkenes, M., van Eerdt, M., Janse, J., Mandryk, M. *et al.* (2018). Pathways for agriculture and forestry to contribute to terrestrial biodiversity conservation: A global scenario-study. *Biological Conservation* 221, 137-150. https://doi.org/10.1016/j.biocon.2018.03.003.

Le Blanc, D. (2015). Towards integration at last? The sustainable development goals as a network of targets. *Sustainable Development* 23(3), 176-187. https://doi.org/10.1002/sd.1582.

Lucas, P., Ludwig, K., Kok, M. and Kruitwagen, S. (2016). *Sustainable Development Goals in the Netherlands: Building Blocks for Environmental Policy for 2030*. The Hague: PBL Netherlands Environmental Assessment Agency. http://www.pbl.nl/sites/default/files/cms/publicaties/pbl-2016-sustainable-development-in-the-Netherlands_1966.pdf.

Lucas P. L. and Wilting H. (2018). *Using Planetary Boundaries to Support National Implementation of Environment-Related Sustainable Development Goals*. The Hague: PBL Netherlands Environmental Assessment Agency. http://www.pbl.nl/sites/default/files/cms/publicaties/Using%20planetary%20boundaries%20to%20support%20national%20implementation%20of%20environment-related%20Sustainable%20Development%20Goals%20-%202748.pdf.

Lucas, P.L., Marcel, T., Kok, J., Nilsson, M. and Alkemade, R. (2014). Integrating biodiversity and ecosystem services in the post-2015 development agenda: Goal structure, target areas and means of implementation. *Sustainability* 6(1), 193-216. http://dx.doi.org/10.3390/su6010193.

Meadows, D.H. (1998). *Indicators and Information Systems for Sustainable Development*. Hartland, VT: The Sustainability Institute. http://donellameadows.org/wp-content/userfiles/IndicatorsInformation.pdf.

Nilsson, M., Griggs, D. and Visbeck, M. (2016). Policy: Map the interactions between sustainable development goals. *Nature* 534(7607). http://dx.doi.org/10.1038/534320a.

Nilsson, M., Lucas, P. and Yoshida, T. (2013). Towards an integrated framework for SDGs: Ultimate and enabling goals for the case of energy. *Sustainability* 5(10), 4124-4151. http://dx.doi.org/10.3390/su5104124.

Nilsson, M., Griggs, D., Visbeck, M. and Ringler, C. (2016). *A Draft Framework for Understanding SDG Interactions*. Paris: International Council for Science. https://www.icsu.org/cms/2017/05/SDG-interactions-working-paper.pdf.

Organisation for Economic Co-operation and Development (2015). *Better Policies for Development 2015: Policy Coherence and Green Growth*. Paris. https://www.oecd-ilibrary.org/better-policies-for-development-2015_5js0bt7443lr.pdf?itemId=%2Fcontent%2Fpublication%2F9789264236813-en&mimeType=pdf.

PBL Netherlands Environmental Assessment Agency (2017). *People and the Earth: International Cooperation for the Sustainable Development Goals in 23 infographics*. Kok, M., Sewell, A., de Blois, F., Warrink, A., Lucas, P. and van Oorschot, M. (eds.). The Hague: PBL Netherlands Environmental Assessment Agency. http://www.pbl.nl/sites/default/files/cms/publicaties/People_and_the_Earth_WEB.pdf.

Pinter, L., Almassy, D., Antonio, E., Hatakeyama, S., Niestroy, I., Olsen, S. *et al.* (2014). *Sustainable Development Goals and Indicators for a Small Planet. Part I: Methodology and Goal Framework*. Singapore: Asia-Europe Foundation. https://www.asef.org/images/stories/publications/ebooks/ASEF_Report_Sustainable-Development-Goals-Indicators_01.pdf.

Raworth, K. (2012). *A Safe and Just Space for Humanity: Can we Live Within the Doughnut*. Oxfam Policy and Practice: Climate Change and Resilience. https://www.oxfam.org/sites/www.oxfam.org/files/dp-a-safe-and-just-space-for-humanity-130212-en.pdf.

Raworth, K. (2017). *Doughnut Economics: Seven Ways to Think Like a 21st-Century Economist*. Chelsea Green Publishing. https://www.chelseagreen.com/product/doughnut-economics/.

Reid, A.J., Brooks, J.L., Dolgova, L., Laurich, B., Sullivan, B.G., Szekeres, P. *et al.* (2017). Post-2015 sustainable development goals still neglecting their environmental roots in the Anthropocene. *Environmental Science and Policy* 77, 179-184. http://dx.doi.org/10.1016/j.envsci.2017.07.006.

Rockström, J., Steffen, W., Noone, K., Persson, Å., Chapin III, F.S., Lambin, E.F. *et al.* (2009). A safe operating space for humanity. *Nature* 461(7263), 472-475. http://dx.doi.org/10.1038/461472a.

Singh, G.G., Cisneros-Montemayor, A.M., Swartz, W., Cheung, W., Guy, J.A., Kenny, T.A. *et al.* (2018). A rapid assessment of co-benefits and trade-offs among Sustainable Development Goals. *Marine Policy* 93, 223-231. http://dx.doi.org/10.1016/j.marpol.2017.05.030.

Stafford-Smith, M., Griggs, D., Gaffney, O., Ullah, F., Reyers, B., Kanie, N. *et al.* (2017). Integration: The key to implementing the sustainable development goals. *Sustainability Science* 12(6), 911-919. http://dx.doi.org/10.1007/s11625-016-0383-3.

Steffen, W., Richardson, K., Rockström, J., Cornell, S.E., Fetzer, I., Bennett, E.M. *et al.* (2015). Planetary boundaries: Guiding human development on a changing planet. *Science* 347(6223), 1259855. http://dx.doi.org/10.1126/science.1259855.

United Nations (1972). *Report of the United Nations Conference on the Human Environment*. A/CONF.48/14/Rev.1. http://www.un-documents.net/aconf48-14r1.pdf

United Nations (2015a). *Transforming Our World: The 2030 Agenda for Sustainable Development*. New York, NY. https://sustainabledevelopment.un.org/content/documents/21252030%20Agenda%20for%20Sustainable%20Development%20web.pdf.

United Nations (2015b). *Global Sustainable Development Report - 2015 Edition Advance Unedited Version*. New York, NY. https://sustainabledevelopment.un.org/content/documents/1758GSDR%202015%20Advance%20Unedited%20Version.pdf.

United Nations (2017). *Global Indicator Framework for the Sustainable Development Goals and Targets of the 2030 Agenda for Sustainable Development*. A/RES/71/313. E/CN.3/2018/2. New York https://unstats.un.org/sdgs/indicators/Global%20Indicator%20Framework%20after%20refinement_Eng.pdf

United Nations (2018). *Tier Classification for Global SDG Indicators*. https://unstats.un.org/sdgs/files/Tier%20Classification%20of%20SDG%20Indicators_15%20October%202018_web.pdf

United Nations Convention to Combat Desertification (2017). *Global Land Outlook*. Bonn. https://www.unccd.int/sites/default/files/documents/2017-09/GLO_Full_Report_low_res.pdf.

United Nations Environment Assembly of the United Nations Environment Programme (2016). *Delivering on the Environmental Dimension of the 2030 Agenda for Sustainable Development: A Concept Note - Information Note by the Executive Director*. UNEP/EA.1/INF/18. http://sdgtoolkit.org/wp-content/uploads/2017/02/Delivering-on-the-Environmental-Dimension-of-the-2030-Agenda-for-Sustainable-Development-%E2%80%93-a-concept-note.pdf

United Nations Environment Programme (2015). *Policy Coherence of the Sustainable Development Goals: A Natural Resource Perspective*. Nairobi. https://wedocs.unep.org/bitstream/handle/20.500.11822/9720/-Policy_Coherence_of_the_Sustainable_Development_Goals_A_Natural_Resource_Perspective-2015Policy_Coherence_of_the_Sustainable_Development_Goals_-_A_N.pdf?sequence=3&isAllowed=y.

United Nations Framework Convention on Climate Change (2015). *Adoption of the Paris Agreement*. FCCC/CP/2015/L.9/Rev.1. 12 December. https://unfccc.int/resource/docs/2015/cop21/eng/l09r01.pdf

van der Esch, S., ten Brink, B., Stehfest, E., Bakkenes, M., Sewell, A., Bouwman, A. *et al.* (2017). *Exploring Future Changes in Land Use and Land Condition and the Impacts on Food, Water, Climate Change and Biodiversity: Scenarios for the UNCCD Global Land Outlook*. The Hague: PBL Netherlands Environmental Assessment Agency. http://www.pbl.nl/sites/default/files/cms/publicaties/pbl-2017-exploring-future-changes-in-land-use-and-land-condition-2076b.pdf.

van Vuuren, D.P., Kok, M., Lucas, P.L., Prins, A.G., Alkemade, R., van den Berg, M. *et al.* (2015). Pathways to achieve a set of ambitious global sustainability objectives by 2050: Explorations using the IMAGE integrated assessment model. *Technological Forecasting and Social Change* 98, 303-323. http://dx.doi.org/10.1016/j.techfore.2015.03.005.

Waage, J., Yap, C., Bel, S., Levy, C., Mace, G., Pegram, T. *et al.* (2015a). Governing the sustainable development goals: Interactions, infrastuctures, and institutions. *The Lancet global health* 3(5), e251–e252. http://dx.doi.org/10.1016/S2214-109X(15)70112-9.

Waage, J., Yap, C., Bell, S.J., Levy, C., Mace, G., Pegram, T. *et al.* (2015b). Governing sustainable development goals: Interactions, infrastructures and institutions. In *Thinking Beyond Sectors for Sustainable Development*. Waage, J. and Yap, C. (eds.). London: Ubiquity Press. 79-88. http://discovery.ucl.ac.uk/1505333/1/Waage_governing-sustainable-development-goals-interactio.pdf

Weitz, N., Carlsen, H., Nilsson, M. and Skånberg, K. (2017). Towards systemic and contextual priority setting for implementing the 2030 Agenda. *Sustainability Science* 13(2), 531-548. http://dx.doi.org/10.1007/s11625-017-0470-0.

Weitz, N., Nilsson, M. and Davis, M. (2014). A nexus approach to the post-2015 agenda: Formulating integrated water, energy, and food SDGs. *SAIS Review of International Affairs* 34(2), 37-50. http://dx.doi.org/10.1353/sais.2014.0022.

World Health Organization (2006). *WHO Air Quality Guidelines for Particulate Matter, Ozone, Nitrogen Dioxide and Sulfur Dioxide: Global Update 2005. Summary of Risk Assessment*. Geneva. http://apps.who.int/iris/bitstream/handle/10665/69477/WHO_SDE_PHE_OEH_06.02_eng.pdf;jsessionid=F232D7F7BDD673090F5BB2309F58CAA8?sequence=1.

Yillia, P.T. (2016). Water-Energy-Food nexus: Framing the opportunities, challenges and synergies for implementing the SDGs. *Österreichische Wasser- und Abfallwirtschaft* 68(3-4), 86-98. http://dx.doi.org/10.1007/s00506-016-0297-4.

Zhou, X. and Moinuddin, M. (2017). *Sustainable Development Goals Interlinkages and Network Analysis: A Practical Tool for SDG Integration and Policy Coherence*. Hayama: Institute for Global Environmental Strategies. https://pub.iges.or.jp/pub_file/igesresearch-reportsdg/download.

Chapter 21

Future Developments Without Targeted Policies

Coordinating Lead Author: Paul Lucas (PBL Netherlands Environmental Assessment Agency), Steve Hedden (Frederick S. Pardee Center for International Futures), Detlef van Vuuren (PBL Netherlands Environmental Assessment Agency)

Lead Author: Katherine V. Calvin (Joint Global Change Research Institute/ PNNL), Serena H. Chung (United States Environmental Protection Agency), Mike Harfoot (United Nations Environment Programme World Conservation Monitoring Centre (UNEP-WCMC)), Alexandre C. Köberle (Universidade Federal do Rio de Janeiro), Jonathan D. Moyer (Frederick S. Pardee Center for International Futures), Yoshihide Wada (International Institute for Applied Systems Analysis)

Contributing Authors: Barry B. Hughes (Frederick S. Pardee Center for International Futures), Fintan Hurley (Institute of Occupational Medicine), Terry Keating (United States Environmental Protection Agency)

GEO Fellow: Katrina Lyne (James Cook University)

Executive summary

Together with other tools, model-based scenario analysis provides a useful method to explore whether environment related targets of the Sustainable Development Goals (SDGs) and related Multilateral Environmental Agreements (MEAs) are going to be achieved *(well established)*. Assessing whether current trends are leading to fulfilment of the selected targets outlined in Chapter 20 is complex: it requires insights into the interactions of different trends and systems, inertia and cross-scale relationships. A combination of qualitative storylines and quantitative scenario tools can help to explore possible futures trends, while taking the many complexities into account. While scenarios can never be forecasts, because surprises may occur, they can inform decision makers of the likely implications of current trends. {21.2}

An assessment of the scenario literature concludes that a continuation of current trends will probably not lead to fulfilment of selected environment related targets of the SDGs and related MEAs *(well established)*. **While some improvement is projected for indicators related to human development – albeit not fast enough to meet the targets – those related to the natural resource base are projected to move further in the wrong direction** *(established, but incomplete)*. Projected ongoing population growth, and economic development imply that the demand for food, water and energy will strongly increase towards 2050. At the same time, business-as-usual scenarios show a clear improvement over time in reducing hunger, increasing access to safe drinking water and adequate sanitation, and increasing access to modern energy services, but not fast enough to meet the related SDG targets by 2030. Furthermore, although projected improvements in resource efficiencies across the board (agricultural yields, nutrient use efficiency, water use efficiency and energy efficiency) will somewhat limit the impacts of resource use on the environment, these improvements are not enough to reduce the pressure on already stressed environmental systems. As a result, trends in environmental degradation are projected to continue at a rapid rate. Related targets are not achieved. {21.4}

Under current trends, environmental pressures related to the agricultural and food system will further increase *(well established)***.** As the global population and per capita incomes are projected to grow, both per capita and total food consumption expected to rise. At the same time, the number of undernourished people is projected to decline to 300-650 million in 2030, a figure that still significantly exceeds the target for ending global hunger. Furthermore, food production and land use are directly related to many environmental problems. Global agricultural demand is projected to increase by 50-60 per cent. Over the last decades, around four-fifth of the increase in food demand was met by agricultural intensification, and one-fifth by an increase in agricultural area. This trend is expected to more or less continue. Together, food production systems will continue to contribute to land expansion, increasing water demand and nutrient runoff, biodiversity loss and land degradation. Related targets are not achieved. {21.3.2}

Without new policies, the objectives of the Paris Agreement are not achieved *(well established)***.** Primary energy supply is projected to grow by 50-70 per cent between 2010 and 2050. Moreover, fossil fuels are expected to remain prominent in the world energy system. As a result, energy use is expected to continue to be the main cause of greenhouse gas (GHG) emissions. In addition, the agricultural systems and land use will continue to contribute to GHG emissions. Current and planned climate policies, as formulated by different countries under the Paris Agreement of the United Nations Framework Convention on Climate Change, are expected to lead, at best, to a sstabilization of emissions. This is considerably less than would be needed to achieve the objectives of the Paris Agreement, i.e. to keep the temperature well below 2°C, and if possible below 1.5°C. Achieving these objectives would require an almost complete decarbonization of the energy system. {21.3.3}

Ambient air pollution is expected to continue to contribute to millions of premature deaths in the coming decades *(established but incomplete)***.** There are different forms of urban and regional air pollution. Exposure to ambient fine particulate matter ($PM_{2.5}$) is estimated to have caused approximately 4 million premature deaths in 2016 and can be used as an indicator of adverse health effects of ambient air pollution. Without stringent policies to control air pollution, ambient $PM_{2.5}$ concentrations are expected to increase. Most trend scenarios assume that past trends of stricter air pollution policies coupled with increasing incomes continue in the future, i.e. that more stringent air pollution policies are applied in developing countries as their incomes increase, thereby projecting a slow decrease in emissions of $PM_{2.5}$ and its precursors in most global regions. However, this trend would still not be sufficient to reduce $PM_{2.5}$ concentrations below the least stringent air quality target of the World Health Organization (WHO) in large parts of Asia, the Middle East and Africa, resulting in 4.5 to 7 million premature deaths globally by mid-century {21.3.3}

Global water scarcity and the population affected by it are expected to increase *(established, but incomplete)*. Global human water demand is projected to increase by around 25 to 40 per cent this century. The rise is primarily driven by rapid population growth and increased industrial activities (higher electricity and energy use) in developing countries. An increase in irrigated area and irrigation intensity is also projected, but its effect will probably be compensated by improvements in irrigation efficiency in regions with strong economic development. Changing rainfall patterns will exert additional pressure on regional water availability. By 2050, the Asian population living in areas exposed to severe water stress is projected to increase by around 50 per cent compared with 2010 levels, putting severe pressure on non-renewable groundwater reserves. {21.3.4}

Oceans are expected to continue to be polluted and overexploited *(established, but incomplete).* Nutrient (nitrogen and phosphorus) flows from freshwater into world oceans exceed sustainable levels and as a result the risks of dead zones and toxic algae blooms in coastal areas are projected to increase. This is largely related to increased fertilizer use in agricultural production and developments in wastewater treatment that are lagging behind improvements in access to sanitation. As a result of an increasing concentration of carbon dioxide (CO_2), oceans are expected to further acidify, negatively affecting marine organisms' ability to create shells and skeletons or even resulting in their dissolution. Acidification is expected to increase most rapidly in polar regions. Finally, under current fishing strategies, the projected increase in demand for fish is expected to reduce the proportion of fish stocks that remain at biologically sustainable levels. {21.3.5}

Preventable environmental health risks are projected to remain prominent in 2030, with related negative impacts on child mortality *(established, but incomplete).* Nearly one-quarter of all deaths globally in 2012 can be attributed to environmental factors, with a greater portion occurring in vulnerable populations (children and the elderly) and in developing countries. Prominent environmental risk factors – i.e. exposure to ambient air pollution, and not having access to clean water, adequate sanitation or modern energy services – together with global hunger are expected to improve towards 2030, but not fast enough to achieve related targets in all countries. Related global child mortality is projected to decline, but not enough to achieve the SDG targets in many developing countries. Especially in sub-Saharan Africa, child mortality rates remain high, with a continued, although smaller, share related to preventable environmental risk factors. {21.3.6}

21.1 Introduction

Chapter 20 provided an overview of environment related targets that the international community committed to support, based on the Sustainable Development Goals (SDGs) and a range of Multilateral Environmental Agreements (MEAs). This chapter examines the international scenario literature to assess to what extent current and long-term trends are in line with achieving these targets, and to understand and highlight potential implementation gaps.

21.2 Global environmental scenarios

Environmental and sustainable development targets are usually formulated for a time period somewhere in the future. In an effort to inform decision makers concerned with global environmental and sustainability challenges, Global Environmental Assessments (GEA) explore possible futures, with a special focus on investigating the consequences of current trends and assessing whether the committed goals and targets are going to be met (Clark, Mitchell and Cash 2006; van Vuuren *et al.* 2012). This is not straightforward: clearly, no one knows which path the world will take in the next 40 years, and world views influence our expectations of this path. Assessments deal with the outlook component and the uncertainty involved in different ways. Some use a reference scenario that captures a likely future state. Others use multiple scenarios reflecting different storylines. In all cases, the scenarios are "plausible descriptions of how the future developments might evolve, based on a coherent and internally consistent set of assumptions ("scenario logic") about the key relationships and driving forces (i.e. the technology, economy, environment interplay)" (Nakicenovic *et al.* 2000). Often a storyline is quantified within a model. While model-based quantification can help to take account of the many relationships that exist across scales, between regions, in time and across various sectors and environmental problems, the storyline elements help to ensure consistency for other elements that are more difficult to quantify. The main purpose of this scenario methodology is to be as scientifically rigorous as possible, while providing policy relevant information (van Vuuren *et al.* 2012).

Over the past few years, a large number of environmental assessment reports has been published. Many of these focus on specific environmental issues, such as climate change (Intergovernmental Panel on Climate Change [IPCC] 2014a), biodiversity loss (Secretariat of the Convention on Biological Diversity [SCBD] 2014) and the management and restoration of land resources (United Nations Convention to Combat Desertification [UNCCD] 2017). These can ensure the input of scientific information into decision-making processes for the three Rio conventions, i.e. UNFCCC, CBD and UNCCD, respectively. Other environmental assessments have a less clear focus on specific decision-making processes. More sectoral environmental assessments address key drivers of environmental change, for example, the global energy system (Global Energy Assessment [GEA] 2012; International Energy Agency [IEA] 2017a) and the global agricultural system (Organisation for Economic Co-operation and Development [OECD] and the Food and Agriculture Organization of the United Nations [FAO] 2017). Finally, some environmental assessments look more closely at the interrelations between environmental issues and how they relate to human development. Examples include the Millennium Ecosystem Assessment (Millennium Ecosystem Assessment [MEA] 2005) and the United Nations Environment Programme's (UNEP) Global Environment Outlook reports (e.g. UNEP 2012).

Interestingly, a limited number of key archetypical scenarios, or scenario families, reappear in many of these assessments (van Vuuren *et al.* 2012). The term 'scenario family' denotes a set of scenarios in the literature that share a similar storyline or logic, resulting in a similar kind of quantification. Based on these key elements, van Vuuren *et al.* (2012) identified six scenario families:

i. economic-technological optimism/conventional markets scenarios;
ii. reformed market scenarios;
iii. global sustainability scenarios;
iv. regional competition/regional markets scenarios;
v. regional sustainable development scenarios; and
vi. business-as-usual or trend scenarios.

None of these scenarios is a prediction or forecast of the future. They are only meant to explore plausible future development pathways.

This chapter focuses on business-as-usual or trend scenarios. This type of scenario assumes that basic socioeconomic mechanisms continue to operate as they did in the past, and that no explicit new policies are introduced to meet specific policy targets. We assess the scenarios literature to analyse to what extent current and long-term trends are in line with achieving the environment-related targets of the SDGs and related MEAs (see selection of targets in Chapter 20), and to understand and highlight potential implementation gaps. The scenario assessment is used as a benchmark against which possible alternative future development pathways, that aim to achieve the selected targets simultaneously, are evaluated (see Chapter 22).

Box 21.1: Waste as an important cause of environmental degradation

Part A of the sixth Global Environment Outlook (GEO-6) identified a number of consistent waste management issues of global concern, especially food waste and marine litter. It also pointed to a growing disparity across regions. Developed regions generate some 56 per cent of food waste, compared with 44 per cent in developing regions. Also, while developed regions increasingly invest in circular economy measures, about 3 billion people lack access to controlled waste disposal facilities, generating both health risks and environmental impacts. Plastic waste is of particular concern. However, there is a general lack of future waste flow studies, and such scenarios are mostly missing from the literature. Chapters 21 and 22 therefore do not discuss waste as a separate issue. Nevertheless, the problem of waste management is central to achieving several SDGs, and policies to address the growing waste problem will need to carefully consider both demand reduction and supply side restructuring.

Box 21.2: The Shared Socioeconomic Pathways

The Shared Socioeconomic Pathways (SSPs) are five distinct global pathways describing the future evolution of key aspects of society that together imply a range of challenges for mitigating and adapting to climate change (O'Neill *et al.* 2017; Riahi *et al.* 2017). They have been developed into storylines and quantitative measures for a broad range of issues, including energy and land-use developments and related GHG emissions, based on scenarios and various other methods. As they are formulated relatively broadly and cover a wide range of possible futures, they are also used extensively for other fields of environmental research and assessment. The five SSPs are:

i. sustainable development (SSP1);
ii. the middle road (SSP2);
iii. a fragmented world (SSP3);
iv. inequality (SSP4); and
v. fossil fuel-based development (SSP5).

The SSPs are not policy-free; they include all kinds of assumptions on policies. However, since they were formulated specifically to support climate change research and assessment, the reference scenario versions of the SSPs are free of climate policy beyond the base year. The SSPs are certainly not free of environmental and sustainable development policies. In fact, key elements in SSP1, for instance, are low population growth, economic convergence, rapid technology development of environmentally friendly technologies and the introduction of environmental policies. By contrast, in SSP3 the fragmented world leads to high population growth, slow economic development and a focus on security issues; and thus, little priority for environmental issues.

Box 21.3: The need for coordination among environmental assessments

There are currently several assessments that include an outlook component and focus on specific SDGs. These include the Global Biodiversity Outlook, the Global Land Outlook, the Intergovernmental Science-Policy Platform on Biodiversity and Ecosystem Services, and the Global Environment Outlook. As highlighted in the 2030 Agenda for Sustainable Development, it is clear that the SDGs cannot be achieved without taking into account important synergies and trade-offs across these goals (see also Section 22.4.2). For this reason, it will become increasingly important to coordinate the work across the various assessments, and check whether the key findings address the SDG agenda as a whole.

Not all relevant issues are addressed equally in the scenario literature. For instance, there are many scenarios published on climate change, while only a few scenario studies focus on water pollution. Other issues of global concern, such as chemicals and waste, are hardly addressed in the scenario literature **(see Box 21.1)**. The available literature thus limits the scope of the analysis. As a result, not all relevant issues that are discussed in Part A of GEO-6, such as chemicals and waste, are covered in the analysis in Chapters 21 and 22.

A widely used set of scenarios are the Shared Socioeconomic Pathways (SSPs; see **Box 21.2**). The SSPs were developed principally to support climate research (van Vuuren *et al.* 2014; Riahi *et al.* 2017) but are also used extensively for other fields of environmental research and assessment. They explore a wide set of possible futures. Within the set, SSP2 is a scenario that represents medium developments for key drivers such as population, economic growth and technology development. Our assessment focuses on business-as-usual or trend scenarios, using the middle-of-the-road SSP2 scenario as a common thread across the different issues discussed. Other trend scenarios are used where relevant. Furthermore, SSP3 scenario results are shown where possible to indicate the risk of higher population growth.

21.3 The achievement of SDGs and related MEAs in trend scenarios

Business-as-usual or trend scenarios, as discussed in Section 21.2, are used to assess to what extent current and long-term trends are in line with achieving selected SDG targets **(see Tables 20.1 and 20.2)**. Achievement of the selected targets is assessed in five distinct clusters of closely related environmental issues **(see Figure 21.1)**. While interlinkages across these clusters exist, presenting them in these five clusters allows a more focused discussion. Four of the five clusters are closely related to the five environmental themes discussed in Part A (air, biodiversity, oceans, land and freshwater). Biodiversity and land are clustered together, as they are both strongly linked to developments in agriculture. In addition, human health is discussed as an individual cluster.

21.3.1 Drivers

There are several key drivers of global environmental change. Here, we focus on population, urbanization and economic development. While climate change is discussed as a driver in Chapter 2, given that it cannot be influenced in the short term, it is assessed here as a global environmental challenge, as part of the energy, climate and air cluster. The role of technology is also discussed within the different clusters, primarily in the context of efficiency of resource use. It should be noted that (un)sustainable consumption and production practices also play a key role (see Section 2.5).

Population

The United Nations' World Population Prospects (United Nations 2017) and the population scenarios underlying the SSPs (Samir and Lutz 2017) are among those mainly used in the literature (see Section 2.2). The scenarios share important characteristics: they project population growth to continue and to reach a level of around 8.5 billion by 2030 and around 9-10 billion by 2050 **(see Figure 21.2)**. The high population growth scenarios, such as SSP3, are typically associated with slow improvements in human development. Low population projections result from a relative rapid drop in fertility rates. More than half the anticipated growth is expected to occur in Africa and around 30 per cent in Asia (mainly South Asia). After 2050, Africa is projected to be the only region experiencing substantial population growth.

Urbanization

On a global scale, more people currently live in urban compared to rural areas (see Section 2.3). Urbanization levels differ significantly across world regions, with more than 80 per cent of the population living in urban areas in Latin America and the Caribbean, and only around 40 per cent in Africa. Urbanization is projected to increase in all regions, with average global urbanization levels growing to around 60 per cent in 2030 and 67 per cent in 2050 under the United Nations' World Urbanization prospects and SSP2 (United Nations 2014; Jiang and O'Neill 2017) **(see Figure 21.2)**. Together with an increasing overall population, the global urban population is projected to grow by more than two-thirds between now and

Figure 21.1: Selected targets and their related clusters as examined in this chapter

	Agriculture, food, land and biodiversity	Energy, air and climate	Fresh water	Oceans	Human health
Human well-being	• End hunger	• Achieve universal access to modern energy services	• Achieve universal access to safe drinking water and adequate sanitation	• End hunger	• End preventable deaths of children under 5
Sustainable consumption and production	• Increase agricultural productivity • Increase nutrient use efficiency	• Increase energy efficiency • Increase the share of renewable energy	• Increase water-use efficiency	• Increase agricultural productivity • Increase nutrient use efficiency	
Natural resource base	• Achieve land degradation neutrality • Halt biodiversity loss	• Limit global warming • Improve air quality in cities	• Reduce water scarcity • Improve water quality	• Sustainably manage ocean resources • Minimize ocean acidification • Reduce marine nutrient pollution	

Figure 21.2: Future projections of the global population (left) and urbanization (right)

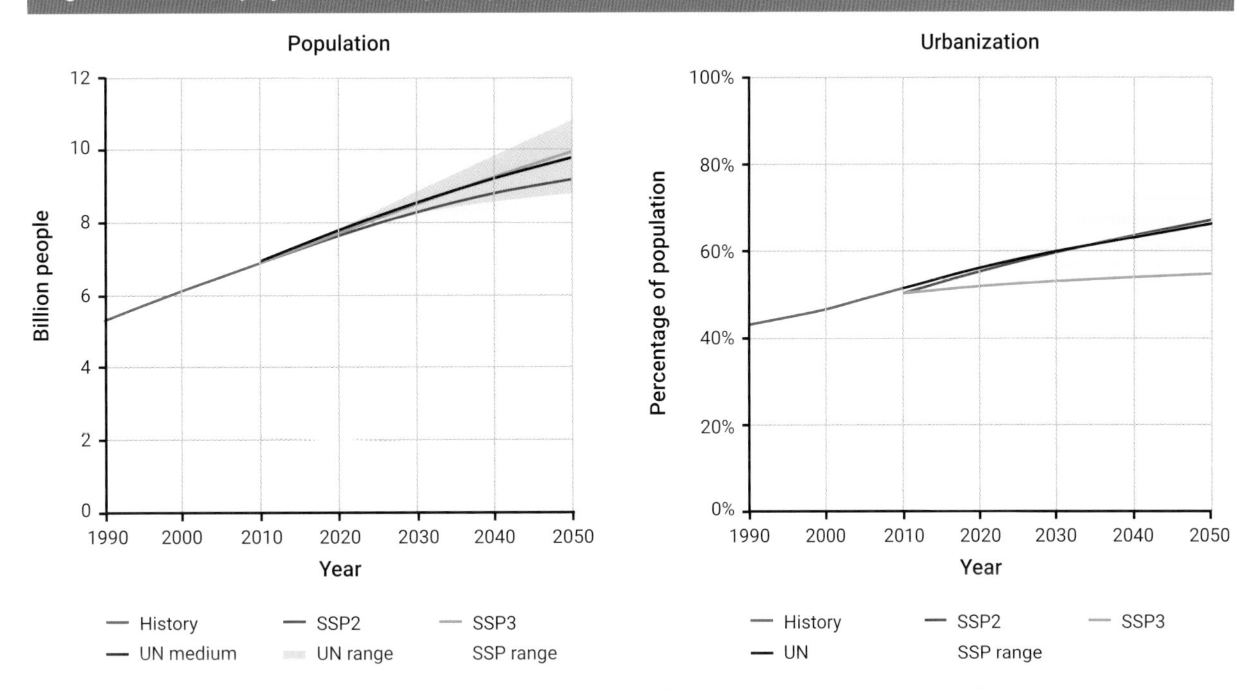

Source: SSP projections (Jiang and O'Neill 2017; Samir and Lutz 2017); UN projections (United Nations 2014; United Nations 2017).

2050, with an increase of nearly 90 per cent in Africa and Asia (United Nations 2014). Under SSP3, assumptions on slow economic development coincide with a slower urbanization rate. Still, as total population growth is much higher, the urban population is projected to increase significantly.

Economic development

In the past 30 years, the world economy grew on average by 3.5 per cent per year, a rate buoyed by strong economic growth in emerging economies in Asia (see Section 2.4). Most economic institutes only publish projections for the coming decade or shorter. In fact, most long-term economic projections are developed as part of environmental

Figure 21.3: Future projections of total GDP per region under SSP2 (left) and global GDP under SSP2 and SSP3 (right)

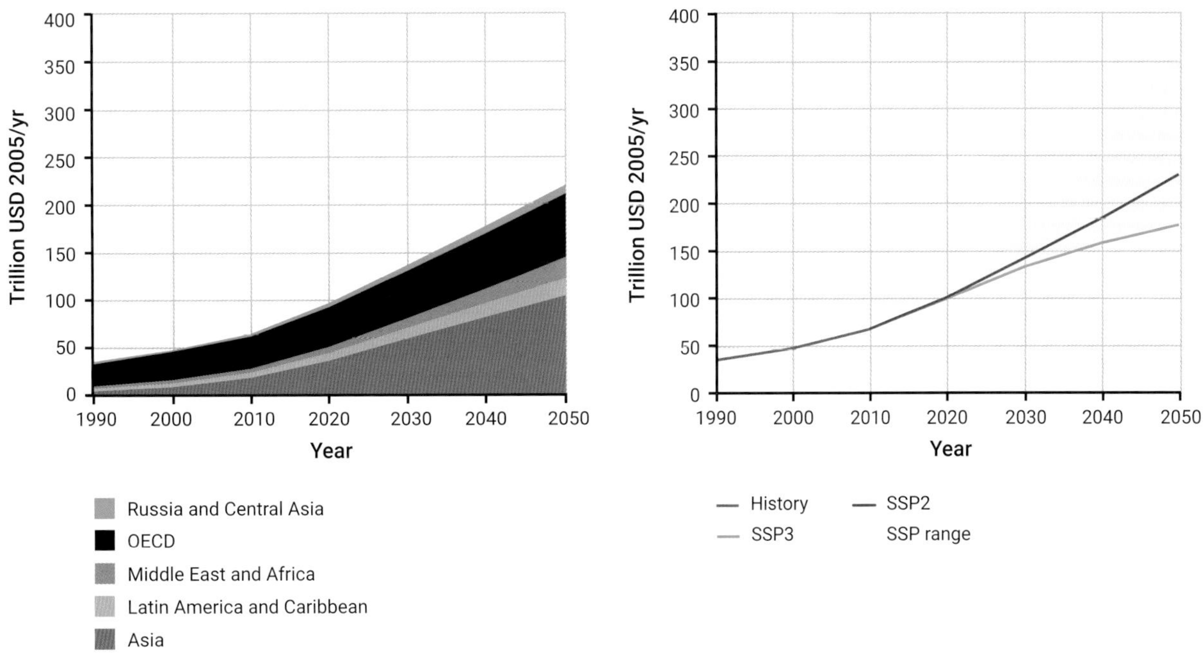

Source: Dellink *et al.* (2017).

assessments. The SSPs project the historic growth to continue, with around 3.1 per cent per year under SSP2 and 2.5 per cent under SSP3 (Dellink *et al.* 2017). For OECD countries, the projected economic growth is somewhat lower than the historical rate (1.7 per cent under SSP2), with annual growth declining over time due to ageing of the population. By contrast, low-income countries are projected to grow by around 3-5 per cent per year. Due to currently scarce capital inflows and high returns on capital investments, the potential for growth in labour and capital productivity is strong. For Asian countries, the projected growth rates are slightly lower than the rapid historical rates, further diminishing over time as a result of the maturing of their economies, with productivity levels approaching those of OECD countries. By contrast, growth rates are higher in Latin America and the Caribbean as well as Middle East and North Africa. While this implies that in relative terms there is some convergence between different parts of the world, the gaps remain significant. In terms of relative shares, these projections imply a strong shift. For instance, the OECD countries' current share of the global economy under SSP2 is projected to fall to less than one-third (compared with around one-half at present), while Asia's share is projected to grow to almost 50 per cent.

The economic projections also have implications for poverty. Historically, absolute poverty, as indicated by the number of people living on less than US$1.90 a day, fell from 1.85 billion in 1990 to fewer than 800 million in 2013 (World Bank 2016a). Projections for 2030 range from 100 million to more than 1 billion, with most studies suggesting a level of 400 to 600 million people (Chandy, Ledlie and Penciakova 2013; Burt, Hughes and Milante 2014). The variation in the scenarios is due to the large differences in assumptions about growth in household consumption and changes in income distribution.

21.3.2 Agriculture, food, land and biodiversity

Food production and land use are directly related to a range of environmental problems (see Chapter 8). As population and incomes are projected to rise, food consumption, both per capita and total, is also expected to rise (Bijl *et al.* 2017; Popp *et al.* 2017). In 2050, estimated total crop production (food, feed and biofuels) ranges from 5,800 million tons per year to 8,300 million tons per year, an increase of 50 to 130 per cent from 2010 levels (Tilman *et al.* 2011; Popp *et al.* 2017), though most projections suggest an increase of 50 to 60 per cent.

Over the last decades, globally, about 80 per cent of the increase in food demand was met by agricultural intensification (increase in crop yields and move towards more intensified animal husbandry systems) and 20 per cent by an increase in agricultural area (Smith *et al.* 2010). These shares are projected to more or less continue into the future. The net result is an increase in total agricultural area (cropland and pasture) in 2050 of 3-9 per cent (Popp *et al.* 2017). Given that, in a number of regions the most productive lands have already been put to use, land expansion takes place on less productive land, requiring more area per ton of production (van der Esch *et al.* 2017). Overall, expansion comes at the expense of forests and savannahs, which are home to important biodiversity hotspots, carbon sinks and other ecosystem services (CBD 2014).

Overall, cereal yields are projected to increase by 0.4 to 0.9 per cent per year between 2010 and 2050 (Alexandratos and Bruinsma 2012; Popp *et al.* 2017), down from 1.9 per cent per year between 1961 and 2007 (Alexandratos and Bruinsma 2012). These future yield increases will be achieved through a combination of changes in fertilizer application, irrigation use and other means (e.g. mechanization, breeding), potentially increasing environmental pressures.

Box 21.4: Climate change impacts on agriculture

In the business-as-usual scenarios, global mean temperature is expected to increase to more than 2°C in 2050 and 2.5-6°C in 2100 (IPCC 2014b). For agriculture, such climate change is expected to pose significant risks by changing seasonal rainfall patterns, increasing peak temperatures, increasing the frequency and severity of droughts, increasing the risk of catastrophic events (storms) and disrupting ecosystem services to agriculture. This could clearly have a negative impact on the ability of the agricultural system to achieve the SDGs with respect to hunger, sustainable agriculture and protection of biodiversity. Projected impacts vary across crops and regions, and for different adaptation scenarios. In general, tropical regions are expected to experience more severe negative impacts than temperate regions – and they do so at lower levels of warming (in the historical period, stronger impacts in temperate regions have been reported according to IPCC). If adaptation to climate change is implemented, yields may increase, particularly in temperate regions, via the combined effect of climate change and CO_2 fertilization. According to the IPCC assessment, after 2050 there is a greater risk of more severe impacts. The combination of the growing demand for food and a temperature rise in the high end of the projections, implies substantial risks to food security at global and regional levels.

Figure 21.4: Future projections of global average crop yield (top left), crop production (top right), agricultural area (bottom left), and forest and other natural land area (bottom right)

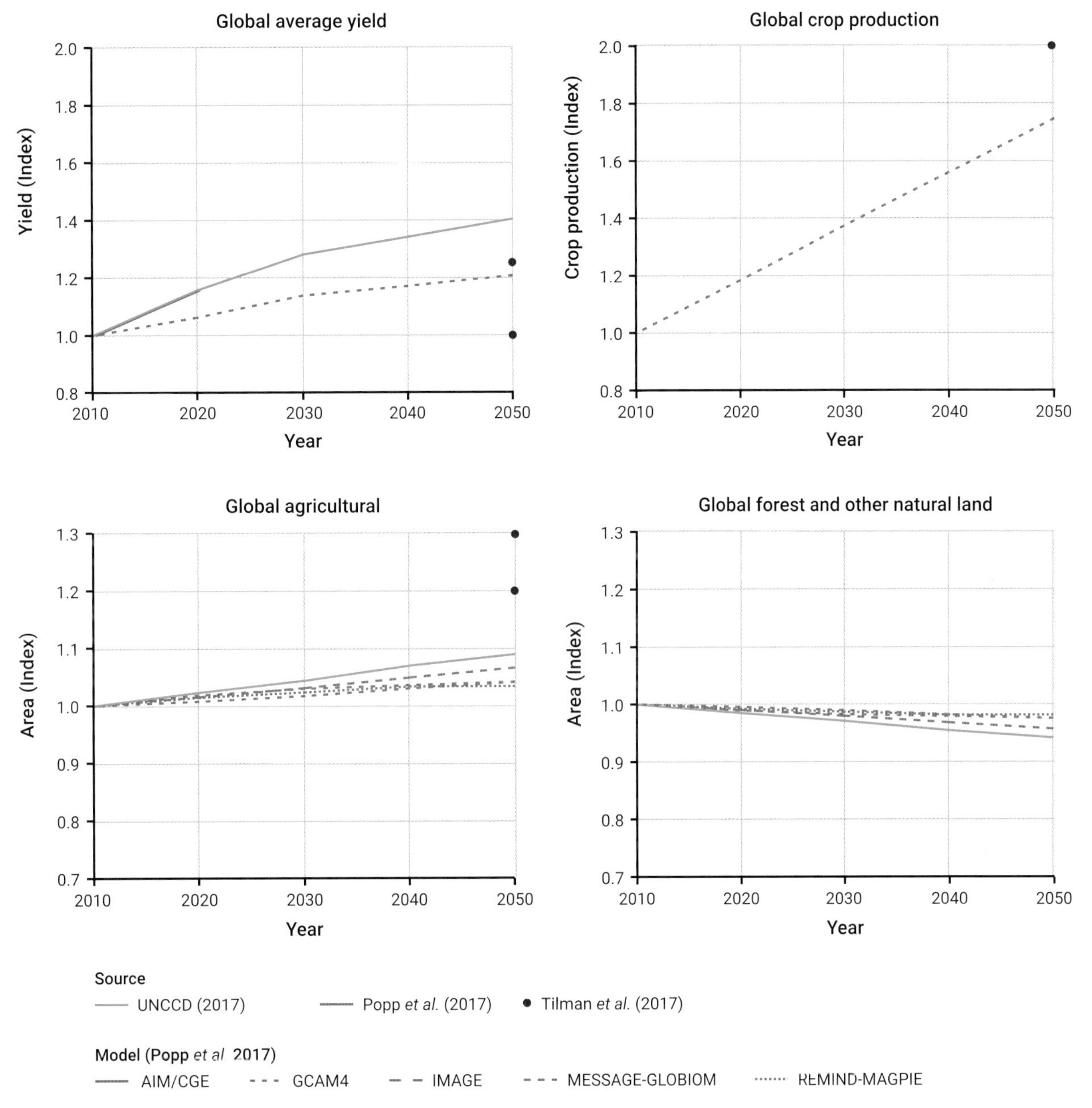

Source: Tilman *et al.* 2011; Popp *et al.* 2017; UNCCD 2017. For Popp *et al.* (2017), only results from the SSP2 scenario are shown. For Tilman *et al.* (2011), only the 'current tech' scenario with fertilizer use greater than or equal to today's values are shown.

Trends in hunger

A key challenge in this cluster is ending hunger by 2030 (SDG target 2.1). Between 2005 and 2014, global hunger has decreased, both in absolute and relative terms. However, since 2014 hunger has been on the rise, with an estimated 815 million people being undernourished in 2016 (FAO *et al.* 2017). Global models project a decrease in the undernourished population towards 2030, driven mostly by an expected income increase in current low-income regions (Alexandratos and Bruinsma 2012; Hasegawa *et al.* 2015; Laborde *et al.* 2016; Bijl *et al.* 2017; FAO *et al.* 2017). These projections are generally based on data from before 2014, and therefore do not include the recent rise in levels of hunger. Differences in historic hunger levels generally relate to the year that the model switches from historic data to model projections (here mostly 2005 or 2010). The number of undernourished people across the different studies is projected to be 300- 650 million people in 2030 and around 100 to 300 million in 2050 **(Figure 21.5)**. While this represents an improvement compared with today's figures, these levels significantly exceed the target of ending hunger by 2030.

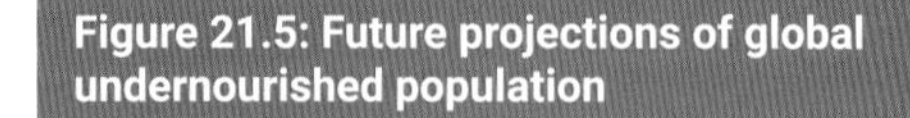

Figure 21.5: Future projections of global undernourished population

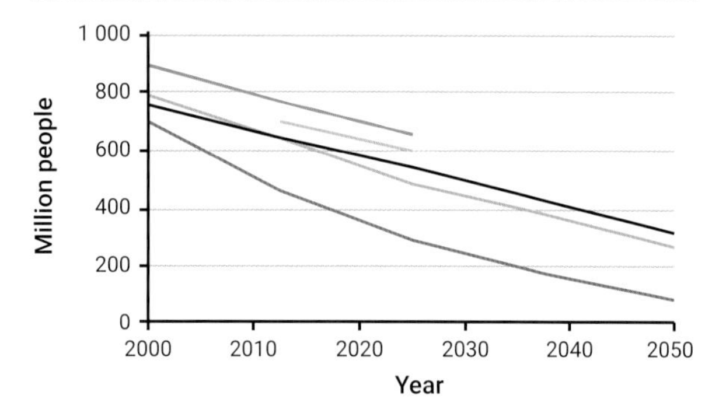

- SSP2 from Hasegawa *et al.* (2015)
- SSP2 from Bijl *et al.* (2017)
- Laborde *et al.* (2016)
- Alexandratos *et al.* (2012)
- FAO/IFAD/WFP (2015)

Source: Alexandratos and Bruinsma 2012; FAO, International Fund for Agricultural Development and World Food Programme 2015; Hasegawa *et. al* 2015; Laborde *et al.* 2016; Bijl *et al.* 2017.

Trends in biodiversity loss

As discussed in Chapter 6, biodiversity loss has a number of causes (SDG target 15.5). Most scenario studies have ascertained that natural habitat loss has been, and still is, the single most important factor causing biodiversity loss (Millennium Ecosystem Assessment 2005; Newbold *et al.* 2015). However, trends in water scarcity, climate change, pollution and disturbance all drive a further decline in biodiversity, while business-as-usual scenarios show that many of these factors are likely to worsen in the future (SCBD 2014; Kok *et al.* 2018). Climate change is projected to become a major cause of biodiversity loss, with species impacted by a range of factors including: temperature increase, altered precipitation patterns and rising sea levels. Climate change is also projected to alter the distribution of biomes over the coming century, with one-tenth to one-half of global land being highly vulnerable to biome shifts under climate change (Gonzalez *et al.* 2010). Boreal forests may be particularly vulnerable (Gonzalez *et al.* 2010; Gauthier *et al.* 2015), but projections also show that 5-6 per cent of land area in Latin America could undergo biome shifts as a result of climate change by the end of the century (Boit *et al.* 2016). The combination of anthropogenic drivers could push some regional social-ecological systems beyond tipping points and transition them to states with severely reduced biodiversity and ecosystem services (Leadley *et al.* 2014).

Studies published after Global Biodiversity Outlook-4 provide further consensus that biodiversity will continue to decline under business-as-usual scenarios. Model projections suggest changes in a number of different dimensions of biodiversity. The Global Land Outlook provides projections of Mean Species Abundance (MSA), a measure of the intactness of ecosystems, projecting a further increase of MSA loss: from 34 per cent in 2010 to 43 and 46 per cent in 2050, under SSP2 and SSP3, respectively (van der Esch *et al.* 2017; **Figure 21.6**). Using scenarios of land-use change consistent with the IPCC's Representative Concentration Pathway (RCP) 8.5, Newbold *et al.* (2015) projected a fall of local species richness by 3.4 per cent by 2100 **(see Figure 21.6)**. The combination of climate and land use change for the same RCP scenario is projected to lead to a cumulative loss of 38 per cent of species from vertebrate communities (Newbold 2018). Many of the effects are concentrated in biodiverse but economically disadvantaged countries (Newbold *et al.* 2015) and in tropical grasslands and savannahs (Newbold 2018). Furthermore, an extrapolation of the Living Planet Index (LPI), a measure of changes in

Figure 21.6: Future projections of relative local species richness for a range of climate stabilisation scenarios and Mean Species Abundance (MSA) for SSP2 and SSP3 land-use

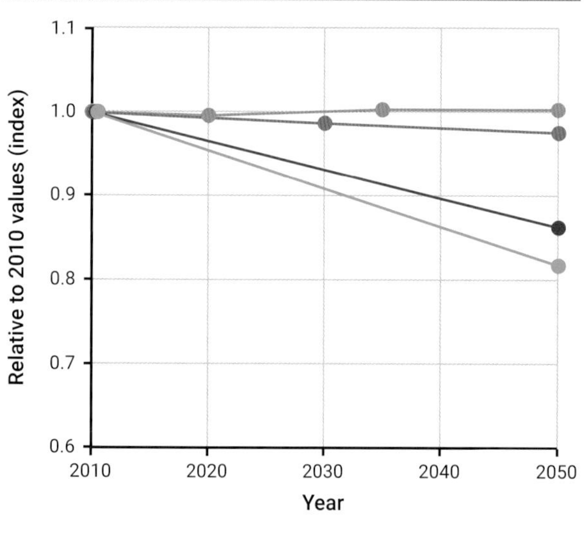

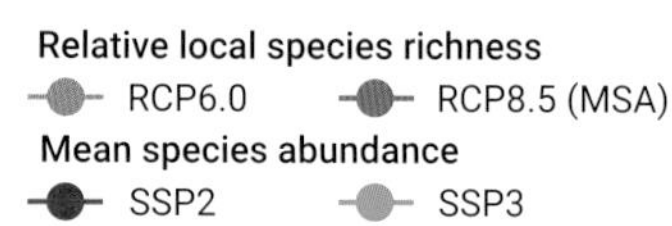

Source: Relative species richness from Newbold *et al.* 2015; Mean species abundance from van der Esch *et al.* 2017

terrestrial, marine and freshwater vertebrate populations, suggests that by 2020 populations will have declined by 67 per cent, on average, compared to their 1970 population size (World Wide Fund for Nature [WWF] 2016). Population sizes of mammalian carnivores and ungulates are also projected to decline under future land use and climate change (Visconti *et al.* 2016). Declines of 18-35 per cent by 2050 were projected in a LPI-like index, base on species dispersal ability. These abundance declines are associated with an 8-23 per cent increase in the risk of extinction (Visconti *et al.* 2016).

Trends in land degradation

Land degradation (SDG target 15.3) is a major problem worldwide, and is linked to food insecurity, vulnerability to climate change and poverty, as well as to mitigation of GHG emissions (UNCCD 2017; van der Esch *et al.* 2017). Estimates of the number of people affected by land degradation vary between a low range of 1.3-1.5 billion (Bai *et al.* 2008; Barbier and Hochard 2016; see Chapter 8) and much higher estimates of 3.2 billion (Intergovernmental Science-Policy Platform on Biodiversity and Ecosystem Services [IPBES] 2018). Although a significant proportion of those affected are poor rural inhabitants living on marginal lands, land degradation also occurs in prime agricultural lands due to mismanagement and/or overgrazing. For example, in Brazil, more than half of all pastures are in an advanced state of degradation, causing significant loss of productivity (Strassburg *et al.* 2014; Assad *et al.* 2015). In fact, expansion and unsustainable practices in agriculture and livestock production are the most important direct drivers of land degradation (IPBES 2018).

Globally, between 1982 and 2010, when correcting for climate effects, a declining trend in net primary productivity (NPP) is observed in about 12 per cent of agricultural land and in about 5 per cent of natural land (Schut *et al.* 2015; van der Esch *et al.* 2017). Regionally, the Russian Federation/Central Asia and sub-Saharan Africa have about double the share of agricultural land with a declining NPP (Schut *et al.* 2015; van der Esch *et al.* 2017). Compared to an undisturbed state, current NPP is estimated to be significantly lower on 28 million km^2, or 23 per cent, of the global terrestrial area, corresponding to about a 5 per cent loss in global NPP (Smith *et al.* 2016; van der Esch *et al.* 2017). Soil degradation involves, among other things, soil erosion and loss of soil organic carbon. Historically, around 176Gt of soil organic carbon (8 per cent) has already been lost as a result of land-use changes, including the conversion of natural land to agriculture, and overgrazing in grasslands (Stoorvogel *et al.* 2017a; Stoorvogel *et al.* 2017b). As a consequence of continued land conversion and unsustainable land management, an additional 27Gt of soil organic carbon is projected to be lost between 2010 and 2050, affecting agricultural yields through reduced water-holding capacity and loss of nutrients (van der Esch *et al.* 2017). Furthermore, losses in soil organic carbon will have wider effects on biodiversity, hydrology and carbon emissions. Further land degradation is expected to occur based on trends in land use **(see Figure 21.4)**, climate change and increasing pressure on land and water resources. Land degradation is especially of concern in drylands, where, by 2050, human populations are projected to increase by 40 to 50 per cent under the SSP2 scenario, which is far greater than the 25 per cent increase projected for non-drylands (van der Esch *et al.* 2017). Overall, these trends show that without targeted policies, a land degradation neutral world will not be achieved.

21.3.3 Energy, air and climate

The energy system plays a crucial role in achieving sustainable development. The use of energy is a prerequisite for human welfare. At the same time, current energy consumption and production patterns contribute strongly to climate change and air pollution. In the next few decades, energy demand

Figure 21.7: Future projections of global primary energy consumption (left panel) and per energy carrier in the SSP2 marker scenario (right panel)

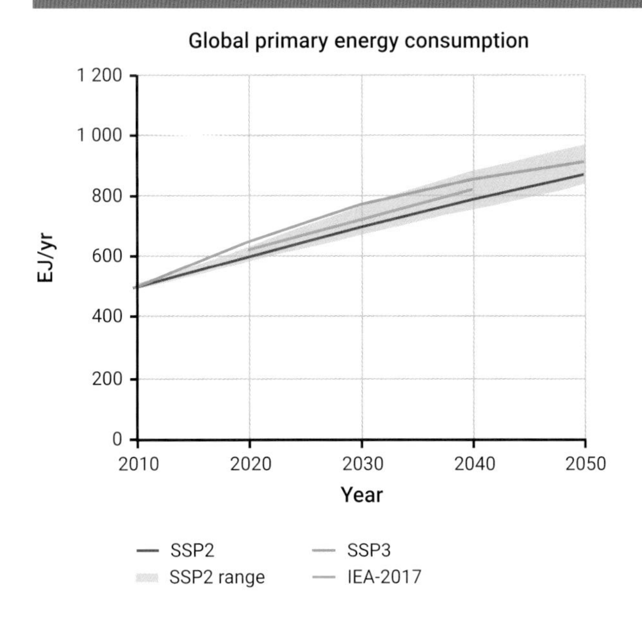

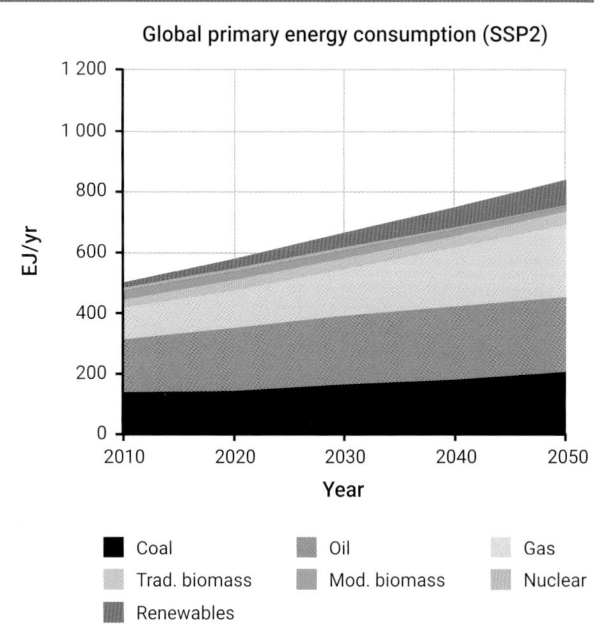

Source: SSP Public Database 2016; IEA 2017a ; Riahi *et al.* 2017.

is expected to increase further, driven by population growth and related human activities. Overall, most scenarios project a 50-70 per cent increase in primary energy demand over the 2015-2050 period. This is despite a projected decrease in energy intensity of around 1-2.5 per cent per year, similar to that achieved historically (e.g. van Vuuren *et al.* 2016; Riahi *et al.* 2017). Although renewables are by far the fastest growing form of energy, fossil fuels continue to contribute the lion's share of total energy supply in scenarios without new policies. In most business-as-usual scenarios, the renewable energy share increases from 15 per cent in 2015 to around 20-30 per cent in 2050 (full range 10-30 per cent) (van Vuuren *et al.* 2016). Renewables started from a small share in 2015 and are mostly successful in replacing fossil fuels in the power sector, which constitutes an important but limited part of total energy consumption. Scenario studies specifically examining the impact of policies formulated by countries as part of their pledges under the Paris Agreement often have a somewhat lower energy demand and faster growth of non-fossil energy, but at best these result in a stabilization of fossil fuel demand.

Trends in access to modern energy services

Access to modern energy services (SDG target 7.1) is an important prerequisite for human development (IEA 2017b). Currently, around 2.8 billion people worldwide rely on traditional biomass, kerosene and coal – fuels not considered to be clean – while 1.1 billion people do not have access to electricity (IEA 2017b). Exposure to household air pollution, caused by the use of traditional biomass in open fires or traditional stoves, can lead to child and adult mortality and morbidity. Overall, in 2015, household air pollution was responsible for almost 3 million deaths worldwide, including 250,000 child deaths (GBD 2016 SDG Collaborators 2017). Since 2000, progress has been made in all regions, especially in Asia, while in sub-Saharan Africa population growth largely outpaced the progress made. These trends are projected to continue towards 2030 (Lucas *et al.* 2015; Dagnachew *et al.* 2017; IEA 2017b; Lucas, Dagnachew and Hof 2017). Overall, the population without access to clean cooking fuels is projected to decrease to around 2.3 billion people by 2030, with larger improvements in urban areas than in rural areas (IEA 2017b). Under SSP2 assumptions, by 2030 around 140,000 children under five per year are projected to die as a result of household air pollution (Lucas *et al.* 2018). Model projections also show a decline in the population without access to electricity to around 700 million in 2030, with most regions, except sub-Saharan Africa, reaching near universal access (IEA 2017b). Although between 2010 and 2030, more than 550 million additional people are projected to gain access to electricity in sub-Saharan Africa, 500 million people would still not have access in 2030, many of them living in rural areas (Dagnachew et al. 2017).

Trends in climate change

Increasing fossil fuel use implies increasing greenhouse gas emissions and related global mean temperature (SDG 13). For the 2010-2050 period, GHG emissions are projected to increase by 30-70 per cent (*Riahi et al.* 2017; IEA 2017a; **Figure 21.8**). Most of this increase is projected for low-income countries. Nevertheless, per capita emissions remain highest in OECD countries. Emissions of greenhouse gases are not only expected to increase due to these energy-related trends; other activities are also expected to contribute. These include CO_2 emissions from land-use change (LUC) (slowly decreasing over time) and non-CO_2 emissions related to energy and agriculture. The sum of non-CO_2 emissions is expected to increase further over time, driven mostly by trends in agriculture.

Figure 21.8: Projected increase in global CO_2 emissions (left) and total GHG emissions (right)

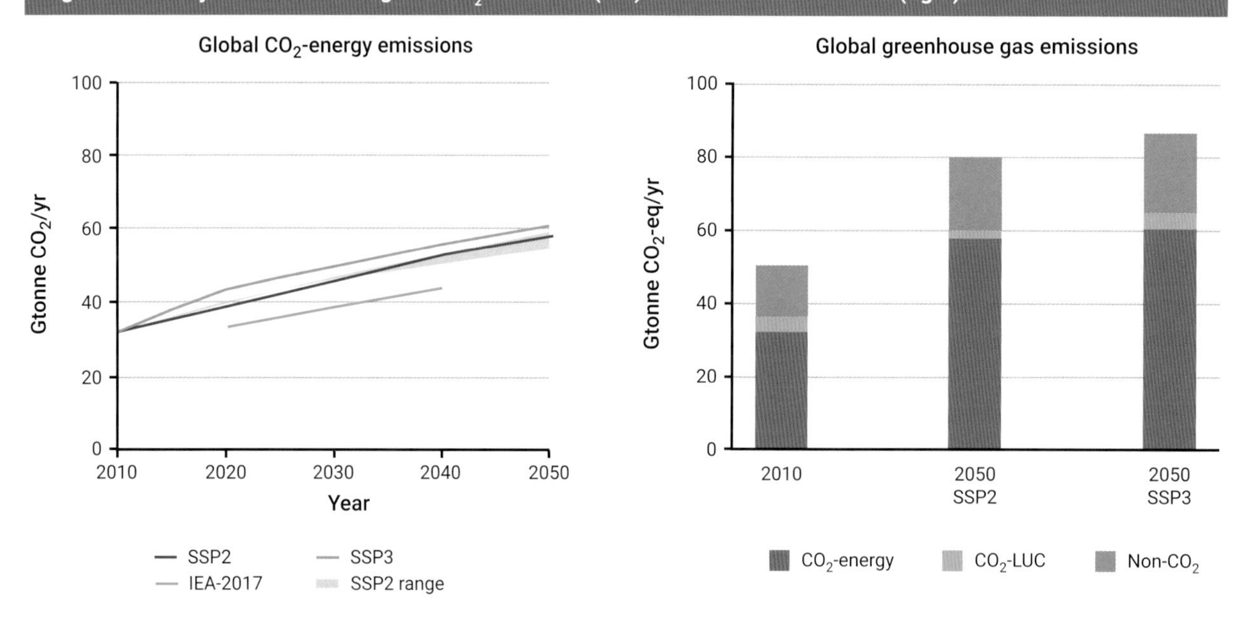

Source: Results are shown for SSP2 (average across the different models elaborating this scenario and the lowest and highest in the range, Riahi *et al.* 2017) and SSP3 (model average; Riahi *et al.* 2017), as well as the IEA scenario (IEA 2017a, reference case).

As a result of this global emission increase, global temperature is expected to rise from around 1°C above pre-industrial levels in 2016 (Visser *et al.* 2018) to around 4°C by 2100 in SSP2, most likely passing the 2°C target of the Paris Agreement before 2050 (IPCC 2014b; see **Figure 21.9**). There are substantial differences in temperature increase across different areas of the world. Typically, temperature increase is greater at higher latitudes, such as the temperate zone and in polar regions. In addition to changes in temperature, considerable changes in precipitation are projected to occur , with some regions becoming drier and others becoming wetter. However, such detailed patterns in climate change variables are still very uncertain. In many places, the projected warming would exceed the global mean temperature increase (which is defined as the increase in temperature above land and oceans). Using projected changes in temperature, the IPCC has assessed the impacts associated with climate change (IPCC 2014b). For a warming as high as in the business-as-usual scenarios projected here the impacts are assessed to be severe, for all categories. This includes sea level rise, negative impacts on agriculture globally (**see Box 21.4**), negative impacts on biodiversity, and the risk of irreversible changes in the complete climate system.

Figure 21.9: Global mean temperature increase

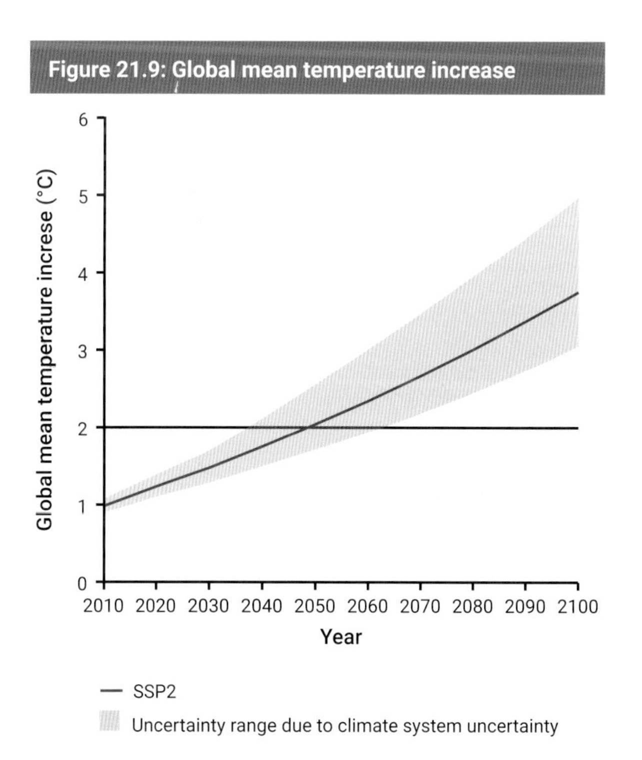

Source: SSP2 database, range taken from baseline scenarios in IPCC (2014b).

Trends in air pollution

From the public health perspective, airborne particles and ground level ozone are the most important air pollutants (SDG target 11.6), with exposure to $PM_{2.5}$ contributing the most to premature deaths (see Chapter 5). The Global Burden of Disease study estimated that exposure to ambient $PM_{2.5}$ was the fifth-ranking mortality risk factor in 2015, contributing to approximately 4 million deaths and 103 million years of healthy life lost (Cohen *et al.* 2017).

Several projections of future air pollution have been made over the past few years (Stohl *et al.* 2015; IEA 2016; OECD 2016; Klimont *et al.* 2017; Rao *et al.* 2017; UNEP 2017). Many projections have been built using emission factors from successive versions of the Greenhouse Gas and Air Pollution Interactions and Synergies (GAINS) model (Amann *et al.* 2011; Klimont *et al.* 2017). In addition to the evolution of these underlying emission factors, future projections differ in their assumptions about energy supply and demand, and the extent to which air pollution control policies will be implemented in the future.

The most pessimistic scenarios examine a situation in which no additional air pollution control policies are implemented. An example is the OECD (2016) study, that projects a significant increase in air pollution emissions by 2060 due to increasing economic activity and energy demand. More realistic scenarios look into expected improvement in air pollution legislation. For instance, the IEA-New Policies Scenario (IEA 2016) projects a more optimistic future in which recently announced policies, including Nationally Determined Contributions (NDCs), under the Paris Agreement, are implemented, leading to the use of emission control technologies and facilitating a shift to cleaner energy sources. These new policies will lead to a slow decline in global air pollutant emissions by 2040, although rapidly developing regions are likely to continue to experience increases in air pollution emissions (IEA 2016; UNEP 2017). In high-income countries, emission decreases are expected as a result of increasingly stringent air pollutant emission standards on power plants and vehicles, shifts to low-carbon sources for electricity generation, and increased energy efficiency. In developing regions, the growth in economic activity and energy demand is still expected to outpace efforts on pollution control until income level reaches a certain point. However, $PM_{2.5}$ emissions (primarily organic carbon (OC) and black carbon (BC)) and exposures may decline due to increased access to clean energy sources for cooking, heating and lighting (IEA 2016). The ECLIPSE project (Stohl *et al.* 2015; Klimont *et al.* 2017), which has been used more broadly by the atmospheric research community, e.g. the 2017 Emissions Gap Report (UNEP 2017), includes a current legislation (CLE) scenario that falls between the OECD and IEA pathways. Air pollution emissions scenarios were developed for each of the SSPs based on sets of assumptions about the stringency and implementation of future air pollution emissions controls consistent with the overall scenario storylines (Rao *et al.* 2017). These scenarios also used the emission factors of the GAINS model. To 2040, the range of the SSP air pollutant emission scenarios captures the more pessimistic scenarios of continued growth without additional policies (OECD 2016) and the more optimistic scenarios of new policies (IEA 2016).

Figure 21.10: Future projections of emissions for air pollutants SO_2, NO_x and BC

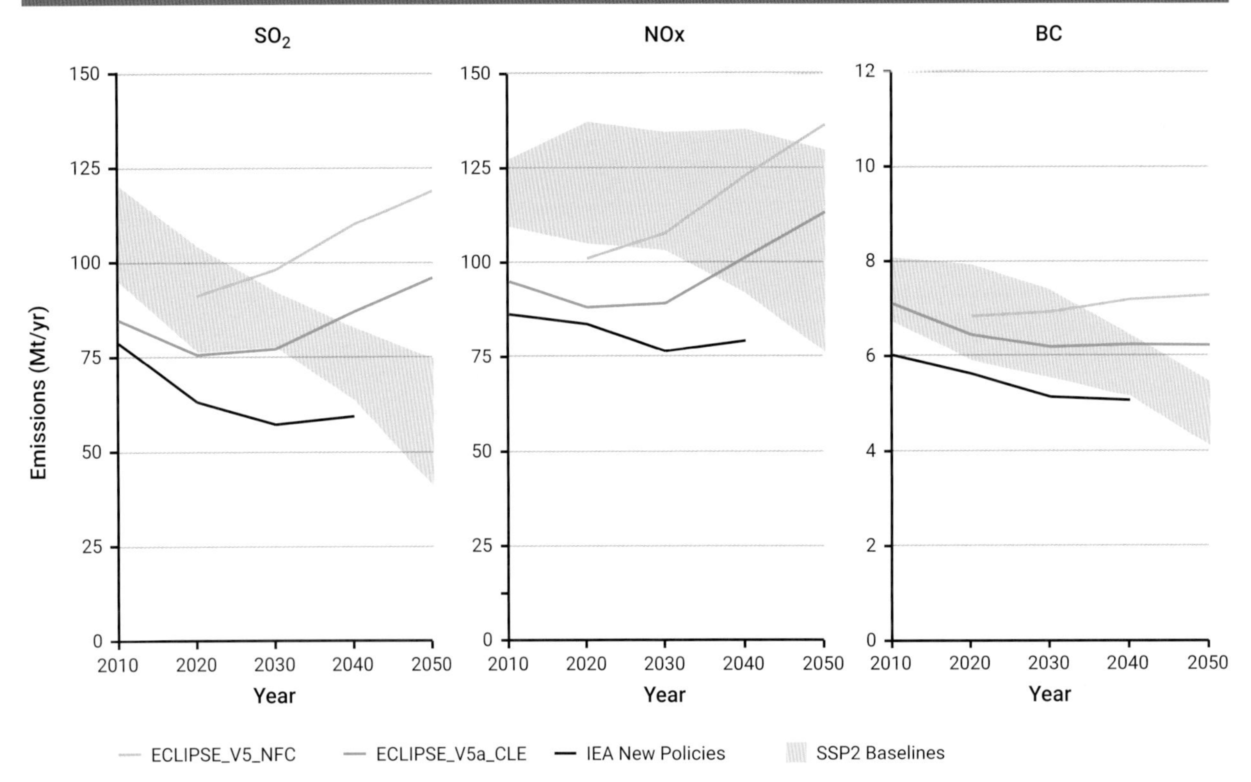

Source: ECLIPSE_V5_NFC and ECLIPSE_V5a_CLE represent ECLIPSE's 'no further control' and 'current legislation' scenarios (Stohl *et al.* 2015; Klimont *et al.* 2017). IEA New Policies represent IEA's new policy scenario that includes NDCs of the Paris Agreement (IEA 2016). For the SSP2 scenario, the shading represents the range for all models (Rao *et al.* 2017).

Overall, scenarios without new policies show a small decrease or increase in air pollutant emissions. Based on data from the Global Burden of Disease (GBD 2016 Risk Factors Collaborators 2017), in 2016 approximately 95 per cent and 58 per cent of the world's population lived in areas where annual mean $PM_{2.5}$ concentrations exceeded the 10μg/m³ guideline and the least stringent interim target of 35μg/m³, respectively, contributing to approximately 4 million premature deaths (Health Effects Institute 2018). Using emissions data as input, the TM5-FASST source-receptor model (van Dingenen *et al.* 2018) can be used to estimate annual $PM_{2.5}$ concentrations, which can be mapped onto future population projections to estimate those exposed to specific $PM_{2.5}$ levels and the number of premature deaths. By 2050, the projected values for populations exposed to $PM_{2.5}$ concentrations above 10μg/m³ and 35μg/m³ are 63 and 9 per cent, respectively, for the SSP2 marker scenario (Rao *et al.* 2017), and 81 and 40 per cent, respectively, for the ECLIPSE V5a 'current legislation' scenario (Stohl *et al.* 2015; Klimont *et al.* 2017). This implies that air pollution will continue to contribute to millions of premature deaths annually, with the IEA New Policies scenario estimate of 4.5 million premature deaths for 2040 (IEA 2016), and the ECLIPSE V5a scenario estimate of 7 million premature deaths for 2050 (Stohl *et al.* 2015; Klimont *et al.* 2017) spanning the range in the literature. Global economic losses due to decreased labour productivity, increased health care costs and decreased crop yields could amount to 1 per cent of global gross domestic product (GDP) by 2060 (OECD 2016). There are large differences across regions, with some experiencing as much as 3 per cent loss in GDP (OECD 2016).

21.3.4 Freshwater

Freshwater - related environmental problems are closely linked to developments in the agriculture, food, land and biodiversity cluster (see Section 21.3.2) and the energy, air and climate cluster (see Section 21.3.3), both as a natural resource and as a sink for pollution. Freshwater is essential for human health (drinking water and sanitation), as well as for agriculture and energy production, while an imbalance of freshwater supply and demand can cause severe water scarcity. Furthermore, excess nutrient losses (nitrogen and phosphorus) to aquatic ecosystems through run-off and erosion can cause the eutrophication of lakes and rivers.

Trends in drinking water and sanitation

In 2015, nearly 2.1 billion people lacked access to safely managed drinking water services (SDG target 6.1), while 0.8 billion of these even lacked an improved source (World Health Organization [WHO] and United Nations Children's Fund [UNICEF] 2017). Furthermore, 4.5 billion people lacked access to safely managed sanitation services (SDG target 6.2), while 2.3 billion of these even lacked an improved source (WHO and UNICEF 2017). Overall, unsafe drinking water, sanitation and hand washing (WASH) were responsible for around 1.5 million deaths, including 410,000 child deaths, mainly due to diarrheal diseases (GBD 2016 SDG Collaborators 2017). Lucas *et al.* (2018) project that, by 2030, more than 30 million children will still live without access to improved drinking water services, and about 150 million will lack access to improved sanitation, under SSP2 assumptions. This translates into 400 million people that live without access to improved drinking water services and about 2 billion without access to improved sanitation. Especially the sanitation challenge is a pressing one. Between 2015 and 2030, around 5.6 billion people will require safely managed sanitation and around 1.3 billion people will need to switch from open to fixed point defecation (Mara and Evans 2018). Improved access to safely managed WASH would lead to a significant reduction in the number of children suffering from related ill health. However, under SSP2 assumptions, by 2030 around 220,000 children under five are projected to die as the result of inadequate drinking water and sanitation facilities (Lucas *et al.* 2018).

Trends in water quality

Freshwater pollution includes different types of chemicals, but also excessive nutrient loading (nitrogen and phosphorus) of aquatic ecosystems through run-off and erosion and declining silica concentrations. Since scenario studies related to chemicals are largely missing from the scientific literature **(see Box 21.3)**, the analysis of trends for this target focuses on nutrient pollution.

Nitrogen (N) and phosphorus (P) fertilizers have played a major role in food production, but they have also found their way into nearly every water body across the globe, causing eutrophication of rivers, lakes and reservoirs (SDG target 6.3). The most important anthropogenic source of nitrogen in freshwater ecosystems is agriculture-related N-fixation (nitrogen fertilizer use and biological crop fixation). For phosphorus, the main anthropogenic sources are phosphorus fertilizer use and wastewater. Current agricultural N-fixation is estimated at 116-127TgN/yr (Bouwman *et al.* 2017). Alexandratos and Bruinsma (2012) project an increase of synthetic N use to 138TgN/yr in 2050, while Mogollón *et al.* (2018a) project an increase to 185 and 260TgN/yr in 2050 under an SSP2 and SSP3 scenario, respectively, illustrating the uncertainty in future projections. Global phosphorus inputs to cropland are projected to increase from 14.5TgP/yr in 2010 (Bouwman *et al.* 2017) to 26 and 27TgP/yr in 2050 under an SSP2 and SSP3 scenario, respectively (Mogollón *et al.* 2018b).

Future projections of N fertilizer use depend strongly on developments in N use efficiency (NUE), which declined from 0.42 in 1970 to 0.35 in the 1980s and then increased again to 0.42 in 2010 (Bouwman *et al.* 2017). The decreasing trend in the 1980s was due to increasing fertilizer use in low-input countries, which initially led to an apparent decline in efficiency, while the later increase was largely the result of improvement in agricultural practices and environmental legislation in developed regions (Bouwman *et al.* 2017; Rao *et al.* 2017). These trends are projected to continue in the future, with projected NUE for SSP2 increasing to 0.55 in 2050 (Mogollón *et al.* 2018a). The P use efficiency (PUE) declined from 0.51 in 1970 to somewhat lower values in the 1980s, and then to 0.6 in 2010 (Mogollón *et al.* 2018b). Future PUE values depend strongly on phosphorus accumulation (low PUE) in residual soil pools or their depletion (high PUE), which can be regarded as a contribution to future production.

It is clear that current and projected nitrogen and phosphorus use in agriculture significantly exceeds the target levels of 62TgN/yr and 6.2TgP/yr. At the same time, construction of dams and the development of reservoirs for water storage and hydropower generation leads to trapping of silica (Si) (e.g. Mavaara, Dürr and van Cappellen2014; Ran *et al.* 2018). The distortion of the nutrient stoichiometry (increasing N:P, increasing N:Si) may lead to the proliferation of harmful algae. The global problem of harmful algae is now on a pathway of more and more frequent blooms, in more places with an increasing extent, and with more toxins (Glibert 2017).

Wastewater is another important source of nutrients in freshwater systems. Improved sanitation is focused on health aspects, and sanitation systems are designed to hygienically separate excreta from human contact. However, without wastewater treatment, sewage systems create direct emissions of nutrients and organic waste to surface water (van Puijenbroek *et al.* 2015). Although access to sanitation is projected to increase, expansion of wastewater treatment will be outpaced by population growth and urbanization trends in developing countries (van Puijenbroek, Beusen and Bouwman 2019). As a result, global nutrient emissions from untreated sewage are projected to increase from 10TgN/yr in 2010 to 17TgN/yr in 2050, and from 1.5TgP to 2.4TgP under SSP2 assumptions (van Puijenbroek, Beusen and Bouwman 2019).

Trends in water scarcity

At present, more than 2 billion people across the globe live in river basins with excess water stress (SDG target 6.4), i.e. the proportion of total freshwater withdrawal to total renewable freshwater above a threshold of 40 per cent (Oki and Kanae 2006; Veldkamp *et al.* 2015; Liu *et al.* 2017). In some countries in Africa and Asia, the proportion extends beyond 70 per cent (Economic and Social Council 2017).

Global human water demand, i.e. water withdrawal, is projected to increase under all trend scenarios. Some scenarios show quite large increases, i.e. from around 4,000km^3 yr^{-1} now to 5,500km^3 yr^{-1} by 2050 (38 per cent) under SSP2 (Wada *et al.* 2016; Satoh *et al.* 2017). Others show a smaller increase, based on expected efficiency improvements. For instance, Bijl *et al.* (2018) project a 26 per cent increase in total water demand by 2050 under an SSP2 scenario. For the high demand scenario, consumptive water use is projected to increase from 2,000km^3 yr^{-1} now to 2,500km^3 yr^{-1} by 2050 (25 per cent) under SSP2 (Wada and Bierkens 2014). An additional 10 per cent increase of water use is expected under SSP3 (Wada *et al.* 2016).

Improvement in water-use efficiency is expected to vary for different sectors (agriculture, industry and households), and ranges between 0.3 and 1.0 per cent per year under the SSP scenario, which mostly follows historical development (Flörke *et al.* 2013; Wada *et al.* 2016). Furthermore, the efficiency improvement is expected to vary substantially across different regions, depending on available infrastructure and economic investments. The greatest increases in total water demand are expected in Africa, many parts of Asia, the western United States of America, Mexico, and Latin America (Hanasaki *et al.* 2013a; Hanasaki *et al.* 2013b; Wada *et al.* 2016) and will largely be the result of rapidly growing population and increasing industrial activities (higher electricity and energy use) in currently developing countries (Hanasaki *et al.* 2013a; Hanasaki *et al.* 2013b; Bijl *et al.* 2016; Wada *et al.* 2016; Satoh *et al.* 2017).

Increases in future agricultural water demand are primarily driven by the expansion of irrigated areas and projected climate change, which enhances evaporative demand for irrigated crops (Hanasaki *et al.* 2013a; Hanasaki *et al.* 2013b; Wada and

Bierkens 2014; Mouratiadou *et al.* 2016). Compared with the domestic and industrial sector, the projected irrigation water demand shows a much lower increase of 20 to 30 per cent by the end of this century (Elliot *et al.* 2014), although some project a doubling of irrigation water withdrawals between 2010 and 2050 (Chaturvedi *et al.* 2015). Although modest changes in global average irrigation efficiency are projected (Hanasaki *et al.* 2013a; Hanasaki *et al.* 2013b), this will probably compensate for the increase in irrigated areas and irrigation intensity (Wada *et al.* 2013), with significant differences in efficiencies across regions (Chaturvedi *et al.* 2015). It should be noted that increasing atmospheric CO_2 concentrations can improve crop growth and reduce crop transpiration, while simultaneously, increased biomass use could potentially offset the gains in crop transpiration (Wada *et al.* 2013).

The trends under both medium and high water demand scenarios (SSP2 and SSP3) imply that water scarcity is expected to increase. For instance, studies project a large increase in water scarcity over 74 to 86 per cent of the total area of Asia under different SSP scenarios, and that at least 20 per cent of the area will probably be subject to severe water stress by the 2050s in Asia (Wada and Bierkens 2014; Satoh *et al.* 2017). It is important to note that a severe reduction in water resources is expected in many parts of arid and semi-arid regions due to climate change (Schewe *et al.* 2014). At present, more than 1 billion people in Asia live in regions with severe water stress, totalling almost one-third of the Asian population (Liu *et al.* 2017). By 2050, the Asian population exposed to severe water stress conditions is expected to increase by 42 to 75 per cent, depending on which scenario is considered (SSP1-3), potentially extending to 2 billion people under the SSP3 high water demand scenario (Satoh *et al.* 2017). Globally, the number of people living in a severe water stress areas show a similar trend, increasing from 2 billion people now to between 2.8 and 3.4 billion people by 2030, under the SSP2 and SSP3 scenarios, respectively (Hanasaki *et al.* 2013a; Hanasaki *et al.* 2013b). Increased water stress can damage renewable freshwater resources to the point where they are unable to sustain human activities and fulfil their ecological functions (Satoh *et al.* 2017; Greve *et al.* 2018). The consequences of water stress also affect agriculture, health and income. Studies show that water stress could lead to a 7-10 per cent reduction in GDP in Central and East Asia by 2050 (World Bank 2016b; Satoh *et al.* 2017).

At the same time, the amount of non-renewable groundwater abstraction is projected to double for almost all major groundwater users under the SSP2 and SSP3 scenarios. The share of non-renewable to total groundwater abstraction is expected to increase from 30 to 40 per cent, indicating a growing reliance of human water use on non-renewable groundwater resources (Elliot *et al.* 2014; Wada and Bierkens 2014). In some areas, the groundwater table may drop too deep, or the aquifer or river may run out of water, increasing concern for food security, energy, cities and ecosystems (Vanham *et al.* 2018).

21.3.5 Oceans

Ocean - related environmental problems are closely linked to developments in the agriculture, food, land and biodiversity cluster (see Section 21.3.2) and the energy, air and climate cluster (see Section 21.3.3), both as a natural resource and as a sink for pollution. Oceans are an important source of food and nutrition for billions of people, and although not discussed in the current section, they are also important for renewable energy, including offshore wind farms and tidal energy. With respect to pollution, excess nitrogen and phosphorus loadings associated with anthropogenic activities, including agriculture and sewage, can cause dead zones and toxic algae blooms in inland and coastal waters, while increasing CO_2 emissions, mostly generated by the energy system, exacerbate ocean acidification.

Trends in marine nutrient pollution

Major threats from nutrient enrichment and changing nutrient ratios are the development of dead zones and toxic algae blooms in inland and coastal waters. The Si:N and Si:P ratios in rivers have declined steadily during the past century (Billen, Lancelot and Meybeck 1991). This is due to elevated N and P loadings associated with anthropogenic activities (Beusen *et al.* 2016), while dissolved Si supply to rivers (primarily from rock weathering) is decreasing due to enhanced Si retention in reservoirs (Conley 2002). As a result community structures change, since siliceous algae (diatoms) require Si in balance with N and P (Si:N ≈ 1; Si:P ≈ 16). Threats from marine nutrient pollution (SDG target 14.1) occur when N and P are present in excess relative to Si, and phytoplankton communities are dominated by non-diatoms and often toxic algae and cyanobacteria proliferate (Anderson, Glibert and Burkholder 2002).

Global river N export estimates range from 37Tg N/yr (Beusen *et al.* 2016) to 43Tg N/yr (Seitzinger *et al.* 2010) for the year 2000. For global river P export estimates range from 4TgN/yr (Beusen *et al.* 2016) to 9Tg N/yr (Seitzinger *et al.* 2010). Increasing inputs from agriculture and wastewater are projected to result in an increase of the global river nitrogen export from 40Tg N/yr in 2006 to 47Tg N/yr in 2050, while P exports are projected to increase from 4Tg P/yr in 2006 (based on Beusen *et al.* 2016) to 5Tg P/yr in 2050, according to the SSP2 scenario (Ligtvoet *et al.* 2018). Although there are considerable uncertainties in historic estimates of phosphorus flow from freshwater systems into the ocean, future trends are moving in the wrong direction.

Trends in ocean acidification

Increases in CO_2 concentrations result in increased ocean acidity (SDG target 14.3) and decreased ocean productivity. Under a high emissions scenario (RCP 8.5) the global average ocean acidity level (pH) is projected to decline by approximately 0.2 in 2060 (Palter *et al.* 2018) and by 0.33 in 2090 (Bopp *et al.* 2013), compared with the 1990s. Lower pH levels (higher acidity) reduce the concentration of carbonate ions, which are required by marine organisms to create shells and skeletons. Higher acidity means that the global average calcium carbonate ($CaCO_3$) saturation state of seawater with respect to aragonite (a type of $CaCO_3$ produced by marine organisms, the saturation state of which is denoted: Ωarg) in the upper water column would decline to levels that are significantly below the selected target level of 2.75Ωarg – from 2.94Ωarg in 2010 to around 1.8Ωarg in 2100 (Zheng and Cao

2014). Declines in carbonate saturation state make it more difficult for marine organisms to form shells and skeletons, can lead to their dissolution, and may increase natural mortality or decrease somatic growth and egg viability (Cattano *et al.* 2018). Regionally, acidification is expected to increase most rapidly in polar areas, with carbonate ion concentrations projected to fall below aragonite saturation levels in the Arctic Ocean beginning in 2048, and in the Southern Ocean in around 2067 (Bopp *et al.* 2013; Ciais *et al.* 2013). It should be noted that nitrogen and phosphorus run-off into the ocean from agriculture and industrial sources can lead to locally enhanced ocean acidification (Billé *et al.* 2013).

Trends in ocean resources

Protecting ocean resources (SDG target 14.4) is critical, as oceans are sources of food and nutrition for billions of people, especially in income-poor coastal zones where significant shares of nutrition and income derive from fisheries. In addition to being a direct source of human food, fish also contribute indirectly to human nutrition when used as fishmeal in aquaculture and livestock feed. Historically, fish demand per capita has risen significantly from 6kg/yr in 1950 to 20.3kg/yr in 2016 (FAO 2018), with other estimates spanning the range 18.8-21.4kg/yr in 2011 (Troell *et al.* 2014; Béné *et al.* 2015). At the same time, there has been a trend towards farmed fish. Since 2014, humans have consumed more farmed fish than wild fish (FAO 2016). Projections from FAO suggest that demand for fish will continue to grow in the future (FAO 2018). However, studies indicate that a sustainable increase in wild fish catch will be difficult under current fishing strategies (Garcia, Rice and Charles 2016; FAO 2018). One important concern is that projections of marine primary productivity, which supports all marine fisheries, and ultimately all marine life, suggest a decline to 2100 under an RCP 8.5 scenario (Bopp *et al.* 2013; Fu, Randerson and Moore 2016), although considerable uncertainties remain (Laufkötter *et al.* 2015). No projections of the 'proportion of fish stocks within biologically sustainable levels' – the official SDG indicator – are available in the literature. As a proxy, projections of global fisheries under an unchanged climate and current management scenario, suggest that, the proportion of fish stocks at or below a target biomass that can undergo recovery would increase from 53 per cent today to 88 per cent in 2050 (Costello *et al.* 2016). However, a wide range of improved management measures already in place. In most countries that are funding science and management adequately (Melnychuk *et al.* 2017), significantly improve the prospects for sustainability (Costello *et al.* 2016). Catch potential is projected to decline by an average of 7.7 per cent by 2050, while revenue might decline by 10.4 per cent over the same period (Lam *et al.* 2016).

21.3.6 Human health

In 2012, 23 per cent of deaths globally were due to modifiable environmental factors– "those reasonably amenable to management or change given current knowledge and technology, resources, and social acceptability," (Prüss-Üstün *et al.* 2016) – with a greater portion occurring in vulnerable populations (children and the elderly) and developing countries (Prüss-Üstün *et al.* 2016). The environment affects human health within households (e.g. through unsafe water, sanitation and hygiene, and indoor air pollution), in communities (e.g. outdoor air pollution), and on a global scale (e.g. climate change) (Smith and Ezzati 2005; Hughes *et al.* 2011).

The proportion of the population with access to safe water, sanitation, and hygiene facilities, as well as clean cooking facilities has been increasing significantly reducing health impacts related to communicable diseases. These trends are projected to continue to 2050 (see Sections 21.3.3 and 21.3.4). For example, global Disability Adjusted Life Years (DALYs), the number of years lost to poor health or early death related to household air pollution due to use of solid fuels, decreased from 9.2 per cent of total DALYs in 1990 to 6.8 per cent in 2016 (Institute for Health Metrics and Evaluation 2016), and is projected to further decline to under 3 per cent by 2024 (Kuhn *et al.* 2016). Hughes *et al.* (2011) also project significant decreases in mortality from communicable diseases, largely related to strong economic development. However, many people are projected to live without proper access to improved drinking water and sanitation and clean cooking facilities by 2030, and the levels of improvement across these risk factors vary widely by region. Furthermore, health risks associated with outdoor air pollution and climate change have been increasing (WHO 2014; Forouzanfar *et al.* 2015; Cohen *et al.* 2017). The impact from ambient particulate matter pollution will continue to contribute to millions of premature deaths annually in the coming decades (see Section 21.3.3). Likewise, climate change is projected to have substantial negative health impacts in the coming decades, among them heat exposure, coastal flooding, diarrhoea, malaria and undernutrition (Hughes *et al.* 2011; WHO 2014).

Environmental risk factors at household level have been declining since 1990, while risk factors at community and global level have been increasing. Global health risks have been shifting away from environmental risks and towards behavioural risks (e.g. smoking, childhood undernutrition, and alcohol use) and metabolic risks (e.g. high blood pressure, and high body mass index) (WHO 2009; Forouzanfar *et al.* 2015). This shift in risk factors is part of a larger epidemiological transition, which has occurred globally over the past two centuries – mortality rates have been decreasing and shifting towards risks that affect people later in life (Murray *et al.* 2015).

Trends in child mortality

Under-five mortality is generally seen as a good indicator of quality of life (see Section 20.4.1). Global child mortality (SDG target 3.2) declined dramatically from 91 deaths per thousand live births in 1990 to 43 per thousand live births in 2015, one of the most successful achievements of the Millennium Development Goal period (You *et al.* 2015). Yet more than 5 million children died in 2016 before reaching their fifth birthday, and 26 per cent of these deaths were due to environmental factors within our control (Prüss-Üstün *et al.* 2016). The five leading environmental risk factors (in order of health impact) are: household air pollution, unsafe drinking water, ambient particulate matter, unsafe sanitation, and insufficient handwashing (WHO 2009; Forouzanfar *et al.* 2015). Furthermore, malnutrition, including fetal growth restriction, child stunting and wasting, micronutrient deficiencies and suboptimal breastfeeding are important health risk factors, related to about 45 per cent of child deaths in 2011 (Black *et al.* 2013). In 1990, these five leading environmental factors accounted for nearly 2.8 million deaths in children under five (30 per cent of total under-five deaths), which decreased to just over 800,000 deaths in 2016 (24 per cent of total under-5 deaths). Currently, 79 countries have under-five mortality rates higher than the SDG target of 25 per 1,000 live births –

Figure 21.11: Projected under-five mortality rate in 2030

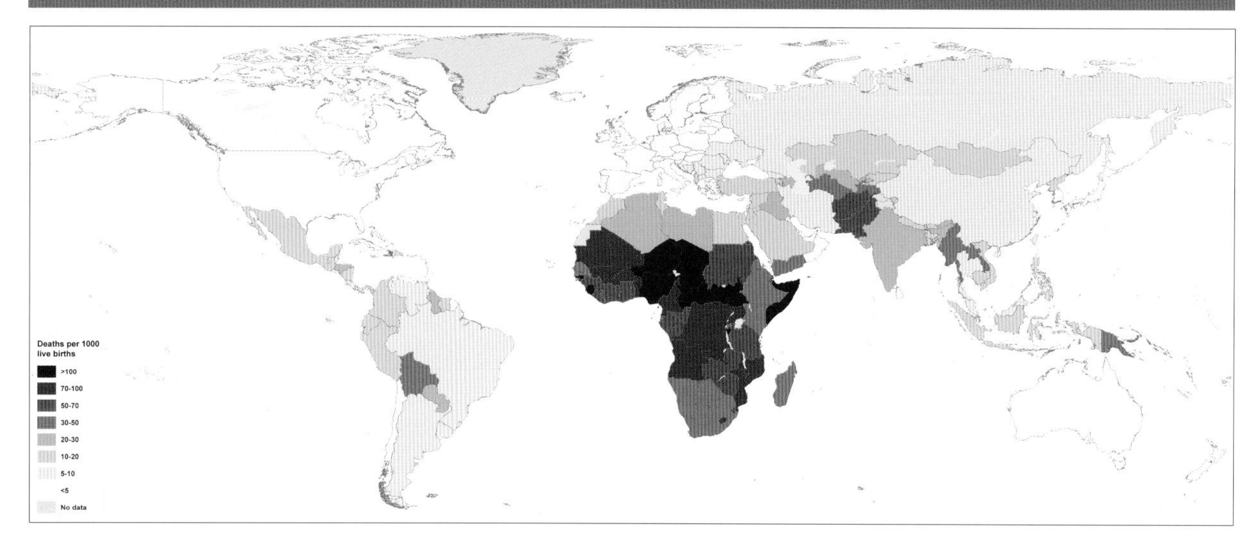

Source: Moyer and Hedden (2018).

partly due to persistent and sometimes even increasing environmental risk factors in low- and middle-income countries (GBD 2015 SDG Collaborators 2016).

Global child mortality is projected to decrease further to around 23 to 39 deaths per thousand live births by 2030, which is not enough to achieve the SDG target (Hughes *et al.* 2011; Liu *et al.* 2015; You *et al.* 2015; GBD 2016 SDG Collaborators 2017). This means that an estimated 47 countries are not on track to achieve the target by 2030, mostly in sub-Saharan Africa (You *et al.* 2015). As a result of decreasing child mortality, global average life expectancy at birth is projected to increase to over 77 by 2050, compared with 71 in 2015 (Samir and Lutz 2017; United Nations 2017). It should be noted that average figures such as these conceal huge differences in life expectancy, especially related to differences in poverty within and across countries.

Child mortality, especially from diarrhoea and pneumonia, is expected to decrease significantly, due to projected improvements in hunger levels (see Sections 21.3.2), access to modern sources of energy (see Section 21.3.3), and access to clean drinking water and sanitation (see Section 21.3.4), as well as to improved overall development levels (Lucas *et al.* 2018). However, by 2030, the five leading environmental risk factors, together with child underweight and malaria, are nevertheless projected to contribute to around 15 per cent of total child deaths, with a greater portion in sub-Saharan Africa (Lucas *et al.* 2018). These health impacts are largely preventable, but require interventions aimed at ensuring cleaner and sustainable access to food, water and energy services. Furthermore, climate change can exacerbate child mortality risks, for example, through impacts on food security and consequent levels of child underweight. Including climate change impacts on child underweight in their base case scenario projection, Hughes *et al.* (2011) project 70,000 additional child deaths in 2050, mostly in southern Asia and sub-Saharan Africa.

It should be noted that household, community and global environmental factors affect many human development indicators beyond child mortality and, in turn, that child mortality is affected by many factors other than modifiable environmental factors. As indicated above, the epidemiological transition implies a change in the balance of risks attributable to the environment, e.g. from water - related communicable diseases, to non-communicable diseases related to, for example, ambient air pollution. Success in dealing with the first of these will show far greater benefits for under-five mortality than success in dealing with the second. Furthermore, other environmental health risk factors affecting health in the under-fives – undernourishment for some, obesity for others – may show as increased risks to health and mortality in later life, rather than (or in addition to) mortality in the under-fives.

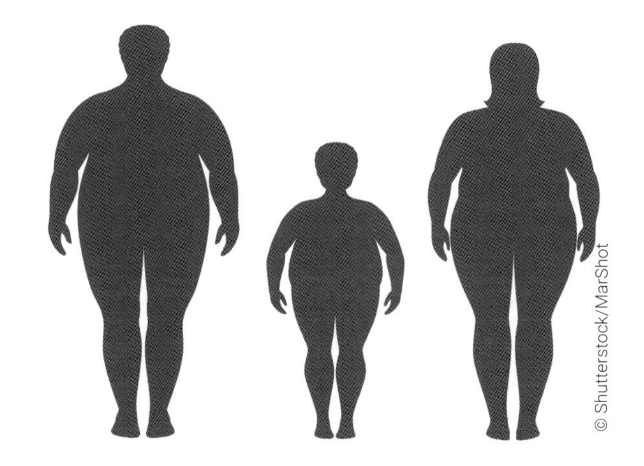

Box 21.5: Country level achievement of selected SDG targets

The 2030 Agenda is a global agenda, to be implemented at national level. This chapter explicitly evaluates future developments of selected environment-related SDG targets at a global level. The analysis concludes that without enhanced policies, the targets are unlikely to be met. However, the level of success differs largely across countries. Moyer and Hedden (2018) explored future country level progress under a SSP2 scenario, for eight SDG targets and nine related indicators. These targets link to selected SDG targets for the human well-being cluster, supplemented by SDG targets addressing poverty eradication and education. The nine indicators, though not comprehensive, represent multiple dimensions of human development.

In line with conclusions in this chapter **(see Table 21.2)**, the study concludes that between 2015 and 2030, the world will make only marginal progress towards achieving the nine SDG indicators. On all country indicator combinations explored (nine indicators for 186 countries = 1,674 indicator values in each year), 43 per cent had already achieved target values in 2015, which is projected to increase to 53 per cent in 2030. Only 17 per cent of the countries are projected to achieve all SDG targets analysed by 2030, while 15 per cent of the countries do not achieve any of the selected SDG targets. Most of these latter countries are in sub-Saharan Africa. The analysis highlights difficulty in achieving target values for access to sanitation, universal lower secondary education and reducing the prevalence of underweight children, representing persistent development issues. The child mortality SDG target is projected to be achieved by 67 per cent of countries in 2030, while only 8 per cent achieve this target in sub-Saharan Africa.

Table 21.1: Percentage of countries by region projected to achieve selected SDG targets in 2030

	Europe and Russian Federation	Latin America and Caribbean	Middle East and North Africa	Non-OECD Asia Pacific	North America	OECD Asia Pacific	South Asia	Sub-Saharan Africa	World
Extreme poverty	100	68	85	70	100	100	79	21	67
Hunger	95	32	70	26	100	100	43	10	48
Underweight children	82	48	30	26	100	100	14	0	37
Child mortality	98	90	90	74	100	50	71	6	67
Primary school completion	100	94	85	78	100	100	86	33	77
Lower secondary school	89	35	40	48	100	100	50	4	45
Access to safe water	98	94	95	70	100	100	93	17	72
Improved sanitation	80	29	65	43	100	100	43	4	44
Access to electricity	100	68	90	48	100	100	71	2	60

Source: Moyer and Hedden (2018).

21.4 Are we achieving the targets?

The results of the scenario assessment are summarized in **Table 21.2** with reference to the selected SDG targets and indicators (see Chapter 20). None of these targets is assessed to be achieved in the business-as-usual scenarios examined, but clear differences exist.

For the targets grouped under human well-being, the projected trends show improvements over time, but not sufficient to achieve them by 2030. However, several targets are projected to be achieved in the longer term, or at least come relatively close to being achieved; for instance, the prevalence of undernourishment is projected to be reduced by two-thirds or more in 2050. While progress is sufficiently rapid to compensate for the growing world population, many people will nevertheless be left without proper access to food, modern energy services or adequate drinking water and sanitation. Linked to this, environmentally related human health impacts are decreasing significantly, but far from enough to achieve the SDG target on under-five mortality. Environmental health risk factors remain especially prominent in sub-Saharan Africa.

Table 21.2: Past and future trends related to selected targets (see Section 20.4)

Cluster	Selected target for GEO-6	Indicator	2010 Level	Target value	Projected value[1]	Trend
Human well-being	End hunger	Prevalence of undernourishment	800-900 million people	0 in 2030	300-500 million people in 2030	↘
	Universal access to modern energy services	People without access to electricity and people without access to clean cooking fuels	2.8 billion and 1.1 billion	0 in 2030	2.3 billion and 700 million in 2030	↘
	Universal access to safe drinking water and adequate sanitation	People who lack access to improved drinking water and people who lack access to improved sanitation	0.8 billion and 2.3 billion	0 in 2030	0.4 billion and 2 billion in 2030	↘
	End preventable deaths of children under 5	Under-five mortality rate	52 deaths per 1,000 live births	<25 deaths per 1,000 live births in 2030	23-39 deaths per 1,000 live births in 2030	↘
Natural resource base	Improve water quality	Nitrogen fertilizer use and biological nitrogen fixation	120TgN/yr	< 62TgN/yr	185TgN/yr in 2050	↗
		Fertilizer use with phosphorus	14.5TgP/yr	< 6.2TgP/yr	26TgP/yr in 2050	↗
	Reduce water scarcity	Population living in water scare areas	2 billion	-	2.8 billion in 2030	↗
	Improve air quality in cities	Percentage population exposed to $PM_{2.5}$ above 35µg/m³	58 per cent	0 per cent in 2050	9-40 per cent in 2050	↘
	Limit global warming	Global mean temperature increase	1°C in 2016	< 2.0 / 1.5°C by 2100	4°C by 2100	↗
	Reduce marine nutrient pollution	P flow from freshwater systems into the ocean	4TgP/yr in 2006	-	5TgP/yr in 2050	↗
	Minimize ocean acidification	Average global surface aragonite saturation level	2.94Ωarg	> 2.75Ωarg	1.8Ωarg in 2100	↗
	Sustainable management of ocean resources	Proportion of fish stocks at or below target biomass that can undergo recovery	53 per cent	-	88 per cent in 2050	↗
	Achieve land degradation neutrality	Loss in soil organic carbon	176Gt historically	-	27GtC between 2010 and 2050	↗
	Halt biodiversity loss	Loss in Mean Species Abundance (MSA)	34 per cent	< 36 per cent from 2030 onwards	43 per cent in 2050	↗

↗ Target projected not to be achieved; trend in opposite direction or no significant improvement;
↘ Target projected not to be achieved; trend in right direction;
↓ Target projected to be achieved (none).

Source: All values are based on Section 21.3. [1] Projected values are mostly for SSP2 scenarios, except for energy access, under five mortality rate, ocean acidification and fish stocks above target biomass. For air pollution, results of the ECLIPSE scenario are also included due to the large uncertainty range.

For the targets grouped under natural resource base, the gap remains relatively large, and for most targets they even become wider. Although resource use efficiency (in terms of yield and nutrient use, water and energy efficiency) is projected to improve, mostly in line with historical trends **(see Table 21.3)**, trends in climate change, biodiversity loss, water scarcity, nutrient pollution and land degradation are projected to continue to move in the wrong direction. Only the proportion of the population exposed to $PM_{2.5}$ concentrations above 35µg/m³ is projected to decrease. Still, under a SSP2 scenario, 63 per cent of the population is projected to be exposed to $PM_{2.5}$ above 10µg/m³ by 2050, concluding that air pollution will continue to contribute to millions of premature deaths in the coming decades.

Overall, the scenario analysis shows that the world is not on track to achieve selected environment - related targets of the SDGs and related MEAs. A significant increase in the rate of improvement with respect to reducing child mortality, air pollution control, hunger eradication and achieving access to clean water, sanitation and modern sources of energy is required. Furthermore, achieving the targets that address the natural resource base, including on climate change, biodiversity loss, land degradation, water scarcity and pollution of freshwater and oceans, requires a clear break with current trends, with absolute decoupling of human development from environmental degradation.

Table 21.3: Historic and business-as-usual trends in resource use efficiency sustainable consumption and production

Selected target for GEO-6	Indicator	Historic development	Trend in business-as-usual scenarios
Increase agricultural productivity (Section 21.3.2)	Yield improvement over time (cereals)	1.9 per cent/yr (1970-2010)	0.4-0.9 per cent/yr (2010-2050)
Increase nutrient-use efficiency (Section 21.3.2)	Total N inputs to the crop N yields	0.42 in 2010	0.55 in 2050
Increase water-use efficiency (Section 21.3.4)	Change in water-use efficiency over time	0.2-1 per cent/yr (1970-2010)	0.3-1 per cent/yr (2010-2050)
Increase the share of renewable energy (Section 21.3.3)	Renewable energy share in total final energy consumption	15 per cent in 2010	20-30 per cent in 2050
Increase energy efficiency (Section 21.3.3)	Reduction in energy intensity over time (measured in terms of primary energy and GDP)	1-2 per cent/yr (1970-2010)	1-2.5 per cent/yr (2010-2050)

References

Alexandratos, N. and Bruinsma, J. (2012). *World Agriculture Towards 2030/2050: The 2012 Revision.* Rome: Food and Agriculture Organization of the United Nations. http://www.fao.org/docrep/016/ap106e/ap106e.pdf.

Amann, M., Bertok, I., Borken-Kleefeld, J., Cofala, J., Heyes, C., Höglund-Isaksson, L. *et al.* (2011). Cost-effective control of air quality and greenhouse gases in Europe: Modeling and policy applications. *Environmental Modelling and Software* 26(12), 1489-1501. https://doi.org/10.1016/j.envsoft.2011.07.012.

Anderson, D.M., Glibert, P.M. and Burkholder, J.M. (2002). Harmful algal blooms and eutrophication: Nutrient sources, composition, and consequences. *Estuaries* 25(4), 704-726. https://doi.org/10.1007/BF02804901.

Assad, E., Pavão, E., Jesus, M.d. and Martins, S.C. (2015). *Invertendo o sinal de carbono da agropecuária brasileira: uma estimativa do potencial de mitigação de tecnologias do Plano ABC de 2012 a 2023.* São Paulo: Observtório ABC. http://medias.canalrural.com.br/resources/pdf/1/5/1435789855051.pdf

Bai, Z.G., Dent, D.L., Olsson, L. and Schaepman, M.E. (2008). Proxy global assessment of land degradation. *Soil Use and Management* 24(3), 223-234. https://doi.org/10.1111/j.1475-2743.2008.00169.x.

Barbier, E.B. and Hochard, J.P. (2016). Does land degradation increase poverty in developing countries? *PLoS One* 11(5), e0152973. https://doi.org/10.1371/journal.pone.0152973.

Béné, C., Barange, M., Subasinghe, R., Pinstrup-Andersen, P., Merino, G., Hemre, G.I. *et al.* (2015). Feeding 9 billion by 2050 – putting fish back on the menu. *Food Security* 7(2), 261-274. https://doi.org/10.1007/s12571-015-0427-z.

Beusen, A.H.W., Bouwman, A.F., Van Beek, L.P.H., Mogollón, J.M. and Middelburg, J.J. (2016). Global riverine N and P transport to ocean increased during the 20th century despite increased retention along the aquatic continuum. *Biogeosciences* 13, 2441-2451. https://doi.org/10.5194/bg-13-2441-2016.

Bijl, D.L., Bogaart, P.W., Kram, T., de Vries, B.J.M. and van Vuuren, D.P. (2016). Long-term water demand for electricity, industry and households. *Environmental Science and Policy* 55, 75-86. https://doi.org/10.1016/j.envsci.2015.09.005.

Bijl, D.L., Bogaart, P.W., Dekker, S.C., Stehfest, E., de Vries, B.J.M. and van Vuuren, D.P. (2017). A physically-based model of long-term food demand. *Global Environmental Change* 45, 47-62. https://doi.org/10.1016/j.gloenvcha.2017.04.003.

Bijl, D.L., Bogaart, P.W., Dekker, S.C. and van Vuuren, D.P. (2018). Unpacking the nexus: Different spatial scales for water, food and energy. *Global Environmental Change* 48, 22-31. https://doi.org/10.1016/j.gloenvcha.2017.11.005.

Billé, R., Kelly, R., Biastoch, A., Harrould-Kolieb, E., Herr, D., Joos, F. *et al.* (2013). Taking action against ocean acidification: A review of management and policy options. *Environmental Management* 52(4), 761-779. https://doi.org/10.1007/s00267-013-0132-7.

Billen, G., Lancelot, C. and Meybeck, M. (1991). N, P, and Si retention along the aquatic continuum from land to ocean. In *Ocean Margin Processes in Global Change.* Mantoura, R.F.C., Martin, J.M. and Wollast, R. (eds.). New York, NY: John Wiley and Sons. 19-44.

Black, R.E., Victora, C.G., Walker, S.P., Bhutta, Z.A., Christian, P., de Onis, M. *et al.* (2013). Maternal and child undernutrition and overweight in low-income and middle-income countries. *The Lancet* 382(9890), 427-451. https://doi.org/10.1016/S0140-6736(13)60937-X.

Boit, A., Sakschewski, B., Boysen, L., Cano-Crespo, A., Clement, J., Garcia-alaniz, N. *et al.* (2016). Large-scale impact of climate change vs. land-use change on future biome shifts in Latin America. *Global Change Biology* 22(11), 3689-3701. https://doi.org/10.1111/gcb.13355.

Bopp, L., Resplandy, L., Orr, J.C., Doney, S.C., Dunne, J.P., Gehlen, M. *et al.* (2013). Multiple stressors of ocean ecosystems in the 21st century: Projections with CMIP5 models. *Biogeosciences* 10(10), 6225-6245. https://doi.org/10.5194/bg-10-6225-2013.

Bouwman, A.F., Beusen, A.H.W., Lassaletta, L., van Apeldoorn, D.F., van Grinsven, H.J.M., Zhang, J. *et al.* (2017). Lessons from temporal and spatial patterns in global use of N and P fertilizer on cropland. *Scientific Reports* 7(40366). https://doi.org/10.1038/srep40366.

Burt, A., Hughes, B. and Milante, G. (2014). *Eradicating Poverty in Fragile States: Prospects of Reaching the "High-Hanging" Fruit by 2030.* Policy Research Working Paper7002. Washington, D.C.: World Bank. http://documents.worldbank.org/curated/en/909761468170347362/pdf/WPS7002.pdf

Cattano, C., Claudet, J., Domenici, P. and Milazzo, M. (2018). Living in a high CO_2 world: A global meta-analysis shows multiple trait-mediated fish responses to ocean acidification. *Ecological Monographs* 88(3), 320-335. https://doi.org/10.1002/ecm.1297.

Chandy, L., Ledlie, N. and Penciakova, V. (2013). *The Final Countdown: Prospects for Ending Extreme Poverty by 2030* Policy Paper 2013-04. Washington, D.C.:The Brookings Institution. https://www.brookings.edu/wp-content/uploads/2016/06/The_Final_Countdown.pdf

Chaturvedi, V., Mohamad Hejazi, M., Edmonds, J., Clarke, L., Kyle, P., Davies, E. *et al.* (2015). Climate mitigation policy implications for global irrigation water demand. *Mitigation and Adaptation Strategies for Global Change* 20(3), 389-407. https://doi.org/10.1007/s11027-013-9497-4.

Ciais, P., Sabine, C., Bala, G., Bopp, L., Brovkin, V., Canadell, J. *et al.* (2013). Carbon and other biogeochemical cycles. In *Climate Change 2013: The Physical Science Basis. Contribution of Working Group I to the Fifth Assessment Report of the Intergovernmental Panel on Climate Change.* Stocker, T.F., Qin, D., Plattner, G.-K., Tignor, M., Allen, S.K., Boschung, J. *et al.* (eds.). Cambridge: Cambridge University Press. chapter 6. https://www.ipcc.ch/pdf/assessment-report/ar5/wg1/WG1AR5_Chapter06_FINAL.pdf

Clark, W.C., Mitchell, R.B. and Cash, D.W. (2006). Evaluating the influence of global environmental assessments. In *Global Environmental Assessments: Information and Influence.* Mitchell, R.B., Clark, W.C., Cash, D.W. and Dickson, N.M. (eds.). Cambridge: MIT Press, chapter 1. https://sites.hks.harvard.edu/gea/pubs/geavol_info_chap_1.pdf.

Cohen, A.J., Brauer, M., Burnett, R., Anderson, H.R., Frostad, J., Estep, K. *et al.* (2017). Estimates and 25-year trends of the global burden of disease attributable to ambient air pollution: An analysis of data from the global burden of diseases study 2015. *The Lancet* 389(10082), 1907-1918. https://doi.org/10.1016/S0140-6736(17)30505-6.

Conley, D. (2002). Terrestrial ecosystems and the global biogeochemical silica cycle. *Global Biogeochemical Cycles* 16(4), 68-61-68-68. https://doi.org/10.1029/2002GB001894

Costello, C., Ovando, D., Clavelle, T., Strauss, C.K., Hilborn, R., Melnychuk, M.C. *et al.* (2016). Global fishery prospects under contrasting management regimes. *Proceedings of the National Academy of Sciences* 113(18), 5125-5129. https://doi.org/10.1073/pnas.1520420113.

Dagnachew, A.G., Lucas, P.L., Hof, A.F., Gernaat, D.E.H.J., de Boer, H.-S. and van Vuuren, D.P. (2017). The role of decentralized systems in providing universal electricity access in Sub-Saharan Africa – A model-based approach. *Energy* 139, 184-195. https://doi.org/10.1016/j.energy.2017.07.144.

Dellink, R., Chateau, J., Lanzi, E. and Magné, B. (2017). Long-term economic growth projections in the shared socioeconomic pathways. *Global Environmental Change* 42, 200-214. https://doi.org/10.1016/j.gloenvcha.2015.06.004.

Economic and Social Council (2017). *Progress towards the Sustainable Development Goals: Report of the Secretary-General. E/2018/64.* https://unstats.un.org/sdgs/files/report/2018/secretary-general-sdg-report-2018--EN.pdf.

Elliott, J., Deryng, D., Müller, C., Frieler, K., Konzmann, M., Gerten, D. *et al.* (2014). Constraints and potentials of future irrigation water availability on agricultural production under climate change. *Proceedings of the National Academy of Sciences* 111(9), 3239-3244. https://doi.org/10.1073/pnas.1222474110.

Flörke, M., Kynast, E., Bärlund, I., Eisner, S., Wimmer, F. and Alcamo, J. (2013). Domestic and industrial water uses of the past 60 years as a mirror of socio-economic development: A global simulation study. *Global Environmental Change* 23(1), 144-156. https://doi.org/10.1016/j.gloenvcha.2012.10.018.

Food and Agriculture Organization of the United Nations (2016). *The State of World Fisheries and Aquaculture 2016: Contributing to Food Security and Nutrition for All.* Rome. http://www.fao.org/3/a-i5555e.pdf.

Food and Agriculture Organization of the United Nations (2018). *The State of World Fisheries and Aquaculture 2018: Meeting the Sustainable Development Goals.* Rome. http://www.fao.org/3/I9540EN/i9540en.pdf.

Food and Agriculture Organization of the United Nations, International Fund for Agricultural Development and World Food Programme (2015). *The State of Food Insecurity in the World: Meeting the 2015 International Hunger Targets: Taking Stock of Uneven Progress.* Rome. http://www.fao.org/3/a-i4646e.pdf.

Food and Agriculture Organization of the United Nations, International Fund for Agricultural Development, United Nations Children's Fund, World Food Programme and World Health Organization (2017). *The State of Food Insecurity in the World 2017. Building Resilience for Peace and Food Security.* Rome. http://www.fao.org/3/a-I7695e.pdf

Forouzanfar, M.H., Alexander, L., Anderson, H.R., Bachman, V.F., Biryukov, S., Brauer, M. *et al.* (2015). Global, regional, and national comparative risk assessment of 79 behavioural, environmental and occupational, and metabolic risks or clusters of risks in 188 countries, 1990–2013: A systematic analysis for the Global Burden of Disease Study 2013. *The Lancet* 386(10010), 2287-2323. https://doi.org/10.1016/S0140-6736(15)00128-2.

Fritsche, U.R., Eppler, U., Iriarte, L., Laaks, S., Wunder, S., Kaphengst, T. *et al.* (2015). *Resource-Efficient Land Use – Towards a Global Sustainable Land Use Standard (GLOBALANDS).* Umweltbundesamt, Dessau-Roßlau https://www.umweltbundesamt.de/sites/default/files/medien/378/publikationen/texte_82_2015_resource_efficient_land_use.pdf

Fu, W., Randerson, J.T. and Moore, J.K. (2016). Climate change impacts on net primary production (NPP) and export production (EP) regulated by increasing stratification and phytoplankton community structure in the CMIP5 models. *Biogeosciences* 13, 5151-5170. https://doi.org/10.5194/bg-13-5151-2016.

Garcia, S.M., Rice, J. and Charles, A. (2016). Bridging fisheries management and biodiversity conservation norms: Potential and challenges of balancing harvest in ecosystem-based frameworks. *ICES Journal of Marine Science* 73(6), 1659-1667. https://doi.org/10.1093/icesjms/fsv230.

Gauthier, S., Bernier, P., Kuuluvainen, T., Shvidenko, A.Z. and Schepaschenko, D.G. (2015). Boreal forest health and global change. *Science* 349(6250), 819-822. https://doi.org/10.1126/science.aaa9092.

GBD 2015 SDG Collaborators (2016). Measuring the health-related Sustainable Development Goals in 188 countries: A baseline analysis from the Global Burden of Disease Study 2015. *The Lancet* 388(10053), 1813-1850. https://doi.org/10.1016/S0140-6736(16)31467-2.

GBD 2016 Risk Factors Collaborators (2017). Global, regional, and national comparative risk assessment of 84 behavioural, environmental and occupational, and metabolic risks or clusters of risks, 1990-2016: A systematic analysis for the Global Burden of Disease Study 2016. *Lancet* 390(10100), 1345-1422. https://doi.org/10.1016/s0140-6736(17)32366-8.

GBD 2016 SDG Collaborators (2017). Measuring progress and projecting attainment on the basis of past trends of the health-related Sustainable Development Goals in 188 countries: An analysis from the global burden of disease study 2016. *The Lancet* 390(10100), 1423-1459. https://doi.org/10.1016/S0140-6736(17)32336-X.

Glibert, P.M. (2017). Eutrophication, harmful algae and biodiversity — Challenging paradigms in a world of complex nutrient changes. *Marine Pollution Bulletin* 124(2), 591-606. https://doi.org/10.1016/j.marpolbul.2017.04.027.

Global Energy Assessment (2012). *Global Energy Assessment: Toward a Sustainable Future.* Cambridge: International Institute for Applied Systems Analysis. http://www.cambridge.org/gb/knowledge/isbn/item6852590/?site_locale=en_GB.

Gonzalez, P., Neilson, R.P., Lenihan, J.M. and Drapek, R.J. (2010). Global patterns in the vulnerability of ecosystems to vegetation shifts due to climate change. *Global Ecology and Biogeography* 19(6), 755-768. https://doi.org/10.1111/j.1466-8238.2010.00558.x.

Hanasaki, N., Fujimori, S., Yamamoto, T., Yoshikawa, S., Masaki, Y., Hijioka, Y. *et al.* (2013a). A global water scarcity assessment under Shared Socio-economic Pathways – Part 1: Water use. *Hydrology and Earth System Sciences* 17, 2375-2391. https://doi.org/10.5194/hess-17-2375-2013.

Hanasaki, N., Fujimori, S., Yamamoto, T., Yoshikawa, S., Masaki, Y., Hijioka, Y. *et al.* (2013b). A global water scarcity assessment under Shared Socio-economic Pathways – Part 2: Water availability and scarcity. *Hydrology and Earth System Sciences* 17(2393-2413). https://doi.org/10.5194/hess-17-2393-2013.

Hasegawa, T., Fujimori, S., Takahashi, K. and Masui, T. (2015). Scenarios for the risk of hunger in the twenty-first century using shared socioeconomic pathways. *Environmental Research Letters* 10(1), 014010. https://doi.org/10.1088/1748-9326/10/1/014010.

Health Effects Institute (2018). *State of Global Air 2018: A Special Report on Global Exposure to Air Pollution and Its Disease Burden.* Boston, MA. https://www.stateofglobalair.org/sites/default/files/soga-2018-report.pdf.

Hughes, B.B., Kuhn, R., Peterson, C.M., Rothman, D.S., Solórzano, J.R., Mathers, C.D. *et al.* (2011). Projections of global health outcomes from 2005 to 2060 using the International Futures integrated forecasting model. *Bulletin of the World Health Organization* 89(7), 478-486. https://doi.org/10.2471/blt.10.083766.

Institute for Health Metrics and Evaluation (2016). *GBD Compare Data Visualization.* https://vizhub.healthdata.org/gbd-compare/.

Intergovernmental Panel on Climate Change (2014a). *Climate Change 2014: Synthesis Report. Contribution of Working Groups I, II and III to the Fifth Assessment Report of the Intergovernmental Panel on Climate Change.* Pachauri, R.K. and Meyer, L.A. (eds.). Geneva: Intergovernmental Panel on Climate Change. http://www.ipcc.ch/pdf/assessment-report/ar5/syr/AR5_SYR_FINAL_All_Topics.pdf

Intergovernmental Panel on Climate Change (2014b). *Climate Change 2014: Impacts, Adaptation, and Vulnerability. Summary for policymakers.* Field, C.B., Barros, V.R., Dokken, D.J., Mach, K.J., Mastrandrea, M.D., Bilir, T.E. *et al.* (eds.). Cambridge: Cambridge University Press. http://www.ipcc.ch/pdf/assessment-report/ar5/wg2/ar5_wgII_spm_en.pdf.

Intergovernmental Science-Policy Platform on Biodiversity and Ecosystem Services (2018). *Land Degradation and Restoration Assessment.* Bonn. https://www.ipbes.net/system/tdf/ipbes_6_inf_1_rev.1_2.pdf?file=1&type=node&id=16514.

International Energy Agency (2016). *World Energy Outlook 2016.* Paris. https://webstore.iea.org/world-energy-outlook-2016.

International Energy Agency (2017a). *World Energy Outlook 2017.* Paris. https://webstore.iea.org/world-energy-outlook-2017.

International Energy Agency (2017b). *Energy Access Outlook 2017: From Poverty to Prosperity.* Paris. https://www.iea.org/publications/freepublications/publication/WEO2017SpecialReport_EnergyAccessOutlook.pdf.

Jiang, L. and O'Neill, B.C. (2017). Global urbanization projections for the shared socioeconomic pathways. *Global Environmental Change* 42, 193-199. https://doi.org/10.1016/j.gloenvcha.2015.03.008.

Klimont, Z., Kupiainen, K., Heyes, C., Purohit, P., Cofala, J., Rafaj, P. *et al.* (2017). Global anthropogenic emissions of particulate matter including black carbon. *Atmospheric Chemistry and Physics* 17(14), 8681-8723. https://doi.org/10.5194/acp-17-8681-2017.

Kok, M.T.J., Alkemade, R., Bakkenes, M., van Eerdt, M., Janse, J., Mandryk, M. *et al.* (2018). Pathways for agriculture and forestry to contribute to terrestrial biodiversity conservation: A global scenario-study. *Biological Conservation* 221, 137-150. https://doi.org/10.1016/j.biocon.2018.03.003.

Kuhn, R., Rothman, D.S., Turner, S., Solórzano, J. and Hughes, B.B. (2016). Beyond attributable burden: Estimating the avoidable burden of disease associated with household air pollution. *PLoS One* 11(3), e0149669. https://doi.org/10.1371/journal.pone.0149669.

Laborde, D., Bizikova, L., Lallemant, T. and Smaller, C. (2016). *Ending Hunger: What Would It Cost?* International Institute for Sustainable Development. http://www.iisd.org/sites/default/files/publications/ending-hunger-what-would-it-cost.pdf.

Lam, V.W.Y., Cheung, W.W.L., Reygondeau, G. and Sumaila, U.R. (2016). Projected change in global fisheries revenues under climate change. *Scientific Reports* 6, 32607. https://doi.org/10.1038/srep32607.

Laufkötter, C., Vogt, M., Gruber, N., Aita-Noguchi, M., Aumont, O., Bopp, L. *et al.* (2015). Drivers and uncertainties of future global marine primary production in marine ecosystem models. *Biogeosciences* 12, 6955-6984. https://doi.org/10.5194/bg-12-6955-2015.

Leadley, P.W., Krug, C.B., Alkemade, R., Pereira, H.M., Sumaila U.R., Walpole, M. *et al.* (2014). *Progress Towards the Aichi Biodiversity Targets: An Assessment of Biodiversity Trends, Policy Scenarios and Key Actions.* CBD Technical Series. Montreal: Secretariat of the Convention on Biological Diversity. https://www.cbd.int/doc/publications/cbd-ts-78-en.pdf.

Ligtvoet, W., Bouwman, A., Knoop, J., de Bruin, S., Nabielek, K., Huitzing, H. *et al.* (2018). *The Geography of Future Water Challenges.* The Hague: PBL Netherlands Environmental Assessment Agency. http://www.pbl.nl/sites/default/files/cms/publicaties/pbl-2018-the-geography-of-future-water-challenges-2920.pdf.

Liu, L., Oza, S., Hogan, D., Perin, J., Rudan, I., Lawn, J.E. *et al.* (2015). Global, regional, and national causes of child mortality in 2000–13, with projections to inform post-2015 priorities: An updated systematic analysis. *The Lancet* 385(9966), 430-440. https://doi.org/10.1016/S0140-6736(14)61698-6.

Liu, J., Yang, H., Gosling, S.N., Kummu, M., Flörke, M., Pfister, S. *et al.* (2017). Water scarcity assessments in the past, present, and future. *Earth's Future* 5(6), 545-559. https://doi.org/10.1002/2016EF000518.

Lucas, P.L., Dagnachew, A.G. and Hof, A.F. (2017). *Towards Universal Electricity Access in Sub-Saharan Africa: A Quantitative Analysis of Technology and Investment Requirements.* The Hague: PBL Netherlands Environmental Assessment Agency. http://www.pbl.nl/sites/default/files/cms/publicaties/pbl-2017-towards-universal-electricity-access-in-sub-saharan-africa-1952.pdf.

Lucas, P.L., Hilderink, H.B.M., Janssen, P., Samir, K.C., Niessen, L.W. and van Vuuren, D.P. (2018). *Future Impacts of Environmental Factors on Achieving The SDG Target On Child Mortality – A Synergistic Assessment.* PBL Working Paper 24. The Hague: PBL Netherlands Environmental Assessment Agency. http://www.pbl.nl/sites/default/files/cms/publicaties/pbl-2018-future-impacts-of-environmental-factors-on-achieving-the-sdg-target-on-child-mortality.pdf.

Lucas, P.L., Nielsen, J., Calvin, K., McCollum, D., Marangoni, G., Strefler, J. *et al.* (2015). Future energy system challenges for Africa: Insights from integrated assessment models. *Energy Policy* 86, 705–717. https://doi.org/10.1016/j.enpol.2015.08.017.

Lucas, P.L., Dagnachew, A.G. and Hof, A.F. (2017). *Towards Universal Electricity Access in Sub-Saharan Africa: A Quantitative Analysis of Technology and Investment Requirements.* http://www.pbl.nl/sites/default/files/cms/publicaties/pbl-2017-towards-universal-electricity-access-in-sub-saharan-africa-1952.pdf

Mara, D. and Evans, B. (2018). The sanitation and hygiene targets of the sustainable development goals: Scope and challenges. *Journal of Water, Sanitation and Hygiene for Development* 8(1), 1-16. https://doi.org/10.2166/washdev.2017.048.

Maavara, T., Dürr, H.H. and Van Cappellen, P. (2014). Worldwide retention of nutrient silicon by river damming: From sparse data set to global estimate. *Global Biogeochemical Cycles* 28(8), 842-855. https://doi.org/10.1002/2014GB004875.

Melnychuk, M.C., Peterson, E., Elliott, M. and Hilborn, R. (2017). Fisheries management impacts on target species status. *Proceedings of the National Academy of Sciences* 114(1), 178-183. https://doi.org/10.1073/pnas.1609915114.

Meybeck, M. (1982). Carbon, nitrogen and phosphorous transport by world rivers. *American Journal of Science* 282, 401-450.

Millennium Ecosystem Assessment (2005). *Ecosystems and Human Well-being: Synthesis.* Washington, D.C: Island Press. https://www.millenniumassessment.org/documents/document.356.aspx.pdf.

Mogollón, J.M., Lassaletta, L., Beusen, A.H.W., van Grinsven, H.J.M., Westhoek, H. and Bouwman, A.F. (2018a). Assessing future reactive nitrogen inputs into global croplands based on the shared socioeconomic pathways. *Environmental Research Letters* 13(4). https://doi.org/10.1088/1748-9326/aab212.

Mogollón, J.M., Beusen, A.H.W., van Grinsven, H.J.M., Westhoek, H. and Bouwman, A.F. (2018b). Future agricultural phosphorus demand according to the shared socioeconomic pathways. *Global Environmental Change* 50, 149-163. https://doi.org/10.1016/j.gloenvcha.2018.03.007.

Mouratiadou, I., Biewald, A., Pehl, M., Bonsch, M., Baumstark, L., Klein, D. *et al.* (2016). The impact of climate change mitigation on water demand for energy and food: An integrated analysis based on the shared Socioeconomic Pathways. *Environmental Science and Policy* 64, 48-58. https://doi.org/10.1016/j.envsci.2016.06.007.

Moyer, J.D. and Hedden, S. (2018). How achievable are human development SDGs on our current path of development?. [in preparation].

Murray, C.J.L., Barber, R.M., Foreman, K.J., Ozgoren, A.A., Abd-Allah, F., Abera, S.F. *et al.* (2015). Global, regional, and national disability-adjusted life years (DALYs) for 306 diseases and injuries and healthy life expectancy (HALE) for 188 countries, 1990-2013: Quantifying the epidemiological transition. *The Lancet* 386(10009), 2145-2191. https://doi.org/10.1016/S0140-6736(15)61340-X.

Nakicenovic, N., Alcamo, J., Grubler, A., Riahi, K., Roehrl, R.A., Roger, H.-H. *et al.* (2000). *Special Report on Emissions Scenarios (SRES), A Special Report of Working Group III of the Intergovernmental Panel on Climate Change.* Cambridge: Cambridge University Press. http://pure.iiasa.ac.at/id/eprint/6101/1/emissions_scenarios.pdf.

Newbold, T., Hudson, L.N., Hil, S.L.L., Contu, S., Lysenko, I., Senior, R.A. *et al.* (2015). Global effects of land use on local terrestrial biodiversity. *Nature* 520, 45-50. https://doi.org/10.1038/nature14324.

Newbold, T. (2018). Future effects of climate and land-use change on terrestrial vertebrate community diversity under different scenarios. *Proceedings of the Royal Society B: Biological Sciences* 285(1881). https://doi.org/10.1098/rspb.2018.0792.

Oki, T. and Kanae, S. (2006). Global hydrological cycles and world water resources. *Science* 313(5790), 1068-1072. https://doi.org/10.1126/science.1128845.

O'Neill, B.C., Kriegler, E., Ebi, K.L., Kemp-Benedict, E., Riahi, K., Rothman, D.S. *et al.* (2017). The roads ahead: Narratives for shared socioeconomic pathways describing world futures in the 21st century. *Global Environmental Change* 42, 169-180. https://doi.org/10.1016/j.gloenvcha.2015.01.004.

Organisation for Economic Co-operation and Development (2016). *The Economic Consequences of Outdoor Air Pollution.* Paris. https://www.oecd-ilibrary.org/the-economic-consequences-of-outdoor-air-pollution_5jlzg2tj7mvf.pdf?itemId=%2Fcontent%2Fpublication%2F9789264257474-en&mimeType=pdf.

Organisation for Economic Co-operation and Development and Food and Agriculture Organization of the United Nations (2017). *OECD-FAO Agricultural Outlook 2017-2026. Special Focus: Southeast Asia.* Paris. https://www.oecd-ilibrary.org/agriculture-and-food/oecd-fao-agricultural-outlook-2017-2026_agr_outlook-2017-en;jsessionid=q8WsoLO97ujTtcppyvzESJNM.ip-10-240-5-90

Palter, J.B., Frölicher, T.L., Paynter, D. and John, J.G. (2018). Climate, ocean circulation, and sea level changes under stabilization and overshoot pathways to 1.5 K warming. *Earth System Dynamics* 9(2), 817-828. https://doi.org/10.5194/esd-9-817-2018.

Popp, A., Calvin, K., Fujimori, S., Havlik, P., Humpenöder, F., Stehfest, E. *et al.* (2017). Land-use futures in the shared socio-economic pathways. *Global Environmental Change* 42, 331-345. https://doi.org/10.1016/j.gloenvcha.2016.10.002.

Prüss-Ustün, A., Wolf, J., Corvalán, C., Bos, R. and Neira, M. (2016). *Preventing Disease Through Healthy Environments: A Global Assessment of the Burden of Disease from Environmental Risks.* Geneva: World Health Organization. http://apps.who.int/iris/bitstream/handle/10665/204585/9789241565196_eng.pdf?sequence=1.

Ran, X., Bouwman, A.F., Yu, Z. and Liu, J. (2018). Implications of eutrophication for biogeochemical processes in the Three Gorges Reservoir, China. *Regional Environmental Change*, 1-9. https://doi.org/10.1007/s10113-018-1382-y.

Rao, S., Klimont, Z., Smith, S.J., Van Dingenen, R., Dentener, F., Bouwman, L. *et al.* (2017). Future air pollution in the shared socio-economic pathways. *Global Environmental Change* 42, 346-358. https://doi.org/10.1016/j.gloenvcha.2016.05.012.

Riahi, K., van Vuuren, D.P., Kriegler, E., Edmonds, J., O'Neill, B.C., Fujimori, S. *et al.* (2017). The shared socioeconomic pathways and their energy, land use, and greenhouse gas emissions implications: An overview. *Global Environmental Change* 42, 153-168. https://doi.org/10.1016/j.gloenvcha.2016.05.009.

Samir, K.C. and Lutz, W. (2017). The human core of the shared socioeconomic pathways: Population scenarios by age, sex and level of education for all countries to 2100. *Global Environmental Change* 42, 181-192. https://doi.org/10.1016/j.gloenvcha.2014.06.004.

Satoh, Y., Kahil, T., Byers, E., Burek, P., Fischer, G., Tramberend, S. *et al.* (2017). Multi-model and multi-scenario assessments of Asian water futures: The Water Futures and Solutions (WFaS) initiative. *Earth's Future* 5(7), 823-852. https://doi.org/10.1002/2016EF000503.

Schewe, J., Heinke, J., Gerten, D., Haddeland, I., Arnell, N.W., Clark, D.B. *et al.* (2014). Multimodel assessment of water scarcity under climate change. *Proceedings of the National Academy of Sciences* 111(9), 3245-3250. https://doi.org/10.1073/pnas.1222460110.

Schut, A.G.T., Ivits, E., Conijn, J.G., ten Brink, B. and Fensholt, R. (2015). Trends in global vegetation activity and climatic drivers indicate a decoupled response to climate change. *PLoS One* 10(10), e0138013. https://doi.org/10.1371/journal.pone.0138013.

Secretariat of the Convention on Biological Diversity (2014). *Global Biodiversity Outlook 4.* Montréal. https://www.cbd.int/gbo/gbo4/publication/gbo4-en-hr.pdf.

Seitzinger, S.P., Mayorga, E., Bouwman, A.F., Kroeze, C., Beusen, A.H.W., Billen, G. *et al.* (2010). Global river nutrient export: A scenario analysis of past and future trends. *Global Biogeochemical Cycles* 24(4). https://doi.org/10.1029/2009GB003587.

Smith, K.R. and Ezzati, M. (2005). How environmental health risks change with development: The epidemiologic and environmental risk transitions revisited. *Annual Review of Environment and Resources* 30(1), 291-333. https://doi.org/10.1146/annurev.energy.30.050504.144424.

Smith, P., Gregory, P.J., van Vuuren, D., Obersteiner, M., Havlík, P., Rounsevell, M. *et al.* (2010). Competition for land. *Philosophical Transactions of the Royal Society B: Biological Sciences* 365(1554), 2941-2957. https://doi.org/10.1098/rstb.2010.0127.

Smith, P., House, J.I., Bustamante, M., Sobocká, J., Harper, R., Pan, G. *et al.* (2016). Global change pressures on soils from land use and management. *Global Change Biology* 22(3), 1008-1028. https://doi.org/10.1111/gcb.13068.

SSP Public Database (2016). *SSP Database (Shared Socioeconomic Pathways) - Version 1.1.* https://tntcat.iiasa.ac.at/SspDb/dsd?Action=htmlpage&page=about2018.

Steffen, W., Richardson, K., Rockström, J., Cornell, S.E., Fetzer, I., Bennett, E.M. *et al.* (2015). Planetary boundaries: Guiding human development on a changing planet. 347(6223). https://doi.org/10.1126/science.1259855 doi: 10.1126/science.1259855.

Stohl, A., Aamaas, B., Amann, M., Baker, L.H., Bellouin, N., Berntsen, T.K. *et al.* (2015). Evaluating the climate and air quality impacts of short-lived pollutants. *Atmospheric Chemistry and Physics* 15(18), 10529-10566. https://doi.org/10.5194/acp-15-10529-2015.

Stoorvogel, J.J., Bakkenes, M., Temme, A.J., Batjes, N.H. and Ten Brink, B.J. (2017a). S-world: A global soil map for environmental modelling. *Land Degradation and Development* 28(1), 22-33. https://doi.org/10.1002/ldr.2656.

Stoorvogel, J.J., Bakkenes, M., Ten Brink, B.J. and Temme, A.J. (2017b). To what extent did we change our soils? A global comparison of natural and current conditions *Land Degradation and Development* 28(7), 1982-1991. https://doi.org/10.1002/ldr.2721.

Strassburg, B.B.N., Latawiec, A.E., Barioni, L.G., Nobre, C.A., da Silva, V.P., Valentim, J.F. *et al.* (2014). When enough should be enough: Improving the use of current agricultural lands could meet production demands and spare natural habitats in Brazil. *Global Environmental Change* 28, 84-97. https://doi.org/10.1016/j.gloenvcha.2014.06.001.

Tilman, D., Balzer, C., Hill, J. and Befort, B.L. (2011). Global food demand and the sustainable intensification of agriculture. *Proceedings of the National Academy of Sciences* 108(50), 20260-20264. https://doi.org/10.1073/pnas.1116437108.

Troell, M., Naylor, R.L., Metian, M., Beveridge, M., Tyedmers, P.H., Folke, C. *et al.* (2014). Does aquaculture add resilience to the global food system? *Proceedings of the National Academy of Sciences* 111(37), 13257-13263. https://doi.org/10.1073/pnas.1404067111.

United Nations (2014). *World Urbanization Prospects: The 2014 Revision (ST/ESA/SER.A/366).* New York, NY. https://esa.un.org/unpd/wup/Publications/Files/WUP2014-Report.pdf.

United Nations (2017). *World Population Prospects: Key Findings and Advance Tables. The 2017 Revision.* New York, NY. https://esa.un.org/unpd/wpp/Publications/Files/WPP2017_KeyFindings.pdf.

United Nations Convention to Combat Desertification (2017). *Global Land Outlook.* Bonn. https://www2.unccd.int/sites/default/files/documents/2017-09/GLO_Full_Report_low_res.pdf.

United Nations Environment Programme (2012). *Global Environment Outlook 5: Environment for the Future We Want.* Nairobi. http://wedocs.unep.org/bitstream/handle/20.500.11822/8021/GEO5_report_full_en.pdf?sequence=5&isAllowed=y.

United Nations Environment Programme (2017). *The Emissions Gap Report 2017: A UN Environment Synthesis Report.* Nairobi. https://wedocs.unep.org/bitstream/handle/20.500.11822/22070/EGR_2017.pdf?sequence=1&isAllowed=y.

van der Esch, S., ten Brink, B., Stehfest, E., Bakkenes, M., Sewell, A., Bouwman, A. *et al.* (2017). *Exploring Future Changes in Land Use and Land Condition and the Impacts on Food, Water, Climate Change and Biodiversity: Scenarios for The UNCCD Global Land Outlook.* The Hague: PBL Netherlands Environmental Assessment Agency. http://www.pbl.nl/sites/default/files/cms/publicaties/pbl-2017-exploring-future-changes-in-land-use-and-land-condition-2076b.pdf.

van Dingenen, R., Dentener, F., Crippa, M., Leitao, J., Marmer, E., Rao, S. *et al.* (2018). TM5-FASST: A global atmospheric source-receptor model for rapid impact analysis of emission changes on air quality and short-lived climate pollutants. *Atmospheric Chemistry and Physics.* https://doi.org/10.5194/acp-2018-112.

van Puijenbroek, P.J., Bouwman, A.F., Beusen, A.H. and Lucas, P.L. (2015). Global implementation of two shared socioeconomic pathways for future sanitation and wastewater flows. *Water Science and Technology* 71(2), 227-233. https://doi.org/10.2166/wst.2014.498.

van Puijenbroek, P.J.T.M., Beusen, A.H.W. and Bouwman, A.F. (2019). Global nitrogen and phosphate in urban waste water based on the Shared Socio-economic Pathways. *Journal of Environmental Management* 231, 446-456. https://doi.org/10.1016/j.jenvman.2018.10.048.

van Vuuren, D.P., Kok, M.T.J., Girod, B., Lucas, P.L. and de Vries, B. (2012). Scenarios in global environmental assessments: Key characteristics and lessons for future use. *Global Environmental Change* 22(4), 884-895. https://doi.org/10.1016/j.gloenvcha.2012.06.001.

van Vuuren, D.P., Kriegler, E., O'Neill, B.C., Ebi, K.L., Riahi, K., Carter, T.R. *et al.* (2014). A new scenario framework for climate change research: Scenario matrix architecture. *Climatic Change* 122(3), 373-386. https://doi.org/10.1007/s10584-013-0906-1.

van Vuuren, D.P., Heleen van, S., Keywan, R., Leon, C., Volker, K., Elmar, K. *et al.* (2016). Carbon budgets and energy transition pathways. *Environmental Research Letters* 11(7), 075002. https://doi.org/10.1088/1748-9326/11/7/075002.

Vanham, D., Hoekstra, A.Y., Wada, Y., Bouraoui, F., de Roo, A., Mekonnen, M.M. *et al.* (2018). Physical water scarcity metrics for monitoring progress towards SDG target 6.4: An evaluation of indicator 6.4.2 "Level of water stress". *Science of the Total Environment* 613-614, 218-232. https://doi.org/10.1016/j.scitotenv.2017.09.056.

Veldkamp, T.I.E., Wada, Y., de Moel, H., Kummu, M., Eisner, S., Aerts, J.C.J.H. *et al.* (2015). Changing mechanism of global water scarcity events: Impacts of socioeconomic changes and inter-annual hydro-climatic variability. *Global Environmental Change* 32, 18-29. https://doi.org/10.1016/j.gloenvcha.2015.02.011.

Visconti, P., Bakkenes, M., Baisero, D., Brooks, T., Butchart, S.H.M., Joppa, L. *et al.* (2016). Projecting global biodiversity indicators under future development scenarios. *Conservation Letters* 9(1), 5-13. https://doi.org/10.1111/conl.12159.

Visser, H., Dangendorf, S., van Vuuren, D.P., Bregman, B. and Petersen, A.C. (2018). Signal detection in global mean temperatures after "Paris": An uncertainty and sensitivity analysis. *Climate of the Past* 14(139-155). https://doi.org/10.5194/cp-14-139-2018.

Wada, Y. and Bierkens, M.F.P. (2014). Sustainability of global water use: Past reconstruction and future projections. *Environmental Research Letters* 9(10). https://doi.org/10.1088/1748-9326/9/10/104003.

Wada, Y., Wisser, D., Eisner, S., Flörke, M., Gerten, D., Haddeland, I. *et al.* (2013). Multimodel projections and uncertainties of irrigation water demand under climate change. *Geophysical Research Letters* 40(17), 4626-4632. https://doi.org/10.1002/grl.50686.

Wada, Y., Flörke, M., Hanasaki, N., Eisner, S., Fischer, G., Tramberend, S. *et al.* (2016). Modeling global water use for the 21st century: The Water Futures and Solutions (WFaS) initiative and its approaches. *Geoscientific Model Development* 9(1), 175-222. https://doi.org/10.5194/gmd-9-175-2016.

World Bank (2016a). *Poverty and Shared Prosperity 2016: Taking on Inequality.* Washington, D.C. https://openknowledge.worldbank.org/bitstream/handle/10986/25078/9781464809583.pdf.

World Bank (2016b). *High and Dry: Climate Change, Water, and the Economy.* Washington, D.C. https://openknowledge.worldbank.org/bitstream/handle/10986/23665/K8517.pdf?sequence=3&isAllowed=y.

World Health Organization (2009). *Global Health Risks: Mortality and Burden of Disease Attributable to Selected Major Risks.* Geneva: World Health Organization. http://apps.who.int/iris/bitstream/handle/10665/44203/9789241563871_eng.pdf?sequence=1&isAllowed=y.

World Health Organization (2014). *Quantitative Risk Assessment of the Effects of Climate Change on Selected Causes of Death, 2030s and 2050s.* Hales, S., Kovats, S., Lloyd, S. and Campbell-Lendrum, D. (eds.). Geneva. http://apps.who.int/iris/bitstream/handle/10665/134014/9789241507691_eng.pdf?sequence=1.

World Health Organization and United Nations Children's Fund (2017). *Progress on Drinking Water, Sanitation and Hygiene: 2017 Update and SDG Baselines.* Geneva. https://www.unicef.org/publications/files/Progress_on_Drinking_Water_Sanitation_and_Hygiene_2017.pdf.

World Wide Fund for Nature (2016). *Living Planet Report 2016: Risk and Resilience in A New Era.* Gland. http://awsassets.panda.org/downloads/lpr_living_planet_report_2016.pdf.

You, D., Hug, L., Ejdemyr, S., Idele, P., Hogan, D., Mathers, C. *et al.* (2015). Global, regional, and national levels and trends in under-5 mortality between 1990 and 2015, with scenario-based projections to 2030: A systematic analysis by the UN inter-agency group for child mortality estimation. *The Lancet* 386(10010), 2275-2286. https://doi.org/10.1016/S0140-6736(15)00120-8.

Zheng, M.-D. and Cao, L. (2014). Simulation of global ocean acidification and chemical habitats of shallow- and cold-water coral reefs. *Advances in Climate Change Research* 5(4), 189-196. https://doi.org/10.1016/j.accre.2015.05.002.

Chapter 22

Pathways Toward Sustainable Development

Coordinating Lead Authors: Detlef van Vuuren (PBL Netherlands Environmental Assessment Agency), Paul Lucas (PBL Netherlands Environmental Assessment Agency)

Lead Authors: Katherine V. Calvin (Joint Global Change Research Institute/ Pacific Northwest National Laboratory), Serena H. Chung (United States Environmental Protection Agency), Mike Harfoot (World Conservation Monitoring Centre [UNEP-WCMC]), Steve Hedden (Frederick S. Pardee Center for International Futures), Alexandre C. Köberle (Universidade Federal do Rio de Janeiro), Yoshihide Wada (International Institute for Applied Systems Analysis)

Contributing Authors: Lex Bouwman (PBL Netherlands Environmental Assessment Agency), Chenmin He (Peking University), Barry B. Hughes (Frederick S. Pardee Center for International Futures), Fintan Hurley (Institute of Occupational Medicine), Terry Keating (United States Environmental Protection Agency), Jonathan D. Moyer (Frederick S. Pardee Center for International Futures), Marco Rieckmann (University of Vechta)

GEO Fellow: Beatriz Rodríguez-Labajos (Universitat Autònoma de Barcelona)

Executive summary

Model-based scenario analysis can help in identifying ways to achieve the environmental targets of the Sustainable Development Goals (SDGs) and related multilateral environmental agreements (MEAs) (*well established*). Target-seeking scenarios provide insight into the required level of effort, promising measures, and possible synergies and trade-offs between these measures and a range of targets. The usefulness of scenarios can be illustrated by the successful use of such scenarios in the literature on climate policy. Scenarios can be used to explore different pathways for achieving long-term targets and provide insights into the costs and benefits of these pathways. There are important interrelations (synergies and trade-offs) between the achievement of the various SDGs and related MEAs.. This means that strategies that aim to achieve sets of targets will have to take account for these interrelations. At the moment, scenarios that explore the fulfilment of a large set of SDG targets simultaneously are mostly lacking. An assessment of possible pathways must therefore rely on more narrowly focused scenarios in the literature. This does lead to a higher level of uncertainty and some clear knowledge gaps. {22.2}

Overall, available scenario literature suggests that different pathways exist for achieving the targets, but that these pathways require transformative changes (*established, but incomplete*). The rate of change in the pathways, required to meet the targets identified in Chapter 20, indicate that incremental environmental policies will not suffice. Significant improvements in resource efficiency with respect to land, water and energy are required. This includes large productivity gains in agriculture, significant improvements in nutrient-use and water-use efficiency, almost a doubling of the energy efficiency improvement rate and a more rapid introduction of 'carbon-free' energy options. Similarly, achieving full access to food, water and energy resources will require a clear break with current trends. {22.3; 22.4.1}

Achieving the sustainability goals will require a broad portfolio of measures based on technological improvements, lifestyle changes and localized solutions (*established, but incomplete*). The pathways emphasize a number of key transitions that are associated with achieving sustainable consumption and production patterns for energy, food and water, in order to provide universal access to these resources, while preventing climate change, air pollution, land degradation, loss of biodiversity, water scarcity, over-exploitation and pollution of the oceans. These transitions include changes in lifestyle, consumption preferences and consumer behaviour on the one hand, and cleaner production processes, resource efficiency and decoupling, and corporate responsibility on the other. {22.3}

Concurrently eliminating hunger, preventing biodiversity loss and halting land degradation is possible by combining measures related to consumption, production and access to food with nature conservation policies (*well established*). Several measures have been identified that together can help minimize the associated trade-offs, including sustainable agricultural intensification (e.g. increased water- and nutrient-use efficiencies), shifts to low-meat diets, reductions in food loss and waste, improved access to food and nutrition management, landscape management and an expansion of protected areas. {22.3.1}

The strong links between biodiversity loss and land use mean that more coordinated international action is needed (*established, but incomplete*). Scenario literature clearly shows that meeting targets to halt biodiversity loss would not be feasible if land use follows projected business-as-usual trajectories. Also, other policies outside the realm of traditional nature conservation policies are urgently needed to protect biodiversity, such as those related to infrastructure development and climate change. Ensuring more coordinated policy action is therefore important at all levels – within national governments, but also internationally - in particular between land-use planning and biodiversity protection. {22.3.1}

There are multiple pathways to reduce greenhouse gas emissions to levels consistent with the Paris Climate Agreement. Each, however, requires transformative changes and needs to be implemented rapidly (*well established*). Measures that reduce greenhouse gas emissions include lifestyle changes (e.g. a shift to low-meat diets and a move to more public modes of transport), a doubling of energy efficiency improvement, a more rapid introduction of low- and zero-carbon technologies (including hydropower, solar and wind, and carbon-capture-and-storage), reduction of non-CO_2 greenhouse gas emissions and the use of land-based mitigation options (e.g. reforestation and bioenergy). Emission reduction measures need to be implemented rapidly, because the carbon budgets for achieving the Paris Agreement are very tight. As a broad guideline, the rate of decoupling CO_2 emissions from gross domestic product (GDP) needs to increase from the historic rate of 1 to 2 per cent per year to between 4 and 6 per cent per year between now and 2050 if the Paris Agreement targets are to be met. {22.3.2}

Air pollution emissions can be reduced significantly, but pathways towards meeting the most stringent air quality guidelines are currently not available (*established, but incomplete*). Introducing air pollution policies alone is often not enough to achieve stringent air quality standards. However, climate change mitigation (e.g. phasing out fossil fuels) also significantly reduces air pollutant emissions. As a result, scenarios that combine climate policies with stringent air pollution policies show strong reductions in emissions of particulate matter with diameter less than 2.5 µm ($PM_{2.5}$), leading to a significant improvement in air quality in all regions. In the best case scenarios, less than 5 per cent of the population is projected to be exposed to $PM_{2.5}$ levels above the World Health Organization's most lenient interim target of 35 µg/m^3, though more than half of the population is still projected to be exposed to levels above the guideline of 10 µg/ m^3. {22.3.2}

Reducing global water stress, including groundwater depletion, requires more efficient water use, increasing water storage and investing in wastewater reuse and desalination capacity (*established, but incomplete*). To maintain or even reduce the global population suffering from water scarcity by 2050 and beyond, water-use efficiency needs to improve by more than 20-50 per cent globally. This includes increasing agricultural water productivity, improving irrigation efficiency and more efficient water use in domestic and industrial sectors. Wastewater reuse and desalination strategies require a large amount of economic investment and modernizing of existing infrastructure, which might not be feasible for many developing countries. Alternatively, nature-based solutions can increase and / or regulate water supply by mitigating water pollution, while limiting economic investments. {22.3.3}

Achieving environmental targets related to oceans requires consistent policies in other sectors (*well established*). Preventing ocean acidification is highly dependent on climate change mitigation (i.e. reduced CO_2 emissions). Reducing marine nutrient pollution, and related hypoxia and harmful algal blooms, requires a significant reduction in nutrient run-off, primarily from fertilizer use and untreated wastewater {22.3.4}

Ending preventable death of children under five years of age requires continued efforts to reduce environmental risk factors, but also increased emphasis on poverty eradication, education of women and girls, and child and maternal health care (*established, but incomplete*). Ending hunger and achieving universal and equitable access to safe drinking water, adequate sanitation and modern energy services would improve health significantly – especially for children under five. However, even if all the environment-related SDG targets were achieved by 2030, the under-five mortality target would not be met. A healthy planet alone is not enough for healthy people. Achieving the SDG target on child mortality also requires addressing non-environmental risk factors, including poverty alleviation, education of women and girls as well as child and maternal health-care. {22.3.5}

Understanding interlinkages between measures and targets is crucial for synergistic implementation and policy coherence (*well established*). Where measures generally aim at achieving specific targets, or clusters of targets, they can also affect other targets. Integrated approaches are needed to grasp the synergies and deal with the potential trade-offs to achieve the environmental targets simultaneously. {22.3; 22.4.2}

Overall, the literature reveals more synergies than trade-offs within and among the SDGs and their targets (*established, but incomplete*). Significant synergies across human well-being and natural resource targets can be harnessed. For example, reducing agricultural demand by changing dietary patterns towards less meat intake and reducing food loss and waste, reduces the pressure on land and water, thereby reducing biodiversity loss and contributing to climate change mitigation. Other examples discussed in the chapter include education and reducing air pollution. Phasing out unabated use of fossil fuels leads to important co-benefits by achieving both climate and air quality targets, the latter having synergies with improving human health, increasing agricultural production and reducing biodiversity loss. {22.4.2}. The chapter also identifies several trade-offs. This could imply that such measures are less attractive or additional policies are needed to mitigate the trade-offs. {22.3}

Yield improvement and bioenergy are important measures to address biodiversity loss and climate change, respectively, but they can conflict with achieving other targets (*well established*). While nearly all scenarios consistent with the Paris agreement rely on land-based mitigation measures, their use increases demand for land, with related biodiversity impacts, and they potentially lead to higher food prices. Increasing agricultural yields can improve overall food availability and reduce pressure on natural land but could also, through higher levels of water, pesticide and fertiliser use and mechanization, lead to land degradation, water scarcity, hypoxia and harmful algal blooms and biodiversity loss. {22.4.2}

Further model development and pathway analysis is needed to cover a wider set of linkages across the SDGs (*well established*). The scenario literature is still patchy with respect to achieving a broad range of targets. Climate change and land-use issues are well covered, while scenarios addressing land degradation and many challenges related to oceans, but also to chemicals and waste, are mostly lacking. Furthermore, many synergies and trade-offs are discussed in the literature, but besides thematic studies, a thorough overview of all relevant interrelations is lacking. More dedicated analyses are required, including systematic reviews of the existing literature and dedicated integrated assessment modelling, with specific attention to interlinkages that are currently underexplored. {22.5.1}

22

22.1 Introduction

The identified targets associated with the environmental dimension of the Sustainable Development Goals (SDGs) and related Multilateral Environmental Agreements (MEAs) from Chapter 20 will not be achieved under current trends (Chapter 21). This chapter assesses the scenario literature for possible pathways that would achieve those targets, thereby closing the implementation gap **(Table 21.2)**. The focus is on the question of what would be needed to achieve these targets – and what are important synergies and trade-offs between different measures and these targets. This chapter does not discuss the social or political feasibility of the pathways. Moreover, the focus is on measures (e.g. energy efficiency improvement or changes in yield) and not on the policies to implement these measures (e.g. taxes or regulation). The latter will be discussed further in Chapter 24.

A range of scenarios can be found in the literature that analyse how to implement specific targets such as those related to climate change or land-use change (e.g. Global Energy Assessment [GEA] 2012; Clarke *et al.* 2014; Obersteiner *et al.* 2016). Scenarios that address achieving multiple environmental and/or development targets at the same time are far more scarce, with only a few exceptions (e.g. van Vuuren *et al.* 2015; The World in 2050 Initiative [TWI2050] 2018). Furthermore, there is no comprehensive study that explores all the key interrelations between a broad set of measures and SDG targets. Such studies are important, as the SDG targets and those related to MEAs depend on each other in different ways, leading to both synergies and trade-offs in response strategies (Nilsson, Griggs and Visbeck 2016). This gap in the literature means that, in our assessment, the required measures and the interrelations between different targets need to be based on interpretation of existing work.

22.2 Pathways definition

A range of different scenarios exist that describe a move towards sustainable development (see van Vuuren *et al.* 2012 for an overview of different scenario types). Some scenarios explore the consequences of introducing a set of assumptions about key drivers (e.g. population, economic development and technology) consistent with an emphasis on sustainable development. These subsequently look at the impacts for human development and the environment. Examples include the SSP1 (Sustainable Development) scenario of the Shared Socio-economic Pathways (SSPs) (Riahi *et al.* 2017; van Vuuren *et al.* 2017a; **Box 21.2**), the TechnoGarden scenario of the Millennium Ecosystem Assessment (Millennium Ecosystem Assessment 2005) and the Great Transition Scenarios of the Global Scenarios Group (Raskin *et al.* 2002). These scenarios all lead to relatively positive developments for environmental problems, although they typically do not reach all the targets introduced in Chapter 20. Other scenarios apply a 'back-casting approach' – showing pathways towards reaching a set of sustainable development objectives (e.g. the Road from Rio+20 scenarios; see **Box 22.1**). Two recent scenarios focus specifically on the role of lifestyle change and the possible implications for climate change mitigation (Grubler *et al.* 2018; van Vuuren *et al.* 2018). Sustainable development scenarios differ from current trend scenarios (see Chapter 21) in many ways – including in the nature of economic activities and personal lifestyles, the availability and performance of technologies, and the interventions, regulations and policies that are applied – leading to differences in associated levels of effort, and synergies and trade-offs that will be required to achieve sustainable development (PBL Netherlands Environmental Assessment Agency 2012).

This chapter assesses available scenarios in the literature. No new scenarios were developed. The scenarios cited here should be seen as illustrations of possible pathways towards sustainable development and not as well-defined blueprints. Where possible, SSP-derived scenarios are used **(see Box 21.2)**. Furthermore, the storylines of the Roads from Rio+20 study are used to show that there are different ways to strengthen and direct, or redirect, technologies, preferences and incentives in society towards sustainable development (van Vuuren *et al.* 2015; **Box 22.1**). As such, the underlying dimensions of the Roads from Rio+20 study can also be used to qualify the measures analysed in this chapter. The first dimension then makes the distinction between options that depend on global cooperation and those that specifically focus on the local situation (mostly related to ensuring heterogeneity and local governance). The second dimension distinguishes between options that focus on introducing more sustainable production patterns versus more sustainable consumption patterns. The Roads from Rio+20 scenarios can also be mapped on these dimensions **(Figure 22.1)**. It should be noted that, so far in model-based scenario analysis, strategies based on making production patterns more sustainable have received more attention than strategies focused on changing consumption patterns.

Figure 22.1: The scenarios from the Roads from Rio+20 study

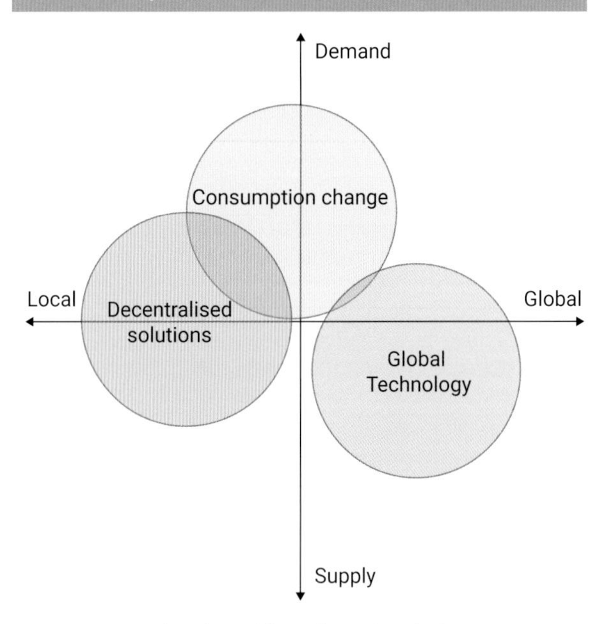

These scenarios are based on a different focus along the dimensions global versus local interventions and production- versus consumption-side orientation. The scenarios are used in this chapter to illustrate that there are different strategies in moving towards sustainable development.

22.3 Pathways towards achieving the targets

A range of measures identified as necessary to achieve the selected targets (see Chapter 20) are listed in **Figure 22.2**. These measures are linked to the five clusters of closely related

Box 22.1: Roads from Rio+20

The Rio+20 study looked into model-based pathways that simultaneously achieve a broad set of long-term environment and development targets (van Vuuren *et al.* 2015). The pathways were developed using the IMAGE integrated assessment model. The targets were based on existing, pre-2012, international agreements (SDGs avant-la-lettre). The study focused on two key sets of related challenges:

1. Eradicating hunger and halting biodiversity loss;
2. Universal access to modern energy and mitigating climate change.

The study further addressed trade-offs with water, nutrients and health. The study introduced three possible pathways towards achieving sustainability targets: (1) global technology, (2) decentralized solutions, and (3) lifestyle change. The different trajectories for the alternative scenarios can be explained by the differences in perceived urgency, economic and institutional effectiveness, and feasible rate of lifestyle changes. The scenarios can be characterized as follows:

- Global technology: In the global technology pathway, international and national decision makers feel an urgency to deal with global sustainability issues and manage to convince most citizens to introduce large-scale, global solutions to resolve these issues. The problems and solutions are primarily perceived and solved as large in scale and global in outreach.
- Decentralized solutions: The belief that a sustainable quality of life can only be realized at the local or regional level gets more priority than the possible impacts of long-term issues. As a result, sustainability problems are primarily seen and resolved in the form of small-scale and decentralized technologies and organizational efforts. Local 'smart' solutions may also fall into this strategy. This is a 'bottom-up' evolving world.
- Consumption change: Partly because there is a growing awareness of sustainability issues, important changes in lifestyle take place that facilitate a transition towards less material- and energy-intensive activities. Targets that still have not been achieved are bridged with additional existing technologies.

Figure 22.2: Selected measures and their related clusters as examined in this chapter

	Agriculture, food, land and biodiversity	Energy, air and climate	Fresh water	Oceans	Human health
Human well-being	• Nutrition management • Improve access to food	• Improve energy access	• Improve access to water, sanitation and hygiene		• Poverty alleviation • Child and maternal healthcare • Education
Sustainable consumption and production	• Reduce food loss and waste • Improve yields • Improve nutrient use efficiency • Dietary change • Land ownership	• Behavioral change • End-use electrification • Low- to zero-emission technologies • Bioenergy • Improve energy efficiency • Air pollution control • Non-CO_2 emission control	• Improve water-use efficiency		
Natural resource base	• Manage soil organic carbon • Protection of terrestrial ecosystems • Forest management • Land use planning • Minimize land damage	• Negative emission technologies	• Wastewater treatment • Water quality standards • Desalination • Integrated water resources management	• Protection of marine ecosystems • Sustainable fisheries • Ocean regulation	

environmental issues and the three groups of SDGs, mirroring the framework of **Figure 22.1**. Note that, in line with Chapter 21, targets associated with production and consumption, such as the rate of yield improvement or energy intensity improvement, are discussed as means to achieve the desired situation. They are discussed within the different clusters and summarized in the synthesis at the end of this chapter.

The following sections review the scenario literature for pathways to achieve the targets within each cluster, discussing the measures required for achieving the targets, and potential synergies and trade-offs between the different measures and targets within each cluster.

In Part A, chemicals and waste, and wastewater were also identified as a major global environmental problem. As explained in **Box 21.1**, there is not a lot of specific scenario literature on these issues. We do discuss reducing food loss and waste in the agriculture, food, land and biodiversity cluster. In the energy, air and climate and freshwater clusters, we pay attention to increasing efficiency – which addresses the issue of wasting energy and water, as well as wastewater treatment.

22.3.1 Agriculture, food, land and biodiversity

The selected targets for the agriculture, food, land and biodiversity cluster can be summarized as ending global hunger, while at the same time halting biodiversity loss and achieving land-degradation neutrality (see Chapter 20). Selected targets that contribute to achieving these endpoint targets include increasing agricultural productivity and increasing nutrient-use efficiency.

Without additional measures, none of these three targets are projected to be met (Chapter 21). While hundreds of millions of people are projected to still be undernourished in 2050, agricultural area is projected to expand by between 150 and 425 million ha between 2010 and 2050, resulting in declines in natural area, including forests. Biodiversity projections suggest a further decline in species richness and abundance, and land degradation is projected to continue. Achieving the targets requires a major transformation of the food production system, the main driver for human-induced land-use change.

With respect to ending hunger, the Food and Agriculture Organization of the United Nations (FAO) definition of food security is used: "Food security exists when all people, at all times, have physical and economic access to sufficient, safe and nutritious food that meets their dietary needs and food preferences for an active and healthy life" (FAO 1996). In practice, not all scenarios include enough information to assess all aspects of this definition. Therefore, we have taken qualitative descriptions of the scenarios to assess whether the target is met. For biodiversity, the target is based on the Convention on Biological Diversity (CBD) strategic plan for biodiversity 2011-2020 (CBD 2010), translated to halt biodiversity loss by 2020 for developed countries and from 2030 onward for developing countries (Kok *et al.* 2018). Halting biodiversity loss is therefore taken to mean preventing further declines in the diversity within species, across species and within ecosystems, as well as the abundance and coverage of these organisms. For achieving land degradation neutrality no quantitative analysis is available.

There are important linkages between this cluster and other cluster targets. For instance, combating climate change might require significant amounts of bioenergy and land devoted to its production. Total land area dedicated to bioenergy production is a major uncertainty in future scenarios, especially those with stringent emissions abatement targets (Popp *et al.* 2014). In addition, increased agricultural production could require increasing inputs of freshwater, nitrogen and phosphorus.

In order to simultaneously end hunger and prevent biodiversity loss and further land degradation, enough food needs to be produced to feed a global population of 9-10 billion people by 2050 without expanding agricultural land (at least on a global scale). At the same, there will also be other demands for land such as biomass production for energy and demand to produce timber. Reducing hunger not only requires sufficient production, but also, much more importantly, issues of access (economic and physical) will need to be addressed in order to ensure that all people receive adequate food. Additionally, this needs to occur with minimal pollution (nitrogen, phosphorus or other). Further land protection and land restoration may be required to prevent biodiversity loss and avoid or reverse land degradation.

There are several scenarios in the literature that achieve these targets in an integrated way. These studies show that that there are multiple routes for achieving the targets, such as via more technology-focused routes, changing demand or focusing more on governance structures, land tenure and creating markets (Tilman *et al.* 2011; Bajželj *et al.* 2014; van Vuuren *et al.* 2015; Obersteiner *et al.* 2016). More recent literature based on the SSPs discusses multiple routes that could lead to zero hunger by 2050 (Hasegawa *et al.* 2015), some of which are achieved without expanding agricultural area (Popp *et al.* 2017). However, it is important to note that food security involves not just security of supply but also demand factors such as access to food, including affordability and distributional concerns (Qureshi, Dixon and Wood 2015), and its nutritional value. However, issues of access, distribution and nutritional value are largely excluded from the scenario literature and thus not discussed in depth in this chapter.

Most scenario studies that discuss prevention of biodiversity loss assume a suite of land-, agriculture- and biodiversity-related measures acting together, including increasing agricultural productivity, reducing consumption of meat, dairy and eggs, reducing food loss and waste, avoiding fragmentation and expanding protected areas. Such measures can reduce biodiversity loss (van Vuuren *et al.* 2015) and extinction risks for birds and mammals (Tilman *et al.* 2017).

Overall, a broad range of measures is discussed in the literature, including measures related to agricultural production, agricultural demand-side measures and measures that aim for protection of terrestrial ecosystems.

Measures related to agricultural production
One option to achieve the targets in the agriculture, food, land and biodiversity cluster is to change agricultural production patterns. This includes yield improvement (to avoid further expansion of agricultural land), but also other efficiency measures, such as for nutrient and water use, to reduce the environmental pressure of agriculture.

Improving yield

In the SSP2 baseline (Fricko *et al.* 2017), between 2010 and 2050, per capita demand for food, feed and energy crops increases by 60 per cent. In the same period, global average aggregate food, feed and energy crop yields (mean tons of agricultural products per hectare) also increases (by around 1.0 per cent per year). As a result, the net effect in SSP2 is an increase in cropland area of about 15 per cent in 2050 (230 million ha) **(Figure 22.3)**. This is in line with the FAO projection for yield improvements and agricultural area expansion through 2050 (Alexandratos and Bruinsma 2012). To limit cropland expansion, yield growth would need to increase from around 1.0 to 1.4 per cent per year. It is thus useful to look into the evidence on the question whether fast yield improvements are possible in the future. First, similar yield improvement rates have been achieved historically (Alexandratos and Bruinsma 2012). Moreover, several scenarios indeed show high future yield increase **(Figure 22.3)**. There is also a large yield gap between the most- and least-productive regions (Global Yield Gap and Water Productivity Atlas 2018),

Figure 22.3: Percentage change in non-energy crop production versus the percentage change in non-energy cropland area from 2010 to 2030 and 2050

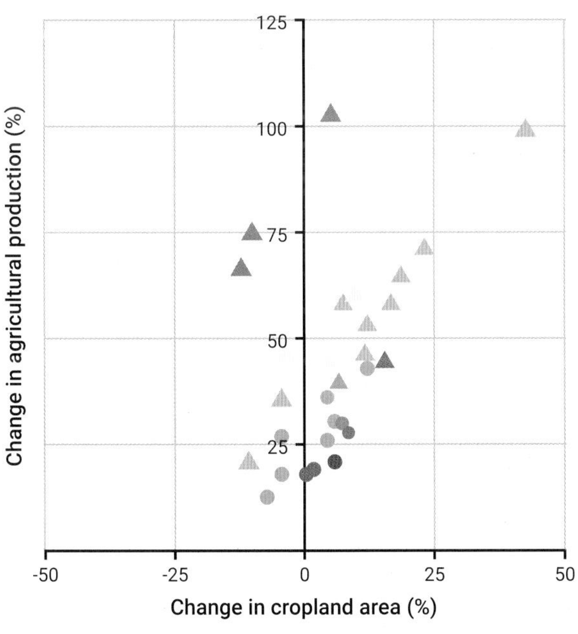

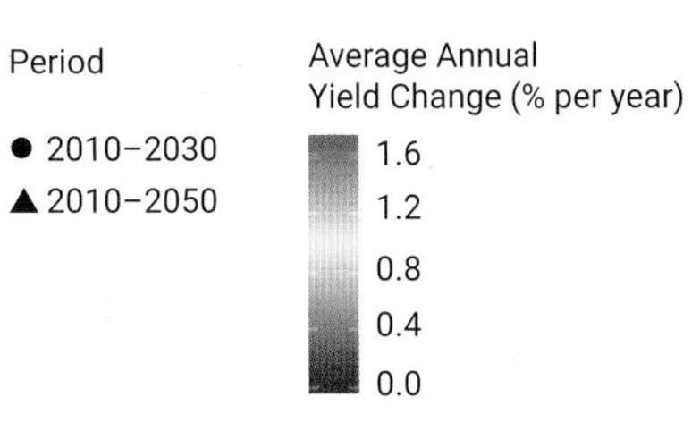

Each marker is a model-scenario-year combination. Colour indicates the annual percentage change in yield over the same time period. Yellow is close to historical trends (about 1 per cent per year between 1960 and today from Alexandratos and Bruinsma 2012); blue indicates yield growth faster than historical trends; red indicates yield growth slower than historical trends. For the SSPs, yield is the global average yield for cereal crops. For the Bajželj *et al.* (2014) scenarios, yield is the global average yield for wheat and data are referenced with respect to 2009.

Sources: SSPs (Popp *et al.* 2017) and Bajželj *et al.* (2014).

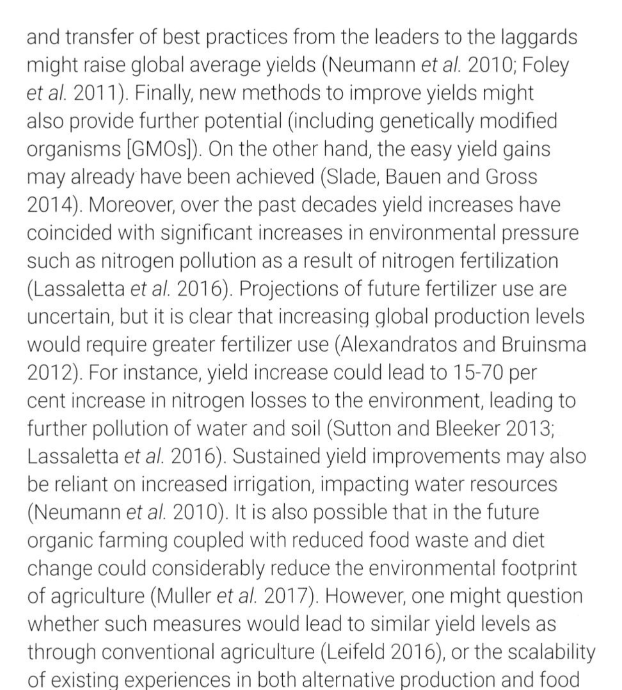

and transfer of best practices from the leaders to the laggards might raise global average yields (Neumann *et al.* 2010; Foley *et al.* 2011). Finally, new methods to improve yields might also provide further potential (including genetically modified organisms [GMOs]). On the other hand, the easy yield gains may already have been achieved (Slade, Bauen and Gross 2014). Moreover, over the past decades yield increases have coincided with significant increases in environmental pressure such as nitrogen pollution as a result of nitrogen fertilization (Lassaletta *et al.* 2016). Projections of future fertilizer use are uncertain, but it is clear that increasing global production levels would require greater fertilizer use (Alexandratos and Bruinsma 2012). For instance, yield increase could lead to 15-70 per cent increase in nitrogen losses to the environment, leading to further pollution of water and soil (Sutton and Bleeker 2013; Lassaletta *et al.* 2016). Sustained yield improvements may also be reliant on increased irrigation, impacting water resources (Neumann *et al.* 2010). It is also possible that in the future organic farming coupled with reduced food waste and diet change could considerably reduce the environmental footprint of agriculture (Muller *et al.* 2017). However, one might question whether such measures would lead to similar yield levels as through conventional agriculture (Leifeld 2016), or the scalability of existing experiences in both alternative production and food waste reduction methods (Schneider *et al.* 2014). For pasture area, the intensification of livestock production could limit the increase in pasture area, and possibly lead to a decrease.

Reducing environmental pressures associated with agriculture

High-yield agricultural systems are usually associated with high levels of nitrogen loss as reported in the previous section. There is evidence, however, that the negative impact of high-yield agriculture on nitrogen loss could be limited by improving nitrogen-use efficiency (Lassaletta *et al.* 2016; Bouwman *et al.* 2017). This can be shown by the large variation in application rates, with excess application in some regions leading to significant environmental impact, especially in China (Zhang *et al.* 2016; Cui *et al.* 2018). In fact, rapidly increasing global nitrogen-use efficiency from the current 40 per cent to close to 70 per cent may lead to a sharp decline in excess nitrogen to 50 Tg N/year, with the added benefit of potentially leading to stabilization of total nitrogen inputs in global crop production (Zhang *et al.* 2015). Mogollon *et al.* (2018) present similar findings but emphasize that this can only happen in optimistic sustainability scenarios (limited increase in demand and high efficiencies). The relationship of crop yield to nitrogen application means there are diminishing returns to higher nitrogen application in regions with high fertilizer application rates and more potential for increased production in regions with low application rates. This means there is room globally to optimize nitrogen application. The trade-off in this case would be an increase in international trade of agricultural commodities.

It is also important to reduce other environmental pressures – such as high levels of water consumption (see Section 22.3.3), the negative impacts of use of herbicides and pesticides, and eutrophication of inland and coastal waters due to excess nutrient use in food production and sewage water discharge. Scientific evidence shows that it is important to maintain agricultural sustainability to ensure services such as natural pest control, pollination and fertility (Oerke 2006; de Vries *et al.* 2013; Garibaldi *et al.* 2017). For instance, except for cereals (which are not insect-pollinated), many important global food

crops depend, at least partly, on animal pollinators (usually insects) for yield and/or quality, and pollinator-dependent crops contribute 35 per cent of the global crop production volume (Klein *et al.* 2007). Reducing negative impacts can to some degree be achieved in high-yield agricultural systems. There is some evidence that organic farming could be an alternative as it may support greater local species richness and higher densities of natural organisms compared with conventional farms (Bengtsson, Ahnström and Weibull 2005; Tuck *et al.* 2014). However, organic farming could also lead to lower yields and thus increased land use (Clark and Tilman 2017). The role of organic farming cannot be really assessed in this chapter as, at present, the issue of organic farming is hardly addressed in scenario studies. In fact, the same goes for strategies to preserve sufficient genetic diversity. While there is some evidence that it is important to maintain diversity as a buffer against all kinds of environmental variability, again this is not really addressed in scenario studies. Such diversity can be encouraged by rotating crops, intercropping and varying crop varieties.

Preventing land degradation

The loss of soil organic carbon and other forms of soil degradation can significantly impact crop yields and the nutritional values of food produced (Godfray *et al.* 2010; Lal 2015; Rojas *et al.* 2016). Therefore, maintaining soil health, through the management of soil organic carbon and preventing land degradation, is important. The recently published Global Land Outlook is one of the few studies that discuss land degradation in the context of different scenarios, but it only discusses trend scenarios and not pathways towards achieving the land degradation neutrality target (United Nations Convention to Combat Desertification [UNCCD] 2017; van der Esch *et al.* 2017). Land restoration and protection targets are projected to increase tree cover by 4 million km^2 in 2050 compared to the area in 2000 and increase forest carbon stocks by 50Gt over the same time period (Wolff *et al.* 2018). However, due to the limited scenarios literature, it is hard to assess the role of avoiding land degradation in achieving the SDGs.

Agricultural demand-side measures

To limit cropland expansion, it is also possible to reduce the food demand that would occur in baseline projections. Reductions in demand could come from reduced food consumption, reduced waste or reduced feed/fuel uses of crops.

Dietary change

Changes in diet are considered an effective measure for reducing land-use impacts of agriculture. Diet changes resulting in less meat consumption would reduce crop use as animal feed, which in turn would reduce demand for land, since direct human consumption of crops requires less land (Stehfest *et al.* 2009). In particular, a reduction in beef consumption would have the most direct positive impact on environmental indicators, as ruminants have the lowest feed and protein conversion rates of all livestock (Béné *et al.* 2015). This implies that reduction of meat consumption to levels consistent with health recommendations in high-income countries could lead to positive impacts in terms of reducing agricultural land-use and increasing human health (Stehfest *et al.* 2009) – as on average current consumption of beef is above this level. Strong reductions in land area for food production as a result of dietary shifts towards more plant-based diets have been reported by Foley *et al.* (2011) and Stehfest *et al.* (2009). Such a shift would also lead to health benefits, according to these studies. Land-efficiency gains can also be gained by eating different meat. Meat from non-ruminant livestock (e.g. pigs) has a lower impact than beef, and the land footprint of their diets can be improved by shifting to more efficient (higher-yielding) fodder crops (Béné *et al.* 2015; van Zanten *et al.* 2018). Thus, diets based on lower shares of ruminants would reduce land demand. In the case of bivalves, aquaculture may even remove nutrient run-off into estuaries through filtration, a potential synergy.

More recent scenarios in the literature have also focused on dietary change, including the SSP1 scenarios (see Popp *et al.* 2017), and the 'consumption change' pathway from Roads from Rio+20 (van Vuuren *et al.* 2015; van Vuuren *et al.* 2018) and others (Bajželj *et al.* 2014; Tilman and Clark 2014). The dietary change ranges from modest shifts towards non-ruminants (the SSP1 scenario) to complete elimination of meat (Tilman and Clark's Vegetarian scenario). Several of these scenarios limit the expansion of cropland area, but these also include enhanced yields, suggesting that dietary change alone is not enough to limit cropland expansion given a growing population. Note that, in addition to changes in yield and diet, these scenarios also have limited expansion of bioenergy cropland (60 and 140 million ha in 2050 in the SSP1 scenario of the IMAGE and GCAM models, respectively). In the end, this means that a combination of yield improvement, diet change and control of bioenergy expansion offers the most likely situation in which expansion of agricultural area can be avoided.

Waste and loss reduction

Global agricultural production in 2010 (about 3,900 kcal of food crops per person per day) was more than enough food to feed the world, yet more than 800 million people were undernourished (Alexandratos and Bruinsma 2012; Kummu *et al.* 2012). One reason is that 25-40 per cent of food produced is wasted, either through supply-chain waste or end-consumption waste (Godfray *et al.* 2010; Kummu *et al.* 2012). Reducing food waste and loss is one way of reducing hunger, while limiting cropland expansion. The amount of food wasted today is enough to feed several hundred million people a year (West *et al.* 2014), with some studies showing that if half of this waste were redistributed to consumers an extra billion people could be fed (Kummu *et al.* 2012). Similarly, Bajželj *et al.* (2014) show that cutting food waste in half would reduce cropland area by 14 per cent. Muller *et al.* (2017) show that, in addition to reducing land demand, dietary change and waste reduction can result in reduced fertilizer and water use. Bijl *et al.* (2017) show that, although significant improvement can be achieved through yield increase, the improvement is less than expected – mostly because meat is, on average, wasted less than other agricultural products. Several of the scenarios that look into waste reduction also report limited cropland expansion (consumption change from van Vuuren *et al.* 2015 and some scenarios of Bajželj *et al.* 2014). Each of these scenarios also assumes enhanced yields leading to the conclusion that waste reduction alone is not enough to limit cropland expansion given an increasing population.

Changes in food distribution
Hunger is to some degree a function of available calories, but more importantly the distribution of these calories. Income distribution plays a key role in food distribution (Wanner *et al.* 2014; Hasegawa *et al.* 2015). In their analysis, Hasegawa *et al.* (2015) conclude that future developments in global hunger are mostly determined by population growth, inequality in food distribution and per capita domestic food production. Improving access to food for the poorest households significantly reduces the required increase in food production to feed the global population in 2050 (van Vuuren *et al.* 2015). Also avoiding food waste reduces demand for cropland and could still allow for meat consumption, albeit at a lower rate than current-trend projections (Röös *et al.* 2017).

In baseline scenarios, childhood stunting and wasting are also projected to decrease, but not enough to achieve the SDG target of elimination by 2030 (Global Burden of Disease [GBD] 2015 SDG Collaborators 2016; GBD 2016 SDG Collaborators (2017). Meanwhile, the prevalence of overweight children has been increasing over the past 15 years (GBD 2015 SDG Collaborators 2016): fewer than 5 per cent of countries are projected to achieve the SDG target for overweight children (GBD 2016 SDG Collaborators 2017). Achieving these targets therefore requires accelerated action on nutrition as well as the more distal drivers of poor health outcomes – poverty, low levels of education and health spending, as well as conflict (GBD 2016 SDG Collaborators 2017; see also Section 22.3.5).

Maintaining terrestrial biodiversity
The baseline scenarios covered in Chapter 21 show a further decline in biodiversity. Some scenarios have been published that specifically look into how to halt biodiversity loss (e.g. van Vuuren *et al.* 2015; Obersteiner *et al.* 2016; Kok *et al.* 2018; Leclere *et al.* 2018). These scenarios show that, in addition to preserving terrestrial biodiversity in protected areas, it will be at least as important to reduce the external drivers that lead to loss of biodiversity such as expansion of land use, climate change and expansion of infrastructure. We briefly discuss some of these elements below. All-in-all, the scenario literature suggests that pathways to halting biodiversity loss exist – but that such scenarios will be difficult to implement.

Protecting terrestrial ecosystems
Protected areas are a key land management conservation tool. Syntheses have demonstrated that, compared with other locations, the diversity of species within protected areas tends to be 10 per cent greater and the abundance of species 15 per cent greater (Coetzee, Gaston and Chown 2014; Gray *et al.* 2016). Also, habitat conversion rates are 7 per cent lower within protected areas (Geldmann *et al.* 2013). While the CBD's Aichi Target 11 suggests a 17 per cent coverage target, in 2016 protected areas occupied 14.6 per cent of the terrestrial land area. As shown in Chapter 21, current trends will lead to a dramatic loss of biodiversity. Therefore, coordinated international action is urgently needed to balance land-use decision-making and biodiversity conservation. Expansion of the protected land area by 5 per cent in a well-designed way could lead to a significant increase in the protection of biodiversity (Pollock, Thuiller and Jetz 2017). Many scenarios in the literature have explicit assumptions on protected area trends. However, protected area expansion should not be the only consideration and should not come at the expense of effective management of current protected areas (Barnes *et al.* 2018). Furthermore, environmental policy outside of the formal protected areas network is of critical importance.

Land ownership
Land ownership has implications for land management and can therefore have implications for biodiversity residing on it. For example, private versus publicly owned lands have different bird species compositions (Maslo, Lockwood and Leu 2015) and private temperate forests contain a greater diversity and density of microhabitats that can support greater biodiversity (Johann and Schaich 2016). Over one-quarter of the whole terrestrial land surface is managed or under the tenure rights of indigenous groups and this land intersects with approximately 40 per cent of protected areas and ecologically intact landscapes (Garnett *et al.* 2018). In addition to public and private land ownership, local committees, and indigenous peoples' land rights and the manner in which they manage that land is therefore likely to be essential to meeting local and global conservation goals. Assessing the role of land ownership in pathways towards sustainability beyond this is difficult, however, because land ownership is seldom incorporated explicitly into scenario exercises.

Land-use planning
Land-use planning involves the systematic assessment of environmental, economic and social impacts of the range of potential uses of land in order to decide on the optimal pattern of land use. Land-use planning and systematic conservation planning has seldom been explored explicitly as a tool in global scenarios. The most noteworthy exceptions are the recent scenarios by Leclere *et al.* (2018) that use the biodiversity value of land areas to determine optimal land use and also can inform GEO assessments in the future. They find that such an approach in land-use planning can indeed contribute to a strategy that aims to halve biodiversity loss.

Forest management
Meta-analysis shows that different categories of forest management types have different implications for biodiversity loss, with selection and retention systems having the least detrimental effect on species diversity, while timber and fuelwood plantations have the worst effect (Chaudhary *et al.* 2016). Although forest management practices are not always explicitly represented in scenario simulations, studies suggest that consistent implementation of any single management regime results in suboptimal biodiversity outcomes compared with an optimal combination of management regimes (Monkkonen *et al.* 2014).

Significant trade-offs across the targets
A number of trade-offs can be identified between specific measures and the various targets within this cluster. Three important ones are as follows.

- Increases in cropland area can help reduce hunger by enabling increased food production. This expansion is included in many of the scenarios in the literature (e.g. Tilman *et al.* 2011; Bajželj *et al.* 2014; Tilman and Clark 2014; Popp *et al.* 2017). However, expansion of cropland area can lead to clearing of natural lands and increased land-use change emissions, which have implications for biodiversity, land degradation and climate change. Note that limiting the expansion of cropland area has implications for crop yields, fertilizer use and energy crop

production as well (see Chapter 5 and Sections 22.3.2). Additionally, limiting cropland expansion could have implications for development (Sandker, Ruiz-Perez and Campbell 2012).

- Increasing fertilizer application rates may help increase agricultural yields in regions with persistent yield gaps but can also have severe consequences for freshwater and coastal ocean eutrophication, and climate change, with excess nitrogen and phosphorus run-off potentially impacting water quality (Beusen *et al.* 2016; Bouwman *et al.* 2017). On the other hand, sustainable intensification of agriculture (e.g. through precision agriculture) can help deliver higher yields while preserving ecosystem services and reducing environmental impacts (Foley *et al.* 2011; Garnett *et al.* 2013; Garbach *et al.* 2017). Increasing global nitrogen-use efficiency can reduce nitrogen run-off to the environment (see Chapter 8).
- Monoculture plantations of exotic, fast-growing trees have been used to maximize carbon sequestration (Chazdon 2008; Hunt 2008), negatively impacting local biodiversity. However, plantations of multiple native species can be an effective alternative (Hulvey *et al.* 2013; Cunningham *et al.* 2015), while also providing greater benefits for biodiversity (Bradshaw *et al.* 2013). Furthermore, natural regrowth is an alternative to plantations that has been shown in tropical forests to be more ecologically beneficial, cost-effective and resilient (Crouzeilles *et al.* 2017).

23.3.2 Energy, climate and air

The selected targets for the energy climate and air cluster can be summarized as the challenge to achieve universal access to modern energy services, while at the same time combating climate change and improving air quality (see Chapter 20). Selected targets that contribute to achieving these endpoint targets include improving energy efficiency and increasing the share of renewable energy.

Under current trends, none of these three targets are projected to be met (Chapter 21). By 2030, more than 2 billion people are still projected to cook on traditional biomass stoves or open fires and around 700 million people do not have access to electricity. The global mean temperature is projected to increase further, while a significant share of the global population is still exposed to concentrations of particulate matter with diameter less than 2.5 μm ($PM_{2.5}$) above 35 μg/m^3. Achieving these targets requires a major transformation of the energy system.

Modern energy services include electricity and clean fuels for cooking, heating and lighting, with 'clean' defined by the emission rate targets and specific fuel recommendations (i.e. compared to unprocessed coal and kerosene) of the World Health Organization (WHO) guidelines for indoor air quality (WHO 2014). Combating climate change means keeping the global mean temperature change well below 2°C and if possible below 1.5°C (United Nations Framework Convention on Climate Change [UNFCCC] 2015). Improving air quality means air pollution levels should, in the long term, be consistent with the WHO guidelines – that is, the interim target of annual mean $PM_{2.5}$ concentration should be below 35 μg/m^3 by 2030 (WHO 2006).

There are important linkages between this cluster and other cluster targets. For instance, most low-carbon pathways that limit global mean temperature to 2°C (or 1.5°C) include significant amounts of bioenergy. The role of land-based ecosystems, both natural and managed, is essential for achieving net-zero and net-negative emissions.

There is a rich literature of scenarios that have looked at the challenge of meeting ambitious climate targets (for an overview, see Clarke *et al.* 2014, and more recent studies including Riahi *et al.* 2017; Rogelj *et al.* 2018; van Vuuren *et al.* 2018). Fewer published scenarios have looked at meeting ambitious energy access targets (e.g. Pachauri *et al.* 2013; International Energy Agency [IEA] 2017) or air pollution targets at a global scale (e.g. Rao *et al.* 2017). A broad range of measures is discussed in the literature, including improving energy access (electricity and clean cooking fuels), reducing greenhouse emissions by addressing both energy demand and production, and air pollution control.

Improving access to energy

Universal access to modern energy services will not be achieved by 2030 in a baseline scenario, particularly not in sub-Saharan Africa (for electricity and clean fuels and technologies) and in Asia (mainly clean fuels and technologies) (see Chapter 21). Achieving universal access to electricity requires further expansion of generation capacity and transmission and distribution networks, as well as access to more efficient and affordable appliances, with a specific focus on poor, remote communities (GEA 2012; IEA 2017; Lucas, Dagnachew and Hof 2017). To achieve universal access to clean fuels and technologies, the affordability, availability and safety of fuels and practices for cooking, heating and lightning should be improved (Modi *et al.* 2006). Improved fuels include liquified petroleum gas (LPG), natural gas and electricity in urban areas, and a range of technologies (including biogas and the use of advanced biomass cookstoves) in rural areas (IEA 2017). Modelling studies have shown that there are different pathways to achieve universal access to modern energy services (Pachauri *et al.* 2013; Dagnachew *et al.* 2017).

The choice of the electrification system – grid-based, mini-grid or off-grid – depends on a range of mostly local factors, including the level of household electricity demand, the distance to the existing grid and local resource availability (Dagnachew *et al.* 2017). Grid-based electrification is attractive for densely populated areas with an expected high demand for electricity and/or within a reasonable distance of existing high voltage power lines, while decentralized electrification systems are key to reaching out to semi-urban areas with low consumption density, and remote rural areas (Dagnachew *et al.* 2017; IEA 2017; Lucas, Dagnachew and Hof 2017). Total annual investments to achieve universal access are estimated at US$52 billion globally (IEA 2017) and US$24-49 billion in sub-Saharan Africa alone (Dagnachew *et al.* 2017; Lucas, Dagnachew and Hof 2017), depending primarily on total household electricity demand and the cost of high-voltage transmission and distribution.

Policies that could encourage a transition to clean fuels and technologies for cooking, heating and lightning include fuel subsidies and grants or microlending facilities to make access to credit easier and lower households' cost of borrowing (Riahi *et al.* 2012). The use of improved or advanced biomass stoves may in fact lead to economic gains instead of costs, as the investments would be countered by the reduction in spending on fuelwood (van Ruijven 2008). Total required investments to achieve universal access to clean fuels for cooking, heating and lighting are projected to be less than 10 per cent of what is needed for achieving universal access to electricity (Pachauri *et al.* 2013; IEA 2017). Improving access to clean fuels can significantly improve health (Pachauri *et al.* 2013; Landrigan *et al.* 2018). Climate policy can induce energy savings, reducing the overall investment required for achieving universal access (Dagnachew *et al.* 2018).

Reducing greenhouse gas emissions

The Paris climate targets set very stringent constraints for the development of future energy systems. Although some recent publications have shown that carbon budgets are subject to considerable uncertainty (Intergovernmental Panel on Climate Change [IPCC] 2018; Rogelj *et al.* 2016; Millar *et al.* 2017), the main message is that they are small compared to current emissions. To meet the Paris climate targets, cumulative CO_2 emissions from now onwards need to be in the order of 1000 -1600 gigatons of CO_2 (2°) or even 300-900 gigatons of CO_2 (1.5°). The current emissions are in the order of 40-42 gigatons CO_2/year (Le Quéré *et al.* 2016; IPCC 2018). Assuming a linear reduction without negative emissions, unabated fossil fuel use thus needs to be phased out somewhere around the middle of the century (van Vuuren *et al.* 2017a). This would require an immediate halt to investments into CO_2-emitting technology, but possibly even a faster retirement of existing fossil fuel infrastructure (Johnson *et al.* 2015; Gambhir *et al.* 2017).

The option, however, also exists to actively remove CO_2 from the atmosphere, for instance by afforestation and bioenergy, combined with carbon-capture-and-storage, direct-air-capture, enhanced weathering and increasing soil carbon (IPCC 2018). However, the amount of CO_2 that can be removed from the atmosphere in this way is not unlimited: both afforestation/reforestation and bioenergy are restricted by the amount of land available, as well as possible impacts on biodiversity and food production (Smith *et al.* 2016). Moreover, the storage potential for CO_2 is limited (Koelbl *et al.* 2013). Among various options for CO_2 removal that have been assessed, under current technologies, only sequestration in geological formations is considered to have the capacity and permanence necessary to store CO_2 at the gigaton level, which is necessary to reduce CO_2 emissions significantly (Benson *et al.* 2012). While the estimated storage capacity is more than enough to meet emissions reduction targets, the estimates do not consider the risks associated with permanent storage (e.g. environmental contamination from leakage, seismic activities) (de Coninck and Benson 2014; Bui *et al.* 2018). Therefore, rapid emissions reduction will be needed in the short term regardless of the availability of negative emissions technologies (van Vuuren *et al.* 2017a). **Figure 22.4** shows the range of scenarios in the SSP database following the SSP2 baseline and those consistent with the Paris targets of well below 2°C (Riahi *et al.* 2017; Rogelj *et al.* 2018). The scenarios depicted here are based on low cost pathways, assuming an immediate response. There are several papers in the literature that show that a delayed response is more expensive and could even make it impossible to reach stringent targets (Riahi *et al.* 2015; Rogelj *et al.* 2018). Such delayed response would, for instance, occur if countries decide to follow the currently formulated climate policies and aim for a rapid implementation of climate policy after 2030.

Globally, energy-related CO_2 emissions would need to be reduced by around 60-70 per cent by the middle of the century in order to meet the Paris target, even when accounting for

Figure 22.4: Global CO_2 emissions and associated global mean temperature increase for the SSP2 baseline and derived scenarios consistent with the Paris target to stay well below 2°C increase

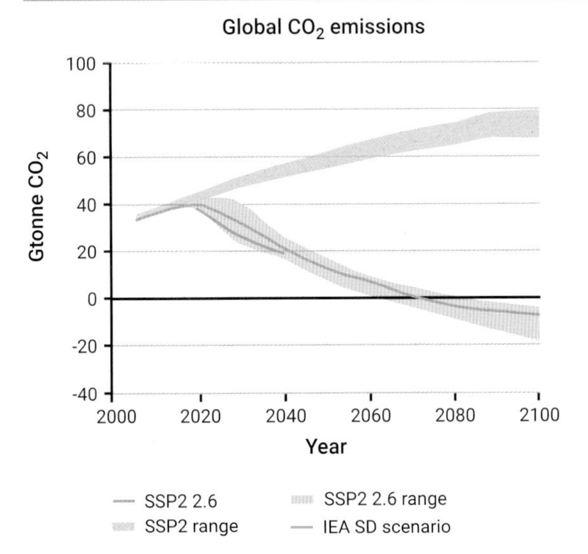

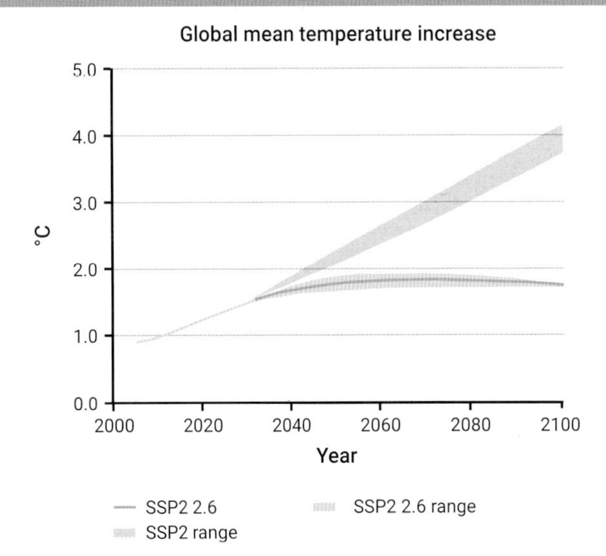

Source: IEA (2017); Riahi *et al.* (2017); Rogelj *et al.* (2018).

negative emissions **(see Figure 22.5)**. There are various ways to reach these targets. While demand-side measures mostly reduce energy intensity, supply-side measures would increase the share of low-carbon options. These two indicators can provide an insight into the challenge that such reductions would pose.

The final energy-intensity (energy divided by GDP) reduction rate in many countries has typically been around 1-2 per cent

Figure 22.5: 2010-2050 energy intensity reduction rate and the 2050 share of low-greenhouse gas technologies in total energy mix of the scenarios included in the SSP database

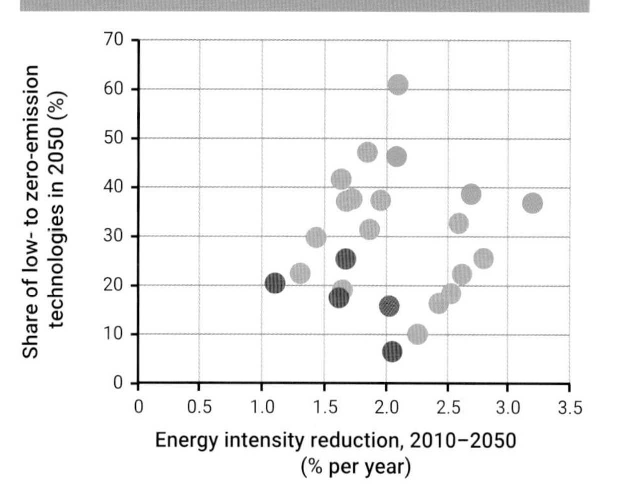

The colours of the dots indicate the projected 2100 temperatures.

Source: Riahi *et al.* (2017); Rogelj *et al.* (2018).

per year in the period since 1970. This has been driven by both increase in energy efficiency and sectoral changes. Relatively high values for energy intensity reduction occurred during the 1973 and 2005 oil crises in response to prices and government policies in Organisation for Economic Co-operation and Development (OECD) countries that aimed to conserve energy (Schippers and Meyers 1992; Sweeney 2016). The share of low-greenhouse gas emitting technologies is at the moment around 20 per cent, consisting mostly of traditional biomass, hydropower and nuclear power. To reach the 2°C target, the combination of energy intensity reduction and increase in the share of low-greenhouse gas technologies would need to be significantly larger than historical values. As shown in **Figure 22.5**, the large scale transformation required for this can be achieved by reducing energy demand (by means of energy efficiency and/or different and lower activity levels) and by decarbonizing energy supply (renewables, carbon-capture-and-storage, nuclear, fuel substitution). Energy efficiency increase, however, would need to be at least 2-3.5 per cent per year. Furthermore, the level of non-CO_2 emitting supply options would need to increase from around 15 per cent today to at least 40-60 per cent by 2050 (for the scenarios included in **Figure 22.5**) or even 50 up to 100 per cent for the most stringent scenarios in the wider literature (van Vuuren *et al.* 2016; IPCC 2018). The low-range value of 40 per cent is only sufficient if combined with a rapid decline in energy demand. The amount of renewables would be around 30-40 per cent **(Figure 22.5)** or up to 60 per cent (full range) for 2 degrees (van Vuuren *et al.* 2016) and 70-85 per cent for 1.5 degrees C (IPCC 2018). It should be noted that the range for renewables largely overlaps with the range of total CO_2-free energy production, as the different options can easily be substituted. All-in-all, the reduction in the carbon intensity of the global economy (rate of change of the ratio of CO_2 over GDP) needs to increase from around 1-2 per cent per year historically to around 4-6 per cent per year towards 2050; for the most stringent scenarios, values up to 8 per cent can be found in the literature (van Vuuren *et al.* 2016).

Emissions of greenhouse gases can be reduced by measures associated with energy demand and decarbonization of energy supply. In addition, it is possible to reduce so-called non-CO_2 emissions from both agricultural and energy systems. In other words, to achieve the Paris targets far-reaching transitions in energy, land, urban infrastructure (including transport and buildings), and industrial systems are needed (IPCC 2018). The contribution of these measures is discussed in more detail in the subsequent paragraphs. **Box 22.2** discusses in more detail the role of land-based mitigation options.

Reducing energy demand

Figure 22.6 presents the aggregated energy use of three different pathways consistent with the 2°C target. The total reduction in energy demand in the pathways is about 25 per cent, compared with the Trend scenario (see also Edelenbosch 2018). Studies focusing on the potential for energy efficiency show even higher possible efficiency improvement rates (Cullen, Allwood and Borgstein 2011; Graus, Blomen and Worrell 2011). Final energy demand is dominated by the industry, transport and residential sectors. Energy consumption in all three sectors would therefore need to be mitigated in order to reach sustainable development targets. Transport is a key sector, as here emissions are increasing most rapidly, driven by increasing emissions from car travel, road freight transport, marine transport and air travel. Different response options exist for decarbonizing the transport sector. For instance, one important option would be an almost complete electrification of most transport modes. This would require a corresponding transition in infrastructure, and its effectiveness in lowering emissions would depend on the carbon intensity of power generation. It should also be noted that, for many parts of the world, such a transition will take a lot of time and, in the meantime, it will be important to minimize emissions, for instance, by promoting car efficiency (Bae and Kim 2017). For modes that cannot be electrified, natural gas (in the short term), fossil fuel with carbon-capture-and-storage (CCS), hydrogen and bioenergy could play a role. Earlier, many studies identified bioenergy use as an effective response strategy for most transport modes. However, because of the possible negative impacts of bioenergy for other targets, the use of bioenergy is assumed to be limited here, restricting bioenergy to those sectors that are hard to abate or that could generate negative emissions. This means that effective measures for transport include electrification, rapid improvement of fuel efficiency and the development of new fuels (hydrogen, synthetic fuels). Alternatively, in a scenario focusing more

Figure 22.6: Different pathways leading to a global mean temperature increase well below 2°C

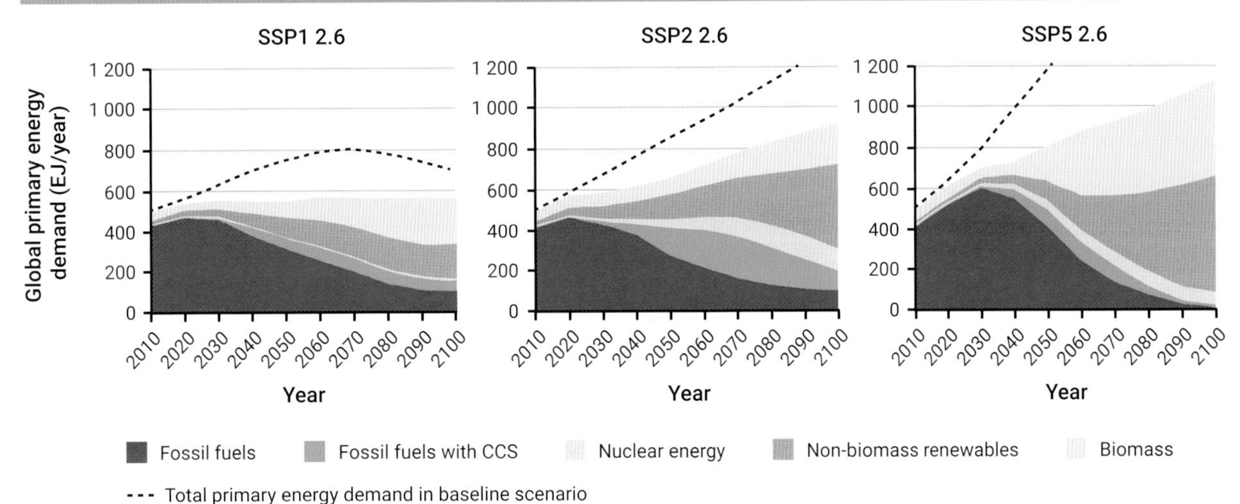

Source: Bauer *et al.* (2017); Riahi *et al.* (2017).

on lifestyle change (e.g. the 'consumption change' pathway), emission reductions occur primarily through a transition away from the use of airplanes and private cars to local electric public transport and fast trains.

Decarbonizing energy supply

A high proportion of the required emission reductions would need to come from supply-side changes **(see Figure 22.6)**. Fossil fuels currently account for around 80 per cent of total primary energy use. This needs to be reduced to a maximum of 20-30 per cent by 2050, depending on the use of negative emission technologies (after 2050) and the ambition of the climate target (Bauer *et al.* 2017; van Vuuren *et al.* 2017b). The fossil fuels need to be replaced by low- to zero-emission technologies, such as bioenergy, other renewables and nuclear energy, and fossil fuel energy combined with carbon-capture-and-storage.

It will be very important to introduce new policies that stimulate the further penetration of renewables. The literature also shows that there is some degree of freedom in choice of technology. For instance, there can be different roles for renewable energy, nuclear power and carbon-capture-and-storage, depending on societal choices and technology development. It should be noted, however, that the size of the overall transformation is – in absolute terms and the period for which it should be sustained – without historical precedent (van der Zwaan *et al.* 2013; van Sluisveld *et al.* 2015). It is in fact well beyond the rate of transitions in the past, highlighting the considerable challenge of meeting the 2°C target (Napp *et al.* 2017). In relative terms (e.g. per cent of investment in new technologies), there are several examples of similar rapid transitions in the past.

There are many ways to decarbonize energy supply in future scenarios (Clarke *et al.* 2014; Kriegler *et al.* 2018; Rogelj *et al.* 2018). One method is fossil fuel energy combined with carbon-capture-and-storage. Most scenarios rely heavily on this option. While the advantage is that it would require relatively far change in energy supply, this option suffers from a limited storage potential and, above all, relatively little societal support. Renewables such as wind and solar power form an important alternative. The costs of these options have decreased rapidly over the last few years, making these technologies a reasonable alternative for fossil fuels even in the absence of stringent climate policy. However, for higher levels of penetration these options suffer from additional costs related to intermittency. This implies that the expansion of renewable energy will require investment in infrastructure to deal with intermittency (e.g. via expanding grid connections and providing storage options). The transition to renewables would also lead to a change in demand for materials (to create solar and wind power plants). Most assessments find the latter not being restrictive (Arvesen *et al.* 2018). Finally, a transition to renewables will also require different operating regimes for the power system. The option of bioenergy could also be attractive as a supply for fuels and, in combination with carbon storage, a pathway to negative emissions. As bioenergy requires large amounts of land it would, however, compete with the targets

Box 22.2: Contribution of land-use-based mitigation options to climate policies

About 20-30 per cent of total greenhouse gas emissions are associated with agricultural activities (Smith *et al.* 2014). In terms of climate policy, the contribution of the land-use sector is very important. First of all, reaching stringent targets would require reducing land-use-related emissions. In addition, it is also possible to contribute to emission reductions by so-called land-use-related mitigation options. This includes, for instance, reforestation and the use of bioenergy. In fact, more than 80 per cent of the nations that are signatories to the Paris Agreement plan to use land-use-related mitigation options to fulfil their Nationally Determined Contributions (NDCs). Analysis has shown that both afforestation and the use of bioenergy in combination with carbon-capture-and-storage are cost-efficient in nearly all scenarios. As a result, the use of land for mitigation might in 2050 be in the order of 25-30 per cent of total cropland in some scenarios (i.e. 10 per cent of total agricultural area). An important challenge, however, is that the use of land-use-based mitigation options could lead to significant trade-offs with the targets to end hunger and preserve biodiversity, due to competition for land (see Section 22.3.1).

mentioned in the previous cluster. This is discussed further in Section 22.4.2. Alternative pathways that rely less on negative emission technologies could be based on stronger changes in lifestyle (van Vuuren *et al.* 2018). Finally, nuclear power can also provide zero-emission energy. However, this technology poses both safety and waste risks and a lack of societal support in many countries.

Reducing non-CO_2 emissions

Although carbon dioxide forms the lion's share of greenhouse gas emissions, non-CO_2 greenhouse gases such as methane, nitrous oxide and fluorinated greenhouse gases also contribute significantly to climate change. Thus, non-CO_2 emissions in pathways that limit global warming to the Paris targets also show deep reductions (IPCC 2018). Some of the non-CO_2 emissions are relatively easy to abate, such as those associated with losses in the energy system. Moreover, these reductions often have high co-benefits including the reduction of methane (also leading to ozone pollution) and soot (leading to climate change and health impacts). In contrast, other sources are relatively hard to abate. For instance, it is hard to imagine how methane emissions from roaming cattle could be reduced to zero. As a result, in most 2°C scenarios, land-use-related emissions are reduced by around 50 per cent compared with current emission levels. Reducing emissions further would typically require reduced meat consumption (see Section 22.3.1).

Controlling air pollution

Future air pollution emissions stemming from human activities, with the energy sector playing a dominant role, require the application of specific measures to reduce air pollutant emissions. Many of the strategies that decrease greenhouse gas emissions, such as increasing energy efficiency, switching fuel types and changing lifestyles, also lower emissions of other air pollutants, resulting in health co-benefits (Markandya *et al.* 2018). Similarly, air pollution policies have climate implications, for example, by affecting emissions of short-term climate forcers such as black carbon.

To explore the limit on what air pollution emission decreases might be possible by introducing air pollution control measures, Stohl *et al.* (2015) defined a maximum technologically feasible reductions scenario by applying the lowest emission rates from known technology regardless of costs. Other scenarios have taken costs and local circumstances into account, such as the 'new policy' and 'clean air' scenarios of the International Energy Agency (IEA 2016). While the 'new policy' scenario considers policies and measures that had been adopted or announced as intended (as of 2015), the 'clean air' scenario includes additional measures that achieve significant reduction of air pollutant emissions. Relative to the 'new policy scenario', the 'clean air' scenario includes an additional US$2.3 trillion invested in advanced air pollution control technologies and a similar amount (US$2.5 trillion) invested in accelerating the transition to cleaner and renewable energy sources. These measures would result in a 50 per cent decrease in SO_2 and NO_X emissions and an almost 75 per cent decrease in particulate matter emissions and would avoid more than 3 million premature deaths per year, with 1.7 million deaths attributable to reduced ambient air pollution and 1.6 million deaths attributable to reduced household air pollution (IEA 2016).

The importance of climate mitigation for air pollution emissions can also be illustrated using SSP results (Rao *et al.* 2017): increasingly stringent climate policy also reduces emissions of air pollutants. The extent to which coal is used for electricity production and manufacturing has a strong influence on CO_2 emissions and largely determines the path of SO_2 emissions. For transportation sector emissions, the level of electrification is important. Electrification, combined with autonomous vehicles and shared mobility services, could lead to dramatic decreases in emissions and associated pollutant exposures (Fulton, Mason and Meroux 2017). Black carbon emissions, associated with diesel engines and residential combustion of traditional biomass fuels, are much less correlated with fossil fuel use (and thus climate policy), but more with use of traditional energy (and thus with the introduction of access to modern energy services); this is reflected in the different black carbon emission levels for the baselines, but also in a much lower response to climate policies (Rao *et al.* 2017).

Short-lived climate pollutants (SLCPs) contribute to atmospheric warming, and include black carbon, tropospheric ozone, methane and hydrofluorocarbons. Among SLCPs, black carbon, methane and tropospheric ozone contribute to air pollution. Reducing emissions of SLCPs can provide near-term climate benefits (Shindell *et al.* 2017; Xu and Ramanathan 2017; Haines *et al.* 2018). For black carbon, measures are available to decrease emissions from diesel engines, biomass cooking fuel, kerosene lighting, and household and small industry coal use. There are opportunities to decrease methane emissions associated with the extraction of coal, oil and natural gas, disposal of waste, switching management of emissions from livestock and manure and rice paddy production. Compliance with the Kigali Amendment (United Nations 2016) will decrease hydrofluorocarbon emissions by 61 per cent from 2018 to 2050 compared to a reference scenario, but substitutions could be made earlier and a 98 per cent decrease is technically possible (Höglund-Isaksson *et al.* 2017). Implementation of such demonstrated technical measures to address SLCP could decrease average global warming, although estimates on the exact level differ by study (United Nations Environment Programme [UNEP] 2017) (see also **Box 22.3**).

Box 22.3: The Climate and Clean Air Coalition

Efforts to simultaneously address air quality and climate impacts of short-lived climate pollutants (SLCPs) include the Climate and Clean Air Coalition (CCAC; http://www.ccacoalition.org), which was launched in 2012 and is a voluntary partnership of governments, intergovernmental organizations, businesses, scientific institutions and civil society organizations committed to improving air quality and mitigating climate change by reducing SLCPs. Approaches for reducing black carbon include clean and efficient household cooking, lighting and heating technologies; modern brick kiln technology for brick production; and clean fuel for heavy-duty diesel vehicles and engines. The focus for reducing methane emissions includes reducing gas leakage from gas distribution systems, improving manure management, using alternative rice farming practices and strategies to reduce enteric fermentation emissions from livestock. As of July 2017, some 178 countries had included methane, 100 had included hydrofluorocarbons, and four had included black carbon in their Nationally Determined Contributions (NDCs) or Intended NDCs for meeting the climate goals of the Paris Agreement. A number of countries are expected to update their NDC to strengthen the inclusion of SLCPs. It is important to note that reducing emission of SLCPs as a complement to reducing greenhouse gas emissions provides opportunities to limit near-term climate warming but is not a substitute for reducing long-lived greenhouse gases to mitigate long-term climate change.

Figure 22.7a: Projected global emissions for SO_2, NO_x and black carbon under different climate and air pollution policies

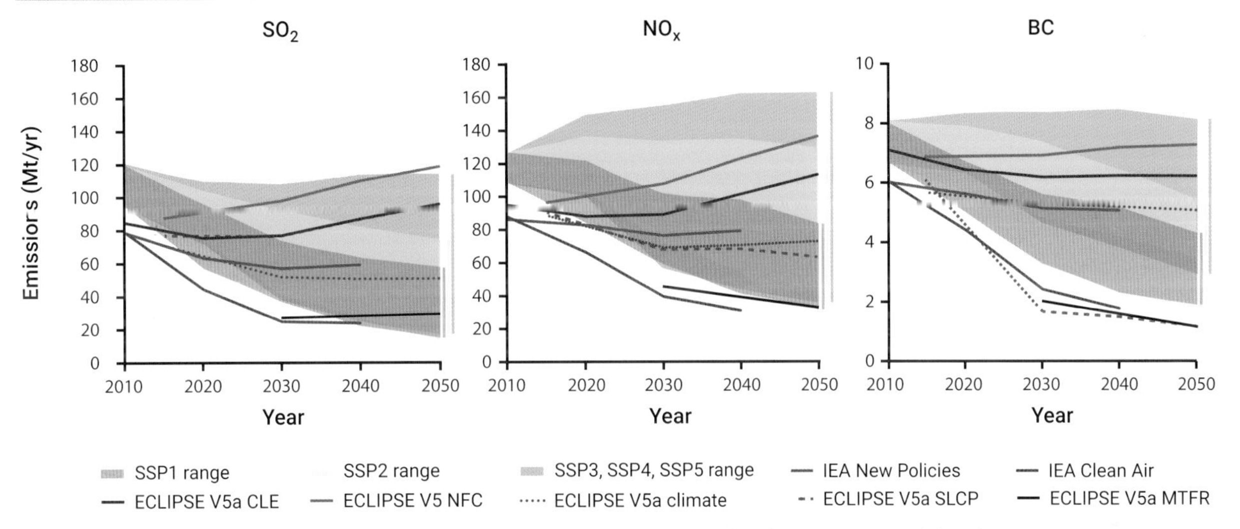

For the SSP baselines the shading represents the ranges over all Integrated Assessment Models (IAMs) included in Rao *et al.* (2017).

Source: SSPs (Rao *et al.* 2017); ECLIPSE (Stohl *et al.* 2015; Klimont *et al.* 2017); IEA (IEA 2016).

Figure 22.7b: Differences in air pollution emissions between various climate mitigation scenarios, and the SSP2 baseline

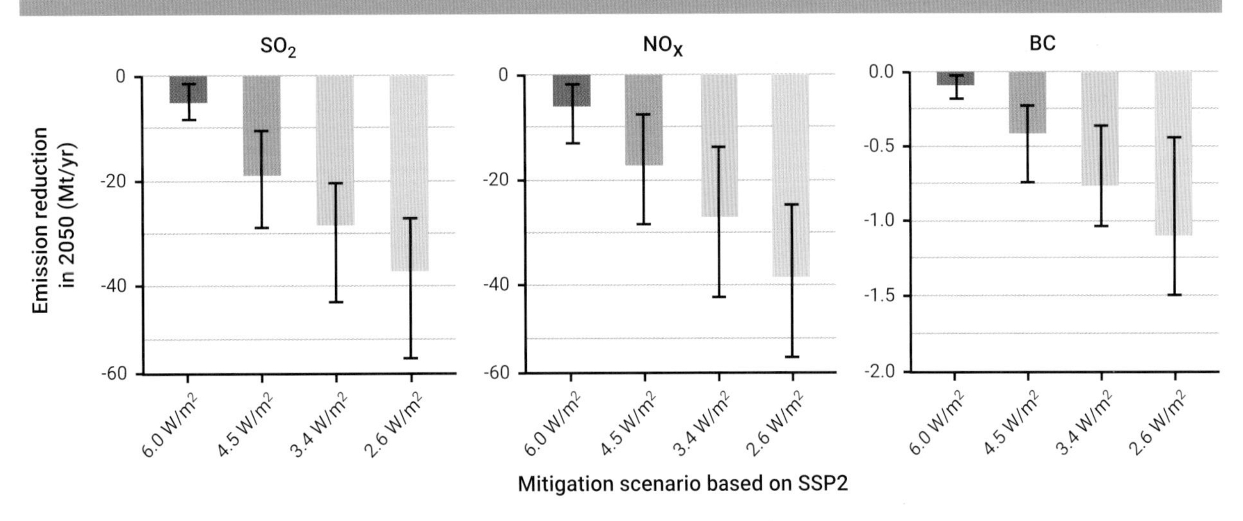

Error bars represent the range of all Integrated Assessment Models (IAMs) included in Rao *et al.* (2017).

Source: Rao et al. (2017).

Although climate policies lead to significant decreases in air pollution in all SSP marker scenarios, these decreases are not sufficient to achieve the WHO air quality guideline of 10 µg/m³ for annual mean $PM_{2.5}$ concentrations by 2050 **(Figure 22.8)**. The ECLIPSE maximum technically feasible reduction (MTFR) scenario without climate mitigation, which has the lowest air pollutant emissions among all scenarios **(Figure 22.7)**, is also insufficient to achieve the WHO guideline. Worldwide, about 60 per cent of the population is projected to be exposed to levels above the standard in the best-case air pollution scenarios (SSP1 or SSP5 with 2.6 W/m² climate mitigation target, or the ECLIPSE MTFR scenario). The worst exposures are projected for Asia and the Middle East and Africa regions. However, by 2050 less than 5 per cent is expected to be above the most lenient interim target of 35 µg/m³ annual mean $PM_{2.5}$ concentrations for SSP2 and SSP5 if climate mitigation is included. These results reflect the air quality benefit of strong air pollution control and the co-benefit of climate mitigation, and for the ECLIPSE MTFR scenario, reflecting the maximum air quality benefit achievable with current air pollution control technologies.

Significant synergies and trade-offs between measures and targets

The measures introduced to achieve universal access to modern energy services, combat climate change or improve air quality in cities can have important synergies and trade-offs (e.g. McCollum *et al.* 2018).

- Most of the climate policies lead to an increase in energy system costs, with potentially increasing energy prices as a result. Higher energy prices, especially for clean fuels for cooking (e.g. electricity, liquified petroleum gas, natural gas), make it more difficult to achieve universal energy access, or to provide affordable energy in general (Daioglou, van Ruijven and van Vuuren 2012; Cameron *et al.* 2016). There are, however, various ways to compensate for this, including targeted subsidies or redistribution of carbon taxes (Cameron *et al.* 2016).
- Policies aiming to increase energy access could lead to an increase in energy consumption and thus impact both climate change and air pollution. These impacts, however, are relatively small (van Vuuren *et al.* 2012) and can, if needed, be mitigated by ensuring that energy access is achieved via low-greenhouse gas energy supply systems. Achieving universal electricity access is estimated to have only a very small increasing effect on global greenhouse gas emissions (Pachauri *et al.* 2013; van Vuuren *et al.* 2015; Dagnachew *et al.* 2018). Furthermore, universal access to clean fuels for cooking could reduce total air pollutant and greenhouse gas emissions, resulting from a switch away from traditional biomass, increased biomass-use efficiency and sustainable harvesting of biomass (Pachauri *et al.* 2013; van Vuuren *et al.* 2015). There are also both synergistic effects and trade-offs between air pollution and climate policy. One example of a possible trade-off is that burning biomass as a low-carbon energy source can lead to more air pollution if appropriate air quality management practices are not put in place (Giuntoli *et al.* 2015). Another is that diesel cars emit less CO_2 than petrol (gasoline) cars but emit more PM (Mazzi and Dowlatabadi 2007; Tanaka *et al.* 2012; O'Driscoll *et al.* 2018). Also, the use of end-of-pipe emission controls may reduce PM emissions of passenger vehicles, but at the cost of reducing fuel efficiency. For petrol vehicles, replacing port fuel injection with direct injection engine technology generally increases fuel efficiency, thus reduces CO_2 emissions, but increases PM and black carbon emissions (Zhu *et al.* 2016; Zimmerman *et al.* 2016; Saliba *et al.* 2017).
- However, in most cases, climate policy reduces air pollution by having an impact on emissions of PM, SO_2 and NO_x. If well designed, air pollution control measures can also limit climate change. This implies that especially for countries currently experiencing high air pollution levels, designing strategies that address both air pollution and climate change can be very attractive (see also **Box 22.4**).
- Geo-engineering (e.g. direct air capture) generally requires additional energy use providing a possible trade-off with air pollution or energy access.

Box 22.4: Possible synergy between climate mitigation and reducing air pollution in China

In response to strong public concerns about air pollution, the China State Council announced in 2013 the 'Action Plan of Prevention and Control of Air Pollution'. The action plan sets specific targets for air pollution. Among others, the 2017 concentration of particulate matter with diameter less than 10 µm (PM_{10}) should fall by at least 10 per cent compared to 2012 concentrations. For some regions, however, more stringent targets are formulated. The plan indicates that one way to implement these targets is through promotion of clean energy, including renewable energy, nuclear power, natural gas, combined with a transition of the energy system, energy conservation and control of coal use. This is completely consistent with low-carbon development in China. Since then, the economic and structural changes together with air pollution control measures resulted in a peak in coal production in 2013/2014. This has also led to a reduction of CO_2 emissions. From 2015 to 2017, there was a rapid expansion of wind, solar and hydro power as well as nuclear power. If, in the future, the increase in energy demand is relatively slow, any expansion can be covered by the increase in renewable energy, nuclear and natural gas, so that the decline in coal capacity can continue. Under these circumstances, a further decline of CO_2 emissions is possible. In the meantime, sustainable development is a basic long-term national strategy in China. China has started to enhance its energy efficiency policies in its Eleventh Five-Year plan and is expected to continue to do so in subsequent Five-Year plans. The main focus for these policies will continue to be on improving energy efficiency in the industrial sector, but new policies are also targeting domestic energy consumption. On this basis, the target is to reduce the share of coal in the energy mix from 64 per cent in 2015 to 58 per cent in 2020. According to the announcement of "Interim measures of the replacement of coal consumption in key regions" from China's National Development and Reform Commission (NDRC *et al.* 2014), 8 provinces and municipalities in key areas, including Beijing-Tianjin-Hebei Region, Shandong province, the Yangtze River Delta and the Pearl River Delta will be required to set up the reduction targets for coal consumption. These policies are meant to reduce air pollutants. All-in-all, it means that current Chinese policies to improve air quality could have a huge benefit for public health, but also lead to a reduction of CO_2 emissions.

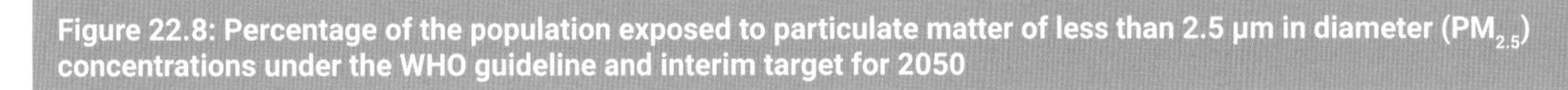

Figure 22.8: Percentage of the population exposed to particulate matter of less than 2.5 μm in diameter ($PM_{2.5}$) concentrations under the WHO guideline and interim target for 2050

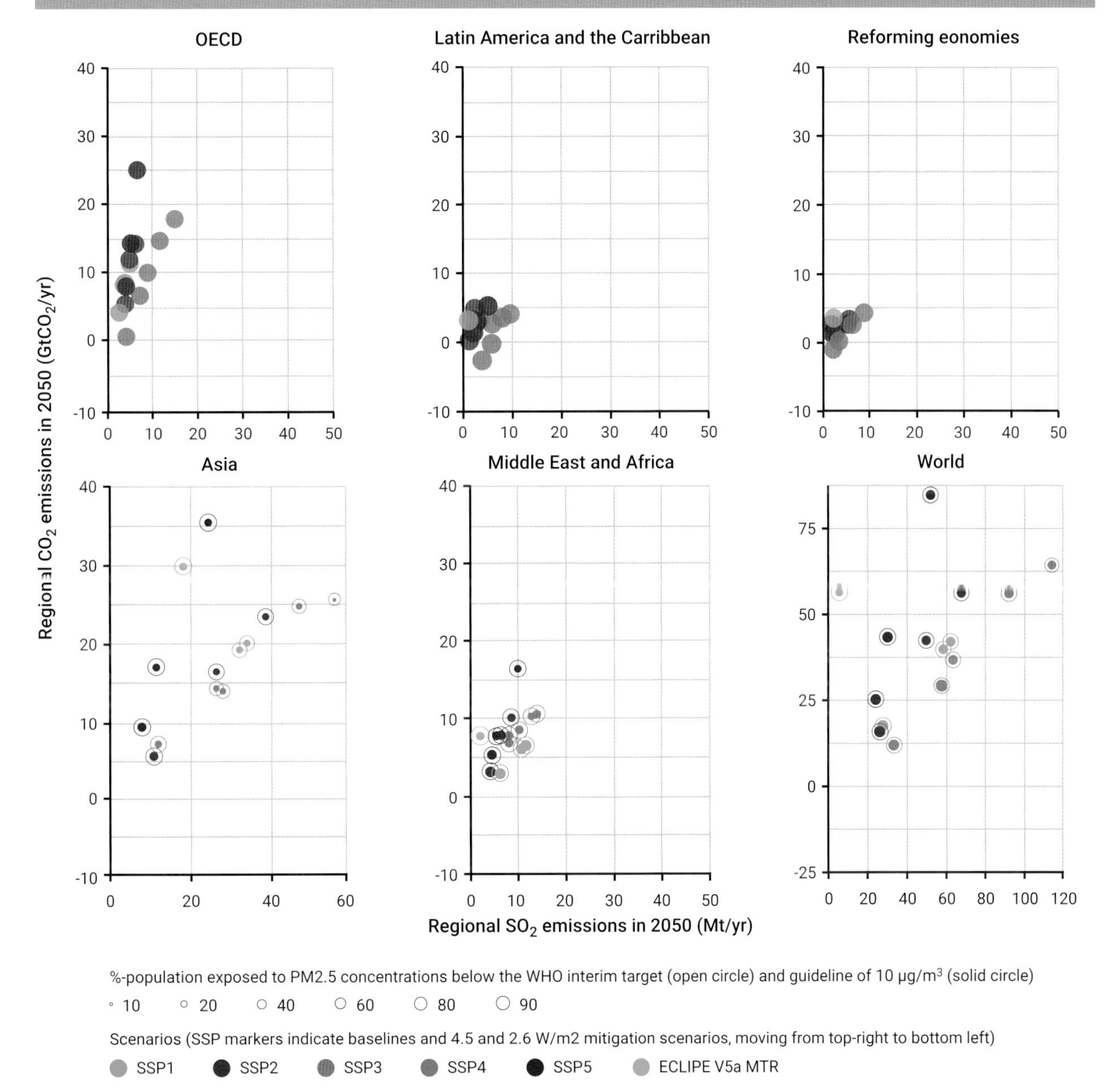

Sources: Rao *et al.* (2017); Population exposure is based $PM_{2.5}$ concentrations determined by applying the TM5-FASST source-receptor model (van Dingenen *et al.* 2018) to marker SSP emission scenarios and the related 4.5 W/m² and 2.6 W/m² climate mitigation scenarios.

22.3.3 Freshwater

The selected targets for the freshwater cluster may be summarized as reducing water scarcity and ensuring water quality, while at the same time providing universal access to safe drinking water and adequate sanitation (Chapter 20). The world is not on track to achieve these targets (see Chapter 21). More than 400 million people are projected to still lack access to at least basic water facilities in 2030 and about 2 billion people still do not have access to at least basic sanitation. Furthermore, the fraction of the global population that lives in water-stressed areas is projected to increase up to about 50 per cent by the end of the century, mostly driven by population growth.

There are important linkages between this cluster and other clusters, especially the agriculture, food, land and biodiversity cluster and the energy, air and climate cluster. Globally, the largest demand for water comes from the agricultural sector (over 70 per cent). Also, many freshwater and ocean pollutants come from agriculture, and agriculture is the dominant source of nitrogen and phosphorus in global watersheds (see Chapter 21).

There are several scenario studies on water scarcity. However, most of these focus on future projections instead of target-based scenarios. In contrast, Wada, Gleeson and Esnault (2014) propose six strategies, or 'water-stress wedges', that collectively lead to a reduction in the water-stressed population by 2050. Bijl *et al.* (2018) discuss some strategies that could

lead to reduced water scarcity, including increased efficiency, other allocation strategies and reducing agricultural water demand via diet change and food waste reduction. Here, we discuss different measures largely linked to the individual targets, addressing increasing access to water, sanitation and hygiene (WASH), decreasing water demand, increasing water supply and reducing water pollution.

Investing in access to water, sanitation and hygiene

Achieving the targets on drinking water and sanitation will require increased investment in infrastructure, especially sanitation (United Nations Conference on Trade and Development 2014; Hutton and Varughese 2016). Due to population growth, an additional 3.4 billion people will require access to sanitation by 2030, or 620,000 per day, 2.5 times the number of people served during the 2001-2015 period (Mara and Evans 2018). The current levels of investment are likely to cover the capital costs of basic service provision for access to WASH by 2030, but not enough for safely managed service provision. To achieve universal access to safely managed WASH services, investment levels will need to increase threefold (Hutton and Varughese 2016). Achieving universal access to safe water and adequate sanitation is as much about changing behaviour as it is about changing infrastructure. This requires better marketing, communication and community-led sanitation (Water and Sanitation Program 2004; Kar and Chambers 2008; Devine and Kullmann 2011).

Increasing water-use efficiency

Water scarcity including groundwater often needs to be managed at the watershed or aquifer level (Scott *et al.* 2014). These can be within one country, but often there are multiple countries involved. In those cases, an international framework is needed to evaluate strategies to reduce water stress and maximize mitigation (Wada, Gleeson and Esnault 2014). Wada, Gleeson and Esnault (2014) conclude that four demand-side measures are required: increasing agricultural water productivity (more crop per drop), improving irrigation efficiency (reducing water losses), more efficient water use in domestic and industrial sectors including reducing water leakage and improving recycling, and limiting the rate of population growth. To maintain or even reduce the global population under water scarcity by 2050 and beyond, water-use efficiency for these demand-side measures needs to improve by more than 20-50 per cent globally (0.5-1.2 per cent improvement per year). Moreover, strategies for water management at the level of watersheds are necessary to deal with competing demands for agricultural production, industrial activities, household water use and ecological services. The precise mix depends on economic, social, legal and political issues such as international or subnational water treaties, rights or disputes (Wada, Gleeson and Esnault 2014). Various scenarios have shown that increased water efficiency in agriculture, households and industry can have a significant impact on reducing water scarcity (e.g. Bijl *et al.* 2017).

Increasing water supply

Increasing water supply can be done using more conventional measures such as building more water storage or dams, by investing more in desalination capacity in coastal regions (Wada, Gleeson and Esnault 2014) or by wastewater reuse. Furthermore, groundwater resources could serve as a buffer during droughts or severe water scarcity because of their ubiquitous presence across the globe.

Increasingly, countries are implementing desalination strategies – for example, in the Middle East, North Africa and the United States of America (e.g. California) (World Water Assessment Programme 2003; Hanasaki *et al.* 2016). The global amount of desalinated water use has been rapidly increasing since the 1990s and it is currently estimated to exceed 10 km^3 annually (Food and Agriculture Organization of the United Nations 2018). Although this amount is important for coastal regions, the global total currently accounts for much less than 1 per cent of water withdrawals worldwide (4,000 km^3). Hanasaki *et al.* (2016) projects that under different SSP scenarios (1-3), the use of seawater desalination will increase 1.4- to 2.1-fold in 2011-2040 compared with the present, and 6.7- to 17.3-fold in 2041-2070. The associated costs are in the order of US$2 billion to US$200 billion. The large spreads in these projections are primarily attributable to substantial socioeconomic variations in the SSP scenarios. To scale up desalination of seawater in coastal water-stressed basins, a 10- to 50-fold increase is projected to be required; however, this would imply significant capital and energy costs, and it would generate wastewater that would need to be disposed of safely (Wada, Gleeson and Esnault 2014; Hanasaki *et al.* 2016).

Wastewater reuse enables upgrading of unsuitable water quality originating from households and industry to sufficient quality for different purposes. The amount of wastewater reuse or recycling has been increasing worldwide especially for agriculture, as small-scale farmers in urban and peri-urban areas of developing countries depend largely on wastewater or wastewater-polluted water sources to irrigate high-value crops for market (Qadir *et al.* 2010). However, higher-quality water is needed for drinking purpose and the establishment of water reuse guidelines is critical (Bixio *et al.* 2006; Bixio *et al.* 2008). Ongoing technological innovations, such as the use of membranes, and dedicated economic instruments are expected to further increase the use of wastewater as a resource in various regions with limited surface- and groundwater resources. In order to reduce water limitations in urban areas or megacities, a similar magnitude of future scaling up is required for wastewater reuse combined with the desalination of seawater (Wada, Gleeson and Esnault 2014).

It should be noted, however, that these two supply-side measures require a large amount of economic investment and modernizing of existing infrastructure, which might not be feasible for many developing countries (Neverre, Dumas and Nassopoulos 2016). Alternatively, nature-based solutions may have high potential to increase and/or regulate water supply by reducing degradation of water quality, while limiting economic investments (Vörösmarty *et al.* 2010). Multiple ecosystem services or sustainable infrastructure can mitigate water pollution and increase water supply for humans and ecosystems (Reddy *et al.* 2015; Liquete *et al.* 2016). These examples highlight an important role for development and deployment of water conservation technologies and practices to achieve water-related SDG targets (Hejazi *et al.* 2014).

Reducing water pollution

Experience in developed countries has shown that it is possible to reduce water pollution. Unfortunately, there is very little scenario literature addressing water pollution problems and ways to achieve future sustainability targets. However, there is some literature discussing reduced nutrient pollution,

for example by wastewater treatment. A global decrease in nutrient discharge is possible only when wastewater treatment plants are extended with at least tertiary treatment in developing countries and with advanced treatment in developed countries. Separate collection systems for urine can reduce nutrient pollution to 15TgN/yr and 1.2TgP/yr (van Puijenbroek, Beusen and Bouwman 2019). When all effluent from sewage systems receive tertiary treatment, global nutrient discharge is projected to decrease to 1990 levels (Ligtvoet *et al.* 2018). For phosphorus, a further decrease could be realized when all laundry and dishwasher detergents are phosphorus-free. This is now mandatory in the European Union, United States of America, Japan and some other countries.

Increasing crop yields and fertilizer-use efficiencies will have a direct effect on the nutrient loading of streams and rivers. However, starting from a situation of low crop yields and minimal nutrient inputs, nutrient loading of watersheds may well increase in scenarios with a shift towards food production systems now prevalent in industrialized countries. Since watersheds retain nitrogen and phosphorus, there may be legacies of past management. As a consequence, nitrogen concentrations in many rivers respond only slowly to increased nitrogen-use efficiency in food production. For example, due to these legacies, European water quality is threatened by rapidly increasing nitrogen-phosphorus ratios (e.g. Romero *et al.* 2013). Developing countries can avoid such problems by managing both nitrogen and phosphorus, accounting for residual soil phosphorus, while avoiding legacies associated with the past and continuing mismanagement of high-income countries.

Significant synergies and trade-offs between measures and targets

A number of synergies and trade-offs can be identified between specific measures and the various targets within this cluster. A few important ones are as follows.

- Increased access to and use of improved and safely managed WASH facilities has direct health benefits and can also improve overall quality of life. Women in developing countries often travel long distances to access water and sanitation facilities, even more so than men because of domestic-related tasks that more often fall to women, and because of menstrual hygiene (Pommells *et al.* 2018). Not only does this leave women more susceptible to health risks from more frequent contact with unsafe facilities, but there is a growing body of literature on the prevalence, and lack of documentation, of assault and rape on these trips (Sorenson, Morssink and Campos 2011; Watt and Chamberlain 2011; Sahoo *et al.* 2015; Sommer *et al.* 2015; Freshwater Action Network South Asia and Water Supply and Sanitation Collaborative Council 2016; Pommells *et al.* 2018).
- Increased levels of access to at least basic safe drinking water and adequate sanitation can drive increased domestic water demand, further contributing to water stress (Hanasaki *et al.* 2013a; Hanasaki *et al.* 2013b; Wada *et al.* 2016).
- Water scarcity negatively affects agriculture and biodiversity and also energy supply. In fact, water stress is one of the five global risks of highest concern according to the World Economic Forum (Wada, Gleeson and Esnault 2014).
- Agriculture is the dominant source of nutrients in global watersheds leading to eutrophication, resulting in hypoxia symptoms in many inland and coastal areas. There is a tendency towards increasing nitrogen-phosphorus ratios and declining silica; this distortion of nutrient ratios leads to the proliferation of harmful algal blooms, both in global watersheds and coastal parts of oceans.
- Improved sanitation facilities without, or with only primary, wastewater treatment are major polluters of freshwater, due to nitrogen and phosphorus discharge (van Puijenbroek *et al.* 2015),
- While the only option for some water-scarce communities, desalination is very energy-intensive, potentially counteracting interventions to reduce industrial water demand (Pinto and Marques 2017).

22.3.4 Oceans

The selected targets for the oceans cluster are limiting ocean acidification, reducing nutrient pollution and sustainably managing ocean resources (see Chapter 20). For all three targets, trends are projected to go in the wrong direction (see Chapter 21). There is strong evidence that the current trend towards declining fish populations and reduced species richness impair the ecological functioning of oceans, including their role in providing food (Worm *et al.* 2006). Nutrients from fertilizers used to increase agricultural yields have also found their way into nearly every water body across the globe where they stimulate aquatic plant production. As a consequence, hypoxia, a growing global problem, occurs where organic matter decay consumes oxygen faster than its diffusion from the oxygen-rich surface. Furthermore, the global problem of harmful algae is now on a pathway of more and more frequent blooms, in more places and with increasing severity, with more toxins (Glibert 2017).

Pathways in this cluster are largely linked to developments in other clusters. With respect to ocean acidification, the scenario literature is linked to climate change (i.e. the reduction of CO_2 emissions; Section 22.3.2), marine nutrient pollution with agricultural production measures (Section 22.3.1) and freshwater pollution (Section 22.3.3). Here, we discuss different measures linked to the individual targets, addressing ocean acidification measures and sustainable ocean management. No scenario studies were found that address the reduction of marine nutrient pollution to stop related hypoxia and harmful algal blooms.

Ocean acidification measures

Ocean acidification is a result of the increased absorption of CO_2 in the oceans, which in turn is a result of an increasing global atmospheric CO_2 concentration. Billé *et al.* (2013) identify three means of preventing ocean acidification:

i. reducing CO_2 concentrations, either by lowering emissions or removing CO_2 from the atmosphere, for example through carbon-capture-and-storage under the seabed (see Section 22.3.2);
ii. limiting ocean warming; and
iii. reducing nutrient run-off into the ocean.

Furthermore, they identify means of reversing acidification after it has occurred, including additives (e.g. alkalinization) and ecological restoration.

Reducing emissions of CO_2 thus reduces ocean acidification directly, while other climate policy measures can have an indirect effect via reducing sea surface temperature. For example, Mora *et al.* (2013) find less reduction in ocean pH and ocean productivity in Representative Concentration Pathway (RCP) 4.5 than in RCP 8.5. Similarly, Bopp *et al.* (2013) find a decline in ocean pH of only 0.07 and an increase in sea surface temperature of only 0.71°C in a stringent climate policy scenario, compared with a decline in pH of 0.33 and an increase in sea temperature of 2.73°C in a high-emission scenario. In fact, carbonate ion concentrations do not fall below saturation levels in the stringent climate policy scenario for any ocean (Bopp *et al.* 2013). Concentrations below saturation level can lead to dissolution of shells and skeletons of marine organisms.

Sustainable ocean management

Currently, fisheries worldwide are severely degraded as a result of overfishing. Several scenarios have looked at the impact of strong fisheries management (among others through the reduction of catch) to find that there could be a decrease in the proportion of exploited fish stocks to close to a recovery target biomass. This would, in the long run, also mean an increase in total global fisheries profit, relative to both the trend scenario and even the present day. Costello *et al.* (2016) analysed data excluding small-scale and artisanal fisheries but representing 78 per cent of global catches and found that applying management policies for returning catch to maximum sustainable yield or even maximum profits through rights-based fisheries management was projected to produce improvements in catch profit, and fish stock biomass relative to the business-as-usual management scenario. By 2050, some 98 per cent of stocks could be biologically healthy under strong fisheries management (Costello *et al.* 2016).

Similarly, under a low-greenhouse gas emissions scenario, Lam *et al.* (2016) projected a smaller decline in catch potential (4 per cent versus 7 per cent in the trend scenario), suggesting that climate policy can limit the impacts of climate change on global fisheries. Also, Cheung, Reygondeau and Frölicher (2016) estimated the benefits to global fisheries from meeting the 1.5°C warming target in the Paris Agreement: every degree of warming above this target resulted in a projected 3 million (metric) tons reduction in potential catch.

Another way to promote more sustainable fisheries and protect biodiversity is by introducing protected areas (Agardy 2000). Marine protected areas tend to increase the biomass of fish (Gill *et al.* 2017), but there is debate about the effectiveness of marine protected areas for biodiversity (Worm *et al.* 2006; Edgar *et al.* 2014). The effectiveness of protected areas regimes depends strongly on their management and enforcement (Edgar *et al.* 2014; Gill *et al.* 2017). In addition, by introducing better strategies for selecting protected areas, their impact can be increased significantly (Davis *et al.* 2017). However, similar to protection of terrestrial biodiversity, it is clear that for preventing biodiversity loss, increasing protected areas will not be enough (Mora and Sale 2011).

Significant synergies and trade-offs between measures and targets

A number of synergies and trade-offs can be identified between specific measures and the various targets within this cluster. A few important ones are as follows.

- Reviving current fish stocks will require a period of reduced catches, therefore potentially reducing the contribution of fish resources in reducing hunger. However, as shown, in the long run this will lead to higher sustainable yields.
- Reduced marine nutrient pollution could make coral reefs less vulnerable to ocean acidification and reduce the predicted shift from net accretion to net erosion (Silbiger *et al.* 2018).
- Reducing ocean acidification by means of limiting CO_2 emissions is also important to conserve marine biodiversity and to secure the availability of fish resources to reduce hunger worldwide.

22.3.5 Human development

The selected target for the human development cluster is ending preventable deaths of children under five years of age (see Chapter 20), with the acknowledgement that other environmental health impacts and age groups are also relevant for human health (see also Section 20.3.1). For example, exposure to ambient $PM_{2.5}$ was the fifth-ranking mortality risk factor in 2015 (Cohen *et al.* 2017; Chapter 5) and the deadliest of any environmental risk factor. More than half of the premature deaths attributed to ambient air pollution occur among those older than 50 years of age, while household air pollution, the second highest environmental risk factor, predominantly affects children and women (GBD 2016 Risk Factors Collaborators 2017; see also Section 5.3.1). Future projections show a reduction in the global child mortality rate, but not enough to achieve the target, while air pollution is projected to continue to contribute to millions of premature deaths annually (Chapter 21).

There are strong links between the child mortality target and several other targets discussed in this chapter. Important health risk factors affecting under-five mortality rates include malnutrition (strongly related to hunger), no access to safe drinking water, adequate sanitation and hygiene (WASH), indoor air pollution and (more indirectly) also climate change.

There are very few studies that look at reducing child mortality in relation to a range of environmental risk factors (e.g. Hughes *et al.* 2011; Lucas *et al.* 2018). Most studies focus on individual risks, most prominently malnutrition (i.e. prevalence of undernourishment) and ambient air pollution. Ending preventable death of children under five, especially with respect to environmental health risks, largely depends on achieving specific targets discussed for the other clusters in this chapter. However, pathway studies suggest that a healthy planet alone is not enough for achieving healthy people (Hughes *et al.* 2011; van Vuuren *et al.* 2015; Lucas *et al.* 2018; Moyer and Bohl 2018). The success of the different pathways in reducing child mortality depends on the degree to which they also address non-environmental risk factors, reducing both wealth inequalities and social inequalities. Here, we discuss four broad measures – reducing exposure to environmental risk factors, poverty alleviation, women and girl's education, and child and maternal health care.

Reducing exposure to environmental risk factors

Preventable risks for children under five include malnutrition (e.g. child underweight), exposure to fine particulate emissions causing pneumonia, and micropathogens and vectors that can transmit infectious diseases such as diarrhoea and malaria. Climate change can negatively impact several of these risk factors, including child underweight (Hughes *et al.* 2011) and malaria (Craig, Snow and le Sueur 1999). Measures for reducing exposure to related risk factors are extensively discussed in Sections 22.3.1 to 22.3.3. Here, we repeat some of these measures and discuss overall impacts on child mortality.

For ending malnutrition (SDG target 2.1), interventions include increased food availability through (for example) yield improvement, diet changes and waste reduction, as well as improving access to food and nutrition management for the poor (Section 22.3.1). Reduced consumption in high-income countries does not necessarily increase availability and access for poor communities and therefore has a low impact on reducing malnutrition and related child mortality (Moyer and Bohl 2018). A combination of availability and access measures are thus required. For reducing air pollution (SDG target 11.6), interventions include introducing effective air pollution controls, cleaner vehicles, better public transport and encouragement of active modes of transport via easily accessible walkways and bicycle paths, and finally reduced household air pollution through improved access to cleaner fuels and cookstoves (SDG target 7.1) (Section 22.3.2). For children under five, improving indoor air pollution through a transition away from traditional biomass on open fires or traditional stoves can result in significant health benefits. Finally, interventions to reduce exposure to microbial pathogens include increased levels of access to and knowledge of safe water, safely managed sanitation and hygiene (SDG targets 6.1 and 6.2) (Landrigan *et al.* 2018, p. 40) (Section 22.3.3).

Through interventions on all three risk factors, the environmental risks of under-five mortality are lessened, leading to reduced mortality from malnutrition, diarrhoea, pneumonia and other common infectious diseases (e.g. malaria). However, even if all the related environmental SDG targets were achieved by 2030, the under-five mortality target would not be met (Hughes *et al.* 2011; van Vuuren *et al.* 2015; Lucas *et al.* 2018; Moyer and Bohl 2018). Lucas *et al.* (2018) show that achieving health-related SDG targets on child nutrition, access to improved drinking water and sanitation, and access to modern energy services can avoid globally around 440,000 child deaths in 2030, reducing projected 2030 global under-five mortality by around 8 per cent. Hughes *et al.* (2011) conclude that, between 2005 and 2060, some 131.6 million cumulative child deaths (23 per cent of total deaths related to communicable diseases) could be avoided by gradually reducing childhood underweight, unsafe water, poor sanitation and hygiene, indoor air pollution and global climate change.

Alleviating poverty

There is considerable overlap between poor health and poverty (Aber *et al.* 1997; Yoshikawa, Aber and Beardslee 2012). In fact, while poverty is generally indicated as a measure of income, it can also be defined in terms of relative deprivation in a range of capabilities, including good health, but also higher levels of education (Hulme and Shepherd 2003; Alkire 2007). Poverty as defined by low income negatively impacts both health and education outcomes driving further deprivation (Hulme and Shepherd 2003). Conversely, eradicating extreme poverty (SDG target 1.1), and thereby improving the income situation of poor households, can improve health, especially of children under five.

Women and girl's education

Inclusive and equitable quality education (SDG 4), especially of women, is highly correlated with reduced child mortality. Furthermore, higher levels of education are associated with better overall health, lower fertility rates, increased economic growth, reduced poverty levels and more democracy (Dickson, Hughes and Irfan 2010; Lutz and Samir 2013; Dickson, Irfan and Hughes 2016). Over half the decline in child mortality from 1970 to 2009 can be attributed to increased education of women of reproductive age (Gakidou *et al.* 2010). Lucas *et al.* (2018) show that through a comprehensive strategy that includes universal female education, piped drinking water, a complete phase-out of biomass use for cooking and advanced malaria control, 777,000 child deaths can be avoided in 2030, reducing the projected 2030 global child mortality rate by around 13 per cent. The largest health gains are projected for sub-Saharan Africa.

Child and maternal health care

Reducing child mortality is inseparable from reducing maternal mortality – a healthy life begins with a healthy mother and a healthy birth. Reducing child mortality thus also requires addressing other SDG targets, including reducing maternal mortality itself (SDG target 3.1), increasing access to family planning and reducing the adolescent birth rate (SDG target 3.7), achieving universal health coverage (SDG target 3.8) and registering all births with a civil authority (SDG target 16.9) (United Nations Children's Fund [UNICEF] 2015; WHO and UNICEF 2017). Increased contraceptive use in developing countries has reduced the maternal mortality ratio by 26 per cent over the last decade by reducing unintended pregnancies and could reduce it by another 30 per cent if the unmet need is met (Cleland *et al.* 2012). Further, access to modern contraception directly reduces child mortality because increasing the interval between pregnancies reduces likelihood of prematurity and low birthweight, and infants with siblings less than two years old have a higher likelihood of death (Cleland *et al.* 2012).

Synergies and trade-offs between measures and socioeconomic developments

Apart from the obvious improvements to quality of life for people across the globe, improving health outcomes can also have significant impacts on demographics (Lee 2003; Hughes *et al.* 2011) and economic development (van Zon and Muysken 2003; Bloom, Canning and Sevilla 2004; Ashraf, Lester and Weil 2008; Suri *et al.* 2011).

- Reductions in child mortality are typically followed by fertility rate reduction, with a lag of about ten years (Angeles 2010; Bohl, Hughes and Johnson 2016). This has transformative implications (i.e. a larger working-age population followed by an ageing population) for the demographic structure of regions such as sub-Saharan Africa and South Asia, which currently have relatively high rates of both under-five mortality and fertility (Bohl, Hughes and Johnson 2016). When the working-age population growth rate exceeds that of the youth population, the growing labour force also creates economic opportunities, called the 'demographic dividend' (Bloom *et al.* 2009; Lee and Mason 2011, and see Chapter 2). During this time, fiscal burdens associated with service provision to youth (and elderly) populations are minimized, while aggregate economic productivity tends to increase (Lee and Mason

2011). However, a growing elderly population can create new budgetary constraints and more intense pressures on health and social services (Tabata 2005; Lee and Mason 2011; Bohl, Hughes and Johnson 2016; Burrows, Bohl and Moyer 2017).

- Reductions in mortality often result in reductions in morbidity of working-age populations (Hughes *et al.* 2011), further increasing aggregate economic productivity and attracting foreign investment into an economy via reduced labour-market uncertainty (Jamison *et al.* 2006; Hughes *et al.* 2011). Improved health outcomes can also lead to increased school attendance, improved cognitive skills and better educational outcomes for students (Baldacci *et al.* 2004; Soares 2006; Ashraf, Lester and Weil 2008), which improves human capital, and results in increased productivity and more healthy economies once these children move into working-age cohorts (Hughes *et al.* 2011).
- Decreased child mortality, especially when combined with female education and access to modern contraception, will likely lead to lower fertility rates in the longer term, curbing population growth, one of the major drivers of environmental degradation (Angeles 2010; Gakidou *et al.* 2010).

22.4 An integrated approach

In the previous sections, we discussed how to achieve a set of environment-related SDG targets (see Chapter 20 for target selection) and showed that, for many targets, pathways can be identified that could lead to meeting the targets by 2030 or 2050 – or at least result in a major improvement. Here we discuss some overall results from the analysis and a more in-depth analysis of key synergies and trade-offs between the different clusters.

22.4.1 Transformative change

The analysis showed that, in all areas, marginal improvements will not suffice; large, transformative changes are needed to realize the different targets, including significant improvements in resource efficiency with respect to yields, and water-, energy- and nitrogen-use efficiency **(see Table 22.1)**. For instance, reaching the targets related to energy access, climate change and air pollution, would imply decoupling of CO_2 emissions from economic growth at a rate of 4-6 per cent a year, over the coming three decades. In comparison, the same ratio only declined by 1-2 per cent a year historically, thus requiring a threefold increase of the historical rate. Furthermore, without demand-side measures, an average increase in productivity of around 1.4 per cent per year in agriculture would be needed to end global hunger, while simultaneously limiting biodiversity loss. While here the required efficiency improvements are comparable to historical improvement rates, it is clear that this will be more difficult to achieve in the future given that, in most cases, easy gains have already been implemented and agricultural production also will have to become more sustainable, including reduced water and nutrient use.

Earlier, we indicated that technological changes, lifestyle changes and multi-scale approaches are available. The measures discussed in this chapter are part of such approaches. However, given the scale of the required transition it seems far more likely that these strategies will have to be combined to achieve the level of transformation that is needed. It can also be concluded that the approaches used to unlock the available potential presented by any of these approaches has, thus far, not been very successful. The existing MEAs have not led to any break with the past (Part A and Chapter 21). It is therefore important to ensure that there is sufficient interest among actors to implement a different set of strategies. This interest is, among other influences, related to the different trade-offs and synergies of the different measures.

22.4.2 Synergies and trade-offs

Sections 22.3.1-22.3.5 discussed interrelations between measures and targets within the five clusters. However, there are also many synergies and trade-offs between these clusters. The SDGs and associated targets form a complicated network of interlinkages, not made explicit in their formulation (International Council for Science and International Social Science Council

Table 22.1: Trends in resource-use efficiency: baseline (Chapter 21) versus pathways towards achieving the targets (this chapter)

Target	Indicator	Baseline (Chapter 21)	More sustainable pathways[a] (Chapter 22)
Increase agricultural productivity (Section 22.3.1)	Yield improvement over time (total)	1 per cent/year (2010-2050)	1.4 per cent/year (2010-2050)
Increase nutrient-use efficiency (Section 22.3.1)	Total N inputs to the crop N yields	0.55 in 2050	0.67 in 2050
Increase water-use efficiency (Section 22.3.3)	Change in water-use efficiency over time	0.3-1 per cent (2010-2050)	0.5-1.2 per cent (2010-2050)
Increase the share of renewable energy (Section 22.3.2)	Renewable energy share in total final energy consumption	20-30 per cent in 2050	30-60 per cent[b] in 2050
Increase energy efficiency (Section 22.3.2)	Reduction in energy intensity over time (measured in terms of primary energy and GDP)	1-2.5 per cent (2010-2050)	2.2-3.5 per cent (2010-2050)

N: nitrogen.

[a] Not for all topics, the pathways found in the literature and discussed in this chapter were able to meet the selected target as presented in Chapter 20 (see Section 22.3.1 to 22.3.5).

[b] Renewable energy includes the full range of renewables and non-CO_2 emission reductions in the mitigation scenarios derived from the SSP scenarios (see Section 3.2.2).

2015; Le Blanc 2015). Understanding the interlinkages, beyond the clusters focused on here, is crucial for synergistic implementation and policy coherence (Nilsson, Griggs and Visbeck 2016; TWI2050 2018). Accounting for interlinkages can help enhance the effectiveness of implementation and, to some extent, also reduce the total burden and cost of achieving targets individually (Elder, Bengtsson and Akenji 2016). Furthermore, it can help with identifying coherent clusters of targets to be pursued together (Weitz *et al.* 2018).

Analysing the integrated nature of the SDGs has been a research area since their agreement in 2015. However, only a few broad studies so far have analysed interrelations across all SDGs (e.g. Prahdan *et al.* 2017; Zhou and Moinuddin 2017). Difficulties with such studies are that they generally do not look at specific measures, do not take into account future developments, and can only conclude correlations between targets, not causality. Studies that do take these elements into account in their analysis generally focus on a subset of SDGs (International Council for Science 2017; van Vuuren *et al.* 2015) or specific themes, such as energy (McCollum *et al.* 2018; Nerini *et al.* 2018), climate mitigation (von Stechow *et al.* 2016), air pollution (Elder and Zusman 2016), land use and food security (Obersteiner *et al.* 2016; Conijn *et al.* 2018), oceans (Singh *et al.* 2017) and ecosystem services (Wood *et al.* 2018). These studies are either based on the existing literature – as is also the case in this chapter – or on dedicated modelling.

Overall, these studies identify more synergies than trade-offs within and among the SDGs and their targets. However, many interrelations are highly context-specific (Nilsson, Griggs and Visbeck 2016; Weitz *et al.* 2018). There are multiple links between two targets, with potentially different and sometimes conflicting interrelations. Furthermore, outcomes depend on the governance and geographical context, as well as the time-horizon taken (Nilsson *et al.* 2018), to name a few. Providing a full analysis of all interrelations across the measures and targets discussed in this chapter thus requires a dedicated, place-based analysis, which is beyond the scope of GEO-6. In this section, we therefore further elaborate on some of the interrelations among measures and targets between the different clusters for which the scenario literature concludes significant interrelations.

Table 22.2 provides a broad overview of measures with strong synergistic effects and measures with strong trade-offs across the targets, based on the scenario assessments of Section 22.3.1 to 22.3.5 and a quick-scan presented in **Box 22.5**. From this set, key measures with respect to strong synergies and trade-offs are selected for a more in-depth discussion.

Table 22.2: Measures with significant synergies or trade-offs across the selected targets

	Synergies	Trade-offs
Discussed here	(Female) education Reducing agricultural demand via loss and waste reduction, changing diets and nutrition management Reducing air pollution	Land-based mitigation, including large-scale bioenergy deployment Agricultural intensification Environmental policy (potentially conflicting with poverty eradication)
Other examples	Improving resource efficiency of energy, land and water resources (although risk of rebound effects exists) Move towards non-biomass renewable energy (e.g. wind and solar power) Ecosystem restoration Integrated water resources management	Competition for scarce resources Economic development (potentially leading to further demand for resources) Desalination

Box 22.5: A snapshot of interrelations between the selected measures and targets

To get an overview of the many interrelations across the selected measures and targets discussed in this chapter, an expert assessment has been conducted under the authors of this chapter. This expert assessment was compared with the literature and the input from authors of Part A of GEO-6. Experts were asked to score the interrelations using the seven-point scale of Nilsson, Griggs and Visbeck (2016). Interrelations were scored from the most positive score (the measure is indivisible to achieve the target) to the most negative score (the measure can cancel achievement of the target). The result (average score over the different expert scores) is presented in **Figure 22.9**.

Some clear patterns emerge from the analysis. Most interlinkages are flagged between the different measures and the targets that address climate change and biodiversity loss. Furthermore, in line with conclusions from earlier interrelation studies, there are more synergies than trade-offs. The strongest synergies are between measures and targets within the same cluster (see the synergies and trade-offs discussion for each of the individual clusters in Section 22.3). Finally, clear trade-offs are identified between the measures for yield improvement, bioenergy use and desalination, and a broad range of targets. However, as the strongest negative score was not given, the experts suggest that these trade-offs could be addressed with extra mitigating measures.

The analysis also concludes that the extent of the interrelations is not always straightforward. For many interrelations, the experts showed some level of disagreement. These stem partly from different assumptions on the overall context in which the measures are taken, but also that several measures can have both synergies and trade-offs requiring some kind of assessment of their strength. From a similar exercise in the literature, focusing on SDGs on health, energy and oceans, it was concluded that interactions depend on key factors such as geographical context, resource endowments, time-horizon and governance (Nilsson *et al.* 2018). **Figure 22.9** thus only presents a first snapshot or quick-scan of key interrelations involved. To draw policy conclusions, a more dedicated analysis is required. This includes systematic reviews, coding existing literature with respect to specific interactions and integrated assessment modelling, with the latter analysing interlinkages within and across a broader range of subsystems than is currently done (see also Nilsson *et al.* 2018).

Figure 22.9: Quick-scan of synergies and trade-offs between selected measures and targets

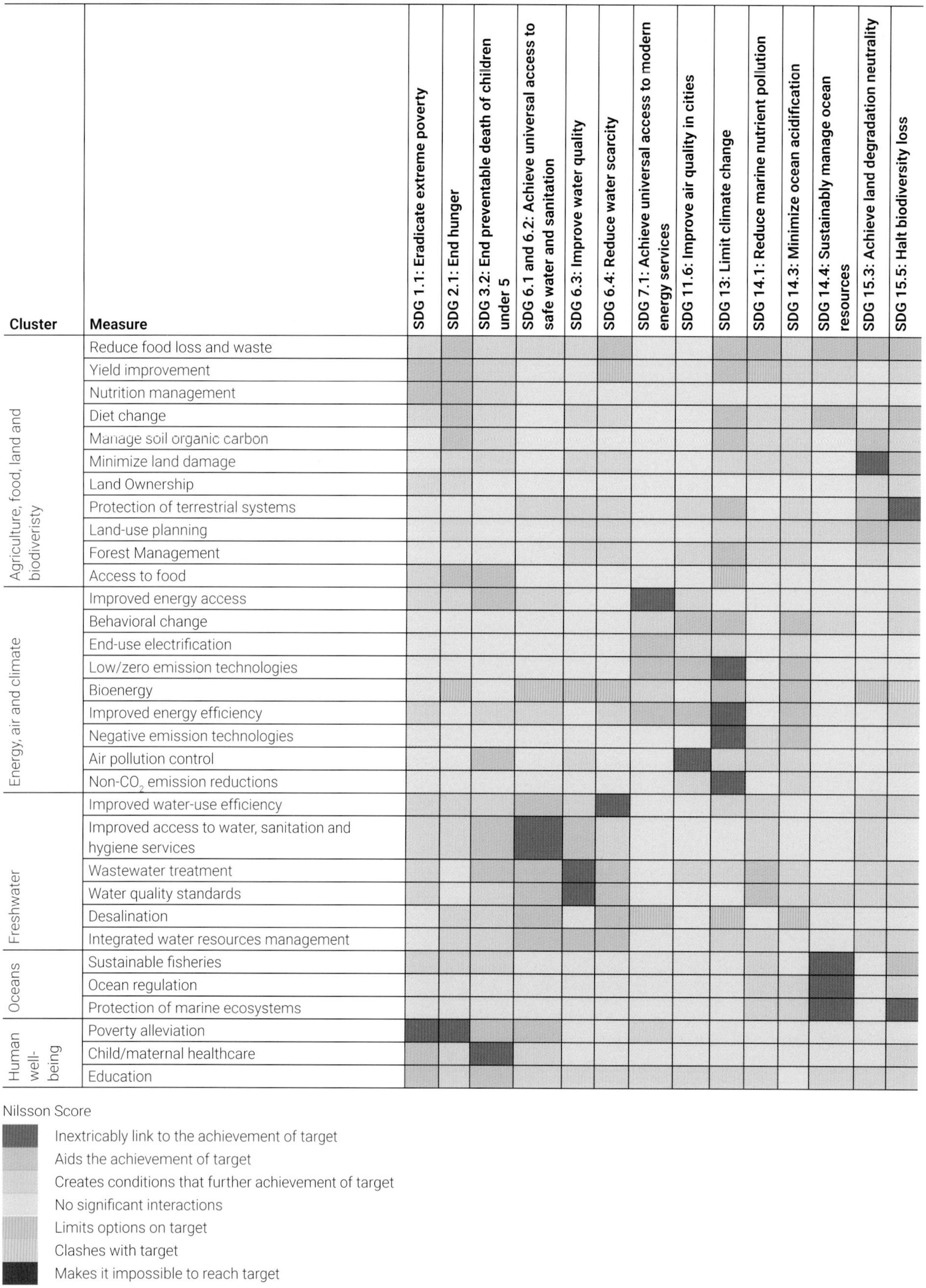

Cluster	Measure	SDG 1.1: Eradicate extreme poverty	SDG 2.1: End hunger	SDG 3.2: End preventable death of children under 5	SDG 6.1 and 6.2: Achieve universal access to safe water and sanitation	SDG 6.3: Improve water quality	SDG 6.4: Reduce water scarcity	SDG 7.1: Achieve universal access to modern energy services	SDG 11.6: Improve air quality in cities	SDG 13: Limit climate change	SDG 14.1: Reduce marine nutrient pollution	SDG 14.3: Minimize ocean acidification	SDG 14.4: Sustainably manage ocean resources	SDG 15.3: Achieve land degradation neutrality	SDG 15.5: Halt biodiversity loss
Agriculture, food, land and biodiveristy	Reduce food loss and waste														
	Yield improvement														
	Nutrition management														
	Diet change														
	Manage soil organic carbon														
	Minimize land damage														
	Land Ownership														
	Protection of terrestrial systems														
	Land-use planning														
	Forest Management														
	Access to food														
Energy, air and climate	Improved energy access														
	Behavioral change														
	End-use electrification														
	Low/zero emission technologies														
	Bioenergy														
	Improved energy efficiency														
	Negative emission technologies														
	Air pollution control														
	Non-CO_2 emission reductions														
Freshwater	Improved water-use efficiency														
	Improved access to water, sanitation and hygiene services														
	Wastewater treatment														
	Water quality standards														
	Desalination														
	Integrated water resources management														
Oceans	Sustainable fisheries														
	Ocean regulation														
	Protection of marine ecosystems														
Human well-being	Poverty alleviation														
	Child/maternal healthcare														
	Education														

Nilsson Score

- Inextricably link to the achievement of target
- Aids the achievement of target
- Creates conditions that further achievement of target
- No significant interactions
- Limits options on target
- Clashes with target
- Makes it impossible to reach target

Source: Scores are based on expert elicitation, using the seven-point scale of Nilsson, Griggs and Visbeck (2016).

22

Selected measures with significant synergies across the selected targets

Education

Education is a basic human right (Universal Declaration of Human Rights, Article 26), an SDG in itself (SDG 4) and, like health, a measure of human development (United Nations Development Programme [UNDP] 2016). Improved education has considerable synergistic effects with both well-being and environment-related targets (United Nations Educational, Scientific and Cultural Organization [UNESCO] 2017). Education, especially for women, has a particularly strong connection with health outcomes. It can significantly affect child health, through reduced malnutrition (Smith and Haddad 2000; Marmot, Allen and Goldblatt 2010) and improved hygiene. Over half the decline in child mortality from 1970 to 2009 can be attributed to increased education in women of reproductive age (Gakidou *et al.* 2010). In addition, higher levels of education are associated with lower fertility rates, increased economic growth, reduced poverty levels and more democracy (Dickson, Hughes and Irfan 2010; Lutz and Samir 2013; Dickson, Irfan and Hughes 2016). The link between improved educational metrics and economic growth and poverty alleviation are well established (Hulme and Shepherd 2003; Verner 2004; Awan *et al.* 2011; Cremin and Nakabugo 2012; UNDP 2016). Improved education also contributes to coping with climate change and coping with the increased occurrence and severity of natural disasters (Cordero, Todd and Abellera 2008; Kagawa and Selby 2012; Chang 2015). Climate change education contributes to capacity-building for decision makers, but also empowers people to implement their own adaptation strategies, among other things, by equipping people to understand complexity and perceive risks (Mochizuki and Bryan 2015). Improving access to safe drinking water, sanitation and hygiene, and sound management of freshwater ecosystems can also benefit from education (Çoban *et al.* 2011; Michelsen and Rieckmann 2015; Karthe *et al.* 2016).

Dietary change

Dietary change, particularly towards reduced ruminant consumption, is synergistic with achieving multiple environmental targets. Furthermore, it can help end hunger and improve human health, with minimal effect on land degradation and biodiversity. In particular, dietary change can reduce cropland expansion (Stehfest *et al.* 2009; Tilman and Clark 2014) and at the same time increase food supply (Foley *et al.* 2011). In addition, dietary change can result in reduced greenhouse gas emissions, reduced pollution, reduced water use and improved health. Dietary change results in reduced emissions of methane from reduced livestock consumption, N_2O and ammonia from reduced fertilizer application, and CO_2 from reduced cropland conversion (Stehfest *et al.* 2009; van Vuuren *et al.* 2017a). The decrease in greenhouse gas emissions associated with dietary change can be significant, with greenhouse gas emission reductions of as much as 70-80 per cent possible (Aleksandrowicz *et al.* 2016). Reducing methane emissions also has positive implications for air quality, as it is a precursor to ozone pollution. Reduction in nitrogen fertilizer use associated with changes in diet has the co-benefit of improving air quality and health by reducing emissions of ammonia and the subsequent formation of fine particulate matter (Zhao *et al.* 2017; Giannadaki *et al.* 2018). Reductions in nitrogen fertilizer use associated with changes in diet also have positive implications for water quality. Reduction in water use can be as much as 50 per cent (Aleksandrowicz *et al.* 2016; Jalava *et al.* 2016; Bijl *et al.* 2017; van Vuuren *et al.* 2017a). Finally, dietary shifts to lower consumption of livestock products yields benefits in all-cause mortality (Milner *et al.* 2015; Aleksandrowicz *et al.* 2016; Springmann *et al.* 2018). It should be noted that some researchers do not find a significant increase of food availability and access for poor communities, resulting from reduced meat consumption in high-income countries (Moyer and Bohl 2018). To be effective, measures to shift diets need to take into account the regional and developmental context (World Economic Forum 2017).

Air pollution control

Reduced air pollution has clear positive impacts on human health. However, there are also synergies with agricultural production, biodiversity and climate change. Ozone is a strong oxidant that can enter plants through the leaves and damage vegetation by affecting photosynthesis and other physiological functions. Several studies have reviewed the links between ozone concentrations, forest productivity and agricultural yields (e.g. Ainsworth *et al.* 2012; Talhelm *et al.* 2014). Averaged over 2010-2012, ozone is estimated to have reduced wheat yield by 9.9 per cent in the Northern Hemisphere and by 6.2 per cent in the Southern Hemisphere (Mills *et al.* 2018). Shindell *et al.* (2012) quantified how measures to reduce black carbon and methane lead to reduced ozone and thus improved agricultural yield, production and value. They found an increase in production of approximately 27 million tons and 24 million tons due to measures to reduce methane and black carbon, respectively. Avnery, Mauzerall and Fiore (2013) report that methane emission controls could increase production of wheat, maize and soybean in North America in 2030 by up to 3.7 million tons. Capps *et al.* (2016) showed that reduction in emissions of nitrogen oxides (NO_x) as a co-benefit of limiting CO_2 emissions from coal power plants in the United States of America could reduce potential productivity loss due to ozone exposure by as much as 16 per cent and 13 per cent for individual crops and tree species, respectively. Reduction in SO_2 and NO_x emissions leads to reductions in acid and nitrogen deposition, and subsequent ecosystem impacts such as eutrophication (Greaver *et al.* 2012).

Selected measures with significant trade-offs across the selected targets

Land-based mitigation

Nearly all climate scenarios consistent with the Paris Agreement rely on significant use of land-use based mitigation (see also **Box 22.2**). This includes the use of bioenergy, avoiding deforestation and afforestation/reforestation. A special case is the role of negative emissions (bioenergy plus carbon-capture-and-storage and afforestation), which seems a requirement for the stringent climate targets – certainly those which allow for higher short-term emissions (Fuss *et al.* 2014; van Vuuren *et al.* 2017a). The use of land-based mitigation options can have important implications for other sustainability targets, in particular food security and protecting terrestrial biodiversity (Wicke 2011; Reilly *et al.* 2012; Calvin *et al.* 2014; Popp *et al.* 2014; Smith *et al.* 2016; Heck *et al.* 2018). For example, pathways with high bioenergy use can negatively impact land degradation and biodiversity, as these pathways would typically lead to higher food prices, reduced forest cover and reduced natural lands. While pathways with significant afforestation could potentially lead to a synergy with reducing biodiversity loss they could still lead to increased competition for land and thus potentially to higher food prices. Bioenergy

use could also lead to higher demand for water and more fertilizer use, with the latter increasing the risk of eutrophication from higher nitrogen and phosphorus run-off (e.g. Gerben-Leenes, Hoekstra and van der Meer 2009; Hejazi *et al.* 2015; Mouratiadou *et al.* 2016).

Although bioenergy is one of several important options for future energy systems, increasing global trade and consumption of bioenergy has been accompanied by a growing concern about the environmental, ecological and social impacts of modern bioenergy production (Wicke 2011). For example, trade-offs between bioenergy and food security, and between the impact of biomass on poverty reduction and on the environment have been widely reported (Wicke 2011; Smith *et al.* 2016). Yamagata *et al.* (2018) report water-food-ecosystem trade-offs for global negative CO_2 emission scenarios. They point to three outstanding conflicts:

- vast conversion of food cropland into rain-fed bio-crop cultivations yields a considerable loss of food production;
- when irrigation is applied to bio-crop production, the bioenergy crop productivity is enhanced – this reduces the area necessary for bio-crop production by half, but water consumption is doubled, increasing water scarcity and groundwater depletion; and
- if conversion of forest land for bioenergy crop cultivation is allowed, large areas of tropical forest could be used for bioenergy crop production, which can cause serious extensive decline in carbon stock and related ecosystem services, leading to increasing CO_2 emissions from land-use change.

More attention needs to be paid to the co-benefit of biodiversity conservation and climatic change mitigation activities for optimizing various sustainability benefits.

Figure 22.10 illustrates that in the majority of the scenarios the use of bioenergy increases for more stringent climate targets. In the SSP scenario database in fact all scenarios consistent with the targets of the Paris Agreement lead to a demand for bioenergy of more than 200 exajoules/year in 2050. Earlier, the Intergovernmental Panel on Climate Change (IPCC) did an assessment of the bioenergy supply in 2050 under different sustainability constraints. It concluded that at least 100 exajoules/year would be available under these constraints. It also concluded that possibly 300 exajoules/year

Figure 22.10: Global mean temperature increase in 2100 versus bioenergy use in various SSP reference scenarios and derived mitigation scenarios

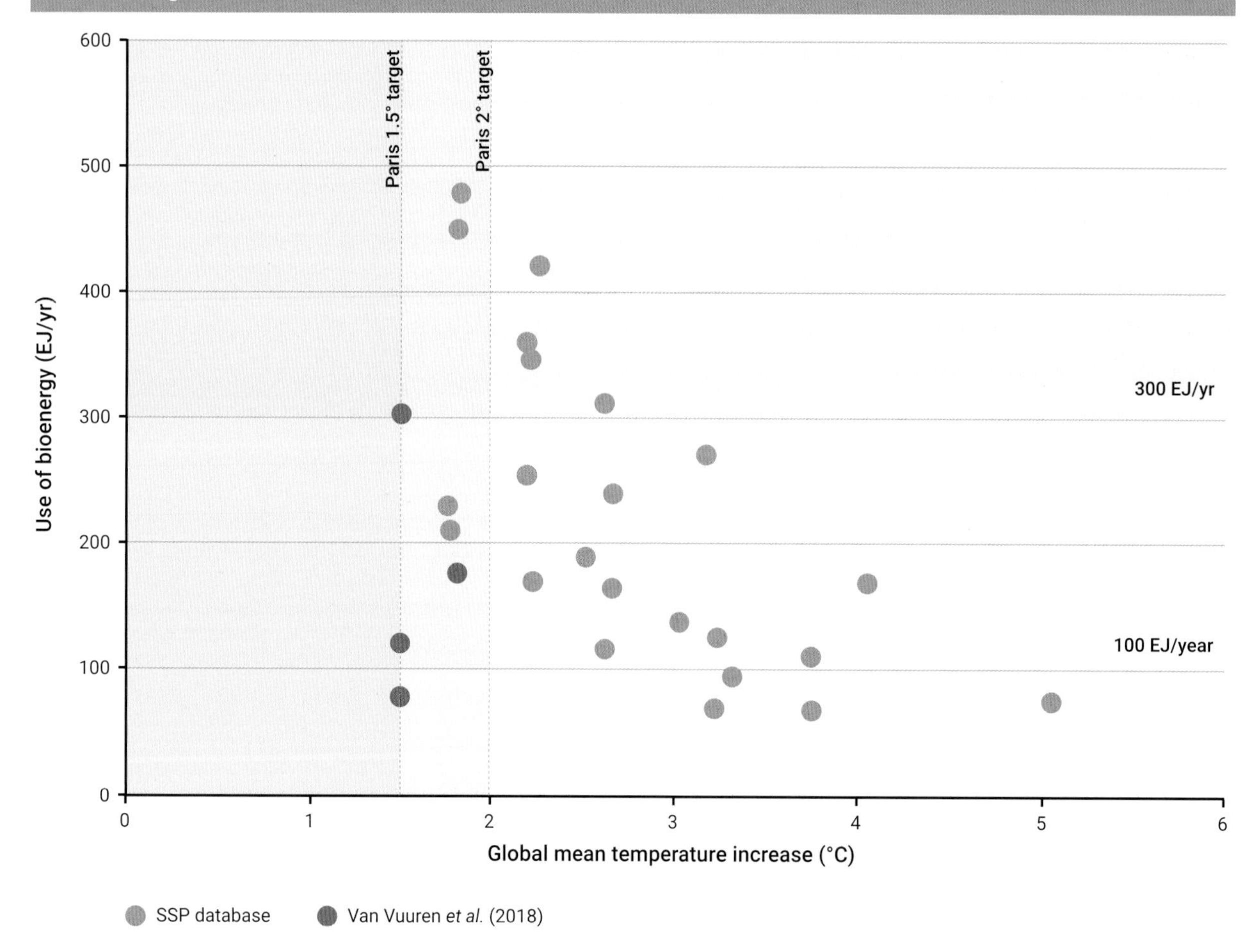

The different background colors indicate the Paris Climate Targets (vertical lines, starting at 1.5° and 2°C) and the range for sustainable biodiversity supply indicated by the Intergovernmental Panel on Climate Change (IPCC) (IPCC indicated 100 exajoules/year was most likely available; 300 exajoules/year could be available).

Source: Riahi *et al.*, 2017; Vuuren *et al.* (2018).

would be available (but with a much higher level of uncertainty). A bioenergy potential of about 100 exajoules/year was found to have high agreement of being sustainable, while values above that threshold had lower levels of agreement as to the sustainability of the bioenergy supply (Creutzig *et al.* 2015). This means that no scenarios in the database would actually be consistent with a stringent interpretation of both the Paris target and the sustainability constraints on bioenergy. Van Vuuren *et al.* (2018) explored different alternative pathways to reach ambitious climate targets that could possibly reduce the need for negative emissions (and thus bioenergy). These scenarios, for instance, assumed diet change towards low-meat diets consistent with health recommendations, ambitious implementation of non-CO_2 emission reduction or alternatively the production of cultivated meat. Such assumptions could lead to a much lower demand for negative emission technologies and thereby bioenergy in combination with carbon-capture-and-storage. Recently, a model comparison study looked into stringent climate policy scenarios with limited bioenergy supply (Bauer *et al.* 2018). Here, some models did find low-bioenergy pathways through careful optimization of their use (e.g. the application of bioenergy in combination with carbon-capture-and-storage only for production of transport fuels).

Agricultural intensification

Improving agricultural yields is seen as a prerequisite for producing enough food and bioenergy to meet future demand while at the same minimizing or completely eliminating the need for agricultural land expansion. The use of fertilizers can potentially deliver yield increases but can also have severe consequences for freshwater and ocean quality and related ecosystems, as well as climate change (Bouwman *et al.* 2017). Impacts on biodiversity largely depend on how higher yields are achieved.

Improving yields can increase overall food availability, especially when these yield improvements are achieved in current low-yield countries and areas with high prevalence of undernourishment. At the same time, it can negatively impact nutrition if high-yield crops contain less micronutrients than average dietary requirements (DeFries *et al.* 2015; Rao *et al.* 2018). Furthermore, when yields are increased without specifically addressing distributional aspects, the increased production does not necessarily reach the communities most in need. At the same time, obesity in high- and middle-income countries could rise as a result of overall decreasing food prices (van Vuuren *et al.* 2015). Finally, when yield increases are accompanied by scale increase, smallholders might be forced to move to cities, which does not necessarily improve their income situation.

Improving agricultural yields reduces land demand for growing crops, reducing pressure on existing natural lands, thus potentially reducing deforestation and biodiversity loss. On the other hand, increasing yields usually demands higher levels of fertilizers, pesticides, and water for irrigation, thus negatively impacting water quality and water scarcity. Use of nitrogen fertilizers also causes higher N_2O emissions, meaning trade-offs with climate change mitigation. Mechanization and monocultures associated with yield increases in the past led to erosion, soil compaction and loss of soil organic carbon, increasing the likelihood of land degradation. This can be further exacerbated by leaching and salinization of land from long-term irrigation. All these factors negatively impact biodiversity.

Poverty alleviation and environmental protection

Higher incomes, decreasing hunger and improved access to water and energy are expected to push up demand for food, water and energy, thereby increasing environmental pressures. In reality, however, both synergies and trade-offs exist – and, while some are important to take into account, others are relatively small. Scenario analysis shows that eradicating hunger and providing universal access to modern energy services (beyond production increases that result from population and economic growth) would not necessarily negatively affect global biodiversity or climate change (e.g. Riahi *et al.* 2012; van Vuuren *et al.* 2015; Dagnachew *et al.* 2018). Although most studies addressing access to modern energy services show that a decrease in biomass use is generally accompanied by an increase in the use of fossil-fuel-based products (e.g. liquified petroleum gas, natural gas, electricity), the increase in global CO_2 emissions is usually small (Dagnachew *et al.* 2018). Furthermore, increasing CO_2 emissions are partly compensated by reduced emissions from deforestation and black carbon. Similarly, the additional demand for food, resulting from the eradication of hunger, is estimated to be relatively small, especially when compared to current production levels and the required increase to keep pace with an increasing and more wealthy global population (van Vuuren *et al.* 2015). If hunger eradication would be facilitated by a redistribution of current consumption levels, the required increase in production would be even less (van Vuuren *et al.* 2015). Obviously, however, further development beyond the minimum levels could be associated with further environmental pressure. Therefore, it is important to add sustainability considerations in policies that aim for higher levels of economic development in order to prevent such a trade-off.

Several studies have emphasized another potential trade-off between achieving environmental targets and ensuring access to basic resources and services. This is because, in many cases, policies for achieving environmental targets could lead to a cost increase. While such cost increases might be relatively unimportant for populations with high income levels, they could have a strong impact on the poor. It has been shown that if implemented without additional compensatory measures, climate policy could lead to negative impacts on access to electricity (Dagnachew *et al.* 2017), access to clean fuels for cooking (Cameron *et al.* 2016) and on food security (Hasegawa *et al.* 2018).

22.5 Conclusions and recommendations

We have assessed the scenario literature to analyse a broad range of measures relevant for achieving the selected environmental targets of the SDGs and related MEAs, with a specific focus on synergies and trade-offs. Overall, the scenario literature provides a broad range of options to move towards achieving these targets, but this knowledge is hampered by the lack of concrete pathways.

22.5.1 Knowledge gaps

The discussion in this chapter shows that model-based scenario analysis can be an effective tool to support integration of knowledge in the effort required to reach the environmental targets of the SDGs and related MEAs and to highlight the linkages across time, scales and issues.

However, from the literature assessment, it can be concluded that the scenario literature is still patchy on analysis to show possible pathways to achieving the SDGs. No fully integrated scenario studies exist. Furthermore, the literature is well advanced in some areas, while for other areas literature is mostly lacking. As a result, it is still difficult to estimate the exact size of different strengths and weaknesses of specific measures. There is extensive literature that discusses pathways to achieve the selected targets in the energy, air and climate cluster and, although to a lesser extent, also in the agriculture, food, land and biodiversity cluster. In the latter case, these studies mostly address hunger and biodiversity, with relatively few scenario studies that aim to meet specific targets and virtually no scenario studies that address how to achieve land degradation neutrality. Ocean acidification is well discussed in the literature, mostly linked to scenarios that address climate targets.

For the freshwater, oceans and human well-being (health) clusters, target-seeking scenarios are much less common in the literature. For the freshwater cluster, scenarios look at water scarcity issues, while the literature around WASH and water quality are sparse. For health (e.g. child mortality), very few target-seeking scenarios were found in the literature. Finally, as already concluded in Chapter 21, quantitative scenario studies on chemicals and waste and wastewater are almost non-existent.

While many synergies and trade-offs are discussed in the literature, besides thematic studies (mostly based on existing scenario literature), a thorough overview of all relevant interrelations across the measures and targets discussed in this chapter is still lacking. This is partly because there are still caveats in the scenario literature and because these interrelations are highly context-specific, making it difficult to provide unambiguous scores. Sectoral studies looking at interlinkages often emphasize the key role of that sector in achieving the overall targets, providing very few options for prioritization. As result, large gaps exist in current understanding of linkages with other sectors or themes.

It should be noted that indirect interlinkages often also exist and that, in many cases, interlinkages can lead to both synergies and trade-offs. For example, fertilizer application could lead to higher yields, requiring less land and thus reducing biodiversity loss and potentially land expansion, while it would also increase nitrogen and phosphorus run-off leading to freshwater and marine nutrient pollution, causing hypoxia and harmful algal blooms, and related biodiversity loss. These complex interrelations and the absence of broad interlinkage studies imply that more dedicated analyses are required, including systematic reviews of the existing literature and dedicated integrated assessment modelling, with specific attention to interlinkages that are currently underexplored.

22.5.2 Policy recommendations

From the scenario analysis, it can be concluded that pathways exist towards achieving a broad range of environmental targets of the SDGs and related MEAs, but they require a clear break with current trends (transformational change). Marginal improvements will not suffice. Large, transformative changes are needed to realize the different targets. Significant improvements in resource efficiency with respect to land, water and energy are required, including an almost 50 per cent increase in agricultural yields compared with current trends, and a doubling of energy efficiency improvement.

Achieving the targets will require a broad portfolio of measures, including a mix of technological improvements, lifestyle changes and localized solutions. The many different challenges require dedicated measures that improve access to, for example, food, water and energy, while at the same time reducing the pressure on environmental resources and ecosystems. A key contribution may come from a redistribution of access to resources. From a production perspective, the changes would include elements such as cleaner production processes and decoupling of resource consumption from economic development. Also changes in demand-side efficiency and consumer behaviour should be considered. The latter may include dietary changes towards reduced ruminant consumption, but also changes in transport moving towards less energy-intensive transport modes.

Understanding interlinkages between measures and targets is crucial for synergistic implementation and policy coherence. Where measures generally aim at achieving specific targets, or a cluster of targets, the analysis showed some clear synergies between measures and targets in other areas. Examples include education, dietary change and air pollution control, with all three having positive impacts on both a Healthy Planet and Healthy People. This chapter also highlights important possible trade-offs, such as the impact of climate policy on the costs of energy and consequently energy access. In many cases, it is also possible to address these trade-offs by introducing mitigation measures (in the example above, specific policies to support energy access for the poor could prevent specific trade-offs).

The economic and technical potential is available to move towards implementation of the targets in Chapter 20. However, a full consideration also needs to account for social feasibility. The feasibility of the transformation processes can only be discussed in the light of current trends and ongoing innovation processes of citizens and businesses worldwide. Chapter 23 will do this. Finally, in Chapter 24, we will discuss how policy measures could induce the transformations presented here. In many cases, social feasibility can be enhanced by ensuring a proper consideration of possible synergies and trade-offs.

References

Aber, J.L., Bennett, N.G., Conley, D.C. and Li, J. (1997). The effects of poverty on child health and development. *Annual Review of Public Health* 18(1), 463-483. https://doi.org/10.1146/annurev.publhealth.18.1.463.

Agardy, T. (2000). Effects of fisheries on marine ecosystems: A conservationist's perspective. *ICES Journal of Marine Science* 57(3), 761-765. https://doi.org/10.1006/jmsc.2000.0721.

Ainsworth, E.A., Yendrek, C.R., Sitch, S., Collins, W.J. and Emberson, L.D. (2012). The effects of tropospheric ozone on net primary productivity and implications for climate change. *Annual Review of Plant Biology* 63, 637-661. https://doi.org/10.1146/annurev-arplant-042110-103829.

Aleksandrowicz, L., Green, R., Joy, E.J.M., Smith, P. and Haines, A. (2016). The impacts of dietary change on greenhouse gas emissions, land use, water use, and health: A systematic review. *PLoS ONE* 11(11), e0165797. https://doi.org/10.1371/journal.pone.0165797.

Alexandratos, N. and Bruinsma, J. (2012). *World Agriculture Towards 2030/2050: The 2012 Revision*. ESA Working paper. Rome: Food and Agriculture Organization of the United Nations. http://www.fao.org/docrep/016/ap106e/ap106e.pdf.

Alkire, S. (2007). The missing dimensions of poverty data: Introduction to the special issue. *Oxford Development Studies* 35(4), 347-359. https://doi.org/10.1080/13600810701701863.

Angeles, L. (2010). Demographic transitions: Analyzing the effects of mortality on fertility. *Journal of Population Economics* 23(1), 99-120. https://doi.org/10.1007/s00148-009-0255-6.

Arvesen, A., Luderer, G., Pehl, M., Bodirsky, B.L. and Hertwich, E.G. (2018). Deriving life cycle assessment coefficients for application in integrated assessment modelling. *Environmental Modelling and Software* 99, 111-125. https://doi.org/10.1016/j.envsoft.2017.09.010.

Ashraf, Q.H., Lester, A. and Weil, D.N. (2008). *When Does Improving Health Raise GDP?* National Bureau of Economic Research Working Papers. Cambridge, MA: National Bureau of Economic Research. http://www.nber.org/papers/w14449.pdf.

Avnery, S., Mauzerall, D.L. and Fiore, A.M. (2013). Increasing global agricultural production by reducing ozone damages via methane emission controls and ozone-resistant cultivar selection. *Global Change Biology* 19(4), 1285–1299. https://doi.org/10.1111/gcb.12118.

Awan, M.S., Malik, N., Sarwar, H. and Waqas, M. (2011). Impact of education on poverty reduction. *International Journal of Academic Research* 3(1), 659-664. https://mpra.ub.uni-muenchen.de/31826/.

Bae, C. and Kim, J. (2017). Alternative fuels for internal combustion engines. *Proceedings of the Combustion Institute* 36(3), 3389-3413. https://doi.org/10.1016/j.proci.2016.09.009.

Bajželj, B., Richards, K.S., Allwood, J.M., Smith, P., Dennis, J.S., Curmi, E. *et al.* (2014). Importance of food-demand management for climate mitigation. *Nature Climate Change* 4, 924-929. https://doi.org/10.1038/nclimate2353.

Baldacci, E., Cui, Q., Clements, M.B.J., Gupta, S. and Cui, Q. (2004). *Social Spending, Human Capital, and Growth in Developing Countries: Implications for Achieving the MDGs*. IMF Working Papers: International Monetary Fund. https://www.imf.org/external/pubs/ft/wp/2004/wp04217.pdf.

Barnes, M.D., Glew, L., Wyborn, C. and Craigie, I.D. (2018). Prevent perverse outcomes from global protected area policy. *Nature Ecology and Evolution* 2, 759–762. https://doi.org/10.1038/s41559-018-0501-y.

Bauer, N., Calvin, K., Emmerling, J., Fricko, O., Fujimori, S., Hilaire, J. *et al.* (2017). Shared socio-economic pathways of the energy sector – quantifying the narratives. *Global Environmental Change* 42, 316-330. https://doi.org/10.1016/j.gloenvcha.2016.07.006.

Bauer, N., K. Rose, S., Fujimori, S., Vuuren, D., Weyant, J., Wise, M. *et al.* (2018). Global energy sector emission reductions and bioenergy use: Overview of the bioenergy demand phase of the EMF-33 model comparison. *Climatic Change*. https://doi.org/10.1007/s10584-018-2226-y.

Bauer, N., Rose, S.K., Fujimori, S., van Vuuren, D.P., Weyant, J., Wise, M. *et al.* (2018). Global energy sector emission reductions and bioenergy use: Overview of the bioenergy demand phase of the EMF 33 model comparison. *Climatic Change*. https://doi.org/10.1007/s10584-018-2226-y.

Béné, C., Barange, M., Subasinghe, R., Pinstrup-Andersen, P., Merino, G., Hemre, G.I. *et al.* (2015). Feeding 9 billion by 2050 – Putting fish back on the menu. *Food Security* 7(2), 261-274. https://doi.org/10.1007/s12571-015-0427-z.

Bengtsson, J., Ahnström, J. and Weibull, A.-C. (2005). The effects of organic agriculture on biodiversity and abundance: A meta-analysis. *Journal of applied ecology* 42(2), 261-269. https://doi.org/10.1111/j.1365-2664.2005.01005.x.

Benson, S.M., Bennaceur, K., Cook, P., Davison, J., de Coninck, H., Farhat, K. *et al.* (2012). Carbon capture and storage. In *Global Energy Assessment - Toward a Sustainable Future*. Gomez-Echeverri, L. and Johansson, T.B. (eds.). Cambridge: Cambridge University Press. chapter 13. 993-1068. http://www.iiasa.ac.at/web/home/research/Flagship-Projects/Global-Energy-Assessment/GEA_Chapter13_CCS_lowres.pdf.

Beusen, A.H.W., Bouwman, A.F., van Beek, L.P.H., Mogollón, J.M. and Middelburg, J.J. (2016). Global riverine N and P transport to ocean increased during the 20th century despite increased retention along the aquatic continuum. *Biogeosciences* 13, 2441-2451. https://doi.org/10.5194/bg-13-2441-2016.

Bijl, D.L., Biemans, H., Bogaart, P.W., Dekker, S.C., Doelman, J.C., Stehfest, E. *et al.* (2018). A global analysis of future water deficit based on different allocation mechanisms. *Water Resources Research* 54(8), 5803-5824. https://doi.org/10.1029/2017WR021688.

Bijl, D.L., Bogaart, P.W., Dekker, S.C., Stehfest, E., de Vries, B.J.M. and van Vuuren, D.P. (2017). A physically-based model of long-term food demand. *Global Environmental Change* 45, 47-62. https://doi.org/10.1016/j.gloenvcha.2017.04.003.

Billé, R., Kelly, R., Biastoch, A., Harrould-Kolieb, E., Herr, D., Joos, F. *et al.* (2013). Taking action against ocean acidification: A review of management and policy options. *Environmental Management* 52(4), 761-779. https://doi.org/10.1007/s00267-013-0132-7.

Bixio, D., Thoeye, C., De Koning, J., Joksimovic, D., Savic, D., Wintgens, T. *et al.* (2006). Wastewater reuse in Europe. *Desalination* 187(1-3), 89-101. https://doi.org/10.1016/j.desal.2005.04.070.

Bixio, D., Thoeye, C., Wintgens, T., Ravazzini, A., Miska, V., Muston, M. *et al.* (2008). Water reclamation and reuse: Implementation and management issues. *Desalination* 218(1), 13-23. https://doi.org/10.1016/j.desal.2006.10.039.

Bloom, D.E., Canning, D., Fink, G. and Finlay, J.E. (2009). Fertility, female labor force participation, and the demographic dividend. *Journal of Economic Growth* 14(2), 79-101. https://doi.org/10.1007/s10887-009-9039-9.

Bloom, D.E., Canning, D. and Sevilla, J. (2004). The effect of health on economic growth: A production function approach. *World development* 32(1), 1-13. https://doi.org/10.1016/j.worlddev.2003.07.002.

Bohl, D.K., Hughes, B.B. and Johnson, S. (2016). *Understanding and Forecasting Demographic Risk and Benefits*. Denver, CO: University of Denver. http://pardee.du.edu/sites/default/files/Demographic%20Risk%20Report%20v44%20%28Final%29.pdf.

Bopp, L., Resplandy, L., Orr, J.C., Doney, S.C., Dunne, J.P., Gehlen, M. *et al.* (2013). Multiple stressors of ocean ecosystems in the 21st century: Projections with CMIP5 models. *Biogeosciences* 10, 6225-6245. https://doi.org/10.5194/bg-10-6225-2013.

Bouwman, A.F., Beusen, A.H.W., Lassaletta, L., van Apeldoorn, D.F., van Grinsven, H.J.M., Zhang, J. *et al.* (2017). Lessons from temporal and spatial patterns in global use of N and P fertilizer on cropland. *Scientific Reports* 7(40366), 40366-40366. https://doi.org/10.1038/srep40366.

Bradshaw, C.J.A., Bowman, D.M.J.S., Bond, N.R., Murphy, B.P., Moore, A.D., Fordham, D.A. *et al.* (2013). Brave new green world – Consequences of a carbon economy for the conservation of Australian biodiversity. *Biological Conservation* 161, 71-90. https://doi.org/10.1016/j.biocon.2013.02.012.

Bui, M., Adjiman, C.S., Bardow, A., Anthony, E.J., Boston, A., Brown, S. *et al.* (2018). Carbon capture and storage (CCS): The way forward. *Energy and Environmental Science* 11(5), 1062-1176. https://doi.org/10.1039/C7EE02342A.

Burrows, M., Bohl, D.K. and Moyer, J.D. (2017). *Our World Transformed: Geopolitical Shocks and Risks*. Zurich Insurance Group, Atlantic Council and University of Denver Pardee Center for International Futures http://www.atlanticcouncil.org/images/publications/Our_World_Transformed_web_0421.pdf.

Calvin, K., Wise, M., Kyle, P., Patel, P., Clarke, L. and Edmonds, J. (2014). Trade-offs of different land and bioenergy policies on the path to achieving climate targets. *Climatic Change* 123(3-4), 691-704. https://doi.org/10.1007/s10584-013-0897-y.

Cameron, C., Pachauri, S., Rao, N.D., McCollum, D., Rogelj, J. and Riahi, K. (2016). Policy trade-offs between climate mitigation and clean cook-stove access in South Asia. *Nature Energy* 1(15010). https://doi.org/10.1038/nenergy.2015.10.

Capps, S.L., Driscoll, C.T., Fakhraei, H., Templer, P.H., Craig, K.J., Milford, J.B. *et al.* (2016). Estimating potential productivity cobenefits for crops and trees from reduced ozone with U.S. coal power plant carbon standards. *Journal of Geophysical Research* 121(24), 14,679-614,690. https://doi.org/10.1002/2016JD025141.

Chang, C.-H. (2015). Teaching climate change – a fad or a necessity? *International Research in Geographical and Environmental Education* 24(3), 181-183. https://doi.org/10.1080/10382046.2015.1043763.

Chaudhary, A., Burivalova, Z., Koh, L.P. and Hellweg, S. (2016). Impact of forest management on species richness: Global meta-analysis and economic trade-offs. *Scientific Reports* 6, 23954. https://doi.org/10.1038/srep23954.

Chazdon, R.L. (2008). Beyond deforestation: Restoring forests and ecosystem services on degraded lands. *Science* 320(5882), 1458-1460. https://doi.org/10.1126/science.1155365.

Cheung, W.W.L., Reygondeau, G. and Frölicher, T.L. (2016). Large benefits to marine fisheries of meeting the 1.5°C global warming target. *Science* 354(6319), 1591-1594. https://doi.org/10.1126/science.aag2331.

China National Development and Reform Commission, China, Ministry of Industry and Information Technology, China, Ministry of Finance, China, Ministry of Environmental Protection, National Bureau of Statistics and China National Energy Administration (2014). Interim measures of the replacement coal consumption in key areas. http://www.ndrc.gov.cn/gzdt/201501/t20150114_660128.html.

Clark, M. and Tilman, D. (2017). Comparative analysis of environmental impacts of agricultural production systems, agricultural input efficiency, and food choice. *Environmental Research Letters* 12(6), 064016. https://doi.org/10.1088/1748-9326/aa6cd5.

Clarke, L., Jiang, K., Akimoto, K., Babiker, M., Blanford, G., Fisher-Vanden, K. *et al.* (2014). Assessing transformation pathways In *Climate Change 2013: Mitigation of Climate Change Contribution of Working Group 3rd to the 5th Assessment Report of the Intergovernmental Panel on Climate Change*. Edenhofer, O., Pichs-Madruga, R., Sokona, Y., Farahani, E., Kadner, S., Seyboth, K. *et al.* (eds.). Cambridge: Cambridge University Press. chapter 6. https://www.ipcc.ch/pdf/assessment-report/ar5/wg3/ipcc_wg3_ar5_chapter6.pdf

Cleland, J., Conde-Agudelo, A., Peterson, H., Ross, J. and Tsui, A. (2012). Contraception and health. *The Lancet* 380(9837), 149-156. https://doi.org/10.1016/S0140-6736(12)60609-6.

Çoban, G.Ü., Akpınar, E., Küçükcankurtaran, E., Yıldız, E. and Ergin, Ö. (2011). Elementary school students' water awareness. *International Research in Geographical and Environmental Education* 20(1), 65-83. https://doi.org/10.1080/10382046.2011.540103.

Coetzee, B.W.T., Gaston, K.J. and Chown, S.L. (2014). Local scale comparisons of biodiversity as a test for global protected area ecological performance: A meta-analysis. *PLoS ONE* 9(8), e105824. https://doi.org/10.1371/journal.pone.0105824.

Cohen, A.J., Brauer, M., Burnett, R., Anderson, H.R., Frostad, J., Estep, K. *et al.* (2017). Estimates and 25-year trends of the global burden of disease attributable to ambient air pollution: An analysis of data from the global burden of diseases study 2015. *The Lancet* 389(10082), 1907-1918. https://doi.org/10.1016/S0140-6736(17)30505-6.

Conijn, J.G., Bindraban, P.S., Schröder, J.J. and Jongschaap, R.E.E. (2018). Can our global food system meet food demand within planetary boundaries? *Agriculture, Ecosystems and Environment* 251, 244-256. https://doi.org/10.1016/j.agee.2017.06.001.

Convention on Biological Diversity (2010). *Decision Adopted by the Conference of The Parties to the Convention on Biological Diversity at its Tenth Meeting. UNEP/CBD/COP/DEC/X/2*. https://www.cbd.int/doc/decisions/cop-10/cop-10-dec-02-en.pdf.

Cordero, E.C., Todd, A.M. and Abellera, D. (2008). Climate change education and the ecological footprint. *Bulletin of the American Meteorological Society* 89(6), 865-872. https://doi.org/10.1175/2007bams2432.1.

Costello, C., Ovando, D., Clavelle, T., Strauss, C.K., Hilborn, R., Melnychuk, M.C. *et al.* (2016). Global fishery prospects under contrasting management regimes. *Proceedings of the National Academy of Sciences* 113(18), 5125-5129. https://doi.org/10.1073/pnas.1520420113.

Craig, M.H., Snow, R.W. and le Sueur, D. (1999). A climate-based distribution model of malaria transmission in Sub-Saharan Africa. *Parasitology Today* 15(3), 105-111. https://doi.org/10.1016/S0169-4758(99)01396-4.

Cremin, P. and Nakabugo, M.G. (2012). Education, development and poverty reduction: A literature critique. *International Journal of Educational Development* 32(4), 499-506. https://doi.org/10.1016/j.ijedudev.2012.02.015.

Creutzig, F., Ravindranath, N.H., Berndes, G., Bolwig, S., Bright, R., Cherubini, F. *et al.* (2015). Bioenergy and climate change mitigation: An assessment. *Global Change Biology: Bioenergy* 7(5), 916-944. https://doi.org/10.1111/gcbb.12205.

Crouzeilles, R., Ferreira, M.S., Chazdon, R.L., Lindenmayer, D.B., Sansevero, J.B.B., Monteiro, L. *et al.* (2017). Ecological restoration success is higher for natural regeneration than for active restoration in tropical forests. *Science Advances* 3(11), e1701345. https://doi.org/10.1126/sciadv.1701345.

Cui, Z., Zhang, H., Chen, X., Zhang, C., Ma, W., Huang, C. *et al.* (2018). Pursuing sustainable productivity with millions of smallholder farmers. *Nature* 555, 363-366. https://doi.org/10.1038/nature25785.

Cullen, J.M., Allwood, J.M. and Borgstein, E.H. (2011). Reducing energy demand: What are the practical limits? *Environmental science and technology* 45(4), 1711-1718. https://doi.org/10.1021/es102641n.

Cunningham, S.C., Cavagnaro, T.R., Mac Nally, R., Keryn, P., Baker, P.J., Beringer, J. *et al.* (2015). Reforestation with native mixed-species plantings in a temperate continental climate effectively sequesters and stabilizes carbon within decades. *Global Change Biology* 21(4), 1552-1566. https://doi.org/10.1111/gcb.12746.

Dagnachew, A.G., Lucas, P.L., Hof, A.F., Gernaat, D.E.H.J., de Boer, H.-S. and van Vuuren, D.P. (2017). The role of decentralized systems in providing universal electricity access in Sub-Saharan Africa – A model-based approach. *Energy* 139, 184-195. https://doi.org/10.1016/j.energy.2017.07.144.

Dagnachew, A.G., Lucas, P.L., Hof, A.F. and van Vuuren, D.P. (2018). Trade-offs and synergies between universal electricity access and climate change mitigation in Sub-Saharan Africa. *Energy Policy* 114, 355-366. https://doi.org/10.1016/j.enpol.2017.12.023.

Daioglou, V., van Ruijven, B.J. and van Vuuren, D.P. (2012). Model projections for household energy use in developing countries. *Energy* 37(1), 601-615. https://doi.org/10.1016/j.energy.2011.10.044.

Davis, K.F., Rulli, M.C., Seveso, A. and D'Odorico, P. (2017). Increased food production and reduced water use through optimized crop distribution. *Nature Geoscience* 10(12), 919–924. https://doi.org/10.1038/s41561-017-0004-5.

de Coninck, H. and Benson, S.M. (2014). Carbon dioxide capture and storage: Issues and prospects. *Annual Review of Environment and Resources* 39, 243-270. https://doi.org/10.1146/annurev-environ-032112-095222.

de Vries, F.T., Thébault, E., Liiri, M., Birkhofer, K., Tsiafouli, M.A., Bjørnlund, L. *et al.* (2013). Soil food web properties explain ecosystem services across European land use systems. *Proceedings of the National Academy of Sciences* 110(35), 14296-14301. https://doi.org/10.1073/pnas.1305198110.

DeFries, R., Fanzo, J., Remans, R., Palm, C., Wood, S. and Anderman, T.L. (2015). Metrics for land-scarce agriculture. *Science* 349(6245), 238-240. https://doi.org/10.1126/science.aaa5766.

Devine, J. and Kullmann, C. (2011). *Introductory Guide to Sanitation Marketing*. Water and Sanitation Program. http://www.wsp.org/sites/wsp.org/files/publications/WSP-Introductory-Guide-Sanitation-Marketing.pdf.

Dickson, J.R., Hughes, B. and Irfan, M.T. (2010). *Advancing Global Education: Patterns of Potential Human Progress*. Denver, BO: Routledge. https://pardee.du.edu/patterns-potential-human-progress.

Dickson, J.R., Irfan, M.T. and Hughes, B.B. (2016). *Usc 2030: Exploring Impacts, Costs, and Financing*. Background Paper for the International Commission on Financing Global Education Opportunity. Denver, BO. https://pardee.du.edu/sites/default/files/Pardee%20for%20Intl%20Com%20Fin%20Ed%20Background%20Final%255b7%255d.pdf.

Edelenbosch, O.Y., van Vuuren, D.P., Blok, K., Calvin, K. and Fujimori, S. (2018). Mitigating energy demand emissions: The integrated modelling perspective. *The 37th Edition of International Energy Workshop*. Gothenburg. PBL Netherlands Environmental Assessment Agency Submitted https://iew2018.org/wp-content/uploads/2018/07/5D_Edelenbosch.pdf.

Edgar, G.J., Stuart-Smith, R.D., Willis, T.J., Kininmonth, S., Baker, S.C., Banks, S. *et al.* (2014). Global conservation outcomes depend on marine protected areas with five key features. *Nature* 506, 216-220. https://doi.org/10.1038/nature13022.

Elder, M., Bengtsson, M. and Akenji, L. (2016). An optimistic analysis of the means of implementation for sustainable development goals: Thinking about goals as means. *Sustainability* 8(9), 962. https://doi.org/10.3390/su8090962.

Elder, M. and Zusman, E. (2016). *Strengthening the Linkages Between Air Pollution and the Sustainable Development Goals*. IGES Policy Briefs. Kanagawa: Institute for Global Environmental Strategies. https://pub.iges.or.jp/system/files/publication_documents/pub/policy/5528/PB_35_0707_2.pdf.

Foley, J.A., Ramankutty, N., Brauman, K.A., Cassidy, E.S., Gerber, J.S., Johnston, M. *et al.* (2011). Solutions for a cultivated planet. *Nature* 478, 337–342. https://doi.org/10.1038/nature10452.

Food and Agriculture Organization of the United Nations (1996). Rome declaration on world food security. *World Food Summit*. Rome, 13-17 November. http://www.fao.org/docrep/003/w3613e/w3613e00.HTM (Accessed: 11/06/2018).

Food and Agriculture Organization of the United Nations (2018). *AQUASTAT database*. http://www.fao.org/nr/water/aquastat/main/index.stm.

Freshwater Action Network South Asia and Water Supply and Sanitation Collaborative Council (2016). *Leave No One Behind: Voices of Women, Adolescent Girls, Elderly and Disabled People, and Sanitation Workers*. Hyderabad. https://www.wsscc.org/wp-content/uploads/2016/03/Leave-No-One-Behind-Report-by-WSSCC-and-FANSA-2016.pdf.

Fricko, O., Havlik, P., Rogelj, J., Klimont, Z., Gusti, M., Johnson, N. *et al.* (2017). The marker quantification of the shared socioeconomic pathway 2: A middle-of-the-road scenario for the 21st century. *Global Environmental Change* 42, 251-267. https://doi.org/10.1016/j.gloenvcha.2016.06.004.

Fulton, L., Mason, J. and Meroux, D. (2017). *Three Revolutions in Urban Transportation*. Institute of Transportation Studies. https://steps.ucdavis.edu/wp-content/uploads/2017/05/STEPS_ITDP-3R-Report-5-10-2017-2.pdf.

Fuss, S., Canadell, J.G., Peters, G.P., Tavoni, M., Andrew, R.M., Ciais, P. *et al.* (2014). Betting on negative emissions. *Nature Climate Change* 4, 850-853. https://doi.org/10.1038/nclimate2392.

Gakidou, E., Cowling, K., Lozano, R. and Murray, C.J.L. (2010). Increased educational attainment and its effect on child mortality in 175 countries between 1970 and 2009: A systematic analysis. *The Lancet* 376(9745), 959-974. https://doi.org/10.1016/S0140-6736(10)61257-3.

Gambhir, A., Drouet, L., McCollum, D., Napp, T., Bernie, D., Hawkes, A. *et al.* (2017). Assessing the feasibility of global long-term mitigation scenarios. *Energies* 10(1), 89. https://doi.org/10.3390/en10010089.

Garbach, K., Milder, J.C., DeClerck, F.A.J., Montenegro de Wit, M., Driscoll, L. and Gemmill-Herren, B. (2017). Examining multi-functionality for crop yield and ecosystem services in five systems of agroecological intensification. *International Journal of Agricultural Sustainability* 15(1), 11-28. https://doi.org/10.1080/14735903.2016.1174810.

Garibaldi, L.A., Gemmill-Herren, B., D'Annolfo, R., Graeub, B.E., Cunningham, S.A. and Breeze, T.D. (2017). Farming approaches for greater biodiversity, livelihoods, and food security. *Trends in Ecology and Evolution* 32(1), 68-80. https://doi.org/10.1016/j.tree.2016.10.001.

Garnett, S.T., Burgess, N.D., Fa, J.E., Fernández-Llamazares, Á., Molnár, Z., Robinson, C.J. *et al.* (2018). A spatial overview of the global importance of Indigenous lands for conservation. *Nature Sustainability* 1(7), 369-374. https://doi.org/10.1038/s41893-018-0100-6.

Garnett, T., Appleby, M.C., Balmford, A., Bateman, I.J., Benton, T.G., Bloomer, P. *et al.* (2013). Sustainable intensification in agriculture: Premises and policies. *Science* 341(6141), 33-34. https://doi.org/10.1126/science.1234485.

Geldmann, J., Barnes, M., Coad, L., Craigie, I.D., Hockings, M. and Burgess, N.D. (2013). Effectiveness of terrestrial protected areas in reducing habitat loss and population declines. *Biological Conservation* 161, 230-238. https://doi.org/10.1016/j.biocon.2013.02.018.

Gerbens-Leenes, W., Hoekstra, A.Y. and van der Meer, T.H. (2009). The water footprint of bioenergy. *Proceedings of the National Academy of Sciences* 106(25), 10219-10223. https://doi.org/10.1073/pnas.0812619106.

Giannadaki, D., Giannakis, E., Pozzer, A. and Lelieveld, J. (2018). Estimating health and economic benefits of reductions in air pollution from agriculture. *Science of the total environment* 622-623, 1304-1316. https://doi.org/10.1016/j.scitotenv.2017.12.064.

Gill, D.A., Mascia, M.B., Ahmadia, G.N., Glew, L., Lester, S.E., Barnes, M. *et al.* (2017). Capacity shortfalls hinder the performance of marine protected areas globally. *Nature* 543, 665-669. https://doi.org/10.1038/nature21708.

Giuntoli, J., Caserini, S., Marelli, L., Baxter, D. and Agostini, A. (2015). Domestic heating from forest logging residues: Environmental risks and benefits. *Journal of Cleaner Production* 99, 206-216. https://doi.org/10.1016/j.jclepro.2015.03.025.

Glibert, P.M. (2017). Eutrophication, harmful algae and biodiversity — Challenging paradigms in a world of complex nutrient changes. *Marine Pollution Bulletin* 124(2), 591-606. https://doi.org/10.1016/j.marpolbul.2017.04.027.

Global Burden of Disease 2015 SDG Collaborators (2016). Measuring the health-related sustainable development goals in 188 countries: A baseline analysis from the global burden of disease study 2015. *Lancet* 388(10053), 1813-1850. https://doi.org/10.1016/S0140-6736(16)31467-2.

Global Burden of Disease 2016 Risk Factors Collaborators (2017). Global, regional, and national comparative risk assessment of 84 behavioural, environmental and occupational, and metabolic risks or clusters of risks, 1990–2016: A systematic analysis for the Global Burden of Disease Study 2016. *The Lancet* 390(10100), 1345-1422. https://doi.org/10.1016/S0140-6736(17)32366-8.

Global Burden of Disease 2016 SDG Collaborators (2017). Measuring progress and projecting attainment on the basis of past trends of the health-related sustainable development goals in 188 countries: An analysis from the global burden of disease study 2016. *The Lancet* 390(10100), 1423-1459. https://doi.org/10.1016/S0140-6736(17)32336-X.

Global Energy Assessment (2012). *Global Energy Assessment: Towards a Sustainable Future*. Cambridge: International Institute of Applied Systems Analysis. http://www.iiasa.ac.at/web/home/research/Flagship-Projects/Global-Energy-Assessment/GEA-Summary-web.pdf.

Global Yield Gap and Water Productivity Atlas (2018). *Food security analysis: From local to global*. http://www.yieldgap.org/ (Accessed: 12 November 2018).

Godfray, H.C.J., Beddington, J.R., Crute, I.R., Haddad, L., Lawrence, D., Muir, J.F. *et al.* (2010). Food security: The challenge of feeding 9 billion people. *Science* 327(5967), 812-818. https://doi.org/10.1126/science.1185383.

Graus, W., Blomen, E. and Worrell, E. (2011). Global energy efficiency improvement in the long term: A demand-and supply-side perspective. *Energy efficiency* 4(3), 435-463. https://doi.org/10.1007/s12053-010-9097-z.

Gray, C.L., Hill, S.L., Newbold, T., Hudson, L.N., Börger, L., Contu, S. *et al.* (2016). Local biodiversity is higher inside than outside terrestrial protected areas worldwide. *Nature Communications* 7(12306). https://doi.org/10.1038/ncomms12306.

Greaver, T.L., Sullivan, T.J., Herrick, J.D., Barber, M.C., Baron, J.S., Cosby, B.J. *et al.* (2012). Ecological effects of nitrogen and sulfur air pollution in the US: What do we know? *Frontiers in Ecology and the Environment* 10(7), 365-372. https://doi.org/10.1890/110049.

Grubler, A., Wilson, C., Bento, N., Boza-Kiss, B., Krey, V., McCollum, D.L. *et al.* (2018). A low energy demand scenario for meeting the 1.5 °C target and sustainable development goals without negative emission technologies. *Nature Energy* 3(6), 515-527. https://doi.org/10.1038/s41560-018-0172-6.

Haines, A., Amann, M., Borgford-Parnell, N., Leonard, S., Kuylenstierna, J. and Shindell, D. (2018). Short-lived climate pollutant mitigation and the sustainable development goals. *Nature Climate Change* 7(12), 863-869. https://doi.org/10.1038/s41558-017-0012-x.

Hanasaki, N., Fujimori, S., Yamamoto, T., Yoshikawa, S., Masaki, Y., Hijioka, Y. *et al.* (2013b). A global water scarcity assessment under shared socio-economic pathways – Part 2: Water availability and scarcity. *Hydrology and Earth System Sciences* 17(7), 2393-2413. https://doi.org/10.5194/hess-17-2393-2013.

Hanasaki, N., Fujimori, S., Yamamoto, T., Yoshikawa, S., Masaki, Y., Hijioka, Y. *et al.* (2013a). A global water scarcity assessment under shared socio-economic pathways – part 1: Water use. *Hydrology and Earth System Sciences* 17(7), 2375–2391. https://doi.org/10.5194/hess-17-2375-2013.

Hanasaki, N., Yoshikawa, S., Kakinuma, K. and Kanae, S. (2016). A seawater desalination scheme for global hydrological models. *Hydrology and Earth System Sciences* 20(10), 4143-4157. https://doi.org/10.5194/hess-20-4143-2016.

Hasegawa, T., Fujimori, S., Havlik, P., Valin, H., Bodirsky, B.L., Doelman, J.C. *et al.* (2018). Risk of increased food insecurity under stringent global climate change mitigation policy. *Nature Climate Change* 8, 699-703. https://doi.org/10.1038/s41558-018-0230-x.

Hasegawa, T., Fujimori, S., Takahashi, K. and Masui, T. (2015). Scenarios for the risk of hunger in the twenty-first century using shared socioeconomic pathways. *Environmental Research Letters* 10(1), 014010. https://doi.org/10.1088/1748-9326/10/1/014010.

Heck, V., Gerten, D., Lucht, W. and Popp, A. (2018). Biomass-based negative emissions difficult to reconcile with planetary boundaries. *Nature Climate Change* 8, 151-155. https://doi.org/10.1038/s41558-017-0064-y.

Hejazi, M., Edmonds, J., Clarke, L., Kyle, P., Davies, E., Chaturvedi, V. *et al.* (2014). Long-term global water projections using six socioeconomic scenarios in an integrated assessment modeling framework. *Technological Forecasting and Social Change* 81(1), 205-226. https://doi.org/10.1016/j.techfore.2013.05.006.

Hejazi, M.I., Voisin, N., Liu, L., Bramer, L.M., Fortin, D.C., Hathaway, J.E. *et al.* (2015). 21st century United States emissions mitigation could increase water stress more than the climate change it is mitigating. *Proceedings of the National Academy of Sciences* 112(34), 10635-10640. https://doi.org/10.1073/pnas.1421675112.

Höglund-Isaksson, L., Purohit, P., Amann, M., Bertok, I., Rafaj, P., Schöpp, W. *et al.* (2017). Cost estimates of the Kigali Amendment to phase-down hydrofluorocarbons. *Environmental Science and Policy* 75, 138-147. https://doi.org/10.1016/j.envsci.2017.05.006.

Hughes, B.B., Kuhn, R., Peterson, C.M., Rothman, D.S., Solorzano, J.R. and Dickson, J. (2011). *Improving Global Health: Patterns in Potential Human Progress*. Boulder, CO: Paradigm Publishers. http://pardee.du.edu/sites/default/files/PPHP3_Full_Volume.pdf.

Hulme, D. and Shepherd, A. (2003). Conceptualizing chronic poverty. *World development* 31(3), 403-423. https://doi.org/10.1016/S0305-750X(02)00222-X.

Hulvey, K.B., Hobbs, R.J., Standish, R.J., Lindenmayer, D.B., Lach, L. and Perring, M.P. (2013). Benefits of tree mixes in carbon plantings. *Nature Climate Change* 3, 869-874. https://doi.org/10.1038/nclimate1862.

Hunt, C. (2008). Economy and ecology of emerging markets and credits for bio-sequestered carbon on private land in tropical Australia. *Ecological Economics* 66(2-3), 309-318. https://doi.org/10.1016/j.ecolecon.2007.09.012.

Hutton, G. and Varughese, M. (2016). *The Costs of Meeting the 2030 Sustainable Development Goal Targets on Drinking Water, Sanitation, and Hygiene*. Water and Sanitation Program: Technical Paper 103171. Washington, D.C: World Bank. http://documents.worldbank.org/curated/en/415441467988938343/pdf/103171-PUB-Box394556B-PUBLIC-EPI-K8543-ADD-SERIES.pdf.

Intergovernmental Panel on Climate Change (2018). *Global Warming of 1.5 °C. An IPCC Special Report on the Impacts of Global Warming Of 1.5 °C Above Pre-Industrial Levels and Related Global Greenhouse Gas Emission Pathways, in the Context of Strengthening the Global Response to the Threat of Climate Change, Sustainable Development, and Efforts to Eradicate Poverty*. Geneva. http://www.ipcc.ch/report/sr15/.

International Council for Science (2017). *A Guide to SDG Interactions: From Science to Implementation*. Griggs, D.J., Nilsson, M., Stevance, A. and McCollum, D. (eds.). Paris: International Council for Science. http://pure.iiasa.ac.at/id/eprint/14591/1/SDGs-Guide-to-Interactions.pdf.

International Council for Science and International Social Science Council (2015). *Review of Targets for the Sustainable Development Goals: The Science Perspective*. Paris: International Council for Science. https://council.science/cms/2017/05/SDG-Report.pdf.

International Energy Agency (2016). *World Energy Outlook 2016. Special Report: Energy and Air Pollution*. Paris. https://webstore.iea.org/world-energy-outlook-2016.

International Energy Agency (2017). *Energy Access Outlook 2017: From Poverty to Prosperity*. Paris. https://www.iea.org/publications/freepublications/publication/WEO2017SpecialReport_EnergyAccessOutlook.pdf.

International Institute for Applied Systems Analysis (2015). *ECLIPSE V5a global emission fields (July 2015)*. http://www.iiasa.ac.at/web/home/research/researchPrograms/air/ECLIPSEv5a.html (Accessed: 16 November 2018).

Jalava, M., Guillaume, J.H.A., Kummu, M., Porkka, M., Siebert, S. and Varis, O. (2016). Diet change and food loss reduction: What is their combined impact on global water use and scarcity? *Earth's Future* 4(3), 62-78. https://doi.org/10.1002/2015EF000327.

Jamison, D.T., Breman, J.G., Measham, A.R., Alleyne, G., Claeson, M., Evans, D.B. *et al.* (eds.) (2006). *Disease Control Priorities in Developing Countries*. Washington, D.C: World Bank. https://www.who.int/management/referralhospitals.pdf.

Johann, F. and Schaich, H. (2016). Land ownership affects diversity and abundance of tree microhabitats in deciduous temperate forests. *Forest Ecology and Management* 380, 70-81. https://doi.org/10.1016/j.foreco.2016.08.037.

Johnson, N., Krey, V., McCollum, D.L., Rao, S., Riahi, K. and Rogelj, J. (2015). Stranded on a low-carbon planet: Implications of climate policy for the phase-out of coal-based power plants. *Technological Forecasting and Social Change* 90, 89-102. https://doi.org/10.1016/j.techfore.2014.02.028.

Kagawa, F. and Selby, D. (2012). Ready for the storm: Education for disaster risk reduction and climate change adaptation and mitigation. *Journal of Education for Sustainable Development* 6(2), 207-217. https://doi.org/10.1177/0973408212475200.

Kar, K. and Chambers, R. (2008). *Handbook on Community-Led Total Sanitation*. London: Plan UK and Institute of Development Studies at the University of Sussex. http://www.communityledtotalsanitation.org/sites/communityledtotalsanitation.org/files/cltshandbook.pdf.

Karthe, D., Reeh, T., Walther, M., Niemann, S. and Siegmund, A. (2016). School-based environmental education in the context of a research and development project on integrated water resources management: experiences from Mongolia. *Environmental Earth Sciences* 75(18), 1286. https://doi.org/10.1007/s12665-016-6036-0.

Klein, A.M., Vaissiere, B.E., Cane, J.H., Steffan-Dewenter, I., Cunningham, S.A., Kremen, C. *et al.* (2007). Importance of pollinators in changing landscapes for world crops. *Proceedings of the Royal Society of London B: Biological Sciences* 274(1608), 303-313. https://doi.org/10.1098/rspb.2006.3721.

Klimont, Z., Kupiainen, K., Heyes, C., Purohit, P., Cofala, J., Rafaj, P. *et al.* (2017). Global anthropogenic emissions of particulate matter including black carbon. *Atmospheric Chemistry and Physics* 17, 8681-8723. https://doi.org/10.5194/acp-17-8681-2017.

Koelbl, B.S., van den Broek, M., van Ruijven, B., van Vuuren, D.P. and Faaij, A.P.C. (2013). A Sensitivity analysis of the global deployment of CCS to the cost of storage and storage capacity estimates. *Energy Procedia* 37, 7537-7544. https://doi.org/10.1016/j.egypro.2013.06.697.

Kok, M.T.J., Alkemade, R., Bakkenes, M., van Eerdt, M., Janse, J., Mandryk, M. *et al.* (2018). Pathways for agriculture and forestry to contribute to terrestrial biodiversity conservation: A global scenario-study. *Biological Conservation* 221, 137-150. https://doi.org/10.1016/j.biocon.2018.03.003.

Kriegler, E., Luderer, G., Bauer, N., Baumstark, L., Fujimori, S., Popp, A. *et al.* (2018). Pathways limiting warming to 1.5°C: A tale of turning around in no time? *Philosophical Transactions of the Royal Society A: Mathematical, Physical and Engineering Sciences* 376(2119). https://doi.org/10.1098/rsta.2016.0457.

Kummu, M., de Moel, H., Porkka, M., Siebert, S., Varis, O. and Ward, P.J. (2012). Lost food, wasted resources: Global food supply chain losses and their impacts on freshwater, cropland, and fertiliser use. *Science of the total environment* 438, 477-489. https://doi.org/10.1016/j.scitotenv.2012.08.092.

Lal, R. (2015). Restoring soil quality to mitigate soil degradation. *Sustainability* 7(5), 5875-5895. https://doi.org/10.3390/su7055875.

Lam, V.W.Y., Cheung, W.W.L., Reygondeau, G. and Sumaila, U.R. (2016). Projected change in global fisheries revenues under climate change. *Scientific Reports* 6, 32607. https://doi.org/10.1038/srep32607.

Landrigan, P.J., Fuller, R., Acosta, N.J.R., Adeyi, O., Arnold, R., Basu, N. *et al.* (2018). The lancet commission on pollution and health. *The Lancet* 391(10119), 462–512. https://doi.org/10.1016/S0140-6736(17)32345-0.

Lassaletta, L., Billen, G., Garnier, J., Bouwman, L., Velazquez, E., Mueller, N.D. *et al.* (2016). Nitrogen use in the global food system: Past trends and future trajectories of agronomic performance, pollution, trade, and dietary demand. *Environmental Research Letters* 11(9), 095007. https://doi.org/10.1088/1748-9326/11/9/095007.

Le Blanc, D. (2015). Towards integration at last? The sustainable development goals as a network of targets. *Sustainable Development* 23(3), 176-187. https://doi.org/10.1002/sd.1582.

Le Quéré, C., Andrew, R.M., Canadell, J.G., Sitch, S., Ivar Korsbakken, J., Peters, G.P. *et al.* (2016). Global carbon budget 2016 data. *Earth System Science Data* 8, 605-649. https://doi.org/10.5194/essd-8-605-2016.

Leclere, D., Obersteiner, M., Alkemade, R., Almond, R., Barrett, M., Bunting, G. *et al.* (2018). *Towards Pathways Bending The Curve Terrestrial Biodiversity Trends Within The 21st Century*. International Institute for Applied Systems Analysis. http://pure.iiasa.ac.at/id/eprint/15241/1/Leclere_et_al_IIASA_2018_TowardsPathwaysBendingTheCurveOfTerrestrialBiodiversityTrendsWithinThe21stCentury.pdf.

Lee, R. (2003). The demographic transition: Three centuries of fundamental change. *Journal of economic perspectives* 17(4), 167-190. https://doi.org/10.1257/089533003772034943.

Lee, R.D. and Mason, A. (2011). *Population Aging and the Generational Economy: A Global Perspective*. Cheltenham: Edward Elgar Publishing. https://idl-bnc-idrc.dspacedirect.org/bitstream/handle/10625/47092/IDL-47092.pdf.

Leifeld, J. (2016). Current approaches neglect possible agricultural cutback under large-scale organic farming. A comment to Ponisio *et al.* *Proceedings of the Royal Society B: Biological Sciences* 283(1824). https://doi.org/10.1098/rspb.2015.1623.

Liquete, C., Udias, A., Conte, G., Grizzetti, B. and Masi, F. (2016). Integrated valuation of a nature-based solution for water pollution control. Highlighting hidden benefits. *Ecosystem Services* 22(Part B), 392-401. https://doi.org/10.1016/j.ecoser.2016.09.011.

Lucas, P.L., Dagnachew, A.G. and Hof, A.F. (2017). *Towards Universal Electricity Access in Sub-Saharan Africa: A Quantitative Analysis of Technology and Investment Requirements*. The Hague: PBL Netherlands Environmental Assessment Agency. http://www.pbl.nl/sites/default/files/cms/publicaties/pbl-2017-towards-universal-electricity-access-in-sub-saharan-africa-1952.pdf.

Lucas, P.L., Hilderink, H.B.M., Janssen, P., Samir, K.C., van Vuuren, D.P. and Niessen, L.W. (2018). *Future Impacts of Environmental Factors on Achieving the SDG Target on Child Mortality – A Synergistic Assessment*. Environmental Health Perspectives. The Hague: PBL Netherlands Environmental Assessment Agency. http://www.pbl.nl/sites/default/files/cms/publicaties/pbl-2018-future-impacts-of-environmental-factors-on-achieving-the-sdg-target-on-child-mortality.pdf.

Lutz, W. and Samir, K.C. (2011). Global human capital: Integrating education and population. *Science* 333(6042), 587-592. https://doi.org/10.1126/science.1206964.

Mara, D. and Evans, B. (2018). The sanitation and hygiene targets of the sustainable development goals: Scope and challenges. *Journal of Water Sanitation and Hygiene for Development* 8(1), 1-16. https://doi.org/10.2166/washdev.2017.048.

Markandya, A., Sampedro, J., Smith, S.J., Van Dingenen, R., Pizarro-Irizar, C., Arto, I. *et al.* (2018). Health co-benefits from air pollution and mitigation costs of the Paris Agreement: A modelling study. *The Lancet Planetary Health* 2(3), e126-e133. https://doi.org/10.1016/S2542-5196(18)30029-9.

Marmot, M., Allen, J. and Goldblatt, P. (2010). A social movement, based on evidence, to reduce inequalities in health. *Social Science and Medicine* 71(7), 1254-1258. https://doi.org/10.1016/j.socscimed.2010.07.011.

Maslo, B., Lockwood, J.L. and Leu, K. (2015). Land ownership patterns associated with declining forest birds: Targeting the right policy and management for the right birds. *Environmental Conservation* 42(3), 216-226. https://doi.org/10.1017/S0376892915000041.

Mazzi, E.A. and Dowlatabadi, H. (2007). Air quality impacts of climate mitigation: UK policy and passenger vehicle choice. *Environmental science and technology* 41(2), 387-392. https://pubs.acs.org/doi/abs/10.1021/es060517w.

McCollum, D.L., Gomez Echeverri, L., Busch, S., Pachauri, S., Parkinson, S., Rogelj, J. *et al.* (2018). Connecting the sustainable development goals by their energy inter-linkages. *Environmental Research Letters* 13(3), 033006. https://doi.org/10.1088/1748-9326/aaafe3.

Michelsen, G. and Rieckmann, M. (2015). The contribution of education for sustainable development in promoting sustainable water use. In *Sustainable Water Use and Management: Examples of New Approaches and Perspectives*. Leal Filho, W. and Sümer, V. (eds.). Cham: Springer International Publishing. 103-117. https://doi.org/10.1007/978-3-319-12394-3_6.

Millar, R.J., Fuglestvedt, J.S., Friedlingstein, P., Rogelj, J., Grubb, M.J., Matthews, H.D. *et al.* (2017). Emission budgets and pathways consistent with limiting warming to 1.5 C. *Nature Geoscience* 10, 741-747. https://doi.org/10.1038/ngeo3031.

Millennium Ecosystem Assessment (2005). *Millennium Ecosystem Assessment: Scenarios Assessment*. Washington, D.C: Island Press. https://www.millenniumassessment.org/en/Scenarios.html.

Mills, G., Sharps, K., Simpson, D., Pleijel, H., Broberg, M., Uddling, J. *et al.* (2018). Ozone pollution will compromise efforts to increase global wheat production. *Global Change Biology* 24(8), 3560-3574. https://doi.org/10.1111/gcb.14157.

Milner, J., Green, R., Dangour, A.D., Haines, A., Chalabi, Z., Spadaro, J. *et al.* (2015). Health effects of adopting low greenhouse gas emission diets in the UK. *BMJ open* 5(4), e007364. https://doi.org/10.1136/bmjopen-2014-007364.

Mochizuki, Y. and Bryan, A. (2015). Climate change education in the context of education for sustainable development: Rationale and principles. *Journal of Education for Sustainable Development* 9(1), 4-26. https://doi.org/10.1177/0973408215569109.

Modi, V., McDade, S., Lallement, D. and Saghir, J. (2006). *Energy Services for the Millenium Development Goals: Achieving the Millennium Development Goals*. New York, NY. http://lutw.org/wp-content/uploads/Energy-services-for-the-millennium-development-goals.pdf.

Mogollón, J.M., Lassaletta, L., Beusen, A.H.W., van Grinsven, H.J.M., Westhoek, H. and Bouwman, A.F. (2018). Assessing future reactive nitrogen inputs into global croplands based on the shared socioeconomic pathways. *Environmental Research Letters* 13(4). https://doi.org/10.1088/1748-9326/aab212.

Mönkkönen, M., Juutinen, A., Mazziotta, A., Miettinen, K., Podkopaev, D., Reunanen, P. *et al.* (2014). Spatially dynamic forest management to sustain biodiversity and economic returns. *Journal of Environmental Management* 134, 80-89. https://doi.org/10.1016/j.jenvman.2013.12.021.

Mora, C. and Sale, P.F. (2011). Ongoing global biodiversity loss and the need to move beyond protected areas: A review of the technical and practical shortcomings of protected areas on land and sea. *Marine Ecology Progress Series* 434, 251-266. https://doi.org/10.3354/meps09214.

Mora, C., Wei, C.-L., Rollo, A., Amaro, T., Baco, A.R., Billett, D. *et al.* (2013). Biotic and human vulnerability to projected changes in ocean biogeochemistry over the 21st century. *PLoS Biology* 11(10), e1001682. https://doi.org/10.1371/journal.pbio.1001682.t001.

Mouratiadou, I., Biewald, A., Pehl, M., Bonsch, M., Baumstark, L., Klein, D. *et al.* (2016). The impact of climate change mitigation on water demand for energy and food: An integrated analysis based on the shared socioeconomic pathways. *Environmental Science and Policy* 64, 48-58. https://doi.org/10.1016/j.envsci.2016.06.007.

Moyer, J.D. and Bohl, D. (2018). Alternative pathways to human development: Assessing trade-offs and synergies in achieving the Sustainable Development Goals. *Futures*. https://doi.org/10.1016/j.futures.2018.10.007.

Muller, A., Schader, C., El-Hage Scialabba, N., Brüggemann, J., Isensee, A., Erb, K.H. *et al.* (2017). Strategies for feeding the world more sustainably with organic agriculture. *Nature Communications* 8(1290). https://doi.org/10.1038/s41467-017-01410-w.

Napp, T., Bernie, D., Thomas, R., Lowe, J., Hawkes, A. and Gambhir, A. (2017). Exploring the feasibility of low-carbon scenarios using historical energy transitions analysis. *Energies* 10(1), 116. https://doi.org/10.3390/en10010116.

Nerini, F.F., Tomei, J., To, L.S., Bisaga, I., Parikh, P., Black, M. *et al.* (2018). Mapping synergies and trade-offs between energy and the sustainable development goals. *Nature Energy* 3(1), 10-15. https://doi.org/10.1038/s41560-017-0036-5.

Neumann, K., Verburg, P.H., Stehfest, E. and Müller, C. (2010). The yield gap of global grain production: A spatial analysis. *Agricultural Systems* 103(5), 316-326. https://doi.org/10.1016/j.agsy.2010.02.004.

Neverre, N., Dumas, P. and Nassopoulos, H. (2016). Large-scale water scarcity assessment under global changes: Insights from a hydroeconomic framework. *Hydrology and Earth System Sciences*. https://doi.org/10.5194/hess-2015-502.

22

Nilsson, M., Chisholm, E., Griggs, D., Howden-Chapman, P., McCollum, D., Messerli, P. *et al.* (2018). Mapping interactions between the sustainable development goals: Lessons learned and ways forward. *Sustainability Science*. https://doi.org/10.1007/s11625-018-0604-z.

Nilsson, M., Griggs, D. and Visbeck, M. (2016). Policy: Map the interactions between sustainable development goals. *Nature* 534(7607), 320-322. https://doi.org/10.1038/534320a.

Obersteiner, M., Walsh, B., Frank, S., Havlík, P., Cantele, M., Liu, J. *et al.* (2016). Assessing the land resource–food price nexus of the sustainable development goals. *Science Advances* 2(9), e1501499. https://doi.org/10.1126/sciadv.1501499.

O'Driscoll, R., Stettler, M.E.J., Molden, N., Oxley, T. and ApSimon, H.M. (2018). Real world CO_2 and NOx emissions from 149 Euro 5 and 6 diesel, gasoline and hybrid passenger cars. *Science of the total environment* 621, 282-290. https://doi.org/10.1016/j.scitotenv.2017.11.271.

Oerke, E.C. (2006). Crop losses to pests. *The Journal of Agricultural Science* 144(1), 31-43. https://doi.org/10.1017/S0021859605005708.

Pachauri, S., van Ruijven, B.J., Nagai, Y., Riahi, K., van Vuuren, D.P., Brew-Hammond, A. *et al.* (2013). Pathways to achieve universal household access to modern energy by 2030. *Environmental Research Letters* 8(2), 024015. https://doi.org/10.1088/1748-9326/8/2/024015.

PBL Netherlands Environmental Assessment Agency (2012). *Roads from Rio+20: Pathways to Achieve Global Sustainability Goals by 2050*. The Hague. http://www.pbl.nl/sites/default/files/cms/publicaties/pbl-2012-roads-from-rio-pathways-to-achieve-global-sustainability-goals-by-2050.pdf.

Pinto, F.S. and Marques, R.C. (2017). Desalination projects economic feasibility: A standardization of cost determinants. *Renewable and Sustainable Energy Reviews* 78, 904-915. https://doi.org/10.1016/j.rser.2017.05.024.

Pollock, L.J., Thuiller, W. and Jetz, W. (2017). Large conservation gains possible for global biodiversity facets. *Nature* 546, 141-144. https://doi.org/10.1038/nature22368.

Pommells, M., Schuster-Wallace, C., Watt, S. and Mulawa, Z. (2018). Gender violence as a water, sanitation, and hygiene risk: Uncovering violence against women and girls as it pertains to poor WaSH access. *Violence Against Women*. https://doi.org/10.1177/1077801218754410.

Popp, A., Calvin, K., Fujimori, S., Havlik, P., Humpenöder, F., Stehfest, E. *et al.* (2017). Land-use futures in the shared socio-economic pathways. *Global Environmental Change* 42, 331-345. https://doi.org/10.1016/j.gloenvcha.2016.10.002.

Popp, A., Rose, S.K., Calvin, K., Van Vuuren, D.P., Dietrich, J.P., Wise, M. *et al.* (2014). Land-use transition for bioenergy and climate stabilization: Model comparison of drivers, impacts and interactions with other land use based mitigation options. *Climatic Change* 123(3-4), 495-509. https://doi.org/10.1007/s10584-013-0926-x.

Pradhan, P., Costa, L., Rybski, D., Lucht, W. and Kropp, J.P. (2017). A systematic study of Sustainable Development Goal (SDG) interactions. *Earth's Future* 5(11), 1169-1179. https://doi.org/10.1002/2017EF000632.

Qadir, M., Wichelns, D., Raschid-Sally, L., McCornick, P.G., Drechsel, P., Bahri, A. *et al.* (2010). The challenges of wastewater irrigation in developing countries. *Agricultural Water Management* 97(4), 561-568. https://doi.org/10.1016/j.agwat.2008.11.004.

Qureshi, M.E., Dixon, J. and Wood, M. (2015). Public policies for improving food and nutrition security at different scales. *Food Security* 7(2), 393-403. https://doi.org/10.1007/s12571-015-0443-z.

Rao, N.D., Min, J., DeFries, R., Ghosh-Jerath, S., Valin, H. and Fanzo, J. (2018). Healthy, affordable and climate-friendly diets in India. *Global Environmental Change* 49, 154-165. https://doi.org/10.1016/j.gloenvcha.2018.02.013.

Rao, S., Klimont, Z., Smith, S.J., Van Dingenen, R., Dentener, F., Bouwman, L. *et al.* (2017). Future air pollution in the shared socio-economic pathways. *Global Environmental Change* 42, 346-358. https://doi.org/10.1016/j.gloenvcha.2016.05.012.

Raskin, P., Banuri, T., Gallopin, G., Gutman, P., Hammond, A., Kates, R. *et al.* (2002). *Great Transition: The Promise and Lure of the Times Ahead. A Report of the Global Scenario Group*. Stockholm Environment Institute http://rwkates.org/pdfs/b2002.01.pdf.

Reddy, S.M.W., McDonald, R.I., Maas, A.S., Rogers, A., Girvetz, E.H., North, J. *et al.* (2015). Finding solutions to water scarcity: Incorporating ecosystem service values into business planning at The Dow Chemical Company's Freeport, TX facility. *Ecosystem Services* 12, 94-107. https://doi.org/10.1016/j.ecoser.2014.12.001.

Reilly, J., Melillo, J., Cai, Y., Kicklighter, D., Gurgel, A., Paltsev, S. *et al.* (2012). Using land to mitigate climate change: Hitting the target, recognizing the trade-offs. *Environmental science and technology* 46(11), 5672-5679. https://doi.org/10.1021/es2034729.

Riahi, K., Dentener, F., Gielen, D., Grubler, A., Jewell, J., Klimont, Z. *et al.* (2012). Energy pathways for sustainable development. In *Global Energy Assessment: Toward a Sustainable Future*. Cambridge: Cambridge University Press and the International Institute for Applied Systems Analysis. chapter 17. 1203-1306. http://www.iiasa.ac.at/web/home/research/Flagship-Projects/Global-Energy-Assessment/GEA_Chapter17_pathways_lowres.pdf.

Riahi, K., Kriegler, E., Johnson, N., Bertram, C., den Elzen, M., Eom, J. *et al.* (2015). Locked into Copenhagen pledges — Implications of short-term emission targets for the cost and feasibility of long-term climate goals. *Technological Forecasting and Social Change* 90, 8-23. https://doi.org/10.1016/j.techfore.2013.09.016.

Riahi, K., Van Vuuren, D.P., Kriegler, E., Edmonds, J., O'neill, B.C., Fujimori, S. *et al.* (2017). The shared socioeconomic pathways and their energy, land use, and greenhouse gas emissions implications: An overview. *Global Environmental Change* 42, 153-168. https://doi.org/10.1016/j.gloenvcha.2016.05.009.

Rogelj, J., Popp, A., Calvin, K.V., Luderer, G., Emmerling, J., Gernaat, D. *et al.* (2018). Scenarios towards limiting global mean temperature increase below 1.5° C. *Nature Climate Change* 8, 325–332 https://doi.org/10.1038/s41558-018-0091-3.

Rogelj, J., Schaeffer, M., Friedlingstein, P., Gillett, N.P., Van Vuuren, D.P., Riahi, K. *et al.* (2016). Differences between carbon budget estimates unravelled. *Nature Climate Change* 6(3), 245. https://doi.org/10.1038/nclimate2868.

Rojas, R.V., Achouri, M., Maroulis, J. and Caon, L. (2016). Healthy soils: A prerequisite for sustainable food security. *Environmental Earth Sciences* 75(3), 180. https://doi.org/10.1007/s12665-015-5099-7.

Romero, E., Garnier, J., Lassaletta, L., Billen, G., Le Gendre, R., Riou, P. *et al.* (2013). Large-scale patterns of river inputs in southwestern Europe: Seasonal and interannual variations and potential eutrophication effects at the coastal zone. *Biogeochemistry* 113(1-3), 481-505. https://doi.org/10.1007/s10533-012-9778-0.

Röös, E., Bajželj, B., Smith, P., Patel, M., Little, D. and Garnett, T. (2017). Greedy or needy? Land use and climate impacts of food in 2050 under different livestock futures. *Global Environmental Change* 47, 1-12. https://doi.org/10.1016/j.gloenvcha.2017.09.001.

Sahoo, K.C., Hulland, K.R.S., Caruso, B.A., Swain, R., Freeman, M.C., Panigrahi, P. *et al.* (2015). Sanitation-related psychosocial stress: A grounded theory study of women across the life-course in Odisha, India. *Social Science and Medicine* 139, 80-89. https://doi.org/10.1016/j.socscimed.2015.06.031.

Saliba, G., Saleh, R., Zhao, Y., Presto, A.A., Lambe, A.T., Frodin, B. *et al.* (2017). Comparison of gasoline Direct-injection (GDI) and port fuel injection (PFI) vehicle emissions: Emission certification standards, cold-start, secondary organic aerosol formation potential, and potential climate impacts. *Environmental Science and Technology* 51(11), 6542-6552. https://doi.org/10.1021/acs.est.6b06509.

Sandker, M., Ruiz-Perez, M. and Campbell, B.M. (2012). Trade-offs between biodiversity conservation and economic development in five tropical forest landscapes. *Environmental management* 50(4), 633-644. https://doi.org/10.1007/s00267-012-9888-4.

Schipper, L. and Meyers, S. (1992). *Energy Efficiency and Human Activity: Past Trends, Future Prospects*. Cambridge: Cambridge University Press. https://www.cambridge.org/vi/academic/subjects/engineering/energy-technology/energy-efficiency-and-human-activity-past-trends-future-prospects?format=PB.

Schneider, M.K., Lüscher, G., Jeanneret, P., Arndorfer, M., Ammari, Y., Bailey, D. *et al.* (2014). Gains to species diversity in organically farmed fields are not propagated at the farm level. *Nature Communications* 5, 4151. https://doi.org/10.1038/ncomms5151.

Scott, C.A., Vicuña, S., Blanco-Gutiérrez, I., Meza, F. and Varela-Ortega, C. (2014). Irrigation efficiency and water-policy implications for river basin resilience. *Hydrology and Earth System Science* 18, 1339-1348. https://doi.org/10.5194/hess-18-1339-2014.

Shindell, D., Borgford-Parnell, N., Brauer, M., Haines, A., Kuylenstierna, J.C.I., Leonard, S.A. *et al.* (2017). A climate policy pathway for near- and long-term benefits. *Science* 356(6337), 493-494. https://doi.org/10.1126/science.aak9521.

Shindell, D., Kuylenstierna, J.C.I., Vignati, E., van Dingenen, R., Amann, M., Klimont, Z. *et al.* (2012). Simultaneously mitigating near-term climate change and improving human health and food security. *Science* 335(6065), 183-189. https://doi.org/10.1126/science.1210026.

Silbiger, N.J., Nelson, C.E., Remple, K., Sevilla, J.K., Quinlan, Z.A., Putnam, H.M. *et al.* (2018). Nutrient pollution disrupts key ecosystem functions on coral reefs. *Proceedings of the Royal Society B: Biological Sciences* 285(1880). https://doi.org/10.1098/rspb.2017.2718.

Singh, G.G., Cisneros-Montemayor, A.M., Swartz, W., Cheung, W., Guy, J.A., Kenny, T.-A. *et al.* (2017). A rapid assessment of co-benefits and trade-offs among Sustainable Development Goals. *Marine Policy* 93, 223-231. https://doi.org/10.1016/j.marpol.2017.05.030.

Slade, R., Bauen, A. and Gross, R. (2014). Global bioenergy resources. *Nature Climate Change* 4(2), 99-105. https://doi.org/10.1038/nclimate2097.

Smith, L.C. and Haddad, L.J. (2000). *Explaining Child Malnutrition in Developing Countries: A Cross-Country Analysis*. Washington, D.C.: International Food Policy Research Institute. https://ageconsearch.umn.edu/bitstream/94515/2/explaining%20child%20malnutrition%20in%20developing%20countries.pdf.

Smith, P., Davis, S.J., Creutzig, F., Fuss, S., Minx, J., Gabrielle, B. *et al.* (2016). Biophysical and economic limits to negative CO_2 emissions. *Nature Climate Change* 6(1), 42-50. https://doi.org/10.1038/NCLIMATE2870.

Smith, P., M. Bustamante., Ahammad, H., Clark, H., Dong, H., Elsiddig, E.A. *et al.* (2014). Agriculture, forestry and other land use. In *Climate Change 2014: Mitigation of Climate Change. Contribution of Working Group III to the Fifth Assessment Report of the Intergovernmental Panel on Climate Change*. Edenhofer, O., Pichs-Madruga, R., Sokona, Y., Farahani, E., Kadner, S., Seyboth, K. *et al.* (eds.). Cambridge: Cambridge University Press. chapter 11. 811-922. https://www.ipcc.ch/pdf/assessment-report/ar5/wg3/ipcc_wg3_ar5_chapter11.pdf.

Soares, R.R. (2006). The effect of longevity on schooling and fertility: Evidence from the Brazilian demographic and health survey. *Journal of Population Economics* 19(1), 71-97. https://doi.org/10.1007/s00148-005-0018-y.

Sommer, M., Ferron, S., Cavill, S. and House, S. (2015). Violence, gender and WASH: Spurring action on a complex, under-documented and sensitive topic. *Environment and Urbanization* 27(1), 105-116. https://doi.org/10.1177/0956247814564528.

Sorenson, S.B., Morssink, C. and Campos, P.A. (2011). Safe access to safe water in low income countries: Water fetching in current times. *Social Science and Medicine* 72(9), 1522-1526. https://doi.org/10.1016/j.socscimed.2011.03.010.

Springmann, M., Wiebe, K., Mason-D'Croz, D., Sulser, T.B., Rayner, M. and Scarborough, P. (2018). Health and nutritional aspects of sustainable diet strategies and their association with environmental impacts: A global modelling analysis with country-level detail. *The Lancet Planetary Health* 2(10), e451-e461. https://doi.org/10.1016/S2542-5196(18)30206-7.

Stehfest, E., Bouwman, L., van Vuuren, D., den Elzen, M., Eickhout, B. and Kabat, P. (2009). Climate benefits of changing diet. *Climatic Change* 95(1-2), 83-102. https://doi.org/10.1007/s10584-008-9534-6.

Stohl, A., Aamaas, B., Amann, M., Baker, L.H., Bellouin, N., Berntsen, T.K. *et al.* (2015). Evaluating the climate and air quality impacts of short-lived pollutants. *Atmospheric Chemistry and Physics* 15, 10529-10566. https://doi.org/10.5194/acp-15-10529-2015.

Suri, T., Boozer, M.A., Ranis, G. and Stewart, F. (2011). Paths to success: The relationship between human development and economic growth. *World development* 39(4), 506-522. https://doi.org/10.1016/j.worlddev.2010.08.020.

Sutton, M.A. and Bleeker, A. (2013). The shape of nitrogen to come. *Nature* 494(7438), 435-437. https://doi.org/10.1038/nature11954.

Sweeney, J. (2016). *Energy Efficiency: Building A Clean, Secure Economy*. Hoover Institution Press. http://www.hooverpress.org/Energy-Efficiency-P626.aspx.

Tabata, K. (2005). Population aging, the costs of health care for the elderly and growth. *Journal of Macroeconomics* 27(3), 472-493. https://doi.org/10.1016/j.jmacro.2004.02.008.

Talhelm, A.F., Pregitzer, K.S., Kubiske, M.E., Zak, D.R., Campany, C.E., Burton, A.J. *et al.* (2014). Elevated carbon dioxide and ozone alter productivity and ecosystem carbon content in northern temperate forests. *Global Change Biology* 20(8), 2492-2504. https://doi.org/10.1111/gcb.12564.

Tanaka, K., Berntsen, T., Fuglestvedt, J.S. and Rypdal, K. (2012). Climate effects of emission standards: The case for gasoline and diesel cars. *Environmental science and technology* 46(9), 5205-5213. https://doi.org/10.1021/es204190w.

The World in 2050 Initiative (2018). *Transformations to Achieve the Sustainable Development Goals: Report Prepared by the World in 2050 Initiative*. Laxenburg: International Institute for Applied Systems Analysis. http://pure.iiasa.ac.at/id/eprint/15347/1/TWI2050_Report_web-071718.pdf.

Tilman, D., Balzer, C., Hill, J. and Befort, B.L. (2011). Global food demand and the sustainable intensification of agriculture. *Proceedings of the National Academy of Sciences* 108(50), 20260-20264. https://doi.org/10.1073/pnas.1116437108.

Tilman, D. and Clark, M. (2014). Global diets link environmental sustainability and human health. *Nature* 515, 518-522. https://doi.org/10.1038/nature13959.

Tilman, D., Clark, M., Williams, D.R., Kimmel, K., Polasky, S. and Packer, C. (2017). Future threats to biodiversity and pathways to their prevention. *Nature* 546(7656), 73-81. https://doi.org/10.1038/nature22900.

Tuck, S.L., Winqvist, C., Mota, F., Ahnström, J., Turnbull, L.A. and Bengtsson, J. (2014). Land-use intensity and the effects of organic farming on biodiversity: A hierarchical meta-analysis. *Journal of applied ecology* 51(3), 746-755. https://doi.org/10.1111/1365-2664.12219.

United Nations (2016). *Amendment to the Montreal Protocol on Substances that Deplete the Ozone Layer, Kigali, 5 October 2016*. https://treaties.un.org/doc/Publication/CN/2016/CN.872.2016-Eng.pdf.

United Nations Children's Fund (2015). *Committing to Child Survival: A Promise Renewed Progress Report 2015*. New York, NY. https://www.unicef.org/publications/files/APR_2015_9_Sep_15.pdf.

United Nations Conference on Trade and Development (2014). *World Investment Report 2014: Investing in the SDGs - An Action Plan*. Geneva. https://unctad.org/en/PublicationsLibrary/wir2014_en.pdf.

United Nations Convention to Combat Desertification (2017). *Global Land Outlook*. Bonn. https://knowledge.unccd.int/sites/default/files/2018-06/GLO%20English_Full_Report_rev1.pdf.

United Nations Development Programme (2016). *Human Development Report 2016. Human Development for Everyone*. New York, NY. http://hdr.undp.org/sites/default/files/2016_human_development_report.pdf.

United Nations Educational, Scientific and Cultural Organization (2017). *Education for Sustainable Development Goals. Learning Objectives*. Paris. http://unesdoc.unesco.org/images/0024/002474/247444e.pdf.

United Nations Environment Programme (2017). *The Emissions Gap Report 2017. A UN Environment Synthesis Report*. Nairobi. https://wedocs.unep.org/bitstream/handle/20.500.11822/22070/EGR_2017.pdf?isAllowed=y&sequence=1.

United Nations Framework Convention on Climate Change (2015). *Adoption of the Paris Agreement. FCCC/CP/2015/L.9/Rev.1*. Paris. https://unfccc.int/resource/docs/2015/cop21/eng/l09r01.pdf.

van der Esch, S., ten Brink, B., Stehfest, E., Bakkenes, M., Sewell, A., Bouwman, A. *et al.* (2017). *Exploring Future Changes in Land Use and Land Condition and the Impacts on Food, Water, Climate Change and Biodiversity: Scenarios for the UNCCD Global Land Outlook*. The Hague: PBL Netherlands Environmental Assessment Agency. http://www.pbl.nl/sites/default/files/cms/publicaties/pbl-2017-exploring-future-changes-in-land-use-and-land-condition-2076b.pdf.

van der Zwaan, B.C.C., Rösler, H., Kober, T., Aboumahboub, T., Calvin, K.V., Gernaat, D.E.H.J. *et al.* (2013). A cross-model comparison of global long-term technology diffusion under a 2°c climate change control target. *Climate Change Economics* 4(4), 1-24. https://doi.org/10.1142/S2010007813400137.

van Dingenen, R., Dentener, F., Crippa, M., Leitao, J., Marmer, E., Rao, S. *et al.* (2018). TM5-FASST: A global atmospheric source-receptor model for rapid impact analysis of emission changes on air quality and short-lived climate pollutants. *Atmospheric Chemistry and Physics Discussions* 18, 16173-16211. https://doi.org/10.5194/acp-18-16173-2018.

van Puijenbroek, P.J.T.M., Beusen, A.H.W. and Bouwman, A.F. (2019). Global nitrogen and phosphorus in urban waste water based on the shared socio-economic pathways. *Journal of Environmental Management* 231, 446-456. https://doi.org/10.1016/j.jenvman.2018.10.048.

van Puijenbroek, P.J.T.M., Bouwman, A.F., Beusen, A.H.W. and Lucas, P.L. (2015). Global implementation of two shared socioeconomic pathways for future sanitation and wastewater flows. *Water Science and Technology* 71(2), 227-233. https://doi.org/10.2166/wst.2014.498.

van Ruijven, B.J. (2008). *Energy and Development - A Modelling Approach*. PhD Thesis, Utrecht University https://dspace.library.uu.nl/bitstream/handle/1874/31562/ruijven.pdf?sequence=2.

van Sluisveld, M.A.E., Harmsen, J.H.M., Bauer, N., McCollum, D.L., Riahi, K., Tavoni, M. *et al.* (2015). Comparing future patterns of energy system change in 2 °C scenarios with historically observed rates of change. *Global Environmental Change* 35, 436-449. https://doi.org/10.1016/j.gloenvcha.2015.09.019.

van Vuuren, D.P., Kok, M.T.J., Girod, B., Lucas, P.L. and de Vries, B. (2012). Scenarios in global environmental assessments: Key characteristics and lessons for future use. *Global Environmental Change* 22(4), 884-895. https://doi.org/10.1016/j.gloenvcha.2012.06.001.

van Vuuren, D.P., Kok, M., Lucas, P.L., Prins, A.G., Alkemade, R., van den Berg, M. *et al.* (2015). Pathways to achieve a set of ambitious global sustainability objectives by 2050: Explorations using the IMAGE integrated assessment model. *Technological Forecasting and Social Change* 98, 303-323. https://doi.org/10.1016/j.techfore.2015.03.005.

van Vuuren, D.P., Stehfest, E., Gernaat, D., Doelman, J.C., Van den Berg, M., Harmsen, M. *et al.* (2017a). Energy, land-use and greenhouse gas emissions trajectories under a green growth paradigm. *Global Environmental Change* 42(1), 237-250. https://doi.org/10.1016/j.gloenvcha.2016.05.008.

van Vuuren, D.P., Hof, A.F., van Sluisveld, M.A.E. and Riahi, K. (2017b). Open discussion of negative emissions is urgently needed. *Nature Energy* 2(12), 902-904. https://doi.org/10.1038/s41560-017-0055-2.

van Vuuren, D.P., Stehfest, E., Gernaat, D.E.H.J., van den Berg, M., Bijl, D.L., de Boer, H.S. *et al.* (2018). Alternative pathways towards the 1.5 degree target focusing on the need of negative emission technologies. *Nature Climate Change* 8(5), 391–397. https://doi.org/10.1038/s41558-018-0119-8.

van Vuuren, D.P., van Soest, H., Riahi, K., Clarke, L., Krey, V., Kriegler, E. *et al.* (2016). Carbon budgets and energy transition pathways. *Environmental Research Letters* 11(7). https://doi.org/10.1088/1748-9326/11/7/075002.

van Zanten, H.H.E., Bikker, P., Meerburg, B.G. and de Boer, I.J.M. (2018). Attributional versus consequential life cycle assessment and feed optimization: Alternative protein sources in pig diets. *The International Journal of Life Cycle Assessment* 23(1), 1-11. https://doi.org/10.1007/s11367-017-1299-6.

van Zon, A. and Muysken, J. (2003). *Health as a Principal Determinant of Economic Growth*. Maastricht Maastricht Economic Research Institute on Innovation and Technology,Maastricht University. http://collections.unu.edu/eserv/UNU:1144/rm2003-021.pdf.

Verner, D. (2004). *Education and Its Poverty-Reducing Effects: The Case Of Paraiba, Brazil*. World Bank Policy Research Working Paper. Washington, D.C: World Bank. https://openknowledge.worldbank.org/bitstream/handle/10986/14083/wps3321.pdf?sequence=1&isAllowed=y.

von Stechow, C., Minx, J.C., Riahi, K., Jewell, J., McCollum, D.L., Callaghan, M.W. *et al.* (2016). 2 °C and SDGs: United they stand, divided they fall? *Environmental Research Letters* 11(3). https://doi.org/10.1088/1748-9326/11/3/034022.

Vörösmarty, C.J., McIntyre, P.B., Gessner, M.O., Dudgeon, D., Prusevich, A., Green, P. *et al.* (2010). Global threats to human water security and river biodiversity. *Nature* 467(7315), 555-561. https://doi.org/10.1038/nature09440.

Wada, Y., Flörke, M., Hanasaki, N., Eisner, S., Fischer, G., Tramberend, S. *et al.* (2016). Modeling global water use for the 21st century: The Water Futures and Solutions (WFaS) initiative and its approaches. *Geoscientific Model Development* 9(1), 175. https://doi.org/10.5194/gmd-9-175-2016.

Wada, Y., Gleeson, T. and Esnault, L. (2014). Wedge approach to water stress. *Nature Geoscience* 7(9), 615. https://doi.org/10.1038/ngeo2241.

Wanner, N., Cafiero, C., Troubat, N. and Conforti, P. (2014). *Refinements to the FAO Methodology for Estimating the Prevalence of Undernourishment Indicator*. Rome: Food and Agriculture Organization of the United Nations. http://www.fao.org/3/a-i4046e.pdf.

Water and Sanitation Program (2004). *The Case for Marketing Sanitation*. http://www.wsp.org/sites/wsp.org/files/publications/af_marketing.pdf.

Watt, S. and Chamberlain, J. (2011). Water, climate change, and maternal and newborn health. *Current Opinion in Environmental Sustainability* 3(6), 491-496. https://doi.org/10.1016/j.cosust.2011.10.008.

Weitz, N., Carlsen, H., Nilsson, M. and Skånberg, K. (2018). Towards systemic and contextual priority setting for implementing the 2030 Agenda. *Sustainability Science* 13(2), 531-548. https://doi.org/10.1007/s11625-017-0470-0.

West, P.C., Gerber, J.S., Engstrom, P.M., Mueller, N.D., Brauman, K.A., Carlson, K.M. *et al.* (2014). Leverage points for improving global food security and the environment. *Science* 345(6194), 325-328. https://doi.org/10.1126/science.1246067.

Wicke, B. (2011). *Bioenergy Production on Degraded and Marginal Land: Assessing its Potentials, Economic Performance, and Environmental Impacts for Different Settings and Geographical Scales*. Utrecht University https://dspace.library.uu.nl/bitstream/handle/1874/203772/wicke.pdf.

Wolff, S., Schrammeijer, E.A., Schulp, C.J.E. and Verburg, P.H. (2018). Meeting global land restoration and protection targets: What would the world look like in 2050? *Global Environmental Change* 52, 259-272. https://doi.org/10.1016/j.gloenvcha.2018.08.002.

Wood, S.L.R., Jones, S.K., Johnson, J.A., Brauman, K.A., Chaplin-Kramer, R., Fremier, A. *et al.* (2018). Distilling the role of ecosystem services in the sustainable development goals. *Ecosystem Services* 29, 70-82. https://doi.org/10.1016/j.ecoser.2017.10.010.

World Economic Forum (2017). *Shaping the Future of Global Food Systems: A Scenarios Analysis* Geneva: World Economic Forum. http://www3.weforum.org/docs/IP/2016/NVA/WEF_FSA_FutureofGlobalFoodSystems.pdf.

World Health Organization (2006). *WHO Air Quality Guidelines for Particulate Matter, Ozone, Nitrogen Dioxide and Sulfur Dioxide: Global Update 2005. Summary of Risk Assessment*. Geneva. http://apps.who.int/iris/bitstream/handle/10665/69477/WHO_SDE_PHE_OEH_06.02_eng.pdf;jsessionid=F232D7F7BDD673090F5BB2309F58CAA8?sequence=1.

World Health Organization (2014). *Who Guidelines for Indoor Air Quality: Household Fuel Combustion*. Geneva. http://apps.who.int/iris/bitstream/handle/10665/141496/9789241548885_eng.pdf?sequence=1.

World Health Organization and United Nations Children's Fund (2017). *Reacing the Every Newborn National 2020 Milestones Country Progress, Plans and Moving Forward*. Geneva. http://apps.who.int/iris/bitstream/handle/10665/255719/9789241512619-eng.pdf?sequence=1.

World Water Assessment Programme (2003). *Water for People: Water for Life, The United Nations World Water Development Report*. Paris: United Nations Educational, Scientific and Cultural Organization. http://unesdoc.unesco.org/images/0012/001297/129726e.pdf.

Worm, B., Barbier, E.B., Beaumont, N., Duffy, J.E., Folke, C., Halpern, B.S. *et al.* (2006). Impacts of biodiversity loss on ocean ecosystem services. *Science* 314(5800), 787-790. https://doi.org/10.1126/science.1132294.

Xu, Y. and Ramanathan, V. (2017). Well below 2 °C: Mitigation strategies for avoiding dangerous to catastrophic climate changes. *Proceedings of the National Academy of Sciences* 114(39), 10315-10323. https://doi.org/10.1073/pnas.1618481114.

Yamagata, Y., Hanasaki, N., Ito, A., Kinoshita, T., Murakami, D. and Zhou, Q. (2018). Estimating water–food–ecosystem trade-offs for the global negative emission scenario (IPCC-RCP2.6). *Sustainability Science* 13(2), 301-313. https://doi.org/10.1007/s11625-017-0522-5.

Yoshikawa, H., Aber, J.L. and Beardslee, W.R. (2012). The effects of poverty on the mental, emotional, and behavioral health of children and youth: Implications for prevention. *American Psychologist* 67(4), 272-284. https://doi.org/10.1037/a0028015.

Zhang, W., Cao, G., Li, X., Zhang, H., Wang, C., Liu, Q. *et al.* (2016). Closing yield gaps in China by empowering smallholder farmers. *Nature* 537(7622), 671-674. https://doi.org/10.1038/nature19368.

Zhang, X., Davidson, E.A., Mauzerall, D.L., Searchinger, T.D., Dumas, P. and Shen, Y. (2015). Managing nitrogen for sustainable development. *Nature* 528, 51-59. https://doi.org/10.1038/nature15743.

Zhao, Z.Q., Bai, Z.H., Winiwarter, W., Kiesewetter, G., Heyes, C. and Ma, L. (2017). Mitigating ammonia emission from agriculture reduces PM2.5 pollution in the Hai River Basin in China. *Science of the total environment* 609, 1152-1160. https://doi.org/10.1016/j.scitotenv.2017.07.240.

Zhou, X. and Moinuddin, M. (2017). *Sustainable Development Goals Interlinkages and Network Analysis: A Practical Tool for SDG Integration and Policy Coherence*. Institute for Global Environmental Strategies. Hayama. https://pub.iges.or.jp/pub_file/igesresearch-reportsdg/download.

Zhu, R., Hu, J., Bao, X., He, L., Lai, Y., Zu, L. *et al.* (2016). Tailpipe emissions from gasoline direct injection (GDI) and port fuel injection (PFI) vehicles at both low and high ambient temperatures. *Environmental Pollution* 216, 223-234. https://doi.org/10.1016/j.envpol.2016.05.066.

Zimmerman, N., Wang, J.M., Jeong, C.-H., Wallace, J.S. and Evans, G.J. (2016). Assessing the climate trade-offs of gasoline direct injection engines. *Environmental science and technology* 50(15), 8385-8392. https://doi.org/10.1021/acs.est.6b01800.

Chapter 23

Bottom-up Initiatives and Participatory Approaches for Outlooks

© Shutterstock/StevenK

Coordinating Lead Author: Laura Pereira (Centre for Complex Systems in Transition, Stellenbosch University)

Lead Authors: Ghassem Asrar (Pacific Northwest National Laboratory), Laur Hesse Fisher (Massachusetts Institute of Technology), Angel Hsu (Yale University), Jeanne Nel (Vrije Universiteit Amsterdam), Nadia Sitas (Council for Scientific and Industrial Research, South Africa), James Ward (University of South Australia), Joost Vervoort (Utrecht University),

Contributing Authors: Odirilwe Selomane (Stockholm Resilience Centre, Stockholm University), Christopher Trisos (National Socio-Environmental Synthesis Center, United States of America), Thomas Malone (Massachusetts Institute of Technology), Yaolin Zhang (Yale-Nus College)

GEO Fellows: Rohan Bhargava (Utrecht University), Mandy Angèl van den Ende (Utrecht University), Amy Weinfurter (Yale University)

Executive summary

The challenge of sustainable development offers the opportunity for more effective integration of global and local scenario approaches in environmental assessments and outlooks to support decision-making for all 17 Sustainable Development Goals (SDGs) at all levels (i.e. local, national, regional and global) (*established, but incomplete*). A bottom-up perspective on the future, which is based on local scenarios and practices offers potential benefits for exploring alternative futures that are grounded in local realities and start with existing practical action that can be appropriately scaled. Linking top-down and bottom-up approaches to multilevel scenario development provides an opportunity for global processes to inform local actions and for taking account of local actions in global agreements. Co-developing approaches with diverse stakeholders will help to overcome the current limitations in scaling innovations up, out and deep, and in transferring valuable lessons and results from local to both regional and global levels, and vice versa {23.1}.

The bottom-up approach engages a broad range of scientific and action-oriented knowledge, perspectives and opinions about a desirable world in the future and the ways to get there, including pathways to achieve long-term sustainability goals (e.g. the SDGs) (*established, but incomplete*). Since there is no single answer to achieving sustainability, having multiple perspectives is essential for defining different desirable futures. Through a combination of crowdsourcing platforms, participatory workshops in different regions of the world, analyses of existing sustainability solutions and an assessment of regional outlooks, novel methods for linking the generic results of global models with complementary information and insight from the local level can be undertaken. The outcome from the implementation of such an innovative framework provides useful and relevant information and knowledge for policymakers and practitioners to make more informed decisions about how to achieve a sustainable future {23.4, 23.6}.

A groundswell of bottom-up efforts to realize the SDGs and other multilateral environmental agreements is currently under way, as are efforts to support and facilitate collaboration among them (*established, but incomplete*). Reviewing platforms of bottom-up initiatives provides a preliminary understanding of the breadth and depth of ideas, actions and programmes that seek to help achieve sustainable development objectives. The clear majority of the platforms have a global level of coverage, drawing on examples and encouraging connections from all over the world. Most of the platforms facilitate knowledge-sharing and the identification of solutions in two ways. First, this is through the collection of examples, solutions and best practices (e.g. United Nations Framework Convention on Climate Change [UNFCCC] Momentum for Change Lighthouse Initiative, PANORAMA Solutions for a Healthy Planet; WOCAT; see Annex 23-1), and, second, by creating forums for sharing technical or regional tools and know-how for on-the-ground activities (e.g. Biofin Knowledge Platform, ClimateTechWiki). Other platforms use contests or crowdsourcing to generate and synthesize solutions to challenging questions (e.g. VertMTL, MIT Climate CoLab). These platforms highlight the importance of involving a wider variety of people to complement government policies and initiatives {23.9}.

The GEO Regional Assessments highlight important global environmental pressures in the future, but also emphasize regional priorities and solutions that are critical in decision-making processes (*well established*). GEO-Africa focused on so-called leapfrogging development and emphasized low-carbon, resilient infrastructure for meeting food, energy, water and housing needs while maintaining the continent's ecological assets. Asia and the Pacific had different regional priorities, including disaster-risk reduction as an important consideration, and smart cities were outlined as potential solutions, given population and urbanization trends. Latin America and the Caribbean focused on decoupling economic growth from the use of natural resources through sustainable management and ecosystem-based resilience. Europe recognized the need for lifestyle and consumption changes to reach sustainability goals. North America did provide a scenario analysis but emphasized technological innovation and the data revolution as mechanisms for achieving sustainable development. West Asia emphasized peace and security and the importance of integrated resource management to manage limited natural resources such as water. While climate change is a driver considered across all the regions, adaptation and mitigation pathways are suggested within framings {23.10}.

Data and knowledge gaps exist in the bottom-up analysis, emphasizing a need to broaden out the participatory approaches across scales (*unresolved*). The gaps associated with these bottom-up processes can be grouped into four broad categories as follows. (1) Gaps to do with interconnections across regions: connections and inter-dependencies across regions were not highlighted in most Regional Assessments. (2) Gaps to do with cross cutting themes such as gender, equity and inequality are absent in all but the assessments for Africa and Latin America and the Caribbean. These are more likely to be addressed through a bottom-up approach. (3) Gaps to do with specific resources: interventions for freshwater and oceans are the least addressed in bottom-up initiatives. The Climate CoLab proposals and initiatives focused on freshwater interventions mostly on WASH (water, sanitation and hygiene) and no bottom-up initiatives addressed desalination or ocean regulation. (4) Gaps to do with human well-being: these include solutions predominantly focused on poverty alleviation, while child and maternal health care was addressed by only one Climate CoLab proposal. This highlights an important area for government interventions to target these specific areas {23.9}.

Participatory approaches to identify and assess transformative solutions and envision pathways towards greater sustainability can provide decision makers with a useful landscape of initiatives and concrete synergistic solutions (*established, but incomplete*). By engaging with stakeholders through global workshops and Climate CoLab, GEO-6 could collect many diverse solutions and visions that can complement, and potentially be fed into, quantitative information in integrated assessment models. These participatory approaches can help to shift the focus of outlooks work from the 'what' to the 'how'. For example, there was a strong emphasis on food systems as critical intervention points to move towards a healthy planet, healthy people. Chapter 22 identifies yield-improvement targets and general

solutions such as diet change and reduced pesticide use. These are complemented by initiatives from the workshops and the Climate CoLab that elaborate on specific campaigns taking place right now that provide examples of how to promote diet change and innovations for more high-yield sustainable farming, e.g. Apps to promote sharing economies to reduce food waste in cities; urban agriculture; aquaculture; indigenous and local knowledge exchanges {23.12}.

Transformations to sustainability require both social and technical innovations as well as an enabling policy environment in which to scale these ideas and solutions appropriately *(established, but incomplete)*. Sustainability transformations refer to the systemic changes that are needed to move from a business-as-usual trajectory to a more sustainable future. Transformation is often broken down into multiple phases with temporal periods related to a problematic status quo, a preparation phase in which innovations begin to develop, a navigation/acceleration phase in which innovations grow and become part of the new system, and an institutionalization phase in which a more desirable system is made sustainable in the longer term. Each of these phases requires strongly enabling governance conditions for transformations to occur successfully. These enabling conditions can best be broken into:

- Supporting conditions for the appropriate scaling of innovations (establishing and supporting markets for innovations; supporting innovation experimentation and learning; financial resource mobilization; human resource mobilization)
- Disrupting conditions for the weakening of existing, problematic structures (control policies; rules reform; reduction in existing regime support; changes in networks and key actors and their relationships) {23.12}.

The combined analysis of bottom-up and regional solutions for achieving a healthy planet, healthy people highlight the need to consider a full range of actors, to enable distributive justice and to ensure fair perceptions about where action should be expected to take place *(established, but incomplete)*. Many solutions offer the opportunity for developing countries to leapfrog onto more sustainable and equitable development trajectories. The use of information and communications technology (ICT) plays a major role as a tool for driving change. Furthermore, the roles of different societal actors are made explicit in bottom-up pathways. For instance, there is an important role for city-level government actors, in many of the initiatives assessed in this report, as well as for global networks, like sustainable cities or energy cooperatives. Based on the experiences of GEO-6, participatory work in the future can be enhanced by engaging globally with stakeholders from a greater diversity of backgrounds, focusing on policy-relevant data collection, such as actor roles and barriers to change, and further refining processes of transformation and the equity implications of proposed interventions {23.13}.

23.1 Introduction

The rapid pace and scale of societal and environmental changes in the Anthropocene, where human activity dominates most of the Earth's processes (Crutzen 2006; Leach *et al.* 2013; Steffen *et al.* 2015) are changing how assessments are carried out. Global environmental assessments (including GEO-6) are moving the focus from current trends (e.g. what is the current state of biodiversity?) towards the required transformations for a more sustainable future, and the means to get there (e.g. what interventions are needed to keep global warming below 1.5°C?) (Kowarsch *et al.* 2017; Minx *et al.* 2017). Decision-makers, scholars and practitioners are demanding a deeper and more explicit focus on response options and policy analysis (Jabbour and Flachsland 2017). This shift in intention and direction is especially relevant in the context of the Sustainable Development Goals (SDGs), where nations have set the ambition to achieve a broad range of globally accepted and integrated social, economic and environmental targets for 2030. However, medium- to long-term decision-making is complicated by the fact that the future is uncertain, and it is often not obvious how existing policies and practices can be transformed to achieve desired future outcomes (Miller 2013; Miller, Poli and Rossel 2013; Bennett *et al.* 2016).

Global environmental assessments distil, synthesize and interpret existing information in ways that are relevant to decision makers and can help governments to achieve consensus when negotiating complex international accords and agreements (e.g. the Paris Climate Agreement and the United Nations 2030 Agenda for Sustainable Development) (Jabbour and Flachsland 2017). However, while global environmental assessments often rely on global-scale quantitative scenarios to assess potential futures and to navigate uncertainty (van Vuuren *et al.* 2012), they struggle to integrate dynamics that can bridge local, regional and global scales (Bennett *et al.* 2003). Furthermore, integrated assessment models like those employed in Chapter 22 to develop quantitative global scenarios, struggle to simulate decisions that engage multiple jurisdictional levels, as well as diverse actors, and therefore cannot capture the impact of trends emerging from subglobal scales. As a result, while such scenarios present archetypal, globally unified futures, it is not always clear to decision makers how national policies can use these in ways that are geared for local decisions and action (Biggs *et al.* 2015; Pereira *et al.* 2018a).

The successful implementation of transformative pathways requires an understanding of:

i. how transformational changes occur at local, national, regional and global levels;
ii. which actors and what disruptive technologies (i.e. those that replace incumbent technologies creating new markets) drive such changes; and
iii. what the consequences of transformative action might be in terms of cross-scale connections (Cash *et al.* 2006; Feola 2015; Patterson *et al.* 2017).

This is where the combination of top-down scenarios and bottom-up analyses is crucial.

This chapter assesses participatory processes and local practices seeking transformed futures and grounds the interventions proposed in Chapter 22 with existing examples. The following sections provide background information on cross-level interactions in sub-global assessments and existing research on aggregating local practices towards effective implementation of the SDGs. The later sections describe the methodology used for the GEO-6 bottom-up analysis, followed by the assessment findings and insights gained from the analysis.

23.2 Integrating global assessments and bottom-up analyses

The assessment of transformation pathways can be conducted from global to local, or from local to global levels. For example, Chapters 21 and 22 present global scenario and pathway analyses, but such analyses can also be conducted at local and regional levels. Additionally, pathways can be formulated from the bottom-up by using existing, potentially transformative initiatives as a starting point (Pereira *et al.* 2018b). As described in Chapter 22, global scenarios integrate models and data at the global scale to project plausible future pathways and outcomes. These methods are used to explore a wide range of possible futures (explorative scenarios), and the impacts of recommended solutions or policy options (target-seeking scenarios) (van Vuuren *et al.* 2012; Intergovernmental Science-Policy Platform on Biodiversity and Ecosystem Services [IPBES] 2016). Most global approaches and integrated assessment models cannot, however, engage effectively with: (1) the roles and behaviour of specific actors and the multilevel political mechanisms that support transformation, (2) disruptive technologies and, (3) geographic disaggregation.

Participatory, local scenario approaches can, in contrast, use existing narratives and initiatives to imagine and observe actor behaviour, consider disruptive change and develop future pathways that are locally contextualized and practical (Merrie *et al.* 2018). However, these local scenarios face the challenge of scaling up and transferring the accumulated knowledge and results from individual cases, from local to regional and global levels. Further, local approaches lack the specificity of model-based approaches since they are often only partially quantified or aggregated, limiting their applicability at higher levels.

From these alternative starting points, multilevel scenarios can be developed in two directions. Global scenarios can be downscaled in a top-down manner for use at regional and local levels; and local scenarios can be aggregated through bottom-up approaches to complement global scenarios by inserting local contexts to address biases and assumptions. The downscaling of global scenarios has been investigated and published widely (Zurek and Henrichs 2007; Mason-D'Croz *et al.* 2016; Palazzo *et al.* 2017). The creation of global scenarios through the aggregation of bottom-up approaches or through other innovative scaling up of local scenarios has, by contrast, received little research attention. This area offers many potential benefits for integrating more imaginative futures across scales in global environmental assessments to provide more useful information for informing policies and decisions (Bennett *et al.* 2016).

23.3 Sub-global assessments in a multilevel context

Regional or sub-global assessments based on top-down scenarios offer useful insights and experience on navigating multi- and cross-scale dynamics. There are significant challenges associated with the creation and connection of scenarios across different scales and levels, but also significant opportunities for greater policy relevance. The existing literature has mostly assumed that higher-level (global) scenarios can serve as a framework for lower-level (regional, national or local) scenarios in five ways (Zurek and Henrichs 2007; Table 1. p.1292):

i. scenarios between different levels are viewed as being *equivalent* in all aspects if what is considered true at the global level is also true at the local level;
ii. they are *consistent* when all the key assumptions that frame global scenarios can be used to constrain local-level scenarios. This is generally how regional GEO assessments were developed prior to GEO-6 (i.e. the Regional Assessments discussed in this chapter);
iii. less directly connected scenarios are considered *coherent* if they share some, but not all, basic assumptions about the future across all levels – with other assumptions typically being specific to each level. An example is a set of regional scenarios created with West African policy concerns in mind but connected in terms of some key assumptions to the global scenarios developed by the Intergovernmental Panel on Climate Change (IPCC) community (See Palazzo *et al.* 2017);
iv. *comparable* scenario sets investigate the same scope of topics and issues, but are not connected in terms of key assumptions about the future. The regional IPBES assessments followed this process to a certain extent by using scenario archetypes as tools for comparison across the different regions (Sitas and Harmáčková *et al.* submitted for publication); and
v. *independent* scenarios may extend this further, based on different concerns and focus.

The scientific literature demonstrates how higher-level scenarios can be integrated with more local scenario sets – with scenario links ranging from those that are close to equivalency (Kok *et al.* 2015) to those having comparable scenario sets (Mason-D'Croz *et al.* 2016). There is a major gap, however, in the existence of studies that use local- and regional-level scenarios to inform global-level scenarios through a bottom-up approach. This is a major new focus for the outlooks presented by GEO-6.

Both GEO and IPBES share an interest in bottom-up future scenarios (IPBES 2016; Rosa *et al.* 2017; Lundquist *et al.* 2017). IPBES regional assessments offer an important point of comparison that include a broad review of sub-global scenarios and pathways efforts – **(see Box 23.1)**. Another highly relevant example of the use of regional pathways and scenarios is the CGIAR Research Program on Climate Change, Agriculture and Food Security (Vervoort *et al.* 2014). This enables understanding of how pathway development can be directly connected to policy formulation across different sub-global regions.

Box 23.1: IPBES and bottom-up scenario processes

The IPBES methodological assessment on scenarios and models explored the basis for how scenarios can be employed as tools for decision-making (IPBES 2016). Like GEO, IPBES aims to link science with policy on a variety of scales through Regional Assessments, which are used as a scientific knowledge base for policy development. Generally, IPBES focuses on the planet's state related to biodiversity, ecosystems and nature's contributions to people, grounded in interactions between the human and non-human world (Pascual *et al.* 2017). Findings of the regional assessments show that ecosystems, and consequently their services, are increasingly degrading, thus there is high need for policies addressing this challenge to be investigated from the local to the global (IPBES 2016). IPBES is undertaking scenario reviews both at global levels (IPBES Global report in preparation) and regional levels (IPBES 2018a; IPBES 2018b; IPBES 2018c; IPBES 2018d), allowing for a more specific focus on how bottom-up futures can contribute to global narratives and assist with better understanding of how to achieve more desirable futures, coupled human-nature systems and sustainable development (Lundquist *et al.* 2017).

There is increasing consensus in the literature that scenarios could be made more useful, especially in the IPBES process, through the creative development of more stakeholder-engaged bottom-up, diverse, multi-scale scenarios that are consistent within a global scenario context (Kok *et al.* 2016; Rosa *et al.* 2017). This has been reinforced in the findings for a need to build capacity in the role of scenarios in decision-making – a key finding in some of the IPBES regional assessments (See IPBES 2018a; IPBES 2018b). In response to this, the IPBES 3c Expert Group on scenarios and modelling decided on a way forward to start filling in the gaps on scenario exercises (Rosa *et al.* 2017).

The expert group recognized that:

1. scenarios fail to incorporate policy objectives related to nature conservation and social-ecological feedbacks
2. scenarios are typically relevant at only a particular spatial level, and
3. nature and its contributions to people are treated as the consequence of human decisions rather than being at the centre of the analysis (Lundquist *et al.* 2017).

To address these issues, the expert group initiated the development of a set of multiscale scenarios for nature futures based on positive visions for human relationships with nature. The first step in this process was a visioning workshop with multi-sectoral stakeholders and experts (4-8 September 2017 in Auckland, New Zealand; see Lundquist *et al.* 2017). Using an adapted Manoa mash-up approach based on Pereira *et al.* (2018a), the workshop resulted in seven visions of positive nature futures based on a bottom-up scenario approach that will be further developed in the workplan of the expert group. The process of refining the visions into scenarios that can have a quantitative element for modelling, as well as for filling in gaps, will involve iterative cycles of visioning, stakeholder consultation and modelling through a variety of different forums (Lundquist *et al.* 2017).

23.4 Bottom-up futures based on existing local practices

The need to consider the contributions of bottom-up initiatives is being recognized formally in global assessments. This demonstrates both political commitment to bottom-up implementation and the potential offered to achieve environmental goals, such as decarbonization by 2050. In the Fifth Assessment Report of the IPCC, Chapter 12 on human settlements, infrastructure and spatial planning acknowledged the role of local actors in global climate mitigation (Seto *et al.* 2014). The United Nations Environment Programme synthesis report of the Emissions Gap Report 2016 included, for the first time, an assessment of multiple studies that quantified the additional contribution of local actors to mitigation (United Nations Environment Programme [UNEP] 2016). This analysis found that subnational and non-state actors could reduce emission by an equivalent 0.4-10.0 gigatons of CO_2 in 2020. These cuts would help to narrow the 12-14 Gt gap in 2030 between national governments' emissions cuts and what global scenarios specify is needed to avoid a 2°C increase in global temperatures, although, the latest IPCC report emphasises the need to garner global action towards a 1.5°C target (IPCC 2018). In September 2018, Jerry Brown, the governor of California and, Michael Bloomberg, the former mayor of New York City, hosted a Global Climate Action Summit that highlighted the role that could be played by diverse actors such as universities, civil society organizations, businesses and local governments through bottom-up and participatory processes to address climate change (Global Climate Action Summit 2018). The critical role of cities in climate adaptation and mitigation has also been identified in a report by the Urban Climate Change Research Network that identifies pathways to sustainable urban transformations (Rosenzweig *et al.* 2018).

Several approaches for bottom-up futures identify local practices and small-scale sustainability initiatives at varying geographic levels and across sectors. At the global level, the Seeds of Good Anthropocenes and Climate CoLab projects are two examples of such initiatives. The Seeds of Good Anthropocenes project is developing a collection of local, social, technological, economic, ecological and social-ecological initiatives to help envision positive environmental futures (Bennett *et al.* 2016). Climate CoLab is an online platform for anyone to submit and discuss climate change solutions (Malone *et al.* 2017). While the Seeds of Good Anthropocenes project focuses on the identification and investigation of the practices of local initiatives, Climate CoLab focuses primarily on the process of initiative identification, development and evaluation through a crowdsourcing mechanism. An example of a sector-specific global database is the World Overview of Conservation Approaches and Technologies (WOCAT). The WOCAT network was established in 1992 to compile, document, evaluate, share, disseminate and apply knowledge for sustainable land management (WOCAT 2018). It was a trendsetter in recognizing the vital importance of sustainable land management and the pressing need for corresponding knowledge management. In early 2014, it was officially recognized by the United Nations Convention to Combat Desertification (UNCCD) as the primary recommended database for best practices in sustainable land management.

Regionally, three European Union projects, namely TESS, TRANSMANGO and PATHWAYS, have also collected local initiatives on a variety of environmental themes. TESS developed a database of small-scale social innovation initiatives in Europe focused on climate change (TESS 2018), while TRANSMANGO focused on food sustainability (TRANSMANGO 2018), and PATHWAYS developed a database on local and regional transitions for a sustainable, low-carbon Europe (PATHWAYS 2018).

The Seeds of Good Anthropocenes initiative calls for "seed-based" scenarios in which collected bottom-up initiatives are scaled up, out and deep (Bennett *et al.* 2016), with the first activities recently completed (Lundquist *et al.* 2017; Pereira *et al.* 2018b). Climate CoLab and TESS do not explore initiatives through scenarios explicitly, but Climate CoLab has conducted experiments in which the public has been invited to integrate local proposals to create national-level climate action plans for many countries and regions of the world (Malone *et al.* 2017). Meanwhile, TRANSMANGO and PATHWAYS have built bottom-up scenarios. The TRANSMANGO project based these on 18 case studies to explore local future pathways to sustainable food systems. By contrast, PATHWAYS integrated knowledge from its database into its development and analysis of transition pathways, but did not base these pathways on combinations of initiatives. While there are a variety of databases of bottom-up initiatives that could be used for building bottom-up or seeds-based scenarios, no global scenarios relevant to all aspects of environmental change are specifically based on such seeds. Methodologies from the Seeds of Good Anthropocenes and TRANSMANGO (Hebinck *et al.* 2016; Hebinck *et al.* 2018; Pereira *et al.* 2018a) provide a starting point for developing such bottom-up global scenarios. The related literature on bottom-up planning and decision-making (Fraser *et al.* 2006; Reed, Fraser and Dougill 2006; Reed 2008; Kuramochi, Wakiyama and Kuriyama 2016; Nemoto and Biazoti 2017) and crowdsourcing (Wiggins and Crowston 2011; Gellers 2015; Vasileiadou, Huijben and Raven 2016), provide useful guidelines for the methods used in this chapter.

23.5 Methodological rationale and approach

Part of the conceptual basis for this chapter is the notion that global integrated assessments and bottom-up processes drawing on innovative practices have complementary benefits, and that their connection offers unique insights **(Table 23.1)**. As outlined in Chapter 22, global, quantitative simulations of pathways towards the SDGs have the benefit of offering a strong numeric understanding of the global changes needed to reach these goals, and of unexpected positive and negative impacts that attempts to create these changes may have. Such global pathways also have the benefit of offering a context whereby global drivers of change- like those captured by the no-intervention scenarios presented in Chapter 21- can be considered. As a complementary approach to these global assessments, this chapter assesses three complementary modes of analysis:

i. an assessment of existing platforms featuring bottom-up sustainability initiatives;
ii. the assessment of local practices through illustrative examples of crowdsourcing and participatory approaches; and

iii. analysing sub-global interventions for shifting to more sustainable futures, as highlighted by sub-global/regional assessments **(Figure 23.1)**.

The analysis of local-level initiatives offers to support global pathways in tangible examples and mechanisms for change – especially when based on initiatives that are already occurring, even if in pilot or niche form. Sub-global assessments offer regional specificity while still providing broadly applicable meso-level context for national and local pathways.

23.6 Investigating the broad landscape of bottom-up initiatives

The broader landscape of bottom-up initiatives not captured in the participatory processes is diverse, but methods for capturing this diversity are limited due to data availability. A range of platforms that collate a variety of environment- and sustainability-related bottom-up initiatives has been identified through an online search and coded. While not exhaustive, Annex 23-1 provides a sample of around 20 bottom-up initiative

Table 23.1: Different types of assessment model

Global integrated assessment model	❖ Global context ❖ Integration of many dimensions of change ❖ Simulation of effects of global interventions ❖ Quantification of magnitude of challenges
Subglobal	❖ Regional contextualization of interventions in terms of physical, economic, political and cultural conditions, challenges, opportunities
Synthesis of local practices	❖ Populating macro-level interventions with the 'who and how' – the many actors and innovations that provide feasibility to global and regional pathways

Figure 23.1: Outline of how this chapter's bottom-up approaches complement the top-down findings of Chapters 21 and 22 and how together they can offer policy insights for Chapter 24

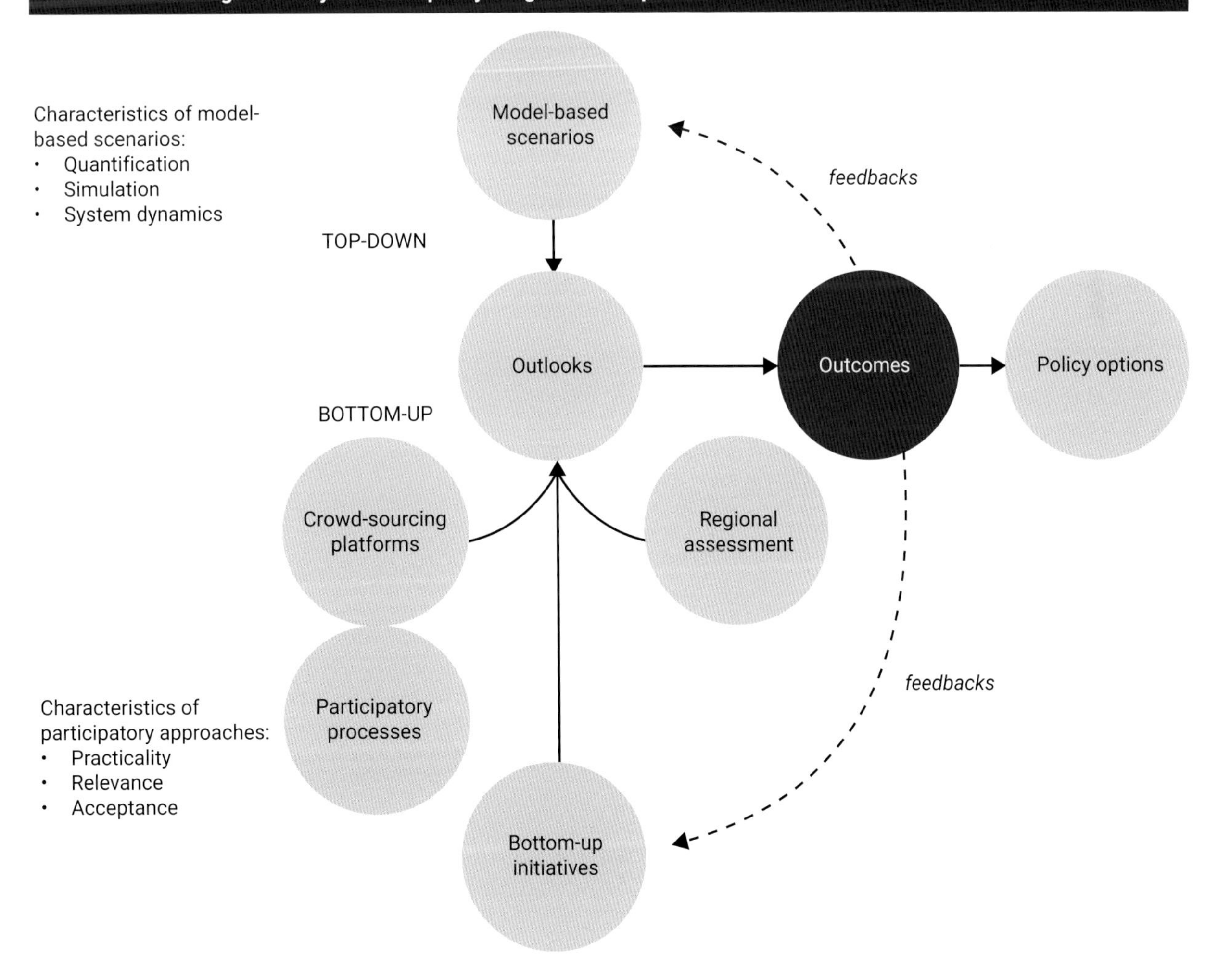

platforms. These were identified through Internet searches using keywords that included "sustainability platform" and "bottom-up environmental initiatives" as well as through prior knowledge of initiatives. These platforms are led by a range of both non-government and government actors and provide a preliminary understanding of the breadth and range of ideas, actions and programmes that seek to implement and to help achieve sustainable development objectives.

23.7 GEO-6 participatory initiatives

Two participatory processes were organized to develop bottom-up pathways focusing on concrete system transformations. These pathways were developed to complement global integrated assessment models and are based on a diversity of potentially transformative on-the-ground practices and knowledge. These pathways also help to connect GEO-6 to stakeholders globally. The first participatory process was a series of workshops held in Bangkok, Guangzhou, Nairobi and Singapore, where local stakeholders were invited to envision specific transformation pathways based on local practices, within the frameworks of the three pathways of Roads from Rio+20 (PBL Netherlands Environmental Assessment Agency 2012): global technology, decentralized solutions, and lifestyle change (UNEP 2017a; UNEP 2017b; UNEP 2017c; UNEP 2018). The second participatory process was an online contest held in conjunction with the Climate CoLab platform (see **Figure 21.9: Global mean temperature increase**; Climate CoLab 2018). The contest asked participants to combine existing proposals within the Climate CoLab platform to build creative combinations of actions that can achieve climate change goals alongside other SDGs.

As a new and innovative aspect of GEO-6, this chapter and the participatory initiatives offer an illustrative assessment of how participatory actions can add stakeholder perspectives and on-the-ground knowledge to integrated assessment models. This analysis therefore has two goals: (1) it helps to link bottom-up and top-down perspectives on transformative systemic change for future GEO reports, and, (2) it provides insights on potentially impactful existing practices that could help to achieve transformative change towards sustainability.

From the four workshops and the Climate CoLab contest, three different types of data were gathered: innovative practices and concepts (called seeds), a combination of seed ideas into larger proposals that focused on specific system changes, and Climate CoLab proposals (these are proposals that combined existing ideas within the platform in new and innovative ways). Seeds are examples of existing, but not yet dominant social initiatives, new technologies, economic tools or social-ecological projects, or organizations, movements or new ways of acting that appear to be making a substantial contribution towards creating a future that is just, prosperous and sustainable (Pereira *et al.* 2018a). The workshops collected seeds and asked participants to build proposals for how to achieve as many SDGs as possible by combining those seeds with one another and exploring how they could interact (UNEP 2017a; UNEP 2017b; UNEP 2017c; UNEP 2018). Both the seeds and the combined proposals were framed around one of the three Roads from Rio+20 pathways mentioned above (See PBL Netherlands Environmental Assessment Agency 2012 and Chapter 22). The four workshops led to 156 seeds and 24 proposals for specific system transformations; and the Climate CoLab competition led to 70 proposals, from which judges selected 34 semi-finalists, 12 finalists and two winners (one selected by public vote and one by the judges see **Figure 21.9: Global mean temperature increase** and **Box 23.4**).

To assess the outcomes of the participatory process, the seeds and the Climate CoLab semi-finalist proposals were coded along the five dimensions summarized in **Table 23.2**. These dimensions were selected to best capture the diversity of results and to integrate results with Chapter 22. As an iterative and participatory process, seeds and proposals were coded based on the availability and quality of the data submitted, so

Box 23.2: Climate CoLab

Climate CoLab is an online contest platform and community run by the Massachusetts Institute of Technology (MIT) Center for Collective Intelligence, with the goal of harnessing the collective intelligence of thousands of people from all around the world to address global climate change. People work on the platform with each other and with over 800 experts on climate change and related topics, to create, analyse and select detailed proposals for what to do about different aspects of the climate change problem. The Climate CoLab site has over 100,000 registered members and has received over 2,500 proposals.

The contest, given the title Exploring Synergistic Solutions for Sustainable Development, began accepting submissions on November 1, 2017, and invited anyone to submit proposals answering the question: "What combinations of Climate CoLab proposals could help achieve multiple SDGs?"

The judges' contest was promoted through a wide range of networks, including through UN Environment, MIT and other organizational partners worldwide, as well as being promoted to the Climate CoLab community itself. The judges selected 12 finalists plus a judges' choice winner out of these. The global public was also invited to vote for the proposal most deserving of a popular choice award out of the 12. These winners were announced on 15 March 2018 **(See Box 23.4)**.

Contest statistics:

- 73 proposals submitted
- 112 proposal authors (individuals or as part of a team)
- 188 proposal comments submitted by experts, authors and other members
- 3,064 valid votes cast

See the contest web pages at: http://www.climatecolab.org/contests/2017/exploring-synergistic-solutions-for-sustainable-development

Source: Climate CoLab (2018).

Table 23.2: Coding dimensions

Dimension	Categories	Description
Initiative benefits	17 SDGs	Coding by SDGs captures the range of benefits for each of the seeds and proposals. Results identify the diversity of outcomes and potential SDG synergies.
Global measures category	41 global "measures" or "interventions" (as referred to in this chapter when describing specific initiatives) broken down into five system-focused clusters	Results were categorized under the 32 measures identified in Chapter 22 along with nine additional interventions, identified during the coding process, that did not fit neatly under any of the existing 32 measures. The clusters for freshwater and for oceans were grouped together due to low representation in the results.
Theory of change	❖ New technology ❖ Decentralization ❖ Design/infrastructure ❖ Monitoring and reporting ❖ Change in production practices ❖ Lifestyle change ❖ New organization/business ❖ Knowledge/data platform ❖ Policy change ❖ Finance/incentives/subsidies ❖ Awareness, knowledge, skills development	The theory of change identifies the type of change or solution of the initiative. These categories are based on an iterative coding process of results to best capture the diversity while minimizing overlapping categories.
Actor	❖ International organizations ❖ Governments (local, regional, national) ❖ Private sector/businesses ❖ Civil society ❖ Academic and research institutions ❖ Households/individuals	The type of actor focuses on their involvement in each of the initiatives.
Geography (only for Climate CoLab semi-finalists)	❖ By country	The country or countries where the initiative would be deployed and where the authors originated.

not all the results could be coded on all dimensions. It should also be noted that some dimensions were self-identified by the contributors, while others were specifically coded by the GEO author team.

23.8 GEO-6 Regional Assessments

Six GEO Regional Assessments were completed in 2016: for Africa, Asia and the Pacific, Europe, Latin America and the Caribbean, North America, and West Asia. These can be viewed as intermediate assessments between global and local assessments. Each Regional Assessment highlights region-specific environmental challenges and the key interventions for addressing them. When viewed together, these assessments provide a global set of overarching challenges and responses for securing a more sustainable future that still retain certain regional priorities. In this chapter, we compare the key interventions identified in the six regions with those identified from the review of the scenario literature (Chapter 22) and the bottom-up processes **(Figure 23.15)**. By doing so, we aim to identify potential gaps in the interventions considered at all three levels of assessment (global, regional and local), and to draw insights that enhance the range of interventions and policy options available to decision makers.

In a similar way to participatory initiatives, key interventions identified in the Outlooks chapters of the GEO Regional Assessments were coded according to the interventions identified by the review of the scenarios literature (Chapter 22, **Table 22.1**). Interventions that were not on the predefined list from the scenario literature review were added to derive an updated list of interventions.

23.9 Findings from a bottom-up approach

23.9.1 Broader bottom-up platforms and the diverse actors needed for change

National governments throughout the world have begun to recognize that sound climate scenario modelling and assessment require contributions from bottom-up sources (Hsu *et al.* in press). Global emissions trajectories modelled from top-down integrated assessment models (van Vuuren *et al.* 2011) do not explicitly incorporate information from bottom-up initiatives and individual contributions from local governments, businesses and civil society organizations. Top-down emission pathways assume that these mitigation efforts are subsumed into national government pledges, but bottom-up actors make climate commitments that could be considered additional to or outside of national climate efforts, complicating the assessment of climate mitigation scenarios (Hsu *et al* 2015; Jordan *et al.* 2015). Compounding this complexity, individual actors frequently form hybrid coalitions, often in cooperation with national governments, building transnational climate governance networks. These partnerships demonstrate the potential additive effects of individual bottom-up climate

actions when actors align targeted goals and coordinate efforts (Andonova, Betsill and Bulkeley 2009).

In December 2014, the United Nations Framework Convention on Climate Change (UNFCCC) launched the Global Climate Action portal (also known as NAZCA after its former name, the Non-state Actor Zone for Climate Action) in an effort to capture and track the diversity of bottom-up actors and commitments pledging climate mitigation, adaptation, financing, capacity-building and other actions to address climate change (UNFCCC 2018; see **Box 23.3**). The Global Climate Action portal was initially developed to illustrate an 'all hands on deck' approach (Hale 2016) to climate governance, and the scientific and analytical community is now moving towards a consistent methodology to account for quantifying bottom-up actor contributions in global climate mitigation scenarios (Initiative for Climate Action Transparency [ICAT] 2018; Hsu *et al.* in press). This effort is intended to serve multiple objectives:

i. quantifying the global aggregation of bottom-up climate efforts and its additional impact in existing climate scenarios will allow for more accurate appraisal of existing emission pathways and gaps.
ii. understanding the mitigation contributions of bottom-up efforts will provide national governments with additional information by which to leverage more ambitious Nationally Determined Contributions to the Paris Climate Agreement in review cycles (UNFCCC 2015). This knowledge of decentralized impacts could also prompt and enable governments to better support and scale up these activities.
iii. incorporating bottom-up initiatives into global climate scenarios will provide recognition of small-scale initiatives or qualitative contributions (e.g. capacity-building) that are critical to advancing lower-carbon trajectories but are difficult to quantify (Chan, Brandi and Bauer 2016).

Results from the analysis of bottom-up platforms

Over 50,000 individual bottom-up actions were identified, but their different structures and goals made comparisons challenging. Evaluating platforms, rather than individual commitments, helped to facilitate comparisons between different kinds of bottom-up action, and also to shed light on the structures in place to enable and support the continued growth and development of these initiatives. The platforms identified through the online search range from the Amazon Vision Coordination and Information Platform, which is based in Colombia and includes more than 200 initiatives that support the implementation of mitigation activities against greenhouse gas emissions (GHG), to Sustainia 100, which has tracked more than 4,500 sustainable solutions being deployed by 188 companies. The aims of these platforms vary, from providing crowdsourcing solutions to listing microfinancing options, to giving information that connects stakeholders **(Figure 23.2)**. Platforms often seek to support or feature initiatives from a wide range of actor types while others have a narrower focus on a particular type of actor, such as business. Drawing examples from all over the world, all but five platforms have a global level of coverage. The five non-global initiatives focus on city (MTLGreen), regional (Amazon Vision Coordination and Information Platform, MACBIO – Pacific) and national (e.g. WorthWild, GreenCrowd) issues.

The majority of the platforms considered, facilitated knowledge-sharing and the identification of solutions in two ways. One was through the collection of examples, solutions and best practices (e.g. UNFCCC Momentum for Change Lighthouse Initiative, PANORAMA Solutions for a Healthy Planet), and the second was by creating forums for sharing technical or regional tools and know-how, to support a wide range of on-the-ground activities (e.g. Biofin Knowledge Platform, ClimateTechWiki). Still other platforms used contests or crowdsourcing to generate and synthesize solutions to challenging questions (e.g. VertMTL, MIT Climate CoLab). Fewer platforms focused on tracking the progress or impacts of activities (e.g. REDDX) or on enabling project implementation by matching projects with funds or other forms of technical or capacity support (e.g. WorthWild, Greencrowd, Divvy, LifeWeb Initiative).

The coding analysis revealed a wide variety of actors working at all scales to implement the SDGs **(Figure 23.3)**. The platforms we identified are convened, curated or led mainly by a range of non-government and government actors, and primarily facilitate knowledge-sharing and the identification of solutions between bottom-up initiatives. These spaces may provide an important route for scaling solutions out, and could lay the foundations to scale solutions up, by collecting and distilling best practices and innovative solutions. Creating forums for collaboration and exchange may also help to facilitate loose coordination and a mutually beneficial division of labour between different actors. Abbott's (2012) research on transnational initiatives, for example, finds that many coalitions perform activities that national governments may be less suited to implement, such as information-sharing and capacity-building.

Box 23.3: The Global Climate Action portal

The Global Climate Action portal, also known as NAZCA, is an online platform currently featuring more than 12,000 commitments to climate change action made by local governments, businesses, civil society organizations, higher education institutions and investors. These range from individual companies adopting internal carbon prices to constrain emissions growth, to city governments pledging carbon neutrality. The Global Climate Action portal also includes initiatives like the World Food Programme's R4 Rural Resilience Initiative (World Food Programme 2018), which aims to increase resilience to climate change through an integrated risk management system for 100,000 farmers. By far the most numerous bottom-up actor types in the portal are subnational and local governments, with close to three-quarters of cities in the platform located in Europe (Hsu *et al.* 2016). This geographic overrepresentation of bottom-up actors in the global North, due to a lack of reported data, is one of the major limitations of efforts to understand the scope of climate action. The vast majority of the climate commitments are focused on emission reduction targets, with 85 per cent of subnational efforts and close to 40 per cent of corporate actions addressing climate mitigation. Most of the actions in the Global Climate Action portal recognize the role of local efforts to promote clean energy production and alter the consumption systems that are responsible for global climate change.

Figure 23.2: The number of initiatives covered in a sample of platforms that feature bottom-up sustainability initiatives (see Annex 23-1 for a brief description of the platforms)

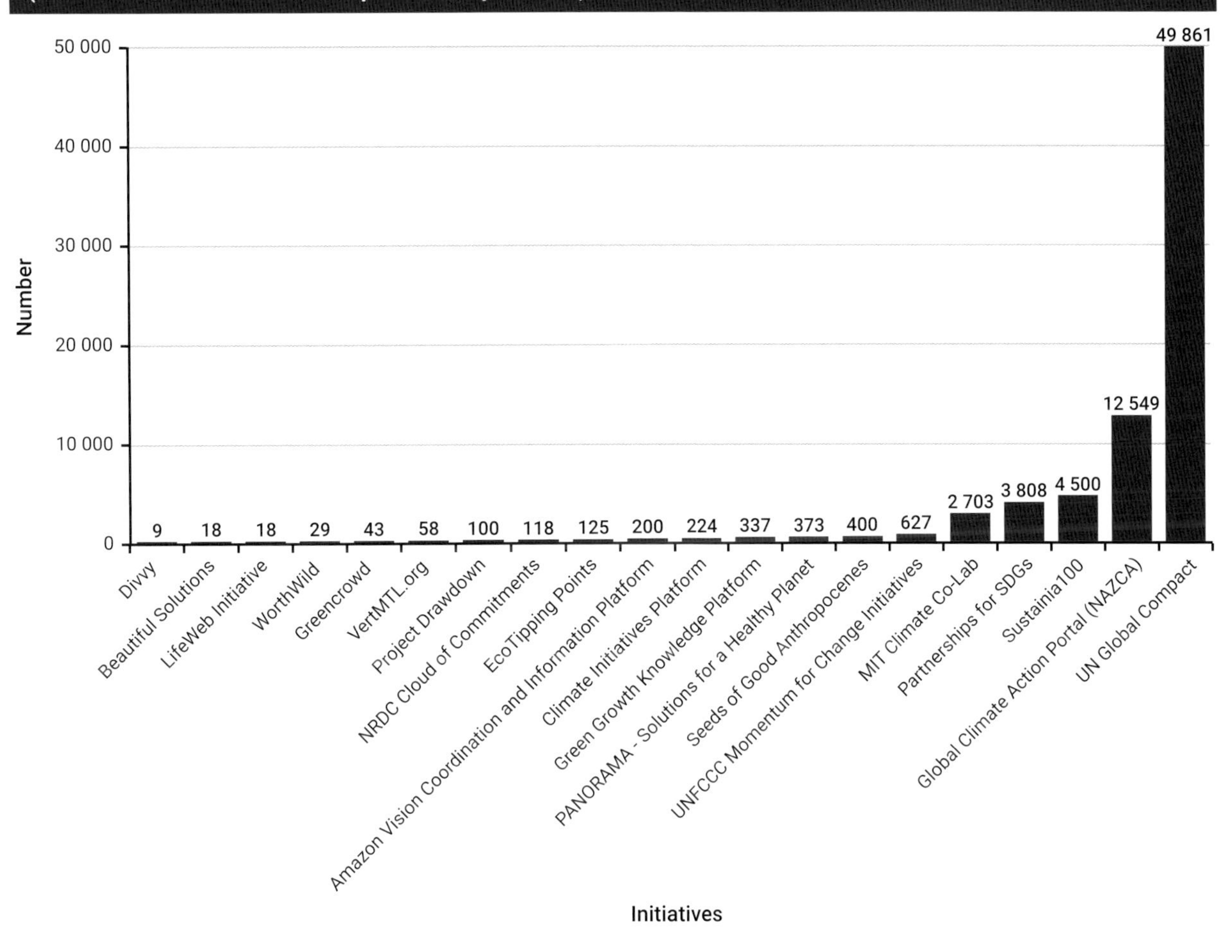

23

Figure 23.3: The SDGs represented proportionally by how they are covered by the selected bottom-up sustainability initiative platforms. Some initiatives are narrower in scope and strictly relate to one, two or three SDGs, while others are diverse and capture a wider range of SDGs (four or more) (see Annex 23-1 for a brief description of the initiative platforms)

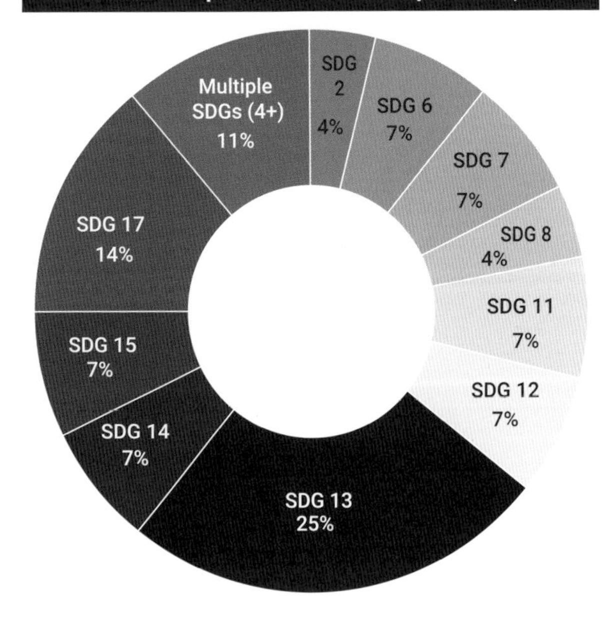

23.9.2 Participatory processes

While Section 23.10 outlines the array of platforms that are already collecting initiatives with the potential to help meet the 2030 Agenda for Sustainable Development, this section presents results from the participatory workshops and Climate CoLab crowdsourcing that further bring to light the diversity of solutions found globally. These initiatives were identified as concrete examples of typical solutions in the measure categories outlined in Chapter 22. They also challenge some of the assumptions of how change happens within top-down models, and highlight the interrelated trends of SDGs, their potential synergies, and the role of diverse actors in achieving the 2030 Agenda – while the top-down models help focus on trade-offs. This section first provides an overview of trends found across all workshop seeds and Climate CoLab proposals before breaking down the results by the four clusters studied in Chapter 22. In addition, a fifth cluster of measure/intervention categories was created based on the solutions found in the bottom-up work that did not neatly fit within those four existing clusters – these are discussed more fully in Section 23.10. The assessment demonstrates the potential of bottom-up initiatives to aid policymakers and top-down analyses, but, due to the limited sample size it does not present a comprehensive overview of all on-the-ground solutions globally.

General landscape of initiatives

SDGs

The workshop seeds and Climate CoLab proposals targeted all 17 SDGs to varying extents. **Figure 23.4** highlights the range of SDGs that were found in the analysis. In the case of the workshop seeds, SDG 12 (responsible consumption and production) and SDG 11 (sustainable cities and communities) were most represented. As for Climate CoLab proposals, SDG 13 (climate action) was targeted by over 80 per cent of proposals, followed by SDG 3 (good health and well-being).

Actors

Workshop seeds focused most on government actors, private sector/business and households/individuals **(Figure 23.5)**. Over 60 per cent of workshop seeds indicated a role for the government, with local government mentioned most, followed by national governments. Similarly, Climate CoLab proposals also emphasized the role of government, with national governments being referenced most. The importance of assessing diverse actors is elaborated further in Section 23.11.

Geography

In submitting proposals on the Climate CoLab platform, contributors were requested to identify up to five countries where their proposals would be active **(Figure 23.6a)**.

Within the 34 Climate CoLab semi-finalists the individual countries that were most covered were the Republic of Kenya (11 proposal mentions), the Republic of India (8) and the United Republic of Tanzania (7). As an open and global crowdsourcing project for solutions, the emphasis on the global South points towards a geographical inequity about where change is perceived to be needed, and highlights a need for transformations to be more equitable across regions (see Section 23.14 for a discussion on distributive justice and equitable transformations). Some of the solutions emanated from the global North for application in the global South making the case for equity particularly relevant. Although not deliberate, this trend can be seen to reinforce the narrative that the North can continue on a business-as-usual trajectory while the South develops more sustainably, and also misses out some of the nuance of how contextual the interpretation of sustainability is and how to achieve it (see Vercoe and Brinkman 2009). However, the high number of suggestions made by contributors from the Global South **(Figure 23.6b)** also points to the innovative thinking that is happening in these parts of the world, where the urgency for action towards meeting Agenda 2030 is greater (Nagendra 2018). By enabling contributions from across the globe, the participatory processes of GEO 6 could capture a range of context-specific solutions for achieving sustainable development.

Figure 23.4: SDGs targeted by the total workshop seeds and the total Climate CoLab proposals

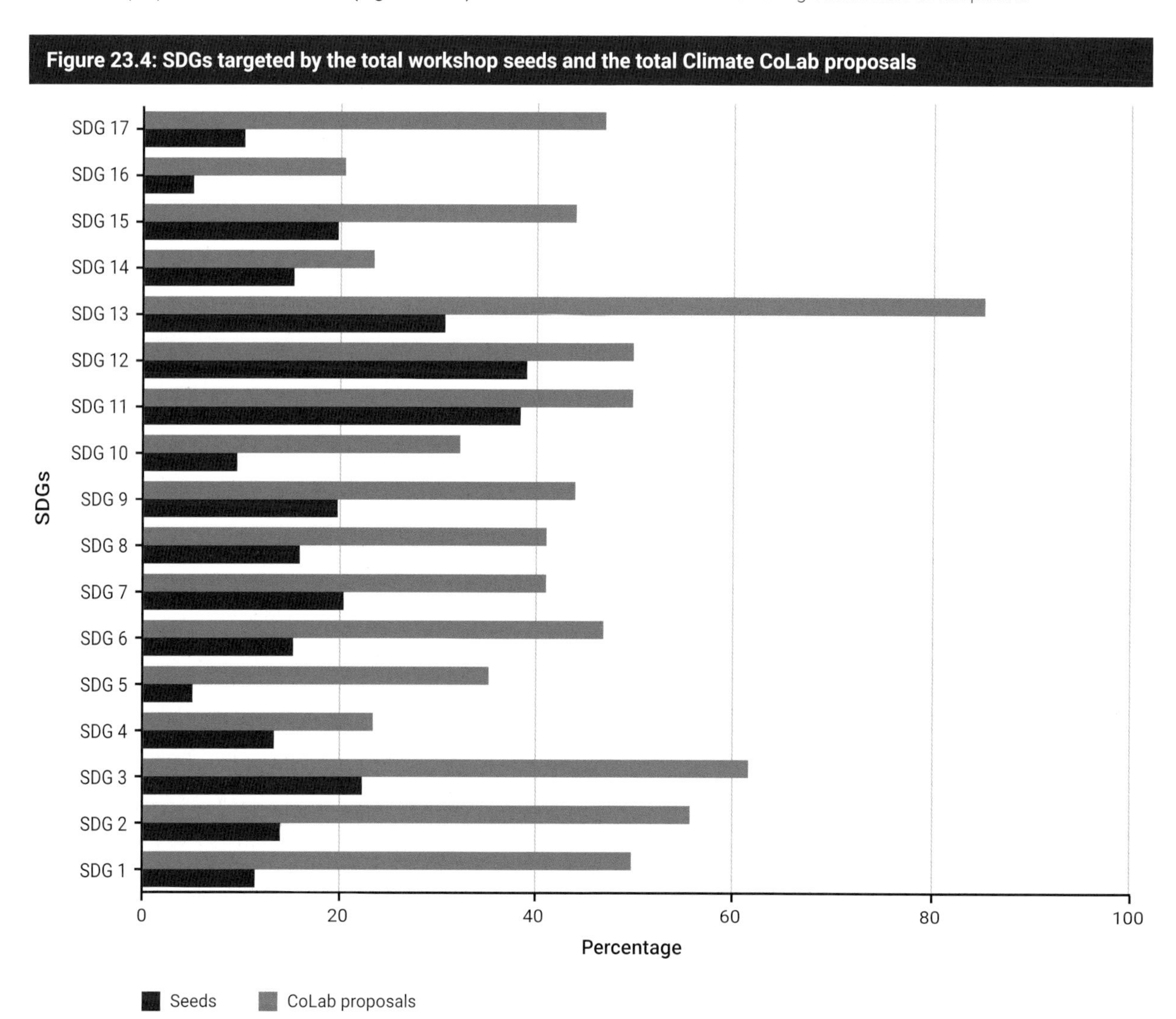

Figure 23.5: Actor types represented by total seeds and total Climate CoLab proposals

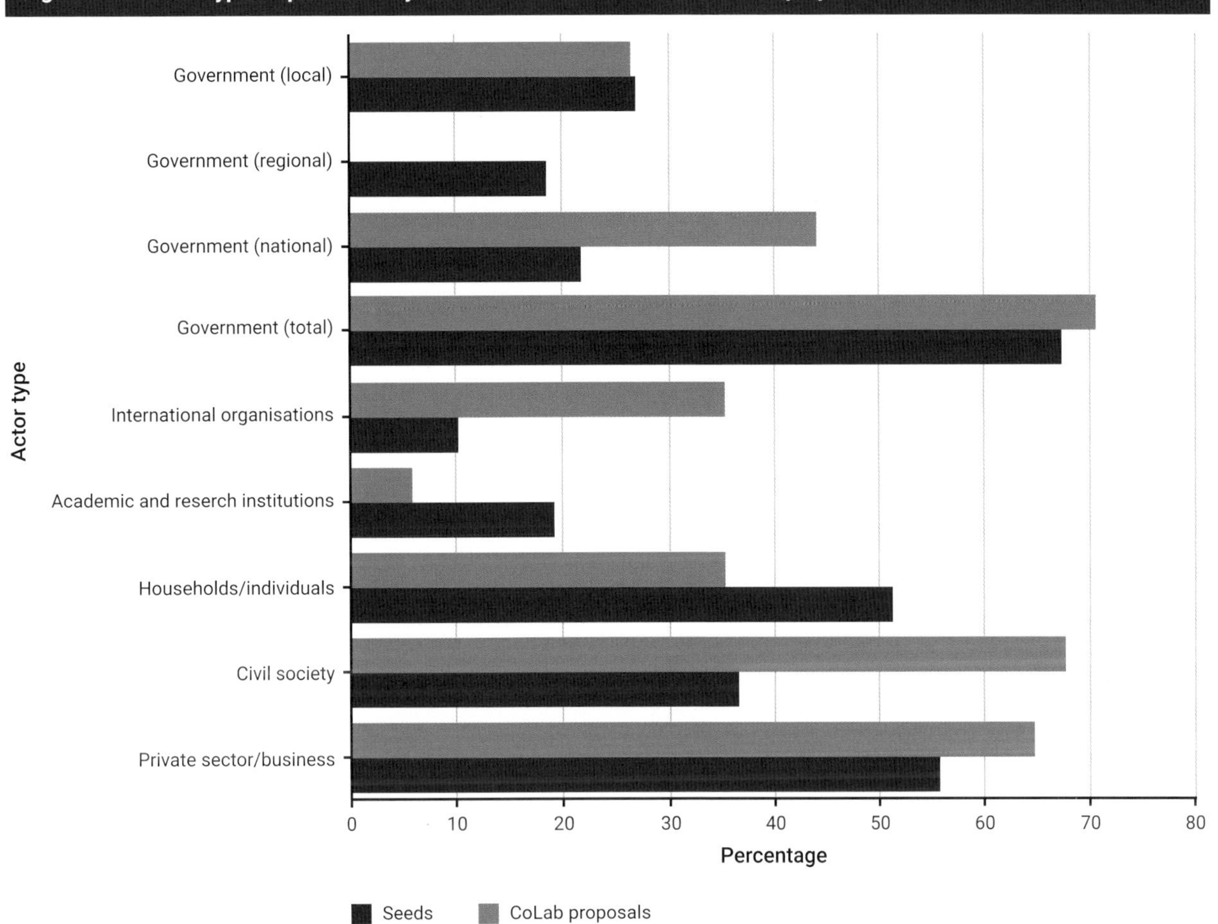

23

Figure 23.6a: Regions covered by Climate CoLab proposals

Figure 23.6b: Regional breakdown of Climate CoLab proposals

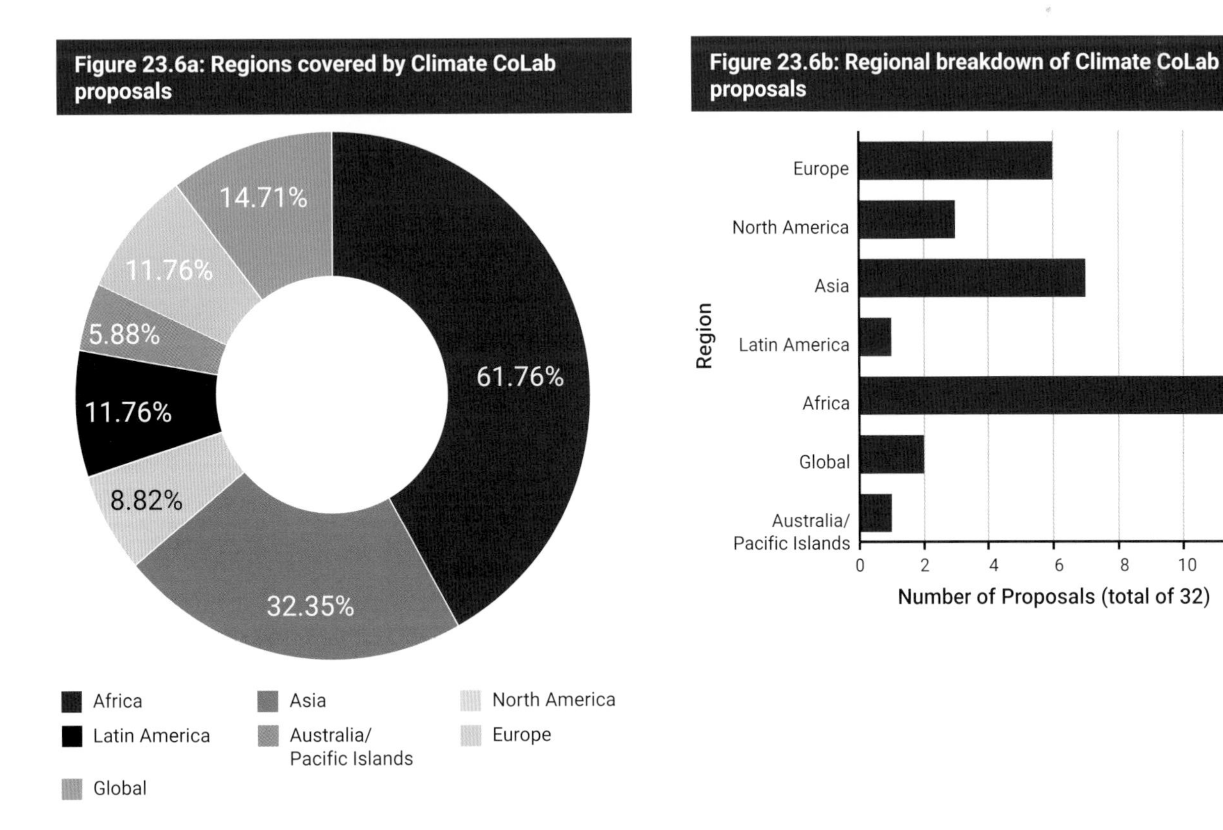

Theory of change
Seeds and Climate CoLab proposals overwhelmingly focused on new technologies to reach their desired goals **(Figure 23.7)**. Climate CoLab proposals also largely emphasized changes in production practices, thus making producers more responsible for sustainability interventions. Seeds focused more on lifestyle change and putting the responsibility on consumers rather than producers. Within the 'new technology category' there was also a large emphasis on app-based solutions. Eleven seeds and one Climate CoLab proposal proposed apps, largely to enable users to monitor and report on sustainability issues and to connect with others over them. These included the Climate CoLab proposal to enable urban dwellers to report on the quality of their environment (C'SQUARE), and seeds like a plastic waste footprint calculator app or apps to report water pollution to relevant authorities, monitor energy consumption, help report and identify plants, and several sharing economy apps related to ride-sharing, waste exchanges, and product borrowing from neighbours. The winning Climate CoLab proposals had technological innovations at their core: ClimateCoop was based on blockhain technology and The Community-Based Framework for Sustainable Development integrated existing technologies to meet multiple sustainability needs holistically (For an example see **Box 23.4**).

Clusters within workshop seeds and Climate CoLab proposals
Workshop seeds and Climate CoLab proposals were coded by types of intervention and broad clusters, according to the categories outlined in **Table 22.13** in Chapter 22. The cluster coding allows for the bottom-up initiatives to complement and reinforce the top-down analysis. As the bottom-up approaches are new to GEO-6, the following discussion is intended to be illustrative of the possibilities offered by these complementary methods. In future assessments, a larger data set could be gathered, and results linked more explicitly to the top-down efforts, and, in turn, the top-down analysis could be enhanced by including some of the findings from the bottom-up analysis. Coding was done by subjectively assigning as many intervention types as were appropriate, based on the description provided for each seed or Climate CoLab proposal; as such, it is common for multiple intervention types across more than one cluster to be represented in a given proposal. Due to low representation in the two clusters of freshwater and oceans, these have been grouped together throughout this chapter for cluster-based analysis and based on the original cluster in Chapter 22. However, it is recommended that freshwater and oceans are considered separately in future assessments.

Figure 23.7: How each theory of change is represented by the total seeds and proposals

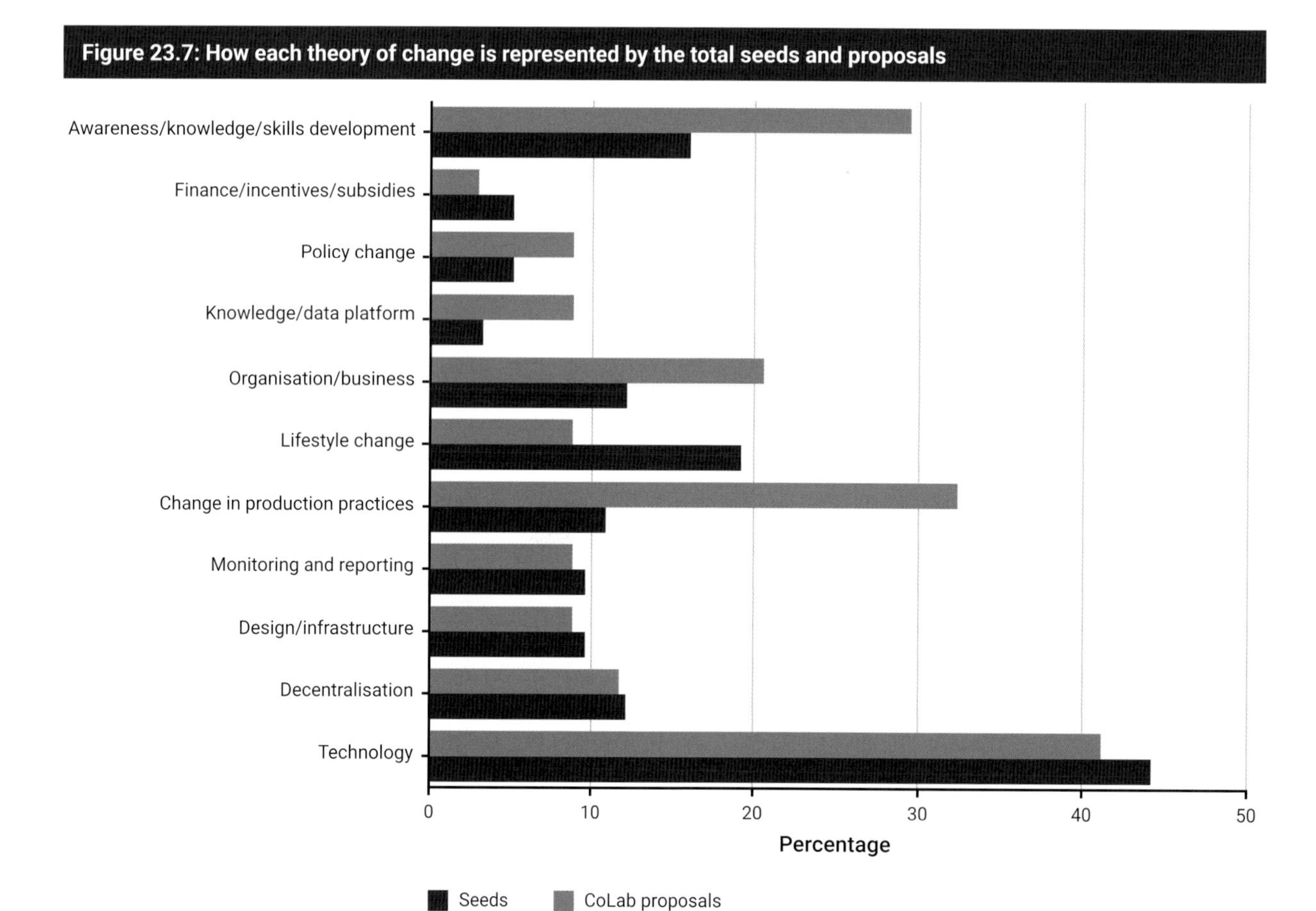

Box 23.4: Climate CoLab Winners

ClimateCoop - The Climate Consortium Blockchain (Judges' Choice Winner)

ClimateCoop is a blockchain-based platform that allows for decentralized, local, and transparent action on SDGs. This distributed platform connects interested parties (e.g. individuals, researchers, sponsors, international organizations, governments, businesses) and facilitates collaborations for new ideas and initiatives. On the platform, initiative creators can update their progress, while accredited members can review and approve future initiatives. The developers of the ClimateCoop Platform believe that their innovation utilizes the best of distributed digital technology, modern social patterns, decentralized matrix governance, and disruptive economic models (e.g. crowdsourcing) to efficiently support bottom-up climate and sustainability action. Their platform empowers individuals and institutions to cooperate and collaborate.

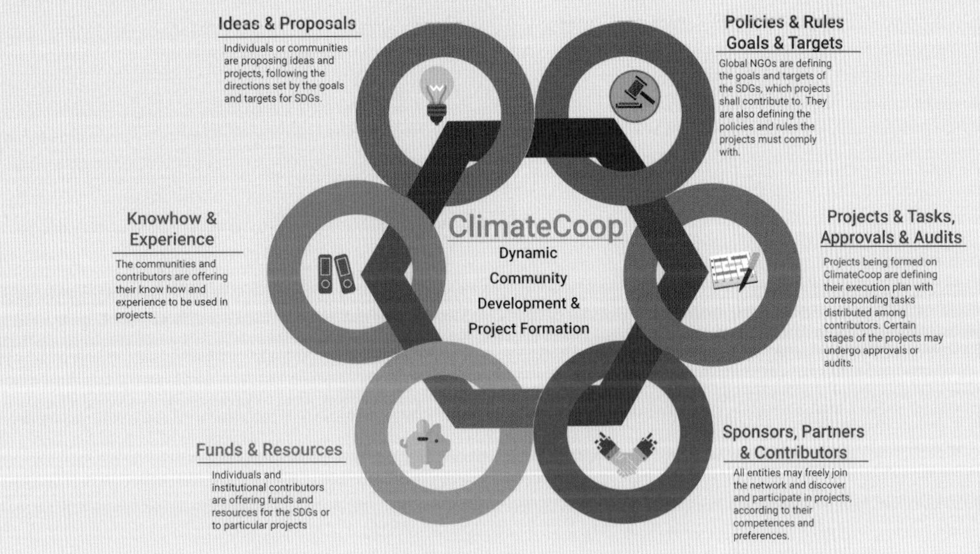

Source: ClimateCoop (2018)

Framework for Community-Based Sustainable Development (Popular Choice Winner)

The Framework for Community-Based Sustainable Development introduces a comprehensive, integrated roadmap for communities to pursue sustainable development. This integrated roadmap builds upon the energy, water/waste, and food sectors to create a holistic approach to community sustainability. By emphasizing the synergistic nature of infrastructure and society, this roadmap helps future development consider the "human factor" within sustainability, ensuring environmental sustainability that is community inclusive. The framework's independent components such as the development of biogas technology, vertical hydroponic farms, and rainwater harvesting are designed to be adaptable to different localities.

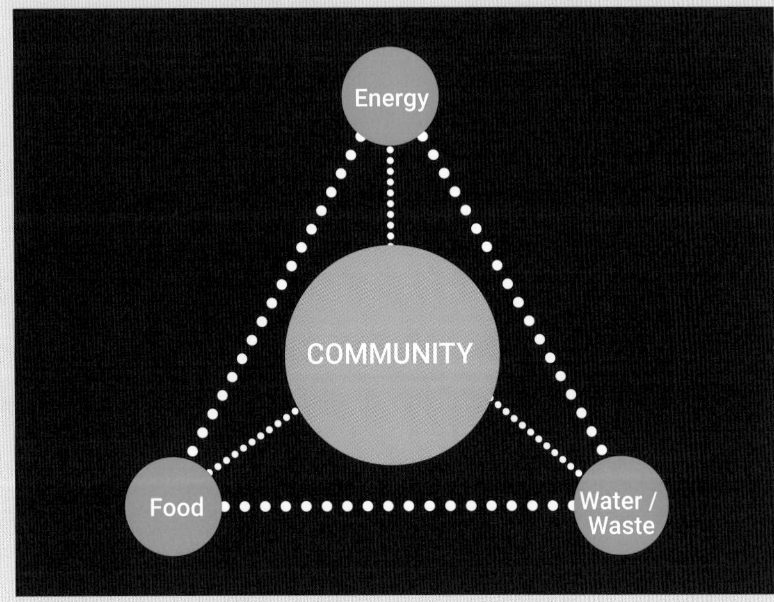

Source: Wright, Yang and Ma (2018).

23

Figure 23.8 shows that in the workshop seeds, there was strong representation of the energy, climate and air cluster, particularly linked to SDGs 7, 11 and 13. Specific interventions within the cluster are detailed below, but popular interventions related to low/zero emissions, behaviour change, energy efficiency and (to a lesser degree) energy access. The seeds showed strong representation of the various categories described as "other" (not part of the main cluster groupings identified in Chapter 22), particularly awareness and skills building, monitoring and reporting, plastics and consumer waste reduction, and circular economy, with the strongest SDG links being to SDGs 11 and 12, with slightly less strong links to SDGs 3 and 13. There was modest interest in the food, agriculture, land and biodiversity cluster, with the strongest interventions relating to diet change and protection of terrestrial ecosystems. Due to the participatory workshops taking place in cities, there was a big focus on how to meet SDGs in an urban context –**Box 23.5** provides some of these key findings from the interventions that emerged in the participatory processes.

© Shutterstock/Sopotnicki

Box 23.5: Urban systems

GEO-6 identifies urbanization as one of five key drivers of environmental change, creating fundamental changes in natural and social systems, as well as one of 12 cross-cutting issues that require urgent and systemic responses (see Chapters 2 and 4). With around 60 per cent of the urban areas yet to be built to accommodate the urban population of 2050, it is critical to ensure that urban systems designed today are made as sustainable as possible.

The participatory results focused to a large extent on improving urban environments, with SDG 11 (sustainable cities and communities) mentioned often, by 38 per cent of all workshop seeds and half of all Climate CoLab proposals. Analysis of these results also showed a variety of SDG synergies, supporting the idea of urbanization being a cross-cutting issue in which solutions can have multiple co-benefits. Seeds addressing SDG 11 had large synergies for addressing SDGs 3, 9, 12 and 13. Climate CoLab proposals also indicated several synergies with SDG 11, including for SDGs 3, 12, 13 and 17. These coding results were further reflected in the descriptions of relevant seeds and proposals, as many spoke of a variety of co-benefits for urban-based solutions.

Urban-related seeds often focused on empowering citizens using online platforms and smartphone applications. Some apps focused on allowing users to monitor and report their energy usage, air and water pollution, to identify plant species (biodiversity), and more. A core aspect of these apps was to enable data-based action in addition to educating users. An app to monitor energy consumption incorporated monetary incentives to change electricity use habits, and an app to monitor water quality connected directly to relevant municipal water agencies. Urban seeds also focused on infrastructure, particularly on developing green infrastructure through green roofs, community gardens and green building standards more generally.

In all four workshops, seeds-based visions often coalesced around sustainable cities or communities. Urban areas were imagined in which buildings are fitted with solar panels and/or green roofs, are built with sustainable materials, and make use of smart technologies to minimize energy usage. Pathways to sustainable futures often included setting aside spaces and providing infrastructure to enable urban agriculture, the products of which could be used for food as well as for sustainable consumer goods such as biodegradable or edible cutlery. One pathway focused specifically on an international cities platform that allows for environmental data and actions to be aggregated internationally, and to be used by citizens to learn and engage in sustainable community actions.

Figure 23.8: Heat map of workshop seeds, showing pairings of specific measures/interventions and SDGs

Cluster	Measure category	No poverty (1)	Zero hunger (2)	Good health and well-being (3)	Quality education (4)	Gender equality (5)	Clean water and sanitation (6)	Affordable and clean energy (7)	Decent work and economic growth (8)	Industry, innovation and infrastructure (9)	Reduced inequalities (10)	Sustainable cities and communities (11)	Responsible consumption and production (12)	Climate action (13)	Life below water (14)	Life on land (15)	Peace, justice and strong institutions (16)	Partnership for the goals (17)
Energy, Climate and Air	Energy access	1	0	0	1	0	0	9	4	3	1	4	1	4	0	0	1	1
	Behavioural change (transport and households)	1	0	5	1	0	2	6	4	5	1	10	6	10	1	1	0	1
	End-use electrification	0	0	0	0	0	0	5	1	3	0	2	0	4	0	0	0	0
	Low/ zero emission technologies (non-biomass)	1	0	5	1	0	2	18	6	7	1	8	5	10	1	1	0	2
	Bioenergy (with and without CCS)	0	0	0	0	0	0	0	0	0	0	0	0	0	0	0	0	0
	Improve energy efficiency	0	1	4	0	0	2	5	4	7	0	7	4	7	0	1	0	0
	Negative emission technologies	0	0	0	0	0	0	0	0	0	0	0	0	1	0	0	0	0
	Air pollution control	0	0	0	0	0	0	0	0	0	0	0	0	0	0	0	0	0
	Non-CO_2 emission reduction	0	1	0	0	0	0	1	0	1	0	0	1	2	0	0	0	0
Agriculture, Food, Land and Biodiversity	Reduce food waste	0	1	1	0	0	1	0	0	0	0	1	3	1	0	0	0	0
	Yield improvement	3	4	2	0	1	0	0	1	1	0	1	2	0	0	0	0	0
	Nutrition management	0	1	2	0	0	0	0	0	0	0	1	1	1	0	1	0	0
	Food access	2	5	3	0	1	0	1	2	1	1	2	1	0	0	1	0	1
	Diet change	0	2	4	1	0	1	1	0	1	0	4	4	3	3	4	0	1
	Manage soil carbon loss	1	1	0	0	0	1	0	1	0	0	0	0	0	0	1	0	1
	Minimize land damage	1	2	1	0	0	2	2	2	0	0	0	0	1	0	2	0	2
	Land ownership	0	0	0	0	0	0	0	0	0	0	0	0	0	0	0	0	0
	Protection of terrestrial ecosystems	2	2	1	0	0	3	2	3	3	0	3	1	4	2	6	0	1
	Land-use planning	1	0	0	0	0	0	0	1	1	0	2	1	1	0	1	0	0
	Forest management	1	3	1	0	0	1	1	1	0	0	0	1	3	0	4	0	0
Human Well-being	Poverty alleviation	2	1	2	1	1	0	3	4	3	3	3	3	2	0	1	1	1
	Child/ maternal healthcare	0	0	0	0	0	0	0	0	0	0	0	0	0	0	0	0	0
	Education	2	1	2	6	0	1	0	3	0	3	1	0	0	0	1	1	2
Freshwater and Oceans	Improve water-use efficiency	0	0	0	0	0	2	0	0	1	0	0	0	1	0	1	0	0
	Blue Carbon	0	0	0	0	0	0	0	0	0	0	0	0	0	0	0	0	0
	WASH	0	1	2	0	0	7	1	0	2	1	3	0	0	1	1	0	0
	Wastewater treatment	0	0	1	0	0	2	1	0	0	0	0	1	0	0	0	0	0
	Water quality standards	0	0	0	0	0	1	0	0	0	0	0	0	0	0	0	0	0
	Desalination	0	0	0	0	0	0	0	0	0	0	0	0	0	0	0	0	0
	Integrated water resource management	0	0	0	0	0	1	0	0	1	0	3	1	0	0	0	0	0
	Sustainable fisheries	0	1	1	0	0	0	0	0	0	0	1	1	0	1	0	0	0
	Ocean regulation	0	0	0	0	0	0	0	0	0	0	0	0	0	0	0	0	0
	Protection of marine ecosystems	0	0	0	0	0	1	0	0	0	0	0	0	1	2	1	0	0
Other	Monitoring and reporting	3	3	10	3	1	3	3	4	6	3	8	5	3	3	5	2	3
	Circular economy	1	0	1	1	0	2	1	2	4	1	6	14	1	2	1	0	1
	Sharing economy	1	0	1	0	0	0	0	0	0	1	4	6	1	2	2	0	1
	Plastics and consumer waste reduction	1	2	3	1	0	2	1	1	2	0	6	16	8	5	5	0	1
	Awareness and skills building	3	4	8	10	4	5	5	6	6	5	12	12	10	8	7	3	6
	Gender equality	0	1	1	1	2	1	0	0	0	0	0	1	0	1	1	0	1
	Smart cities for sustainability	1	0	2	0	0	1	1	2	1	0	6	3	5	1	1	0	0
	Ecosystem restoration	0	0	0	0	0	1	0	0	0	0	0	0	0	1	0	0	0
	Effective governance	0	0	0	0	0	0	0	0	0	0	0	0	0	0	0	0	0

Numbers indicate the count of proposals coded with the specific pairing of intervention (row) and SDG (column). 'Other' is described more in Section 23.11

Figure 23.9: Heat map of Climate CoLab proposals showing pairings of measures/interventions and SDGs

Cluster	Measure category	No poverty (1)	Zero hunger (2)	Good health and well-being (3)	Quality education (4)	Gender equality (5)	Clean water and sanitation (6)	Affordable and clean energy (7)	Decent work and economic growth (8)	Industry, innovation and infrastructure (9)	Reduced inequalities (10)	Sustainable cities and communities (11)	Responsible consumption and production (12)	Climate action (13)	Life below water (14)	Life on land (15)	Peace, justice and strong institutions (16)	Partnership for the goals (17)
Energy, Climate and Air	Energy access	2	2	2	1	1	2	4	2	2	1	0	1	4	1	1	0	2
	Behavioural change (transport and households)	3	3	3	1	2	2	4	3	3	2	3	3	5	3	3	2	4
	End-use electrification	1	1	1	0	0	2	2	1	1	0	1	0	1	1	1	0	0
	Low/zero emission technologies (non-biomass)	3	4	4	1	2	3	5	4	2	2	2	0	5	1	2	1	3
	Bioenergy (with and without CCS)	0	0	0	0	0	0	1	0	1	0	0	0	1	0	0	0	0
	Improve energy efficiency	2	2	2	1	1	2	1	2	1	1	2	0	3	1	1	0	1
	Negative emission technologies	0	0	0	0	0	0	0	0	0	0	0	0	0	0	0	0	0
	Air pollution control	0	0	0	0	0	0	0	0	0	0	0	0	0	0	0	0	0
	Non-CO_2 emission reduction	0	1	1	0	0	1	0	0	1	0	0	1	1	0	1	0	0
Agriculture, Food, Land and Biodiversity	Reduce food waste	2	2	2	1	2	1	1	1	1	1	2	1	2	1	1	1	2
	Yield improvement	3	3	2	0	2	1	0	1	1	1	1	1	3	0	1	0	2
	Nutrition management	0	0	0	0	0	0	0	0	0	0	0	0	0	0	0	0	0
	Food access	7	10	10	4	6	4	6	8	3	5	4	6	10	3	6	2	8
	Diet change	0	1	1	0	0	1	0	0	1	0	0	1	1	0	1	0	0
	Manage soil carbon loss	3	3	2	1	2	3	1	1	1	1	2	1	3	1	1	1	2
	Minimize land damage	5	8	8	3	6	7	6	7	5	5	4	6	10	4	7	3	6
	Land ownership	0	0	0	0	0	0	0	0	0	0	0	0	0	0	0	0	0
	Protection of terrestrial ecosystems	3	5	5	2	3	5	4	4	3	2	3	5	6	3	5	1	3
	Land-use planning	1	2	2	1	0	1	0	0	1	0	0	1	2	0	1	0	1
	Forest management	2	3	2	1	1	4	3	2	1	0	1	3	4	2	3	0	1
Human Well-being	Poverty alleviation	8	9	9	3	3	5	5	7	4	4	3	5	10	3	5	1	5
	Child/ maternal healthcare	1	1	1	1	1	1	1	1	1	1	1	1	1	1	1	1	1
	Education	1	1	1	1	1	1	1	1	1	1	1	1	1	1	1	1	1
Freshwater and Oceans	Improve water-use efficiency	1	1	0	0	0	1	0	0	0	0	0	0	1	0	0	0	0
	Blue carbon	0	0	0	0	0	0	0	0	0	0	0	0	0	0	0	0	0
	WASH	0	0	0	0	0	0	0	0	0	0	0	0	0	0	0	0	0
	Wastewater treatment	0	0	0	0	0	0	0	0	0	0	0	0	0	0	0	0	0
	Water quality standards	0	0	0	0	0	0	0	0	0	0	0	0	0	0	0	0	0
	Desalination	0	0	0	0	0	0	0	0	0	0	0	0	0	0	0	0	0
	Integrated water resource management	1	3	3	2	3	2	2	3	0	2	2	0	3	0	1	1	3
	Sustainable fisheries	0	0	0	0	0	0	0	0	0	0	0	0	0	0	0	0	0
	Ocean regulation	0	0	0	0	0	0	0	0	0	0	0	0	0	0	0	0	0
	Protection of marine ecosystems	1	1	1	1	1	1	1	1	1	1	1	1	1	1	1	1	1
Other	Monitoring and reporting	1	1	2	1	0	2	1	1	2	1	1	1	2	0	1	2	2
	Circular economy	3	3	5	1	3	1	1	1	2	2	4	4	4	1	2	1	2
	Sharing economy	2	2	2	1	1	1	1	1	1	1	1	1	2	1	1	1	2
	Plastics and consumer waste reduction	2	1	3	0	2	1	0	0	1	1	4	4	4	1	3	0	2
	Awareness and skills building	7	8	8	5	7	7	7	8	5	7	9	7	13	5	7	5	6
	Gender equality	5	6	7	2	7	3	3	5	1	5	6	3	7	1	4	2	6
	Smart cities for sustainability	0	0	0	0	0	0	0	0	0	0	0	0	0	0	0	0	0
	Ecosystem restoration	1	1	1	0	0	1	1	1	0	0	0	1	1	1	1	0	0
	Effective governance	0	0	0	0	0	0	0	0	0	0	1	1	1	1	1	0	1

Numbers indicate the count of proposals coded with the specific pairing of intervention (row) and SDG (column). 'Other' is described more in Section 23.11

The cluster groupings were quite different between the seeds and the Climate CoLab proposals. In the latter, agriculture, food, land and biodiversity emerged as a very strong cluster, far more so than in the workshop seeds, with many Climate CoLab proposals targeting food access and minimizing land damage **(Figure 23.9)**. Climate CoLab proposals also focused heavily on poverty alleviation. The added intervention type, awareness and skills building, was strongly represented in both the seeds and Climate CoLab proposals. SDGs 1, 2, 3 and 13 emerge as strongly linked across many proposals. Comparatively few Climate CoLab proposals had interventions relating to energy, climate and air despite strong representation of SDG 13 (climate action). Gender equality emerged as a strong intervention in Climate CoLab proposals compared with the seeds, but it was not strongly related to any other SDG. Neither the seeds nor the Climate CoLab produced any substantial focus on the merged cluster for freshwater and oceans, although this gap is partially addressed in the analysis of the Regional Assessments.

Figure 23.10 shows the number of seeds/proposals that sit across multiple clusters. The Climate CoLab proposals were more likely to be relevant in more than one cluster, whereas the seeds tended to stay within one cluster. This makes sense because the seeds were typically single initiatives rather than a combination of interventions into one proposal. Seeds show

Figure 23.10: Inter-cluster pairings across the seeds and Climate CoLab proposals

Workshop

Cluster	Agriculture, Food, Land and Biodiversity	Energy, Climate and Air	Freshwater and Oceans	Human well-being	Others
Agriculture, Food, Land and Biodiversity	30	4	4	5	6
Energy, Climate and Air	4	43	5	4	13
Freshwater and Oceans	4	5	17	0	3
Human well-being	5	4	0	16	4
Others	6	13	3	4	84

Climate CoLab

Cluster	Agriculture, Food, Land and Biodiversity	Energy, Climate and Air	Freshwater and Oceans	Human well-being	Others
Agriculture, Food, Land and Biodiversity	18	6	5	9	12
Energy, Climate and Air	6	12	2	7	6
Freshwater and Oceans	5	2	5	2	4
Human well-being	9	7	2	14	9
Others	12	6	4	9	23

Numbers indicate the count of seeds/proposals with at least one intervention from each of the intersecting cluster groups

23

a tendency to pair energy, climate and air with the "other" cluster, while in the Climate CoLab proposals, this pairing is one of the least common. The Climate CoLab proposals are far more likely to show pairings between various clusters and human well-being due to the strong representation of poverty alleviation across the Climate CoLab proposals. The key conclusion from this figure is that, when looking at real-world examples, it is possible for interventions to work across clusters. It is therefore also possible to give specific example of how to achieve the synergies described in Chapter 22.

Agriculture, food, land and biodiversity
Seeds and Climate CoLab proposals within the cluster for agriculture, food, land and biodiversity were most related to food access, protection of terrestrial ecosystems, and minimizing land damage **(Figure 23.11)**. No solutions targeted land ownership, and only two addressed nutrition management.

Some key trends emerging from this cluster are the decentralization and localization of food production (e.g. community-supported agriculture, urban farming innovations) to improve food access, minimize land damage and potentially improve yields. These types of solution could potentially address the yield-improvement trade-offs that were identified by Chapter 22, for example against addressing climate change and water scarcity.

Energy, climate and air
Seeds and proposals that fit within the energy, climate and air cluster were most related to low- or zero-emission technologies, behavioural change in the use of transport and household energy, energy access, and improved energy efficiency **(Figure 23.12)**. Bioenergy, negative-emissions technologies, and air-pollution control were addressed very sparsely. One of the Climate CoLab proposals "Adapting the indigenous approach to climate change adaptation and mitigation" makes clear the importance of not relying only on technological fixes, but recognising the relevance of local innovations that draw on a variety of knowledge sources.

Freshwater and oceans
The clusters for freshwater and oceans, combined for the analysis, were among the least-addressed ones, especially in Climate CoLab proposals **(Figure 23.13)**. Seeds within this combined cluster focused most on WASH (water, sanitation and hygiene) while no seeds or proposals addressed desalination or ocean regulation.

Figure 23.11: Total number of workshop seeds and Climate CoLab proposals addressing each intervention in the agriculture, food, land and biodiversity cluster (seeds and proposals are double counted when they meet multiple measures)

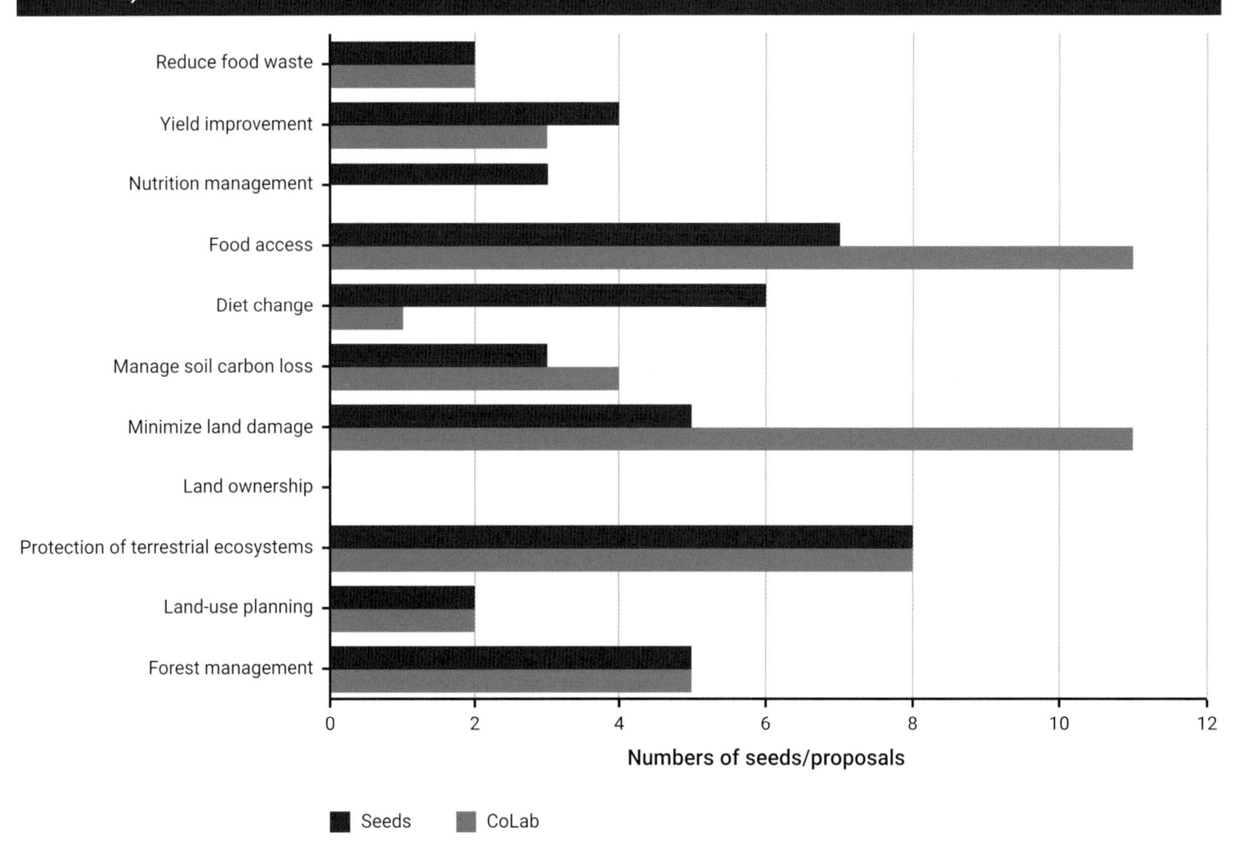

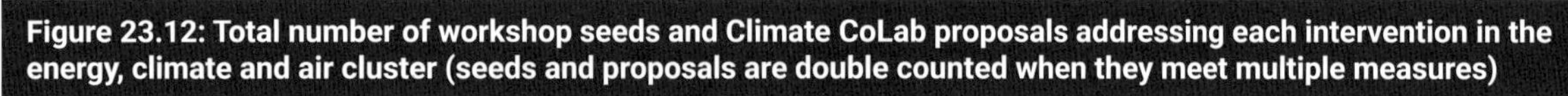

Figure 23.12: Total number of workshop seeds and Climate CoLab proposals addressing each intervention in the energy, climate and air cluster (seeds and proposals are double counted when they meet multiple measures)

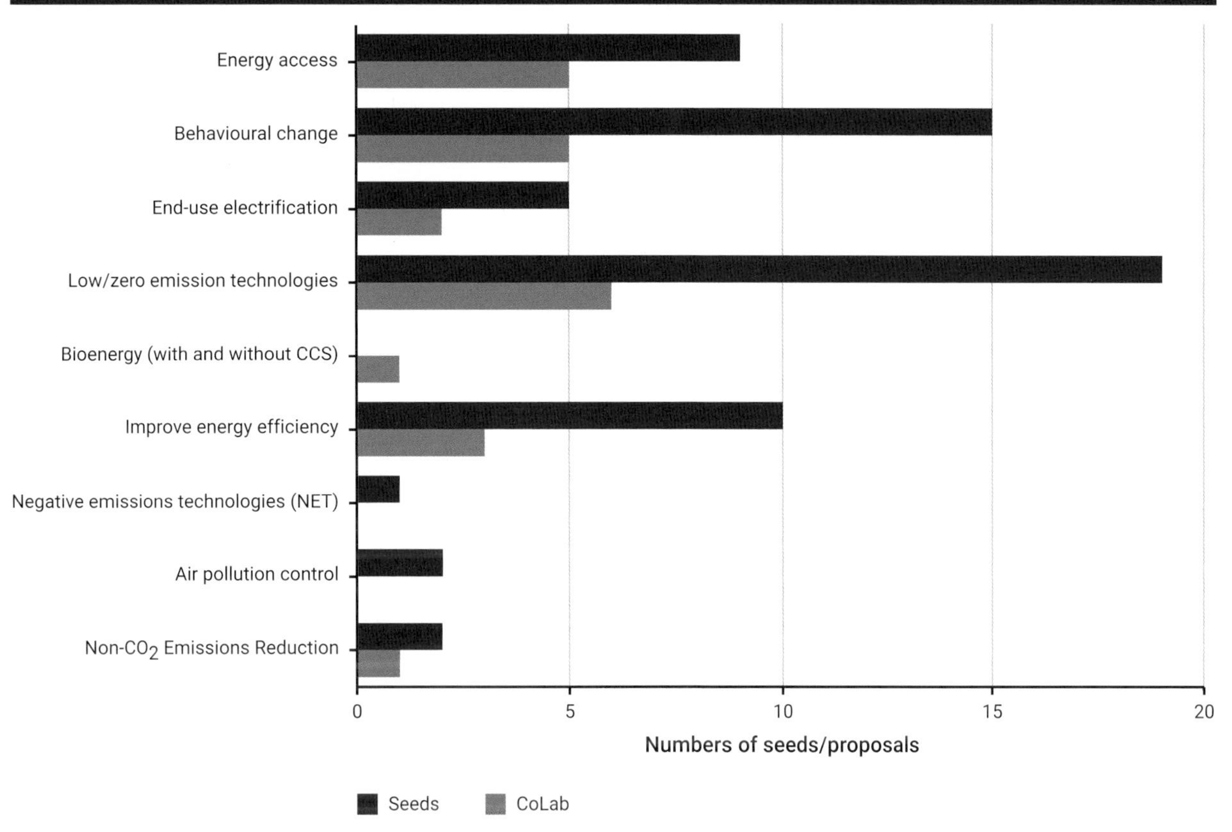

Figure 23.13: Total number of workshop seeds and Climate CoLab proposals addressing each intervention in the combined clusters for freshwater and oceans (seeds and proposals are double counted when they meet multiple measures)

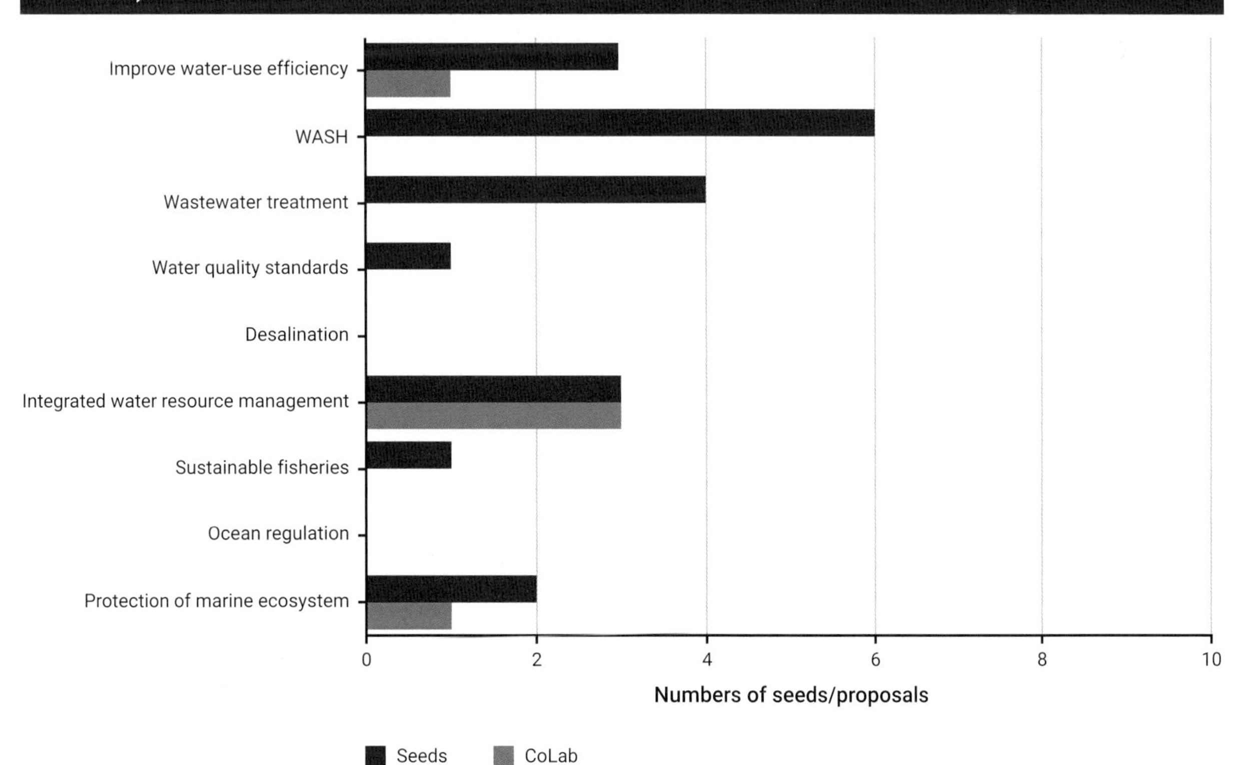

Human well-being
Solutions related to human well-being focused predominantly on the alleviation of poverty while child and maternal health-care was addressed by only one Climate CoLab proposal **(Figure 23.14)**. This could highlight an important area for government interventions to specifically target these areas.

23.10 GEO Regional Assessment synthesis

The additional interventions highlighted by the GEO-6 Regional Assessments are presented below, followed by an outline of the main regional emphasis of different clusters of interventions, and a comparison with the prevalent top-down and bottom-up interventions.

23.10.1 Relevance of additional interventions for different regions

Nine additional interventions were highlighted in the Regional Assessments (see Section 23.11 for a more in-depth discussion). Two of these – effective governance, and awareness and skills building – were highlighted as important interventions across all six Regional Assessments. The Regional Assessments indicate the need to involve a diverse range of actors in seeking transformative solutions to achieve sustainable development, and all of the regional assessments emphasize the development of new collaborations between business, government and civil society. In addition to these commonalities, the assessments strongly reflected region-specific issues, which emphasizes the need for considering bottom-up initiatives. In North America, the identified governance and capacity-building needs focused on integrated forward-looking approaches that leveraged new technologies and citizen science in monitoring and reporting that would ultimately internalize environmental costs in the economy. Africa and Latin America and the Caribbean emphasized effective implementation and regulation to prevent further habitat loss and land degradation, focusing strongly on policies that strengthen equitable landownership and sustainable use of natural resources. Europe, and Asia and the Pacific strongly emphasized regional policy integration and cooperation, although the outlook for Europe focused its policy coordination around encouraging sustainable lifestyles, while Asia and the Pacific emphasized coordination as an adaptation response in disaster risk reduction. In West Asia, the dominant governance issue was peace and security. Only three assessments (Africa, Europe, and Latin America and the Caribbean) emphasized the need for global governance in addressing tele-coupling aspects that transfer the impacts of production and consumption to other regions. This limited consideration of interregional impacts, particularly from major regions of consumption such as North America and parts of Asia and the Pacific, is concerning and should be included as an explicit criterion in future Regional Assessments.

Monitoring and reporting, plastic and consumer waste reduction, and ecosystem restoration were also prevalent regional interventions that were not originally emphasized in the review of the scenario literature. Monitoring and reporting was emphasized by all regions except Europe, and the focus was on the use of new technologies and citizen science to monitor future trends and report on sustainable development. Plastic and consumer waste reduction was emphasized by most regions – except Africa, and Latin America and the Caribbean – and focused primarily on solutions against landfill being used for solid waste management. Ecosystem restoration was emphasized by Europe, North America and West Asia, but the focus differed in each region. In North America, restoration was considered important for improved water-quality management, while in West Asia restoration was strongly focused on restoring coastal marine ecosystems as a strategy to reduce disaster risk. In Europe, restoration was an integrative pathway to realizing multiple goals for biodiversity conservation, the rewilding of abandoned farmlands, a reduction of nitrogen and GHG emissions, and the mental and physical health benefits of restoring blue-green infrastructure.

Circular economies and smart cities for sustainability were highlighted as interventions by only some of the Regional Assessments **(Figure 23.15)** Nevertheless, at least two regions identified these as priority interventions, and there are indications from the bottom-up initiatives that these interventions represent emerging opportunities that can be leveraged as integrated and synergistic approaches to achieve sustainable futures.

Figure 23.14: Total number of workshop seeds and Climate CoLab proposals addressing each intervention in the human well-being cluster (seeds and proposals are double counted when they meet multiple measures)

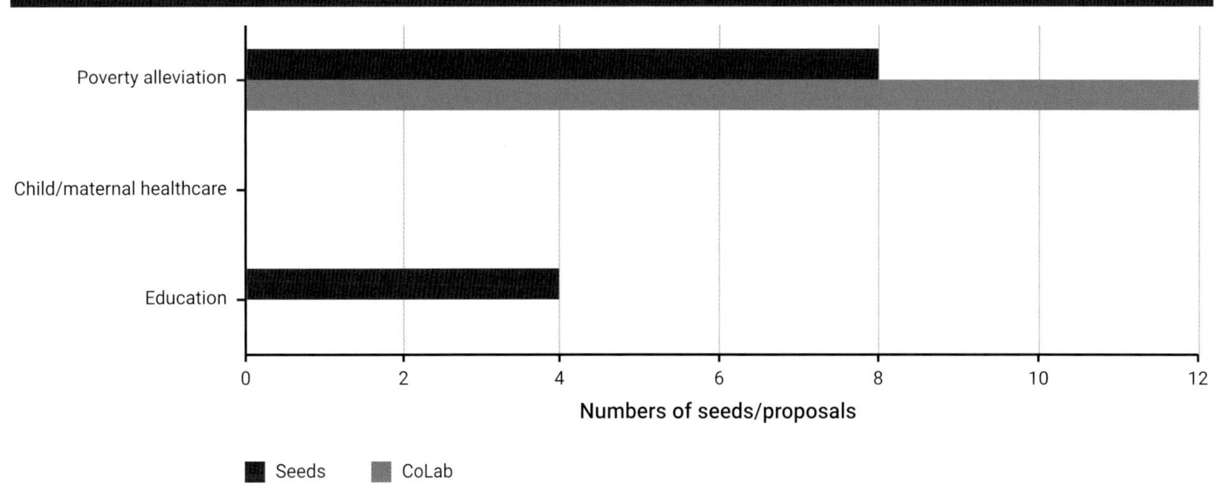

Figure 23.15: The interventions highlighted by the outlook chapters of the GEO Regional Assessments

Cluster	Measure Category	North America	Latin America and the Caribbean	Africa	Europe	Asia Pacific	West Asia
Energy, air and climate	Energy access			■			
	Behavioural change (transport and households)	■		■	■	■	
	End-use electrification						
	Low/zero emission technologies (non-biomass)	■	■	■	■	■	■
	Bioenergy (with and without CCS)	■	■		■		
	Improve energy efficiency	■			■		■
	Negative emission technologies				■		
	Air pollution control			■	■		
	Non-CO_2 emission reduction				■		
Agriculture, food, land and biodiversity	Reduce food waste	■			■		■
	Yield improvement	■		■	■		
	Nutrition management						
	Food access			■			
	Diet change						
	Manage soil carbon loss	■		■			
	Minimize land damage	■	■	■			
	Land ownership			■			
	Protection of terrestrial ecosystems	■	■	■	■		■
	Land-use planning		■	■	■		
	Forest management		■	■	■		
	Improve water-use efficiency	■		■		■	
Freshwater and Oceans	Blue carbon						
	WASH			■		■	■
	Wastewater treatment	■					
	Water quality standards	■					■
	Desalination						
	Integrated water resource management	■		■			■
	Sustainable fisheries			■			■
	Ocean regulation					■	
	Protection of marine ecosystems			■		■	
Human well-being	Poverty alleviation						
	Child/maternal health care					■	
	Education		■				
Other regional and bottom-up interventions	Effective governance	■	■	■	■	■	■
	Awareness and skills building	■	■	■	■	■	■
	Monitoring and reporting	■	■	■		■	■
	Plastics and consumer waste reduction	■			■	■	■
	Ecosystem restoration	■			■		■
	Smart cities for sustainability	■				■	
	Circular economy				■	■	■
	Sharing economy						
	Gender equality						

Blue indicates that the intervention was highlighted by the associated regional assessment for moving towards a more sustainable development trajectory; white indicates absence of the intervention. The interventions are clustered similarly to the grouping used in Chapter 22. Interventions that were not on the predefined list from the scenario literature review (those labelled "Other regional and bottom-up interventions") were added to derive an updated list of interventions (see Section 23.8).

23.10.2 Regional emphasis of different clusters of interventions

The most frequently occurring interventions across regions were low/zero-emission technologies, the protection of terrestrial biodiversity, effective governance, skills and awareness building, and monitoring and reporting. The Regional Assessments highlighted roughly similar proportions of interventions in the energy, climate and air cluster and in the agriculture, food, land and biodiversity cluster, with interventions in the combined cluster for freshwater and oceans showing only slightly less prevalence **(Figure 23.16)**. There was a marked absence of interventions that directly addressed the human well-being cluster (unlike the Climate CoLab proposals in which this cluster was emphasized heavily) Below the emphasis the regions place on the interventions within the clusters identified in Chapter 22 is discussed.

Energy, climate and air

A positive finding, also emphasized in Chapter 22, is that renewable energies are on the agendas of all regions. All six assessments – no matter whether primarily comprising developed or developing economies – emphasize renewable energies in their key interventions. In Africa, this is not only viewed as a way of improving air quality and GHG emissions, but also as a means of improving access to basic services by providing off-grid development in rural areas. In West Asia, renewable energies are viewed as a fundamental consideration for food and water security. Even though the GHG emissions in Latin America and the Caribbean are currently the lowest globally, the region places strong emphasis on renewable energies as a means of curbing current trends, which are expected to increase dramatically in the scenarios in which there is no investment in low-carbon futures.

Although renewable energies are on the agendas of all regions, there are clear gaps in interventions dealing with emissions reductions, with Europe being the only region to emphasize the full range of energy, climate and air interventions. This lack of direct action for climate mitigation is concerning. In addressing climate-change adaptation there is a definite regional difference: both Africa and North America emphasize food and water security; Asia and the Pacific, and West Asia emphasize disaster-risk reduction; Europe emphasizes air quality and health; and Latin America and the Caribbean focuses on ecosystem-based resilience and the need systematically to consider alternative sustainability framings that can be found in indigenous and local knowledge.

Agriculture, food, land and biodiversity

This cluster of interventions reflects the management of the land system, which has conventionally been dominated by ecological and biophysical perspectives. While protection of terrestrial ecosystems still dominates the interventions in this cluster, there are indications that conventional approaches are broadening in scope to include more integrated social-ecological initiatives, such as reduced food waste, yield improvement, agro-biodiversity, and forest and land management **(Figure 23.16)**. In Africa and Latin America and the Caribbean, this shift from a protection approach towards more sustainable land management has been reflected in the concept of ecological infrastructure and the complementary benefits it has for built infrastructure. In all regions, the lack of nutrition management and diet change, however, are notable gaps in the interventions highlighted by the Regional Assessments, indicating that the more behavioural aspects of the social-ecological spectrum have not yet been fully entrenched into this cluster in the regions. Similar gaps in socioeconomic interventions are prevalent in the human well-being cluster.

The interventions in this cluster also reflected region-specific environmental issues. Africa, and Latin America and the Caribbean had a very strong emphasis on protection of terrestrial ecosystems and sustainable land management, reflecting the need to address the enormous pressures these regions face around large-scale land conversion for agriculture. The Africa region, in its focus for leapfrogging to more sustainable development, also highlighted the potential for investment in agricultural intensification to increase efficiencies and improve agricultural yield simultaneously, and thereby minimize further habitat loss. Europe and North America placed strong emphasis on yield improvement and reduced food waste, with Europe also focusing attention on land abandonment and rewilding. Food access was another social intervention that was highlighted in this cluster, and this pertained to providing improved opportunities to smallholder farmers in Africa and West Asia.

Freshwater and oceans

Like the workshop seeds and Climate CoLab proposals, the Regional Assessments emphasized proportionally fewer interventions in the freshwater and oceans cluster compared with the previous two clusters in this section. The outlooks for Europe, and Latin America and the Caribbean were particularly scant on emphasizing interventions in this cluster. Both regions show signs of improvement in their key freshwater challenges (improved water quality in Europe; improved water supply and sanitation in Latin America and the Caribbean), so there may be more important regional challenges, such as production and consumption changes in Europe, and sustainable land management in Latin America and the Caribbean.

The most frequently emphasized freshwater interventions were integrated water resource management, improved water use efficiency, and water and sanitation. The first two of these are often bundled together, with the predominant narrative being around integrated water resource management to address water scarcity and water allocation issues. This was emphasized by Africa, Asia and the Pacific, and North America (the latter after recent droughts and under climate change projections). Interestingly, West Asia did not emphasize water scarcity in itself, but rather the investment costs of groundwater abstraction and desalination for continued water supply and sanitation of rapidly expanding cities. This indicates that at least one region is explicitly emphasizing diversification of water sources as a feasible response to water supply challenges. Water quality issues – both in terms of safe wastewater treatment and water supply quality – were addressed separately from integrated water resource management. Water quality interventions were emphasized in the Outlooks presented for North America and West Asia, where both regions highlighted issues with wastewater treatment as well as chemical contaminants.

Figure 23.16: Number of regions emphasizing interventions within the clusters identified in Chapter 22

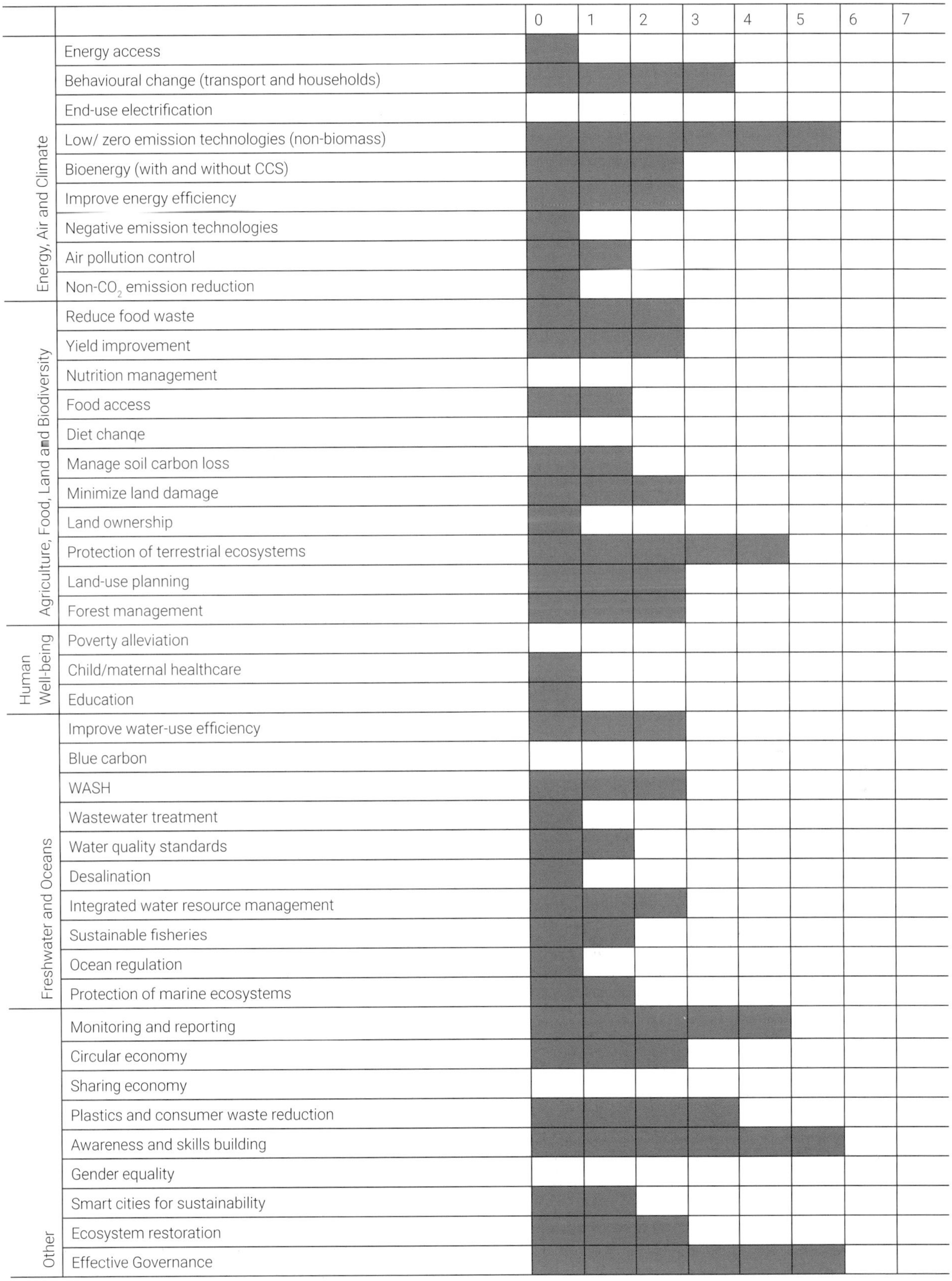

Cluster	Intervention
Energy, Air and Climate	Energy access
	Behavioural change (transport and households)
	End-use electrification
	Low/ zero emission technologies (non-biomass)
	Bioenergy (with and without CCS)
	Improve energy efficiency
	Negative emission technologies
	Air pollution control
	Non-CO_2 emission reduction
Agriculture, Food, Land and Biodiversity	Reduce food waste
	Yield improvement
	Nutrition management
	Food access
	Diet change
	Manage soil carbon loss
	Minimize land damage
	Land ownership
	Protection of terrestrial ecosystems
	Land-use planning
	Forest management
Human Well-being	Poverty alleviation
	Child/maternal healthcare
	Education
Freshwater and Oceans	Improve water-use efficiency
	Blue carbon
	WASH
	Wastewater treatment
	Water quality standards
	Desalination
	Integrated water resource management
	Sustainable fisheries
	Ocean regulation
	Protection of marine ecosystems
Other	Monitoring and reporting
	Circular economy
	Sharing economy
	Plastics and consumer waste reduction
	Awareness and skills building
	Gender equality
	Smart cities for sustainability
	Ecosystem restoration
	Effective Governance

Clusters identified in Chapter 22 (0 = none, 6 = all regions)

The regional Outlooks for Africa and Asia and the Pacific were the only ones that highlighted interventions for the oceans. For Africa, this was mainly around protecting marine ecosystems for sustainable fisheries. In Asia and the Pacific, protecting marine ecosystems was viewed as a strategy both for sustainable fisheries management and disaster risk reduction, particularly in relation to protection and restoration of mangroves.

Human well-being

There was a distinct lack of emphasis placed on the interventions in the human well-being cluster. Only one regional outlook (for Latin America and the Caribbean) identified one intervention (education) as a key intervention for transforming to a sustainable future. The more socially oriented interventions in other clusters were either poorly emphasized (e.g. energy access, food access, smart cities for sustainability), or not highlighted at all (e.g. nutrition management, diet change, poverty alleviation, sharing economy, gender equity and equality). This is not to say that human well-being interventions are ignored throughout the Regional Assessments or even in the chapter presenting the outlook. Indeed, in many cases, the synergies with human well-being SDGs are discussed, and in detail in some cases (e.g. Africa, and Latin America and the Caribbean). However, these are not emphasized as interventions in and of themselves. Instead, the Regional Assessments regarded interventions in this cluster as the fortunate spin-offs of managing the previous three clusters, rather than explicitly planning for synergistic target achievement. Future Regional Assessments could strive for more integrative strategies through explicitly addressing and planning for this cluster of interventions.

23.11 Regional outlook interventions and bottom-up initiatives

23.11.1 Additional categories of intervention

A large portion of solutions did not fit neatly into any of the categories of measures in Chapter 22 in the process of collecting and assessing the seeds and proposals, and reviewing the emphasized regional interventions (see "other" in **Figures 23.15** and **23.16**).

As a result, nine new categories were developed and coded as part of the analysis:

- **Monitoring and reporting**: Innovations to improve the monitoring and reporting of environmental conditions, including citizen science initiatives.
- **Circular economy:** Innovations that involve the increased efficiency of resource use, specifically through new business models that better engage with the issue of waste products of other production processes (See Ghisellini, Cialani and Ulgiati 2016).
- **Sharing economy**: Innovations related to the peer-to-peer sharing of goods and services, primarily through information and communications technology (ICT) platforms (See Hamari, Sköklint and Ukkonen 2016).
- **Plastic and solid waste reduction:** Innovations that help to reduce plastic and solid waste.
- **Awareness and skills building:** Education related to sustainability and environmental issues to improve public awareness and build relevant skills.
- **Gender equality:** Solutions that promote the fair treatment of all genders, including female empowerment and considerations of gender equity.
- **Smart cities for sustainability:** Smart cities use modern digital technologies, such as apps for mobile phones, to engage and connect citizens in addressing their key sustainability challenges, such as city transportation, consumption patterns, energy, nutrition, water and waste.
- **Ecosystem restoration:** The process of assisting the recovery of an ecosystem that has been degraded, damaged or destroyed. Although this category would fit well under the agriculture, food, land and biodiversity cluster, it is considered as a separate category here due to the emphasis on this intervention in the reports. In future assessments, it could be adapted to refer to nature-based solutions, encapsulating those relevant innovations that draw on indigenous knowledge and ecological infrastructure.

Figure 23.17: Seeds and proposals by cluster

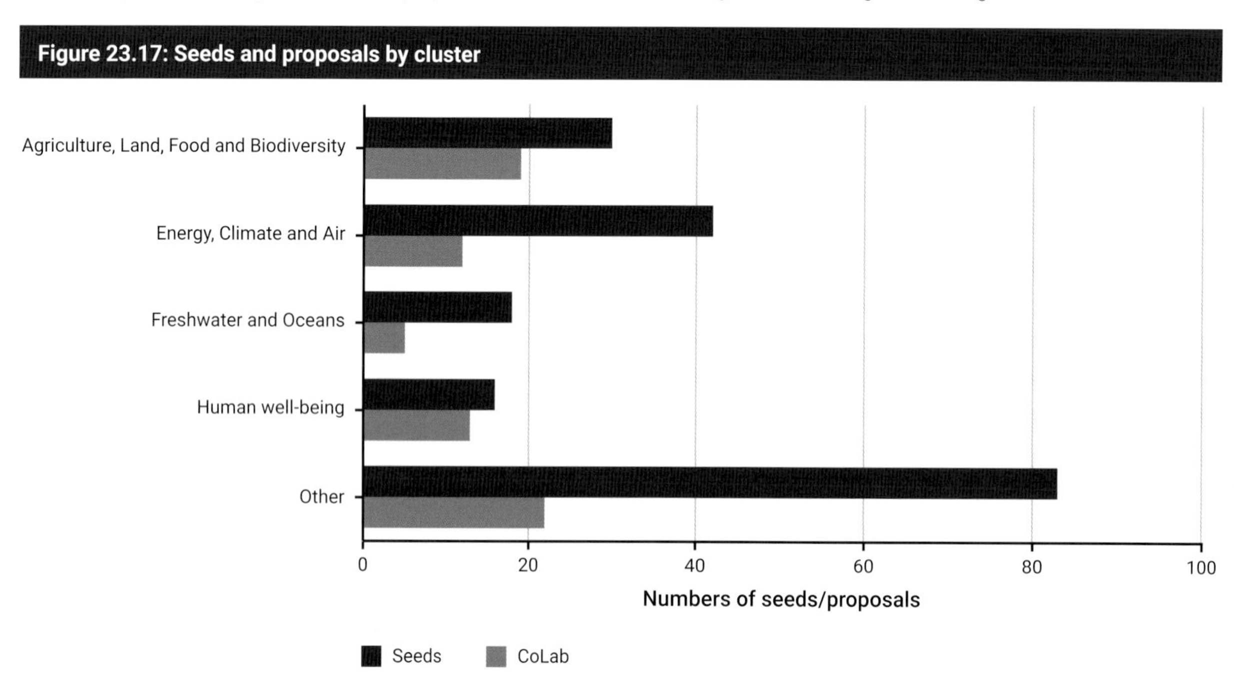

- **Effective governance**: Solutions to improve regional cooperation, and harmonization across scales, including to improve the management of interlinkages and tele-coupling between systems to reduce interregional inequalities.

Sixty out of the 157 workshop seeds, and seven out of the 34 Climate CoLab proposals, were coded against interventions exclusively from this new set of categories grouped as "other" **(Figure 23.18)**. For seeds and proposals with measures that were coded across both "other" and at least one of the four clusters, some preliminary patterns emerged, although the sample sizes were small. For seeds, the most common cluster to be paired with "other" measures was energy, climate and air, with seeds linking this cluster to monitoring and reporting, smart cities, and awareness and skills building. Gender equality appeared in only two seeds and neither of these was coded against any of the four main clusters. In contrast, in the Climate CoLab proposals, gender equality, and awareness and skills building emerged as the strongest intervention categories and appeared in various proposals paired with all of the four main clusters. These proposals ranged in their suggestions from a mentoring network for women to female economic empowerment through activities like beekeeping. Agriculture food, land and biodiversity emerged as the strongest cluster paired with various "other" interventions. While monitoring and reporting was a strongly represented measure in seeds, it was far less prevalent in Climate CoLab proposals.

Two interventions are highlighted in the platforms of bottom-up initiatives that are not included in the global assessment: sharing economies and circular economies. These show

Figure 23.18: Count of the number of pairings of "other" measures with at least one intervention from a main cluster group

Workshop Seeds

"Other" category	Agriculture, Food, Land and Biodiversity	Energy, Climate and Air	Freshwater and Oceans	Human Well-being
Monitoring and reporting	2	5	0	1
Circular economy	2	1	2	2
Sharing economy	0	0	0	0
Plastics and consumer waste reduction	2	1	0	2
Awareness and skills building	1	3	0	0
Gender equality	0	0	0	0
Smart cities for sustainability	0	4	0	0
Ecosystem restoration	0	0	1	0
Effective governance	0	0	0	0

Climate CoLab

"Other" category	Agriculture, Food, Land and Biodiversity	Energy, Climate and Air	Freshwater and Oceans	Human Well-being
Monitoring and reporting	0	0	0	1
Circular economy	3	0	1	2
Sharing economy	2	1	0	1
Plastics and consumer waste reduction	2	1	0	0
Awareness and skills building	5	5	2	4
Gender equality	6	3	3	2
Smart cities for sustainability	0	0	0	0
Ecosystem restoration	1	0	0	1
Effective governance	0	1	0	0

innovations that would boost the energy cluster, and also address production and consumption challenges in the agriculture, food, land and biodiversity cluster **(see Box 23.6)**.

23.11.2 Implications for future assessments

The global review of the scenario provided a useful overview for synthesizing the range of potential interventions available for moving to a more sustainable future. Furthermore, by having concrete examples, it was possible to analyse the likely synergies and trade-offs between these interventions. However, the nine additional interventions that were uncovered in the bottom-up analysis should be considered in future global Outlooks **(Figure 23.18)**. Smart cities, for example, were emphasized in the regional Outlooks as a means of achieving integrated responses to sustainability that capture many interventions towards transformative change. Exploring these urban opportunities, and the role they have in shifting urban-rural dynamics, should be a strong focus in global assessments given current population and urbanization trends. The bottom-up initiatives highlight sharing economies and circular economies as fast-evolving, and region-specific emerging interventions. Future global assessments should aim to factor the impact of such interventions into their outlook. Although the important role of indigenous and local knowledge in sustainability innovations not added in as a separate category, this has been captured as an important aspect in similar participatory processes undertaken by IPBES (See Lundquist 2017; IPBES 2018e) and could be highlighted in the next iteration of assessments.

Regional emphasis for the same intervention, or clusters of interventions, can differ enormously across and within regions. Collecting, piloting and scaling a diverse range of bottom-up initiatives that are relevant to the local context can therefore be extremely useful in providing tangible examples to policymakers of otherwise generic pathways. Effective governance, and awareness and skills building were two interventions that all Regional Assessments emphasized. By comparing the interventions identified in the chapters presenting the Outlooks from the Regional Assessments with interventions identified from the review of the scenario literature, we identified several gaps, which should be noted and explicitly considered in future Regional Assessments. The most notable gaps were in the human well-being cluster, and in the inclusion of more social and behavioural interventions in the other clusters (e.g. nutrition management, diet change, energy access).

The review of the global scenario literature showed clearly that some interventions towards sustainable development could achieve synergies across multiple targets, while others may lead to trade-offs with specific targets. **Table 22.1** provides a template for understanding which interventions trade off against each other or provide co-benefits. This systematic consideration of synergies and trade-offs between interventions would ensure an integrated approach that links top-down and bottom-up visioning.

23.12 Enabling conditions for transformations

The literature argues that transformations for sustainability require innovation – both technological and institutional (Olsson *et al.* 2017). Chapter 24 elaborates more fully on the relationship between policy and enabling transformative change towards achieving specific future goals. This chapter concludes with a discussion of what types of conditions are

Box 23.6: Case study: food systems

GEO-6 identifies the food system as a key cross-cutting issue due to its wide-ranging environmental impacts (water, land and GHG emissions) (see Chapters 4, 8 and 17). In the stakeholder engagement and crowdsourcing initiatives throughout the GEO-6 process, 27 out of the 156 workshop-collected seeds related directly to food, and 11 out of the 34 Climate CoLab finalists' proposals did as well. There was a willingness demonstrated by participants to embrace a more sustainable food system, with a large diversity of proposals including dietary change (e.g. eating less meat), reduction of waste in the food distribution system, and alternative production systems. Some workshop seed proposals did not address environmental impacts explicitly, such as those relating to food waste; however, given that an estimated one-third of food produced globally is wasted (see Chapter 8), reducing this would make more effective use of the natural resources consumed by agricultural production.

Several of the workshop seed proposals related to dietary change, specifically advocating increased uptake of – and support for – vegetarian and vegan diets. Such diets are widely understood to demand less land, water and energy than meat-based diets (Pimentel and Pimentel 2003), although regionally appropriate livestock rearing on pasture can be sustainable (Eisler *et al.* 2014). Others related to alternative farming methods (e.g. urban agriculture, rooftop farms, agroforestry) that could potentially have a positive impact on food security while reducing dependence on land and/or water resources. The Climate CoLab proposals contained more detail than the seed initiatives collected during the face-to-face stakeholder workshops. While the dominant focus of these proposals was obviously climate change, about one-third were related to the food system. Proposed solutions ranged from very broad-scope, global interventions such as a sustainability network involving "tens of thousands of food forests" through to more targeted interventions such as improving the moisture-retention capacity of agricultural soils in drought-affected parts of Africa. Notwithstanding the challenge of demonstrating effectiveness, the bottom-up scenarios show a clear willingness to embrace changes in the food system, suggesting a degree of public awareness of the necessary changes identified in the modelled pathways in Chapter 22.

Some of the proposed interventions, both from the seed workshops and from the Climate CoLab platform, could represent game-changers that – subject to further, rigorous examination – have the potential to fundamentally alter the way to develop model-based food-production scenarios in the future. The modelled links between population, meat consumption, average agricultural yields and resultant land use could be substantially reimagined in light of, for instance, widespread reuse of food waste for nutrient recovery (Cordell *et al.* 2011), combined with regenerative, ecological and multifunctional agriculture systems that have the potential to both increase and diversify yields (Horlings and Marsden 2011). In addition, radical models of optimized hypothetical diets have also been presented in the literature (Schramski *et al.* 2011; Ward *et al.* 2014), which could play a role in altering the conventional views in scenarios, of a rigid relationship between humans and land use.

required to enable bottom-up initiatives to scale and achieve potentially transformative change. There are many existing and ongoing initiatives that aim to achieve the SDGs and other global multilateral environmental agreements. Although these initiatives may be the potential building blocks of a more desirable future for people and the planet, higher-level enabling governance conditions will be crucial to their scalability (Moore, Riddell and Vocisano 2015).

A large amount of literature exists regarding sustainability transformations that provides a useful framework to understand the governance conditions needed to transform unsustainable systems and scale the innovations mentioned by workshop seeds and Climate CoLab proposals. Sustainability transformations are often broken down into multiple phases, with temporal periods related to a problematic status quo, a preparation phase in which innovations begin to develop, a navigation/acceleration phase in which innovations grow and become part of the new system, and an institutionalization phase in which a more desirable system is made sustainable in the long term (Olsson *et al.* 2006; Moore *et al.* 2014; Pereira *et al.* 2018a). For transformations to occur successfully, each of these phases requires governance conditions that are strongly enabling. These enabling conditions can best be broken into supporting conditions for the scaling innovations appropriately and disrupting conditions for the weakening of existing, problematic structures.

To connect the theory to the bottom-up results, **Table 23.3** introduces the enabling and disruptive conditions for the transformations identified by the existing literature and provides examples that connect back to the workshop seeds and Climate CoLab proposals.

23.13 Key messages

The analysis of potential bottom-up and regional solutions for achieving a healthy planet, healthy people highlighted the need to do the following:

1. Integrate top-down and bottom-up approaches to developing scenarios.
2. Consider the full range of actors involved in achieving sustainability.
3. Recognize the need for distributive justice when setting expectations about where action should take place.

Table 23.3: Summary of enabling and disruptive conditions for the appropriate scaling up, out and deep of potentially transformative innovations

ENABLING CONDITIONS	
Establishing and supporting markets for innovations *Governance for transformations should involve establishing and supporting new markets for innovations. This consists of policies like regulations, tax exemptions, deployment subsidies and labelling*	Some seeds and proposals mentioned creating and expanding markets such as an ethical fashion industry, and many others looked at innovations related to new and growing markets within the circular and sharing economies. These changes may require market-supporting policies like the labelling of fashion projects that meet certain standards, and subsidies that make niche innovations (e.g. in reusing waste) more affordable for consumers. More generally, policymakers and stakeholders should constantly explore how more sustainable markets related to identified innovations can be supported until they become the norm.
Supporting innovation experimentation and learning *Learning and experimentation support includes support for research and development, deployment and demonstration, policies that stimulate entrepreneurship, incubators, low-interest loans, venture capital and supportive regulatory conditions*	Not many seeds and proposals specifically addressed experimentation and learning support. The most relevant seed was an innovation lab focused on sustainable innovations at the local level. However, given that the seeds and proposals are new innovations predominantly in their prototype or early stages of development, support for innovation experimentation and learning is needed to ensure continued growth. Governance related to all seeds and proposals should strive for continuing improvements to make the solutions viable in the long term.
Financial resource mobilization *Financial resource support is the mobilization of financial capital through funding mechanisms, low-interest loans and venture capital*	A large number of seeds and Climate CoLab proposals identified a need for greater financial mobilization including the mobilization of domestic funds; the Inga Foundation's proposal seeks international funding to help fund its projects; Govardhan Ecovillage proposes a Green Innovations Fund; and "Framework for Community-based Sustainable Development" mentions a need for developed countries to transfer financial resources (and technological expertise) to less developed countries. Related to supporting markets, supporting experimentation and learning, and financial resource mobilization is the emphasis on subsidies and incentives to support new innovations. Workshop pathways, particularly those developed in the Singapore workshop, emphasized the need for subsidies to promote renewable energy development, green urban infrastructure, and sustainable farming. Climate CoLab proposals went into further depth. The proposal "Climate protection by the elderly" called for incentives for the elderly to work, incentives for developing carbon sinks, and education subsidies for children involved in the programme. A proposal submitted by the Govardhan Ecovillage suggested subsidies for organic farmers. Another example, "Business plan for production and marketing of compost from urban solid wastes", suggests incentives and subsidies for individuals, cooperatives, businesses, etc.

Enabling/disruptive condition and description (Adapted from Kivimaa and Kern 2016)	Connection to the workshop seeds and Climate CoLab proposals
Human resource mobilization *Human resource support is the mobilization of human capital through education and labour policies*	Human resource mobilization was a salient theme within the seeds and proposals, particularly the role of educating and engaging people on environmental issues. There was a large number of awareness, knowledge, and skills development solutions, all of which help to mobilize people towards transformations. Seeds-based visions from all four workshops also listed public awareness as a key component of realizing the participants' imagined sustainable futures. Some unique and exciting examples of human resource mobilization include educating the youth to work on climate issues through the 'Youth Climate Leaders' and "Youth Informing Communities on Climate Change Adaptation through building homes" Climate CoLab proposals, and the many app-based solutions that make environmental engagement accessible. More broadly, for significant scaling up of solutions, labour policies will need to promote and reflect the same development priorities as the solutions. Considering that many seeds referred to the development of solar power, there will be a need for labour and training policies, for example, to help promote education and skills development to meet scaling up needs.
DISRUPTIVE CONDITIONS	
Control policies *Control policies are taxes, trade restrictions and regulations that can be instituted by government actors to make existing processes less profitable or more sustainable*	Seeds and proposals related to control policies included introducing limits on plastic, cutting red meat from diets, and bans and taxes on plastic packaging. Control policies like taxes that internalise social and environmental costs and restrictions appeared less often in the bottom-up initiatives than many other enabling conditions as they are related to dealing with existing structures rather than innovating for new solutions. It is important to acknowledge that for all seeds and proposals, transformations usually have winners and losers (Meadowcroft 2011; Geels 2014). As such, for every new innovation there are displacements that can be promoted through control policies (and should be explored), although such policies should consider their wider implications as they can have unintended consequences.
Rules reform *Rules reform consisting of radical policy reforms and changes in overarching rule structures*	A few seeds and proposals suggested entirely new rule structures to promote sustainability, such as embracing the concept of a well-being economy. These included lowering the age of decision makers (e.g. to vote) and policymakers, introducing new financial systems that incorporate the value of the environment, and expanding the circular economy with extended producer responsibility.
Reduction in existing regime support *The removal of supporting conditions that have allowed for the existing, problematic structures to be successful*	Solutions that tackled the conditions that make existing systems successful mostly focused on informing and engaging people on why the existing structures are problematic and how to do things differently. For example, many apps looked at teaching users how their lifestyle was environmentally unfriendly and how to improve, and programmes such as 'No Straw Tuesdays' aimed to challenge the excessive use of straws and plastics more broadly. This can be extended to include the removal of environmentally and socially perverse subsidies.
Changes in networks and key actors *The replacement of incumbent actors and the breaking of powerful actor-network structures in favour of new actors and networks more favourable to the desired transformations*	Several workshop pathways and Climate CoLab finalists referenced changing current actor relations, specifically through building collaborative environments and new, involved networks of stakeholders. Decentralized power and action in large networks was a key component of many seeds. One Climate CoLab proposal, 'C'SQUARE' reflected the trend found in workshop pathways and mentioned the need to empower and mobilize citizens in order to gather their opinions to improve urban areas. Its success was dependent on strong partners and collaborations. The 'Organic Monetary Fund' and "Framework for Community based Sustainable Development" Climate CoLab proposals focused on engaging stakeholders at all levels, including the national governments, international organizations, local communities and relevant experts.

Ongoing efforts to incorporate the impact of bottom-up climate action into existing climate scenarios illustrate how including bottom-up activities can do the following.

- Create a more accurate understanding of existing sustainability pathways and where there are gaps.
- Help national governments to support and account for bottom-up activities in their own agenda setting.
- Identify small-scale initiatives that could provide functions (e.g. capacity-building, piloting of innovative solutions) that may be difficult to quantify but can be critical to achieving the transition to a low-carbon society (Chan, Brandi and Bauer 2016). The concentration of Climate CoLab proposals in the global South suggests that these activities could, for instance, fill a key data gap in current records of sustainability innovations beyond the global North.

23.13.1 Methodological learnings

This lack of bottom-up futures in the context of sustainability poses major challenges. In terms of legitimacy, large-scale global or regional futures that do not represent the diversity of many different lived experiences, world views and discourses risks giving insufficient space for the concerns and needs of different societal actors. It is difficult to imagine transformative change if large-scale sustainability futures do not draw on

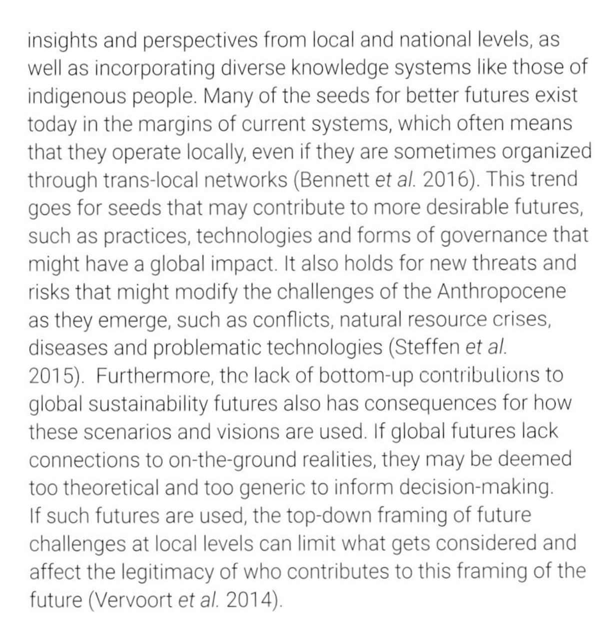

insights and perspectives from local and national levels, as well as incorporating diverse knowledge systems like those of indigenous people. Many of the seeds for better futures exist today in the margins of current systems, which often means that they operate locally, even if they are sometimes organized through trans-local networks (Bennett *et al.* 2016). This trend goes for seeds that may contribute to more desirable futures, such as practices, technologies and forms of governance that might have a global impact. It also holds for new threats and risks that might modify the challenges of the Anthropocene as they emerge, such as conflicts, natural resource crises, diseases and problematic technologies (Steffen *et al.* 2015). Furthermore, the lack of bottom-up contributions to global sustainability futures also has consequences for how these scenarios and visions are used. If global futures lack connections to on-the-ground realities, they may be deemed too theoretical and too generic to inform decision-making. If such futures are used, the top-down framing of future challenges at local levels can limit what gets considered and affect the legitimacy of who contributes to this framing of the future (Vervoort *et al.* 2014).

The top-down scenarios based on integrated assessment models, and the participant-based bottom-up initiatives both have strengths and weaknesses as tools to chart a course towards sustainability. If used well, both approaches have the potential to complement and mutually reinforce one another, as shown in **Figure 23.19**.

The seeds workshops and Climate CoLab proposals represent a small sample, but they show that some solutions are highly synergistic in terms of the SDGs addressed, extremely diverse in scope, and multidimensional in ways that make categorization by any single dimension challenging. The initiatives targeted all SDGs, but were most focused mainly on SDGs 2, 3, 11, 12 and 13. The domains addressed by the initiatives were diverse, and – beyond the expected focus on climate change by Climate CoLab proposals– both the seeds and the Climate CoLab proposals focused in a cross-sectoral manner on the food, energy, water, and waste sectors and their interconnections. Seeds and Climate CoLab proposals envisioned changing systems largely through new technologies, but they also envisioned change occurring through lifestyle shifts, enabled by improving environmental awareness through education, skills development and knowledge generation. Climate CoLab proposals differed slightly and looked at changes in production practices and proposed new organizations and businesses as well as proposing the development of awareness, knowledge and skills. Finally, in examining the Climate CoLab proposals, an overwhelming focus was put on solutions for the global South, particularly for countries in Africa and Asia.

At the same time as quantitative, top-down approaches can be used to inform and strengthen the physical basis for bottom-up initiatives, those bottom-up ideas can in turn challenge overly rigid or outdated assumptions in top-down models. Using bottom-up approaches, it can be possible to identify game-changing concepts that fundamentally restructure the way we view future scenarios. One tangible example is the development of small-scale, decentralized renewable energy systems. The rapid pace of technological development and the associated decrease in the cost of, among others, solar photovoltaics and battery storage, coupled with ICT, makes microgrids a new possibility for areas not yet served by conventional electricity from fossil fuels. This has already become a reality in Kenya since the establishment of M-KOPA, a mobile-enabled payment system for Solar Home Systems in 2013. These technologies – and the public demand to embrace them – mean that the types of energy transition characterizing the past (coal to oil, oil to gas, gas to large-scale renewables) may not necessarily characterize the leapfrog development of energy supplies in the future.

There are many similarities between the macro-level pathways in Chapter 22 and the bottom-up interventions in this chapter. Interventions discussed in both have significant co-benefits for several SDGs. There is a prominent focus on urban sustainability and on food waste and diet change in both analyses **(see Boxes 23.5 and 23.6)**. A crucial complementarity that becomes clear is that the macro-level pathways in the global models allow for an integrative analysis of many contextual drivers and interventions, while the bottom-up pathways provide information about the theories of change underlying the ways of scaling of high-potential practices to achieve the SDGs. The complementary insights provided by the bottom-up and the macro-level pathway analyses demonstrate that further integration of these approaches has much potential. For instance, global modelling results could be used

Figure 23.19: Conceptual framework for mutually beneficial feedbacks between top-down and bottom-up approaches to generating sustainable scenarios

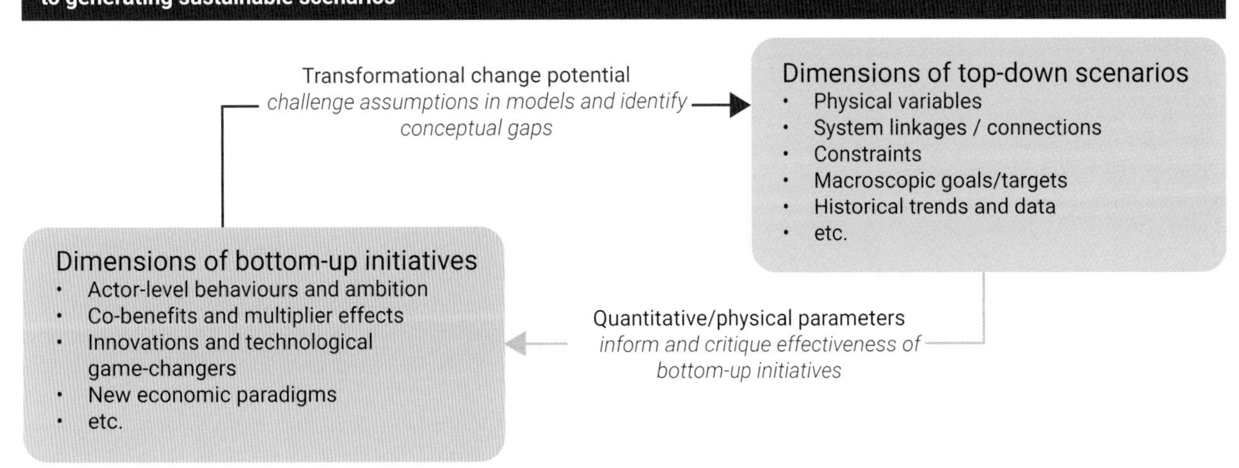

to provide direct global contexts for stakeholders developing bottom-up pathways; and bottom-up pathways can provide directions for future model extensions.

The platforms pioneered in GEO-6 represent an opportunity – if adopted in future assessments – for the top-down scenario-development community to receive feedback on the public acceptance of the various interventions and their trade-offs adopted. To meet the requirement of an increased food supply, for example, pathways include the expansion of agricultural land for rain-fed agriculture (at the expense of biodiversity), or increased use of fertilizer and irrigation to improve yields on the land already in use (at the expense of water resources and pollution). Stakeholders could be consulted to gain insights into the relative acceptance of different options, as well as to identify blind spots in the modelling approach that may mean alternative, synergistic solutions are being overlooked. Similarly, gaps in actual interventions that could help to achieve SDG targets can also be revealed – as is the case with interventions specifically aimed at drivers like population growth that present an important challenge to sustainability, as identified in Chapter 22 and across the chapters of Part A.

Longer-term possibilities for integration could include quantitative aggregation of local scenarios and seed initiatives with direct links to model inputs and outputs; and model integration with online crowdsourcing of bottom-up pathway elements.

23.14 Key interventions and a critical need to recognize distributive justice given global inequities and inequality

The analysis of the Climate CoLab proposals, where an overwhelming focus was put on solutions for the global South, particularly countries in Africa and Asia, highlights existing inequities in the perceptions of where interventions are necessary for transformation, and of who needs to act. While our analysis was of a small subset of studies, if it is indicative of broader perceptions, the burden placed on the global South to transform and implement development initiatives or solutions exacerbates current power inequities in global governance structures (Nagendra 2018; Newell 2005; Parks and Roberts 2008; United Nations Research Institute for Social Development [UNRISD] 2016). This imbalance can obscure or ignore the role of the global North in current development trajectories (e.g. focusing only on poverty alleviation and not discussing wealth redistribution). While the GEO Regional Assessment for Europe did highlight trade-offs and tensions associated with tele-coupling, the limited emphasis on tele-couplings generally is of concern and requires concerted effort (tele-couplings highlight consumption patterns in one region driving environmental concerns related to production in another region) (Liu *et al.* 2013; Seaquist, Johansson and Nicholas 2014). Here, incorporating principles of distributive justice – normative principles designed to guide the allocation of the benefits and burdens of economic activity based on fair distribution (Lamont and Favor 2008) – can help to construct a development agenda based on principles of equity and equality. Such an equality-based and equity-focussed framework can help to account for the disparate developmental conditions of the global South and global North (Rosales 2008; Pelletier 2010; Nagendra *et al.* 2018). This process can provide more equitable options for where and how to implement the solutions with the most transformative potential to achieve sustainable development; for example, in reforming consumption and production patterns or in instituting market mechanisms such as caps in emission-trading schemes, carbon taxes and offsetting schemes. Addressing these global inequities is a means through which to achieve the global goal of equality.

Many of the solutions presented in this chapter do offer the opportunity for developing countries to leapfrog onto more sustainable and equitable development trajectories. The use of ICT plays a major role in driving change in the bottom-up pathways – a result of a stronger focus on theories of change and on *how* change processes are facilitated. There are already many good examples of how this is being leveraged for change in the global South (Karpouzoglou, Pereira and Doshi 2017; Ockwell *et al* 2018). The roles of different societal actors and diverse knowledge systems are made explicit in bottom-up pathways. There is an important role, for instance, for city-level government actors in many proposals. The proposals also include a role for global networks of, for instance, sustainable cities or energy cooperatives. Similarly, diverse higher-level enabling conditions like international agreements, again tied to specific actors, are discussed as part of the bottom-up pathways and their seed initiatives (see Byrne *et al.* 2018 for a discussion on the need for international agreements to enable niches for achieving global energy and climate ambitions).

Chapter 22 identifies trade-offs in the balance between yield improvements and a set of human and environmental goals that include preventing nutrient pollution, limiting climate change, improving child health, providing universal access to clean water and sanitation, and neutralizing land degradation. The present chapter has offered some potential solutions for minimizing such trade-offs and maximizing the synergies. There was a large emphasis on food systems being a critical intervention point for moving towards a healthier planet as well as healthier people. Many seeds and proposals addressed current challenges in the food system by referencing examples that are taking place right now –examples of urban agriculture, aquaculture, diet-change initiatives, and indigenous and local knowledge exchanges (see Annex 23-1).

Chapter 22 also identified a challenge in promoting economic development while reducing emissions. The many initiatives for sharing and circular economies seen in the bottom-up pathways help towards a transformative shift to a well-being economy that no longer presents trade-offs. These pathways offer grounded methods to address global trade-offs.

This analysis has highlighted the specific interventions that governments could facilitate in the shift towards a healthier planet with healthier people, and has highlighted how these interventions differ across different locations. It has also offered some specific examples of where and how change is starting to happen. These are further developed in Chapter 24.

References

Abbott, K.W. (2012). The transnational regime complex for climate change. *Environment and Planning C: Government and Policy* 30(4), 571-590. https://doi.org/10.2139/ssrn.1813198.

Andonova, L.B., Betsill, M.M. and Bulkeley, H. (2009). Transnational climate governance. *Global Environmental Politics* 9(2), 52-73. https://doi.org/10.1162/glep.2009.9.2.52.

Bennett, E.M., Carpenter, S.R., Peterson, G.D., Cumming, G.S., Zurek, M. and Pingali, P. (2003). Why global scenarios need ecology. *Frontiers in Ecology and the Environment* 1(6). https://doi.org/10.1890/1540-9295(2003)001[0322:WGSNE]2.0.CO;2.

Bennett, E.M., Solan, M., Biggs, R., McPhearson, T., Norström, A.V., Olsson, P. *et al.* (2016). Bright spots: Seeds of a good Anthropocene. *Frontiers in Ecology and the Environment* 14(8), 441-448. https://doi.org/10.1002/fee.1309.

Biggs, R., Raudsepp-Hearne, C., Atkinson-Palombo, C., Bohensky, E., Boyd, E., Cundill, G. *et al.* (2007). Linking futures across scales: A dialog on multiscale scenarios. *Ecology and Society* 12(1), 17. https://doi.org/10.5751/ES-02051-120117.

Byrne, R., Mbeva, K. and Ockwell, D. (2018). A political economy of niche-building: Neoliberal-developmental encounters in photovoltaic electrification in Kenya. *Energy Research and Social Science* 44, 6-16. https://doi.org/10.1016/j.erss.2018.03.028.

Cash, D., Adger, W.N., Berkes, F., Garden, P., Lebel, L., Olsson, P. *et al.* (2006). Scale and Cross-Scale Dynamics: Governance and Information in a Multilevel World. *Ecology and Society* 11(2), 8. https://doi.org/10.5751/ES-01759-110208.

Chan, S., Brandi, C. and Bauer, S. (2016). Aligning transnational climate action with international climate governance: The road from Paris. *Review of European Community and International Environmental Law* 25(2), 238-247. https://doi.org/10.1111/reel.12168

ClimateCoop (2018). Climatecoop – The climate consortium blockchain [MIT Center for Collective Intelligence https://www.climatecolab.org/contests/2017/exploring-synergistic-solutions-for-sustainable-development/c/proposal/1334268 (Accessed: 27 December 2018).

Climate CoLab (2018). *Exploring synergistic solutions for sustainable development 2018.* MIT Center for Collective Intelligence https://www.climatecolab.org/contests/2017/exploring-synergistic-solutions-for-sustainable-development (Accessed: 8 October 2018).

Cordell, D., Rosemarin, A., Schröder, J.J. and Smit, A.L. (2011). Towards global phosphorus security: A systems framework for phosphorus recovery and reuse options. *Chemosphere* 84(6), 747-758. https://doi.org/10.1016/j.chemosphere.2011.02.032.

Crutzen, P.J. (2006). The "Anthropocene". In *Earth System Science the Anthropocene*. Ehlers, E. and Krafft, T. (eds.). Berlin: Springer. 13-18. https://link.springer.com/chapter/10.1007/3-540-26590-2_3

Eisler, M.C., Lee, M.R., Tarlton, J.F., Martin, G.B., Beddington, J., Dungait, J.A. *et al.* (2014). Agriculture: Steps to sustainable livestock. *Nature* 507(7490), 32-34. https://doi.org/10.1038/507032a.

Feola, G. (2015). Societal transformation in response to global environmental change: A review of emerging concepts. *Ambio* 44(5), 376-390. https://doi.org/10.1007/s13280-014-0582-z.

Fraser, E.D.G., Dougill, A.J., Mabee, W.E., Reed, M. and McAlpine, P. (2006). Bottom-up and top down: Analysis of participatory processes for sustainability indicator identification as a pathway to community empowerment and sustainable environmental management. *Journal of Environmental Management* 78(2), 114-127. https://doi.org/10.1016/j.jenvman.2005.04.009.

Geels, F.W. (2014). Regime resistance against low-carbon transitions: introducing politics and power into the multi-level perspective. *Theory, Culture and Society* 31(5), 21-40. https://doi.org/10.1177/0263276414531627.

Gellers, J.C. (2015). *Crowdsourcing Sustainable Development Goals from Global Civil Society: A Content Analysis*. https://papers.ssrn.com/sol3/Delivery.cfm/SSRN_ID2562122_code1560115.pdf?abstractid=2562122&mirid=1.

Ghisellini, P., Cialani, C. and Ulgiati, S. (2016). A review on circular economy: the expected transition to a balanced interplay of environmental and economic systems. *Journal of cleaner production* 114, 11-32. https://doi.org/10.1016/j.jclepro.2015.09.007.

Global Climate Action Summit (2018). *Global Climate Action Summit.* https://globalclimateactionsummit.org (Accessed: 12 June 2018).

Hale, T. (2016). "All hands on deck": The Paris agreement and nonstate climate action. *Global Environmental Politics* 16(3), 12-22. https://doi.org/10.1162/GLEP_a_00362.

Hamari, J., Sjöklint, M. and Ukkonen, A. (2016). The sharing economy: Why people participate in collaborative consumption. 67(9), 2047-2059. https://doi.org/10.1002/asi.23552.

Hebinck, A., Vervoort, J.M., Hebinck, P., Rutting, L. and Galli, F. (2018). Imagining transformative futures: Participatory foresight for food systems change. *Ecology and Society* 23(2). https://doi.org/10.5751/ES-10054-230216.

Hebinck, A., Villarreal Herrera, G., Oostindië, H.A., Hebinck, P.G.M., Zwart, T.A., Vervoort, J.M. *et al.* (2016). *Urban Agriculture Policy-Making: Proeftuin040. TRANSMANGO Scenario Workshop Report, The Netherlands*. Wageningen University. https://transmango.files.wordpress.com/2018/04/hebinck-et-al-2016-ua-policy-making-proeftuin040-workshop-report.pdf.

Horlings, L.G. and Marsden, T.K. (2011). Towards the real green revolution? Exploring the conceptual dimensions of a new ecological modernisation of agriculture that could 'feed the world'. *Global Environmental Change* 21(2), 441-452. https://doi.org/10.1016/j.gloenvcha.2011.01.004.

Hsu, A., Cheng, Y., Weinfurter, A., Xu, K. and Yick, C. (2016). Track climate pledges of cities and companies. *Nature* 532(7599), 303-306. https://doi.org/10.1038/532303a.

Hsu, A., Höhne, N., Kuramochi, T., Roelfsema, M., Weinfurter, A., Xie, Y. *et al.* (in press). Defining a research roadmap for quantifying non-state and subnational climate action.

Hsu, A., Schwartz, J.D., Moffat, A.S. and Weinfurter, A.J. (2015). Towards a new climate diplomacy. *Nature Climate Change* 5(6), 501-503. https://doi.org/10.1038/nclimate2594.

Initiative for Climate Action Transparency (ICAT) (2018). *Non-state and Subnational Action Guidance*. https://climateactiontransparency.org/wp-content/uploads/2018/08/ICAT-Non-State-and-Subnational-Action-Guidance-July-2018.pdf.

Intergovernmental Panel on Climate Change (2018). *Global Warming of 1.5 °C: An IPCC Special Report on the Impacts of Global Warming Of 1.5 °C Above Pre-Industrial Levels and Related Global Greenhouse Gas Emission Pathways, in the Context of Strengthening the Global Response to the Threat of Climate Change, Sustainable Development, and Efforts to Eradicate Poverty.* http://www.ipcc.ch/report/sr15/.

Intergovernmental Science-Policy Platform on Biodiversity and Ecosystem Services (2016). *The Methodological Assessment Report on Scenarios and Models of Biodiversity and Ecosystem Services.* Ferrier, S., Ninan, K.N., Leadley, P., Alkemade, R., Acosta, L.A., Akçakaya, H.R. *et al.* (eds.). Bonn: Intergovernmental Science-Policy Platform on Biodiversity and Ecosystem Service. https://www.ipbes.net/sites/default/files/downloads/pdf/2016.methodological_assessment_report_scenarios_models.pdf.

Intergovernmental Science-Policy Platform on Biodiversity and Ecosystem Services (2018a). *The Regional Assessment Report on Biodiversity and Ecosystems Services for Africa.* Bonn. https://www.ipbes.net/system/tdf/ipbes_6_inf_3_rev.1_final.pdf?file=1&type=node&id=16516.

Intergovernmental Science-Policy Platform on Biodiversity and Ecosystem Services (2018b). *The Regional Assessment Report on Biodiversity and Ecosystems Services for the Americas.* Bonn. https://www.ipbes.net/system/tdf/ipbes-6-inf-4-rev.1.pdf?file=1&type=node&id=16517.

Intergovernmental Science-Policy Platform on Biodiversity and Ecosystem Services (2018c). *The Regional Assessment Report on Biodiversity and Ecosystems Services for Asia and the Pacific.* Bonn. https://www.ipbes.net/system/tdf/ipbes_6_inf_5_rev.1_1.pdf?file=1&type=node&id=16518.

Intergovernmental Science-Policy Platform on Biodiversity and Ecosystem Services (2018d). *The Regional Assessment Report on Biodiversity and Ecosystems Services for Europe and Central Asia.* Bonn. https://www.ipbes.net/system/tdf/ipbes_6_inf_6_rev.1.pdf?file=1&type=node&id=16519.

Intergovernmental Science-Policy Platform on Biodiversity and Ecosystem Services (2018e). *Summary for Policymakers of the Regional Assessment Report on Biodiversity and Ecosystem Services for Africa of the Intergovernmental Science-Policy Platform on Biodiversity and Ecosystem Services.* E. Archer, L. E. Dziba, K. J. Mulongoy, M. A. Maoela, M. Walters, R. Biggs *et al.* (eds.). Bonn. https://www.ipbes.net/system/tdf/spm_africa_2018_digital.pdf?file=1&type=node&id=28397.

Jabbour, J. and Flachsland, C. (2017). 40 years of global environmental assessments: A retrospective analysis. *Environmental Science Policy* 77, 193-202. https://doi.org/10.1016/j.envsci.2017.05.001.

Jordan, A.J., Huitema, D., Hildén, M., van Asselt, H., Rayner, T.J., Schoenefeld, J.J. *et al.* (2015). Emergence of polycentric climate governance and its future prospects. *Nature Climate Change* 5(11), 977-982. https://doi.org/10.1038/nclimate2725.

Karpouzoglou, T., Pereira, L.M. and Doshi, S. (2018). Bridging ICTs with governance capabilities for food-energy-water sustainability. *Food, Energy and Water Sustainability*, 222-238. https://doi.org/10.9774/GLEAF.9781315696522_13.

Kivimaa, P. and Kern, F. (2016). Creative destruction or mere niche support? Innovation policy mixes for sustainability transitions. *Research Policy* 45(1), 205-217. https://doi.org/10.1016/j.respol.2015.09.008.

Kok, K., Pedde, S., Jäger, J. and Harrison, P. (2015). *European Shared Socioeconomic Pathways.* Wageningen: Wageningen University and Research. https://climate-adapt.eea.europa.eu/metadata/publications/european-shared-socioeconomic-pathways/download.pdf.

Kok, M.T.J., Kok, K., Peterson, G.D., Hill, R., Agard, J. and Carpenter, S.R. (2016). Biodiversity and ecosystem services require IPBES to take novel approach to scenarios. *Sustainability Science* 12(1), 177-181. https://doi.org/10.1007/s11625-016-0354-8.

Kowarsch, M., Jabbour, J., Flachsland, C., Kok, M.T.J., Watson, R., Haas, P.M. *et al.* (2017). A road map for global environmental assessments. *Nature Climate Change* 7, 379-382. https://doi.org/10.1038/nclimate3307.

Kuramochi, T., Asuka, J., Fekete, H., Tamura, K. and Höhne, N. (2016). Comparative assessment of Japan's long-term carbon budget under different effort-sharing principles. *Climate Policy* 16(8), 1029-1047. https://doi.org/10.1080/14693062.2015.1064344.

Lamont, J. and Favor, C. (2008). Distributive Justice Zalta, E.N. *The Stanford Encyclopedia of Philosophy*. Stanford, CA https://philpapers.org/rec/LAMDJ

Leach, M., Raworth, K. and Rockström, J. (2013). Between social and planetary boundaries: Navigating pathways in the safe and just space for humanity. In *World Social Science Report 2013: Changing Global Environments.* Paris: United Nations Educational, Scientific and Cultural Organization. chapter 6. 84-90. https://www.oecd-ilibrary.org/docserver/9789264203419-en.pdf?expires=1533653701&id=id&accname=ocid54015570&checksum=CC551D1F3C5B6C1E71F36A23BDEBC407

Liu, J., Hull, V., Batistella, M., DeFries, R., Dietz, T., Fu, F. *et al.* (2013). Framing Sustainability in a Telecoupled World. *Ecology and Society* 18(2). https://doi.org/10.5751/ES-05873-180226.

Lundquist, C.J., Pereira, H.M., Alkemade, R., den Belder, E., Carvalho Ribeiro, S., Davies, K. *et al.* 83 (2017). Visions for Nature and Nature's Contributions to People for the 21st Century *NIWA Science and Technology Series*. Intergovernmental Science-Policy Platform on Biodiversity and Ecosystem Services, Auckland, 123 https://www.niwa.co.nz/files/IPBES-Nature-Futures-report_2017_ExecSum.pdf

Malone, T.W., Nickerson, J.V., Laubacher, R.J., Fisher, L.H., de Boer, P., Han, Y. *et al.* (2017). Putting the pieces back together again: Contest webs for large-scale problem solving. *Proceedings of the 2017 ACM Conference on Computer Supported Cooperative Work and Social Computing*. Portland, OR. 2998343 1661-1674. https://doi.org/10.1145/2998181.2998343

Mason-D'Croz, D., Vervoort, J., Palazzo, A., Islam, S., Lord, S., Helfgott, A. *et al.* (2016). Multi-factor, multi-state, multi-model scenarios: Exploring food and climate futures for Southeast Asia. *Environmental Modelling and Software* 83, 255-270. https://doi.org/10.1016/j.envsoft.2016.05.008.

Meadowcroft, J. (2011). Engaging with the politics of sustainability transitions. *Environmental Innovation and Societal Transitions* 1(1), 70-75. https://doi.org/10.1016/j.eist.2011.02.003.

Merrie, A., Keys, P., Metian, M. and Österblom, H. (2018). Radical ocean futures-scenario development using science fiction prototyping. *Futures* 95, 22-32. https://doi.org/10.1016/j.futures.2017.09.005.

Miller, R. (2013). Changing the conditions of change by learning to use the future differently. In *World Social Science Report 2013: Changing Global Environments.* Paris: United Nations Educational, Scientific and Cultural Organization. chapter 10. 107-112. https://www.oecd-ilibrary.org/world-social-science-report-2013_5k43jt1wr17d.pdf?itemId=%2Fcontent%2Fpublication%2F9789264203419-en&mimeType=pdf

Miller, R., Poli, R. and Rossel, P. (2013). *The Discipline of Anticipation: Exploring Key Issues.* UNESCO Foresight. Paris: United Nations Educational, Scientific and Cultural Organization. http://filer.fumee.dk/5/The%20Discipline%20of%20Anticipation%20-%20Miller,%20Poli,%20Rossel.pdf.

Minx, J.C., Lamb, W.F., Callaghan, M.W., Bornmann, L. and Fuss, S. (2017). Fast growing research on negative emissions. *Environmental Research Letters* 12(3), 035007. https://doi.org/10.1088/1748-9326/aa5ee5.

Moore, M.-L., Riddell, D. and Vocisano, D. (2015). Scaling out, scaling up, scaling deep: Strategies of non-profits in advancing systemic social innovation. *The Journal of Corporate Citizenship* (58), 67-84. https://doi.org/10.9774/GLEAF.4700.2015.ju.00009.

Moore, M.-L., Tjornbo, O., Enfors, E., Knapp, C., Hodbod, J., Baggio, J.A. *et al.* (2014). Studying the complexity of change: toward an analytical framework for understanding deliberate social-ecological transformations. *Ecology and Society* 19(4). https://doi.org/10.5751/ES-06966-190454.

Nagendra, H. 557 (2018). The global south is rich in sustainability lessons that students deserve to hear *Nature*. 485-488 https://www.nature.com/articles/d41586-018-05210-0

Nagendra, H., Bai, X., Brondizio, E.S. and Lwasa, S. (2018). The urban south and the predicament of global sustainability. *Nature Sustainability* 1(7), 341-349. https://doi.org/10.1038/s41893-018-0101-5.

Nemoto, E.H. and Biazoti, A.R. (2017). Urban agriculture: How bottom-up initiatives are impacting space and policies in São Paulo. *Future of Food: Journal on Food, Agriculture and Society* 5(3), 21-34. https://kobra.uni-kassel.de/bitstream/handle/123456789/2017110153670/fofjVol5No3S21.pdf?sequence=1&isAllowed=y

Newell, P. (2005). Race, Class and the Global Politics of Environmental Inequality. 5(3), 70-94. https://doi.org/10.1162/1526380054794835.

Ockwell, D., Byrne, R., Hansen, U.E., Haselip, J. and Nygaard, I. (2018). The uptake and diffusion of solar power in Africa: Socio-cultural and political insights on a rapidly emerging socio-technical transition. *Energy Research and Social Science* 44, 122-129. https://doi.org/10.1016/j.erss.2018.04.033.

Olsson, P., Gunderson, L.H., Carpenter, S.R., Ryan, P., Lebel, L., Folke, C. *et al.* (2006). Shooting the Rapids: Navigating Transitions to Adaptive Governance of Social-Ecological Systems. *Ecology and Society* 11(1). https://doi.org/10.5751/ES-01595-110118.

Palazzo, A., Vervoort, J.M., Mason-D'Croz, D., Rutting, L., Havlík, P., Islam, S. *et al.* (2017). Linking regional stakeholder scenarios and shared socioeconomic pathways: Quantified West African food and climate futures in a global context. *Global Environmental Change* 45, 227-242. https://doi.org/10.1016/j.gloenvcha.2016.12.002.

Parks, B.C. and Roberts, J.T. (2008). Inequality and the global climate regime: breaking the north-south impasse. *Cambridge Review of International Affairs* 21(4), 621-648. https://doi.org/10.1080/09557570802452979.

Pascual, U., Balvanera, P., Díaz, S., Pataki, G., Roth, E., Stenseke, M. *et al.* (2017). Valuing nature's contributions to people: The IPBES approach. *Current Opinion in Environmental Sustainability* 26-27, 7-16. https://doi.org/10.1016/j.cosust.2016.12.006.

PATHWAYS (2018). *Participation to healthy workplaces and inclusive strategies in the work sector: Welcome to PATHWAYS project.* Neurological Institute Carlo Besta IRCCS Foundation https://www.path-ways.eu/ (Accessed: 15 May 2018).

Patterson, J., Schulz, K., Vervoort, J., van der Hel, S., Widerberg, O., Adler, C. *et al.* (2017). Exploring the governance and politics of transformations towards sustainability. *Environmental Innovation and Societal Transitions* 24, 1-16. https://doi.org/10.1016/j.eist.2016.09.001.

PBL Netherlands Environmental Assessment Agency (2012). *Roads from Rio+20. Pathways to Achieve Global Sustainability Goals by 2050.* van Vuuren, D. and Kok, M. (eds.). The Hague: PBL Netherlands Environmental Assessment Agency. http://www.pbl.nl/sites/default/files/cms/publicaties/PBL_2012_Roads%20from%20Rio_500062001.pdf.

Pelletier, N. (2010). Environmental sustainability as the first principle of distributive justice: Towards an ecological communitarian normative foundation for ecological economics. *Ecological Economics* 69(10), 1887-1894. https://doi.org/10.1016/j.ecolecon.2010.04.001.

Pereira, L.M., Hichert, T., Hamann, M., Preiser, R. and Biggs, R. (2018a). Using futures methods to create transformative spaces: Visions of a good Anthropocene in southern Africa. *Ecology and Society* 23(1). https://doi.org/10.5751/ES-09907-230119.

Pereira, L.M., Bennett, E.M., Biggs, R.O., Peterson, G.D., McPhearson, T., Norström, A.V. *et al.* (2018b). Seeds of the future in the present: Exploring pathways for navigating towards "Good Anthropocenes". In *Urban Planet: Knowledge towards Sustainable Cities.* Elmqvist, T., Bai, X., Frantzeskaki, N., Griffith, C., Maddox, D., McPhearson, T. *et al.* (eds.). Cambridge: Cambridge University Press. 327-350. http://openaccess.city.ac.uk/19567/1/seeds_of_the_future_in_the_present.pdf

Pimentel, D. and Pimentel, M. (2003). Sustainability of meat-based and plant-based diets and the environment. *The American Journal of Clinical Nutrition* 78(3), 660S-663S. https://doi.org/10.1093/ajcn/78.3.660S.

Reed, M.S. (2008). Stakeholder participation for environmental management: A literature review. *Biological Conservation* 141(10), 2417-2431. https://doi.org/10.1016/j.biocon.2008.07.014.

Reed, M.S., Fraser, E.D.G. and Dougill, A.J. (2006). An adaptive learning process for developing and applying sustainability indicators with local communities. *Ecological Economics* 59(4), 406-418. https://doi.org/10.1016/j.ecolecon.2005.11.008.

Rosa, I.M.D., Pereira, H.M., Ferrier, S., Alkemade, J.R.M., Acosta, L.A., Resit Akcakaya, H. *et al.* (2017). Multiscale scenarios for nature futures. *Nature Ecology and Evolution* 1, 1416–1419. https://doi.org/10.1038/s41559-017-0273-9.

Rosales, J. (2008). The politics of equity: Precedent for post-Kyoto per capita schemes Grover, V.I. *Global Warming and Climate Change: Ten Years After Kyoto And Still Counting.* Science Publishers, Enfield, NH, 87-106 5 https://www.taylorfrancis.com/books/e/9781439843444/chapters/10.1201%2Fb11007-9

Rosenzweig, C., Solecki, W., Romero-Lankao, P., Mehrotra, S., Dhakal, S., Bowman, T. *et al.* (2018). Climate Change and Cities: Second Assessment Report of the Urban Climate Change Research Network. In *Climate Change and Cities: Second Assessment Report of the Urban Climate Change Research Network.* Rosenzweig, C., Romero-Lankao, P., Mehrotra, S., Dhakal, S., Ali Ibrahim, S. and Solecki, W.D. (eds.). Cambridge: Cambridge University Press. xvii-xlii. https://www.cambridge.org/core/books/climate-change-and-cities/climate-change-and-cities-second-assessment-report-of-the-urban-climate-change-research-network/BE242A59BEA99C3DB5E663BAF5FD480F

Schramski, J.R., Rutz, Z.J., Gattie, D.K. and Li, K. (2011). Trophically balanced sustainable agriculture. *Ecological Economics* 72, 88-96. https://doi.org/10.1016/j.ecolecon.2011.08.017.

Seaquist, J.W., Johansson, E.L. and Nicholas, K.A. (2014). Architecture of the global land acquisition system: applying the tools of network science to identify key vulnerabilities. *Environmental Research Letters* 9(11), 114006. https://doi.org/10.1088/1748-9326/9/11/114006.

Seto, K.C., Dhakal, S., Bigio, A., Blanco, H., Delgado, G.C., Dewar D. *et al.* (2014). Human settlements, infrastructure, and spatial planning. In *Climate Change 2014: Mitigation of Climate Change. Contribution of Working Group III to the Fifth Assessment Report of the Intergovernmental Panel on Climate Change.* Edenhofer, O., Pichs-Madruga, R., Sokona, Y., Farahani, E., Kadner, S. and Seyboth, K. (eds.). Cambridge: Cambridge University Press. chapter 12. 923-1000. https://www.ipcc.ch/pdf/assessment-report/ar5/wg3/ipcc_wg3_ar5_chapter12.pdf

Sitas, N. and Harmáčková, Z. (submitted for publication). *Exploring the utility of scenario archetypes in science-policy processes: A cross-regional comparison from the Intergovernmental science-policy platform for biodiversity and ecosystem services.* Intergovernmental Science-Policy Platform on Biodiversity and Ecosystem Services.

Steffen, W., Deutsch, L., Ludwig, C., Broadgate, W. and Gaffney, O. (2015). The trajectory of the Anthropocene: The great acceleration. *Anthropocene Review* 2(1), 81-98. https://doi.org/10.1177/2053019614564785.

TRANSMANGO (2018). *The transmango game jam tour.* https://transmango.wordpress.com/ (Accessed: 22 June 2018).

United Nations Environment Programme (2016). *The Emissions Gap Report 2016: A UNEP Synthesis Report.* Nairobi. http://wedocs.unep.org/bitstream/handle/20.500.11822/10016/emission_gap_report_2016.pdf?isAllowed=y&sequence=1.

United Nations Environment Programme (2017a). First GEO-6 Innovative Scenarios and Policy Pathways Stakeholder Visioning Workshop. *The Second Global Authors Meeting of the Sixth Global Environment Outlook (GEO-6).* Bangkok, Thailand, 22 - 26 May 2017. 21 http://wedocs.unep.org/bitstream/handle/20.500.11822/21463/Outlooks_Meeting_First_GEO-6_Workshop%20Report_v5_final.pdf?sequence=1&isAllowed=y

United Nations Environment Programme (2017b). Second GEO-6 Innovative Scenarios and Policy Pathways Stakeholder Visioning Workshop. *The Third Global Authors Meeting of the Sixth Global Environment Outlook (GEO-6).* Guangzhou, 9 - 14 October 2017.

United Nations Environment Programme (2017c). Third GEO-6 Innovative Scenarios and Policy Pathways Stakeholder Visioning Workshop. *Innovative Outlooks Visioning Workshop.* Nairobi, 1 December 2017. United Nations Environment Programme

United Nations Environment Programme (2018). Fourth GEO-6 Innovative Scenarios and Policy Pathways Stakeholder Visioning Workshop. Singapore, 19-23 February 2018. United Nations Environment Programme https://wedocs.unep.org/bitstream/handle/20.500.11822/25512/Summary_Scenarios%20and%20Policy%20Pathways%20Stakeholder%20Visioning%20Workshop_Final.pdf?sequence=1&isAllowed=y

United Nations Framework Convention on Climate Change (2015). *The Paris Agreement.* United Nations Framework Convention on Climate Change. https://unfccc.int/sites/default/files/english_paris_agreement.pdf

United Nations Research Institute for Social Development (2016). *Policy Innovations for Transformative Change: Implementing the 2030 Agenda for Sustainable Development.* Geneva. http://www.unrisd.org/80256B42004CCC77/(httpInfoFiles)/2D9B6E61A43A7E87C125804F003285F5/$file/Flagship2016_FullReport.pdf.

van Vuuren, D., Edmonds, J., Kainuma, M., Riahi, K., Thomson, A., Hibbard, K. *et al.* (2011). The representative concentration pathways: An overview. *Climatic Change* 109(5). https://doi.org/10.1007/s10584-011-0148-z.

van Vuuren, D., Kok, M., Girod, B., Lucas, P.L. and de Vries, B.J.M. (2012). Scenarios in global environmental assessments: Key characteristics and lessons for future use. *Global Environmental Change* 22(4). https://doi.org/10.1016/j.gloenvcha.2012.06.001.

Vasileiadou, E., Huijben, J.C.C.M. and Raven, R.P.J.M. (2016). Three is a crowd? Exploring the potential of crowdfunding for renewable energy in the Netherlands. *Journal of Cleaner Production* 128, 142-155. https://doi.org/10.1016/j.jclepro.2015.06.028.

Vercoe, R. and Brinkmann, R. (2012). A tale of two sustainabilities: Comparing sustainability in the global north and south to uncover meaning for educators. *Journal of Sustainability Education* http://www.susted.com/wordpress/content/a-tale-of-two-sustainabilities-comparing-sustainability-in-the-global-north-and-south-to-uncover-meaning-for-educators_2012_03/ (Accessed: 06 February 2018).

Vervoort, J.M., Thornton, P.K., Kristjansson, P., Foerch, W., Ericksen, P.J., Kok, K. *et al.* (2014). Challenges to scenario-guided adaptive action on food security under climate change. *Global Environmental Change: Human and Policy Dimensions* 28, 383-394. https://doi.org/10.1016/j.gloenvcha.2014.03.001.

Ward, J.D., Ward, P.J., Mantzioris, E. and Saint, C. (2014). Optimising diet decisions and urban agriculture using linear programming. *Food Security* 6(5), 701-718. https://doi.org/10.1007/s12571-014-0374-0.

Wiggins, A. and Crowston, K. (2011). From conservation to crowdsourcing: A typology of citizen science. *44th Hawaii International Conference on System Sciences.* Kauai, HI, 4-7 January 2011. https://doi.org/10.1109/HICSS.2011.207

World Food Programme (2018). *The R4 rural resilience initiative.* http://www1.wfp.org/r4-rural-resilience-initiative (Accessed: 8 October 2018).

World Overview of Conservation Approaches and Technologies (2018). *Welcome to WOCAT – The world overview of conservation approaches and technologies.* https://www.wocat.net/en/ (Accessed: 8 October 2018).

Wright, C., Yang, C. and Ma, D. (2018). Framework for community-based sustainable development [MIT Center for Collective Intelligence https://www.climatecolab.org/contests/2017/exploring-synergistic-solutions-for-sustainable-development/c/proposal/1334291 (Accessed: 27 December 2018).

Zurek, M.B. and Henrichs, T. (2007). Linking scenarios across geographical scales in international environmental assessments. *Technological Forecasting and Social*

Chapter 24

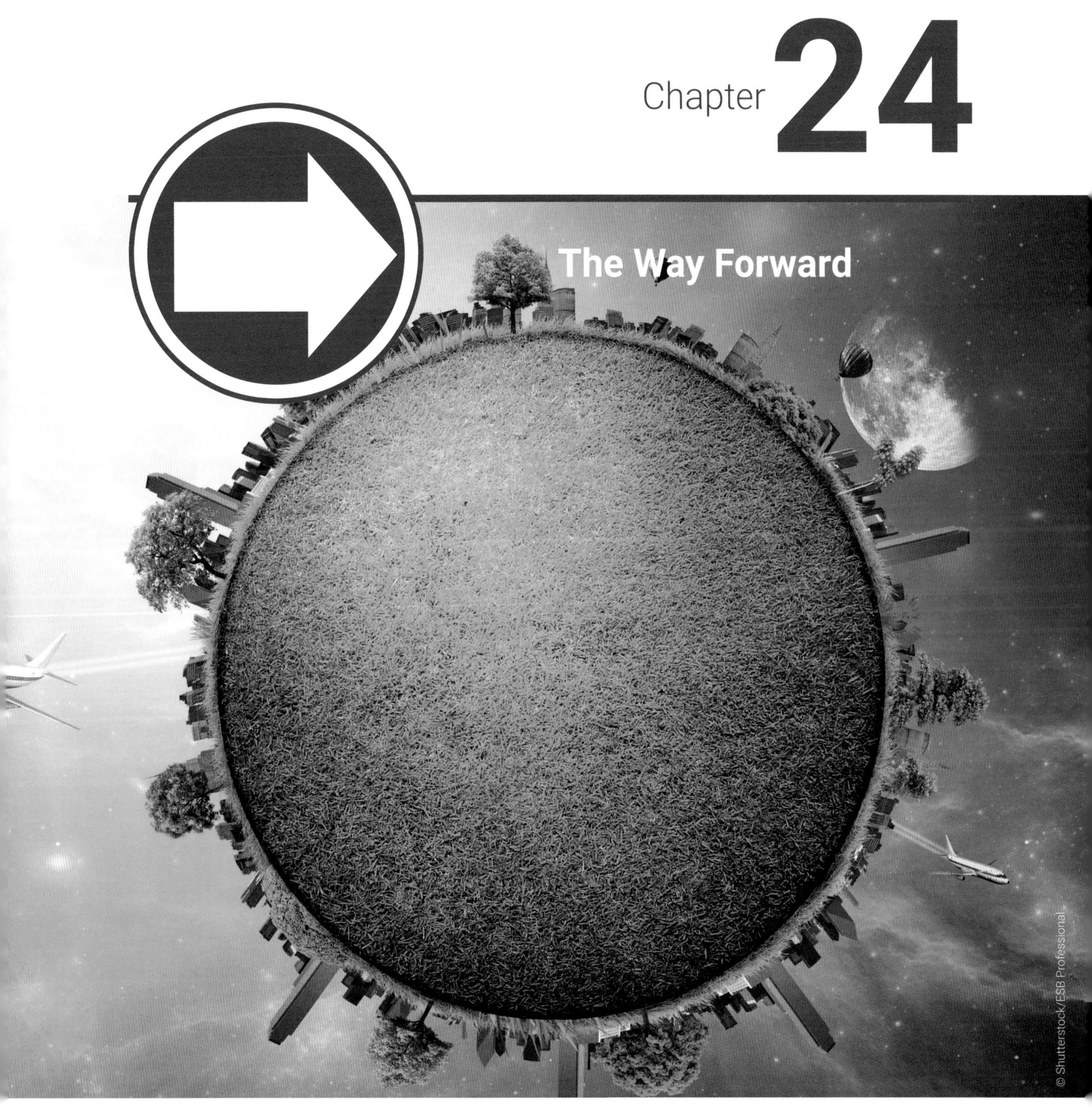

The Way Forward

© Shutterstock/ESB Professional

Coordinating Lead Authors: Mikiko Kainuma (Institute for Global Environmental Strategies [IGES]), Diana Mangalagiu (University of Oxford and Neoma Business School), Ghassem R. Asrar (Pacific Northwest National Laboratory (PNNL), Klaus Jacob (Freie Universität Berlin)

Lead Authors: Laura Pereira (Centre for Complex Systems in Transition, Stellenbosch University), Alexis Rocamora (Institute for Global Environmental Strategies), Detlef van Vuuren (PBL Netherlands Environmental Assessment Agency), Fintan Hurley (Institute of Occupational Medicine), Steve Hedden (Frederick S. Pardee Center for International Futures, University of Denver), Paul Lucas (PBL Netherlands Environmental Assessment Agency), Peter King (Institute for Global Environmental Strategies)

Contributing Authors: Kei Gomi (National Institute for Environmental Studies [NIES]), Robyn Lucas (National Centre for Epidemiology and Population Health, The Australian National University)

GEO Fellow: Mandy Angèl van den Ende (Utrecht University)

Executive summary

Systemic and transformative policies, technologies and social practices, if used together and holistically, have the potential for achieving the Sustainable Development Goals (SDGs) and other Multilateral Environmental Agreements (MEAs) (*established but incomplete*). Transformation is a disruptive process that goes beyond the mere incremental improvement of existing technologies and practices to serve human needs, in an innovative manner. Its origin could be technology, policy or social norms and practices, but to be transformative it should be all encompassing (i.e. holistic). Transformations do not necessarily result from top-down approaches. They emerge from the co-evolution of multiple interdependent factors and the active engagement of diverse stakeholders. {24.2}

Transformative pathways to sustainable development require (1) visions to guide systemic innovation towards sustainability, (2) social and policy innovation, (3) the phasing out of unsustainable practices, (4) policy experimentation and, (5) engaging and enabling actors and stakeholders (*established but incomplete*)**.** Innovative solutions are required to link policies to SDGs, to promote viable business models, to finance the support and management of investment risks, to support international cooperation, and to address the concerns of citizens and stakeholders and ensure their active participation in the entire process. {24.3}

The promotion of systemic innovation is key to socioeconomic development (*established but incomplete*). Many countries are struggling to develop, adopt and diffuse innovative technologies due to the perceived high costs associated with them and, in some cases, technical or regulatory barriers to implementation. For example, in some countries, low-carbon technologies have been adopted by industries only to the extent that they have been successful in market competition. However, the development of policies and governance – including financial mechanisms, policy innovation and the relevant human capacities – at local, subnational and national levels to create an enabling environment, is crucial for wide-scale diffusion. {24.3.1}

Transformative environmental policies have the potential to complement existing ones (*established but incomplete*)**.** The potential of the environmental policies developed and implemented over the past decades is far from realized. Some strategically important environmental policies that address technologies, reduce emissions and improve resource use efficiency lack effective implementation. For example, sectoral policies often lack a consideration of environmental concerns. Transformative policies do have considerable potential to go beyond these measures, but it is less certain that experimental and systemic innovation will succeed in the short term. Accordingly, both approaches, with a focus on more effective implementation of strategically important existing and transformative policies, should be pursued together. {24.1, 24.4}

A healthy planet is the ultimate foundation for supporting all life forms and human well-being, which depend on the viability of Earth's life-support system (*well established*). The Healthy Planet, Healthy People perspective recognizes that human activities have transformed Earth's natural systems and disrupted its self-regulatory mechanisms and life-support system. Economic growth has come at the cost of ecosystem health. The resulting environmental degradation has increased the burden of disease through exposure to harmful pollutants as well as through reduced access to the ecosystem services that we enjoy (e.g. clean air, biodiverse ecosystems, healthy food, clean oceans, land and freshwater). The Healthy Planet, Healthy People approach will be central to global efforts to promote the stewardship of resources from air, biodiversity, land, oceans and freshwater to support human well-being, and the sustainability of the Earth system. For example, the global health savings from reduced air pollution are estimated to be 1.4-2.5 times greater than the costs of mitigating climate change. The proposed strategy to reach the less than 2°C warming target by the end of this century is projected to have the highest benefit-to-cost ratio – where the global health savings (US$54.1 trillion) are estimated to be more than double the global policy costs (US$22.1 trillion). {24.4}

24.1 Approaches for environmental policy: strategic and transformative

The 2030 Agenda for Sustainable Development, together with a range of Multilateral Environmental Agreements (MEAs), set an ambitious long-term vision for the universal pursuit of sustainable development through economic, social, environmental and institutional transformation (Chapter 20). Although progress has been made in managing some environmental problems (e.g. ozone depletion, acid rain), overall global agreements and associated policies have not been able to bend the unsustainable trajectory. Without new policies and effective actions, the ambitious sustainable development vision will not be met (Chapter 21).

Options for bending the prevailing trends do exist. Moving to a sustainable path requires a mix of technological innovations, lifestyle changes and local, regional, global and decentralized solutions with stakeholder engagement, at an unprecedented pace of change (Chapters 22 and 23). The potential from efficiency improvements and emission reductions are far from fully exploited, yet because of rebound and growth effects, it is questionable that they will be sufficient. More disruptive and transformative changes, including new social practices, seem necessary. This chapter discusses promising innovative approaches and transformative, effective policies that will help to attain the goal of a Healthy Planet, Healthy People.

Part A of this report provides the evidence that the current pace of change is inadequate to reverse the environmental harm we are already experiencing. Without a fundamental redirection, most environmental domains will continue to degrade, threatening the economic and social progress achieved to date and the fate of the multiple species that share planet Earth. Part B concludes that, despite a proliferation of policy innovation, often only second-best and small-scale solutions are being observed, rarely going beyond technological fixes. Moreover, potentially effective and ambitious environmental policies are not getting traction. The future projections and potential pathways in Part C suggest that new policies and measurable actions are required at all levels (i.e. local, national, regional and global) to attain the Sustainable Development Goals (SDGs) and targets by 2030, and beyond. The analysis shows that for most environmental goals, the projected conditions appear to worsen, e.g.,

- More and more people will be living in water-stressed areas (Hejazi *et al.* 2014).
- Increasing greenhouse gas emissions will result in a large overshoot of the "well-below-2°C" target of the Paris Agreement on climate change (Iyer *et al.* 2015; United Nations Framework Convention on Climate Change [UNFCCC] 2015).
- The rapid decline in biodiversity will continue (Intergovernmental Science-Policy Platform on Biodiversity and Ecosystem Services [IPBES] 2018).
- Stressed food systems will continue to result in persistent malnourishment, affecting both human well-being and planetary health (Whitmee *et al.* 2015).

Some pathways for change are assessed in various sustainability scenarios in Chapter 22, and through the potential seeds of change in Chapter 23. The sustainability challenge, however, requires new strategies that will stretch humanity's collective imagination, and the current knowledge and action. Incremental steps are insufficient.

Figure 24.1 illustrates the sustainability trajectories for integrated and transformative approaches compared with business as usual. Business as usual, with unambitious environmental policies, lacks effective implementation and holistic integration in other sectoral policies and therefore will not contribute to safeguarding the environment and meeting the sustainable development goals. Stronger environmental policies, including those that provide economic incentives for reducing emissions and improving the efficiency of resource use, do have considerable potential. A transformative approach, based on experimentation and consideration of social practices may be more open-ended and less certain in its direction and chance of success, but it offers greater potential for higher impact and achieving sustainability goals. Both policy approaches could be pursued in parallel to ensure a greater chance of success in both the short and long term.

Figure 24.1: Different policy approaches

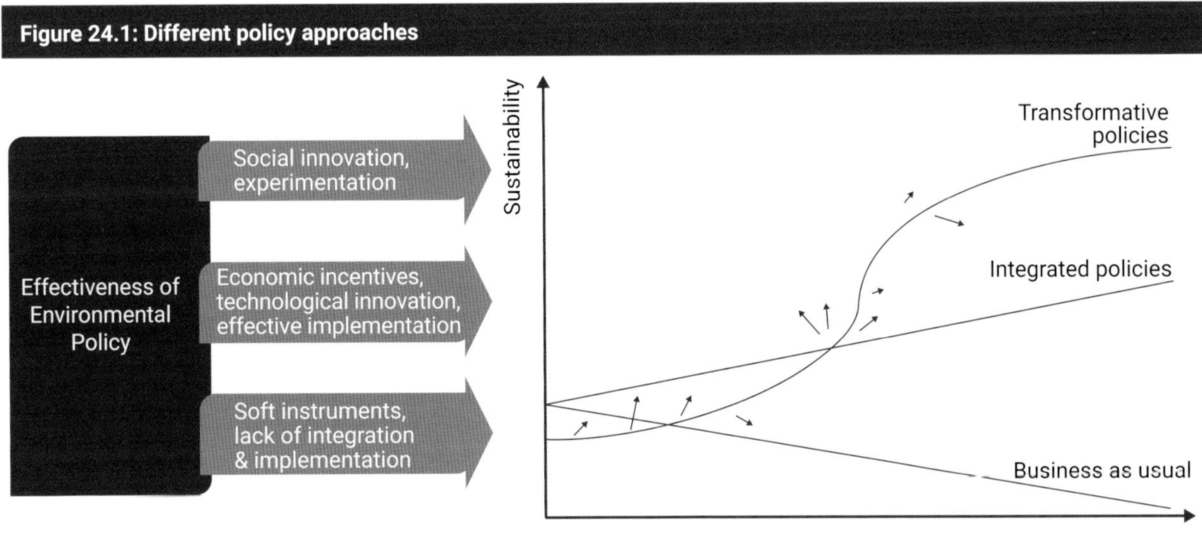

24.2 Transformative change

Our needs for nutrition, health, energy, housing mobility, and so on, are met by a range of social-ecological, socio-technical and socioeconomic systems (Folke *et al.* 2011; Geels and Schot 2017: Díaz *et al.* 2018). Such systems provide their services to society not only by a single technology or service, but are embedded in infrastructure, markets, institutions and social practices, including norms and values (Grießhammer and Brohmann 2015). The different elements of these systems mutually reinforce and stabilize each other, but they are viewed relatively independently of each other, making it difficult for environmental policies to fundamentally change the structure of the systems and organizations involved, let alone their interlinkages and interactions.

Environmental policies have triggered innovation in many sectors through strategies and actions such as ecological modernization, green economy, the valuation of ecosystem services, and the potential for further innovation remains considerable. There is significant potential for improving resource productivity by factors of four to ten (i.e. with one unit of resource, four to ten times more goods are produced) (Schmidt-Bleek 2008; von Weizsäcker *et al.* 2009). Improved resource productivity is necessary but not sufficient because it does not change the underlying systems adequately to achieve the required transformation towards a sustainable society. Therefore, a fundamental reconfiguration of societal systems, including mental models and thought processes, institutions, and norms and values, is necessary (Westley *et al.* 2013; Olsson, Galaz and Boonstra 2014; Bennett *et al.* 2016). Such transformations do not necessarily result from top-down approaches. They emerge from the co-evolution of multiple interdependent factors and the active engagement of diverse stakeholders (Chapter 23). It is important to coordinate actors and resources, guided by a vision of a dramatically different future.

Fostering the ability to transform could enable new development trajectories for social-ecological systems that are more sustainable and have more space for dynamic innovation (Folke *et al.* 2010; Jacob *et al.* 2018). Transformations start from niches defined as small, protected spaces in which new practices can develop, thus causing changes from local to regional and global scales (Loorbach and Raak 2006; Olsson *et al.* 2006; Jänicke and Rennings 2011; Olsson, Galaz and Boonstra 2014). Once feedback mechanisms have reached a critical mass, however, transformative change can be abrupt, and existing technologies and their supporting infrastructure, knowledge, capital and institutions are de-legitimized, and the transformative change is ultimately well integrated into norms and practices (Arthur 2011).

Historical transformation has followed this pattern, starting from innovation in niches, and challenging prevailing practices, with a co-evolutionary and emergent character (Diamond 1997; Arthur 2011; Westley, McGowan and Tjörnbo eds. 2017). In many cases, these transformations were unguided processes that led to increased resource use, emissions and environmental degradation rather than the sustainable use and stewardship of resources and the environment. Hence, there is a need to navigate and guide transformations onto more desirable trajectories (Olsson *et al.* 2006; Jacob *et al.* 2018).

Transformative approaches may differ country by country. Moreover, while current policies have been insufficient to address environmental problems, they need to continue in terms of pollution control, efficiency improvements, planning for the environment and so on. Some countries could achieve transformative changes by leapfrogging to best practices, whereas others may need incremental changes in their policies and practices before reaching transformative stages. Deploying instruments such as economic incentives for innovation and changes in existing economic frameworks, including internalization of external costs, eliminating environmental subsidies, promoting the valuation of ecosystem services, reforming green budget investments, could all play key roles in bringing about transformative changes.

There is no simple recipe for enabling transformative change towards sustainability, but recent methodological innovations emphasize the need for different actors to come together and to experiment with innovations that have the potential for systemic transformation (Frantzeskaki, Wittmayer and Loorbach 2014: Pereira *et al.* 2015). Many of these processes are dubbed lab-based processes. The features of these real-world labs that contribute to transformation include experimental methods, a transdisciplinary mode of research, and the scalability and transferability of results as well as scientific and societal learning and reflexivity (Schapke *et al.* 2018). Examples include social-innovation labs (Westley *et al.* 2012), resilience labs (Frantzeskaki *et al.* 2018), transformation labs (Charli-Joseph *et al.* 2018; Zgambo 2018; van Zwanenberg *et al.* 2018), living labs (Budweg *et al.* 2011; Hooli *et al.* 2016), including urban living labs (Cosgrave *et al.* 2013; Voytenko *et al.* 2016) and transition arenas (Loorbach 2010). Other related processes draw on fields such as foresight – an approach that covers a wide range of methods to systematically investigate the future across systems like the food system (Hebinck *et al.* 2018), urban systems (Potjer, Hajer and Pelzer 2018) or energy systems (Hajer and Pelzer 2018). Some processes refer to new ways of thinking about how change needs to happen, from the individual level through ideas like "inscaping", where individuals surface their inner experiences (Nilsson and Paddock 2014), to how groups can undergo change using concepts like "Theory U" (Scharmer 2007), and drawing more on stories and lived experiences to create real connections with people and their environments in the future (Galafassi *et al.* 2018). These system interventions have been defined as transformative spaces, safe collaborative environments in which experimentation with new configurations of social-ecological systems, crucial for transformation, can occur (see Charli-Joseph *et al.* 2018; Drimie *et al.* 2018; Dye 2018; Galafassi *et al.* 2018; Hebinck *et al.* 2018; Marshall *et al.* 2018; Moore *et al.* 2018; Pereira *et al.* 2018; van Zwanenberg *et al.* 2018). These approaches can be an important step in navigating onto a more sustainable trajectory.

24.3 Building blocks for transformation

Five key approaches to guide, shape and enable transformation can be identified:

i. visions to guide systemic innovation towards sustainability;
ii. social and policy innovation;
iii. the phasing out of unsustainable practices;
iv. policy experimentation; and
v. engaging and enabling actors and stakeholders.

These necessary ingredients are discussed and illustrated through the examples given in the sections that follow.

24.3.1 Visions to guide systemic innovation towards sustainability

An increasing number of governments, cities, companies and communities are expressing compelling visions of a more sustainable future and sharing their strategies and plans for achieving those visions. Many of these visions realize that new ways of measuring progress are also needed (Midgley and Lindhult 2017).

The concept of gross national happiness (GNH) as an alternative to monetary values to measure societal progress was introduced in Bhutan's 1999 strategy for sustainable development (Niestroy, Schmidt and Esche 2013; Jacob, Kannen and Niestroy 2014). Since then it has been evolved as the core vision for Bhutan's governmental and economic activities. Policies and investments are assessed against their contribution to increased GNH instead of their monetary cost and benefits. GNH is key for Bhutan's five-year plans and is included in its Constitution. A GNH commission monitors the implementation. GNH is based on four pillars:

i. equitable socioeconomic development (equity between individuals, communities and regions to provide social harmony and stability);
ii. conservation of the environment;
iii. preservation and promotion of culture (appreciation of the country's cultural heritage and the preservation of spiritual and emotional values); and
iv. promotion of good governance (developing institutions and human resources and providing opportunities for participation).

In response to a regrettable history of deforestation and environmental degradation (Food and Agriculture Organization of the United Nations 2016), Costa Rica has developed a vision of modernity that gives environmental quality a prime place (Silva 2002; Johnson 2016). The 1994 Constitution of Costa Rica provides for "the right to a healthy and ecologically balanced environment" (United Nations, General Assembly 2014). Some recent policy approaches to attaining that vision include payment for ecosystem services, forest preservation for carbon credits, forest credit certificates, legal protection and preservation of iconic species, a ban on open pit mining and, most recently, a pledge to become carbon neutral by 2021. Although challenges remain in relation to water quality and marine protection, significant environmental improvements have stemmed from this overarching vision. For example, forest cover has improved from 26 per cent in the 1980s to 52 per cent in 2010 (United Nations, General Assembly 2014).

An increasing number of cities, communities and regions worldwide aim to reduce their carbon footprint and aspire to become zero-emission or carbon-neutral places (Yamanoshita and Aamano 2012). A clear definition for the scope of emissions (e.g. internal emissions based on the geographic boundary, or external emissions directly caused by municipal activities) addressed by such labels at the city level is under development worldwide (Kennedy and Sgouridis 2011; Straatman *et al.* 2018). Globally, 19 cities have committed to making net-zero-carbon buildings and infrastructure a central piece of their investment strategy by 2030, and to revisit their current planning policies and regulations for existing buildings infrastructure to make them net-zero carbon by 2050 (C40 Cities 2018). Zero-emission city prototypes have been attempted by using renewable energy, cutting-edge technology, innovative urban planning and an emphasis on total reuse (Premalatha *et al.* 2013). Other initiatives focus on helping existing cities to get on a pathway towards net zero emissions (e.g. World Business Council for Sustainable Development 2017) in which municipalities work together with businesses to jointly reduce CO_2 emissions, while focusing on sustainability priorities (Zadek 2004; Moore, Riddell and Vocisano 2015).

ProjectZero (2016) in the Sønderborg region (77,000 inhabitants) in the south of the Kingdom of Denmark has the declared vision of becoming CO_2-neutral by 2029, based on sustainable growth resulting in new green jobs. This vision is being implemented by a public-private partnership involving the municipality and major businesses in the region. A milestone of a 25 per cent reduction in CO_2 emissions in 2015 was exceeded (at 35 per cent) (World Future Council 2016). Technological initiatives are taking place in cities and regions worldwide, such as expanded district heating networks, the conversion of supplies to CO_2-neutral sources and the installation of onshore wind turbines and photovoltaic facilities, coupled with programmes that involve citizens and industries, such as the ZEROhousing and ZEROcompany programmes (Bulkeley and Betsill 2005; Betsill and Bulkeley 2006; Frantzeskaki, Wittmayer and Loorbach 2014; Fujino and Asakawa 2017; City of Melbourne 2018).

Iskandar Regional Development Authority (IRDA), a Malaysian federal government agency overseeing the country's economic and physical development, formulated a vision known as the Low-carbon Society Blueprint 2025. IRDA developed the Green Economy Guideline Manual as a means to implement this vision with the active participation of the business operators in the region, where there is significant domestic and foreign investment (Ho *et al.* 2013; Iskandar Regional Development Authority [IRDA] 2014).

24.3.2 Social and policy innovation

There is no single blueprint for the achievement of these visions, as they are all socially and ecologically embedded in national and local contexts, historical developments, cultural norms and values, and so on. Accordingly, transformation encourages massive social and policy innovation with no guarantees about which forms will ultimately prove successful and worthy of emulation in other domains. One emerging approach that is finding multiple applications is the concept of the sharing economy (e.g. shared accommodation and mobility systems), helping to move societies away from wasteful consumption of both renewable and non-renewable resources (see Section 23.3; Frenken 2017). Sharing accommodation and mobility to reduce environmental impacts is potentially transformative. Private vehicle ownership and solo use, with the high running costs of insurance, parking, maintenance, fuel, and so on, may be reduced by as much as 80 per cent within a decade if sound regulations and incentive schemes are implemented (Arbib and Seba 2017). Trust is no longer based on personal ties but on mechanisms such as peer ratings, business and liability regulations and third-party verification (Lan *et al.* 2017).

Some cities are contemplating making all public transport free. In some cities of Switzerland, for example, hotels provide guests with free passes to use public transport and avoid the traffic and parking congestion. Since 2013, permanent residents of Tallinn, the capital of the Republic of Estonia, have been entitled to use public transport after registration and the purchase of a green card for just two euros, after which all transport is free of charge. The motivations to introduce the scheme were:

i. to promote a modal shift from private cars to public transport;
ii. to improve accessibility for people on low incomes; and
iii. to stimulate the registration of Tallinn residents and so increase the returns from income taxes (Cats, Susilo and Reimal 2017).

As more and more people gravitate to cities, the urban footprint on the hinterland becomes increasingly detrimental to the environment. One promising policy approach to minimizing these impacts, addressing climate change and strengthening community bonds is to create the necessary enabling conditions for increased urban agriculture – green rooftops, vertical farms and community gardens, for example. Of course, for many developing countries, urban agriculture has been a way of life (Orsini *et al.* 2013) with 11 per cent (Indonesia) to almost 70 per cent (Viet Nam and Nicaragua) of urban households earning income from urban agriculture. What has changed has been the increasing sophistication of urban agriculture, such as vertical farming (Association of Vertical Farming 2018) and green rooftops (City of Melbourne 2018), predominantly in more developed countries.

24

Promoting a circular economy is another potential opportunity for reducing CO_2 emissions and other waste and preserving natural resources and ecosystems (see Chapter 17). This concept is captured in the approach to managing the consumption of natural resources and to addressing related environmental and socioeconomic challenges that has been taken by the European Commission Circular Economy Action Plan, published in December 2015 (Wilts 2017; European Commission 2018). If materials are preserved in high-quality products or recycled and used as high-quality secondary raw materials, the circular economy can reduce industries' demands for primary raw materials (Wilts 2017). The concept of circular economy also promotes a decentralized approach to sharing, to providing services and to businesses' dematerializing innovations. For example, a decentralized mode of service provision, which is not necessarily dependent on product and material ownership, is rapidly becoming possible through the development of information and communication technologies and new business models (Kishita *et al.* 2018).

24.3.3 The phasing out of unsustainable practices

A commitment to changing the current, unsustainable socioeconomic and environmental trajectory offers great opportunities in all aspects of daily life, with a high potential to generate the required transformations. The banning of single-use plastics provides one of the most recent examples, where the initial phase-out of lightweight plastic bags has moved into a much broader policy response at all levels, addressing the use of all kinds plastics (Onyanga-Omara 2013; European Commission 2018; United Nations Environment Programme [UNEP] 2018). Developing countries are leading this transformation. In 2002, the People's Republic of Bangladesh became the first country in the world to completely ban thin plastic bags after it realized that around 80 per cent of the waterlogging in cities during floods was being caused by polyethylene bags blocking drains and increasing standing water. This also produced a breeding ground for mosquitoes, increasing the incidence of diseases such as dengue and malaria. Several other countries joined with similar initiatives, including the State of Eritrea in 2005 and the Republic of Kenya in 2017 (Njugunah 2017). Scaling out from tackling plastic bags, the European Commission (2018) made the ground-breaking announcement of banning around ten single-use plastic items (e.g. cutlery, straws, cotton buds, plates, coffee cups and stirrers) that account for 70 per cent of garbage in regional waters and beaches. This example was immediately followed by India, marking a historic breakthrough.

In some circumstances, natural materials may provide alternatives to plastics. For example, the Republics of Indonesia, India, Philippines and Kenya are using water hyacinth, which is among the most effective plants for removing carbon dioxide from the atmosphere, as a source of hard wearing fibre or to produce paper and paper products, with the potential to reduce the demand for conventional plastic products. UN Environment is leading the information-sharing and education process at a global level through its Clean Seas campaign and, most recently, by making the theme for World Environment Day 2018 to beat plastic pollution (Dris *et al.* 2015; Ocean Care 2017). The policies to replace plastics with alternative materials will fail, however, to reduce marine debris if the disposal of the new alternative materials is not considered prior to their introduction. Better collection, recycling and waste management will help to reduce debris on land and in the ocean (Trucost 2016).

24.3.4 Policy experimentation

Transformative policy can often be judged as successful only with the benefit of hindsight and careful monitoring and evaluation. Policy mistakes directly introduced at a national level may have long-lasting implications, such as some of the regrettable policies in the past for controlling population growth (Zhang 2017). Accordingly, the precautionary principle suggests that policy experimentation at smaller scales, combined with national support and continuous evaluation, may be a more sensible choice (Heilman 2008; Husain 2017; Shin 2018).

Policy experimentation at a local scale followed by scaling up is a hallmark of China's policy success (Heilmann 2008). This approach of deliberate experimentalism dates back to early land reforms and addressing agricultural production in the 1940s (Husain 2017). Local-scale policy experiments provide a space for tailoring and innovating policies that are closely monitored; if successful, they are subsequently scaled up, or if unsuccessful, halted. Shin (2018) refers to this approach as experimentation under hierarchy, complemented by performance incentives for local officials.

Experimental governance differs from traditional governance in that it emphasizes learning processes based on public-private partnerships. Experimentation is goal-oriented and seeks to overcome gaps between top-down policies and the challenges at a grass-roots level (Antikainen, Alhola and Jaaskelainen 2017). These types of policy experiments have been practised for climate adaptation in the Kingdom of the Netherlands (McFagen and Huitema 2018). Climate adaptation experiments have also been practised in cities in developing countries, where experiments rely on community-based strategies that involve concerned community members and professionals, and that gain support from external agents. These adaptation experiments need to be in coherence with their urban political economic contexts to ensure transformative change (Broto and Bulkeley 2013; Chu 2016). Policy experimentation works well when the processes are more iterative and more participatory, reflecting both a long-term goal formulation and interactive strategy (Hilden, Jordan and Huitema 2017).

24.3.5 Engaging and enabling actors and stakeholders

Transformation, by definition, will change existing social-economic systems and create winners and losers. Such changes should not be feared, as the continuation of business as usual involves even greater disruption and larger numbers of losers. While it is not possible here to describe all the actors who need to thrive in the context of these emerging visions of sustainable development, many new opportunities will be created and need to be supported. Participatory approaches to engaging decision makers and actors in all phases of transformative change ensure greater acceptance and significantly reduce the time to adoption and produce greater ownership of such changes (Mitchell, Agle and Wood 1997; Umaemiya, Rametsteiner and Kraxner 2010; Smith, Ansett and Erez 2011; Asrar, Ryabinin and Detemmerman 2012; Asrar, Hurrell and Busalacchi 2013; IRDA 2014; Vallentin 2016). Such approaches are widely recognized in a number of international agreements stemming from Principle 10 of the Rio Declaration (United Nations Educational, Scientific and Cultural Organization 1992): "Environmental issues are best handled with the participation of all concerned citizens" (e.g. the Aarhus Convention, Escazu Convention, Talanoa Dialogue).

Innovative finance represents a key breakthrough in the complex pathway to achieving the SDGs. Business as usual does not present any option to close the estimated gap between current and required spending on the SDGs of US$2.5 trillion per year in developing countries (United Nations Conference on Trade and Development 2014). Innovative finance not only aims to establish new financial instruments but also refers to doing business in the future through more inclusive processes (SDG 17) (Porter and Kramer 2006; Ritzén and Sandströma 2017). Instruments that are complementary to grants or financial stimuli may help to unlock the additional capital needed to support sustainable investments. Examples that could form part of a smarter funding mix include loans, equity, quasi-equity and guarantees, and green, blue and social bonds (Venugopal and Srivastava 2012; International Capital Market Association 2018). Global companies are being encouraged to, not only deliver financial performance, but also show how their businesses make a positive contribution to society (Porter and Kramer 2006; Downie 2017). A number of institutional investors, banks and other private-sector financial institutions have joined this appeal, reframing their strategies in asset management and shifting investment capital to companies that incorporate environmental, social and governance considerations into fundamental financial analyses (Noguer and Houillier 2010; Enright, McElrath and Taylor 2016).

Cooperative arrangements between governments and the private sector to create new financial instruments are also beginning to emerge. For example, the Federal Republic of Nigeria, Africa's biggest oil producer, in December 2017 became the first country on the continent to issue a green bond to finance projects and programmes on renewable energy micro-utilities and afforestation. The success of the first issuance of N10.69 billion pushed the government to target an additional N150 billion green bonds in 2018. This bond issuance aims to reduce Nigeria's CO_2 emissions by 40 per cent by 2030. Assessing the progress, evaluating the impact and sharing the lessons learned and experiences gained from such initiatives are key to successful transformative change in policies and practices (Asrar and Hurrell 2013; Premalatha *et al.* 2013).

The transformative potential of engagement and cooperation between businesses, governments and non-governmental organizations (NGOs) is also important to highlight. The Southern Africa Food Lab (SAFL) is a platform established to provide a space for diverse stakeholders from across the food system to engage in dialogue, paying particular attention to the relationship between dialogue and action (Drimie *et al.* 2018). One of these processes involved creating transformative scenarios for the future of the Republic of South Africa's food system, at the same time as the policy on national food and nutrition security was being approved (See Freeth and Drimie 2016). The scenario process brought together a diverse group of interested stakeholders across the food system, including government officials, big business and civil-society activists and legal organizations, who all navigated through their different perspectives to build the meaningful relationships that are fundamental to policy engagement and ultimately to policy change (Freeth and Drimie 2016). SAFL has also become a rallying point for partnerships between NGOs, researchers and small businesses to engage around transformative change in the food system. Many of these partnerships involve the World Wide Fund for Nature (WWF) as a boundary organization for transformative change towards sustainability (Cockburn *et al.* 2018), by increasing consumer awareness of sustainability challenges like overfishing (WWF 2014) and transcending the partisan biases that sometimes hamstring innovative interventions (Drimie and Pereira 2016). The role of NGOs as actors enabling positive change is well documented and needs to be leveraged in order to achieve the sustainable development agenda.

24.4 Healthy Planet, Healthy people: challenge and opportunity

A healthy planet is the ultimate foundation for supporting all life forms, including the health and well-being of humans, which depend on the viability of this life-support system. This principle is captured in the 2030 Agenda for Sustainable Development and related multilateral environmental agreements. Improving human health and well-being, food security and nutrition, social justice and economic prosperity and environmental stewardship through sustainable development is the major theme of GEO-6.

Human activities have already transformed Earth's natural systems and disrupted their self-regulatory mechanisms, with irreversible consequences for the planetary system and human well-being (Millennium Ecosystem Assessment

2005; Rockström *et al.* 2009; Intergovernmental Panel on Climate Change 2014; Steffen *et al.* 2015; Whitmee *et al.* 2015; Ceballosa, Ehrlichb and Dirzob 2017; IPBES 2018; see Part A of this report).

The Healthy Planet, Healthy People approach is key to promoting stewardship of the air, biodiversity, oceans, land and freshwater that are essential for supporting human well-being and the sustainability of Earth systems for current and future generations. Central to this approach is taking a holistic and systemic approach, whereby the identified challenges for all aspects of Earth's life-support system (e.g. clean air, freshwater, food production from oceans and land, habitats for species) are pursued together with the socioeconomic and health dimensions (e.g. gender, equity, poverty) (Commission on Social Determinants of Health [CSDH] 2008; Gordon *et al.* 2017; Dye 2018). The complex interlinkages between the different aspects of environmental change are illustrated by the 12 selected cross-cutting issues described in Chapter 4 and the synergies and trade-offs analysed in Section 22.4.2.

About a quarter of annual deaths globally are caused by modifiable environmental factors (Prüss-Ustün *et al.* 2016). Human health depends on much more than a healthy planet though. Even if it were desirable and feasible to attain a healthy, sustainable planet without addressing socioeconomic issues and the associated determinants of health, it would still leave humanity far short of the goal of healthy people (see also Section 22.2.5, on achieving the SDG target on child mortality). Socioeconomic and cultural factors have significant health impacts, through lifestyle choices, inequalities and damaging practices such as war, violence, unsafe working conditions and child labour (CSDH 2008; see Section 4.1). Therefore, the social determinants of health, including social and wealth inequalities, must also be addressed effectively (Camfield, Møller and Rojas 2015; Donkin *et al.* 2017).

As reported in Section 4.1, human health is mediated by multiple factors in the natural, social and built environments, including our perceptions of equity and safety as well as equitable access to environmental resources and human contact with nature (CSDH 2008). This perspective complements the classical definition of human health as "a state of complete physical, mental and social well-being and not merely the absence of disease or infirmity" (World Health Organization 1948), and the practice of using well-being (Camfield Møller and Rojas 2015; Maggino 2015) together with health to incorporate the psychological, emotional and social dimensions. The multiple relationships between planetary and human systems link health and well-being directly and indirectly to the majority of the SDGs. As such, the SDGs offer the opportunity to approach human health systemically, unlike other major health initiatives that are often focused on a given disease or pandemic event.

Several frameworks have been developed in recent years to help ensure that research and policy development take account of the complex interrelations between health, socioeconomic and environmental factors (Buse *et al.* 2018). However, much of the scientific evidence about the effect of the environment on human health has a narrower focus, on pollution and disease (i.e. mortality and morbidity), with limited attention to the wider concept of well-being or to the social determinants of health. Within this narrower classical framework of environmental health, the commission on pollution and health of the journal *The Lancet* (Landrigan *et al.* 2017) estimated that environmental pollution caused about 9 million premature deaths in 2015; mainly from outdoor and indoor air pollution, which together caused 6.4 million deaths (Cohen *et al.* 2017). Also, environmental pressures and their impacts on health and well-being are not equitably distributed (see Part A). They especially hit groups that are already vulnerable or disadvantaged, such as younger, older and female demographic groups, poor people, those with chronic health conditions, indigenous people and those targeted by racial profiling (Solomon *et al.* 2016; Landrigan *et al.* 2017).

The cost of failing to address the challenges of poor environmental conditions must be examined and communicated widely (Haines 2017; see the example in **Box 24.1**). Such costs are pervasive, through the loss of life and property; disability; the costs incurred from cardiovascular and respiratory diseases; the costs of health damages due to the multiple stresses of extreme weather events, to conflicts over food and water insecurity; gross inequality and poverty; and the tragic plight of refugees around the world.

Box 24.1: The health benefits outweigh the costs of implementing the Paris Agreement

The costs of implementing the Paris Agreement (UNFCCC 2015) between 2020 and 2050 could be outweighed by the health benefits of reductions in air pollution-related diseases and deaths alone, according to one modelling study (Markandya *et al.* 2018). The study modelled emission levels under various scenarios and estimated the costs of the consequent air pollution-related deaths (as a result of respiratory diseases ranging from acute lower respiratory tract infections to chronic obstructive pulmonary disease and heart disease, stroke and lung cancer), and compared this with the costs of climate-change mitigation by country or region (the People's Republic of China, the European Union, the Republic of India, the United States of America and the rest of the world). The scenarios include doing nothing, continuing current country-level policies, and three different strategies for implementing and funding the agreement towards the 2°C and 1.5°C warming limits.

Depending on the scenario used, the health benefits from reduced air pollution were estimated to be, at the global level, 1.4 to 2.5 times greater than the costs of mitigation. The highest benefit-to-cost ratio was for the emission strategy to reach the 2°C target: global health savings were estimated to be US$54.1 trillion, dwarfing the global policy costs of US$22.1 trillion.

Under all the scenarios examined, the countries likely to see the biggest health savings from improved emission-reduction measures were China and India. The cost of implementing climate-mitigation policies in China and India would be fully compensated for by the health savings under most scenarios, and the added costs of pursuing a 1.5°C target instead of 2°C could generate substantial benefits (for India, about US$3.3-8.4 trillion and for China, about US$0.3-2.3 trillion). For the European Union and the United States, the health savings would be large, but not enough to fully compensate the costs.

All of these contribute to inequality and instability and they are all far less expensive to prevent than to react to, in an attempt to manage consequences.

No nation is isolated from the impact of poor environmental conditions. To successfully advance policies, practices and financial investment in global development as well as address environmental challenges, justifications must be framed holistically based on how they can improve the security, prosperity and well-being of citizens and nations globally; policy options and sound solutions should be backed by economic analysis and data to demonstrate the savings and/or the new sources of revenue (Haines 2017; Markandya *et al.* 2018).

In the view of the public health authors of the joint commission on planetary health of the Rockefeller Foundation and *The Lancet*, "solutions [to the environmental crisis] lie within reach and should be based on the redefinition of prosperity to focus on the enhancement of quality of life and delivery of improved health for all, together with respect for the integrity of natural systems" (Whitmee *et al.* 2015).

The changes needed to ensure a Healthy Planet, Healthy People are on such a scale and are so complex and extensive that it would be presumptuous to claim that they could be foreseen in full. Nevertheless, investing in the global environment, development and human health through multilateral agreements and actions, and building the wide coalitions that are necessary for transformative change, are certainly elements of an effective path to holistically addressing these transboundary challenges. The theme of Healthy Planet, Healthy People embodies this integrated approach to the contemporary environment and to addressing the socioeconomic and health challenges faced by current and future generations wanting a sustainable planet for themselves, their children and for all life on Earth.

References

Antikainen, R., Alhola, K. and Jaaskelainen, T. (2017). Experiments as a means towards sustainable societies - Lessons learnt and future outlooks from a Finnish perspective. *Journal of Cleaner Production* 169, 216-224. https://doi.org/10.1016/j.jclepro.2017.06.184.

Arbib, J. and Seba, T. (2017). *Rethinking Transportation 2020-2030: The Disruption of Transportation and the Collapse of the Internal-Combustion Vehicle and Oil Industries; A RethinkX Sector Disruption Report*. RethinkX. https://static1.squarespace.com/static/585c3439be65942f022bbf9b/t/591a2e4be6f2e1c13df930c5/1494888038959/RethinkX+Report_051517.pdf.

Arthur, B. (2011). *The Nature of Technology: What it is and how it Evolves*. http://www.simonandschuster.com/books/The-Nature-of-Technology/W-Brian-Arthur/9781416544067.

Asrar, G.R., Hurrell, J. and Busalacchi, A. (2013). A need for "actionable" climate science and information: Summary of WCRP open science conference outcomes. *World Climate Research Program Open Science Conference*. Denver, CO, 24–28 October 2011. http://journals.ametsoc.org/doi/pdf/10.1175/BAMS-D-12-00011.1.

Asrar, G.R. and J. W. Hurrell. (eds.) (2013). *Climate Science for Serving Society: Research, Modeling and Prediction Priorities*: Springer. https://www.springer.com/gp/book/9789400766914.

Asrar, G.R., Ryabinin, V. and Detemmerman, V. (2012). Climate science and services: Providing climate information for adaptation, sustainable development and risk management. *Current Opinion in Environmental Sustainability* 4(1), 88-100. https://doi.org/10.1016/j.cosust.2012.01.003.

Association of Vertical Farming (2018). *Introducing the Vertical Farming Global Sustainability Registry (SURE) Network*. Association for Vertical Farming https://sure.vertical-farming.net/ (Accessed: 26 January 2018).

Bennett, E.M., Solan, M., Biggs, R., McPhearson, T., Norström, A.V., Olsson, P. *et al.* (2016). Bright spots: Seeds of a good Anthropocene. *Frontiers in Ecology and the Environment* 14(8), 441-448. https://doi.org/10.1002/fee.1309.

Betsill, M.M. and Bulkeley, H. (2006). Cities and the multilevel governance of global climate change. *Global Governance* 12(2), 141-159. http://journals.rienner.com/doi/pdf/10.5555/ggov.2006.12.2.141.

Broto, V.C. and Bulkeley, H. (2013). A survey of urban climate change experiments in 100 cities. *Global Environmental Change* 23(1), 92-102. https://doi.org/10.1016/j.gloenvcha.2012.07.005.

Budweg, S., Schaffers, H., Ruland, R., Kristensen, K. and Prinz, W. (2011). Enhancing collaboration in communities of professionals using a Living Lab approach. *Production Planning and Control* 22(5-6). https://doi.org/10.1080/09537287.2010.536630.

Bulkeley, H. and Betsill, M. (2005). Rethinking sustainable cities: Multilevel governance and the 'urban' politics of climate change. *Environmental Politics* 14(1), 42-63. https://doi.org/10.1080/0964401042000310178.

Buse, C.G., Oestreicher, J.S., Ellis, N.R., Patrick, R., Brisbois, B., Jenkins, A.P. *et al.* (2018). Public health guide to field developments linking ecosystems, environments and health in the Anthropocene. *Journal of Epidemiology and Community Health* 72(5), 420-425. https://doi.org/10.1136/jech-2017-210082.

C40 Cities (2018). *C40 cities*. https://www.c40.org/ (Accessed: 24 February 2018).

Camfield, L., Møller, V. and Rojas, M. (eds.) (2015). *Global handbook of quality of life: Exploration of well-being of nations and continents*: Springer. https://www.springer.com/gp/book/9789401791779.

Cats, O., Susilo, Y.O. and Reimal, T. (2017). The prospects of fare-free public transport: Evidence from Tallinn. *Transportation* 44(5), 1083–1104. https://doi.org/10.1007/s11116-016-9695-5.

Ceballosa, G., Ehrlichb, P.R. and Dirzob, R. (2017). Biological annihilation via the ongoing sixth mass extinction signaled by vertebrate population losses and declines. *Proceedings of the National Academy of Sciences*. https://doi.org/10.1073/pnas.1704949114.

Charli-Joseph, L., Siqueiros-Garcia, J.M., Eakin, H., Manuel-Navarrete, D. and Shelton, R. (2018). Promoting agency for social-ecological transformation: A transformation-lab in the Xochimilco social-ecological system. *Ecology and Society* 23(2), 46. https://doi.org/10.5751/ES-10214-230246.

Chu, E.K. (2016). The governance of climate change adaptation through urban policy experiments. *Environmental Policy and Governance* 26(6), 439-451. https://doi.org/10.1002/eet.1727.

City of Melbourne (2018). *Green Rooftop Project*. http://www.melbourne.vic.gov.au/building-and-development/sustainable-building/Pages/rooftop-project.aspx (Accessed: 22 March 2018).

Cockburn, J., Koopman, V., Pereira, L.M. and van Niekerk, J. (2018). Institutional bricolage to address sustainability challenges in the South African sugarcane industry. In *Food, Energy and Water Sustainability*. Routledge in association with GSE Research. 133-151. https://www.taylorfrancis.com/books/e/9781317446194/chapters/10.4324%2F9781315696522-14.

Cohen, A.J., Brauer, M., Burnett, R., Anderson, H.R., Frostad, J., Estep, K. *et al.* (2017). Estimates and 25-year trends of the global burden of disease attributable to ambient air pollution: An analysis of data from the Global Burden of Diseases Study 2015. *The Lancet* 389(10082), 1907-1918. https://doi.org/10.1016/S0140-6736(17)30505-6.

Commission on Social Determinants of Health (2008). *Closing the Gap in a Generation: Health Equity through Action on the Social Determinants of Health*. Geneva: World Health Organization. http://apps.who.int/iris/bitstream/handle/10665/43943/9789241563703_eng.pdf?sequence=1.

Cosgrave, E., Arbuthnot, K. and Tryfonas, T. (2013). Living labs, innovation districts and information marketplaces: A systems approach for smart cities. *Procedia Computer Science* 16, 668-677. https://doi.org/10.1016/j.procs.2013.01.070.

Diamond, J. (1997). *Guns, Germs, and Steel*. W. W. Norton and Company, Inc. http://books.wwnorton.com/books/978-0-393-35432-4/.

Díaz, S., Pascual, U., Stenseke, M., Martín-López, B., Watson, R.T., Molnár, Z. *et al.* (2018). Assessing nature's contributions to people: Recognizing culture, and diverse sources of knowledge, can improve assessments. *Science* 359(6373), 270-272. https://doi.org/10.1126/science.aap8826.

Donkin, A., Goldblatt, P., Allen, J., Nathanson, V. and Marmot, M. (2017). Global action on the social determinants of health. *BMJ Global Health* 3(1). https://doi.org/10.1136/bmjgh-2017-000603.

Downie, C. (2017). Business actors, political resistance, and strategies for policymakers. *Energy Policy* 108, 583-592. https://doi.org/10.1016/j.enpol.2017.06.018.

Drimie, S., Hamann, R., Manderson, A.P. and Mlondobozi, N. (2018). Creating transformative spaces for dialogue and action: Reflecting on the experience of the Southern Africa Food Lab. *Ecology and Society* 23(3). https://doi.org/10.5751/ES-10177-230302.

Dris, R., Imhof, H., Sanchez, W., Gasperi, J., Galgani, F., Tassin, B. *et al.* (2015). Beyond the ocean: Contamination of freshwater ecosystems with (micro-) plastic particles. *Environmental Chemistry* 12(5), 539-550. https://doi.org/10.1071/EN14172.

Dye, C. (2018). Expanded health system for sustainable development: Advance transformative research for 2030 agenda. *Science* 359(6381), 1337-1339. https://doi.org/10.1126/science.aaq1081.

Enright, S., McElrath, R. and Taylor, A. (2016). *The Future of Stakeholder Engagement: Transformative Engagement for Inclusive Business*. San Francisco, CA: Business for Social Responsibility (BSR). https://www.bsr.org/reports/BSR_Future_of_Stakeholder_Engagement_Report.pdf.

European Commission (2018). *Implementation of the circular economy action plan*. [European Union http://ec.europa.eu/environment/circular-economy/index_en.htm (Accessed: 04 April 2018).

Folke, C., Carpenter, S.R., Walker, B., Scheffer, M., Chapin, T. and Rockström, J. (2010). Resilience thinking: Integrating resilience, adaptability and transformability. *Ecology and Society* 15(4), 20. https://doi.org/10.5751/es-03610-150420.

Folke, C., Jansson, Å., Rockström, J., Olsson, P., Carpenter, S.R., Chapin, F.S. *et al.* (2011). Reconnecting to the biosphere. *Ambio* 40(7), 719–738. https://doi.org/10.1007/s13280-011-0184-y.

Food and Agriculture Organisation of the United Nations (2016). *The Global Forest Resources Assessment 2015: How are the World's Forests Changing?* Rome: Food and Agriculture Organisation. http://www.fao.org/3/a-i4793e.pdf.

Frantzeskaki, N., van Steenbergen, F. and Stedman, R.C. (2018). Sense of place and experimentation in urban sustainability transitions: The resilience lab in Carnisse, Rotterdam, The Netherlands. *Sustainability Science* 13(4), 1045-1059. https://doi.org/10.1007/s11625-018-0562-5.

Frantzeskaki, N., Wittmayer, J. and Loorbach, D. (2014). The role of partnerships in 'realising' urban sustainability in Rotterdam's City Ports Area, The Netherlands. *Journal of Cleaner Production* 65, 406-417. https://doi.org/10.1016/j.jclepro.2013.09.023.

Freeth, R. and Drimie, S. (2016). Participatory scenario planning: from scenario 'stakeholders' to scenario 'owners'. *Environment: Science and Policy for Sustainable Development* 58(4), 32-43. https://doi.org/10.1080/00139157.2016.1186441.

Frenken, K. (2017). Sustainability perspectives on the sharing economy. *Environmental Innovation and Societal Transitions* 23, 1-2. https://doi.org/10.1016/j.eist.2017.04.004.

Fujino, J. and Asakawa, K. (2017). *Taking actions on the SDGs in Japanese Cities: The "Future City" Initiative and its Achievement on the SDGs*. Kamiyaguchi: Institute for Global Environmental Strategies. https://pub.iges.or.jp/pub_file/iges-dp-sdgs-city-en-1pdf/download.

Galafassi, D., Daw, T.M., Thyresson, M., Rosendo, S., Chaigneau, T., Bandeira, S. *et al.* (2018). Stories in social-ecological knowledge cocreation. *Ecology and Society* 23(1). https://doi.org/10.5751/ES-09932-230123.

Geels, F.W. and Schot, J. (2017). Typology of sociotechnical transition pathways. *Research Policy* 36(3), 399-417. https://doi.org/10.1016/j.respol.2007.01.003.

Gordon, L.J., Bignet, V., Crona, B., Henriksson, P.J.G., Van Holt, T., Jonell, M. *et al.* (2017). Rewiring food systems to enhance human health and biosphere stewardship. *Environmental Research Letters* 12(10). https://doi.org/10.1088/1748-9326/aa81dc.

Grießhammer, R. and Brohmann, B. (2015). *Wie Transformationen und Gesellschaftliche Innovationen Gelingen Können: Transformationsstrategien und Models of Change für Nachhaltigen Gesellschaftlichen Wandel*. Umweltbundesamt. https://www.umweltbundesamt.de/sites/default/files/medien/376/publikationen/wie_transformationen_und_gesellschaftliche_innovationen_gelingen_koennen.pdf.

Haines, A. (2017). Health co-benefits of climate action. *The Lancet Planetary Health* 1(1), e4-e5. https://doi.org/10.1016/S2542-5196(17)30003-7.

Hajer, M.A. and Pelzer, P. (2018). 2050 - An energetic odyssey: Understanding 'Techniques of Futuring' in the transition towards renewable energy. *Energy Research and Social Science* 44, 222-231. https://doi.org/10.1016/j.erss.2018.01.013.

Hebinck, A., Vervoort, J.M., Hebinck, P., Rutting, L. and Galli, F. (2018). Imagining transformative futures: Participatory foresight for food systems change. *Ecology and Society* 23(2). https://doi.org/10.5751/ES-10054-230216.

Heilmann, S. (2008). Policy experimentation in China's economic rise. *Studies in Comparative International Development* 43(1), 1-26. https://doi.org/10.1007/s12116-007-9014-4.

Hejazi, M., Edmonds, J., Clarke, L., Kyle, P., Davies, E., Chaturvedi, V. *et al.* (2014). Long-term global water projections using six socioeconomic scenarios in an integrated assessment modeling framework. *Technological Forecasting and Social Change* 81, 205-226. https://doi.org/10.1016/j.techfore.2013.05.006.

Hildén, M., Jordan, A. and Huitema, D. (2017). Special issue on experimentation for climate change solutions editorial: The search for climate change and sustainability solutions - The promise and the pitfalls of experimentation. *Journal of Cleaner Production* 169, 1-7. https://doi.org/10.1016/j.jclepro.2017.09.019.

Ho, C.S., Matsuoka, Y., Chau, L.W., Teh, B.T., Simson, J.J. and Gomi, K. (2013). Blueprint for the development of low carbon society scenarios for Asian regions: Case study of Iskandar Malaysia. *IOP Conference Series: Earth and Environmental Science* 16, 012125. https://doi.org/10.1088/1755-1315/16/1/012125.

Hooli, L.J., Jauhiainen, J.S. and Lähde, K. (2016). Living labs and knowledge creation in developing countries: Living labs as a tool for socio-economic resilience in Tanzania. *African Journal of Science, Technology, Innovation and Development* 8(1), 61-70. https://doi.org/10.1080/20421338.2015.1132534.

Husain, L. (2017). Policy experimentation and innovation as a response to complexity in China's management of health reforms. *Globalization and Health* 13(54). https://doi.org/10.1186/s12992-017-0277-x.

Intergovernmental Panel on Climate Change (2014). *Climate Change 2014: Impacts, Adaptation, and Vulnerability. Part B: Regional Aspects. Contribution of Working Group II to the Fifth Assessment Report of the Intergovernmental Panel on Climate Change*. Barros, V.R., Field, C.B., Dokken, D.J., Mastrandrea, M.D., Mach, K.J., Bilir, T.E. *et al.* (eds.). Cambridge: Cambridge University Press. https://www.cambridge.org/core/books/climate-change-2014-impacts-adaptation-and-vulnerability-part-b-regional-aspects/036A899BD52861D61B0D519C5F2B9334.

Intergovernmental Science-Policy Platform on Biodiversity and Ecosystem Services (2018). *Summary for Policymakers of the Regional Assessment Report on Biodiversity and Ecosystem Services for Africa of the Intergovernmental Science-Policy Platform on Biodiversity and Ecosystem Services*. Archer, E., Dziba, L.E., Mulongoy, K.J., Maoela, M.A., Walters, M., Biggs, R. *et al.* (eds.). Bonn: Intergovernmental Science-Policy Platform on Biodiversity and Ecosystem Services. https://www.ipbes.net/system/tdf/spm_africa_2018_digital.pdf?file=1&type=node&id=28397.

International Capital Market Association (2018). *Green and Social Bonds: A High-Level Mapping to the Sustainable Development Goals*. Zurich: International Capital Market Association. https://www.icmagroup.org/assets/documents/Regulatory/Green-Bonds/Mapping-SDGs-to-Social-and-Sustainability-Bonds-Final-030818.pdf.

Iskandar Regional Development Authority (2014). *Green Economy Guideline Manual*. Iskandar Regional Development Authority (IRDA). http://www.greengrowthknowledge.org/sites/default/files/learning-resources/action/IRDA%20GEG%20Manual%20-%20Tourism.pdf.

Iyer, G.C., Edmonds, J.A., Clarke, L.E., Asrar, G.R., Hultman, N.E., Jeong, M. *et al.* (2015). The contribution of Paris to limit global warming to 2°C. *Environmental Research Letters* 10(12). https://doi.org/10.1088/1748-9326/10/12/125002

Jacob, K., Graaf, L., Wolff, F. and Heyen, D.A. (2018). *Transformative Umweltpolitik: Ansätze zur Förderung Gesellschaftlichen Wandels*. Berlin: Federal Ministry for the Environment. https://www.oeko.de/fileadmin/oekodoc/Impulspapier_Transformative_Umweltpolitik.pdf.

Jacob, K., Kannen, H. and Niestroy, I. (2014). Nachhaltigkeitsstrategien im internationalen Vergleich. In *Nachhaltigkeitsstrategien erfolgreich entwickeln: Strategien für eine nachhaltige Zukunft in Deutschland, Europa und der Welt*. Stiftung, B. (ed.). Gütersloh. https://www.bertelsmann-stiftung.de/fileadmin/files/BSt/Publikationen/GrauePublikationen/Studie_Nachhaltigkeitsstrategien_erfolgreich_entwickeln-de_NW.pdf

Jänicke, M. and Rennings, K. (2011). Ecosystem dynamics: The principle of co-evolution and success stories from climate policy. *International Journal of Technology, Policy and Management* 11(3-4), 198-219. https://doi.org/10.1504/IJTPM.2011.042084.

Johnson, N. (2016). *Costa Rica modernized without wrecking the environment. Here's how*. Grist. https://grist.org/food/costa-rica-modernized-without-wrecking-the-environment-heres-how/ (Accessed: 06 April 2018).

Kennedy, S. and Sgouridis, S. (2011). Rigorous classification and carbon accounting principles for low and Zero Carbon Cities. *Energy Policy* 39(9), 5259-5268. https://doi.org/10.1016/j.enpol.2011.05.038.

Kishita, Y., Kuroyama, S., Matsumoto, M., Kojima, M. and Umeda, Y. (2018). Designing future visions of sustainable consumption and production in Southeast Asia. *Procedia CIRP* 69, 66-71. https://doi.org/10.1016/j.procir.2017.11.150.

Lan, J., Ma, Y., Zhu, D., Mangalagiu, D. and Thornton, T.F. (2017). Enabling value co-creation in the sharing economy: The case of Mobike. *Sustainability Science* 9(9), 1-20. https://doi.org/10.3390/su9091504.

Landrigan, P.J., Fuller, R., Acosta, N.J.R., Adeyi, O., Arnold, R., Basu, N.N. *et al.* (2017). The Lancet Commission on pollution and health. *The Lancet* 391(10119), 1-57. https://doi.org/10.1016/S0140-6736(17)32345-0.

Loorbach, D. and van Raak, R. (2006). *Strategic Niche Management and Transition Management: Different but Complementary Approaches*. Erasmus University. https://repub.eur.nl/pub/37247.

Maggino, F. (2015). Assessing the subjective wellbeing of nations. In *Global Handbook of Quality of Life*. Glatzer, W., Camfield, L., Møller, V. and Rojas, M. (eds.). Dordrecht: Springer. chapter 10. 803-822. https://link.springer.com/chapter/10.1007/978-94-017-9178-6_37#enumeration

Markandya, A., Sampedro, J., Smith, S.J., Van Dingenen, R., Pizarro-Iriza, C., Arto, I. *et al.* (2018). Health co-benefits from air pollution and mitigation costs of the Paris agreement: A modelling study. *The Lancet Planetary Health* 2(3), e126-e133. https://doi.org/10.1016/S2542-5196(18)30029-9.

Marshall, F., Dolley, J. and Priya, R. (2018). Transdisciplinary research as transformative space making for sustainability: Enhancing propoor transformative agency in periurban contexts. *Ecology and Society* 23(3), 8. https://doi.org/10.5751/ES-10249-230308

McFadgen, B. and Huitema, D. (2018). Experimentation at the interface of science and policy: A multi-case analysis of how policy experiments influence political decision-makers. *Policy Science* 51(2), 161-187. https://doi.org/10.1007/s11077-017-9276-2.

Midgley, G. and Lindhult, E. (2017). *What is Systemic Innovation?* Research Memorandum 99. https://www.researchgate.net/profile/Gerald_Midgley/publication/315692364_What_is_Systemic_Innovation/links/58dbe0fda6fdcc7c9f191ff6/What-is-Systemic-Innovation.pdf?origin=publication_detail.

Millennium Ecosystem Assessment (2005). *Ecosystems and Human Well-being: Synthesis*. Washington, D.C: Island Press. https://www.millenniumassessment.org/documents/document.356.aspx.pdf.

Mitchell, R.K., Agle, B.R. and Wood, D.J. (1997). Toward a theory of stakeholder identification and salience: Defining the principle of who and what really counts. *The Academy of Management Review* 22(4), 853–886. https://doi.org/10.5465/amr.1997.9711022105.

Moore, M.L., Olsson, P., Nilsson, W., Rose, L. and Westley, F.R. (2018). Navigating emergence and system reflexivity as key transformative capacities: Experiences from a Global Fellowship program. *Ecology and Society* 23(2), 38. https://doi.org/10.5751/ES-10166-230238.

Moore, M.L., Riddell, D. and Vocisano, D. (2015). Scaling out, scaling up, scaling deep: Strategies of non-profits in advancing systemic social innovation. *Journal of Corporate Citizenship* 58, 67–84. https://doi.org/10.9774/GLEAF.4700.2015.ju.00009.

Niestroy, I., Schmidt, A.G. and Esche, A. (2013). *Bhutan: Paradigm Matters. Case Study for the Reinhard Mohn Prize 2013*. Stiftung, B. (ed.). Gütersloh: Bertelsmann Foundation. https://www.bertelsmann-stiftung.de/fileadmin/files/Projekte/31_Nachhaltigkeitsstrategien/Case-Study-Bhutan_Reinhard-Mohn-Prize-2013_20131016.pdf.

Nilsson, W. and Paddock, T. (2014). Social Innovation from the Inside Out. *Stanford Social Innovation Review*, Winter 2014. https://ssir.org/pdf/Social_Innovation_from_the_Inside_out.pdf.

Njugunah, M. (2017). List of countries that have banned plastic paper bags. *Capital Business*, Capital Group Limited. https://www.capitalfm.co.ke/business/2017/08/list-of-countries-that-have-banned-plastic-paper-bags/.

Noguer, S.N. and Houillier, S. (2010). *Minding your Stakeholders' Business: The Key to Sustainability*. Deloitte. http://globaldialogue.ca/doc/ca_consulting_minding_your_stakeholders_business.pdf.

Ocean Care (2017). Marine debris and the sustainable development goals. *UN Ocean Conference*. New York, NY, 5-9 June. United Nations https://www.oceancare.org/wp-content/uploads/2017/05/Marine_Debris_neutral_2018_web.pdf.

Olsson, P., Galaz, V. and Boonstra, W.J. (2014). Sustainability transformations: A resilience perspective. *Ecology and Society* 19(4). https://doi.org/10.5751/ES-06799-190401.

Olsson, P., Gunderson, L.H., Carpenter, S.R., Ryan, P., Lebel, L., Folke, C. *et al.* (2006). Shooting the rapids: Navigating transitions to adaptive governance of social-ecological systems. *Ecology and Society* 11(1). http://www.ecologyandsociety.org/vol11/iss1/art18/.

Onyanga-Omara, J. (2013). Plastic bag backlash gains momentum. *BBC News*, British Broadcasting Corporation. https://www.bbc.com/news/uk-24090603.

Orsini, F., Kahane, R., Nono-Womdim, R. and Gianquinto, G. (2013). Urban agriculture in the developing world: A review. *Agronomy for Sustainable Development* 33(4), 695–720. https://doi.org/10.1007/s13593-013-0143-z.

Pereira, L., Karpouzoglou, T., Doshi, S. and Frantzeskaki, N. (2015). Organizing a safe space for socio-ecological transformation to sustainability. *International Journal of Environmental Research and Public Health* 12(6), 6027-6044. https://doi.org/10.3390/ijerph120606027.

Pereira, L.M., McElroy, C.A., Littaye, A. and Girard, A.M. (eds.) (2018). *Food, Energy and Water Sustainability: Emergent Governance Strategies*: Routledge. https://www.routledge.com/Food-Energy-and-Water-Sustainability-Emergent-Governance-Strategies/Pereira-McElroy-Littaye-Girard/p/book/9781138904095.

Porter, M. and Kramer, M. (2006). Strategy and society: The link between competitive advantage and corporate social responsibility. *Harvard Business Review* 84(12), 78–92. https://doi.org/10.1108/sd.2007.05623ead.006.

Potjer, S., Hajer, M. and Pelzer, P. (2018). *Learning to Experiment: Realising the Potential of Urban Agenda for the EU*. Utrecht: Urban Futures Studio. https://www.docdroid.net/99DbF6c/research-urbanfuturesstudio-web-def-1.pdf#page=2.

Premalatha, M., Tauseef, S.M., Abbasi, T. and Abbasi, S.A. (2013). The promise and the performance of the world's first two zero carbon eco-cities. *Renewable and Sustainable Energy Reviews* 25, 660-669. https://doi.org/10.1016/j.rser.2013.05.011.

ProjectZero (2016). *35% less CO_2 in Sonderborg*. http://brightgreenbusiness.com/en-GB/News/Archive/2016/35-less-CO2-in-Sonderborg.aspx (Accessed: 10 January 2018).

Prüss-Ustün, A., Wolf, J., Corvalán, C., Bos, R. and Neira, M. (2016). *Preventing Disease through Healthy Environments: A Global Assessment of the Burden of Disease from Environmental Risks*. Geneva: World Health Organization. http://apps.who.int/iris/bitstream/10665/204585/1/9789241565196_eng.pdf?ua=1.

Ritzén, S. and Sandströma, G.Ö. (2017). Barriers to the circular economy: Integration of perspectives and domains. *Procedia CIRP* 64, 7-12. https://doi.org/10.1016/j.procir.2017.03.005.

Rockström, J., Steffen, W., Noone, K., Persson, Å., Stuart Chapin III, F., Lambin, E.F. *et al.* (2009). A safe operating space for humanity. *Nature* 461, 472–475. https://doi.org/10.1038/461472a.

Schäpke, N., Wagner, F., Parodi, O. and Meyer-Soylu, S. (2018). Strengthening the transformative impulse while mainstreaming real-world labs: Lessons learned from three years of BaWü-Labs. *GAIA - Ecological Perspectives for Science and Society* 26(4), 262-264. https://doi.org/10.14512/gaia.27.2.19.

Scharmer, O. (2007). *Theory U: Leading from the Future as it Emerges*. Berrett-Koehler Publishers. http://www.ottoscharmer.com/publications/executive-summaries.

Schmidt-Bleek, F. (2008). Factor 10: The future of stuff. *Sustainability: Science, Practice and Policy* 4(1), 1-4. https://doi.org/10.1080/15487733.2008.11908009.

Shin, K. (2018). Environmental policy innovations in China: A critical analysis from a low-carbon city. *Environmental Politics* 27(5), 830-851. https://doi.org/10.1080/09644016.2018.1449573.

Silva E. (2002). National environmental policies: Costa Rica. In *Capacity Building in National Environmental Policy*. Weidner H. and Jänicke M. (eds.). Berlin: Springer. 147-175. https://link.springer.com/chapter/10.1007/978-3-662-04794-1_7.

Smith, N.C., Ansett, S. and Erez, L. (2011). *What's at Stake? Stakeholder Engagement Strategy as the Key to Sustainable Growth*. Fontainebleau: INSEAD. https://sites.insead.edu/facultyresearch/research/doc.cfm?did=47212.

Solomon, G.M., Morello-Frosch, R., Zeise, L. and Faust, J.B. (2016). Cumulative environmental impacts: Science and policy to protect communities. *Annual Review of Public Health* 37, 83-96. https://doi.org/10.1146/annurev-publhealth-032315-021807.

Steffen, W., Richardson, K., Rockström, J., Cornell, S.E., Fetzer, I., Bennett, E.M. *et al.* (2015). Planetary boundaries: Guiding human development on a changing planet. *Science* 347(6223), 1259855. https://doi.org/10.1126/science.1259855.

Straatman, B., Boyd, B., Mangalagiu, D., Rathje, P., Eriksen, C., Madsen, B. *et al.* (2018). A consumption-based, regional input-output analysis of greenhouse gas emissions and the carbon regional index. *International Journal of Environmental Technology and Management* 21(1-2). https://doi.org/10.1504/IJETM.2018.10013804.

Trucost (2016). *Plastics and Sustainability: A Valuation of Environmental Benefits, Costs and Opportunities for Continuous Improvement*. Trucost. https://plastics.americanchemistry.com/Plastics-and-Sustainability.pdf.

Umemiya, C., Rametsteiner E. and Kraxner, F. (2010). Quantifying the impacts of the quality of governance on deforestation. *Environmental Science and Policy* 13(8), 695-701. https://doi.org/10.1016/j.envsci.2010.07.002.

United Nations, General Assembly (2014). *Report of the Independent Expert on the issue of human rights obligations relating to the enjoyment of a safe, clean, healthy and sustainable environment, John H. Knox - Addendum: Mission to Costa Rica*. Human Rights Council: Promotion and protection of all human rights, civil, political, economic, social and cultural rights, including the right to development, Twenty-fifth session. A/HRC/25/53/Add.1. https://www.ecoi.net/en/file/local/1247651/1930_1399473512_a-hrc-25-53-add-1-en.doc.

United Nations Conference on Trade and Development (2014). *World Investment Report 2014: Investing in the SDGs; An Action Plan*. Geneva: United Nations Conference on Trade and Development https://unctad.org/en/PublicationsLibrary/wir2014_en.pdf.

United Nations Environment Programme (2018). *Exploring the Potential for Adopting Alternative Materials to Reduce Marine Plastic Litter*. Nairobi: United Nations Environment Programme. http://wedocs.unep.org/bitstream/handle/20.500.11822/25485/plastic_alternative.pdf?sequence=1&isAllowed=y.

United Nations Framework Convention on Climate Change (2015). *The Paris Agreement*. United Nations Framework Convention on Climate Change. https://unfccc.int/process-and-meetings/the-paris-agreement/the-paris-agreement.

Vallentin, D. (2016). *North Rhine-Westphalia's Industry in Transition – Great Achievements, Great Challenges Ahead*. Energy Transition Platform. https://www.stiftung-mercator.de/media/downloads/3_Publikationen/The_Climate_Group_Briefing_Energy_Transiton_Platform_November_2016.pdf

van Zwanenberg, P., Cremaschi, A., Obaya, M., Marin, A. and Lowenstein, V. (2018). Seeking unconventional alliances and bridging innovations in spaces for transformative change: The seed sector and agricultural sustainability in Argentina. *Ecology and Society* 23(3), 11. https://doi.org/10.5751/ES-10033-230311.

Venugopal, S. and Srivastava, A. (2012). *Glossary of Financing Instruments*. Washington, D.C: World Resources Institute. http://pdf.wri.org/glossary_of_financing_instruments.pdf.

von Weizsacker, E.U., Hargroves, C., Smith, M.H., Desha, C. and Stasinopoulos, P. (2009). *Factor Five: Transforming the Global Economy through 80% Improvements in Resource Productivity*. 1st edn. London: Routledge https://www.taylorfrancis.com/books/9781136545801.

Voytenko, Y., McCormick, K., Evans, J. and Schliwa, G. (2016). Urban living labs for sustainability and low carbon cities in Europe: Towards a research agenda. *Journal of Cleaner Production* 123, 45-54. https://doi.org/10.1016/j.jclepro.2015.08.053.

Westley, F., McGowan, K. and Tjörnbo, O. (eds.) (2017). *The Evolution of Social Innovation: Building Resilience through Transitions*. Cheltenham: Edward Elgar. https://www.e-elgar.com/shop/eep/preview/book/isbn/9781786431158/.

Westley, F.R., Laban, S., Rose, C., McGowan, K., Robinson, K., Tjornbo, O. *et al.* (2012). *Social Innovation Lab Guide*. Waterloo: Waterloo Institute for Social Innovation and Resilience. https://uwaterloo.ca/waterloo-institute-for-social-innovation-and-resilience/sites/ca.waterloo-institute-for-social-innovation-and-resilience/files/uploads/files/10_silabguide_final.pdf.

Westley, F.R., Tjornbo, O., Schultz, L., Olsson, P., Folke, C., Crona, B. *et al.* (2013). A theory of transformative agency in linked social-ecological systems. *Ecology and Society* 18(3). https://doi.org/10.5751/ES-05072-180327.

Whitmee, S., Haines, A., Beyrer, C., Boltz, F., Capon, A.G., de Souza Dias, B.F. *et al.* (2015). Safeguarding human health in the Anthropocene epoch: Report of The Rockefeller Foundation–Lancet Commission on planetary health. *The Lancet* 386(10007), 1973-2028. https://doi.org/10.1016/S0140-6736(15)60901-1.

Wilts, H. (2017). Key challenges for transformations towards a circular economy: The status quo in Germany. *International Journal of Waste Resources* 7(1). https://doi.org/10.4172/2252-5211.1000262.

World Business Council for Sustainable Development (2017). *Zero emissions cities*. [World Business Council for Sustainable Development https://www.wbcsd.org/Programs/Cities-and-Mobility/Zero-Emissions-Cities (Accessed: 12 November 2017).

World Future Council (2016). *Mobilizing Actors for the Local Energy Transition*. Hamburg: World Future Council. https://www.worldfuturecouncil.org/wp-content/uploads/2016/05/PZ-2016.09.26-ProjectZero-World-Future-Council-visit-to-Sonderborg-english.pdf.

World Health Organization (1948). *Constitution of the World Health Organisation*. World Health Organization. http://apps.who.int/gb/bd/PDF/bd47/EN/constitution-en.pdf?ua=1.

World Wide Fund for Nature (2014). *Ten Years of Being SASSI: A Documentation of the Sustainable Seafood Movement in South Africa*. Cape Town. http://www.ee.uct.ac.za/sites/default/files/image_tool/images/258/Papers/WWF-SA%2010%20years%20of%20SASSI.pdf.

Yamanoshita, M. and Aamano, M. (2012). Capability development of local communities for project sustainability in afforestation/reforestation clean development mechanism. *Mitigation and Adaptation Strategies for Global Change* 17(4), 425-440. https://doi.org/10.1007/s11027-011-9334-6.

Zadek, S. (2004). The path to corporate responsibility. *Harvard Business Review* 82(12), 125–132. https://doi.org/10.1007/978-3-540-70818-6_13.

Zgambo, O. (2018). *Exploring food system transformation in the greater Cape Town area*. Master of Philosophy in Sustainable Development, Stellenbosch University http://scholar.sun.ac.za/bitstream/handle/10019.1/103445/zgambo_exploring_2018.pdf?sequence=1&isAllowed=y.

Zhang, J. (2017). The evolution of China's one-child policy and its effects on family outcomes. *Journal of Economic Perspectives* 31(1), 141-160. https://doi.org/10.1257/jep.31.1.141.

PART D

Remaining Data and Knowledge Gaps

Chapter 25

Future Data and Knowledge Needs

© Shutterstock/kan_chana

Coordinating Lead Authors: Florence Daguitan (Tebtebba, Indigenous Peoples' International Centre for Policy Research and Education), Charles Mwangi (Global Learning and Observations to Benefit the Environment [GLOBE] Program), Michelle Tan (ADEC Innovations)

Lead Authors: Graeme Clark (University of New South Wales), Daniel Cooper (University of Oxford), James Donovan (ADEC Innovations), Sheryl Gutierrez (ADEC Innovations), Nina Kruglikova (University of Oxford), Pali Lehohla (Pan African Institute for Evidence), Joni Seager (Bentley University), William Sonntag (Group on Earth Observations)

GEO Fellow: Amit Patel (Planned Systems International, Inc.)

Executive summary

Citizen science is providing unprecedented opportunities for engaging the public in collecting and analysing vast amounts of environmental data (*well established*). The potential for massively dispersed teams of observers, coupled with new technologies such as smart sensors, mobile telephony, Internet and computing capabilities, is offering new approaches for research and engaging the public on environmental issues. As well as collecting large volumes of data, the advancement of new technologies has also enhanced the quality and veracity of the data collected. Key opportunities presented by citizen science include greater frequency of data from dispersed sources, the ability to address large knowledge and funding deficits, the ability to educate the public about environmental policy issues, and the use of local knowledge. {25.2.1}

Big data is one of the world's emerging valuable resources, shifting the landscape of environmental assessment at global, national and local scales (*well established*). Traditional processing techniques cannot handle the volume, velocity, variety and veracity of big data, demanding new algorithms, programming and statistical methods to derive information and draw evidence-based conclusions. There is enormous potential for advancing environmental knowledge if big data can be effectively harnessed and interrogated. {25.2.2}

Governments, organizations, academia and the private sector have initiatives seeking opportunities to tap the potential of big data for sustainability and development (*well established*). Current initiatives include the establishment of the United Nations Pulse Labs for pilot studies on big data, the formation of the United Nations Global Working Group on Big Data in monitoring the Sustainable Development Goals (SDGs), and the availability of repository sites and open data sources from multilateral organizations, research centres and government collaborations. Big data from web-based and geospatial mapping technologies, remote sensing and statistical visualization provide a basis for environmental assessment. {25.2.2}

Challenges for using big data in environmental assessments include its accessibility, quality, varying scale and context, and incomplete time series (*well established*). Despite efforts to generate globally acceptable and available big data, capacities are limited by resources and funding, especially in developing countries. Much real-time big data are controlled and held by the private sector, though many data products are made freely available for public good in a process known as data philanthropy. Recommendations for building a holistic system for big data include the establishment of leadership and data governance; collaborations among governments, institutions and the private sector; and institutionalizing legal frameworks with safeguards on information. {25.2.2}

Strengthening the ability to gather, interpret and use data for effective planning, policymaking, management and evaluation could provide countries with a comprehensive view of environmental impacts (*well established*). Governments and society need to adapt to the evolving data landscape, including the possible use of artificial intelligence to manage environmental concerns. Coping with the shift in the data landscape entails new information-technology skills and a holistic approach in utilizing emerging and existing data and knowledge tools. {25.3}

Traditional knowledge held by indigenous peoples and local communities is increasingly seen as a valuable resource for environmental assessment and sustainable development (*well established*). This revaluation is evidenced by the increase in discussions and studies on traditional knowledge, and its inclusion in global policy agreements. In order to address current and future challenges such as climate change, research suggests that the best approaches may be characterized by the coordination of modern science and technology with traditional knowledge. While cooperation between local and global communities and knowledge systems has proven to be successful for the health of individuals and the planet, certain challenges remain. {25.2.3}

25.1 Introduction

This chapter discusses emerging areas of environmental information and statistics, including citizen science, big data and traditional knowledge. It aims to summarize the gaps and opportunities for improving the environmental knowledge base.

The global landscape is changing, technology is advancing and more and more data are available. These new data sources will not override the need for traditional means of data collection but will provide additional opportunities for environmental monitoring and assessment. This chapter analyses these new and emerging means of data collection and presents a perspective for the future of environmental monitoring and assessment.

25.2 Emerging tools for environmental assessment

Citizen science, big data and traditional knowledge are not new sources of information; what is new is their regular and systematic use in environmental assessments. This section highlights some current experiences and the need to use these innovative sources of information to fill data gaps.

25.2.1 Citizen Science

Citizen science entails the engagement of volunteers in science and research. Volunteers are commonly involved in data collection, but can also be involved in initiating questions, designing projects, disseminating results and interpreting data (Blaney *et al.* 2016). Coupling Citizen Science with new emergent technologies is providing unprecedented opportunities for doing research and sensitization of the public on environmental issues (Newman *et al.* 2012, p. 298).

The possibility of tapping into a massive, dispersed team of observers in different regions of the world has created opportunities for collating and analyzing data at unprecedented spatial and temporal scales. Citizen Science projects have the potential to gather large amounts of scientific data but this is only helpful if data collected is utilized in one way or the other (Dickinson, Zuckerberg and Bonter 2010; Kim *et al.* 2011; Dickinson *et al.* 2012).

Citizen science has numerous benefits, the main one being the opportunity to collect data over wider spatial coverage and longer periods at lower cost. Additional benefits include the creation of jobs, increased scientific literacy, citizen engagement in local and environmental issues, cost effectiveness for governments and benefits to the environment being monitored. Citizen Science also allows the expertise of scientists to be brought to the public while at the same time exposing the scientists to the indigenous knowledge and expertise available within the local community (Conrad and Hilchey 2011; Blaney *et al.* 2016). Some of the key benefits of citizen science are highlighted in **Figure 21.1: Selected targets and their related clusters as examined in this chapter**.

The fields of astronomy and ornithology have led the charge for citizen science. In 1900, Frank Chapman, an ornithologist with the American Museum of Natural History initiated the Christmas Bird Count (CBC). This project has survived thanks to the enthusiasm of citizen scientists over the years and is currently being run by the National Audubon Society (Dickinson, Zuckerberg and Bonter 2010). Since then, there have been many citizen science projects over the years at local, regional and global scales, covering different areas of interest.

More recently, citizen science projects have included a wide variety of initiatives, ranging from building collaborative knowledge (e.g. Wikipedia, OpenStreetMap), volunteer computing (e.g. CitizenGrid, climateprediction.net), and pattern classification (e.g. Galaxy Zoo, eyewire), to the community collection of observations (e.g. bird counting, air sensor toolbox) (Mathieu *et al.* 2016).

Many environmental interests that transcend government boundaries, such as pollution and bird migration, have increased the engagement of citizen scientists to monitor these issues of concern. More innovative projects include the use of Google's reCAPTCHA, which has facilitated the digitization of books and millions of articles by turning words that cannot be read by computers into CAPTCHAs for people to solve (Conrad and Hilchey 2011; Google 2018).

There are two main approaches used in the organization of citizen science projects; top-down or bottom-up. These approaches are similar to the concepts in Chapter 10 on evaluation of policy effectiveness.

Figure 25.1: Some of the benefits of citizen science

Individual Citizen

- Learn observational and analytical skills
- Gain a better understanding of the natural world
- Job opportunities
- Capacity building

Governments

- Lower cost of data collection
- Wider spatial and temporal coverage of data
- Promote environmental stewardship

Communities

- Monitor the health of the environment
- Increased interaction of the community
- Promote environmental stewardship

Scientists and Researchers

- Large numbers of participants reduce workload
- Scientists are able to build connection with community
- Teach people how to research

The top-down approach is mostly driven by scientists who train volunteers on the procedures and the research to be undertaken. Based on this approach, the volunteers play limited roles mostly in data collection. The bottom-up approach is driven by the community. More often than not, this is driven by the need of the community to understand or gather evidence of a concern. The community can then approach scientists for support and guidance during the process (Roelfsema *et al.* 2016; Shirk *et al.* 2012).

The level of engagement, skills and knowledge needed by volunteers to participate in citizen science projects varies depending on the scope of the research. Some projects require basic data collection knowledge requiring minimal or no training of volunteers, while others require intensive training (Haklay 2013; Shirk *et al.* 2012). **Figure 21.4: Future projections of global average crop yield (top left), crop production (top right), agricultural area (bottom left), and forest and other natural land area (bottom right)**, illustrates the various levels of engagement of volunteers in citizen science projects.

Citizen scientists can help to uncover critical information about our environment which could possibly take scientists years to discover by themselves. An example is illustrated by the infographic shown in **Figure 25.3** where rivers need a citizen science movement for monitoring, and how the collected data and findings are used to maintain ecosystem services and human wellbeing (Pottinger 2012). The figure also demonstrates a step-by-step procedure for conducting citizen science. This data collection and analysis procedure can be replicated across the Drivers (Chapter 2) and the various environmental themes (Chapters 5 to 9).

Figure 25.2: Levels of citizen science by increasing depth of the participation

Level 4
Extreme Citizen Science
- Collaborative science - problem definition, data collection and analysis

Level 3

Participatory Science
- Participation in problem definition and data collection

Level 2

Distributed Intelligence
- Citizens as basic interpreters
- Volunteered thinking

Level 1

Crowdsourcing
- Citizens as sensors
- Volunteered computing

Source: Haklay (2013).

Trends in citizen science

The technology revolution has heralded multiple novel ways of collecting, archiving, analyzing, and transmitting data. The emergence of the internet-of-things (IoT), miniaturized smart sensors with geo-location functions, ease of accessing internet and data as well as the potential of cloud storage and computing has expanded the possibilities and opportunities for data collection and analysis. This rapid advancement in technology coupled with greater exposure and sensitization of the public, have led to an explosion in uptake of projects based on citizen science (Mathieu *et al.* 2016).

The availability of internet and geographic information system (GIS)-enabled web applications has enabled citizen scientists to collect large volumes of geographically-referenced data and submit them electronically to centralized databases. An example of such a system is the Global Learning for Observation to Benefit the Environment (GLOBE) program, which uses students to collect environmental data and archive it in the GLOBE program database (Dickinson *et al.* 2012; GLOBE 2018).

The expansion in the use of smartphones, the possibility of digital photo validation of observations, and the capability of creating simple online data-entry systems is revolutionizing the process of initiating citizen science projects while ensuring data accuracy at minimal cost. Currently, it is now possible to create mobile phone apps for collecting different types of datasets and automatically geo-locating the data, using the in-built GPS receiver chip on most mobile phones (Dickinson, Zuckerberg and Bonter 2010; Dickinson *et al.* 2012).

Scientists are now increasingly using citizen scientists to collect geo-referenced *in-situ* data which can be used to support the calibration and validation of Earth Observation satellite data products. Citizen scientists are also involved in the interpretation and digitization of Earth Observation (EO) data sets (Mathieu *et al.* 2016; See *et al.* 2016). Tomnod is such an example of using crowdsourcing and citizen scientists to identify objects and places in satellite images. Tomnod was used in trying to locate the missing Malaysian Airlines flight MH370 aircraft using satellite imagery. Approximately 2.3 million Internet users submitted 18 million tags for over 745,000 satellite images clearly illustrating the potential of citizen science (Mazumdar *et al.* 2017).

Another example of the use of citizen scientists to validate satellite data is the partnership between NASA's Global Precipitation Measurement (GPM) satellite mission and the GLOBE program. The GLOBE program is an environmental

Figure 25.3: An example of citizen science that demonstrates how it is needed and can be replicated

Case Study: Why our rivers need a citizen science movement

Most of our decisions are based on incomplete or inadequate data/information - in the abscence of professional scientists to fill these gaps, citizen science can step in to help uncover information and findings. This case study explores opportunities where citizen scientists have filled this gap.

TOO MUCH WORK, TOO FEW SCIENTISTS

- There is too much research that needs to be done yet hardly enough scientists to undertake it all by themselves.
- Volunteers in the US state of Oregon are helping scientists survey 146 miles of streams by locating and counting salmon and native trout species and helping restore habitat.
- Hundreds of volunteers with The Nature Conservancy annually survey how much desert land is made wet by the San Pedro River; they cover more than 250 miles.

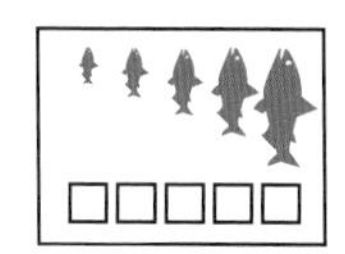

BECAUSE THEY'RE OUR RIVERS

- Volunteers for the Mystic River Watershed Association in the Eastern US, with support of scientists, take monthly samples at 15 locations along the river to monitor water quality.
- Advocacy based on their results has helped improve the river's cleanliness and has enabled the residents get involved in their natural environment through hands-on science.

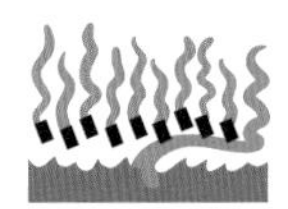

Just 1 in 10 rivers now reach the sea

8/10 people depend on river resources

Just 2/3 of Earth's major rivers are dammed

WHAT WE DON'T KNOW CAN HURT US

- Citizen science can be used to document basic information about a river system, as well as changes over time to its flow, sediment load, species and water quality.
- In China, the South-North Water Transfer Project was envisioned to have massive impacts on many key waterways, volunteers were recruited for a 4 year assessment of 10,000km of China's western rivers.

UNCOVERING RIVER MYSTERIES

- Citizen scientists can fill in gaps in crucial baseline knowledge about a river's species or general health.
- The Mekong River in Southeast Asia supports several species of giant fish and very little is known about them. More information is needed on where they spawn, what natural cues drive them to spawn, population estimates, and maps of their life-cycle territory.
- A step-by-step guide on Mekong River citizen study illustrates how this can be realized.

STEP-BY-STEP CITIZEN SCIENCE GUIDE: CASE STUDY OF FISHING VILLAGERS DOCUMENTING MEKONG'S RIVER'S NATURAL WEALTH

IDENTIFY THE QUESTIONS YOU WANT TO ANSWER

- In 1994, Thailand built Pak Mun Dam on the largest tributary of the Mekong, destroying local fisheries and harming river-based communities.
- Information on local fisheries was scant.
- In 2001, the Thai government opened the dam's floodgates 1-year study of its impacts to fisheries.

FORM A RESEARCH TEAM

- South East Asia Rivers Network (SEARIN) and Assembly of the Poor teamed up to monitor the changes caused by the dam.
- Their innovative citizens' science research method, called Thai Bahn (Thai Villager) research, relied on local fishers to gather information.

DEVELOP A PLAN OF ACTION

- Methods, areas of study, and research team members were all decided by the local villagers.
- SEARIN helped develop a plan of action, write up their findings and increase international awareness.

DOCUMENT YOUR FINDINGS

- The natural flows of the one-year trial period allowed people to resume traditional ways of life and eased resource conflicts among river communities.
- Local fish species not seen for eight years came back; researchers found a total of 156 fish species had returned to the Mun River.

ANALYZE YOUR DATA

- SEARIN helped create a report on the team's findings, in two languages.
- The report is considered one of the most thorough documentations of Mekong fisheries produced for that area.

SHARE YOUR FINDINGS, AND USE THEM FOR ACTION

- Thanks to this citizen science effort, the villagers succeeded in getting the Thai government to open the dam gates for four months each year to allow for fish migration.
- Subsequent governments have not implemeneted this agreement.
- The project has inspired many other citizen science projects to protect rivers in the region.

Source: International Rivers (2012).

educational program for primary and secondary schools, where students from schools across the world collect precipitation data using rain gauges as shown in **Figure 25.4**. The collected data, as well as data collected from other sources, is used by NASA to calibrate and validate GPM precipitation measurement data (United States National Aeronautics and Space Administration [NASA] 2018).

Figure 25.4: GLOBE Students in St. Scholastica Catholic School in Nairobi collecting and recording the amount of precipitation for the GPM Satellite Mission field campaign

Source: © GLOBE Program (Kenya).

Automated and autonomous equipment such as drones, remotely operated sensors, autonomous underwater vehicles (AUV's) and underwater gliders are predicted to play an increasing role in citizen science. These autonomous systems can be a primary source of data or complement data collected *in situ*, provide high resolution data nearly in real time, be deployed on a need basis and often enable access to remote or extreme locations such as observation of marine environments. In addition, they are low cost compared to satellites and are thus offering alternative and credible sources of EO data (Macauley and Brennan 2016; Garcia-Soto 2017).

Citizen science, as well as other data sources, contributes to Big Data collection and these huge volumes of data need processing. Numerous approaches have been explored to involve the huge numbers of citizen scientists to assist in analyzing these huge volumes of data, one of which is the development of game-like systems (gamification). Citizen participation in these games help to speed up the data analysis and allow science to advance more rapidly (Van Vliet and Moore 2016; Spitz *et al.* 2017; McCallum *et al.* 2018).

An example of gamification is Cropland Capture, a game version of the GeoWiki project, which engaged citizen scientists in global land cover research, helping researchers identify farmland around the world. The game managed to collect 4 million classifications from over 3,000 players identifying images with and without cropland present (See *et al.* 2013).

Table 25.1 shows some of the global and regional projects dedicated to citizen science.

The potential of citizen science should not be limited to engaging volunteers to collect and collate scientific data as illustrated in **Figure 25.5**. Citizen science can be used to sensitize and engage the community on issues related to their natural environment, to better understand them and allow them to take charge, and provide an avenue for showcasing the need

Table 25.1: A selection of citizen-science projects and websites

Programme	Region	Description	Website
UNEP Environment Live	Global	UN open access platform of global, regional and national environmental data	https://environmentlive.unep.org
SciStarter	Global	Aggregates information, video and blogs about citizen-science projects	www.scistarter.com
Data Observation Network for Earth	Global	Provides a framework to access data from multiple data sources (including citizen science data)	www.dataone.org
CitSci.org	Global	Provides tools for citizen scientists to guide them on the entire research process such as: process of initiating research projects, managing the process of data collection, and analysis	www.citsci.org
iSpot	Global	Website aimed at helping anyone identify anything in nature by connecting citizen scientists with experts in species identification	www.ispotnature.org
eBird	Global	Online database of bird observations with real-time data about bird distribution and abundance	www.ebird.org

Figure 25.5: Citizen scientists collecting environmental data

Source: © GLOBE Program (Kenya).

to maintain and conserve our ecosystems given the increasing pressures on the environment (Roelfsema *et al.* 2016).

Challenges of citizen science

Challenges in citizen science mostly revolve around three main issues: *organizational* issues, *data-collection* issues and *data-use* issues. At the organizational level, the challenges include the process of recruiting volunteers, motivating and providing incentives for their participation and ensuring sustainability of the initiative as well as funding. On data collection, the issues that arise include: data fragmentation, data representativeness, data quality (for example data intentionally flawed by the data collector) and/or lack of essential metadata. In data use, the challenges include: differences in protocols and standards, legal issues, data-privacy concerns and the question of allowing open access (Conrad and Hilchey 2011; Hochachka *et al.* 2012; Rotman *et al.* 2012; See *et al.* 2016)

Due to misunderstandings and lack of technical knowledge and skills to handle such data, concerns have emerged over the credibility, comparability, completeness of, and lack of metadata, as well as challenges in data access and sharing, and these have resulted in these data not being seriously considered by policy and decision makers. In most cases, perception of poor data quality, rather than the actual data quality and fitness for use, have influenced the value and use of citizen science data (University of the West of England, Science Communication Unit 2013; Storksdieck *et al.* 2016).

The key opportunities presented by citizen science, mainly include:

i. use of local knowledge;
ii. timely data from dispersed sources;
iii. capability to address large knowledge and funding deficits;
iv. ability to educate the public about environmental policy issues; and
v. enhance participatory democracy.

For citizen science to be widely accepted, there is a need for appropriate training and support for citizen science project coordinators and those that use the data that emerge from it. Careful design of citizen science projects and application of appropriate quality assurance methods, as illustrated in **Figure 25.3**, can ensure that the effort of citizen scientists is not wasted (University of the West of England, Science Communication Unit 2013; Storksdieck *et al.* 2016).

There are on-going initiatives, such as the Public Participation in Scientific Research (PPSR)-Core data model framework as illustrated in **Figure 25.6**, to establish data and metadata standards to facilitate international collaboration and improve data standardization, interoperability, integration, accessibility, and dissemination of citizen science data (Bowser *et al.* 2017). Citizen science has the potential to provide credible data to bridge the data gaps highlighted in Chapter 3 and to provide data to enable the monitoring of SDG environmental indicators.

25.2.2 Big data and data analytics

Big data can be defined as "datasets whose size is beyond the ability of typical database software tools to capture, store, manage and analyze" (Manyika *et al.* 2011).

Data are one of the world's valuable resources, shifting the landscape of environmental assessment across global, national and local scales ("The world's most valuable resource is no longer oil, but data" 2017). From 1.8 zettabytes (1.8 trillion gigabytes) of data generated in 2011 (International Data Corporation [IDC] 2012), the total amount of data is expected to reach 40 zettabytes (40 trillion gigabytes) by 2020 (Dell EMC and IDC 2014). With this influx, traditional processing applications will be unable to cope with the quantity of data from multiple sources. Big data is characterized by the four Vs of large storage capacity (volume), speed at which data are generated and transmitted (velocity), the complexity of unstructured data types (variety), and the uncertainty of data sources (veracity) **(Figure 25.7)**. A fifth V (value) is achieved through the application of data analytics (International Business Machines [IBM] 2017).

The science of data analytics is needed to create patterns from intricate data sets and find correlations (e.g. chemical pollution and locations in aerial photographs) by using algorithms, programming, and mechanical and statistical methods to draw evidence-based conclusions and obtain information that is useful for decision-making purposes (Monnappa 2017). Examples of insights drawn from big data analytics include those from projects in the United Nations Global Pulse initiative, such as:

i. urban dynamics drawn from mobile data used to improve transportation in Sao Paulo and Abidjan;
ii. campaign developments based on a survey of perceptions of HIV on social media; and
iii. and a support-services location plan based on the spatial epidemiology of Dengue fever (Kirkpatrick 2016).

Current trends and initiatives in big data

United Nations member states, in partnership with the academic and research communities, non-governmental organizations and the private sector, are seeking out innovations and looking for opportunities to tap into the optimum potential of big data for sustainability and development.

Innovation for public good

The United Nations Global Pulse initiative was founded in 2009 to progressively establish a global network of Pulse Labs to collect digital data for decision-making purposes (United Nations 2018a). Pulse Labs continue to innovate machines and to conduct pilot studies on the scalability of the capture and analytics of big data for sustainable development – some examples are presented in **Table 25.2.**

Figure 25.6: The PPSR-Core data-model framework

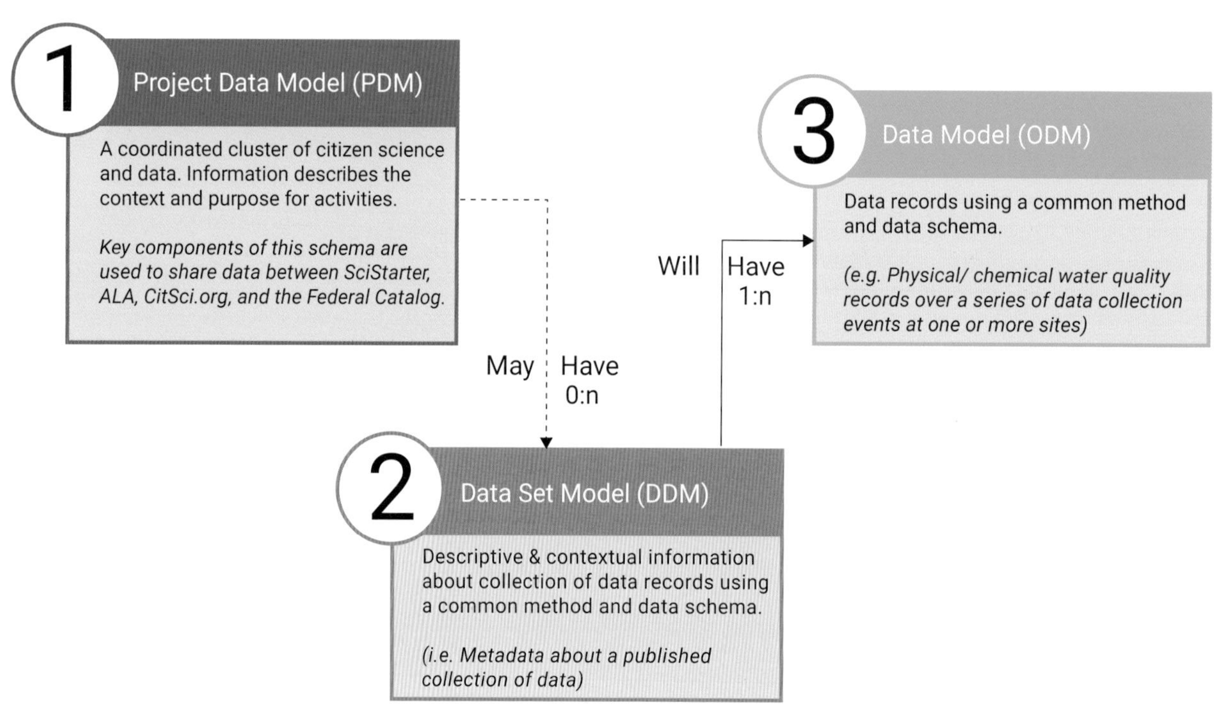

Source: Bowser *et al.* (2017).

25

Figure 25.7: Characteristics of big data and the role of analytics

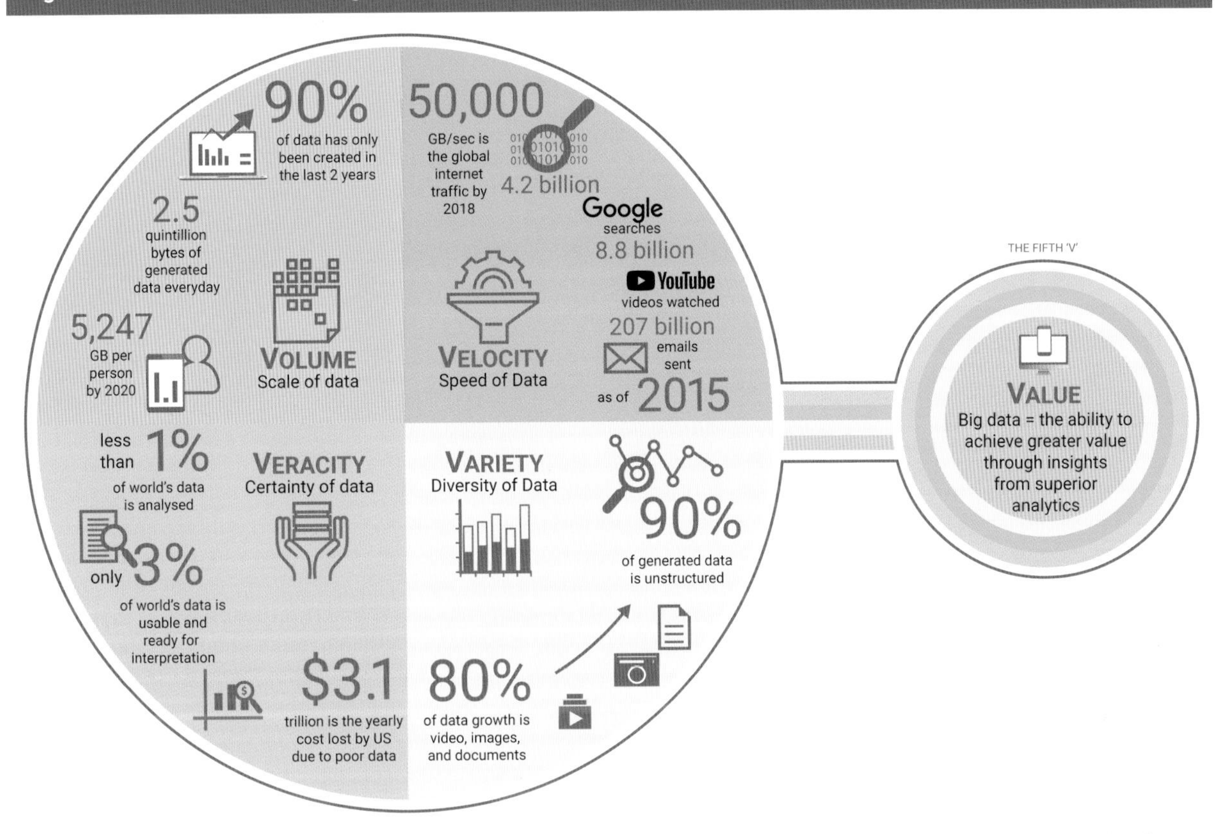

Source: Adapted and recreated the infographics of IBM, with information from World Bank (2016a), IBM (2017); IDC (2012); Harvard Business Review (2016).

Table 25.2: Pulse Lab research and studies

United Nations Global Pulse partner	Project description	Insights and results
Stellenbosch University Pulse Lab Kampala (2017)	Radio content analysis, prototype speech-to-text software that converts public radio content into categorized texts	Searchable topics of interest related to SDGs and development
Office of the United Nations High Commissioner for Refugees (UNHCR) Vacarelu (2017)	Understanding forced displacement of European refugees by utilizing Twitter data	Real-time social media monitoring system relevant to humanitarian actions
World Food Programme Webb and Usher (2017)	Determining the extent of drought in Indonesia, its impact on food market prices, and the resilience of affected areas through a vulnerability monitoring platform	Real-time information platform in support of climate-impacted populations
UNHCR Hoffman (2017)	Gaining insights on the displacement patterns from Libya to Italy and Malta, and the magnitude of rescue operations using vessel data	Revealed rescue activity patterns, capacity of rescue vessels, and patterns of distress signals. Optimized rescue operations by studying migration patterns in the Mediterranean

Source: Blog posts at United Nations Global Pulse (United Nations 2018b).

In recognition of the significance of big data for official statistics, the UN Statistical Commission (UNSC) established the UN Global Working Group (GWG) on Big Data in 2014 to tap the potential of big data in monitoring the SDGs. Various collaborations, research and projects addressing the quality, collection, accessibility, management and feasibility of big data have been developed over the years. These have included task-force teams focusing on the relevance of large volumes of information coming from mobile phones, satellite imagery, social media, virtual platforms and technological applications (United Nations 2018c).

Other initiatives of the United Nations include the UN Environment Live platform **(see Table 25.1)**, which in addition to a repository related to citizen science has data from official national, regional and global statistical and geospatial data series on different thematic areas (freshwater, forests, climate etc.). Another example is the Sustainable Development Goals Interface Ontology (SDGIO), developed by UN Environment to harmonize the relationships across different SDGs through a taxonomy and semantic framework, for SDG monitoring purposes (Jensen 2017).

Data collaboratives

Governments, leading technology companies, innovators, academia, research institutions and non-governmental organizations are convening to understand the challenges around big data and to search for development solutions through collaborative learning. The UNSC Global Working Group believes that a multifaceted approach to data collection, leading to timely delivery of trusted information, can be made possible through close relationships among the private and public sectors, including civil society (United Nations 2018c). Such dialogues open doors for the co-creation of more innovation hubs, allowing capacity-building and skills transfer from countries with more experience in big data to those entering the field more recently (e.g. Vacarelu 2017).

Open data access

Access to open data is essential to harnessing big data's potential for sustainability and development. The global non-profit network Open Knowledge International and the World Wide Web Foundation's Open Data Barometer promote open data as accessible, readily available and free of charge for universal use (World Wide Web Foundation 2017). Open access to valuable and timely data from the outputs of surveys, field experiments and scientific research provides a powerful resource for presenting the state of the environment, validating our knowledge of the anthropogenic climate-change impacts, and towards proposing feasible solutions.

Box 25.1 presents a selection of open-data initiatives at global and national levels.

Environmental assessments and evaluation

Big data analytics enable illustrations of trends and progress over time (e.g. comparing current with historical data at a specific spatial site), reaches more end users beyond geographical boundaries, and allows a predictive analysis of the future using models and comparisons with historical data. Web-based and geospatial mapping technologies, remote sensing and statistical visualization provide a basis for analysis of environmental implications and raise issues on the scalability of data collected, as reflected in the sample of assessments and evaluations in **Box 25.2**. Big data from satellite imagery and sensors make the environmental indicators measurable (Uitto 2016).

The Data-Pop Alliance describes big data as a socio-technological phenomenon, emerging from a novel data ecosystem that defines the complexity of human behaviours and beliefs – generated and captured by digital devices, computational and analytical tools, and the active correspondence of communities (Anttila-Hughes *et al.* 2015). With this paradigm shift, technology has been used to survey public insights, leveraging big data to improve the environmental process.

Box 25.1: Examples of open-data systems

National Bioscience Database Center

National database of Japan, containing global life sciences data that can be readily accessed by users.

Air Now

A global database that presents the daily nationwide forecast of the Air Quality Index (AQI) of over 400 cities, providing visual representation of the air quality condition relevant to health interpretation.

Open Data for Business Assessment and Engagement Tool

Launched by World Bank in 2016 to establish the use of open government data for industrial or business purposes.

Estonia's X Road

Online e-government system, serving as a platform of data exchange among participating institutions and private companies, with 70 per cent composed of government agencies.

Source: Estonia Digital Society (n.d.); United States Environmental Protection Agency (n.d.); Japan Science and Technology Agency (2011); Manley (2015).

Box 25.2: Examples of web-based and geospatial technologies using big data

Big data use for environmental Assessments and Evaluation

- **Participatory web-based GIS tool** used in the Strategic Environmental Assessment (SEA) process, complementing the traditional public consultation by developing a user-friendly and comprehensible system in Ireland
- Reporting of factors affecting **environmental litigation through GIS-generated graphics and location** (e.g. extent of pollution and potential contaminants)
- **Geospatial relations in Environmental Epidemiology,** investigation of disease in relation to the subject's location (e.g. disease mapping, cluster analyses and geographic correlation studies). Epidemiology studies increased from 43 in 1990 to 934 in 2014, based on the PubMed publication index.

Source: Gozales *et.al.* (2012); Rominger and Ikeda (2015).

Box 25.3: Comprehensive air-quality forecasting in India using big data

There are several hundred manuals and about a hundred regulatory air-monitoring stations in operation in India. These are limited to urban agglomerations, so second-tier towns and rural areas do not have access to any on-ground monitoring data. Before being able to manage and improve air quality effectively, citizens and policymakers need to know the status of air quality and to have information on the sources of pollution. The system, known as India air-quality forecasts, developed by *urbanemissions.info*, uses a modelling approach to predict, for the next three days, the estimated pollution levels and source contributions for all of the 640 districts in India. This is not a substitute for a robust monitoring system, but the estimates can be used to support informed decision-making while more monitoring capacity and systems are built.

While the methodology is continually improving, the key challenge in this approach is using a detailed emissions inventory and its spatial and temporal granularity. The programme currently uses information from official reports, academic publications, and survey analysis, and the following open data (dynamic feeds), which are updated every day.

- Remote-sensing satellite data (from NASA's Visible Infrared Imaging Radiometer Suite) for location of open fires, which when overlaid with land-use imagery, builds a dynamic emissions inventory for agricultural and forest fires at 1 km resolution.
- Meteorological data at 1 km resolution linked to emissions from multiple sectors. For example: (a) a surface temperature profile is used to trigger space heating in the residential sector (b) grids with precipitation over 1 mm/h are adjusted for lesser vehicle movement (c) grids with precipitation are adjusted for dust resuspension on the roads and dust at the construction sites (d) the dynamic calculations within the meteorological model estimation of likely dust storms, sea salt emissions, lightning, dry deposition rates and wet scavenging rates by grid.
- Google provides a wealth of information on traffic movement. Over cities, transit speeds are extracted at 1 km resolution, which is used as a proxy to dynamically allocate vehicle exhaust emissions and estimate road dust resuspension. For example, during peak times, if a grid shows speeds under 5 km/h, the emissions profile is adjusted to increase the exhaust emissions due to idling, and the road dust resuspension is zeroed.
- Google Earth imagery is used to generate spatial data on brick kilns, power plants, industrial zones, and mining and quarrying areas.
- Load dispatch centres across India report information on demand and supply by power grid, which is used as a proxy to dynamically adjust the use of diesel generator sets in cities.

These dynamic feeds are making air-quality forecasting more robust, allowing the model to capture trends, and they help to understand the source contributions better. Multiple microsensor networks are being tested and evaluated in several cities, promising to further improve the forecasting process and strengthen the on-ground data availability.

Figure 25.8: Forecasting air quality for Indian districts

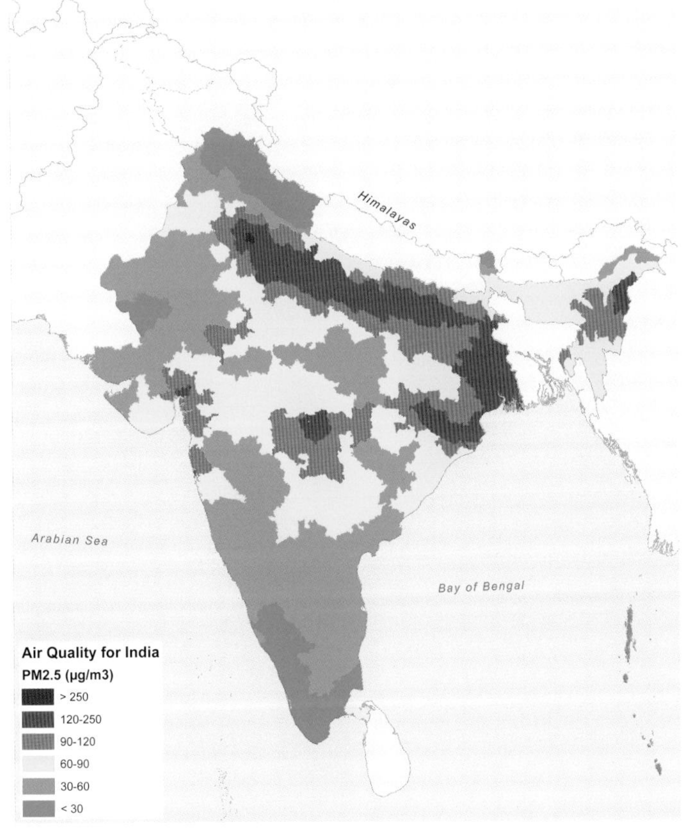

Source: UrbanEmissions.info (2018).

Challenges in Big Data
Gaps in the collection, monitoring, analysis and interpretation of data identified in GEO-5 continue to challenge the reliability of Big Data as a tool in environmental assessment today in GEO-6. Issues include the accessibility, quality and sparsity of data for varying scales, contexts and time series (United Nations Environment Programme [UNEP] 2012). Constraints, generally encountered across the thematic areas covered in GEO-6, are summarized in **References**. These challenges on scope, privacy and potential for misinterpretation of data have not been sufficiently addressed over the years. Efforts to generate globally acceptable and available Big Data are pursued, but the actual capacities are limited by scarce resources and funding, especially in developing countries.

Although the private sector has been pioneering Big Data strategies, increasingly governments and global initiatives are exploring the benefits of Big Data for transparency, market analysis, research, education and environmental protection. In South-East Asia, six countries have formed an open-government partnership to advance their vision of scaling up public services and leveraging Big Data for development (Bhunia 2017).

Environmental agreements provide United Nations member states with guidelines on how to make environmental data publicly accessible, and provide open, geographically referenced data, together with opportunities for public participation in decision-making, multi-stakeholder involvement and the promotion of government transparency and accountability. These agreements include the Aarhus Convention on Access to Information, Public Participation in Decision-Making and Access to Justice in Environmental Matters (United Nations 1998) and the Kyiv *Protocol on Pollutant Release and Transfer Registers* (2003). The Data-Pop Alliance is a global coalition for people-centred data pioneered by the Harvard Humanitarian Initiative, Massachusetts Institute of Technology Media Lab, and the Overseas Development Institute. This alliance, funded by the United Nations Economic Commission for Latin America and the World Bank Group among others, creates a collaborative workspace for researchers, experts, practitioners and activists to overcome foreseen challenges on Big Data (i.e. technological biases, lack of access to an online knowledge-sharing facility, and limited technical capacity development) through research, capacity-building and community engagement (Data-Pop Alliance n.d.). A 2015 report by the alliance explores the opportunities for leveraging Big Data to monitor climate change hazards, mitigate the impacts, guide disaster response and increase the resilience of vulnerable countries (Anttila-Hughes *et al.* 2015).

Real-time Big Data are often controlled and held by the private sector (Kirkpatrick 2016). Therefore, collaborations are needed where both parties benefit without sacrificing the economic value of data, and at the same time maintain fair competition among businesses. The private sector has been providing the public sector, including research institutions and industry practitioners, with access to data through what Robert Kirkpatrick (2016) describes as data philanthropy. This collaboration has been in existence within the United Nations system. In pursuit of companies' contributions to SDGs, data scientists at the firms interpret private data for public good and well-being, which, in return, reduces the risks to business. Another form of collaboration is the public-private partnership where resources and capabilities on Big Data are shared between governments, National Statistical offices, research institutions, and the private sector, including leading technology and data companies across the globe. **Table 25.3** presents some examples of economic improvement achieved through public-private partnerships.

For Big Data to become an effective tool for environmental assessment and development, this emerging form of data and knowledge should be seen as a valuable asset. Big-data analytics involve not only compiling information but also creating a comprehensible view of the environment and its social attributes as a basis for proposing solutions and drafting policies. Factors that contribute to establishing a holistic data system include leadership and data governance, including the appointment of a chief data officer in national and local government agencies; partnerships among governments, institutions and the private sector; and institutionalizing legal frameworks with safeguards on information.

Box 25.4: Some challenges of using Big Data

Big Data Challenges

Accessibility

Reluctance of governments and private sector from sharing information lead to data disintegration. Most data in the private domain are bounded by privacy and security limits-intellectual property rights, thus entailing a price for accessibility. Data sharing and availability are not only part of the legal issues, but of the political concern as well-like in China where the government place restrictions in the release of valuable environmental information.

Quality

There is no assurance in the reliability of data due to inconsistencies in the methodologies used by different countries and insufficient technical capability on data interpretation and analysis. Baseline studies are paralyzed by insufficient valuable data (e.g. biodiversity, hydrometeorological, wastewater treatment).

Sparsity

The availability of data varies across temporal and spatial scale. To fill the gaps, researchers utilize secondary information as proxy data and model estimates to fill the gaps, making global data incomparable.

Table 25.3: Example public-private partnerships

Partnership	Project description	Source
Government of Nigeria and Cellulant	**Wide-scale mobile e-wallet system** that directly coordinates the distribution of seeds, and transfers subsidized fertilizers to farmers, thus streamlining public-service delivery.	World Bank (2016b), p. 94
Organisation for Economic Co-operation and Development and various governments	**Global earthquake model** that communicates earthquake risks through open-access to catastrophe models across the globe.	Thomas and McSharry (2015)
Willis Research Network and various partners	**Willis Research Network** provides and supports the scientific research and development of applications for universities, modelling companies, governments and non-governmental organizations	

25.2.3 Traditional knowledge

Many terms are used to describe the knowledge held by indigenous peoples and local communities. Some refer to the term traditional ecological knowledge, which is "a cumulative body of knowledge, practice, and belief, evolving by adaptive processes and handed down through generations by cultural transmission, about the relationship of living beings (including humans) with one another and with their environment" (Berkes, Colding and Folke 2000, p. 1,252). Others prefer to use the terms indigenous knowledge, folk knowledge, local knowledge or traditional knowledge. There is no universally accepted definition of the diversity of expressions within this epistemic landscape; however, to include the widest understanding, this section uses the term traditional knowledge to include ecological, local, indigenous and folk knowledge.

According to Article 8j of the Convention on Biological Diversity, traditional knowledge includes cultural values, beliefs, rituals, community laws, local language and knowledge related to practical fields such as agriculture, fishing, hunting, medicine, horticulture, forestry and environmental management in general (Secretariat of the Convention on Biodiversity n.d.). The following definition, meanwhile, was published by the World Intellectual Property Organization (WIPO) in 2010 and remains the description given by the body (WIPO n.d.): "knowledge, know-how, skills and practices that are developed, sustained and passed on from generation to generation within a community, often forming part of its cultural or spiritual identity".

Perceived by some in the mainstream as superstitious and anecdotal, traditional knowledge has been historically marginalized. In the past 20 years, however, it has been acknowledged as a valuable resource for sustainable development. Its promotion and protection has been expressed in several United Nations agreements and agencies (e.g. Article 8(j) mentioned above, the United Nations University Institute for the Advanced Study of Sustainability, and WIPO). The foundational operating principles of the Intergovernmental Science-Policy Platform on Biodiversity and Ecosystem Services, established in 2012, include recognition and respect for the contribution of indigenous and local knowledge to the conservation and sustainable use of biodiversity. Moreover, the United Nations Convention to Combat Desertification (UNCCD) in its Decision 20/COP.12 adopted "improvement of knowledge dissemination including traditional knowledge, best practices, and success stories" (UNCCD 2016, p. 57).

An increasing amount of research concludes that traditional knowledge developed through direct interaction with local ecosystems is of equal value to that of Western scientific knowledge. Both knowledge systems have commonalities, but each has distinct features **(see Figure 25.9)** that complement each other to better understand the natural world (Agrawal 1995; Tsuji and Ho 2002).

Figure 25.9: Comparing indigenous/traditional knowledge and Western science

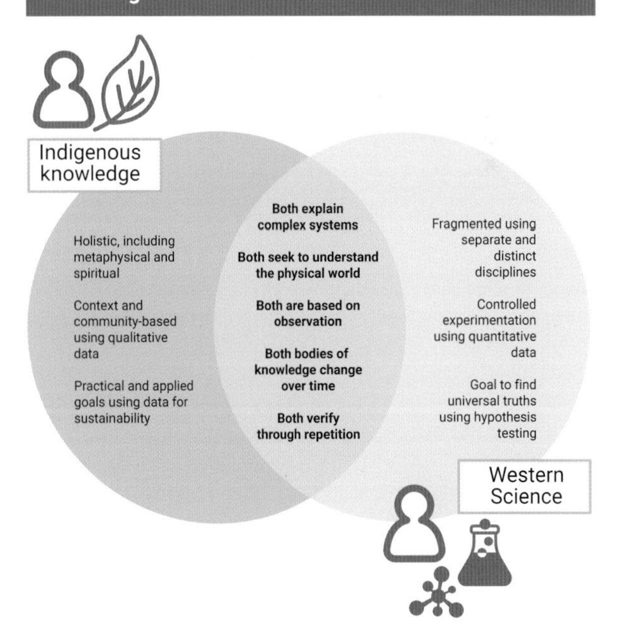

Source: Baker, Rayner and Wolowic (2011).

Traditional knowledge and Western scientific knowledge
GEO-5 called for coordination in the realm of knowledge-building, which implies that cooperation is necessary between Western scientists and the various holders of traditional knowledge. Recent progress has enabled the development of new tools and approaches in measurement, reporting and verification. Among others, these include the community-based monitoring and information systems, and the multiple evidence-base system, both of which foster knowledge co-creation between indigenous and Western systems, as an important way to help advance the recognition of the leadership role of indigenous peoples in stewarding their lands and waters (Raygorodetsky 2017). See more detail about these and other approaches in **Box 25.5** and **Table 25.4.**

These approaches combine traditional community monitoring systems with modern software (e.g. GIS, Google Maps, GPS, Microsoft Excel) and hardware (e.g. drones, remote sensing devices, trackers, smartphones, electronic tablets) to generate data and develop information on trends in ecosystems in order to assess development interventions. These modern tools and monitoring systems enable individuals and communities to decide on what actions to take by providing informed, improved and timely decision-making.

As traditional knowledge is based on *in situ* data generation, it is effectively used to give on-the-ground truth to Big Data, and like citizen science, could involve all sectors in the community. **Table 25.5** outlines new partnerships between communities and researchers that have produced innovative new approaches to the documentation and analysis of traditional knowledge.

Traditional knowledge and the SDGs
Indigenous peoples' engagement in the development of the 2030 Agenda for Sustainable Development (United Nations 2015) resulted in the inclusion of six references to

Box 25.5: Complementary uses of traditional knowledge and Western science

Community-based monitoring and information systems refer to initiatives by indigenous peoples and local communities/organizations to monitor their community's well-being and the state of their territories and natural resources, applying a mix of traditional knowledge and innovative tools and approaches (Ferrari, de Jong and Belohrad 2015).

The multiple evidence-base system generates new insights and innovations through complementarities between indigenous peoples, local communities and Western scientific knowledge systems. The system emphasizes that the evaluation of knowledge occurs primarily within, rather than across, knowledge systems (Tengö *et al.* 2014).

Indigenous peoples make and use maps to: assert their rights to lands and waters; manage their territory; preserve knowledge of their own history, culture and environment; and communicate some of this knowledge to others (Tebtebba Foundation 2015).

Successful integration of traditional knowledge with modern science, technology and innovation can be seen in the example of a recent tech start-up called Indigital. This Aboriginal-owned and operated social enterprise, based in the Kakadu World Heritage Area in the Northern Territory of Australia, uses digital technology to showcase local sacred sites, knowledge and stories in augmented and virtual realities, contributing to the preservation of heritage while creating jobs in the digital economy (Cooper and Kruglikova 2018).

Table 25.4: Studies that combine traditional knowledge with Western scientific knowledge

Study	Summary
Genome-wide association study to identify the genetic base of smallholder farmer preferences of durum wheat traits (Biodiversity International 2017; Kidane *et al.* 2017)	The study demonstrates that researchers using modern scientific tools (i.e. genetic analysis), and holders of traditional knowledge using cultural practices in crop selection can work together to advance crop breeding to cope with the changing climate.
Arbediehtu pilot project on documentation and protection of Sami traditional knowledge (Porsanger and Guttorm eds. 2011)	The study highlights community work, legal questions and legislation, ethics of documentation, institutional relationships, history and identity, information technologies, transmission, management, and legitimacy.
Traditional knowledge and nutritive value of indigenous foods in the Oraon tribal community of Jharkhand: an exploratory cross-sectional study (Ghosh-Jerath *et al.* 2015)	The study identifies more than 130 varieties of indigenous foods, many of which are rich sources of micronutrients and medicinal properties, and explains how they can be leveraged to address malnutrition in tribal communities.
Cree traditional ecological knowledge and science: a case study of the sharp-tailed grouse (Tsuji 1996)	The study shows that Cree traditional knowledge is factual and often quantitative in nature. While limitations exist in the distinction between observations and interpretations, this knowledge can be added to databases to facilitate resource co-management.
Collaborative Partnership on Forests (2018)	The project shows that the integration of traditional knowledge into forest management practices is a prerequisite for achieving sustainable forest management because it strengthens the rights and participation of indigenous peoples and local communities, and clarifies land tenure.

Figure 25.10: Recognition of indigenous peoples in the 2030 Agenda for Sustainable Development

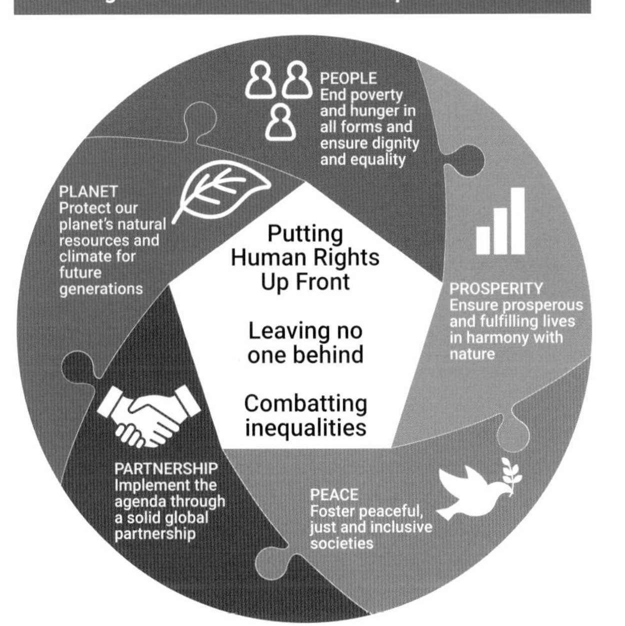

Source: United Nations (n.d.).

Figure 25.11: Lands/territories of indigenous peoples are the base of their knowledge

© Tebtebba Foundation 2008.

indigenous peoples. Aside from these six, many other SDGs address indigenous peoples indirectly, through the principles of human rights, equality, non-discrimination, sustainability, and participation by right-holders. Sustainability, as the banner of Agenda 2030, is underpinned by environmental health.

Having evolved after years of observation and experience from a holistic relationship between people and nature, traditional knowledge sustains life and landscapes. Meanwhile, current development strategies are largely based on Western science and technology, which are often detrimental to the environment and our well-being in many ways. Therefore, there is a need to harness scientific knowledge, technology, and traditional knowledge to solve many of the issues related to sustainable natural resource management and biodiversity conservation. Some examples of studies on the potential of traditional knowledge for sustainability are presented in **Table 25.5**.

The promotion and development of traditional knowledge is hampered by contemporary bias in existing power relations. In many places, holders of traditional knowledge continue to face exploitation when trying to defend their territories against further abuse. As a result, their lands – which constitute the basis of their knowledge systems – have been subjected to resource development projects such as mono-crop plantations for rubber, timber and palm oil, large hydroelectric dams, mineral extraction activities (Asia Indigenous Peoples Pact 2015), and conservation projects (Vidal 2016).

To address historical and continuing injustices adequately, indigenous peoples and local communities, their land, and the knowledge systems that hold these together must be included in the development process. As expressed in the United Nations

Table 25.5: Studies on the potential of traditional knowledge for sustainable development

Study	Summary
Local biodiversity outlooks: indigenous peoples' and local communities' contributions to the implementation of the strategic plan for biodiversity 2011-2020 (Forest Peoples Programme 2016).	This study presents snapshots of on-the-ground initiatives by indigenous peoples and local communities and demonstrates that they are making vital contributions to the implementation of the five strategic goals and the 20 Aichi biodiversity targets, although many challenges remain. It outlines the way forward, highlighting key potential actions to accelerate progress in the implementation of the strategic plan for biodiversity as it relates to indigenous peoples and local communities.
Sustaining and enhancing forests through traditional resource management (Enchaw and Njobdi 2014)	This study highlights women's roles in forest management and sustainable farming within the context of indigenous knowledge.
United Nations Educational Scientific and Cultural Organization (UNESCO) World Heritage: Agricultural Landscape (UNESCO 2013)	This study deals with communities that demonstrate rich cultural and landscape diversity and sustainable land-use systems. It also highlights how some people struggle for daily survival under extreme climatic and environmental conditions.

Declaration on the Rights of Indigenous Peoples (United Nations, General Assembly 2007), these peoples have the right to self-determination, as well as to their land, resources and the freedom to pursue their own way of life. These rights are necessary for peoples to sustain, innovate and develop traditional knowledge systems and customary practices. Granting and implementing these rights is crucial for ensuring the maintenance of a balance between the economic, social and environmental dimensions of sustainable development.

Traditional knowledge for a Healthy Planet, Healthy People

Traditional knowledge is an invaluable resource for sustaining a healthy population and planet. Indigenous territories constitute up to 22 per cent of the world's land surface and sustain 80 per cent of the planet's biodiversity (Food and Agriculture Organization of the United Nations [FAO] 2017). The vast majority of the world's genetic resources and a considerable part of global biodiversity survives within indigenous and community-conserved areas. This correlation is not coincidental but is due to the application of traditional knowledge and the customary sustainable use of biological resources over centuries (Independent Expert Advisory Group Secretariat 2014). Moreover, the indigenous ethics and values, plus culture and identity, related to land and wildlife stewardship, hold great promise for more effective resource management as well as for more effective risk reduction in human health (Houde 2007). **Figure 25.12** is just one illustration of the wealth and potential of traditional knowledge and practices for environmental management that need to be optimized in their contribution to improving the health of both land and people.

Further examples of benefit include the many advantages of maintaining a well-functioning food web – including the enhanced diversification of wild and cultivated food systems, enhanced nutrition, and a healthy environment (Kuhnlein *et al.* eds. 2013).

However, traditional knowledge remains underutilized in environmental assessment and management. While advances have been made, there are still challenges that need to be addressed. As cited by Genetic Resources Action International [GRAIN] and Kalpavriksh (2002), these include:

i. the continued loss of indigenous peoples' lands, making it challenging for indigenous communities to sustain their knowledge;
ii. the risk of misappropriating traditional knowledge or patenting life forms for commercial purposes without sharing the benefits with knowledge holders, as demonstrated by the case of turmeric in India;
iii. the proliferation and all-out promotion of so-called modern medicine and agriculture, which replace diverse plants and crops that would have been able to resist pests, diseases, and changing climatic and economic conditions (GRAIN and Kalpavriksh 2002);
iv. the co-production of knowledge processes that do not always guarantee fairness, equal standing, or power symmetries (Williams and Hardison 2013); and
v. the rapid erosion of linguistic diversity is accompanied by the loss of indigenous ways of knowing and understanding of the natural world (UNESCO 2017).

Other challenges relate to the full and effective implementation of free prior and informed consent (United Nations, General Assembly 2007). To address these challenges in an efficient way, policies need to be formulated and implemented on the basis of "further interdisciplinary action research that brings together indigenous knowledge holders and scientists, both natural and social, to build mutual understanding and reinforce dialogue" (Nakashima *et al.* 2012, p. 97).

25.3 Environmental monitoring for the future

25.3.1 Measuring what matters

There is a maxim that what gets measured gets done. Its origin is debatable, but the message is clear: measuring something gives us the information we need to make sure we actually achieve what we set out to do.

As noted in Section 3.3.1, the data requirements for the SDG indicators are almost as unprecedented as the SDGs themselves, and constitute a tremendous challenge to all countries. Unfortunately, much of the data required to monitor the SDGs are unavailable. Issues relating to quality, timeliness, human and financial capacity, and the lack of standardized methodologies all hamper our ability to comprehensively track this important agenda. As highlighted in the United Nations (2016) report on the SDGs, tracking their progress will require a shift in how data are collected, processed, analysed and disseminated, including a move to using data from new, diverse and innovative sources. More than ever, this demands that

Figure 25.12: Indigenous peoples as stewards of the environment

In forests in the Amazon, ecosystems improve when indigenous peoples inhabit them.

In mountains indigenous peoples have designed agricultural systems that protect the soil, reduce erosion, conserve water and reverse the risk of disasters.

In range lands indigenous pastoralist communities manage cattle grazing and cropping in sustainable ways that allow the preservation of rangeland diversity.

Source: FAO (2016).

we explore other data streams – citizen science, Big Data and traditional knowledge – to complement conventional, official statistics and Earth observations.

25.3.2 Translating local information into national data

A shift in data generation by electronic devices, data modelling, cloud computing and other technologies has produced unprecedented volumes of information. The availability of these data and information varies, however, between developed and less developed countries, between and within social groups – by gender, ethnicity, and social and income status – and markedly between the global and local levels. Data and information gathering at national and local levels is given lower priority, especially in developing countries. But to solve global problems, actions for solutions must emanate from the local and national level. Policy effectiveness should be measured according to its implementation and fulfilment at the local to national level (see Part B).

Over time, there has been an accumulation of knowledge on good and bad practices of environmental management, enabling some conclusions as to what needs to be done. However, there remains a need for making traditional knowledge more accessible and to integrate traditional knowledge with other sources of information.

On the global level, states and civil society organizations, academia, indigenous peoples and activists all agree that for the promotion of environmental protection and restoration, biological loss and climate change need to be addressed. Yet at the country level, positions about environmental protection vary, and often even contradict each other.

The SDGs aim to reduce environmental degradation while at the same time upholding basic human rights and promoting economic empowerment. This will require knowledge and technology to be shared between communities, businesses and governments – from local, through national, to global levels.

25.3.3 Open data and reproducible research

The open-data movement has gained significant traction in recent years and is expected to continue to grow. The concept of open data is that data resulting from publicly funded research should be freely available to all, for equity, transparency, and to catalyse the advancement of science. The principles of open data have been proposed in a number of different variations, including the Open Data Charter (2015), which stipulates that data should be:

i. open by default;
ii. timely and comprehensive;
iii. accessible and usable;
iv. comparable and interoperable;
v. for improved governance and citizen engagement; and
vi. for inclusive development and innovation.

There are clearly demonstrated benefits of open data, particularly in the health sector (Kostkova *et al.* 2016). The principles of open science, open innovation, open access and open source adopted by the malaria research community, for example, have allowed it to achieve more progress than would otherwise be possible (Wells *et al.* 2016). Additionally, many countries now have open government data portals that include environmental data.

There is a growing call for reproducible research alongside open data, and the two are often considered in tandem by the open-science movement. To be reproducible, research should be reported in a manner that allows it to be replicated precisely (Mesirov 2010). There are three aspects of research reproducibility – it should be:

i. empirical – based on scientific experiments and observations;
ii. computational – code, software, hardware and implementation details are made available; and
iii. statistical – based on statistical tests and model parameters (Stodden 2014).

Computational and statistical reproducibility are most pertinent to data practices. In theory, the publication of code and data together means that users can understand and critique the entire process of analysis and inference, including details of the techniques used and any assumptions made. Publication of code is now an essential prerequisite of many scientific journals, and this trend is increasing such that it is expected to become the norm in coming years.

Reproducible data analyses are especially important in an era of open data, since users of data will become increasingly detached from those who collect and curate the data. Open-access data increases the risk of data being misused or misinterpreted, but the publication of code circumvents this problem in that the treatment of data is transparent and can be scrutinized by readers. Reproducibility not only improves the quality of scientific output, but also increases trust in the results, and therefore uptake (Laine *et al.* 2007).

25.3.4 Coping with the changing data landscape

Strengthening the ability to gather, interpret and use data for effective planning, policymaking, management and evaluation is necessary for providing countries with a comprehensive view of environmental impacts, ranging from geo-political perspectives through to industrial operations, naturally occurring or anthropogenic environmental change or a combination of all these. The challenge escalates as the magnitude and types of generated data, both structured and unstructured, grow over time. Management information systems alone are insufficient to draw the full value from the exponential growth of potential data assets. It is imperative that governments and society learn to cope with the evolving data landscape, to shift from mere reporting and conventional data repository functions towards a predictive and prescriptive analysis for both modelling different environmental scenarios and creating appropriate policies to address these foreseen challenges.

The information landscape is changing as the technologies for harnessing data evolve from data-as-a-service (DaaS) or software-as-a-service (SaaS) to insights-as-a-service (IaaS), which uses prescriptive analytics **(Figure 25.13)**. With DaaS, data could reach several users beyond geographical limits and organizational segmentation, bringing together data in a central repository (Olson 2010). DaaS and SaaS have the capacity to present both historical and current states, reporting what happened, why it happened, and what is happening now with regard to both environment and society. From cleaning and consolidating Big Data, analytics-as-a-service (AaaS) shifts the demand from internal manual services to web-delivered technologies, outsourcing the needs from the Internet of things or IoT (Atos 2013) and providing virtual services. AaaS accompanies users throughout their experience, imparting knowledge through artificial intelligence to find expertise and support throughout the entire journey (Takahashi 2017). AaaS and knowledge-as-a-service (KaaS) apply predictive analytics and modelling to interpret Big Data from multiple sources and project the future. Tapping the potential of artificial intelligence could develop strategies not only to foresee future environmental and social challenges, but also to advance solutions by predicting the outcomes of the countries' efforts, responding to the question, how do we make it happen?

The exponential growth of Big Data, technological solutions, complex algorithms and open data sources propels and integrates artificial intelligence into our everyday lives, cities and world networks each year (Herweijer 2018). Artificial intelligence is a powerful tool for countries to navigate in managing environmental concerns and furthering the SDGs, but risks around privacy, biases, declining human intervention and autonomy have to be considered. A road map is needed for how artificial intelligence could transform traditional systems and add value in delivering services, combating the impacts of climate change, building sustainable and liveable cities, and protecting environmental and social welfare. As the scale of economic and health impacts broadens as a result of environmental degradation, strategic measures have to be developed to establish not only human-friendly but also Earth-friendly artificial intelligence.

The new forms of data and knowledge, coupled with conventional tools, will dramatically influence the way solutions are created and delivered. Coping with the shift in the data landscape will need new information technology skills and a holistic approach to utilize emerging and existing data and knowledge tools – thus making data on healthy people and a healthy planet more accessible for environmental assessment and other purposes.

25.3.5 Crucial assets of technology

The collection of data in statistical operations follows well-founded methodological approaches, such as sample survey methods and designs. These include well-identified and defined sources, on which systematic methods are applied to transform the data into statistics that lend themselves to time series – the fundamental value of any statistic. New technologies such as remote sensing, transactional data, block chain and artificial intelligence algorithms have the potential to create a wealth of information that is useful for environmental purposes. However, there is a challenge to harness these new technologies to produce time series data, to achieve real-time monitoring and to bring this information into the scope of official statistics.

Only with time series do statistics transform into information, and thereby a fundamental knowledge system. The roles and responsibilities of those who count and those who are counted are also clear cut. The governance of official statistics is predicated on a set of fundamental principles laid down by the United Nations (2014). There are ten principles, including covering the fact that statistical practice should be impartial, should ensure the protection of the privacy of individuals, should be transparent, and should work to ensure the quality of the information produced. Technology has crucial assets which are founded in their development in compliance with standards. Beyond compliance is replication and scaling up. In a way, statistical operations and technological advancement can speed up optimization and give value to visibility, application and transformation of, and from, systems of data, statistics, and can give information to verifiable knowledge systems.

Figure 25.13: The evolution of the data landscape

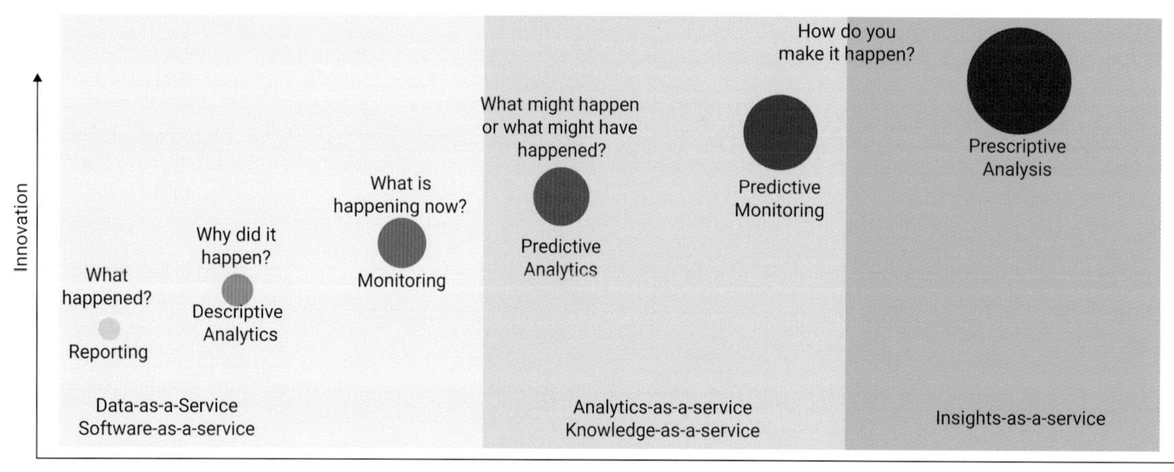

Source: Chartered Property Causality Underliner [CPCU] (2015); ASI Business Solutions (2016).

But are these fundamental principles still relevant in the changed data environment? Maintaining the principles of official statistics involves many challenges in the new data landscape. These include definitional considerations such as concerning the selection of statistical units, data items and their associated spatio-temporal nature. The major paradigm leap for official statistics is how established methodologies of data collection and statistical time series can be adapted in the context of a deluge of unstructured, private (and potentially anonymized) data recorded on electronic devices. Phones, computers and other devices lend themselves readily to standardization and the possibility of replication and scale. For statistics to function optimally, though, standards are needed. Statisticians, technology gurus and data purveyors will need to navigate this space carefully as the hard and laborious slog of carefully designed statistical operations faces a real existential challenge of easy-to-use and fashionable observational tools. Statistics – and its essence, time series – can be greatly challenged in this new environment. The truth is, in the absence of time series, all data can be rendered useless. If a zealous adaptation of technology results in the loss of time series data, it will become impossible to track trends over time. The intersection of technology, data, statistics, knowledge, finance and governance needs to be found.

25.3.6 Data assurance and quality practices

With the increasing use of complementary data alongside traditional statistics to support environmental and sustainability policy, questions of how data quality, pedigree and provenance can be assured will need to be systematically answered to determine data that are credible and fit for purpose. Environmental data may come to include digital sources, incorporating Earth observations, citizen science, environmental monitoring, development data and statistics, administrative data sets, and population- and survey-derived data.

In addition to the fundamental principles of statistical practice mentioned above, references and standard practice documents are also emerging for complementary data sources. For example, metadata standards and practices now serve as a basis for data description, including methodological description and data quality. Metadata – the documentation of data – serves the purpose of making data discoverable, usable and understandable. Many discipline-specific or community-specific metadata standards have been developed to support systems of data management and data discovery, and to capture and convey information to users. Examples include directory interchange format (DIF), ecological metadata language (EML), sensor model language (SensorML), climate science modelling language (CSML), and netCDF markup language (NcML). Additionally, the International Organization for Standardization (ISO) has developed a series of standards to describe geographical information – ISO 19115 and ISO 11179.

Data-quality assurance for citizen-science data is in its formative stages, engaging a variety of digital-platform techniques for quality checking field observations contributed by citizens. Examples include the Local Environmental Observer (LEO) Network in the North American Arctic region, using a smartphone app that uploads observations for expert checks before being used in graphic displays on maps and in tabular data sets (LEO Network 2017). Recent workshops convened by the European Commission's Joint Research Centre are also beginning the process of establishing principles for mobile apps and platforms.

The Group on Earth Observations has propagated its Global Earth Observation System of Systems (GEOSS) data management principles, which are being widely adopted by Earth Observation entities (Group on Earth Observations 2014). Included in the coverage of the principles are: discoverability, accessibility, usability, preservation and curation.

As open-source analytics, community-sourced query codes and custom data-integration methods advance, there will be more community curation of data sets, exchange standards and application-programming interfaces. Code sets incorporating analytics and queries are now routinely community-curated on open collaboration platforms such as the GitHub development platform.

The Research Data Alliance sets registry standards for long-term curation and for defining the parameters of data sets in Earth science and other research domains. Complementary data sources will be judged to be credible as far as they conform with these registry standards.

The ultimate test for the data revolution of open and accessible digital data will be user satisfaction and integrated platform requirements for aligning with an array of recognized standards, practices and open source community-driven testing for methodology and data quality from across environmental, natural resource and development data.

25.4 Conclusion: Challenges, gaps and opportunities

The challenges, gaps and opportunities related to environmental data and statistics are presented below. Data and knowledge are valuable assets that need to be shared.

25.4.1 Data disaggregation

The SDGs call for a data revolution that leaves no one behind, incorporating disaggregated data and reporting at all levels of the 17 goals. As highlighted in Chapter 3 of this report, assessing the nexus between society and the environment can be done only if there is disaggregated information on different populations because not all people have the same level of dependence on the environment, nor the same impact on it. To tease out these differences, then, there is a need for information that can be disaggregated by income, gender, age, ethnicity, migratory status, disability, geographical location and other characteristics relevant in national contexts. Unfortunately, there is currently a dearth of environment-related information that can be disaggregated, and data from household surveys on access to water, energy and other natural resources is available only at the household level, which makes understanding differences at other levels difficult.

In addition to disaggregation by socioeconomic variables there is also a need for geospatial disaggregation of environmental information. Biological ecosystems do not follow national boundaries, so to understand both the state of particular ecosystems and the interactions between them and people

and the economy requires spatially disaggregated information. Disaggregation will require bringing together Earth observation data, data from traditional censuses and surveys, as well as citizen science, traditional knowledge and Big Data into an integrated data ecosystem.

25.4.2 Open data access

Many developing countries will need access to unrestricted open data. In terms of data governance, the overall management of the availability, usability, integrity and security of the data should be made transparent. The continental leaders of vast, validated data sets – North America and Europe – are currently the data stewards making their valuable resources readily available to the rest of the world as public goods. What may be needed is a sound data-governance programme that includes a governing body or council and a defined set of procedures, with a plan for their execution. Governments should ensure that there are legal frameworks in place to promote good data governance. A documented, transparent data-governance policy will establish a set of easy-to-follow guidelines for ensuring the proper management of information, helping to answer questions revolving around sovereignty, security, data quality and privacy. In particular, protecting the privacy of individual people is an essential component of data governance. To help shepherd this governance programme, multilateral data stewards, responsible for the management and fitness of data elements, should be assigned to ensure that data exchanges are executed precisely and consistently between computer systems and between different collection methods (see 'Institutional arrangements' below).

25.4.3 Data and information governance

Governance of data involves managing and leveraging all data assets (dispersed among large external open data sources, central databases, and other existing and emerging data and knowledge sources from governments, institutions, research, and studies). Information technology alone is insufficient to draw full value from the exponential growth of potential data assets. Policy makers generally lack an understanding of the distinct issues that information technology has to deal with, and the need for active organizational involvement and institutional ownership. Addressing the challenges posed by Big Data means the integration of valuable information through a combined process of 'data semantics' and the intervention of a data-governance specialist.

Dealing with Big Data involves the storage of all data resources, sorting them out and identifying which data make sense beyond local use, and accessing the disparate data and finding their relationships. Having the right tools is not enough to sift through Big Data and find any relevant patterns and relationships. Like any valuable asset, information needs to be managed and secured through careful practices of designing, creating, developing and deploying data. With proper governance, the meaning, use and integrity of the data are preserved over time, increasing the value of data as an asset.

25.4.4 Public-private partnerships

The challenges of scarce resources and lack of financial capability in progressing the use of big-data analytics, particularly in developing countries, could be helped through public-private partnerships. The public sector could exercise the social implementation of big-data analytics as private entities finance and build on the technical know-how (Groff 2017). Public-private projects provide opportunities to strengthen statistical and information systems, develop innovative data-collection methods and initiate co-creation through people-first partnerships. Partnerships between the government and the private sector that involve civil society may result in the adoption of best practices in data collection, analytical assessment and monitoring for SDG-related information.

25.4.5 Institutional arrangements

The institutional arrangements between different authorities and agencies producing and holding environmental information are in their early stages (e.g. official statistics, map-based information on natural resources, hydrometeorological, geospatial monitoring, environmental and geospatial portals, open-data governmental portals, etc.). Any future measures should aim to improve institutional arrangements to facilitate the effective integration, sharing and reuse of data across barriers, including those of sovereignty, mandates, bureaucracy, knowledge gaps, standards and standardization, digitization and digitalization. These challenges need to be resolved. Open data and instruments that facilitate its use, such as the Statistical Data and Metadata eXchange (SDMX), hold promise for seamless data sharing. However, many countries face challenges in terms of implementing open-data instruments. National governments need to commit to common global goods, and intergovernmental arrangements are necessary to facilitate solutions to some of the political problems arising from government limitations.

25.4.6 Capacity-building

Sustainable capacity is achieved when the entire value chain works – where data from collation and analysis through to use are dealt with competently to deliver the value of the information, disseminate it through the most effective and efficient channels, and ensure that those for whom its value is intended are able to use it. To reach this stage of competency, environmental education should be implemented at all levels. Indigenous knowledge systems, including emerging opportunities offered by technological advances to democratize the participation of citizens in the exercise of science and scientific discovery should be facilitated.

25.4.7 Traditional knowledge

The challenge around the continued use and development of traditional knowledge is the lack of effective implementation of rights to indigenous knowledge, resulting in the loss of lands and the misappropriation of traditional knowledge. The inclusion of traditional knowledge does not guarantee the equality of benefit sharing.

The opportunities for traditional knowledge to contribute to sustainable development include the improved effort in documenting traditional knowledge and the increasing network building partnerships to develop innovative tools and methodologies for a better understanding of traditional knowledge to be used in strategic plans and interventions and to assess changes in ecosystems. The commitment of UN member states for sustainable development also matters when discussing traditional knowledge.

25.4.8 Integrated data systems

The physical and social systems, hand in hand, comprise the real state of the environment. Without both, the picture of where we are now and where we are heading would be incomplete. A combination of conventional methods with emerging forms of data and knowledge, and with the social data, could provide a holistic view of the environment, encompassing social, physical and economic perspectives.

Societies often notice change before measurable environmental change occurs – but, we also need to look into how data, information and knowledge are used and understood. In the era of information technology, the problem may not be the lack of knowledge or data on what is happening, but rather on the question of which available data should be acted on, what kind of information, based on which assumptions, is considered legitimate, and which reality counts (Davies 1994).

A shift in thought processes on environmental assessments would bring data, including from Big Data, citizen science and traditional knowledge to the forefront with Earth observation and official statistics. This would allow the environmental challenges to be identified, the progress in addressing the SDGs to be monitored, and the solutions to be drafted for sound and evidence-based policy-making. This would require not only turning the challenges of environmental data into opportunities, but also improving data integration, utilizing both existing and emerging tools for environmental assessment, and making data openly available for use.

References

Agrawal, A. (1995). Dismantling the divide between indigenous and scientific knowledge. *Development and Change* 26(3), 413-439. https://doi.org/10.1111/j.1467-7660.1995.tb00560.x.

Anttila-Hughes, J., Dumas, M., Jones, L., Pestre, G., Qiu, Y., Levy, M. *et al.* (2015). *Big Data for Climate Change and Disaster Resilience: Realising the Benefits for Developing Countries*. Data-Pop Alliance. http://datapopalliance.org/wp-content/uploads/2015/11/Big-Data-for-Resilience-2015-Report.pdf.

ASI Business Solutions Inc (2016). *Evolution of analytics: Where does your company stand?* http://asi-solutions.com/2016/12/evolution-of-analytics-where-does-your-company-stand/ (Accessed: 06 February 2018).

Asia Indigenous Peoples Pact (2015). *Recognition of Indigenous Peoples' Customary Land Rights in Asia. Asia Indigenous Peoples Pact.* Chiang Mai: Asia Indigenous Peoples Pact. https://aippnet.org/wp-content/uploads/2015/07/iva.aippnet.org_wp-content_uploads_2015_07_CLR-AIPP-Corrected-2.pdf.

Atos (2013). *Data Analytics as a Service: Unleashing the Power of Cloud and Big Data*. Atos. https://atos.net/wp-content/uploads/2017/10/01032013-AscentWhitePaper-DataAnalyticsAsAService.pdf.

Baker, J., Rayner, A. and Wolowic, J. (2011). *Native Science: A Primer for Science Teachers*. https://ctabobandung.files.wordpress.com/2011/11/ns-primer.pdf.

Berkes, F., Colding, J. and Folke, C. (2000). Rediscovery of traditional ecological knowledge as adaptive management. *Ecological Applications* 10(5), 1251-1262. https://doi.org/10.1890/1051-0761(2000)010[1251:ROTEKA]2.0.CO;2.

Bhunia, P. (2017). *Brief look at open government data in 6 ASEAN countries*. [OpenGov https://www.opengovasia.com/brief-look-at-open-government-data-in-6-asean-countries/ (Accessed: 02 January 2018).

Biodiversity International (2017). *Modern science meets traditional knowledge to improve crop breeding*. https://www.bioversityinternational.org/news/detail/modern-science-meets-traditional-knowledge-to-improve-crop-breeding/ (Accessed: 11 October 2018).

Blaney, R.J.P., Philippe, A.C.V., Pocock, M.J.O. and Jones, G.D. (2016). *Citizen science and Environmental Monitoring: Towards a Methodology for Evaluating Opportunities, Costs and Benefits*. Wiltshire: UK Environmental Observation Framework. http://www.ukeof.org.uk/resources/citizen-science-resources/Costbenefitcitizenscience.pdf.

Bowser, A., Brenton, P., Stevenson, R., Newman, G., Schade, S., Bastin, L. *et al.* (2017). *Citizen science Association Data and Metadata Working Group: Report from CSA 2017 and Future Outlook*. Washington, DC.: Woodrow Wilson International Center for Scholars. https://www.wilsoncenter.org/sites/default/files/wilson_171204_meta_data_f2.pdf.

Chartered Property Casualty Underliner (2015). *Big Data and Analytics Webinar*. 7 April. https://www.slideshare.net/PatriciaSaporito/cpcu-big-data-analytics-webinar-april-7-2015sedslideshare.

Collaborative Partnership on Forests (2018). *About the collaborative partnership on forests*. http://www.cpfweb.org/73947/en/ (Accessed: 11 October 2018).

Conrad, C.C. and Hilchey, K.G. (2011). A review of citizen science and community-based environmental monitoring: Issues and opportunities. *Environmental Monitoring and Assessment* 176(1-4), 273-291. https://doi.org/10.1007/s10661-010-1582-5.

Cooper, D. and Kruglikova, N. (2018). *Augmented Realities: The Digital Economy of Indigenous Knowledge*. International Labour Organization.

Data Pop Alliance (n.d.). *Vision and members*. Data Pop Alliance. http://datapopalliance.org/about/vision-and-members/ (Accessed: 02 July 2018).

Davies, S. (1994). Introduction: Information, Knowledge and Power. *IDS Bulletin* 25(2), 1-13. https://doi.org/10.1111/j.1759-5436.1994.mp25002001.x.

Dell EMC and International Data Corporation (2014). *IDC study: Digital universe in 2020*. https://www.emc.com/leadership/digital-universe/index.htm (Accessed: 05 May 2018).

Dickinson, J.L., Shirk, J., Bonter, D., Bonney, R., Crain, R.L., Martin, J. *et al.* (2012). The current state of citizen science as a tool for ecological research and public engagement. *Frontiers in Ecology the Environment* 10(6), 291-297. https://doi.org/10.1890/110236.

Dickinson, J.L., Zuckerberg, B. and Bonter, D.N. (2010). Citizen science as an ecological research tool: Challenges and benefits. *Annual Review of Ecology, Evolution, and Systematics* 41, 149-172. https://doi.org/10.1146/annurev-ecolsys-102209-144636.

Enchaw, G.B. and Njobdi, I. (2014). *Sustaining and Enhancing Forests through Traditional Resource Management*. Tebtebba Foundation. http://www.tebtebba.org/index.php/content/276-sustaining-a-enhancing-forests-through-traditional-resource-management-volume-2.

Estonia Digital Society (n.d.). *x-road*. https://e-estonia.com/solutions/interoperability-services/x-road/ (Accessed: 20 November 2018).

Ferrari, M., de Jong, C. and Belohrad, V.S. (2015). Community-based monitoring and information systems (CBMIS) in the context of the Convention on Biological Diversity (CBD). *Biodiversity* 16(2-3), 57-67. https://doi.org/10.1080/14888386.2015.1074111.

Food and Agriculture Organization of the United Nations (2016). *Indigenous Peoples can feed the world*. Rome. http://www.fao.org/3/a-c0386e.pdf

Food and Agriculture Organization of the United Nations (2017). *6 ways indigenous peoples are helping the world achieve #ZeroHunger*. http://www.fao.org/indigenous-peoples/news-article/en/c/1029002 (Accessed: 11 October 2018).

Forest Peoples' Programme (2016). *Local Biodiversity Outlooks: Indigenous Peoples and Local Communities Contributions to the Implementation of the Strategic Plan for Biodiversity 2011-2010. A complement to the Fourth Edition of the Global Biodiversity Outlook*. Development Southern Africa. Moreton-in-Marsh. https://www.cbd.int/gbo/gbo4/publication/lbo-en.pdf.

Garcia-Soto, C., van der Meeren, G.I., Busch, J.A., Delany, J., Domegan, C., Dubsky, K. *et al.* (2017). *Advancing Citizen Science for Coastal and Ocean Research: Position Paper 23*. French, V., Kellett, P., Delany, J. and McDonough, N. (eds.). Ostend: European Marine Board. http://www.marineboard.eu/sites/marineboard.eu/files/public/publication/EMB_PP23__Citizen_Science_web_4.pdf.

Genetic Resources Action International and Kalpavriksh (2002). *Traditional Knowledge of Biodiversity in Asia-Pacific: Problems of Piracy and Protection*. New Delhi. https://www.grain.org/article/entries/81-traditional-knowledge-of-biodiversity-in-asis-pacific-problems-of-piracy-and-protection.pdf.

Ghosh-Jerath, S., Singh, A., Kamboj, P., Goldberg, G. and Magsumbol, M.S. (2015). Traditional knowledge and nutritive value of indigenous foods in the oraon tribal community of jharkhand: An exploratory cross-sectional study. *Ecology of Food and Nutrition* 54(5), 493-519. https://doi.org/10.1080/03670244.2015.1017758.

Global Learning and Observations to Benefit the Environment (2018). *About GLOBE program*. https://www.globe.gov/about/overview (Accessed: 20 July 2018).

Gonzales, A., Gilmer, A., Foley, R., Sweeney, J. and Fry, J. (2007). Developing and applying a user-friendly web-based GIS for participative environmental assessment. *Proceedings of the Geographical Information Science Research UK Conference*. Maynooth, 11-13 April 2007. http://eprints.maynoothuniversity.ie/1428/1/FoleyDevelopingGISRUK.pdf

Google (2018). *Google reCAPTCHA: The new way to stop bots*. https://www.google.com/recaptcha/intro/v3beta.html# (Accessed: 26 May 2018).

Groff, S.P. (2017). 'Better data is the key for developing Asia to successfully implement the SDGs by 2030'. https://blogs.adb.org/blog/data-matters-meet-sdgs (Accessed: 30 June 2017)

Group on Earth Observations (2014). *The GEOSS data sharing principles post-2015*. http://www.earthobservations.org/documents/dswg/10_GEOSS%20Data%20Sharing%20Principles%20post%202015.pdf (Accessed: 11 October 2018).

Haklay, M. (2013). Citizen science and volunteered geographic information: Overview and typology of participation. In *Crowdsourcing Geographic Knowledge*. Berlin: Springer. 105-122.

Harvard Business Review (2016). *The Explainer: Big Data and Analytics*. https://hbr.org/video/3633937151001/the-explainer-big-data-and-analytics (Accessed: 22 January 2018).

Herweijer, C. (2018). *8 Ways AI can help save the planet*. [World Economic Forum https://www.weforum.org/agenda/2018/01/8-ways-ai-can-help-save-the-planet/ (Accessed: 21 February 2018).

Hochachka, W.M., Fink, D., Hutchinson, R.A., Sheldon, D., Wong, W.-K. and Kelling, S. (2012). Data-intensive science applied to broad-scale citizen science. *Trends in Ecology Evolution* 27(2), 130-137. https://doi.org/10.1016/j.tree.2011.11.006.

Hoffmann, K. (2017). Using vessel data to study rescue patterns in the Mediterranean Sea. 20 September 2017 https://www.unglobalpulse.org/news/using-vessel-data-study-rescue-patterns-mediterranean-sea

Houde, N. (2007). The six faces of traditional ecological knowledge: challenges and opportunities for Canadian co-management arrangements. *Ecology Society* 12(2). https://doi.org/10.5751/ES-02270-120234

Independent Expert Advisory Secretariat (2014). *A World that Counts Mobilising the Data Revolution for Sustainable Development*. New York, NY: United Nations. http://wedocs.unep.org/bitstream/handle/20.500.11822/20065/ieag_world.pdf?sequence=1&isAllowed=y.

International Business Machines (2017). *Extracting business value from the 4 V's of big data*. International Business Machines Corporation http://www.ibmbigdatahub.com/sites/default/files/infographic_file/4Vs_Infographic_final.pdf

International Data Corporation (2012). *Worldwide Big Data Technology and Services 2012-2015 Forecast*. International Data Corporation. http://ec.europa.eu/newsroom/dae/document.cfm?doc_id=6242.

International Rivers (2012). Citizen science Supports a Healthy Mekong. *World Rivers Review* 4. https://www.internationalrivers.org/sites/default/files/attached-files/wwrdec_web.pdf.

Japan Science and Technology Agency (2011). *National bioscience database center*. https://biosciencedbc.jp/en/ (Accessed: 20 November 2018).

Jensen, M. (2017). *UNEP Sustainable Development Goals Initiatives Ontology*. http://ncgia.buffalo.edu/OntologyConference/Abstract/Jensen.pdf.

Kidane, Y.G., Mancini, C., Mengistu, D.K., Frascaroli, E., Fadda, C., Pè, M.E. *et al.* (2017). Genome wide association study to identify the genetic base of smallholder farmer preferences of durum wheat traits. *Frontiers in Plant Science* 8(1230). https://doi.org/10.3389/fpls.2017.01230.

Kim, S., Robson, C., Zimmerman, T., Pierce, J. and Haber, E.M. (2011). Creek watch: Pairing usefulness and usability for successful citizen science. *Proceedings of the SIGCHI Conference on Human Factors in Computing Systems*. Vancouver, 7-12 May 2011. ACM 2125-2134 doi: ttps://doi.org/10.1145/1978942.1979251

Kirkpatrick, R. (2016). 'The importance of big data partnerships for sustainable development'. 31 May 2016 https://www.unglobalpulse.org/big-data-partnerships-for-sustainable-development

Kostkova, P., Brewer, H., de Lusignan, S., Fottrell, E., Goldacre, B., Hart, G. *et al.* (2016). Who owns the data? Open data for healthcare. *Frontiers in Public Health* 4(7). https://doi.org/10.3389/fpubh.2016.00007.

Kuhnlein, H.V., Erasmus, B., Spigelski, D. and Burlingame, B. (eds.) (2013). *Indigenous Peoples' Food Systems and Well-being: Interventions and Policies for Healthy Communities*. Rome: Food and Agriculture Organization of the United Nations. http://www.fao.org/docrep/018/i3144e/i3144e.pdf.

Laine, C., Goodman, S.N., Griswold, M.E. and Sox, H.C. (2007). Reproducible research: Moving toward research the public can really trust. *Annals of Internal Medicine* 146(6), 450-453. https://doi.org/10.7326/0003-4819-146-6-200703200-00154.

Local Environmental Observer Network (2017). *LEO reporter mobile app*. https://www.leonetwork.org/en/docs/about/mobile (Accessed: 11 October 2018).

Macauley M. and Brennan T. (2016). Data from drones: A new way to see the natural world. *Resources for the Future*, 192. 25 May 2016. http://www.rff.org/files/document/file/RFF-Resources-192_DataFromDrones.pdf.

Manley, L. (2015). 'Open data for business tool: Learning from initial pilots'. 24 August 2015 http://blogs.worldbank.org/ic4d/open-data-business-tool-learning-initial-pilots

Manyika, J., Chui, M., Brown, B., Bughin, J., Dobbs, R., Roxburgh, C. *et al.* (2011). *Big Data: The Next Frontier for Innovation, Competition, and Productivity*. McKinsey Global Institute. https://www.mckinsey.com/~/media/McKinsey/Business%20Functions/McKinsey%20Digital/Our%20Insights/Big%20data%20The%20next%20frontier%20for%20innovation/MGI_big_data_full_report.ashx.

Mathieu, P., Borgeaud, M., Desnos, Y., Rast, M. and Brockmann, C. (2016). *ESAEarth Observation Open Science: Accelerating Data-Intensive Research in the Digital Age*. IEEE Magazine.

Mazumdar, S., Wrigley, S. and Ciravegna, F. (2017). Citizen science and crowdsourcing for earth observations: An analysis of stakeholder opinions on the present and future. *Remote Sensing* 9(1), 87. https://doi.org/10.3390/rs9010087.

McCallum, I., See, L., Sturn, T., Salk, C., Perger, C., Duerauer, M. *et al.* (2018). Engaging citizens in environmental monitoring via gaming. *International Journal of Spatial Data Infrastructures Research* 13, 15-23 https://doi.org/10.2902/1725-0463.2018.13.art3.

Mesirov, J.P. (2010). Accessible reproducible research. *Science* 327(5964), 415–416. https://doi.org/10.1126/science.1179653.

Monnappa, A. (2017). *Data science vs. big data vs. data analytics*. https://www.simplilearn.com/data-science-vs-big-data-vs-data-analytics-article (Accessed: 27 April 2018).

Nakashima, D., McLean, K.G., Thulstrup, H.D., Castillo, A.R. and Rubis, J.T. (2012). *Weathering Uncertainty: Traditional Knowledge for Climate Change Assessment and Adaptation*. Paris: United Nations Educational, Scientific, and Cultural Organization and United Nations University. http://unesdoc.unesco.org/images/0021/002166/216613e.pdf.

Newman, G., Wiggins, A., Crall, A., Graham, E., Newman, S. and Crowston, K. (2012). The future of citizen science: Emerging technologies and shifting paradigms. *Frontiers in Ecology and the Environment* 10(6), 298-304. https://doi.org/10.1890/110294.

Olson, J.A. (2010). Data as a service: Are we in the clouds? *Journal of Map and Geography Libraries* 6(1), 76–78. https://doi.org/10.1080/15420350903432739.

Open Data Charter (2015). *International Open Data Charter*. https://opendatacharter.net/wp-content/uploads/2015/10/opendatacharter-charter_F.pdf.

Porsanger, J. and Guttorm, G. (2011). *Working with Traditional Knowledge: Communities, Institutions, Information Systems, Law and Ethics*. Trondheim: Sámi University College. https://brage.bibsys.no/xmlui/handle/11250/177065.

Pottinger, L. (2012). *Why our rivers need a citizen science movement*. https://www.internationalrivers.org/resources/why-our-rivers-need-a-citizen-science-movement-7764.

Protocol on Pollutant Release and Transfer Registers (2003). Signed 21 May 2003. https://www.unece.org/fileadmin/DAM/env/documents/2003/pp/ch_XXVII_13_ap.pdf.

Pulse Lab Kampala (2017). Using machine learning to analyse radio content in Uganda: Opportunities for sustainable development and humanitarian action. 11 September 2017 https://www.unglobalpulse.org/news/using-machine-learning-accelerate-sustainable-development-solutions-uganda-0 (Accessed: 6 November 2017).

Raygorodetsky, G. (2017). 'Braiding science together with indigenous knowledge'. *Observations*, 21 December 2017 https://blogs.scientificamerican.com/observations/braiding-science-together-with-indigenous-knowledge/

Roelfsema, C., Thurstan, R., Beger, M., Dudgeon, C., Loder, J., Kovacs, E. *et al.* (2016). A citizen science approach: A detailed ecological assessment of subtropical reefs at Point Lookout, Australia. *PLoS ONE* 11(10), e0163407. https://doi.org/10.1371/journal.pone.0163407.

Rominger, J. and Ikeda, S. (2015). The role of big data in solving environmental problems. *Gradient Trends Risk Science and Application*, Fall 2015, 64. https://gradientcorp.com/pdfs/newsletter/Trends64.pdf.

Rotman, D., Preece, J., Hammock, J., Procita, K., Hansen, D., Parr, C. *et al.* (2012). Dynamic changes in motivation in collaborative citizen-science projects. *Proceedings of the ACM 2012 conference on Computer Supported Cooperative Work*, 217-226. https://doi.org/10.1145/2145204.2145238

Secretariat of the Convention on Biodiversity (n.d.). *Article 8(j) -Traditional Knowledge, Innovations and Practices*. https://www.cbd.int/traditional/ (Accessed: 14 March 2018).

See, L., Fritz, S., Dias, E., Hendriks, E., Mijling, B. and Snik, F. (2016). Supporting earth-observation calibration and validation: A new generation of tools for crowdsourcing and citizen science. *IEEE Geoscience and Remote Sensing Magazine* 4(3), 38-50. https://doi.org/10.1109/MGRS.2015.2498840.

Shirk, J., Ballard, H., Wilderman, C., Phillips, T., Wiggins, A., Jordan, R. *et al.* (2012). Public participation in scientific research: A framework for deliberate design. *Ecology and Society* 17(2), 29. https://doi.org/10.5751/ES-04705-170229.

Spitz, R., Pereira, C., Leite, L.C., Ferranti, M.P., Kogut, R., Oliveira, W. *et al.* (2017). Gamification, citizen science and civic engagement: In search of the common good. *Balance-Unbalance 2017*. Plymouth, 21-23 August 2017. http://balance-unbalance2017.org/events/gamification-citizen-science-and-civic-engagement-in-search-of-the-common-good/

Stodden, V. (2014). *What scientific idea is ready for retirement?* [Edge Foundation https://www.edge.org/responses/what-scientific-idea-is-ready-for-retirement (Accessed: 05 June 2018).

Storksdieck, M., Shirk, J.L., Cappadonna, J.L., Domroese, M., Göbel, C., Haklay, M. *et al.* (2016). Associations for citizen science: Regional knowledge, global collaboration. *Citizen science: Theory and Practice* 1(2), 1-10. https://doi.org/10.5334/cstp.55.

Takahashi, D. (2017). *Got It debuts knowledge-as-service that uses AI to help you find human experts*. [VentureBeat https://venturebeat.com/2017/06/06/got-it-debuts-knowledge-as-a-service-that-uses-ai-to-find-you-human-experts/ (Accessed: 22 February 2018).

Tebtebba Foundation (2015). Mapping our lands and waters: Protecting our future. *Global Conference on Community Participatory Mapping in Indigenous Peoples' Territories*. North Sumatra, 25-27 August 2013. Baguio City http://www.iapad.org/wp-content/uploads/2017/01/mapping-our-lands-waters-protecting-our-future.pdf

Tengö, M., Brondizio, E.S., Elmqvist, T., Malmer, P. and Spierenburg, M. (2014). Connecting diverse knowledge systems for enhanced ecosystem governance: The multiple evidence base approach. *Ambio* 43(5), 579-591. https://doi.org/10.1007/s13280-014-0501-3.

The world's most valuable resource is no longer oil, but data (2017). *The Economist*, 6 May. https://www.economist.com/news/leaders/21721656-data-economy-demands-new-approach-antitrust-rules-worlds-most-valuable-resource.

Thomas, R. and McSharry, P. (2015). *Big Data Revolution: What Farmers, Doctors and Insurance Agents Teach Us About Discovering Big Data Patterns*. Wiley. https://www.wiley.com/en-us/Big+Data+Revolution%3A+What+farmers%2C+doctors+and+insurance+agents+teach+us+about+discovering+big+data+patterns-p-9781118943717.

Tsuji, L.J.S. (1996). Cree traditional ecological knowledge and science: A case study of the sharp-tailed grouse, *Tympanuchus Phasianllus Phasianllus*. *The Canadian Journal of Native Studies* XVI (1), 67-79. http://www3.brandonu.ca/cjns/16.1/tsuji.pdf.

Tsuji, L.J.S. and Ho, E. (2002). Traditional environmental knowledge and western science: In search of common ground. *Canadian Journal of Native Studies* 22(2), 327-360. http://www3.brandonu.ca/cjns/22.2/cjnsv.22no.2_pg327-360.pdf.

Uitto, J. (2016). Use of big data in environmental evaluation: A focus session on use of new technologies in M&E and implications for evaluation. *19th Meeting of the DAC Network on Development Evaluation*. Paris, 26-27 April 2016. https://www.gefieo.org/sites/default/files/ieo/ieo-documents/Presentation-big-data.pdf

United Nations (1998). *Convention on Access to Information, Public Participation in Decision-Making and Access to Justice in Environmental Matters*. https://treaties.un.org/pages/viewdetails.aspx?src=ind&mtdsg_no=xxvii-13&chapter=27&clang=_en (Accessed: 24 January 2018).

United Nations (2014). Resolution adopted by the General Assembly on 29 January 2014. Fundamental Principles of Official Statistics. A/RES/68/261 New York, NY https://unstats.un.org/unsd/dnss/gp/FP-New-E.pdf

United Nations (2015). *Transforming Our World: The 2030 Agenda for Sustainable Development*. New York, NY. https://sustainabledevelopment.un.org/content/documents/21252030%20Agenda%20for%20Sustainable%20Development%20web.pdf.

United Nations (2016). *The Sustainable Development Goals Report 2016*. New York, NY. https://unstats.un.org/sdgs/report/2016/The%20Sustainable%20Development%20Goals%20Report%202016.pdf.

United Nations (2018a). *About*. https://www.unglobalpulse.org/about-new (Accessed: 11 October 2018).

United Nations (2018b). *News*. https://www.unglobalpulse.org/blog (Accessed: 11 October 2018).

United Nations (2018c). *Big data: News*. https://unstats.un.org/bigdata (Accessed: 11 October 2018).

United Nations, General Assembly (2007). *United Nations Declaration on the Rights of Indigenous Peoples*. New York. https://www.un.org/esa/socdev/unpfii/documents/DRIPS_en.pdf.

United Nations (n.d.). *Indigenous Peoples: 2030 Agenda for Sustainable Development*. https://www.un.org/esa/socdev/unpfii/documents/2016/Docs-updates/Indigenous-Peoples-and-the-2030-Agenda-with-indicators.pdf

United Nations Convention to Combat Desertification (2016). *Report of the Conference of the Parties on its Twelfth Session, held in Ankara from 12 to 23 October 2015 Part two: Action Taken by the Conference of the Parties at its Twelfth Session - Addendum* 12-23 October. ICCD/COP (12)/20/Add.1. https://www.unccd.int/sites/default/files/sessions/documents/ICCD_COP12_20_Add.1/20add1eng.pdf.

United Nations Educational Scientific and Cultural Organization (2013). *Agricultural Landscapes*. October 2013 (69). United Nations Educational, Scientific and Cultural Organization http://es.calameo.com/read/00332997247675cccaf1e

United Nations Educational Scientific and Cultural Organization (2017). *Local Knowledge, Global Goals*. Paris. http://unesdoc.unesco.org/images/0025/002595/259599e.pdf.

United Nations Environment Programme (2012). *Global Environment Outlook-5: Environment for the Future We Want*. Nairobi: United Nations Environment Programme. http://wedocs.unep.org/bitstream/handle/20.500.11822/8021/GEO5_report_full_en.pdf?sequence=5&isAllowed=y.

United States Environmental Protection Agency (n.d.). *AirNow*. https://airnow.gov/ (Accessed: 20 November 2018).

United States National Aeronautics and Space Administration (2018). *Engaging citizen scientists with GPM*. https://pmm.nasa.gov/articles/engaging-citizen-scientists-gpm (Accessed: 20 July 2018).

University of the West of England, Science Communication Unit, (2013). *Science for Environment Policy In-depth Report: Environmental Citizen Science*. Report produced for the European Commission. http://ec.europa.eu/environment/integration/research/newsalert/pdf/IR9_en.pdf.

UrbanEmissions.info (2016). *Air quality forecasting for all India*. http://www.urbanemissions.info/ (Accessed: 11 October 2018).

Vacarelu, F. (2017). 'New paper from UN Global Pulse and UNHCR explores use of digital data for insights into forced displacement'. https://www.unglobalpulse.org/news/new-paper-un-global-pulse-and-unhcr-explores-use-digital-data-insights-forced-displacement (Accessed: 6 November 2017)

Van Vliet, K. and Moore, C. (2016). Citizen science initiatives: Engaging the public and demystifying science. *Journal of Microbiology and Biology Education* 17(1), 13-16. https://doi.org/10.1128/jmbe.v17i1.1019.

Vidal, J. (2016). The tribes paying the brutal price of conservation. *The Guardian*. https://www.theguardian.com/global-development/2016/aug/28/exiles-human-cost-of-conservation-indigenous-peoples-eco-tourism.

Webb, A. and Usher, D. (2017). 'New vulnerability monitoring platform to assist drought-affected populations in Indonesia'. https://www.unglobalpulse.org/news/new-vulnerability-monitoring-platform-assist-drought-affected-populations-indonesia

Wells, T.N.C., Willis, P., Burrows, J.N. and van Huijsduijnen, R.H. (2016). Open data in drug discovery and development: Lessons from malaria. *Nature Reviews Drug Discovery* 15(10), 661-662. https://doi.org/10.1038/nrd.2016.154.

Williams, T. and Hardison, P. (2013). Culture, law, risk and governance: Contexts of traditional knowledge in climate change adaptation. *Climate Change*, 531–544. https://doi.org/10.1007/s10584-013-0850-0.

World Bank (2016a). *The World Bank Annual Report 2016*. Washington, D.C. https://openknowledge.worldbank.org/bitstream/handle/10986/24985/9781464808524.pdf?sequence=3&isAllowed=y.

World Bank (2016b). *World Development Report 2016: Digital Dividends*. Washington, D.C: World Bank. https://openknowledge.worldbank.org/bitstream/handle/10986/23347/9781464806711.pdf.

World Intellectual Pr

Annexes

Annex 1-1: Mission of the sixth Global Environment Outlook

Within UN Environment's mandate to keep the environment under review, Member States have requested that UN Environment continue to review the environmental dimension of the SDGs, which are at the core of the 2030 Agenda for Sustainable Development (United Nations 2015a). GEO-6 is a powerful tool to strengthen UN Environment's role within the science-policy interface with multiple functions:

- Support UN Environment's pivotal role in providing assessments, policy analysis, integrative analytics, and approaches to deliver on the environmental dimension of the SDGs, including the follow-up and review process;
- Be UN Environment's instrument to support member states, major groups, stakeholders, and UN system entities' implementation of the 2030 Agenda through the UN System-wide Strategies on the Environment adopted in 2016;
- Help UN Environment align its strategic planning to the 2030 Agenda and strengthen collaboration with the rest of the UN system, and in doing so, embed the environment in global normative frameworks, and address emerging environmental issues (UNEP 2016d).

The GEO-6 assessment also supports UN Environment's core principles on delivering the environmental dimension of the 2030 Agenda, including the principles of:

a. Universality [all people – beyond borders – collective action]: The 2030 Agenda is global, applying to all people in all countries. It is a shared agenda that requires a collective response from the international community, governments, businesses and citizens' groups.
b. Human rights and equity [pathway to a fair, just and sustainable world]: The 2030 Agenda encourages a more-even distribution of wealth and resources, equitable access to opportunities, information and the rule of law; including the development of new approaches that build capacities at all levels of society.
c. Integration [acting as a harmonious whole]: Past approaches treated the social, environmental and economic dimensions of sustainable development as disconnected pillars, but the new agenda integrates and balances all three.
d. Innovation [invention is the master key to progress]: The acceleration and transfer of technological innovations is key to delivering the 2030 Agenda. The world will need new innovation pathways that draw on formal science, traditional knowledge and citizens' common sense (UNEP 2015b).

Compared with previous GEOs, the sixth edition provides the first integrative baseline in light of global megatrends supported by various sources of open and accessible data and information, and a pluralistic knowledge base, with due consideration given to gender and youth, indigenous knowledge, and cultural dimensions. Also new in this edition is the integration and discussion of economic aspects of sustainable development and the dimension of social equity throughout the assessment, not only to strengthen overall policy relevance, but also to highlight that environmental change and degradation cause tremendous pressure on global economic prosperity, social justice and overall human well-being. GEO-6 also reflects on the impact of economic prosperity and social justice on environmental degradation.

The GEO process laid the foundation for continued and intensified socio-economic-environmental assessment across relevant scales, with a thematic as well as an integrative focus, enabling and informing societal transitions and the tracking of SDG goals and targets, as well as previously internationally agreed environmental goals. Therefore, GEO-6 aims to assist Member States, international organizations, and Major Groups (like non-governmental organizations) to position themselves on the most effective pathway for transitions towards a sustainable future over various time frames (2030/2050), considering the extensive inter-dependencies between the environment and people's well-being (e.g. Healthy Planet, Health People).

Annex 1-2: Range of integrated environmental assessments which the sixth Global Environment Outlook draws from

GEO-6 draws and integrates findings from major global environmental assessments, including IPCC, IPBES, etc.

Table A.1: Examples of Global Environmental Assessments and their links to GEO-6

Assessment	Lead Organization	link	Objectives	Links to GEO-6
Assessment Reports of the International Panel of Climate Change (IPCC)	UNEP, WMO	http://www.ipcc.ch/	To provide policymakers with regular assessments of the scientific basis for climate change, its impacts and future risks, and options for adaptation and mitigation.	Results were used as a key reference by addressing climate change as a cross-cutting issue, affecting all other themes, including policy responses and outlooks.
Global and Regional Assessments on Biodiversity and Ecosystem Services (IPBES)	UNEP, UNESCO, FAO, UNDP	http://www.ipbes.net/	To assess the state and trends of biodiversity and of the ecosystem services it provides to society, in response to requests from decision makers. Strengthen the science-policy interface for biodiversity and ecosystem services for the conservation and sustainable use of biodiversity, long-term human well-being and sustainable development.	Results were used as a key reference for the state of the environment chapters on biota, land, freshwater and oceans. Results were used as a baseline also in Part B (policy and governance) and Part C (Outlooks).
Global Biodiversity Outlook (GBO) IV	Convention on Biological Diversity (CBD)	https://www.cbd.int/gbo4/	To periodically assess and summarize the latest data on the state of biodiversity and draw conclusions relevant to the further implementation of the Convention.	Used as a key reference in the thematic chapter on biodiversity, including policy responses and outlooks.
World Water Assessment Programme (WWAP)	UNESCO, UN-Inter-agency (UN-Water)	http://www.unesco.org/new/en/natural-sciences/environment/water/wwap	Comprehensive review that gives an overall picture of the state of the world's freshwater resources and aims to provide decision-makers with the tools to implement sustainable use of water resources. To provide a mechanism for monitoring changes in the resource and its management, while tracking progress towards achieving targets, particularly those of the MDGs/SDGs. To offer best practices as well as in-depth theoretical analyses to help stimulate ideas and actions for better stewardship in the water sector.	Reports within the WWAP and their results served as a baseline for the thematic chapter on freshwater, and related cross-cutting issues, including policy responses and outlooks.
World Ocean Assessment I (2015)	Group of Experts of the Regular Process/UN General Assembly	http://www.worldoceanassessment.org/	The global mechanism for reviewing the state of the marine environment, including socioeconomic aspects, on a continual basis by providing regular assessments at the global and supraregional levels and an integrated view of environmental, economic and social aspects.	Results of the World Ocean Assessment I served as a baseline for the thematic chapter on oceans and coasts and relevant cross-cutting issues, including policy response and outlooks.
Global Land Degradation Assessment/Global Soil Health Assessment (2015)	FAO, UNEP	http://www.fao.org/soils-portal/soil-degradation-restoration/global-soil-health-indicators-and-assessment/jp/	To provide a global scientific assessment of current and projected soil conditions built on regional data analysis and expertise; to explore the implications of these soil conditions for food security, climate change, water quality and quantity, biodiversity, and human health and wellbeing; and to conclude with a series of recommendations for action by policymakers and other stakeholders.	Results of these assessments served as a baseline for the thematic chapter on land and relevant cross-cutting issues, like food security, including policy responses and outlooks.

Assessment	Lead Organization	link	Objectives	Links to GEO-6
Global Land Outlook (GLO) (2017)	UNCCD	http://www2.unccd.int/publications/global-land-outlook	The GLO presents an overview of the status of land and a clear set of responses to optimize land use, management, and planning, and thereby create synergies across sectors in the provision of land-based goods and services. This integrated approach is the basis of the conceptual framework for land degradation neutrality, a target which is seen as the driving vehicle for the implementation of the United Nations Convention to Combat Desertification (UNCCD) and an important part of the 2030 Agenda for Sustainable Development.	Results of the GLO were used for the thematic chapter on land and soil and relevant cross-cutting issues, like food security, including policy responses and outlooks.
Global Waste Management Outlook (2015)	UNEP, International Association of Solid Waste Management	http://www.unep.org/ourplanet/september-2015/unep-publications/global-waste-management-outlook	To assess the global state of waste management. Develop a holistic approach towards waste management and recognizing waste and resource management as a significant contributor to sustainable development and climate change mitigation. To complement the Sustainable Development Goals of the Post-2015 Development Agenda/SDGs, the Outlook sets forth Global Waste Management Goals and a Global Call to Action to achieve those goals.	Results of the Global Waste Management Outlook were used as a core reference to address key challenges of waste and resource management as cross-cutting issues within GEO-6, including policy responses and outlooks
Global Chemicals Outlook I	UNEP	http://www.unep.org/chemicalsandwaste/what-we-do/policy-and-governance/global-chemicals-outlook	Develop a comprehensive environmental understanding and up to date assessment of the trends and changes affecting the production and use of chemicals, their health and environmental effects, economic implications, and policy options throughout their life cycle. The GCO I is meant to be informative so as to illustrate both the economic interest and the necessity to invest in the sound management of chemicals.	Results of the Global Chemicals Outlook I were used as core reference to address key challenges of chemicals as cross-cutting issues within GEO-6, including policy responses and outlooks
Global Mercury Assessment (2002/2008/2013/2018)	UNEP	http://web.unep.org/chemicalsandwaste/what-we-do/technology-and-metals/mercury/global-mercury-assessment	The Global Mercury Assessment provides the most recent information available for the worldwide emissions, releases, and transport of mercury in atmospheric and aquatic environments. The Global Mercury Assessment is intended as a basis for decision making, emphasis is given to anthropogenic emissions (mercury going into the atmosphere) and releases (mercury going into water and land), that is, those associated with human activities.	Results of the latest Global Mercury Assessment (2013/2018) were used as reference within the air thematic chapter, the cross-cutting theme of chemicals, including policy responses and outlooks.
Global Gender and Environment Outlook (GGEO) (2016)	UNEP	http://web.unep.org/ggeo	The GGEO for the first time provides a comprehensive global overview of the linkages between gender and environment in the contexts of SDGs and 2030 Development agenda. Its objectives are to enable better understanding of the environment through a gender lens, to support better integration of gender perspectives in development and implementation of environmental policies at international and national levels, and to drive impact through partnerships.	For chapter 4 and 17 on Cross-cutting issues the GGEO has been instrumental as a basis for the gender-related language. As the GGEO is specifically looking into gender aspects of diverse environmental areas, policies, data and approaches, the insights of GGEO are also integrated into several other GEO6 chapters and sections.

Annex 1-3: Theory of Change for the sixth Global Environment Outlook (GEO-6)

Based on the principles of integrated environmental assessments, the theory of change in GEO-6 is embedded in its structure and purpose and based on a social process that moves a community of institutions and people towards a new way of (strategic) thinking and (goal-oriented) acting. Through this social process, the evidence presented in the GEO-6 assessment is considered legitimate, credible and relevant (salient) to the community, which facilitates its acceptance as an input to improved environmental policy, which in turn helps make progress towards sustainable development.

GEO-6 aims to create change through a process that encompasses data, science and experimental and participatory approaches. It uses multidisciplinary perspectives to generate knowledge-based conclusions. GEO-6 also aims to create change by highlighting the benefits and opportunities to citizens and communities from achieving change, even disruptive change. New earth observation and other technologies have revolutionized our ability to understand environmental change and its impacts on human well-being and vice versa. GEO-6 aims to communicate the results of the assessment in a way that can influence action by stakeholders and policymakers. This, in turn, facilitates the development of more appropriate, equitable, and effective policy responses, including shifting investment, production, distribution, and consumption in more sustainable directions, as well as better governance capacities at multiple scales.

Figure A.1 shows how GEO has impacts through its influence on people's actions:

Activities and process

The Global Environment Outlook process is designed through consultations with governments and other stakeholders. From these consultations, nominations of government officials, stakeholders and experts who will be involved in the process create the community that will follow, and be influenced by the process. Regular meetings and conference calls are necessary to keep this community engaged in the process and also to obtain their advice so that they feel ownership of the process and the product. Peer review and intergovernmental review processes allow a broader community of experts, governmental officials and stakeholders to contribute their advice and expertise and experience a higher level of engagement. The community members, motivated by a sense of ownership through their participation in the process, become ambassadors for GEO's messages.

Figure A.1: Theory of Change of GEO-6

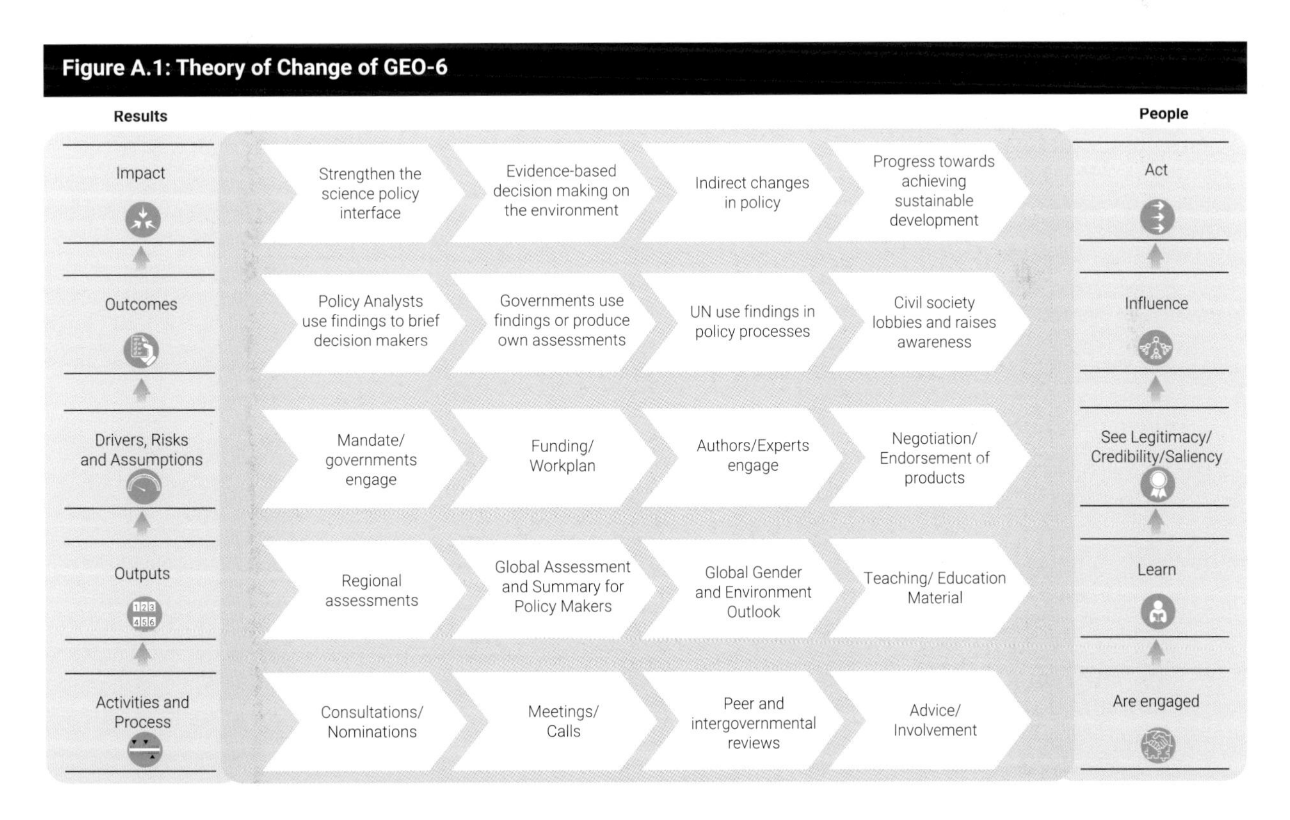

Outputs

The four main outputs of the GEO process document the evidence and rationale for the findings that will influence the future path of environmental and sustainable development policy. These outputs include:

- Six regional environmental assessments which present policy relevant information which is actionable either regionally and/or nationally. These regional assessments allow for a deeper ownership of the findings at a level where governments can act. They also engage different geographic groupings in the process of implementing GEO's findings;
- Following the publication of the regional assessments, a global environmental assessment is produced, which can look at broader issues, such as the state of the world's environment, presenting the findings at a level where governments, together, can act.
- One of the main equity issues related to environmental analysis is how environmental impacts and actions are differentiated by gender. For this reason, there is a systemic link with the Global Gender and Environment Outlook.
- The findings and new knowledge of GEO should be translated and disseminated through the production of teaching and outreach material to stimulate further capacity building. Capacity building makes GEO accessible to a much broader audience such as youth and educators and enables its findings and new recommended directions to be promoted more widely, thereby enhancing GEO's influence on environmental change over a longer period and strengthening the long-term durability of these changes.

Drivers, Assumptions and Risks

A main assumption of the assessment process is that the findings will be considered legitimate, scientifically credible and relevant by the intended audiences, especially governments. These three criteria are defined as:

- Legitimacy: considered unbiased and respecting different stakeholder perspectives and conforming to law or authoritative rules.
- Credibility: considered scientifically sound and authoritative. Producing information that can be believed and trusted.
- Relevant: considered timely and related to the needs of the end users.

The GEO achieves legitimacy, credibility, and relevance through several avenues, based on certain assumptions, and subject to some risks, including:

- A clear mandate is provided by governments to produce the GEO and governments help to define some of the key parameters, such as the timeframe, organizational structure, work plan, outline for the analysis, etc.
- Sufficient funding is made available, following a clear work plan which is developed and approved by the Secretariat, in consultation with appropriate advisory bodies that are part of the process.
- A sufficiently large and diverse cohort of authors and experts is engaged to produce the report such that they ensure the scientific credibility of the assessment and can devote the appropriate amount of time to the analysis and drafting of chapters during the approved period of the work programme. This includes peer reviewers and other experts working outside the main drafting process.
- A robust process for negotiating related products meant for policy makers (e.g. Summary for Policymakers) is undertaken, which is transparent and considers the views of all Member States. These products help increase the legitimacy of the assessment process and, with appropriate endorsement, can lend credibility to these products.

Outcomes

The outcomes of the GEO process focus on increased awareness of the current state of the environment, knowledge of the possible policy solutions that could be used to address these, including the future implications of not acting and the future benefits of following particular pathways to achieve environmental goals. In order to achieve these goals, the findings of the assessment must be understood and/or used by various actors and applied in their daily work and personal lives:

- Policy analysts must access and understand GEO's findings, using them appropriately to inform decision makers;
- More broadly, governments (and potentially other non-state actors) should understand the findings in order to use them to advance their policy work. Governments can also use the GEO methodology to prepare their own regional, national or sub-national assessments if desirable.
- United Nations and other international organizations should be able to understand and apply GEO's findings in their own assessment, policy work, and practice.
- Civil society and non-governmental organizations should be able to understand and apply the findings of GEO in their own work, e.g. by influencing the policy and decision-making processes on the environment.

Impact

The impact of the GEO will be judged by the responses and actions that governments, institutions, and people take in their work arenas and daily lives. To increase the impact of GEO, UN Environment facilitates actions in the following areas:

- Helping countries strengthen the science-policy interface through the promotion of the GEO findings and process;
- Promoting the use of evidence-based decision making based on the findings of the GEO, its various derivative products, and other scientific sources;
- Encouraging, directly and indirectly, changes at the regional and national policy level that are in line with the GEO reports and process.

The theory of change for GEO supports various actors, including national governments, to make progress towards achieving the Sustainable Development Goals. This can be facilitated by incorporating the findings of the GEO into the Agenda 2030 policy process and implementation.

Annex 1-4: Structure and rationale for confidence statements used in the sixth Global Environment Outlook

Guidance from the Scientific Advisory Panel

This document is adapted from guidance developed by the Intergovernmental Science-Policy Platform on Biodiversity and Ecosystem Services (IPBES), IPBES/5/INF/6.

Developing and applying confidence terms

Characterizing and communicating the confidence and uncertainty in findings is essential to ensure the scientific credibility of the assessment process, help stakeholders and decision-makers understand the strength and weight of the underlying evidence base and lead to more informed decision-making. This guidance note is intended to assist authors of the Global Environment Outlook (GEO-6) to describe, in a consistent and transparent manner, the confidence and uncertainty associated with their findings. The note suggests a common approach and calibrated language that can be used broadly for developing expert judgments and for evaluating and communicating the degree of certainty.

What is confidence?

The use of confidence statements in assessments reflects how assured authors are about the findings (data and information) presented within their chapters. Low confidence describes a situation where we have incomplete knowledge and therefore cannot fully explain an outcome or reliably predict a future outcome, whereas high confidence conveys that we have extensive knowledge and are able to explain an outcome or predict a future outcome with much greater certainty.

Confidence terms should always be used in three key parts of an assessment:

1. They should be assigned to the key findings in **Executive Summaries** of the technical chapters in an assessment report.
2. They should be used for the key findings in any **Technical Summary** produced from the main report.
3. They should be used within the **Summary for Policymakers.**

It is not mandatory to apply confidence terms throughout the main text of the assessment report. However, in some parts of the main text, in areas where there are a range of views that need to be described, confidence terms may be applied where considered appropriate by the author team. In no case should the terms be used colloquially or casually to avoid confusing readers. Only use these terms if you have followed the recommended steps for assessing confidence.

Assessing confidence

As they develop their key findings, author teams should evaluate the associated evidence and agreement within the evidence base. Depending on the nature of the evidence evaluated, teams may either use a qualitative level of confidence or quantify the uncertainty in the finding probabilistically. Qualitative assessments of confidence reflect expert judgment about agreement and evidence. Quantitative assessments of confidence are estimates of the likelihood (probability) that a well-defined outcome will occur in the future. Probabilistic estimates are based on statistical analysis of observations or model results, or both, combined with expert judgment. However, it may be that quantitative assessments of confidence are not possible for all findings due to the nature of the evidence available.

In order to ensure consistency in communication, specific phrases or terms will be used to describe the level of confidence or the extent of uncertainty. The choice of the term used will be based on the author team's expert judgement on the quantity and quality of the supporting evidence and the level of scientific agreement.

The sixth Global Environment Outlook uses a four-box model of confidence **(see Figure A.2)** based on evidence and agreement that gives four main confidence terms for the qualitative assessment of confidence: "well established" (much evidence and high agreement), "unresolved" (much evidence but low agreement), "established but incomplete" (limited evidence but good agreement) and "inconclusive" (limited or no evidence and little agreement).

Qualitative assessment of confidence

This section discusses the process and language that all author teams must apply to evaluate and communicate confidence qualitatively. The following factors should be considered while assessing the confidence in a message or finding: the type, quantity, quality and consistency of evidence (the existing peer-reviewed literature and grey literature etc.), and the level of agreement (the level of concurrence in the data, literature and amongst experts, not just across the author team). The author team's expert judgement on the level

Figure A.2: The four-box model for the qualitative communication of confidence

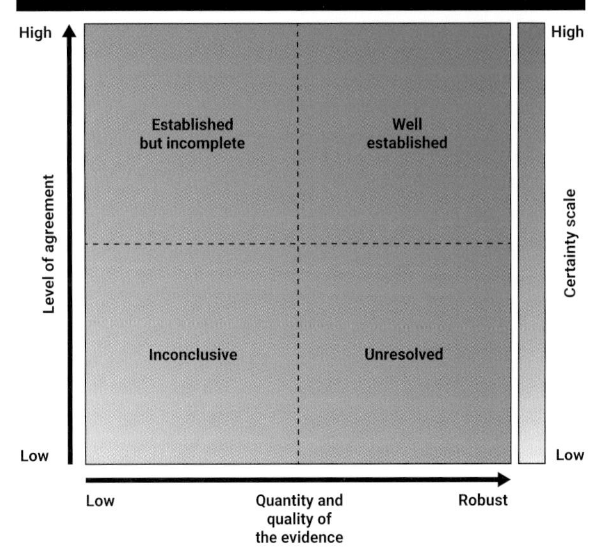

Confidence increases towards the top-right corner as suggested by the increasing strength of shading.

Source: IPBES (2017).

of evidence and agreement should then be used to apply a confidence term **(Figure A.2)**:

- ***Inconclusive*** – existing as or based on a suggestion or speculation; no or limited evidence and no clear consensus in the evidence.
- ***Unresolved*** – multiple independent studies exist but conclusions do not agree.
- ***Established but incomplete*** – general agreement although only a limited number of studies exist but no comprehensive synthesis and, or the studies that exist imprecisely address the question.
- ***Well established*** – comprehensive meta-analysis or other synthesis or multiple independent studies that agree.

The ***well-established*** box in **Figure A.2** can be further subdivided in order to give author teams the flexibility to emphasise key messages and findings that the author team have very high confidence in:

- ***Very well established*** – very comprehensive evidence base and very low amount of disagreement.
- ***Virtually certain*** –very robust evidence base covering multiple temporal and spatial scales and almost no disagreement.

Note that the term "virtually certain" above still reflects a qualitative assessment of confidence. It should not be interpreted probabilistically and does not convey any level of "statistical significance". These sub-classifications of the "well established" box provide authors flexibility to emphasize findings that may be considered as fact or reflecting scientific consensus.

The degree of confidence in findings that are conditional on other findings should be evaluated and reported separately.

When evaluating the level of evidence and agreement for a statement, it is important to standardise the use of the terms within and across the author teams, and when possible, across the assessment, to ensure their consistent use. The use of the above confidence terms can be standardised by taking key messages and findings in the **Executive Summaries** and discussing, as an author team, what terms should be applied and the reasons why. When appropriate, teams may consider using formal elicitation methods to organise and quantify the selection of confidence terms.

Teams should be aware of the tendency for a group to converge on an expressed view and become over confident in it. One method to avoid this would be to ask each member of the author team to write down his or her individual assessment of the level of confidence before entering into a group discussion. If this is not done before group discussion, important views and ranges of confidence may be inadequately discussed and assessed. It is important to recognize when individual views are adjusting as a result of group interactions and allow adequate time for such changes in viewpoint to be reviewed (Mastrandrea *et al.* 2010). Whichever approach is taken, traceable accounts should be produced and recorded to demonstrate how confidence was evaluated (see section on Traceability).

It is important to carefully consider how the sentences in the key messages and findings are structured because it will influence the clarity with which we communicate our understanding of the level of confidence. For example, sometimes the key finding combines an element that is ***well-established*** with one that is ***established but incomplete***. In this case it can be helpful to arrange the phrasing so that the ***well-established*** element comes first, and the ***established but incomplete*** element comes second, or as a separate sentence. Where possible avoid the use of the ***unresolved*** and ***established but incomplete*** by writing or rewording key messages and findings in terms of what is known rather than unknown. Author teams should focus on presenting what is ***well-established*** as far as possible in order to make it clear to decision makers what is known. Assigning confidence terms to our key findings will therefore often require that we re-write sentences, rather than simply adding the terms to existing text.

Quantitative assessment of confidence

In many cases it may be possible to quantitatively assess the uncertainty in an outcome or event. This section discusses the process and language that author teams may wish to apply in order to evaluate and communicate the confidence that an outcome will occur quantitatively. Likelihood expresses a probabilistic estimate of the occurrence of a single event or of an outcome within a given range. Probabilistic estimates are based on statistical analysis of observations or model results, or both, combined with expert judgment.

When sufficient probabilistic information is available, consider ranges of outcomes and their associated probabilities with attention to outcomes of potential high consequence. The author team's expert judgement on the magnitude of the probability should then be used to apply a likelihood term from **Figure A.3**.

Figure A.3: Likelihood scale for the quantitative communication of the probability of an outcome occurring

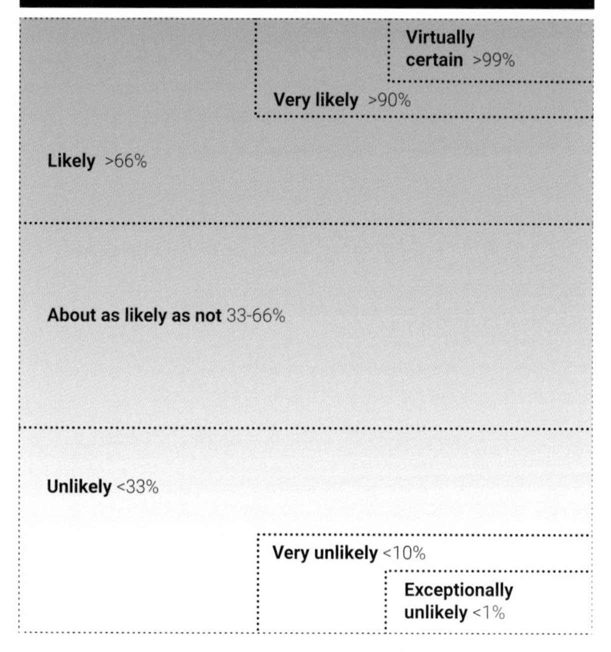

Note that the extreme levels of probability are nested within the broader levels of "likely" and "unlikely".

Source: Adapted from Mastrandrea *et al.* (2010).

Categories in **Figure A.3** can be considered to have nested boundaries. For example, describing an outcome as ***likely*** or ***very likely*** conveys in both cases that the probability of this outcome could fall within the range of 95 per cent to 100 per cent probability, but in the case of ***likely***, the larger range (66-100 per cent) indicates a higher degree of confidence than ***very likely*** (90-100 per cent). In making their expert judgement, author teams should start at ***about as likely as not*** and consider whether there is sufficient quantitative information available to assign either a ***likely*** or ***unlikely*** probability range. Only after thinking about this initial range should the author teams consider whether there is sufficient evidence to move to more extreme levels of probability.

Author teams should note that using a likelihood term for a specific outcome implies that alternative outcomes have the inverse likelihood e.g., if an outcome is ***likely*** (a range of 66-100 per cent) than that would imply that other outcomes are ***unlikely*** (0-33 per cent probability).

If the author team consider that sufficiently robust information is available with which to make a 'best estimate' of the probability of the occurrence of an event, then it is preferable to specify the full probability range (e.g. 90-95 per cent) in the text without using the terms in **Figure A.3**. Also, ***about as likely as not*** should not be used to communicate a lack of knowledge, only an estimate of probability based on the available information.

Author teams should be aware of the way in which key messages and findings are phrased. The way in which a statement is framed will have an effect on how it is interpreted e.g., a 10 per cent chance of dying is interpreted more negatively than a 90 per cent chance of surviving. Consider reciprocal statements to avoid value-laden interpretations e.g., report chances both of dying and of surviving (Mastrandrea *et al.* 2010).

Finally, author teams should try not to avoid controversial events, such as impacts or events with high consequence but extremely low probability, in their effort to achieve consensus within an author team.

How to present confidence terms - Presenting confidence using the four-box model

Confidence terms are communicated as part of the key findings of an assessment. The key findings are set out in the **Executive Summaries** for each of the assessment's chapters in the full technical report. The key findings are the facts and information drawn directly from the chapter. It is recommended that key findings should be set out as follows.

The first sentence of the finding should be bolded and contain a confidence term from the four-box model in italics and brackets at the end of the sentence. This first sentence is followed by two to four sentences which then supports the information contained in this first sentence. Subsequent sentences may contain confidence terms within brackets where appropriate. It is not necessary to include confidence terms with each sentence if the whole paragraph falls under the same confidence term.

The words that make up the four-box model and likelihood scale should not be used in the text of the assessment except when formally assigning confidence**.** If, for example, there was a sentence that used the word "likely" but not with the intended meaning from the likelihood scale, then the word should be replaced with another (e.g. probably).

Presenting confidence using the likelihood scale

In some instances, as above, author teams may wish to complement the use of the ***well-established*** confidence term with a term from the likelihood scale. If terms from the likelihood scale are used then they should be incorporated into the text and italicised prior to the impact or outcome the probability of which they are describing.

Traceability

The author team's expert judgment of their confidence in the key messages and findings should be explained by providing a clear traceable account. A traceable account is a description in the chapter of the evaluation of the type, quantity, quality and consistency of the evidence and level of agreement that forms the basis for the given key message or finding (Mastrandrea *et al.* 2010). Where possible, the description should identify and discuss the sources of confidence. In order to ensure consistency in how the author teams classify sources of confidence within and across Global Environment Outlook assessments, author teams should use the typology shown in **Table A.2** below.

A key statement in the **Summary for Policymakers** should be readily traceable back to an **Executive Summary** statement(s) that in turn should be readily traceable back to a section(s) of the chapter text, which in turn should be traceable where appropriate to the primary literature through references.

References to the relevant **Executive Summary** statement should be included in curly brackets (e.g. {1.2}).

Summary of Steps for applying confidence terms

The steps recommended for assessing and communicating confidence for Executive Summaries and Summaries for Policymakers.

1. Identify the chapter's key messages and findings.
2. Evaluate the supporting evidence and the level of scientific agreement.
3. Establish whether the evidence is probabilistic or not (e.g. from model predictions).
4. Where the evidence is qualitative instead or probabilistic, select a confidence term from the four-box model **(Figure A.2)** to communicate the author team's confidence in the key message or finding.
   (a) Assess the quantity and quality of evidence and the level of agreement in the scientific community.
   (b) Establish how confident the author team is and select the appropriate term.
5. Where quantitative estimates of the probability of an outcome or impact occurring are available (e.g. from model predictions), select a likelihood term from the likelihood scale **(Figure A.3)** to communicate the author teams' expert judgement of the range of the probability of occurrence.
6. Ensure that there is always a 'traceable account' in the main text describing how the author team adopted the specific level of confidence, including the important lines of evidence used, standard of evidence applied and approaches to combine/reconcile multiple lines of evidence.
7. OPTIONAL: Consider using formal frameworks for assessing expert judgement for each author team.

Table A.2: Sources of low confidence

Sources of low confidence	Definition and examples	Qualities	Means of dealing with low confidence
Imprecise meanings of words (Linguistic uncertainty)	Vagueness and ambiguity of terms EXAMPLE: When terms such as human welfare, risks, plant reproductive success, pollination deficits are central to the finding.	Reducible Not quantifiable	❖ Clear, common definition of terms (IPBES Common Glossary). ❖ Protocols as used in agent-based modelling to deal with context dependence.
Inherently unpredictable systems (Stochastic uncertainty)	Low confidence due to the chaotic nature of complex natural, social or economic systems (sometimes known as 'aleatory' uncertainty). Findings that depend on weather or climate variables, or market prices, will be subject to this low confidence. EXAMPLE: Pollination deficits and values measured at local scales.	Not reducible Quantifiable	❖ Clear communication. ❖ Using probabilistic approaches. ❖ Support large scale, long term multi-site studies to quantify the variation over space and time to characterise the low confidence. ❖ Evidence synthesis. ❖ Capacity building for researchers and decision makers.
Limits of methods and data (Scientific uncertainty)	Where there is insufficient data to fully answer the question, due to unsatisfactory methods, statistical tools, experimental design or data quality (also referred to as epistemic uncertainty). EXAMPLE: Impacts of pesticides on pollinator populations in the field, trends in pollinator abundance, estimations of ecosystem service delivery.	Reducible Quantifiable	❖ Acknowledge differences in conceptual frameworks (within and between knowledge systems). ❖ Improve experimental design. ❖ Expand data collection. ❖ Support detailed, methodological research. ❖ Knowledge quality assessment. ❖ Evidence synthesis. ❖ Capacity building for scientists.
Differences in understanding of the world (Decision uncertainty)	Low confidence that is caused by variation in subjective human judgments, beliefs, world views and conceptual frameworks (sometimes called epistemic uncertainty). In terms of policy decisions, low confidence is due to preferences and attitudes that may vary with social and political contexts. This can mean a finding looks different in different knowledge systems that cannot easily be aligned. EXAMPLES: Effects of organic farming look different if you take the view that wild nature beyond farmland has a higher value than farmland biodiversity, and overall food production at a large scale is more important than local impacts. There are divergent interpretations/perceptions of well-being.	Sometimes reducible Not quantifiable	❖ Acknowledge differences in conceptual frameworks (within and between knowledge systems). ❖ Document, map and integrate where possible. ❖ Acknowledge existence of biases. ❖ Multi-criteria analysis, decision support tools. ❖ Capacity building for decision makers.

* Adapted from the IPBES guide on the production of assessments

Annex 4-1: Towards monitoring the environmental dimension of the SDGs

Introduction

The Sustainable Development Goals, the Multilateral Environment Agreement indicators and other indicators related to the environmental drivers, state, pressures, impacts and responses are useful for conducting environmental assessments, including on particular aspects of the environment, and multiple levels (global, regional and national). Additionally, socio-economic indicators can be combined with environmental indicators to better contextualized the environment and to understand the linkages between the environment, people and the economy.

The Sustainable Development Goals (SDGs) are a framework which elaborates the global development agenda toward achieving a better and more sustainable future for all. The Sustainable Development Goals are a call for action by all countries – poor, rich and middle-income – to promote prosperity while protecting the planet. They recognize that ending poverty must go hand-in-hand with strategies that build economic growth and address a range of social needs including education, health, social protection, and job opportunities, while tackling climate change and increasing environmental protection. A monitoring framework of 244 indicators has been agreed for monitoring the SDGs; however, this indicator framework does not represent a complete list of all information that is needed to understand the planet.

The current GEO publication is accompanied by a statistical annex https://environmentlive.unep.org/media/global_assessment/review_documents/annex4_1.pdf which includes tabular information to be used by technical experts to better understand the environment and the nexus between the environment, society and the economy. Additionally, this annex extracts certain indicators from the Statistical Annex in order to highlight the current state of progress towards achieving the environment-related SDGs. Note that the Statistical Annex does not include any analysis or figures. An analysis of the information in the Statistical Annex will be forthcoming in a GEO derivate product entitled: Measuring Progress, which is a follow-up to a publication that was produced for GEO-5.

This annex has taken the data in the Statistical Annex which directly links to particular environmental SDGs and extrapolated information in order to provide a summary of the current state of the environmental dimension of the SDGs.

Statistical Methods

The phrase, the environmental dimension of the SDGs, does not have a precise definition and there are many different views on what the environmental dimension of development should include (should it include only indicators related to the state of the environment, what about indicators related to access to natural resources such as water or perhaps it should include all indicators, since every aspect of life is related to the environment). For the purpose of this analysis, the list of environment-related indicators from the perspective of the UN Environment Programme will be used. A list of SDG indicators which are considered to be part of the environmental dimension of the SDGs was established by the UN Environment Programme Secretariat and was presented to the UN Environment Assembly Committee of Permanent Representatives at the sub-committee meeting on 20 September 2018 (see: https://www.unenvironment.org/events/subcommittee-meetings/committee-permanent-representatives-subcommittee-meeting-14) - it is also included at the end of this document.

The data in the Statistical Annex and in this paper are based on data which are included in the UN Environment Live Global database (https://uneplive.unep.org). The data in the database come from a variety of international databases and other sources, UN Environment maintains strict criteria for the information in the UN Environment Live Global database which include:

1. data must be published by a UN agency or other reputable global entity;
2. data must have transparent methodologies and metadata which is publicly available;
3. data must be compiled at the global level (data which is only available for a single country or region is not included);
4. only data with a timeseries which includes more than 2 timepoints is included; and
5. the most recent point in the timeseries must be no more than 10 years old.

The UN Environment Live Global database also uses a statistical methodology for aggregating national data to produce global, regional, sub-regional and special country groupings; information on aggregation procedures can be found at: https://uneplive.unep.org/media/docs/graphs/aggregation_methods.pdf.

This annex uses simple extrapolation procedures to estimate if the SDG targets would be met based on the current state of the SDG indicators (i.e. no efforts were made to change the current data trend). Thus progress in the next 15 years was estimated to be identical to the progress in the last 15 years at a global level. UN Environment extrapolated the aggregated data using the exponential regression model based on available data points from year to year. The cut-off used for data extrapolation and analysis is the year 2030. We determined if the target will be met or not by comparing the 2030 data to the indicator target. For example, if there is an increase in a target by 5% by 2030, it is considered as a positive progress, a change in condition based on this indicator is shown as a positive direction between 2000 and 2030. The same is applied for any decrease higher than -5 per cent. Any per cent change between +5 per cent and -5 per cent has been considered as representing very little negative or positive change in this indicator between 2000 and 2030.

An indicator is considered to have no data if there is not enough data for global aggregation. To determine this we have followed the global aggregation model explained on Environment Live. Where sufficient data are available, aggregations are performed for all indicators which share a common unit and are believed to be internationally comparable. Indicators which are expressed in national currency or another national unit are not aggregated.

Progress toward the SDGs

Of the 93 environment-related SDG indicators, there are 20 for which good progress has been made over the last 15 years and if this progress continues then it is likely that these SDGs will be met. However, many of these indicators include particular reporting or funding efforts. For example, there has been an increase in terrestrial, mountain and marine protected areas; the effort to combat invasive species has increased; there has been significant progress toward grid-connected renewable energy; sustainability reporting and mainstreaming in policy has increased; and development assistance for climate change and the environment has increased. For 8 of the environment-related SDG indicators the progress has been relatively flat and for 7 of the SDG indicators, additional efforts will be needed. In particular, many of the indicators related to the state of the environment show a negative trend (these include indicators related to forests, sustainable fisheries, endangered species, domestic material consumption and material footprint). Unfortunately, this is a very incomplete picture because there is too little data to formally assess the status of 58 of the 93 environment-related SDG indicators – however, scientific research and the current GEO has shown that many of these areas have shown a particular negative trend. A snapshot of the progress toward these indicators is shown in the graphic below and an overall table of progress is also shown.

Figure A.4: Relative progress on SDG indicators

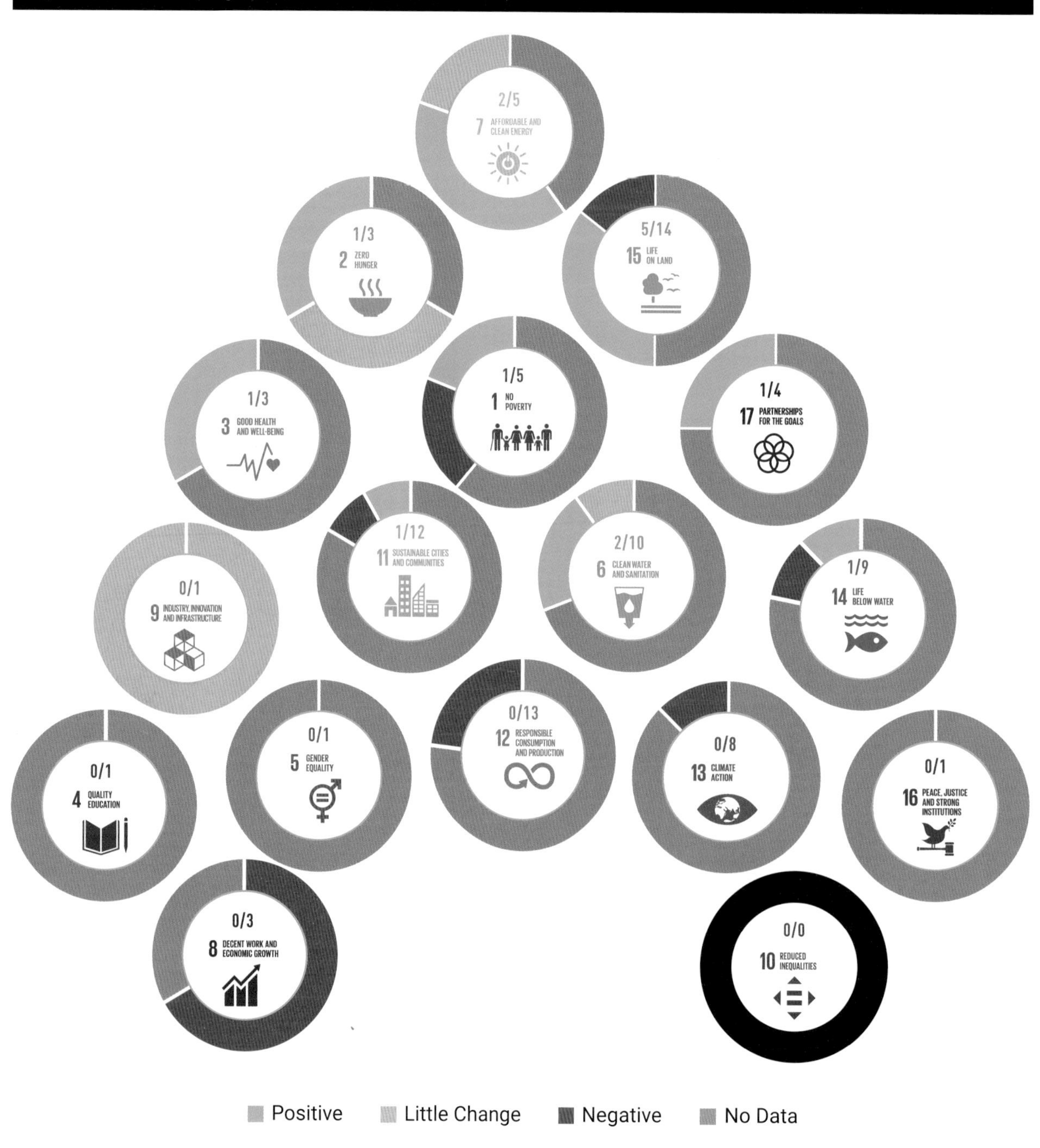

Figure A.5: Environmental Dimensions of the SDGs – Score Card

SDG 1: End Poverty

Indicator	+ change	Little change	Negative change	Some data	No data
Land tenure (SDG 1.4.2)					●
Disasters: persons affected (SDG 1.5.1)			●		
Disasters: economic loss (SDG 1.5.2)	●				
Disaster risk reduction strategies (SDG 1.5.3)				●	
Disaster risk reduction for local government (SDG 1.5.4)				●	

SDG 2: Food Security

Indicator	+ change	Little change	Negative change	Some data	No data
Sustainable agricultural practices (SDG 2.4.1)					●
Secure genetic resources for food (SDG 2.5.1)	●				
Local breeds for agriculture (SDG 2.5.2)		●			

SDG 3: Health

Indicator	+ change	Little change	Negative change	Some data	No data
Air pollution mortality (SDG 3.9.1)				●	
Water-related mortality (SDG 3.9.2)				●	
Unintentional poisoning mortality (SDG 3.9.3)	●				

SDG 4: Education

Indicator	+ change	Little change	Negative change	Some data	No data
Environmental education (SDG 4.7.1)					●

SDG 5: Gender

Indicator	+ change	Little change	Negative change	Some data	No data
Women agricultural land owners (SDG 5.a.1)					●

SDG 6: Water

Indicator	+ change	Little change	Negative change	Some data	No data
Safe drinking water (SDG 6.1.1)	●				
Wastewater treatment (SDG 6.3.1)					●
Water quality (SDG 6.3.2)				●	
Water efficiency (SDG 6.4.1)					●
Water stress (SDG 6.4.2)				●	
Water resource management (SDG 6.5.1)				●	
Water cooperation (SDG 6.5.2)				●	
Water ecosystems (SDG 6.6.1)		●			
Investment in water and sanitation (SDG 6.a.1)	●				
Local water management (SDG 6.b.1)				●	

SDG 7: Energy

Indicator	+ change	Little change	Negative change	Some data	No data
Reliance on clean fuels (SDG 7.1.2)	●				
Renewable energy (SDG 7.2.1)		●			
Energy intensity (SDG 7.3.1)	●				
Clean energy research and technology (SDG 7.a.1)					●
Investment in energy efficiency (SDG 7.b.1)					●

SDG 8: Decent Work and Economic Growth

Indicator	+ change	Little change	Negative change	Some data	No data
Material footprint (SDG 8.4.1)			●		
Domestic material consumption (SDG 8.4.2)			●		
Employment in sustainable tourism (SDG 8.9.2)					●

SDG 9: Industry, Innovation and Infrastructure

Indicator	+ change	Little change	Negative change	Some data	No data
CO_2 Emissions (SDG 9.4.1)		●			

SDG 10: Reduced Inequalities

The environmental dimension is not represented in Goal 10

SDG 11: Cities and Communities

Indicator	+ change	Little change	Negative change	Some data	No data
Access to public transport (SDG 11.2.1)					●
Land consumption (SDG 11.3.1)					●
Urban planning (SDG 11.3.2)					●
Investment in cultural and natural heritage (SDG 11.4.1)					●
Disasters: persons affected (SDG 11.5.1)			●		
Disasters: economic loss (SDG 11.5.2)	●				
Urban solid waste management (SDG 11.6.1)				●	
Ambient air pollution (SDG 11.6.2)				●	
Public land in cities (SDG 11.7.1)					●
Local disaster risk reduction strategies (SDG 11.b.1)				●	
National disaster risk reduction strategies (SDG 11.b.2)				●	
Financial ass. to least developed countries (SDG 11.c.1)					●

SDG 12: Responsible Lifestyles

Indicator	+ change	Little change	Negative change	Some data	No data
Action plans for sustainability (SDG 12.1.1)				●	
Material footprint (SDG 8.4.1)			●		
Domestic material consumption (SDG 8.4.2)			●		
Food loss and waste (SDG 12.3.1)					●
Chemicals convention reporting (SDG 12.4.1)			●		
Hazardous waste generation (SDG 12.4.2)					●
Recycling (SDG 12.5.1)					●
Corporate sustainability reporting (SDG 12.6.1)					●
Sustainable public procurement (SDG 12.7.1)					●
Education for sustainable lifestyles (SDG 12.8.1)					●
Research for sustainable lifestyles (SDG 12.a.1)					●
Sustainable tourism strategies (SDG 12.b.1)					●
Fossil fuel subsidies (SDG 12.c.1)				●	

SDG 13: Climate Action

Indicator	+ change	Little change	Negative change	Some data	No data
Disasters: persons affected (SDG 13.1.1)			●		
Disaster risk reduction strategies (SDG 13.1.2)				●	
Disaster risk reduction for local government (SDG 13.1.3)				●	
Climate change action plans (SDG 13.2.1)					●
Climate change education (SDG 13.3.1)					●
Community based approaches for CC (SDG 13.3.2)					●
Resources mobilized for climate action (SDG 13.a.1)					●
Climate action support for LDCs (SDG 13.b.1)					●

SDG 14: Oceans

Indicator	+ change	Little change	Negative change	Some data	No data
Marine litter and coastal eutrophication (SDG 14.1.1)					●
Management of marine areas (SDG 14.2.1)					●
Marine acidification (SDG 14.3.1)					●
Sustainable fish stocks (SDG 14.4.1)			●		
Marine protected areas (SDG 14.5.1)	●				
Fishing regulation (SDG 14.6.1)					●
Fishing contribution to GDP (SDG 14.7.1)					●
Research on sustainable marine technology (SDG 14.a.1)				●	
Ocean conservation instruments (SDG 14.c.1)					●

SDG 15: Land and biodiversity

Indicator	+ change	Little change	Negative change	Some data	No data
Forest area (SDG 15.1.1)			●		
Protection of key biodiversity areas (SDG 15.1.2)	●				
Sustainable forest management (SDG 15.2.1)	●				
Land degradation (SDG 15.3.1)					●
Mountain protected areas (SDG 15.4.1)	●				
Mountain green cover (SDG 15.4.2)				●	
Endangered species (SDG 15.5.1)			●		
Strategies for sharing biodiversity benefits (SDG 15.6.1)	●				
Trade in poached or illicitly trafficked wildlife (SDG 15.7.1)					●
Strategies for preventing invasive alien species (SDG 15.8.1)					●
Progress towards Aichi Biodiversity Target 2 (SDG 15.9.1)					●
Investment in biodiversity and ecosystems (SDG 15.a.1)	●				
Investment in sustainable forests (SDG 15.b.1)					●
Protection against poaching, trafficking and trade (15.c.1)					●

SDG 16: Peace and Justice

Indicator	+ change	Little change	Negative change	Some data	No data
Participation in global governance (SDG 16.8.1)				●	

SDG 17: Partnerships and means of implementation

Indicator	+ change	Little change	Negative change	Some data	No data
Science and technology cooperation (SDG 17.6.1)					●
Funding for environmentally sound technologies (SDG17.7.1)					●
Funding for capacity building (SDG 17.9.1)	●				
Mechanisms enhancing policy coherence (SDG 17.14.1)					●

Represents a change in condition based on this indicator in a + direction between 2000-2017 (does not represent that the SDG target will be achieved).
Represents very little negative or positive change in this indicator between 2000-2017.
Represents a change in condition based on this indicator in a negative direction between 2000-2017.
Some data is available, but not enough to analyze changes over time.
No data is available.

Table A.3: Description of environment relevant SDG targets and indicators in the SDG Global Indicator Framework

Goal	70 Targets	93 Indicators
Goal 1. End poverty in all its forms everywhere	1.4 By 2030, ensure that all men and women, in particular the poor and the vulnerable, have equal rights to economic resources, as well as access to basic services, ownership and control over land and other forms of property, inheritance, natural resources, appropriate new technology and financial services, including microfinance	1.4.2 Proportion of total adult population with secure tenure rights to land, with legally recognized documentation and who perceive their rights to land as secure, by sex and by type of tenure
	1.5 By 2030, build the resilience of the poor and those in vulnerable situations and reduce their exposure and vulnerability to climate-related extreme events and other economic, social and environmental shocks and disasters	1.5.1 Number of deaths, missing persons and directly affected persons attributed to disasters per 100,000 population
		1.5.2 Direct economic loss attributed to disasters in relation to global gross domestic product (GDP)
		1.5.3 Number of countries that adopt and implement national disaster risk reduction strategies in line with the Sendai Framework for Disaster Risk Reduction 2015-2030
		1.5.4 Proportion of local governments that adopt and implement local disaster risk reduction strategies in line with national disaster risk reduction strategies
Goal 2. End hunger, achieve food security and improved nutrition and promote sustainable agriculture	2.4 By 2030, ensure sustainable food production systems and implement resilient agricultural practices that increase productivity and production, that help maintain ecosystems, that strengthen capacity for adaptation to climate change, extreme weather, drought, flooding and other disasters and that progressively improve land and soil quality	2.4.1 Proportion of agricultural area under productive and sustainable agriculture
	2.5 By 2020, maintain the genetic diversity of seeds, cultivated plants and farmed and domesticated animals and their related wild species, including through soundly managed and diversified seed and plant banks at the national, regional and international levels, and promote access to and fair and equitable sharing of benefits arising from the utilization of genetic resources and associated traditional knowledge, as internationally agreed	2.5.1 Number of plant and animal genetic resources for food and agriculture secured in either medium or long-term conservation facilities
		2.5.2 Proportion of local breeds classified as being at risk, not-at-risk or at unknown level of risk of extinction
Goal 3. Ensure healthy lives and promote well-being for all at all ages	3.9 By 2030, substantially reduce the number of deaths and illnesses from hazardous chemicals and air, water and soil pollution and contamination	3.9.1 Mortality rate attributed to household and ambient air pollution
		3.9.2 Mortality rate attributed to unsafe water, unsafe sanitation and lack of hygiene (exposure to unsafe Water, Sanitation and Hygiene for All (WASH) services)
		3.9.3 Mortality rate attributed to unintentional poisoning
Goal 4. Ensure inclusive and equitable quality education and promote lifelong learning opportunities for all (1/1/0)	4.7 By 2030, ensure that all learners acquire the knowledge and skills needed to promote sustainable development, including, among others, through education for sustainable development and sustainable lifestyles, human rights, gender equality, promotion of a culture of peace and non-violence, global citizenship and appreciation of cultural diversity and of culture's contribution to sustainable development	4.7.1 Extent to which (i) global citizenship education and (ii) education for sustainable development, including gender equality and human rights, are mainstreamed at all levels in: (a) national education policies, (b) curricula, (c) teacher education and (d) student assessment
Goal 5. Achieve gender equality and empower all women and girls	5.a Undertake reforms to give women equal rights to economic resources, as well as access to ownership and control over land and other forms of property, financial services, inheritance and natural resources, in accordance with national laws	5.a.1 (a) Proportion of total agricultural population with ownership or secure rights over agricultural land, by sex; and (b) share of women among owners or rights-bearers of agricultural land, by type of tenure
Goal 6. Ensure availability and sustainable management of water and sanitation for all	6.1 By 2030, achieve universal and equitable access to safe and affordable drinking water for all	6.1.1 Proportion of population using safely managed drinking water services
	6.3 By 2030, improve water quality by reducing pollution, eliminating dumping and minimizing release of hazardous chemicals and materials, halving the proportion of untreated wastewater and substantially increasing recycling and safe reuse globally	6.3.1 Proportion of wastewater safely treated
		6.3.2 Proportion of bodies of water with good ambient water quality
	6.4 By 2030, substantially increase water-use efficiency across all sectors and ensure sustainable withdrawals and supply of freshwater to address water scarcity and substantially reduce the number of people suffering from water scarcity	6.4.1 Change in water-use efficiency over time
		6.4.2 Level of water stress: freshwater withdrawal as a proportion of available freshwater resources
	6.5 By 2030, implement integrated water resources management at all levels, including through transboundary cooperation as appropriate	**6.5.1** Degree of integrated water resources management implementation (0-100)
		6.5.2 Proportion of transboundary basin area with an operational arrangement for water cooperation
	6.6 By 2020, protect and restore water-related ecosystems, including mountains, forests, wetlands, rivers, aquifers and lakes	**6.6.1** Change in the extent of water-related ecosystems over time
	6.a By 2030, expand international cooperation and capacity-building support to developing countries in water- and sanitation-related activities and programmes, including water harvesting, desalination, water efficiency, wastewater treatment, recycling and reuse technologies	**6.a.1** Amount of water- and sanitation-related official development assistance that is part of a government-coordinated spending plan
	6.b Support and strengthen the participation of local communities in improving water and sanitation management	**6.b.1** Proportion of local administrative units with established and operational policies and procedures for participation of local communities in water and sanitation management

Note: Indicators for which UN Environment is Custodian Agency are marked in blue font

Goal	70 Targets	93 Indicators
Goal 7. Ensure access to affordable, reliable, sustainable and modern energy for all	7.1.2 Proportion of population with primary reliance on clean fuels and technology	7.1.2 Proportion of population with primary reliance on clean fuels and technology
	7.2 By 2030, increase substantially the share of renewable energy in the global energy mix	7.2.1 Renewable energy share in the total final energy consumption
	7.3 By 2030, double the global rate of improvement in energy efficiency	7.3.1 Energy intensity measured in terms of primary energy and GDP
	7.a By 2030, enhance international cooperation to facilitate access to clean energy research and technology, including renewable energy, energy efficiency and advanced and cleaner fossil-fuel technology, and promote investment in energy infrastructure and clean energy technology	7.a.1 International financial flows to developing countries in support of clean energy research and development and renewable energy production, including in hybrid systems
	7.b By 2030, expand infrastructure and upgrade technology for supplying modern and sustainable energy services for all in developing countries, in particular least developed countries, small island developing States and landlocked developing countries, in accordance with their respective programmes of support	7.b.1 Investments in energy efficiency as a proportion of GDP and the amount of foreign direct investment in financial transfer for infrastructure and technology to sustainable development services
Goal 8. Promote sustained, inclusive and sustainable economic growth, full and productive employment and decent work for all	8.4 Improve progressively, through 2030, global resource efficiency in consumption and production and endeavour to decouple economic growth from environmental degradation, in accordance with the 10Year Framework of Programmes on Sustainable Consumption and Production, with developed countries taking the lead	**8.4.1** Material footprint, material footprint per capita, and material footprint per GDP
		8.4.2 Domestic material consumption, domestic material consumption per capita, and domestic material consumption per GDP
	8.9 By 2030, devise and implement policies to promote sustainable tourism that creates jobs and promotes local culture and products	8.9.2 Proportion of jobs in sustainable tourism industries out of total tourism jobs
Goal 9. Build resilient infrastructure, promote inclusive and sustainable industrialization and foster innovation	9.4 By 2030, upgrade infrastructure and retrofit industries to make them sustainable, with increased resource-use efficiency and greater adoption of clean and environmentally sound technologies and industrial processes, with all countries taking action in accordance with their respective capabilities	9.4.1 CO2 emission per unit of value added
Goal 11. Make cities and human settlements inclusive, safe, resilient and sustainable	11.2 By 2030, provide access to safe, affordable, accessible and sustainable transport systems for all, improving road safety, notably by expanding public transport, with special attention to the needs of those in vulnerable situations, women, children, persons with disabilities and older persons	11.2.1 Proportion of population that has convenient access to public transport, by sex, age and persons with disabilities
	11.3 By 2030, enhance inclusive and sustainable urbanization and capacity for participatory, integrated and sustainable human settlement planning and management in all countries	11.3.1 Ratio of land consumption rate to population growth rate
		11.3.2 Proportion of cities with a direct participation structure of civil society in urban planning and management that operate regularly and democratically
	11.4 Strengthen efforts to protect and safeguard the world's cultural and natural heritage	11.4.1 Total expenditure (public and private) per capita spent on the preservation, protection and conservation of all cultural and natural heritage, by type of heritage (cultural, natural, mixed and World Heritage Centre designation), level of government (national, regional and local/municipal), type of expenditure (operating expenditure/investment) and type of private funding (donations in kind, private non-profit sector and sponsorship)
	11.5 By 2030, significantly reduce the number of deaths and the number of people affected and substantially decrease the direct economic losses relative to global gross domestic product caused by disasters, including water-related disasters, with a focus on protecting the poor and people in vulnerable situations	11.5.1 Number of deaths, missing persons and directly affected persons attributed to disasters per 100,000 population
		11.5.2 Direct economic loss in relation to global GDP, damage to critical infrastructure and number of disruptions to basic services, attributed to disasters
	11.6 By 2030, reduce the adverse per capita environmental impact of cities, including by paying special attention to air quality and municipal and other waste management	11.6.1 Proportion of urban solid waste regularly collected and with adequate final discharge out of total urban solid waste generated, by cities
		11.6.2 Annual mean levels of fine particulate matter (e.g. PM2.5 and PM10) in cities (population weighted)
	11.7 By 2030, provide universal access to safe, inclusive and accessible, green and public spaces, in particular for women and children, older persons and persons with disabilities	11.7.1 Average share of the built-up area of cities that is open space for public use for all, by sex, age and persons with disabilities
	11.b By 2020, substantially increase the number of cities and human settlements adopting and implementing integrated policies and plans towards inclusion, resource efficiency, mitigation and adaptation to climate change, resilience to disasters, and develop and implement, in line with the Sendai Framework for Disaster Risk Reduction 2015-2030, holistic disaster risk management at all levels	11.b.1 Number of countries that adopt and implement national disaster risk reduction strategies in line with the Sendai Framework for Disaster Risk Reduction 2015-2030
		11.b.2 Proportion of local governments that adopt and implement local disaster risk reduction strategies in line with national disaster risk reduction strategies
	11.c Support least developed countries, including through financial and technical assistance, in building sustainable and resilient buildings utilizing local materials	11.c.1 Proportion of financial support to the least developed countries that is allocated to the construction and retrofitting of sustainable, resilient and resource-efficient buildings utilizing local materials

Goal	70 Targets	93 Indicators
Goal 12. Ensure sustainable consumption and production patterns	12.1 Implement the 10-Year Framework of Programmes on Sustainable Consumption and Production Patterns, all countries taking action, with developed countries taking the lead, taking into account the development and capabilities of developing countries	**12.1.1** Number of countries with sustainable consumption and production (SCP) national action plans or SCP mainstreamed as a priority or a target into national policies
	12.2 By 2030, achieve the sustainable management and efficient use of natural resources	**12.2.1** Material footprint, material footprint per capita, and material footprint per GDP
		12.2.2 Domestic material consumption, domestic material consumption per capita, and domestic material consumption per GDP
	12.3 By 2030, halve per capita global food waste at the retail and consumer levels and reduce food losses along production and supply chains, including post-harvest losses	**12.3.1** Global food loss index
	12.4 By 2020, achieve the environmentally sound management of chemicals and all wastes throughout their life cycle, in accordance with agreed international frameworks, and significantly reduce their release to air, water and soil in order to minimize their adverse impacts on human health and the environment	**12.4.1** Number of parties to international multilateral environmental agreements on hazardous waste, and other chemicals that meet their commitments and obligations in transmitting information as required by each relevant agreement
		12.4.2 Hazardous waste generated per capita and proportion of hazardous waste treated, by type of treatment
	12.5 By 2030, substantially reduce waste generation through prevention, reduction, recycling and reuse	**12.5.1** National recycling rate, tons of material recycled
	12.6 Encourage companies, especially large and transnational companies, to adopt sustainable practices and to integrate sustainability information into their reporting cycle	**12.6.1** Number of companies publishing sustainability reports
	12.7 Promote public procurement practices that are sustainable, in accordance with national policies and priorities	**12.7.1** Number of countries implementing sustainable public procurement policies and action plans
	12.8 By 2030, ensure that people everywhere have the relevant information and awareness for sustainable development and lifestyles in harmony with nature	12.8.1 Extent to which (i) global citizenship education and (ii) education for sustainable development (including climate change education) are mainstreamed in (a) national education policies; (b) curricula; (c) teacher education; and (d) student assessment
	12.a Support developing countries to strengthen their scientific and technological capacity to move towards more sustainable patterns of consumption and production	**12.a.1** Amount of support to developing countries on research and development for sustainable consumption and production and environmentally sound technologies
	12.b Develop and implement tools to monitor sustainable development impacts for sustainable tourism that creates jobs and promotes local culture and products	12.b.1 Number of sustainable tourism strategies or policies and implemented action plans with agreed monitoring and evaluation tools
	12.c Rationalize inefficient fossil-fuel subsidies that encourage wasteful consumption by removing market distortions, in accordance with national circumstances, including by restructuring taxation and phasing out those harmful subsidies, where they exist, to reflect their environmental impacts, taking fully into account the specific needs and conditions of developing countries and minimizing the possible adverse impacts on their development in a manner that protects the poor and the affected communities	**12.c.1** Amount of fossil-fuel subsidies per unit of GDP (production and consumption) and as a proportion of total national expenditure on fossil fuels
Goal 13. Take urgent action to combat climate change and its impacts	13.1 Strengthen resilience and adaptive capacity to climate-related hazards and natural disasters in all countries	13.1.1 Number of deaths, missing persons and directly affected persons attributed to disasters per 100,000 population
		13.1.2 Number of countries that adopt and implement national disaster risk reduction strategies in line with the Sendai Framework for Disaster Risk Reduction 2015-2030
		13.1.3 Proportion of local governments that adopt and implement local disaster risk reduction strategies in line with national disaster risk reduction strategies
	13.2 Integrate climate change measures into national policies, strategies and planning	13.2.1 Number of countries that have communicated the establishment or operationalization of an integrated policy/strategy/plan which increases their ability to adapt to the adverse impacts of climate change, and foster climate resilience and low greenhouse gas emissions development in a manner that does not threaten food production (including a national adaptation plan, nationally determined contribution, national communication, biennial update report or other)
	13.3 Improve education, awareness-raising and human and institutional capacity on climate change mitigation, adaptation, impact reduction and early warning	13.3.1 Number of countries that have integrated mitigation, adaptation, impact reduction and early warning into primary, secondary and tertiary curricula
		13.3.2 Number of countries that have communicated the strengthening of institutional, systemic and individual capacity-building to implement adaptation, mitigation and technology transfer, and development actions
	13.a Implement the commitment undertaken by developed-country parties to the United Nations Framework Convention on Climate Change to a goal of mobilizing jointly $100 billion annually by 2020 from all sources to address the needs of developing countries in the context of meaningful mitigation actions and transparency on implementation and fully operationalize the Green Climate Fund through its capitalization as soon as possible	13.a.1 Mobilized amount of United States dollars per year between 2020 and 2025 accountable towards the $100 billion commitment
	13.b Promote mechanisms for raising capacity for effective climate change-related planning and management in least developed countries and small island developing States, including focusing on women, youth and local and marginalized communities	13.b.1 Number of least developed countries and small island developing States that are receiving specialized support, and amount of support, including finance, technology and capacity-building, for mechanisms for raising capacities for effective climate change-related planning and management, including focusing on women, youth and local and marginalized communities

Goal	70 Targets	93 Indicators
Goal 14. Conserve and sustainably use the oceans, seas and marine resources for sustainable development	14.1 By 2025, prevent and significantly reduce marine pollution of all kinds, in particular from land-based activities, including marine debris and nutrient pollution	**14.1.1** Index of coastal eutrophication and floating plastic debris density
	14.2 By 2020, sustainably manage and protect marine and coastal ecosystems to avoid significant adverse impacts, including by strengthening their resilience, and take action for their restoration in order to achieve healthy and productive oceans	**14.2.1** Proportion of national exclusive economic zones managed using ecosystem-based approaches
	14.3 Minimize and address the impacts of ocean acidification, including through enhanced scientific cooperation at all levels	14.3.1 Average marine acidity (pH) measured at agreed suite of representative sampling stations
	14.4 By 2020, effectively regulate harvesting and end overfishing, illegal, unreported and unregulated fishing and destructive fishing practices and implement science-based management plans, in order to restore fish stocks in the shortest time feasible, at least to levels that can produce maximum sustainable yield as determined by their biological characteristics	14.4.1 Proportion of fish stocks within biologically sustainable levels
	14.5 By 2020, conserve at least 10 per cent of coastal and marine areas, consistent with national and international law and based on the best available scientific information	**14.5.1** Coverage of protected areas in relation to marine areas
	14.6 By 2020, prohibit certain forms of fisheries subsidies which contribute to overcapacity and overfishing, eliminate subsidies that contribute to illegal, unreported and unregulated fishing and refrain from introducing new such subsidies, recognizing that appropriate and effective special and differential treatment for developing and least developed countries should be an integral part of the World Trade Organization fisheries subsidies negotiation	14.6.1 Progress by countries in the degree of implementation of international instruments aiming to combat illegal, unreported and unregulated fishing
	14.7 By 2030, increase the economic benefits to small island developing States and least developed countries from the sustainable use of marine resources, including through sustainable management of fisheries, aquaculture and tourism	14.7.1 Sustainable fisheries as a proportion of GDP in small island developing States, least developed countries and all countries
	14.a Increase scientific knowledge, develop research capacity and transfer marine technology, taking into account the Intergovernmental Oceanographic Commission Criteria and Guidelines on the Transfer of Marine Technology, in order to improve ocean health and to enhance the contribution of marine biodiversity to the development of developing countries, in particular small island developing States and least developed countries	14.a.1 Proportion of total research budget allocated to research in the field of marine technology
	14.c Enhance the conservation and sustainable use of oceans and their resources by implementing international law as reflected in the United Nations Convention on the Law of the Sea, which provides the legal framework for the conservation and sustainable use of oceans and their resources, as recalled in paragraph 158 of "The future we want"	14.c.1 Number of countries making progress in ratifying, accepting and implementing through legal, policy and institutional frameworks, ocean-related instruments that implement international law, as reflected in the United Nation Convention on the Law of the Sea, for the conservation and sustainable use of the oceans and their resources
Goal 15. Protect, restore and promote sustainable use of terrestrial ecosystems, sustainably manage forests, combat desertification, and halt and reverse land degradation and halt biodiversity loss	15.1 By 2020, ensure the conservation, restoration and sustainable use of terrestrial and inland freshwater ecosystems and their services, in particular forests, wetlands, mountains and drylands, in line with obligations under international agreements	15.1.1 Forest area as a proportion of total land area
		15.1.2 Proportion of important sites for terrestrial and freshwater biodiversity that are covered by protected areas, by ecosystem type
	15.2 By 2020, promote the implementation of sustainable management of all types of forests, halt deforestation, restore degraded forests and substantially increase afforestation and reforestation globally	15.2.1 Progress towards sustainable forest management
	15.3 By 2030, combat desertification, restore degraded land and soil, including land affected by desertification, drought and floods, and strive to achieve a land degradation-neutral world	15.3.1 Proportion of land that is degraded over total land area
	15.4 By 2030, ensure the conservation of mountain ecosystems, including their biodiversity, in order to enhance their capacity to provide benefits that are essential for sustainable development	**15.4.1** Coverage by protected areas of important sites for mountain biodiversity
		15.4.2 Mountain Green Cover Index
	15.5 Take urgent and significant action to reduce the degradation of natural habitats, halt the loss of biodiversity and, by 2020, protect and prevent the extinction of threatened species	15.5.1 Red List Index
	15.6 Promote fair and equitable sharing of the benefits arising from the utilization of genetic resources and promote appropriate access to such resources, as internationally agreed	15.6.1 Number of countries that have adopted legislative, administrative and policy frameworks to ensure fair and equitable sharing of benefits
	15.7 Take urgent action to end poaching and trafficking of protected species of flora and fauna and address both demand and supply of illegal wildlife products	15.7.1 Proportion of traded wildlife that was poached or illicitly trafficked
	15.8 By 2020, introduce measures to prevent the introduction and significantly reduce the impact of invasive alien species on land and water ecosystems and control or eradicate the priority species	15.8.1 Proportion of countries adopting relevant national legislation and adequately resourcing the prevention or control of invasive alien species
	15.9 By 2020, integrate ecosystem and biodiversity values into national and local planning, development processes, poverty reduction strategies and accounts	**15.9.1** Progress towards national targets established in accordance with Aichi Biodiversity Target 2 of the Strategic Plan for Biodiversity 2011-2020
	15.a Mobilize and significantly increase financial resources from all sources to conserve and sustainably use biodiversity and ecosystems	15.a.1 Official development assistance and public expenditure on conservation and sustainable use of biodiversity and ecosystems
	15.b Mobilize significant resources from all sources and at all levels to finance sustainable forest management and provide adequate incentives to developing countries to advance such management, including for conservation and reforestation	15.b.1 Official development assistance and public expenditure on conservation and sustainable use of biodiversity and ecosystems
	15.c Enhance global support for efforts to combat poaching and trafficking of protected species, including by increasing the capacity of local communities to pursue sustainable livelihood opportunities	15.c.1 Proportion of traded wildlife that was poached or illicitly trafficked

Goal	70 Targets	93 Indicators
Goal 16. Promote peaceful and inclusive societies for sustainable development, provide access to justice for all and build effective, accountable and inclusive institutions at all levels	16.8 Broaden and strengthen the participation of developing countries in the institutions of global governance	16.8.1 Proportion of members and voting rights of developing countries in international organizations
Goal 17. Strengthen the means of implementation and revitalize the Global Partnership for Sustainable Development	17.6 Enhance North-South, South-South and triangular regional and international cooperation on and access to science, technology and innovation and enhance knowledge-sharing on mutually agreed terms, including through improved coordination among existing mechanisms, in particular at the United Nations level, and through a global technology facilitation mechanism	17.6.1 Number of science and/or technology cooperation agreements and programmes between countries, by type of cooperation
	17.7 Promote the development, transfer, dissemination and diffusion of environmentally sound technologies to developing countries on favourable terms, including on concessional and preferential terms, as mutually agreed	**17.7.1** Total amount of approved funding for developing countries to promote the development, transfer, dissemination and diffusion of environmentally sound technologies
	17.9 Enhance international support for implementing effective and targeted capacity-building in developing countries to support national plans to implement all the Sustainable Development Goals, including through North-South, South-South and triangular cooperation	17.9.1 Dollar value of financial and technical assistance (including through North-South, South-South and triangular cooperation) committed to developing countries
	17.14 Enhance policy coherence for sustainable development	**17.14.1** Number of countries with mechanisms in place to enhance policy coherence of sustainable development
Total	**72**	**93**

Annex 6-1: The Principal Biodiversity-related Conventions

Convention on Biological Diversity

Convention on Biological Diversity

The objectives of the CBD are the conservation of biological diversity, the sustainable use of its components, and the fair and equitable sharing of the benefits arising from commercial and other utilization of genetic resources. The agreement covers all ecosystems, species, and genetic resources.

Convention on International Trade in Endangered Species of Wild Fauna and Flora (CITES)

The CITES aims to ensure that international trade in specimens of wild animals and plants does not threaten their survival. Through its three Appendices, the Convention accords varying degrees of protection to more than 36,000 plant and animal species.

Convention on the Conservation of Migratory Species of Wild Animals

The CMS, or the Bonn Convention aims to conserve terrestrial, marine and avian migratory species throughout their range. Parties to the CMS work together to conserve migratory species and their habitats by providing strict protection for the most endangered migratory species, by concluding regional multilateral agreements for the conservation and management of specific species or categories of species, and by undertaking co-operative research and conservation activities. www.cms.int

The International Treaty on Plant Genetic Resources for Food and Agriculture

The objectives of the International Treaty are the conservation and sustainable use of plant genetic resources for food and agriculture and the fair and equitable sharing of the benefits arising out of their use, in harmony with the Convention on Biological Diversity, for sustainable agriculture and food security. The International Treaty covers all plant genetic resources for food and agriculture, while its Multilateral System of Access and Benefit-sharing covers a specific list of 64 crops and forages. The Treaty also includes provisions on Farmers' Rights.

Convention on Wetlands (also known as the Ramsar Convention)

The Ramsar Convention is the only international treaty focused on wetlands. It provides a platform of 170 Contracting Parties working together for wetland conservation and wise use, and to develop the best available data, advice and policy recommendations to realize the benefits of fully functional wetlands to nature and society. The Convention recognized wetlands as ecosystems which are extremely important for biodiversity conservation. Parties to the Convention have already committed to maintaining the ecological character of over 2,300 Wetlands of International Importance covering nearly 250 million hectares, 13-18 per cent of global wetlands.

United Nations Educational, Scientific and Cultural Organization · World Heritage Convention

World Heritage Convention (WHC)

The primary mission of the WHC is to identify and conserve the world's cultural and natural heritage, by drawing up a list of sites whose outstanding values should be preserved for all humanity and to ensure their protection through a closer co-operation among nations.

International Plant Protection Convention

International Plant Protection Convention (IPPC)

The IPPC aims to protect world plant resources, including cultivated and wild plants by preventing the introduction and spread of plant pests and promoting the appropriate measures for their control. The convention provides the mechanisms to develop the International Standards for Phytosanitary Measures (ISPMs), and to help countries to implement the ISPMs and the other obligations under the IPPC, by facilitating the national capacity development, national reporting and dispute settlement. The Secretariat of the IPPC is hosted by the Food and Agriculture Organization of the United Nations (FAO).

Annex 9-1: Water Contaminants and Occurrences

Pathogens	Human and livestock excrement (bacterial)	Inadequate treatment of sewer effluents; sewer and storm water overflows into rivers, lakes and wetlands	90 per cent of child deaths are caused by diarrheal diseases (WHO and UNICEF 2012).	1/3 of all rivers in Africa, Asia-Pacific and Latin America regions (UNEP 2016a)
Parasites (non-bacterial)	Human and livestock excrement (non-bacterial)	Human and livestock excrement; septic leakage into surface and groundwater	Approximately half the deaths of children under the age of five years.	Cities and rural communities in Africa, Asia-Pacific, Latin America, India, Pakistan, China, Nigeria, and Democratic Republic of Congo
Viruses (non-bacterial)		Treated drinking water (Bergeron *et al.* 2015); natural occurrence in range of concentrations (Kümmerer 2009)		
Antibiotic/ antimicrobial compounds	Human excretion; intensive agriculture and aquaculture practices	Sewer effluents; agricultural and urban runoff	Human illness and death due to antimicrobial- and antibiotic-resistant infections	Projected to become a major cause of death worldwide by 2050 (O'Neill Commission 2014)
Nutrients	Agricultural inorganic fertilizer (UNEP 2016a); human and livestock excrement	Inadequately/untreated sewage discharges; urban and agricultural runoff; aquaculture	Eutrophication and algal blooms (OECD 1982; Research Center for Sustainability and Environment-Shiga University and International Lake Environment Committee Foundation 2014)	All five UNEP regions; rural areas in China, India, Thailand and Philippines also affected by excessive chemical fertilizer application (Novotny *et al.* 2010)
		River contributions of total nutrients to coastal areas increased by approximately 80 per cent during 1970-2000	Effects of harmful algal blooms (HABs) can impact ecosystem functions Aquaculture; livestock and human health via bioaccumulation of toxicity. (O'Neil *et al.* 2012)	Thirty-seven Latin American transboundary rivers are highly polluted with wastewaters and agricultural runoff nutrients at the basin level
Sediment	Deforestation; poor agricultural practices; livestock overgrazing; intensive fuelwood harvesting; sand mining; unplanned settlements causing exposed soil surfaces and erosion	Storm-generated runoff can carry sediments, nutrients, heavy metals, pesticides and other pollutants into rivers, lakes and wetlands, particularly in agricultural areas	Sediment-associated pollutants interfere with human water use Can have health impacts, degrade aquatic organism metabolism and habitats	In Asia-Pacific, some rivers carry high loads of sediment-associated heavy metals
	Changes in sediment flow paths (dykes, channels, urban drainage, dams) can lead to erosion and high sediment loads		Sediment loading to oceans and coastal ecosystems (river deltas, wetlands, beaches, etc.)	50 per cent of upslope-eroded soil is deposited in White Volta sub-basin in West Africa Human-induced erosion has affected approximately 2.2 million km^2 of land in Latin America
Biodegradable Organic Pollutants	Process characterized by a high Biological Oxygen Demand (BOD) from microbial decomposition of human and livestock wastes, and eutrophication-associated algal blooms, particularly in lakes and wetlands	Industrial and domestic wastewater discharges	Bacterial-mediated decomposition of algae/ aquatic plants can cause hypoxia/oxygen depletion in waterbodies, resulting in fish kills and facilitating release of heavy metals from bottom sediments back into water column	Increasing in African, Asia-Pacific and Latin America (UNEP 2016a) in contrast, decreasing in developed countries with enhanced wastewater treatment
	Industrial and agricultural applications and operations			Rapidly-urbanizing and industrializing countries (e.g., China India; Ethiopia; Mexico) and rivers downstream of major Central Asian cities

Water Contaminant	Contaminant Sources	Pathways into Waterbody	Impacts of Contaminated Water and inadequate Sanitation/Hygiene	Examples of Occurrence
Persistent Organic Pollutants (POPs, including organic pesticides; industrial chemicals and organic neonicotinoid insecticides solvents)	DDT (produced globally); Neonicotinoid insecticides (introduced in 1990s); Organic chemicals and solvents in manufacturing processes	Agricultural and urban runoff; Industrial and domestic wastewater discharges	Accumulates and persists in fatty tissues of humans, fish and other aquatic organisms, damaging their health if toxic; Neonicotinoid insecticides toxic to aquatic invertebrates and biodiversity; DDT has human carcinogenic and teratogenic risks (e.g., elevated DDT levels found in Lake Kariba ecosystem, and in breast milk of women living in area);	DDT still used in many developing countries to control malaria; Neonicotinoid insecticides most-widely used insecticides in the world; Estimated 40 per cent of world land area affected by insecticide runoff Wide range of industrial chemical processes involving organic solvents; Reducing DDT use had some positive results (e.g., recovery of eagles and other birds in North America)
	Neonicotinoid insecticides		Contaminate freshwater resources, wetlands, estuarine habitats and marine systems globally Poses serious threat to pollinators such as bees (IPBES 2017); contaminate food chains Significantly increases human exposure to synthetic chemicals (Kim *et al.* 2017).	
Heavy Metals	Industrial, agricultural, medical, technological and mining wastes; stormwater runoff (e.g., highways);	Untreated industrial and municipal wastewater discharges into rivers, lakes, wetlands; land runoff; sedimentation	Can affect human health directly via ingestion of drinking water Can bioaccumulate in vegetables, rice and other edible plants irrigated with contaminated irrigation water (Arunakumara, Walpola and Yoon 2013; Lu *et al.* 2015) Mercury, lead, chromium, cadmium and arsenic have toxic effects on humans and other organisms	Asia-Pacific river waters and sediments contain high heavy metal levels from untreated tannery and metal-finishing operation discharges, and from highway runoff (e.g., zinc in West Java; lead in Erdenet, Mongolia; chromium in some Bangladeshi and Japanese rivers) (Sikder *et al.* 2013); Chinese urban rivers (Qu and Fan 2010). South American urban areas In contrast, heavy metal contamination has generally diminished in EU countries
	Natural contamination (e.g., arsenic in groundwater)			Widespread in Bangladeshi and Indian groundwater; some parts of China, Iran, Mongolia, Pakistan and Nepal (Rahman, Ng and Naidu 2009)
Salinity	Agricultural irrigation drainage; lake and wetland evaporation	High evaporation rates	Most freshwater organisms and ecosystems have limited salinity tolerance (UNEP 2016a); salinization impairs agricultural and industrial water uses Salt water intrusion can result in salinized water in coastal aquifers	Salinity problems affect one-tenth of all rivers in Africa, Asia-Pacific and Latin America; surface water salinization is major issue in Central Asia
	Intensive agricultural practices; domestic and industrial sewer effluents	Production of salinized soils		
	seawater intrusion	Over-abstraction of groundwater		

Water Contaminant	Contaminant Sources	Pathways into Waterbody	Impacts of Contaminated Water and inadequate Sanitation/Hygiene	Examples of Occurrence
Contaminants of Emerging Concern (CEC's)	Veterinary and human pharmaceuticals; insect repellents; antimicrobial disinfectants; fire retardants; detergent metabolites	Municipal and industrial sewer effluents	Hormone imbalances contributing to reduced human fertility and feminization of male fish (Gross-Sorokin, Roast and Brighty 2006); Increasing evidence of antibiotic-resistant organisms in water sources, possibly altering aquatic microbial ecosystems	US Geological Survey detected these contaminants in 80 per cent of streams sampled in US; also detected in all pan-European seas
	Microplastics and nanoparticles (Kolpin *et al.* 2002)		Impacts both freshwater and marine ecosystems Microplastics known to contain and absorb toxic chemicals	Global issue (Dris *et al.* 2015)
Additional Water Quality Concerns	Groundwater pollution associated with oil and gas fracking activities	Discharge of large volumes of "produced water" and associated chemicals enter waterways	Fracking-associated pollutants are being researched (Osborn *et al.* 2011)	The Americas (Vengosh *et al.* 2014)
	Lake acidification from atmospheric deposition of fossil fuel emissions	Acid rain	Impacts freshwater ecosystems, including fish and other aquatic organisms	Lake acidification remains problematic in areas lacking soils or bedrock capable of buffering acid rain Situation is improving where SOx and NOx emissions have decreased (e.g., affected lakes in New York Adirondack region recovering at different rates) (Driscoll *et al.* 2016)

Annex 13-1 Biodiversity Conservation and International Environmental Agreements (IEAs)

The University of Oregon has built the most comprehensive database on IEAs to-date. We searched the iea.uoregon.edu database for IEAs related to biodiversity conservation. The search terms "biodiversity" AND "conservation" were used to search for multilateral and bilateral agreements, which returned 45 IEAs in total. Amendments to agreements already deemed relevant were also excluded here to prevent double counting. After the screening process, 33 IEAs were identified that concerned biodiversity conservation. They were signed within three decades (between 1985 and 2015). Four were bilateral (signed by only 2 countries) and the rest were multilateral (signed by 3 or more countries). Of the multilateral agreements, the number of signatories ranged from 3 to 196 (median = 7). Twenty-eight IEAs focussed on specific geographic regions while six had a global scope. Of those agreements on terrestrial regions, seven focused on conservation of ecosystems or species within North America, five in Europe, six in Asia, and three in Africa. Seven IEAs also focussed on biodiversity conservation in a non-terrestrial context (Indian, Atlantic, and Pacific Ocean, and the Baltic sea).

Table A.4: List of International Environmental Agreements signed between 2010 and 2015

IEA	Year signed	Themes
Agreement on the Protection and Sustainable Development of the Prespa Park Area	2010	4
Nagoya Protocol on Access to Genetic Resources and the Fair and Equitable Sharing of Benefits Arising from their Utilization to the Convention on Biological Diversity	2010	3
Agreement between the Governments of the Member States of the Association of Southeast Asian Nations and the Republic of Korea on Forest Cooperation	2011	2
Protocol on Sustainable Forest Management to the Framework Convention on the Protection and Sustainable Development of the Carpathians	2011	2
Protocol on Sustainable Tourism to the Framework Convention on the Protection and Sustainable Development of the Carpathians	2011	2
Agreement on the establishment of the Global Green Growth Institute	2012	2
Convention on the Conservation and Management of High Seas Fisheries Resources in the North Pacific Ocean	2012	2
Protocol Amending the Agreement Between The United States And Canada On Great Lakes Water Quality	2012	1
Benguela Current Convention	2013	4
Protocol on Sustainable Transport to the Framework Convention on the Protection and Sustainable Development of the Carpathians	2014	2
Paris Agreement under the United Nations Framework Convention on Climate Change	2015	3

The IEAs have been categorised into four themes: pollution prevention (1), sustainable use of biodiversity (2), environmental process (3) and protection of ecosystem/species/genes (4) based on the predominant context they fall under.

Source: Mukherjee *et al.* (2018).

Annex 13-2: Overview of Key Policy Developments and Governance Responses at a Global Level

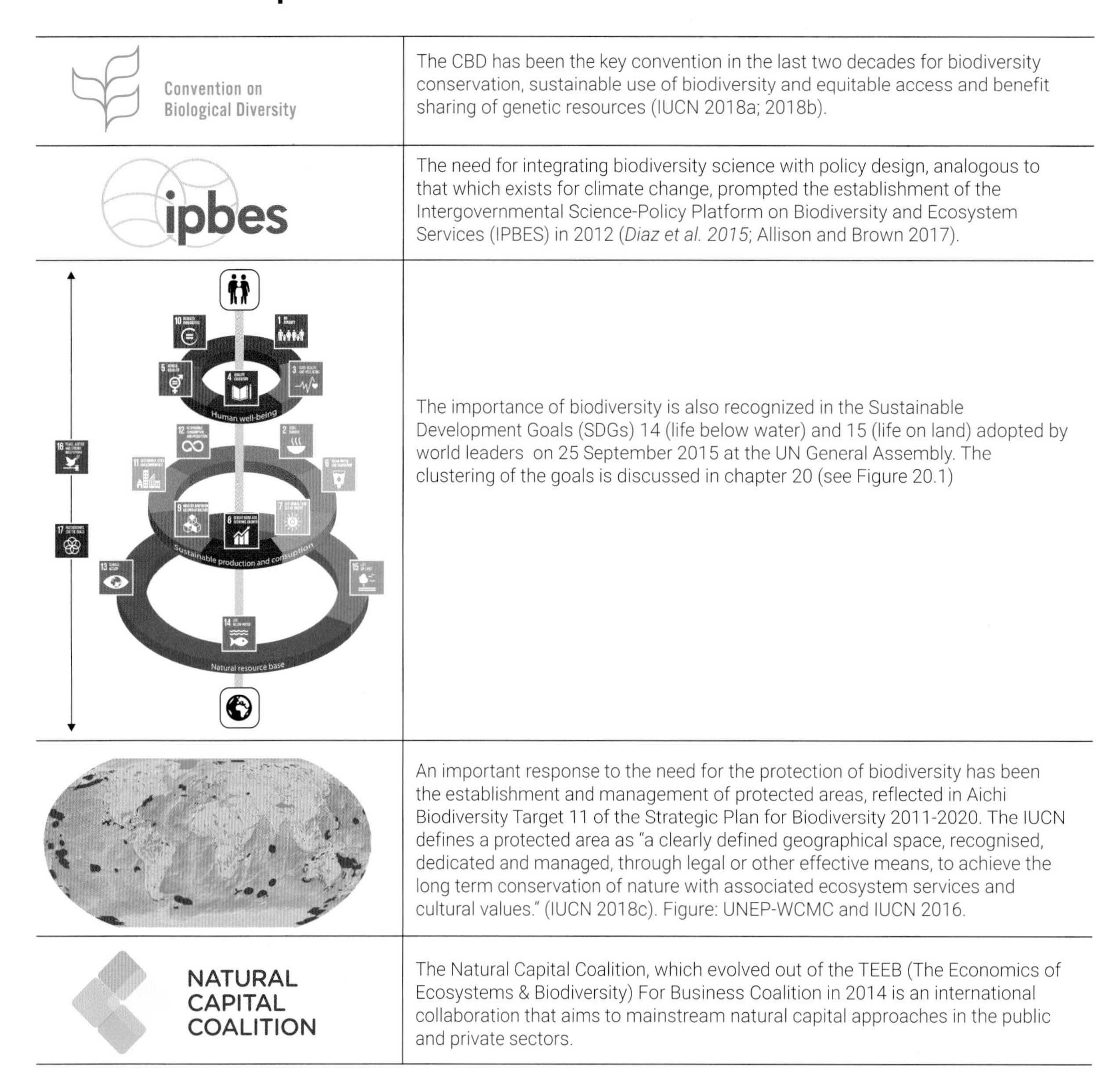

Convention on Biological Diversity	The CBD has been the key convention in the last two decades for biodiversity conservation, sustainable use of biodiversity and equitable access and benefit sharing of genetic resources (IUCN 2018a; 2018b).
ipbes	The need for integrating biodiversity science with policy design, analogous to that which exists for climate change, prompted the establishment of the Intergovernmental Science-Policy Platform on Biodiversity and Ecosystem Services (IPBES) in 2012 (*Diaz et al. 2015*; Allison and Brown 2017).
Human well-being Sustainable production and consumption Natural resource base	The importance of biodiversity is also recognized in the Sustainable Development Goals (SDGs) 14 (life below water) and 15 (life on land) adopted by world leaders on 25 September 2015 at the UN General Assembly. The clustering of the goals is discussed in chapter 20 (see Figure 20.1)
	An important response to the need for the protection of biodiversity has been the establishment and management of protected areas, reflected in Aichi Biodiversity Target 11 of the Strategic Plan for Biodiversity 2011-2020. The IUCN defines a protected area as "a clearly defined geographical space, recognised, dedicated and managed, through legal or other effective means, to achieve the long term conservation of nature with associated ecosystem services and cultural values." (IUCN 2018c). Figure: UNEP-WCMC and IUCN 2016.
NATURAL CAPITAL COALITION	The Natural Capital Coalition, which evolved out of the TEEB (The Economics of Ecosystems & Biodiversity) For Business Coalition in 2014 is an international collaboration that aims to mainstream natural capital approaches in the public and private sectors.

Annex 23-1: Bottom-up Initiative Platforms and Results

Initiative	Location	Platform Secretariat	Platform Participants	Description and Projects	Theory of Change	Initiatives	SDGs	Potential Pathway	Link
Amazon Vision Coordination and Information Platform (CIP)	Regional (Colombia)	Civil Society, local government	All government, public, civil society, private sector, international organizations	By providing a structure to pool knowledge, increase capacity, and facilitate the deployment of financial resources, CIP aims to support the implementation of GHG mitigation activities, and ensure that the results of such activities (e.g. information, tools, programs and incentives) reach the right beneficiaries.	Knowledge/data platform; Monitoring and reporting; Finance/incentives/ subsidies	200	13	Decentralized Solutions	http://www.pidamazonia.com
Blue Solutions/ Marine and Coastal Biodiversity Management in Pacific Island Countries (MACBIO - Pacific)	Regional (Pacific Island Region)	National government, international organizations	Practitioners, national governments, regional governments	Collate, document and share successful approaches for addressing marine and coastal challenges. By strengthening institutional and individual capacity, to manage and conserve biodiversity in marine and coastal ecosystems, MACBIO supports sustainable economies and livelihoods of Pacific Island Countries.	Knowledge/data platform		14	Decentralized solutions	http://macbio-pacific.info/qbook/
Climate Initiatives Platform	Global	International organization	All government, private sector, civil society, international organizations	Online portal for collecting, sharing and tracking information on international cooperative climate initiatives among city, region, company, investor, civil society and national government participants.	Knowledge/data platform; Monitoring and reporting	224	13	Decentralized solutions	http://climateinitiativesplatform.org/index.php/Welcome
Green Growth Knowledge Platform	Global	Civil society, international organizations	Practitioners, international organizations, academic research, civil society, all government	The Green Growth Knowledge Platform (GGKP) is a global network of international organizations and experts, that identifies and addresses major knowledge gaps in green growth theory and practice. By encouraging widespread collaboration and world-class research, the GGKP offers practitioners and policymakers the policy guidance, good practices, tools, and data necessary to support the transition to a green economy.	Knowledge/data platform; Awareness, knowledge, skills development; New organization/business	337	12	Decentralized solutions	http://www.greengrowthknowledge.org/
LifeWeb Initiative	Global	International organization	All government, international organizations, civil society	The LifeWeb Initiative highlights biodiversity, conservation and adaptation to climate change needs to donors who are able to create development assistance partnerships, through an online clearing-house and participation in roundtable meetings.	Knowledge/data platform; Finance/incentives/ subsidies	18	14, 15, 17	Decentralized solutions	https://lifeweb.cbd.int/
Non-State Actor Zone for Climate Action (NAZCA)	Global	International organization	All government, private sector, civil society, international organizations, academic research	NAZCA captures commitments to climate action made by companies, cities, subnational, regions, investors, and civil society organizations. NAZCA aims to track the mobilization and action that are helping countries achieve and exceed their national commitments to address climate change.	Knowledge/data platform; Monitoring and reporting	12549	11, 13, 17	Decentralized solutions	http://climateaction.unfccc.int/

Initiative	Location	Platform Secretariat	Platform Participants	Description and Projects	Theory of Change	Initiatives	SDGs	Potential Pathway	Link
UN Global Compact	Global	International organization	Private sector	The UN Global Compact helps companies align their strategies and operations with Ten Principles on human rights, labour, environment and anti-corruption. It takes strategic actions to advance broader societal goals, such as the UN Sustainable Development Goals, with an emphasis on collaboration and innovation.	Awareness, knowledge, skills development	49861	6, 7, 8, 11	Decentralized solutions	www.unglobalcompact.org/
NRDC Cloud of Commitments	Global	Civil society	All government, private sector, civil society, international organizations	A platform to highlight the emerging coalitions, networks, partnerships, and other initiatives taking action on energy, water, cities, and other key sustainability challenges discussed at the UN Conference on Sustainable Development (UNCSD, or Rio+20). It is expected to evolve into a platform working toward transparency, engagement, assessments, and accountability for the commitments highlighted.	Knowledge/data platform; Monitoring and reporting	118	Various	Decentralized solutions	www.cloudofcommitments.org/
SEEDS of the Anthropocene	Global	International organization	All government, private sector, civil society, international organizations	This initiative gathers "SEEDS," which are existing initiatives that are not widespread. They can be social initiatives, new technologies, economic tools, or social-ecological projects, or organizations, movements or new ways of acting that have that appear to be making a substantial contribution towards creating a future that is just, prosperous, and sustainable.	Knowledge/data platform; Awareness, knowledge, skills development	400	13, 17	Decentralized Solutions, Consumption Change, Global Technology	https://goodanthropocenes.net/
Project Drawdown	Global	Public, practitioners, civil society, private sector, policymakers, governments, academic research	Practitioners, civil society, private sector, all governments, academic research, international organizations	Project Drawdown is facilitating a broad coalition of researchers, scientists, graduate students, PhDs, post-docs, policy makers, business leaders and activists to assemble and present the best available information on climate solutions in order to describe their beneficial financial, social and environmental impact over the next thirty years.	Knowledge/data platform; Awareness, knowledge, skills development	100	13	Decentralized solutions	http://www.drawdown.org/
MIT Climate Co-Lab	Global	Academic research	Public, civil society	Climate CoLab is an open problem-solving platform where a growing community of over 90,000 people - including hundreds of the world's leading experts on climate change and related fields - work on and evaluate plans to reach global climate change goals.	Knowledge/data platform; Awareness, knowledge, skills development; New organization/business	2703	Various	Decentralized solutions	https://climatecolab.org/contests
VertMTL.org	Local (Montreal, Canada)	Local government	Public, civil society	As part of its consultation on reducing Montrealers' reliance on fossil fuels, the Montreal Public Advisory Board is challenging the innovation community by asking them to prototype innovative approaches to meeting the needs of Canadians.	Knowledge/data platform; Awareness, knowledge, skills development; New organization/business	58	6, 13, 2, 12	Decentralized solutions	https://marathoncreatif.sparkboard.com/

Initiative	Location	Platform Secretariat	Platform Participants	Description and Projects	Theory of Change	Initiatives	SDGs	Potential Pathway	Link
Sustainia100	Global (188 countries in 2016)	Civil society	Private sector, all government, civil society, international organizations, academic research	Now in its fifth year, the Sustainia100 has tracked more than 4,500 solutions to date from all over the world that respond to interconnected global challenges and help achieve the Sustainable Development Goals.	Knowledge/data platform; Awareness, knowledge, skills development	4500	Various	Decentralized solutions	http://www.sustainia.me/solutions/
Beautiful Solutions	Global	Civil society	Public, civil society, practitioners	The Beautiful Solutions Gallery and Lab is an interactive space for sharing the stories, solutions and big ideas needed to build new institutional power and point the way toward a just, resilient, and democratic future. It is an online platform, book, and training program designed to give people tools to create the world we want.	Knowledge/data platform; Awareness, knowledge, skills development	18	13	Decentralized solutions	https://solutions.thischangeseverything.org/
EcoTipping Points	Global	Civil society	Public, civil society, practitioners	The EcoTipping Points Project's pragmatic goal is to help people identify "tipping point" levers right at home - concrete actions that they and their community can act upon. The EcoTipping Points Project is dedicated to making the stories and their lessons known through the media, workshops, and direct collaboration with community groups.	Knowledge/data platform; Awareness, knowledge, skills development	125	15, 17	Decentralized solutions	http://www.ecotippingpoints.org
PANORAMA - Solutions for a Healthy Planet	Global	Civil society, international organizations, national government	Public, civil society, practitioners, all governments, civil society, academic research, international organizations, private sector	A partnership initiative to document and promote examples of inspiring, replicable solutions across a range of conservation and development topics, enabling cross-sectoral learning and inspiration. PANORAMA allows practitioners to share their stories, get recognized for successful work, and learn how others have tackled problems across the globe, by encouraging reflection on and learning from proven approaches.	Knowledge/data platform; Awareness, knowledge, skills development	373	Various	Decentralized Solutions	http://www.panorama.solutions/
UNFCCC Momentum for Change Initiatives	Global	International organizations	Private sector, civil soceity, all governments, international organizations	Momentum for Change is an initiative spearheaded by the UN Climate Change, that recognizes innovative and transformative solutions that address both climate change and wider economic, social and environmental challenges. The initiative features practical, scalable and replicable examples of what people, businesses, governments and industries are doing to tackle climate change.	New organizationorganization/business; Finance/incentives/subsidies	627	Various	Global Technology	http://unfccc.int/secretariat/momentum_for_change/items/6214.php

Initiative	Location	Platform Secretariat	Platform Participants	Description and Projects	Theory of Change	Initiatives	SDGs	Potential Pathway	Link
WorthWild	Regional (United States)	Private sector	Public, civil society, private sector, practitioners	WorthWild is a crowdfunding platform for environmentally-conscious businesses, nonprofits, and individuals who want to raise money to fund projects that protect and sustain the planet. WorthWild utilizes technology to rally behind green ideas, guiding first-time crowdfunders and experienced philanthropists through the process of building effective campaigns from start to finish.	New organizationorganization/ business; Finance/ incentives/subsidies	29	7	Decentralized Solutions	http://www.worthwild.com/
Greencrowd	Regional (Netherlands)	Private sector	Public, civil society, private sector, practitioners	Netherlands-based Greencrowd is founded to accelerate the realization of sustainable energy projects. Investors in Greencrowd projects will realize an environmental impact as well as a financial profit. Greencrowd thoroughly evaluates the risks involved in the project and assures there are guarantees (e.g. insurances, real estate as collateral) to mitigate the potential losses.	New organizationorganization/ business; Finance/ incentives/subsidies	43	Various	Decentralized Solutions; Consumption Change	https://greencrowd.nl/
Divvy	Global	Civil society	Public, civil society, private sector, practitioners	Divvy is a crowdfunding platform for community sustainability projects. Our platform makes it super-easy for green-minded community leaders to turn their sustainability ideas into reality. If needed, we even help project initiators get competitive quotes from local professionals so that they can focus on engaging their community.	New organization/business; Finance/incentives/ subsidies	9	All	Decentralized Solutions; Global Technology	http://divvygreen.com/
Partnerships for SDGs	Global	International organization	Public, practitioners, civil society, all governments, international organizations, private sector, academic research	The Partnerships for SDGs online platform is United Nations' global registry of voluntary commitments and multi-stakeholder partnerships, facilitating global engagement of all stakeholders in support of the implementation of the Sustainable Development Goals.	Knowledge/data platform; Monitoring and reporting	3808	All	Decentralized Solutions	https://sustainabledevelopment.un.org/partnerships/

Contributed Bottom-up Initiatives

Workshop Seeds

Workshop	Seed Name	Description
Singapore [MM81]	Renewable energy microgrids	Renewable energy microgrids implemented in climate and disaster vulnerable areas to strengthen the energy security, resilience and the ability for remote or secluded communities to bounce back from climate events.
Singapore	Celebrating Singapore Shores	Platform to bring together marine groups to celebrate the International Year of the Reefs. This occurs every 10 years.
Singapore	Plastic waste footprint calculator	Similar to carbon footprint calculation for individuals, an app/website that approximates the plastic waste footprint of individuals based on their daily lifestyle. The app can then summarize or extrapolate how much the person generates in a week/month/year and provide suggestions on how they can personally tweak their lifestyle to reduce plastic waste.
Singapore	Blockchain open source reporting	Using blockchain as a tool to aid in CSOs reporting. Using technology to ease reporting and measuring data and impact so as it could help cso to report and raise awareness.
Singapore	Solar Light Cooperative	Decentralized community and cooperative based power grid. Prioritizes poor and underserved communities. Communities manage their own systems and operates like a cooperative.
Singapore	Implementation of biomimetics for lifestyle change	To keep the linkage between industry and biodiversity conservation. A project to promote biomimicry - technology inspired by nature. Biomimetics as cultural services inspired by ecosystem services. For example, we have swimsuits that are inspired by the skin of the shark, city planning model inspired by the ecosystem itself.
Singapore	Marina barrage	Using barrage as a case study to alleviate flooding issues auto flood gates
Singapore	Using Drone to environmental assessment before development	Collect of Real time environmental informations such as air quality and forest structure
Singapore	Repair Kopitiam	Conducting monthly repair workshops to teach residents in different areas to repair appliances and reduce e-waste. Also involves uploading a series of videos so that anyone can conduct their own workshops. A strength is that everyone can gain repair skills and this also brings back a culture of repairing items.
Singapore	Drainage water level sensors	Sensors to monitor water level of drains and canals to provide real-time site conditions update during heavy storms, to improve response times to floods. This would be really useful for urban cities with dense canal system.
Singapore	Melbourne open data platform for environmental management	Collate information on all the trees in the city of Melbourne
Singapore	Satellite imagery to detect the palms health	Ability to detect the affected trees without a need to destructive methods.
Singapore	Urban Farming by using traditional vegetables	Urban Farming in Tokyo Metropolitan area of traditional and indigenous vegetables. Agricultural cooperatives that keep the traditional seeds to promote regional culture and identity. (for example, white radish)
Singapore	Citizen pollution monitoring	Giving citizens the tools to report on local pollution (especially vehicles). One strength is that is leverages pre-existing technology and another strength is that it makes pollution visible.
Singapore	China Black and Smelly Waters app	Citizens can report instances of foul or smelly water in urban areas through a smartphone app that connects to WeChat. Local government officials have to respond within 7 days to the complaint.
Singapore	Repair Kopitiam	"Kopitiam" is a neighborhood coffee shop. This project is run by SL2 (Sustainable Living Lab). Underneath public housing the organization sets up an area where residents can bring down broken items to get fixed. The volunteers teach the residents how to fix electronics, clothing, household and consumer items etc. Typically items of emotional value. Afterwards, pictures are taken. This is set up somewhere around the city every Sunday.
Singapore	Citizen science reporting	Shared data collected by individuals helping the different cause.
Singapore	Carbon dioxide capture for decarbonisation of atmosphere	Direct air capture of CO2 from ambient air through engineered chemical reactions. The plant sits on top of a waste heat recovery facility that powers the process. Fans push air through a filter system that collects CO2. When the filter is saturated, CO2 is separated at temperatures above 100 degrees Celsius. The gas is then sent through an underground pipeline to a greenhouse.
Singapore	Seed water	You can drink water without plastic bottle water
Singapore	Seaweed farming for livestock feed	Seaweed as a feed substitute for livestock and dairy cattle. Seaweed has been shown to to reduce the amount of methane produce by ruminating animals.
Singapore	Transport Network Vehicle Systems	Big data-based transport system. Users hail rides in the most convenient way using smart phone apps. Offers lower ownership costs per passenger-km and reduces environmental impacts of transport. It is a potential solution to reduce private car use. People get to share rides, while still having the comfort and convenience of private transport.
Singapore	Underwater reporting	Using a single video recording equipment and live share the video image for different people are the world to observe without being there
Singapore	Versatile solar panel	Having solar panels everywhere to increase renewable energy use
Singapore	Solar farming	Solar farming, large scale transition to renewable energy. Located on Low yielding agricultural land solar farming has provided Farmers experiencing reducing yields and effects of climate change with an alternative source of income. Additional solar farming supports the reductions of emissions and the transition to renewable energy. Solar farming has also created a new job market and income for many rural communities. Livestock as sheep can still be grazed under and around the panels.
Singapore	Ecological Mangrove Restoration - Restore Ubin Mangroves Initiative	A community project to restore mangroves in abandoned Aquaculture ponds at Pulau Ubin. Mangrove restoration without planting. Community-based effort involving academics, fish farmers, nature enthusiasts, fishermen, marine advocates. Technology - based on scientific geographical mapping of the site to be restored (mangroves)
Singapore	Swapping resources	To reduce buying and encourage swapping existing resources
Singapore	Blue SG - Electric Carsharing	Similar to bike-sharing, 3-4 electric cars with charging stations are placed in heartland carparks. Residents can use it any time. Better than owning their own car.

Workshop	Seed Name	Description
Singapore	Smart Farm	Smart farm uses a variety of technologies to monitor the state of vegetables. It helps farmers to know how much the vegetables and fruits need the sun and waters. It helps to make sure the food security.
Singapore	Karthavyam (Dutiful citizens for SDG)	A hands on student diploma on public problem solving through the Sustainable Development Goals. The diploma is a 6 month programme with 4 pathways where children are taught via films, podcasts, and other media that can be easily shared. Classes involve experiential learning, designing localized initiatives, writing story books, watching films, and engaging with the community. A strength is that it uses visual mediums (filmmaking) to create a decentralized platform for knowledge sharing.
Singapore	Heka Leka	bring together for social cohesion, community building through education.
Singapore	Repair Kopitiam	Bring residents together to learn how to repair broken appliances in order to reduce waste. It also aims to reduce consumption, and also equip residents with employable skills. At the same time, it also preserves legacies of twilight industries (such as cobblers, etc.)
Singapore	Great British Bee Count	Individuals are encouraged to engage with bees (and biodiversity) by taking photos of bees and sharing it on a centralized platform. Individuals can also buy bee saving materials to plant flowers in neighbourhoods in exchange for photos of bees. A strength of the initiative is that it allows for data collection on bees. Additionally, it leverages pre-existing technology.
Singapore	Global Circular Economy Database	A database that captures circular economy initiatives and sharing the information on a central platform. A strength is that it can mobilize many different groups to contribute to the database without much effort. This database can then act as a learning platform for others.
Singapore	Edible cutlery	Cutlery made from wheat, rice, and sorghum. There are over 160 million tons of plastic cutlery used in India every year. An initiative sponsored by the Govt of India.
Singapore	horticulture along the banks of perennial rivers in India	The idea would be to replace plants by trees along the banks of perennial rivers in order to prevent soil erosion and to promote economic growth for farmers.
Singapore	Trash to Treasure (Free flea market)	A free flea market where personal items and belongings that are unwanted can be redistributed to those who need them or could better utilize them. A strength is that it's free for everyone, can be organised anywhere, and doesn't require any technology.
Singapore	Green Roofs	Planned and built green roofs and not just those green roofs that are imposed upon existing buildings. These green roofs can reduce energy use from air conditioning. Well built green roofs also collect rainwater and reduce the flow fast-flowing water, and also reduces the risk of flooding in flood-prone areas. Provides habitat for biodiversity.
Singapore	carpooling	encourage students and faculty members of a given university to use carpooling as much as they can
Singapore	Setting up of Wormery	The setting up of wormeries allows for the decentralization of food waste collection. Food waste is collected locally and composted locally using earthworms. The castings (worm poop) are used as fertilizers and they are sent to the community gardens.
Singapore	Smart Solar Charging	Electric car sharing initiative in Utrecht
Singapore	Skillsfuture SG	Decentralized education through multiple course providers and institutions, conducted on a governmental platform with government funding. A strength is that it leverages government budgets, which are much larger. Additionally, it sponsors and increases the educational level of the country.
Singapore	regulate diet	The aim is to encourage people to change their diet by promoting vegetarian options
Singapore	Environmentalist foundation of India	Volunteering opportunities for individuals to restore urban lakes and rural water bodies through community action. A strength is that it is simple and connects with people's volunteering aspirations.
Singapore	Sustainable Alternative Lighting (SALt) lamp	It is an environment-friendly and sustainable alternative light source that runs on saltwater
Singapore	Community in Bloom	Setting up of localized community gardens in Singapore. Around 2000 of such gardens have already been set up all over Singapore. Gardens are also managed by their own communities.
Singapore	Electricity productions from ocean currents	Production of electricity by the use of underwater turbines based on the difference of temperatures in water.
Singapore	Intel Make Tomorrow	Skills development in using microControllers and IoT for vocational institution students.
Singapore	Refugee crisis management	Using technology to help refugees like a message alert that can reach out to them
Singapore	Fresh Direct Container Farms	A Nigerian entrepreneur turned shipping containers into indoor governments, and employ mainly needy women
Singapore	Smartphone app to monitor energy consumption	Powershop is a company that provides an app platform that enable consumers to track the energy consumption of their home. The tracking is live and accounts for energy inputs from solar PV. The app also provides monetary incentives to reduce consumption by displaying $ values supported from solar energy input which supports conscious consumption and transition to renewable energy.
Singapore	Solar-powered Water Purification	Used by local communities for water filtration and sanitation. Filtration system within a bottle which allows communities to use the water.
Singapore	Green Building Standards	Setting standards for new buildings construction and renovations
Singapore	App for plant identification	The app helps to identify trees, plants, and flowers. People who see unknown plants, they can take a photo using the app. It plays a significant role in educating people.
Singapore	AI driverless electric cars	To integrate AI INTO our transport system
Singapore	Palm oil-targeting activist organizations like People's Movement to Stop Haze	The seed initiative tries to promote the use of sustainable palm oil in the Singapore market by engaging both F&B sector (supply side) and consumers (demand side).
Singapore	Precious Plastic	A startup that provides open-source guides and designs for communities to create plastic recycling machines and tools. The startup provides support and guidance for anyone interested in creating such machines. A strength is that all the information is open sourced and it allows decentralized recycling initiatives to emerge.
Singapore	Youth Ki Awaaz	A decentralized online platform for people to write stories on social issues. It allows anyone to create a campaign and facilitate change. Additionally, stories are powerful in tackling global challenges and this platform allows anyone to participate.
Singapore	Wearable devices	Wearable devices for people's health

Workshop	Seed Name	Description
Singapore	Gaia Grid	An off grid farming community that uses crowdfunding and social media to create a self sustaining community. It co-ops tribal villages, weeds out social problems, and encourages organic farming.
Singapore	Gravity Light	Using gravity to create electricity
Singapore	Safe spaces for deep conservations on climate change	A NGO that trains facilitators to help facilitate home-based, friend networked conversations on climate change. This activates individuals to lead community actions.
Singapore	Street Feeders of KL	A regular gathering of volunteers to distribute perishable and non-perishable food to the homeless. This also helps increase the understanding on the background of the homeless. The homeless can also be linked to job opportunities. A strength is that this facilitates face-to-face conversations that help connect communities with the homeless. It also offers hope to the underprivileged.
Singapore	UN REDD+ Carbon Credit System in a Quirino Protected	To restore fragmented landscapes and promote planting of fruit trees (for food security as well) and provide subsidy to Farmers for being advocates of the protected area
Singapore	Sharks Fin Database	A centralized database that allows citizens to share the location and names of restaurants that serve sharks fin. This creates awareness of these restaurants and allows citizens to boycott or engage with the restaurants that serve sharks fin. A strength is that this is citizen sourced data that leverages existing technology and is low cost.
Singapore	No Straw Tuesdays	Plastic-lite started 1.5 years ago as a way for people to reduce and be mindful of their plastic consumption. Volunteer-run, self-funded group. Rolling out initiatives among communities to promote lifestyle changes. Taps into the power of social media to galvanise participants. 'No Straw Tuesdays', rolled out in schools one day a week.
Singapore	GrabHitch	A technology platform that connects non-taxi drivers and riders to facilitate carpooling in order to reduce the number of cars and fuel demands. This leverages existing technology and apps to reduce the number of cars on the road.
Singapore	Amsterdam Rain Proof programme	Harnessing urban water runoff for alternative products eg. beer and closing the water loop.
Singapore	Bitcoin Mining Heater for Homeless	Bitcoin mining releases lot of heat energy.
Singapore	Dog Poo Bag Station	Provide self sustainable and convenient way to encourage do owners to clean up after their pets, for pet owner to share their unused dog poo plastic bag with fellow pet owners, to clean up dog poo.
Singapore	Making of pet plastic bottles into t-shirts	Tzu Chi charity employs the use of disposed pet plastic bottles and upscale them into t-shirts and blankets, which are then donated to victims of natural disasters
Singapore	Lendor (app)	P2P library of things (e.g. household objects) that users can borrow, instead of buying for one-time use
Singapore	Innisfree Empty Bottle Recycling Campaign	Customers can bring used containers back to stall (up to 50 points redeemed), get discount on future purchases. Campaign uses statistics on how many bottles have been recycled and repurposed. Appeals to consumers, "feel-good" aspect.
Singapore	500 Women Scientists	Improve openness, equality in science in Latin America. Goal to create scientific culture, promote scientific literacy, embrace technology and sciences. Grassroots movement - get people to recognise the presence of female scientists in particular. Host social events on a monthly basis, invite individuals to chat, speed-dating style. Mentoring, going to schools and talking to girls about S&T, policy in government, does not seem accessible, sense of cultural inferiority (for the Old White Man), means of decolonizing academia and science.
Singapore	Swapaholic	Online clothes swapping platform. Participants bring in pre-loved, quality clothing in exchange for points that can be spent at clothes swapping events hosted around Singapore on a regular basis.
Singapore	Plastic Footprint Calculator	http://whatismycarbonfootprint.com/plastic-footprint - Calculates an individuals's plastic footprint, aiming to use information to educate and reduce usage of plastics.
Singapore	Plastic Bank (app)	https://www.plasticbank.org/what-we-do/ - Turns waste into currency by incentivising individuals to collect plastics in exchange for rewards which are distributed and authenticated through the Plastic Bank app which uses Blockchain technology.Transfers values into the hands of those who collect plastic.
Singapore	Vegan/Vegetarian UN Environment Meals	Provide vegan/vegetarian options during meals at UN/INGO conferences and events to showcase that vegan/vegetarian meals.
Singapore	Local Water Commissions	Community-organizing in the event of a drought/water rationing, to pool resources, help less abled members collect water, draw on connections. Can also be applied to energy sharing and food security.
Singapore	Grab (app)	Ride-hailing app, expanded to GrabShare, GrabHitch, incentivise passengers by using cheaper prices as opposed to riding individually. Reduces fuel consumption, company can mobilize clean energy vehicles (e.g. electric cars).
Singapore	First climate change course in a Costa Rican University	Addresses lack of existing climate change education and communication in the country
Singapore	SECMDL	An alternative learning school for youth established by local youth in a community. A strength of this is that it is a decentralized, replicable, and self-sustaining project that can be transferred to other communities.
Singapore	Plant Diet/Veganism	include and promote more plant-based menus
Singapore	Sustainable Aquaculture	Integrates multi-trophic systems, using outputs (e.g. waste) of a species as inputs (e.g. food) for species up the chain. Also sources for local/indigenous species to breed in Singapore, to encourage Singapore's heritage, change tastes and preferences to reduce carbon footprint from food imports.
Singapore	Superwomarket	A supermarket/cafe designed by women scientists filled with products they selected. It will have: - Products showing carbon/water footprint and relevant SDGs - Products with minimum wrapping - A breastfeeding and expressing space, which will also have a sit-in nurse who can check/give guidance for breast cancer - A communication space (cafe) for women to network, conduct events, etc. - A childcare space with a sitter where young children can play while the mothers are shopping or networking It will provide a "safe environment" for women to share their expertise.

Workshop	Seed Name	Description
Singapore	Greening the GEO Conference	The next GEO meeting will be more green. We must practice what we preach. It could potentially include: vegetarian/vegan meal options, remote conferencing options using conferencing robots (e.g. "Double") (This will be an inclusive option for persons who cannot travel, such as those like myself who cannot travel due to childcare, or persons with mobility issues), paperless, less air conditioning, sustainable hotel practices, smaller carbon footprint (less plastic), sourced by renewable energy The existing 2009 UN Environment guidelines (http://www.greeningtheblue.org/sites/default/files/GreenMeetingGuide.pdf) could be updated through online consultations with the GEO authors (e.g. "what do you want to see in the next GEO conference?"), and then e-published together with GEO-6 as a spin-off product.
Guangzhou	Reduce the consumption of wildlife	- Reduce consumption of wildlife (e.g. do not eat wildlife) - Reduce the purchase of wildlife products
Guangzhou	Light and Shadow Ocean Pavilion	- The public should not visit the aquarium of the captive cetacean; (reduce the number of captive cetacean because captive conditions are not suitable for their growth) - The public could visit the Light and Shadow Ocean Pavilion (where physical visit could be replaced by Light and shadow technology)
Guangzhou	Museum of environmental photography	- To build environment protection photography museum
Guangzhou	Sharing Community and say "NO" to waste	- Propose potential mechanism for sharing to promote sustainable development in cities.
Guangzhou	Cellphone sharing	- Frequent replace of cellphones is not encouraged - Cellphone recycling is encouraged - Cellphone sharing is encouraged
Guangzhou	Reduce the use of solid wood furniture	- Reduce the use of solid wood furniture
Guangzhou	Development and utilization of natural gas hydrate	- Natural gas hydrate resource is very rich, which could be used by humans for 1000 years. Currently, lots of resource spots are found and exploration tech is greatly improved; How to encourage all countries to explore natural gas hydrate in a clean manner should be a priority.
Guangzhou	Intelligent Green Building	- Intelligent Green Building is able to utilize natural spontaneous process (e.g. air convection) to reduce energy consumption.
Guangzhou	Anhydrous aluminum radiator	Technology which emphasizes the use of electricity instead of coal is an efficient approach to increase energy efficiency and decrease carbon emission. E.g. Anhydrous aluminum radiator
Guangzhou	Use of Big Data tech to change citizen purchase behavior	Big Data has significant impact on environment awareness promotion and consuming behavior by normal citizens.
Guangzhou	Distributed intelligent energy storage technology	- Distributed intelligent energy storage, energy internet, intelligent energy community
Guangzhou	Online intelligent detection of drinking water purification system	This initiative consists of following elements: Internet, online monitor, artificial intelligence, artificial manufacture, purification tech (e.g. physical filtration and chemical decomposition), big data and cloud computing
Guangzhou	Reduce water consumption in daily life	Residential wastewater could be recycled for other utilization (such as toilet and car washing)
Guangzhou	Tableware made of sorghum flour as substitution of disposable tableware	Tableware could be made of sorghum flour; Such tableware could be utilized as substitution of disposable tableware.
Guangzhou	Reuse and recycling of daily necessity packages	Package of daily necessities (e.g. make-up, shampoo etc.) could be reused and recycled.
Guangzhou	Self drink container	When customer brings drink container themselves at drink shop, they could enjoy discount. This could reduce the number of disposable container.
Guangzhou	Plastic limit and environmental court	To establish environment court
Guangzhou	Plastic limit in universities	University prohibit the use of disposable tableware and thin-plastic bag.
Guangzhou	Simplify packaging of express delivery	If packages of express delivery could be simplified, it is a way to reduce amount of waste.
Guangzhou	Reduce use of plastic, use environmental friendly makeup and packaging materials	This initiative recommends to reduce use of makeup which contains plastic molecules. It could be implemented in a similar way as the plastic bag limit in supermarket in previous years.
Guangzhou	Environment education	All the responsible people take efforts to promote environment education in regional areas with clear objective.
Guangzhou	Transfer air into fresh and clean water	After transferring, such water is fresh and clean, which could meet highest standards.
Guangzhou	Brain-computer interface	Collect EEG (brain language) in human brain and build corresponding database; Once there is brain wave, there will be corresponding computer language to enable robot action.
Guangzhou	MR Disaster Prevention	To promote disaster prevention education
Guangzhou	Decentralized distribution	Community Supported Agriculture, Farmers market
Bangkok	Global CEO alliance	The initiative is to get to the core of private sector engagement/establishing the value-proposition from the Sustainable Development Goals (what's in it for private sector)
Bangkok	Initiative on sharing economy	For example, platforms such as Uber, AirBnB, clothes swapping etc. There is growing movement where under-utilized resources are being used more efficiently – i.e., most cars sitting idle; this is expanding into all sorts of new areas and gets to the heart of SCP.

Workshop	Seed Name	Description
Bangkok	Innovation lab that functions as an incubator for ideas	To help scale-up small-scale innovation/ technological entrepreneurial ideas (i.e., recycling innovation idea for cans)
Bangkok	Green rooftops in urban spaces	Use to grow food, clean water, ...application of green infrastructure; these efforts could up hugely up-scaled
Bangkok	Rain water harvesting	Particularly in the urban context where there are fewer and fewer permeable surfaces...
Bangkok	Ethical fashion industry	-Use discarded fabrics and textiles from the fashion industry; - Use circular economy concept and applying it to the design, production, retail, and purchasing and of fashion products: addressing a range of issues including exploitation, fair trade etc. while tackling sustainable production and environmental protection.
Bangkok	Solar panel windows for skyscrapers	Massive renewable energy potential for the urban env; vast amount of glass in skyscrapers represents enormous potential for an emerging technology that turns windows into solar panels. (Yale 360: Transforming Buildings into Energy Producers)
Bangkok	Box-type solar cookers for rooftops	Relatively simple, low-tech, low cost
Bangkok	Big data and business intelligence	At scale to tackle Zero discharge of illegal chemicals/ dyes in the supply chain
Bangkok	Low-carbon initiatives	Climate Change Asia initiative launched at AIT – a pioneer initiatives in the region- helping to understand how vulnerable habitats can be restored.
Bangkok	Climate smart agriculture and community forestry	
Bangkok	Intelligent transportation systems	For major cities to tackle air pollution, resource efficiency, safety... fixed route software integrated in all cars, integrated scheduling systems, fully integrated CAD/AVL system,
Bangkok	Global public awareness campaigns	To counter some of the rhetoric that some government leaders are spreading regarding climate denial
Bangkok	Urban green infrastructure - urban parks connectivity	Deliberate urban planning and design that focuses maximizing connectivity of urban green space in including inner city parks; softening park edges and better connections to the peri-urban fridge
Bangkok	Green infrastructure for urban heat stress reduction	Encourage capital infrastructure improvement projects (such as more regular street-upgrades, community level heat-reducing practices like tree plantings, etc.)
Bangkok	Lowering the age of decision-makers	Tackle social barriers, address countries that have age limits. (Italy, France etc.)
Bangkok	Small scale renewable energy projects	Examples include residential solar panel projects, smaller hydropower plants
Bangkok	Innovating and strengthening traditional agricultural knowledge	Counter balance to the forces that are downgrading TKL; seeing soil as a living matter that needs to be cared for
Bangkok	Food systems approach - from upstream to downstream	Multi-sector engagement at every stage
Bangkok	Natural capital accounting	To link nature conservation and development impact and catalyze technological services (i.e., e-waste tracking)
Bangkok	Resource-oriented sanitation	Convert wastes in the waste chain back to agricultural inputs/food systems
Bangkok	Circular economy and extended producer responsibility	Reuse of e-waste, old-phones, etc.
Bangkok	Technology in renewable hydrogen	as an element of the circular economy
Bangkok	DIY waste management systems	Use recycled materials for furniture
Bangkok	Knowledge-sharing strategies	Use digital platforms to share ideas
Nairobi	Smart Energy microgrids	
Nairobi	Smart H20	Smart H2OSM is a user-friendly app that is freely available to all citizens world-wide. With just one-click, it allows them to report leakages, violations and water waste very easily and in very little time. Through Smart H2OSM, citizens can partner with their water utility and be proactively involved in saving water.
Nairobi	Smart agriculture (productivity crops)	
Nairobi	Biogas (waste)	
Nairobi	Earth Observation	"Sanivation" and "locate it." Task force on high resolution spatial data to track decofly in school, health, transport, and energy. The seed requires technology, evidence (??), and communication. Very simple app, upload to Airbus platform. The seed is just starting up in Norok County, Nairobi. The seed is innovative as it is a low-cost application and a global platform. Addresses SDGs 3, 4, 10
Nairobi	Education - connectivity of schools	Use technology to produce quality education. Use kindles instead of wastes and reduce paper use. Addresses SDG 4.
Nairobi	Affordable air quality monitoring	Low cost sensor devices
Nairobi	Crowdsourcing behavior with smart apps	
Nairobi	Products from plastics	Produce petroleum products from plastics, reuse plastics. Innovation from producing threads of 3D printing. This seed has a "microeconomy approach."
Nairobi	Awareness building	

Workshop	Seed Name	Description
Nairobi	Ecoflame - toilet that separates solid/liquid waste	Solid waste management system that recycles waste to produce biogas. This technology produces energy from compost. Portal toilets fit within houses. Waste is baked in a 90 degree parabolic mirror and turned into charcoal. Ecoflame has been in use for a while and has over 10,000 users in Naivasha (??). The product does not smell
Nairobi	Residential rooftop solar energy	Catalyst for creating social awareness for energy issues. Addresses SDGs 13, 9, 7, 11. This seed is gaining momentum globally. Example from Pham Binh (Vietnam), Durga Prasad Dawadi (Nepal), and Peter Mburu (Kenya)
Nairobi	Smart energy	
Nairobi	Rooftop rainwater harvesting	
Nairobi	Water/agriculture management in Iran (small grant program by GEF)	Water management Iran to address local community and health. The seed has many small projects to engage communities and farms to prevent dust and sand storms. Addresses SDGs 1, 6, 8, 17. The seed is already established and engaging with government and local communities. This seed is currently operating around the border of Iran and could scale up to neighbouring countries to address desertification. The seeds main weaknesses are its difficulty to link national government and local communities' civil society. Several actors are involved including the media, government, local communities, private sector, banks/investment.
Nairobi	Protected areas in Madagascar	Protected areas in the east of Madagascar for the Credit Carbon, reducing deforestation, and promoting smart climate agriculture. The seed addresses SDGS 2, 3, 6, 7, 8, 13, 15. Deforestation is prevented through alternative economic opportunities. The seed's main weaknesses include the need to promote across the rest of the country the idea of climate smart agriculture, and it is unclear if people will accept new ideas/approaches. Actors involved include the rural population and local government.
Nairobi	Electric cars/Tesla and trucks	This is an established seed that addresses SDGs 3, 7, 8, 9, 11, 12, 13. It is based in EU, US, China, and some ME countries. The seed replaces diesel/gasoline vehicles. With the Tesla app you can get real-time info on the vehicle through the software integration in cars. Particularly promising is that Beijing sells more EVs than regular vehicles with a 40 per cent subsidy of the cost from the government. Three main weaknesses can be identified: (1) energy infrastructure issues if electricity generation comes from coal or fossil fuels, (2) congestion problems can just get worse, (3) rebound effect, (4) competition with e-bikes. Actors involved include business, consumers, and individuals. Superchargers, hydrogen batters (and other tech breakthroughs) could make EVs faster and easier.
Nairobi	Innovative public transport	This includes e-ships for public transport and tourism. The seed addresses SDGs 3, 7, 8, 9, 11, 12, 13. It currently in start-up phase and based in Iraq. The ships are to be cheaper for consumers, energy efficient, faster, can alleviate road accidents, reduce congestion, and improve safety. Weaknesses include congestion, potential conflict with water, could conflict with pollution of air, water, waste. Actors involved include business, govt, and individuals.
Nairobi	Car and bicycle sharing	This addresses SDGs 3, 7, 8, 11, 12, 13. It is established and in use in at least hundreds of cities (Mobike in 110 cities). The seed uses clean energy, is a low cost solution, and easy to implement/upscale. Major weaknesses include the need for an app/smart phone, too many bikes lead to crowding and accidents. Enabling conditions include infrastructure and critical mass; proper education and respect for cultural issues, many co-benefits for integration with other smart-city initiatives.
Nairobi	E-pay initiatives/cashless systems	This seed addresses SDGs 5, 9, 8, 12. It is established in some countries and a start-up in others, but examples can be found globally. It is innovative as it adds convenience, reduces the need for cash, and allows for easier exchanges. There is a global megatrend towards this seed and can help with sustainable consumption and production. The seed's main weaknesses are waste and and the increased consumption of solid waste.
Nairobi	Reduce packaging	Optional packaging when purchasing goods, package free shops, plastic bag bans, taxes. This seed addresses SDGs 2, 6, 9, 11, 12, 13. It is at the start-up stage and examples can be found in Dresden, Germany and other places. It is innovative and beneficial as it includes: (1) biopackaging from agricultural waste which closes system loops, (2) reduces food waste, (3) incentive alternatives, (4) visible and a good communication tool. It is a very scalable solution. There are three main weaknesses: (1) bulk/wholesale - health/sanitary issues, (2) alternatives to plastics could be worse and potentially increase the amount of bags used, (3) compostable/biopackaging needs inputs that add stress to agriculture. Actors involved include citizens and governments, business, urban shops and retailers.

Workshop Seeds-Based Visions

Workshop	Pathway	Description
Bangkok	Smart Communities (Decentralized Solutions focused)	This alliance proposed a vision and an approach for developing what they described as "smart communities" – a new and radically different vision for future built environments. The basic premise of the envisaged future presented here, is to challenge the conventional model and principles of urbanism and the traditional processes through which existing cities grow, evolve and function. The idea which builds on the so-called New Urbanism concept seeks to address the disconnection between the current models of urban/ city-planning and interactions at the peri-urban interface that characterize today's-built environments. For example, the group aimed to address the inefficiencies with urban and suburban sprawl, simplistic and counterproductive patterns of metropolitan growth, perverse incentives around infrastructure investment, rural-urban migration etc. The proposed Smart Communities Alliance brings together several common elements and mutually reinforcing attributes that lead to development of smarter and more sustainable communities. The main 'seeds' or game-changers ideas that are part of this alliance include: circular economy, sustainable peri-urban agriculture, microfinancing, intelligent and sustainable transportation systems, and public/ community awareness. This alliance proposes to address all five regional environmental challenges identified. The group felt that the Smart Communities vision was relevant to all 17 Sustainable Development Goals but in particular those related to 1, 3, 6, 7, 11, 12, 13 and 17 with an emphasis on the following synergies: smart changes, behavioural choices, sustained investments in R&D, innovation and clean technology, political and social adaptability (and adaptive governance). Finally, as a point of clarification, the group indicated that the intention of the Smart Communities was not to convert existing large cities, but rather, to shape future build environments and areas that are currently in the early stages of urbanizing. The Smart Communities Alliance received a final score of 17 and succeeded in addressing a number of Sustainable Development Goals and leveraging synergies between urban sustainability objectives and sustainable (or eco-centric) urban infrastructure investments. One of the most important enabling conditions for bottom-up approaches to succeed is sustainability, and the need for gap analysis. Here, the alliance was able partially successful, however a major shortcoming was a lack of discussion on the need for social and political acceptability.
Bangkok	Smart Future (Lifestyle Change focused)	The second alliance in the visioning exercise, proposed a holistic approach to bringing together and catalyzing large-scale behavioural changes through a process of "influencing the influencers". Here, the alliance stressed the importance of finding a new delivery mechanism to identify who the main private and public sector leaders (or influencers) are to communicate a single value proposition about the Sustainable Development Goals. As a secondary approach, the alliance discussed the need to target consumers, and to leverage the opportunities brought about through big data/ data revolution. The alliance suggested that the Smart Futures vision was relevant to all 17 Sustainable Development Goals and addressed all five regional challenges. The common attributes of this alliance include: disruptive innovation technologies (e.g., smartphone applications, cloud computing, social networking), data-driven decision systems, sustainable/ smart cities, agro-economy solutions, highly inclusive/ people-centric initiatives, integration, meaningful private-public partnerships, and results-based performance to improve decision-making processes. The Smart Futures Alliance received a final score of 9 from the SAP and despite some promising game-changing ideas, and several areas of convergence, the alliance was ultimately unsuccessful as they rolled a higher dice. Reflecting on the process, the group found that the principal barrier was that their ideas were too broad and that their main inputs were spread across too many competing (and sometimes mutually exclusive) objectives.
Bangkok	Planet Tech (Global tech focused)	The third and final alliance was the 'Planet Tech' group presented a futuristic, hyper technology rich vision of the future with a focus on planet altering technologies and of Earth systems including: geoengineering/ carbon capture storage technologies, mesopelagic exploration, planetary tech, and artificial intelligence. The proposed vision was predominantly geared towards addressing macro/ planetary scale environmental challenges including climate change, biodiversity and complex atmospheric-ocean related issues. The common thread for this alliance was the potential for plenary harm and conversely opportunities for transformational 'planet-alerting' solutions. The main Sustainable Development Goals that the Planet Tech Alliance was targeting include 12, 14 and 17. Several institutional obstacles and gaps were identified including mechanisms to circumvent conflict, intergovernmental and global governance issues (e.g., UN Security Council issues). The Planet Tech Alliance received a final score of 8 and struggled with a scenario that was overly complex, far too doomsday oriented and ultimately not inspiring or compelling enough. Their high dice roll meant that their scenario also did not succeed. The group acknowledged that the overall concept was not conducive and/or accessible enough to attract meaningful political engagement and that they needed to refine their technology dominant strategy.

Workshop	Pathway	Description
Guangzhou	Proposal 1	**How are the seeds combined?** The habitat is essential for survival. In order to ensure coexistence between humans and animals, new and appropriate links must be developed. **How does the proposal help realize the pathway – toward what SDGs?** To provide better habitat condition for animals, which do good for biodiversity. **What policy changes are needed to help realize the proposal in this pathway? How can policies help deal with any trade-offs related to other SDGS?** 1. Public awareness of animal protection 2. Laws and regulations for animal protection
Guangzhou	Proposal 2	**How are the seeds combined?** These seeds have similar features, which could share similar support on policy and addressing mechanism. **How does the proposal help realize the pathway – toward what SDGs?** Nowadays online shopping produces large amount of plastic waste, which is difficult to be either recycled or degraded and also causes significant environmental burden. By limiting use of plastic and solid wood furniture, negative environment impact caused by these could be reduced. What policy changes are needed to help realize the proposal in this pathway? How can policies help deal with any trade-offs related to other SDGS? 1. Implementation of Plastic Limit. 2. Waste management and categorization. 3. Concrete policies and regulations need to be developed under framework of environment protection law. 4. Public awareness of environment protection
Guangzhou	Proposal 3	**How are the seeds combined?** 1. Constructions and facilities are everywhere in the community, advanced energy techs can be applied. 2. Model sustainable community is easy to scale up, copy and paste. 3. Communication and cooperation between communities, cities and different countries are helpful to combine all the seeds. 4. Community could be the "lab" for green tech. **How does the proposal help realize the pathway – toward what SDGs?** This proposal could help achieve following goals: Goal 3: Ensure Health and Well-being Goal 4: Quality Education Goal 6: Clean Water and Sanitation Goal 7: Affordable and Clean Energy Goal 9: Industry, Innovation and Infrastructure Goal 11: Sustainable Cities and Communities Goal 12: Responsible Consumption and Production Goal 13: Climate Action Goal 17: Partnerships for the Goals **What policy changes are needed to help realize the proposal in this pathway? How can policies help deal with any trade-offs related to other SDGS?** 1. Campaigns to raise public awareness 2. Cut land and tax policy 3. Public fundings from government These aspects could help increase energy efficiency.
Nairobi	Behavior change	Use seeds for ISO-like framework to standardize sustainability for national governments (i.e., standards for waste, recycling, packaging, etc.) Policy changes needed/trade-offs related to other SDGs: - UN resolution - national level legislation - how to sell to national govts - competitive advantage, better business opportunities for leading companies
Nairobi	Global Tech	Proposals: 1. healthy people 2. smart cities conference

Workshop	Pathway	Description
Nairobi	Decentralized Solutions	Consists of climate smart agriculture, protection of forests, healthcare and promotion of gender equality Seeds are combined with guidance from national/local committees. SDGs addressed: 1,3,10,11,13,15,17,16 Challenges: linkages between national and local community/ potential risk to marine life, and sending messages to high-level policy makers and media to increase public awareness Policy changes needed/trade-offs related to other SDGs: Potential risk to marine life is there is an increase in agricultural fertilizer. Need for strengthening link between local governments and communities. Inviting investment from private sector. Need to work across silos between healthcare, agriculture, forest protection. Need to identify and engage all stakeholders to increase public awareness. Important for high-level policymakers to understand local activities - GEO can help facilitate this.
Singapore	Alternative energy solutions to promote mixed land use	Ppps lock in guaranteed customers for the solar farm; drones can be used for solar farm citing; citizens can help monitor operations through iterative feedback loops
Singapore	Smart ag systems for sustainable development	The tech seeds we came up with can support the urban ag and community indigenous knowledge in the existing seeds.
Singapore	Appification for Everyday Lifestyle Changes	Cover different aspects of daily life
Singapore	Straw-lite Campaign 'Same Taste, Less Waste'	Builds on Straw-lite Campaign to extend to local businesses and eateries, getting eateries in Singapore to not give out straws as a default, working on various zones, to approach eateries to reduce straw usage.
Singapore	Sustainable Urban Living	Green buildings are used for urban farming. These urban farms produce edible cutlery as well as food. Biomimicy technology and ride sharing further promote the community's sustainability
Singapore	No impact on the environment	They can be implemented in the same institution, namely the university or a private company
Singapore	Change from Consumers to Community	All relate to sharing rather than consuming and building communities
Singapore	Energy efficient community	Green building standards require a wide set of sustainability building and renovation rules, and require roofs to be used by solar cooperatives.
Singapore	Sustainable Urban food production and consumption	Both seeds reduce waste from food consumption, and address sustainable production of food.
Singapore	Community resilient gardens	Goals: SDG 10, 2, 12, 13, 3, 8, 6. The proposal aims to create synergies between the different seeds proposed by incorporating different metrics and initiatives of environmental sustainability with the fair employment of employees that are mentally challenged.
Singapore	A Containerized, Modular, Sustainable City	Each of our seed in this proposal address a specific urban city challenge, with the function/technology of the seed being transformed into to create a transportable container module which makes up the building blocks of a sustainable city.
Singapore	Urbanites (engaging citizens in community environment action)	Create an international cities platform online where data and environmental action is aggregated. Multiple features are included on this platform, including environmental education, citizen information reporting, enforcement, skills education embedded in circular economy concepts and also logs/pins where activities are for the nearby communities to participate in. SDG 17 is fulfilled through multiple partnerships 1. Safe space conversations to tie sustainability to very local impacts (SDG 13) 2. After activating these citizens, they contribute by engaging with this online platform. They can choose to be active citizen information providers, signing up for skills education etc... (SDG 12, 11, 4) 3. The platform is not a passive platform but actively engages experts and practitioners. E.g., citizen reporting of biodiversity can be linked to Researchers; citizen reporting of vehicle or water pollution is linked to regulatory enforcement officers, after finishing skills education they can provide paid services (SDG 3, 4, 8) 4. Education arm of the app provides both environmental and skills based education (aka udemy but specialized in environment such as circular economy, repair, waste management, composting skills etc) (SDG 4, 9) 5. Aggregation of multiple community initiatives, including waste reduction, plastic pollution, biodiversity, poverty and hunger (SDG 1, 2, 15, 14) learn from different cities and communities as everyone will upload their initiatives to this platform 6. App enables easy access to action

Workshop	Pathway	Description
Singapore	500 Científicas (500 Women Scientists)	Getting people to know female scientists (perhaps due to our family-oriented, traditional culture female scientists will be perceived as more approachable, as most people identify with the "advice of a mother or a sister". This will help create a local science culture and improve scientific literacy. By working with other seeds we can increase the reach of female scientists, motivate women to incur in traditionally male-dominated fields, and improve equality and access.
Singapore	Off Grid Rural Development	Gaia Grid is the foundation of the seed proposal - crowdfunding from social media to purchase degraded non-forested land. Regional farmers are mobilized to do organic farming on the land. Volunteers from the community that need jobs and skills development are invited to help. Farm uses off-grid clean energy. Volunteer programme is set up to develop framing skills that can be used for future integration back into society and explicitly addresses social issues such as alcoholism and drugs that plague the neighbouring community. (SDG 1, 2, 3, 7, 8, 12) By connecting to additional seeds we achieve extra SDGs synergistically: - Environmentalist foundation of India: explicitly uses land and water restoration projects and tools (SDG 6, 15) - SECMDL; Karthavyam; Youth Ki Awaaz: Alternative education which provides bottom-up experiential learning and skills development through mediums like story-telling, filmmaking and using local issues and indigenous local knowledge and perspectives (SDG 4, 18) - Safe Spaces for Deep Conversations on Climate Change: Introducing climate change perspectives into community conversations using issues that are very relevant to the community to bring abstract global debates into a local context (SDG 13, 17) - Watly: Bringing access to technology to rural areas in order to give communities access to global networks (SDG 9)

Climate CoLab Semi-Finalists

Proposal	Description
C'SQUARE - Urban Comfort Rating	C'SQUARE is an online-based application with the objective of empowering citizens by rating places in urban areas concerning their comfort level (thermal comfort, air and noise pollution, security, safety and beauty/cleanliness) and proposing solutions for solving trade-offs to minimize its impact on the environment.
Youth Informing Communities on Climate Change Adaptation through building homes	We are putting forward a strategy encouraging youth to lead an environmentally efficient, educational strategy through the expansion of housing. The solution will promote environmental education with the skill to adapt to future issues in First Nations communities in Canada.
An innovative model harmonizing ecology, emotions, and economy of the villages	A model Ecovillage showcasing fulfilment of the village community needs of food, water, waste, and energy while meeting the maximum SDGs
Local organization for the promotion of environmental data analysis	This proposal proposes the implementation of local agencies that address this problem in a systemic and proactive way, working in conjunction with government agencies, academic institutions and different groups of civil society.
Global complementary currency pegged to the production of organic living biomass	The Organic Monetary Fund, set up as a division of the International Monetary Fund, will help to achieve almost all SDGs in record time
Adding value to all waste at source through blockchain technology and IoT	Recyclebot is a blockchain-based waste management platform that allows any waste buyer to source specific waste directly from the producer, giving you the ability to buy, sell and reduce waste wherever you are, when you want. Unlike landfills, Recyclebot provides an affordable, fast and safe way to add value to all waste right where it is generated.
Synergistic Solutions Global Network	Networking existing resources and engaging women and youth to deliver proven solutions to address ten Sustainable Development Goals.
Leo Leo Eco-Transport	Leo Leo Eco-Transport aims to foster economic empowerment and promote food security while ensuring sustainability in Arusha city. Leo Leo Eco-Transport is proposing a visionary and innovative synergistic solution to sustainable development of food security, reliable access to solar energy utilization, economic empowerment, employment opportunities and climate change that affects the surrounding environment.
Beekeeping as mechanism of biodiversity conservation and livelihoods improvement	Alohen, partnering with Barefoot College, will establish adoption of beekeeping practice in Tanzania. Following its expansion strategy, Barefoot will further develop the project in Madagascar, Liberia, India and Pacific Islands, through its network of training centers worldwide. The main objective of the project is to create conservation awareness, empowering on the modern beekeeping methods, targeting women from rural areas and supporting them by creating local social enterprises based on bees' byproducts. Such project was initiated by BC in Zanzibar in 2015, creating the B.Barefoot Honey brand to provide trainings and markets for local women beekeepers. Replication was initiated in India in 2016.
New clean energy tech will reverse climate change, eliminate world poverty	Disruptive 0E energy tech works better than fossil/nuke at a better value, based on a new, previously undocumented aspect of physics.
Media houses and institutions for communication awareness on the 17 SDGs	Media Groups are mainly associated with local political issues and propaganda, rather than ensuring communication awareness on the 17 SDGs.
Binder-less board made from coconut husk and sugarcane bagasse	Ecovon are developing sustainably sourced, newly engineered wood made from coconut husk and sugar cane bagasse better for people and planet.
Culturing the future with permaculture	Build a physical Permaculture Resource Hub. A space where people can network with individuals and projects, as well as catalyze community organization and action. Our mission is to build resilience at the personal, household and community levels while creating thriving examples of abundance based on ecological wisdom.
Climate protection by the elderly	A brief, the elderly plants a fast growing fruit bearing tree rich in proteins and minerals. Once the tree starts bearing fruits, these fruits are then distributed in the market by either his children or son in laws. His children will be given an incentive of Rs 100 for the work of transporting the commodities to the government market. Once the government procures the fruits it distributes it to the anganwadis, orphanages and old age homes free of cost under ministry of social welfare scheme.
ClimeDoc to achieve SDGs	A documentary on impacts of climate change significantly contributes promotion of appropriate actions to achieve SDGs in developing nations
Cut out red meat to pause climate for 25 years	Eliminating grazed animal consumption will buy 25 years to fix emissions and will profoundly benefit 7 SDGs and 5 Planetary Boundaries
Biochar electrode for the recovery of acid mine drainage produced by Tailings	Tailings have a real potential for contamination of ecosystem told the relaxation of metals and metalloids that should manage adequately.
ClimateCoop	Blockchain based collaboration and governance platform enabling dynamic community development and project formation for SDGs, e.g. ClimateAction
Global waste reduction and climate change slowdown through systemic approach	We're intended to create a holistic systemic approach and processes on waste reduction which is easily adaptable to any country.
Transforming plastic garbage into plastic timbers to preserve forests	A solar powered extrude machine that uses innovative technology to transform plastic garbage into plastic timber to reduce deforestation.
System approach for greening the conventional agribusiness	Alliancing Green producers and green consumers for green future
Acacia: Answer to climate change, economic empowerment and food security in ASALs	The adoption of acacia Senegal tree species/Licorice as an agroforestry system for the benefit of environment and populations in ASAL areas.
Business plan for production and marketing of compost from urban solid wastes	Converting urban solid wastes into compost forms will have social, economic and environmental benefits; and helps to reduce GHG emissions
Youth Climate Leaders (YCL)	Young people traveling the world and working together to learn more about climate change and start their careers as climate leaders
Framework for community-based sustainable development	A modular sustainable development framework for communities that is cognizant of the synergistic nature of infrastructure and society.

Proposal	Description
Waste, a source of resource-learn to empower	This project utilizes poor women affected by dumped waste by providing earning source at the same time preventing production greenhouse gas.
Living Energy - connecting science, design and nature to light up our world	Living Light is an atmospheric lamp which harvests its energy from the plant living next to the lamp itself. The light of the lamp is produced by 'plant microbial fuel cell technology ': energy generated by bacteria in the soil which release electrons while breaking down organic compounds of the plant. These collected electrons, for which we have created a storage mechanism, produce enough energy to light up the LED lights for about an hour in the current phase of development. Because the Living Light symbolysis and extra-ordinary lamp, we have given it an extra-ordinary switch: by softly stroking the leaves transforms this ordinary plant into a Living Light.
An environmental conservation approach to handling oil and gas in Uganda	Development of oil deposits in an environmentally conscious manner
Land for life: an alternative to slash-and-burn in the world's rain forests	"Guama", an integrated agroforestry model, is transforming family livelihoods, saving rain forests; restoring degraded soils and landscapes
Population control	Population control solves EVERY identified problem facing humanity. You don't get more synergistic that that.
Potential for zeolites for conserving moisture in drought prone areas of Africa	Moisture conserving ability of Zeolite minerals can be exploited to enhance crop and tree production in drought prone areas of Africa.
Leveraging telecommunication for decoupling SDGs using a whole systems approach	Building an ICT-led citizen movement to address some of the Sustainable Development Goals and climate action from the bottom up.
Adapting the indigenous approach to climate change adaptation and mitigation	The project is dedicated to emphasizing the indigenous adaptation and mitigation techniques adapted by local communities in responding to climate change issues.
Solar based poly-generation system that can provide power, heat, and clean water	Our system uses Solar concentrators to provide Power, Heat and clean water at 1/3rd the capital cost of SolarPV and has no disposable costs.

The GEO-6 Process

"We all share one planet and are one humanity;
there is no escaping this reality."

Wangari Maathai (1940-2011), Nobel Lauriate

Objectives, Scope and Process

The Mandate for the sixth Global Environment Outlook was obtained from Member States at the first UN Environment Assembly (resolution 1/4, operative paragraph 8). More information on this mandate can be found in Annex 1-1 of this report. The objectives, scope and process for GEO-6 were defined and adopted in a Final Statement by the Global Intergovernmental and Multi-Stakeholder Consultation that took place in October 2014. It was attended by more than 133 delegates with more than 100 governments represented.

Objectives

The consultation reaffirmed the UNEA-1 mandate by identifying the following objectives for the assessment:

- provide a comprehensive, integrated and scientifically credible global environmental assessment to support decision-making processes at appropriate levels;
- facilitate broader participation by major groups and stakeholders, in particular from the private sector and NGOs and to increase outreach to target audiences;
- The analysis should draw on diverse knowledge systems, including by using accepted guidelines for the use of peer reviewed scientific literature, grey literature, data and indigenous and local knowledge;
- A clear process and organizational structure is needed to ensure credibility, legitimacy and relevance;
- "The assessment should build on and be consistent with previous GEOs, as well as the work of other relevant intergovernmental organizations and processes, including Multilateral Environmental Agreements, in order to maintain its branding and role in keeping the environmental situation under review";[1]
- inform, as appropriate, the strategic directions of UNEP and other relevant UN bodies;
- strengthen the policy relevance of GEO-6 by including an analysis of case studies of policy options, that incorporates environmental, economic, social and scientific data and information and their indicative costs and benefits to identify promising policy options to speed up achievement of the internationally agreed goals such as the Sustainable Development Goals and other multilateral environmental agreements;
- identify data gaps in the thematic issues considered by GEO-6.

Scope

GEO-6 builds on previous GEO reports and continues to provide an analysis of the state of the global environment, the global, regional and national policy response as well as the outlook for the foreseeable future. It differs from previous GEO reports in its emphasis on Sustainable Development Goals and in providing possible means of accelerating achievement of these goals. GEO-6 is made up of four distinct but closely linked parts.

- **Part A** assesses the state of the global environment in relation to key internationally agreed goals such as the Sustainable Development Goals and those of various multilateral environmental agreements. The assessment is based on national, regional and global analyses and datasets.

- **Part B** provides an analysis of the effectiveness of the policy response to these environmental challenges as well as an analysis of progress towards achieving specific environmentals goals.
- **Part C** reviews the scenarios literature and assesses pathways towards achieving Agenda 2030 as well as achieving a truly sustainable world in 2050.
- **Part D** identifies future data and knowledge necessary to improve our ability to assess environmental impacts and pathways for achieving sustainability.

The GEO-6 also considers key policy questions. These include:

- What are the primary drivers of environmental change?
- What is the current state of the environment and why?
- How successful have we been in achieving our internationally agreed environmental goals?
- Have there been successful environmental policies?
- What are the policy lessons learned and possible solutions?
- Is the current policy response enough?
- What are the business as usual scenarios and what does a sustainable future look like?
- What are the emerging issues and megatrends including their possible impacts?
- What are the possible pathways to achieving Agenda 2030 and other internationally agreed environmental goals?

Process

The October 2014 consultation also provided direction for strengthening the process of the GEO-6 assessment, including:

- The assessment process shall be supported by two main advisory bodies: the High-level Intergovernmental and Stakeholder Advisory Group (HLG) and the Scientific Advisory Panel (SAP);
- Advice shall also be obtained from an Assessment Methodologies, Data and Information Working Group;
- Other GEO-6 roles would include: Coordinating Lead Authors (CLAs); Lead Authors; up to 20 GEO-6 Fellows and Coordinators; global experts; regional experts; community of practice moderators; review editors; and reviewers;
- the CLAs will provide technical summaries of the GEO 6 and prepare the negotiating drafts of the Summary for Policymakers in close collaboration with and under the leadership of the HLG, ensuring that the technical aspects of GEO-6 are reflected in the draft. The SPM would be negotiated at a dedicated intergovernmental and stakeholder meeting;
- Relevant MEAs, international organizations and scientific institutions will be invited to actively contribute to the GEO-6 process.
- The GEO-6 will ensure scientific credibility, policy relevance and legitimacy of the assessment by engaging a wide range of stakeholders;
- The assessment will be subjected to extensive scientific expert peer review and government review;

1 Outcomes document of the Intergovernmental and Multi-stakeholder consultation, 21-23 October, 2014, Berlin, Germany. The full text of the Outcomes document can be found in the Appendix to this section.

- The assessment process will continue to target institutional capacity building by engaging developing country experts;
- The assessment should strive to communicate key messages and findings to target audiences in an accessible manner.

TIMELINE

The sixth Global Environment Outlook process was characterized by 4 larger authors meetings, two smaller drafting meetings on Outlooks and Policy, as well as face to face meetings of the advisory bodies, Review Editors and Member States. The meeting and drafting schedule followed 3 basic principles established by the advisory bodies:

- There should be coherence across the different Parts of GEO-6 and the 12 cross-cutting issues should be drafted in tandem with the assessment of the 5 environmental themes.
- There should be opportunities for robust interaction between the authors and the advisory bodies to ensure both policy relevance and scientific integrity are maintained throughout the process.
- The author teams should be kept small since the regional assessments contain much of the information that is needed in the global assessment and they should form the foundation of the global assessment.

To ensure robust interaction with the advisory bodies, 3 of the 4 larger authors meetings had participation from the High-level Group and the Scientific Advisory Panel. To ensure coherence across the assessment, the larger meetings were used to allow for 'speed dating' between the thematic chapter authors and the cross-cutting issue authors. This 'speed dating' allowed for 1 hour of discussion between the authors teams where issues were discussed and writing assignments given. To ensure that the author teams were kept small, a core of coordinating lead authors were first selected into the process and then skills gaps were identified. From the skills gap analysis, invitations were sent to lead authors to complement the drafting of the chapters.

As the work programme evolved it became clear that additional authors meetings for the Policy and Outlooks chapters would be needed. The Secretariat proceeded to organize these during the months of May and June, 2018. In addition, the Scientific Advisory Panel requested to meet one last time in order to formulate their opinion on the scientific credibility of the GEO process. This meeting was organized back-to-back with the final Review Editors meeting in October, 2018. This allowed for the two groups to share information about the peer review processes and their overall rigour.

The drafting meeting for the Summary for Policymakers involved the High-level Group, Coordinating Lead Authors as well as the Co-chairs of the assessment. The Co-chairs of the Scientific Advisory Panel also participated as observers and provided some of their experience with drafting of Summaries in other assessment processes.

The final meeting of the GEO process was the meeting of Member States to finalize and adopt the Summary for Policymakers. This meeting was held at UN Environment headquarters which allowed for a broad participation of

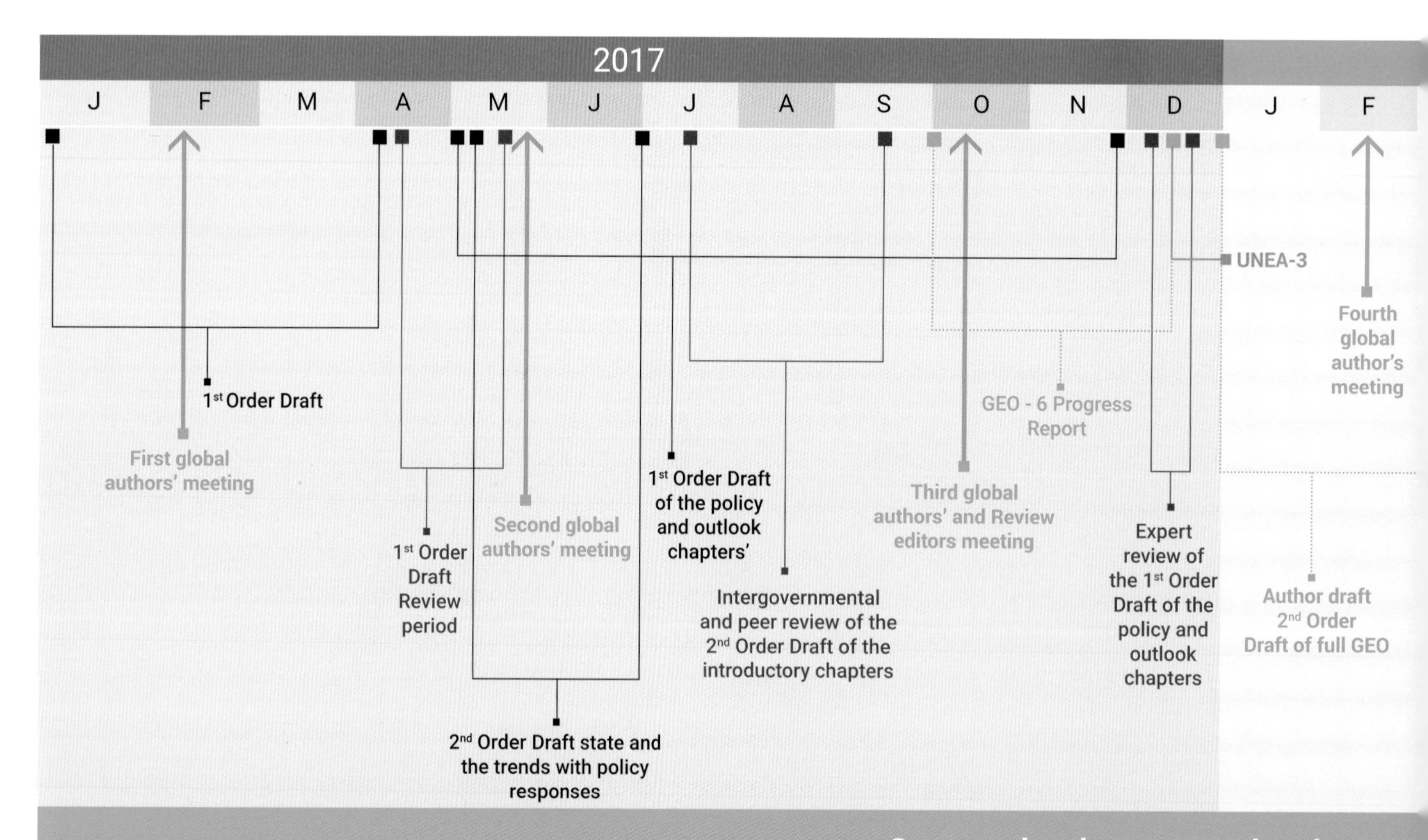

Member States. The 4-day meeting sought to review the text of the Summary and make changes that would allow for its adoption by all Member States present. The final adopted document was submitted to the fourth UN Environment Assembly for endorsement.

PARTNERSHIPS AND COLLABORATION

The development of GEO-6 involved extensive collaboration both within UN Environment and between UN Environment and a network of multidisciplinary experts and research institutions, all of whom made their valuable time and knowledge available to the process.

The consultation requested that experts for content development, including reviewers and advisory groups, be nominated by governments and other main stakeholders based on their expertise and using a transparent nomination process. The nominated experts were then convened by the UN Environment Secretariat based on their expertise with due consideration of gender and regional balance.

Chapter expert groups

The GEO-6 report contains 25 chapters. An expert author group was established for each chapter to conceptualize, research, draft, revise and finalize each chapter. More than 150 authors and fellows were involved in content development. Each chapter expert was under the leadership of three or four coordinating lead authors and supported by a UN Environment chapter coordinator. Other members of the chapter expert groups comprised lead authors and contributing authors.

GEO-6 fellows

GEO-6 continued to pursue the Fellowship initiative established during the GEO-4 process in 2005. This engages early career professionals in the GEO process so that they can gain experience from participating in a major global environmental assessment. A total of 27 fellows from 15 countries participated in GEO-6.

REVIEW PROCESS

The GEO-6 assessment underwent five rounds of review involving more than 1000 experts. In total the GEO-6 assessment was reviewed five times at different stages of its development and the process yielded more than 14,000 comments. Due to this process, the draft chapters have been re-written, adjusted and edited to improve the quality while the drafting process has been adjusted to improve its effectiveness.

The first nine introductory chapters of the assessment: introduction, drivers of environmental change, state of our data and knowledge, the crosscutting chapter as well as the state of the global environment, across 5 main thematic areas: air, biodiversity , oceans, land and freshwater were reviewed earlier in the process than the policy and outlooks chapters. At the end of the review process all chapters were provided for review by technical experts then for a longer intergovernmental and expert review. For the final review the chapters were provided as individual chapters (25 chapters separately) and as a complete assessment report (all chapters as a single document). This offered reviewers an opportunity to either

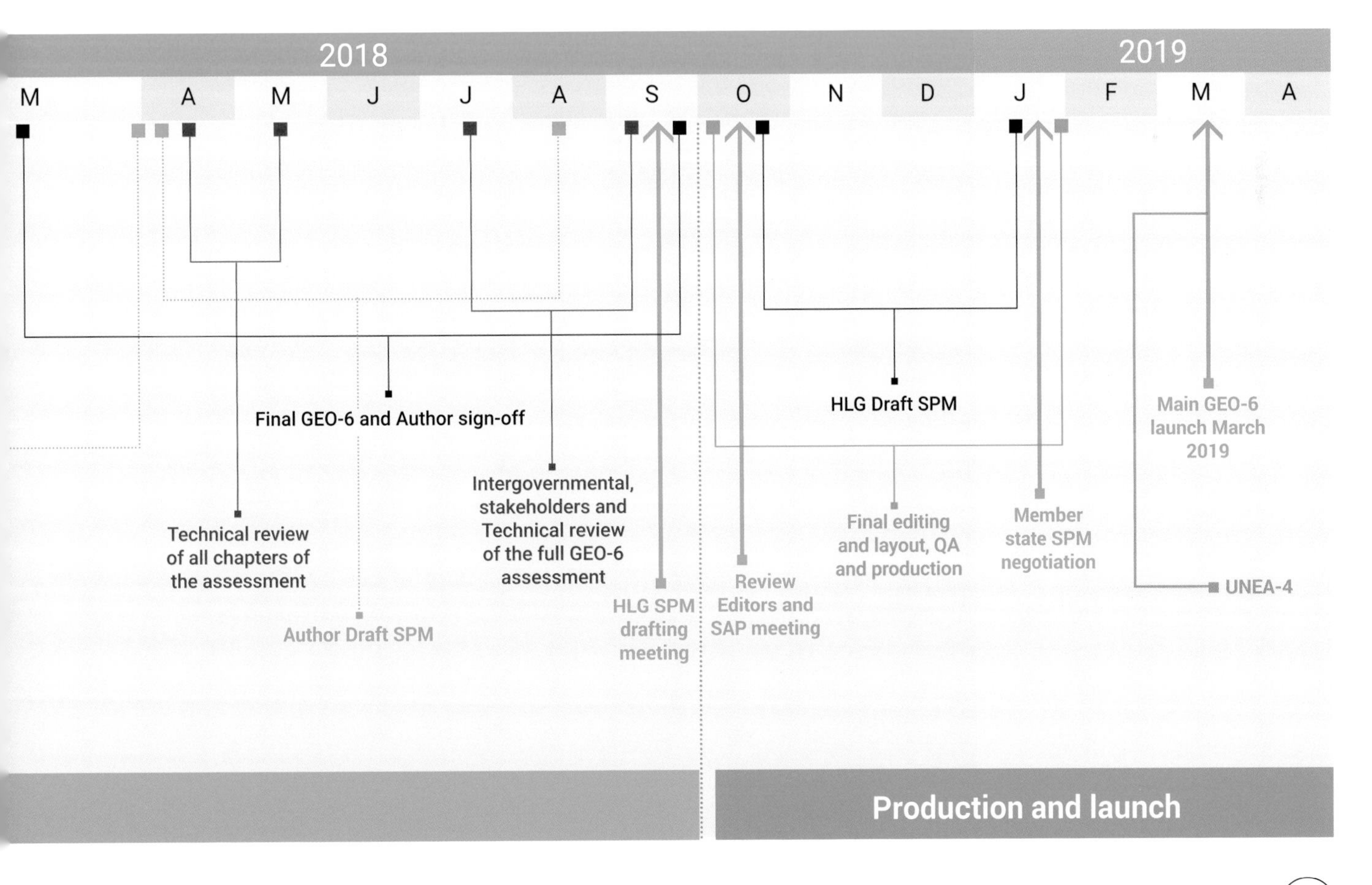

review specific chapters that were directly related to their areas of expertise or review the whole assessment report to comment on the report's coherence.

For all review periods the secretariat offered a 'service desk' where all reviewers with questions or concerns were supported in this task. Virtual meetings were organized for all reviewers coordinated by the secretariat to first orient the review team before the start of the review process and then to check on progress as well as answer questions. These virtual meetings were conducted by the secretariat with support from the lead review editors, who listened in and provided advice on any issues. The preparatory review material/tools were discussed during these meetings with the concentration being on the reviewer's guidelines. Terms of reference for the reviewers were developed and updated for each review period, including the ethical responsibilities of GEO-6 Reviewers. During the review period the secretariat conducted follow-up calls for all available reviewers to assess the progress and review important deadlines. All review call recordings were shared with the whole review team to ensure that other reviewers were aware of the tasks and the plan for moving forward.

GEO-6 ADVISORY BODIES

Three external specialized advisory bodies were established to support the assessment process.

High-Level Intergovernmental and Stakeholder Advisory Group

The panel included 33 high-level government representatives from all six UN Environment regions as well as 8-10 key stakeholders. The High-level Group assessed and formulated strategic advice to GEO-6 authors and other groups to assist them in their assessment work. They also provided initial guidance on the structure and content of the GEO-6 Summary for Policymakers and further guidance to the experts in finalizing the draft Summary, in preparation for the final intergovernmental negotiation. In addition, ad-hoc guidance was provided to UN Environment throughout the assessment process to align the GEO-6 process with other relevant global assessments. The High-level Group met face-to-face seven times between 2015 and 2018. The Advisory Group also met virtually on a monthly basis throughout the preparation of the global assessment, from May 2016 to September 2018.

Science Advisory Panel

The Panel included 22 distinguished scientists who met face-to-face five times. The Panel was responsible for providing advice on the scientific credibility of the assessment process. The Panel provided scientific advice; standards and guidelines for the assessment and review process; and reviewed the findings of the mid-term evaluation of the assessment process. The Panel met virtually on a monthly basis throughout the preparation of the global assessment, from June 2016 to October 2018.

Assessment Methodologies, Data and Information Working Group

The working group comprised of 12 professionals who met face-to-face three times between 2015 and 2018 and provided support to the assessment process and provide guidance on the use of core datasets and indicators. They consulted with experts to review the methods used in GEO-6, identify priority environmental indicators as well as data gaps and related issues. The Working Group met virtually as needed throughout the process.

CONSULTATION PROCESS

UNEP organized panel discussions at all authors meetings throughout the assessment process. These panel discussions were intended to delve into specific environmental issues that were relevant to the region and location of the meeting. The following are some of the key meetings convened since the inception of the GEO-6 process.

GEO-6 planning meetings

Two planning meetings were convened with the High-level Group and the Scientific Advisory Panel May and June 2016. The meetings produced a final annotated outline for the global assessment and a list of recommended co-chairs and coordinating lead authors.

Global Intergovernmental and Multi-stakeholder Consultation

This consultation defined and adopted the scope, objectives and process for GEO-6 in October 2014. Participants at the Intergovernmental and Multi-Stakeholder Consultation concluded that GEO-6 would be an integrated environmental assessment using the Drivers – Pressures – State – Impacts – Response (DPSIR) approach. The report would build on regional assessments and include an inter-governmentally negotiated Summary for Policymakers. The analysis would aim to present findings and deliver products to targeted audiences including decision makers, across the public and private sectors, such as businesses and the youth.

Outlooks expert meeting

In May 2018, an outlooks expert group, was convened to move the policy chapters to third order draft quality by addressing all comments from the science editors, as well as comments received from the second order draft technical review period.

Policy expert meeting

In June 2018, a policy expert group, was convened to move the policy chapters to third order draft quality by addressing all comments from the science editors, as well as comments received from the second order draft technical review period.

Global authors' meetings

Four global production and authors' meetings were convened in February 2017, May 2017, October 2017 and in February 2018 to discuss and develop GEO-6-chapter content and outlines, to address review comments, and to harmonize different approaches and presentation styles.

Chapter working group meetings

Hundreds of virtual chapter meetings were convened to prepare, review and revise the drafts for individual chapters.

Summary for Policymakers intergovernmental meeting

A final open-ended intergovernmental meeting was convened from January 21-24, 2019 in Nairobi, Kenya to negotiate and adopt the GEO-6 Summary for Policymakers (SPM). The meeting attended by 95 Governments adopted the summary, which presents the policy-relevant findings of GEO-6 and is published as a separate document. The GEO-6 Summary for Policymakers was presented to the fourth UN Environment Assembly for endorsement.

The launch of GEO-6 will coincide with the fourth United Nations Environmental Assembly. GEO-6 highlights the current state, trends and outlook for the planet and its people, and showcases more than 35 case studies of policies that have been assessed for their effectiveness.

GEO-6 highlights not just the perils of delaying action, but the options for transforming our economic, environmental and social systems to achieve a truly sustainable world.

Further information is available at https://www.unenvironment.org/global-environment-outlook

Appendix

Statement by the Global Intergovernmental and Multi-stakeholder Consultation on the Sixth Global Environment Outlook held in Berlin from 21 – 23 October 2014

UNEP/IGMC.2 Rev.2

Strengthening the Science Policy Interface:

Building the Evidence Base for the Post-2015 Agenda

23 October 2014

Organisation of work

The Global Intergovernmental Multi-stakeholder Consultation (IGMS) met in Berlin from 21-23 October 2014. It was attended by 133 delegates, with more than 100 governments represented.

The meeting was opened by Achim Steiner, Executive Director of UNEP.

The election of officials followed. Idunn Eidheim (Norway) and Dr. Majid Shafie-Pour (Iran) were elected Co-Chairs. Dr. Peter Denton (Major Groups and Stakeholders) was elected Rapporteur.

Background

Reference was made to the Rio+20 outcome document, earlier Governing Council decisions, and specifically to UNEA Resolutions 4 and 10.

The Secretariat presented the recommendations of the independent evaluation of GEO-5 which stated the need to:

"(1) *facilitate stakeholder engagement*; (2) *enhance capacity building*; (3) *increase the use of grey literature and indigenous knowledge*; (4) *promote relevance at all scales*; (5) *increase developing country participation*; (6) *facilitate access to information*; (7) *use results based management and evidence for evaluations*; and (8) *improve financial planning and funding*."

Participants at the IGMS noted the findings of the evaluation and expressed the need to facilitate broader participation by major groups and stakeholders, in particular from the private sector and NGOs and to increase outreach to target audiences. The analysis should draw on diverse knowledge systems, including by using accepted guidelines for the use of peer reviewed scientific literature, grey literature, data and indigenous and local knowledge. A clear process and organizational structure is needed to ensure credibility, legitimacy and relevance. The assessment should build on and be consistent with previous GEOs, as well as the work of other relevant intergovernmental organizations and processes including such as MEAs, in order to maintain its branding and role in keeping the environmental situation under review.

Under Agenda Item Four, the meeting participants discussed options and timing for GEO-6.

Structure for the content of GEO-6

Participants at the IGMS supported that GEO-6 would be an integrated environmental assessment, using the Drivers – Pressures – State – Impacts – Response (DPSIR) approach in the GEO conceptual framework. The Report will build on regional assessments and include an inter-governmentally negotiated Summary for Policymakers. The analysis will aim at presenting findings and delivering products to targeted audiences among decision makers, across the public and private sectors at global to local levels.

GEO-6 will reflect three broad, analytical components.

Global Environment: State and Trends

The first component will include an analysis of the environmental state and trends for air, biota, land and water and their multiple contributions to environment and human well-being. This will be achieved through an analysis of interactions with cross-cutting issues such as climate change; environmental disasters; food; energy; human health; economic development; resource use; chemicals and waste; and culture and society, and relevant policies.

Environmental Policies, Goals and Objectives: A Review of Policy Responses and Options

The second component will provide a policy analysis of the links between the state and trends in the environment and global and regional environmental goals and objectives, including those reflected in national policy responses, and an assessment of progress towards them.

Global Environment Outlook

The third component will be comprised of an integrated analysis of megatrends and environmental change, and refer to the outputs of modeling, scenarios and regional outlooks. The analysis will take into account the Global Sustainable Development Report and provide support to the environmental components of the post-2015 agenda.

Timing of GEO-6

Participants expressed broad support for the following delivery dates: GEO-6 regional assessments to be delivered by early 2016 and the complete GEO-6 including its Summary for Policymakers to be delivered not later than 2018, at an appropriate event to be determined in consultation between UNEP and governments. Regional assessments will be undertaken during 2015. The Executive Director of UNEP will report on progress to UNEA 2 in 2016.

Process and operational structure of GEO-6

Participants also voiced support for the establishment of two advisory bodies: the High-level Intergovernmental and Stakeholder Advisory Group (HLG) and the Scientific Advisory Panel (SAP). There will also be an Assessment Methodologies, Data and Information Working Group. The HLG will include five representatives from each UN region, plus five representatives from the Major Groups and Stakeholders. The SAP will be comprised of three representatives from each UNEP region, plus up to six global experts. The Assessment Methodologies, Data and Information Working Group will be comprised of three representatives from each UNEP region, plus up to six global experts. Participants expressed a wish to include individuals with indigenous and local knowledge.

Other GEO-6 roles would include: Coordinating Lead Authors (CLAs); Lead Authors; up to 20 GEO-6 Fellows; Global Experts; Regional Experts; Community of Practice Moderators; Review Editors; and Reviewers.

The participants discussed the terms of reference for the operational structure as set out in the Annex 1.

Based on practice from earlier GEOs and other international scientific assessments, the CLAs will provide technical summaries of the GEO 6 and preparing the negotiating drafts of the Summary for Policymakers in close collaboration with and under the leadership of the HLG, ensuring that the technical aspects of GEO-6 are reflected in the draft. The SPM would be negotiated at a dedicated intergovernmental and stakeholder meeting.

It was noted that UNEP Live will be used by the Secretariat to enhance capacity development and to support GEO-6 by providing the platform for the GEO-6 Communities of Practice and the Nominations Portal. UNEP Live will also support the global and regional analyses through relevant data collection related to *inter alia* UNSEEA and indicator development; encouraging sharing and access to national data and information; linking to peer-reviewed literature from various language domains; providing access to indigenous and local knowledge and information drawn from attributable, public sources. Information should also be provided on the benefits of UNEP Live for countries; the roles of MEAs in UNEP Live and on the UNEP Live programme of work.

Support was given for the GEO-6 to use Communities of Practice to encourage sharing of knowledge amongst the various groups, increase stakeholder engagement and support capacity development. CoPs will be established for the major areas of GEO-6 and regional assessments. Capacity development would be supported through the fellowship programme, the implementation of national reporting systems, along with participation in regional environmental information networks and regional assessments.

Relevant MEAs, international organizations and scientific institutions will be invited to actively contribute to the GEO-6 process.

Support was given for the multi -stage peer review, based on the following principles. First the best possible scientific and technical advice should be included to ensure that the assessment represents the latest scientific, technical and socioeconomic findings and is as comprehensive as possible. Second, a broad circulation process ensuring representation of experts not involved in the preparation of the parts they are reviewing, with particular emphasis on involving as many experts from developing countries as possible. Third, the peer-review by governments will include both technical and policy aspects with due respect to the independence of the reviewers. Finally, the multi -stage review process to be balanced, open and transparent. Conflicts of interest will be identified through a process based on those used by IPBES and IPCC.

Nomination process

Participants emphasized the need for an open and transparent nomination process for all the GEO-6 roles, using the GEO-6 Nominations Portal in UNEP Live. The experts will be nominated using the criteria outlined in Annex II, and be selected by UNEP in a transparent manner with due consideration of the need to ensure geographic, disciplinary and gender balance. The nomination period will run until January 31, 2015. The selection process will be completed by the end of February 28, 2015. Late nominations will be accepted under mitigating circumstances. The selected experts and nominees for the advisory bodies will be sent to governments for review. The list of selected experts will be published on-line.

Governmental representatives for the HLG must be nominated by their respective governments and will act in this capacity. The selection process for the stakeholder representatives will be overseen by the UNEP Major Groups and Stakeholders Branch. The selection procedure for the HLG will be determined within the UN regional groups.

The nomination process will be initiated by a letter from the Secretariat to be sent to governments and Major Groups and Stakeholders. This correspondence will be in the relevant UN language and append details of the GEO-6 processes, including remuneration of experts and a GEO-6 timetable.

Acronyms and Abbreviations

AaaS Analytics as a Service
ABNJ areas beyond national jurisdiction
ADB Asian Development Bank
AEM Agri-environment measures
AIDS Acquired Immune Deficiency Syndrome
AGGI Annual Greenhouse Gas Index
ALRTI acute lower respiratory infections
AMAP Arctic Monitoring and Assessment Programme
AMCEN African Ministerial Conference on the Environment
AMD acid mine drainage
AOC Areas of Concern
ASEAN Association of Southeast Asian Nations
ASGM artisanal and small scale gold mining
BACT best available control technology
BaP benzo[a]pyrene
BAT best available techniques
BAU business-as-usual
BC black carbon
BECCS Bioenergy with crabon capture and storage
BEV battery electric vehicles
CA conservation agriculture
CaCO3 calcium carbonate
CAP Common Agricultural Policy (EU)
CAS Chemical Abstract Service
CBMIS community-based monitoring and information systems
CBD Convention on Biological Diversity
CBMIS Community-based monitoring and information systems
CBO Congressional Budget Office
CCAC Climate and Clean Air Coalition for the Reduction of Short-Lived Climate Pollutants
CCAK Clean Cookstoves Association of Kenya
CCAMLR Conservation for the Antarctic Marine Living Resources
CCB China Construction Bank Corporation
CCE climate change education
CCFSC Central Committee for Flood and Storm Control
CCP command and control policies
CFC chlorofluorocarbon
CCS carbon capture and storage
CD compact disk
CDC Centre for Disease Control and Prevention (United States)
CEC 1) contaminants of emerging concern, or 2) Commission for Environmental Cooperation (under NAFTA)
CEDS Community Emissions Data System
CENESTA Centre for Sustainable Development and Environment
CFC-11 trichlorofluoromethane
CH_4 methane
CITES Convention on International Trade in Endangered Species of Wild Fauna and Flora
CLRTAP Convention on Long-range Transboundary Air Pollution
CLTS community-led total sanitation
CMM cutaneous malignant melanoma
CMS Convention on the Conservation of Migratory Species of Wild Animals
CNS central nervous system
CO carbon monoxide
CO_2 carbon dioxide
COMEAP Committee on the Medical Effects of Air Pollutants
CONAFOR National Forestry Commission of Mexico
COP Conference of the Parties
COPD chronic obstructive pulmonary disease
CORSIA Carbon Offsetting and Reduction Scheme for International Aviation
CSDH Commission on Social Detriments of Health
CSIRO Commonwealth Scientific and Industrial Research Organisation (Australia)
CSML Climate Science Model Language
CSO civil society organization
DaaS Data as a Service
DALY disability adjusted life year
DDT dichlorodiphenyltrichloro-ethane
DEFRA Department for Environment, Food and Rural Affairs (United Kingdom of Great Britian and Northern Ireland)
DESD UN Decade of Education for Sustainable Development (UNCCD)
DFO Department of Fisheries and Oceans (Canada)
DIF Directory Interchanged Format
DLDD desertification, land degradation and drought (UNCCD)
DPSIR drivers, pressures, state, impacts, responses
DRR disaster risk reduction
DSI dust storm intensity
DSF desert storm frequency
DWAF Department of Water Affairs and Forestry (South Africa)
EAP Environmental Action Programme (EU)
EBA ecosystem-based adaptation
EBAFOSA Ecosystem Based Adaptation for Food Security Assembly
EC European Commission
ECLAC United Nations Economic Commission for Latin America and the Caribbean
EDC endocrine-disrupting chemicals
EEA European Environment Agency
EGA Environmental Goods Agreement

EIA 1) Energy Information Administration (United States of America), or 2) environmental impact assessment
EID emerging infectious diseases
EIP economic incentive policies
ELD Economics of Land Degradation Initiative
EML Ecological Metadata Language
ENGO Environmental Non-Governmental Organizations
EPA 1) environmental performance assessment, or 2) Environmental Protection Agency (United States)
EPI environmental policy integration
ESA 1) environmentally sensitive area, or 2) European Space Agency
ESD Education for Sustainable Development
ESDIS Earth Science Data and Information System
EU European Union
FAO Food and Agriculture Organization of the United Nations
FBS food balance sheet (FAO)
FBSP Free Basic Sanitation Policy
FBWP free basic water policy
FIT feed-in tariff
FSC Forest Stewardship Council
G7 Group of Seven (Canada, France, Germany, Italy, Japan, United Kingdom, United States)
GACC Global Alliance for Clean Cookstoves
GAEC Good Agricultural and Environmental Conditions (EU)
GAP Global Action Programme on Education for Sustainable Development
GAWSiS Global Atmospheric Watch Station Information System
GBD Global Burden of Disease
GBR Great Barrier Reef
GBRMPA Great Barrier Reef Marine Park Authority
GCM 1) general circulation model, or 2) Global Climate Model
GDI Gasoline direct injection
GDP gross domestic product
GEA Global Environmental Assessment
GEF Global Environment Facility
GEMI UN-Water's Integrated Monitoring Initiative
GEO Global Environment Outlook
GEOSS Global Earth Observation System of Systems
GESAMP Joint Group of Experts on the Scientific Aspects of Marine Environmental Protection
GGEO Global Gender and Environment Outlook
GGW Great Green Wall (China)
GHG greenhouse gas
GIS geographical information systems
GLADIS Global Land Degradation Information System
GLASOD Global Assessment of Human-Induced Soil Degradation
GLOBE Global Learning and Observation to Benefit the Environment
GLRI Great Lakes Restoration Initiative
GM genetically modified
GMACC Global Military Advisory Council on Climate Change
GMO genetically modified organism
GMSL Global Mean Sea Level
GNH Gross National Happiness
GPA Global Program of Action for the Protection of the Marine Environment from Land-based Activities
GRI Global Reporting Initiative
GSP Generalized Scheme of Preferences
$GtCO_2$ gigatonne of carbon dioxide
GTP Global Temperature Potential
GW gigawatt
GWG Global Working Group
GWP 1) Global Water Partnership, or 2) global warming potential
ha hectares
HALE health adjusted life expectancy
HCFC hydrochlorofluorocarbon
HEI Health Effects Institute
HFC hydrofluorocarbon
Hg mercury
HIV human immunodeficiency virus
HLPF High-level Political Forum on Sustainable Development
HS Harmonized System
IAEG 1) Inter-agency and Expert Group, or 2) Internationally Agreed Environmental Goals
IAM integrated assessment model
IAS invasive alien species
ICAO International Civil Aviation Organization
ICCA indigenous and community- conserved areas
ICCT International Council on Clean Transportation
ICMM International Council of Mining and Metals
ICP International Cooperative Program
ICPDR International Commission for the Protection of the Danube River
ICS improved cookstove
ICT information and communication technology
IDMC Internal Displacement Monitoring Centre
IDP's internally displaced people
IEA 1) International Energy Agency, or 2) integrated environmental assessment
IFAW International Fund for Animal Welfare
IFPRI International Food Policy Research Institute
IGRAC International Groundwater Resource Assessment Centre
IIED International Institute for Environment and Development
IJC International Joint Commission
IK indigenous knowledge

ILBM	Integrated Lake Basin Management
ILO	International Labour Organization
IMO	International Maritime Organization
INBO	International Network of Basin Organizations
INDC	Intended Nationally Determined Contribution
INTERPOL	International Criminal Police Organization
IPLC	Indigenous Peoples and Local Communities
IPAT	Impact = Population x Affluence x Technology
IPBES	Intergovernmental Science-Policy Platform on Biodiversity and Ecosystem Services
IPCC	Intergovernmental Panel on Climate Change
IRDA	Iksandar Regional Development Authority
ISA	International Seabed Authority
ISO	International Organization for Standardization
ITF	International Transport Forum
ITPS	Intergovernmental Technical Panel on Soils
ITQ	Individual Transferable Quota
IUCN	International Union for the Conservation of Nature and Natural Resources
IUU	illegal, unreported and unregulated fishing
IWRM	integrated water resources management
JMP	Joint Monitoring Programme for Water Supply and Sanitation of WHO/UNICEF
JRC	Joint Research Centre (European Commission)
LA	local authorities
LAC	Latin America and the Caribbean
LAER	lowest achievable emission rate
LANCE	Land, Atmosphere near real-time Capability for Earth Observing System
LCT	low carbon technology
LDN	Land Degradation Neutrality (UNCCD)
LMO	living modified organism
LMMA	Locally Managed Marine Areas
LPG	liquefied petroleum gas
LPI	Living Planet Index
LSF	large-scale fisheries
LTEMP	Glen Canyon Dam Long-term Experimental and Management Plan
MAP	Mediterranean Action Plan for the Barcelona Convention
MARPOL	International Convention for the Prevention of Pollution from Ships
MCA	Minerals Council of Australia
MDG	Millennium Development Goal
MEA	1) Multilateral Environmental Agreement, or 2) Millennium Ecosystem Assessment
MEB	multiple evidence base
MPA	marine protected area
MRV	measurement, reporting, and verification
MSY	maximizing sustainable yield
MTFR	maximum technologically feasible reduction
MUDP	Environmental Technology Development and Demonstration Program (Denmark)
MUFPP	Milan Urban Food Policy Pact
N_2O	nitrous oxide
NAAQS	National Ambient Air Quality Standards (United States)
NAP	National Adaptation Plan
NASA	National Aeronautics Space Administration (United States)
NAZCA	Non-State Actor Zone for Climate Action
NBS	Nature-based solutions
NBSAP	National Biodiversity Strategies and Action Plans
NcML	netCDF Markup Language
NCP	nature's contribution to people
NCSDs	National Councils for Sustainable Development
NDC	nationally determined contribution
NDVI	Normalized Difference Vegetation Index
NEPA	National Environment Policy Act (United States)
NFCP	Natural Forest Conservation Program (China)
NGO	non-governmental organization
NH_3	ammonia
NIP	National Implementation Plan
NMVOC	non-methane volatile organic compounds
NO_2	nitrogen dioxide
NOAA	National Oceanic and Atmospheric Administration (United States)
NOWPAP	Action Plan for the Protection, Management and Development of the Marine and Coastal Development of the Northwest Pacific Region
NO_x	nitrogen oxides
NPP	net primary productivity
NSPS	New Source Performance Standards
NT	no-tillage
O_3	ozone
OC	organic carbon
OCP	organochlorine pesticides
ODA	official development assistance
ODS	ozone-depleting substance
OECD	Organisation for Economic Co-operation and Development
OSCAR	Observing Systems Capability Analysis and Review
OSPAR	Convention for the Protection of the Marine Environment of the North-East Atlantic
PA	protected area
PACC	Pacific Adaptation to Climate Change
PAME	protected area management effectiveness
PAH	polycyclic aromatic hydrocarbons
PAR	protect, accommodate and retreat
PAWS	Protection of Asian Wildlife Species
Pb	lead
PBDE	polybrominated diphenyl ethers
PBT	persistent, bioaccumulative toxic chemicals
PCB	Polychlorinated biphenyls
PCFV	Partnership for Clean Fuel and Vehicles
PCN	Polychlorinated napthalenes

PEER	Publishing and the Ecology of European Research
PES	payment for ecosystem services
PFAS	per- and polyfluoroakyl substances
PFC	prefluorinated chemicals
PFI	port fuel injection
PHE	phenanthrene
PM	particulate matter
PM_{10}	particulate matter with a diameter of 10 micrometres (0.01) millimetre) or less
$PM_{2.5}$	particulate matter with a diameter of 2.5 micrometres (0.0025 millimetre) or less
POPs	persistent organic pollutants
PoWPA	Programme of Work on Protected Areas
PPP	1) purchasing power parity or, 2) public private partnership
PPSR	Public Participation in Scientific Research
PRTR	Pollutant Release and Transfer Registers
R&D	research and development
RACT	Reasonably Available Control Technology
RBFM	rights-based fisheries management
RBG	Royal Botanic Garden, Kew
RBM	resilience-based management
RBO	River Basin Organizations
RCP	Representative Concentration Pathways
RD&D	research, development and demonstration
REACH	Registration, Evaluation, Authorisation and Restriction of Chemical (EU)
REC	renewable energy credits
REDD	Reducing Emissions from Deforestation and Forest Degradation
RES	renewable energy systems
RFMO	regional fisheries management organizations
RLI	Red List Index
ROD	Record of Decision
RPS	renewable portfolio standard
SaaS	Software as a Service
SADC	Southern African Development Community
SAICM	Strategic Approach to International Chemicals Management
SBI	Subsidiary Body for Implementation
SBSTTA	Subsidiary Body on Scientific, Technical and Technological Advice
SCBD	Secretariat of the Convention on Biological Diversity
SDG	Sustainable Development Goal
SDGIO	Sustainable Development Goals Interface Ontology
SDMX	Statistical Data and Metadata Exchange
SDS	sand and dust storms
SDS-WAS	Sand and Dust Storm Warning Advisory and Assessment System
SEA	strategic environmental assessment
SEEA	System of Environmental-Economic Accounting
SensorML	Sensor Model Language
SFA	State Forestry Administration (China)
SFM	sustainable forest management
SGSV	Svalbard Global Seed Vault
SIA	Sustainability Impact Assessment
SIDS	small island developing states
SLCP	short-lived climate pollutant
SLM	sustainable land management
SO_2	sulphur dioxide
SPC	1) Secretariat of the Pacific Community, or 2) South Pacific Community
SPREP	Secretariat of the Pacific Regional Environment Programme
SSA	Sub-Saharan Africa
SSF	small-scale fisheries
SSP	Shared Socio-economic Pathway
SST	sea surface temperature
TAC	total allowable catch
TCDD	Tetra-chlorodibenzo-dioxin
TEEB	The Economics of Ecosystems and Biodiversity
TEK	traditional ecological knowledge
TK	traditional knowledge
TNRSF	Three Northern Regions Shelter Forest (China)
TOAR	Tropospheric O_3 Assessment Report
TSD	Trade and Sustainable Development
TURF	Territorial Use Rights for Fishing
TWAP	Transboundary Waters Assessment Programme
U5MR	under-five mortality rate
UK	United Kingdom
UN	United Nations
UNCCD	United Nations Convention to Combat Desertification
UNCDF	United Nations Capital Development Fund
UNCLOS	United Nations Convention on the Law of the Sea
UNCSD	United Nations Commission on Sustainable Development
UNCTAD	United Nations Conference on Trade and Development
UNDP	United Nations Development Programme
UNDRIP	United Nations Declaration on the Rights of Indigenous People
UNECE	United Nations Economic Commission for Europe
UNECLAC	United Nations Economic Commission for Latin America and the Caribbean
UNEP	United Nations Environment Programme
UNEP-DHI	United Nations Environment Programme - Institute for Water and Environment
UN-GGIM	United Nations Committee of Experts on Global Geospatial Information Management
UNEP-UNECE	United Nations Environment Programme – Economic Commission for Europe
UNEP-WCMC	United Nations Environment Programme – World Conservation Monitoring Centre
UNESCAP	United Nations Economic and Social Commission for Asia and the Pacific

UNESCO United Nations Educational, Scientific and Cultural Organization
UNFCCC United Nations Framework Convention on Climate Change
UNFPA United Nations Population Fund
UNGA United Nations General Assembly
UNHCR United Nations High Commissioner for Refugees
UNICEF United Nations International Children's Emergency Fund
UNIDO United Nations Industrial Development Organization
UNISDR United Nations Office for Disaster Risk Reduction
UNODC UN Office on Drugs and Crime
UNSCN United Nations System Standing Committee on Nutrition
UNSDSN United Nations Sustainable Development Solutions Network
US EPA United States Environmental Protection Agency
USA United States of America
USAID United States Agency for International Development
UV ultraviolet
VGGT FAO Voluntary Guidelines on the Responsible Governance of Tenure
VME vulnerable marine ecosystems
VOC volatile organic compounds
VWE vulnerable marine ecosystems
WAD World Atlas of Desertification
WASH water, sanitation and hygiene
WAVES Wealth Accounting and the Valuation of Ecosystem Services
WBCSD World Business Council for Sustainable Development
WDPA World Database on Protected Areas
WEF World Economic Forum
WFD Waste Framework Directive of the EU
WfW Working for Water programme
WHC World Heritage Convention
WHO World Health Organization
WIPO World Intellectual Property Organization
WMO World Meteorological Organization
WOCAT World Overview of Conservation Approaches and Technologies
WRI World Resources Institute
WTO World Trade Organization
WWAP World Water Assessment Programme
WWF World Wide Fund for Nature
WWF-SA World Wide Fund for Nature, South Africa
ZSL Zoological Society of London

Contributors

GEO-6 Author teams

Co- chairs: Paul Ekins [University College London, United Kingdom of Great Britian and Northern Ireland]; Joyeeta Gupta [University of Amsterdam, The Netherlands].

Vice- chairs: Jane Bemigisha [ESIPPS International Ltd, Uganda]; Kejun Jiang [Energy Research Institute, China].

Chapter 1 Introduction and Context:

Mark Elder [Institute for Global Environmental Strategies, Japan]; Christian Loewe [German Environment Agency, Germany].

Chapter 2 Drivers of Environmental Change:

Tariq Banuri [University of Utah, United States of America]; Matthew Kosko (GEO Fellow) [University of Utah, United States of America]; Diego Martino [AAE Asesoramiento Ambiental Estratégico and ORT University, Uruguay]; Indu K. Murthy [Indian Institute of Science, India]; Jacob Park [Green Mountain College, United States of America]; Fernando Filgueira Prates [Centro de Informaciones y Estudios del Uruguay (CIESU), Uruguay]; Maria Jesus Iraola Trambauer (GEO Fellow) [University College London (UCL), United Kingdom of Great Britian and Northern Ireland]; Dimitri Alexis Zenghelis [London School of Economics, United Kingdom of Great Britian and Northern Ireland].

Chapter 3 The State of Our Data and Knowledge:

Graeme Clark [University of New South Wales, Australia]; Florence Mayocyoc-Daguitan [Tebtebba (Indigenous Peoples' International Centre for Policy Research and Education), Philippines]; James Donovan [ADEC Innovations, United Kingdom of Great Britian and Northern Ireland]; Pali Lehohla [Pan African Institute for Evidence, South Africa]; Sheryl Joy Anne S. Gutierrez [ADEC Innovations, Philippines]; Charles Mwangi [GLOBE ProgramKenya]; Amit R. Patel (GEO Fellow) [Planned Systems International Inc., United States of America]; Joni Seager [Bentley University, United States of America]; William Sonntag [The Group on Earth Observation Secretariat, United States of America]; Michelle Tan [ADEC Innovations, Kenya].

Chapter 4 Cross-cutting Issues:

Babatunde Joseph Abiodun [University of Cape Town, South Africa]; Giovanna Armiento [ENEA - Italian National Agency for New Technologies, Energy and Sustainable Economic Development, Italy]; Rob Bailey [Chatham House, The Royal Institute of International Affairs, United Kingdom of Great Britian and Northern Ireland]; Rajasekhar Balasubramanian [National University of Singapore, Singapore]; Ricardo Barra [University of Concepcion, Chile]; Kathryn Jennifer Bowen [Australian National University, Australia]; John Crump [GRID-Arendal, Norway]; Irene Dankelman [Radboud University, Netherlands]; Kari De Pryck (GEO Fellow) [SciencesPo, France]; Riyanti Djalante [United Nations University – Institute for the Advanced Study of Sustainability, Japan]; Monica Dutta [The Energy and Resources Institute (TERI), India]; Francois Gemenne [The Hugo Observatory, Université de Liège, Belgium]; Linda Godfrey [Council for Scientific and Industrial Research (CSIR), South Africa]; James Grellier [University of Exeter, United Kingdom of Great Britian and Northern Ireland]; Maha Halalsheh [University of Jordan, Jordan]; Fintan Hurley [Institute of Occupational Medicine, United Kingdom of Great Britian and Northern Ireland]; Richard King [Chatham House, The Royal Institute of International Affairs, United Kingdom of Great Britian and Northern Ireland]; Andrei P. Kirilenko [University of Florida, United States of America]; Peter Lemke [Alfred-Wegener-Institut, Germany]; Daniela Liggett [University of Canterbury, New Zealand]; Robyn M. Lucas [National Centre for Epidemiology and Population Health, The Australian National University, Australia]; Oswaldo Lucon [Sao Paulo State Environment Secretariat, Brazil]; Katrina Lyne (GEO Fellow) [James Cook University, Australia]; Diego Martino [AAE Asesoramiento Ambiental Estratégico and ORT University, Uruguay]; Ritu Mathur [The Energy and Resources Institute (TERI), India]; Shanna N. McClain [Environmental Law Institute, United States of America]; Catherine P. McMullen, Stockholm Environment Institute -Asia Centre, Thailand]; Emma Gaalaas Mullaney [Bucknell University, United States of America]; Unai Pascual [Ikerbasque, the Basque Foundation for Science, Spain]; Leisa N. Perch [SAEDI Consulting, Trinidad and Tobago]; Marco Rieckmann [University of Vechta, Germany]; Fülöp Sándor [National University of Public Services, Hungary]; Atilio Savino [ARS, Argentina]; Heinz Schandl [Commonwealth Scientific and Industrial Research Organisation (CSIRO), Australia]; Joeri Scholtens [University of Amsterdam, Netherlands]; Patricia Nayna Schwerdtle (GEO Fellow) [Monash University, Australia]; Joni Seager [Bentley University, United States of America]; Lei Shi [Tsinghua University, China]; Frank Thomalla [Stockholm Environment Institute - Asia Centre, Thailand]; Maria Jesus Iraola Trambauer (GEO Fellow) [University College London (UCL), United Kingdom of Great Britian and Northern Ireland]; Laura Wellesley [Chatham House, The Royal Institute of International Affairs, United Kingdom of Great Britian and Northern Ireland]; Caradee Y. Wright [South African Medical Research Council, South Africa]; Dan Wu [Sun Yat-Sen University, China]; Dimitri Alexis Zenghelis [London School of Economics, United Kingdom of Great Britian and Northern Ireland]; Caroline Zickgraf [The Hugo Observatory, Université de Liège, Belgium].

Chapter 5 State of the Global Environment: Air:

Babatunde Joseph Abiodun [University of Cape Town, South Africa]; Kathryn Jennifer Bowen [Australian National University, Australia]; Serena H. Chung [U.S. Environmental Protection Agency, United States of America]; Phillip Dickerson [U.S. Environmental Protection Agency, United States of America]; Riyanti Djalante [United Nations University – Institute for the Advanced Study of Sustainability, Japan]; Cristina de B. B. Guerreiro [Norwegian Institute for Air Research – NILU, Portugal]; Chenmin He (GEO Fellow) [Peking University, China]; Fintan Hurley [Institute of Occupational Medicine, United Kingdom of Great Britian and Northern Ireland]; Terry Keating [U.S. Environmental Protection Agency, United States of America]; Andrei P. Kirilenko [University of Florida, United States of America]; Robyn M. Lucas [National Centre for Epidemiology

and Population Health, The Australian National University, Australia]; John Muthama Nzioka [University of Nairobi, Kenya]; Stefan Reis [Centre for Ecology and Hydrology, United Kingdom of Great Britian and Northern Ireland]; Caradee Y. Wright [South African Medical Research Council, South Africa].

Chapter 6 Biodiversity:

Rob Bailey [Chatham House, The Royal Institute of International Affairs, United Kingdom of Great Britian and Northern Ireland]; Colin Butler [University of Canberra, Australia]; Irene Dankelman [Radboud University]; Jonathan Davies [University of British Columbia, United Kingdom of Great Britian and Northern Ireland]; Linda Godfrey [Council for Scientific and Industrial Research (CSIR), South Africa]; Jeremy Hills [The University of the South Pacific, United Kingdom of Great Britian and Northern Ireland]; Andrei P. Kirilenko [University of Florida, United States of America]; Daniela Liggett [University of Canterbury, New Zealand]; Louise McRae [Institute of Zoology, Zoological Society of London, United Kingdom of Great Britian and Northern Ireland]; Gavin Mudd [RMIT University, Australia]; Dolors Armenteras Pascual [Universidad Nacional de Colombia, Colombia]; Joni Seager [Bentley University, United States of America]; Peter Stoett [University of Ontario Institute of Technology, Canada]; Carol Zastavniouk (GEO Fellow) [Golder Associates, Canada]; Caroline Zickgraf [The Hugo Observatory, Université de Liège, Belgium].

Chapter 7 Oceans and Coasts:

AlAnoud Alkhatlan (GEO Fellow) [Arabian Gulf University, Bahrain]; Elaine Baker [GRID-Arendal at the University of Sydney, Australia]; James Grellier [University of Exeter, United Kingdom of Great Britian and Northern Ireland]; Peter Harris [GRID-Arendal, Norway]; Adelina Mensah [Institute for Environment and Sanitation Studies - University of Ghana, Ghana]; Jake Rice [Department of Fisheries and Oceans Canada – Emeritus, Canada].

Chapter 8 Land and Soil:

Nicolai Dronin [Moscow State University, Russian Federation]; Andrés Guhl [Universidad de los Andes, Colombia]; Gensuo Jia [Chinese Academy of Sciences, China]; Javier Naupari [Universidad Nacional Agraria La Molina, Peru]; Darshini Ravindranath (GEO Fellow) [University College London (UCL), United Kingdom of Great Britian and Northern Ireland]; Hung Vo (GEO Fellow) [Harvard Graduate School of Design, United States of America]; Ying Wang (GEO Fellow) [Tongji University, China].

Chapter 9 Freshwater:

Erica Gaddis [Utah Department of Environmental Quality, United States of America]; Anna Maria Grobicki [Food and Agriculture Organization, Italy]; Rowena Hay [Umvoto, South Africa]; Gavin Mudd [RMIT University, Australia]; Walter Rast [Meadows Center for Water and Environment -Texas State University, United States of America]; Jaee Sanjay Nikam [Arizona State University, United States of America]; Beatriz Rodríguez-Labajos (GEO Fellow) [Universitat Autònoma de Barcelona, Spain]; Ying Wang [Tongji University, China].

Chapter 10 Approach to Assessment of Policy Effectiveness:

Klaus Jacob [Freie Universität Berlin, Germany]; Peter King [Institute for Global Environmental Strategies, Thailand]; Diana Mangalagiu [University of Oxford and Neoma Business School, United Kingdom of Great Britian and Northern Ireland]; Beatriz Rodríguez-Labajos (GEO Fellow) [Universitat Autònoma de Barcelona, Spain].

Chapter 11 Policy Theory and Practice:

Pedro Fidelman [Centre for Policy Futures, The University of Queensland, Australia]; Leandra Regina Gonçalves [University of Campinas/Center for Environmental Studies and Research (NEPAM), Portugal]; Chenmin He (GEO Fellow) [Peking University, China]; James Hollway [Graduate Institute of International and Development Studies, Switzerland]; Klaus Jacob [Freie Universität Berlin, Germany]; Peter King [Institute for Global Environmental Strategies, Thailand]; Sebastian Sewerin [Swiss Federal Institute of Technology Zurich (ETH Zurich), Switzerland].

Chapter 12 Overview of Air Policy Instruments:

Frederick Ato Armah [University of Cape Coast, Ghana]; Kari De Pryck (GEO Fellow) [SciencesPo, France]; Phillip Dickerson [U.S. Environmental Protection Agency, United States of America]; Cristina de B. B. Guerreiro [Norwegian Institute for Air Research – NILU, Norway]; Terry Keating [U.S. Environmental Protection Agency, United States of America]; Peter King [Institute for Global Environmental Strategies, Thailand]; Oswaldo Lucon [Sao Paulo State Environment Secretariat, Brazil]; Asami Miyazaki [Kumamoto Gakuen University, Japan]; Amit R. Patel (GEO Fellow) [Planned Systems International Inc., United States of America]; Stefan Reis [Centre for Ecology and Hydrology, United Kingdom of Great Britian and Northern Ireland].

Chapter 13 Biodiversity Policy:

Irene Dankelman [Radboud University, Netherlands]; Jonathan Davies [University of British Columbia, United Kingdom of Great Britian and Northern Ireland]; Leandra Regina Gonçalves [University of Campinas/Center for Environmental Studies and Research (NEPAM), Portugal]; Souhir Hammami (GEO Fellow) [Freie Universität Berlin, Germany]; Jeremy Hills [The University of the South Pacific, Fiji]; Diana Mangalagiu [University of Oxford and Neoma Business School, United Kingdom of Great Britian and Northern Ireland]; Louise McRae [Institute of Zoology, Zoological Society of London, United Kingdom of Great Britian and Northern Ireland]; Nibedita Mukherjee [University of Cambridge, United Kingdom of Great Britian and Northern Ireland]; Dolors Armenteras Pascual [Universidad Nacional de Colombia, Colombia]; Peter Stoett [University of Ontario Institute of Technology, Canada]; Caradee Y. Wright [South African Medical Research Council, South Africa]; Carol Zastavniouk (GEO Fellow) [Golder Associates, Canada].

Chapter 14 Oceans and Coastal Policy:

AlAnoud Alkhatlan (GEO Fellow) [Arabian Gulf University, Bahrain]; Elaine Baker [GRID-Arendal at the University of Sydney, Australia]; Pedro Fidelman [Centre for Policy Futures, The University of Queensland, Australia]; Leandra Regina Gonçalves

[University of Campinas/Center for Environmental Studies and Research (NEPAM), Portugal]; Peter Harris [GRID-Arendal, Norway]; James Hollway [Graduate Institute of International and Development Studies, Switzerland]; Rakhyun E. Kim [Utrecht University, Netherlands]; Diana Mangalagiu [University of Oxford and Neoma Business School, United Kingdom of Great Britian and Northern Ireland]; Jake Rice [Department of Fisheries and Oceans Canada – Emeritus, Canada].

Chapter 15 Land and Soil Policy:

Katharina Helming [Leibniz Centre for Agricultural Landscape Reseaerch (ZALF), Germany]; Klaus Jacob [Freie Universität Berlin, Germany]; Peter King [Institute for Global Environmental Strategies, Thailand]; Diana Mangalagiu [University of Oxford and Neoma Business School, United Kingdom of Great Britian and Northern Ireland]; Andrew Onwuemele [Nigerian Institute of Social and Economic Research (NISER), Nigeria]; Darshini Ravindranath (GEO Fellow) [University College London (UCL), United Kingdom of Great Britian and Northern Ireland]; Hung Vo (GEO Fellow) [Harvard Graduate School of Design, United States of America]; Leila Zamani (GEO Fellow) [Department of Environment Islamic Republic of Iran, Iran (Islamic Republic of)]; Pandi Zdruli [Mediterranean Agronomic Institute of Bari (CIHEAM), Italy].

Chapter 16 Freshwater Policy:

Erica Gaddis [Utah Department of Environmental Quality, United States of America]; James Grellier [University of Exeter, United Kingdom of Great Britian and Northern Ireland]; Anna Maria Grobicki [Food and Agriculture Organization, Italy]; Rowena Hay [Umvoto, South Africa]; Peter King [Institute for Global Environmental Strategies, Thailand]; Naho Mirumachi [King's College London, United Kingdom of Great Britian and Northern Ireland]; Gavin Mudd [RMIT University, Australia]; Farhad Mukhtarov [International Institute of Social Studies, Erasmus University Rotterdam, Netherlands]; Jaee Sanjay Nikam [Arizona State University, United States of America]; Walter Rast [Meadows Center for Water and Environment -Texas State University, United States of America]; Beatriz Rodríguez-Labajos (GEO Fellow) [Universitat Autònoma de Barcelona, Spain]; Patricia Nayna Schwerdtle (GEO Fellow) [Monash University, Australia].

Chapter 17 Systemic Policy Approaches for Cross-cutting Issues:

Babatunde Joseph Abiodun [University of Cape Town, South Africa]; Giovanna Armiento [ENEA - Italian National Agency for New Technologies, Energy and Sustainable Economic Development, Italy]; Rob Bailey [Chatham House, The Royal Institute of International Affairs, United Kingdom of Great Britian and Northern Ireland]; Elaine Baker [GRID-Arendal at the University of Sydney, Australia]; Kathryn Jennifer Bowen [Australian National University, Australia];John Crump [GRID-Arendal, Norway]; Irene Dankelman [Radboud University, Netherlands]; Riyanti Djalante [United Nations University – Institute for the Advanced Study of Sustainability, Japan]; Monica Dutta [The Energy and Resources Institute (TERI), India]; Fintan Hurley [Institute of Occupational Medicine, Ireland]; Klaus Jacob [Freie Universität Berlin, Germany]; Rakhyun E. Kim [Utrecht University, Netherlands]; Peter King [Institute for Global Environmental Strategies, Thailand]; Richard King [Chatham House, The Royal Institute of International Affairs, United Kingdom of Great Britian and Northern Ireland]; Andrei P. Kirilenko [University of Florida, United States of America]; Oswaldo Lucon [Sao Paulo State Environment Secretariat, Brazil]; Diana Mangalagiu [University of Oxford and Neoma Business School, United Kingdom of Great Britian and Northern Ireland]; Diego Martino [AAE Asesoramiento Ambiental Estratégico and ORT University, Uruguay]; Ritu Mathur [The Energy and Resources Institute (TERI), India]; Gavin Mudd [RMIT University, Australia]; Joni Seager [Bentley University, United States of America]; Sebastian Sewerin [Swiss Federal Institute of Technology Zurich (ETH Zurich), Switzerland]; Tim Stephens [University of Sydney, Australia]; Patricia Schwerdtle [Monash University, Australia];Maria Jesus Iraola Trambauer (GEO Fellow) [University College London (UCL), United Kingdom of Great Britian and Northern Ireland]; Laura Wellesley [Chatham House, The Royal Institute of International Affairs, United Kingdom of Great Britian and Northern Ireland]; Caradee Y. Wright [Medical Research Council of South Africa, South Africa].

Chapter 18 Conclusions on Policy Effectiveness:

Klaus Jacob [Freie Universität Berlin, Germany]; Peter King [Institute for Global Environmental Strategies, Thailand]; Diana Mangalagiu [University of Oxford and Neoma Business School, United Kingdom of Great Britian and Northern Ireland]; Beatriz Rodríguez-Labajos (GEO Fellow) [Universitat Autònoma de Barcelona, Spain].

Chapter 19 Outlooks in GEO-6:

Ghassem R. Asrar [Pacific Northwest National Laboratory's (PNNL), United States of America]; Rohan Bhargava (GEO Fellow) [Utrecht University, Netherlands]; Paul Lucas [PBL Netherlands Environmental Assessment Agency, Netherlands]; Laura Pereira [Centre for Complex Systems in Transition (CST), Stellenbosch University, South Africa]; Detlef van Vuuren [PBL Netherlands Environmental Assessment Agency, Netherlands]; Joost Vervoort [Utrecht University, Netherlands].

Chapter 20 A long-term vision for 2050:

Mark Elder [Institute for Global Environmental Strategies, Japan]; Fintan Hurley [Institute of Occupational Medicine, United Kingdom of Great Britian and Northern Ireland]; Paul Lucas [PBL Netherlands Environmental Assessment Agency, Netherlands]; Maryam Meftahi (GEO Fellow) [Tehran Provincial Department of Environment, Iran (Islamic Republic of)]; Detlef van Vuuren [PBL Netherlands Environmental Assessment Agency, Netherlands].

Chapter 21 Future developments without targeted policies:

Katherine V. Calvin [Joint Global Change Research Institute, Pacific Northwest National Laboratory's (PNNL), United States of America]; Serena H. Chung [U.S. Environmental Protection Agency, United States of America]; Mike Harfoot [World Conservation Monitoring Centre (UNEP-WCMC), United Kingdom of Great Britian and Northern Ireland]; Steve Hedden [Frederick S. Pardee Center for International Futures, University of Denver, United States of America]; Barry B. Hughes

[Frederick S. Pardee Center for International Futures, University of Denver, United States of America]; Fintan Hurley [Institute of Occupational Medicine, United Kingdom of Great Britian and Northern Ireland]; Alexandre C. Köberle [Universidade Federal do Rio de Janeiro, Brazil]; Paul Lucas [PBL Netherlands Environmental Assessment Agency, Netherlands]; Katrina Lyne (GEO Fellow) [James Cook University, Australia]; Jonathan D. Moyer [Frederick S. Pardee Center for International Futures, University of Denver, United States of America]; Detlef van Vuuren [PBL Netherlands Environmental Assessment Agency, Netherlands]; Yoshihide Wada [International Institute for Applied Systems Analysis (IIASA), Austria].

Chapter 22 Pathways Toward Sustainable Development:

Lex Bouwman [PBL Netherlands Environmental Assessment Agency, Netherlands]; Katherine V. Calvin [Joint Global Change Research Institute, Pacific Northwest National Laboratory's (PNNL), United States of America]; Serena H. Chung [U.S. Environmental Protection Agency, United States of America]; Mike Harfoot [World Conservation Monitoring Centre (UNEP-WCMC), United Kingdom of Great Britian and Northern Ireland]; Chenmin He (GEO Fellow) [Peking University, China]; Steve Hedden [Frederick S. Pardee Center for International Futures, University of Denver, United States of America]; Barry B. Hughes [Frederick S. Pardee Center for International Futures, University of Denver, United States of America]; Fintan Hurley [Institute of Occupational Medicine, United Kingdom of Great Britian and Northern Ireland]; Alexandre C. Köberle [Universidade Federal do Rio de Janeiro, Brazil]; Paul Lucas [PBL Netherlands Environmental Assessment Agency, Netherlands]; Jonathan D. Moyer [Frederick S. Pardee Center for International Futures, University of Denver, United States of America]; Marco Rieckmann [University of Vechta, Germany]; Beatriz Rodríguez-Labajos (GEO Fellow) [Universitat Autònoma de Barcelona, Spain]; Detlef van Vuuren [PBL Netherlands Environmental Assessment Agency, Netherlands]; Yoshihide Wada [International Institute for Applied Systems Analysis (IIASA), Austria].

Chapter 23 Bottom-up Initiatives and Participatory Approaches for Outlooks:

Ghassem R. Asrar [Pacific Northwest National Laboratory's (PNNL), United States of America]; Rohan Bhargava (GEO Fellow) [Utrecht University, Netherlands]; Laur Hesse Fisher [Massachusetts Institute of Technology (MIT), United States of America]; Angel Hsu [Yale University, United States of America]; Thomas Malone [Massachusetts Institute of Technology (MIT), United States of America]; Jeanne Nel [Vrije Universiteit Amsterdam, Netherlands]; Laura Pereira [Centre for Complex Systems in Transition (CST), Stellenbosch University, South Africa]; Odirilwe Selomane [Stockholm Resilience Centre, Stockholm University, Sweden]; Nadia Sitas [Council for Scientific and Industrial Research (CSIR), South Africa]; Christopher Trisos [National Socio-Environmental Synthesis Center (SESYNC), University of Maryland, United States of America]; Mandy Angèl van den Ende (GEO Fellow) [Utrecht University, Netherlands]; Joost Vervoort [Utrecht University, Netherlands]; James Ward [University of South Australia, Australia]; Amy Weinfurter (GEO Fellow) [Data-Driven Yale, United States of America]; Yihao Xie [Yale-NUS College, Singapore]; Yaolin Zhang [Yale-NUS College, Singapore].

Chapter 24 The Way Forward:

Ghassem R. Asrar [Pacific Northwest National Laboratory's (PNNL), United States of America]; Kei Gomi [National Institute for Environmental Studies, Japan]; Steve Hedden [Frederick S. Pardee Center for International Futures, University of Denver, United States of America]; Fintan Hurley [Institute of Occupational Medicine, United Kingdom of Great Britian and Northern Ireland]; Klaus Jacob [Freie Universität Berlin, Germany]; Mikiko Kainuma [Institute for Global Environmental Strategies, Japan]; Peter King [Institute for Global Environmental Strategies, Thailand]; Diana Mangalagiu [University of Oxford and Neoma Business School, United Kingdom of Great Britian and Northern Ireland]; Paul Lucas [PBL Netherlands Environmental Assessment Agency, Netherlands]; Robyn M. Lucas [National Centre for Epidemiology and Population Health, The Australian National University, Australia]; Laura Pereira [Centre for Complex Systems in Transition (CST), Stellenbosch University, South Africa]; Alexis Rocamora [Institute for Global Environmental Strategies (IGES), Japan]; Mandy Angèl van den Ende (GEO Fellow) [Utrecht University, Netherlands]; Detlef van Vuuren [PBL Netherlands Environmental Assessment Agency, Netherlands].

Chapter 25 Future Data and Knowledge Needs:

Graeme Clark [University of New South Wales, Australia]; Daniel Cooper [University of Oxford, United Kingdom of Great Britian and Northern Ireland]; Florence Mayocyoc-Daguitan [Tebtebba (Indigenous Peoples' International Centre for Policy Research and Education), Philippines]; James Donovan [ADEC Innovations, United Kingdom of Great Britian and Northern Ireland]; Pali Lehohla [Pan African Institute for Evidence, South Africa]; Sheryl Joy Anne S. Gutierrez [ADEC Innovations, Philippines]; Nina Kruglikova [University of Oxford, United Kingdom of Great Britian and Northern Ireland]; Charles Mwangi [GLOBE Program, Kenya]; Amit R. Patel (GEO Fellow) [Planned Systems International Inc., United States of America]; Joni Seager [Bentley University, United States of America]; William Sonntag [The Group on Earth Observation Secretariat, United States of America]; Michelle Tan [ADEC Innovations, Kenya].

Fellows:

AlAnoud Alkhatlan [Arabian Gulf University, Bahrain]; Rohan Bhargava [Utrecht University, Netherlands]; Kari De Pryck [Sciences Po Paris, France]; Priyanka DeSouza [Massachusetts Institute of Technology, United States of America]; Souhir Hammami [Freie Universität Berlin, Germany]; Chenmin He [Peking University, China]; Matthew D. Kosko [University of Utah, United States of America]; Katrina Lyne [James Cook University, Australia]; Maryam Meftahi [Tehran Provincial Department of Environment, Iran (Islamic Republic of)]; Semie Memuna [Centre for International Governance Innovation, Canada]; Emma Gaalaas Mullaney [Bucknell University, United States of America]; Jaee Sanjay Nikam [Arizona State University, United States of America]; Amit R. Patel [Planned Systems International, Inc., United States of America]; Darshini Ravindranath [University College London (UCL), United Kingdom of Great Britian and Northern Ireland]; Beatriz Rodríguez-Labajos [Universitat Autònoma de Barcelona, Spain]; Mayar Sabet [CEDARE, Egypt]; Joeri Scholtens [University of

Amsterdam, Netherlands]; Patricia Nayna Schwerdtle [Monash University, Australia]; Maria Jesus Iraola Trambauer [University College London (UCL), United Kingdom of Great Britian and Northern Ireland]; Natalie Unterstell [Brazil]; Mandy Angèl van den Ende [Utrecht University, Netherlands]; Hung Vo [Harvard Graduate School of Design, United States of America]; Ying (Grace) Wang [Tongji University, China]; Amy Weinfurter [Data-Driven Yale, United States of America]; ChangXia Wu [Dalhousie University, Canada]; Leila Zamani [Department of Environment, Iran (Islamic Republic of)]; Carol Zastavniouk [Golder Associates, Canada].

High-Level Intergovernmental and Stakeholder Advisory Group:

Nassir S. Al-Amri, [King Abdulaziz University, Saudi Arabia]; Hæge Andenæs [Ministry of Climate and Environment, Norway]; Juan Carlos Arredondo [Secretariat de Medio Ambiente y Recursos Naturales, Mexico]; Julio Baena (alternate) [Ministry of the Environment, Brazil]; Sara Baisai Feresu [University of Zimbabwe, Zimbabwe]; Benon Bibbu Yassin [Ministry of Natural Resources, Energy and Environment, Malawi]; Simon Birkett [Clean Air in London, United Kingdom of Great Britian and Northern Ireland]; Gillian Bowser [Colorado State University, United States of America]; Joji Carino [Forest Peoples Programme, England]; Fernando E.L.S. Coimbra [Embassy of the Federative Republic of Brazil, Brazil]; Pascale Collas [Environment and Climate Change, Canada]; Marine Collignon (alternate) [Ministry of Foreign Affairs and International Development, France]; Victoria de Higa Rodriguez [Ministry of the Environment and Sustainable Development, Argentina]; Laksmi Dhewanthi [Ministry of Environment and Forestry, Indonesia]; Noasilalaonomenjahary Ambinintsoa Lucie [Ministry of Environment Ecology and Forest, Madagascar]; Arturo Flores Martinez (alternate) [Ministry of Environment and Natural Resources, Mexico]; Sascha Gabizon [WECF International, Germany]; Prudence Galega [Ministry of Environment, Protection of Nature and Sustainable Development, Cameroon]; Edgar Gutiérrez Espeleta [University of Costa Rica , Costa Rica]; Keri Holland (alternate) [U.S. Department of State, United States of America]; Pascal Valentin Houénou (vice-chair) [Université Nangui Abrogoua, Côté d'ivoire]; Yi Huang (co-chair) [Peking University, China]; Mork-Knutsen Ingeborg (alternate) [Ministry of Climate and Environment, Norway]; Melinda Kimble [United Nations Foundation, United States of America]; Asdaporn Krairapanond [Office of Natural Resources and Environmental Policy and Planning, Thailand]; Yaseen M. Khayyat [Minister of Environment, Jordan]; Pierluigi Manzione [Ministry of Environment Land and Sea, Italy]; Veronica Marques (alternate) [Ministry of the Environment, Brazil]; Jock Martin [European Environment Agency, Denmark]; John M. Matuszak [U.S. Department of State, United States of America]; Megan Meaney [ICLEI – Local Governments for Sustainability, Canada]; Naser Moghaddasi [Department of Environment, Iran (Islamic Republic of)]; Bedrich Moldan [Charles University, Czech Republic]; Roger Roberge [Environment and Climate Change, Canada]; Najib Saab [General Authority for Meteorology and for Meteorology and Environment Protection, Saudi Arabia]; Mohammed Salahuddin [Ministry of Environment, Forests and Climate Change, India]; Jurgis Sapijanskas (alternate) [Ministry for the Ecological and Inclusive Transition of France, France]; Paolo Soprano (co-chair) [The Ministry for Environment, Land and Sea Protection of Italy, Italy]; Xavier Sticker [Ministry for the Environment, France]; Sibylle Vermont (vice-chair) [Swiss Federal Office for the Environment, Switzerland]; Andrea Vincent (alternate) [University of Costa Rica, Costa Rica]; Terry Yosie [World Environment Center, United States of America].

Scientific Advisory Panel:

Asma Abahussain [Arabian Gulf University, Bahrain]; John B.R Agard [The University of the West Indies, Jamaica]; Odeh Al-Jayyousi [Arabian Gulf University, Bahrain]; Paulo Eduardo Artaxo Netto [University of São Paolo, Brazil]; Rosina M. Bierbaum [University of Michigan, United States of America]; Enrico Giovannini [Università di Roma "Tor Vergata", Italy]; Sarah Green (co-chair) [Michigan Technological University, United States of America]; Torkil Jønch Clausen [World Water Council, France]; Ahmed Khater [National Water Research Center, Egypt]; Nicholas King (co-chair) [Independent, South Africa]; Paolo Laj [Institut des Géosciences de l'Environnement, France]; Byung-Kook Lee [Korea Environment Institute, Republic of Korea]; Alastair Charles Lewis [University of York, United Kingdom of Great Britian and Northern Ireland]; Franklyn Lisk [University of Warwick and HEART, United Kingdom of Great Britian and Northern Ireland]; Majid Shafiepour Motlagh [University of Tehran, Iran (Islamic Republic of)]; Carlos Afonso Nobre [National Institute of S&T for Climate Change, Brazil]; Toral Patel-Weynand [US Forest Service, United States of America]; Anand Patwardhan [University of Maryland School of Public Policy, United States of America]; N.H Ravindranath [Indian Institute of Science, India]; Wendelin Stark [ETH Zurich, Switzerland]; Danling Tang [Chinese Academy of Sciences, China]; Maria del Mar Viana Rodriguez (vice-chair) [Spanish National Research Council, Spain]; Naohiro Yoshida [Tokyo Institute of Technology, Japan].

Assessment Methodologies, Data and Information Working Group

Maria Andrzejewska [UNEP/GRID-Warsaw, Poland]; Ousséni Arouna [Université Nationale des Sciences, Technologies, Ingénierie et Mathématiques, Bénin]; Sandra De Carlo (co-chair) [Presidency (Brazil)]; Rosario Gomez, Universidad del Pacifico, Peru]; Wabi Marcos [Ministère de l'Environnement Chargé de la Gestion des Changements Climatiques, Bénin]; Reza Maknoon [Amirkabir University of Technology, Iran (Islamic Republic of)]; Graciela Metternicht [University of New South Wales, Australia]; Thy Nguyen Van [Vietnam Environment Administration, Viet Nam]; Nicolas Perritaz (co-chair) [Federal Office for the Environment FOEN, Switzerland]; Qurat ul Ain Ahmad [Global Change Impact Studies Center, Pakistan]; Mathis Wackernagel [Global Footprint Network, United States of America]; Fei Wang [Northwest A&F University, China].

UNEP Extended Team:

Misha Alberizzi; Neville Ash; Jennifer Bailey; Matthew Billot; Peter Bjornsen; Oli Brown; Alex Caldas; Kilian Christ; Thierry De Oliveira; Fanny Demassieux; Francesco Gaetani; Tessa Goverse; Alexander Juras; Thomas Koetz; Pushpam Kumar; Monika MacDevette; Tomas Marques; Jacqueline McGlade; Abdelmenam Mohamed; Pascal Peduzzi; Corli Pretorius; Rula Qalyoubi; Tatiana Terekhova; Frank Turyatunga; Dirk Wagener; Clarice Wilson; Jinhua Zhang; Laetitia Zobel; Jochem Zoetelief; Sheeren Zorba.

Reviewers from Other UN Bodies and Partners Requested to Review:

Maher Amer [PERSGA]; Joseph Appiott [CBD]; Regina Asariotis [UNCTAD]; Alfonso Ascencio-Herrera [ISA]; Julian Barbière [UNESCO]; Uwe Barg [FAO]; Stefano Belfiore [WMO]; Maija Bertule [UNEP-DHI]; Marie Bourrel-McKinnon [ISA]; Edgard Cabrera [WMO]; Michele Cavinato [UNHCR]; Isabel Chavez [UNESCO]; Nishikawa Chihiro [UNESCO]; Genevieve Connors [The World Bank]; Rey Da Silva [UNESCO]; Mario Abel Diaz Anzueto [IPBES]; Fanny Douvere [UNESCO]; Milen F. Dyoulgerov [The World Bank]; Paul Egerton [WMO]; Kim Friedman [FAO]; Dirk Glaesser [UNWTO]; Paul Glennie; Sarah Grimes [WMO]; Ulrike Guerin [UNESCO]; Fredrik Haag [IMO]; Valerie Hikey [The World Bank]; Jan Hladik [UNESCO]; Andrew Hudson [UNDP]; Byonug-Hwa Hwang [The World Bank]; Peter Koefoed Bjørnsen [UNEP-DHI]; Neno Kukuric [UN-IGRAC]; Juhyun Lee [CBD]; Annukka Lipponen [UNECE]; Gareth James Lloyd [UNEP-DHI]; Michael Lodge [ISA]; Warren Lee Long [SPREP]; Robert Masters [WMO]; Arni Mathiesen [FAO]; Chris McOwen [UNEP-WCMC]; Kate Medlicott [WHO]; Stefan Micalle [IMO]; Hassan Mohammadi [ROPME]; Wahid Mouffadal; Audrey Nepveu [IFAD]; David Osborn [IAEA]; Sivaji Patra [SACEP]; Manzoor Qadir [UNU], Mechtild Rössler [UNESCO]; Vladimir Ryabinin [UNESCO]; Susana Salvador [OSPAR]; Zita Sebesvari [UNU-EHS]; Cameron Shilton [UNHCR]; Monika Stankiewicz [Helcom]; Christian Susan [UNIDO]; Peter Wolfgang Swarzenski [IAEA]; Xu Tang [WMO]; Tumi Tómasson [UNU-FTP]; Brandt Wagner [ILO]; Sara Walker [WRI]; Marcus Wijnen [The World Bank]; Andrew Wright [CCAMLR]; Joseph Zelasney [FAO]; Wenjian Zhang [WMO].

External Reviewers[2]:

Magdi Tawfik Abdelhamid [Egypt]; Mohamed Abdel-Monem [Egypt]; Ahmed Abdelrehim [Egypt]; Anwar Abdo [Bahrain]; Amani Abdou [Niger]; Maisharou Abdou [Niger]; Abdulkader Abed [Jordan]; Mohamed Jamil Saleh Anbdulrazzak [Saudi Arabia]; Ehsan Abedualemer Jassem Abbas [Iraq]; Mohammad Abido [Syrian Arab Republic]; Tamiru Alemayehu Abiye [Ethiopia]; Iyad Aburdeineh [State of Palestine]; Khaled Abu-Zeid [Egypt]; Priscilla Mbarumun Achakpa [Nigeria]; David Acosta [Colombia]; Mange Ram Adhana [India]; Alphonse Adite [Benin]; Carolina Adler [Chile]; Jean Paul Brice Affana [Cameroon]; John B.R. Agard [Trinidad and Tobago]; Maxime Agossou [Benin]; Christer Ågren; Qurat ul Ain Ahmad [Pakistan]; Emmanyel Adegboyega Ajao [Nigeria]; Afif Akel [Jordan]; Hajime Akimoto; Thabit Zahran Salim Al Abdulsalaam [Oman]; Abdulwali Al-Aghbari [Syrian Arab Republic]; Mohammad Al Ahmad [Kuwait]; Amani Abdullah Al-Assaf [Jordan]; Seyed Kazem Alavipanah [Iran (Islamic Republic of)]; Amr Osama Al-Aziz [Egypt]; Susan Al Banaa [Iraq]; Khaldoun Al-Bassam [Iraq]; Pedro Manuel Alcolado-Menendez [Cuba]; Nourah Alenezi [Kuwait]; Meshari Al-Harbi [Kuwait]; Belal Al-Hayek [Syrian Arab Republic]; Suzan Al-Ajjawi [Bahrain]; Lilian Alessa [Canada]; Björn Alfthan [Canada]; Fatima Alhemyani [United States of America]; Israa Jassim Mohamed Ali [Iraq]; Thamer Ali [Iraq]; Mohammed Al-Kalbani [Oman]; Al-Anoud Al-Khatlan [Kuwait]; Mukdad Al-Khateeb [Iraq]; Hussien Al-Kisswani [Jordan]; Myles Allen [United Kingdom of Great Britain and Northern Ireland]; Ismail Almadani [Bahrain]; Mazen Almalkawi [Jordan]; Mouza Al Mansouri [United Arab Emirates]; Dora Almassy [Hungary]; Reem AlMealla [Bahrain]; Khawla Al Muhannadi [Bahrain]; Mubarak Aman Al-Noaimi [Kingdom of Bahrain]; Savas Alpay [Turkey]; Israa Jasim Al-Rubaye [Iraq]; Khalid Al-Rwis [Saudi Arabia]; Yaser Al-Sharif [Jordan]; Omran Alshibabi [Syrian Arab Republic]; Afaf Sayed Ali Al-Shoala [Bahrain]; Wasan Alaa A-Deen Mahmood Al-Tai'e [Iraq]; Shubar Ebrahim Al-Widae [Bahrain], Ibrahim Al Zu'bi [Jordan]; Farshad Amiraslani [Iran (Islamic Republic of)]; Soudabeh Amiri [Iran (Islamic Republic of)]; Patila Malua Amosa [Samoa]; Joseph Armathé Amougou [Cameroon]; Koffi Gautier Amoussou [Benin]; Martin Andriamahafehiarivo [Madagascar]; Luciano Andriamaro [Madagascar]; Rivoniony Andrianasolo [Madagascar]; Maria Andrzejewska [Poland]; Muhammad Rehan Anis [Pakistan]; Marina Antonopoulou [Greece]; Ken Anthony; Lawrence Anukam [Nigeria]; Chika Aoki-Suzuki [Japan]; Chandani Appadoo [Mauritius]; Bernadette Arakwiye [United States of America]; Mojtaba Ardestani [Iran (Islamic Republic of)]; Herto Dwi Ariesyady [Indonesia]; Maria Teresa Armijosburneo [United Kingdom and Northern Ireland]; Hyacinth Armstrong-Vaughn [Barbados]; Luca Arnold [Switzerland]; Ousséni Arouna [Benin]; Awadis Arslan [Syrian Arab Republic]; Gulaiym Ashakeeva [Kyrgyzstan]; Hamed Assaf [Jordan];Nibal Assaly [Jordan]; Nabegh Ghazal Asswad [Syrian Arab Republic]; Fakher Aukour [Jordan]; Hassan Awad [Egypt]; Katia Awaujo [United States of America]; Mouina Badran [Syrian Arab Republic]; Marc Baeta [Spain]; Festus D. Kibiri Bagoora [Uganda]; Kenneth Bagstad [United States of America]; Alkiviadis F. Bais [Greece]; Malini Balakrishnan [India]; Rajasekhar Balasubramanian [Singapore]; Robert Baldwin [United Kingdom of Great Britain and Northern Ireland]; Bhawna Bali [India]; Samjwal Ratna Bajracharya [Nepal]; Jamal Ali Bamaileh [Saudi Arabia]; Jayanta Bandyopadhyay; Manjushree Banerjee [India]; Abderrazak Bannari [Canada]; Grazia Barberio [Italy]; Francisco José Barbosa de Oliveira Filho [Brazil]; Garfield Barnwell [Guyana]; Ana Flávia Barros-Platiau; Edwin A. Barry [United States of America]; Christian Barthod [France]; Ferdo Basic [Croatia]; Andrea Bassi [Italy]; Vidya Batra [India]; Maarten Bavinck [Netherlands]; Yannick Beaudoin [Canada]; Sarah Bell [Great Britain [Australia]; Jane Bemigisha [Uganda]; Magnus Bengtsson [Sweden]; Mirta Estela Benítez Herrera [Panama]; Abdelaziz Benjouad [Morocco]; Thomas Bernauer [Switzerland]; Luis Berríos-Negrón [Puerto Rico]; Suress Bhagwant [Mauritius]; Souvik Bhattacharjya [India]; Inogwabini Bila-Isia [Congo]; Peter Koefoed Bjørnsen; Dylan Blake [South Africa]; Gabriel Blanco [Argentina]; Raimund Bleischwitz [Germany]; Ivan Blinkov [The Former Yugoslav Republic of Macedonia]; Rizaldi Boer [Indonesia]; Chandradeo Bokhoree [Mauritius]; Jariya Boonjawat; Helvecia María Bonilla Delgado [Panama]; Jared Bosire [Kenya]; Zalia Yacouba Boubacar [Niger]; Nouzha Bouchareb [Morocco]; Philippe Bourdeau [Belgium]; Kerry W. Bowman [Canada]; Hans Brauch [Germany]; Jean-Jacques Gabriel Marie Braun [France]; Bernard Brillet [France]; Stefan Bringezu [Germany]; Ravina Brizmohun [Mauritius]; Lluis Brotons; Bradford Brown [United States of America]; Carl Bruch; Claudia Brunori [Italy]; Neil Burgess; Reginald Burke [Barbados]; Monday Businge [Uganda]; Thomas Butler [United States of America]; Isabella Buttino [Italy]; Enrico Cabras [Italy]; Jialiang Cai [China]; Edison Calderón [Ecuador]; Pedro Lando Bumba Canga [Angola]; Anthony Capon [New Zealand]; Rene Pablo Capote- Lopez [Cuba]; Wilfredo M. Carandang; Felipe Carazo Ortiz [Costa Rica]; Beatriz Cárdenas; Jose Carlos Orihuela [Colombia]; María José Carroquino Saltó [Spain]; Guillermo Castro [Panama]; Ben Cave [United Kingdom of Great Britian and Northern Ireland]; Alexander Ceron [Colombia]; Farid

2 The external reviewers listed include those who reviewed or were invited to review the GEO-6 report.

Chaaban [Lebanon]; Vanda Chan Ting [Samoa]; Alvin Chandra [Fiji]; Hoon Chang [Republic of Korea]; Vasantha Chase [Saint Lucia]; Rajiv Kumar Chaturvedi [India]; Deliang Chen [Sweden]; Norma Cherry-Fevrier [Saint Lucia]; Mariano Cherubini [Italy]; Sosten Chiotha [Malawi]; Irene G. Lungu Chipili [Zambia]; Victoria Chomo [United States of America]; Nee Sun Choong Kwet Yive [Mauritius]; Liu Chuang [China]; Alistair Clark [United Kingdom of Great Britain and Northern Ireland]; Suani Coelho; Augustin Collette; Maria Cordeiro [Portugal]; Dana Cordell [Australia]; Robert (Bob) Corell [United States of America]; Cosmin Corendea [Romania]; Maria Teresa Cornide - Hernandez [Cuba]; Robert (Bob) Costanza [United States of America]; Tim Coulborn [United Kingdom of Great Britian and Northern Ireland]; Barbara Cremaschi [Italy]; Yiyun (Ryna) Cui; Philippe Cullet [Switzerland]; Laura Cutaia [Italy]; Saed Dababneh [Jordan]; Arthur Dahl [United States of America / Switzerland]; Allan Dale; Salvatore D'Angelo [Italy]; Karine Danielyan [Armenia]; Hy Dao [Switzerland]; Aliou Mohamed Daouda [Benin]; Sir Partha Dasgupta; Lésan Etiennette Florence Dassi [Benin]; Divya Datt [India]; Liliana Dávalos; Eric A. Davidson [United States of America]; John Day [United States of America]; Sandra De Carlo [Brazil]; Fábio De Castro [Brazil]; Francesca De Crescenzo [Italy]; Sabino De Gisi [Italy]; Roberto Bonilla De La Lastra [Panama]; Genoveva Clara de Mahieu [Argentina]; Carlos Alberto de Mattos Scaramuzza [Brazil]; Tom De Meulenaer [Belgium]; Laura De Simone Borma [Brazil]; Elizabete de Souza Cândido [Brazil]; Roberto De Vogli [Italy]; Cassandra De Young [United States of America]; Dimitry D Deheyn [Belgium]; Alex Dehgan [United States of America]; Rosario Del C. Oberto G. [Panama]; Getahun Demissie Gemeda [Ethiopia]; Andriy Demydenko [Ukraine]; Manfred Denich [Germany]; Nickolai Denisov [Russian Federation]; Peter Denton [Canada]; Michael Depledge [United Kingdom of Great Britain and Northern Ireland]; Shobhakar Dhakal [Nepal]; Yakhya Aicha Diagne [Senegal]; Sandra Myrna Diaz [Argentina]; Susana Beatriz Diaz [Argentina]; Robert Didham [United States of America]; Yihun Dile [Ethiopia]; Guglielmina Diolaiuti [Italy]; Salif Diop [Senegal]; Rodolfo Dirzo [Mexico]; Pierre Francois Djocgoue [Cameroon]; Gordana Djurovic [Serbia]; Isaac Gcina Dladla [Swaziland]; Edward J. Dlugokencky [United States of America]; Tomoko Doko [Japan]; Kumar Dookhitram [Mauritius]; William Dougherty [United States of America]; Marra Dourma [Togo]; Stephen Dovers; Ousmane Drame [Senegal]; R Driejana [Indonesia]; Paul Dumble [United Kingdom of Great Britain and Northern Ireland]; Anton Earle [South Africa]; Jonas Ebbesson; Kristie L. Ebi; François Edwards; Blaise Efendene [Cameroon]; Ehab Eid [Jordan]; Akram Eissa Darwich [Syrian Arab Republic]; Hossam El Din Elalkamy [Egypt]; Manal Elewah [Egypt]; Yomn El Hamky [Egypt]; Essam El-Hinnawi [Egypt]; Nagwa El Karawy El Karawy [Egypt]; Abdelfattah El Kassab [Morocco]; Ahmed El-Kholei [Egypt]; Lorraine Elliott; Asim El Moghraby [Sudan]; Kassem El Saddik [Lebanon]; Amr Abdel-Aziz El-Sammak [Egypt]; Hany Gaber El Shaer [Egypt]; Mohamed Eltayeb [Sudan]; Wael El Zerey [State of Palestine]; Lisa Emberson [United Kingdom of Great Britain and Northern Ireland]; Tareq Emtairah [Egypt]; Francois Alwyn Engelbrecht [South Africa]; Jonathan Ensor [United Kingdom of Great Britain and Northern Ireland]; Mamaa Entsua-Mensah [Ghana]; Velasco Saldana Hector Erik; Kevin Erwin [United States of America]; Carlos Ariel Escudero Nuñez [Panama]; Lima Euloge [Benin]; Jaén Núñez Eustorgio [Panama]; Olivier Evrard [Belgium]; Joan Fabres [Spain]; Sunita Facknath [Mauritius]; Hilde Fagerli; Marco Falconi [Italy]; Amy Fallon [United Kingdom of Great Britian and Northern Ireland]; Nadim Farajalla [Lebanon]; Zilda Maria Faria Veloso [Brazil]; Akhmad Fauzi [Indonesia]; Benjamin Fayomi [Benin]; Asghar Mohammadi Fazel [Iran (Islamic Republic of)]; Daniel Feldman [United States of America]; Fabio Feldmann; Maurizio Ferrari [Italy]; Beatrice Ferreira [Brazil]; Francisco Ferreira; Manoel Ferreira Cardoso [Brazil]; Christian Flachsland [Germany]; Martina Floerke [Germany]; Arturo Flores Martinez [Mexico]; Cheikh Fofana [Senegal]; Gary Foley [United States of America]; DDM Fonollera [Philippines]; Jaume Fons-Esteve [Spain]; Patrick Forghab Mbomba [Cameroon]; Eric Fotsing [Cameroon]; Ulrich Franck [Germany]; Niki Frantzeskaki [Greece]; Naoya Furuta [Japan]; Françoise Gaill [France]; Samia Galal [Egypt]; Elsa Patricia Galarza Contreras [Peru]; François Galgani [France]; Easter Catherine Galuvao [Samoa]; Edson Gandiwa [Zimbabwe]; Nadezhda Gaponenko [Russian Federation]; Jennifer Garard [Canada]; Dida Gardera [Indonesia]; Luca Garibaldi [Italy]; Hathairatana Garivait; Jean-Marc Garreau [France]; Domenico Gaudioso [Italy]; Jose Marcelo Gaviño Novillo [Argentina]; Chazhong Ge [China]; Louis Géli [France]; Ibrahim Abdel Gelil [Egypt]; Giorgos Georgiadis [Greece]; Nesreen Ghaddar [Lebanon]; Nadia Abdul Ghaffar [Saudi Arabia]; Razieh Ghayuomi [Iran (Islamic Republic of)]; Shahina Ghazanfar [United Kingdom of Great Britain and Northern Ireland]; Fereidoon Ghazban [Iran (Islamic Republic of)]; Kidane Giday Gebremedhin [Ethiopia]; Vladimir Gil Ramon [Peru]; Aidan Gilligan [Ireland]; Hector Ginzo [Argentina]; Naituli Gitile [Kenya]; Jane Glavan [Canada]; Biljana Gligoric [Serbia and Montenegro]; Kissao Gnandi [Togo]; William Godfrey [United States of America]; Khatuna Gogaladze [Georgia]; Jose Gómez [Spain]; Carlos Gómez [Panama]; Rosario Gómez [Peru]; Tania Merino Gómez [Cuba]; Paulo Rogério Gonçalves [Brazil]; Andy Gonzalez; Rianna Gonzales [Trinidad and Tobago]; Chris Gordon; Alexander Gorobets [Ukraine]; Zhou Goumei [China]; Edwin Grandcourt [United Kingdom of Great Britain and Northern Ireland]; Gilles Grandjean [France]; Marco Grasso [Italy]; Julie Greenwalt [United States of America]; Christophe Grenier [France]; Renáta Grófová [Slovakia]; Sergey Gromov [Russian Federation]; Cisse Gueladio [Côte d'Ivoire]; Katharina Gugerell [Austria]; Richard Guldin [United States of America]; Jing Guo [China]; Eshita Gupta [India]; Joyeeta Gupta [Netherlands]; Jeannette Denholm Gurung [United States of America and United Kingdom of Great Britian and Northern Ireland]; David I. Gustafson [United States of America]; Ayma Abou Hadid [Egypt]; Joanna Haigh [United Kingdom of Great Britain and Northern Ireland]; Muki Haklay [Israel]; Catherine Hallmich [Canada]; David Halpern [United States of America]; Shadi Hamadeh [Lebanon]; Muhannad Hamed [Jordan]; Garba Hamissou [Niger]; Waleed Hamza; Quentin Hanich; Muhammad Hanif [Pakistan]; James Hansen [United States of America]; Rikke Munk Hansen [Denmark]; Fahad Hareb [United Arab Emirates]; Khaled Allam Harhash [Egypt]; Stuart L. Hart [United States of America]; Kristopher Hartley; Chris Hartnady [South Africa]; Muhamm Zia Ur Rahman Hashmi [Pakistan]; Amna Ibrahim Hassan [Sudan]; Rashed Abdul Karim Hassan [Bahrain]; Tareq Ahmed Abdo Hassan [Yemen]; Imad Hassoun [Syrian Arab Republic]; Christophe Häuser [Germany]; Marcus Haward; Charlie Heaps [United States of America]; Lisa Hebbelmann [South African]; Anhar Hegazi [Egypt]; Gabriele Clarissa Hegerl [United Kingdom of Great Britain and Northern Ireland]; Sherry Heileman [Trinidad and Tobago]; Alan Hemmings; Yves Henocque [France]; Sunil Herath [Australia]; Gladys Hernandez-Pedraza [Cuba]; Jeffrey Herrick; Mark Hibberd; Kevin Hicks [United Kingdom of Great Britain and Northern Ireland]; Ivonne Higuero [Panama]; Colin D. Hills [United Kingdom of Great Britain and Northern Ireland]; Denise Hills [Brazil]; Alistair

Hobday; Ove Hoegh-Guldberg; Holger Hoff [Germany]; Ron N. Hoffer [United States of America]; Niklas Höhne [Germany]; Jose Holguin-Veras [Costa Rica]; Katherine Homewood [United Kingdom of Great Britain and Northern Ireland]; Yasuhiko Hotta [Japan]; Christophe Sègbè Houssou [Benin]; Solomon Hsiang [United States of America]; Jinhui Jeanne Huang [Canada]; Marc Hufty [Switzerland]; Carol Hunsberger [Canada]; Natalıia Husieva [Ukraine]; Raja Imran Hussain [Austria]; Malaki Iakopo [Samoa]; Karen Hussey; Anastasiya Idrisova [Tajikistan]; Taema Imo-Seuoti [Samoa]; David Inouye [United States of America]; Leilani Duffy Iosefa [Samoa]; Roger Noel Iroume [Cameroon]; Douglas Irwin; Abdullaev Iskandar; Toko Imorou Ismaïla [Benin]; Yuyun Ismawati [Indonesia]; Mirjana Ivanov [Montenegro]; Maria Ivanova [Bulgaria]; Gokul Iyer; Richard J.T. Klein; Rima Jabado [Canada]; Tronczynski Jacek [France]; Mark Z. Jacobson [United States of America]; Joy Jadam [Lebanon]; Anita James [Saint Lucia]; Chubamenla Jamir [India]; Sadik Bakir Jawad [Iraq]; Ljubomir Jeftic [Croatia]; Seongwoo Jeon [Republic of Korea]; Kejun Jiang [China]; Zhigang Jiang [China]; Prisca Roselyne Sènami Jimaja [Benin]; Luz Adriana Jimenez [Colombia]; Refiloe Joala [South Africa]; Lyndon John [Saint Lucia]; Francis Johnson [United States of America]; Alirou Yedidia Jonas [Nigeria]; Julia Jones]; Richard Jordan [United States of America]; Omar Jouzdan [Syrian Arab Republic]; Kupiainen Kaarle; Pavel Kabat]; Adel Abdel Kader [Egypt]; Thoko Kaime [United Kingdom of Great Britain and Northern Ireland]; Sankwe Michael Kambole [Zambia]; Anurag Kandya [India]; Paula Kankaanpää [Finland]; Shilpi Kapur [India]; Ghada Kassab [Jordan]; Bronwyn Keatley [Canada]; Bibi Nasreen Khadun [Mauritius]; Talib Khalaf [Iraq]; Ahmed Khaled Mostafa Abdel Wahid [Egypt]; Ziad Khalifa [Egypt]; Ahmed Khalil [Sudan]; Shaker Khamdan [Kingdom of Bahrain]; Ahmed S. Khan [Canada and Sierra Leone]; Azmat Hayat Khan [Pakistan]; Muhammad Ajmal Khan; Imad Khatib [Jordan]; Sayed Khalil Khattari [Jordan]; Charles Kihampa [United Republic of Tanzania]; Jeong In Kim [Republic of Korea]; Danielson Kisanga [United Republic of Tanzania]; Leo Klasinc [Croatia]; Carlos Augusto Klink [Brazil]; Zoran Kljajic [Montenegro]; Stefan Knights [Guyana]; John Knox; Reto Knutti [Switzerland]; Lilja Dóra Kolbeinsdóttir [Iceland]; Richard Kock [United Kingdom of Great Britain and Northern Ireland]; Marcel Kok [Netherlands]; Souleymane Konate [Côte d'Ivoire]; Peter Kouwenhoven [Netherlands]; Martin Kowarsch [Germany]; Nawarat Krairapanond [Thailand]; Tom Kram [Netherlands]; Pavel V. Krasilnikov [Russian Federation]; Prabhakar Sivapuram Venkata Rama Krishna [India]; Indu Krishnamurthy [India]; Jürgen P. Kropp [Germany]; Nina Kruglikova [Russian Federation]; Ida Kubiszweski [United States of America]; Michael Kuhndt [Germany]; Tiina Kurvits [Canada]; Sigrid Kusch [Germany]; Johan Kuylenstierna [Sweden]; Hammou Laamrani [Morocco]; Jean-Philippe Lagrange [France]; Elton Laisi [Malawi]; Annamaria Lammel [France]; Johan Larsson [Sweden]; Jonatan Lassa; Márton László [Hungary]; Mojib Latif [Germany]; Edwin Laurent [Saint Lucia]; Roberto Lava [Italy]; Kai Po Jenny Law [China]; Yoon Lee [Republic of Korea]; Enrique Lendo Fuentes [Mexico]; Louis Lengrendre [Canada]; Cuauhtemoc Leon [Mexico]; Vanessa Leonardi [Italy]; David Lesolle [Botswana]; Marc Levy [United States of America]; Xia Li [China]; Mweemba Liberty; Hanlie Liebenberg-Enslin; Zuzana Lieskovská [Slovakia]; Willem Ligtvoet [Netherlands]; Bundit Limmeechokchai [Thailand]; Rosilena Lindo [Panama]; Mark Little [United States of America]; Yu Liya E [Singapore]; Josep Enric Llebot [Spain]; Ivana Logar [Croatia]; Francesco Loro [Italy]; Andreas Löschel [Germany]; Heila Lotz-Sisitka [South Africa]; Ronald Loughland [Australia]; Gordon Lovegrove [Canada]; Naglaa M. Loufty [Egypt]; L. Hunter Lovins [United States of America]; Shengji Luan [China]; Jesada Luangjame; André Lucena [Brazil]; Shuaib Lwasa [Uganda]; Patricia Maccagno [Argentina]; Mary MacDonald [Canada]; Georgina Mace; Masego Madzwamuse [South Africa]; Clever Mafuta [Zimbabwe]; Flora John Magige [United Republic of Tanzania], Robin Mahon [Barbados]; Juliette Maitre [France]; Nada Majdalani [State of Palestine]; Anna Makarova [Russian Federation]; Majid Makhdoum [Iran (Islamic Republic of)]; Reza Maknoon [Iran (Islamic Republic of)]; Malayang III [Philippines]; Sri Ramachandra Murthy Manchiraju [India]; Makoala Marake [Lesotho]; Ney Maranhão [Brazil]; Wabi Marcos [Benin]; Sergio Margulis [Brazil]; Adama Mariko [Mali]; Marina Markovic [Montenegro]; Prasad Modak [India]; Eric Martin [France]; Miguel Martìnez [Guatemala]; Maria Amparo Martinez Arroyo [Mexico]; Olena Maslyukivska [Ukraine]; Mohammad Masnavi [Iran (Islamic Republic of)]; Rania Masri [Lebanon]; Vlado Matevski [The former Yugoslav Republic of Macedonia]; Jörg Matschullat [Germany]; Vedast Max Makota [United Republic of Tanzania]; Simone Maynard [Australia]; Hermann Désiré Mbouobda [Cameroon]; Kezia Mwanga Mbwambo; Patrick Adrian McConney [Barbados]; Bruce McCormack [South Africa]; Michael McGrady [United States of America]; Liana Mcmanus; Victor Makarius Mdemu [United Republic of Tanzania]; Shahbaz Mehmood [Pakistan]; Antonio Augusto Melo Malard [Brazil]; Graciela Metternicht [Argentina]; Karina Miglioranza [Argentina]; Piotr Mikolajczyk [Poland]; Richard Mills; Ziad Mimi [Jordan]; Emmanuel Charles Mkomwa [Malawi]; Jennifer Mohamed-Katerere [Zimbabwe]; Tšepo Mokuku [Lesotho]; Luisa T. Molina; Giuseppina Montanari [Italy]; Lourenço Monteiro de Jesus [Sao Tome and Principe]; Iliana Monterroso [Guatemala]; Felipe Montoya-Greenheck [Costa Rica]; Adam Moolna [United Kingdom of Great Britian and Northern Ireland]; Claudio Morana [Italy]; Ana Rosa Moreno [Mexico]; Tiffany Morrison; Ozore Mossana [Central African Republic]; Pargol Ghavam Mostafavi [Iran (Islamic Republic of)]; Stanley Mubako [Zimbabwe]; Ackmez Mudhoo [Mauritius]; Prisca Mugabe [Zimbabwe]; Ijaz Muhammad [Pakistan]; Arif Goheer Muhammad [Pakistan]; Dusko Mukaetov [The former Yugoslav Republic of Macedonia]; Rupa Mukerji; Yacob Mulugetta; Olegario Pablo Muniz-Ugarte [Cuba]; Kevi Murphy [United States of America]; Radhika Murti [Fiji]; Josephine Kaviti Musango [Kenya]; Patience Mutopo [Zimbabwe]; Iyngararasan Mylvakanam; Nora Mzavanadze [Lithuania]; Etien N'Dah [Côte d'Ivoire]; Mohamed Nabil Chalabi [Syrian Arab Republic]; Cuthbert L. Nahonyo [United Republic of Tanzania]; M P Sukumaran Nair [India]; Adil Najam; Evelyn Namubiru-Mwaura [Kenya]; Stephen Nanthambwe [Malawi]; Humood Abdulla Naser [Bahrain]; Nabil Z. Nasr [United States]; Shahida Nasreen Zakir [Pakistan]; Nabil Nassif [Egypt]; Nilwala Nayanananda [Sri Lanka]; Mzime Ndebele- Murisa [Zimbabwe]; Admire Ndhlovu [Zimbabwe]; Ousmane Ndiaye [Senegal]; Jacques Andre Ndione [Senegal]; Cecile Ndjebet [Cameroon]; Nakicenovic Nebojsa; Filomena Nelson [Samoa]; Robin L. Newmark [United States of America]; Robert Njilla Mengnjo Ngalim [Republic of Cameroon]; Martha Raymond Ngalowera [United Republic of Tanzania]; Tatiana Ngangoum Nana [Cameroon]; Édouard Kouakou N'guessan [Côte d'Ivoire]; Thang Nguyen Trung [Viet Nam]; Lars Nordberg; Barbara Ntombi [Ngwenya [Botswana]; Kimberly Nicholas [United States of America]; Mark Nieuwenhuijsen [Netherlands]; Maeve Nightingale [United Kingdom of Great Britain and Northern Ireland]; Geert-Jan Nijsten [Netherlands]; Ian Noble; William Nordhaus; Pascal Ntahompagaze [Burundi]; Ernst-August Nuppenau [Germany];

Dieudonné Nwaga [Cameroon]; Julius William Nyahongo [United Republic of Tanzania]; Kamwenje Nyalugwe [Zambia]; Deogratius Paul Nyangu [United Republic of Tanzania]; Douglas Nychka [United States of America]; Tarcisius Nyobe [Cameroon]; Nguyen Thi Kim Oanah [Viet Nam]; Joseph O'Brien [United States of America]; Kenneth Ochoa [Colombia]; Karen T. Odhiambo [Kenya]; Washington Odongo Ochola [Kenya]; Patrick O'Farrell; Ibrahim Oanda Ogachi [Kenyan]; Philip Gbenro Oguntunde [Nigeria]; Krzysztof Olendrzynski; Lennart Olsson [Sweden]; Alice Oluoko-odingo [Kenya]; Jean Pierre H.B. Ometto [Brazil]; Jean Michel Onana [Cameroon]; Choon Nam Ong [Singapore]; James J. Orbinski [Canada]; Alexander Orlov [United Kingdom of Great Britain and Northern Ireland]; Jean-Nicolas Ormsby [France]; Isis Karinna Alvarez Ortiz [Colombia]; Ahmad Osman [Lebanon]; Eugene Otaigbe Itua [Nigeria]; Yasser Othman [Egypt]; Dorcas Otieno [Kenya]; Begüm Ozkaynak [Turkey]; Jon Padgham [United States of America]; Emilio Padoa-Schioppa [Italy]; Amber Pairis [United States of America]; Jean Palutikof; Arnico K Panday [Nepal]; Ruchi Pant [India]; Samuel Pare [Burkina Faso]; Kwang Kook Park [Republic of Korea]; Kemraj Parsram [Guyana]; Trista Patterson [United States of America]; Jose Paula [Portugal]; Gunter Pauli [Belgium]; Rosália Marta Pedro [Mozambique]; Tony Penikett [Canada]; Renat Perelet [Russian Federation]; Nicolas Perritaz [Switzerland]; Linn Persson [Sweden]; Marcello Petitta [Italy]; Rohan Pett Pethiyagoda [Australian]; Freddy Picado Trana [Nicaragua]; Stefano Picchi [Italy]; Ramon Pichs-Madruga [Cuba]; Kate Pickett [United Kingdom of Great Britain and Northern Ireland]; Michael D. Pido [Philippines]; Kevin Pietersen [South Africa]; Patrícia Pinheiro Beck Eichler [Brazil]; László Pintér [Hungary]; Gilles Pipien [France]; Were Pitala [Togo]; Andrius Plepys [Lithuania]; Jan Plesnik [Czech Republic]; Erika Podest; Katherine Pond [United Kingdom of Great Britian and Northern Ireland]; Siwatt Pongpiachan [Thailand]; Daniele Ponzi [Italy]; Felix Preston [United Kingdom of Great Britain and Northern Ireland]; Emilia Noel Ptak [Denmark]; Muhammad Qasim [Pakistan]; Florian Rabitz; Kareff Rafisura [Philippines]; Kristin Vala Ragnarsdóttir [Iceland]; David Anthony Raitzer [United States of America]; Jean Roger Rakotoarijaona [Madagascar]; Elysé Odon Rakotonirainy [Madagascar]; Frederic Joel Ramarolahivonjitiana [Madagascar]; Paul Randrianarisoa [Madagascar]; Mohamed Abdel Raouf [Egypt]; Adel Abdul Rasheed]; Harunur Rashid [Bangladesh]; Yousef Rashidi [Iran (Islamic Republic of)]; Anne Rasmussen [Samoa]; Jacquis Rasoanaina [Madagascar]; Valentina Rastelli [Italy]; Jerry Ratsimandresy [Madagascar]; Akkihebbal Ramaiah Ravishankara [United States of America]; Brian K. Ray [Canada]; Hanitriniaina Razafindramboa [Madagascar]; Keith Reid [Australia]; Françoise Breton Renard [Spain]; Yuri Resnichenko [Uruguay]; Lorena Aguilar Revelo [Costa Rica]; Markus Reuter [Germany]; Frances Brown Reupena [Samoa]; Keywan Riahi; Kornelius Riemann [Germany]; Ntep Rigobert [Cameroon]; Sandy Rikoon [United States of America]; Callum Roberts; Debra Roberts [South Africa]; Johan Rockström; Jose Manuel Mateo Rodriguez [Cuba]; Cesar Edgardo Rodriguez Ortega [Mexico]; Jenny Roe [United Kingdom of Great Britain and Northern Ireland]; Dilys Roe; Dannely Romano [Dominican Republic]; Jaime Romero [Colombia]; Espen Ronnenberg [Norway]; Marina Rosales Benites [Peru]; Antoni Rosell Melé [Spain]; Cynthia Rosenzweig; Jean Rosete; Ariana Rossen [Argentina]; Laurence Rouil; Ximena Rueda Fajardo; Romano Ruggeri [Italy]; Blanca Ruiz Franco [Spain]; Ernest Rukangira [Rwanda]; Markku Rummukainen [Finland]; Federico Sabetta [Italy]; Hounada Sadat [Syrian Arab Republic]; David Saddington [United Kingdom of Great Britain and Northern Ireland]; Tarek Mohie El-Din Sadek [Egypt]; Abdul-Karim Sadik [Kuwait]; Edwin Safari [Iran (Islamic Republic of)]; Donna-May Sakura-Lemessy [Trinidad and Tobago]; Hilmi Salem [State of Palestine]; Samira Omar Salem [Kuwait]; Jon Samseth [Norway]; Sergio Sánchez; Roberto Sánchez-Rodríguez [Mexico]; Komla Sanda [Togo]; Simone Sandholz [Germany]; Roberto San Jose; Salieu Kabba Sankoh [Sierra Leone]; Shilpanjali Deshpande Sarma [India]; Makiko Sato [United States of America]; Elsa Sattout [Lebanon]; Geoffrey B. Saxe; Roberto Schaeffer; Rüdiger Markus Holger Schaldach [Germany]; Pedro Manuel Scheel Monteiro [South Africa]; Michael Schlesinger [United States of America]; Alexander J. Schmidt [Germany]; Andreas Schmittner [United States of America]; Laura Schneider; Thomas Schneider von Deimling [Germany]; Roland Scholz [Switzerland]; Tina Schoolmeester [Belgium]; Dieter Schwela [Germany]; William Scott; Jamilla Sealy [Barbados]; Sedigheh sedigheh [Iran (Islamic Republic of)]; Gita Sen [India]; Kanyinke Sena [Kenya]; Sonia I. Seneviratne [Switzerland]; Mazen M. Senjab [Syrian Arab Republic]; Daniel Sertvije [United Kingdom of Great Britain and Northern Ireland]; Sunny Seuseu [Samoa/New Zealand]; Ali Seydou Moussa [Niger]; Kalim Shah [Trinidad and Tobago]; Jeremy D. Shakun [United States of America]; Merab Sharabidze [Georgia]; Constantine Shayo [United Republic of Tanzania]; Charles Sheppard [United Kingdom of Great Britain and Northern Ireland]; Mohamed Yasser Sherif [Egypt]; John Shilling [United States of America]; Binaya Raj Shivakoti [Nepal]; Arun Bhaka Shrestha [Nepal]; Abdou Salami Amadou Siako [Benin]; Susana Siar [Philippines]; Fethi Silajdzic [Bosnia and Herzegovina]; Riziki Silas Shemdoe [United Republic of Tanzania]; Óscar F. Silvarcampos [Peru]; Alan Simcock; Ramesh P. Singh [India]; Sunita Singh [India]; Amrikha Singh [Guyana]; Asha Singh [Guyana]; Nigel Sizer [United Kingdom of Great Britain and Northern Ireland]; Posa A. Skelton [Samoa]; Risa Smith [Canada]; Lars Tov Søftestad [Norway]; Santiago Solda [Argentina]; Anama Solofa [Samoa]; Pamela Soltis; Andrea Sonnino [Italy]; Viriato Soromenho-marques [Portugal]; Edmond Sossoukpe [Benin]; Doris Soto [Chile]; Jeffrey Soule [United States of America]; Aboubacar Souley [Niger]; Ousmane Sow [Senegal]; Clive Spash [Austria]; Olga Speranskaya [Russian Federation]; Simon Spooner [United Kingdom of Great Britain and Northern Ireland]; Mark Stafford Smith; Trajce Stafilov [The former Yugoslav Republic of Macedonia]; Julia A. Stegemann [Canada]; Martin Steinbacher [Germany]; Rolf Steinhilper [Germany]; PJ Stephenson [United Kingdom of Great Britain and Northern Ireland]; Wendy Stephenson [United Kingdom of Great Britain and Northern Ireland]; Josephine Stowers Fiu [Samoa]; Nina Stoyanova [Bulgaria]; Tepa Suaesi [Samoa]; Avelino Suarez-Rodriguez [Cuba]; Laura Suazo [Hondurus]; Parita Sureshchandrashah [Kenya]; Enid J. Sullivan Graham [United States of America]; Riad Sultan [Mauritius]; Vanisa F. Surapipith [Thailand]; Lawrence Surendra [India]; Dinesh Surroop [Mauritius], William J. Sutherland]; Chakkaphan Sutthirat [Thailand]; Paul Sutton [United States of America]; Darren Swanson [Canada]; Mark Swilling; Ian R. Swingland [United Kingdom of Great Britain and Northern Ireland]; Marc Sydnor [United States of America]; Mouhamadou Bamba Sylla [Burkina Faso]; Elemér Szabo [Hungary]; John Robert Stephen Tabuti [Uganda]; Hippolyte Tapamo [Cameroon]; Jaume Targa [Spain]; Vikash Tatayah [Mauritius]; Azadeh Tavakoli [Iran (Islamic Republic of)]; Mohamed Tawfic Ahmed; Egline Tawuya [Zimbabwe]; Anders Telenius [Sweden]; Agossou Brice Hugues Tente [Benin]; Anyai Thomas [Trinidad and Tobago]; Wilfried Thuiller; Donatha Damian Tibuhwa [United Republic of

Tanzania]; Virginie Tilot [France]; Mulipola Tainau Ausetalia Titimaea [Samoa]; Eisaku Toda [Japan]; Amir Tolouei [Iran (Islamic Republic of)]; Javier Tomasella [Argentina]; Elham Tomeh [Syrian Arab Republic]; Masui Toshihiko [Japan]; Tibor Tóth [Hungary]; Yongyut Trisurat [Thailand]; George Tsolakis [Greece]; Joy Tukahirwa [Uganda]; Arnold Tukker [Netherlands]; Bishnunarine Tulsie [Saint Lucia]; Leonardo Tunesi [Italy]; Carol Turley [United Kingdom of Great Britain and Northern Ireland]; Gemedo Dalle Tussie [Ethiopia]; Hector Tuy [Guatemala]; Natalie Unterstell [Brazil]; Haman Unusa [Cameroon]; Nathan M. Urban [United States of America]; Diana Urge-Vorsatz [Hungary]; Sybille van den Hove [Belgium]; Emma Archer van Garderen [South Africa]; Eric van Praag [Venezuela]; Nguyen Van Thuy [Viet Nam]; Marco Vattano [Italy]; Karen Vella; Joberto Veloso de Freitas [Brazil]; Joost Vervoort [Netherlands]; Sonja Vidic; Petteri Vihervaara [Finland]; Joanna Vince; Johannes Vogel [Germany]; John Vogler; Graham von Maltitz [South Africa]; Vladimir Vulic [Montenegro]; Nikola Vulic [Montenegro]; Mathis Wackernagel [Switzerland]; Takako Wakiyama [Japan]; Fei Wang [China]; Supat Wangwongwatana; Mostafa Warith [Canada]; Robin Warner; Meriel Watts; Kenneth Webster [United Kingdom of Great Britian and Northern Ireland]; Rathnadeera Weddikkara Kankanamge [Sri Lanka]; Judith Weis [United States of America]; Kadmiel Wekwete [Zimbabwe]; Chris West [United Kingdom of Great Britain and Northern Ireland]; James West [Australia]; Henk Westhoek [Netherlands]; Florian Wetzel [Germany]; Daniel R. Wildcat [United State of America]; Richard Wilkinson [United Kingdom of Great Britain and Northern Ireland]; Meryl J Williams; John R. A. WILSON [Barbados]; Simon Wilson [United Kingdom of Great Britain and Northern Ireland]; Nicholas Winfield [Canada]; Ron Witt [United States of America]; Poh Poh Wong [Singapore]; Jeremy Woods; Lukasz Wyrowski [Poland]; Ran Xie [China]; Ibouraïma Yabi [Benin]; Salissou Yahouza [Niger]; Changrong Yan [China]; Naama Raz Yaseef [Israel]; Bullat Yessekin [Kazakhstan]; Emmanuel Dieudonné Kam Yogo [Cameroon]; Anthony Young [Canada]; Abourabi Yousra [Morocco]; Liya Yu [Taiwan Province of China]; Sha Yu [China]; Yuqing Yu [China]; Abduljalil M. Zainal [Bahrain]; Bushra M. Zalloom [Jordan]; Pandi Zdruli [Albania]; Irina Safitri Zen [Malaysia]; Saltanat Zhakenova [Kazakhstan]; Frank Zimmerman [Germany]; Siphamandla Zondi [South Africa]; Waleed Zubari [Bahrain]; Claudio Zucca [Italy]; Rami Zurayk [Lebanon]; Eric Zusman.

Intergovernmental Reviewers:

Janine van Aalst [Netherlands]; Mohammed Abdelraouf; Aisha Al Abdooli [United Arab Emirates]; G.A.U.P. Abeypala [Sri Lanka]; Fábio Abreu [Brazil]; Mary Beth Adams [United States of America]; Henry A. Adornado [Republic of the Philippines]; Wills Agricole [Seychelles]; Aji Awa Kaira [Gambia]; Jasim Ali Al-Amaadi [Qatar]; Gustavo Induni Alfaro [Costa Rica]; Gudi Alkemade [Netherlands]; Ahmed Falah Al-Remithi [Qatar]; Travis Ancelet [New Zealand]; Mojtaba Ardestani [Iran (Islamic Republic of)]; Robert Argent [Australia]; K. Arulananthan [Sri Lanka]; A.M.A.S. Attanayake [Sri Lanka]; Miak Aw [Singapore]; Fátima Azevedo [Portugal]; Mevr. Stephanie Baclin [Belgium]; Julio Cesar Baena [Brazil]; Bhumika Bakshi [Canada]; Nyada Yoba Baldeh [Gambia]; Felipe Barbosa [Brazil]; Nathan Bartlett [Australia]; Viviane Bartlett [Canada]; Julian Bauer [Stakeholder]; Elias Begnini [Brazil]; Thijs van den Berg [Netherlands]; Carmen Terry Berro [Cuba]; Brianna Besch [United States of America]; Medani P. Bhandari [Stakeholder]; Meena Bilgi [India]; Patrick Newton Bondo [Stakeholder]; Deborah Bossio; Valerie Brachya [Israel]; Francis Brancart [Belgium]; Ben ten Brink [Netherlands]; Vitória Adail Brito [Brazil]; Hermien Busschbach [Netherlands]; João Batista Drummond Câmara [Brazil]; Odalys C.Goicochea Cardoso [Cuba]; Dr. Edin J. Castellanos [Guatemala]; Yan Changrong [China]; Ge Chazhong [China]; Marion Cheatle [Stakeholder]; Nino Chikovani [Georgia]; Ga Youn Cho [Republic of Korea]; Wacharee Chuaysri [Thailand]; Lorenzo Ciccarese [Italy]; Fernando E. L. de S. Coimbra [Brazil]; Marine Collignon [France]; Sarah R. Cooley [Stakeholder]; María Verónica Cordova [Ecuador]; Sylvie Cote [Canada]; Carlos Alberto Coury [Brazil]; Zeljko Crnojevic [Croatia]; LI Daoji [China]; Samir Kaumar Das [Stakeholder]; Jeff Davis [Canada]; Alain Decomarmond [Seychelles]; Paul Deogratius [United Republic of Tanzania]; Jonathan Derham [Ireland]; Brigitte Dessing-Peerbooms [Netherlands]; Alvaro Aguilar Díaz [Costa Rica]; Ana Lúcia Lima Barros Dolabella [Brazil]; Jiang Dong [China]; Ariuntuya Dorjsuren [Mongolia]; Aljosa Duplic [Croatia]; Ralalaharisoa Christine Edmee [Madagascar]; Efransjah [Republic of Indonesia]; Arthur Eijs [Netherlands]; Pedro Faria [Stakeholder]; Parvin Farshchi [Iran (Islamic Republic of)]; Daniel Favrat [Switzerland]; Asghar Mohammadi Fazel [Iran (Islamic Republic of)]; Wang Fei [China]; George Porto Ferreira [Brazil]; MA. Lourdes G. Ferrer [Republic of the Philippines]; Liz Fox-Tucker [United Kingdom of Great Britian and Northern Ireland]; Blanca Ruiz Franco [Spain]; Keondra Freemyn [Stakeholder]; Meridith Fry [United States of America]; Marcus André Fuckner [Brazil]; Lourdes Coya de la Fuente [Cuba]; Janet Gamble [United States of America]; Sylla Sékou Gaoussou; Mirela Garaventta [Brazil]; Garcia [Peru]; Réka Gaul [Hungary]; Réka Orsolya Gaul [Hungary]; Zita Géller [Hungary]; Jennifer Gleed [Stakeholder]; Geraldo Sandoval Góes [Brazil]; Nino Gokhelashvili [Georgia]; Elise Golan [United States of America]; Verónica Gordillo [Ecuador]; A.J.M. Gunasekera [Sri Lanka]; Zhou Guomei [China]; Gillian Guthrie [Jamaica]; Aysun Demet *Güvendiren* [Turkey]; Hayo Haanstra [Netherlands]; Mohamed Salem Hamouda [Libya]; Dai Hancheng [China]; David Hanrahan [Stakeholder]; Katalin Hargitai [Hungary]; Radhiya Al Hashimi [Stakeholder]; Hasnawir [Indonesia]; Chris Heartley [United States of America]; Guadalupe Heras [Ecuador]; Francisco Heras Hernández [Spain]; Astrid Hilgers [Netherlands]; Elizabeth Hess [Canada]; Vincent V. Hilomen; John van Himbergen [Netherlands]; Keri Holland [United States of America]; Sung Chul Hong [Republic of Korea]; Wang Hongtao [China]; Soonwhan Hwang [Republic of Korea]; Sang-il Hwang [Republic of Korea]; Mohamed Abdi Ibrahim [Qatar]; Caroline Icaza [Ecuador]; Mork-Knutsen Ingeborg [Norway]; Adriana Jácome [Ecuador]; Darren Janzen [Canada]; Ehssan A. Jasim [Iraq]; Maia Javakhishvili [Georgia]; Kulasekara Jayantha [Sri Lanka]; S.M.D.P. Anura Jayatilake [Sri Lanka]; D.S. Jayaweera [Sri Lanka]; Liu Jianguo [China]; Xu Jianhua [China]; Zhang Jieqing [China]; Jaime Camps Saiz Junior [Brazil]; Claudia Kabel [Germany]; Bangoura Abdel Kader; Shurooq Saad Kasim [Iraq]; Patrick Kavanagh [New Zealand]; Melih Kayal [Turkey]; Julio Thadeu da Silva Kettelhut [Brazil]; Kevin Khng [Singapore]; Joe Kiesecker; Francis Kihumba [Kenya]; R.P.P. Kjayasinghe [Sri Lanka]; Andrew Klekociuk [Australia]; Dr.Suranga Kodithuwkku [Sri Lanka]; Lamin Komma [Gambia]; Tom Kompier [Netherlands]; Sasha Koo-Oshima [United States of America]; Rene Korenromp [Netherlands]; Nataša Kova [Slovenia]; Jasna Kufrin [Croatia]; Lei Kun [China]; Budi Kurniawan [Indonesia]; Felipe Rodrigo Cortes Labra [Chile]; T.J. Lah [Republic of Korea]; David Lapp [Canada]; Henrik Larsson [Sweden]; Sang Hee Lee [Republic of Korea]; Byoungyoon Lee [Republic of Korea]; George Leonard [Stakeholder]; Tampushi Leonard [Kenya]; Ruomei Li [China];

Ephraim Leibtag [United States of America]; Régis Pinto de Lima [Brazil]; Martin Lok [Netherlands]; Ulises Lovera [Stakeholder]; Cecilia Loya [Portugal]; Gabriel Henrique Lui [Brazil]; Carol L. MacCurdy [United States of America]; Vincent Madadi [Kenya]; Jaqueline Leal Madruga [Brazil]; Salomar Mafaldo [Brazil]; Salomar Mafaldo [Brazil]; Mahfudz [Indonesia]; Enikő Zita Majoros [Hungary]; Mariam Makarova [Georgia]; Ghulam Mohd Malikyar [Afghanistan]; Anna Mampye [South Africa]; Kätlin Mandel [Estonia]; Cai Mantang [China]; Molnárné Galambos Mária [Hungary]; Caitrin Martin [United States of America]; Jock Martin; Magaly Torres Martínez [Cuba]; John Matuszak [United States of America]; Susannah Mayhew [Stakeholder]; Andrew McCartor [Stakeholder]; Rob McDonald; Noe Megrelishvili [Georgia]; Hans Meijer [Netherlands]; Agustín Gómez Méndez [Costa Rica]; Dan Metcalfe [Australia]; Onana Jean Michael [Cameroon]; Dr. (Mrs.) Andjelka Mihajlov [Stakeholder]; Jason Minor [Canada]; Abhay Sagar Minz [Stakeholder]; Antônio Calazans Reis Miranda [Brazil]; Andrés Mogro [Ecuador]; Philip More [United States of America]; Cristóbal Díaz Morejón [Cuba]; Emilio Canda Moreno [Spain]; Helen Murphy [Australia]; Patricia Murphy [United States of America]; Richard Mwendandu [Kenya]; Ashley Nelson [United States of America]; Martha Ngalowera [United Republic of Tanzania]; Lucy Nganga [Kenya]; Wu Ning [China]; Robert Njilla [Stakeholder];Dr. Saad Al Numairi [United Arab Emirates]; Erica L. Nunez [United States of America]; Engr Hubert Ibezim Nwobi [Nigeria]; Stephen Mutuku Nzika [Kenya]; Peter O.Otieno [Kenya]; Pacifica Ogola [Kenya]; Kahraman Öğuz [Turkey]; Mirian de Oliveira [Brazil]; Kennedy Ondimu [Kenya]; Alexander R. O'Neill [United States of America]; Segundo Onofa [Ecuador]; Laivao Orner [Madagascar]; Sylvie Ote [Canada]; Mark Overman [Netherlands]; Osman Özdemir [Turkey]; Sule Ozkal [Turkey]; GloriaGómez Pais [Cuba]; Nirmalie Pallewatta [Sri Lanka]; Toral Patel-Weynand [United States of America]; Rungnapar Pattanavibool [Thailand]; Lakshman Peiris [Sri Lanka]; Pro. Athula Perera [Sri Lanka]; Nicholas Perritaz [Switzerland]; Stephen Stec J.D. M. Phil [Stakeholder]; Yadira Pilco [Ecuador]; Anita Pirc-Velkavrh [European Environment Agency]; Anabelle Rosalina E. Plantilla [Biodiversity Management Bureau]; Pokorny [Czech Republic]; Lukas Pokorny [Czech Republic]; Sharon Polishuk [Australia]; Tereza Ponocna [Czech Republic]; Hugh Possingham [The Nature Conservancy]; B.H.J. Premathilaka [Sri Lanka]; Luciana Melchert Saguas Presas [Brazil]; Christopher Prins [United States of America]; Justin Prosper [Seychelles]; Eric Rabenasolo[Madagascar]; Kamal Kumar Rai [Stakeholder]; Indrika Rajapaksha [Sri Lanka]; Jean Roger Rakotoarljaona [Madagascar]; Ranto Rakotoaridera [Madagascar]; Paul Ralison [Madagascar]; Joel Frederic Ramarolahivonjtlana [Madagascar]; Carlota de Azevedo Bezerra Vitor Ramos [Brazil]; Yvette Ramos [Switzerland]; Sampath Aravinda Ranasinghe [Sri Lanka]; Yousef Rashidi [Iran (Islamic Republic of)]; Jacquis Rasoaniaina [Madagascar]; Jolanta Rawska-Olejniczak [Poland]; Omer van Renterghem [Netherlands]; Caroline Ridley [United States of America]; Monsieur Ntep Rigobert [Cameroon]; Rabemananjara Rivomalala [Madagascar]; David Guimarães Rocha [Brazil]; Rene Rollon [Stakeholder]; Micah Rosenblum [United States of America]; Bernarda Rozman [Croatia]; Danny Rueda [Ecuador]; Omar Ruiz [Peru]; Liselotte Säll [Sweden]; Oscar Arturo Lücke Sanchez [Costa Rica]; Marcos Oliveira Santana [Brazil]; Raquel Breda dos Santos [Brazil]; Orlando Rey Santos [Cuba]; Jurgis Sapilanskas [France]; Teresa Cruz Sardiñas [Cuba]; Carlos Alberto Scaramuzza [Brazil]; Andreas Benjamin Schei [Norway]; Kees Schotten [European Environment Agency]; Marcos Serrano [Chile]; Xie Shuguang [China]; Wang Shuxiao [China]; Debora Pereira da Silva [Brazil]; Dharani Thanuja de Silva [Sri Lanka]; Gina Sinclair [Canada]; Ashbindu Singh [Stakeholder]; Aldo Sirotic [United States of America]; Virana Sonnasinh [Lao People's Democratic Republic]; Nonglak Sopakayoung [Thailand]; Pedro Tiê Candido Souza [Brazil]; Nicola Speranza [Brazil]; Jorden Splinter [Netherlands]; Anna Stabrawa [Stakeholder]; Andreja Steinberger [Croatia]; Andrew Stott [United Kingdom of Great Britian and Northern Ireland]; Ana Strbenac [Croatia]; Mariam Sulkhanishvili [Georgia]; W.L. Sumathipala [Sri Lanka]; Momodou J. Suwareh [Gambia]; Edyta Sysło [Poland]; Elemer Szabó [Hungary]; Marcel Taal [Netherlands]; Eiji Tanaka [Japan]; Veronica Marques Tavares [Brazil]; Alexandre Lima de Figueiredo Teixeira [Brazil]; Andrina Crnjak Thavenet [Croatia]; Mads Thelander [Denmark]; Osman Tikansak [Turkey]; Carolina de la torre [Ecuador]; Vinícius Fox Drummond Cançado Trindade [Brazil]; R. Talbor Trotter [United States of America]; Nathalie Trudeau [Canada]; İrfan UYSAL [Turkey]; César Vaca [Ecuador]; Lisa-Marie Vaccaro [Canada]; Daksha Vaja [Stakeholder]; Eddy López Valdés [Cuba]; Freddy Valencia [Ecuador]; Martin van Veelen [South Africa]; Henrique Veiga [Brazil]; Jean Venables [United Kingdom of Great Britian and Northern Ireland]; Mevr. Veronique Verbeke [Belgium]; Véronique Verbeke [Belgium]; Marielle Verret [Canada]; Maria Tereza Viana [Brazil]; Nina Vik [Norway]; Larissa Carolina Loureiro Villarroel [Brazil]; Hanitra Viviane [Madagaşçar]; Niels Vlaanderen [Netherlands]; Jan Voet [Belgium]; Rahanitriniaina Volatiana [Madagascar]; Barbara Bernard Vukadin [Slovenia]; Brendan Wall [Ireland]; Margaret Walsh [United States of America]; See Wan [Singapore]; Chris Weaver [United States of America]; Devaka Weerakoon [Sri Lanka]; Mona Mejsen Westergaard [Denmark]; Henk Westhoek [Netherlands]; Leers Wiebke [Germany]; Piet de Wildt [Netherlands]; Scott Wilson [Canada]; Mary Omble Wuya [Nigeria]; Huang Yi [China]; Jeongki Yoon [Republic of Korea]; Rafaralahy Tovoharison Zakaria [Madagascar]; Holger Zambrano [Ecuador]; Nicolás Zambrano [Ecuador]; Jiang Zhigang [China]; Mira Zovko [Croatia]; Nina Zovko [Croatia]; Shepard Zvigadza [Zimbabwe].

Review Editors:

Amr Osama Abdel-Aziz [Egypt]; Ahmed Abdelrehim [Egypt]; Majdah Aburas [Saudi Arabia]; Mohammad Al Ahmad [Kuwait]; Chandani Appadoo [Mauritius]; Michael Brody [United States of America]; Louis Cassar [Malta]; William W. Dougherty [United States of America]; Manal Elewah [Egypt]; Amr El-Sammak [Egypt]; Elsa Patricia Galarza Contreras [Peru]; Jose Holguin-Veras [Costa Rica]; Muhammad Ijaz [Pakistan]; Joy Jadam [Lebanon]; Emmanuel Dieudonné Kam Yogo [Cameroon]; Yoon Lee [Republic of Korea]; Clever Mafuta [Zimbabwe]; Simone Maynard [Australia]; Joan Momanyi [Kenya]; Jacques André Ndione [Senegal]; Washington Odongo Ochola [Kenya]; Renat Perelet [Russian Federation]; Linn Persson [Sweden]; Jan Plesnik [Czech Republic]; Ariana Rossen [Argentina]; Mayar Sabet [Egypt]; John Shilling [United States of America]; Binaya Raj Shivakoti [Nepal]; Asha Singh [Guyana]; Asha Sitati [Kenya]; Lawrence Surendra [India]; Paul C. Sutton [United States of America]; Khulood Abdul Razzaq Tubaishat [Jordan]; Emma Archer van Garderen [South Africa]; Lei Yu [China]; Samy Mohamed Zalat [Egypt].

Contributing Institutions and Organizations:

Arabian Gulf University [Bahrain]; Charles University [Czech Republic]; Chinese Academy of Sciences [China]; Clean Air in London [United Kingdom of Great Britain and Northern Ireland]; Colorado State University [United States of America]; Department of Environment [Iran (Islamic Republic of)]; Embassy of the Federative Republic of Brazil [Brazil]; Environment and Climate Change [Canada]; ETH Zurich [Switzerland]; European Environment Agency [Denmark]; Forest Peoples Programme [England]; General Authority for Meteorology and for Meteorology and Environment Protection [Saudi Arabia]; ICLEI – Local Governments for Sustainability [Canada]; Indian Institute of Science [India]; Institut des Géosciences de l'Environnement [France]; King Abdulaziz University [Saudi Arabia]; Korea Environment Institute [Republic of Korea]; Michigan Technological University [United States of America]; Ministry for the Ecological and Inclusive Transition of France [France]; Ministry for the Environment [France]; Ministry of Climate and Environment [Norway]; Ministry of Environment and Forestry [Indonesia]; Ministry of Environment and Natural Resources [Mexico]; Ministry of Environment Ecology and Forest [Madagascar]; Ministry of Environment Land and Sea [Italy]; Ministry of Environment, Forests and Climate Change, India]; Ministry of Environment [Jordan]; Ministry of Environment, Protection of Nature and Sustainable Development [Cameroon]; Ministry of Natural Resources, Energy and Environment [Malawi]; Ministry of the Environment and Sustainable Development [Argentina]; National Institute of S&T for Climate Change [Brazil]; National Water Research Center [Egypt]; Office of Natural Resources and Environmental Policy and Planning [Thailand]; Peking University [China]; Secretariat de Medio Ambiente y Recursos Naturales [Mexico]; Spanish National Research Council [Spain]; Swiss Federal Office for the Environment [Switzerland]; The University of the West Indies [Jamaica]; Tokyo Institute of Technology [Japan];U.S. Department of State [United States of America]; United Nations Foundation [United States of America]; Università di Roma "Tor Vergata" [Italy]; Université Nangui Abrogoua [Côté d'ivoire]; University of Costa Rica [Costa Rica]; University of Maryland School of Public Policy [United States of America]; University of Michigan [United States of America]; University of São Paolo [Brazil]; University of Tehran [Iran (Islamic Republic of)]; University of Warwick and HEART [United Kingdom of Great Britian and Northern Ireland]; University of York [United Kingdom of Great Britian and Northern Ireland]; University of Zimbabwe [Zimbabwe]; US Forest Service [United States of America]; WECF International [Germany]; World Environment Center [United States of America];World Water Council [France].

Glossary

This glossary is compiled from citations in different chapters, and draws from glossaries and other resources available on the websites of the following organizations, networks and projects:

American Academy of Opthamology; American Meteorological Society; Asian Development Bank ; Biodiversity Journal; Business Dictionary; Business Dictionary ; Cambridge Dictionary; Center for Transportation Excellence (United States); Centers for Disease Control and Prevention; Charles Darwin University(Australia); Collins Dictionary; Consultative Group on International Agricultural Research; Convention on Biological Diversity; Convention on Wetlands of International Importance especially as Waterfowl Habitat (Ramsar); Department of Agriculture (United States); Department of the Interior (United States); Department of Transportation (United States); Deutsche Gesellschaft für Internationale Zusammenarbeit, GmbH, GiZ; Edwards Aquifer Website (United States); Encyclopaedia Britannica; Encyclopedia of Earth; Energy Information Administration (United States); Environmental Protection Agency (United States); Environmental Science and Pollution Research; Europe's Information Society; European Commission; European Environmental Agency; European Nuclear Society; Farlex Free; Food and Agriculture Organization of the United Nations, Foundation for Research; Gender GEO; Global Earth Observation System of Systems; Global Environment Outlook Sixth Edition; Global Footprint Network ; Global Land Outlook; Glossary of Environment Statistics; GreenFacts Glossary; Hayes' Handbook of Pesticide Toxicology; Healthline; IGI Global; Illinois Clean Coal Institute (United States); Illuminating Engineering Society of North America; Industrial Organisation Economics and Competition Law; Intellectual Property Organization; Intergovernmental Panel on Climate Change; Intergovernmental Science-Policy Platform on Biodiversity and Ecosystem Services; International Centre for Research in Agroforestry; International Comparison Program; International; Federation of Organic Agriculture Movements; International Research Institute for Climate and Society at Columbia University (United States); International Strategy for Disaster Reduction; International Union for Conservation of Nature; Journal of Pharmaceutical Microbiology; Journal of the Association for Information Science and Technology; Lyme Disease Foundation (United States); Manual Práctico de Ecodiseño; Medical Dictionary; Merriam-Webster Dictionary; Millennium Ecosystem Assessment; Ministerial Conference on the Protection of Forests in Europe; Ministry of Environment New Zealand; Ministry of Rural Development (Malaysia); MIT Press; National Aeronautics Space Administration (United States); National Bureau of Economic Research; National Cancer Institute (United States); National Center for Biotechnology Information (United States); National Geographic; National Heart, Lung and Blood Institute (United States); National Oceanic and Atmospheric Administration (United States); National Safety Council (United States); National Snow and Ice Data Centre (United States); Natsource (United States); Organisation for Economic Co-operation and Development; Organisation for Economic Co-operation and Development; Oxford Dictionary; PPP Knowledge Lab; Professional Development for Livelihoods (United Kingdom of Great Britian and Northern Ireland); RadioPaedia; Redefining Progress (United States); SafariX eTextbooks Online; Science and Technology (New Zealand); Science Dictionary; SDG Knowledge platform; Semanticscolar.org; SER Primer; The IUP Journal of Applied Economics; TheFreeDictionary.com; Tirana Declaration; UN Environment; UN-Habitat; United Nations Convention to Combat Desertification; United Nations Development Group; United Nations Development Programme; United Nations Development Programme; United Nations Economic and Social Commission for Asia and the Pacific; United Nations Educational, Scientific and Cultural Organization; United Nations Framework Convention on Climate Change ; United Nations Industrial Development Organization; United Nations International Strategy for Disaster Reduction; United Nations Statistics Division; United Nations Water; United Nations Women; United State Geogical Survey; University of Sydney; USLegal.com; Water Footprint Network, (Netherlands); Water Quality Association (United States); Wikipedia; World Bank; World Health Organization; World Health Organization; World Meteorological Organization; World Wide Fund for Nature

Abundance
The number of individuals or related measure of quantity (such as biomass) in a population, community or spatial unit.

Abrupt change
The change that takes place so rapidly and unexpectedly that human or natural systems have difficulty adapting to it.

Acidification
Change in natural chemical balance caused by an increase in the concentration of acidic elements.

Acidity
A measure of how acid a solution may be. A solution with a pH of less than 7.0 is considered acidic.

Adaptation
Adjustment in natural or human systems to a new or changing environment, including anticipatory and reactive adaptation, private and public adaptation, and autonomous and planned adaptation.

Adaptive capacity
The ability of a system to adjust to climate change (including climate variability and extremes) to moderate potential damages, to take advantage of opportunities, or to cope with the consequences.

Adaptive governance
A governance approach that incorporates methods of adaptive management, adaptive policy making and transition management for addressing complex, uncertain and dynamic issues. Adaptive governance relies on polycentric institutional arrangements for decision making at multiple scales. Spanning the local and global levels, this form of governance provides for collaborative, flexible, learning-based approaches to ecosystem management.

Aeroponics
A plant-cultivation technique in which the roots hang suspended in the air while nutrient solution is delivered to them in the form of a fine mist climate geoengineering

Afforestation
Establishment of forest plantations on land that is not classified as forest.

Aflatoxin
Aflatoxins are poisonous substances produced by certain kinds of fungi (moulds) that are found naturally all over the world; they can contaminate food crops and pose a serious health threat to humans and livestock. Aflatoxins also pose a significant economic burden, causing an estimated 25 per cent or more of the world's food crops to be destroyed annually.

Agglomeration economies
The benefi ts that come when firms and people locate near one another together in cities and industrial clusters. These benefi ts all ultimately come from transport costs savings: the only real difference between a nearby firm and one across the continent is that it is easier to connect with a neighbor.

Agricultural Intensification
Agricultural intensification can be technically defined as an increase in agricultural production per unit of inputs (which may be labour, land, time, fertilizer, seed, feed or cash). For practical purposes, intensification occurs when there is an increase in the total volume of agricultural production that results from a higher productivity of inputs, or agricultural production is maintained while certain inputs are decreased (such as by more effective delivery of smaller amounts of fertilizer, better targeting of plant or animal protection, and mixed or relay cropping on smaller fields). Intensification that takes the form of increased production is most critical when there is a need to expand the food supply, for example during periods of rapid population growth. Intensification that makes more efficient use of inputs may be more critical when environmental problems or social issues are involved. In either case, changes caused by intensification are to be understood conceptually in contrast to extensive adjustments, which involve increases or decreases in the amount of inputs used. Historically, the most common and effective extensive adjustment in agricultural production has been to increase or decrease the area of land planted.

Agroecology
An ecological approach to agriculture that views agricultural areas as ecosystems and is concerned with the ecological impact of agricultural practices.

Agroecosystems
Organisms and environment of an agricultural area considered as an ecosystem.

Agrotechnology
The application of technology in agriculture.

Albedo
The fraction of solar energy that is diffusely reflected from the Earth back into space. It shows how reflective earth's surface is.

Alienation
Unlawfully transferring records or losing custody of them to an unauthorized organization or person.

Alien species (also non-native, non-indigenous, foreign, exotic)
Species accidentally or deliberately introduced outside its normal distribution.

Alkalinisation
A process that lowers the amount of acid in a solution. In medicine, an alkali, such as sodium bicarbonate, may be given to patients to lower high levels of acid in the blood or urine that can be caused by certain medicines or conditions.

All-cause mortality
All of the deaths that occur in a population, regardless of the cause. It is measured in clinical trials and used as an indicator of the safety or hazard of an intervention.

Anthropocene
A term used by scientists to name a new geologic epoch (following the most recent Holocene) characterized by significant changes in the Earth's atmosphere, biosphere and hydrosphere due primarily to human activities.

Antimicrobial resistance
The ability of a microorganism (like bacteria, viruses, and some parasites) to stop an antimicrobial (such as antibiotics, antivirals and antimalarials) from working against it. As a result, standard treatments become ineffective, infections persist and may spread to others.

Aquatic ecosystem
Basic ecological unit composed of living and non-living elements interacting in water.

Aquifer
An aquifer is an underground layer of water-bearing rock. Water-bearing rocks are permeable, meaning they have openings that liquids and gases can pass through. Sedimentary rock such as sandstone, as well as sand and gravel, are examples of water-bearing rock. The top of the water level in an aquifer is called the water table.

Arable land
Land under temporary crops (double-cropped areas are counted only once), temporary meadows for mowing or pasture, land under market and kitchen gardens, and land temporarily fallow (less than five years). The abandoned land resulting from shifting cultivation is not included in this category.

Asymptote
A line that continually approaches a given curve but does not meet it at any finite distance.

Benthic
Of, relating to, or occurring at the bottom of a body of water.

Billion
10^9 (1 000 000 000).

Bioaccumulation
The increase in concentration of a chemical in organisms. Also used to describe the progressive increase in the amount of a chemical in an organism resulting from rates of absorption of a substance in excess of its metabolism and excretion.

Biocapacity
The capacity of ecosystems to produce useful biological materials and to absorb waste materials generated by humans, using current management schemes and extraction technologies. The biocapacity of an area is calculated by multiplying the actual physical area by the yield factor and the appropriate equivalence factor. Biocapacity is usually expressed in units of global hectares.

Biochemical Oxygen Demand
A measure of the organic pollution of water: the amount of oxygen , in mg per litre of water, absorbed by a sample kept at 20°C for five days.

Biodiversity (a contraction of biological diversity)
The variety of life on Earth, including diversity at the genetic level, among species and among ecosystems and habitats. It includes diversity in abundance, distribution and behavior, as well as interaction with socio-ecological systems. Biodiversity also incorporates human cultural diversity, which can both be affected by the same drivers as biodiversity, and itself has impacts on the diversity of genes, other species and ecosystems.

Bioenergy
Renewable energy produced by living organisms.

Biofuel
Fuel produced from dry organic matter or combustible oils from plants, such as alcohol from fermented sugar or maize, and oils derived from oil palm, rapeseed or soybeans.

Biogas
Gas, rich in methane, which is produced by the fermentation of animal dung, human sewage or crop residues in an airtight container.

Biogeochemical cycles
The flow of chemical elements and compounds between living organisms (biosphere) and the physical environment (atmosphere, hydrosphere, lithosphere).

Biomass
Organic material, above and below ground and in water, both living and dead, such as trees, crops, grasses, tree litter and roots.

Biomagnification
The build-up of certain substances in the bodies of organisms at higher trophic levels of food webs. Organisms at lower trophic levels accumulate small amounts. Organisms at the next higher level of the food chain eat many of these lower-level organisms and hence accumulate larger amounts. The tissue concentration increases at each trophic level in the food web when there is efficient uptake and slow elimination.

Biome
The largest unit of ecosystem classification that is convenient to recognize below the global level. Terrestrial biomes are typically based on dominant vegetation structure (such as forest or grassland). Ecosystems within a biome function in a broadly similar way, although they may have very different species composition. For example, all forests share certain properties regarding nutrient cycling, disturbance and biomass that are different from the properties of grasslands.

Biosphere
The part of the Earth and its atmosphere in which living organisms exist or that is capable of supporting life.

Black carbon
Operationally defined aerosol based on measurement of light absorption and chemical reactivity and/or thermal stability. Black carbon is formed through the incomplete combustion of fossil fuels, biofuel and biomass, and is emitted as part of anthropogenic and naturally occurring soot. It consists of pure carbon in several linked forms. Black carbon warms the Earth by absorbing sunlight and re-emitting heat to the atmosphere and by reducing albedo (the ability to reflect sunlight) when deposited on snow and ice.

Bleaching (of coral reefs)
A phenomenon occurring when corals under stress expel their mutualistic microscopic algae, called zooxanthellae. This results in a severe decrease or even total loss of photosynthetic pigments. Since most reef-building corals have white calcium carbonate skeletons, these then show through the corals' tissue and the coral reef appears bleached.

Blue water
Fresh surface and groundwater, in other words, the water in freshwater lakes, rivers and aquifers. The blue water footprint is the volume of surface and groundwater consumed as a result of the production of a good or service. Blue water consumption refers to the volume of freshwater used and then evaporated or incorporated into a product. It also includes water abstracted from surface or groundwater in a catchment and returned to another catchment or the sea. It is the amount of water abstracted from groundwater or surface water that does not return to the catchment from which it was withdrawn.

Bottom-up
From the lowest level of a hierarchy or process to the top.

By-catch
The unwanted fish and other marine creatures caught during commercial fishing for a different species.

Cadastre
A register of property showing the extent, value, and ownership of land for taxation.

Capacity development
The process through which individuals, organizations and societies obtain, strengthen and maintain the capabilities to set and achieve their own development objectives over time.

Cap and trade (system)
A regulatory or management system that sets a target level for emissions or natural resource use, and, after distributing shares in that quota, lets trading in those permits determine their price.

Capital
Resource that can be mobilized in the pursuit of an individual's goals. Thus, natural capital (natural resources such as land and water), physical capital (technology and artefacts), social capital (social relationships, networks and ties), financial capital (money in a bank, loans and credit), human capital (education and skills).

Carbon dioxide equivalent (CO2-equivalent or CO2e)
The universal unit of measurement used to indicate the global warming potential of the different greenhouse gases. Carbon dioxide – a naturally occurring gas that is a byproduct of burning fossil fuels and biomass, land-use changes and other industrial processes – is the reference against which other greenhouse gases are measured.

Carbon fertilization
The CO_2 fertilization effect begins with enhanced photosynthetic CO_2 fixation. Non-structural carbohydrates tend to accumulate in leaves and other plant organs as starch, soluble carbohydrates or polyfructosans, depending on species. In some cases, there may be feedback inhibition of photosynthesis associated with accumulation of non-structural carbohydrates. Increased carbohydrate accumulation, especially in leaves, may be evidence that crop plants grown under CO_2 enrichment may not be fully adapted to take complete advantage of elevated CO_2. This may be because the CO_2-enriched plants do not have an adequate sink (inadequate growth capacity), or lack capacity to load phloem and translocate soluble carbohydrates. Improvement of photoassimilate utilization should be one goal of designing cultivars for the future.

Carbon sequestration
The process of increasing the carbon content of a reservoir other than the atmosphere.

Carbon stock
The quantity of carbon contained in a "pool," meaning a reservoir or system which has the capacity to accumulate or release carbon.

Cataracts
A cloudiness or opacity in the normally transparent crystalline lens of the eye. This cloudiness can cause a decrease in vision and may lead to eventual blindness.

Catchment (area)
The area of land from which precipitation drains into a river, basin or reservoir. See also Drainage basin.

Chikungunya
Chikungunya is a viral disease transmitted to humans by infected mosquitoes. It causes fever and severe joint pain. Other symptoms include muscle pain, headache, nausea, fatigue and rash.

Circular economy
A circular economy is a systems approach to industrial processes and economic activity that enables resources used to maintain their highest value for as long as possible. Key considerations in implementing a circular economy are reducing and rethinking research use, and the pursuit of longevity, renewability, reusability, reparability, replaceability, upgradability for resources and products that are used.

Citizen science
The collection and analysis of data relating to the natural world by members of the general public, typically as part of a collaborative project with professional scientists.

Citizen scientist
A member of the general public who collects and analyses data relating to the natural world, typically as part of a collaborative project with professional scientists.

Civil society
The aggregate of non-governmental organizations and institutions representing the interests and will of citizens.

Clean Development Mechanism (CDM)
The mechanism provided by Article 12 of the Kyoto Protocol, designed to assist developing countries achieve sustainable development by permitting industrialized countries to finance projects for reducing greenhouse gas emissions in developing countries and receive carbon credits for doing so.

Climate change
The UN Framework Convention on Climate Change defines climate change as "a change of climate which is attributed directly or indirectly to human activity that alters the composition of the global atmosphere and which is in addition to natural climate variability observed over comparable time periods."

Climate proofing
A shorthand term for identifying risks to a development project, or any other specified natural or human asset, as a consequence of climate variability and change, and ensuring that those risks are reduced to acceptable levels through long-lasting and environmentally sound, economically viable, and socially acceptable changes implemented at one or more of the following stages in the project cycle: planning, design, construction, operation and decommissioning.

Climate variability
Variations in the mean state and other statistics (such as standard deviations and the occurrence of extremes) of the climate on all temporal and spatial scales beyond that of individual weather events. Variability may be due to natural internal processes in the climate system (internal variability), or to variations in natural or anthropogenic external forcing (external variability).

Chlorofluorocarbons (CFCs)
A group of chemicals, consisting of chlorine, fluorine and carbon, highly volatile and of low toxicity, widely used in the past as refrigerants, solvents, propellants and foaming agents. Chlorofluorocarbons have both ozone depletion and global warming potential.

Community-based monitoring and information systems (CBMIS)
This term refers to initiatives by indigenous peoples and local community organisations to monitor their community's well-being and the state of their territories and natural resources, applying a mix of traditional knowledge and innovative tools and approaches.

Cross-cutting issue
An issue that cannot be adequately understood or explained without reference to the interactions of several of its dimensions that are usually defined separately.

Crowdsourcing
A problem-solving and production process that involves outsourcing tasks to a network of people, also known as the crowd. This process can occur both online and offline.

Conjunctival melanoma
A pigmented lesion of the ocular surface. It is an uncommon but potentially devastating tumor that may invade the local tissues of the eye, spread systemically through lymphatic drainage and hematogenous spread, and recur in spite of treatment.

Conservation
The protection, care, management and maintenance of ecosystems, habitats, wildlife species and populations, within or outside of their natural environments, in order to safeguard the natural conditions for their long-term permanence.

Crop
(The total amount collected of) a plant such as a grain, fruit or vegetable grown in large amounts.

Cultural services
In the context of ecosystems, the non-material benefits for people, including spiritual enrichment, cognitive development, recreation and aesthetic experience.

Custodian agencies
United Nations bodies (and in some cases, other international organizations) responsible for compiling and verifying country data and metadata, and for submitting the data, along with regional and global aggregates, to the United Nations Statistics Division (UNSD). Furthermore, custodian agencies are expected to take the lead in developing missing indicators.

Cutaneous malignant melanoma
The most common subtype of malignant melanoma, a malignant neoplasm that arises from melanocytes. Melanocytes predominantly occur in the basal layer of the epidermis but do occur elsewhere in the body. Primary cutaneous melanoma is by far the most common type of primary melanoma, although it may occur in other tissues, e.g. primary uveal malignant melanoma.

Dataset
A collection of data on a particular issue.

DDT (dichlorodiphenyltrichloroethane)
A synthetic organochlorine insecticide, one of the persistent organic pollutants listed for control under the Stockholm Convention on Persistent Organic Pollutants.

Decarbonization
Remove carbon or carbonaceous deposits from (an engine or other metal object).

Deforestation
Conversion of forested land to non-forest areas.

Dengue
An infectious diseases caused by any one of four related viruses transmitted by mosquitoes. The dengue virus is a leading cause of illness and death in the tropic and subtropics. As many as 400 million people are infected yearly.

Desertification
Land degradation in arid, semi-arid and dry sub-humid areas resulting from various factors, including climatic variations and human activities. It involves crossing thresholds beyond which the underpinning ecosystem cannot restore itself, but requires ever-greater external resources for recovery.

Detoxification
The process of removing toxic substances or qualities.

Disability-adjusted life years (DALYS)
The sum of years of potential life lost due to premature mortality and the years of productive life lost due to disability.

Disaggregation
To separate into component parts.

Disaster risk management
The application of disaster risk reduction policies and strategies, to prevent new disaster risks, reduce existing disaster risks, and manage residual risks, contributing to the strengthening of resilience and reduction of losses. Disaster risk management actions can be categorized into; prospective disaster risk management, corrective disaster risk management and compensatory disaster risk management (also referred to as residual risk management).

Disaster risk reduction
The conceptual framework of elements intended to minimize vulnerability to disasters throughout a society, to avoid (prevention) or limit (mitigation and preparedness) the adverse impacts of hazards, within the broad context of sustainable development.

DPSIR Framework
UNEP adopted the DPSIR causal framework approach for the GEO assessments. This represents a systems-analysis view in which the driving forces of social and economic development exert pressures on the environment, which change the state of the environment. The changing state of the environment leads to impacts on, for example, human well-being and ecosystem health, which then produces human responses to remedy these impacts, such as social controls, redirecting investments, and/or policies and political interventions to influence human activity. Finally, these responses influence the state of the environment, either directly or indirectly, through the driving forces or the pressures. Existing policies increasingly need to be assessed in terms of how they address the drivers and impacts of environmental challenges.

Drainage basin
(Also called watershed, river basin or catchment)
Land area where precipitation runs off into streams, rivers, lakes and reservoirs. It is a land feature that can be identified by tracing a line along the highest elevations between different areas, often a ridge.

Drip irrigation
Sometimes called trickle irrigation and involves dripping water onto the soil at very low rates (2-20 litres/hour) from a system of small diameter plastic pipes fitted with outlets called emitters or drippers. Water is applied close to plants so that only part of the soil in which the roots grow is wetted (Figure 60), unlike surface and sprinkler irrigation, which involves wetting the whole soil profile. With drip irrigation water, applications are more frequent (usually every 1-3 days) than with other methods and this provides a very favourable high moisture level in the soil in which plants can flourish.

Driver
The overarching socio-economic forces that exert pressures on the state of the environment.

Drylands
Areas characterized by lack of water, which constrain two major, linked ecosystem services: primary production and nutrient cycling. Four dryland sub-types are widely recognized: dry sub-humid, semi-arid, arid and hyper-arid, showing an increasing level of aridity or moisture deficit.

E-waste (electronic waste)
A generic term encompassing various forms of electrical and electronic equipment that has ceased to be considered of value and is disposed of.

Early warning
The provision of timely and effective information, through identified institutions, that allows individuals exposed to a hazard to take action to avoid or reduce their risk and prepare an effective response.

Earth System
The Earth System is a complex social-environmental system of interacting physical, chemical, biological and social components and processes that determine the state and evolution of the planet and life on it.

Eco-design
The integration of environmental aspects into product design and development with the aim of reducing adverse environmental impacts throughout a product's life cycle.

Ecological footprint
A measure of the area of biologically productive land and water an individual, population or activity uses to produce all the resources it consumes and to absorb the corresponding waste (such as carbon dioxide emissions from fossil fuel use), using prevailing technology and resource management practices. The ecological footprint is usually measured in global hectares.

Ecoregion
A major ecosystem defined by distinctive geography and receiving uniform solar radiation and moisture.

Ecosystem
A dynamic complex of plant, animal and micro-organism communities and their non-living environment, interacting as a functional unit.

Ecosystem approach
A strategy for the integrated management of land, water and living resources that promotes conservation and sustainable use in an equitable way. An ecosystem approach is based on the application of appropriate scientific methods, focused on levels of biological organization that encompass the essential structure, processes, functions and interactions among and between organisms and their environment. It recognizes that humans, with their cultural diversity, are an integral component of many ecosystems.

Ecosystem boundaries
Ecosystem boundaries are zones of transitions between two adjacent habitats. They occur naturally in all biomes but the extent of boundaries has been greatly increased by anthropogenic habitat modification. Transition zones are characterized by a profound change in the composition of plant and animal communities and that transition may be abrupt, gradual or even occur via a series of intermediate habitat types.

Ecosystem collapse
The endpoint of ecosystem decline, and occurs when all occurrences of an ecosystem have moved outside the natural range of spatial and temporal variability in composition, structure and/or function.

Ecosystem function
An intrinsic ecosystem characteristic related to the set of conditions and processes whereby an ecosystem maintains its integrity (such as primary productivity, food chain and biogeochemical cycles). Ecosystem functions include such processes as decomposition, production, nutrient cycling, and movements of nutrients and energy.

Ecosystem health
The degree to which ecological factors and their interactions are reasonably complete and function for continued resilience, productivity and renewal of the ecosystem.

Ecosystem management
An approach to maintaining or restoring the composition, structure, function and delivery of services of natural and modified ecosystems for the goal of achieving sustainability. It is based on an adaptive, collaboratively developed vision of desired future conditions that integrates ecological, socio-economic, and institutional perspectives, applied within a geographic framework, and defined primarily by natural ecological boundaries.

Ecosystem resilience
The level of disturbance that an ecosystem can withstand without crossing a threshold to become a different structure or deliver different outputs. Resilience depends on ecological dynamics as well as human organizational and institutional capacity to understand, manage and respond to these dynamics.

Ecosystem restoration
The process of assisting the recovery of an ecosystem that has been degraded, damaged or destroyed.

Ecosystem-based adaptation
The use of biodiversity and ecosystem services as part of an overall strategy to help people adapt to the adverse effects of climate change.

Ecotourism
Travel undertaken to witness the natural or ecological quality of particular sites or regions, including the provision of eco-friendly services to facilitate such travel.

Effluent
In issues of water quality, refers to liquid waste (treated or untreated) discharged to the environment from sources such as industrial process and sewage treatment plants.

El Niño (also El Niño-Southern Oscillation (ENSO))
In its original sense, it is a warm water current that periodically flows along the coast of Ecuador and Peru, disrupting the local fishery. This oceanic event is associated with a fluctuation of the inter-tropical surface pressure pattern and circulation in the Indian and Pacific Oceans, called the Southern Oscillation. This atmosphere-ocean phenomenon is collectively known as El Niño-Southern Oscillation. During an El Niño event, the prevailing trade winds weaken and the equatorial countercurrent strengthens, causing warm surface waters in the Indonesian area to flow eastward to overlie the cold waters of the Peru current off South America. This event has great impact on the wind, sea surface temperature and precipitation patterns in the tropical Pacific. It has climatic effects throughout the Pacific region and in many other parts of the world. The opposite of an El Niño event is called La Niña.

Electrification
The action or process of charging something with electricity.

Emission inventory
Details the amounts and types of pollutants released into the environment.

Endangered species
A species is endangered when the best available evidence indicates that it meets any of the criteria A to E specified for the endangered category of the IUCN Red List, and is therefore considered to be facing a very high risk of extinction in the wild.

Endocrine disruptor
An external substance that interferes (through mimicking, blocking, inhibiting or stimulating) with function(s) of the hormonal system and consequently causes adverse health effects in an intact organism, or its progeny, or (sub) populations.

Energy intensity
Ratio of energy consumption to economic or physical output. At the national level, energy intensity is the ratio of total domestic primary energy consumption or final energy consumption to gross domestic product or physical output. Lower energy intensity shows greater efficiency in energy use.

Environment statistics
Statistics that describe the state of and trends in the environment, covering the media of the natural environment (air/ climate, water, land/soil), the living organisms within the media, and human settlements.

Environmental assessment
The entire process of undertaking an objective evaluation and analysis of information designed to support environmental decision making. It applies the judgement of experts to existing knowledge to provide scientifically credible answers to policy-relevant questions, quantifying where possible the level of confidence. It reduces complexity but adds value bysummarizing, synthesizing and building scenarios, and identifies consensus by sorting out what is known and widely accepted from what is not known or not agreed. It sensitizes the scientific community to policy needs and the policy community to the scientific basis for action.

Environmental degradation
Environmental degradation is the deterioration in environmental quality from ambient concentrations of pollutants and other activities and processes such as improper land use and natural disasters.

Environmental education
The process of recognizing values and clarifying concepts in order to develop skills and attitudes necessary to understand and appreciate the interrelatedness of humans, their culture and biophysical surroundings. Environmental education also entails practice in decision-making and self-formulation of a code of behaviour about issues concerning environmental quality.

Environmental flows
Quantity, timing and quality of water flows required to sustain freshwater and estuarine ecosystems and the human livelihoods and well-being that depend on these ecosystems. Through implementation of environmental flows, water managers strive to achieve a flow regime, or pattern, that provides for human uses and maintains the essential processes required to support healthy river ecosystems.

Environmental footprint
The effect that a person, company, activity, etc. has on the environment, for example the amount of natural resources that they use and the amount of harmful gases that they produce.

Environmental governance
Environmental Governance is the means by which society determines and acts on goals and priorities related to the management of natural resources. This includes the rules, both formal and informal, that govern human behavior in decision-making processes as well as the decisions themselves. Appropriate legal frameworks on the global, regional, national and local level are a prerequisite for good environmental governance.

Environmental health
Those aspects of human health and disease that are determined by factors in the environment. It also refers to the theory and practice of assessing and controlling factors in the environment that can potentially affect health. Environmental health includes both the direct pathological effects of chemicals, radiation and some biological agents, and the effects, often indirect, on health and well-being of the broad physical, psychological, social and aesthetic environment. This includes housing, urban development, land use and transport.

Environmental impact assessment (EIA)
An analytical process or procedure that systematically examines the possible environmental consequences of a given activity or project. The aim is to ensure that the environmental implications are taken into account before the decisions are made.

Environmental justice
A mechanism of accountability for the protection of rights and the prevention and punishment of wrongs related to the disproportionate impacts of growth on the poor and vulnerable in society from rising pollution and degradation of ecosystem services, and from inequitable access to and benefits from the use of natural assets and extractive resources.

Environmental monitoring
Regular, comparable measurements or time series of data on the environment.

Environmental policy
A policy aimed at addressing environmental problems and challenges.

Environmental pressure
Pressure resulting from human activities which bring about changes in the state of the environment.

Environmental refugees and internally displaced people (IDPs)
People who have been forced to leave their traditional habitat temporarily or permanently, because of a marked environmental disruption (natural or triggered by people) that jeopardizes their existence and/or seriously affected the quality of their life. (Science for Peace) belong to a larger group of immigrants known as environmental refugees. Environmental refugees include immigrants forced to flee because of natural disasters, such as volcanoes and tsunamis.

Epidemiology
The branch of medicine which deals with the incidence, distribution, and possible control of diseases and other factors relating to health.

Equity
Fairness of rights, distribution and access. Depending on context, this can refer to access to resources, services or power.

Estuary
Water passage where the tide meets a river current.

Eutrophication
The degradation of water or land quality due to enrichment by nutrients, primarily nitrogen and phosphorous, which results in excessive plant (principally algae) growth and decay. Eutrophication of a lake normally contributes to its slow evolution into a bog or marsh and ultimately to dry land. Eutrophication may be accelerated by human activities that speed up the ageing process.

Evapotranspiration
Combined loss of water by evaporation from the soil or surface water, and transpiration from plants and animals.

External cost (also externality)
A cost that is not included in the market price of the goods and services produced. In other words, a cost not borne by those who create it, such as the cost of cleaning up contamination caused by discharge of pollution into the environment.

Feed-in tariff
A feed-in tariff is an energy policy focused on supporting the development and dissemination of renewable power generation. In a feed-in tariff scheme, providers of energy from renewable sources, such as solar, wind or water, receive a price for what they produce based on the generation costs. This purchase guarantee is offered generally on a long-term basis, ranging from 5 to 20 years, but most commonly spanning 15–20 years.1 The cost of the tariff payments are typically shared with the electricity consumers.

Feedback
Where non-linear change is driven by reactions that either dampen change (negative feedbacks) or reinforce change (positive feedbacks).

Fipronil systemic insecticides
Phenyl-pyrazole fipronil are insecticides with systemic properties. Their physicochemical characteristics, mainly assessed in terms of their octanol water partition coefficient (*Kow*) and dissociation constant (pKa), enable their entrance into plant tissues and their translocation to all its parts. Regardless of the manner of application and route of entry to the plant, they translocate throughout all plant tissues making them toxic to any insects (and potentially other organisms) that feed upon the plant. This protects the plant from direct damage by herbivorous (mainly sap feeding) insects and indirectly from damage by plant viruses that are transmitted by insects.

Floods (river, flash and storm surge)
Usually classified into three types: river flood, flash flood and storm surge. River floods result from intense and/or persistent rain over large areas. Flash floods are mostly local events resulting from intense rainfall over a small area in a short period of time. Storm surge floods occur when flood water from the ocean or large lakes is pushed on to land by winds or storms.

Food security
Physical and economic access to food that meets people's dietary needs as well as their food preferences.

Food system
1) Food systems are usually conceived as a set of activities ranging from production to consumption. It is a broad concept encompassing food security and its components – availability, access and utilization – and including the social and environmental outcomes of these activities. Food systems in developing countries have been largely transformed by globalization. This change offers tremendous opportunities for food workers to access new and better employments. Yet, small scale food producers and other food workers are still too often excluded from the benefits generated by food businesses.

Food-water-energy nexus
The water-food-energy nexus is central to sustainable development. Demand for all three is increasing, driven by a rising global population, rapid urbanization, changing diets and economic growth. Agriculture is the largest consumer of the world's freshwater resources, and more than one-quarter of the energy used globally is expended on food production and supply. The inextricable linkages between these critical domains require a suitably integrated approach to ensuring water and food security, and sustainable agriculture and energy production worldwide.

Forest
Land spanning more than 0.5 hectares with trees higher than 5 metres and a canopy cover of more than 10 per cent, or trees able to reach these thresholds *in situ*. It does not include land that is predominantly under agricultural or urban use.

Forest degradation
Changes within the forest that negatively affect the structure or function of the stand or site, and thereby lower the capacity to supply products and/or services.

Forest management
The processes of planning and implementing practices for the stewardship and use of forests and other wooded land aimed at achieving specific environmental, economic, social and/or cultural objectives.

Forest plantation
Forest stands established by planting and/or seeding in the process of afforestation or reforestation. They are either of introduced species (all planted stands), or intensively managed stands of indigenous species, which meet all the following criteria: contain one or two species, are of similar age and regularly spaced. "Planted forest" is another term used for plantation.

Fossil fuel
Coal, natural gas and petroleum products (such as oil) formed from the decayed bodies of animals and plants that died millions of years ago.

Free-riding
Free riding occurs when one firm (or individual) benefits from the actions and efforts of another without paying or sharing the costs. For example, a retail store may initially choose to incur costs of training its staff to demonstrate to potential customers how a particular kitchen appliance works. It may do so in order to expand its sales. However, the customers may later choose to buy the product from another retailer selling at a lower price because its business strategy is not to incur these training and demonstration costs. This second retailer is viewed as "free riding" on the efforts and the costs incurred by the first retailer. If such a situation persists, the first retailer will not have the incentive to continue.

Gender
Gender refers to the roles, behaviors, activities, and attributes that a given society at a given time considers appropriate for men and women. In addition to the social attributes and opportunities associated with being male and female and the relationships between women and men and girls and boys, gender also refers to the relations between women and those between men. These attributes, opportunities and relationships are socially constructed and are learned through socialization processes. They are context/ time-specific and changeable. Gender determines what is expected, allowed and valued in a woman or a man in a given context. Gender is part of the broader socio-cultural context, as are other important criteria for socio-cultural analysis including class, race, poverty level, ethnic group, sexual orientation, age, etc.

Gender analysis
Gender analysis is a critical examination of how differences in gender roles, activities, needs, opportunities and rights/ entitlements affect men, women, girls and boys in certain situation or contexts. Gender analysis examines the relationships between females and males and their access to and control of resources and the constraints they face relative to each other. A gender analysis should be integrated into all sector assessments or situational analyses to ensure that gender-based injustices and inequalities are not exacerbated by interventions, and that where possible, greater equality and justice in gender relations are promoted.

Gender equality (Equality between women and men)
This refers to the equal rights, responsibilities and opportunities of women and men and girls and boys. Equality does not mean that women and men will become the same but that women's and men's rights, responsibilities and opportunities will not depend on whether they are born male or female. Gender equality implies that the interests, needs and priorities of both women and men are taken into consideration, recognizing the diversity of different groups of women and men. Gender equality is not a women's issue but should concern and fully engage men as well as women. Equality between women and men is seen both as a human rights issue and as a precondition for, and indicator of, sustainable people-centered development.

Gender gap
The term gender gap refers to any disparity between women and men's condition or position in society. It is often used to refer to a difference in average earnings between women and men, e.g. "gender pay gap." However, gender gaps can be found in many areas, such as economic participation and opportunity, educational attainment, health and survival and political empowerment.

Gender mainstreaming
Gender mainstreaming is the chosen strategy of the United Nations system for implementing greater equality for women and girls in relation to men and boys. Mainstreaming a gender perspective is the process of assessing the implications for women and men of any planned action, including legislation, policies or programs, in all areas and at all levels. It is a way to make women's as well as men's concerns and experiences an integral dimension of the design, implementation, monitoring and evaluation of policies and programs in all political, economic and societal spheres so that women and men benefit equally and inequality is not perpetuated. The ultimate goal is to achieve gender equality.

Gender-disaggregated data
Information collected and presented separately according to people's gender. It typically includes the state of being masculine or feminine based on social or cultural identities, constructs and differences.

Genetic diversity
The variety of genes within a particular species, variety or breed.

GEO Data Portal (now Environmental Data Explorer)
The source for datasets used by UNEP and its partners in the Global Environment Outlook report and other integrated environmental assessments. Its online database holds more than 500 different variables, including national, sub-regional, regional and global statistics as well as geospatial datasets (maps), covering themes such as freshwater, population, forests, emissions, climate, disasters, health and GDP.

Geomorphology
The study of the physical features of the surface of the earth and their relation to its geological structures.

Geospatial
Relating to or denoting data that is associated with a particular location.

Geostationary orbit
Circular orbit 35,785 km (22,236 miles) above Earth's Equator in which a satellite's orbital period is equal to Earth's rotation period of 23 hours and 56 minutes. A spacecraft in this orbit appears to an observer on Earth to be stationary in the sky. This particular orbit is used for meteorological and communications satellites. The geostationary orbit is a special case of the geosynchronous orbit, which is any orbit with a period equal to Earth's rotation period.

Geothermal energy
The word geothermal comes from the Greek words geo (earth) and therme (heat). Geothermal energy is heat within the earth. People can use this heat as steam or as hot water to heat buildings or to generate electricity. Geothermal energy is a renewable energy source because heat is continuously produced inside the earth.

Glacial periods
A period in the earth's history when polar and mountain ice sheets were unusually extensive across the earth's surface.

Glacier retreat
Glaciers retreaz when their terminus does not extend as far downvalley as it previously did. Glaciers may retreat when their ice melts or ablates more quickly than snowfall can accumulate and form new glacial ice. Higher temperatures and less snowfall have been causing many glaciers around the world to retreat recently.

Global (international) environmental governance
The assemblage of laws and institutions that regulate society-nature interactions and shape environmental outcomes.

Global commons
Natural un-owned assets such as the atmosphere, oceans, outer space and the Antarctic.

Global Earth Observation System of Systems (GEOSS)
A network aiming to link existing and planned Earth observing systems (e.g., satellites and networks of weather stations and ocean buoys) around the world, support the development of new systems where gaps currently exist, and promote common technical standards so that data from the thousands of different instruments can be combined into coherent datasets. It aims to provide decision support tools to policy makers and other users in areas such as health, agriculture and disasters.

Global hectare
A hypothetical hectare with world-average ability to produce resources and absorb wastes.

Global observation system
A set of coordinated monitoring activities that would collect much needed data at a global scale on a variety of indicators such as biodiversity, water quality and quantity, atmospheric pollution, land degradation and chemical release.

Global public good
Public goods that have universal benefits, covering multiple groups of countries and all populations.

Global warming
Increase in surface air temperature, referred to as the global temperature, induced by emissions of greenhouse gases into the air.

Globalization
The increasing integration of economies and societies around the world, particularly through trade and financial flows, and the transfer of culture and technology.

Governance
The act, process, or power of governing for the organization of society/ies. For example, there is governance through the state, the market, or through civil society groups and local organizations. Governance is exercised through institutions: laws, property-rights systems and forms of social organization.

Green economy
There is no internationally agreed definition of green economy and at least eight separate definitions were identified in recent publications. For example, UNEP has defined the green economy as "one that results in improved human well-being and social equity, while significantly reducing environmental

risks and ecological scarcities. It is low carbon, resource efficient, and socially inclusive" (UNEP, 2011). This definition has been cited in a number of more recent reports, including by the UNEMG and the OECD. Another definition for green economy offered by the Green Economy Coalition (a group of NGOs, trade union groups and others doing grassroots work on a green economy) succinctly defines green economy as "a resilient economy that provides a better quality of life for all within the ecological limits of the planet."

Greenhouse gases (GHGs)
Gaseous constituents of the atmosphere, both natural and anthropogenic, that absorb and emit thermal radiation. This property causes the greenhouse effect. Water vapour (H2O), carbon dioxide (CO2), nitrous oxide (N2O), methane (CH4) and ozone (O3) are the primary greenhouse gases in the Earth's atmosphere. There are human-made greenhouse gases in the atmosphere, such as halocarbons and other chlorine- and bromine-containing substances. Beside CO_2, N_2O and CH_4, the Kyoto Protocol deals with sulphur hexafluoride (SF6), hydrofluorocarbons (HFCs), perfluorocarbons (PFCs) and nitrogen trifluoride (NF3).

Grey water
Water the quality of which has been adversely affected by human use, in industrial, agriculture or domestically. The grey water footprint of a product is an indicator of freshwater pollution that can be associated with the production of a product over its full supply chain. It is defined as the volume of freshwater that is required to assimilate the load of pollutants based on natural background concentrations and existing ambient water quality standards. It is calculated as the volume of water that is required to dilute pollutants to such an extent that the quality of the water remains above agreed water quality standards.

Gross domestic product (GDP)
The value of all final goods and services produced in a country in one year. GDP can be measured by adding up all of an economy's incomes – wages, interest, profits, and rents – or expenditures – consumption, investment, government purchases, and net exports (exports minus imports).

Ground truthing
A process by which the content of satellite images, aerial photographs – or maps based on them – is compared with the reality on the ground through site visits and field surveys. It is used to verify the accuracy of the images or the way they have been interpreted to produce maps.

Groundwater
Water that flows or seeps downward and saturates soil or rock, supplying springs and wells. The upper surface of the saturated zone is called the water table.

Gyres
A large system of rotating ocean currents, primarily driven by wind movement. Large gyres exist in the Indian Ocean, North Atlantic, North Pacific, South Atlantic and South Pacific.

Habitat fragmentation
Alteration of habitat resulting in spatial separation of habitat units from a previous state of greater continuity.

Habitat
(1) The place or type of site where an organism or population occurs naturally.
(2) Terrestrial or aquatic areas distinguished by geographic, living and non-living features, whether entirely natural or semi-natural.

Hadley cell
A large-scale atmospheric convection cell in which air rises at the equator and sinks at medium latitudes, typically about 30° north or south.

Hazard
A potentially damaging physical event, phenomenon or human activity that may cause the loss of life or injury, property damage, social and economic disruption or environmental degradation.

Hazardous waste
A used or discarded material that can damage human health and the environment. Hazardous wastes may include heavy metals, toxic chemicals, medical wastes or radioactive material.

Heavy metals
A subset of elements that exhibit metallic properties, including transitional metals and semi-metals (metalloids), such as arsenic, cadmium, chromium, copper, lead, mercury, nickel and zinc, that have been associated with contamination and potential toxicity.

Helminth
Worm-like parasites.

Heterogeneities
The quality or sate of being diverse in character or content.

High seas
The oceans outside national jurisdictions, lying beyond each nation's exclusive economic zone or other territorial waters.

Human footprint
The impact of human activities measured in terms of the area of biologically productive land and water required to produce the goods consumed and to assimilate the wastes generated.

Human health
Health is a state of complete physical, mental and social well-being and not merely the absence of disease or infirmity.

Human well-being
The extent to which individuals have the ability to live the kinds of lives they have reason to value; the opportunities people have to pursue their aspirations. Basic components of human well-being include: security, meeting material needs, health and social relations.

Hybridization
The process of an animal or plant breeding with an individual of another species or variety.

Hydraulic fracturing
The gas-fired plants come courtesy of the revolution in hydraulic fracturing ("fracking"), which has delivered a vast supply of low-cost natural gas to an electricity market that has struggled with steadily rising coal prices since 2001.

Hydrochlorofluorocarbons (HCFCs)
Organic and human-made substances composed of hydrogen, chlorine, fluorine and carbon atoms. As the ozone-depleting potential of HCFCs is much lower than that of CFCs, HCFCs were considered acceptable interim substitutes for CFCs.

Hydrological cycle
Succession of stages undergone by water in its passage from the atmosphere to the Earth's surface and its return to the atmosphere. The stages include evaporation from land, sea or inland water, condensation to form clouds, precipitation, accumulation in the soil or in water bodies, and re-evaporation.

Hydrometeorology
A branch of meteorology that deals with water in the atmosphere especially as precipitation.

Hydroponics
The process of growing plants in sand, gravel, or liquid, with added nutrients but without soil.

Hypoxia
Lack of oxygen. In the context of eutrophication and algal blooms, hypoxia is the result of a process that uses up dissolved oxygen in the water. Algal blooms cause water to become more opaque, thereby reducing light availability to submerged aquatic vegetation, and interfering with beneficial human water uses. When the bloom dies off, algae sink to the bottom and are decomposed by bacteria using up the available oxygen. Hypoxia is particularly severe in the late summer, and can be so severe in some areas that they are referred to as "dead zones" because only bacteria can survive there.

IAS Invasive alien species
Invasive alien species are plants, animals, pathogens and other organisms that are non-native to an ecosystem, and which may cause economic or environmental harm or adversely affect human health. In particular, they impact adversely upon biodiversity, including decline or elimination of native species - through competition, predation, or transmission of pathogens - and the disruption of local ecosystems and ecosystem functions.

Improved drinking water
"Improved" sources of drinking water include piped water into dwellings; piped water into yards/plots; public taps or standpipes; tube wells or boreholes; protected dug wells; protected springs; and rainwater.

Improved sanitation
"Improved" sanitation includes flush lavatories; piped sewer systems; septic tanks; flush/pour flush to pit latrines; ventilated improved pit latrines (VIP); pit latrines with slab; composting lavatories.

In vitro
(Of a process) performed or taking place in a test tube, culture dish, or elsewhere outside a living organism.

Inertial forces
Any force invoked by an observer to maintain the validity of Isaac Newton's second law of motion in a reference frame that is rotating or otherwise accelerating at a constant rate. For specific inertial forces, see centrifugal force; Coriolis force; d'Alembert's principle .

Institutions
Regularized patterns of interaction by which society organizes itself: the rules, practices and conventions that structure human interaction. The term is wide and encompassing, and could be taken to include law, social relationships, property rights and tenurial systems, norms, beliefs, customs and codes of conduct as much as multilateral environmental agreements, international conventions and financing mechanisms. Institutions could be formal (explicit, written, often having the sanction of the state) or informal (unwritten, implied, tacit, mutually agreed and accepted).

Integrated coastal zone management
Approaches that integrate economic, social and ecological perspectives for the management of coastal resources and areas.

Integrated water resources management (IWRM)
A process which promotes the coordinated development and management of water, land and related resources, in order to maximize the resultant economic and social welfare in an equitable manner without compromising the sustainability of vital ecosystems.

Intersectionality
The understanding that social roles and identities overlap and have intertwined effects. The identity of any individual reflects and is shaped by a range of social and cultural categories such as race, class, gender, sexual orientation, and religion (among others). Oppressions within society are enacted through these multiple and linked identities.

Invasive species
Introduced species that have spread beyond their area of introduction (and, rarely, native species that have recently expanded their populations), and which are frequently associated with negative impacts on the environment, human economy or human health.

Jetstream
A narrow variable band of very strong predominantly westerly air currents encircling the globe several miles above the earth. There are typically two or three jet streams in each of the northern and southern hemispheres.

Keratinocyte
Cells found in the epidermis. Keratinocytes at the outer surface of the epidermis are dead and form a tough protective layer. The cells underneath divide to replenish the supply.

Kyoto Protocol
A protocol to the 1992 United Nations Framework Convention on Climate Change (UNFCCC) adopted at the Third Session of the Conference of the Parties to the UNFCCC in 1997 in Kyoto, Japan. It contains legally binding commitments in addition to those included in the UNFCCC. Countries included in Annex B of the protocol (most OECD countries and countries

with economies in transition) agreed to control their national anthropogenic emissions of greenhouse gases (CO2, CH4, N2O, HFCs, PFCs, SF6 and NF3) so that the total emissions from these countries would be at least 5 per cent below 1990 levels in the commitment period, 2008 to 2012.

Land cover
The physical coverage of land, usually expressed in terms of vegetation cover or lack of it. Influenced by but not synonymous with land use.

Land degradation neutrality
A state whereby the amount and quality of land resources, necessary to support ecosystem functions and services and enhance food security, remains stable or increases within specified temporal and spatial scales and ecosystems.

Land degradation neutrality
A state whereby the amount and quality of land resources, necessary to support ecosystem functions and services and enhance food security, remains stable or increases within specified temporal and spatial scales and ecosystems.

Land degradation
A long-term loss of ecosystem function and services, caused by disturbances from which the system cannot recover unaided.

Land grabbing
Large-scale land grabbing is defined as "acquisitions or concessions that are one or more of the following: (i) in violation of human rights, particularly the equal rights of women; (ii) not based on free, prior and informed consent of affected land-users; (iii) not based on a thorough assessment, or in disregard of social, economic and environmental impacts including the way those impacts are gendered; (iv) not based on transparent contracts that specify clear and binding commitments about activities, employment and benefits sharing; and (v) not based on effective democratic planning, independent oversight and meaningful participation."

Land Tenure
The relationship, whether legally or customarily defined, among people, as individuals or groups, with respect to land. (For convenience, "land" is used here to include other natural resources such as water and trees.) Land tenure is an institution, i.e., rules invented by societies to regulate behaviour. Rules of tenure define how property rights to land are to be allocated within societies. They define how access is granted to rights to use, control, and transfer land, as well as associated responsibilities and restraints. In simple terms, land tenure systems determine who can use what resources for how long, and under what conditions.

Land use planning
The systematic assessment of land and water potential, alternative patterns of land use and other physical, social and economic conditions, for the purpose of selecting and adopting land-use options which are most beneficial to land users.

Land use
The functional dimension of land for different human purposes or economic activities. Examples of land use categories include agriculture, industrial use, transport and protected areas.

Land-use planning
Land-use planning involves the systematic assessment of environmental, economic and social impacts of the range of potential uses of land in order to decide on the optimal pattern of land use. Land-use planning and systematic conservation planning has seldom been explored explicitly as a tool in global scenarios.

Legitimacy
Measure of political acceptability or perceived fairness. State law has its legitimacy in the state; local law and practices work on a system of social sanction, in that they derive their legitimacy from a system of social organization and relationships.

Leverage point
A place in a system's structure where a relatively small amount of force can effect change. It is a low leverage point if a small amount of force causes a small change in system behaviour, or a high leverage point if a small amount of force causes a large change.

Life-cycle analysis
A technique to assess the environmental impacts associated with all the stages of the life of a product – from raw material extraction through materials processing, manufacture, distribution, use, repair and maintenance, and disposal or recycling (cradle-to-grave).

Lifetime (in the atmosphere)
The approximate amount of time it takes for concentrations of an atmospheric pollutant to return to the background level (assuming emissions cease) as a result of either being converted to another chemical compound or being taken out of the atmosphere through a sink. Atmospheric lifetimes can vary from hours or weeks (sulphate aerosols) to more than a century (CFCs).

Livelihood
(The way someone earns) the money people need to pay for food, a place to live, clothing, etc.

Mainstreaming
Taking into consideration as an integral part of the issue in question.

Mangrove
A tree or shrub that grows in chiefly tropical coastal swamps that are flooded at high tide. Mangroves typically have numerous tangled roots above ground and form dense thickets.

Marginalization
Treatment of a person, group, or concept as insignificant or peripheral.

Mariculture
The cultivation of marine organisms in their natural environment.

Marine protected area (MPA)
A geographically defined marine area that is designated or regulated and managed to achieve specific conservation objectives.

Market-based instrument
Market-based instruments span a range of measures and approaches. Fundamentally, they are policy measures that influence outcomes through their effect on costs and profits. In the hands of policymakers, they can affect the operation of established markets or create new ones. They are commonly also referred to as 'economic' instruments because they attribute value to assets and directly affect decisions based on considerations of price and income.

Market-based/ Economic incentives
Market-based approaches or incentives provide continuous inducements, monetary and near-monetary, to encourage polluting entities to reduce releases of harmful pollutants. As a result, market-based approaches create an incentive for the private sector to incorporate pollution abatement into production or consumption decisions and to innovate in such a way as to continually search for the least costly method of abatement.

Material flow accounting
The quantification of all materials used in economic activities. It accounts for the total material mobilized during the extraction of materials and for the materials actually used in economic processes measured in terms of their mass.

Megacities
Urban areas with more than 10 million inhabitants.

Merit goods
Goods or services (such as education and vaccination) provided free for the benefit of the entire society by a government, because they would be under-provided if left to the market forces or private enterprise.

Merkel cell carcinoma
A very rare disease in which malignant (cancer) cells form in the skin. Merkel cells are found in the top layer of the skin. These cells are very close to the nerve endings that receive the sensation of touch.

Methemoglobinemia
A condition in which a higher-than-normal amount of methemoglobin is found in the blood. Methemoglobin is a form of hemoglobin that cannot carry oxygen. In methemoglobinemia, tissues cannot get enough oxygen. Symptoms may include headache, dizziness, fatigue, shortness of breath, nausea, vomiting, rapid heartbeat, loss of muscle coordination, and blue-colored skin. Methemoglobinemia can be caused by injury or being exposed to certain drugs, chemicals, or foods. It can also be an inherited condition.

Microbeads
A tiny sphere of plastic (such as polyethylene or polypropylene).

Microbial and non-microbial contaminants
Microbiological contamination refers to the non-intended or accidental introduction of microbes such as bacteria, yeast, mould, fungi, virus, prions, protozoa or their toxins and by-products. Prominent changes for product contamination include: loss of viscosity and sedimentation due to depolymerisation of suspending agents, pH changes, gas production, faulty smell, shiny viscous masses etc.

Microhabitat
A habitat which is of small or limited extent and which differs in character from some surrounding more extensive habitat.

Microplastics
Small plastic pieces, less than five millimeters long which can be harmful to our ocean and aquatic life.

Millennium Development Goals (MDGs)
The eight Millennium Development Goals – which range from halving extreme poverty to halting the spread of HIV/AIDS and providing universal primary education, all by the target date of 2015 – formed a blueprint agreed to by all the world's countries and all the world's leading development institutions.

Monocultural farming systems
The cultivation or growth of a single crop or organism especially on agricultural or forest land.

Morphology
(1) The physical characteristics of living organisms.
(2) The branch of biology that deals with the form of living organisms, and with relationships between their structures.

Multilateral environmental agreements (MEAs)
Treaties, conventions, protocols and contracts between several states regarding specific environmental problems.

Mycotoxin
Mycotoxins are toxic compounds that are naturally produced by certain types of moulds (fungi). Moulds that can produce mycotoxins grow on numerous foodstuffs such as cereals, dried fruits, nuts and spices. Mould growth can occur either before harvest or after harvest, during storage, on/in the food itself often under warm, damp and humid conditions. Most mycotoxins are chemically stable and survive food processing.

Nanomaterial
A natural, incidental or manufactured material containing particles, in an unbound state, as an aggregate or as an agglomerate and where, for 50 per cent or more of the particles in the number size distribution, one or more external dimension is in the size range 1–100 nanometres (a nanometre is one billionth of a metre). Such particles/materials are generally termed as nanoparticles, nanochemicals or nanomaterials.

Natural capital
Natural assets in their role of providing natural resource inputs and environmental services for economic production. Natural capital includes land, minerals and fossil fuels, solar energy, water, living organisms, and the services provided by the interactions of all these elements in ecological systems.

Natural infrastructure
Strategically planned and managed network of natural lands, such as forests and wetlands, working landscapes, and other open spaces that conserves or enhances ecosystem values and functions and provides associated benefits to human populations.

Natural resources
Materials or substances such as minerals, forests, water, and fertile land that occur in nature and can be used for economic gain.

Nature's Contribution to People
Nature's contribution to people (NCP are all the contributions, both positive and negative, of living nature (i.e. diversity of organisms, ecosystems and their associated ecological and evolutionary processes) to the quality of life for people. Beneficial contribution from nature include such things as food provision, water purification, flood control and artistic inspiration, whereas detrimental contributions include disease transmission and predation that damages people or their assets. Many NCP may be perceived as benefits or detriment depending on the cultural, temporal or spatial context.

Neonicotinoid
Neonicotinoids are an acetylcholine-interfering neurotoxic class of insecticides that are utilized in a variety of venues ranging from veterinary medicine, urban landscaping, and use in many agricultural systems as agents of crop protection. They can be applied by multiple methods as foliar sprays to above-ground plants, as root drenches to the soil, or as trunk injections to trees. However, it is estimated that approximately 60 per cent of all neonicotinoid applications globally are delivered as seed/soil treatments.

Net primary production (NPP)
The rate at which all the plants in an ecosystem produce net useful chemical energy. Some net primary production goes toward growth and reproduction of primary producers, while some is consumed by herbivores.

Neurotoxin
A poison which acts on the nervous system.

Nitrogen deposition
The input of reactive nitrogen, mainly derived from nitrogen oxides and ammonia emissions, from the atmosphere into the biosphere.

Non-Hodgkins lymphoma
Any of a large group of cancers of lymphocytes (white blood cells). Non-Hodgkin lymphomas can occur at any age and are often marked by lymph nodes that are larger than normal, fever, and weight loss. There are many different types of non-Hodgkin lymphoma. These types can be divided into aggressive (fast-growing) and indolent (slow-growing) types, and they can be formed from either B-cells or T-cells.

Non-state actors
Non-state actors are categorized as entities that (i) participate or act in the sphere of international relations; organizations with sufficient power to influence and cause change in politics which (ii) do not belong to or exist as a state-structure or established institution of a state; do not have the characteristics of this, these being legal sovereignty and some measure of control over a country's people and territories.

Normalized Difference Vegetation Index
To determine the density of green on a patch of land, researchers must observe the distinct colors (wavelengths) of visible and near-infrared sunlight reflected by the plants. As can be seen through a prism, many different wavelengths make up the spectrum of sunlight. When sunlight strikes objects, certain wavelengths of this spectrum are absorbed and other wavelengths are reflected. The pigment in plant leaves, chlorophyll, strongly absorbs visible light (from 0.4 to 0.7 μm) for use in photosynthesis. The cell structure of the leaves, on the other hand, strongly reflects near-infrared light (from 0.7 to 1.1 μm). The more leaves a plant has, the more these wavelengths of light are affected, respectively.

No-till (zero tillage)
A technique of drilling (sowing) seed with little or no prior land preparation, which has a positive impact on soil erosion.

Nutrient pollution
Contamination of water resources by excessive inputs of nutrients.

Nutrients
The approximately 20 chemical elements known to be essential for the growth of living organisms, including nitrogen, sulphur, phosphorus and carbon.

Ocean Acidification
Term used to describe significant changes to the chemistry of the ocean. It occurs when carbon dioxide gas (or CO2) is absorbed by the ocean and reacts with seawater to produce acid. Although CO_2 gas naturally moves between the atmosphere and the oceans, the increased amounts of CO_2 gas emitted into the atmosphere, mainly as a result of human activities (e.g. burning fossil fuels), has been increasing the amount of CO_2 absorbed by the ocean, which results in seawater that is more acidic.

Ocean eutrophication
A process driven by the enrichment of water by nutrients, especially compounds of nitrogen and/or phosphorus, leading to: increased growth, primary production and biomass of algae; changes in the balance of organisms; and water quality degradation. The consequences of eutrophication are undesirable if they appreciably degrade ecosystem health and biodiversity and/or the sustainable provision of goods and services. Nitrogen and phosphorous are the primary inorganic nutrients responsible for the eutrophication of marine waters. Nitrogen and phosphorous occur naturally in marine waters, transferred from land via streams, rivers and runoff of rainwater and also from degradation of organic material within the water.

Oceanography
The branch of science that deals with the physical and biological properties and phenomena of the sea.

Organic agriculture
A production system that sustains the health of soils, ecosystems and people. It relies on ecological processes, biodiversity and cycles adapted to local conditions, rather than the use of synthetic inputs.

Organic carbon (OC)
Organic carbon, as used in climate research, usually refers to the carbon fraction of the aerosol that is not black. This term is an oversimplification because organic carbon may contain hundreds or thousands of different organic compounds with varying atmospheric behaviour. It is the quantity that results from thermal analysis of carbon aerosols.

Organizations
Bodies of individuals with a specified common objective. Organizations could be political organizations, political parties, governments and ministries; economic organizations, federations of industry; social organizations (non-governmental organizations (NGOs) and self-help groups) or religious organizations (church and religious trusts). The term organizations should be distinguished from institutions.

Organochlorine compounds
Any of a class of organic chemical compounds containing carbon, hydrogen and chlorine, such as dioxins, poly-chlorinated-biphenyls (PCBs) and some pesticides such as DDT.

Outmigration
The action of leaving one place to settle in another, especially within a country.

Overexploitation
The excessive extraction of raw materials without considering the long- term ecological impacts of such use.

Overgrazing
Excessive grazing (feeding of livestock) which causes damage to grassland.

Overshoot
The situation that occurs when humanity's demand on the biosphere exceeds supply or regenerative capacity. At the global level, ecological deficit and overshoot are the same, since there is no net import of resources to the planet.

Oxidant
An oxidizing agent.

Ozone layer
A region of the atmosphere situated at an altitude of 10–50 km above the Earth's surface (called the stratosphere) which contains diluted ozone.

ozone-depleting substances (ODSs)
Volatile organic compounds (VOC) are organic chemicals that when released into the atmosphere can react with sunlight and nitrogen oxides (NOx) to form tropospheric (ground-level) ozone. Two general classes of pesticide products contribute the vast majority of pesticidal VOC emissions: fumigants and emulsifiable concentrates.

Panacea
A solution or remedy for all difficulties or diseases.

Participatory approach
Securing an adequate and equal opportunity for people to place questions on an agenda and to express their preferences about a final outcome during decision making to all group members. Participation can occur directly or through legitimate representatives. Participation may range from consultation to the obligation of achieving a consensus.

Particulate matter (PM)
Tiny solid particles or liquid droplets suspended in the air.

Pastoralism
The husbandry of domestic animals as a primary means of obtaining resources.

Pasture
Ground covered with grass or herbage, used or suitable for the grazing of livestock.

Pathogen
A bacterium, virus, or other microorganism that can cause disease.

Payment for environmental services/payment for ecosystem services (PES)
Appropriate mechanisms for matching the demand for environmental services with incentives for land users whose actions modify the supply of those environmental services.

Peatland
Peatlands are a type of wetlands that occur in almost every country on Earth, currently covering 3 per cent of the global land surface. The term 'peatland' refers to the peat soil and the wetland habitat growing on its surface.

Per- and polyfluoroalkyl substances
Per- and polyfluoroalkyl substances (PFAS) are a group of man-made chemicals that includes PFOA, PFOS, GenX, and many other chemicals. PFAS have been manufactured and used in a variety of industries around the globe, including in the United States since the 1940s. PFOA and PFOS have been the most extensively produced and studied of these chemicals. Both chemicals are very persistent in the environment and in the human body – meaning they don't break down and they can accumulate over time. There is evidence that exposure to PFAS can lead to adverse human health effects.

Perennial
Lasting or existing for a long or apparently infinite time; enduring or continually recurring.

Peri-urban
(Especially in Africa) denoting or located in an area immediately adjacent to a city or urban area.

Permafrost
Soil, silt and rock located in perpetually cold areas, and that remains frozen year-round for two or more years.

Pernicious
Having a harmful effect, especially in a gradual or subtle way.

Persistent organic pollutants (POPs)
Chemical substances that persist in the environment, bioaccumulate through the food web, and pose a risk of causing adverse effects to human health and the environment.

Phenology
The study of cyclic and seasonal natural phenomena, especially in relation to climate and plant and animal life.

Photoconjuctivitis
Inflammation of the conjunctiva of the eye caused by exposure to UV.

Photokeratitis
Painful eye condition that occurs when your eye is exposed to invisible rays of energy called ultraviolet (UV) rays, either from the sun or from a man-made source.

Phytoplankton
Microscopically small plants that float or swim weakly in fresh or saltwater bodies.

Planetary boundaries
A framework designed to define a safe operating space for humanity for the international community, including governments at all levels, international organizations, civil society, the scientific community and the private sector, as a precondition for sustainable development.

Plasticizers
A substance (typically a solvent) added to a synthetic resin to produce or promote plasticity and flexibility and to reduce brittleness.

Pneumonia
Pneumonia is a bacterial, viral, or fungal infection of one or both sides of the lungs that causes the air sacs, or alveoli, of the lungs to fill up with fluid or pus. Symptoms can be mild or severe and may include a cough with phlegm (a slimy substance), fever, chills, and trouble breathing. Many factors affect how serious pneumonia is, such as the type of germ causing the lung infection, your age, and your overall health. Pneumonia tends to be more serious for children under the age of five, adults over the age of 65, people with certain conditions such as heart failure, diabetes, or COPD (chronic obstructive pulmonary disease), or people who have weak immune systems due to HIV/AIDS, chemotherapy (a treatment for cancer), or organ or blood and marrow stem cell transplant procedures.

Policy diffusion
The process of a policy being taken up, copied, implemented in other areas, fields, regions or sectors.

Policy
Any form of intervention or societal response. This includes not only statements of intent, but also other forms of intervention, such as the use of economic instruments, market creation, subsidies, institutional reform, legal reform, decentralization and institutional development. Policy can be seen as a tool for the exercise of governance. When such an intervention is enforced by the state, it is called public policy.

Policymaker
A member of a government department, legislature, or other organization who is responsible for making new rules, laws, etc.

Pollutant
Any substance that causes harm to the environment when it mixes with soil, water or air.

Pollution
The presence of minerals, chemicals or physical properties at levels that exceed the values deemed to define a boundary between good or acceptable and poor or unacceptable quality, which is a function of the specific pollutant.

Polycentric
Having many centres, especially of authority or control.

Poverty
The state of one who lacks a defined amount of material possessions or money. Absolute poverty refers to a state of lacking basic human needs, which commonly include clean and freshwater, nutrition, health care, education, clothing and shelter.

Precautionary approach/principle
The precautionary approach or precautionary principle states that if an action or policy has a suspected risk of causing harm to the public or to the environment, in the absence of scientific consensus that the action or policy is harmful, the burden of proof that it is not harmful falls on those taking the action.

Prediction
The act of attempting to produce a description of the expected future, or the description itself, such as "it will be 30°C tomorrow, so we will go to the beach."

Premature deaths
Deaths occurring earlier due to a risk factor than would occur in the absence of that risk factor.

Primary energy
Energy embodied in natural resources (such as coal, crude oil, sunlight or uranium) that has not undergone any anthropogenic conversion or transformation.

Private sector
The private sector is part of a country's economy which consists of industries and commercial companies that are not owned or controlled by the government.

Projection
The act of attempting to produce a description of the future subject to assumptions about certain preconditions, or the description itself, such as "assuming it is 30°C tomorrow, we will go to the beach."

Protected area
A clearly defined geographical space, recognized, dedicated and managed, through legal or other effective means, to achieve the long-term conservation of nature with associated ecosystem services and cultural values.

Provisioning services
The products obtained from ecosystems, including, for example, genetic resources, food and fibre, and freshwater.

Pterygium
Growth of the conjunctiva or mucous membrane that covers the white part of your eye over the cornea. The cornea is the clear front covering of the eye. This benign or noncancerous growth is often shaped like a wedge. A pterygium usually doesn't cause problems or require treatment, but it can be removed if it interferes with your vision.

Precision agriculture
It involves the observation, impact assessment and timely strategic response to fine-scale variation in causative components of an agricultural production process. Therefore, precision agriculture may cover a range of agricultural enterprises, from dairy herd management through horticulture to field crop production. The philosophy can be also applied to pre- and post-production aspects of agricultural enterprises.

Public sector
The portion of society that comprises the general government sector plus all public corporations including the central bank.

Public-private partnership
A contractual agreement between a public agency (federal, state or local) and a private sector entity. Through such an agreement, the skills and assets of each sector (public and private) are shared in delivering a service or facility.

Quasi-equity
A form of company debt that could also be considered to possess some traits of equity, such as being non-secured by any collateral.

Radiative forcing
A measure of the net change in the energy balance of the Earth with space, that is, the change in incoming solar radiation minus outgoing terrestrial radiation.

REDD/REDD+
Reducing Emissions from Deforestation and Forest Degradation in Developing Countries. REDD+ involves enhancing existing forests and increasing forest cover. In order to meet these objectives, policies need to address enhancement of carbon stocks by providing funding and investments in these areas.

Reforestation
Planting of forests on lands that have previously contained forest, but have since been converted to some other use.

Regulating services
The benefits obtained from the regulation of ecosystem processes, including, for example, the regulation of climate, water and some human diseases.

Remote sensing
Collection of data about an object from a distance. In the environmental field, it normally refers to aerial or satellite data for meteorology, oceanography or land cover assessment.

Renewable energy source
An energy source that does not rely on finite stocks of fuels. The most widely known renewable source is hydropower; other renewable sources are biomass, solar, tidal, wave and wind.

Resilience (Ecological)
The capacity of a system to absorb disturbance and reorganize while undergoing change so as to still retain essentially the same function, structure, identity, and feedbacks.

Resilience-based management
The focus on the processes that are essential to the ability of corals to withstand the effects of climate-related stress (resistance), and to recover (recruitment, growth survival) after major impacts.

Resistance
The capacity of a system to withstand the impacts of drivers without displacement from its present state.

Resource management activities
Activities related with the management of natural resources (monitoring, control, surveys, administration and actions for facilitating structural adjustments of the sectors concerned).

Riparian
Related to or located on the bank of a natural watercourse, usually a river, but sometimes a lake, tidewater or enclosed sea.

River fragmentation
Degree to which river connectivity and flow regimes have been altered, usually by dams and reservoirs.

Riverine
Relating to or situated on a river or riverbank; riparian.

Run-off
A portion of rainfall, melted snow or irrigation water that flows across the ground's surface and is eventually returned to streams. Run-off can pick up pollutants from air or land and carry them to receiving waters.

Sahel
A loosely defined strip of transitional vegetation that separates the Sahara desert from the tropical savannahs to the south. The region is used for farming and grazing, and because of the difficult environmental conditions at the border of the desert, the region is very sensitive to human-induced land-cover change. It includes parts of Senegal, the Gambia, Mauritania, Mali, Niger, Nigeria, Burkina Faso, Cameroon and Chad.

Salinisation/salination
The process by which water-soluble salts accumulate in the soil. Salinization may occur naturally or because of conditions resulting from management practices.

Sand and dust storms
Sand and dust storms are common meteorological hazards in arid and semi-arid regions. They are usually caused by thunderstorms – or strong pressure gradients associated with cyclones – which increase wind speed over a wide area. These strong winds lift large amounts of sand and dust from bare, dry soils into the atmosphere, transporting them hundreds to thousands of kilometres away. Some 40 per cent of aerosols in the troposphere (the lowest layer of Earth's atmosphere) are dust particles from wind erosion. The main sources of these mineral dusts are the arid regions of Northern Africa, the Arabian Peninsula, Central Asia and China. Comparatively, Australia, America and South Africa make minor, but still important, contributions. Global estimates of dust emissions, mainly derived from simulation models, vary between one and three Gigatons per year.

Scale
The spatial, temporal (quantitative or analytical) dimension used to measure and study any phenomena. Specific points on a scale can thus be considered levels (such as local, regional, national and international).

Scenario
A description of how the future may unfold based on if-then propositions, typically consisting of a representation of an initial situation, a description of the key drivers and changes that lead to a particular future state. For example, "given that we are on holiday at the coast, if it is 30°C tomorrow, we will go to the beach."

Seagrass bed
Profusion of grass-like marine plants, usually on shallow, sandy or muddy areas of the seabed.

Seamounts
Underwater mountain formed by volcanic activity.

Secondary pollutant
Not directly emitted as such, but forms when other pollutants (primary pollutants) react in the atmosphere.

Security
Relates to personal and environmental security. It includes access to natural and other resources, and freedom from violence, crime and war, as well as security from natural and human-caused disasters.

Sediment
Solid material that originates mostly from disintegrated rocks and is transported by, suspended in or deposited from water, wind, ice and other organic agents.

Sedimentation
Strictly, the act or process of depositing sediment from suspension in water or ice. Broadly, all the processes whereby particles of rock material are accumulated to form sedimentary deposits. Sedimentation, as commonly used, involves transport by water, wind, ice and organic agents.

Sequestration
In GEO-5, sequestration refers to the capture of carbon dioxide in a manner that prevents it from being released into the atmosphere for a specified period of time.

Sex-disaggregated data
Sex-disaggregated data is data that is cross-classified by sex, presenting information separately for men and women, boys and girls. Sex-disaggregated data reflect roles, real situations, general conditions of women and men, girls and boys in every aspect of society. For instance, the literacy rate, education levels, business ownership, employment, wage differences, dependants, house and land ownership, loans and credit, debts, etc. When data is not disaggregated by sex, it is more difficult to identify real and potential inequalities. Sex-disaggregated data is necessary for effective gender analysis.

Sharing economy
The peerto-peer-based activity of obtaining, giving, or sharing the access to goods and services, coordinated through community-based online services.

Short-term climate forcers
Substances such as methane, black carbon, tropospheric ozone, and many hydrofluorocarbons, which have a significant impact on climate change, and a relatively short lifespan in the atmosphere compared to carbon dioxide and other longer-lived gases.

Siltation
The deposition of finely divided soil and rock particles on the bottom of stream and riverbeds and reservoirs.

Silvopastoral production systems
The integration of trees and shrubs in pastures with animals for economic, ecological and social sustainability.

Smart cities
A smart city is a designation given to a city that incorporates information and communication technologies (ICT) to enhance the quality and performance of urban services such as energy, transportation and utilities in order to reduce resource consumption, wastage and overall costs. The overarching aim of a smart city is to enhance the quality of living for its citizens through smart technology.

Social amenities
Refer to places, buildings or infrastructural facilities which are to be shared and to become convergence spots for the local and surrounding communities. It has become a basic necessity for villages and settlement areas to have well-built and complete social amenities for the benefit of the local and surrounding communities, so as to facilitate them in conducting social functions and activities, which in turn would help shape a united, harmonious, advanced, dynamic and progressive society.

Social ecological systems
Complex adaptive systems composed of many diverse human and non-human entities that interact. They adapt to changes in their environment and their environment changes as a result.

Social network
A social structure made up of a set of actors, such as individuals or organizations, and the ties between these actors, such as relationships, connections or interactions.

Socioeconomic
Of, relating to, or involving a combination of social and economic factors.

Soft law
Rules that are neither strictly binding in nature nor completely lacking legal significance. They are weakened along one or more of the dimensions of obligation, precision and delegation. In the context of international law, soft law refers to guidelines, policy declarations or codes of conduct which set standards of conduct. However, they are not directly enforceable.

Spawning (fisheries)
To deposit or fertilize spawn; to produce young especially in large numbers.

Species (biology)
An interbreeding group of organisms that is reproductively isolated from all other organisms, although there are many partial exceptions to this rule. A generally agreed fundamental taxonomic unit that, once described and accepted, is associated with a unique scientific name.

Species diversity
Biodiversity at the species level, often combining aspects of species richness, their relative abundance and their dissimilarity.

Species richness
The number of species within a given sample, community or area.

Spillover effect
The trickle down of growth from one region to another.

Stewardship
The job of supervising or taking care of something, such as an organization or property.

Strategic environmental assessment (SEA)
A range of analytical and participatory approaches that aim to integrate environmental considerations into policies, plans and programmes and evaluate the links with economic and social considerations. An SEA is undertaken for plans, programmes and policies. It helps decision makers reach a better understanding of how environmental, social and economic considerations fit together.

Stratospheric ozone depletion
Chemical destruction of the stratospheric ozone layer, particularly by substances produced by human activities.

Surface water
All water naturally open to the atmosphere, including rivers, lakes, reservoirs, streams, impoundments, seas and estuaries. The term also covers springs, wells or other collectors of water that are directly influenced by surface waters.

Sustainability
A characteristic or state whereby the needs of the present population can be met without compromising the ability of future generations or populations in other locations to meet their needs.

Sustainable agriculture
Sustainable Agriculture puts the emphasis on methods and processes that improve soil productivity while minimising harmful effects on the climate, soil, water, air, biodiversity and human health. It aims to minimise the use of inputs from nonrenewable sources and petroleum-based products and replace them with those from renewable resources. It Focuses on local people and their needs, knowledge, skills, socio-cultural values and institutional structures. It ensures that the basic nutritional requirements of current and future generations are met in both quantity and quality terms. It provides long-term employment, an adequate income and dignified and equal working and living conditions for everybody involved in agricultural value chains. It educes the agricultural sector's vulnerability to adverse natural conditions (e.g. climate), socioeconomic factors (e.g. strong price fluctuations) and other risks.

Sustainable development
Development that meets the needs of the present generation without compromising the ability of future generations to meet their own needs.

Sustainable forest management (SFM)
The stewardship and use of forests and forest lands in a way, and at a rate, that maintains their biodiversity, productivity, regeneration capacity, vitality and potential to fulfill, now and in the future, relevant ecological, economic and social functions, at local, national and global levels, and that does not cause damage to other ecosystems.

Synergies
These arise when two or more processes, organizations, substances or other agents interact in such a way that the outcome is greater than the sum of their separate effects.

System
A system is a collection of component parts that interact with one another within some boundary.

Taxonomy
A system of nested categories (taxa) reflecting evolutionary relationships or morphological similarities.

TechnoGarden
The TechnoGarden scenario depicts a globally connected world relying strongly on technology and highly managed, often engineered ecosystems, to deliver ecosystem services.

Technology transfer
A broad set of processes covering the flows of know-how, experience and equipment among different stakeholders.

Technology
Physical artefacts or the bodies of knowledge of which they are an expression. Examples are water extraction structures, such as tube wells, renewable energy technologies and traditional knowledge. Technology and institutions are related. Any technology has a set of practices, rules and regulations surrounding its use, access, distribution and management.

Temperate region
The region in which the climate undergoes seasonal change in temperature and moisture. Temperate regions of the Earth lie primarily between 30° and 60° latitude in both hemispheres.

Theory of change
A theory of change is a method that explains how a given intervention, or set of interventions, is expected to lead to specific development change, drawing on a causal analysis based on available evidence.

Thermohaline circulation
Large-scale density-driven circulation in the ocean, caused by differences in temperature and salinity. In the North Atlantic, the thermohaline circulation consists of warm surface water flowing northward and cold deep water flowing southward, resulting in a net poleward transport of heat. The surface

water sinks in highly restricted sinking regions located in high latitudes. Also referred to as the (global) ocean conveyor belt or the meridional overturning circulation.

Threshold
The level of magnitude of a system process at which sudden or rapid change occurs. A point or level at which new properties emerge in an ecological, economic or other system, invalidating predictions based on mathematical relationships that apply at lower levels.

Tipping point
The critical point in an evolving situation that leads to a new and sometimes irreversible development.

Top-down
Used to refer to a situation in which decisions are made by a few people in authority rather than by the people who are affected by the decisions.

Topography
The study or detailed description of the surface features of a region.

Traditional or local ecological knowledge
A cumulative body of knowledge, know-how, practices or representations maintained or developed by peoples with extended histories of interaction with the natural environment.

Transformation
State of being transformed. In the context of GEO-5, transformation refers to a series of actions that explores opportunities to stop doing the things that pull the Earth System in the wrong direction and at the same time provide resources, capacity and an enabling environment for all that is consistent with the sustainable-world vision.

Transformational change
The process whereby positive development results are achieved and sustained over time by institutionalizing policies, programmes and projects within national strategies. It should be noted that this embodies the concept of institutionally sustained results – consistency of achievement over time. This is in order to exclude short-term, transitory impact.

Transformative pedagogy
A progressive educational approach that includes democratic constructivist-based pedagogy for the promotion of social justice and democratic ideals to transform students and society. Transformative pedagogy empowers learners to engage in dialogue to co-construct meaning from educational material and experiences through an inquiry-based approach (as opposed to what Paulo Freire calls a "banking" orientation). It also promotes personal experiences, dialogical pedagogy, and aligning education with social justice.

Transitions
Non-linear, systematic and fundamental changes of the composition and functioning of a societal system with changes in structures, cultures and practices.

Transpiration
The loss of water vapour from parts of plants, especially in leaves but also in stems, flowers and roots.

Trillion
10^{12} (1 000 000 000 000).

Trophic level
Successive stages of nourishment as represented by the links of the food chain. The primary producers (phytoplankton) constitute the first trophic level, herbivorous zooplankton the second and carnivorous organisms the third trophic level.

Tropospheric ozone
Ozone at the bottom of the atmosphere, and the level at which humans, crops and ecosystems are exposed. Also known as ground-level ozone.

Urban agglomeration
The population contained within the contours of a contiguous territory inhabited at urban density levels without regard to administrative boundaries." In other words, it integrates the 'City Proper' plus suburban areas that are part of what can be considered as city boundaries; a term that in itself is controversial.

Urban sprawl
The decentralization of the urban core through the unlimited outward extension of dispersed development beyond the urban fringe, where low density residential and commercial development exacerbates fragmentation of powers over land use.

Urbanism
An integration of urban and rural development in terms of sustainable resource use and the convergence of human well-being.

Urbanization
An increase in the proportion of the population living in urban areas.

Venture capital
Venture capital is capital that is invested in projects that have a high risk of failure, but that will bring large profits if they are successful.

Virtual water trade
The idea that when goods and services are traded, the water needed to produce them (embedded) is traded as well.

Volatile Organic Compounds (VOCs)
Volatile organic compounds (VOC) means any compound of carbon, excluding carbon monoxide, carbon dioxide, carbonic acid, metallic carbides or carbonates and ammonium carbonate, which participates in atmospheric photochemical reactions, except those designated by EPA as having negligible photochemical reactivity.

Vulnerability
An intrinsic feature of people at risk. It is a function of exposure, sensitivity to impacts of the specific unit exposed (such as a watershed, island, household, village, city or country), and the ability or inability to cope or adapt. It is multi-dimensional, multi-disciplinary, multi-sectoral and dynamic. The exposure is to hazards such as drought, conflict or extreme price fluctuations, and also to underlying socio-economic, institutional and environmental conditions.

Wastewater treatment
Any of the mechanical, biological or chemical processes used to modify the quality of wastewater in order to reduce pollution levels.

Water column
An imaginary column extending through a water body from its floor to its surface.

Water quality
The chemical, physical and biological characteristics of water, usually in respect to its suitability for a particular purpose.

Water scarcity
Occurs when annual water supplies drop below 1 000 m3 per person, or when more than 40 per cent of available water is used.

Water security
A term that broadly refers to the sustainable use and protection of water systems, the protection against water related hazards (floods and droughts), the sustainable development of water resources and the safeguarding of (access to) water functions and services for humans and the environment.

Water stress
Occurs when low water supplies limit food production and economic development, and affect human health. An area is experiencing water stress when annual water supplies drop below 1 700 m^3 per person.

Wetland
Area of marsh, fen, peatland, bog or water, whether natural or artificial, permanent or temporary, with water that is static or flowing, fresh, brackish or salt, including areas of marine water to a depth, at low tide, that does not exceed 6 metres.

Whole-genome sequencing
A laboratory process that is used to determine nearly all of the approximately 3 billion nucleotides of an individual's complete DNA sequence, including non-ending sequence.

Wildlife
Wild animals collectively; the native fauna (and sometimes flora) of a region.

Woodland
Wooded land, which is not classified as forest, spanning more than 0.5 hectares, with trees higher than 5 metres and a canopy cover of 5–10 per cent, or trees able to reach these thresholds *in situ*, or with a combined cover of shrubs, bushes and trees above 10 per cent. It does not include areas used predominantly for agricultural or urban purposes.

Zettabyte
A unit of information equal to one sextillion (10^{21}) or, strictly, 2^{70} bytes.

Zika
A mosquito-borne virus of the genus Flavivirus (family Flaviviridae), found in parts of Africa and in Malaysia; it causes Zika fever.

Zoonotic disease
(Also known as zoonosis) An infection or disease that is transmissible from animals to humans under natural conditions.